TEACHER'S WRAPAROUND EDITION

Glencoe

Algebra 1

Integration
Applications
Connections

GLENCOE
McGraw-Hill

New York, New York   Columbus, Ohio   Mission Hills, California   Peoria, Illinois

Glencoe/McGraw-Hill

A Division of The McGraw-Hill Companies

Send all inquiries to:
Glencoe/McGraw-Hill
8787 Orion Place
Columbus, OH 43240-4027

ISBN: 0-02-824846-5 (Student Edition)
          0-02-824871-6 (Teacher's Wraparound Edition)

4 5 6 7 8 9 10 071/043 04 03 02 01 00

TEACHER'S
WRAPAROUND
EDITION

Table of Contents

Algebra 1

INTEGRATION ● APPLICATIONS ● CONNECTIONS

Glencoe Brings Algebra ❶ to Life!

INTEGRATION ● **APPLICATIONS** ● **CONNECTIONS**

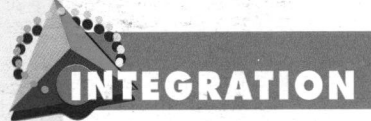
Students relate and apply algebraic concepts to geometry, statistics, data analysis, probability, and discrete mathematics.

Realistic and relevant applications help answer the question "When am I ever going to use this stuff?" Sports, space, world cultures, and consumerism are just a few of the real-life problem settings that students explore.

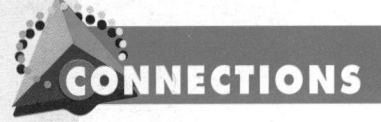
Students connect mathematics to other topics they are studying, like biology, geography, art, history, and health, through problems that are rich in algebraic content.

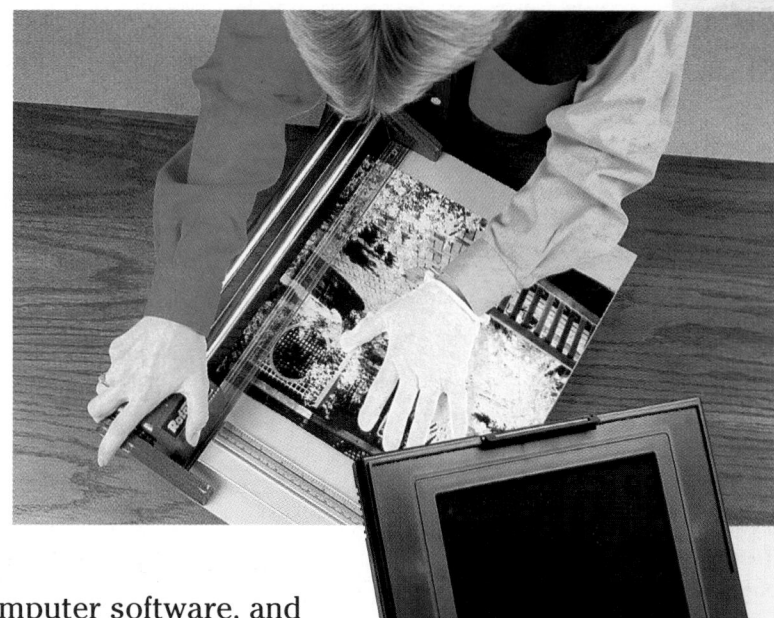

TECHNOLOGY

Graphing calculators, interactive computer software, and CD-ROM multimedia technology provide tools for both problem solving and discovery. The variety of integrated technology supports diverse learning styles.

HANDS-ON ACTIVITIES

Modeling Mathematics activities and multi-day Investigations engage students with manipulatives and mathematical models. Both college-bound and tech-prep students find meaningful hands-on activities that relate to life and work.

ASSESSMENT

A broad spectrum of evaluation and assessment tools, ranging from short-response questions to open-ended problems, lets all students demonstrate what they have learned.

ACCESSIBILITY

Practical strategies meet the needs of students with differing ability levels, rates of learning, and learning styles. Modeling, journal writing, cooperative learning projects, and technology tools help all students experience success.

NCTM STANDARDS

The sequence of topics, the content emphasis, integration with other mathematics areas, hands-on activities, and technology combine to form a program that reflects the NCTM Curriculum, Teaching, and Assessment Standards.

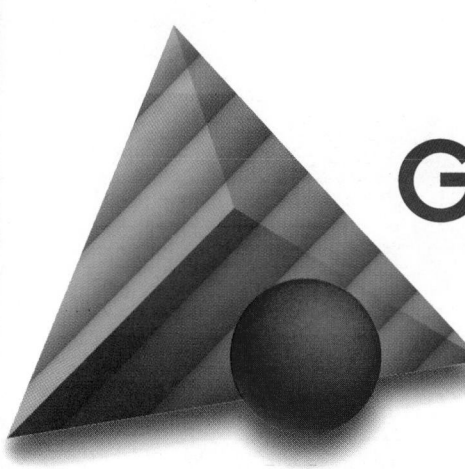

Glencoe's Dynamic Lesson Structure

Each lesson follows a straightforward format —

> **Objective**
> **Application/Connection/Integration**
> **Examples**
> **Check for Understanding**
> **Exercises**

— that focuses on concepts and applications.

OBJECTIVE

The lesson objectives help students quickly see what performance is indicated and expected. "After studying this lesson, you should be able to...."

APPLICATION/CONNECTION/INTEGRATION

A motivating lesson opener introduces the concept by posing a real-world application, an interdisciplinary connection, or a mathematical integration.

EXAMPLES

A set of completely worked-out examples with clear explanations supports the objectives and mirrors the exercises in the Guided Practice and Practice sections that follow.

MODELING MATHEMATICS/ EXPLORATION

Activities with hands-on manipulatives and technology are fully integrated into appropriate lessons.

3-4 Integration: Geometry
Angles and Triangles

Objectives
After studying this lesson, you should be able to:
* find the complement and supplement of an angle, and
* find the measure of the third angle of a triangle given the measures of the other two angles.

The symbol for angle is ∠.

APPLICATION
Football

On fourth down in a football game, the team with the ball usually *punts*. A punt is a kick in which the ball is dropped by the kicker and is kicked before it hits the ground.

The longest punt ever kicked in a college football game was kicked by Pat Brady of the University of Nevada-Reno in their 1950 game against Loyola University. The punt was 99 yards long—just 1 yard short of spanning the entire length of the football field.

How far a football will travel once it is punted depends not only on the strength of the punter, but also on the angle at which the ball is kicked. If the angle is too small, the ball will travel close to the ground, then drop; if the angle is too large, the ball will travel high, but not very far. What do you think would be the best angle at which to kick the ball?

You can use a *protractor* to measure angles, as shown below.

Angle ABC (denoted ∠ABC) measures 60°. However, where ray BC (denoted $\vec{BC}$) intersects the curve of the protractor, there are two readings, 60° and 120°. The measure of ∠DBC is 120°. What is the sum of the measures of ∠ABC and ∠DBC?

Supplementary Angles > Two angles are supplementary if the sum of their measures is 180°.

162 Chapter 3 Solving Linear Equations

Example 1 The measure of an angle is three times the measure of its supplement. Find the measure of each angle.

Let x = the lesser measure. Then $3x$ = the greater measure.

$x + 3x = 180$ *The angles are supplementary.*
$4x = 180$ *Add like terms.*
$\frac{4x}{4} = \frac{180}{4}$ *Divide each side by 4.*
$x = 45$

The measures are 45° and 3 · 45° or 135°. *Check this result.*

Complementary Angles Two angles are complementary if the sum of their measures is 90°.

Example 2
APPLICATION
Golf

Lowest winning scores in the U.S. Masters Tournament
1. 271, Raymond Floyd, 1976
1. 271, Jack Nicklaus, 1965
3. 274, Ben Hogan, 1953
4. 275, Severiano Ballesteros, 1980
4. 275, Fred Couples, 1992

The backward slant of the face of a golf club head is called the *loft*. It is designed to drive the ball in a high arc. Assuming that the angle made by the ground and the face of the club is 79°, what is the measure of the loft?

Let x = the measure of the loft.

$79° + x = 180°$ *The angles are complementary.*
$79° - 79° + x = 90° - 79°$ *Subtract 79° from each side.*
$x = 11°$

The club has an 11° loft.

Example 3 The measure of an angle is 34° greater than its complement. Find the measure of each angle.

Let x = the lesser measure. Then $x + 34$ = the greater measure.

$x + (x + 34) = 90$ *The angles are complementary.*
$2x + 34 = 90$
$2x + 34 - 34 = 90 - 34$ *Subtract 34 from each side.*
$2x = 56$
$\frac{2x}{2} = \frac{56}{2}$ *Divide each side by 2.*
$x = 28$

The measures are 28° and 28° + 34° or 62°. *Check this result.*

A **triangle** is a polygon with three sides and three angles. What is the sum of the measures of the three angles of a triangle?

Lesson 3-4 Geometry Angles and Triangles 163

MODELING MATHEMATICS   Angles of a Triangle

Materials: 3 sheets of construction paper   protractor   scis

Copy each triangle below onto a sheet of construction paper. If possible, you may want to enlarge the triangles on a photocopier.

b. Cut out one of the triangles. Label the A, B, and C. Tear off the angles and arrange shown.
c. When the angles were put together, what measure of the new angle they formed?
d. Try this activity with another triangle. Whi you think is true about the measures of the of any triangle?

Your Turn
a. Use a protractor to measure the angles of each triangle above. What is the sum of the measures of the three angles of each triangle?

Sum of the Angles of a Triangle   The sum of the measures of the angles in any triangle is 180

In an **equilateral triangle**, each angle has the same measure. We sa the angles are **congruent**. The sides of an equilateral triangle are also congruent. What is the measure of each angle of an equilateral triangle

Let x = the measure of each angle.

$x + x + x = 180$
$3x = 180$ *Add like terms.*
$\frac{3x}{3} = \frac{180}{3}$ *Divide each side by 3*
$x = 60$ Each angle measures 60°.

In an **isosceles triangle**, at least two angles have the same measure. Generally, the two congruent angles are the base angles.

Example 4 What are the measures of the base angles of an isosceles triangle which the vertex angle measures 45°?

Let x = the measure of each base angle.

$x + x + 45 = 180$
$2x + 45 = 180$ *Add like terms.*
$2x + 45 - 45 = 180 - 45$ *Subtract 45 from each side.*
$2x = 135$
$\frac{2x}{2} = \frac{135}{2}$ *Divide each side by 2*
$x = \frac{135}{2}$ or $67\frac{1}{2}$ The base angles each measure $67\frac{1}{2}$°.

164 Chapter 3 Solving Linear Equations

CHECK FOR UNDERSTANDING

This section ensures that all students are engaged in the lesson and understand the concepts.

Communicating Mathematics
Students work in small groups or as a whole class to define, explain, describe, make lists, draw models, use symbols, or create graphs.

Modeling Mathematics and **Math Journal** activities, which further strengthen communication skills, appear frequently.

Guided Practice
Keyed to the lesson objectives, the Guided Practice exercises present a representative sample of the exercises in the Practice section.

EXERCISES

The Exercises always include **Practice, Critical Thinking, Applications and Problem Solving,** and **Mixed Review.**

Practice
The Practice exercises are separated into A, B, and C sections, indicated only in the Teacher's Wraparound Edition. The A and B sections generally match the Guided Practice exercises in a 3:1 ratio. The C section offers challenging, thought-provoking questions.

Critical Thinking
Each lesson contains critical thinking exercises, in which students explain, evaluate, and justify mathematical concepts and relationships.

Applications and Problem Solving
Students find numerous opportunities to apply concepts to both real-life and mathematical problem situations.

Mixed Review
This spiraled, cumulative review comprises about 15% of the total number of exercises in each lesson and includes at least one application, connection, or integration problem.

Glencoe's Algebra ① Exemplifies the NCTM Standards

CONTENT EMPHASIS

Algebra 1: Integration, Applications, and Connections implements the shift from just manipulative skills to algebra as a means of representation.

> *"The proposed algebra curriculum will move away from a tight focus on manipulative facility to include a greater emphasis on conceptual understanding, on algebra as a means of representation, and on algebraic methods as a problem-solving tool." NCTM Standards, page 150, 1989.*

In each chapter, students explore the language of algebra in verbal, tabular, graphical, and symbolic forms. Problem-solving activities and applications encourage students to model patterns and relationships with variables and functions.

> *"The primary role of algebra at the school level is to develop confidence and facility in using variables and functions to model numerical patterns and quantitative relations." NCTM Statement on Algebra, February 1994.*

Each lesson opener motivates students to master the content they need to solve the application, connection, or integration. Additional applications, connections, and integration in the exercises enable students to apply what they have learned.

Glencoe's ***Algebra 1*** incorporates graphing-calculator and computer-software activities for discovery, problem solving, and modeling. The program deemphasizes paper-and-pencil methods of transforming, simplifying, and solving symbolic expressions, while retaining those procedures that students need for algebraic understanding.

> *"Development of algebra should use technology to explore ideas and methods from at least three connected perspectives—graphic, numeric, and symbolic." NCTM Statement on Algebra, February 1994.*

Examples, exercises, and assessment activities in Glencoe's ***Algebra 1*** integrate algebra topics with statistics, geometry, and discrete mathematics.

> *"Refocusing algebra on modeling...serves to break down artificial barriers between algebra and statistics, between algebra and geometry, and between algebra and discrete mathematics." NCTM Statement on Algebra, February 1994.*

Applications, modeling activities, and open-ended projects encourage a diversity of approaches and engage today's students in Algebra 1.

ASSESSMENT

The wide and varied collection of assessment tools and materials that accompany Glencoe's *Algebra 1* support the NCTM Assessment Standards. *[Note: The bulleted headings and quotations below are taken from the Assessment Standards for School Mathematics, NCTM, 1995, pages 11, 13, 15, 17, 19, and 21.]*

● **The Mathematics Standard** — "Assessment should reflect the mathematics that all students need to know and be able to do."

    Skills, procedural knowledge, factual knowledge, reasoning, applying mathematics, and problem solving are assessed through a broad range of short-response exercises and longer-term, open-ended activities.

● **The Learning Standard** — "Assessment should enhance mathematics learning."

    Assessment is an integral part of the Communicating Mathematics, Guided Practice, and Exercises sections of each Glencoe's *Algebra 1* lesson. Students are able to evaluate their own learning by writing in their math journals and selecting portfolio items.

● **The Equity Standard** — "Assessment should promote equity."

    The wide variety of assessment tools enable all students to demonstrate what they know and can do. Performance Assessment tasks for each chapter include a scoring guide.

● **The Openness Standard** — "Assessment should be an open process."

    Learning objectives tell students what they are expected to learn in each lesson. The examples, activities, and assessment match those objectives. Guided Practice prepares students to successfully complete the Exercises.

● **The Inferences Standard** — "Assessment should promote valid inferences about mathematics learning."

    The assortment of assessment tools provide many sources of evidence. Exercises, Mixed Review, and Chapter Tests are all correlated to the learning objectives and practice activities.

● **The Coherence Standard** — "Assessment should be a coherent process."

    The numerous, varied assessment tools are carefully designed to complement each other. They enable you to easily construct the assessment program that best fits your needs.

Learning Objectives & NCTM Standards Correlation

Lesson	Lesson Objectives	NCTM Standards
1-1	Translate verbal expressions into mathematical expressions and vice versa.	1-5
1-2	Solve problems by looking for a pattern.	1-5
1-3	Use the order of operations to evaluate real number expressions.	1-5, 8
1-4	Display and interpret data on a stem-and-leaf plot.	1-5, 10
1-5	Solve open sentences by performing arithmetic operations.	1-5
1-6	Recognize and use the properties of identity and equality. Determine the multiplicative inverse of a number.	1-5
1-7A	Model the distributive property.	1-5
1-7	Use the distributive property to simplify expressions.	1-5, 8
1-8	Recognize and use the commutative and associative properties when simplifying expressions.	1-5
1-9	Interpret graphs in real-world settings. Sketch graphs for given functions.	1-6, 8
2-1	State the coordinate of a point on a number line. Graph integers on a number line. Add integers by using a number line.	1-5, 14
2-2	Interpret numerical data from a table. Display and interpret statistical data on a line plot.	1-4, 10
2-3A	Use counters to add and subtract integers.	1-5
2-3	Find the absolute value of a number. Add and subtract integers.	1-5
2-4	Compare and order rational numbers. Find a number between two rational numbers.	1-5
2-5	Add and subtract rational numbers. Simplify expressions that contain rational numbers.	1-5
2-6A	Use counters to multiply integers.	1-5
2-6	Multiply rational numbers.	1-5
2-7	Divide rational numbers.	1-5
2-8A	Use base-ten tiles to model square roots.	1-5
2-8	Find square roots. Classify numbers. Graph solutions of inequalities on number lines.	1-5, 14
2-9	Explore problem situations. Translate verbal sentences and problems into equations or formulas and vice versa.	1-5
3-1A	Solve equations by using cups and counters to model variables and integers.	1-5
3-1	Solve equations by using addition and subtraction.	1-5
3-2	Solve equations by using multiplication and division.	1-5
3-3A	Solve equations by using cups and counters to model variables and integers.	1-5
3-3	Solve equations involving more than one operation. Solve problems by working backward.	1-5
3-4	Find the complement and supplement of an angle. Find the measure of the third angle of a triangle given the measures of the other two angles.	1-5, 7

Lesson	Lesson Objectives	NCTM Standards
3-5	Solve equations with the variable on both sides. Solve equations containing grouping symbols.	1-5, 7
3-6	Solve equations and formulas for a specified variable.	1-5, 7
3-7	Find and interpret the mean, median, and mode of a set of data.	1-5, 10
4-1A	Collect data to determine ratios.	1-5
4-1	Solve proportions.	1-5
4-2	Find the unknown measures of the sides of two similar triangles.	1-5, 7
4-3	Use trigonometric ratios to solve right triangles.	1-5, 9
4-4	Solve percent problems. Solve problems involving simple interest.	1-5
4-5	Solve problems involving percent of increase or decrease. Solve problems involving discounts or sales tax.	1-5
4-6	Find the probability of a simple event. Find the odds of a simple event.	1-5, 11
4-7	Solve mixture problems. Solve problems involving uniform motion.	1-5
4-8	Solve problems involving direct and inverse variation.	1-5
5-1	Graph ordered pairs on a coordinate plane. Solve problems by making a table.	1-5, 8
5-2A	Use a graphing calculator to graph relations.	1-5, 8
5-2	Identify the domain, range, and inverse of a relation. Show relations as sets of ordered pairs, tables, mappings, and graphs.	1-5
5-3A	Use a graphing calculator to investigate relations and determine the ranges when the domains are given.	1-5
5-3	Determine the range for a given domain. Graph the solution set for the given domain.	1-5
5-4A	Use a graphing calculator to graph linear relations and functions.	1-5, 8
5-4	Graph linear equations.	1-5, 8
5-5	Determine whether a given relation is a function. Find the value of a function for a given element of the domain.	1-6
5-6	Write equations to represent relations, given some of the solutions for the equation.	1-6
5-7A	Use a graphing calculator to calculate measures of variation.	1-4, 10
5-7	Calculate and interpret the range, quartiles, and interquartile range of sets of data.	1-4, 10
6-1A	Use a geoboard to model a line segment and calculate its slope.	1-5, 8
6-1	Find the slope of a line, given the coordinates of two points on the line.	1-5, 8
6-2	Write linear equations in point-slope form. Write linear equations in standard form.	1-5, 8
6-3	Graph and interpret points on a scatter plot. Draw and write equations for best-fit lines, and make predictions by using those equations. Solve problems by using models.	1-5, 10
6-4	Determine the x- and y-intercepts of linear graphs from their equations. Write equations in slope-intercept form. Write and solve direct variation equations.	1-5
6-5A	Use a graphing calculator to determine whether a group of graphs forms a family.	1-5
6-5	Graph a line given any linear equation.	1-5
6-6	Determine if two lines are parallel or perpendicular by their slopes. Write equations of lines passing through a given point, parallel or perpendicular to the graph of a given equation.	1-5, 8

Lesson	Lesson Objectives	NCTM Standards
6-7	Find the coordinates of the midpoint of a line segment in the coordinate plane.	1-5, 8
7-1	Solve inequalities by using addition and subtraction.	1-5
7-2A	Use cups and counters to solve inequalities.	1-5
7-2	Solve inequalities by using multiplication and division.	1-5
7-3	Solve linear inequalities involving more than one operation. Find the solution set for a linear inequality when replacement values are given for the variables.	1-5
7-4	Solve problems by making a diagram. Solve compound inequalities and graph their solution sets. Solve problems that involve compound inequalities.	1-5
7-5	Find the probability of a compound event.	1-4, 11
7-6	Solve open sentences involving absolute value and graph the solutions.	1-5
7-7	Display and interpret data on box-and-whisker plots.	1-5, 10
7-7B	Use a graphing calculator to compare two sets of data by using a double box-and-whisker plot.	1-5, 10
7-8A	Use the "Shade C" command on a graphing calculator to graph inequalities in two variables.	1-5
7-8	Graph inequalities in the coordinate plane.	1-5
8-1A	Use a graphing calculator to solve systems of equations.	1-5, 8
8-1	Solve systems of equations by graphing. Determine whether a system of equations has one solution, no solutions, or infinitely many solutions by graphing.	1-5, 8
8-2	Solve systems of equations by the substitution method. Organize data to solve problems.	1-5
8-3	Solve systems of equations by the elimination method using addition or subtraction.	1-5
8-4	Solve systems of equations by the elimination method using multiplication and addition.	1-5
8-5	Solve systems of inequalities by graphing.	1-5, 8
9-1	Multiply monomials. Simplify expressions involving powers of monomials. Solve problems by looking for a pattern.	1-5
9-2	Simplify expressions involving quotients of monomials. Simplify expressions containing negative exponents.	1-5
9-3	Express numbers in scientific and standard notation. Find products and quotients of numbers expressed in scientific notation.	1-5
9-4A	Use algebra tiles to model polynomials.	1-5
9-4	Find the degree of a polynomial. Arrange the terms of a polynomial so that the powers of a variable are in ascending or descending order.	1-5
9-5A	Use algebra tiles to add and subtract polynomials.	1-5
9-5	Add and subtract polynomials.	1-5
9-6A	Use algebra tiles to model the product of simple polynomials.	1-5
9-6	Multiply a polynomial by a monomial. Simplify expressions involving polynomials.	1-5
9-7A	Use algebra tiles to find the product of binomials.	1-5
9-7	Use the FOIL method to multiply two binomials. Multiply any two polynomials by using the distributive property.	1-5
9-8	Use patterns to find $(a + b)2$, $(a - b)2$, and $(a + b)(a - b)$.	1-5

Lesson	Lesson Objectives	NCTM Standards
10-1	Find the prime factorization of integers. Find the greatest common factors (GCF) for sets of monomials.	1-5
10-2A	Use algebra tiles to factor binomials.	1-5
10-2	Use the GCF and the distributive property to factor polynomials. Use grouping techniques to factor polynomials with four or more terms.	1-5
10-3A	Use algebra tiles to factor simple trinomials.	1-5
10-3	Solve problems by using guess and check. Factor quadratic trinomials.	1-5
10-4	Identify and factor binomials that are the differences of squares.	1-5
10-5	Identify and factor perfect square trinomials.	1-5
10-6	Use the zero product property to solve equations.	1-5
11-1A	Use a graphing calculator to graph a quadratic function and find the coordinates of its vertex.	1-6
11-1	Find the equation of the axis of symmetry and the coordinates of the vertex of a parabola. Graph quadratic functions.	1-6, 13
11-1B	Use a graphing calculator to study the characteristics of families of parabolas.	1-6
11-2	Use estimation to find roots of quadratic equations. Find roots of quadratic equations by graphing.	1-6, 13
11-3	Solve quadratic equations by using the quadratic formula.	1-6
11-4A	Use a graphing calculator to graph exponential functions.	1-6, 13
11-4	Graph exponential functions. Determine if a set of data displays exponential behavior. Solve exponential equations.	1-6, 13
11-5	Solve problems involving growth and decay.	1-6
12-1	Simplify rational expressions. Identify values excluded from the domain of a rational expression.	1-5
12-1B	Use a graphing calculator to check the simplification of rational expressions.	1-6
12-2	Multiply rational expressions.	1-5
12-3	Divide rational expressions.	1-5
12-4	Divide a polynomial by a binomial.	1-5
12-5	Add and subtract rational expressions with like denominators.	1-5
12-6	Add and subtract rational expressions with unlike denominators. Solve problems by making an organized list of possibilities.	1-5
12-7	Simplify mixed expressions and complex fractions.	1-5
12-8	Solve rational equations.	1-5
13-1A	Use a geoboard to find areas by using the Pythagorean theorem.	1-5, 7
13-1	Use the Pythagorean theorem to solve problems.	1-5, 7
13-2	Simplify square roots. Simplify radical expressions.	1-5
13-3A	Use a graphing calculator to simplify and approximate values of expressions containing radicals.	1-5
13-3	Simplify radical expressions involving addition and subtraction.	1-5
13-4	Solve radical equations.	1-5
13-5	Find the distance between two points in the coordinate plane.	1-5, 8
13-6A	Use algebra tiles as a model for completing the square.	1-5
13-6	Solve problems by identifying subgoals. Solve quadratic equations by completing the square.	1-5

Planning Your Algebra 1 Course

40- TO 50-MINUTE CLASS PERIODS

● One Year

The chart below gives suggested pacing guides for two options, Standard and Honors, for four 9-week grading periods. The Standard option covers Chapters 1-12 while the Honors option covers Chapters 1-13. *A pacing guide that addresses both of these options is provided in each lesson of the Teacher's Wraparound Edition.*

The total days suggested for each option is about 160 days. This allows for teacher flexibility in planning due to school cancellation or shortened class periods.

Grading Period	Standard		Honors	
	Chapter	Days	Chapter	Days
1	1	15	1	14
	2	15	2	15
	3	12	3	12
2	4	15	4	8
	5	14	5	7
	6	13	6	7
3	7	14	7	13
	8	9	8	8
	9	15	9	14
			10-1 to 10-3	5
4	10	12	10-4 to end	6
	11	11	11	10
	12	14	12	13
			13	11

● Two Years

The suggested pacing guide for teaching Glencoe's **Algebra 1** in two years is based on teaching Chapters 1-7 the first year preceded by a 2-week review of pre-algebra. The second year begins with a 6-week review of the first year and then Chapters 8-13. *The pacing guide in each lesson of the Teacher's Wraparound Edition also addresses teaching Algebra 1 in two years.*

For day-by-day lesson plans and other supplemental materials, please refer to Glencoe's Algebra 1 in Two Years Instructional Resources.

Grading Period	Year One		Year Two	
	Chapter	Days	Chapter	Days
1	Review	10	Review	30
	1	22	8-1 to 8-4	9
	2-1 to 2-3	8		
2	2-4 to end	20	8-5 to end	5
	3	19	9	26
			10-1 to 10-3	10
3	4	20	10-4 to end	12
	5	21	11	19
			12-1 to 12-3	
4	6	18	12-4 to end	18
	7	21	13	22

BLOCK SCHEDULE, 90-MINUTE CLASS PERIODS

The charts below give suggested pacing guides for teaching Glencoe's
Algebra 1 using block scheduling. The guides are based on teaching Chapters
1-13. *The pacing guide in each lesson of the Teacher's Wraparound Edition also
addresses block scheduling.*

● 90 Days (1 Semester or 1 Year)

The suggested pacing guide below can be used for block scheduling in
1 semester (classes meet every day) or 1 year (classes meet about 85 days).

Chapter	Class Periods
1	7
2	9
3	7
4	7
5	7
6	7
7	7
8	5
9	8
10	6
11	5
12	6
13	7

● 180 Days (1 or 2 Years)

The suggested pacing guide below can be used for block scheduling in
1 year (classes meet every day) or 2 years (classes meet about 85 days
per year).

One Year		Two Years	
Chapter	Class Periods	Chapter	Class Periods
Review	7	Review	5
1	12	1	11
2	15	2	14
3-1 to 3-3	4	3	10
3-4 to end	6	4	10
4	11	5	11
5	12	6	9
6	10	7	11
7	12	Review	15
8	8	8	7
9	14	9	13
10-1 to 10-3	5	10-1 to 10-4	6
10-4 to end	7	10-5 to end	5
11	10	11	10
12	14	12	14
13	12	13	11

For day-by-day lesson plans, please refer to
Glencoe's Algebra 1 Block Scheduling Booklet.

Classroom Vignettes

Glencoe's unique **Classroom Vignettes** are classroom-tested teaching suggestions made by experienced mathematics educators. Each vignette was written by a teacher who uses Glencoe's *Algebra 1* or by a Glencoe reviewer, consultant, or author. Glencoe thanks these outstanding mathematics educators for their unique contributions.

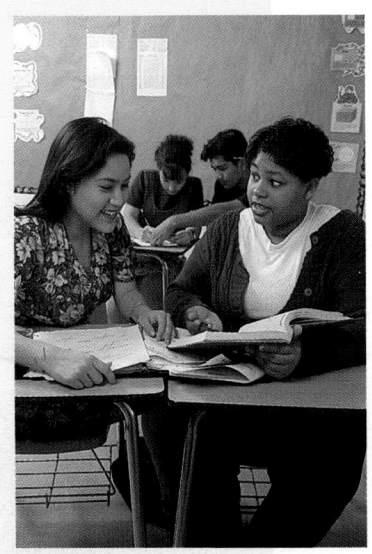

Frank Arcuri (p. 508)
Croton Harmon High School
Croton-on-Hudson, New York

Betsy Barron (p. 74)
Rochester High School
Rochester, Indiana

Cindy Boyd (p. 465)
Abilene High School
Abilene, Texas

Jenny Brawner (p. 87)
Lee-Scott Academy
Auburn, Alabama

William Brinkman (p. 235)
Paynesville High School
Paynesville, Minnesota

John Cline (p. 241)
Greenfield Central High School
Greenfield, Indiana

Emma Lou Crawford (p. 223)
Roanoke High School
Robersonville, North Carolina

Linda Crawford (p. 121)
Brown County Junior High School
Nashville, Indiana

Lara Dillinger (p. 219)
Rock Creek High School
St. George, Kansas

Barbara Ferguson (p. 243)
Enumclaw High School
Enumclaw, Washington

Eva Gates (pp. 143 & 346)
Independent Mathematics Consultant
Pearland, Texas

Joan Gell (p. 637)
Palos Verdes High School
Palos Verdes Estates, California

Linda Glover (p. 537)
Conway Junior High School
Conway, Arkansas

Harvey A. Johnecheck (p. 505)
Gaylord High School
Gaylord, Michigan

Nancy Kinard (p. 25)
Palm Beach Gardens High School
Palm Beach Gardens, Florida

Barbara Lasko (p. 130)
Roosevelt-Wilson Middle School
Clarksburg, West Virginia

Joyce Anne MacKenzie (p. 662)
Camdenton R-III High School
Camdenton, Missouri

Bea Moore-Harris (p. 56)
Educational Specialist, Region IV
Education Service Center
Houston, Texas

Stan Musgrave (p. 254)
Rochester High School
Rochester, Indiana

Barbara Owens (p. 710)
Davidson Middle School
Davidson, North Carolina

Karen C. Painter (p. 585)
Smoky Mountain High School
Sylva, North Carolina

Deborah Phillips (pp. 258 & 272)
St. Vincent-St. Mary High School
Akron, Ohio

Donna Rehling (p. 145)
Elder High School
Cincinnati, Ohio

Carolyn Stanley (p.128)
Tidewater Academy
Wakefield, Virginia

Linda Weissinger (p. 421)
Rochester High School
Rochester, Indiana

People in the News

Glencoe's unique **People in the News** features young people from across the United States who have appeared "in the news" because of a noteworthy accomplishment. Glencoe congratulates these exceptional citizens.

Jorge Aguilar (p. 323)
Texas

Liz Alvarez (p. 609)
Florida

Ayrris Layug Aunario (p. 193)
Illinois

Julie & Tony Goskowicz (p. 451)
Wisconsin

Shakema Hodge (p. 383)
New York

Ryan Holladay (p. 711)
Virginia

Wendy Isdell (p. 5)
Virginia

Jamie Knight (p. 253)
Pennsylvania

Joe Maktima (p. 659)
Arizona

Marianne Ragins (p. 141)
Georgia

Robert Rodriquez (p. 557)
Texas

Danny Seo (p. 71)
Pennsylvania

Jenny Slabaugh & Amy Gusfa (p. 495)
Michigan

Internet Connections

Glencoe's unique **Internet Connection** is a source for more information related to the **Chapter Project** and the interests of the person featured in **People in the News.**

The sites referenced in Glencoe's **Internet Connections** are not under the control of Glencoe. Therefore, Glencoe can make no representation concerning the content of these sites. Extreme care has been taken to list only reputable links by using educational and government sites whenever possible. Internet searches have been used that return only sites that contained no content apparently intended for mature audiences. The following list of mathematics Internet sites may prove helpful. Useful search tools include http://webcrawler.com/, http://www.yahoo.com/search.html, and http://www.microsoft.com/search.

Internet Site	Comments
http://www.yahoo.com/science/mathematics	An ideal starting point that has links to other math sites.
http://www.tc.cornell.edu/Edu/MathSciGateway	Provides links to math and science resources; for grades 9-12.
http://forum.swarthmore.edu/dr.math/	Students in grades K-12 can ask Dr. Math their own questions.
http://www.enc.org/	This is the Eisenhower clearinghouse for K-2 math and science instructional materials.
http://forum.swarthmore.edu/mathmagic	K-12 telecommunications project in Texas that uses computers to increase problem-solving and communication skills.

Glencoe's Algebra ❶ Research Activities

Glencoe's **Algebra 1: Integration, Applications, and Connections**, as well as the entire Glencoe Mathematics Series, is the product of ongoing, classroom-oriented research that involves students, teachers, curriculum supervisors, administrators, parents, and college-level mathematics educators.

The programs that make up the Glencoe Mathematics Series are currently being used by millions of students and tens of thousands of teachers. The key reason that Glencoe Mathematics programs are so successful in the classroom is the fact that each Glencoe author team is a mix of practicing classroom teachers, curriculum supervisors, and college-level mathematics educators. Glencoe's balanced author teams help ensure that Glencoe Mathematics programs are both practical and progressive.

Prior to publication of a Glencoe program, typical research activities include:

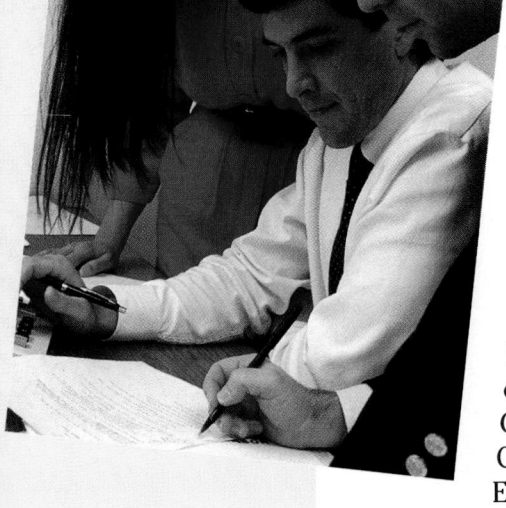

- a review of educational research and recommendations made by groups such as NCTM
- mail surveys of mathematics educators
- discussion groups involving mathematics teachers, department heads, and supervisors
- focus groups involving mathematics educators
- face-to-face interviews with mathematics educators
- telephone surveys of mathematics educators
- in-depth analyses of manuscript by a wide range of reviewers and consultants
- field tests in which students and teachers use pre-publication manuscript in the classroom

Feedback from teachers, curriculum supervisors, and even students who currently use Glencoe mathematics programs is also incorporated as Glencoe plans and publishes new and revised programs. For example, Classroom Vignettes, which are printed in the Teacher's Wraparound Editions, are one result of this feedback.

All of this research and information is used by Glencoe's authors and editors to publish the best instructional resources possible.

WHY IS ALGEBRA IMPORTANT?

Why do I need to study algebra? When am I ever going to have to use algebra in the real world?

Many people, not just algebra students, wonder why mathematics is important. *Algebra 1* is designed to answer those questions through **integration, applications,** and **connections.**

Geometry

Did you know that algebra and geometry are closely related? Topics from all branches of mathematics, like geometry and statistics, are integrated throughout the text.

You'll learn how to find the complement and supplement of an angle and how to find the measure of the third angle of a triangle. (Lesson 3–4, pages 162–164)

Nutrition

I can't believe that a double cheeseburger has that much fat! Real-world uses of mathematics are presented.

The number of grams of fat in a double cheeseburger is determined by solving an open sentence. (Lesson 1–5, page 32)

Seymour, Wisconsin
"The Home of the Hamburger"

Biology

What does biology have to do with mathematics? Mathematical topics are connected to other subjects that you study.

Punnett squares, which are models that show the possible ways that genes combine, are connected to squaring a binomial. (Lesson 9–8, page 542)

Authors

WILLIAM COLLINS teaches mathematics at James Lick High School in San Jose, California. He has served as the mathematics department chairperson at James Lick and Andrew Hill High Schools. Mr. Collins received his B.A. in mathematics and philosophy from Herbert H. Lehman College in Bronx, New York, and his M.S. in mathematics education from California State University, Hayward. Mr. Collins is a member of the Association of Supervision and Curriculum Development and the National Council of Teachers of Mathematics, and is active in several professional mathematics organizations at the state level. He is also currently serving on the Teacher Advisory Panel of the *Mathematics Teacher*.

"In this era of educational reform and change, it is good to be part of a program that will set the pace for others to follow. This program integrates the ideas of the NCTM Standards with real tools for the classroom, so that algebra teachers and students can expect success every day."

GILBERT CUEVAS is a professor of mathematics education at the University of Miami in Miami, Florida. Dr. Cuevas received his B.A. in mathematics and M.Ed. and Ph.D., both in educational research, from the University of Miami. He also holds a M.A.T. in mathematics from Tulane University. Dr. Cuevas is a member of many mathematics, science, and research associations on the local, state, and national levels and has been an author and editor of several National Council of Teachers of Mathematics (NCTM) publications. He is also a frequent speaker at NCTM conferences, particularly on the topics of equity and mathematics for all students.

ALAN G. FOSTER is a former mathematics teacher and department chairperson at Addison Trail High School in Addison, Illinois. He obtained his B.S. from Illinois State University and his M.A. in mathematics from the University of Illinois. Mr. Foster is a past president of the Illinois Council of Teachers of Mathematics (ICTM) and was a recipient of the ICTM's T.E. Rine Award for Excellence in the Teaching of Mathematics. He also was a recipient of the 1987 Presidential Award for Excellence in the Teaching of Mathematics for Illinois. Mr. Foster was the chairperson of the MATHCOUNTS question writing committee in 1990 and 1991. He frequently speaks and conducts workshops on the topic of cooperative learning.

BERCHIE GORDON is the mathematics/science coordinator for the Northwest Local School District in Cincinnati, Ohio. Dr. Gordon has taught mathematics at every level from junior high school to college. She received her B.S. in mathematics from Emory University in Atlanta, Georgia, her M.A.T. in education from Northwestern University in Evanston, Illinois, and her Ph.D. in curriculum and instruction at the University of Cincinnati. Dr. Gordon has developed and conducted numerous inservice workshops in mathematics and computer applications. She has also served as a consultant for IBM, and has traveled throughout the country making presentations on graphing calculators to teacher groups.

"Using this textbook, you will learn to think mathematically for the 21st century, solve a variety of problems based on real-world applications, and learn the appropriate use of technological devices so you can use them as tools for problem solving."

BEATRICE MOORE-HARRIS is an educational specialist at the Region IV Education Service Center in Houston, Texas. She is also the Southwest Regional Director of the Benjamin Banneker Association. Ms. Moore-Harris received her B.A. from Prairie View A&M University in Prairie View, Texas. She has also done graduate work there, at Texas Southern University in Houston, Texas, and at Tarleton State University in Stephenville, Texas. Ms. Moore-Harris is a consultant for the National Council of Teachers of Mathematics (NCTM) and serves on the Editorial Board of the NCTM's *Mathematics Teaching in the Middle School.*

"This program will bring algebra to life by engaging you in motivating, challenging, and worthwhile mathematical tasks that mirror real-life situations. Opportunities to use technology, manipulatives, language, and a variety of other tools are an integral part of this program, which allows all students full access to the algebra curriculum."

JAMES RATH has 30 years of classroom experience in teaching mathematics at every level of the high school curriculum. He is a former mathematics teacher and department chairperson at Darien High School in Darien, Connecticut. Mr. Rath earned his B.A. in philosophy from The Catholic University of America and his M.Ed. and M.A. in mathematics from Boston College. He has also been a Visiting Fellow in the mathematics department at Yale University in New Haven, Connecticut.

DORA SWART is a mathematics teacher at W.F. West High School in Chehalis, Washington. She received her B.A. in the mathematics education at Eastern Washington University in Cheney, Washington, and has done graduate work at Central Washington University in Ellensburg, Washington, and Seattle Pacific University in Seattle, Washington. Ms. Swart is a member of the National Council of Teachers of Mathematics, the Western Washington Mathematics Curriculum Leaders, and the Association of Supervision and Curriculum Development. She has developed and conducted numerous inservices and presentations to teachers in the Pacific Northwest.

"Glencoe's algebra series provides the best opportunity for you to learn algebra well. It explores mathematics through hands-on learning, technology, applications, and connections to the world around us. Mathematics can unlock the door to your success— this series is the key."

LESLIE J. WINTERS is the former secondary mathematics specialist for the Los Angeles Unified School District and is currently supervising student teachers at California State University, Northridge. Mr. Winters received bachelor's degrees in mathematics and secondary education from Pepperdine University and the University of Dayton, and master's degrees from the University of Southern California and Boston College. He is a past president of the California Mathematics Council-Southern Section, and received the 1983 Presidential Award for Excellence in the Teaching of Mathematics and the 1988 George Polya Award for being the Outstanding Mathematics Teacher in the state of California.

v

Consultants, Writers, and Reviewers

Consultants

Cindy J. Boyd
Mathematics Teacher
Abilene High School
Abilene, Texas

Gail Burrill
National Center for Research/
    Mathematical & Science Education
University of Wisconsin
Madison, Wisconsin

David Foster
Glencoe Author and Mathematics
    Consultant
Morgan Hill, California

Eva Gates
Independent Mathematics
    Consultant
Pearland, Texas

Joan Gell
Mathematics Department Chairman
Palos Verdes High School
Palos Verdes Estates, California

Daniel Marks
Consultant, Real-World Applications
Associate Professor of Mathematics
Auburn University at Montgomery
Montgomery, Alabama

Melissa McClure
Consultant, Tech Prep
Mathematics Consultant
Teaching for Tomorrow
Fort Worth, Texas

Dr. Luis Ortiz-Franco
Consultant, Diversity
Associate Professor of Mathematics
Chapman University
Orange, California

Writers

David Foster
Writer, Investigations
Glencoe Author and Mathematics
    Consultant
Morgan Hill, California

Jeri Nichols-Riffle
Writer, Graphing Technology
Assistant Professor
Teacher Education/Mathematics and
    Statistics
Wright State University
Dayton, Ohio

Reviewers

Susan J. Barr
Mathematics Department Chairperson
Dublin Coffman High School
Dublin, Ohio

William Biernbaum
Mathematics Teacher
Platteview High School
Springfield, Nebraska

John R. Blickenstaff
Mathematics Teacher
Martinsville High School
Martinsville, Indiana

Wayne Boggs
Mathematics Supervisor
Ephrata High School
Ephrata, Pennsylvania

Donald L. Boyd
Mathematics Teacher
South Charlotte Middle School
Charlotte, North Carolina

Judith B. Brigman
Mathematics Teacher
South Florence High School
Florence, South Carolina

William A. Brinkman
Mathematics Department Chairperson
Paynesville High School
Paynesville, Minnesota

Louis A. Bruno
Mathematics Department Chairperson
Somerset Area Senior High School
Somerset, Pennsylvania

Luajean Nipper Bryan
Mathematics Teacher
McMinn County High School
Athens, Tennessee

Kenneth Burd, Jr.
Mathematics Teacher
Hershey Senior High School
Hershey, Pennsylvania

Todd W. Busse
Mathematics/Science Teacher
Wenatchee High School
Wenatchee, Washington

Esther Corn
Mathematics Department Chairperson
Renton High School
Renton, Washington

Kimberly C. Cox
Mathematics Teacher
Stonewall Jackson High School
Manassas, Virginia

Janis Frantzen
Mathematics Department Chairperson
McCullough High School
The Woodlands, Texas

Vicki Fugleberg
Mathematics Teacher
May-Port CG School
Mayville, North Dakota

Sabine Goetz
Mathematics Teacher
Hewitt-Trussville Junior High
Trussville, Alabama

Dee Dee Hays
Mathematics Teacher
Henry Clay High School
Lexington, Kentucky

Ralph Jacques
Mathematics Department Chairperson
Biddeford High School
Biddeford, Maine

Dixie T. Johnson
Mathematics Department Chairperson
Paul Blazer High School
Ashland, Kentucky

Jane Housman Jones
Mathematics/Technology Teacher
du Pont Manual High School
Louisville, Kentucky

Rebecca A. Luna
Mathematics Teacher
Austin Junior High School
San Juan, Texas

Sandra A. Nagy
Mathematics Resource Teacher
Mesa Schools
Mesa, Arizona

Linda Palmer
Mathematics Teacher
Clark High School
Plano, Texas

Cynthia H. Ray
Mathematics Teacher
Ravenswood High School
Ravenswood, West Virginia

Hazel Russell
Mathematics Teacher
Riverside Middle School
Fort Worth, Texas

Mabel S. Sechrist
Mathematics Teacher
Mineral Springs Middle School
Winston-Salem, North Carolina

Diane Baish Stamy
Mathematics Chairperson
Big Spring School District
Newville, Pennsylvania

Robert L. Thompson
Mathematics Teacher
Raines High School
Jacksonville, Florida

John J. Turechek
Mathematics Department Chairperson
Bunnell High School
Stratford, Connecticut

Jan S. Wessell
Mathematics Supervisor
New Hanover County Schools
Wilmington, North Carolina

Kathy Zwanzig
Middle School Algebra Resource
    Teacher
Jefferson County Public Schools
Louisville, Kentucky

Bernita Jones
Mathematics Department Chairperson
Turner High School
Kansas City, Kansas

Nancy W. Kinard
Mathematics Teacher
Palm Beach Gardens High School
Palm Beach Gardens, Florida

Frances J. MacDonald
Area Mathematics Coordinator, K–12
Pompton Lakes School District
Pompton Lakes, New Jersey

Roger O'Brien
Mathematics Supervisor
Polk County Schools
Bartow, Florida

Deborah Patonai Phillips
Mathematics Department Head
St. Vincent-St. Mary High School
Akron, Ohio

Carrol W. Rich
Mathematics Teacher
High Point Central
High Point, North Carolina

Judith Ryder
Mathematics Teacher
South Charlotte Middle School
Charlotte, North Carolina

Richard K. Sherry
Mathematics Department Chairman
Roosevelt Junior High School
Altoona, Pennsylvania

Carolyn Stanley
Mathematics Department Chairperson
Tidewater Academy
Wakefield, Virginia

Linda K. Tinga
Mathematics Teacher
Laney High School
Wilmington, North Carolina

Tonya Dessert Urbatsch
Mathematics Curriculum Coordinator
Davenport Community Schools
Davenport, Iowa

Ted Wiecek
Mathematics Department Chairperson
Wells Community Academy
Chicago, Illinois

Cheryl M. Jones
Mathematics Department Coordinator
Winnetonka High School
Kansas City, Missouri

Margaret M. Kirk
Mathematics Teacher
Welborn Middle School
High Point, North Carolina

George R. Murphy
Mathematics Department Chairperson
Pine-Richland High School
Gibsonia, Pennsylvania

Barbara S. Owens
Mathematics Teacher
Davidson Middle School
Davidson, North Carolina

Kristine Powell
Mathematics Teacher
Lehi Junior High School
Lehi, Utah

Mary Richards
Mathematics Teacher
Clyde A. Erwin High School
Asheville, North Carolina

Donna Schmitt
Reviewer, Technology
Mathematics and Computer Science
    Teacher
Dubuque Senior High School
Dubuque, Iowa

Catherine C. Sperando
Mathematics Teacher
Shades Valley High School
Birmingham, Alabama

Richard P. Strausz
Math/Computer Coordinator
Farmington Schools
Farmington, Michigan

Kimberly Troise
Mathematics Teacher
A. C. Reynolds High School
Asheville, North Carolina

Anthony J. Weisse
Mathematics Teacher
Central High School
LaCrosse, Wisconsin

Rosalyn Zeid
Mathematics Director
Union High School
Union, New Jersey

Table of Contents

CHAPTER **2** **Exploring Rational Numbers** **70**

CHAPTER **3** **Solving Linear Equations** **140**

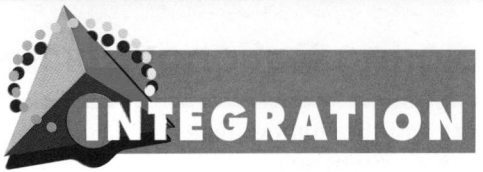

INTEGRATION

Content Integration

What does geometry have to do with algebra? Believe it or not, you can study most math topics from more than one point of view. Here are some examples.

Music Sales by Categories

Sales (millions of dollars)

— = Rock
— = Rap
— = Country

Year

Source: The Recording Industry Association of America

◀ **Probability** You'll model the roll of two dice with relations and line plots. (Lesson 5–2, Exercise 45)

▲ **Problem Solving** You'll interpret graphs and learn how to sketch graphs of real-world situations. (Lesson 1–9, page 62)

LOOK BACK

You can refer to Lesson 1-7 for information on the distributive property.

Look Back features refer you to skills and concepts that have been taught earlier in the book.

Source: Lesson 9–6, page 529

Discrete Mathematics ▶ You'll investigate patterns and sequences and use them to solve a variety of problems. (Lesson 1–2, pages 12–18)

Statistics You'll learn how to ▶ display data about the speeds of the fastest animals on a line plot. (Lesson 2–2, page 80)

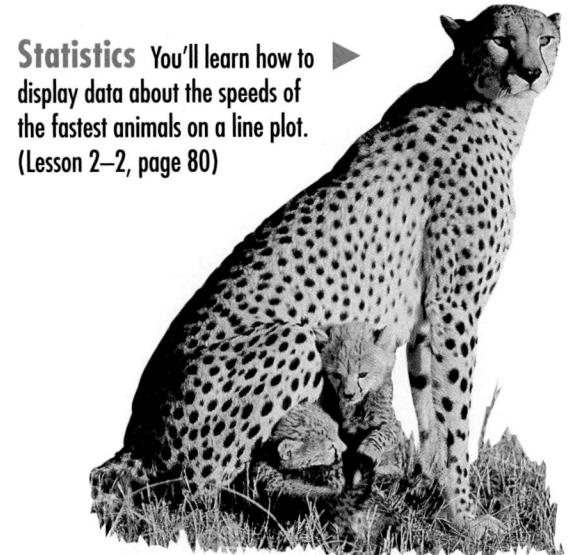

Geometry You'll use your algebra skills to find the missing measures of similar triangles. (Lesson 4–2, pages 201–205)

APPLICATIONS

Real-Life Applications

Have you ever wondered if you'll ever actually use math? Every lesson in this book is designed to help show you where and when math is used in the real world. Since you'll explore many interesting topics, we believe you'll discover that math is relevant and exciting. Here are some examples.

Top Five List, FYI, and **Fabulous Firsts** contain interesting facts that enhance the applications.

Selling Prices of Paintings

1. *Portrait du Dr. Gachet* by van Gogh, $75,000,000
2. *Au Moulin de la Galette* by Renoir, $71,000,000
3. *Les Noces de Pierrette* by Picasso, $51,700,000
4. *Irises* by van Gogh, $49,000,000
5. *Yo Picasso* by Picasso, $43,500,000

Source: Lesson 9–4, page 514

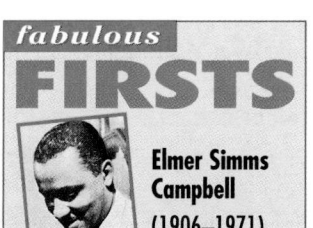

Elmer Simms Campbell

(1906–1971)

Elmer Simms Campbell was the first African-American cartoonist to work for national publications. He contributed cartoons and other art work to *Esquire, Cosmopolitan,* and *Redbook,* as well as syndicated features in 145 newspapers.

Source: Lesson 8–3, page 469

Football You'll see how punting a football is related to angle measure. (Lesson 3–4, page 162)

Carpentry You'll study an industrial technology application that involves computing with rational numbers. (Lesson 2–7, Exercise 47)

F Y I

In September 1995, blue M&M's® completely replaced the tan ones. The ratios of colors for plain M&M's are as follows.

brown	30%
red	20%
yellow	20%
orange	10%
green	10%
blue	10%

Source: Lesson 2–2, page 79

Mathematics and SOCIETY

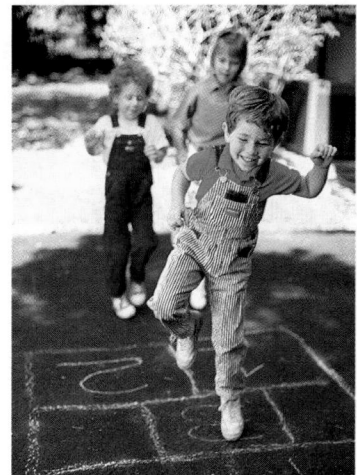

◀ Recreation Games from several world cultures that are similar to hopscotch will illustrate multiplying a polynomial by a monomial. (Lesson 9–6, page 529)

air shaft

grand gallery

King's chamber

ascending corridor

Queen's chamber

entrance

escape shaft

unfinished chamber

descending corridor

▲ World Cultures The Great Pyramid of Khufu is the setting for an application involving systems of linear equations. (Lesson 8–3, Exercise 42)

Space Science You'll discuss the extreme temperatures encountered during a walk in space as you learn about absolute value. (Lesson 2–3, page 85)

Teen Talk Barbies

What do Barbie dolls have to do with mathematics? Actual reprinted articles illustrate how mathematics is a part of our society. (Lesson 3–7, page 183)

CHAPTER **8** # Solving Systems of Linear Equations and Inequalities **450**

Long-Term Investigation
Ready, Set, Drop! **448–449**
• Working on the Investigation **461**
Technology
Explorations **471, 484**
Modeling Mathematics
Activities **464**
Mathematics and Society
High-Tech Checkout Lanes **481**

CHAPTER **9** # Exploring Polynomials **494**

Long-Term Investigation
Ready, Set, Drop!
• Working on the Investigation **519, 541**
• Closing the Investigation **548**
Technology
Explorations **502, 507**
Modeling Mathematics
Activities **545**
Mathematics and Society
Measurements Great and Small **512**

xv

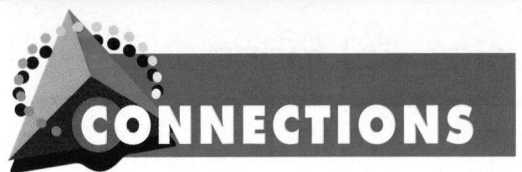

Interdisciplinary Connections

Did you realize that mathematics is used in biology? in history? in geography? Yes, it may be hard to believe, but mathematics is frequently connected to other subjects that you are studying.

GLOBAL CONNECTIONS

About 15,000 years ago, an important application of the lever appeared as hunters used an *atlatl*, the Aztec word for spear-thrower. This simple device was a handle that provided extra mechanical advantage by adding length to the hunter's arm. This enabled the hunter to use less motion to get greater results when hunting.

Source: Lesson 7–2, page 392

◀ **Global Connections** features introduce you to a variety of world cultures.

Health You'll write inequalities to model target heart rates and exercise. (Lesson 7–8, Exercise 47)

Number of Manatees Found Dead

▲ **Biology** You'll study the mortality rates of Florida's manatees, an endangered species. (Lesson 5–2, page 262)

CAREER CHOICES

A **zoologist** studies animals, their origin, life process, behavior, and diseases. They are usually identified by the animal group in which they specialize, for example herpetology (reptiles), ornithology (birds), and ichthyology (fish).

Zoology requires a minimum of a bachelor's degree. Many college positions in this field require a Ph.D. degree.

For more information, see:

Occupational Outlook Handbook, 1994–95, U.S. Department of Labor.

Source: Lesson 5–2, page 264

◀ **Career Choices** features include information on interesting careers.

Geography In 1994, the population of New York was surpassed by Texas. This situation is modeled as a system of linear equations. (Lesson 8–2, page 462)

▲ **Art** A Mondrian painting entitled *Composition with Red, Yellow, and Blue* is used to model polynomials. (Lesson 9–3, page 514)

Math Journal exercises give you the opportunity to assess yourself and write about your understanding of key math concepts. (Lesson 7–1, Exercise 6)

MATH JOURNAL

6. Sometimes statements we make can be translated into inequalities. For example, *In some states, you have to be at least 16 years old to have a driver's license* can be expressed as $a \geq 16$, and *Tomás cannot lift more than 72 pounds* can be translated into $w \leq 72$. Following these examples, write three statements that deal with your everyday life. Then translate each into a corresponding inequality.

CHAPTER **12** Exploring Rational Expressions
and Equations   **658**

Long-Term Investigation
A Growing Concern **656–657**
• Working on the Investigation **665, 695**
Technology   Explorations **662, 678**
Modeling Mathematics   Activities **676**
Mathematics and Society   Reading the Labels **689**

CHAPTER **13** Exploring Radical Expressions
and Equations   **710**

Long-Term Investigation
A Growing Concern
• Working on the Investigation **718, 741**
• Closing the Investigation **748**
Technology   Explorations **720, 733**
Mathematics and Society   Nonlinear Math **736**

xvii

TECHNOLOGY

Do you know how to use computers and graphing calculators? If you do, you'll have a much better chance of being successful in today's high-tech society and workplace.

GRAPHING CALCULATORS

There are several ways in which graphing calculators are integrated.

- **Introduction to Graphing Calculators**   On pages 2–3, you'll get acquainted with the basic features and functions of a graphing calculator.

- **Graphing Technology Lessons**   In Lesson 5–2A, you'll learn how to plot points using a graphing calculator.

- **Graphing Calculator Explorations**   You'll learn how to use a graphing calculator to find the mean and median in Lesson 3–7.

- **Graphing Calculator Programs**   In Lesson 3–5, Exercise 39 includes a graphing calculator program that can be used to solve equations of the form $ax + b = cx + d$.

- **Graphing Calculator Exercises**   Many exercises are designed to be solved using a graphing calculator. For example, see Exercises 40–42 in Lesson 7–8.

COMPUTER SOFTWARE

- **Spreadsheets**   On page 297 of Lesson 5–6, a spreadsheet is used to help write an equation that models a relation.

- **BASIC Programs**   The program on page 8 of Lesson 1–1 is designed to evaluate expressions.

- **Graphing Software**   The graphing software Exploration on page 469 in Lesson 8–3 involves graphing and solving systems of linear equations.

Technology Tips, such as this one on page 216 of Lesson 4–4, are designed to help you make more efficient use of technology through practical hints and suggestions.

TECHNOLOGY *Tip*

Most calculators have a key labeled %. To find 28% of 58.4 on a TI-34, enter:

28 2nd % ×

58.4 =   16.352.

SYMBOLS AND MEASURES

Symbols

$=$	is equal to		π	pi
$\neq$	is not equal to		$\%$	percent
$>$	is greater than		$0.1\overline{2}$	decimal 0.12222...
$<$	is less than		$^\circ$	degree
$\geq$	is greater than or equal to		$f(x)$	f of x, the value of f at x
$\leq$	is less than or equal to		(a, b)	ordered pair a, b
$\approx$	is approximately equal to		$\overline{AB}$	line segment $\overline{AB}$
$\sim$	is similar to		$\overset{\frown}{AB}$	arc AB
$\times$ or $\cdot$	times		$\overrightarrow{AB}$	ray AB
$\div$	divided by		$\overleftrightarrow{AB}$	line AB
$-$	negative or minus		AB	measure of $\overline{AB}$
$+$	positive or plus		$\angle$	angle
$\pm$	positive or negative		Δ	triangle
$-a$	opposite or additive inverse of a		$\cos A$	cosine of A
$|a|$	absolute value of a		$\sin A$	sine of A
$a \overset{?}{=} b$	Does a equal b?		$\tan A$	tangent of A
$a : b$	ratio of a to b		$(\ )$	parentheses; *also* ordered pairs
$\sqrt{a}$	square root of a		$[\ ]$	brackets; *also* matrices
$P(A)$	probability of A		$\{\ \}$	braces; *also* sets
O	origin		$\varnothing$	empty set

Measures

mm	millimeter		in.	inch
cm	centimeter		ft	foot
m	meter		yd	yard
km	kilometer		mi	mile
g	gram		in^2 or sq in.	square inch
kg	kilogram		s	second
mL	milliliter		min	minute
L	liter		h	hour

GETTING ACQUAINTED WITH THE GRAPHING CALCULATOR

What is it?
What does it do?
How is it going to help me learn math?

These are just a few of the questions many students ask themselves when they first see a graphing calculator. Some students may think, "Oh, no! Do we *have* to use one?", while others may think, "All right! We get to use these neat calculators!" There are as many thoughts and feelings about graphing calculators as there are students, but one thing is for sure: a graphing calculator *can* help you learn mathematics.

So what is a graphing calculator? Very simply, it is a calculator that draws graphs. This means that it will do all of the things that a "regular" calculator will do, *plus* it will draw graphs of simple or very complex equations. In algebra, this capability is nice to have because the graphs of some complex equations take a lot of time to sketch by hand. Some are even considered impossible to draw by hand. This is where a graphing calculator can be very useful.

But a graphing calculator can do more than just calculate and draw graphs. You can program it, work with matrices, and make statistical graphs and computations, just to name a few things. If you need to generate random numbers, you can do that on the graphing calculator. If you need to find the absolute value of numbers, you can do that, too. It's really a very powerful tool—so powerful that it is often called a pocket computer. But don't let that intimidate you. A graphing calculator can save you time and make doing mathematics easier.

As you may have noticed, graphing calculators have some keys that other calculators do not. The Texas Instruments TI-82 will be used throughout this text. The keys located on the bottom half of the calculator are probably familiar to you as they are the keys found on basic scientific calculators. The keys located just below the screen are the graphing keys. You will also notice the up, down, left, and right arrow keys. These allow you to move the cursor around on the screen and to "trace" graphs that have been plotted. The other keys located on the top half of the calculator access the special features such as statistical and matrix computations.

There are some keystrokes that can save you time when using the graphing calculator. A few of them are listed below.

- Any light blue commands written above the calculator keys are accessed with the 2nd key, which is also blue. Similarly, any gray characters above the keys are accessed with the ALPHA key, which is also gray.

- 2nd ENTRY copies the previous calculation so you can edit and use it again.

- Pressing ON while the calculator is graphing stops the calculator from completing the graph.

- 2nd QUIT will return you to the home (or text) screen.

- 2nd A-LOCK locks the ALPHA key, which is like pressing "shift lock" or "caps locks" on a typewriter or computer. The result is that all caps will be typed and you do not have to hold the shift key down. (This is handy for programming.)

- 2nd OFF turns the calculator off.

Some commonly used mathematical functions are shown in the table below. As with any scientific calculator, the graphing calculator observes the order of operations.

Mathematical Operation	Examples	Keys	Display
evaluate expressions	Find 2 + 5.	2 + 5 ENTER	2+5 7
exponents	Find 3^5.	3 ⌃ 5 ENTER	3^5 243
multiplication	Evaluate $3(9.1 + 0.8)$.	3 × ⌈ 9.1 + .8 ⌉ ENTER	3(9.1+.8) 29.7
roots	Find $\sqrt{14}$.	2nd √ 14 ENTER	√ 14 3.741657387
opposites	Enter -3.	(−) 3	−3

Graphing on the TI–82

Before graphing, we must instruct the calculator how to set up the axes in the coordinate plane. To do this, we define a **viewing window**. The viewing window for a graph is the portion of the coordinate grid that is displayed on the **graphics screen** of the calculator. The viewing window is written as [left, right] by [bottom, top] or [Xmin, Xmax] by [Ymin, Ymax]. A viewing window of [−10, 10] by [−10, 10] is called the **standard viewing window** and is a good viewing window to start with to graph an equation. The standard viewing window can be easily obtained by pressing ZOOM 6. Try this. Move the arrow keys around and observe what happens. You are seeing a portion of the coordinate plane that includes the region from -10 to 10 on the x-axis and from -10 to 10 on the y-axis. Move the cursor, and you can see the coordinates of the points for the current position of the cursor.

Any viewing window can be set manually by pressing the WINDOW key. The window screen will appear and display the current settings for your viewing window. First press ENTER . Then, using the arrow and ENTER keys, move the cursor to edit the window settings. Xscl and Yscl refer to the x-scale and y-scale. This is the number of tick marks placed on the x- and y-axes. Xscl=1 means that there will be a tick mark for every unit of one along the x-axis. The standard viewing window would appear as follows.

$$\text{Xmin} = -10$$
$$\text{Xmax} = 10$$
$$\text{Xscl} = 1$$
$$\text{Ymin} = -10$$
$$\text{Ymax} = 10$$
$$\text{Yscl} = 1$$

Graphing equations is as simple as defining a viewing window, entering the equations in the Y= list, and pressing GRAPH . It is often important to view enough of a graph so you can see all of the important characteristics of the graph and understand its behavior. The term **complete graph** refers to a graph that shows all of the important characteristics such as intercepts or maximum and minimum values.

Example: Graph $y = x - 14$ in the standard viewing window.

Enter:

We see only a small portion of the graph. Why?

The graph of $y = x - 14$ is a line that crosses the x-axis at 14 and the y-axis at -14. The important features of the graph are plotted off the screen and so the graph is not complete. A better viewing window for this graph would be $[-20, 20]$ by $[-20, 20]$, which includes both intercepts. This is considered a complete graph.

Programming on the TI–82

The TI–82 has programming features that allow us to write and execute a series of commands to perform tasks that may be too complex or cumbersome to perform otherwise. Each program is given a name. Commands begin with a colon (:), followed by an expression or an instruction. Most of the features of the calculator are accessible from program mode.

When you press PRGM , you see three menus: EXEC, EDIT, and NEW. EXEC allows you to execute a stored program by selecting the name of the program from the menu. EDIT allows you to edit or change an existing program and NEW allows you to create a new program. To break during program execution, press ON . The following example illustrates how to create and execute a new program that stores an expression as Y and evaluates the expression for a designated value of X.

1. Enter PRGM ▶ ▶ ENTER to create a new program.

2. Type EVAL ENTER to name the program. (Make sure that the caps lock is on.) You are now in the program editor, which allows you to enter commands. The colon (:) in the first column of the line indicates that this is the beginning of the command line.

3. The first command lines will ask the user to designate a value for x. Enter PRGM ▶ 3 2nd A-LOCK " ENTER THE VALUE FOR X " ALPHA ENTER PRGM ▶ 1 X,T,θ ENTER .

4. The expression to be evaluated for the value of x is $x - 7$. To store the expression as Y, enter X,T,θ $-$ 7 STO▶ ALPHA Y ENTER .

5. Finally, we want to display the value for the expression. Enter PRGM ▶ 3 ALPHA Y ENTER .

6. Now press 2nd QUIT to return to the home screen.

7. To execute the program, press PRGM . Then press the down arrow to locate the program name and press ENTER , or press the number or letter next to the program name. The program asks for a value for x. You will input any value for which the expression is defined and press ENTER . To immediately re-execute the program, simply press ENTER when Done appears on the screen.

While a graphing calculator cannot do everything, it can make some things easier. To prepare for whatever lies ahead, you should try to learn as much as you can. The future will definitely involve technology, and using a graphing calculator is a good start toward becoming familiar with technology. Who knows? Maybe one day you will be designing the next satellite, building the next skyscraper, or helping students learn mathematics with the aid of a graphing calculator!

Exploring Expressions, Equations, and Functions

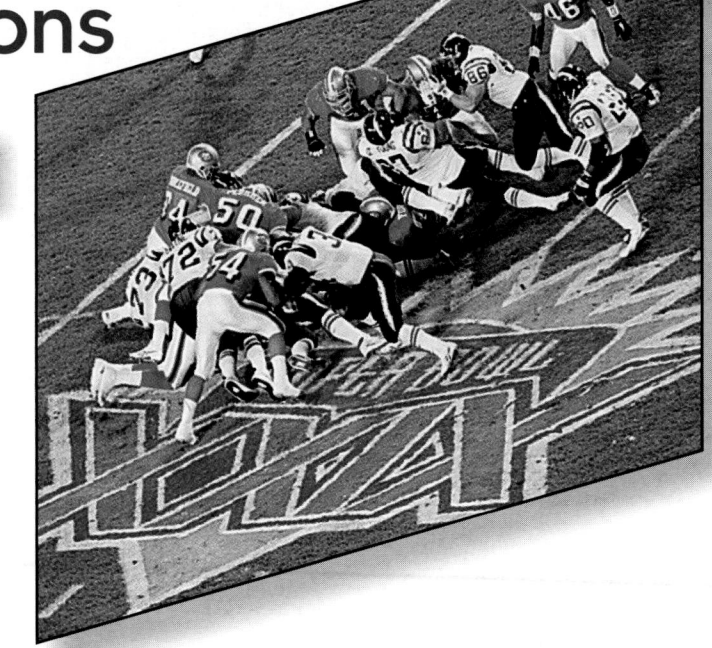

PREVIEWING THE CHAPTER

This chapter provides the necessary introduction and practice to prepare students for the successful study of algebra. Students apply the concept of variables as they write and simplify algebraic expressions. Patterns and sequences are introduced to enhance problem-solving skills, and stem-and-leaf plots are integrated into the chapter to provide students with another tool for data representation. Students connect relevant number rules and properties to the organization and structure of algebra. The distributive property is emphasized because of its vital role in performing algebraic functions. The chapter concludes with students using graphs to interpret and organize information.

Lesson (Pages)	Lesson Objectives	NCTM Standards	State/Local Objectives
1-1 (6–11)	Translate verbal expressions into mathematical expressions and vice versa.	1–5	
1-2 (12–18)	Solve problems by looking for a pattern.	1–5	
1-3 (19–24)	Use the order of operations to evaluate real number expressions.	1–5, 8	
1-4 (25–31)	Display and interpret data on a stem-and-leaf plot.	1–5, 10	
1-5 (32–36)	Solve open sentences by performing arithmetic operations.	1–5	
1-6 (37–43)	Recognize and use the properties of identity and equality. Determine the multiplicative inverse of a number.	1–5	
1-7A (44)	Model the distributive property.	1–5	
1-7 (45–50)	Use the distributive property to simplify expressions.	1–5, 8	
1-8 (51–55)	Recognize and use the commutative and associative properties when simplifying expressions.	1–5	
1-9 (56–62)	Interpret graphs in real-world settings. Sketch graphs for given functions.	1–6, 8	

ORGANIZING THE CHAPTER

You may want to refer to the **Course Planning Calendar** on page T12 for detailed information on pacing.
PACING: Standard—15 days; **Honors**—14 days; **Block**—7 days; **Two Years**—22 days

LESSON PLANNING CHART

| Lesson (Pages) | Materials/ Manipulatives | Extra Practice (Student Edition) | BLACKLINE MASTERS | | | | | | | | | Real-World Applications | Interactive Mathematics Tools Software | Teaching Transparencies |
			Study Guide	Practice	Enrichment	Assessment and Evaluation	Modeling Mathematics	Multicultural Activity	Tech Prep Applications	Graphing Calculator	Science and Math Lab Manual			
1-1 (6–11)	graphing calculator	p. 756	p. 1	p. 1	p. 1								1-1	1-1A 1-1B
1-2 (12–18)	string scissors calculator	p. 756	p. 2	p. 2	p. 2									1-2A 1-2B
1-3 (19–24)	graphing calculator	p. 756	p. 3	p. 3	p. 3	p. 16	p. 72		p. 1				1-3	1-3A 1-3B
1-4 (25–31)		p. 757	p. 4	p. 4	p. 4									1-4A 1-4B
1-5 (32–36)		p. 757	p. 5	p. 5	p. 5	pp. 15, 16		p. 1			pp. 1–4	1		1-5A 1-5B
1-6 (37–43)		p. 757	p. 6	p. 6	p. 6									1-6A 1-6B
1-7A (44)	algebra tiles* product mat*						p. 19						1-7A	
1-7 (45–50)	algebra tiles*	p. 758	p. 7	p. 7	p. 7	p. 17			p. 2					1-7A 1-7B
1-8 (51–55)		p. 758	p. 8	p. 8	p. 8									1-8A 1-8B
1-9 (56–62)		p. 758	p. 9	p. 9	p. 9	p. 17			p. 2	p. 1		2		1-9A 1-9B
Study Guide/ Assessment (63–67)						pp. 1–14, 18–20								

*Included in Glencoe's Student Manipulative Kit and Overhead Manipulative Resources.

ORGANIZING THE CHAPTER

OTHER CHAPTER RESOURCES

Student Edition
Chapter Opener, pp. 4–5
Mathematics and Society, p. 11

Teacher's Classroom Resources
Investigations and Projects Masters, pp. 25–28
 Algebra and Geometry Overhead Manipulative Resources, p. 1

Technology
Test and Review Software (IBM and Macintosh)
CD-ROM Interactions (Windows and Macintosh)

Professional Publications
Block Scheduling Booklet
Glencoe Mathematics Professional Series

OUTSIDE RESOURCES

Books/Periodicals
A Gebra Named Al, Free Spirit Publishing
Assessment Standards for School Mathematics, NCTM
Curriculum and Evaluation Standards for School Mathematics, NCTM
Everybody Counts: A Report to the Nation on the Future of Mathematics Education, National Academy Press
Mathematics: The Science of Patterns, W. H. Freeman and Co.

 Software
Algebraic Patterns, Sunburst/Wings for Learning
Algebra Concepts, Volume 1, Ventura Educational Systems

Videos/CD-ROMs
Getting Acquainted with Algebra/Order of Operations, The Wisconsin Foundation for Vocational, Technical, and Adult Education, Inc.
How to Take the Mystery Out of Algebra, Learning Forum

ASSESSMENT RESOURCES

Student Edition
Math Journal, pp. 28, 34, 41, 53, 59
Mixed Review, pp. 18, 24, 31, 36, 43, 50, 55, 62
Self Test, p. 36
Chapter Highlights, p. 63
Chapter Study Guide and Assessment, pp. 64–66
Alternative Assessment, p. 67
 Portfolio, p. 67

Teacher's Wraparound Edition
5-Minute Check, pp. 6, 12, 19, 25, 32, 37, 45, 51, 56
Check for Understanding, pp. 9, 15, 22, 28, 34, 40, 48, 52, 59
Closing Activity, pp. 11, 17, 23, 30, 36, 43, 50, 55, 61
Cooperative Learning, pp. 27, 34

Assessment and Evaluation Masters
Multiple Choice Tests, Forms 1A (Honors), 1B (Average), 1C (Basic), pp. 1–6
Free Response Tests, Forms 2A (Honors), 2B (Average), 2C (Basic), pp. 7–12
Calculator-Based Test, p. 13
Performance Assessment, p. 14
Mid-Chapter Test, p. 15
Quizzes A–D, pp. 16–17
Standardized Test Practice, p. 18
Cumulative Review, pp. 19–20

ENHANCING THE CHAPTER

Examples of some of the materials for enhancing Chapter 1 are shown below.

DIVERSITY

Multicultural Activity Masters, pp. 1, 2

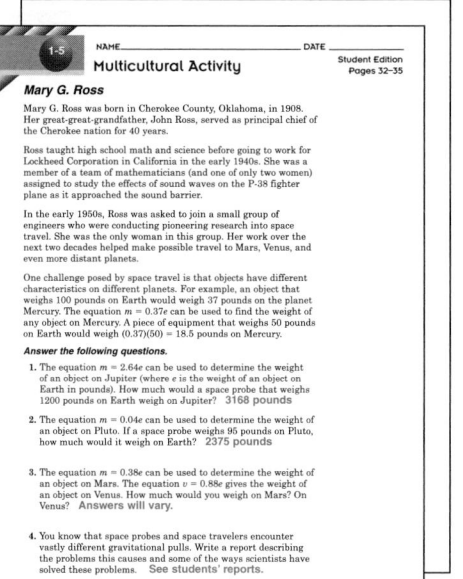

Multicultural Activity

1-5
NAME _____ DATE _____
Student Edition
Pages 32–35

Mary G. Ross

Mary G. Ross was born in Cherokee County, Oklahoma, in 1908. Her great-great-grandfather, John Ross, served as principal chief of the Cherokee nation for 40 years.

Ross taught high school math and science before going to work for Lockheed Corporation in California in the early 1940s. She was a member of a team of mathematicians (and one of only two women) assigned to study the effects of sound waves on the P-38 fighter plane as it approached the sound barrier.

In the early 1950s, Ross was asked to join a small group of engineers who were conducting pioneering research into space travel. She was the only woman in this group. Her work over the next two decades helped make possible travel to Mars, Venus, and even more distant planets.

One challenge posed by space travel is that objects have different characteristics on different planets. For example, an object that weighs 100 pounds on Earth would weigh 37 pounds on the planet Mercury. The equation $m = 0.37e$ can be used to find the weight of any object on Mercury. A piece of equipment that weighs 50 pounds on Earth would weigh (0.37)(50) = 18.5 pounds on Mercury.

Answer the following questions.

1. The equation $m = 2.64e$ can be used to determine the weight of an object on Jupiter (where e is the weight of an object on Earth in pounds). How much would a space probe that weighs 1200 pounds on Earth weigh on Jupiter? 3168 pounds

2. The equation $m = 0.04e$ can be used to determine the weight of an object on Pluto. If a space probe weighs 95 pounds on Pluto, how much would it weigh on Earth? 2375 pounds

3. The equation $m = 0.38e$ can be used to determine the weight of an object on Mars. The equation $v = 0.88e$ gives the weight of an object on Venus. How much would you weigh on Mars? On Venus? Answers will vary.

4. You know that space probes and space travelers encounter vastly different gravitational pulls. Write a report describing the problems this causes and some of the ways scientists have solved these problems. See students' reports.

APPLICATIONS

Real-World Applications, 1, 2

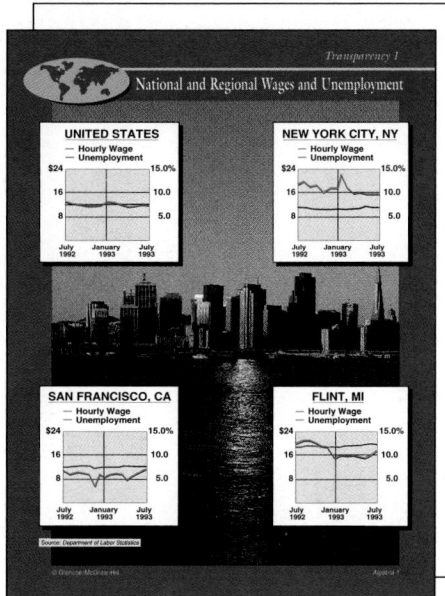

Transparency 1

National and Regional Wages and Unemployment

TECHNOLOGY

Graphing Calculator Masters, p. 1

Graphing Calculator Activity

1-9
NAME _____ DATE _____
Student Edition
Pages 56–62

Make a Line Graph

The value of collectibles such as antiques and works of art increases over time. This is called *appreciation*. The table shows what happened to the value of a painting in the first five years after it was purchased for $550.

The procedure for graphing this information on a TI-82 graphing calculator is shown below.

Number of Years Owned	Value
0	$550
1	600
2	650
3	700
4	770
5	850

a. Press STAT 1. Under L₁, enter the numbers from the left-hand column of the table. Press ENTER to move from one line to the next.

b. Press ▶ to move the cursor under L₂. Then enter the values from the right-hand side of the table.

c. Press 2nd Y= 1. The screen should look like this:

If necessary, use the direction keys and ENTER to change selections.

d. Press ZOOM 9.

Use a graphing calculator for the following exercises.

1. Press TRACE. Then describe what happens when you press the ▶ key. The cursor moves from point (0, 0) to point (1, 600).

2. The value of the painting increased every year. However, the increase was greater in some years than in others. How is this shown on the graph? A greater increase is shown by a steeper line segment.

3. Depreciation occurs when the value of an item decreases over time. The table shows how the value of a certain car depreciates. Use steps a–d above to graph this information on a TI-82 graphing calculator. Then draw a sketch of your graph.

Number of Years Owned	Value
0	$17,000
1	13,500
2	11,000
3	8,000
4	6,750
5	4,500

4. Between what two years did the car depreciate the least? How is this indicated on the graph? between year 3 and year 4; graph is flatter between these two points

TECH PREP

Tech Prep Applications Masters, pp. 1, 2

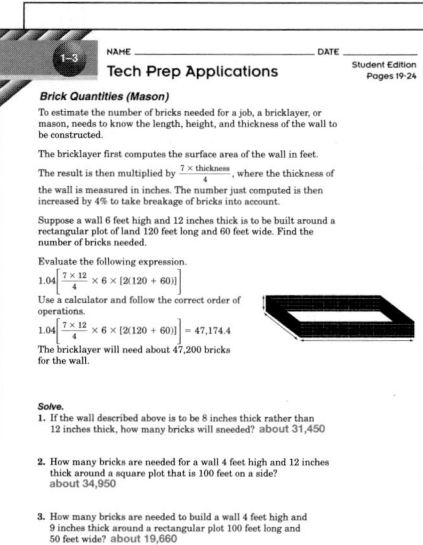

Tech Prep Applications

1-3
NAME _____ DATE _____
Student Edition
Pages 19–24

Brick Quantities (Mason)

To estimate the number of bricks needed for a job, a bricklayer, or mason, needs to know the length, height, and thickness of the wall to be constructed.

The bricklayer first computes the surface area of the wall in feet.

The result is then multiplied by $\frac{7 \times \text{thickness}}{4}$, where the thickness of the wall is measured in inches. The number just computed is then increased by 4% to take breakage of bricks into account.

Suppose a wall 6 feet high and 12 inches thick is to be built around a rectangular plot of land 120 feet long and 60 feet wide. Find the number of bricks needed.

Evaluate the following expression.

$$1.04\left[\frac{7 \times 12}{4} \times 6 \times [2(120 + 60)]\right]$$

Use a calculator and follow the correct order of operations.

$$1.04\left[\frac{7 \times 12}{4} \times 6 \times [2(120 + 60)]\right] = 47,174.4$$

The bricklayer will need about 47,200 bricks for the wall.

Solve.

1. If the wall described above is to be 8 inches thick rather than 12 inches thick, how many bricks will sneeded? about 31,450

2. How many bricks are needed for a wall 4 feet high and 12 inches thick around a square plot that is 100 feet on a side? about 34,950

3. How many bricks are needed to build a wall 4 feet high and 9 inches thick around a rectangular plot 100 feet long and 50 feet wide? about 19,660

4. If the height of a wall to be built is doubled, will the number of bricks needed to build it double? yes

CONNECTIONS

Science and Math Lab Manual, pp. 1–4

Science and Math Lab 1

1
NAME _____ DATE _____
Student Edition
Pages 32–35

Using Physical Properties

Introduction

Suppose you've been collecting pennies in a huge milk jug. You're curious to know what your collection is worth, but you don't have the time—or the energy—to count each coin. How can algebra and data about physical properties save you time?

Objectives

- Measure thickness and mass of pennies using a metric ruler and a balance.

- Compare relationships among thickness, mass, and number of pennies.

- Write verbal descriptions and algebraic equations for calculating mass, length, or number of pennies given two other measurements.

Materials

- pennies (10)
- balance
- metric ruler
- roll of pennies

Figure 1

Procedure

A. Measuring Thickness

1. Use the metric ruler to find the thicknesses of 1 penny, 2 pennies, 3 pennies, 4 pennies, 6 pennies, 8 pennies, and 10 pennies. See Figure 1. Measure each thickness to the nearest 0.5 mm. Record the thicknesses in Table 1.

2. Record in the table the number of pennies in the roll. Measure the length of the roll. Record that value in the table.

B. Measuring Mass

3. Use the balance to determine the mass of 1 penny, 2 pennies, 3 pennies, 4 pennies, 6 pennies, 8 pennies, and 10 pennies to the nearest 0.1 g. Record the masses in Table 1.

4. Use the balance to find the mass of the roll of coins. Record the values in Table 1.

5. Find the average thickness of one penny. Record the value in Table 1.

PROBLEM SOLVING

Problem of the Week Cards, 1, 2, 3

Puzzling Pentomino Paths

Problem-of-the-Week
Card 1

The Problem

A pentomino is an arrangement of 5 squares placed together so that each square shares at least one side with another square. In this problem, pretend that ants always crawl to the right and down as they go from point A to point X. If each ant must take a different path, determine which pentomino shape below will permit the most ants to arrive at point X.

Strategies and Hints

1. Be certain that you understand the conditions of the problem. Which two directions can ants travel? Which directions can they not travel?

2. Try solving the problem using "easier" pentominoes such as these.

3. Simplify the original problem by having the ants always begin by following paths for which the beginning two moves are down.

4. Number the corners as you go from 1 to the end with the number of ways it is possible to get to each corner. Many of the outside points will have the number 1.

CHAPTER

1

Exploring Expressions, Equations, and Functions

Objectives

In this chapter, you will:
- translate verbal expressions into mathematical expressions,
- solve problems by looking for a pattern,
- use mathematical properties to evaluate expressions,
- solve open sentences, and
- use and interpret stem-and-leaf plots, tables, graphs, and functions.

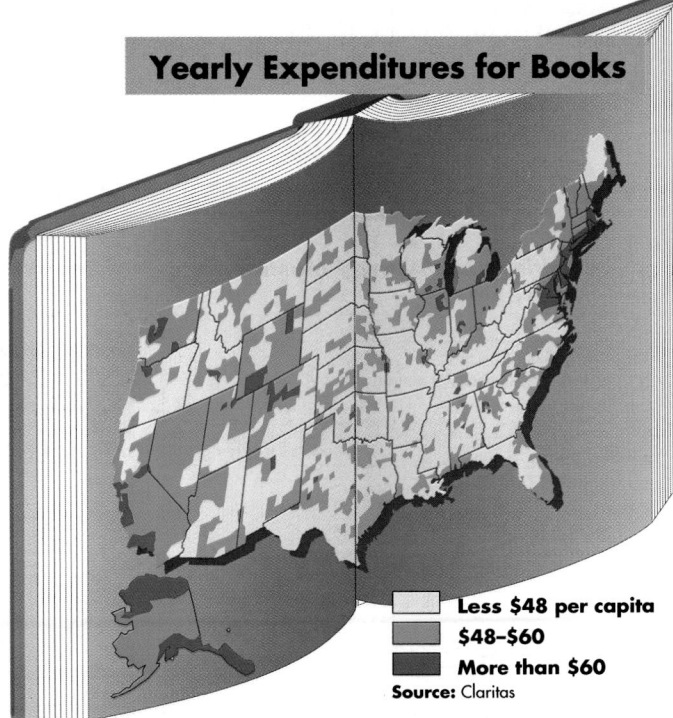

Yearly Expenditures for Books

- Less $48 per capita
- $48–$60
- More than $60

Source: Claritas

The writing and publishing of books is an important part of our economy. Do you like to express yourself? Do you share your thoughts by writing poems or short stories? Would you like to eventually make a living putting words on paper?

TIME *Line*

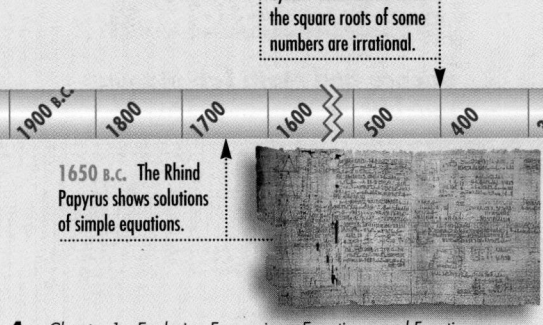

400 B.C. Theodorus of Cyrene demonstrates that the square roots of some numbers are irrational.

1650 B.C. The Rhind Papyrus shows solutions of simple equations.

A.D. 786 The Grand Mosque of Cordoba, Spain, is completed.

4 Chapter 1 Exploring Expressions, Equations, and Functions

TIME *Line*

Ask students to conduct research and write a one-page report about both Egyptian and Greek mathematics. Point out that applied mathematics began in Egypt, while pure mathematics began in Greece. Have students compare and contrast the two.

inter**NET** CONNECTION

Resources for writers (career planning, advice from professional writers, writing interactive multimedia, and more) are available on-line!

World Wide Web
http://www.seanet.com/Users/warlock/writers1.html

When **Wendy Isdell** was in her eighth grade Algebra 1 class, she got an idea for a short story. Later, she wrote an outline and completed a first draft. The following year, she entered the finished story in the Virginia Young Authors' Contest and won first place in the state competition. Wendy sent her story to Free Spirit Publishing in Minneapolis, and her book *A Gebra Named Al* was published during her senior year in high school.

Wendy's Algebra 1 and 2, geometry, trigonometry, chemistry, Earth science, and physical science classes have provided background for the information in her book. She hopes to make a living as a professional writer.

Chapter Project

Choose a poet and read 10 of his or her poems.

- Describe some patterns that you see in these poems.

- Count the number of words in each poem. Make a stem-and-leaf plot of the number of words in the poems. Does the poet seem to write poems with about the same number of words? Are one or more of the poems exceptionally long or short for this poet?

- Make a graph that models one of the poems. Write a paragraph that describes the relationship between the graph and the poem.

1663 The first known work on probability is published.

1922 Irish novelist James Joyce publishes *Ulysses*.

1958 American vocalist Billie Holiday records *Lady in Satin*.

| 900 | 1000 | 1100 | 1200 | 1300 | 1400 | 1500 | 1600 | 1700 | 1800 | 1900 | 2000 |

1455 The Gutenberg Bible is printed in Mainz, Germany.

1881 French impressionist painter Auguste Renoir completes the *Boating Party Lunch*.

1993 Toni Morrison becomes the first African-American to win the Nobel Prize for literature.

Chapter 1 **5**

Mathematical ideas offer a wealth of ideas for creative fiction. Working imaginatively with the concept of space and time has led to many science fiction stories. One is the book, *Flatland*, by A. Square (real name: Edwin A. Abbott) in 1884. In this story, the characters are points, lines, and two-dimensional figures like squares, triangles, and so on. Abbott illustrates the concept of dimension through this tale. It also contains biting social commentary about the class structure of the time.

Chapter Project

Cooperative Learning You may choose to have students pair off or arrange them in cooperative groups to create the graph. Students need to set up the horizontal and vertical axes. Since there are decreases as well as increases, the vertical axis must extend below the horizontal axis. One student can construct the axis system, being careful to allow for an appropriate range of values. The other student(s) can plot the data on the graph.

Investigations and Projects Masters, p. 25

Alternative Chapter Projects

Two other chapter projects are included in the *Investigations and Projects Masters.* In Chapter 1 Project A, pp. 25–26, students extend the topic in the chapter opener. In Chapter 1 Project B, pp. 27–28, students analyze sports statistics.

1

NAME_____ DATE_____

Chapter 1 Project A   Student Edition Pages 6–62

The "Write" Way

1. Conduct a survey in which you ask at least 20 working people to rank the importance of good writing skills from 1 to 3, with 1 being "very important for my job," 2 being "somewhat important for my job," and 3 being "not important for my job."

2. Using the data you gathered in step 1, calculate the percentage of responses in each category.

3. Go to the library and find a book that contains descriptions of various jobs and the skills that are necessary to do them. Make a list of jobs that require good writing skills.

4. Research the kind of writing that is involved in each career on your list. You might want to interview people who do a lot of writing on the job. Find out about the purpose of the writing each person does. Is it meant to inform? To entertain? To persuade? You might also want to ask about the audience for each piece of writing.

5. Make a poster or brochure that helps younger students understand the importance of having strong writing skills.

NCTM Standards: 1–5

Instructional Resources

- Study Guide Master 1-1
- Practice Master 1-1
- Enrichment Master 1-1

 Transparency 1-1A contains the 5-Minute Check for this lesson; **Transparency 1-1B** contains a teaching aid for this lesson.

Recommended Pacing	
Standard Pacing	Days 1 & 2 of 15
Honors Pacing	Day 1 of 14
Block Scheduling*	Day 1 of 7
Alg. 1 in Two Years*	Days 1 & 2 of 22

 *For more information on pacing and possible lesson plans, refer to the *Block Scheduling Booklet* and *Algebra 1 in Two Years*.

1 FOCUS

5-Minute Check

Write an arithmetic expression that represents each situation. Then find its value.

1. the sum of 18, 23, and 51
   18 + 23 + 51; 92
2. the quotient of 2,910,000 and 550,000 **2,910,000 ÷ 550,000; 5.291**
3. the square of 9 **9^2; 81**
4. the product of 4, 5, and 6
   $4 \times 5 \times 6$; 120
5. 10 decreased by the cube of 2 **$10 - 2^3$; 2**

Motivating the Lesson

Questioning Ask students the following questions concerning the protection of the skin from the sun.

1. Do you use sunscreen when you go to the beach or pool?
2. If you use sunscreen, do you know its SPF number?
3. Do you know how long you can be in the sun without burning?

Objective

After studying this lesson, you should be able to:

- translate verbal expressions into mathematical expressions and vice versa.

10×5 and 10×15 are numerical expressions.

CAREER CHOICES

A registered nurse (RN) cares for the injured and sick, administers medications, and provides for the physical and emotional needs of his or her patients.

Nursing requires graduation from an accredited nursing school and passing a national licensing examination. About 20% of the 3.8 million new jobs in health care will be for RNs.

For more information, contact:
National League for Nursing
250 Hudson St.
New York, NY 10014.

The symbol "×" as shown in x × y is often avoided because it may be confused with the variable "x."

1-1

Variables and Expressions

 APPLICATION
Health

Many people enjoy going to the beach in the summer. Unfortunately, the ultraviolet radiation in direct sunlight causes skin cancer. In recent years, the number of skin cancer cases has increased. Many sunbathers use sunscreen lotions to protect themselves in the sun.

The Sun Protection Factor (SPF) scale gives numbers that represent the length of time you can stay in the sun without burning if you have lotion on. Let's say you can stay in the sun with no sunscreen for 10 minutes without burning. If you put on SPF 5 lotion, you can stay in the sun for 10 minutes × 5 or 50 minutes, or a little less than one hour. Use the table below to see the pattern.

No Sunscreen	With Sunscreen	
Minutes in Sun Without Burning	SPF Number	Minutes in Sun Without Burning
10	5	10×5
10	15	10×15
5	4	5×4
15	8	15×8
m	s	$m \times s$

The letters m and s are called **variables,** and $m \times s$ is an **algebraic expression.** In algebra, variables are symbols that are used to represent unspecified numbers. Any letter may be used as a variable. We selected m because it is the first letter of the word "minutes" and s because it is the first letter of "SPF".

An algebraic expression consists of one or more numbers and variables along with one or more arithmetic operations. Here are some other examples of algebraic expressions.

$$x - 2 \qquad \frac{a}{b} + 3 \qquad t \times 2s \qquad 7mn \div 3k$$

In algebraic expressions, a raised dot or parentheses are often used to indicate multiplication. Here are ways to represent the product of x and y.

$$xy \qquad x \cdot y \qquad x(y) \qquad (x)(y) \qquad (x)y \qquad x \times y$$

In each of the multiplication expressions, the quantities being multiplied are called **factors,** and the result is called the **product.**

CAREER CHOICES

Encourage students to discuss the possible uses of mathematics in nursing and other related health care fields. For example, standard dosages are given for a certain body weight.

It is often necessary to translate verbal expressions into algebraic expressions.

Verbal Expression	Algebraic Expression
7 less than the product of 3 and a number x	$3x - 7$
the product of 7 and s divided by the product of 8 and y	$7s \div 8y$
four years younger than Sarah (S = Sarah's age)	$S - 4$
half as big as last night's crowd (c = size of last night's crowd)	$\frac{c}{2}$

Example Write an algebraic expression for each verbal expression.

    a. three times a number x subtracted from 24

    $24 - 3x$

    b. 5 greater than half of a number t

    $\frac{t}{2} + 5$ or $\frac{1}{2}t + 5$

Another important skill is translating algebraic expressions into verbal expressions.

Example **2** Write a verbal expression for each algebraic expression.

    a. $(3 + b) \div y$

    the sum of 3 and b divided by y

    b. $5y + 10x$

    the product of 5 and y plus the product of 10 and x

An expression like x^n is called a **power**. The variable x is called the **base**, and n is called the **exponent**. The exponent indicates the number of times the base is used as a factor.

    x^2 means $x \cdot x$. 12^4 means $12 \cdot 12 \cdot 12 \cdot 12$.

Symbols	Words	Meaning
5^1	5 to the first power	5
5^2	5 to the second power or 5 squared	$5 \cdot 5$
5^3	5 to the third power or 5 cubed	$5 \cdot 5 \cdot 5$
5^4	5 to the fourth power	$5 \cdot 5 \cdot 5 \cdot 5$
$3a^5$	three times a to the fifth power	$3 \cdot a \cdot a \cdot a \cdot a \cdot a$
x^n	x to the nth power	$\underbrace{x \cdot x \cdot x \cdot \ldots \cdot x}_{n \text{ factors}}$
	An expression of the form x^n is read as "x to the nth power."	

Teaching Tip Letters in algebra are called *variables* because their values can *vary*.

Teaching Tip Note that for a product of a number and a variable, the number is written first. For example, the product of x and 5 is $5x$, not $x5$.

In-Class Examples

For Example 1
Write an algebraic expression for each verbal expression.

a. m increased by 5 $m + 5$
b. the difference of x and 9 $x - 9$
c. 7 times the product of x and t $7xt$

For Example 2
Write two different verbal expressions for each algebraic expression.

a. $9t$ 9 times t, product of 9 and t
b. $8 + a$ 8 plus a, sum of 8 and a
c. $7 - 3y$ difference of 7 and 3 times y, 7 minus the product of 3 and y

---

GLENCOE *Technology*

Interactive Mathematics Tools Software

This multimedia software provides an interactive lesson for finding the relationship between the number of cricket chirps per minute and the degrees Fahrenheit. A **Computer Journal** gives students an opportunity to write about what they have learned.

For Windows & Macintosh

Alternative Teaching Strategies

Student Diversity Review lesson vocabulary with students. If necessary, write a variety of expressions on the board and have students identify the variables, factors, products, bases, and exponents within the expressions.

Example **3** **Write a power that represents the number of smallest squares in the large square.**

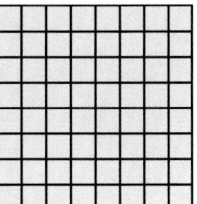

There are 8 squares on each side.

The total number of squares is $8^2 = 8 \cdot 8$ or 64.

You can use the x^2 key on a calculator to square a number.

Enter: 8 $\boxed{x^2}$ *64*

To **evaluate** an expression means to find its value.

Example **4** **Evaluate 3^4.**

$$3^4 = 3 \cdot 3 \cdot 3 \cdot 3$$

$$= 81$$

You can use the $\boxed{y^x}$ key on a calculator to raise a number to a power.

Enter: 3 $\boxed{y^x}$ 4 $\boxed{=}$ *81*

EXPLORATION   PROGRAMMING

BASIC is a computer language. The symbols used in BASIC are similar to those used in algebra.

+ means add.	/ means divide.
− means subtract.	↑ means exponent.
* means multiply.	

Numerical variables in BASIC are represented by capital letters. The program at the right can be used to add, subtract, multiply, divide, and find powers.

```
10 INPUT A,B
20 PRINT "A + B ="; A + B
30 PRINT "A − B ="; A − B
40 PRINT "A * B ="; A * B
50 PRINT "A / B ="; A / B
60 PRINT "A ↑ B ="; A ↑ B
```

Your Turn

a. Let $a = 10$ and $b = 3$. Use the program to find $10 + 3$, $10 − 3$, 10×3, $10 \div 3$, and 10^3.   **13; 7; 30; 3.333; 1000**

b. Let $a = 9$ and $b = 3$. Use the program to find $9 + 3$, $9 − 3$, 9×3, $9 \div 3$, and 9^3.   **12; 6; 27; 3; 729**

c. Let $a = 5.2$ and $b = 2$. Use the program to find $5.2 + 2$, $5.2 − 2$, 5.2×2, $5.2 \div 2$, and 5.2^2.   **7.2; 3.2; 10.4; 2.6; 27.04**

Communicating Mathematics

Study the lesson. Then complete the following.

1. **Describe** how you would compute the length of time you could sunbathe using an SPF 20 sunscreen lotion if you can stay in the sun for 15 minutes without burning. **Multiply 15 times 20.**

2. Can you find the volume of a cube if you only know the length of a side, s? How? **Yes; evaluate s^3 where s is the measure of a side.**

3. **Describe** what x^7 means. **x used as a factor 7 times**

4. **Explain** the difference between numerical expressions and algebraic expressions. **Algebraic expressions include variables.**

5. **You Decide** Darcy and Tonya are studying for an exam. Darcy says, "The square of a number is always greater than the number." Tonya disagrees. Who is correct? Explain your answer.
Tonya; Sample explanation: $0.5^2 = 0.25$ and $0.5 > 0.25$.

Guided Practice

Write an algebraic expression for each verbal expression.

6. the product of the fourth power of a and the second power of b **$a^4 \cdot b^2$**

7. six less than three times the square of y **$3y^2 - 6$**

Write a verbal expression for each algebraic expression.

8. 7^5 **7 to the fifth power**

9. $3x^2 + 4$ **3 times x squared increased by 4**

Write each expression as an expression with exponents.

10. $4 \cdot 4 \cdot 4$ **4^3**

11. $a \cdot a \cdot a \cdot a \cdot a \cdot a \cdot a$ **a^7**

Evaluate each expression.

12. 6^2 **36**

13. 2^5 **32**

14. **Geometry** The area of a circle can be found by multiplying the number π times the square of the radius. If the radius of the circle is r, write an expression that represents the area of the circle. **πr^2**

Practice

Ⓐ

Write an algebraic expression for each verbal expression.

15. the sum of k and 20 **$k + 20$**

16. the product of 16 and p **$16p$**

17. a to the seventh power **a^7**

18. 49 increased by twice a number **$49 + 2x$**

Ⓑ

19. two thirds of the square of a number **$\frac{2x^2}{3}$ or $\frac{2}{3}x^2$**

20. the product of 5 and m plus half of n

20. **$5m + \frac{n}{2}$ or $5m + \frac{1}{2}n$**

21. 8 pounds heavier than his brother (b = brother's weight) **$b + 8$**

22. 3 times as many wins as last season (w = last season's wins) **$3w$**

Write a verbal expression for each algebraic expression.

23. $4m^5$ **4 times m to the fifth power**

24. $\frac{x^2}{2}$ **x squared divided by 2**

25. $c^2 + 23$ **c squared plus 23**

26. $\frac{1}{2}n^3$ **one half the cube of n**

27. **2 times 4 times 5 squared**

28. **3 times x squared minus 2 times x**

27. $2(4)(5)^2$

28. $3x^2 - 2x$

Lesson 1–1 Variables and Expressions **9**

Reteaching

Translating Expressions Have students answer each question with an algebraic expression. Let d represent Dawn's age now.

1. How old will Dawn be in two years? **$d + 2$**

2. How old is Ellen now if she is five years younger than Dawn? **$d - 5$**

3. How old is Frank now if he is three times as old as Dawn? **$3d$**

Check for Understanding

Exercises 1–14 are designed to help you assess your students' understanding through reading, writing, speaking, and modeling. You should work through Exercises 1–5 with your students and then monitor their work on Exercises 6–14.

Error Analysis

Students often translate expressions involving subtraction in the wrong order. Point out that "the difference of 5 and t" and "5 decreased by t" are translated as $5 - t$ while "5 less than t" and "5 subtracted from t" are translated as $t - 5$.

Assignment Guide
Core: 15–49 odd **Enriched:** 16–46 even, 47–50

For **Extra Practice,** see p. 756.

The red A, B, and C flags, printed only in the Teacher's Wraparound Edition, indicate the level of difficulty of the exercises.

Study Guide Masters, p. 1

1-1 NAME_____ DATE _____
Study Guide Student Edition Pages 6–11

Variables and Expressions

Any letter used to represent an unspecified number is called a variable. You can use variables to translate verbal expressions into algebraic expressions.

Words	Symbols
4 more than a number	$x + 4$
a number decreased by 8	$b - 8$
the product of 5 and a number	$5c$
a number divided by 8	$h \div 8$ or $\frac{h}{8}$
a number squared	y^2

The algebraic expression x^n represents a product in which each factor is the same. The small raised n is the exponent and it tells how many times the base, x, is used as a factor.

Example: Evaluate 3^4.
$3^4 = 3 \cdot 3 \cdot 3 \cdot 3$
$= 81$

Answers may vary. Sample answers are given.

Write a verbal expression for each algebraic expression.

1. $w - 1$ **one less than w**

2. $\frac{1}{3}a^3$ **one third the cube of a**

3. $81 + 2x$ **eighty-one increased by twice x**

Write an algebraic expression for each verbal expression.

4. a number decreased by 5 **$n - 5$**

5. four times a number **$4n$**

6. 8 less than a number **$n - 8$**

7. a number divided by 6 **$n \div 6$**

8. a number multiplied by 37 **$37n$**

9. the sum of a number and 9 **$n + 9$**

10. 3 less than 5 times a number **$5n - 3$**

11. twice the sum of 15 and a number **$2(15 + n)$**

12. 7 more than the product of 6 and a number **$6n + 7$**

13. 30 increased by 3 times the square of a number **$30 + 3n^2$**

Write each expression as an expression with exponents.

14. $7 \cdot 7 \cdot 7$ **7^3**

15. $3 \cdot p \cdot p$ **$3p^2$**

16. $9(b)(b)(b)(b)(b)$ **$9b^5$**

Evaluate each expression.

17. 2^3 **8**

18. 10^5 **100,000**

19. 4^4 **256**

Write each expression as an expression with exponents.

29. $8 \cdot 8$ 8^2

30. $10 \cdot 10 \cdot 10 \cdot 10 \cdot 10$ 10^5

31. $4 \cdot 4 \cdot 4 \cdot 4 \cdot 4 \cdot 4 \cdot 4$ 4^7

32. $t \cdot t \cdot t$ t^3

33. $z \cdot z \cdot z \cdot z \cdot z$ z^5

34. $d \cdot d \cdot d \cdot d \cdot d \cdot d \cdot d \cdot d \cdot d \cdot d$ d^{10}

Evaluate each expression.

35. 7^2 49

36. 9^2 81

37. 4^3 64

38. 5^3 125

39. 2^4 16

40. 3^5 243

Write an algebraic expression for each verbal expression.

41. triple the difference of 55 and the cube of w $3(55 - w^3)$

$42.\ 4(r + s) + 2(r - s)$

42. four times the sum of r and s, increased by twice the difference of r and s

43. the sum of a and b, increased by the quotient of a and b $a + b + \dfrac{a}{b}$

44. the perimeter of a square if the length of a side is s $4s$

45. the amount of money in Karen's savings account if she has y dollars and adds x dollars per week for 10 weeks $y + 10x$

46. Geometry Write an expression that represents the total number of small cubes in the large cube shown at the right. Then evaluate the expression. 5^3; 125 cubes

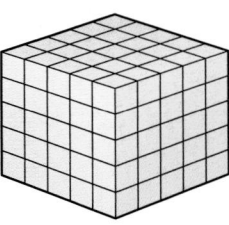

Critical Thinking

47. Evaluate 4^2 and 2^4.   16; 16   47a. They are equal; no; no.
  a. What do you notice? From this example, can you say that the same relationship exists between 2^5 and 5^2? What about between 3^4 and 4^3?
  b. In general, do you think that $a^b = b^a$ for positive whole numbers a and b? Support your answer.   no; for example, $2^3 \neq 3^2$

Applications and Problem Solving

48. Babysitting According to *Smart Money* magazine, in New York City, the average teenage babysitter makes $7 per hour for a Saturday night. In Atlanta, the average teenage babysitter makes only $3 per hour.
  a. Write an expression representing the amount of money that the average teenage babysitter in New York City makes for working x hours on a Saturday night. $7x$
  b. Write an expression representing the amount of money that the average teenage babysitter in Atlanta makes for working x hours on a Saturday night. $3x$
  c. Write an expression representing the difference between the amount that the New York babysitter makes and the Atlanta babysitter makes for working x hours on a Saturday night. $7x - 3x$

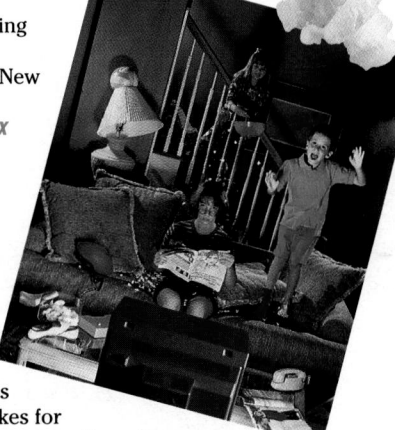

Practice Masters, p. 1

Extension

Reasoning Have students solve: Carolyn received d dollars on Sunday. She spent half of this money on Monday, one-third of what was left on Tuesday, and one-fourth of what was left on Wednesday. Write an algebraic expression to represent how much money she spent each day.

a. Monday $\dfrac{d}{2}$

b. Tuesday $\dfrac{1}{3}\left(d - \dfrac{d}{2}\right)$

c. Wednesday $\dfrac{1}{4}\left[\dfrac{d}{2} - \dfrac{1}{3}\left(d - \dfrac{d}{2}\right)\right]$

49. Recycling The alarming amount of garbage produced by our society requires that everyone work to recycle. According to *Vitality* magazine, each person in the United States produces an average of 3.5 pounds of trash a day.

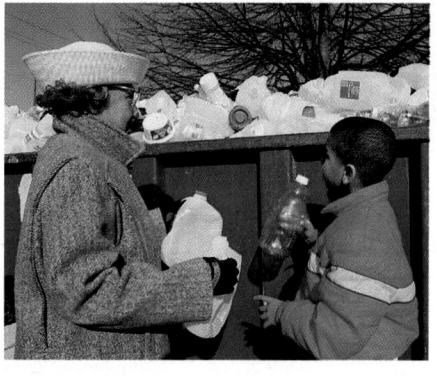

   a. Write an expression representing the pounds of trash produced in a day by a family that has *x* members. **3.5*x***

   b. Write an expression representing the pounds of trash produced in a day by a family that has *y* members. **3.5*y***

   c. Write an expression representing the pounds of trash produced in one day by both of the families in parts a and b. **3.5*x* + 3.5*y***

50. Geometry The surface area of a rectangular prism is the sum of the product of twice the length ℓ and the width w and the product of twice the length and the height h and the product of twice the width and the height. Write an expression that represents the surface area of this type of prism. **$2\ell w + 2\ell h + 2wh$**

Mathematics and SOCIETY

International Mathematical Olympiad

The excerpt below appeared in an article in *TIME* magazine on August 1, 1994.

IF THE UNITED STATES SENDS SIX KIDS TO THE International Mathematical Olympiad in Hong Kong, where a perfect individual score is 42, and together they score 252, does the country have reason to cheer? If you can't answer that one, then you desperately need a remedial course in arithmetic (or perhaps just new batteries in your calculator). The U.S. team members, all public high-school students, started out by competing against 350,000 of their peers on the American High School Mathematics Examination, aced two tougher exams, and prepped for a month at the U.S. Naval Academy. Only then did they board a plane and became the first squad in the Math Olympiad's 35-year history to get perfect scores across the board, outstripping 68 other nations to win the competition. . . . "They showed the world!" suggests the justifiably proud U.S. coach, Walter Mientka, a math professor at the University of Nebraska at Lincoln. ■

1. Do you think this type of international contest is a good idea? Why or why not? **1–3. See students' work.**

2. Would you like to represent your country in this competition? Why or why not?

3. How do you feel about the math courses you have had in school? What factors have influenced the way you feel?

Lesson 1–1 Variables and Expressions **11**

Chapter 1 **11**

Instructional Resources

- Study Guide Master 1-2
- Practice Master 1-2
- Enrichment Master 1-2

 Transparency 1-2A contains the 5-Minute Check for this lesson; **Transparency 1-2B** contains a teaching aid for this lesson.

Recommended Pacing	
Standard Pacing	Day 3 of 15
Honors Pacing	Day 2 of 14
Block Scheduling*	Day 2 of 7 (along with Lesson 1-3)
Alg. 1 in Two Years*	Days 3 & 4 of 22

 *For more information on pacing and possible lesson plans, refer to the *Block Scheduling Booklet* and *Algebra 1 in Two Years.*

1 FOCUS

 ### 5-Minute Check
(over Lesson 1-1)

1. Write a verbal expression for $2x - x^2$. **twice a number less the square of the number**
2. Evaluate 3^4. **81**

Write an algebraic expression for each verbal expression.

3. a number increased by 17 **$n + 17$**
4. twice a number decreased by 45 **$2n - 45$**
5. the square of a number decreased by twice the number **$n^2 - 2n$**

Motivating the Lesson

Questioning Write the following sequence on the board: *a, abb, cabbc, cabbcdd,* . . . . Have students continue the sequence. Then have them explain the steps they followed to continue the sequence.

Patterns and Sequences

Objective

After studying this lesson, you should be able to:
- extend sequences.

 APPLICATION
Handicrafts

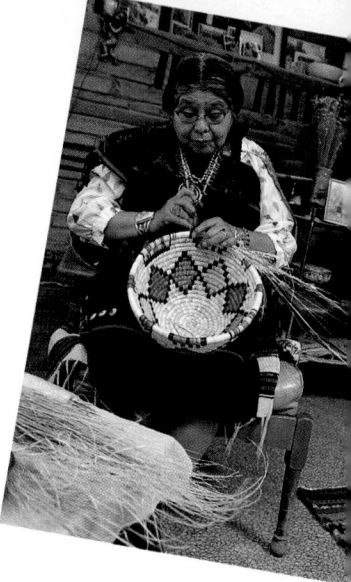

The photograph at the right shows a Hopi woman completing a basket. Baskets are important in the Hopi culture since they can serve either a functional or a spiritual purpose. Looking at the inside of the basket, we can see a pattern. Following this pattern in a clockwise direction, what color would you expect to find on the first diamond shape that cannot be seen in the photo? **green**

GLOBAL CONNECTIONS

Many American Indians play traditional games passed down through the generations. In one such game, markers made from bone, pottery, or shell are marked and tumbled in a small basket. The array of the markers is the basis for scoring. Five or six of a kind would win.

Example Study the pattern below.

a. **Draw the next three figures in the pattern.**
b. **Draw the 35th square in the pattern.**

a. The pattern consists of squares with one corner shaded. The corner that is shaded is rotated in a clockwise direction. The next three figures are drawn below.

You may find a different pattern. For example, the seventh square may be the same as the fifth square, the eighth square may be the same as the fourth square, and the ninth square may be the same as the third square.

b. The pattern repeats after every 4 designs. Since 32 is the greatest number less than 35 that is divisible by 4, the 33rd square will be the same as the first square.

  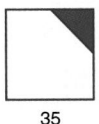

The 35th square will have its upper right corner shaded.

GLOBAL CONNECTIONS

These traditional games required an intuitive understanding of probability. Six of a kind is less likely than five of a kind. Why do three spotted and two striped beat three of a kind? **Because the former is less likely to occur than the latter.**

The numbers 2, 4, 6, 8, 10, and 12 form a pattern called a **sequence**. A sequence is a set of numbers in a specific order. The numbers in the sequence are called **terms**.

Example **Find the next three terms in each sequence.**

a. 7, 13, 19, 25, . . .

Study the pattern in the sequence.

7,   13,   19,   25,...
   + 6   + 6   + 6

Each term is 6 more than the term before it.

$25 + 6 = 31$

$31 + 6 = 37$

$37 + 6 = 43$

The next three terms are 31, 37, and 43.

b. 243, 81, 27, 9, . . .

Study the pattern in the sequence.

243,   81,   27,   9,...
   $\times \frac{1}{3}$ $\times \frac{1}{3}$ $\times \frac{1}{3}$

Each term is $\frac{1}{3}$ of the term before it.

$9 \times \frac{1}{3} = 3$

$3 \times \frac{1}{3} = 1$

$1 \times \frac{1}{3} = \frac{1}{3}$

The next three terms are 3, 1, and $\frac{1}{3}$.

One of the most-used strategies in problem solving is **look for a pattern**. When using this strategy, you will often need to make a table to organize the information.

Example **What is the number of diagonals in a 10-sided polygon?**

INTEGRATION
Geometry

Drawing a 10-sided polygon and all its diagonals would be difficult. Another way to solve the problem is to study the number of diagonals for polygons with fewer sides and then look for a pattern.

triangle	quadrilateral	pentagon	hexagon
3 sides	4 sides	5 sides	6 sides
0 diagonals	2 diagonals	5 diagonals	9 diagonals

(continued on the next page)

Lesson 1-2 Patterns and Sequences **13**

Alternative Learning Styles

Visual Share with students that most numerical sequences can be represented by points with lines joining the points. The sequence 3, 6, 12. . . can be visualized as follows.

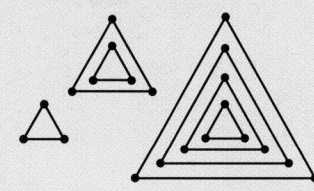

The sequence can also be visualized as follows.

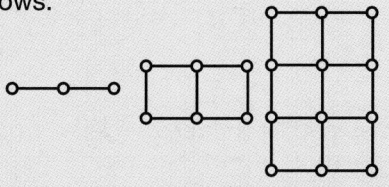

Since the visual representation is not unique, have students use grid paper and creativity to draw their own representations for this sequence.

In-Class Examples

For Example 1
Study the pattern below.

a. Draw the next three figures in the pattern.

b. Draw the 26th triangle in the pattern.

For Example 2
Find the next three numbers in each sequence.

a. 51, 43, 35, 27 19, 11, 3
b. 3072, 768, 192 48, 12, 3

For Example 3
How many points does the sixth figure in this sequence have?
63

Teaching Tip In Example 2, suggest that students begin looking for a pattern by determining if the second term is derived from the first term by addition, subtraction, multiplication, or division.

Teaching Tip Point out to students that in some cases it looks as if there is a pattern, but later on the pattern is broken. For example, show how the pattern is broken in the eighth digit of $111,111,111 \div 9 = 12,345,67\underline{9}$.

Use a chart to see the pattern.

Number of Sides	3	4	5	6	7	8	9	10
Number of Diagonals	0	2	5	9	?	?	?	?

+ 2   + 3   + 4

Use the pattern to complete the chart.

Number of Sides	3	4	5	6	7	8	9	10
Number of Diagonals	0	2	5	9	14	20	27	35

+ 2   + 3   + 4   + 5   + 6   + 7   + 8

A 10-sided polygon has 35 diagonals.

Recognizing patterns is important, but it does not always guarantee finding the correct answer. You may wish to check this answer by drawing a 10-sided polygon and its diagonals.

Sometimes a pattern can lead to a general rule that can be written as an algebraic expression.

Example **Study the following pattern. Write an algebraic expression for the perimeter of the pattern consisting of *n* trapezoids.**

1 trapezoid	2 trapezoids	3 trapezoids	4 trapezoids
P = 5 units	P = 8 units	P = 11 units	P = 14 units

Use a chart to look for a pattern.

Number of Trapezoids	1	2	3	4	5	6	n
Perimeter	5	8	11	14	?	?	?

+ 3   + 3   + 3

Notice that each trapezoid adds 3 units to the perimeter.
 For 1 trapezoid, the perimeter is $3 \cdot 1 + 2$, or 5 units.
 For 2 trapezoids, the perimeter is $3 \cdot 2 + 2$, or 8 units.
 For 3 trapezoids, the perimeter is $3 \cdot 3 + 2$, or 11 units.
 For 4 trapezoids, the perimeter is $3 \cdot 4 + 2$, or 14 units.

Extend this pattern.
 For 5 trapezoids, the perimeter should be $3 \cdot 5 + 2$, or 17 units.
 For 6 trapezoids, the perimeter should be $3 \cdot 6 + 2$, or 20 units.
 For *n* trapezoids, the perimeter should be $3 \cdot n + 2$, or $3n + 2$ units.

The algebraic expression $3n + 2$ represents the perimeter of this pattern with *n* trapezoids.

MODELING MATHEMATICS

Looking for Patterns

Materials: string ✂ scissors

If you use a pair of scissors to cut a piece of string in the normal way, you will have 2 pieces of string. What happens if you loop the string around one of the cutting edges of the scissors and cut?

0 loop   1 loop

Your Turn

a. Make a piece of string loop once around your scissors as shown above. Cut the string. How many pieces do you have? **3 pieces**

b. Make 2 loops and cut. How many pieces do you have? **4 pieces**

c. Add 2 to the number of loops; 3, 4, 5, 6, 7, …

c. Continue making loops and cutting until you see a pattern. Describe the pattern and write the sequence.

d. How many pieces would you have if you made 20 loops? **22 pieces**

e. Now tie the ends of the string together before you loop the string around the scissors. Investigate to determine how many pieces you would have if you made 10 loops with this string. **22 pieces**

0 loop   1 loop

CHECK FOR UNDERSTANDING

Communicating Mathematics

1. See margin.
2. Sample answer: 7, 12, 17, 22

Study the lesson. Then complete the following.

1. **Explain** how looking for a pattern can help to solve some problems.

2. **Write** a sequence that has 22 as its fourth term.

3. **You Decide** Chi-Yo studies the sequence 1, 2, and 4. She notices that the second number is 1 more than the first, and the third number is 2 more than the second. She concludes that the next number in the sequence is $4 + 3$ or 7. Alonso disagrees. He notices that each number is twice the number before it. He says the next number in the sequence is 8. Who is correct? Explain. **See margin.**

MODELING MATHEMATICS

4. Refer to the Modeling Mathematics activity above. Suppose the ends of a string are *not* tied together.

    a. The string is looped around the scissors 50 times, and the string is cut. How many pieces will you have? **52 pieces**

    b. The string is looped around the scissors y times, and the string is cut. How many pieces will you have? $y + 2$

Guided Practice

5–7. Sample answers are given.

7. $5x + 1, 6x + 1$

Give the next two items for each pattern.

5.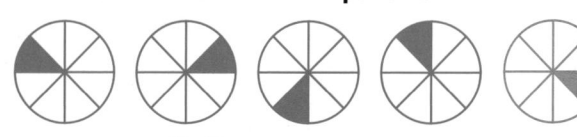

6. 85, 76, 67, 58,… **49, 40**

7. $1x + 1, 2x + 1, 3x + 1, 4x + 1,…$

---

3 PRACTICE/APPLY

Check for Understanding

Exercises 1–9 are designed to help you assess your students' understanding through reading, writing, speaking, and modeling. You should work through Exercises 1–4 with your students and then monitor their work on Exercises 5–9.

Error Analysis

Students will sometimes find a pattern in the first three terms and not notice that the fourth term does not fit the pattern.

Additional Answers

1. When the problem is difficult, you can solve simpler problems and look for a pattern that could help you solve the more difficult problem.

3. Either person could be right. You would need to know the fourth term to decide which pattern is correct.

Reteaching

Using Patterns Find the next two terms of this sequence.

				5 3	5 3
		3	3		
1	212	212	42124	42124	6421246
		3	3		
				3 5	3 5

For **Extra Practice,** see p. 756.

The red A, B, and C flags, printed only in the Teacher's Wraparound Edition, indicate the level of difficulty of the exercises.

Additional Answers

8b.

12.

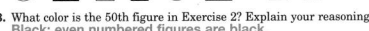
---

8. Consider the following pattern.

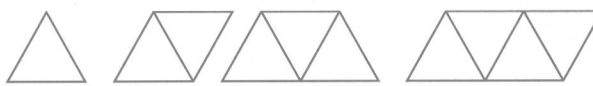

8b. See margin; 7 units.

 a. Suppose the length of each side of the triangles is 1 unit. What is the perimeter of each figure in the pattern? 3 units; 4 units; 5 units; 6 units
 b. Draw the next figure in the pattern. What is the perimeter of this figure?
 c. What is the perimeter of the tenth figure in this pattern? 12 units
 d. What is the perimeter of the *n*th figure in this pattern? n + 2 units

9. a. Copy and complete the following table.

4^1	4^2	4^3	4^4	4^5
4	16	64	256	1024

 b. If you found the value of 4^6, what number do you think will be in the ones place? Check your conjecture. 6; $4^6 = 4096$
 c. Find the number in the ones place for the value of 4^{225}. Explain your reasoning. 4; When the exponent is odd, 4 is in the ones place.

EXERCISES

Practice
A

Give the next two items for each pattern.

10.

11.

12. See margin.

12.

13. 3, 6, 12, 24, ... 48, 96   14. 4, 5.5, 7, 8.5, ... 10, 11.5

B

15. 1, 4, 9, 16, ... 25, 36   16. 9, 7, 10, 8, 11, 9, 12, ... 10, 13

18. $x - 4y, x - 5y$

17. $a + 1, a + 3, a + 5, ...$ $a + 7, a + 9$ 18. $x - 1y, x - 2y, x - 3y, ...$

19. a. Draw the next three figures in the following pattern.

19b. White; even-numbered figures are white.

19c. 12 sides; For reasoning, see margin.

 b. What is the color of the 38th figure? Explain your reasoning.
 c. How many sides will the 19th figure have? Explain your reasoning.

20. a. Copy and complete the following table.

3^1	3^2	3^3	3^4	3^5	3^6
3	9	27	81	243	729

b. Write the sequence of numbers representing the numbers in the ones place. **3, 9, 7, 1, 3, 9, ...**

c. What are the next six numbers in this sequence? **7, 1, 3, 9, 7, 1**

d. Find the number in the ones place for the value of 3^{100}. Explain your reasoning. **1; See margin for reasoning.**

21. a. Copy and find each sum.

$1 = 1$

$1 + 3 = ?$ **4**

$1 + 3 + 5 = ?$ **9**

$1 + 3 + 5 + 7 = ?$ **16**

$1 + 3 + 5 + 7 + 9 = ?$ **25**

21b. See margin.

b. Pythagoras is credited with a discovery about the sum of consecutive odd numbers such as those shown in part a. What did he discover?

c. What is the sum of the first 100 odd numbers? **10,000**

d. What is the sum of the first x odd numbers? x^2

22. If y represents the counting numbers 1, 2, 3, 4, ..., then the algebraic expression $2y$ represents the terms of a sequence. To find the first term of the sequence, replace y with 1 and find $2 \cdot 1$.

22a. Replace y with 2 and find $2 \cdot 2$.

22b. 2, 4, 6, 8, 10, 12, 14, 16, 18, 20

a. How could you find the second term of the sequence?

b. Write the first ten terms of the sequence.

c. Describe the terms of this sequence. **even numbers**

23. a. Use a calculator to find each product.

i. $999,999 \times 2$ **1,999,998**

ii. $999,999 \times 3$ **2,999,997**

iii. $999,999 \times 4$ **3,999,996**

iv. $999,999 \times 5$ **4,999,995**

b. Without using a calculator, find the product of 999,999 and 9. **8,999,991**

 24. a. Copy and complete the following table.

10^1	10^2	10^3	10^4	10^5
10	100	1000	10,000	100,000

b. Use the pattern to find the value of 10^0. **1**

25. If n represents the counting numbers 1, 2, 3, 4, ..., then the algebraic expression $3n + 1$ represents the terms of a sequence. Write the first five terms of the sequence. **4, 7, 10, 13, 16**

Critical Thinking

26. A special sequence is called the *Fibonacci sequence*. It is named after Leonardo Fibonacci from Italy who presented it in 1201. This sequence is very intriguing because the numbers in the sequence are often found in nature. The first six terms of the sequence are listed below. Give the next six numbers in the sequence. 1, 1, 2, 3, 5, 8, ... **13, 21, 34, 55, 89, 144**

Applications and Problem Solving

27. Transportation Olga has part of a bus schedule. She wishes to take the bus to go to the mall, but she cannot leave until after 1:00 P.M. What is the earliest time Olga can catch the bus? **1:13 P.M.**

Bus Schedule
Departures
8:25 A.M.
9:13 A.M.
10:01 A.M.
10:49 A.M.

Closing Activity

Modeling Have students work in pairs. One student should use toothpicks to create a pattern. The other student should determine the pattern and continue it.

Additional Answers

19c. The shapes come in pairs. Since $19 \div 2 = 9.5$, the 19th figure will be part of the 10th pair. The 10th pair will have $10 + 2$ or 12 sides.

20d. 100 is divisible by 4. According to the pattern, all powers with exponents divisible by 4 have 1 in the ones place.

21b. The sums are perfect square numbers; that is $1^2, 2^2, 3^2, 4^2, 5^2, \ldots$.

Practice Masters, p. 2

top layer

second layer

third layer

28. Sales Paula needs to make a tower of soup cans as a display in a supermarket. Each layer of the tower will be in the shape of a rectangle as shown at the left. The length and width of each layer will be one less than the layer below it.

a. How many cans will be needed for the fourth layer? **30 cans**

b. What is the total number of cans needed for an 8-layer tower? **328 cans**

29a. 100 cards

29b–d. See students' work.

29. Recreation The house of cards at the right used 26 cards to build four stories.

a. How many cards are needed to build a similar house that is eight stories high?

b. Plan another way to construct a house of cards. Make a drawing showing four stories of your house.

c. Write a sequence that represents the number of cards needed to build 1, 2, 3, and 4 stories of your house.

d. How many cards will you need to build eight stories of your house?

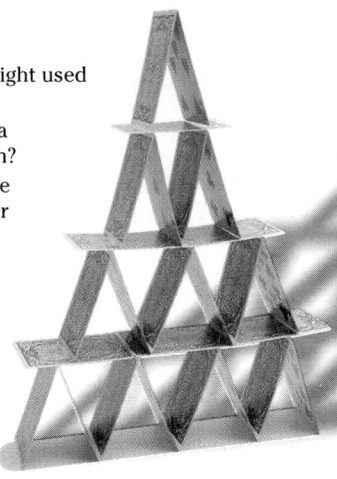

Mixed Review

30. Write an algebraic expression for *eight less than the square of q*. (Lesson 1–1) $q^2 - 8$

31. *x* cubed divided by 9

31. Write a verbal expression for $\frac{x^3}{9}$. (Lesson 1–1)

32. Write $m \cdot m \cdot m \cdot m \cdot m \cdot m \cdot m \cdot m \cdot m$ as an algebraic expression with exponents. (Lesson 1–1) m^9

33. Geometry Write an expression that represents the total number of small cubes in the large cube at the right. Then evaluate the expression. (Lesson 1–1) 4^3; **64 cubes**

34. Write an algebraic expression for the amount of mileage on Seth's car if the car initially has 20,000 miles and Seth adds an average of *x* miles per month for 2 years. (Lesson 1–1) $20,000 + 24x$

35. $x + \frac{1}{11}x$

35. Science When water freezes, its volume is increased by one eleventh. In other words, the volume of the ice would equal the sum of the volume of the water and the product of $\frac{1}{11}$ and the volume of the water. Suppose *x* cubic inches of water is placed in a freezer. Write an expression for the volume of the ice that is formed. (Lesson 1–1)

18 Chapter 1 *Exploring Expressions, Equations, and Functions*

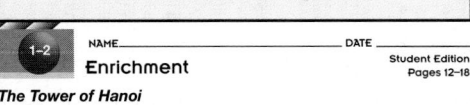
Extension

Reasoning For Exercise 29a, describe the number of cards needed to construct level *n* when levels are counted from top to bottom and *n* is not the bottom level. **3n** Describe the number of cards needed when *n* is the bottom level. **3n − n**

Order of Operations

Objective

After studying this lesson, you should be able to:

- use the order of operations to evaluate real number expressions.

APPLICATION
Investments

Bobbie Jackson is investing money in stocks to help pay for her child's college education. She buys one share of Nike stock at $16. She also purchases five shares of Disney stock at $35 each. The expression below represents the amount of money Ms. Jackson spends for the stock purchase.

cost of 1 share of Nike stock ⌐ ⌐ *cost of 1 share of Disney stock*

$$16 + 5 \cdot 35$$

number of Disney stocks

Numerical and algebraic expressions often contain more than one operation. A rule is needed to let you know which operation to perform first. This rule is called the **order of operations.**

To find the total amount of money Ms. Jackson invests, evaluate the expression $16 + 5 \cdot 35$. Which of the following methods is correct?

Method 1		**Method 2**	
$16 + 5 \cdot 35 = 16 + 175$	*Multiply first.*	$16 + 5 \cdot 35 = 21 \cdot 35$	*Add first.*
$= 191$	*Then add.*	$= 735$	*Then multiply.*

The answers are not the same because a different order of operations was used in each method. Since numerical expressions must have only one value, the following order of operations has been established.

Order of Operations	1. **Simplify the expressions inside grouping symbols, such as parentheses, brackets, and braces, and as indicated by fraction bars.** 2. **Evaluate all powers.** 3. **Do all multiplications and divisions from left to right.** 4. **Do all additions and subtractions from left to right.**

Based on the context of the problem and the order of operations, Method 1 is correct. Therefore, Ms. Jackson invested $16 + 5 \cdot 35$ or $191 in the stock market.

NCTM Standards: 1–5, 8

Instructional Resources

- Study Guide Master 1-3
- Practice Master 1-3
- Enrichment Master 1-3
- Assessment and Evaluation Masters, p. 16
- Modeling Mathematics Masters, p. 72
- Tech Prep Applications Masters, p. 1

Transparency 1-3A contains the 5-Minute Check for this lesson; **Transparency 1-3B** contains a teaching aid for this lesson.

Recommended Pacing

Standard Pacing	Day 4 of 15
Honors Pacing	Day 3 of 14
Block Scheduling*	Day 2 of 7 (along with Lesson 1-2)
Alg. 1 in Two Years*	Days 5 & 6 of 22

*For more information on pacing and possible lesson plans, refer to the *Block Scheduling Booklet* and *Algebra 1 in Two Years*.

1 FOCUS

5-Minute Check
(over Lesson 1-2)

Write the next three numbers in each sequence.

1. 2, 4, 10, 28 . . . 82, 244, 730
2. 5, 15, 35, 75 . . . 155, 315, 635
3. 10, 90, 890, 8890 . . . 88,890; 888,890; 8,888,890
4. 5, 17, 53, 161 . . . 485, 1457, 4373
5.

Other expressions can be evaluated by using the order of operations.

Example **Evaluate $5 \times 7 - 6 \div 2 + 3^2$.**

$$
\begin{aligned}
\text{Evaluate } 5 \times 7 - 6 \div 2 + 3^2 &= 5 \times 7 - 6 \div 2 + 9 &&\textit{Evaluate } 3^2. \\
&= 35 - 6 \div 2 + 9 &&\textit{Multiply 5 by 7.} \\
&= 35 - 3 + 9 &&\textit{Divide 6 by 2.} \\
&= 32 + 9 &&\textit{Subtract 3 from 35.} \\
&= 41 &&\textit{Add 32 and 9.}
\end{aligned}
$$

In mathematics, grouping symbols such as parentheses ( ), brackets [ ], and braces { } are used to clarify or change the order of operations. They indicate that the expression within the grouping symbol is to be evaluated first. When more than one grouping symbol is used, start evaluating within the innermost grouping symbols.

Example **Evaluate $8[6^2 - 3(2 + 5)] \div 8 + 3$.**

$$
\begin{aligned}
8[6^2 - 3(2 + 5)] \div 8 + 3 &= 8[6^2 - 3(7)] \div 8 + 3 &&\textit{Add 2 + 5, the innermost group.} \\
&= 8[36 - 3(7)] \div 8 + 3 &&\textit{Evaluate } 6^2. \\
&= 8[36 - 21] \div 8 + 3 &&\textit{Multiply 3 by 7.} \\
&= 8[15] \div 8 + 3 &&\textit{Subtract 21 from 36.} \\
&= 120 \div 8 + 3 &&\textit{Multiply 8 by 15.} \\
&= 15 + 3 &&\textit{Divide 120 by 8.} \\
&= 18 &&\textit{Add 15 and 3.}
\end{aligned}
$$

Algebraic expressions can be evaluated when the values of the variables are known. First, replace the variables by their values. Then, calculate the value of the numerical expression.

Example **The figure at the right is a rectangle.**

INTEGRATION
Geometry

a. Find the perimeter of the rectangle when $s = 5$.
b. Find the area of the rectangle.

s + 3 mm

s
mm

a. The perimeter of the rectangle is the sum of 2 times the measure of the width (s) and 2 times the measure of the length ($s + 3$).

$$
\begin{aligned}
P &= 2s + 2(s + 3) \\
&= 2(5) + 2(5 + 3) &&\textit{Replace the variable s with 5.} \\
&= 2(5) + 2(8) &&\textit{Add 5 and 3.} \\
&= 10 + 2(8) &&\textit{Multiply 2 and 5.} \\
&= 10 + 16 &&\textit{Multiply 2 and 8.} \\
&= 26 &&\textit{Add 10 and 16.}
\end{aligned}
$$

The perimeter is 26 mm.

20 Chapter 1 *Exploring Expressions, Equations, and Functions*

b. The area is the product of the width (s) and the length ($s + 3$).

$$A = s(s + 3)$$
$$= 5(5 + 3) \quad \textit{Replace the variable s with 5.}$$
$$= 5(8) \quad \textit{Add 5 and 3.}$$
$$= 40 \quad \textit{Multiply 5 and 8.}$$

The area is 40 mm². *Area is expressed in units squared.*

EXPLORATION

GRAPHING CALCULATORS

You can use a graphing calculator to evaluate algebraic expressions. Use a calculator to evaluate $\frac{0.25x^2}{7x^3}$ when $x = 0.75$.

Enter: .75 [STO▶] [X,T,θ] [2nd] [:] [(] .25 [X,T,θ] [x²] [)] [÷] [(] 7 [X,T,θ] [∧] 3 [)] [ENTER] 0.0476190476

Your Turn
a. Evaluate the expression when $x = 24.076$. 0.0014833978
b. Evaluate $\frac{2x^2}{(x^2 - x)}$ when $x = 27.89$. 2.074377092
c. Work through some of the examples in this lesson using a graphing calculator.

The fraction bar is another grouping symbol. It indicates that the numerator and denominator should each be treated as a single value.

$$\frac{2 \times 3}{1 + 2} \text{ means } (2 \times 3) \div (1 + 2) \text{ or } 2.$$

Example **④** Evaluate $\frac{x^3 + y^3}{x^2 - y^2}$ when $x = 4.2$ and $y = 1.8$.

$$\frac{x^3 + y^3}{x^2 - y^2} = \frac{(4.2)^3 + (1.8)^3}{(4.2)^2 - (1.8)^2}$$

Estimate: $\frac{4^3 + 2^3}{4^2 - 2^2} = \frac{64 + 8}{16 - 4} = \frac{72}{12}$ *or 6*

Use a scientific calculator.

Enter: [(] 4.2 [yˣ] 3 [+] 1.8 [yˣ] 3 [)] [÷] [(] 4.2 [x²] [−] 1.8 [x²] [)] [=] 5.55

Is the answer reasonable?

Check for Understanding

Exercises 1–12 are designed to help you assess your students' understanding through reading, writing, speaking, and modeling. You should work through Exercises 1–3 with your students and then monitor their work on Exercises 4–12.

Error Analysis

Students may tend to evaluate expressions from left to right, ignoring the correct order of operations. For example, they may evaluate $24 - 8 \div 2$ as $16 \div 2$ or 8 and not $24 - 4$ or 20. Emphasize the importance of looking at the entire expression to determine the correct order of operations before evaluating.

Assignment Guide

Core: 13–41 odd, 42, 43, 45–53
Enriched: 14–40 even, 42–53

For **Extra Practice,** see p. 756.

The red A, B, and C flags, printed only in the Teacher's Wraparound Edition, indicate the level of difficulty of the exercises.

Study Guide Masters, p. 3

Communicating Mathematics

Study the lesson. Then complete the following. 1. Multiply 7 and 2.

1. When evaluating the expression $4 + 7 \cdot 2$, what would you do first?
2. **Name** two types of grouping symbols and explain how you may use them. 2–3. See margin.
3. **Explain** how you would evaluate the expression $2[5 + (30 \div 6)]^2$.

Guided Practice

Evaluate each expression.

4. $15 + 3 \cdot 2$ 21
5. $5^3 + 3(4^2)$ 173
6. $\frac{38 - 12}{2 \cdot 13}$ 1
7. $12 \div 3 \cdot 5 - 4^2$ 4

Evaluate each expression when $w = 12$, $x = 5$, $y = 6$, and $z = 4$.

8. $wx - yz$ 36
9. $2w + x^2 - yz$ 25

10. Sample answer:
$xz - \frac{w}{y}$

10. Use two or more of the variables in Exercises 8–9 and their assigned values to write an expression whose value is 18.
11. If $a = 1.5$, which is less, $a^2 - a$ or 1? $a^2 - a$
12. **Geometry** Find the area of the rectangle when $t = 6$ inches. 78 in^2

Practice

Evaluate each expression.

A

13. $3 + 2 \cdot 3 + 5$ 14
14. $(4 + 5)7$ 63
15. $29 - 3(9 - 4)$ 14
16. $50 - (15 + 9)$ 26
17. $15 \div 3 \cdot 5 - 4^2$ 9
18. $4(11 + 7) - 9 \cdot 8$ 0

B

19. $\frac{(4 \cdot 3)^2 \cdot 5}{(9 + 3)}$ 60
20. $[7(2) - 4] + [9 + 8(4)]$ 51
21. $\frac{6 + 4^2}{3^2(4)}$ $\frac{11}{18}$
22. $(5 - 1)^3 + (11 - 2)^2 + (7 - 4)^3$ 172
23. $\frac{2 \cdot 8^2 - 2^2 \cdot 8}{2 \cdot 8}$ 6
24. $7(0.2 + 0.5) - 0.6$ 4.3

Evaluate each expression when $v = 5$, $x = 3$, $a = 7$, and $b = 5$.

25. $(v + 1)(v^2 - v + 1)$ 126
26. $\frac{x^2 - x + 6}{x + 3}$ 2
27. $[a + 7(b - 3)]^2 \div 3$ 147
28. $v^2 - (x^3 - 4b)$ 18
29. $(2v)^2 + ab - 3x$ 126
30. $\frac{a^2 - b^2}{v^3}$ $\frac{24}{125}$

Reteaching

Using Guess and Check Have students insert parentheses so that each expression has the indicated value.

1. $56 \div 6 + 2 \div 1$; 7
   $56 \div (6 + 2) \div 1$
2. $2 \cdot 8 - 2 + 6 \div 3$; 14
   $2 \cdot (8 - 2) + 6 \div 3$
3. $2 \cdot 8 - 2 + 6 \div 3$; 0
   $2 \cdot (8 - (2 + 6)) \div 3$
4. $20 - 2 \cdot 3^2 + 12 \div 6$; 0
   $20 - (2 \cdot 3^2 + 12 \div 6)$

Additional Answers

2. parentheses, brackets, braces, fraction bar; to show what operation should be completed first

3. First, divide 30 by 6, which equals 5. Second, add 5 and 5, which equals 10. Third, square 10, which equals 100. Fourth, multiply 100 times 2, which equals 200.

Write an algebraic expression for each verbal expression. Then, evaluate the expression when $r = 2$, $s = 5$, and $t = \frac{1}{2}$.

31. the square of r increased by $3s$ $r^2 + 3s$; 19
32. t times the sum of four times s and r $t(4s + r)$; 11
33. the sum of r and s times the square of t $(r + s)t^2$; $\frac{7}{4}$
34. r to the fifth power decreased by t $r^5 - t$; $31\frac{1}{2}$

Geometry

A formula for the perimeter of each figure is given. Find the perimeter when $a = 5$, $b = 6$, and $c = 8.5$.

35. triangle

$P = a + b + c$

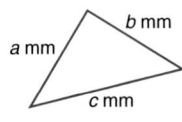

19.5 mm

36. square

$P = 4c$

c yd

34 yd

37. parallelogram

$P = 2(a + b)$

22 in.

Graphing Calculator

Use a graphing calculator to evaluate each expression to the nearest hundredth.

38. $5(2)^4 + 3$ 83

39. $\frac{(5 \cdot 7)^2 + 5}{(9 \cdot 3^2) - 7}$ 16.62

40. $4.79\,(0.05)^2 + 0.375\,(6.34)^3$ 95.58

41. $1 - 2u + 3u^2$ when $u = 1.35$ 3.77

Critical Thinking

42. **Patterns** Consider the value of the expression $\frac{2x - 1}{2x}$ for various values of x.

a. Copy and complete the chart below.

x	$\frac{1}{2}$	1	10	50	100
$\frac{2x-1}{2x}$	0	$\frac{1}{2}$	$\frac{19}{20}$	$\frac{99}{100}$	$\frac{199}{200}$

b. As the value of x gets larger, the value of the expression approaches 1.

b. Describe the pattern that is emerging.
c. What would the approximate value of the expression be if x were a very large number? 1

43. Consider the following sequence of numbers.

4 2 5 3 2

a. Insert operation symbols and parentheses in the sequence so that its value is 2. Sample answer: $(4 - 2)5 \div (3 + 2)$
b. Insert operation symbols and parentheses in the sequence so that the answer is the largest possible number. $4 \times 2 \times 5 \times 3 \times 2$

Applications and Problem Solving

44. **Accounting** Alicia and Travis are selling tickets for a school talent show. Bleacher seats cost $3.00, and floor seats cost $4.00. Alicia sells 30 bleacher seat tickets and 25 floor seat tickets. Travis sells 65 bleacher seat tickets.

a. Write an expression to show how much money Alicia and Travis have collected for the tickets. $30(3) + 25(4) + 65(3)$
b. How much money have they collected? $385

Lesson 1–3 Order of Operations **23**

4 ASSESS

Closing Activity

Writing Have students work as teams writing numerical expressions using only the four digits from the four-digit number that represents the current year. For example, 1997 can be represented as

$1 \cdot (9 - 9) \cdot 7 = 0$.

Compare answers to see what values were generated and how many different expressions were written for particular values.

Chapter 1, Quiz A (Lessons 1-1 through 1-3), is available in the *Assessment and Evaluation Masters*, p. 16.

Practice Masters, p. 3

1-3
NAME_____ DATE_____
Student Edition
Pages 19–24
Practice

Order of Operations
Evaluate each expression.

1. $(5 + 3) \div 2 + 2$ 6
2. $7 \cdot 5 - 3 \cdot 4$ 23
3. $3 \cdot 6 + 9 \div 3 - 6$ 15
4. $5^3 - 8 \cdot 5 + 6$ 91
5. $(3 + 6) \div 3^2$ 1
6. $6 \cdot (12 - 7.5) - 7$ 20
7. $(8 - 3)(12 \div 4) - 5$ 10
8. $2.1 \div (0.5 + 0.2)$ 3
9. $20 \div 4 \cdot 5 \cdot 2 \div 10$ 5
10. $125 \div [5(2 + 3)]$ 5
11. $(6 + 8)(8 - 3) \div (9 + 3 - 2)^2$ 0.7
12. $3(2^5 + 4^5)$ 72
13. $50 - \frac{1}{2}(17 + 5)$ 39
14. $6(0.2 + 0.3) - 0.25$ 2.75
15. $\frac{(6 + 2)^2}{16} + 3 \cdot 9$ 31
16. $\frac{8^2 - 6(4)}{2(5)} - 4$ 0
17. $[6^2 - (2 + 4)2]3$ 72
18. $\frac{6^3 - 4^2}{2(3 - 2)} - 2^3$ 2
19. $5[\frac{1}{2} + (\frac{3}{5} \cdot \frac{5}{6}) \div \frac{5}{8}]$ $6\frac{1}{2}$
20. $[\frac{3}{4} \cdot \frac{2}{3} - (\frac{1}{2} - \frac{1}{3})]12$ 4
21. $6 - [\frac{2 + 7}{3} - (2 \cdot 3 - 5)]$ 4
22. $12[10 - \frac{(5^2 - 6)3}{6}]$ 6

45. World Landmarks The Great Pyramid of Cheops in Egypt is considered to be one of the Seven Wonders of the World. It is also the largest pyramid in the world. The area of its base is 4050 square meters. The volume of any pyramid is one-third the product of the area of the base B and its height h.

Pyramid	Height (meters)
Great Pyramid of Cheops in Egypt	147
Bent Pyramid in Egypt	101
Inca Pyramid in Peru	75
Pyramid of the Sun in Mexico	60
Step Pyramid of Djoser in Egypt	60
Pyramid of the Sun in Peru	50

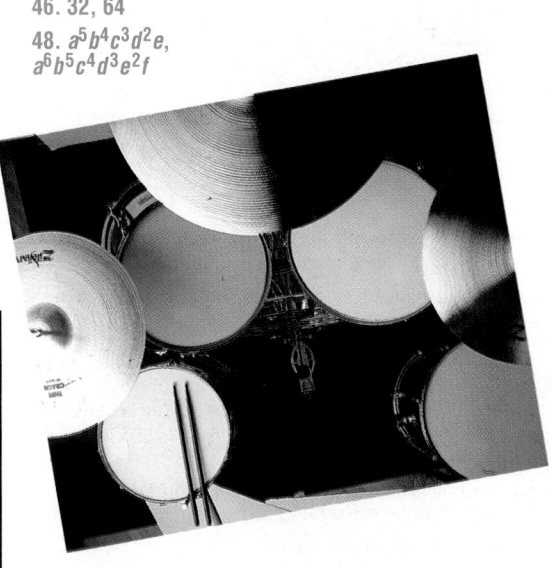

a. Write an expression that represents the volume of a pyramid. $\frac{1}{3}Bh$

b. Find the volume of the Great Pyramid of Cheops. **198,450 m³**

Mixed Review

46. 32, 64

48. $a^5b^4c^3d^2e$, $a^6b^5c^4d^3e^2f$

46. Find the next two terms in the sequence 2, 4, 8, 16, ... (Lesson 1–2)

47. Find the next two terms in the sequence 2, 5.5, 9, 12.5, ... (Lesson 1–2) **16, 19.5**

48. Find the next two items in the pattern a, a^2b, a^3b^2c, $a^4b^3c^2d$, ... (Lesson 1–2)

49. Ishi wants to save her money to buy a drum set from her uncle. Her uncle is willing to sell her the drums for $525. She has already saved $357, and she plans to add $1 on June 1, $2 on June 2, $3 on June 3, and so on. On what day will Ishi be able to buy the drums? (Lesson 1–2) **June 18**

50. Write an algebraic expression for h to the fifth power. (Lesson 1–1) h^5

51. The price of tickets for the spring concert is $3 more than last year. If last year's tickets cost t dollars, write an expression for the cost of the tickets this year. (Lesson 1–1) $t + 3$

52. Evaluate 11^2. (Lesson 1–1) **121**

53. Write a verbal expression for $9 + 2y$. (Lesson 1–1) **9 more than two times y**

24 Chapter 1 Exploring Expressions, Equations, and Functions

Enrichment Masters, p. 3

NAME_____ DATE_____

Student Edition Pages 19–24

1-3

Enrichment

Puzzling Primes

A *prime number* has only two factors, itself and 1. The number 6 is not prime because it has 2 and 3 as factors; 5 and 7 are prime. The number 1 is not considered to be prime.

1. Use a calculator to help you find the 25 prime numbers less than 100.

   2, 3, 5, 7, 11, 13, 17, 19, 23, 29, 31, 37 41, 43, 47, 53, 59, 61,

   67, 71, 73, 79, 83, 89, 97

Prime numbers have interested mathematicians for centuries. They have tried to find expressions that will give all the prime numbers, or only prime numbers. In the 1700s, Euler discovered that the expression $x^2 + x + 41$ will yield prime numbers for values of x from 0 through 39.

2. Find the prime numbers generated by Euler's formula for x from 0 through 7.

   41, 43, 47, 53, 61, 71, 83, 97

3. Show that the expression $x^2 + x + 31$ will not give prime numbers for very many values of x.

   It works for $x = 0, 2, 3, 5,$ and 6 but not for $x = 1, 4,$ and 7.

4. Find the largest prime number generated by Euler's formula. ___1601

Goldbach's Conjecture is that every nonzero even number greater than 2 can be written as the sum of two primes. No one has ever proved that this is always true. No one has disproved it, either.

5. Show that Goldbach's Conjecture is true for the first 5 even numbers greater than 2.

   $4 = 2 + 2, 6 = 3 + 3, 8 = 3 + 5, 10 = 3 + 7, 12 = 5 + 7$

6. Give a way that someone could disprove Goldbach's Conjecture.

   Find an even number that cannot be written as the sum of

   two primes.

Extension

Reasoning Determine all the possible values for the numerical expression $8 - 2 \cdot 3 + 1$ if you ignore the rules for order of operation.

There are five possible values: 0, 1, 3, 19, and 24.

Integration: Statistics
Stem-and-Leaf Plots

Objective

After studying this lesson, you should be able to:

- display and interpret data on a stem-and-leaf plot.

APPLICATION
Consumerism

Carmela Perez sells artificial joints to hospitals. She does a lot of highway driving and frequently works with the doctors who perform operations to replace joints. Her present car is old, and she needs to buy a new one. Before deciding which car to buy, she wants to find out the miles-per-gallon (MPG) ratios of the cars she likes. This information will help her decide which cars to investigate in detail before she chooses one to buy. The MPG ratios for 25 cars are listed below.

31	30	28	26	22	31	26	34	47
32	18	33	26	23	18	29	13	40
31	42	17	22	50	12	41		

A stem may have one or more digits. A leaf usually has just one digit.

Each day when you read newspapers or magazines, watch television, or listen to the radio, you are bombarded with numerical information about food, sports, the economy, politics, and so on. Interpreting this numerical information, or **data,** is important to your understanding of the world around you. A branch of mathematics called **statistics** helps provide you with methods of collecting, organizing, and interpreting data. One way to organize the data above clearly is by using a **stem-and-leaf plot.**

In a stem-and-leaf plot, the greatest common place value of the data is used to form the *stems.* The numbers in the next greatest place-value position are then used to form the *leaves.* In the list above, the greatest place value is tens. Thus, 31 miles per gallon would have stem 3 and leaf 1.

Classroom Vignette

"I use a version of stem-and-leaf plots all the time to show grades from each test."

Example:

10	0
9	5 6 9
9	0 3 4
8	9
8	0 3 3

Nancy Kinard
Nancy N. Kinard Palm Beach Gardens High School
Palm Beach Gardens, Florida

1-4 LESSON NOTES

NCTM Standards: 1–5, 10

Instructional Resources

- Study Guide Master 1-4
- Practice Master 1-4
- Enrichment Master 1-4

Transparency 1-4A contains the 5-Minute Check for this lesson; **Transparency 1-4B** contains a teaching aid for this lesson.

Recommended Pacing	
Standard Pacing	Days 5 & 6 of 15
Honors Pacing	Days 4 & 5 of 14
Block Scheduling*	Day 3 of 7
Alg. 1 in Two Years*	Days 7 & 8 of 22

*For more information on pacing and possible lesson plans, refer to the *Block Scheduling Booklet* and *Algebra 1 in Two Years.*

1 FOCUS

5-Minute Check
(over Lesson 1-3)

Evaluate each expression.

1. $5 + 6 \cdot 3 + 7$ 30
2. $3(10 + 6) - 8 \cdot 6$ 0
3. $5(3^2 + 2^3)$ 85

Evaluate if $x = 4$, $y = 2$, and $z = 5$.

4. $3xz - y^3$ 52
5. $6x + y^2 - z^2$ 3

Motivating the Lesson

Hands-On Activity Have students pair off and measure one another's height, recording the data in inches rounded to the nearest whole number. Students should write their partner's height on the board in no particular order. Then ask students to think of ways to organize and present the data for the entire class.

In-Class Example

For Example 1
Use the stem-and-leaf plot of the history test scores given below to answer each question.

Stem	Leaf
9	1 5
8	0 2 6 8
7	3 8
6	5 9

7|3 = 73

a. In which interval did most students score? **80–89**

b. What were the highest and lowest scores? **95, 65**

c. How many scores were between 90 and 99?
2 scores

To make the stem-and-leaf plot, first make a vertical list of the stems. Since the mileage data range from 12 to 50, the stems range from 1 to 5. Then, plot each number by placing the units digit (leaf) to the right of its correct stem. Thus, the mileage 31 is plotted by placing leaf 1 to the right of stem 3. Include a key with the plot. The complete stem-and-leaf plot is shown below.

Stem	Leaf
1	8 8 3 7 2
2	8 6 2 6 6 3 9 2
3	1 0 1 4 2 3 1
4	7 0 2 1
5	0 3|1 = 31 ← Key

A second stem-and-leaf plot can be made to arrange the leaves in numerical order from least to greatest as shown below. This will make it easier for Ms. Perez to analyze the data.

Stem	Leaf
1	2 3 7 8 8
2	2 2 3 6 6 6 8 9
3	0 1 1 1 2 3 4
4	0 1 2 7
5	0 3|1 = 31

Example ❶

APPLICATION
Consumerism

Use the information in the stem-and-leaf plot shown above to answer each question.

a. Which MPG ratio did Ms. Perez plot most frequently?
26 and 31 (each three times)

b. What are the highest and lowest MPG ratios?
50 and 12

c. Each line of the stem-and-leaf plot represents an interval of the data. In which mileage interval did Ms. Perez find the most cars?
20–29 miles per gallon (8 cars)

d. How many cars have a ratio of between 20 and 39 miles per gallon?
15 cars

e. If you were Ms. Perez, which cars might you investigate more closely? Why?
Ms. Perez might want to investigate the cars represented in the last two lines of the stem-and-leaf plot, since she should get more miles per gallon with these cars than the other cars.

Sometimes the data for a stem-and-leaf plot are numbers that mostly begin with the same digit. In this case, use the digits in the first two places to form the stems.

Example 2

APPLICATION

Health

The weights (in pounds) of the students in a health and nutrition class are listed below. Make a stem-and-leaf plot of the student weights and answer the questions.

102	117	119	147	135	148	122	137	103
116	147	152	117	149	108	123	130	123
147	112	133	99	101	135	138	155	118
142	103	159	131	137	156	149	120	98

Since the data range from 159 to 98 pounds, the stems range from 15 to 9.

Notice that the numbers in the stem can be placed in ascending or descending order.

Stem	Leaf
15	9 6 5 2
14	9 9 8 7 7 7 2
13	8 7 7 5 5 3 1 0
12	3 3 2 0
11	9 8 7 7 6 2
10	8 3 3 2 1
9	9 8 *12|3 = 123*

a. What does 14|8 represent on the plot?
It represents 148 pounds.

b. Which interval has the most students in it?
130–139 pounds (8 students)

c. What is the difference between the lowest and highest weight?
159−98 or 61 pounds

d. Which weight occurred most frequently?
147 pounds (3 students)

A *back-to-back stem-and-leaf plot* can also be used to compare two related sets of data.

Example 3

Kyle and Mikito wanted to compare boys' and girls' heights. They measured the height (in inches) of every student in their class. The data they collected and stem-and-leaf plot they made are shown below.

Boys' Heights (in.)		Girls' Heights (in.)	
65	60	72	64
63	70	57	60
69	72	61	63
71	66	65	62
73	71	59	61
59	58	61	71

(continued on the next page)

Cooperative Learning

Co-op Co-op Separate the class into 4 or 5 small groups. Tell each group it is to poll at least 20 individuals regarding the number of hours a week that they watch TV. Assign each group a particular age range for the individuals they are to poll. Have each group summarize its data in a stem-and-leaf plot. Display the plots to the class to determine whether a relationship exists between age and amount of time watching TV. For more information on the co-op co-op strategy, see *Cooperative Learning in the Mathematics Classroom,* one of the titles in the Glencoe Mathematics Professional Series, page 32.

For Example 3
Measurements were taken of a stand of oak and maple trees. Below is a table summarizing heights rounded to the nearest foot. Construct a stem-and-leaf plot and answer the questions.

Oaks		Maples	
35	51	28	27
42	28	35	28
38	35	38	31
35	48	27	22
47	42	22	32
28	38	41	35
38	36	32	29
30	42	35	27

Oak	Stem	Maple
1	5	
22278	4	1
05556888	3	122558
88	2	22777889

$3 \mid 1 = 31$

a. What is the height of the shortest oak? The shortest maple? **28, 22**

b. What is the difference in height between the shortest oak and the shortest maple? **6**

c. What does $3 \mid 5$ represent in each plot? **35 feet**

d. What is the greatest number of oaks that are the same height? What is the greatest number of maples that are the same height? **3, 3**

e. How many maples are over 35 feet tall? How many oaks are over 40 feet tall? **2, 6**

Teaching Tip Notice that all of the leaves in the examples are in numerical order. Encourage students to automatically order the leaves.

3 PRACTICE/APPLY

Check for Understanding
Exercises 1–10 are designed to help you assess your students' understanding through reading, writing, speaking, and modeling. You should work through Exercises 1–4 with your students and then monitor their work on Exercises 5–10.

In this case, the heights of the boys and the heights of the girls are to be compared. To compare these numbers most effectively, use a back-to-back stem-and-leaf plot.

Boys	Stem	Girls
3 2 1 1 0	7	1 2
9 6 5 3 0	6	0 1 1 1 2 3 4 5
9 8	5	7 9 $5 \mid 7 = 57$

a. **What is the height of the shortest boy? the shortest girl?**
The shortest boy is 58 inches tall. The shortest girl is 57 inches tall.

b. **What is the difference in height between the shortest boy and the tallest girl?**
$72 - 58$ or 14 inches

c. **What does $6 \mid 3$ represent in each plot?**
63 inches

d. **What is the greatest number of boys who are the same height? What is the greatest number of girls who are the same height?**
There are 2 boys who are 71 inches tall and 3 girls who are 61 inches tall.

e. **What patterns, if any, do you see in the data?**
The boys appear to be slightly taller than the girls. Seven boys are 65 inches or taller, while only 2 of the girls are that tall.

CHECK FOR UNDERSTANDING

Communicating Mathematics

Study the lesson. Then complete the following.

1. **Describe** the information you can determine by looking at a stem-and-leaf plot. **See margin.**

2. **Show** how the number 478.1 would be plotted if the number 254.3 is plotted using stem 25 and leaf 4. **stem 47, leaf 8 (47 | 8)**

3. **List** the steps used to make a stem-and-leaf plot. **See margin.**

 MATH JOURNAL

4. **Assess Yourself** Write a paragraph about your use of stem-and-leaf plots. Begin the paragraph by completing this sentence: "I can use stem-and-leaf plots to _____ ." Think of the ways in which you deal with numbers: earnings from a job, number of assignments turned in, test scores, attendance at club meetings, and so on. **See margin.**

Guided Practice

Suppose the number 25,678 is rounded to 25,700 and plotted using stem 25 and leaf 7. Write the stem and leaf for each number below if the numbers are part of the same set of data. **7. stem 126, leaf 9**

5. 12,221 **stem 12, leaf 2** 6. 6323 **stem 6, leaf 3** 7. 126,896

8. Write the stems that would be used for a plot of the following set of data.

57, 43, 34, 12, 29, 8 **0, 1, 2, 3, 4, 5**

Reteaching

Using Discussion Have students discuss the advantages and disadvantages of stem-and-leaf plots. Have them compare a stem-and-leaf plot to a simple list of data. Emphasize that while original data must be reconstructed from a stem-and-leaf plot, it is a more concise representation of the data and displays the data in order.

Additional Answers

1. **Answers will vary. Responses should include least and greatest values, frequency in categories, distribution, clusters, and gaps.**

3a. **Decide on stems.**

3b. **Place stems and leaves on a chart.**

3c. **Arrange leaves in numerical order.**

9. Use the stem-and-leaf plot below to answer each question.

Number of Prom Tickets Sold During a 20-Day Period

Stem	Leaf
3	1 3 5
4	5 5 6 7
5	0 0 1 1 2 3 3
6	3 4 4 4
7	1 5 *5|1 = 51*

a. What does the entry 3│5 represent?   **35 tickets sold one day**
b. What was the least number of tickets sold?   **31 tickets**

9c. 75 tickets

c. What was the greatest number of tickets sold during the 20-day period?
d. What was the total number of tickets sold?   **1043 tickets**

10. Architecture   The *World Almanac* lists 23 tall buildings in Denver, Colorado. Each number below represents the number of floors for one of these buildings.

54	52	43	41	40	36	35	31
32	34	42	32	29	26	28	33
31	30	30	29	27	26	56	

a. Make a stem-and-leaf plot of these data.   **See margin.**
b. What was the least number of floors?   **26**
c. What was the greatest number of floors?   **56**
d. How many buildings had 35 floors or more?   **9**
e. How many buildings were in the 20–29 floor range?   **6**

EXERCISES

Practice

Suppose the number 178,651 is rounded to 179,000 and plotted using stem 17 and leaf 9. Write the stem and leaf for each number below if the numbers are part of the same set of data.

11. stem 13, leaf 3

11. 133,271   **12.** 44,589   **13.** 442,672

12. stem 4, leaf 5

13. stem 44, leaf 3

14. 99,278   **15.** 1,112,750   **16.** 8443
  stem 9, leaf 9   stem 111, leaf 3   stem 0, leaf 8

Suppose the number 0.0478 is rounded to 0.048 and plotted using stem 4 and leaf 8. Write the stem and leaf for each number below if the numbers are part of the same set of data.

17. 0.14278   **18.** 0.00997   **19.** 1.114
  stem 14, leaf 3   stem 1, leaf 0   stem 11, leaf 4

Write the stems that would be used for a plot of each set of data.   **20–22. See margin.**

20. 123, 436, 507, 449, 278, 489, 134, 770, 98, 110, 398
21. 12,367; 24,003; 27,422; 9447; 39,550; 38,045; 40,196
22. 37.2, 8.9, 12.74, 33.5, 27, 17.001, 13.5, 29.6

Additional Answers

4. Sample answer: I can use stem-and-leaf plots to keep a record of my math test scores. From this plot, I can determine my lowest and highest scores. I can also determine the interval where most of my scores fall.

10a.
Stem	Leaf
2	6 6 7 8 9 9
3	0 0 1 1 2 2 3 4 5 6
4	0 1 2 3
5	2 4 6 *4|3 = 43*

20. 9, 11, 12, 13, 27, 39, 43, 44, 48, 50, 77
21. rounded to the nearest hundred 9, 12, 24, 27, 38, 39, 40
22. 8, 12, 13, 17, 27, 29, 33, 37

Assignment Guide

Core: 11–25 odd, 27–31
Enriched: 12–22 even, 23–31

For **Extra Practice,** see p. 757.

The red A, B, and C flags, printed only in the Teacher's Wraparound Edition, indicate the level of difficulty of the exercises.

Study Guide Masters, p. 4

NAME_____   DATE_____
1-4   **Study Guide**   Student Edition Pages 25–31

Integration: Statistics
Stem-and-Leaf Plots

Mrs. Corrigan interviewed some students who were willing to help her with some typing. The number of words typed per minute by each student is listed below.

20	19	18	16	17	17	14

This data can be organized on a **stem-and-leaf plot.** The greatest common place value of each piece of data is used to form the **stem.** The next greatest place value is used to form the **leaves.**

Stem	Leaf
1	9 8 6 7 7 4
2	0

1|8 represents a rate of 18 words per minute.

Data with more than two digits may be rounded [13.5 ⇒ 14] or truncated [13.5 ⇒ 13].

The rounded and truncated values of the data at the right can be compared in a back-to-back stem-and-leaf plot.

61.5	51.9	60.0
46.0	38.5	39.1
56.2	33.0	61.5
65.4	51.0	52.3

Rounded	Stem	Truncated
9 9 3	3	3 8 9
6	4	6
6 2 2 1	5	1 1 2 6
5 2 2 0	6	0 1 1 5

In a back-to-back stem-and-leaf plot, the same stem is used for the leaves of both plots.

Solve each problem.

1. The stem-and-leaf plot at the right shows the height, in feet, of buildings in San Francisco that are over 500 ft tall.

Stem	Leaf
5	2 3 3 4 5 6 7 7 7
6	
7	8
8	5

a. How tall is the tallest building?   850 ft
b. What is the height of the shortest building that is taller than 500 ft?   520 ft
c. What building height occurs most frequently?   570 ft

5|6 represents a height of 560 feet.

2. Lori sells tickets at a movie theater. Her sales for week 1 were $173, $194, $160, $182, $183, and $247. Sales for week 2 were $137, $182, $151, $193, $199, and $194.
a. Make a stem-and-leaf plot of two weeks' sales. Round the data to the nearest $10.

Stem	Leaf
1	3 5 6 7 8 8 8 9 9 9 9
2	4

b. What is the most amount of money made in one day during the two week period? $247

1|6 represents $160-$169.

Closing Activity

Speaking Have students explain the steps to follow in making a stem-and-leaf plot.

Additional Answers

23a. Sample answer: Both teens and young adults had similar distribution of responses.

23b. Sample answer: The market research shows that the new game is equally appealing to teens and young adults. Therefore, we should concentrate our marketing efforts toward both teens and young adults.

Critical Thinking

23. Manufacturing A video game company tests its products before selling them in stores. Its latest product is an inexpensive space detective adventure. The company would like to know whether to market the game to teens or young adults. A group of 25 teens and 25 young adults rated the game after playing it. The players rated the game for features like graphics, level of difficulty, and player interaction. The ratings consisted of a score between 1 (Trash it, man!) and 60 (Excellent, dude!). The players' ratings are listed below.

Teens (13–19 years old)	19 45 22 44 30 35 41 43 18 35 21 43 30
	57 17 27 20 35 15 41 57 22 33 41 55
Young Adults (20–25 years old)	51 43 36 51 42 33 27 48 42 31 48 26 48
	31 38 26 45 13 37 25 52 12 44 12 37

a. Should this game be sold to teens or young adults (or to both groups)?

b. Write a memo to the company with a recommendation. Be sure to give reasons for your decision.   a–b. See margin for sample answers.

Applications and Problem Solving

24. Geology According to the *Universal Almanac,* there were 40 major earthquakes from June, 1990, to June, 1994. The stem-and-leaf plot below shows the magnitude of each earthquake in Richter scale units.

Stem	Leaf
5	1 4 4 5 6 9 9
6	0 0 1 2 2 2 2 4 4 5 8 8 8 8 8 8
7	0 0 1 2 2 2 2 3 4 5 5 5 6 7 7
8	0 7|3 = 7.3

24a. 6.8

a. What was the most frequent magnitude of the earthquakes?

b. What was the highest Richter scale value recorded for an earthquake? **8.0**

c. How many more earthquakes had magnitudes in the 6.0–6.9 range than in the 5.0–5.9 range? **9 more**

d. What was the magnitude of the weakest earthquake? **5.1**

Practice Masters, p. 4

NAME_____ DATE _____

1-4 Practice

Student Edition Pages 25–31

Integration: Statistics
Stem-and-Leaf Plots

1. The stem-and-leaf plot at the right below shows the scores for a 100-point unit exam in a science class.

a. What were the highest and lowest scores on this test? 96, 52

b. Which test score occurred most frequently? 74

c. In which 10-point interval did most of the students score? 70-80

d. How many students received a score of 70 or better? 13

e. How many students received a score less than 70? 7

Stem	Leaf
5	2 4 6 6
6	1 5 8
7	2 4 4 4 5 7
8	2 3 3
9	0 2 5 6

7|2 = 72

2. Each number below represents the age of a worker at Fred's Fast Food.

20 52 21 39 40 58 27 48 36 20 51 26
45 30 49 22 59 50 33 35 28 43 55 20

a. Make a stem-and-leaf plot of the data. Use the space provided.

b. How many people work at Fred's Fast Food? 24

c. What is the difference in the ages between the oldest and youngest workers at Fred's? 39 years

d. What is the most common age for a worker? 20 years

e. Which age group is most widely represented? people in their 20s

f. How many workers are older than 35 years? 13

Stem	Leaf
2	0 0 0 1 2 6 7 8
3	0 3 5 6 9
4	0 3 5 8 9
5	0 1 2 5 8 9

2|1 = 21

3. Each number below represents round lots of XXYZ stock traded each day over a 28-day period.

814 722 144 413 812 118 214 533 219 254
183 733 624 121 325 166 722 766 814 723
325 445 388 357 746 537 155

a. Make a stem-and-leaf plot of the data in the space provided. Round numbers of shares to the nearest ten.

b. How many times were more than 300 round lots of stock traded? 18 times

Stem	Leaf
1	2 2 4 6 7 8
2	1 2 4 5
3	3 3 6 9
4	1 5
5	3 4
6	2
7	2 2 2 3 5 7
8	1 1 1

2|4 = 235–244

Tech Prep

Geologist Students who are interested in geology may wish to do further research on the statistics given in Exercise 24 and explore the potential growth of this career. For more information on tech prep, see the *Teacher's Handbook.*

25. Economics The table at the right shows the average acreage per farm for six western states in 1980 and 1993 according to the *Universal Almanac*.

State	1980	1993
Arizona	5080	4557
Colorado	1358	1286
Montana	2601	2445
Nevada	3100	3708
New Mexico	3467	3274
Wyoming	3846	3742

a. Round each number to the nearest hundred. Then make a back-to-back stem-and-leaf plot of the average farm sizes for 1980 and 1993 for these six states.   **See margin.**

b. Which interval shows the most common farm size found in 1980 for these six states? in 1993?   **3000 to 3900; same**

c. Write a statement about the information the stem-and-leaf plot presents.
Sample answer: Farm sizes seem to be shrinking.

26. Geography The numbers below show the lengths (in miles) of 30 major rivers in North America as listed in the *World Almanac*.

424	313	444	301	659	652	314	538
377	800	883	525	360	512	500	722
865	360	390	425	309	336	430	692
540	610	800	350	300	420		

a. Round each number to the nearest ten. Then make a stem-and-leaf plot of the data.   **See margin.**

b. What is the difference in length between the shortest and longest rivers?   **580 miles**

c. How many rivers are less than 400 miles long?   **11**

Mixed Review

27. Evaluate $3 \cdot 6 - \frac{12}{4}$.   (Lesson 1–3)   **15**

28. Evaluate $9a - 4^2 + b^2 \div 2$ when $a = 3$ and $b = 6$.   (Lesson 1–3)   **29**

29. Weather The time between seeing lightning flash and hearing the bang of thunder it produces can be used to approximate the distance from the lightning. The distance in miles from lightning can be approximated by dividing the number of seconds between seeing the lightning and hearing the thunder by 5. Lightning within three miles can be dangerous and is a warning to take shelter.

a. Suppose there are s seconds between seeing the lightning and hearing the thunder. Write an algebraic expression for the distance from the lightning.   (Lesson 1–1) $\frac{s}{5}$

b. Fernando counted 10 seconds between seeing lightning and hearing thunder. How far away was the lightning? Is Fernando in danger?
(Lesson 1–3)   **2 miles; yes**

30. Find the next two terms in the sequence $\frac{1}{2}, \frac{3}{4}, \frac{5}{8}, \frac{7}{16}, \frac{9}{32}, \dots$ $\frac{11}{64}, \frac{13}{128}$
(Lesson 1–2)

31. Give the next two items for the following pattern.   (Lesson 1–2)

Lesson 1–4   **INTEGRATION**   *Statistics*   *Stem-and-Leaf Plots*   **31**

Extension

Reasoning The table below shows the percent of households in two major cities that own the appliances listed. Explain why a stem-and-leaf plot of the data would not be an adequate representation.

Even though a back-to-back stem-and-leaf plot would distinguish between the data for City A and City B, it would not distinguish between the data for TVs and dishwashers.

Appliance	City A	City B
TV	88	94
Dishwasher	28	43

Additional Answers

25a.

1980	Stem	1993
4	1	3
6	2	4
8 5 1	3	3 7 7
	4	6
1	5	

$3|7 = 3700$

26a.

Stem	Leaf
3	0 0 1 1 1 4 5 6 6 8 9
4	2 2 3 3 4
5	0 1 3 4 4
6	1 5 6 9
7	2
8	0 0 7 8

$6|5 = 650$

Enrichment Masters, p. 4

Instructional Resources

- Study Guide Master 1-5
- Practice Master 1-5
- Enrichment Master 1-5
- Assessment and Evaluation Masters, pp. 15–16
- Multicultural Activity Masters, p. 1
- Real-World Applications, 1
- Science and Math Lab Manual, pp. 1–4

 Transparency 1-5A contains the 5-Minute Check for this lesson; **Transparency 1-5B** contains a teaching aid for this lesson.

Recommended Pacing

Standard Pacing	Day 7 of 15
Honors Pacing	Day 6 of 14
Block Scheduling*	Day 4 of 7 (along with Lesson 1-6)
Alg. 1 in Two Years*	Days 9 & 10 of 22

 *For more information on pacing and possible lesson plans, refer to the *Block Scheduling Booklet* and *Algebra 1 in Two Years.*

1 FOCUS

 ### 5-Minute Check
(over Lesson 1-4)

1. Use a stem-and-leaf plot to represent the following test data: 56, 75, 73, 89, 92, 87, 64, and 82.

Stem	Leaf	
9	2	
8	2 7 9	
7	3 5	
6	4	
5	6	9|2 = 92

2. Based on the plot, in what interval did the most grades fall? **80–89**
3. Given the data 125, 128, 134, 158, 164, 144, 168, 172, and 165, what stems would you use to represent the data? **12, 13, 14, 15, 16, 17**

Objective
After studying this lesson, you should be able to:
- solve open sentences by performing arithmetic operations.

The largest hamburger in the world was made in Seymour, Wisconsin, on August 2, 1989. It was 21 feet in diameter and weighed 5520 pounds.

APPLICATION
Nutrition

The grams of fat found in cheeseburgers sold by various fast food chains are listed in the stem-and-leaf plot at the right.

Stem	Leaf
1	3 3 3 6 9
2	0 5 8 9
3	6 9
4	6
6	3

4\|6 = 46

Obviously, some of the fast food chains are selling burgers with more than the usual amount of fat. Some students at Middletown High School have formed a consumer awareness club and are concerned about the high content of fat in this particular cheeseburger. They decide to write a letter to the company recommending a way to reduce the fat of their cheeseburger to 37 grams.

The students know that a one-ounce slice of American cheese has about 7 grams of fat and that ground beef has about 6 grams of fat per ounce. The students need to determine the number of ounces of ground beef that can be used to make a cheeseburger with a fat content of 37 grams.

Let b represent the number of ounces of beef that will go into the cheeseburger. This problem can be represented by the equation below.

$$6b + 7 = 37$$

The number 6 represents the amount of fat in one ounce of ground beef. The 7 represents the number of grams of fat in one slice of cheese. The 37 represents the total number of grams of fat for a cheeseburger.

Mathematical statements with one or more variables, or unknown numbers, are called **open sentences.** An open sentence is neither true nor false until the variable has been replaced by a value. Finding a replacement for the variable that results in a true sentence is called **solving the open sentence.** This replacement is called a **solution** of the open sentence.

32 Chapter 1 Exploring Expressions, Equations, and Functions

A little arithmetic will show that each square foot of the hamburger weighed 15.9 lbs. Have students use the current price of hamburger to determine the cost of this giant burger.

Replace b in $6b + 7 = 37$ with the values 3, 4, 5, and 6. Then see whether each replacement results in a true or false sentence.

Replace b with:	$6b + 7 = 37$	True or False?
3	$6(3) + 7 \stackrel{?}{=} 37 \rightarrow 25 \neq 37$	false
4	$6(4) + 7 \stackrel{?}{=} 37 \rightarrow 31 \neq 37$	false
5	$6(5) + 7 \stackrel{?}{=} 37 \rightarrow 37 = 37$	true
6	$6(6) + 7 \stackrel{?}{=} 37 \rightarrow 43 \neq 37$	false

Since $b = 5$ makes the sentence $6b + 7 = 37$ true, the solution for $6b + 7 = 37$ is 5. The students can write a letter suggesting that the chain reduce the amount of fat to 37 grams by using 5 ounces of ground beef in their cheeseburger.

A set of numbers from which replacements for a variable may be chosen is called a **replacement set**. A **set** is a collection of objects or numbers. Sets are often shown by using braces { }. Each object or number in a set is called an **element**, or member. Sets are usually named by capital letters. Set A has three elements; they are 1, 3, and 5.

$$A = \{1, 3, 5\} \qquad B = \{2, 4, 5\} \qquad C = \{1, 2, 3, 4, 5\}$$

The **solution set** of an open sentence is the set of all replacements for the variable that make the sentence true.

Example **Find the solution set for $y + 5 \leq 7$ if the replacement set is $\{0, 1, 2, 3, 4\}$.**

Replace y with:	$y + 5 \leq 7$	True or False?
0	$0 + 5 \stackrel{?}{\leq} 7 \rightarrow 5 \leq 7$	true
1	$1 + 5 \stackrel{?}{\leq} 7 \rightarrow 6 \leq 7$	true
2	$2 + 5 \stackrel{?}{\leq} 7 \rightarrow 7 \leq 7$	true
3	$3 + 5 \stackrel{?}{\leq} 7 \rightarrow 8 \nleq 7$	false
4	$4 + 5 \stackrel{?}{\leq} 7 \rightarrow 9 \nleq 7$	false

The symbol $\leq$ means "less than or equal to." The symbol $\geq$ means "greater than or equal to."

Therefore, the solution set for $y + 5 \leq 7$ is $\{0, 1, 2\}$.

A sentence that contains an equals sign, $=$, is called an **equation**. A sentence having the symbols $<$, $\leq$, $>$, or $\geq$ is called an **inequality**. Which of the open sentences below are equations? Which ones are inequalities?

Mathematical Sentence	Equation or Inequality?
$2x + 10 = 50$	equation
$3a \leq 43$	inequality
$y - 6 > 12$	inequality

Sometimes you can solve an equation by simply applying the order of operations.

Lesson 1–5 Open Sentences **33**

Motivating the Lesson

Situational Problem Ramon, Kesha, Larry, and Maria are going to the movies together. After buying her ticket, Kesha has $1.75. She wants to buy a bucket of popcorn, which costs $2.90. Ramon has $1.25, Larry has $1.10, and Maria has $1.18. How can Kesha determine which of her friends does not have enough money to help her buy the popcorn? **She should add the money that she has ($1.75) to the money that each of the friends has ($1.25, $1.10, and $1.18) and determine which sum is less than $2.90. Since $1.75 + $1.10 < $2.90, Larry is the one who does not have enough money.**

2 TEACH

Teaching Tip In mathematics, sentences that can be classified as true or false are called *statements*. In general, *open sentences* are true for some variable replacements and false for others. The original sentence is neither true nor false.

In-Class Example

For Example 1
Find the solution set for each open sentence if the replacement set is $\{0, 2, 4, 6, 8\}$.

a. $8 - x < 7$ $\{2, 4, 6, 8\}$

b. $\dfrac{x + 3}{4} > 3$ $\{ \}$ or $\emptyset$

Teaching Tip Point out to students that an open sentence in mathematics, like that in Example 1, is like an English sentence with a blank in it. The sentence _____ *is a color* is true if *Blue* is placed in the blank, false if *Hard* is placed in the blank, and neither true nor false if the blank is left empty.

3 PRACTICE/APPLY

Check for Understanding
Exercises 1–12 are designed to help you assess your students' understanding through reading, writing, speaking, and modeling. You should work through Exercises 1–5 with your students and then monitor their work on Exercises 6–12.

34 Chapter 1

CHECK FOR UNDERSTANDING

Communicating Mathematics

1–4. See Solutions Manual.

Study the lesson. Then complete the following.

1. **Explain** why an open sentence always has at least one variable.
2. **Explain** the difference between an expression and an open sentence.
3. **Define** in your own words the phrase *solution set of an open mathematical sentence*.
4. **Explain** how to find the solution set for $7 + 2n > 31$ if the replacement set for n is $\{10, 11, 12, 13\}$.

 MATH JOURNAL

5. Make up an inequality with a replacement set. Find the solution set and explain how you obtained it.   See student's work.

Guided Practice

State whether each equation is *true* or *false* for the value of the variable given.

6. $7(x^2) - 15 \div 5 = 25, x = 2$ true
7. $\dfrac{7a + a}{(9 \cdot 3) - 7} = 4, a = 5$ false

Find the solution set for each inequality, given the replacement set.

8. $3x + 2 > 2; \{0, 1, 2\}$ $\{1, 2\}$
9. $2y^2 - 1 > 0; \{1, 3, 5\}$ $\{1, 3, 5\}$

Solve each equation for y if x is replaced with 6.

10. $x - 2 = y$ 4
11. $2x^2 + 3 = y$ 75

34   *Chapter 1   Exploring Expressions, Equations, and Functions*

Reteaching ▬▬▬

Decision Making State whether each sentence is *true, false,* or *open.* If the sentence is open, replace the variable with a value to make the sentence true.

1. $2 + 3 \cdot 5 = 30$ false
2. $2 + 3 \cdot 5 = x$ open; 17
3. $40 - 8 \div 2^2 = 38$ true
4. $x - 8 \div 2^2 = 24$ open; 26

 ## Cooperative Learning

Group Discussion Separate the class into small groups. Have them fill in the blanks in the open sentence, "The green ____ is bigger than the blue ____" in as many ways as possible. Discuss how students might have filled in this open sentence to make it false. For more information on the group discussion strategy, see *Cooperative Learning in the Mathematics Classroom,* one of the titles in the Glencoe Mathematics Professional Series, page 31.

12. Diet During a lifetime, the average American drinks 15,579 glasses of milk, 6220 glasses of fruit juice, and 18,995 glasses of soft drinks.
    a. Write an equation for the total number of glasses of milk, juice, and soft drinks that the average American drinks in a lifetime.
    b. What is the total number of glasses of milk, juice, and soft drinks consumed by the average American in a lifetime? **40,794 glasses**
    12a. $g = 15{,}579 + 6220 + 18{,}995$

EXERCISES

Practice

State whether each equation is *true* or *false* for the value of the variable given.

 A

13. $a + \frac{3}{4} = \frac{3}{2} + \frac{1}{4}$, $a = \frac{1}{2}$ false   **14.** $\frac{3 + 15}{x} = \frac{1}{2}(x)$, $x = 6$ true

15. $y^6 = 4^3$, $y = 2$ true   **16.** $3x^2 - 4(5) = 6$, $x = 3$ false

17. $\frac{5^2 - 2y}{5^2 - 6} \leq 1$, $y = 3$ true   **18.** $a^5 \div 8 \div a^2 \div a < \frac{1}{2}$, $a = 2$ false

Find the solution set for each inequality if the replacement sets are $x = \left\{ \frac{1}{2}, \frac{3}{4}, 1, \frac{5}{4} \right\}$ and $y = \{5, 10, 15, 20\}$.

20. $\{10, 15, 20\}$

21. $\left\{ \frac{1}{2}, \frac{3}{4} \right\}$

 B

19. $y - 2 < 6$ $\{5\}$ **20.** $y + 2 > 7$ **21.** $8x + 1 < 8$

22. $2x > 1$ $\left\{ \frac{3}{4}, 1, \frac{5}{4} \right\}$ **23.** $\frac{y}{5} \geq 2$ $\{10, 15, 20\}$ **24.** $3x \leq 4$ $\left\{ \frac{1}{2}, \frac{3}{4}, 1, \frac{5}{4} \right\}$

Solve each equation.

25. $y = \frac{14 - 8}{2}$ 3   **26.** $4(6) + 3 = a$ 27   **27.** $\frac{21 - 3}{12 - 3} = x$ 2

C

28. $d = 3\frac{1}{2} \div 2$ $1\frac{3}{4}$ **29.** $s = 4\frac{1}{2} + \frac{1}{3}$ $4\frac{5}{6}$ **30.** $x = 5^2 - 2^3$ 17

Critical Thinking

31. Find five pairs of values for p and q such that the open sentence $q + 2 > 3p$ is true. **See margin.**

Applications and Problem Solving

32. Weather Forecasting The graph at the right shows the states with the most tornadoes.

Source: National Severe Storm Forecast Center

    a. Write an equation that estimates the number of tornadoes Texas will have in the next three years. Justify why your equation will provide a good estimate. **See margin.**
    b. At this rate, how many tornadoes will Texas have in the next three years? **432 tornadoes**
    c. Make up another problem using the data in the graph. **See margin.**

33a. $c = \frac{3500 \cdot 4}{14}$

33b. 1000 calories

33. Nutrition A person must burn 3500 calories to lose one pound of weight.
    a. Write an equation that represents the number of calories a person would have to burn a day to lose 4 pounds in two weeks.
    b. How many calories would the person have to burn each day?

Lesson 1–5 Open Sentences   **35**

 Tech Prep

Meteorologist Meteorologists must often look to data like that in Exercise 32 as they prepare long-range weather forecasts. Students who are interested in meteorology may wish to find out more about how meteorologists use that information in the science of weather forecasting. For more information on tech prep, see the *Teacher's Handbook*.

Extension

Communication Write a verbal statement for each open sentence.
1. $2 + 6x = 26$ Two more than six times a number is equal to 26.
2. $5(r - 4) = 20$ Five times the difference of a number and 4 is 20.
3. $a^3 > 10$ The cube of a number is greater than 10.
4. $7(1 + 3z)^2 \leq \frac{z}{2}$ Seven times the quantity one plus three times z squared is less than or equal to half of z.

Assignment Guide

Core: 13–33 odd, 35–40
Enriched: 14–30 even, 31–40
All: Self Test, 1–10

For **Extra Practice,** see p. 757.

The red A, B, and C flags, printed only in the Teacher's Wraparound Edition, indicate the level of difficulty of the exercises.

Additional Answers

31. Sample answer: $p = 1$ and $q = 2$, $p = 2$ and $q = 10$, $p = 3$ and $q = 8$, $p = 4$ and $q = 20$, and $p = 5$ and $q = 15$

32a. $t = 3(144)$; Texas should have about 3 times their yearly average

32c. Sample answer: Write an equation that estimates the number of tornadoes that Oklahoma will have in the next five years.
    $t = 5(45)$

Practice Masters, p. 5

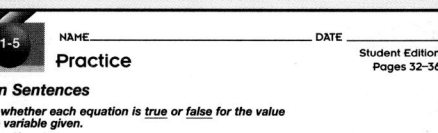

4 ASSESS

Closing Activity

Modeling Have students explore the classroom for pairs of objects that they can handle such that each pair satisfies the open sentence "_____ is heavier than _____." Lay the objects out in such pairs.

Chapter 1, Quiz B (Lessons 1-4 and 1-5), is available in the *Assessment and Evaluation Masters,* p. 16.

Mid-Chapter Test (Lessons 1-1 through 1-5) is available in the *Assessment and Evaluation Masters,* p. 15.

Answers to the Self Test

7.
Stem	Leaf
4	8
5	4
6	7
7	7
8	5 9

$6 | 7 = 67$

8.
Stem	Leaf
1	0
2	4 5
3	5 9
4	5
7	5 6

$3 | 5 = 350$

Enrichment Masters, p. 5

NAME_____ DATE _____
Enrichment   Student Edition Pages 32–35

Solution Sets

Consider the following open sentence.

It is the tallest building in the world.

You know that a replacement for the variable *It* must be found in order to determine if the sentence is true or false. If *It* is replaced by either the Empire State Building or the Sears Tower, the sentence is true.

The set {Empire State Building, Sears Tower} is called the *solution set* of the open sentence given above. This set includes all replacements for the variable that make the sentence true.

Write the solution set of each open sentence.

1. It is the name of a state beginning with the letter A.   1. {Alabama, Alaska, Arizona, Arkansas}

2. It is a primary color.   2. {red, yellow, blue}

3. Its capital is Harrisburg.   3. {Pennsylvania}

4. It is a New England state.   4. {Maine, New Hamp., Vermont, Mass., Rhode Is., Conn.}

5. $x + 4 = 10$ 5. {6}

6. It is the name of a month that contains the letter *r*.   6. {Jan, Feb, Mar, Apr, Sept, Oct, Nov, Dec}

7. During the 1970s, she was the wife of a U.S. President.   7. {Pat Nixon, Betty Ford, Rosalyn Carter}

8. It is an even number between 1 and 13.   8. {2, 4, 6, 8, 10,12}

9. $31 = 72 - k$ 9. {41}

10. It is the square of 2, 3, or 4.   10. {4, 9, 16}

Write an open sentence for each solution set.

11. {A, E, I, O, U}   11. It is a vowel.

12. {1, 3, 5, 7, 9}   12. It is an odd number between 0 and 10.

13. {June, July, August}   13. It is a summer month.

14. {Atlantic, Pacific, Indian, Arctic}   14. It is an ocean.

---

34. **Biology** Insects are the largest class of animals in the world with at least 800,000 species. The fastest-moving insect is the large tropical cockroach. It scurries at speeds of up to 3.36 miles per hour, which is about 2.3 feet per second.

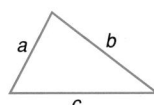

   a. Write an equation that represents how many miles the roach can travel in 1.5 hours. $d = (3.36)(1.5)$
   b. How many miles can the roach travel in 1.5 hours? **5.04 mi**
   c. Can the roach travel 100 feet in less than a minute? Explain.
   Yes; it can travel 100 ft in $100 \div 2.3$ or 43.5 s

Mixed Review

35. Write the stems that would be used to form a stem-and-leaf plot for the following set of data. (Lesson 1–4)

   2.7, 5.9, 2.0, 7.7, 5.2, 6.0, 5.4, 9.9, 5.4   **2, 5, 6, 7, 9**

36. **Hockey** The stem-and-leaf plot at the right shows the greatest number of goals scored by a National Hockey League player during any one season. Mario Lemieux is credited with the fourth greatest number of goals in a season. How many goals did Lemieux score during that season? (Lesson 1–4) **85**

Stem	Leaf
7	1 2 3 6 6 6
8	5 6 7
9	2

$7 | 2 = 72$

37. Evaluate $5(13 - 7) - 22$. (Lesson 1–3) **8**

38. Write the next two expressions for the pattern $2a + 1, 4a + 3, 6a + 5, 8a + 7, \ldots$ (Lesson 1–2) **$10a + 9, 12a + 11$**

39. Write a verbal expression for $x^5 - 5$. (Lesson 1–1) **5 less than x to the fifth power**

40. **Geometry** Write an algebraic expression for the perimeter of the triangle at the right. (Lesson 1–1) **$a + b + c$**

SELF TEST

Write an algebraic expression for each verbal expression. (Lesson 1–1) **1. $3a + b^2$**

1. the sum of three times a and the square of b 2. w to the fifth power minus 37 $w^5 - 37$

3. **Patterns** Each Tuesday morning at Central High School, the first seven bells ring at 8:00, 8:43, 8:47, 9:30, 9:34, 10:17, and 10:21 A.M. Give the times of the other morning bells. (Lesson 1–2)
11:04, 11:08, 11:51, 11:55

Evaluate each expression. (Lesson 1–3)

4. $5(8 - 3) + 7 \cdot 2$ **39**   5. $6(4^3 + 2^2)$ **408**   6. $(9 - 2 \cdot 3)^3 - 27 + 9 \cdot 2$ **18**

Make a stem-and-leaf plot for each set of data. (Lesson 1–4) **7–8. See margin.**

7. 67, 85, 54, 48, 89, 77   8. 236, 450, 748, 254, 755, 347, 97, 386

Find the solution set for each open sentence if the replacement set is {4, 5, 6, 7, 8}.
(Lesson 1–5)

9. $x + 2 > 7$ **{6, 7, 8}**   10. $9x - 20 = x^2$ **{4, 5}**

36   Chapter 1 Exploring Expressions, Equations, and Functions

---

SELF TEST

The Self Test provides students with a brief review of the concepts and skills in Lessons 1-1 through 1-5. Lesson numbers are given to the right of exercises or instruction lines so students can review concepts not yet mastered.

Identity and Equality Properties

Objectives

After studying this lesson, you should be able to:

- recognize and use the properties of identity and equality, and
- determine the multiplicative inverse of a number.

Football

According to the graph at the right, the price of Super Bowl tickets increased by $0 from 1975 to 1977. Also, from 1981 to 1983, ticket prices stayed at $40. The open sentences below represent each situation.

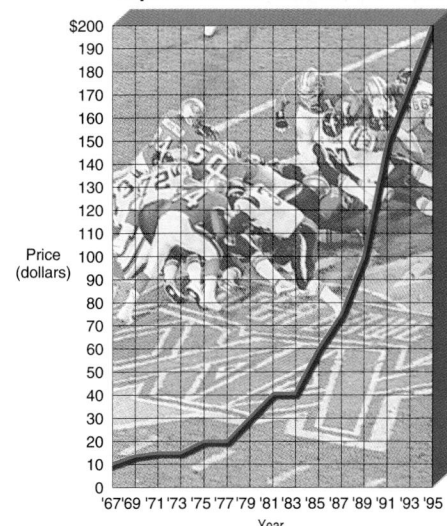

Super Bowl Ticket Prices 1967–1995

Price (dollars) / Year

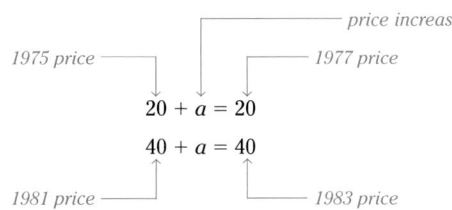

1975 price —— *price increase* —— *1977 price*

$$20 + a = 20$$

$$40 + a = 40$$

1981 price —— *1983 price*

The solution for each equation is 0. Super Bowl ticket prices increased $0 during each time period.

Equations such as these can be summarized in algebraic terms. The sum of any number and 0 is equal to that number. Zero is called the **additive identity**.

Additive Identity Property	For any number a, $a + 0 = 0 + a = a$.

The equations below also represent the ticket prices from 1975 to 1977 and from 1981 to 1983. The variable m represents the number of times of increase.

1975 price —— *number of times of increase* —— *1977 price*

$$20 \cdot m = 20$$

$$40 \cdot m = 40$$

1981 price —— *1983 price*

Lesson 1-6   Identity and Equality Properties   **37**

1-6 LESSON NOTES

NCTM Standards: 1–5

Instructional Resources

- Study Guide Master 1-6
- Practice Master 1-6
- Enrichment Master 1-6

Transparency 1-6A contains the 5-Minute Check for this lesson; **Transparency 1-6B** contains a teaching aid for this lesson.

Recommended Pacing	
Standard Pacing	Day 8 of 15
Honors Pacing	Day 7 of 14
Block Scheduling*	Day 4 of 7 (along with Lesson 1-5)
Alg. 1 in Two Years*	Days 11 & 12 of 22

*For more information on pacing and possible lesson plans, refer to the *Block Scheduling Booklet* and *Algebra 1 in Two Years*.

1 FOCUS

5-Minute Check
(over Lesson 1-5)

Solve each equation.

1. $b = \dfrac{10 + 6}{4}$ **4**

2. $15.2 - 4.25 = s$ **10.95**

3. $k = \dfrac{8.6 - 1.16}{(1.2)(2)}$ **3.1**

4. $t = \dfrac{3 \cdot 9 - 5^2}{2^3 + 3^2}$ $\dfrac{2}{17}$

5. Find the solution set for $4x + 1 \geq 3(x + 1)$ if the replacement set is {2, 3, 4, 5, 6}. **{2, 3, 4, 5, 6}**

Motivating the Lesson

Situational Problem To attract customers to a car dealership, the manager decides to sell one of the models at cost so she will make $0.00 in profit on each car sold. How much profit will she make if she sells 2 cars? If she sells 10 cars? if she sells n cars? **$0.00, $0.00, $0.00**

Teaching Tip To help explain why it must be assumed that no variable equals zero in Example 1, have students explore division by zero on a calculator.

Enter: 6 ⊕ 0 ⊜ ERROR

Why does the display show "ERROR"?
Division by zero is undefined.

In-Class Example

For Example 1
Name the multiplicative inverse of each number or variable. Assume that no variable equals zero.

a. 3 $\frac{1}{3}$

b. $\frac{x}{3}$ $\frac{3}{x}$

c. xy $\frac{1}{xy}$

The solution for each equation is 1. Since the product of any number and 1 is equal to the number, 1 is called the **multiplicative identity.**

Multiplicative Identity Property	**For any number a, $a \cdot 1 = 1 \cdot a = a$.**

Suppose you purchase a number of Super Bowl tickets for $200 each. If you sell three of them at face value ($200), your profit is $0. The following equation describes the situation.

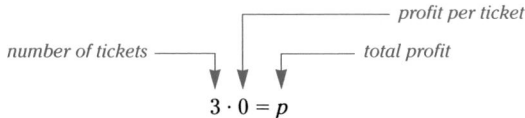

$$3 \cdot 0 = p$$

In this equation, one of the factors is 0 and the value of p is 0. This equation suggests the following property.

Multiplicative Property of Zero	**For any number a, $a \cdot 0 = 0 \cdot a = 0$.**

Two numbers whose product is 1 are called **multiplicative inverses** or **reciprocals.** Zero has no reciprocal because any number times 0 is 0.

Multiplicative Inverse Property	**For every nonzero number $\frac{a}{b}$, where $a, b \neq 0$, there is exactly one number $\frac{b}{a}$ such that $\frac{a}{b} \cdot \frac{b}{a} = 1$.**

Example **Name the multiplicative inverse of each number or variable. Assume that no variable equals zero.**

a. 5

Since $5 \cdot \frac{1}{5} = 1$, $\frac{1}{5}$ is the multiplicative inverse of 5.

b. x

Since $x \cdot \frac{1}{x} = 1$, the multiplicative inverse is $\frac{1}{x}$.

$x \neq 0$; why?

c. $\frac{2}{3}$

Using the property, the multiplicative inverse is $\frac{3}{2}$.

At the beginning of algebra class, Ms. Escalante gave each of her students a single strip of paper 8 inches long. She instructed the students to divide their paper strips any way they wished.

Staci left her strip as one 8-inch strip.

Amad cut his strip to form a 6-inch strip and a 2-inch strip.

Liam cut his strip to form a 5-inch strip and a 3-inch strip.

Using the strips of paper, we know the following to be true.

$$8 = 8 \qquad 6 + 2 = 6 + 2 \qquad 3 + 5 = 3 + 5$$

The **reflexive property of equality** says that any quantity is equal to itself.

Reflexive Property of Equality	For any number a, $a = a$.

Using the paper strips, we can show the following statements are true.

If $8 = 6 + 2$, then $6 + 2 = 8$.

If $3 + 5 = 6 + 2$, then $6 + 2 = 3 + 5$.

The **symmetric property of equality** says that if one quantity equals a second quantity, then the second quantity also equals the first.

Symmetric Property of Equality	For any numbers a and b, if $a = b$, then $b = a$.

A third property can also be shown using the paper strips.

If $3 + 5 = 8$ and $8 = 6 + 2$, then $3 + 5 = 6 + 2$.

If $8 = 3 + 5$ and $3 + 5 = 6 + 2$, then $8 = 6 + 2$.

The **transitive property of equality** says that if one quantity equals a second quantity and the second quantity equals a third quantity, then the first and third quantities are equal.

Transitive Property of Equality	For any numbers a, b, and c, if $a = b$ and $b = c$, then $a = c$.

We know that $5 + 3 = 6 + 2$. Since $5 + 3$ is equal to 8, we can substitute 8 for $5 + 3$ to get $8 = 6 + 2$. The **substitution property of equality** says that a quantity may be substituted for its equal in any expression.

Substitution Property of Equality	If $a = b$, then a may be replaced by b in any expression.

3 PRACTICE/APPLY

Check for Understanding

Exercises 1–19 are designed to help you assess your students' understanding through reading, writing, speaking, and modeling. You should work through Exercises 1–5 with your students and then monitor their work on Exercises 6–19.

Error Analysis
In such large and complex expressions, students must be careful to follow the rules of the order of operations. One suggestion is to number the operations first and then evaluate the expression.

Example:

$$3 (8 \cdot 2^3 - 7 \cdot 3)$$

You can use the properties of identity and equality to justify each step when evaluating an expression.

Example

APPLICATION

Sales

The pep club at Roosevelt High School is selling submarine sandwiches, lemonade, and apples at the district swim meet. Each sandwich costs $2.00 to make and sells for $3.00. Each glass of lemonade costs $0.25 to make and sells for $1.00. Each apple costs the club $0.25, and the members have decided to sell apples for $0.25 each. Write an expression that represents the profit for 80 sandwiches, 150 glasses of lemonade, and 40 apples. Evaluate the expression, indicating the property used in each step.

sandwiches   *lemonade*   *apples*
sold   *profit*   *sold*   *profit*   *sold*   *profit*

$$80(3.00 - 2.00) + 150(1.00 - 0.25) + 40(0.25 - 0.25)$$

$= 80(1) + 150(0.75) + 40(0)$	*Substitution* $(=)$
$= 80 + 150(0.75) + 40(0)$	*Identity* $(\times)$
$= 80 + 112.50 + 40(0)$	*Substitution* $(=)$
$= 80 + 112.50 + 0$	*Multiplicative prop. of 0*
$= 192.50 + 0$	*Substitution* $(=)$
$= 192.50$	*Identity* $(+)$

The club would make a profit of $192.50.

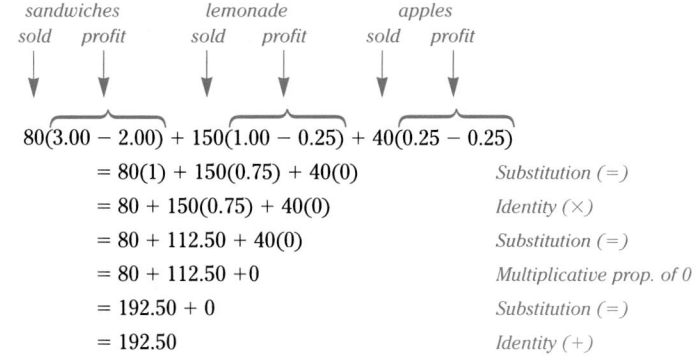

1. two or more things that are exactly alike

2. No; because $a + 1 \neq a$.

3. You cannot divide by 0.

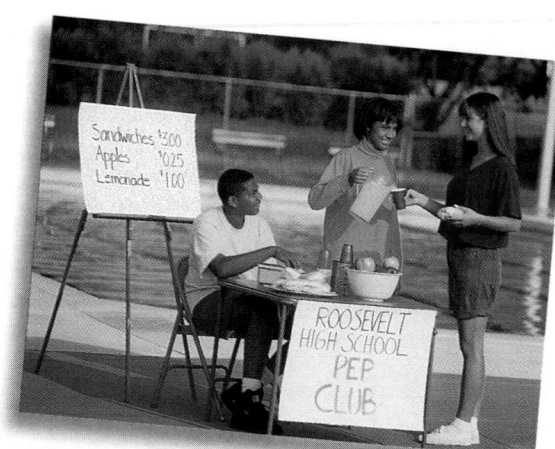

CHECK FOR UNDERSTANDING

Communicating Mathematics

Study the lesson. Then complete the following.

1. **Define** the term *identity* in your own words.
2. **Explain** whether or not 1 can be the additive identity.
3. **Explain** why 0 does *not* have a multiplicative inverse.
4. **Name** the multiplicative inverse of 1.   1

40   Chapter 1   Exploring Expressions, Equations, and Functions

Reteaching

Using Guess and Check Replace the variable to make each open sentence true.

1. $(2 + x)^2 = 5^2$ 3
2. $(2 + x)^2 = 2^2$ 0
3. $3(2 - a) = 3$ 1
4. $3(2 - a) = 0$ 2
5. $m(1) + m(0) = 5$ 5

MATH JOURNAL

5. Write a paragraph explaining how to determine the multiplicative inverse of a number.
   Reverse the numerator and the denominator.

Guided Practice

Name the multiplicative inverse of each number or variable. Assume that no variable represents zero.

6. 7 $\frac{1}{7}$ 7. $\frac{9}{2}$ $\frac{2}{9}$ 8. c $\frac{1}{c}$

Match the expressions in the left-hand column with the properties in the right-hand column.

9. $0 \cdot 36 = 0$ c   **a.** Additive identity property
10. $1(68) = 68$ b   **b.** Multiplicative identity property
11. $14 + 16 = 14 + 16$ e   **c.** Multiplicative property of 0
12. $(9 - 7)(5) = 2(5)$ g   **d.** Multiplicative inverse property
13. $\frac{3}{4} \times \frac{4}{3} = 1$ d   **e.** Reflexive property (=)
14. $0 + g = g$ a   **f.** Symmetric property (=)
15. If $8 + 1 = 9$,   **g.** Substitution property (=)
    then $9 = 8 + 1$.   f

Name the property used in each step.

16. $(14 \cdot \frac{1}{14} + 8 \cdot 0) \cdot 12 = (1 + 8 \cdot 0) \cdot 12$ multiplicative inverse
    $= (1 + 0) \cdot 12$ multiplicative prop. of 0
    $= 1 \cdot 12$ additive identity
    $= 12$ multiplicative identity

17–18. For properties, see Solutions Manual.

Evaluate each expression. Name the property used in each step.

17. $6(12 - 48 \div 4) + 9 \cdot 1$ 9   18. $3 + 5(4 - 2^2) - 1$ 2

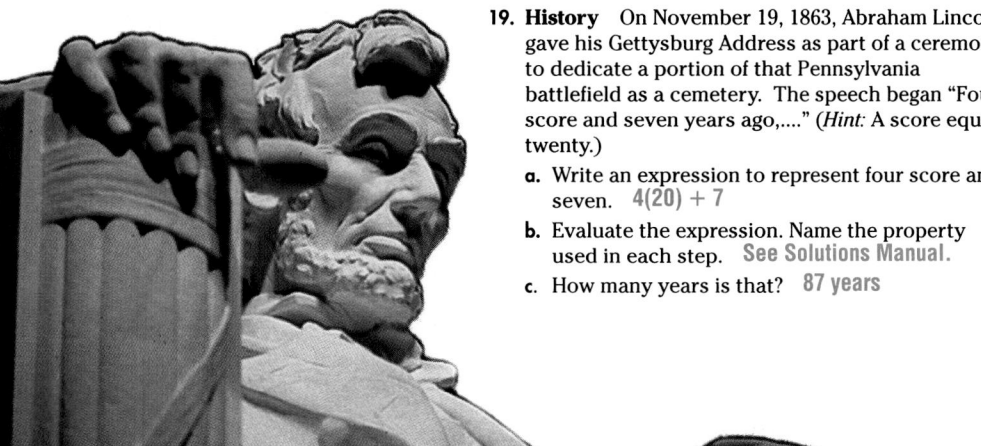

19. **History** On November 19, 1863, Abraham Lincoln gave his Gettysburg Address as part of a ceremony to dedicate a portion of that Pennsylvania battlefield as a cemetery. The speech began "Four score and seven years ago,...." (*Hint:* A score equals twenty.)

    a. Write an expression to represent four score and seven. $4(20) + 7$

    b. Evaluate the expression. Name the property used in each step.   See Solutions Manual.

    c. How many years is that?   87 years

26. symmetric (=)
27. substitution (=)
28. substitution (=)
29. multiplicative identity
30. multiplicative inverse
31. multiplicative inverse, multiplicative identity
32. additive identity
33. symmetric (=)
34. multiplicative property of 0
35. reflexive (=)

Practice Masters, p. 6

EXERCISES

Practice Name the multiplicative inverse of each number or variable. Assume that no variable represents zero.

20. 9 $\frac{1}{9}$
21. $\frac{1}{9}$ 9
22. $\frac{1}{4}$ 4
23. p $\frac{1}{p}$
24. $\frac{2}{a}$ $\frac{a}{2}$
25. $1\frac{1}{2}$ $\frac{2}{3}$

Name the property or properties illustrated by each statement.

26. If $7 \cdot 2 = 14$, then $14 = 7 \cdot 2$.
27. $8 + (3 + 9) = 8 + 12$
28. $(10 - 8)(5) = 2(5)$
29. $mnp = 1mnp$
30. $\left(\frac{3}{4}\right)\left(\frac{4}{3}\right) = 1$
31. $3\left(5^2 \cdot \frac{1}{25}\right) = 3$
32. $0 + 23 = 23$
33. If $6 = 9 - 3$, then $9 - 3 = 6$.
34. $5(0) = 0$
35. $32 + 21 = 32 + 21$
36. If $4 \cdot 2 = 8$ and $8 = 6 + 2$, then $4 \cdot 2 = 6 + 2$. transitive (=)

Name the property used in each step.

37. $2(3 \cdot 2 - 5) + 3 \cdot \frac{1}{3} = 2(6 - 5) + 3 \cdot \frac{1}{3}$ substitution (=)
$= 2(1) + 3 \cdot \frac{1}{3}$ substitution (=)
$= 2 + 3 \cdot \frac{1}{3}$ multiplicative identity
$= 2 + 1$ multiplicative inverse
$= 3$ substitution (=)

38. $26 \cdot 1 - 6 + 5(12 \div 4 - 3) = 26 \cdot 1 - 6 + 5(3 - 3)$ substitution (=)
$= 26 \cdot 1 - 6 + 5(0)$ substitution (=)
$= 26 - 6 + 5(0)$ multiplicative identity
$= 26 - 6 + 0$ multiplicative prop. of 0
$= 20 + 0$ substitution (=)
$= 20$ additive identity

39. $7(5 \cdot 3^2 - 11 \cdot 4) = 7(5 \cdot 9 - 11 \cdot 4)$ substitution (=)
$= 7(45 - 44)$ substitution (=)
$= 7 \cdot 1$ substitution (=)
$= 7$ multiplicative identity

Evaluate each expression. Name the property used in each step.

40–45. See Solutions Manual for properties.

40. $4(16 \div 4^2)$ 4
41. $(15 - 8) \div 7 \cdot 25$ 25
42. $(8 \cdot 3 - 19 + 5) + (3^2 + 8 \cdot 4)$ 51
43. $(2^5 - 5^2) + (4^2 - 2^4)$ 7
44. $8[6^2 - 3(11)] \div 8 \cdot \frac{1}{3}$ 1
45. $5^3 + 9\left(\frac{1}{3}\right)^2$ 126

46. Think about the relationship "is less than," represented by the symbol $<$. Does $<$ work with each of the following? Explain why or why not. Give examples that support your reasoning.
  a. reflexive property   no
  b. symmetric property   no
  c. the transitive property   yes

See margin for explanations.

Applications and Problem Solving

47. Entertainment   The students in Mr. Toshio's class are planning an ice cream party. They have selected the ice cream flavors shown in the table below. The prices listed are for $\frac{1}{2}$-cup servings. A survey of the students shows that 12 students want vanilla, 15 want chocolate, and 10 want chocolate chip cookie dough.

Flavor	Brand	Price
Vanilla	Breyers	21¢
Chocolate	Edy's/Dreyer's Grand	23¢
Chocolate Chip Cookie Dough	Ben & Jerry's	67¢

  a. Write an expression that represents how much it will cost to purchase ice cream for the class if each student gets a 1-cup serving.   See margin.
  b. Evaluate the expression. Name the property used in each step.   See margin.
  c. What is the total cost? $25.34

48. Postal Rates   Patricia wants to mail a package to her cousin in Los Angeles. The cost of first class mail is $0.32 for the first ounce and $0.23 for each additional ounce or fraction of an ounce. Patricia's package weighs 14.4 ounces.

48a. $0.32 + 0.23(14)$
48b. See margin.

  a. Write an expression that represents the cost to mail the package.
  b. Evaluate the expression. Name the property used in each step.
  c. How much will Patricia need to pay in postage? $3.54

Mixed Review

49. *True or False*: $15 \div 3 + 7 < 13$.   (Lesson 1–5)   true
50. Solve $m = (18 - 3) \div (3^2 - 2^2)$.   (Lesson 1–5)   3
51. *True or False*: $(2n^2 + 6) \div 4 < 5$, when $n = 3$.   (Lesson 1–5)   false

52. when there are 2 sets of data
53. See students' work.

52. When is a back-to-back stem-and-leaf plot used?   (Lesson 1–4)
53. Write the first and last names of ten of your classmates. Make a stem-and-leaf plot showing the number of letters in these ten names. (Lesson 1–4)
54. Evaluate $5(7 - 2) - 3^2$.   (Lesson 1–3)   16
55. Evaluate $xy - 2y$ when $x = 6$ and $y = 9$.   (Lesson 1–3)   36
56. Find the next two terms in the sequence 1, 4, 7, 10, ...   (Lesson 1–2)   13, 16
57. Write an algebraic expression for the number of months in y years. (Lesson 1–1)   12y

Extension

Problem Solving Dentmaster Car Rental charges $21.95 per day and $0.11 a mile to rent one of its cars. Bryan rented a car for one day to visit friends in Altton, 85 miles away, Latton, 80 miles away, or Talton, 75 miles away. If he can spend only $40.00 on his car rental, which friends could he visit?
the friends in Latton or Talton

Additional Answers

47a. $[21(12 \cdot 2)] + [23(15 \cdot 2)] + [67(10 \cdot 2)]$
47b. $[21(12 \cdot 2)] + [23(15 \cdot 2)] + [67(10 \cdot 2)]$
   $= [21(24)] + [23(30)] + [67(20)]$
                          Substitution (=)
   $= 504 + 690 + 1340$
                          Substitution (=)
   $= 2534$ Substitution (=)
48b. $0.32 + 0.23(14) = 0.32 + 3.22$
                          Substitution (=)
   $= 3.54$ Substitution (=)

---

4 ASSESS

Closing Activity
Speaking Discuss the following relationships between two people to determine which ones work with the

a) reflexive property
b) symmetric property, or
c) transitive property.

1. is taller than   no, no, yes
2. sits next to   no, no, yes
3. has same color eyes as   yes, yes, yes
4. is the father of   no, no, no
5. lives within one mile of   yes, yes, no

Additional Answers
46a. A number is never less than itself. Sample example: 4 is *not* less than 4.
46b. If one number is less than a second number, then the second can never be less than the first number. Sample example: 4 is less than 5, but 5 is *not* less than 4.
46c. If one number is less than a second number and a second number is less than a third number, then the first number will be less than the third number. Sample example: 4 < 5 and 5 < 6, so 4 < 6.

Enrichment Masters, p. 6

NCTM Standards: 1–5

Objective
Model the distributive property.

Recommended Time
Demonstration and discussion: 15 minutes; Exercises: 30 minutes

Instructional Resources
For each student or group of students
Student Manipulative Kit
• algebra tiles
Modeling Mathematics Masters
• p. 15 (product mat)
• p. 19 (worksheet)
For teacher demonstration
Algebra and Geometry Overhead Manipulative Resources

1 FOCUS

Motivating the Lesson
Addition can be modeled by the length and width of a rectangle, and multiplication can be modeled by its area. The distributive property relates these two operations. How could the relationship between adding and multiplying be represented geometrically? **Answers will vary.**

2 TEACH

Teaching Tip Watch for students who may try to give the x-tile a numerical value by comparing the length of the x-tile. Reinforce the idea that the tiles are models, not exact quantities, and that x represents an unknown quantity.

3 PRACTICE/APPLY

Assignment Guide

Core: 1–7
Enriched: 1–7

1-7A The Distributive Property

A Preview of Lesson 1–7

Materials: algebra tiles   product mat

Throughout your study of mathematics, you have used rectangles to model multiplication. For example, the figure below shows the multiplication $2(3 + 1)$ as a rectangle that is 2 units wide and $3 + 1$ units long. The model shows that the expression $2(3 + 1)$ is equal to $2 \cdot 3 + 2 \cdot 1$. The sentence $2(3 + 1) = 2 \cdot 3 + 2 \cdot 1$ illustrates the distributive property.

You can use special tiles called **algebra tiles** to form rectangles that model multiplication. A 1-tile is a square that is 1 unit long and 1 unit wide. Its area is 1 square unit. An x-tile is a rectangle that is 1 unit wide and x units long. Its area is x square units.

Activity **Find the product $3(x + 2)$ by using algebra tiles. First, model $3(x + 2)$ as the area of a rectangle.**

Step 1 The rectangle has a width of 3 units and a length of $x + 2$ units. Use your algebra tiles to mark off the dimensions on a product mat.

Step 2 Using the marks as a guide, make the rectangle with algebra tiles.

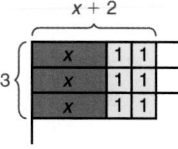

The rectangle has 3 x-tiles and 6 1-tiles. The area of the rectangle is $x + 1 + 1 + x + 1 + 1 + x + 1 + 1$, or $3x + 6$. Thus, $3(x + 2) = 3x + 6$.

Model **Find each product by using algebra tiles.**

1. $2(x + 1)$ 2. $5(x + 2)$ 3. $2(2x + 1)$ 4. $2(3x + 3)$
   $2x + 2$ $5x + 10$ $4x + 2$ $6x + 6$

Draw **Tell whether each statement is true or false. Justify your answer with algebra tiles and a drawing.** 5–6. See Solutions Manual for drawings.

5. $3(x + 3) = 3x + 3$ false   6. $x(3 + 2) = 3x + 2x$ true

Write **7. You Decide** Helen says that $3(x + 4) = 3x + 4$, but Adita says that $3(x + 4) = 3x + 12$.

   a. Which classmate is correct? Adita

   b. Use words and/or models to give an explanation that demonstrates the correct equation. See Solutions Manual.

44 Chapter 1 *Exploring Expressions, Equations, and Functions*

4 ASSESS

Observing students working in cooperative groups is an excellent method of assessment.

GLENCOE Technology

Interactive Mathematics Tools Software

In this interactive computer lesson, students use algebra tiles and a product mat to model the distributive property. A **Computer Journal** gives students an opportunity to write about what they have learned.

For Windows & Macintosh

1-7

The Distributive Property

Objective

After studying this lesson, you should be able to:

• use the distributive property to simplify expressions.

APPLICATION
Food Service

In the school cafeteria, each student can pick one hot or cold entree and one salad from the following in order to get a "Meal Deal" lunch at a special low price.

Hot Lunch	Cold Lunch	Salads
• Tacos • Meat loaf and mashed potatoes • Steamed veggies and rice • Spaghetti	• Submarine sandwich • Turkey club sandwich • Tuna salad with tomato	• Tossed salad • Fruit salad

Jenine wants to know how many different lunches are possible. She can make a table to represent all the possible combinations.

Hot Lunch

Salads	Entree			
	Tacos	Meat Loaf	Veggies	Spaghetti
Tossed	x	x	x	x
Fruit	x	x	x	x

Cold Lunch

Salads	Entree		
	Submarine	Club Sandwich	Tuna Salad
Tossed	x	x	x
Fruit	x	x	x

There are 2 · 4 or 8 possible hot lunches.
There are 2 · 3 or 6 possible cold lunches.
There are (2 · 4) + (2 · 3) possible lunches from which students can choose.
Note that (2 · 4) + (2 · 3) = 8 + 6 or 14.

Lesson 1–7   The Distributive Property   **45**

Instructional Resources

• Study Guide Master 1-7
• Practice Master 1-7
• Enrichment Master 1-7
• Assessment and Evaluation Masters, p. 17
• Multicultural Activity Masters, p. 2

Transparency 1-7A contains the 5-Minute Check for this lesson; **Transparency 1-7B** contains a teaching aid for this lesson.

Recommended Pacing	
Standard Pacing	Day 10 of 15
Honors Pacing	Day 9 of 14
Block Scheduling*	Day 5 of 7
Alg. 1 in Two Years*	Days 14 & 15 of 22

*For more information on pacing and possible lesson plans, refer to the *Block Scheduling Booklet* and *Algebra 1 in Two Years*.

1 FOCUS

5-Minute Check
(over Lesson 1-6)

Solve each equation.

1. $0 + r = 5$ **5**
2. $b \cdot 1 = 3$ **3**

Name the property illustrated by each statement.

3. If $3 + 0 = 3$, then $3 = 3 + 0$. **symmetric (=)**
4. $3(0) = 3(0)$ **reflexive (=)**
5. If $8 + 7 = 15$, then $15 = 8 + 7$. **symmetric (=)**

Motivating the Lesson

Questioning Which of the following expressions represent the area, in square meters, of the large rectangle at the left? **2, 3, and 7**

1. $30 + x$ 2. $30 + 6x$
3. $6(5 + x)$ 4. $6(5x)$
5. $30 + x^2$ 6. $11x$
7. $(5 + x)6$ 8. $30x$
9. $5 = 6x$ 10. $22 + 2x$

Chapter 1   **45**

Teaching Tip A common mistake is for students to evaluate an expression such as 2(5)(8) as 2(5) + 2(8), getting 26 instead of 80.

In-Class Example

For Example 1
Evaluate each expression in two different ways.

a. 11(8 + 2)
$$11(8 + 2) = 11(10)$$
$$= 110$$
$$11(8 + 2) = 11(8) + 11(2)$$
$$= 110$$

b. 6(7 − 3)
$$6(7 − 3) = 6(4)$$
$$= 24$$
$$6(7 − 3) = 6(7) − 6(3)$$
$$= 24$$

Jenine can also represent this problem by using the following table.

	Entree						
	Hot Lunch				Cold Lunch		
Salads	Taco	Meat Loaf	Veggie	Spaghetti	Sub	Club Sand.	Tuna
Tossed	x	x	x	x	x	x	x
Fruit	x	x	x	x	x	x	x

According to this table, there are 2 types of salads times the total number of entrees, 4 + 3, or 7.

$$2(4 + 3) = 2 \cdot 7 \text{ or } 14$$

Either way you look at it, there are 14 possible lunches. That's because the following is true.

$$2(4 + 3) = 2 \cdot 4 + 2 \cdot 3$$

This is an example of the **distributive property**.

Distributive Property	For any numbers a, b, and c, $a(b + c) = ab + ac$ and $(b + c)a = ba + ca$; $a(b − c) = ab − ac$ and $(b − c)a = ba − ca$.

Notice that it doesn't matter whether a is placed on the right or the left of the expression in parentheses.

The symmetric property of equality allows the distributive property to be written as follows.

If $a(b + c) = ab + ac$, then $ab + ac = a(b + c)$.

Example

INTEGRATION
Geometry

The Oak Grove High School Spirit Club wants to make a banner for the championship football game. The students have two large pieces of paper to use for the banner. One piece is 5 feet by 13 feet, and the other is 5 feet by 10 feet. They plan to use both pieces for the banner. Find the total area of the banner.

The total area of the banner can be found in two ways.

Method 1: Add the areas of the smaller rectangles.
$$A = w\ell_1 + w\ell_2$$
$$= 5(13) + 5(10)$$
$$= 65 + 50$$
$$= 115$$

Using Manipulatives

Use a rectangular piece of paper as a geometric model to illustrate the distributive property.

(1) $A = 8(5 + 4) = 8(9)$ or 72
(2) $A = 8(5) + 8(4) = 40 + 32$ or 72

Method 2: Multiply the width by the length.

$A = w\ell$

$\quad = 5(13 + 10) \quad$ *w = 5, l = 13 + 10*

$\quad = 5(23)$

$\quad = 115$

13 ft   10 ft

5 ft

The area is 115 square feet.

You can use the distributive property to multiply in your head.

Example **Use the distributive property to find each product.**

   a. 7 · 98

$7 \cdot 98 = 7(100 - 2)$

$\qquad = 700 - 14$

$\qquad = 686$

   b. 8(6.5)

$8(6.5) = 8(6 + 0.5)$

$\qquad = 48 + 4$

$\qquad = 52$

A **term** is a number, a variable, or a product or quotient of numbers and variables. Some examples of terms are x^3, $\frac{1}{4}a$, and $4y$. The expression $9y^2 + 13y^2 + 3$ has three terms.

Like terms are terms that contain the same variables, with corresponding variables having the same power. In the expression $8x^2 + 2x^2 + 5a + a$, $8x^2$ and $2x^2$ are like terms, and $5a$ and a are also like terms.

We can use the distributive property and properties of equality to show that $3x + 8x = 11x$. In this expression, $3x$ and $8x$ are like terms.

$3x + 8x = (3 + 8)x \qquad$ *Distributive property*

$\qquad\quad = 11x \qquad$ *Substitution (=)*

The expressions $3x + 8x$ and $11x$ are called **equivalent expressions** because they denote the same number. An expression is in **simplest form** when it is replaced by an equivalent expression having no like terms and no parentheses.

Example **Simplify $\frac{1}{4}x^2 + 2x^2 + \frac{11}{4}x^2$.**

In this expression, $\frac{1}{4}x^2$, $2x^2$, and $\frac{11}{4}x^2$ are like terms.

$\frac{1}{4}x^2 + 2x^2 + \frac{11}{4}x^2 = \left(\frac{1}{4} + 2 + \frac{11}{4}\right)x^2 \qquad$ *Distributive property*

$\qquad\qquad\qquad\quad = 5x^2 \qquad$ *Substitution (=)*

The **coefficient** of a term is the numerical factor. For example, in $23ab$, the coefficient is 23. In xy, the coefficient is 1 since, by the multiplicative identity property, $1 \cdot xy = xy$. Like terms may also be defined as terms that are the same or that differ only in their coefficients.

In-Class Examples

For Example 2
Use the distributive property to find each product.

a. 16(101) **1616**
b. 9(10.6) **95.4**

For Example 3
Simplify $\frac{6xy^2}{5} + \frac{7xy^2}{5}$. $\frac{13xy^2}{5}$

Teaching Tip When the text refers to a coefficient, it will be referring to the numerical coefficient of the term.

Teaching Tip Emphasize that in Example 5a, $5w^4$ and w^2 cannot be combined into one term because their exponents are different.

Teaching Tip Emphasize that in Example 5b, the coefficient of $\frac{a^3}{4}$ is $\frac{1}{4}$. This is often difficult for students to understand.

3 PRACTICE/APPLY

Check for Understanding
Exercises 1–23 are designed to help you assess your students' understanding through reading, writing, speaking, and modeling. You should work through Exercises 1–4 with your students and then monitor their work on Exercises 5–23.

Study Guide Masters, p. 7

Example ④ **Name the coefficient in each term.**

a. $145x^2y$ The coefficient is 145.

b. ab^2 The coefficient is 1 since $ab^2 = 1ab^2$.

c. $\frac{4a^2}{5}$ The coefficient is $\frac{4}{5}$ because $\frac{4a^2}{5}$ can be written as $\frac{4}{5} \cdot a^2$.

Example ⑤ **Simplify each expression.**

a. $4w^4 + w^4 + 3w^2 - 2w^2$

Remember that $w^4 = 1w^4$.

$4w^4 + w^4 + 3w^2 - 2w^2$
$= (4 + 1)w^4 + (3 - 2)w^2$
$= 5w^4 + 1w^2$
$= 5w^4 + w^2$

b. $\frac{a^3}{4} + 2a^3$

Remember that $\frac{a^3}{4} = \frac{1}{4}a^3$.

$\frac{a^3}{4} + 2a^3 = \frac{1}{4}a^3 + 2a^3$
$= \left(\frac{1}{4} + 2\right)a^2$
$= 2\frac{1}{4}a^3$

CHECK FOR UNDERSTANDING

Communicating Mathematics

1. See margin.

Study the lesson. Then complete the following.

1. **Explain** why the equation $2(a - 3) = 2a - 3$ is *not* a true sentence.

2. **Write** an expression that meets the following conditions.
   a. It has four terms,   Sample answer: $5a + 3a + a + 2b$
   b. three terms are like terms, and
   c. one has a coefficient of 1.

3. **Describe** how you would simplify $3(2x - 4)$.   See margin.

MODELING MATHEMATICS

4. **Draw** a rectangular model or use tiles to represent $4(x + 1)$.   See margin.

Guided Practice

Match an expression in the left-hand column with the equivalent expression in the right-hand column.

5. $8(10 + 4)$ b
6. $(12 - 3)6$ e
7. $(4 \cdot 2) + (x \cdot 2)$ a
8. $2(x + 5)$ d
9. $2(x - 1)$ c

a. $(4 + x)2$
b. $8 \cdot 10 + 8 \cdot 4$
c. $2x - 2$
d. $2x + 10$
e. $(12)(6) - (3)(6)$

Use the distributive property to rewrite each expression without parentheses.

10. $3(2x + 6)$ $6x + 18$
11. $2(a - b)$ $2a - 2b$

Use the distributive property to find each product.

12. $15 \cdot 99$ 1485
13. $28\left(2\frac{1}{7}\right)$ 60

Reteaching

Using Alternative Methods Explain how the multiplication algorithm is just an example of the distributive property. Students should demonstrate this by using an example like the following.

$\begin{array}{r} 14 \\ \times\ 76 \\ \hline 84 \\ +980 \\ \hline 1064 \end{array}$ $\begin{array}{l} = 14 \times 6 \\ = 14 \times 70 \\ = 14(70 + 6) \end{array}$

Additional Answers

1. According to the distributive property, $2(a - 3) = 2a - 6$.

3. Multiply 3 times $2x$ and 3 times 4. Subtract the product of 3 and 4 from the product of 3 and $2x$.

4.

$x + 1$

4	x	1
	x	1
	x	1
	x	1

Name the coefficient of each term.

14. $2.5cd$ **2.5**   **15.** $7a^2b$ **7**   **16.** $\frac{3b}{5}$ $\frac{3}{5}$

Name the like terms in each expression.

17. $4y^4 + 3y^3 + y^4$ $4y^4, y^4$ **18.** $3a^2 + 4c + a + 3b + c + 9a^2$
                                               $3a^2, 9a^2; 4c, c$

Simplify each expression, if possible. If not possible, write *in simplest form*.

19. $t^2 + 2t^2 + 4t$ $3t^2 + 4t$ **20.** $25x^2 + 5x$ **in simplest form**

21. $23a^2b + 3ab^2$

21. $16a^2b + 7a^2b + 3ab^2$ **22.** $7p + q - p + \frac{2q}{3}$ $6p + 1\frac{2}{3}q$

23. $5.35(24) + 5.35(32)$;
$5.35(24 + 32)$

23. Employment   Maria and Mark are salesclerks at a local department store. Each earns \$5.35 per hour. Maria works 24 hours per week, and Mark works 32 hours per week. Write two expressions to represent the amount the two of them earn per week.

EXERCISES

Practice   **Use the distributive property to rewrite each expression without parentheses.**

24. $2(4 + t)$ $8 + 2t$ **25.** $(g - 9)5$ $5g - 45$

26. $5(x + 3)$ $5x + 15$ **27.** $8(3m + 6)$ $24m + 48$

28. $28\left(y - \frac{1}{7}\right)$ $28y - 4$ **29.** $a(5 - b)$ $5a - ab$

Use the distributive property to find each product.

30. $5 \cdot 97$ 485 **31.** $\left(3\frac{1}{17}\right) \times 17$ 52

32. $16(102)$ 1632 **33.** $24(2.5)$ 60

34. $999 \cdot 6$ 5994 **35.** 3×215 645

Simplify each expression, if possible. If not possible, write *in simplest form*.

37. in simplest form

36. $15x + 18x$ $33x$ **37.** $14a^2 + 13b^2 + 27$

38. $10n + 3n^2 + 9n^2$ $10n + 12n^2$ **39.** $5a + 7a + 10b + 5b$ $12a + 15b$

40. $21x^2y - 28xy^2 + 7xy$

40. $7(3x^2y - 4xy^2 + xy)$ **41.** $13p^2 + p$ **in simplest form**

44. $\frac{5}{3}c - \frac{1}{2} + cb$

42. $5(6a + 4b - 3b)$ $30a + 5b$ **43.** $3(x + 2y) - 2y$ $3x + 4y$

44. $\frac{2}{3}\left(c - \frac{3}{4}\right) + c(1 + b)$ **45.** $a + \frac{a}{5} + \frac{2}{5}a$ $1\frac{3}{5}a$

46. $4(3g + 2) + 2(g + 3)$ $14g + 14$ **47.** $3(x + y) + 2(x + y) + 4x$ $9x + 5y$

Programming

48. The program at the right tests values of *A*, *B*, and *C* to determine if the distributive property holds for division. That is, does
$$\frac{A + B}{C} = \frac{A}{C} + \frac{B}{C}?$$

a. Run the program for 10 different sets of values for *A*, *B*, and *C*. Does the property hold?   **yes, if $C \neq 0$**

48b. Change the / in the If statement to $\wedge$; only for $C = 1$.

b. How could you change the program to test if $(A + B)^C = A^C + B^C$? Does the distributive property hold for powers?

```
Program: DISTPROP
: Prompt A, B, C
: If(A + B)/C = A/C + B/C
: Then
: Disp "YES, IT HOLDS."
: Else
: Disp "TRY AGAIN."
: END
```

Lesson 1–7   The Distributive Property   **49**

Closing Activity

Writing Have students list other possible "distributive" properties and state whether these properties are always true. Examples might include addition distributing over multiplication, division distributing over subtraction, and raising to a power distributing over multiplication. They should include counterexamples to show properties that are not always true.

Additional Answer

56.

Stem	Leaf
2	9
3	6 6 7
4	5 5
5	1 5 8

$3|7 = 37$

Chapter 1, Quiz C (Lessons 1-6 and 1-7), is available in the *Assessment and Evaluation Masters*, p. 17.

Enrichment Masters, p. 7

NAME_____ DATE _____

Enrichment

Student Edition
Pages 45–50

Closure

A *binary operation* matches two numbers in a set to just one number. Addition is a binary operation on the set of whole numbers. It matches two numbers such as 4 and 5 to a single number, their sum.

If the result of a binary operation is always a member of the original set, the set is said to be *closed* under the operation. For example, the set of whole numbers is not closed under subtraction because $3 - 6$ is not a whole number.

Is each operation binary? Write yes or no.

1. the operation ↩, where a ↩ b means to choose the lesser number from a and b yes

2. the operation ©, where a © b means to cube the sum of a and b yes

3. the operation sq, where $sq(a)$ means to square the number a no

4. the operation exp, where $exp(a, b)$ means to find the value of a^b yes

5. the operation ⇑, where a ⇑ b means to match a and b to any number greater than either number   no

6. the operation ⇒, where a ⇒ b means to round the product of a and b up to the nearest 10   yes

Is each set closed under addition? Write yes or no. If your answer is no, give an example.

7. even numbers   yes

8. odd numbers   no; $3 + 7 = 10$

9. multiples of 3   yes

10. multiples of 5   yes

11. prime numbers   no; $3 + 5 = 8$

12. nonprime numbers   no; $22 + 9 = 31$

Is the set of whole numbers closed under each operation? Write yes or no. If your answer is no, give an example.

13. multiplication: $a \times b$ yes

14. division: $a \div b$ no; $4 \div 3$ is not a whole number

15. exponentiation: a^b yes

16. squaring the sum: $(a + b)^2$ yes

---

49. If $2(b + c) = 2b + 2c$, does $2 + (b \cdot c) = (2 + b)(2 + c)$? Choose values for b and c to show that these may be true or find *counterexamples* to show that they are not.   No; sample counterexample: $2 + (4 \cdot 5) \neq (2 + 4)(2 + 5)$

50a. $4(16.15 + 32.45)$;
$4(16.15) + 4(32.45)$

50. **Economics**   According to *American Demographics,* the average 13- to 15-year-old male received $16.15 per week for allowance in 1993. The average 16- to 19-year-old female received $32.45 per week. Toni was 17 years old, and her brother Carlos was 14 in 1993. Assume that they got the average amount for their weekly allowance.

   a. Write two expressions to represent the amount their parents paid in allowance during the month of February.

   b. What was the amount their parents paid in allowance during that month? $194.40

51. **Geometry**   The largest permanently installed theater screen has an area of 6768 square feet. It is at the Keong Emas Imax Theater in Jakarta, Indonesia. The movie screen below represents a typical movie screen in U.S. movie theaters.

   a. Write an expression for the perimeter of the movie screen below and then simplify the expression.
   $2[x + (x + 14)] = 4x + 28$

Keong Emas Imax Theater

x feet

$(x + 14)$ feet

   b. Evaluate the expression if $x = 17$.   96
   c. Find the area of the movie screen if $x = 17$.
   527 ft^2

53. multiplicative prop. of 0

52. Name the property illustrated by the following statement: *If $19 - 3 = 16$, then $16 = 19 - 3$.*   (Lesson 1–6)   symmetric (=)

53. Name the property illustrated by $9 \times 0 = 0$.   (Lesson 1–6)

54. Find the solution set for inequality $3x - 5 > 7$ if the replacement set is $\{2, 3, 4, 5, 6\}$.   (Lesson 1–5) $\{5, 6\}$

55. **Physics**   Sound travels 1129 feet per second through air.   (Lesson 1–5)

   a. Write an equation that represents how many feet sound can travel in 2 seconds when it is traveling through air. $d = (1129)(2)$

   b. How many feet can sound travel in 2 seconds when traveling through air?   2258 ft

56. Make a stem-and-leaf plot for the following set of data.   (Lesson 1–4)
   37, 45, 36, 51, 55, 29, 45, 58, 36   See margin.

57. Evaluate $\dfrac{4^2 - 2^3}{24 - 2(10)}$.   (Lesson 1–3)   2

58. What are the next two expressions for the pattern $5a, 10a^2, 15a^3, 20a^4, ...$?   (Lesson 1–2) $25a^5, 30a^6$

59. **Culture**   Each year in the Chinese calendar is named for one of 12 animals. Every 12 years, the same animal is named. If 1992 was the Year of the Monkey, how many years in the 20th century were Years of the Monkey?   (Lesson 1–2)   8 years

60. Write an algebraic expression for *37 less than 2 times a number k.*   (Lesson 1–1) $2k - 37$

Extension

Connections   Carita placed six 2-cm by 2-cm squares flat on her desk so that each square had one side in common with at least one other square. What are the possible perimeters for all the figures she can make? (*Hint:* Make drawings for all the possible figures.)
20 cm, 24 cm, 28 cm

Commutative and Associative Properties

Objective

After studying this lesson, you should be able to:

- recognize and use the commutative and associative properties to simplify expressions.

APPLICATION
Networks

The map below shows the location of Leticia's house and school. It also shows the time in minutes that it takes Leticia to walk around her neighborhood.

Yesterday, Leticia walked to school by way of Lincoln Street and Washington Avenue. She took $3 + 3 + 3 + 2 + 2 + 2$ or 15 minutes to walk to school. Today, Leticia walked to school by way of Madison Avenue and Wilson Street. She took $2 + 2 + 2 + 3 + 3 + 3$ or 15 minutes. The time it took Leticia to walk to school is the same for both days.

$$3 + 3 + 3 + 2 + 2 + 2 = 2 + 2 + 2 + 3 + 3 + 3$$

In this equation, the addends are the same, but their order is different. The **commutative property** says that the order in which you add or multiply two numbers does not change their sum or product.

Commutative Property	For any numbers a and b, $a + b = b + a$ and $a \cdot b = b \cdot a$.

One easy way to find the sum or product of numbers is to group or *associate* the numbers. The **associative property** says that the way you group three numbers when adding or multiplying does not change their sum or product.

Associative Property	For any numbers a, b, and c, $(a + b) + c = a + (b + c)$ and $(ab)c = a(bc)$.

Lesson 1-8 Commutative and Associative Properties **51**

NCTM Standards: 1–5

Instructional Resources

- Study Guide Master 1-8
- Practice Master 1-8
- Enrichment Master 1-8

Transparency 1-8A contains the 5-Minute Check for this lesson; **Transparency 1-8B** contains a teaching aid for this lesson.

Recommended Pacing

Standard Pacing	Day 11 of 15
Honors Pacing	Day 10 of 14
Block Scheduling*	Day 6 of 7 (along with Lesson 1-9)
Alg. 1 in Two Years*	Days 16 & 17 of 22

*For more information on pacing and possible lesson plans, refer to the *Block Scheduling Booklet* and *Algebra 1 in Two Years*.

1 FOCUS

5-Minute Check
(over Lesson 1-7)

Simplify each expression.

1. $2a(4a^2 - 2ab)$ $8a^3 - 4a^2b$
2. $a(2a + 3b) - ab$ $2a^2 + 2ab$
3. $3m + 2n + 4p + 3n - m$ $2m + 5n + 4p$
4. $4(3x + 2y) + 6(3y - 2x)$ $26y$
5. $2(7x + y) + 4(3x + 2y)$ $26x + 10y$

Motivating the Lesson

Situational Problem Have students list the steps they would follow, in order, when wrapping a gift. Then have them indicate on the list what steps cannot come after other steps. For example, tissue paper cannot be placed in the package after ribbon has been wrapped around the package.

For Example 1
Which of the following is the greatest?

a. $6 \times 4\frac{1}{2} \times 3$

b. $5\frac{1}{2} \times 5 \times 4$

c. $7 \times 3 \times 2\frac{1}{2}$

$b = 110$

For Example 2

a. Write an algebraic expression for the verbal expression, *the difference of x cubed and three increased by the difference of 5 and x cubed.* $(x^3 - 3) + (5 - x^3)$

b. Simplify the algebraic expression in part a, indicating all of the properties used. **2, associative, commutative**

Teaching Tip Ask students to explain the difference between the commutative and associative properties. **The commutative properties allow a different ordering of numbers while the associative properties allow a different grouping of numbers.**

3 PRACTICE/APPLY

Check for Understanding
Exercises 1–15 are designed to help you assess your students' understanding through reading, writing, speaking, and modeling. You should work through Exercises 1–5 with your students and then monitor their work on Exercises 6–15.

Error Analysis
Students often confuse the symmetric property of equality and the commutative properties. Emphasize that expressions are interchanged with respect to an equal sign for the symmetric property, while expressions are interchanged with respect to either an addition or a multiplication sign for the commutative properties.

You can group $11 + 12 + 7 + 9 + 7 + 6$ to make mental addition easy.

$$11 + 12 + 7 + 9 + 7 + 6 = 11 + 9 + 7 + 6 + 7 + 12 \quad \text{Commutative } (+)$$
$$= (11 + 9) + (7 + 6 + 7) + 12 \quad \text{Associative } (+)$$
$$= 20 + 20 + 12$$
$$= 52$$

The commutative and associative properties can be used with the other properties you have studied when evaluating and simplifying expressions.

Example ❶

APPLICATION
Marketing

Why has the order of the factors changed?

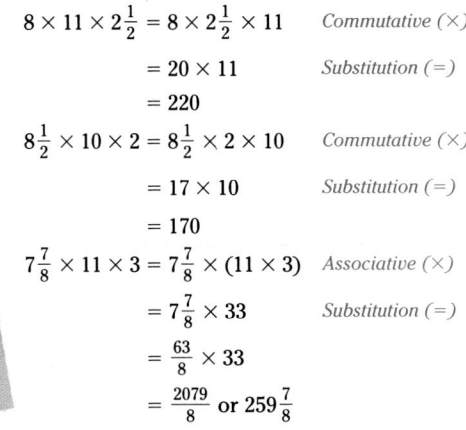

Juan Martinez is a product manager for a cereal company. Part of his job is to determine what size box should be used to package the company's Toasty Oatsies cereal. Juan can choose from among the following sizes: 8″ by 11″ by $2\frac{1}{2}$″, $8\frac{1}{2}$″ by 10″ by 2″, or $7\frac{7}{8}$″ by 11″ by 3″. Juan wants to know which package will hold the most cereal.

The box with the greatest volume will hold the most cereal. To find the volume of each box, multiply the length times the width times the height.

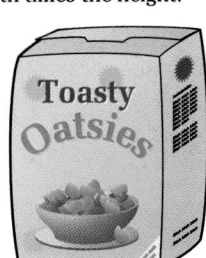
Toasty Oatsies

$$8 \times 11 \times 2\frac{1}{2} = 8 \times 2\frac{1}{2} \times 11 \quad \text{Commutative } (\times)$$
$$= 20 \times 11 \quad \text{Substitution } (=)$$
$$= 220$$
$$8\frac{1}{2} \times 10 \times 2 = 8\frac{1}{2} \times 2 \times 10 \quad \text{Commutative } (\times)$$
$$= 17 \times 10 \quad \text{Substitution } (=)$$
$$= 170$$
$$7\frac{7}{8} \times 11 \times 3 = 7\frac{7}{8} \times (11 \times 3) \quad \text{Associative } (\times)$$
$$= 7\frac{7}{8} \times 33 \quad \text{Substitution } (=)$$
$$= \frac{63}{8} \times 33$$
$$= \frac{2079}{8} \text{ or } 259\frac{7}{8}$$

The box that is $7\frac{7}{8}$″ by 11″ by 3″ has the greatest volume.

The chart below summarizes the properties you can use to simplify expressions.

The following properties are true for any numbers *a*, *b*, and *c*.		
	Addition	**Multiplication**
Commutative	$a + b = b + a$	$ab = ba$
Associative	$(a + b) + c = a + (b + c)$	$(ab)c = a(bc)$
Identity	0 is the identity. $a + 0 = 0 + a = a$	1 is the identity. $a \cdot 1 = 1 \cdot a = a$
Zero		$a \cdot 0 = 0 \cdot a = 0$
Distributive	$a(b + c) = ab + ac$ and $(b + c)a = ba + ca$	
Substitution	If $a = b$, then a may be substituted for b.	

Example ❷   **a.** Write an algebraic expression for the verbal expression *the sum of two and the square of t increased by the sum of t squared and 3.*

b. Then simplify the algebraic expression, indicating all of the properties used.

a. *the sum of two and the square of t*	*increased by*	*the sum of t squared and 3*
$2 + t^2$	$+$	$t^2 + 3$

b. $2 + t^2 + t^2 + 3 = t^2 + t^2 + 2 + 3$ *Commutative (+)*

$= (t^2 + t^2) + (2 + 3)$ *Associative (+)*

$= (1 \cdot t^2 + 1 \cdot t^2) + (2 + 3)$ *Multiplicative identity*

$= (1 + 1)t^2 + (2 + 3)$ *Distributive property*

$= 2t^2 + 5$ *Substitution (=)*

CHECK FOR UNDERSTANDING

Communicating Mathematics

Study the lesson. Then complete the following.   1–4. See margin.

1. **Illustrate** the associative property of multiplication with a numerical equation.

2. **Explain** how a property or properties can be used to find the product of $5(6.5)(2)$ without using a calculator or paper and pencil.

3. **Explain** the difference between the commutative and associative properties.

4. **Write** a short explanation as to whether or *not* there is a commutative property of division. Consider these examples.

$$5 \div 3 = 1\frac{2}{3} \qquad\qquad 3 \div 5 = \frac{3}{5}$$

 MATH JOURNAL

5. **Assess Yourself** Do you believe that you understand the properties described in this chapter? Which properties were easiest for you to understand? List any properties that give you difficulty and explain why. **See students' work.**

Guided Practice

Name the property illustrated by each statement.   6–9. See margin.

6. $(6 + 4) + 2 = 6 + (4 + 2)$ 7. $7 + 6 = 6 + 7$

8. $(2 + 5) + x = 7 + x$ 9. $7(ab) = (7a)b$

10. Name the property used in each step.
    a. $ab(a + b) = (ab)a + (ab)b$ **distributive property**
    b. $\qquad\qquad = a(ab) + (ab)b$ **commutative ($\times$)**
    c. $\qquad\qquad = (a \cdot a)b + a(b \cdot b)$ **associative ($\times$)**
    d. $\qquad\qquad = a^2b + ab^2$ **substitution (=)**

Simplify.

11. $4a + 2b + a$ $5a + 2b$ 12. $3p + 2q + 2p + 8q$ $5p + 10q$

13. $3(4x + y) + 2x$ $14x + 3y$ 14. $6(0.4x + 0.2y) + 0.5x$ $2.9x + 1.2y$

15. Write an algebraic expression for the verbal expression *the product of six and the square of z, increased by the sum of seven, z^2, and 6.* Then simplify, indicating the properties used.   **See margin.**

Reteaching

Logical Thinking For each situation, determine whether "followed by" is commutative.

a. Putting mustard on your hamburger, followed by putting ketchup on your hamburger. **yes**

b. Opening the car door, followed by starting the car's engine. **no**

c. Buying tickets for the homecoming dance, followed by buying an outfit to wear to the homecoming dance. **yes**

d. Have students make up their own example of a situation in which "followed by" is commutative and a situation in which "followed by" is not commutative. **Answers may vary.**

Assignment Guide

Core: 17–47 odd, 48–56
Enriched: 16–44 even, 45–56

For **Extra Practice**, see p. 758.

The red A, B, and C flags, printed only in the Teacher's Wraparound Edition, indicate the level of difficulty of the exercises.

Additional Answers

16. commutative (+)
17. multiplicative identity
18. distributive property
19. associative (×)
20. multiplicative property of zero
21. distributive property
22. substitution (=)
23. commutative (×)
24. associative (+)
25. associative (+)
39. $2(s + t) - s$
$= 2s + 2t - s$
      distributive property
$= 2t + 2s - s$
      commutative (+)
$= 2t + (2s - s)$
      associative (+)
$= 2t + (2s - 1s)$
      multiplicative identity
$= 2t + (2 - 1)s$ distributive
$= 2t + 1s$ substitution (=)
$= 2t + s$
      multiplicative identity

Practice Masters, p. 8

54 Chapter 1

EXERCISES

Practice A

Name the property illustrated by each statement. 16–25. See margin.

16. $67 + 3 = 3 + 67$
17. $1 \cdot b^2 = b^2$
18. $10x + 10y = 10(x + y)$
19. $(5 \cdot m) \cdot n = 5 \cdot (m \cdot n)$
20. $(3x^2) \cdot 0 = 0$
21. $4(a + 5b) = 4a + 20b$
22. $(2 + 3)a + 7 = 5a + 7$
23. $fh + 2g = hf + 2g$
24. $7a + \left(\frac{1}{2}b + c\right) = \left(7a + \frac{1}{2}b\right) + c$
25. $(5x^2 + x + 3) + 15 = (5x^2 + x) + (3 + 15)$

B

26. Name the property used in each step.
   a. $3c + 5(2 + c) = 3c + 5(2) + 5c$ distributive property
   b. $\qquad = 3c + 5c + 5(2)$ commutative (+)
   c. $\qquad = (3c + 5c) + 5(2)$ associative (+)
   d. $\qquad = (3 + 5)c + 5(2)$ distributive property
   e. $\qquad = 8c + 10$ substitution (=)

Simplify. 37. $\frac{3}{4} + \frac{5}{3}m + \frac{4}{3}n$ 38. $\frac{23}{10}p + \frac{6}{5}q$

27. $4x + 5y + 6x$ $10x + 5y$
28. $8a + 3b + a$ $9a + 3b$
30. 4 + 8ac + 12b
29. $5x + 3y + 2x + 7y$ $7x + 10y$
30. $4 + 6(ac + 2b) + 2ac$
31. $2(3x + y) + 4x$ $10x + 2y$
32. $4y^4 + 3y^2 + y^4$ $5y^4 + 3y^2$
33. $16a^2 + 16 + 16a^2$ $32a^2 + 16$
34. $3.2(x + y) + 2.3(x + y) + 4x$
34. 9.5x + 5.5y

C

35. $\frac{1}{4}x + 2x + 2\frac{3}{4}x$ $5x$
36. $0.5[3x + 4(3 + 2x)]$ $5.5x + 6$
37. $\frac{3}{4} + \frac{2}{3}(m + 2n) + m$
38. $\frac{3}{5}\left(\frac{1}{2}p + 2q\right) + 2p$

Write an algebraic expression for each verbal expression. Then simplify, indicating the properties used. 39–42. See margin.

39. twice the sum of s and t decreased by s
40. half of the sum of p and $2q$ increased by three-fourths q
41. five times the product of x and y increased by $3xy$
42. four times the sum of a and b increased by twice the sum of a and $2b$
43. Find the following product.
   $\frac{1}{2} \cdot \frac{2}{3} \cdot \frac{3}{4} \cdot \ldots \cdot \frac{98}{99} \cdot \frac{99}{100}$
   Recall that an ellipsis "..." means "to continue the pattern." Write a sentence or two explaining how you found the product. See margin.

Critical Thinking

44. Is subtraction commutative? Write a short explanation that includes examples to support your answer. See margin.
45. Suppose the operation ✳ is defined for all numbers a and b as $a ✳ b = a + 2b$. Is the operation ✳ commutative? Give examples to support your answer. See margin.

Applications and Problem Solving

46. **Chemistry** Chemists use water to dilute acid. One important rule of chemistry is to always pour the acid into the water. Pouring water into acid could produce spattering and burns.
   a. Would you say that combining acid and water is commutative? no
   b. Give an example from your life that is commutative. See margin.
   c. Give an example from your life that is not commutative. See margin.

54 Chapter 1 Exploring Expressions, Equations, and Functions

Additional Answers

40. $\frac{1}{2}(p + 2q) + \frac{3}{4}q = \frac{1}{2}p + q + \frac{3}{4}q$
      distributive property
$= \frac{1}{2}p + \left(q + \frac{3}{4}q\right)$ associative (+)
$= \frac{1}{2}p + \left(1q + \frac{3}{4}q\right)$ multiplicative identity
$= \frac{1}{2}p + \left(1 + \frac{3}{4}\right)q$ distributive property
$= \frac{1}{2}p + \frac{7}{4}q$ substitution (=)

41. $5xy + 3xy = (5 + 3)xy$
      distributive property
$= 8xy$ substitution (=)
42. $4(a + b) + 2(a + 2b)$
$= 4a + 4b + 2a + 4b$
      distributive property
$= 4a + 2a + 4b + 4b$
      commutative (+)
$= (4 + 2)a + (4 + 4)b$
      distributive property
$= 6a + 8b$ substitution (=)

47. Physics The pressure you experience as you ride a roller coaster is produced by the fast changes in direction. This pressure is what pins you down to the back or bottom of your seat. This pressure is known as *g-force*. The formula used to evaluate the g-force is, $G = \frac{(speed)^2}{32.2 \times radius\ of\ the\ curve}$, where the speed is in feet per second, and the radius of the curve is in feet. The acceleration of gravity in $\frac{ft}{sec^2}$ is 32.2.

a. If you are riding the Steel Phantom at 60 feet per second (about 40 mph) and the car is turning a curve with a radius of 30 feet, what g-force will you experience? *G = 3.73*

b. The g-force has an effect on your weight. For example, if you weigh 100 pounds and experience a g-force of 2 g's, your weight at that moment feels like 2×100 or 200 pounds. What will your weight feel like while taking the curve at 60 feet per second?
See students' work.

Mixed Review

48. Name the coefficient of $\frac{4m^2}{5}$.
(Lesson 1–7) $\frac{4}{5}$

49. Environment According to the U.S. Environmental Protection Agency, a typical family of four uses 100 gallons of water flushing the toilet, 80 gallons of water showering and bathing, and 8 gallons of water using the bathroom sink each day. Write two expressions that represent the amount of water a typical family of four uses for these purposes in *d* days.
(Lesson 1–7) *100d + 80d + 8d, 188d*

50. Name the property illustrated by the following statement.
If a + b = c and a = b, then a + a = c. (Lesson 1–6) **substitution (=)**

51. Name the multiplicative inverse of $\frac{2}{3}$. (Lesson 1–6) $\frac{3}{2}$

52. *True or False:* $\frac{3k-3}{7} + 13 < 17$, when *k* = 8. (Lesson 1–5) **true**

53. Weather The following data represents the average daily temperature in Fahrenheit for Santiago, Chile, for two weeks during the month of February. Organize the data using a stem-and-leaf plot. (Lesson 1–4)

101, 95, 100, 100, 99, 101, 97,

99, 97, 100, 103, 101, 101, 99 **See margin.**

54. Evaluate $(25 - 4) \div (2^2 - 1^3)$. (Lesson 1–3) **7**

55. Find the next two items in the sequence. (Lesson 1–2)

56. Write an algebraic expression for *five times p to the sixth power*.
(Lesson 1–1) *5p⁶*

Extension

Reasoning Define the operation × as follows: For any numbers *a* and *b*, *a* × *b = a + b + ab*. Use the commutative and associative properties to show that the operation × is commutative.

$$a \times b = b \times a$$
$$a + b + ab = b + a + ba$$
$$a + b + ab = a + b + ba$$

---

4 ASSESS

Closing Activity

Writing Have students list the different properties discussed in Lessons 1-6, 1-7, and 1-8 along with an example of each property.

Additional Answers

43. $\frac{1}{100}$; Each denominator and the following numerator represent the number 1. The resulting expression is 1 in the numerator and 100 in the denominator multiplied by numerous 1s. Since 1 is the multiplicative identity, the product is $\frac{1}{100}$.

44. No; answers will vary. Sample answer: $5 - 3 \neq 3 - 5$.

45. No; Sample example: Let *a* = 1 and *b* = 2; then 1 * 2 = 1 + 2(2) or 5 and 2 * 1 = 2 + 2(1) or 4.

46b. Sample answer: practicing piano and doing homework

46c. Sample answer: putting clothes into the washer and dryer

53.

Stem	Leaf
10	0 0 0 1 1 1 1 3
9	5 7 7 9 9 9

10|3 = 103

Enrichment Masters, p. 8

NCTM Standards: 1–6, 8

Instructional Resources

- Study Guide Master 1-9
- Practice Master 1-9
- Enrichment Master 1-9
- Assessment and Evaluation Masters, p. 17
- Graphing Calculator Masters, p. 1
- Real-World Applications, 2
- Tech Prep Applications Masters, p. 2

Transparency 1-9A contains the 5-Minute Check for this lesson; **Transparency 1-9B** contains a teaching aid for this lesson.

Recommended Pacing	
Standard Pacing	Days 12 & 13 of 15
Honors Pacing	Days 11 & 12 of 14
Block Scheduling*	Day 6 of 7 (along with Lesson 1-8)
Alg. 1 in Two Years*	Days 18 & 19 of 22

*For more information on pacing and possible lesson plans, refer to the *Block Scheduling Booklet* and *Algebra 1 in Two Years*.

1 FOCUS

5-Minute Check
(over Lesson 1-8)

Name the property illustrated by each statement.

1. $by + 3a = yb + 3a$
   commutative property of multiplication
2. $4b + 5c = 5c + 4b$
   commutative property of addition
3. $(2 \cdot a) \cdot b = 2 \cdot (a \cdot b)$
   associative property of multiplication

Simplify.

4. $4x + 5y + 3z + 6x$
   $10x + 5y + 3z$
5. $4(3a + b) + 3a$ $15a + 4b$

Objectives

After studying this lesson, you should be able to:
- interpret graphs in real-world settings, and
- sketch graphs for given functions.

1-9 A Preview of Graphs and Functions

APPLICATION
Economics

The dollar value of a car begins to decrease immediately after it is sold. This is called *depreciation*. Mathematically, depreciation can be defined by the following open sentence.

depreciation = original cost of car − value of car when sold

The table below shows how a typical $15,000 car depreciates over a period of five years. From the table, you can see that as the age of the car increases, the value decreases. Why do you think this happens?

Age of Car (year)	Approximate Value of Car
0	$15,000
1	11,000
2	8000
3	6000
4	5000
5	4000

This information can also be presented in a graph. The graph below shows the relationship between the current value of the car and the age of the car.

The value of the car is said to be a **function** of the age of the car. A function is a relationship between input and output. In a function, the output depends on the input. There is exactly one output for each input.

There are two number lines on the graph. The line that represents the current value of the car is the **vertical axis.** The line that represents the age of the car is the **horizontal axis.**

Car Depreciation Over a Five-Year Period

56   Chapter 1   *Exploring Expressions, Equations, and Functions*

Classroom Vignette

"Ask students to research (using Blue Books, newspaper ads, etc.) the price of several makes of cars to determine the better buy. Insurance costs, maintenance, and fuel efficiency should also be considered. They should then prepare graphs to show the depreciation of the cars researched as part of their findings. This can serve a dual role of addressing interdisciplinary instruction and performance assessment."

Bea Moore-Harris
Educational Specialist,
Region IV Education Service Center
Houston, Texas

Notice that there are six points shown on the graph. Each point represents the dollar value of the car (in thousands) at the end of each year. For example, after year 1, the value is $11,000. You can express this relationship as (1, 11). What do points *A* and *B* represent?

A(2, 8) means that at the end of year 2 the value of the car is about $8000.

B(3, 6) means that at the end of year 3 the value of the car is about $6000.

(1, 11), (2, 8), and (3, 6) are called **ordered pairs.** The order in which the pair of numbers is written is important. Ordered pairs are used to locate points on the graph. The first number in an ordered pair corresponds to the horizontal axis, and the second number in an ordered pair corresponds to the vertical axis. The ordered pair (0, 0) corresponds to the **origin.**

Many real-world situations can be modeled using functions.

Example **1**

APPLICATION

Sales

Shim owns a farm market. The amount a customer pays for sweet corn depends on the number of ears that are purchased. Shim sells a dozen ears of corn for $3.00.

a. **Make a table showing the price of various purchases of sweet corn.**

Number of Dozen	Ears of Corn	Price
$\frac{1}{2}$	6	$1.50
1	12	$3.00
$1\frac{1}{2}$	18	$4.50
2	24	$6.00

b. **Write four ordered pairs that represent the number of ears of corn and the price of the corn.**

Four ordered pairs can be determined from the chart. These ordered pairs are (6, 1.50), (12, 3.00), (18, 4.50), and (24, 6.00).

c. **Describe a set of axes that could be used to graph the number of ears of corn and the price of the corn.**

Let the horizontal axis represent the number of ears of corn, and let the vertical axis represent the price of the corn.

d. **Draw a graph that shows the relationship between the number of ears of corn and the price.**

Graph the four points described by the ordered pairs.

e. **As you read the graph from left to right, describe the trend you see. Explain.**

The graph goes upward, because the price increases as the number of ears increases. The price is a function of the number of ears purchased.

Lesson 1–9   A Preview of Graphs and Functions   **57**

2 TEACH

In Example 1, the price a customer pays depends on the number of ears purchased. Therefore, the number of ears purchased is called the **independent variable,** and the price is called the **dependent variable.** Usually the independent variable is graphed on the horizontal axis and the dependent variable is graphed on the vertical axis.

You can use a graph without scales on either axis to show the general shape of the graph that models a situation.

Example **2** Lucinda jogs up a hill at a steady speed. Then she runs down the hill and picks up her speed.

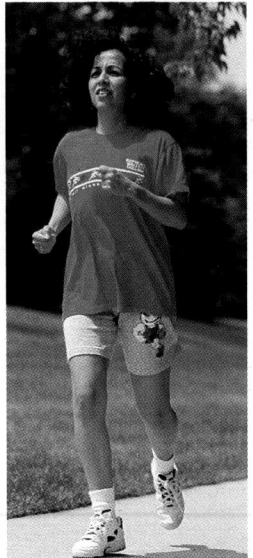

a. What happens to her speed when Lucinda jogs at a steady pace?

The speed remains the same because her jogging pace is steady.

b. What happens to her speed when Lucinda runs down the hill?

The speed increases when she runs downhill.

c. Match the graph to the situation.

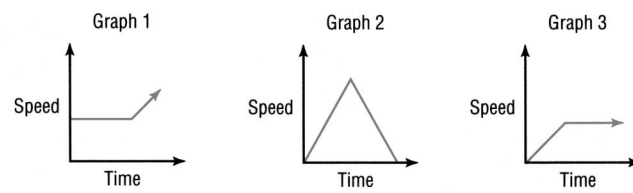

To select the graph that best represents this situation, find the graph that shows a steady speed (a horizontal line), followed by an increase in speed (a line that goes up) the longer she exercises. Graph 1 matches this description.

Example **3** Mitchell is not feeling well and has missed a day of school. The graph below shows Mitchell's temperature as a function of the time of day. Write a story about Mitchell's illness.

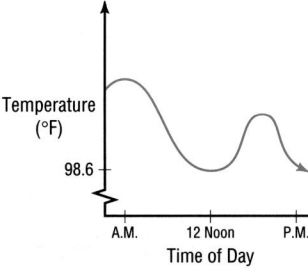

Here is a possible story.

When Mitchell got up in the morning, he had a high fever. As the morning went on, his temperature went down. By noon it was almost normal, but his temperature went up again after lunch. Then his mom gave him some medication. By evening his fever was gone, and his temperature was back to normal.

Communicating Mathematics

1. See margin.

2. Sample answers: $3000; $800

Study the lesson. Then complete the following.

1. **Explain** why the order of the numbers in an ordered pair is important.

2. Refer to the application at the beginning of the lesson. Extend the graph to predict the value of the car after 6 years and after 10 years.

3. Use the graph at the right to answer the following questions. a–e. See margin.

   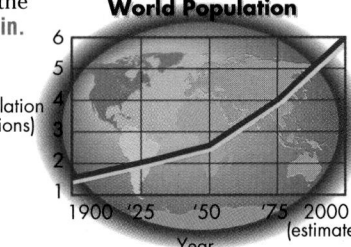

   World Population

   a. **Describe** the information that the graph shows.

   b. **Identify** the variable represented along each axis.

   c. **Identify** two points on the graph and write the ordered pair for each point. Explain what each point represents.

   d. **Explain** what the ordered pair (2000, 6.2) represents.

   e. **Write** a statement that explains the function in the graph.

HERMAN

"I really look forward to your cheery little visits."

4. **Write** a story that explains what the graph in the cartoon at the left might represent.
   Sample answer: Every day, the patient's condition becomes worse.

Guided Practice

5. Match each description with the most appropriate graph.

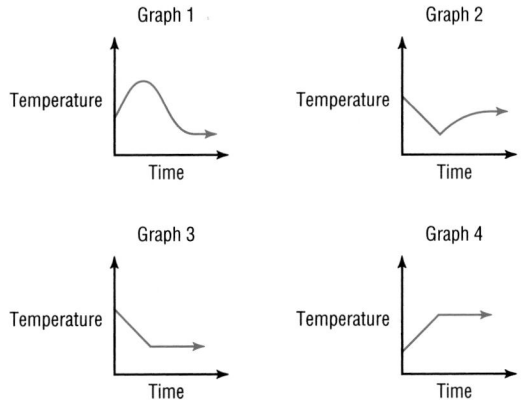

Graph 1

Graph 2

Graph 3

Graph 4

5a. Graph 3

5b. Graph 4

5c. Graph 2

5d. Graph 1

a. In August, you enter a hot house and turn on the air conditioner.

b. In January, you enter a cold house and turn up the thermostat to 68° F.

c. You put ice cubes in your fruit punch and then drink it slowly.

d. You put a cup of water in the microwave and heat it for one minute.

Lesson 1–9   A Preview of Graphs and Functions   **59**

---

3 PRACTICE/APPLY

Check for Understanding
Exercises 1–7 are designed to help you assess your students' understanding through reading, writing, speaking, and modeling. You should work through Exercises 1–4 with your students and then monitor their work on Exercises 5–7.

Error Analysis
Students may mix up the axes. Remind them that it is customary to represent the independent variable along the horizontal axis and the dependent variable along the vertical axis.

Additional Answers

1. The numbers represent different values. The first number represents the number on the horizontal axis and the second represents the number on the vertical axis.

3a. world population in billions of people from 1900 to the year 2000

3b. horizontal axis: time in years; vertical axis: population in billions

3c. Sample answer: (1925, 2): In 1925 the world's population was about 2 billion people. (1950, 2.5): In 1950 the world's population was about 2.5 billion people.

3d. The estimated world population for the year 2000 is 6.2 billion.

3e. Sample answer: The world's population has increased since 1900.

Reteaching

Using Data Have groups of students think of data that could be portrayed using a graph and construct their own graph using their data. Also ask students to write five questions that pertain to the graph. Make copies of each group's graph and questions and distribute them to the other groups to answer the questions.

Additional Answers

6.

7a. *A* is the younger player, but *B* runs the mile in less time.

7b. *A* made more 3-point shots, but *B* made more 2-point shots.

7c. *B* is the older player and *B* made more 2-point shots.

7d. *A* is the younger player and *A* made more free-throw shots.

8. b; The bus gains speed, and then travels at a level speed. Later, it slows to a stop. Then the pattern continues.

9. a; An average person makes no money as a child, then his or her income rises for several years, and finally levels off.

Study Guide Masters, p. 9

---

6. Sketch a reasonable graph for the following statement. *How hard you hit your thumb with a hammer impacts on how much it hurts.* **See margin.**

7. The basketball coach at City College is recruiting basketball players. The graphs below describe two of the players being recruited.

7b. false

Determine whether each statement is *true* or *false*. Explain your reasoning. **a–d. See margin for explanations.**

a. The younger player is faster than the older player. **false**

b. One player made more 2-point and more 3-point shots than the other.

c. The older player made more 2-point shots. **true**

d. The younger player tried and made more free-throw shots. **true**

EXERCISES

Practice
A

8. Identify the graph that matches the following statement. Explain your answer. **See margin.**

A bus frequently stops to pick up passengers.

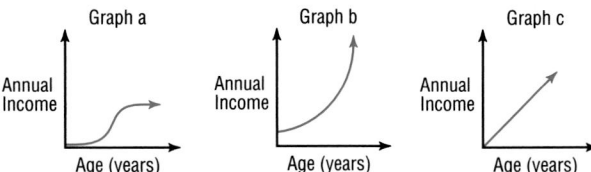

9. Identify the graph that represents annual income as a function of age. Explain your answer. **See margin.**

Graph a   Graph b   Graph c

10. See margin.
B

10. The graph at the right shows the amount of money in Jorge's savings account as a function of time. Describe what is happening to Jorge's money. Be sure to explain why the graph rises and falls at particular points.

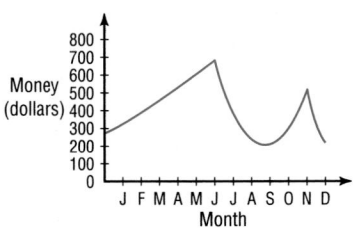

Additional Answer

10. Sample answer: Jorge saved steadily from January to June. In July, he withdrew money to go on vacation. He started saving again in September. Then in November he withdrew money for holiday presents.

11. Match each table of data to the most appropriate graph.

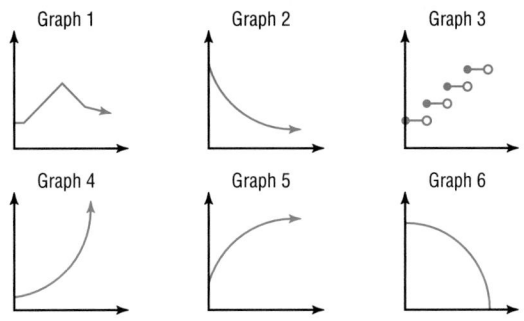

Graph 1   Graph 2   Graph 3

Graph 4   Graph 5   Graph 6

a. Primary Seismic Wave   **Graph 5**

Distance from Epicenter (km)	1000	2000	3000	4000	5000	6000	7000
Time (minutes)	2	4	5.8	7.2	8.5	9.7	10.8

b. Cost of First-Class Postage Stamps   **Graph 3**

Year	1983	1985	1987	1989	1991	1993	1995
Cost of Stamp	22¢	25¢	25¢	25¢	29¢	29¢	32¢

c. Immigration into the United States   **Graph 1**

Year	1987	1988	1989	1990	1991	1992	1993
Legal Admissions (millions)	0.6	0.6	1.1	1.5	1.8	1.0	0.9

d. Development of Human Fetus   **Graph 4**

Time (weeks)	4	8	12	16	20	24	28
Mass (grams)	0.5	1	28	110	300	650	1200

e. Height of Falling Object   **Graph 6**

Time (seconds)	0	0.5	1.0	1.5	2.0	2.5	3.0
Height (feet)	200	196	184	164	136	100	56

f. Cooling of Boiling Water   **Graph 2**

Time (minutes)	0	1	2	3	4	5	6
Temperature (°F)	100	50	25	12.5	6.25	3.125	1.5625

Sketch a reasonable graph for each situation. 12–15. See margin.

12. Rashaad likes to trade basketball cards. During the first two weeks of the month, he added lots of cards to his collection. Then, he lost some cards and sold others.

13. A radio-controlled car moves along and then crashes against a wall.

14. Phyllis is riding her bike at a steady pace. Then she has a flat tire. She walks to a gas station to have the tire repaired. When the tire is fixed, she continues her ride.

15. The height of a football that was thrown into the end zone for a touchdown compared to the distance the football is from the end zone.

Lesson 1–9   A Preview of Graphs and Functions   **61**

Additional Answers

12.

Number of Cards / Time

13.

Speed / Time

14.

Speed / Time

15.

Height / Distance

4 ASSESS

Closing Activity

Speaking Encourage a discussion of the main terms introduced in this lesson: *vertical axis, horizontal axis, origin, ordered pair, independent variable,* and *dependent variable.* Encourage students to find a sentence definition for each term.

Chapter 1, Quiz D (Lessons 1-8 and 1-9), is available in the *Assessment and Evaluation Masters,* p. 17.

Teaching Tip The graphs in Exercise 11 are previews of the step functions, quadratic functions, and exponential functions students will study in this text.

Practice Masters, p. 9

1. Identify the graph that matches the following statement. Explain your answer. *A train stops to let off passengers.*

a. Speed / Time   b. Speed / Time   c. Speed / Time

c; The train gains speed, and then travels at a level speed. Later it slows to a stop. Then the pattern continues.

2. Sketch a reasonable graph for the situation. *A football is thrown straight up and then curves downward upon its descent.*

The circle graphs below show how Kelly and Lynn spend their day. Use the graphs to answer each question.

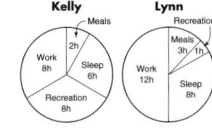

3. How does Kelly's time for recreation compare to Lynn's time for recreation? Kelly's time for recreation is 8 times greater than Lynn's.

4. Which activity takes most of Lynn's time? work

5. How much more time does Lynn spend working than Kelly does? 4 hours

The graphs below show the results of two surveys on favorite pastas. Compare the graphs and answer each question.

6. In Survey 1, the number of votes for spaghetti is twice the number of votes for which pasta in Survey 2? Linguini

7. How many more people preferred spaghetti in Survey 2 than preferred spaghetti in Survey 1? 10 people

8. In Survey 1, which pasta received the most votes? the fewest votes? spaghetti; linguini

Chapter 1   **61**

62 Chapter 1

Critical Thinking

16. Which of the sports listed will produce a graph like the one below? Explain why you chose the sport you did and give reasons why you did not choose the other sports. **See students' work.**

Archery   High Jumping
Biking   Javelin Throwing
Fishing   Pole Vaulting
Golf   Skateboarding
High Diving   Sky Diving

Applications and Problem Solving

17a–e. See margin.

17. **Charities** The graph at the right shows the amount of money raised by the Muscular Dystrophy Association over a 14-year period.

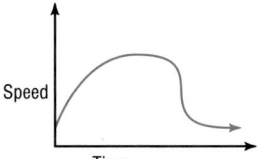
Money Raised by MDA Telethon

Source: Muscular Dystrophy Association

a. Name the variables represented along each axis.

b. What does the ordered pair (1991, 45) represent?

c. Write a statement about the amount of money raised from 1981 to 1986.

d. Write a statement about the amount of money raised from 1987 to 1994.

e. Use the information to predict the amount of money raised each year since 1994. Research to see if your prediction is correct.

18. **Music Business** The graph at the right shows the sales in millions of dollars for different music categories. Write two conclusions you can make from the information presented in the graph. **See margin.**

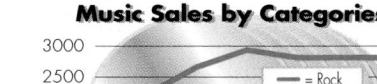
Music Sales by Categories

= Rock
= Rap
= Country

Source: The Recording Industry Association of America

Mixed Review

19. Simplify $5p + 7q + 9 + 4q + 2p$. (Lesson 1–8) $7p + 11q + 9$

20. Simplify $9a + 14(a + 3)$. (Lesson 1–7) $23a + 42$

21. Evaluate $14(1) − 27(0)$. (Lesson 1–6) **14**

22. *True or False:* $5m + 6^2 = 56$, when $m = 4$. (Lesson 1–5) **true**

23. $\frac{1}{3}Bh$

23. **Geometry** The volume of a cone equals one-third the area of the base B times the height h. Write an algebraic expression for the volume of a cone. (Lesson 1–1)

Extension

Reasoning Markham starts running quickly, slows down, stops, runs quickly again, then stops. Draw a rough graph of speed against time, and a graph of distance against time.

VOCABULARY

After completing this chapter, you should be able to define each term, property, or phrase and give an example or two of each.

Algebra
additive identity (p. 37)
algebraic expression (p. 6)
algebra tiles (p. 44)
associative property (p. 51)
base (p. 7)
coefficient (p. 47)
commutative property (p. 51)
dependent variable (p. 58)
distributive property (p. 46)
equation (p. 33)
equivalent expressions (p. 47)
evaluate (p. 8)
exponent (p. 7)
factors (p. 6)
function (p. 56)
horizontal axis (p. 56)
independent variable (p. 58)

inequality (p. 33)
like terms (p. 47)
multiplicative identity (p. 38)
multiplicative inverse (p. 38)
multiplicative property of zero (p. 38)
open sentence (p. 32)
order of operations (p. 19)
ordered pairs (p. 57)
power (p. 7)
product (p. 6)
reciprocal (p. 38)
reflexive property of equality (p. 39)
replacement set (p. 33)
simplest form (p. 47)
solution (p. 32)
solution set (p. 33)
solving the open sentence (p. 32)
substitution property of equality (p. 39)

symmetric property of equality (p. 39)
term (pp. 13, 47)
transitive property of equality (p. 39)
variable (p. 6)
vertical axis (p. 56)

Discrete Mathematics
element (p. 33)
set (p. 33)
sequence (p. 13)

Statistics
data (p. 25)
statistics (p. 25)
stem-and-leaf plots (p. 25)

Problem Solving
look for a pattern (p. 13)

UNDERSTANDING AND USING THE VOCABULARY

Choose the letter of the term that best matches each statement or phrase.

1. For any number a, $a + 0 = 0 + a = a$.   **a**
2. For any number a, $a \cdot 1 = 1 \cdot a = a$.   **c**
3. For any number a, $a \cdot 0 = 0 \cdot a = 0$.   **e**
4. For any number a, there is exactly one number $\frac{1}{a}$ such that $a \cdot \frac{1}{a} = \frac{1}{a} \cdot a = 1$.   **d**
5. For any number a, $a = a$.   **f**
6. For any numbers a and b, if $a = b$, then $b = a$.   **h**
7. For any numbers a and b, if $a = b$, then b may be substituted for a.   **g**
8. For any numbers a, b, and c, if $a = b$ and $b = c$, then $a = c$.   **i**
9. For any numbers a, b, and c, $a(b + c) = ab + ac$.   **b**

a. additive identity property
b. distributive property
c. multiplicative identity property
d. multiplicative inverse property
e. multiplicative property of zero
f. reflexive property
g. substitution property
h. symmetric property
i. transitive property

Chapter 1 Highlights **63**

Using the STUDY GUIDE AND ASSESSMENT

Skills and Concepts Encourage students to refer to the objectives and examples on the left as they complete the review exercises on the right.

Assessment and Evaluation Masters, pp. 9–10

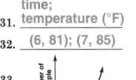

SKILLS AND CONCEPTS

OBJECTIVES AND EXAMPLES	REVIEW EXERCISES

Upon completing this chapter, you should be able to:

Use these exercises to review and prepare for the chapter test.

• translate verbal expressions into mathematical expressions and vice versa (Lesson 1–1)

Write an algebraic expression for the sum of twice a number x and fifteen.

$$2x + 15$$

Write a verbal expression for $4x^2 - 13$.
four times a number x squared minus thirteen

15. three times a number m to the fifth power

16. the sum of one half a number x and 2

17. the difference of four times m squared and twice m

Write an algebraic expression for each verbal expression. 12. $3x - 8$

10. the sum of a number x and twenty-one $x + 21$

11. a number x to the fifth power x^5

12. the difference of three times a number x and 8

13. five times the square of a number x $5x^2$

Write a verbal expression for each algebraic expression. 14. twice a number p squared

14. $2p^2$ 15. $3m^5$

16. $\frac{1}{2}x + 2$ 17. $4m^2 - 2m$

• solve problems by extending sequences (Lesson 1–2)

Give the next two items for each sequence.

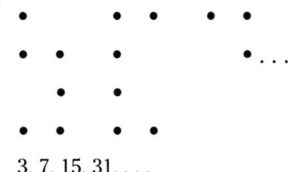

$$3, 7, 15, 31, \ldots$$

3   7   15   31   63   127

+4 +8 +16 +32 +64

The next two terms are 63 and 127.

Give the next two items for each pattern.

18.

19. $2, 4, 8, 16, \ldots$ **32, 64** 20. $1, 1, 2, 3, 5, \ldots$ **8, 13**

21. $x + y, 2x + y, 3x + y, \ldots$ **$4x + y, 5x + y$**

22. $100{,}000, 1{,}000{,}000$

Consider the following sequence.
$$10, 100, 1000, 10{,}000, \ldots$$

22. Give the next two terms of the sequence.

23. Express the first four terms in the sequence as a power of 10. $10^1, 10^2, 10^3, 10^4$

24. What are the 9th and 13th terms? $10^9, 10^{13}$

• use the order of operations to evaluate real number expressions (Lesson 1–3)

$5 \times 2^2 - 15 \div 3 + 4$	$[3(6) - 4^2]^3 \times 15 \div 5$
$= 5 \times 4 - 15 \div 3 + 4$	$= [3(6) - 16]^3 \times 15 \div 5$
$= 20 - 15 \div 3 + 4$	$= [18 - 16]^3 \times 15 \div 5$
$= 20 - 5 + 4$	$= [2]^3 \times 15 \div 5$
$= 15 + 4$	$= 8 \times 15 \div 5$
$= 19$	$= 120 \div 5$
	$= 24$

Evaluate each expression.

25. $3 + 2 \times 4$ **11**   26. $(10 - 6) \div 8$ $\frac{1}{2}$

27. $18 - 4^2 + 7$ **9**   28. $0.8 (0.02 + 0.05) - 0.006$

29. $(3 \times 1)^3 - (4 + 6) \div (9.1 + 0.9)$ **26**   **28. 0.05**

Evaluate each expression when $x = 0.1$, $t = 3$, and $y = 2$.

30. $t^2 + 3y$ **15**   31. xty^3 **2.4**   32. $ty \div x$ **60**

33. the product of y and x plus t raised to the second power **9.2**

GLENCOE Technology

Test and Review Software

You may use this software, a combination of an item generator and item bank, to create your own tests or worksheets. Types of items include free response, multiple choice, short answer, and open ended.

For IBM & Macintosh

OBJECTIVES AND EXAMPLES

• display and interpret data on a stem-and-leaf plot (Lesson 1–4)

Stem	Leaf
2	6 8
3	3 5 6 8 9
4	0 2 3 4 4 4 5 6 7 7 8 8 9
5	0 1 2 2 2 3 3 4 6 7
6	1 3 5 9
7	0 3 2|6 = 26

The highest entry is 73. The lowest entry is 26. The total number of entries is 36. Most of the data is in the interval 40–49. **36. 46 38. 60–70**

• solve open sentences by performing arithmetic operations (Lesson 1–5)

State whether each sentence is true or false.

$x + 3 \geq 5$, $x = 1$ $5^2 - 3y \leq 1$, $y = 8$

$1 + 3 \geq 5$ $5^2 - 3(8) \leq 1$

$\quad 4 \geq 5$ $25 - 24 \leq 1$

$\quad$ false $1 \leq 1$

$\qquad\qquad$ true

• recognize and use the properties of identity and equality (Lesson 1–6)

$36 + 7 \times 1 + 5(2 - 2)$

$= 36 + 7 \times 1 + 5(0)$ *Substitution (=)*

$= 36 + 7 + 5(0)$ *Multiplicative identity*

$= 36 + 7$ *Multiplicative prop. of zero*

$= 42$ *Substitution (=)*

48. multiplicative identity   49. additive identity

• use the distributive property to simplify expressions (Lesson 1–7)

$5(t + 3)$ $2x^2 + 4x^2 + 7x$

$= 5(t) + 5(3)$ $= (2 + 4)x^2 + 7x$

$= 5t + 15$ $= 6x^2 + 7x$

51. $2 \times 4 + 2 \times 7$ 52. $4x + 4$ 53. $1 - 3p$

57. $8m + 8n$ 58. $2p + 32pr$

REVIEW EXERCISES

The ages (in years) at the time of death for the presidents of the United States are listed below.

67, 90, 83, 85, 73, 80, 78, 79, 68, 71, 53, 65, 74, 64, 77, 56, 66, 63, 70, 49, 56, 67, 71, 58, 60, 72, 67, 57, 60, 90, 63, 88, 78, 46, 64, 81   **34. See margin.**

34. Make a stem-and-leaf plot of the data.

35. How old was the oldest president at death? **90**

36. How old was the youngest president at death?

37. Does the total number of entries match the number of presidents? Explain. **See margin.**

38. In which interval did most of the ages fall?

State whether each sentence is *true* or *false* for the value of the variable given.

39. $x + 13 = 22$, $x = 8$ **false**

40. $2b + 2 < b^3$, $b = 2$ **true**

41. $(y + 4) \div (y + 2) \leq 2$, $y = 2$ **true**

42. Find the solution set for $3x + 1 \leq 13$ if the replacement set is {2, 4, 6, 8}. **{2, 4}**

Solve each equation.

43. $y = 4\frac{1}{2} + 3^2$ **$13\frac{1}{2}$**   44. $y = 5[2(4) - 1^3]$ **35**

Name the multiplicative inverse of each number or variable.

45. 3 **$\frac{1}{3}$**   46. y **$\frac{1}{y}$**   47. 0.2 **5**

Name the property illustrated by each statement.

48. $xy = 1xy$ 49. $0 + 22x = 22x$

50. Evaluate $2[3 \div (19 - 4^2)]$ and name the property used in each step. **2; See margin.**

Use the distributive property to rewrite each expression without parentheses.

51. $2(4 + 7)$ 52. $4(x + 1)$ 53. $3\left(\frac{1}{3} - p\right)$

Use the distributive property to find each product.

54. 6×103 55. 3×98 56. $12(1.5)$ **18**
       618   **294**

Simplify each expression.

57. $3m + 5m + 12n - 4n$ 58. $2p(1 + 16r)$

Additional Answers

34.

Stem	Leaf
4	6 9
5	3 6 6 7 8
6	0 0 3 3 4 4 5 6 7 7 7 8
7	0 1 1 2 3 4 7 8 8 9
8	0 1 3 5 8
9	0 0 7|2 = 72

37. No, because some presidents are still alive.

Applications and Problem Solving Encourage students to work through the exercises in the Applications and Problem Solving section to strengthen their problem-solving skills.

OBJECTIVES AND EXAMPLES

• when simplifying expressions recognize and use the commutative and associative properties (Lesson 1–8)

$3x + 7xy + 9x$

$= 3x + 9x + 7xy$ *Commutative (+)*

$= (3 + 9)x + 7xy$ *Distributive*

$= 12x + 7xy$ *Substitution (=)*

64–65. See Solutions Manual for algebraic expressions and properties used.

REVIEW EXERCISES

Name the property illustrated by each statement.

59. $7(p + q) = 7(q + p)$ **commutative (+)**

60. $3 + (x + y) = (3 + x) + y$ **associative (+)**

61. $(b + a)c = c(b + a)$ **commutative (×)**

62. $2(mn) = (2m)n$ **associative (×)**

63. Simplify $2x + 2y + 3x + 3y$.   **5x + 5y**

Write an algebraic expression for each verbal expression. Then simplify, indicating the properties used.   **64.** $3x + 5y$ **65.** $3pq$

64. five times the sum of x and y decreased by $2x$

65. twice the product of p and q increased by pq

• interpret graphs in real-world settings (Lesson 1–9)

Sketch a graph for the following statement.

As a band gets more popular, their prices increase to a point and then level off.

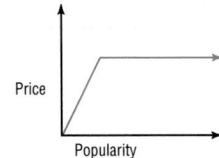

Match each description with the most appropriate graph.

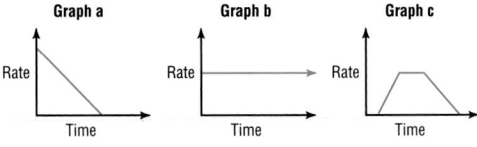

66. the speed of light   **Graph b**

67. an airplane taking off, then landing   **Graph c**

68. a car approaching a stop sign   **Graph a**

71a. $3 + 4 > 5, 3 + 5 > 4, 4 + 5 > 3$ **71b.** $x > 3$ ft

APPLICATIONS AND PROBLEM SOLVING

69. Food and Nutrition There are 80 Calories in one serving of skim milk and 8 servings in a half gallon. (Lesson 1–1)

   a. Write an expression that describes how many Calories you get if you drink s servings of skim milk.   **80s**

   b. How many Calories do you consume if you drink 4 servings of skim milk a day?   **320**

   c. How many Calories are in a half gallon of skim milk?   **640**

70. Investments The equation $I = prt$ describes simple interest on a savings account, where I is the amount of interest, p is the amount deposited, r is the annual interest rate, and t is the time in years. (Lesson 1–3)

   a. Solve for I if $p = 100$, $r = 0.05$, and $t = 2$.   **$10**

   b. If you deposit $200 in a savings account that earns 6% interest (6% = 0.06), how much money will be in your account after the first year?   **$212**

71. Geometry The triangle inequality states that if you have a triangle, then the sum of any two of its sides must be greater than its third side. (Lesson 1–5)

   a. Verify that a triangle with sides of length 3, 4, and 5 units satisfies the triangle inequality.

   b. Suppose you want to construct a triangle such that two of the sides are 3 feet and 6 feet long. What is the minimum length of the third side? (Use x for the length of the third side.)

A practice test for Chapter 1 is provided on page 787.

ALTERNATIVE ASSESSMENT

COOPERATIVE LEARNING PROJECT

Operating a Music Store In this chapter, you learned how to model verbal sentences with algebraic sentences. The basic principles of modeling verbal sentences involve being able to determine whether or not you have an equality or an inequality and determining what your variables are.

In this project, imagine that you and your friends plan to open a new music store that will sell only compact discs. After some research, you find that the monthly rent for your store will be $500 and that you can expect to spend about $200 a month for utilities. Assume that your inventory of compact discs must be at least 400 different titles with 4 discs per title and that the wholesale cost for compact discs is $8.00 per disc.

Outline the costs for starting your business and plan how much in CD sales you would need to show a profit each month.

Follow these steps to organize your business.

- Determine the cost to open your new business.
- Determine your monthly overhead costs.
- Find a price for which you will sell your CDs.
- Write an algebraic model that describes the number of CDs you have to sell in order to break even each month.
- Draw a graph to show that if sales increase over time, the profits will also increase.
- Write several paragraphs describing your plan and incorporate your algebraic models and graphs to help support it.

THINKING CRITICALLY

- Does the distributive property also apply to multiplication? In other words, does the product $a(bc) = (ab)(ac)$? Why or why not?
- Create a sequence of numbers in which there is more than one pattern. Explain why this is possible.

PORTFOLIO

Do you organize your work so that you and anyone else reading it can tell what you were thinking? Find an example of your work that you have done that is well organized and list the qualities that make it so. Then find an example of your work that isn't as well organized and list what you could have done to make it more so. Place both of these in your portfolio.

SELF EVALUATION

Are you confident with your knowledge of mathematics? Most math students would rather not try to solve a math problem than try to and get the wrong answer.

Assess yourself. How confident are you? Would you rather not attempt a problem, or are you willing to give your best effort and learn from the outcome? Describe how you can become more confident in your problem solving both in mathematics and in your daily life.

Assessment and Evaluation Masters, pp. 14, 25

1

NAME_____ DATE_____

Chapter 1 Performance Assessment

Instructions: *Demonstrate your knowledge by giving a clear, concise solution to each problem. Be sure to include all relevant drawings and justify your answers. You may show your solution in more than one way or investigate beyond the requirements of the problem.*

1. Some important area formulas are given below.

   Area of a circle = πr^2, where r is the radius of the circle

   Area of a trapezoid = $\frac{(b_1 + b_2)}{2} \cdot h$, where b_1 and b_2 are the lengths of

   the bases and h is the height of the trapezoid

   a. Write a formula for the area of the figure shown at the right.

   b. Simplify the formula in part a. Justify each step in your simplification by naming the property used.

   c. State the order of operations indicated by the formula in part b. Then find the area when $a = 2$.

2. In each sentence, circle key words or phrases that indicate a mathematical operation and write the corresponding mathematical symbol above each. Then write an equation for each sentence.

   a. Three times a number decreased by four is equal to twenty-five.

   b. The sum of the square of a number and a second number is eighteen.

   c. A number multiplied by 7 and divided by three gives five more than the number.

   d. One-half of a number added to itself equals twice the difference of the number and five.

3. Think of a situation that could be modeled by this graph. Then label the axes of the graph and write several sentences describing the situation.

Scoring Guide
Chapter 1
Performance Assessment

Level	Specific Criteria
3 Superior	• Shows thorough understanding of the concepts of *evaluating and simplifying mathematical expressions, translating verbal expressions into mathematical expressions,* and *interpreting graphs.* • Uses appropriate strategies to solve problems. • Computations are correct. • Written explanations are exemplary. • Goes beyond requirements of some or all problems.
2 Satisfactory, with Minor Flaws	• Shows understanding of the concepts of *evaluating and simplifying mathematical expressions, translating verbal expressions into mathematical expressions,* and *interpreting graphs.* • Uses appropriate strategies to solve problems. • Computations are mostly correct. • Written explanations are effective. • Satisfies all requirements of problems.
1 Nearly Satisfactory, with Serious Flaws	• Shows understanding of most of the concepts of *evaluating and simplifying mathematical expressions, translating verbal expressions into mathematical expressions,* and *interpreting graphs.* • May not use appropriate strategies to solve problems. • Computations are mostly correct. • Written explanations are satisfactory. • Satisfies most requirements of problems.
0 Unsatisfactory	• Shows little or no understanding of the concepts of *evaluating and simplifying mathematical expressions, translating verbal expressions into mathematical expressions,* and *interpreting graphs.* • May not use appropriate strategies to solve problems. • Computations are incorrect. • Written explanations are not satisfactory. • Does not satisfy requirements of problems.

Alternative Assessment

The Alternative Assessment section provides students with the opportunity to assess their own work by thinking critically, working with others, keeping a portfolio, and honestly evaluating their own progress. For more information on alternative forms of assessment, see *Alternative Assessment in the Mathematics Classroom,* one of the titles in the Glencoe Mathematics Professional Series.

Performance Assessment

Performance Assessment tasks for this chapter are included in the *Assessment and Evaluation Masters.* A scoring guide is also provided.

NCTM Standards: 1–5

This Investigation is designed to be completed over several days or weeks. It may be considered optional. You may want to assign the Investigation and the follow-up activities to be completed at the same time.

Objective

Use mathematics to analyze aspects of the greenhouse effect and report your findings.

Mathematical Overview

This Investigation will use the following mathematical skills and concepts from Chapters 2 and 3.

• making charts and graphs
• adding and subtracting integers
• looking for patterns

Recommended Time		
Part	**Pages**	**Time**
Investigation	68–69	1 class period
Working on the Investigation	83, 111, 154, 177	20 minutes each
Closing the Investigation	184	1 class period

Instructional Resources

Investigations and Projects Masters, pp. 1–4

A recording sheet, teacher notes, and scoring guide are provided for each Investigation in the *Investigations and Projects Masters.*

1 MOTIVATION

News reports often refer to the threat of global warming. Ask students what is meant by *global warming* or the *greenhouse effect* and what dangers the situation poses for Earth.

the Greenhouse Effect

MATERIALS NEEDED

lamp with a 100-watt bulb

stopwatch

thermometer

plastic zipper-style sandwich bag

ruler

The *greenhouse effect* refers to the warming of Earth by the sun. The main culprit in the greenhouse effect is carbon dioxide (CO_2). Carbon dioxide in the atmosphere works like the panes of glass on a greenhouse. Glass is transparent to visible light, allowing the sun's rays to warm Earth's surface. But when the surface gives off excess heat, the hot air stays in the greenhouse, which continues to keep the air warm. Similarly, CO_2 in the atmosphere absorbs the sun's infrared rays, allowing some of the excess heat to stay in the atmosphere rather than escaping into space. How much heat is retained depends on how much CO_2 is in the air.

Over the last 200 years, the amount of CO_2 in our atmosphere has increased, raising Earth's average temperature. Some scientists predict that if global warming continues and Earth's average temperature goes up an additional 3° to 8°F,

we could see a marked increase in the number of weather-related disasters like heat waves, droughts, floods, and hurricanes.

In this Investigation, you will use mathematics to analyze aspects of the greenhouse effect and report your findings. As a scientist commissioned to explore the greenhouse effect, you will conduct two experiments. The first experiment is a *control experiment* that will be used to compare data with the second experiment, called the *greenhouse experiment*. Both experiments will involve temperature as a function of distance.

Make an Investigation Folder in which you can store all of your work on this Investigation for future use.

 ## Cooperative Learning

This Investigation offers an excellent opportunity for using cooperative learning groups. For more information on cooperative learning strategies and group management, see *Cooperative Learning in the Mathematics Classroom,* one of the titles in the Glencoe Mathematics Professional Series.

CONTROL EXPERIMENT

Test	Distances Between Thermometer and Bulb	First Temperature	Second Temperature	Difference in Temperatures
1				
2				
3				
4				

CONTROL EXPERIMENT

1 Begin by copying the chart above onto a sheet of paper.

2 Set your lamp about 5 inches away from the thermometer. Make sure the lamp is turned off.

3 Measure and record the temperature and the distance between the thermometer and the bulb.

4 Turn the lamp on and leave it on for five minutes. Use the stopwatch to keep accurate time.

5 Record the new temperature and find the difference of the temperatures.

6 Let the thermometer and bulb cool to room temperature and repeat the process, moving the lamp a little farther from the thermometer. Conduct the test a total of four times at four different distances. Make sure you let the thermometer cool to room temperature between each test.

GREENHOUSE EXPERIMENT

7 Begin by making a chart like the one you made for the Control Experiment.

8 Make sure that the thermometer has cooled to room temperature. Then record the temperature.

9 Put the thermometer in the plastic bag and close the bag. Make sure that the bulb of the thermometer does not touch the bag. Then place the thermometer the same distance from the lamp as you did in the control experiment (about 5 inches). Record the distance between the thermometer and the lamp bulb. Then turn the lamp on and leave it on for five minutes.

10 Record the new temperature. Then find the difference of the temperatures.

11 Conduct four tests, using the same distances you used in the control experiment. Make sure you take the thermometer out of the bag and let it cool to room temperature between each test. Be sure to take accurate readings and measurements.

You will continue working on this Investigation throughout Chapters 2 and 3.

Be sure to keep your chart and materials in your Investigation Folder.

The Greenhouse Effect Investigation

Working on the Investigation
Lesson 2–2, p. 83

Working on the Investigation
Lesson 2–6, p. 111

Working on the Investigation
Lesson 3–2, p. 154

Working on the Investigation
Lesson 3–6, p. 177

Closing the Investigation

End of Chapter 3, 184

Investigation: The Greenhouse Effect **69**

2 SETUP

You may wish to have a student read the first two paragraphs of the Investigation to provide information about why the greenhouse effect is an important issue. You may then wish to read the third paragraph, which introduces the activity. Discuss the activity with the class. Then separate the class into groups of four.

3 MANAGEMENT

Each group member should be responsible for a specific task.

Recorder Collects data.
Measurer Sets and measures the distances between the lamp and the thermometer.
Timer Times different aspects of the activity and turns the lamp on and off at the appropriate times.
Weatherperson Reports temperature readings to the recorder and makes sure that the thermometer is adequately prepared for each aspect of the activity.

At the end of the activity, each member should turn in his or her respective equipment.

Investigations and Projects Masters, p. 4

2, 3

NAME _____ DATE _____

Investigation, Chapters 2 and 3 Student Edition Pages 68–69, 83, 111, 154, 177, 184

The Greenhouse Effect

Work with your group to add to the list of questions to be considered.

- What was the difference in the way the two experiments were designed?
- Do you think life could exist on Earth without CO_2?
- Do you think life could exist on Venus with CO_2?
- How much CO_2 was in the atmosphere in 1800 and in 1950?
- How many degrees on the Celsius scale would 15°F be?
- What is the difference between the boiling point and the freezing point on each of the scales?

_____
_____
_____
_____
_____

Reread the information about the greenhouse effect on pages 68–69 to find significant information. You may wish to look in some other sources. List your findings below.

_____
_____
_____
_____
_____
_____
_____

Please keep this page and any other research in your Investigation folder.

Exploring Rational Numbers

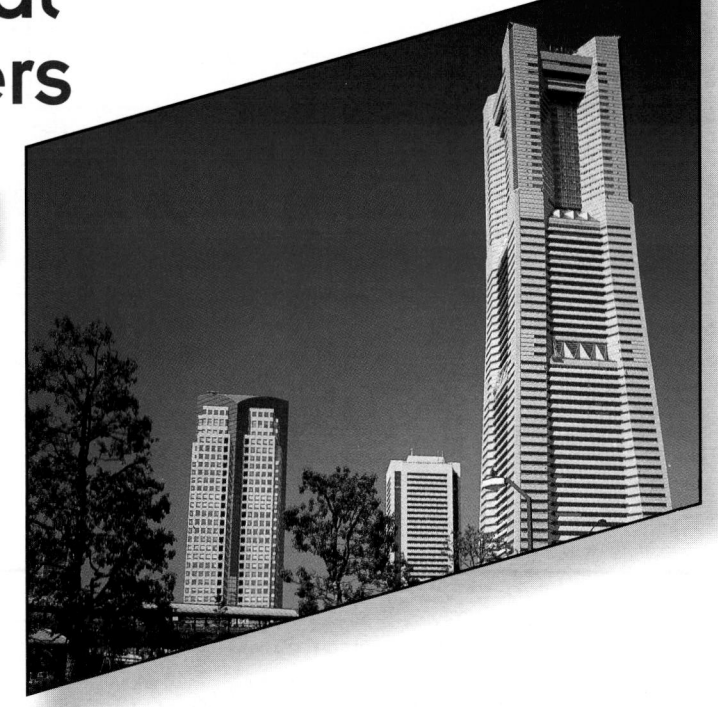

PREVIEWING THE CHAPTER

This chapter explores the basic operations as applied to integers and rational numbers. The number line is used as a mathematical model to develop the rules for the addition of integers. Students apply their knowledge of the rules for addition and the additive inverse property to develop rules for the subtraction of integers. Then the connection is established between these rules for integers and the rules for rational numbers. Rules for multiplying and dividing rational numbers are also explored. Finally, students examine the information presented in verbal problems and translate that information into algebraic expressions, as well as reversing the process and writing descriptions of the problems from algebraic expressions.

Lesson (Pages)	Lesson Objectives	NCTM Standards	State/Local Objectives
2-1 (72–77)	State the coordinate of a point on a number line. Graph integers on a number line. Add integers by using a number line.	1–5, 14	
2-2 (78–83)	Interpret numerical data from a table. Display and interpret statistical data on a line plot.	1–4, 10	
2-3A (84)	Use counters to add and subtract integers.	1–5	
2-3 (85–92)	Find the absolute value of a number. Add and subtract integers.	1–5	
2-4 (93–99)	Compare and order rational numbers. Find a number between two rational numbers.	1–5	
2-5 (100–104)	Add and subtract rational numbers. Simplify expressions that contain rational numbers.	1–5	
2-6A (105)	Use counters to multiply integers.	1–5	
2-6 (106–111)	Multiply rational numbers.	1–5	
2-7 (112–117)	Divide rational numbers.	1–5	
2-8A (118)	Use base-ten tiles to model square roots.	1–5	
2-8 (119–125)	Find square roots. Classify numbers. Graph solutions of inequalities on number lines.	1–5, 14	
2-9 (126–132)	Explore problem situations. Translate verbal sentences and problems into equations or formulas and vice versa.	1–5	

ORGANIZING THE CHAPTER

You may want to refer to the **Course Planning Calendar** on page T12 for detailed information on pacing.
PACING: Standard—15 days; **Honors**—15 days; **Block**—9 days; **Two Years**—28 days

LESSON PLANNING CHART

Lesson (Pages)	Materials/ Manipulatives	Extra Practice (Student Edition)	BLACKLINE MASTERS									Real-World Applications	Interactive Mathematics Tools Software	Teaching Transparencies
			Study Guide	Practice	Enrichment	Assessment and Evaluation	Modeling Mathematics	Multicultural Activity	Tech Prep Applications	Graphing Calculator	Science and Math Lab Manual			
2-1 (72–77)	graphing calculator	p. 759	p. 10	p. 10	p. 10						pp. 5–8		2-1	2-1A 2-1B
2-2 (78–83)	small packages of plain M & Ms	p. 759	p. 11	p. 11	p. 11							3		2-2A 2-2B
2-3A (84)	counters* integer mat*						p. 20						2-3A	
2-3 (85–92)	computer calculator	p. 759	p. 12	p. 12	p. 12	p. 44	p. 73		p. 3			4		2-3A 2-3B
2-4 (93–99)	calculator	p. 760	p. 13	p. 13	p. 13					p. 2			2-4	2-4A 2-4B
2-5 (100–104)		p. 760	p. 14	p. 14	p. 14	pp. 43, 44		p. 3					2-5	2-5A 2-5B
2-6A (105)	counters* integer mat*						p. 21							
2-6 (106–111)		p. 760	p. 15	p. 15	p. 15				p. 4			5		2-6A 2-6B
2-7 (112–117)	calculator	p. 761	p. 16	p. 16	p. 16	p. 45	pp. 36–38							2-7A 2-7B
2-8A (118)	base-ten tiles*						p. 22							
2-8 (119–125)	calculator	p. 761	p. 17	p. 17	p. 17									2-8A 2-8B
2-9 (126–132)	rectangular box scissors	p. 761	p. 18	p. 18	p. 18	p. 45	pp. 39–41	p. 4				6	2-9	2-9A 2-9B
Study Guide/ Assessment (133–137)						pp. 29–42, 46–48								

*Included in Glencoe's Student Manipulative Kit and Overhead Manipulative Resources.

ORGANIZING THE CHAPTER

OTHER CHAPTER RESOURCES

Student Edition
Investigation, pp. 68–69
Chapter Opener, pp. 70–71
Mathematics and Society, p. 77
Working on the Investigation,
   pp. 83, 111

Teacher's Classroom Resources
Investigations and Projects Masters,
   pp. 29–32
Algebra and Geometry Overhead
   Manipulative Resources, pp. 2–5

Technology
Test and Review Software (IBM
   and Macintosh)
CD-ROM Interactions (Windows
   and Macintosh)

Professional Publications
Block Scheduling Booklet
Glencoe Mathematics Professional
   Series

OUTSIDE RESOURCES

Books/Periodicals
Gray, Virginia, *The Write Tool to Teach Algebra*,
   Dale Seymour Publications
Taylor, Harold and Loretta Taylor, *Developing
   Skills in Algebra One*, Dale Seymour
   Publications

Software
Tobbs Learns Algebra, WINGS for Learning,
   Sunburst

Videos/CD-ROMs
Algebra for Everyone, NCTM

ASSESSMENT RESOURCES

Student Edition
Math Journal, pp. 97, 122
Mixed Review, pp. 77, 83, 92,
   99, 104, 111, 117, 125, 132
Self Test, p. 99
Chapter Highlights, p.133
Chapter Study Guide and
   Assessment, pp. 134–136
Alternative Assessment, p. 137
       Portfolio, p. 137

Cumulative Review, pp. 138–139

Teacher's Wraparound Edition
5-Minute Check, pp. 72, 78, 85,
   93, 100, 106, 112, 119, 126
Check for Understanding, pp. 75,
   80, 89, 96, 102, 109, 115,
   122, 129
Closing Activity, pp. 76, 82, 91,
   99, 104, 111, 117, 125, 132
Cooperative Learning, pp. 113,
   127

Assessment and Evaluation Masters
Multiple-Choice Tests, Forms 1A
   (Honors), 1B (Average), 1C
   (Basic), pp. 29–34
Free-Response Tests, Forms 2A
   (Honors), 2B (Average), 2C
   (Basic), pp. 35–40
Calculator-Based Test, p. 41
Performance Assessment, p. 42
Mid-Chapter Test, p. 43
Quizzes A–D, pp. 44–45
Standardized Test Practice, p. 46
Cumulative Review, pp. 47–48

ENHANCING THE CHAPTER

Examples of some of the materials for enhancing Chapter 2 are shown below.

DIVERSITY

Multicultural Activity Masters, pp. 3, 4

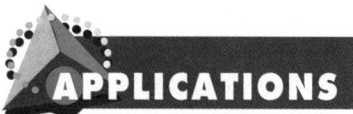

APPLICATIONS

Real-World Applications, 3, 4, 5, 6

TECHNOLOGY

Graphing Calculator Masters, p. 2

TECH PREP

Tech Prep Applications Masters, pp. 3, 4

CONNECTIONS

Science and Math Lab Manual, pp. 5–8

PROBLEM SOLVING

Problem of the Week Cards, 4, 5, 6

This two-page introduction to the chapter provides students with an opportunity to explore contemporary topics and their applications to mathematics.

Background Information
Veggie Lovers Unite! There are many types of vegetarians. Vegetarians eat only vegetables, fruits, nuts, seeds, grains, and legumes. A lacto-vegetarian will also consume milk and milk products. Lacto-ovo-vegetarians include eggs in their diets. Pesco-vegetarians will eat fish, while pollo-vegetarians will include poultry in their diets. All vegetarians abstain from eating red meat.

CHAPTER

2

Exploring Rational Numbers

Veggie Lovers Unite!

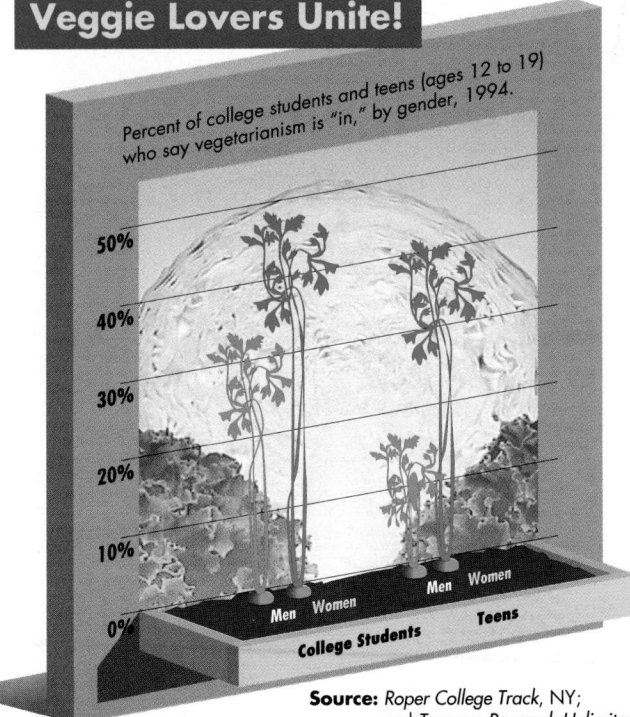

Percent of college students and teens (ages 12 to 19) who say vegetarianism is "in," by gender, 1994.

Source: *Roper College Track, NY; and Teenage Research Unlimited*

Objectives

In this chapter, you will:
* display and interpret statistical data on line plots,
* add, subtract, multiply, and divide rational numbers,
* find square roots, and
* write equations and formulas.

Hamburgers and pepperoni pizzas have been teen diet staples for decades. But lately there is a growing trend toward meatless meals. Why do teens switch to "veggie values" and go for tofu and eggplant? Are there benefits to a diet without meat? Are there dangers to a diet without meat?

TIME *Line*

1000 B.C. Chinese counting boards originate.

1812 French chef Nicolas Appert invents canning, starting the first commercial cannery.

A.D. 1565 The first potatoes from the New World arrive in Spain.

1790 Metric system developed by group of French scientists.

TIME *Line*

The process of counting primarily deals with the positive integers. Students might find it interesting to investigate and report on the Chinese counting boards that were developed as far back as 1000 B.C.

*inter*NET
CONNECTION
You can learn about vegetarian news, recipes, and environmental and biological concerns on-line.

World Wide Web
http://www.cs.unc.edu/~barman/
vegetarian.html

ChapterProject

What would you eat if you didn't eat meat? Research nutrition and dietary resource materials at the library to help you devise a healthy vegetarian diet and a healthy meat diet for an average teenager.

- Plan two diets. One should be for a no animal meat vegetarian, and one should be for a meat-eating person.

- Devise a seven-day menu for each diet, listing all nutrition facts such as calories, fat,

cholesterol, sodium, carbohydrates, proteins, and vitamins.

- Can a healthy diet be maintained without animal products? Is an increase in fruit and vegetable consumption healthy? Explain your reasoning.

- Draw some conclusions about individual diets and food production from your research. Share your findings with the class.

The concern for animals is one reason teenagers give up eating meat. Groups like Earth 2000, a teens-only advocacy group founded by **Danny Seo,** want the focus of their generation to be "no animal cruelty, no meat." Danny and members of Earth 2000 cite teen celebrities like Michael Stipe of R.E.M., Eddie Vedder of Pearl Jam, and Jennie Garth of *Beverly Hills 90210* as vegetarian role models.

Danny also cites health reasons for maintaining a meatless diet, such as cutting back on the intake of fat. In a balanced vegetarian diet, only about one third of the calories come from fat.

1930 Ruth Wakefield invents the Toll House cookie.

1950 Luis Buñuel, Spanish filmmaker, directs his Mexican film *Los Olvidados,* "The Forgotten."

1890 | 1900 | 1910 | 1920 | 1930 | 1940 | 1950 | 1960 | 1970 | 1980 | 1990 | 2000

1925 Clarence Birdseye pioneers the development of packaged frozen foods.

1991 Carolyn Napoli co-invented the technology that can be applied to block genes in plants to reduce sugar and oil content.

Chapter 2 **71**

Alternative Chapter Projects

Two other chapter projects are included in the *Investigations and Projects Masters.* In Chapter 2 Project A, pp. 29–30, students extend the topic in the chapter opener. In Chapter 2 Project B, pp. 31–32, students design and produce a board game.

The following menu lists what a healthy 13-year-old vegetarian might eat in a day. Of the 2260 calories, 32% come from fat.

Breakfast orange; oatmeal with soy milk; toast with jam

Lunch miso soup with tofu; pita-bread sandwich with hummus, tomato, and spinach; glass of carrot juice; dried figs

Dinner rice, bean, squash, onion, and broccoli casserole with safflower oil; cooked kale

Snacks slice of watermelon; strawberries; trail mix

ChapterProject

Cooperative Learning Have students work in groups of three to investigate levels of fat, protein, and carbohydrates in cereal. One group member should investigate a sugary children's cereal, another should look at a cereal that is promoted as healthful, and the third group member should investigate a cereal that does not obviously fall into either category. Have group members compare their findings and then discuss them with the class.

Investigations and Projects Masters, p. 29

---

2 NAME_____ DATE_____
Chapter 2 Project A   Student Edition
                                     Pages 72–132

Good for You!

1. Choose at least five different kinds of breakfast cereal. Copy the information given in the "Nutrition Facts" chart on each package and note the price. (Be sure each package contains the same amount of cereal.)

2. For each cereal find the total number of vitamins with a Percent Daily Value of at least 25% in one serving. (Use the information given for the cereal alone—without milk.)

3. Rank the cereals in order from least sugar to most sugar per serving.

4. Rank the cereals in order from lowest price to highest price.

5. Present your findings to the class by doing one of the following activities.
   - Make a poster that summarizes what you discovered about the cereals.
   - Write an article for the school newspaper about nutrition in breakfast cereals.
   - Create a magazine or television ad for the cereal you think is most nutritious.

NCTM Standards: 1–5, 14

Instructional Resources

- Study Guide Master 2-1
- Practice Master 2-1
- Enrichment Master 2-1
- Science and Math Lab Manual, pp. 5–8

 Transparency 2-1A contains the 5-Minute Check for this lesson; **Transparency 2-1B** contains a teaching aid for this lesson.

Recommended Pacing	
Standard Pacing	Day 1 of 15
Honors Pacing	Day 1 of 15
Block Scheduling*	Day 1 of 9 (along with Lesson 2-2)
Alg. 1 in Two Years*	Days 1, 2, & 3 of 28

 *For more information on pacing and possible lesson plans, refer to the *Block Scheduling Booklet* and *Algebra 1 in Two Years*.

1 FOCUS

 ### 5-Minute Check
(over Chapter 1)

Simplify.

1. $6r + 2s + 3r + s$ $9r + 3s$
2. $9(a + b) + 3a$ $12a + 9b$

Evaluate each expression if $a = 5$, $b = 4$, and $c = 3$.

3. $3ac - bc$ 33
4. $b - c + 2ab$ 41

Name the property shown.
5. $5 + (x + y) = 5 + (y + x)$
   commutative property (+)

Motivating the Lesson

Situational Problem An outdoor thermometer shows a temperature of 10°C. Several hours later, the temperature has dropped 13 degrees. What temperature does the thermometer show now? −3°C

Objectives

After studying this lesson, you should be able to:

- state the coordinate of a point on a number line,
- graph integers on a number line, and
- add integers by using a number line.

Most Landed-on Squares in Monopoly®
1. Illinois Ave.
2. Go
3. B&O Railroad
4. Free Parking
5. Tennessee Ave.

 APPLICATION
Entertainment

Integers and the Number Line

2-1

Have you ever played Monopoly®? Most likely you have at one time or another. Parker Brothers has sold over 100 million sets of the game of Monopoly since it was invented in 1933 by Charles Darrow.

In order to play Monopoly, you must know how to move playing pieces around the board. To do this, you add the numbers rolled on the dice and then move your playing piece the corresponding number of spaces forward. You also need to know how to move your playing piece *backward* when directed to do so. Understanding how to add and subtract integers can help you to play.

Nidi, Lynn, Michael, and Domingo are playing a game of Monopoly. The chart below shows a turn for each player. How can you represent Nidi's moves using numerals?

Player	Turn 1
Nidi	Rolls 2; lands on Chance — go back 3 spaces.
Lynn	Rolls double 3s; rolls 7.
Michael	Rolls 6 — go directly to jail (back 20).
Domingo	Rolls 9; lands on Community Chest — advance to next utility (forward 11).

You can solve problems like the one above by using a **number line.** A number line is drawn by choosing a starting position, usually 0, and marking off equal distances from that point. The set of **whole numbers** is often represented on a number line. This set can be written {0, 1, 2, 3, ...}, where "..." means that the set continues indefinitely.

Although only a portion of the number line is shown, the arrowhead indicates that the line and the set of numbers continue.

We can use the expression $2 - 3$ to represent Nidi's moves.

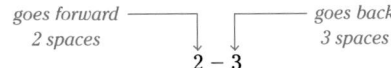
goes forward 2 spaces   *goes back 3 spaces*
$2 - 3$

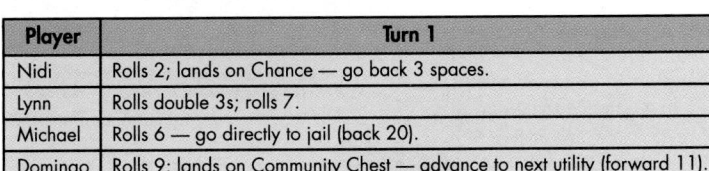
Mr. Darrow's Monopoly became so popular among his family and friends that he began making them their own sets. He then began selling sets for $4.00 each. When first offered the game, Parker Brothers turned it down, believing the game would never be accepted by the public.

The number line below shows that the value of $2 - 3$ should be 1 less than 0. However, there is no whole number that corresponds to 1 less than 0. You can write the number 1 *less than* 0 as -1. This is an example of a **negative number**.

Any nonzero number written without a sign is understood to be positive.

To include negative numbers on the number line, extend the number line to the left of zero and mark off equal distances. The points to the right of zero are named using the *positive sign* (+). The points to the left of zero are named using the *negative sign* (−). Zero is neither positive nor negative.

Read "−5" as *negative* 5.   Read "+5" as *positive* 5.

The set of numbers used on the number line above is called the set of **integers**. This set can be written $\{..., -6, -5, -4, -3, -2, -1, 0, 1, 2, 3, 4, 5, 6, ...\}$.

Venn diagrams are figures often used to represent sets of numbers.

Sets	Examples
Natural numbers	1, 2, 3, 4, 5, ...
Whole numbers	0, 1, 2, 3, 4, ...
Integers	..., −2, −1, 0, 1, 2, ...

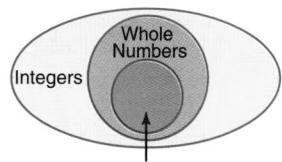

Natural Numbers

Notice that the natural numbers are a subset of the whole numbers and the whole numbers are a subset of the integers.

To **graph** a set of numbers means to draw, or plot, the points named by those numbers on a number line. The number that corresponds to a point on a number line is called the **coordinate** of that point.

Example ① **Name the set of numbers graphed.**

a.

The bold arrow means that the graph continues indefinitely in that direction.

The set is $\{-5, -4, -3, -2, -1, 0, 1, 2, ...\}$.

b.

The set is $\{-4, -2, -1, 1, 3\}$.

***GLENCOE* Technology**

Interactive Mathematics Tools Software

This multimedia software provides an interactive lesson that uses a number line to determine the rules for adding and subtracting positive and negative integers. A **Computer Journal** gives students an opportunity to write about what they have learned.

For Windows & Macintosh

Alternative Teaching Strategies

Reading Algebra After studying the concept of positive and negative numbers, ask students to discuss what business terms such as *negative worth* and *negative cash flow* might mean.

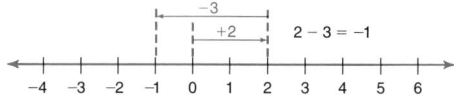

There are many varieties of Domino games. In China, one finds Tiu-ü (fishing), Tau Ngau (bull fighting), Pai Kow, Tien Kow, and Mah Jong. In Europe, one finds the Block Game, Cyprus, Matador, Bergen, Tiddlywinks, and Blind Hughie.

Teaching Tip In a Venn diagram, ovals, circles, and squares or rectangles are used to represent the universal, or largest, set and its subsets. However, the size and shape of a Venn diagram is unimportant.

In-Class Example

For Example 1
Name the set of numbers graphed.

$\{-5, -3, -1, 1, 3, 5\}$

MODELING MATHEMATICS

Adding Integers

To find the sum of -7 and -5, follow the steps below.

Step 1
Draw an arrow starting at 0 to -7, the first addend.

Step 2
Starting at -7, draw an arrow and go to the left 5 units long. This represents the second addend, -5.

Step 3
The arrow ends at the sum, -12.

So, $-7 + (-5) = -12$.

Notice that parentheses are used in the equation so that the sign of the number is not confused with a subtraction sign.

Your Turn a–c. See margin.

a. What addition sentence is modeled on the number line shown at the right?

b. Draw a number line to show the addition sentence $-3 + (-2) = -5$.

c. Use a number line to find $-4 + 5$.

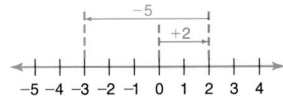

Example ②

APPLICATION
Meteorology

At 4:08 A.M., the temperature in Casper, Wyoming, was $-10°$F. By 1:30 P.M., the temperature had risen $17°$ to the daytime high. What was the high temperature?

Draw and label a number line. Draw an arrow from 0 to -10. Then draw an arrow to the right 17 units long.

The high temperature of the day was $7°$F.

Communicating Mathematics

3. See Solutions Manual.

Study the lesson. Then complete the following. 1–2. See margin.

1. **Explain** how you would find the sum $-4 + 6$ on a number line.

2. **Draw** a number line to show that $-17 < 0$, $-3 < 1$, and $-5 > -10$.

3. Refer to the application at the beginning of the lesson. Draw number lines to show the first Monopoly® moves for Lynn, Michael, and Domingo.

4. **Research** instances where integers are used in real-life situations by collecting newspaper clippings that show integers. See students' work.

Left margin column

MODELING MATHEMATICS When showing addition on a number line, emphasize that negative numbers are indicated by arrows pointing to the left, and positive numbers are indicated by arrows pointing to the right. The first arrow must start at the origin. The second arrow starts at the arrowhead of the first arrow. The sum is at the head of the second arrow.

Answers for Modeling Mathematics

a. $2 + (-5) = -3$

b.

c.

Teaching Tip Point out that a number line is not always horizontal. A thermometer is an instrument with a vertical number line.

In-Class Example

For Example 2
The next day in Casper, Wyoming, the low temperature was $-8°$F. The high was $21°$ higher than the low temperature. What was the high temperature? **13°F**

Teaching Tip Point out that since addition is commutative, the sum of $(2) + (-4)$ can be found either by finding $+2$ on the number line and then moving 4 units left, or by first finding -4 on the number line and then moving 2 units to the right.

Additional Answers

1. An arrow would be drawn, starting at 0 and going to -4. Starting at -4, an arrow would be drawn to the right 6 units long. The arrow ends at the sum of 2.

2.

Classroom Vignette

"We actually put a number line on the floor and left it there for weeks. We used it to demonstrate problems throughout this chapter. It really seemed effective, and the students enjoyed physically adding and subtracting on the number line."

Mrs. Betsy A. Barron

Betsy Barron
Rochester High School
Rochester, Indiana

MODELING
MATHEMATICS

5. Use a number line to find $5 + (-6)$.   **See margin.**

Guided
Practice

Name the set of numbers graphed.

6.
$\{-3, -2, -1\}$

7.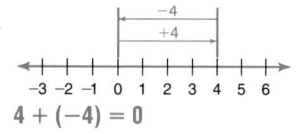
$\{-1, 0, 1, 2, ...\}$

Graph each set of numbers on a number line.

8. $\{0, 2, 4, 6\}$

9. $\{-1, 0, 1, 2, 3, ...\}$

10. {integers greater than -3}   **8–10. See Solutions Manual.**

Write a corresponding addition sentence for each diagram.

11.
$-4 + (-3) = -7$

12.
$4 + (-4) = 0$

Find each sum. If necessary, use a number line.

13. $-8 + 3$ -5

14. $-7 + (-15)$ -22

15. Football   The Barnesville Bruins' offense lined up on their 20-yard line. They gained 5 yards on the first down, lost 3 yards on the second down, gained 7 yards on the third down, and punted on the fourth down. What was the net gain or loss after the third down?   **9-yard gain**

EXERCISES

Practice

16. $\{..., -2, -1, 0,$
1, 2, 3\}

17. $\{-4, -3, -2, -1\}$

18. $\{0, 2, 5, 6, 8\}$

19. $\{-7, -3\}$

20. $\{-2, -1, 1, 2\}$

21. $\{..., -5, -4, -3,$
$-2, -1, 0\}$

22–30. See Solutions
Manual.

Name the set of numbers graphed.

16.

17.

18.

19.

20.

21.

Graph each set of numbers on a number line.

22. $\{-4, -3, -1, 3\}$

23. $\{..., -2, -1, 0, 1\}$

24. {integers between -6 and 10}

25. {integers less than 0}

26. {integers between -2 and 5}

27. {integers less than $-3 + (-1)$}

28. {whole numbers greater than $-4 + 4$}

29. {integers less than 0 but greater than -6}

30. {integers less than or equal to -3 and greater than or equal to -10}

Find each sum. If necessary, use a number line.

31. $8 + 5$ **13**

32. $-6 + (-7)$ **-13**

33. $7 + (-12)$ **-5**

34. $-4 + (-9)$ **-13**

35. $0 + (-12)$ **-12**

36. $5 + (-3)$ **2**

37. $-3 + 9$ **6**

38. $-13 + 13$ **0**

39. $-14 + (-9)$ **-23**

3 PRACTICE/APPLY

Check for Understanding

Exercises 1–15 are designed to help you assess your students' understanding through reading, writing, speaking, and modeling. You should work through Exercises 1–5 with your students and then monitor their work on Exercises 6–15.

Error Analysis

The "$-$" is used in three ways in math. Students may interchange "minus," "negative," and "opposite of."

Expression	Meaning
$-x$	opposite of x (relationship)
-3	negative (direction)
$5 - 4$	5 minus 4 (operation)

The context determines how the "$-$" is being used. Consider $-x - 3 - (-2) - y$. This means the opposite of x, minus 3, minus negative 2, minus y.

Assignment Guide

Core: 17–39 odd, 40, 41, 43, 45–52
Enriched: 16–38 even, 40–52

For **Extra Practice,** see p. 759.

Study Guide Masters, p. 10

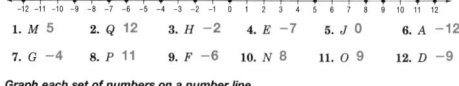

2-1   NAME_____   DATE_____
Study Guide   Student Edition Pages 72–77

Integers and the Number Line

The figure at the right is part of a number line. On a number line, the distances marked to the right of 0 are named by members of the set of **whole numbers**.

The set of numbers used to name the points marked on the number line at the right is called the set of **integers**.

To graph a set of numbers means to locate the points named by those numbers on the number line. The number that corresponds to a point on the number line is called the **coordinate** of the point.

Name the coordinate of point G.

The coordinate of G is -2.

A number line is often used to show addition of integers. For example, to find the sum of 3 and -5, follow the steps at the right.

Step 1 Draw an arrow, starting at 0 and going to 3.
Step 2 Start at 3. Draw an arrow 5 units to the left.
Step 3 The second arrow points to the sum, -2.

Name the coordinate of each point.

1. M 5   2. Q 12   3. H -2 4. E -7 5. J 0   6. A -9

7. G -4 8. P 11   9. F -6 10. N 8   11. O 9   12. D -9

Graph each set of numbers on a number line.

13. $\{-3, -1, 1, 3\}$ 14. $\{-5, -2, 1, 4\}$ 15. {integers less than 0}

16. $\{0, 1, 3, 5\}$ 17. $\{-3, -2, 2\}$ 18. $\{..., -2, -1, 0, 1\}$

Find each sum. If necessary, use a number line.

19. $2 + 3$ 5   20. $9 + 1$ 10   21. $-5 + (-1)$ -6

22. $-10 + 6$ -4 23. $9 + (-9)$ 0   24. $0 + (-4)$ -4

25. $-8 + (-3)$ -11 26. $6 + (-10)$ -4 27. $-6 + 6$ 0

Reteaching ─────

Act It Out "Walk out" addition problems by dividing a large strip of paper into squares labeled -9 to 9, with positives on the right and negatives on the left. To "walk out" $2 + (-5)$, start at zero and walk 2 places to the right ($+2$). Then walk 5 places to the left (-5) ending on -3. Have students "walk out" $2 + 5$, $(-2) + 5$, and $(-2) + (-5)$.

Additional Answer

5.

Closing Activity

Writing Have students determine the relationship that exists between two addends with the same sign and the sign of their sum. Then have students identify the relationship between addends with different signs and the sign of their sum. Students should write a short paragraph describing their findings.

Additional Answers

42a. 11 P.M. the same day
42b. 7 A.M. the same day
42c. 12 noon the same day
42d. 8:30 P.M. the same day

Practice Masters, p. 10

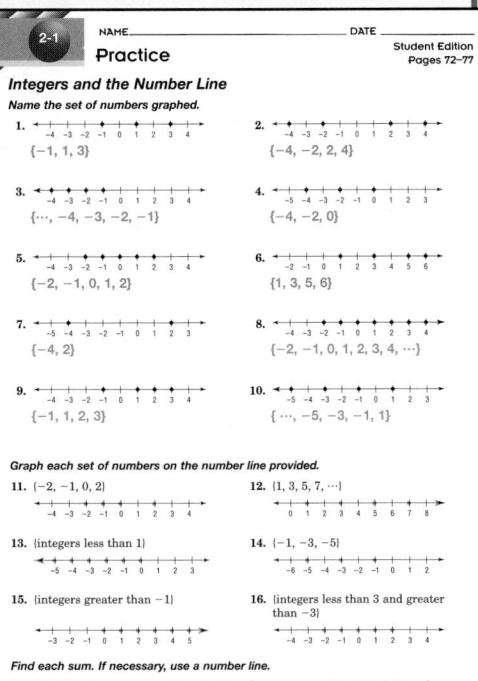

Critical Thinking

40. Write three different equations using integers that have a sum of -14.
See students' work.

Applications and Problem Solving

For Exercises 41-42, write an open sentence using addition. Then solve each problem.

41. Meteorology On February 10, the low temperature in Houlton, Maine, and Fort Yukon, Alaska, was -17°. On the same date, the high temperature in Lajitas, Texas, was 82° higher. What was the temperature in Lajitas, Texas, on February 10? -17 + 82 = 65; 65°F

42. Geography Use the map below to determine what time it is in each city if it is 10:00 A.M. in New York City. a-d. See margin.

   a. Hong Kong   **b.** Los Angeles
   c. Rio de Janeiro   **d.** Bombay

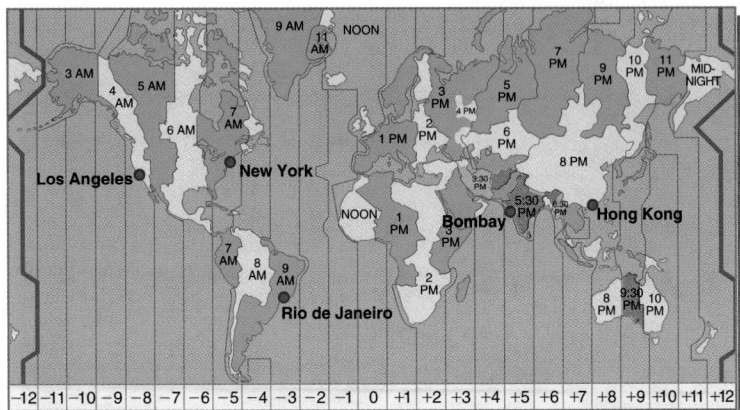

43. Meteorology *Windchill factor* is an estimate of the cooling effect the wind has on a person in cold weather. Meteorologists use a chart like the one below to predict the windchill factor. Use the chart below to find each windchill factor.

43a. 4°F   **a.** 20°F, 10 mph
43b. -31°F   **b.** 0°F, 15 mph
43c. -15°F   **c.** -10°F, 5 mph
43d. 9°F   **d.** 30°F, 15 mph

Wind speed (mph)	Actual temperature (°Fahrenheit)						
	30	20	10	0	-10	-20	-30
	Equivalent temperature (°Fahrenheit)						
0	30	20	10	0	-10	-20	-30
5	27	16	6	-5	-15	-26	-36
10	16	4	-9	-21	-33	-46	-58
15	9	-4	-18	-31	-45	-58	-72

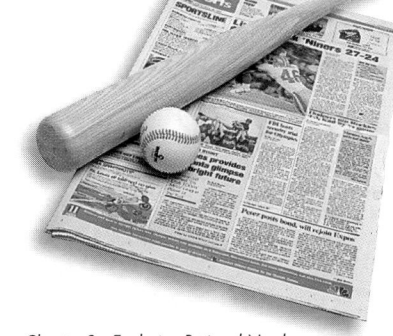

44. Baseball Sometimes baseball team owners develop incentive clauses for hitters in an effort to win more games. Suppose an owner decided to pay each hitter $500 for each run scored, $400 for each hit, and $500 for each run batted in. Look in the sports section of a newspaper. Choose a team's box score and calculate the value of the incentive clause for the first five hitters in a game. See students' work.

Extension

Reasoning Have students study the following number pattern.
              -15, -12, -9, -6, . . .
Ask students to identify the pattern that exists in this sequence. Each number is 3 more than the number that preceded it. Have students identify the tenth term in the number pattern. 12

45. Entertainment Juanita has the volume on her stereo turned up because her parents are not home. Suddenly, she sees her father's car pull into the driveway. She runs to her room to turn down the volume. Juanita's father comes in, grabs an umbrella, and leaves, so Juanita returns the volume to its previous level. Sketch a reasonable graph to show the volume of Juanita's stereo during this time. (Lesson 1–9) **See margin.**

46. Physical Fitness Mitchell likes to exercise regularly. On Mondays, he likes to walk two miles, run three miles, sprint one-half of a mile, and then walk for another mile. Identify the graph that best represents Mitchell's heart rate as a function of time. (Lesson 1–9) **b**

a.

Heart
Rate

Time

b.

Heart
Rate

Time

c.

Heart
Rate

Time

47. commutative (×)

47. Name the property illustrated by $8(2 \cdot 6) = (2 \cdot 6)8$. (Lesson 1–8)

48. Simplify $16a + 21a + 30b - 7b$. (Lesson 1–7) **$37a + 23b$**

49. substitution (=)

49. Name the property illustrated by $(12 - 9)(4) = 3(4)$. (Lesson 1–6)

50. Solve $5(7) + 6 = x$. (Lesson 1–5) **41**

51. Evaluate $9(0.4 + 1.2) - 0.5$. (Lesson 1–3) **13.9**

52. Patterns Find the next three numbers in the pattern 5, 6.5, 8, 9.5, …. (Lesson 1–2) **11, 12.5, 14**

Mayan Mathematics

The excerpt below is from an article that appeared in *The Columbus Dispatch* on January 15, 1995. It describes the number system developed by the Mayan Indians, an ancient civilization that lived in southern Mexico and Central America until the 16th century.

THE MAYAS USED A COUNTING SYSTEM based on 20. And while we add digits to the left of numbers to show their increased size, the Mayas piled them upwards. In our decimal system, as you look at a number from right to left, each digit tells us how much value it has, based on the number 10. The number 326, for instance, is really 3 hundreds, 2 tens and 6 ones….The Mayas used the same principle, but used the vigesimal system (based on 20). Units were represented by dots. Five dots were represented by a horizontal line. ∎

Upper Level

= 1
= 5
= 5
= 5

Lower Level

$16 \times 20 = 320$

= 1
= 5

$6 \times 1 = 6$

1. See margin.   2–3. See students' work.

1. What combinations of dots and horizontal lines (bars) are needed to write the number 252 using the Mayan system? (*Hint:* Two levels are required.)

2. What uses would the Mayas and other ancient civilizations have had for the number systems they developed?

3. Would a number system based on a number other than 10 be as accurate as the decimal system? Would it be as easy to use? Explain your answers.

Additional Answer

45.

Volume

Time

Answer for Mathematics and Society

1. Upper Level

○ = 1
○ = 1
▭ = 5
▭ = 5

$12 \times 20 = 240$

Lower Level

○ = 1
○ = 1
▭ = 5
▭ = 5

$12 \times 1 = 12$

Enrichment Masters, p. 10

 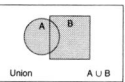

NCTM Standards: 1–4, 10

Instructional Resources

- Study Guide Master 2-2
- Practice Master 2-2
- Enrichment Master 2-2
- Real-World Applications, 3

 Transparency 2-2A contains the 5-Minute Check for this lesson; **Transparency 2-2B** contains a teaching aid for this lesson.

Recommended Pacing

Standard Pacing	Day 2 of 15
Honors Pacing	Day 2 of 15
Block Scheduling*	Day 1 of 9 (along with Lesson 2-1)
Alg. 1 in Two Years*	Days 4 & 5 of 28

*For more information on pacing and possible lesson plans, refer to the *Block Scheduling Booklet* and *Algebra 1 in Two Years.*

1 FOCUS

5-Minute Check
(over Lesson 2-1)

1. Graph {2, 3, 4} on a number line.

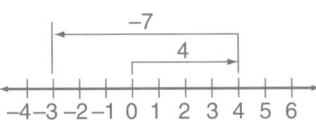

−2 −1 0 1 2 3 4 5 6

2. Show the addition 4 + (−7) on a number line. −3

−4 −3 −2 −1 0 1 2 3 4 5 6

Find each sum.

3. 9 + (−24) −15
4. (−18) + (−47) −65
5. −27 + 45 18

Teaching Tip Have students discuss and write what information can be gained from the line plot in the Example. You might ask, in general, how many wins were necessary to make the playoffs? between 40 and 60

Objectives

After studying this lesson, you should be able to:
- interpret numerical data from a table, and
- display and interpret statistical data on a line plot.

Integration: Statistics
Line Plots

APPLICATION
Television

Are you a fan of Will on the television show *Fresh Prince of Bel-Air*? The Nielsen ratings are used to determine which shows are popular and which shows are not. They are also used to determine which shows should be continued or canceled. The chart at the right shows the Nielsen ratings for Monday, February 6, 1995.

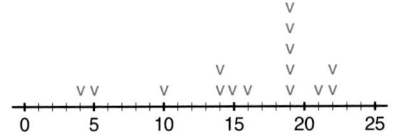

Nielsen Ratings

Monday, February 6, 1995		
TIME	PROGRAM	VIEWERS (rounded to nearest million)
8:00	Fresh Prince of Bel-Air (NBC)	22
	The Nanny (CBS)	19
	Melrose Place (Fox)	14
	Coach (ABC)	15
	Star Trek: Voyager (UPN)	14
8:30	Dave's World (CBS)	21
	Blossom (NBC)	19
	Sneakers (ABC)	16
9:00	Murphy Brown (CBS)	22
	Serving in Silence (NBC)	19
	Models, Inc. (Fox)	10
	Platypus Man (UPN)	5
9:30	Cybill (CBS)	19
	Pig Sty (UPN)	4
10:00	Chicago Hope (CBS)	19

In some cases, data can be presented on a number line. Numerical data displayed on a number line is called a **line plot.** The data in the table above can be presented in a line plot as follows.

Step 1 Draw and label a number line. You can see that the data in the table ranges from 4 million to 22 million viewers. In order to represent the data on a number line, a *scale* must be used that includes this range of values. You can use a scale from 0 to 25 with *intervals* of five.

Step 2 Draw the line plot. Write a "v" for each TV show above its share of viewers. A completed line plot for the Nielsen ratings is shown below.

```
                              v
                              v
                              v
                  v     v     v
        v v    v  v v v   v v v
        +--------+--------+--------+--------+--------+
        0     5    10    15    20    25
```

Notice that some data values are located between marked intervals on the number line.

Example
APPLICATION
Basketball

In 1994 and 1995, the Houston Rockets won back-to-back NBA championships. The table at the right shows the standings for each of the 27 NBA teams for the 1994–1995 season on April 28, 1995.

a. Make a line plot to show the number of wins by each playoff team.

Final Regular-Season Standings

EASTERN CONFERENCE	W	L	WESTERN CONFERENCE	W	L
ATLANTIC			**MIDWEST**		
y-Orlando	57	25	y-San Antonio	62	20
x-New York	55	27	x-Utah	60	22
x-Boston	35	47	x-Houston	47	35
Miami	32	50	x-Denver	41	41
New Jersey	30	52	Dallas	36	46
Philadelphia	24	58	Minnesota	21	61
Washington	21	61			
CENTRAL			**PACIFIC**		
z-Indiana	52	30	z-Phoenix	59	23
x-Charlotte	50	32	x-Seattle	57	25
x-Chicago	47	35	x-L.A. Lakers	48	34
x-Cleveland	43	39	x-Portland	44	38
x-Atlanta	42	40	Sacramento	39	43
Milwaukee	34	48	Golden State	26	56
Detroit	28	54	L.A. Clippers	17	65

x–clinched playoff berth; y–clinched conference; z–clinched division

Source: *USA Today,* 4-28-95

The teams with an x, y, or z next to their names made the playoffs. The number of wins by the playoff teams ranges from 35 to 62. You can use a "w" to represent the number of each team's wins.

b. **How many playoff teams won fewer than 50 games?**
   From the line plot, you can see that 8 teams won fewer than 50 games. These teams were Boston, Chicago, Cleveland, Atlanta, Houston, Denver, L.A. Lakers, and Portland.

c. **Which playoff team had the best record? the worst record?**
   San Antonio had the best record and Boston had the worst record of the playoff teams.

d. **How many teams made the playoffs?**
   Count the number of w's on the line plot. Sixteen teams made the playoffs.

The data in the application at the beginning of the lesson was the result of an actual survey. The data in the example was collected by checking NBA records. Data can be collected by taking actual measurements, by conducting surveys or polls, by using questionnaires, by simulation, or by consulting reference materials.

It is important that you know how the data were obtained. For example, would you want to draw conclusions about changing the name of your school mascot based on a result of a survey of seniors only? Why or why not?

In September 1995, blue M&M's® completely replaced the tan ones. The ratios of colors for plain M&M's are as follows.

brown   30%
red   20%
yellow 20%
orange 10%
green   10%
blue   10%

MODELING MATHEMATICS

Line Plots

Materials: small packages of plain M&M's®

In 1994 and 1995, the Mars Company conducted surveys to determine whether a new color should be added to M&M's®. The choices were blue, pink, purple, or leave them as they are. The color blue was chosen.

Your Turn   a–e. See students' work.

a. Open a package of M&M's. Separate the candies by color. Find the total number of each color.

b. Make a line plot to show the number of each color. Use b for brown, r for red, y for yellow, o for orange, g for green, and bl for blue.

c. Do the colors cluster around any number?

d. Make a class line plot of your data. Are the data in the class line plot different or the same as yours? Explain.

e. Make a class line plot showing the total number of M&M's in each of your packages. Do the packages have the same number in them?

Lesson 2–2   INTEGRATION Statistics   Line Plots   **79**

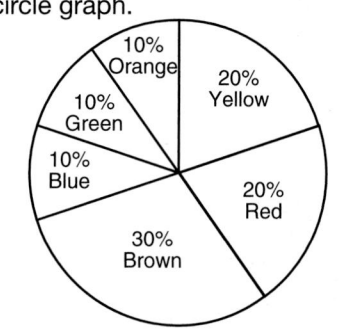

3 PRACTICE/APPLY

Check for Understanding

Exercises 1–6 are designed to help you assess your students' understanding through reading, writing, speaking, and modeling. You should work through Exercises 1–3 with your students and then monitor their work on Exercises 4–6.

Error Analysis

In order to answer questions using a line plot, the scales should be accurate enough to read right off the plot. They should also be as precise as possible.
Example:

Additional Answers

4. from 0 to 70

5. from 35 to 80

6a.

6b. $1950 to $11, 475

6c.

7a.

Communicating Mathematics

Study the lesson. Then complete the following.

1. **Describe** a situation in which it would be valuable to obtain data in order to make a decision.   1–3. See students' work.

2. **Compare and contrast** line plots and tables as a means of reporting data. What are the advantages and disadvantages of each?

3. **Develop** a survey and collect data on the Monday night television viewing habits of your classmates. Make a line plot of the data. Compare your class findings to those of another class and to a Nielsen ratings chart from a magazine or newspaper. What conclusions can you draw?

Guided Practice

State the scale you would use to make a line plot for the following data. Then draw the line plot.   4–5. See margin.

6. See margin.

4. 50, 50, 30, 30, 20, 10, 60, 40   5. 52, 43, 67, 69, 37, 76

6. **Retail Sales**   The advertisement below lists the cars for sale at Texas Motors.

a. Make a line plot to show how many cars from each year are in stock.

b. What is the price range for the used cars?

c. Make a line plot to show the costs of the cars. Round each price to the nearest thousand dollars. Let each number on the number line represent thousands of dollars.

Applications and Problem Solving

7b. 70 mph; cheetah

7. **Animals**   The speeds of 20 of the fastest animals in miles per hour according the *The World Almanac*, 1995, are listed below.

40	61	50	50	32	70	35	30	50	45
43	40	30	30	35	45	42	32	40	30

a. Make a line plot of the data.   See margin.

b. What is the fastest speed of any animal? Research to find the name of the fastest animal.

c. What is the slowest speed of the 20 animals?   30 mph

d. Which speed occurred most frequently?   30 mph

e. How many animals had a speed of at least 40 mph?   12

f. How many animals had speeds greater than 30 and less than 40 mph?   4

Reteaching

Using Discussion Have students compare and contrast how data are presented in a table and on a line plot.

8. History The table below lists the fifty United States and the years they entered the Union.

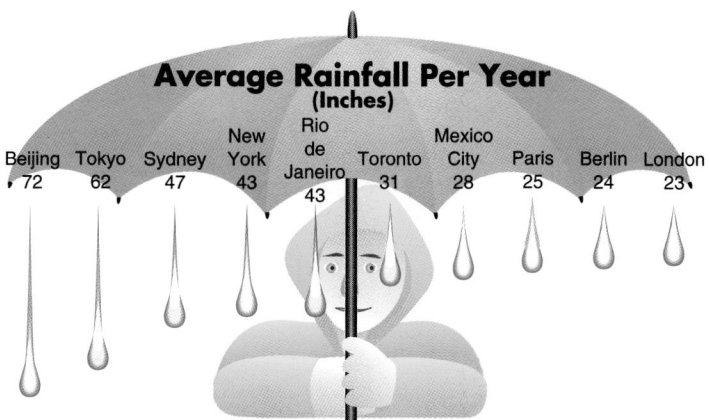

Years States Entered the Union

State	Year	State	Year	State	Year	State	Year
AL	1819	IN	1816	NE	1867	SC	1788
AK	1959	IA	1846	NV	1864	SD	1889
AZ	1912	KS	1861	NH	1788	TN	1796
AR	1836	KY	1792	NJ	1787	TX	1845
CA	1850	LA	1812	NM	1912	UT	1896
CO	1876	ME	1820	NY	1788	VT	1791
CT	1788	MD	1788	NC	1789	VA	1788
DE	1787	MA	1788	ND	1889	WA	1889
FL	1845	MI	1837	OH	1803	WV	1863
GA	1788	MN	1858	OK	1907	WI	1848
HI	1959	MS	1817	OR	1859	WY	1890
ID	1890	MO	1821	PA	1787		
IL	1818	MT	1889	RI	1790		

Source: *The World Almanac, 1995*

a, c. See margin.

a. What is the range of years for states entering the Union?

8b. See Solutions Manual.

b. Make a line plot that shows the year each state joined the Union.

c. Do the years cluster around a certain number? If so, which years are those? What would explain this?

9. Weather The average rainfall per year, rounded to the nearest inch, for ten cities around the world is shown on the map below.

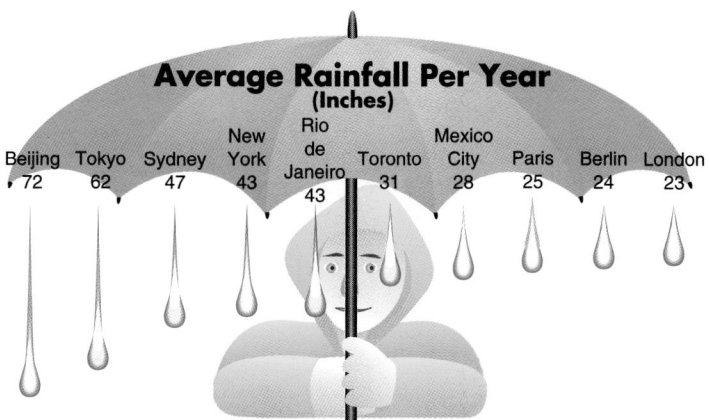

Average Rainfall Per Year
(Inches)

Beijing	Tokyo	Sydney	New York	Rio de Janeiro	Toronto	Mexico City	Paris	Berlin	London
72	62	47	43	43	31	28	25	24	23

a. Make a line plot of the data. What scale did you use? See margin.

9b. yes, 25 and 42

b. Do the average amounts of rain cluster about a certain number?

c. Do you think that the average rainfall for your community is close to that of any of the cities above? Why or why not? c–d. See students' work.

d. Look up the average rainfall for your community. How does it compare to the average rainfall for London? for Tokyo?

Assignment Guide

Core: 7–11 odd, 13–18
Enriched: 8–12 even, 13–18

For **Extra Practice**, see p. 759.

The red A, B, and C flags, printed only in the Teacher's Wraparound Edition, indicate the level of difficulty of the exercises.

Additional Answers

8a. from 1787 to 1959
8c. Yes, 1788; states ratified constitution and became official United States of America.
9a. from 20 to 75 by 5s

Study Guide Masters, p. 11

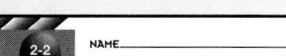

2-2
Study Guide

NAME_____ DATE_____
Student Edition
Pages 78–84

Integration: Statistics
Line Plots

The table to the right shows the final results of the Great Frog Competition in Dickerson County. The frog jumps were measured in inches.

	Jumps		
Frogs	**1**	**2**	**3**
Slippery	61.5	51.9	60.0
Spots	46.0	38.5	39.1
Inky	56.2	33.0	61.5
Popper	65.4	51.0	52.3

Numerical information displayed on a number line is called a **line plot.** The line plot below is another way to show the data for the frog jumping contest.

1. Make a table comparing the highest mountains in the United States: Mt. McKinley, 20,633 ft; Mt. Elbert, 14,433 ft; Mt. Rainier, 14,410 ft; Humphrey Peak, 12,633 ft; Kings Peak, 13,528 ft.

Mountain	Height (feet)
Mt. McKinley	20,633
Mt. Elbert	14,433
Mt. Rainier	14,410
Humphrey Peak	12,633
Kings Peak	13,528

Use the table at the right for Exercises 2–4.

2. Make a line plot representing the weights of junior varsity wrestlers. Insert labels and "x"s on the number line below.

Weights of Junior Varsity Wrestlers (pounds)				
170	160	135	135	160
122	188	154	108	135
140	122	103	190	154

3. How many wrestlers are in weight classes over 140 lb? 7

4. What is the largest weight class? 190 lb

Closing Activity

Modeling Have students obtain data from the newspaper that would be good to use on a line plot. Have them cut out the data, make a line plot, and then present it to the class.

Additional Answer

11b. No; in Ms. Martinez's class, the hours that the students talked on the phone were close to each other. (4 ± 2 hours). In Mr. Thomas's class, the students had a wider range of hours spent talking on the phone. The range was from 0 to 8 hours, with no hours clustered around a certain number.

Practice Masters, p. 11

B **10. Tornadoes** A tornado is a powerful, twisting windstorm. The winds of a tornado are the most dangerous winds that occur on Earth, with speeds of more than 200 miles per hour. The table below lists the average number of tornadoes that occurred per year in each state for a 30-year period.

TORNADO OCCURRENCE BY STATE, 1962–1991

STATE	AVG.	STATE	AVG.	STATE	AVG.	STATE	AVG.
AL	22	IN	20	NE	37	SC	10
AK	0	IA	36	NV	1	SD	29
AZ	4	KS	40	NH	2	TN	12
AR	20	KY	10	NJ	3	TX	139
CA	5	LA	28	NM	9	UT	2
CO	26	ME	2	NY	6	VT	1
CT	1	MD	3	NC	15	VA	6
DE	1	MA	3	ND	21	WA	2
FL	53	MI	19	OH	15	WV	2
GA	21	MN	20	OK	47	WI	21
HI	1	MS	26	OR	1	WY	12
ID	3	MO	26	PA	10		
IL	27	MT	6	RI	0		

Source: National Severe Storm Forecast Center

10a. See Solutions Manual.
10b. yes; 1, 2, and 3

a. Make a line plot of the average number of tornado occurrences.
b. Do any of the numbers appear clustered? If so, which ones?
c. Which state had the highest average? Why do you think it had the most tornadoes? Texas, because of weather patterns and geographical location.

11. School Mr. Thomas and Ms. Martinez each asked 10 students from their algebra classes how many hours they spent talking on the telephone last week. The results are shown in the line plot below. Use this plot to answer each question.

Number of Hours on the Telephone

0 1 2 3 4 5 6 7 8

x – Mr. Thomas' class
● – Ms. Martinez's class

a. Which group of 10 students talked on the telephone the most? Ms. Martinez's
b. Does the pattern for the number of hours spent talking on the telephone appear to be the same for both groups? Explain. See margin.

Extension

Problem Solving The enrollment per class in each of the three high schools in Brunswick is given in the table. Using the enrollment for each high school as a separate set of data, make a line plot that differentiates between each set.

	9th	10th	11th	12th
North	500	450	432	405
Central	648	640	563	552
South	425	405	372	354

350 400 450 500 550 600 650

12. Winter Olympics The table below lists the medal standings for the 1994 Winter Olympics.

Winter Olympics, Medal Standings

Nation	Gold	Silver	Bronze	Total
Norway	10	11	5	26
Germany	9	7	8	24
Russia	11	8	4	23
Italy	7	5	8	20
United States	6	5	2	13
Canada	3	6	4	13
Switzerland	3	4	2	9
Austria	2	3	4	9
South Korea	4	1	1	6
Finland	0	1	5	6

a. Make a line plot that shows the number of gold medals won by the countries. **a–b. See margin.**

b. Write three questions that can be answered using the line plot.

Critical Thinking

13. Use the line plot in Exercise 11. Find the average number of hours that the students in Mr. Thomas and Ms. Martinez's classes spent on the telephone. Do the values support your answer to Exercise 11b? Explain. **See margin.**

Mixed Review

14. Graph {...,−5, −4, −3} on a number line. (Lesson 2–1) **See margin.**

15. Name the property illustrated by $6(jk) = (6j)k$. (Lesson 1–8)

15. associative (×)

16. Use the distributive property to find 15(124). (Lesson 1–7) **1860**

17. Solve $m = \frac{22 - 8}{7}$. (Lesson 1–5) **2**

18. Evaluate $np + st$ when $n = 7$, $p = 6$, $s = 4$, and $t = 5$. (Lesson 1–3) **62**

In·ves·ti·ga·tion

Refer to the Investigation on pages 68–69.

the Greenhouse Effect

1 Examine the data from the control experiment. Look for patterns and relationships in the data. Analyze how the change in temperature is a function of the distance. Describe your analysis in writing. Be sure to explain any patterns you see in the data.

2 Examine the data from the greenhouse experiment. Look for patterns and relationships in the data. Analyze how the change in temperature

is a function of the distance. Describe your analysis in writing. Be sure to explain any patterns you see in the data.

3 Compare the two experiments. List the similarities and differences in the data and explain the factors that might cause them.

4 Think about the design of the two experiments. How were they different? Why do you think they were designed the way they were? How do they relate to the greenhouse effect?

Add the results of your work to your Investigation Folder.

Lesson 2-2 **INTEGRATION** *Statistics Line Plots* **83**

In·ves·ti·ga·tion

Working on the Investigation

The Investigation on pages 68–69 is designed to be a long-term project that is completed over several days or weeks. Encourage students to keep their materials in their Investigation Folder as they work on the Investigation.

Additional Answers

12a.

x
x x x x   x x   x x x
0 1 2 3 4 5 6 7 8 9 10 11

12b. Answers will vary. Sample answer:
1. Did any countries win the same number of gold medals?
2. How many countries won more than three gold medals?
3. How many gold medals did the countries win altogether?

13. Mr. Thomas's, 3.6 hours; Ms. Martinez's, 3.7 hours; yes

14.

−5 −4 −3 −2 −1 0 1 2

Enrichment Masters, p. 11

NAME_____ DATE_____

2-2   **Enrichment**   Student Edition Pages 78–83

The Digits of π

The number π (pi) is the ratio of the circumference of a circle to its diameter. It is a nonrepeating and nonterminating decimal. No block of the digits of π ever repeats. Here are the first 201 digits of π including 200 digits that follow the decimal point.	3.14159 26535 89793 23846
	69399 37510 58209 74944
	86280 34825 34211 70679
	09384 46095 50582 23172
	84102 70193 85211 05559
	26433 83279 50288 41971
	59230 78164 06286 20899
	82148 08651 32823 06647
	53594 08128 34111 74502
	64462 29489 54930 38196

Solve each problem.

1. If each of the digits appeared with equal frequency, how many times would each digit appear in the first 200 places following the decimal point? **20**

2. Complete this frequency distribution table for the first 200 digits of π that follow the decimal point.

Digit	Frequency (Tally Marks)	Frequency (Number)	Cumulative Frequency
0	ᵀᴴᴸ ᵀᴴᴸ ᵀᴴᴸ ////	19	19
1	ᵀᴴᴸ ᵀᴴᴸ ᵀᴴᴸ ᵀᴴᴸ	20	39
2	ᵀᴴᴸ ᵀᴴᴸ ᵀᴴᴸ ᵀᴴᴸ ////	24	63
3	ᵀᴴᴸ ᵀᴴᴸ ᵀᴴᴸ ᵀᴴᴸ	20	83
4	ᵀᴴᴸ ᵀᴴᴸ ᵀᴴᴸ ᵀᴴᴸ //	22	105
5	ᵀᴴᴸ ᵀᴴᴸ ᵀᴴᴸ ᵀᴴᴸ	20	125
6	ᵀᴴᴸ ᵀᴴᴸ ᵀᴴᴸ /	16	141
7	ᵀᴴᴸ ᵀᴴᴸ //	12	153
8	ᵀᴴᴸ ᵀᴴᴸ ᵀᴴᴸ ᵀᴴᴸ ////	24	177
9	ᵀᴴᴸ ᵀᴴᴸ ᵀᴴᴸ ᵀᴴᴸ ///	23	200

3. Explain how the cumulative frequency column can be used to check a project like this one. **The last number should be 200, the number of items being counted.**

4. Which digit(s) appears most often? **2 and 8**

5. Which digit(s) appears least often? **7**

NCTM Standards: 1–5

Objective
Use counters to add and subtract integers.

Recommended Time
Demonstration and discussion: 15 minutes; Exercises: 30 minutes

Instructional Resources
For each student or group of students
Student Manipulative Kit
• 20 red/yellow 2-sided counters
Modeling Mathematics Masters
• p. 2 (counters)
• p. 13 (integer mat)
• p. 20 (worksheet)
For teacher demonstration
Algebra and Geometry Overhead Manipulative Resources

1 FOCUS

Motivating the Lesson
Ask students how much money they would really have if they just cashed a check for $40, but owe their parents $25.

2 TEACH

Teaching Tip Caution students that $-1 - (-4)$ presents a problem. Having put down one negative counter, four counters cannot be removed. This can be solved by first putting down four zero pairs.

3 PRACTICE/APPLY

Assignment Guide

Core: 1–13
Enriched: 1–13

Additional Answer

13. Answers should include using the number line.

MODELING MATHEMATICS

2-3A Adding and Subtracting Integers

A Preview of Lesson 2–3

Materials: ⬤ counters ▭ integer mat

You can use counters to help you understand addition and subtraction of integers. In these activities, yellow counters represent positive integers, and red counters represent negative integers.

Rules for Integer Models	
A zero pair is formed by pairing one positive counter with one negative counter.	⊕ ⊖
You can remove or add zero pairs to a set because removing or adding zero does not change the value of the set.	⊕ ⊖ ⊖ → ⊖

Activity 1   Use counters to find the sum $-2 + 3$.

Step 1 Place 2 negative counters and 3 positive counters on the mat.

Step 2 Remove the 2 zero pairs. Since 1 positive counter remains, the sum is 1. Therefore, $-2 + 3 = 1$.

Activity 2   Use counters to find the difference $3 - (-2)$.

Step 1 Place 3 positive counters on the mat. There are no negative counters, so you cannot remove 2 negatives. Add 2 zero pairs to the mat. Adding zero pairs does not change the value of the set.

Step 2 Now remove 2 negative counters. Since 5 positive counters remain, the difference is 5. Therefore, $3 - (-2) = 5$.

 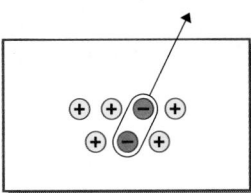

Model Find each sum or difference by using counters.

1. $4 + 2$ 6
2. $4 + (-2)$ 2
3. $-4 + 2$ -2
4. $-4 + (-2)$ -6
5. $4 - 2$ 2
6. $-4 - (-2)$ -2
7. $4 - (-2)$ 6
8. $-4 - 2$ -6

Draw Tell whether each statement is *true* or *false*. Justify your answer with a drawing.

9. $5 - (-2) = 3$
   false
10. $-5 + 7 = 2$
    true
11. $2 - 3 = -1$
    true
12. $-1 - 1 = 0$
    false

Write 13. Write a paragraph explaining how to find the sum of two integers without using counters. Be sure to include all possibilities. See margin.

---

4 ASSESS

Observing students working in cooperative groups is an excellent method of assessment.

GLENCOE Technology

Interactive Mathematics Tools Software

This multimedia software provides an interactive lesson using a game in which students find the differences between two positive and/or negative integers. A **Computer Journal** gives students an opportunity to write about what they have learned.

For Windows & Macintosh

Adding and Subtracting Integers

Objectives

After studying this lesson, you should be able to:

- find the absolute value of a number, and
- add and subtract integers.

APPLICATION

Space Science

On February 10, 1995, Dr. Bernard A. Harris, Jr. became the first African-American to walk in space. During the walk, he and Dr. Michael Foale, a British-born astronaut who is a U.S. citizen, were exposed to temperatures as cold as −125°F. The scheduled five-hour spacewalk to test thermal improvements to NASA spacesuits was cut short by 30 minutes because the astronauts experienced icy fingers.

Drs. Michael Foale and Bernard Harris in space

Usually, spacewalkers spend some of their time basking in the sun's rays where temperatures can be as high as 200°F. But throughout the spacewalk, Drs. Harris and Foale remained in the shadow of the shuttle Discovery and Earth, the coldest possible spot.

As you can see, spacewalkers must be able to survive in extreme temperatures. You can use a number line to determine the range of the temperature extremes. −125 and 200 are graphed on the number line below. Notice that −125°F is 125 units from 0 and 200°F is 200 units from 0. So, the total number of units from −125 to 200 is 125 + 200 or 325.

The temperature range from −125°F to 200°F is 325°F.

You used the idea of **absolute value** to find the range from −125°F to 200°F.

Definition of Absolute Value	The absolute value of a number is its distance from zero on a number line.

Since distance is always positive, absolute value is always positive.

The symbol for the absolute value of a number is two vertical bars around the number.

Note that the absolute value bars can serve as grouping symbols.

$|-125| = 125$ is read *The absolute value of −125 equals 125.*

$|200| = 200$ is read *The absolute value of 200 equals 200.*

$|-7 + 6| = 1$ is read *The absolute value of the quantity −7 + 6 equals 1.*

Lesson 2-3   Adding and Subtracting Integers   **85**

NCTM Standards: 1–5

Instructional Resources

- Study Guide Master 2-3
- Practice Master 2-3
- Enrichment Master 2-3
- Assessment and Evaluation Masters, p. 44
- Modeling Mathematics Masters, p. 73
- Real-World Applications, 4
- Tech Prep Applications Masters, p. 3

Transparency 2-3A contains the 5-Minute Check for this lesson; **Transparency 2-3B** contains a teaching aid for this lesson.

Recommended Pacing	
Standard Pacing	Day 4 of 15
Honors Pacing	Day 4 of 15
Block Scheduling*	Day 2 of 9
Alg. 1 in Two Years*	Days 7 & 8 of 28

*For more information on pacing and possible lesson plans, refer to the *Block Scheduling Booklet* and *Algebra 1 in Two Years*.

1 FOCUS

5-Minute Check
(over Lesson 2-2)

True or False:

1. A line plot helps you see the distribution and clustering of data. true
2. A line plot would be more useful than a table when reviewing your test grades in algebra over the past semester. false

Use the line plot below to answer each question.

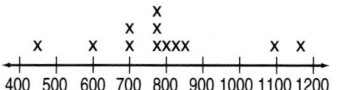

3. What was the greatest attendance? almost 1200
4. What appears to be the average attendance, rounded to the nearest hundred? 800

Hands-On Activity A scuba diver swimming at a depth of 30 meters descends another 10 meters. Draw a diagram that shows the path of the diver. At what depth is the diver now swimming?
40 meters

2 TEACH

Teaching Tip Emphasize that the absolute value of a number is always nonnegative. Absolute value will be used in the rules for adding signed numbers.

In-Class Examples

For Example 1
Evaluate each expression.

a. $|r + 2|$ if $r = -5$ **3**
b. $-|4 + t|$ if $t = -8$ **−4**
c. $-|7 + p|$ if $p = 3$ **−10**

For Example 2
Find each sum.

a. $-6 + (-2)$ **−8**
b. $-56 + 42$ **−14**
c. $38 + (-36)$ **2**

Teaching Tip Students may not remember the meaning of the term *addend*. Remind them that an addend is one of the numbers or terms being added.

You can evaluate expressions involving absolute value.

Example ❶ Evaluate $-|x + 6|$ if $x = -10$.

$$-|x + 6| = -|-10 + 6| \quad \text{\textit{Substitution property of equality}}$$
$$= -|-4|$$
$$= -(4) \quad \text{\textit{The absolute value of −4 is 4.}}$$
$$= -4$$

You can use absolute value to add integers.

Same Signs

a. $4 + 3 = 7$ \qquad *Notice that the sign of each addend is positive. The sum is positive.*

b. $-4 + (-3) = -7$ \quad *Notice that the sign of each addend is negative. The sum is negative.*

Different Signs

c. $6 + (-8) = -2$ \quad *Notice that $8 - 6 = 2$. Since the integer −8 has the greater absolute value, the sign of the sum is also negative.*

d. $-6 + 8 = 2$ \qquad *Notice that $8 - 6 = 2$. Since the integer 8 has the greater absolute value, the sign of the sum is also positive.*

These examples suggest the following rules.

Adding Integers	To add integers with the *same sign,* add their absolute values. Give the result the same sign as the integers.
	To add integers with *different signs,* subtract the lesser absolute value from the greater absolute value. Give the result the same sign as the integer with the greater absolute value.

Example ❷ Find each sum.

a. $-10 + (-17)$

$$-10 + (-17) = -(|-10| + |-17|) \quad \text{\textit{Both numbers are negative,}}$$
$$= -(10 + 17) \quad \text{\textit{so the sum is negative.}}$$
$$= -27$$

b. $39 + (-22)$

$$39 + (-22) = +(|39| - |-22|) \quad \text{\textit{Subtract the absolute values. Since the}}$$
$$= +(39 - 22) \quad \text{\textit{number with the greater absolute value,}}$$
$$= 17 \quad \text{\textit{39, is positive, the sum is positive.}}$$

c. $-28 + 16$

$$-28 + 16 = -(|-28| - |16|) \quad \text{\textit{Subtract the absolute values. Since the}}$$
$$= -(28 - 16) \quad \text{\textit{number with the greater absolute value,}}$$
$$= -12 \quad \text{\textit{28, is negative, the sum is negative.}}$$

Notice that both +2 and −2 are 2 units away from 0. That is, −2 and 2 have the same absolute value.

Every positive integer can be paired with a negative integer. These pairs are called **opposites.** For example, the opposite of +2 is −2.

A number and its opposite are called **additive inverses** of each other. Look at the sum of each number and its additive inverse below. What pattern do you notice?

$$2 + (−2) = 0 \qquad −67 + 67 = 0 \qquad 409 + (−409) = 0$$

These examples suggest the following rule.

Additive Inverse Property	For every number a, $a + (−a) = 0$.

Additive inverses can be used when you subtract numbers.

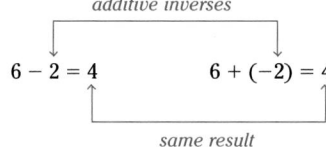

It appears that subtracting a number is equivalent to adding its additive inverse.

Subtracting Integers	To subtract a number, add its additive inverse. For any numbers a and b, $a − b = a + (−b)$.

Example Find each difference.

a. **6 − 14**

Method 1
Use the rule.

To subtract 14, add −14.
$$6 − 14 = 6 + (−14)$$
$$= −8$$

Method 2
Use models.

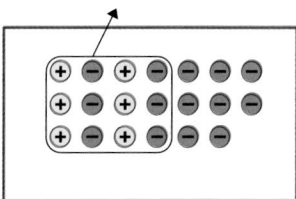

Place 6 positive counters on the mat. Add 14 negative counters. Form and remove six zero pairs.
$$6 − 14 = −8$$

In-Class Example

For Example 3
Find each difference.

a. $−9 − (−4)$ **−5**
b. $−3 − 7$ **−10**
c. $12 − (−48)$ **60**

Classroom Vignette

"I put integers in the five triangles at the bottom of the large triangle and ask students to add or subtract integers as they progress up the triangle. The example shows an addition triangle."

Jenny Brawner

Jenny Brawner
Lee-Scott Academy
Auburn, Alabama

For Example 4

a. Add the matrices.

$$\begin{bmatrix} 800 & 650 & 240 \\ 1200 & 880 & 310 \end{bmatrix} + \begin{bmatrix} 650 & 410 & 170 \\ 800 & 600 & 200 \end{bmatrix}$$

$$\begin{bmatrix} 1450 & 1060 & 410 \\ 2000 & 1480 & 510 \end{bmatrix}$$

b. Subtract the matrices.

$$\begin{bmatrix} 800 & 650 & 240 \\ 1200 & 880 & 310 \end{bmatrix} - \begin{bmatrix} 650 & 410 & 170 \\ 800 & 600 & 200 \end{bmatrix}$$

$$\begin{bmatrix} 150 & 240 & 70 \\ 400 & 280 & 110 \end{bmatrix}$$

For Example 5
Simplify $4a - 9a$. $-5a$

---

3 PRACTICE/APPLY

Check for Understanding

Exercises 1–24 are designed to help you assess your students' understanding through reading, writing, speaking, and modeling. You should work through Exercises 1–6 with your students and then monitor their work on Exercises 7–24.

Error Analysis
Students often interpret $-x$ to be a negative number. Use a table of values to show otherwise.

x	$-x$	$\lvert x \rvert$	$-\lvert x \rvert$
3	−3	3	−3
−2	2	2	−2
−1.5	1.5	1.5	−1.5
0	0	0	0
2	−2	2	−2
−7	7	7	−7

On a TI-82, enter the following keystrokes.

(−) 12 −
(−) 8 ENTER   −4

The plural of matrix is matrices.

---

b. $-12 - (-8)$

Method 1

Use the rule.

To subtract -8, add its inverse, $+8$.

$-12 - (-8) = -12 + (+8)$
$\qquad\qquad\quad = -4$

Method 2

Use a scientific calculator.

The +/− key on a scientific calculator, called the *change-sign key*, changes the sign of the number on the display. You can find $-12 - (-8)$ by using this key.

Enter: 12 +/− − 8 +/− =   −4

c. $43 - (-26)$

$43 - (-26) = 43 + 26$
$\qquad\qquad\quad = 69$

Check by using a calculator.

Enter: 43 − 26 +/− =   69 ✓

A **matrix** is a rectangular arrangement of elements in rows and columns. Although matrices are sometimes used as a problem-solving tool, their importance extends to another branch of mathematics called **discrete mathematics**. Discrete mathematics deals with finite or discontinuous quantities. The distinction between continuous and discrete quantities is one that you have encountered throughout your life. Think of a staircase. You can slide your hand up the banister, but you have to climb the stairs one by one. The banister represents a continuous quantity, like the graph of {all numbers greater than 3}. However, each step represents a discrete quantity, like an element in a matrix, or a point in the graph of {2, 4, 7, 12}.

When entries in corresponding positions of two matrices are equal, the matrices are equal.

$$\begin{bmatrix} 4 & 2 \\ -2 & 5 \end{bmatrix} = \begin{bmatrix} 4 & 2 \\ -2 & 5 \end{bmatrix} \qquad \begin{bmatrix} 4 & 2 \\ -2 & 5 \end{bmatrix} \neq \begin{bmatrix} 2 & 4 \\ -2 & 1 \end{bmatrix}$$

You can add or subtract matrices only if they have the same number of rows and columns. You add and subtract matrices by adding or subtracting corresponding entries.

Example 4

APPLICATION
Business

The Paw Print at Gahanna Lincoln High School in Gahanna, Ohio, is a school store run by seniors in a yearlong marketing education class. The store is stocked with school items such as sweatshirts, T-shirts, school supplies, and snacks. The sales, cost of goods, and expenses for each semester in 1994 and 1995, rounded to the nearest dollar, are shown below.

Semester 1	Sales	Cost of Goods	Expenses
1994	$47,981	$32,627	$12,999
1995	$70,018	$49,013	$12,705

Semester 2	Sales	Cost of Goods	Expenses
1994	$31,988	$21,752	$8666
1995	$30,008	$21,005	$5445

a. Find the total sales, cost of goods, and expenses for each year at The Paw Print.

To find the total sales, cost of goods, and expenses for each year, add the matrices for semester 1 and semester 2.

$$\begin{bmatrix} \$47{,}981 + \$31{,}988 & \$32{,}627 + \$21{,}752 & \$12{,}999 + \$8666 \\ \$70{,}018 + \$30{,}008 & \$49{,}013 + \$21{,}005 & \$12{,}705 + \$5445 \end{bmatrix}$$

$$= \begin{bmatrix} \$79{,}969 & \$54{,}379 & \$21{,}665 \\ \$100{,}026 & \$70{,}018 & \$18{,}150 \end{bmatrix}$$

b. Find the store's gross income, or profit, by year. The gross income is the difference between the sales and the sum of the cost of goods and expenses.

Profit for 1994 = $79,969 − (54,379 + 21,665)
            = $3925

Profit for 1995 = $100,026 − (70,018 + 18,150)
            = $11,858

The rule for subtracting integers and the distributive property can be used to combine like terms.

Example ⑤ **Simplify $2x - 5x$.**

$2x - 5x = 2x + (-5x)$ *To subtract 5x, add its additive inverse, −5x.*

$\qquad = [2 + (-5)]x$ *Use the distributive property.*

$\qquad = -3x$

CHECK FOR UNDERSTANDING

Communicating Mathematics

1. See margin.

2–3. See margin for sample answers.

Study the lesson. Then complete the following.

1. **Explain** how a zero pair of counters illustrates the additive inverse property.

2. **Write** an integer on a piece of paper. Have a friend explain how to find its absolute value without a number line and without knowing the integer.

3. **Explain** to a friend how you would find each sum or difference below.
   a. $-3 + (-1)$ **b.** $5 + (-7)$ **c.** $4 - 9$ **d.** $6 - (-5)$

4. **a.** Find a value for x if $x = -x$.   **0**
   b. Find a value for x if $|x| = -|x|$.   **0**

MODELING MATHEMATICS

5. Write a statement that represents the model shown below. $-3 + 5 = 2$

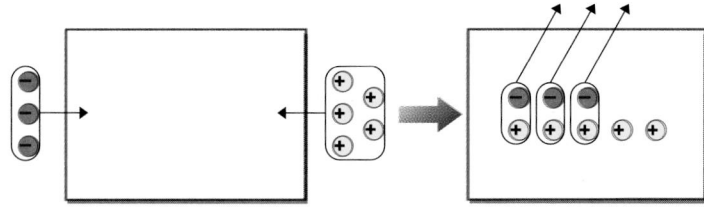

6. Find $-9 - (-4)$ by using counters.   **−5**

Reteaching

Using Discussion Ask students the following questions.

1. What integer is its own additive inverse?   **0**

2. A number x is added to the absolute value of x. If the sum is 0, what conclusion can you draw about the number x?   **It is a negative integer.**

Additional Answers

1. A zero pair of counters has equal amounts of positive and negative counters, therefore there are equal pairs of opposites and they will add to zero, which shows the additive inverse property.

2. Sample answer: −2; Start at zero. The absolute value is two places to the right.

3a. Sample answers: On a number line, start at zero. Move left three spaces to −3. Then move left one space to add −1. The sum is −4.

3b. On a number line, start at zero. Move right five spaces to +5. Then move left seven spaces to add −7. The sum is −2.

3c. On a number line, start at zero. Move right four spaces to +4. Then move nine spaces to the left to subtract 9. The difference is −5.

3d. On a number line, start at zero. Move right six spaces to +6. Subtracting −5 is the same as adding +5. So move right five more spaces to add +5. The sum is 11.

Assignment Guide

Core: 25–61 odd, 62, 63, 65, 67–78
Enriched: 26–60 even, 62–78

For **Extra Practice,** see p. 759.

The red A, B, and C flags, printed only in the Teacher's Wraparound Edition, indicate the level of difficulty of the exercises.

Additional Answers

22. $\begin{bmatrix} 7 & 0 \\ 4 & 6 \end{bmatrix}$

23. $\begin{bmatrix} -1 & 4 \\ 6 & -6 \end{bmatrix}$

58. $\begin{bmatrix} 3 & 7 \\ 1 & -5 \end{bmatrix}$

59. $\begin{bmatrix} 0 & 3 \\ 2 & 2 \end{bmatrix}$

60. $\begin{bmatrix} 4 & 12 \\ -6 & -1 \end{bmatrix}$

61. $\begin{bmatrix} -4 & -5 \\ -4 & -5 \\ 10 & -3 \end{bmatrix}$

Study Guide Masters, p. 12

NAME_____ DATE _____

Student Edition
Pages 85–92

Study Guide

Adding and Subtracting Integers
Use the following definitions, rules, and properties when adding or subtracting integers.

Definition, Rule, or Property		Example
Definition of Absolute Value	For any real number a: if $a > 0$, then $|a| = a$, and if $a < 0$, then $|a| = -a$. $|0| = 0$	$|2| = 2$ $|-2| = 2$
Adding Integers with the Same Sign	To add integers with the same sign, add their absolute values. Give the sum the same sign as the addends.	$3 + 2 = 5$ $-3 + (-2) = -5$
Adding Integers with Different Signs	To add integers with different signs, subtract the lesser absolute value from the greater absolute value. Give the result the same sign as the addend with the greater absolute value.	$-7 + 6 = -1$ $8 + (-4) = 4$
Additive Inverse Property	For every number a, $a + (-a) = 0$.	$-9 + 9 = 0$
Subtraction Rule	To subtract a number, add its additive inverse. For any numbers a and b, $a - b = a + (-b)$.	$8 - (-2) = 8 + 2$ $= 10$

You can use the distributive property and the addition and subtraction rules for integers to simplify expressions with like terms.

Example: Simplify $-6x - x + 9x$.
$-6x - x + 9x = -6x + (-1x) + 9x$
$= [-6 + (-1) + 9]x$
$= (-7 + 9)x$
$= 2x$

Find each sum or difference.

1. $-17 + (-16)$ -33
2. $107 + (-40)$ 67
3. $75 + 86$ 161
4. $11 - 41$ -30
5. $15 - (-21)$ 36
6. $-33 - (-17)$ -16
7. $3m + (-15m) - 11m$ $-23m$
8. $-6a + 15a + (-11a)$ $-2a$
9. $-9y + 20y - (-6y)$ $17y$

Evaluate each expression if $x = -4$, $y = 3$, and $z = -7$.

10. $456 + |z|$ 463
11. $z + (-71) + |y|$ -75
12. $-11 - |x|$ -15
13. $31 - y - |x|$ 24

Guided Practice

State the additive inverse and absolute value of each integer.

7. $+7$ $-7, 7$
8. -20 $20, 20$
9. 0 $0, 0$

Find each sum or difference.

10. $-12 + (-8)$ -20
11. $-11 - (-7)$ -4
12. $-36 + 15$ -21
13. $18 - 29$ -11
14. $|-6 - 4|$ 10
15. $-|5 - 5|$ 0

Simplify each expression.

16. $-16y - 5y$ $-21y$
17. $32c - (-8c)$ $40c$
18. $-9d + (-6d)$ $-15d$

Evaluate each expression if $x = -5$, $y = 3$, and $z = -6$.

19. $y - 7$ -4
20. $16 + z$ 10
21. $|8 + x|$ 3

Let $A = \begin{bmatrix} 3 & 2 \\ 5 & 0 \end{bmatrix}$ and $B = \begin{bmatrix} 4 & -2 \\ -1 & 6 \end{bmatrix}$.   22–23. See margin.

22. Find $A + B$.
23. Find $A - B$.

24. **Space Science** Refer to the application at the beginning of the lesson. The Discovery astronauts were working in temperatures ranging from $-125°$F to $200°$F.
   a. Which is colder, $-125°$F or $200°$F? $-125°$F
   b. Write an equation that uses addition or subtraction to show how much colder one temperature is than the other. $200 - (-125) = 325$

EXERCISES

Practice

A **State the additive inverse and absolute value of each integer.**

25. $+12$ $-12, 12$
26. -45 $45, 45$
27. -302 $302, 302$

Find each sum or difference.

28. $18 + (-16)$ 2
29. $47 + (-47)$ 0
30. $-9 + 32$ 23
31. $0 - 32$ -32
32. $16 - (-23)$ 39
33. $-5 + (-11)$ -16
34. $-104 + 16$ -88
35. $-9 + (-61)$ -70
36. $-18 - 4$ -22

B 37. $9 - 24$ -15
38. $-21 - (-24)$ 3
39. $-13 - (-8)$ -5

Simplify each expression.

40. $29t - 17t$ $12t$
41. $17b - (-23b)$ $40b$
42. $-6w + (-13w)$ $-19w$
43. $6p + (-35p)$ $-29p$
44. $54y - 47y$ $7y$
45. $-5d + 31d$ $26d$

Evaluate each expression if $b = -4$, $d = 2$, and $p = -5$.

46. $b + 14$ 10
47. $d - 6$ -4
48. $15 + p$ 10
49. $d + (-17)$ -15
50. $d + 7$ 9
51. $|p|$ 5
52. $|6 + b|$ 2
53. $|p - 3|$ 8
54. $-|-22 + p|$ -27

Find each sum or difference.   57. -59 58–61. See margin.

C

INTEGRATION

Discrete Mathematics

55. $|-58 + (-41)|$ 99
56. $|-93| - (-43)$ 136
57. $-|-345 - (-286)|$

58. $\begin{bmatrix} 3 & 2 \\ 4 & -1 \end{bmatrix} + \begin{bmatrix} 0 & 5 \\ -3 & -4 \end{bmatrix}$

59. $\begin{bmatrix} 2 & 4 \\ -1 & 0 \end{bmatrix} + \begin{bmatrix} -2 & -1 \\ 3 & 2 \end{bmatrix}$

60. $\begin{bmatrix} 1 & 6 \\ -4 & -5 \end{bmatrix} - \begin{bmatrix} -3 & -6 \\ 2 & -4 \end{bmatrix}$

61. $\begin{bmatrix} 2 & 3 \\ 4 & 2 \\ 5 & -5 \end{bmatrix} - \begin{bmatrix} 6 & 8 \\ 8 & 7 \\ -5 & -2 \end{bmatrix}$

Alternative Learning Styles

Visual Use the graph below to complete the following table.

Average Temperature

Month

Month	Average Temp.	Increase from Previous Month
Jan.	-15	—
Feb.	-3	$+12$
March	$+20$	$+23$
April	$+46$	$+26$
May	$+63$	$+17$
June	$+72$	$+9$
July	$+56$	-12
Aug.	$+46$	-10
Sept.	$+37$	-9
Oct.	$+18$	-19
Nov.	$+8$	-10
Dec.	-7	-15

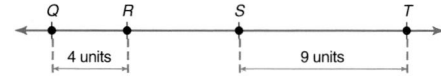

Critical Thinking

62. Geometry   Use the number line below to find the length of segment *QS* if segment *RT* has a length of 15 units.   **10 units**

Q   R   S   T

|← 4 units →| |← 9 units →|

Applications and Problem Solving

Yokohama Landmark Tower

63. Elevators   The 70-floor Yokohama Landmark Tower houses the world's fastest passenger elevators. Passengers travel from the second floor to the 69th floor in 40 seconds at an average speed of 28 miles per hour. If an employee rode the elevator up to his office on the 67th floor, then came down 43 floors to deliver a report, which floor is he on now?   **24th floor**

64. Puzzles   The sum of the numbers in each row, column, and diagonal of a magic square is the same. In the magic square at the right, the sum is 0. New magic squares can be formed by adding the same number to each entry. Complete each magic square.

1	2	−3
−4	0	4
3	−2	−1

a.
3	4	−1
−2	2	6
5	0	1

b.
−2	−1	−6
−7	−3	1
0	−5	−4

c.
−6	−5	−10
−11	−7	−3
−4	−9	−8

65. Business   The manager of The Best Bagel Shop keeps a record of each type of bagel sold each day at their two stores. Two days of sales are shown below.

Day	Store	Sesame	Poppy	Blueberry	Plain
Monday	East	120	80	64	75
	West	65	105	77	53
Tuesday	East	112	76	56	74
	West	69	95	82	50

a. Write a matrix for each day's sales.   **a–c. See margin.**

b. Find the sum of the two days' sales using matrix addition.

c. Subtract Tuesday's sales from Monday's sales by using matrix subtraction. What does this matrix represent?

66. Golf   In golf, scores are based on *par*. Par 72 means that a golfer should hit the ball 72 times to complete 18 holes of golf. A score of 68, or 4 under par, is written as −4. A score of 2 over par is written as +2. At the Oldsmobile Classic LPGA tournament in June 1995, at a par-72 course in Michigan, Helen Alfredsson shot 65, 74, 69, and 71 during four rounds of golf.

a. Use integers to write Ms. Alfredsson's score for each round as over or under par.   **−7, +2, −3, −1**   **b. −7 + 2 + (−3) + (−1) = −9**

b. Add the integers to find Ms. Alfredsson's overall score.

66c. Under; yes, it is better than par 72.

c. Was Ms. Alfredsson's score under or over par? Would you want to have her score? Explain.

Lesson 2–3   Adding and Subtracting Integers   **91**

Tech Prep

Retail Manager   Students who are interested in a career in retail management may wish to do further research into the use of mathematics in that occupation, as mentioned in Exercise 65. For more information on tech prep, see the *Teacher's Handbook*.

Closing Activity

Modeling Have students "walk out" addition and subtraction problems, using a large strip of squares labeled −9 to 9 with positives to the right and negatives to the left. Adopt the conventions shown below.

Direction	Face
positive	right (R)
negative	left (L)

Operation	Walk
addition	forward (F)
subtraction	backward (B)

Problem	Start	Face	Walk	Steps	End
5 − (+2)	5	R	B	2	3
5 − (−2)	5	L	B	2	7
5 + (+2)	5	R	F	2	7
−4 − (−6)	−4	L	B	6	2

Chapter 2, Quiz A (Lessons 2-1 through 2-3), is available in the *Assessment and Evaluation Masters,* p. 44.

Enrichment Masters, p. 12

67. Trees The table below lists twenty different species of American trees and their highest recorded height in feet. (Lesson 2–2)

American Trees

Type	Height (feet)	Type	Height (feet)	Type	Height (feet)
American Beech	130	Bitter Cherry	104	Mockernut Hickory	125
American Chestnut	110	Black Maple	118	Pecan	143
American Elm	100	Black Walnut	105	Shellback Hickory	105
American Sweetgum	136	Cherrybark Oak	124	Siberian Elm	146
Arizona Sycamore	114	Common Hackberry	111	Virginia Pine	120
Balsam Poplar	138	Digger Pine	114	Weeping Willow	114
Bishop Pine	112	Honeylocust	115		

Source: *The World Almanac, 1995*

a. What is the range of heights for the trees? **from 100 to 146**
b. Make a line plot showing the height of the trees. **See Solutions Manual.**
c. Do the heights cluster around a certain number? **yes; 114**
d. Which tree is the tallest? the shortest? **See margin.**

68. Write a corresponding addition sentence for the diagram at the right. (Lesson 2–1) **4 + (−6) = −2**

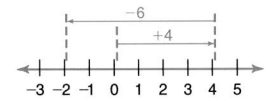

69. Sketch a reasonable graph for the following situation. The water level in Craig Pond is extremely high at the beginning of the summer. Throughout the summer, the water level lowers steadily. By the end of the summer, the pond is almost dry. (Lesson 1–9) **See margin.**

70. commutative (+)

70. Name the property illustrated by $3 + 4 = 4 + 3$. (Lesson 1–8)

71. Simplify $23y^2 + 32y^2$. (Lesson 1–7) **$55y^2$**

Name the property illustrated by each statement. (Lesson 1–6)

72. multiplicative property of 0

73. symmetric (=)

72. $3abc \cdot 0 = 0$

73. If $12xy − 3 = 4$, then $4 = 12xy − 3$.

74. Solve $p = 6\frac{1}{4} + \frac{1}{2}$. (Lesson 1–5) **$6\frac{3}{4}$**

75. Statistics Write the stems that would be used for a plot of the following data: 29.1, 34.7, 20.05, 74.9, 64, 14.2, 84.9, 38.26, 20.9, 34, 59.42, 37.107, and 43.676. (Lesson 1–4) **See margin.**

76. Evaluate $19 + 5 \cdot 4$. (Lesson 1–3) **39**

77. 24

77. Patterns Find the sixth term in the sequence 4, 8, 12, 16, ... (Lesson 1–2)

78. $n + 33$

78. Write an algebraic expression for the sum of n and 33. (Lesson 1–1)

Extension

Reasoning Is $|a| + |b|$ always equal to $|a + b|$? Prove your answer with an example. **No. Examples will vary. Sample is $|5| + |−2|$ and $|5 + (−2)|$.**

Additional Answers

67d. Siberian Elm; American Elm

69.

75. 14, 20, 29, 34, 37, 38, 43, 59, 64, 74, 84

Rational Numbers

Objectives

After studying this lesson, you should be able to:

- compare and order rational numbers, and
- find a number between two rational numbers.

APPLICATION
Recycling

Many people all over the world recycle their aluminum cans in order to help our environment. The graph below shows the fraction of aluminum cans that have been recycled over the years.

Recycling
Fraction of aluminum cans that are recycled:

$\frac{1}{6}$ $\frac{1}{4}$ $\frac{11}{20}$ $\frac{1}{2}$ $\frac{16}{25}$ $\frac{13}{20}$

'74   '78   '82   '86   '90   '94

Source: The Aluminum Association

The numbers shown in the graph are examples of **rational numbers.**

Definition of a Rational Number	A rational number is a number that can be expressed in the form $\frac{a}{b}$, where a and b are integers and b is not equal to 0.

Examples of rational numbers expressed in the form $\frac{a}{b}$ are shown below.

Notice that all integers are rational numbers.

Rational Numbers	4	$-3\frac{3}{4}$	0.250	0	$0.33\overline{3}$
Form $\frac{a}{b}$	$\frac{4}{1}$	$-\frac{15}{4}$	$\frac{1}{4}$	$\frac{0}{1}$	$\frac{1}{3}$

Rational numbers can be graphed on a number line in the same manner as integers. The number line below is separated into fourths to show the graphs of some common fractions and decimals.

-1 -0.75 $-\frac{1}{2}$ -0.25 0 $\frac{1}{4}$ 0.5 $\frac{3}{4}$ 1

You can compare rational numbers by graphing them on a number line. Recall that a mathematical sentence that uses $<$ and $>$ to compare two expressions is called an *inequality.*

NCTM Standards: 1–5

Instructional Resources

- Study Guide Master 2-4
- Practice Master 2-4
- Enrichment Master 2-4
- Graphing Calculator Masters, p. 2

 Transparency 2-4A contains the 5-Minute Check for this lesson; **Transparency 2-4B** contains a teaching aid for this lesson.

Recommended Pacing	
Standard Pacing	Day 5 of 15
Honors Pacing	Day 5 of 15
Block Scheduling*	Day 3 of 9
Alg. 1 in Two Years*	Days 9 & 10 of 28

*For more information on pacing and possible lesson plans, refer to the *Block Scheduling Booklet* and *Algebra 1 in Two Years.*

1 FOCUS

 5-Minute Check
(over Lesson 2-3)

Find each sum or difference.

1. $18 - 25$ **−7**
2. $-6 + 29$ **23**
3. $9 + (-24)$ **−15**
4. $(-18) + -(47)$ **−65**

Evaluate each expression if $b = -4$, $j = 2$, and $n = -7$.

5. $b + 12$ **8**
6. $j - n$ **9**
7. $-|b + (-9)|$ **−13**

Motivating the Lesson

Questioning Display a picture of an appliance such as a refrigerator from a sales flyer. Note the sale price on the back of the picture. Have students take turns asking questions about the sale price of the item using the terms "greater than" and "less than." Students should use responses from one another's questions to guess the price.

Teaching Tip Students may benefit from making concrete comparisons involving height, weight, or length to prove that exactly one of the three comparison sentences is true. For example, use three students in the classroom whose heights are different to demonstrate the comparison sentences.

Teaching Tip The comparison property is sometimes called the *trichotomy property.*

In-Class Example

For Example 1
Replace each _?_ with <, >, or = to make each sentence true.

a. -11 _?_ $-20 + 9$ =
b. $-15 + 6$ _?_ 0 <
c. 5 _?_ $8 + (-12)$ >

Teaching Tip You may wish to introduce the symbols for "is not less than" (≮) and "is not greater than" (≯) to your students.

The following statements can be made about the graphs of -5, -2, $1\frac{1}{2}$ and 3.5 shown on the number line below.

a. The graph of -2 is to the left of the graph of 3.5. $-2 < 3.5$
b. The graph of $1\frac{1}{2}$ is to the right of the graph of -5. $1\frac{1}{2} > -5$

These examples suggest the following rule.

Comparing Numbers on the Number Line	If a and b represent any numbers and the graph of a is to the left of the graph of b, then $a < b$. If the graph of a is to the right of the graph of b, then $a > b$.

If $<$, $>$, and $=$ are used to compare two numbers, then the following property applies.

Comparison Property	For any two numbers a and b, exactly one of the following sentences is true. $a < b$ $a = b$ $a > b$

The symbols $\neq$, $\leq$, and $\geq$ can also be used to compare numbers. The chart below shows several inequality symbols and their meanings.

Symbol	Meaning
$<$	is less than
$>$	is greater than
$\neq$	is not equal to
$\leq$	is less than or equal to
$\geq$	is greater than or equal to

Example **Replace each _?_ with <, >, or = to make each sentence true.**

a. -75 _?_ 13

Since any negative number is less than any positive number, the true sentence is $-75 < 13$.

Note that the closed ends of $<$ and $>$ always point to the lesser number.

b. -14 _?_ $-22 + 9$

-14 _?_ $-22 + 9$

-14 _?_ -13 *Simplify.*

Since -14 is less than -13, the true sentence is $-14 < -22 + 9$.

c. $\frac{3}{8}$ _?_ $-\frac{7}{8}$

Since 3 is greater than -7, the true sentence is $\frac{3}{8} > -\frac{7}{8}$.

You can use **cross products** to compare two fractions with different denominators. When two fractions are compared, the cross products are the products of the terms on the diagonals.

GLENCOE Technology

Interactive Mathematics Tools Software

This multimedia software provides an interactive lesson that uses a number line to compare positive and negative rational numbers. A **Computer Journal** gives students an opportunity to write about what they have learned.

For Windows & Macintosh

Comparison Property for Rational Numbers	For any rational numbers $\frac{a}{b}$ and $\frac{c}{d}$, with $b > 0$ and $d > 0$: 1. if $\frac{a}{b} < \frac{c}{d}$, then $ad < bc$, and 2. if $ad < bc$, then $\frac{a}{b} < \frac{c}{d}$.

This property also holds if $<$ is replaced by $>$, $\leq$, $\geq$, or $=$.

Example ②

Replace each __?__ with $<$, $>$, or $=$ to make each sentence true.

a. $\frac{7}{13}$ _?_ $\frac{4}{15}$

$7(15)$ _?_ $13(4)$
$105 > 52$

The true sentence is $\frac{7}{13} > \frac{4}{15}$.

b. $\frac{7}{8}$ _?_ $\frac{8}{9}$

$7(9)$ _?_ $8(8)$
$63 < 64$

The true sentence is $\frac{7}{8} < \frac{8}{9}$.

Every rational number can be expressed as a terminating or repeating decimal. You can use a calculator to write rational numbers as decimals.

Example ③

It is helpful to know these commonly used fraction-decimal equivalencies by memory.

$\frac{1}{2} = 0.5$, $\frac{1}{3} = 0.\overline{3}$,

$\frac{1}{4} = 0.25$, $\frac{1}{5} = 0.2$,

$\frac{1}{8} = 0.125$

Use a calculator to write the fractions $\frac{3}{8}$, $\frac{4}{5}$, and $\frac{1}{6}$ as decimals. Then write the fractions in order from least to greatest.

$\frac{3}{8} = 0.375$ This is a terminating decimal.

$\frac{4}{5} = 0.8$ This is also a terminating decimal.

$\frac{1}{6} = 0.16666\ldots$ or $0.1\overline{6}$ This is a repeating decimal.

In order from least to greatest, the decimals are $0.1\overline{6}$, 0.375, 0.8. So, the fractions in order from least to greatest are $\frac{1}{6}$, $\frac{3}{8}$, $\frac{4}{5}$.

You can use a calculator to compare the cost per unit, or **unit cost**, of two similar items. Many people comparison shop to find the best buys at the grocery store. The item that has the lesser unit cost is the better buy.

unit cost = total cost ÷ number of units

Example ④

Erica and Gabriella are running in a 5-kilometer race to raise money for The Multiple Sclerosis Foundation. They want to buy a sports beverage to drink after they finish the race. Which is the better buy?

Use a calculator to find the unit cost of each brand. In each case, the unit cost is expressed in cents per ounce.

unit cost of 20-ounce bottle: 1.09 ÷ 20 =   *0.0545*

unit cost of 32-ounce bottle: 1.69 ÷ 32 =   *0.0528*

Since $0.053 < 0.055$, the 32-ounce All Sport® drink is the better buy.

Teaching Tip To further emphasize the ease of using the comparison property, change the fractions to equivalent fractions with a common denominator.

$\frac{6}{11} = \frac{72}{132}$ $\frac{7}{12} = \frac{77}{132}$

Notice that $\frac{72}{132} < \frac{77}{132}$. Thus, $\frac{6}{11} < \frac{7}{12}$.

Alternative Learning Styles

Kinesthetic Use items available at students' desks, such as pencils, paper, or backpacks, to demonstrate the comparison and transitive properties by comparing lengths of pencils, weights of backpacks, and thicknesses of paper. Point out to students that the transitive property applies to inequalities also. If $a \geq b$ and $b \geq c$, then $a \geq c$.

For Example 5
Find a rational number between $\frac{1}{9}$ and $\frac{3}{5}$. Sample answer: $\frac{16}{45}$

3 PRACTICE/APPLY

Check for Understanding
Exercises 1–13 are designed to help you assess your students' understanding through reading, writing, speaking, and modeling. You should work through Exercises 1–6 with your students and then monitor their work on Exercises 7–13.

Additional Answers

3. One way is to represent these numbers as decimals and then find three decimal values between 0.2 and 0.25, such as 0.21 or 0.24, and then convert these back into fraction form. Another way is to use the calculator to find the mean of the two decimal equivalents. That is one in-between value. Then find the mean of the first decimal and the new value for the second value and the mean of the second decimal and the new value for the third value.

4. A fraction cannot have zero in the denominator because that would make the fraction undefined.

5. Yes; since $\frac{a}{b} < \frac{c}{d}$ then what needs to be shown is that $\frac{a}{b} < \frac{a+c}{b+d} < \frac{c}{d}$.
If $\frac{a}{b} < \frac{c}{d}$, then $ad < bc$ and then adding ab to both sides, $ad + ab < bc + ab$. Then $a(d + b) < b(c + a)$ and $\frac{a}{b} < \frac{a+c}{b+d}$.
The same method can be done to show the other side also. Have students show that side.

Can you always find another rational number between two rational numbers? One point that lies between any two points is their midpoint. Consider $\frac{1}{3}$ and $\frac{1}{2}$.

To find the coordinate of the midpoint for $\frac{1}{3}$ and $\frac{1}{2}$, find the average, or mean, of the two numbers.

To find the average of two numbers, add the numbers and then divide by 2.
Multiplying by $\frac{1}{2}$ is the same as dividing by 2.

average: $\frac{1}{2}\left(\frac{1}{3} + \frac{1}{2}\right) = \frac{1}{2}\left(\frac{5}{6}\right)$ or $\frac{5}{12}$

This process can be continued indefinitely. The pattern suggests the **density property.**

Density Property for Rational Numbers	Between every pair of distinct rational numbers, there are infinitely many rational numbers.

Example 5 Find a rational number between $-\frac{2}{5}$ and $-\frac{3}{8}$.

Find the average of the two rational numbers.

Method 1
Use a pencil and paper.

$\frac{1}{2}\left[-\frac{2}{5} + \left(-\frac{3}{8}\right)\right] = \frac{1}{2}\left[-\frac{16}{40} + \left(-\frac{15}{40}\right)\right]$

$= \frac{1}{2}\left(-\frac{31}{40}\right)$

$= -\frac{31}{80}$

Method 2
Use a scientific calculator.

Enter: (2 ÷ 5 +/–
+ 3 ÷ 8 +/–
) ÷ 2 =   –0.3875

A rational number between $-\frac{2}{5}$ and $-\frac{3}{8}$ is $-\frac{31}{80}$ or -0.3875.

CHECK FOR UNDERSTANDING

Communicating Mathematics

1. Sample answers: $\frac{1}{2}$, $\frac{10}{3}$, 0.32

2. none

3–5. See margin.

Study the lesson. Then complete the following.

1. **Write** three examples of rational numbers.

2. **State** any assumptions you can make about y and z if $x < y$ and $x < z$.

3. **Show** two ways to find three rational numbers between $\frac{1}{5}$ and $\frac{1}{4}$.

4. **Explain** why the definition of a rational number states that b cannot be zero.

5. **You Decide** To find a number between $\frac{3}{4}$ and $\frac{6}{7}$, John added as follows.
$\frac{3+6}{4+7} = \frac{9}{11}$; $\frac{9}{11}$ is between $\frac{3}{4}$ and $\frac{6}{7}$.
He claimed that this method will always work. Do you agree? Explain.

96   Chapter 2 Exploring Rational Numbers

Reteaching

Translating Expressions Write an algebraic expression for each verbal expression.

1. The cost is under $195. $c < 195$
2. x is positive. $x > 0$
3. x is nonnegative. $x \geq 0$
4. Tom is at least 18 years old. $T \geq 18$
5. The stream is no more than 50 feet across. $s \leq 50$

MATH JOURNAL

6. Assess Yourself Describe two situations in which you would prefer to use decimals instead of fractions. See students' work.

Guided Practice

Replace each _?_ with <, >, or = to make each sentence true.

7. -4 _?_ 8 <

8. -5 _?_ $0 - 3$ <

9. $\frac{5}{14}$ _?_ $\frac{25}{70}$ =

Write the numbers in each set in order from least to greatest.

10. $\frac{2}{3}, \frac{1}{6}, \frac{1}{2}$ $\frac{1}{6}, \frac{1}{2}, \frac{2}{3}$

11. $2.5, \frac{3}{4}, -0.5, \frac{7}{8}$ $-0.5, \frac{3}{4}, \frac{7}{8}, 2.5$

12. Which is a better buy: a 6-ounce can of tuna for $1.59 or a package of 3 three-ounce cans for $2.19?

13. Find a number between $\frac{1}{2}$ and $\frac{5}{7}$.   See students' work.

EXERCISES

Practice

A

Replace each _?_ with <, >, or = to make each sentence true.

14. -3 _?_ 5 <

15. -1 _?_ -4 >

16. $-6 - 3$ _?_ -9 =

17. 5 _?_ $8.4 - 1.5$ <

18. 4 _?_ $\frac{16}{3}$ <

19. $\frac{8}{15}$ _?_ $\frac{9}{16}$ <

B

20. $\frac{14}{5}$ _?_ $\frac{25}{13}$ >

21. $\frac{4}{3}(6)$ _?_ $4\left(\frac{3}{2}\right)$ >

22. $\frac{0.4}{3}$ _?_ $\frac{1.2}{8}$ <

Write the numbers in each set in order from least to greatest.

23–28. See margin.

C

23. $\frac{6}{7}, \frac{2}{3}, \frac{3}{8}$

24. $-\frac{4}{15}, -\frac{6}{17}, -\frac{3}{16}$

25. $\frac{4}{14}, \frac{3}{23}, \frac{8}{42}$

26. $6.7, -\frac{5}{7}, \frac{6}{13}$

27. $0.2, -\frac{2}{5}, -0.2$

28. $\frac{4}{5}, \frac{9}{10}, 0.7$

Which is the better buy?

29. a 16-ounce drink for $0.59 or a 20-ounce drink for $0.89

30. a 32-ounce bottle of shampoo for $3.59 or a 64-ounce bottle for $6.99

31. a package of 48 paper plates for $2.39 or a package of 75 paper plates for $3.29

Find a number between the given numbers.

32–37. Sample answers are given.

32. $\frac{2}{5}$ and $\frac{7}{2}$ $\frac{1}{2}$

33. $\frac{19}{30}$ and $\frac{31}{45}$ $\frac{2}{3}$

34. $-\frac{2}{15}$ and $-\frac{6}{3}$ -1

35. Name a fraction between $\frac{1}{4}$ and $\frac{1}{2}$ whose denominator is 20. $\frac{7}{20}$

36. Name a fraction between $\frac{1}{3}$ and $\frac{5}{6}$ whose denominator is 12. $\frac{6}{12}$

37. Name a fraction between -0.5 and $\frac{1}{3}$ whose denominator is 6. $\frac{1}{6}$

Lesson 2–4 Rational Numbers **97**

Assignment Guide

Core: 15–41 odd, 43–49
Enriched: 14–36 even, 38–49
All: Self Test, 1–10

For **Extra Practice,** see p. 760.

The red A, B, and C flags, printed only in the Teacher's Wraparound Edition, indicate the level of difficulty of the exercises.

Additional Answers

23. $\frac{3}{8}, \frac{2}{3}, \frac{6}{7}$

24. $-\frac{6}{17}, -\frac{4}{15}, -\frac{3}{16}$

25. $\frac{3}{23}, \frac{8}{42}, \frac{4}{14}$

26. $-\frac{5}{7}, \frac{6}{13}, 6.7$

27. $-\frac{2}{5}, -0.2, 0.2$

28. $0.7, \frac{4}{5}, \frac{9}{10}$

Study Guide Masters, p. 13

2-4 NAME_____ DATE_____

Study Guide   Student Edition Pages 93–99

Rational Numbers

| Definition of a Rational Number | A rational number is a number that can be expressed in the form $\frac{a}{b}$, where a and b are integers and $b \neq 0$. |

You can compare rational numbers by graphing them on a number line.

| Comparing Numbers on the Number Line | If a and b represent any numbers and the graph of a is to the left of the graph of b, then $a < b$. If the graph of a is to the right of the graph of b, then $a > b$. |
| Comparison Property | For any two numbers a and b, exactly one of the following sentences is true. $a < b$ $a = b$ $a > b$ |

Example 1: $-3\frac{1}{2} < -\frac{1}{2}$ The graph of $-3\frac{1}{2}$ is to the left of the graph of $-\frac{1}{2}$.

Example 2: $-2\frac{1}{4} > -3\frac{1}{4}$ The graph of $-2\frac{1}{4}$ is to the right of the graph of $-3\frac{1}{4}$.

Example 3: Replace _?_ with <, >, or = to make the sentence true.
-15 _?_ -3
$-15 < -3$ Since -15 is to the left of -3 on a number line, -15 is less than -3.

The symbols $\neq$, $\leq$ and $\geq$ can also be used to compare numbers and are called **inequality symbols.**

You can use **cross products** to compare two fractions with different denominators.

| Comparison Property for Rational Numbers | For any rational numbers $\frac{a}{b}$ and $\frac{c}{d}$, with $b > 0$ and $d > 0$: 1. if $\frac{a}{b} < \frac{c}{d}$ then $ad < bc$, and 2. if $ad < bc$, then $\frac{a}{b} < \frac{c}{d}$. |

This property also holds if < is replaced by >, $\leq$, $\geq$, or =.

A property that is true for rational numbers but is not true for integers is the **density property.**

| Density Property for Rational Numbers | Between every pair of distinct rational numbers, there is another rational number. |

Replace each _?_ with <, >, or = to make each sentence true.
1. -4 _?_ 10 <
2. $\frac{-29}{2}$ _?_ $-28.5 + 14$ =
3. $-5 - 6$ _?_ $-12 - 1$ >

Write the numbers in each set in order from least to greatest.
4. $3\frac{1}{3}, \frac{5}{8}, 0.4$ $0.4, \frac{5}{8}, 3\frac{1}{3}$
5. $-\frac{3}{2}, \frac{1}{4}, 0.2$ $-\frac{3}{2}, 0.2, \frac{1}{4}$

Find a number between the given numbers. See students' work.
6. $\frac{1}{2}$ and $\frac{7}{9}$ $\frac{23}{36}$
7. $\frac{7}{6}$ and $\frac{9}{8}$ $\frac{55}{48}$
8. $\frac{9}{17}$ and $\frac{2}{5}$ $\frac{79}{170}$

Teaching Tip
You may wish to have students measure their own tibia or radius and apply the appropriate formula to determine if it provides a reasonably accurate measurement of their height.

Critical Thinking

39. $E = \frac{4}{14}$ or $\frac{2}{7}$;

$G = \frac{10}{14}$ or $\frac{5}{7}$;

$H = \frac{13}{14}$

38. Three numbers a, b, and c satisfy the following conditions: $b - c < 0$, $a - b > 0$, and $c - a < 0$. Which one is the greatest? *a*

39. Find the coordinates of E, G, and H if the distances between the points are equal.

Applications and Problem Solving

40. Personal Computers Study the graph at the right. Write an inequality that compares the total number of PCs shipped for home use in 1995 to those shipped in 1996. **8.2 < 9.8**

41. Aviation When building or servicing airplanes, an airplane mechanic sometimes needs to drill holes. Drill bit sizes are given in fractions, yet many blueprints list hole diameters in decimals. One blueprint calls for a hole of 0.391 inches. Find the fractional drill size required if the hole must be drilled $\frac{1}{64}$-inch undersize. **0.375 inch**

Computers Shipped for Home Use

(millions)

Year	
1990	4.0
1991	3.9
1992	4.9
1993	5.9
1994	6.9
1995	8.2
1996	9.8

Note: 1994-96 are forecasts.

Sources: Dataquest, LINK

42. Archaeology To calculate a person's height, archaeologists measure the lengths of certain bones. The bones measured are the femur or thigh bone F, the tibia or leg bone T, the humerus or upper arm bone H, and the radius or lower arm bone R. When the length of one of these bones is known, scientists can use one of the formulas below to determine the person's height h in centimeters. **a. 172.4 cm   b. 179.6 cm**

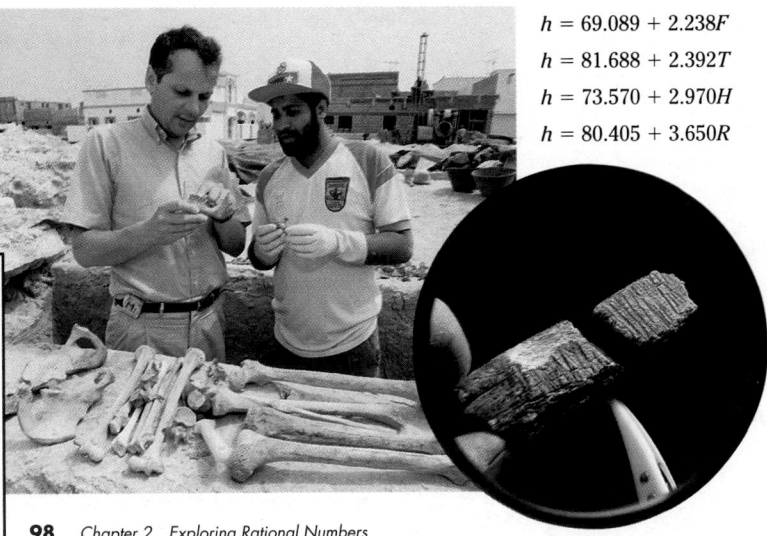

Males	Females
$h = 69.089 + 2.238F$	$h = 61.412 + 2.317F$
$h = 81.688 + 2.392T$	$h = 72.572 + 2.533T$
$h = 73.570 + 2.970H$	$h = 64.977 + 3.144H$
$h = 80.405 + 3.650R$	$h = 73.502 + 3.876R$

a. The femur of a 37-year old woman measured 47.9 cm. Use a calculator to find the height of the woman to the nearest tenth of a centimeter.

b. The humerus of a 49-year old man measured 35.7 cm. Use a calculator to find the height of the man to the nearest tenth of a centimeter.

98   Chapter 2   Exploring Rational Numbers

Practice Masters, p. 13

2-4 NAME_____ DATE_____

Practice

Student Edition Pages 93–99

Rational Numbers

Replace each ? with <, >, or = to make each sentence true.

1. $\frac{2}{3}\ ?\ \frac{5}{8}$ >
2. $\frac{3}{4}\ ?\ \frac{5}{7}$ >
3. $\frac{4}{2}\ ?\ \frac{5}{2}$ <
4. $\frac{7}{11}\ ?\ \frac{2}{3}$ <
5. $\frac{3}{14}\ ?\ \frac{15}{70}$ =
6. $\frac{4}{9}\ ?\ \frac{1}{2}$ <
7. $\frac{0}{1}\ ?\ \frac{0}{2}$ =
8. $\frac{10}{10}\ ?\ \frac{100}{100}$ =
9. $\frac{9}{7}\ ?\ \frac{7}{4}$ <
10. $\frac{5}{7}\ ?\ \frac{7}{9}$ <
11. $\frac{33}{40}\ ?\ \frac{17}{24}$ >
12. $\frac{12}{6}\ ?\ \frac{18}{9}$ =
13. $\frac{2}{3}\ ?\ \frac{9}{16}$ >
14. $\frac{3}{16}\ ?\ \frac{5}{12}$ <
15. $\frac{1}{5}\ ?\ \frac{1}{2}$ <
16. $\frac{7}{8}\ ?\ \frac{15}{16}$ <
17. $\frac{100}{10}\ ?\ \frac{40}{4}$ =
18. $\frac{3}{4}\ ?\ \frac{11}{16}$ >
19. $\frac{1.1}{3}\ ?\ \frac{3.3}{9}$ =
20. $\frac{1}{8}\ ?\ \frac{2}{3}$ <
21. $\frac{10}{29}\ ?\ \frac{15}{43}$ <

Write the numbers in order from least to greatest.

22. $\frac{18}{29}, \frac{24}{39}, \frac{3}{5}, \frac{24}{5}, \frac{18}{29}$ $\frac{3}{5}, \frac{18}{29}, \frac{24}{39}$

23. $\frac{10}{29}, \frac{5}{14}, \frac{15}{43}$ $\frac{10}{29}, \frac{15}{43}, \frac{5}{14}$

24. $\frac{3}{4}, \frac{5}{8}, \frac{5}{6}$ $\frac{5}{8}, \frac{3}{4}, \frac{5}{6}$

25. $0.25, \frac{7}{8}, \frac{3}{8}$ $0.25, \frac{3}{8}, \frac{7}{8}$

26. $0.85, 0.84, \frac{19}{22}$ $0.84, 0.85, \frac{19}{22}$

27. $\frac{16}{19}, \frac{18}{21}, \frac{17}{20}$ $\frac{16}{19}, \frac{17}{20}, \frac{18}{21}$

28. $\frac{5}{6}, 0.8, \frac{2}{3}$ $\frac{2}{3}, 0.8, \frac{5}{6}$

29. $\frac{6}{6}, 0, \frac{1}{2}$ $0, \frac{1}{2}, \frac{6}{6}$

Find a number between the given numbers. Sample answers are given.

30. $\frac{1}{5}$ and $\frac{2}{3}$ $\frac{13}{30}$
31. $\frac{9}{10}$ and $\frac{1}{4}$ $\frac{23}{40}$
32. $\frac{18}{5}$ and $\frac{6}{7}$ $\frac{78}{35}$
33. $\frac{1}{2}$ and $\frac{1}{9}$ $\frac{11}{36}$
34. $\frac{2}{3}$ and $\frac{3}{2}$ $\frac{13}{12}$
35. $\frac{6}{1}$ and $\frac{2}{2}$ $\frac{7}{2}$

Tech Prep

Airplane Mechanic Students who are interested in a career in aviation mechanics may wish to do further research into the use of mathematics in that occupation, as mentioned in Exercise 41. For more information on tech prep, see the *Teacher's Handbook*.

43. State the additive inverse and absolute value of +9. (Lesson 2–3) **−9, 9**

44. Statistics State the scale you would use to make a line plot for the following data. Then draw the line plot. (Lesson 2–2)
145, 130, 135, 150, 145, 145, 140, 130, 145, 150, 130

45. Graph $\{-3, -2, 2, 3\}$ on a number line. (Lesson 2–1) **44–45. See margin.**

46. Soccer Bryce, Maria, and Holly are three offensive players for the Spazmatics soccer team. The graph at the right describes the players' offensive skills in their last season. (Lesson 1–9)

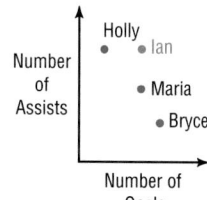

a. Who scored the most goals? **Bryce**

b. Who helped the team most with goals and assists overall? **Maria**

c. Another player, Ian, had the same number of goals as Maria and the same number of assists as Holly. Describe where on the graph a point should go that would show his offensive skills. **See students' work.**

47. Simplify $\frac{2}{5}m + \frac{1}{5}(6n + 3m) + \frac{1}{10}(8n + 15)$. (Lesson 1–6)
$m + 2n + \frac{3}{2}$

48. Patterns Find the next three terms in the sequence 1, 3, 9, 27, ... (Lesson 1–2) **81, 243, 729**

49. Write a verbal expression for $2x^2 + 6$. (Lesson 1–1)
two times x squared plus six

SELF TEST

Graph each set of numbers on a number line. (Lesson 2–1)

1. $\{-3, 0, 2\}$
2. {numbers greater than −2}
1–2. See margin.

Find each sum or difference. (Lesson 2–3)

3. $-9 + (-8)$ **−17**
4. $6 + (-15)$ **−9**
5. $23 - (-32)$ **55**
6. $-12 - 4$ **−16**

7. Mountain Bikes The table at the right shows prices of several mountain bikes. (Lesson 2–2)

a. Make a line plot of the data. **See Solutions Manual.**

b. Do the numbers cluster around any certain price? **yes; $270**

Bike	Price
Trek 820	$325
Mongoose Threshold	270
Giant Yukon	370
Giant Rincon	300
Bianchi Timber Wolf	270
Schwinn Clear Creek	320
Schwinn Sidewinder	260
Specialized Hardrock	270
Raleigh M40GS	290
Trek 800	240

Replace each ? with <, >, or = to make each sentence true. (Lesson 2–4)

8. $-7 - 5 \underline{?} -11$ **<**
9. $\frac{7}{16} \underline{?} \frac{8}{15}$ **<**
10. $\frac{5}{4}(4) \underline{?} 8\left(\frac{1}{2}\right)$ **>**

Lesson 2–4 Rational Numbers **99**

SELF TEST

The Self Test provides students with a brief review of the concepts and skills in Lessons 2-1 through 2-4. Lesson numbers are given to the right of exercises or instruction lines so students can review concepts not yet mastered.

Extension

Connections Draw two rectangles with equal dimensions. Then show that $\frac{3}{5} > \frac{1}{3}$.

---

4 ASSESS

Closing Activity

Speaking Have students determine how many whole numbers satisfy the inequality $x < 5$. **5** Then ask students how many rational numbers satisfy the inequality. **an infinite number**

Additional Answers

44. from 130 to 150 by 5s

45.

Answers for the Self Test

1.

2.

Enrichment Masters, p. 13

NCTM Standards: 1–5

Instructional Resources

- Study Guide Master 2-5
- Practice Master 2-5
- Enrichment Master 2-5
- Assessment and Evaluation Masters, pp. 43–44
- Multicultural Activity Masters, p. 3

 Transparency 2-5A contains the 5-Minute Check for this lesson; **Transparency 2-5B** contains a teaching aid for this lesson.

Recommended Pacing	
Standard Pacing	Day 6 of 15
Honors Pacing	Day 6 of 15
Block Scheduling*	Day 4 of 9
Alg. 1 in Two Years*	Days 11, 12, & 13 of 28

 *For more information on pacing and possible lesson plans, refer to the *Block Scheduling Booklet* and *Algebra 1 in Two Years.*

1 FOCUS

 5-Minute Check
(over Lesson 2-4)

Write the fractions in order from least to greatest.

1. $\frac{3}{15}, \frac{5}{17}, \frac{2}{7}$ $\frac{3}{15}, \frac{2}{7}, \frac{5}{17}$

2. $\frac{7}{12}, \frac{6}{11}, \frac{8}{15}$ $\frac{8}{15}, \frac{6}{11}, \frac{7}{12}$

Find a rational number between the given numbers.

3. $-\frac{3}{4}, \frac{5}{6}$ 0

4. $\frac{1}{2}, \frac{6}{7}$ $\frac{19}{28}$

5. Find the unit price if 8 ounces of coffee sells for $3.84. $0.48

Adding and Subtracting Rational Numbers

Objectives

After studying this lesson, you should be able to:

- add and subtract rational numbers, and
- simplify expressions that contain rational numbers.

 APPLICATION
Employment

In 1982, the average number of weekly overtime hours worked by a U.S. worker was $2\frac{1}{4}$ hours. This average increased $1\frac{1}{4}$ hours by 1989 and then decreased $\frac{1}{2}$ hour by 1991.

What was the average number of weekly overtime hours worked in 1991?

To answer this question, you need to find the following sum.

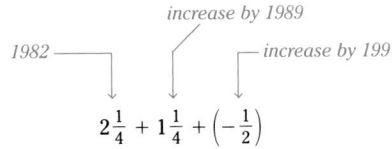

$$2\frac{1}{4} + 1\frac{1}{4} + \left(-\frac{1}{2}\right)$$

To add rational numbers such as these, use the same rules you used to add integers. When adding three or more rational numbers, you can use the commutative and associative properties to rearrange the addends.

LOOK BACK

You can refer to Lesson 1–8 for information on the commutative and associative properties.

Method 1: Use the rules.

$$2\frac{1}{4} + 1\frac{1}{4} + \left(-\frac{1}{2}\right) = 3\frac{2}{4} + \left(-\frac{1}{2}\right) \quad \text{\textit{Add the first two addends.}}$$

$$= 3\frac{1}{2} + \left(-\frac{1}{2}\right) \quad \text{\textit{Simplify.}}$$

$$= +\left(\left|3\frac{1}{2}\right| - \left|-\frac{1}{2}\right|\right) \quad \text{\textit{$3\frac{1}{2}$ has the greater absolute value,}}$$

$$= +\left(3\frac{1}{2} - \frac{1}{2}\right) \quad \text{\textit{so the sign of the sum is positive.}}$$

$$= 3$$

Method 2: Use a number line.

As with integers, you can also add rational numbers by using a number line.

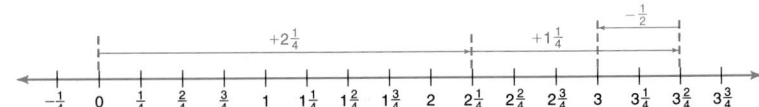

The average number of weekly overtime hours worked in 1991 was 3.

 GLENCOE Technology

CD-ROM Interaction

A multimedia simulation allows students to see how math is used in operating a pizza parlor. A blackline master activity with teacher's notes provides a follow-up to the CD-ROM simulation.

For Windows & Macintosh

Example **Find each sum.**

   a. $4\frac{1}{8} + \left(-1\frac{1}{2}\right)$ *Estimate:* $4 + (-2) = 2$

$$4\frac{1}{8} + \left(-1\frac{1}{2}\right) = 4\frac{1}{8} + \left(-1\frac{4}{8}\right)$$ *The LCD is 8. Replace* $-1\frac{1}{2}$ *with* $-1\frac{4}{8}$.

$$= +\left(\left|4\frac{1}{8}\right| - \left|-1\frac{4}{8}\right|\right)$$ *The sum is positive. Why?*

$$= +\left(4\frac{1}{8} - 1\frac{4}{8}\right)$$

$$= +\left(3\frac{9}{8} - 1\frac{4}{8}\right)$$ *Replace* $4\frac{1}{8}$ *with* $3\frac{9}{8}$.

$$= 2\frac{5}{8}$$ *Compare to the estimate.*

   b. $-1.34 + (-0.458)$ *Estimate:* $-1 + (-0.5) = -1.5$

$$-1.34 + (-0.458) = -(|-1.34| + |-0.458|)$$ *The numbers have the same*

$$= -(1.34 + 0.458)$$ *sign. Their sum is negative.*

$$= -1.798$$

   c. $-\frac{2}{5} + 1\frac{1}{2} + \left(-\frac{2}{3}\right)$ *Estimate:* $-\frac{1}{2} + 1\frac{1}{2} + \left(-\frac{1}{2}\right) = \frac{1}{2}$

$$-\frac{2}{5} + 1\frac{1}{2} + \left(-\frac{2}{3}\right) = \left[-\frac{2}{5} + \left(-\frac{2}{3}\right)\right] + 1\frac{1}{2}$$ *Group the negative numbers together.*

$$= \left[-\frac{12}{30} + \left(-\frac{20}{30}\right)\right] + \frac{45}{30}$$ *The LCD is 30.*

$$= -\frac{32}{30} + \frac{45}{30}$$

$$= \frac{13}{30}$$ *Compare to the estimate.*

To subtract rational numbers, use the same process you used to subtract integers.

Example ② **Find each difference.**

   a. $-\frac{3}{5} - \left(-\frac{4}{7}\right)$

$$-\frac{3}{5} - \left(-\frac{4}{7}\right) = -\frac{3}{5} + \frac{4}{7}$$ *To subtract* $-\frac{4}{7}$, *add its inverse,* $+\frac{4}{7}$.

$$= -\frac{21}{35} + \frac{20}{35}$$ *The LCD is 35.*

$$= -\frac{1}{35}$$

   b. $-6.24 - 8.52$

$$-6.24 - 8.52 = -6.24 + (-8.52)$$

$$= -14.76$$

Lesson 2–5 Adding and Subtracting Rational Numbers   **101**

Motivating the Lesson

Questioning Have students read the stock report from a recent newspaper. Discuss the meanings of such headings as "High," "Low," and "Chg." Have each class member select and monitor a particular stock during a one-week period.

2 TEACH

Teaching Tip Students are often confused with the various sets of numbers. Emphasize that a rational number is a ratio of two integers.

Teaching Tip In Example 1, although the positive sign is not necessary in the step that reads

$$+\left(\left|4\frac{1}{8}\right| - \left|-1\frac{4}{8}\right|\right),$$

encourage students to, at first, place the sign outside the parentheses. Emphasize this step in analyzing what sign the sum will have.

In-Class Examples

For Example 1
Find each sum.

a. $5\frac{3}{7} + \left(-6\frac{1}{2}\right)$ $-1\frac{1}{14}$

b. $-8.007 + (-5.755)$ -13.762

For Example 2
Find each difference.

a. $-\frac{5}{9} - \left(-\frac{3}{5}\right)$ $\frac{2}{45}$

b. $-\frac{1}{2} - \left(-\frac{6}{7}\right)$ $\frac{5}{14}$

c. $6.32 - (-3.42)$ 9.74

3 PRACTICE/APPLY

Example ③

APPLICATION
Stock Market

The Intel Corporation produces the Pentium™ processor for personal computers. In November of 1994, Intel announced that there was a flaw in the processor that could cause errors in complex mathematical calculations. In one week, Intel's stock dropped $2\frac{1}{4}$ points. By December 13, the stock had dropped another $2\frac{1}{8}$ points. How many points did the stock drop in this time period?

We can represent the drops in the price of the stock with negative numbers.

$-2\frac{1}{4} - 2\frac{1}{8} = -2\frac{1}{4} + \left(-2\frac{1}{8}\right)$ *To subtract $2\frac{1}{8}$, add its additive inverse.*

$= -2\frac{2}{8} + \left(-2\frac{1}{8}\right)$ *The LCD is 8.*

$= -4\frac{3}{8}$

Intel's stock dropped $4\frac{3}{8}$ points during this time period.

You can also evaluate expressions involving the addition and subtraction of rational numbers.

Example ④

a. Evaluate $p - 3.5$ if $p = 2.8$.

$p - 3.5 = 2.8 - 3.5$ *Replace p with 2.8.*

$= -0.7$

b. Evaluate $\frac{11}{4} - x$ if $x = \frac{27}{8}$.

$\frac{11}{4} - x = \frac{11}{4} - \frac{27}{8}$ *Replace x with $\frac{27}{8}$.*

$= \frac{22}{8} - \frac{27}{8}$ *The LCD is 8.*

$= -\frac{5}{8}$

CHECK FOR UNDERSTANDING

Communicating Mathematics

1. See Solutions Manual.

Study the lesson. Then complete the following.

1. **Describe** two ways to find $-2.4 + 5.87 + (-2.87) + 6.5$.

2. **Find** an example of addition or subtraction of rational numbers in a newspaper or magazine. **See students' work.**

MODELING MATHEMATICS

3. **Draw** and use a number line to show the sum of $-\frac{4}{5}$ and $\frac{3}{5}$. **See margin.**

4. **Use** a number line to solve the problem in Example 3. **See Solutions Manual.**

Guided Practice

Find each sum or difference.

5. $\frac{7}{9} - \frac{8}{9}$ **$-\frac{1}{9}$** 6. $-\frac{7}{12} + \frac{5}{6}$ **$\frac{1}{4}$** 7. $-\frac{1}{8} + \left(-\frac{5}{2}\right)$ **$-2\frac{5}{8}$**

8. $-69.5 - 82.3$ **-151.8** 9. $4.57 + (-3.69)$ **0.88** 10. $-\frac{2}{5} + \frac{3}{15} + \frac{3}{5}$ **$\frac{2}{5}$**

Evaluate each expression.

11. $a - (-5)$, if $a = 0.75$ **5.75**

12. $\frac{11}{2} - b$ if $b = -\frac{5}{2}$ **8**

13. **Business** For the week of July 22, 1995, the following day-to-day changes were recorded in the Dow Jones Industrial Average:

Monday, $+2\frac{3}{8}$; Tuesday, $-\frac{1}{4}$; Wednesday, $-\frac{3}{8}$;

Thursday, $+2\frac{1}{4}$; Friday, $-3\frac{1}{2}$.

  a. What was the net change for the week? $+\frac{1}{2}$

  b. Draw a number line that shows the change each day.
    See Solutions Manual.

EXERCISES

Practice   **A**   **Find each sum or difference.**

14. $\frac{5}{8} + \left(-\frac{3}{8}\right)$ $\frac{1}{4}$

15. $-\frac{7}{8} - \left(-\frac{3}{16}\right)$ $-\frac{11}{16}$

16. $-1.6 - 3.8$ -5.4

17. $3.2 + (-4.5)$ -1.3

18. $\frac{1}{2} + \left(-\frac{8}{16}\right)$ 0

19. $\frac{2}{3} + \left(-\frac{2}{9}\right)$ $\frac{4}{9}$

20. $-38.9 + 24.2$ -14.7

21. $-0.0007 + (-0.2)$ -0.2007

22. $-\frac{3}{7} + \frac{1}{4}$ $-\frac{5}{28}$

23. $-5\frac{7}{8} - 2\frac{3}{4}$ $-8\frac{5}{8}$

24. $79.3 - (-14.1)$ 93.4

25. $-0.0015 + 0.05$ 0.0485

B   26. $-9.16 - (-10.17)$ 1.01

27. $-5.6 + (-9.45) + (-7.89)$ -22.94

28. $\frac{1}{4} + 2 + \left(-\frac{3}{4}\right)$ $1\frac{1}{2}$

29. $-0.87 + 3.5 + (-7.6) + 2.8$ -2.17

30. $\frac{3}{4} + \left(-\frac{4}{5}\right) + \frac{2}{5}$ $\frac{7}{20}$

31. $-4\frac{1}{4} + 6\frac{1}{3} + 2\frac{2}{3} + \left(-3\frac{3}{4}\right)$ 1

Evaluate each expression.

32. $-3\frac{1}{3} - b$, if $b = \frac{1}{2}$ $-3\frac{5}{6}$

33. $r - 2.7$, if $r = -0.8$ -3.5

34. $n - 0.5$, if $n = -0.8$ -1.3

35. $t - (-1.3)$, if $t = -18$ -16.7

C   36. $\frac{4}{3} - p$, if $p = \frac{4}{5}$ $\frac{8}{15}$

37. $-\frac{12}{7} - s$, if $s = \frac{16}{21}$ $-\frac{52}{21}$

 INTEGRATION
Discrete Mathematics

Find each sum or difference.   38–42. See margin.

38. $\begin{bmatrix} 3 & -2 \\ 5 & 6.2 \end{bmatrix} + \begin{bmatrix} 2.4 & 5 \\ -4 & 1 \end{bmatrix}$

39. $\begin{bmatrix} -1.3 & 4.2 \\ 0 & -3.4 \end{bmatrix} - \begin{bmatrix} 2.5 & 4.3 \\ -1.7 & -6.3 \end{bmatrix}$

40. $\begin{bmatrix} 1 & -2 \\ 2 & 1 \\ 7 & 4 \end{bmatrix} + \begin{bmatrix} 3 & -5 \\ 2 & 6 \\ 3 & \end{bmatrix}$

41. $\begin{bmatrix} \frac{1}{2} & 6 \\ -1 & 5 \\ 4 & -7 \end{bmatrix} + \begin{bmatrix} -\frac{1}{4} & 5 \\ -4 & 3 \\ -3\frac{1}{2} & -5 \end{bmatrix}$

42. $\begin{bmatrix} 3.2 & 6.4 & -4.3 \\ 5 & -4 & 0.4 \end{bmatrix} - \begin{bmatrix} 1.4 & 3.7 & 5 \\ -1 & 5 & 0.4 \end{bmatrix}$

Critical Thinking

43. **Discrete Mathematics** Create four 2×2 matrices. Use the matrices to support or disprove the following statement: *Matrix addition is associative and commutative.* See margin.

Additional Answer

43. Answers will vary. Sample answer:

$\begin{bmatrix} 5 & 3 \\ 1 & 7 \end{bmatrix}, \begin{bmatrix} 4 & -3 \\ 1 & 2 \end{bmatrix}, \begin{bmatrix} 9 & 0 \\ 2 & 9 \end{bmatrix}, \begin{bmatrix} 18 & 0 \\ 4 & 18 \end{bmatrix}$

$\begin{bmatrix} 5 & 3 \\ 1 & 7 \end{bmatrix} + \begin{bmatrix} 4 & -3 \\ 1 & 2 \end{bmatrix} = \begin{bmatrix} 9 & 0 \\ 2 & 9 \end{bmatrix}$

$= \begin{bmatrix} 4 & -3 \\ 1 & 2 \end{bmatrix} + \begin{bmatrix} 5 & 3 \\ 1 & 7 \end{bmatrix}$

$\left(\begin{bmatrix} 5 & 3 \\ 1 & 7 \end{bmatrix} + \begin{bmatrix} 4 & -3 \\ 1 & 2 \end{bmatrix}\right) + \begin{bmatrix} 9 & 0 \\ 2 & 9 \end{bmatrix} = \begin{bmatrix} 18 & 0 \\ 4 & 18 \end{bmatrix}$

$= \begin{bmatrix} 5 & 3 \\ 1 & 7 \end{bmatrix} + \left(\begin{bmatrix} 4 & -3 \\ 1 & 2 \end{bmatrix} + \begin{bmatrix} 9 & 0 \\ 2 & 9 \end{bmatrix}\right)$

Assignment Guide

Core: 15–45 odd, 47–54
Enriched: 14–42 even, 43–54

For **Extra Practice,** see p. 760.

The red A, B, and C flags, printed only in the Teacher's Wraparound Edition, indicate the level of difficulty of the exercises.

Additional Answers

38. $\begin{bmatrix} 5.4 & 3 \\ 1 & 7.2 \end{bmatrix}$

39. $\begin{bmatrix} -3.8 & -0.1 \\ 1.7 & 2.9 \end{bmatrix}$

40. $\begin{bmatrix} \frac{7}{2} & -7 \\ \frac{23}{3} & \frac{25}{4} \end{bmatrix}$

41. $\begin{bmatrix} \frac{1}{4} & 11 \\ -5 & 8 \\ \frac{1}{2} & -12 \end{bmatrix}$

42. $\begin{bmatrix} 1.8 & 2.7 & -9.3 \\ 6 & -9 & 0 \end{bmatrix}$

Practice Masters, p. 14

Closing Activity

Writing Ask students to write a problem involving a real-life situation in which rational numbers are either added or subtracted. Have students exchange problems with classmates and solve each other's creation.

Chapter 2, Quiz B (Lessons 2-4 and 2-5), is available in the *Assessment and Evaluation Masters,* p. 44.

Mid-Chapter Test (Lessons 2-1 through 2-5) is available in the *Assessment and Evaluation Masters,* p. 43.

Additional Answer

49a.

```
       X       X         X X
  X X  XX XX XX  XXXX X XXX  X  X  X X  X X
++++++++++++++++++++++++++++++++++++++++++++++
  10  15  20  25  30  35  40  45  50  55  60
```

Enrichment Masters, p. 14

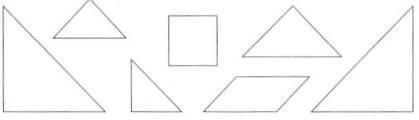

2-5 · NAME_____ DATE _____
Enrichment
Student Edition
Pages 100–104

Tangram Puzzles

These seven geometric figures are called **tans.** They are used in a very old Chinese puzzle called **tangrams.**

Glue the seven tans on heavy paper and cut them out. Use all seven pieces to make each shape shown. Record your solutions.

1.   2.   3.

4.   5.

6. Each of the two figures is made from all seven tans. They seem to be exactly alike, but one has a foot and the other does not. Where does the second figure get his foot?
In the left figure, the body is made from 4 pieces rather than 3. The extra piece becomes the foot in the right figure.

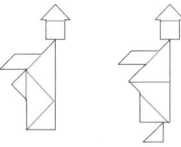

Applications and Problem Solving

44. Book Binding A textbook like this one is made up of sets of 32 pages called *signatures.* A signature is made from one piece of large paper that is folded several times. Signatures are bound together to make a book. Suppose when this book was printed, there were $1\frac{1}{2}$ rolls of paper available in the warehouse. The printer had three jobs to do that utilized this type of paper. This textbook required $\frac{1}{4}$ roll, another book required $\frac{2}{5}$ roll, and the third book needed $\frac{1}{2}$ roll. Is there enough paper available for all three jobs? Explain. **Yes; $\frac{7}{20}$ of a roll is left over.**

45. Discrete Mathematics An arithmetic sequence is a sequence in which the difference between any two consecutive terms is the same.
   a. Write the next three terms in the arithmetic sequence 4.53, 5.65, 6.77, 7.89, ... **9.01, 10.13, 11.25**
   b. What is the common difference? **1.12**
   c. Write the first five terms of an arithmetic sequence in which the first term is -2 and the common difference is $\frac{3}{4}$. Find the sum of the first five terms. **$-2, -\frac{5}{4}, -\frac{1}{2}, \frac{1}{4}, 1; -2\frac{1}{2}$**

46. Track and Field In 1995, Sheila Hudson-Strudwick set the U.S. women's triple jump record. Suppose her first jump was 46 feet 9 inches and her next jump was $1\frac{3}{4}$ inches longer. Her record-setting jump was 1 foot $2\frac{1}{2}$ inches longer than her second jump. What was the length of her record-setting jump? **48 feet $1\frac{1}{4}$ inches**

Mixed Review

47. Replace __?__ with $<$, $>$, or $=$ to make -3.833 __?__ -3.115 true. (Lesson 2–4) **$<$**

48. Evaluate $-|-9 + 2x|$ when $x = 7$. (Lesson 2–3) **-5**

49. Statistics The points scored by the winning teams in the first 29 Super Bowls are listed below. (Lesson 2–2)

35	33	16	23	16	24	14	24
16	21	32	27	35	31	27	26
27	38	38	46	39	42	20	55
20	37	52	30	49			

   a. Make a line plot of the data. **See margin.**
   b. Do the numbers cluster around any number? **no**

50. Restaurant Sales Rosa's Deli made 3 cheesecakes, each having 12 slices, on Sunday. On Monday, they sold 3 slices. On Tuesday, they sold one entire cheesecake and 2 additional slices. The baker made another cheesecake on Wednesday and sold nine slices. If Rosa's Deli sold 11 slices of cheesecake on Thursday, how many slices were left on Friday? (Lesson 2–1) **11 slices**

51. Name the property illustrated by $7(0) = 0$. (Lesson 1–6) **mult. prop. of 0**

52. Evaluate $7[4^3 - 2(4 + 3)] \div 7 + 2$. (Lesson 1–3) **52**

53. Patterns Find the next term in the sequence $\frac{1}{3}, \frac{3}{6}, \frac{5}{12}, ...$ (Lesson 1–2) **$\frac{7}{24}$**

54. Evaluate 8^4. (Lesson 1–1) **4096**

Extension

Reasoning Insert one set of parentheses so that the value of the expression shown becomes $-\frac{37}{60}$.

$$\left(-\frac{3}{4}\right) - \left(-\frac{1}{5}\right) - \frac{3}{5} - \left(-\frac{2}{3}\right)$$

$$\left(-\frac{3}{4}\right) - \left(\left(-\frac{1}{5}\right) - \frac{3}{5} - \left(-\frac{2}{3}\right)\right)$$

MODELING MATHEMATICS

A Preview of Lesson 2–6

2-6A Multiplying Integers

Materials: counters ▭ integer mat

You can use counters to model multiplication of integers. Remember that yellow counters represent positive integers and red counters represent negative integers. When multiplying whole numbers, 2 × 4 means *two sets of four items*. When multiplying integers with counters, (+2)(+4) means *to put in two sets of four positive counters*. (−2)(+4) means to take out two sets of four positive counters.

Activity 1 **Use counters to find the product (+2)(−4).**

Step 1 Place, or *put in,* two sets of four negative counters on the mat.

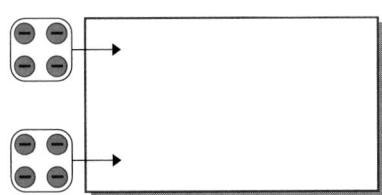

Step 2 Since there are eight negative counters on the mat, the product is −8. Thus, (+2)(−4) = −8.

Activity 2 **Use counters to find the product (−2)(+4).**

Step 1 Add enough zero pairs so that you can *take out* two sets of four positive counters.

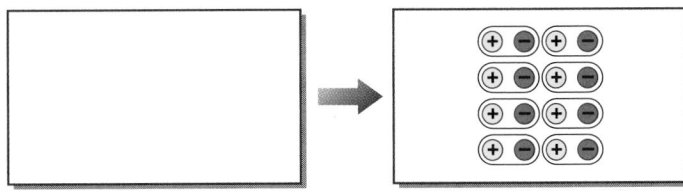

Step 2 Now take out two sets of four positive counters from the mat.

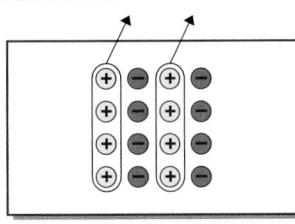

Step 3 Since there are eight negative counters on the mat, the product is −8. Thus, −2(+4) = −8.

Write 1. Model −2(−4). Write a short paragraph explaining what (−2)(−4) means. See students' work; take out two sets of negative counters.

Model **Use counters to find each product.** 2–7. See Solutions Manual for models.

2. 2(−5) −10   3. −2(5) −10   4. −2(−5) 10
5. 5(−2) −10   6. −5(2) −10   7. −5(−2) 10

Write 8. How are the operations −2(5) and 5(−2) the same? How do they differ? See margin.

Additional Answer

8. The product is the same. The order of multiplication using counters is different.

4 ASSESS

Observing students working in cooperative groups is an excellent method of assessment. You may wish to ask a student from each group to explain how to model and solve a problem. Also watch for and acknowledge students who are helping others understand the concept being taught.

NCTM Standards: 1–5

Objective
Use counters to model multiplication of integers.

Recommended Time
Demonstration and discussion: 15 minutes; Exercises: 30 minutes

Instructional Resources
For each student or group of students
Student Manipulative Kit
• 20 red/yellow 2-sided counters
Modeling Mathematics Masters
• p. 2 (counters)
• p. 13 (integer mat)
• p. 21 (worksheet)
For teacher demonstration
Algebra and Geometry Overhead Manipulative Resources

1 FOCUS

Motivating the Lesson
Remind students that multiplication is simply repeated addition. For example, 3 × 6 = 6 + 6 + 6. 3 × 6 is 6 added together 3 times, or 3 added together 6 times. Ask students to solve 3(−6) in the same way. What is the sign of the answer? **negative**

2 TEACH

Teaching Tip Point out that in this activity, the − symbol can be viewed in two different ways. When it occurs outside the parentheses, it symbolizes the act of "taking out" counters. When it occurs inside the parentheses, it symbolizes negative counters.

3 PRACTICE/APPLY

Assignment Guide
Core: 1–8
Enriched: 1–8

NCTM Standards: 1–5

Instructional Resources

- Study Guide Master 2-6
- Practice Master 2-6
- Enrichment Master 2-6
- Real-World Applications, 5
- Tech Prep Applications Masters, p. 4

 Transparency 2-6A contains the 5-Minute Check for this lesson; **Transparency 2-6B** contains a teaching aid for this lesson.

Recommended Pacing	
Standard Pacing	Day 8 of 15
Honors Pacing	Day 8 of 15
Block Scheduling*	Day 5 of 9
Alg. 1 in Two Years*	Days 15 & 16 of 28

 *For more information on pacing and possible lesson plans, refer to the *Block Scheduling Booklet* and *Algebra 1 in Two Years.*

1 FOCUS

 ### 5-Minute Check
(over Lesson 2-5)

Find each sum or difference.

1. $3.1 + (-4.56)$ **−1.46**

2. $-\frac{23}{3} + 2\frac{1}{5}$ **−5$\frac{7}{15}$**

3. $9.16 - 10.17$ **−1.01**

Evaluate each expression.

4. $-1\frac{2}{3} - b$, if $b = \frac{1}{3}$ **−2**

5. $k - (-5.8)$, if $k = 6.3$ **12.1**

Motivating the Lesson
Questioning Ask students to name examples from everyday life in which they use fractions. They might also suggest instances requiring the use of fractions from their science classes.

2-6

Objective
After studying this lesson, you should be able to:
- multiply rational numbers.

Multiplying Rational Numbers

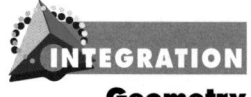 **Geometry**

Once called "the world's largest office building," the Pentagon, the five-story, five-sided defense building in Washington, D.C., was designed by G.E. Bergstrom to maximize space and efficiency. No two offices are more than a seven-minute walk from one another.

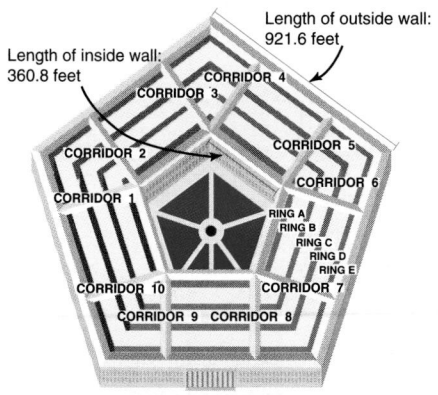

Length of outside wall: 921.6 feet
Length of inside wall: 360.8 feet
CORRIDOR 1, CORRIDOR 2, CORRIDOR 3, CORRIDOR 4, CORRIDOR 5, CORRIDOR 6, CORRIDOR 7, CORRIDOR 8, CORRIDOR 9, CORRIDOR 10
RING A, RING B, RING C, RING D, RING E

THE PENTAGON

The building is made up of 10 corridors and five rings. There are 230 restrooms, 150 stairways, and 7748 windows. Each outside wall of the building is 921.6 feet in length, slightly longer than three football fields. The inner walls, also in the shape of a pentagon, are each 360.8 feet in length. What is the outer perimeter of the Pentagon?

One way to solve this problem is to use repeated addition.

$$921.6 + 921.6 + 921.6 + 921.6 + 921.6 = 4608$$

An easier method would be to multiply 921.6 by 5.

$$5(921.6) = 4608$$

The perimeter of the Pentagon is 4608 feet.

Since this method would not work if you wanted to find the product of $\frac{2}{3}$ and $-\frac{2}{5}$ or the product of -5 and -0.3, you can use the following patterns to discover a rule for multiplying rational numbers.

$$\frac{2}{3} \times \frac{2}{5} = \frac{4}{15}$$

$$\frac{2}{3} \times \frac{1}{5} = \frac{2}{15}$$

$$\frac{2}{3} \times \frac{0}{5} = \frac{0}{15}$$

$$\frac{2}{3} \times \left(-\frac{1}{5}\right) = -\frac{2}{15}$$

$$\frac{2}{3} \times \left(-\frac{2}{5}\right) = -\frac{4}{15}$$

$$-5 \cdot 0.3 = -1.5$$

$$-5 \cdot 0.2 = -1.0$$

$$-5 \cdot 0.1 = -0.5$$

$$-5 \cdot 0 = 0$$

$$-5 \cdot (-0.1) = 0.5$$

$$-5 \cdot (-0.2) = 1.0$$

$$-5 \cdot (-0.3) = 1.5$$

The examples on the previous page suggest the following rules.

| **Multiplying Two Rational Numbers** | **The product of two numbers having the *same sign* is positive. The product of two numbers having *different signs* is negative.** |

Example **Find each product.**

a. $(-9.8)4$ *Estimate:* $(-10)4 = -40$

$(-9.8)4 = -39.2$ *Since the factors have different signs, the product is negative.*

b. $\left(-\frac{3}{4}\right)\left(-\frac{2}{3}\right)$ *Estimate:* $(-1)\left(-\frac{1}{2}\right) = \frac{1}{2}$

$\left(-\frac{3}{4}\right)\left(-\frac{2}{3}\right) = \frac{6}{12}$ or $\frac{1}{2}$ *Since the factors have the same sign, the product is positive.*

Sometimes you need to evaluate expressions that contain rational numbers.

Example **Evaluate** $a\left(\frac{5}{6}\right)^2$ **if** $a = 2$.

$a\left(\frac{5}{6}\right)^2 = 2\left(\frac{5}{6} \cdot \frac{5}{6}\right)$ *Replace a with 2.*

$= 2\left(\frac{25}{36}\right)$ *Multiply.*

$= \frac{\overset{1}{\cancel{2}}}{1}\left(\frac{25}{\underset{18}{\cancel{36}}}\right)$ or $\frac{25}{18}$

You may need to simplify expressions by multiplying rational numbers.

Example **3** **Simplify each expression.**

a. $(2b)(-3a)$

$(2b)(-3a) = 2(-3)ab$ *Commutative and associative properties of multiplication*

$= -6ab$

b. $3x(-3y) + (-6x)(-2y)$

$3x(-3y) + (-6x)(-2y) = -9xy + 12xy$ *Multiply.*

$= 3xy$ *Combine like terms.*

Notice that multiplying a number or expression by -1 results in the opposite of the number or expression.

$$-1(4) = -4 \qquad (1.5)(-1) = -1.5 \qquad (-1)(-3m) = 3m$$

| **Multiplicative Property of −1** | **The product of any number and −1 is its additive inverse.** $-1(a) = -a$ **and** $a(-1) = -a$ |

To find the product of three or more numbers, you may want to first group the numbers in pairs.

Lesson 2–6 Multiplying Rational Numbers **107**

2 TEACH

In-Class Examples

For Example 1
Find each product.

a. $20\left(-\frac{4}{5}\right)$ -16

b. $(-1.4)7$ -9.8

For Example 2
Evaluate $a\left(\frac{2}{3}\right)^3$ if $a = 3$. $\frac{8}{9}$

For Example 3
Simplify each expression.

a. $(-2a)(3b) + (4a)(-6b)$ $-30ab$

b. $(5x)(-3y) + (-7x)(4y)$ $-43xy$

108 Chapter 2

Teaching Tip The product of any pair of negative numbers is positive. An even number of negative factors can be paired exactly to produce a positive result. An odd number of negative factors cannot be paired exactly, so a negative result is produced.

In-Class Examples

For Example 4
Multiply.

a. $\left(-\frac{2}{3}\right)\left(-\frac{1}{6}\right)\left(\frac{9}{5}\right)\left(-\frac{1}{2}\right)$ $-\frac{1}{10}$

b. $\left(-\frac{5}{6}\right)(-4)\left(7\frac{1}{5}\right)\left(\frac{11}{12}\right)$ 22

For Example 5
Find the new fares if American lowers its fares by 40% or $\frac{4}{10}$.

$$\begin{bmatrix} \$118.80 & \$118.80 & \$728.40 \\ \$147.60 & \$734.40 & \$805.20 \end{bmatrix}$$

Additional Answers

1a. ab is positive if a and b have the same signs, either both positive or both negative.

1b. ab is negative if a and b have opposite signs, one is negative and the other is positive.

1c. ab is equal to 0 if either a or b is 0, or both a and b are 0.

2a. If a^2 is positive, then a is either negative or positive.

2b. If a^3 is positive, then a is positive.

2c. If a^3 is negative, then a is negative.

Example **4** Find $\left(-\frac{3}{4}\right)\left(-4\frac{1}{3}\right)\left(3\frac{2}{5}\right)(4)(-1)$.

$$\left(-\frac{3}{4}\right)\left(-4\frac{1}{3}\right)\left(3\frac{2}{5}\right)(4)(-1) = \left[\left(-\frac{3}{4}\right)(4)\right]\left[\left(-4\frac{1}{3}\right)(-1)\right]\left(3\frac{2}{5}\right) \quad \text{Commutative and associative properties}$$

$$= (-3)\left(4\frac{1}{3}\right)\left(3\frac{2}{5}\right)$$

$$= \left(-\frac{\overset{1}{\cancel{3}}}{1}\right)\left(\frac{13}{\underset{1}{\cancel{3}}}\right)\left(\frac{17}{5}\right)$$

$$= -\frac{221}{5} \text{ or } -44\frac{1}{5}$$

You can multiply any matrix by a constant. This is called **scalar multiplication.** When scalar multiplication is performed, each element is multiplied by that constant and a new matrix is formed.

Scalar Multiplication of a Matrix	$m\begin{bmatrix} a & b & c \\ d & e & f \end{bmatrix} = \begin{bmatrix} ma & mb & mc \\ md & me & mf \end{bmatrix}$

Example **5**

APPLICATION
Airlines

From July 3 through July 6, 1995, American Airlines quoted round-trip coach fares, in dollars, to and from the selected cities below.

	Chicago	Dallas	Las Vegas
Atlanta	$198.00	$198.00	$1214.00
New York	$246.00	$1224.00	$1342.00

Suppose the major airlines lower fares and begin an airfare war. To increase their air travel, American Airlines decides to lower their fares by 30% or $\frac{3}{10}$. Find the new fares for travel to the cities above.

To find the new fares, you can use scalar multiplication. If fares are reduced by $\frac{3}{10}$, travelers will pay $1 - \frac{3}{10}$ or $\frac{7}{10}$ of the original price. Multiply to find the discount prices.

$$\frac{7}{10}\begin{bmatrix} 198.00 & 198.00 & 1214.00 \\ 246.00 & 1224.00 & 1342.00 \end{bmatrix} = \begin{bmatrix} 138.60 & 138.60 & 849.80 \\ 172.20 & 856.80 & 939.40 \end{bmatrix}$$

The new fares are shown below.

	Chicago	Dallas	Las Vegas
Atlanta	$138.60	$138.60	$849.80
New York	$172.20	$856.80	$939.40

CHECK FOR UNDERSTANDING

Communicating Mathematics

Study the lesson. Then complete the following.   1–2. See margin.

1. **List** the conditions under which each statement is true. Then give an example that verifies your conditions.
   a. ab is positive.   **b.** ab is negative.   **c.** ab is equal to 0.

2. **Explain** what you can conclude about a if each statement is true.
   a. a^2 is positive.   **b.** a^3 is positive.   **c.** a^3 is negative.

3. If $a = -2$, which is greater, $a + a^2$ or -4? $a + a^2$

 Alternative Teaching Strategies

Student Diversity Many countries use the Celsius system for measuring temperatures. The United States still uses the Fahrenheit system for weather reports for the general public. The formula used to convert degrees Fahrenheit to degrees Celsius is

$$C = \frac{5}{9}(F - 32).$$

4. Write a multiplication sentence for the model at the right.
$2 \times (-3) = -6$

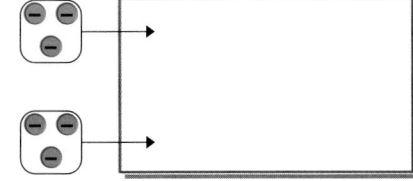

Guided Practice

Find each product.

5. $6(-3)$ -18

6. $(-4)(-8)$ 32

7. $(-4)(2)(-3)$ 24

8. $\left(\frac{7}{3}\right)\left(\frac{7}{3}\right)$ $\frac{49}{9}$

9. $\left(-\frac{4}{5}\right)\left(-\frac{1}{5}\right)(-5)$ $-\frac{4}{5}$

10. $\left(\frac{3}{5}\right)\left(-\frac{4}{7}\right)$ $-\frac{12}{35}$

Evaluate each expression if $x = \frac{1}{2}$ **and** $y = -\frac{2}{3}$.

11. $3y - 4x$ -4

12. $x^2 y$ $-\frac{1}{6}$

Simplify.

13. $5s(-6t) + 2s(-8t)$ $-46st$

14. $6x(-7y) + (-3x)(-5y)$ $-27xy$

Find each product. 15–16. See margin.

15. $3\begin{bmatrix} -2 & 4 \\ -1 & 5 \end{bmatrix}$

16. $-5\begin{bmatrix} -1 & 0 \\ 4.5 & 8 \\ 3.2 & -4 \end{bmatrix}$

17. **Business** Employees of Glencoe/McGraw-Hill are reimbursed for the "wear and tear" that occurs to their car while on company business. One employee's odometer read 19,438.6 at the beginning of the day and 19,534.1 at the end of the day. How much should the employee be reimbursed if the rate is $0.30 per mile? $28.65

EXERCISES

Practice
A

Find each product.

18. $6(13)$ 78

19. $(-5)(12)$ -60

20. $(-7)(-6)$ 42

21. $\left(-\frac{8}{9}\right)\left(\frac{9}{8}\right)$ -1

22. $-\frac{5}{6}\left(-\frac{2}{5}\right)$ $\frac{1}{3}$

23. $(-5)\left(-\frac{2}{5}\right)$ 2

24. $(-100)(-3.6)$ 360

25. $(-2.93)(-0.003)$

26. $(-5)(3)(-4)$ 60

27. $\left(\frac{2}{3}\right)\left(\frac{3}{5}\right)(-3)$ $-\frac{6}{5}$

28. $\left(-\frac{7}{12}\right)\left(\frac{6}{7}\right)\left(-\frac{3}{4}\right)$ $\frac{3}{8}$

29. $\frac{6}{11}\left(-\frac{33}{34}\right)$ $-\frac{9}{17}$

30. $(-0.075)(-5.5)$

31. $(-5.8)(-6.425)(2.3)$

32. $(-4)(0)(-2)(-3)$ 0

33. $\frac{3}{5}(5)(-2)\left(-\frac{1}{2}\right)$ 3

34. $(3)(-4)(-1)(-2)$

35. $\frac{2}{11}(-11)(-4)\left(-\frac{3}{4}\right)$

25. 0.00879

30. 0.4125

31. 85.7095

34. -24

35. -6

37. $\frac{13}{12}$ 38. $-\frac{17}{16}$ **B**

Evaluate each expression if $m = -\frac{2}{3}$, $n = \frac{1}{2}$, $p = -3\frac{3}{4}$, **and** $q = 2\frac{1}{6}$.

36. $6m$ -4

37. nq

38. $2m - 3n$

39. $pq - m$ $-\frac{179}{24}$

40. $m^2\left(-\frac{1}{4}\right)$ $-\frac{1}{9}$

41. $n^2(q + 2)$ $\frac{25}{24}$

Lesson 2–6 Multiplying Rational Numbers **109**

Reteaching

Using Reasoning

1. Write the sign of each product.
   a. $(-)(-)$ $+$
   b. $(-)(-)(-)(+)(+)$ $-$
   c. product of seven negative and three positive factors $-$

2. Find each product.
   a. $(-1)^8$ $+1$
   b. $(-1)^{19}$ -1
   c. $\left(\frac{3}{4}\right)\left(\frac{2}{3}\right)\left(-\frac{1}{2}\right)$ $-\frac{1}{4}$
   d. $(-2)(-12)(0)(-6)$ 0

---

3 PRACTICE/APPLY

Check for Understanding

Exercises 1–17 are designed to help you assess your students' understanding through reading, writing, speaking, and modeling. You should work through Exercises 1–4 with your students and then monitor their work on Exercises 5–17.

Assignment Guide
Core: 19–57 odd, 59–68
Enriched: 18–52 even, 54–68

For **Extra Practice,** see p. 760.

The red A, B, and C flags, printed only in the Teacher's Wraparound Edition, indicate the level of difficulty of the exercises.

Additional Answers

15. $\begin{bmatrix} -6 & 12 \\ -3 & 15 \end{bmatrix}$

16. $\begin{bmatrix} 5 & 0 \\ -22.5 & -40 \\ -16 & 20 \end{bmatrix}$

Study Guide Masters, p. 15

2-6
NAME_____ DATE_____
Study Guide
Student Edition
Pages 106–111

Multiplying Rational Numbers

You can use the rules below when multiplying rational numbers.

	Rule or Property	Example
Multiplying Two Numbers with Different Signs	The product of two numbers that have different signs is negative.	$(-2x)5y = -(2)(5)xy$ $= -10xy$
Multiplying Two Numbers with the Same Sign	The product of two numbers that have the same sign is positive.	$(-4)(-7) = 28$
Multiplicative Property of -1	The product of any number and -1 is its additive inverse. $-1(a) = -a$ and $a(-1) = -a$	$(-3)(-4)(-1)(2) = 12(-1)(2)$ $= -12(2)$ $= -24$

To find the product of two or more numbers, first group the numbers in pairs.

Example: Multiply $(-2.3)(5.6)(-0.7)(-0.2)$
$(-2.3)(5.6)(-0.7)(-0.2) = [(-2.3)(5.6)][(-0.7)(-0.2)]$ Associative property (×)
$= (-12.88)(0.14)$ Substitution property (=)
$= -1.8032$

Find each product.

1. $(-24)(-2)$ 48

2. $(6.0)(-0.3)$ -1.8

3. $(-2)(-3)(-4)$ -24

4. $\left(\frac{1}{2}\right)(-10)(5)$ -25

5. $(-22)(-3)\left(-\frac{2}{3}\right)$ -44

6. $\left(\frac{4}{5}\right)(-5)(0)(4)$ 0

Simplify.

7. $-6(5) + (2)(4)$ -22

8. $\left(-\frac{3}{4}\right)\left(\frac{1}{8}\right) - \left(\frac{1}{4}\right)\left(\frac{1}{8}\right)$ $-\frac{1}{8}$

9. $-5(2x + x) - 3(-xy)$ $-15x + 3xy$

10. $(-5ab)6 - ab(4 + 1)$ $-35ab$

11. $2.3(4c - d) - 0.1(-0.5c + 8d)$ $9.25c - 3.1d$

12. $(-0.9)(1.1) - (-5.1)(0.6)$ 2.07

48. $\begin{bmatrix} 7 & -57.4 & 0 \\ -28 & -39.2 & 7 \\ -22.4 & -49 & -49 \end{bmatrix}$

49. $\begin{bmatrix} 2 & 6 & 3 \\ \frac{5}{2} & 5 & 1 \end{bmatrix}$

50. $\begin{bmatrix} 5.2 & -8 & -16 \\ 2 & -1.2 & 20 \\ 26.4 & 8.4 & -32 \end{bmatrix}$

51. $\begin{bmatrix} -9 & 22.68 \\ -22.4 & -10 \\ 28.8 & -11.12 \end{bmatrix}$

52. $\begin{bmatrix} -1.6 & -36 \\ 11.2 & 24 \\ -24 & -19.2 \\ 56 & 25.6 \end{bmatrix}$

53. $\begin{bmatrix} 6 & 18 & 4 \\ 0 & 2 & \frac{8}{3} \end{bmatrix}$

56a. $\frac{1}{9}, \frac{1}{27}, \frac{1}{81}$

56b. $\frac{1}{3}$

56c. $-6, -3, -1.5, -0.75,$
$-0.375, -11.625$

Practice Masters, p. 15

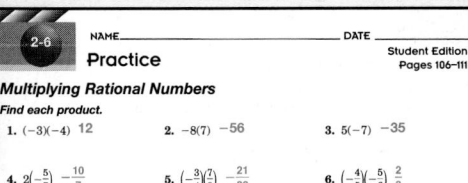

110 Chapter 2

Simplify.

42. $6ac - 36ry$

42. $-2a(-3c) + (-6y)(6r)$
43. $(5t)(-6r) - (-4s)$ $-30rt + 4s$
44. $7m(-3n) + 3m(-4n)$ $-33mn$
45. $5(2x - x) + 4(x + 3x)$ $21x$
46. $(-6b)(-3c) - (-9a)(7b)$ $18bc + 63ab$
47. $3.2(5x - y) - 0.3(-1.6x + 7y)$ $16.48x - 5.3y$

INTEGRATION
Discrete
Mathematics

Find each product. 48–53. See margin.

48. $-7\begin{bmatrix} -1 & 8.2 & 0 \\ 4 & 5.6 & -1 \\ 3.2 & 7 & 7 \end{bmatrix}$

49. $\frac{1}{2}\begin{bmatrix} 4 & 12 & 6 \\ 5 & 10 & 2 \end{bmatrix}$

50. $4\begin{bmatrix} 1.3 & -2 & -4 \\ 0.5 & -0.3 & 5 \\ 6.6 & 2.1 & -8 \end{bmatrix}$

51. $-4\begin{bmatrix} 2.25 & -5.67 \\ 5.6 & 2.5 \\ -7.2 & -2.78 \end{bmatrix}$

52. $-8\begin{bmatrix} 0.2 & 4.5 \\ -1.4 & -3 \\ 3 & 2.4 \\ -7 & -3.2 \end{bmatrix}$

53. $\frac{2}{3}\begin{bmatrix} 9 & 27 & 6 \\ 0 & 3 & 4 \end{bmatrix}$

Critical Thinking

54. If a product has an even number of negative factors, what must be true of the product? **It is positive.**

55. If a product has an odd number of negative factors, what must be true of the product? **It is negative.**

Applications and Problem Solving

56. **Discrete Mathematics** A *geometric sequence* is a sequence in which the ratio of any term divided by the term before it is the same for any two terms.
a. Write the next three terms in the geometric sequence $9, 3, 1, \frac{1}{3}, \ldots$.
b. What is the common ratio? **a–c. See margin.**
c. Write the first five terms in a geometric sequence in which the first term is -6 and the common ratio is 0.5. Then find the sum of the five terms.

57. **Construction** Ryduff Builders is building homes in a new residential development. The building code for the development states that lots must have a minimum of 1250 square feet and no dimension can be less than 20 feet. **a–c. See margin.**
a. Determine whether a plot plan would be approved if it measured 32 feet by 38 feet. Explain your reasoning.
b. Determine whether a plot plan would be approved if it measured 19 feet by 70 feet. Explain your reasoning.
c. If one dimension of a plot plan was 42 feet, find another dimension that would satisfy the building code conditions.

58. **Civics** The length of a flag is called the *fly*, and the width is called the *hoist*. The blue rectangle in the United States flag is called the *union*. The length of the union is $\frac{2}{5}$ of the fly, and the width is $\frac{7}{13}$ of the hoist. If the fly of a United States flag is 3 feet, how long is the union?

58. $1\frac{1}{5}$ feet

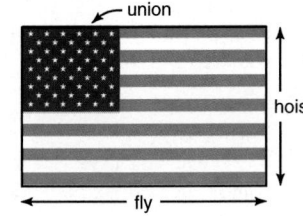
union
hoist
fly

110 Chapter 2 *Exploring Rational Numbers*

57a. No; it meets the requirement that no dimensions can be less than 20 feet, but the minimum of 1250 square feet is not met because this yard would have 1216 square feet.

57b. No; it meets the requirement that yards must have a minimum of 1250 square feet because it would have 1330 square feet, but the requirement that no dimension can be less than 20 feet is not met because it has a side of length 19 feet.

57c. Answers will vary. Sample answer: 30 feet

Mixed Review

59. Find $5.7 + (-7.9)$. (Lesson 2–5) **−2.2**

60. Replace _?_ with $<$, $>$, or = to make 12 _?_ $15 - 7$ true. (Lesson 2–4) **>**

61. Evaluate $a - 12$ if $a = -8$. (Lesson 2–3) **−20**

62. Statistics State the scale you would use to make a line plot for the following data. Then draw the line plot. (Lesson 2–2) **See margin.**

4, 9, 2, 2, 15, 7, 6, 6, 9, 12, 1, 3, 2, 11, 10, 2, 6, 4, 12, 13

63. Find $6 + (-13)$. If necessary, use a number line. (Lesson 2–1) **−7**

64. Write an algebraic expression for *five times the sum of x and y decreased by z*. (Lesson 1–8) $5(x + y) - z$ **or** $5(x + y - z)$

65. Find the solution set for $y + 3 > 8$ if the replacement set is {3, 4, 5, 6, 7}. (Lesson 1–5) **{6, 7}**

66. Statistics Suppose the number 27,878 is rounded to 27,900 and plotted using stem 27 and leaf 9. Write the stem and leaf for each number below if the numbers are part of the same set of data. (Lesson 1–4) **See margin.**
  a. 13,245   **b.** 35,684   **c.** 153,436

67. Evaluate $6(4^3 + 2^2)$. (Lesson 1–3) **408**

68. Geometry Write an expression that represents the total number of small cubes in the large cube shown at the right. Then evaluate the expression. (Lesson 1–1) 4^3; **64 cubes**

WORKING ON THE
In·ves·ti·ga·tion

Refer to the Investigation on pages 68–69.

the Greenhouse Effect

Without a heat-trapping blanket of naturally occurring CO_2, it is believed that Earth would have an average surface temperature of $-17°C$, instead of its current average, $15°C$. There is evidence that Mars has little CO_2 in its atmosphere, and its temperature never exceeds $-31°C$. At the other extreme, Venus, with lots of CO_2, has an average temperature of $454°C$.

1 What would the difference in temperature be between Earth with CO_2 and Earth without CO_2? Do you think life could exist on Earth without CO_2? Explain.

2 What is the difference in temperature between Mars and Earth without CO_2? Explain what other factors contribute to the difference in temperature.

3 What is the difference in temperature between Venus and Mars? Explain what other factors contribute to the difference in temperature.

4 Make a chart that shows the three planets and their temperatures with CO_2 in their atmospheres and with little or no CO_2. Assume that the differences in the temperatures for the other planets would be the same as the difference in Earth's temperatures with CO_2 and without CO_2. You will use this information later on as you work on the Investigation.

Add the results of your work to your Investigation Folder.

Lesson 2–6 Multiplying Rational Numbers **111**

Extension

Problem Solving Negative three is added to a certain number 27 times. If the result is −29, what was the original number? **52**

In·ves·ti·ga·tion

Working on the Investigation
The Investigation on pages 68–69 is designed to be a long-term project that is completed over several days or weeks. Encourage students to keep their materials in their Investigation Folder as they work on the Investigation.

---

4 ASSESS

Closing Activity

Speaking Without finding each product, have students indicate whether the product will be positive, negative, or 0.

1. $(-4)(-3)$ **positive**

2. $(-1.3)(4)(-2.54)$ **positive**

3. $\left(-\frac{3}{4}\right)\left(-\frac{1}{2}\right)(0)\left(-\frac{7}{10}\right)$ **0**

4. $\left(\frac{2}{5}\right)\left(-\frac{3}{8}\right)\left(\frac{5}{9}\right)$ **negative**

Additional Answers

62. from 1 to 15 by 1s

66a. stem 13 and leaf 2
66b. stem 35 and leaf 7
66c. stem 153 and leaf 4

Enrichment Masters, p. 15

2-6 NAME_____ DATE_____
Enrichment Student Edition Pages 106–111

Convergence, Divergence, and Limits

Imagine that a runner runs a mile from point A to point B. But, this is not an ordinary race! In the first minute, he runs one-half mile, reaching point C. In the next minute, he covers one-half the remaining distance, or $\frac{1}{4}$ mile, reaching point D. In the next minute he covers one-half the remaining distance, or $\frac{1}{8}$ mile, reaching point E.

In this strange race, the runner approaches closer and closer to point B, but never gets there. However close he is to B, there is still some distance remaining, and in the next minute he can cover only half of that distance.

This race can be modeled by the infinite sequence
$\frac{1}{2}, \frac{3}{4}, \frac{7}{8}, \frac{15}{16}, \ldots$

The terms of the sequence get closer and closer to 1. An infinite sequence that gets arbitrarily close to some number is said to **converge** to that number. The number is the limit of the sequence.

Not all infinite sequences converge. Those that do not are called **divergent**.

Write C if the sequence converges and D if it diverges. If the sequence converges, make a reasonable guess for its limit.

1. 2, 4, 6, 8, 10, … **D**   **2.** 0, 3, 0, 3, 0, 3, … **D**

3. $1, \frac{1}{2}, \frac{1}{3}, \frac{1}{4}, \frac{1}{5}, \ldots$ **C, 0**   **4.** 0.9, 0.99, 0.999, 0.9999, … **C, 1**

5. −5, 5, −5, 5, −5, 5, … **D**   **6.** 0.1, 0.2, 0.3, 0.4, … **D**

7. $2\frac{1}{4}, 2\frac{3}{8}, 2\frac{7}{16}, 2\frac{15}{16}, \ldots$ **C, 3**   **8.** $6, 5\frac{1}{2}, 5\frac{1}{3}, 5\frac{1}{4}, 5\frac{1}{5}, \ldots$ **C, 5**

9. 1, 4, 9, 16, 25, … **D**   **10.** $1, -\frac{1}{2}, \frac{1}{3}, -\frac{1}{4}, \frac{1}{5}, -\frac{1}{6}, \ldots$ **C, 0**

11. Create one convergent sequence and one divergent sequence. Give the limit for your convergent sequence. **Answers will vary.**

Chapter 2 **111**

Instructional Resources

- Study Guide Master 2-7
- Practice Master 2-7
- Enrichment Master 2-7
- Assessment and Evaluation Masters, p. 45
- Modeling Mathematics Masters, pp. 36–38

 Transparency 2-7A contains the 5-Minute Check for this lesson; **Transparency 2-7B** contains a teaching aid for this lesson.

Recommended Pacing	
Standard Pacing	Day 9 of 15
Honors Pacing	Day 9 of 15
Block Scheduling*	Day 6 of 9
Alg. 1 in Two Years*	Days 17, 18, & 19 of 28

 *For more information on pacing and possible lesson plans, refer to the *Block Scheduling Booklet* and *Algebra 1 in Two Years.*

1 FOCUS

 ### 5-Minute Check
(over Lesson 2-6)

Find each product.

1. $\left(-\frac{2}{3}\right)\left(\frac{5}{4}\right)$ $-\frac{5}{6}$

2. $(0.1)(-1.1)(-0.01)$ **0.0011**

Evaluate each expression if $x = -3$ and $y = \frac{4}{3}$.

3. $2xy + 4x$ **−20**

4. $-2(x - 6xy)$ **−42**

5. Find $\frac{3}{4}\begin{bmatrix} -\frac{3}{4} & 4 & 16 \\ 2 & 0 & -1\frac{1}{3} \end{bmatrix}$.

$\begin{bmatrix} -\frac{9}{16} & 3 & 12 \\ \frac{3}{2} & 0 & -1 \end{bmatrix}$

2-7

Dividing Rational Numbers

Objective
After studying this lesson, you should be able to:
- divide rational numbers.

 ### APPLICATION
Pilot

Vicki Van Meter had a lot to write in her report about her summer vacation of 1993. At 12 years old, inspired by Amelia Earhart, she became the youngest female to pilot a plane across the Atlantic Ocean. Vicki did all of the flying, navigation, and communication for her 3200-kilometer flight aboard a single-engine Cessna 210. She also had to be concerned about how supplies and fuel were loaded onto the plane.

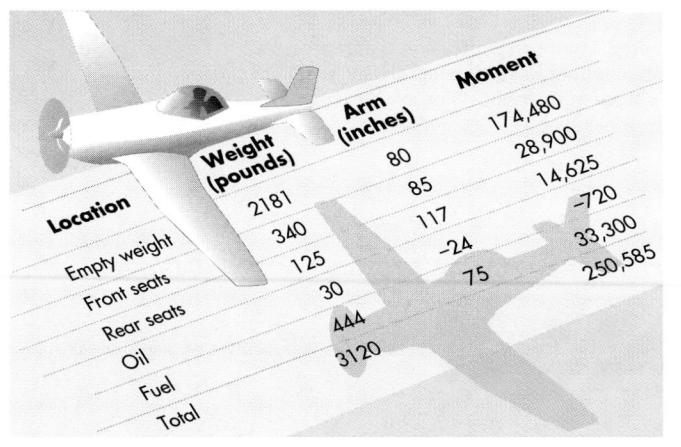

Before any small aircraft can take off, the pilot must be sure that the aircraft is loaded so that the center of gravity is within certain safe limits. If it is, the pilot is ready for takeoff. If not, the weight must be rebalanced.

You can use the table at the left to find the center of gravity for a particular aircraft. The safe limit for this aircraft is 82.1. For each location, the weight and the arm are multiplied to find the moment.

Now add the moments, divide by the total weight, and round to the nearest tenth to find the center of gravity.

Total Moments		Total Weight		Center of Gravity
250,585	÷	3120	=	80.3

Since 80.3 is less than 82.1, the aircraft is safe and ready for takeoff.

You already know that the quotient of two positive numbers is positive. But how do you determine the sign of the quotient when negative numbers are involved? Since multiplication and division are inverse operations, the rule for finding the sign of the quotient of two numbers is similar to the rule for finding the sign of the product. Study these patterns.

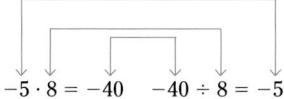

$$-5 \cdot 8 = -40 \qquad -40 \div 8 = -5$$

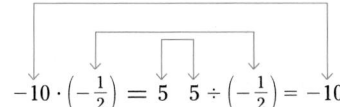

$$-10 \cdot \left(-\frac{1}{2}\right) = 5 \qquad 5 \div \left(-\frac{1}{2}\right) = -10$$

These examples suggest the following rules.

Dividing Two Rational Numbers	The quotient of two numbers having the *same sign* is positive. The quotient of two numbers having *different signs* is negative.

Teaching Tip To find the center of gravity of an aircraft, the pilot multiplies the weight of all of the items on the plane by its *arm,* or the distance from a certain point called the *datum.* This product is called the *moment* of each item. The moments are then added together and divided by the total weight of the items. The quotient equals the distance of the center of gravity from the datum. The pilot then checks this number to see if the center of gravity is within safe limits.

Example 1 Find each quotient.

a. $-75 \div (-15)$

This division problem may also be written as $\frac{-75}{-15}$.

$\frac{-75}{-15} = 5$ *The fraction bar indicates division.*

Since the signs are the same, the quotient is positive.

b. $\frac{72}{-8}$

$\frac{72}{-8} = -9$ *Since the signs are different, the quotient is negative.*

Recall that you can change any division expression to an equivalent multiplication expression. To divide by any nonzero number, multiply by the reciprocal of that number.

Example 2 Find each quotient.

a. $\frac{1}{2} \div 5$

$\frac{1}{2} \div 5 = \frac{1}{2} \cdot \frac{1}{5}$ *Multiply by $\frac{1}{5}$, the reciprocal of 5.*

$= \frac{1}{10}$ *The signs are the same, so the product is positive.*

b. $-\frac{6}{7} \div 3$

$-\frac{6}{7} \div 3 = -\frac{6}{7} \cdot \frac{1}{3}$ *Multiply by $\frac{1}{3}$, the reciprocal of 3.*

$= -\frac{6}{21}$ or $-\frac{2}{7}$ *The signs are different, so the product is negative.*

Example 3

APPLICATION
Nursing

Three of the measurements nurses commonly use are cubic centimeters (cc), drops, and grains. Use this information to solve the following.

a. A doctor orders $\frac{1}{400}$ of a grain of medicine to be given to a patient. The nurse has a vial labeled $\frac{1}{200}$ grains per cc. How many cc of the medicine should the nurse give the patient?

b. The doctor also prescribes a 1000-cc intravenous (IV) bottle to be given to the patient over an 8-hour period. If there are 15 drops in 1 cc, for how many drops per minute should the nurse set the IV?

a. How many cc are in $\frac{1}{400}$ of a grain? Divide $\frac{1}{400}$ grain by $\frac{1}{200}$ grains/cc.

$$\frac{1}{400} \text{ grain} \div \frac{1}{200} \text{ grain/cc} = \frac{1 \text{ grain}}{400} \cdot \frac{200 \text{ cc}}{1 \text{ grain}}$$

$$= \frac{1 \text{ grain}}{\underset{2}{400}} \cdot \frac{\overset{1}{200 \text{ cc}}}{1 \text{ grain}}$$

$$= \frac{1}{2} \text{ cc}$$

The nurse should give the patient $\frac{1}{2}$ cc of the medicine.

(continued on the next page)

Lesson 2–7 Dividing Rational Numbers **113**

(continued on the next page)

Motivating the Lesson
Hands-On Activity Have students look up the term *inverse* in the dictionary. Develop student understanding of the term by having a volunteer mime a sequence of three actions. Ask another volunteer to mime the inverse of these actions. Then discuss the meaning of a multiplicative inverse, or reciprocal.

2 TEACH

In-Class Examples

For Example 1
Find each quotient.

a. $-32 \div -8$ **4**

b. $\frac{81}{-9}$ **−9**

For Example 2
Find each quotient.

a. $\frac{5}{6} \div 3$ $\frac{5}{18}$

b. $-\frac{3}{4} \div \frac{5}{8}$ $-1\frac{1}{5}$

c. $-\frac{1}{5} \div \frac{3}{10}$ $-\frac{2}{3}$

For Example 3

a. If the doctor orders $\frac{1}{50}$ of a grain to be given to a patient, how many cc should the nurse administer? **4cc**

b. If the doctor prescribes an 800-cc intravenous bottle to be given over a 6-hour period, for how many drops per minute should the nurse set the IV?

$33\frac{1}{3}$

Teaching Tip Remind students that division by 0 is not possible. In addition, $\frac{0}{n} = 0$ regardless of the sign of n if $n \neq 0$.

Cooperative Learning

Trade-A-Problem Have students work in pairs. Each pair should develop ten problems dealing with division of fractions, with both students agreeing on the answers. Pairs of students should trade sets of problems, solve them, and then join to check answers.

For more information on the trade-a-problem strategy, see *Cooperative Learning in the Mathematics Classroom*, one of the titles in the Glencoe Mathematics Professional Series, pages 25–26.

In-Class Examples

For Example 4
Write each fraction in simplest form.

a. $\dfrac{\frac{3}{4}}{\frac{9}{4}}$ $\dfrac{4}{3}$

b. $-\dfrac{6}{\frac{3}{2}}$ -4

For Example 5
Simplify.

a. $\dfrac{-8x - 30}{2}$ $-4x - 15$

b. $\dfrac{-25y - 4}{5}$ $-5y - \dfrac{4}{5}$

3 PRACTICE/APPLY

Check for Understanding

Exercises 1–13 are designed to help you assess your students' understanding through reading, writing, speaking, and modeling. You should work through Exercises 1–4 with your students and then monitor their work on Exercises 5–13.

Error Analysis

Since division is not commutative, determining the major division bar in a complex fraction is crucial.

Consider $\dfrac{\frac{3}{4}}{8}$. Is this $\dfrac{\frac{3}{4}}{8}$ or $\dfrac{3}{\frac{4}{8}}$?

Their values are $\dfrac{3}{32}$ and 6, respectively. The author of the fraction should indicate the major fraction bar by some emphasis (bolder, longer, or wider). If no major fraction bar is indicated, the lowest one is the major one by default.

Example: $\dfrac{\frac{64}{8}}{4}$ defaults to $\dfrac{\frac{64}{8}}{4}$, which

is $64 \div 8 \div 4$ or 2.

---

b. $\dfrac{1000 \text{ cc}}{8 \text{ h}} = \left(\dfrac{\overset{125}{\cancel{1000 \text{ cc}}}}{\underset{1}{\cancel{8 \text{ h}}}}\right)\left(\dfrac{\overset{1}{\cancel{15 \text{ drops}}}}{1 \text{ cc}}\right)\left(\dfrac{1 \text{ h}}{\underset{4}{\cancel{60 \text{ min}}}}\right)$ *There are 15 drops in 1 cc and*

                                                   60 minutes in 1 hour.

$\phantom{\dfrac{1000 \text{ cc}}{8 \text{ h}}} = \dfrac{125 \text{ drops}}{4 \text{ min}}$

$\phantom{\dfrac{1000 \text{ cc}}{8 \text{ h}}} = 31\dfrac{1}{4}$ drops/min

The nurse should set the IV for $31\dfrac{1}{4}$ drops per minute.

If a fraction has one or more fractions in the numerator or denominator, it is called a **complex fraction**. To simplify a complex fraction, rewrite it as a division sentence.

Example (4) **Write each fraction in simplest form.**

a. $\dfrac{\frac{2}{3}}{8}$

Rewrite the fraction as $\dfrac{2}{3} \div 8$, since fractions indicate division.

$\dfrac{2}{3} \div 8 = \dfrac{2}{3} \cdot \dfrac{1}{8}$ *Multiply by $\frac{1}{8}$, the reciprocal of 8.*

$\phantom{\dfrac{2}{3} \div 8} = \dfrac{2}{24}$ or $\dfrac{1}{12}$

b. $\dfrac{-5}{\frac{3}{7}}$

Rewrite the fraction as $-5 \div \dfrac{3}{7}$.

$-5 \div \dfrac{3}{7} = -5 \times \dfrac{7}{3}$ *Multiply by $\frac{7}{3}$, the reciprocal of $\frac{3}{7}$.*

$\phantom{-5 \div \dfrac{3}{7}} = -\dfrac{35}{3}$ or $-11\dfrac{2}{3}$ *The signs are different, so the product is negative.*

You can use the distributive property to simplify fractional expressions.

Example (5) **Simplify** $\dfrac{-3a + 16}{4}$.

Method 1

$\dfrac{-3a + 16}{4} = (-3a + 16) \div 4$

$\phantom{\dfrac{-3a + 16}{4}} = (-3a + 16)\left(\dfrac{1}{4}\right)$ *To divide by 4, multiply by $\frac{1}{4}$.*

$\phantom{\dfrac{-3a + 16}{4}} = -3a\left(\dfrac{1}{4}\right) + 16\left(\dfrac{1}{4}\right)$ *Distributive property*

$\phantom{\dfrac{-3a + 16}{4}} = -\dfrac{3}{4}a + 4$

Method 2

$\dfrac{-3a + 16}{4} = -\dfrac{3a}{4} + \dfrac{16}{4}$

$\phantom{\dfrac{-3a + 16}{4}} = -\dfrac{3}{4}a + 4$

Communicating Mathematics

Study the lesson. Then complete the following. 1, 4. See margin.

1. **Compare** multiplying rational numbers to dividing rational numbers. How are the two operations similar?

2. **Complete** the sentence: Dividing by any number, except zero, is the same as ___?___. **multiplying by its reciprocal**

3. **Find** a value for x if $\frac{1}{x} > x$. **Sample answer: $x = \frac{1}{2}$**

4. **You Decide** Simone says that $-\frac{4}{5}$ is equal to $\frac{-4}{-5}$. Miguel says that $-\frac{4}{5}$ is equal to $\frac{-4}{5}$ or $\frac{4}{-5}$. Which one is correct, and why?

Guided Practice

Simplify.

5. $\frac{32}{-8}$ -4

6. $\frac{-77}{11}$ -7

7. $-\frac{3}{4} \div 8$ $-\frac{3}{32}$

8. $\frac{2}{3} \div 9$ $\frac{2}{27}$

9. $\frac{-\frac{5}{6}}{8}$ $-\frac{5}{48}$

10. $\frac{54s}{6}$ $9s$

11. $\frac{-300x}{50}$ $-6x$

12. $\frac{6b + 12}{6}$ $b + 2$

13. **Money** On October 10, 1994, *USA Today* reported that the United States government spends about $168 million per hour. Stanley Newberg, who came to the United States from Austria in 1906, died in 1994 at age 81 and left the government $5.6 million. How long did it take the government to spend Mr. Newberg's money? **about 2 minutes**

Practice

Simplify.

14. $\frac{-36}{4}$ -9

15. $\frac{-96}{-16}$ 6

16. $-9 \div \left(-\frac{10}{17}\right)$ $\frac{153}{10}$

17. $-\frac{2}{3} \div 12$ $-\frac{1}{18}$

18. $-64 \div (-8)$ 8

19. $-\frac{3}{4} \div 12$ $-\frac{1}{16}$

20. $-18 \div 9$ -2

21. $78 \div (-13)$ -6

22. $-108 \div (-9)$ 12

23. $-\frac{2}{3} \div 8$ $-\frac{1}{12}$

24. $\frac{-1}{3} \div (-4)$ $\frac{1}{12}$

25. $-9 \div \left(-\frac{10}{27}\right)$ $\frac{243}{10}$

26. $\frac{\frac{5}{6}}{-10}$ $-\frac{1}{12}$

27. $-\frac{7}{\frac{3}{5}}$ $-\frac{35}{3}$

28. $\frac{-5}{\frac{2}{7}}$ $-\frac{35}{2}$

29. $\frac{-650m}{10}$ $-65m$

30. $\frac{81c}{-9}$ $-9c$

31. $\frac{8r + 24}{8}$ $r + 3$

33. $-20a - 25b$

32. $\frac{6a + 24}{6}$ $a + 4$

33. $\frac{40a + 50b}{-2}$

34. $\frac{-5x + (-10y)}{-5}$ $x + 2y$

37. $-a - 4b$

35. $\frac{42c - 18d}{-3}$ $-14c + 6d$

36. $\frac{-8f + (-16g)}{8}$ $-f - 2g$

37. $\frac{-4a + (-16b)}{4}$

Evaluate if $a = 5$, $b = -6$, and $c = -1.5$.

38. $\frac{b}{c}$ 4

39. $b \div a$ -1.2

40. $(a + b) \div c$ $0.\overline{6}$

41. $-0.8\overline{3}$

41. $(a + b + c) \div 3$

42. $\frac{c}{a}$ -0.3

43. $\frac{ab}{ac}$ 4

Reteaching

Using Alternative Methods Here is an easy way to use signed numbers to find an average A of a set of n numbers. Guess an average G. Subtract G from each number to find its deviation. Find the total of the deviations and then divide by n to get the average deviation D.

$A = G + D$

Scores	Deviations for	
	$G = 80$	$G = 75$
92	12	17
60	-20	-15
84	4	9
73	-7	-2
81	1	6
TOTAL	-10	15

$D = -2$ $D = 3$
$A = 80 + (-2)$ $A = 75 + 3$

Additional Answers

1. Both multiplying and dividing rational numbers involve multiplication. In division, you multiply by the reciprocal of the second number.

4. Miguel is correct.

$-\frac{4}{5} = (-1)\frac{4}{5} = \frac{(-1)4}{5} =$

$\left(\frac{1}{-1}\right)\frac{4}{5} = \frac{4}{-5}$

$\frac{-4}{-5} = \frac{(-1)4}{(-1)5} = \left(\frac{-1}{-1}\right)\left(\frac{4}{5}\right) =$

$(1)\frac{4}{5} = \frac{4}{5}$

Assignment Guide

Core: 15–43 odd, 44, 45, 47, 48–55
Enriched: 14–42 even, 44–55

For **Extra Practice**, see p. 761.

The red A, B, and C flags, printed only in the Teacher's Wraparound Edition, indicate the level of difficulty of the exercises.

Study Guide Masters, p. 16

2-7 Study Guide

NAME_____ DATE_____

Student Edition Pages 112–117

Dividing Rational Numbers
Use the following rules to divide rational numbers.

	Rule or Property	Example
Dividing Rational Numbers	The quotient of two numbers is positive if the numbers have the same sign. The quotient of two numbers is negative if the numbers have different signs.	$-60 \div (-10) = 6$ $-48 \div 4 = -12$
Multiplicative Inverse Property	For every nonzero number a, there is exactly one number $\frac{1}{a}$, such that $(a)\frac{1}{a} = \frac{1}{a}(a) = 1$.	$\frac{1}{3} \cdot \frac{3}{1} = \frac{1}{3} \cdot \frac{3}{1}$ $= \frac{4}{9}$
Division Rule	For all numbers a and b, with $b \neq 0$, $a \div b = \frac{a}{b} = a\left(\frac{1}{b}\right) = \frac{1}{b}(a)$.	$6 \div 2 = \frac{6}{2}$ $= (6)\frac{1}{2}$ $= \frac{1}{2}(6) = 3$

Since the fraction bar indicates division, you can use the division rules and the distributive property to simplify rational expressions.

Example: Simplify $\frac{-20a + 15}{5}$.

$\frac{-20a + 15}{5} = (-20a + 15)\left(\frac{1}{5}\right)$
$= (-20a)\left(\frac{1}{5}\right) + (15)\left(\frac{1}{5}\right)$
$= -5a + 3$

Simplify.

1. $\frac{-44a}{4}$ $-11a$

2. $\frac{16x}{2}$ $8x$

3. $\frac{80}{5}$ 16

4. $\frac{81}{-27}$ -3

5. $\frac{-144a}{6}$ $-24a$

6. $\frac{-30}{-10} \div \frac{30}{10}$ 1

7. $\frac{57y}{3}$ $19y$

8. $-\frac{1}{2} \div 8$ $-\frac{1}{16}$

9. $\frac{18a - 6b}{-3}$ $-6a + 2b$

10. $\frac{12x}{3} \div \frac{1}{12x} + xyz$ $48x^2 + xyz$

11. $\frac{36a - 12}{12}$ $3a - 1$

12. $\frac{-\frac{5}{8}}{5}$ $-\frac{1}{8}$

44. The 1/x key on a scientific calculator is called the *reciprocal key*. When this key is pressed, the calculator replaces the number on the display with its reciprocal. **a–c. See margin.**

  a. Enter 0 and then press the reciprocal key. What happened? Explain.

  b. Enter a number and then press the reciprocal key twice. What happened? Predict what will happen if you press the key *n* times.

  c. Enter 6.435 1/x × 6.435 = . What is the result? Why?

45. Drafting Sofia Fernandez is making a dresser for her little sister's nursery. The dresser will be 30 inches high and will have a 4-inch-thick base and a $1\frac{1}{2}$-inch-thick top. Four equal-sized drawers are to fit in the remaining space, with $\frac{3}{4}$ inch between each drawer.

  a. Sketch the dresser, labeling each part. **See margin.**

  b. What is the height of each drawer?
  $5\frac{9}{16}$ inches

46. Economics Kim Lee's shoe store has a buy one, get one half-price sale. You must pay full price for the more expensive pair of shoes in order to receive half off the second pair. If you buy a single pair of shoes, you get $\frac{1}{5}$ off. Together, Sharon and LaShondra find five pair of shoes that cost $45.99, $23.88, $36.99, $19.99, and $14.99. How should they pay for the shoes in order to get the best buy? **See margin.**

47. Carpentry The students in the industrial technology class at Monroe High School are making hat racks to donate to local nursing homes. They need to drill six equally-spaced holes for the pegs in a plywood strip that is $24\frac{1}{4}$ inches long. There must also be 2 inches from each end of the strip to the center of the first hole on each end. Find the distance from center to center of two consecutive holes.

$4\frac{1}{20}$ inches

Mixed Review

48. Find $\left(-\frac{1}{5}\right)\left(\frac{3}{2}\right)(-2)$. (Lesson 2–6) $\frac{3}{5}$

49. Evaluate $\frac{9}{4} + \frac{x}{6}$ if $x = -7$. (Lesson 2–5) $\frac{13}{12}$

50. Sample answer: $\frac{19}{24}$

50. Name a fraction between $\frac{2}{3}$ and $\frac{7}{8}$ whose denominator is 24. (Lesson 2–4)

51a. $\begin{bmatrix} -2 & 4 \\ 3 & 12 \end{bmatrix}$

51b. $\begin{bmatrix} 4 & 4 \\ 7 & 2 \end{bmatrix}$

51. Let $A = \begin{bmatrix} 1 & 4 \\ 5 & 7 \end{bmatrix}$ and $B = \begin{bmatrix} -3 & 0 \\ -2 & 5 \end{bmatrix}$. (Lesson 2–3)

    a. Find $A + B$.   **b.** Find $A - B$.

52. Statistics The table below shows the mean scores for the mathematics portion of the Scholastic Assessment Test (SAT) by state for 1994. (Lesson 2–2)

SAT Mean Mathematics Scores, 1994									
AL	529	HI	480	MA	475	NM	528	SD	548
AK	477	ID	508	MI	537	NY	472	TN	535
AZ	496	IL	546	MN	562	NC	455	TX	474
AR	518	IN	466	MS	528	ND	559	UT	558
CA	482	IA	574	MO	532	OH	510	VT	472
CO	513	KS	550	MT	523	OK	537	VA	469
CT	472	KY	523	NE	543	OR	491	WA	488
DE	464	LA	530	NV	484	PA	462	WV	482
FL	466	ME	463	NH	486	RI	462	WI	557
GA	446	MD	479	NJ	475	SC	443	WY	521

Source: College Entrance Examination Board

    a. What is the range of scores for the fifty states? **131**
    b. Make a line plot of the data. **See Solutions Manual.**
    c. Do the scores cluster around a certain number? If so, which ones? **yes, 472**

53. Simplify $8b + 12(b + 2)$. (Lesson 1–7) **$20b + 24$**

54. Name the multiplicative inverse of $\frac{4}{5}$. (Lesson 1–6) $\frac{5}{4}$

55. Statistics The list below shows the prices of several 1995 models of sport utility vehicles, rounded to the nearest hundred dollars. (Lesson 1–4)

    $22,000 $19,400 $29,000 $13,000 $22,200 $17,300
    $25,400 $25,100 $33,000 $20,000 $30,700 $15,400
    $27,900 $34,600 $30,400 $24,500 $52,500 $17,200

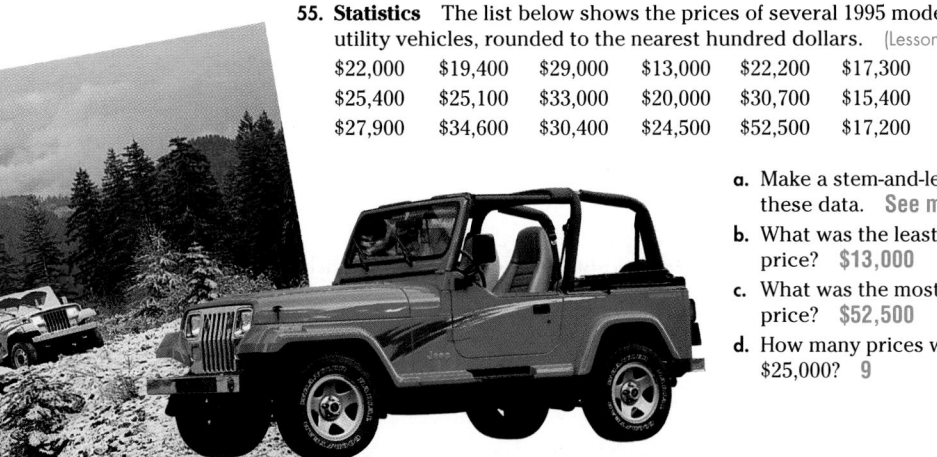

    a. Make a stem-and-leaf plot of these data. **See margin.**
    b. What was the least expensive price? **$13,000**
    c. What was the most expensive price? **$52,500**
    d. How many prices were over $25,000? **9**

Extension

Reasoning Matt had an average of 72 on three 100-point tests. What score would he need on his next test to raise his average to 75? **84**

NCTM Standards: 1–5

Objective
Use base-ten tiles to model square roots.

Recommended Time
Demonstration and discussion: 15 minutes; Exercises: 30 minutes

Instructional Resources
For each student or group of students
Modeling Mathematics Masters
• p. 16 (base-ten models)
• p. 22 (worksheet)
For teacher demonstration
Algebra and Geometry Overhead Manipulative Resources

1 FOCUS

Motivating the Lesson
Ask students to suggest a number whose square root is greater than 2 but less than 3. Point out that $2^2 = 4$ and $3^2 = 9$, so 5, 6, 7, and 8 are all correct answers to the question. Explain that students can use base-ten tiles to find such estimates of squares roots.

2 TEACH

Teaching Tip One strategy is to arrange the first four tiles into a two-by-two square. Fill in two adjacent sides to form a three-by-three square, and so on.

3 PRACTICE/APPLY

Assignment Guide
Core: 1–6
Enriched: 1–6

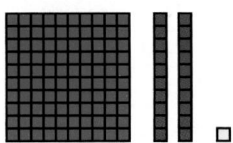

A Preview of Lesson 2–8

2–8A Estimating Square Roots

Materials: 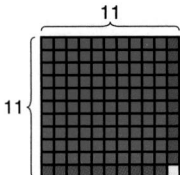 base-ten tiles

You can use base-ten tiles to model square roots. A **square root** is one of two identical factors of a number. For example, a square root of 144 is 12 since $12^2 = 144$.

Activity 1   Use base-ten tiles to find the square root of 121.

Step 1 Model 121 with base-ten tiles.

Step 2 Arrange the tiles into a square. The square root of 121 is 11 because $11^2 = 121$.

Activity 2   Use base-ten tiles to estimate the square root of 151.

Step 1 Model 151 with base-ten tiles.

Step 2 Arrange the tiles into a square. The largest square possible has 144 tiles, with 7 tiles left over.

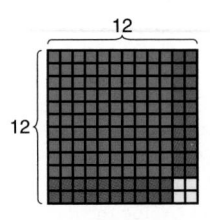

Trade a 10-tile for 10 1-tiles when necessary.

Step 3 Add tiles until you have the next larger square. You need to add 18 tiles. Since 151 is between 144 and 169, the square root of 151 is between 12 and 13.

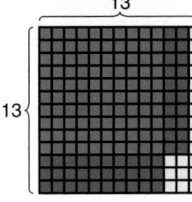

....

Model   Use base-ten tiles to estimate the square root of each number.

1. 20 4–5   2. 450 21–22   3. 180 13–14   4. 200 14–15   5. 2 1–2

Write   6. If you list all of the factors of a number in numerical order, the square root of the number either is the middle number or lies between the two middle numbers. How can you use tiles to show this? See margin.

4 ASSESS

Observing students working in cooperative groups is an excellent method of assessment. You may wish to ask a student from each group to explain how to model and solve a problem. Also watch for and acknowledge students who are helping others understand the concept being taught.

Additional Answer

6. Begin with a $1 \times n$ rectangle made of blocks, where *n* is the number being factored. As you form new rectangles with the same blocks, the first factor increases in value while the second factor decreases. When you find the square root, both dimensions are the same.

Square Roots and Real Numbers

Objectives

After studying this lesson, you should be able to:

- find square roots,
- classify numbers, and
- graph solutions of inequalities on number lines.

Biology

Swedish botanist Carolus Linnaeus (1707–1778) developed the system we use today to classify every kind of living thing according to common characteristics. For example, an African elephant is from the Mammalia Class and also from the Animalia Kingdom.

In mathematics, we classify numbers that have common characteristics. So far in this text, we have classified numbers as natural numbers, whole numbers, integers, and rational numbers.

Finding a square root of 81 is the same as finding a number whose square is 81.

The square roots of perfect squares are classified as rational numbers. A **square root** is one of two equal factors of a number. For example, one square root of 81 is 9 since $9 \cdot 9$ or 9^2 is 81. A rational number like 81, whose square root is a rational number, is called a **perfect square**.

Kingdom: Animalia
Phylum: Chordata
Class: Mammalia
Order: Proboscidea
Family: Elephantidae
Genus: Loxodonta
Species: Loxodonta africana

African elephant

It is also true that $-9 \cdot (-9) = 81$. Therefore, -9 is another square root of 81.

$$9^2 = 9 \cdot 9$$
$$= 81$$

9^2 is read "nine squared" and means that 9 is used as a factor two times.

$$(-9)^2 = (-9)(-9)$$
$$= 81$$

-9 is used as a factor two times.

Definition of Square Root	If $x^2 = y$, then x is a square root of y.

The symbol $\sqrt{}$, called a **radical sign**, is used to indicate a nonnegative or **principal square root** of the expression under the radical sign.

$$\sqrt{81} = 9 \qquad \sqrt{81} \text{ indicates the } principal \text{ square root of 81.}$$
$$-\sqrt{81} = -9 \qquad \sqrt{-81} \text{ indicates the } negative \text{ square root of 81.}$$
$$\pm\sqrt{81} = \pm9 \qquad \pm\sqrt{81} \text{ indicates } both \text{ square roots of 81.}$$

$\pm\sqrt{81}$ is read "plus or minus the square root of 81."

Lesson 2-8 Square Roots and Real Numbers **119**

Teaching Tip Encourage students to check square roots by squaring.

Example **1** Find each square root.

a. $\sqrt{25}$

Some square roots can be found mentally.

The symbol $\sqrt{25}$ represents the principal square root of 25.
Since $5^2 = 25$, you know that $\sqrt{25} = 5$.

b. $-\sqrt{144}$

The symbol $-\sqrt{144}$ represents the negative square root of 144.
Since $12^2 = 144$, you know that $-\sqrt{144} = -12$.

c. $\pm\sqrt{0.16}$

The symbol $\pm\sqrt{0.16}$ represents both square roots of 0.16.
Since $(0.4)^2 = 0.16$, you know that $\pm\sqrt{0.16} = \pm0.4$.

Most scientific calculators have a *square root* key labeled $\boxed{\sqrt{}}$ or $\boxed{\sqrt{x}}$. When you press this key, the number in the display is replaced by its principal square root.

Example **2** Use a scientific calculator to evaluate each expression if $x = 2401$, $a = 147$, and $b = 78$.

a. $\sqrt{x}$

$\sqrt{x} = \sqrt{2401}$ *Replace x with 2401.*

Enter: 2401 $\boxed{\text{2nd}}$ $\boxed{\sqrt{x}}$ *49*

Therefore, $\sqrt{2401} = 49$.

b. $\pm\sqrt{a + b}$

$\pm\sqrt{a + b} = \pm\sqrt{147 + 78}$ *Replace a with 147 and b with 78.*

Enter: $\boxed{(}$ 147 $\boxed{+}$ 78 $\boxed{)}$ $\boxed{\text{2nd}}$ $\boxed{\sqrt{x}}$ *15*

Therefore, $\pm\sqrt{a + b}$ is ± 15.

Numbers such as $\sqrt{5}$ and $\sqrt{13}$ are the square roots of numbers that are *not* perfect squares. Notice what happens when you find these square roots with your calculator.

Enter: 5 $\boxed{\text{2nd}}$ $\boxed{\sqrt{x}}$ *2.236067978...*

Enter: 13 $\boxed{\text{2nd}}$ $\boxed{\sqrt{x}}$ *3.605551275...*

These numbers continue indefinitely without any pattern of repeating digits. These numbers are not rational numbers since they are not repeating or terminating decimals. Numbers like $\sqrt{5}$ and $\sqrt{13}$ are called **irrational numbers.**

Definition of an Irrational Number	An irrational number is a number that *cannot* be expressed in the form $\frac{a}{b}$, where a and b are integers and $b \neq 0$.

The set of rational numbers and the set of irrational numbers together form the set of **real numbers**. The Venn diagram at the right shows the relationships among natural numbers, whole numbers, integers, rational numbers, irrational numbers, and real numbers.

Real Numbers

Rationals
Integers
Whole Numbers
Natural Numbers
Irrationals

Example **Name the set or sets of numbers to which each real number belongs.**

a. **0.8333333...**   This repeating decimal is a rational number since it is equivalent to $\frac{5}{6}$.   *This number can also be expressed as $0.8\overline{3}$.*

b. **$-\sqrt{16}$**   Since $-\sqrt{16} = -4$, this number is an integer and a rational number.

c. **$\frac{14}{2}$**   Since $\frac{14}{2} = 7$, this number is an integer, a natural number, a whole number, and a rational number.

d. **$\sqrt{120}$**   Since $\sqrt{120} = 10.95445115...$, which is not a repeating or terminating decimal, this number is irrational.

The solutions to many real-world problems are irrational numbers.

Example **The area of a square is 325 square inches. Find its perimeter to the nearest hundredth.**

INTEGRATION
Geometry

First find the length of each side. Since the area of a square is the length of the side squared, find the square root of 325.

325 square inches

Enter: 325 [2nd] [√x] *18.02775638*

The length of each side is about 18.02775638 inches. Use the formula $P = 4s$ to find the perimeter.

$P = 4s$

$ = 4 \cdot 18.02775638$ *Replace s with 18.02775638.*

Enter: 18.02775638 [×] 4 [=] *72.111026551*

The perimeter is about 72.11 inches.

You have graphed rational numbers on number lines. Yet, if you graphed all of the rational numbers, the number line would still not be complete. The irrational numbers complete the number line. The graph of all real numbers is the entire number line. This is illustrated by the **completeness property**.

Completeness Property for Points on the Number Line	Each real number corresponds to exactly one point on the number line. Each point on the number line corresponds to exactly one real number.

Classroom Vignette

"To show the relationship between the various sets of numbers (whole, natural, rational, integers, irrational, and real), I made a set of nested boxes. In each box, I included small slips of paper with representative numbers of that set."

Linda Crawford
Brown County Junior High School
Nashville, Indiana

Linda L. Crawford

In-Class Examples

For Example 3
Name the set or sets of numbers to which each number belongs. Use *N* for natural numbers, *W* for whole numbers, *Z* for integers, *Q* for rational numbers, and *I* for irrational numbers.

a. -9 *Z, Q*
b. 0.101101110...   *I*
c. $\sqrt{81}$ *N, W, Z, Q*
d. 0.33333...   *Q*

For Example 4
The area of a square is 827 square inches. Find its perimeter.
approximately 115 inches

3 PRACTICE/APPLY

Recall that equations like $x - 5 = 11$ are open sentences. Inequalities like $x < 6$ are also considered to be open sentences. To solve $x < 6$, determine what replacements for x make $x < 6$ true. All numbers less than 6 make the inequality true. This can be shown by the solution set {real numbers less than 6}. Not only does this include integers like 3, 0, and −4, but it also includes all rational numbers less than 6 such as $\frac{1}{2}$, $-5\frac{3}{8}$, and −3 and all irrational numbers less than 6 such as $\sqrt{5}$, $\sqrt{3}$, and π.

Example **5** **Graph each solution set.**

a. $y \geq -7$

The heavy arrow indicates that all numbers to the right of −7 are included. The *dot* indicates that the point corresponding to −7 is included in the graph of the solution set.

b. $p \neq \frac{3}{4}$

The heavy arrows indicate that all numbers to the left and to the right of $\frac{3}{4}$ are included in the graph of the solution set. The *circle* indicates that the point corresponding to $\frac{3}{4}$ is not included in the graph.

CHECK FOR UNDERSTANDING

Communicating Mathematics

MATH JOURNAL

Study the lesson. Then complete the following. 1–2. See margin.

1. You have studied the following sets of numbers: integers, irrational, natural, rational, real, and whole numbers. Draw a number line and label at least one number from each set of numbers. Indicate which number is from each set.

2. Study the comic at the left. Explain why it is humorous.

3. **Determine** whether 36 is a perfect square. If it is, to what set of numbers does $\sqrt{36}$ belong? Yes, because its square root is a rational number, 6; rationals, integers, whole numbers, natural numbers.

4. **Explain** why 3 and −3 are both square roots of 9. 3(3) = 9 and (−3)(−3) = 9. Therefore, 3 and −3 are both square roots of 9.

5. **Write** an inequality for the graph below. $x \leq 0$

6. **Write** a paragraph explaining the difference between rational and irrational numbers to a classmate. See margin.

Guided
Practice

10. ± 0.28

Find each square root. Use a calculator if necessary. Round to the nearest hundredth if the result is not a whole number.

7. $\sqrt{64}$ 8   8. $-\sqrt{36}$ -6 9. $\sqrt{122}$ 11.05   10. $\pm\sqrt{0.08}$

Evaluate each expression. Use a calculator if necessary. Round to the nearest hundredth if the result is not a whole number.

11. $\sqrt{x}$, if $x = 256$ 16   12. $\sqrt{y}$, if $y = 151$ 12.29

Name the set or sets of numbers to which each real number belongs. Use N for natural numbers, W for whole numbers, Z for integers, Q for rational numbers, and I for irrational numbers.

13. $-\frac{3}{4}$ Q   14. $\frac{8}{4}$ Q, Z, W, N   15. $0.6666\ldots$ Q   16. $\sqrt{13}$ I

Graph the solution set of each inequality on a number line.

17–19. See margin.

17. $p < 7$ 18. $r \geq -3$ 19. $x \neq 2$

20. **Discrete Mathematics** In the geometric sequence 5, 15, $\underline{\ ?\ }$, 135, 405, the missing number is called the *geometric mean* of 15 and 135. It can be found by evaluating $\sqrt{ab}$, where a and b are the numbers on either side of the geometric mean. Find the missing number.   45

EXERCISES

Practice

Find each square root. Use a calculator if necessary. Round to the nearest hundredth if the result is not a whole number or simple fraction.

A

21. $\sqrt{169}$ 13   22. $\sqrt{0.0049}$ 0.07   23. $\sqrt{\frac{4}{9}}$ $\frac{2}{3}$

24. $-\sqrt{289}$ -17 25. $\sqrt{420}$ 20.49   26. $\sqrt{\frac{25}{64}}$ $\frac{5}{8}$

27. $\sqrt{225}$ 15   28. $\pm\sqrt{1158}$ ± 34.03 29. $-\sqrt{625}$ -25

30. $\sqrt{1.96}$ 1.4   31. $\sqrt{\frac{9}{25}}$ $\frac{3}{5}$ 32. $-\sqrt{5.80}$ -2.41

Evaluate each expression. Use a calculator if necessary. Round to the nearest hundredth if the result is not a whole number.

33. $\sqrt{x}$, if $x = 87$ 9.33   34. $\pm\sqrt{t}$, if $t = 529$ ± 23

35. $-\sqrt{m}$, if $m = 2209$ -47 36. $\sqrt{c + d}$, if $c = 23$ and $d = 56$ 8.89

B

37. $-\sqrt{np}$, if $n = 16$ and $p = 25$ -20 38. $\pm\sqrt{\frac{a}{b}}$, if $a = 64$ and $b = 4$ ± 4

Name the set or sets of numbers to which each real number belongs. Use N for natural numbers, W for whole numbers, Z for integers, Q for rational numbers, and I for irrational numbers.

39. $-\sqrt{49}$ Z, Q   40. 0 W, Z, Q   41. 0.4583 Q

42. $0.\overline{3}$ Q   43. $-\frac{1}{2}$ Q   44. $\sqrt{49}$ N, W, Z, Q

45. 0.6666 Q   46. $\frac{10}{5}$ N, W, Z, Q   47. $\sqrt{37}$ I

48. 3.14 Q   49. $\frac{3}{5}$ Q   50. 5 N, W, Z, Q

Lesson 2–8   Square Roots and Real Numbers   **123**

For **Extra Practice**, see p. 761.

The red A, B, and C flags, printed only in the Teacher's Wraparound Edition, indicate the level of difficulty of the exercises.

Additional Answers

6. Sample answer: A rational number is a number that can be expressed as a common fraction. An irrational number is a number that cannot be expressed as a common fraction.

17.

18.

19.

Study Guide Masters, p. 17

Chapter 2   **123**

124 Chapter 2

51–59. See margin.

Graph the solution set of each inequality on a number line.

51. $y > -2$
52. $x < 1$
53. $p \geq -4$
54. $n \neq 6$
55. $c > -12$
56. $r \leq 4.5$
57. $b \geq -5.2$
58. $y \neq \frac{3}{4}$
59. $s \leq 5\frac{1}{2}$

60. Geometry The volume of a rectangular solid is 100 cm³. Its height is the product of its length and width. If the base of the solid is a square, find the dimensions of the solid. Use a calculator as needed. Round decimal answers to the nearest hundredth.
3.16 cm × 3.16 cm × 10 cm; 10 cm × 10 cm × 1 cm

Critical Thinking

61–62. See margin.

61. Determine whether $\sqrt{733}$ is an irrational number. If it is, name two consecutive integers between which its graph lies on the number line.

62. Find all numbers of the form $\sqrt{n}$ such that n is a natural number and the graph of $\sqrt{n}$ lies between each pair of numbers on the number line.
a. 3 and 4
b. 5.25 and 5.5

Applications and Problem Solving

63. Military The formula to determine the distance d in miles that an object can be seen on a clear day on the surface of the ocean is $d = 1.4\sqrt{h}$, where h is the height in feet the viewer's eyes are above the surface of the water. About how many miles can the pilot of a U.S. Coast Guard plane see if he is flying at 1275 feet? **about 50 miles**

64. Traffic Safety When investigating a traffic accident, police officers often need to estimate how fast a car was traveling by measuring the length of its skid marks. The formula $s = \sqrt{24d}$ can be used to estimate the speed on a dry, concrete road. In the formula, s is the speed in miles per hour, and d is the distance in feet the car skidded after its brakes were applied. What was the approximate speed of a car that left skid marks 40 feet long?
about 31 miles per hour

65. Math History One of the great mathematical challenges in early mathematics was to find an exact value for π. Today, 3.14 and $\frac{22}{7}$ are two generally accepted approximations for π. In the third century, Chinese mathematician Liu Hui used the number $3\frac{7}{50}$ as an approximation of π. The Italian mathematician Leonardo Fibonacci used $\frac{864}{275}$ as an approximation of π. Which mathematician's approximation was closer to the generally-accepted approximation of π, 3.141592654...? **Fibonacci**

66. Business The Spanish Club at Alexander Manor High School is selling chocolate bars to raise money for a trip to Mexico. The bars are sold only by the case, with 24 bars per case. Each bar costs $1.00. The club's goal is to raise at least $2450 in gross sales.
a. How many cases of chocolate bars must the club sell in order to reach or exceed their goal? **a–b. See margin.**
b. Graph the solution set.

Mixed Review

67. Evaluate $(-3a)(4)\left(\frac{2}{3}\right)\left(\frac{1}{6}a\right)$ if $a = 3$. (Lesson 2–6) **−12**

68. Find $-\frac{5}{12} + \frac{3}{8}$. (Lesson 2–5) $-\frac{1}{24}$

69. Which is the better buy, <u>a 1-pound package of lunch meat for $1.95</u> or a 12-ounce package for $1.80? (Lesson 2–4)

70. Find $-5 - (-6)$. (Lesson 2-3) **1**

71. Statistics State the scale you would use to make a line plot for the following data: 504, 509, 520, 500, 513, 517, 517, 508, 502, 518, 509, 520, 504, 509, 517, 511, 520. Then draw the line plot. (Lesson 2–2) **See margin.**

72. Graph $\{0, 2, 6\}$ on a number line. (Lesson 2–1) **See margin.**

73. Suppose a basketball was dropped from a tall building onto a sidewalk. Identify the graph below that best represents this situation. (Lesson 1–9) **b**

a.

b.

c.

74. Name the property illustrated by $a + (2b + 5c) = (a + 2b) + 5c$. (Lesson 1–8) **associative (+)**

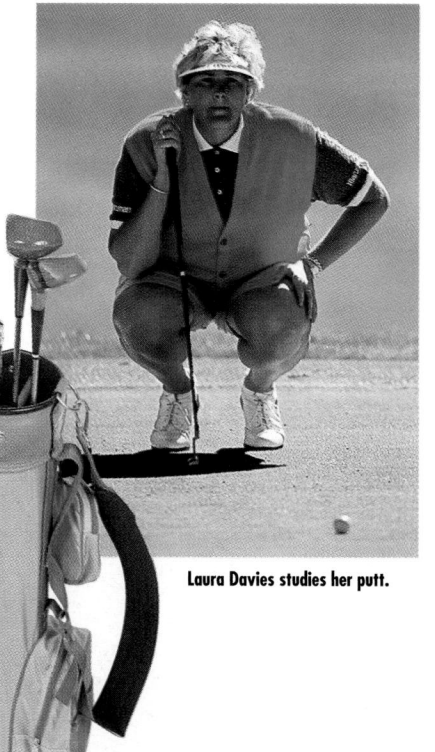

Laura Davies studies her putt.

75. Golf The stem-and-leaf plot below shows the earnings of the top 25 U.S. women professional golfers for 1994 according to the *Sports Almanac*. (Lesson 1–4)

Stem	Leaf
6	6 9
5	0
4	0 1 2 3 7
3	2 3 4 5 9
2	0 0 1 1 3 4 4 5 6 7 7 8 $4\,|\,3 = \$430,000$

a. to the nearest ten-thousand
a. How have these numbers been rounded?
b. How many of these golfers made between $300,000 and $490,000? **10**
c. Laura Davies had the highest earnings of all women golfers in 1994. How much money did she earn?
$690,000

76. Evaluate $12(19 - 15) - 3 \cdot 8$. (Lesson 1–3) **24**

77. Find the next two terms in the sequence 2, 8, 14, 20,...
(Lesson 1–2) **26, 32**

78. Write an algebraic expression for p to the sixth power.
(Lesson 1–1) p^6

Lesson 2–8 Square Roots and Real Numbers **125**

Extension

Connections The area of a triangle is 38 m². If the base is twice as long as the height, what is the base of the triangle? (Use the formula $A = \frac{1}{2}bh$.)

$2\sqrt{38}$ m or 12.3 m

---

4 ASSESS

Closing Activity
Speaking Have students give examples of natural numbers, whole numbers, integers, rational numbers, and irrational numbers.

Additional Answers
71. from 500 to 520 by 5s

72.

Enrichment Masters, p. 17

2-8	NAME_____ DATE _____
	Enrichment Student Edition Pages 119–125

Squares and Square Roots From a Graph

The graph of $y = x^2$ can be used to find the squares and square roots of numbers.

To find the square of 3, locate 3 on the x-axis. Then find its corresponding value on the y-axis.

The arrows show that $3^2 = 9$.

To find the square root of 4, first locate 4 on the y-axis. Then find its corresponding value on the x-axis. Following the arrows on the graph, you can see that $\sqrt{4} = 2$.

A small part of the graph at $y = x^2$ is shown below. A 1:10 ratio for unit length on the y-axis to unit length on the x-axis is used.

Example: Find $\sqrt{11}$.
The arrows show that $\sqrt{11} = 3.3$ to the nearest tenth.

Use the graph above to find each of the following to the nearest whole number.

1. 1.5^2 2   **2.** 2.7^2 7   **3.** 0.9^2 1
4. 3.6^2 13   **5.** 4.2^2 18   **6.** 3.9^2 15

Use the graph above to find each of the following to the nearest tenth.

7. $\sqrt{15}$ 3.9   **8.** $\sqrt{8}$ 2.8   **9.** $\sqrt{3}$ 1.7
10. $\sqrt{5}$ 2.2   **11.** $\sqrt{14}$ 3.7   **12.** $\sqrt{17}$ 4.1

Instructional Resources

- Study Guide Master 2-9
- Practice Master 2-9
- Enrichment Master 2-9
- Assessment and Evaluation Masters, p. 45
- Modeling Mathematics Masters, pp. 39–41
- Multicultural Activity Masters, p. 4
- Real-World Applications, 6

Transparency 2-9A contains the 5-Minute Check for this lesson; **Transparency 2-9B** contains a teaching aid for this lesson.

Recommended Pacing	
Standard Pacing	Days 12 & 13 of 15
Honors Pacing	Days 12 & 13 of 15
Block Scheduling*	Day 8 of 9
Alg. 1 in Two Years*	Days 23, 24, & 25 of 28

*For more information on pacing and possible lesson plans, refer to the *Block Scheduling Booklet* and *Algebra 1 in Two Years.*

1 FOCUS

5-Minute Check
(over Lesson 2-8)

Find each square root to the nearest hundredth.

1. $\sqrt{0.0256}$ 0.16
2. $-\sqrt{280}$ −16.73

Name the set of numbers to which each real number belongs.

3. $\sqrt{7}$ irrational
4. $\sqrt{0.04}$ rational
5. Graph the solution set of $x \geq -3\frac{1}{2}$ on a number line.

Objectives
After studying this lesson, you should be able to:
- explore problem situations, and
- translate verbal sentences and problems into equations or formulas and vice versa.

CONNECTION
Biology

Bugs, bugs, everywhere! That could be the slogan for James Fujita, an avid bug collector. Fujita has been fascinated with bugs since he was bitten by a cricket at the age of three. His collection includes hundreds of bugs—beetles, spiders, millipedes, scorpions, just to name a few—both living and dead.

Fujita has studied bugs as far away as Japan and Costa Rica, as well as in his own backyard in Oxnard, California. He has given bug presentations and workshops, donated some of his bugs to museums and zoos, and even appeared on a television talk show with some of his larger bugs. Fujita is a big believer in reading and exploring all you can about bug collecting—if you get the itch!

You can explore problems by asking and answering questions. In this text, we will use a four-step plan to solve problems. All of the steps are listed below.

Problem-Solving Plan
1. **Explore the problem.**
2. **Plan the solution.**
3. **Solve the problem.**
4. **Examine the solution.**

CAREER CHOICES

An **entomologist** is a scientist who studies the origin, development, anatomy, and distribution of insects and their effects on humans.

An inquisitive mind, and a knowledge of biology, chemistry, and mathematics are necessary are required for this career.

For more information, contact:

The Entomological Society of America
9301 Annapolis Road
Lanham, MD 20706

The first step in solving a problem is to read and explore it until you completely understand the relationships in the given information.

Step 1: Explore the Problem
To solve a verbal problem, first read the problem carefully and explore what the problem is about.

- Identify what information is given.
- Identify what you are asked to find.

Questions	Answers
a. How old was James when he became interested in bugs?	**a.** 3 years old
b. In what countries other than the U.S. has James studied bugs?	**b.** Japan and Costa Rica
c. If James is n years old now, for how many years has he been a collector of bugs?	**c.** $n - 3$
d. If James has b bugs and finds 14 new bugs, how many bugs are in his collection now?	**d.** $b + 14$

CAREER CHOICES

Entomology contains many subfields: anatomy, physiology, insect behavior, bionomics and ecology, systematics, economic entomology, and medical entomology. One of the main concerns of medical entomology is the role insects play in the transmission of diseases.

GLENCOE Technology

Interactive Mathematics Tools Software

This multimedia software provides an interactive lesson that allows students to use guess-and-check as a means of finding the formulas for converting degrees Fahrenheit to Celsius and degrees Celsius to Fahrenheit. A **Computer Journal** gives students an opportunity to write about what they have learned.
For Windows & Macintosh

Other strategies are:
- *look for a pattern*
- *solve a simpler problem*
- *act it out*
- *guess and check*
- *draw a diagram*
- *make a table or chart*
- *work backward*

Step 2: Plan the Solution
One strategy you can use to solve a problem is to write an equation. Choose a variable to represent one of the unspecified numbers in the problem. This is called **defining the variable.** Then use the variable to write expressions for the other unspecified numbers in the problem.

Step 3: Solve the Problem
Use the strategy you chose in Step 2 to solve the problem.

Step 4: Examine the Solution
Check your answer within the context of the original problem. Does your answer make sense? Does it fit the information in the problem?

Example ❶

APPLICATION
Computers

Gregory Arakelian of Herndon, Virginia, set a speed record for typing the most words per minute, with no errors, on a personal computer in the KeyTronic World Invitational Type-Off on September 24, 1991. Suppose his closest competitor typed 10 fewer words per minute than Arakelian. If his closest competitor typed 148 words per minute, how many words did Arakelian type per minute?

Explore • How many words can Arakelian type in relation to his closest competitor?   10 more words

• How many words per minute did his competitor type? 148 words

Plan Write an equation to represent the situation. Let *w* represent the words per minute that Arakelian typed.

$$\underbrace{w}_{\substack{\text{words}\\ \text{Arakelian typed}}} \quad -10 = \quad \underbrace{148}_{\substack{\text{words}\\ \text{competitor typed}}}$$

Solve $w - 10 = 148$ *Solve mentally by asking,*
       $w = 158$ *"What number minus 10 is 148?"*

Arakelian typed 158 words per minute.

Examine The problem asks how many words Arakelian typed per minute. His competitor typed 10 fewer words per minute. Since $158 - 10 = 148$, the answer makes sense.

Gregory Arakelian

In Lesson 2–4, you studied meanings for the symbols $<, >, \neq, \leq,$ *and* $\geq$.

When solving problems, many sentences can be written as equations. Use variables to represent the unspecified numbers or measures referred to in the sentence or problem. Then write the verbal expressions as algebraic expressions. Some verbal expressions that suggest the *equals sign* are listed below.

• is   • is equal to   • is as much as
• equals   • is the same as   • is identical to

For Example 2
Translate the following sentences into an equation.
One number is 12 greater than a second number. The sum of these two numbers is 86.
$n + (n + 12) = 86$

For Example 3
Translate each sentence into a formula.

a. The distance d is equal to the product of the rate r and the time t. $d = rt$
b. The volume of a cone V is equal to the area of its base a times its height h divided by 3. $V = \frac{ah}{3}$

For Example 4
Translate each equation into a verbal sentence.

a. $(x + 5)^2 = 100$
The square of the sum of a number and five equals 100.
b. $\frac{n^2 - 15}{3} > 32$ The difference of a number squared and fifteen divided by three is greater than 32.

For Example 5
Write a problem based on the given information.
w = weight in pounds of fully packed box
$w - 30$ = weight in pounds of partially packed box
$3w + (w - 30) = 270$
Sample problem: A partially packed box weighs 30 pounds less than a fully packed box. The total weight of three fully packed boxes and one partially packed box is 270 pounds. How much does a fully packed box weigh?

 MODELING MATHEMATICS Students might find an additional challenge by trying to cut the shape out of paper as one continuous cut-out. This will also help them to see physical connections between the different dimensions.

Example **Translate each sentence into an algebraic sentence.**

a. *Six times a number x is equal to 7 times the sum of z and y.*

Six times a number x is equal to 7 times the sum of z and y.
$6 \times x = 7 \times (z + y)$

The algebraic sentence is $6x = 7(z + y)$.

b. *A number is less than or equal to 5.*

A number is less than or equal to 5.
$n \leq 5$

The algebraic sentence is $n \leq 5$.

Another strategy you can use to solve a problem is to write a formula. A **formula** is an equation that states a rule for the relationship between certain quantities. Sometimes you can develop a formula by making a model.

MODELING MATHEMATICS

Surface Area

Materials: ▱ rectangular box   ✂ scissors

You need to find the surface area of a rectangular box.

- Cut up a box like the one shown at the right.
- Find the area of each face.

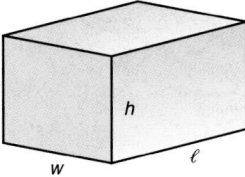

- Write the formula by adding the areas.

$S = wh + wh + \ell h + \ell h + \ell w + \ell w$
$\quad = 2wh + 2\ell h + 2\ell w$
$\quad = 2(wh + \ell h + \ell w)$

Your Turn   a–c. See students' work.

a. Find a box that is the shape of a cube. Take the box apart. Draw and label each side. Then write a formula for finding the surface area of the cube.

b. How are the box (rectangular prism) and cube the same? How are they different?

c. Find a formula for the surface area of the triangular prism below. Use S, h, w, and ℓ in your formula.

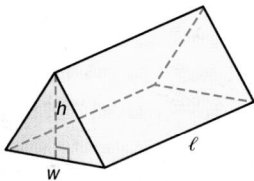

128   Chapter 2   *Exploring Rational Numbers*

Classroom Vignette

"I gave my students a soup can and asked them to find the surface area. It took a while, but finally one group discovered that if they took the label off, it was a rectangle, and the rest fell into place."

Carolyn Stanley
Tidewater Academy
Wakefield, Virginia

Example ③ Translate the sentence into a formula.

The area of a circle equals the product of π and the square of the radius r.

The area of a circle equals π times the square of the radius r.

$$\underbrace{A}_{} = \underbrace{\pi}_{} \times \underbrace{r^2}_{}$$

The formula is $A = \pi r^2$.

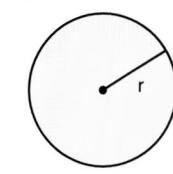

INTEGRATION
Geometry

You can also translate equations into verbal sentences or make up your own verbal problem if you are given an equation or two.

Example ④ Translate $x^2 + 6 = 39$ into a verbal sentence.

$$\underbrace{x^2}_{} \quad + \quad \underbrace{6}_{} = \underbrace{39}_{}$$

The sum of the square of a number and six equals 39.

Example ⑤ Write a problem based on the given information.

ℓ = **Lawana's height in inches**
$\ell + 5$ = **Tatewin's height in inches**
$2\ell + (\ell + 5) = 194$

Here's a sample problem.

Tatewin is 5 inches taller than Lawana. The sum of Tatewin's height and twice Lawana's is 194 inches. How tall is Lawana?

CHECK FOR UNDERSTANDING

Communicating Mathematics

Study the lesson. Then complete the following.   1–3. See margin.

1. **Write** three questions you would ask yourself in order to understand the problem below.
   Consuela's target heart rate is 140 beats per minute. Her normal heart rate is 60 beats per minute. After running for 10 minutes, her heart rate rises by 65 beats. It drops by 32 beats after a 5-minute rest period and then rises by 50 beats after another 15-minute run. Did she reach her target heart rate?

2. **Explain** whether an equation is a type of formula or a formula is a type of equation.

3. **Write** a problem that uses the formula for the area of a rectangle, $A = \ell w$.

MODELING MATHEMATICS

4. **Find** any gift box. Measure each side. Use the formula $S = 2(wh + \ell h + \ell w)$ to find its surface area, including the lid.   **See students' work.**

Lesson 2–9 Problem Solving: Write Equations and Formulas   **129**

---

3 PRACTICE/APPLY

Check for Understanding
Exercises 1–12 are designed to help you assess your students' understanding through reading, writing, speaking, and modeling. You should work through Exercises 1–4 with your students and then monitor their work on Exercises 5–12.

Additional Answers

1. Answers will vary. Sample answers:
   How much does her normal heart rate have to increase to reach her target heart rate? During the entire workout, how much does her heart rate decrease? During the entire workout, how much does her heart rate increase?
2. A formula is a type of equation that states a rule for the relationship between certain quantites.
3. Answers will vary. Sample answer: If José has a garden with the dimensions of 20 feet by 30 feet, what is the area?

Reteaching

Using Variables Define the variable and then write an equation for each problem. Do not solve.

1. Four times Zina's age 3 years ago is 56. How old is Zina now? *x = Zina's age now;* $4(x - 3) = 56$

2. Carrie is 3 years older than Subir. The sum of their ages in 6 years will be 41. How old is Subir? *x = Subir's age now;* $(x + 6) + (x + 3 + 6) = 41$

3. The sum of Sherry's age and her dad's age is 60. In 12 years, Sherry will be half as old as her dad. Find Sherry's age. *x = Sherry's age now;* $x + 12 = \frac{1}{2}(60 - x + 12)$

4. Emilio is 5 years younger than Teri. In 8 years the sum of their ages will be 55. Find Teri's age. *x = Teri's age;* $(x + 8) + (x - 5 + 8) = 55$

Additional Answers

9. Let s = average speed;
$s = \frac{189}{3} = 63$

10. Let q = number of quarters;
$q + (q + 4) + (q + 4 - 7) = 28$

11. The product of a times the sum of y and 1 is b.

12. Answers will vary. Sample answer: Your mother works 8 less than 5 times the number of hours that you work during the summer. How many hours did your mother work if you worked 20 hours this week?

Study Guide Masters, p. 18

Guided Practice

5. Answer the related questions for the verbal problem below.
   Each question in the computation section of a mathematics test is worth 4 points. Each question in the problem-solving section is worth 6 points. Madison needs to answer 15 questions correctly to have a total of 86 points and get a B on the test. How many problem-solving questions must she answer correctly?
   a. How many points are the computation questions worth?   **4 points**
   b. How many questions does Madison need to answer correctly?
   c. What score does Madison need to make on the test?   **86**
   d. How many points are the problem-solving questions worth?   **6 points**
   e. If n represents the number of correct computation questions, how many problem-solving questions does she need to answer correctly?   **15 − n**
   f. If n represents the number of correct answers in the computation section of the test, how does Madison compute the number of points in this section of the test?   **4n**

5b. 15 questions

Translate each sentence into an equation, inequality, or formula.

6. The sum of twice x and three times y is equal to thirteen. $2x + 3y = 13$

7. The sum of a number and 5 is at least 48. $n + 5 \geq 48$

8. The perimeter P of a parallelogram is twice the sum of the lengths of two adjacent sides, a and b. $P = 2(a + b)$

Define a variable, and then write an equation for each problem. Do *not* try to solve.   9–10. See margin.

9. Olivia drove 189 miles from her home in Fort Worth, Texas, to Austin in three hours. What was her average speed?

10. Shane has 4 more dimes than quarters and 7 fewer nickels than dimes. He has 28 coins in all. How many quarters does Shane have?

11. Translate $a(y + 1) = b$ into a verbal sentence.   **See margin.**

12. Write a problem based on the given information.
    Let h = the number of hours you can work during the summer.
    $5h - 8$ **See margin.**

EXERCISES

Practice A

13. Answer the related questions for the verbal problem below.
    The Johnson Sisters' Construction Company is planning to build a playground for a housing development. The blueprints allow an area of 20,000 square feet for the playground. The width of the rectangular playground is 80 feet less than its length, and its perimeter is 640 feet. Is there enough space to build the playground as designed?
    a. What is the perimeter of the playground?   **640 feet**
    b. If y represents the length of the playground, what is its width?   **y − 80**
    c. What is the formula for the perimeter of a rectangular shape?
    d. What is the length of the playground?   **200 ft**
    e. What is the area of the playground?   **24,000 ft²**
    f. What does the problem ask?

13c. $\ell + \ell + w + w$ or $2\ell + 2w$

13f. Is there enough space to build the playground as designed?

Classroom Vignette

"I liken 'translating' into Algebra to translating from English to French. I encourage students to literally translate statements such as '5 less than a number is 7' to an algebraic equation. In this case, '5 less than' means 'subtract 5 from something,' x represents 'a number,' and 'is' means 'equals.' Therefore, the statement is $x - 5 = 7$."

Barbara Lasko
Roosevelt-Wilson Middle School
Clarksburg, West Virginia

14. $V = \frac{1}{3}bh$

15. $a^2 + b^3 = 25$

16. $\frac{x}{y} \geq 18 + 5(x + y)$

17. $x < (a - 4)^2$

18. $C = 2\pi r$

19. $\frac{7}{8}(a + b + c^2) = 48$

Translate each sentence into an equation, inequality, or formula.

14. The volume V of a pyramid is equal to one third the product of the area of the base b and its height h.

15. The sum of the square of a and the cube of b is equal to twenty-five.

16. The quotient of x and y is at least 18 more than five times the sum of x and y.

17. The quantity x is less than the square of the difference of a and 4.

18. The circumference C of a circle is the product of 2, π, and the radius r.

19. Seven-eighths of the sum of a, b, and the square of c is the same as 48.

Define a variable, and then write an equation for each problem. Do *not* try to solve.   20–23. See margin.

20. Nicole has 26 CDs and tapes altogether. When she tries to pair them up, she has 4 CDs left over. How many tapes does Nicole have?

21. One number is 25 more than a second number. The sum of these two numbers is 106. Find the numbers.

22. Anna is twice as old as her brother Dennis, who is 4 years older than their brother Curtis. The sum of their ages is 32. How old is Anna?

23. Hector works for a lawn care service. He gets paid twice a month and saves $50 of each pay. After 7 months, how much will he have saved?

Translate each equation or formula below into a verbal sentence.

24. $(mn)^2 = p$ 25. $V = \frac{ah}{3}$ 26. $\frac{(a + 2)}{5} > 10$

24–26. See margin.

Write a problem based on the given information.   27–29. See margin.

27. x = the number of miles from home to school
$3x = 36$

28. p = the cost of a suit
$p - 25 = 150$

29. d = the number of appointments a doctor has scheduled
$6d + 5 = 47$

Write a formula for each shaded area.   30–32. See margin.

30.

31.

32.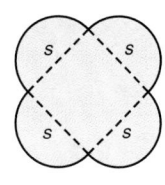

Critical Thinking

33. **Geometry**   The two line segments below are the same length.

Write an equation that describes this situation. Then, use guess and check or any other strategy to find x.   See margin.

Lesson 2-9   Problem Solving: Write Equations and Formulas   **131**

Additional Answers

29. Sample answer: The number of minutes spent waiting for a doctor in the waiting room is 5 minutes more than 6 times the number of appointments she has. How long would you have to wait for a doctor with d appointments?

30. $a^2 - b^2$

31. $ab + cd + 2d - 8$

32. $s^2 + \pi \cdot \frac{s^2}{2}$

33. Sample answer:
$4.5 + x + x = 24 + x$
$x = 19.5$

Additional Answers

20. Sample answer:
t = how many tapes Nicole has. $2t + 4 = 26$

21. Sample answer:
x = first number
$x + 25$ = second number
$25 + 2x = 106$

22. Sample answer:
$2D$ = Anna's age
D = Dennis' age
$D - 4$ = Curtis' age
$2D + D + D - 4 = 32$

23. Sample answer:
y = how much money Hector saves
$y = 2(7)(50)$

24. The square of the product of m and n is equal to p.

25. V is equal to the quotient of a times h and three.

26. The quantity of the sum of a and two, divided by five is greater than ten.

27. Sample answer: Three times the number of miles home from school is 36. How many miles away from home is the school?

28. Sample answer: Manny's Suits for Men is having a $25.00 off sale. If the original cost of a suit is p, find the original cost of the suit if the sale price is $150.00.

Closing Activity

Writing Have students each write a problem using the following information. Let h = Chris's height now, $h - 6$ = Ryan's height. Let students switch problems with a classmate and solve.

Chapter 2, Quiz D (Lessons 2-8 and 2-9), is available in the *Assessment and Evaluation Masters*, p. 45.

Additional Answers

34a. What is the area of two 7-inch pizzas and what is the area of one 12-inch pizza?

34b. See students' work.

34c. $A = \pi r^2$
(two 7-inch pizzas)
$A = \pi(3.5)^2$
$A = 12.25\pi$ or ≈ 38.5 in^2 each
2 pizzas ≈ 77 in^2
(one 12-inch pizza)
$A = \pi(6)^2$
$A = 36\pi$ or ≈ 113.1 in^2
One 12-inch pizza is the best buy.

34d. If crust were a factor, you would have to take circumference into consideration and figure out how the circumferences compare.

Enrichment Masters, p. 18

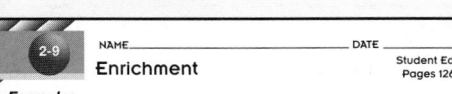

Applications and Problem Solving

34. See margin.

34. Food Aaron and Ellie went out for pizza. After looking at the menu, they couldn't decide whether to order two 7-inch round pizzas or one 12-inch round pizza because the cost was the same. Which is the better buy?

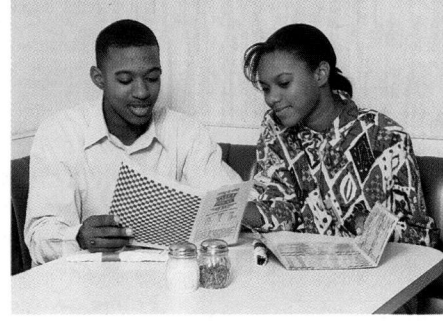

  a. What questions do you need to answer?
  b. Draw and label diagrams that describe the situation.
  c. Write the formulas or equations needed to solve the problem, then solve.
  d. Ellie's favorite part of a pizza is the crust. Explain how knowing this might affect your answer.

35a. $A = \frac{1}{2}h(a + b)$

35b. $A = 108\frac{1}{2}$ ft^2

36a. See margin.

36b. $P = x + 2y + 3x + y + 2x + y$ or $P = 6x + 4y$

36c. $A = xy + 3xy$ or $A = 4xy$

36d. $P - 6(5.5) + 4(4) = 49$ cm; $A = 4(5.5)(4) = 88$ cm^2

35. Geometry The area of a trapezoid equals one half the product of its height and the sum of the lengths of its two bases.
  a. Write a formula for the area of a trapezoid. Let A represent the area, let a and b represent the bases, and let h represent the height.
  b. Find the area for a trapezoid with a base of 20 feet, another base of 11 feet, and a height of 7 feet.

36. Geometry Study the diagram at the right.
  a. Based on the dimensions given, label the remaining sides. Assume that all angles are right angles.
  b. Write a formula to represent the perimeter.
  c. Write a formula to represent the area.
  d. If x is 5.5 cm and y is 4 cm, find the perimeter and area.

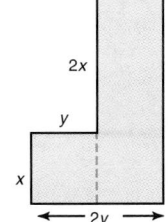

Mixed Review

37. Find $-\sqrt{64}$. (Lesson 2–8) **-8**

38. Simplify $-72 \div (-6)$. (Lesson 2–7) **12**

39. Find $-4\begin{bmatrix} -2 & 0.4 \\ -3 & 5 \end{bmatrix}$. (Lesson 2–6) $\begin{bmatrix} 8 & -1.6 \\ 12 & -20 \end{bmatrix}$

40. Sample answer: $\frac{20}{42}$

40. Find a fraction between $\frac{1}{3}$ and $\frac{4}{7}$ whose denominator is 42. (Lesson 2–4)

41. State the additive inverse and absolute value of $+25$. (Lesson 2–3) **$-25, 25$**

42. Identify the graph below that best represents the following situation. Brandon has a deflated balloon. He slowly fills the balloon up with air. Without tying the balloon, he lets it go. (Lesson 1–9) **C**

43. Find the solution set for the inequality $4y + 2 > 9$ if the replacement set is $\{1, 2, 3, 4, 5\}$. (Lesson 1–5) **$\{2, 3, 4, 5\}$**

Extension

Communication Have students write and then compare the algebraic expressions for each verbal expression.

1. square of the sum of x and a; sum of the square of x and the square of a
$(x + a)^2; x^2 + a^2$

2. product of m, n, and the square of p; product of the squares of m, n, and p; square of the product of m, n, and p $mnp^2, m^2n^2p^2, (mnp)^2$

Additional Answer

36a.

VOCABULARY

After completing this chapter, you should be able to define each term, property, or phrase and give an example or two of each.

Algebra
absolute value   (p. 85)
adding integers   (p. 86)
additive inverses   (p. 87)
additive inverse property   (p. 87)
comparison property   (p. 94)
completeness property   (p. 121)
complex fraction   (p. 114)
cross products   (p. 94)
defining the variable   (p. 127)
density property   (p. 96)
dividing rational numbers   (p. 112)
formula   (p. 128)
integer   (p. 73)

irrational number   (p. 120)
multiplicative property of −1   (p. 107)
multiplying rational numbers   (p. 107)
negative number   (p. 73)
number line   (p. 72)
opposites   (p. 87)
perfect square   (p. 119)
principal square root   (p. 119)
radical sign   (p. 119)
rational number   (p. 93)
real number   (p. 121)
square root   (pp. 118, 119)
subtracting integers   (p. 87)
unit cost   (p. 95)

Venn diagram   (p. 73)
whole number   (p. 72)

Geometry
coordinate   (p. 73)
graph   (p. 73)

Statistics
line plot   (p. 78)

Problem Solving
problem-solving plan   (p. 126)

Discrete Mathematics (p. 88)
matrix   (p. 88)
scalar multiplication   (p. 108)

UNDERSTANDING AND USING THE VOCABULARY

State whether each sentence is *true* or *false*. If false, replace the underlined word or number to make a true sentence.

1. The absolute value of −26 is <u>26</u>.   true
2. The <u>multiplicative inverse</u> of 2 is −2.   false, additive inverse
3. Terminating decimals are <u>rational</u> numbers.   true
4. The square root of 144 is <u>12</u>.   true
5. $2\frac{1}{2}$ is a complex fraction.   false, sample answer: $\frac{\frac{1}{2}}{5}$
6. $-\sqrt{576}$ is an <u>irrational number</u>.   false, integer
7. 225 is a <u>perfect square</u>.   true
8. <u>−3.1</u> is an integer.   false, sample answer: −3
9. <u>0.666</u> is a repeating decimal.   true
10. $\frac{10}{5}$ is a whole number.   true

Chapter 2   Highlights   **133**

Instructional Resources

Three multiple-choice tests and three free-response tests are provided in the *Assessment and Evaluation Masters*. Forms 1A and 2A are for honors pacing, and Forms 1B, 1C, 2B, and 2C are for average pacing. Chapter 2 Test, Form 1B is shown at the right, Chapter 2 Test, Form 2B is shown on the next page.

Chapter 2 Test, Form 1B

Write the letter for the correct answer in the blank at the right of each problem.

1. What set of numbers is graphed below?   1. __B__
   A. {negative integers}   B. {· · ·, −3, −2, −1, 0}
   C. {−4, −3, −2, −1, 0, · · ·}   D. none of these

2. Which of the following is the graph of {integers greater than −2}?   2. __D__

3. You are making a line plot. Each unit on the line stands for $100. Where will you plot $5.16?   3. __A__

4. Use a number line to find the sum of −9 and (−13).   4. __A__
   A. −22   B. 22   C. −4   D. 4

5. What is the value of −387 + 432?   5. __B__
   A. −45   B. 45   C. −9   D. 9

6. Find the value of −|w + 12| if w = −82.   6. __A__
   A. −70   B. 70   C. −94   D. 94

7. Find −0.007 + 0.02.   7. __D__
   A. −0.09   B. 0.13   C. −0.027   D. 0.013

8. Simplify 16a + (−7a) + (−12a) + a.   8. __C__
   A. 0   B. −20a   C. −2a   D. −3a

9. Which fractions are written in order from least to greatest?   9. __D__
   A. $\frac{4}{25}, \frac{3}{19}, \frac{5}{27}$ B. $\frac{3}{19}, \frac{5}{27}, \frac{4}{25}$ C. $\frac{5}{27}, \frac{4}{25}, \frac{3}{19}$ D. $\frac{3}{19}, \frac{4}{25}, \frac{5}{27}$

10. Which of the following is a rational number between $\frac{2}{9}$ and $\frac{11}{40}$?   10. __D__
    A. $\frac{2}{11}$ B. $\frac{9}{31}$ C. $\frac{179}{360}$ D. $\frac{179}{720}$

Chapter 2 Test, Form 1B (continued)

Exercises 11 through 13: What symbol can replace the __?__ to make each sentence true?

11. −4 __?__ $\frac{4}{3}$ 11. __B__
    A. =   B. <   C. >

12. (2.1)(−1) __?__ −5   12. __C__
    A. =   B. <   C. >

13. $\frac{3}{4}$ __?__ $\frac{5}{7}$ 13. __C__
    A. =   B. <   C. >

14. Find $\begin{bmatrix} 3 & -4 \\ 2.1 & 1.7 \end{bmatrix} + \begin{bmatrix} 6.6 & -1 \\ -5 & 0.5 \end{bmatrix}$.   14. __D__
    A. $\begin{bmatrix} 2 & 2.6 \\ 2.6 & -3.3 \end{bmatrix}$ B. $\begin{bmatrix} 2.6 & 2 \\ -3.3 & 2.6 \end{bmatrix}$ C. $\begin{bmatrix} -2 & -3.5 \\ 8.7 & 0.7 \end{bmatrix}$ D. $\begin{bmatrix} 9.6 & -5 \\ -2.9 & 2.2 \end{bmatrix}$

15. Find $\left(-\frac{1}{3}\right)\left(-\frac{6}{7}\right)$.   15. __B__
    A. $\frac{7}{10}$ B. $\frac{2}{7}$ C. $-\frac{2}{7}$ D. $-\frac{7}{10}$

16. Find $-\frac{1}{4} \div 6$.   16. __A__
    A. $-\frac{1}{24}$ B. $-\frac{3}{2}$ C. $\frac{1}{24}$ D. $\frac{3}{2}$

17. Simplify $\frac{24k - 6}{-6}$.   17. __B__
    A. −4k − 6   B. −4k + 1   C. −24k − 1   D. 24k − 1

18. Simplify $\sqrt{\frac{81}{256}}$.   18. __A__
    A. $\frac{9}{16}$ B. $\pm\frac{9}{16}$ C. $\frac{3}{4}$ D. $\pm\frac{3}{4}$

19. Which number is irrational?   19. __B__
    A. $\sqrt{\frac{4}{9}}$ B. 9.737737773· · ·
    C. 8.888· · ·   D. They are all irrational.

20. Which equation can be used to solve the following problem? The Sears Tower in Chicago is 40 feet taller than the Empire State Building in New York. The sum of their heights is 2868 feet. How tall is the Empire State Building?   20. __C__
    A. x + 40 = 2868   B. x = 2868 − 40
    C. x + (x + 40) = 2868   D. x − 40 = 2868

Bonus What is the opposite of the reciprocal of the multiplicative inverse of the additive inverse of 5?   Bonus __5__

Skills and Concepts Encourage students to refer to the objectives and examples on the left as they complete the review exercises on the right.

Assessment and Evaluation Masters, pp. 37–38

CHAPTER **2** **STUDY GUIDE AND ASSESSMENT**

SKILLS AND CONCEPTS

OBJECTIVES AND EXAMPLES	REVIEW EXERCISES

Upon completing this chapter, you should be able to:

• graph integers on a number line   (Lesson 2–1)

Graph $\{ \ldots, -5, -4, -3 \}$.

Use these exercises to review and prepare for the chapter test.   **11–13. See margin.**

Graph each set of numbers on a number line.

11. $\{5, 3, -1, -3\}$

12. $\{$integers greater than $-3\}$

13. $\{$integers less than 4 and greater than or equal to $-2\}$

• add integers by using a number line (Lesson 2–1)

$4 + (-3) = 1$

Find each sum. If necessary, use a number line.

14. $4 + (-4)$	**0**	15. $2 + (-7)$	**−5**
16. $-8 + (-12)$	**−20**	17. $-9 + 5$	**−4**
18. $-14 + (-8)$	**−22**	19. $6 + (-11)$	**−5**

• display and interpret statistical data on a line plot   (Lesson 2–2)

Make a line plot for the set of data.

78, 74, 86, 88, 99, 63, 85, 85, 85

The following table lists the percent of 18-year-olds with high school diplomas for each of 9 years.

Year	'50	'55	'60	'65	'70	'75	'80	'85	'90
%	56	65	72	71	77	74	72	74	72

20. Make a line plot of the data.   **See margin.**

21. What was the lowest percent of 18-year-olds with high school diplomas?   **56%**

22. In how many years was the percent with diplomas between 70 and 75%?   **6 years**

22. 3%, from 1970–1975

• add and subtract integers   (Lesson 2–3)

$-14 + (-9) = -(|-14| + |-9|)$
$= -(14 + 9)$
$= -23$

$7 - 9 = 7 + (-9)$ *To subtract 9, add −9.*
$= -2$

Find each sum and difference.

23. $17 + (-9)$	**8**	24. $14 - 36$	**−22**
25. $-10 + 8$	**−2**	26. $18 - (-5)$	**23**
27. $-7 - (-11)$	**4**	28. $-17 + (-31)$	**−48**
29. $-12 + 7$	**−5**	30. $-54 - (-34)$	**−20**

134   *Chapter 2   Study Guide and Assessment*

Additional Answers

11.

12.

13.

20.

OBJECTIVES AND EXAMPLES	REVIEW EXERCISES

compare and order rational numbers (Lesson 2–4)

$$\frac{5}{9} \;?\; \frac{3}{5}$$

$5 \cdot 5 \;?\; 9 \cdot 3$ *Find the cross products.*

$$25 < 27$$
$$\frac{5}{9} < \frac{3}{5}$$

Replace each ? with <, >, or = to make each sentence true.

31. $-8 \;?\; -14$ $>$
32. $\frac{3}{8} \;?\; \frac{4}{11}$ $>$
33. $-5.6 \;?\; -4.5$ $<$
34. $\frac{-3.6}{0.6} \;?\; -7$ $>$

35–36. Sample answers are given.
Find a number between the given numbers.

35. $-\frac{3}{5}$ and $\frac{7}{12}$ 0
36. $-\frac{2}{9}$ and $-\frac{5}{8}$ $-\frac{20}{45}$

add and subtract rational numbers (Lesson 2–5)

$-0.37 + 0.812 = +(|0.812| - |0.37|)$
$= 0.442$

$7\frac{3}{10} - \left(-4\frac{1}{5}\right) = 7\frac{3}{10} + 4\frac{1}{5}$
$= 7\frac{3}{10} + 4\frac{2}{10}$
$= 11\frac{5}{10}$ or $11\frac{1}{2}$

Find each sum or difference.

37. $\frac{6}{7} + \left(-\frac{13}{7}\right)$ -1
38. $-0.0045 + 0.034$ 0.0295
39. $3.72 - (-8.65)$ 12.37
40. $-\frac{4}{3} + \frac{5}{6} + \left(-\frac{7}{3}\right)$ $-\frac{17}{6}$
41. $-4.57 - 8.69$ -13.26
42. $-4.5y - 8.1y$ $-12.6y$

multiply rational numbers (Lesson 2–6)

$-4(-2) + 6(-3) = 8 + (-18)$
$= -10$

$-2\frac{1}{7}\left(3\frac{2}{3}\right) + \left(-5\frac{5}{7}\right) = \left(-\frac{15}{7}\right)\left(\frac{11}{3}\right) + \left(-\frac{40}{7}\right)$
$= -\frac{55}{7} + \left(-\frac{40}{7}\right)$
$= -\frac{95}{7}$

Find each product.

43. $(-11)(9)$ -99
44. $(-7)(12)(-3)$ 252
45. $\left(\frac{3}{5}\right)\left(-\frac{5}{7}\right)$ $-\frac{3}{7}$
46. $(-5.733)(-2.43)(-3.6)$ -50.152284
47. $-2(45)$ -90
48. $-4\left(\frac{7}{12}\right)$ $-\frac{7}{3}$

divide rational numbers (Lesson 2–7)

$\frac{-12}{-\frac{2}{3}} = -12 \div \left(-\frac{2}{3}\right)$
$= -12\left(-\frac{3}{2}\right)$
$= 18$

Simplify.

49. $\frac{-54}{6}$ -9
50. $-15 \div \left(\frac{3}{4}\right)$ -20
51. $\frac{\frac{4}{5}}{-7}$ $\frac{-4}{35}$
52. $\frac{-575x}{5}$ $-115x$
53. $218 \div (-2)$ -109
54. $-78 \div (-6)$ 13

Applications and Problem Solving Encourage students to work through the exercises in the Applications and Problem Solving section to strengthen their problem-solving skills.

Additional Answer

59a.

OBJECTIVES AND EXAMPLES

• find square roots (Lesson 2–8)

Evaluate $\sqrt{a+b}$ if $a = 489$ and $b = 295$.

$$\sqrt{489 + 295} = \sqrt{784}$$
$$= 28$$

REVIEW EXERCISES

Evaluate each expression. Use a calculator if necessary. Round to the nearest hundredth if the result is not a whole number.

55. $\sqrt{y}$, if $y = 196$ **14**

56. $\pm\sqrt{t}$, if $t = 112$ **±10.58**

57. $-\sqrt{ab}$, if $a = 36$ and $b = 25$ **−30**

58. $\pm\sqrt{\dfrac{c}{d}}$, if $c = 169$ and $d = 16$ **±3.25**

---

62. 4.6 ohms and 1500 watts or 5.2 ohms and 1200 watts

63a. They both saw a gain of $\frac{1}{4}$ for the week.

63c. It changed $-\frac{3}{4}$ from Wednesday to Thursday.

APPLICATIONS AND PROBLEM SOLVING

59. Consumerism   The cost per cup, in cents, of 28 different liquid laundry detergents are listed below. (Lesson 2–2)

28	17	16	18
19	21	26	15
19	19	16	14
21	12	26	17
30	17	13	18
14	22	20	12
19	9	15	12

a. Make a line plot of the data.   **See margin.**

b. How many detergents cost at most 17¢ per cup?   **14 detergents**

60. Aquatics   A submarine descended to a depth of 432 meters and then rose 189 meters. How far below the surface of the water is the submarine now? (Lesson 2–3)
243 meters below the surface

61. Consumerism   Which is the better buy: 0.75 liter of soda for 89¢ or 1.25 liters of soda for $1.31? (Lesson 2–4)

62. Electricity   A circuit is designed with two resistance settings R, 4.6 ohms and 5.2 ohms, and two power settings P, 1200 watts and 1500 watts. Which settings can be used so that the voltage of the circuit is between 75 volts and 85 volts? Use $V = \sqrt{PR}$. (Lesson 2–8)

63. Stock   The changes in two separate stocks for one week are listed below. (Lesson 2–5)

	M	T	W	TH	F
CompNet	$+\frac{3}{8}$	$+\frac{1}{8}$	0	$-\frac{3}{4}$	$+\frac{1}{2}$
AccuFirm	$-\frac{1}{4}$	$+\frac{1}{8}$	$+\frac{1}{8}$	$-\frac{1}{4}$	$+\frac{1}{2}$

a. Which stock saw the largest gain for the week?

b. Which stock had the greatest change from one day to the next?   **CompNet**

c. What was the change and on what days?

Define a variable, and then write an equation for each problem. Do not try to solve. (Lesson 2–9)

64. Minal weighs 8 pounds less than Claudia. Together they weigh 182 pounds. How much does Minal weigh?   **Let w = Minal's weight; $w + (w + 8) = 182$.**

65. Three times a number decreased by 21 is 57. Find the number.
Let n = the number; $3n - 21 = 57$.

66. Four years ago, three times Cecile's age was 42, her father's age now. How old is Cecile now?
Let c = Cecile's age now; $3(c - 4) = 42$.

A practice test for Chapter 2 is provided on page 788.

ALTERNATIVE ASSESSMENT

COOPERATIVE LEARNING PROJECT

Cooking   In this project, you are hosting a party for the cast members of the school play. You must prepare enough chili and peanut butter brownies to serve 100 individuals. (Recipes are given below.) You will also need bowls, napkins, and spoons. You start out with $50 in your checking account. You write a check to a party goods store for the bowls, napkins, and spoons. You then go to the grocery and buy everything except the ground beef because you decide to buy the ground beef at the local meat market.

Follow these steps to prepare a grocery list and a checkbook register.

- Determine the amount of each ingredient needed for the chili and the peanut butter brownies.
- Write a grocery list including the total amount of ingredients needed.
- Go to the grocery and record the size of the container you would need to buy, the quantity you would need to buy, and the price of the container.
- Keep a running total with entries for your checkbook register.
- Write several paragraphs incorporating your data and your money situation for a report to be given to the school play director.

Every time you run a negative balance in your checking account your bank deducts $10. Did you ever run a negative balance? If so, when would have been a good time to deposit more money? How much more money should you have deposited in order to cover all your checks?

THINKING CRITICALLY

- Write a sentence explaining why terminating and repeating decimals are rational numbers.
- A number is both multiplied and divided by the same rational number n, where $0 < n < 1$. Which is greater, the product or the quotient? Explain your reasoning.

PORTFOLIO

Make a grocery list containing ten items. Go to a grocery store and record the price of two different size containers for each of these items. Be sure to also include the brand names and the different sizes in your record. Make a chart of this data.

When you have completed the chart, calculate the unit cost and include it in your chart. Write a new grocery list of the ten items containing the exact brand name and size of container that should be bought in order to insure the best buy. Place these items in your portfolio.

SELF EVALUATION

Comparing numbers is often useful when solving a problem. It can be beneficial when estimating and in decision-making. When comparing numbers or items, be sure to consider other factors.

Assess yourself. When do you use comparison? Do you evaluate new situations based on comparing that situation to another situation that was similar? Describe two ways in which you have used or could use comparison to solve a problem or situation in mathematics and in your daily life.

Chili

1 lb. ground beef, lean	3 1/2 tsp. chili powder
1 medium onion, chopped	2 tbsp. flour
	1 6-oz. can tomato paste
1 1/2 tsp. salt	1 cup water

Brown beef in large skillet. Drain. Add onion and salt. Cook slowly for 5 minutes. Stir in chili powder, flour, tomato paste, and water. Cover and cook on low heat for 1 hour. Stir occasionally. Serves 8.

Peanut Butter Brownies

1/3 cup butter or margarine	1 cup sifted flour
1/2 cup peanut butter	1 tsp. baking powder
1 cup sugar	1/4 tsp. salt
1/4 cup brown sugar	6 oz. chocolate chips
2 eggs	1/2 tsp. vanilla

Beat butter and peanut butter together until light. Add sugars. Add eggs, one at a time, beating after each. Add flour, baking powder, and salt. Add chocolate chips and vanilla last. Spread into greased 9 x 9-in. pan. Bake 30 minutes at 350°. Cool. Serves 10.

Assessment and Evaluation Masters, pp. 42, 53

2   NAME_____   DATE_____

Chapter 2 Performance Assessment

Instructions: *Demonstrate your knowledge by giving a clear, concise solution to each problem. Be sure to include all relevant drawings and justify your answers. You may show your solution in more than one way or investigate beyond the requirements of the problem.*

1. Draw a diagram to show each sum or difference.
   a. $4 + (-7)$ **b.** $-3 - 4$ **c.** $3\frac{1}{2} - \left(-2\frac{3}{4}\right)$ **d.** $|5| - |-8|$

2. Ching, Alex, Ellen, and Ashley are playing in a local golf tournament. For the first two days of the tournament, both of Ching's scores were under -2. Alex had scores under par, or zero. Ellen's scores were less than Ching's scores and Ashley's scores were par or greater.
   a. Draw a graph showing possible scores for Ching.
   b. Are Ashley's scores greater than Ching's? Explain your reasoning.
   c. Were Alex's scores greater than Ching's? Explain your reasoning.
   d. Were Ellen's scores less than Ashley's? Explain your reasoning.

3. Replace each ● with <, >, or = to make a true sentence. Explain your reasoning.
   a. $\left(-\frac{3}{4}\right)\left(\frac{7}{8}\right)$ ● $\left(-\frac{4}{5}\right)\left(-\frac{5}{6}\right)$
   b. $(-5)(7)(-3) - \left(\frac{2}{3}\right)(360)$ ● $\frac{3}{8}(-24)(5)(3)$
   c. $-8 \div \left(-\frac{9}{14}\right)$ ● $\frac{7}{5}$
   d. $\frac{-15x + 21}{3}$ ● $\frac{6 - 10x}{2}$

4. Write a short paragraph describing the set of data shown in the line plot below.

Scoring Guide
Chapter 2
Performance Assessment

Level	Specific Criteria
3 Superior	• Shows thorough understanding of the concepts of *adding, subtracting, and comparing rational numbers; absolute value; graphing inequalities;* and *line plot.* • Uses appropriate strategies to solve problems. • Computations are correct. • Written explanations are exemplary. • Graphs and diagrams are accurate and appropriate. • Goes beyond requirements of some or all problems.
2 Satisfactory, with Minor Flaws	• Shows understanding of the concepts of *adding, subtracting, and comparing rational numbers; absolute value; graphing inequalities;* and *line plot.* • Uses appropriate strategies to solve problems. • Computations are mostly correct. • Written explanations are effective. • Graphs and diagrams are mostly accurate and appropriate. • Satisfies all requirements of problems.
1 Nearly Satisfactory, with Serious Flaws	• Shows understanding of most of the concepts of *adding, subtracting, and comparing rational numbers; absolute value; graphing inequalities;* and *line plot.* • May not use appropriate strategies to solve problems. • Computations are mostly correct. • Written explanations are satisfactory. • Graphs and diagrams are mostly accurate and appropriate. • Satisfies most requirements of problems.
0 Unsatisfactory	• Shows little or no understanding of the concepts of *adding, subtracting, and comparing rational numbers; absolute value; graphing inequalities;* and *line plot.* • May not use appropriate strategies to solve problems. • Computations are incorrect. • Written explanations are not satisfactory. • Graphs are not accurate or appropriate. • Does not satisfy requirements of problems.

Alternative Assessment

The Alternative Assessment section provides students with the opportunity to assess their own work by thinking critically, working with others, keeping a portfolio, and honestly evaluating their own progress. For more information on alternative forms of assessment, see *Alternative Assessment in the Mathematics Classroom,* one of the titles in the Glencoe Mathematics Professional Series.

Performance Assessment

Performance Assessment tasks for this chapter are included in the *Assessment and Evaluation Masters.* A scoring guide is also provided.

2

NAME_____ DATE _____

Chapter 2 Cumulative Review

Write an algebraic expression for each verbal expression. (Lesson 1-1)

1. 7 more than the square of a number   1. _____$n^2 + 7$_____

2. 5 less than 4 times a number   2. _____$4n - 5$_____

Evaluate each expression. (Lesson 1-3)

3. $6 - 2^2 + 14 \cdot 3 \div 7$ 3. _____8_____

4. $3a^2(a^2 - b^2)$ if $a = 4$ and $b = 3$ 4. _____336_____

Solve each equation. (Lesson 1-5)

5. $a = \frac{3}{5} \cdot 15$ 5. _____9_____

6. $\frac{84 \div 14}{27 \div 9} = m$ 6. _____2_____

Name the property illustrated by each statement. (Lessons 1-6, 1-7, and 1-8)

7. If $x = 7 + 2$ and $7 + 2 = w$, then $x = w$.   7. ____transitive____

8. $6 + (7 \cdot 5) = 6 + 35$ 8. ___substitution___

9. $3(4a + b) = 12a + 3b$ 9. ___distributive___

10. $(7 + x) + a = (x + 7) + a$ 10. ___commutative of add.___

Simplify. (Lesson 1-7)

11. $6a^2 + 11a^2 - 3$ 11. ____$17a^2 - 3$____

12. $2(2x + y) + 6(x + 2y)$ 12. ____$10x - 14y$____

Translate each sentence into an equation, inequality, or formula. (Lesson 2-9)

13. y is the product of a and b, increased by c.   13. ___$y = ab + c$___

14. The number m equals the square of the difference of n and p.   14. ___$m = (n - p)^2$___

Replace each __?__ with <, >, or = to make each sentence true. (Lesson 2-4)

15. -3 __?__ -6 15. _____>_____

16. $\frac{4}{13}$ __?__ $\frac{3}{11}$ 16. _____>_____

2

NAME_____ DATE _____

Chapter 2 Cumulative Review (continued)

17. Name the set of numbers graphed below. (Lesson 2-1)   17. {integers less than 3}

18. Graph $\{-3, -1, 0, 2\}$ on the number line provided. (Lesson 2-1)   18. (number line)

19. Graph the solution set of $w \le 3$ on the number line. (Lesson 2-1)   19. (number line)

Find each sum or difference. (Lessons 2-3 and 2-5)

20. $-15 + a$ if $a = -10$ 20. _____-25_____

21. $-13 + 6 + (-8)$ 21. _____-15_____

22. $4m + (-3m) + (-12m)$ 22. _____$-11m$_____

23. $\begin{bmatrix} -3 & -\frac{2}{7} \\ 2.7 & 12 \end{bmatrix} + \begin{bmatrix} -18 & \frac{11}{14} \\ -5.4 & 6 \end{bmatrix}$ 23. $\begin{bmatrix} -21 & \frac{2}{2} \\ -2.7 & 18 \end{bmatrix}$

24. Make a line plot for the following data. (Lesson 2-2) 21, 20, 29, 29, 22, 20, 24, 27, 29, 26.   24. (line plot)

Find each scalar product. (Lesson 2-6)

25. $4\begin{bmatrix} -3 & -3 \\ 2.6 & 12 \end{bmatrix}$ 25. $\begin{bmatrix} -12 & -3 \\ 10.4 & 48 \end{bmatrix}$

26. $-\frac{2}{3}\begin{bmatrix} 9.6 & \frac{9}{4} & -12 \end{bmatrix}$ 26. $\begin{bmatrix} -6.4 & -\frac{3}{2} & 8 \end{bmatrix}$

27. Find $\sqrt{38}$ to the nearest hundredth. (Lesson 2-8)   27. _____6.16_____

Simplify. (Lessons 2-6 and 2-7)

28. $5(-7) + (-3)(6)$ 28. _____-53_____

29. $3(5t - t) + 4(2t + t)$ 29. _____$24t$_____

30. $\left(-\frac{1}{3}\right)\left(\frac{3}{7}\right) + \left(\frac{5}{6}\right)\left(\frac{12}{35}\right)$ 30. _____$\frac{1}{7}$_____

31. $\frac{5a - 10}{5}$ 31. _____$a - 2$_____

Define a variable, then write an equation for each problem. Do not try to solve. (Lesson 2-9)

32. Wayne bought a suit on sale and saved $15.35. The regular price was $139.29. What was the sale price?   Let s = sales price; 32. $s = 139.29 - 15.35$

33. Lisa is 5 years older than Susan. The sum of their ages is 25. How old is Susan?   Let s = Susan's age; 33. $s + (s + 5) = 25$

SECTION ONE: MULTIPLE CHOICE

There are eight multiple-choice questions in this section. After working each problem, write the letter of the correct answer on your paper.

1. Patterns Find the sixth term in the sequence $-6, -3\frac{1}{2}, -1, \ldots.$ **D**

A. 0

B. $2\frac{1}{2}$

C. 4

D. $6\frac{1}{2}$

2. Travel Travel agencies often give information on a state's temperature during the year in order to prepare their customers for their vacation. The stem-and-leaf plot shows daily high temperatures recorded for April in Ohio. Which statement best describes the data in order to prepare travelers for their visit to Ohio in April?   **B**

Stem	Leaf
8	0 2 2 5
7	0 3 3 4 5 7 7 7 8 9 9
6	1 2 3 5 5 6 7 8 8 9
5	4 7 8 8
4	9 $5 \mid 7 = 57$

A. April temperatures can range from 49 to 85 degrees.

B. April is mild, with temperatures usually between 60 and 80 degrees.

C. April's temperatures can soar as high as 85 degrees.

D. April is a cold month with temperatures as low as 49 degrees.

3. Evaluate $3t^2 - g(w - t)$ if $t = 3, g = 5$, and $w = -2$.   **D**

A. -110

B. 20

C. 43

D. 52

---

4. Tyrone is 3 years older than Bill. In 5 years, twice the sum of their ages will equal 5 times Bill's age now. Choose the equation that should be used to solve this problem.   **C**

A. $b + (b + 3) = 5b$

B. $2b + (b + 3) + 5 = 5b$

C. $2[(b + 5) + (b + 3 + 5)] = 5b$

D. $2[(b + 3) + (b - 3 + 5)] = 5b$

5. Quan is working as a mailroom clerk in a busy downtown office building for the summer. His morning route takes him from the mailroom up two floors to the payroll office, up four more floors to the attorney's office, down three floors to the cafeteria, up seven floors to the executive offices, and down twelve floors to the security desk on the first floor. On what floor is the mailroom?   **B**

A. second floor

B. third floor

C. sixth floor

D. top floor

6. Stephen bought a 5-foot-long submarine sandwich to serve at his Super Bowl party. How many pieces would he have to serve if he cut the sub into $1\frac{1}{2}$-inch pieces?   **D**

A. 3 pieces

B. 8 pieces

C. 30 pieces

D. 40 pieces

7. Simplify $10a^3 + 7 - 2(2a^3 + a) + 3$.   **C**

A. $-2a^4 + 10a^3 + 10$

B. $8a^3 + 2a + 10$

C. $6a^3 - 2a + 10$

D. $6a^3 + a + 10$

8. Choose the expression that has a value of 28.   **D**

A. $4 + 3 \cdot 4$

B. $\frac{75}{3} - 2$

C. $8 \cdot 4 - 8$

D. $(5 + 3) \cdot 7 \div 2$

Standardized Test Practice Questions are also provided in the *Assessment and Evaluation Masters,* p. 46.

Additional Answers

11.

13.

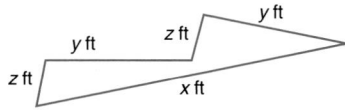

85 86 87 88 89 90 91 92 93 94 95

The scores ranged from a low of 85 to a high of 95 with a cluster around the score of 91.

SECTION TWO: FREE RESPONSE

This section contains seven questions for which you will provide short answers. Write your answer on your paper.

9. Rewrite the expression $7 + 15 \div 2 + 4 - 2$ using grouping symbols so that the value of the expression is $6\frac{1}{2}$. $7 + 15 \div 2 + 4\,(-2)$

10. Three integers x, y, and z satisfy the following conditions:

$y - z < 0$, $x - y > 0$, and $z - x < 0$.

Arrange the integers in order from least to greatest.   *y, z, x*

11. Graph the solution set of $k \neq 1$ on a number line.   **See margin.**

12. Geometry Find the perimeter of the figure below if $x = 7$, $y = 3$, and $z = 1\frac{1}{2}$.   **16 ft**

13. School The students in Miss Stickler's algebra class scored the following points on their last test: 88, 89, 85, 92, 91, 86, 90, 95, 91, 86, 90, 92, 91, 89, 91, and 90. Make a line plot of the scores and write a sentence to describe the data shown.   **See margin.**

14. The average hourly rate of a teenage baby-sitter on a Saturday night is $5 in Los Angeles and $2 in Pittsburgh. Corey, who lives in Pittsburgh, baby-sits his little brother every Saturday night for 5 hours. Ella lives in Los Angeles and baby-sits her cousin for 3 hours every other Saturday night. How many Saturday nights will each have to baby-sit to earn the same amount of money?   **Corey, 3; Ella, 2**

15. What is the absolute value of a if $-a > a$? $-a$

16. Write an algebraic expression to represent *nine decreased by the product of y and 3*. $9 - 3y$

17. Patterns Find the next five terms of the sequence 3, 4.5, 6, ...   **Sample answer: 7.5, 9, 10.5, 12, 13.5**

18. Replace ___?___ with $>$, $<$, or $=$ to make a true statement.

$-2 - 1$ ___?___ $-2(-1)$ $<$

SECTION THREE: OPEN-ENDED

This section contains two open-ended problems. Demonstrate your knowledge by giving a clear, concise solution to each problem. Your score on these problems will depend on how well you do the following.

- Explain your reasoning.
- Show your understanding of the mathematics in an organized manner.
- Use charts, graphs, and diagrams in your explanation.
- Show the solution in more than one way or relate it to other situations.
- Investigate beyond the requirements of the problem.

19. Latisha and Joia are planning to purchase a CD player for their dorm room. Latisha works at the campus book store and earns $5.35 per hour. Joia has a job off campus as a waitress and earns $2.00 per hour, plus tips. Each student works 20 hours per week.

A. How long will they have to work to earn enough money to buy a CD player worth $350?   **about $2\frac{1}{2}$ weeks, not counting tips**

B. What other factors may affect when they can buy the CD player?   **other expenses they have, amount of tips Joia makes**

20. If $2(b + c) = 2b + 2c$, does $2 + (b \cdot c) = (2 + b)(2 + c)$? Use values for b and c to justify your answer.   **No; see students' work.**

3

Solving Linear Equations

PREVIEWING THE CHAPTER

This chapter addresses the crucial skills and concepts involved in solving equations. Cups and counters are used as models to develop an understanding of equations, and then, in carefully sequenced lessons, the addition, subtraction, multiplication, and division properties of equality are presented and applied to solving equations. Then students use their knowledge of equations to integrate geometry skills by studying properties of angles and triangles. They also learn to use more than one property to solve equations with the variable on either side or on both sides. The chapter concludes with students learning to find the mean, median, and mode.

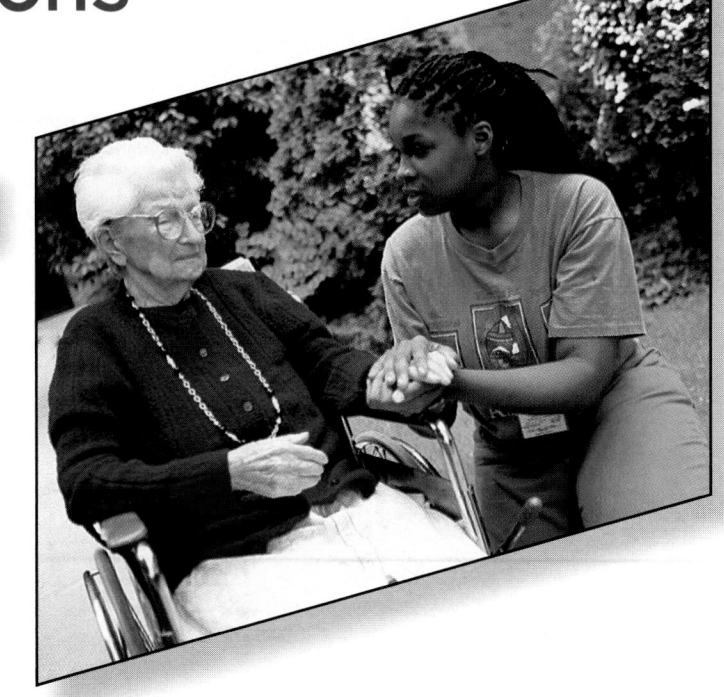

Lesson (Pages)	Lesson Objectives	NCTM Standards	State/Local Objectives
3-1A (142–143)	Solve equations by using cups and counters to model variables and integers.	1–5	
3-1 (144–149)	Solve equations by using addition and subtraction.	1–5	
3-2 (150–154)	Solve equations by using multiplication and division.	1–5	
3-3A (155)	Solve equations by using cups and counters to model variables and integers.	1–5	
3-3 (156–161)	Solve equations involving more than one operation. Solve problems by working backward.	1–5	
3-4 (162–167)	Find the complement and supplement of an angle. Find the measure of the third angle of a triangle given the measures of the other two angles.	1–5, 7	
3-5 (168–172)	Solve equations with the variable on both sides. Solve equations containing grouping symbols.	1–5, 7	
3-6 (173–177)	Solve equations and formulas for a specified variable.	1–5, 7	
3-7 (178–183)	Find and interpret the mean, median, and mode of a set of data.	1–5, 10	

ORGANIZING THE CHAPTER

You may want to refer to the **Course Planning Calendar** on page T12 for detailed information on pacing.
PACING: **Standard**—14 days; **Honors**—13 days; **Block**—6 days; **Two Years**—19 days

LESSON PLANNING CHART

| Lesson (Pages) | Materials/ Manipulatives | Extra Practice (Student Edition) | BLACKLINE MASTERS | | | | | | | | | | Real-World Applications | Interactive Mathematics Tools Software | Teaching Transparencies |
			Study Guide	Practice	Enrichment	Assessment and Evaluation	Modeling Mathematics	Multicultural Activity	Tech Prep Applications	Graphing Calculator	Science and Math Lab Manual			
3-1A (142–143)	cups and counters* equation mat*						p. 23							3-1A	
3-1 (144–149)	calculator cups and counters* equation mat*	p. 762	p. 19	p. 19	p. 19		p. 74								3-1A 3-1B
3-2 (150–154)	cups and counters* equation mat*	p. 762	p. 20	p. 20	p. 20	p. 72			p. 5	p. 3		7		3-2	3-2A 3-2B
3-3A (155)	cups and counters* equation mat*						p. 24								
3-3 (156–161)	graphing calculator cups and counters* equation mat*	p. 762	p. 21	p. 21	p. 21		pp. 42–44	p. 5	p. 6			8			3-3A 3-3B
3-4 (162–167)	construction paper protractor* scissors	p. 763	p. 22	p. 22	p. 22	pp. 71, 72		p. 6			pp. 9–12			3-4	3-4A 3-4B
3-5 (168–172)		p. 763	p. 23	p. 23	p. 23										3-5A 3-5B
3-6 (173–177)		p. 763	p. 24	p. 24	p. 24	p. 73									3-6A 3-6B
3-7 (178–183)	graphing calculator	p. 764	p. 25	p. 25	p. 25	p. 73						9		3-7	3-7A 3-7B
Study Guide/ Assessment (185–189)						pp. 57–70, 74–76									

*Included in Glencoe's Student Manipulative Kit and Overhead Manipulative Resources.

ORGANIZING THE CHAPTER

OTHER CHAPTER RESOURCES

Student Edition
Chapter Opener, pp. 140–141
Mathematics and Society, p. 183
Working on the Investigation,
    pp. 154, 177
Closing the Investigation, p. 184

Teacher's Classroom Resources
Investigations and Projects Masters,
    pp. 33–36
Algebra and Geometry Overhead
    Manipulative Resources,
    pp. 5–12

Technology
Test and Review Software (IBM
    and Macintosh)
CD-ROM Interactions (Windows
    and Macintosh)

Professional Publications
Block Scheduling Booklet
Glencoe Mathematics Professional
    Series

OUTSIDE RESOURCES

Books/Periodicals
Codes, Ciphers, and Secret Writing, Dover
Mathematics and Humor, NCTM

Software
Data Insights, Sunburst
Geometric Connectors: Coordinates, Sunburst
Teasers by Tobbs with Whole Numbers, Sunburst

Videos/CD-ROMs
Becoming Successful Problem Solvers, HRM Video
Math ... Who Needs It?! FASE Productions

ASSESSMENT RESOURCES

Student Edition
Math Journal, pp. 170, 175, 181
Mixed Review, pp. 149, 154,
    161, 167, 172, 177, 183
Self Test, p. 161
Chapter Highlights, p. 185
Chapter Study Guide and
    Assessment, pp. 186–188
Alternative Assessment, p. 189
 Portfolio, p. 189

Teacher's Wraparound Edition
5-Minute Check, pp. 144, 150,
    156, 162, 168, 173, 178
Check for Understanding, pp. 148,
    152, 159, 165, 170, 175, 181
Closing Activity, pp. 149, 154,
    161, 167, 172, 177, 183
Cooperative Learning, pp. 164, 180

Assessment and Evaluation Masters
Multiple-Choice Tests, Forms 1A
    (Honors), 1B (Average), 1C
    (Basic), pp. 57–62
Free-Response Tests, Forms 2A
    (Honors), 2B (Average), 2C
    (Basic), pp. 63–68
Calculator-Based Test, p. 69
Performance Assessment, p. 70
Mid-Chapter Test, p. 71
Quizzes A–D, pp. 72–73
Standardized Test Practice, p. 74
Cumulative Review, pp. 75–76

ENHANCING THE CHAPTER

Examples of some of the materials for enhancing Chapter 3 are shown below.

DIVERSITY

Multicultural Activity Masters, pp. 5, 6

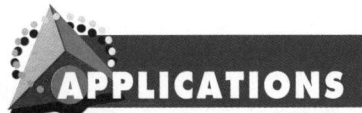

APPLICATIONS

Real-World Applications, 7, 8, 9

TECHNOLOGY

Graphing Calculator Masters, p. 3

TECH PREP

Tech Prep Applications Masters, pp. 5, 6

CONNECTIONS

Science and Math Lab Manual, pp. 9–12

PROBLEM SOLVING

Problem of the Week Cards, 7, 8, 9

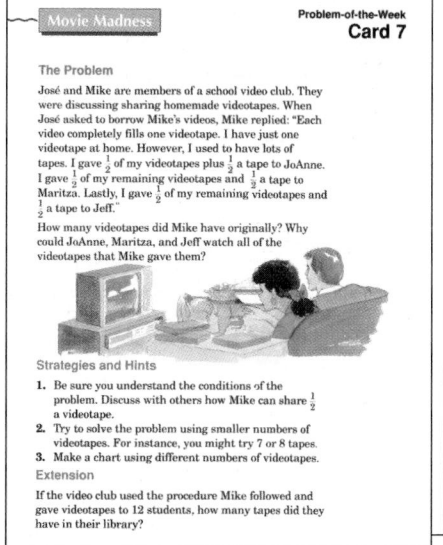

Background Information

The Rising Cost of College It's never too early to start saving for college. MoneySense, of Dallas, Texas, determined in 1994 that if the parent of a newborn started placing $214 each month into a savings plan earning 8% interest, he or she would accumulate the projected $115,615 required for four years of college in 18 years.

Students should be aware that there are many kinds of grants, loans, and scholarships available for college. There are currently seven federal programs:

- Federal Pell Grant—up to $2300
- Federal Supplemental Educational Opportunity Grant (FSEOG)—up to $4000
- Subsidized Stafford Loan—up to $5500
- Unsubsidized Stafford Loan—up to $5500
- Perkins Loan
- Parent Loan for Undergraduate Students (PLUS)
- Federal Work Study

In addition, thousands of organizations, foundations, and businesses offer scholarships. Almost every college also offers academic, art, athletic, drama, music, and science scholarships. Finally, most colleges offer on-campus part-time employment, regardless of financial status.

Solving Linear Equations

Objectives

In this chapter, you will:

- solve equations using one or more operations,
- solve problems that can be represented by equations,
- work backward to solve problems,
- define and study angles and triangles, and
- find measures of central tendency.

The Rising Cost of College

Annual Cost (in 1992–93 dollars)

Private 4-year colleges

Public 4-year colleges

School Year Ending

Source: U.S. Department of Education

Over the last thirty years, tuition at public 4-year colleges has increased by about 50%, and tuition at private 4-year colleges has increased by a whopping 110%! How do you and your family intend to fund your college experience?

TIME Line

A.D. 240 In *Arithmetica*, Diophantus obtains negative numbers as solutions to equations, but dismisses them as "absurd."

575 B.C. Early standard coins, Lydia, Turkey

TIME Line

Ask students to conduct research and write a one-page report about the changes in computer technology since the invention of ENIAC in 1946. Point out that they could focus on cost, size, power, speed of calculations, use of electricity, and/or modern usage.

interNET CONNECTION

Use Fast WEB (a searchable database) to find information about financial aid available for college expenses.

World Wide Web
http://www.cs.cmu.edu/afs/cs/user/mkant/Public/FinAid/finaid.html

When **Marianne Ragins** was a high-school student in her hometown of Macon, Georgia, she applied for scholarship funds from several sources so she could go to college. By the time Marianne was a freshman at Florida A & M University, she had won more than $400,000 in scholarship offers.

When she graduates from college, Marianne will have used $120,000 of those scholarships for her college expenses. Marianne, who is now 21, is the author of *Winning Scholarships for College,* and she now conducts The Scholarship Workshop across the country.

Chapter Project

Juan Ramirez has been offered two college scholarships. Under the terms of the first scholarship, University A will pick up one-half of all his college expenses. Under the terms of the second scholarship, University B will pick up $30,000 of his college expenses. There is $27,500 in the college fund his parents set up for him when he was born.

- Suppose college fees for one year at University A, a state university, are $9500 per year and fees for one year at University B, a private university, are $19,000.

- Write an equation to represent the cost for Juan to attend University A for four years. Write another equation to represent the cost for Juan to attend University B for four years.

- Which school should Juan attend, and why? Be sure to completely explain your reasoning.

780 Muhammed ibn Musa Al-Khowarizmi, Arabian mathematician, born; from his name we get the word *algebra*.

1636 Harvard College, the first college in the United States, is founded.

1946 ENIAC, the first all-purpose electronic computer is completed; it uses so much electricity that the lights dim in a nearby town.

| 900 | 1000 | 1100 | 1200 | 1300 | 1400 | 1500 | 1600 | 1700 | 1800 | 1900 |

1158 The University of Bologna, Italy, is founded.

1848 Maria Mitchell is the first woman elected to the American Academy of Science.

Alternative Chapter Projects

Two other chapter projects are included in the *Investigations and Projects Masters.* In Chapter 3 Project A, pp. 33–34, students extend the topic in the chapter opener. In Chapter 3 Project B, pp. 35–36, students research numeration in other civilizations.

Sample Answers

$$f_A = \frac{4(9500)}{2}; f_B = 4(19,000) - 30,000$$

University A would be more economical and his parents could easily pay the remainder of the fees. University B may provide other benefits such as a smaller campus, more individualized instruction, and so on. Juan's family would have to come up with the remaining fees.

Marianne began her research by gathering information at her local library. She contacted several corporations and applied for all the scholarships for which she was eligible. The first two she received were from Coca-Cola Corporation ($20,000) and from Rhodes College ($60,000).

Marianne has three tips that she shares with students in her workshop.

1. Don't let your parents do the footwork; you have to really want the scholarships to get them.
2. Write to local banks, as well as to radio and television stations, for information on available scholarships.
3. Attend college fairs. Contact your guidance counselor about fair dates and locations.

Chapter Project

Cooperative Learning You may have students pair off or arrange them in cooperative groups to discuss possible solutions for this activity. This activity allows students to make comparisons between two beneficial plans and choose the one that best fits the situation.

Investigations and Projects Masters, p. 33

3
NAME_____ DATE_____
Chapter 3 Project A
Student Edition Pages 142–184

Higher Learning

1. What do you think the ideal college would be like? Would it be large or small? Would it be located in a big city or a small town? Would the weather be warm or cold? What would the professors, classes, and sports teams be like? Think about your answers to these questions. Make up your own questions. Then write a description of your ideal college.

2. Go to the library and find a book with descriptions of colleges. Make a list of schools that fit the description that you wrote for exercise 1.

3. Find the estimated annual cost of attending each school.

4. Research the financial aid that is available to students attending three of the schools on your list. What scholarships do the colleges offer? Will the schools assist you in finding a part-time job? How much money can you earn on such a job? Can you get a loan from the colleges, the federal government, or some other agency?

5. Draw up a financial plan for meeting your college expenses.

6. Work with a partner to complete one of the projects listed below, based on the information you gathered in exercises 1–5.

- Conduct an interview.
- Write a book report.
- Write a proposal.
- Write an article for the school paper.
- Stage a debate.
- Write a report.
- Make a display.
- Make a graph or chart.
- Plan an activity.
- Design a checklist.

NCTM Standards: 1–5

Objective
Solve equations by using cups and counters to model variables and integers.

Recommended Time
Demonstration and discussion: 30 minutes; Exercises: 30 minutes

Instructional Resources
For each student or group of students
Student Manipulative Kit
• 20 red/yellow 2-sided counters
Modeling Mathematics Masters
• pp. 2–3 (cups and counters)
• p. 13 (equation mat)
• p. 23 (worksheet)
For teacher demonstration
Algebra and Geometry Overhead Manipulative Resources

1 FOCUS

Motivating the Lesson
Since this is the first time students will see the process used to solve equations using cups and counters, you may want to allow students some time for exploration and discovery. Introduce the terms and manipulatives and invite students to try to work out the methodology.

2 TEACH

Teaching Tip Another way to solve the equation in Activity 1 is to add 3 positive integers to each side of the mat and then remove the zero-pairs.

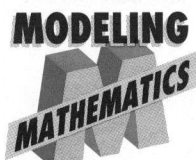

3–1A Solving One-Step Equations

Materials: cups & counters   equation mat

A Preview of Lesson 3–1

You can use cups and counters as a model for solving equations. After you model the equation, the goal is to get the cup by itself on one side of the mat by using the rules stated below.

Rules for Equation Models
You can remove or add the same number of identical counters to each side of the equation mat without changing the equation.
You can remove or add zero pairs to either side of the equation mat without changing the equation.

Activity 1   Use an equation model to solve $x + (-3) = -5$.

Step 1 Model the equation $x + (-3) = -5$ by placing 1 cup and 3 negative counters on one side of the mat. Place 5 negative counters on the other side of the mat. The two sides of the mat represent equal quantities.

$$x + (-3) = -5$$

Step 2 Remove 3 negative counters from each side to get the cup by itself.

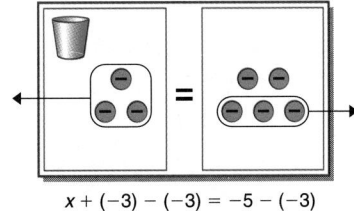

$$x + (-3) - (-3) = -5 - (-3)$$

Step 3 The cup on the left side of the mat is matched with 2 negative counters. Therefore, $x = -2$.

$$x = -2$$

 Cooperative Learning

This lesson offers an excellent opportunity for using cooperative learning groups. For more information on cooperative learning strategies and group management, see *Cooperative Learning in the Mathematics Classroom,* one of the titles in the Glencoe Mathematics Professional Series.

GLENCOE Technology

 Interactive Mathematics Tools Software

This multimedia software provides an interactive lesson that uses counters to solve one-step equations with addition or subtraction. A **Computer Journal** gives students an opportunity to write about what they have learned.

For Windows & Macintosh

Activity 2 Use an equation model to solve $2p = -6$.

Step 1 Model the equation $2p = -6$ by placing 2 cups on one side of the mat. Place 6 negative counters on the other side of the mat.

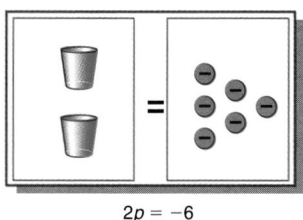

$2p = -6$

Step 2 Separate the counters into 2 equal groups to correspond to the 2 cups. Each cup on the left is matched with 3 negative counters. Therefore, $p = -3$.

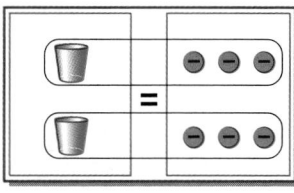

$p = -3$

Activity 3 Use an equation model to solve $r - 2 = 3$.

Step 1 Write the equation in the form $r + (-2) = 3$. Place 1 cup and 2 negative counters on one side of the mat. Place 3 positive counters on the other side of the mat. Notice that it is not possible to remove the same kind of counters from each side. Add 2 positive counters to each side.

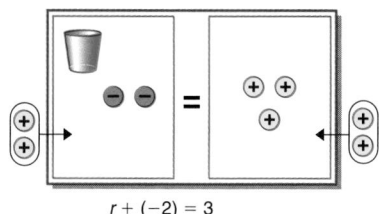

$r + (-2) = 3$

Step 2 Group the counters to form zero pairs. Then remove all the zero pairs. The cup on the left is matched with 5 positive counters. Therefore, $r = 5$.

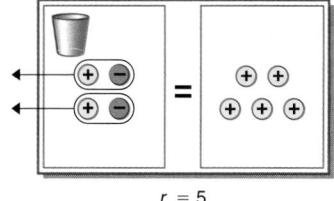

$r = 5$

Model Use equation models to solve each equation.

1. $x + 4 = 5$ 1
2. $y + (-3) = -1$ 2
3. $y + 7 = -4$ -11
4. $3z = -9$ -3
5. $m - 6 = 2$ 8
6. $-2 = x + 6$ -8
7. $8 = 2a$ 4
8. $w - (-2) = 2$ 0

Draw Tell whether each number is a solution of the given equation. Justify your answer with a drawing.

9. $-3; x + 5 = -2$ no
10. $-1; 5b = -5$ yes
11. $-4; y - 4 = -8$ yes

Write 12. Write a paragraph explaining how you use zero pairs to solve an equation like $m + 5 = -8$.   See margin.

Teaching Tip In Activity 3, remind students that subtracting a number is the same as adding its opposite. So subtracting 2 from x is the same as adding -2 to x.

Teaching Tip Remind students to always check their solutions algebraically.

3 PRACTICE/APPLY

Assignment Guide
Core: 1–12
Enriched: 1–12

Additional Answer

12. Add 5 negative counters to each side. Group and remove the counters that form zero pairs. The cup will be matched with 13 negative counters. Thus, $m = -13$.

4 ASSESS

Observing students working in cooperative groups is an excellent method of assessment.

Classroom Vignette

"Before I use tiles or cups and counters to demonstrate solving equations, I use a balance scale, some #2 unsharpened pencils, and centimeter cubes to model the concept of equality and transformation of equations. By building an equation using a visual model, students have a conceptual understanding of equality."

Eva Gates
Independent Mathematics Consultant
Pearland, Texas

Instructional Resources

- Study Guide Master 3-1
- Practice Master 3-1
- Enrichment Master 3-1
- Modeling Mathematics Masters, p. 74

 Transparency 3-1A contains the 5-Minute Check for this lesson; **Transparency 3-1B** contains a teaching aid for this lesson.

Recommended Pacing	
Standard Pacing	Days 2 & 7 of 14
Honors Pacing	Day 2 of 13
Block Scheduling*	Day 1 of 6
Alg. 1 in Two Years*	Days 2 & 3 of 19

 *For more information on pacing and possible lesson plans, refer to the *Block Scheduling Booklet* and *Algebra 1 in Two Years.*

1 FOCUS

5-Minute Check
(over Chapter 2)

Simplify.

1. $-\frac{38}{2}$ −19

2. $\frac{30}{-5}$ −6

3. $\frac{-200x}{50}$ −4x

Write an equation for each problem.

4. A number decreased by 14 is 72. What is the number?
   $x - 14 = 72$

5. Jon is twice as old as Peter. The sum of their ages is 51. Find their ages. $x + 2x = 51$

Objective

After studying this lesson, you should be able to:
- solve equations by using addition and subtraction.

F Y I

As winners of the Super Bowl in 1993 and 1994, the Dallas Cowboys became the fifth team to win the Super Bowl for two consecutive years. The Pittsburgh Steelers did it twice, winning in 1975, 1976, 1979, and 1980.

APPLICATION
Football

According to the graph below, the Oakland Raiders and the Dallas Cowboys have each appeared on *Monday Night Football* a total of 41 times through the 1992–1993 season.

Most *Monday Night Football* Appearances

Source: Capital Cities/ABC

If during the next five seasons, each team appears an average of two times per season, the two teams would still have an equal number of *Monday Night Football* appearances.

$$41 = 41$$

$$41 + 5(2) = 41 + 5(2)$$ *Two appearances per season for five seasons can be represented as 5(2).*

This example illustrates the **addition property of equality.**

Addition Property of Equality	For any numbers *a*, *b*, and *c*, if $a = b$, then $a + c = b + c$.

Note that *c* can be positive, negative, or 0. In the equation on the left below, $c = 5$. In the equation on the right, $c = -5$.

$$18 + 5 = 18 + 5 \qquad\qquad 18 + (-5) = 18 + (-5)$$

If the same number is added to each side of an equation, then the result is an **equivalent equation.** Equivalent equations are equations that have the same solution.

$y + 7 = 13$ *The solution to this equation is 6.*

$y + 7 + 3 = 13 + 3$ *Using the addition property of equality, add 3 to each side.*

$y + 10 = 16$ *The solution to this equation is also 6.*

F Y I

The Dallas Cowboys were defeated in their quest to become the first team to win the Super Bowl for three consecutive years. After defeating the Cowboys in the NFC championship game, the San Francisco 49ers won the Super Bowl in 1995. The Buffalo Bills, on the other hand, are the only team to have lost the Super Bowl for four consecutive years.

Recall that an equation mat is a model of an equation. The sides of the mat represent the sides of an equation. When you add counters to each side, the result is an equivalent equation, as shown below.

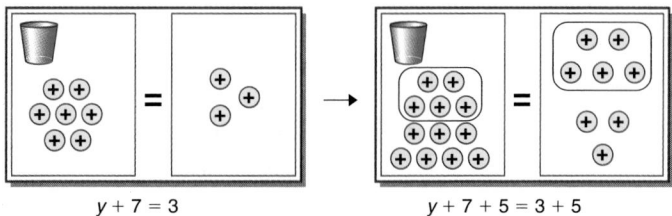

$$y + 7 = 3 \qquad\qquad y + 7 + 5 = 3 + 5$$

Remember, x means 1 · x. The coefficient of x is 1.

To **solve an equation** means to isolate the variable having a coefficient of 1 on one side of the equation. You can do this by using the addition property of equality.

Example ① Solve $23 + t = -16$.

$$23 + t = -16$$
$$23 + t + (-23) = -16 + (-23) \quad \textit{Add } -23 \textit{ to each side.}$$
$$t + 0 = -39 \qquad\qquad \textit{The sum of 23 and } -23 \textit{ is } 0.$$
$$t = -39$$

To check that -39 is the solution, substitute -39 for t in the original equation.

Check:
$$23 + t = -16$$
$$23 + (-39) \stackrel{?}{=} -16$$
$$-16 = -16 \checkmark$$

The solution is -39.

You can and will use equations to solve many real-world problems.

Example ②

APPLICATION
Football

Refer to the application at the beginning of the lesson. The San Francisco 49ers appeared fewer times than the Washington Redskins through the 1992–1993 season. How many more times did the Redskins appear?

Explore   Read the problem to find out what is asked. Then define a variable.

The problem asks how many more times the Washington Redskins appeared on *Monday Night Football* than the San Francisco 49ers through the 1992–1993 season. From the chart, we can see that the Redskins appeared 43 times.

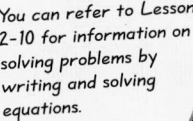
LOOK BACK

You can refer to Lesson 2-10 for information on solving problems by writing and solving equations.

Let n = the difference in the number of times each team appeared. Then $43 - n$ = the number of times the 49ers appeared.

Plan   Write an equation.

You know that $43 - n$ represents the number of times the 49ers appeared. From the chart, we can see that the 49ers appeared 35 times. Thus, $43 - n = 35$.

(continued on the next page)

Lesson 3-1   *Solving Equations with Addition and Subtraction*   **145**

Classroom Vignette

"I ask one student to represent a parent and two other students to represent children. The parent stands between the two children and acts like an equals sign. The parent then divides a box or bag of candy between the two children. Whatever the parent does to one side, he or she must do to the other side to keep things fair."

Donna Rehling
Elder High School
Cincinnati, Ohio

Motivating the Lesson

Hands-On Activity Display a scale to the class. Place two classroom objects such as a pencil and a piece of chalk on the sides of the scale. Have students explain the result. Develop the idea that the scale is balanced when its sides hold equal weights. Challenge students to find two different objects that are equal in weight. Check the validity of student predictions by placing the objects on the scale.

2 TEACH

In-Class Examples

For Example 1
Solve.

a. $21 + r = -2$ -23
b. $5 + p = -10$ -15
c. $13 + t = -14$ -27

For Example 2
Jeff Simons sold 27 cars this month. He sold 63 cars last month. How many fewer cars did he sell this month?   36 cars

Teaching Tip Emphasize the importance of utilizing the answer check. For Example 1, show how $t = 39$ produces a false statement in the last line of the check.
$$23 + t = -16$$
$$23 + 39 = -16$$
$$62 \neq -16$$

Teaching Tip Explain the symbol $\stackrel{?}{=}$ to students. For example, $a \stackrel{?}{=} b$ means "Is a equal to b?"

Teaching Tip Review the four steps in the problem-solving plan: explore, plan, solve, and examine. Emphasize the importance of checking the answers to all problems as in Example 2.

Teaching Tip It is helpful to learn a concept with basic examples and then extend to more complex ones.

Teaching Tip Point out that the minus sign preceding the variable in Example 3 can be read as "the opposite of." For example, $-k = 21$ means "the opposite of k is positive 21." Therefore, k is negative 21; that is, $k = -21$.

Solve   Solve the equation and answer the problem.

$$43 - n = 35$$

$$43 + (-43) - n = 35 + (-43) \quad \textit{Add} -43 \textit{ to each side.}$$

$$-n = -8 \qquad \textit{The opposite of n is negative 8.}$$

$$n = 8 \qquad \textit{Therefore, n is positive 8.}$$

Thus, the Washington Redskins appeared 8 times more than the San Francisco 49ers.

Examine   Check to see if the answer makes sense.

If the Redskins appeared 43 times and the 49ers appeared 35 times, then the difference is $43 - 35$ or 8 times.

In addition to the addition property of equality, there is a **subtraction property of equality** that may also be used to solve equations.

Subtraction Property of Equality	For any numbers a, b, and c, if $a = b$, then $a - c = b - c$.

Example **Solve $190 - x = 215$.**

$$190 - x = 215$$

$$190 - x - 190 = 215 - 190 \quad \textit{Subtract 190 from each side.}$$

$$-x = 25 \qquad \textit{The opposite of x is 25.}$$

$$x = -25$$

Check:
$$190 - x = 215$$
$$190 - (-25) \stackrel{?}{=} 215$$
$$190 + 25 \stackrel{?}{=} 215$$
$$215 = 215 \quad ✔$$

The solution is -25.

Most equations can be solved in two ways. Remember that subtracting a number is the same as adding its inverse.

Example **Solve $a + 3 = -9$ in two ways.**

Method 1: Use the subtraction property of equality.

$$a + 3 = -9$$

$$a + 3 - 3 = -9 - 3 \quad \textit{Subtract 3 from each side.}$$

$$a = -12$$

Check:
$$a + 3 = -9$$
$$-12 + 3 \stackrel{?}{=} -9$$
$$-9 = -9 \quad ✔$$

Alternative Learning Styles

Auditory Read the following problem to your students. Have them solve the problem by writing and solving an equation. You may have to read the problem more than once.

Todd went to the store and bought three rolls of tape for $0.89 each, two packs of notebook paper for $1.39 each, and a notebook. If he spent $9.34, not including tax, how much did he spend on the notebook? $3.89

Method 2: Use the addition property of equality.

$$a + 3 = -9$$
$$a + 3 + (-3) = -9 + (-3) \quad \text{Add } -3 \text{ to each side.}$$
$$a = -12 \quad \text{The answers are the same.}$$

The solution is -12.

Sometimes equations can be solved more easily if they are first rewritten in a different form.

Example **5** Solve $y - \left(-\frac{3}{7}\right) = -\frac{4}{7}$.

This equation is equivalent to $y + \frac{3}{7} = -\frac{4}{7}$. *Why?*

$$y + \frac{3}{7} = -\frac{4}{7}$$
$$y + \frac{3}{7} - \frac{3}{7} = -\frac{4}{7} - \frac{3}{7} \quad \text{Subtract } \frac{3}{7} \text{ from each side.}$$
$$y = -\frac{7}{7} \text{ or } -1$$

Check: $y - \left(-\frac{3}{7}\right) = -\frac{4}{7}$

$$-\frac{7}{7} - \left(-\frac{3}{7}\right) \stackrel{?}{=} -\frac{4}{7}$$
$$-\frac{7}{7} + \frac{3}{7} \stackrel{?}{=} -\frac{4}{7}$$
$$-\frac{4}{7} = -\frac{4}{7} \checkmark$$

The solution is -1.

EXPLORATION

CALCULATORS

You can use a scientific or graphing calculator to solve equations that involve decimals.

 If you're using a scientific calculator like the TI-34, you can use the *plus/minus* key to input negative numbers or to change the sign of any number.

 If you're using a graphing calculator like the TI-81 or TI-82, you can use the *negative* key to input negative numbers.

 On all calculators, use the *subtraction* key to subtract two numbers.

Your Turn a–d. See margin.

a. Take some time to work through the examples in this lesson using your calculator.

b. Describe the difference between the key you use to indicate a negative number on your calculator and the subtraction key.

c. Does the order in which you use these keys matter? Why or why not?

d. How would you use your calculator to solve $x + (-9.016) = 5.14$?

Lesson 3–1 Solving Equations with Addition and Subtraction **147**

Answers for the Exploration

b. The plus/minus or negative key is used to indicate a negative number. For example, you would use the plus/minus key to add -3 and 5. The subtraction key is used only for subtraction. You would use both keys to subtract -7 and 2.

c. Yes, the order in which you use the keys does matter. If you are using the plus/minus key, you press it *after* entering the number. If you are using the negative key, you press it *before* entering the number.

d. 5.14 $\boxed{-}$ 9.016 $\boxed{+/-}$ $\boxed{=}$

Teaching Tip The equation in Example 5 can be solved by adding -11 to each side, that is, by performing the inverse operation. However, most students can solve it more easily by first writing the equation in a simpler form, as shown. That is also the case with Example 6, which could be solved by adding 7.2 to each side.

EXPLORATION

After you give students some time to explore how their own calculators work, have several students demonstrate solving an equation that involves a negative number on their calculators. Make sure they understand that the order in which they use the keys does matter, as well as the difference between a negative or plus/minus key and the subtraction key.

Study Guide Masters, p. 19

3 PRACTICE/APPLY

Check for Understanding
Exercises 1–13 are designed to help you assess your students' understanding through reading, writing, speaking, and modeling. You should work through Exercises 1–5 with your students and then monitor their work on Exercises 6–13.

Error Analysis
Students may be tempted to solve word problems without translating them into open sentences. With simple word problems, students may solve and "undo" the problem as they read and interpret it. To help students focus on translation, assign some word problems with these instructions: "Translate into open sentences. Do not solve the problems."

Practice Masters, p. 19

Practice
3-1

NAME _____ DATE _____

Student Edition
Pages 144–149

Solving Equations by Using Addition and Subtraction
Solve each equation. Then check your solution.

1. $13 + c = -38$ -51 2. $w + 14 = -8$ -22

3. $4.2 = n + 7.3$ -3.1 4. $y + (-10) = 6$ 16

5. $y - 7 = 8$ 15 6. $-13 = 5 + x$ -18

7. $19 - k = 32$ -13 8. $x - 4 = 6$ 10

9. $-12 = 7 - h$ 19 10. $-15 = u + (-5)$ -10

11. $j - (-17) = 36$ 19 12. $-10 - a = -17$ 7

13. $y - (-1.5) = 0.5$ -1 14. $k + (-4.21) = -19.3$ -15.09

15. $52 = 41 - u$ -11 16. $s + (-28) = 0$ 28

17. $\frac{9}{10} = k + \frac{7}{10}$ $\frac{1}{5}$ 18. $-\frac{7}{8} - x = -\frac{7}{12}$ $-\frac{7}{24}$

Define a variable, write an equation, and solve each problem. Then check your solution.

19. A number decreased by 14 is -46. Find the number. -32

20. Thirteen subtracted from a number is -5. Find the number. 8

21. A number increased by -56 is -82. Find the number. -26

22. The sum of a number and 67 is -34. Find the number. -101

Example **Solve** $b + (-7.2) = -12.5$.

This equation is equivalent to $b - 7.2 = -12.5$.   *Why?*

$$b - 7.2 = -12.5$$
$$b - 7.2 + 7.2 = -12.5 + 7.2 \quad \textit{Add 7.2 to each side.}$$
$$b = -5.3$$

Method 1	**Method 2**
Check: $b + (-7.2) = -12.5$	Use a calculator.
$-5.3 + (-7.2) \overset{?}{=} -12.5$	5.3 [+/−] [+] 7.2
$-12.5 = -12.5$ ✔	[+/−] [=] −12.5 ✔

The solution is -5.3.

CHECK FOR UNDERSTANDING

Communicating Mathematics

Study the lesson. Then complete the following.

1. Choose the equation that is equivalent to $2.3 - w = 7.8$.   **b**
   a. $w - 2.3 = 7.8$ **b.** $w - 2.3 = -7.8$ **c.** $-w = -5.5$

2. **Write** three equivalent equations.   **See students' work.**

3. See margin.

3. **Explain** how you can prove that 4.5 is the solution of $x - 1.2 = 3.3$.

4. **Complete:** If $x - 6 = 21$, then $x + 7 = $ _?_ . Explain your reasoning.   **34**

MODELING MATHEMATICS

5. Write an equation for the model at the right. Then use cups and counters to solve the equation. $x - 5 = -10$; $x = -5$

Guided Practice

Solve each equation. Then check your solution.

6. $m + 10 = 7$ -3 7. $a - 15 = -32$ -17 8. $5.7 + a = -14.2$ -19.9

9. $y + (-7) = -19$ -12 10. $\frac{1}{6} - n = \frac{2}{3}$ $-\frac{1}{2}$ 11. $d - (-27) = 13$ -14

Define a variable, write an equation, and solve each problem. Then check your solution.   12. $n - 13 = -5$; 8   13. $n + (-56) = -82$; -26

12. Thirteen subtracted from a number is -5. Find the number.

13. A number increased by -56 is -82. Find the number.

EXERCISES

Practice
A

Solve each equation. Then check your solution.

14. $k + 11 = -21$ -32 15. $41 = 32 - r$ -9 16. $-12 + z = -36$ -24

17. $2.4 = m + 3.7$ -1.3 18. $-7 = -16 - k$ -9 19. $0 = t + (-1.4)$ 1.4

20. $r + (-8) = 7$ 15 21. $h - 26 = -29$ -3 22. $-23 = -19 + n$ -4

23. $-11 = k + (-5)$ -6 24. $r - 6.5 = -9.3$ -2.8 25. $t - (-16) = 9$ -7

26. 2.32
28. −3.6

B

26. $-1.43 + w = 0.89$ 27. $m - (-13) = 37$ 24 28. $-4.1 = m + (-0.5)$

29. $-\frac{5}{8} + w = \frac{5}{8}$ $1\frac{1}{4}$ 30. $x - \left(-\frac{5}{6}\right) = \frac{2}{3}$ $-\frac{1}{6}$ 31. $g + \left(-\frac{1}{5}\right) = -\frac{3}{10}$ $-\frac{1}{10}$

148   Chapter 3   Solving Linear Equations

Reteaching

Using Substeps State the operation and the number you would apply to each side to solve each equation. Then solve. Check each solution.

1. $p + 12 = -4$ -12 or $+(-12)$; -16
2. $-12 + b = 12$ $+12$ or $-(-12)$; 24
3. $-14 - a = -21$ $+14$ or $-(-14)$; 7
4. $k + (-13) = 21$ $+13$ or $-(-13)$; 34
5. $b - (-8) = 23$ $+(-8)$ or -8; 15
6. $z + (-17) = 0$ $+17$ or $-(-17)$; 17

Additional Answer

3. $x - 1.2 = 3.3$
   $4.5 - 1.2 \overset{?}{=} 3.3$
       $3.3 = 3.3$

Define a variable, write an equation, and solve each problem. Then check your solution.

32. Twenty-three minus a number is 42. Find the number. $23 - n = 42; -19$

33. A number increased by 5 is equal to 34. Find the number. $n + 5 = 34; 29$

34. What number decreased by 45 is −78? $n - 45 = -78; -33$

35. The difference of a number and −23 is 35. Find the number.

36. A number increased by −45 is 77. Find the number. $n + (-45) = 77; 122$

37. The sum of a number and −35 is 98. Find the number. $n + (-35) = 98; 133$

35. $n - (-23) = 35; 12$

Critical Thinking

38. Suppose a solution of an equation is a number n. Is it possible for $-n$ to also be a solution of this equation? If not, explain why not. If so, give an example of such an equation and its solutions n and $-n$.
Yes; a typical answer is $n^2 = 25; n = \pm 5$.

Applications and Problem Solving

Write two different equations to represent each situation. Then solve the problem.

39. **Telecommunications** The graph at the right shows the growth in the number of people who own and use cellular telephones.

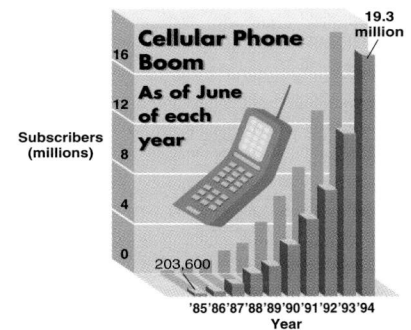

Cellular Phone Boom
As of June of each year

Subscribers (millions)

19.3 million

203,600

'85 '86 '87 '88 '89 '90 '91 '92 '93 '94
Year

Source: Cellular Telecommunications Industry Association

a. How many more cellular phone subscribers were there in 1994 than there were in 1985?
b. Predict how many cellular phone subscribers there will be in the year 2000. Write a justification for your prediction.

40. **Consumerism** According to Runzheimer International, in 1994, a tube of lipstick cost $5.94 in Los Angeles, California. The same tube cost $4.89 more than that in London, England. In Sao Paulo, Brazil, the lipstick cost $26.54.

a. How much did the lipstick cost in London?
b. How much more did the lipstick cost in Sao Paulo than in Los Angeles?

Mixed Review

39a. $203{,}600 + s = 19{,}300{,}000; 19{,}096{,}400$ subscribers

39b. Sample answer: 25 million

40a. $t - 5.94 = 4.89;$ $10.83

40b. $5.94 + t = 26.54;$ $20.60

41e. 24 years

41. Alejandra Salazar said, "I am 24 years younger than my mom and the sum of our ages is 68 years." (Lesson 2–9)
a. How old was Mrs. Salazar when Alejandra was born? **24 years old**
b. How much older than Alejandra is Mrs. Salazar now? **24 years**
c. What will be the sum of their ages in 5 years? **78 years**
d. How old was Mrs. Salazar when Alejandra was 10 years old? **34 years old**
e. In ten years, how much younger than Mrs. Salazar will Alejandra be?

42. Find the principal square root of 256. (Lesson 2–8) **16**

43. Simplify $65 \div (-13)$. (Lesson 2–7) **−5**

44. **Statistics** Odina's algebra average dropped three fourths of a point for each of eight consecutive months. If her average was originally 82, what was her average at the end of eight months? (Lesson 2–6) **76**

45. Find the sum. $-0.23x + (-0.5x)$ (Lesson 2–5) **−0.73x**

46. Simplify $6(5a + 3b - 2b)$. (Lesson 1–7) **30a + 6b**

47. Evaluate $12 \div 4 + 15 \cdot 3$. (Lesson 1–3) **48**

Extension

Reasoning Have students work through the following problem in groups.

Would you rather have 100 pounds of dimes or a ton of pennies?

Students will need to do some research to determine the weights of a dime and a penny to solve the problem.

Since there are 11 dimes per ounce and 9 pennies per ounce, 100 pounds of dimes equals $1760 while 2000 pounds of pennies equals $2880.

Assignment Guide

Core: 15–37 odd, 38, 39, 41–47
Enriched: 14–36 even, 38–47

For **Extra Practice**, see p. 762.

The red A, B, and C flags, printed only in the Teacher's Wraparound Edition, indicate the level of difficulty of the exercises.

4 ASSESS

Closing Activity

Writing Have students make up a problem for which they can write an equation and use addition or subtraction to solve. Then have them write an equation using addition and one using subtraction to solve their problem. Finally, have them solve both equations and compare their answers.

Enrichment Masters, p. 19

3-1   NAME_____ DATE_____

Enrichment   Student Edition Pages 144–149

Equations With No Solutions

Not every equation has a solution. Watch what happens when we try to solve the following equation.

$$8 - (3 - 2x) = 5x - 3x$$
$$8 - 3 + 2x = 2x$$
$$5 + 2x = 2x$$
$$5 = 0$$

Since the equation is equivalent to the false statement $5 = 0$, it has no solution. There is no value of x that will make the equation true.

Write a false statement that shows each equation has no solution. Answers may vary. Sample answers are given.

1. $1 - 2t = 2(1 - t)$
   $1 = 2$

2. $11y - 7y = 5 + 4y - 6$
   $0 = -1$

3. $-7x^2 + 5 + 6x^2 = 12 - (2 + x^2)$
   $5 = 10$

4. $2(2 - y^2) = 5 - (5 + 2y^2)$
   $4 = 0$

5. $\frac{3}{2} + \frac{2}{3}p - 1 = \frac{1}{3}(1 + 2p)$
   $\frac{1}{2} = \frac{1}{3}$

6. $0.5(1 + 3m) = 1.05 - (1 - 1.5m)$
   $0.5 = 0.05$

Solve each equation if possible.

7. $5(3 - m) = 15m + 15$
   $m = 0$

8. $-9x + 12x = 3(2 - x)$
   $x = 1$

9. $10(0.2 + 0.4c) = 10c + 0.2 - 6c$
   no solution

10. $13 - (3 - n) = 5(n + 2)$
    $n = 0$

11. $2(1 + 4t) = 8 - (3 - 8t)$
    no solution

12. $3(d - 1) + 2 = 3(d + 2) - 5$
    no solution

Instructional Resources

- Study Guide Master 3-2
- Practice Master 3-2
- Enrichment Master 3-2
- Assessment and Evaluation Masters, p. 72
- Graphing Calculator Masters, p. 3
- Real-World Applications, 7
- Tech Prep Applications Masters, p. 5

 Transparency 3-2A contains the 5-Minute Check for this lesson; **Transparency 3-2B** contains a teaching aid for this lesson.

Recommended Pacing	
Standard Pacing	Day 3 of 14
Honors Pacing	Day 3 of 13
Block Scheduling*	Day 2 of 6
Alg. 1 in Two Years*	Days 4 & 5 of 19

 *For more information on pacing and possible lesson plans, refer to the *Block Scheduling Booklet* and *Algebra 1 in Two Years*.

1 FOCUS

 ### 5-Minute Check
(over Lesson 3-1)

Solve.

1. $m + 10 = 7$ -3
2. $18 + j = 54$ 36
3. $t - 11 = 46$ 57
4. $x - (-24) = -37$ -61
5. $-24 - g = -74$ 50
6. The sum of two integers is -42. One integer is 12. Find the other integer. -54

3-2 Solving Equations with Multiplication and Division

Objective
After studying this lesson, you should be able to:
- solve equations by using multiplication and division.

 **APPLICATION**
Construction

In 1990, Congress passed into law the Americans with Disabilities Act. One of the provisions of that law has to do with ramps installed on buildings to give people with disabilities access to those buildings. The law states that for a *rise* of 1″, there should be at least 12″ of *run*. The maximum rise is 30″.

Rick Hansen, wheelchair athlete

If a contractor wants to build a ramp that has a 180-inch run, what is the greatest rise that the ramp can have?

In the table at the right, the pattern suggests that the run is always 12 times the rise. Let x represent the number of inches in the rise. Then $12x$ represents the number of inches in the run. Write an equation to represent the situation.

Rise	Run
1	12
2	24
3	36
4	48
x	$12x$

In this situation, run is a function of rise. Thus, the equation $12x = 180$ is a function.

$12x = 180$ *This equation will be solved in Example 3.*

To solve equations with multiplication and division, you will need new tools. Equations of the form $ax = b$, where a and/or b are fractions, are generally solved by using the **multiplication property of equality**.

Multiplication Property of Equality	For any numbers a, b, and c, if $a = b$, then $ac = bc$.

Example ❶ Solve $\frac{g}{24} = \frac{5}{12}$.

$$\frac{g}{24} = \frac{5}{12}$$

$$24\left(\frac{g}{24}\right) = 24\left(\frac{5}{12}\right) \quad \text{Multiply each side by 24.}$$

$$g = 2(5) \text{ or } 10$$

Check: $\frac{g}{24} = \frac{5}{12}$

$$\frac{10}{24} \overset{?}{=} \frac{5}{12} \quad \text{Replace } g \text{ with 10.}$$

$$\frac{5}{12} = \frac{5}{12} \quad \text{✔} \qquad \text{The solution is 10.}$$

Example 2 Solve each equation.

a. $\left(3\frac{1}{4}\right)p = 2\frac{1}{2}$

$$\left(3\frac{1}{4}\right)p = 2\frac{1}{2}$$

$$\frac{13}{4}p = \frac{5}{2} \qquad \text{\textit{Rewrite the mixed numbers as improper fractions.}}$$

$$\frac{4}{13}\left(\frac{13}{4}\right)p = \frac{4}{13}\left(\frac{5}{2}\right) \qquad \text{\textit{Multiply each side by }} \frac{4}{13}, \text{ \textit{the reciprocal of }} \frac{13}{4}.$$

$$p = \frac{20}{26} \text{ or } \frac{10}{13} \qquad \text{\textit{Check this result.}}$$

The solution is $\frac{10}{13}$.

b. $40 = -5d$

$$40 = -5d$$

$$-\frac{1}{5}(40) = -\frac{1}{5}(-5d) \qquad \text{\textit{Multiply each side by }} -\frac{1}{5}.$$

$$-8 = d \qquad \text{\textit{Check this result.}}$$

The solution is -8.

The equation in Example 2b, $40 = -5d$, was solved by multiplying each side by $-\frac{1}{5}$. The same result could have been obtained by dividing each side by -5. This method uses the **division property of equality**. It is often easier to use than the multiplication property of equality.

Division Property of Equality	For any numbers a, b, and c, with $c \neq 0$, if $a = b$, then $\frac{a}{c} = \frac{b}{c}$.

Example 3 **APPLICATION** **Construction**

Refer to the application at the beginning of the lesson. Solve $12x = 180$.

$$12x = 180$$

$$\frac{12x}{12} = \frac{180}{12} \qquad \text{\textit{Divide each side by 12.}}$$

$$x = 15$$

12x or 180"

Check:
$$12x = 180$$
$$12(15) \stackrel{?}{=} 180$$
$$180 = 180 \checkmark$$

The rise of the ramp could be at most 15 inches.

Motivating the Lesson

Situational Problem Read the following to the class.
Mr. Kwan bought 15 computer disks and a carrying case for $28.50. If the carrying case cost $6.75, what was the cost of each disk?
Then have students identify the procedure they would use to solve the problem. Determine the procedure identified that involves the least number of steps.

2 TEACH

In-Class Examples

For Example 1
Solve.

a. $\frac{3}{4} = \frac{a}{12}$ 9

b. $\frac{2}{7} = \frac{b}{14}$ 4

c. $\frac{3}{5} = \frac{c}{30}$ 18

For Example 2
Solve.

a. $\left(3\frac{1}{2}\right)t = 5\frac{1}{4}$ $1\frac{1}{2}$

b. $\left(1\frac{3}{4}\right)m = 4\frac{1}{2}$ $2\frac{4}{7}$

c. $\left(-5\frac{2}{5}\right)y = -4\frac{1}{20}$ $\frac{3}{4}$

For Example 3
Solve.

a. $-5t = 60$ -12

b. $15 = 6n$ $2\frac{1}{2}$

c. $-3v = -129$ 43

Teaching Tip Point out that mixed numbers are less difficult to work with if they are changed to improper fractions or decimals, as in Example 2.

Teaching Tip Be sure to help students develop the habit of checking their solutions, as in Examples 2 and 3.

Teaching Tip You may wish to show Example 3 using the division property of equality.
$$40 = -5d$$
$$\frac{40}{-5} = \frac{-5d}{-5}$$
$$-8 = d$$
The solution is -8.

3 PRACTICE/APPLY

Check for Understanding

Exercises 1–13 are designed to help you assess your students' understanding through reading, writing, speaking, and modeling. You should work through Exercises 1–5 with your students and then monitor their work on Exercises 6–13.

Additional Answers

1. Doralina is correct. There is no value for x that would make the statement $0x = 8$ true.
2. Rise is the number of units in a vertical direction, and run is the number of units in a horizontal direction.
4.

	17 m
x m	$A = 51$ m^2

Let x = the width. Then $17x = 51$, and $x = 3$.

Study Guide Masters, p. 20

Example ④ **Solve** $3x = -9$.

$3x = -9$

$\frac{3x}{3} = \frac{-9}{3}$ *Divide each side by 3.*

$x = -3$

Recall that an equation mat is a model of an equation. You can group each cup on the left side of the mat with the same number of counters on the right side of the mat.

$3x = -9$ $x = -3$

The solution is -3.

CHECK FOR UNDERSTANDING

Communicating Mathematics

Study the lesson. Then complete the following. 1–2. See margin.

1. **You Decide** Kezia says that the equation $0x = 8$ has a solution. Doralina says it doesn't. Which one is correct, and why?

2. **Define** the terms *rise* and *run* in your own words.

3. **Refer** to the application at the beginning of the lesson. If the maximum rise is 30″, what is the maximum run? **360 inches or 30 feet**

4. **Draw** a model and explain how to find the width of a rectangle if its area is 51 m^2 and its length is 17 m. **See margin.**

MODELING MATHEMATICS

5. Use cups and counters to model $4x = 16$. Then solve the equation. **See margin.**

Guided Practice

Solve each equation. Then check your solution.

6. $-8t = 56$ -7
7. $-5s = -85$ 17
8. $42.51x = 8$ 0.19
9. $\frac{k}{8} = 6$ 48
10. $-10 = \frac{b}{-7}$ 70
11. $-5x = -3\frac{2}{3}$ $\frac{11}{15}$

Define a variable, write an equation, and solve the problem. Then check your solution.

12. Eight times a number is 216. What is the number? $8n = 216$; 27

13. The product of -7 and a number is 1.477. What is the number? $-7n = 1.477$; -0.211

Reteaching

Using Substeps State the operation and the number you would apply to each side to solve each equation. Then solve. Check each solution.

1. $4n = -20$ $\times \frac{1}{4}$ or $\div 4$; -5
2. $12 = -3a$ $\times \left(-\frac{1}{3}\right)$ or $\div (-3)$; -4
3. $1.2x = 6$ $\times \left(\frac{1}{1.2}\right)$ or $\div 1.2$; 5
4. $\frac{x}{3} = 6$ $\times 3$ or $\div \frac{1}{3}$; 18
5. $\frac{m}{-6} = 12$ $\times (-6)$ or $\div \left(-\frac{1}{6}\right)$; -72
6. $\frac{c}{2.4} = 12$ $\times 2.4$ or $\div \frac{1}{2.4}$; 28.8

Additional Answer

5.

$x = 4$

Practice

16. 4.44 or $4\frac{4}{9}$

17. −17.33 or $-17\frac{1}{3}$

18. −3.67 or $-3\frac{2}{3}$

Solve each equation. Then check your solution.

14. $-4r = -28$ 7   15. $5x = -45$ −9   16. $9x = 40$

17. $-3y = 52$ 18. $3w = -11$ 19. $434 = -31y$ −14

20. $1.7b = -39.1$ −23   21. $0.49x = 6.277$ 12.81   22. $-5.73c = 97.41$ −17

23. $-0.63y = -378$ 600   24. $11 = \frac{x}{5}$ 55   25. $\frac{h}{11} = -25$ −275

26. $\frac{c}{-8} = -14$ 112   27. $\frac{2}{5}t = -10$ −25   28. $-\frac{11}{8}x = 42$ $-30\frac{6}{11}$

29. $-\frac{13}{5}y = -22$ $8\frac{6}{13}$ 30. $3x = 4\frac{2}{3}$ $1\frac{5}{9}$ 31. $\left(-4\frac{1}{2}\right)x = 36$ −8

Define a variable, write an equation, and solve each problem. Then check your solution.

32. Six times a number is −96. Find the number. $6n = -96$; −16

33. Negative twelve times a number is −156. What is the number?

34. One fourth of a number is −16.325. What is the number?

35. Four thirds of a number is 4.82. What is the number? $\frac{4}{3}n = 4.82$; 3.615

36. Seven eighths of a number is 14. What is the number? $\frac{7}{8}n = 14$; 16

33. $-12n = -156$; 13
34. $\frac{1}{4}n = -16.325$; −65.3

Complete.

37. If $3x = 15$, then $9x = \underline{\ ?\ }$.   45   38. If $10y = 46$, then $5y = \underline{\ ?\ }$.   23

39. If $2a = -10$, then $-6a = \underline{\ ?\ }$.   30   40. If $12b = -1$, then $4b = \underline{\ ?\ }$. $-\frac{1}{3}$

41. If $7k - 5 = 4$, then $21k - 15 = \underline{\ ?\ }$.   12

Critical Thinking

42. If $-x$ and $-1 \cdot x$ always represent the same number, does $-x$ always stand for a negative number? Explain your reasoning.   No; if x is negative, then $-x$ is positive.

Applications and Problem Solving

Write an equation to represent each situation. Then solve the problem.

43. **Sociology** Research conducted by *USA Today* has shown that about 1 in 7 people in the world are left-handed.
   a. About how many left-handed people are there in a group of 350 people? 583 people? $7\ell = 350$, 50 people; $7\ell = 583$, ≈ 83 people
   b. If there are 65 left-handed people in a group, about how many people are in that group? $\frac{p}{7} = 65$; 455 people

44. **Demographics** In the year 2000, there will be about $1\frac{1}{2}$ times the number of 15- to 19-year-olds in the U.S than there were in 1960. If the U.S. Census Bureau projects that there will be 19,819,000 15- to 19-year-olds in the year 2000, about how many were there in 1960?

44. $1\frac{1}{2}t = 19{,}819{,}000$; 13,212,667

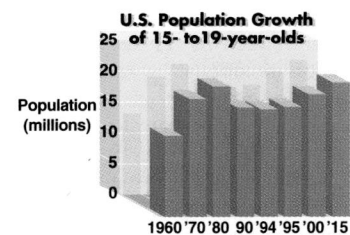

U.S. Population Growth of 15- to19-year-olds

Population (millions)

1960 '70 '80 90 '94 '95 '00 '15
Year (est.)

Source: U.S.Census Bureau

45. **Telecommunications** In 1995, the long-distance company Sprint introduced Sprint Sense, a plan in which long-distance calls placed on weekends cost only $0.10 per minute.
   a. How long could you talk for $2.30? $0.10m = 2.30$; 23 minutes
   b. What would be the cost of an 18-minute call? $0.10(18) = c$; $1.80

4 ASSESS

Closing Activity

Writing Have students identify and write about specific instances in which algebra is used in geometry. As they write, have them create problems to illustrate their answers.

Chapter 3, Quiz A (Lessons 3-1 and 3-2), is available in the *Assessment and Evaluation Masters*, p. 72.

Additional Answer

50.

$$-1\ 0\ 1\ 2\ 3\ 4\ 5$$

Enrichment Masters, p. 20

Mixed Review

47. $17 + 1\frac{1}{2}y = 33\frac{1}{2}$; **11 years**

51a. The number of bags increased during the day.

51b. The machine was refilled.

51c. See students' work.

46. Solve $-11 = a + 8$. Then check your solution. (Lesson 3–1) **−19**

47. Define a variable, then write an equation for the problem below. (Lesson 2–9) Ponderosa pines grow about $1\frac{1}{2}$ feet each year. If a pine tree is now 17 feet tall, about how long will it take for the tree to become $33\frac{1}{2}$ feet tall?

48. **World Records** The longest loaf of bread ever baked was 2132 feet $2\frac{1}{2}$ inches. If this loaf were cut into $\frac{1}{2}$-inch slices, how many slices of bread would there have been? (Lesson 2–7) **51,173 slices**

49. Find a number between $\frac{1}{2}$ and $\frac{6}{7}$. (Lesson 2–4) **$\frac{19}{28}$**

50. Graph $\{-1, 1, 3, 5\}$ on a number line. (Lesson 2–1) **See margin.**

51. The graph at the right shows the number of bags of potato chips in the snack machine at Lee High School two times during an average day. (Lesson 1–9)
   a. What can you learn from looking at the graph?
   b. What do you think happened between the times indicated by the dots?
   c. If lunch at Lee is between 11:45 A.M. and 1:15 P.M., estimate the number of bags you think would be in the machine at 1:30 P.M. Justify your estimate.

Number of Bags

10 11 12 1 2
A.M. A.M. NOON P.M. P.M.
Time of Day

52. Simplify $5(3x + 2y - 4y)$. (Lesson 1–7) **$15x - 10y$**

53. Evaluate $a + b^2 + c^2$ if $a = 6$, $b = 4$, and $c = 3$. (Lesson 1–1) **31**

WORKING ON THE In·ves·ti·ga·tion

Refer to the Investigation on pages 68–69.

the Greenhouse Effect

The changes in the levels of carbon dioxide (CO_2) on Earth happened naturally until nearly 200 years ago. During the Industrial Revolution of the early 1800s, a new factor was thrown into the equation. When wood and fossil fuels like coal, oil, and natural gas are burned, CO_2 is released in large quantities. Oceans and vegetation absorb the gas. But due to the widespread cutting of trees to produce goods in the 1900s, few remained to soak up excess CO_2. In 1920, atmospheric CO_2 stood at about 280 parts per million (ppm). By 1996, it had risen to 356 ppm. **1–4. See margin.**

1 Make a chart that shows the changes in the levels of CO_2 in our atmosphere. Assume that the levels of CO_2 grow at a constant rate over the years. Use the headings *Year, Amount of CO_2 (ppm)*, and *Change*. Start the chart with 1920, and end with the present year.

2 Examine the chart that you made. Look for patterns and write a paragraph about the change in ppm as a function of time.

3 Based on the relationships you found in the previous question, determine how much CO_2 was in the atmosphere in 1800 and 1950. Explain how you found those amounts.

4 Predict the amounts of CO_2 that will be in the atmosphere in the years 2000, 2020, 2050, and 3000. Explain how you made your predictions.

Add the results of your work to your Investigation Folder.

154 *Chapter 3 Solving Linear Equations*

Extension ▬

Problem Solving JoAnn bought 18 cupcakes for a party. She had writing put on seven of the cupcakes at an extra cost of 20¢ each. If the total cost of the cupcakes was $11.30, how much did each plain cupcake cost? **55¢**

Working on the Investigation

The Investigation on pages 68–69 is designed to be a long-term project that is completed over several days or weeks. Encourage students to keep their materials in their Investigation Folder as they work on the Investigation.

3–3A Solving Multi-Step Equations

Materials: cups & counters ⬚ equation mat

You can use an equation model to solve equations with more than one operation or equations with a variable on each side.

A Preview of Lesson 3–3

Activity 1   Use an equation model to solve 2x + 2 = −4.

Step 1 Model the equation by placing 2 cups and 2 positive counters on one side of the mat. Place 4 negative counters on the other side of the mat. Add 2 negative counters to each side to form zero pairs on the left.

Step 2 Group the counters to form zero pairs and remove the zero pairs. Separate the remaining counters into 2 equal groups to match the 2 cups. Each cup is paired with 3 negative counters. Therefore, *x* = −3.

2x + 2 = −4

x = −3

Activity 2   Use an equation model to solve w − 3 = 2w − 1.

Step 1 Model the equation by placing 1 cup and 3 negative counters on one side of the mat. Place 2 cups and 1 negative counter on the other side of the mat. Remove 1 negative counter from each side of the mat.

Step 2 You can remove the same number of cups from each side of the mat. In this case, remove 1 cup from each side. The cup on the right is matched with 2 negative counters. Therefore, *w* = −2.

w − 3 = 2w − 1

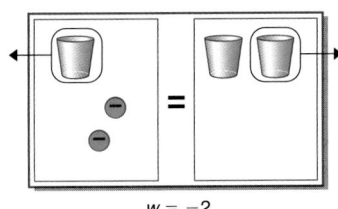

w = −2

Model   Use equation models to solve each equation.

1. 2x + 3 = 13   5
2. 2y − 2 = −4   −1
3. −4 = 3a + 2   −2
4. 3m − 2 = 4   2
5. 3x + 2 = x + 6   2
6. 3x + 7 = x + 1   −3
7. 3x − 2 = x + 6   4
8. y + 1 = 3y − 7   4
9. 2b + 3 = b + 1   −2

Cooperative Learning

This lesson offers an excellent opportunity for using cooperative learning groups. For more information on cooperative learning strategies and group management, see *Cooperative Learning in the Mathematics Classroom,* one of the titles in the Glencoe Mathematics Professional Series.

4 ASSESS

Observing students working in cooperative groups is an excellent method of assessment.

NCTM Standards: 1–5

Objective
Solve equations by using cups and counters to model variables and integers.

Recommended Time
Demonstration and discussion: 15 minutes; Exercises: 30 minutes

Instructional Resources
For each student or group of students
Student Manipulative Kit
• 20 red/yellow 2-sided counters
Modeling Mathematics Masters
• pp. 2–3 (cups and counters)
• p. 13 (equation mat)
• p. 24 (worksheet)
For teacher demonstration
Algebra and Geometry Overhead Manipulative Resources

1 FOCUS

Motivating the Lesson
Remind students that this is not a new procedure, but a combination of the two types of equations they solved in Lesson 3-1A. Help students to see the connection between the processes used to solve one-step equations and those used to solve multi-step equations.

2 TEACH

Teaching Tip Encourage students to check their solutions algebraically by replacing the variable in the original equation with the solution to see if the equation holds.

3 PRACTICE/APPLY

Assignment Guide
Core: 1–9
Enriched: 1–9

NCTM Standards: 1–5

Instructional Resources

- Study Guide Master 3-3
- Practice Master 3-3
- Enrichment Master 3-3
- Modeling Mathematics Masters, pp. 42–44
- Multicultural Activity Masters, p. 5
- Real-World Applications, 8
- Tech Prep Applications Masters, p. 6

Transparency 3-3A contains the 5-Minute Check for this lesson; **Transparency 3-3B** contains a teaching aid for this lesson.

Recommended Pacing	
Standard Pacing	Days 5 & 6 of 14
Honors Pacing	Day 5 of 13
Block Scheduling*	Day 3 of 6
Alg. 1 in Two Years*	Days 7 & 8 of 19

*For more information on pacing and possible lesson plans, refer to the *Block Scheduling Booklet* and *Algebra 1 in Two Years.*

1 FOCUS

5-Minute Check
(over Lesson 3-2)

Solve.

1. $\frac{k}{8} = -24$ -192
2. $39 = -3b$ -13
3. $\left(1\frac{1}{3}\right)p = 2\frac{1}{6}$ $1\frac{5}{8}$
4. $-65.8 = 8.27t$ -7.96
5. $\frac{4x}{5} = -32$ -40
6. Suzanne bought 7 books for $55.23. If all the books cost the same amount, what was the price of 1 book? $7.89

3-3

Solving Multi-Step Equations

Objectives

After studying this lesson, you should be able to:

- solve equations involving more than one operation, and
- solve problems by working backward.

APPLICATION
World Cultures

On June 9, 1993, Japan's Crown Prince Naruhito wed former diplomat Masako Owada. It took Ms. Owada about $2\frac{1}{2}$ hours to put on her wedding kimono, a 12-layered silk garment that weighed about 30 pounds. The prince wore a kimono of bright orange—the "sunrise color"—that can only be worn by the heir to the throne.

After the wedding, the crown prince changed into a tuxedo and the princess into a bridal gown to formally announce their marriage to the Emperor and Empress. To change into the bridal gown, Princess Masako and her attendants had to remove the wedding kimono, a process that was similar to putting on the garment, only in reverse order. You will solve certain kinds of problems by working in reverse order, or **working backward.** Work backward is one of many *problem-solving strategies* that you can use to solve problems. Here are some other problem-solving strategies.

F Y I

The *kimono* has been worn by Japanese men and women since the 7th Century A.D. It is an ankle-length gown with long wide sleeves and is secured by a sash called an *obi*.

Problem-Solving Strategies	
draw a diagram	solve a simpler (or a similar) problem
make a table or chart	eliminate the possibilities
make a model	look for a pattern
guess and check	act it out
check for hidden assumptions	list the possibilities
use a graph	identify subgoals

Example ❶ **Due to melting, an ice sculpture loses one-half its weight every hour. After 8 hours, it weighs $\frac{5}{16}$ of a pound. How much did it weigh in the beginning?**

Make a table to show weight as a function of time. Work backward to find the original weight.

Recall that the sculpture loses one-half its weight every hour. Multiply the current weight of the sculpture by 2 to find its weight an hour before. Continue multiplying until you reach the original weight.

The original weight of the sculpture was 80 pounds.

Hour	Weight
8	$\frac{5}{16}$
7	$2\left(\frac{5}{16}\right) = \frac{5}{8}$
6	$2\left(\frac{5}{8}\right) = \frac{5}{4}$
5	$2\left(\frac{5}{4}\right) = \frac{5}{2}$
4	$2\left(\frac{5}{2}\right) = 5$
3	$2(5) = 10$
2	$2(10) = 20$
1	$2(20) = 40$
0	$2(40) = 80$

F Y I

The wedding of Crown Prince Naruhito and Masako Owada was the first to take advantage of high-definition television (HDTV). In keeping with tradition, the groom's parents did not attend the wedding ceremony. They were, however, able to watch a computer simulation of the private ceremony on their HDTV set.

To solve equations with more than one operation, often called **multi-step equations,** you will undo the operations by working backward.

Example 2

Mrs. Guzman needs a repairperson to fix her washing machine. Since her machine is pretty old, she doesn't want to spend more than $100 for repairs. The owner of Albie's Appliances told her that a service call would cost $35 and the labor would be an additional $20 per hour. What is the maximum number of hours that the repairperson can work and keep the total cost at $100?

Explore   Read the problem and define the variable.
Let h represent the maximum number of hours the repair can take.

Plan   Write an equation.

$\underbrace{\$35}\ \underbrace{plus}\ \underbrace{\$20\ per\ hour\ for\ h\ hours}\ \underbrace{equals}\ \underbrace{\$100.}$
$\ \ \ \ 35\ \ \ \ \ +\ \ \ \ \ \ \ \ \ \ \ \ \ 20h\ \ \ \ \ \ \ \ \ \ \ \ =\ \ \ \ 100$

Solve   Work backward to solve the equation.

$35 + 20h = 100$ *Undo the addition first. Use the subtraction*
$35 - 35 + 20h = 100 - 35$ *property of equality.*
$20h = 65$
$\dfrac{20h}{20} = \dfrac{65}{20}$ *Then undo the multiplication. Use the*
division property of equality.
$h = \dfrac{13}{4}\ or\ 3\dfrac{1}{4}$

The repairperson has up to $3\dfrac{1}{4}$ hours to fix the washing machine.

Examine   Check to see if the answer makes sense.

$3\dfrac{1}{4} \times \$20 = \$65$ $\$35 + \$65 = \$100$

EXPLORATION   GRAPHING CALCULATORS

You can use a graphing calculator, like the TI-82, to solve multi-step equations. The *solve(* function will solve an equation if it is rewritten as an expression that equals zero, or most commonly, in the form $ax + b - c = 0$. This function also requires that you include a guess about the solution to the equation. The format is as follows.

solve(*expression,variable,guess*)

To solve the equation $3x - 4 = 2$, access the math function by pressing the [MATH] key. Then choose 0, for solve. Enter the expression $3x - 4 - 2$, the variable x, and a guess about its solution. If we guess -1, then this appears on the calculator screen: solve(3X-4-2,X,-1). When you press [ENTER], the solution, 2, appears on the screen.

Your Turn

a. Work through the examples in this lesson by using a graphing calculator.

b. Describe the process of using a graphing calculator to solve equations in your own words.

b–d. See margin.

c. Why did we input the equation $3x - 4 = 2$ as $3x - 4 - 2$?

d. How would you use a graphing calculator to solve the equation $3x - 4 = -2x + 6$?

For Example 3
Solve.

a. $\frac{t}{3} + 11 = 29$ 54

b. $\frac{x}{5} - 17 = 3$ 100

c. $\frac{2}{3}a - 14 = 61$ $112\frac{1}{2}$

d. $12 = \frac{2 - 3x}{4}$ $-15\frac{1}{3}$

e. $\frac{3b + 1}{2} = -25$ -17

f. $\frac{5 + 2x}{8} = 14$ $53\frac{1}{2}$

Teaching Tip You may wish to show an alternate method for solving Example 3a.

$$\frac{y}{5} + 9 = 6$$
$$5\left(\frac{y}{5} + 9\right) = 5(6)$$
$$y + 45 = 30$$
$$y + 45 - 45 = 30 - 45$$
$$y = -15$$

The solution is -15.

Additional Answers

2. Let $2n + 2$ represent consecutive even numbers and let $2n + 1$ and $2n + 3$ represent consecutive odd numbers.
$$2n + 2n + 2 = (2n + 1) + (2n + 3)$$
$$4n + 2 = 4n + 4$$
$$2 = 4$$
Since this is not a true statement, the sum of two consecutive even numbers can never equal the sum of two consecutive odd numbers.

4. $$4x + 3x - 5 = 27$$
$$4(-2) + 3(-2) - 5 \stackrel{?}{=} 27$$
$$-8 + (-6) - 5 \stackrel{?}{=} 27$$
$$-19 \neq 27$$

8. Add 5 to each side, then divide each side by 4.

9. Subtract 7 from each side, then divide each side by 3.

10. Add 6 to each side, then multiply each side by $\frac{9}{2}$.

11. Subtract 3 from each side, then multiply each side by -7.

12. Multiply each side by 7, then add 3 to each side.

13. Rewrite numerator as $p + 5$, multiply each side by -2, and then subtract 5.

You have seen a multi-step equation in which the first, or *leading*, coefficient is an integer. You can use the same steps if the leading coefficient is a fraction.

Example 3 Solve each equation.

a. $\frac{y}{5} + 9 = 6$

$\frac{y}{5} + 9 = 6$ *The leading coefficient is $\frac{1}{5}$.*

$\frac{y}{5} + 9 - 9 = 6 - 9$ *First, subtract 9 from each side. Why?*

$\frac{y}{5} = -3$

$5\left(\frac{y}{5}\right) = 5(-3)$ *Then, multiply each side by 5.*

$y = -15$

Check: $\frac{y}{5} + 9 = 6$

$\frac{-15}{5} + 9 \stackrel{?}{=} 6$

$-3 + 9 \stackrel{?}{=} 6$

$6 = 6$ ✔

The solution is -15.

b. $\frac{d - 2}{3} = 7$

$\frac{d - 2}{3} = 7$

$3\left(\frac{d - 2}{3}\right) = 3(7)$ *Multiply each side by 3. Why?*

$d - 2 = 21$

$d - 2 + 2 = 21 + 2$ *Add 2 to each side.*

$d = 23$

Check: $\frac{d - 2}{3} = 7$

$\frac{23 - 2}{3} \stackrel{?}{=} 7$

$\frac{21}{3} \stackrel{?}{=} 7$

$7 = 7$ ✔

The solution is 23.

Consecutive integers are integers in counting order, such as 3, 4, 5. Beginning with an even integer and counting by two will result in *consecutive even integers*. For example, -6, -4, -2, 0, and 2 are consecutive even integers. Beginning with an odd integer and counting by two will result in *consecutive odd integers*. For example, -1, 1, 3, and 5 are consecutive odd integers.

The study of numbers and the relationships between them is called **number theory**. Number theory involves the study of odd and even numbers.

Alternative Learning Styles

Kinesthetic Have students work in groups of four. Give each group a set of cards numbered 1 through 8. Each student should hold two cards. The group's task is to arrange its members in the formation shown at the right so that no two consecutive integers occupy neighboring squares horizontally, vertically, or diagonally.

Once all of the groups have reached a solution, have one member from each group record and explain the group's solution to the class. There are several solutions. One possible solution is provided.

Example **4** **Find three consecutive odd integers whose sum is -15.**

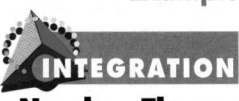
INTEGRATION
Number Theory

Let n = the least odd integer.
Then $n + 2$ = the next greater odd integer,
and $n + 4$ = the greatest of the three odd integers.

$$n + (n + 2) + (n + 4) = -15$$
$$3n + 6 = -15 \qquad \textit{Simplify.}$$
$$3n + 6 - 6 = -15 - 6 \qquad \textit{Subtract 6 from each side.}$$
$$3n = -21$$
$$\frac{3n}{3} = \frac{-21}{3} \qquad \textit{Divide each side by 3.}$$
$$n = -7$$

$n + 2 = -7 + 2 \qquad\qquad n + 4 = -7 + 4$
$n + 2 = -5 \qquad\qquad\quad n + 4 = -3$

The consecutive odd integers are -7, -5, and -3.

Explain why this answer makes sense.

CHECK FOR UNDERSTANDING

Communicating Mathematics

Study the lesson. Then complete the following.

1. a. **Explain** how you would solve $2p + 10 = 42$ if you had to undo the multiplication first.   **Divide each term by 2.**
   b. **Explain** why undoing the multiplication first would be inconvenient for solving the equation $7x - 4 = 24$.   **Fractions would be introduced.**

2. See margin.

2. **Determine** whether the sum of two consecutive even numbers can ever equal the sum of two consecutive odd numbers. Justify your answer.

3. Subtract 2; $n - 2$.

3. If n is an even integer, explain how to find the even integer just before it.

4. See margin.

4. **Explain** why -2 is not the solution of the equation $4x + 3x - 5 = 27$.

5. $-22, -20, -18$

5. **List** three consecutive even integers if the greatest one is -18.

6. **Complete:** If $2x + 1 = 5$, then $3x - 4 = \underline{\ ?\ }$. Explain your reasoning.   **2**

MODELING MATHEMATICS

7. Write an equation for the model at the right. Then use cups and counters to solve the equation.
   $3x + 2 = 8$; 2

Guided Practice

8–13. See margin for explanations.

Explain how to solve each equation. Then solve.

8. $4a - 5 = 15$ **5**
9. $7 + 3c = -11$ **-6**
10. $\frac{2}{9}v - 6 = 14$ **90**
11. $3 - \frac{a}{7} = -2$ **35**
12. $\frac{x-3}{7} = -2$ **-11**
13. $\frac{p-(-5)}{-2} = 6$ **-17**

Define a variable, write an equation, and solve each problem. Then check your solution.   15. $n + (n + 2) + (n + 4) = 21$; 5, 7, 9

14. $n + (n + 1) = -31$; $-16, -15$

14. Find two consecutive integers whose sum is -31.

15. Find three consecutive odd integers whose sum is 21.

16. $29 = 13 + 4n$; $n = 4$

16. Twenty-nine is 13 added to 4 times a number. Find the number.

Lesson 3-3 Solving Multi-Step Equations **159**

Reteaching

Using Alternative Methods Solve the equations by undoing the multiplication or division first. Be sure to multiply or divide each side completely (all terms) by the number.

1. $42 = 10 + 2p$ **16**
2. $\frac{x}{4} + 9 = 6$ **-12**
3. $\frac{d-4}{3} = 5$ **19**
4. Two germs live in a tin can. At 4:00 P.M. they begin to double in

number every second. At 5:00 P.M. the can is completely full of germs.

a. When was the can
   i. $\frac{1}{2}$ full? **4:59**
   ii. $\frac{1}{4}$ full? **4:58**
   iii. $\frac{1}{8}$ full? **4:57**

b. How many germs does the can hold? **2^{61}**

In-Class Example

For Example 4
Find four consecutive odd integers whose sum is 72.
15, 17, 19, 21

Teaching Tip In Example 4, be sure students understand that zero is even and that both positive and negative integers can be odd or even.

3 PRACTICE/APPLY

Check for Understanding
Exercises 1–16 are designed to help you assess your students' understanding through reading, writing, speaking, and modeling. You should work through Exercises 1–7 with your students and then monitor their work on Exercises 8–16.

Error Analysis
Students may choose to undo multiplication before undoing addition. Emphasize that the whole side (all terms) must be divided. The distributive property shows why using the multiplicative property of equality is equivalent to multiplying each term by the number.

Study Guide Masters, p. 21

NAME_____ DATE _____
3-3
Study Guide
Student Edition
Pages 156–161

Solving Multi-Step Equations

When solving some equations you must perform more than one operation on both sides. First, determine what operations have been done to the variable. Then undo these operations in the reverse order.

Example 1: How would you solve $\frac{n}{3} - 7 = 28$?

$\frac{n}{3} - 7 = 28$ First, n was divided by 3. } To solve, first add 7 to each side.
  Then 7 was subtracted. } Then multiply each side by 3.

| Procedure for Solving a Two-Step Equation | 1. Undo any indicated additions or subtractions. |
| | 2. Undo any indicated multiplications or divisions involving the variable. |

Example 2: $5x + 3 = 23$ Addition of 3 is indicated.   **Check:**
$5x + 3 - 3 = 23 - 3$ Therefore, subtract 3 from each side. $5x + 3 = 23$
$5x = 20$ Multiplication by 5 is also indicated. $5(4) + 3 \stackrel{?}{=} 23$
$\frac{5x}{5} = \frac{20}{5}$ Therefore, divide each side by 5. $20 + 3 \stackrel{?}{=} 23$
$x = 4$ $23 = 23$ ✔

Solve each equation. Then check your solution.

1. $5z + 16 = 51$ 7
2. $14n - 8 = 34$ 3
1. $0.6x - 1.5 = 1.8$ 5.5
4. $\frac{4b+8}{-2} = 10$ -7
5. $16 = \frac{d-12}{14}$ 236
6. $8 + \frac{3n}{12} = 13$ 20
7. $\frac{7}{8}p - 4 = 10$ 16
8. $\frac{g}{-5} + 3 = -13$ 80
9. $-4 = \frac{7x-(-1)}{-8}$ $\frac{31}{7}$, or $4\frac{3}{7}$

Define a variable, write an equation, and solve each problem. Then check your solution.

10. Find three consecutive integers whose sum is 96.
   $n + (n+1) + (n+2) = 96$; 31, 32, 33
11. Find two consecutive odd integers whose sum is 176.
   $n + (n+2) = 176$; 87, 89

Chapter 3 **159**

160 Chapter 3

EXERCISES

Practice

Solve each equation. Then check your solution.

17. $3x - 2 = -5$ −1
18. $-4 = 5n + 6$ −2
19. $5 - 9w = 23$ −2
20. $17 = 7 + 8y$
21. $-2.5d - 32.7 = 74.1$
22. $0.2n + 3 = 8.6$ 28
23. $\frac{3}{2}a - 8 = 11$ $12\frac{2}{3}$
24. $5 = -9 - \frac{p}{4}$ −56
25. $7 = \frac{c}{-3} + 5$ −6
26. $\frac{m}{-5} + 6 = 31$ −125
27. $\frac{g}{8} - 6 = -12$ −48
28. $6 = -12 + \frac{h}{-7}$ −126
29. $\frac{b + 4}{-2} = -17$ 30
30. $\frac{z - 7}{5} = -3$ −8
31. $-10 = \frac{17 - s}{4}$ 57
32. $\frac{4t - 5}{-9} = 7$ −14.5
33. $\frac{7n + (-1)}{6} = 5$ $4\frac{3}{7}$
34. $\frac{-3j - (-4)}{-6} = 12$ $25\frac{1}{3}$

INTEGRATION
Number Theory

Define a variable, write an equation, and solve each problem. Then check your solution. 36–37. See margin.

35. Twelve decreased by twice a number is -7. $12 - 2n = -7; 9\frac{1}{2}$
36. Find three consecutive integers whose sum is -33.
37. Find four consecutive integers whose sum is 86.
38. Find two consecutive odd integers whose sum is 196.
    $n + (n + 2) = 196$; 97, 99

INTEGRATION
Geometry

Make a diagram to represent each situation. Then define a variable, write an equation, and solve each problem. Check your solution.

39. The lengths of the sides of a triangle are consecutive odd integers. The perimeter is 39 meters. What are the lengths of the sides?
40. The lengths of the sides of a quadrilateral are consecutive even integers. Twice the length of the shortest side plus the length of the longest side is 120 inches. Find the lengths of the four sides.
    $2n + (n + 6) = 120$; 38 in., 40 in., 42 in., 44 in.

Graphing Calculator

Solve each equation. Then use a graphing calculator to verify your solution.

41. $0.2x + 3 = 8.6$ 28
42. $4.91 + 7.2x = 39.4201$ 4.793
43. $\frac{3 + x}{7} = -5$ −38
44. $\frac{-3n - (-4)}{-6} = -9$ −16.67 or $-16\frac{2}{3}$

Critical Thinking

45. Write an expression for the sum of three consecutive even integers if $3n - 1$ is the least integer. $(3n - 1) + (3n + 1) + (3n + 3)$

Applications and Problem Solving

For Exercises 46–47, write an equation to represent each situation. Then solve the problem.

46. **Health** According to the American Medical Association, the average birth weight of a baby in the United States is 7.5 pounds. After birth, most babies lose an average of 1 ounce a day for the first 5 days and then gain an average of 1 ounce a day for the next 13 weeks.
    a. Suppose a baby weighs 7.5 pounds at birth. After how many days will the baby weigh 12 pounds? $(120 - 5) + (d - 5) = 192$; 82 days
    b. Draw a graph that shows the expected weight of a 7.5-pound baby for each of its first 13 weeks of life. **See margin.**

47. **World Cultures** Hawaii became our 50th state in 1959. It's the only state that is not on the North American continent. The English alphabet contains 2 more than twice as many letters as the Hawaiian alphabet. How many letters are there in the Hawaiian alphabet? $2a + 2 = 26$; 12 letters

160 Chapter 3 Solving Linear Equations

Sidebar answers:
20. 1.25 or $1\frac{1}{4}$
21. −42.72
39. $n + (n + 2) + (n + 4) = 39$; 11 m, 13 m, 15 m

648. Wilson, 16 bags;
Martinez, 8 bags;
Brightfeather, 4 bags;
Wimberly, 2 bags

48. Work Backward Four families went to a baseball game. A vendor selling bags of popcorn came by. The Wilson family bought half of the bags of popcorn plus one. The Martinez family bought half of the remaining bags of popcorn plus one. The Brightfeather family bought half of the remaining bags of popcorn plus one. The Wimberly family bought half of the remaining bags of popcorn plus one, leaving the vendor with no bags of popcorn. If the Wimberlys bought 2 bags of popcorn, how many bags did each of the four families buy?

Mixed Review

49a–b. See students' work.

49. Health One way to calculate your suggested daily intake of fat grams is to multiply your ideal weight in pounds times 0.454. (Lesson 3–2)
  a. Write an equation to represent your suggested daily intake.
  b. What is your suggested daily intake of fat grams?

50. Complete: If $x + 4 = 15$, then $x - 2 = \underline{\ ?\ }$. (Lesson 3–1) **9**

51. Name the set or sets of numbers to which $\sqrt{11}$ belongs. Use N for natural numbers, W for whole numbers, Z for integers, Q for rational numbers, or I for irrational numbers. (Lesson 2–8) **I**

52. Simplify $\frac{-200x}{50}$. (Lesson 2–7) **$-4x$**

53. Simplify $4(7) - 3(11)$. (Lesson 2–6) **-5**

54. Name the property illustrated by $9 + (2 + 10) = 9 + 12$. (Lesson 1–6) **substitution (=)**

55. Solve $a = \frac{12 + 8}{4}$. (Lesson 1–5) **5**

56. Evaluate $\frac{3}{4}(6) + \frac{1}{3}(12)$. (Lesson 1–3) **$8\frac{1}{2}$**

SELF TEST

Solve each equation. (Lessons 3–1, 3–2, and 3–3)

1. $-10 + k = 34$ **44**
2. $y - 13 = 45$ **58**
3. $20.4 = 3.4y$ **6**
4. $-65 = \frac{f}{29}$ **-1885**
5. $-3x - 7 = 18$ **$-8\frac{1}{3}$**
6. $5 = \frac{m - 5}{4}$ **25**

Write an equation and solve. Then check your solution.

7. Twenty-three minus a number is 42. Find the number. (Lesson 3–1) **$23 - n = 42$; -19**

8. **Geometry** The length of one side of a rectangle is 34 cm. The area of the rectangle is 68 cm². Find the width of the rectangle. (Lesson 3–2) **$34w = 68$; 2 cm**

9. **Number Theory** Find two consecutive even integers whose sum is 126. (Lesson 3–3) **$n + (n + 2) = 126$; 62, 64**

10. **Economics** According to the *Baltimore Sun*, if inflation continues at the current rate, the price of a movie ticket in 40 years will be $3.50 less than 10 times the current average price of $5.05. What will be the price of a movie ticket in 40 years? (Lesson 3–3) **$p = 10(5.05) - 3.50$; $47**

Extension

Connections A rectangular garden is fenced on all four sides. If the garden is 6 feet longer than it is wide, what are its dimensions if 156 feet of fencing is needed to enclose it?
42 feet by 36 feet

SELF TEST

The Self Test provides students with a brief review of the concepts and skills in Lessons 3-1 through 3-3. Lesson numbers are given to the right of exercises or instruction lines so students can review concepts not yet mastered.

NCTM Standards: 1–5, 7

Instructional Resources

- Study Guide Master 3-4
- Practice Master 3-4
- Enrichment Master 3-4
- Assessment and Evaluation Masters, pp. 71–72
- Multicultural Activity Masters, p. 6
- Science and Math Lab Manual, pp. 9–12

 Transparency 3-4A contains the 5-Minute Check for this lesson; **Transparency 3-4B** contains a teaching aid for this lesson.

Recommended Pacing

Standard Pacing	Day 7 of 14
Honors Pacing	Day 6 of 13
Block Scheduling*	Day 4 of 6 (along with Lesson 3-5)
Alg. 1 in Two Years*	Days 9 & 10 of 19

 *For more information on pacing and possible lesson plans, refer to the *Block Scheduling Booklet* and *Algebra 1 in Two Years*.

1 FOCUS

 ### 5-Minute Check
(over Lesson 3-3)

Solve.

1. $\frac{y}{4} + 7 = -4$ **−44**

2. $-3 + \frac{2r}{3} = 13$ **24**

3. $\frac{(t-5)}{4} = 19$ **81**

4. A rectangle is 5 times as long as it is wide. The difference of its length and width is 9 inches. Find its perimeter. **27 in.**

5. Find three consecutive integers whose sum is 54. **17, 18, 19**

Objectives

After studying this lesson, you should be able to:

- find the complement and supplement of an angle, and
- find the measure of the third angle of a triangle given the measures of the other two angles.

The symbol for angle is ∠.

Integration: Geometry
Angles and Triangles

 ### APPLICATION
Football

On fourth down in a football game, the team with the ball usually *punts*. A punt is a kick in which the ball is dropped by the kicker and is kicked before it hits the ground.

The longest punt ever kicked in a college football game was kicked by Pat Brady of the University of Nevada-Reno in their 1950 game against Loyola University. The punt was 99 yards long—just 1 yard short of spanning the entire length of the football field.

How far a football will travel once it is punted depends not only on the strength of the punter, but also on the angle at which the ball is kicked. If the angle is too small, the ball will travel close to the ground, then drop; if the angle is too large, the ball will travel high, but not very far. What do you think would be the best angle at which to kick the ball?

You can use a *protractor* to measure angles, as shown below.

Angle ABC (denoted $\angle ABC$) measures 60°. However, where ray BC (denoted $\overrightarrow{BC}$) intersects the curve of the protractor, there are two readings, 60° and 120°. The measure of $\angle DBC$ is 120°. What is the sum of the measures of $\angle ABC$ and $\angle DBC$?

Supplementary Angles	Two angles are supplementary if the sum of their measures is 180°.

Example The measure of an angle is three times the measure of its supplement. Find the measure of each angle.

Let x = the lesser measure. Then $3x$ = the greater measure.

$x + 3x = 180$ *The angles are supplementary.*

$4x = 180$ *Add like terms.*

$\frac{4x}{4} = \frac{180}{4}$ *Divide each side by 4.*

$x = 45$

The measures are 45° and 3 · 45° or 135°. *Check this result.*

Complementary Angles	Two angles are complementary if the sum of their measures is 90°.

Example The backward slant of the face of a golf club head is called the *loft*. It is designed to drive the ball in a high arc. Assuming that the angle made by the ground and the face of the club is 79°, what is the measure of the loft?

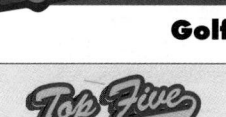

APPLICATION

Golf

Lowest winning scores in the U.S. Masters Tournament
1. 271, Raymond Floyd, 1976
1. 271, Jack Nicklaus, 1965
3. 274, Ben Hogan, 1953
4. 275, Severiano Ballesteros, 1980
4. 275, Fred Couples, 1992

Let x = the measure of the loft.

$79° + x = 90°$ *The angles are complementary.*

$79° - 79° + x = 90° - 79°$ *Subtract 79° from each side.*

$x = 11°$

The club has an 11° loft.

Example The measure of an angle is 34° greater than its complement. Find the measure of each angle.

Let x = the lesser measure. Then $x + 34$ = the greater measure.

$x + (x + 34) = 90$ *The angles are complementary.*

$2x + 34 = 90$

$2x + 34 - 34 = 90 - 34$ *Subtract 34 from each side.*

$2x = 56$

$\frac{2x}{2} = \frac{56}{2}$ *Divide each side by 2.*

$x = 28$

The measures are 28° and 28° + 34° or 62°. *Check this result.*

A **triangle** is a polygon with three sides and three angles. What is the sum of the measures of the three angles of a triangle?

 Alternative Teaching Strategies

Reading Algebra Knowing the Latin derivation of a nonmathematical English word may help students understand and remember a mathematical definition of the same word. For example, the word *complement* comes from the Latin word *complere*, which means to "make full" or "to complete." A nonmathematical meaning in music is an interval that completes a given octave.

Have students look up the meanings of the Latin roots of the following words, as well as a nonmathematical definition.

a. supplement
b. acute
c. congruent

Questioning Draw the angles below on the chalkboard. Ask students to name each angle in two ways.

1.
∠ XYZ
∠ Y

2.
∠ MNP
∠ N

3.
∠ RST
∠ S

4.
∠ BCD
∠ C

2 TEACH

Teaching Tip Point out that on the protractor shown, we can see that angle *ABC* is the supplement of angle *DBC* and vice versa.

In-Class Examples

For Example 1
The measure of an angle is 20° less than three times its supplement. Find the measure of each angle. **130°, 50°**

For Example 2
The measure of an angle is 26° greater than its complement. Find the measure of each angle. **32°, 58°**

For Example 3
An angle measures 42° less than its complement. Find the measure of each angle. **24°, 66°**

Point out that the winner of a golf game is the player who has the lowest score. The next lowest scores on the list were shared by many players: 276, by Arnold Palmer (1964), Jack Nicklaus (1975), and Tom Watson (1977); 277, by Bob Goalby (1968), Johnny Miller (1975), Gary Player (1978), Ben Crenshaw (1984), Ian Woosnam (1991), Raymond Floyd (1992), and Bernhard Langer (1993).

Instead of tearing the corners of the triangle, students can fold the vertices together, as shown below.

Help students to see that since the three angles form a straight angle, the sum of their measures is 180°.

In-Class Example

For Example 4
What are the measures of the base angles of an isosceles triangle in which the vertex angle measures 20°? **80°**

Teaching Tip Remind students that the word *congruent* means "equal measures."

 Angles of a Triangle

Materials: 3 sheets of construction paper   protractor   scissors

Copy each triangle below onto a sheet of construction paper. If possible, you may want to enlarge the triangles on a photocopier.

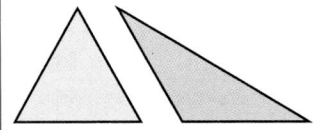

b. Cut out one of the triangles. Label the angles *A*, *B*, and *C*. Tear off the angles and arrange as shown.

c. When the angles were put together, what is the measure of the new angle they formed? **180°**

d. Try this activity with another triangle. What do you think is true about the measures of the angles of any triangle? **The sum of the interior angles is 180°.**

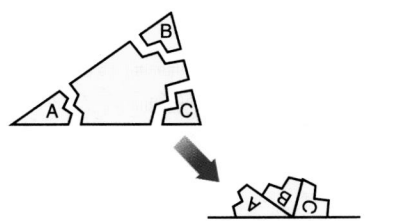

Your Turn

a. Use a protractor to measure the angles of each triangle above. What is the sum of the measures of the three angles of each triangle? **180°**

| Sum of the Angles of a Triangle | The sum of the measures of the angles in any triangle is 180°. |

In an **equilateral triangle,** each angle has the same measure. We say that the angles are **congruent.** The sides of an equilateral triangle are also congruent. What is the measure of each angle of an equilateral triangle?

Let x = the measure of each angle.

$x + x + x = 180$

$3x = 180$ *Add like terms.*

$\dfrac{3x}{3} = \dfrac{180}{3}$ *Divide each side by 3.*

$x = 60$ Each angle measures 60°.

In an **isosceles triangle,** at least two angles have the same measure. Generally, the two congruent angles are the base angles.

Example **4** What are the measures of the base angles of an isosceles triangle in which the vertex angle measures 45°?

Let x = the measure of each base angle.

$x + x + 45 = 180$

$2x + 45 = 180$ *Add like terms.*

$2x + 45 - 45 = 180 - 45$ *Subtract 45 from each side.*

$2x = 135$

$\dfrac{2x}{2} = \dfrac{135}{2}$ *Divide each side by 2.*

$x = \dfrac{135}{2}$ or $67\frac{1}{2}$ The base angles each measure $67\frac{1}{2}°$.

 GLENCOE Technology

Interactive Mathematics Tools Software

This multimedia software provides an interactive lesson by helping students find the sum of the interior angles of any triangle by manipulating a triangle into various shapes and sizes. A **Computer Journal** gives students an opportunity to write about what they have learned.

For Windows & Macintosh

 Cooperative Learning

Formations You may wish to have students work in cooperative groups to model the various types of triangles. For more information on the formations strategy, see *Cooperative Learning in the Mathematics Classroom*, one of the titles in the Glencoe Mathematics Professional Series, pages 30–31.

Triangles are often classified in one of three ways. A **right triangle** has one angle that measures 90°. An **obtuse triangle** has one angle with measure greater than 90°. In an **acute triangle,** all of the angles measure less than 90°.

Example **5** The measures of the angles of a triangle are given as $x°$, $2x°$, and $3x°$.

a. What are the measures of each angle?
b. Classify the triangle.

a. The sum of the measures of the angles of a triangle is 180°.

$$x + 2x + 3x = 180$$
$$6x = 180 \quad \textit{Add like terms.}$$
$$\frac{6x}{6} = \frac{180}{6} \quad \textit{Divide each side by 6.}$$
$$x = 30$$

The measures are 30°, 2(30°) or 60°, and 3(30°) or 90°.

b. Since the triangle has one angle that measures 90°, it is a right triangle.

CHECK FOR UNDERSTANDING

Communicating Mathematics

Study the lesson. Then complete the following. 1–2. See margin.

1. The words *complement* and *compliment* sound very much alike.
   a. What are their meanings?
   b. Which word has the mathematical meaning?

2. **Compare and contrast** the standard definition and the mathematical meaning of the word *supplement.*

3. acute, obtuse, and right, respectively

3. **Classify** each of the three triangles in the Modeling Mathematics activity on page 164 as right, obtuse, or acute. Justify your classifications.

4. **Make up a problem** in which you must use an equation to find the measures of the angles of a triangle. Then solve the problem, explaining each step. See students' work.

5. Draw a right triangle on a separate piece of paper and cut it out. Cut or tear off the two acute angles and arrange as shown. What seems to be true about the two acute angles? The sum of their measures is 90°.

Guided Practice

Find both the complement and the supplement of each angle measure.

6. 130° none, 50°   7. 11° 79°, 169°   8. 24° 66°, 156°

9. (90 − 3x)°, (180 − 3x)°   9. 3x°   10. (2x + 40)°   11. (x − 20)°

10. (50 − 2x)°, (140 − 2x)°
11. (110 − x)°, (200 − x)°

14. (160 − 2x)°

Find the measure of the third angle of each triangle in which the measures of two angles of the triangle are given.

12. 16°, 42° 122°   13. 50°, 45° 85°   14. $x°$, $(x + 20)°$

Lesson 3–4 **INTEGRATION** Geometry Angles and Triangles **165**

Reteaching

Using Substeps The measure of an angle is 5 times its complement. If x is the lesser measure, then $5x$ is the greater measure. Fill in the blanks to solve the problem.

1. The angles are complementary.
$$\frac{?}{5x} + \frac{?}{x} = \frac{?}{90°}$$

2. Simplify.
$$\frac{?}{6x} = \frac{?}{90°}$$

3. Divide each side by the same number.
$$\frac{?}{?} = \frac{?}{?} \qquad \frac{6x}{6} = \frac{90°}{6}$$

4. What are the measures of the angles?
$$x = ?, 5x = ?$$
$$x = 15°, 5x = 75°$$

Chapter 3 **165**

Practice Masters, p. 22

Make a diagram to represent each situation. Then define a variable, write an equation, and solve each problem. Check your solution.

15. An angle measures 38° less than its complement. Find the measures of the two angles. $(c − 38) + c = 90; 26°, 64°$

16. $x + (x + 5) + (2x + 3) = 180; 43°, 48°, 89°$

16. The measures of the angles of a triangle are given as $x°$, $(x + 5)°$, and $(2x + 3)°$. What are the measures of each angle?

EXERCISES

Practice

A

24. $(90 − y)°$, $(180 − y)°$
25. $(90 − 3a)°$, $(180 − 3a)°$

Find both the complement and the supplement of each angle measure.

17. 42° 48°, 138°   18. 87° 3°, 93°   19. 125° none, 55°
20. 90° none, 90°   21. 21° 69°, 159°   22. 174° none, 6°
23. 99° none, 81°   24. $y°$ 25. $3a°$
26. $(x + 30)°$ 27. $(b − 38)°$ 28. $(90 − z°)$
$(60 − x)°$, $(150 − x)°$ $(128 − b)°$, $(218 − b)°$ $z°$, $(90 + z)°$

The measures of two angles of a triangle are given. Find the measure of the third angle.

34. $180 − (x + y)°$ or $(180 − x − y)°$

B

29. 40°, 70° 70°   30. 90°, 30° 60°   31. 63°, 12° 105°
32. 43°, 118° 19°   33. 4°, 38° 138°   34. $x°$, $y°$
35. $p°$, $(p − 10)°$ 36. $c°$, $(2c + 1)°$ 37. $y°$, $(135 − y)°$ 45°
$(190 − 2p)°$ $(179 − 3c)°$

Draw a diagram to represent each situation. Then define a variable, write an equation, and solve each problem. Check your solution.

38–43. See students' diagrams.

39. $x + x − 30 = 180; 75°$

38. One of the congruent angles of an isosceles triangle measures 37°. Find the measures of the other angles. $37 + 37 + x = 180; 37°, 106°$

39. Find the measure of an angle that is 30° less than its supplement.

40. One of the angles of a triangle measures 53°. Another angle measures 37°. What is the measure of the third angle? $53 + 37 + x = 180; 90°$

41. One of two complementary angles measures 30° more than three times the other. Find the measure of each angle. $x + (30 + 3x) = 90; 15°, 75°$

C

42. Find the measure of an angle that is one-half the measure of its supplement. $x + 2x = 180; 60°$

43. The measures of the angles of a triangle are given as $x°$, $(3x)°$, and $(4x)°$. What are the measures of each angle? $x + 3x + 4x = 180; 22.5°, 67.5°, 90°$

Critical Thinking

44. Draw a diagram that shows the relationships among right, isosceles, equilateral, acute, scalene, and obtuse triangles. See students' work.

Applications and Problem Solving

45. **Carpentry** A *stringer* is a triangular piece of wood on which staircases are based. If the stringer makes a 30° angle with the floor, what is the measure of the angle that the stringer makes with the vertical? 60°

 Alternative Learning Styles

Visual Students will need grid paper to solve the following problem. Have students draw the path described in the problem and label their drawings appropriately. Then have them draw and describe three different paths Marta could have taken. Have them identify the most direct path.

Marta walked 8 blocks north and 3 blocks east from her home to school. After school, she walked 2 blocks south and 1 block west to the library. From the library, she traveled 1 block south and 2 blocks east to visit a friend. How far does Marta have to walk to return home? **5 blocks south and 4 blocks west**

46. Aeronautics One of the newer space shuttles is *Endeavour.* Its first flight left Earth on May 7, 1992. The oldest shuttle is *Columbia,* which flew its first flight on April 12, 1981. A space shuttle lands at an angle six times as great as that of the average commercial jet.

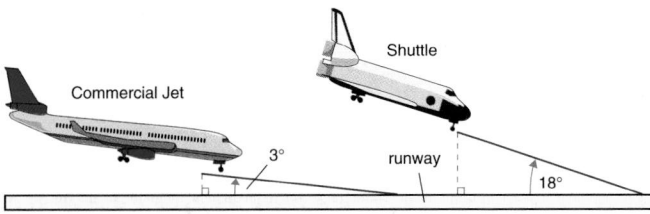

a. What is the measure of the angle the path of the jet makes with the vertical? **87°**

b. What is the measure of the angle the path of the shuttle makes with the vertical? **72°**

c. Which craft lands at a steeper angle? **shuttle**

47. Mountain Climbing *Rapelling* is a technique used by climbers to make a difficult descent. Climbers back to the edge and then spring off. To avoid slipping, the climber's legs should remain at a 90° angle with the side of the ledge. The ledge at the right is 50° off the vertical. What is the measure of the climber's angle with the vertical? **40°**

Mixed Review

48. Solve $4 + 7x = 39$. (Lesson 3–3) **5**

49. Consumerism When stores go out of business, there is generally a period when all of the items in the store can be purchased for a fraction of the original price. When Central Hardware went out of business, all of the items in the store began at $\frac{1}{4}$ off. (Lesson 3–2)

a. What was the original price of a circular saw if you save $20.00 during the sale? **$80**

b. How much would you save on patio furniture that originally cost $299.00? **$74.75**

50. Solve $h + (-13) = -5$. (Lesson 3–1) **8**

51. Simplify $5(3t - 2t) + 2(4t - 3t)$. (Lesson 2–6) **7t**

52. Find the sum $5y + (-12y) + (-21y)$. (Lesson 2–5) **−28y**

53. American History Each number below represents the age of a U.S. president on his first inauguration. (Lesson 2–2) **53a–b. See margin.**

57 61 57 57 58 57 61 54 68 51 49 64 50 48

65 52 56 46 54 49 50 47 55 55 54 42 51 56

55 51 54 51 60 62 43 55 56 61 62 69 64 46

a. Make a number line plot of the ages of the presidents.

b. Do any of the ages appear to be clustered? If so, which ones?

54. comm. (+)

54. Name the property illustrated by $5a + 2b = 2b + 5a$. (Lesson 1–8)

55. See margin.

55. Statistics Make a stem-and-leaf plot of the data in Exercise 53. (Lesson 1–4)

Extension

Reasoning Show that when the measure of a given angle is added to twice the measure of the complement of the angle, the result equals the measure of the supplement of the angle.

$$a + 2(90 - a) = a + 180 - 2a$$
$$= 180 - a$$

Tech Prep

Aeronautical Technician For students who are interested in aeronautics, you may wish to point out that an aeronautical technician is a vital part of the landing operations of an aircraft (see Exercise 46). For more information on tech prep, see the *Teacher's Handbook.*

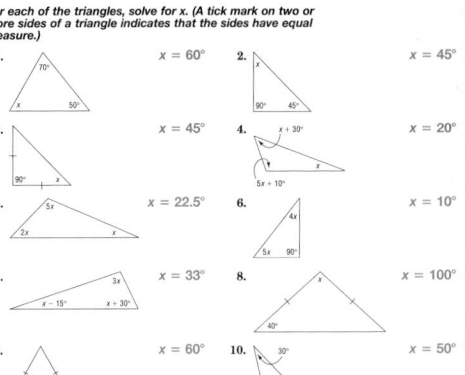

Solving Equations with the Variable on Both Sides

NCTM Standards: 1–5, 7

Instructional Resources

- Study Guide Master 3-5
- Practice Master 3-5
- Enrichment Master 3-5

Transparency 3-5A contains the 5-Minute Check for this lesson; **Transparency 3-5B** contains a teaching aid for this lesson.

Recommended Pacing	
Standard Pacing	Days 8 & 9 of 14
Honors Pacing	Days 7 & 8 of 13
Block Scheduling*	Day 4 of 6 (along with Lesson 3-4)
Alg. 1 in Two Years*	Days 11 & 12 of 19

*For more information on pacing and possible lesson plans, refer to the *Block Scheduling Booklet* and *Algebra 1 in Two Years.*

1 FOCUS

5-Minute Check
(over Lesson 3-4)

Find the measure of the third angle of each triangle in which the measure of two angles of the triangle are given.

1. 15°, 86° **79°**
2. 22°, 110° **48°**
3. 66°, 52° **62°**

Find the complement and supplement of each angle measure.

4. 24° **66°, 156°**
5. 67° **23°, 113°**
6. 80° **10°, 100°**

Objectives
After studying this lesson, you should be able to:
- solve equations with the variable on both sides, and
- solve equations containing grouping symbols.

fabulous
FIRSTS

Wilma Rudolph (1940–1994)

Wilma Rudolph was the first woman to win three gold medals at a single Olympic games. She won the 100-meter sprint, the 200-meter dash, and the 4 × 100-meter relay at the 1960 Olympic Games.

APPLICATION
Track and Field

In the 1928 Olympics, the winner of the men's 800-meter run, Douglas Lowe of Great Britain, won the race in 1 minute 51.8 seconds, or 111.8 seconds. In that same year, the winner of the women's 800-meter run, Lina Radke of Germany, won the race in 2 minutes 16.8 seconds, or 136.8 seconds. Over the next 64 years, the men's winning times decreased an average of 0.127 seconds per year, and the women's winning times decreased an average of 0.332 seconds per year. Suppose the times continued to decrease at these rates. When will men and women have the same winning times in the 800-meter run?

After x years, the men's winning times can be represented by $111.8 - 0.127x$. After x years, the women's winning times can be represented by $136.8 - 0.332x$. The times would be the same when the two expressions are equal.

$$111.8 - 0.127x = 136.8 - 0.332x$$

Many equations contain variables on each side. To solve these types of equations, first use the addition or subtraction property of equality to write an equivalent equation that has all of the variables on one side. Then solve the equation.

Method 1

$$111.8 - 0.127x = 136.8 - 0.332x \qquad \textit{Use a calculator.}$$
$$111.8 - 0.127x + 0.332x = 136.8 - 0.332x + 0.332x \qquad \textit{Add 0.332x to each side.}$$
$$111.8 + 0.205x = 136.8$$
$$111.8 - 111.8 + 0.205x = 136.8 - 111.8 \qquad \textit{Subtract 111.8 from each side.}$$
$$0.205x = 25$$
$$\frac{0.205x}{0.205} = \frac{25}{0.205} \qquad \textit{Divide each side by 0.205.}$$
$$x \approx 122$$

At these rates, the men's and women's winning times will be equal about 122 years after 1928, or in the year 2050.

fabulous
FIRSTS

Wilma Rudolph overcame many obstacles on her way to becoming a champion. At 4, she was stricken with double pneumonia and scarlet fever. The illnesses crippled her left leg. When she was 8 she began walking with a corrective shoe. At 16, in 1956, she won a bronze medal as a member of the U.S. Olympic 4 × 100-meter relay team.

Method 2

Since two of the decimals involve thousandths, another way to solve the equation would be to multiply each side by 1000 to clear the decimals.

$$111.8 - 0.127x = 136.8 - 0.332x$$

Multiply each side by 1000. $\quad 1000(111.8) - 1000(0.127x) = 1000(136.8) - 1000(0.332x)$

$$111,800 - 127x = 136,800 - 332x$$

Add 332x to each side. $\quad\quad\quad\quad 111,800 + 205x = 136,800$

Subtract 111,800 from each side. $\quad\quad\quad 205x = 25,000$

Divide each side by 205. $\quad\quad\quad\quad\quad\quad x \approx 122$

Example ❶ Solve $\frac{3}{8} - \frac{1}{4}x = \frac{1}{2}x - \frac{3}{4}$.

TECHNOLOGY
Tip

Look back to Lesson 3–3 to see how you would solve an equation with the variable on both sides by using a graphing calculator.

$$\frac{3}{8} - \frac{1}{4}x = \frac{1}{2}x - \frac{3}{4} \quad\quad \textit{The least common denominator is 8.}$$

$$8\left(\frac{3}{8} - \frac{1}{4}x\right) = 8\left(\frac{1}{2}x - \frac{3}{4}\right) \quad\quad \textit{Multiply each side by 8.}$$

$$8\left(\frac{3}{8}\right) - 8\left(\frac{1}{4}x\right) = 8\left(\frac{1}{2}x\right) - 8\left(\frac{3}{4}\right) \quad\quad \textit{Use the distributive property.}$$

$$3 - 2x = 4x - 6 \quad\quad \textit{The fractions are eliminated.}$$

$$3 = 6x - 6 \quad\quad \textit{Add 2x to each side.}$$

$$9 = 6x \quad\quad \textit{Add 6 to each side.}$$

$$\frac{3}{2} = x \quad\quad \textit{Divide each side by 6.}$$

The solution is $\frac{3}{2}$. *Check this result.*

When solving equations that contain grouping symbols, first use the distributive property to remove the grouping symbols.

Example ❷

INTEGRATION
Geometry

One angle of a triangle measures 10° more than the second. The measure of the third angle is twice the sum of the first two angles. Find the measure of each angle.

Make a drawing. Let y = the measure of one angle. Let $y + 10$ = the measure of another angle. Then $2[y + (y + 10)]$ = the measure of the third angle. Recall that the sum of the measures of the angles in any triangle is 180°.

$2[y + (y + 10)]°$

$y°$

$(y + 10)°$

$$y + (y + 10) + 2[y + (y + 10)] = 180$$

$$2y + 10 + 2(2y + 10) = 180 \quad\quad \textit{Simplify.}$$

$$2y + 10 + 4y + 20 = 180 \quad\quad \textit{Use the distributive property.}$$

$$6y + 30 = 180 \quad\quad \textit{Simplify.}$$

$$6y + 30 - 30 = 180 - 30 \quad\quad \textit{Subtract 30 from each side.}$$

$$6y = 150$$

$$y = 25 \quad\quad \textit{Divide each side by 6.}$$

The measures of the three angles are 25°, $(25 + 10)°$ or 35°, and $2(25 + 35)°$ or 120°. *Check this result.*

Motivating the Lesson

Hands-On Activity Separate the class into two groups. Think of a trait, such as people wearing red sneakers, and count the number of students in each half of the class who exhibit the trait. Tell the class that you are thinking of a trait and that they need to identify it. Also tell them how many students in each half show the trait.

Students should examine each half of the class and use the clue to guess the trait. They can check the validity of their guess by having students move between the halves. Each time a student moves to a different half of the class, you must again announce how many students in each half show the trait. Continue in this manner until the trait has been identified.

Point out how the nature of the task changed as students moved between the halves. Show that the nature of an equation can also change if variables are not properly handled when solving equations.

2 TEACH

In-Class Examples

For Example 1
Solve.

a. $\frac{2}{5}n - 9 = 7 - \frac{3}{5}n$ 16

b. $\frac{3}{7}t + 4 = \frac{1}{7}t - 6$ −35

c. $8 - \frac{1}{2}p = \frac{1}{4}p - 7$ 20

For Example 2
The perimeter of a rectangle is 24 inches. Find its dimensions if its length is 3 inches greater than its width.
$4\frac{1}{2}$ in. by $7\frac{1}{2}$ in.

Teaching Tip Students usually have the least difficulty with equations similar to the one in Example 2 if they apply the distributive property first. Then they can finish solving the equation by applying properties of equality.

Alternative Teaching Strategies

Student Diversity Review with students the different meanings of the equal sign. $3 = 1 + 2$ means that the left side is always equal to the right side. This is also true of $7x = 7x$. However, $2x = 8$ is not always true. For this to have a solution means that there is at least one value for x that makes the two sides equal.

3 PRACTICE/APPLY

170   *Chapter 3*

Some equations with the variable on both sides may have no solution. That is, there is no value of the variable that will result in a true equation.

Example **3**   Solve $5n + 4 = 7(n + 1) - 2n$.

$$5n + 4 = 7(n + 1) - 2n$$
$$5n + 4 = 7n + 7 - 2n \qquad \textit{Distributive property}$$
$$5n + 4 = 5n + 7$$
$$4 = 7 \quad \text{Since } 4 = 7 \text{ is a false statement, this equation has no solution.}$$

An equation that is true for every value of the variable is called an **identity**.

Example **4**   Solve $7 + 2(x + 1) = 2x + 9$.

$$7 + 2(x + 1) = 2x + 9$$
$$7 + 2x + 2 = 2x + 9 \qquad \textit{Distributive property}$$
$$2x + 9 = 2x + 9 \qquad \textit{Reflexive property of equality}$$

Since the expressions on each side of the equation are the same, this equation is an identity. The statement $7 + 2(x + 1) = 2x + 9$ is true for all values of x.

CHECK FOR UNDERSTANDING

Communicating Mathematics

1. See Solutions Manual.

5. $-\frac{1}{2}$ 6. $\frac{12}{11}$

12. $6x + (x - 3) + (3x + 7) = 180$; $105.6°$, $14.6°$, $59.8°$

13. $\frac{1}{2}n + 16 = \frac{2}{3}n - 4$;

120

M<small>ATH</small> J<small>OURNAL</small>

Study the lesson. Then complete the following.   2–3. See margin.

1. a. Explain why you think winning times in the 800-meter run for men and women are growing closer.

   b. Use a table to show how women's winning times in the 800-meter run could catch up to men's winning times in the same race.

   c. Do you think the times will ever be the same? Why or why not?

2. Describe the difference between an identity and an equation with no solution.

3. You Decide   Lauren says that to solve the equation $2[x + 3(x - 1)] = 18$, the first step should be to multiply 2 by $[x + 3(x - 1)]$. Carmen says the first step should be to multiply 3 by $(x - 1)$. Which one is correct, and why?

4. Assess Yourself   Describe your favorite sport and explain how math is used in it.   **See students' work.**

Guided Practice

Explain how to solve each equation. Then solve each equation and check your solution.   5–10. See margin for explanations.

5. $3 - 4x = 10x + 10$ **6.** $8y - 10 = -3y + 2$ **7.** $\frac{3}{5}x + 3 = \frac{1}{5}x - 7$ **−25**

8. $5.4y + 8.2 = 9.8y - 2.8$ **9.** $5x - 7 = 5(x - 2) + 3$ **10.** $4(2x - 1) = -10(x - 5)$
2.5   **all numbers**   **3**

Define a variable, write an equation, and solve each problem. Then check your solution.

11. Twice the greater of two consecutive odd integers is 13 less than three times the lesser. Find the integers. $2(n + 2) = 3n - 13$; 17, 19

12. Geometry   The measures of the angles of a triangle are given as $6x°$, $(x - 3)°$, and $(3x + 7)°$. What are the measures of each angle?

13. One half of a number increased by 16 is four less than two thirds of the number. Find the number.

Reteaching

Using Substeps   List the first two steps you would take to solve each equation. Do *not* solve.

1. $6x + 1 = 15 - x$ **Add x to each side. Then subtract 1 from each side.**

2. $2x + 3 = 5x - 9$ **Subtract $2x$ from each side. Then add 9 to each side.**

3. $x + 36 = 1 - 4(x - 5)$ **Multiply 4 and $x - 5$. Then subtract x from each side.**

4. $3(2x + 4) = 5x + 12 + x$ **Multiply 3 and $2x + 4$. Then add $5x$ and x.**

5. $3(2x - 5) = 6x + 7$ **Multiply 3 and $2x - 5$. Then subtract $6x$ from each side.**

Practice

Solve each equation. Then check your solution.

14. $6x + 7 = 8x - 13$ **10**

15. $17 + 2n = 21 + 2n$ **no solution**

22. 3.57 or $3\frac{4}{7}$

16. $\frac{3n - 2}{5} = \frac{7}{10}$ **1.83 or $1\frac{5}{6}$**

17. $\frac{7 + 3t}{4} = -\frac{t}{8}$ **-2**

27. 2.6 or $2\frac{3}{5}$

18. $13.7b - 6.5 = -2.3b + 8.3$ **0.925**

19. $18 - 3.8x = 7.36 - 1.9x$ **5.6**

30. 3.6 or $3\frac{3}{5}$

20. $\frac{3}{2}y - y = 4 + \frac{1}{2}y$ **no solution**

21. $\frac{3}{4}n + 16 = 2 - \frac{1}{8}n$ **-16**

22. $-7(x - 3) = -4$

23. $4(x - 2) = 4x$ **no solution**

24. $28 - 2.2y = 11.6y + 262.6$ **-17**

25. $1.03x - 4 = -2.15x + 8.72$ **4**

26. $7 - 3x = x - 4(2 + x)$ **no solution**

27. $6 = 3 + 5(y - 2)$

28. $6(y + 2) - 4 = -10$ **-3**

29. $5 - \frac{1}{2}(b - 6) = 4$ **8**

B

30. $-8(4 + 9x) = 7(-2 - 11x)$

31. $2(x - 3) + 5 = 3(x - 1)$ **2**

33. all numbers

32. $4(2a - 8) = \frac{1}{7}(49a + 70)$ **42**

33. $-3(2n - 5) = \frac{1}{2}(-12n + 30)$

Define a variable, write an equation, and solve each problem. Then check your solution.

34. Three times the greatest of three consecutive even integers exceeds twice the least by 38. Find the integers. $3(n + 4) = 2n + 38$; 26, 28, 30

35. $\frac{1}{5}n + 5n = 7n - 18$;

10

35. One fifth of a number plus five times that number is equal to seven times the number less 18. Find the number.

36. The difference of two numbers is 12. Two fifths of the greater number is six more than one third of the lesser number. Find both numbers.
$\frac{2}{5}(n + 12) = 6 + \frac{1}{3}(n)$; 18, 30

Geometry

Make a diagram to represent each situation. Then define a variable, write an equation, and solve each problem. Check your solution.

38. $x + (x + 30) + 3[x + (x + 30)] = 180$; 7.5°, 37.5°, 135°

37. The measures of the angles of a certain triangle are consecutive even integers. Find their measures. $n + (n + 2) + (n + 4) = 180$; 58°, 60°, 62°

C

38. One angle of a triangle measures 30° more than another angle. The measure of the third angle is three times the sum of the first two angles. Find the measure of each angle.

Programming

39. The graphing calculator program at the right can help you solve equations of the form $ax + b = cx + d$.

39a. $4x + 6 = 4x + 6$; identity

39b. $5x - 7 = x + 3$; 2.5

39c. $-3x + 6 = 3x - 6$; 2

39d. $5.4x + 6.8 = 4.6x + 2.8$; -5

39e. $2x - 8 = 2x - 6$; no solution

Write each equation in the form $ax + b = cx + d$. Then run the program to find the solution.

a. $2(2x + 3) = 4x + 6$

b. $5x - 7 = x + 3$

c. $6 - 3x = 3x - 6$

d. $6.8 + 5.4x = 4.6x + 2.8$

e. $5x - 8 - 3x = 2(x - 3)$

```
PROGRAM:SOLVE
: Disp "ENTER A, B, C, D"
: Input A: Input B: Input C:
  Input D
: If A - C ≠ 0
: Goto 2
: If D - B ≠ 0
: Goto 1
: Disp "THIS IS AN"
: Disp "IDENTITY"
: Goto 3
: Lbl 1
: Disp "NO SOLUTION"
: Goto 3
: Lbl 2
: Disp "X = ", (D - B)/(A - C)
: Lbl 3
```

Additional Answers

5. Add $4x$ to each side, subtract 10 from each side, then divide each side by 14.

6. Add $3y$ to each side, add 10 to each side, then divide each side by 11.

7. Subtract $\frac{1}{5}x$ from each side, subtract 3 from each side, then multiply by $\frac{5}{2}$.

8. Subtract $5.4y$ from each side, add 2.8 to each side, then divide each side by 4.4.

10. Multiply 4 by $(2x - 1)$ and -10 by $(x - 5)$. Subtract $8x$ from each side, and subtract 50 from each side. Then divide each side by -18.

Closing Activity

Speaking Ask students to explain the difference between an equation whose solution set is the empty set and an equation that is an identity.

Critical Thinking

40. Mathematics History Diophantus of Alexandria was one of the greatest mathematicians of the Greek civilization. Little is known of his life except for the following problem, which was first printed in a work called the *Greek Anthology*. Although the problem was not written by Diophantus, it is believed to accurately describe his life.

Diophantus passed one sixth of his life in childhood, one twelfth in youth, and one seventh more as a bachelor. Five years after his marriage, there was born a son who died four years before his father, at half his father's (final) age. How old was Diophantus when he died? **84 years old**

Applications and Problem Solving

For Exercises 41–42, assume that the rates of change continue indefinitely.

41. Sales According to the Association of Home Appliance Manufacturers, sales of room air conditioners were 4.6 million units in 1988, and sales of window fans were 0.975 million units in 1988. Since 1988, sales of room air conditioners have decreased about 0.425 million units per year, and sales of window fans have increased about 0.106 million units per year.

a. If the trend continues, after how many years will sales of room air conditioners and window fans be equal? **6.827 years**

b. How would you explain these trends? **Air conditioner sales are decreasing while fan sales are increasing.**

42. Pet Ownership According to the American Veterinary Medical Association, in 1987, 34.7 million households owned a dog, and 27.7 million households owned a cat. Since 1987, dog ownership has decreased by about 0.025 million households per year, and cat ownership has increased by about 0.375 million households per year. If the trend continues, after how many years will the number of households that own dogs and cats be equal? **17.5 years**

Mixed Review

43. Find the supplement of a 32° angle. (Lesson 3–4) **148°**

44. Work Backward A number is decreased by 35, then multiplied by 6, then added to 87, then divided by 3. The result is 67. What is the number? (Lesson 3–3) **54**

45. Solve $x + 4.2 = 1.5$. Check your solution. (Lesson 3–1) **−2.7**

46. Define a variable, then write an equation for the following problem. (Lesson 2–9)

Karen is 10, and she has always admired her 15-year-old sister Kristy. Karen wants to be exactly like Kristy. She knows that she will never catch up with Kristy in age, but she has heard that when one person is 0.9 times as old as another, you can't really tell them apart. How long will it be before Karen is 0.9 times as old as Kristy? **$10 + x = 0.9(15 + x)$, 35 years**

47. Which is greater, $\frac{11}{9}$ or $\frac{12}{10}$? (Lesson 2–4) **$\frac{11}{9}$**

48. As students spend more time watching TV, their test scores go down.

48. Explain what the graph at the right tells you about test scores as a function of hours spent watching television. (Lesson 1–9)

49. Solve $14.8 − 3.75 = t$. (Lesson 1–5) **11.05**

Enrichment Masters, p. 23

NAME_____ DATE _____

3-5 Enrichment

Student Edition
Pages 168–172

Using Equations

Use what you have learned about equations to solve each problem.

1. In many universities, grade point averages are figured out using a table like this.

grade	A	B	C	D	F
point value	4	3	2	1	0

Different classes often meet for a different number of hours each week. If you had the following schedule and grades, your average would be 2.45.

Course	Hours	Grade
Chemistry	3	C
Chemistry Lab	2	F
Calculus	5	A
Phys. Ed.	1	C
Total	11	

$$\frac{3(2) + 2(0) + 5(4) + 1(2)}{11} = \frac{28}{11} = 2.54$$

Suppose you are taking the following courses and you know all your grades but your math grade. What is the lowest math grade you could receive and still have a grade point average of over 3.00? **B**

Course	Hours	Grade
Biology	5	A
Spanish	3	B
Math	4	?
English	4	B
Phys. Ed.	1	D
Total	17	

2. Read the information below the soccer standings at the right to understand how a team's points are determined. A shootout is a form of tie-breaker used when the score is still tied after an overtime period.

a. A team had 5 wins, 2 shootout victories and 23 bonus points. Show that the number of points must be 61.
6(5) + 4(2) + 23 = 61

b. To find the number of shootout victories for Golden Bay, let x = number of shootout victories. Then, 5 − x = number of regulation or overtime wins. 30 − 6x + 4x + 32 = 60
6(5 − x) + 4x + 32 = 60   −2x = −2
Solve this equation.   x = 1

c. Find the number of shootout victories for Chicago.
6(10 − x) + 4x + 42 = 94
60 − 6x + 4x + 42 = 94
−2x = −8
x = 4

Soccer
NASL

Tech Prep

Veterinary Technician Students who are interested in veterinary medicine may wish to do further research on the statistics given in Exercise 42 and explore the potential growth of this career. For more information on tech prep, see the *Teacher's Handbook*.

Extension

Connections The lengths of the sides of a seven-sided figure are consecutive odd integers. The sum of the shortest and longest sides is 34 cm. Find the perimeter of the figure. **119 cm**

3-6

Solving Equations and Formulas

Objective

After studying this lesson, you should be able to:

• solve equations and formulas for a specified variable.

Geometry

Some equations contain more than one variable. At times you will need to solve these equations for one of the variables. For example, suppose the variables x, y, and z were all in the same equation. To *solve for x* would mean to get x by itself on one side of the equation, with no x's on the other side. In the same way, to *solve for y* would mean to get y by itself on one side of the equation, with no y's on the other side.

The formula for the area of a triangle is $A = \frac{1}{2}bh$ where b represents the length of the base and h represents the height of the triangle. Suppose you know the areas and the lengths of the bases of several triangles and you want to find the height of each triangle. Rather than solve the formula over and over for different values of A and b, it would be easier to solve the formula for h before substituting the values for the other variables.

$$A = \frac{1}{2}bh$$

$$2A = 2\left(\frac{1}{2}bh\right) \quad \text{Multiply each side by 2.}$$

$$2A = bh$$

$$\frac{2A}{b} = \frac{bh}{b} \quad \text{Divide each side by b.}$$

$$\frac{2A}{b} = h$$

When you divide by a variable in an equation, remember that division by 0 is undefined. For example, in the formula above, b cannot equal zero.

Example Solve the equation $-5x + y = -56$
a. **for y, and**
b. **for x.**

a.
$$-5x + y = -56$$
$$-5x + 5x + y = -56 + 5x \quad \text{Add 5x to each side.}$$
$$y = -56 + 5x$$

b.
$$-5x + y = -56$$
$$-5x + y - y = -56 - y \quad \text{Subtract y from each side.}$$
$$-5x = -56 - y$$
$$\frac{-5x}{-5} = \frac{-56 - y}{-5} \quad \text{Divide each side by –5.}$$
$$x = \frac{-56 - y}{-5} \text{ or } \frac{56 + y}{5}$$

LOOK BACK

You can refer to Lesson 2-9 for information on writing equations and formulas.

NCTM Standards: 1–5, 7

Instructional Resources

• Study Guide Master 3-6
• Practice Master 3-6
• Enrichment Master 3-6
• Assessment and Evaluation Masters, p. 73

 Transparency 3-6A contains the 5-Minute Check for this lesson; **Transparency 3-6B** contains a teaching aid for this lesson.

Recommended Pacing	
Standard Pacing	Days 10 & 11 of 14
Honors Pacing	Days 9 & 10 of 13
Block Scheduling*	Day 5 of 6 (along with Lesson 3-7)
Alg. 1 in Two Years*	Days 13 & 14 of 19

 *For more information on pacing and possible lesson plans, refer to the *Block Scheduling Booklet* and *Algebra 1 in Two Years.*

1 FOCUS

 ### 5-Minute Check
(over Lesson 3-5)

Solve.
1. $8m + 3 = 5m + 15$ **4**
2. $\frac{2}{3}t - 7 = 5 - \frac{2}{3}t$ **9**
3. $\frac{3}{4}x + \frac{1}{2} = \frac{1}{3}x - \frac{1}{4}$ $-\frac{9}{5}$
4. $4(x + 3) - 5 = 7(x - 1) + 9$ $\frac{5}{3}$
5. $6(4 + y) - 3 = 4(y - 3) + 2y$
   no solution

Motivating the Lesson

Questioning Begin the lesson by having students draw a number of geometric figures such as a parallelogram, trapezoid, and triangle. Then they should write the formulas used to calculate the perimeter and area of each figure.

Teaching Tip Point out that solving equations for one of two variables will help students in their future work. For example, when writing linear equations in slope-intercept form, students will have to solve the equations for *y*.

In-Class Examples

For Example 1
Solve $2x - 4y = 7$ for x. Then solve $2x - 4y = 7$ for y.
$x = \frac{4y + 7}{2}$; $y = \frac{2x - 7}{4}$

For Example 2
Solve for *x*.

a. $3ax - b = d - 4cx$ $\frac{b + d}{3a + 4c}$

b. $4x - 3m = 2mx - 5$ $\frac{3m - 5}{4 - 2m}$

c. $x + 3ab = ax + n$ $\frac{n - 3ab}{1 - a}$

For Example 3
Solve for *w* in the formula for the perimeter of a rectangle, $P = 2\ell + 2w$. $\frac{P - 2\ell}{2}$

Teaching Tip Emphasize that although there are more variables in the equation in Example 2, the process of solving for a single variable does not change. Operations being performed on that variable still must be undone.

CAREER CHOICES

Computer programming is only one career choice in a wide and varied field. Other career options for people with degrees in computer science include systems analysis, management of information systems (MIS), data processing, and local area network (LAN) administration.

Example **2**   Solve for *y* in $3y + z = am - 4y$.

$$3y + z = am - 4y$$
$$3y + z - z = am - 4y - z \qquad \textit{Subtract z from each side.}$$
$$3y = am - 4y - z$$
$$3y + 4y = am + 4y - 4y - z \qquad \textit{Add 4y to each side.}$$
$$3y + 4y = am - z$$
$$7y = am - z$$
$$(3 + 4)\, y = am - z \qquad \textit{Use the distributive property.}$$
$$\frac{7y}{7} = \frac{am - z}{7} \qquad \textit{Divide each side by 7.}$$
$$y = \frac{am - z}{7}$$

CAREER CHOICES

A **computer programmer** writes a detailed plan for solving a problem with an ordered sequence of instructions, often including equations or formulas.

Programming requires a degree in computer science with a working knowledge of computer languages and mathematics.

For more information, contact:

Association for Computing
  Machinery
11 W. 42nd St.
3rd Floor
New York, NY 10036

Many real-world problems require the use of formulas. When using formulas, you may need to use **dimensional analysis.** This is the process of carrying units throughout a computation. You may also be asked to solve the formula for a specific variable. This may make it easier to use certain formulas.

Example **3**   The formula $P = \frac{1.2W}{H^2}$ represents the amount of pressure exerted on the floor by the heel of a shoe. In this formula, *P* represents the pressure in pounds per square inch (lb/in²), *W* represents the weight of the person wearing the shoe in pounds, and *H* is the width of the heel of the shoe in inches.

  a. Find the amount of pressure exerted if a 130-pound person wore shoes with heels $\frac{1}{2}$ inch wide.

  b. Solve the formula for *W*.

  c. Find the weight of the person if the heel is 3 inches wide and the pressure exerted is 40 lb/in².

a. $P = \frac{1.2W}{H^2}$

$= \frac{1.2(130 \text{ lb})}{\left(\frac{1}{2}\text{ in}\right)^2}$ $W = 130\ lb,$
                $H = \frac{1}{2}\ in.$

$= \frac{156 \text{ lb}}{\frac{1}{4}\text{ in}^2}$

$= 624 \text{ lb/in}^2$

624 lb/in² of pressure are exerted.

b. $P = \frac{1.2W}{H^2}$

$H^2P = H^2\left(\frac{1.2W}{H^2}\right)$ *Multiply each side by H^2.*

$H^2P = 1.2W$

$\frac{H^2P}{1.2} = \frac{1.2W}{1.2}$ *Divide each side by 1.2.*

$\frac{H^2P}{1.2} = W$

GLENCOE Technology

 CD-ROM Interaction

A multimedia simulation connects tables, graphs, and equations to business decisions in a fireworks factory. A blackline master activity with teacher's notes provides a follow-up to the CD-ROM simulation.

For Windows & Macintosh

c.

$$\frac{H^2 P}{1.2} = W$$

$$\frac{(3 \text{ in.})^2 (40 \text{ lb/in}^2)}{1.2} = W$$ *Estimate:* $[(3)^2 - 40] \div 1 = 360$
Will the actual answer be greater or less? Why?

$$\frac{360 \text{ lb}}{1.2} = W$$ *(9 in²)(40 lb/in²) = 360 lb*

$$300 \text{ lb} = W$$ The person weighs 300 pounds.

CHECK FOR UNDERSTANDING

Communicating Mathematics

1. $\frac{7y + z}{a} = m$

 MATH JOURNAL

Guided Practice

4. $\frac{7 + 4y}{3}$ 5. $\frac{3x - 7}{4}$

6. $\frac{b}{a} - 1$ 7. $\frac{c - b}{2}$

Study the lesson. Then complete the following.

1. **Show** how you would solve the equation in Example 2 for m.

2. **Compare and contrast** the amount of pressure being exerted in the case of the two people described in Example 3.   **See Solutions Manual.**

3. Find a formula in a newspaper or magazine. Define all of the variables and explain what the purpose of the formula is.   **See students' work.**

Solve each equation or formula for the variable specified.

4. $3x - 4y = 7$, for x 5. $3x - 4y = 7$, for y

6. $a(y + 1) = b$, for y 7. $4x + b = 2x + c$, for x

8. $F = G\left(\frac{Mm}{d^2}\right)$, for M $\frac{Fd^2}{Gm}$ 9. $S = \frac{n}{2}(A + t)$, for A $\frac{2S - nt}{n}$

Write an equation and solve for the variable specified.

10. Twice a number x and 12 is 31 less than three times another number y. Solve for y. $2x + 12 = 3y - 31$; $\frac{2x + 43}{3}$

EXERCISES

Practice

 A

Solve each equation or formula for the variable specified.

11. $-3x + b = 6x$, for x $\frac{b}{9}$ 12. $ex - 2y = 3z$, for x $\frac{3z + 2y}{e}$

13. $\frac{y + a}{3} = c$, for y $3c - a$ 14. $\frac{3}{5}y + a = b$, for y $\frac{5(b - a)}{3}$

15. $v = r + at$, for a $\frac{v - r}{t}$ 16. $y = mx + b$, for m $\frac{y - b}{x}$

17. $I = prt$, for r $\frac{I}{pt}$ 18. $\frac{by + 2}{3} = c$, for y $\frac{3c - 2}{b}$

19. $H = (0.24)I^2 Rt$, for R $\frac{H}{0.24I^2 t}$ 20. $P = \frac{E^2}{R}$, for R $\frac{E^2}{P}$

B

21. $4b - 5 = -t$, for b $\frac{-t + 5}{4}$ 22. $km + 5x = 6y$, for m $\frac{6y - 5x}{k}$

23. $c = \frac{3}{4}y + b$, for y $\frac{4}{3}(c - b)$ 24. $p(t + 1) = -2$, for t $\frac{-2 - p}{p}$

25. $\frac{5x + y}{a} = 2$, for a $\frac{5x + y}{2}$ 26. $\frac{3ax - n}{5} = -4$, for x $\frac{-20 + n}{3a}$

Lesson 3–6 Solving Equations and Formulas   **175**

Reteaching

Checking Solutions To check the solution to solving an equation for a specific variable, substitute a set of values for all other variables in both the original equation and its equivalent that has been solved for a given variable. Find the value of the given variable in each and compare.

3 PRACTICE/APPLY

Check for Understanding

Exercises 1–10 are designed to help you assess your students' understanding through reading, writing, speaking, and modeling. You should work through Exercises 1–3 with your students and then monitor their work on Exercises 4–10.

Assignment Guide

Core: 11–33 odd, 35–41
Enriched: 12–30 even, 31–41

For **Extra Practice,** see p. 763.

The red A, B, and C flags, printed only in the Teacher's Wraparound Edition, indicate the level of difficulty of the exercises.

Study Guide Masters, p. 24

Write an equation and solve for the variable specified.

C **27.** Twice a number x increased by 12 is 31 less than three times another number y. Solve for x. $2x + 12 = 3y - 31; \dfrac{3y - 43}{2}$

28. Five eighths of a number x is three more than one half of another number y. Solve for y. $\dfrac{5}{8}x = \dfrac{1}{2}y + 3; \dfrac{5}{4}x - 6$

29. Five more than two thirds of a number x is the same as three less than one half of another number y. Solve for x. $\dfrac{2}{3}x + 5 = \dfrac{1}{2}y - 3; \dfrac{3}{4}y - 12$

Critical Thinking **30. Weather** Here's a conversation between a mother and her teenage daughter, adapted from the pages of *Games* magazine.

> Mrs. Weatherby: "This thermometer's no good. It's in Celsius, and I want to know the temperature in Fahrenheit."
>
> Mercuria: "You're so old-fashioned, Mother. Just use the formula $\dfrac{9}{5}C + 32 = F$, where C is the temperature in Celsius and F is the temperature in Fahrenheit. If that's too hard for you, then just double the number on the thermometer and add 30. Of course, you won't get a precise answer."

30. $\dfrac{9}{5}C + 32 = 2C + 30$; 10°C or 50°F

The two women each computed the temperature in degrees Fahrenheit, Mercuria with her formula and Mrs. Weatherby by the approximation. Surprisingly, they got exactly the same answer. Just how hot was it? Justify your reasoning.

Applications and Problem Solving **31. Business** The formula in Exercise 17 is the formula for computing simple interest, where I is the interest, p is the principal or amount invested, r is the interest rate, and t is the time in years. Find the amount of interest earned if you were to invest $5000 at 6% interest (use 0.06), for 3 years. **$900**

32. Physics The formula $d = vt + \dfrac{1}{2}at^2$ is the formula for distance d, given the initial velocity v, the time t, and the acceleration a. Suppose a biker going 4.5 meters per second (m/s) passes a lamppost at the top of a hill and then accelerates down the hill at a constant rate of 0.4 m/s^2 for 12 seconds. How far does she go down the hill during this time? **82.8 meters**

33. Work The formula $s = \dfrac{w - 10e}{m}$ is often used by placement services to find keyboarding speeds. In the formula, w represents the number of words typed, e represents the number of errors, m represents the number of minutes typed, and s represents the speed in words per minute.

a. On a 10-minute keyboarding test, Sally typed 420 words with 6 errors. Find Sally's speed in words per minute. **36 words per minute**

33b. Clarence, with 74 words per minute

b. Who has the greater speed on a 5-minute test: Isabel, who types 500 words with 14 errors, or Clarence, who types 410 words with 4 errors?

176 Chapter 3 Solving Linear Equations

Extension

Problem Solving Experiment with a sheet of $8\dfrac{1}{2} \times 11$-inch paper to make a figure with the greatest number of sides possible. Then use another sheet of paper to create a figure with the least number of sides possible.

Practice Masters, p. 24

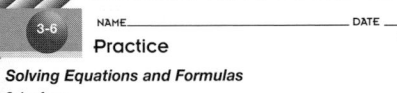

NAME_____ DATE _____

3-6 **Practice**

Student Edition Pages 173–177

Solving Equations and Formulas

Solve for x.

1. $7x = t$ $x = \dfrac{t}{7}$

2. $\dfrac{x - c}{2} = d$ $x = c + 2d$

3. $ax - c = b$ $x = \dfrac{b + c}{a}$

4. $fx + 3y = 2z$ $x = \dfrac{2z - 3y}{f}$

5. $e = rx$ $x = \dfrac{e}{r}$

6. $2p = kx - q$ $x = \dfrac{2p + q}{k}$

Solve for y.

7. $ry + s = tx - m$ $y = \dfrac{tx - m - s}{r}$

8. $x + 3y = 1$ $y = \dfrac{1 - x}{3}$

9. $x - 2y = 1$ $y = \dfrac{x - 1}{2}$

10. $\dfrac{2}{3}y + k = j$ $y = \dfrac{3j - 3k}{2}$

11. $5(2a + y) = 3b$ $y = \dfrac{3b - 10a}{5}$

12. $\dfrac{2}{3}y + a = a + b$ $\dfrac{3}{2}b$

Write an equation and solve for the variable specified.

13. Five more than a number x is six less than twice a number y. Solve for x.
$x + 5 = 2y - 6$; $x = 2y - 11$

14. One fourth of a number x is three more than three times a number y. Solve for x.
$\dfrac{1}{4}x = 3y + 3$; $x = 12y + 12$

15. Five times a number x minus 9 is two thirds of a number y. Solve for y.
$5x - 9 = \dfrac{2}{3}y$; $y = \dfrac{15x - 27}{2}$

176 Chapter 3

Mixed Review

34. Solve $2(x - 2) = 3x - (4x - 5)$. (Lesson 3–5) **3**

35. Find the complement of an 85° angle. (Lesson 3–4) **5°**

36. Work Backward Greta Moore's plane is scheduled to leave for Dallas at 9:30 A.M. She is to pick up her tickets at the boarding gate thirty minutes before departure. It takes Ms. Moore fifty minutes to drive to the airport, ten minutes to park her car in the overnight lot, and ten more minutes to get to the gate. On her way to the airport, Ms. Moore needs to pick up some materials from her office. This detour will take about fifteen minutes. If it takes her an hour to get dressed and out the door, for what time should Ms. Moore set her alarm? (Lesson 3–3) **6:35 A.M.**

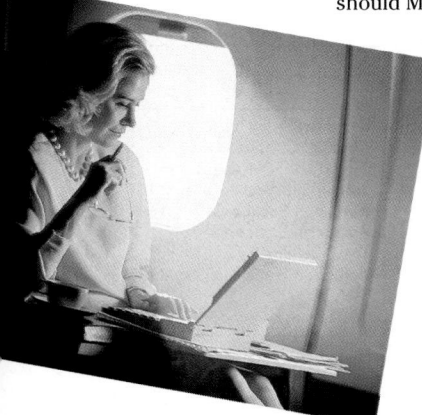

37. Write an inequality for the graph shown below. (Lesson 2–4) $x \neq 2$

$$-2 \quad -1 \quad 0 \quad 1 \quad 2 \quad 3 \quad 4$$

38. Use a number line to find the sum of 9 and −5. (Lesson 2–1) **4**

39. Find the solution set for $x - 3 > \frac{x + 1}{2}$ if the replacement set is {4, 5, 6, 7, 8}. (Lesson 1–5) **{8}**

40. Health Your optimum exercise heart rate per minute is given by the expression $0.7(220 - a)$, where a represents your age. Find your optimum exercise heart rate. (Lesson 1–1) **See students' work.**

WORKING ON THE
In·ves·ti·ga·tion

Refer to the Investigation on pages 68–69.

the Greenhouse Effect

Two scales are used to measure temperature. The Celsius scale, part of the metric system, was invented in 1741 by Swedish astronomer Anders Celsius (1701–1744). The Celsius scale has the freezing point of water at 0°C and the boiling point of water at 100° C. The scale most frequently used in the United States is the Fahrenheit scale, invented in 1714 by German physicist Gabriel Daniel Fahrenheit (1686–1736). On the Fahrenheit scale, the boiling point of water is 212°F, and the freezing point is 32°F.

To convert Celsius temperatures to Fahrenheit, use the formula $F = \frac{9}{5}C + 32$.

1 Using the formula, convert the average temperatures of Mars, Earth, and Venus from Celsius to Fahrenheit. How do the temperatures compare?

2 The greenhouse effect may raise the average temperature of Earth 8°F by the year 2050. What will be the average temperature of Earth in degrees Celsius at that time?

3 Change the formula for converting Celsius to Fahrenheit to a formula for converting Fahrenheit to Celsius. **1–3. See margin.**

Add the results of your work to your Investigation Folder.

Lesson 3-6 Solving Equations and Formulas **177**

In·ves·ti·ga·tion

Working on the Investigation

The Investigation on pages 68–69 is designed to be a long-term project that is completed over several days. Encourage students to keep their materials in their Investigation Folder as they work on the Investigation.

Sample Answers

Planet	Earth	Mars	Venus
Temp. with CO_2	15°C 59°F	1°C 33.8°F	454°C 849.2°F
Temp. without CO_2	−17°C 1.4°F	−31°C −23.8°F	422°C 791.6°F

−13.3°C change in temperature; $\frac{5}{9}(F - 32) = C$

Chapter 3 **177**

NCTM Standards: 1–5, 10

Instructional Resources
- Study Guide Master 3-7
- Practice Master 3-7
- Enrichment Master 3-7
- Assessment and Evaluation Masters, p. 73
- Real-World Applications, 9

 Transparency 3-7A contains the 5-Minute Check for this lesson; **Transparency 3-7B** contains a teaching aid for this lesson.

Recommended Pacing

Standard Pacing	Day 12 of 14
Honors Pacing	Day 11 of 13
Block Scheduling*	Day 5 of 6 (along with Lesson 3-6)
Alg. 1 in Two Years*	Days 15 & 16 of 19

 *For more information on pacing and possible lesson plans, refer to the *Block Scheduling Booklet* and *Algebra 1 in Two Years.*

1 FOCUS

5-Minute Check
(over Lesson 3-6)

Solve for *n*.

1. $\frac{1}{2}n + b = n$ $2b$

2. $ab - bn = 1$ $a - \frac{1}{b}$

3. $cn - 7 = 4$ $\frac{11}{c}$

4. $\frac{a}{c} = n + bn$ $\frac{a}{c + bc}$

5. $an - bn + cn = 5$ $\frac{5}{a - b + c}$

Motivating the Lesson

Situational Problem Reports in newspapers and on TV often refer to the "average." The "average" American earns about $25,000 per year, the "average" cost of housing is about $10,000 per year, and so on. Ask students how averages are determined. Are they determined in the same manner for all types of situations?

Teaching Tip You may wish to suggest that students use a calculator to aid in computation.

3-7

Integration: Statistics
Measures of Central Tendency

Objective
After studying this lesson, you should be able to:
- find and interpret the mean, median, and mode of a set of data.

Charities

The table at the right shows the top 10 charities in amount of contributions for 1993.

When analyzing data, it is helpful to have one number that describes the entire set of data. Numbers known as **measures of central tendency** are often used to describe sets of data because they represent a centralized, or *middle,* value. Three of the most commonly used measures of central tendency are the **mean, median,** and **mode.**

Charity	Millions of Dollars
Salvation Army	$726
Catholic Charities USA	411
United Jewish Appeal	407
Second Harvest	407
American Red Cross	395
American Cancer Society	355
YMCA	317
American Heart Association	235
YWCA	218
Boy Scouts of America	211

Source: *The Chronicle of Philanthropy, Nov., 1993*

Definition of Mean	The mean of a set of data is the sum of the numbers in the set divided by the number of numbers in the set.

Until 1994, the Cardwell sisters of Sweetwater, Texas, were the oldest living set of triplets. However, on October 2, 1994, the firstborn triplet, Faith, died at the age of 95. She was survived by her sisters Hope and Charity.

To find the mean of the charity data, find the sum of the dollar values and divide by 10, the number of numbers in the set. *Estimate: Is an answer of $500 reasonable?*

$$\text{mean} = \frac{726 + 411 + 407 + 407 + 395 + 355 + 317 + 235 + 218 + 211}{10}$$

$$= \frac{3682}{10}$$

$$= 368.2$$

The mean is $368.2 million. Notice that the amount collected by the Salvation Army, $726 million, is far greater than the amounts collected by the other charities. Because the mean is an average of several numbers, a single number that is so much greater than the others can affect the mean a great deal. In extreme cases, the mean becomes less representative of the values in a set of data.

When the Cardwell sisters were born, their doctor suggested they be named Lillie, Lola, and Lula. Their parents did not like those names or any others, so for six months, the sisters went unnamed. Finally, Frances Cleveland, the wife of then-President Grover Cleveland, suggested that the babies be named Faith, Hope, and Charity, and thus, they were.

The median is another measure of central tendency.

Definition of Median	> The median of a set of data is the middle number when the numbers in the set are arranged in numerical order.

To find the median of the charity data, arrange the dollar values in order.

726   411   407   407   395   355   317   235   218   211

If there were an odd number of dollar values, the middle one would be the median. However, since there is an even number of dollar values, the median is the average of the two middle values, 395 and 355.

$$\text{median} = \frac{395 + 355}{2}$$

$$= \frac{750}{2}$$

$$= 375$$

The median is $375 million. Notice that the number of values that are greater than the median is the same as the number of values that are less than the median.

A third measure of central tendency is the mode.

Definition of Mode	> The mode of a set of data is the number that occurs most often in the set.

To find the mode of the charity data, look for the number that occurs most often.

726   411   407   407   395   355   317   235   218   211

In this set, 407 appears twice. Thus, $407 million is the mode of the data.

It is possible for a set of data to have more than one mode. For example, the set of data {2, 3, 3, 4, 6, 6} has two modes, 3 and 6.

Based on our results, the charity data has mean $368.2 million, median $375 million, and mode $407 million. As you can see, the mean, median, and mode are rarely the same value.

Example ❶

CONNECTION

History

When the Declaration of Independence was signed in 1776, George III was king of Great Britain. After George III, Great Britain had seven monarchs before Queen Elizabeth II was crowned queen in 1952. The table at the right shows the number of years that each monarch reigned. Find the mean, median, and mode of the data.

Monarch	Reign (years)
George III	59
George IV	10
William IV	7
Victoria	63
Edward VII	9
George V	25
Edward VIII	1
George VI	15

$$\text{mean} = \frac{59 + 10 + 7 + 63 + 9 + 25 + 1 + 15}{8}$$

$$= \frac{189}{8}$$

$$= 23.625 \quad \text{The mean is 23.625 years.} \quad \textit{(continued on the next page)}$$

Lesson 3–7 *Statistics   Measures of Central Tendency*   **179**

In-Class Example

For Example 1
The heights of players on South High School's basketball team are 72", 74", 70", 77", 75", and 70". Find the mean, median, and mode of the heights.
Mean is 73", median is 73", and mode is 70".

GLENCOE Technology

Interactive Mathematics Tools Software

This multimedia software provides an interactive lesson that uses histograms to observe the changes in the relationships between the mean, mode, and median as the data change. A **Computer Journal** gives students an opportunity to write about what they have learned.

For Windows & Macintosh

The median is $\frac{10 + 15}{2}$ or $12\frac{1}{2}$ years. Why do you think the mean and median are so different? Which value, mean or median, do you think is more representative of the data?

mode   Since each number appears exactly once, there is no mode.

You can also determine the mean, median, and mode of a set of data by examining stem-and-leaf plots.

Example **2**

APPLICATION
Demographics

Since it would be inconvenient to list every stem, we have used dots to indicate that some stems are missing.

The stem-and-leaf plot at the right shows population data for the 20 largest cities in the United States to the nearest ten thousand. Find the mean, median, and mode of the data.

Stem	Leaf
73	2
⋮	⋮
34	9
⋮	⋮
27	8
⋮	⋮
16	9 3
⋮	⋮
10	3 1
⋮	⋮
8	8
7	9 9 4 3
6	8 4 4 3 3 3 1
5	7

$8 \mid 8 = 880,000$ people

mean   Add the 20 values and then divide by 20. Since the sum of the 20 values is 2791, the mean is $\frac{2791}{20}$ or 139.55. This represents a mean population of 1,395,500.

median   Since there are 20 values, the median is the average of the 10th and 11th values. Counting from the top down, you will find that the 10th and the 11th values are 79 and 74. The average of 79 and 74 is $\frac{79 + 74}{2}$ or 76.5. Thus, the median population is 765,000.

mode   There are three entries for 63. Thus, the mode is 630,000.

Communicating Mathematics

Study the lesson. Then complete the following.

1. **Explain** why you think the mean, median, and mode are rarely the same value. **See margin.**

2. **Discuss** which of the measures of central tendency you would use to describe the data in Example 2. Explain your reasoning. **See margin.**

3. **Describe** the steps you would take to find the mean, median, and mode of the data represented by the line plot shown at the right. **See margin.**

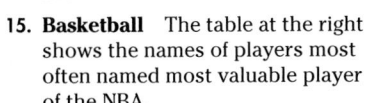

```
                 ×
          × ×          ×
  ×   × × × ×     × × × ×
  +---+---+---+---+---+---+---+---+---+---+---+
  0   1   2   3   4   5   6   7   8   9   10
```

4. Suppose you conducted a survey in which you asked your classmates, "What's your favorite rock group?" Which measure of central tendency could you use to analyze the data, and why? **See margin.**

MATH JOURNAL

5. Find some data that interests you in a newspaper or magazine. Find the mean, median, and mode and discuss which of these measures you think best describes the data. **See students' work.**

Complete.

6. Measures of central tendency represent __?__ values of a set of data. **middle**

7. If the numbers in a set of data are arranged in numerical order, then the __?__ of the set is the middle number. **median**

8. Extremely high or low values affect the __?__ of a set of data. **mean**

9. If all the numbers in a set of data occur the same number of times, then the set has no __?__ . **mode**

Guided Practice

Find the mean, median, and mode for each set of data.

10. 4, 6, 12, 5, 8   **7; 6; none**

11. 8, 8, 8, 8, 9   **8.2; 8; 8**

12. **93.77; 94; 82**

Stem	Leaf
7	3 5
8	2 2 4
9	0 4 7 9
10	5 8
11	4 6

$9|4 = 94$

13–14. Sample answers are given.

List six numbers that satisfy each set of conditions.

13. The mean is 50, the median is 40, and the mode is 20.   **20, 20, 30, 50, 90, 90**

14. The mean is 70, the median is 70, and the modes are 65 and 70.   **65, 65, 70, 70, 73, 77**

15. **Basketball** The table at the right shows the names of players most often named most valuable player of the NBA.

   a. Find the mean, median, and mode of the data.   **3.857; 3; 3**

   b. Which measure of central tendency best represents these data, and why?   **mean**

Player	MVP Wins
Kareem Abdul-Jabbar	6
Bill Russell	5
Wilt Chamberlain	4
Larry Bird	3
Magic Johnson	3
Moses Malone	3
Michael Jordan	3

Source: *World Almanac, 1995*

Reteaching ▬▬▬▬

Using Data Give students a list of the scores from their last test. Have them find the mean, median, and mode of these scores and then discuss their results.

Additional Answer

4. Mode; since the data are nonnumeric, the mean and median would have no meaning in this context.

Check for Understanding

Exercises 1–15 are designed to help you assess your students' understanding through reading, writing, speaking, and modeling. You should work through Exercises 1–9 with your students and then monitor their work on Exercises 10–15.

Additional Answers

1. **The mean is affected by extreme values, where median is not. Mode is affected by the amount of repetition of values.**

2. **The mean, since it finds the average amount of the contributions. The mean is affected by extreme values.**

3. **Count the total number of x's. To find the mean, multiply the number by the number of x's in that column. Then divide by the total number of x's (4.8). To find the median, divide the total number of x's by 2 (13 ÷ 2 or ≈ 7) and count from left to right. The seventh x represents the median (4). The mode is the number with the greatest number of x's in its column (2).**

Study Guide Masters, p. 25

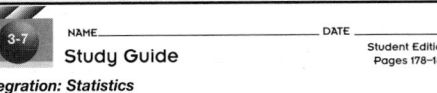

3-7   NAME_____   DATE_____

Study Guide   Student Edition Pages 178–183

Integration: Statistics
Measures of Central Tendency

In working with statistical data, it is often useful to have one value represent the complete set of data. For example, **measures of central tendency** represent centralized values of the data. Three measures of central tendency are the **mean, median,** and **mode.**

	Definitions	Examples
Mean	Sum of the elements in the set divided by the number of elements in the set.	Data: 24, 36, 21, 30, 21, 30 $\frac{24 + 36 + 21 + 30 + 21 + 30}{6} = 27$
Median	The middle of a set of data when the numbers are arranged in numerical order. In an even number of elements, the median is halfway between the two middle elements.	Data: 21, 21, 25, 30, 31, 42 $\frac{25 + 30}{2} = 27.5$
Mode	The number that occurs most often in a set of data.	Data: 21, 21, 24, 30, 30, 36 There are two modes, 21 and 30.

Find the mean, median, and mode for each set of data.

1.
Month	Days above 90°F
May	4
June	7
July	14
August	12
September	8

9; 8; none

2.
```
              ×
        ×     ×  × ×
  × ×   × ×   × × ×
  +--+--+--+--+--+--+--+
  1  2  3  4  5  6  7
```
3.7; 4; 1 and 5

3. 3, 6, 6, 3, 6, 6, 3, 3
   4.5, 4.5, 3 and 6

4. 19, 3, 0, 1
   5.75; 2; none

5. $\frac{1}{4}, \frac{2}{5}, \frac{2}{8}, \frac{1}{3}$
   $\frac{37}{120}; \frac{7}{24}; \frac{1}{4}$

6. $1, \frac{1}{2}, 2, \frac{1}{3}, 3, \frac{1}{4}, 4, \frac{1}{4}$
   $\frac{269}{192}; \frac{3}{4}; $ none

Find the median and mode(s) of the data shown in each stem-and-leaf plot.

7.
Stem	Leaf
2	4 7 7
3	1 2 6 6 6 9
4	0
5	8 8 9

$\frac{312}{10} = 32$
36; 36

8.
Stem	Leaf
9	0 0 1 3 9
10	2 2 5
11	
12	0 3 3 8 8 9

$\frac{1015}{10} = 105$
103.5; 102, 123, 128

Assignment Guide

Core: 17–31 odd, 32–37
Enriched: 16–26 even, 27–37

For **Extra Practice,** see p. 764.

The red A, B, and C flags, printed only in the Teacher's Wraparound Edition, indicate the level of difficulty of the exercises.

Additional Answers

27b. Yes, because the article indicates that the median value is in the middle of the data set.

27c. No, he should have used the term *mean* since the mean can be affected by extremely high values, causing the majority to be below the mean value.

Practice Masters, p. 25

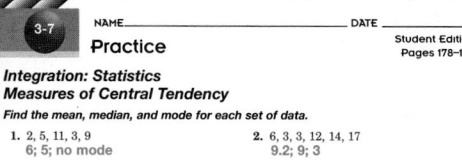

```
3-7   NAME_____ DATE _____
      Practice                    Student Edition
                                  Pages 178–183
Integration: Statistics
Measures of Central Tendency
Find the mean, median, and mode for each set of data.
1. 2, 5, 11, 3, 9           2. 6, 3, 3, 12, 14, 17
   6; 5; no mode               9.2; 9; 3
3. 1, 6, 7, 6, 8, 10, 10    4. 5, 0, 11, 1, 14, 3, 4, 7, 11
   6.9; 7; 6 and 10            6.2; 5; 11
```

Find the median and mode of the data shown in each line plot.

```
5. 5; 4                     6. 50; 59
```

Find the median and mode of the data shown in each stem-and-leaf plot.

```
7. Stem | Leaf          8. Stem | Leaf         9. Stem | Leaf
   2    | 1 3 7 7 8        5    | 5 7 8           5    | 2 4 6
   3    | 0 1 9            6    | 1 3 7 9         6    | 4 5
   4    | 2 4 4 4 6 8      7    | 0 4 6 6         7    | 0 2 6 6 9
   2|3 = 23               5|7 = 57               8    | 1 5 7 7 7
   35; 44                 67; 76                 9    | 8
                                                 5|2 = 5.2
                                                 7.6; 8.7
```

```
10. The prices of a sweater in six       11. Jack's scores on 20-point quizzes are
    different stores are $31.95, $26.30,     16, 18, 10, 15, 19, 16, 15, 19, 20, 16,
    $29.00, $32.50, $25.25, and $38.00.      12, 17, 16, 15, 18, 14, and 16. Find
    Find the mean and median prices of       the mean, median, and mode of his
    the sweater.                             quiz scores.
    $30.50; $30.48                           16, 16, 16
```

182 Chapter 3

EXERCISES

Practice

Find the mean, median, and mode for each set of data.

16. 2, 4, 7, 9, 12, 15
  $8\frac{1}{6}$; 8; none

17. 300, 34, 40, 50, 60   96.8; 50; none

18. 23, 23, 23, 12, 12, 12
  17.5; 17.5; 23, 12

19. 10, 3, 17, 1, 8, 6, 12, 15   9; 9; none

20. 7, 19, 9, 4, 7, 2   8; 7; 7

21. 2.1, 7.4, 13.9, 1.6, 5.21, 3.901
  5.69; 4.56; none

22.

Stem	Leaf
5	3 6 8
6	5 8
7	0 3 7 7 9
8	1 4 8 8 9
9	9 6|8 = 68

75.3; 77; 77, 88

23.

Stem	Leaf
19	3 5 5
20	2 2 5 8
21	5 8 8 9 9 9
22	0 1 7 8 9
	21|5 = 215

212.94; 218; 219

24. List ten numbers that satisfy each set of conditions.

Sample answers are given.
a. 6, 6, 6, 6, 6, 6, 6, 6, 6, 0
b. 5, 5, 5, 5, 5, 5, 5, 5, 5, 25.
d. No, because it is always in the middle of the set of data.

  a. The mean is less than all but one of the numbers.
  b. The mean is greater than all but one of the numbers.
  c. The mean is greater than all of the numbers.   not possible
  d. Would you be able to complete parts a, b, and c if the word *mean* were replaced by the word *median*? Why or why not?

25. The mean of a set of ten numbers is 5. When the greatest number in the set is eliminated, the mean of the new set of numbers is 4. What number was eliminated from the original set of numbers?   14

26. The first of three consecutive odd integers is $2n + 1$.
  a. What is the mean of the three integers? $\frac{6n + 9}{3}$ or $2n + 3$
  b. What is the median? $2n + 3$

Critical Thinking

27. This paragraph appeared in the *San Diego Union-Tribune* on August 8, 1993.

> Lawyers at the top of their profession earn more than $1 million a year, while the typical attorney makes almost $67,000—far more than the average American. . . . According to census figures, lawyers and judges earned a median pay of $66,784 in 1991, which means that half of them made more than that amount and half made less. . . . However R. Wilson Montjoy II of Jackson, Mississippi, said, "We have lots more people at or below the median than above it."

27a. mean

27b–c. See margin.

  a. To which measure of central tendency do you think the word *average* refers?
  b. Has *median* been defined correctly in this passage? Why or why not?
  c. In the statement by R. Wilson Montjoy II, did he use the term *median* correctly? If not, which term should he have used? Justify your answer.

Applications and Problem Solving

28a. 29.4; 16; none

28b. The mean is most representative.

28. Demographics   The table at the right shows population data for 10 South American countries, according to the 1990 Census.
  a. Find the mean, median, and mode of the data.
  b. If you were reporting these data in a newspaper article, which measure would you use, and why?

Country	Population (millions)
Argentina	32
Bolivia	7
Brazil	150
Chile	13
Colombia	33
Ecuador	10
Paraguay	5
Peru	22
Uruguay	3
Venezuela	19

29. Geography The areas of each of the South American countries listed in the table on the previous page are as follows (in millions of square miles): 1085, 139, 7778, 670, 3166, 1068, 177, 1135, 154, and 1234, respectively. Find the mean, median, and mode area of the countries. $1660\frac{3}{5}$; $1076\frac{1}{2}$; none

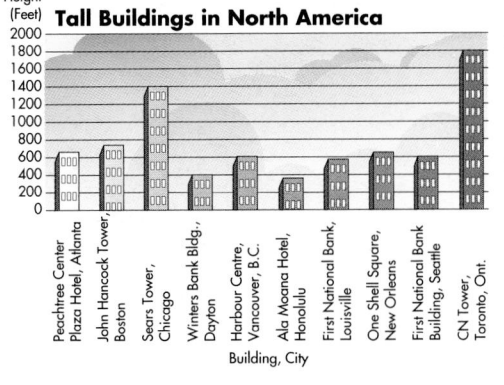

Height
(Feet)
Tall Buildings in North America

30. American Landmarks The graph at the left shows the height in feet of several tall buildings in North America. Find the mean and median height of the buildings. **about 798; about 625**

31. Commuting On her way home from work, Luisa drives 10 miles through the city at about 30 mph and then 10 miles on the highway at about 50 mph. What is her average speed? Use $d = rt$. (*Hint:* The answer is *not* 40 mph.) **37.5 mph**

Mixed Review

32. Solve $\frac{a+5}{3} = 7x$ for x. (Lesson 3–6) $\frac{a+5}{21}$

33. Solve $\frac{3}{4}n - 3 = 9$. Then check your solution. (Lesson 3–3) **16**

34. Evaluate $|k| + |m|$ if $k = 3$ and $m = -6$. (Lesson 2–3) **9**

35. Use a number line to add $-6 + (-14)$. (Lesson 2–1) **−20**

36. Evaluate $\left(13 + \frac{2}{5} \cdot 5\right)(3^2 - 2^3)$. Indicate the property used in each step. (Lesson 1–8) **15**

37. Simplify $5 + 7(ac + 2b) + 2ac$. (Lesson 1–7) $5 + 9ac + 14b$

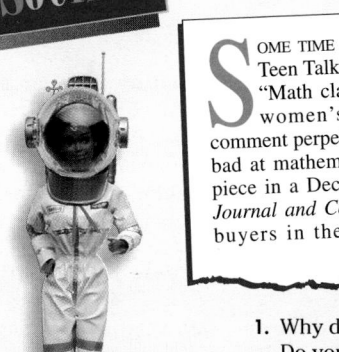

Mathematics and SOCIETY

Teen Talk Barbies

The article below appeared in *Antique Week* on August 30, 1993.

SOME TIME AGO, MATTEL RECALLED ANY Teen Talk Barbies programmed to say "Math class is tough" because some women's groups complained the comment perpetuated the view that girls are bad at mathematics. According to a news piece in a December issue of *The Atlanta Journal and Constitution,* only five Barbie buyers in the United States turned in their dolls because of Mattel's offer. While some talking Barbies may have found math tough, the consumers/collectors who own the doll won't have trouble adding up the profits.... Savvy collectors were tearing open Teen Talk boxes trying to find a Barbie who talked about math. Mike Huen, a professional toy collector, estimates the doll, which sold for $35, will be worth several hundred dollars now. ∎

1. Why do you think some women's groups objected to what the dolls say? Do you think that what the dolls say has any influence on the girls who play with them? Why or why not? **1–3. See students' work.**

2. Why do you think only five Barbie buyers have returned their dolls? Explain.

3. Do you think it is right for people to make money on this doll? Justify your reasoning.

Extension

Problem Solving Find the mean, median, and mode of the hourly wages of 20 workers. Ten workers earn $4.75 per hour, one earns $5.50 per hour, one earns $6.75 per hour, two earn $4.80 per hour, and six earn $5.25 per hour. **mean = $5.04, median = $4.78, mode = $4.75**

Mathematics and SOCIETY

Point out that there are articles and information that relate to mathematics in many popular magazines and journals. For example, this article appeared in *Antique Week.*

Discuss the feelings and opinions of those who wanted the dolls recalled and those who opposed the recall.

Closing the Investigation

This activity provides students an opportunity to bring their work on the Investigation to a close. For each Investigation, students should present their findings to the class. Here are some ways students can display their work.

- Conduct and report on an interview or survey.
- Write a letter, proposal, or report.
- Write an article for the school or local paper.
- Make a display, including graphs and/or charts.
- Plan an activity.

Assessment

To assess students' understanding of the concepts and topics explored in the Investigation and its follow-up activities, you may wish to examine students' Investigation Folders.

The scoring guide provided in the *Investigations and Projects Masters,* p. 3, provides a means for you to score students' work on the Investigation.

Investigations and Project Masters, p. 3

Scoring Guide
Chapters 2 and 3
Investigation

Level	Specific Criteria
3 Superior	· Shows thorough understanding of the concepts of *the greenhouse effect, global warming, data analysis, subtraction of rational numbers, equations,* and *formulas.* · Uses appropriate strategies to solve problems. · Computations are correct. · Written explanations are exemplary. · Charts, graph, and letter are appropriate and sensible. · Goes beyond requirements of all or some problems.
2 Satisfactory, with Minor Flaws	· Shows understanding of the concepts of *the greenhouse effect, global warming, data analysis, subtraction of rational numbers, equations,* and *formulas.* · Uses appropriate strategies to solve problems. · Computations are mostly correct. · Written explanations are effective. · Charts, graph, and letter are appropriate and sensible. · Satisfies all requirements of problems.
1 Nearly Satisfactory, with Obvious Flaws	· Shows understanding of most of the concepts of *the greenhouse effect, global warming, data analysis, subtraction of rational numbers, equations,* and *formulas.* · May not use appropriate strategies to solve problems. · Computations are mostly correct. · Written explanations are satisfactory. · Charts, graph, and letter are appropriate and sensible. · Satisfies most requirements of problems.
0 Unsatisfactory	· Shows little or no understanding of the concepts of *the greenhouse effect, global warming, data analysis, subtraction of rational numbers, equations,* and *formulas.* · Does not use appropriate strategies to solve problems. · Computations are incorrect. · Written explanations are not satisfactory. · Charts, graph, and letter are not appropriate or sensible. · Does not satisfy requirements of problems.

---

the Greenhouse Effect

Refer to the Investigation on pages 68–69.

For more than a decade, scientists have warned that cars and factories spew so many gases into the atmosphere that Earth could soon be affected by disastrous climatic changes. The loss of rain forests is reducing the number of trees to offset the large increases in carbon dioxide in our atmosphere. The possible consequences are so frightening that it makes sense to slow the buildup of CO_2 through preventive measures, such as encouraging energy conservation, developing alternatives to fossil fuels, and preventing the destruction of the rain forests.

Analyze

You have conducted experiments and done research on this important problem. It is now time to analyze your findings and state your conclusions.

PORTFOLIO ASSESSMENT

You may want to keep your work on this Investigation in your portfolio.

1 Create a graph of the data from your experiments. Let the vertical axis represent the changes in temperature. Let the horizontal axis represent the distances between the lamp and the thermometer. Use different colors to indicate the control experiment and the greenhouse experiment.

2 Use the data from your experiments to draw a conclusion about the relationship between the temperature and the distance between the thermometer and the lamp. Describe this relationship as a function, and justify your reasoning.

3 Use the data from Working on the Investigation in Lesson 2–6 to make a chart.

Use the headings *Planet, Temperature Without CO₂, Temperature With CO₂,* and *Distance from the Sun.*

4 Draw a conclusion about the relationship between the distance from the sun and the temperature of Venus and Earth with CO_2. Describe this relationship as a function, and justify your reasoning.

5 Draw a conclusion about the relationship between the distance from the sun and the temperature of Mars and Earth with CO_2. Describe this relationship as a function, and justify your reasoning.

6 How does your experiment explore the temperatures of the planets? How are the relationships similar and different?

7 Predict what the temperature of Mars would be if it had the same amount of CO_2 that Earth has. Explain your calculations.

Write

Some scientists predict that by the year 2050, Earth's atmosphere will contain between 500 and 700 ppm of CO_2. Write a one-page paper explaining the situation to a group of concerned citizens. Consider some of the following questions to address in your paper.

8 What factors will cause this increase in CO_2?

9 How different is this estimate from the amount of CO_2 calculated earlier?

10 In your opinion, what will the approximate average temperature on Earth be at that time?

11 What might occur if the temperature reaches that point? What are some of the dangers involved?

Sample Answers

Planet	Earth	Mars	Venus
Temp. with CO_2	15°C	1°C	454°C
Temp. without CO_2	−17°C	−31°C	422°C
Distance from the sun	150 million km	228 million km	108 million km

Answers will vary based on the data collected. However, general observations should be that the farther the object is from the light source, the colder the object becomes. The plastic bag over the thermometer holds in the heat in much the same way that CO_2 does for a planet.

VOCABULARY

After completing this chapter, you should be able to define each term, property, or phrase and give an example or two of each.

Algebra
addition property of equality   (p. 144)
consecutive integers   (p. 158)
dimensional analysis   (p. 174)
division property of equality   (p. 151)
equivalent equation   (p. 144)
identity   (p. 170)
multi-step equations   (p. 157)
multiplication property of equality   (p. 150)
number theory   (p. 158)
solve an equation   (p. 145)
subtraction property of equality   (p. 146)

Statistics
mean   (p. 178)
measures of central tendency   (p. 178)
median   (p. 178)
mode   (p. 178)

Geometry
acute triangle   (p. 165)
complementary angles   (p. 163)
congruent   (p. 164)
equilateral triangle   (p. 164)
isosceles triangle   (p. 164)
obtuse triangle   (p. 165)
right triangle   (p. 165)
supplementary angles   (p. 162)
triangle   (p. 163)

Problem Solving
work backward   (p. 156)

UNDERSTANDING AND USING THE VOCABULARY

Choose the letter of the term that best matches each statement or phrase.

1. If $a = b$, then $a + c = b + c$.   b
2. If $a = b$, then $a - c = b - c$.   l
3. If $a = b$, then $ac = bc$.   i
4. If $a = b$ and $c \neq 0$, then $\frac{a}{c} = \frac{b}{c}$.   c
5. a triangle in which each angle has the same measure   d
6. a triangle in which two angles have the same measure   e
7. a triangle in which one angle measures 90°   k
8. a triangle in which one angle measures more than 90°   j
9. a triangle in which all of the angles measure less than 90°   a
10. the sum of the numbers divided by the number of numbers   f
11. the middle number when data is in numerical order   g
12. the number that occurs most often in a set of data   h

a. acute triangle
b. addition property of equality
c. division property of equality
d. equilateral triangle
e. isosceles triangle
f. mean
g. median
h. mode
i. multiplication property of equality
j. obtuse triangle
k. right triangle
l. subtraction property of equality

Chapter 3   Highlights   **185**

Instructional Resources

Three multiple-choice tests and three free-response tests are provided in the *Assessment and Evaluation Masters.* Forms 1A and 2A are for honors pacing, and Forms 1B, 1C, 2B, and 2C are for average pacing. Chapter 3 Test, Form 1B is shown at the right. Chapter 3 Test, Form 2B is shown on the next page.

Assessment and Evaluation Masters, pp. 59–60

---

3   NAME_____ DATE_____

Chapter 3 Test, Form 1B

Write the letter for the correct answer in the blank at the right of each problem.

1. What is the solution of $m - (-4) = 7$?   1. __A__
   A. 3   B. 23   C. 11   D. 211

2. What is the solution of $-11 = n - (-11)$?   2. __C__
   A. 0   B. 22   C. -22   D. none of these

3. The sum of two integers is -46. The greater integer is 13. What is the lesser integer?   3. __B__
   A. 59   B. -59   C. -33   D. 33

4. What is the solution of $-26y = 884$?   4. __D__
   A. 910   B. 858   C. 34   D. -34

5. What is the solution of $\frac{3}{5}x = 15$?   5. __C__
   A. 45   B. 5   C. 25   D. 75

6. What is the solution of $5x + 3 = 23$?   6. __A__
   A. 4   B. -4   C. 5   D. -5

7. What is the solution of $\frac{n}{3} - 8 = -2$?   7. __D__
   A. -30   B. 30   C. -18   D. 18

8. What is the solution of $-14 = \frac{c + 12}{-6}$?   8. __B__
   A. -72   B. 72   C. 96   D. -96

9. What is the supplement of an angle whose measure is $(x - 28)°$?   9. __C__
   A. $(62 - x)°$ B. $(118 - x)°$ C. $(208 - x)°$ D. $(152 - x)°$

10. What is the measure of an angle that is twice its complement?   10. __B__
    A. 30°   B. 60°   C. 90°   D. 120°

11. The measures of two angles of a triangle are 44° and 32°. What is the measure of the third angle?   11. __D__
    A. 76°   B. 14°   C. 44°   D. 104°

---

3   NAME_____ DATE_____

Chapter 3 Test, Form 1B (continued)

12. What is the solution of $2x + 7 = 5x + 16$?   12. __A__
    A. -3   B. $\frac{2}{3}$ C. $-\frac{23}{3}$ D. 3

13. What is the solution of $\frac{2}{3}(6x - 3) = x + 5(x + 8) - 2x$?   13. __B__
    A. identity   B. no solution   C. 6   D. 0

14. What is the solution of $-3(h - 6) = 5(2h + 3)$?   14. __B__
    A. $-\frac{3}{13}$ B. $\frac{3}{13}$ C. $-\frac{9}{13}$ D. $\frac{9}{13}$

15. Let x represent the lesser of two consecutive even integers. Which equation would you use to find the two integers if their sum is 126?   15. __A__
    A. $x + (x + 2) = 126$ B. $x + 2x = 126$
    C. $(x + 2) + (x + 4) = 126$ D. $x(x + 2) = 126$

16. A number is added to 9. The result is then multiplied by 4 to give a new result of 120. What is the number?   16. __D__
    A. $27\frac{3}{4}$ B. 489   C. 39   D. 21

17. What is the solution for x of $2x - y = y$?   17. __C__
    A. $2y - 2$ B. $y - 2$ C. y D. 0

18. Find the mean of the set of data 4, 6, 7, 7, 9, 11, 17, 21, 25 rounded to the nearest tenth.   18. __C__
    A. 7   B. 9   C. 11.9   D. 11.5

19. Find the median of the set of data 5.11, 7.07, 8.95, 6.89, 8.63, 8.95.   19. __A__
    A. 7.85   B. 8.95   C. 7.60   D. no median

20. Which is affected by extremely high or low values in a set of data?   20. __C__
    A. median   B. mode   C. mean   D. All are affected.

Bonus   **Bonus**

A high school team had just won a conference mathematics contest. A reporter from the local paper asked the team captain, "How many gold, silver, and bronze medals did the team win?" The captain replied, "The product of the three numbers is 72 and their sum is the same as today's date." The reporter thought for a while and said, "I know the date but I still can't find the answer." The captain exclaimed, "I'm sorry. I forgot to tell you that the smallest number of medals was of bronze and the greatest number of medals was gold." "Thank you," replied the reporter, who then left to write the story. How many medals of each type were won?

9 gol
8 silv
1 bron

Using the STUDY GUIDE AND ASSESSMENT

Skills and Concepts Encourage students to refer to the objectives and examples on the left as they complete the review exercises on the right.

Assessment and Evaluation Masters, pp. 65–66

NAME _____ **DATE** _____

Chapter 3 Test, Form 2B

Solve each equation.

1. $12 + r = 3$
2. $-12 = 7 - p$
3. $n - (-8) = 3$
4. $d + (-11) = -4$
5. $-\frac{3}{4} - p = \frac{1}{2}$
6. $-7b = -35$
7. $31 = -\frac{1}{6}n$
8. $-\frac{5}{8}w = -9$
9. $-3a + 4 = -14$
10. $3.2n + 2.6 = -23$
11. $\frac{x}{-4} + 5 = 1$
12. $2 = \frac{-2y - (-6)}{13}$
13. $6y - 3 = 6y + 8$
14. $1.4f + 1.1 = 8.3 - f$
15. $-4(p + 2) + 8 = 2(p - 1) - 7p + 15$
16. $\frac{n}{3} + \frac{3}{4} = \frac{5}{6}n - 1$
17. Find the complement of an angle whose measure is 33°.
18. Find the supplement of an angle whose measure is 85°.
19. Find the supplement of an angle whose measure is $(y - 13)°$.
20. Find the complement of an angle whose measure is $(40 + a)°$.
21. Find the measure of the third angle of the triangle.

Solve for x.

22. $\frac{a}{b}x - c = 0$
23. $\frac{x + a}{b} = c$

1.	-9
2.	19
3.	-5
4.	$\frac{7}{-5}$
5.	$-\frac{5}{4}$
6.	5
7.	-186
8.	$\frac{72}{5}$
9.	6
10.	-8
11.	16
12.	-10
13.	no soln.
14.	3
15.	13
16.	$\frac{7}{2}$
17.	$57°$
18.	$95°$
19.	$(193 - y)°$
20.	$(50 - a)°$
21.	$135°$
22.	$x = \frac{(cb)}{a}$
23.	$x = cb - a$

NAME _____ **DATE** _____

Chapter 3 Test, Form 2B (continued)

Define a variable, write an equation, and solve each problem.

24. A number is subtracted from -6. If the answer is 55, find the number.
   24. $n = $ the number
   $-6 - n = 55$
   $n = -61$

25. In 1989, the Chicago Cubs pitching staff had 50 saves. Williams had 36 saves while Lancaster had 8 saves. How many saves did the rest of the pitching staff have?
   25. $s = $ no. of saves
   $36 + 8 + s = 50$
   $s = 6$

26. A number is divided by 3 and the result added to 1. The new result is doubled to give a final result of 80. What is the number?
   26. $n = $ the number
   $(1 + \frac{n}{3})2 = 80$
   $n = 117$

27. Find two consecutive even integers whose sum is 98.
   27. $n = $ lesser number
   $n + (n + 2) = 98$
   $n = 48, 50$

28. Find three consecutive integers such that the sum of the first and third is 84.
   28. $n = $ least number
   $n + (n + 2) = 84$
   $n = 41, 42, 43$

29. Four times a number decreased by 6 is 150. What is the number?
   29. $n = $ the number
   $4n - 6 = 150$
   $n = 39$

30. The length of a rectangle is 5 less than twice the width. The perimeter of the rectangle is 80. Find the dimensions of the rectangle.
   30. $w = $ width
   $2w + 2(2w - 5) = 80$
   width: 15
   length: 25

The table below gives the average monthly temperatures over nine months for San Francisco.

Month	Jan.	Feb.	Mar.	Apr.	May	June	July	Aug.	Sept.
SF(°F)	49	52	53	55	58	61	62	63	64

31. Find the mean of the average monthly temperatures.
   31. about 57.4°F
32. Find the median of the average monthly temperatures.
   32. 58°F
33. Find the mode of the average monthly temperatures.
   33. none

Bonus
$\frac{97}{9 + 7} = 6.06$ $\frac{42}{4 + 2} = 7$ $\frac{61}{6 + 1} = 8.714$
Find the greatest two-digit whole number that when divided by the sum of its digits gives the greatest result.

Bonus: 90

OBJECTIVES AND EXAMPLES

Upon completing this chapter, you should be able to:

• solve equations by using addition or subtraction (Lesson 3-1)

$$x - 13 = 45$$
$$x - 13 + 13 = 45 + 13$$
$$x = 58$$

$$y - (-33) = 14$$
$$y + 33 = 14$$
$$y + 33 - 33 = 14 - 33$$
$$y = -19$$

• solve equations by using multiplication or division (Lesson 3-2)

$$\frac{4}{9}t = -72$$
$$\frac{9}{4}\left(\frac{4}{9}t\right) = \frac{9}{4}(-72)$$
$$t = -162$$

$$-5c = -8$$
$$\frac{-5c}{-5} = \frac{-8}{-5}$$
$$c = \frac{8}{5}$$

30. $\frac{9}{10}$

• solve equations involving more than one operation (Lesson 3-3)

$$34 = 8 - 2t$$
$$34 - 8 = 8 - 8 - 2t$$
$$26 = -2t$$
$$\frac{26}{-2} = \frac{-2t}{-2}$$
$$-13 = t$$

$$\frac{d + 5}{3} = -9$$
$$3\left(\frac{d + 5}{3}\right) = 3(-9)$$
$$d + 5 = -27$$
$$d = -32$$

REVIEW EXERCISES

Use these exercises to review and prepare for the chapter test.

Solve each equation. Then check your solution. 18. -8.5

13. $r - 21 = -37$ -16
14. $14 + c = -5$ -19
15. $-27 = -6 - p$ 21
16. $b + (-14) = 6$ 20
17. $r + (-11) = -21$ -10
18. $d - (-1.2) = -7.3$
19. A number decreased by 14 is -46. Find the number. -32
20. The sum of two numbers is -23. One of the numbers is 9. What is the other number? -32
21. Eighty-two increased by some number is -34. Find the number. -116

Solve each equation. Then check your solution.

22. $6x = -42$ -7
23. $-7w = -49$ 7
24. $\frac{3}{4}n = 30$ 40
25. $-\frac{3}{5}y = -50$ $83\frac{1}{3}$
26. $\frac{5}{2}x = -25$ -10
27. $\frac{5}{12} = \frac{r}{24}$ 10
28. Negative seven times a number is -56. Find the number. 8
29. Three fourths of a number is -12. What is the number? -16
30. One and two thirds of a number equals one and a half. What is the number?

Solve each equation. Then check your solution.

31. $4t - 7 = 5$ 3
32. $6 = 4n + 2$ 1
33. $\frac{y}{3} + 6 = -45$ -153
34. $\frac{c}{-4} - 8 = -42$ 136
35. $\frac{4d + 5}{7} = 7$ 11
36. $\frac{7n + (-1)}{8} = 8$ $9\frac{2}{7}$
37. Four times a number decreased by twice the number is 100. What is the number? 50
38. Find four consecutive integers whose sum is 130. $31, 32, 33, 34$
39. Find two consecutive integers such that twice the lesser integer increased by the greater integer is 49. $16, 17$

GLENCOE Technology

Test and Review Software

You may use this software, a combination of an item generator and item bank, to create your own tests or worksheets. Types of items include free response, multiple choice, short answer, and open ended.

For IBM & Macintosh

OBJECTIVES AND EXAMPLES

• find the complement and supplement of an angle (Lesson 3–4)

Find the complement and supplement of an angle with measure 70°.

complement: $90° - 70°$ or $20°$

supplement: $180° - 70°$ or $110°$

REVIEW EXERCISES

Find both the complement and supplement of each angle measure.

40. 28°   **62°; 152°**   41. 69°   **21°; 111°**
42. $5x°$ 43. $(y + 20)°$
   $(90 - 5x)°$; $(180 - 5x)°$ **$(70 - y)°$; $(160 - y)°$**
44. Find the measure of an angle that is 10° more than its supplement.   **95°**
45. Find the measure of an angle that is one-half the measure of its complement.   **30°**

• find the measure of the third angle of a triangle given the measures of the other two angles (Lesson 3–4)

The measures of two angles of a triangle are 89° and 90°. The measure of the third angle is $180° - (89° + 90°)$ or $1°$.

49. $180° - (z + z - 30°)$ or $(210° - 2z)$

Find the measure of the third angle of each triangle in which the measures of two angles of the triangle are given.

46. 16°, 47°   **117°**
47. 45°, 120°   **15°**
48. $y°, x°$ **$180° - (x + y)°$**
49. $(z - 30)°, z°$

• solve equations with the variable on both sides and solve equations containing grouping symbols   (Lesson 3–5)

$$-3(x + 5) = 3(x - 1)$$
$$-3x - 15 = 3x - 3$$
$$-3x - 15 + 15 = 3x - 3 + 15$$
$$-3x = 3x + 12$$
$$-3x - 3x = 3x - 3x + 12$$
$$\frac{-6x}{-6} = \frac{12}{-6}$$
$$x = -2$$

Solve each equation. Then check your solution.

50. $4(3 + 5w) = -11$ **$-1\frac{3}{20}$**
51. $\frac{2}{3}n + 8 = \frac{1}{3}n - 2$ **-30**
52. $3x - 2(x + 3) = x$ **no solution**
53. $\frac{4 - x}{5} = \frac{1}{5}x$ **2**
54. The sum of two numbers is 25. Twelve less than four times one of the numbers is 16 more than twice the other number. Find both numbers.   **12, 13**

• solve equations and formulas for a specified variable   (Lesson 3–6)

Solve $\frac{x + y}{b} = c$ for x.

$$b\left(\frac{x + y}{b}\right) = b(c)$$
$$x + y = bc$$
$$x = bc - y$$

Solve each equation for the variable specified.

55. $5x = y$, for x 55. $\frac{y}{5}$
56. $ay - b = c$, for y 56. $\frac{c + b}{a}$
57. $yx - a = cx$, for x 57. $\frac{a}{y - c}$
58. $\frac{2y - a}{3} = \frac{a + 3b}{4}$, for y **$\frac{7a + 9b}{8}$**

CHAPTER 3   STUDY GUIDE AND ASSESSMENT

OBJECTIVES AND EXAMPLES

• find and interpret the mean, median, and mode of a set of data (Lesson 3–7)

Find the mean, median, and mode of 1.5, 3.4, 5.4, 5.6, 5.7, 6.2, 6.8, 7.1, 7.1, 8.4, and 9.9.

mean: The sum of the 11 values is 67.1.

$\frac{67.1}{11} = 6.1$

median: The 6th value is 6.2.

mode: 7.1

REVIEW EXERCISES

59. **School**   Marisa's scores on the 25-point quizzes in her English class are 20, 21, 18, 21, 22, 22, 24, 21, 20, 19, and 23. Find the mean, median, and mode of her scores.   **21; 21; 21**

60. **Business**   Of the 42 employees at Pirate Printing, sixteen make $4.75 an hour, four make $5.50 an hour, three make $6.85 an hour, six make $4.85 an hour, and thirteen make $5.25 an hour. Find the mean, median, and mode of the hourly wages.   **$5.14; $4.85; $4.75**

APPLICATIONS AND PROBLEM SOLVING

61. **Food and Nutrition**   According to the National Eating Trends Service, in 1984, the average person ate 71 meals in a restaurant and purchased 20 meals to go. The number of meals eaten in a restaurant decreased 0.7 meals per year, and the number of meals purchased to go increased 1.3 meals per year. (Lesson 3–5)

   a. After how many years will people eat the same number of meals in restaurants as they order to go and eat at home?   **25.5 years**

   b. How would you explain these trends?
      Answers will vary.

62. **Health**   You can estimate your ideal weight with the following formulas.

Male	Female
$w = 100 + 6(h - 60)$	$w = 100 + 5(h - 60)$

   In each formula, w is your ideal weight in pounds and h represents your height in inches.
   (Lesson 3–6)

   a. Find your ideal weight.   **See students' work.**

   b. Solve each formula for h.

   c. How tall is a man that is at his ideal weight of 170 pounds?

   b. $\frac{w + 260}{6}$; $\frac{w}{5} + 40$ c. $71\frac{2}{3}$ in.

63. **Home Economics**   According to the *Dallas Morning News,* the average refrigerator lasts $2\frac{1}{4}$ years less than twice as long as the average color television. The average color television lasts for 8 years. How long does the average refrigerator last? (Lesson 3–3)

   $13\frac{3}{4}$ years

64. **Postal Service**   The graph below shows the cost of mailing a first-class letter in several countries in 1994. (Lesson 3–7)

   a. Find the mean, median, and mode cost.

   b. Round each cost to the nearest cent. Then find the mean, median, and mode cost and compare your two answers.

   a. 40.4¢; 32.2¢; 32¢   b. 40.6¢; 32¢; 32¢

Cost of mailing a first-class letter

Japan 73.6¢
Germany 56.8¢
Norway 43.7¢
Australia 32.2¢
Canada 32¢
U.S. 32¢
Mexico 12.5¢

Source: *USA TODAY* research

A practice test for Chapter 3 is provided on page 789.

ALTERNATIVE ASSESSMENT

COOPERATIVE LEARNING PROJECT

Balances and Solving Equations In this chapter, you learned the addition, subtraction, multiplication, and division properties of equality. Each definition begins with an equation in *balance*, $a = b$, and states that if something is done to one side of the equation, the equation can be kept in balance by doing the same thing to the other side.

A basic principle is that two quantities in balance will remain in balance until something is done to only one of the quantities. To bring the quantities back into balance, you must do the same thing to the other quantity or undo what was done to the first quantity.

In this project, you will construct a device to demonstrate the principle of balance and how it relates to solving equations. All balances have certain features in common. A horizontal bar is balanced on a vertical support. Pans are suspended from the ends of the bar. A pointer attached to the bar moves along a scale on the support, indicating when the pans are balanced.

Follow these steps to design and build your balance.

- Outline a plan you can follow.
- Use materials you can easily find. The precision of your balance depends more on the care and accuracy with which you build it than it does on the materials you use.
- Carry out your plan.
- Determine how you might use your balance to represent equations. Include ways to solve equations by using the balance.
- Write several paragraphs describing how you can solve equations using the balance. Be sure to give examples of equations and their solutions.

THINKING CRITICALLY

- Write an equation that has no solution. Explain why there is no solution.
- Write an equation that has an infinite number of solutions. Explain why.

PORTFOLIO

Select one of the assignments from this chapter that you found especially challenging. Revise your work if necessary and place it in your portfolio. Explain why you found it to be a challenge.

SELF EVALUATION

One characteristic of a good problem solver is persistence. If you don't succeed the first time you attempt something, you should learn from your mistakes and try again.

Assess yourself. How persistent are you? How do you react when you fail at something? List two or three ways that you can make a conscious effort to be more persistent in solving problems both in your mathematics studies and in your daily life.

Assessment and Evaluation Masters, pp. 70, 81

3 NAME_____ DATE_____

Chapter 3 Performance Assessment

Instructions: *Demonstrate your knowledge by giving a clear, concise solution to each problem. Be sure to include all relevant drawings and justify your answers. You may show your solution in more than one way or investigate beyond the requirements of the problem.*

1. Use the equation $10x + 18 = 48$ for parts a–d.

   a. Add 4 to each side of the equation. Tell, in your own words, how you know that the two sides of the equation are still equal.

   b. How are the solutions of the original equation and the equation in part a related? What does the term *equivalent equations* mean?

   c. Use subtraction, multiplication, and division to form three equations equivalent to the original equation.

   d. Solve the original equation. Explain each step in finding the solution.

2. Use the equation $\frac{1}{2}(x - 3) = \frac{1}{3}x + 1$ for parts a–c.

   a. How can you eliminate the fraction from each side of the equation to make finding the solution easier?

   b. Write a word problem for the equation.

   c. Solve the equation. Explain each step and give the meaning of the answer.

3. a. Solve $ry + s = tx - m$ for y. Explain each step in your solution.

   b. Would there be any limitations for the value of each variable? If so, explain the limitation.

4. The opening day scores in a local golf tournament are 78, 83, 70, 84, 89, 67, 84, 92, 78, 91, 85, 77, 68, 80, 71, 78, 99, 81, 75, 88, 90, 71, 73. Find the mean, median, and mode for the set of scores. Give a reason for finding each of these values.

Scoring Guide
Chapter 3
Performance Assessment

Level	Specific Criteria
3 Superior	• Shows thorough understanding of the concepts of *solving equations, solving formulas for a given variable,* and *finding measures of central tendency.* • Uses appropriate strategies to solve problems. • Computations are correct. • Written explanations are exemplary. • Word problem concerning equation is appropriate and makes sense. • Goes beyond requirements of all or some problems.
2 Satisfactory, with Minor Flaws	• Shows understanding of the concepts of *solving equations, solving formulas for a given variable,* and *finding measures of central tendency.* • Uses appropriate strategies to solve problems. • Computations are mostly correct. • Written explanations are effective. • Word problem concerning equation is appropriate and makes sense. • Satisfies all requirements of problems.
1 Nearly Satisfactory, with Serious Flaws	• Shows understanding of most of the concepts of *solving equations, solving formulas for a given variable,* and *finding measures of central tendency.* • May not use appropriate strategies to solve problems. • Computations are mostly correct. • Written explanations are satisfactory. • Word problem concerning equation is mostly appropriate and sensible. • Satisfies most requirements of problems.
0 Unsatisfactory	• Shows little or no understanding of the concepts of *solving equations, solving formulas for a given variable,* and *finding measures of central tendency.* • May not use appropriate strategies to solve problems. • Computations are incorrect. • Written explanations are not satisfactory. • Word problem concerning equation is not appropriate or sensible. • Does not satisfy requirements of problems.

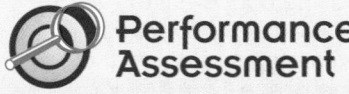

Alternative Assessment

The Alternative Assessment section provides students with the opportunity to assess their own work by thinking critically, working with others, keeping a portfolio, and honestly evaluating their own progress. For more information on alternative forms of assessment, see *Alternative Assessment in the Mathematics Classroom*, one of the titles in the Glencoe Mathematics Professional Series.

Performance Assessment

Performance Assessment tasks for this chapter are included in the *Assessment and Evaluation Masters*. A scoring guide is also provided.

NCTM Standards: 1–5

This Investigation is designed to be completed over several days or weeks. It may be considered optional. You may want to assign the Investigation and the follow-up activities to be completed at the same time.

Objective

Use the capture-recapture method and proportions to estimate the number of fish in a large lake.

Mathematical Overview

This Investigation will use the following mathematical skills and concepts from Chapters 4 and 5.

- Solve proportions.
- Find the probability of a simple event.
- Show relations.
- Determine and interpret the range, quartiles, and interquartile range of a set of data.

Recommended Time		
Part	**Pages**	**Time**
Investigation	190–191	1 class period
Working on the Investigation	200, 227, 269, 302	20 minutes each
Closing the Investigation	314	1 class period

Instructional Resources

Investigations and Projects Masters, pp. 5–8

A recording sheet, teacher notes, and scoring guide are provided for each investigation in the *Investigations and Projects Masters*.

1 MOTIVATION

This Investigation uses common household items to investigate fish population in a lake. Ask students if they are familiar with the capture-recapture method used to study animal populations. Discuss reasons why it is important to estimate populations of animals and ways to do this without harming the animals.

MATERIALS NEEDED

paper lunch bag

2 bags of dry beans, each a different color

5-oz. paper cup

Imagine that you are asked to determine the number of fish in a nearby pond. To count the fish one by one, you could remove the fish from the pond and stack them to one side, or mark each fish so you would not count them over and over again. Counting like this could be hazardous to a fish's health!

To determine the number of animals in a population, scientists often use the *capture-recapture* method. A number of animals are captured, carefully tagged, and returned to their native habitat. Then a second group of animals is captured and counted, and the number of tagged animals is noted. Scientists then use proportions to estimate the number in the entire population.

In this Investigation, you will work in pairs to model the process used by scientists to estimate the number of fish in a large lake. The lunch bag will represent the lake, the beans will represent fish, and the paper cup will represent a net.

Make an Investigation Folder in which you can store all of your work on this Investigation for future use.

 ## Cooperative Learning

This Investigation offers an excellent opportunity for using cooperative learning groups. For more information on cooperative learning strategies and group management, see *Cooperative Learning in the Mathematics Classroom*, one of the titles in the Glencoe Mathematics Professional Series.

CASTING	A Total Number of Tagged Fish	B Total Number of Fish in Sample	C Number of Tagged Fish in Sample	D Estimate
1				
2				
3				
4				
5				

CAPTURE

1 Begin by copying the chart above onto a sheet of paper.

2 Write your name on the paper bag. Empty one bag of beans into the bag.

3 Use your net to remove a sample of fish. Count the number of fish you netted. Since this number will remain constant for all casts, you can record this number in each row of column A.

4 Replace all of the beans you counted with beans of a different color. Put these "tagged fish" back into the lake and gently shake your bag to mix the fish.

RECAPTURE

5 For your first casting, use your net to remove a sample of fish. Count the total number of fish in your sample and record this number in column B. Then count the number of tagged fish in this sample and record this number in column C of your chart. Return these fish to the lake and gently shake your bag to mix the fish.

6 Cast your net a second time and record your findings. Continue casting and recording until you have counted five samples.

You will continue working on this Investigation throughout Chapters 4 and 5.

Be sure to keep your chart and materials in your Investigation Folder.

Go Fish! Investigation

Working on the Investigation
Lesson 4–1, p. 200

Working on the Investigation
Lesson 4–5, p. 227

Working on the Investigation
Lesson 5–2, p. 269

Working on the Investigation
Lesson 5–6, p. 302

Closing the Investigation
End of Chapter 5, p.314

Investigation: Go Fish! **191**

You may wish to have a student read the first two paragraphs of the Investigation to provide information about how we can study animal populations without threatening the animals' health. You may then wish to read the next two paragraphs, which introduce the activity. Discuss the activity with your students. Then separate the class into pairs.

3 MANAGEMENT

Each group member should be responsible for a specific task.

Recorder Collects data.
Measurer Puts beans into bag. Uses net to remove sample fish from bag.

At the end of the activity, each member should turn in his or her respective equipment.

Sample Answers

Answers will vary as they are based on the number of fish in the sample.

Investigations and Projects Masters, p. 8

NAME _____ DATE _____

4, 5 Investigation, Chapters 4 and 5 — Student Edition Pages 190–191, 200, 227, 269, 302, 314

Go Fish!

Work with your group to add to the list of questions to be considered.

1. Find the population of fish in a pond given the following information:
   Number of fish tagged: 35
   Number of tagged fish from cast sample 1: 7
   Number of untagged fish from cast sample 2: 48
   Number of tagged fish from cast sample 2: 9
   Number of untagged fish from cast sample 2: 65

2. If you knew that 20% of a population of fish were tagged, could you determine how many total tagged fish there were? Explain how you could find that number or, if you couldn't, what else you would need to know.

3. Why is it important for scientists to keep track of animal population data?

Use this chart to record your calculations from page 227 of the Investigation.

Actual Population	
Estimated Population	
Percent in Lake Tagged	
Percent in Sample Tagged	
Probability of Catching a Tagged Fish	
Reasoning	

Use this diagram for the mapping on page 269.

4

Using Proportional Reasoning

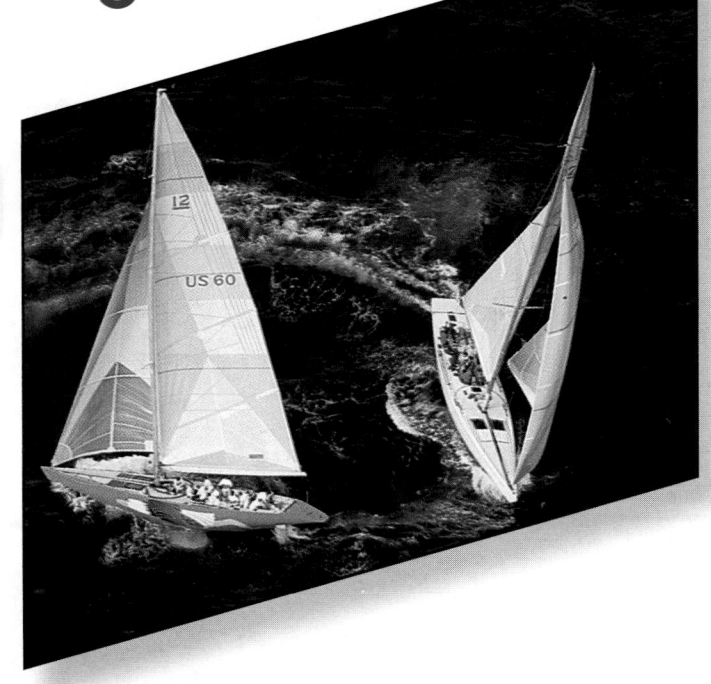

PREVIEWING THE CHAPTER

In this chapter, students apply the process of mathematical modeling to real-world problem situations, making the connection with the equation-solving skills and concepts that they mastered in the previous chapter. Students solve proportions in the first lesson. Then, they study relationships between similar triangles and use similarity properties to find the measures of corresponding parts of similar triangles. This background leads to the study of the sine, cosine, and tangent ratios associated with an acute angle in a right triangle. Next, students solve problems involving percent, simple interest, percent of increase or decrease, discount, and sales tax. Students learn to find the probability and the odds of a simple event. The chapter concludes with lessons involving mixture problems, uniform motion problems, and direct and inverse variation problems.

Lesson (Pages)	Lesson Objectives	NCTM Standards	State/Local Objectives
4-1A (194)	Collect data to determine ratios.	1–5	
4-1 (195–200)	Solve proportions.	1–5	
4-2 (201–205)	Find the unknown measures of the sides of two similar triangles.	1–5, 7	
4-3 (206–214)	Use trigonometric ratios to solve right triangles.	1–5, 9	
4-4 (215–221)	Solve percent problems. Solve problems involving simple interest.	1–5	
4-5 (222–227)	Solve problems involving percent of increase or decrease. Solve problems involving discounts or sales tax.	1–5	
4-6 (228–232)	Find the probability of a simple event. Find the odds of a simple event.	1–5, 11	
4-7 (233–238)	Solve mixture problems. Solve problems involving uniform motion.	1–5	
4-8 (239–244)	Solve problems involving direct and inverse variation.	1–5	

A complete, 1-page lesson plan is provided for each lesson in the *Lesson Planning Guide*. Answer keys for each lesson are available in the *Answer Key Masters*.

ORGANIZING THE CHAPTER

You may want to refer to the **Course Planning Calendar** on page T12 for detailed information on pacing.
PACING: Standard—15 days; **Honors**—15 days; **Block**—7 days; **Two Years**—20 days

LESSON PLANNING CHART

Lesson (Pages)	Materials/ Manipulatives	Extra Practice (Student Edition)	BLACKLINE MASTERS										Real-World Applications	Interactive Mathematics Tools Software	Teaching Transparencies
			Study Guide	Practice	Enrichment	Assessment and Evaluation	Modeling Mathematics	Multicultural Activity	Tech Prep Applications	Graphing Calculator	Science and Math Lab Manual			
4-1A (194)	tape measure*						p. 25								
4-1 (195–200)	calculator	p. 764	p. 26	p. 26	p. 26								4-1	4-1A 4-1B
4-2 (201–205)		p. 764	p. 27	p. 27	p. 27	p. 100						10		4-2A 4-2B
4-3 (206–214)	calculator protractor* straws tape paper clips graphing calculator string	p. 765	p. 28	p. 28	p. 28								4-3.1 4-3.2 4-3.3	4-3A 4-3B
4-4 (215–221)	calculator compass* protractor*	p. 765	p. 29	p. 29	p. 29	pp. 99, 100		p. 7	p. 7					4-4A 4-4B
4-5 (222–227)	graphing calculator	p. 765	p. 30	p. 30	p. 30		p. 75					11		4-5A 4-5B
4-6 (228–232)	graph paper	p. 766	p. 31	p. 31	p. 31	p. 101							4-6	4-6A 4-6B
4-7 (233–238)		p. 766	p. 32	p. 32	p. 32		pp. 45–47					12		4-7A 4-7B
4-8 (239–244)		p. 766	p. 33	p. 33	p. 33	p. 101		p. 8	p. 8	p. 4	pp. 13–16	13	4-8	4-8A 4-8B
Study Guide/ Assessment (245–249)						pp. 85–98, 102–104								

*Included in Glencoe's Student Manipulative Kit and Overhead Manipulative Resources.

ORGANIZING THE CHAPTER

OTHER CHAPTER RESOURCES

Student Edition
Investigation, pp. 190–191
Chapter Opener, pp. 192–193
Mathematics and Society, p. 214
Working on the Investigation, pp. 200, 227

Teacher's Classroom Resources
Investigations and Projects Masters, pp. 37–40
Algebra and Geometry Overhead Manipulative Resources, pp. 13–16

Technology
Test and Review Software (IBM and Macintosh)
CD-ROM Interactions (Windows and Macintosh)

Professional Publications
Block Scheduling Booklet
Glencoe Mathematics Professional Series

OUTSIDE RESOURCES

Books/Periodicals
Chazan, Daniel, *Similarity*, Sunburst
Lovell, Robert, *Probability Activities*, Dale Seymour Publications

Software
A Chance Look, Sunburst
Taking Chances, Sunburst

Videos/CD-ROMs
Mathmedia, ITSCO
Mathwise, ITSCO

ASSESSMENT RESOURCES

Student Edition
Math Journal, pp. 224, 236
Mixed Review, pp. 200, 205, 214, 220, 227, 232, 238, 244
Self Test, p. 221
Chapter Highlights, p. 245
Chapter Study Guide and Assessment, pp. 246–248
Alternative Assessment, p. 249
 Portfolio, p. 249

Cumulative Review, pp. 250–251

Teacher's Wraparound Edition
5-Minute Check, pp. 195, 201, 206, 215, 222, 228, 233, 239
Check for Understanding, pp. 198, 203, 210, 218, 224, 230, 236, 242
Closing Activity, pp. 200, 205, 214, 220, 227, 232, 238, 244
Cooperative Learning, pp. 197, 217

Assessment and Evaluation Masters
Multiple-Choice Tests, Forms 1A (Honors), 1B (Average), 1C (Basic), pp. 85–90
Free-Response Tests, Forms 2A (Honors), 2B (Average), 2C (Basic), pp. 91–96
Calculator-Based Test, p. 97
Performance Assessment, p. 98
Mid-Chapter Test, p. 99
Quizzes A–D, pp. 100–101
Standardized Test Practice, p. 102
Cumulative Review, pp. 103–104

ENHANCING THE CHAPTER

Examples of some of the materials for enhancing Chapter 4 are shown below.

DIVERSITY

Multicultural Activity Masters, pp. 7, 8

APPLICATIONS

Real-World Applications, 10, 11, 12, 13

TECHNOLOGY

Graphing Calculator Masters, p. 4

TECH PREP

Tech Prep Applications Masters, pp. 7, 8

CONNECTIONS

Science and Math Lab Manual, pp. 13–16

PROBLEM SOLVING

Problem of the Week Cards, 10, 11, 12

4

Using Proportional Reasoning

Objectives

In this chapter, you will:

- solve proportions,
- find the unknown measures of the sides of two similar triangles,
- use trigonometric ratios to solve right triangles,
- solve percent problems,
- find the probability and odds of a simple event, and
- solve problems involving direct and inverse variation.

Incoming Immigrants

Country	Immigrants
United Kingdom	18,783
El Salvador	26,818
Poland	27,846
India	40,121
Dominican Republic	45,420
Former Soviet Union	58,571
Vietnam	59,614
Philippines	63,457
China	65,576
Mexico	126,561

Source: U.S. Immigration and Naturalization Service

Some people blame immigrants, both legal and illegal, for many of the economic and social problems in the United States. Others believe immigrants bring vitality, diversity, and economic strength to the United States. What effect does immigration have on American society? Do you have any personal experiences with new immigrants?

TIME Line

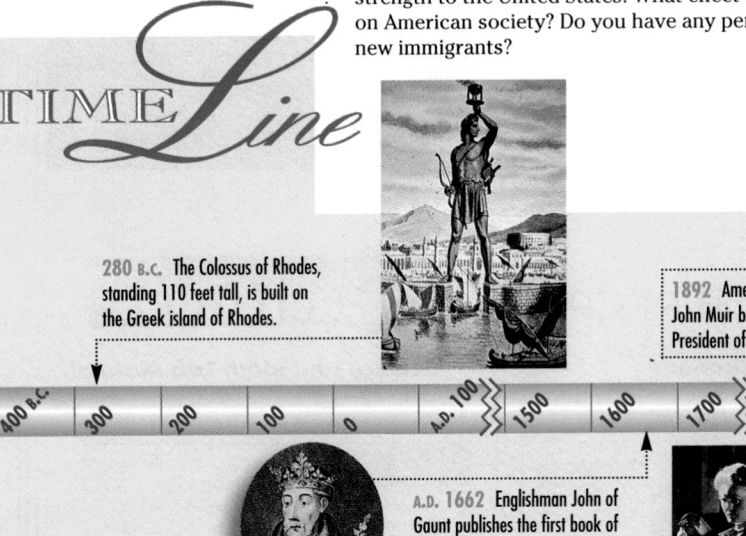

280 B.C. The Colossus of Rhodes, standing 110 feet tall, is built on the Greek island of Rhodes.

1892 American naturalist John Muir becomes first President of the Sierra Club.

| 400 B.C. | 300 | 200 | 100 | 0 | A.D. 100 | 1500 | 1600 | 1700 | 1885 | 1890 | 1895 |

A.D. 1662 Englishman John of Gaunt publishes the first book of statistics.

1903 Marie Sklodowska Curie becomes the first woman to win a Nobel Prize in science.

TIME Line

Have students pick a country and research one famous person from that country. How did that person contribute to our society in the United States?

interNET CONNECTION

Explore the Immigration Issues Menu for articles on the cost of immigration, the quota system, and legal and illegal immigration.

World Wide Web
http://www.fairus.org/issues/04001604.htm

Chapter Project

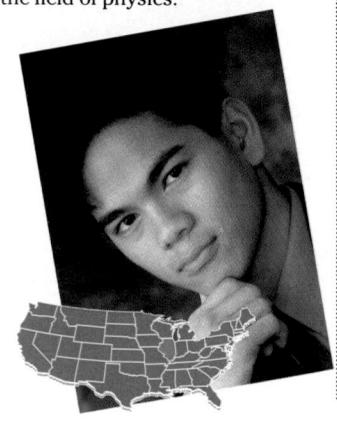

Ayrris Layug Aunario won the grand prize in *Filipinas* magazine's 1994 National Essay Contest with an essay describing what it means to be Filipino-American. The 16-year-old from Waukegan, Illinois, is a student at the Illinois Mathematics and Science Academy, a public school for gifted students in science and mathematics.

Ayrris' essay explains what it was like to be a 10-year-old immigrant to the U.S., arriving in Chicago and learning to live with people of many different cultures and beliefs. He had to master English quickly and soon learned the American virtues of independence, self-reliance, and assertiveness. Ayrris plans to attend college in the field of physics.

Work in cooperative groups to take a heritage survey of students in your school.

- Determine the geographic birthplace of each student and the birthplace of their parents or guardians. If possible, gather information about the birthplace of each student's grandparents and great-grandparents.

- Use a world map to pinpoint each geographic location. Connect the generations of each student using colored yarn.

- Draw some conclusions about the birthplaces of the students you surveyed. Share your findings with the class.

Once, Ayrris was competing in a spelling bee. He was asked to spell the word *illusion*. He spelled it correctly, but the pronouncer insisted it was spelled "illusian." He respectfully sat down, but checked a dictionary and showed it to the judges. He rejoined the competition and won first place.

Chapter Project

Cooperative Learning Students might also want to do library research on the immigration profile for the general population of the United States. Compare the findings for the school with the general population of the U.S.

Investigations and Projects Masters, p. 37

1920 German Expressionist painter Emil Nolde paints his watercolor *Red Poppies*.

1963 British artist Barbara Hepworth creates her sculpture *Hollow Form with White Interior*.

1954 Ellis Island in New York Bay closes after 62 years and the processing of 20 million immigrants.

1992 Physicists in Germany fuse two isotopes of bismuth and iron to make Element 109, the heaviest known element in the universe.

Chapter 4 **193**

4 Chapter 4 Project A

NAME _____ DATE _____
Student Editi
Pages 194–2

A World of Names

1. For this project, you will work in a small group to research names and naming customs of different ethnic groups. In some groups, for example, the family name comes before the other names. Sometimes, a person's full family name is made up of both parents' names. Use reference books to find out more about such customs. You can also ask friends from different ethnic backgrounds about customs connected with their names. Find out what some common family names mean.

2. Do research to find out the most common family names of students at your school. You may be able to obtain lists of students' last names from classroom teachers or from the principal's office.

3. Make a list of the five most common names at your school. Determine what percent of the students at your school has each of these names.

4. Do research to find out the most common family names in your community. Use one or more telephone directories and the list of the most common family names at your school. Make a list of the five most common family names in your community and determine what percent of the population has each name.

5. If possible, interview an expert in the field of demographics, the study of human populations. Ask how the results of your research would have been different 25, 50, and 100 years ago. Also, ask the interviewee what a list of common last names might look like 100 years from now.

6. Write an article for your school newspaper about names and naming customs and the results of your investigation.

Alternative Chapter Projects

Two other chapter projects are included in the *Investigations and Projects Masters*. In Chapter 4 Project A, pp. 37–38, students extend the topic in the chapter opener. In Chapter 4 Project B, pp. 39–40, students investigate sales and discounts.

NCTM Standards: 1–5

Objective
Collect data to determine ratios of the length of parts of the human body to the length of the head.

Recommended Time
Demonstration and discussion: 15 minutes; Exercises: 30 minutes

Instructional Resources
For each student or group of students
Student Manipulative Kit
• tape measure
Modeling Mathematics Masters
• p. 25 (worksheet)
For teacher demonstration
Algebra and Geometry Overhead Manipulative Resources

1 FOCUS

Motivating the Lesson
This modeling activity presents the idea that there are certain ratios that exist in nature. Such constants are fundamental to science. Students will also see that, in the case of the human body, nothing is exact.

2 TEACH

Teaching Tip Make it clear to students that the unit of measure is not the inch, but their heads.

3 PRACTICE/APPLY

Assignment Guide
Core: 1–2
Enriched: 1–2

MODELING MATHEMATICS

A Preview of Lesson 4–1

4-1A   Ratios

Materials: tape measure

You can collect data to determine whether there is a relationship between the length of an individual's head and an individual's body.

Activity 1   **Use a tape measure to find the length of your head.**

Step 1 Measure your head from the top of your skull to the bottom of your chin. Since the top of your head is round, hold something flat, such as a piece of cardboard, across the top of your head to get an accurate measurement.

Step 2 Make a measuring "stick" with the length of your head as one unit to find the following measurements.
• total height
• chin to waist
• waist to hip
• knee to ankle
• ankle to bottom of bare heel
• underarm to elbow
• elbow to wrist
• wrist to tip of finger
• shoulder to tip of finger

Step 3 Make a table to record the relationship, or *ratio*, of each measurement to the measurement of your head.

Body Proportions of an Average Adult

Draw   1. The average adult is seven and one-half heads tall. Use the information in your table to draw an outline of the proportions of your body. How does your drawing compare to the one above?   **See students' work.**

Write   2. Compare your ratios to those of your classmates. Find the number of heads tall of a random student in your class. Write a paragraph explaining how you compare to this average.   **See students' work.**

4 ASSESS

Observing students working in cooperative groups is an excellent method of assessment.

Cooperative Learning

This lesson offers an excellent opportunity for using cooperative learning groups. For more information on cooperative learning strategies and group management, see *Cooperative Learning in the Mathematics Classroom*, one of the titles in the Glencoe Mathematics Professional Series.

Ratios and Proportions

Objective

After studying this lesson, you should be able to:
- solve proportions.

F Y I

Almost 500 new candy and chocolate products were introduced in the first four months of 1995.

APPLICATION

Food

How would you like to eat chocolate for a living? That's exactly what Carl Wong does every day as associate director of product development for Hershey Foods. His job is to create new candy bars as well as taste test existing candy bars in order to improve them.

Hershey Foods keeps their candy recipes confidential, but the basic ingredients found in a batch of Hershey's chocolate bars are sugar, cocoa beans, milk, and flavorings. The ingredients for a batch of candy are shown in the table below.

Ingredient	Parts Per Batch
Sugar	10
Cocoa Beans	5
Milk	4
Flavorings	1

A **ratio** is a comparison of two numbers by division. The ratio of x to y can be expressed in the following ways.

$$x \text{ to } y \qquad\qquad x{:}y \qquad\qquad \frac{x}{y}$$

Ratios are often expressed as fractions in simplest form. A ratio that is equivalent to a whole number is written with a denominator of 1.

In the application at the beginning of the lesson, the table shows that for every 10 parts sugar in a batch of chocolate bars, there are 5 parts cocoa beans. The ratio of sugar to cocoa beans is $\frac{10}{5}$. Suppose Hershey uses 30 pounds of sugar and 15 pounds of cocoa beans in a batch of chocolate bars. This ratio is $\frac{30}{15}$. Is this ratio different than the first ratio of $\frac{10}{5}$? When simplified, both ratios are equivalent to $\frac{2}{1}$.

$$\begin{array}{cc} \overset{\div 5}{\frac{10}{5} = \frac{2}{1}} & \overset{\div 15}{\frac{30}{15} = \frac{2}{1}} \\ {}_{\div 5} & {}_{\div 15} \end{array}$$

An equation stating that two ratios are equal is called a **proportion**. So, $\frac{10}{5} = \frac{30}{15}$ is a proportion.

F Y I

Chocolate is made from cocoa beans. It also contains small amounts of two stimulants—theobromine and caffeine.

GLENCOE Technology

CD-ROM Interaction

A multimedia simulation links the use of ratios, graphs, and estimation to bicycling. A blackline master activity with teacher's notes provides a follow-up to the CD-ROM simulation.

For Windows & Macintosh

Teaching Tip For Example 2, you may wish to explain that this type of problem can be solved using the following method.

$$\frac{5}{4} = \frac{75}{m}$$
$$m\left(\frac{5}{4}\right) = m\left(\frac{75}{m}\right)$$
$$\frac{5}{4}m = 75$$
$$\frac{4}{5}\left(\frac{5m}{4}\right) = \frac{4}{5}(75)$$
$$m = 60$$

Teaching Tip The proportion $\frac{a}{b} = \frac{c}{d}$ is generally read "a is to b as c is to d."

Teaching Tip A proportion may also be written as $a:b = c:d$. In this form, it is easy to see that a and d are the extremes, since they are at the extreme ends. The means are the middle terms.

To solve a proportion, *cross multiply.*

One way to determine if two ratios form a proportion is to check their cross products. In the proportion at the right, the cross products are $10 \cdot 15$ and $5 \cdot 30$. In this proportion, 10 and 15 are called the **extremes,** and 5 and 30 are called the **means.**

$$\frac{10}{5} = \frac{30}{15}$$
$$10(15) = 5(30)$$
$$\text{extremes} = \text{means}$$
$$150 = 150$$

The cross products of a proportion are equal.

Means-Extremes Property of Proportions	In a proportion, the product of the extremes is equal to the product of the means. If $\frac{a}{b} = \frac{c}{d}$, then $ad = bc$.

Example 1

Use cross products to determine whether each pair of ratios forms a proportion.

a. $\frac{2}{3}, \frac{12}{18}$

$$2 \cdot 18 \stackrel{?}{=} 3 \cdot 12$$
$$36 = 36$$
So, $\frac{2}{3} = \frac{12}{18}$.

This is a proportion.

b. $\frac{2.5}{6}, \frac{3.4}{5.2}$

Enter: 2.5 $\boxed{\times}$ 5.2 $\boxed{=}$ *13*

Enter: 6 $\boxed{\times}$ 3.4 $\boxed{=}$ *20.4*

Since $13 \neq 20.4$, $\frac{2.5}{6} \neq \frac{3.4}{5.2}$.

This is not a proportion.

You can write proportions that involve a variable and then use cross products to solve the proportion.

Example 2

APPLICATION
Food

Refer to the application at the beginning of the lesson. Suppose Hershey makes a batch of chocolate with 75 pounds of cocoa beans. How many gallons of milk will they use?

You know that the ratio of cocoa beans to milk is 5:4. Let m represent the gallons of milk.

$$\frac{5 \text{ parts cocoa beans}}{4 \text{ parts milk}} = \frac{75 \text{ pounds of cocoa beans}}{m \text{ gallons of milk}}$$

$$\frac{5}{4} = \frac{75}{m}$$

$$5m = 4(75) \quad \text{\textit{Find the cross products.}}$$

$$5m = 300$$

$$m = 60$$

In a batch of chocolate with 75 pounds of cocoa beans, Hershey needs to use 60 gallons of milk.

A ratio called a **scale** is used when making a model to represent something that is too large or too small to be conveniently drawn at actual size. The scale compares the size of the model to the actual size of the object being modeled.

Example ③

In the movie *Jurassic Park*, the dinosaurs were scale models and so was the sport utility vehicle that the T-Rex overturned. The vehicle was made to the scale of 1 inch to 8 inches. The actual vehicle was about 14 feet long. What was the length of the model sport utility vehicle?

First, change 14 feet to 168 inches. Then let ℓ represent the length of the model vehicle.

$$\begin{array}{c} scale \rightarrow \\ actual \rightarrow \end{array} \frac{1}{8} = \frac{\ell}{168}$$

$168 = 8\ell$ *Find the cross products.*

$21 = \ell$

The model sport utility vehicle was 21 inches long.

You can solve proportions by using a calculator.

Example ④ **Solve each proportion.**

a. $\frac{5}{4.25} = \frac{11.32}{m}$

Method 1
Use paper and a pencil.

$\frac{5}{4.25} = \frac{11.32}{m}$

$5m = 4.25(11.32)$

$5m = 48.11$

$m = 9.622$

Method 2
Use a calculator.

By the means-extremes property, $5m = 4.25(11.32)$. Multiply the means. Then divide by 5.

Enter: 4.25 $\boxed{\times}$ 11.32 $\boxed{\div}$ 5 $\boxed{=}$ *9.622*

Rounded to the nearest hundredth, the solution is 9.62.

b. $\frac{x}{3} = \frac{x+5}{15}$

$\frac{x}{3} = \frac{x+5}{15}$

$15x = 3(x + 5)$ *Means-extremes property*

$15x = 3x + 15$ *Distributive property*

$12x = 15$

$x = \frac{5}{4}$ *Check this result.*

The solution is $\frac{5}{4}$.

The ratio of two measurements having different units of measure is called a **rate**. For example, 30 miles per gallon is a rate. Proportions are often used to solve problems involving rates.

Cooperative Learning

Group Discussion Separate students into groups of three to solve this problem.
In the proportion $\frac{e}{f} = \frac{g}{h}$, e is less than g, and e, f, g, and h are nonzero. What can be said about the relationship between f and h? *h > f*

For more information on the group discussion strategy, see *Cooperative Learning in the Mathematics Classroom,* one of the titles in the Glencoe Mathematics Professional Series, page 31.

3 PRACTICE/APPLY

Check for Understanding

Exercises 1–13 are designed to help you assess your students' understanding through reading, writing, speaking, and modeling. You should work through Exercises 1–4 with your students and then monitor their work on Exercises 5–13.

Additional Answers

1. If the cross products of two ratios in a proportion are equal, then the two ratios are equivalent.

2. Sample answer: In the proportion $\frac{x}{2.5} = \frac{3}{4}$, multiply 2.5 and 3, then divide by 4.

APPLICATION

State Fair

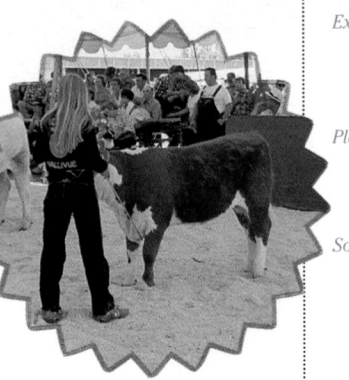

Example 5

In the first 30 minutes of the opening day of the Texas State Fair, 1252 people entered the gates. If this attendance rate continued, how many people visited the fair during the operating hours of 8:00 A.M. and 12:00 midnight the first day?

Explore   Let p represent the number of people attending the fair on opening day.

Plan   Write a proportion for the problem.

$$\frac{1252}{0.5} = \frac{p}{16}$$ *Notice that both ratios compare the number of people per hour.*

Solve
$$\frac{1252}{0.5} = \frac{p}{16}$$
$$1252(16) = 0.5p$$
$$40{,}064 = p$$

If the attendance rate continued, 40,064 people visited the fair on opening day.

Examine   Use estimation to check your answer. About 1250 people entered the fair every 30 minutes. This means that about 2500 people entered every hour. The fair was open for 16 hours each day. Therefore, about 16×2.5 thousand or 40,000 people entered, and the answer is reasonable.

CHECK FOR UNDERSTANDING

Communicating Mathematics

Study the lesson. Then complete the following.   1–2. See margin.

1. **Explain** how to determine whether two ratios are equivalent.

2. **Explain** how to use a calculator to solve a proportion.

3. **Find** three examples of ratios in a newspaper or magazine.
3–4. See students' work.

4. **Draw** an outline of an infant's body if the average infant is three heads long.

Guided Practice

Use cross products to determine whether each pair of ratios forms a proportion.

5. $\frac{3}{2}, \frac{21}{14}$ $=$

6. $\frac{2.3}{3.4}, \frac{0.3}{3.6}$ $\neq$

Solve each proportion.

7. $\frac{2}{3} = \frac{8}{x}$ 12

8. $\frac{4}{w} = \frac{2}{10}$ 20

9. $\frac{3}{15} = \frac{1}{y}$ 5

10. $\frac{5.22}{13.92} = \frac{b}{48}$ 18

11. $\frac{1.1}{0.6} = \frac{8.47}{n}$ 4.62

12. $\frac{x}{1.5} = \frac{2.4}{1.6}$ 2.25

13. **Travel** A 96-mile trip requires 6 gallons of gasoline. At that rate, how many gallons would be required for a 152-mile trip? **9.5 gallons**

198   Chapter 4   *Using Proportional Reasoning*

Reteaching

Using Alternative Methods Solve these proportions by multiplying each side by the LCD to clear fractions. First tell by what number each side can be multiplied to clear fractions. Then solve.

1. $\frac{x}{5} = \frac{12}{15}$ $15; x = 4$

2. $\frac{6}{9} = \frac{10}{x}$ $9x; x = 15$

Practice

Use cross products to determine whether each pair of ratios forms a proportion.

14. $\frac{6}{8}, \frac{22}{28}$ $\neq$

15. $\frac{4}{5}, \frac{16}{20}$ $=$

16. $\frac{4}{11}, \frac{12}{33}$ $=$

17. $\frac{8}{9}, \frac{16}{17}$ $\neq$

18. $\frac{2.1}{3.6}, \frac{5}{7}$ $\neq$

19. $\frac{0.4}{0.8}, \frac{0.7}{1.4}$ $=$

Solve each proportion.

22. $-\frac{63}{16}; -3.938$

27. $-\frac{149}{6}; -24.8$

28. $\frac{3}{5}; 0.6$

34. 63.368

20. $\frac{3}{4} = \frac{x}{8}$ 6

21. $\frac{a}{45} = \frac{3}{15}$ 9

22. $\frac{y}{9} = \frac{-7}{16}$

23. $\frac{3}{5} = \frac{x+2}{6}$ $\frac{8}{5}$; 1.6

24. $\frac{w+2}{5} = \frac{7}{5}$ 5

25. $\frac{x}{8} = \frac{0.21}{2}$ 0.84

26. $\frac{5+y}{y-3} = \frac{14}{10}$ 23

27. $\frac{m+9}{5} = \frac{m-10}{11}$

28. $\frac{r+7}{-4} = \frac{r-12}{6}$

29. $\frac{85.8}{t} = \frac{70.2}{9}$ 11

30. $\frac{z}{33} = \frac{11.75}{35.25}$ 11

31. $\frac{0.19}{2} = \frac{0.5x}{12}$ 2.28

32. $\frac{2.405}{3.67} = \frac{g}{1.88}$ 1.232

33. $\frac{x}{4.085} = \frac{5}{16.33}$ 1.251

34. $\frac{3t}{9.65} = \frac{21}{1.066}$

Critical Thinking

35. Mariah is exactly eight years older than her cousin Louis.
   a. Copy and complete the table below.

Louis' age	1	2	3	6	10	20	30
Mariah's age	9	10	11	14	18	28	38

35c. $r = \frac{y+8}{y}$

35e. No; if the ratio equaled 1, Mariah and Louis would be the same age.

   b. Find the ratio in decimal form of Mariah's age to Louis' age for each pair of ages listed in the table. $9, 5, 3.\overline{6}, 2.\overline{3}, 1.8, 1.4, 1.2\overline{6}$
   c. Write a formula for the ratio of Mariah's age to Louis' age when Louis is y years old.
   d. As Mariah and Louis grow older, explain what happens to the ratio of their ages. The ratio gets smaller.
   e. Explain whether the ratio of their ages will ever equal 1.

Applications and Problem Solving

36. **Biology** A flea, usually less than an eighth of an inch long, can long jump about thirteen inches and high jump about eight inches. If humans could jump in proportion to the flea, it would take only nine leaps to go one mile! At this rate, how many leaps would it take a human to travel 693 miles from Louisville, Kentucky to Norfolk, Virginia? **6237 leaps**

38. 50 pounds

37. **Movies** When rating movies in 1994, critic Gene Siskel gave four thumbs up to every five thumbs up given by his partner Roger Ebert. If Mr. Siskel gave thumbs up to 68 movies, how many movies did Mr. Ebert rate favorably? **85 movies**

38. **Recycling** When a pair of blue jeans is made, the leftover denim scraps can be recycled to make stationery, pencils, and more denim. One pound of denim is left after making every five pairs of jeans. How many pounds of denim would be left from 250 pairs of jeans?

Assignment Guide

Core: 15–37 odd, 39–46
Enriched: 14–34 even, 35–46

For **Extra Practice,** see p. 764.

The red A, B, and C flags, printed only in the Teacher's Wraparound Edition, indicate the level of difficulty of the exercises.

Practice Masters, p. 26

4-1

NAME_____ DATE _____

Practice Student Edition
Pages 195–200

Ratios and Proportions

Solve each proportion.

1. $\frac{x}{4} = \frac{3}{2}$ 6
2. $\frac{3}{a} = \frac{1}{6}$ 18
3. $\frac{-3}{1} = \frac{12}{y}$ −4

4. $\frac{2}{7} = \frac{x}{-42}$ −12
5. $\frac{x}{-3} = \frac{3}{-2}$ $\frac{9}{2}$
6. $\frac{4}{5} = \frac{3}{w}$ $\frac{15}{4}$

7. $\frac{-5}{c} = \frac{1}{3}$ −15
8. $\frac{k}{8} = \frac{-2}{5}$ $\frac{-16}{5}$
9. $\frac{6}{z} = \frac{-3}{5}$ −10

10. $\frac{5}{12} = \frac{x+1}{4}$ $\frac{2}{3}$
11. $\frac{r+2}{7} = \frac{5}{7}$ 3
12. $\frac{-6}{13} = \frac{8}{x+1}$ $\frac{55}{3}$

13. $\frac{x-1}{-4} = \frac{2}{3}$ $\frac{5}{3}$
14. $\frac{3}{7} = \frac{x-2}{6}$ $\frac{32}{7}$
15. $\frac{7}{5} = \frac{1}{x+3}$ $\frac{16}{7}$

16. $\frac{x-1}{3} = \frac{-1}{5}$ $\frac{2}{5}$
17. $\frac{1}{3} = \frac{2}{x+2}$ 4
18. $\frac{x+2}{-3} = \frac{9}{-7}$ $\frac{13}{7}$

19. $\frac{x+2}{3} = \frac{x-1}{6}$ −5
20. $\frac{z-1}{7} = \frac{z-2}{3}$ $\frac{11}{4}$
21. $\frac{n+8}{-5} = \frac{n-3}{-1}$ $\frac{23}{4}$

22. $\frac{x+5}{3} = \frac{x-2}{-4}$ −2
23. $\frac{x+1}{x-1} = \frac{2}{3}$ −5
24. $\frac{5-t}{3+t} = \frac{13}{12}$ $\frac{21}{25}$

Closing Activity

Speaking Discuss with students the similarities and differences among proportions, fractions, and ratios.

Mixed Review

39. **Statistics** Find the mean, median, and mode for the following set of data. (Lesson 3–7) **23; 19; 18**

    19, 21, 18, 22, 46, 18, 17

40. Solve $a = \frac{v}{t}$ for t. (Lesson 3–6) $\quad t = \frac{v}{a}$

41. Define a variable, write an equation, and then solve the following problem. (Lesson 3–2) $\quad 4x - 2x = 100; 50$
    Four times a number decreased by twice the number is 100. What is the number?

42. Solve $-15 + d = 13$. (Lesson 3–1) **28**

43. **Sales** Tanya is a sales representative for Incredible Universe. A CD and tape player manufacturer is having a sales competition. Anyone who sells more than 40 CD players in one day wins a trip to Hawaii. Tanya sold 93 CD and tape players in one day for a total of $9695. The CD players sold for $135 and the tape players for $80. (Lesson 2–9) **43c. more than 40**

    a. How many CD and tape players were sold? **93**

    b. What was the total amount of Tanya's CD and tape player sales? **$9695**

    c. How many CD players does Tanya need to sell for her trip award?

    d. If p represents the number of CD players sold, what was the number of tape players sold? $\quad 93 - p$

    e. Will Tanya win the trip to Hawaii? **yes**

44. **Golf** In four rounds of a recent golf tournament, Heather shot 3 under par, 2 over par, 4 under par, and 1 under par. What was her score for the tournament? (Lesson 2–5) **6 under par, or −6**

45. Evaluate $|m - 4|$ if $m = -6$. (Lesson 2–3) **10**

46. Simplify $x^2 + \frac{7}{8}x - \frac{x}{8}$. (Lesson 1–7) $\quad x^2 + \frac{3}{4}x$

WORKING ON THE In·ves·ti·ga·tion

Refer to the Investigation on pages 190–191.

Since it is impossible to count all the fish in the lake, you can use your samples to estimate the population in the lake.

1. Write a proportion that relates the numbers in your chart and the estimated number of fish in the lake.

2. Imagine that the only information you had was the data you recorded after your first casting.

What would have been your estimate of the number of fish in the lake? Justify your reasoning. Record your estimate in the last column of your chart.

3. Make estimates for each of your castings and record them in the last column of your chart.

4. Taking all of your castings into consideration, if you had to give an official estimate of the number of fish in the lake, what would it be? Justify your reasoning.

Add the results of your work to your Investigation Folder.

200 Chapter 4 *Using Proportional Reasoning*

Enrichment Masters, p. 26

Extension

Reasoning Solve each proportion for x.

1. $\dfrac{x + a}{x} = \dfrac{b + a}{2b} \quad \dfrac{2ab}{a - b}$

2. $\dfrac{3 - x}{x} = \dfrac{4 - b}{2a} \quad \dfrac{6a}{4 - b + 2a}$

In·ves·ti·ga·tion

Working on the Investigation

The Investigation on pages 190–191 is designed to be a long-term project that is completed over several days or weeks. Encourage students to keep their materials in their Investigation Folder as they work on the Investigation.

Integration: Geometry
Similar Triangles

Objective

After studying this lesson, you should be able to:

- find the unknown measures of the sides of two similar triangles.

APPLICATION
Architecture

"Rock n' roll is here to stay," is what you might hear after visiting the Rock and Roll Hall of Fame and Museum in Cleveland, Ohio. Memorabilia from musicians such as the Beatles, U2, and Carlos Santana cover the walls inside. The automatic teller machines in the building are even shaped like jukeboxes.

The building itself, designed by architect I.M. Pei, is made up of several geometric forms—a triangle, a cylinder, a rectangle, and a trapezoid. The triangular section of the building is similar to the one Pei designed for the Louvre in Paris, France. Both are made up of similar triangles.

Two figures are **similar** if they have the same shape, but not necessarily the same size.

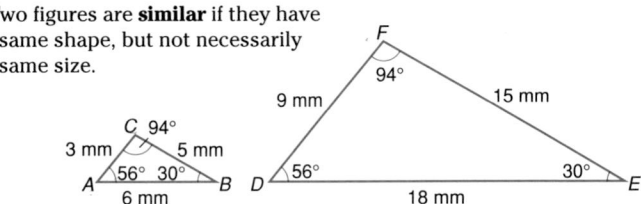

If **corresponding angles** of two triangles have equal measures, the triangles are similar. The two triangles above are similar. We write this as $\triangle ABC \sim \triangle DEF$. The order of the letters indicates the angles that correspond.

corresponding angles	corresponding sides
$\angle A$ and $\angle D$	$\overline{AB}$ and $\overline{DE}$
$\angle B$ and $\angle E$	$\overline{BC}$ and $\overline{EF}$
$\angle C$ and $\angle F$	$\overline{AC}$ and $\overline{DF}$

The sides opposite corresponding angles are called **corresponding sides**. Compare the measures of the corresponding sides. Note that AC means the measure of $\overline{AC}$, DF means the measure of $\overline{DF}$ and so on.

$$\frac{AB}{DE} = \frac{6}{18} = \frac{1}{3} \qquad \frac{BC}{EF} = \frac{5}{15} = \frac{1}{3} \qquad \frac{AC}{DF} = \frac{3}{9} = \frac{1}{3}$$

When the measures of the corresponding sides form equal ratios, the measures are said to be *proportional*.

Similar Triangles	If two triangles are similar, the measures of their corresponding sides are proportional, and the measures of their corresponding angles are equal.

4-2 LESSON NOTES

NCTM Standards: 1–5, 7

Instructional Resources

- Study Guide Master 4-2
- Practice Master 4-2
- Enrichment Master 4-2
- Assessment and Evaluation Masters, p. 100
- Real-World Applications, 10

Transparency 4-2A contains the 5-Minute Check for this lesson; **Transparency 4-2B** contains a teaching aid for this lesson.

Recommended Pacing	
Standard Pacing	Day 3 of 15
Honors Pacing	Day 3 of 15
Block Scheduling*	Day 2 of 7 (along with Lesson 4-3)
Alg. 1 in Two Years*	Days 4 & 5 of 20

*For more information on pacing and possible lesson plans, refer to the *Block Scheduling Booklet* and *Algebra 1 in Two Years.*

1 FOCUS

5-Minute Check
(over Lesson 4-1)

Use cross products to determine whether each pair of ratios forms a proportion.

1. $\frac{9}{19}, \frac{8}{18}$ $\neq$

2. $\frac{16}{24}, \frac{10}{15}$ $=$

Solve each proportion.

3. $\frac{x}{20} = \frac{3}{5}$ 12

4. $\frac{8}{y} = \frac{4}{7}$ 14

5. $\frac{z-5}{8} = \frac{3}{4}$ 11

Hands-On Activity Before class, draw a triangle with side lengths 4 cm, 10 cm, and 8 cm on an overhead transparency. Project the triangle on the chalkboard and determine how far from the chalkboard the projector must be to obtain a triangle with sides 8, 20, and 16 cm. Then move the projector to the right and back until the side lengths are 12, 30, and 24 cm and again note the position of the projector. In class, move the projector to the predetermined locations, trace the triangles with chalk, and have students give you the measures of the angles and sides of the two triangles. Show that the corresponding angles are congruent and the corresponding sides are proportional.

2 TEACH

In-Class Examples

For Example 1
A tree 9 feet tall casts a shadow 5 feet long. How high is a flagpole that casts a shadow 15 feet long? **27 ft**

For Example 2

a. Find the distance, *UV*, across the pond shown below. $\triangle STU$ is similar to $\triangle WVU$. **The pond is 105 meters across.**

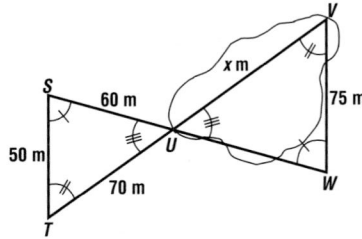

b. $\triangle ABC$ is similar to $\triangle ADE$. Find *x*. **5**

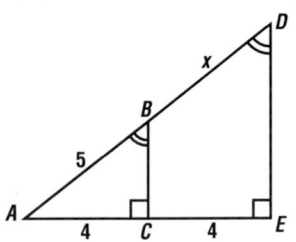

Proportions can be used to find the measures of the sides of similar triangles when some measurements are known.

Example **1**

Surveyors normally use instruments to measure objects that are too large or too far away to measure by hand. They also can use the shadows that objects cast to find the height of the objects without measuring them directly. How can a surveyor use a telephone pole that is 25 feet tall and casts a shadow 20 feet long to find the height of a building that casts a shadow 52 feet long?

$\triangle NOP$ is similar to $\triangle MRQ$.

$\dfrac{RQ}{OP} = \dfrac{QM}{PN}$ *Corresponding sides of similar triangles are proportional.*

$\dfrac{25}{x} = \dfrac{20}{52}$

$20x = 25(52)$ *Find the cross products.*

$20x = 1300$

$x = 65$

The building is 65 feet high.

Similar triangles may be positioned so that the corresponding parts are not obvious.

Example 2

Find the missing measures of the sides if each pair of triangles below are similar.

Remember that the sum of the measures of the angles in a triangle is 180°.

a. The measure of $\angle Q$ is $180° - (117° + 27°)$ or 36°.

The measure of $\angle T$ is $180° - (117° + 36°)$ or 27°.

Since the corresponding angles have equal measures, $\triangle QRS \sim \triangle VUT$. This means that the lengths of the corresponding sides are proportional.

$\dfrac{SQ}{TV} = \dfrac{QR}{VU}$ $\dfrac{SQ}{TV} = \dfrac{SR}{TU}$

$\dfrac{33}{12} = \dfrac{x}{6}$ $\dfrac{33}{12} = \dfrac{y}{8}$

$12x = 33(6)$ $12y = 33(8)$

$12x = 198$ $12y = 264$

$x = 16.5$ $y = 22$

The missing measures are 16.5 and 22.

b. $\triangle NRT$ is similar to $\triangle MST$.

$\dfrac{RN}{SM} = \dfrac{NT}{MT}$

$\dfrac{9}{x} = \dfrac{12}{28}$

$12x = 9(28)$ *Find the cross products.*

$12x = 252$

$x = 21$

The measure of $\overline{SM}$ is 21 meters.

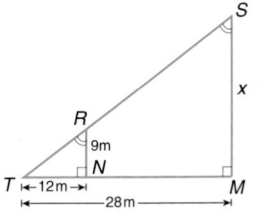

Alternative Teaching Strategies

Student Diversity Have students draw a variety of triangles on plain pieces of paper. Use a copy machine to enlarge and reduce the triangles. Then have students cut them out and match similar triangles.

CHECK FOR UNDERSTANDING

Communicating Mathematics

Study the lesson. Then complete the following. 2–3. See margin.

1. **Name** the pairs of corresponding angles in Example 2a. ∠R and ∠U, ∠S and ∠T, ∠Q and ∠V

2. **List** what you know about the angles and sides of two triangles that are similar.

3. **List** the corresponding angles and corresponding sides if △RED ~ △SOX. Then write three proportions that correspond to these triangles.

 MODELING MATHEMATICS

4. **Model** a triangle using toothpicks or straws. Then make a similar triangle using more toothpicks or straws. Write a proportion to show that the two triangles are similar. See students' work.

Guided Practice

5. For the pair of similar triangles at the right, name the triangle that is similar to △XYZ. Make sure you have the letters in the correct order. △DEF

Determine whether each pair of triangles is similar. Justify your answer.

6. no

7. yes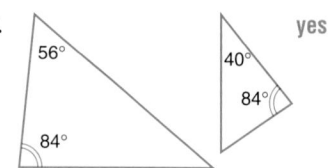

△**KLM** and △**NOP** are similar. For each set of measures given, find the measures of the remaining sides.

8. o = 20, p = 10
9. ℓ = 12, m = 6
10. k = 3, o = 8

8. k = 24, ℓ = 30, m = 15, n = 16
9. k = 9, n = 6, o = 8, p = 4
10. n = 6, p = 2.5, ℓ = 4, m = 1.25

11. If a tree 6 feet tall casts a shadow 4 feet long, how high is a flagpole that casts a shadow 18 feet long at the same time of day? 27 feet high

Lesson 4-2 INTEGRATION Geometry Similar Triangles **203**

Reteaching

Using Models Have students move a mirror along the ground between an observer and an object until the top of the observer's line of sight is reflected in the mirror. Then use similar triangles to find the height of the object.

---

3 PRACTICE/APPLY

Check for Understanding
Exercises 1–11 are designed to help you assess your students' understanding through reading, writing, speaking, and modeling. You should work through Exercises 1–4 with your students and then monitor their work on Exercises 5–11.

Error Analysis
In identifying similar triangles, students are sometimes careless in making sure corresponding vertices are given in the same order. To check, make sure the angles at vertices given in corresponding order are congruent.

Additional Answers

2. Sample answer: The measures of their corresponding sides are proportional, and the measures of their corresponding angles are equal.

3. ∠R and ∠S, ∠E and ∠O, ∠D and ∠X
$\frac{RE}{SO} = \frac{ED}{OX}, \frac{RE}{SO} = \frac{RD}{SX}, \frac{ED}{OX} = \frac{RD}{SX}$

Study Guide Masters, p. 27

Chapter 4 **203**

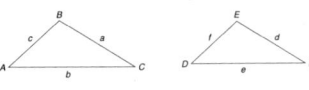
204 Chapter 4

EXERCISES

Practice

For each pair of similar triangles, name the triangle that is similar to △**WXY**.

12. △PNY

13. △DFE

14. △RQY

Determine whether each pair of triangles is similar. Justify your answer.

15. no

16. yes

17. yes

18. no

19. no

20. no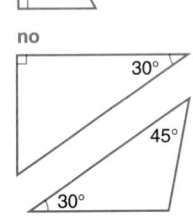

21. $a = \frac{55}{6}$, $b = \frac{22}{3}$

22. $b = \frac{25}{7}$, $c = \frac{30}{7}$

23. $d = \frac{51}{5}$, $c = 9$

24. $d = \frac{112}{13}$, $f = \frac{84}{13}$

△**ABC** and △**DEF** are similar. For each set of measures given, find the measures of the remaining sides.

21. $c = 11$, $f = 6$, $d = 5$, $e = 4$

22. $a = 5$, $d = 7$, $f = 6$, $e = 5$

23. $a = 17$, $b = 15$, $c = 10$, $f = 6$

24. $a = 16$, $e = 7$, $b = 13$, $c = 12$

25. $d = 2.1$, $b = 4.5$, $f = 3.2$, $e = 3.4$ $a = 2.78$, $c = 4.24$

26. $f = 12$, $d = 18$, $c = 18$, $e = 16$ $a = 27$, $b = 24$

27. $c = 5$, $a = 12.6$, $e = 8.1$, $f = 2.5$ $b = 16.2$, $d = 6.3$

28. $f = 5$, $a = 10.5$, $b = 15$, $c = 7.5$ $d = 7$, $e = 10$

29. $a = 4\frac{1}{4}$, $b = 5\frac{1}{2}$, $e = 2\frac{3}{4}$, $f = 1\frac{3}{4}$ $c = \frac{7}{2}$, $d = \frac{17}{8}$

Critical Thinking

30. △*ABC* and △*DEF* are similar. If the ratio of *AC* to *DF* is 2 to 3, what is the ratio of the area of △*ABC* to the area of △*DEF*? **4:9**

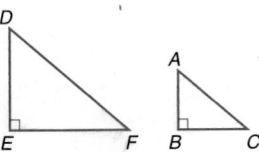

31. See margin.

31. Draw an equilateral triangle. Try to cut it into three similar parts with no two parts that are congruent. Describe the process you used.

204 Chapter 4 *Using Proportional Reasoning*

Applications and Problem Solving

32. Billiards Meda is playing billiards on a table like the one show at the right. If she can make her next shot, she figures she will have no trouble winning. She wants to strike the cue ball at *D*, bank it at *C*, and hit another ball at the mouth of pocket *A*. Use similar triangles to find where Meda's cue ball should strike the rail. **24 inches from pocket B**

33. Tennis When serving in tennis, a player hits the ball standing at point E as shown in the figure at the right. To be a good serve, the ball must go over the net and land in the shaded area. Assume that all serves travel in a straight line and that all serves just clear the net. The figure below shows the view from the sideline.

Let $\overline{EY}$ represent the flight of the ball. **a.** $\frac{3}{9} = \frac{d}{d+39}$; $d = 19.5$ ft; yes

a. Suppose you hit the ball when it is 9 feet above the ground. Write a proportion and solve for *d* to determine whether the serve is good.

b. Repeat the calculation of the distance *d* if the ball is hit from 8 feet above the ground and from 10 feet above the ground. Use the results to help you determine which players have an easier time hitting a good serve, tall players or short players. Explain. **See margin.**

Mark Clark Expressway

34. Bridges Haylee is making a model of the Mark Clark Expressway, a continuous truss bridge that spans the Cooper River in Charleston, South Carolina. Parts of the trusses are made up of triangles. Haylee plans to make the model bridge to the following scale: 1 inch to 12 feet. If the height of the triangle on the actual bridge is 40 feet, what will the height be on the model? $3\frac{1}{3}$ **inches**

Mixed Review

35. Drafting The scale on the blueprint for a house is 1 inch to 3 feet. If the living room on the blueprint is $5\frac{1}{2}$ inches by 7 inches, what are the dimensions of the actual room? (Lesson 4–1) $16\frac{1}{2}$ **feet by 21 feet**

36. Solve $7 = 3 - \frac{n}{3}$. (Lesson 3–3) **−12**

37. Table Tennis Each Ping-Pong™ ball weighs about $\frac{1}{10}$ of an ounce. How many Ping-Pong balls weigh 1 pound altogether? (Lesson 3–2) **160**

38. Solve $x + (-8) = -31$. (Lesson 3–1) **−23**

39. 60 million; 420 million

39. Ecology Americans use about 2.5 million plastic bottles every hour. How many do they use in one day? How many in one week? (Lesson 2–6)

40. Replace __?__ with <, >, or = to make $\frac{8}{15}$ __?__ $\frac{9}{16}$ true. (Lesson 2–4) **<**

41. Evaluate $(19 - 12) \div 7 \cdot 23$. Name the property used in each step. (Lesson 1–6) **23; See margin.**

Lesson 4–2 **INTEGRATION** Geometry Similar Triangles **205**

Extension

Problem Solving Suppose a man is 180 centimeters tall. His image on the film of a camera is 1.5 centimeters tall. If the film is 2 centimeters from the lens of the camera, how far is the man from the camera? **240 cm**

---

4 ASSESS

Closing Activity

Speaking Have students explain how they can use proportions to find the missing measures of similar triangles.

Chapter 4, Quiz A (Lessons 4-1 and 4-2), is available in the *Assessment and Evaluation Masters,* p. 100.

Additional Answers

33b. When the ball is served from 8 feet, $d = 23.4$ ft, and the serve is a fault. When the ball is served from 10 feet, $d = 16.7$ ft. Therefore, tall players have an easier time serving.

41. $(19 - 12) \div 7 \times 23$
  $= 7 \div 7 \times 23$
  Substitution property of equality
  $= 1 \times 23$
  Substitution property of equality
  $= 23$
  Multiplicative identity property

Enrichment Masters, p. 27

4-2 NAME_____ DATE_____
Enrichment Student Edition Pages 201–205

A Curious Construction

Many mathematicians have been interested in ways to construct the number π. Here is one such geometric construction.

In the drawing, triangles *ABC* and *ADE* are right triangles. The length of $\overline{AD}$ equals the length of $\overline{AC}$ and *FB* is parallel to $\overline{EG}$.

The length of $\overline{BG}$ gives a decimal approximation of the fractional part of π to six decimal places.

Follow the steps to find the length of $\overline{BG}$.

1. Use the length of $\overline{BC}$ and the Pythagorean theorem to find the length of $\overline{AC}$.
  $AC = \sqrt{1^2 + \left(\frac{7}{8}\right)^2} = 1.3287682$

2. Find the length of $\overline{AD}$.
  $AD = AC = 1.3287682$

3. Use the length of $\overline{AD}$ and the Pythagorean theorem to find the length of $\overline{AE}$.
  $AE = \sqrt{(AD)^2 + \left(\frac{1}{2}\right)^2} = 1.4197271$

4. The sides of the similar triangles *FED* and *DEA* are in proportion. So, $\frac{FE}{0.5} = \frac{0.5}{AE}$. Find the length of $\overline{FE}$.
  $FE = \frac{1}{4(AE)}$ $FE = 0.1760902$

5. Find the length of $\overline{AF}$.
  $AF = AE - FE = 1.2436369$

6. The sides of the similar triangles *AFB* and *AEG* are in proportion. So, $\frac{AF}{AE} = \frac{AB}{AG}$. Find the length of $\overline{AG}$.
  $AG = \frac{AB \cdot AE}{AF} = \frac{AE}{AF} = 1.1415929$

7. Now, find the length of $\overline{BG}$.
  $BG = AG - AB = AG - 1 = 0.1415929$

8. The value of π to seven decimal places is 3.1415927. Compare the fractional part of π with the length of $\overline{AG}$.
  $0.1415929 - 0.1415927 = 0.0000002$, an error of less than 1 part in a million

Chapter 4 **205**

Integration: Trigonometry
Trigonometric Ratios

Instructional Resources

- Study Guide Master 4-3
- Practice Master 4-3
- Enrichment Master 4-3

Transparency 4-3A contains the 5-Minute Check for this lesson; **Transparency 4-3B** contains a teaching aid for this lesson.

Recommended Pacing	
Standard Pacing	Days 4 & 5 of 15
Honors Pacing	Days 4 & 5 of 15
Block Scheduling*	Day 2 of 7 (along with Lesson 4-2)
Alg. 1 in Two Years*	Days 6 & 7 of 20

*For more information on pacing and possible lesson plans, refer to the *Block Scheduling Booklet* and *Algebra 1 in Two Years*.

1 FOCUS

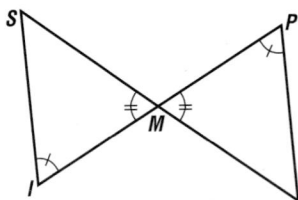

5-Minute Check
(over Lesson 4-2)

1. If △NBC ∼ △FOX, then $\frac{BC}{?} = \frac{CN}{?}$. **OX; XF**

2. Name the triangle that is similar to △SIM. **△LPM**

3. What is the value of *x*? **35**

4. A tree five feet tall casts a shadow two feet long. At the same time, a flagpole casts a shadow nine feet long. How tall is the flagpole?
22.5 ft tall

Objective
After studying this lesson, you should be able to:
- use trigonometric ratios to solve right triangles.

APPLICATION
Architecture

The Leaning Tower of Pisa, a bell tower in Pisa, Italy, currently leans 16.5 feet off center. The tower was closed in 1990 in order for workers to attempt to stabilize the tower's foundation to prevent it from eventually collapsing. By 1994, the tower's lean had decreased about $\frac{2}{5}$ of an inch. No date has been scheduled for reopening the tower.

If you were to draw a line from the top of the tower to the ground, a right triangle would be formed with the ground. If enough is known about a right triangle, certain ratios can be used to find the measures of the remaining parts of the triangle. These ratios are called **trigonometric ratios.** *Trigonometry is an area of mathematics that studies angles and triangles.*

A typical right triangle is shown at the right.

a is the measure of the side *opposite* ∠A, $\overline{BC}$.

b is the measure of the side *opposite* ∠B, $\overline{AC}$.

c is the measure of the side *opposite* ∠C, $\overline{AB}$.

a is *adjacent* to ∠B and ∠C.

b is *adjacent* to ∠A and ∠C.

c is *adjacent* to ∠A and ∠B.

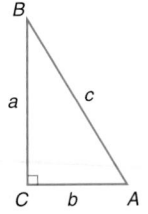

The side opposite ∠C, the right angle, is called the **hypotenuse.** The other two sides are called **legs.**

Three common trigonometric ratios are defined as follows.

Definition of Trigonometric Ratios	
	sine of ∠**A** = $\frac{\text{measure of leg opposite } \angle A}{\text{measure of hypotenuse}}$
	sin A = $\frac{a}{c}$
	cosine of ∠**A** = $\frac{\text{measure of leg adjacent to } \angle A}{\text{measure of hypotenuse}}$
	cos A = $\frac{b}{c}$
	tangent of ∠**A** = $\frac{\text{measure of leg opposite } \angle A}{\text{measure of leg adjacent to } \angle A}$
	tan A = $\frac{a}{b}$

Notice that sine, cosine, and tangent are abbreviated as sin, cos, and tan, respectively.

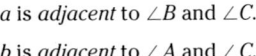

Alternative Teaching Strategies

Reading Algebra A *polygon* is a many-sided figure. A *trigon* is a three-sided figure. This was also the name of a stringed musical instrument in ancient Greece. The suffix *-metry* means "measure." So, *trigonometry* means "the study of finding measures of three-sided figures, or triangles."

Example 1

Find the sine, cosine, and tangent of each acute angle. Round your answers to the nearest thousandth.

$$\sin J = \frac{\text{opposite leg}}{\text{hypotenuse}}$$

$$= \frac{7}{25} \text{ or } 0.280$$

$$\cos J = \frac{\text{adjacent leg}}{\text{hypotenuse}}$$

$$= \frac{24}{25} \text{ or } 0.960$$

$$\tan J = \frac{\text{opposite leg}}{\text{adjacent leg}}$$

$$= \frac{7}{24} \text{ or about } 0.292$$

$$\sin L = \frac{\text{opposite leg}}{\text{hypotenuse}}$$

$$= \frac{24}{25} \text{ or } 0.960$$

$$\cos L = \frac{\text{adjacent leg}}{\text{hypotenuse}}$$

$$= \frac{7}{25} \text{ or } 0.280$$

$$\tan L = \frac{\text{opposite leg}}{\text{adjacent leg}}$$

$$= \frac{24}{7} \text{ or about } 3.429$$

Consider triangles QTV and MSO. Since corresponding angles have the same measure, $\triangle QTV \sim \triangle MSO$. Recall that in similar triangles, the corresponding sides are proportional.

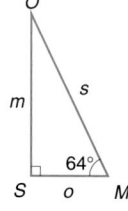

$$\frac{m}{q} = \frac{s}{t}$$

$$\frac{q}{s} \cdot \frac{m}{q} = \frac{q}{s} \cdot \frac{s}{t} \quad \textit{Multiply each side by } \frac{q}{s}.$$

$$\frac{m}{s} = \frac{q}{t}$$

$$\sin M = \sin Q$$

In general, the sine of a 64° angle of a right triangle will be the same number no matter how large or small the triangle is. A similar result holds for cosine and tangent.

You can use a calculator to find the values of trigonometric functions or to find the measure of an angle.

Example 2

Find the value of sin 64° to the nearest ten thousandth.

Use a scientific calculator.

Enter: 64 [SIN] *0.898794046*   *The calculator must be in degree mode.*

Rounded to the nearest ten thousandth, sin 64° ≈ 0.8988.

Example 3

Find the measure of ∠P to the nearest degree.

Since the length of the opposite and adjacent sides are known, use the tangent ratio.

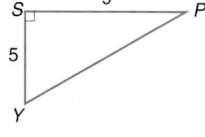

$$\tan P = \frac{\text{opposite leg}}{\text{adjacent leg}}$$

$$= \frac{5}{9}$$

Use a scientific calculator to find the angle measure with a tangent of $\frac{5}{9}$.

Enter: 5 [÷] 9 [=] [2nd] [TAN⁻¹] *29.0546041*   *The TAN⁻¹ key "undoes" the original function.*

To the nearest degree, the measure of ∠P is 29°.

Lesson 4–3   **INTEGRATION** Trigonometry   Trigonometric Ratios   **207**

Motivating the Lesson
Situational Problem The steepest streets in the U.S. are Filbert Street, 22nd Street, and Russian Hill in San Francisco. They rise 1 foot for every 3.17 feet. Have students use trigonometry to find the angle these streets form with the horizontal. **approximately 17.5°**

2 TEACH

Teaching Tip SOH-CAH-TOA is a helpful mnemonic device for remembering these ratios.

$$\sin = \frac{\text{opposite}}{\text{hypotenuse}} \text{ or } s = \frac{o}{h}$$

$$\cos = \frac{\text{adjacent}}{\text{hypotenuse}} \text{ or } c = \frac{a}{h}$$

$$\tan = \frac{\text{opposite}}{\text{adjacent}} \text{ or } t = \frac{o}{a}$$

In-Class Examples

For Example 1
Find the sine, cosine, and tangent of each acute angle to the nearest thousandth.

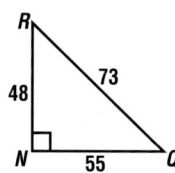

sin Q = 0.658; cos Q = 0.753; tan Q = 0.873; sin R = 0.753; cos R = 0.658; tan R = 1.146

For Example 2
Find the value of cos 72° to the nearest ten thousandth. **0.3090**

For Example 3
Find the measure of ∠A to the nearest degree. **18°**

 Alternative Learning Styles

Visual Given a trigonometric ratio of an angle, find the other two ratios for the angle by using a drawing of the relative lengths of a representative right triangle. For example, given $\sin A = \frac{3}{5}$, then the triangle at the right would be a representative right triangle.

$\cos A = \frac{4}{5}$; $\tan A = \frac{3}{4}$

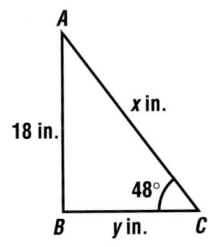

In-Class Examples

For Example 4
Guy wires are attached to a radio tower. They are 52 yards long and are anchored 30 yards from the base of the tower. Find the angle the wires make with the base of the tower to the nearest degree. **about 55°**

For Example 5
Solve $\triangle ABC$.
$x \approx 24.2$, $y \approx 16.2$

Many real-world applications involve trigonometry.

Example 4

Architecture

Refer to the application at the beginning of the lesson. Engineers working to correct the lean of the Leaning Tower of Pisa could check the angle the Tower makes with the ground to find any increase in the lean. Use the measurements to find the angle to the nearest degree that the Tower made with the ground.

Since the length of adjacent side and hypotenuse are known, use the cosine.

$$\cos x = \frac{\text{adjacent leg}}{\text{hypotenuse}}$$

$$\cos x = \frac{16.5}{179}$$

$$\cos x = 0.09217877 \quad \textit{Use a calculator to find x.}$$

$$x \approx 85°$$

179 ft

16.5 ft

To the nearest degree, the measure of the angle is 85°.

You can find the missing measures of a right triangle if you know the measure of two sides of the triangle or the measure of one side and one acute angle. Finding all of the measures of the sides and the angles in a right triangle is called **solving the triangle**.

Example 5 **Solve $\triangle ABC$.**

The measure of $\angle C$ is $90° - 42°$ or $48°$.

$$\sin 42° = \frac{x}{22}$$
$$0.6691 \approx \frac{x}{22}$$
$$14.7 \approx x$$

$$\cos 42° = \frac{y}{22}$$
$$0.7431 \approx \frac{y}{22}$$
$$16.3 \approx y$$

Thus, $\overline{BC}$ is about 14.7 inches long.

Thus, $\overline{AB}$ is about 16.3 inches long.

In the real world, trigonometric ratios are often used to find distances or lengths that cannot be measured directly. In these applications, you will sometimes use an angle of elevation or an angle of depression. An **angle of elevation** is formed by a horizontal line of sight and a line of sight above it. An **angle of depression** is formed by a horizontal line of sight and a line of sight below it.

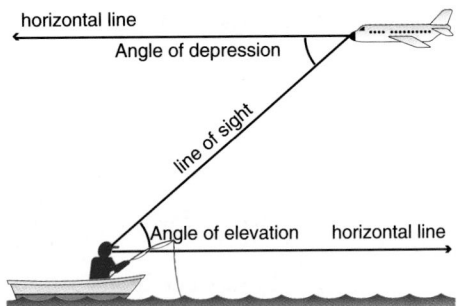

horizontal line
Angle of depression
line of sight
Angle of elevation   horizontal line

Make a Hypsometer

Materials: protractor straw tape paper clip

~~ string

You can make a hypsometer to find the angle of elevation of an object that is too tall to measure.

Straw

180
170
160
150
140
130
120
110
100
90
10 20 30 40 50 60 70 80

String

Paper Clip

Your Turn

a. Tie one end of a piece of string to the middle of a straw. Tie the other end of string to a paper clip.

b. Tape a protractor to the side of the straw. Make sure that the string hangs freely to create a vertical or plumb line.

c. Find an object outside that is too tall to measure directly, such as a basketball hoop, a flagpole, or the school building.

line of sight

horizontal line

x

d. Look through the straw to the top of the object you are measuring. Find a horizontal line of sight as shown above. Find the measurement where the string and protractor intersect. Determine the angle of elevation by subtracting this measurement from 90°.

e. Use the equation tan (angle sighted) = $\frac{\text{height of object} - x}{\text{distance of object}}$, where x represents the distance from the ground to your eye level, to find the height of the object.

f. Compare your answer with someone who measured the same object. Did your heights agree? Why or why not? **See students' work.**

Example **6**

APPLICATION

Balloons

What would the Macy's Thanksgiving Day Parade be without balloons? In 1994, Dr. Suess' The Cat in the Hat made its first appearance. It took 36 people to maneuver the cat along the parade route. Suppose you were watching the parade. You noticed that two of the handling lines formed a right triangle with the ground. You estimated the angle of elevation to the bottom of the balloon to be 60°, and the handlers to be about 35 feet away from each other. Find the altitude of the bottom of the balloon to the nearest foot.

Explore You know the distance from one handler to another and the approximate angle of elevation. You need to find the height from the handlers to the bottom of the balloon.

(continued on the next page)

In this modeling activity, students build a simple version of a sextant. Since the instrument in the activity uses a counterweight, it can measure only vertical angles.

In-Class Example

For Example 6
If the angle of elevation to the top of the balloon is about 65° and the handler is about 45 feet away from the bottom of the balloon, what is the height of the balloon to the nearest foot?
97 ft

Check for Understanding

Exercises 1–20 are designed to help you assess your students' understanding through reading, writing, speaking, and modeling. You should work through Exercises 1–6 with your students and then monitor their work on Exercises 7–20.

Error Analysis
Students sometimes take the terms *opposite leg* and *adjacent leg* to be absolute rather than relative. In a right triangle with acute angles *A* and *B*, the opposite leg of *A* is the adjacent leg of *B*. These labels are relative to the acute angle named.

Additional Answer

3. Use the sine when the opposite side and hypotenuse are known, use the cosine when the adjacent side and hypotenuse are known, and use the tangent when the opposite side and adjacent side are known.

Plan	Let *h* represent the distance to the bottom of the balloon in feet. Use the tangent ratio to solve this problem.

Solve

$$\tan 60° = \frac{\text{opposite leg}}{\text{adjacent}}$$

$$\tan 60° = \frac{h}{35}$$

$$1.7321 \approx \frac{h}{35}$$

$$1.7321(35) \approx h$$

$$60.6235 \approx h$$

The altitude of the bottom of the balloon is about 61 feet above the handlers.

Examine   You can examine the solution by determining the angle of elevation in the equation $\tan x = \frac{61}{35}$. Since $x = 60.15°$, the solution is reasonable.

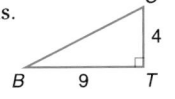

CHECK FOR UNDERSTANDING

Communicating Mathematics

Study the lesson. Then complete the following.

1. **Use** the figure at the right to answer the following questions.
   a. What is the measure of the leg adjacent to ∠C?   4
   b. What is the measure of the leg opposite ∠C?   9

2. **List** the trigonometric ratio or ratios that involve the measure of the hypotenuse.   sine and cosine

3. **Explain** how to determine which trigonometric ratio to use when solving for an unknown measure of a right triangle.   See margin.

4. **Name** the angles of elevation and depression in the drawing at the right.   ∠QRP; ∠SPR

5. State which trigonometric ratio you could use to find *x*.

a.   12 ft   x°   13 ft   **cosine**

b.   x m   39°   12 m   **tangent**

c.   46°   x ft   15 ft   **sine**

MODELING MATHEMATICS

6. **Use** a hypsometer to find the height of a tree in your community. Then write a paragraph explaining how you found the height.   See students' work.

Reteaching

Using Alternative Methods   Have students make flash cards with the name of a trigonometric ratio on one side and its definition on the other side. Encourage students to practice with these until they know them well.

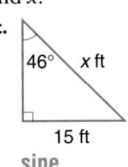

GLENCOE Technology

Interactive Mathematics Tools Software

A multimedia software provides an interactive lesson by helping students explore the relationship of the sine ratio to the length of the hypotenuse of the right triangle. A **Computer Journal** gives students an opportunity to write about what they have learned.

For Windows & Macintosh

Guided Practice

For each triangle, find sin _Y_, cos _Y_, and tan _Y_ to the nearest thousandth. 7–8. See margin.

7.

8.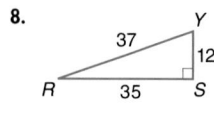

Use a calculator to find the value of each trigonometric ratio to the nearest ten thousandth.

9. cos 35° **0.8192**   10. sin 63° **0.8910**   11. tan 7° **0.1228**

Use a calculator to find the measure of each angle to the nearest degree.

12. cos _C_ = 0.9613 **16°**   13. sin _X_ = 0.7193 **46°**   14. tan _W_ = 2.4752 **68°**

For each triangle, find the measure of the marked acute angle to the nearest degree.

15. **36°**

16. **10°**
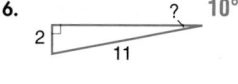

Solve each right triangle. State the side lengths to the nearest tenth and the angle measures to the nearest degree. 17–19. See margin.

17.

18.

19.

20. **Transportation** Find the angle of elevation of a train track if a train in the mountains rises 15 feet for every 250 feet it moves along the track.
about 3.4°

EXERCISES

Practice

For each triangle, find sin _G_, cos _G_, and tan _G_ to the nearest thousandth. 21–23. See margin.

A

21.

22.

23.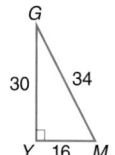

Assignment Guide

Core: 21–57 odd, 58–69
Enriched: 22–54 even, 55–69

For **Extra Practice,** see p. 765.

The red A, B, and C flags, printed only in the Teacher's Wraparound Edition, indicate the level of difficulty of the exercises.

Additional Answers

7. sin _Y_ = 0.600, cos _Y_ = 0.800, tan _Y_ = 0.750
8. sin _Y_ ≈ 0.946, cos _Y_ ≈ 0.324, tan _Y_ ≈ 2.917
17. ∠_B_ = 50°, _AC_ ≈ 12.3 m, _BC_ ≈ 10.3 m
18. ∠_B_ = 20°, _AB_ ≈ 9.6 cm, _AC_ ≈ 3.3 cm
19. ∠_A_ = 30°, _AC_ ≈ 13.9 m, _BC_ = 8 m
21. sin _G_ = 0.600, cos _G_ = 0.800, tan _G_ = 0.750
22. sin _G_ ≈ 0.246, cos _G_ ≈ 0.969, tan _G_ ≈ 0.254
23. sin _G_ ≈ 0.471, cos _G_ ≈ 0.882, tan _G_ ≈ 0.533

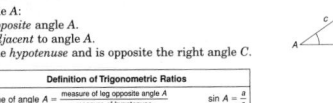
For each triangle, find sin G, cos G, and tan G to the nearest thousandth.   24–26. See margin.

24.

25.

26.
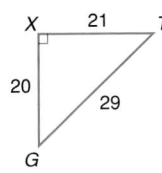

Use a calculator to find the value of each trigonometric ratio to the nearest ten thousandth.

27. sin 21°   0.3584
28. cos 15°   0.9659
29. sin 76°   0.9703
30. tan 56°   1.4826
31. cos 68°   0.3746
32. tan 30°   0.5774

Use a calculator to find the measure of each angle to the nearest degree.

B

33. tan A = 0.6473   33°
34. cos B = 0.7658   40°
35. sin F = 0.3823   22°
36. cos Q = 0.2993   73°
37. sin R = 0.8827   62°
38. tan J = 8.8988   84°

For each triangle, find the measure of the marked acute angle to the nearest degree.

C

39. 77°

40. 33°

41. 58°

42.

17°

43.

18°

44.
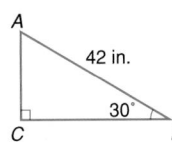
42°

Solve each right triangle. State the side lengths to the nearest tenth and the angle measures to the nearest degree.   45–53. See margin.

45.

46.

47.

48.

49.

50.

51.

52.

53.

Programming

To run the program repeatedly, press ENTER *after each run is completed and the prompt will appear again for new values of a and b.*

54. Suppose ∠*ABC* has a right angle *C*, and the measures of its legs are *a* and *b*. The graphing calculator program at the right asks you to enter the lengths of the legs of the triangle. It then calculates the measure of the hypotenuse *c*, the measures of the acute angles, and the trigonometric ratios for each acute angle. The PAUSE commands in the program stop the calculator's computations and allow you to record the information shown on the screen. Press ENTER to continue the program. Make sure your calculator is in degree mode before running the program. **a–d. See margin.**

```
PROGRAM: LEGS
: Disp "ENTER LENGTHS OF",
  "LEGS A AND B"
: Prompt A, B
: √(A² + B²)→C
: Disp "SIDE C =",C
: Disp "ANGLE A =", tan⁻¹ (A/B)
: Disp "ANGLE B =", tan⁻¹ (B/A)
: Pause
: Disp "sin A=", A/C
: Disp "sin B=", B/C
: Disp "cos A=", B/C
: Pause
: Disp "cos B=", A/C
: Disp "tan A=", A/B
: Disp "tan B=", B/A
```

Use the program for each pair of values for *a* and *b*.

a. 3, 4   **b.** 5, 12

c. 20, 40   **d.** 125, 100

Critical Thinking

55. Use the triangle below to determine which of the following statements are true. Justify your answers. **a, d, e; See margin.**

a. $\sin C = \cos B$

b. $\cos B = \dfrac{1}{\sin B}$

c. $\tan C = \dfrac{\cos C}{\sin C}$

d. $\tan C = \dfrac{\sin C}{\cos C}$

e. $\sin B = (\tan B)(\cos B)$

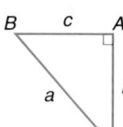

Applications and Problem Solving

56. Archaeology The largest of the pyramids of Egypt has a square base with sides 755 feet long. ∠*PQR* has a measure of 52°. The top of the pyramid is no longer there. What was the pyramid's original height ($\overline{RP}$) to the nearest foot? **483 ft**

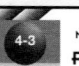
755 ft

57. Make a Drawing On September 30, 1995, the Ohio State Buckeyes defeated the Notre Dame Fighting Irish in football as the Goodyear blimp relayed aerial views from above. Suppose the blimp was directly over the 50-yard line. A photographer on the ground estimated the angle of elevation to be 85° when she was 65 yards away from the point on the ground directly under the blimp. How high is the blimp? Round your answer to the nearest foot. **2229 ft**

Lesson 4–3   **INTEGRATION** Trigonometry   Trigonometric Ratios   **213**

Extension

Connections The length in miles ℓ along the surface of Earth of one degree longitude is given by the formula $\ell = 2\pi \dfrac{(4000 \cos x)}{360}$, where *x* is the number of degrees latitude. Find the length to the nearest mile of one degree longitude at each latitude.

a. 30° latitude   **60 miles**

b. 60° latitude   **35 miles**

c. 90° latitude   **0 miles**

Using the Programming Exercises The program given in Exercise 54 is for use with a TI-82 graphing calculator. For other programmable calculators, have students consult their owner's manual for commands similar to those presented here.

Additional Answers

54a. 5; 36.9°, 53.1°; sin *A* = 0.600, sin *B* = 0.800, cos *A* = 0.800, cos *B* = 0.600, tan *A* = 0.750, tan *B* = 1.333

54b. 13; 22.6°, 67.4°; sin *A* = 0.385, sin *B* = 0.923; cos *A* = 0.923, cos *B* = 0.385, tan *A* = 0.417, tan *B* = 2.400

54c. 44.7; 26.6°, 63.4°; sin *A* = 0.447, sin *B* = 0.894, cos *A* = 0.894, cos *B* = 0.447, tan *A* = 0.500, tan *B* = 2.000

54d. 160.1; 51.3°, 38.7°; sin *A* = 0.781, sin *B* = 0.625, cos *A* = 0.625, cos *B* = 0.781, tan *A* = 1.250, tan *B* = 0.800

55a. $\sin C = \dfrac{c}{a}$, $\cos B = \dfrac{c}{a}$

55b. $\cos B = \dfrac{c}{a}$, $\dfrac{1}{\sin B} = \dfrac{a}{b}$

55c. $\tan C = \dfrac{c}{b}$, $\dfrac{\cos C}{\sin C} = \dfrac{b}{a} \div \dfrac{c}{a} = \dfrac{b}{c}$

Practice Masters, p. 28

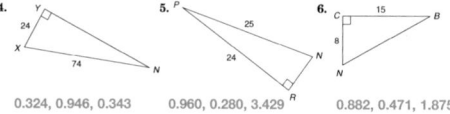

4-3   NAME _____   DATE _____

Practice   Student Edition Pages 206–214

Integration: Trigonometry
Trigonometric Ratios

For each triangle, find sin N, cos N, and tan N to the nearest thousandth.

1. 0.385, 0.923, 0.417   **2.** 0.600, 0.800, 0.750   **3.** 0.385, 0.923, 0.417

4. 0.324, 0.946, 0.343   **5.** 0.960, 0.280, 3.429   **6.** 0.882, 0.471, 1.875

Use a calculator to find the measure of each angle to the nearest degree.

7. sin *A* = 0.3779   22°   **8.** cos *B* = 0.8931   27°   **9.** cos *G* = 0.4545   63°

10. tan *C* = 1.4213   55°   **11.** sin *K* = 0.9898   82°   **12.** tan *X* = 0.3785   21°

For each triangle, find the measure of the marked acute angle to the nearest degree.

13. 21°   **14.** 27°   **15.** 71°

Additional Answers

55d. $\tan C = \dfrac{c}{b}$, $\dfrac{\sin C}{\cos C} = \dfrac{c}{a} \div \dfrac{b}{a} = \dfrac{c}{b}$

55e. $\sin B = \dfrac{b}{a}$, $(\tan B)(\cos B) = \dfrac{b}{c} \cdot \dfrac{c}{a} = \dfrac{b}{a}$

Chapter 4   **213**

4 ASSESS

Closing Activity

Speaking Ask students to give the definition of trigonometry ratios in their own words without using terms such as *hypotenuse*. If students have to use simpler definitions, students are more likely to remember them.

Enrichment Masters, p. 28

NAME _____ DATE _____

Enrichment   Student Edition Pages 206–214

4-3

Modern Art

The painting below, aptly titled, "Right Triangles," was painted by that well-known artist Two-loose La-Rectangle. Using the information below, find the dimensions of this masterpiece. (*Hint*: The triangle that includes ∠*F* is isosceles.)

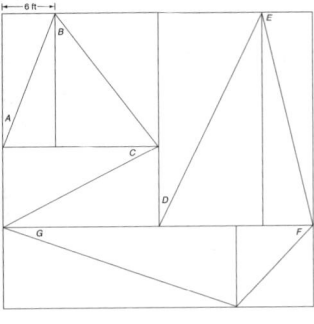

$\tan A = \frac{2}{5}$ $\tan B = \frac{4}{5}$ $\tan C = \frac{1}{2}$

$\tan D = \frac{1}{2}$ $\tan E = 4$ $\tan F = 1$

$\tan G = \frac{1}{3}$

1. What is the length of the painting? **33 ft**
2. What is the width of the painting? **36 ft**

Mixed Review

58. 121m

58. Civil Engineering The city of Mansfield plans to build a bridge across Spring Lake. Use the information in the diagram at the right to find the distance across Spring Lake. (Lesson 4–2)

59. Solve $\frac{x}{5} = \frac{x+3}{10}$. (Lesson 4–1) **3**

60. Baseball The batting averages for 10 players on a baseball team are 0.234, 0.253, 0.312, 0.333, 0.286, 0.240, 0.183, 0.222, 0.297, and 0.275. Find the median batting average for these players. (Lesson 3–7) **0.264**

61. Solve $\frac{2}{5}y + \frac{y}{2} = 9$. (Lesson 3–5) **10**

62. Write an equation to represent the situation below. Then solve the problem. (Lesson 3–2) **5x = $47.50; $9.50**

Joyce Conners paid $47.50 for five football tickets. What was the cost per ticket?

63. Name the sets of numbers to which $\sqrt{36}$ belongs. Use N for natural numbers, W for whole numbers, Z for integers, Q for rational numbers, and I for irrational numbers. (Lesson 2–8) **N, W, Z, Q**

64. Simplify $\frac{3a+9}{3}$. (Lesson 2–7) **a + 3**

65. Find $\left(-\frac{2}{3}\right)\left(-\frac{1}{5}\right)$. (Lesson 2–6) **$\frac{2}{15}$**

66. Find $-13 + (-8)$. (Lesson 2–3) **−21**

67. Simplify $8x + 2y + x$. (Lesson 1–8) **9x + 2y**

68. Evaluate $(9 - 2 \cdot 3)^3 - 27 + 9 \cdot 2$. (Lesson 1–6) **18**

69. Evaluate 5^3. (Lesson 1–1) **125**

Mathematics and SOCIETY

Morgan's Conjecture

The excerpt below appeared in an article in *The Columbus Dispatch* on December 21, 1994.

RYAN MORGAN WOULD HAVE GOTTEN an "A" in geometry even if he hadn't unearthed a mathematical treasure. But the persistent Baltimore high school sophomore pushed a hunch into a theory. He calls it "Morgan's Conjecture" and is hoping it will soon be "Morgan's Theorem." In geometric circles, developing a theorem is a big deal—especially if you're only 15....What did Ryan see?...When the sides of a triangle are divided by an odd number larger than one, and when lines are drawn from the division points on the sides of the triangle to the vertices, there always will be a hexagon in the interior of the triangle. And the area of that hexagon always will be a predictable fraction of the area of the larger triangle. The fraction is determined by a complex formula that Ryan worked out. ■

1. In talking about his work, the student said, "It's mostly just luck in playing around with it." What other factors besides luck might have been involved? **See margin.**

2. Are you surprised to learn that students can discover theorems? Why or why not? **See students' work.**

3. Is there an area of mathematics that interests you that you would like to explore further? How would you do it? **See students' work.**

214 *Chapter 4 Using Proportional Reasoning*

Mathematics and SOCIETY

Mathematics has always been a field for young prodigies. The great mathematicians have usually been most productive in their twenties and thirties. One reason may be that success in mathematics does not depend on knowledge and experience of the real world, but on intellectual insight.

Answer for Mathematics and Society

1. Answers will vary. Sample answers: curiosity, determination, a systematic approach, willingness to try new methods, looking for patterns, learning from mistakes, and keeping good records.

Percents

4-4

Objectives

After studying this lesson, you should be able to:

- solve percent problems, and
- solve problems involving simple interest.

APPLICATION
Cosmetics

According to *Good Housekeeping*, 3 out of 5 teenage girls wear mascara. What is the rate of teenage girls wearing mascara per 100 girls?

You can solve this problem by using a proportion. The ratio of teenage girls who wear mascara to the total number of teenage girls is $\frac{3}{5}$. Write a proportion that sets $\frac{3}{5}$ equal to a ratio with a denominator of 100.

$$\frac{3}{5} = \frac{n}{100}$$

$$60 = n$$

Teenage girls wear mascara at a rate of 60 per 100, or 60 percent. A **percent** is a ratio that compares a number to 100. Percent also means *per hundred*, or *hundredths*. Percents can be expressed with a percent symbol (%), as fractions, or as decimals.

$$60\% = \frac{60}{100} = 0.60$$

Example **1** Write $\frac{3}{4}$ as a percent.

Method 1
Use a proportion.

$$\frac{3}{4} = \frac{n}{100}$$

$$300 = 4n$$

$$75 = n$$

Thus, $\frac{3}{4}$ is equal to $\frac{75}{100}$ or 75%.

Method 2
Use a calculator.

$$\frac{3}{4} = \frac{n}{100}$$

$$100\left(\frac{3}{4}\right) = n$$

Enter: 100 ⊠ 3 ÷ 4 ⊟ 75

Proportions are often used to solve percent problems. One of the ratios in these proportions is always a comparison of two numbers called the **percentage** and the **base**. The other ratio, called the **rate**, is a fraction with a denominator of 100.

$$\begin{array}{l} percentage \rightarrow \\ base \rightarrow \end{array} \quad \frac{3}{4} = \frac{75}{100} \Big\} \leftarrow Rate$$

Percent Proportion	$\dfrac{\text{Percentage}}{\text{Base}} = \text{Rate}$ or $\dfrac{\text{Percentage}}{\text{Base}} = \dfrac{r}{100}$

The $\frac{percentage}{base}$ represents $\frac{part}{whole}$.

4-4 LESSON NOTES

NCTM Standards: 1–5

Instructional Resources

- Study Guide Master 4-4
- Practice Master 4-4
- Enrichment Master 4-4
- Assessment and Evaluation Masters, pp. 99–100
- Multicultural Activity Masters, p. 7
- Tech Prep Applications Masters, p. 7

Transparency 4-4A contains the 5-Minute Check for this lesson; **Transparency 4-4B** contains a teaching aid for this lesson.

Recommended Pacing	
Standard Pacing	Days 6 & 7 of 15
Honors Pacing	Days 6 & 7 of 15
Block Scheduling*	Day 3 of 7 (along with Lesson 4-5)
Alg. 1 in Two Years*	Days 8 & 9 of 20

*For more information on pacing and possible lesson plans, refer to the *Block Scheduling Booklet* and *Algebra 1 in Two Years*.

1 FOCUS

5-Minute Check
(*over Lesson 4-3*)

For the triangle below, express each trigonometric ratio as a fraction in simplest form.

1. sin A $\frac{8}{17}$
2. sin B $\frac{15}{17}$
3. cos B $\frac{8}{17}$
4. cos A $\frac{15}{17}$
5. tan A $\frac{8}{15}$

Example   **a. 30 is what percent of 50?**

Use the percent proportion.
$$\frac{\text{Percentage}}{\text{Base}} = \frac{r}{100}$$

The percentage is 30.
The base is 50. $\frac{30}{50} = \frac{r}{100}$

$$3000 = 50r$$
$$60 = r$$

Thus, 30 is 60% of 50.

b. 20 is what percent of 30?

Use a calculator.
$$\frac{\text{Percentage}}{\text{Base}} = \frac{r}{100}$$

$$\frac{20}{30} = \frac{r}{100}$$

Enter: 20 ÷ 30 × 100 =
66.66666667

Thus, 20 is 66.$\overline{6}$% of 30. This can also be written as $66\frac{2}{3}$%.

You can also write equations to solve problems with percents.

Example   **a. 60% of what number is 54?**

60% of what number is 54?

$$\frac{60}{100} \cdot x = 54$$
$$0.6x = 54$$
$$x = 90$$

Thus, 60% of 90 is 54.

b. What number is 40% of 37.5?

What number is 40% of 37.5?

$$x = \frac{40}{100} \cdot 37.5$$
$$x = 15$$

Thus, 15 is 40% of 37.5.

TECHNOLOGY Tip

Most calculators have a key labeled %. To find 28% of 58.4 on a scientific calculator, enter:

28 [2nd] [%] [×]
58.4 [=] 16.352.

MODELING MATHEMATICS

Making Circle Graphs

Materials: 📐 compass 📐 protractor

You can use a circle graph to compare parts of a whole. Follow the steps to display the data below in a circle graph. **a–e. See margin.**

Where's the Remote?	
Number of Times TV Remote Misplaced Per Week	Number of People Responding
Never	220
1–5	190
5 or more	85
Don't Know	5

b. Find the ratio that compares the number of people that responded for each category to the total number of people that responded.

c. Since there are 360° in a circle, multiply each ratio by 360 to find the number of degrees for each section of the graph. Round to the nearest degree.

d. Use a compass to draw a circle and a radius as shown at the right.

e. Use a protractor to draw angles. Start with the least number of degrees. Repeat for the remaining sections. Label each section and give the graph a title.

Your Turn

a. Find the total number of people surveyed.

f. What is the sum of the percents in your graph? **f. 100%**

Answers for Modeling Mathematics

a. 500

b. $\frac{220}{500}, \frac{190}{500}, \frac{85}{500}, \frac{5}{500}$

c. 158°, 137°, 61°, 4°

d. See students' work.

e. Number of Times Remote Misplaced Per Week

Percents are also used in simple interest problems. **Simple interest** is the amount paid or earned for the use of money. The formula $I = prt$ is used to solve problems involving simple interest. In the formula, I represents the interest, p represents the amount of money invested, called the *principal*, r represents the annual interest rate, and t represents time in years.

Example ④

APPLICATION

Finance

a. **Luis Hernandez has some money he has saved over the summer from mowing lawns. He wants to put some of it in a 6-month certificate of deposit (CD) account at BancOne that would pay 6% annual interest. He doesn't want to put all of his money in the CD because he wants some spending money. He is hoping to earn $45 in interest to buy a new video game. How much money should Luis put in the CD?**

Explore   Let p = the amount of money Luis should deposit.

Plan $I = prt$
             $45 = p(0.06)(0.5)$ *Write 6% as 0.06 and 6 months as 0.5 years.*

Solve $45 = 0.03p$ *Multiply 0.06 by 0.5.*
             $1500 = p$ *Divide each side by 0.03.*

Examine   When p is 1500, $I = (1500)(0.06)(0.5)$ or $45. Luis should deposit $1500.

b. **Whitney Williamson has $30,000 she would like to invest. She has a choice of two interest-paying bonds, one offering a 6% annual interest and the other paying 7.5% annual interest. She would like to earn $2100 in interest in one year. If she earns any more than $2100, she would need to pay a higher rate of income tax. How much money should she invest in each bond?**

Explore   Let n = the amount of money invested at 6%.
             Then $30,000 - n$ = the amount of money invested at 7.5%.

Plan   interest on 6% investment

             $n(0.06)(1)$ or $0.06n$

             interest on 7.5% investment

             $(30,000 - n)(0.075)(1)$ or $2250 - 0.075n$

Solve   interest at 6% + interest at 7.5% = total interest

             $0.06n$ + $(2250 - 0.075n) =$ 2100
                          $-0.015n + 2250 = 2100$ *Add 0.06n and*
                                 $-0.015n = -150$ *$-0.075n.$*
                                       $n = 10,000$

             Whitney should invest $10,000 at 6% and $30,000 - $10,000 or $20,000 at 7.5%.

Examine $(10,000)(0.06)(1) + (20,000)(0.075)(1) = 2100$

             The answer checks.

Teaching Tip Emphasize that when a problem asks for the percent, the students must find the value of r.

MODELING MATHEMATICS This modeling activity shows students how the circle graph displays parts of a whole. Each part is a percent of the total area of the circle.

In-Class Example

For Example 4
a. Luis sees that another bank has a 3-month certificate of deposit that would pay 8% annual interest. Which would be better for Luis? **BancOne; his $1500 would earn only $30 at the other bank.**

b. If Whitney had invested part of her money at 5% and the rest at 8%, how much had she invested at each rate if the interest after 1 year was $1995? **$13,500 at 5%; $16,500 at 8%**

Teaching Tip Remind students that interest rate is annual. Therefore, time should be expressed in years when working with the simple interest formula.

Cooperative Learning

Categorizing Separate students into groups of five for this library research project. Have students list five occupations in which salary is dependent upon commission. Have students research these occupations to find the usual rate of commission. Challenge students to analyze their lists to find a relationship between the dollar value of the item sold and the size of the commission awarded an employee. For more information on the categorizing strategy, see *Cooperative Learning in the Mathematics Classroom*, one of the titles in the Glencoe Mathematics Professional Series, pages 31–32.

Check for Understanding

Exercises 1–15 are designed to help you assess your students' understanding through reading, writing, speaking, and modeling. You should work through Exercises 1–5 with your students and then monitor their work on Exercises 6–15.

Additional Answer

2. Enter: 57
    = 23.94

CHECK FOR UNDERSTANDING

Communicating Mathematics

1. They are the same.

3. The investment at the higher rate may be too risky.

4. The sum of the percents is greater than 100.

Study the lesson. Then complete the following.   2. See margin.

1. **Compare** $\frac{r}{100}$ in the percent proportion and r in the simple interest formula. How are they the same? How are they different?

2. **Explain** how to find 57% of 42 using a scientific calculator.

3. **Explain** why an investor might invest a greater amount of money in a lower rate of interest instead of a higher rate.

4. **Read** the following excerpt from an article about the America's Cup yacht race in *Motor Boating & Sailing* magazine. Explain what is incorrect in this excerpt and why.

> With a Ph.D. from MIT, (Dr. Koch) takes a scientific view of things, and quantifies the effort involved in winning the Cup as 55 percent boat speed, 20 percent tactics, 20 percent crew work, 5 percent luck, and less than 5 percent physical strength.

MODELING MATHEMATICS

5. **Draw** a circle graph to represent the number of students in your school by grade level.   See students' work.

Guided Practice

Write each ratio as a percent and then as a decimal.

6. $\frac{3}{4}$ 75%, 0.75
7. $\frac{43}{100}$ 43%, 0.43
8. $\frac{2}{25}$ 8%, 0.08

Use a proportion to answer each question.

10. $56\frac{1}{4}$%; 56.25%

9. Eleven is what percent of 20?   55%
10. What percent of 80 is 45?

Use an equation to answer each question.

12. $21\frac{1}{3}$%; 21.$\overline{3}$%

11. What number is 30% of 50?   15
12. What percent of 75 is 16?

Use $I = prt$ to find the missing quantity.

13. Find I if $p = \$1500$, $r = 7\%$, and $t = 6$ months. $52.50
14. Find p if $I = \$196$, $r = 10\%$, and $t = 7$ years. $280
15. **Finance**   Kaylee Richardson invested $7200 for one year, part at 10% annual interest and the rest at 14% annual interest. Her total interest for the year was $960. How much money did she invest at each rate? $1200 at 10%, $6000 at 14%

EXERCISES

Practice
A

Write each ratio as a percent and then as a decimal.

16. $\frac{67}{100}$ 67%, 0.67
17. $\frac{6}{20}$ 30%, 0.30
18. $\frac{5}{8}$ $62\frac{1}{2}$%, 0.625

19. $\frac{7}{10}$ 70%, 0.70
20. $\frac{5}{6}$ $83\frac{1}{3}$%, 0.8$\overline{3}$
21. $\frac{9}{5}$ 180%, 1.8

22. $\frac{2}{3}$ $66\frac{2}{3}$%, 0.$\overline{6}$
23. $\frac{25}{40}$ $62\frac{1}{2}$%, 0.625
24. $\frac{20}{8}$ 250%, 2.5

Reteaching

Using Models Have students go through the business sections of newspapers. They can compare different terms and rates of certificates of deposit for different banks. They can also compare interest rates on mortgages offered through different companies.

25. 50%
27. 12½%; 12.5%

Use a proportion to answer each question.

25. Thirty-five is what percent of 70?
26. What percent of 60 is 18? **30%**
27. Eight is what percent of 64?
28. Six is what percent of 15? **40%**
29. What percent of 2 is 8? **400%**
30. What percent of 14 is 4.34? **31%**

Use an equation to answer each question.

B

31. What percent of 160 is 4? **2.5%**
32. What number is 25% of 56? **14**

33. 72.3
34. 40%
35. 702.4

33. Twelve is 16.6% of what number?
34. Thirty-two is what percent of 80?
35. 17.56 is 2.5% of what number?
36. What percent of 75 is 30? **40%**

Solve.

C

37. $64.93 is what percent of $231.90? **28%**
38. Find 112% of $500. **$560**
39. Find 81% of 32. **25.92**

Use I = prt to find the missing quantity.

40. Find r if $I = \$5920$, $p = \$4000$, and $t = 3$ years. **49.3%**
41. Find r if $I = \$780$, $p = \$6500$, and $t = 1$ year. **12%**
42. Find I if $p = \$3200$, $r = 9\%$, and $t = 18$ months. **$432**
43. Find I if $p = \$5000$, $r = 12\frac{1}{2}\%$, and $t = 5$ years. **$3125**

44. 4½ years

44. Find t if $I = \$2160$, $p = \$6000$, and $r = 8\%$.
45. Find p if $I = \$756$, $r = 9\%$, and $t = 3\frac{1}{2}$ years. **$2400**

Critical Thinking

46. If x is 225% of y, then y is what percent of x? **44.4̄%**

Applications and Problem Solving

47. **Health** Physicians in the United States treated 4.1 million broken bones in 1992. The chart at the right shows the fractures by age. Find the percentage of fractures by age. **39%, 34%, 15%, 12%**

Age	Fractures
Under 18	1.6 million
18–44	1.4 million
45–64	600,000
65 and older	500,000

Source: American Academy of Orthopedic Surgeons

48. **Finance** Melanie Morgan invested $5000 for one year, part at 9% annual interest and the rest at 12% annual interest. The interest from the investment at 9% was $198 more than the interest from the investment at 12%. How much money did she invest at 9%? **$3800**

49. **Finance** Juan and Rosita Diaz have invested $2500 at 10% annual interest. They have $6000 more to invest. At what rate must they invest the $6000 to have a total of $9440 at the end of the year? **11.5%**

50. **Statistics** According to The Alden Group, desserts in the home are not always eaten in the kitchen or dining room. In fact, only 30% of desserts are eaten in the kitchen and 14% in the dining room. Eighteen percent are most often eaten in the living room, 10% in the den, and an amazing 28% are eaten in the bedroom. Make a circle graph showing these data. **See margin.**

For **Extra Practice,** see p. 765.

The red A, B, and C flags, printed only in the Teacher's Wraparound Edition, indicate the level of difficulty of the exercises.

Additional Answer

50. **Desserts Eaten in the Home**

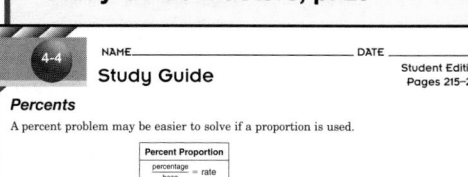

kitchen 30%
bedroom 28%
den 10%
living room 18%
dining room 14%

Study Guide Masters, p. 29

Classroom Vignette

"We looked at the Nutritional Information on food boxes and used proportions to find the percentage of fat calories in one serving. For example, a muffin mix has 2 grams of fat and 140 calories in one serving. One gram of fat is about 9 calories.

Fat calories → $\frac{18}{140}$ = % ← percentage of fat
Total calories → 140 = 100 ← total percentage

So, one serving of the muffin mix is about 13% fat calories. I encourage students to find foods that have less than 25% fat calories."

Lara Dillinger
Rock Creek High School
St. George, Kansas

4 ASSESS

Closing Activity

Modeling The form below can be used for all percent problems.

$$\frac{part}{base\ or\ whole} = \frac{\#\ of\ percent}{100}$$

Percent is based on 100. A certain part (percentage) out of the whole or base is examined. The base follows the word *of*. Percent problems give two of the three quantities: part, base, and number of percent. So, there are three types of percent problems:

1. find the part (percentage),
2. find the base, or
3. find the number of percent.

Go through the exercises to determine the type of each problem.

Practice Masters, p. 29

51. Consumerism Luke's father used his credit card to purchase a $330 television. The bank that issued the credit card charges an annual interest rate of 19.8%. Each month they charge interest on the unpaid balance of the account. If Luke's father does not make any more purchases and makes no payments, how much interest will he be charged next month? **$5.45**

52. Nutrition The table at the right lists the number of calories and grams of fat in eight popular pizza toppings. The estimates are based on two slices of a 14-inch pizza. To find the percentage of calories from fat, multiply the fat grams by 9 and divide the result by the number of calories. Which of the toppings gets the highest percentage of its calories from fat? **black olives**

Topping	Calories	Fat Grams
Extra Cheese	168	8
Sausage	97	8
Pepperoni	80	7
Black Olives	56	5
Ham	41	2
Onion	11	<1
Green Pepper	5	0
Mushrooms	5	0

53. Data Analysis A survey was taken recently to determine what Americans would serve President Clinton and his family for dinner. The survey results are shown at the right.

a. Suppose 258 people were surveyed. How many people said they would serve chicken marsala? How many said pizza?

b. The sum of the percentages in the survey is not 100%. Why do you think this is the case? **Respondents could choose more than one entree.**

What Would You Serve?	
steak and potatoes	39%
lasagna	25%
chicken marsala	21%
poached salmon	18%
hamburgers or hot dogs	15%
spaghetti and meatballs	14%
pasta primavera	13%
chef salad	10%
pizza	9%
spaghetti with tomato sauce	8%
soup and sandwiches	5%

Source: The Ragu Corporation

53a. about 54 people; about 23 people

54. Money The number of coins minted in 1994 is shown in the table at the right. Make a circle graph of the data. **See margin.**

U.S. Coins Minted in 1994	
Pennies	14,120,000,000
Nickels	1,480,000,000
Dimes	2,600,000,000
Quarters	1,760,000,000
Half-Dollars	40,000,000

Mixed Review

General Sherman

55. Trees While visiting the Sequoia National Park in California, Tyler stops near a tree called the "General Sherman." The guide tells him it is the world's largest tree according to its volume of wood. Tyler is standing about 160 feet from the tree. If he is looking at the top of the tree at a 60° angle, about how tall is the tree? (Lesson 4–3) **about 277 feet**

56. Geometry $\triangle ABC$ and $\triangle RQP$ are similar. If $a = 10$, $r = 5$, $q = 3$, and $p = 4$, find the measures of the remaining sides. (Lesson 4–2) **b = 6; c = 8**

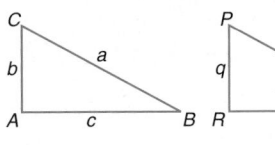

57. Solve $\frac{9}{x-8} = \frac{4}{5}$. (Lesson 4–1) $\frac{77}{4}$; **19.25**

Extension

Problem Solving At Joe's Clothing, the manager increased the price of a coat *x* percent and then reduced the price *x* percent. At the local department store, the same coat that originally sold for the same price was reduced *x* percent and then the price was increased by *x* percent. If you purchased the coat now, which store would have the cheaper price? **Both are the same.**

58. Statistics In a girl's basketball game between East High School and Monroe High School, the East players' individual scores were 12, 4, 5, 3, 11, 23, 4, 6, 7, and 8. Find the mean, median, and mode of the individual points. (Lesson 3–7) **8.3, 6.5, 4**

59. Solve $y + (-7.5) = -12.2$. (Lesson 3-1) **−4.7**

60. Diving Greg Louganis of the United States has won a record of five world titles in diving. He won the highboard title in 1978, the highboard and springboard in 1982 and 1986, and four Olympic gold medals in 1984 and 1988. The total score for a dive is the sum of the scores of each dive times the degree of difficulty of the dive. His final dive in 1988 had a degree of difficulty of 3.4. If his scores were 9, 8, and 8.5, find his total score for the dive. (Lesson 2–6) **86.7**

61. Find $5y + (-12y) + (-21y)$. (Lesson 2–5) **−28y**

62. Statistics Mr. Crable asked his students how many hours of sleep they got on the average each night for a week. He labeled the data with G for girls and B for boys. Use Gs and Bs to make a line plot of these data. (Lesson 2–2) **See margin.**

10-G	7.5-B	9-B	9-G	8-G	8.5-G	9.5-B	8.5-B	8.5-G
9.5-B	8-G	8-B	10-B	10-G	9.5-G	7.5-G	9-G	8.5-B

63. Solve $\frac{21-3}{12-3} = x$. (Lesson 1–5) **2**

64. Patterns Find the next three terms in the sequence 3, 8.5, 14, 19.5.... (Lesson 1–2) **25, 30.5, 36**

SELF TEST

Solve each proportion. (Lesson 4–1)

1. $\frac{2}{10} = \frac{1}{a}$ **5**

2. $\frac{3}{5} = \frac{24}{x}$ **40**

3. $\frac{y}{4} = \frac{y+5}{8}$ **5**

4. $\triangle WYQ$ and $\triangle MVT$ shown at the right are similar. Find the measures of the missing sides. (Lesson 4–2) **x = 15, y = 25**

Solve each right triangle. State the side lengths to the nearest tenth and the angle measures to the nearest degree. (Lesson 4–3)

5.

$\angle B = 34°, b = 11.5, c = 20.5$

6.

$\angle A = 37°, \angle B = 53°, c = 20$ mm

7. Refer to the triangles in Exercises 5 and 6. Determine whether the triangles are similar. (Lesson 4–2) **7. no**

8. Thirty-six is 45% of what number? (Lesson 4–4) **80**

9. Fifty-five is what percent of 88? (Lesson 4–4) **$62\frac{1}{2}\%$, 62.5%**

10. Research Suppose 6% of 8000 people polled regarding an election expressed no opinion. How many people had an opinion? (Lesson 4–4) **7520 people**

SELF TEST

The Self Test provides students with a brief review of the concepts and skills in Lessons 4-1 through 4-4. Lesson numbers are given to the right of exercises or instruction lines so students can review concepts not yet mastered.

Chapter 4, Quiz B (Lessons 4-3 and 4-4), is available in the *Assessment and Evaluation Masters,* p. 100.

Mid-Chapter Test (Lessons 4-1 through 4-4) is available in the *Assessment and Evaluation Masters,* p. 99.

Additional Answer

62.

Enrichment Masters, p. 29

NCTM Standards: 1–5

Instructional Resources

- Study Guide Master 4-5
- Practice Master 4-5
- Enrichment Master 4-5
- Modeling Mathematics Masters, p. 75
- Real-World Applications, 11

Transparency 4-5A contains the 5-Minute Check for this lesson; **Transparency 4-5B** contains a teaching aid for this lesson.

Recommended Pacing	
Standard Pacing	Day 8 of 15
Honors Pacing	Day 8 of 15
Block Scheduling*	Day 3 of 7 (along with Lesson 4-4)
Alg. 1 in Two Years*	Days 10 & 11 of 20

*For more information on pacing and possible lesson plans, refer to the *Block Scheduling Booklet* and *Algebra 1 in Two Years*.

1 FOCUS

5-Minute Check
(over Lesson 4-4)

Solve.

1. What number increased by 40% equals 14? **10**
2. What number increased by $33\frac{1}{3}$% equals 52? **39**
3. Twenty is 20% less than what number? **25**
4. Which earns more interest, $1000 at 8% interest for one year or $2000 at 2% interest for two years? **They both earn the same amount.**
5. Which earns more interest, $1500 at 10% interest for one year or $500 at 4% interest for ten years? **$500 at 4%**

4-5

Percent of Change

Objectives

After studying this lesson, you should be able to:

- solve problems involving percent of increase or decrease, and
- solve problems involving discounts or sales tax.

APPLICATION
Food

mallow plant

If you've ever bought a package of marshmallows, you've mostly bought air. That's right, marshmallows are 80% air! Today, marshmallows come in a variety of flavors such as vanilla, chocolate, toasted coconut, and fruit flavors like laser lime and astroberry. The marshmallow was invented in ancient Egypt and made with honey and the sap of the mallow plant that grows in swamps, or marshes. Since then, the marshmallow has gone through many changes.

Today, marshmallows are mass produced and made from cornstarch and gelatin. In 1955, the height of marshmallow making in the United States, there were 30 marshmallow companies. Today there are only a handful.

Let's say there are only six marshmallow companies today. You can write a ratio that compares the amount of decrease in marshmallow companies to the original number of companies in 1955. This ratio can be written as a percent. First, subtract to find the amount of change: $30 - 6 = 24$. Then divide by the original number of companies.

$$\begin{array}{c}\text{amount of decrease} \rightarrow \\ \text{original amount} \rightarrow\end{array} = \frac{r}{100}$$

$$24 \cdot 100 = 30 \cdot r \quad \textit{Find the cross products.}$$

$$80 = r \quad \textit{Solve for r.}$$

The amount of decrease is $\frac{80}{100}$ or 80% of the original number of companies. So, we can say that the **percent of decrease** is 80%.

The **percent of increase** can be found in a similar way.

Example ❶

CONNECTION
Biology

In 1982, the California condor was on the verge of extinction. The population had dwindled to a total of 21, making the condor the rarest bird in North America. According the Refuge Department of the U.S. Fish and Wildlife Service, however, there were 64 condors in existence in 1992 and 103 as of October 1995. Find the percent of increase in the population of California condors from 1992 to 1995.

Method 1
First, subtract to find the amount of change.

$$103 - 64 = 39$$

$$\frac{\text{amount of increase}}{\text{original number}} = \frac{39}{64}$$

$$\frac{39}{64} = \frac{r}{100}$$

$$60.9375 = r$$

The percent of increase is about 61%.

Method 2
Divide the new amount by the original amount.

$$103 \div 64 = 1.609375$$

Subtract 1 from the result and write the decimal as a percent.

$$1.609375 - 1 = 0.609375 \text{ or } 60.9\%$$

The percent of increase is about 61%.

Sometimes an increase or decrease is given as a percent, rather than an amount. Two applications of percent of change are discounts and sales tax.

Alternative Learning Styles

Auditory Have students conduct a telephone survey aimed at gathering information about how merchants determine selling prices of items. Challenge the class to review the results of the survey to determine general markup percentages used in various industries.

Example ②

APPLICATION

Sales

A discount of 20% means that the price is decreased by 20%.

Ayita received a coupon in the mail offering her a 20% discount on the price of a pair of jeans at The Limited. Ayita visited the store and found a pair of jeans she wants to buy that cost $48. What will be the discounted price?

Explore The original price is $48.00, and the discount is 20%.

Plan You want to find the amount of discount, then subtract that amount from $48.00. The result is the discounted price.

Estimate: *20% of $48 → 2(10% of $50) = 2($5) or $10.*

Solve 20% of $\$48.00 = 0.20(48.00)$ *Note that $20\% = \frac{20}{100} = 0.20$.*
$$= 9.60$$

Subtract this amount from the original price.
$$\$48.00 - \$9.60 = \$38.40$$
The discounted price will be $38.40.

Examine Solve the problem another way. The discount was 20%, so the discounted price will be 80% of the original price.   *$100\% - 20\% = 80\%$*

Find 80% of $48.00. $0.80(48.00) = 38.40$

This method produces the same discounted price, $38.40.
How does this compare to the estimate?

Example ③

APPLICATION

Sales

Sales tax of $5\frac{3}{4}\%$ means that the price is increased by $5\frac{3}{4}\%$.

Karen Danko, a sophomore at Westerville North High School, is ordering a class ring. The ring she has chosen costs $169.99. She also needs to pay a sales tax of $5\frac{3}{4}\%$. What is the total price?

Explore The price is $169.99 and the tax rate is $5\frac{3}{4}\%$.

Plan First find $5\frac{3}{4}\%$ of $169.99. Then add the result to $169.99.

Estimate: $5\frac{3}{4}\%$ *of $169.99 → $\frac{1}{2}$(10% of 170) =*
$\frac{1}{2}$ *(17) or $8.50. The total will be about*
$170 + $8.50 or $178.50.

Solve $5\frac{3}{4}\%$ of $\$169.99 = 0.0575(169.99)$ *Note that $5\frac{3}{4}\% = 0.0575$.*

$= 9.774425$ *Round 9.774425 to 9.78, since tax is always rounded up.*

Add this amount to the original price.
$$\$169.99 + \$9.78 = \$179.77$$
The total price is $179.77.

Examine The tax rate is $5\frac{3}{4}\%$, so the total price was $100\% + 5\frac{3}{4}\%$ or 105.75% of the purchase price. Find 105.75% of $169.99.
$$(1.0575)(169.99) = 179.76442$$
Thus, the total price of $179.77 is correct.
How does this compare to the estimate?

Motivating the Lesson

Situational Problem Have students consider the following. A pair of sneakers is reduced 10%. The sales tax on the sneakers is 6%. Is the final price of the sneakers equal to the original price less 4%? **No, because the sales tax is calculated on the sale price, not on the original price of the sneakers.**

2 TEACH

Teaching Tip Point out that the difference is always compared to the original price.

In-Class Examples

For Example 1
A retailer increased the price of ball-point pens from $0.40 to $0.45 each. What was the percent of increase? **12.5%**

For Example 2
Mieko received a 5% discount on a calculator. The original price was $15.00. What was the sale price? **$14.25**

For Example 3
Tim bought a pair of running shoes for $35. There was a 4% sales tax. Find the tax and the total price Tim paid.
tax = $1.40, total = 36.40

Teaching Tip In Example 2, you may wish to explain that since there is no monetary unit that represents $\frac{1}{10}$ of a cent, amounts of money are generally rounded to the nearest whole cent. If the original price of the jeans was $48.99, the discount would be $9.798, which would be rounded to $9.80.

3 PRACTICE/APPLY

Check for Understanding

Exercises 1–11 are designed to help you assess your students' understanding through reading, writing, speaking, and modeling. You should work through Exercises 1–4 with your students and then monitor their work on Exercises 5–11.

Additional Answers

1. Subtract old from new. Divide the result by old. Move decimal point 2 places to the right.
2. Multiply tax rate by original price.
3. $\frac{1773}{17,972} = \frac{r}{100}$

EXPLORATION   PROGRAMMING

Suppose a newspaper ad states that a discount store's microwave ovens are discounted 20% from the manufacturer's suggested retail price (MSRP). The store is running a special sale that says all kitchen appliances have been discounted an additional 15%. What is the final sale price of a microwave oven whose MSRP is $250?

There are two ways you might interpret the ad in the paper.

• The discounts can be successive. That is, 20% is taken off the MSRP and then 15% is taken off the resulting price.

• The discounts can be combined. So the price is 35% off the MSRP.

The graphing calculator program at the right finds the cost of an item for each interpretation of the ad. You must enter the original price and each discount amount (as a decimal) when prompted.

```
PROGRAM:DISCOUNT
: Disp "ORIG. PRICE"
: Input P
: Disp "1ST DISCOUNT?"
: Input A
: Disp "2ND DISCOUNT?"
: Input B
: P(1-A)(1-B)→C
: P(1-(A+B))→D
: Disp "SUCCESSIVE",
  "DISCOUNT",C
: Disp "COMBINED",
  "DISCOUNT",D
```

e. The combined discount always yields a lower price. Stores often use successive discounts.

Your Turn

Copy the following table and use the program to complete it.

	Price	First Discount	Second Discount	Sale Price Successive Discount	Sale Price Combined Discount
a.	$49.00	20%	10%	$35.28	$34.30
b.	$185.00	25%	10%	$124.88	$120.25
c.	$12.50	30%	12.5%	$7.66	$7.19
d.	$156.95	30%	15%	$93.39	$86.32

e. What is the relationship between the sale price using successive discounts and the sale price using combined discounts? Which one is usually used in stores?

CHECK FOR UNDERSTANDING

Communicating Mathematics

Study the lesson. Then complete the following.   1–3. See margin.

1. **Explain** how to find a percent of increase.

2. **Explain** how to find the sales tax if you know the rate of tax.

3. **Write** the equation you would use to find the percent of increase on a car that cost $17,972 in 1995 and $19,705 in 1996.

4. **Assess Yourself**   Find an advertisement or article in the newspaper that shows a percent of change. Determine whether it is a percent of increase or decrease. Explain your reasoning.   **See students' work.**

Reteaching

Using Substeps Make certain that students realize that two steps are required to solve problems involving percent of change. First, they must subtract to find the amount of change. Then, they must compare that amount, increasing or decreasing, to the *original*. In each exercise at the right, require students to show both steps to find the percent of increase or decrease.

1. old: $74.00
   new: $79.18 $5.18, 7%
2. old: $62.00
   new: $65.72 $3.72, 6%
3. old: $139.40
   new: $164.49 $25.09, 18%
4. old: $65.48
   new: $60.24 $5.24, 8%

Guided Practice

State whether each percent of change is a percent of increase or a percent of decrease. Then find the percent of increase or decrease.

5. original: $50
   new: $70   **I; 40%**

6. original: $200
   new: $172   **D; 14%**

7. original: 72 ounces
   new: 36 ounces

Find the final price of each item. When there is a discount and sales tax, first compute the discount price and then compute the sales tax and final price.

8. CD player: $149
   discount: 15%   **$126.65**

9. athletic shoes: $89.99
   discount: 10%
   sales tax: 6%   **$85.85**

10. sweater: $45
    sales tax: 6.5%
    $47.93

11. **Sales** Kevin Mason paid $205.80 for his senior class photographs. This included 5% sales tax. What was the cost of the pictures before taxes?   **$196**

EXERCISES

Practice

State whether each percent of change is a percent of increase or a percent of decrease. Then find the percent of increase or decrease. Round to the nearest whole percent.   **14. I; 162%**

A

12. original: $100
    new: $59   **D; 41%**

13. original: 324 people
    new: 549 people   **I; 69%**

14. original: 58 homes
    new: 152 homes

15. original: 66 dimes
    new: 30 dimes   **D; 55%**

16. original: $53
    new: $75   **I; 42%**

17. original: 15.6 liters
    new: 11.4 liters   **D; 27%**

B

18. original: $3.78
    new: $2.50   **D; 34%**

19. original: 231.2 mph
    new: 236.4 mph   **I; 2%**

20. original: 124 tons
    new: 137 tons   **I; 10%**

Find the final price of each item. When there is a discount and sales tax, first compute the discount price and then compute the sales tax and final price.

21. VCR: $219
    sales tax: 6.5%
    $233.24

22. jeans: $39.99
    discount: 15%
    sales tax: 4%   **$35.35**

23. book: $19.95
    discount: 5%
    sales tax: 5%

C

24. concert tickets: $52.50
    sales tax: 7%
    $56.18

25. in-line skates: $99.99
    discount: 20%
    sales tax: 6.75%   **$85.39**

26. hiking boots: $59
    discount: 10%
    sales tax: 5.5%

27. compact disc: $15.88
    sales tax: 4.5%
    $16.59

28. amusement park
    tickets: $37.50
    sales tax: 6%   **$39.75**

29. software: $29.99
    discount: 6%
    sales tax: 6.75%

Graphing Calculator

30. Use a graphing calculator to find the sale price of each item using successive discounts and then one combined discount.

    a. price, $89; discounts, 25% and 10%   **$60.08; $57.85**

    b. price, $254; discounts 30% and 14.5%   **$152.02; $140.97**

Critical Thinking

31. An amount is increased by 15%. The result is decreased by 15%. Is the final result equal to the original amount? Explain your reasoning with an example.   **No; see students' work.**

Assignment Guide

Core: 13–39 odd, 40–46
Enriched: 12–30 even, 31–46

For **Extra Practice,** see p. 765.

The red A, B, and C flags, printed only in the Teacher's Wraparound Edition, indicate the level of difficulty of the exercises.

Teaching Tip Before students work on Exercises 21–29, you may wish to discuss why it is necessary to deduct an amount of discount from an item before computing the sales tax on the item.

Study Guide Masters, p. 30

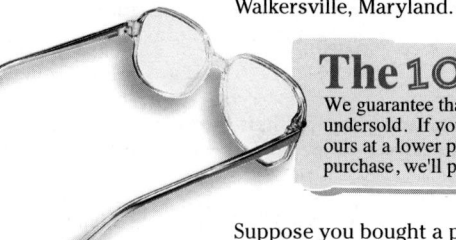

32. Sales The advertisement below was used by a local optician's office in Walkersville, Maryland.

The 100% Price Pledge!

We guarantee that we will simply not be undersold. If you find the same eyewear as ours at a lower price within 30 days of your purchase, we'll pay you **110%** of the difference.

Suppose you bought a pair of glasses from this optician for $150. You then found the same glasses from a competitor for $125. What should the optician do to honor the pledge?   pay the customer $27.50

Applications and Problem Solving

33. 133%

33. Predictions The World Future Society predicts that by the year 2020, airplanes will be able to carry 1400 passengers. Today's biggest jets can carry 600 people. What will be the percent of increase of airplane passengers?

34. Computers In 1995, America Online had about 3,000,000 users. Over the next decade, users are expected to increase from a few million to the tens of millions. Suppose the number of users increases by 150% by the year 2000. How many users will there be in the year 2000?   7,500,000 users

35. 8.6%

35. Taxes As the saying goes, time is money. The graph at the right shows a function of how much time it takes out of each eight-hour day to earn enough money to pay a day's worth of taxes. What was the percent of increase from 1970 to 1994?

36. Sales Music Systems, Inc. allows a 10% discount if a purchase is paid for within 30 days. An additional 5% discount is given if the purchase is paid for within 15 days. Brent Goodson buys a sound system that originally cost $360. If he pays the entire amount at the time of purchase, how much does he pay for his system after the successive discounts? $307.80

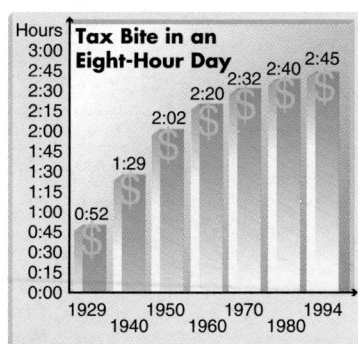

Tax Bite in an Eight-Hour Day

Hours: 3:00, 2:45, 2:30, 2:15, 2:00, 1:45, 1:30, 1:15, 1:00, 0:45, 0:30, 0:15, 0:00

1929: 0:52
1940: 1:29
1950: 2:02
1960: 2:20
1970: 2:32
1980: 2:40
1994: 2:45

Source: *The Universal Almanac*, 1995

37. Health The excerpt below is from the March 1994 issue of *Runner's World* magazine.   an increase of more than 22%

When you first get up in the morning, your muscles and soft tissues are tight. In fact, at that time your muscles are generally about 10 percent shorter than their normal resting length. As you move around, they stretch to their normal length. Then when you start to exercise, your muscles stretch even more, to about 10 percent longer than resting length. This means you have a 20 percent change in muscle length from the time you get out of bed until your muscles are well warmed up.

According to the basic laws of physics, muscles work more efficiently when they are longer; they can exert more force with less effort. This means, too, that longer muscles are much less prone to injury.

What is the percent change in muscle length from the time a runner gets up in the morning until after he or she has exercised for a while?

226   Chapter 4   *Using Proportional Reasoning*

Tech Prep

Health and Fitness Expert Students who are interested in health and fitness may wish to do further research on the data given in Exercise 37 and explore the potential growth of this career. For more information on tech prep, see the *Teacher's Handbook*.

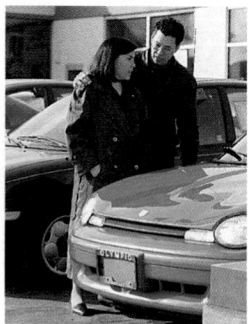

38. **Consumer Awareness** A windsuit that costs $75 is discounted 25%. The sales tax is 6%.
   a. Would it be better for the customer to have the discount taken off before adding the sales tax or after the sales tax has been added? Explain your reasoning.   **no difference**
   b. Do you think stores have a choice in which order to calculate discounts and sales tax? Explain your reasoning.   **See students' work.**

39. **Automobiles** As soon as a new car is purchased and driven away from the dealership, it begins to lose its value, or depreciate. Alonso bought a 1994 Plymouth Neon for $9559. One year later, the value of the car was $8500. What is the percent of decrease of the value of the car?   **11%**

Mixed Review

40. **Entertainment** A theater was filled to 75% capacity. How many of the 720 seats were filled?   (Lesson 4–4)   **540 seats**

41. **Geometry** A chimney casts a shadow 75 feet long when the angle of elevation of the sun is 41°. How tall is the chimney?   (Lesson 4–3)   **65 ft**

42. **Construction** To paint his house, Lonnie needs to purchase an extension ladder that reaches at least 24 feet off the ground. Ladder manufacturers recommend the angle formed by the ladder and the ground be no more than 75°. What is the shortest ladder he could buy to reach 24 feet safely? (Lesson 4–3)   **at least 25 feet long**

43. Solve $0.2x + 1.7 = 3.9$.   (Lesson 3–6)   **11**

44. **Geometry** The measures of two angles of a triangle are 38° and 41°. Find the measure of the third angle.   (Lesson 3–4)   **101°**

45. heavy, 571;
light, 286

45. **Dietetics** A hospitalized diabetic is on a carefully controlled 2000-calorie diet distributed over five feedings a day. There are two heavy feedings consisting each of $\frac{2}{7}$ of the total calories. There are also three light feedings consisting each of $\frac{1}{7}$ of the total calories. Determine the number of calories for a heavy feeding and for a light feeding.   (Lesson 2–6)

46. **Statistics** State the scale you would use to make a line plot for the following data. Then draw the line plot. 4.2, 5.3, 7.6, 9.6, 7.3, 6.7   (Lesson 2–2)
See margin.

WORKING ON THE
In·ves·ti·ga·tion

Refer to the Investigation on pages 190–191.

1 Count all of the fish in your lake. Record the actual population and your estimates on a class chart.

2 How close were your estimates? How would you account for the difference between your estimates and the actual number of fish in the lake?

3 What percent of the fish in the lake was tagged? What percent of the fish in your samples was tagged?

4 Suppose you went fishing at that lake and caught one fish. What is the probability, or chance, that the fish you caught would be tagged? Justify your reasoning.

5 Write a paragraph or two that relates proportions, percents, and probability in this situation.
You will learn more about probability in the next lesson.

Add the results of your work to your Investigation Folder.

Lesson 4–5   Percent of Change   **227**

Extension

Reasoning A $75 blazer is marked down 25%. The buyer has to pay sales tax on the regular price. How much could have been saved if the sale price was taxed instead?   **$1.12**

In·ves·ti·ga·tion

Working on the Investigation
The Investigation on pages 190–191 is designed to be a long-term project that is completed over several days or weeks. Encourage students to keep their materials in their Investigation Folder as they work on the Investigation.

Chapter 4   **227**

Instructional Resources

- Study Guide Master 4-6
- Practice Master 4-6
- Enrichment Master 4-6
- Assessment and Evaluation Masters, p. 101

Transparency 4-6A contains the 5-Minute Check for this lesson; **Transparency 4-6B** contains a teaching aid for this lesson.

Recommended Pacing

Standard Pacing	Days 9 & 10 of 15
Honors Pacing	Days 9 & 10 of 15
Block Scheduling*	Day 4 of 7
Alg. 1 in Two Years*	Days 12 & 13 of 20

*For more information on pacing and possible lesson plans, refer to the *Block Scheduling Booklet* and *Algebra 1 in Two Years.*

5-Minute Check
(over Lesson 4-5)

Solve.

1. What is 30% more than 30?
   39
2. What is 15% less than 40?
   34
3. A price decreased from $36 to $27. What is the percent of decrease? **25%**
4. A price increased from $68 to $85. What is the percent of increase? **25%**
5. An item sells for $48 after a 20% discount. What was the original price of the item? **$60**

Motivating the Lesson

Hands-On Activity Select a student randomly from the class. Ask the student, "What is the probability that you would have been selected?" Separate the class in half. Select a student from one half. Ask this student the same question.

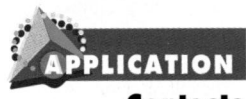

Integration: Probability
Probability and Odds

Objectives

After studying this lesson, you should be able to:
- find the probability of a simple event, and
- find the odds of a simple event.

APPLICATION
Contests

Have you ever heard anyone say that one man's trash is another man's treasure? For garbage truck driver Craig Randall of East Bridgewater, Massachusetts, the treasure was a Wendy's soft-drink cup. By peeling off a contest sticker on the discarded cup, Randall won the grand prize of $200,000 toward the purchase of a home! Since there was only one grand prize available and 24,059,900 cups were distributed, the probability of winning the grand prize was $\frac{1}{24,059,900}$.

You can calculate the chance, or **probability,** that a certain event, such as winning a particular prize, will happen. The **probability of an event** is a ratio of the number of ways a certain event can occur to the number of possible outcomes. The numerator is the number of favorable outcomes, and the denominator is the total number of possible outcomes. *The probability of an event may be written as a percent, a fraction, or a decimal.*

For example, suppose you want to know the probability of getting a 2 on one roll of a die. When you roll a die, there are six possible outcomes. Of these outcomes, only one is favorable, a 2. Therefore, the probability is $\frac{1}{6}$. We write $P(2)$ to represent *the probability of getting a 2 on one roll of a die.*

Definition of Probability	$P(\text{event}) = \dfrac{\text{number of favorable outcomes}}{\text{total number of possible outcomes}}$

Example **1**

APPLICATION
Games

Letters of Alphabet with Most Entries in *Oxford English Dictionary*

1. S
2. C
3. P
4. M
5. A

Lauren Slocum is representing her sophomore class in a *Scrabble* **tournament. She is the first player to select her seven tiles out of the 100 available tiles. The distribution of tiles is shown at the right. Find the probability of each selection below.**

Letters	Number of Tiles
J, K, Q, X, Z	1
B, C, F, H, M, P, V, W, Y, blank	2
G	3
D, L, S, U	4
N, R, T	6
O	8
A, I	9
E	12

a. an O on the very first selection
   There are eight O tiles and 100 tiles in all. $P(\text{selecting an O}) = \frac{8}{100}$ or $\frac{2}{25}$
   The probability that Lauren selects an O is $\frac{2}{25}$ or 8%.

b. an E if 5 tiles have been selected and 2 of them were Es
   There are 12 E tiles. If 2 have been chosen, 10 remain. There are $100 - 5$ or 95 tiles from which to select.
   $P(\text{choosing an E}) = \frac{10}{95}$ or $\frac{2}{19}$
   The probability of selecting an E is $\frac{2}{19}$ or about 10%.

Top Five Languages Spoken in the U.S.

	Language	Speakers
1.	English	224,900,000
2.	Spanish	19,640,000
3.	French	1,930,000
4.	German	1,750,000
5.	Italian	1,480,000

The probability that an event will occur is somewhere between 0 and 1.

- A probability of 0 means that it is impossible for an event to occur.
- A probability of 1 means that an event is certain to occur.
- A probability between 0 and 1 means that an event is neither impossible nor certain.

Based on the situations above, the probability of any event can be expressed as $0 \leq P(\text{event}) \leq 1$.

Some outcomes have an equal chance of occurring. We say that such outcomes are **equally likely.** When an outcome is chosen without any preference, we say that the outcome occurs at **random.**

 Probability

Materials: grid paper

What are the chances that two of the Oscar-winning actors or actresses from 1983 to 1993 share the same birthday? Considering that there are only 24 people in this group, you would probably say the chances are slim.

You may be surprised that in a group of only 24 people, the likelihood that two of them have the same birthday is just about 50%, or one out of two.

For example, in this group of Academy Award winners, Holly Hunter and William Hurt were both born on March 20.

Your Turn   a–d. See students' work.

a. Survey a group of 24 people in your school. Tally the results.

b. Does your group contain a shared birthday? Display your findings in a chart or graph.

c. Combine your results with those of your classmates. Does your class graph differ from your individual graph? Explain.

d. Use the graph below to find the likelihood of birthdays shared by Presidents of the United States. Research to find any shared Presidential birthdays. Compare your findings. Do you agree or disagree with the percentages? Explain your reasoning.

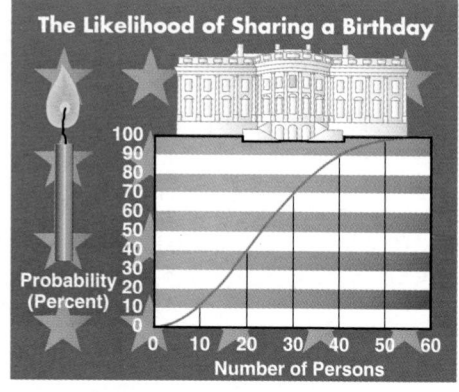

Another way to express the chance of an event's occurring is with **odds.** The odds of an event is the ratio that compares the number of ways an event can occur to the number of ways it *cannot* occur.

Definition of Odds	The odds of an event occurring is the ratio of the number of ways the event can occur (successes) to the number of ways the event cannot occur (failures).

Odds = number of successes : number of failures

Lesson 4–6   INTEGRATION *Probability   Probability and Odds*   **229**

---

2 TEACH

In-Class Example

For Example 1
Lynn collects stamps from different countries. He has five from Canada, two from France, one from Russia, four from Great Britain, and one from Germany. If he accidentally loses one stamp, what is the possibility that it is the stamp from Russia? $\frac{1}{13}$

MODELING MATHEMATICS Encourage students to follow this graphing method with other probability exercises. Survey and graph other characteristics such as height, eye color, or foot size. What is the likelihood of each characteristic?

Teaching Tip Develop student understanding of probability and odds by asking the following questions.

1. What is the meaning of the word *random*?

2. What are examples of situations that have a probability of 0? of 1?

GLENCOE *Technology*

Interactive Mathematics Tools Software

This multimedia software provides an interactive lesson that uses histograms to observe the results when considering various chance outcome activities. Students will compare the results of the mean, mode, and median of these activities. A **Computer Journal** gives students an opportunity to write about what they have learned.

For Windows & Macintosh

For Example 2

The door prize at a party with 25 people is given by writing numbers 1 through 25 on the bottom of the paper plates used.

a. What is the probability that an individual had the winning plate? **1:25**

b. What are the odds of winning the door prize? **1:24**

c. What are the odds of not winning the door prize? **24:1**

For Example 3

The probability that Jim Jackson will make his next free throw is 75% or $\frac{3}{4}$. Find the odds that he will not make his next free throw. **1:3**

---

3 PRACTICE/APPLY

Check for Understanding

Exercises 1–12 are designed to help you assess your students' understanding through reading, writing, speaking, and modeling. You should work through Exercises 1–5 with your students and then monitor their work on Exercises 6–12.

Study Guide Masters, p. 31

NAME _____ DATE _____
Study Guide   Student Edition
                               Pages 228–232

Integration: Probability
Probability and Odds

The **probability** of an event is a ratio that tells how likely it is that the event will take place.

Definition of Probability
$P(\text{event}) = \frac{\text{number of favorable outcomes}}{\text{number of possible outcomes}}$

Example 1: Mr. Babcock picks 5 of the 25 students in his algebra class at random for a special project. What is the probability of being picked?

$$P(\text{being picked}) = \frac{\text{number of students picked}}{\text{total number of students}}$$

The probability of being picked is $\frac{5}{25}$ or $\frac{1}{5}$.

The probability of any event has a value from 0 to 1. If the probability of an event is 0, it is impossible for the event to occur. An event that is certain to occur has a probability of 1. This can be expressed as $0 \le P(\text{event}) \le 1$.

The odds of an event occurring is the ratio of the number of ways an event can occur (successes) to the number of ways the event cannot occur (failures).

Definition of Odds
Odds = $\frac{\text{number of successes}}{\text{number of failures}}$

Example 2: Find the odds that a member of Mr. Babcock's class will be picked for the special project.

Number of successes: 5   Number of failures: 20
Odds of being picked = number of successes : number of failures
= 5:20, or 1:4

Solve each problem.

1. There were 2 colas, 1 ginger ale, 5 cherry sodas, and 4 root beers in the cooler. What is the probability of pulling out a cola? $\frac{1}{6}$

2. It will rain 8 times in November and snow 3 times. The other days it will be sunny. What is the probability of sun? the odds of sun? $\frac{19}{30}$; 19:11

There is a bowl of money at the carnival. The bowl contains 50 quarters, 75 dimes, 100 nickels, and 125 pennies.

3. If one coin is chosen, what is the probability that a quarter will be chosen? $\frac{1}{7}$

4. What would be the odds of choosing a quarter if all the dimes were removed first? 2:9

5. What are the odds that a dime will not be chosen? 11:3

6. What are the odds of choosing a penny? 5:9

Example **2**

Jersey Mike's Submarine Shop has a business card drawing for a free lunch every Tuesday. Four coworkers from Invo Accounting put their business cards in the bowl for the drawing. If 80 cards were in the bowl, what are the odds that one of the coworkers will win a free lunch?

The coworkers have 4 of the 80 business cards. Thus, there are $80 - 4$ or 76 business cards that will not be winning cards for the coworkers.

Odds of winning = $\frac{\text{number of chances of}}{\text{drawing winning card}} : \frac{\text{number of chances of}}{\text{drawing other cards}}$
= 4:76 or 1:19   *1:19 is read "1 to 19".*

Example **3**

Do you have a better chance of winning a prize or not winning a prize?

Refer to the application at the beginning of the lesson. The probability of winning any prize or food discount in Wendy's peel-off sticker contest was 20% or $\frac{1}{5}$. Find the odds of winning a prize or discount.

If the probability of winning a prize or discount is $\frac{1}{5}$, then the number of successes (prize) is 1, while the total number of outcomes is 5. This means that the number of failures (no prize) must be $5 - 1$ or 4.

Odds of prize = number of successes : number of failures
= 1:4   *This is read "1 to 4."*

CHECK FOR UNDERSTANDING

Communicating Mathematics

Study the lesson. Then complete the following.   4–5. See students' work.

1. **Refer** to Example 1. Find the probability of selecting a Z if 25 tiles have been selected and one of them was a Z.   **0**

2. 7; a natural number less than 7

2. **Give** examples of an impossible event and a certain event when rolling a die.

3. **Tell** what the odds would be for an event not occurring if the odds for the event occurring are 3:5.   **5:3**

4. **Write** a problem in which the answer will be a probability of $\frac{3}{4}$.

5. Choose another group of 24 people to survey. Record their birthdays. Does your group contain a shared birthday?

Guided Practice

Determine the probability of each event.

6. This is an algebra book.   **1**   7. A coin will land tails up. $\frac{1}{2}$

Find the probability of each outcome if a die is rolled.

8. a 6 $\frac{1}{6}$ 9. a number greater than 2 $\frac{2}{3}$

Find the odds of each outcome if a die is rolled.

10. a number greater than 2   **2:1**   11. not a 3   **5:1**

12. If the probability that an event will occur is $\frac{3}{7}$, what are the odds that the event will *not* occur?   **4:3**

Reteaching

Using Probability Have students work in pairs. Have one student present an experiment and ask the probability of a certain event happening. Have the other student find the probability and then ask the first student the probability of another event happening in that same experiment, and so on. Have students record their questions and answers.

Practice

Determine the probability of each event.

13. A coin will land heads up. $\frac{1}{2}$

14. There is a January 1 this year. 1

15. A baby will be a boy. $\frac{1}{2}$

16. Pigs will fly. 0

17. $\frac{1}{2}$ 18. $\frac{1}{12}$

17. You will roll a multiple of 2 on a die.

18. A person's birthday is in June.

Find the probability of each outcome if a computer randomly chooses a letter in the word "probability."

19. the letter b $\frac{2}{11}$

20. $P(not\ i)$ $\frac{9}{11}$

21. the letter e 0

22. $P(vowel)$ $\frac{4}{11}$

23. *not b* or *y* $\frac{8}{11}$

24. the letters *a* or *t* $\frac{2}{11}$

Find the odds of each outcome if you randomly select a coin from a jar containing 80 pennies, 100 nickels, 70 dimes, and 50 quarters.

25. selecting a quarter 1:5

26. selecting a nickel 1:2

27. selecting a penny 4:11

28. *not* selecting a nickel 2:1

29. *not* selecting a dime 23:7

30. selecting a penny or dime 1:1

A card is selected at random from a standard deck of 52 cards.

31. What is the probability of selecting a red card? $\frac{1}{2}$

32. What is the probability of selecting a queen? $\frac{1}{13}$

33. What are the odds of selecting a heart? 1:3

34. What are the odds of *not* selecting a black 6? 25:1

35. $\frac{8}{13}$

35. If the odds that an event will occur are 8:5, what is the probability that the event will occur?

36. If the probability that an event will occur is $\frac{2}{3}$, what are the odds that the event will occur? 2:1

Critical Thinking

37. Geometry If a point inside the figure at the right is chosen at random, what is the probability that it will be in the shaded region? $\frac{1}{7}$

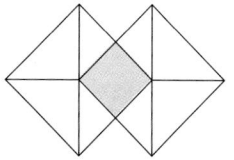

38. Geometry If a point inside the figure at the right is chosen at random, what is the probability that it will *not* be in the shaded region? $\frac{6}{19}$

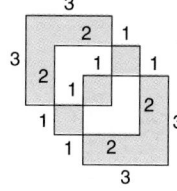

Assignment Guide

Core: 13–41 odd, 42–49
Enriched: 14–36 even, 37–49

For **Extra Practice,** see p. 766.

The red A, B, and C flags, printed only in the Teacher's Wraparound Edition, indicate the level of difficulty of the exercises.

Practice Masters, p. 31

4-6 NAME_____ DATE_____
Practice Student Edition Pages 228–232

Integration: Probability
Probability and Odds

Find the probability of each outcome if a die is rolled.

1. a 1 $\frac{1}{6}$

2. a number less than 3 $\frac{1}{3}$

3. an odd number $\frac{1}{2}$

4. a number greater than 6 0

5. an even number less than 6 $\frac{1}{3}$

6. a number greater than 0 1

Find the odds of each outcome if a die is rolled.

7. a multiple of 2 1:1

8. a number less than 2 1:5

9. a factor of 6 2:1

10. a 5 1:5

11. not a 1 5:1

12. a number greater than 5 1:5

A card is selected at random from a standard deck of 52 cards.

13. What is the probability of selecting a red card? $\frac{1}{2}$

14. What is the probability of selecting an ace? $\frac{1}{13}$

15. What are the odds of selecting a diamond? 1:3

16. What are the odds of not selecting a black 2? 25:1

The number of male and female doctors in Camron are listed by age in the table at the right. Use this data to answer the following questions.

17. What is the probability that a doctor chosen is in the 45-54 age group? $\frac{1}{6}$

18. What is the probability that a doctor chosen is a female? $\frac{31}{132}$

Camron Doctors		
Age	Male	Female
Under 35	29	18
35–44	36	8
45–54	19	3
55–64	17	2

19. What are the odds that a doctor chosen is a female under 35 years? 3:19

20. What are the odds that a doctor chosen is a male between the ages of 35 and 54 years? 5:7

Closing Activity

Speaking Suppose you have seven notebooks, three of which are blue, one red, two green, and one yellow. If you select one at random, do you have a greater chance of choosing any one color? What two colors do you have an equal chance of selecting? **yes, blue; red and yellow**

Chapter 4, Quiz C (Lessons 4-5 and 4-6), is available in the *Assessment and Evaluation Masters*, p. 101.

Additional Answer

48a.

Stem	Leaf
1	6 7 7 8 8 8 8 8 9 9 9 9 9 9
2	0 0 2 3 4 4 5 5 6 7 7
3	0 2 3 3 5 6
4	5 8
5	5

$3|5 = 35$

Enrichment Masters, p. 31

Geometric Probability

If a dart, thrown at random, hits a triangular board shown at the right, what is the probability that it will hit the shaded region? This probability can be determined by comparing the area of the shaded region to the area of the board. This ratio indicates what fraction of the tosses should hit in the shaded region.

$$\frac{\text{area of shaded region}}{\text{area of triangular board}} = \frac{\frac{1}{2}(4)(6)}{\frac{1}{2}(8)(6)}$$

$$= \frac{12}{24} \text{ or } \frac{1}{2}$$

In general, if S is a subregion of some region R, then the probability, $P(S)$, that a point, chosen at random, belongs to subregion S is given by the following.

$$P(S) = \frac{\text{area of subregion } S}{\text{area of region } R}$$

Find the probability that a point, chosen at random, belongs to the shaded subregions of the following regions.

1. $\frac{1}{2}$ 2. $\frac{5}{9}$ 3. $\frac{\pi}{4}$

4. $\frac{1}{2}$ 5. $\frac{\sqrt{3}}{4}$ 6. $\frac{2}{3}$

7.   8.   9. $\frac{3}{4}$

Applications and Problem Solving

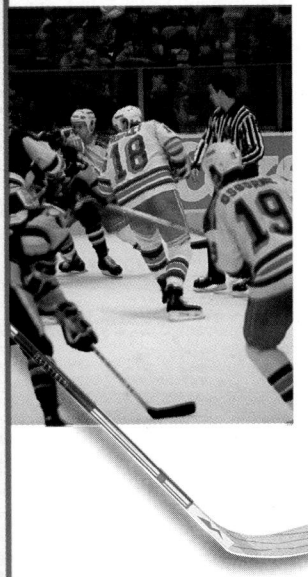

39a. 1:14   39b. $\frac{1}{27}$

39. Games The game board on *Jeopardy!* is divided into 30 squares. There are six categories with five answers in each category. In the Double Jeopardy! round, 2 Daily Double squares are hidden among the 30 squares.
  a. What are the odds of choosing a Daily Double square on the first selection?
  b. If you are making the fourth selection of the Double Jeopardy! round, what is the probability of choosing a Daily Double square if one has been chosen?

40. School Have you ever had to make an oral presentation before your classmates? Did the teacher tell everyone to be prepared and then choose the first day's presenters at random?
  a. If there are 35 students in your class, and four students will be chosen to give a presentation each day until everyone has made a presentation, what is the probability that you will be chosen to make your presentation the first day?
  b. If you have not been chosen by the fourth day, what is the probability that you will be chosen on the fourth day?

41. Statistics The stem-and-leaf plot at the right shows the total points earned by each of the 26 teams in the National Hockey League for the 1993–1994 season.
  a. What is the probability that a team earned 97 points?
  b. What is the probability that a team earned less than 80 points?
  c. What are the odds in favor of a team scoring more than 100 points? **3:23**

Stem	Leaf
11	2
10	0 1 6
9	1 5 6 7 7 7 8
8	0 2 3 4 5 7 8
7	1 1 6
6	3 4 6
5	7
4	
3	7 $9|1 = 91$

Mixed Review

40a. $\frac{4}{35}$ **40b.** $\frac{4}{23}$

41a. $\frac{3}{26}$ **41b.** $\frac{4}{13}$

42. Chocolate Illinois is the number one candy-producing state. The Little Chocolatier shop in Sterling, Illinois, is known for its pecan dumplings. The owners used to make nine pounds of pecan dumplings at one time. Now they make 300 pounds in the same amount of time! Find the percent of change. (Lesson 4–5) **3233%**

43. Geometry In a parking garage, there are 20 feet between each level. Each ramp to a level is 130 feet long. Find the measure of the angle of elevation of each ramp. (Lesson 4–3) **9°**

44. Solve $\frac{5}{8}x + \frac{3}{5} = x$. (Lesson 3–5) $\frac{8}{5}$; **1.6**

45. Work Backward Kristen spent one fifth of her money for gasoline. Then she spent half of what was left for a haircut. She bought lunch for $7. When she got home, she had $13 left. How much did Kristen have originally? (Lesson 3–3) **$50**

46. Simplify $\pm\sqrt{1764}$. (Lesson 2–8) **±42**

47. Find $\frac{17}{21} + \left(-\frac{13}{21}\right)$. (Lesson 2–5) $\frac{4}{21}$

48. School Each number below represents the age of a student in Ms. Wallace's evening calculus class at DeSantis Community College. (Lesson 1–4)
22 17 25 24 19 27 33 16 35 26 20 18 24 33 18 19 48
36 19 23 55 18 18 19 27 18 19 25 17 32 19 45 19 20 30
  a. Make a stem-and-leaf plot of the data. **See margin.**
  b. How many people attend Ms. Wallace's class? **35 people**
  c. What is the difference in ages between the oldest and youngest person in class? **39 years**
  d. What is the most common age for a student in the class? **19 years**
  e. Which age group is most widely represented in the class? **teens**

49. Evaluate $8(a - c)^2 + 3$ if $a = 6$, $b = 4$, and $c = 3$. (Lesson 1–3) **75**

Extension

Connections Twenty-one Little League baseball teams each played seven games. The stem-and-leaf plot shows the sum of the runs scored by each team in the seven games. Use this to answer each question.

Stem	Leaf
6	0 2 3
7	1 5 8 9
8	2 4 5 7 9
9	5 7 8 8 8
10	1 4 6 7

$7|5 = 75$

1. What is the probability that a team scored 98 runs? $\frac{1}{7}$

2. What is the probability that a team scored fewer than 80 runs? $\frac{1}{3}$

3. What are the odds in favor of a team scoring more than 100 runs? **4:17**

4-7

Weighted Averages

Objectives

After studying this lesson, you should be able to:

- solve mixture problems, and
- solve problems involving uniform motion.

APPLICATION
School

In Mr. Calloway's American History class, semester grades are based on five unit exams, one semester exam, and a long-term project. Each unit exam is worth 15% of the grade, the semester exam is worth 20% of the grade, and the project is worth 5% of the grade.

Semester Grades	
Item	Score
Unit Exam A	79%
Unit Exam B	83
Unit Exam C	96
Unit Exam D	91
Unit Exam E	89
Semester Exam	90
Project	95

Parker's scores for the semester are given in the table at the right. If 100% is a perfect score for each item, find Parker's average for the semester.

In order to find Parker's average for the semester, you may want to add the percentages and then divide by the number of items, 7. But this assumes that each has the same weight. So, you need to find the **weighted average** of the scores.

Definition of Weighted Average	The weighted average *M* of a set of data is the sum of the product of each number in the set and its weight divided by the sum of all the weights.

You can find the weighted average for the application above by multiplying each score by the percentage of the semester grade that it represents and then dividing by the sum of the weights, or 100.   *5(15) + 1(20) + 1(5) = 100*

$$M = \frac{15(79 + 83 + 96 + 91 + 89) + 20(90) + 5(95)}{100}$$

$$= \frac{8845}{100} \text{ or } 88.45$$

So, Parker's average for the semester is about 88%.

Mixture problems involve weighted averages. In a mixture problem, the weight is usually a price or a percentage of something.

Example ❶

PROBLEM SOLVING
Make a Chart

Suppose the Central Perk coffee shop sells a cup of espresso for $2.00 and a cup of cappuccino for $2.50. On Friday, Rachel sold 30 more cups of cappuccino than espresso, and she sold $178.50 worth of espresso and cappuccino. How many cups of each were sold?

Explore   Let *e* represent the number of cups of espresso sold. Then *e* + 30 represents the number of cups of cappuccino sold.

(continued on the next page)

Lesson 4–7   Weighted Averages   **233**

4-7 LESSON NOTES

NCTM Standards: 1–5

Instructional Resources

- Study Guide Master 4-7
- Practice Master 4-7
- Enrichment Master 4-7
- Modeling Mathematics Masters, pp. 45–47
- Real-World Applications, 12

 Transparency 4-7A contains the 5-Minute Check for this lesson; **Transparency 4-7B** contains a teaching aid for this lesson.

Recommended Pacing	
Standard Pacing	Days 11 & 12 of 15
Honors Pacing	Days 11 & 12 of 15
Block Scheduling*	Day 5 of 7
Alg. 1 in Two Years*	Days 14 & 15 of 20

*For more information on pacing and possible lesson plans, refer to the *Block Scheduling Booklet* and *Algebra 1 in Two Years*.

  ### 5-Minute Check
(over Lesson 4-6)

Find the probability of each event when a single die is rolled.

1. 3 $\frac{1}{6}$
2. even number $\frac{1}{2}$
3. less than 3 $\frac{1}{3}$

Find the probability of each event when 5 coins are tossed.

4. 5 heads $\frac{1}{32}$
5. 1 head $\frac{5}{32}$
6. The odds of an event not occurring are 7:3. What is the probability that the event will occur? $\frac{3}{10}$

Chapter 4   **233**

Questioning Open the lesson by asking what units of measure should be used for each rate described below.

1. the speed of a person running
2. the speed of a caterpillar
3. the rate of growth of a child
4. the speed of light

2 TEACH

In-Class Examples

For Example 1
Barrett's bookstore sells pencils for $0.10 each and erasers for $0.15 each. Last Tuesday, the store sold 17 more pencils than erasers for a total of $23.45. How many of each item were sold? **104 pencils, 87 erasers**

For Example 2
Zina has 48 mL of a solution that is 25% acid. How much water should she add to obtain a solution that is 15% acid? **32 mL of water**

Teaching Tip For Examples 1 and 2, emphasize to students that in mixture problems they are trying to find the perfect blend or balance of several parts to achieve the desired result.

Plan   Make a chart of the information.

	Number of Cups	Price Per Cup	Total Price
espresso	e	$2.00	$2e$
cappuccino	$e + 30$	$2.50	$2.5(e + 30)$

Solve

$$\underbrace{\text{total sales of espresso}} + \underbrace{\text{total sales of cappuccino}} = \underbrace{\text{total sales}}$$

$$2e + 2.5(e + 30) = 178.50$$
$$2e + 2.5e + 75 = 178.50$$
$$4.5e + 75 = 178.50$$
$$4.5e = 103.50$$
$$e = 23$$

There were 23 cups of espresso sold. There were $23 + 30$, or 53 cups of cappuccino sold.

Examine If 23 cups of espresso were sold, the total sales of those cups of coffee would be $23(\$2.00)$ or $46.00. If 53 cups of cappuccino were sold, the total sales of those cups of coffee would be $53(\$2.50)$ or $132.50.

Since $46.00 + 132.50 = 178.50$, the solution is correct.

Sometimes mixture problems are expressed in terms of percents.

Example

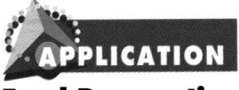
Food Preparation

An advertisement for an orange drink claims that the drink contains 10% orange juice. Jamel needs 6 quarts of the drink to serve at a party and he wants the drink to contain 40% orange juice. How much of the 10% drink and pure orange juice should Jamel mix to obtain 6 quarts of a mixture that contains 40% orange juice?

Explore Let p represent the amount of pure juice to be added.

Plan   Make a chart of the information.

	Quarts	Amount of Orange Juice
10% juice	$6 - p$	$0.10(6 - p)$
Pure juice	p	$1.00p$
40% juice	6	$0.40(6)$

Solve

$$\underbrace{\begin{array}{c}\text{amount of orange}\\\text{juice in 10\% juice}\end{array}} + \underbrace{\begin{array}{c}\text{amount of orange}\\\text{juice in pure juice}\end{array}} = \underbrace{\begin{array}{c}\text{amount of orange}\\\text{juice in 40\% juice}\end{array}}$$

$$0.10(6 - p) + 1.00p = 0.40(6)$$
$$0.6 - 0.1p + 1.00p = 2.4$$
$$0.9p = 1.8$$
$$p = 2$$

Jamel needs to combine 2 quarts of pure orange juice with $6 - 2$ or 4 quarts of 10% juice to obtain a 6-quart mixture that is 40% orange juice. *Examine this solution.*

Motion problems are another application of weighted averages. When an object moves at a constant speed, or rate, it is said to be in **uniform motion**. The formula $d = rt$ is used to solve uniform motion problems. In the formula, d represents distance, r represents rate, and t represents time. You can also use equations and charts when solving motion problems.

Example ③

APPLICATION
Travel

On Friday, Shenae and her brother Rafiel went to visit their grandparents for the weekend. Luckily, traffic was light and they were able to make the 50-mile trip in exactly one hour. On Sunday, they weren't so lucky. The trip home took exactly two hours. What was their average speed for the round trip?

To find the average speed for each leg of the trip, rewrite $d = rt$ as $r = \frac{d}{t}$.

Going	Returning
$r = \frac{d}{t}$	$r = \frac{d}{t}$
$= \frac{50 \text{ miles}}{1 \text{ hour}}$ or 50 miles per hour	$= \frac{50 \text{ miles}}{2 \text{ hours}}$ or 25 miles per hour

You may think that the average speed of the trip would be $\frac{50 + 25}{2}$ or 37.5 miles per hour. However, Shenae did not drive at these speeds for equal amounts of time. You can find the weighted average for their trip.

Round Trip

$$M = \frac{50(1) + 25(2)}{3}$$

$$= \frac{100}{3} \text{ or } 33\frac{1}{3}$$

Their average speed was $33\frac{1}{3}$ miles per hour.

Example ④

APPLICATION
Transportation

Two city buses leave their station at the same time, one heading east and the other heading west. The eastbound bus travels at 35 miles per hour, and the westbound bus travels at 45 miles per hour. In how many hours will they be 60 miles apart?

Explore   Draw a diagram to help analyze the problem.

60 miles

Westbound bus   Bus Station   Eastbound bus
45 mph   35 mph

Plan   Organize the information in a chart. Let t represent the number of hours until they are 60 miles apart. Remember that $rt = d$.

Bus	r	t	d	
Eastbound	35	t	35t	*The eastbound bus travels 35t miles.*
Westbound	45	t	45t	*The westbound bus travels 45t miles.*

$$\underbrace{\text{eastbound distance}}_{35t} + \underbrace{\text{westbound distance}}_{45t} = \underbrace{\text{total distance}}_{60}$$

(continued on the next page)

In-Class Examples

For Example 3
Lisa and Jamaal traveled 20 miles in $\frac{1}{2}$ hour. What was their average speed?   **40 mph**

For Example 4
Suppose John and Gina leave at the same time traveling in the same direction. Gina drives at a rate of 85 km/h and John at 70 km/h. How long until they are 90 km apart?   **6 hours**

Teaching Tip For Examples 3 and 4, remind students that the actual solution of a uniform motion problem is a number with an appropriate unit of measure, not the equation containing the isolated variable.

Classroom Vignette

"For Example 4, have two students stand back to back, headed away from each other. Start them walking at the same time. Clap loudly when they are "60 miles" apart. They will quickly come up with the fact that the time is the same and that the *eastbound distance + westbound distance = total distance*."

William Brinkman
Paynesville High School
Paynesville, Minnesota

Check for Understanding

Exercises 1–7 are designed to help you assess your students' understanding through reading, writing, speaking, and modeling. You should work through Exercises 1–3 with your students and then monitor their work on Exercises 4–7.

Additional Answers

1. d = distance, r = rate, t = time
2. to organize information and model the situation

Study Guide Masters, p. 32

Solve $35t + 45t = 60$
$80t = 60$
$t = \frac{3}{4}$

In $\frac{3}{4}$ hour, or 45 minutes, the buses will be 60 miles apart.

Examine To check the answer, find the distance each bus could travel in $\frac{3}{4}$ hour and see if it totals 60 miles.

$\left(\frac{3}{4}h\right)(35 \text{ mph}) + \left(\frac{3}{4}h\right)(45 \text{ mph}) \overset{?}{=} 60 \text{ mi}$

$26\frac{1}{4} \text{ mi} + 33\frac{3}{4} \text{ mi} \overset{?}{=} 60 \text{ mi}$

$60 \text{ mi} = 60 \text{ mi}$ ✔

CHECK FOR UNDERSTANDING

Communicating Mathematics

Study the lesson. Then complete the following.   1–2. See margin.

1. **Tell** what the d, r, and t represent in the formula $d = rt$.
2. **Explain** why it is sometimes helpful to use charts and diagrams.

MATH JOURNAL

3. **Describe** your favorite mode of transportation. Then write a problem using the formula $d = rt$ and your favorite mode.   **See students' work.**

Guided Practice

4. **Chemistry** Joshua is doing a chemistry experiment that calls for a 30% solution of copper sulfate. He has 40 mL of 25% solution. How many milliliters of 60% solution should Joshua add to obtain the required 30% solution?   **about 6.7 mL**

	Amount of Solution (mL)	Amount of Copper Sulfate
25% solution	40	0.25(40)
60% solution	x	$0.60x$
30% solution	$40 + x$	$0.30(40 + x)$

5. 226 dozen chocolate chip; 311 dozen peanut butter

5. **Sales** The Cookie Crumbles Company sells two kinds of cookies daily: peanut butter at $6.50 per dozen and chocolate chip at $9.00 per dozen. Yesterday, Cookie Crumbles sold 85 dozen more peanut butter than chocolate chip cookies. The total sales for both were $4055.50. How many dozen of each were sold?

6. **Air Travel** An airplane flies 1000 miles due east in 2 hours and 1000 miles due south in 3 hours. What is the average speed of the airplane?   **400 mph**

7. **Boating** *The Yankee Clipper* leaves the pier at 9:00 A.M. at 8 knots (nautical miles per hour). A half hour later, *The Riverboat Rover* leaves the same pier in the same direction traveling at 10 knots. At what time will *The Riverboat Rover* overtake *The Yankee Clipper*?   **11:30 A.M.**

EXERCISES

Applications and Problem Solving

A

B

C

14. 5 adult tickets, 3 student tickets

8. **Sales** The Madison Local High School marching band sold gift wrap to earn money for a band trip to Orlando, Florida. The gift wrap in solid colors sold for $4.00 per roll, and the print gift wrap sold for $6.00 per roll. The total number of rolls sold was 480, and the total amount of money collected was $2340. How many rolls of each kind of gift wrap were sold? **270 solid; 210 print**

	Number of Rolls	Price Per Roll	Total Price
Solid	r	$4	$4r$
Print	$480 - r$	$6	$6(480 - r)$

9. **Money** Rochelle has $2.55 in dimes and quarters. She has eight more dimes than quarters. How many quarters does she have? **5 quarters**

10. **Sales** The Nut House sells walnuts for $4.00 a pound and cashews for $7.00 a pound. How many pounds of cashews should be mixed with 10 pounds of walnuts to obtain a mixture that sells for $5.50 a pound? **10 pounds**

11. **Travel** Ryan and Jessica Wilson leave their home at the same time, traveling in opposite directions. Ryan travels at 57 miles per hour and Jessica travels at 65 miles per hour. In how many hours will they be 366 miles apart? **3 hours**

12. **Travel** At 7:00 A.M., Brooke leaves home to go on a business trip driving 35 miles per hour. Fifteen minutes later, Bart discovers that Brooke forgot her presentation materials. He drives 50 miles per hour to catch up with her. If Bart is delayed 30 minutes with a flat tire, when will he catch up with Brooke? **9:30 A.M.**

13. **Cycling** Two cyclists begin traveling in the same direction on the same bike path. One travels at 20 miles per hour, and the other at 14 miles per hour. After how many hours will they be 15 miles apart? **2.5 hours**

14. **Entertainment** A local radio station is having a drawing to win tickets to see En Vogue in concert. The winners will be driven to the concert in a limousine that holds eight passengers. Concert tickets cost $22 for adults and $19 for students. After renting the limousine, the radio station has $167 left in their giveaway budget to buy the eight tickets. How many adult tickets and student tickets can the station buy?

15. **Travel** Pablo is driving 40 miles per hour. After he has driven 30 miles, his brother Ricardo starts driving in the same direction. At what rate must Ricardo drive to catch up with Pablo in 5 hours? **46 mph**

16. **Automotives** A car radiator has a capacity of 16 quarts and is filled with a 25% antifreeze solution. How much must be drained off and replaced with pure antifreeze to obtain a 40% antifreeze solution? **3.2 quarts**

17. **Law Enforcement** A state patrol officer is chasing a car he believes is speeding. The officer is traveling 70 miles per hour to try to catch up with the car. But he can't say later that the other car was going 70 mph because the driver could ask, "If we were going the same speed, how did you catch up? Obviously I was going slower than you. I wasn't speeding." The officer sees the driver pass a mile marker, one-fourth of a mile ahead. From the marker, it takes the officer five minutes $\left(\frac{1}{12}\text{ of an hour}\right)$ to catch up. How fast was the driver going? **67 mph**

Lesson 4-7 Weighted Averages **237**

Reteaching

Using Discussion Have students discuss each problem in groups of two or three. Have them talk through what information is given and what is needed. Have each group develop a plan of attack; then have students solve the exercises individually.

Tech Prep

Law Enforcement Students who are interested in law enforcement may wish to do further research on the data provided in Exercise 17 and explore the potential growth of this career. For more information on tech prep, see the *Teacher's Handbook*.

Assignment Guide

Core: 9–21 odd, 23–29
Enriched: 8–20 even, 21–29

For **Extra Practice,** see p. 766.

The red A, B, and C flags, printed only in the Teacher's Wraparound Edition, indicate the level of difficulty of the exercises.

Practice Masters, p. 32

4-7 NAME_____ DATE _____

Practice Student Edition Pages 233–238

Weighted Averages

Solve.

1. **Entertainment** At a spring concert, tickets for adults cost $4.00 and tickets for students cost $2.50. How many of each kind of tickets were purchased if 125 tickets were bought for $413? **67 adult tickets; 58 student tickets**

2. Two trains leave Raleigh at the same time, one traveling north, the other south. The first train travels at 50 miles per hour and second at 60 miles per hour. In how many hours will the trains be 275 miles apart? **2.5 h**

3. A train leaves a city heading west and travels at 50 miles per hour. Three hours later, a second train leaves from the same place and travels in the same direction at 65 miles per hour. How long will it take for the second train to overtake the first train? **10 h**

4. **Sales** Susan wants to mix 10 pounds of Virginia peanuts that cost $3.50 a pound with Spanish peanuts that cost $3.00 a pound to obtain a mixture that costs $3.40 a pound. How many pounds of Spanish peanuts should she use? **2.5 lb**

5. A ship travels from New York to Southampton, England at 33 miles per hour. A plane traveling 605 miles per hour takes 104 hours less time to make the same trip. How far apart are the two cities? **3630 mi**

6. **Advertising** An advertisement for a pineapple drink claims that the drink contains 15% pineapple juice. How much pure pineapple juice would have to be added to 8 quarts of the drink to obtain a mixture containing 50% pineapple juice? **5.6 qt**

Chapter 4 **237**

4 ASSESS

Closing Activity

Modeling Have each student create a diagram that presents a motion problem. An example might be drawing two cars moving in opposite directions with labels showing the starting time of each as well as their rates. Tell students to switch drawings and then create and solve a motion problem from their classmate's diagram.

Additional Answer

21. Sample answer: How much pure antifreeze must be mixed with a 20% solution to produce 40 quarts of a 28% solution?

---

18. **Travel** An express train travels 80 kilometers per hour from Ironton to Wildwood. A local train, traveling at 48 kilometers per hour, takes 2 hours longer for the same trip. How far apart are Ironton and Wildwood? **240 km**

19. **Colleges** Emilio is trying to decide what college he would like to attend. He has determined his five most important factors in choosing a school. A score of 1 is the lowest and 4 is the highest. The table at the right shows his ratings for one school he is considering.

	Importance Factor (1 to 4)	Rating (1 to 4)	Total
Distance from home	4	4	16
Academic reputation	3	2	6
Social climate	3	1	3
Quality of dorms	1	4	4
Community	2	3	6

a. Find the weighted average of his ratings for this school. **7**
b. What is the highest rating that one of Emilio's schools can receive? **16**

20. **School** Many schools base a student's grade point average, or GPA, on the grade a student receives as well as on the number of credits received for the class. Mercedes' grades for this semester are listed in the table at the right. Find Mercedes' GPA if a grade of A equals 4 and a B equals 3. **3.56**

Grade Card		
Class	Credits Rating	Grade
Honors Algebra	1	A
Biology	1	B
English 1	1	A
Spanish 2	1	B
Phys. Ed.	$\frac{1}{2}$	A

Critical Thinking

21. See margin.

21. Write a mixture problem for the equation $1.00x + 0.20(40 - x) = 0.28(40)$.

22. Monica Morrison drove to work on Monday at 40 miles per hour and arrived one minute late. She left at the same time on Tuesday, drove at 45 miles per hour, and arrived one minute early. How far does Monica drive to work? **12 miles**

Mixed Review

23. **Probability** Find the probability of getting a number less than 1 if a die is rolled. (Lesson 4–6) **0**

24. **Baseball** In 1994, DeMarini Sports Inc. sold 7500 aluminum softball bats to softball enthusiasts. In 1995, the company expected to increase sales by 60%. How many bats did the company expect to sell in 1995? (Lesson 4–5)

24. 12,000 bats
26. 58 feet

25. **Finance** Selena Cruz wants to invest a portion of her $16,000 savings at 8% annual interest and the balance at a safer 5% annual interest. She hopes to earn $1130 in the next year to pay for a Caribbean cruise. How much money should she invest at 8%? (Lesson 4–4) **$11,000**

26. **Model Trains** Model trains are scaled-down replicas of real trains that come in a variety of sizes, called scales. One of the most popular-sized models is called the HO. Every dimension of the HO model measures $\frac{1}{87}$ that of a real engine. The HO model of a modern diesel locomotive is about 8 inches long. About how many feet long is the real locomotive? (Lesson 4–1)

27. Solve $5.3 - 0.3x = -9.4$. (Lesson 3–3) **49**

28. Solve $24 = -2a$. (Lesson 3–2) **−12**

29. associative prop. (+)

29. Name the property illustrated in $(a + 3b) + 2c = a + (3b + 2c)$. (Lesson 1–8)

Extension

Connections Ben, who is bicycling at the rate of x mph, has a head start of 2 miles on Rachel, who is bicycling at the rate of y mph ($y > x$). In how many hours will Rachel overtake Ben?

$\frac{2}{y - x}$ hours

4-8 Direct and Inverse Variation

Objective

After studying this lesson, you should be able to:

- solve problems involving direct and inverse variation.

APPLICATION
Water Usage

Do you find yourself singing in the shower? If you are like the average American, you have enough time to sing several songs. The national average for time spent in the shower is 12.2 minutes. A standard shower head uses about 6 gallons of water per minute. That means, you use 73.2 gallons of water for each shower you take. Taking one shower per day for one year would use 26,718 gallons of water!

The gallons of water used depends *directly* on the amount of time spent in the shower. The table below shows the gallons of water used y as a function of time in the shower x.

x (minutes)	3	6	9	12	15
y (gallons)	18	36	54	72	90

The relationship between the number of minutes in the shower and the gallons of water used is shown by the equation $y = 6x$. This type of equation is called a **direct variation**. We say that y *varies directly as x* or y *is directly proportional to x*. This means that as x increases in value, y increases in value, or as x decreases in value, y decreases in value.

Definition of Direct Variation	A direct variation is described by an equation of the form $y = kx$, where $k \neq 0$.

In the equation $y = kx$, k is called the **constant of variation**. To find the constant of variation, divide each side by x.

$$\frac{y}{x} = k$$

Example 1

APPLICATION
Employment

Julio's wages vary directly as the number of hours that he works. If his wages for 5 hours are $29.75, how much will they be for 30 hours?

First, find Julio's hourly pay. Let x = number of hours Julio works, and let y = Julio's pay. Find the value of k in the equation $y = kx$. The value of k is the amount of money Julio is paid per hour.

$$k = \frac{y}{x}$$

$$k = \frac{29.75}{5}$$

$$k = 5.95$$

Julio is paid $5.95 per hour.

(continued on the next page)

---

1 FOCUS

5-Minute Check
(over Lesson 4-7)

Solve.

1. Lois has 27 coins in nickels and dimes. In all she has $1.90. How many of each coin does she have?
   16 nickels, 11 dimes

2. How much pure copper must be added to 50.255 kg of an alloy containing 12% copper to raise the copper content to 21%? 5.725 kg

3. Juan traveled 270 miles. What is his rate if he made the trip in 9 hours? 30 mph

4. David rode his bicycle 36 kilometers. What is his rate if he began riding 4 hours ago? 9 km/h

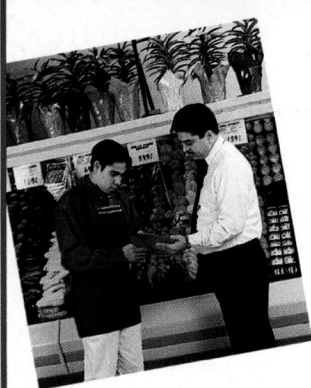

Motivating the Lesson

Hands-On Activity Put 400 mL of water in a 500-mL beaker. Then add an object such as a spoon or glass stirring rod to the beaker. Ask students to describe what happens. **The water level rises.** Add additional spoons or rods to the beaker and observe. Challenge students to identify the relationship between the water level and the objects. **The rise of water level is directly proportional to the number of spoons or rods added to the beaker.**

2 TEACH

In-Class Examples

For Example 1
A car uses 8 gallons of gasoline to travel 290 miles. How much gasoline will the car use to travel 400 miles?
11.034 gallons

For Example 2
Solve each problem. Assume that y varies directly as x.

a. If $y = 6$ when $x = 8$, find y when $x = 12$. **9**

b. If $y = 8$ when $x = 20$, find x when $y = 6$. **15**

c. If $y = 3$ when $x = 9$, find y when $x = 40.5$. **13.5**

d. If $x = -5$ when $y = 6$, find y when $x = -8$. **9.6**

Teaching Tip For Example 2, point out that the phrase y *varies directly as* x can also be phrased y *is directly proportional to* x. Both phrases mean that as y increases or decreases in value, x increases or decreases in value, respectively.

Next find out how much Julio's wages will be for 30 hours.

$$y = kx$$
$$y = 5.95(30)$$
$$y = 178.50$$

Thus, Julio's wages will be \$178.50 for 30 hours of work.

Using the table in the application at the beginning of the lesson, many proportions can be formed. Two examples are shown below.

number of minutes
↓ ↓
$$\frac{3}{18} = \frac{9}{54} \qquad \text{number of minutes} \; \langle \frac{3}{9} = \frac{18}{54} \rangle \; \text{gallons of water}$$
↑ ↑
gallons of water

x_1 is read "x sub 1."

Two general forms for proportions like these can be derived from the equation $y = kx$. Let (x_1, y_1) be a solution for $y = kx$. Let a second solution be (x_2, y_2). Then $y_1 = kx_1$ and $y_2 = kx_2$.

$y_1 = kx_1$ *This equation describes a direct variation.*

$\dfrac{y_1}{y_2} = \dfrac{kx_1}{kx_2}$ *Use the division property of equality. Since y_2 and kx_2 are equivalent, you can divide the left side by y_2 and the right side by kx_2.*

$\dfrac{y_1}{y_2} = \dfrac{x_1}{x_2}$ *Simplify.*

Another proportion can be derived from this proportion.

$x_2y_1 = x_1y_2$ *Find the cross products of the proportion above.*

$\dfrac{x_2y_1}{y_1y_2} = \dfrac{x_1y_2}{y_1y_2}$ *Divide each side by y_1y_2.*

$\dfrac{x_2}{y_2} = \dfrac{x_1}{y_1}$ *Simplify.*

You can use either of these forms to solve problems involving direct proportion.

Example **If y varies directly as x, and $y = 28$ when $x = 7$, find x when $y = 52$.**

Use $\dfrac{y_1}{y_2} = \dfrac{x_1}{x_2}$ to solve the problem.

$\dfrac{28}{52} = \dfrac{7}{x_2}$ *Let $y_1 = 28$, $x_1 = 7$, and $y_2 = 52$.*

$28x_2 = 52(7)$ *Find the cross products.*

$28x_2 = 364$

$x_2 = 13$

Thus, $x = 13$ when $y = 52$.

GLENCOE Technology

Interactive Mathematics Tools Software

This multimedia software provides an interactive lesson that uses a manipulative resistor to determine the relationship between current and resistance on a circuit. A **Computer Journal** gives students an opportunity to write about what they have learned.

For Windows & Macintosh

The reverse of direct variation is **inverse variation**. We say that *y varies inversely as x*. This means that as *x* increases in value, *y* decreases in value, or as *y* decreases in value, *x* increases in value. For example, the more miles you drive, the less gasoline you have in the tank.

Definition of Inverse Variation	An inverse variation is described by an equation of the form $xy = k$, where $k \neq 0$.

Sometimes inverse variations are written in the form $y = \frac{k}{x}$.

Example ❸

APPLICATION

Music

The length of a violin string varies inversely as the frequency of its vibrations. A violin string 10 inches long vibrates at a frequency of 512 cycles per second. Find the frequency of an 8-inch string.

Let ℓ represent the length in inches and f represent the frequency in cycles per second. Find the value of k.

$$\ell f = k$$

$(10)(512) = k$ *Substitute 10 for ℓ and 512 for f.*

$5120 = k$ *The constant of variation is 5120.*

Next, find the frequency, in cycles per second, of the 8-inch string.

$\ell f = k$ *Use the same inverse variation equation.*

$8 \cdot f = 5120$ *Replace ℓ with 8 and k with 5120.*

$f = \frac{5120}{8}$ *Divide each side by 8.*

$f = 640$

The frequency of an 8-inch string is 640 cycles per second.

Let (x_1, y_1) be a solution of an inverse variation, $xy = k$. Let (x_2, y_2) be a second solution. Then $x_1 y_1 = k$ and $x_2 y_2 = k$.

$x_1 y_1 = k$

$x_1 y_1 = x_2 y_2$ *You can substitute $x_2 y_2$ for k because $x_2 y_2 = k$.*

The equation $x_1 y_1 = x_2 y_2$ is called the product rule for inverse variations. Study how it can be used to form a proportion.

$x_1 y_1 = x_2 y_2$

$\frac{x_1 y_1}{x_2 y_1} = \frac{x_2 y_2}{x_2 y_1}$ *Divide each side by $x_2 y_1$.*

$\frac{x_1}{x_2} = \frac{y_2}{y_1}$ *Notice that this proportion is different from the proportion for direct variation.*

Classroom Vignette

"I actually bring a 12-foot board to the classroom. I allow students to find the center of the board and then move on the board until they are balanced. The students measure the distances and weights to find the constant of variation."

John Cline
Greenfield Central High School
Greenfield, Indiana

In-Class Examples

For Example 4
Solve each problem. Assume that y varies inversely as x.

a. If $y = 6$ when $x = 2$, find x when $y = 4$. **3**

b. If $y = 15$ when $x = 21$, find x when $y = 27$. $\dfrac{35}{3}$

For Example 5
A 200-pound weight is located 5 feet from the fulcrum. How far from the fulcrum should 125 pounds be placed to balance the lever? **8 ft**

---

3 PRACTICE/APPLY

Check for Understanding

Exercises 1–11 are designed to help you assess your students' understanding through reading, writing, speaking, and modeling. You should work through Exercises 1–4 with your students and then monitor their work on Exercises 5–11.

Study Guide Masters, p. 33

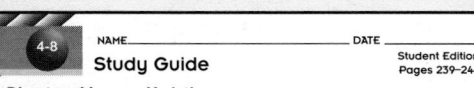

4-8 NAME_____ DATE_____

Study Guide

Student Edition
Pages 239–244

Direct and Inverse Variation

If two variables x and y are related by the equation $y = kx$, where k is a nonzero constant, then the equation is called a **direct variation**, and k is called the **constant of variation**.
If two variables x and y are related by the equation $xy = k$, where $k \neq 0$, then the equation is called an **inverse variation**.

Example 1: If y varies directly as x, and $y = 12$ when $x = 4$, find y when $x = 7$.

$y = kx$
$12 = 4k$
$\dfrac{12}{4} = \dfrac{4k}{4}$
$3 = k$

$y = kx$
$y = 3(7)$
$y = 21$

Example 2: If y varies inversely as x, and $y = 3$ when $x = 12$, find x when $y = 4$.

$xy = k$
$12(3) = k$
$36 = k$

$xy = k$
$4x = 36$
$\dfrac{4x}{4} = \dfrac{36}{4}$
$x = 9$

Solve. Assume that y varies directly as x.

1. If $y = 4$ when $x = 2$, find y when $x = 16$.
32

2. If $y = 9$ when $x = -3$, find x when $y = 6$.
-2

3. If $y = \dfrac{2}{5}$ when $x = \dfrac{1}{3}$, find y when $x = \dfrac{1}{4}$.
$\dfrac{3}{10}$

4. If $y = \dfrac{1}{4}$ when $x = \dfrac{1}{8}$, find x when $y = \dfrac{3}{16}$.
$\dfrac{3}{32}$

Solve. Assume that y varies inversely as x.

5. If $y = 9$ when $x = 7$, find y when $x = 2$.
$\dfrac{63}{2}$

6. If $y = 4.3$ when $x = 12.9$, find y when $x = 15.8$.
$\dfrac{5547}{1580}$

7. If $x = \dfrac{1}{2}$ when $y = \dfrac{1}{3}$, find y when $x = \dfrac{1}{4}$.
$\dfrac{2}{3}$

8. If $y = -6$ when $x = \dfrac{1}{4}$, find y when $x = \dfrac{3}{16}$.
-8

242 Chapter 4

You can use either the product rule or the proportion rule to solve problems involving inverse variation.

Example ❹ **If y varies inversely as x, and $y = 5$ when $x = 15$, find x when $y = 3$.**

Let $x_1 = 15$, $y_1 = 5$, and $y_2 = 3$. Solve for x_2.

Method 1	**Method 2**
Use the product rule.	Use the proportion.
$x_1 y_1 = x_2 y_2$	$\dfrac{x_1}{x_2} = \dfrac{y_2}{y_1}$
$15 \cdot 5 = x_2 \cdot 3$	$\dfrac{15}{x_2} = \dfrac{3}{5}$
$\dfrac{75}{3} = x_2$	$75 = 3x_2$
$25 = x_2$	$25 = x_2$

Thus, $x = 25$ when $y = 3$.

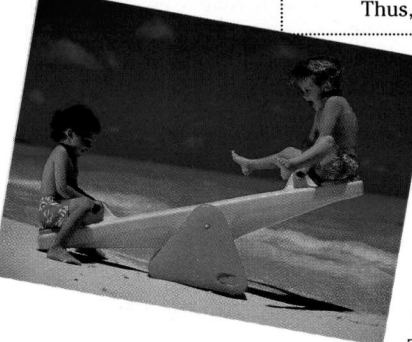

If you have observed people on a seesaw, you may have noticed that the heavier person must sit closer to the fulcrum (pivot point) to balance the seesaw. A seesaw is a type of *lever,* and all lever problems involve inverse variation.

Suppose weights w_1 and w_2 are placed on a lever at distances d_1 and d_2, respectively, from the fulcrum. The lever is balanced when $w_1 d_1 = w_2 d_2$. This property of levers is illustrated at the right.

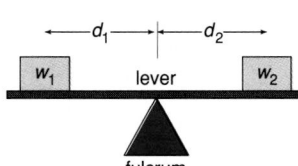

Example ❺

APPLICATION
Physics

The fulcrum is placed in the middle of a 20-foot seesaw. Cholena, who weighs 120 pounds, is seated 9 feet from the fulcrum. How far from the fulcrum should Antonio sit if he weighs 135 pounds?

Let $w_1 = 120$, $d_1 = 9$, and $w_2 = 135$. Solve for d_2.

$$w_1 d_1 = w_2 d_2$$
$$120 \cdot 9 = 135 \cdot d_2$$
$$1080 = 135 d_2$$
$$d_2 = 8$$

Antonio should sit 8 feet from the fulcrum.

CHECK FOR UNDERSTANDING

Communicating Mathematics

2. $y = kx$, where $k \neq 0$

3. Divide each side of $y = kx$ by x.

Study the lesson. Then complete the following.

1. **Tell** whether an equation of the form $xy = k$, where $k \neq 0$, represents a direct or inverse variation. **inverse**

2. **Write** the equation used when describing a direct variation.

3. **Explain** how you find the constant of variation in a direct variation.

4. **You Decide** Morgan says that a person's height varies directly as his or her age. Do you agree? Explain. **No, because growth rates vary widely.**

Reteaching

Using Manipulatives Have a small group of students use a triple-beam balance and 4 stacks of uniform washers. Each stack should contain a different number of washers tied together with strong thread so students cannot weigh just one washer at a time.

# of Washers (*n*)	Weight (*W*)

Count and record the number of washers in each stack. Weigh the first stack. Use this result to predict the weights of each of the three remaining stacks. Then weigh these stacks. Write and solve a proportion for the first stack to find the *constant of variation k* in $W = kn$. What does it represent as a rate? **weight of 1 washer**

Guided Practice

Determine which equations represent inverse variations and which represent direct variations. Then find the constant of variation.

5. $mn = 5$ I, 5

6. $a = -3b$ D, -3

Solve. Assume that y varies directly as x.

7. If $y = 27$, when $x = 6$, find x when $y = 45$. 10

8. If $y = -7$ when $x = -14$, find y when $x = 20$. 10

Solve. Assume that y varies inversely as x.

10. $\frac{12}{5}$ or 2.4

9. If $y = 99$ when $x = 11$, find x when $y = 11$. 99

10. If $y = -6$ when $x = -2$, find y when $x = 5$.

11. **Physics** An 8-ounce weight is placed at one end of a yardstick. A 10-ounce weight is placed at the other end. Where should the fulcrum be placed to balance the yardstick? 20 inches from the 8-ounce weight, 16 inches from the 10-ounce weight

EXERCISES

Practice

Determine which equations represent inverse variations and which represent direct variations. Then find the constant of variation.

 A

12. $c = 3.14d$ D, 3.14

13. $15 = rs$ I, 15

14. $\frac{35}{p} = q$ I, 35

15. $s = \frac{9}{t}$ I, 9

16. $\frac{1}{3}x = z$ D, $\frac{1}{3}$

17. $4a = b$ D, 4

Solve. Assume that y varies directly as x.

18. $2\frac{1}{4}$ 19. $26\frac{1}{4}$

18. If $y = -8$ when $x = -3$, find x when $y = 6$.

19. If $y = 12$ when $x = 15$, find x when $y = 21$.

B

20. If $y = 2.5$ when $x = 0.5$, find y when $x = 20$. 100

21. If $y = 4$ when $x = 12$, find y when $x = -24$. -8

22. If $y = -6$ when $x = 9$, find y when $x = 6$. -4

23. If $y = 2\frac{2}{3}$ when $x = \frac{1}{4}$, find y when $x = 1\frac{1}{8}$. 12

Solve. Assume that y varies inversely as x.

C

24. If $y = 9$ when $x = 8$, find y when $x = 6$. 12

25. If $x = 2.7$ when $y = 8.1$, find y when $x = 3.6$. 6.075

26. If $y = 24$ when $x = -8$, find y when $x = 4$. -48

27. If $x = 6.1$ when $y = 4.4$, find x when $y = 3.2$. 8.3875

28. $\frac{2}{3}$ 29. $\frac{1}{4}$

28. If $y = 7$ when $x = \frac{2}{3}$, find y when $x = 7$.

29. If $x = \frac{1}{2}$ when $y = 16$, find x when $y = 32$.

Critical Thinking

30. Assume that y varies inversely as x. 30a. It is halved. 30b. It is divided by 3.

a. If the value of x is doubled, what happens to the value of y?

b. If the value of y is tripled, what happens to the value of x?

Lesson 4–8 Direct and Inverse Variation **243**

Classroom Vignette

"After the initial introduction, examples, and discussion, I have students work in groups of four to solve the problems in this section. Each student completes the assignment, and one paper is collected at random from each group."

Barbara Ferguson
Enumclaw High School
Enumclaw, Washington

Barbara K. Ferguson

Assignment Guide

Core: 13–29 odd, 30, 31, 33–40
Enriched: 12–28 even, 30–40

For **Extra Practice,** see p. 766.

The red A, B, and C flags, printed only in the Teacher's Wraparound Edition, indicate the level of difficulty of the exercises.

Practice Masters, p. 33

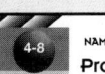

NAME_____ DATE _____

4-8 Practice

Student Edition Pages 239–244

Direct and Inverse Variation

Solve. Assume that y varies directly as x.

1. If $y = -4$ when $x = 2$, find y when $x = -6$. 12

2. If $y = 16$ when $x = 4$, find y when $x = 6$. 24

3. If $y = -5$ when $x = 12.5$, find x when $y = 15$. -37.5

4. If $y = 7$ when $x = 4$, find y when $x = 12$. 21

5. If $y = 80$ when $x = 32$, find x when $y = 100$. 40

6. If $y = 198$ when $x = 22$, find y when $x = 182$. 1638

7. If $y = 28$ when $x = 168$, find y when $x = 108$. 18

8. If $y = 24$ when $x = 6$, find x when $y = -4$. -16

Solve. Assume that y varies inversely as x.

9. If $y = -4$ when $x = 2$, find y when $x = -6$. $\frac{4}{3}$

10. If $y = 16$ when $x = 4$, find y when $x = 6$. $\frac{32}{3}$

11. If $y = -5$ when $x = 12.5$, find x when $y = 15$. -4.16

12. If $y = 7$ when $x = 4$, find y when $x = 12$. $\frac{7}{3}$

13. If $y = 27$ when $x = 12$, find x when $y = -12$. -27

14. If $y = 6$ when $x = -4$, find x when $y = \frac{12}{5}$. -10

15. If $y = 60$ when $x = 80$, find x when $y = -20$. -240

16. If $y = 40$ when $x = 16$, find y when $x = 10$. 64

Applications and Problem Solving

31. 18 pounds

31. **Space** The weight of an object on the moon varies directly as its weight on Earth. With all of his gear on, Neil Armstrong weighed 360 pounds on Earth. When he became the first person to step on the moon on July 20, 1969, he weighed 60 pounds. Tara weighs 108 pounds on Earth. What would she weigh on the moon?

32. **Physics** Pam and Adam are seated on the same side of a seesaw. Pam is 6 feet from the fulcrum and weighs 115 pounds. Adam is 8 feet from the fulcrum and weighs 120 pounds. Kam is seated on the other side of the seesaw, 10 feet from the fulcrum. If the seesaw is balanced, how much does Kam weigh? **165 pounds**

33. **Music** The pitch of a musical tone varies inversely as its wavelength. If one tone has a pitch of 440 vibrations per second and a wavelength of 2.4 feet, find the wavelength of a tone that has a pitch of 660 vibrations per second. **1.6 feet**

Mixed Review

34. **Travel** At 8:00 A.M., Alma drove west at 35 miles per hour. At 9:00 A.M., Reiko drove east from the same point at 42 miles per hour. At what time will they be 266 miles apart? (Lesson 4–7) **12:00 noon**

35. **Probability** If the probability that an event will occur is $\frac{2}{3}$, what are the odds that the event will occur? (Lesson 4–6) **2:1**

Neil Armstrong on the moon

36. **Retail Sales** A department store buys clothing at wholesale prices and then marks the clothing up 25% to sell at retail price to customers. If the retail price of a jacket is $79, what was the wholesale price? (Lesson 4–5) **$63.20**

37. **Finance** Hiroko invested $11,700, part at 5% interest and the balance at 7% interest. If her annual earnings from both investments is $733, how much is invested at each rate? (Lesson 4–4) **$4300 at 5%, $7400 at 7%**

38. Solve $\frac{2x}{5} + \frac{x}{4} = \frac{26}{5}$ for *x*. (Lesson 3–6) **8**

39. Solve $\frac{x}{4} + 9 = 6$. (Lesson 3–3) **−12**

40. Simplify $\frac{4a + 32}{4}$. (Lesson 2–7) **a + 8**

Extension

Problem Solving Six feet of steel wire weigh 0.7 kilogram. How much do 100 feet of steel wire weigh? **11.6 kg**

VOCABULARY

After completing this chapter, you should be able to define each term, property, or phrase and give an example or two of each.

Algebra
base (p. 215)
constant of variation (p. 239)
direct variation (p. 239)
extremes (p. 196)
inverse variation (p. 241)
means (p. 196)
odds (p. 229)
percent (p. 215)
percentage (p. 215)
percent of decrease (p. 222)
percent of increase (p. 222)
percent proportion (p. 215)
proportion (p. 195)

rate (pp. 197, 215)
ratio (p. 195)
scale (p. 197)
simple interest (p. 217)
uniform motion (p. 235)
weighted average (p. 233)

Geometry
angle of depression (p. 208)
angle of elevation (p. 208)
corresponding angles (p. 201)
corresponding sides (p. 201)
hypotenuse (p. 206)
legs (p. 206)

similar (p. 201)
similar triangles (p. 201)
trigonometric ratios (p. 206)

Probability
equally likely (p. 229)
probability (p. 228)
probability of an event (p. 228)
random (p. 229)

Trigonometry
cosine (p. 206)
sine (p. 206)
solving the triangle (p. 208)
tangent (p. 206)

UNDERSTANDING AND USING THE VOCABULARY

Choose the correct term to complete each sentence.

1. The angle formed by a person in a radio tower looking up at an airplane in flight is called the (*angle of depression*, *angle of elevation*).

2. The equation $\frac{3}{5} = \frac{9}{15}$ is a (*proportion*, *ratio*).

3. If $\triangle ABC \sim \triangle DEF$, then their corresponding angles are (*equal*, *proportional*), and their corresponding sides are (*equal*, *proportional*).

4. A ratio (*can*, *cannot*) be expressed in the following ways: $\frac{4}{5}$, 4:5, and 4 to 5.

5. In a right triangle, the sides that form the right angle are called the (*hypotenuse*, *legs*).

6. In $\frac{2}{x} = \frac{3}{9}$, 3 and x are called the (*extremes*, *means*).

7. The (*odds*, *probability*) that a certain event can occur is the ratio of the number of ways it can occur to the number of ways it cannot occur.

8. In the ratio $\frac{5}{28}$, 5 is the (*base*, *percentage*), and 28 is the (*base*, *percentage*).

9. A(n) (*direct variation*, *inverse variation*) is described by an equation of the form $xy = k$, where $k \neq 0$.

10. The tangent of an angle is defined as the measure of the (*adjacent*, *opposite*) side divided by the measure of the (*adjacent side*, *hypotenuse*).

Instructional Resources

Three multiple-choice tests and three free-response tests are provided in the *Assessment and Evaluation Masters*. Forms 1A and 2A are for honors pacing, and Forms 1B, 1C, 2B, and 2C are for average pacing. Chapter 4 Test, Form 1B is shown at the right. Chapter 4 Test, Form 2B is shown on the next page.

Using the
STUDY GUIDE AND ASSESSMENT

Skills and Concepts Encourage students to refer to the objectives and examples on the left as they complete the review exercises on the right.

Assessment and Evaluation Masters, pp. 93–94

NAME _____ DATE _____

Chapter 4 Test, Form 2B

Solve each proportion.

1. $\frac{x}{6} = \frac{2}{9}$ 2. $\frac{c+2}{5} = \frac{3}{7}$

1. _____ $1\frac{1}{3}$

2. _____ $\frac{1}{7}$

3. If 500 sheets of paper weigh 7 pounds, how much do 2 sheets weigh?   3. _____ 0.028 lb

4. Kathy can type 13 business letters in 2 hours. At that same rate, how many letters can she type in 8 hours?   4. _____ 52 letters

Triangles MET and CUB are similar. Use this information for exercises 5 and 6.

5. If $t = 3$, $m = 7$, $e = 9$, and $b = 2$, find c and u.   5. _____ $c = \frac{14}{3}$; $u = 6$

6. If $m = 10$, $c = 7$, $b = 4$, and $e = 15$, find t and u.   6. _____ $t = \frac{40}{7}$; $u = 10\frac{1}{2}$

7. Find the height of the tree.

7. _____ $40\frac{1}{2}$ ft

Use the triangle below for exercises 8 and 9.

8. If $r = 16$ and $s = 23$, find the measure of $\angle R$ to the nearest degree.   8. _____ 35°

9. If $s = 5$, $r = 12$, and $t = 13$, find cos S.   9. _____ $\frac{12}{13}$

10. Find the height of a kite to the nearest yard if the length of the string is 140 yards and the angle of elevation to the kite is 54 degrees. Assume the string is straight.   10. _____ about 113 yd

11. What number is 23% of 86?   11. _____ 19.78

12. Fifty-five is what percent of 220?   12. _____ 25%

13. Of those children adopted in the United States, 6.5% were over 10 years old. If 8400 were adopted, how many were over 10 years old?   13. _____ 546

14. Johnny paid $24.36 including 5% sales tax for a pair of pants. What was the price of the pants without sales tax?   14. _____ $23.20

15. A price increased from $145 to $156.60. What was the percent of increase?   15. _____ 8%

NAME _____ DATE _____

Chapter 4 Test, Form 2B (continued)

16. A television normally selling for $423.00 is discounted 22%. Find the sale price.   16. _____ $329.94

17. If the probability in favor of an event is $\frac{5}{12}$, what are the odds of that event occurring?   17. _____ 5:7

18. A computer randomly selects a letter of the alphabet. What is the probability that the letter is one of the letters in the word "probability"?   18. _____ $\frac{9}{26}$

19. The results of a survey of 284 randomly chosen students is shown below. What is the probability that a student chosen at random favors strawberry?   19. _____ $\frac{121}{284}$

Preference for Frozen Yogurt Flavors			
Students	Strawberry	Chocolate	No Preference
Male	25	87	20
Female	96	34	22

20. Joe and Janna leave home at the same time, traveling in opposite directions. Joe drives 45 mi/h and Janna drives 40 mi/h. In how many hours will they be 510 miles apart?   20. _____ 6 h

21. A factory has an order for 23,500 paper clips. Machine A can make 1500 paper clips per hour while Machine B can make 2500 paper clips per hour. Machine A starts at 8:00 A.M. and Machine B starts at 1:00 P.M. At what time will the two machines complete the job?   21. _____ 5 P.M.

22. A chemist has 20 liters of a 10% acid solution. How many liters of pure acid must be added to produce a 25% acid solution?   22. _____ 4 liters

23. If y varies directly as x and $y = 25$ when $x = 10$, find y when $x = 4$.   23. _____ 10

24. If y varies inversely as x and $y = 9$ when $x = 5$, find y when $x = 6$.   24. _____ $\frac{15}{2}$, or $7\frac{1}{2}$

25. Suppose that in a certain restaurant the number of hamburgers sold varies inversely as the price. If 70 burgers are sold for $1.50 each, how many burgers would be sold at $1.25 each?   25. _____ 84 burgers

Bonus The intensity of illumination, I, on a movie screen varies inversely as the square of its distance from the projector. If the distance between the projector and the screen is changed from 200 feet to 100 feet, by what factor is the intensity of illumination changed?   Bonus _____ 4

OBJECTIVES AND EXAMPLES

Upon completing this chapter, you should be able to:

- solve proportions   (Lesson 4–1)

$$\frac{x}{3} = \frac{x+1}{2}$$

$2(x) = 3(x+1)$ *Find the cross products.*

$2x = 3x + 3$

$-3 = x$

- find the unknown measures of the sides of two similar triangles   (Lesson 4–2)

$$\frac{10}{5} = \frac{6}{a}$$

$10a = 5(6)$

$10a = 30$

$a = 3$

5 cm

a cm

6 cm

10 cm

19. $d = \frac{45}{8}$, $e = \frac{27}{4}$ 20. $d = \frac{48}{5}$, $e = \frac{36}{5}$

21. $b = \frac{44}{3}$, $d = 6$ 22. $a = \frac{35}{2}$, $e = 8$

- use trigonometric ratios to solve right triangles (Lesson 4–3)

$$\sin A = \frac{\text{measure of leg opposite } \angle A}{\text{measure of hypotenuse}}$$

$$\cos A = \frac{\text{measure of leg adjacent } \angle A}{\text{measure of hypotenuse}}$$

$$\tan A = \frac{\text{measure of leg opposite } \angle A}{\text{measure of leg adjacent } \angle A}$$

- use trigonometric ratios to solve right triangles (Lesson 4–3)

$\cos M = 0.3245$

Enter: 0.3245 [2nd] [COS⁻¹]   71.064715

The measure of $\angle M$ is about 71°.

REVIEW EXERCISES

Use these exercises to review and prepare for the chapter test.

Use cross products to determine whether each pair of ratios forms a proportion.

11. $\frac{10}{3}$, $\frac{150}{45}$ =   12. $\frac{8}{7}$, $\frac{2}{1.75}$ =

13. $\frac{30}{12}$, $\frac{5}{4}$ ≠   14. $\frac{2.7}{3.1}$, $\frac{8.1}{9.3}$ =

Solve each proportion.

15. $\frac{6}{15} = \frac{n}{45}$ 18   16. $\frac{x}{11} = \frac{35}{55}$ 7

17. $\frac{y+4}{y-1} = \frac{4}{3}$ 16   18. $\frac{z-7}{6} = \frac{z+3}{7}$ 67

△ABC and △DEF are similar. For each set of measures given, find the measures of the remaining sides.

19. $c = 16$, $b = 12$, $a = 10$, $f = 9$

20. $a = 8$, $c = 10$, $b = 6$, $f = 12$

21. $c = 12$, $f = 9$, $a = 8$, $e = 11$

22. $b = 20$, $d = 7$, $f = 6$, $c = 15$

For △ABC, find each value to the nearest thousandth.

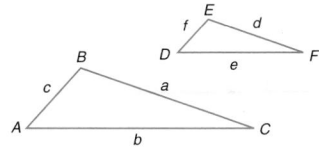

B

53 ft

28 ft

C   45 ft   A

23. cos B 0.528

24. tan A 0.622

25. sin B 0.849

26. cos A 0.849

27. tan B 1.607

28. sin A 0.528

Use a calculator to find the measure of each angle to the nearest degree.

29. tan $M = 0.8043$ 39°

30. sin $T = 0.1212$ 7°

31. tan $Q = 5.9080$ 80°

32. cos $F = 0.7443$ 42°

246 Chapter 4 Study Guide and Assessment

GLENCOE Technology

Test and Review Software

You may use this software, a combination of an item generator and item bank, to create your own tests or worksheets. Types of items include free response, multiple choice, short answer, and open ended.

For IBM & Macintosh

OBJECTIVES AND EXAMPLES	REVIEW EXERCISES

OBJECTIVES AND EXAMPLES

● use trigonometric ratios to solve right triangles
(Lesson 4–3)

Solve right $\triangle ABC$ if $m\angle B = 40°$ and $c = 6$.

$m\angle A = 180° - (90° + 40°)$ or $50°$

$\cos 40° = \dfrac{a}{6}$ $\sin 40° = \dfrac{b}{6}$

$0.7660 \approx \dfrac{a}{6}$ $0.6428 \approx \dfrac{b}{6}$

$0.7660(6) \approx a$ $0.6428(6) \approx b$

$4.596 \approx a$ $3.857 \approx b$

REVIEW EXERCISES

Solve each right triangle. State the side lengths to the nearest tenth and the angle measures to the nearest degree.

33.

34.

33. $m\angle B = 45°$, $AB = 8.5$, $BC = 6$

35.

36.

34. $m\angle B = 12°$, $AB = 134.7$, $BC = 131.7$
35. $m\angle A = 67°$, $AB = 9.8$, $m\angle B = 23°$
36. $m\angle B = 60°$, $AC = 22.5$, $BC = 13°$

● solve percent problems   (Lesson 4–4)

Nine is what percent of 15?

$\dfrac{9}{15} = \dfrac{r}{100}$ *Use a proportion.*

$9(100) = 15r$ *Find the cross products.*

$900 = 15r$

$60 = r$

Solve.

37. What number is 60% of 80?   **48**
38. Twenty-one is 35% of what number?   **60**
39. Eighty-four is what percent of 96?   **87.5%**
40. What percent of 17 is 34?   **200%**
41. What number is 0.3% of 62.7?   **0.1881**
42. Find 0.12% of $5200.   **6.24**

● solve problems involving percent of increase or decrease   (Lesson 4–5)

original: $120   new: $114

amount of decrease: $120 − $114 = $6

$\dfrac{\text{amount of decrease}}{\text{original}} = \dfrac{r}{100}$

$\dfrac{6}{120} = \dfrac{r}{100}$

$5 = r$

The percent of decrease is 5%.

43. decrease, 13%
44. increase, 19%
45. increase, 6%
46. increase, 8%
47. decrease, 10%
48. decrease, 6%

State whether each percent of change is a percent of increase or a percent of decrease. Then find the percent of increase or decrease. Round to the nearest whole percent.

43. original: $40
    new: $35

44. original: 97 cases
    new: 115 cases

45. original: $35
    new: $37.10

46. original: $50
    new: $88

47. original: 1500
    employees
    new: 1350 employees

48. original: 12,500
    students
    new: 11,800 students

● solve problems involving discounts or sales tax
(Lesson 4–5)

running shoes: $74; discount: 15%; sales tax: 6%

15% of $74 = (0.15)(74) or $11.10

$74 − $11.10 = $62.90   ← *sale price*

6% of $62.90 = (0.06)(62.90) or $3.77

$62.90 + $3.77 = $66.67   ← *price with tax*

The total price is $66.67 for the shoes.

Find the final price of each item. When there is a discount and sales tax, compute the discount price first.

49. calculator: $81
    sales tax: 5.75%
    $85.66

50. gasoline: $21.50
    discount: 2%
    $21.07

51. used car: $8,690
    sales tax: 6.7%
    $9272.23

52. dress: $89
    discount: 25%
    sales tax: 7%   **$71.42**

CHAPTER 4 STUDY GUIDE AND ASSESSMENT

OBJECTIVES AND EXAMPLES	REVIEW EXERCISES

• find the probability of a simple event and find the odds of a simple event (Lesson 4–6)

Find the probability of a computer randomly choosing the letter I in the word MISSISSIPPI.

$$\frac{\text{number of favorable outcomes}}{\text{number of possible outcomes}} = \frac{4}{11}$$

Find the odds that you will randomly *not* select the letter S in the word MISSISSIPPI.

number of successes:number of failures = 7:4

53. $\frac{1}{12}$ 54. $\frac{1}{4}$ 55. $\frac{5}{6}$ 56. $\frac{1}{4}$

57. 10:39   58. 18:31   59. 34:15   60. 25:24

Find the probability of each outcome if a computer randomly chooses a letter in the word REPRESENTING.

53. the letter S   54. the letter E

55. $P(\text{not N})$ 56. the letters R or P

Find the odds of each outcome if you randomly select a coin from a jar containing 90 pennies, 75 nickels, 50 dimes, and 30 quarters.

57. selecting a dime   58. selecting a penny

59. *not* selecting a nickel

60. selecting a nickel or a dime

• solve problems involving direct and inverse variation (Lesson 4–8)

If x varies directly as y and $x = 15$ when $y = 1.5$, find x when $y = 9$.	If y varies inversely as x and $y = 24$ when $x = 30$, find x when $y = 10$.
$\frac{1.5}{9} = \frac{15}{x}$	$\frac{30}{x} = \frac{10}{24}$
$1.5x = 9(15)$	$(30)(24) = 10x$
$1.5x = 135$	$720 = 10x$
$x = 90$	$72 = x$

Solve. Assume that *y* varies directly as *x*.

61. If $y = 15$ when $x = 5$, find y when $x = 7$.   **21**

62. If $y = 35$ when $x = 175$, find y when $x = 75$.   **15**

63. If $y = 10$ when $x = 0.75$, find x when $y = 80$.   **6**

64. If $y = 3$ when $x = 99.9$, find y when $x = 522.81$.   **15.7**

Solve. Assume that *y* varies inversely as *x*.

65. If $y = 28$ when $x = 42$, find y when $x = 56$.   **21**

66. If $y = 15$ when $x = 5$, find y when $x = 3$.   **25**

67. If $y = 18$ when $x = 8$, find x when $y = 3$.   **48**

68. If $y = 35$ when $x = 175$, find y when $x = 75$.   **81.$\overline{6}$**

APPLICATIONS AND PROBLEM SOLVING

69. **Travel** Two airplanes leave Dallas at the same time and fly in opposite directions. One airplane travels 80 miles per hour faster than the other. After three hours, they are 2940 miles apart. What is the rate of each airplane? (Lesson 4–7)   **450 mph and 530 mph**

A practice test for Chapter 4 is provided on page 790.

70. **Coffee** Anne Leibowitz owns "The Coffee Pot," a specialty coffee store. She wants to create a special mix using two coffees, one priced at $8.40 per pound and the other at $7.28 per pound. How many pounds of the $7.28 coffee should she mix with 9 pounds of the $8.40 coffee to sell the mixture for $7.95 per pound? (Lesson 4–7)   **6 pounds**

ALTERNATIVE ASSESSMENT

COOPERATIVE LEARNING PROJECT

Letter Occurrence In this project, you will determine percentages, percent of change, and probability and odds.

There is an uneven distribution of letters which occur in words. Christofer Sholes, the inventor of the typewriter, may not have been aware of this unevenness. He gave the left hand 56 percent of all the strokes, and the two most agile fingers on the right hand hit two of the least often used letters of the alphabet, j and k.

The table below shows the percentages of the occurrence of letters in large samples.

E	12.3%	R	6.0%	F	2.3%	K	1.5%
T	9.6%	H	5.1%	M	2.2%	Q	0.2%
A	8.1%	L	4.0%	W	2.0%	X	0.2%
O	7.9%	D	3.7%	Y	1.9%	J	0.1%
N	7.2%	C	3.2%	B	1.6%	Z	0.1%
I	7.2%	U	3.1%	G	1.6%		
S	6.6%	P	2.3%	V	0.9%		

Use the second paragraph on this page to count the number of times that e, s, c, w, and k occur and compute their percentages. Compare your answers with the table. What is the probability that a vowel was used in that paragraph? Are the odds better that a vowel or a consonant was used in the above paragraph? What are these odds?

Write a paragraph describing similar triangles. Then determine the percentages of the occurrence of the above mentioned letters in your paragraph. Do they coincide with the percentages in the table? Determine the percent of change for the above letters in the paragraph above and the paragraph you wrote.

Follow these guidelines.

- Determine what type of data you will need in order to calculate your answers.
- Devise a chart that will be helpful in organizing the data you will need.
- Write a paragraph describing what your answers mean.
- Using the percentages from the above table, draw your idea of where the letters should be on a typewriter and explain why.

THINKING CRITICALLY

- Use an example to describe the difference between odds and probability. Explain why some events are described in terms of odds and others in terms of probability.
- Use the definitions of the trigonometric ratios to explain why, in any right $\triangle ABC$ with right angle C, $\sin A = \cos B$, and $\cos A = \sin B$.

PORTFOLIO

Even after working through a chapter of material, you can get to the end and still feel uncertain about a concept or a certain type of problem. Find a problem that you still cannot solve or that you are worried you might not be able to solve on a test. Write out the question and as much of the solution as you can until you get to the hard part. Then explain what it is that keeps you from solving the problem. Be clear and precise. Place this in your portfolio.

SELF EVALUATION

Theme learning occurs when the subject matter being studied is all related to the same theme, such as water, animals, and so on. When the course material is presented, it is all applied to that theme.

Assess yourself. Are you a person who would enjoy theme learning? Would you like to relate all your course work to one area or do you prefer having a variety of application themes? Describe a theme that would be of interest to you and tell how you would use that theme to introduce and apply the contents of a specific math subject.

Assessment and Evaluation Masters, pp. 98, 109

4

NAME_____ DATE _____

Chapter 4 Performance Assessment

Instructions: *Demonstrate your knowledge by giving a clear, concise solution to each problem. Be sure to include all relevant drawings and justify your answers. You may show your solution in more than one way or investigate beyond the requirements of the problem.*

1. The following questions are related to the formula $d = rt$, where d represents distance, r represents rate, and t represents time.

   a. Write what is meant by direct variation.

   b. Write what is meant by inverse variation.

   c. Every day at lunchtime Kelli walks 30 minutes before returning to work. On cooler days she walks more briskly than on warm days. Does the distance she walks vary directly or inversely with her rate of walk? Justify your answer.

   d. At the Indianapolis 500 auto race, each car finishing the race travels 500 miles. Does the average speed of the cars finishing the race vary directly or inversely with the time taken to complete the race? Justify your answer.

2. A landscaper wishes to make a scale drawing of a rectangular garden. She makes the measurements indicated.

   a. Find the measurements labeled a, b, c, and d. Show your work and explain each step.

   b. Make a scale drawing based on the length of the garden being represented by 10 inches on your drawing. Show your work in determining each length in your drawing.

3. Explain how to find the sale price when you know the original price and the percent of discount.

4. Explain how to find the odds that an event will occur if you know the probability that it will occur.

Scoring Guide
Chapter 4
Performance Assessment

Level	Specific Criteria
3 Superior	• Shows thorough understanding of the concepts of *proportion; direct and inverse variation; similar triangles; sine, cosine, and tangent; percent of change;* and *probability.* • Uses appropriate strategies to solve problems. • Computations are correct. • Written explanations are exemplary. • Diagrams are accurate and appropriate. • Goes beyond requirements of some or all problems.
2 Satisfactory, with Minor Flaws	• Shows understanding of the concepts of *proportion; direct and inverse variation; similar triangles; sine, cosine, and tangent; percent of change;* and *probability.* • Uses appropriate strategies to solve problems. • Computations are mostly correct. • Written explanations are effective. • Diagrams are mostly accurate and appropriate. • Satisfies all requirements of problems.
1 Nearly Satisfactory, with Serious Flaws	• Shows understanding of most of the concepts of *proportion; direct and inverse variation; similar triangles; sine, cosine, and tangent; percent of change;* and *probability.* • May not use appropriate strategies to solve problems. • Computations are mostly correct. • Written explanations are satisfactory. • Diagrams are mostly accurate and appropriate. • Satisfies most requirements of problems.
0 Unsatisfactory	• Shows little or no understanding of the concepts of *proportion; direct and inverse variation; similar triangles; sine, cosine, and tangent; percent of change;* and *probability.* • May not use appropriate strategies to solve problems. • Computations are incorrect. • Written explanations are not satisfactory. • Diagrams are not accurate or appropriate. • Does not satisfy requirements of problems.

Alternative Assessment

The Alternative Assessment section provides students with the opportunity to assess their own work by thinking critically, working with others, keeping a portfolio, and honestly evaluating their own progress. For more information on alternative forms of assessment, see *Alternative Assessment in the Mathematics Classroom,* one of the titles in the Glencoe Mathematics Professional Series.

Performance Assessment

Performance Assessment tasks for this chapter are included in the *Assessment and Evaluation Masters.* A scoring guide is also provided.

Using the CUMULATIVE REVIEW

These two pages review the skills and concepts presented in Chapters 1–4. This review is formatted to reflect new trends in standardized testing.

A more traditional cumulative review is provided in the *Assessment and Evaluation Masters,* pp. 103–104.

Assessment and Evaluation Masters, pp. 103–104

SECTION ONE: MULTIPLE CHOICE

There are nine multiple-choice questions in this section. After working each problem, write the letter of the correct answer on your paper.

1. Solve the equation $7(3b + x) = 5a$ for x.   **D**

   A. $5a - 21b = x$ **B.** $\frac{5a}{7} - 21b = x$

   C. $\frac{5a - 3b}{7} = x$ **D.** $\frac{5a - 21b}{7} = x$

2. **Sales** A Tampa Bay Buccaneers sweatshirt decreased in price from $40 to $35 during a clearance sale. Find the percent of decrease.   **B**

   A. 5%   **B.** 12.5%

   C. 14.3%   **D.** 114.3%

3. A caterpillar was trying to crawl out of a bucket into which it had fallen. It crawled up $1\frac{1}{4}$ feet before sliding down $\frac{3}{8}$ feet. It then crawled up $\frac{5}{6}$ feet before resting. How much farther must the caterpillar crawl to reach the top of the 2-foot bucket?   **A**

   A. $\frac{7}{24}$ feet   **B.** $\frac{7}{8}$ feet

   C. $1\frac{1}{8}$ feet   **D.** $1\frac{17}{24}$ feet

4. **Geometry** Choose the correct proportion to find the distance across the lake (AB) given $\triangle ABC$ is similar to $\triangle EDC$.   **C**

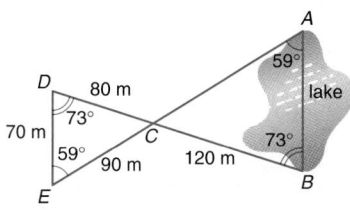

   A. $\frac{70}{AB} = \frac{90}{120}$ **B.** $\frac{80}{70} = \frac{AB}{120}$

   C. $\frac{70}{AB} = \frac{80}{120}$ **D.** $\frac{80}{90} = \frac{120}{AB}$

Standardized Test Practice Questions are also provided in the *Assessment and Evaluation Masters,* p. 102.

---

5. Kassim read 36 pages of a novel in 2 hours. At the same rate, find the number of hours it will take him to read the remaining 135 pages.   **B**

   A. $3\frac{3}{4}$ **B.** $7\frac{1}{2}$

   C. $9\frac{1}{2}$ **D.** $10\frac{1}{2}$

6. Choose an equivalent equation to represent "h is the product of g and the difference of the square of b and t".   **D**

   A. $h - t = gb^2$

   B. $h = g(b^2 - t)$

   C. $h = g(b - t)^2$

   D. $h + t = gb^2$

7. **Probability** The freshman class of Perry High School is planning an end-of-the-year dance. The dance committee decides to award door prizes. They want the odds of winning a prize to be 1:5. If they plan for 180 tickets to be sold for the dance, how many prizes will be needed?   **C**

   A. 6 prizes   **B.** 36 prizes

   C. 30 prizes   **D.** 150 prizes

8. **Sales** Raven sold her concert ticket to Sandra for 75% of the original ticket price. If Sandra paid $26.25 for the ticket, choose the proportion that could be used to find the ticket's original price.   **A**

   A. $\frac{75}{100} = \frac{26.25}{x}$ **B.** $\frac{x}{26.25} = \frac{75}{100}$

   C. $\frac{x}{100} = \frac{75}{26.25}$ **D.** $\frac{x}{75} = \frac{26.25}{100}$

9. **Statistics** Carita's bowling scores for the first four games of a five-game series are $b + 2$, $b + 3$, $b - 2$, and $b - 1$. What must her score be for the last game to have an average of $b + 2$?   **D**

   A. $b - 2$ **B.** $b + 2$

   C. $b - 8$ **D.** $b + 8$

SECTION TWO: FREE RESPONSE

This section contains nine questions for which you will provide short answers. Write your answer on your paper.

10. Describe the steps required to solve the equation $4t - 5 = 7t - 23$.   **See students' work.**

11. Geometry   Find the dimensions of a rectangle whose width is 5 inches less than its length and has a perimeter of 70 inches.   **15 in. × 20 in.**

12. Finance   In the beginning of the month, the balance in Marisa Fuente's checking account was $428.79. After writing checks totaling $1097.31, depositing 2 checks of $691.53 each, and withdrawing $100 from a bank machine, what was the new balance in Marisa's account?   **$614.54**

13. Finance   Maria Cruz wants to invest part of her $8000 savings in a bond paying 12% annual interest, and the rest of it in a savings account paying 8%. If she wants to earn a total of $760, how much money should Maria invest at each rate?   **$5000 at 8%, $3000 at 12%**

14. Weather   The time t in hours, that a storm will last is given by the formula $t = \sqrt{\dfrac{d^3}{216}}$, where d is the diameter of the storm in miles. Suppose the umpires in a baseball game declared rain delay at 4:00 P.M. The storm causing the rain delay has a diameter of 12 miles. After the rain has stopped, can the game be restarted before 6:00 P.M.?   **No, since the rain will end about 6:50 P.M.**

15. Statistics   Find the mean, median, and mode of the data represented in the line plot below to the nearest tenth.   **47.5, 46, 35 and 63**

16. Running   Raul ran a distance of 385 yards in $1\frac{3}{4}$ minutes. His younger sister, Luisa, estimated it would take her $3\frac{1}{2}$ minutes to run the same distance. If Luisa ran at a rate of 94 yards per minute, was her estimation correct? Verify your answer.   **See students' work.**

17. Government   The term of office for a United States Senator is 150% as long as the term of office for the President. How long is the term of office for a United States Senator?   **6 years**

18. Physics   Shannon weighs 126 pounds and Minal weighs 154 pounds. They are seated at opposite ends of a seesaw. Shannon and Minal are 16 feet apart, and the seesaw is balanced. How far is Shannon from the fulcrum?   **8.8 ft**

SECTION THREE: OPEN-ENDED

This section contains two open-ended problems. Demonstrate your knowledge by giving a clear, concise solution to each problem. Your score on these problems will depend on how well you do the following.

- Explain your reasoning.
- Show your understanding of the mathematics in an organized manner.
- Use charts, graphs, and diagrams in your explanation.
- Show the solution in more than one way or relate it to other situations.
- Investigate beyond the requirements of the problem.

19. Geometry   One angle of a triangle measures 15° more than the second. The measure of the third angle is the sum of the first two angles tripled. Find the measure of each angle.   **15°, 30°, 135°**

20. Biology   An adult African elephant can weigh up to 12,000 pounds. An adult blue whale can weigh as much as 37 adult African elephants. A newborn blue whale weighs one-third of an adult African elephant's weight. The newborn whale will gain 200 pounds per day until it reaches its adult weight. How many days will it take for the newborn to reach adult weight?   **2200 days**

Cumulative Review   Chapters 1–4   **251**

Graphing Relations and Functions

PREVIEWING THE CHAPTER

This chapter makes the connection between algebraic equations and their geometric models. Students identify coordinates and locate points on the coordinate plane, identify the domain, range, and inverse of a relation, and solve linear equations for a specific variable or a given domain. They graph relations from a table of ordered pairs and determine whether the relation is a function. Graphing technology is integrated throughout the chapter. Finally, students examine measures of variation and their connection to the distribution of data.

Lesson (Pages)	Lesson Objectives	NCTM Standards	State/Local Objectives
5-1 (254–259)	Graph ordered pairs on a coordinate plane. Solve problems by making a table.	1–5, 8	
5-2A (260–261)	Use a graphing calculator to graph relations.	1–5, 8	
5-2 (262–269)	Identify the domain, range, and inverse of a relation. Show relations as sets of ordered pairs, tables, mappings, and graphs.	1–5	
5-3A (270)	Use a graphing calculator to investigate relations and determine the ranges when the domains are given.	1–5	
5-3 (271–277)	Determine the range for a given domain. Graph the solution set for the given domain.	1–5	
5-4A (278–279)	Use a graphing calculator to graph linear relations and functions.	1–5, 8	
5-4 (280–286)	Graph linear equations.	1–5, 8	
5-5 (287–294)	Determine whether a given relation is a function. Find the value of a function for a given element of the domain.	1–6	
5-6 (295–302)	Write equations to represent relations, given some of the solutions for the equation.	1–6	
5-7A (303–304)	Use a graphing calculator to calculate measures of variation.	1–4, 10	
5-7 (305–313)	Calculate and interpret the range, quartiles, and interquartile range of sets of data.	1–4, 10	

ORGANIZING THE CHAPTER

You may want to refer to the **Course Planning Calendar** on page T12 for detailed information on pacing.
PACING: Standard—14 days; **Honors**—13 days; **Block**—7 days; **Two Years**—21 days

LESSON PLANNING CHART

| Lesson (Pages) | Materials/ Manipulatives | Extra Practice (Student Edition) | Blackline Masters | | | | | | | | | | Real-World Applications | Interactive Mathematics Tools Software | Teaching Transparencies |
			Study Guide	Practice	Enrichment	Assessment and Evaluation	Modeling Mathematics	Multicultural Activity	Tech Prep Applications	Graphing Calculator	Science and Math Lab Manual			
5-1 (254–259)	geoboards graphing calculator	p. 767	p. 34	p. 34	p. 34			p. 9				14	5-1.1 5-1.2	5-1A 5-1B
5-2A (260–261)	graphing calculator									pp. 14, 15				
5-2 (262–269)	grid paper straightedge colored pencils	p. 767	p. 35	p. 35	p. 35	p. 128			p. 9		pp. 17–22		5-2	5-2A 5-2B
5-3A (270)	graphing calculator									pp. 16, 17				
5-3 (271–277)	graphing calculator	p. 767	p. 36	p. 36	p. 36			p. 10						5-3A 5-3B
5-4A (278–279)	graphing calculator									pp. 18, 19				
5-4 (280–286)	graphing calculator	p. 768	p. 37	p. 37	p. 37	pp. 127, 128				p. 5				5-4A 5-4B
5-5 (287–294)		p. 768	p. 38	p. 38	p. 38	pp. 48–50, 76			p. 10			15	5-5	5-5A 5-5B
5-6 (295–302)		p. 768	p. 39	p. 39	p. 39	p. 129								5-6A 5-6B
5-7A (303–304)	graphing calculator									pp. 20, 21				
5-7 (305–313)		p. 769	p. 40	p. 40	p. 40	p. 129						16		5-7A 5-7B
Study Guide/ Assessment (315–319)						pp. 113–126, 130–132								

ORGANIZING THE CHAPTER

OTHER CHAPTER RESOURCES

Student Edition
Chapter Opener, pp. 252–253
Mathematics and Society, p. 313
Working on the Investigation,
pp. 269, 302
Closing the Investigation, p. 314

Teacher's Classroom Resources
Investigations and Projects Masters,
pp. 41–44
Algebra and Geometry Overhead
Manipulative Resources,
pp. 16–18

Technology
Test and Review Software (IBM
and Macintosh)
CD-ROM Interactions (Windows
and Macintosh)

Professional Publications
Block Scheduling Booklet
Glencoe Mathematics Professional
Series

OUTSIDE RESOURCES

Books/Periodicals
Algebra Experiments I: Exploring Linear Functions,
Addison-Wesley
Cartesian Cartoons, Dale Seymour Publications

Software
Coordinate Math, MECC
Visualizing Algebra: The Function Analyzer,
Sunburst

Videos/CD-ROMs
Solving Equations 1, Coronet, MTI

ASSESSMENT RESOURCES

Student Edition
Math Journal, pp. 274, 283,
291
Mixed Review, pp. 259, 269,
277, 285, 294, 302, 312
Self Test, p. 286
Chapter Highlights, p. 315
Chapter Study Guide and
Assessment, pp. 316–318
Alternative Assessment, p. 319
 Portfolio, p. 319

Teacher's Wraparound Edition
5-Minute Check, pp. 254, 262,
271, 280, 287, 295, 305
Check for Understanding, pp. 257,
266, 274, 283, 291, 298, 309
Closing Activity, pp. 259, 269,
276, 285, 294, 302, 312
Cooperative Learning, pp. 281,
307

Assessment and Evaluation Masters
Multiple-Choice Tests, Forms 1A
(Honors), 1B (Average), 1C
(Basic), pp. 113–118
Free-Response Tests, Forms 2A
(Honors), 2B (Average), 2C
(Basic), pp. 119–124
Calculator-Based Test, p. 125
Performance Assessment, p. 126
Mid-Chapter Test, p. 127
Quizzes A–D, pp. 128–129
Standardized Test Practice, p. 130
Cumulative Review, pp. 131–132

ENHANCING THE CHAPTER

Examples of some of the materials for enhancing Chapter 5 are shown below.

DIVERSITY

Multicultural Activity Masters, pp. 9, 10

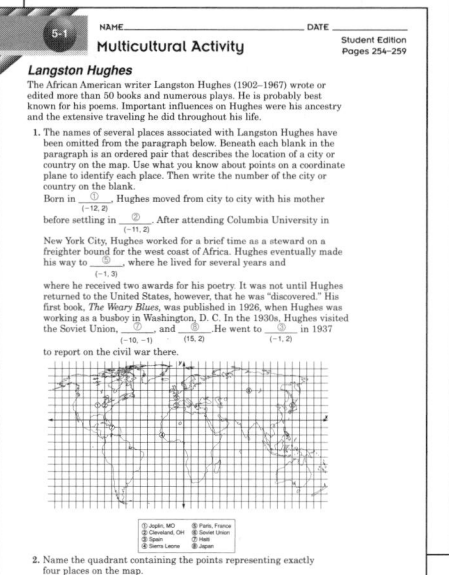

5-1 NAME_____ DATE_____
Multicultural Activity
Student Edition Pages 254–259

Langston Hughes

The African American writer Langston Hughes (1902–1967) wrote or edited more than 50 books and numerous plays. He is probably best known for his poems. Important influences on Hughes were his ancestry and the extensive traveling he did throughout his life.

1. The names of several places associated with Langston Hughes have been omitted from the paragraph below. Beneath each blank in the paragraph is an ordered pair that describes the location of a city or country on the map. Use what you know about points on a coordinate plane to identify each place. Then write the number of the city or country on the blank.

Born in ___①___, Hughes moved from city to city with his mother
(−12, 2)
before settling in ___②___. After attending Columbia University in
(−11, 2)
New York City, Hughes worked for a brief time as a steward on a freighter bound for the west coast of Africa. Hughes eventually made his way to ___⑤___, where he lived for several years and
(−1, 3)
where he received two awards for his poetry. It was not until Hughes returned to the United States, however, that he was "discovered." His first book, *The Weary Blues*, was published in 1926, when Hughes was working as a busboy in Washington, D. C. In the 1930s, Hughes visited the Soviet Union, ___⑦___, and ___⑧___. He went to ___⑥___ in 1937
(−10, −1)   (15, 2)   (−1, 3)
to report on the civil war there.

① Joplin, MO   ② Paris, France
③ Cleveland, OH ⑥ Soviet Union
④ Spain   ⑦ Haiti
⑤ Sierra Leone ⑧ Japan

2. Name the quadrant containing the points representing exactly four places on the map.

APPLICATIONS

Real-World Applications, 14, 15, 16

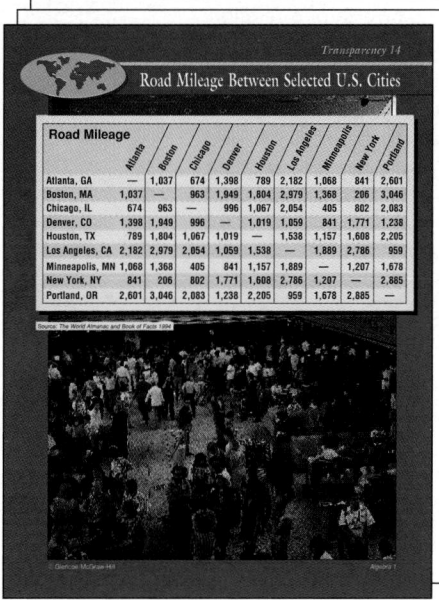

Transparency 14

Road Mileage Between Selected U.S. Cities

Road Mileage	Atlanta	Boston	Chicago	Denver	Houston	Los Angeles	Minneapolis	New York	Portland
Atlanta, GA	—	1,037	674	1,398	789	2,182	1,068	841	2,601
Boston, MA	1,037	—	963	1,949	1,804	2,979	1,368	206	3,046
Chicago, IL	674	963	—	996	1,067	2,054	405	802	2,083
Denver, CO	1,398	1,949	996	—	1,019	1,059	841	1,771	1,238
Houston, TX	789	1,804	1,067	1,019	—	1,538	1,157	1,608	2,205
Los Angeles, CA	2,182	2,979	2,054	1,059	1,538	—	1,889	2,786	959
Minneapolis, MN	1,068	1,368	405	841	1,157	1,889	—	1,207	1,678
New York, NY	841	206	802	1,771	1,608	2,786	1,207	—	2,885
Portland, OR	2,601	3,046	2,083	1,238	2,205	959	1,678	2,885	—

Source: The World Almanac and Book of Facts 1994

©Glencoe/McGraw-Hill   Algebra 1

TECHNOLOGY

Graphing Calculator Masters, p. 5

5-4 NAME_____ DATE_____
Graphing Calculator Activity
Student Edition Pages 280–285

The WINDOW Key

When you draw a graph on grid paper, you must first decide how large to make the graph and what numbers to mark on the x- and y-axes. Often you draw your graphs with the origin (0, 0) at the center of the graph. However, this is not always convenient. How to mark the scale depends on the coordinates of the points that need to be displayed. The [WINDOW] key is used to determine what portion of the graph will be displayed on the screen.

Tell whether the given points will fit on the screen if the given settings are used.

1. Points: (−5, 12), (4, −14), (8, 20)   2. Points: (−10, 0), (−5, 15), (0, 8)
yes; yes; no   yes; no; no

	x-axis	y-axis
Minimum	−5	−15
Maximum	5	15
Scale size	1	1

	x-axis	y-axis
Minimum	−12	0
Maximum	−6	20
Scale size	2	2

For each problem, adjust the window so that a square with the given coordinates can be drawn centered on the screen. Then draw the square. (Hint: See Example 3 in Lesson 5-2A.) Record the values that you use for the window settings. (Note: Your square may look like a rectangle when it is graphed.) Answers may vary. Sample answers are given, in the following order: min (x), max (x), scale size (x); min (y), max (y), scale size (y).

3. (12, 3), (18, 3), (18, −3), (12, −3)   4. (7, 9), (9, 9), (9, 7), (7, 7)
10, 20, 1; −5, 5, 1   5, 11, 1; 5, 11, 1

5. (−5, −10), (−5, 2), (7, 2), (7, −10)   6. (−10, 10), (−10, −10), (10, −10), (10, 10)
−7, 9, 1; −12, 4, 1   −12, 12, 1; −12, 12, 1

For each problem, the coordinates of three of the four vertices of a quadrilateral are given. Plot the points and connect them in order. Find the coordinates of the fourth vertex so that the figure will have a line of symmetry. Finish drawing the quadrilateral. Answers may vary. Sample answers are given.

7. (0, 7), (−5, 12), (0, 0)   8. (8, 2), (4, 7), (−6, −1)
(−5, −5)   (−2, −6)

TECH PREP

Tech Prep Applications Masters, pp. 9, 10

5-2 NAME_____ DATE_____
Tech Prep Applications
Student Edition Pages 262–269

Traffic Patterns (Traffic Technician)

Highway planners often find a need to expand an existing highway system. Another bridge over a bay may need to be built. An existing roadway may need to be widened. What was once a country road may need to become a super highway.

One way to estimate how many cars and trucks use a particular route is to count them. Traffic technicians can actually count them and record results on paper. They can also use automatic devices to achieve the same goal. In any case, a statistical record of traffic patterns is necessary to determine whether a highway system needs to be expanded.

The table at the right shows the actual count of cars and trucks that use a particular roadway during rush hour on one weekday selected at random.

Time	Cars	Trucks
6:00 A.M.–6:30 A.M.	133	65
6:30 A.M.–7:00 A.M.	200	60
7:00 A.M.–7:30 A.M.	210	63
7:30 A.M.–8:00 A.M.	210	63
8:00 A.M.–8:30 A.M.	250	55
8:30 A.M.–9:00 A.M.	222	50

What relations does the table suggest? What are the domains and ranges of the relations?

There is a relationship between time intervals and the count of cars.

Domain: set of half-hour time intervals between 6:00 A.M. and 9:00 A.M.
Range: {133, 200, 210, 222, 250}

There is a relationship between time intervals and the count of trucks.

Domain: set of half-hour time intervals between 6:00 A.M. and 9:00 A.M.
Range: {50, 55, 60, 63, 65}

Solve.

1. In what time interval is the count of cars the greatest?
8:00 A.M.–8:30 A.M.

2. Is there a relationship between the number of cars using the road at one time and the number of trucks using the road at the same time?
Yes; the relation is the set of ordered pairs (count of cars, count of trucks).

3. Is there a trend in the count of cars as rush hour progresses?
Yes; the count increases until it reaches a peak of 250 between 8:00 A.M.–8:30 A.M. and then decreases.

CONNECTIONS

Science and Math Lab Manual, pp. 17–22

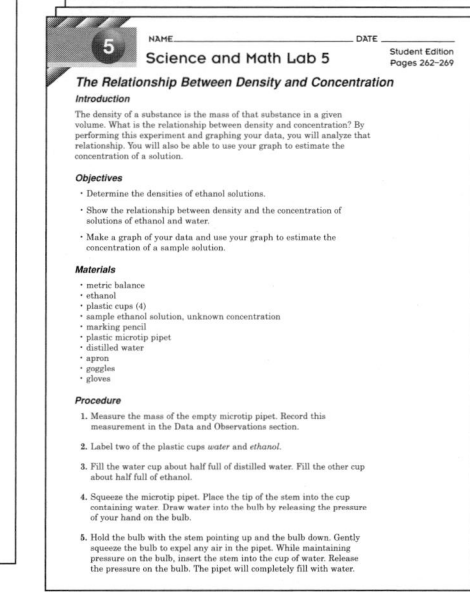

5 NAME_____ DATE_____
Science and Math Lab 5
Student Edition Pages 262–269

The Relationship Between Density and Concentration

Introduction

The density of a substance is the mass of that substance in a given volume. What is the relationship between density and concentration? By performing this experiment and graphing your data, you will analyze that relationship. You will be able to use your graph to estimate the concentration of a solution.

Objectives

- Determine the densities of ethanol solutions.
- Show the relationship between density and the concentration of solutions of ethanol and water.
- Make a graph of your data and use your graph to estimate the concentration of a sample solution.

Materials

- metric balance
- ethanol
- plastic cups (4)
- sample ethanol solution, unknown concentration
- marking pencil
- plastic microtip pipet
- distilled water
- apron
- goggles
- gloves

Procedure

1. Measure the mass of the empty microtip pipet. Record this measurement in the Data and Observations section.

2. Label two of the plastic cups *water* and *ethanol*.

3. Fill the water cup about half full of distilled water. Fill the other cup about half full of ethanol.

4. Squeeze the microtip pipet. Place the tip of the stem into the cup containing water. Draw water into the bulb by releasing the pressure of your hand on the bulb.

5. Hold the bulb with the stem pointing up and the bulb down. Gently squeeze the bulb to expel any air in the pipet. While maintaining pressure on the bulb, insert the stem into the cup of water. Release the pressure on the bulb. The pipet will completely fill with water.

PROBLEM SOLVING

Problem of the Week Cards, 13, 14, 15

Divide and Conquer
Problem-of-the-Week
Card 13

The Problems

A. George and Raphael work at a kennel. One day they are told to move gates into a 20 ft × 20 ft pen so that 9 dogs can be separated. The dogs must also stay as far apart from each other as possible (see diagram). When they check their supplies, they find four gate sections 10 feet long and four other fence sections that are a little more than 14 feet long. If they are not allowed to bend or cut the gates, and if some dogs can have more space than others, how can they fence in the dogs if they decide to build square fences?

B. If x = 2, evaluate the expressions and then draw four straight lines to separate the square area into 11 smaller areas such that the sum of the expressions in each area is 14.

Strategies and Hints

Problem A

1. Be sure to keep the dogs in their approximate positions.

2. Think about how squares could be drawn inside the total area.

3. Try to solve the problem ignoring the given lengths.

Problem B

1. Find ways to divide a plane into 11 parts using three straight lines.

2. Think about the location of lines in relationship to the value of the algebraic expressions.

©Glencoe/McGraw-Hill   Algebra 1

	Country/ Western	Rock	Jazz
TOTAL	51.7%	43.5%	34.2%
AGE			
18 to 24	39	70	30
25 to 34	50	59	41
35 to 44	53	57	39
45 to 54	61	39	33
55 to 64	58	14	30
65 to 74	54	9	27
75 to 96	46	7	21
EDUCATION			
Grade school	48	12	10
Some high school	59	27	15
High school graduate	57	42	28
Some college	50	54	42
College graduate	42	54	50
Graduate school	46	53	54
FAMILY INCOME			
Under $5,000	43	36	27
$5,000 to $9,999	52	32	21
$10,000 to $14,999	55	33	25
$15,000 to $24,999	57	39	29
$25,000 to $49,999	54	50	36
$50,000 or more	48	55	47
Not ascertained	42	35	35

CHAPTER 5

Graphing Relations and Functions

Objectives

In this chapter, you will:
- graph ordered pairs, relations, and equations,
- solve problems by making a table,
- identify the domain, range, and inverse of a relation,
- determine if a relation is a function,
- write an equation to represent a relation, and
- calculate and interpret the range, quartiles, and interquartile range of a set of data.

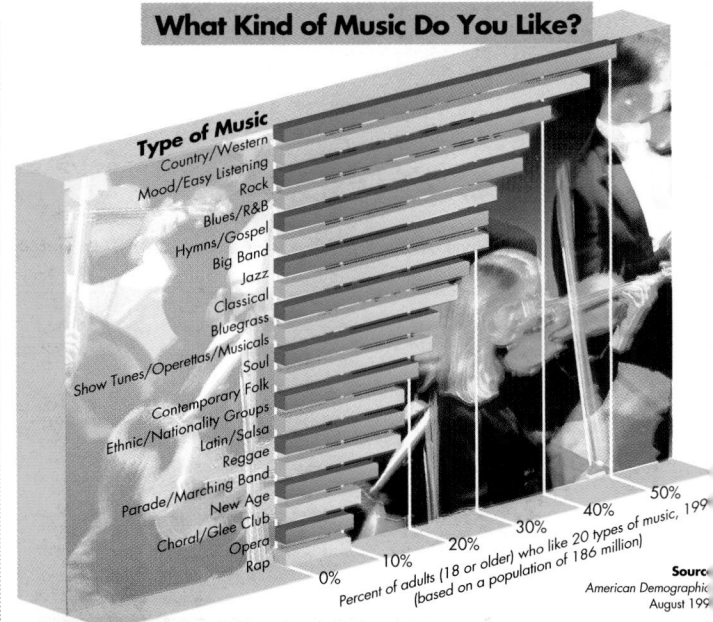

What Kind of Music Do You Like?

Type of Music

Country/Western, Mood/Easy Listening, Rock, Blues/R&B, Hymns/Gospel, Big Band, Jazz, Classical, Bluegrass, Show Tunes/Operettas/Musicals, Soul, Contemporary Folk, Ethnic/Nationality Groups, Latin/Salsa, Reggae, Parade/Marching Band, New Age, Choral/Glee Club, Opera, Rap

0% 10% 20% 30% 40% 50%

Percent of adults (18 or older) who like 20 types of music, 199
(based on a population of 186 million)

Source
American Demographic
August 199

What kinds of music do you listen to most often? Most teenagers have predictable tastes in music—listening to and buying popular rock and country/western favorites. As you become an adult, you may learn to enjoy listening to many other types of music. Who knows, you may become a fan of 12th century Gregorian chants, or perhaps even Zydeco!

TIME Line

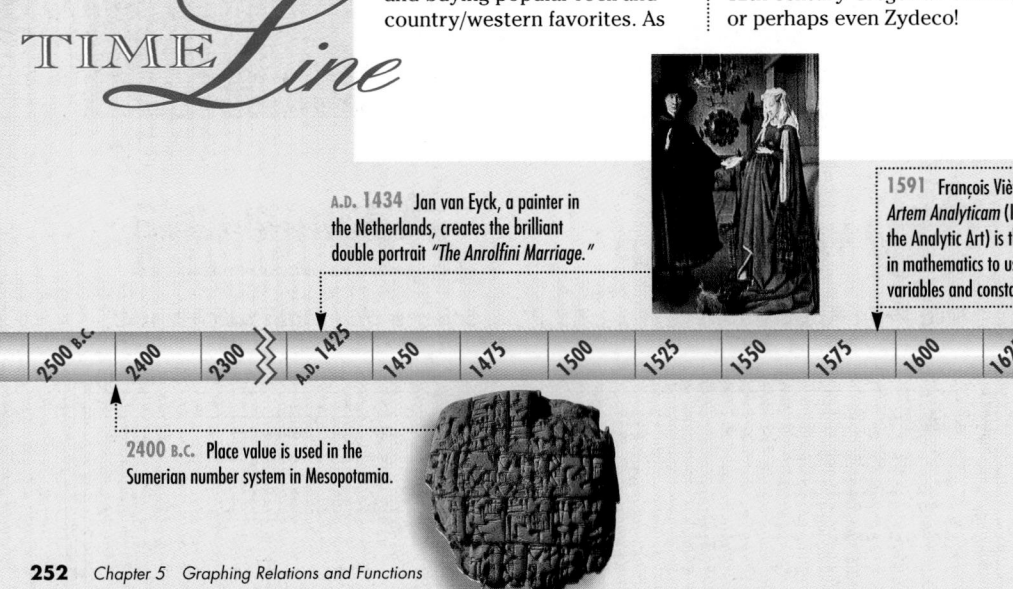

A.D. 1434 Jan van Eyck, a painter in the Netherlands, creates the brilliant double portrait "The Anrolfini Marriage."

1591 François Viète's *Isago Artem Analyticam* (Introductic the Analytic Art) is the first w in mathematics to use letters variables and constants.

2500 B.C. 2400 2300 A.D. 1425 1450 1475 1500 1525 1550 1575 1600 1625

2400 B.C. Place value is used in the Sumerian number system in Mesopotamia.

TIME Line

The Sumerian number system, which is based on 60, still survives today. Our hour is divided into 60 minutes, and our minute is divided into 60 seconds. Students might want to look further into the mathematics and applications of the base 60 system.

interNET CONNECTION

The Jazz Fan Attic site contains information on blues, ragtime, Dixieland, contemporary jazz, and general interest.

World Wide Web
http://www2.magmacom.com/~rbour/

Chapter Project

By the young age of 10, **Jamie Knight** was already an accomplished jazz singer. She has even performed at New York City's Carnegie Hall, following in the footsteps of Ella Fitzgerald and Mel Torme. Jamie is from Philadelphia, Pennsylvania, and has been performing since the age of 6. She has appeared at jazz festivals and on *Good Morning America.* She also has her own CD.

Work in groups of three to conduct your own music survey.

- Each person in the group should survey 20 people of varying ages to determine their musical tastes. Use the graph on the previous page for ideas.

- Combine your results and compare them to the graph. Does your survey match the national results?

- Create a bar graph to show your results for each type of music. Also create bar graphs

for the five most popular types of music you surveyed showing the age distribution for each type of music. Then create a line graph comparing male and female musical tastes in relation to age groups.

- Compare your graphs, make some conclusions, and report your findings.

- Do you think certain types of music are associated with different cultures? You may want to include your thoughts on this in your report.

1846 Adolphe Sax, attempting to improve on the bass clarinet, invents a new instrument, the saxophone.

1965 Engineer Robert Moog, with the help of composers Herbert Deutsch and Walter Carlos, creates the Minimoog, the first electronic synthesizer.

1993 *Jurassic Park* was the top money-making movie and a best-selling paperback.

| 1725 | 1750 | 1775 | 1800 | 1825 | 1850 | 1875 | 1900 | 1925 | 1950 | 1975 | 2000 |

1750 The great Chinese work of fiction *The Dream of the Red Chamber* is written by Tsao Hsueh Chin.

1943 Duke Ellington composes his concert work *Black, Brown, and Beige.*

Chapter 5 **253**

Alternative Chapter Projects

Two other chapter projects are included in the *Investigations and Projects Masters.* In Chapter 5 Project A, pp. 41–42, students extend the topic in the chapter opener. In Chapter 5 Project B, pp. 43–44, students investigate fundraising.

Jamie has performed at Carnegie Hall, with her own rendition of "It Don't Mean a Thing If It Ain't Got That Swing." While she likes all kinds of music, Jamie says, "Jazz is different. It has its own flavor."

Chapter Project

Cooperative Learning Students might want to organize into groups of four and then work in pairs within their groups. They can compile the results from the two teams to create a larger sample for their data.

Investigations and Projects Masters, p. 41

NAME_____ DATE____
Student Editic
Pages 254–3

5 Chapter 5 Project A

Venues and Revenues

1. Work with a partner to make a list of places in your area that feature live performances by musicians. You may need to look in a newspaper for ideas.

2. Do research to find out the seating capacity of each place, or venue, you listed in exercise 1. Also, find out what kind(s) of music is (are) usually performed in each venue. Present your findings in the form of a graph.

3. Write an equation that can be used to estimate the amount of money that would be generated by a sold-out show at each venue.

4. Find the average ticket price for upcoming concerts at the venues you researched. You may be able to do this by looking at newspaper ads.

5. Use the information about ticket prices from exercise 4 and the equation from exercise 3 to determine how much money different concerts would make at each venue. Then find out how the money from ticket sales is divided among a venue's owners, the musicians, the promoter, the company selling the tickets, and others.

6. Use what you've learned to write a newspaper article about the business of performing live music. Be sure to include any graphs you've drawn.

NCTM Standards: 1–5, 8

Instructional Resources

- Study Guide Master 5-1
- Practice Master 5-1
- Enrichment Master 5-1
- Multicultural Activity Masters, p. 9
- Real-World Applications, 14

 Transparency 5-1A contains the 5-Minute Check for this lesson; **Transparency 5-1B** contains a teaching aid for this lesson.

Recommended Pacing

Standard Pacing	Day 1 of 14
Honors Pacing	Day 1 of 13
Block Scheduling*	Day 1 of 7
Alg. 1 in Two Years*	Days 1 & 2 of 21

 *For more information on pacing and possible lesson plans, refer to the *Block Scheduling Booklet* and *Algebra 1 in Two Years.*

1 FOCUS

 ### 5-Minute Check
(over Chapter 4)

Solve.

1. $\frac{6}{15} = \frac{x}{45}$ **18**
2. What number is 75% of 200? **150**
3. Bill is driving at 50 kilometers per hour. How far will he travel in 2 hours? In h hours? **100 km; 50h km**
4. If y varies directly as x and $y = -13$ when $x = -6$, find y when $x = 3$. **6.5**
5. If $\triangle NBC \sim \triangle FOX$, then $\frac{BC}{?} = \frac{CN}{?}$. **OX; XF**

5-1

The Coordinate Plane

Objectives
After studying this lesson, you should be able to:
- graph ordered pairs on a coordinate plane, and
- solve problems by making a table.

APPLICATION
Landmarks

In 1988, as part of Cincinnati's 200th birthday celebration, the city's Bicentennial Commission created parks along the Ohio River. They sold over 33,800 personalized bricks that were placed in the pathways in the parks. However, people had difficulty finding their bricks. So, in 1989, Mr. Frank Albi of Business Information Storage (BIS) developed a master directory, in which each patron's name was listed along with the section of the park and the column in which his or her brick was laid. BIS then added locator bricks to identify the columns in each section.

Perpendicular lines are lines that meet to form 90° angles.

Unless otherwise designated, you can assume that each division on the axes represents 1 unit.

The system used in Cincinnati helped to locate the specific site of each brick. In mathematics, points are located in reference to two perpendicular number lines called **axes.** The axes intersect at their zero points, a point called the **origin.** The horizontal number line, called the **x-axis,** and the vertical number line, called the **y-axis,** divide the plane into four **quadrants.** The quadrants are numbered as shown at the right. Notice that neither the axes nor any point on the axes are located in any quadrant. The plane containing the x- and y-axes is called the **coordinate plane.**

Points in the coordinate plane are named by *ordered pairs* of the form (x, y). The first number, or **x-coordinate,** corresponds to the numbers on the x-axis. The second number, or **y-coordinate,** corresponds to the numbers on the y-axis. The ordered pair for the origin is $(0, 0)$.

A point can be referred to by just a letter. Thus, Z can be used to mean point Z.

To find the ordered pair for point Z, shown at the right, first follow along a vertical line through Z to find the x-coordinate on the x-axis and along a horizontal line through Z to find the y-coordinate on the y-axis. The number on the x-axis that corresponds to Z is 4. The number on the y-axis that corresponds to Z is -6. Thus, the ordered pair for point Z is $(4, -6)$. This ordered pair can also be written as $Z(4, -6)$. The x-coordinate of Z is 4, and the y-coordinate of Z is -6. Point Z is located in Quadrant IV.

Classroom Vignette

"I give each student a slip of paper as he or she enters the room. The paper has a coordinate pair written on it. The abscissa is a letter and the ordinate is a whole number. The room is laid out like a coordinate grid. They use the coordinates to locate their new seats."

Stan Musgrave
Rochester High School
Rochester, Indiana

Stan Musgrave

A table is often used when working with ordered pairs. It is a helpful organizational tool in problem solving.

Example 1

Write ordered pairs for points *E, F, G,* and *H* shown at the right. Name the quadrant in which each point is located.

Use a table to help find the coordinates of each point.

LOOK BACK
You can review more about using tables in Lesson 4-7.

Point	x-coordinate	y-coordinate	Ordered Pair	Quadrant
E	4	–2	(4, –2)	IV
F	0	–5	(0, –5)	none
G	–3.5	–4	(–3.5, –4)	III
H	–6	3	(–6, 3)	II

To **graph** an ordered pair means to draw a dot at the point on the coordinate plane that corresponds to the ordered pair. This is sometimes called *plotting a point.* When graphing an ordered pair, start at the origin. The *x*-coordinate indicates how many units to move right (positive) or left (negative). The *y*-coordinate indicates how many units to move up (positive) or down (negative).

Example 2

Plot the following points on a coordinate plane.

a. *N*(−3, −5)

Start at the origin, *O.* Move left 3 units since the *x*-coordinate is −3. Then move down 5 units since the *y*-coordinate is −5. Draw a dot and label it *N.*

b. *K*(−4.5, 8)

Start at the origin, *O.* Move left 4.5 units and up 8 units. Draw a dot and label it *K.*

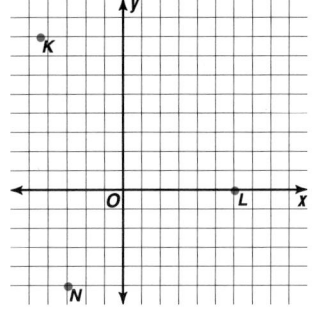

c. *L*(6, 0)

Start at the origin, *O.* Move right 6 units. Since the *y*-coordinate is 0, the point will be located on the *x*-axis. Draw the dot and label it *L.*

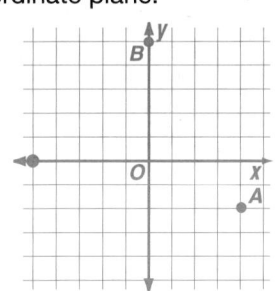

Teaching Tip Students often confuse horizontal and vertical. Emphasize that a horizontal line runs in the same direction as the *horizon*. A vertical line runs up and down, or perpendicular to the horizontal line.

In-Class Example

For Example 3
In which sector is each landmark located?

a. San Antonio International Airport (3, *A*)
b. Stinson Airport (3, *D*)
c. St. Mary's University (2, *B*)
d. The University of Texas at San Antonio (1, *A*)

Cassini published his major work on cartography, *Geometric Description of the Earth,* in 1775.

3 PRACTICE/APPLY

Check for Understanding
Exercises 1–12 are designed to help you assess your students' understanding through reading, writing, speaking, and modeling. You should work through Exercises 1–3 with your students and then monitor their work on Exercises 4–12.

Error Analysis
A common mistake that students make is to plot the first coordinate of an ordered pair on the vertical axis and the second coordinate on the horizontal axis. Picture graphing as in Exercises 37 and 38 is a fun, effective way to practice graphing points and provides a self-check for students.

Each point in the coordinate plane can be named by exactly one ordered pair, and each ordered pair names exactly one point in the plane. Thus, there is a one-to-one correspondence between points and ordered pairs.

Completeness Property for Points in the Plane	1. Exactly one point in the plane is named by a given ordered pair of numbers.
	2. Exactly one ordered pair of numbers names a given point in the plane.

On city and state maps, letters and numbers are usually used to locate landmarks on the map. The number representing the horizontal coordinate is usually listed first followed by the letter representing the vertical coordinate. However, unlike the coordinate plane where a specific point is denoted by the ordered pair, the coordinates on a map define a region or sector in which the landmark is located.

Example 3

APPLICATION
Cartography

F Y I
César-François Cassini, who was born in Thury, France, in 1714, was the first cartographer to create a map according to modern principles. In 1744, he directed the first National Geographic survey, the triangulation of France.

Use ordered pairs to name all of the sectors that Interstate Highway 35 passes through on the map of San Antonio, Texas, shown below.

I-35 begins in the southwest corner and extends through San Antonio to the northeast corner. Part of it shares the same path as I-410.

I-35 goes through (1, *D*), continuing through (2, *D*), across the southeast corner of (2, *C*), turning upward through (3, *C*), across the southern portion of (3, *B*), sharing the path with I-410 through (4, *B*), and exiting the map at the eastern edge of (4, *A*).

256 *Chapter 5   Graphing Relations and Functions*

GLENCOE *Technology*

 Interactive Mathematics Tools Software

This multimedia software provides an interactive lesson by graphing ordered pairs on a coordinate plane and observing what occurs when the *x* and *y* values are exchanged. A **Computer Journal** gives students an opportunity to write about what they have learned.

For Windows & Macintosh

 Alternative Teaching Strategies

Student Diversity

• Review lesson vocabulary words. Make sure students can identify the origin, *x*-axis, *y*-axis, quadrants, and the *x*- and *y*-coordinates within an ordered pair.

• Use a variety of maps with different scales (city, state, etc.) to help students better understand the concept of using coordinates.

Communicating Mathematics

Study the lesson. Then complete the following.

1. **Draw** a coordinate plane. Label the origin, *x*-axis, *y*-axis, and the quadrants. **See coordinate plane on page 254.**

2. **Explain** how you can tell which quadrant a point is in by just looking at the signs of its coordinates. $(+, +)$, I; $(-, +)$, II; $(-, -)$, III; $(+, -)$, IV

MODELING MATHEMATICS

3. **Copy or sketch** a map of the major thoroughfares in your community. Using a different color, draw a coordinate system on the map, making your home the origin. Find the coordinates of your school, a library, a shopping center or department store, and your favorite restaurant. **See students' work.**

Guided Practice

Write the ordered pair for each point shown at the right. Name the quadrant in which the point is located.

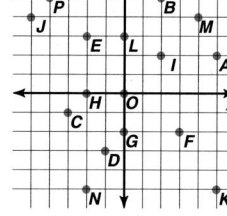

5. $(-3, -1)$; III

7. $(0, -2)$; none

4. A $(5, 2)$; I

5. C

6. E $(-2, 3)$; II

7. G

Graph each point. 8–11. See margin.

8. $W(-5, 0)$

9. $X(-3, -4)$

10. $Y(3, -5)$

11. $Z(0, 4)$

12. Write the ordered pair that describes a point 15 units up from and 13 units to the left of the origin. $(-13, 15)$

Practice

Refer to the coordinate plane for Exercises 4–7. Write the ordered pair for each point. Name the quadrant in which the point is located.

 A

13. D $(-1, -3)$; III
14. H $(-2, 0)$; none
15. L $(0, 3)$; none
16. P $(-4, 5)$; II
17. O $(0, 0)$; none
18. K $(5, -5)$; IV
19. F $(3, -2)$; IV
20. B $(2, 5)$; I
21. I $(2, 2)$; I
22. M $(4, 4)$; I
23. J $(-5, 4)$; II
24. N $(-2, -5)$; III

Graph each point. 25–36. See margin.

 B

25. $A(2, -1)$
26. $B(-3, -3)$
27. $C(0, 3.5)$
28. $D(5, -2)$
29. $E(-3, 0)$
30. $F(3, 5)$
31. $G(-3, -4)$
32. $H(-5, -2)$
33. $I(0, -2)$
34. $J(-\frac{1}{2}, 1)$
35. $K(4, 0)$
36. $L(4, 2)$

Graph each point. Then connect the points in alphabetical order and identify the figure. 37–38. See margin for graphs.

37. $A(2, 0)$, $B(2, 3)$, $C(1, 3)$, $D(-0.5, -1)$, $E(-2, 3)$, $F(-3, 3)$, $G(-3, -3)$, $H(-2, -3)$, $I(-2, 0)$, $J(-0.5, -3)$, $K(1, 0)$, $L(1, -3)$, $M(2, -3)$, $N(2, 0)$ **M**

 C

38. $A(9, 0.5)$, $B(5, 1)$, $C(2, 1)$, $D(-1, 5)$, $E(-2, 5)$, $F(-1, 1)$, $G(-4, 1)$, $H(-5, 3)$, $I(-6, 3)$, $J(-5.5, 1)$, $K(-5.5, 0)$, $L(-6, -2)$, $M(-5, -2)$, $N(-4, 0)$, $P(-1, 0)$, $Q(-2, -4)$, $R(-1, -4)$, $S(2, 0)$, $T(5, 0)$, $U(9, 0.5)$ **airplane**

Lesson 5–1   The Coordinate Plane   **257**

Reteaching ▬▬▬

Using Visual Models To practice reading coordinates of points, have students make a picture using straight lines on a coordinate grid. List the coordinates of the points in order. Exchange ordered lists of points, and plot and connect the points in the order given as in dot-to-dot puzzles.

Additional Answer

38.

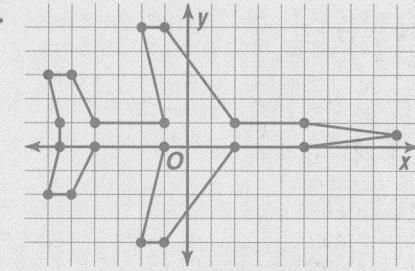

Additional Answers

8–11.

25–36.

37.

Study Guide Masters, p. 34

Practice Masters, p. 34

Programming

39. The graphing calculator program at the right will plot a group of points on the same coordinate plane. You will need to set an appropriate window so that you will be able to see all of the points. After each point is plotted, press ENTER to plot another point. To end the program, press 2nd QUIT while you are at the home screen. Use this program to check the points you plotted in Exercises 37 and 38. **See students' work.**

```
PROGRAM:PLOTPTS
:ClrDraw
:Lbl 1
:Prompt X,Y
:Pt-On(X,Y)
:Pause
:Goto 1
```

Critical Thinking

40. Describe the possible locations, in terms of quadrants or axes, for the graph of (x, y) if x and y satisfy the given conditions.
   a. $xy > 0$ **I or III**   **b.** $xy < 0$ **II or IV**   **c.** $xy = 0$ **x- or y-axis**

Applications and Problem Solving

41. **Geography** On the map of the United States shown below, latitude and longitude can be used to form ordered pairs that name the sectors on the map. The longitude is represented along the horizontal axis and should be the first coordinate of the ordered pair. The latitude is represented along the vertical axis and should be the second coordinate.

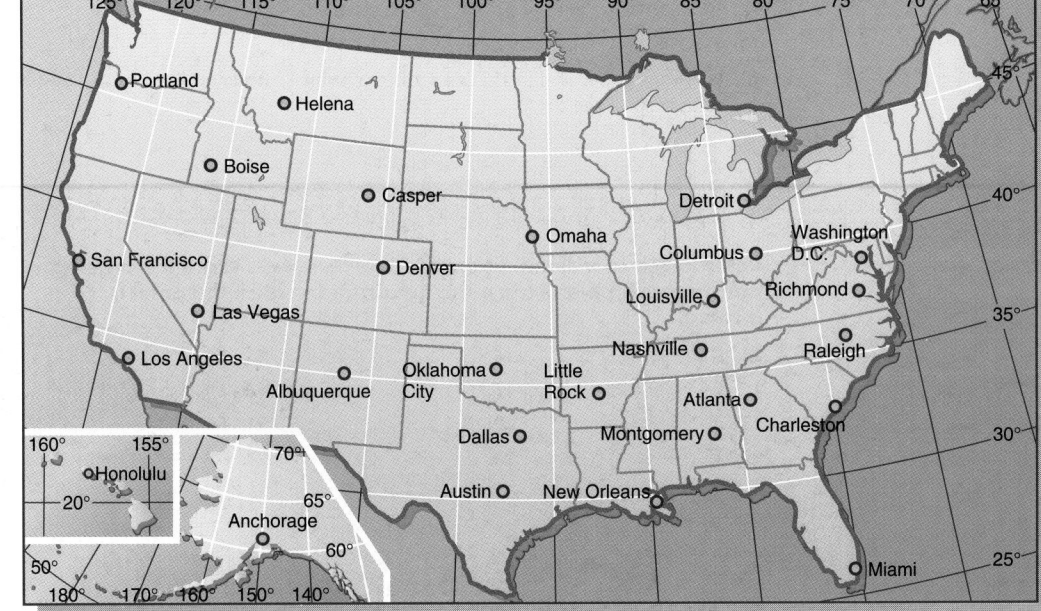

a. Name the city at (90°, 30°). **New Orleans**
b. Name the state in which (120°, 45°) is located. **Oregon**
c. Estimate the latitude and longitude of our nation's capital to the nearest 5°. **Answers may vary. Sample answer: (75°, 40°)**
d. What state capital has its location at (157°, 21°)? **Honolulu, Hawaii**
e. What is the approximate latitude and longitude of your community?
f. Look at the longitude and latitude lines on a globe and compare them to the ones shown on this map. What do you find?

41e. See students' work.
41f. Sample answer: The longitude lines are not the same distance apart; they meet at the poles.

Classroom Vignette

"I use the idea in Exercise 43 every fall to make up my seating chart. I arrange the desks in rows and columns, and as the students enter the class, I hand them a card with an ordered pair on it—then they find their seats!"

Deborah Phillips
St. Vincent–St. Mary High School
Akron, Ohio

Deborah Patonai Phillips

Opa-Locka Airport
North South Expwy.
Miami Int'l. Airport
East West Expressway
Coral Gables
John Kennedy Causeway
Julia Tuttle Causeway
Miami Beach
MacArthur Causeway

Miami, Florida
AND VICINITY
Scale in Miles

42. Tourism Diana works for a travel agency and has been asked to highlight key landmarks on a map of the Miami area for a customer. Use the map of Miami at the left to answer the following questions.

   a. What causeway is in sector (3, M)?

   b. In what sector is the city of Coral Gables? **(1, 0)**

   c. Which interstate goes from sector (2, L) to (2, O)? **I-95**

   d. Name all of the sectors through which Highway 826 passes.

43. Draw a Diagram Make a seating chart of the desks in your mathematics classroom. Create a coordinate plane on your diagram so that each desk corresponds to an ordered pair in the plane. **See students' work.**

   a. What are the coordinates of your desk in the room?

   b. Name another student in your class. What is the coordinate of this person's desk?

42a. John Kennedy 42d. (1, P), (1, O), (1, N), (1, M), (1, L), (2, L), (3, L)

Mixed Review

44. If *y* varies directly as *x*, and *y* = 3 when *x* = 15, find *y* when *x* = −25. (Lesson 4–8) **−5**

45. Two trains leave York at the same time, one traveling north, the other south. The first train travels at 40 miles per hour and the second at 30 miles per hour. In how many hours will the trains be 245 miles apart? (Lesson 4–7) **$3\frac{1}{2}$ hours**

46. Use a calculator to find the measure of ∠*A* to the nearest degree if sin *A* = 0.2756. (Lesson 4–3) **16°**

47. Find the mean, median, and mode for {5, 9, 1, 2, 3}. (Lesson 3–7) **4; 3; none**

Simplify each expression.

48. $\sqrt{\dfrac{16}{25}}$ (Lesson 2–8) **$\dfrac{4}{5}$**

49. $\dfrac{7a + 35}{-7}$ (Lesson 2–7) **$-a - 5$**

50. Agriculture Agribusiness technicians assist in the organization and management of farms. They may prepare graphs of weather data for a given region. For example, the graph at the right displays normal maximum, minimum, and mean temperatures for the region around Des Moines, Iowa. (Lesson 1–9)

50a. 65°
50b. 51° to 72°

   a. What is the normal maximum temperature for October?

   b. What is the normal range of temperatures for May?

51. Simplify 7(5*a* + 3*b*) − 4*a*. (Lesson 1–8) **31*a* + 21*b***

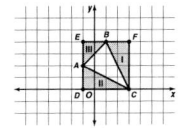

Normal Temperature Range for Des Moines

Temperature (°F)

— normal maximum/minimum
● normal mean

J F M A M J J A S O N D
Month

Extension ▬▬▬▬

Connections A parallelogram has three of its vertices at (0, 0), (2, 2), and (7, 1). Find the possible ordered pairs that could correspond to its fourth vertex. **(9, 3) and (5, −1)**

 Tech Prep

Travel Agent Students who are interested in a career in tourism may wish to do further research on the use of maps by travel agents, as in Exercise 42. For more information on tech prep, see the *Teacher's Handbook*.

Enrichment Masters, p. 34

5-1   NAME_____   DATE____

Enrichment   Student Edition Pages 254–259

Coordinate Geometry and Area

How would you find the area of a triangle whose vertices have the coordinates *A* (−1, 2), *B* (1, 4), and *C* (3, 0)?

When a figure has no sides parallel to either axis, the height and base are difficult to find.

One method of finding the area is to enclose the figure in a rectangle and subtract the area of the surrounding triangles from the area of the rectangle.

Area of rectangle *DEFC*
= 4 × 4
= 16 square units

Area of triangle I = $\frac{1}{2}$(2)(4) = 4

Area of triangle II = $\frac{1}{2}$(2)(4) = 4

Area of triangle III = $\frac{1}{2}$(2)(2) = 2
               Total = 10 square units

Area of triangle *ABC* = 16 − 10, or 6 square units

Find the areas of the figures with the following vertices.

1. *A*(−4, −6), *B*(0, 4), *C*(4, 2)
24 square units

2. *A*(6, −2), *B*(8, −10), *C*(12, −6)
20 square units

3. *A*(0, 2), *B*(2, 7), *C*(6, 10), *D*(9, −2)
55 square units

5-2A Graphing Technology
Relations

A Preview of Lesson 5–2

NCTM Standards: 1–5, 8

Objective
Use a graphing calculator to graph relations.

Recommended Time
25 minutes

Instructional Resources
Graphing Calculator Masters, pp. 14 and 15

These masters provide keystroking instruction for this lesson for the TI-81 and Casio graphing calculators.

1 FOCUS

Motivating the Lesson
Review with students and have them practice methods for setting standard windows and customizing a window by specifying Xmin, Xmax, Ymin, and Ymax.

2 TEACH

Teaching Tip When plotting a relation on the graphing calculator, the window must be chosen carefully. One way to do this is to set Xmin to 1 less than the smallest *x* value for all the pairs given and Xmax at 1 greater than the largest *x* value. Then set Ymin and Ymax in the same way.

A **relation** is a set of ordered pairs. You can graph a relation using a graphing calculator. Points can be plotted on the coordinate plane using the DRAW menu.

Example ① **Plot the following points on a coordinate plane:**
M(4, 7), *A*(−10, 25), *T*(−17, −17), and *H*(23, −11).

Begin with a viewing window of [−47, 47] by [−31, 31] using a scale of 10 for each axis. This can be done quickly by entering ZOOM 6 ZOOM 8 ENTER. *This sets the scale to integer values.*

Clear the coordinate plane by turning off all stat plots and clearing the Y= list.

If you press 2nd DRAW ▶ *1 before pressing* GRAPH*, the prompt Pt-On(appears. To plot a point from this prompt, enter the coordinates of the point, a closing parenthesis, and* ENTER*. Then proceed with the cursor for the other points.*

- To plot the first point, enter GRAPH 2nd DRAW ▶ 1. *The cursor appears at the origin and X = 0 and Y = 0 appears at the bottom of the screen.*

- Now use the arrow keys to move the cursor to the coordinates of the point you wish to graph. *The coordinates of each position of the cursor appear at the bottom of the viewing screen.*

- Press ENTER to put a dot at that point.

- Continue to move the cursor and press ENTER to graph the other three points.

- When finished, you can clear the screen by pressing 2nd DRAW 1.

If you do not need a range as great as the one you get by pressing ZOOM 8, you can set the WINDOW values in the usual manner or press ZOOM 6. Once you set your new window and try the method shown in Example 1, you will find that you cannot plot points at exact integer values by using the cursor. However, there is another way to plot exact points.

Example ② **Plot *G*(−4, 5), *R*(1, 6), *A*(−3, −6), *P*(5, −8), and *H*(0, 7) in the standard viewing window.**

- Set the standard viewing window. **Enter:** ZOOM 6

- Return to the home screen by pressing 2nd QUIT.

- You can graph all of the points by linking commands into a series using a colon.

Enter: 2nd DRAW ▶ 1 *Pt-On(appears on the screen.*

[(-)] 4 [,] 5 [)] [2nd] [:]

This enters the first point. Now repeat the steps without pressing ENTER to enter all the other points. Finally, press [ENTER] to graph the points.

Enter: [2nd] [DRAW] [▶] 1 1 [,]
6 [)] [2nd] [:] [2nd] [DRAW]
[▶] 1 [(-)] 3 [,] [(-)] 6 [)]
[2nd] [:] [2nd] [DRAW] [▶] 1
5 [,] [(-)] 8 [)] [2nd] [:]
[2nd] [DRAW] [▶] 1 0 [,] 7 [)] [ENTER]

You can also use the Line(feature on the DRAW menu to connect the points. This can help you analyze the points in a relation. You must first clear the screen by pressing [2nd] [DRAW] 1 [ENTER] before beginning a new graph.

Example 3 Determine if the graphs of (3, 5), (−3, −5), and (0, 7) are on the same line.

If you connect the points and only a segment appears, the three points are probably on the same line.

Enter: [2nd] [DRAW] 2 3 [,] 5
[,] [(-)] 3 [,] [(-)] 5 [)]
[2nd] [:] [2nd] [DRAW] 2 [(-)]
3 [,] [(-)] 5 [,] 0 [,] 7 [)]
[2nd] [:] [2nd] [DRAW] 2
3 [,] 5 [,] 0 [,] 7 [)] [ENTER]

Because the resulting figure is a triangle, the three points are not on the same line.

EXERCISES

1–6. See margin.

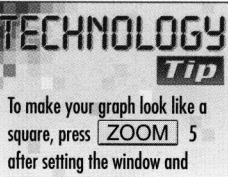

TECHNOLOGY *Tip*

To make your graph look like a square, press [ZOOM] 5 after setting the window and before using DRAW.

Use a graphing calculator to plot each point in the viewing window [−47, 47] by [−31, 31].

1. $X(10, 10)$ 2. $Y(0, -6)$ 3. $Z(-26, 11)$
4. $Q(-17, -19)$ 5. $T(-21, 4)$ 6. $P(-20, 0)$

7. Explain why, when you graph the points in Exercises 2 and 6, you do not see them on the screen.   **They are points on the axes.**

8. **a.** Plot $H(9, 9)$ and $J(29, 9)$.   **8a–c. See margin.**
   b. Find two additional points E and F so that these four points form the vertices of a square. Are there other possible coordinates of E and F?
   c. Plot E and F to complete the square.

3 PRACTICE/APPLY

Assignment Guide

Core: 1–8
Enriched: 1–8

4 ASSESS

Observing students working with technology is an excellent method of assessment.

Additional Answers

1–6.

8a.
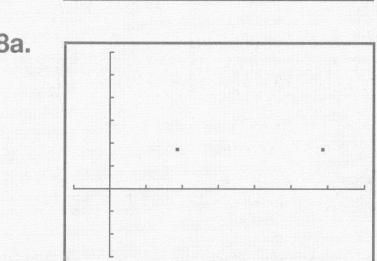

8b. $E(9, -11), F(29, -11);$
      $E(9, 29), F(29, 29)$

8c.

NCTM Standards: 1–5

Instructional Resources

- Study Guide Master 5-2
- Practice Master 5-2
- Enrichment Master 5-2
- Assessment and Evaluation Masters, p. 128
- Science and Math Lab Manual, pp. 17–22
- Tech Prep Applications Masters, p. 9

Transparency 5-2A contains the 5-Minute Check for this lesson; **Transparency 5-2B** contains a teaching aid for this lesson.

Recommended Pacing	
Standard Pacing	Day 3 of 14
Honors Pacing	Day 3 of 13
Block Scheduling*	Day 2 of 7
Alg. 1 in Two Years*	Days 4 & 5 of 21

*For more information on pacing and possible lesson plans, refer to the *Block Scheduling Booklet* and *Algebra 1 in Two Years*.

1 FOCUS

5-Minute Check
(over Lesson 5-1)

Name the quadrant in which each point is located.

1. $(-3, 2)$ **Quadrant II**
2. $(6, 2)$ **Quadrant I**
3. $(-8, -1)$ **Quadrant III**
4. $(4, -7)$ **Quadrant IV**
5. $(9, 0)$ **It lies on the *x*-axis.**

Motivating the Lesson

Questioning Tell students you are going to say two pairs of numbers. The numbers in each pair are related in the same way. The students need to identify the relationship and then name an additional pair that is related in the same way. For example, you might say "7 and 2, 4 and −1." The students might respond with "difference of the numbers in each pair is 5." Continue the activity with different number pairs.

262 *Chapter 5*

Relations

Objectives

After studying this lesson, you should be able to:
- identify the domain, range, and inverse of a relation, and
- show relations as sets of ordered pairs, tables, mappings, and graphs.

F Y I

The U.S. Congress passed the Rare and Endangered Species Act in 1966, setting up rules by which the Fish and Wildlife Service can list and protect species in danger of extinction.

What is the trend over time?

CONNECTION
Biology

There are 1.4 million classified species of microorganisms, invertebrates, plants, fish, birds, reptiles, amphibians, and mammals. Currently, 930 species worldwide have been listed as *endangered*. A species becomes endangered when its numbers are so low that the species is in danger of extinction.

In the United States, the manatee, an aquatic mammal, is considered to be endangered. Although manatees were once common throughout the coastal areas of North America, they now live mainly in Florida. In the table below, you will find the number of manatees that have been found dead since 1976.

Manatee Mortality In Florida									
Year	1976	1977	1978	1979	1980	1981	1982	1983	1984
Number of Manatees	62	114	84	77	63	116	114	81	128
Year	1985	1986	1987	1988	1989	1990	1991	1992	1993
Number of Manatees	119	122	114	133	168	206	174	163	145

Source: Florida Marine Research Institute

The manatee data could also be represented by a set of ordered pairs, as shown in the list below. Each first coordinate is the year, and the second coordinate is the number of manatees. Each ordered pair can then be graphed.

{(1976, 62), (1977, 114), (1978, 84), (1979, 77), (1980, 63), (1981, 116), (1982, 114), (1983, 81), (1984, 128), (1985, 119), (1986, 122), (1987, 114), (1988, 133), (1989, 168), (1990, 206), (1991, 174), (1992, 163), (1993, 145)}

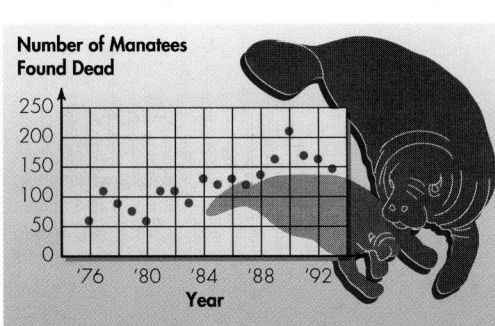

262 *Chapter 5   Graphing Relations and Functions*

F Y I

As of June 1993, the following animals were also on the list of endangered species in the United States: American alligator, American crocodile, brown bear, ocelot, Carolina northern flying squirrel.

A **relation** is a set of ordered pairs, like the one shown on the previous page. The set of first coordinates of the ordered pairs is called the **domain** of the relation. The domain usually contains the *x*-coordinates. The set of second coordinates is called the **range** of the relation. It usually contains the *y*-coordinates.

Definition of the Domain and Range of a Relation	**The domain of a relation is the set of all first coordinates from the ordered pairs in the relation. The range of the relation is the set of all second coordinates from the ordered pairs.**

For the relation representing the manatee data, the domain and range are as follows.

Domain = {1976, 1977, 1978, 1979, 1980, 1981, 1982, 1983, 1984, 1985, 1986, 1987, 1988, 1989, 1990, 1991, 1992, 1993}

Range = {62, 114, 84, 77, 63, 116, 81, 128, 119, 122, 133, 168, 206, 174, 163, 145}

An element is a member of a set.

In addition to ordered pairs, tables, and graphs, a relation can be represented by a **mapping**. A mapping illustrates how each element of the domain is paired with an element in the range. For example, the relation {(3, 3), (−1, 4), (0, −4)} can be modeled in each of the following ways.

Ordered Pairs	Table	Graph	Mapping
(3, 3) (−1, 4) (0, −4)		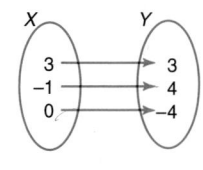	

Example 1 Represent the relation shown in the graph at the right as

a. a set of ordered pairs,
b. a table, and
c. a mapping.
d. Then determine the domain and range.

a. The set of ordered pairs for the relation is {(−3, 3), (−1, 2), (1, 1), (1, 3), (3, −2), (4, −2)}.

b.

x	y
−3	3
−1	2
1	1
1	3
3	−2
4	−2

c.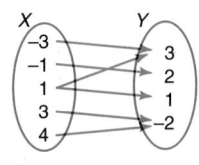

When giving the domain and range, if a value is repeated, you list it only once.

d. The domain for this relation is {−3, −1, 1, 3, 4}, and the range is {−2, 1, 2, 3}.

 Alternative Learning Styles

Visual Graph the relation {(2, 4), (−1, 3), (0, 2), (1, −2)} and indicate the domain and range by marking those values on the *x*-axis and the *y*-axis.

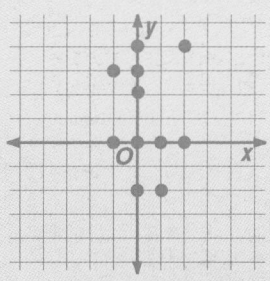

In-Class Example

For Example 1
Represent the relation shown in the graph below as

a. a set of ordered pairs,
b. a table, and
c. a mapping.

Then determine the domain and range.

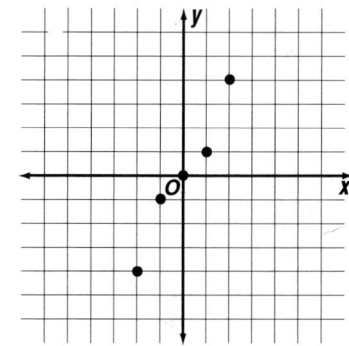

a. (−2, −4), (−1, −1), (0, 0), (1, 1), (2, 4)

b.

x	y
−2	−4
−1	−1
0	0
1	1
2	4

c.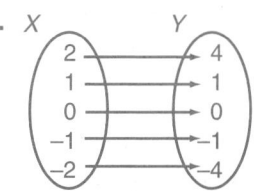

The domain is {−2, −1, 0, 1, 2} and the range is {−4, −1, 0, 1, 4}.

Teaching Tip The word *relation* simply indicates a set of ordered pairs. There may or may not be a reason for the pairing. Stress that a relation can be shown in many different ways. Remind students that the domain is a set of numbers, not a set of ordered pairs. The same is true of the range.

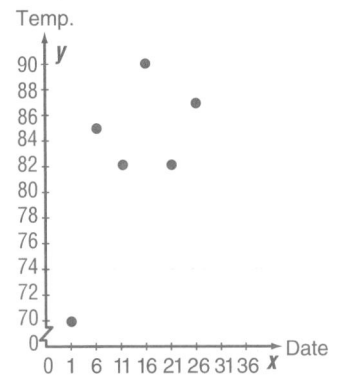

In-Class Example

For Example 2
The following table lists the high temperatures for certain days in June.

Date (June)	1	6	11	16	21	26
Temperature (°F)	70	85	82	90	82	87

a. Determine the domain and range of the relation.
domain = {1, 6, 11, 16, 21, 26}, range = {70, 82, 85, 87, 90}

b. Graph the data.

CAREER CHOICES

Median annual earnings for jobs in the biological and life sciences were about $34,500 in 1992. The middle 50 percent earned between $26,000 and $46,800. Ten percent earned less than $20,400 and ten percent earned over $56,900.

In real-life situations, you may need to select a range of values for the *x*- or *y*-axis that does not begin with 0 and does not have units of 1.

Example 2

CONNECTION
Biology

CAREER CHOICES

A **zoologist** studies animals, their origin, life process, behavior, and diseases. They are usually identified by the animal group in which they specialize, for example herpetology (reptiles), ornithology (birds), and ichthyology (fish).

Zoology requires a minimum of a bachelor's degree. Many college positions in this field require a Ph.D. degree.

For more information, contact:

Occupational Outlook Handbook, 1994–95, U.S. Department of Labor.

In order to protect endangered species, there is a limit to the number of animals that can be exported to other countries. The table below shows yearly quotas for the number of wild crocodiles that can be exported from Indonesia.

a. Determine the domain and range of the relation.
b. Graph the data.
c. What conclusions might you make from the graph of the data?

Maximum Exportation of Crocodiles from Indonesia									
Year	1986	1987	1988	1989	1990	1991	1992	1993	1994
Number of Crocodiles	2000	2000	4000	4000	3000	3000	2700	1500	1500

Source: *Traffic USA*, World Wildlife Fund, August 1992

a. The domain is {1986, 1987, 1988, 1989, 1990, 1991, 1992, 1993, 1994}. The range is {1500, 2000, 2700, 3000, 4000}.
Values are usually listed in numerical order.

b. The values for the *x*-axis need to go from 1986 to 1994. It is not efficient to begin the scale at 0. The values on the *y*-axis need to include values from 1500 to 4000. You can include 0 and use units of 500.

c. While the number of crocodiles that could be exported rose in 1988, it has decreased since 1991. This may indicate that crocodiles in Indonesia are even more endangered.

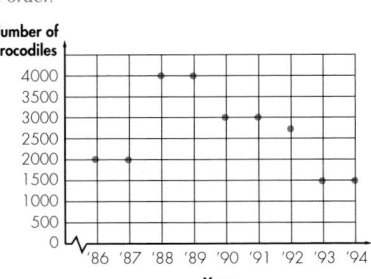

A jagged line is used to represent the values prior to 1986 that have been omitted.

The **inverse** of any relation is obtained by switching the coordinates in each ordered pair.

Relation	Inverse
(1, 4)	(4, 1)
(−3, 2)	(2, −3)
(7, −9)	(−9, 7)

Notice that the domain of the relation becomes the range of the inverse. Similarly, the range of the relation becomes the domain of the inverse.

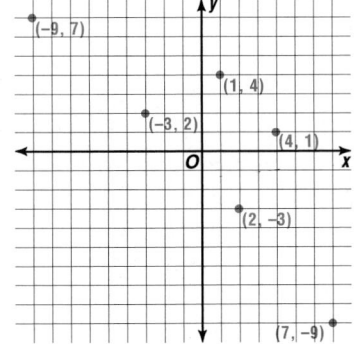

Alternative Teaching Strategies

Reading Algebra Have students look up the various definitions of *inverse*. Use these to help students develop a better understanding of the meaning of *inverse relation*.

| Definition of the Inverse of a Relation | Relation Q is the inverse of relation S if and only if for every ordered pair (a, b) in S, there is an ordered pair (b, a) in Q. |

Example 3

Express the relation shown in the mapping as a set of ordered pairs. Write the inverse of the relation and draw a mapping to model the inverse.

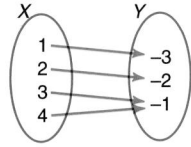

The mapping shows the relation $\{(1, -3), (2, -2), (3, -1), (4, -1)\}$.

The inverse of the relation is $\{(-3, 1), (-2, 2), (-1, 3), (-1, 4)\}$.

The mapping at the right shows the inverse.

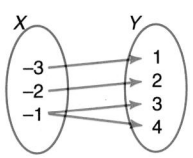

Relations and their inverses display some interesting characteristics, as you will discover below.

MODELING MATHEMATICS

Relations and Inverses

Materials: grid paper straightedge colored pencils

There is a special relationship between a relation and its inverse, which you can discover by paper folding.   c. Sample ordered pairs: $(-2, -2), (-1, -1), (0, 0), (1, 1)$.
For each (x, y), $x = y$.

Step 1 Graph the relation $\{(4, 5), (-3, 6), (-5, -3), (4, -7), (5, 0), (0, -3)\}$ on a coordinate plane using one color of pencil.

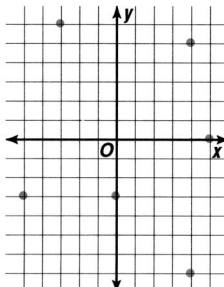

Connect the points in order using the same color pencil.

Step 2 Use a different color pencil to graph the inverse of the relation, connecting its points in order.

Step 3 Fold the graph paper through the origin so that the positive y-axis lies on top of the positive x-axis. Hold the paper up to a light so that you can see all of the points you graphed.

Your Turn

a. What do you notice about the location of the points you graphed when you looked at the folded paper? **See margin.**

b. Unfold the paper. Describe the pattern formed by the lines connecting the points in the relation and its inverse. **They are mirror images of each other.**

c. What do you think are the ordered pairs that represent the points on the fold line? Describe these in terms of x and y.

d. How could you graph the inverse of a function without writing ordered pairs first? **See margin.**

In-Class Example

For Example 3
Express the relation shown in the mapping below as a set of ordered pairs. Then write the inverse of this relation.

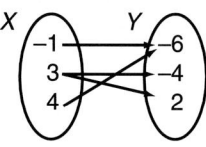

Relation = $\{(3, 2), (4, -6), (3, -4), (-1, -6)\}$
Inverse = $\{(2, 3), (-6, 4), (-4, 3), (-6, -1)\}$

MODELING MATHEMATICS This modeling activity introduces the idea that a relation and its inverse are symmetric with respect to each other about the line $y = x$, the diagonal.

Answers for Modeling Mathematics

a. The inverse of each point matches the point.

d. Draw the line through the ordered pairs (x, y) where $x = y$. Count the squares up/down until you meet the line. From that point, count the same number of squares right. Plot the point.

Check for Understanding

Exercises 1–15 are designed to help you assess your students' understanding through reading, writing, speaking, and modeling. You should work through Exercises 1–4 with your students and then monitor their work on Exercises 5–15.

Additional Answers

1. The domain values are the *x* values and the range values are the *y* values. If you do not know which is which, your plots will be incorrect.

2. After a high in 1990, the number of deaths seems to be decreasing.

4.

7. {(1, 3), (2, 4), (3, 5), (5, 7)};
D = {1, 2, 3, 5}; R = {3, 4, 5, 7}; Inv = {(3, 1), (4, 2), (5, 3), (7, 5)}

8. {(1, 4), (3, −2), (4, 4), (6,−2)};
D = {1, 3, 4, 6}; R = {−2 , 4}; Inv = {(4, 1), (−2, 3), (4, 4), (−2, 6)}

9. {(1, 3), (2, 2), (4, 9), (6, 5)};
D = {1, 2, 4, 6}; R = {2, 3, 5, 9}; Inv = {(3, 1), (2, 2), (9, 4), (5, 6)}

10. {(−3, −2), (−2, −1), (0, 0), (1, 1)}; D = {−3, −2, 0, 1};
R = {−2, −1, 0, 1}; Inv = {(−2, −3), (−1, −2), (0, 0), (1, 1)}

11. {(−2, 2), (−1, 1), (0, 1), (1, 1), (1, −1), (2, −1), (3, 1)};
D = {−2, −1, 0, 1, 2, 3};
R = {−1, 1, 2}; Inv = {(2, −2), (1, −1), (1, 0), (1, 1), (−1, 1), (−1, 2), (1, 3)}

15c. Except for the first 6 months of 1992, the unemployment rate seems to be decreasing.

CHECK FOR UNDERSTANDING

Communicating Mathematics

3. The domain of the relation becomes the range of the inverse. The range becomes the domain of the inverse.

4. See margin.

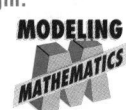

MODELING MATHEMATICS

Guided Practice

Study the lesson. Then complete the following.   1–2. See margin.

1. **Explain** why it is important to identify the domain and range values of a relation when graphing the relation.

2. Refer to the application at the beginning of the lesson. Explain what the manatee graph suggests about patterns in the rate of manatee deaths.

3. **State** the relationship between the domain and range of a relation and the domain and range of its inverse.

4. Graph the relation {(0, 5), (2, 3), (1, −4), (−3, 3), (−1, −2)}. Draw a line that goes through the points (−3, −3) and (3, 3). Use this line to graph the inverse of the relation without writing the ordered pairs of the inverse.

State the domain and range of each relation.

5. {(0, 2), (1, −2), (2, 4)}
D = {0, 1, 2}; R = {2, −2, 4}

6. {(−4, 2), (−2, 0), (0, 2), (2, 4)}
D = {−4, −2, 0, 2}; R = {2, 0, 4}

Express the relation shown in each table, mapping, or graph as a set of ordered pairs. Then state the domain, range, and inverse of the relation.   7–11. See margin.

7.

x	y
1	3
2	4
3	5
5	7

8.

x	y
1	4
3	−2
4	4
6	−2

9.

10.

11.

12–14. Graph the relations shown in Exercises 7–9.   See Solutions Manual.

15. **Economics**   The graph at the right shows the unemployment rate in the United States from January 1992 to December 1994.

   a. Name three ordered pairs shown by the graph.

   b. Estimate the least value in the range.   5.6%

   c. What conclusions might you make from the graph?   See margin.

   15a. See students' work.

U.S. Unemployment Rate 1992–1994

Source: *Wall Street Journal*, Dec. 5, 1994

Reteaching

Using Real-World Examples
Relations in which elements of a first set are linked to elements of a second set are common in the real world. For example, names of students → homeroom, names of U.S. citizens → Social Security number. Ask students to name other relations and identify the domain and range of each.

Practice

State the domain and range of each relation.

16. {(1, 3), (2, 5), (1, −7), (2, 9), (3, 3)}   D = {1, 2, 3}; R = {−7, 3, 5, 9}

17. {(1, 7), (−2, 7), (3, 7), (−5, 7)}   D = {−5, −2, 1, 3}, R = {7}

18. {(3.1, −1), (−4.7, 3.9), (2.4, −3.6), (−9, 2)}

19. $\left\{\left(\frac{1}{2}, \frac{1}{4}\right), \left(1\frac{1}{2}, -\frac{2}{3}\right), \left(-3, \frac{2}{5}\right), \left(-5\frac{1}{4}, -6\frac{2}{7}\right)\right\}$

18. D = {−9, −4.7, 2.4, 3.1}; R = {−3.6, −1, 2, 3.9}

19. D = $\left\{-5\frac{1}{4}, -3, \frac{1}{2}, 1\frac{1}{2}\right\}$; R = $\left\{-6\frac{2}{7}, -\frac{2}{3}, \frac{1}{4}, \frac{2}{5}\right\}$

Express the relation shown in each table, mapping, or graph as a set of ordered pairs. Then state the domain, range, and inverse of the relation. 20–28. See margin.

20.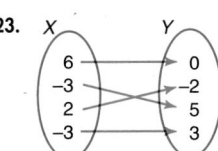

x	y
0	4
1	5
2	6
3	6

21.
x	y
6	4
4	−2
3	4
1	−2

22.
x	y
−4	2
−2	0
0	2
2	4

23.

24.

25.

26. temperature of boiled water as it cools

time (minutes)	0	5	10	15	20	25	30
temperature (°C)	100	90	81	73	66	60	55

27. cost of car repairs

time (hours)	0	1	2	3
cost (dollars)	25	50	75	100

28. distance traveled at a rate of 55 mph

time (hours)	1.25	3.75	4.5	5.5	6
distance (miles)	68.75	206.25	247.5	302.5	330

29–34. See Solutions Manual.

29.

30.

31.

32.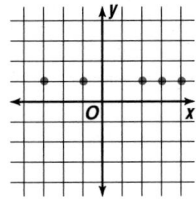

33.

34.

Lesson 5-2   Relations   **267**

Additional Answers

26. {(0, 100), (5, 90), (10, 81), (15, 73), (20, 66), (25, 60), (30, 55)}; D = {0, 5, 10, 15, 20, 25, 30}; R = {55, 60, 66, 73, 81, 90, 100}; Inv = {(100, 0), (90, 5), (81, 10), (73, 15), (66, 20), (60, 25), (55, 30)}

27. {(0, 25), (1, 50), (2, 75), (3, 100)}; D = {0, 1, 2, 3}; R = {25, 50, 75, 100}; Inv = {(25, 0), (50, 1), (75, 2), (100, 3)}

28. {(1.25, 68.75), (3.75, 206.25), (4.5, 247.5), (5.5, 302.5), (6, 330)}; D = {1.25, 3.75, 4.5, 5.5, 6}; R = {68.75, 206.25, 247.5, 302.5, 330}; Inv = {(68.75, 1.25), (206.25, 3.75), (247.5, 4.5), (302.5, 5.5), (330, 6)}

Assignment Guide

Core: 17–41 odd, 42, 43, 45, 46–53
Enriched: 16–40 even, 42–53

For **Extra Practice,** see p. 767.

The red A, B, and C flags, printed only in the Teacher's Wraparound Edition, indicate the level of difficulty of the exercises.

Additional Answers

20. {(0, 4), (1, 5), (2, 6), (3, 6)}; D = {0, 1, 2, 3}; R = {4, 5, 6}; Inv = {(4, 0), (5, 1), (6, 2), (6, 3)}

21. {(6, 4), (4, −2), (3, 4), (1, −2)}; D = {1, 3, 4, 6}; R = {−2, 4}; Inv = {(4, 6), (−2, 4), (4, 3), (−2, 1)}

22. {(−4, 2), (−2, 0), (0, 2), (2, 4)}; D = {−4, −2, 0, 2}; R = {0, 2, 4}; Inv = {(2, −4), (0, −2), (2, 0), (4, 2)}

23. {(6, 0), (−3, 5), (2, −2), (−3, 3)}; D = {−3, 2, 6}; R = {−2, 0, 3, 5}; Inv = {(0, 6), (5, −3), (−2, 2), (3, −3)}

24. {(5, 2), (−3, 1), (2, 2), (1, 7)}; D = {−3, 1, 2, 5}; R = {1, 2, 7}; Inv = {(2, 5), (1, −3), (2, 2), (7, 1)}

25. {(3, 4), (3, 2), (2, 9), (5, 4), (5, 8), (−7, 2)}; D = {−7, 2, 3, 5 }; R = { 2, 4, 8, 9}; Inv = {(4, 3), (2, 3), (9, 2), (4, 5), (8, 5), (2, −7)}

Study Guide Masters, p. 35

Draw a mapping and a graph for each relation.

35–37. See Solutions
Manual.

35. {(8, 2), (4, 2), (8, −9), (7, 5), (−3, 2)}

36. {(3, 3), (2, 7), (−3, 3), (1, 3), (4, 1)}

37. {(6, 0), (6, −4), (4, −3), (5, −3)}

Graphing Calculator

Graph each relation on a graphing calculator.

a. State the WINDOW settings that you used.

b. Write the coordinates of the inverse. Then graph the inverse.

c. Name the quadrants in which each point of the relation and its inverse lies.

38–41. See Solutions
Manual.

38.

x	y
1992	77
1993	200
1994	550
1995	880

39.

x	y
0	10
2	−8
6	6
9	−4

40.

x	y
−1	18
−2	23
−3	28
−4	33

41. Review your answers to part c of Exercises 38–40. What conclusions can you draw about the quadrant in which the inverse of a point will lie?

Critical Thinking

42. Write a relation with five ordered pairs such that the relation is its own inverse. Then describe the graph of the relation and its inverse.
Relations will vary. Sample answer: {(5, 3), (3, 5), (2, 2), (−4, −2) (−2, −4)}; the graphs are the same.

Applications and Problem Solving

43. Sample answers:

43a. 157 billion, 191 billion

43b. Retail sales have increased from 1992 to 1994.

43c. As unemployment decreases, retail sales increase because people have more money to spend.

43. Retail Sales The graph at the right shows U.S. retail sales in billions of dollars. The amounts for each month have been adjusted to account for seasonal increases in sales.

a. Approximate the least and greatest values in the range.

b. In your own words, describe what the pattern formed by the points represents.

c. Compare this graph with the one in Exercise 15. How do you think unemployment rates might be compared with retail spending trends? Explain.

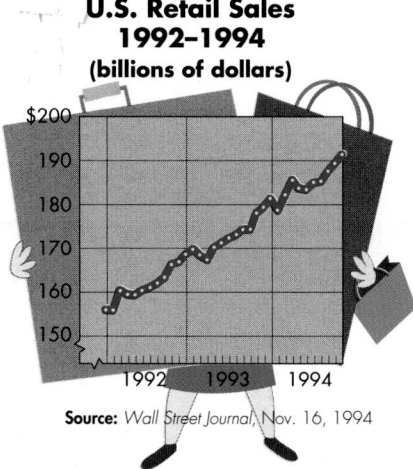

U.S. Retail Sales 1992–1994 (billions of dollars)

Source: *Wall Street Journal,* Nov. 16, 1994

44. Make a Table Mandy Chin is saving money to pay for the first year of her car insurance, which will cost about $1200. She already has $500 in her savings account. She gets a job working at a fast-food restaurant and plans to save $45 per week.

a. Make a table to show the number of weeks and how much money Mandy will have in her account after each week. For example, on Week 1, she will have $545. Graph the results. **See Solutions Manual.**

44b. after 16 weeks

b. When will she have enough money to pay her insurance premium?

Extension

Reasoning A relation has domain A = {2, 4, 6} and range B = {1, 3}. The Cartesian product A × B is defined as the relation consisting of all ordered pairs (a, b), where a is a member of A and b is a member of B.

1. What is the relation A × B? {(2, 1), (2, 3), (4, 1), (4, 3), (6, 1), (6, 3)}

2. How many elements are contained in A × B? 6

3. Suppose the domain A contains 213 elements and range B contains 432 elements. How many elements are contained in A × B?
213 × 432 = 92,016 ordered pairs

45. Probability There are 36 different possible outcomes when you roll two dice. The outcomes can be expressed as ordered pairs, such as (4, 3).

   a. Graph all 36 possible outcomes. What is the domain and range? **See margin.**

   b. What is the domain and range of the inverse? What do you notice about the relation and its inverse?

   c. How many different sums are possible when two dice are rolled? Draw a line plot showing how many ways each sum can be rolled. Describe your line plot. **See margin.**

   d. Graph the outcomes that have a sum of 7. **See margin.**

   e. What is the probability of rolling a sum of 7? Explain.

   45b. D = R = {1, 2, 3, 4, 5, 6}; relation = inverse

   45e. $\frac{6}{36}$; There are 6 out of 36 ways to roll a sum of 7.

Mixed Review

46. Graph $A(6, 2)$, $B(-3, 6)$, and $C(-5, -4)$ on the same coordinate plane. (Lesson 5-1) **See margin.**

47. If y varies inversely as x, and $y = 8$ when $x = 24$, find y when $x = 6$. (Lesson 4-8) **32**

48. 4 adult, 12 children

48. Entertainment At Backintime Cinema, tickets for adults cost $5.75 and tickets for children cost $3.75. How many of each kind of ticket were purchased if 16 tickets were bought for $68? (Lesson 4-7)

49. Use $I = prt$ to find I if $p = \$8000$, $r = 6\%$, and $t = 1$ year. (Lesson 4-4) **$480**

50. Solve $E = mc^2$ for m. (Lesson 3-6) $m = \dfrac{E}{c^2}$

51. Space Science Halley's Comet flashes through the sky every 76.3 years. It last appeared in 1986. In what year of the 23rd century is Halley's Comet expected to reappear? (Lesson 2-6) **2214, 2290**

52. Agriculture Refer to the graph in Exercise 50 on page 259. (Lesson 1-9)
   a. What is the normal minimum temperature for March? **27°F**
   b. What happens to the temperature after July? **It continually decreases.**

53. Simplify $4[1 + 4(5x + 2y)]$. (Lesson 1-8) $4 + 80x + 32y$

WORKING ON THE

Refer to the Investigation on pages 190–191.

In tracking populations of fish, the Department of Natural Resources and Wildlife often chart their findings in a graph. They use these graphs to monitor the populations of the different types of fish. In this way they track which species are on the increase and which are in decline. This can be especially important in monitoring species that are rare or in danger of extinction.

1 Use the information in your chart to make a table of ordered pairs (x, y) such that x represents the number of the casting and y represents your estimate for the number of fish in the lake.

2 Draw a mapping for this relation.

3 Graph the ordered pairs on a coordinate grid. Describe the graph.

Add the results of your work to your Investigation Folder.

Lesson 5-2 Relations **269**

Working on the Investigation

The Investigation on pages 190–191 is designed to be a long-term project that is completed over several days or weeks. Encourage students to keep their materials in their Investigation Folder as they work on the Investigation.

4 ASSESS

Closing Activity

Speaking Ask students to describe the terms *relation*, *domain*, *range*, and *inverse* in their own words. Then have them identify any three ordered pairs and apply each term.

Chapter 5, Quiz A (Lessons 5-1 and 5-2), is available in the *Assessment and Evaluation Masters*, p. 128.

Additional Answers

46.

Enrichment Masters, p. 35

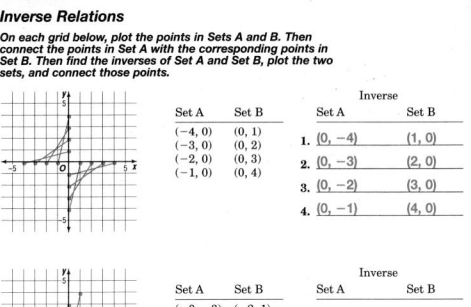

NCTM Standards: 1–5

Objective
Use a graphing calculator to investigate relations and determine the ranges when the domains are given.

Recommended Time
15 minutes

Instructional Resources
Graphing Calculator Masters, pp. 16 and 17

These masters provide keystroking instruction for this lesson for the TI-81 and Casio graphing calculators.

1 FOCUS

Motivating the Lesson
You may want to have students review the use of the GRAPH feature on the calculator.

2 TEACH

Teaching Tip Encourage students to check their solutions by replacing the variable with the *x* values and computing the *y* values by hand.

3 PRACTICE/APPLY

Assignment Guide

Core: 1–4
Enriched: 1–4

4 ASSESS

Observing students working with technology is an excellent method of assessment.

5–3A Graphing Technology
Equations
A Preview of Lesson 5–3

Ordered pairs can be used to represent solutions to equations in two variables. Thus, the solution set of an equation for a defined domain is a relation. We can use a graphing calculator to investigate the relation and determine the range when the domain is given.

Example

Solve $y = 2.5x + 4$ if the domain is $\{-8, -6, -4, -2, 0, 2, 4, 6, 8\}$.

Method 1

Store the elements of the domain as a list of values in L1.

Enter: [2nd] [{] [(−)] 8 [,] [(−)] 6 [,] [(−)] 4 [,] [(−)] 2 [,]
0 [,] 2 [,] 4 [,] 6 [,] 8 [2nd] [}] [STO▸] [2nd]

[L1] [ENTER]   {-8 -6 -4 -2 0...

To calculate the range values, compute the right-hand side of the equation, substituting L1 for *x*.

Enter: 2.5 [2nd] [L1] [+] 4 [ENTER]   {-16 -11 -6 -1...

Press the arrow keys to scroll right or left through the display to see all the values in the range.

The solution set is $\{(-8, -16), (-6, -11), (-4, -6), (-2, -1), (0, 4), (2, 9), (4, 14), (6, 19), (8, 24)\}$.

Method 2

Store the equation in the Y= menu of the calculator and view the table of values it creates.

Enter: [Y=] 2.5 [X,T,θ] [+] 4 [2nd] [TblSet] [(−)] 8 [ENTER] 2

[2nd] [TABLE]

ΔTbl means the increments in which the x values will appear in the table.

A table with columns X and Y_1 appears on the screen. The values in the Y_1 column represent the corresponding range values for each *x* value in the domain. This table helps you find the ordered pairs that are solutions for the equation. The solution set is $\{(-8, -16), (-6, -11), (-4, -6), (-2, -1), (0, 4), (2, 9), (4, 14), (6, 19), (8, 24)\}$, which agrees with the list in Method 1.

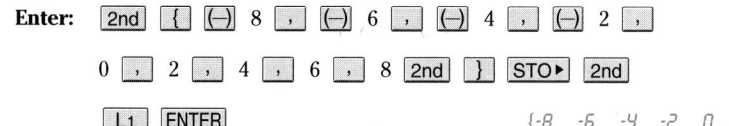

EXERCISES

Use a graphing calculator to solve each equation if the domain is $\{-3, -2, -1, 0, 1, 2, 3\}$. 1–4. See margin.

1. $y = 4x - 7$ 2. $y = x^2 + 11$ 3. $1.2x - y = 6.8$ 4. $6x + 2y = 12$

Additional Answers

1. $\{(-3, -19), (-2, -15), (-1, -11), (0, -7), (1, -3), (2, -1), (3, 5)\}$
2. $\{(-3, 20), (-2, 15), (-1, 12), (0, 11), (1, 12), (2, 15), (3, 20)\}$
3. $\{(-3, -10.4), (-2, -9.2), (-1, -8), (0, -6.8), (1, -5.6), (2, -4.4), (3, -3.2)\}$
4. $\{(-3, 15), (-2, 12), (-1, 9), (0, 6), (1, 3), (2, 0), (3, -3)\}$

Using Technology
This lesson offers an excellent opportunity for using technology in your algebra classroom. For more information on using technology, see *Graphing Calculators in the Mathematics Classroom,* one of the titles in the Glencoe Mathematics Professional Series.

Equations as Relations

CONNECTION
Physical Science

Objectives

After studying this lesson, you should be able to:
- determine the range for a given domain, and
- graph the solution set for the given domain.

Are you as "light" as a feather or as "heavy" as a rock? The *density* of an object is the measure of its "lightness" or "heaviness." Density is defined as the mass per unit of volume. The chart at the right shows the densities of some common substances.

Substance	Density (g/cm³)
air	0.0013
aluminum	2.7
gasoline	0.7
gold	19.3
lead	11.3
mercury	13.5
silver	10.5
steel	7.8
water	1.0

F Y I

Water has a density of 1.0. Any substance with a density less than 1.0 will float in water. Any substance with a density greater than 1.0 will sink in water.

An equation relating density, mass, and volume is $m = DV$, where m is the mass of a substance, D is the density, and V is the volume. For example, the equation for gold is $m = 19.3V$ because the density of gold is 19.3 g/cm³. This equation is called an **equation in two variables,** m and V.

No one knows exactly when gold was first discovered. However, there have been gold artifacts dug up in Ur in Mesopotamia (modern day Iraq) that date as early as 3500 B.C. Suppose we wanted to know the mass of four gold artifacts with volumes of 10 cm³, 50 cm³, 100 cm³, and 150 cm³. We can use a table to find ordered pairs (V, m) that satisfy the equation.

V	D · V	m	Ordered Pair
10	19.3(10)	193	(10, 193)
50	19.3(50)	965	(50, 965)
100	19.3(100)	1930	(100, 1930)
150	19.3(150)	2895	(150, 2895)

These four ordered pairs are graphed at the right. Since each ordered pair satisfies the equation, each ordered pair is a solution of the equation.

Definition of the Solution of an Equation in Two Variables	If a true statement results when the numbers in an ordered pair are substituted into an equation in two variables, then the ordered pair is a solution of the equation.

Since the solutions of an equation in two variables are ordered pairs, such an equation describes a relation. In an equation involving x and y, the set of x values is the domain of the relation. The set of corresponding y values is the range of the relation.

Lesson 5-3 Equations as Relations **271**

5-3 LESSON NOTES

NCTM Standards: 1–5

Instructional Resources

- Study Guide Master 5-3
- Practice Master 5-3
- Enrichment Master 5-3
- Multicultural Activity Masters, p. 10

Transparency 5-3A contains the 5-Minute Check for this lesson; **Transparency 5-3B** contains a teaching aid for this lesson.

Recommended Pacing

Standard Pacing	Day 5 of 14
Honors Pacing	Day 5 of 13
Block Scheduling*	Day 3 of 7
Alg. 1 in Two Years*	Days 7 & 8 of 21

*For more information on pacing and possible lesson plans, refer to the *Block Scheduling Booklet* and *Algebra 1 in Two Years*.

1 FOCUS

5-Minute Check
(over Lesson 5-2)

State the domain and range of each relation.

1. {(0, −3), (1, 4), (1, −3)}
   D = {0, 1}, R = {−3, 4}
2. {(1, 5), (2, 5), (3, 5), (4, 5)}
   D = {1, 2, 3, 4}, R = {5}
3. Show the inverse of the relation in Exercise 1 by using a table.

x	y
−3	0
4	1
−3	1

4. A relation contains five ordered pairs. If its domain is {2, 4, 6, 8, 10} and each y-coordinate is three more than the x-coordinate, then what are the ordered pairs in the relation? {(2, 5), (4, 7), (6, 9), (8, 11), (10, 13)}

272 Chapter 5

Motivating the Lesson

Hands-On Activity Have students work in pairs. Provide each pair with an ample amount of play money. One member of each pair should give a random amount of money to the other member. The second student computes 15% of the amount and returns the original amount along with the interest to the first student. Ask students to think about the equation involved in the transaction.

2 TEACH

In-Class Examples

For Example 1

a. Solve $y = -\frac{1}{3}x$ if the domain is $\{-3, 0, 2, 6, 8\}$. The solution set is $\left\{(-3, 1), (0, 0), \left(2, -\frac{2}{3}\right), \left(6, -2\right), \left(8, -\frac{8}{3}\right)\right\}$.

b. Solve $y = -2x + 1$ if the domain is $\{-4, -2, 0, 2, 4\}$. The solution set is $\{(-4, 9), (-2, 5), (0, 1), (2, -3), (4, -7)\}$.

For Example 2

Solve $2x - 3y = 12$ if the domain is $\{-9, -6, -3, 0, 3, 6, 9\}$. The solution set is $\{(-9, -10), (-6, -8), (-3, -6), (0, -4), (3, -2), (6, 0), (9, 2)\}$.

Teaching Tip Transforming equations, as in Example 2, provides good practice in applying the properties of solving equations.

Teaching Tip Make sure students realize that the domain and range are not x's and y's. The domain and range are sets of numbers from which values for a variable are chosen. Letters other than x and y can be used for the variables.

Example **Solve each equation for the given domain values. Graph the solution set.**

a. $y = 4x$ if the domain is $\{-3, -2, 0, 1, 2\}$

Make a table. The values of x come from the domain. Substitute each value of x into the equation to determine the corresponding values of y.

Domain x	$4x$	Range y	Ordered Pair (x, y)
-3	$4(-3)$	-12	$(-3, -12)$
-2	$4(-2)$	-8	$(-2, -8)$
0	$4(0)$	0	$(0, 0)$
1	$4(1)$	4	$(1, 4)$
2	$4(2)$	8	$(2, 8)$

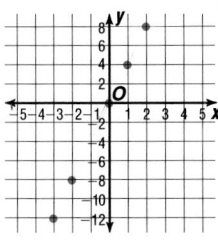

Then graph the solution set $\{(-3, -12), (-2, -8), (0, 0), (1, 4), (2, 8)\}$.

b. $y = x + 6$ if the domain is $\{-4, -3, -1, 2, 4\}$

Make a table. Find the values of y in the range by substituting the values of x from the domain into the equation.

x	$x + 6$	y	(x, y)
-4	$-4 + 6$	2	$(-4, 2)$
-3	$-3 + 6$	3	$(-3, 3)$
-1	$-1 + 6$	5	$(-1, 5)$
2	$2 + 6$	8	$(2, 8)$
4	$4 + 6$	10	$(4, 10)$

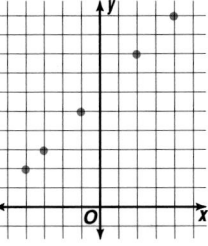

Then graph the solution set $\{(-4, 2), (-3, 3), (-1, 5), (2, 8), (4, 10)\}$.

It is often helpful to solve an equation for y before substituting each domain value into the equation. This makes creating a table of values easier.

Example **Solve $8x + 4y = 24$ if the domain is $\{-2, 0, 5, 8\}$.**

First solve the equation for y in terms of x.

$$8x + 4y = 24$$
$$4y = 24 - 8x \quad \text{Subtract 8x from each side.}$$
$$y = 6 - 2x \quad \text{Divide each side by 4.}$$

Now substitute each value of x from the domain to determine the corresponding values of y in the range.

x	$6 - 2x$	y	(x, y)
-2	$6 - 2(-2)$	10	$(-2, 10)$
0	$6 - 2(0)$	6	$(0, 6)$
5	$6 - 2(5)$	-4	$(5, -4)$
8	$6 - 2(8)$	-10	$(8, -10)$

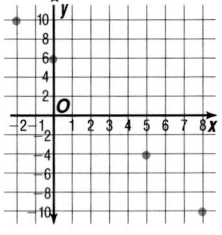

Then graph the solution set $\{(-2, 10), (0, 6), (5, -4), (8, -10)\}$.

Classroom Vignette

"When I teach my students to solve for y, I like the x value to be first, in slope-intercept form. So, in Example 2, when we solve for y, we would write $y = -2x + 6$. That way, when we get to that, the students do not have to relearn how to solve for y. It's already in the correct form."

Deborah Phillips
St. Vincent–St. Mary High School
Akron, Ohio

You can enter selected x values in the TABLE feature of the graphing calculator, and it will calculate the corresponding y values for a given equation.

- Solve the equation for y and enter the equation into the Y= list.
- Press [2nd] [TblSet] and use the arrow keys to go to Indpnt:. Highlight Ask.
- Press [2nd] [TABLE]. Scroll up to the top of the X list. Enter the first value of the domain and press [ENTER]. The corresponding Y value appears in the second column. Enter the other values of the domain and record the range values.

Your Turn a. {(−4, 32), (−2, 22), (0, 12), (1, 7), (3, −3), (5, −13)}

a. Use a graphing calculator to find the corresponding y values for $y = -5x + 12$ if $x = \{-4, -2, 0, 1, 3, 5\}$. Write the solutions as ordered pairs.

b. Why do you think the Ask feature on the TABLE menu is helpful in finding the ordered pairs that are solutions for equations? **See margin.**

EXPLORATION

The object of this Exploration is to introduce students to the GRAPH and ASK features on the TABLE menu.

Answer for the Exploration

b. You can use it to find y values for specific x values that may not appear in the general table setup. It also saves time from scrolling through the TABLE list of values.

In-Class Example

For Example 3
Suppose a rectangle has a perimeter of 42 mm.

a. Solve the perimeter formula for w. $w = 21 - \ell$
b. State the independent and dependent variables. w is the dependent variable, and ℓ is the independent variable.
c. Choose five values for ℓ and find the corresponding values of w. Answers will vary.

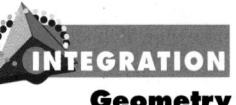

LOOK BACK

You can look back to Lesson 1-9 to review independent and dependent variables.

Variables other than x and y are often used in equations representing real situations. The domain contains values represented by the *independent variable*. Graph the domain values on the horizontal axis. The range contains the corresponding values represented by the *dependent variable*, determined by the given equation. Graph the range values on the vertical axis.

When you solve an equation for a given variable, that variable becomes the dependent variable. That is, its value depends upon the domain values chosen for the other variable.

Example ③

INTEGRATION
Geometry

The equation for the perimeter of a rectangle is $2w + 2\ell = P$. Suppose the perimeter of a rectangle is 24 centimeters.

a. Solve the equation for ℓ.
b. State the independent and dependent variables.
c. Choose five values for w and find the corresponding values of ℓ.

a. First substitute 24 for P in the equation. Then solve for ℓ.

$$2w + 2\ell = P$$
$$2w + 2\ell = 24$$
$$2\ell = 24 - 2w \qquad \textit{Subtract 2w from each side.}$$
$$\frac{2\ell}{2} = \frac{24 - 2w}{2} \qquad \textit{Divide each side by 2.}$$
$$\ell = 12 - w$$

b. Since the value of ℓ depends on the value of w, ℓ is the dependent variable, and w is the independent variable.

(continued on the next page)

Check for Understanding

Exercises 1–12 are designed to help you assess your students' understanding through reading, writing, speaking, and modeling. You should work through Exercises 1–6 with your students and then monitor their work on Exercises 7–12.

Additional Answers

6. The independent variable's values are chosen or assigned. The dependent variable's value varies with the value of the independent variable. The independent is graphed along the horizontal axis and the dependent along the vertical axis.

11.

19. {(−3, −12), (−2, −8), (0, 0), (3, 12), (6, 24)}
20. {(−3, −10), (−2, −7), (0, −1), (3, 8), (6, 17)}
21. {(−3, 10), (−2, 9), (0, 7), (3, 4), (6, 1)}
22. {(−3, −7), (−2, −6), (0, −4), (3, −1), (6, 2)}
23. {(−3, 11), (−2, 9.5), (0, 6.5), (3, 2), (6, −2.5)}
24. {(−3, 19), (−2, 14), (0, 4), (3, −11), (6, −26)}
25. {(−3, −12), (−2, −7), (0, 3), (3, 18), (6, 33)}
26. {(−3, −5.5), (−2, −5), (0, −4), (3, −2.5), (6, −1)}
27. {(−3, 1.8), (−2, 1.4), (0, 0.6), (3, −0.6), (6, −1.8)}

---

c. When choosing values for w, you can choose any number greater than zero. Suppose we choose the domain {1, 2, 3, 4, 5}. Make a table of values.

w	$12 - w$	ℓ	(ℓ, w)
1	12 − 1	11	(1, 11)
2	12 − 2	10	(2, 10)
3	12 − 3	9	(3, 9)
4	12 − 4	8	(4, 8)
5	12 − 5	7	(5, 7)

The solution set for the domain we chose is {(1, 11), (2, 10), (3, 9), (4, 8), (5, 7)}. *Check to see if each solution satisfies the equation $2w + 2\ell = 24$.*

CHECK FOR UNDERSTANDING

Communicating Mathematics

1a. air, gasoline; air, aluminum, gasoline, lead, silver, steel, water
1b. $D = \{10, 50, 100, 150\}$; $R = \{193, 965, 1930, 2895\}$
4. Length of a side cannot be zero or a negative number.

Study the lesson. Then complete the following.

1. Refer to the connection at the beginning of the lesson.
   a. Which of the substances listed will float on water? on mercury?
   b. State the domain and range of the graphed relation.
2. **Show** why $(1, -1)$ is or is not a solution of $x + 2y = 3$.   not; $1 + 2(-1) \neq 3$
3. **Determine** which of the points shown in the graph at the right represent ordered pairs that are solutions for $y = 2x + 1$.   A, C, F, G

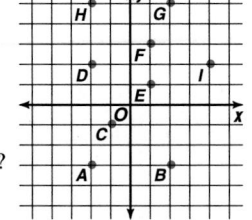

4. **Explain** why, in Example 3, you only consider values greater than zero.
5. Why should you solve for y before finding ordered pairs that are solutions for an equation? It makes calculating the y value easier.

MATH JOURNAL

6. In your own words, explain the difference between an independent variable and a dependent variable and how you know on which axis to graph each of them.   See margin.

Guided Practice

Solve each equation for the domain given in the table.

7. $y = 2x + 3$

x	y	(x, y)
−2	−1	(−2, −1)
−1	1	(−1, 1)
0	3	(0, 3)
1	5	(1, 5)
2	7	(2, 7)
3	9	(3, 9)

8. $a = \dfrac{3b - 5}{2}$

b	a	(b, a)
−5	−10	(−5, −10)
−2	$-\frac{11}{2}$	$\left(-2, -\frac{11}{2}\right)$
0	$-\frac{5}{2}$	$\left(0, -\frac{5}{2}\right)$
2	$\frac{1}{2}$	$\left(2, \frac{1}{2}\right)$
5	5	(5, 5)

Reteaching

Using Different Approaches
Demonstrate two approaches to solving an equation given its domain. First, substitute each value of the domain for x and solve for the corresponding y value. Second, solve for y in general. Then substitute each value of the domain for x into the phrase found equal to y.

9. a,d

10. b,c

11. {(−2, 8), (−1, 7.5), (0, 7), (1, 6.5), (2, 6)}; See margin for graph.

Which ordered pairs are solutions of each equation?

9. $1 + 5y = 2x$ **a.** $(−7, −3)$ **b.** $(7, 3)$ **c.** $(2, 1)$ **d.** $(−2, −1)$

10. $11 − 2y = 3x$ **a.** $(1, 3)$ **b.** $(3, 1)$ **c.** $(5, −2)$ **d.** $(−1, 4)$

11. Find the solution set for $x + 2y = 14$ if the domain is {−2, −1, 0, 1, 2}. Then graph the solution set.

12. Geometry The formula for the area of a rectangle is $A = \ell w$. Suppose the area of a rectangle is 36 square meters.
   a. Solve the equation for ℓ. $\ell = \dfrac{A}{w}$ or $\ell = \dfrac{36}{w}$
   b. Choose five values for w and find the corresponding values of ℓ.
   Sample answer: (1, 36), (2, 18), (3, 12), (4, 9), (6, 6)

Practice

14. b,c

18. a,b,c,d

19–27. See margin.

Which ordered pairs are solutions of each equation?

13. $3a + b = 8$ **a.** $(4, −4)$ **b.** $(8, 0)$ **c.** $(2, 2)$ **d.** $(3, 1)$ a,c

14. $y = 3x$ **a.** $(6, 2)$ **b.** $(−2, −6)$ **c.** $(0, 0)$ **d.** $(−15, −5)$

15. $3a − 8b = −4$ **a.** $(0, 0.5)$ **b.** $(4, 2)$ **c.** $(2, 0.75)$ **d.** $(2, 4)$ a,b

16. $x = 3y − 7$ **a.** $(2, −1)$ **b.** $(2, 4)$ **c.** $(−1, 2)$ **d.** $(2, 3)$ c,d

17. $2y + 4x = 8$ **a.** $(0, 2)$ **b.** $(−3, 0.5)$ **c.** $(2, 0)$ **d.** $(−2, 1)$ c

18. $3x + 3y = 0$ **a.** $(1, −1)$ **b.** $(2, −2)$ **c.** $(−1, 1)$ **d.** $(−2, 2)$

Solve each equation if the domain is {−3, −2, 0, 3, 6}.

19. $y = 4x$ **20.** $y = 3x − 1$ **21.** $2x + 2y = 14$

22. $x = 4 + y$ **23.** $3x = 13 − 2y$ **24.** $y = 4 − 5x$

25. $5x + 3 = y$ **26.** $5x − 10y = 40$ **27.** $2x + 5y = 3$

Make a table and graph the solution set for each equation and domain.

28. $y = 3x$ for $x = \{−3, −2, −1, 0, 1, 2, 3\}$ 28–31. See Solutions Manual.

29. $y = 2x + 1$ for $x = \{−5, −3, 0, 1, 3\, 6\}$

30. $3x − 2y = 5$ for $x = \{−3, −1, 2, 4, 7\}$

31. $5x = 8 − 4y$ for $x = \{−2, −1, 0, 1, 3, 4, 5\}$

32. Geometry A regular pentagon has five angles that have equal measures. Suppose the measure of one angle is $(a + b)°$. The sum of the measures of the angles of any pentagon is 540°.
   a. Write an equation for the sum of the angle measures of a regular pentagon. $5a + 5b = 540$
   b. Solve the equation for a. $a = 108 − b$
   c. Choose five values for b and find the corresponding values of a.

32c. Sample answer: (1, 107), (2, 106), (3, 105), (4, 104), (5, 103)

33b. $y = 45 − \dfrac{3}{4}x$

33c. Sample answer: (1, 44.25), (2, 43.5), (3, 42.75), (4, 42), (5, 41.25)

33. Geometry Two angles that are supplementary have measures whose sum is 180°. Suppose the measures of two supplementary angles are $(x + y)°$ and $(2x + 3y)°$.   **a.** $3x + 4y = 180$
   a. Write an equation for the sum of the measures of these two angles.
   b. Solve the equation for y.
   c. Choose five values for y and find the corresponding values of x.

36. $\left\{ -\dfrac{5}{6}, -\dfrac{2}{3}, -\dfrac{1}{2}, -\dfrac{1}{6}, 0 \right\}$

37. $\left\{ -\dfrac{7}{4}, -\dfrac{1}{2}, 2, \dfrac{13}{4}, \dfrac{9}{2} \right\}$

Find the domain for each equation if the range is {−2, −1, 0, 2, 3}.

34. $y = x + 7$ {−9, −8, −7, −5, −4}   **35.** $y = 3x$ $\left\{ -\dfrac{2}{3}, -\dfrac{1}{3}, 0, \dfrac{2}{3}, 1 \right\}$

36. $6x − y = −3$ **37.** $5y = 8 − 4x$

Core: 13–49 odd, 51–59
Enriched: 14–44 even, 45–59

For **Extra Practice,** see p. 767.

The red A, B, and C flags, printed only in the Teacher's Wraparound Edition, indicate the level of difficulty of the exercises.

Study Guide Masters, p. 36

NAME_____ DATE _____

5-3   **Study Guide**   Student Edition Pages 271–277

Equations as Relations

An equation in two variables describes a relation. It is often easier to determine the solution of such an equation by solving for one of the variables.

Example: Solve $3y + 2x = 10$ if the domain is {−7, −1, 8}.
First solve for y in terms of x.
$$3y + 2x = 10$$
$$3y = 10 − 2x$$
$$y = \dfrac{10 − 2x}{3}$$
Then substitute values of x.

x	$\dfrac{10-2x}{3}$	y	(x, y)
−7	$\dfrac{10-2(-7)}{3}$	8	$(-7, 8)$
−1	$\dfrac{10-2(-1)}{3}$	4	$(-1, 4)$
8	$\dfrac{10-2(8)}{3}$	−2	$(8, -2)$

Which ordered pairs are solutions of each equation?

1. $y = 3x + 1$ **a.** $(0, 1)$ **b.** $\left(\frac{1}{3}, 2\right)$ **c.** $\left(-1, -\frac{2}{3}\right)$ **d.** $(-1, -2)$
a, b, d

2. $2a = 5 − b$ **a.** $(5, 0)$ **b.** $(5, -5)$ **c.** $\left(\frac{5}{2}, 0\right)$ **d.** $(1, -3)$
b, c

Solve each equation if the domain is {−4, −2, 0, 2, 4}.

3. $x + y = 4$
{(−4, 8), (−2, 6), (0, 4), (2, 2), (4, 0)}

4. $y = -4x − 6$
{(−4, 10), (2, 2), (0, −6), (2, −14), (4, −22)}

5. $5a − 3b = 15$
{(−4, −11⅔), (−2, −8⅓), (0, −5), (2, −1⅔), (4, 1⅔)}

6. $3x − 5y = 8$
{(−4, −4), (−2, −14/5), (0, −8/5), (2, −2/5), (4, −4/5)}

7. $6x + 3y = 18$
{(−4, 14), (−2, 10), (0, 6), (2, 2), (4, −2)}

8. $4x + 8 = 6y$
{(−4, −4/3), (−2, 0), (0, 4/3), (2, 8/3), (4, 4)}

Closing Activity

Writing For each equation, find the range for the given domain.

1. $5y - 3x = 8$, where the domain is $\{-2, 0, 5\}$. $R = \left\{\frac{2}{5}, 1\frac{3}{5}, 4\frac{3}{5}\right\}$

2. $3z - 6w = 1$, where the domain is $\{-1, 0, 2\}$. $R = \left\{-1\frac{2}{3}, \frac{1}{3}, 4\frac{1}{3}\right\}$

Additional Answers

41. $\{(-2.5, -4.26),$
$(-1.75, -3.21), (0, -0.76),$
$(1.25, 0.99), (3.33, 3.902)\}$

42. $\{(-125, -425.5),$
$(-37, -117.5), (-6, -9),$
$(12, 54), (57, 211.5),$
$(150, 537)\}$

43. $\{(-100, 350), (-30, 116.\overline{6}),$
$(0, 16.6), (120, -383.\overline{3}),$
$(360, -1183.3),$
$(720, -2383.3)\}$

44. $\{(-10, 4.\overline{6}), (-5, 3), (0, 1.\overline{3}),$
$(5, -0.\overline{3}), (10, -2), (15, -3.\overline{6})\}$

46a. The other graphs have points that seem to lie in a straight line. These points do not.

46b. $y = x^2 - 3x - 10$ is U-shaped; $y = x^3$ is both an upward and downward curve; $y = 3^x$ is J-shaped.

48a. North America, 19; South America, 35; Europe, 54; Asia, 7; Africa, 5

Practice Masters, p. 36

Make a table and graph the solution set for each equation and domain. 38–40. See Solutions Manual.

38. $y = x^2 - 3x - 10$ if the domain is $\{-3, -1, 0, 1, 3, 5\}$
39. $y = x^3$ if the domain is $\{-2, -1, 0, 1, 2\}$
40. $y = 3^x$ if the domain is $\{1, 2, 3, 4\}$

Graphing Calculator

Use a graphing calculator to find the solution set for the given domain. 41–44. See margin.

41. $y = 1.4x - 0.76$ for $x = \{-2.5, -1.75, 0, 1.25, 3.33\}$
42. $y = 3.5x + 12$ for $x = \{-125, -37, -6, 12, 57, 150\}$
43. $3.6y + 12x = 60$ for $x = \{-100, -30, 0, 120, 360, 720\}$
44. $75y + 25x = 100$ for $x = \{-10, -5, 0, 5, 10, 15\}$

Critical Thinking

45a. $\{-6, -4, 0, 4, 6\}$
45b. $\{-13, -8, -4, 4, 8, 13\}$
45c. $\{-5, 0, 4, 8, 13\}$

45. Find the domain values of each relation if the range is $\{0, 16, 36\}$.
 a. $y = x^2$ **b.** $y = |4x| - 16$ **c.** $y = |4x - 16|$

46. Compare the graphs you drew in Exercises 38–40 with the other graphs in this lesson. See margin.
 a. What is different about the pattern of the points for these relations?
 b. Study the equations associated with the graphs. How do you think you could predict the pattern of the points by looking at the equation?

Applications and Problem Solving

48b. 354,234,275; 504,531,323; 718,597,477

Growth Rates of Continents

North America	3.6
South America	2.1
Europe	1.3
Asia	9.7
Africa	15.1

0   4   8   12   16
Percent

47. Physical Science Refer to the connection at the beginning of the lesson. Suppose you have an unknown substance with a mass of 378 grams and a volume of 36 cm^3 and a second unknown substance with a mass of 87.5 grams and a volume of 125 cm^3.
 a. Solve the formula $m = DV$ for D. $D = \frac{m}{V}$
 b. Identify the unknown substances. silver and gasoline

48. Demographics The time T (in years) needed for a population to double is calculated using the formula $T = \frac{70}{R}$, where R represents the percent growth rate of the population. a. See margin.
 a. Use the formula to approximate the doubling time to the nearest year for the continents shown in the graph at the left.
 b. The population of the United States in 1990 was 248,709,873. If the U.S. maintains the same growth rate as North America, what will be the population of the U.S. in the year 2000? in 2010? in 2020?

49. Anatomy The formula for relating the shoe size S and the length of a woman's foot in inches L is $S = 3L - 22$. The formula for relating a man's shoe size with his foot length is $S = 3L - 26$. Copy and complete each table below to determine shoe sizes given lengths of feet.

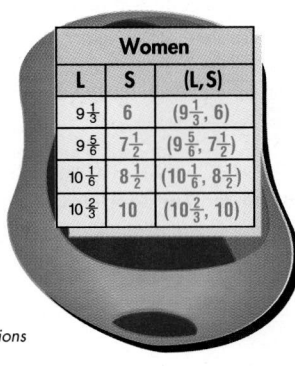

Women		
L	S	(L, S)
$9\frac{1}{3}$	6	$(9\frac{1}{3}, 6)$
$9\frac{5}{6}$	$7\frac{1}{2}$	$(9\frac{5}{6}, 7\frac{1}{2})$
$10\frac{1}{6}$	$8\frac{1}{2}$	$(10\frac{1}{6}, 8\frac{1}{2})$
$10\frac{2}{3}$	10	$(10\frac{2}{3}, 10)$

Men		
L	S	(L, S)
$11\frac{1}{3}$	8	$(11\frac{1}{3}, 8)$
$11\frac{5}{6}$	$9\frac{1}{2}$	$(11\frac{5}{6}, 9\frac{1}{2})$
$12\frac{1}{3}$	11	$(12\frac{1}{3}, 11)$
$12\frac{5}{6}$	$12\frac{1}{2}$	$(12\frac{5}{6}, 12\frac{1}{2})$

50. **Make a Table** Winona Brownsman is doing small jobs for her rich aunt. Her aunt has asked Winona how she wishes to be paid for her services. Winona has two choices.

- *Plan A:* Winona can be paid one dollar the first day of the month and an additional dollar for each day for a month. For example, she will be paid $1 on Day 1, $2 on Day 2, $3 on Day 3, and so on.
- *Plan B:* Winona will be paid one cent on the first day and for each succeeding day, double the amount of the previous day. For example she will be paid 1¢ on Day 1, 2¢ on Day 2, 4¢ on Day 3, 8¢ on Day 4, and so on.   **a–b. See margin.**
  a. Use two tables to determine which plan would be more profitable.
  b. On which day will the better plan become greater than the other?

Mixed Review

51. **Basketball** The table at the right shows the number of years that specific NBA basketball players have been playing and the average number of points each player made during the 1993–1994 season. (Lesson 5–2)

Player	Years	Points
Charles Barkley	10	21.6
Glen Rice	5	21.1
Chris Mullin	9	16.8
Jamal Mashburn	1	19.2
Scottie Pippen	7	22.0
Dominique Wilkins	12	26.0

Source: *Hawes Fantasy Basketball Guide,* 1994–1995

  a. Plot the points and describe the graph.   **a–b. See margin.**
  b. What do you think the graph suggests about the relationship of the number of years played and the point average?

52. Graph each point below. Then connect the points in alphabetical order and identify the figure.   (Lesson 5–1)   **star**

$$A(0, 5), B(4, -3), C(-5, 2), D(5, 2), E(-4, -3), F(0, 5)$$

53. **Employment** Hugo's wages vary directly with the time he works. If his wages for 4 days are $110, how much will they be for 17 days? (Lesson 4–8)   **$467.50**

54. What number decreased by 80% is 14?   (Lesson 4–5)   **70**

55. **Finance** If Li Fong had earned one fourth of a percent more in annual interest on an investment, the interest for one year would have been $45 greater. How much did he invest at the beginning of the year?   (Lesson 4–4)

56. Solve $\frac{1}{3}a - 2b = -9c$ for a.   (Lesson 3–6) $a = -27c + 6b$

57. **Geometry** The perimeter of a rectangle is 148 inches.
  a. Write an equation to represent this situation. $2w + 2\ell = 148$
  b. Find its dimensions if the length is 17 inches greater than three times the width.   (Lesson 3–5) w, **14.25 in.**; ℓ, **59.75 in.**

58. Find a number between $-\frac{8}{17}$ and $\frac{1}{9}$.   (Lesson 2–4)   **Sample answer: 0**

59. Evaluate $3x^3 - 2y$ if $x = 0.2$ and $y = 4$.   (Lesson 1–3) -7.976

55. $18,000

Lesson 5–3   Equations as Relations **277**

Extension

Reasoning A relation is called transitive if whenever (a, b) and (b, c) are in the relation, then so is the ordered pair (a, c).
a. Is "is taller than" a transitive relation?   yes
b. Is "is the father of" a transitive relation? no

Additional Answers

51a.

(graph of Point Avg. vs. Years)

51b. The more years played, the higher the point-per-game average of the player.

Chapter 5 **277**

NCTM Standards: 1–5, 8

Objective
Use a graphing calculator to graph linear relations and functions.

Recommended Time
25 minutes

Instructional Resources
Graphing Calculator Masters, pp. 18 and 19

These masters provide keystroking instruction for this lesson for the TI-81 and Casio graphing calculators.

1 FOCUS

Motivating the Lesson
Review with students the method for converting a linear equation from standard form to slope-intercept form.

2 TEACH

Teaching Tip Students might find it helpful to first plot a few points on graph paper.

3 PRACTICE/APPLY

Assignment Guide

Core: 1–10
Enriched: 1–10

4 ASSESS

Observing students working with technology is an excellent method of assessment.

5-4A Graphing Technology
Linear Relations and Functions
A Preview of Lesson 5–4

The graphing calculator is a powerful tool that can be used to study a wide variety of graphs. In many cases, equations are graphed in the standard viewing window, $[-10, 10]$ by $[-10, 10]$. To set the standard viewing window, press ZOOM 6. You do not need to press GRAPH after using ZOOM 6.

The examples below illustrate how you can use the calculator to graph linear relations and functions.

Example ❶ **Graph $2y - 6x = 8$ in the standard viewing window.**

First solve the equation for *y*.
$$2y - 6x = 8$$
$$2y = 6x + 8$$
$$y = 3x + 4$$

Note that the graph of this equation is a line.

Then enter the equation into the calculator, set the standard viewing window, and graph.

Enter: Y= 3 X,T,θ + 4 *Enters the equation.*

ZOOM 6 *Sets the standard window.*

Example ❷ **Graph $y = -x + 20$ in the standard viewing window.**

Enter: Y= (−) X,T,θ + 20

ZOOM 6

An equation whose graph is a non-vertical line is called a <u>linear function</u>.

What happened? None of the graph is shown in the standard viewing window. Pressing the WINDOW key and changing the viewing window settings to $[-5, 25]$ by $[-5, 25]$ with scale settings of 5 for each axis will allow you to see where the graph crosses both the *x*- and *y*-axes.

When the equation in Example 2 is graphed in the $[-5, 25]$ by $[-5, 25]$ viewing window, it is considered a **complete graph**. A complete graph shows all of the important characteristics of the graph on the screen. These include the origin and the points at which the graph crosses the *x*- and *y*-axes.

278   Chapter 5   *Graphing Relations and Functions*

Additional Answers

1.

2.

3.

Example ❸ Use a graphing calculator to draw a graph that represents the solutions to this problem. Then list three of the solutions.

A second number is two more than the opposite of the first.

Let x represent the first number, and let y represent the second number. The equation $y = -x + 2$ describes the relationship between the numbers for any first number x. Select the standard viewing window, enter the equation, and press GRAPH to view the graph of the equation. A complete graph is displayed.

You can find sample solutions for the equation in two ways.

Method 1

Press the TRACE key. The cursor appears as a flashing square and the approximate coordinates of the location of the cursor appears at the bottom of the screen. Use the right and left arrows to move the cursor along the line. New coordinates appear for each location of the cursor. These ordered pairs are approximate solutions for the equation. Sample solutions are $(0, 2)$, $(0.42553191, 1.5744681)$, and $(-0.6382979, 2.6382979)$.

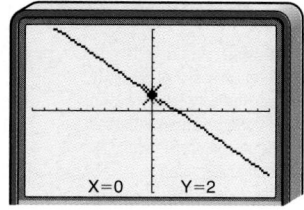

Method 2

You can obtain exact solutions by pressing 2nd TABLE . A table of x and y values for the equation appears.

Use the up and down arrow keys to scroll through the list of values. Sample solutions are $(-3, 5)$, $(0, 2)$, and $(3, -1)$.

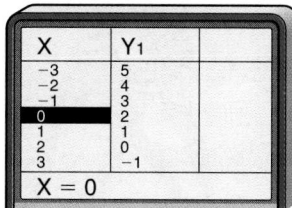

EXERCISES

Use a graphing calculator to graph each equation in the standard viewing window. Sketch each graph on a sheet of paper.

1–6. See margin.

1. $y = 3x - 6$
2. $y = 0.5x - 4$
3. $2x - 3y = 5$
4. $5x + y = 8$
5. $2y - 2x = 3$
6. $-10x - 2y = -14$

Graph each linear function.
a. Determine a viewing window that shows a complete graph.
b. Sketch the graph on a sheet of paper, noting the scale used for each axis.
c. Name three solutions for the equation.

7–10. Answers will vary. See margin for sample answers.

7. $0.1x + y = 10$
8. $y = 3x + 17$
9. $y = 0.01x + 0.02$
10. $200x + y = 150$

Using Technology

This lesson offers an excellent opportunity for using technology in your algebra classroom. For more information on using technology, see *Graphing Calculators in the Mathematics Classroom,* one of the titles in the Glencoe Mathematics Professional Series.

Additional Answers

10b. Xscl: 0.1, Yscl: 10

10c. $(0, 150)$, $(1, -50)$, $(2, -250)$

Additional Answers

4.

5.

6.

7–10. Sample answers are given.
7a. $[-10, 110]$ by $[-5, 15]$
7b. Xscl: 10, Yscl: 1

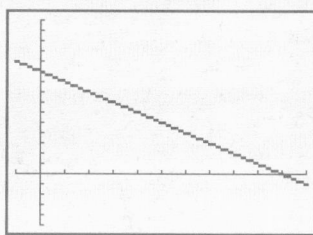

7c. $(0, 10)$, $(100, 0)$, $(10, 9)$
8a. $[-10, 10]$ by $[-5, 20]$
8b. Xscl: 1, Yscl: 1

8c. $(0, 17)$, $(1, 20)$, $(2, 23)$
9a. $[-2.5, 0.5]$ by $[-0.05, 0.05]$
9b. Xscl: 0.5, Yscl: 0.01

9c. $(-2, 0)$, $(0, 0.02)$, $(1, 0.03)$
10a. $[-1, 1]$ by $[-10, 160]$

Instructional Resources

- Study Guide Master 5-4
- Practice Master 5-4
- Enrichment Master 5-4
- Assessment and Evaluation Masters, pp. 127–128
- Graphing Calculator Masters, p. 5

Transparency 5-4A contains the 5-Minute Check for this lesson; **Transparency 5-4B** contains a teaching aid for this lesson.

Recommended Pacing

Standard Pacing	Day 7 of 14
Honors Pacing	Day 7 of 13
Block Scheduling*	Day 4 of 7
Alg. 1 in 2 Years*	Days 10 & 11 of 21

*For more information on pacing and possible lesson plans, refer to the *Block Scheduling Booklet* and *Algebra 1 in Two Years*.

1 FOCUS

5-Minute Check
(over Lesson 5-3)

Which ordered pairs are solutions of each equation?

1. $2m - 5n = 1$
   a. $(-2, -1)$ b. $(2, 1)$
   c. $(7, 3)$ d. $(-7, -3)$ a, d
2. $2x + 3y = 11$
   a. $(3, 1)$ b. $(1, 3)$
   c. $(-2, 5)$ d. $(4, -1)$ b, c
3. $4m + n = 7$
   a. $(1, 3)$ b. $(2, 1)$
   c. $(-1, 11)$ d. $(4, 7)$ a, c

Solve each equation if the domain is $\{-2, -1, 0, 1, 2\}$.

4. $y = 4x - 1$ $\{(-2, -9),$ $(-1, -5), (0, -1), (1, 3),$ $(2, 7)\}$
5. $3r + 2s = 4$ $\{(-2, 5), \left(-1, \frac{7}{2}\right),$ $(0, 2), \left(1, \frac{1}{2}\right), (2, -1)\}$

Objective

After studying this lesson, you should be able to:
- graph linear equations.

CONNECTION
Health

The manner in which you burn Calories depends on your weight and the activity you are doing. The chart at the right shows the number of Calories burned per minute per kilogram (C/min/kg).

Manuel weighs 70 kilograms and wants to know how many Calories he burns playing football. The formula $C = wtr$, where w is his weight in kilograms, t is the time in minutes, and r is C/min/kg, can be used to determine how many Calories he burns.

$$C = wtr$$
$$C = 70 \cdot t \cdot 0.132$$
$$C = 9.24t$$

The number of Calories burned is dependent upon how long he exercises. So, t is the independent variable and C is the dependent variable.

t	$9.24t$	C	(t, C)
0	9.24(0)	0	(0, 0)
10	9.24(10)	92.4	(10, 92.4)
20	9.24(20)	184.8	(20, 184.8)
30	9.24(30)	277.2	(30, 277.2)
40	9.24(40)	369.6	(40, 369.6)

When you graph the ordered pairs, a pattern begins to form. The points seem to lie in a line.

Suppose the domain of $C = 9.24t$ is the set of real numbers. There would be an infinite number of ordered pairs that are solutions for the equation. If you graphed all the solutions, they would form a line. The line shown in the graph at the right represents all the solutions for $C = 9.24t$.

Since the graph of $C = 9.24t$ is a line, $C = 9.24t$ is called a **linear equation**.

Activity	C/min/kg
Basketball	0.138
Cycling (leisure)	0.064
Cycling (racing)	0.169
Dancing (aerobic)	0.135
Dancing (normal)	0.075
Drawing	0.036
Eating	0.023
Football	0.132
Free weights	0.086
Golf	0.085
Gymnastics	0.066
Jumping rope	0.162
Playing drums	0.066
Playing flute	0.035
Playing horn	0.029
Playing piano	0.040
Playing trumpet	0.031
Nautilus® training	0.092
Running (7.2- min km)	0.135
Running (5.0- min km)	0.208
Running (3.7- min km)	0.252
Sitting Quietly	0.021
Swimming	0.156
Walking	0.080
Writing	0.029
Word processing	0.027

Linear equations may contain one or two variables with no variable having an exponent other than 1.

Definition of a Linear Equation in Standard Form	A linear equation is an equation that can be written in the form $Ax + By = C$, where A, B, and C are any real numbers, and A and B are not both zero.

Example Determine whether each equation is a linear equation. If so, identify A, B, and C.

a. $4x = 7 + 2y$

First, rewrite the equation so that both variables are on the same side of the equation.

$$4x = 7 + 2y$$

$4x - 2y = 7$ *Subtract 2y from each side.*

The equation is now in the form $Ax + By = C$, where $A = 4$, $B = -2$, and $C = 7$. This is a linear equation.

b. $2x^2 - y = 7$

The exponents of the variables in a linear equation must be 1. Since the exponent of x is 2, this is not a linear equation.

c. $x = 12$

This equation can be rewritten as $x + 0y = 12$. Therefore, it is a linear equation in the form $Ax + By = C$, where $A = 1$, $B = 0$, and $C = 12$.

To graph a linear equation, it is often helpful to make a table of ordered pairs that satisfy the equation. Then graph the ordered pairs and connect the points with a line.

Example Graph each equation.

a. $y = 8x - 4$

Select five values for the domain and make a table. Then graph the ordered pairs and connect them to draw the line.

x	$8x - 4$	y	(x, y)
-2	$8(-2) - 4$	-20	$(-2, -20)$
-1	$8(-1) - 4$	-12	$(-1, -12)$
0	$8(0) - 4$	-4	$(0, -4)$
1	$8(1) - 4$	4	$(1, 4)$
2	$8(2) - 4$	12	$(2, 12)$

From the graph, you can find other ordered pairs that are solutions; for example, (2.5, 16).

The range of this relation is the set of real numbers.

(continued on the next page)

Cooperative Learning

Think-Pair-Share Have students work in pairs to solve Example 2b. One student in each pair should follow the procedure in the text. The other student should graph the equation on a graphing calculator, and then use the TRACE feature to verify the values in the table. Have pairs discuss their findings with the class. For more information on the think-pair-share strategy, see *Cooperative Learning in the Mathematics Classroom,* one of the titles in the Glencoe Mathematics Professional Series, pages 24–25.

Motivating the Lesson

Questioning Ask students to identify applications from previous lessons in which there appears to be a linear relationship between the two variables in a relation. Some examples might include a salary that is dependent upon the number of hours worked and commission based on the amount of sales.

2 TEACH

Teaching Tip Make sure students understand the definition of a linear equation. The phrase "can be written" does not mean the equation has to appear in that form. However, the equation must be equivalent to an equation in the form $Ax + By = C$.

In-Class Examples

For Example 1
Determine whether each equation is a linear equation. If so, identify A, B, and C.

a. $9y = x + 16$ linear; $A = 1$, $B = -9$, $C = -16$
b. $2x^2 + 3y = 4$ not linear, has x^2 term
c. $x = -5$ linear; $A = 1$, $B = 0$, $C = -5$

For Example 2
Graph each equation.

a. $y = 3x - 4$

b. $2x + 5y = 8$

Teaching Tip In Example 2a, students should be aware that any value can be chosen for x.

In-Class Example

For Example 3
Yukari averages 40 miles per hour when she drives from Los Angeles to San Francisco.

a. What equation relates the distance traveled to the number of hours traveled?
$d = 40t$

b. Graph the relation by letting the horizontal axis represent the time and the vertical axis represent the distance.

---

3 PRACTICE/APPLY

Check for Understanding
Exercises 1–15 are designed to help you assess your students' understanding through reading, writing, speaking, and modeling. You should work through Exercises 1–5 with your students and then monitor their work on Exercises 6–15.

Error Analysis
Students often confuse the graphs of $y = a$ and $x = b$. Invite students to reconstruct which line, horizontal or vertical, corresponds to which graph by making (x, y) tables that fit the equations. To differentiate these similar concepts, focus on just one of the ideas.

---

b. $2x + 5y = 10$

In order to find values for y more easily, solve the equation for y.

$2x + 5y = 10$

$5y = 10 - 2x$ *Subtract 2x from each side.*

$y = \dfrac{10 - 2x}{5}$ *Divide each side by 5.*

Now make a table and draw the graph.

x	$\dfrac{10 - 2x}{5}$	y	(x, y)
-10	$\dfrac{10 - 2(-10)}{5}$	6	$(-10, 6)$
-5	$\dfrac{10 - 2(-5)}{5}$	4	$(-5, 4)$
0	$\dfrac{10 - 2(0)}{5}$	2	$(0, 2)$
5	$\dfrac{10 - 2(5)}{5}$	0	$(5, 0)$
10	$\dfrac{10 - 2(10)}{5}$	-2	$(10, -2)$

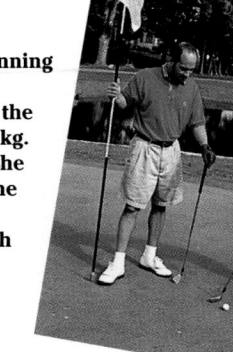

Using graphs is a good way to make comparisons in real-life situations.

Example ③

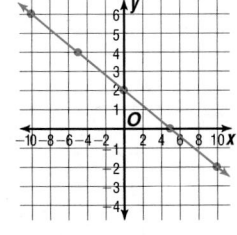

Refer to the connection at the beginning of the lesson. Carmen Delgado is running in a cross-country meet at the rate of 5 min/km. Her weight is 50 kg. How does the number of Calories she is burning compare to those that she would be burning if she were at home word processing her research paper or if she were playing golf with her dad?

CONNECTION
Health

Explore   Look up the Calories burned for each of the three types of activities in the chart on page 280.

TECHNOLOGY
Tip
Graphing calculators are helpful when comparing equations that are not easily graphed with pencil and paper.

Plan   Write the equation for each activity using the formula $C = wtr$ and 50 for w.

Running	$C = 50(0.208)t$ or $C = 10.4t$
Word processing	$C = 50(0.027)t$ or $C = 1.35t$
Golfing	$C = 50(0.085)t$ or $C = 4.25t$

Solve   Use a graphing calculator to graph all three equations. Enter the equations as Y_1, Y_2, and Y_3. Let x represent the t in each equation. Make a sketch of each graph and label each with its activity.

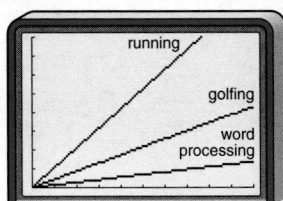

Reteaching

Using Examples To convince students that different samples from a domain of real numbers will produce the same line for a linear relation, use two or three different sets of numbers for x.

For each value of t, Carmen burns many more Calories running than she does golfing or word processing. Carmen would burn more Calories golfing than she would word processing.

Examine Think about the physical movement required for each activity and how much energy that would take. Does the result make sense?

CHECK FOR UNDERSTANDING

Communicating Mathematics

Study the lesson. Then complete the following. 1–2. See margin.

1. **Explain** why the ordered pair $(-1, 3)$ is a solution of $y = -2x + 1$.

2. **Describe** the graph of a linear equation in the form $Ax + By = C$ for which
   a. $A = 0$ b. $B = 0$ c. $C = 0$

3. The first graph is a set of points; the second is a line; see margin for graphs.

3. **Show** how the graph of $y = 2x + 1$ for the domain $\{-1, 0, 2, 3\}$ differs from the graph of $y = 2x + 1$ for the domain of all real numbers.

4. **Explain** why the x and y values used to set the viewing window in Example 3 do not include negative numbers. Time and Calories are nonnegative quantities.

5. **Assess Yourself** During the next 48 hours, record three activities that you perform and the length of each activity. Determine how many Calories you burned in each activity. In which activity did you burn the most Calories? Explain why you think this activity burned the most Calories. See students' work.

Guided Practice

Determine whether each equation is a linear equation. If an equation is linear, rewrite it in the form $Ax + By = C$.

6. $3x - 5y = 0$ yes; $3x - 5y = 0$ 7. $2x = 6 - y$ yes; $2x + y = 6$
8. $3x^2 + 3y = 4$ no   9. $3x = 7 - 2y$ yes; $3x + 2y = 7$

Graph each equation. 10–15. See margin.

10. $3x + y = 4$ 11. $4x + 3y = 12$ 12. $\frac{1}{2}x = 8 - y$
13. $x = 6$ 14. $y = -5$ 15. $x - y = 0$

EXERCISES

Practice

Determine whether each equation is a linear equation. If an equation is linear, rewrite it in the form $Ax + By = C$.

17. $\frac{3}{5}x - \frac{2}{3}y = 5$ **A**
20. $3y = -2$
21. $3x - 2y = 8$
22. $5x = 7$
23. $7x - 7y = 0$
25. $3m - 2n = 0$
26. $\frac{1}{2}x - \frac{2}{3}y = 10$
27. $6a - 7b = -5$

16. $\frac{3}{x} + \frac{4}{y} = 2$ no
18. $x + y^2 = 25$ no
20. $3y + 2 = 0$ yes
22. $5x - 7 = 0$ yes
24. $4x^2 - 3x = y$ no
26. $\frac{x}{2} = 10 + \frac{2y}{3}$ yes

17. $\frac{3}{5}x - \frac{2}{3}y = 5$ yes
19. $x + \frac{1}{y} = 7$ no
21. $2y = 3x - 8$ yes
23. $2x + 5x = 7y$ yes
25. $3m = 2n$ yes
27. $8a - 7b = 2a - 5$ yes

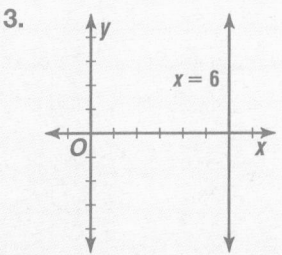

Assignment Guide

Core: 17–59 odd, 61–70
Enriched: 16–56 even, 57–70
All: Self Test, 1–10

For **Extra Practice,** see p. 768.

The red A, B, and C flags, printed only in the Teacher's Wraparound Edition, indicate the level of difficulty of the exercises.

Additional Answers

57a. Sample answer: Parallel lines that slant upward and intersect the x-axis at −7, −2.5, 0, and 4.5.

57b. Sample answer: Parallel lines that slant downward and intersect the x-axis at 0, $1\frac{1}{3}$, $2\frac{1}{3}$, and $-3\frac{1}{3}$.

Study Guide Masters, p. 37

Graph each equation. 28–45. See Solutions Manual.

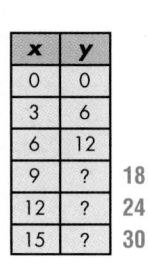

28. $x + 6 = -5$
29. $y = 3x + 1$
30. $6x + 7 = -14y$
31. $2x + 7y = 9$
32. $y + 3 = 4$
33. $x - 6 = -\frac{1}{3}y$
34. $8x - y = 16$
35. $3x + 3y = 12$
36. $6x = 24 - 6y$
37. $x - \frac{7}{2} = 0$
38. $3x - 4y = 60$
39. $4x - \frac{3}{8}y = 1$
40. $2.5x + 5y = 7.5$
41. $x + 5y = 16$
42. $y + 0.25 = 2$

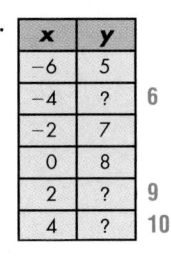

43. $\frac{4x}{3} = \frac{3y}{4} + 1$
44. $y + \frac{1}{3} = \frac{1}{4}x - 3$
45. $\frac{3x}{4} + \frac{y}{2} = 6$

Each table below represents points on a linear graph. Copy and complete each table.

46.

x	y
0	?	4
1	5
2	6
3	7
4	8
5	?	9

47.

x	y
10	?	−5
5	−2.5
0	0
−5	2.5
−10	5
−15	?	7.5

48.

x	y
0	0
3	6
6	12
9	?	18
12	?	24
15	?	30

49.

x	y
−6	5
−4	?	6
−2	7
0	8
2	?	9
4	?	10

Graphing Calculator

Graph each equation using a graphing calculator. Determine the WINDOW setting to use so that a complete graph is shown. Make a sketch of each graph noting the scale used on each axis.

50. $y = 2x + 4$
51. $4x - 9y = 45$
52. $27x + 75y = 100$
53. $17y = 22$
54. $0.2x - 9.7y = 8.9$
55. $\frac{1}{2}x - \frac{2}{3}y = 10$

50–55. See Solutions Manual.

56. Clear the screen. Press [2nd] [DRAW] 4 3 [ENTER]. Describe the graph. Explain how you could use this function to graph $x = -25$. Graph of $x = 3$; use the same procedure but change 3 to −25.

Critical Thinking

57a–b. See margin.

57. The graphs of each group of equations form a family of graphs. Graph each family on the same coordinate plane. Then write an explanation of the similarities and differences that exist in the graphs of each family.
  a. $y = 2x$ $y = 2x + 5$ $y = 2x - 9$ $y = 2x + 14$
  b. $y = -3x$ $y = -3x + 4$ $y = -3x - 10$ $y = -3x + 7$

Applications and Problem Solving

58b. about 14 seconds

58. **Science** As a thunderstorm approaches, you see lightning as it occurs, but you hear the accompanying sound of the thunder a short time afterward. The distance y, in miles, that sound travels in t seconds is given by the equation $y = 0.21t$.
  a. Draw a graph of this equation. See margin.
  b. Use the graph to estimate how long it will take you to hear the thunder from a storm that is 3 miles away.

284 Chapter 5 Graphing Relations and Functions

Additional Answers

57b. cont.

58a.

59. Employment Elva Duran works as a sales representative for Quasar Electronics. She receives a salary of $1800 per month plus a 6% commission on monthly sales over her target. Her sales in July will be $800, $1300, or $2000 over target, depending on when her orders are filled by the company's distribution center.
 a. Graph ordered pairs that represent each of her possible incomes y for the three sales figures x. **See margin.**
 b. Will her total income for July be more than $1850 if the equation that represents her income is $y = 1800 + 0.06x$? Explain.

59b. yes, but only if her sales are $1300 or $2000 over target

60. Weight Training Hector Farentez, who weighs 68.2 kg, wanted to increase his body size by beginning a weight-training program. A local fitness club suggests that he begin his program using Nautilus® equipment before he attempts the free weights. The Nautilus® equipment includes various machines, each of which exercises a certain group of muscles. After several weeks of training, Hector has expanded his workout time from 30 minutes to 1 hour. **a.** $C = 0.092(68.2)t$; **See Solutions Manual for graph.**

 a. Use the information in the table at the beginning of the lesson to determine a formula for determining how many Calories Hector is burning. Graph the equation showing each 5-minute interval.
 b. Later in his training at a weight of 72.3kg, Hector plans to spend one-half hour on Nautilus® and one-half hour on free weights. Will he use more or fewer Calories in this type of workout than an hour of Nautilus®? Explain.
 He will burn more calories in the new routine because $0.092(30)(72.3) + 0.132(30)(72.3) > 0.092(60)(72.3)$.

Mixed Review

62. See Solutions Manual.

61. Solve $8x + 2y = 6$ for y. (Lesson 5–3) $y = 3 - 4x$

62. Draw a mapping for the relation $\{(2, 7), (-3, 7), (2, 4)\}$. (Lesson 5–2)

63. Geometry (Lesson 5–1)
 a. Graph (3, 2), (6, 2), and (6, 5). **See Solutions Manual.**
 b. If these three points represent the vertices of a square, what would be the coordinates of the fourth point? **(3, 5)**
 c. Assume that each unit on each axis represents one inch. What is the perimeter of the square? **12 inches**

64. Probability A gumball machine has 24 cherry gumballs, 5 apple gumballs, 18 grape gumballs, 14 orange gumballs, and 9 licorice gumballs. What is the probability of getting an orange gumball? (Lesson 4–6) $\frac{1}{5}$

65. Jimmy got a discount of $4.50 on a new radio. The discounted price was $24.65. What was the percent of discount to the nearest percent? (Lesson 4–5) **15%**

Lesson 5–4 Graphing Linear Equations **285**

Extension

Reasoning Write a linear equation for y in terms of x for the following ordered pairs: (1, 4), (2, 9), (3, 14), (5, 24), (8, 39).
$y = 5x - 1$

Answers for the Self Test

1.

2. {(1, 3), (2, 7), (4, 1), (−3, 3), (−3, −3)}; D = {−3, 1, 2, 4}; R = {−3, 1, 3, 7}; Inv = {(3, 1), (7, 2), (1, 4), (3, −3), (−3, −3)}

3. {(−5, −3), (−1, 4), (4, 4), (4, −3)}; D = {−5, −1, 4}; R = {−3, 4}; Inv = {(−3, −5), (4, −1), (4, 4), (−3, 4)}

Enrichment Masters, p. 37

66. **Running** One member of a cross-country team placed fourth in a meet. The next four finishers on the team placed in consecutive order, but farther behind. The team score was 70. In what places did the other members finish? (Lesson 3–5) **15th, 16th, 17th, 18th**

67. **Miniature Golf** In miniature golf, balls bounce off the walls of the course at the same angle at which they hit. If a ball strikes the wall at 30°, what is the measure of the angle between the two paths of the ball? (Lesson 3–4) **120°**

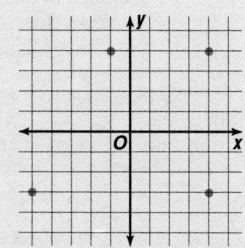

30°   30°

68. Two meters of copper tubing weigh 0.25 kilograms. How much do 50 meters of the same tubing weigh? (Lesson 3–3) **6.25 kg**

69. Simplify $\dfrac{\frac{-3}{4}}{-36}$. (Lesson 2–7) $\dfrac{1}{48}$

70. Find $-21 + 52$. (Lesson 2–3) **31**

SELF TEST

1. Graph $A(-6, 6)$, $B(3, -5)$, $C(0, 2)$, and $D(-3, -1)$. (Lesson 5–1) **1–3. See margin.**

Express each relation as a set of ordered pairs. Then state the domain, the range, and the inverse of the relation. (Lesson 5–2)

2.

3.
(graph)

Solve each equation if the domain is {−2, −1, 0, 1, 3}. Then graph the solution set.
(Lesson 5–3) **4–6. See Solutions Manual for graphs.**

4. $3y + 6x = 12$ for y $y = 4 - 2x$
5. $2a + 3b = 9$ for b $b = 3 - \dfrac{2}{3}a$
6. $5r = 8 - 4s$ for s $s = \dfrac{8 - 5r}{4}$

Graph each equation. (Lesson 5–4) **7–9. See Solutions Manual.**

7. $y = x - 1$ 8. $y = 2x - 1$ 9. $3x + 2y = 4$

10. **Geometry** The formula for the area of a trapezoid is $A = \dfrac{1}{2}h(b_1 + b_2)$, where b_1 and b_2 are the lengths of the bases and h is the height. Find the height of the trapezoid shown at the right if its area is 40 m². (Lesson 5–3) **about 5.33 m**

5 m
h
10 m

Functions

Objectives

After studying this lesson, you should be able to:

- determine whether a given relation is a function, and
- find the value of a function for a given element of the domain.

APPLICATION
Air Travel

During certain times of the year, airlines offer lower rates to fly to selected cities. The advertisement below shows discount ticket fares. The mileage between the cities is also listed for those people who are enrolled in frequent flier programs.

Plan Your Vacation Now!

From	To	Regular Fare	Discount Fare	Mileage
Atlanta	New York	$ 89	$ 79	892
Baltimore	Houston	149	129	1454
Boston	Greensboro	109	79	786
Cleveland	Philadelphia	94	79	441
Dayton	Houston	109	99	1178
Greensboro	Miami	84	79	814
Houston	Jacksonville	104	99	875
Houston	Miami	109	99	1231
Los Angeles	San Antonio	139	129	1277

Sample sale prices. Coach class each way. Round-trip purchase required. Seats are limited and may not be available on all flights/days.

Suppose we let r represent the regular fare, d the discount fair, and m the mileage. The relation whose ordered pairs are of the form (r, d) is graphed below on the left. The relation whose ordered pairs are of the form (m, d) is graphed below on the right.

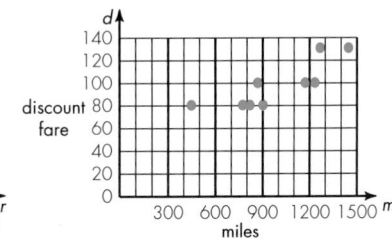

Notice that in the graph on the left, when $r = 109$, there is more than one value of d, 79 and 99. However, in the other graph, there is exactly one value of d for each value of m. Relations with this characteristic are called **functions**.

Definition of Function	A function is a relation in which each element of the domain is paired with *exactly* one element of the range.

NCTM Standards: 1–6

Instructional Resources

- Study Guide Master 5-5
- Practice Master 5-5
- Enrichment Master 5-5
- Modeling Mathematics Masters, pp. 48–50, 76
- Real-World Applications, 15
- Tech Prep Applications Masters, p. 10

 Transparency 5-5A contains the 5-Minute Check for this lesson; **Transparency 5-5B** contains a teaching aid for this lesson.

Recommended Pacing

Standard Pacing	Days 8 & 9 of 14
Honors Pacing	Day 8 of 13
Block Scheduling*	Day 5 of 7 (along with Lesson 5-6)
Alg. 1 in Two Years*	Days 12 & 13 of 21

 *For more information on pacing and possible lesson plans, refer to the *Block Scheduling Booklet* and *Algebra 1 in Two Years*.

1 FOCUS

 5-Minute Check
(over Lesson 5-4)

Determine whether each equation is a linear equation. If so, identify *A*, *B*, and *C*.

1. $2a - 3b = 7$ **linear; $A = 2$, $B = -3$, $C = 7$**

2. $y = \frac{1}{x}$ **not linear**

3. $3x = 2y - 14$ **linear; $A = 3$, $B = -2$, $C = -14$**

4. Graph $2x - y = 8$.

Example **Determine whether each relation is a function. Explain your answer.**

a. {(2, 3), (3, 0), (5, 2), (−1, −2), (4, 1)}
Since each element of the domain is paired with exactly one element of the range, this relation is a function.

b.

This mapping represents a function since, for each element of the domain, there is *only one* corresponding element in the range.
It does not matter if two elements of the domain are paired with the same element in the range.

c.

x	y
4	−1
5	2
5	3
6	6
−1	1

This table represents a relation that is not a function. The element 5, in the domain, is paired with both 2 and 3 in the range.

There are several ways to determine whether an equation represents a function.

Example **Determine whether $x - 4y = 12$ is a function.**

Method 1: Make a table of solutions.

First solve for y.

$x - 4y = 12$

$-4y = -x + 12$ *Subtract x from each side.*

$y = \frac{1}{4}x - 3$ *Divide each side by −4.*

Next make a table of values like the one at the right.

It appears that for any given value of x, there is only one value for y that will satisfy the equation. Therefore, the equation $x - 4y = 12$ is a function.

x	y
−8	−5
−4	−4
−2	−3.5
0	−3
2	−2.5
4	−2
8	−1

Method 2: Graph the equation.

Since the equation is in the form $Ax + By = C$, the graph of the equation will be a line. Graph the ordered pairs from Method 1 and connect them with a line.

Now place your pencil at the left of the graph to represent a vertical line. Slowly move the pencil to the right across the graph.

For each value of x, this vertical line passes through no more than one point on the graph. Thus, the line represents a function.

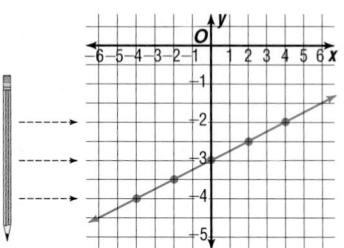

Using a pencil to see if a graph represents a function is one way to perform the **vertical line test.**

Vertical Line Test for a Function	If any vertical line passes through no more than one point of the graph of a relation, then the relation is a function.

Example Use the vertical line test to determine if each relation is a function.

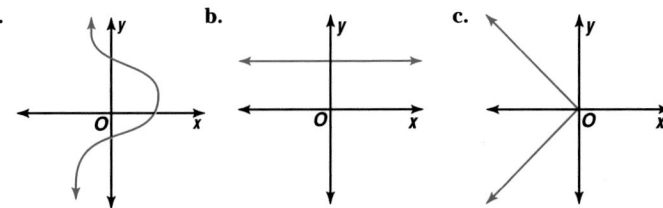

a.   b.   c.

Graph b is the only relation to pass the vertical line test. Thus, it is the only function. With graphs a and c, a vertical line intersects the graph in more than one point. Thus, they are *not* functions.

Letters other than f are also used for names of functions. For example, g(x) and h(x) are sometimes used.

Equations that are functions can be written in a form called **functional notation.** For example, consider the equation $y = 3x - 7$.

equation	functional notation
$y = 3x - 7$	$f(x) = 3x - 7$
	The symbol f(x) is read "f of x."

The ordered pair (4, f(4)) is a solution of the function f.

In a function, x represents the elements of the domain and $f(x)$ represents the elements of the range. Suppose you want to find the value in the range that corresponds to the element 4 in the domain. This is written $f(4)$ and is read "f of 4." The value of $f(4)$ is found by substituting 4 for x in the equation. So, $f(4) = 3(4) - 7$ or 5.

Example If $f(x) = 2x - 9$, find each value.

a. $f(6)$
$f(6) = 2(6) - 9$
$= 12 - 9$
$= 3$

b. $f(-2)$
$f(-2) = 2(-2) - 9$
$= -4 - 9$
$= -13$

c. $f(k + 1)$
$f(k + 1) = 2(k + 1) - 9$
$= 2k + 2 - 9$
$= 2k - 7$

The functions we have studied thus far have been linear functions. Many functions are not linear. However, you can find values of functions in the same way.

Example If $h(z) = z^2 - 4z + 9$, find each value.

a. $h(-3)$
$h(-3) = (-3)^2 - 4(-3) + 9$
$= 9 + 12 + 9$ or 30

(continued on the next page)

Lesson 5–5 *Functions* **289**

Teaching Tip Have students perform the vertical line test with the graphs in the application at the beginning of the lesson.

Teaching Tip Have students look at the graphs in Lesson 5-2 and determine which of the graphs are functions.

In-Class Examples

For Example 3
Use the vertical line test to determine if the relation is a function. **It is a function.**

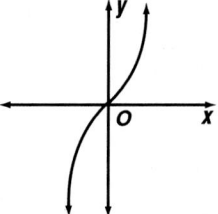

For Example 4
Find each value.

a. $f(1)$ if $f(x) = 3x - 2$ 1
b. $f(2)$ if $f(x) = 7 - 4x$ −1

Teaching Tip It may be helpful for some students to think of $f(x)$ as "f at x" since it represents the value of f at x.

b. $h(5c)$

$h(5c) = (5c)^2 - 4(5c) + 9$ *Substitute 5c for z.*
$= 5 \cdot 5 \cdot c \cdot c - 4 \cdot 5 \cdot c + 9$
$= 25c^2 - 20c + 9$

c. $5[h(c)]$

$5[h(c)] = 5[(c)^2 - 4(c) + 9]$ *5[h(c)] means 5 times the value of h(c).*
$= 5 \cdot c^2 - 5 \cdot 4c + 5 \cdot 9$ *Distributive property*
$= 5c^2 - 20c + 45$

Notice that when comparing parts b and c, $5[h(c)] \neq h(5c)$.

Functions are often used in solving real-life problems.

 Example 6

CONNECTION
Health

The normal systolic blood pressure S is a function of the age a of the individual. That is, a person's normal blood pressure depends on how old the person is. To determine the normal systolic blood pressure of an individual, you can use the equation $S = 0.5a + 110$, where a represents age in years.

a. Write the equation in functional notation.

b. Find $S(10)$, $S(30)$, $S(50)$, and $S(70)$.

c. Graph the function.

d. Use the graph of the function to estimate whether blood pressure increases or decreases with age. Then estimate the blood pressure of an 80-year-old person.

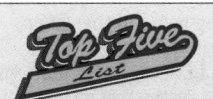

Best Selling Prescription Drugs in the World

Brand Name	Used for
1. Zantac	ulcers
2. Vasotec	hypertension*
3. Capoten	hypertension*
4. Voltaren	arthritis
5. Tenormin	hypertension*

*high blood pressure

a. Let $S(a)$ represent the function. The equation becomes $S(a) = 0.5a + 110$.

b. Make a table with 10, 30, 50, and 70 as values of a.

a	$S(a) = 0.5a + 110$	$S(a)$	$(a, S(a))$
10	$S(10) = 0.5(10) + 110$	115	(10, 115)
30	$S(30) = 0.5(30) + 110$	125	(30, 125)
50	$S(50) = 0.5(50) + 110$	135	(50, 135)
70	$S(70) = 0.5(70) + 110$	145	(70, 145)

c. Use the ordered pairs from the table to graph the function.

d. The graph indicates that as you age, your blood pressure is expected to rise. An 80-year-old person would expect to have a systolic pressure of about 150.

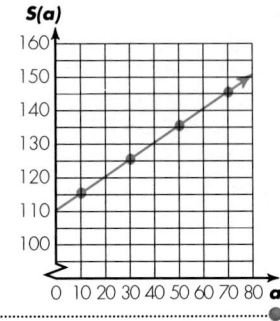

Communicating Mathematics

Study the lesson. Then complete the following. 1–2. See margin.

1. **Explain** any differences between a relation and a function.

2. **Describe** how the phrase "is a function of" relates to the mathematical definition of function. For example, the diameter of a tree is a function of its age.

3. Substitute 1 for *x* in the equation and evaluate; $g(1) = 15$.

3. **Write** how you would find $g(1)$ if $g(x) = 3x + 12$.

4. False; $x = 4$ is not a function.

4. **You Decide** Is the statement "All linear equations are functions" true? Support your answer with examples or counterexamples.

MATH JOURNAL

5. In your own words, explain how and why the vertical line test works. See margin.

Guided Practice

Determine whether each relation is a function.

6.

yes

7.

x	y
−2	3
5	3
5	4
4	0

no

8.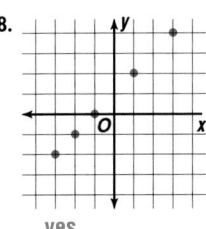

yes

9. $\{(-3, 1), (-1, 3), (1, -2), (3, 2)\}$ yes

10. $\{(3, 1), (-2, 2), (1, -1), (1, 2), (-3, 1), (-2, 5)\}$ no

11. $y + 5 = 7x$ yes

12. $2x^2 + 3y^2 = 36$ no

13. The graph of $y^2 = x + 4$ is shown at the right. Use the vertical line test to determine if this relation is a function. no

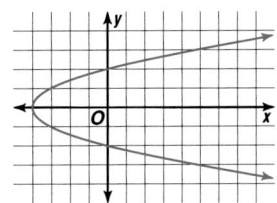

14. Which of the following graphs represent a function? a, c

a. $y = |x - 1|$

b. $x = |y| + 1$

c. $y = |x| + 1$

d. $x = 1 - |y|$

LOOK BACK

You can refer to Lesson 2-3 for more information on absolute value. $|x|$.

If $h(x) = 3x + 2$, find each value.

15. $h(-4)$ −10   16. $h(2)$ 8   17. $h(w)$ $3w + 2$ 18. $h(r - 6)$ $3r - 16$

Lesson 5–5 Functions **291**

---

3 PRACTICE/APPLY

Check for Understanding
Exercises 1–18 are designed to help you assess your students' understanding through reading, writing, speaking, and modeling. You should work through Exercises 1–5 with your students and then monitor their work on Exercises 6–18.

Additional Answers

1. A relation is a set of ordered pairs. A function is a relation in which each member of the domain is paired with only one member of the range.
2. The value of one variable is dependent upon the value of the other variable. That is, if *w* is a function of *f*, the value of *w* is dependent upon the value of *f* in the expression defining *w*.
5. Sample answer: If you can draw a vertical line through the graph at any point and it intersects the graph more than once, the graph is not the graph of a function.

Reteaching

Using Alternatives Demonstrate how functional notation $f(x)$ is easier to use than *y*.

$y = 2x - 3$ vs. $f(x) = 2x - 3$
Find *y* when   Find $f(4)$.
x is 4.
$y = 2(4) - 3$ $f(4) = 2(4) - 3$
$y = 5$ when $x = 4$ $f(4) = 5$

EXERCISES

Practice

Determine whether each relation is a function.

19.

a	b
−3	3
−2	3
0	4
2	4
4	4

yes

20.

r	s
−4	3
−4	2
−4	1
4	0
4	−1
4	−2

no

21.

no

22.

yes

23.

yes

24.

no

25. yes   26. no
27. yes   28. yes

25. {(6, 3), (5, −2), (2, 3), (12, −12)}   26. {(4, 5), (3, −2), (−2, 5), (4, 7)}

27. {(5, −1), (6, −1), (−8, −1), (0, −1)}   28. {(4, −2), (−4, −2), (9, −2), (0, −2)}

B

29. $y = -15$ yes   30. $x = 13$ no   31. $y = 3x - y$ yes

32. $y = |x|$ yes   33. $x = |y|$ no   34. $yx = 36$ yes

41. −18
42. 3.84
43. $9b^2 - 6b$
46. $12w^2 - 12w$
47. $5a^4 - 10a^2$
48. $4c + 14$
49. $24p - 36$
50. $-60w - 30$

If $f(x) = 4x + 2$ and $g(x) = x^2 - 2x$, find each value.

35. $f(-4)$ −14   36. $g(4)$ 8   37. $g\left(\frac{1}{5}\right)$ $-\frac{9}{25}$ 38. $f\left(\frac{3}{4}\right)$ 5

39. $g(3.5)$ 5.25   40. $f(6.2)$ 26.8   41. $3[f(-2)]$ 42. $-6[g(0.4)]$

43. $g(3b)$ 44. $f(2y)$ $8y + 2$ 45. $-3[g(1)]$ 3   46. $3[g(2w)]$

C

47. $5[g(a^2)]$ 48. $f(c + 3)$ 49. $6[f(p - 2)]$ 50. $-3[f(5w + 2)]$

51. **a.** Create and graph a relation that is a function and has an inverse that is also a function. Answers will vary; a sample answer is $f(x) = x$.

b. Create and graph a relation that is a function and has an inverse that is not a function. Answers will vary; a sample answer is $f(x) = x^2$.

Choose the graph that best represents the information given. Explain your choice. Determine whether the graph represents a function.

52. **Communication** A call to London, England costs $1.78 for the first minute and $1.00 for each additional minute. Fractions of a minute are charged as an entire minute. b; Each step represents a minute or any part thereof.

a.
Cost ($)
Time (minutes)

b.
Cost ($)
Time (minutes)

c.
Cost ($)
Time (minutes)

53. **Manufacturing** A company charges $9.95 for each custom-printed T-shirt. If you order 8 or more, but less than 15, the charge is $8.25 each. If you order 15 or more, the charge is $6.50 each. **See margin.**

a. Price ($)   b. Price ($)   c. Price ($)
Number Ordered   Number Ordered   Number Ordered

Critical Thinking

54c. The inverse is a reflection of the function with $y = x$ acting as the mirror.

54. Let the domain of $f(x) = x^2 + 3x + 2$ be $\{-5, -4, -3, -2, -1, 0, 1, 2\}$.
   a. Graph the function, the ordered pairs that make up the inverse of the function, and the line $y = x$ on the same coordinate plane. **See margin.**
   b. Is the inverse of this function a function? **no**
   c. How are the graphs of the function and its inverse related to the line $y = x$?

Applications and Problem Solving

55. **Metallurgy** Since pure gold is very soft, other metals are often added to gold to make an alloy that is stronger and more durable. The relative amount of gold in a piece of jewelry is measured in karats. The formula for the relationship is $g = \frac{25k}{6}$, where k represents the number of karats and g is the percent of gold in the piece. **a. See margin.**
   a. Suppose the domain for the function includes the set $\{0, 6, 12, 18, 24\}$. Make a table using these values for k and graph the function.
   b. Describe the graph of the function. **It is a line.**
   c. How many karats are in a ring made of pure gold? **24 karats**

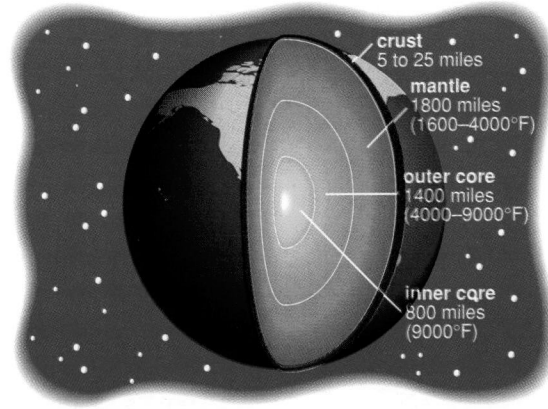

crust
5 to 25 miles
mantle
1800 miles
(1600–4000°F)
outer core
1400 miles
(4000–9000°F)
inner core
800 miles
(9000°F)

56c. Because the crust is thinnest when measured from the bottom of the ocean and thicker from the tops of mountains.

56. **Earth Science** Earth's interior is composed of four levels: the crust, the mantle, the outer core, and the inner core. The inner core is believed to be solid while the outer core is molten. The temperature in the first 62 miles below Earth's surface can be approximated with the formula $T = 35d + S$, where T is the temperature in degrees Fahrenheit, d is the distance in miles, and S is the temperature (°F) on the surface.
   a. Suppose the temperature at the surface is 75°F. Find the temperature at the lower edge of the crust if the crust is 30 miles deep at that point. **1125°F**
   b. Russian scientists are credited with drilling the deepest hole ever on the Kola Peninsula, which is located near the Arctic Circle in northwest Russia about 250 miles from Finland. Drilling began in 1970, but is only 7 miles deep thus far. What is the temperature at the bottom of this hole if the outside temperature is 0°F? **245°F**
   c. Why do you think that there is a range of values given for the depth of the crust?

Lesson 5–5 Functions **293**

Extension

Problem Solving If $h(x) = 4x^3 + 3x^2 + x - 12$, find $3[h(2)]$. **102**

Chapter 5 **293**

Closing Activity

Modeling Using materials on hand, construct a simple fulcrum-and-lever device. With objects of different weights, demonstrate for students how varying amounts of pressure are required to lift the different objects using the lever. Explain that the amount of pressure applied to lift the object is a function of the weight of the object.

Additional Answer

58.

$y - x = -5$

Enrichment Masters, p. 38

Enrichment — Student Edition Pages 287–294

Composite Functions

Three things are needed to have a function—a set called the *domain*, a set called the *range*, and a *rule* that matches each element in the domain with only one element in the range. Here is an example.

Rule: $f(x) = 2x + 1$

$f(x) = 2x + 1$
$f(1) = 2(1) + 1 = 2 + 1 = 3$
$f(2) = 2(2) + 1 = 4 + 1 = 5$
$f(-3) = 2(-3) + 1 = -6 + 1 = -5$

Suppose we have three sets A, B, and C and two functions described as shown below.

Rule: $f(x) = 2x + 1$ Rule: $g(y) = 3y - 4$

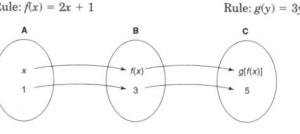

$g(y) = 3y - 4$
$g(3) = 3(3) - 4 = 5$

Let's find a rule that will match elements of set A with elements of set C without finding any elements in set B. In other words, let's find a rule for the **composite function** $g[f(x)]$.

Since $f(x) = 2x + 1$, $g[f(x)] = g(2x + 1)$.
Since $g(y) = 3y - 4$, $g(2x + 1) = 3(2x + 1) - 4$, or $6x - 1$.
Therefore, $g[f(x)] = 6x - 1$.

Find a rule for the composite function $g[f(x)]$.

1. $f(x) = 3x$ and $g(y) = 2y + 1$
$g[f(x)] = 6x + 1$

2. $f(x) = x^2 + 1$ and $g(y) = 4y$
$g[f(x)] = 4x^2 + 4$

3. $f(x) = -2x$ and $g(y) = y^2 - 3y$
$g[f(x)] = 4x^2 + 6x$

4. $f(x) = \frac{1}{x-3}$ and $g(y) = y^{-1}$
$g[f(x)] = x - 3$

5. Is it always the case that $g[f(x)] = f[g(x)]$? Justify your answer.
No. For example, in Exercise 1, $f[g(x)] = f(2x + 1) = 3(2x + 1) = 6x + 3$, not $6x + 1$

294 Chapter 5

57a. \$260.87

57b. Equations may vary; Sample equation:
$B = \frac{3}{23}P$

Mixed Review

59. $\{-16, -13, 5, 11\}$

60. $\{(1, -1), (9, -5), (6, 4)\}$

61. $\frac{3}{13}$

---

57. Accounting The following is adapted from an article by Gail A. Eisner, CPA (certified public accountant), in *The Mathematics Teacher*.

> Early in my accounting career, the partner who was directly supervising me, a very bright, young CPA, called me into his office and asked how a person could possibly calculate a bonus if the company's formula required that the bonus be 15 percent of the net profits *after* the bonus has been deducted? I showed him the linear equation $B = 0.15(P - B)$, where B is the bonus, and P (a known quantity) represents the profits before subtracting the bonus.

a. Suppose the profits are \$2000. Determine the amount of the bonus.
b. Solve the equation for B and graph the resulting equation.

58. Graph $y - x = -5$. (Lesson 5–4) **See margin.**

59. Solve $3a - b = 7$ if the domain is $\{-3, -2, 4, 6\}$. (Lesson 5–3)

60. State the inverse of the relation $\{(-1, 1), (-5, 9), (4, 6)\}$. (Lesson 5–2)

61. Probability A card is selected at random from a deck of 52 cards. What is the probability of selecting a jack, queen, or king? (Lesson 4–6)

62. 44 is what percent of 89? (Lesson 4–4) **49.4%**

63. For what angle are the sine and cosine equal? (Lesson 4–3) **45°**

64. Trigonometry Find the measure of the third angle of a triangle if the other angles have measures 167° and 4°. (Lesson 3–4) **9°**

65. Solve $\frac{z}{-4} - 9 = 3$. (Lesson 3–3) **−48**

66. Budgeting The total of Jon Young's gas and electric bills was \$210.87. His electric bill was \$95.25. How much was his gas bill? (Lesson 3–2) **\$115.62**

67. Sports The players with the most runs batted in (RBI) for the National League are listed below. (Lesson 2–2)

Year	Name	RBI	Year	Name	RBI
1971	Joe Torre	137	1983	Dale Murphy	121
1972	Johnny Bench	125	1984	Mike Schmidt	106
1973	Willie Stargell	119		Gary Carter	
1974	Johnny Bench	129	1985	Dave Parker	125
1975	Greg Luzinski	120	1986	Mike Schmidt	119
1976	George Foster	121	1987	Andre Dawson	137
1977	George Foster	149	1988	Will Clark	109
1978	George Foster	120	1989	Kevin Mitchell	125
1979	Dave Winfield	118	1990	Matt Williams	122
1980	Mike Schmidt	121	1991	Howard Johnson	117
1981	Mike Schmidt	91	1992	Darren Daulton	109
1982	Dale Murphy	109	1993	Barry Bonds	123
	Al Oliver		1994	Jeff Bagwell	116

Source: *The World Almanac, 1995*

a. Make a line plot of the data. **See margin.**
b. What was the greatest number of RBIs during one season? **149**
c. What was the least number of RBIs during one season? **91**
d. What was the most frequently-occurring number of RBIs during one season? **121 and 125**
e. How many of the players had from 119 to 129 RBIs? **13 players**

294 Chapter 5 Graphing Relations and Functions

Additional Answer

67a.

Writing Equations from Patterns

Objective

After studying this lesson, you should be able to:

• write equations to represent relations, given some of the solutions for the equations.

APPLICATION
Scuba Diving

As scuba divers descend, the pressure of the water increases. Scuba divers can determine their depth by the pressure. Pressure can be expressed in atmospheres. An atmosphere is equivalent to 14.7 psi (pounds per square inch) of pressure. The table below shows the relationship between atmospheres of pressure and ocean depth.

Pressure (atmospheres)	1	2	3	4	5
Depth of Ocean (feet)	0	33	66	99	132

Suppose we let p represent the atmospheres of pressure, and let d represent the ocean depth. The relation shown in the table can also be represented by a graph. Graph the points and observe the pattern.

Notice that when the ordered pairs are graphed, they form a linear pattern. You can use a straightedge to draw the line through these points. Points that lie in a linear pattern can be described by an equation. Look at the relationship between the domain and range to find a pattern that can be described by an equation.

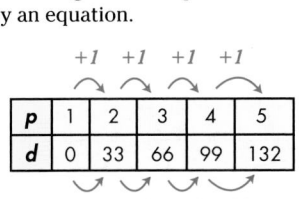

$$\frac{range\ differences}{domain\ differences} = \frac{33}{1}$$

Notice that the differences in the d values are 33 times the differences of the corresponding p values. It seems that this relationship can be described by the equation $d = 33p$. Check to see if this equation is correct by substituting values of p into the equation.

p	1	2	3	4	5
$33p$	33	66	99	132	165

These values do not match those in the first chart. However, note that each corresponding value for d using the equation is 33 more than the d in the relation. Thus, we need to adjust our equation to compensate for this difference. The equation that describes this relation is $d = 33p - 33$. Since this relation is also a function, we can write this equation as $d(p) = 33p - 33$.

Lesson 5–6 Writing Equations from Patterns **295**

In 1943, Jacques-Yves Cousteau and E. Gagnon developed the Aqua-Lung™, to free divers from hoses connected to air sources on the surface. It is called *scuba* gear, which is an acronym for *self-contained underwater breathing apparatus*.

Jacques-Yves Cousteau is known to many for his adventures aboard his research vessel, *The Calypso*, and for his three widely read books, *The Silent World* (1953), *The Living Sea* (1962), and *World Without Sun* (1964).

NCTM Standards: 1–6

Instructional Resources

• Study Guide Master 5-6
• Practice Master 5-6
• Enrichment Master 5-6
• Assessment and Evaluation Masters, p. 129

Transparency 5-6A contains the 5-Minute Check for this lesson; **Transparency 5-6B** contains a teaching aid for this lesson.

Recommended Pacing	
Standard Pacing	Day 10 of 14
Honors Pacing	Day 9 of 13
Block Scheduling*	Day 5 of 7 (along with Lesson 5-5)
Alg. 1 in Two Years*	Days 14 & 15 of 21

*For more information on pacing and possible lesson plans, refer to the *Block Scheduling Booklet* and *Algebra 1 in Two Years*.

1 FOCUS

5-Minute Check
(over Lesson 5-5)

1. Is the relation {(1, 3), (2, 4), (3, 5), (−1, 3), (−2, 4), (−3, 5)} a function? **yes**
2. Is the inverse of the relation shown in Exercise 1 a function? **no**
3. Is the relation described by $y = -2$ a function? **yes**

If $f(x) = 3x^2 - 4$, find each value.

4. $f(-2)$ **8**
5. $f(0)$ **−4**

Motivating the Lesson

Situational Problem Present the following number sequences to students. Ask them to determine what number is added to each term as well as what the missing terms are.

1. 3, 7, 11, 15, __, __
   Add 4; 19, 23
2. 6, 4, 2, 0, −2, __, __
   Add −2; −4, −6
3. 3, −1, −5, −9, __, __
   Add −4; −13, −17

2 TEACH

Teaching Tip In Example 1, it may help students to consider the question, "What must I do to x to get y?"

In-Class Examples

For Example 1
Write an equation in functional notation for the relation given in the chart.

x	1	2	3	4	5
y	6	12	18	24	30

$f(x) = 6x$

For Example 2
Write an equation in functional notation for the relation given in the chart.

x	1	2	3	4	5	6
y	−1	1	3	5	7	9

$f(x) = 2x - 3$

Example **Plot the points in the relation shown in the table. Then write an equation in functional notation for the relation.**

x	−4	−8	−12	−16	−20
y	1	2	3	4	5

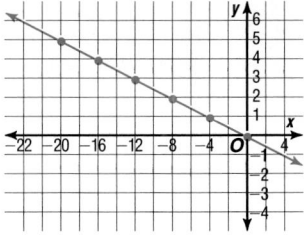

Since the points form a linear pattern, we know that there is a linear equation that describes this relation.

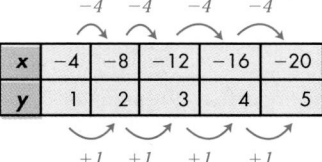

$$\frac{range\ differences}{domain\ differences} = \frac{1}{-4}$$

The differences in the y values are one-fourth the differences in the x values. This pattern implies that the equation $y = -\frac{1}{4}x$ may describe the relation. Check the equation for the domain values in the relation.

Check: If $x = -4$, then $y = -\frac{1}{4}(-4)$ or 1. ✔

If $x = -8$, then $y = -\frac{1}{4}(-8)$ or 2. ✔

The equation that represents the relation is $y = -\frac{1}{4}x$. Since this relation is also a function, we can write this equation in functional notation as $f(x) = -\frac{1}{4}x$.

Sometimes you may need to find the equation for a graph.

Example **Write an equation in functional notation for the relation graphed at the right.**

First make a table of ordered pairs for several points on the graph.

k	1	2	4	5	6	9
m	8	6	2	0	−2	−8

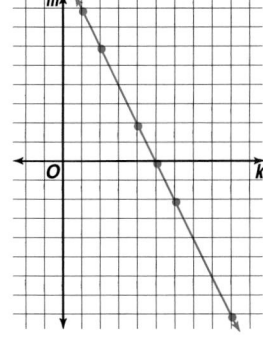

Now find the common differences of the domain and range.

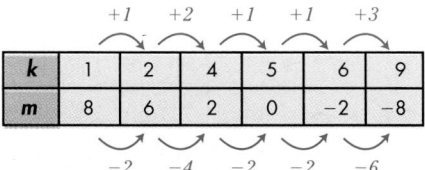

Notice that there is not a consistent change in the domain and range values. However, you can still find the ratio of the range difference to the domain difference for each ordered pair. There are three possible ratios.

GLENCOE Technology

 CD-ROM Interaction

A multimedia simulation allows students to use data, graphs, and equations to analyze ocean depth and pressure. A blackline master activity with teacher's notes provides a follow-up to the CD-ROM simulation.

For Windows & Macintosh

$$\frac{\text{range differences}}{\text{domain differences}} \quad \frac{-2}{1} = -2 \quad \frac{-4}{2} = -2 \qquad \frac{-6}{3} = -2$$

Since all of the ratios are the same, it appears that the differences in the m values are -2 times that of the k values. This pattern suggests that $m = -2k$. Check the equation to see if it is correct.

Check: If $k = 2$, then $m = -2(2)$ or -4. But the range value for $k = 2$ is 6, a difference of 10. Try some other values in the domain to see if the same difference occurs.

k	1	4	5	6	9
$-2k$	-2	-8	-10	-12	-18
m	8	2	0	-2	-8

$\}$ *m is 10 more than $-2k$.*

This pattern suggests that 10 should be added to one side of the equation in order to correctly describe the relation. Thus, the equation for this relation is $m = -2k + 10$. Check this equation.

Check: If $k = 2$, then $m = -2(2) + 10$ or 6. ✔
        If $k = 9$, then $m = -2(9) + 10$ or -8. ✔

Therefore, $m = -2k + 10$ describes this relation. Since this relation is also a function, we can write this equation in functional notation as $f(k) = -2k + 10$.

EXPLORATION

Students might check their conjecture by testing it with paper and pencil on the first few ordered pairs.

Many times you can generalize the relationships among data you have collected with an equation.

EXPLORATION

SPREADSHEETS

In the relation $\{(-2, 2), (-1, 5), (0, 8), (1, 11), (2, 14)\}$, the range difference is 3 when the domain difference is 1. You might suggest that the equation of the relation is $y = 3x$. Let's use spreadsheet software to check this equation.

- Each cell of a spreadsheet is named by a letter and number. The letter refers to the column and the number of the row. Enter each x value into cells A1 through A5. Enter the y values into cells B1 through B5.

- In cell C1, enter the formula A1*3. This formula means take the value in cell A1 and multiply it by 3. Copy this formula to cells C2 through C5. The spreadsheet will automatically change the formula so that the appropriate cell in column A will be used.

- In cell D1, enter the formula B1 − C1. This formula will subtract the range value of the equation from the range value of the relation. Copy this formula to cells D2 through D5.

Column D tells what number to add to the equation so that it is correct. In this case, the entries in column D are 8s. The correct equation should be $y = 3x + 8$.

Your Turn   a. A1 * −2; See students' work.

a. What formula would you use in cell C1 to test the first formula found in Example 2? Use a spreadsheet to check all the values in the relation.

b. What number would you expect to appear in column D if the equation you found is correct for a given relation?   0

In-Class Example

For Example 3

Akbar notices that the counter on his VCR advances by 12 units every minute. Letting c = counter reading and t = time in minutes, find the equation for the counter as a function of time if Akbar begins the tape at $c = 0$ and $t = 0$.
$c(t) = 12t$

The penny terminology has been largely replaced today by more precise designations. A 6-common or a 6-finish nail gives information about length and usage.

3 PRACTICE/APPLY

Check for Understanding

Exercises 1–10 are designed to help you assess your students' understanding through reading, writing, speaking, and modeling. You should work through Exercises 1–4 with your students and then monitor their work on Exercises 5–10.

Additional Answers

2. Test the values of the domain in the equation. If the resulting values match the range, the equation is correct.

3. Yes; she can use points on the line since the line represents all solutions to the equation.

Example 3

F Y I

This pattern for nails does not hold for all penny sizes. For example, a 60-penny nail is only 6 inches long.

Pennies are units for measuring the lengths of nails. Nails measured in pennies can range from 2-penny nails to 60-penny nails. The graph at the right illustrates the lengths of several penny nails.

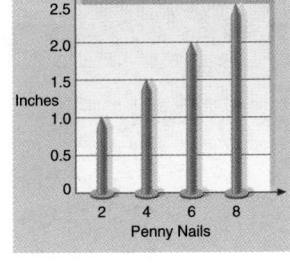

a. The length of a nail as a function of its penny status can be modeled by a linear function for 2- through 10-penny nails. Write an equation in functional notation for this relationship.

b. Find the lengths of a 3-penny nail, a 5-penny nail, and a 10-penny nail.

a. We can see from the graph that the points form a linear pattern. Thus, there is a linear equation that describes the relation for 2- through 10-penny nails.

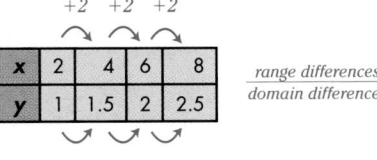

$$\frac{range\ differences}{domain\ differences} = \frac{0.5}{2}\ or\ 0.25$$

This pattern suggests that the equation $f(x) = 0.25x$ might model this situation. However, if you check the domain values with this equation, you will find that the values of $f(x)$ are off by 0.5. The correct equation is $f(x) = 0.25x + 0.5$.

b. *3-penny nail*   *5-penny nail*   *10-penny nail*

$f(3) = 0.25(3) + 0.5$ $f(5) = 0.25(5) + 0.5$ $f(10) = 0.25(10) + 0.5$
$\quad = 1.25$ $\quad = 1.75$ $\quad = 3$

A 3-penny nail is 1.25 inches long, a 5-penny nail is 1.75 inches long, and a 10-penny nail is 3 inches long.

CHECK FOR UNDERSTANDING

Communicating Mathematics

4. Sample answer: $2x + y = 3$; No, $4x + 2y = 6$ and other equivalent equations also have these solutions.

Study the lesson. Then complete the following.

1. **Analyze** the graphs in Examples 1 and 2 along with the ratios of the range differences to the domain differences. Then complete these sentences.

   a. If the ratio is positive, then the line slants ___?___. upward

   b. If the ratio is negative, then the line slants ___?___. downward

2. **Explain** how you can determine if an equation correctly represents a relation given as a table. See margin.

3. **You Decide** When the teacher was explaining Example 2, Narissa said she had another way to find the equation without calculating three ratios. She said she found other points that lie on the line so the differences were common. Is Narissa's method valid? Explain. See margin.

4. **Write** a linear equation in functional notation that has both (1, 1) and (0, 3) as solutions. Is this the only linear equation that has these two solutions? Explain.

Reteaching

Using Review Games For a fun way to practice finding equations from relations, play the function rule game. A person thinks of an equation like $y = 10 - 2x$. Students give input values and are given the functional output values. For example, $1 \rightarrow 8$, $3 \rightarrow 4$. The goal is to guess the function rule.

7. $y = \frac{1}{2}x - \frac{3}{2}$

8. $y = x$

10b. $f(w) = \frac{12}{11} w$

Guided Practice

Write an equation in functional notation for each relation.

5.

x	3	4	5	6	7
f(x)	12	14	16	18	20

$f(x) = 2x + 6$

6.

x	2	4	6	8	10
f(x)	−4	−3	−2	−1	0

$f(x) = \frac{1}{2}x - 5$

Write an equation for each graphed relation.

7.

8.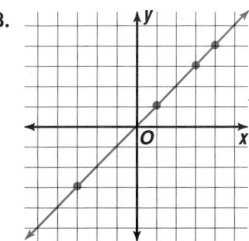

9. The table at the right represents values for the function $N(m)$. Copy and complete the table at the right. Explain how you determined the missing value.   −6; See students' work.

m	8	4	2	0
N(m)	−10	−8	−7	?

10. **Chemistry**   Most substances contract when they freeze. However, water expands in volume when it freezes. Eleven cubic feet of water becomes 12 cubic feet of ice, 33 cubic feet of water becomes 36 cubic feet of ice, and 66 cubic feet of water becomes 72 cubic feet of ice.
   a. Make a graph of these data.   See margin.
   b. Write a functional equation for the relationship between the volume of water and the corresponding volume of ice.

Calvin and Hobbes   by Bill Watterson

EXERCISES

Practice

Copy and complete the table for each function.

11.

x	1	2	3	4	5
f(x)	12	24	36	48	60

12.

x	−4	−2	0	2	4
g(x)	−2	−1	0	1	2

13.

x	−3	−1	1	2	4
h(x)	18	8	−2	−7	−17

14.

x	−2	0	2	4	6
p(x)	0	1	2	3	4

15.
x	1	2	3	4	5
f(x)	5	10	15	20	25

$f(x) = 5x$

16.
n	1	2	3	4	5
f(n)	1	4	7	10	13

$f(n) = 3n - 2$

17.
x	−2	−1	1	2	4
g(x)	13	12	10	9	7

$g(x) = 11 - x$

18.
n	−4	−2	0	2	4
m(n)	−11	−3	5	13	21

$m(n) = 4n + 5$

B

19.
x	0	6	12	18	24
h(x)	−2	0	2	4	6

$h(x) = \frac{1}{3}x - 2$

20.
n	−4	0	4	6	8
m(n)	26	18	10	6	2

$m(n) = 18 - 2n$

21.

$y = -3x$

22.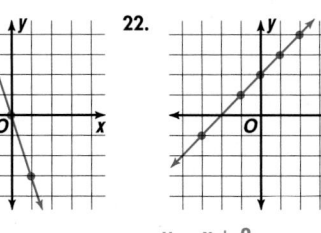

$y = x + 2$

23.

$y = \frac{1}{2}x$

24.

$y = 6 - x$

25.

$y = 2x - 10$

26.

$y = 12 - 3x$

C
27. $\{(6, -4), (2, -12), (1, -24), (-3, 8), (-6, 4)\}$ $xy = -24$

28. $\{(-3, 10), (-2, 5), (-1, 2), (0, 1), (1, 2), (2, 5), (3, 10)\}$ $y = x^2 + 1$

29. $\{(-3, -27), (-1, -1), (2, 8), (3, 27), (10, 1000)\}$ $y = x^3$

30. $\{2, 12), (1, 48), (-1, 48), (-2, 12), (-4, 3)\}$ $y = \frac{48}{x^2}$

Critical Thinking

31. The y- and x-intercepts are those points at which a graph intersects the y- and x-axes, respectively. Use functional notation to describe these points. y-intercept: $f(0)$, x-intercept: $f(x) = 0$

32. See margin.

32. Suppose the ratio of range differences to domain differences in a linear function $g(x)$ is $\frac{2}{3}$ and the graph of $g(x)$ goes through the point $(-3, 2)$. Draw the graph of $g(x)$ and explain why you drew it the way you did.

Applications and Problem Solving

33a. $f(x) = 34x - 34$

33b. You must go deeper in fresh water to get the same pressure as in ocean water.

33. Aquatics The table below illustrates the relationship between atmospheres of pressure and the depth of fresh water.

Pressure (atmospheres)	1	2	3	4	5
Depth of Fresh Water (ft)	0	34	68	102	136

a. Write an equation in functional notation for the relation.

b. Refer to the application at the beginning of the lesson. How does the pressure in fresh water compare to the pressure in ocean water?

34. Archaeology Archaeologists sometimes use lengths of heavy-duty wire to hold together the bones that compose the spinal columns of various vertebrates. Suppose a 70-foot length of wire weighs 23 pounds.

  a. If this sample of wire is representative of all lengths of this wire, draw a graph showing the weight of the wire as a function of its length.

  b. Use your graph to estimate the weight of 50 feet of wire to the nearest pound. **about 16 lb**

  c. Write an equation to describe the relationship between the length of a wire and its weight.

  d. Evaluate the equation in part c for $\ell = 50$. How does this result compare with your estimate in part b? **16.4 lb, slightly more than estimate**

35. Health Alex Knockelmann is concerned about the number of Calories that she needs to intake during the basketball season so she will not lose weight. She wants to eat as many Calories before practice as she will use during practice. The table below shows some fast foods from restaurants near her school and their Calories.

Fast Food	Calories
Arby's® Ham and Cheese	380
Arby's® Roast Beef	350
Burger King's® Whopper™ with Cheese	760
Burger King's® Regular Fries	240
Burger King's® Chocolate Milk Shake	380
Wendy's® Cheeseburger	577
Wendy's® Double Cheeseburger	797
Wendy's® Fries	327
Wendy's® Frosty™	390
Pepsi® (20 oz)	241
7-Up® (20 oz)	219

  a. Before practice, Alex went to Wendy's. She ate a double cheeseburger, fries, and a Frosty™. What was her Calorie intake for this meal? **1514 C**

  b. In Lesson 5–4, you learned that playing basketball burns 0.138 C/min/kg. At Alex's weight of 80 kg, determine how many Calories she would burn each minute. **11.04 C/min**

  c. Make a table to show the number of Calories that Alex burns each minute for the first 10 minutes of practice. Graph the points. **See margin.**

  d. Let C represent the number of Calories burned and t represent the number of minutes. Write a functional equation to represent the relation. Is it a linear function? $C(t) = 11.04t$; yes

35e. 1324.8 C burned; yes, 189.2 C remaining

  e. Use the graph and your equation to predict the number of Calories Alex would use during two hours of practice. Did she eat enough food so that she had Calories remaining at the end of practice?

Lesson 5–6   Writing Equations from Patterns   **301**

Extension

Problem Solving Tables are given at the right for two linear functions. If the domain of each of these functions is the set of all numbers, find the value(s) of x that make $f(x) = g(x)$. $x = 3$

x	2	4	6	8
f(x)	5	9	13	17

$y = 2x + 1$

x	2	4	6	8
g(x)	4	10	16	22

$y = 3x - 2$

Additional Answer

35c.

Minute	Calories
1	11.04
2	22.08
3	33.12
4	44.16
5	55.2
6	66.24
7	77.28
8	88.32
9	99.36
10	110.4

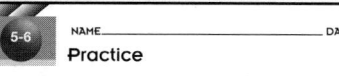

Practice Masters, p. 39

Closing Activity

Writing The table below shows a linear relation between *x* and *y*. Find the values of *a*, *b*, *c*, and *d*. Then write an equation that relates *x* and *y*.

x	8	9	10	d
y	a	5	b	c

$a = 3, b = 7, c = 9, d = 11;$
$y = 2x - 13$

Chapter 5, Quiz C (Lessons 5-5 and 5-6), is available in the *Assessment and Evaluation Masters*, p. 129.

Additional Answers

41.
$$-2\ -1\ 0\ 1\ 2\ 3\ 4\ 5\ 6\ 7\ 8\ 9$$

42.

Stem	Leaf
2	3
3	2
4	3 5 7
5	3 7 8 9 9 9
6	0 0 2 2 4 4 5
•	5 6 6 6 6 6
•	6 7 8 8 8 8 9
7	0 2 3 4 4 4
•	4 4 4 5 5 5
•	5 5 5 5 5 5

$2 | 3 = 23$¢

Enrichment Masters, p. 39

5-6
NAME_____ DATE _____
Enrichment
Student Edition
Pages 295–302

Polynomial Functions

Suppose a linear equation such as $-3x + y = 4$ is solved for *y*. Then an equivalent equation, $y = 3x + 4$, is found. Expressed in this way, *y* is a function of *x*, or $f(x) = 3x + 4$. Notice that the right side of the equation is a binomial of degree 1.

x	y
$-1\frac{1}{2}$	$-2\frac{2}{8}$
-1	0
0	1
1	2
$1\frac{1}{2}$	$4\frac{3}{8}$

$f(x) = x^3 + 1$

Higher-degree polynomials in *x* may also form functions. An example is $f(x) = x^3 + 1$, which is a polynomial function of degree 3. You can graph this function using a table of ordered pairs.

For each of the following polynomial functions, make a table of values for x and y = f(x). Then draw the graph on the grid.

1. $f(x) = 1 - x^2$

2. $f(x) = x^3 - 5$

3. $f(x) = x^2 + 4x - 1$

4. $f(x) = x^3$

Mixed Review

36. **Biology** The function $f(t) = \frac{t}{0.2} - 32$ shows that the number of times a cricket chirps in an hour is a function of the temperature in degrees Celsius. (Lesson 5–5) **36a. 28 times 36b. 10°C**

   a. If it is 12°C outside, how many times will a cricket chirp in an hour?

   b. If a cricket chirps 54 times in 3 hours, what is the outside temperature?

37. D = {1, 3, 5}; R = {2, 4, 6}

37. State the domain and range of {(1, 6), (3, 4), (5, 2)}. (Lesson 5–2)

38. What number is 16% of 50? (Lesson 4–4) **8**

39. **Trigonometry** Given the similar triangles at the right, find *s*. (Lesson 4–2) **5**

40. Solve $x + (-7) = 36$. (Lesson 3–1) **43**

41. See margin.

41. Graph {0, 3, 7} on a number line. (Lesson 2–1)

42. **Personal Finance** Many credit cards offer a protection plan so you can pay off your balance if you lose your job or have an accident and cannot work. The table below shows the monthly premium (in cents per $100 outstanding balance) for the plan on a Discover® card in the United States and its territories. Round each amount to the nearest cent and make a stem-and-leaf plot of these fees. (Lesson 1–4) **See margin.**

Area	Fee	Area	Fee	Area	Fee	Area	Fee	Area	Fee	Area	Fee
AK	68.2	FL	75	LA	74	NC	64.6	OK	75	UT	68.9
AL	66	GA	75	MA	63.5	ND	64.8	OR	69.8	VA	59.3
AR	66	HI	57.5	MD	75	NE	68	PA	22.5	VI	66
AZ	73.7	IA	58.8	ME	62	NH	46.9	PR	74	VT	56.7
CA	75	ID	60	MI	66	NJ	64.2	RI	67	WA	60
CO	68.1	IL	75	MN	32.3	NM	74	SC	71.9	WI	59
CT	53.2	IN	62.1	MO	75	NV	74	SD	66	WV	74
DC	66	KS	68.3	MS	74	NY	45.3	TN	75	WY	73.1
DE	66	KY	75	MT	74	OH	75	TX	42.7		

WORKING ON THE In·ves·ti·ga·tion

Refer to the Investigation on pages 190–191.

Many of the procedures from one area of science can be applied to other areas of science. Estimating populations is not reserved just for fish. Conservationists also keep track of species of birds, insects, plants, and other wildlife in a similar way.

1 Design a different experiment that illustrates the capture-recapture method. Complete your experiment a few times to see if it is as accurate as the one you have completed.

2 What was the range of your estimates using the method described in the Investigation? What was the range of your estimates using your method? Which do you think is the most reliable method?

Add the results of your work to your Investigation Folder.

302 Chapter 5 *Graphing Relations and Functions*

In·ves·ti·ga·tion

Working on the Investigation

The Investigation on pages 190–191 is designed to be a long-term project that is completed over several days or weeks. Encourage students to keep their materials in their Investigation. Folder as they work on the Investigation.

5–7A Graphing Technology
Measures of Variation

A Preview of Lesson 5–7

The graphing calculator is a powerful tool for observing patterns and analyzing data that come from research or experimentation. Many of the measures of variation you will learn about in Lesson 5–7 can be calculated by using a graphing calculator.

Before analyzing any data, you must know how to enter it into a list.

Example **①** **The Glencoe Publishing Golf Tournament is held every September to raise money for a scholarship fund. The list below includes the scores for the employees of the manufacturing department. Enter the data and find the range of these scores.**

139, 99, 105, 115, 88, 91, 105, 80, 102, 101, 103, 95, 99, 77, 112

First enter the data.

Method 1   Assign the data to List 1 (L1) from the home screen.

Enter: [2nd] [{] 139 [,] 99 [,] 105 [,] 115 [,] 88 [,] 91

[,] 105 [,] 80 [,] 102 [,] 101 [,] 103 [,] 95 [,] 99

[,] 77 [,] 112 [2nd] [}] [STO▶] [2nd] [L1] [ENTER]

Method 2   Enter the data directly into the list display.

Before entering your data, you must clear the L1 list of previously entered data.

Enter: [STAT] 4 [2nd] [L1] [ENTER]

Enter each score into the L1 list.

Enter: [STAT] 1

The cursor appears in the L1 list. Enter the data one by one pressing [ENTER] after each entry, including the last one.

Then find the range.

The range of the scores is the difference of the maximum and minimum values in the list.

Enter: [2nd] [QUIT] [2nd] [LIST] [▶] 2 [2nd] [L1] [ )] [−]

[2nd] [LIST] [▶] 1 [2nd] [L1] [ )] [ENTER]   *62*

The range of the data is 62.

LOOK BACK
You can review another method for using the graphing calculator to find the median of a set of data in Lesson 3-7.

You can use the calculator to sort the data from least to greatest by pressing
[STAT] 2 [2nd] [L1]
[ENTER] *. You can view the list by pressing* [STAT]
[ENTER] *.*

If you do not use [2nd]
[QUIT] *to leave the LIST menu, the range will be added to L1, which alters your data.*

NCTM Standards: 1–4, 10

Objective
Use a graphing calculator to determine measures of variation.

Recommended Time
25 minutes

Instructional Resources
Graphing Calculator Masters, pp. 20 and 21

These masters provide keystroking instruction for this lesson for the TI-81 and Casio graphing calculators.

1 FOCUS

Motivating the Lesson
Ask students what the first step would be in finding the median for a long list of values such as a basketball player's points per game for an entire season. Students should suggest that arranging the values in increasing order would have to be done first. Explain that on a calculator, the data could be input in any order to find the median.

2 TEACH

Teaching Tip Review with students the difference between median and mean.

4 ASSESS

Observing students working with technology is an excellent method of assessment.

The data can be arranged into four groups that have approximately the same number of data in each group. The points separating these groups are called **quartiles** and are symbolized by Q_1, Q_2, and Q_3. The median of the set is Q_2. The **interquartile range** is $Q_3 - Q_1$. Half of the data lie in this range. You can use a graphing calculator to help you find each of these measures.

Example **Find the quartiles and the interquartile range of the golf data in Example 1.**

Before finding the measures you want, you must first make sure the calculator is set to refer to the correct set of data. Our data is in L_1.

Enter: [STAT] [▶] 3 and highlight L_1 under 1-Var Stats. Make sure that the frequency is 1.

Enter: [STAT] [▶] 1 [ENTER]

Use the down arrow key to scroll to the end of the statistics. Your screen should look like the one at the right. This tells you there are 15 items of data, the least value (minX) is 77, the greatest value (maxX) is 139, the lower quartile (Q_1) is 91, the median is 101, and the upper quartile (Q_3) is 105.

```
1–Var Stats
↑ n=15
  minX=77
  Q₁=91
  Med=101
  Q₃=105
  maxX=139
```

Use Q_3 and Q_1 to find the interquartile range: $Q_3 - Q_1 = 105 - 91$ or 14.

EXERCISES

Remember to change the Setup when calculating the measures for each set of data.

Enter each set of data into the list given. Then find the lower quartile, median, upper quartile, range, and interquartile range of the data.

1. 12, 17, 16, 23, 18 in L_1 Q_1, 14; Med, 17; Q_3, 20.5; R, 11; IQR, 6.5

2. 56, 45, 37, 43, 10, 34 in L_2 Q_1, 34; Med, 40; Q_3, 45; R, 46; IQR, 11

3. 77, 78, 68, 96, 99, 84, 65 in L_3 Q_1, 68; Med, 78; Q_3, 96; R, 34; IQR, 28

4. 30, 90, 40, 70, 50, 100, 80, 60 in L_4 Q_1, 45, Med, 65; Q_3, 85; R, 70; IQR, 40

5. Q_1, 3.4; Med, 5.3; Q_3, 21; R, 77; IQR, 17.6

5. 3, 3.2, 6, 45, 7, 26, 1, 3.4, 4, 5.3, 5, 78, 8, 21, 5 in L_5

6. Q_1, 62; Med, 73; Q_3, 77; R, 37; IQR, 15

6. 85, 77, 58, 69, 62, 73, 55, 82, 67, 77, 59, 92, 75, 69, 76 in L_6

7. At least 50% of the data is clustered around the median.

7. Look at the range and the interquartile range of each set of data. What do you think it means if the range is great but the interquartile range is small?

8. Suppose you entered a set of data and the interquartile range was 0. What does that mean?   Each item of data from Q_1 to Q_3 is the same number.

Using Technology

This lesson offers an excellent opportunity for using technology in your algebra classroom. For more information on using technology, see *Graphing Calculators in the Mathematics Classroom*, one of the titles in the Glencoe Mathematics Professional Series.

Integration: Statistics
Measures of Variation

Objective

After studying this lesson, you should be able to:

- calculate and interpret the range, quartiles, and interquartile range of sets of data.

LOOK BACK

You can review finding the mean and median of a set of data in Lesson 3-7.

APPLICATION
Climate

Justin Williams has been promoted and has the option of relocating either to Columbus, Ohio, or San Francisco, California. The Williams family lives in Tampa, Florida, where the average temperature is about 72°F and it is usually warm all year long. His family's wardrobe includes basically all lightweight apparel. He wondered if moving would mean they would have to get lots of new clothes. He looked up the average high temperature for the two cities and found the following information in a table of average high temperatures of U.S. cities.

Mr. Williams calculated the mean and median of the temperatures of each city. The mean high temperature for Columbus was 63.5°, and for San Francisco, it was 62.4°. The median high temperature for Columbus was 63.5°, and for San Francisco, it was 63°. He figured that meant the climate in the two cities was just about the same. But then he decided to take another look at the temperatures and find another way to analyze them.

	Columbus	San Francisco
Jan.	35°	
Feb.	38°	56°
Mar.	49°	59°
Apr.	62°	60°
May	73°	61°
June	81°	63°
July	84°	64°
Aug.	83°	64°
Sept.	77°	65°
Oct.	65°	69°
Nov.	51°	68°
Dec.	39°	63°
		57°

Source: U.S. National Oceanic and Atmospheric Administration

The months and temperatures for each city form a relation. Mr. Williams decided to graph the relation for each city. He used different colors for each city.

5-7 LESSON NOTES

NCTM Standards: 1–4, 10

Instructional Resources

- Study Guide Master 5-7
- Practice Master 5-7
- Enrichment Master 5-7
- Assessment and Evaluation Masters, p. 129
- Real-World Applications, 16

Transparency 5-7A contains the 5-Minute Check for this lesson; **Transparency 5-7B** contains a teaching aid for this lesson.

Recommended Pacing	
Standard Pacing	Day 12 of 14
Honors Pacing	Day 11 of 13
Block Scheduling*	Day 6 of 7
Alg. 1 in Two Years*	Days 17 & 18 of 21

*For more information on pacing and possible lesson plans, refer to the *Block Scheduling Booklet* and *Algebra 1 in Two Years.*

1 FOCUS

5-Minute Check
(over Lesson 5-6)

Complete the table for the function given.

1.

x	1	3	5	7	9
f(x)	4	12	20	28	36

Write an equation for each relation.

2.

x	3	6	9	12	15
f(x)	1	2	3	4	5

$f(x) = \dfrac{1}{3x}$

3.

$y = x - 2$

He noticed that the pattern in the temperatures was very different. This case shows that measures of central tendency may not give an accurate enough description of the data. Often **measures of variation** are also used to describe the distribution of the data.

One of the most common measures of variation is the **range**. Unlike the other definition of range in this chapter, the range of a set of data is a measure of the spread of the data.

Definition of Range	The range of a set of data is the difference between the greatest and the least values of the set.

Let's find the range for each set of temperatures. Make a table to help organize the data.

City	Greatest Temperature	Least Temperature	Range
Columbus	84°	35°	84° − 35° or 49°
San Francisco	69°	56°	69° − 56° or 13°

The range of temperatures for Columbus is greater than that for San Francisco. This means that temperatures in Columbus vary more than temperatures in San Francisco. If his family moves to Columbus, he will definitely need some new clothes for their colder winters.

Another commonly used measure of variation is called the **interquartile range.** In a set of data, the **quartiles** are values that divide the data into four equal parts. Statisticians often use Q1, Q2, and Q3 to represent the three quartiles. Remember that the median separates the data into two equal parts. Q2 is the median. Q1 is the **lower quartile.** It divides the lower half of the data into two equal parts. Likewise, Q3 is the **upper quartile.** It divides the upper half of the data into two equal parts. The difference between the upper and lower quartiles is the interquartile range (IQR).

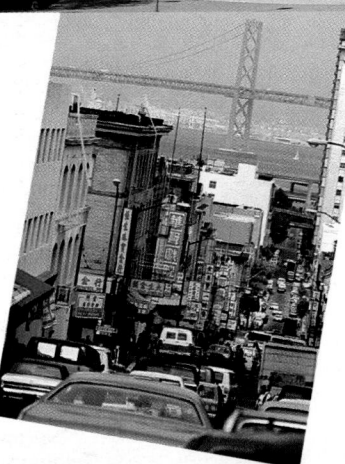

The abbreviations LQ and UQ are often used to represent the lower quartile and upper quartile.

Definition of Interquartile Range	The difference between the upper quartile and the lower quartile of a set of data is called the interquartile range. It represents the middle half, or 50%, of the data in the set.

Example ① The table below shows the heights and weights of 13 veteran players on the 1994 Dallas Cowboys football team. Find the median, upper and lower quartiles, and the interquartile range for the weights of the players.

APPLICATION
Football

Player	Height	Weight (lb)
Aikman	6'4"	228
Daniel	5'11"	192
Gaines	5'11"	228
Harper	6'3"	208
Holmes	5'10"	181
Irvin	6'2"	205
Jones	6'2"	237
Marlon	5'11"	189
Newton	6'3"	325
Price	6'3"	247
Smith	5'11"	180
Tolbert	6'6"	263
Williams	5'9"	192

Source: *Gridiron 1994 Pro Football Yearbook*

Order the 13 weights from least to greatest. Then find the median.

180   181   189   192   192   205   **208**   228   228   237   247   263   325
↑
median

Remember that the median is the middle number when the data are arranged in numerical order.

The lower quartile is the median of the lower half of the data, and the upper quartile is the median of the upper half of the data. If the median is an item in the set of data, it is not included in either half.

180   181   189   192   192   205   **208**   228   228   237   247   263   325
$Q_1 = 190.5$ $Q_2 =$ median $Q_3 = 242$

The interquartile range is $242 - 190.5$ or 51.5. Therefore, the middle half, or 50% of the weights of the football players, varies within 51.5 pounds.

In Example 1, one weight, 325 pounds, is much greater than the others. In a set of data, a value that is much greater or much less than the rest of the data can be called an **outlier.** An outlier is defined as any element of a set of data that is at least 1.5 interquartile ranges greater than the upper quartile or less than the lower quartile.

An outlier will not affect the quartiles, but it will affect the mean.

The interquartile range of the football data is 51.5. So 1.5(51.5) is 77.25.

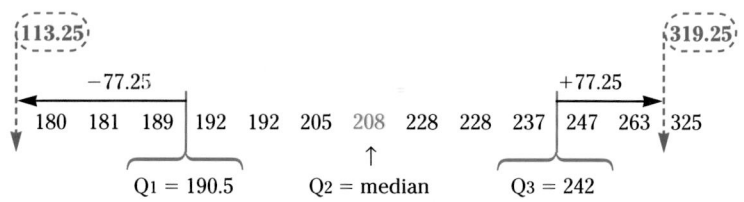

(113.25)   (319.25)
      -77.25 $+77.25$
180   181   189 │ 192   192   205   **208**   228   228   237 │ 247   263 │325
$Q_1 = 190.5$ $Q_2 =$ median $Q_3 = 242$

From the diagram above, you can see that the only item of data occurring beyond the dashed lines is 325. So, this is the only outlier.

Cooperative Learning

Trade-A-Problem Separate the class into two groups. Have group members measure and record one another's heights in inches or centimeters and then work together to determine the median, upper and lower quartiles, interquartile range, and any outliers for the values. Then have groups trade sets of values and determine the same

information for the other group's heights. Have the whole class discuss their findings and methods. For more information on the trade-a-problem strategy, see *Cooperative Learning in the Mathematics Classroom,* one of the titles in the Glencoe Mathematics Professional Series, pages 25–26.

In-Class Example

For Example 1
Mrs. Kollar gave a quiz in her statistics class. The scores were 23, 30, 22, 20, 20, 14, 15, 19, 19, 20, 23, 20, 22, and 24. Find the median, the upper and lower quartiles, the interquartile range, and any outliers.
median = 20, UQ = 23, LQ = 19, interquartile range = 4, outlier = 30

Example 2

APPLICATION

Commerce

LOOK BACK

You can review stem-and-leaf plots in Lesson 1-4.

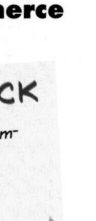

The double stem-and-leaf plot below represents the tonnage of domestic and foreign imports being delivered at the top 19 busiest ports in the United States during 1992.

Domestic	Stem	Foreign
[5	10	
	9	[5
	8	
5	7	3
5	6	
	5	
8⬚0	4	1 4
8 1	3	6⬚6
8 5]△[1	2	0⬚0 3 4 5 5]△[5 9 9
9 9 7 ⬚5 2 0	1	8 8
9 8]	0	8]

$5 \mid 2 = 25,000,000$ tons $7 \mid 3 = 73,000,000$ tons

Source: Army Corps of Engineers, U.S. Dept. of Defense

a. The brackets group the values in the lower half and the values in the upper half. What do the triangles contain? What do the boxes contain?

b. Find the interquartile ranges of the domestic and the foreign tonnage.

c. Which type of imports had a more consistent tonnage—the domestic or the foreign?

d. Find any outliers.

a. The triangles contain the medians.
The boxes contain the lower and upper quartiles.

b. **Domestic**

Q_1 is 15 and Q_3 is 40.

$IQR = Q_3 - Q_1$

$= 40 - 15$ or 25

Foreign

Q_1 is 20 and Q_3 is 36.

$IQR = Q_3 - Q_1$

$= 36 - 20$ or 16

c. The foreign imports had more consistent tonnage because it had the smaller interquartile range.

d. Outliers are those values that lie 1.5 IQRs above Q_3 or below Q_1.

Domestic

$1.5(IQR) + Q_3 = 1.5(25) + 40$ or 77.5

$Q_1 - 1.5 (IQR) = 15 - 1.5(25)$ or -22.5

Since $105 > 77.5$, 105 is an outlier. There are no outliers below -22.5.

Foreign

$1.5(IQR) + Q_3 = 36 + (1.5)(16)$ or 60

$Q_1 - 1.5 (IQR) = 20 - (1.5)(16)$ or -4

Since $95 > 60$ and $73 > 60$, 95 and 73 are outliers.

There are no outliers below -4.

Communicating Mathematics

1. San Francisco, 5; Columbus, 35

2. Outliers may make the mean much higher or lower than the mean of the data excluding the outliers.

Study the lesson. Then complete the following.

1. Refer to the application at the beginning of the lesson. Compare the interquartile ranges of the temperatures in San Francisco and Columbus.

2. **Describe** how the mean is affected by an outlier.

3. **Predict** the following.   Sample answers are given.
   a. If you were to measure the height of all of the students in your school, in which group of students might you find outliers?   the basketball teams
   b. If you were to measure the weight of all of the students in your school, in which group of students might you find outliers?   the football team

4. **Explain** the difference between range as defined in Lesson 5–2 and range as defined in this lesson.   The range in Lesson 5–2 is a set of values that correspond to a set of domain values. Range in this lesson describes the spread of data.

MODELING MATHEMATICS

5. Along with your classmates, line up according to your heights.
   a. "Fold" the line in half by having the shortest person meet the tallest person. What is the median height?   See students' work.
   b. Do the same with each half of the line. What is lower quartile and upper quartile of the heights of the students in your classroom?   See students' work.

Guided Practice

Find the range, median, upper quartile, lower quartile, and interquartile range of each set of data. Identify any outliers.

6. 16, 24, 11, 17, 19   13, 17, 21.5, 13.5, 8

7. 43, 45, 56, 37, 11, 34

8.

Top 10 Jobs of the Future	
Job	Average Salary
Civil engineer	$55,800
Computer system analyst	42,700
Electrical engineer	59,100
Geologist	50,800
High school teacher	32,500
Pharmacist	47,500
Physical therapist	37,200
Physician	148,000
Psychologist	53,000
School principal	57,300

Source: *American Careers,* Fall 1994

7. 45, 40, 45, 34, 11; 11
8. 115,500; 51,900; 57,300; 42,700; 14,600; $148,000

9.
Stem	Leaf
0	0 2 3
1	1 7 9
2	2 3 5 6
3	3 4 4 5 9
4	0 7 8 8

$2|2 = 22$ 48, 26, 39, 17, 22

10. med, 6'2"; Q₃, 6'3.5"; Q₁, 5'11"; IQR, 4.5"

10. **Football**   Refer to the data in Example 1. Find the median, upper and lower quartiles, and the interquartile range for the heights of the players.

Check for Understanding

Exercises 1–10 are designed to help you assess your students' understanding through reading, writing, speaking, and modeling. You should work through Exercises 1–5 with your students and then monitor their work on Exercises 6–10.

Error Analysis

In order to find the mean and mode, the data do not necessarily have to be in order. However, watch that the students do order the data to find the median, upper and lower quartiles, and the interquartile range.

Reteaching

Using Alternative Methods Have students work the processes they have just learned in reverse. Provide students with a median and a range and have them develop sets of data to satisfy these measures. You might also want to specify the number of data points within the set.

EXERCISES

Assignment Guide

Core: 11–21 odd, 22, 23–29 odd, 30–34
Enriched: 12–20 even, 22–34

For **Extra Practice**, see p. 769.

The red A, B, and C flags, printed only in the Teacher's Wraparound Edition, indicate the level of difficulty of the exercises.

Additional Answers

21b. No; the libraries would have accumulated books throughout the years.
22a. Sample answer: {26, 27, 28, 29, 30, 36, 37, 38, 39, 40, 41, 42, 43, 45, 46, 50, 55, 58, 86}
22b. Sample answer: {29, 31, 32, 33, 34, 36, 37, 38, 39, 40, 41, 42, 43, 44, 45, 46, 47, 48, 49}

Practice

Ⓐ

11. 34, 78, 96, 68, 28
12. 29, 19, 22, 18, 4
13. 10, 5, 8, 2, 6
14. 36, 78.5, 88, 68.5, 19.5

Find the range, median, upper quartile, lower quartile, and interquartile range for each set of data.

11. 77, 78, 68, 96, 99, 84, 65
12. 17°, 46°, 18°, 22°, 18°, 21°, 19°
13. 2, 4, 6, 8, 2, 4, 6, 8, 10, 0
14. 89, 56, 75, 82, 64, 73, 87, 92
15. 30.8, 29.9, 30.0, 31.0, 30.1, 30.5, 30.7, 31.0   1.1, 30.6, 30.9, 30.05, 0.85
16. 78, 2, 3.4, 4, 45, 7, 5.3, 5, 3, 1, 3.2, 6, 26, 8, 5   77, 5, 8, 3.2, 4.8
17. 1050, 1175, 835, 1075, 1025, 1145, 1100, 1125, 975, 1005, 1125, 1095, 1075, 1055   340, 1075, 1125, 1025, 100

Ⓑ

18.

Stem	Leaf
5	3 6 8
6	5 8
7	0 3 7 7 9
8	1 4 8 8 9
9	9

$9|9 = 9900$

4600, 7700, 8600, 6650, 1950

19.

Stem	Leaf
19	3 5 5
20	2 2 5 8
21	5 8 8 9 9 9
22	0 1 7 8 9
23	2

$19|3 = \$193$

39, 218, 221, 202, 19

20.

Stem	Leaf
5	0 3 7 9
6	1 3 4 5 5 6
7	1 5 6 6 9
8	1 2 3 5 8
9	2 5 6 9

$5|0 = 5.0\ cm$

4.9, 7.55, 8.4, 6.35, 2.05

Graphing Calculator

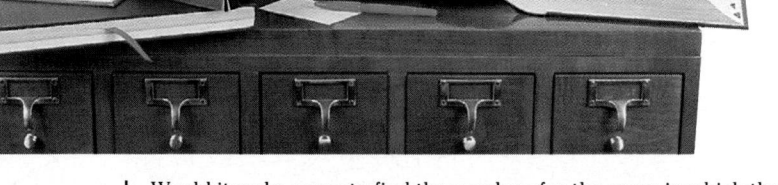

21. **a.** Use a graphing calculator to find the median, upper and lower quartiles, range, and interquartile range of the number of books given in the table below.   9,198,630; 11,750,000; 5,700,000; 24,000,000; 6,050,000

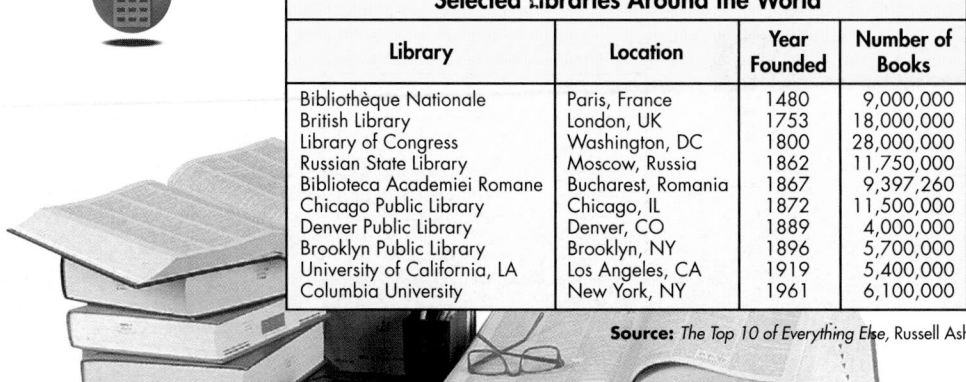

Selected Libraries Around the World

Library	Location	Year Founded	Number of Books
Bibliothèque Nationale	Paris, France	1480	9,000,000
British Library	London, UK	1753	18,000,000
Library of Congress	Washington, DC	1800	28,000,000
Russian State Library	Moscow, Russia	1862	11,750,000
Biblioteca Academiei Romane	Bucharest, Romania	1867	9,397,260
Chicago Public Library	Chicago, IL	1872	11,500,000
Denver Public Library	Denver, CO	1889	4,000,000
Brooklyn Public Library	Brooklyn, NY	1896	5,700,000
University of California, LA	Los Angeles, CA	1919	5,400,000
Columbia University	New York, NY	1961	6,100,000

Source: *The Top 10 of Everything Else*, Russell Ash

b. Would it make sense to find these values for the years in which these libraries were founded? Why or why not?   See margin.

Critical Thinking

22. **a.** Write an example of a set of 19 numbers that has a range of 60, a median of 40, an interquartile range of 16, and one outlier.   See margin.
   b. Write an example of a set of 19 numbers that has a range of 20, a median of 40, an interquartile range of 11, and no outliers.   See margin.

 Tech Prep

Librarian Students who are interested in library science may wish to do further research on the world's great libraries, as mentioned in Exercise 21, and explore the potential growth of this career. For more information on tech prep, see the *Teacher's Handbook*.

Applications and Problem Solving

23. 3760, 3224, 4201.5, 2780.5, 1421
24. 145,687; 262,306.5; 299,737; 219,683; 80,054

Phantom of the Opera

Find the range, median, upper quartile, lower quartile, and interquartile range for each set of data.

23. **Entertainment**

Long-Running Broadway Plays (as of 7/17/94)	
Play	Performances
42nd Street	3,486
Annie	2,377
Cats	4,917
Chorus Line	6,137
Fiddler on the Roof	3,242
Grease	3,388
Hello Dolly	2,844
Les Miserables	3,005
Life with Father	3,224
My Fair Lady	2,717
Oh, Calcutta	5,959
Phantom of the Opera	2,717
Tobacco Road	3,182

Source: *Variety*, 1994

24. **Transportation**

Popular Passenger Cars in the U.S. (1993)	
Name of Car	Number Sold
Chevrolet Cavalier	273,617
Chevrolet Lumina	219,683
Ford Escort	269,034
Ford Taurus	360,448
Ford Tempo	217,644
Honda Accord	330,030
Honda Civic	255,579
Pontiac Grand Am	214,761
Saturn	229,356
Toyota Camry	299,737

Source: American Automobile Manufacturers Association

25. **Education**

Enrollment at Selected Colleges in the U.S. (1993–1994)	
College	Enrollment
Baylor University	12,194
Brown University	7,655
California Institute of Technology	1,977
Columbia University	3,441
Dartmouth College	5,475
Georgia State University	23,651
Howard University	10,736
Massachusetts Institute of Technology	9,790
Princeton University	6,444
Stanford University	14,002
Yale University	10,844

Source: *Peterson's Guides,*© 1994

21,674; 9790; 12,194; 5475; 6719

26. **Nutrition**

Calories in Foods, per serving	
Food	Calories
Apple	100
Banana	130
Bread	60
Cupcake	200
Donut	120
Ice cream	185
Orange juice	85
Peach	50
Peanut butter	190
Potato	100
Soft taco	225
Whole milk	133

Source: U.S. Dept. of Agriculture

175, 125, 187.5, 92.5, 95

Hello Dolly

Lesson 5–7 **INTEGRATION** Statistics Measures of Variation **311**

Closing Activity

Writing Have students briefly define each of the following terms: range, quartile, upper quartile, lower quartile, interquartile range, and outlier.

Chapter 5, Quiz D (Lesson 5-7), is available in the *Assessment and Evaluation Masters*, p. 129.

27a. males: 36, 29, 37.5, 44, 15; females: 23, 29, 33, 39, 10

27c. The ages of the top female golfers are less varied than those of the top male golfers.

28a. $67.8 million; $50.5 million; $74.1 million; $29.05 million; $45.05 million

Practice Masters, p. 40

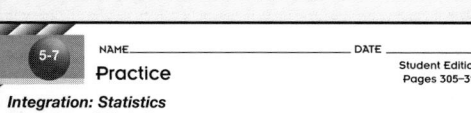

5-7

NAME_____ DATE_____

Practice   Student Edition Pages 305–313

Integration: Statistics
Measures of Variation

Find the range, median, upper quartile, lower quartile, and interquartile range for each set of data.

1. 13, 15, 25, 22, 18, 19
12; 18.5; 22, 15; 7

2. 2, 7, 3, 6, 10, 8, 9
8; 7; 9, 3; 6

3. 8, 9, 10, 6, 15, 6, 10, 5
10; 8.5; 10, 6; 4

4. 107, 57, 47, 40, 34, 20, 25, 37, 46, 57, 69
87; 46; 57, 34; 23

5. 200, 250, 230, 180, 160, 140
110; 190; 230, 160; 70

6. 2.1, 3.2, 4.6, 7.1, 2.9
5; 3.2; 5.85; 2.5; 3.35

7.
Stem	Leaf
2	0 1
3	2 4 6
4	3 3 8 9
5	2 2 4
6	3 6

4|3 = 43
46; 45.5; 52; 34; 18

8.
Stem	Leaf
1	0 0 1
2	4 6
3	2 3 3
4	7 8 9
5	1 2 2 3

2|6 = 26
43; 33; 51; 24; 27

9.
Stem	Leaf
7	0 2 5
8	0 0
9	3 5 9
10	1 1 4 6
11	6 7 7 8

10|1 = 101
48; 100; 111; 80; 31

10.
Stem	Leaf
8	1 9
9	2 3 6
10	1 9
11	2 4 8 9
12	3 7

10|1 = 101
46; 109; 118.5, 92.5; 26

---

27. Sports The double stem-and-leaf plot below represents the ages of the top 20 male and female golfers for the 1994 season, according to *Golfer's Almanac,* 1995.

Male	Stem	Female
8	1	
9 9 8 8 5	2	3 4 5 5 9 9
8 8 7 6 6 5	3	1 2 2 2 4 4 5 8 9 9 9
6 2 0 0	4	0 4 6
4 3 1 0	5	

4|5 = 54 years old   *4|6 = 46 years old*

a. Find the ranges, quartiles, and interquartile ranges of the average golfer's age for male and female golfers.

b. Identify any outliers.   There are no outliers.

c. Compare the ranges and interquartile ranges for male and female golfers. What can you conclude from these statistics?

28. Entertainment The total sales of movie rentals as of April 1994 for selected Academy Award winners are listed below.

Title	Total Rental Sales ($ million)	Title	Total Rental Sales ($ million)
Schindler's List	38.7	Ordinary People	23.1
Silence of the Lambs	59.8	Kramer vs. Kramer	60.0
Dances with Wolves	81.5	The Deer Hunter	27.5
Driving Miss Daisy	50.5	Annie Hall	19.0
Rain Man	86.8	Rocky	56.5
Platoon	70.0	One Flew Over the	
Out of Africa	43.5	Cuckoo's Nest	60.0
Amadeus	23.0	The Godfather, Part II	30.6
Terms of Endearment	50.25	The Sting	78.2
Ghandi	25.0	The Godfather	86.3
Chariots of Fire	30.6	The Sound of Music	80.0

Source: *Variety,* May 1994

a. Find the range, median, upper and lower quartiles, and interquartile range for the total rental sales.

b. Identify any outliers.   There are none.

c. The least value on the list of all-time top 20 movie rentals as of April, 1994, is $96.3 million. What can you conclude about Academy Award winners in relation to this data?   None of these are on the top 20 list.

29. a. List the grades you have made during this grading period in mathematics. Determine the range, the quartiles, and the interquartile range of the scores.   See students' work.

b. List the grades you have made during this grading period in English. Determine the range, the quartiles, and the interquartile range of the scores.   See students' work.

c. In which of the two courses are your grades more consistent? Why do you think this is the case?   See students' work.

Mixed Review

30. Geology The underground temperature of rocks varies with their depth below the surface. The temperature at the surface is about 20°C. At a depth of 2 km, the temperature is about 90°C, and at a depth of 10 km, the temperature is about 370°C.   (Lesson 5–6)

a. Write an equation to describe this relationship. $y = 35x + 20$

b. Use the equation to predict the temperature at a depth of 13 km.   475°C

Extension

Reasoning A set of data contains 7 elements. The median is 8900, the interquartile range is 1200, and there is one outlier above the upper quartile. Find a set of data that satisfies all the conditions. Compile the student responses to investigate how different the data might be. Answers will vary.

31. Which ordered pairs are solutions to $5 - 1.5x = 2y$? (Lesson 5–3) **c, d**

   a. (0, 1)   **b.** (8, 2)   **c.** $\left(4, -\dfrac{1}{2}\right)$ **d.** (2, 1)

32. Ecology For every 20 grams of smog produced by a typical automobile in an hour, gasoline-powered lawn mowers produce 50 grams. If it takes Jodi 1.5 hours to mow her lawn, how long would she be able to drive her car and produce the same amount of smog? (Lesson 4–1) **3.75 hours**

33. Sports The table at the right shows how many millions of people participated in in-line skating and mountain biking in 1991 and 1992. Write an equation to represent each situation. Then solve. (Lesson 3–1)

   a. How many more people were participating in in-line skating in 1992 than in 1991?

   b. How many more people were mountain biking in 1992 than in 1991? **6.0 + p = 6.9; about 0.9 million people**

33a. **6.2 + p = 9.4; about 3.2 million people**

Year	1991	1992
In-Line Skating	6.2	9.4
Mountain Biking	6.0	6.9

Source: *Men's Fitness*

34. Simplify $3(-4) + 2(-7)$. (Lesson 2–6) **−26**

This is the right sidebar.

Mathematics and SOCIETY

Meteorology makes extensive use of computer/mathematical models to represent the changes in the atmosphere. Dynamic meteorology deals with air flow in the atmosphere. Mathematics' chaos theory developed from research carried out in dynamic meteorology.

Mathematics and SOCIETY

Record Cold Temperatures

The excerpt below appeared in an article in *Mother Earth News* in January, 1995. All of the temperatures are given in °F.

How cold does it get in Alaska? The all-time low temperature is −80 degrees at Prospect Creek Camp on January 23, 1971. Think of it: that temperature is to 0 degrees what 0 degrees is to a typical summer afternoon. Still, you might say, Alaska is up there near the North Pole. Surely, down here in the "lower 48," we never approach such levels of incredible frigidity. Yes we do. On January 20, 1954, the temperature in Rogers Pass, Montana, nudged −70 degrees.... Is there anywhere that temperatures can get colder than in Alaska? Siberia can get a little colder. But there is one place on Earth that is much more frigid: Antarctica. Earth's all-time record low was set at Vostok Station in Antarctica in 1983: −128 degrees. ∎

Antarctica

2. No; the wind-chill factor applies to people and animals, but not to inanimate objects.

1. How much colder is Earth's record low temperature than a temperature of 80°F in summer? **208°F**

2. The wind-chill factor is a measure of how cold we feel when we experience wind along with a low temperature. Do you think that the wind-chill factor also applies to nonliving objects such as bicycles and cars? Why or why not?

3. The hottest temperature recorded on Earth was 136°F at Azizia, Tripolitaina, in northern Africa, on September 13, 1922. If every country's record high and low temperatures for the 20th century were entered in a computer, what would be the range of the data? **264°F**

4. If the interquartile range of temperatures in a country is very small, what does that tell you about the climate of that country? **The temperature stays the same about half of the time.**

Lesson 5–7 **INTEGRATION** *Statistics Measures of Variation* **313**

Enrichment Masters, p. 40

Enrichment worksheet reproduction.

5-7

NAME_____ DATE _____

Enrichment

Student Edition
Pages 305–314

Standard Deviation

The most commonly used measure of variation is called the **standard deviation**. It shows how far the data are from their mean. You can find the standard deviation using the steps given below.

a. Find the mean of the data.
b. Find the difference between each value and the mean.
c. Square each difference.
d. Find the mean of the squared differences.
e. Find the square root of the mean found in Step **d.** The result is the standard deviation.

Example: Calculate the standard deviation of the test scores 82, 71, 63, 78, and 66.

mean of the data $(m) = \dfrac{82 + 71 + 63 + 78 + 66}{5} = \dfrac{360}{5} = 72$

x	$x - m$	$(x - m)^2$
82	$82 - 72 = 10$	$10^2 = 100$
71	$71 - 72 = -1$	$(-1)^2 = 1$
63	$63 - 72 = -9$	$(-9)^2 = 81$
78	$78 - 72 = 6$	$6^2 = 36$
66	$66 - 72 = -6$	$(-6)^2 = 36$

mean of the squared differences $= \dfrac{100 + 1 + 81 + 36 + 36}{5} = \dfrac{254}{5} = 50.8$

standard deviation $= \sqrt{50.8} \approx 7.13$

Use the test scores 94, 48, 83, 61, and 74 to complete Exercises 1-3.

1. Find the mean of the scores. **72**

2. Show that the standard deviation of the scores is about 16.2.

x	$x - m$	$(x - m)^2$
94	$94 - 72 = 22$	484
48	$48 - 72 = -24$	576
83	$83 - 72 = 11$	121
61	$61 - 72 = -11$	121
74	$74 - 72 = 2$	4

mean of $(x - m)^2 = 261.2$
S.D. $= \sqrt{261.2} \approx 16.2$

3. Which had less variations, the test scores listed above or the test scores in the example? **the test scores in the example**

Chapter 5 **313**

Closing the Investigation

This activity provides students an opportunity to bring their work on the Investigation to a close. For each Investigation, students should present their findings to the class. Here are some ways students can display their work.

- Conduct and report on an interview or survey.
- Write a letter, proposal, or report.
- Write an article for the school or local paper.
- Make a display, including graphs and/or charts.
- Plan an activity.

Assessment

To assess students' understanding of the concepts and topics explored in the Investigation and its follow-up activities, you may wish to examine students' Investigation Folders.

The scoring guide provided in the *Investigations and Projects Masters,* p. 7, provides a means for you to score students' work on the Investigation.

Investigations and Projects Masters, p. 7

Scoring Guide
Chapters 4 and 5
Investigation

Level	Specific Criteria
3 Superior	· Shows thorough understanding of the concepts of *the capture-recapture method, data collection and analysis, proportions, percents,* and *probability*. · Uses appropriate strategies to solve problems. · Computations are correct. · Written explanations are exemplary. · Charts, graphs, and speech are appropriate and sensible. · Goes beyond the requirements of some or all problems.
2 Satisfactory, with Minor Flaws	· Shows understanding of the concepts of *the capture-recapture method, data collection and analysis, proportions, percents,* and *probability*. · Uses appropriate strategies to solve problems. · Computations are mostly correct. · Written explanations are effective. · Charts, graphs, and speech are appropriate and sensible. · Satisfies the requirements of problems.
1 Nearly Satisfactory, with Obvious Flaws	· Shows understanding of most of the concepts of *the capture-recapture method, data collection and analysis, proportions, percents,* and *probability*. · May not use appropriate strategies to solve problems. · Computations are mostly correct. · Written explanations are satisfactory. · Charts, graphs, and speech are appropriate and sensible. · Satisfies most requirements of problems.
0 Unsatisfactory	· Shows little or no understanding of the concepts of *the capture-recapture method, data collection and analysis, proportions, percents,* and *probability*. · Does not use appropriate strategies to solve problems. · Computations are incorrect. · Written explanations are not satisfactory. · Charts, graphs, and speech are not appropriate or sensible. · Does not satisfy the requirements of problems.

Go Fish!

Refer to the Investigation on pages 190–191.

Sampling is a research method frequently used in many fields. Manufacturers use sampling to determine the quality of their products. Surveys are a way of sampling the opinions of the public about items they encounter in everyday life such as what television programs they watch or what food is their favorite. Capture-recapture is only one of many methods used in sampling a population.

Analyze

You have conducted experiments and organized your data in various ways. It is now time to analyze your findings and state your conclusions.

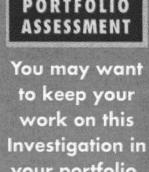

PORTFOLIO ASSESSMENT

You may want to keep your work on this Investigation in your portfolio.

1 What are the strengths and weaknesses of the capture-recapture method of sampling?

2 How many samples do you think give the best results? Why?

3 Write a one-page report on how the capture-recapture method is used in real life. Be sure to include references for your report.

Write

Imagine that the fish population in Moonlit Lake had been estimated each year for nine years and that these estimates are reflected in the following graph.

Fish Population in Moonlit Lake

Estimated Number of Fish — Year ('90–'98): 235, 347, 358, 427, 489, 569, 604, 678, 775

Notice that the fish population is increasing over time. Suppose city council member Aubrey Howard wanted to show other members of the city council that the growth in the fish population has been rapid. Mr. Howard might choose to find a best-fit line for the graph.

4 Plot the information from the graph as ordered pairs on a coordinate plane.

5 Use two vertical lines to separate the data into three sets of equal size.

6 Locate a point in each set whose x- and y-coordinates represent the medians of the x- and y-coordinates of all the points in that set. In other words, locate the middle point of the three sets.

7 Now take a straightedge and align it with the outer two median points. Then slide it one third of the distance to the median point of the center set of points. Then draw the line. This is a *best-fit line*.

8 What might Mr. Howard's argument be? Write a one-page speech for the council member, using the graph as evidence for your opinion.

314 *Chapter 5 Graphing Relations and Functions*

Using the
CHAPTER HIGHLIGHTS

The Chapter Highlights begins with a listing of the new terms, properties, and phrases that were introduced in this chapter. Have students define each term and provide an example or two of it, if appropriate.

Assessment and Evaluation Masters, pp. 115–116

VOCABULARY

After completing this chapter, you should be able to define each term, property, or phrase and give an example or two of each.

Algebra

axes (p. 254)

complete graph (p. 278)

completeness property for points in the plane (p. 256)

coordinate plane (p. 254)

domain (p. 263)

equation in two variables (p. 271)

function (p. 287)

functional notation (p. 289)

graph (p. 255)

inverse of a relation (p. 264)

linear equation (p. 280)

linear equation in standard form (p. 281)

linear function (p. 278)

mapping (p. 263)

origin (p. 254)

quadrants (p. 254)

range (p. 263)

relation (pp. 260, 263)

solution of an equation in two variables (p. 271)

vertical line test (p. 289)

x-axis (p. 254)

x-coordinate (p. 254)

y-axis (p. 254)

y-coordinate (p. 254)

Statistics

interquartile range (pp. 304, 306)

lower quartile (p. 306)

measures of variation (p. 306)

outlier (p. 307)

quartiles (pp. 304, 306)

range (p. 306)

upper quartile (p. 306)

Problem Solving

use a table (p. 255)

UNDERSTANDING AND USING THE VOCABULARY

Choose the letter of the term that best matches each statement or phrase.

1. In the coordinate plane, the axes intersect at the __?__. e

2. A __?__ is a set of ordered pairs. g

3. __?__ are graphed on a coordinate plane. d

4. In a coordinate system, the __?__ is a horizontal line. h

5. In the ordered pair, A(2, 7), 7 is the __?__. j

6. The coordinate axes separate a plane into four __?__. f

7. An equation whose graph is a non-vertical straight line is called a __?__. c

8. In the relation $\{(4, -2), (0, 5), (6, 2), (-1, 8)\}$, the __?__ is the set $\{-1, 0, 4, 6\}$. a

9. The domain contains values represented by the __?__. b

a. domain
b. independent variable
c. linear function
d. ordered pairs
e. origin
f. quadrants
g. relation
h. x-axis
i. y-axis
j. y-coordinate

Chapter 5 Highlights **315**

Instructional Resources

Three multiple-choice tests and three free-response tests are provided in the *Assessment and Evaluation Masters*. Forms 1A and 2A are for honors pacing, and Forms 1B, 1C, 2B, and 2C are for average pacing. Chapter 5 Test Form 1B is shown at the right. Chapter 5 Test Form 2B is shown on the next page.

Skills and Concepts Encourage students to refer to the objectives and examples on the left as they complete the review exercises on the right.

Assessment and Evaluation Masters, pp. 121–122

SKILLS AND CONCEPTS

OBJECTIVES AND EXAMPLES	REVIEW EXERCISES

Upon completing this chapter, you should be able to:

Use these exercises to review and prepare for the chapter test. **10–13. See Solutions Manual.**

• graph ordered pairs on a coordinate plane (Lesson 5–1)

Graph $T(3, -2)$ and state the quadrant in which the point is located.

$T(3, -2)$ is located in Quadrant IV.

Graph each point.

10. $A(4, 2)$ **11.** $B(-1, 3)$

12. $C(0, -5)$ **13.** $D(-3, -2)$

Write the ordered pair for each point shown at the right. State the quadrant in which the point is located.

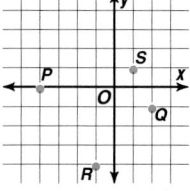

14. P $(-4, 0)$; none
15. Q $(2, -1)$; IV
16. R $(-1, -4)$; III
17. S $(1, 1)$; I

• identify the domain, range, and inverse of a relation (Lesson 5–2)

Determine the domain, range, and inverse of $\{(6, 6), (4, -3), (6, 0)\}$.

The domain is $\{4, 6\}$.

The range is $\{-3, 0, 6\}$.

The inverse is $\{(6, 6), (-3, 4), (0, 6)\}$.

State the domain and range of each relation.

18. $\{(4, 1), (4, 6), (4, -1)\}$ $D = \{4\}, R = \{-1, 1, 6\}$
19. $\{(-3, 5), (-3, 6), (4, 5), (4, 6)\}$
20. $\{(-2, 1), (-5, 1), (-7, 1)\}$
21. $\{(-3, 1), (-2, 0), (-1, 1), (0, 2)\}$

19. $D = \{-3, 4\}, R = \{5, 6\}$
20. $D = \{-7, -5, -2\}, R = \{1\}$
21. $D = \{-3, -2, -1, 0\}, R = \{0, 1, 2\}$
22. $\{(4, 4), (5, -3), (-1, 4), (3, 0)\}$
23. $\{(2, 0), (-1, 3), (2, 2), (-1, -2)\}$

22–23. See Solutions Manual for mappings and graphs.
Draw a mapping and a graph for each relation. State the inverse of each relation.

22. $\{(4, 4), (-3, 5), (4, -1), (0, 3)\}$
23. $\{(0, 2), (3, -1), (2, 2), (-2, -1)\}$

• determine the range for a given domain (Lesson 5–3)

Solve $2x + y = 8$ if the domain is $\{3, 2, 1\}$.

Solve for y.

$2x + y = 8$
$y = 8 - 2x$

x	$8 - 2x$	y	(x, y)
3	$8 - 2(3)$	2	$(3, 2)$
2	$8 - 2(2)$	4	$(2, 4)$
1	$8 - 2(1)$	6	$(1, 6)$

Solve each equation if the domain is $\{-4, -2, 0, 2, 4\}$.   **24–27. See margin.**

24. $y = 4x + 5$ **25.** $x - y = 9$
26. $3x + 2y = 9$ **27.** $4x - 3y = 0$

28–29. See Solutions Manual.
Make a table and graph the solution set for each equation and domain.

28. $y = 7 - 3x$ for $x = \{-3, -2, -1, 0, 1, 2, 3\}$
29. $5x - y = -3$ for $x = \{-2, 0, 2, 4, 6\}$

Chapter 5 Test, Form 2B (inset)

5   NAME_____ DATE_____

Chapter 5 Test, Form 2B

Use the table to answer questions 1 and 2.

1. Write the relation as a set of ordered pairs.

x	y
6	-2
-3	4
0	5

1. $\{(6, -2), (-3, 4), (0, 5)\}$

2. State the domain, range, and inverse of the relation.

2. D: $\{6, -3, 0\}$; R: $\{-2, 4, 5\}$; I: $\{(-2, 6), (4, -3), (5, 0)\}$

3. Graph these points on the coordinate plane provided.
$A(3, 2)$ $B(-3, 0)$ $C(0, -2)$ $D(3, -4)$

3.

4. Name the quadrant in which each point is located.
$A(-5, 7)$ $B(2, -4)$ $C(6, 0)$

4. II; IV; none

5. Draw a mapping for the relation $\{(1, 4), (-5, 0), (7, 4)\}$.

5.

Use the graph to answer questions 6 and 7.

6. Write the relation in the graph as a set of ordered pairs.

6. $\{(-3, 2), (-3, -1), (0, 1), (4, 3)\}$

7. State the domain, range, and inverse of the relation.

7. D: $\{-3, 0, 4\}$; R: $\{-1, 1, 2, 3\}$; I: $\{(-1, -3), (1, 0), (2, -3), (3, 4)\}$

8. Solve $3x + y = 6$ for y.

8. $y = 6 - 3x$

9. Solve the equation $m - 2n = 9$ if the domain is $\{-1, 0, 5\}$.

9. $\{(-1, -5), (0, -\frac{9}{2}), (5, -2)\}$

10. Which of the following equations are linear equations? Write the letter for each linear equation in the answer blank.
a. $xy = 6$ b. $y = 0$ c. $4x = 2y$
d. $x - y = 0$ e. $y^2 - 4x = 0$

10. b, c, d

11. Graph $x - 4y = 2$. Use the coordinate plane provided.

11.

5   NAME_____ DATE_____

Chapter 5 Test, Form 2B (continued)

12. Graph $2x = -3y$. Use the coordinate plane provided.

12.

13. Determine whether each relation is a function. Then write the letter for each function in the answer blank.
a. $\{(5, 3), (2, -1), (-1, 3)\}$ e.
b. $\{(2, 1), (2, -4), (-3, 0)\}$
c. $x - 8 = 0$
d. $4y = 9x$

13. a, d, e

14. Given $f(x) = x^2 + 3x - 2$, find $f(2)$.

14. 8

15. Given $h(x) = 4x + 8$, find $4[h(-\frac{3}{4})]$.

15. 20

Write an equation for the relation given in each chart.

16.
x	1	2	3	4	5
y	5	7	9	11	13

16. $y = 2x + 3$

17.
x	-2	0	2	4
y	10	2	-6	-14

17. $y = -4x + 2$

18. Li Teng is ordering graduation announcements for the senior class. If 100 announcements are ordered, the cost is $225. The cost for 175 announcements is $375, and 200 announcements cost $425. Write an equation to describe this relationship and use it to find the cost of 250 announcements.

18. $c = 25 + 2a$; $525

19. Find the median and upper and lower quartiles for this set of data: 200, 250, 230, 180, 160, 140.

19. median: 190; UQ: 230; LQ: 160

20. Find the interquartile range for the data in question 19.

20. interq. range: 70

Bonus Graph $x = 3$, $y = -1$, and $2x - 2y = 0$ on the coordinate plane provided. Name the geometric figure formed. Then state the coordinates of each vertex.

Bonus triangle; (3, 3), (3, -1), (-1, -1)

GLENCOE Technology

Test and Review Software

You may use this software, a combination of an item generator and item bank, to create your own tests or worksheets. Types of items include free response, multiple choice, short answer, and open ended.

For IBM & Macintosh

24. $\{(-4, -11), (-2, -3), (0, 5),$
    $(2, 13), (4, 21)\}$

25. $\{(-4, -13), (-2, -11), (0, -9),$
    $(2, -7), (4, -5)\}$

26. $\left\{\left(-4, 10\frac{1}{2}\right), \left(-2, 7\frac{1}{2}\right), \left(0, 4\frac{1}{2}\right),$
    $\left(2, 1\frac{1}{2}\right), \left(4, -1\frac{1}{2}\right)\right\}$

27. $\left\{\left(-4, -5\frac{1}{3}\right), \left(-2, -2\frac{2}{3}\right), (0, 0),$
    $\left(2, 2\frac{2}{3}\right), \left(4, 5\frac{1}{3}\right)\right\}$

OBJECTIVES AND EXAMPLES

• graph linear equations (Lesson 5–4)

Graph $y = 3x - 4$.

x	3x − 4	y	(x, y)
0	3(0) − 4	−4	(0, −4)
1	3(1) − 4	−1	(1, −1)
2	3(2) − 4	2	(2, 2)
x	3x − 4	y	(x, y)

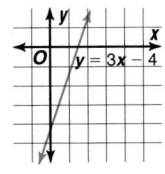

REVIEW EXERCISES

Graph each equation. 30–35. See margin.

30. $y = -x + 2$
31. $x + 5y = 4$
32. $2x - 3y = 6$
33. $5x + 2y = 10$
34. $\frac{1}{2}x + \frac{1}{3}y = 3$
35. $y - \frac{1}{3} = \frac{1}{3}x + \frac{2}{3}$

30.

31.

• determine whether a given relation is a function (Lesson 5–5)

Is $\{(3, 2), (5, 3), (4, 3), (5, 2)\}$ a function?

Because there are two values of y for one value of x, 5, the relation is *not* a function.

Determine whether each relation is a function.

36.

a	b
−2	6
3	−2
3	0
4	6

no

37.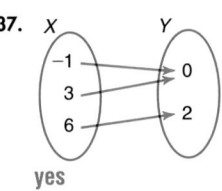

yes

38. $\{(3, 8), (9, 3), (-3, 8), (5, 3)\}$ yes
39. $x - y^2 = 4$ no
40. $xy = 6$ yes
41. $3x - 4y = 7$ yes

32.

• find the value of a function for a given element of the domain (Lesson 5–5)

Given $g(x) = 2x - 1$, find $g(-6)$.

$g(-6) = 2(-6) - 1$
$\quad\quad = -12 - 1$
$\quad\quad = -13$

If $g(x) = x^2 - x + 1$, find each value.

42. $g(2)$ 3
43. $g(-1)$ 3
44. $g\left(\frac{1}{2}\right)$ $\frac{3}{4}$
45. $g(a + 1)$ $a^2 + a + 1$
46. $g(-2a)$ $4a^2 + 2a + 1$
47. $2g(a - 3)$ $2a^2 - 14a + 26$

33.

• write an equation to represent a relation, given some of the solutions for the equation (Lesson 5–6)

Write an equation for the relation given in the chart below.

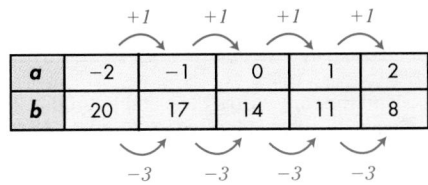

The equation is $b = 14 - 3a$.

Write an equation for each relation.

48.

x	0	1	2	3	4
y	5	8	11	14	17

49.

x	2	4	5	7	10
y	−2	0	1	3	6

50.

x	3	6	9	12	15
y	−1	−3	−5	−7	−9

48. $y = 3x + 5$ 49. $y = x - 4$ 50. $y = -\frac{2}{3}x + 1$

34.

35.

Applications and Problem
Solving Encourage students to
work through the exercises in the
Applications and Problem Solving
section to strengthen their
problem-solving skills.

Additional Answer

55a.

CHAPTER 5   STUDY GUIDE AND ASSESSMENT

OBJECTIVES AND EXAMPLES

• calculate and interpret the range, quartiles, and
interquartile range of sets of data   (Lesson 5–7)

Find the range, median, upper quartile, lower
quartile, and interquartile range for the set of
data below.

$$25, 20, 30, 24, 22, 26, 28, 29, 19$$

Order the set of data from least to greatest.

$$19 \quad 20 \quad 22 \quad 24 \quad 25 \quad 26 \quad 28 \quad 29 \quad 30$$

The range is $30 - 19 = 11$.

The median is the middle number, 25.

The lower quartile is $\frac{20 + 22}{2}$ or 21.

The upper quartile is $\frac{28 + 29}{2}$ or 28.5.

The interquartile range is $28.5 - 21$ or 7.5.

REVIEW EXERCISES

**Find the range, median, upper quartile,
lower quartile, and interquartile range for
each set of data.**

51. 30, 90, 40, 70, 50, 100, 80, 60   **70, 65, 85, 45, 40**

52. 3, 3.2, 45, 7, 2, 1, 3.4, 4, 5.3, 5, 78, 8, 21, 5

53. 85, 77, 58, 69, 62, 73, 55, 82, 67, 77, 59, 92, 75, 69, 76

54. The average annual snowfall, in inches, for
each of 12 northeastern cities is listed below.

111.5	70.7	59.8	68.6	63.8	254.8
64.3	82.3	91.7	88.9	110.5	77.1

52. 77, 5, 8, 3.2, 4.8

53. 37, 73, 77, 62, 15

54. 195, 79.7, 101.1, 66.45, 34.65

APPLICATIONS AND PROBLEM SOLVING

55. Finance To save for a new bicycle, Ralph
begins a savings plan. Ralph's savings are
described by the equation $s = 5w + 56$,
where s represents his total savings in dollars
and w represents the number of weeks since
the start of the savings plan.   (Lesson 5–4)

   a. Draw a graph of this equation.   **See margin.**

   b. Use the graph to determine how much
   money Ralph had already saved when the
   plan started.   **$56**

56. Car Rental The
cost of a one-day car
rental from A-1 Car
Rental is $41 if
you drive 100
miles, $51.80 if
you drive 160
miles, and
$63.50 if you
drive 225 miles.
(Lesson 5–6)

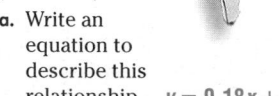

   a. Write an
   equation to
   describe this
   relationship. $y = 0.18x + 23$

   b. Use the equation to determine the per-mile
   charge.   **$0.18 per mile**

57. Entertainment The ratings of the top 20
favorite syndicated television programs for
1993–94 are listed below.   (Lesson 5–7)

Program	Rating
Action Pack Network	6.1
Baywatch	6.5
Cops	5.7
Current Affair	6.6
Donahue	5.1
Entertainment Tonight	8.4
Family Matters	5.9
Hard Copy	6.7
Inside Edition	7.3
Jeopardy!	12.6
Married With Children	6.9
National Geographic on Assignment	7.5
Oprah Winfrey	9.7
Roseanne	7.9
Sally Jessy Raphael	5.2
Star Trek	10.8
Star Trek: Deep Space Nine	8.2
Wheel of Fortune	14.7
Wheel of Fortune-Weekend	7.2
World Wrestling Federation	5.7

   a. Find the range, quartiles, and interquartile
   range for the ratings.   **9.6, 6, 7.05, 8.3, 2.3**

   b. Identify any outliers.   **12.6, 14.7**

**A practice test for Chapter 5 is provided on
page 791.**

ALTERNATIVE ASSESSMENT

COOPERATIVE LEARNING PROJECT

Graphs and Business In this chapter, you learned how to graph relations and determine whether the relation is a linear equation. Once a relation has been graphed, the process of analyzing it to determine more applicable information is necessary in order to completely understand the problem or situation.

In this project, imagine that your family owns a catering business and you want to look at the amount that you charge for banquets. You have been asked to present a report to the other members of your family that shows recommended charges and how they will affect your profit. After some research, you find that the initial basic charge for a banquet should be $250 plus $4 per person. Your food costs per person are $2.50, and your other costs such as labor, time, etc. total $1.00 per person.

Graph the various charges determined by the number of persons. Graph the various costs determined by the number of persons.

Follow these steps to organize your presentation.

- Determine the charges for various banquet sizes.
- Determine the costs for various banquet sizes.
- Determine how to plot and display your graphs appropriately.
- Write an algebraic model that describes the total charge for a banquet.
- Write an algebraic model that describes the total cost for a banquet.
- Write several paragraphs for your presentation and incorporate your algebraic models and graphs to help your presentation.

THINKING CRITICALLY

- Use one set of data to construct two graphs in which each represents the data, but portrays different meanings.
- Create a relation that is a function and one that is not a function. Compare and contrast the two relations.

PORTFOLIO

You have been asked to show another student how to graph an equation, but it must be completely written and nonverbal. Select one of the graphing assignments from this chapter and list the steps involved in graphing this equation. Be sure to include your steps in the appropriate order.

When you have completed this paper, give it to another student in your class and have him or her read it and follow the written steps to graph a different equation. Collect his or her graph and check for accuracy. Place both of these in your portfolio.

SELF EVALUATION

Graphs can be used to analyze functions or data. To *analyze* means to separate or distinguish all the parts of something in order to discover more information about the complex whole element.

Assess yourself. How analytical are you? Do you delve deeper into a problem and look for smaller components that will help you understand the larger problem, or do you look at the big picture and attempt to solve the problem within the larger view of things? Describe two or three ways in which you could divide a complex problem or situation in mathematics and/or in your daily life in order to analyze it.

Assessment and Evaluation Masters, pp. 126, 137

5 NAME_____ DATE_____

Chapter 5 Performance Assessment

Instructions: *Demonstrate your knowledge by giving a clear, concise solution to each problem. Be sure to include all relevant drawings and justify your answers. You may show your solution in more than one way or investigate beyond the requirements of the problem.*

1. The members of sets X and Y are shown at the right.

   a. Draw a mapping of X to Y that shows a function. Tell how you know it is a function.

   b. Draw a mapping of X to Y that represents a relation but not a function. Tell how you know it is a relation. Tell how you know it is not a function.

   c. Graph the relation $\{(-1, 3), (3, 5), (3, -2), (4, 0), (-3, 2)\}$. Use the vertical line test to determine if the relation is a function. Write a sentence to explain your reasoning.

   d. Express the relation shown in the mapping for part a as a set of ordered pairs. Then write the inverse of the relation. Is the inverse a function? Tell why or why not.

2. a. Complete the pattern in the chart.

x	-2	-1	0	1	2	3		
y	-9	-5	-1	3	7			

   b. Write an equation for the relation shown in the chart.

   c. Is the equation in part b a linear equation? Explain your answer.

   d. Graph the equation in part b.

   e. Is the inverse of the relation a linear function? Explain your answer.

3. The opening day scores in a local golf tournament are 78, 83, 70, 84, 89, 67, 84, 92, 78, 91, 85, 77, 68, 80, 71, 78, 99, 81, 75, 88, 90, 71, 73.

   a. Find the range and interquartile range of the scores. Tell in your own words the purpose of finding the range or interquartile range.

   b. Is the range or the interquartile range more affected by outliers?

   c. Display the results in a stem-and-leaf plot.

Scoring Guide
Chapter 5
Performance Assessment

Level	Specific Criteria
3 Superior	• Shows thorough understanding of the concepts of *relation, function, inverse function, domain, range, linear equation,* and *measures of variation.* • Computations are correct. • Written explanations are exemplary. • Graphs and diagrams are accurate and appropriate. • Goes beyond requirements of some or all problems.
2 Satisfactory, with Minor Flaws	• Shows understanding of the concepts of *relation, function, inverse function, domain, range, linear equation,* and *measures of variation.* • Computations are mostly correct. • Written explanations are effective. • Graphs and diagrams are accurate and appropriate. • Satisfies all requirements of problems.
1 Nearly Satisfactory, with Serious Flaws	• Shows understanding of most of the concepts of *relation, function, inverse function, domain, range, linear equation,* and *measures of variation.* • Computations are mostly correct. • Written explanations are satisfactory. • Graphs and diagrams are mostly accurate and appropriate. • Satisfies most requirements of problems.
0 Unsatisfactory	• Shows little or no understanding of the concepts of *relation, function, inverse function, domain, range, linear equation,* and *measures of variation.* • May not use appropriate strategies to solve problems. • Computations are incorrect. • Written explanations are not satisfactory. • Graphs and diagrams are not accurate or appropriate. • Does not satisfy requirements of problems.

 Alternative Assessment

The Alternative Assessment section provides students with the opportunity to assess their own work by thinking critically, working with others, keeping a portfolio, and honestly evaluating their own progress. For more information on alternative forms of assessment, see *Alternative Assessment in the Mathematics*

Classroom, one of the titles in the Glencoe Mathematics Professional Series.

Performance Assessment

Performance Assessment tasks for this chapter are included in the *Assessment and Evaluation Masters.* A scoring guide is also provided.

NCTM Standards: 1–5

This Investigation is designed to be completed over several days or weeks. It may be considered optional. You may want to assign the Investigation and the follow-up activities to be completed at the same time.

Objective
Use mathematics to collect and analyze data regarding the effects of secondhand smoke.

Mathematical Overview
This Investigation will use the following mathematical skills and concepts from Chapters 6 and 7.

- making charts and graphs
- using the formulas for radius and volume

Recommended Time

Part	Pages	Time
Investigation	320–321	1 class period
Working on the Investigation	338, 361, 398, 426	20 minutes each
Closing the Investigation	442	1 class period

Instructional Resources
Investigations and Projects Masters, pp. 9–12

A recording sheet, teacher notes, and scoring guide are provided for each Investigation in the *Investigations and Projects Masters.*

1 MOTIVATION

In this Investigation, students use simple materials and background information to investigate the effects of secondhand smoke. Ask students to discuss why this problem might be harmful for our society.

Smoke Gets In Your Eyes

MATERIALS NEEDED

- stopwatch
- grid paper
- balloons
- tape measure
- string

In the 1950s and '60s, scientists piled up a mountain of evidence on the life-threatening health consequences of smoking. Recently there have been studies done on the effects of secondhand smoke. Secondhand smoke is the smoke others breathe when they are close to a person who is smoking. In 1993, the Environmental Protection Agency (EPA) estimated that secondhand smoke is responsible for several thousands of cases of lung cancer among nonsmokers each year in the U.S. Passive smoke joins a select company of only about a dozen other environmental pollutants in this risk category.

As an aware citizen and health-conscious teenager, you understand this problem and want to start a campaign to limit the effects of secondhand smoke. In this Investigation, you will gather evidence about how much air is inhaled when you breathe and compare that to smoke in the room. You will also examine the financial and health aspects of this problem. Work in groups of three. Make an Investigation Folder in which you can store all of your work on this Investigation for future use.

Adult Cigarette Use in the United States

Source: U.S. Department of Health and Human Resources

320 *Investigation: Smoke Gets In Your Eyes*

Cooperative Learning

This Investigation offers an excellent opportunity for using cooperative learning groups. For more information on cooperative learning strategies and group management, see *Cooperative Learning in the Mathematics Classroom,* one of the titles in the Glencoe Mathematics Professional Series.

GROUP MEMBER	1	2	3
Number of breaths			
Circumference of balloon			
Radius of balloon			
Volume of balloon			
Volume of each breath			

COLLECT THE DATA

1 Begin by copying the chart above onto a sheet of paper.

2 Stretch each balloon. Have each member of the group blow up at least three round balloons to full size and let the air out.

3 Give each team member time to rest.

4 Cut a piece of string that is approximately 30 inches long. Tie the string so that it can form a circle that is approximately 8 inches in diameter when laid on a flat surface.

5 One member of the group should hold the string hoop around the balloon and another member should blow the balloon up again. The remaining group member should count and record the number of puffs it takes to blow up each balloon to the size of the hoop. Each group member should blow up one balloon.

6 Measure each inflated balloon to estimate its circumference at the fullest part. Find the approximate radius of the balloon by using the formula $r = \frac{C}{2\pi}$. Record the radius.

ANALYZE THE DATA

7 Display your data on a graph.

8 Calculate the average number of breaths it takes to fill a balloon.

9 The volume of the spherical balloon can be estimated by the formula $V = \frac{4}{3}\pi r^3$. Compute the volume of each balloon. Then estimate the volume of each breath.

You will continue working on this Investigation throughout Chapters 6 and 7.

Be sure to keep your chart and materials in your Investigation Folder.

Smoke Gets In Your Eyes Investigation

Working on the Investigation
Lesson 6–2, p. 338

Working on the Investigation
Lesson 6–5, p. 361

Working on the Investigation
Lesson 7–2, p. 398

Working on the Investigation
Lesson 7–6, p. 426

Closing the Investigation
End of Chapter 7, p. 442

Investigation: Smoke Gets In Your Eyes **321**

You may wish to have a student read the first paragraph of the Investigation to provide information about why secondhand smoke is an important issue. You may then wish to read the last paragraph, which introduces the activity. Discuss the activity with students. Then separate the class into groups of three.

3 MANAGEMENT

Each group member should be responsible for a specific task.

Recorder Collects data.
Measurer Sets, measures, and ties the string and holds the string hoop around the balloon.
Timer Counts and records number of puffs that it takes to blow up each balloon to the size of the hoop.

At the end of the activity, each member should turn in his or her respective equipment.

Sample Answers

Answers will vary based on the efforts of each team member.

Investigations and Projects Masters, p. 12

6, 7 | NAME_____ DATE____
Student Edition
Investigation, Chapters 6 and 7 Pages 320–321
338, 361, 398, 426, 442

Smoke Gets In Your Eyes. . .

Work with your group to answer the following questions and add others to be considered.

• If you were in a closed room with someone who smoked two packs of cigarettes (there are 20 cigarettes in a pack), what estimated volume of smoke would you breathe? How many cigarettes does that equal?

• Based on what you have found in the Investigation, would you support or oppose laws that restrict people from smoking in public buildings and enclosed spaces? Explain why you think so.

_____
_____
_____
_____
_____

Reread the information about secondhand smoke on pages 320 and 442. Then do research to find out more about this topic. Record your findings in the chart below.

Topic	Cancer	Heart Disease	Other Respiratory Diseases
Risks to children in smokers' households			
Risks to nonsmoking adults in smokers' households			
Risks to general population encountering secondhand smoke in public places			

Analyzing Linear Equations

In this chapter, students begin by finding the slope of a line given the coordinates of two points on the line. Then, students study the point-slope, standard, and slope-intercept forms of linear equations. The concepts of scatter plots and best-fit lines are integrated into the chapter to provide students with another tool for data representation and problem solving. Next, students use a graphing calculator and the concepts of slopes and intercepts to graph equations. Students apply the slope-intercept form to write the equation of a line and then learn how to write equations for parallel or perpendicular lines. The chapter concludes with a lesson in which students find the coordinates of the midpoint of a line given the coordinates of the endpoints.

Lesson (Pages)	Lesson Objectives	NCTM Standards	State/Local Objectives
6-1A (324)	Use a geoboard to model a line segment and calculate its slope.	1–5, 8	
6-1 (325–331)	Find the slope of a line, given the coordinates of two points on the line.	1–5, 8	
6-2 (332–338)	Write linear equations in point-slope form. Write linear equations in standard form.	1–5, 8	
6-3 (339–345)	Graph and interpret points on a scatter plot. Draw and write equations for best-fit lines, and make predictions by using those equations. Solve problems by using models.	1–5, 10	
6-4 (346–353)	Determine the x- and y-intercepts of linear graphs from their equations. Write equations in slope-intercept form. Write and solve direct variation equations.	1–5	
6-5A (354–355)	Use a graphing calculator to determine whether a group of graphs forms a family.	1–5	
6-5 (356–361)	Graph a line given any linear equation.	1–5	
6-6 (362–368)	Determine if two lines are parallel or perpendicular by their slopes. Write equations of lines that pass through a given point, parallel or perpendicular to the graph of a given equation.	1–5, 8	
6-7 (369–374)	Find the coordinates of the midpoint of a line segment in the coordinate plane.	1–5, 8	

A complete, 1-page lesson plan is provided for each lesson in the *Lesson Planning Guide*. Answer keys for each lesson are available in the *Answer Key Masters*.

You may want to refer to the **Course Planning Calendar** on page T12 for detailed information on pacing.
PACING: Standard—13 days; **Honors**—12 days; **Block**—7 days; **Two Years**—18 days

LESSON PLANNING CHART

Lesson (Pages)	Materials/ Manipulatives	Extra Practice (Student Edition)	Study Guide	Practice	Enrichment	Assessment and Evaluation	Modeling Mathematics	Multicultural Activity	Tech Prep Applications	Graphing Calculator	Science and Math Lab Manual	Real-World Applications	Interactive Mathematics Tools Software	Teaching Transparencies
						BLACKLINE MASTERS								
6-1A (324)	geoboard* rubber bands*						p. 26						6-1A	
6-1 (325–331)	spreadsheet	p. 769	p. 41	p. 41	p. 41			p. 11				17		6-1A 6-1B
6-2 (332–338)		p. 769	p. 42	p. 42	p. 42	p. 156								6-2A 6-2B
6-3 (339–345)	graphing calculator	p. 770	p. 43	p. 43	p. 43					pp. 23–28			6-3	6-3A 6-3B
6-4 (346–353)		p. 770	p. 44	p. 44	p. 44	pp. 155, 156		p. 12	p. 11			18		6-4A 6-4B
6-5A (354–355)	graphing calculator									pp. 22, 23				
6-5 (356–361)	graphing calculator	p. 770	p. 45	p. 45	p. 45			p. 12		p. 6			6-5.1 6-5.2	6-5A 6-5B
6-6 (362–368)	grid paper scissors*	p. 771	p. 46	p. 46	p. 46	p. 157	pp. 51–53, 77						6-6	6-6A 6-6B
6-7 (369–374)	grid paper ruler* colored pencils	p. 771	p. 47	p. 47	p. 47	p. 157								6-7A 6-7B
Study Guide/ Assessment (375–379)						pp. 141–154, 158–160								

*Included in Glencoe's Student Manipulative Kit and Overhead Manipulative Resources.

ORGANIZING THE CHAPTER

OTHER CHAPTER RESOURCES

Student Edition
Investigation, pp. 320–321
Chapter Opener, pp. 322–323
Mathematics and Society, p. 374
Working on the Investigation, pp. 338, 361

Teacher's Classroom Resources
Investigations and Projects Masters, pp. 45–48
Algebra and Geometry Overhead Manipulative Resources, pp. 18–22

Technology
Test and Review Software (IBM and Macintosh)
CD-ROM Interactions (Windows and Macintosh)

Professional Publications
Block Scheduling Booklet
Glencoe Mathematics Professional Series

OUTSIDE RESOURCES

Books/Periodicals
Taylor, Harold, *Developing Skills in Algebra One,* Dale Seymour Publications
Winter, Mary Jean and Ronald J. Carlson, *Algebra Experiments I,* Dale Seymour Publications

Software
Algebra Concepts, Vol. I, Ventura Educational Systems
Topics in Discrete Mathematics: Computer-Supported Problem Solving, Dale Seymour Publications

Videos/CD-ROMs
Algebra for Everyone, Videotape and Discussion Guide, NCTM

ASSESSMENT RESOURCES

Student Edition
Math Journal, pp. 343, 350, 359
Mixed Review, pp. 331, 337, 345, 352, 361, 368, 374
Self Test, p. 353
Chapter Highlights, p. 375
Chapter Study Guide and Assessment, pp. 376–378
Alternative Assessment, p. 379
 Portfolio, p. 379

Cumulative Review, pp. 380–381

Teacher's Wraparound Edition
5-Minute Check, pp. 325, 332, 339, 346, 356, 362, 369
Check for Understanding, pp. 328, 335, 343, 349, 359, 366, 371
Closing Activity, pp. 331, 338, 345, 352, 361, 368, 373
Cooperative Learning, pp. 340, 347

Assessment and Evaluation Masters
Multiple-Choice Tests, Forms 1A (Honors), 1B (Average), 1C (Basic), pp. 141–146
Free-Response Tests, Forms 2A (Honors), 2B (Average), 2C (Basic), pp. 147–152
Calculator-Based Test, p. 153
Performance Assessment, p. 154
Mid-Chapter Test, p. 155
Quizzes A–D, pp. 156–157
Standardized Test Practice, p. 158
Cumulative Review, pp. 159–160

ENHANCING THE CHAPTER

Examples of some of the materials for enhancing Chapter 6 are shown below.

DIVERSITY

Multicultural Activity Masters, pp. 11, 12

APPLICATIONS

Real-World Applications, 17, 18

TECHNOLOGY

Graphing Calculator Masters, p. 6

TECH PREP

Tech Prep Applications Masters, pp. 11, 12

CONNECTIONS

Science and Math Lab Manual, pp. 23–28

PROBLEM SOLVING

Problem of the Week Cards, 16, 17, 18

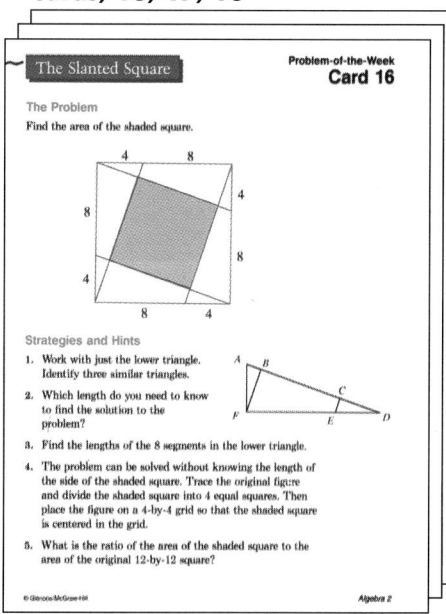

Analyzing Linear Equations

Objectives

In this chapter, you will:

- find the slope of a line, given the coordinates of two of its points,
- write linear equations in point-slope, standard, and slope-intercept forms,
- draw a scatter plot and find the equation of a best-fit line for the data,
- solve problems by using models,
- graph linear equations, and
- use slope to determine if two lines are parallel or perpendicular.

Immigration on the Rise

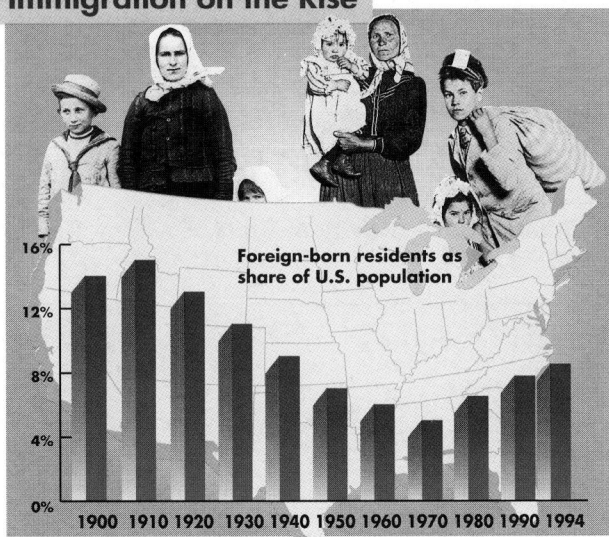

Foreign-born residents as share of U.S. population

1900 1910 1920 1930 1940 1950 1960 1970 1980 1990 1994

Source: *U.S. News and World Report,* September 25, 1995

Almost $20 billion a year is spent on bilingual programs in U.S. schools. These programs were launched in 1968 to help immigrant children learn English. The percentage of foreign-born Americans is rising. More than 22 states have enacted laws proclaiming English as the official language. What are your views on making English the official language of the U.S.?

TIME Line

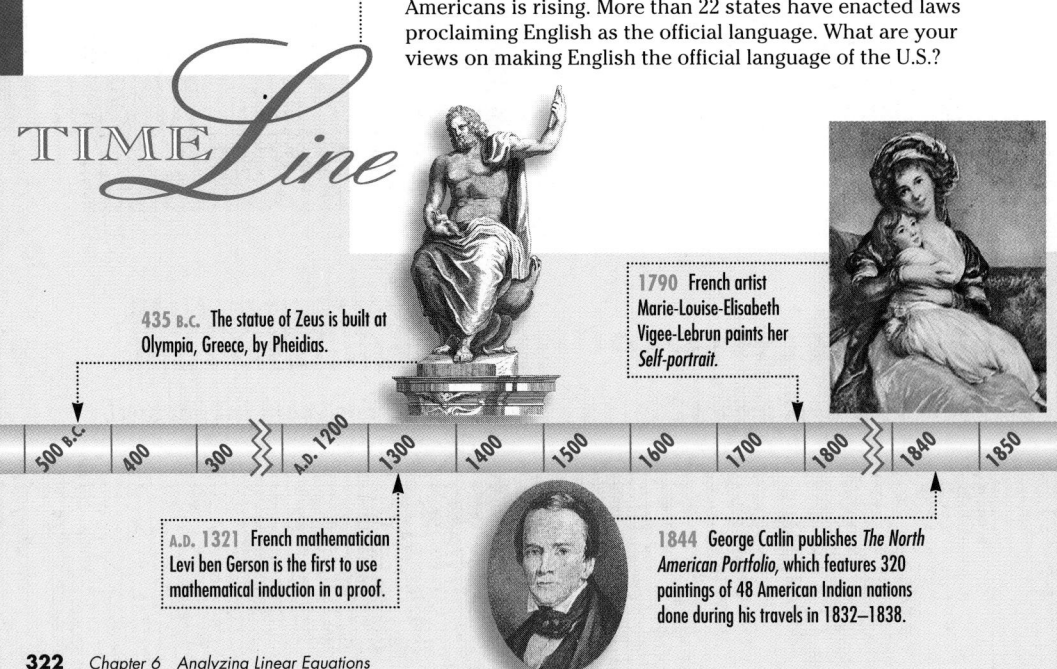

435 B.C. The statue of Zeus is built at Olympia, Greece, by Pheidias.

1790 French artist Marie-Louise-Elisabeth Vigee-Lebrun paints her *Self-portrait.*

500 B.C. 400 300 A.D. 1200 1300 1400 1500 1600 1700 1800 1840 1850

A.D. 1321 French mathematician Levi ben Gerson is the first to use mathematical induction in a proof.

1844 George Catlin publishes *The North American Portfolio,* which features 320 paintings of 48 American Indian nations done during his travels in 1832–1838.

TIME Line

Students should investigate the Sieve of Eratosthenes, developed by the Libyan mathematician in 230 B.C. During their investigation, students should notice that the method does not generate prime numbers, but tests each integer to determine if it is prime.

interNET CONNECTION

Learn about languages of the world through The Human-Languages Page.

World Wide Web
http://www.willamette.edu/~tjones/
Language-Page.html

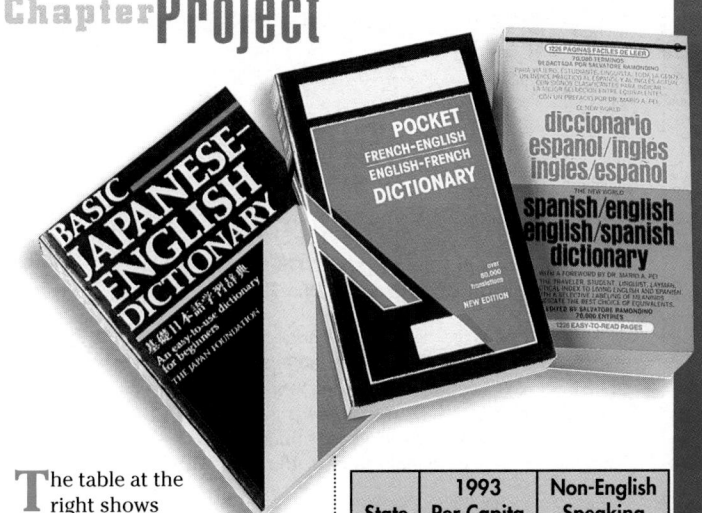

Chapter Project

One student from Webster, Texas, **Jorge Arturo Pineda Aguilar,** is a member of the new immigrant minority. He and his family are from Monterrey, Mexico, and recently moved to Texas. At age 13, Jorge is one of the recent winners of the 1995 Youth Honor Awards presented by *Skipping Stones,* a multicultural children's magazine. His 1995 winning essay, written in both Spanish and English, deals with racial and ethnic prejudice.

Jorge regrets that so many people have ill-feelings about someone who doesn't speak English. He writes, "All of us are human beings and the color or nationality of a person should not be an issue. What should matter are the feelings of each person. We have to be together—work together for a common well being."

The table at the right shows 15 selected states, the per capita income, and the total non-English speaking population (ages 5 years or older). The population is given according to the 1990 U.S. Census.

Round the data to the nearest thousand. Graph the data for each state by plotting the per capita income data along the horizontal axis and the non-English-speaking population data along the vertical axis.

• What pattern, if any, do you notice?

• Can a best-fit line be drawn? If so, what might be the equation of this line?

State	1993 Per Capita Income ($)	Non-English Speaking Population
AL	17,234	107,866
AR	16,143	60,781
FL	20,857	2,098,315
IL	22,582	1,499,112
IN	19,203	245,826
KY	17,173	86,482
LA	16,667	391,994
NC	17,488	240,866
NM	16,297	493,999
NY	24,623	3,908,720
OK	17,020	145,798
TN	17,666	131,550
TX	19,189	3,970,304
VA	21,634	418,521
WV	16,209	44,203

• Can a valid relationship exist between these data? Why or why not?

A quote from Jorge's essay follows.
"All of us are human beings and the color or nationality of a person should not be an issue. What should matter are the feelings of each person. . . . We should always find a place for peace in our hearts."

Chapter Project

Cooperative Learning In setting up cooperative teams, try to include in each group at least one student from a bilingual family. Students might also want to compare the results of their community with a profile for the entire country.

1965 The national origins quota system of immigration abolished.

1973 Marion Wright Edelman founds the Children's Defense Fund.

1890 | 1900 | 1910 | 1920 | 1930 | 1940 | 1950 | 1960 | 1970 | 1980 | 1990 | 2000

1926 Mexican artist Diego Rivera paints *The Tortilla Maker.*

1995 Twins Chris and Courtney Salthouse from Chamblee, Georgia *both* score a perfect 1600 on their SATs.

Chapter 6 **323**

Alternative Chapter Projects

Two other chapter projects are included in the *Investigations and Projects Masters*. In Chapter 6 Project A, pp. 45–46, students extend the topic in the chapter opener. In Chapter 6 Project B, pp. 47–48, students investigate the growth of plants under various conditions.

Investigations and Projects Masters, p. 45

NCTM Standards: 1–5, 8

Objective
Use a geoboard to model a line segment and calculate its slope.

Recommended Time
Demonstration and discussion: 15 minutes; Exercises: 30 minutes

Instructional Resources
For each student or group of students
Student Manipulative Kit
• geoboard
• red rubber band, green rubber band, blue or yellow rubber band
Modeling Mathematics Masters
• p. 6 (geoboard pattern)
• p. 26 (worksheet)
For teacher demonstration
Algebra and Geometry Overhead Manipulative Resources

1 FOCUS

Motivating the Lesson
Make certain that the desks in the room are arranged in rows and columns. Select two seated students. Count the number of rows one must walk, then the number of columns one must walk, to get from one student to the other. Do this for several pairs of students.

2 TEACH

Teaching Tip Point out to students that they must be careful to divide the vertical distance by the horizontal distance. Doing the opposite gives an incorrect answer.

3 PRACTICE/APPLY

Assignment Guide
Core: 1–5
Enriched: 1–5

MODELING MATHEMATICS

6-1A Slope

Materials: geoboard   rubber bands

A Preview of Lesson 6–1

The **slope** of a line segment is the ratio of the change in the vertical distance between the endpoints of the segment to the change in the horizontal distance. You can use a geoboard to model a line segment and calculate its slope.

Activity **Model line segment *AB* whose endpoints are *A*(1, 2) and *B*(3, 5). Then find the slope of segment *AB*.**

Step 1 Suppose the pegs of the geoboard represent ordered pairs on a coordinate plane. Let's define the lower left peg as the point representing the ordered pair (1, 1). Locate the pegs that represent (1, 2) and (3, 5). Place a rubber band around these pegs to model segment *AB*.

Step 2 Use a different color rubber band to show the horizontal distance from the *x* value of *A* to the *x* value of *B*. Use another color rubber band to show the vertical distance from the *y* value of *A* to the *y* value of *B*.

Step 3 As you move from point *A* to point *B*, you can go right and then up. These are positive directions on the coordinate plane. So, the red rubber band represents 2 units. The green rubber band represents 3 units. The slope of segment *AB* is the ratio of the *y* distance to the *x* distance or $\frac{3}{2}$.

1. $-\frac{1}{2}$

2. $\frac{2-5}{1-3} = \frac{3}{2}$; $\frac{4-2}{1-5} = \frac{2}{-4}$ or $-\frac{1}{2}$; It is the same ratio as the slope of the segments.

Model 1. Use a geoboard to find the slope of a segment whose endpoints are at (5, 2) and (1, 4).

Write 2. For the activity above and Exercise 1, write ratios that compare the differences in the *y* values of the ordered pairs to the differences in the *x* values of the ordered pairs. How do these compare with the slopes of the respective segments?

3. What value could you assign to the lower left peg if you were modeling segment *CD* with endpoints *C*(3, 5) and *D*(6, 7)? **Sample answer: (3, 3)** **4. See Solutions Manual.**

4. Write a rule for finding the slope of any line segment shown on a coordinate plane.

5. Do you think the rule applies to ordered pairs that involve negative numbers? Explain your answer and give examples. **See Solutions Manual.**

324 *Chapter 6 Analyzing Linear Equations*

---

4 ASSESS

Observing students working in cooperative groups is an excellent method of assessment.

GLENCOE *Technology*

Interactive Mathematics Tools Software

This multimedia software provides an interactive lesson by students observing the effect of the changes of the coefficients of *x*-terms on the slope of the line as well as when the lines have the same slope. A **Computer Journal** gives students an opportunity to write about what they have learned.
For Windows & Macintosh

Slope

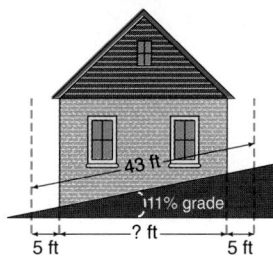

Objective

After studying this lesson, you should be able to:

* find the slope of a line, given the coordinates of two points on the line.

Construction

Alan and Mabel Wong bought a lot in San Francisco on which to build a house. The lot is located on a street having an 11% grade. The length of the sidewalk along the lot is 43 feet. City ordinances state that there must be a 5-foot clearance on each side of the house to the property line. The house they would like to build is 32 feet wide. Will they be able to build their house on the lot they bought? *This problem will be solved in Example 4.*

What do you think of when you hear the word *slope*? You might think of how something slants either uphill or downhill. Suppose you placed a ladder against a wall. Then you moved the base of the ladder out about a foot. What happens to the slant of the ladder? Suppose you move it again. What happens? Observe the pattern in the diagram below.

Notice that the top of the ladder moves down the wall when the bottom of the ladder moves out. As the top moves down the wall, the slant, or slope, becomes less steep. If you continued to move the ladder out, eventually it would lie flat on the ground, and there would be no slant at all.

The steepness of the line representing the ladder is called the **slope** of the line. It is defined as the ratio of the **rise,** or vertical change, to the **run,** or horizontal change, as you move from one point on the line to another. The graph at the right shows a line that passes through the origin and the point at (5, 4).

$$\text{slope} = \frac{\text{change in } y \text{ (rise)}}{\text{change in } x \text{ (run)}}$$

$$= \frac{4}{5}$$

So, the slope of this line is $\frac{4}{5}$.

6-1 LESSON NOTES

NCTM Standards: 1–5, 8

Instructional Resources

* Study Guide Master 6-1
* Practice Master 6-1
* Enrichment Master 6-1
* Multicultural Activity Masters, p. 11
* Real-World Applications, 17

Transparency 6-1A contains the 5-Minute Check for this lesson; **Transparency 6-1B** contains a teaching aid for this lesson.

Recommended Pacing	
Standard Pacing	Day 2 of 13
Honors Pacing	Day 2 of 12
Block Scheduling*	Day 1 of 7
Alg. 1 in Two Years*	Day 2 of 18

*For more information on pacing and possible lesson plans, refer to the *Block Scheduling Booklet* and *Algebra 1 in Two Years.*

1 FOCUS

5-Minute Check
(over Chapter 5)

State the domain and range of each relation.

1. {(5, 2), (0, 0), (−9, −1)}
   Domain = {5, 0, −9};
   Range = {2, 0, −1}
2. {(1, 1), (2, 4), (−2, 4), (3, 9)}
   Domain = {1, 2, −2, 3};
   Range = {1, 4, 9}

Determine whether each relation is a function. Then state the inverse relation and determine whether it is a function.

3. {(3, 1), (5, 1), (7, 1)} yes;
   {(1, 3), (1, 5), (1, 7)}; no
4. {(5, 4), (−6, 5), (4, 5), (0, 4)}
   yes; {(4, 5), (5, −6), (5, 4), (4, 0)}; no
5. If $f(x) = 3x − 5$, determine the value of $f(2)$. 1

Motivating the Lesson

Questioning Introduce this lesson on slope by discussing the pitch of a roof. Ask why the pitch varies in different parts of the country. What determines the pitch of a roof when an architect draws up building plans? Develop the idea that a relationship exists between the amount of precipitation an area generally receives and the roof structures used in the region.

2 TEACH

Teaching Tip In Example 1, point out that a line with a slope of 0 is horizontal (parallel to the *x*-axis) and a line with an undefined slope is vertical (parallel to the *y*-axis).

In-Class Examples

For Example 1
Determine the slope of the line through the origin and the point named.

A	3
B	2
C	1
D	$\frac{1}{2}$
E	$\frac{1}{3}$
F	$-\frac{1}{3}$
G	$-\frac{1}{2}$
H	−1
I	−2
J	−3

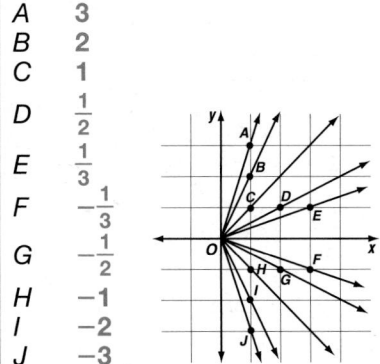

For Example 2
Determine the slope of the line that passes through each pair of points.

a. (−5, 3), (2, 1) $-\frac{2}{7}$
b. (4, 3), (1, −1) $\frac{4}{3}$

Teaching Tip In Example 2, students should be aware that the *y*-coordinates can be subtracted in any order as long as the corresponding *x*-coordinates are subtracted in that same order.

$$\frac{y_1 - y_2}{x_1 - x_2} \text{ or } \frac{y_2 - y_1}{x_2 - x_1}, \text{ not } \frac{y_2 - y_1}{x_1 - x_2}.$$

Definition of Slope	The slope *m* of a line is the ratio of the change in the *y*-coordinates to the corresponding change in the *x*-coordinates.

For simplicity, we will refer to the change in the y-coordinates as the y change and the corresponding change in the x-coordinates as the x change.

Example ❶ Determine the slope of each line.

a.   b.   c.   d.

a. $\frac{y \text{ change}}{x \text{ change}} = \frac{3}{2}$

$m = \frac{3}{2}$

b. $\frac{y \text{ change}}{x \text{ change}} = \frac{4}{-3}$

$m = \frac{4}{-3}$

c. $\frac{y \text{ change}}{x \text{ change}} = \frac{0}{2}$

$m = 0$

d. $\frac{y \text{ change}}{x \text{ change}} = \frac{3}{0}$

Since division by 0 is undefined, the slope is undefined.

Let's analyze the slopes of the graphs in Example 1.
- Graph a slopes upward from left to right and has a positive slope.
- Graph b slopes downward from left to right and has a negative slope.
- Graph c is a horizontal line and has a slope of 0.
- Graph d is a vertical line and its slope is undefined.

These observations are true of other lines that have the same characteristics.

Since a line is made up of an infinite number of points, you can use any two points on that line to find the slope of the line. So, we can generalize the definition of slope for any two points on the line.

Determining Slope Given Two Points	Given the coordinates of two points, (x_1, y_1) and (x_2, y_2), on a line, the slope *m* can be found as follows: $m = \frac{y_2 - y_1}{x_2 - x_1}$, where $x_1 \neq x_2$.

y_2 is read "y sub 2." The 2 is called a <u>subscript</u>.

Example ❷ Determine the slope of the line that passes through (2, −5) and (7, −10).

Method 1	**Method 2**
Let $(2, -5) = (x_1, y_1)$ and $(7, -10) = (x_2, y_2)$.	Let $(7, -10) = (x_1, y_1)$ and $(2, -5) = (x_2, y_2)$.
$m = \frac{y_2 - y_1}{x_2 - x_1}$	$m = \frac{y_2 - y_1}{x_2 - x_1}$
$= \frac{-10 - (-5)}{7 - 2}$	$= \frac{-5 - (-10)}{2 - 7}$
$= \frac{-5}{5}$ or -1	$= \frac{5}{-5}$ or -1

The slope of the line is −1.

As you can see in Example 2, it does not matter which ordered pair is selected as (x_1, y_1).

 Alternative Learning Styles

Kinesthetic Have students determine the slope of their driveway or curb ramp. Ask them to describe the procedure they used in calculating the value of this slope.

You can use a spreadsheet to calculate the slopes of several lines very quickly.

A spreadsheet is a table of cells that can contain text (labels) or numbers and formulas. Each cell is named by the column and row in which it is located. In the spreadsheet below, cells A1, B1, C1, D1, E1, and F1 contain labels.

a. Enter the formula (E2–C2)/(D2–B2) into cell F2.

b. Copy the formula to all of the cells in the F column. When you do, the spreadsheet automatically changes the formula to correspond to the values in that row. That is, for F3 the formula becomes (E3–C3)/(D3–B3), for F4 it becomes (E4–C4)/(D4–B4), and so on.

	===A===	==B===	===C===	==D===	===E===	===F===
1	LINE	X1	Y1	X2	Y2	SLOPE
2	A	2	3	4	6	1.5
3						
4						
5						

Your Turn b. Slopes are given for each line in part a.

a. Enter the following ordered pairs from each line in the spreadsheet.

line a $(5, 5), (11, 11)$ **1**   line e $(-1, 5), (6, 3)$ **−0.28571**
line b $(4, -4), (3, 5)$ **−9**   line f $(5, 3), (4, 0)$ **3**
line c $(6, -1), (4, -1)$ **0**   line g $(2, 3), (11, 14)$ **1.22222**
line d $(11, 3), (1, 1)$ **0.2**   line h $(10, -5), (10, 3)$ **undefined**

b. Use the calculate command to have the spreadsheet compute the slope.

c. Use the ordered pairs to graph each line on grid paper. Label each line.

d. From the results of your spreadsheet, make lists of lines with positive slope, negative slope, and 0 slope. **c−e. See margin.**

e. Compare the lists with the graphs of the lines. Write a statement to summarize your observations.

If you know the slope of a line and the coordinates of one of the points on a line, you can find the coordinates of other points on that line.

Example 3 Determine the value of r so the line through $(r, 6)$ and $(10, -3)$ has a slope of $-\frac{3}{2}$.

$$m = \frac{y_2 - y_1}{x_2 - x_1}$$

$$-\frac{3}{2} = \frac{-3 - 6}{10 - r} \quad (x_1, y_1) = (r, 6)$$
$$\quad\quad\quad\quad\quad (x_2, y_2) = (10, -3)$$

$$-\frac{3}{2} = \frac{-9}{10 - r}$$

$$-3(10 - r) = -9(2) \quad \text{Find the cross products.}$$

$$-30 + 3r = -18 \quad \text{Solve for r.}$$

$$3r = 12$$

$$r = 4$$

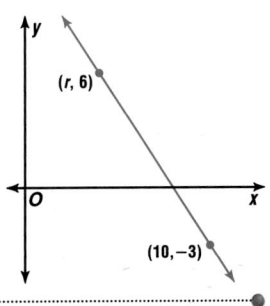

Lesson 6–1 Slope **327**

EXPLORATION

This Exploration illustrates the use of spreadsheet software to calculate a large number of slopes. Point out to students that since a graph is created, input errors are harder to detect.

Answers for the Exploration

c.

d. positive slopes: *a, d, f,* and *g;* negative slopes: *b* and *e;* zero slope: *c;* undefined slope: *h*

e. Sample answer: Positive slope means that the graphs slant upward to the right. Negative slope means that the graphs slant downward to the right. Zero slope means the lines are horizontal.

In-Class Example

For Example 3
Determine the value of *r* so the line through the two given points has the indicated slope.

a. $(-2, 4), (r, 5), m = \frac{1}{5}$ **3**

b. $(3, 4), (-1, r), m = -\frac{3}{4}$ **7**

Teaching Tip Have students graph the line through $(9, -2)$ with slope $-\frac{3}{2}$ to verify that the point $(5, 4)$ is on this line. They can think of slope as directions from one point to another on a line. That is, from $(9, -2)$, go 2 units right and 3 units down to locate another point on the line.

In-Class Example

For Example 4
How many feet does a road with a 4% grade drop in 5 miles? **1056 feet**

3 PRACTICE/APPLY

Check for Understanding

Exercises 1–14 are designed to help you assess your students' understanding through reading, writing, speaking, and modeling. You should work through Exercises 1–6 with your students and then monitor their work on Exercises 7–14.

Additional Answers

Sample answers:

4a.

4b.

4c.

4d.
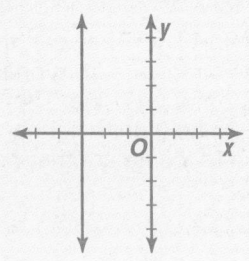

You can use slope with other mathematical skills to solve real-world problems like the one presented at the beginning of this lesson.

Example **4**

APPLICATION
Construction

LOOK BACK

You can refer to Lesson 4-3 for more information on trigonometric functions.

Refer to the application at the beginning of the lesson. Will the Wongs be able to build their house on their lot?

The width of the house and the two 5-foot clearances equal 32 + 2(5) or 42 feet. Thus, the horizontal measure between the property lines must be at least 42 feet.

We can use trigonometric ratios to determine if there is adequate room for the house. A grade of 11 percent means that the rise is 11 feet while the run is 100 feet. We can use this information to find the measure of $\angle A$.

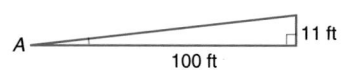

$\tan A = \frac{11}{100}$ ← *opposite*
                          ← *adjacent*

$\tan A = 0.11$

$A \approx 6.3°$ *Use a calculator.*

We can use the measure of $\angle A$ to find the horizontal measure x of the property.

$\cos A = \frac{x}{43}$ ← *adjacent*
                        ← *hypotenuse*

$43(\cos 6.3°) = x$ *Multiply each side by 43.*

$42.74 = x$ *Use a calculator.*

The lot is 42.74 feet wide. The house and clearances will fit the 43-foot lot. So, the Wongs can build their house on the lot they bought.

CHECK FOR UNDERSTANDING

Communicating Mathematics

1. From one point to another, go down 3 units and right 5 units. The slope is $-\frac{3}{5}$.

2. It means that as you travel 100 ft horizontally, your altitude increases 8 ft.

3. Yes, it would affect the sign of the slope.

MODELING MATHEMATICS

Study the lesson. Then complete the following.

1. **Explain** how you would find the slope of the line in the graph at the right.

2. **Describe** what it means if a road has an 8% grade.

3. Would it matter if you subtracted the x-coordinates in the opposite order from the way you subtracted the y-coordinates?

4. **Draw** the graph of a line having each slope.
   a. a positive slope
   b. a negative slope
   c. a slope of 0
   d. undefined slope   a–d. See margin.

5. **Explain** why the formula for determining the slope using two points does not apply to vertical lines.   The difference in the x values is always 0, and division by 0 is undefined.

6. Use a geoboard to model line segments with each slope. Make a sketch of your model.   a–e. See margin.
   a. 4   **b.** $\frac{3}{4}$ **c.** 0   **d.** undefined   **e.** $-\frac{3}{2}$

Reteaching

Using Alternative Methods Review the concept of slope by looking at four lines that pass through the origin: (1) a horizontal line, (2) a line that goes upward from left to right, (3) a vertical line, and (4) a line that goes downward from left to right. Emphasize the signs of the slopes.

Guided Practice

Determine the slope of each line.

7. $\frac{5}{3}$

8. 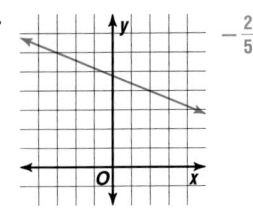 $-\frac{2}{5}$

Determine the slope of the line that passes through each pair of points.

9. $(7, -4), (9, -1)$ $\frac{3}{2}$

10. $(5, 7), (-2, -3)$ $\frac{10}{7}$

11. $(0.75, 1), (0.75, -1)$ undefined

Determine the value of r so the line that passes through each pair of points has the given slope.

12. $(6, -2), (r, -6), m = -4$ 7

13. $(9, r), (6, 3), m = -\frac{1}{3}$ 2

14a. Sample answer: rise, 23 mm; run, 15 mm

14. **Architecture** The slope of the roof line of a building is often referred to as the *pitch* of the roof.

a. Use a millimeter ruler to find the rise and run of the roof of the tower on the horse barn in Versailles, Kentucky, shown at the right.

b. Then write the pitch as a decimal. based on sample answer, about 1.5

EXERCISES

Practice

A

Determine the slope of each line.

15. a

16. b

17. c 0

18. d undefined

19. e 1

20. f -6

15. $-\frac{1}{5}$

16. $\frac{1}{4}$

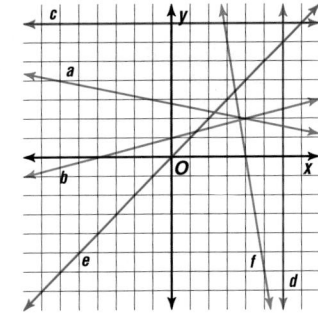

Determine the slope of the line that passes through each pair of points.

B

21. $(2, 3), (9, 7)$ $\frac{4}{7}$

22. $(-3, -4), (5, -1)$ $\frac{3}{8}$

23. $(2, -1), (5, -3)$ $-\frac{2}{3}$

24. $(2, 6), (-1, 3)$ 1

25. $(-5, 4), (-5, -1)$

26. $(-2, 3), (8, 3)$ 0

27. $(4, -5), (4, 2)$

28. $\left(2\frac{1}{2}, -1\frac{1}{2}\right), \left(-\frac{1}{2}, \frac{1}{2}\right)$

29. $\left(\frac{3}{4}, 1\frac{1}{4}\right), \left(-\frac{1}{2}, -1\right)$ $\frac{9}{5}$

25. undefined

27. undefined

28. $-\frac{2}{3}$

Additional Answers

6c.

6d.

6e.

Assignment Guide

Core: 15–39 odd, 40, 41, 43, 45–52
Enriched: 16–38 even, 40–52

For **Extra Practice**, see p. 769.

The red A, B, and C flags, printed only in the Teacher's Wraparound Edition, indicate the level of difficulty of the exercises.

Additional Answers

6a.

6b.

Study Guide Masters, p. 41

6-1 NAME_____ DATE_____
Study Guide Student Edition Pages 325–331

Slope

The ratio of *rise* to *run* is called **slope**. The slope of a line describes its steepness, or rate of change.

Definition of Slope
The slope m of a line is the ratio of the change in the y-coordinates to the corresponding change in the x-coordinates.
Slope $= \frac{\text{change in } y}{\text{change in } x}$ or $m = \frac{\text{change in } y}{\text{change in } x}$

On a coordinate plane, a line extending from lower left to upper right has a positive slope. A line extending from upper left to lower right has a negative slope. The slope of a horizontal line is zero. A vertical line has *no slope*.

The slope of a nonvertical line can be determined from the coordinates of any two points on the line.

Determining Slope Given Two Points
Given the coordinates of two points, (x_1, y_1) and (x_2, y_2), on a line, the slope m can be found as follows:
$m = \frac{y_2 - y_1}{x_2 - x_1}$, where $x_1 \neq x_2$.

Example: Determine the slope of the line that passes through $(-1, 5)$ and $(4, -2)$.

$m = \frac{y_2 - y_1}{x_2 - x_1}$

$= \frac{-2 - 5}{4 - (-1)}$

$= \frac{-7}{5} = -\frac{7}{5}$

Determine the slope of the line that passes through each pair of points.

1. $(2, 1), (8, 9)$ $\frac{4}{3}$

2. $(4, 9), (1, 6)$ 1

3. $(7, -8), (14, -6)$ $\frac{2}{7}$

4. $(-10, 7), (-20, 8)$ $-\frac{1}{10}$

5. $(3, 11), (-12, 18)$ $-\frac{7}{15}$

6. $(-4, -1), (-2, -5)$ -2

Determine the value of r so the line that passes through each pair of points has the given slope.

7. $(10, r), (3, 4), m = -\frac{2}{7}$ 2

8. $(-1, -3), (7, r), m = \frac{3}{4}$ 3

9. $(-2, r), (10, 4), m = -\frac{1}{2}$ 10

10. $(12, r), (r, 6), m = 2$ 10

11. $(6, 8), (r, -2), m = -3$ $\frac{28}{3}$

12. $(r, 9), (7, 5), m = 6$ $\frac{23}{3}$

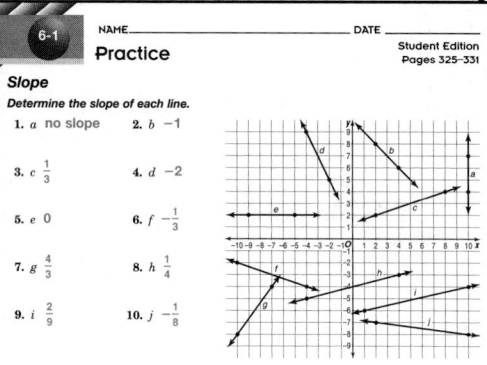
Determine the value of *r* so the line that passes through each pair of points has the given slope.

30. (5, r), (2, −3), m = $\frac{4}{3}$ 1
31. (−2, 7), (r, 3), m = $\frac{4}{3}$ −5
32. (4, −5), (3, r), m = 8 −13
33. (6, 2), (9, r), m = −1 −1
34. (4, r), (r, 2), m = $-\frac{5}{3}$ 7
35. (r, 5), (−2, r), m = $-\frac{2}{9}$ 7

Draw a line through the given point that has the given slope.

36–38. See margin.

36. (3, −1), m = $\frac{1}{3}$
37. (−2, −3), m = $\frac{4}{3}$
38. (4, −2), m = $-\frac{2}{5}$

39. The points A(12, −4) and B(6, 8) lie on a line. Find the coordinates of a third point that lies on line AB. Describe how you determined the coordinates of this point. **See margin.**

Critical Thinking

40. Carpentry A carpenter was a member of a crew building a roof that is 30 feet long at the base. The roof had a pitch of 4 inches for every foot of length along the base. His task on the crew was to put in vertical supports every 16 inches along the base. He would climb up the ladder, measure 16 inches horizontally, and then measure the vertical height to the roof. He then climbed down the ladder, sawed the piece he needed, and went back up the ladder to put it in place. He wondered if there wasn't some way he could figure out how long each support would be ahead of time so he wouldn't have to climb the ladder so many times. Explain how you can use what you have learned in this chapter to help him out. **See margin.**

30 ft

Applications and Problem Solving

41. about 1477 feet

41. Road Construction The Castaic Grade in Southern California has a grade of 5%. The length of the roadway is 5.6 miles. What is the change in elevation from the top of the grade to the bottom of the grade in feet?

42. Carpentry Julio Mendez is a carpenter. He is building a staircase between the first and second floors of a house, a height of 9 feet. The *tread*, or depth of each step, must be 10 inches. The slope of the staircase cannot exceed $\frac{3}{4}$.

riser
tread
9 ft

a. How many steps should he plan? **15 steps**
b. What is the measure of the riser, or height of each step? **7.2 in.**

43. Driving The Eisenhower Tunnel in Colorado was completed in 1973 and is the fourth longest road tunnel in the United States. The eastern entrance of the Eisenhower Tunnel is at an elevation of 11,080 feet. The tunnel is 8941 feet long and has an upgrade of 0.895% toward the western end. What is the elevation of the western end of the tunnel? **11,160 feet**

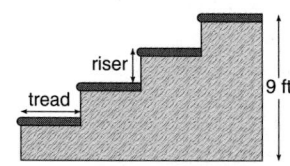

Longest Road Tunnels in the World

Tunnel	Length (miles)
1. St. Gotthard (Switzerland)	10.14
2. Arlberg (Austria)	8.69
3. Fréjus (France/Italy)	8.02
4. Mont-Blanc (France/Italy)	7.21
5. Gudvangen (Norway)	7.08

The very first tunnels were built by cave dwellers, who cut through clay and soft rock to connect adjacent caves.

Extension

Problem Solving The highway approaching Castaic Lake comes down a 5% grade for a distance of 5.2 miles. If the elevation of the lake is 1440 feet, calculate the drop from the top of the grade to the lake. What is the elevation of the top of the grade? The drop is about 1373 feet. The top of the grade is about 2813 feet.

44. Architecture Use a millimeter ruler to estimate the pitch, or slope, of each object.

a.

Uro Indian huts of Peru
1

b.

S. Miniato, Florence, Italy
$\frac{3}{2}$

c.

Great Pyramid at Giza, Egypt
1

Mixed Review

45. Statistics Find the range, median, upper and lower quartiles, and interquartile range for the set of data. (Lesson 5–7)
3, 3.2, 6, 45, 7, 26, 2, 3.4, 4, 5.3, 5, 78, 8, 1, 5 **77; 5; 8; 3.2; 4.8**

46. Patterns Copy and complete the table below. (Lesson 5–5)

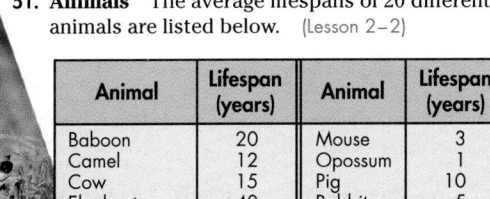

n	1	2	3	4	5
f(n)	14	13	12	11	10

47. Use the formula $I = prt$ to find r if $I = \$2430$, $p = \$9000$, and $t = 2$ years 3 months. (Lesson 4–5) **12%**

48. Solve $\frac{6}{14} = \frac{7}{x-3}$. (Lesson 4–1) $\frac{58}{3}$

Solve each equation. Check your solution.

49. $\frac{2}{3}x + 5 = \frac{1}{2}x + 4$ (Lesson 3–5) **−6** **50.** $3x − 8 = 22$ (Lesson 3–3) **10**

51. Animals The average lifespans of 20 different animals are listed below. (Lesson 2–2)

Animal	Lifespan (years)	Animal	Lifespan (years)
Baboon	20	Mouse	3
Camel	12	Opossum	1
Cow	15	Pig	10
Elephant	40	Rabbit	5
Fox	7	Sea Lion	12
Gorilla	20	Sheep	12
Hippopotamus	25	Squirrel	10
Kangaroo	7	Tiger	16
Lion	15	Wolf	5
Monkey	15	Zebra	15

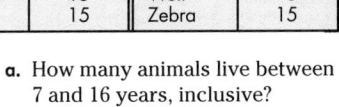

a. How many animals live between 7 and 16 years, inclusive?

b. Make a line plot of the average lifespans of the animals from part a. **See margin.**

c. Which number occurred most frequently? **15 years**

d. How many animals lived at least 20 years? **4 animals**

51a. **12 animals**

52. Evaluate $29 − 3(9 − 4)$. (Lesson 1–3) **14**

Lesson 6–1 Slope **331**

Additional Answer

51b.

Closing Activity

Speaking Ask students to explain how they would determine the slope of a line containing two given points.

Additional Answer

40. The slope is $\frac{120\text{ in.}}{30\text{ ft}} = \frac{10\text{ ft}}{30\text{ ft}}$ or $\frac{1}{3}$. Use the slope to figure the length of the support.

distance (in.)	16	32	64	96
length (in.)	$\frac{16}{3}$	$\frac{32}{3}$	$\frac{64}{3}$	$\frac{96}{3}$

Enrichment Masters, p. 41

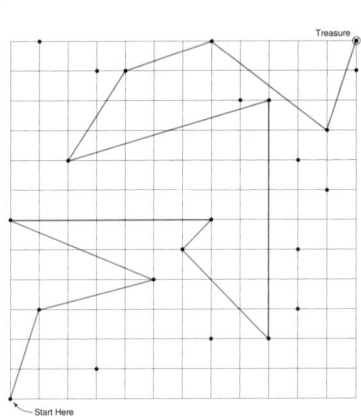
Chapter 6 **331**

Instructional Resources

- Study Guide Master 6-2
- Practice Master 6-2
- Enrichment Master 6-2
- Assessment and Evaluation Masters, p. 156

Transparency 6-2A contains the 5-Minute Check for this lesson; **Transparency 6-2B** contains a teaching aid for this lesson.

Recommended Pacing	
Standard Pacing	Days 3 & 4 of 13
Honors Pacing	Day 3 of 12
Block Scheduling*	Day 2 of 7
Alg. 1 in Two Years*	Days 3 & 4 of 18

*For more information on pacing and possible lesson plans, refer to the *Block Scheduling Booklet* and *Algebra 1 in Two Years.*

1 FOCUS

5-Minute Check
(over Lesson 6-1)

Determine the slope of the line that passes through each pair of points.

1. $(5, -2), (-2, 1)$ $-\frac{3}{7}$

2. $(-3, -8), (4, -2)$ $\frac{6}{7}$

3. $(8, 3), (-1, 3)$ 0

Find r so the given line through each pair of points has the given slope.

4. $(5, r), (-2, 4), m = -1$ -3

5. $(2, 6), (r, -3), m = \frac{3}{4}$ -10

Objectives

After studying this lesson, you should be able to:

- write linear equations in point-slope form, and
- write linear equations in standard form.

6-2

Writing Linear Equations in Point-Slope and Standard Forms

CONNECTION
Geography

If you lived in Miami, Florida, and moved to Denver, Colorado, what adjustments do you think you might have to make? Obviously, the weather is different, but did you know you would also have to adjust to living at a higher altitude? As the altitude increases, the oxygen in the air decreases. This can affect your breathing and cause dizziness, headache, insomnia, and loss of appetite.

Over time, people who move to higher elevations experience acclimatization, or the process of getting accustomed to the new climate. Their bodies develop more red blood cells to carry oxygen to the muscles. Generally, long-term acclimatization to an altitude of 7000 feet takes about 2 weeks. After that, it takes 1 week for each additional 2000 feet of altitude. The graph at the right shows the acclimatization for altitudes greater than 7000 feet.

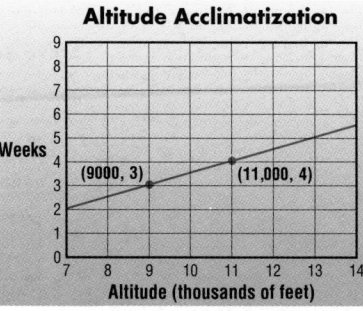
Altitude Acclimatization

We can use any two points on the graph to find the slope of the line. For example, let $(9000, 3)$ and $(11,000, 4)$ represent (x_1, y_1) and (x_2, y_2), respectively.

$$m = \frac{y_2 - y_1}{x_2 - x_1}$$

$$= \frac{4 - 3}{11,000 - 9000} \text{ or } \frac{1}{2000}$$

Suppose we let (x, y) represent any other point on the line. We can use the slope and one of the given ordered pairs to write an equation for the line.

$$m = \frac{y_2 - y_1}{x_2 - x_1}$$

$$\frac{1}{2000} = \frac{y - 3}{x - 9000} \quad m = \frac{1}{2000}, (x_1, y_1) = (9000, 3), (x_2, y_2) = (x, y)$$

$$\frac{1}{2000}(x - 9000) = y - 3 \quad \textit{Multiply each side by } (x - 9000).$$

$$y - 3 = \frac{1}{2000}(x - 9000) \quad \textit{Reflexive property } (=)$$

Since this form of the equation was generated using the coordinates of a known point and the slope of the line, it is called the **point-slope form.**

Point-Slope Form of a Linear Equation	For a given point (x_1, y_1) on a nonvertical line having slope m, the point-slope form of a linear equation is as follows. $$y - y_1 = m(x - x_1)$$

You can write an equation in point-slope form for the graph of any nonvertical line if you know the slope of the line and the coordinates of one point on that line.

Example Write the point-slope form of an equation for each line.

a. a line that passes through $(-3, 5)$ and has a slope of $-\frac{3}{4}$

$$y - y_1 = m(x - x_1)$$
$$y - 5 = -\frac{3}{4}[x - (-3)] \quad \textit{Replace } x_1 \textit{ with } -3,$$
$$\textit{y}_1 \textit{ with 5, and m with } -\frac{3}{4}.$$
$$y - 5 = -\frac{3}{4}(x + 3)$$

An equation of the line is $y - 5 = -\frac{3}{4}(x + 3)$.

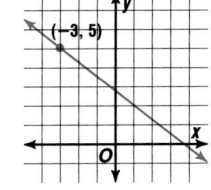

b. a horizontal line that passes through $(-6, 2)$

Horizontal lines have a slope of 0.

$$y - y_1 = m(x - x_1)$$
$$y - 2 = 0[x - (-6)]$$
$$y - 2 = 0$$
$$y = 2$$

An equation of the line is $y = 2$.

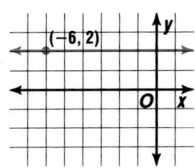

You can also write the equations of horizontal lines by using the y-coordinate as in Example 1b.

A vertical line has a slope that is undefined, so you cannot use the point-slope form of an equation. However, you can write the equation of a vertical line by using the coordinates of the points through which it passes. Suppose the line passes through $(3, 5)$ and $(3, -2)$. The equation of the line is $x = 3$, since the x-coordinate of every point on the line is 3.

Any linear equation can also be expressed in the form $Ax + By = C$, where A, B, and C are integers and A and B are not both zero. This is called the **standard form** of a linear equation. *Usually, $A > 0$.*

Standard Form of a Linear Equation	The standard form of a linear equation is $Ax + By = C$, where A, B, and C are integers, $A \geq 0$, and A and B are not both zero.

Lesson 6–2 Writing Linear Equations in Point-Slope and Standard Forms **333**

Motivating the Lesson
Hands-On Activity Students need to develop an understanding of approximate slopes of lines. Have them make a paper cut-out of a straight line. Draw a grid on paper and place the cut-out in various positions. Students should estimate a reasonable slope for each position.

2 TEACH

In-Class Example

For Example 1
Write the point-slope form of an equation of the line that passes through the given point and has the given slope.

a. $(3, 5)$, $\frac{4}{3}$ $y - 5 = \frac{4}{3}(x - 3)$

b. $(-2, 0)$, $-\frac{3}{2}$ $y = -\frac{3}{2}(x + 2)$

c. $(-3, 2)$, $-\frac{1}{2}$
$y - 2 = -\frac{1}{2}(x + 3)$

d. $(0, 5)$, -3 $y - 5 = -3x$

Teaching Tip The point-slope form of a linear equation is not unique. This form is used as an intermediate step when finding the standard form or slope-intercept form. Ask students why the definition of the point-slope form includes the phrase "a nonvertical line."

Alternative Teaching Strategies

Student Diversity Have students make a chart with the following headings: *Standard Form, Point-Slope Form, Slope, x-Intercept, y-Intercept, Points.* On the first line, have them write $3x - 2y = 12$ under *Standard Form.* On the second line, have them write 4 under *x-Intercept* and -1 under *y-Intercept.* On the third line, write $(4, 1)$ and $(-3, 5)$ under *Points.* Then have students complete the chart.

334 *Chapter 6*

Linear equations that are written in point-slope form can be rewritten in standard form.

Example Write $y + 5 = -\frac{5}{4}(x - 2)$ in standard form.

$$y + 5 = -\frac{5}{4}(x - 2)$$

$4(y + 5) = 4 \cdot \left(-\frac{5}{4}\right)(x - 2)$ *Multiply each side by 4 to eliminate the fraction.*

$4y + 20 = -5x + 10$ *Distributive property*

$4y = -5x - 10$ *Subtract 20 from each side.*

$5x + 4y = -10$ *Add 5x to each side.*

The standard form of the equation is $5x + 4y = -10$.

You can write an equation of a line if you know the coordinates of two points on that line.

Example ③ **Write the point-slope form and the standard form of an equation of the line that passes through $(-8, 3)$ and $(4, 5)$.**

First determine the slope of the line.

$m = \frac{5 - 3}{4 - (-8)}$ or $\frac{1}{6}$

You can use either point for (x_1, y_1) in the point-slope form.

Method 1: Use $(-8, 3)$.

$y - 3 = \frac{1}{6}(x - (-8))$

$y - 3 = \frac{1}{6}(x + 8)$

Method 2: Use $(4, 5)$.

$y - 5 = \frac{1}{6}(x - 4)$

Both $y - 3 = \frac{1}{6}(x + 8)$ and $y - 5 = \frac{1}{6}(x - 4)$ are valid equations for the point-slope form of an equation for the line passing through $(-8, 3)$ and $(4, 5)$. Now let's find the standard form of each equation.

Method 1

$y - 3 = \frac{1}{6}(x + 8)$

$6(y - 3) = 6 \cdot \frac{1}{6}(x + 8)$ *Multiply each side by 6.*

$6y - 18 = x + 8$

$6y = x + 26$ *Add 18 to each side.*

$-x + 6y = 26$ *Subtract x from each side.*

$x - 6y = -26$ *Multiply by −1.*

Method 2

$y - 5 = \frac{1}{6}(x - 4)$

$6(y - 5) = 6 \cdot \frac{1}{6}(x - 4)$ *Multiply each side by 6.*

$6y - 30 = x - 4$

$6y = x + 26$ *Add 30 to each side.*

$-x + 6y = 26$ *Subtract x from each side.*

$x - 6y = -26$ *Multiply by −1.*

Regardless of which point was used to find the point-slope form, the standard form of the equation, when all common factors are removed, always results in the same equation.

Example **4**

Write the equations of the lines containing the sides of parallelogram ABCD in point-slope and standard forms.

Geometry

Since the lines containing segments AB and CD are horizontal, the slope of each line is 0.

The equation of $\overleftrightarrow{AB}$ is $y = 3$.

The equation of $\overleftrightarrow{CD}$ is $y = -1$.

Now use the coordinates of the points to find the slopes and equations of the lines containing the other two sides.

Use $A(-1, 3)$ and $D(-3, -1)$ to find the slope of $\overline{AD}$.

$$m = \frac{3 - (-1)}{-1 - (-3)}$$

$$= \frac{4}{2} \text{ or } 2$$

Let's use $A(-1, 3)$ for (x_1, y_1) in the point-slope form.

$$y - y_1 = m(x - x_1)$$
$$y - 3 = 2[x - (-1)]$$
$$y - 3 = 2(x + 1)$$

Now find the standard form of this equation.

$$y - 3 = 2(x + 1)$$
$$y - 3 = 2x + 2$$
$$y = 2x + 5$$
$$2x - y = -5$$

Use $B(6, 3)$ and $C(4, -1)$ to find the slope of $\overline{BC}$.

$$m = \frac{3 - (-1)}{6 - 4}$$

$$= \frac{4}{2} \text{ or } 2$$

Let's use $B(6, 3)$ for (x_1, y_1) in the point-slope form.

$$y - y_1 = m(x - x_1)$$
$$y - 3 = 2(x - 6)$$

Now find the standard form of this equation.

$$y - 3 = 2(x - 6)$$
$$y - 3 = 2x - 12$$
$$y = 2x - 9$$
$$2x - y = 9$$

The equations of the lines containing the sides of the parallelogram $ABCD$ are $y = 3$, $y = -1$, $2x - y = -5$, and $2x - y = 9$.

CHECK FOR UNDERSTANDING

Communicating Mathematics

1. x_1 and y_1 are the coordinates of a point through which the line passes.

2. $x = 6$ is a vertical line and $y = 6$ is a horizontal line.

Study the lesson. Then complete the following.

1. **Explain** what x_1 and y_1 in the slope-intercept form of an equation represent.

2. **Compare and contrast** the graphs of $x = 6$ and $y = 6$.

3. **Explain** why A and B in the standard form of a linear equation cannot both be zero.   There would be no variable in the equation.

4. **You Decide**   A line contains the points $P(-3, 2)$ and $Q(2, 5)$. Elena thinks that $R(12, 12)$ is also on the line. Use the point-slope form to determine if she is correct.   See margin.

Guided Practice

7. 3; $(-7, 1)$

State the slope and the coordinates of a point through which the line represented by each equation passes.   5. 4; $(2, -3)$ 6. $-\frac{2}{3}$; $(-5, -6)$

5. $y + 3 = 4(x - 2)$ 6. $y - (-6) = -\frac{2}{3}(x + 5)$ 7. $3(x + 7) = y - 1$

Reteaching

Using Cooperative Groups Have students name two points and then graph them. Determine the slope of the line through the points from the graph. Have half the class use one of the points and the slope to find the standard form of the equation and the other half of the class use the other point and the slope to find the standard form of the equation.

In-Class Examples

For Example 4
Write the equations of the lines containing the sides of trapezoid *ABCD* in point-slope and standard forms.

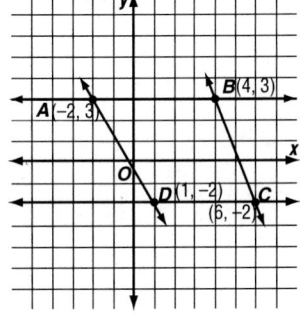

$y = 3, y = -2$

$y - 3 = -\frac{5}{3}(x + 2),$

$5x + 3y = -1$

$y - 3 = -\frac{5}{2}(x - 4),$

$5x + 2y = 26$

3 PRACTICE/APPLY

Check for Understanding

Exercises 1–17 are designed to help you assess your students' understanding through reading, writing, speaking, and modeling. You should work through Exercises 1–4 with your students and then monitor their work on Exercises 5–17.

Error Analysis

Have students write their equation in exact point-slope form to alleviate the problems of using the wrong signs and forgetting about any zeros.
Example: $y + 1 = -2x$ should be rewritten as $y - (-1) = -2(x - 0)$.

Additional Answer

4. slope of $\overleftrightarrow{PQ} = \frac{5 - 2}{2 - (-3)} = \frac{3}{5}$.

An equation of the line is

$y - 5 = \frac{3}{5}(x - 2)$. Try point R in this equation.

$$y - 5 = \frac{3}{5}(x - 2)$$
$$12 - 5 = \frac{3}{5}(12 - 2)$$
$$7 \neq 6$$

Point R does not lie on $\overleftrightarrow{PQ}$.

For **Extra Practice,** see p. 769.

The red A, B, and C flags, printed only in the Teacher's Wraparound Edition, indicate the level of difficulty of the exercises.

Additional Answers

33. $y - 2 = -\frac{1}{3}(x + 5)$ or

$y + 1 = -\frac{1}{3}(x - 4)$

34. $y - 1 = -5(x - 6)$ or

$y + 4 = -5(x - 7)$

35. $y + 1 = \frac{3}{7}(x + 8)$ or

$y - 5 = \frac{3}{7}(x - 6)$

36. $y - 3 = -\frac{2}{3}(x - 2)$ or

$y - 1 = -\frac{2}{3}(x - 5)$

37. $y + 2 = 0(x - 4)$ or

$y + 2 = 0 \, (x - 8)$

38. $y - 3 = 2.5(x - 2.5)$ or

$y + 4.5 = 2.5(x + 0.5)$

Study Guide Masters, p. 42

NAME_____ DATE _____

6-2

Study Guide

Student Edition
Pages 332–338

Writing Linear Equations in Point-Slope and Standard Forms

If you know the slope of a line and the coordinates of one point on the line, you can write an equation of the line by using the **point-slope form.** For a given point (x_1, y_1) on a nonvertical line with slope m, the point-slope form of a linear equation is $y - y_1 = m(x - x_1)$.

Any linear equation can be expressed in the form $Ax + By = C$ where A, B, and C are integers and A and B are not both zero. This is called the **standard form.** An equation that is written in point-slope form can be changed to standard form.

Example 1: Write the point-slope form of an equation of the line that passes through $(6, 1)$ and has a slope of $-\frac{5}{2}$.

$y - y_1 = m(x - x_1)$
$y - 1 = -\frac{5}{2}(x - 6)$

Example 2: Write $y + 5 = 3(x - 4)$ in standard form.

$y + 5 = 3(x - 4)$
$y + 5 = 3x - 12$
$-3x + y = -17$
$3x - y = 17$

You can also find an equation of a line if you know the coordinates of two points on the line. First, find the slope of the line. Then write an equation of the line by using the point-slope form or the standard form.

Write the standard form of an equation of the line that passes through the given point and has the given slope.

1. $(2, 1), 4$
$4x - y = 7$

2. $(-7, 2), 6$
$6x - y = -44$

3. $(\frac{1}{2}, 3), 5$
$10x - 2y = -1$

4. $(4, 9), \frac{3}{4}$
$3x - 4y = -24$

5. $(-6, 7), 0$
$y = 7$

6. $(8, 3), 1$
$x - y = 5$

Write the point-slope form of an equation of the line that passes through each pair of points.

7. $(6, 3), (-8, 5)$
$y - 3 = -\frac{1}{7}(x - 6)$

8. $(-1, 9), (10, 7)$
$y - 7 = -\frac{2}{11}(x - 10)$

9. $(8, 5), (0, -4)$
$y - 5 = \frac{9}{8}(x - 8)$

10. $(-3, -4), (5, -6)$
$y + 6 = -\frac{1}{4}(x - 5)$

11. $(2, 9), (9, 2)$
$y - 2 = -(x - 9)$

12. $(-1, -4), (-6, -10)$
$y + 10 = \frac{6}{5}(x + 6)$

Write the point-slope form of an equation of the line that passes through the given point and has the given slope.

8. $(9, 1), m = \frac{2}{3}$

$y - 1 = \frac{2}{3}(x - 9)$

9. $(-2, 4), m = -3$

$y - 4 = -3(x + 2)$

10. $(-3, 6), m = 0$ $y = 6$

Write each equation in standard form.

11. $y + 3 = -\frac{3}{4}(x - 1)$
$3x + 4y = -9$

12. $y - \frac{1}{2} = \frac{5}{6}(x + 2)$
$5x - 6y = -13$

13. $y - 3 = 2(x + 1.5)$
$2x - y = -6$

Write the point-slope form and the standard form of an equation of the line that passes through each pair of points.

14. $(-6, 1), (-8, 2)$

15. $(-1, -2), (-8, 2)$

16. $(4, 8), (-2.5, 8)$

14. $y - 1 = -\frac{1}{2}(x + 6)$ or $y - 2 = -\frac{1}{2}(x + 8)$; $x + 2y = -4$

17. **Geometry** Write the standard form of the equation of the line containing the hypotenuse of the right triangle shown at the right.
$3x - 5y = -2$

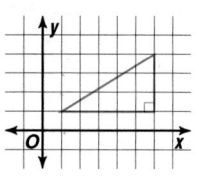

15. $y + 2 = -\frac{4}{7}(x + 1)$ or $y - 2 = -\frac{4}{7}(x + 8)$;

$4x + 7y = -18$

16. $y - 8 = 0$; $y = 8$

EXERCISES

Practice

18. $y - 8 = 2(x - 3)$
19. $y - 5 = 3(x - 4)$
20. $y + 3 = x + 4$
21. $y - 1 = -4(x + 6)$
22. $y - 5 = 0$
23. $y - 3 = -2(x - 1)$
24. $y - 5 = \frac{2}{3}(x - 3)$
25. $y + 3 = \frac{3}{4}(x - 8)$
26. $y - 3 = -\frac{2}{3}(x + 6)$

Write the point-slope form of an equation of the line that passes through the given point and has the given slope.

18. $(3, 8), m = 2$

19. $(4, 5), m = 3$

20. $(-4, -3), m = 1$

21. $(-6, 1), m = -4$

22. $(0, 5), m = 0$

23. $(1, 3), m = -2$

24. $(3, 5), m = \frac{2}{3}$

25. $(8, -3), m = \frac{3}{4}$

26. $(-6, 3), m = -\frac{2}{3}$

Write the standard form of an equation of the line that passes through the given point and has the given slope.

27. $(2, 13), m = 4$ $4x - y = -5$

28. $(-5, -3), m = 4$ $4x - y = -17$

29. $(-4, 6), m = \frac{3}{2}$ $3x - 2y = -24$

30. $(-2, -7), m = 0$ $y = -7$

31. $(8, 2), m = -\frac{2}{5}$ $2x + 5y = 26$

32. $(-5, 5), m = $ undefined $x = -5$

Write the point-slope form of an equation of the line that passes through each pair of points. 33–38. See margin.

33. $(-5, 2), (4, -1)$

34. $(6, 1), (7, -4)$

35. $(-8, -1), (6, 5)$

36. $(2, 3), (5, 1)$

37. $(4, -2), (8, -2)$

38. $(2.5, 3), (-0.5, -4.5)$

Write the standard form of an equation of the line that passes through each pair of points.

39. $(6, 5), (12, -3)$ $4x + 3y = 39$
40. $(-2, -7), (1, 2)$ $3x - y = 1$

42. $4x + y = 1.5$

41. $(-5, 9), (3, -2)$ $11x + 8y = 17$
42. $(0.7, -1.3), (-0.4, 3.1)$

43. $36x - 102y = 61$

43. $\left(-\frac{2}{3}, -\frac{5}{6}\right), \left(\frac{3}{4}, -\frac{1}{3}\right)$
44. $\left(-2, 7\frac{2}{3}\right), \left(-2, \frac{16}{5}\right)$ $x = -2$

Critical Thinking

45. A line contains the points at $(9, 1)$ and $(5, 5)$. Write a convincing argument that the same line intersects the x-axis at $(10, 0)$.   **See margin.**

Applications and Problem Solving

46b. $7x - 2y = -22$;
$7x + 4y = 2$; $y = -3$

46. Geometry Three lines intersect to form triangle ABC, as shown at the right.

a. Write the equation for each line in point-slope form.
b. Write the equation for each line in standard form.

46a. AC: $y - 4 = \frac{7}{2}(x + 2)$ or $y + 3 = \frac{7}{2}(x + 4)$

AB: $y - 4 = -\frac{7}{4}(x + 2)$ or $y + 3 = -\frac{7}{4}(x - 2)$

BC: $y + 3 = 0$

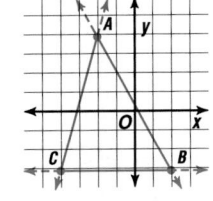

47. Draw a Diagram The Americans with Disabilities Act (ADA) of 1990 states that ramps should have at least a 12-inch run for each rise of 1 inch, with a maximum rise of 30 inches. The post office in Meyersville is changing one of its entrances to accommodate a wheelchair ramp. The distance from the street to the building is 18 feet. The current sidewalk has steps up to the entrance. The entranceway is 30 inches above ground level.

47a. No, for a rise of 30 inches the ramp must be 30 feet long, but there is only 18 feet available.

a. Determine if a wheelchair ramp that meets ADA standards can be placed over the existing sidewalk. Explain why or why not.
b. Draw a diagram of an alternative plan for the ramp.   **See margin.**

Mixed Review

48. Aviation An airplane flying over Albuquerque at an elevation of 33,000 feet begins its descent to land at Santa Fe, 50 miles away. If the elevation of Santa Fe is 7000 feet, what should be the approximate slope of descent, expressed as a percent? (Lesson 6–1)   **9.85%**

49. Statistics The stem-and-leaf plot at the right represents the cost per cup of various brands of coffee. (Lesson 5–7)

a. Find the range and interquartile range for the costs.   **$0.26, $0.09**
b. Identify any outliers.   **$0.32**

Stem	Leaf
0	6 6 6 6 8 9 9 9 9
1	0 2 3 4 5 7 7 8 8
2	4 8 9
3	0 2 $1|2 = \$0.12$

50. Draw a mapping for $\{(-6, 0), (-1, 2), (-3, 4)\}$. (Lesson 5–2)
See margin.
51. If $y = 12$ when $x = 3$, find y when $x = 7$. Assume that y varies directly as x. (Lesson 4–8)   **28**

Extension

Connections The vertices of a triangle are $A(2, 1)$, $B(-3, 4)$, and $C(-5, -2)$. Write the standard form of the equations of the lines that form the sides of the triangle.

side $\overline{AB}$: $3x + 5y = 11$
side $\overline{AC}$: $3x - 7y = -1$
side $\overline{BC}$: $3x - y = -13$

Closing Activity

Writing Place several lines on a coordinate grid and have the students estimate their slopes and *y*-intercepts. Have them write a linear equation in point-slope form for each line.

Chapter 6, Quiz A (Lessons 6-1 and 6-2), is available in the *Assessment and Evaluation Masters*, p. 156.

Enrichment Masters, p. 42

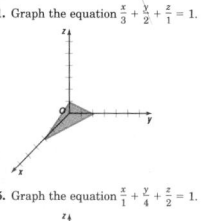

52. **Motion** At 1:30 P.M., an airplane leaves Tucson for Baltimore, a distance of 2240 miles. The plane flies at 280 miles per hour. A second airplane leaves Tucson at 2:15 P.M., and is scheduled to land in Baltimore 15 minutes before the first airplane. At what rate must the second airplane travel to arrive on schedule? (Lesson 4–7) **320 mph**

53. No; because there could be 2 pink and 1 white or 1 pink and 2 white.

53. **Work Backward** In Lupita's garden, all of the flowers are either pink, yellow, or white. Given any three of the flowers, at least one of them is pink. Given any three of the flowers, at least one of them is white. Can you say that given any three of the flowers, at least one of them must be yellow? Why? (Lesson 3–3)

54. **Printing** A customer orders a job requiring 2500 sheets of paper. He has a choice of ordering a full carton of 3000 sheets at $27.50 per thousand or "breaking" a carton (ordering exactly the number of sheets required) and paying $38.40 per thousand. Which would be least expensive in this case? (Lesson 2–4) **full; $82.50 vs. $96**

55. Name the property illustrated by $(a + 3b) + 2c = a + (3b + 2c)$. (Lesson 1–8) **associative (+)**

WORKING ON THE

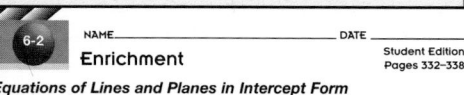

Refer to the Investigation on pages 320–321.

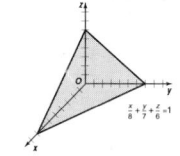

Smoke Gets In Your Eyes

The amount of smoke exhaled into a room on each puff of a cigarette is comparable to the amount of air you exhale in a deep breath. The average smoker takes 10 puffs per cigarette.

1 Suppose that each breath into a balloon approximates the smoke from each exhale during smoking. How many cigarettes are represented by the air in the balloon?

2 Make a graph showing the number of breaths and the cumulative smoke in the air from smoking a cigarette. Describe the graph, and find an equation that would approximate these data.

3 Graph the relationship between the radius of the balloon and the circumference of the balloon. Write an equation to represent this relationship.

4 Find the volume of each balloon in milliliters if $1 \text{ in}^3 \approx 16.4$ milliliters. The average at-rest exhale is approximately 500 milliliters. How does the volume of the balloon compare with the volume of air you would expect from the number of breaths it took to inflate the balloon? What accounts for the difference?

Add the results of your work to your Investigation Folder.

338 *Chapter 6 Analyzing Linear Equations*

In·ves·ti·ga·tion

Working on the Investigation

The Investigation on pages 320–321 is designed to be a long-term project that is completed over several days or weeks. Encourage students to keep their materials in their Investigation Folder as they work on the Investigation.

Integration: Statistics
Scatter Plots and Best-Fit Lines

APPLICATION
Education

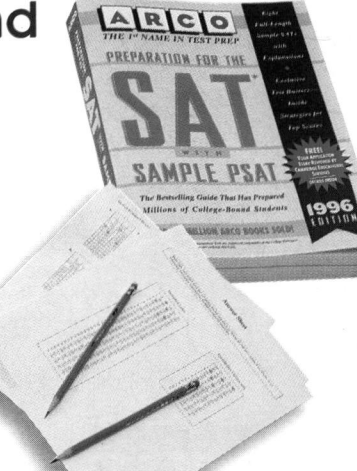

Objectives
After studying this lesson, you should be able to:
- graph and interpret points on a scatter plot,
- draw and write equations for best-fit lines and make predictions by using those equations, and
- solve problems by using models.

If you get a good score on your SAT does that mean you have a good chance of graduating from college? The table below shows the average SAT (Scholastic Assessment Test) scores for freshmen at selected universities and the graduation rate of those students.

School (State)	Average SAT	Graduation Rate
Baylor University (TX)	1045	69%
Brandeis University (MA)	1215	81%
Case Western Reserve University (OH)	1235	65%
College of William and Mary (VA)	1240	90%
Colorado School of Mines (CO)	1200	78%
Georgia Institute of Technology (GA)	1240	68%
Lehigh University (PA)	1140	88%
New York University (NY)	1145	69%
Penn State University, Main Campus (PA)	1096	61%
Pepperdine University (CA)	1070	64%
Rensselaer Polytechnic Institute (NY)	1190	68%
Rutgers at New Brunswick (NJ)	1110	74%
Tulane University (LA)	1168	71%
University of Florida (FL)	1135	64%
University of North Carolina at Chapel Hill (NC)	1045	81%
University of Texas at Austin (TX)	1135	62%
Wake Forest University (NC)	1250	86%

Source: *America's Best Colleges,* June, 1995

To determine if there is a relationship between SAT scores and graduation rates, we can display the data points in a graph called a **scatter plot.** In a scatter plot, the two sets of data are plotted as ordered pairs in the coordinate plane. In this example, the independent variable is the average SAT score, and the dependent variable is the graduation rate. The scatter plot is shown at the left. The graph indicates that higher SAT scores do not necessarily result in higher graduation rates.

One way to solve real-world problems is to **use a model.** Scatter plots are an excellent way to model real-life data to observe patterns and trends.

Game	Yardage
1	348
2	350
3	300
4	295
5	280
6	250
7	225

The yardage per game is decreasing.

6-3 LESSON NOTES

NCTM Standards: 1–5, 10

Instructional Resources

- Study Guide Master 6-3
- Practice Master 6-3
- Enrichment Master 6-3
- Science and Math Lab Manual, pp. 23–28

 Transparency 6-3A contains the 5-Minute Check for this lesson; **Transparency 6-3B** contains a teaching aid for this lesson.

Recommended Pacing	
Standard Pacing	Day 5 of 13
Honors Pacing	Day 4 of 12
Block Scheduling*	Day 3 of 7
Alg. 1 in Two Years*	Days 5 & 6 of 18

 *For more information on pacing and possible lesson plans, refer to the *Block Scheduling Booklet* and *Algebra 1 in Two Years.*

1 FOCUS

 5-Minute Check
(over Lesson 6-2)

Write the standard form of an equation of the line that passes through the given point and has the given slope.

1. $(-2, 6)$, 0 $y = 6$
2. $(4, -3)$, 2 $2x - y = 11$
3. $(5, 7)$, 0 $y = 7$

Write the standard form of an equation of the line that passes through each pair of points.

4. $(5, 3)$, $(-6, 3)$ $y = 3$
5. $(9, 1)$, $(8, 2)$ $x + y = 10$

Motivating the Lesson
Situational Problem A local football team has gained the following yardage in its first seven games. Have students analyze the data in the table on the left to identify a pattern. Develop the idea that neither a stem-and-leaf plot nor a box-and-whisker plot would show this pattern.

In-Class Example

For Example 1
The table below lists the number of shots attempted and the number of shots made by each of eleven members of a basketball team. Make a scatter plot of the data, and answer the questions below.

Player	Shots Attempted	Shots Made
A	240	160
B	200	120
C	135	50
D	80	50
E	75	70
F	35	20
G	20	10
H	12	5
I	9	7
J	5	5
K	2	0

a. Do the points graphed seem to fall on a line? **Yes, with the exception of a few, the points appear to form a line slanting upward.**
b. Is there a negative, a positive, or no correlation between the variables? **positive**

Teaching Tip Point out that with real-world data, rarely will all the data points fall on the line.

Look for a relationship between x and y in the graphs below.

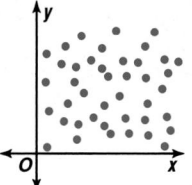

In this graph, x and y have a **positive correlation**. That is, the values are related in the same way. As x increases, y increases.

In this graph, x and y have a **negative correlation**. That is, the values are related in opposite ways. As x increases, y decreases.

In this graph, x and y have *no correlation*. In this case, x and y are not related and are said to be *independent*.

Example ❶ The table below shows 13 of the fastest-growing cities in the United States and their latitude and longitude.

PROBLEM SOLVING
Use a Model

Fastest-Growing Cities	Ranking	North Latitude	West Longitude
Austin, TX	9	30°	98°
Bakersfield, CA	1	35°	119°
Colorado Springs, CO	13	39°	105°
Durham, NC	8	36°	79°
Fresno, CA	2	37°	120°
Laredo, TX	10	28°	100°
Las Vegas, NV	3	36°	115°
Raleigh, NC	6	36°	79°
Reno, NV	12	40°	120°
Sacramento, CA	11	39°	121°
San Bernardino, CA	7	34°	117°
Stockton, CA	5	38°	121°
Tallahassee, FL	4	30°	84°

Source: Bureau of the Census, National Oceanic and Atmospheric Administration

a. Draw one scatter plot to represent the correlation between the rankings and each city's latitude, and another to represent the correlation between the rankings and each city's longitude.
b. Is there a correlation between the cities' locations and their popularity?

a. The scatter plots are shown below.

b. Both of the correlations are fairly weak. However, the correlation between the rankings and latitude is stronger than the correlation between the rankings and longitude because the data approximate a line a little more. It is impossible to tell whether either correlation is positive or negative. Thus, we could say that there is a weak correlation between the cities' location and popularity.

Cooperative Learning

Pairs Check Have students work in pairs for this activity. One student is to come up with several relationships of the following type: (1) miles per gallon and speed (2) hourly wage and weekly pay. Then the other student should determine which value is determined by the other value. For more information on the pairs check strategy, see *Cooperative Learning in the Mathematics Classroom*, one of the titles in the Glencoe Mathematics Professional Series, page 12.

The best-fit line may not pass through any of the data points.

You can use a scatter plot to make predictions. To help you do this, you can draw a line, called a **best-fit line**, that passes close to most of the data points. Use a ruler to draw a line that is close to most or all of the points. Then use the ordered pairs representing points on the line to make predictions.

A best-fit line shows if the correlation between two variables is *strong* or *weak*. The correlation is strong if the data points come close to, or lie on, the best-fit line. The correlation is weak if the data points do not come close to the line.

Example ②

APPLICATION
Animals

The best-fit line drawn in this example is arbitrary. You may draw another best-fit line that is equally as valid.

Dogs age differently than humans do. You may have heard someone say that a dog ages 1 year for every 7 human years. However, that is not the case. The table at the right shows the relationship between dog years and human years.

Dog Years	Human Years
1	15
2	24
3	28
4	32
5	37
6	42
7	47

Source: *National Geographic World*, January, 1995

a. Draw a scatter plot to model the data and determine what relationship, if any, exists in the data.

b. Draw a best-fit line for the scatter plot.

c. Find an equation for the best-fit line.

d. Use the equation to determine how many human years are comparable to 13 dog years.

a. Let the independent variable d be dog years, and let the dependent variable h be human years. The scatter plot is shown below.

The plot seems to indicate that there is nearly a linear relationship between dog years and human years.

There also appears to be a positive correlation between the two variables. As x increases, y increases.

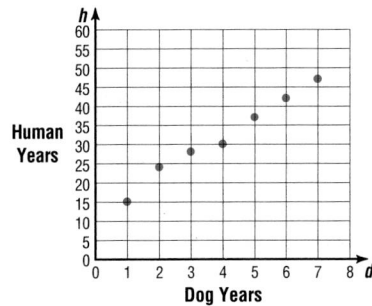

b. No line will pass through all of the data points. However, the best-fit line describes the trend of the data. Draw a line that seems to represent the data.

The slope of the best-fit line is positive. Thus, the correlation between the two variables is also positive.

c. As you can see, the best-fit line we drew passes through three of the data points: (4, 32), (5, 37), and (6, 42). Use two of these points to write the equation of the line.

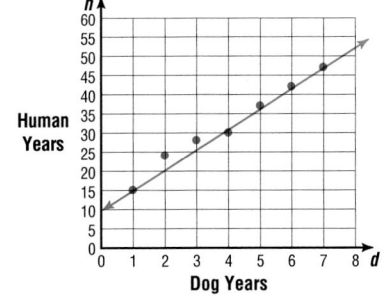

(continued on the next page)

Alternative Teaching Strategies

Reading Algebra A dictionary defines *scatter* using the words, *separate*, *irregular*, *widely*, *diffusely*, and *randomly*. This suggests that the points on a scatter plot are unrelated. This may be true where there is no correlation. Remind students that the point of correlation and regression is to discover some relationship among the data.

GLENCOE Technology

Interactive Mathematics Tools Software

This multimedia software provides an interactive lesson by helping students observe positive and negative relationships of points on a scatter plot. A **Computer Journal** gives students an opportunity to write about what they have learned.

For Windows & Macintosh

First, find the slope.

$$\frac{y_2 - y_1}{x_2 - x_1} = \frac{37 - 32}{5 - 4} \text{ or } 5 \quad (x_1, y_1) = (4, 32) \text{ and } (x_2, y_2) = (5, 37)$$

Then, use the point-slope form.

$h - y_1 = m(d - x_1)$

$h - 32 = 5(d - 4)$ *Let (4, 32) represent (x_1, y_1).*

$h - 32 = 5d - 20$ *Distributive property*

$5d - h = -12$

d. To determine how many human years are comparable to 13 dog years, let $d = 13$ and solve for h.

$5d - h = -12$

$5(13) - h = -12$

$\qquad h = 77$

A dog that is 13 years old is comparable to a human that is 77 years old.

A **regression line** is the most accurate best-fit line for a set of data, and can be determined with a graphing calculator or computer. A graphing calculator assigns each regression line an r value. This value ($-1 \le r \le 1$) measures how closely the data are related. -1 indicates a strong negative correlation and 1 indicates a strong positive correlation.

LOOK BACK

Refer to Lesson 5-7A for more information on entering data into lists on a graphing calculator.

c. The data are more closely related.

EXPLORATION

GRAPHING CALCULATORS

The table at the right shows the years of rate increases and the fares charged to ride the subway in New York City.

a. Use the Edit option on the STAT menu to enter the year data in L1 and the fare data into L2.

b. Use the window [1950, 1995] with a scale factor of 5 and [0, 1.5] with a scale factor of 0.25. Make sure any equations are deleted from the Y= list. Then press [2nd] [STAT PLOT] 1. Make sure the following items are highlighted: On, the first type of graph (scatter plot), L1 as the Xlist, and L2 as the Ylist. Press [GRAPH], and describe the graph.

c. Then find an equation for the regression line.

Enter: [STAT] [▶] 5 [2nd] [L1] [,] [2nd] [L2] [ENTER].

The screen displays the equation $y = ax + b$ and gives values for a, b, and r. Record these values on your paper.

d. Now graph the best-fit line.

Enter: [Y=] [VARS] 5 [▶] [▶] 7 [GRAPH]

Year	Fare
1953	$0.15
1966	0.20
1970	0.30
1972	0.35
1975	0.50
1981	0.75
1984	0.90
1986	1.00
1990	1.15
1992	1.25

Source: *New York Times, July 25, 1993*

Your Turn

a. How well do you feel the graph of the equation fits the data? Justify your answer.   See students' work.

b. Do any of the data points lie on the best-fit line? If so, name them.   no

c. Remove the ordered pair (1953, $0.15) from the set of data and repeat the process. What impact has removing this ordered pair had on the fit of the line to the data? (*Hint:* Compare the r values.)

d. Use the equation and graph to predict the fares for the year 2000. $1.40

Communicating Mathematics

1. If the slope is positive (or negative), the correlation is positive (or negative).

2–3. See Solutions Manual.

Math Journal

Study the lesson. Then complete the following.

1. **Explain** how to determine whether a scatter plot has a positive or negative correlation.

2. **Draw** sketches of scatter plots that have each type of correlation.
   a. very strong positive   b. very strong negative   c. no correlation

3. **Describe** a situation that illustrates each type of plot in Exercise 2.

4. What does the *r* value on a graphing calculator tell you? Give examples.

5. What situations in your own life would present a negative correlation?
   Sample answer: distance driven vs. gallons of gasoline in tank

4. how strongly the data are correlated; −1 is very negative; 1 is very positive

Guided Practice

9b. yes; negative

9c. (7, 18) and (8, 16)

Explain whether a scatter plot for each pair of variables would probably show a *positive*, *negative*, or *no* correlation between the variables.

6. money earned by a restaurant worker and time spent on the job   positive

7. heights of fathers and sons   positive

8. grades in school and how many times you miss class   negative

9. **Archery**   The graph at the right shows the results when three student archers shot at a target nine times each.

Distance from Bull's-eye (in.)

Shot Number

   a. As the archers take more and more shots, what happens to the accuracy of the shots?   It gets better.
   b. Is there a correlation between the variables? Is it positive or negative?
   c. Which data points seem to be extremes from the rest of the data?

10. **Biology**   The table below shows the average body temperature of 14 insects at a given air temperature.

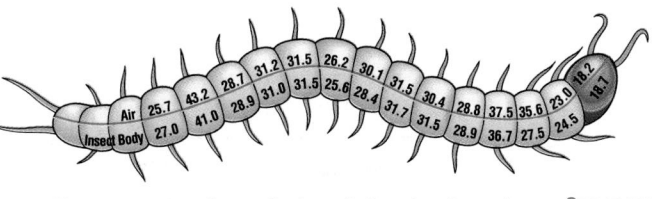

10c. An insect's body temperature closely approximates that of the air around it, and thus, is cold-blooded.

   a. Draw a scatter plot and a best-fit line for these data.   See margin.
   b. Write an equation for the best-fit line.   Sample answer: $y = x$
   c. What conclusion might you draw from these data?

Practice

Explain whether a scatter plot for each pair of variables would probably show a *positive*, *negative*, or *no* correlation between the variables.

A

11. your age and weight from ages 1 to 20   positive

12. temperature of a cup of coffee and the time it sits on a table   negative

13. the amount a mail carrier earns each day and the weight of the mail delivered each day   no

Lesson 6-3   **INTEGRATION** Statistics   Scatter Plots and Best-Fit Lines   **343**

Reteaching

Using Modeling Separate the class into groups and give each group a different age level, but give all groups the same survey question. Have them collect data on the question and make a scatter plot of the data. Show the different age level scatter plots to the whole class to see what happens to each plot as age is varied. Discuss their findings.

Additional Answer

10a.

Check for Understanding

Exercises 1–10 are designed to help you assess your students' understanding through reading, writing, speaking, and modeling. You should work through Exercises 1–5 with your students and then monitor their work on Exercises 6–10.

Error Analysis

It does not matter which data are placed on which axis, but because the data are in ordered pairs, the order in which they are graphed is important. Have students write the data in ordered pair form before graphing.

Assignment Guide

Core: 11–29 odd, 31–35
Enriched: 12–24 even, 25–35

For **Extra Practice,** see p. 770.

The red A, B, and C flags, printed only in the Teacher's Wraparound Edition, indicate the level of difficulty of the exercises.

Study Guide Masters, p. 43

NAME _____   DATE _____

6-3   Study Guide   Student Edition Pages 339–345

Integration: Statistics
Scatter Plots and Best-Fit Lines

A **scatter plot** is a graph that shows the relationship between paired data. The scatter plot may reveal a pattern, or association, between the paired data. This association can be negative or positive. The association is said to be positive when a line suggested by the points slants upward.

The scatter plot at the right represents the relationship between the amount of money Carmen earned each week and the amount she deposited to her savings account. Since the points suggest a line that slants upward, there seems to be a positive relationship between the paired data. In general, the scatter plot seems to show that the more Carmen earned, the more she saved.

Dollars Saved

Dollars Earned

Solve each problem.

1. The table below shows the number of bull's-eyes attempted and the number of bull's-eyes made during a few dart games.
   a. Draw a scatter plot at the right from the data in the table.

Bull's-eyes		
Name	Attempted	Made
Darlene	5	4
Chris	7	7
Mark	5	1
Kathy	6	2

Bull's-eyes Made

Bull's-eyes Attempted

   b. What are the paired data?
      the number of bull's-eyes attempted and the number of bull's-eyes made
   c. Is there a relationship between attempts and successes?
      There is a weak suggestion of a positive correlation.

2. Anna's running speed after 5 minutes was 10 mi/h; at 10 minutes, 8 mi/h; at 15 minutes, 5 mi/h; and at 20 minutes, 4 mi/h.
   a. Make a scatter plot pairing time run with running speed.
   b. How is the data related, positively, negatively, or not at all?   negatively

Miles per Hour

Minutes

Explain whether a scatter plot for each pair of variables would probably show a *positive*, *negative*, or *no* correlation between the variables. 16. negative   17. positive

14. the distance traveled and the time driving   positive

15. a person's height and their birth month   no

16. the amount of snow on the ground and the daily temperature

17. the amount of time you exercise and the amount of calories you burn

18. the time water boils and the amount of water in the pot   positive

19. the number of files stored on a disk and the amount of memory left on the disk   negative

20. Sample reason: Dots lie in a horizontal pattern.

 B

21. Sample reason: Dots are grouped in an upward diagonal pattern.

22. Sample reason: Dots are everywhere, no linear pattern exists.

 C

23. c; if 1 is correct then 19 are wrong, if 2 are correct, then 18 are wrong, and so on. Graph c shows these pairs of numbers.

Determine whether a best-fit line should be drawn for each set of data graphed below. Explain your reasoning.

20. yes

21. yes

22. no
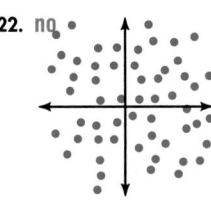

23. A test contains 20 true-false questions. Which scatter plot below best illustrates all the possible combinations of the numbers of correct and incorrect answers if the horizontal axis represents the number of responses that are correct and the vertical axis represents the number that are incorrect? Explain your choice.

a.

b.

c.

Graphing Calculator

24. The average January and July temperatures (°F) for 14 cities in various parts of the United States are listed below.

Jan.	21	35	42	32	51	7	30	21	20	40	28	45	56	39
July	71	79	79	77	82	70	73	73	70	86	81	90	69	82

a. Use a graphing calculator to enter the January temperatures as L1 and the July temperatures as L2 and create a scatter plot.   a–d. See margin.

b. Describe the pattern of the points in the scatter plot.

c. Record the equation of the regression line and the *r* value.

d. Write a sentence that generally describes the relationship between the January temperature in a city and its July temperature.

Critical Thinking

25a–d. See margin for sample answers.

25. Different correlation values are acceptable for different situations. Give an example for each situation. Explain your reasoning.

a. When is a strongly positive correlation a bad thing?

b. When is a mildly positive correlation an acceptable outcome?

c. When would a strongly negative correlation be the desired outcome?

d. When would no correlation be acceptable?

344   Chapter 6   Analyzing Linear Equations

26. Many verbal expressions can be translated into algebraic expressions or equations. How do you think you could illustrate the expression "The bigger they are, the harder they fall" as a scatter plot? **See students' work.**

Applications and Problem Solving

27. Education Refer to the application at the beginning of the lesson.
  a. Describe the correlation of the SAT scores to the graduation rate.
  b. Copy the graph and draw a best-fit line. What is the equation of the best-fit line you drew? **a–b. See margin for sample answers.**

28. Biology Scientists use samples to generalize relationships they observe in nature. The table at the right shows the length and weight of several humpback whales. A long ton is about 2240 pounds.

Length (feet)	Weight (long tons)
40	25
42	29
45	34
46	35
50	43
52	45
55	51

  a. Make a scatter plot of these data. Does a relationship exist? If so, what is the relationship? **See margin.**
  b. Use your scatter plot as a model to predict what you think a humpback whale 51 feet long might weigh. **about 43–45 long tons**

28d. Yes; sample answer: the point (46, 36) falls in the same general area as the other data.

  c. How long might a whale that weighs between 30 long tons and 35 long tons be? **43–46 feet**
  d. A 46-foot long humpback whale weighing 36 long tons was observed. Does this alter your conclusion about the correlation of the data? Explain.

29c. Sample answer:
$y = \frac{2}{3}x + 22$

29. Test Scores The scores received on the first two tests of the grading period for 30 students in algebra class are listed below as ordered pairs.

30, 43	58, 57	55, 61	4, 71	32, 27	68, 59
54, 47	38, 27	56, 47	72, 63	50, 23	70, 67
60, 53	73, 79	55, 68	19, 58	38, 41	71, 59
68, 75	74, 83	58, 67	66, 73	72, 67	42, 57
71, 69	70, 89	94, 59	8, 84	84, 71	82, 73

  a. Make a scatter plot of the data. **See margin.**
  b. What correlation, if any, do you observe in the data? **positive**
  c. Draw a best-fit line and write the equation of the line.

30. Use a Model Collect 10 round objects around your home. Treat each of them as a circle and measure the diameter and circumference in millimeters.
  a. Use a scatter plot as a model to relate the diameter and circumference of each object. **a–c. See students' work.**
  b. Write an equation for the best-fit line.
  c. The actual formula that relates the circumference and diameter of a circle is $C = 2\pi d$. How does this formula compare with your equation?

Mixed Review

31a. $7500x - y = 120,000$
31b. 15,000 feet
31c. No, it only describes the plane's path in that part of the flight.
32. $\{(-2, 5), (-1, 5\frac{1}{6}), (0, 5\frac{1}{3}), (2, 5\frac{2}{3}), (5, 6\frac{1}{6})\}$

31. Air Travel A Boeing 747 takes off from the runway, and 20 minutes after liftoff, the plane is 30,000 feet from the ground. At 22 minutes after liftoff, the plane is 45,000 feet from the ground. (Lesson 6–2)
  a. Write an equation in standard form to represent the flight of the plane during this time, with x = time in minutes, and y = altitude in feet.
  b. Use the equation to find the plane's altitude 18 minutes after takeoff.
  c. Can this equation be used to determine the height of the plane at any time during the flight? If so, explain how.

32. Solve $6b - a = 32$ if the domain is $\{-2, -1, 0, 2, 5\}$. (Lesson 5–3)

33. Trigonometry Use a calculator to solve $\cos W = 0.2598$. (Lesson 4–3) **75°**

34. Simplify $1.2(4x - 5y) - 0.2(-1.5x + 8y)$. (Lesson 2–6) **$5.1x - 7.6y$**

35. Solve $8.2 - 6.75 = m$. (Lesson 1–5) **1.45**

Lesson 6–3 **INTEGRATION** Statistics *Scatter Plots and Best-Fit Lines* **345**

Extension

Reasoning The daily high temperatures in Tampa were recorded for the month of November 1994. The data are shown in the form of a scatter plot, with the horizontal scale representing the day of the month and the vertical scale representing the temperature.

1. What information can be determined from the plot? **weather trends that occurred during the month**

2. What information cannot be determined from the plot? **how the daily temperatures in 1994 compared with average temperatures over an extended period of time**

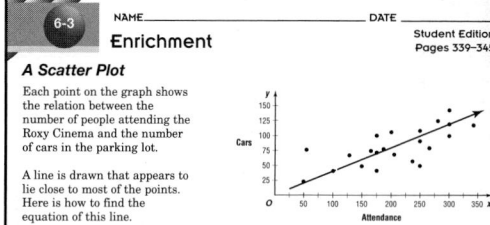
Chapter 6   **345**

Instructional Resources

- Study Guide Master 6-4
- Practice Master 6-4
- Enrichment Master 6-4
- Assessment and Evaluation Masters, pp. 155–156
- Multicultural Activity Masters, p. 12
- Real-World Applications, 18
- Tech Prep Applications Masters, p. 11

 Transparency 6-4A contains the 5-Minute Check for this lesson; **Transparency 6-4B** contains a teaching aid for this lesson.

Recommended Pacing	
Standard Pacing	Day 6 of 13
Honors Pacing	Day 5 of 12
Block Scheduling*	Day 4 of 7
Alg. 1 in Two Years*	Days 7 & 8 of 18

 *For more information on pacing and possible lesson plans, refer to the *Block Scheduling Booklet* and *Algebra 1 in Two Years.*

1 FOCUS

 ## 5-Minute Check
(over Lesson 6-3)

Use the scatter plot shown below to answer the questions.

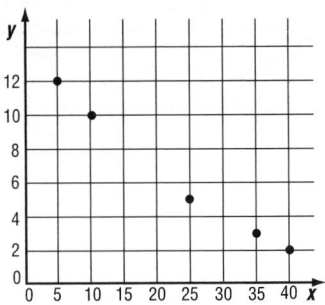

1. What ordered pairs are represented? {(5, 12), (10, 10), (25, 5), (35, 3), (40, 2)}

2. Does the plot indicate a negative, a positive, or no correlation? negative

---

Writing Linear Equations in Slope-Intercept Form

6-4

Objectives

After studying the lesson, you should be able to:
- determine the *x*- and *y*-intercepts of linear graphs from their equations,
- write equations in slope-intercept form, and
- write and solve direct variation equations.

Notice that the x-coordinate of the ordered pair for the y-intercept and the y-coordinate of the ordered pair for the x-intercept are both 0.

APPLICATION
Banking

Have you seen advertisements like the one at the right in the newspaper? The bank is offering free checking. However, in the fine print of this ad, you will find that the free checking only occurs if you maintain a minimum daily balance of at least $2000.

Suppose this bank charges a monthly service fee of $3 plus 10¢ for each check or withdrawal transaction if you fail to maintain a minimum balance of $2000. The graph at the right shows the line that represents this situation. *You will write an equation for this line in Exercise 57.*

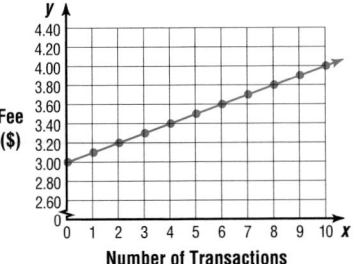

The coordinates at which a graph intersects the axes are known as the **y-intercept** and the **x-intercept**. Since this graph intersects the *y*-axis at (0, 3), the *y*-intercept is 3. If you were to extend the graph to the left, you would find that it crosses the *x*-axis at (−30, 0), so the *x*-intercept is −30.

Number of Transactions

Example **Find the *x*- and *y*-intercepts of the graph of $3x + 4y = 6$.**

Remember that the *x*-intercept is the point at which $y = 0$.

$$3x + 4y = 6$$
$$3x + 4(0) = 6 \qquad \text{\textit{Let y = 0.}}$$
$$3x = 6$$
$$x = \frac{6}{3} \text{ or } 2 \qquad \text{\textit{Divide each side by 3.}}$$

Now let $x = 0$ to find the *y*-intercept.

$$3x + 4y = 6$$
$$3(0) + 4y = 6 \qquad \text{\textit{Let x = 0.}}$$
$$4y = 6$$
$$y = \frac{6}{4} \text{ or } \frac{3}{2} \qquad \text{\textit{Divide each side by 4.}}$$

The *x*-intercept is 2, and the *y*-intercept is $\frac{3}{2}$. This means that the graph crosses the *x*-axis at (2, 0) and the *y*-axis at $\left(0, \frac{3}{2}\right)$. You can use these points to graph the equation $3x + 4y = 6$.

Classroom Vignette

"I usually give students real-world examples to help them understand this concept.
Example: You have 4 gallons of gas in the tank of your car. You pump gas at the rate of 5 gallons/minute.
Students make a table showing time (minutes) and gallons pumped for each minute. They then graph these ordered pairs and draw the graph to determine its *y*-intercept and equation."

Eva Gates
Independent Mathematics Consultant
Pearland, Texas

Consider the graph at the right. The line crosses the y-axis at $(0, b)$, so its y-intercept is b. Write the point-slope form of an equation for this line using $(0, b)$ as (x_1, y_1).

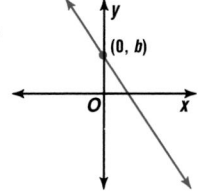

$$y - y_1 = m(x - x_1) \quad \text{\textit{Point-slope form}}$$
$$y - b = m(x - 0) \quad \text{\textit{Replace } } (x_1, y_1) \text{ \textit{with} } (0, b).$$
$$y - b = mx$$
$$y = mx + b \quad \text{\textit{Add b to each side.}}$$

This form of an equation is called the **slope-intercept form** of a linear equation, because in this form the slope and y-intercept are easily identified.

$$y = \underset{\underset{slope}{\uparrow}}{m}x + \underset{\underset{y\text{-}intercept}{\uparrow}}{b}$$

Slope-Intercept Form of a Linear Equation	**Given the slope m and the y-intercept b of a line, the slope-intercept form of an equation of the line is $y = mx + b$.**

If you know the slope and y-intercept of a line, you can write an equation of the line in slope-intercept form. This equation can also be written in standard form.

Example Write an equation of a line in slope-intercept form if the line has a slope of $\frac{2}{3}$ and a y-intercept of 6. Then write the equation in standard form.

$$y = mx + b$$
$$y = \frac{2}{3}x + 6 \quad b = 6, m = \frac{2}{3}$$

Now rewrite the equation in standard form.

$$y = \frac{2}{3}x + 6$$
$$3y = 2x + 18 \quad \text{\textit{Multiply by 3 to eliminate the fraction.}}$$
$$-2x + 3y = 18 \quad \text{\textit{Subtract 2x from each side.}}$$
or
$$2x - 3y = -18 \quad \text{\textit{Multiply by -1 so that A is positive.}}$$

Let's compare the slope-intercept form with the standard form of a linear equation. First solve $Ax + By = C$ for y.

$$Ax + By = C$$
$$Ax - Ax + By = C - Ax \quad \text{\textit{Subtract Ax from each side.}}$$
$$By = -Ax + C \quad \text{\textit{Commutative property of addition}}$$
$$y = \underset{\underset{slope}{\uparrow}}{-\frac{A}{B}}x + \underset{\underset{y\text{-}intercept}{\uparrow}}{\frac{C}{B}} \quad \text{\textit{Divide each side by B.}}$$

Thus, you can identify the slope and y-intercept from an equation written in standard form by using $m = -\frac{A}{B}$ and $b = \frac{C}{B}$.

Lesson 6–4 Writing Linear Equations in Slope-Intercept Form **347**

Chapter 6 **347**

Teaching Tip For Example 3, explain that if students forget $-\frac{A}{B}$ and $\frac{C}{B}$, the slope and y-intercept can still be found. Have them solve the standard-form equation for y as shown below.

$$3x + 2y = 6$$
$$2y = -3x + 6$$
$$y = -\frac{3}{2}x + 3$$

slope y - intercept

In-Class Examples

For Example 3
Find the slope and y-intercept of the graph of each equation.

a. $y = 3x - 7$ $3, -7$

b. $y = \frac{2}{3}x$ $\frac{2}{3}, 0$

c. $y = 5$ $0, 5$

d. $y = -\frac{3}{4}x + 11$ $-\frac{3}{4}, 11$

For Example 4
Write the slope-intercept and standard forms of the equation for a line that passes through $(-2, -1)$ and $(4, 2)$.
$y = \frac{1}{2}x; \; x - 2y = 0$

Teaching Tip After completing Example 4, point out that vertical lines have no y-intercept, while horizontal lines have no x-intercept.

Example ③ Find the slope and y-intercept of the graph of $5x - 3y = 6$.

Method 1
In $5x - 3y = 6$, which is in standard form, $A = 5$, $B = -3$, and $C = 6$.
Find the slope.
$$m = -\frac{A}{B}$$
$$= -\frac{5}{-3} \text{ or } \frac{5}{3}$$

Find the y-intercept.
$$b = \frac{C}{B}$$
$$= \frac{6}{-3} \text{ or } -2$$

Method 2
Solve for y to find the slope-intercept form.
$$5x - 3y = 6$$
$$5x - 5x - 3y = 6 - 5x$$
$$-3y = -5x + 6$$
$$\frac{-3y}{-3} = \frac{-5x}{-3} + \frac{6}{-3}$$
$$y = \frac{5}{3}x + (-2)$$
$$m = \frac{5}{3}, b = -2$$

In Lesson 6–2, you learned how to use the point-slope form to find an equation of a line passing through two given points. Now you have another tool that you can use for the same situation.

Example ④ Write the slope-intercept and standard forms of the equation for a line that passes through $(-3, -1)$ and $(6, -4)$.
First find the slope.
$$m = \frac{y_2 - y_1}{x_2 - x_1}$$
$$= \frac{-4 - (-1)}{6 - (-3)} \quad (x_1, y_1) = (-3, -1) \text{ and } (x_2, y_2) = (6, -4)$$
$$= \frac{-3}{9} \text{ or } -\frac{1}{3}$$

Method 1
In $Ax + By = C$, the slope is $-\frac{A}{B}$.
If $-\frac{A}{B} = -\frac{1}{3}$, then $A = 1$ and $B = 3$.
So $Ax + By = C$ becomes $1x + 3y = C$.
To find the value of C, substitute either ordered pair into the equation. For example, choose $(6, -4)$.

$$x + 3y = C$$
$$6 + 3(-4) = C \quad (x, y) = (6, -4)$$
$$-6 = C$$

An equation in standard form for the line is $x + 3y = -6$.

Solve the equation for y to find the slope-intercept form.

$$x + 3y = -6$$
$$3y = -x - 6$$
$$y = -\frac{1}{3}x - \frac{6}{3}$$
$$y = -\frac{1}{3}x - 2$$

Method 2
Use one of the points in the slope-intercept form to find b.

$$y = mx + b$$
$$-4 = -\frac{1}{3}(6) + b \quad (x, y) = (6, -4)$$
$$-4 = -2 + b$$
$$-2 = b$$

An equation in slope-intercept form for the line is $y = -\frac{1}{3}x - 2$.

Now write the equation in standard form.

$$y = -\frac{1}{3}x - 2$$
$$3y = 3\left(-\frac{1}{3}x - 2\right)$$
$$3y = -x - 6$$
$$x + 3y = -6$$

Check: Make sure that both points satisfy the equation.
$$x + 3y = -6 \qquad\qquad x + 3y = -6$$
$$-3 + 3(-1) \overset{?}{=} -6 \quad (x, y) = (-3, -1) \qquad 6 + 3(-4) \overset{?}{=} -6 \quad (x, y) = (6, -4)$$
$$-6 = -6 \;\checkmark \qquad\qquad -6 = -6 \;\checkmark$$

Alternative Learning Styles

Auditory Discuss the following questions. Is it possible to write an equation of a line given its graph? How? Yes; find the coordinates of two points on the line.

A special case of the slope-intercept form occurs when the y-intercept is 0. When $m = k$ and $b = 0$, $y = mx + b$ becomes $y = kx$. You may recognize this as the equation for *direct variation*. Another way to write this equation is $\frac{y}{x} = k$.

Example ⑤

APPLICATION
Swimming

Penny Dean holds the record for the fastest time swimming the English Channel. The 23-year-old Californian swam the 21-mile distance in $7\frac{2}{3}$ hours on July 29, 1978. At the same rate, how long would it take her to swim the Catalina Channel, a 26-mile distance near Los Angeles?

Explore Make a table to relate the information given in the problem.

Channel	Distance	Time
English	21 miles	$7\frac{2}{3}$ hours
Catalina	26 miles	? hours

Plan The formula that relates time t and distance d is $d = rt$, where r is the rate at which she travels. This formula is an example of direct variation. Use the information from the English Channel swim to find her rate. Then apply that rate to the Catalina Channel swim.

Solve For her English Channel swim, $d = 21$ miles, and $t = 7\frac{2}{3}$ or $\frac{23}{3}$ hours. Find the rate r.

$$d = rt$$
$$21 = r\left(\frac{23}{3}\right) \quad d = 21, r = \frac{23}{3}$$
$$\frac{3}{23}(21) = \left(\frac{23}{3}\right)\left(\frac{3}{23}\right)r$$
$$\frac{63}{23} = r$$

Now use the same formula and the rate r to find the time for the Catalina Channel swim.

$$d = rt$$
$$26 = \frac{63}{23}t$$
$$\frac{23}{63}(26) = \frac{23}{63}\left(\frac{63}{23}\right)t$$
$$9.5 \approx t \quad \textit{Use a calculator.}$$

At the same rate that she swam the English Channel, Ms. Dean could swim the Catalina Channel in 9.5 hours.

Examine You can also use a proportion to solve this problem since the rate r will be the same in each situation. If $d = rt$, then $r = \frac{d}{t}$, a form of the direct variation equation.

$$\frac{d_1}{d_2} = \frac{t_1}{t_2}$$
$$\frac{21}{26} = \frac{\frac{23}{3}}{t}$$
$$21t = \frac{23}{3}(26) \quad \textit{Find the cross products.}$$
$$t = \frac{23 \cdot 26}{3 \cdot 21} \quad \textit{Divide each side by 21.}$$
$$t \approx 9.5 \quad \textit{Use a calculator.}$$

The answer checks.

fabulous FIRSTS

Gertrude Ederle (1907–)

The first woman to swim the English Channel was an American, Gertrude Ederle. She was only 19 years old and made the swim in 14 hours, 39 minutes on August 6, 1926.

CHECK FOR UNDERSTANDING

Communicating Mathematics

Study the lesson. Then complete the following. 1–3. See margin.

1. **Explain** how to find the intercepts of a line given its equation. Then find the intercepts of the standard form $Ax + By = C$.

2. **Write** a sentence to explain how direct variation and the slope-intercept form of a linear equation are related.

3. **Describe** two ways in which a direct variation problem may be solved.

Lesson 6–4 Writing Linear Equations in Slope-Intercept Form **349**

In-Class Example

For Example 5
At the same rate, how long would it take Penny Dean to swim a distance of 40 miles?
14.6 hours

3 PRACTICE/APPLY

Check for Understanding
Exercises 1–18 are designed to help you assess your students' understanding through reading, writing, speaking, and modeling. You should work through Exercises 1–6 with your students and then monitor their work on Exercises 7–18.

Additional Answers

1. Let $x = 0$ to find the y-intercept; let $y = 0$ to find the x-intercept; x: $\frac{C}{A}$; y: $\frac{C}{B}$.

2. Sample answer: Direct variation is the slope-intercept form when $b = 0$.

3. Direct variation may be solved by using the formulas $y = kx$ or $\frac{y}{x} = k$, where $k = m$, which represents slope.

Reteaching

Checking Data Have students find the x- and y-intercepts of a linear equation and graph them. Then write the equation in slope-intercept form. Using the graph, find another point on the line and substitute the coordinates of that point into the original equation to check for equality. Also, find the slope from the graph and check with the slope-intercept form equation.

fabulous FIRSTS

Gertrude Ederle's success in swimming led to financial success. She earned $2000 a week traveling the country giving swimming exhibitions. In 1953 she received the Helms Foundation Hall of Fame Award, and in 1965 she was inducted into the International Swimming Hall of Fame.

4. Because its slope is undefined.

 M*ATH* J*OURNAL*

Guided Practice

11. $y = \frac{2}{3}x - 10$, $2x - 3y = 30$

18a. $y = \frac{1}{4}x + 12$

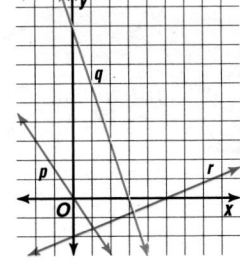

4. Why can't you write the equation of a vertical line in slope-intercept form?

5. **You Decide** Taka wrote $y = 3.5x - 8$ as an equation of a line having a slope of 3.5 and a y-intercept of −8. Chuma wrote $7x - 2y = 16$ as an equation of the same line. Who is correct, and why? **Both; Chuma's equation is the standard form of Taka's slope-intercept form.**

6. List all the types of information you can use to write an equation of a line. Be sure to include examples of each. **See margin for sample answer.**

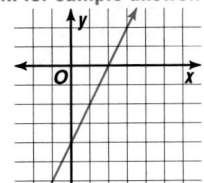

7. For the line shown in the graph at the right:
   a. state the slope, **2**
   b. state the x- and y-intercepts, and **2, −4**
   c. write an equation in slope-intercept form.
   $y = 2x - 4$

Find the x- and y-intercepts of the graph of each equation.

8. $3x + 4y = 24$ **8, 6**

9. $\frac{3}{4}x - 2y + 7 = 0$ $-\frac{28}{3}, \frac{7}{2}$

Write an equation in slope-intercept form of a line with the given slope and y-intercept. Then write the equation in standard form.

10. $m = -4, b = 5$
    $y = -4x + 5, 4x + y = 5$

11. $m = \frac{2}{3}, b = -10$

Find the slope and y-intercept of the graph of each equation.

12. $y = -3x + 7$ **−3, 7**
13. $2x + y = -4$ **−2, −4**
14. $x = 3y - 2$ $\frac{1}{3}, \frac{2}{3}$

Write an equation in standard form for a line that passes through each pair of points.

15. (8, 1), (−2, 3) $x + 5y = 13$
16. (5, 7), (1, 9) $x + 2y = 19$

17. Write a direct variation equation that has (3, 11) as a solution. Then find y when $x = 12$. $y = \frac{11}{3}x; 44$

18. **Construction** The roof line of a house starts 12 feet above the ground and has a slope of $\frac{1}{4}$.
    a. If the outer wall of the house represents the y-axis and the floor represents the x-axis, find the equation, in slope-intercept form, of the line that represents the roof.
    b. If you are standing inside the house 6 feet from the wall, how high is the roof at that point? **13.5 feet**

12 ft

EXERCISES

Practice

For each line graphed at the right:
a. state the slope,
b. state the x- and y-intercepts, and
c. write an equation in slope-intercept form.

A 19. p $-\frac{3}{2}; 0, 0; y = -\frac{3}{2}x$

20. q $-3; 3, 9; y = -3x + 9$

21. r $\frac{2}{5}; 5, -2, y = \frac{2}{5}x - 2$

22. $-\frac{12}{5}, 4$

Find the x- and y-intercepts of the graph of each equation.

22. $5x - 3y = -12$ 23. $4x + 7y = 8$ $2, \frac{8}{7}$ 24. $5y - 2 = 2x$ $-1, \frac{2}{5}$

25. $-\frac{5}{6}, 5$

25. $y - 6x = 5$ 26. $4y - x = 3$ 27. $3y = 18$ **none, 6**

26. $-3, \frac{3}{4}$

Write an equation in slope-intercept form of a line with the given slope and y-intercept. Then write the equation in standard form.

28. $y = 3x + 5$,
$3x - y = -5$

 B

28. $m = 3, b = 5$ 29. $m = 7, b = -2$ 30. $m = -6, b = 0$

31. $m = -1.5, b = 3.75$ 32. $m = \frac{1}{4}, b = -10$ 33. $m = 0, b = -7$

29. $y = 7x - 2$,
$7x - y = 2$

30. $y = -6x$, $6x + y = 0$

Find the slope and y-intercept of the graph of each equation.

31. $y = -1.5x + 3.75$,
$6x + 4y = 15$

34. $3y = 2x - 9$ 35. $5x + 4y = 10$

32. $y = \frac{1}{4}x - 10$,
$x - 4y = 40$

36. $4x - \frac{1}{3}y = -2$ **12, 6**   37. $\frac{2}{3}x + \frac{1}{6}y = 2$ **−4, 12**

33. $y = -7, y = -7$

38. $5(x - 3y) = 2(x + 3)$ $\frac{1}{5}, -\frac{2}{5}$ 39. $4(3x + 9) - 3(5y + 7) = 11$ $\frac{4}{5}, \frac{4}{15}$

34. $\frac{2}{3}, -3$ 35. $-\frac{5}{4}, \frac{5}{2}$

Write an equation in standard form for a line that passes through each pair of points.   40. $10x - 7y = 5$ 41. $y = -2$ 42. $3x + 10y = -8$

44. $5x + 12y = 83$

40. $(-3, -5), (4, 5)$ 41. $(7, -2), (-4, -2)$ 42. $(-6, 1), (4, -2)$

46. $y = -\frac{8}{5}x, -\frac{55}{8}$

43. $(3, 5), (3, -6)$ $x = 3$ 44. $(7, 4), (-5, 9)$ 45. $(2, 9), (-5, 9)$ $y = 9$

47. $y = \frac{11}{24}x, \frac{33}{2}$

Write a direct variation equation that has each ordered pair as a solution. Then find the missing value.

48. $y = -\frac{3}{17}x, -85$

46. $(-5, 8), (?, 11)$ 47. $(24, 11), (36, ?)$ 48. $(-17, 3), (?, 15)$

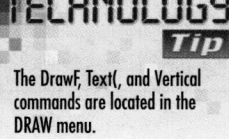 **C**

49. Write an equation in slope-intercept form of a line with slope $\frac{2}{3}$ and an x-intercept of 4.

50. Find the coordinates of a point on the graph of $5x - 3y = 10$ in which the x-coordinate is 5 greater than the y-coordinate.

49. $y = \frac{2}{3}x - \frac{8}{3}$

51. Find the coordinates of a point on the graph of $3x + 7y = -16$ in which the x-coordinate is 3 times the y-coordinate. $(-3, -1)$

50. $\left(-\frac{5}{2}, -\frac{15}{2}\right)$

52. Write the equation in standard form of the line with an x-intercept of 7 and a y-intercept of -2. $2x - 7y = 14$

Programming

53. The program below finds the slope of a line, given the coordinates of two of the points through which it passes. Then it displays the graph of the line and gives the equation of the line in slope-intercept form.

```
PROGRAM:SLOPE
: ClrDraw                : DrawF AX+B
: Input "X1=",Q          : Text(2,5,"EQUATION IS"
: Input "Y1=",R          : Text(8,5,"Y=",A,"X+",B)
: Input "X2=",S          : Else
: Input "Y2=",T          : ClrHome
: If Q-S≠0               : Text(2,5,"SLOPE UNDEFINED")
: Then                   : Text(8,5,"EQUATION OF LINE")
: (R-T)/(Q-S)→A          : Text(15,5,"X=",Q)
: R-AQ→B                 : Vertical Q
```

TECHNOLOGY Tip

The DrawF, Text(, and Vertical commands are located in the DRAW menu.

a. Use the program to write an equation of a line that passes through (8.57, −3.82) and (11.09, 1.31). Round numbers to the nearest hundredth. $y = 2.04x - 21.32$

b. Use the program to check your equations in Exercises 40–45.

Using the Programming Exercises The program given in Exercise 53 is for use with a TI-82 graphing calculator. For other programmable calculators, have students consult their owner's manual for commands similar to those presented here.

Technology Tip
Remind students to use the Clr Home command to clear the screen and return to home position if earlier work is still on the screen.

Study Guide Masters, p. 44

Closing Activity

Speaking Ask students to explain the procedure they would use to determine the slope and y-intercept of the equation $3x + 4y = 12$. Then have them use the procedure described to find each. $-\frac{3}{4}, 3$

Chapter 6, Quiz B (Lessons 6-3 and 6-4), is available in the *Assessment and Evaluation Masters*, p. 156.

Mid-Chapter Test (Lessons 6-1 through 6-4) is available in the *Assessment and Evaluation Masters,* p. 155.

Additional Answers

56c–56d.

Practice Masters, p. 44

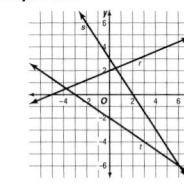

Critical Thinking

54. The x-intercept of a line is p, and the y-intercept is q. Write an equation of the line. $y = -\frac{qx}{p} + q$

Applications and Problem Solving

55. **Chemistry** Charles' Law states that when the pressure is constant, the volume of a gas is directly proportional to the temperature on the Kelvin scale. Write an equation for each situation and solve. 55a. 42.24 ft^3
    a. If the volume of a gas is 35 ft^3 at 290 K, what is the volume at 350 K?
    b. If the volume of a gas is 200 ft^3 at 300 K, at what temperature is the volume 180 ft^3? **270 K**

56. **Life Expectancy** The life expectancies at birth of the average male and female born between 1970 and 2000 (estimated) are shown in the table at the right.

    56a. Sample answer: $y = 0.20x - 326.9$

    56b. Sample answer: $y = 0.18x - 279.9$

Life Expectancy		
Year Born	Males (years)	Females (years)
1970	67.1	74.7
1975	68.8	76.6
1980	70.4	77.8
1985	71.1	78.2
1990	71.8	78.8
1995	72.8	79.7
2000 (est.)	73.2	80.2

    a. Write an equation for the line that passes through (1970, 67.1) and (2000, 73.2).
    b. Write an equation for the line that passes through (1970, 74.7) and (2000, 80.2).
    c. Graph the ordered pairs for male life expectancy and the equation from part a. **c–e. See margin.**
    d. On the same graph, use a different color pencil to graph the ordered pairs for female life expectancy and the equation from part b.
    e. Write a paragraph to describe the relationship between the points and lines you graphed.
    f. Use the data to predict the life expectancy for males and females in the year 2100. **based on sample equations: males, 93.2; females, 98.2**

57. **Banking** Refer to the application at the beginning of the lesson.
    a. Write an equation to represent the line that shows the service fees charged by the bank for an account that maintains a daily balance that is less than $2000. $y = 0.1x + 3$
    b. Suppose you are not able to maintain $2000 in an account and you write about 25 checks each month. Use the equation from part a to determine whether you should open an account at this bank or go to the bank across the street that charges a flat $5 each month. **Go to the other bank, since this one would charge you $5.50.**

Mixed Review

58. **Statistics** The table below shows the keyboarding speeds of 12 students in words per minute (wpm) and their weeks of experience. (Lesson 6–3)

Experience (weeks)	4	7	8	1	6	3	5	2	9	6	7	10
Keyboarding Speed (wpm)	33	45	46	20	40	30	38	22	52	44	42	55

    a. Make a scatter plot of these data. **a–b. See Solutions Manual.**
    b. Draw a best-fit line for the data. Find the equation of the line.
    c. Use the equation to predict the keyboarding speed of a student after a 12-week course. **about 65 wpm**
    d. Why can't this equation be used to predict the speed for any number of weeks of experience? **There's a limit as to how fast one can type.**

Extension

Problem Solving Write an equation in slope-intercept form of the line with y-intercept 6 that passes through the point (3, 4).
$y = -\frac{2}{3}x + 6$

Additional Answer

56e. The life expectancy of females is higher than the life expectancy of males and, as time passes, the life expectancy of both increases. According to the graph, at approximately the year 2350, men and women both will have a life expectancy of about 136 years.

59. Determine the slope of the line that passes through $(14, 3)$ and $(-11, 3)$. (Lesson 6–1) **0**

60. Write an equation for the function shown in the chart below. (Lesson 5–6)

m	1	2	4	5	6	9
n	9	6	0	−3	−6	−15

$n = 12 - 3m$

62. $\frac{1}{2}$

63. 13 ft 11 in.

61. Graph $A(5, -2)$ on a coordinate plane. (Lesson 5–1) **See margin.**

62. **Probability** A card is selected at random from a deck of 52 cards. What is the probability of selecting a black card? (Lesson 4–6)

63. **Construction** When building a roof, a 5-foot support is to be placed at point B as shown on the diagram at the right. Find the length of the support that is to be placed at point A. (Lesson 4–2)

64. Solve $8x + 2y = 6$ for y. (Lesson 3–6) $y = 3 - 4x$

65. Solve $-36 = 4z$. (Lesson 3–2) **−9**

66. Evaluate $|a + k|$ if $a = -5$ and $k = 3$. (Lesson 2–3) **2**

67. Simplify $4(3x + 2) + 2(x + 3)$. (Lesson 1–7) $14x + 14$

SELF TEST

Determine the slope of the line that passes through each pair of points. (Lesson 6–1)

1. $(-7, 10)$ and $(-2, 5)$ **−1**
2. $(-6, 3)$ and $(-12, 3)$ **0**
3. $(-5, 7)$ and $(-5, -15)$
4. Determine the value of r so the line through $(r, 3)$ and $(6, -2)$ has a slope of $-\frac{5}{2}$. (Lesson 6–1) **4**

3. **undefined**

Write the point-slope form of an equation of the line that passes through the given point and has the given slope. (Lesson 6–2)

5. $(-6, 4)$, $m = \frac{1}{2}$ $y - 4 = \frac{1}{2}(x + 6)$
6. $(-12, 12)$, $m = 0$ $y - 12 = 0$

7. Write the standard form of an equation of the line that passes through $(-3, 4)$ and $(2, 3)$. (Lesson 6–2) $x + 5y = 17$

8. **Computers** A newspaper article stated that with advancements in technology, computers would become cheaper and cheaper. Is this true? The table at the right shows the average cost of a computer system for each year from 1991 to 1994. (Lesson 6–3) **a–b. See margin.**
   a. Make a scatter plot of these data. **d–e. See margin.**
   b. Draw a best-fit line and write an equation that describes it.
   c. Does the scatter plot show a positive, negative, or no correlation between the variables? **positive**
   d. Write what this correlation means in words.
   e. With each year, computer systems become faster at handling more and more data. How can you use this fact to substantiate the newspaper article claim?

Year	Average Cost ($)
1991	1100
1992	1143
1993	1183
1994	1219

9. Find the x- and y-intercepts of the graph of $7x + 3y = -42$. (Lesson 6–4) **−6, −14**

Source: *Vitality,* 1995

10. Write the slope-intercept form of an equation for the line that passes through $(0, -3)$ and $(6, 0)$. (Lesson 6–4) $y = \frac{1}{2}x - 3$

Tech Prep

Roofing Contractor Students who are interested in construction may wish to do further research on the data presented in Exercise 63 and explore the potential growth of this career. For more information on tech prep, see the *Teacher's Handbook.*

SELF TEST

The Self Test provides students with a brief review of the concepts and skills in Lessons 6-1 through 6-4. Lesson numbers are given to the right of exercises or instruction lines so students can review concepts not yet mastered.

Additional Answer

61.

Answers for the Self Test

8a, b.

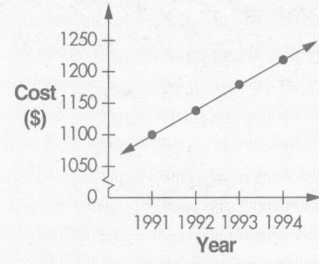

8b. Sample equation:
$y = 40x - 78{,}537$

8d. Each year the cost of a computer increases.

8e. The number of features on a computer increases more quickly than the cost. Thus, for specific features on a computer, it costs less to buy those features now than it did a year ago.

Enrichment Masters, p. 44

Objective
Use a graphing calculator to determine if a group of graphs forms a family.

Recommended Time
25 minutes

Instructional Resources
Graphing Calculator Masters, pp. 22 and 23

These masters provide keystroking instruction for this lesson for the TI-81 and Casio graphing calculators.

1 FOCUS

Motivating the Lesson
Have students graph, with paper and pencil, the equations $y = \frac{x}{2}$, $y = x$, $y = 2x$, and $y = 5x$. This lesson shows students one of the benefits of the graphing calculator.

2 TEACH

Teaching Tip Emphasize that *same slope* means "parallel" and *same intercept* means "spokes of a wheel."

3 PRACTICE/APPLY

Assignment Guide
Core: 1–11
Enriched: 1–11

4 ASSESS

Observing students working with technology is an excellent method of assessment.

6-5A Graphing Technology
Families of Linear Graphs
A Preview of Lesson 6–5

What is a family? Generally, a family is a group of people that are related either by birth, marriage, or adoption. Graphs can also form families. A **family of graphs** includes graphs and equations of graphs that have at least one characteristic in common. That characteristic differentiates the group of graphs from other groups.

Families of linear graphs often fall into two categories—those with the same slope or those with the same intercept. A graphing calculator is a useful tool in studying a group of graphs to determine if they form a family.

Example 1 Graph $y = x$, $y = 2x$, and $y = 4x$ in the standard viewing window. Describe any similarities and differences among the graphs. Write a description of the family.

Clear the Y= list of all other equations.

Enter: $\boxed{Y=}$ $\boxed{X,T,\theta}$ $\boxed{ENTER}$ 2
$\boxed{X,T,\theta}$ $\boxed{ENTER}$ 4 $\boxed{X,T,\theta}$
$\boxed{ENTER}$ $\boxed{ZOOM}$ 6

The slope of each graph is positive and each graph goes through the origin. However, each graph has a different slope. The graph of $y = 4x$ is the steepest, and the graph of $y = x$ is the least steep.

You might describe this family of graphs as lines that pass through the origin. Another description might be lines that have 0 as their y-intercept.

You can use braces to enter equations that have a common characteristic. Since all of the equations in Example 1 are of the form $y = mx$, you can use braces to enter the different values of m using one step as follows.

Enter: $\boxed{Y=}$ $\boxed{2nd}$ $\boxed{\{}$ 1 $\boxed{,}$ 2 $\boxed{,}$ 4 $\boxed{2nd}$ $\boxed{\}}$ $\boxed{X,T,\theta}$ $\boxed{GRAPH}$

This tells the calculator to graphs equations for which the coefficients of x are 1, 2, and 4 and the y-intercept is 0.

Example 2 Graph $y = x$, $y = x + 3$, and $y = x - 3$ in the standard viewing window. Describe any similarities and differences among the graphs. Write a description of the family.

You can enter each of the graphs into the Y= list as you did in Example 1 or use braces to enter the equations in one step. Notice that all three equations can be written in the form $y = x + b$. The braces can be used to enter the different values of b.

Enter: $\boxed{Y=}$ $\boxed{X,T,\theta}$ $\boxed{+}$ $\boxed{2nd}$ $\boxed{\{}$ 0 $\boxed{,}$
3 $\boxed{,}$ $\boxed{(-)}$ 3 $\boxed{2nd}$ $\boxed{\}}$ 6 $\boxed{ZOOM}$

The slope of each graph is positive. However, each graph has a different *y*-intercept.

The graph of $y = x$ has a *y*-intercept of 0. The graph of $y = x + 3$ has a *y*-intercept of 3. The graph of $y = x - 3$ has a *y*-intercept of -3.

The graph of $y = x + 3$ is the same as the graph of $y = x$, but is moved 3 units up the *y*-axis. Likewise, the graph of $y = x - 3$ is the same as that of $y = x$, but is 3 units down the *y*-axis. The slope of each of these lines is 1.

This family of graphs can be described as linear graphs whose slope is 1.

Sometimes a common characteristic is not enough to determine that a group of equations describes a family of graphs.

Example ❸ Graph $y = -2x$, $y = -2x + 3$, and $y = -x + 3$ on the same screen. Describe any similarities and differences among the graphs.

The slope of each graph is negative. However, each graph has a different slope and *y*-intercept. The downward slope is not a distinctive characteristic that would differentiate these graphs from others. Therefore, this group of equations does not describe a family of graphs.

EXERCISES

Graph each set of equations on the same screen. Describe any similarities or differences among the graphs. State what the graphs have in common. 1–4. See margin.

1. $y = -x$
   $y = -2x$
   $y = -4x$

2. $y = x$
   $y = 2x$
   $y = 0.5x$

3. $y = -x$
   $y = -x + 2$
   $y = -x - 3$

4. Write a sentence comparing the graphs of equations with a positive coefficient of *x* and a negative coefficient of *x*.

Sketch a graph on paper to represent how you think the graph of each equation will appear. Describe any similarities to the graphs in Examples 1 and 2 and Exercises 1–3. Then use a graphing calculator to confirm your prediction and describe how accurate your sketch was.

5–8. See Solutions Manual.

9. Sample answer: $y = -3x + 2$

5. $y = x - 5$
6. $y = -x + 6$
7. $y = 0.1x$
8. $y = \frac{1}{3}x + 4$

9. Write an equation of a line whose graph has a negative slope.

10. Write an equation of a line whose graph lies between the graphs of $y = -x + 2$ and $y = -x + 3$. Sample answer: $y = -x + 2.5$

11. Write a paragraph explaining how the values of *m* and *b* in the slope-intercept form affect the graph of the equation. Include several graphs with your paragraph. See margin.

Lesson 6–5A Graphing Technology: Families of Linear Graphs **355**

Using Technology
This lesson offers an excellent opportunity for using technology in your algebra classroom. For more information on using technology, see *Graphing Calculators in the Mathematics Classroom,* one of the titles in the Glencoe Mathematics Professional Series.

Additional Answers

1. All graphs are of the family $y = -ax + 0$, where *a* represents different negative slopes.

2. All graphs are of the family $y = ax + 0$, where *a* represents different positive slopes.

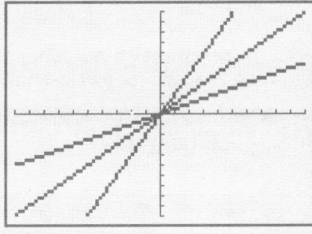

3. All graphs have the same slope, -1, but have different *y*-intercepts.

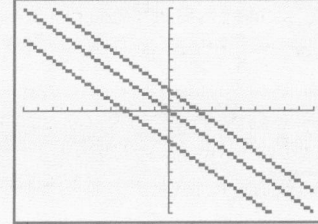

4. Graphs with positive *x* coefficients slant upward from left to right. Graphs with negative *x*-coefficients slant downward from left to right.

11. Sample answer: The value of *m* tells how the line will slant—upward or downward—and how steep it will be. The value of *b* tells where the line will intersect the *y*-axis. See students' work for graphs.

Instructional Resources

- Study Guide Master 6-5
- Practice Master 6-5
- Enrichment Master 6-5
- Graphing Calculator Masters, p. 6
- Tech Prep Applications Masters, p. 12

 Transparency 6-5A contains the 5-Minute Check for this lesson; **Transparency 6-5B** contains a teaching aid for this lesson.

Recommended Pacing

Standard Pacing	Day 8 of 13
Honors Pacing	Day 7 of 12
Block Scheduling*	Day 5 of 7
Alg. 1 in Two Years*	Days 10 & 11 of 18

 *For more information on pacing and possible lesson plans, refer to the *Block Scheduling Booklet* and *Algebra 1 in Two Years*.

1 FOCUS

 5-Minute Check
(over Lesson 6-4)

Find the slope, *x*-intercept, and *y*-intercept for the graph of each equation.

1. $4x - y = 16$ **4, 4, −16**
2. $2x + 3y = 8$ **$-\frac{2}{3}, 4, \frac{8}{3}$**
3. $y = 2x + 7$ **2, $-\frac{7}{2}$, 7**
4. $3x = y - 1$ **3, $-\frac{1}{3}$, 1**
5. $y = \frac{3}{4}x - 7$ **$\frac{3}{4}, \frac{28}{3}, -7$**

Motivating the Lesson

Questioning Introduce this lesson with a review of graphing a linear equation using two ordered pairs that satisfy the equation. Ask students to describe the procedure they would use to graph $x + 3y = 3$. Then have them graph the equation.

Graphing Linear Equations

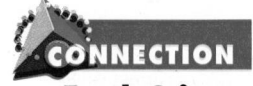
6–5

Objective

After studying this lesson, you should be able to:

- graph a line given any linear equation.

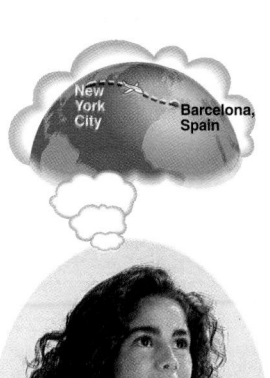

CONNECTION
Earth Science

Solome was flying from New York to Spain to be an exchange student for the spring quarter. After climbing to cruising level, the pilot announced that they were at an altitude of 13,700 meters and the outside temperature was about −76°C. You may think that since you get closer to the sun as you get farther from Earth, you might get warmer as altitude increases. However, as altitude increases, the air gets thinner and colder.

The air temperature above Earth on a day when the temperature at ground level is 15°C can be calculated using the formula $a + 150t = 2250$, where a is the altitude in meters and t is the temperature in degrees Celsius.

Suppose you wanted to graph $a + 150t = 2250$. In Chapter 5, you learned that you could graph a linear equation by finding and then graphing several ordered pairs that satisfy the equation.

Using the x- and y-intercepts is a convenient way to find ordered pairs to graph a linear equation. In $a + 150t = 2250$, these can be called the a-intercept and the t-intercept.

Find the a-intercept.

$$a + 150t = 2250$$
$$a + 150(0) = 2250 \quad \textit{Let } t = 0.$$
$$a = 2250$$

Find the t-intercept.

$$a + 150t = 2250$$
$$0 + 150t = 2250 \quad \textit{Let } a = 0.$$
$$t = 15$$

Now graph (2250, 0) and (0, 15), the ordered pairs for the intercepts. Draw the line representing the equation by connecting the points. Thus, you can graph a line if you know two points on that line.

Use the equation to confirm the pilot's announcement that the temperature was −76°C at 13,700 meters altitude.

$$a + 150t = 2250$$
$$13,700 + 150t = 2250$$
$$150t = -11,450$$
$$t = -76.33$$

So, the pilot was correct in saying that the outside temperature was about −76°C.

You can also graph a line if you know its slope and a point on the line. The forms of linear equations you have learned in this chapter can help you find the slope and a point on the line.

GLENCOE *Technology*

 CD-ROM Interaction

A multimedia simulation connects slopes and equations with family vacations. A blackline master activity with teacher's notes provides a follow-up to the CD-ROM simulation.

For Windows & Macintosh

Example **Graph $y - 1 = 3(x + 2)$.**

This equation is in point-slope form. The slope is 3, and a point on the line is $(-2, 1)$.

Graph the point $(-2, 1)$. Remember that the slope represents the change in y and x.

$$\frac{3}{1} = \frac{\text{change in } y}{\text{change in } x}$$

Thus, from the point $(-2, 1)$, you can move up 3 and right 1. Draw a dot. You can repeat this process to find another point on the graph. Draw a line connecting the points. This line is the graph of $y - 1 = 3(x + 2)$.

Check: Check $(-1, 4)$ to make sure that it satisfies the equation.

$$y - 1 = 3(x + 2)$$
$$4 - 1 \overset{?}{=} 3(-1 + 2)$$
$$3 = 3 \quad \checkmark$$

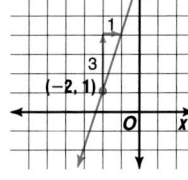

You can also use the slope-intercept form to help you graph an equation.

Example **Graph $\frac{3}{4}x + \frac{1}{2}y = 4$.**

Solve the equation for y to find the slope-intercept form.

$$\frac{3}{4}x + \frac{1}{2}y = 4$$

$$\frac{1}{2}y = -\frac{3}{4}x + 4 \quad \textit{Subtract } \tfrac{3}{4}x \textit{ from each side.}$$

$$y = -\frac{3}{2}x + 8 \quad \textit{Multiply each side by 2.}$$

The slope-intercept form of this equation tells us that the slope is $-\frac{3}{2}$ and the y-intercept is 8. Graph the y-intercept, and then use the slope to find another point on the line. Draw the line.

Be sure to check your second point to make sure it satisfies the *original* equation.

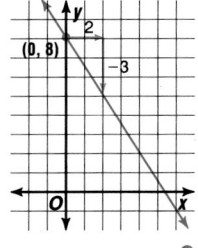

You frequently need to rewrite equations before you can determine the best way to graph them.

Example **Graph $\frac{4}{5}(2x - y) = 6x + \frac{2}{5}y - 10$.**

First, simplify the equation.

$$\frac{4}{5}(2x - y) = 6x + \frac{2}{5}y - 10$$

$$4(2x - y) = 30x + 2y - 50 \quad \textit{Multiply each side by 5 to eliminate the fractions.}$$

$$8x - 4y = 30x + 2y - 50$$

$$0 = 22x + 6y - 50 \quad \textit{Add } -8x \textit{ and } 4y \textit{ to each side.}$$

$$50 = 22x + 6y \quad \textit{Add 50 to each side.}$$

$$25 = 11x + 3y \quad \textit{Divide each side by 2.}$$

$$11x + 3y = 25 \quad \textit{Standard form}$$

(continued on the next page)

---

2 TEACH

Teaching Tip The method shown in Example 1 for checking graphs should be used only when a point on the graph appears to have integral coordinates. Point out that if the coordinates are not integers, the equation probably will not check. The best way to check such graphs is to find another ordered pair that satisfies the equation and then determine whether the point for the ordered pair lies on the line.

In-Class Examples

For Example 1
Graph $2x + 5y = 20$ using the x- and y-intercepts.

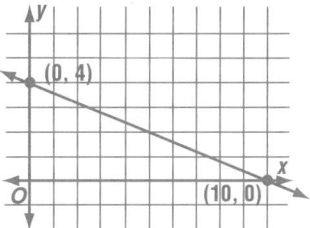

For Example 2
Graph $y = 4x - 1$ using the slope and y-intercept.

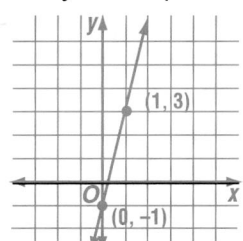

For Example 3
Graph $\frac{2}{3}(6x - 9y) = -\frac{8}{3}x - 4y - 4$.

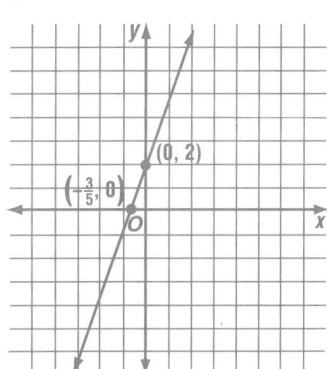

For Example 4

Graph the equation
$y = 2.3 + 7.413x$ and find
y for $x = 1.5$.

The food chain consists of a
chain of predator-prey
relationships. The mule deer is
the predator while the
bitterbrush plant is its prey.

The equation is now in standard form. You can choose one of several
methods to graph this equation.

ordered pairs	intercepts	slope-intercept form
If $x = 2$, $y = 1$.	If $x = 0$, $y = \frac{25}{3}$.	$11x + 3y = 25$
If $x = -1$, $y = 12$.	If $y = 0$, $x = \frac{25}{11}$.	$3y = -11x + 25$
Graph $(2, 1)$ and		$y = -\frac{11}{3}x + \frac{25}{3}$
$(-1, 12)$.	Graph $\left(0, \frac{25}{3}\right)$ and	The y-intercept is $\frac{25}{3}$,
	$\left(\frac{25}{11}, 0\right)$.	and the slope is $-\frac{11}{3}$.

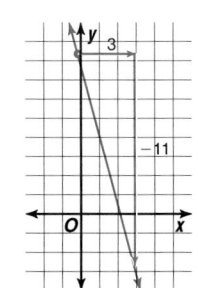

Notice that the result is the same regardless of the method you choose.
Choose the method that is easiest for you.

It is often helpful to make a sketch of the graph of an equation when you are
trying to observe patterns and make predictions.

Example 4

CONNECTION

Biology

The bitterbrush plant can grow
as high as 10 feet. However, a
plant this tall is very rare since
it is often the sole source of food
in many areas in the winter. Its
growth is constantly stunted by
grazing animals.

The mule deer likes to feed upon the bitterbrush plant, which is native
to the dry regions of western North America. The diameter of each
bitterbrush twig is related to the length of the twig by the function
$\ell = 1.25 + 8.983d$, where ℓ is the length (inches) of the twig, d is its
diameter in inches, and $d \geq 0.1$.

a. Graph the function.

b. Describe the relationship between a twig's diameter and length.

c. What would be the length of a twig if its diameter is 0.5 inch?

a. In this equation, d is the independent variable, and ℓ is the
dependent variable. Since d must be a nonnegative value, it may be
easier to choose two values of d and find the corresponding value
of ℓ to graph. A calculator may be helpful.

Let $d = 0.1$
$\ell = 1.25 + 8.983(0.1)$ or
2.1483

Let $d = 1$.
$\ell = 1.25 + 8.983(1)$ or
10.233

Graph $(0.1, 2.148)$ and
$(1, 10.233)$. Connect
them with a line.

Twig Length (in.)

Twig Diameter (in.)

GLENCOE Technology

**Interactive Mathematics
Tools Software**

This multimedia software provides
an interactive lesson by having
students graph an equation, given
the formula of the equation. A
Computer Journal gives students
an opportunity to write about what
they have learned.

For Windows & Macintosh

b. As the diameter of the twig increases, the length of the twig increases.

c. Let $d = 0.5$ and evaluate the equation.

$$\ell = 1.25 + 8.983(0.5)$$
$$= 5.7415$$

A 0.5-inch diameter twig is about 5.74 inches long.

Check to see if this point is on the graph.

CHECK FOR UNDERSTANDING

Communicating Mathematics

1. You can graph a line if you know the slope and a point on the line.

2. A twig cannot have a diameter of 0 and have any length at all.

MATH JOURNAL

Study the lesson. Then complete the following.   3. See margin.

1. **Write** a counterexample to the statement "You must always know two points that satisfy an equation in order to graph the equation."

2. **Explain** why the value of d cannot be zero in Example 4.

3. **Illustrate** how you graph a line if you know a point on the line and its slope.

4. **Describe** two ways in which you can interpret the slope $-\frac{3}{4}$ when using it to find another point on the graph of a line.   down 3, right 4; up 3, left 4

5. **Assess Yourself**   List all the possible ways you know to graph the equation of a line. Which of these methods do you prefer and why? See students' work.

Guided Practice

6–12. See Solutions Manual.

6. Graph $3x - 8y = 12$ by using the x- and y-intercepts.

7. Graph $y + 5 = -2(x + 1)$ by using the slope and a point on the line.

8. Graph $\frac{2}{3}x + \frac{1}{2}y = 3$ by using the slope and y-intercept.

Graph each equation.

9. $y = 2x - 3$ 10. $y = \frac{2}{5}x - 4$ 11. $6y + 12 = 18$ 12. $5x = 9$

13. **Health**   Did you know that your fingernails grow twice as fast as your toenails? Your fingernails grow about $\frac{1}{8}$ inch each month. Suppose the nail on your index finger is $\frac{1}{4}$ inch long.   a, d. See margin.

13b. $1\frac{3}{4}$ inches

13c. $\frac{1}{16}$ inch

a. Graph the line representing your fingernail's growth for a year.

b. How long would your nail be at the end of one year?

c. How much do your toenails grow each month?

d. Graph the line representing your toenail's growth for one year.

e. What does the slope of each line represent?   the rate of growth per month

EXERCISES

Practice

14–16. See Solutions Manual.

A

Graph each equation by using the x- and y-intercepts.

14. $y = 5x - 10$ 15. $6x - y = 9$ 16. $\frac{1}{2}x - \frac{2}{3}y = -6$

Reteaching

Using Cooperative Groups   Separate the class into groups consisting of 2 or 3 students each. Give each group a grid containing 4 different lines. Give a different grid to each group and have these groups determine the x-intercept, y-intercept, and the slope for each line on the graph. Rotate the different graphs around to each of the groups.

3 PRACTICE/APPLY

Check for Understanding

Exercises 1–13 are designed to help you assess your students' understanding through reading, writing, speaking, and modeling. You should work through Exercises 1–5 with your students and then monitor their work on Exercises 6–13.

Additional Answers

3. Sample answer: For (2, 3) and $m = \frac{2}{3}$, you graph (2, 3) and move 2 units up and 3 units right to plot the second point. You can repeat the movements to find additional points.

13a, d.

Study Guide Masters, p. 45

Core: 15–35 odd, 36, 37, 39–45
Enriched: 14–36 even, 37–45

For **Extra Practice,** see p. 770.

The red A, B, and C flags, printed only in the Teacher's Wraparound Edition, indicate the level of difficulty of the exercises.

Additional Answers

34.

35.

Practice Masters, p. 45

6-5

NAME_____ DATE_____

Practice

Student Edition
Pages 356–361

Graphing Linear Equations

Graph each equation by using the x- and y-intercepts.

1. $3x - y = 3$

2. $3y - 2x = 6$

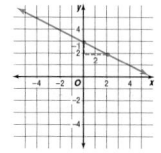

3. $4y + x = -2$

4. $x + y = -3$

Graph each equation by using the slope and y-intercept.

5. $y = -\frac{1}{2}x + 3$

6. $3y + 6 = 2x$

Graph each equation by using the slope and a point on the line.

17–33. See Solutions Manual.

17. $y - 2 = 3(x - 5)$
18. $y + 6 = -\frac{3}{2}(x + 5)$
19. $2(x - 3) = y + \frac{3}{2}$

Graph each equation by using the slope and y-intercept.

B 20. $y = -\frac{3}{4}x + 4$
21. $2x - 3y = -7$
22. $5y + 3 = -5$

Graph each equation.

23. $y = 3x - 5$
24. $6y + 5 = 5y + 3$
25. $3(y + 4) = -2x$
26. $y = \frac{2}{3}x + 1$
27. $5x + 2y = 20$
28. $\frac{2}{3}x - y = 4$
29. $6x - y = 8$
30. $15x - 29y = 429$
31. $y = 0.17x + 1.75$

C 32. $y - 3 = \frac{2}{3}(x - 6)$
33. $y + 1 = -2(x + 3)$

Graphing Calculator

Graph each group of equations on the same screen. Describe any similarities and differences among the graphs. 34–35. See margin.

34. $y = 2x + 4$
$2x - 3y = -12$
$y + 5 = 3(x + 3)$

35. $12x - 3y = 15$
$y = 4x - 11$
$y + x = 5x$

Critical Thinking

36. An equation belongs to the same family of graphs as $y = 3x + 5$, but it has a y-intercept of 7. How could you graph the equation without finding the equation first? It has the same slope as $y = 3x$. Graph the intercept and use the slope, 3.

Applications and Problem Solving

37. **Health** In cardiopulmonary resuscitation (CPR) on an adult, you initially give two breaths before you begin a cycle of one breath for every five pumps on the heart per minute. Each pump of the heart takes about a second. Thus, in a minute (60 seconds), you give about 60 ÷ 5 or 12 breaths. The number of breaths can be described by the equation $B = 2 + 12t$, where t represents the amount of time in minutes and B is the total number of breaths after pumping starts.

37a. See margin.

a. Graph the equation.

b. It is recommended that CPR be sustained until an emergency squad can get to the patient. If the squad takes 10 minutes to get to the patients, how many breaths should the patient have received? 122 breaths

Extension

Problem-Solving Graph each pair of equations on the same coordinate system, and determine whether the lines intersect. If they intersect, determine the coordinates of the point of intersection and check the coordinates in the original equations.

1. $2x - y - 5 = 0$
$x + 2y + 5 = 0$ yes; $(1, -3)$

2. $2x - 3y + 5 = 0$
$4x + 3y - 17 = 0$ yes; $(2, 3)$

3. $4x - 6y = -16$
$2x - 3y + 6 = 0$ no

38. Physical Science The length of an object's shadow depends on its height. At a certain time of day on a particular day of the year, the lengths of shadows can be calculated by the formula $S = 1.5h$, where h is the height and S is the length of the shadow.

a. Graph the equation.   **See margin.**

38b. 20 feet

b. At the time of day for which the formula is valid, the oak tree in your back yard is casting a 30-foot shadow. What is the height of the tree?

39. $y = \frac{2}{5}x + 12$

Mixed Review

39. Write an equation in slope-intercept form of the line with y-intercept 12 and slope the same as the line whose equation is $2x - 5y - 10 = 0$.   (Lesson 6–4)

40. See margin.

40. Draw a mapping for the relation {(1, 3), (2, 3), (2, 1), (3, 2)}.   (Lesson 5–2)

41. Travel Art leaves home at 10:00 A.M., driving at 50 miles per hour. At 11:30 A.M., Jennifer leaves home, going in the same direction, at 45 miles per hour. When will they be 100 miles apart?   (Lesson 4–7)   **4:30 P.M.**

42. Geometry Find the supplement of 87°.   (Lesson 3–4)   **93°**

43. Find the square root of 3.24.   (Lesson 2–8)   **1.8**

44. Find $-13 + (-9)$.   (Lesson 2–1)   **−22**

45. Write a verbal expression for $m - 1$.   (Lesson 1–1)   **a number *m* minus 1**

WORKING ON THE Investigation

Refer to the Investigation on pages 320–321.

Smoke Gets In Your Eyes

Which of your group members was closest to the average in number of breaths it took to blow up a balloon? Use that group member to conduct another experiment. For this activity, assume that each breath into a balloon approximates the exhaled smoke from one puff of a cigarette.

1 Have that member simulate the amount of smoke from each exhale during smoking by blowing into a balloon. Use a tape measure to determine the circumference of the balloon after each breath. Record each measurement.

2 Calculate the volume after each breath. Graph the relationship of the volume to each breath.

Describe the graph. Is it linear? Explain your answer.

3 There are 20 cigarettes in each pack. If a smoker smokes an entire pack of cigarettes, how many balloons would be filled? What would be the volume of smoke-filled air?

4 Graph the data relating the cumulative smoke in the air for each cigarette smoked in a pack of cigarettes.

5 If a smoker smokes an entire carton of cigarettes (10 packs/carton), how many balloons would be filled? What would be the volume of smoke-filled air?

Add the results of your work to your Investigation Folder.

Lesson 6–5 Graphing Linear Equations   **361**

Investigation

Working on the Investigation

The Investigation on pages 320–321 is designed to be a long-term project that is completed over several days or weeks. Encourage students to keep their materials in their Investigation Folder as they work on the Investigation.

---

4 ASSESS

Closing Activity

Modeling Using a cut-out paper line and grid, model the equation $y = 2x + 3$. Students should place the line so that it intersects the y-axis at 3, then tilt the line to a slope of 2.

Additional Answers

37a.

38a.

40.

Enrichment Masters, p. 45

6-5 NAME_____ DATE_____

Enrichment   Student Edition Pages 356–361

Formulas For a Line

You can think of the slope-intercept form of an equation for a line, $y = mx + b$, as a formula for the line given m and b.

Example: Write the formula for a line given two points (x_1, y_1) and (x_2, y_2).

Use the idea that the slope of a line is the same between any two points.

$$\frac{y - y_1}{x - x_1} = \frac{y_2 - y_1}{x_2 - x_1}$$

Multiply each side by $(x - x_1)$.

$$y - y_1 = \frac{y_2 - y_1}{x_2 - x_1}(x - x_1)$$ *Two-point formula*

If you know two points of a line, you can write an equation for the line directly using this formula.

Use the two-point formula to write an equation for the line that passes through the given points.   After exercise 1, only one of the two answers is given.

1. (3, 1) and (2, −1)
$y - 1 = 2(x - 3)$,
$y + 1 = 2(x - 2)$

2. (−3, 1) and (6, 7)
$y - 1 = \frac{2}{3}(x + 3)$

3. (−4, 4) and (0, 3)
$y - 4 = -\frac{1}{4}(x + 4)$

4. (−5, −3) and (7, −3)
$y = -3$

5. (10, 0) and (0, 9)
$y = -\frac{9}{10}(x - 10)$

6. $(a, 0)$ and $(0, b)$
$y = -\frac{b}{a}(x - a)$

Write the standard form of each equation in Exercises 1–6.

7. Exercise 1
$2x - y = 5$

8. Exercise 2
$2x - 3y = -9$

9. Exercise 3
$x + 4y = 12$

10. Exercise 4
$y = -3$

11. Exercise 5
$9x + 10y = 90$

12. Exercise 6
$bx + ay = ab$

Solve each problem.

13. Write a formula for an equation of a line given x-intercept a and y-intercept b.
$bx + ay = ab$, or $\frac{x}{a} + \frac{y}{b} = 1$

14. Write a formula for an equation of a line given the slope m and one point (x_1, y_1).
$y - y_1 = m(x - x_1)$
or $mx - y = mx_1 - y_1$

Chapter 6   **361**

NCTM Standards: 1–5, 8

Instructional Resources

- Study Guide Master 6-6
- Practice Master 6-6
- Enrichment Master 6-6
- Assessment and Evaluation Masters, p. 157
- Modeling Mathematics Masters, pp. 51–53, 77

 Transparency 6-6A contains the 5-Minute Check for this lesson; **Transparency 6-6B** contains a teaching aid for this lesson.

Recommended Pacing	
Standard Pacing	Days 9 & 10 of 13
Honors Pacing	Days 8 & 9 of 12
Block Scheduling*	Day 6 of 7 (along with Lesson 6-7)
Alg. 1 in Two Years*	Days 12 & 13 of 18

 *For more information on pacing and possible lesson plans, refer to the *Block Scheduling Booklet* and *Algebra 1 in Two Years.*

1 FOCUS

 5-Minute Check
(over Lesson 6-5)

Graph each equation.

1. $4x + y = 6$

$(0, 6)$
$(\frac{3}{2}, 0)$

2. $y = -2x + 5$

$(0, 5)$
$(\frac{5}{2}, 0)$

Integration: Geometry
Parallel and Perpendicular Lines

6-6

 World Cultures

Objectives

After studying this lesson, you should be able to:

- determine if two lines are parallel or perpendicular by their slopes, and
- write equations of lines that pass through a given point, parallel or perpendicular to the graph of a given equation.

Kite-making has been a national sport in some Far East countries since ancient times. In September, Chinese families celebrate the Ninth Day of the Ninth Moon as Kites' Day. Entire families go outside and fly kites. These kites may be simple diamond shapes or more complex shapes like fish, birds, dragons, and colorfully-dressed people.

The outline of a simple kite is shown on the coordinate plane at the right. The ordered pairs for the tips of the kite are labeled. We know that the two slats that support that kite meet at right angles. Lines that intersect at right angles are called **perpendicular lines.** *You will learn more about these lines later in the lesson.*

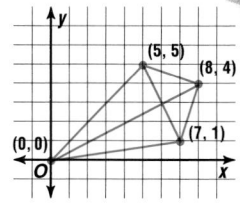
$(5, 5)$
$(8, 4)$
$(0, 0)$
$(7, 1)$

Lines in the same plane that never intersect are called **parallel lines.** Parallel lines can form families of graphs. In Lesson 6–5A, you learned that families of graphs can either have the same slope or the same intercept. The graph at the right shows a family of graphs represented by the following equations.

$$y = 2x$$
$$y = 2x + 3$$
$$y = 2x - 1$$

The slope of each line is 2. Notice that the lines seem to be parallel. In fact, they are parallel.

Definition of Parallel Lines in a Coordinate Plane	If two nonvertical lines have the same slope, then they are parallel. All vertical lines are parallel.

A **parallelogram** is a quadrilateral in which opposite sides are parallel. We can use slope to determine if quadrilaterals graphed on a coordinate plane are parallelograms. *The slope of a segment is the same as the slope of the line containing that segment.*

Example **1** Determine whether quadrilateral *ABCD* is a parallelogram if its vertices are $A(-5, -3)$, $B(5, 3)$, $C(7, 9)$, and $D(-3, 3)$.

Explore Graph the four vertices. Connect the vertices to form the quadrilateral.

Plan Use each pair of vertices to find the slope of each segment.

Use $m = \frac{y_2 - y_1}{x_2 - x_1}$.

 $\overline{AB}$ means a line segment whose endpoints are A and B.

Solve $\overline{AB}$: $m = \frac{3 - (-3)}{5 - (-5)} = \frac{6}{10}$ or $\frac{3}{5}$

$\overline{BC}$: $m = \frac{9 - 3}{7 - 5} = \frac{6}{2}$ or 3

$\overline{CD}$: $m = \frac{3 - 9}{-3 - 7} = \frac{-6}{-10}$ or $\frac{3}{5}$

$\overline{AD}$: $m = \frac{3 - (-3)}{-3 - (-5)} = \frac{6}{2}$ or 3

$\overline{AB}$ and $\overline{CD}$ have the same slope, $\frac{3}{5}$. $\overline{BC}$ and $\overline{AD}$ have the same slope, 3. Both pairs of opposite sides are parallel. Thus, *ABCD* is a parallelogram.

Examine Another definition of a parallelogram is that its opposite sides are congruent, or equal in length. Use a millimeter ruler to measure each side. $\overline{AB}$ and $\overline{CD}$ are approximately 37 millimeters long. $\overline{BC}$ and $\overline{AD}$ are approximately 20 millimeters long. The figure is probably a parallelogram.

You can write the equation of a line parallel to another line if you know a point on the line and the equation of the other line.

Example **2** Write an equation in slope-intercept form of the line that passes through (4, 0) and is parallel to the graph of $4x - 3y = 2$.

First find the slope by rewriting the equation in slope-intercept form.

$4x - 3y = 2$
$-3y = -4x + 2$
$y = \frac{4}{3}x - \frac{2}{3}$

The slope is $\frac{4}{3}$.

Method 1
Use point-slope form.
$y - y_1 = m(x - x_1)$
$y - 0 = \frac{4}{3}(x - 4)$ $m = \frac{4}{3}, x_1 = 4,$ and $y_1 = 0$
$y = \frac{4}{3}x - \frac{16}{3}$

This equation is in slope-intercept form.

An equation for the line is $y = \frac{4}{3}x - \frac{16}{3}$.

Method 2
Use the slope-intercept form.
$y = mx + b$
$0 = \frac{4}{3}(4) + b$
$0 = \frac{16}{3} + b$
$-\frac{16}{3} = b$

Now substitute the values for *m* and *b* into the slope-intercept form.
$y = \frac{4}{3}x - \frac{16}{3}$

Lesson 6-6 INTEGRATION *Geometry   Parallel and Perpendicular Lines*   **363**

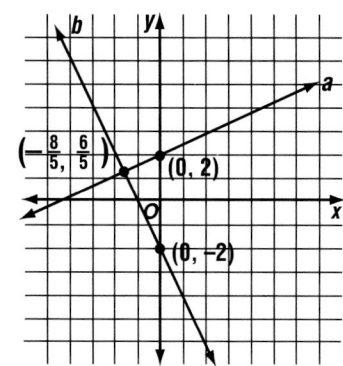
Sometimes graphs can be drawn in a way that leads you to the wrong conclusion. Mathematics can help you determine if a graph is misleading.

Example 3

CONNECTION
Health

The American Heart Association recommends that men keep their weight within a certain "healthy weight band" as shown in the graph at the right. Are the lines parallel as they appear in the graph?

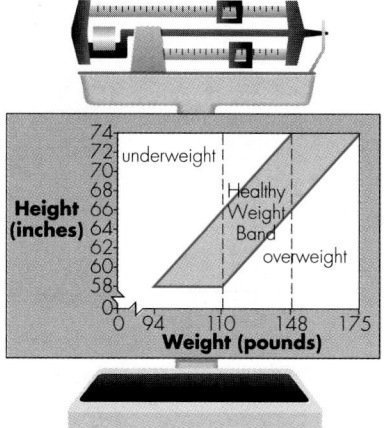

From the graph, we can determine that the endpoints of the minimum edge are (94, 58) and (148, 74). The endpoints of the maximum edge are (110, 58) and (175, 74). Find the slopes of each segment.

$m_1 = \frac{74 - 58}{148 - 94} = \frac{16}{54}$ or about 0.30

$m_2 = \frac{74 - 58}{175 - 110} = \frac{16}{65}$ or about 0.25

The segments do not have the same slope. Therefore, they are not parallel as they appear in this graph.

We have seen that the slopes of parallel lines are equal. What is the relationship of the slopes of perpendicular lines?

Example 4

APPLICATION
Kite-Making

Refer to the application at the beginning of the lesson.

a. **Write equations of the lines containing the slats of the kite in slope-intercept form.**

b. **Determine the relationship between the slopes of perpendicular lines.**

a. The endpoints of the longer slat are (0, 0) and (8, 4). The endpoints of the shorter slat are (5, 5) and (7, 1). Find the slope of each slat and then use the point-slope form to find an equation for each slat.

long slat

$m = \frac{4 - 0}{8 - 0} = \frac{4}{8}$ or $\frac{1}{2}$

Let $(x_1, y_1) = (0, 0)$.
$y - y_1 = m(x - x_1)$
$y - 0 = \frac{1}{2}(x - 0)$
$y = \frac{1}{2}x$

short slat

$m = \frac{1 - 5}{7 - 5} = \frac{-4}{2}$ or -2

Let $(x_1, y_1) = (5, 5)$.
$y - y_1 = m(x - x_1)$
$y - 5 = -2(x - 5)$
$y - 5 = -2x + 10$
$y = -2x + 15$

Equations of the slats are $y = \frac{1}{2}x$ and $y = -2x + 15$.

b. The slats of the kite are perpendicular. The slopes of the slats are $\frac{1}{2}$ and -2. The two slopes are negative reciprocals of each other. Note that their product is $\frac{1}{2}(-2)$ or -1.

These results suggest the following definition.

| Definition of Perpendicular Lines in a Coordinate Plane | If the product of the slopes of two lines is −1, then the lines are perpendicular. In a plane, vertical lines and horizontal lines are perpendicular. |

The slopes of perpendicular lines are negative reciprocals of each other.

You can use a right triangle and a coordinate grid to model the slopes of perpendicular lines.

MODELING MATHEMATICS

Perpendicular Lines on a Coordinate Plane

Materials: grid paper   scissors

a. A scalene triangle is one in which no two sides are equal. Cut out a scalene right triangle ABC so that $\angle C$ is the right angle. Label the vertices and the sides as shown at the right.

b. Draw a coordinate plane on the grid paper. Place $\triangle ABC$ on the coordinate plane so that A is at the origin and side b lies along the positive x-axis.

c. Name the coordinates of B. (b, a)

d. What is the slope of side c? $\dfrac{a}{b}$

Your Turn

a. Rotate the triangle 90° counterclockwise so that A is still at the origin and side b is along the positive y-axis.

b. Name the coordinates of B. $(-a, b)$

c. What is the slope of c? $-\dfrac{b}{a}$

d. What is the relationship between the first position of c and the second? **perpendicular lines**

e. What is the relationship between the slopes of c in each position? **negative reciprocals**

You can use your knowledge of perpendicular lines to write equations of lines perpendicular to a given line.

Example 5 Write the slope-intercept form of an equation that passes through $(8, -2)$ and is perpendicular to the graph of $5x - 3y = 7$.

First find the slope of the given line.

$$5x - 3y = 7$$
$$-3y = -5x + 7$$
$$y = \frac{5}{3}x - \frac{7}{3}$$

The slope of the line is $\dfrac{5}{3}$. The slope of the line perpendicular to this line is the negative reciprocal of $\dfrac{5}{3}$, or $-\dfrac{3}{5}$.

Use the point-slope form to find the equation.

$$y - y_1 = m(x - x_1)$$
$$y - (-2) = -\frac{3}{5}(x - 8) \quad \text{$m = -\frac{3}{5}$, $(x_1, y_1) = (8, -2)$}$$
$$y + 2 = -\frac{3}{5}x + \frac{24}{5} \quad \text{Distributive property}$$
$$y = -\frac{3}{5}x + \frac{14}{5} \quad \text{Subtract 2 or $\frac{10}{5}$ from each side.}$$

An equation for the line is $y = -\dfrac{3}{5}x + \dfrac{14}{5}$.

MODELING MATHEMATICS This hands-on exercise uses the hypotenuse of a scalene right triangle to generate perpendicular lines. Using the lengths of the adjacent sides enables students to visualize slopes.

In-Class Example

For Example 5
Write the slope-intercept form of an equation that passes through each point and is perpendicular to the graph of each equation.

a. $7x - 2y = 3$, $(4, -1)$
$$y = -\frac{2}{7}x + \frac{1}{7}$$

b. $2x + 5y = -3$, $(2, -3)$
$$y = \frac{5}{2}x - 8$$

Check for Understanding

Exercises 1–17 are designed to help you assess your students' understanding through reading, writing, speaking, and modeling. You should work through Exercises 1–5 with your students and then monitor their work on Exercises 6–17.

Additional Answers

3. Example 2

Example 5

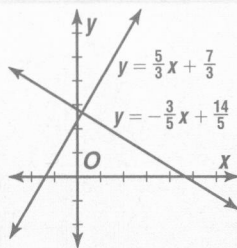

5b. The relationship would not work because there would be no perpendicular lines.

Study Guide Masters, p. 46

6-6

NAME_____ DATE _____
Student Edition
Study Guide Pages 362–368

Integration: Geometry
Parallel and Perpendicular Lines

When you graph two lines, you may encounter the two special types of graphs described at the right.

Parallel Lines and Perpendicular Lines
If two nonvertical lines have the same slope, then they are **parallel**. All vertical lines are parallel.
If the product of the slope of two lines is −1, then the lines are **perpendicular**. In a plane, vertical lines and horizontal lines are perpendicular.

Example: Write an equation in slope-intercept form of the line that passes through (0, 6) and is parallel to the graph of $2x - y = -12$.

The slope of the graph is 2. $\quad -\frac{A}{B} = -\frac{2}{-1}$, or 2

The slope-intercept form of an equation whose graph is parallel to the original graph is $y = 2x + b$. Substitute (0, 6) into the equation and solve for b.
$6 = 0x + b$
$b = 6 \quad$ **The y-intercept is b.**

The equation of the line is $y = 2x + 6$.

Since $2 \cdot \left(-\frac{1}{2}\right) = -1$, any line that is perpendicular to the line in the Example has an equation of the form $y = -\frac{1}{2}x + b$. If the line includes the point (−4, 3), then $3 = -\frac{1}{2}(-4) + b$ and thus $b = 1$. The equation of the line is $y = -\frac{1}{2}x + 1$.

Write an equation in slope-intercept form of the line that passes through the given point and is parallel to the graph of each equation.

1. $y = -\frac{2}{3}x + 4$; (1, −3)
$y = -\frac{2}{3}x - \frac{7}{3}$

2. $y = \frac{1}{2}x + 1$; (4, 2)
$y = \frac{1}{2}x$

3. $2x + y = 5$; (4, −6)
$y = -2x + 2$

4. $3x - y = 0$; (7, 2)
$y = 3x - 19$

5. $x - 8y = 10$; (0, 5)
$y = \frac{1}{8}x + 5$

6. $7x + 2y = 3$; (1, 6)
$y = -\frac{7}{2}x + \frac{19}{2}$

Write an equation in slope-intercept form of the line that passes through the given point and is perpendicular to the graph of each equation.

7. $4x - 3y = 7$; (5, −2)
$y = -\frac{3}{4}x + \frac{7}{4}$

8. $6x + 16y = 8$; (0, 4)
$y = \frac{8}{3}x + 4$

9. $y = 7x + 1$; (6, 4)
$y = -\frac{1}{7}x + \frac{34}{7}$

Communicating Mathematics

Study the lesson. Then complete the following.

1. **Describe** the relationship between the slopes of each of the following.
   a. two parallel lines   The slopes are equal.
   b. two perpendicular lines   The slopes are negative reciprocals.

2. Negative reciprocals are two numbers whose product is −1; −2 and $\frac{1}{2}$.

2. **Explain** what negative reciprocals are. Give an example.

3. **Check** the results of Examples 2 and 5 by graphing the original equation and the answer equation for each example on the same coordinate plane. Describe what happens.   See margin.

MODELING MATHEMATICS

4. Refer to the Modeling Mathematics activity. Why do you think the right triangle should be scalene?   If it were not, a and b would be equal.

5. Repeat the activity with a different size right triangle.
   a. Are the results different?   Slopes are different; relationship is not.
   b. What do you think would happen if you used a triangle that was not a right triangle?   See margin for sample answer.

Guided Practice

State the slopes of the lines parallel to and perpendicular to the graph of each equation.

6. $6x - 5y = 11$ $\quad \frac{6}{5}, -\frac{5}{6}$
7. $y = \frac{2}{3}x - \frac{4}{5}$ $\quad \frac{2}{3}, -\frac{3}{2}$
8. $3x = 10y - 3$ $\quad \frac{3}{10}, -\frac{10}{3}$

Determine whether the graphs of each pair of equations are parallel, perpendicular, or neither.

9. $3x - 7y = 1$, $7x + 3y = 4$
   perpendicular

10. $5x - 2y = 6$, $4y - 10x = -48$
    parallel

Write an equation in slope-intercept form of the line that passes through the given point and is parallel to the graph of each equation.

11. $(9, -3), 5x - 6y = 2$
$y = \frac{5}{6}x - \frac{21}{2}$

12. $(0, 4), 2y = 5x - 7$
$y = \frac{5}{2}x + 4$

13. $(7, -2), x - y = 0$
$y = x - 9$

Write an equation in slope-intercept form of the line that passes through the given point and is perpendicular to the graph of each equation.

14. $(8, 5), 7x + 4y = 23$
$y = \frac{4}{7}x + \frac{3}{7}$

15. $(0, 0), 9y = 3 - 5x$
$y = \frac{9}{5}x$

16. $(-2, 7), 2x - 5y = 3$
$y = -\frac{5}{2}x + 2$

17. **Geometry**   The diagonals of a square are segments that connect the opposite vertices. Use what you've learned in this lesson to determine the relationship between the diagonals $\overline{AC}$ and $\overline{BD}$ of the square graphed at the right.   perpendicular

Reteaching

Using Alternative Methods Put a graph of a line on an overhead and ask each row of students to pick a point not on the line. Students should then find the line that is perpendicular to the original line and that passes through the point they chose. Then they should find the line that is parallel to the original line and that passes through the point they chose. Use slope-intercept form.

EXERCISES

Practice

Determine whether the graphs of each pair of equations are *parallel, perpendicular,* or *neither.*

A

18. $y = -2x + 11$
$y + 2x = 23$
parallel

19. $3y = 2x + 14$
$2x - 3y = 2$
parallel

20. $y = -5x$
$y = 5x - 18$
neither

21. $y = 0.6x + 7$
$3y = -5x + 30$
perpendicular

22. $y = 3x + 5$
$y = 5x + 3$
neither

23. $y + 6 = -5$
$y + x = y + 7$
perpendicular

Write an equation in slope-intercept form of the line that passes through the given point and is parallel to the graph of each equation.

26. $y = -\frac{2}{3}x - \frac{22}{3}$

27. $y = \frac{8}{7}x + \frac{2}{7}$

24. $(2, -7), y = x - 2$
$y = x - 9$

25. $(2, 3), y = x + 5$
$y = x + 1$

26. $(-5, -4), 2x + 3y = -1$

27. $(5, 6), 8x - 7y = 23$

28. $(1, -2), y = -2x + 7$
$y = -2x$

29. $(0, -5), 5x - 2y = -7$
$y = 2.5x - 5$

30. $(5, -4), x - 3y = 8$
$y = \frac{1}{3}x - \frac{17}{3}$

31. $(-3, 2), 2x - 3y = 6$
$y = \frac{2}{3}x + 4$

32. $(2, -1), y = -0.5x + 2$
$y = -0.5x$

Write an equation in slope-intercept form of the line that passes through the given point and is perpendicular to the graph of each equation.

33. $y = -\frac{9}{2}x + 14$ **B**

34. $y = -3x - 8$

35. $y = 3x - 19$

36. $y = -\frac{5}{3}x + 8$

37. $y = -\frac{1}{5}x - 1$

38. $y = -\frac{3}{5}x + \frac{14}{5}$ **C**

43. $y = \frac{5}{4}x$

44. $y = -\frac{3}{5}x + \frac{9}{5}$

33. $(6, -13), 2x - 9y = 5$ **34.** $(-3, 1), y = \frac{1}{3}x + 2$ **35.** $(6, -1), 3y + x = 3$

36. $(6, -2), y = \frac{3}{5}x - 4$ **37.** $(0, -1), 5x - y = 3$ **38.** $(8, -2), 5x - 7 = 3y$

39. $(4, -3), 2x - 7y = 12$ **40.** $(3, 7), y = \frac{3}{4}x - 1$ **41.** $(3, -3), 3x + 7 = 2x$
$y = -\frac{7}{2}x + 11$ $\qquad$ $y = -\frac{4}{3}x + 11$ $\qquad$ $y = -3$

Write an equation of the line having the following properties.

42. passes through $(-5, 3)$ and is perpendicular to the x-axis $\quad x = -5$

43. is parallel to the graph of $y = \frac{5}{4}x - 3$ and passes through the origin.

44. has an x-intercept of 3 and is perpendicular to the graph of $5x - 3y = 2$

45. has a y-intercept of -6 and is parallel to the graph of $x - 3y = 8$
$y = \frac{1}{3}x - 6$

Critical Thinking

46. Lines a, b, and c lie on the same coordinate plane. Line a is perpendicular to line b and intersects it at $(3, 6)$. Line b is perpendicular to line c, which is the graph of $y = 2x + 3$.
 a. Write the equation of lines a and b. $\quad y = 2x, y = -\frac{1}{2}x + \frac{15}{2}$
 b. Graph the three lines. See margin.
 c. What is the relationship of lines a and c? parallel

47. No, because the slope of $\overline{AC}$ is $\frac{6}{7}$ and the slope of $\overline{BD}$ is $-\frac{2}{3}$. These are not negative reciprocals, so the diagonals are not perpendicular.

Applications and Problem Solving

47. Geometry A rhombus is a parallelogram that has perpendicular diagonals. Determine if quadrilateral $ABCD$ with $A(-2, 1)$, $B(3, 3)$, $C(5, 7)$, and $D(0, 5)$ is a rhombus. Explain.

Error Analysis
Watch that students do not just use the coefficient of x for the slope. They must solve for y first.
Example: $-x - 4y = 2$
The slope is not -1. Solve for y.
$$-4y = x + 2$$
$$y = -\frac{1}{4}x - \frac{1}{2}$$
The slope is $-\frac{1}{4}$.

Assignment Guide

Core: 19–45 odd, 46, 47, 49–58
Enriched: 18–44 even, 46–58

For **Extra Practice,** see p. 771.

The red A, B, and C flags, printed only in the Teacher's Wraparound Edition, indicate the level of difficulty of the exercises.

Additional Answer

46b.

Practice Masters, p. 46

NAME_____ DATE _____

6-6

Practice

Student Edition
Pages 362–368

Integration: Geometry
Parallel and Perpendicular Lines

Write an equation in slope-intercept form of the line that passes through the given point and is parallel to the graph of each equation.

1. $2x + y = 5$; (3, 1)
$y = -2x + 7$

2. $3x - y = 5$; (-1, -2)
$y = 3x + 1$

3. $5x - 4y = 1$; (-8, 2)
$y = \frac{5}{4}x + 12$

4. $9x + 3y = 8$; (-1, -4)
$y = -3x - 7$

5. $y = \frac{4}{3}x + 5$; (12, 3)
$y = \frac{4}{3}x - 13$

6. $y = -\frac{3}{4}x + \frac{1}{4}$; (4, -2)
$y = -\frac{3}{4}x + 1$

Write an equation in slope-intercept form of the line that passes through the given point and is perpendicular to the graph of each equation.

7. $x - 6y = 2$; (2, 4)
$y = -6x + 16$

8. $3x + 2y = -7$; (1, 1)
$y = \frac{2}{3}x + \frac{1}{3}$

9. $5x + 4y = 8$; (10, 5)
$y = \frac{4}{5}x - 3$

10. $4x + 3y = -6$; (2, 1)
$y = \frac{3}{4}x - \frac{1}{2}$

11. $y = \frac{1}{4}x - 4$; (-2, 3)
$y = -4x - 5$

12. $2x + 10y = 3$; (2, 3)
$y = 5x - 7$

13. $x = 2y - 1$; (0, 0)
$y = -2x$

14. $4x + 7y = 6$; (-4, 1)
$y = \frac{7}{4}x + 8$

Closing Activity

Speaking Ask students to explain the relationship of the line represented by $ax + by = c$ and all the lines parallel and perpendicular to it.

Chapter 6, Quiz C (Lessons 6-5 and 6-6), is available in the *Assessment and Evaluation Masters*, p. 157.

Additional Answers

50.

(0, 3.5)

(−1, 0)

51.

Fuel Used (gal)

Miles Driven

Enrichment Masters, p. 46

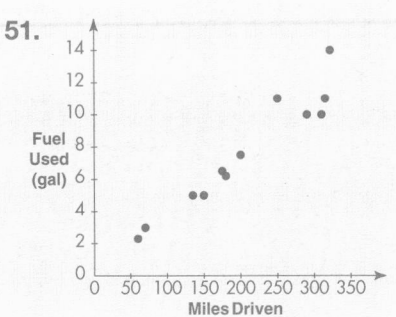

48. **Construction** A large roller is being used to flatten new asphalt on a hill that has a slope of $-\frac{1}{10}$. The steering mechanism for the front roller is perpendicular to the hill.
   a. What is the slope of the steering mechanism? **10**
   b. If the bottom of the hill has coordinates (9, 3) and the point directly below the steering mechanism is at $(-1, 4)$, write the equations of the lines representing the hill and the steering mechanism.

48b. $y = -\frac{x}{10} + \frac{39}{10}$, $y = 10x + 14$

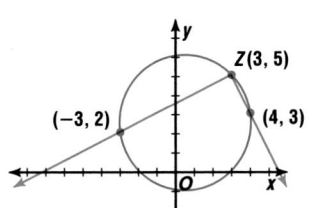
steering mechanism

49. **Geometry** An angle inscribed in a circle is one in which the vertex of the angle lies on the circle and the two sides intersect the circle, as shown at the right.
   a. Find the equations of the lines that contain the sides of $\angle Z$.
   b. What type of angle is $\angle Z$? **right or 90° angle**

49a. $y = \frac{1}{2}x + \frac{7}{2}$, $y = -2x + 11$

Z(3, 5)
(−3, 2)
(4, 3)

Mixed Review

50. Graph $7x - 2y = -7$ by using the x- and y-intercepts. (Lesson 6–5)

50. See margin.

51. **Statistics** Make a scatter plot of the data below. (Lesson 6–3) **See margin.**

Miles Driven	200	322	250	290	310	135	60	150	180	70	315	175
Fuel Used (gallons)	7.5	14	11	10	10	5	2.3	5	6.2	3	11	6.5

52. **Patterns** Copy and complete the table below. (Lesson 5–4)

x	−1	2	5	8	11	14
f(x)	5	−1	−7	−13	−19	−25

55a. $\frac{5}{12}$; $\frac{7}{12}$

53. Find 4% of $6070. (Lesson 4–4) **$242.80**

54. Solve $16 = \frac{s-8}{-7}$. Check your solution. (Lesson 3–3) **−104**

55. **Metallurgy** The gold content of jewelry is given in karats. For example, 24-karat gold is pure gold, and 18-karat gold is $\frac{18}{24}$ or 0.75 gold. (Lesson 2–7)
   a. What fraction of 10-karat gold is pure gold? What fraction is *not* gold?
   b. If a piece of jewelry is $\frac{2}{3}$ gold, how would you describe it using karats? **16-karat gold**

56. Find $-0.0005 + (-0.3)$. (Lesson 2–5) **−0.3005**

57. Name the property illustrated by $1(87) = 87$. (Lesson 1–6) **identity (×)**

58. Solve $x = 6 + 0.28$. (Lesson 1–5) **6.28**

Extension

Communication Have students explain the method for finding an equation in standard form when they know one point and a line to which it is parallel.

Tech Prep

Metallurgist Students who are interested in metallurgy may wish to do further research on the data provided in Exercise 55 and explore the potential growth of this career. For more information on tech prep, see the *Teacher's Handbook.*

Integration: Geometry
Midpoint of a Line Segment

Objective

After studying this lesson, you should be able to:

- find the coordinates of the midpoint of a line segment in the coordinate plane.

APPLICATION
Woodworking

Maria Hernandez made a frame for one of her watercolor paintings in a class offered at the community college. Because her canvas was heavy and the wooden frame was light, her instructor suggested that she stabilize her frame by adding crossbars connecting the consecutive sides. One of the books she used as a reference showed the crossbars inserted so that the bars are placed on the sides of the frame at the **midpoint** of each side. The midpoint of a line segment is the point that is halfway between the endpoints of the segment.

Maria decided to use some grid paper to model her frame and the crossbars she is going to insert. Each square of the paper represented 2 inches of her frame. To find the midpoint of the interior vertical edge, she counted how many units it was from one corner to the next and divided by 2. The inside of the frame is 16 inches high, so the midpoint is 8 inches from the corner. She counted 4 units on the grid and placed a bullet on her drawing. She used a similar method to find the midpoint of the interior horizontal edge. The inside is 20 inches wide, so the midpoint is 10 inches from the end.

Suppose we place Maria's frame drawing on a coordinate plane. Let's list the coordinates of the corners of the frame and the midpoints of the sides. What pattern do you notice?

$A(2, 2)$
$B(22, 2)$
$C(22, 18)$
$D(2, 18)$

Midpoint of $\overline{AB}$: (12, 2)
Midpoint of $\overline{BC}$: (22, 10)
Midpoint of $\overline{CD}$: (12, 18)
Midpoint of $\overline{AD}$: (2, 10)

Notice that the coordinates of the midpoint of $\overline{AB}$ are the mean of the corresponding coordinates of A and B.

Midpoint of $\overline{AB}$:　　(12, 2)

mean of x-coordinates   *mean of y-coordinates*
of A(2, 2) and B(22, 2)   *of A(2, 2) and B(22, 2)*

$$\frac{2+22}{2} = 12 \qquad\qquad \frac{2+2}{2} = 2$$

The woodworking example uses the midpoints of horizontal and vertical line segments. The Modeling Mathematics activity on the next page applies to any line segment.

6-7 LESSON NOTES

NCTM Standards: 1–5, 8

Instructional Resources

- Study Guide Master 6-7
- Practice Master 6-7
- Enrichment Master 6-7
- Assessment and Evaluation Masters, p. 157

Transparency 6-7A contains the 5-Minute Check for this lesson; **Transparency 6-7B** contains a teaching aid for this lesson.

Recommended Pacing	
Standard Pacing	Day 11 of 13
Honors Pacing	Day 10 of 12
Block Scheduling*	Day 6 of 7 (along with Lesson 6-6)
Alg. 1 in Two Years*	Days 14 & 15 of 18

*For more information on pacing and possible lesson plans, refer to the *Block Scheduling Booklet* and *Algebra 1 in Two Years.*

1 FOCUS

5-Minute Check
(over Lesson 6-6)

Write equations in standard form of the lines that pass through the given point and are parallel and perpendicular to the graph of each equation.

1. $5x - 3y = 8$, $(4, -2)$
$5x - 3y = 26$, $3x + 5y = 2$
2. $7x + 2y = 4$, $(-6, 1)$
$7x + 2y = -40$,
$2x - 7y = -19$
3. $x = 7$, $(4, -3)$
$x = 4$, $y = -3$
4. $x - y = 3$, $(4, 6)$
$x - y = -2$, $x + y = 10$

Motivating the Lesson

Hands-On Activity Students are to hang a framed painting. The hook needs to be placed at a point one third the distance from the top to the bottom and at the midpoint of the horizontal dimension. They can use only a ruler to locate the point.

2 TEACH

Teaching Tip Point out that the midpoint is computed using the average of the x-coordinates and the y-coordinates.

In-Class Example

For Example 1
If *P* is the midpoint of line segment *AB,* find the coordinates of the missing point *A* or *B.*

a. A(3, 2), P(6, 3)   B(9, 4)
b. A(6, 3), P(3, 2)   B(0, 1)
c. A(3, 2), P(1, 5)   B(−1, 8)
d. P(6, 3), B(1, 5)   A(11, 1)
e. P(1, 5), B(6, 3)   A(−4, 7)

MODELING MATHEMATICS

Midpoint of a Line Segment

Materials: grid paper   ruler   colored pencils

- Create a coordinate plane on your grid paper.

- Use a ruler to draw any line segment on the coordinate plane. The segment should not be vertical or horizontal. Label the endpoints of the segment *A* and *B* with their coordinates.

- Hold the paper up to the light and fold it so that *A* and *B* coincide. Crease the paper at the fold.

- Unfold the paper. Label the point *M* at which the crease meets the segment and write the coordinates of *M*.

- Use a different color pencil to draw a vertical line through the lower endpoint of your segment. Then draw a horizontal line through the upper endpoint

of your segment. Label the point of intersection *P*.

- Write the coordinates of the midpoints of $\overline{AP}$ and $\overline{BP}$.

- How do the x-coordinate of the vertical segment and the y-coordinate of the horizontal segment compare with the coordinates of *M*?

Your Turn   a–c. See student's work.

a. Create another coordinate plane and draw a segment with a different slope from the first one you drew.

b. Repeat the activity with this segment. What are your results?

c. Write a general rule for finding the midpoint of any segment.

This activity suggests the following rule for finding the midpoint of any segment, given the coordinates of its endpoints.

Midpoint of a Line Segment on a Coordinate Plane	The coordinates of the midpoint of a line segment whose endpoints are at (x_1, y_1) and (x_2, y_2) are given by $\left(\dfrac{x_1 + x_2}{2}, \dfrac{y_1 + y_2}{2}\right)$.

Example **1**   If the vertices of parallelogram *WXYZ* are W(3, 0), X(9, 3), Y(7, 10), and Z(1, 7), prove that the diagonals bisect each other. That is, prove that they intersect at their midpoints.

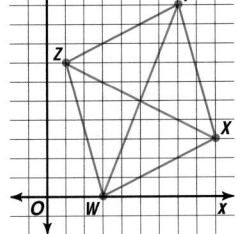

Explore   Graph the vertices and draw the parallelogram and its diagonals.

Plan   Find the coordinates of the midpoints of the diagonals to see if they are the same point.

Solve   Find the midpoints of $\overline{WY}$ and $\overline{XZ}$.

$W(x_1, y_1) = W(3, 0)$ $X(x_1, y_1) = X(9, 3)$
$Y(x_2, y_2) = Y(7, 10)$ $Z(x_2, y_2) = Z(1, 7)$

midpoint of $\overline{WY} = \left(\dfrac{3+7}{2}, \dfrac{0+10}{2}\right)$ midpoint of $\overline{XZ} = \left(\dfrac{9+1}{2}, \dfrac{3+7}{2}\right)$

$\qquad\qquad = \left(\dfrac{10}{2}, \dfrac{10}{2}\right)$ $\qquad\qquad = \left(\dfrac{10}{2}, \dfrac{10}{2}\right)$

$\qquad\qquad = (5, 5)$ $\qquad\qquad = (5, 5)$

Examine   Since the midpoint of $\overline{XZ}$ has the same coordinates as the midpoint of $\overline{WY}$, the diagonals bisect each other.

Alternative Teaching Strategies

Reading Algebra In non-mathematical uses, the word *middle* and the prefix *mid-* both mean "between." But in mathematics, they mean "exactly halfway between."

If you know the midpoint and one endpoint of a segment, you can find the other endpoint.

Example The center of a circle is $M(0, 2)$, and the endpoint of one of its radii is $A(-6, -4)$. If $\overline{AB}$ is a diameter of the circle, what are the coordinates of B?

Use the midpoint formula and substitute the values you know.

$$(x, y) = \left(\frac{x_1 + x_2}{2}, \frac{y_1 + y_2}{2}\right)$$

$$(0, 2) = \left(\frac{-6 + x_2}{2}, \frac{-4 + y_2}{2}\right) \quad \begin{array}{l} (x, y) = (0, 2), \\ (x_1, y_1) = (-6, -4) \end{array}$$

The x- and y-coordinates of the ordered pairs are equal since the ordered pairs are equal. So, two equations can be formed.

Find the x-coordinate.

$$0 = \frac{-6 + x_2}{2}$$
$$0 = -6 + x_2$$
$$6 = x_2$$

Find the y-coordinate.

$$2 = \frac{-4 + y_2}{2}$$
$$4 = -4 + y_2$$
$$8 = y_2$$

The coordinates of endpoint $B(x_2, y_2)$ are $(6, 8)$.

Check: Copy the graph of the circle and points A and M. Extend the radius through M to intersect the circle. This segment is the diameter of the circle. The intersection point is at $(6, 8)$, which are the coordinates of point B.

CHECK FOR UNDERSTANDING

Communicating Mathematics

Study the lesson. Then complete the following.

1. **Explain** how you can use the midpoint formula to find the midpoint of each of the following. a–b. See margin.
   a. a vertical line segment
   b. a horizontal line segment

2. **Demonstrate** how to find the coordinates of vertex C of rectangle $ABCD$ if its two diagonals meet at $M(-3, 4)$ and the coordinates of A are $(0, -1)$. $C(-6, 9)$

3. Draw a rectangle on a coordinate plane. a–c. See students' work.
   a. Find the midpoint of each side of the rectangle.
   b. Find the midpoint of the diagonals.
   c. How do the coordinates of the midpoint of the diagonal relate to the coordinates of the midpoints of the sides?

In-Class Example

For Example 2
The endpoints of a diameter of a circle are at (4, 2) and (8, 26). Find the coordinates of the center of the circle. (6, 22)

3 PRACTICE/APPLY

Check for Understanding
Exercises 1–12 are designed to help you assess your students' understanding through reading, writing, speaking, and modeling. You should work through Exercises 1–3 with your students and then monitor their work on Exercises 4–12.

Additional Answers

1a. The x-coordinates will be the same, so calculate the average of the y-coordinates.
1b. The y-coordinates will be the same, so calculate the average of the x-coordinates.

Reteaching

Using Problem Solving Have students find the midpoint of a given segment, then the midpoint of half of the original segment, and so on. For example, given points $A(-4, -5)$ and $B(12, 3)$, find the coordinates for X, Y, Z if X is the midpoint of $\overline{AB}$, Y is the midpoint of $\overline{AX}$, and Z is the midpoint of $\overline{AY}$.

Guided Practice

Find the coordinates of the midpoint of a segment with each pair of endpoints.

4. $A(5, -2), B(5, 8)$ $(5, 3)$
5. $C(-5, 6), D(8, 6)$ $(1.5, 6)$
6. $E(6, 5), F(14, 7)$ $(10, 6)$
7. $G(-9, -6), H(-3, 8)$ $(-6, 1)$

Find the coordinates of the other endpoint of a segment given one endpoint and the midpoint **M.**

8. $A(3, 6), M(-1.5, 5)$ $(-6, 4)$
9. $B(-3, 7), M(-7, 7)$ $(-11, 7)$
10. $C(8, 11), M\left(5, \frac{17}{2}\right)$ $(2, 6)$
11. $D(8, 4.5), M(8, 7.15)$ $(8, 9.8)$

12. **Archery** The target for archery competition is composed of 10 rings and 5 colored circles. Hitting the bull's-eye is worth ten points, and hitting the outer ring is worth one point. A target is shown on a coordinate plane at the right. The two labeled points are the endpoints of a diameter of the target. Find the coordinates of the center of the bull's-eye. $(-3, 5)$

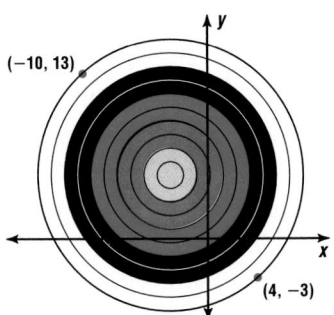

EXERCISES

Practice

Find the coordinates of the midpoint of a segment with each pair of endpoints.

13. $J(6, 6), K(19, 6)$ $(12.5, 6)$
14. $L(5, 9), M(-7, 3)$ $(-1, 6)$
15. $P(-5, 9), Q(7, 1)$ $(1, 5)$
16. $R(7, 4), S(11, -10)$ $(9, -3)$
17. $T(8, -7), U(-2, 11)$ $(3, 2)$
18. $V(-8, 1.2), W(-8, 7.4)$ $(-8, 4.3)$
19. $X(-3, 9), Y(4, -7)$
20. $B(9, -4), C(2, 7)$
21. $D(4.7, -2.9), E(-3.1, 8.3)$
22. $F(a, b), G(c, d)$
23. $Y(6x, 14y), Z(2x, 4y)$ $(4x, 9y)$
24. $M(-2w, -7v), N(6w, 2v)$ $\left(2w, -\frac{5}{2}v\right)$

19. $\left(\frac{1}{2}, 1\right)$

20. $\left(\frac{11}{2}, \frac{3}{2}\right)$

21. $(0.8, 2.7)$

22. $\left(\frac{a+c}{2}, \frac{b+d}{2}\right)$

Find the coordinates of the other endpoint of a segment given one endpoint and the midpoint **M.**

25. $E(-7, 8), M(-7, 4)$ $(-7, 0)$
26. $F(4, 2), M(2, 1)$ $(0, 0)$
27. $G(3, -6), M(12, -6)$ $(21, -6)$
28. $H(5, 3), M(6, 4)$ $(7, 5)$
29. $L(-8, 4), M\left(\frac{1}{2}, 7\right)$ $(9, 10)$
30. $N(5, -9), M(8, -7.5)$ $(11, -6)$
31. $R\left(\frac{1}{6}, \frac{1}{3}\right), M\left(\frac{1}{2}, \frac{1}{3}\right)$ $\left(\frac{5}{6}, \frac{1}{3}\right)$
32. $S(a, b), M\left(\frac{x+a}{2}, \frac{y+b}{2}\right)$ (x, y)

33. $B(2.3, 6.8)$
34. $A(-2.9, 7.1)$
35. $P(6.65, -1.85)$
36. $B(3.8, 4.5)$

If **P** is the midpoint of segment **AB,** find the coordinates of the missing point.

33. $A(6.5, -8.2), P(4.4, -0.7)$
34. $P(1.2, 4.5), B(5.3, 1.9)$
35. $A(9.7, -5.4), B(3.6, 1.7)$
36. $A(-5.9, 7.2), P(-1.05, 5.85)$

372 Chapter 6 *Analyzing Linear Equations*

37. The coordinates of the endpoints of a diameter of a circle are (5, 8) and (−7, 2). Find the coordinates of the center. **(−1, 5)**

38. Find the coordinates of the point on $\overline{AB}$ that is one-fourth the distance from A to B for $A(8, 3)$ and $B(10, −5)$. **(8.5, 1)**

39. The endpoints of $\overline{AB}$ are $A(−2, 7)$ and $B(6, −5)$. Find the coordinates of P if P lies on $\overline{AB}$ and is $\frac{3}{8}$ the distance from A to B. $\left(1, \frac{5}{2}\right)$

Critical Thinking

40. Points W, X, Y, and Z are the midpoints of the sides of quadrilateral $QUAD$. Write a convincing argument that quadrilateral $WXYZ$ is a parallelogram. **The vertices are $W(2, −3)$, $X(6, 2)$, $Y(2, 4)$, and $Z(−2, −1)$. The slope of $\overline{XW}$ = slope of $\overline{YZ} = \frac{5}{4}$. The slope of $\overline{WZ}$ = slope of $\overline{XY} = −\frac{1}{2}$. The opposite sides are parallel, so $WXYZ$ is a parallelogram.**

Applications and Problem Solving

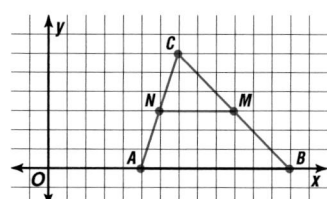

41. Geometry N is the midpoint of $\overline{AC}$ in the triangle shown at the left. M is the midpoint of $\overline{BC}$. **a. $M(6, 3)$, $M(10, 3)$**
 a. Write the coordinates of M and N.
 b. Compare $\overline{MN}$ and $\overline{AB}$. Verify your findings. **parallel, $MN = \frac{1}{2}AB$**

42. Technology The screen of a TI-82 graphing calculator is composed of tiny dots called *pixels* that are turned on or off during a display. Each pixel is named by the row and column in which it lies.

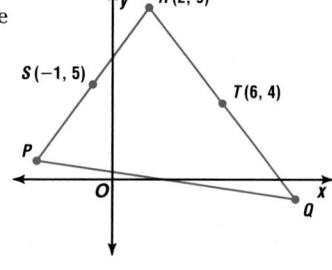

 a. If a segment drawn on the screen has one endpoint at pixel (20, 43) and its midpoint is at pixel (30, 60), what is the pixel name for the other endpoint of the segment? **(40, 77)**
 b. What is the pixel name of the origin? **(31, 47)**
 c. How does this coordinate system differ from the coordinate system we usually use for graphing? **Ordered pairs are (y, x) and the origin is in a different place.**

43. Geometry S and T are the midpoints of their respective sides in the triangle shown at the right.
 a. Find the coordinates of P and Q. **$P(−4, 1)$, $Q(10, −1)$**
 b. If the area of triangle RST is 15.5 square units, what is the area of triangle RPQ? Explain how you arrived at your answer. **62 square units; See margin for sample explanation.**

Extension

Connections The diameter of a circle has its endpoints at (8, 6) and (2, −2). Find the center of the circle and at least two other points on the circle. **Center is at (5, 2); answers may vary for other points on the circle. Possible answers include (2, 6), (8, −2), (9, 5), (9, −1), (1, 5), (1, −1), (10, 2), and (0, 2).**

---

4 ASSESS

Closing Activity

Speaking Ask students to explain the procedure they would use to determine the midpoint of a line segment.

Chapter 6, Quiz D (Lesson 7), is available in the *Assessment and Evaluation Masters*, p. 157.

Additional Answer

43b. Sample answer: The area of the smaller triangle is $\frac{1}{2}bh$. Since the base of the larger triangle is twice that of the smaller one and the height is also twice the length of the small one, the area of the larger is $\frac{1}{2}(2b)(2h)$, or $2bh$. This is 4 times the area of the small one.

Practice Masters, p. 47

Enrichment Masters, p. 47

NAME_____ DATE _____
6-7 Enrichment
Student Edition
Pages 369–374

Celsius and Kelvin Temperatures

If you blow up a balloon and put it in the refrigerator, the balloon will shrink as the temperature of the air in the balloon decreases.

The volume of a certain gas is measured at 30° Celsius. The temperature is decreased and the volume is measured again.

Temperature (t)	Volume (v)
30°C	202 mL
21°C	196 mL
0°C	182 mL
−12°C	174 mL
−27°C	164 mL

1. Graph this table on the coordinate plane provided below.

2. Find the equation of the line that passes through the points you graphed in Exercise 1. $y = \frac{2}{3}x + 182$ or $v = \frac{2}{3}t + 182$

3. Use the equation you found in Exercise 2 to find the temperature that would give a volume of zero. This temperature is the lowest one possible and is called "absolute zero." −273°C

4. In 1848 Lord Kelvin proposed a new temperature scale with 0 being assigned to absolute zero. The size of the degree chosen was the same size as the Celsius degree. Change each of the Celsius temperatures in the table above to degrees Kelvin. 303°, 294°, 273°, 261°, 246°

Mixed Review

44. $y = -\frac{3}{5}x + \frac{14}{5}$

45. $y = -\frac{7}{9}x - \frac{8}{3}$

44. Write an equation in slope-intercept form of the line that passes through (8, −2) and is perpendicular to the graph of $5x - 3y = 7$. (Lesson 6–6)

45. Write an equation in slope-intercept form of the line that passes through (−6, 2) and (3, −5). (Lesson 6–5)

46. Given $g(x) = x^2 - x$, find $g(4b)$. (Lesson 5–5) $16b^2 - 4b$

47. Determine whether $9x + (-7) = 6y$ is a linear equation. If it is, rewrite it in standard form. (Lesson 5–4) **yes; $9x - 6y = 7$**

48. **Trigonometry** If $\sin N = 0.6124$, use a calculator to find the measure of angle N to the nearest degree. (Lesson 4–3) **38°**

49. Solve $-27 - b = -7$. (Lesson 3–1) **−20**

50. In Mr. Tucker's algebra class, students can get extra points for finding the correct solution to the "Riddle of the Week." One week, Mr. Tucker posed this riddle. *Josh is 10 years older than his brother. Next year, he will be three times as old as his brother.* Answer the following questions. (Lesson 2–9)
   a. How old was Josh when his brother was born? **10 years old**
   b. If his brother's age is represented by a, what is Josh's age? **$a + 10$**
   c. India says that Josh is 12 years old and his brother is 4. Does India get the extra points for her answer? Why or why not? **no; $4 + 10 \neq 12$**

51. Simplify $0.2(3x + 0.2) + 0.5(5x + 3)$. (Lesson 1–8) **$3.1x + 1.54$**

3b. Sample answer: No, this does not take into account deaths and births that may occur during the immigration.

Mathematics and SOCIETY

People Around the World

This excerpt is from an article in *The Amicus Journal*, Winter 1994.

THE LAST FORTY YEARS SAW THE FASTEST rise in human numbers in all previous history, from only 2.5 billion people in 1950 to 5.6 billion today....The second half of the 1990s will add an additional 94 million people per year. That is equivalent to a new United States every thirty-three months, another Britain every seven months, a Washington (DC) every six days. A whole Earth of 1800 was added in just one decade, according to United Nations Population Division statistics. After 2000, annual additions will slow, but by 2050 the United Nations expects the human race to total just over 10 billion—an extra Earth of 1980 on top of today's. ∎

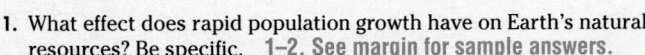

1. What effect does rapid population growth have on Earth's natural resources? Be specific. **1–2. See margin for sample answers.**

2. What changes would you expect to see if the community in which you live doubles in population during your lifetime?

3. A village is deserted, but its buildings are intact and there are ample resources. Suppose 1000 refugees per day flock to this village.
   a. At this rate, how long would it take to reach the midpoint to a population of five million? **about 6.85 years**
   b. Is the proposition of a village growing at this rate realistic? Why or why not?

VOCABULARY

After completing this chapter, you should be able to define each term, property, or phrase and give an example or two of each.

Algebra
family of graphs   (p. 354)
point-slope form   (p. 333)
rise   (p. 325)
run   (p. 325)
slope   (pp. 324, 325)
slope-intercept form   (p. 347)
standard form   (p. 333)
x-intercept   (p. 346)
y-intercept   (p. 346)

Problem Solving
use a model   (p. 339)

Geometry
midpoint   (p. 369)
parallel lines   (p. 362)
parallelogram   (p. 362)
perpendicular lines   (p. 362)

Statistics
best-fit line   (p. 341)
negative correlation   (p. 340)
positive correlation   (p. 340)
regression line   (p. 342)
scatter plot   (p. 339)

UNDERSTANDING AND USING THE VOCABULARY

Choose the correct term to complete each sentence.

1. The lines with equations $y = -2x + 7$ and $y = -2x - 6$ are (*parallel, perpendicular*) lines.

2. The equation $y - 2 = -3(x - 1)$ is written in (*point-slope, slope-intercept*) form.

3. If the endpoints of $\overline{AB}$ are $A(6, -3)$ and $B(2, 4)$, then the (*midpoint, slope*) of $\overline{AB}$ is at $\left(4, \frac{1}{2}\right)$.

4. The (*x-intercept, y-intercept*) of the equation $2x + 3y = -1$ is $-\frac{1}{2}$.

5. The lines with equations $y = \frac{1}{3}x + 1$ and $y = -3x - 5$ are (*parallel, perpendicular*) lines.

6. The slope of a line is defined as the ratio of the (*rise, run*), or vertical change, to the (*rise, run*), or horizontal change, as you move from one point on the line to another.

7. The equation $y = -\frac{1}{2}x + 3$ is written in (*slope-intercept, standard*) form.

8. The (*x-intercept, y-intercept*) of the equation $-x - 4y = 2$ is $-\frac{1}{2}$.

9. The (*midpoint, slope*) of the line with equation $-3y = 2x + 3$ is $-\frac{2}{3}$.

10. The equation $3x - 4y = -5$ is written in (*point-slope, standard*) form.

Instructional Resources

Three multiple-choice tests and three free-response tests are provided in the *Assessment and Evaluation Masters*. Forms 1A and 2A are for honors pacing, and Forms 1B, 1C, 2B, and 2C are for average pacing. Chapter 6 Test, Form 1B is shown at the right. Chapter 6 Test, Form 2B is shown on the next page.

CHAPTER **6** STUDY GUIDE AND ASSESSMENT

Skills and Concepts Encourage students to refer to the objectives and examples on the left as they complete the review exercises on the right.

Assessment and Evaluation Masters, pp. 149–150

NAME_____ DATE_____

Chapter 6 Test, Form 2B

Determine the slope of the line passing through each pair of points. If the slope is undefined, write "no slope."

1. $(6, -4)$ and $(-3, 7)$ 1. $-\frac{11}{9}$

2. $(-2, -6)$ and $(7, -6)$ 2. 0

3. Determine the value of r so that the line through $(-4, 8)$ and $(r, -6)$ has a slope of $\frac{2}{3}$.   3. -25

Write an equation in standard form of the line satisfying the given conditions.

4. has no slope and passes through $(-6, 4)$ 4. $x = -6$

5. passes through $(1, -4)$ and $(-6, 8)$ 5. $12x + 7y = -16$

6. passes through $(7, -3)$ and has a y-intercept of 2   6. $5x + 7y = 14$

The average SAT math and verbal scores for selected years from 1967 to 1985 are listed in the table below. Use the data for problems 7–9.

Year	Verbal Score	Math Score
1967	466	492
1970	460	488
1975	434	472
1980	424	466
1985	431	475

7. Make a scatter plot of the data with the verbal score on the horizontal axis and the math score on the vertical axis.   7.

8. Is the correlation positive or negative?   8. positive

9. Predict the mathematics score corresponding to a verbal score of 445.   9. about 478

10. Graph the line with x-intercept -3 and y-intercept 2. Use the coordinate plane at the right.   10.

NAME_____ DATE_____

Chapter 6 Test, Form 2B (continued)

11. Graph the line with y-intercept 3 and slope $-\frac{3}{4}$. Use the coordinate plane at the right.   11.

12. Graph the equation $4x - 3y = -9$. Use the coordinate plane at the right.   12.

Write an equation in slope-intercept form of the line satisfying the given conditions.

13. passes through $(5, 4)$ and $(6, -1)$ 13. $y = -5x + 29$

14. slope $\frac{1}{3}$ and passes through $(-2, 8)$ 14. $y = \frac{1}{3}x + \frac{26}{3}$

15. passes through $(10, -3)$ and has a y-intercept of 2   15. $y = -\frac{1}{2}x + 2$

16. parallel to the graph of $9x + 3y = 6$ and passes through $(5, 3)$ 16. $y = -3x + 18$

17. perpendicular to the graph of $4x - y = 12$ and passes through $(8, 2)$ 17. $y = -\frac{1}{4}x + 4$

18. parallel to the x-axis and passes through $(4, 2)$ 18. $y = 2$

19. Find the coordinates of the midpoint of the segment whose endpoints are $(-2, 14)$ and $(3, -9)$.   19. $\left(\frac{1}{2}, 2\frac{1}{2}\right)$

20. The center of a circle is $M(-3, 0)$. If $\overline{AB}$ is a diameter, and $A(-7, 3)$ is one endpoint on the circle, find the coordinates of B.   20. $(1, -3)$

Bonus In a certain lake, a 1-year-old bluegill fish is 3 inches long, while a 4-year-old bluegill is 6.6 inches long. Assuming the growth rate can be approximated by a linear equation, write an equation in slope-intercept form for the length of a bluegill fish in inches (y) after x years. Use this equation to estimate the length of a 10-year-old bluegill. About how old is a 9-inch-long bluegill?   Bonus $y = 1.2x + 1.8$; 13.8 in.; 6 yr

---

SKILLS AND CONCEPTS

OBJECTIVES AND EXAMPLES

Upon completing this chapter, you should be able to:

• find the slope of a line, given the coordinates of two points on the line (Lesson 6–1)

Determine the slope of the line that passes through $(-6, 5)$ and $(3, -2)$.

$$m = \frac{y_2 - y_1}{x_2 - x_1}$$
$$= \frac{-2 - 5}{3 - (-6)}$$
$$= -\frac{7}{9}$$

The slope is $-\frac{7}{9}$.

• write linear equations in point-slope form (Lesson 6–2)

Write the point-slope form of an equation of the line that passes through the points at $(-4, 7)$ and $(-2, 3)$.

$$m = \frac{3 - 7}{-2 - (-4)} \text{ or } -2$$
$$y - y_1 = m(x - x_1)$$
$$y - 7 = -2[x - (-4)]$$
$$y - 7 = -2(x + 4)$$

An equation of the line is $y - 7 = -2(x + 4)$.

• write linear equations in standard form (Lesson 6–2)

Write the standard form of an equation of the line that passes through the points at $(6, -4)$ and $(-1, 5)$.

$$m = \frac{5 - (-4)}{-1 - 6} \text{ or } -\frac{9}{7}$$

$$y - y_1 = m(x - x_1) \qquad\qquad y + 4 = -\frac{9}{7}(x - 6)$$
$$y - (-4) = -\frac{9}{7}(x - 6) \qquad 7(y + 4) = 7 \cdot -\frac{9}{7}(x - 6)$$
$$y + 4 = -\frac{9}{7}(x - 6) \qquad 7y + 28 = -9(x - 6)$$
$$\qquad\qquad\qquad\qquad\qquad 7y + 28 = -9x + 54$$
$$\qquad\qquad\qquad\qquad\qquad 9x + 7y = 26$$

An equation of the line is $9x + 7y = 26$.

REVIEW EXERCISES

Use these exercises to review and prepare for the chapter test.

Determine the slope of the line that passes through each pair of points. 12. undefined

11. $(8, 3), (2, 5)$ $-\frac{1}{3}$ 12. $(-2, 5), (-2, 9)$

13. $(-3, 6), (-8, 4)$ $\frac{2}{5}$ 14. $(4, 3), (-5, 3)$ 0

Determine the value of r so the line that passes through each pair of points has the given slope.

15. $(r, 4), (7, 3), m = \frac{3}{4}$ $\frac{25}{3}$

16. $(4, -7), (-2, r), m = \frac{8}{3}$ -23

17. $y + 3 = -2(x - 4)$ 18. $y - 5 = 5(x - 8)$ 19. $y - 7 = 0$ 20. $y - 2 = \frac{1}{2}(x - 6)$ 21. $y - 3 = \frac{3}{5}x$

Write the point-slope form of an equation of the line that passes through the given point and has the given slope.

17. $(4, -3), m = -2$ 18. $(8, 5), m = 5$

19. $(-5, 7), m = 0$ 20. $(6, 2), m = \frac{1}{2}$

Write the point-slope form of an equation of the line that passes through each pair of points.

21. $(0, 3), (-5, 0)$ 22. $(5, 4), (6, 3)$

23. $(4, 1), (-3, 7)$ 24. $(2, -5), (0, 4)$

22. $y - 3 = -(x - 6)$ 23. $y - 1 = -\frac{6}{7}(x - 4)$ 24. $y + 5 = -\frac{9}{2}(x - 2)$

Write the standard form of an equation of the line that passes through the given point and has the given slope. 25–28. See margin.

25. $(4, -6), m = 3$ 26. $(1, 5), m = 0$

27. $(6, -1), m = \frac{3}{4}$ 28. $(8, 3), m = $ undefined

Write the standard form of an equation of the line that passes through each pair of points. 29–32. See margin.

29. $(-2, 5), (9, 5)$ 30. $(0, 5), (-2, 0)$

31. $\left(-2, \frac{2}{3}\right), \left(-2, \frac{2}{7}\right)$ 32. $\left(-5, 7\right), \left(0, \frac{1}{2}\right)$

GLENCOE Technology

Test and Review Software

You may use this software, a combination of an item generator and item bank, to create your own tests or worksheets. Types of items include free response, multiple choice, short answer, and open ended.

For IBM & Macintosh

Additional Answers

25. $3x - y = 18$
26. $y = 5$
27. $3x - 4y = 22$
28. $x = 8$
29. $y = 5$
30. $5x - 2y = -10$
31. $x = -2$
32. $13x + 10y = 5$

Additional Answers

33c.

42.

43.

44.

45.

46.

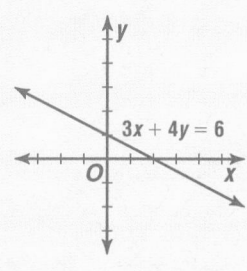

OBJECTIVES AND EXAMPLES

• graph and interpret points on a scatter plot; draw and write equations for best-fit lines and make predictions by using those equations (Lesson 6-3)

The scatter plot below shows a negative relationship since the line suggested by the points would have a negative slope.

Mistakes made

Hours studied

33b. Sample answer: 35 stories
33c. See margin for graph.

REVIEW EXERCISES

33. Draw a scatter plot of the data below. Let the independent variable be height.

Buildings in Oklahoma City	Height (feet)	Number of Stories
Liberty Tower	500	36
First National Center	493	28
City Place	440	33
First Oklahoma Tower	425	31
Kerr-McGee Center	393	30
Mid-America Tower	362	29

a. Does a relationship exist? If so, what is the relationship?   **Yes; it is positive.**

b. Use your scatter plot as a model to predict how many stories you might think a building 475 feet high would be.

c. Draw a best-fit line and write the equation of the line. $y = 0.03x + 21$

• determine the x- and y-intercepts of linear graphs from their equations and write equations in slope-intercept form   (Lesson 6-4)

Determine the slope and x- and y-intercepts of the graph of $3x - 2y = 7$.

Solve for y to find the slope and y-intercept.	Let $y = 0$ to find the x-intercept.
$3x - 2y = 7$	$3x - 2(0) = 7$
$-2y = -3x + 7$	$3x = 7$
$y = \frac{3}{2}x - \frac{7}{2}$	$x = \frac{7}{3}$

The slope is $\frac{3}{2}$, the y-intercept is $-\frac{7}{2}$, and the x-intercept is $\frac{7}{3}$.

Write an equation in slope-intercept form of a line with the given slope and y-intercept.

34. $m = 2, b = 4$ $y = 2x + 4$

35. $m = -3, b = 0$ $y = -3x$

36. $m = -\frac{1}{2}, b = -9$ $y = -\frac{1}{2}x - 9$

37. $m = 0, b = 5.5$ $y = 5.5$

Find the slope and y-intercept of the graph of each equation.

38. $x = 2y - 7$ $\frac{1}{2}, \frac{7}{2}$ **39.** $8x + y = 4$ $-8, 4$

40. $y = \frac{1}{4}x + 3$ $\frac{1}{4}, 3$ **41.** $\frac{1}{2}x + \frac{1}{4}y = 3$ $-2, 12$

• graph a line given any linear equation   (Lesson 6-5)

Graph $-2x - y = 5$.

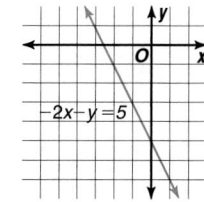

Graph each equation.

42. $3x - y = 9$ **43.** $5x + 2y = 12$

44. $y = \frac{2}{3}x + 4$ **45.** $y = -\frac{3}{2}x - 6$

46. $3x + 4y = 6$ **47.** $5x - \frac{1}{2}y = 2$

48. $y - 4 = -2(x + 1)$ **49.** $y + 5 = -\frac{3}{4}(x - 6)$

42–49. See margin for graphs.

Additional Answers

47.

48.

49.

Additional Answers

50. $y = 4x - 9$

51. $y = -\frac{7}{2}x - 14$

52. $y = -\frac{1}{3}x + 1$

53. $y = -\frac{3}{8}x + \frac{13}{2}$

54. $y = 2x - 16$

55. $y = 5x - 15$

56. $(6, 1)$

57. $\left(1, -\frac{5}{2}\right)$

58. $(8, 2)$

59. $\left(5, \frac{11}{2}\right)$

60. $(3, 2)$

61. $\left(\frac{7}{2}, -\frac{3}{2}\right)$

62. $(19, 9)$

63. $(13, 11)$

64. $(6, -7)$

65. $(-5, 20)$

Applications and Problem Solving Encourage students to work through the exercises in the Applications and Problem Solving section to strengthen their problem-solving skills.

OBJECTIVES AND EXAMPLES

• write equations of lines that pass through a given point, parallel or perpendicular to the graph of a given equation (Lesson 6–6)

Write an equation of the line that is perpendicular to the graph of $2x + y = 6$ and passes through $(2, 3)$.

$2x + y = 6$

$\qquad y = -2x + 6$

The slope of this line is -2. Therefore, the slope of the line perpendicular to it is $\frac{1}{2}$.

$y - 3 = \frac{1}{2}(x - 2) \quad m = \frac{1}{2}, (x_1, y_1) = (2, 3)$

$\qquad y = \frac{1}{2}x + 2$

• find the coordinates of the midpoint of a line segment in the coordinate plane (Lesson 6–7)

The endpoints of a segment are at $(11, 4)$ and $(9, 2)$. Find its midpoint.

$(x, y) = \left(\frac{x_1 + x_2}{2}, \frac{y_1 + y_2}{2}\right)$

$\qquad = \left(\frac{11+9}{2}, \frac{4+2}{2}\right)$

$\qquad = (10, 3)$

56–65. See margin.

REVIEW EXERCISES

Write an equation of the line having the following properties 50–55. See margin.

50. parallel to $4x - y = 7$ and passes through $(2, -1)$

51. perpendicular to $2x - 7y = 1$ and passes through $(-4, 0)$

52. parallel to $3x + 9y = 1$ and passes through $(3, 0)$

53. perpendicular to $8x - 3y = 7$ and passes through $(4, 5)$

54. parallel to $-y = -2x + 4$ and passes through $(5, -6)$

55. perpendicular to $5y = -x + 1$ and passes through $(2, -5)$

Find the coordinates of the midpoint of a segment with each pair of endpoints.

56. $A(3, 5)$, $B(9, -3)$
57. $A(-6, 6)$, $B(8, -11)$
58. $A(14, 4)$, $B(2, 0)$
59. $A(2, 7)$, $B(8, 4)$
60. $A(2, 5)$, $B(4, -1)$
61. $A(10, 4)$, $B(-3, -7)$

Find the coordinates of the other endpoint of a segment given one endpoint and the midpoint M.

62. $A(3, 5)$, $M(11, 7)$
63. $A(5, 3)$, $M(9, 7)$
64. $A(4, -11)$, $M(5, -9)$
65. $A(11, -4)$, $M(3, 8)$

APPLICATIONS AND PROBLEM SOLVING

66. **Skiing** A course for cross-country skiing is regulated so that the slope of any hill cannot be greater than 0.33. Suppose a hill rises 60 meters over a horizontal distance of 250 meters.
   a. What is the slope of the hill? $\frac{6}{25}$
   b. Does the hill meet the requirements?
   (Lesson 6–1) yes, $\frac{6}{25} < \frac{1}{3}$

67. **Travel** Jon Erlanger is taking a long trip. In the first two hours, he drives 80 miles. After that, he averages 45 miles per hour. Write an equation in slope-intercept form relating distance traveled and time. (Lesson 6–4)
   $d = 45t - 10$

A practice test for Chapter 6 is provided on page 792.

68. **Entertainment** Carolyn Parks owns stock in Star Gazer Motion Picture Company. Every other week, she graphs the closing value of a share of the stock. What is the midpoint between the highest and lowest values of the stock? (Lesson 6–7) (7, \$5.50)

ALTERNATIVE ASSESSMENT

COOPERATIVE LEARNING PROJECT

Video Rental   In this chapter, you investigated graphing linear equations. You used slope and data points to graph equations. These graphs represented information that could be analyzed and extended in order to interpret further information than what was given.

In this project, you will analyze two video-store rental policies. There are two stores in Joe's neighborhood where he can rent videos: Video World and Mega Video. At one of the two stores, you have to buy a membership card first. The other one does not have these cards; you simply pay per video. Both have a daily charge (rate per day). Mega Video has a membership card that costs $15 per year and then each video that is rented costs $2 per day. Video World simply rents each video for $3.25 per day.

Which is the most economical video store for Joe to use? If he rarely rents videos, which store should he use? If he often rents videos, which store should he use?

Follow these steps to determine the store he should use.

- Determine a way to graph this information.
- Develop a linear equation for each store that describes the amount of money it costs to rent a video.
- Determine whether the graphs should be drawn separately or on the same grid.
- Investigate the graphs and determine what each graph represents.
- What factors other than price might Joe want to consider before he chooses a store?
- Write a paragraph describing various situations for Joe and the appropriate solutions.

THINKING CRITICALLY

- Give an example of an equation of a line that has the same x- and y-intercepts. What is the slope of the line?

- Give an example of an equation of a line whose x- and y-intercepts are opposites. What is the slope of the line?

- Make a conjecture about the slopes of lines whose x- and y-intercepts are equal or opposites.

- The x-intercept of a line is s, and the y-intercept is t. Write the equation of the line.

PORTFOLIO

In mathematics, there is often more than one way to solve a problem. In order to graph a line for example, various information can be used. Sometimes you use two points, sometimes you use a point and a slope, and sometimes you use x- and y-intercepts. Find an equation from your work in this chapter and describe at least three ways in which to graph it. Place this in your portfolio.

SELF EVALUATION

Checking your answer is the final step in solving a problem. This step is the most critical, but it is also the step that is most often forgotten. When an answer is finally reached, it should be checked to see whether it makes sense and also whether it can be verified.

Assess yourself. Do you use all four steps when you solve a problem? Do you reach an answer, assume that it is the correct answer, and then move on without checking it? Do you ask yourself questions about your answer to verify that it is correct and that it makes sense? Can you think of an example of a problem in your daily life in which you didn't ask yourself questions about your solution to the problem and later you discovered that the solution didn't work?

Assessment and Evaluation Masters, pp. 154, 165

6   NAME _____   DATE _____

Chapter 6 Performance Assessment

Instructions: *Demonstrate your knowledge by giving a clear, concise solution to each problem. Be sure to include all relevant drawings and justify your answers. You may show your solution in more than one way or investigate beyond the requirements of the problem.*

1. a. Write a linear equation in standard form. Tell how you know the equation is linear.

   b. Use the coefficients and constant term to name the slope and y-intercept of the graph. Then write the equation in slope-intercept form.

   c. Use the y-intercept and slope to graph the equation. Explain how you graphed the equation.

   d. Write an equation whose graph is parallel to the graph in part c. Tell how you know they are parallel.

   e. Write an equation of a line perpendicular to the graph in part c. Tell how you know they are perpendicular.

2. a. Describe the pattern of the points in the scatter plot at the right.

   b. Give at least two examples of real-life situations that, if graphed, would result in a correlation like the one shown in this scatter plot.

3. The graphs of each group of equations have at least one characteristic in common. Name the characteristic(s) and then graph each group of equations on the same axes to verify your answer.

   a. $y = x$, $y = 3x$, $y = 5x$

   b. $y = 2x$, $y = 2x + 1$, $y = 2x - 3$

   c. $y = \frac{1}{2}x$, $y = \frac{1}{3}x$, $y = \frac{2}{5}x$

   d. $2y = 6x$, $4y - 12x = 0$, $3x - y = 0$

   e. $2x - 4y = 12$, $3y + 2x = -2$, $x + y = 0$

Scoring Guide
Chapter 6
Performance Assessment

Level	Specific Criteria
3 Superior	• Shows thorough understanding of the concepts of *standard form, linear equation, slope, y-intercept, slope-intercept form, parallel and perpendicular lines, scatter plot,* and *correlation.* • Uses appropriate strategies to solve problems. • Computations are correct. • Written explanations are exemplary. • Word problem concerning a linear function is appropriate and makes sense. • Graphs are accurate and appropriate. • Goes beyond requirements of some or all problems.
2 Satisfactory, with Minor Flaws	• Shows understanding of the concepts of *standard form, linear equation, slope, y-intercept, slope-intercept form, parallel and perpendicular lines, scatter plot,* and *correlation.* • Uses appropriate strategies to solve problems. • Computations are mostly correct. • Written explanations are effective. • Word problem concerning a linear function is appropriate and makes sense. • Graphs are mostly accurate and appropriate. • Satisfies all requirements of problems.
1 Nearly Satisfactory, with Serious Flaws	• Shows understanding of most of the concepts of *standard form, linear equation, slope, y-intercept, slope-intercept form, parallel and perpendicular lines, scatter plot,* and *correlation.* • May not use appropriate strategies to solve problems. • Computations are mostly correct. • Written explanations are satisfactory. • Word problem concerning a linear function is appropriate and sensible. • Graphs are mostly accurate and appropriate. • Satisfies most requirements of problems.
0 Unsatisfactory	• Shows little or no understanding of the concepts of *standard form, linear equation, slope, y-intercept, slope-intercept form, parallel and perpendicular lines, scatter plot,* and *correlation.* • May not use appropriate strategies to solve problems. • Computations are incorrect. • Written explanations are not satisfactory. • Word problem concerning a linear function is not appropriate or sensible. • Graphs are not accurate or appropriate. • Does not satisfy requirements of problems.

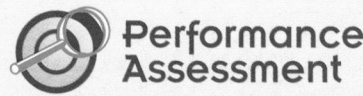

Alternative Assessment

The Alternative Assessment section provides students with the opportunity to assess their own work by thinking critically, working with others, keeping a portfolio, and honestly evaluating their own progress. For more information on alternative forms of assessment, see *Alternative Assessment in the Mathematics Classroom,* one of the titles in the Glencoe Mathematics Professional Series.

Performance Assessment

Performance Assessment tasks for this chapter are included in the *Assessment and Evaluation Masters.* A scoring guide is also provided.

CUMULATIVE REVIEW

CHAPTERS 1–6

SECTION ONE: MULTIPLE CHOICE

There are eight multiple-choice questions in this section. After working each problem, write the letter of the correct answer on your paper.

1. Name the quadrant in which point $P(x, y)$ is located if it satisfies the condition $x < 0$ and $y = -3$. **C**

A. I

B. II

C. III

D. IV

2. Raini made a map of his neighborhood using a coordinate grid. His school has coordinates $(3, 8)$, where 3 is the number of blocks east he must walk and 8 is the number of blocks north. From school, he walks 4 blocks east and 2 blocks south to get to his friend's house. Find the coordinates of the midpoint of the line segment that could be drawn to connect his school and his friend's house. **A**

A. $(5, 7)$

B. $(8, -2)$

C. $(7, 6)$

D. $(5, -6)$

3. Determine the slope of the line that passes through the points at $(-3, 6)$ and $(-5, 9)$. **C**

A. $\frac{3}{2}$

B. $\frac{2}{3}$

C. $-\frac{3}{2}$

D. none

4. Probability A shipment of 100 CDs was just received at Tunes R Us. There is a 4% probability that one of the CDs was damaged during shipment, even though the package may not be cracked. If Craig buys a CD from this shipment, what are the odds that he is buying a damaged CD? **D**

A. 1:25

B. $\frac{4}{100}$

C. 96:4

D. 1:24

5. Choose the graph that represents a function. **B**

A.   B.

C.   D.

6. Choose the property illustrated by the statement *The quantity g plus h times b equals g times b plus h times b.* **A**

A. distributive property

B. commutative property of multiplication

C. multiplicative identity property

D. associative property of addition

7. Choose the line that represents the graph of the equation $y = -\frac{2}{3}x + 2$. **D**

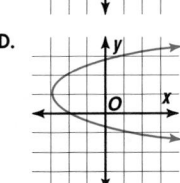

A. line A

B. line B

C. line C

D. line D

8. Geometry The formula Area $= \frac{1}{2}bc \sin A$ can be used to find the area of any triangle. Find the area of the triangle shown below to the nearest tenth of a square centimeter if $b = 16$, $c = 9$, and $\angle A$ is a 36° angle. **B**

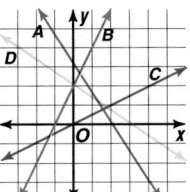

A. 72 cm^2

B. 42.3 cm^2

C. 84.6 cm^2

D. 144 cm^2

Standardized Test Practice Questions are also provided in the *Assessment and Evaluation Masters,* p. 158.

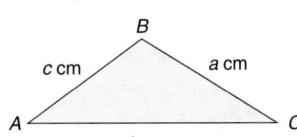

6   NAME_____   DATE_____

Chapter 6 Cumulative Review

1. Write the expression $2 \cdot r \cdot r \cdot s \cdot s$ using exponents. (Lesson 1-1)   1. ___$2r^2s^2$___

2. Evaluate $2xy - y^2$ if $x = 6$ and $y = 12$. (Lesson 1-3)   2. ___0___

Simplify. (Lessons 2-5, 2-6, and 2-7)

3. $4\frac{7}{8} - 2\frac{5}{8}$ 3. ___$2\frac{1}{4}$___

4. $\frac{3}{8} \times 2\frac{7}{18}$ 4. ___$\frac{43}{48}$___

5. $-\frac{2}{3} + \frac{3}{4}$ 5. ___$\frac{1}{12}$___

6. $-3.9 + (-2.5) + (-8.7)$ 6. ___-15.1___

7. $4(2y + y) - 6(4y + 3y)$ 7. ___$-30y$___

8. $\frac{12a - 18b}{-6}$ 8. ___$-2a + 3b$___

Solve each equation. (Lessons 3-1 and 3-2)

9. $13 - m = 21$ 9. ___-8___

10. $\frac{3}{4}x = \frac{2}{3}$ 10. ___$\frac{8}{9}$___

11. Find three consecutive even integers whose sum is 132. (Lesson 3-3)   11. ___42, 44, 46___

12. Which is a better buy: a liter of milk for 59¢ or 1.5 liters for 81¢?   12. ___1.5 liters for 81¢___

13. Six is what percent of 80?   13. ___7.5%___

Find the final price of each item. When there is a discount and sales tax, compute the discount price first.

14. calculator: $90   sales tax: 8%   14. ___$97.20___

15. magazine: $3.95   discount: 10%   sales tax: 6.5%   15. ___$3.79___

---

6   NAME_____   DATE_____

Chapter 6 Cumulative Review (continued)

16. A can contains two different kinds of nuts, weighs 5 pounds, and cost $16. One type of nut costs $3.50 a pound. The other type costs $2.75 a pound. How many pounds of $2.75 nuts are there? (Lesson 4-1)   16. ___2 pounds at $2.75___

17. Solve $x - 2y = 12$ if the domain is $\{-3, -1, 0, 2, 5\}$. (Lesson 5-3)   17. ___$\{(-3, -7.5), (-1, -6.5), (0, -6), (2, -5), (5, -3.5)\}$___

18. Graph $3x - y = 1$. Use the coordinate plane provided. (Lesson 5-4)   18. [graph]

19. State whether the following relation is a function. (Lesson 5-5) $\{(1, 4), (2, 6), (3, 7), (4, 4)\}$ 19. ___yes___

20. Write an equation for the relationship between the variables in the chart at the right. (Lesson 5-6)

x	0	2	4	6
y	2	5	8	11

20. ___$y = \frac{3}{2}x + 2$___

21. Determine the slope of the line passing through (2, 7) and (−5, 2). (Lesson 6-1)   21. ___$\frac{5}{7}$___

22. Write an equation in standard form for the line passing through (2, 6) and having a slope of −3. (Lesson 6-2)   22. ___$3x + y = 12$___

23. Write an equation in slope-intercept form for the line in exercise 22. (Lesson 6-4)   23. ___$y = -3x + 12$___

24. Write an equation for the line that is parallel to the graph of $5x - 3y = 1$ and passes through (0, −4). (Lesson 6-6)   24. ___$y = \frac{5}{3}x - 4$___

25. Graph $y = 2x - 3$. Use the coordinate plane at the right. (Lesson 6-5)   25. [graph]

Additional Answers

9.

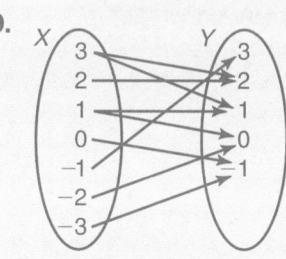

12. {(−2, 14), (−1, 9), (0, 4), (2, −6), (5, −21)}

15.

17a.

17b. The more years of education you have, the higher your income.

SECTION TWO: SHORT ANSWER

This section contains seven questions for which you will provide short answers. Write your answer on your paper.

9. Express the relation shown in the graph as a set of ordered pairs. Then draw a mapping of the relation. *See margin.*

10. Write an equation of the line that passes through the point at (2, −1) and is perpendicular to $4x − y = 7$. $y = -\frac{1}{4}x - \frac{1}{2}$

11. The average person can live for eleven days without water, assuming a mean temperature of 60° F. Determine if survival is possible given daily high temperatures of 59°, 70°, 49°, 62°, 46°, 63°, 71°, 64°, 55°, 68°, and 54° F. Verify your answer. *yes*

12. Write the solution set for $5x + y = 4$, if the domain is {−2, −1, 0, 2, 5}. *See margin.*

13. Peanuts sell for $3.00 per pound. Cashews sell for $6.00 per pound. How many pounds of cashews should be mixed with 12 pounds of peanuts to obtain a mixture that sells for $4.20 per pound? *8 pounds*

14. At Videoville, Eshe earns a weekly salary of $150 plus $0.30 for each video over 100 that she sells each week. If Eshe sells v videos in a week, then her total weekly salary is $C(v) = 150 + 0.30(v − 100)$ for $v > 100$. If she wants to earn $225 each week to save up for a stereo system, how many videos must she sell each week? *350 videos*

15. Graph the solution set {integers greater than or equal to −4} on a number line. *See margin.*

SECTION THREE: OPEN ENDED

This section contains two open-ended problems. Demonstrate your knowledge by giving a clear, concise solution to each problem. Your score on these problems will depend on how well you do the following.

- Explain your reasoning.
- Show your understanding of the mathematics in an organized manner.
- Use charts, graphs, and diagrams in your explanation.
- Show the solution in more than one way or relate it to other situations.
- Investigate beyond the requirements of the problem.

16. For long-distance phone calls, a telephone company charges $1.72 for a 4-minute call, $2.40 for a 6-minute call, and $5.46 for a 15-minute call. Determine the charge for a 1-minute call. *$0.70*

17. The table below shows the annual income and the number of years of college education for eleven people. *a–b. See margin.*

Income (thousands)	College Education (years)
$23	3.0
20	2.0
25	4.0
47	6.0
19	2.5
48	7.5
35	6.5
10	1.0
39	5.5
26	4.5
36	4.0

a. Make a scatter plot of the data.

b. Based on this plot, how does the number of years of college affect income?

7

Solving Linear Inequalities

PREVIEWING THE CHAPTER

This chapter develops the properties for solving inequalities. Students learn the addition and subtraction properties for solving inequalities, as well as the multiplication and division properties which, in some cases, require the reversal of the inequality sign. Students solve compound inequalities, drawing pictures or diagrams when appropriate, as well as open sentences involving absolute value. Students apply their knowledge in lessons involving the probability of compound events and the statistical tool known as a box-and-whisker plot. The chapter concludes with a lesson on graphing inequalities in two variables.

Lesson (Pages)	Lesson Objectives	NCTM Standards	State/Local Objectives
7-1 (384–390)	Solve inequalities by using addition and subtraction.	1–5	
7-2A (391)	Use cups and counters to solve inequalities.	1–5	
7-2 (392–398)	Solve inequalities by using multiplication and division.	1–5	
7-3 (399–404)	Solve linear inequalities involving more than one operation. Find the solution set for a linear inequality when replacement values are given for the variables.	1–5	
7-4 (405–412)	Solve problems by making a diagram. Solve compound inequalities and graph their solution sets. Solve problems that involve compound inequalities.	1–5	
7-5 (413–419)	Find the probability of a compound event.	1–4, 11	
7-6 (420–426)	Solve open sentences involving absolute value and graph the solutions.	1–5	
7-7 (427–432)	Display and interpret data on box-and-whisker plots.	1–4, 10	
7-7B (433–434)	Use a graphing calculator to compare two sets of data by using a double box-and-whisker plot.	1–5, 10	
7-8A (435)	Use the "Shade(" command on a graphing calculator to graph inequalities in two variables.	1–6	
7-8 (436–441)	Graph inequalities in the coordinate plane.	1–5	

A complete, 1-page lesson plan is provided for each lesson in the *Lesson Planning Guide*. Answer keys for each lesson are available in the *Answer Key Masters*.

ORGANIZING THE CHAPTER

You may want to refer to the **Course Planning Calendar** on page T12 for detailed information on pacing.
PACING: **Standard**—14 days; **Honors**—13 days; **Block**—7 days; **Two Years**—21 days

LESSON PLANNING CHART

| Lesson (Pages) | Materials/ Manipulatives | Extra Practice (Student Edition) | BLACKLINE MASTERS | | | | | | | | | Real-World Applications | Interactive Mathematics Tools Software | Teaching Transparencies |
			Study Guide	Practice	Enrichment	Assessment and Evaluation	Modeling Mathematics	Multicultural Activity	Tech Prep Applications	Graphing Calculator	Science and Math Lab Manual			
7-1 (384–390)	graphing calculator	p. 771	p. 48	p. 48	p. 48		p. 78		p. 13				7-1	7-1A 7-1B
7-2A (391)	equation mat* cups and counters* self-adhesive note					p. 27							7-2A.1 7-2A.2	
7-2 (392–398)		p. 772	p. 49	p. 49	p. 49	p. 184			p. 14					7-2A 7-2B
7-3 (399–404)	graphing calculator	p. 772	p. 50	p. 50	p. 50			p. 13				19		7-3A 7-3B
7-4 (405–412)	graphing calculator	p. 772	p. 51	p. 51	p. 51	pp. 183, 184	pp. 54–56			p. 7		20		7-4A 7-4B
7-5 (413–419)		p. 773	p. 52	p. 52	p. 52									7-5A 7-5B
7-6 (420–426)		p. 773	p. 53	p. 53	p. 53	p. 185								7-6A 7-6B
7-7 (427–432)		p. 773	p. 54	p. 54	p. 54						pp. 29–34		7-7	7-7A 7-7B
7-7B (433–434)	graphing calculator									pp. 24, 25				
7-8A (435)	graphing calculator									pp. 26, 27				
7-8 (436–441)	graphing calculator	p. 774	p. 55	p. 55	p. 55	p. 185			p. 14					7-8A 7-8B
Study Guide/ Assessment (443–447)						pp. 169 –182, 186– 188								

*Included in Glencoe's Student Manipulative Kit and Overhead Manipulative Resources.

ORGANIZING THE CHAPTER

OTHER CHAPTER RESOURCES

Student Edition
Chapter Opener, pp. 382–383
Mathematics and Society, p. 419
Working on the Investigation,
   pp. 398, 426
Closing the Investigation,
   p. 442

Teacher's Classroom Resources
Investigations and Projects Masters,
   pp. 49–52
Algebra and Geometry Overhead
   Manipulative Resources,
   pp. 22–25

 Technology
Test and Review Software (IBM
   and Macintosh)
CD-ROM (Windows and
   Macintosh)

Professional Publications
Block Scheduling Booklet
Glencoe Mathematics Professional
   Series

OUTSIDE RESOURCES

Books/Periodicals
*Algebra Experiments I: Exploring Linear
   Functions,* Dale Seymour Publications
Developing Skills in Algebra 1: Book C,
   Tricon Publishing

 Software
Math Connections: Algebra I, Sunburst

Videos/CD-ROMs
The Laws of Algebra/Solving Equations I,
   Coronet/MTI

ASSESSMENT RESOURCES

Student Edition
Math Journal, pp. 388, 415, 439
Mixed Review, pp. 390, 397,
   404, 412, 418, 425, 432, 441
Self Test, p. 412
Chapter Highlights, p. 443
Chapter Study Guide and
   Assessment, pp. 444–446
Alternative Assessment, p. 447
 Portfolio, p. 447

Teacher's Wraparound Edition
5-Minute Check, pp. 384, 392,
   399, 405, 413, 420, 427, 436
Check for Understanding, pp. 387,
   395, 402, 409, 415, 423, 429,
   439
Closing Activity, pp. 390, 398,
   404, 412, 419, 426, 432, 441
Cooperative Learning, pp. 408,
   425

Assessment and Evaluation Masters
Multiple-Choice Tests, Forms 1A
   (Honors), 1B (Average), 1C
   (Basic), pp. 169–174
Free-Response Tests, Forms 2A
   (Honors), 2B (Average), 2C
   (Basic), pp. 175–180
Calculator-Based Test, p. 181
Performance Assessment, p. 182
Mid-Chapter Test, p. 183
Quizzes A–D, pp. 184–185
Standardized Test Practice, p. 186
Cumulative Review, pp. 187–188

ENHANCING THE CHAPTER

Examples of some of the materials for enhancing Chapter 7 are shown below.

DIVERSITY

Multicultural Activity Masters, pp. 13, 14

7-3 NAME_____ DATE_____
Student Edition
Multicultural Activity Pages 399–404

Carlos Montezuma

During his lifetime, Carlos Montezuma (1865?–1923) was one of the most influential Native Americans in the United States. He was recognized as a prominent physician and was also a passionate advocate of the rights of Native American peoples. The exercises that follow will help you learn some interesting facts about Dr. Montezuma's life.

Solve each inequality. The word or phrase next to the equivalent inequality will complete the statement correctly.

1. $-2k > 10$
   Montezuma was born in the state of __?__.
   a. $k < -5$ Arizona
   b. $k > -5$ Montana
   c. $k > 12$ Utah

2. $5 \geq r - 9$
   He was a Native American of the Yavapais, who are a __?__ people.
   a. $r \leq -4$ Navajo
   b. $r \geq -4$ Mohawk
   c. $r \leq 14$ Mohave-Apache

3. $-y \leq -9$
   Montezuma received a medical degree from __?__ in 1889.
   a. $y \geq 9$ Chicago Medical College
   b. $y \geq -9$ Harvard Medical School
   c. $y \leq 9$ Johns Hopkins University

4. $-3 + q > 12$
   As a physician, Montezuma's field of specialization was __?__.
   a. $q > -4$ heart surgery
   b. $q > 15$ internal medicine
   c. $q < -15$ respiratory diseases

5. $5 + 4x - 14 \leq x$
   For much of his career, he maintained a medical practice in __?__.
   a. $x \leq 9$ New York City
   b. $x \leq 3$ Chicago
   c. $x \geq 9$ Boston

6. $7 - t < 7 + t$
   In addition to maintaining his medical practice, he was also a(n) __?__.
   a. $t > 7$ director of a blood bank
   b. $t > 0$ instructor at a medical college
   c. $t < -7$ legal counsel to physicians

7. $3a + 8 \geq 4a - 10$
   Montezuma founded, wrote, and edited __?__, a monthly newsletter that addressed Native American concerns.
   a. $a \leq -2$ Yavapai
   b. $a \geq -18$ Apache
   c. $a \leq 18$ Wassaja

8. $6n > 8n - 12$
   Montezuma testified before a committee of the United States Congress concerning his work in treating __?__.
   a. $n < 6$ appendicitis
   b. $n > -6$ asthma
   c. $n > -10$ heart attacks

APPLICATIONS

Real-World Applications, 19, 20

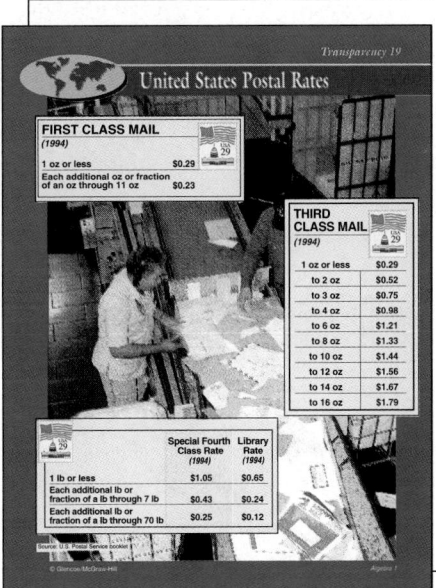

Transparency 19

United States Postal Rates

FIRST CLASS MAIL (1994)

1 oz or less	$0.29
Each additional oz or fraction of an oz through 11 oz	$0.23

THIRD CLASS MAIL (1994)

1 oz or less	$0.29
to 2 oz	$0.52
to 3 oz	$0.75
to 4 oz	$0.98
to 6 oz	$1.21
to 8 oz	$1.33
to 10 oz	$1.44
to 12 oz	$1.56
to 14 oz	$1.67
to 16 oz	$1.79

	Special Fourth Class Rate (1994)	Library Rate (1994)
1 lb or less	$1.05	$0.85
Each additional lb or fraction of a lb through 7 lb	$0.43	$0.24
Each additional lb or fraction of a lb through 70 lb	$0.25	$0.12

Source: U.S. Postal Service booklet
© Glencoe/McGraw-Hill Algebra 1

TECHNOLOGY

Graphing Calculator Masters, p. 7

7-4 NAME_____ DATE_____
Student Edition
Graphing Calculator Activity Pages 405–411

Solving Inequalities

Jenny had test scores of 84, 79, 91, and 76. What is the minimum score she needs on the next test to have an average of at least 80?

Let $x =$ the fifth test score. $\frac{(84 + 79 + 91 + 76 + x)}{5} \geq 80$

Add 84, 79, 91, and 76 to get 330. Place the calculator in computation mode.

a. Use a [0, 100] by [0, 5] viewing window. Set Xscl = 10 and Yscl = 1.
b. **Graph** $((330 + x)/5) \geq 80$.
c. **Trace** the coordinates of the graph, a ray.
d. **Zoom in** to trace the coordinates of the ray more accurately.

Round the x-value of the endpoint of the ray to the nearest integer, 70. Therefore, if Jenny scores a 70 or higher on her next test, her average will be 80 or higher.

Suppose Jenny also wants to calculate the highest test score that would keep her in the 80–85 range. The solution is found by solving the following compound inequality.

$$85 \geq \frac{(84 + 79 + 91 + 76 + x)}{5} \geq 80$$

Change part b above to "**Graph** $85 \geq ((330 + x) / 5)$ and $((330 + x) / 5) \geq 80$." Trace the coordinates of the endpoints of the solution graph to get $70 \leq x \leq 95$. Therefore, Jenny's next test score must be in the 70–95 range to give her an average in the 80–85 range.

Use a graphing calculator to graph each inequality. Use the trace feature to determine the solution.

1. $125 \leq 2x + 80$
   $x > 22.5$
2. $4x + 30 \leq 2x + 80$
   $x < 25$
3. $90 - 3x > 2x - 10$
   $x < 20$
4. $100 < x + 30 < 120$
   $70 < x < 90$
5. $110 < 2x + 20 < 140$
   $45 < x < 60$
6. $-50 < 4x - 150 < -40$
   $25 < x < 27.5$

TECH PREP

Tech Prep Applications Masters, pp. 13, 14

7-1 NAME_____ DATE_____
Student Edition
Tech Prep Applications Pages 384–390

Tolerance (Machinist)

A piston inside an automobile cylinder can be neither too tight nor too loose. The piston must be machined to a very high tolerance. This means that the diameter of the piston must be very close to the diameter of the cylinder in which it slides.

A machinist often uses a vernier caliper to measure an exterior diameter, such as the diameter of a piston, or an interior diameter, such as the diameter of a cylinder.

Suppose that a machinist wants to make a piston with an ideal diameter of 5 inches. The actual piston may vary no more than 0.002 inch from the ideal. Find the greatest and least acceptable diameters.

Let x represent the diameter of the piston. To find the maximum allowable diameter, solve $x - 0.002 \leq 5$.

$$x - 0.002 \leq 5$$
$$x - 0.002 + 0.002 \leq 5 + 0.002$$
$$x \leq 5.002$$

To find the minimum allowable diameter, solve $x + 0.002 \geq 5$.

$$x + 0.002 \geq 5$$
$$x + 0.002 - 0.002 \geq 5 - 0.002$$
$$x \geq 4.998$$

The piston cannot have a diameter less than 4.998 inches or more than 5.002 inches.

Solve.

1. The actual diameter of a piston may vary no more than 0.0025 inch from the ideal diameter of 6.4 inches. Find the greatest and least acceptable diameters.
   6.4025 in. and 6.3975 in.

2. The actual length of a metal rod may vary no more than 0.001 inch from the ideal length of 12.5 inches. Find the greatest and least acceptable lengths.
   12.501 in. and 12.499 in.

3. With a vernier caliper, a machinist determines that a metal rod has a length that is within 0.001 inch of 12 inches. What are the possible lengths of the rod?
   between 11.999 in. and 12.001 in.

Cylinder
Piston

CONNECTIONS

Science and Math Lab Manual, pp. 29–34

7 NAME_____ DATE_____
Student Edition
Science and Math Lab 7 Pages 427–432

Wind Power and Box-and-Whisker Plots

Introduction

Some of the sun's energy combines with the rotation of the Earth to produce wind. Sometimes, people can use wind power to turn turbines and produce electrical energy. In order to use the wind as a source of energy, there must be a steady source of wind, usually of a constant speed of at least 12.8 kilometers per hour.

Objectives

- Construct a device to measure wind speed.
- Measure the wind speed at different times during the day for a week.
- Display measurements in a box-and-whisker plot.
- Determine if wind is a good source of energy in your area.

Materials

- stiff cardboard
- glue or paste
- sheet of grid paper (1)
- magic marker, any color (1)
- needle, long enough to go through ball (1)
- nylon line (30 cm)
- table tennis ball (1)
- scissors

Procedure

1. Cut out the protractor in Figure 1 and glue it to the cardboard.

Figure 1

PROBLEM SOLVING

Problem of the Week Cards, 19, 20, 21

Calculated Cuts
Problem-of-the-Week
Card 19

The Problem

Susan has a very long piece of string. She wants to cut it into pieces, and she wonders whether the way the string is laid out will affect the number of pieces she will get. She is considering laying the piece in a straight line, in a circle, in a half circle, or in the shape of an 8. Each cut Susan makes will be a vertical cut across the shape she makes, and it will be parallel to the other cuts. If she does not care about the size of the pieces, how many pieces will be made from each configuration when she makes 100 cuts in each? For the circle arrangement, you may assume the first cut passes through the point where the ends of the string meet.

Strategies and Hints

1. Think about making the problem simpler. What if Susan makes fewer than 100 cuts?
2. Record what you know in a table.
3. Do you see patterns in the numbers in the table? Can you express these patterns in general, using a variable?

Extensions

1. Can the same number of pieces be made from both a line arrangement and from a circular arrangement? Why or why not?
2. If you want to make exactly 31 pieces, which arrangement(s) could you use? Which ones(s) could you not use?

This two-page introduction to the chapter provides students with an opportunity to explore contemporary topics and their applications to mathematics.

Background Information

Increase Your Earnings Potential The percentage of Americans who have completed college has been increasing steadily. But so has the cost of a college education. Average yearly costs at a public university in 1995 were slightly over $10,000. Still, the data presented here show that college tuition is a wise investment. While a bachelor's degree at a state university averages $44,000, the return on that investment is a $600,000 increase in earnings over a lifetime.

CHAPTER 7

Solving Linear Inequalities

Increase Your Earnings Potential

Objectives

In this chapter, you will:

- solve inequalities,
- graph solutions of inequalities,
- graph solutions of open sentences that involve absolute value,
- solve problems by drawing a diagram, and
- use box-and-whisker plots to display and analyze data.

Degree	Lifetime Earnings
Professional degree	3013
Doctorate	2142
Master's degree	1619
Bachelor's degree	1421
2-year degree	1082
Some college	993
High school graduate	821
Not a high school graduate	609

0 500 1000 1500 2000 2500 3000 3500
Lifetime Earnings (thousands of dollars)

Source: *Chicago Tribune, 1995*

There is a direct relationship between lifetime earnings and educational attainment. College graduation and advanced degrees really do make a difference. Set your sights on a career and aim to be the best you can be.

TIME Line

260 B.C. Archimedes develops a mathematical description of the lever and other simple machines.

1598 Spanish missionaries introduce theater to the residents of what is now New Mexico.

300 B.C. 200 100 A.D. 800 900 1000 1100 1200 1300 1400 1500 1600

A.D. 1142 Adelard of Bath translates Euclid's *Elements* from the Arabic.

1514 Dutch mathematician Vander Hoecke is the first to use plus and minus signs as we do today in algebra.

TIME Line

Students might find it interesting to write a short report on Archimedes. He contributed greatly both to the sciences and to mathematics.

interNET CONNECTION

You can learn about medical, micro-biological, and molecular biological sciences and use links to other sites of such material on the Internet.

World Wide Web
http://www.qmw.ac.uk/~rhbm001/

Chapter Project

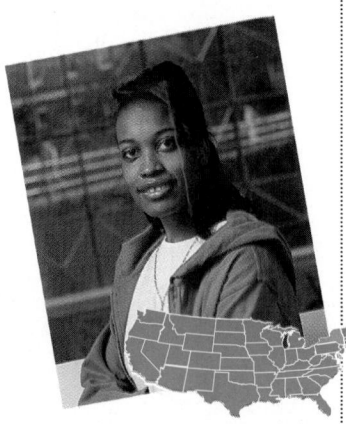

Shakema Hodge is aiming for an advanced degree in microbiology. The young scientist from St. Thomas, Virgin Islands, earned her bachelor's degree in biology at the University of the Virgin Islands. Now she is enrolled in a 5-year Ph.D. program in molecular microbiology at the University of Rochester in New York. She would love to teach high school students, believing that "to capture a young person's mind, you have to expose them to the sciences at an early age."

Kimana plans to study either botany, genetics, or microbiology at a nearby university. She hasn't ruled out teaching as a possible career choice, but is interested in finding other businesses in which she can apply the science she will be studying.

• Visit the public library or local college to investigate careers that relate to Kimana's interests.

• Which careers might offer Kimana an opportunity to earn an amount greater than that presented in the graph for a bachelor's degree?

• Which careers require a degree for which the years studied are greater than those for a bachelor's degree? Choose a career. Write an inequality that might express the time needed to study and train for that career.

1824 German composer Ludwig van Beethoven finished op 125 (choral).

1953 Massey Hall in Toronto, Ontario, was the site of a famous jazz concert featuring Charlie Parker, Dizzy Gillespie, Bud Powell, Max Roach, and Charles Mingus.

...125 | 1750 | 1775 | 1800 | 1825 | 1850 | 1875 | 1900 | 1925 | 1950 | 1975 | 2000

1709 Daniel Gabriel Fahrenheit invents the first accurate thermometer.

1995 Juliet Garcia, president of the University of Texas at Brownsville, becomes the first Hispanic woman to be named the president of a major U.S. university.

Chapter 7 **383**

Alternative Chapter Projects

Two other chapter projects are included in the *Investigations and Projects Masters.* In Chapter 7 Project A, pp. 49–50, students extend the topic in the chapter opener. In Chapter 7 Project B, pp. 51–52, students compare the heights, heart rates, and temperatures of their classmates with the national norm.

Even as a child in St. Thomas, Shakema was interested in "weird" science. She spent much of her play time in creative activities, such as making perfume out of flowers and leaves. Shakema was heavily recruited by a number of American graduate schools, but chose Rochester because of its research in the area of molecular biology.

Chapter Project

Cooperative Learning In groups of two or three, have students investigate careers. Students should prepare a 10-minute class presentation concerning the career, education required, and earning potential.

Investigations and Projects Masters, p. 49

7 NAME_____ DATE_____
Chapter 7 Project A
Student Edition
Pages 384–442

Choose a Career

1. What career do you want to pursue after you graduate from high school? You may be considering several different careers; if so, which of these are you most likely to choose? Do research to find out more about this career. Are there any courses you can take in high school that will help you prepare for your career?

2. What are the requirements for entering your chosen field? Do research to find out about where you can get the necessary education and/or training. Be sure to find out about the money and time you will need to invest to receive such education and/or training.

3. Think about how you will pay for your education and/or training. Write a plan for saving the money required. Consider the answers to the following questions as you work on your savings plan.
   • Can you start saving money now?
   • Are your parents or guardians saving money for you?
   • Are there scholarships for which you might be eligible?

   Write an inequality that expresses the minimum amount you plan to save from all possible sources.

4. Share your plan with your classmates. Ask them if they think the plan is realistic and if they can offer any suggestions for improving it.

NCTM Standards: 1–5

Instructional Resources

- Study Guide Master 7-1
- Practice Master 7-1
- Enrichment Master 7-1
- Modeling Mathematics Masters, p. 78
- Tech Prep Applications Masters, p. 13

 Transparency 7-1A contains the 5-Minute Check for this lesson; **Transparency 7-1B** contains a teaching aid for this lesson.

Recommended Pacing	
Standard Pacing	Day 1 of 14
Honors Pacing	Day 1 of 13
Block Scheduling*	Day 1 of 7
Alg. 1 in Two Years*	Days 1 & 2 of 21

 *For more information on pacing and possible lesson plans, refer to the *Block Scheduling Booklet* and *Algebra 1 in Two Years.*

1 FOCUS

 5-Minute Check
(over Chapter 6)

State the slopes of the lines parallel and perpendicular to the graph of each equation.

1. $4r = 2p - 7$ $\frac{1}{2}$, -2

2. $t = 3m - 7$ 3, $-\frac{1}{3}$

3. $5x - y = 7$ 5, $-\frac{1}{5}$

Find the coordinates of the midpoint of the line segment whose endpoints are given.

4. (9, 5), (17, 3) (13, 4)

5. (11, 4), (9, 2) (10, 3)

Objective

After studying this lesson you should be able to:
- solve inequalities by using addition and subtraction.

CAREER CHOICES

Nutritionists plan dietary programs and supervise the preparation and serving of meals for hospitals, nursing homes, health maintenance organizations, and individuals. A bachelor's degree in dietetics, foods, or nutrition is required that includes studies in biology, physiology, mathematics, and chemistry.

For more information, contact:
The American Dietetic Association
216 West Jackson Blvd.
Chicago, IL 60606

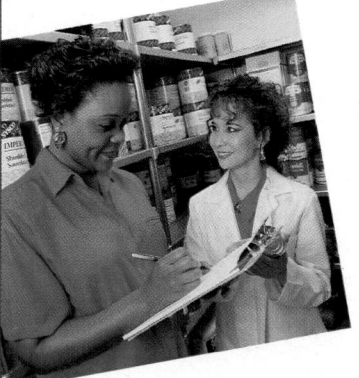

APPLICATION
Nutrition

In 1990, the U.S. Department of Agriculture issued new dietary guidelines. These guidelines recommend that people greatly reduce their fat intake. Your recommended calorie intake depends on your height, desired weight, and physical activity. The average 14-year-old is 5 feet 2 inches tall and weighs 107 pounds. Boys should consume about 2434 calories per day and girls 2208 calories per day to maintain this weight.

Snack	Quantity	Fat (grams)
Cheese/peanut butter crackers	9 crackers	15
Whole roasted almonds	10 almonds	8
Candy bars	2 bars	4
Apples	2 medium	2

Oliana learned in health class that no more than 30% of her calorie intake should come from fat. For her 2030-calorie-a-day diet, that means no more than 68 grams of fat. She keeps track of her snacks for one day and records their fat content, as shown in the table above. How many grams of fat can Oliana have in the other foods she eats that day and stay within the guidelines?

Let's write an inequality to represent the problem. Let g represent the remaining grams of fat that Oliana can eat that day.

grams of fat in snacks	plus	grams of fat remaining	is less than or equal to	total grams
15 + 8 + 4 + 2	+	g	≤	68

That is, $29 + g \leq 68$.

The symbol ≤ indicates *less than or equal to*. It is used in this situation because the total number of grams of fat in Oliana's daily diet should be no greater than 68. If this were an equation, we would subtract 29 from (or add −29 to) each side. Can the same procedure be used in an inequality? *This problem will be solved in Example 1.*

CAREER CHOICES

Nutritionists with less than 5 years experience earned a median annual salary of $28,500 in 1991. Those with 6–10 years experience earned a median salary of $32,900; 11–15 years, $36,000; 16–20 years, $38,400; and more than 20 years, $40,000.

Let's explore what happens if inequalities are solved in the same manner as equations. We know that $7 > 2$. What happens when you add or subtract the same quantity to each side of the inequality? We can use number lines to model the situation.

Add 3 to each side.

$$7 > 2$$
$$7 + 3 \overset{?}{>} 2 + 3$$
$$10 > 5$$

Subtract 4 from each side.

$$7 > 2$$
$$7 - 4 \overset{?}{>} 2 - 4$$
$$3 > -2$$

In each case, the inequality holds true. These examples illustrate two properties of inequalities.

Addition and Subtraction Properties for Inequalities	**For all numbers a, b, and c, the following are true.** 1. If $a > b$, then $a + c > b + c$ and $a - c > b - c$. 2. If $a < b$, then $a + c < b + c$ and $a - c < b - c$.

These properties are also true when $>$ and $<$ are replaced by $\geq$ and $\leq$. So, we can use these properties to obtain a solution to the application at the beginning of the lesson.

Example 1

APPLICATION

Nutrition

Top Five List

Preferred snack foods and
% of calories from fat

1. potato chips	58%
2. tortilla chips	47%
3. popcorn	45%
4. pretzels	8%
5. mixed nuts	80%

Refer to the application at the beginning of the lesson. Solve $29 + g \leq 68$.

$$29 + g \leq 68$$
$$29 - 29 + g \leq 68 - 29 \quad \textit{Subtract 29 from each side.}$$
$$g \leq 39 \qquad \textit{This means all numbers less than or equal to 39.}$$

The solution set can be written as {all numbers less than or equal to 39}.

Check: To check this solution, substitute 39, a number less than 39, and a number greater than 39 into the inequality.

Let $g = 39$.	Let $g = 20$.	Let $g = 40$.
$29 + g \leq 68$	$29 + g \leq 68$	$29 + g \leq 68$
$29 + 39 \overset{?}{\leq} 68$	$29 + 20 \overset{?}{\leq} 68$	$29 + 40 \overset{?}{\leq} 68$
$68 \leq 68$ true	$49 \leq 68$ true	$69 \leq 68$ false

So, Oliana can have 39 or fewer grams of fat in other foods that day and stay within the dietary guidelines.

The solution to the inequality in Example 1 was expressed as a set. A more concise way of writing a solution set is to use **set-builder notation**. The solution in set-builder notation is $\{g \mid g \leq 39\}$. This is read *the set of all numbers g such that g is less than or equal to 39.*

Lesson 7-1 Solving Inequalities by Using Addition and Subtraction **385**

Alternative Learning Styles

Kinesthetic Use the concept of an unbalanced scale to solve inequalities. Use a ruler, algebra tiles or counters, and a triangular piece of cardboard. Place the ruler on the triangle so that the greater side of the inequality is up. The representation of $x - 5 > -3$ is shown at the right.

To solve, place 5 positive tiles on each side of the scale. Thus, the solution is $x > 2$.

Motivating the Lesson
Questioning Introduce the lesson by having students name some values for a and b that make all three of the following inequalities true.
$$a < 0, \, b < 0, \, a - b < 0$$
Then ask them to describe the relationship between a and b for the above inequalities to be true.
$a < b$

In-Class Example

For Example 1
Solve each inequality.

a. $x + 14 < 16$ $\{x \mid x < 2\}$
b. $y + (-21) > 7$ $\{y \mid y > 28\}$
c. $(-11) + t > 5$ $\{t \mid t > 16\}$

Teaching Tip Emphasize that as in Example 1, the solutions of open sentences are numbers. The set of all solutions for a sentence is the solution set.

Top Five List

Suggest that students go to a grocery store and examine the labels on five different items that come in both a regular version and a "lite" version. Have them compare the grams of fat in each version and write down the results. Discuss their findings in class.

Teaching Tip In Example 2, make sure students understand that $52 < z$ and $z > 52$ are equivalent inequalities. Students may find it easier to determine the solution set by writing the inequality with the variable on the left side.

In-Class Examples

For Example 2
Solve each inequality. Then graph the solution.

a. $3r - 17 > 2r + 14$ $\{r \mid r > 31\}$

b. $9t + 5 < 10t - 3$ $\{t \mid t > 8\}$

For Example 3
A stove and a freezer together weigh at least 260 kg. The stove weighs 115 kg. What is the weight of the freezer? **at least 145 kg**

LOOK BACK

You can refer to Lesson 1-5 for information on solution sets.

In Lesson 2–4, you learned that you can show the solution to an inequality on a graph. The solution to Example 1 is shown on the number line below.

28 29 30 31 32 33 34 35 36 37 38 39 40 41 42

The closed circle at 39 tells us that 39 is included in the inequality. The heavy arrow pointing to the left shows that it also includes all numbers less than 39. *If the inequality was $<$, the circle would be open.*

Example ❷ **Solve $13 + 2z < 3z - 39$. Then graph the solution.**

$$13 + 2z < 3z - 39$$
$$13 + 2z - 2z < 3z - 2z - 39 \quad \textit{Subtract 2z from each side.}$$
$$13 < z - 39$$
$$13 + 39 < z - 39 + 39 \quad \textit{Add 39 to each side.}$$
$$52 < z$$

Since $52 < z$ is the same as $z > 52$, the solution set is $\{z \mid z > 52\}$.

The graph of the solution contains an open circle at 52 since 52 is not included in the solution, and the arrow points to the right.

49 50 51 52 53 54 55 56

Verbal problems containing phrases like *greater than* or *less than* can often be solved by using inequalities. The following chart shows some other phrases that indicate inequalities.

Inequalities			
$<$	$>$	$\le$	$\ge$
• less than • fewer than	• greater than • more than	• at most • no more than • less than or equal to	• at least • no less than • greater than or equal to

Example ❸

APPLICATION
Budgeting

Alvaro, Chip, and Solomon have earned $500 to buy equipment for their band. They have already spent $275 on a used guitar and a drum set. They are now considering buying a $125 amplifier. What is the most they can spend on promotional materials and T-shirts for the band if they buy the amplifier?

Explore *At most* means that they cannot go over what is left of their budget of $500. They have spent $275, so they have $225 left. Let $m =$ the amount of money for promotional materials and T-shirts.

GLENCOE Technology

Interactive Mathematics Tools Software

This multimedia software provides an interactive lesson that uses counters to solve and compare inequalities. A **Computer Journal** gives students an opportunity to write about what they have learned.

For Windows & Macintosh

Alternative Teaching Strategies

Student Diversity To help students relate the language used in verbal problems to algebraic symbols, read the phrases on page 386 that indicate inequalities. Have students write the symbol that is used in each expression.

Plan	Total to spend	is at most	$225.
	$125 + m$	$\leq$	225

Solve

$$125 + m \leq 225$$
$$125 - 125 + m \leq 225 - 125 \quad \text{Subtract 125 from each side.}$$
$$m \leq 100$$

The members of the band can spend $100 or less on promotional materials and T-shirts.

Examine Since $275 + $125 + $100 = $500, Alvaro, Chip, and Solomon can spend $100 or less on promotional materials and T-shirts.

When solving problems involving equations, it is often necessary to write an equation that represents the words in the problem. This is also true of inequalities.

Example **4** **Write an inequality for the sentence below. Then solve the inequality and check the solution.**

Three times a number is more than the difference of twice that number and three.

Three times a number	is more than	twice the number	minus	three
$3x$	$>$	$2x$	$-$	3

$$3x > 2x - 3$$
$$3x - 2x > 2x - 2x - 3 \quad \text{Subtract 2x from each side.}$$
$$x > -3$$

The solution set is $\{x \mid x > -3\}$.

CHECK FOR UNDERSTANDING

Communicating Mathematics

1. Sample answer: $x + 5 < -5$, $x - 5 < -15$, $2x < x - 10$

2. The set of all numbers w such that w is greater than -3.

Study the lesson. Then complete the following.

1. **Write** three inequalities that are equivalent to $x < -10$.

2. **Explain** what $\{w \mid w > -3\}$ means.

3. **Explain** the difference between the solution sets for $x + 24 < 17$ and $x + 24 \leq 17$.   One ($\leq$) includes -7; the other does not.

4. **Describe** how you would graph the solution to an inequality. Include examples and graphs in your explanation.   See margin.

5. Is it possible for the solution set of an inequality to be the empty set? If so, give an example.   Yes, sample answer: $x + 6 > x + 8$.

Lesson 7–1 Solving Inequalities by Using Addition and Subtraction **387**

In-Class Example

For Example 4
Write an inequality for the sentence below. Then solve the inequality.
One-fourth a number is less than five minus the number.
$\frac{1}{4}x < 5 - x; \{x \mid x < 4\}$

3 PRACTICE/APPLY

Check for Understanding

Exercises 1–16 are designed to help you assess your students' understanding through reading, writing, speaking, and modeling. You should work through Exercises 1–6 with your students and then monitor their work on Exercises 7–16.

Error Analysis
Students may assume that they have solved an inequality correctly by unknowingly substituting incorrect representative values into the original inequality. For example, suppose they solve $x + 3 < 7$ incorrectly as $x < 1$. Substituting 3 into the original inequality makes it true. However, 3 is not a solution of the inequality.

Additional Answer

4. Answers and graphs will vary. Sample answer: The circle on the graph will be filled if the inequality involves $\leq$ or $\geq$. The circle will be open if it involves $>$ or $<$. The arrow goes to the left if it involves $<$ or $\leq$. It goes to the right for $>$ or $\geq$.

Reteaching

Using Diagrams Use a two-pan balance to demonstrate how to solve inequalities. If no balance is handy, draw a large balance on the chalkboard. Boxes have identical weights. What can be concluded about the weight of one box? Explain.

Additional Answers

17.
13 14 15 16 17 18 19 20 21

18.
−19−18−17−16−15−14−13−12

19.
−4−3−2−1 0 1 2 3 4

20.
13 14 15 16 17 18 19 20 21

21.
1 2 3 $\frac{11}{3}$ 5 6 7 8 9

22.
−3−2 −0.15 1 2 3 4 5

23.
−1 0 1 2 3 4 5 6 7

24.
−4−3−2−1 0 1 2 3 4

Study Guide Masters, p. 48

7-1

Study Guide

NAME_____ DATE_____

Student Edition
Pages 384–390

Solving Inequalities by Using Addition and Subtraction

You can use the addition and subtraction properties for inequalities when solving problems involving inequalities.

Addition and Subtraction Properties for Inequalities
For all numbers a, b, and c,
1. if $a > b$, then $a + c > b + c$ and $a - c > b - c$;
2. if $a < b$, then $a + c < b + c$ and $a - c < b - c$.

Example: Solve $3a + 5 > 4 + 2a$.

$3a + 5 > 4 + 2a$
$3a - 2a + 5 > 4 + 2a - 2a$
$a + 5 > 4$
$a + 5 - 5 > 4 - 5$
$a > -1$

The solution set is {all numbers greater than −1}.

To check your solution, choose two numbers, one greater than −1, and one less than −1. Substitute both numbers in the original inequality. Only those numbers greater than −1 should yield a true statement.

The solution set in the above example, written in set-builder notation is {$a \mid a > -1$}. This is read "The set of all numbers a such that a is greater than −1."

Solve each inequality. Then check your solution.

1. $a + 4 < 14$
   {$a \mid a < 10$}

2. $9k - 12 > 80 + 8k$
   {$k \mid k > 92$}

3. $-19 + x < 2x - 33$
   {$x \mid x > 14$}

4. $6y > 14 - 2 + 7y$
   {$y \mid y < -12$}

5. $3n + 17 < 4n - 6$
   {$n \mid n > 23$}

6. $\frac{3}{2}q - \frac{25}{5} \geq \frac{2q}{4}$
   {$q \mid q \geq 5$}

7. $h + \frac{2}{3} \leq 2 - \frac{2}{3}$
   {$h \mid h \leq \frac{2}{3}$}

8. $4p - 3.2 \geq 3p + 0.7$
   {$p \mid p \geq 3.9$}

9. $-2\frac{1}{2}z \leq 3\frac{1}{3} + 2\frac{1}{3} - 3\frac{1}{2}z$
   {$z \mid z \leq 5\frac{2}{3}$}

10. $\frac{1}{b} + 4 \leq 10 + \frac{1}{b}$, $b \neq 0$
    all numbers

11. $6r > 10r - r - 3r$
    no solution

12. $3.2x < 2x - (9 - 1.2x)$
    no solution

388 Chapter 7

---

6. Sometimes statements we make can be translated into inequalities. For example, *In some states, you have to be at least 16 years old to have a driver's license* can be expressed as $a \geq 16$, and *Tomás cannot lift more than 72 pounds* can be translated into $w \leq 72$. Following these examples, write three statements that deal with your everyday life. Then translate each into a corresponding inequality. **See students' work.**

Guided Practice

Match each inequality with the graph of its solution.

7. $b - 18 > -3$ c

8. $10 \geq -3 + x$ a

9. $x + 11 < 6$ d

10. $4c - 3 \leq 5c$ b

a.
10 11 12 13 14 15 16 17

b.
−6 −5 −4 −3 −2 −1 0 1

c.
10 11 12 13 14 15 16 17

d.
−7 −6 −5 −4 −3 −2 −1 0

Solve each inequality. Then check your solution.

11. $x + 7 > 2$ {$x \mid x > -5$}

12. $10 \geq x + 8$ {$x \mid x \leq 2$}

13. $y - 7 < -12$ {$y \mid y < -5$}

14. $-81 + q > 16 + 2q$ {$q \mid q < -97$}

Define a variable, write an inequality, and solve each problem. Then check your solution.

15. A number decreased by 17 is less than −13. $x - 17 < -13$, {$x \mid x < 4$}

16. A number increased by 4 is at least 3. $x + 4 \geq 3$, {$x \mid x \geq -1$}

EXERCISES

Practice

Solve each inequality. Then check your solution, and graph it on a number line. 17–24. See margin for graphs.

A

17. $a - 12 < 6$ {$a \mid a < 18$}

18. $m - 3 < -17$ {$m \mid m < -14$}

19. $2x \leq x + 1$ {$x \mid x \leq 1$}

20. $-9 + d > 9$ {$d \mid d > 18$}

21. $x + \frac{1}{3} > 4$ {$x \mid x > \frac{11}{3}$}

22. $-0.11 \leq n - (-0.04)$ {$n \mid n \geq -0.15$}

23. $2x + 3 > x + 5$ {$x \mid x > 2$}

24. $7h - 1 \leq 6h$ {$h \mid h \leq 1$}

Solve each inequality. Then check your solution.

B

25. $x + \frac{1}{8} < \frac{1}{2}$ {$x \mid x < \frac{3}{8}$}

26. $3x + \frac{4}{5} \leq 4x + \frac{3}{5}$ {$x \mid x \geq \frac{1}{5}$}

27. $3x - 9 \leq 2x + 6$ {$x \mid x \leq 15$}

28. $6w + 4 \geq 5w + 4$ {$w \mid w \geq 0$}

29. {$x \mid x < 0.98$}

29. $-0.17x - 0.23 < 0.75 - 1.17x$

30. $0.8x + 5 \geq 6 - 0.2x$ {$x \mid x \geq 1$}

31. $3(r - 2) < 2r + 4$ {$r \mid r < 10$}

32. $-x - 11 \geq 23$ {$x \mid x \leq -34$}

388 Chapter 7 Solving Linear Inequalities

Define a variable, write an inequality, and solve each problem. Then check your solution.

33. A number decreased by -4 is at least 9. $x - (-4) \geq 9$, $\{x \mid x \geq 5\}$

34. The sum of a number and 5 is at least 17. $x + 5 \geq 17$, $\{x \mid x \geq 12\}$

35. $3x < 2x + 8$

36. $21 \geq x + (-2)$

35. Three times a number is less than twice the number added to 8. $\{x \mid x < 8\}$

36. Twenty-one is no less than the sum of a number and -2. $\{x \mid x \leq 23\}$

37. The sum of two numbers is less than 53. One number is 20. What is the other number? $20 + x < 53$, $\{x \mid x < 33\}$

▷ C

38. The sum of four times a number and 7 is less than 3 times that number.

38. $4x + 7 < 3x$, $\{x \mid x < -7\}$

39. $2x > x - 6$, $\{x \mid x > -6\}$

39. Twice a number is more than the difference of that number and 6.

40. The sum of two numbers is 100. One number is at least 16 more than the other number. What are the two numbers? $100 - x \geq x + 16$, $\{x \mid x \leq 42\}$

If $3x \geq 2x + 5$, then complete each inequality.

41. $3x + 7 \geq 2x + \underline{\ ?\ }$ 12

42. $3x - 10 \geq 2x - \underline{\ ?\ }$ 5

43. $3x + \underline{\ ?\ } \geq 2x + 3$ -2

44. $\underline{\ ?\ } \leq x$ 5

Programming

45. Geometry For three line segments to form a triangle, the sum of the lengths of any two sides must exceed the length of the third side. Let the lengths of the possible sides be a, b, and c. Then these three inequalities must be true: $a + b > c$, $a + c > b$, and $b + c > a$. The graphing calculator program at the right uses these inequalities to determine if the three lengths can be measures of the sides of a triangle.

```
PROGRAM: TRIANGLE
: Disp "ENTER THREE LENGTHS"
: Prompt A, B, C
: If C≥A+B
: Then
: Goto 1
: End
: If B≥A+C
: Then
: Goto 1
: End
: If A≥B+C
: Then
: Goto 1
: End
: Disp "THIS IS A TRIANGLE."
: Stop
: Lbl 1
: Disp "NOT A TRIANGLE"
```

Use the program to determine whether segments with the given lengths can form a triangle.

Tip: To run the program again after trying one set of numbers, press ENTER .

a. 10 in., 12 in., 27 in. no
b. 3 ft, 4 ft, 5 ft yes
c. 125 cm, 140 cm, 150 cm yes
d. 1.5 m, 2.0 m, 2.5 m yes

47. The value of x falls between -2.4 and 3.6.

Critical Thinking

46. Using an example, show that even though $x > y$ and $t > w$, $x - t > y - w$ may be false. **Answers will vary. Sample answer:** $x = 4$, $y = 2$, $t = 3$, $w = 0$

47. What does the sentence $-2.4 < x < 3.6$ mean?

Lesson 7–1 Solving Inequalities by Using Addition and Subtraction **389**

Using the Programming Exercises The program given in Exercise 45 is for use with a TI-82 graphing calculator. For other programmable calculators, have students consult their owner's manual for commands similar to those presented here.

Practice Masters, p. 48

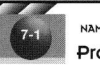

4 ASSESS

Closing Activity

Writing Without solving, rewrite each inequality in a form that has the same meaning.

1. $17 < x + 4$ $x + 4 > 17$
2. $-3 + x > 2x - 4$
   $2x - 4 < -3 + x$
3. $4 + x < 8$ $8 > 4 + x$

Additional Answer

48b.
70 71 72 73 74 75 76 77 78

Enrichment Masters, p. 48

390 *Chapter 7*

Applications and Problem Solving

Define a variable, write an inequality, and solve each problem.

48. **Academics**   Josie must have at least 320 points in her math class to get a B. She needs a B or better to maintain her grade-point average so she can play on the basketball team. The grade is based on four 50-point tests, three 20-point quizzes, two 20-point projects, and a final exam worth 100 points. Josie's record of her grades are shown in the table below.

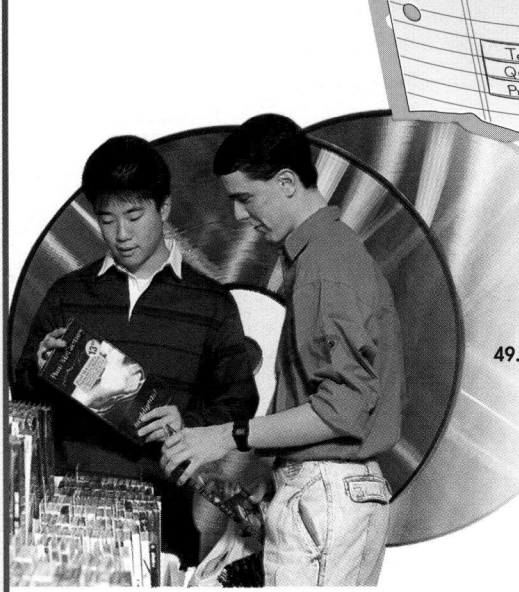

	Points	Total Points
Tests	40, 42, 41, 45	168
Quizzes	15, 12, 19	46
Projects	15, 18	33

   a. Write an inequality to represent the range of scores on the final exam needed in order for Josie to make a B.   **$247 + x \geq 320$**
   b. Solve the inequality and graph its solution.   **$x \geq 73$; See margin for graph.**

49. **Personal Finances**   Tanaka had $75 to buy presents for his family. He bought a $21.95 shirt for his dad, a $23.42 necklace for his mother, and a $16.75 CD for his sister. He still has to buy a present for his brother.
   a. How much can he spend on his brother's present?   **$x \leq \$12.88$**
   b. What factors not stated in the problem may affect how much money he can spend?   **Sample answer: There may be sales tax on his purchases.**

Mixed Review

50. Find the midpoint of the line segment whose endpoints are at $(-1, 9)$ and $(-5, 5)$.   (Lesson 6–7)   **$(-3, 7)$**

Write an equation in slope-intercept form for a line that satisfies each condition.   (Lesson 6–6)

51. perpendicular to $x + 7 = 3y$ that passes through $(1, 0)$ **$y = -3x + 3$**

52. parallel to $\frac{1}{5}y - 3x = 2$ that passes through $(0, -3)$ **$y = 15x - 3$**

53. **Statistics**   Find the range, quartiles, and interquartile range for the set of data at the right.   (Lesson 5–7)   **42, 131, 145, 159, 28**

54. $\left\{(-2,8), \left(-1, \frac{20}{3}\right), \left(0, \frac{16}{3}\right), \left(2, \frac{8}{3}\right), \left(5, -\frac{4}{3}\right)\right\}$

54. Solve $4x + 3y = 16$ if the domain is $\{-2, -1, 0, 2, 5\}$.   (Lesson 5–3)

55. If y varies inversely as x and $y = 32$ when $x = 3$, find y when $x = 8$.   (Lesson 4–8)   **12**

56. Solve $y - \frac{7}{16} = -\frac{5}{8}$ (Lesson 3–1)   **$-\frac{3}{16}$**

57. Replace the ___?___ with $<$, $>$, or $=$ to make $\frac{6}{13}$ ___?___ $\frac{1}{2}$ true.   (Lesson 2–4)   **$<$**

58. **Look for a Pattern**   How many triangles are shown at the right? Count only the triangles pointing upward.   (Lesson 1–2)   **120 triangles**

Stem	Leaf
12	4 6 7 7
13	1 1 6 9
14	0 5 7
15	0 3 9 9 9
16	5 6 6

$14 \mid 0 = \$140$

390   *Chapter 7   Solving Linear Inequalities*

Extension

Problem Solving   Find the least prime number greater than 80.   **83**

7-2A Solving Inequalities

Materials: equation mat   cups and counters

self-adhesive note

A Preview of Lesson 7–2   You can use an equation model to solve inequalities.

Activity   Model the solution for −2x < 4.

Step 1 Use the note to cover the equals sign on the equation mat. Then write a < symbol on the note. Label 2 cups with a negative sign, and place them on the left side. Place 4 positive counters on the right.

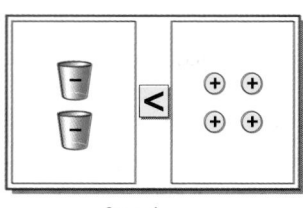

$$-2x < 4$$

Step 2 Since we cannot solve for a negative cup, we must eliminate the negative cups by adding 2 positive cups to each side. Remove the zero pairs.

$$-2x + 2x < 4 + 2x$$

Step 3 Add 4 negative counters to each side to isolate the cups. Remove the zero pairs.

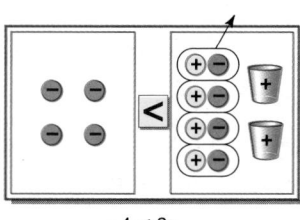

$$-4 < 2x$$

Step 4 Separate the counters into 2 groups.

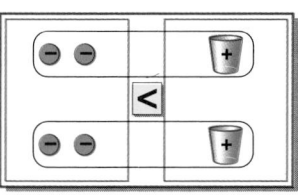

$$-2 < x \text{ or } x > -2$$

Model

1. Compare the symbol and location of the variable in the original problem with those in the solution. What do you find?   **1–4. See Solutions Manual.**

2. Model the solution for 3x > 12. What do you find? How is this different from solving −2x < 4?

Write

3. Write a rule for solving inequalities involving multiplication.

4. Do you think the rule applies to inequalities involving division?
   Remember that dividing by a number is the same as multiplying by its reciprocal.

NCTM Standards: 1–5

Objective
Use cups and counters to solve inequalities.

Recommended Time
Demonstration and discussion: 15 minutes; Exercises: 30 minutes

Instructional Resources
For each student or group of students
Student Manipulative Kit
• 20 red/yellow 2-sided counters
• 5 cups
Modeling Mathematics Masters
• pp. 2–3 (cups and counters)
• p. 13 (equation mat)
• p. 27 (worksheet)
For teacher demonstration
Algebra and Geometry Overhead
Manipulative Resources

1 FOCUS

Motivating the Lesson
Write the inequality −5 < −2 on the board. Ask students if the inequality would still be true if the negative signs were removed. Ask what would have to be done with the inequality symbol to make the sentence true again.

2 TEACH

Teaching Tip Point out that there is an alternative method for Activity 1. After Step 1, they could replace every cup and every counter with its opposite and reverse the direction of the inequality symbol.

3 PRACTICE/APPLY

Assignment Guide
Core: 1–4
Enriched: 1–4

4 ASSESS

Observing students working in cooperative groups is an excellent method of assessment.

Solving Inequalities by Using Multiplication and Division

1 FOCUS

 ### 5-Minute Check
(over Lesson 7-1)

Solve each inequality.

1. $y + 3 > 11$ $\{y \mid y > 8\}$
2. $t - 9 < 6$ $\{t \mid t < 15\}$
3. $3t + 11 < 4$ $\left\{t \mid t < -\frac{7}{3}\right\}$
4. $2y - 10 < y + 6$ $\{y \mid y < 16\}$
5. $0.9 - p < -3.4$ $\{p \mid p > 4.3\}$

Objective

After studying this lesson you should be able to:

- solve inequalities by using multiplication and division.

CONNECTION
Physical Science

A lever can be used to multiply the effort force you exert when trying to move something. The fixed point or *fulcrum* of a lever separates the length of the lever into two sections—the *effort arm* on which the effort force is applied and the *resistance arm* that exerts the resistance force. The *mechanical advantage* of a lever is the number of times a lever multiplies that effort force.

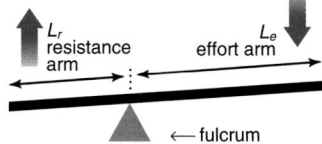

The formula for determining the mechanical advantage *MA* of a lever can be expressed as $MA = \frac{L_e}{L_r}$, where L_e represents the length of the effort arm and L_r represents the length of the resistance arm.

Suppose a group of volunteers is clearing hiking trails at Yosemite National Park. They need to position a lever so that a mechanical advantage of at least 7 is achieved in order to remove a boulder blocking the trail. The volunteers place the lever on a rock so they can use the rock as a fulcrum. They will need the resistance arm to be 1.5 ft long so that it is long enough to get under the boulder. What should be the length of the lever in order to move the boulder?

We need to find the length of the effort arm to find the total length. Let L_e represent the length of the effort arm. We know that 1.5 feet is the length of the resistance arm L_r. Since the mechanical advantage must be at least 7, we can write an inequality using the formula.

$$MA \geq 7$$

$$\frac{L_e}{L_r} \geq 7 \quad \textit{Replace MA with } \tfrac{L_e}{L_r}.$$

$$\frac{L_e}{1.5} \geq 7 \quad \textit{Replace } L_r \textit{ with 1.5.} \qquad \textit{You will solve this problem in Example 1.}$$

If you were solving the equation $\frac{L_e}{1.5} = 7$, you would multiply each side by 1.5. Will this method work when solving inequalities? Before answering this question, let's explore how multiplying (or dividing) an inequality by a positive or negative number affects the inequality. Consider the inequality $10 < 15$, which we know is true.

Multiply by 2.

$10 < 15$

$10(2) < 15(2)$

$20 < 30$ true

Multiply by -2.

$10 < 15$

$10(-2) < 15(-2)$ false

$-20 < -30$ false

$-20 > -30$ true

Divide by 5.

$10 < 15$

$\frac{10}{5} < \frac{15}{5}$

$2 < 3$ true

Divide by -5.

$10 < 15$

$\frac{10}{-5} < \frac{15}{-5}$ false

$-2 < -3$ false

$-2 > -3$ true

These results suggest the following.

• If each side of a true inequality is multiplied or divided by the same positive number, the resulting inequality is also true.

• If each side of a true inequality is multiplied or divided by the same negative number, the direction of the inequality symbol must be *reversed* so that the resulting inequality is also true.

Multiplication and Division Properties for Inequalities

> **For all numbers, a, b, and c, the following are true.**
>
> **1.** **If c is positive and $a < b$, then $ac < bc$ and $\frac{a}{c} < \frac{b}{c}$, and**
>
> **if c is positive and $a > b$, then $ac > bc$ and $\frac{a}{c} > \frac{b}{c}$.**
>
> **2.** **If c is negative and $a < b$, then $ac > bc$ and $\frac{a}{c} > \frac{b}{c}$, and**
>
> **if c is negative and $a > b$, then $ac < bc$ and $\frac{a}{c} < \frac{b}{c}$.**

These properties also hold for inequalities involving $\leq$ and $\geq$.

Example ❶

CONNECTION
Physical Science

Refer to the connection at the beginning of the lesson. What should the minimum length of the lever be?

$$\frac{L_e}{1.5} \geq 7$$

$$1.5 \cdot \frac{L_e}{1.5} \geq 1.5(7) \quad \textit{Multiply each side by 1.5.}$$

$$L_e \geq 10.5$$

The effort arm must be at least 10.5 feet long.

In order to find the length of the lever, add the lengths of the effort arm and the resistance arm. The lever should be at least $10.5 + 1.5$ or 12 feet long.

Motivating the Lesson
Hands-On Activity Have students draw a number line and graph the following points: $A = 0.5$, $B = -1.5$. Then have students consider the following.

1. If the coordinates of A and B are multiplied by 2, where are the corresponding points located on the number line?
$A = 1, B = -3$

2. If the coordinates of A and B are multiplied by -2, where are the corresponding points located on the number line?
$A = -1, B = 3$

2 TEACH

In-Class Example

For Example 1
Suppose a mechanical advantage of 10 is desired and the resistance arm is 20 centimeters long. What should be the minimum length of the lever?
at least 220 centimeters

Example 2 Solve $\frac{x}{12} \le \frac{3}{2}$.

$$\frac{x}{12} \le \frac{3}{2}$$

$$12 \cdot \frac{x}{12} \le 12 \cdot \frac{3}{2} \qquad \textit{Multiply each side by 12.}$$

$$x \le 18 \qquad \textit{Since we multiplied by a positive number, the inequality symbol stays the same.}$$

The solution set is $\{x \mid x \le 18\}$.

Since dividing is the same as multiplying by the reciprocal, there can be two methods to solve an inequality that involves multiplication.

Example 3 Solve $-3w > 27$.

Method 1	Method 2
$-3w > 27$	$-3w > 27$
$\frac{-3w}{-3} < \frac{27}{-3}$ *Divide each side by -3 and change $>$ to $<$.*	$\left(-\frac{1}{3}\right)(-3w) < \left(-\frac{1}{3}\right)(27)$ *Multiply each side by $-\frac{1}{3}$ and change $>$ to $<$.*
$w < -9$	$w < -9$

Check: Let w be any number less than -9.

$$-3w > 27$$

$$-3(-10) \overset{?}{>} 27 \qquad \textit{Suppose we select -10.}$$

$$30 > 27 \qquad \text{true}$$

Numbers less than -9 compose the solution set. The solution set is $\{w \mid w < -9\}$.

Example 4

APPLICATION
Business

Angelica Moreno is a sales representative for an appliance distributor. She needs at least $5000 in weekly sales of a particular TV model to qualify for a sales competition to win a trip to the Bahamas. If the TVs sell for $250 each, how many TVs will Ms. Moreno have to sell to qualify?

Explore Let t represent the number of TVs to be sold. At least $5000 means greater than or equal to $5000.

Plan The price of one TV times the number of sets sold must be greater than or equal to the total amount of sales needed.

The price of one TV	times	the number of TVs sold	is at least	$5000.
$250	$\times$	t	$\ge$	$5000

Solve
$$250t \ge 5000$$

$$\frac{250t}{250} \ge \frac{5000}{250} \qquad \textit{Divide each side by 250.}$$

$$t \ge 20$$

Examine Ms. Moreno must sell a minimum of 20 TVs to qualify for the contest. The amount of money from the sale of 20 TVs is $250(20) or $5000.

 Tech Prep

Real Estate Associate Students who are interested in a career in real estate, as mentioned in Example 4, may wish to do further research into the use of math in that occupation. For more information on tech prep, see the *Teacher's Handbook*.

Example 5

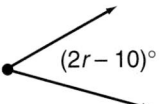

INTEGRATION
Geometry

Triangle *XYZ* is not an acute triangle. The greatest angle in the triangle has a measure of $(6d)°$. What are the possible values of *d*?

Since $\triangle XYZ$ is not acute, the measure of the greatest angle must be 90° or larger, but less than 180°. Thus, $6d \geq 90$ and $6d < 180$.

$6d \geq 90$		$6d < 180$	
$\frac{6d}{6} \geq \frac{90}{6}$	*Divide each side by 6.*	$\frac{6d}{6} < \frac{180}{6}$	*Divide each side by 6.*
$d \geq 15$		$d < 30$	

The value of *d* must be greater than or equal to 15 but less than 30.

CHECK FOR UNDERSTANDING

Communicating Mathematics

Study the lesson. Then complete the following.

1. **Classify** each statement as *true* or *false*. If false, explain how to change the inequality to make it true.

   a. If $x > 9$, then $-3x > -27$.   **False; change $>$ to $<$ in the second inequality.**

   b. If $x < 4$, then $3x < 12$.   **true**

2. **Complete** each statement.

   a. If each side of an inequality is multiplied by the same ___?___ number, the direction of the inequality symbol must be reversed so that the resulting inequality is true.   **negative**

   b. Multiplying by ___?___ is the same as dividing by -6. $-\frac{1}{6}$

   c. An acute triangle has angles whose measures are all less than ___?___.   **90°**

3. **You Decide**   Utina and Paige are discussing the rules for changes in the direction of the inequality symbol when solving inequalities. Paige says the rule is "Whenever you have a negative sign in the problem, the direction of the inequality symbol will change." Utina says that is not always true. Decide who is correct and give examples to support your answer.   **See margin.**

MODELING MATHEMATICS

Use models to solve each inequality. Write the answer in set-builder notation.

4. $3x < 15$ $\{x \mid x < 5\}$

5. $-6x < 18$ $\{x \mid x > -3\}$

6. $2x + 6 > x - 7$ $\{x \mid x > -13\}$

7. $-4x + 8 \geq 14$ $\{x \mid x \leq -1.5\}$

8. Use models to determine the appropriate symbol ($>$, $<$, or $=$) to complete each comparison.

   a. $x + 5$ ___?___ $x - 7$ $>$

   b. $x + 3 + (x - 6)$ ___?___ $(2x + 7) - 4$ $<$

   c. $2x + 8$ ___?___ $2x + 6 - x$ **cannot be determined**

For Example 5
The acute angle shown below has a measure greater than 0° and less than 90°. What is the set of all possible values for *r*?

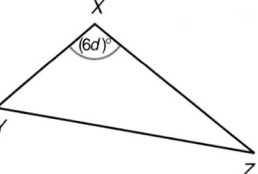

$(2r - 10)°$

$r > 5$ and $r < 50$

---

3 PRACTICE/APPLY

Check for Understanding
Exercises 1–19 are designed to help you assess your students' understanding through reading, writing, speaking, and modeling. You should work through Exercises 1–8 with your students and then monitor their work on Exercises 9–19.

Error Analysis
Some students may add an extra step when solving inequalities that involve dividing or multiplying.
Example:
$-3x \leq 12$
   $x \leq -4$ *Divide by -3.*
   $x \geq -4$ *Reverse the sign.*
Remind students that the sign is reversed in the same step as the multiplying or dividing.

Reteaching

Using Manipulatives Use a two-pan balance to demonstrate how to solve these inequalities.

1.

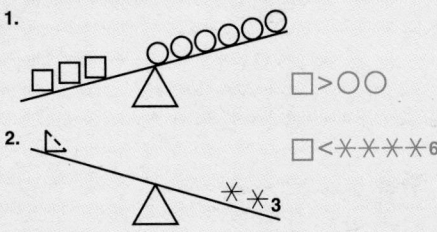

   $\square > \bigcirc\bigcirc$

2.

   $\square < \text{✶✶✶✶✶}6$

Additional Answer

3. Utina is correct because the only time the inequality symbol changes in solving an inequality is when you must multiply or divide by a negative to solve for the variable. For example, in $-3x > 12$, the inequality symbol changes. However, in $3x > -12$, the inequality symbol does not change.

Guided Practice

9. $\times -\frac{1}{6}$, $\div -6$; yes;
$\{y \mid y \le 4\}$

12. $\times -\frac{7}{2}$; yes;
$\{z \mid z \le 42\}$

State the number by which you multiply or divide to solve each inequality. Indicate whether the direction of the inequality symbol reverses. Then solve.

9. $-6y \ge -24$
10. $10x > 20$ $\times \frac{1}{10}$, $\div 10$; no; $\{x \mid x > 2\}$
11. $\frac{x}{4} < -5$ $\times 4$; no; $\{x \mid x < -20\}$
12. $-\frac{2}{7}z \ge -12$

Solve each inequality. Then check your solution.

13. $\frac{4}{5}x < 24$ $\{x \mid x < 30\}$
14. $-\frac{v}{3} \ge 4$ $\{v \mid v \le -12\}$
15. $-0.1t \ge 3$ $\{t \mid t \le -30\}$
16. $5y > -25$ $\{y \mid (y > -5\}$

Define a variable, write an inequality, and solve each problem. Then check your solution.

17. $\frac{1}{5}x \le 4.025$;
$\{x \mid x \le 20.125\}$
18. $-6x < 216$;
$\{x \mid x > -36\}$

17. One fifth of a number is at most 4.025.
18. The opposite of six times a number is less than 216.
19. **Geometry** Determine the value of s so that the area of the square is at least 144 square feet. $s \ge 12$

s ft

s ft

EXERCISES

Practice

22. $\{w \mid w < 25\}$
23. $\{x \mid x \ge -44\}$
25. $\{r \mid r < -6\}$
26. $\{b \mid b \ge -36\}$
28. $\{w \mid w > -33\}$
30. $\left\{b \mid b > \frac{28}{15}\right\}$
31. $\{x \mid x \ge -0.7\}$
33. $\left\{r \mid r < -\frac{1}{20}\right\}$
35. $\{x \mid x < -27\}$
37. $\{m \mid m \ge -24\}$
41. $\frac{3}{4}x \le -24$;
$\{x \mid x \le -32\}$

Solve each inequality. Then check your solution.   21. $\{b \mid b > -12\}$

A
20. $7a \le 49$ $\{a \mid a \le 7\}$
21. $12b > -144$
22. $-5w > -125$
23. $-x \le 44$
24. $4 < -x$ $\{x \mid x < -4\}$ 25. $-102 > 17r$
26. $\frac{b}{-12} \le 3$
27. $\frac{t}{13} < 13$ $\{t \mid t < 169\}$ 28. $\frac{2}{3}w > -22$

B
29. $6 \le 0.8g$ $\{g \mid g \ge 7.5\}$ 30. $-15b < -28$
31. $-0.049 \le 0.07x$
32. $\frac{3}{7}h < \frac{3}{49}$ $\left\{h \mid h < \frac{1}{7}\right\}$ 33. $\frac{12r}{-4} > \frac{3}{20}$
34. $\frac{3b}{4} \le \frac{2}{3}$ $\left\{b \mid b \le \frac{8}{9}\right\}$
35. $-\frac{1}{3}x > 9$
36. $\frac{y}{6} \ge \frac{1}{2}$ $\{y \mid y \ge 3\}$ 37. $\frac{-3m}{4} \le 18$

Define a variable, write an inequality, and solve each problem. Then check your solution.

38. Four times a number is at most 36. $4x \le 36$; $\{x \mid x \le 9\}$
39. Thirty-six is at least one half of a number. $36 \ge 0.5x$; $\{x \mid x \le 72\}$
40. The opposite of three times a number is more than 48. $-3x > 48$; $\{x \mid x < -16\}$
41. Three fourths of a number is at most -24.
42. Eighty percent of a number is less than 24. $0.80x < 24$; $\{x \mid x < 30\}$

C
43. The product of two numbers is no greater than 144. One of the numbers is -8. What is the other number? $-8x \le 144$; -18 or greater

44. Geometry Determine the value of x so that the area of the rectangle at the right is at least 918 square feet. $x \geq 8.5$ feet

36 feet

$3x$ feet

45. Geometry Determine the value of y so that the perimeter of the triangle at the right is less than 100 meters. $y < 7.14$ meters

$3y$ m $5y$ m

$6y$ m

Complete.

46. If $24m \geq 16$, then ___?___ ≥ 12. $18m$ **47.** If $-9 \leq 15b$, then $25b$ ___?___ -15. $\geq$

48. If $5y < -12$, then $20y < $ ___?___. -48 **49.** If $-10a > 21$, then $30a$ ___?___ -63. $<$

Critical Thinking

50. Use an example to show that if $x > y$, then $x^2 > y^2$ is not necessarily true. Answers will vary. Sample answer: $x = -1$, $y = -2$

Applications and Problem Solving

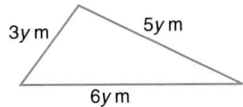

Define a variable, write an inequality, and solve each problem.

51. Travel The charge per mile for a compact rental car at 4-D Rentals is $0.12. Mrs. Rodriguez is on a business trip and must rent a car to attend various meetings. She has a budget of $50 per rental for mileage charges. What is the greatest number of miles Mrs. Rodriguez can travel without going over her budget? up to 416 miles

52. Physics Refer to the application at the beginning of the lesson. A city worker needs to raise a utility access cover. She has an iron bar to use as a lever. When using a lever, the resistance force is equal to the effort force multiplied by the mechanical advantage, or $F_r = MA \cdot F_e$. The worker weighs 120 pounds, so she can supply that much effort force. All utility covers in the city weigh at least 360 pounds. What mechanical advantage does she need to lift the cover? at least 3

53. at least 5883 signatures

53. Political Science A candidate needs 5000 signatures on a petition before she can run for a township office. Experience shows that 15% of the signatures on petitions are not valid. What is the smallest number of signatures the candidate should get to end up with 5000 valid signatures?

Mixed Review

54. Define a variable, write an inequality, and solve the following problem. The difference of five times a number less four times that number plus seven is at most 34. (Lesson 7–1) $5a - 4a + 7 \leq 34$; $a \leq 27$

Lesson 7–2 Solving Inequalities by Using Multiplication and Division **397**

Extension

Problem Solving A plumber charges $36 for the first half hour of work and $16 for every half hour or any part of a half hour thereafter. Find the longest amount of time this plumber can work without having the bill exceed $150.
4 hours

Closing Activity

Speaking Discuss the following.
If $-2a < b$, is it possible that

1. a and b are both positive? **yes**
2. a is positive and b is negative? **yes**
3. a is negative and b is positive? **yes**
4. a and b are both negative? **no**

Chapter 7, Quiz A (Lessons 7-1 and 7-2), is available in the *Assessment and Evaluation Masters,* p. 184.

55. $(-1, 1)$

56. $-3; \frac{1}{3}$

55. **Geometry** Three vertices of a square are at $(-5, -3)$, $(-5, 5)$, and $(3, -3)$. Suppose you were to inscribe a circle in this square. What would the coordinates of the center of the circle be? (Lesson 6–7)

56. **Geometry** Determine the slopes of the lines parallel and perpendicular to the graph of $3x - 6 = -y$. (Lesson 6–6)

57. Determine the slope of the line that passes through $(4, -9)$ and $(-2, 3)$. (Lesson 6–1) **-2**

58. Write an equation in functional notation for the relation at the right. (Lesson 5–6) $b = a - 3$

a	-2	0	2	4	6	8	10
b	-5	-3	-1	1	3	5	7

59. **Budgeting** JoAnne Paulsen's take-home pay is $1782 per month. She spends $325 on rent, $120 on groceries, and $40 on gas. She allows herself 12% of the remaining amount for entertainment. How much can she spend on entertainment each month? (Lesson 4–4) **$155.64**

60. What number is 47% of 27? (Lesson 4–4) **12.69**

61. **Soccer** A soccer field is 75 yards shorter than 3 times its width. Its perimeter is 370 yards. Find its dimensions. (Lesson 3–5) **65 yd by 120 yd**

62. Evaluate $5y + 3$ if $y = 1.3$. (Lesson 1–3) **9.5**

WORKING ON THE
In·ves·ti·ga·tion

Refer to the Investigation on pages 320–321.

Smoke Gets In Your Eyes

Currently, about 50 million Americans smoke, and each smoker averages about 30 cigarettes a day. Consider the amount of smoke produced by a single cigarette. Make some projections on the volume of cigarette smoke generated by all 50 million smoking Americans.

1 Find the volume of smoke produced by the average smoker.

2 Project the amount of smoke generated by 50 million smokers in a year.

3 Estimate the volume of air in your room at home. Write an inequality that would estimate the number of puffs p it would take to fill your room. Solve the inequality.

4 Suppose a smoker was locked in your room with several cartons of cigarettes. Would it be possible for the smoker to fill the room completely with smoke in the same concentration as the smoke exhaled? Explain your answer.

Add the results of your work to your Investigation Folder.

398 *Chapter 7 Solving Linear Inequalities*

Enrichment Masters, p. 49

7-2	NAME_____	DATE _____
	Enrichment	Student Edition Pages 392–398

Traceable Figures

Try to trace over each of the figures below without tracing the same segment twice.

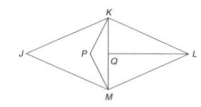

The figure at the left cannot be traced, but the one at the right can. The rule is that a figure is traceable if it has no points, or exactly two points where an odd number of segments meet. The figure at the left has three segments meeting at each of the four corners. However, the figure at the right has exactly two points, L and Q, where an odd number of segments meet.

Determine whether each figure can be traced. If it can, then name the starting point and number the sides in the order in which they should be traced.

1. Yes E

2. No

3. Yes X

4. 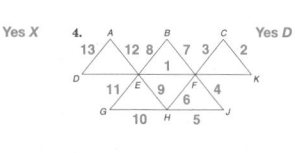 Yes D

In·ves·ti·ga·tion

Working on the Investigation

The Investigation on pages 320–321 is designed to be a long-term project that is completed over several days or weeks. Encourage students to keep their materials in their Investigation Folder as they work on the Investigation.

7-3 Solving Multi-Step Inequalities

Objectives

After studying this lesson you should be able to:

- solve linear inequalities involving more than one operation, and
- find the solution set for a linear inequality when replacement values are given for the variables.

APPLICATION
Engineering

Rosa Whitehair is a partner in an engineering consulting firm. Her fee for consulting on large construction projects is $1000 plus 10% of the design fee that the company charges its clients. Ms. Whitehair is considering two construction projects: a 50-story office building and the design of an airport terminal. She is interested in both projects, but decides to choose the one for which her fee is higher. The company that is designing the office building has agreed to pay Ms. Whitehair a flat fee of $5000. How much does the design company's fee need to be for her to choose the terminal?

Let x represent the design fee for the airport terminal.

The flat fee	plus	10% of the design fee	is more than	the flat fee for the office building.
1000	+	0.10x	>	5000

> **LOOK BACK**
>
> You can review solving multi-step equations in Lesson 3-3.

This inequality involves more than one operation. It can be solved by undoing the operations in reverse of the order of operations in the same way you would solve an equation with more than one operation.

$$1000 + 0.10x > 5000$$
$$1000 - 1000 + 0.10x > 5000 - 1000 \quad \textit{Subtract 1000 from each side.}$$
$$0.10x > 4000$$
$$\frac{0.10x}{0.10} > \frac{4000}{0.10} \quad \textit{Divide each side by 0.10.}$$
$$x > 40,000$$

If the company charges design fees higher than $40,000 for the terminal, Ms. Whitehair will choose them since her consulting fee is higher.

Lesson 7–3 Solving Multi-Step Inequalities **399**

GLENCOE Technology

CD-ROM Interaction

A multimedia simulation links graphing inequalities to determining safety constraints for a fireworks factory. A blackline master activity with teacher's notes provides a follow-up to the CD-ROM simulation.

For Windows & Macintosh

7-3 LESSON NOTES

NCTM Standards: 1–5

Instructional Resources

- Study Guide Master 7-3
- Practice Master 7-3
- Enrichment Master 7-3
- Multicultural Activity Masters, p. 13
- Real-World Applications, 19

Transparency 7–3A contains the 5-Minute Check for this lesson; **Transparency 7–3B** contains a teaching aid for this lesson.

Recommended Pacing

Standard Pacing	Day 4 of 14
Honors Pacing	Day 4 of 13
Block Scheduling*	Day 3 of 7 (along with Lesson 7-4)
Alg. 1 in Two Years*	Days 5 & 6 of 21

*For more information on pacing and possible lesson plans, refer to the *Block Scheduling Booklet* and *Algebra 1 in Two Years*.

1 FOCUS

> ### 5-Minute Check
> (over Lesson 7-2)
>
> **Solve each inequality.**
>
> 1. $\frac{a}{4} > 16$ $\{a \mid a > 64\}$
> 2. $15 < \frac{x}{3}$ $\{x \mid x > 45\}$
> 3. $\frac{5}{7}p > -20$ $\{p \mid p > -28\}$
> 4. $-9v \leq -108$ $\{v \mid v \geq 12\}$
> 5. $-13x \geq 208$ $\{x \mid x \leq -16\}$

Motivating the Lesson

Questioning Write the following equations on the chalkboard.
$-3x + 5 = -10$
$-3(a + 2) = 8(a - 4)$
Ask students to identify the steps needed to solve each equation. Then replace each = with >. Challenge students to compare the steps needed to solve the inequalities with those for equations.

In-Class Examples

For Example 1
Solve each inequality.

a. $5m - 8 > 12$ $m > 4$
b. $12 + 3a > 18$ $a > 2$
c. $20 - 6c < 44$ $c > -4$

For Example 2
Solve each inequality.

a. $5m - 4 < 2m + 11$ $m < 5$
b. $2r - 18 < 5r + 3$ $r > -7$
c. $26p - 20 > 14p + 64$ $p > 7$

For Example 3
Determine the solution set for $2x + 5 < 17$ if the replacement set for x is $\{-10, -6, -2, 2, 6, 10\}$. $\{-10, -6, -2, 2\}$

Teaching Tip In Example 2, suggest to students that they check several numbers in the solution set to determine whether true statements result.

INTEGRATION
Geometry

Example Determine the value of x so that $\angle A$ is acute.

Assume that x is positive.

For $\angle A$ to be acute, its measure must be less than 90°.
Thus, $3x - 15 < 90$.

$$3x - 15 < 90$$
$$3x - 15 + 15 < 90 + 15 \quad \text{Add 15 to each side.}$$
$$3x < 105$$
$$\frac{3x}{3} < \frac{105}{3} \quad \text{Divide each side by 3.}$$
$$x < 35$$

For $\angle A$ to be acute, x must be less than 35.

Sometimes inequalities, like equations, involve variables on each side of the inequality.

Example Solve $-4w + 9 \leq w - 21$.

$$-4w + 9 \leq w - 21$$
$$-w - 4w + 9 \leq w - 21 - w \quad \text{Subtract } w \text{ from each side.}$$
$$-5w + 9 \leq -21$$
$$-5w + 9 - 9 \leq -21 - 9 \quad \text{Subtract 9 from each side.}$$
$$-5w \leq -30$$
$$\frac{-5w}{-5} \geq \frac{-30}{-5} \quad \text{Divide each side by } -5 \text{ and change} \leq \text{ to } \geq .$$
$$w \geq 6$$

The solution set is $\{w \mid w \geq 6\}$.

When we solve an inequality, the solution set usually includes all numbers for a certain criteria, such as $\{x \mid x > 4\}$. Sometimes a replacement set is given from which the solution set can be chosen.

Example 3 Determine the solution set for $3x + 6 > 12$ if the replacement set for x is $\{-2, -1, 0, 1, 2, 3, 4, 5\}$.

Method 1

Substitute values into the inequality to find the values that satisfy the inequality. Try -2.

$$3x + 6 > 12$$
$$3(-2) + 6 > 12$$
$$0 > 12 \quad \text{false}$$

From this trial, we can estimate that the value of x must be much greater than -2 to make the inequality true. Try 2.

$$3x + 6 > 12$$
$$3(2) + 6 > 12$$
$$12 > 12 \quad \text{false}$$

 Alternative Learning Styles

Visual Have students find solutions to the inequality $2x - 3 \leq 5$ using trial-and-error substitution. Each time the substituted value works, have them plot the value on a number line. After students have tried about ten values, they should guess at the solution.

From this trial, we see that values greater than 2 must be in the solution set. Try 3.

$$3x + 6 > 12$$
$$3(3) + 6 > 12$$
$$15 > 12 \quad \text{true}$$

The solution set is {3, 4, 5}.

Method 2

Solve the inequality for all values of x. Then determine which values from the replacement set belong to the solution set.

$$3x + 6 > 12$$
$$3x + 6 - 6 > 12 - 6$$
$$3x > 6$$
$$\frac{3x}{3} > \frac{6}{3}$$
$$x > 2$$

The solution set is those numbers from the replacement set that are greater than 2. Thus, the solution set is {3, 4, 5}.

EXPLORATION

GRAPHING CALCULATORS

You can use the inequality symbols in the TEST menu on the TI-82 graphing calculator to find the solution to an inequality in one variable.

Your Turn a. You see part of the graph of $y = 1$.

a. Clear the [Y=] list. Enter $3x + 6 > 4x + 9$ as Y1. (The symbol $>$ is item 3 on the TEST menu.) Press [GRAPH]. Describe what you see.

b. Use the TRACE function to scan the values along the graph. What do you notice about the values of y on the graph?

c. Solve the inequality algebraically. How does your solution compare to the pattern you noticed in part **b**?

b. $y = 1$ until $x = -3$. Then $y = 0$.

c. $x < -3$; $y = 1$ if the inequality is true for those values of x; $y = 0$ for those values of x for which the inequality is not true.

When solving some inequalities that contain grouping symbols, remember to first use the distributive property to remove the grouping symbols.

Example ④ Solve $5(k + 4) - 2(k + 6) \geq 5(k + 1) - 1$. Then graph the solution.

$$5(k + 4) - 2(k + 6) \geq 5(k + 1) - 1$$
$$5k + 20 - 2k - 12 \geq 5k + 5 - 1 \quad \textit{Distributive property.}$$
$$3k + 8 \geq 5k + 4 \quad \textit{Combine like terms.}$$
$$3k - 5k + 8 \geq 5k - 5k + 4 \quad \textit{Subtract 5k from each side.}$$
$$-2k + 8 \geq 4$$
$$-2k + 8 - 8 \geq 4 - 8 \quad \textit{Subtract 8 from each side.}$$
$$-2k \geq -4$$
$$\frac{-2k}{-2} \leq \frac{-4}{-2} \quad \textit{Divide each side by -2 and change $\geq$ to $\leq$.}$$
$$k \leq 2$$

The solution set is $\{k \mid k \leq 2\}$.
The graph of $k \leq 2$ is shown at the right.

−4 −3 −2 −1 0 1 2 3 4 5

In-Class Example

For Example 4
Solve each inequality. Then graph the solution.

a. $\frac{1}{3}(3t + 6) < \frac{1}{3}(6t - 9)$ $t > 5$

3 4 5 6 7 8 9 10

b. $0.5(n - 2) > 0.4(n + 6)$
$n > 34$

32 33 34 35 36 37 38 39

Check for Understanding

Exercises 1–15 are designed to help you assess your students' understanding through reading, writing, speaking, and modeling. You should work through Exercises 1–6 with your students and then monitor their work on Exercises 7–15.

Additional Answers

1. Sample answer:
$$-3x + 7 < 4x - 5$$
$$3x - 3x + 7 < 4x - 5 + 3x$$
$$7 < 7x - 5$$
$$7 + 5 < 7x - 5 + 5$$
$$12 < 7x$$
$$\frac{12}{7} < \frac{7x}{7}$$
$$\frac{12}{7} < x \text{ or } x > \frac{12}{7}$$

4. Add $5w$ to each side. Subtract 29 from each side. Divide each side by 5. The solution is $\left\{w \mid w < -\frac{13}{5}\right\}$.

Study Guide Masters, p. 50

Communicating Mathematics

Study the lesson. Then complete the following.

1. **Explain** each step in solving $-3x + 7 < 4x - 5$. See margin.

2. **Write** an inequality that expresses the fact that when you add 3 feet to the perimeter of a square of sides with length s, the resulting perimeter does not exceed 50 feet. $4s + 3 \leq 50$

3. Answers will vary. Sample answer: $x + 1 < x - 1$.

3. **Write** an inequality that has no solution.

4. **Describe** how you would solve $16 - 5w > 29$ without dividing by -5 or multiplying by $-\frac{1}{5}$. See margin.

5. Sample answer: Method 2, because it is shorter.

5. Refer to Example 3. Which of the two methods seems more efficient in solving the inequality for the given replacement set? Explain your selection.

MODELING MATHEMATICS

6. Use the method shown in Lesson 7–2A to model the solutions for each inequality.
   a. $3 - 4x \geq 15$ $x \leq -3$
   b. $6x - 1 < 5 + 3x$ $x < 2$

Guided Practice

Choose the correct solution for each inequality.

7. Solve $2m + 5 \leq 4m - 1$. c
   a. $m > -3$ b. $m < 3$ c. $m \geq 3$ d. $m \leq -3$

8. Solve $13r - 11 \geq 7r + 37$. b
   a. $r < 8$ b. $r \geq 8$ c. $r \leq -8$ d. $r > -8$

Solve each inequality. Then check your solution.

9. $9x + 2 > 20$ $\{x \mid x > 2\}$

10. $-4h + 7 > 15$ $\{h \mid h < -2\}$

11. $-2 - \frac{d}{5} < 23$ $\{d \mid d > -125\}$

12. $6a + 9 < -4a + 29$ $\{a \mid a < 2\}$

Find the solution set of each inequality given the replacement set.

13. $3x - 1 > 4$, $\{-1, 0, 1, 2, 3\}$ $\{2, 3\}$

14. $-7a + 6 \leq 48$, $\{-10, -9, -8, -7, -6, -5, -4, -3\}$ $\{-6, -5, -4, -3\}$

15. **Number Theory** Consider the sentence *The sum of two consecutive even integers is greater than 75.*
   a. Write an inequality for this statement. $x + (x + 2) > 75$
   b. Solve the inequality. $x > 36.5$
   c. Name two consecutive even integers that meet the requirements of the statement. Answers will vary; sample answer: 38 and 40.

Practice

18. $\{-10, -9, -8..., 8, 9, 10\}$

19. $\{-10, -9, ..., -5, -4\}$

Find the solution set of each inequality if the replacement set for each variable is $\{-10, -9, -8, ..., 8, 9, 10\}$.

A

16. $n - 3 \geq \frac{n+1}{2}$ $\{7, 8, 9, 10\}$

17. $\frac{2(x + 2)}{3} < 4$ $\{-10, -9, ..., 2, 3\}$

18. $1.3y - 12 < 0.9y + 4$

19. $-20 \geq 8 + 7k$

Reteaching

Using Manipulatives Use a two-pan balance to demonstrate how to solve these inequalities. □ □ 3 represents 2 identical objects of unknown weight and a weight of 3 units.

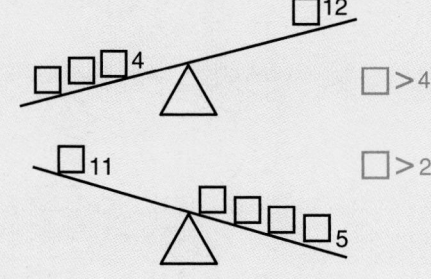

Solve each inequality. Then check your solution.   27. $\{m \mid m < 15\}$

20. $2m + 7 > 17$ $\{m \mid m > 5\}$

21. $-3 > -3t + 6$ $\{t \mid t > 3\}$

22. $-2 - 3x \geq 2$ $\left\{x \mid x \leq -\frac{4}{3}\right\}$

23. $\frac{2}{3}w - 3 \leq 7$ $\{w \mid w \leq 15\}$

24. $7x - 1 < 29 - 2x$ $\left\{x \mid x < \frac{10}{3}\right\}$

25. $8n + 2 - 10n < 20$ $\{n \mid n > -9\}$

26. $2x + 5 < 3x - 7$ $\{x \mid x > 12\}$

27. $5 - 4m + 8 + 2m > -17$

▶B

28. $\frac{2x - 3}{5} < 7$ $\{x \mid x < 19\}$

29. $x < \frac{2x - 15}{3}$ $\{x \mid x < -15\}$

30. $9r + 15 \geq 24 + 10r$ $\{r \mid r \leq -9\}$

31. $6p - 2 \leq 3p + 12$ $\left\{p \mid p \leq \frac{14}{3}\right\}$

32. $4y + 2 < 8y - (6y - 10)$ $\{y \mid y < 4\}$ 33. $3(x - 2) - 8x < 44$ $\{x \mid x > -10\}$

34. $3.1q - 1.4 > 1.3q + 6.7$ $\{q \mid q > 4.5\}$ 35. $-5(k + 4) \geq 3(k - 4)$ $\{k \mid k \leq -1\}$

36. $\{h \mid h < -79\}$

36. $5(2h - 6) - 7(h + 7) > 4h$

37. $\{y \mid y < -1\}$

37. $7 + 3y > 2(y + 3) - 2(-1 - y)$

Define a variable, write an inequality, and solve each problem. Then check your solution.

38. $\frac{2}{3}x - 27 \geq 9$; $\{x \mid x \geq 54\}$

38. Two thirds of a number decreased by 27 is at least 9.

39. Three times the sum of a number and 7 is greater than 5 times the number less 13. $3(x + 7) > 5x - 13$; $\{x \mid x < 17\}$

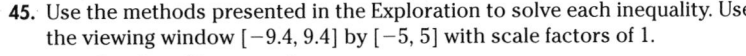
INTEGRATION
Number Theory

▶C

40. The sum of two consecutive odd integers is at most 123. Find the pair with the greatest sum. $2x + 2 \leq 123$; 59 and 61

41. Find all sets of two consecutive positive odd integers whose sum is no greater than 18. $2x + 2 \leq 18$ for $x > 0$; 7 and 9; 5 and 7; 3 and 5; 1 and 3

42. Find all sets of three consecutive positive even integers whose sum is less than 40. $3x + 6 < 40$; 2, 4, 6; 4, 6, 8; 6, 8, 10; 8, 10, 12; 10, 12, 14

Solve each inequality. Then check your solution.

43. $3x + 4 > 2(x + 3) + x$
no solution $(\varnothing)$

44. $3 - 3(y - 2) < 13 - 3(y - 6)$
$\{y \mid y \text{ is a real number.}\}$

Graphing Calculator

45. Use the methods presented in the Exploration to solve each inequality. Use the viewing window $[-9.4, 9.4]$ by $[-5, 5]$ with scale factors of 1.

a. $-5 - 8x \geq 59$ $x \leq -8$

b. $13x - 11 > 7x + 37$ $x > 8$

c. $8x - (x - 5) > x + 17$ $x > 2$

d. $-5(x + 4) \geq 3(x - 4)$ $x \leq -1$

Critical Thinking

46. We can write the expression *the numbers between* -3 *and* 4 as $-3 < x < 4$. What does the algebraic sentence $-3 < x + 2 < 4$ represent?   the numbers between -5 and 2 or $\{x \mid -5 < x < 2\}$

Applications and Problem Solving

47. **Personal Finances**   A couple does not want to have a charge of more than $50 for dinner at a restaurant. A sales tax of 4% is added to the bill, and they plan to tip 15% after the tax has been added. What can the couple spend on the meal? $x + 0.04x + 0.15(x + 0.04x) \leq \$50$, $x \leq \$41.80$

48. **Recreation**   The admission fee to a video game arcade is $1.25 per person, and it costs $0.50 for each game played. Latoya and Donnetta have a total of $10.00 to spend. What is the greatest number of games they will be able to play?   15 games

For **Extra Practice,** see p. 772.

The red A, B, and C flags, printed only in the Teacher's Wraparound Edition, indicate the level of difficulty of the exercises.

Practice Masters, p. 50

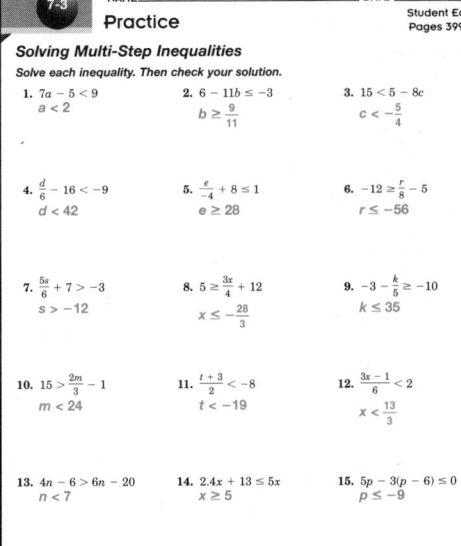

Closing Activity

Writing Have students supply the property or operation applied in each step of the solution to the problem below.

$3(x - 7) - 14 > 2(x + 5) - 5$
$3x - 21 - 14 > 2x + 10 - 5$
distributive property
$\qquad 3x - 35 > 2x + 5$
substitution property
$\qquad\qquad 3x > 2x + 40$
addition property for inequalities
$\qquad\qquad\qquad x > 40$
subtraction property for inequalities

Enrichment Masters, p. 50

 7-3

NAME_____ DATE _____

Enrichment

Student Edition
Pages 399–404

Some Properties of Inequalities

The two expressions on either side of an inequality symbol are sometimes called the *first* and *second* members of the inequality.

If the inequality symbols of two inequalities point in the same direction, the inequalities have the same sense. For example, $a < b$ and $c < d$ have the same sense; $a < b$ and $c > d$ have opposite senses.

In the problems on this page, you will explore some properties of inequalities.

Three of the four statements below are true for all numbers a and b (or a, b, c, and d). Write each statement in algebraic form. If the statement is true for all numbers, prove it. If it is not true, give an example to show that it is false.

1. Given an inequality, a new and equivalent inequality can be created by interchanging the members and reversing the sense.
If $a > b$, then $b < a$.
$a > b$, $a - b > 0$, $-b > -a$, $(-1)(-b) < (-1)(-a)$, $b < a$

2. Given an inequality, a new and equivalent inequality can be created by changing the signs of both terms and reversing the sense.
If $a > b$, then $-a < -b$.
$a > b$, $a - b > 0$, $-b > -a$, $-a < -b$

3. Given two inequalities with the same sense, the sum of the corresponding members are members of an equivalent inequality with the same sense.
If $a > b$ and $c > d$, then $a + c > b + d$.
$a > b$ and $c > d$, so $(a - b)$ and $(c - d)$ are positive numbers, so the sum $(a - b) + (c - d)$ is also positive.
$a - b + c - d > 0$, so $a + c > b + d$.

4. Given two inequalities with the same sense, the difference of the corresponding members are members of an equivalent inequality with the same sense.
If $a > b$ and $c > d$, then $a - c > b - d$. The statement is false. $5 > 4$ and $3 > 2$, but $5 - 3 \not> 4 - 2$.

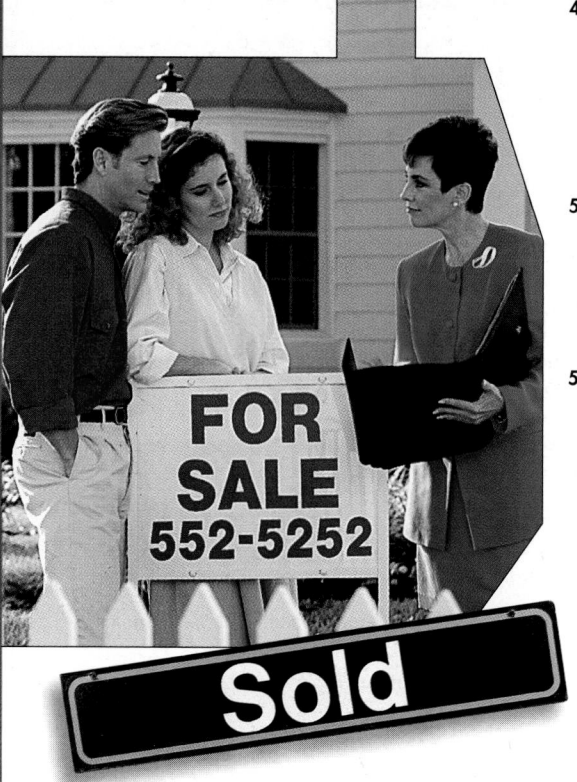

FOR SALE 552-5252

Sold

49. **Fund-raising** A university is running a drive to raise money. A corporation has promised to match 40% of whatever the university can raise from other sources. How much must the school raise from other sources to have a total of at least $800,000 after the corporation's donation? at least $571,428.57

50. **Real Estate** A homeowner is selling her house. She must pay 7% of the selling price to her real estate agent after the house is sold. To the nearest dollar, what must be the selling price of her house to have at least $90,000 after the agent is paid? at least $96,774

51. **Labor** A union worker is currently making $400 per week. His union is seeking a one-year contract. If there is a strike, the new contract will provide for a 6% raise; if there is no strike, the worker expects no increase in salary.
a. Assuming the worker will not be paid during the strike, how many weeks could he go on strike and still make at least as much for the year as he would have made without a strike?
b. How would your answer change if the worker was currently making $575 per week?
c. How would your answer to part a change if the worker's union provided him with $120 per week during the strike? at most 4.1 weeks
51a. at most 2.9 weeks   51b. no change

Mixed Review

Solve each inequality. 52. $\{r \mid r < -6.6\}$ 53. $\{y \mid y > 10\}$

52. $2r - 2.1 < -8.7 + r$ (Lesson 7–2) 53. $7 - 2y < -y - 3$ (Lesson 7–1)

54. $y = -\frac{1}{2}x + 6$

54. Write the slope-intercept form of an equation for the line that passes through $(-12, 12)$ and $(-2, 7)$. (Lesson 6–4)

55. $3x + 2y = 14$

55. Write the standard form of an equation of the line that passes through $(2, 4)$ with slope $-\frac{3}{2}$. (Lesson 6–2)

56. **Business** The owner of No Spots City Car Wash found that if c cars were washed in a day, the average daily profit $P(c)$ was given by the formula $P(c) = -0.027c^2 + 8c - 280$. Find the values of $P(c)$ for various values of c to determine the least number of cars that must be washed each day for No Spots City to make a profit. (Lesson 5–5) **41 cars**

57. $\{-5, -3, -2, 4, 16\}$

57. Find the domain for $3y - 2 = x$ if the range is $\left\{-1, 6, 0, -\frac{1}{3}, 2\right\}$. (Lesson 5–3)

58. **Advertising** For many years, the slogan of Crest® toothpaste has been "Four out of five dentists recommend that their patients use Crest." What are the odds that your dentist will *not* recommend you use Crest? (Lesson 4–6) **1:4**

59. Jason scored the following points for his basketball team during the past ten games: 18, 32, 20, 21, 34, 9, 33, 37, 22, 25. Find the mean, median, and mode of his total points. (Lesson 3–7) **25.1; 23.5; no mode**

60. Find $18 - (-34)$. (Lesson 2–3) **52**

Extension

Communication Give an example of an inequality involving more than one operation for which the solution set is the empty set. **Possible answers include $3x - 6 > 3(x + 2)$.**

Solving Compound Inequalities

Objectives

After studying this lesson you should be able to:

- solve problems by making a diagram,
- solve compound inequalities and graph their solution sets, and
- solve problems that involve compound inequalities.

CONNECTION
Chemistry

The largest fish that spends its whole life in fresh water is the rare Pla beuk, found in the Mekong River in China, Laos, Cambodia, and Thailand. The largest specimen was reportedly 9 feet $10\frac{1}{4}$ inches long and weighed 533.5 pounds.

Such rare fish are sometimes displayed in aquariums. Aquariums can house freshwater or marine life and must be closely monitored to maintain the correct temperature and pH for the animals to survive. pH is a measure of acidity. To determine pH, a scale with values from 0 to 14 is used. One such scale is shown in the diagram below.

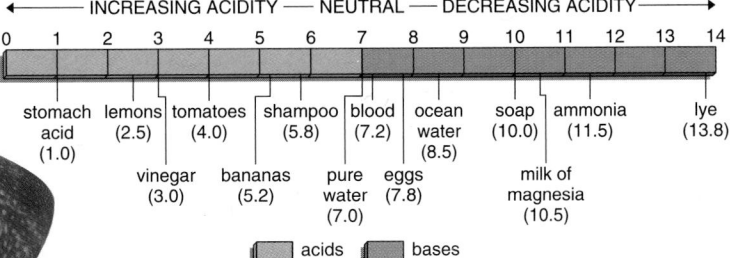

pH SCALE

← INCREASING ACIDITY — NEUTRAL — DECREASING ACIDITY →

0 1 2 3 4 5 6 7 8 9 10 11 12 13 14

stomach acid (1.0) lemons (2.5) tomatoes (4.0) shampoo (5.8) blood (7.2) ocean water (8.5) soap (10.0) ammonia (11.5) lye (13.8)

vinegar (3.0) bananas (5.2) pure water (7.0) eggs (7.8) milk of magnesia (10.5)

acids bases

whale shark

dwarf goby

These statements can also be read p is between 0 and 7 or equal to 0.

If we let p represent the value of the pH scale, we can express the different pH levels by using inequalities. For example, an acid solution will have a pH level of $p \geq 0$ and $p < 7$. When considered together, these two inequalities form a **compound inequality.** This compound inequality can also be written without using *and* in two ways.

$$0 \leq p < 7 \quad \text{or} \quad 7 > p \geq 0$$

The statement $0 \leq p < 7$ can be read *0 is less than or equal to p, which is less than 7.* The statement $7 > p \geq 0$ can be read *7 is greater than p, which is greater than or equal to 0.*

The pH levels of bases could be written as follows.

$$p > 7 \text{ and } p \leq 14 \quad \text{or} \quad 7 < p \leq 14 \quad \text{or} \quad 14 \geq p > 7$$

You will graph this inequality in Exercise 6.

7-4 LESSON NOTES

NCTM Standards: 1–5

Instructional Resources

- Study Guide Master 7-4
- Practice Master 7-4
- Enrichment Master 7-4
- Assessment and Evaluation Masters, pp. 183–184
- Graphing Calculator Masters, p. 7
- Modeling Mathematics Masters, pp. 54–56
- Real-World Applications, 20

 Transparency 7-4A contains the 5-Minute Check for this lesson; **Transparency 7-4B** contains a teaching aid for this lesson.

Recommended Pacing

Standard Pacing	Days 5 & 6 of 14
Honors Pacing	Day 5 of 13
Block Scheduling*	Day 3 of 7 (along with Lesson 7-3)
Alg. 1 in Two Years*	Days 7 & 8 of 21

 *For more information on pacing and possible lesson plans, refer to the *Block Scheduling Booklet* and *Algebra 1 in Two Years.*

1 FOCUS

 5-Minute Check
(over Lesson 7-3)

Solve each inequality.

1. $3x - 15 < 45$ $\{x \mid x < 20\}$
2. $2p - 22 \geq 4p + 14$ $\{p \mid p \leq -18\}$
3. $7y + 12 > 46 - 10y$ $\{y \mid y > 2\}$
4. $3(m + 2) \leq 6(2m - 5)$ $\{m \mid m \geq 4\}$
5. $5(r + 3) > 2(4r - 6)$ $\{r \mid r < 9\}$

You can **draw a diagram** to help solve many problems. Sometimes a picture will help you decide how to work the problem. Other times the picture will show you the answer to the problem.

Example 1

PROBLEM SOLVING
Draw a Diagram

On May 6, 1994, President Francois Mitterrand of France and Queen Elizabeth II of England officially opened the Channel Tunnel connecting England and France. After the ceremonies, a group of 36 English and French government officials had dinner at a restaurant in Calais, France, to celebrate the occasion. Suppose the restaurant staff used small tables that seat four people each, placed end to end, to form one long table. How many tables were needed to seat everyone?

Draw a diagram to represent the tables placed end to end. Use Xs to indicate where the people are sitting. Let's start with a guess, say 10 tables.

Ten tables will seat 22 people. If we use an extra table, we can seat 2 more people. Now, let's look for a pattern.

Number of tables	10	11	12	13	14	15	16	17
Number of people seated	22	24	26	28	30	32	34	36

This pattern shows that the restaurant needed 17 tables to seat all 36 officials.

The logic symbol for <u>and</u> is $\wedge$. You can write $x > 7$ and $x < 10$ as $(x > 7) \wedge (x < 10)$.

A compound inequality containing *and* is true only if *both* inequalities are true. Thus, the graphs of a compound inequality containing *and* is the **intersection** of the graphs of the two inequalities. The intersection can be found by graphing the two inequalities and then determining where these graphs overlap. In other words, draw a diagram to solve the inequality.

Example 2 Graph the solution set of $x \geq -2$ and $x < 5$.

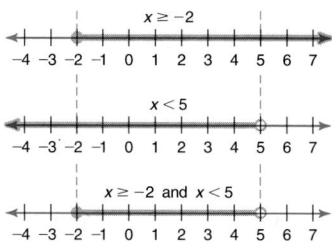

The solution set, shown in the bottom graph, is $\{x \mid -2 \leq x < 5\}$. Note that the graph of $x \geq -2$ includes the point −2. The graph of $x < 5$ does not include 5.

Example ③ **Solve $-1 < x + 3 < 5$. Then graph the solution set.**

First express $-1 < x + 3 < 5$ using *and*. Then solve each inequality.

$$-1 < x + 3 \qquad \text{and} \qquad x + 3 < 5$$
$$-1 - 3 < x + 3 - 3 \qquad\qquad x + 3 - 3 < 5 - 3$$
$$-4 < x \qquad\qquad\qquad x < 2$$

Now graph each solution and find the intersection of the solutions.

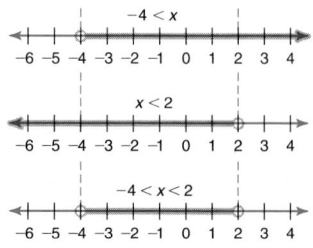

The solution set is $\{x \mid -4 < x < 2\}$.

The following example shows how you can solve a problem by using geometry, a diagram, and a compound inequality.

Example ④

PROBLEM SOLVING
Draw a Diagram

Mai and Luis hope someday to compete in the Olympics in pairs ice skating. Each day they travel from their homes to an ice rink to practice before going to school. Luis lives 17 miles from the rink, and Mai lives 20 miles from it. If this were all the information you were given, determine how far apart Mai and Luis live.

Explore   Mai lives 20 miles from the rink and Luis lives 17 miles from the rink. We do not know the relative positions of Mai's and Luis's homes in relation to the rink.

Plan   Draw a diagram of the situation. Let S be the location of the skating rink. Since Luis lives 17 miles from the rink, we can draw a circle with a radius of 17. Luis will live somewhere on that circle. Likewise, we can draw another circle with radius of 20 for the location of Mai's home.

(continued on the next page)

In-Class Examples

For Example 3
Solve $-3 < x - 2 \le 0$.
Then graph the solution set.
$\{x \mid -1 < x \le 2\}$

For Example 4
Suppose that Luis' family moves and now lives 24 miles from the rink. Determine how far apart Mai and Luis now live.
$4 \le d \le 44$

Teaching Tip In Example 4, point out to students that the triangle inequality is needed only to establish an inequality. It is not needed to compute exact distances.

For Example 5
Graph the solution set of $x \geq 3$ or $x < 2$.

Solve | Let's examine three possibilities for the locations of Mai's and Luis's homes.

(1) Mai and Luis live along the same radius from the skating rink.

(2) Mai and Luis live on opposite radii from the skating rink.

(3) Mai and Luis live somewhere other than the locations described in (1) and (2).

Let L and M represent where Luis and Mai live, respectively.

(1) the same radius

(2) opposite radii

(3) somewhere other than (1) or (2)

  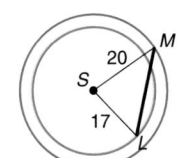

The distance is 20 − 17 or 3 miles.

The distance is 20 + 17 or 37 miles.

By the triangle inequality theorem, the distance must be less than 37 miles and greater than 3 miles.

LOOK BACK
You can find more information about the triangle inequality theorem in Exercise 45 (Programming) of Lesson 7–1.

The distance d can be described by the inequality $3 \leq d \leq 37$.

Examine | Diagrams (1) and (2) show the least and greatest possiblities. To convince yourself that the statement with diagram (3) is always true, you may want to sketch a different triangle.

Another type of compound inequality contains the word *or* instead of *and*. A compound inequality containing *or* is true if one or more of the inequalities is true. The graph of a compound inequality containing *or* is the **union** of the graphs of the two inequalities.

Example **Graph the solution set of $x \geq -1$ or $x < -4$.**

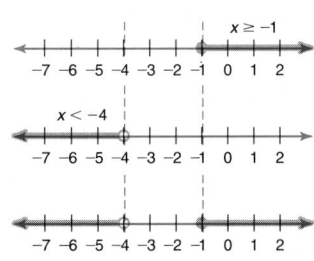

The last graph shows the solution set, $\{x \mid x \geq -1 \text{ or } x < -4\}$.

 Cooperative Learning

Numbered Heads Together Have students work in groups of four to solve the following problem. Make sure that each person in the group understands the solution and is able to explain it to the class.
When changing a dollar bill, you can give one coin (silver dollar), two coins (2 half-dollars), three coins (2 quarters and 1 half-dollar), and so forth. What is the least number of coins that is impossible to give as change for a dollar bill? **77 coins** For more information on the numbered heads together strategy, see *Cooperative Learning in the Mathematics Classroom,* one of the titles in the Glencoe Mathematics Professional Series, pages 9–12.

Example **6** Solve $3w + 8 < 2$ or $w + 12 > 2 - w$. Graph the solution set.

$$3w + 8 < 2 \qquad \text{or} \qquad w + 12 > 2 - w$$
$$3w + 8 - 8 < 2 - 8 \qquad\qquad w + w + 12 > 2 - w + w$$
$$3w < -6 \qquad\qquad 2w + 12 > 2$$
$$\frac{3w}{3} < \frac{-6}{3} \qquad\qquad 2w + 12 - 12 > 2 - 12$$
$$w < -2 \qquad\qquad 2w > -10$$
$$\qquad\qquad \frac{2w}{2} > \frac{-10}{2}$$
$$\qquad\qquad w > -5$$

Now graph each solution and find the union of the two.

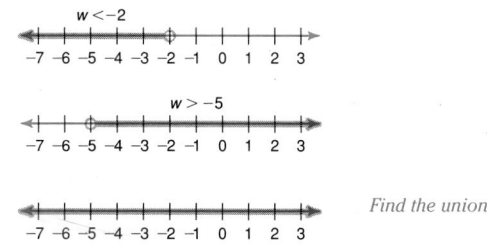

The last graph shows the solution set, $\{x \mid x \text{ is a real number}\}$.

In-Class Example

For Example 6
Solve $3m > m + 4$ or $-2m > m - 6$. Graph the solution set.
$\{m \mid m > 2 \text{ or } m < 2\}$

3 PRACTICE/APPLY

Check for Understanding
Exercises 1–17 are designed to help you assess your students' understanding through reading, writing, speaking, and modeling. You should work through Exercises 1–6 with your students and then monitor their work on Exercises 7–17.

Additional Answers

3. Sample answers: Draw a diagram (helps you to see the solution), make a table (helps you to see pattern), look for a pattern (determines how the tables relate to the number of people seated).

4. The one containing *and* is true when both inequalities are true; the one containing *or* is true if either inequality is true.

6.

7.

8.

CHECK FOR UNDERSTANDING

Communicating Mathematics

1. Answers will vary. Sample answer: The price of an item is at most $18.50 but more than $7.50.

Study the lesson. Then complete the following.

1. **Write** a statement that could be represented by $\$7.50 < p \le \$18.50$.

2. **Write** a compound inequality that describes the graph below.

$x < -4$ or $x \ge 1$

3. **List** two problem-solving strategies used in Example 1. Describe how each strategy helped in the solution of the problem. **See margin.**

4. **Describe** the difference between a compound inequality containing *and* and a compound inequality containing *or*. **See margin.**

5. **Name** two ways in which a picture or diagram can help you solve a problem more easily. **Helps you plan the solution; provides the solution.**

6. **Refer** to the connection at the beginning of the lesson. Graph the inequality that represents the pH levels of bases. **See margin.**

Guided Practice

Write each compound inequality without using *and*. Then graph the solution set.   7–8. See margin for graphs.

7. $x < 9$ and $0 \le x$ $0 \le x < 9$

8. $x > -2$ and $x < 3$ $-2 < x < 3$

Write a compound inequality for each solution set shown below.

9. $-3 < x \le 1$

10. $x < -2$ or $x > 0$

9.

10.

Reteaching

Using Graphing Graph sets A, B, and C.
$A = \{x \mid x > 3\}$ $B = \{x \mid x \le 1\}$
$C = \{x \mid x < 6\}$
Use these graphs to graph the intersections and unions.

1. $A \cap C$

2. $A \cup C$
all numbers

3. $B \cap C$

4. $B \cup C$

5. $A \cup B$

Additional Answers

12.
-6-5-4-3-2-1 0 1 2 3 4

13.
-10-8-6-4-2 0 2 4 6

14.
-3-2-1 0 1 2 3 4 5

15.
-7-6-5-4-3-2-1 0 1 2 3

18.
-8-6-4-2 0 2 4 6

19. ∅

20.
-4-3-2-1 0 1 2 3 4

21.
-7-6-5-4-3-2-1 0 1

22.
-5-4-3-2-1 0 1 2 3

23.
-10-8-6-4-2 0 2 4

410 *Chapter 7*

12–15. See margin for graphs.
12. $\{y \mid -4 \le y < 2\}$
13. $\{h \mid h \le -7 \text{ or } h \ge 1\}$
15. $\{w \mid 1 > w \ge -5\}$

11. Graph the solution set of $y > 5$ and $y < -3$. Describe what the solution set means.   **The solution is the empty set. There are no numbers greater than 5 but less than −3.**

Solve each compound inequality. Then graph the solution set.

12. $2 \le y + 6$ and $y + 6 < 8$
13. $4 + h \le -3$ or $4 + h \ge 5$
14. $b + 5 > 10$ or $b \ge 0$ $\{b \mid b \ge 0\}$
15. $2 + w > 2w + 1 \ge -4 + w$

16. Write a compound inequality without using *and* for the following situation. Solve the inequality and then check the solution.
It costs the same to register mail containing articles with values from $0 to $100. $0 \le c \le \$100$

17. **Draw a Diagram** You can cut a pizza into seven pieces with only three straight cuts as shown at the right. Draw a diagram to show the greatest number of pieces you can make with five straight cuts.   **Drawings will vary; 16 pieces.**

EXERCISES

Practice
A

28–43. See margin for graphs.
28. $\{m \mid m > 3 \text{ or } m < -1\}$
29. $\{x \mid -1 < x < 5\}$
30. $\{y \mid -7 < y < 6\}$
31. $\{x \mid x < -2 \text{ or } x > 3\}$
32. $\{p \mid p < -1\}$
33. $\{c \mid c < 7\}$
36. $\{q \mid -1 < q < 6\}$
37. $\{x \mid x \text{ is a real number}\}$
38. $\{n \mid n \le 4\}$
39. $\{y \mid y > 3 \text{ and } y \ne 6\}$
40. $\{z \mid z \ge -\frac{2}{3}\}$
41. $\{x \mid x \text{ is a real number.}\}$
42. $\{y \mid y < \frac{3}{2}\}$
44. Sample answer: $x > -2$ or $x < 4$
45. Sample answer: $x > 5$ and $x < -4$

Graph the solution set of each compound inequality.

18. $m \ge -5$ and $m < 3$
19. $p < -8$ and $p > 4$
20. $s < 3$ or $s \ge 1$
21. $n \le -5$ or $n \ge -1$
22. $w > -3$ and $w < 1$
23. $x < -7$ or $x \ge 0$
18–23. See margin.

Write a compound inequality for each solution set shown below.

24.
-6-5-4-3-2-1 0 1 2 3 4 5 6
$-3 < x < 1$

25.
-6-5-4-3-2-1 0 1 2 3 4 5 6
$-4 \le x \le 5$

26.
-6-5-4-3-2-1 0 1 2 3 4 5 6
$x < 0$ or $x > 2$

27.
-6-5-4-3-2-1 0 1 2 3 4 5 6
$x \le -2$ or $x > 1$

Solve each compound inequality. Then graph the solution set.

28. $4m - 5 > 7$ or $4m - 5 < -9$
29. $x - 4 < 1$ and $x + 2 > 1$
30. $y + 6 > -1$ and $y - 2 < 4$
31. $x + 4 < 2$ or $x - 2 > 1$
32. $10 - 2p > 12$ and $7p < 4p + 9$
33. $6 - c > c$ or $3c - 1 < c + 13$
34. $4 < 2x - 2 < 10$ $\{x \mid 3 < x < 6\}$
35. $14 < 3h + 2 < 2$ ∅
36. $8 > 5 - 3q$ and $5 - 3q > -13$
37. $-1 + x \le 3$ or $-x \le -4$
38. $3n + 11 \le 13$ or $2n \ge 5n - 12$
39. $3y + 1 > 10$ and $y \ne 6$

B
40. $4z + 8 \ge z + 6$ or $7z - 14 \ge 2z - 4$
41. $5x + 7 > 2x + 4$ or $3x + 3 < 24 - 4x$
42. $2 - 5(2y - 3) > 2$ or $3y < 2(y - 8)$
43. $5w > 4(2w - 3)$ and $5(w - 3) + 2 < 7$ $\{w \mid w < 4\}$

Write a compound inequality for each solution set shown below.

44.
-6-5-4-3-2-1 0 1 2 3 4 5 6

45.
-6-5-4-3-2-1 0 1 2 3 4 5 6

410 *Chapter 7 Solving Linear Inequalities*

Additional Answers

28.
-3-2-1 0 1 2 3 4 5

29.
-3-2-1 0 1 2 3 4 5 6 7

30.
-12-9-6-3 0 3 6 9 12

31.
-4-3-2-1 0 1 2 3 4 5

32.
-7-6-5-4-3-2-1 0 1

33.
1 2 3 4 5 6 7 8 9

34.
1 2 3 4 5 6 7 8

35. ∅

36.
-4-2 0 2 4 6 8 10

37.
0 1 2 3 4 5 6 7 8

Define a variable, write a compound inequality, and solve each problem. Then check your solution.

46. $50 < 3d + 5 < 89$; $\{d \mid 15 < d < 28\}$

47. $n + 2 \le 6$ or $n + 2 \ge 10$; $\{n \mid n \le 4$ or $n \ge 8\}$

48. $7 < 2n + 5 < 11$; $\{n \mid 1 < n < 3\}$

50–52. See margin for graphs.

Graphing Calculator

Critical Thinking

Applications and Problem Solving

46. When three times the distance to the finish line is increased by 5 km, the total kilometers covered in the race will be between 50 and 89 km.

47. The sum of a number and 2 is no more than 6 or no less than 10.

48. The sum of twice a number and 5 lies between 7 and 11.

49. Five less than 6 times a number is at most 37 and at least 31.
$31 \le 6n - 5 \le 37$; $\{n \mid 6 \le n \le 7\}$

Solve each inequality. Then graph the solution set.

50. $3 + y > 2y > -3 - y$
$\{y \mid -1 < y < 3\}$

51. $m > 2m - 1 > m - 5$
$\{m \mid -4 < m < 1\}$

52. $\dfrac{5}{x} + 3 > 0$
$\{x \mid x < -\dfrac{5}{3}$ or $x > 0\}$

53. In Lesson 7–3, you learned how to use a graphing calculator to find graphically which values of x make a given inequality true. You can also use this method to test compound inequalities. The words *and* and *or* can be found in the LOGIC submenu of the TEST menu. Use this method to solve each of the following using your graphing calculator.
   a. $3 + x < -4$ or $3 + x > 4$
   $\{x \mid x < -7$ or $x > 1\}$
   b. $-2 \le x + 3$ and $x + 3 < 4$
   $\{x \mid -5 \le x < 1\}$

54. For what values of a does the compound inequality $-a < x < a$ have no solution? $a \le 0$

55. Write a compound inequality for the solution set shown below.
$-4 \le x \le -1.5$ or $x \ge 2$

```
←+ + + ●━━● + + + ●━━━→
 -6-5-4-3-2-1 0 1 2 3 4 5 6
```

56. **Draw a Diagram** There are eight houses on McArthur Street, all in a row. These houses are numbered from 1 to 8. Allison, whose house number is greater than 2, lives next door to her best friend, Adrienne. Belinda, whose house number is greater than 5, lives two doors away from her boyfriend, Benito. Cheri, whose house number is greater than Benito's, lives three doors away from her piano teacher, Mr. Crawford. Darryl, whose house number is less than 4, lives four doors away from his teammate, Don. Who lives in each house? 1-Darryl, 2-Adrienne, 3-Allison, 4-Mr. Crawford, 5-Don, 6-Benito, 7-Cheri, 8-Belinda

Define a variable, write a compound inequality, and solve each problem.

57. **Physics** According to Hooke's Law, the force F (in pounds) required to stretch a certain spring x inches beyond its natural length is given by $F = 4.5x$. If forces between 20 and 30 pounds, inclusive, are applied to the spring, what will be the range of the increased lengths of the stretched spring? $4.4 < x < 6.7$

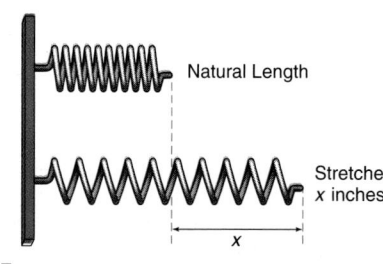

Natural Length

Stretched x inches

x

58. **Statistics** Clarissa must have an average of 92, 93, or 94 points to receive a grade of A– in social studies. She earned 92, 96, and 88 on the first three tests of the grading period. What range of scores on the fourth test will give her an A–? $92 \le s \le 100$

Lesson 7–4 Solving Compound Inequalities **411**

Extension

Reasoning Points A, B, C, and D are located on the same number line. If A is to the left of B, A is to the right of C, and C is to the right of D, what is the order of the points from left to right?
D, C, A, B

Chapter 7 **411**

4 ASSESS

Closing Activity

Modeling Have each student create a diagram to model a word problem. Models can resemble those used in Examples 1 and 4. Then have students switch models and write a word problem that could be solved using their classmate's model.

Chapter 7, Quiz B (Lessons 7-3 and 7-4), is available in the *Assessment and Evaluation Masters*, p. 184.

Mid-Chapter Test (Lessons 7-1 through 7-4) is available in the *Assessment and Evaluation Masters*, p. 183.

Additional Answer

63.

Enrichment Masters, p. 51

 7-4
NAME_____ DATE _____
Enrichment

Student Edition
Pages 405–411

Precision of Measurement

The precision of a measurement depends both on your accuracy in measuring and the number of divisions on the ruler you use. Suppose you measured a length of wood to the nearest one-eighth of an inch and got a length of $6\frac{5}{8}$ in.

The drawing shows that the actual measurement lies somewhere between $6\frac{9}{16}$ in. and $6\frac{11}{16}$ in. This measurement can be written using the symbol ±, which is read "plus or minus." It can also be written as a compound inequality.

$6\frac{5}{8} \pm \frac{1}{16}$ in. $6\frac{9}{16}$ in. $\leq m \leq 6\frac{11}{16}$ in.

In this example, $\frac{1}{16}$ in. is the absolute error. The absolute error is one-half the smallest unit used in a measurement.

Write each measurement as a compound inequality. Use the variable m.

1. $3\frac{1}{2} \pm \frac{1}{4}$ in.
   $3\frac{1}{4}$ in. $\leq m \leq 3\frac{3}{4}$ in.

2. 9.78 ± 0.005 cm
   9.775 cm $\leq m \leq$ 9.785 cm

3. 2.4 ± 0.05 g
   2.35 g $\leq m \leq 2.45$ g

4. $28 \pm \frac{1}{2}$ ft
   $27\frac{1}{2}$ ft $\leq m \leq 28\frac{1}{2}$ ft

5. 15 ± 0.5 cm
   14.5 cm $\leq m \leq$ 15.5 cm

6. $\frac{11}{16} \pm \frac{1}{64}$ in.
   $\frac{43}{64}$ in. $\leq m \leq \frac{45}{64}$ in.

For each measurement, give the smallest unit used and the absolute error.

7. 12.5 cm $\leq m \leq 13.5$ cm
   1 cm, 0.5 cm

8. $12\frac{1}{8}$ in. $\leq m \leq 12\frac{3}{8}$ in.
   $\frac{1}{4}$ in., $\frac{1}{8}$ in.

9. $56\frac{1}{2}$ in. $\leq m \leq 57\frac{1}{2}$ in.
   1 in., $\frac{1}{2}$ in.

10. 23.05 mm $\leq m \leq 23.15$ mm
    0.1 mm, 0.05 mm

412 Chapter 7

Mixed Review

59. $\left\{ m \mid m \geq \frac{44}{3} \right\}$

62. $\frac{3}{7}$

Define a variable, write an inequality, and solve each problem.

59. Three fourths of a number decreased by 8 is at least 3. (Lesson 7–3)

60. A number added to 23 is at most 5. (Lesson 7–1) $\{x \mid x \leq -18\}$

61. If $7x - 2 \leq 9x + 3$, then $2x \geq \underline{\ ?\ }$. (Lesson 7–1) -5

62. **Geometry** Line a is perpendicular to a line that is perpendicular to a line whose equation is $3x - 7y = -3$. If all lines are in the same plane, what is the slope of line a? (Lesson 6–6)

63. Graph $x = 2y - 4$ using the x- and y-intercepts. (Lesson 6–5) **See margin.**

64. Write an equation in functional notation for the relation graphed at the right. (Lesson 5–6)
$f(x) = -2x + 5$

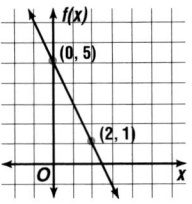

65. Sarah is visiting a friend about two miles away from her house. She sees a hot-air balloon directly above her house. Sara estimates that the angle of elevation formed by the hot-air balloon is about 15°. About how high is the hot-air balloon? (Lesson 4–3) **a little more than half a mile**

66. Solve $\frac{4m - 3}{-2} = 12$. (Lesson 3–3) $-\frac{21}{4}$

67. Simplify $-17px + 22bg + 35px + (-37bg)$. (Lesson 2–5) $18px - 15bg$

68. Evaluate $3ab - c^2$ if $a = 6$, $b = 4$, and $c = 3$. (Lesson 1–3) **63**

SELF TEST

Solve each inequality. Then check your solution. (Lessons 7–1, 7–2, and 7–3)

1. $y + 15 \geq -2$ $\{y \mid y \geq -17\}$

2. $-102 > 17r$ $\{r \mid r < -6\}$

3. $5 - 6n > -19$ $\{n \mid n < 4\}$

4. $\frac{11 - 6w}{5} > 10$ $\{w \mid w < -6.5\}$

5. $7(g + 8) < 3(g + 12)$ $\{g \mid g < -5\}$

6. $0.1y - 2 \leq 0.3y - 5$ $\{y \mid y \geq 15\}$

7. Choose an equivalent statement for $4 \geq x - 1 \geq -3$. (Lesson 7–4) **c**
   a. $5 \geq x \geq -4$ **b.** $3 \geq x \geq -4$ **c.** $5 \geq x \geq -2$ **d.** $3 \geq x \geq -2$

8. Solve $8 + 3t < 2$ or $-12 < 11t - 1$. Then graph the solution set. (Lesson 7–4)
   $\{t \mid t < -2$ or $t > -1\}$; See margin for graph.

9. **Sports** Jennifer has scored 18, 15, and 30 points in her last three starts on the junior varsity girls basketball team. How many points must she score in her next start so that her four-game average is greater than 20 points? (Lesson 7–3) **more than 17 points**

10. **Draw a Diagram** Suppose you roll a die two times. (Lesson 7–4)
    a. Draw a diagram to show the possibilities of rolling an even number the first time and an odd number the second time. **See margin for the nine results.**
    b. How many of the possibilities have a sum of 7? **3**

412 Chapter 7 Solving Linear Inequalities

SELF TEST

The Self Test provides students with a brief review of the concepts and skills in Lessons 7-1 through 7-4. Lesson numbers are given to the right of exercises or instruction lines so students can review concepts not yet mastered.

Answers for Self Test

8.

$-5\ -4\ -3\ -2\ -1\ 0\ 1\ 2$

10a. (2, 1), (2, 3), (2, 5), (4, 1), (4, 3), (4, 5), (6, 1), (6, 3), (6, 5)

Integration: Probability
Compound Events

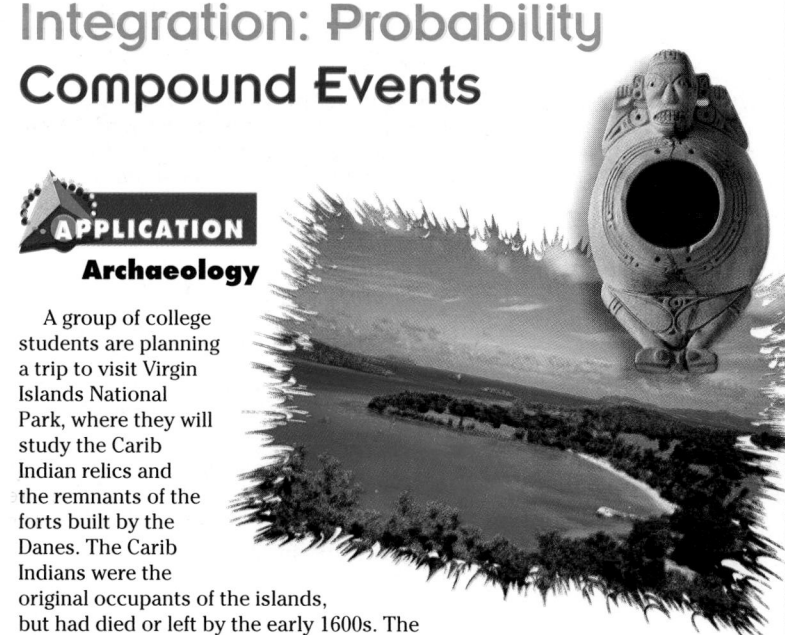

Objective

After studying this lesson you should be able to:

- find the probability of a compound event.

APPLICATION

Archaeology

A group of college students are planning a trip to visit Virgin Islands National Park, where they will study the Carib Indian relics and the remnants of the forts built by the Danes. The Carib Indians were the original occupants of the islands, but had died or left by the early 1600s. The Danes formally claimed the islands in 1666, and they remained under Danish control until 1917.

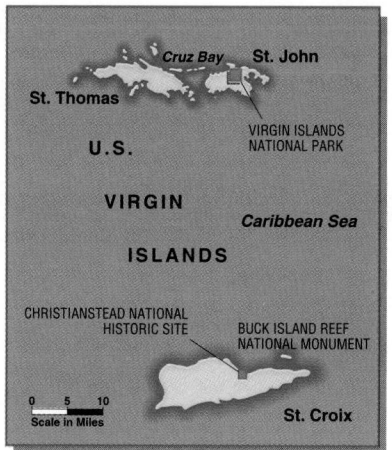

The group's advisor must plan how the group will get to the Virgin Islands. From their college, they will travel to Miami, Florida, by car, bus, train, or plane. Then to travel to St. Thomas in the Virgin Islands, they could take a plane or a ship. Suppose the advisor picks a mode of transportation at random. What is the probability that they will travel by car first and then fly?

To calculate this probability, you need to know all of the possible ways to get to St. Thomas. One method for finding this out is to draw a **tree diagram.** The tree diagram below shows how to get from the college to Miami and then to St. Thomas. The last column details all the possible combinations or **outcomes** of transportation.

Note that the number of outcomes is the product of the number of ways to Miami (4) and the number of ways to the Virgin Islands (2).

To Miami	To St. Thomas	Outcomes
car	ship	car → ship
	plane	car → plane
bus	ship	bus → ship
	plane	bus → plane
train	ship	train → ship
	plane	train → plane
plane	ship	plane → ship
	plane	plane → plane

college

Alternative Learning Styles

Auditory Have students close their books. Read aloud the archaeology application at the beginning of the lesson. Have students construct the tree diagram as you read the passage.

7-5 LESSON NOTES

NCTM Standards: 1–4, 11

Instructional Resources

- Study Guide Master 7-5
- Practice Master 7-5
- Enrichment Master 7-5

 Transparency 7-5A contains the 5-Minute Check for this lesson; **Transparency 7-5B** contains a teaching aid for this lesson.

Recommended Pacing	
Standard Pacing	Day 7 of 14
Honors Pacing	Day 6 of 13
Block Scheduling*	Day 4 of 7 (along with Lesson 7-6)
Alg. 1 in Two Years*	Days 9 & 10 of 21

 *For more information on pacing and possible lesson plans, refer to the *Block Scheduling Booklet* and *Algebra 1 in Two Years.*

1 FOCUS

 5-Minute Check
(over Lesson 7-4)

Write the compound inequality whose solution set is graphed.

1. $x \geq -2$ and $x \leq 5$

2. $m < -7$ or $m \geq 0$

3. $b > 5$ or $b \leq 0$

4. Answers may vary. Sample answer: $n \leq -5$ and $n \geq -1$ or $\varnothing$

5. Answers may vary. Sample answer: $t \leq 8$ or $t \geq 3$

Situational Problem A car manufacturer offers sedan and hatchback versions of a certain model. Either version can be bought with or without a sunroof. All of the versions offered may be purchased in either white or red. Have students make a list to determine all the possible combinations of car options for this model. How many combinations are there? **8**

2 TEACH

In-Class Examples

For Example 1
Three coins are tossed once.
a. Draw a tree diagram to show all of the possible outcomes.

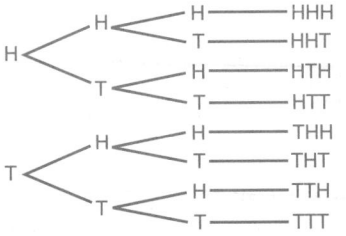

1st Coin 2nd Coin 3rd Coin Outcomes

b. What is the probability of all heads? $\frac{1}{4}$ or 0.25

For Example 2
If three coins are tossed, what is the probability of getting at most two heads? $\frac{7}{8} = 0.875$

How is a compound event similar to a compound inequality?

LOOK BACK

You can find more information on probability in Lesson 4-6.

Since the mode of transportation is chosen at random, you can assume that all combinations of travel are equally likely. Since there are 8 outcomes, the probability of traveling by car first and then flying is $\frac{1}{8}$ or 0.125.

This problem is an example of finding the probability of a **compound event.** A compound event consists of two or more **simple events.** Choosing a car, plane, train, or bus for the first part of the trip is a simple event. Then, selecting a plane or ship for the second stage of the trip is another simple event. The selection of a mode of transportation for each part of the trip is a compound event.

Example Use the tree diagram for the application at the beginning of the lesson to answer each question.

    a. What is the probability that the group will take a ship to get to St. Thomas?

    b. What is the probability that the group will travel by plane for both parts of the trip?

    a. Since 4 of the 8 outcomes involve taking a ship, the probability is $\frac{4}{8}$ or 0.5.

    b. Since only 1 outcome out of 8 involves taking a plane for both parts of the trip, the probability is $\frac{1}{8}$ or 0.125.

Probability is often a factor in determining the possibility of winning a game.

Example ❷ Booker T. Washington High School is having its annual Spring Carnival. The ninth grade class has decided to have a game booth. To win a small stuffed animal, a player will have to draw 2 marbles of the same color from a box containing 3 marbles—1 red, 1 white, and 1 yellow. First a marble is drawn, put back in the box, and then a second marble is drawn.

    a. What are the possible outcomes?

    b. What is the probability of winning the game?

    a. First, let's draw a tree diagram to see the possibilities of drawing 2 marbles of the same color.

1st selection	2nd selection	Outcomes
red	red	red, red
	white	red, white
	yellow	red, yellow
white	red	white, red
	white	white, white
	yellow	white, yellow
yellow	red	yellow, red
	white	yellow, white
	yellow	yellow, yellow

There are 9 outcomes.

b. There are 3 ways out of 9 to draw the same color twice. The probability for each draw is equally likely, so the probability is $\frac{3}{9}$ or $0.\overline{3}$.

Probability can also play an important role in determining the possible outcomes of a sporting event.

Example ③

The Houston Rockets and the New York Knicks are going to play a best two out of three exhibition game series.

a. What are the possible outcomes of the series?

b. Assuming that the teams are equally matched, what is the probability that the the series will end in two games?

a. The tree diagram below shows the possible outcomes of the first two games.

1st game winner	2nd game winner	Outcomes
Rockets	Rockets	Rockets win in two games.
	Knicks	3rd game is required.
Knicks	Rockets	3rd game is required.
	Knicks	Knicks win in two games.

There are 4 possible outcomes

b. The series can end in two games in two ways. The probability is $\frac{2}{4}$ or 0.5.

CHECK FOR UNDERSTANDING

Communicating Mathematics

Study the lesson. Then complete the following. 1, 3, 4. See margin.

1. **Draw** a tree diagram to represent the outcomes of tossing a coin twice.

2. **Describe** a compound event that involves three stages. Draw a tree diagram for such an event. **See students' work.**

3. **Explain** the difference between a simple event and a compound event.

4. **Compare and contrast** the tree diagram used in Example 2 and the one used in Example 3.

 MATH JOURNAL

5. Write a paragraph describing how probabilities are used in real life. Give specific examples. **See students' work.**

Guided Practice

6. **Games** In one turn in the game of Yahtzee®, there are 13 different categories you try to complete by rolling five dice. On the first try of your turn, you roll all five dice. Then you may pick up any or all of the dice and roll again on a second and third try. To get a Yahtzee, you must have the same number appearing on all five dice. What is the probability that you roll a Yahtzee on the first try of your turn?

6. $\frac{1}{7776}$ or 0.0001286
$\frac{5}{6}$

In-Class Example

For Example 3
Karen draws three cards from the ace of hearts, the king of hearts, the queen of hearts, and the jack of hearts. If she draws the ace, what is the probability that she also draws the king and the queen? $\frac{1}{3} \approx 0.333$

3 PRACTICE/APPLY

Check for Understanding

Exercises 1–7 are designed to help you assess your students' understanding through reading, writing, speaking, and modeling. You should work through Exercises 1–5 with your students and then monitor their work on Exercises 6–7.

Additional Answers

1.

3. A compound event consists of two or more simple events. Tossing a coin is a simple event while tossing two coins is a compound event.

4. Answers may vary. Sample answer: The one in Example 2 involves two events while the one in Example 3 involves as many as three events.

Reteaching

Using Diagrams Have students draw a tree diagram for various compound event problems. Once this step is mastered, the problem of finding the probability of a compound event is simplified.

For **Extra Practice,** see p. 773.

The red A, B, and C flags, printed only in the Teacher's Wraparound Edition, indicate the level of difficulty of the exercises.

Additional Answers

7a. See Solutions Manual for tree diagram. Possible combinations are: burger, soup, lemonade; burger, soup, soft drink; burger, salad, lemonade; burger, salad, soft drink; burger, French fries, lemonade; burger, French fries, soft drink; sandwich, soup, lemonade; sandwich, soup, soft drink; sandwich, salad, lemonade; sandwich, salad, soft drink; sandwich, French fries, lemonade; sandwich, French fries, soft drink; taco, soup, lemonade; taco, soup, soft drink; taco, salad, lemonade; taco, salad, soft drink; taco, French fries, lemonade; taco, French fries, soft drink; pizza, soup, lemonade; pizza, soup, soft drink; pizza, salad, lemonade; pizza, salad, soft drink; pizza, French fries, lemonade; pizza, French fries, soft drink.

8a. Let 1–5 represent the roads, and let A and B represent the doors.

$1 \begin{cases} A\text{———}1A \\ B\text{———}1B \end{cases}$

$2 \begin{cases} A\text{———}2A \\ B\text{———}2B \end{cases}$

$3 \begin{cases} A\text{———}3A \\ B\text{———}.3B \end{cases}$

$4 \begin{cases} A\text{———}4A \\ B\text{———}4B \end{cases}$

$5 \begin{cases} A\text{———}5A \\ B\text{———}5B \end{cases}$

7. **Dining**   For lunch at the 66 Diner, you can select one item from each of the following categories for express service guaranteed in 15 minutes.

Entree	Side Dish	Beverage
Burger	Soup	Lemonade
Sandwich	Salad	Soft Drink
Taco	French Fries	
Pizza		

7a. See margin.

7b. $\frac{1}{3}$ or $0.\overline{3}$

7c. $\frac{1}{12}$ or $0.08\overline{3}$

7d. $\frac{1}{24}$ or $0.041\overline{6}$

a. Draw a tree diagram showing all of the meal combinations.
b. What is the probability that a customer will have soup with the meal?
c. What is the probability of selecting a burger with French fries?
d. What is the probability of having pizza with a salad and a soft drink?

EXERCISES

Applications and Problem Solving

8a. See margin.

8. **Video Games**   In a computer video game, you have a choice of five roads to collect an important clue to solve a mystery. At the end of each road, there are two doors. The clue can only be found behind one door.
  a. Draw a tree diagram to show the possibilities you have to find the clue.
  b. What is the probability that you will find the clue? $\frac{1}{10}$ or 0.1

9b. $\frac{1}{5}$ or 0.2

9. **Travel**   Three different airlines fly from Bowling Green to Lexington. Those same three airlines and two other airlines fly from Lexington to Louisville. There are no direct flights from Bowling Green to Louisville.
  a. How many ways can a traveler book flights from Bowling Green to Louisville?   15
  b. What is the probability that flights booked at random from Bowling Green to Louisville use the same airline?

Louisville   Lexington

KENTUCKY

Bowling Green

10. $\frac{1}{6}$ or $0.1\overline{6}$

10. **Business**   On Wednesday, Ralph and Linda were talking about what they were going to wear on Friday since their company had declared it T-Shirt Day. Ralph said he only had three T-shirts he could wear to work. One was a solid color, one was an Earth Day shirt, and the other was a shirt with the company name and logo on it. He also said he would either wear tan pants or jeans. Linda is going to wear jeans and her company T-shirt. If Ralph selects a pair of pants and a T-shirt at random, what is the probability that he, too, will be wearing jeans and the company T-shirt?

416   *Chapter 7   Solving Linear Inequalities*

Alternative Teaching Strategies

Reading Algebra   Students must be cautious not to read too much into the phrase, "The probability of event *E*." The fact that event *E* is more probable than an alternative event E_2 does not mean that it will happen. In fact, unless the probability is 1, it may not happen.

11. $\frac{1}{3}$ or $0.\overline{3}$

B

11. **Food** Kita and Jason are working on the school newspaper's final layout for October. They decide that after they finish, they will go get pizza. They have a coupon for a large pizza with three toppings for $8.99. In discussing what the three toppings should be, they find they have four favorite toppings, and choosing three that both agreed upon was going to be difficult. After some discussion, they decided that pepperoni was the one topping they both agreed upon and they would put the names of the other three toppings in a bag and choose two at random. If the other three toppings are mushrooms, olives, and sausage, what is the probability they will get a pizza with mushrooms?

12. **Games** Twister® is a game composed of a mat with four rows, each having six circles of the same color. One person uses a spinner and calls out *hand* or *foot, right* or *left,* and a color (*blue, red, green,* or *yellow*). Each player must then place that particular body part on a circle of that color. The caller continues to call out body parts and colors with the players moving in the appropriate manner. If a player falls down in the process, he or she is disqualified. The winner is the last player still in position.

a. Draw a tree diagram to show the possibilities for body parts and colors. See margin.

b. What is the probability that the caller will spin and say *"right hand yellow?"* $\frac{1}{16}$ or 0.0625

13. **Playing Cards** The deck of cards for Rook® is composed of 57 cards. There are four suits (red, yellow, black, and green), each containing the numbers 1 to 14, and a rook card. Suppose there are two piles of five cards each with the cards turned face down. This first pile contains a red 3, a black 3, a red 5, a red 14, and a yellow 10. The second pile contains a green 5, a red 10, a black 10, a green 1, and a yellow 14. You are asked to draw a card from each pile.

a. List all the possibilities of drawing two cards. See margin.

13b. $\frac{3}{25}$ or 0.12

b. What is the probability that both cards are red?

13c. $\frac{2}{25}$ or 0.08

c. What is the probability that both cards are 10s?

d. What is the probability that both cards are green? 0

e. What is the probability that the sum of the cards is at least 15?

13e. $\frac{14}{25}$ or 0.56

14a. See margin.   14b. $\frac{13}{16}$ or 0.8125

C

14. **Testing** Marty is taking his final exam in algebra. He takes too long answering the calculation part of the exam to properly evaluate the last five questions, which are true-false. To keep from getting points off for not answering these questions at all, he guesses at them just as the bell rings.

a. Draw a tree diagram to show the possible answer outcomes.

b. What is the probability that he got at least two of the questions correct if the answers were T, F, F, T, F?

c. Suppose Marty answered all the questions as true. Is this a good strategy to try to get the most correct without reading the question? Explain. Answers will vary. See students' work.

Lesson 7–5 *Probability Compound Events* **417**

Additional Answers

12a. See Solutions Manual for tree diagram. Possible combinations are: blue, right, hand; blue, right, foot; blue, left, hand; blue, left, foot; red, right, hand; red, right, foot; red, left, hand; red, left, foot; green, right, hand; green, right, foot; green, left, hand; green, left, foot; yellow, right, hand; yellow, right, foot; yellow, left, hand; and yellow, left, foot.

13a. R3-G5, R3-R10, R3-B10, R3-G1, R3-Y14, B3-G5, B3-R10, B3-B10, B3-G1, B3-Y14, R5-G5, R5-R10, R5-B10, R5-G1, R5-Y14, R14-G5, R14-R10, R14-B10, R14-G1, R14-Y14, Y10-G5, Y10-R10, Y10-B10, Y10-G1, Y10-Y14.

14a. See Solutions Manual for tree diagram. Possible combinations are: TTTTT, TTTTF, TTTFT, TTTFF, TTFTT, TTFTF, TTFFT, TTFFF, TFTTT, TFTTF, TFTFT, TFTFF, TFFTT, TFFTF, TFFFT, TFFFF, FTTTT, FTTTF, FTTFT, FTTFF, FTFTT, FTFTF, FTFFT, FTFFF, FFTTT, FFTTF, FFTFT, FFTFF, FFFTT, FFFTF, FFFFT, and FFFFF.

Study Guide Masters, p. 52

7-5   NAME_____ DATE_____

Study Guide   Student Edition Pages 413–419

Integration: Probability
Compound Events

A **compound event** consists of two or more simple events. The probability of a compound event can be calculated if its outcomes are equally likely. To calculate probability, determine the possible outcomes. A tree diagram is a good way of doing this.

Example: A die is rolled and a coin is tossed. What is the probability of getting a three on the die and heads on the coin?

To calculate the probability, you need to know how many outcomes are possible. There are 12 possible outcomes. The probability of getting a three on the die and a heads on the coin is $\frac{1}{12}$.

Solve each problem.

1. Max makes three piles of cards, each consisting of only a Queen and a Jack. Lisa draws one card from each pile. What is the probability that Lisa draws two Queens and a Jack in any order? Make a tree diagram of all the choices. $\frac{3}{8}$

2. Phyllis drops three pennies in a wishing pond. What is the probability that only one lands with tails showing? Make a tree diagram of all the choices. $\frac{3}{8}$

3. George has 1 pair of red socks and 1 pair of white socks in a drawer. What is the probability of drawing 2 red socks? $\frac{1}{6}$

4. In Exercise 3, what is the probability of picking a red sock and a white sock in that order? $\frac{1}{3}$

tree for Exercises 3 and 4:

15. **Babies** According to the National Center for Health Statistics, each time a baby is born, the probability that it is a boy is always 51.3% and a girl is 48.7%, regardless of the sex of any previous births in the family. Suppose a couple has four children. **15b. about 26.3%**
    a. What is the probability they are all girls? **about 5.6%**
    b. What is the probability there are one girl and three boys?
    c. How many children are there in your family? Find the probability of your family situation. **See students' work.**

16. **Computers** The results of a survey in which students were asked how they used their computers are shown in the Venn diagram at the right. Based on these findings, what is the probability that a student, chosen at random, will use a computer only for word processing, and a second student will use it only for games? **14%**

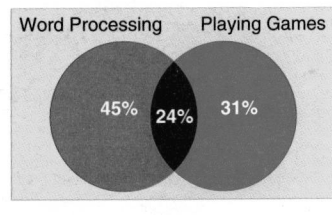

Word Processing		Playing Games
45%	24%	31%

Critical Thinking

17. Every working day Adrianna either bicycles or drives to work. If she oversleeps, she drives to work 80% of the time. If she does not oversleep, she bicycles 80% of the time. She oversleeps 20% of the time. What is the probability that she will drive to work on any particular day? **32%**

18. A bag of marbles contains 2 yellow marbles, 1 blue marble, and 3 red marbles. Suppose you select one marble at random and *do not* put it back in the bag. Then you select another marble.

18a. See students' work; 30 outcomes.

18b. $\frac{1}{5}$

   a. Draw a tree diagram to model this situation. How many possible outcomes are there?
   b. What is the probability of selecting a red marble first and then a yellow one?

Mixed Review

19. **Statistics** Amy's scores on the first three of four 100-point biology tests were 88, 90, and 91. To get an A− in the class, her average must be between 88 and 92, inclusive, on all tests. What score must she receive on the fourth test to get an A− in biology? (Lesson 7–4) **between 83 and 99, inclusive**

20. $\frac{2}{3}n > 99$, $\{n \mid n > 148.5\}$

20. Define a variable, write an inequality, and solve the following. Two thirds of a number is more than 99. (Lesson 7–2)

21. 3; −9

21. Determine the slope and y-intercept of the graph of $3x − y = 9$. (Lesson 6–4)

22. Solve $8r − 7t + 2 = 5(r + 2t) − 9$ for r. (Lesson 5–3) $r = \dfrac{17t − 11}{3}$

Extension

Connections A four-sided (tetrahedral) die has faces numbered 1, 2, 5, 6. If the die is tossed twice, what is the probability that the sum of the numbers facing down is 7? $\frac{1}{4}$

Practice Masters, p. 52

NAME _____ DATE _____

7-5   **Practice**   Student Edition Pages 413–419

tegration: Probability
ompound Events

Draw a tree diagram to show the possibilities for boys and girls in a family of 3 children. Assume that the probabilities for girls and boys being born are the same.

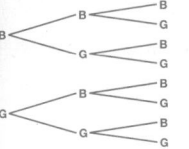

a. What is the probability that the family has exactly 3 boys? $\frac{1}{8}$
b. What is the probability that the family has more boys than girls? $\frac{1}{2}$
c. What is the probability that the family has at most two girls? $\frac{7}{8}$
d. What is the probability that the family has at least two girls? $\frac{1}{2}$

Use a tree diagram to find the probability of getting at most two heads when three fair coins are tossed. $\frac{7}{8}$

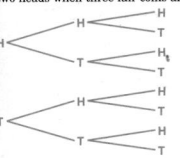

Compare the tree diagram used for Exercise 2 with the one used for Exercise 1.
Both tree diagrams show 3 simple events and 8 outcomes.
Box A contains one blue and two green marbles. Box B contains one green and two red marbles. Box C contains one green marble and two white marbles. A marble is drawn at random from each box.

a. What is the probability that all of the marbles are green? $\frac{2}{27} \approx 0.074$
b. What is the probability that at least one marble is not green? $\frac{25}{27} \approx 0.926$

23. Optics Peripheral vision is the ability to see around you when you keep your head and eyes pointed straight ahead. The average person with good peripheral vision can see about 180° when they are still. For a person in a moving vehicle, peripheral vision decreases as speed increases, as shown in the illustration below. (Lesson 3–4)

a. If Seth Lytle is driving at 65 mph, approximately how large is his angle of vision? **50°**

b. About how large an angle is he not seeing compared to his stationary angle of vision? **130°**

c. If there were a deer standing 50° to the right of Seth's center view and he was driving at 45 mph, would he be able to see the deer? **yes**

Speed of Car
- 40 mph
- 50 mph
- 60 mph
- 70 mph
- 80 mph

Mathematics and SOCIETY

Computers and Chess

The excerpt below appeared in an article in the January 1995 issue of *Discover* magazine.

HOW SMART HAVE COMPUTERS BECOME? Very smart, at least by one measure: this past August, a computer defeated the human world champion at chess for the first time. Russian grand master Garry Kasparov was outwitted by Genius 2, a computer program designed by an English physicist, Richard Lang, who calls himself a mediocre chess player.... Chess programs like Lang's work by focusing on what computers do best: long and intricate calculations. Before deciding on a move, the program examines each of the roughly 36 possibilities in an average game situation. For each of these 36 "branches,"

it looks 16 moves ahead, determining its opponent's possible countermoves, its own possible responses to each countermove, and so on. Sixteen moves ahead, the number of possible board configurations is huge. The genius of Genius 2 lies in the way it prunes the possibility tree, ruling out bad moves from the start. But the software also had help from the hardware; Lang's computer uses a speedy Intel Pentium microprocessor that can execute 166 million instructions per second. "We wouldn't have won (the computer world championship) had we not had this processor," Lang concedes. ∎

1. Many chess-playing programs are written by people who are not top-level chess players. How then can these programs defeat very good players?

2. If you want to improve your chess game by practicing against a computer, would you choose a program that is below your playing level, at your level, a little above your level, or far above your level? Why?

3. Do you think it's fair to match human chess players against computers? Why or why not? **See students' work.**

4. How might a computer use probability in determining the proper move to make? **See margin.**

1–2. See margin.

Answers for Mathematics and Society

1. The computer's speed can evaluate several options for this and future moves better than the best of players.

2. First choose a level a little above your level to stimulate your playing skills. Next choice would be at your level to improve your technique and speed. Playing at a level far above your level might be too frustrating.

4. The computer tests all combinations of possible moves now and in the future. It is programmed to take into consideration the likelihood of certain combinations of moves. Those combinations that are least likely can probably be ignored as possibilities for your opponent's next move.

Closing Activity

Modeling Place one red and one green marble in one jar, one blue and one white marble in another jar, and one black, one pink, and one yellow marble in a third jar. Have students reenact each of the twelve possible outcomes of the experiment of selecting one marble from each jar.

Mathematics and SOCIETY

Students might find it helpful to begin writing a very simple game tree for the game tic-tac-toe. Play the game for a short while, such as five moves. Now, from this configuration, construct a tree for all possible games from this point. Students will quickly see how large a tree would be for a complete tree for all possible tic-tac-toe games.

Enrichment Masters, p. 52

7-5 NAME_____ DATE_____
Enrichment Student Edition Pages 413–419

Conditional Probability

The probability of an event given the occurrence of another event is called **conditional probability.** The conditional probability of event A given event B is denoted $P(A|B)$.

Example: Suppose a pair of number cubes is rolled. It is known that the sum is greater than seven. Find the probability that the number cubes match.

There are 15 sums greater than seven and there are 36 possible pairs altogether.

There are three matching pairs greater than seven, (4, 4), (5, 5), and (6, 6).

$$P(B) = \frac{15}{36}$$

$$P(A \text{ and } B) = \frac{3}{36}$$

$$P(A|B) = \frac{P(A \text{ and } B)}{P(B)}$$

$$= \frac{\frac{3}{36}}{\frac{15}{36}} \text{ or } \frac{1}{5}$$

The conditional probability is $\frac{1}{5}$.

A card is drawn from a standard deck of 52 cards and is found to be red. Given that event, find each of the following probabilities.

1. $P(\text{heart})$ $\frac{1}{2}$
2. $P(\text{ace})$ $\frac{1}{13}$
3. $P(\text{face card})$ $\frac{3}{13}$
4. $P(\text{jack or ten})$ $\frac{2}{13}$
5. $P(\text{six of spades})$ 0
6. $P(\text{six of hearts})$ $\frac{1}{26}$

A sports survey taken at Stirers High School shows that 48% of the respondents liked soccer, 66% liked basketball, and 38% liked hockey. Also, 30% liked soccer and basketball, 22% liked basketball and hockey and 28% liked soccer and hockey. Finally, 12% liked all three sports.

7. Find the probability that Meg likes soccer if she likes basketball. $\frac{5}{11}$
8. Find the probability that Juan likes basketball if he likes soccer. $\frac{5}{8}$
9. Find the probability that Mieko likes hockey if she likes basketball. $\frac{1}{3}$
10. Find the probability that Greg likes hockey if he likes soccer. $\frac{7}{12}$

Instructional Resources

- Study Guide Master 7-6
- Practice Master 7-6
- Enrichment Master 7-6
- Assessment and Evaluation Masters, p. 185

 Transparency 7-6A contains the 5-Minute Check for this lesson; **Transparency 7-6B** contains a teaching aid for this lesson.

Recommended Pacing

Standard Pacing	Day 8 of 14
Honors Pacing	Day 7 of 13
Block Scheduling*	Day 4 of 7 (along with Lesson 7-5)
Alg. 1 in Two Years*	Days 11 & 12 of 21

 *For more information on pacing and possible lesson plans, refer to the *Block Scheduling Booklet* and *Algebra 1 in Two Years.*

1 FOCUS

 ### 5-Minute Check
(over Lesson 7-5)

Two red marbles and four green marbles are placed in a jar. Two marbles are to be selected from the jar at random. A marble is drawn, put back in the jar, and then a second marble is drawn.

1. What is the probability that both are red? $\frac{1}{9}$

2. What is the probability that both are green? $\frac{4}{9}$

3. What is the probability that at least one is green? $\frac{8}{9}$

4. What is the probability that one is green and one is red? $\frac{4}{9}$

5. Which is more likely, at least one is red or at least one is green? **at least one is green**

Objective
After studying this lesson you should be able to:
- solve open sentences involving absolute value and graph the solutions.

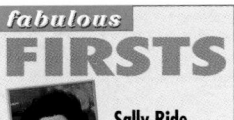 ### fabulous FIRSTS

Sally Ride (1951–)

Sally Ride was the first American woman in space. She was a member of the *Challenger STS-7* crew that was launched June 18, 1983. She has a Ph.D. in physics from Stanford University.

The solution set can also be written as $\{x \mid x = -4 \text{ or } x = 4\}$.

 ### APPLICATION
Space Exploration

On Tuesday, February 7, 1995, the space shuttle *Discovery* maneuvered within 37 feet of the Russian space station *Mir,* 245 miles above Earth. To accomplish this feat, *Discovery* had to be launched within 2.5 minutes of a designated time (12:45 A.M.). This time period is known as the *launch window.* If *t* represents the time elapsed since the launch countdown began and there are 300 minutes scheduled from the beginning of the countdown to blast-off, you can write the following inequality to represent the launch window.

$|300 - t| \le 2.5$ *The difference between 300 minutes and the actual time elapsed since the countdown began must be less than or equal to 2.5 minutes.*

We use absolute value because $300 - t$ cannot be negative.

There are three types of open sentences that can involve absolute value. They are as follows, when *n* is nonnegative.

$$|x| = n \qquad |x| < n \qquad |x| > n$$

First let's consider the case of $|x| = n$.

If $|x| = 4$, this means that the difference from 0 to *x* is 4 units.

Therefore, if $|x| = 4$, then $x = -4$ or $x = 4$. The solution set is $\{-4, 4\}$. So, if $|x| = n$, then $x = -n$ or $x = n$.

Equations involving absolute value can be solved by graphing them on a number line or by writing them as a compound sentence and solving it.

Example ❶ Solve $|x - 3| = 5$.

Method 1: Graphing

$|x - 3| = 5$ means the distance between *x* and 3 is 5 units. To find *x* on the number line, start at 3 and move 5 units in either direction.

The solution set is $\{-2, 8\}$.

 ### fabulous FIRSTS

On her first shuttle flight, Ride helped launch communications satellites, conducted experiments on the production of pharmaceuticals, and tested the space shuttle's remote manipulator arm. Ride flew a second shuttle mission in October 1984.

Method 2: Compound Sentence

$|x - 3| = 5$ also means $x - 3 = 5$ or $-(x - 3) = 5$.

$$x - 3 = 5 \qquad \text{or} \qquad -(x - 3) = 5$$
$$x - 3 + 3 = 5 + 3 \qquad\qquad x - 3 = -5 \qquad \textit{Multiply each side by } -1.$$
$$x = 8 \qquad\qquad x - 3 + 3 = -5 + 3$$
$$x = -2$$

This verifies the solution set.

Now let's consider the case of $|x| < n$. Inequalities involving absolute value can also be represented on a number line or as compound inequalities. Let's examine $|x| < 4$.

$|x| < 4$ means that the distance from 0 to x is less than 4 units.

When the inequality is $<$ or $\leq$, the compound sentence uses and.

Therefore, $x > -4$ and $x < 4$. The solution set is $\{x \mid -4 < x < 4\}$. So, if $|x| < n$, then $x > -n$ and $x < n$.

Example ② Solve $|3 + 2x| < 11$ and graph the solution set.

$|3 + 2x| < 11$ means $3 + 2x < 11$ and $3 + 2x > -11$.

$$3 + 2x < 11 \qquad \text{and} \qquad 3 + 2x > -11$$
$$3 - 3 + 2x < 11 - 3 \qquad\qquad 3 - 3 + 2x > -11 - 3$$
$$2x < 8 \qquad\qquad 2x > -14$$
$$\frac{2x}{2} < \frac{8}{2} \qquad\qquad \frac{2x}{2} > \frac{-14}{2}$$
$$x < 4 \qquad\qquad x > -7$$

The solution set is $\{x \mid x > -7 \text{ and } x < 4\}$, which can be written as $\{x \mid -7 < x < 4\}$.

Now graph the solution set.

Finally let's examine $|x| > 4$. This means that the distance from 0 to x is greater than 4 units.

When the inequality is $>$ or $\geq$, the compound sentence uses or.

Therefore, $x < -4$ or $x > 4$. The solution set is $\{x \mid x < -4 \text{ or } x > 4\}$. So, if $|x| > n$, then $x < -n$ or $x > n$.

Lesson 7-6 Solving Open Sentences Involving Absolute Value **421**

Motivating the Lesson
Situational Problem Jim's average is less than 5 points from Sam's. Sam's average is 74. What conclusion can be drawn about Jim's average? His average is less than 79 (74 + 5) and greater than 69 (74 − 5).

2 TEACH

Teaching Tip When presenting Example 1, remind students that $|x| = 5$ means $x = 5$ or $-x = 5$. Therefore, $x = 5$ or $x = -5$. $|x - 3| = 5$ means $x - 3 = 5$ or $-(x - 3) = 5$. Therefore, $x - 3 = 5$ or $x - 3 = -5$.

In-Class Examples

For Example 1
Solve each equation.

a. $|4p - 3| = 7$
$$\left\{p \mid p = \frac{5}{2} \text{ or } p = -1\right\}$$

b. $|4k + 6| = 5$
$$\left\{k \mid k = -\frac{1}{4} \text{ or } k = -\frac{11}{4}\right\}$$

For Example 2
Solve $|2x + 1| < 8$ and graph the solution set.
$$\left\{x \mid x > -\frac{9}{2} \text{ and } x < \frac{7}{2}\right\}$$

Teaching Tip When presenting Example 2, point out to students that the open sentence $|3 + 2x| < 11$ says that the distance from 0 to $3 + 2x$ is less than 11 units. This means $3 + 2x < 11$ and $-(3 + 2x) < 11$, or $3 + 2x > -11$.

In-Class Examples

For Example 3
Solve $|4x - 5| \geq 3$ and graph the solution set.
$$\left\{x \mid x \leq \frac{1}{2} \text{ or } x \geq 2\right\}$$

For Example 4
If the price of a stock varies more than 3.4 points from its opening price, Mr. Witter buys ten more shares of the stock. If the opening price is 15.8, for what range of values will he buy stock? $\{x \mid x > 19.2 \text{ or } x < 12.4\}$

Teaching Tip If students are having difficulty determining when they should use *and* or *or* in an inequality involving an absolute value, explain that an alternate approach is available.
Solve $|3 + 2x| < 11$.
Solve the inequality as if it were an equation.
$$\begin{array}{ll} 3 + 2x = -11 & 3 + 2x = 11 \\ 2x = -14 & 2x = 8 \\ x = -7 & x = 4 \end{array}$$
Plot these points with open circles on the number line.
The two points separate the number line into three regions, $x < -7$, $-7 < x < 4$, and $x > 4$. Have students test a point from each region in the original inequality, for example, -10, 0, and 5. The region that contains 0 is the region that is the solution. This region is $-7 < x < 4$. This represents a compound sentence involving *and.* Draw a line between the two open circles to represent the answer.

Example **Solve $|5 + 2y| \geq 3$ and graph the solution set.**

$|5 + 2y| \geq 3$ means $5 + 2y \leq -3$ or $5 + 2y \geq 3$.

$$\begin{array}{lcr} 5 + 2y \leq -3 & \text{or} & 5 + 2y \geq 3 \\ 5 - 5 + 2y \leq -3 - 5 & & 5 - 5 + 2y \geq 3 - 5 \\ 2y \leq -8 & & 2y \geq -2 \\ \dfrac{2y}{2} \leq \dfrac{-8}{2} & & \dfrac{2y}{2} \geq \dfrac{-2}{2} \\ y \leq -4 & & y \geq -1 \end{array}$$

The solution set is $\{y \mid y \leq -4 \text{ or } y \geq -1\}$.

Organizations such as OSHA set standards for buildings to meet the needs of those using the building. Building code standards are often written as maximums or minimums that must be met. These standards can be written as inequalities.

Example **4**

There are a number of specifications in the building industry that address the needs of physically-challenged persons. For example, hallways in hospitals must have handrails. The handrails must be placed within a range of 2 inches from a height of 36 inches.

a. Write an open sentence that involves absolute value to represent the range of acceptable heights for hallway handrails.

b. Find and graph the corresponding compound sentence.

a. Let h be an acceptable height for the handrail. Then h can differ from 36 inches by no more than 2 inches. Write an open sentence that represents the range of acceptable heights.

$$\underbrace{h \text{ differs from } 36}_{|h - 36|} \quad \underbrace{\text{by less than or equal to}}_{\leq} \quad \underbrace{2.}_{2}$$

Now, solve $|h - 36| \leq 2$ to find the compound sentence.

$$\begin{array}{lcr} h - 36 \geq -2 & \text{and} & h - 36 \leq 2 \\ h - 36 + 36 \geq -2 + 36 & & h - 36 + 36 \leq 2 + 36 \\ h \geq 34 & & h \leq 38 \end{array}$$

The compound sentence is $34 \leq h \leq 38$.

 Tech Prep

Construction Manager Students who are interested in a career in construction may wish to do further research into the use of math as it relates to building codes, as mentioned in Example 4. For more information on tech prep, see the *Teacher's Handbook*.

b. The graph of this sentence is shown below.

So, the handrail must be placed between 34 and 38, inclusive, inches from the floor.

CHECK FOR UNDERSTANDING

Communicating Mathematics

Study the lesson. Then complete the following. 1–2. See margin.

1. **Describe** two ways you can solve an open sentence involving an absolute value.

2. **Explain** the difference in the solution for $|x + 7| > 4$ and the solution for $|x + 7| < 4$.

3. If $x < 0$ and $|x| = n$, describe n in terms of x. $n = -x$

4. **You Decide** Jamila's teacher said they should work out each problem and test points to determine whether their solutions are correct. Jamila thinks that she knows the solution for problems like $|w + 5| < 0$, where the absolute value quantity is always less than 0, without testing points. Is she correct? Why or why not? Yes, it will always be ∅.

Guided Practice

Choose the replacement set that makes each open sentence true.

5. $|x - 7| = 2$ c
   a. {9, −5}   b. {5, −9}   c. {5, 9}   d. ∅

6. $|x - 2| > 4$ d
   a. {x | −2 < x < 6}   b. {x | x = 6 or x = 2}
   c. {x | x > 6 or x > −2}   d. {x | x < −2 or x > 6}

State which graph below matches each open sentence in Exercises 7–10.

7. $|y| = 3$ c

8. $|y| > 3$ a

9. $|y| < 3$ d

10. $|y| \leq 3$ b

a.
b. (number line −5 to 5)
c. (number line −5 to 5)
d. (number line −5 to 5)

Solve each open sentence. Then graph the solution set.

11–14. See margin for graphs.

11. $|m| \geq 5$ {m | m ≤ −5 or m ≥ 5}   12. $|n| < 6$ {n | −6 < n < 6}

13. $|r + 3| < 6$ {r | −9 < r < 3}   14. $|8 - t| \geq 3$ {t | t ≤ 5 or t ≥ 11}

For each graph, write an open sentence involving absolute value.

15. $|x| = 2$
16. $|x - 1| \leq 3$

15.

16.

Lesson 7–6 Solving Open Sentences Involving Absolute Value **423**

Additional Answers

17.
−6−4−2 0 2 4 6 8 10

18.
−3−2−1 0 1 2 3 4 5

19. ∅

20.
−12−11−10−9−8−7−6−5−4

21.
−2−1 0 1 2 3 4 5 6

22.
−9−8−7−6−5−4−3−2−1 0 1

23. ∅

24.
−5 −4 −3 −2 −1 0

25.
1 2 3 4

26.
−6−5−4−3−2−1 0 1 2 3 4

27.
−4−3−2−1 0 1 2 3 4

Study Guide Masters, p. 53

Practice

17–34. See margin for graphs.

20. $\{w | w \leq −9$ or $w \geq −7\}$

24. $\{x | −\frac{9}{2} < x < −\frac{1}{2}\}$

25. $\{e | \frac{5}{3} < e < 3\}$

26. $\{x | −4 < x < \frac{4}{3}\}$

29. $\{w | 0 \leq w \leq 18\}$

32. $\{r | r \leq −1.\overline{6}$ or $r \geq 2\}$

33. $\{x | x \leq −\frac{8}{3}$ or $x \geq 4\}$

34. $\{p | −1 < p < \frac{4}{3}\}$

38. $|x − 2| = 4$
39. $|x + 1| = 3$
40. $|x| \geq 2$
41. $|x − 1| \leq 1$
42. $|x + 1| < 3$
43. $|x − 8| \geq 3$

Critical Thinking

48. $\{\frac{1}{2}\}$

Applications and Problem Solving

51. $\frac{8}{13}$ or 0.61

Solve each open sentence. Then graph the solution set.

17. $|y − 2| = 4$ $\{−2, 6\}$

18. $|3 − 3x| = 0$ $\{1\}$

19. $|7x + 2| = −2$ ∅

20. $|w + 8| \geq 1$

21. $|2 − y| \leq 1$ $\{y | 1 \leq y \leq 3\}$

22. $|t + 4| \geq 3$ $\{t | t \leq −7$ or $t \geq −1\}$

23. $|4y − 8| < 0$ ∅

24. $|2x + 5| < 4$

25. $|3e − 7| < 2$

26. $|3x + 4| < 8$

27. $|1 − 3y| > −2$ {all numbers}

28. $3 + |x| > 3$ $\{x | x \neq 0\}$

29. $|8 − (w − 1)| \leq 9$

30. $|6 − (11 − b)| = −3$ ∅

31. $|2.2y − 1.1| = 5.5$ $\{−2, 3\}$

32. $|3r − 0.5| \geq 5.5$

33. $|\frac{2 − 3x}{5}| \geq 2$

34. $|\frac{1}{2} − 3p| < \frac{7}{2}$

Express each statement in terms of an inequality involving absolute value. Do _not_ try to solve.

35. The diameter of the lead in a pencil p must be within 0.01 millimeters of 1 millimeter. $|p − 1| \leq 0.01$

36. The cruise control of a car set at 55 mph should keep the speed s within 3 mph of 55 mph. $|s − 55| \leq 3$

37. A liquid at 50°C will change to a gas or liquid if the temperature t increases or decreases more than 50°C. $|t − 50| > 50$

For each graph, write an open sentence involving absolute value.

38.
−3−2−1 0 1 2 3 4 5 6 7

39.
−6−5−4−3−2−1 0 1 2 3 4

40.
−5−4−3−2−1 0 1 2 3 4 5

41.
−4 −3 −2 −1 0 1 2 3 4 5

42.
−6−5−4−3−2−1 0 1 2 3 4

43.
3 4 5 6 7 8 9 10 11 12 13

44. Find all integer solutions of $|x| < 4$. $\{−3, −2, −1, 0, 1, 2, 3\}$

45. Find all integer solutions of $|x| \leq 2$. $\{−2, −1, 0, 1, 2\}$

46. If $a > 0$, how many integer solutions exist for $|x| < a$? $2a − 1$

47. If $a > 0$, how many integer solutions exist for $|x| \leq a$? $2a + 1$

48. Solve $|y − 3| = |2 + y|$.

49. Under what conditions is $−|a|$ negative? positive? $a \neq 0$; never

50. Suppose $8 \leq x \leq 12$. Write an absolute value inequality that is equivalent to this compound inequality. $|x − 10| \leq 2$

51. Probability Suppose $|x| \leq 6$ and x is an integer. Find the probability of $|x|$ being a factor of 18.

52. Chemistry For hydrogen to be a liquid, its temperature must be within 2°C of −257°C. What is the range of temperatures for this substance to remain a liquid? $−259°C < t < 255°C$

Additional Answers

28.
−4−3−2−1 0 1 2 3 4

29.
0 2 4 6 8 10 12 14 16 18

30. ∅

31.
−4−3−2−1 0 1 2 3 4

32.
−4−3−2−1 0 1 2 3 4

33.
−4−3−2−1 0 1 2 3 4 5 6

34.
−1 0 1 2

53. Law Enforcement A radar gun used to determine the speed of passing cars must be within 7 mph of the actual speed of a selected car. If a highway patrol officer reads a speed of 59 mph for a car, does he have irrefutable evidence that the car was speeding in a 55 mph zone? Explain your reasoning and include a graph. **no; $52 \leq s \leq 66$**

54. Space Exploration Refer to the application at the beginning of the lesson. Find how much time can elapse from the beginning of the countdown to remain within the launch window.

54. $297.5 \leq t \leq 302.5$ minutes

55. Entertainment Luis Gomez is a contestant on the *Price is Right*. He must guess within $1500 of the actual price of a Jeep Cherokee without going over in order to win the vehicle. The actual price of the Jeep is $18,000. What is the range of guesses in which Luis can win the vehicle? **$16,500 \leq p \leq $18,000**

56. Spending The graph below shows the spending power of kids aged 3 to 17.

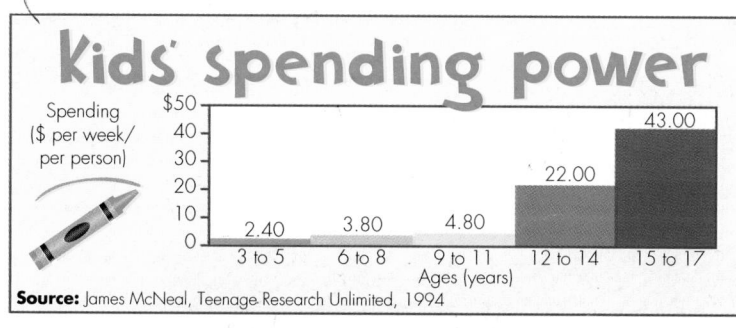

kids' spending power

Spending ($ per week/ per person)

2.40 — 3 to 5
3.80 — 6 to 8
4.80 — 9 to 11
22.00 — 12 to 14
43.00 — 15 to 17

Ages (years)

Source: James McNeal, Teenage Research Unlimited, 1994

a. Write an inequality that represents the spending power of 3- to 17-year olds. Then write an absolute value inequality that describes their spending. $2.40 \leq s \leq 43.00$; $|s - 22.70| \leq 20.30$

b. Write an inequality that represents the spending power of 12- to 17-year olds. Then write an absolute value inequality that describes their spending. $22 \leq s \leq 43$; $|s - 32.50| \leq 10.50$

c. Keep a record of how much money you spend in a week. How does your spending compare with the data in this graph? **See students' work.**

Mixed Review

57. Draw a Diagram The Sanchez family acts as a host family for foreign exchange students during each quarter of the year. Suppose it is equally likely that they get a boy or a girl each quarter. (Lesson 7-5)

a. Draw a tree diagram to represent the possible orders of boys (B) and girls (G) during the four quarters of the year. List the possible outcomes. **See margin.**

57b. $\frac{1}{16}$ or 0.0625, regardless of gender

57c. $\frac{3}{8}$ or 0.375

b. From the tree diagram, what is the probability that all of the students will be girls? boys?

c. From the tree diagram, what is the probability that they will host two boys and two girls?

Lesson 7-6 Solving Open Sentences Involving Absolute Value **425**

Practice Masters, p. 53

7-6
Practice
NAME_____ DATE_____
Student Edition
Pages 420–426

Solving Open Sentences Involving Absolute Value

Solve each open sentence. Then graph the solution set.

1. $|x - 3| < 1$ $2 < x < 4$
2. $|3 - y| > 2$ $y < 1$ or $y > 5$
3. $|2z - 9| \leq 1$ $4 \leq z \leq 5$
4. $|3 + e| > 0$ $e \neq -3$
5. $3 = |1 - 4p|$ $p = -\frac{1}{2}$ or $p = 1$
6. $|2 + n| \geq 1$ $n \leq -3$ or $n \geq -1$

Express each statement in terms of an inequality involving absolute value. Do not try to solve.

7. Mary's bowling score was within 10 pins of her average score of 105. $|x - 105| < 10$
8. The temperature was at least 7 Fahrenheit degrees from 0°F. $|t| \geq 7$

For each graph, write an open sentence involving absolute value.

9. $|x| = 3$
10. $|x + 2| \geq 1$

Cooperative Learning

Send-A-Problem Separate the class into four or five small groups. Have each group member create an inequality and then graph its solution set. Have groups switch graphs and identify the inequality that each solution set solves. Have groups return papers and verify each other's answers. For more information on the send-a-problem strategy, see *Cooperative Learning in the Mathematics Classroom*, one of the titles in the Glencoe Mathematics Professional Series, pages 23–24.

Closing Activity

Speaking Have students explain what must be true of $|m|$ if

a. $m > 5$. $|m| > 5$

b. $m < -5$. $|m| > 5$

Chapter 7, Quiz C (Lessons 7-5 and 7-6), is available in the *Assessment and Evaluation Masters*, p. 185.

Additional Answer

63.

---

58. $\$1932 < p < \$2500$

60. $\{t \mid t > -36\}$

61. $\{k \mid k \geq -15\}$

---

58. Peter wants to buy Crystal an engagement ring. He wants to spend between $1700 and $2200. If he goes shopping during a 12%-off sale to celebrate the store's 12th anniversary, what would his price range be? (Lesson 7–4)

59. Solve $10x - 2 \geq 4(x - 2)$. (Lesson 7–3) $\{x \mid x \geq -1\}$

Solve each inequality.

60. $396 > -11t$ (Lesson 7–2)

61. $-11 \leq k - (-4)$ (Lesson 7–1)

62. Find the coordinates of the midpoint of the line segment whose endpoints are at $(-4, 1)$ and $(10, 3)$. (Lesson 6–7) **(3, 2)**

63. Graph $2x - 9 = 2y$. (Lesson 5–4) **See margin.**

64. **Architecture** Julie is making a model of a house for her drafting class. One part of the house has a triangle like the one below. She wants the base of the triangle to measure 2 inches. What will be the measures of the other two sides? (Lesson 4–2) **1.2 in.; 1.6 in.**

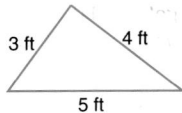

65. Solve $2m + \frac{3}{4}n = \frac{1}{2}m - 9$ for m. (Lesson 3–6) $m = -6 - \frac{n}{2}$

66. Simplify $\frac{-42r + 18}{3}$. (Lesson 2–7) $-14r + 6$

WORKING ON THE **In·ves·ti·ga·tion** Refer to the Investigation on pages 320–321.

Smoke Gets In Your Eyes

Smoking increases the chance of health problems. Constant exposure to secondhand smoke also puts the nonsmoker at risk. According to a statement from a tobacco company, a nonsmoker working among smoking coworkers inhales the smoke of 1.25 cigarettes per month. A restaurant worker in the smoking section breathes just 2 cigarettes per month. The Environmental Protection Agency (EPA) disagrees with these data.

1 Write an equation to express the equivalent number of cigarettes one would breathe if he/she worked in the smoking section of a restaurant for one year.

2 A lawsuit involving the tobacco industry and the EPA states that the EPA did not use the standard significance level of 5% in evaluating their data from various studies. A 5% level means that their numerical conclusions have a 5% margin of error. How does the margin of error relate to absolute value?

3 In January, 1995, the average price of a pack of cigarettes was $2.06, of which 56¢ is federal and state excise tax. The price of cigarettes is on the rise. Write an inequality to represent how much is spent by the average smoker in a year. Solve the inequality.

4 How much tax revenue is generated by the sale of cigarettes to the estimated number of smokers in the United States?

Add the results of your work to your Investigation Folder.

Extension

Problem Solving Find all integer values of x for which $|x - 2| < x$. $x > 1$

In·ves·ti·ga·tion

Working on the Investigation

The Investigation on pages 320–321 is designed to be a long-term project that is completed over several days or weeks. Encourage students to keep their materials in their Investigation Folder as they work on the Investigation.

Enrichment Masters, p. 53

Integration: Statistics
Box-and-Whisker Plots

Objective

After studying this lesson you should be able to:

- display and interpret data on box-and-whisker plots.

APPLICATION
Olympics

In 1992, the site of the Summer Games of the XXV Olympiad was Barcelona, Spain. More than 14,000 athletes from 172 nations competed for medals in 257 events. The table at the right shows the number of gold medals won by the top 16 medal-winning teams.

We can describe these data using the mean, median, and mode. We can also use the median, along with the quartiles and interquartile range, to obtain a graphic representation of the data. A type of diagram, or graph, that shows quartiles and extreme values of data is called a **box-and-whisker plot.**

1992 SUMMER OLYMPICS Top 16 Medal Winners		
Team	**Number of Medals Won**	
	Gold	**Total**
Unified Team (formerly USSR)	45	112
USA	37	108
Germany	33	82
China	16	54
Cuba	14	31
Hungary	11	30
South Korea	12	29
France	8	29
Australia	7	27
Spain	13	22
Japan	3	22
Britain	5	20
Italy	6	19
Poland	3	19
Canada	6	18
Romania	4	18

Source: *The World Almanac, 1995*

Box-and-whisker plots are sometimes called box plots.

LOOK BACK

You can refer to Lesson 3-7 for information on measures of central tendency and Lesson 5-7 for measures of variation.

The median is not included in either half of the data.

Suppose we wanted to make a box-and-whisker plot of the numbers of gold medals won by each of the nations in the table. First, arrange the data in numerical order. Next, compute the median and quartiles. Also, identify the extreme values.

$$3 \quad 3 \quad 4 \quad 5 \quad 6 \quad 6 \quad 7 \quad \boxed{8 \quad \uparrow \quad 11} \quad 12 \quad 13 \quad 14 \quad 16 \quad 33 \quad 37 \quad 45$$

median (Q2)

The median for this set of data is the average of the eighth and ninth values.

$$\text{median} = \frac{8 + 11}{2} \text{ or } 9.5$$

Recall that the *lower quartile* (Q1) is the median of the lower half of the distribution of values. The *upper quartile* (Q3) is the median of the upper half of the data.

$$\boxed{3 \quad 3 \quad 4 \quad 5 \uparrow 6 \quad 6 \quad 7 \quad 8} \quad \boxed{11 \quad 12 \quad 13 \quad 14 \uparrow 16 \quad 33 \quad 37 \quad 45}$$

Q1   Q3

$$Q1 = \frac{5 + 6}{2} \text{ or } 5.5 \qquad\qquad Q3 = \frac{14 + 16}{2} \text{ or } 15$$

The **extreme values** are the least value (LV), 3, and the greatest value (GV), 45.

Lesson 7-7 *Statistics* *Box-and-Whisker Plots* **427**

NCTM Standards: 1–4, 10

Instructional Resources

- Study Guide Master 7-7
- Practice Master 7-7
- Enrichment Master 7-7
- Science and Math Lab Manual, pp. 29–34

 Transparency 7-7A contains the 5-Minute Check for this lesson; **Transparency 7-7B** contains a teaching aid for this lesson.

Recommended Pacing	
Standard Pacing	Day 9 of 14
Honors Pacing	Day 8 of 13
Block Scheduling*	Day 5 of 7
Alg. 1 in Two Years*	Days 13 & 14 of 21

*For more information on pacing and possible lesson plans, refer to the *Block Scheduling Booklet* and *Algebra 1 in Two Years.*

1 FOCUS

5-Minute Check
(over Lesson 7-6)

Solve each open sentence.

1. $|e - 5| = -3$ $\varnothing$
2. $|a - 1| < 4$ $\{a \mid -3 < a < 5\}$
3. $|b + 8| \geq 1$
   $\{b \mid b \leq -9 \text{ or } b \geq -7\}$
4. $|9 - c| \geq 13$
   $\{c \mid c \geq 22 \text{ or } c \leq -4\}$
5. $|3d - 12| < 12$ $\{d \mid 0 < d < 8\}$

Motivating the Lesson

Hands-On Activity Reinforce student understanding of the terms *median, upper quartile,* and *lower quartile* with the following activity. Ask students to arrange themselves in a row in ascending height. Have them determine which class member has the median height. Then ask students to identify the class member who represents the upper quartile and the class member who represents the lower quartile.

Teaching Tip Remind students that items of data that occur more than once must appear the same number of times when arranging the data in ascending order.

Technology Tip
Point out that graphing calculators do not identify outliers.

Now we have the information we need to draw a box-and-whisker plot.

Step 1 Draw a number line. Assign a scale to the number line that includes the extreme values. Plot dots to represent the extreme values (LV and GV), the upper and lower quartile points (Q3 and Q1), and the median (Q2).

The median line will not always divide the box into equal parts.

Step 2 Draw a box to designate the data falling between the upper and lower quartiles. Draw a vertical line through the point representing the median. Draw a segment from the lower quartile to the least value and one from the upper quartile to the greatest value. These segments are the **whiskers** of the plot.

TECHNOLOGY *Tip*

You can use your graphing calculator to find the values you need to make a box-and-whisker plot. Refer to Lesson 5–7A for instructions on finding these values.

Even though the whiskers are different lengths, each whisker contains at least one fourth of the data while the box contains one half of the data. Compound inequalities can be used to describe the data in each fourth. Assume that the replacement set for x is the set of data.

1st fourth	$\{x \mid x < 5.5\}$
2nd fourth	$\{x \mid 5.5 < x < 9.5\}$
3rd fourth	$\{x \mid 9.5 < x < 15\}$
4th fourth	$\{x \mid x > 15\}$

Step 3 Before finishing the box-and-whisker plot, check for outliers. In Lesson 5–7, you learned that an outlier is any element of the set of data that is at least 1.5 interquartile ranges above the upper quartile or below the lower quartile. Recall that the *interquartile range* (IQR) is the difference between the upper and lower quartiles, or in this case, $15 - 5.5$ or 9.5.

$x \geq Q_3 + 1.5(\text{IQR})$ or $x \leq Q_1 - 1.5(\text{IQR})$

$x \geq 15 + 1.5(9.5)$ $x \leq 5.5 - 1.5(9.5)$

$x \geq 15 + 14.25$ $x \leq 5.5 - 14.25$

$x \geq 29.25$ $x \leq -8.75$

Step 4 If x is an outlier in this set of data, then the outliers can be described as $\{x \mid x \leq -8.75 \text{ or } x \geq 29.5\}$. In this case, there are no data less than -8.75. However, 45, 37, and 33 are greater than 29.25, so they are outliers. We now need to revise the box-and-whisker plot. Outliers are plotted as isolated points, and the right whisker is shortened to stop at 16.

APPLICATION
Sports

Refer to the application at the beginning of the lesson. Use the box-and-whisker plot for the gold medals to answer each question.

a. What percent of the teams won between 6 and 15 gold medals?

b. What does the box-and-whisker plot tell us about the upper half of the data compared to the lower half?

a. The box in the plot indicates 50% of the values in the distribution. Since the box goes from 5.5 to 15, we know that 50% of the teams won between 6 and 15 gold medals.

b. The upper half of the data is spread out while the lower half is fairly clustered together.

CHECK FOR UNDERSTANDING

Communicating Mathematics

Study the lesson. Then complete the following.

1. **Explain** how to determine the scale of the number line in a box-and-whisker plot. Scale must be large enough to include the least and greatest values.

2. **Describe** which two points the two whiskers of a box-and-whisker plot connect. LV and Q_1; Q_3 and GV

3. What does Q_2 represent? the median

4. Refer to the box-and-whisker plot at the right. Assume that LV, Q_1, Q_2, Q_3, and GV are whole numbers.

80   90   100   110   120   130

a. What percent of the data is between 85 and 90? 25%

b. Between what two values does the middle 50% of the data lie? 90–120

c. What outliers are represented in the box-and-whisker plot? none

5. Sample answer: quartiles, interquartile range, outliers, whether the data are clustered or diverse; individual point of data, number of data

5. **Describe** what characteristics of a set of data you can gather from its box-and-whisker plot. What are some things you cannot gather from a box-and-whisker plot?

Guided Practice

6a. 1985: 14.0, 18.0, 11.9, 6.1; 1992: 13.1, 15.7, 11.9, 3.8

6. The table at the right shows the birthrates for 15 selected countries.

a. Find the median, upper quartile, lower quartile, and interquartile range for each year's data.

b. Are there any outliers? If so, name them. 21.5, 23.3

c. Draw a box-and-whisker plot for each set of data on the same number line. See margin.

d. Which set of data seems to be more clustered? Why? 1992; because the plot is not as wide.

Birth Rates for Selected Countries
(per 1000 population)

Country	1985	1992
Australia	15.7	15.1
Cuba	18.0	14.5
Denmark	10.6	13.1
France	13.9	12.9
Hong Kong	14.0	11.9
Israel	23.5	21.5
Italy	10.1	9.9
Japan	11.9	9.7
Netherlands	12.3	13.0
Panama	26.6	23.3
Poland	18.2	13.4
Portugal	12.8	11.4
Singapore	16.6	17.7
Switzerland	11.6	12.6
United States	15.7	15.7

Source: United Nations, *Monthly Bulletin of Statistics,* May 1994

For the Example
Identify the extreme values and quartiles and draw a box-and-whisker plot for the following data: 26, 42, 57, 12, 38, 63, 45, 18.
LV = 12, GV = 63, median = 40, Q_3 = 51, Q_1 = 22

Box Plot

10   20   30   40   50   60

3 PRACTICE/APPLY

Check for Understanding
Exercises 1–7 are designed to help you assess your students' understanding through reading, writing, speaking, and modeling. You should work through Exercises 1–5 with your students and then monitor their work on Exercises 6–7.

Additional Answer

6c.

Reteaching

Using Class Discussion Give each row of students a different set of data for a box-and-whisker plot. Have each student draw his or her own plot on an overhead transparency. Put the plots on the overhead so that the class can see them. Ask questions about the plots to see if they were drawn correctly. Discuss problems and good points of the plots.

Assignment Guide

Core: 9–15 odd, 16–20
Enriched: 8–20

For **Extra Practice,** see p. 773.

The red A, B, and C flags, printed only in the Teacher's Wraparound Edition, indicate the level of difficulty of the exercises.

Additional Answers

9c.

10.

7a. A: 25, 65, 30, 60, 40;
B: 20, 70, 40, 60, 45

7. Refer to box-and-whisker plots A and B.

   a. Estimate the least value, greatest value, lower quartile, upper quartile, and median for each plot. Assume that these values are whole numbers.
   b. Which set of data contains the least value? B
   c. Which plot has the greatest interquartile range? A
   d. Which plot has the greatest range? B

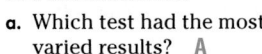

EXERCISES

Applications and Problem Solving

8. **Manufacturing** The box-and-whisker plots at the right show the results of testing the useful life of 10 light bulbs each from two manufacturers.

Hours Burned

   a. Which test had the most varied results? A
   b. Were there any outliers? If so, for which brand? yes; A
   c. How would you compare the medians of the tests? about the same
   d. Based on this plot, from which manufacturer would you buy your light bulbs? Why? B; The lives of the bulbs were more consistent.

9. **Meteorology** Meteorologists keep track of temperatures for four 90-day periods during the year to predict the trends in weather for future years. The following low temperatures were recorded during a 2-week cold snap in Indianapolis during 1993.   9a. $Q_2 = 6.5$, $Q_3 = 16$, $Q_1 = 5$, IQR = 11

   30°, 20°, 2°, 12°, 5°, 4°, 17°, 7°, 6°, 16°, 5°, 0°, 5°, 16°

   a. Find the median, upper quartile, lower quartile, and interquartile range.
   b. Are there any outliers? If so, name them. no
   c. Draw a box-and-whisker plot of the data. See margin.

10. **Football** The table below shows the American Football Conference's leading quarterbacks in touchdowns for the 1993 season. Make a box-and-whisker plot for the data. See margin.

1993 AFC Individual Leaders in Passing		
Player	**Team**	**Touchdowns**
Steve DeBerg	Miami	7
John Elway	Denver	25
Boomer Esiason	N.Y. Jets	16
John Friesz	San Diego	6
Jeff George	Indianapolis	8
Jeff Hostetler	L.A. Raiders	14
Jim Kelly	Buffalo	18
Scott Mitchell	Miami	12
Joe Montana	Kansas City	13
Warren Moon	Houston	21
Neil O'Donnell	Pittsburgh	14
Vinny Testaverde	Cleveland	14

Source: *The World Almanac,* 1995

11. **Demographics** According to the 1990 Census, the American Indian population in the United States is 1.959 million. Many American Indian people live on reservations or trust lands. The stem-and-leaf plot shows the number of reservations in the 34 states that have them.

Stem	Leaf
0	1 1 1 1 1 1 1 1 1 1 1 1 2
•	3 3 3 3 3 4 4 4 4 7 7 8 8 9
1	1 4 9
2	3 5 7
9	6 9|6 = 96

11c. There are four western states that have more American Indian people than other states.

a. Make a box-and-whisker plot of these data.   See margin.
b. Describe the distribution of the data.   clustered with lots of outliers
c. Why do you think there are so many outliers?
d. The mean of these data is about 8.8. How does this compare with the median?   It is greater than the median.

12. **Environment** The table below shows the number of hazardous waste sites in 25 states.

State	No.	State	No.	State	No.	State	No.	State	No.
AL	13	FL	57	LA	13	OH	38	TN	17
AR	12	GA	13	MD	13	OK	11	TX	30
CA	96	IL	37	MI	77	OR	12	VA	25
CO	18	IN	33	NY	85	PA	101	WA	56
CT	16	KY	20	NC	22	SC	24	WV	6

Hazardous Waste Sites in the United States (selected states)

Source: Environmental Protection Agency, May 1994

a. Make a box-and-whisker plot of the data.   See margin.
b. What is the median number of waste sites for the states listed?   22
c. Which states, if any, are outliers?   PA
d. Which half of the data is more widely dispersed?   the upper half

13. **Education** The graph below shows the average American College Testing (ACT) Program mathematics scores for students from 1985–1993.

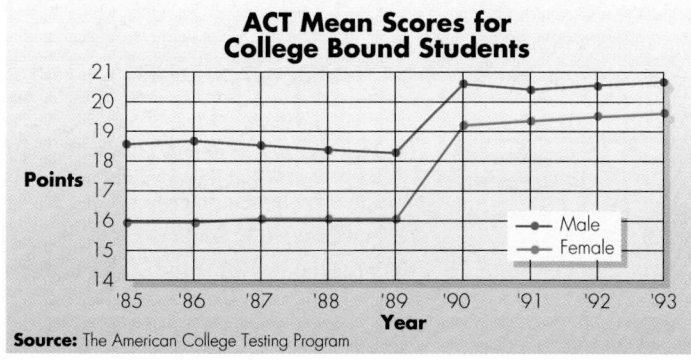

ACT Mean Scores for College Bound Students

Points / Year

— Male
— Female

Source: The American College Testing Program

(continued on the next page)

Lesson 7–7 INTEGRATION Statistics Box-and-Whisker Plots **431**

Practice Masters, p. 54

7-7   NAME_____   DATE_____
Practice   Student Edition Pages 427–43

Integration: Statistics
Box-and-Whisker Plots

Compare box-and-whisker plots A and B to answer each question.

1. What is the median of each set of data?
   Both are 70
2. Which plot has the lesser range? A
3. Which plot has the greater interquartile range? B
4. What is the upper quartile of each set of data? A: 80; B: 85
5. What is the lower quartile of each set of data? A: 65; B: 60
6. What is the least value in plot A? 50
7. What is the greatest value in plot B? 100
8. Which plot illustrates the larger range of data? B
9. What percent of the data in plot B is between 60 and 85? 50%
10. What percent of the data in plot A is greater than 80? 25%
11. What percent of the data in plot A is less than 65? 25%
12. The numbers below represent forty eighteen-hole scores of a golfer at Crystal Springs Golf Club. Make a box-and-whisker plot of this data in the space provided.

    88 80 80 77 84 74 80 79 83 77
    75 75 78 79 77 81 72 85 75 76
    78 76 77 72 75 70 78 77 72 79
    78 75 79 77 77 73 77 78 82 86

13. What is the median score? 77
14. What percent of the scores were higher than 75? 75%

Closing Activity

Speaking Ask students to describe the steps to follow when making a box-and-whisker plot.

Additional Answers

13a.

Male data are more condensed and generally higher than female data.

14a–b.

19.

Enrichment Masters, p. 54

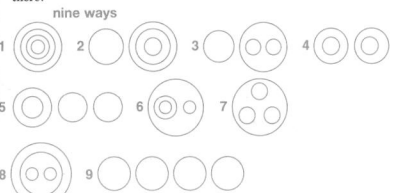
---

a. Make a box-and-whisker plot of the scores for the male students and another for the female students using the same scale. Compare the plots. **See margin.**

b. In a particular year, an entirely new ACT Assessment was given that emphasized rhetorical skills, advanced mathematics items, and a new reading test. From the data on the graph, in which year do you think this occurred? Why? **1990; See students' work.**

14. **History** Did you know that there have been more U.S. vice-presidents than presidents? As of 1995, there have been 41 presidents and 45 vice-presidents. Some presidents had more than one vice-president, and some had none. The line plot below shows the ages of vice-presidents on their inauguration days.

Source: *The World Almanac, 1995*

14a–b. See margin.

a. Make a box-and-whisker plot of these data.

b. The ages for presidents on their inauguration days have the following statistics: LV = 42, Q1 = 51, Q2 = 55, Q3 = 58, GV = 68, and 69 is an outlier. Make a box-and-whisker plot to represent these ages.

14c. The presidents seem more clustered because the box portion, representing 50% of the data, is narrower.

c. Which set of data is more clustered? Explain your answer.

d. Fourteen vice-presidents went on to become presidents. Which ages of vice-presidents were definitely not those who went on to become president? **ages 66–72, assuming no vice-president age 66 or 67 succeeded a president during the president's term**

Critical Thinking

15. The box-and-whisker plots shown below picture the distribution of test scores in two algebra classes taught by two different teachers. If you could select one of the classes to be in, which one would it be, and why?

15. Sample answer: Class A appears to be a more difficult class than B because the students don't do as well.

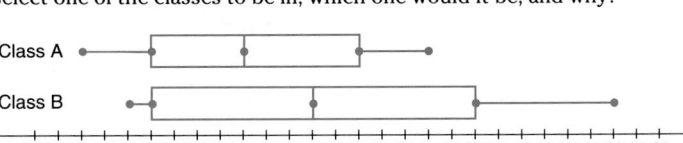

Mixed Review

16. **Travel** Greg's car gets between 18 and 21 miles per gallon of gasoline. If his car's tank holds 15 gallons, what is the range of distances that Greg can drive his car on one tank of gas? (Lesson 7–6) **270 to 315 miles**

17. Solve $2m - 3 > 7$ or $2m + 7 > 9$. (Lesson 7–4) $\{m \mid m > 1\}$

18. Write the standard form of an equation of the line that passes through $(4, 7)$ and $(1, -2)$. (Lesson 6–2) $3x - y = 5$

19. Graph $(0, 6)$, $(8, -1)$, and $(-3, 2)$. (Lesson 5–1) **See margin.**

20. Solve $\frac{6}{x - 3} = \frac{3}{4}$. (Lesson 4–1) **11**

Extension

Connections Have students research the median ages of men and women at the time of their first marriage for each decade from 1900 to the present time. Have students determine the mean, mode, quartiles, range, interquartile range, and any outliers for each set of data, and display the data using box-and-whisker plots. Then compare the plots.

7–7B Graphing Technology
Box-and-Whisker Plots

An Extension of Lesson 7–7

You can use a graphing calculator to compare two sets of data by using a double box-and-whisker plot. The plots that the calculator draws, however, do not account for outliers. If you want to use a graphing calculator to help you sketch a plot, it will be necessary for you to check for outliers and adjust the graph as needed.

In an experiment, the ability of boys and girls to identify objects held in their left hands versus those held in their right hands was tested. The left side of the body is controlled by the right side of the brain and vice versa. The results of the experiment found that the boys did not identify objects with their right hand as well as with their left. The girls could identify objects equally as well with either hand.

Texas Instruments decided to test this premise by conducting the same tests with 12 male and 10 female employees, chosen at random. Thirty small objects were selected and separated into two groups—one for the right hand and one for the left hand. Blindfolded employees felt each of the objects with the prescribed hand and tried to identify them. The results are presented in the chart below.

LOOK BACK

For more information on entering data into lists on the graphing calculator, see Lesson 5-7A.

Each employee had a score for left and right hands.

Correct Responses			
Female Left Hand	Female Right Hand	Male Left Hand	Male Right Hand
8	4	7	12
9	3	8	6
12	7	7	12
11	12	5	12
10	11	7	7
8	11	8	11
12	13	11	12
7	12	4	8
9	11	10	12
11	12	14	11
		13	9
		5	9

Source: *TI-82 Graphics Calculator Guidebook*

NCTM Standards: 1–5, 10

Objective
Use a graphing calculator to compare two sets of data by using a double box-and-whisker plot.

Recommended Time
25 minutes

Instructional Resources
Graphing Calculator Masters, pp. 24 and 25

These masters provide keystroking instruction for this lesson for the TI-81 and Casio graphing calculators.

1 FOCUS

Motivating the Lesson
Have students pair off. Select several objects that can be held in one hand. One student is to be blindfolded. The other student will place the objects in the blindfolded student's hand, one object at a time, alternating hands. The blindfolded student will try to guess what each object is. As a class, discuss whether there were any marked differences for the number of correct responses for each hand or for the results between boys and girls. Ask students what a box-and-whisker plot comparing left-hand and right-hand results might look like.

2 TEACH

Teaching Tip Point out to students that for such an experiment to be valid, many factors must be controlled—for example, the number of left-handed individuals, objects that are equally difficult to recognize, and so on.

Observing students working with technology is an excellent method of assessment.

1. women identifying objects with left hand

2. With the left hand; the data are clustered.

3. The left hand data are more clustered.

6. It does not agree. Sample reason: Adults may have better recognition skills or there may have been a large number of left-handed people in sample.

Make a double box-and-whisker plot of the data to compare the results of Female Left Hand with Female Right Hand using a graphing calculator.

Step 1 Clear lists L1, L2, L3, and L4. Enter the data from each column of the table into lists L1, L2, L3, and L4, respectively.

Step 2 Select the box-and-whisker plot and define which list will be used.

Enter: [2nd] [STAT PLOT] 1 [ENTER]   *Turns plot on.*

[▼] [▶] [▶] [ENTER]   *Selects box-and-whisker plot.*

If L1 is not highlighted in the Xlist, use the down arrow and [ENTER] to highlight it and make sure the frequency is set for 1.

Repeat the process to assign Plot2 as a box-and-whisker plot using L2.

Step 3 Clear the Y= list. Set the WINDOW settings for Xscl = 1, Ymin = 0, and Yscl = 0. Ignore the other settings. Press [ZOOM] 9 to select ZoomStat. This sets the other settings and displays the box-and-whisker plots. It will only display those plots that you have turned on.

Step 4 Use [TRACE] to examine the minX (least value), Q1 (lower quartile), Med (median), Q3 (upper quartile), and maxX (greatest value).

EXERCISES

1. Which set of data does the upper plot represent?

2. Observe the two graphs. Does it appear that the females guessed correctly more often with the left hand or the right? How do you know?

3. Reset your calculator to define Plot1 as L3 and Plot 2 as L4 to examine the males' data. What do you observe in these plots?

4. Reset the calculator to compare the left-hand results of males and females. Were the males or females better at guessing with their left hands?   **females**

5. Reset the calculator to compare the right-hand results of males and females. Which group seemed more adept at identifying objects with their right hands?   **males**

6. How do the results of this experiment compare with the study of boys and girls mentioned at the beginning of this lesson? What reasons may account for any discrepancies?

Using Technology
This lesson offers an excellent opportunity for using technology in your algebra classroom. For more information on using technology, see *Graphing Calculators in the Mathematics Classroom*, one of the titles in the Glencoe Mathematics Professional Series.

7-8A Graphing Technology
Graphing Inequalities

A Preview of Lesson 7-8

Inequalities in two variables can be graphed on a graphing calculator using the "Shade(" command, which is option 7 on the DRAW menu. You must enter *two* functions to activate the shading since the calculator always shades between two specified functions. The first function entered defines the lower boundary of the region to be shaded. The second function defines the upper boundary of the region. The calculator graphs both functions and shades between the two.

NCTM Standards: 1–6

Objective
Use a graphing calculator to graph inequalities in two variables.

Example **Graph $y \geq 2x - 3$ in the standard viewing window.**

The inequality refers to points at which y is *greater than or equal to* $2x - 3$. This means we want to shade above the graph of $y = 2x - 3$. Since the calculator screen shows only part of the coordinate plane, we can use the top of the screen, Ymax or 10, as the upper boundary and $2x - 3$ as the lower boundary.

Before using the "Shade(" option, be sure to clear any equations stored in the [Y] list, and press [ZOOM] 6 for the standard viewing window.

Enter: [2nd] [DRAW] 7 2 [X,T,θ]

[–] 3 [,] 10 [)] [ENTER]

Since both the x- and y-intercepts of the graph and the origin are within the current viewing window, the graph of the inequality is complete.

Recommended Time
15 minutes

Instructional Resources
Graphing Calculator Masters, pp. 26 and 27

These masters provide keystroking instruction for this lesson for the TI-81 and Casio graphing calculators.

Example **Graph $y - x \leq 1$ in the standard viewing window.**

First solve the inequality for y: $y \leq x + 1$. This inequality refers to points where y is *less than or equal to* $x + 1$. This means we want to shade below the graph of $y = x + 1$. We can use the bottom of the screen, Ymin or -10, as the lower boundary and $x + 1$ as the upper boundary.

Enter: [2nd] [DRAW] 7 [(–)] 10 [,]

[X,T,θ] [+] 1 [)] [ENTER]

Don't forget to clear the screen when finished.

Motivating the Lesson
Which of the following ordered pairs, (3, 1), (5, 3), (−2, −6), (−1, −8), satisfy the inequality $2x - y \geq 5$?
(3, 1), (5, 3), (−1, −8)

Teaching Tip Remind students to carefully enter the function at the appropriate time. To shade above a graph, the function must be entered as the first function. To shade below a graph, the function must be entered as the second function.

EXERCISES

Use a graphing calculator to graph each inequality. Sketch each graph on a sheet of paper. 1–9. See Solutions Manual.

1. $y \geq x + 2$	**2.** $y \leq -2x - 4$	**3.** $y + 1 \leq 0.5x$
4. $y \geq 4x$	**5.** $x + y \leq 0$	**6.** $2y + x \geq 4$
7. $3x + y \leq 18$	**8.** $y \geq 3$	**9.** $0.2x + 0.1y \leq 1$

Using Technology
This lesson offers an excellent opportunity for using technology in your algebra classroom. for more information on using technology, see *Graphing Calculators in the Mathematics Classroom,* one of the titles in the Glencoe Mathematics Professional Series.

4 ASSESS

Observing students working with technology is an excellent method of assessment.

Assignment Guide
Core: 1–9
Enriched: 1–9

NCTM Standards: 1–5

Instructional Resources

- Study Guide Master 7-8
- Practice Master 7-8
- Enrichment Master 7-8
- Assessment and Evaluation Masters, p. 185
- Multicultural Activity Masters, p. 14

 Transparency 7-8A contains the 5-Minute Check for this lesson; **Transparency 7-8B** contains a teaching aid for this lesson.

Recommended Pacing

Standard Pacing	Day 12 of 14
Honors Pacing	Day 11 of 13
Block Scheduling*	Day 6 of 7
Alg. 1 in Two Years*	Days 17 & 18 of 21

 *For more information on pacing and possible lesson plans, refer to the *Block Scheduling Booklet* and *Algebra 1 in Two Years.*

1 FOCUS

 5-Minute Check
(over Lesson 7-7)

7-8

Graphing Inequalities in Two Variables

Objective

After studying this lesson you should be able to:
- graph inequalities in the coordinate plane.

 **APPLICATION**
Manufacturing

Rapid Cycle, Inc. is a manufacturer and distributor of racing bicycles. It takes 3 hours to assemble a bicycle and 1 hour to road test a bicycle. Each technician in the company works no more than 45 hours a week. How many racing bikes can one technician assemble, and how many can he or she road test in one week?

Let x represent the number of bikes that are assembled in a week, and let y represent the number of bikes that are road-tested. Then the following inequality can be used to represent the solution.

Total time to assemble x bikes	plus	Total time to road test y bikes	is no more than	45 hours.
$3x$	$+$	y	$\leq$	45

There are an infinite number of ordered pairs that are solutions to this inequality. The easiest way to show all of these solutions is to draw a *graph* of the inequality. Before doing this, let's consider some simpler inequalities. *This problem will be solved in Example 3.*

Example **1** **From the set {(3, 4), (0, 1), (1, 4), (1, 1)}, which ordered pairs are part of the solution set for $4x + 2y < 8$?**

Let's use a table to substitute the x and y values of each ordered pair into the inequality.

x	y	4x + 2y < 8	True or False?
3	4	$4(3) + 2(4) < 8$ $20 < 8$	false
0	1	$4(0) + 2(1) < 8$ $2 < 8$	true
1	4	$4(1) + 2(4) < 8$ $12 < 8$	false
1	1	$4(1) + 2(1) < 8$ $6 < 8$	true

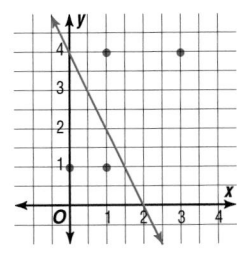

The ordered pairs {(0, 1), (1, 1)} are part of the solution set of $4x + 2y < 8$. The graph above shows the four ordered pairs of the replacement set and the equation $4x + 2y = 8$. Notice the location of the two ordered pairs that are solutions for $4x + 2y < 8$ in relation to the graph of the line.

You can use the following graphing calculator program to find out if a given ordered pair (x, y) is a solution for the inequality $5x - 3y \geq 15$.

```
PROGRAM: XYTEST
: Disp "IS (X, Y) A ","SOLUTION?"
: Prompt X,Y
: If 5X-3Y ≥ 15
: Then                  To run the program for other
: Disp "YES"            ordered pairs, simply press
: Else                  ENTER and the program will
: Disp "NO"             begin again.
```

c. {(−1, −2), (0, −2), (0, −1), (0, 0), (1, −2), (1, −1), (1, 0) (1, 1)}

Your Turn   a. See students' work.

a. Try the program for ten ordered pairs (x, y). Keep a list of which ordered pairs you tried and which ones were solutions.

b. How do you think you could change this program to test the inequality $2x + y \geq 2y$?   **Change the sentence in the If statement.**

c. Use your changed program to find the solution set if $x = \{-1, 0, 1\}$ and $y = \{-2, -1, 0, 1\}$.

The solution set for an inequality contains many ordered pairs when the domain and range are the set of real numbers. The graphs of all of these ordered pairs fill an area on the coordinate plane called a **half-plane.** An equation defines the **boundary** or edge for each half-plane. For example, suppose you wanted to graph the inequality $y > 5$ on the coordinate plane.

First determine the boundary by graphing $y = 5$.

If the inequality contains ≥ or ≤, the graph of the boundary equation would be drawn as a solid line.

Since the inequality involves only $>$, the line should be dashed. The boundary divides the coordinate plane into two half-planes.

To determine which half-plane contains the solution, choose a point from each half-plane and test it in the inequality.

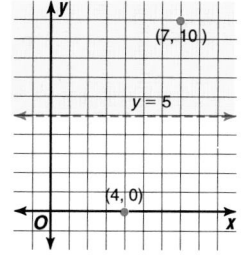

Try (7, 10).

$y > 5$ $y = 10$

$10 > 5$ true

Try (4, 0).

$y > 5$ $y = 0$

$0 > 5$ false

This graph is called an open half-plane because the boundary is not part of the graph.

The half-plane that contains (7, 10) contains the solution. Shade that half-plane.

Example **2**   **Graph $y + 2x \leq 3$.**

First solve for y in terms of x.

$$y + 2x \leq 3$$
$$y + 2x - 2x \leq 3 - 2x \quad \text{Subtract 2x from each side.}$$
$$y \leq 3 - 2x$$

(continued on the next page)

Motivating the Lesson
Situational Problem Have students consider each of the following situations.

1. Your height must be less than 4 feet to go on a kiddy ride at the amusement park.
2. The center must score at least 24 points to set a new record.
3. The cost of five pencils and three pens is at most $4.25.

Ask students to show a relation to represent each of the above sentences.

2 TEACH

In-Class Example

For Example 1
From the set {(1, 1), (2, 8), (5, 0), (8, 1)}, which ordered pairs are part of the solution set for $2x - 3y < 10$? **(1, 1), (2, 8)**

EXPLORATION

This program has a typical if-then-else structure. The *else* case handles the negation of $5x - 3y \geq 15$; that is, $5x - 3y < 15$.

Teaching Tip Point out that the graphs shown are graphs of relations but not of functions. Every x value is paired with many y values. Have students consider the vertical line test to verify that these relations are not functions.

In-Class Examples

For Example 2
Graph $3y - 2x \leq 6$.

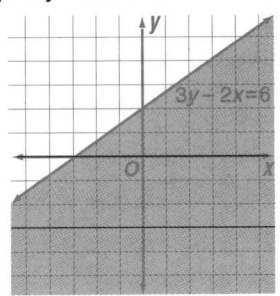

For Example 3
The drama club would like to attend a play at the local playhouse. Tickets for the play cost $15 and $20. If the club can spend no more than $240 on tickets, how many of each ticket can the club purchase? Show the inequality used to represent the problem, and graph the answer.
$15x + 20y \leq 240$

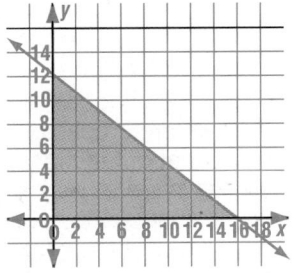

The origin is often used as a test point because the values are easy to substitute into the inequality.

Graph $y = 3 - 2x$. Since $y \leq 3 - 2x$ means $y < 3 - 2x$ or $y = 3 - 2x$, the boundary is included in the graph and should be drawn as a solid line.

Select a point in one of the half-planes and test it. For example, use the origin $(0, 0)$.

The half-plane that contains the origin should be shaded.

Check: Test a point in the other half-plane, for example $(3, 3)$.

Since the statement is false, the half-plane containing $(3, 3)$ is not part of the solution.

This graph is called a closed half-plane because the boundary line is included.

When solving real-life inequalities, the domain and range of the inequality are often restricted to nonnegative numbers or whole numbers.

Example ③

APPLICATION
Manufacturing

F Y I
The world speed record for bicycles over a 200-meter course using a flying start is 65.484 mph by Fred Markham at Mono Lake, California, set in 1986.

Refer to the application at the beginning of the lesson. How many racing bikes will the technician be able to assemble and road test?

First, solve for y in terms of x.

$3x + y \leq 45$
$3x - 3x + y \leq 45 - 3x$
$y \leq 45 - 3x$

Since the open sentence includes the equation, graph $y = 45 - 3x$ as a solid line. Test a point in one of the half-planes, for example $(0, 0)$. The half-plane containing $(0, 0)$ represents the solution since $3(0) + 0 \leq 45$ is true.

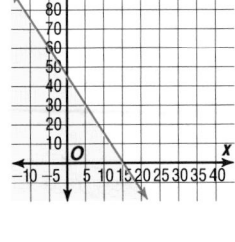

Let's examine what the solution means. The technician cannot complete negative numbers of racing bikes. So, any point in the half-plane whose x- and y-coordinates are whole numbers is a possible solution. That is, only the portion of the shading in the first quadrant is the solution for this problem. One solution is $(5, 30)$. This represents 5 bicycles assembled and 30 road tested by the technician in a 45-hour week.

F Y I

The fastest speed ever achieved on a bicycle was 152.284 mph, by John Howard at Bonneville Salt Flats, Utah, on July 20, 1985. The bicycle followed a car that was pulling a wind barrier, thus reducing air resistance.

Communicating Mathematics

Study the lesson. Then complete the following.

1. a. **Graph** $y \geq x + 1$. **See margin.**
   b. **Identify** the boundary and indicate whether it is included or not.
   c. **Identify** the half-plane that is part of the graph.
   d. **Write** the coordinates of a point not on the boundary that satisfy the inequality.

2. **Explain** how you would check whether a point is part of the graph of an inequality. **See margin.**

 MATH JOURNAL

3. **Assess Yourself** What do you think was the most challenging concept you learned in this chapter? Give an example of that concept and tell why you thought it was challenging. **See students' work.**

Guided Practice

Match each inequality with its graph.

4. $y \geq \frac{1}{2}x - 2$ **c**

5. $y \leq 0.5x - 2$ **a**

6. $y \geq \frac{2}{3}x + 2$ **d**

7. $y \leq \frac{2}{3}x + 2$ **b**

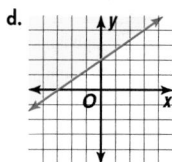

Determine which ordered pairs are solutions to the inequality. State whether the boundary is included in the graph. **8. b, c; yes**

8. $y \leq x$ a. $(-3, 2)$ b. $(1, -2)$ c. $(0, -1)$

9. $y > x - 1$ a. $(0, 0)$ b. $(2, 0)$ c. $(1, 3)$ **a, c; no**

10. Find which ordered pairs from the set $\{(-2, 2), (-2, 3), (2, 2), (2, 3)\}$ are part of the solution set for $a + b < 1$. $\{(-2, 2)\}$

Graph each inequality.   **11–14. See Solutions Manual.**

11. $y > 3$ 12. $x + y > 1$ 13. $2x + 3y \geq -2$ 14. $-x < -y$

Practice

A

Copy each graph. Shade the appropriate half-plane to complete the graph of the inequality.

15. $x > 4$ 16. $3y < x$ 17. $3x + y > 4$ 18. $2x - y \leq -2$

   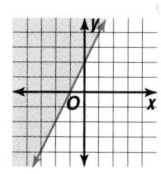

Reteaching ▬▬▬▬

Using Alternative Methods An alternate strategy in determining which half-plane contains the solution to an inequality in two variables starts by transforming the inequality so y is the left number. Shade the lower half-plane for the "$y <$" form or shade the upper half-plane for the "$y >$" form.

---

3 PRACTICE/APPLY

Check for Understanding
Exercises 1–14 are designed to help you assess your students' understanding through reading, writing, speaking, and modeling. You should work through Exercises 1–3 with your students and then monitor their work on Exercises 4–14.

Additional Answers

1a.

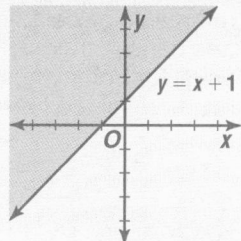

1b. The boundary is the graph of $y = x + 1$ and it is included.
1c. The half-plane above the line is shaded.
1d. Sample answer: $(-3, 3)$
2. Replace the x and y in the inequality with the values of the ordered pair. If the inequality holds true, then the ordered pair is part of the solution set.

Study Guide Masters, p. 55

Assignment Guide

Core: 15–45 odd, 46, 47, 49–56
Enriched: 16–46 even, 47–56

For **Extra Practice,** see p. 774.

The red A, B, and C flags, printed only in the Teacher's Wraparound Edition, indicate the level of difficulty of the exercises.

Using the Programming Exercises The program given in Exercises 43–45 is for use with a TI-82 graphing calculator. For other programmable calculators, have students consult their owner's manual for commands similar to those presented here.

Additional Answers

40.

41.

Practice Masters, p. 55

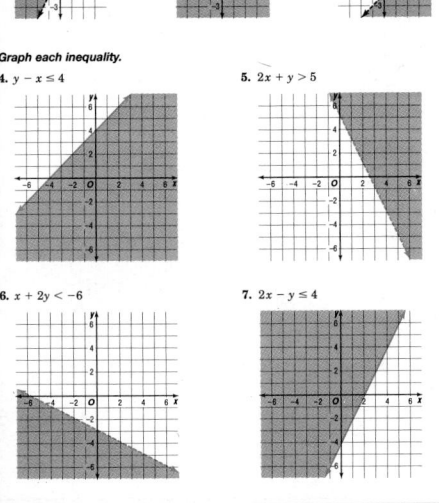

7-8   NAME_____ DATE_____
Practice   Student Edition
                                      Pages 436–441

Graphing Inequalities in Two Variables

Shade the appropriate half-plane to complete the graph of each inequality.

1. $y > 2x$ 2. $y \leq 2$ 3. $y - x < -2$

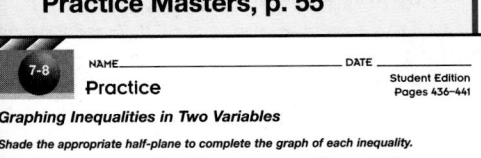

Graph each inequality.

4. $y - x \leq 4$ 5. $2x + y > 5$

6. $x + 2y < -6$ 7. $2x - y \leq 4$

Find which ordered pairs from the given set are part of the solution set for each inequality.

19. $y < 3x$, $\{(-3, 1), (-3, 2), (1, 1), (1, 2)\}$ $\{(1, 1), (1, 2)\}$

20. $y - x > 0$, $\{(1, 1), (1, 2), (4, 1), (4, 2)\}$ $\{(1, 2)\}$

B 21. $2y + x \geq 4$, $\{(-1, -3), (-1, 0), (-2, -3), (-2, 0)\}$ $\varnothing$

Graph each inequality.   22–39. See Solutions Manual.

22. $x > -5$	23. $y < -3$	24. $3y + 6 > 0$
25. $4x + 8 < 0$	26. $y \leq x + 1$	27. $x + y > 2$
28. $x + y < -4$	29. $3x - 1 \geq y$	30. $3x + y < 1$
31. $x - y \geq -1$	32. $x \leq y$	33. $-y > x$
C 34. $2x - 5y \leq -10$	35. $8y + 3x < 16$	36. $|y| \geq 2$
37. $y > |x + 2|$	38. $y > 2$ and $x < 3$	39. $y \leq -x$ and $x \geq -3$

Graphing Calculator   **Use a graphing calculator to graph each inequality. Make a sketch of the graph.**   40–42. See margin.

40. $y > x - 1$ 41. $4y + x < 16$ 42. $x - 2y < 4$

Programming   **Use the program in the Exploration to determine which pairs are solutions for each inequality.**

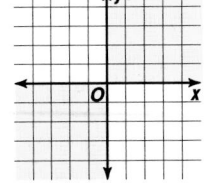

43. $x + 2y \geq 3$ **a.** $(-2, 2)$ **b.** $(4, -1)$ **c.** $(3, 1)$ **d.** $(0, 0)$ **c**

44. $2x - 3y \leq 1$ **a.** $(2, 1)$ **b.** $(5, -1)$ **c.** $(1, 1)$ **d.** $(0, 0)$

45. $-2x < 8 - y$ **a.** $(5, 10)$ **b.** $(3, 6)$ **c.** $(-4, 0)$ **d.** $(0, 0)$
a, b, d   44. a, c, d

Critical Thinking   46. What compound inequality is described by the graph at the right? Find a simple inequality that also describes this graph. $\{x \geq 0$ and $y \geq 0\}$ or $\{x \leq 0$ and $y \leq 0\}$; $xy \geq 0$

Applications and Problem Solving

47. **Health**   The graph below shows the effective heart rate ranges for each type of exercise goal.

Workout Goals for Exercise

Goals
Boost performance as a competitive athlete
Improve cardiovascular conditioning
Lose weight
Improve overall health and reduce risk of heart attack

40% 50% 60% 70% 80% 90% 100%
Target Heart Rate Range
Source: *Vitality,* May 1994

Arrio is 35 years old and just beginning a bench-stepping aerobics class. In the orientation at the beginning of the first class, he learned that during exercise an effective minimum heart rate (beats/minute) should be 70% of the difference of 220 and his age. A maximum heart rate should be 80% of the difference of 220 and his age.

Additional Answer

42.

a. Write a compound inequality that expresses the effective rate zone for a person a years of age. $\quad 0.7(220 - a) \le z \le 0.8(220 - a)$

b. In class, the participants take a break and count their heart beats for 15 seconds. What should be Arrio's effective heart rate zone for that 15-second count? $\quad 32 \le z \le 37$

c. According to the graph and the heart rate range given, what is the goal of the bench-stepping class? $\quad$ improve cardiovascular conditioning

Reported Snowmobile Accidents in Minnesota

Year	Fatal	Water-related	Nonfatal	Total
1987–1988	14	3	261	278
1988–1989	8	1	313	322
1989–1990	10	4	246	260
1990–1991	11	1	354	366
1991–1992	15	0	386	401
1992–1993	19	4	546	569
1993–1994	21	1	531	553

Source: Minnesota Department of Natural Resources

48. Snowmobiling Although snowmobiling is an exhilarating sport, it can be very dangerous. The chart below shows the reported accidents in Minnesota involving snowmobiles.

a. A study suggests that the average number of nonfatal snowmobile accidents per year nationwide is about 350. Suppose x represents the total number of fatal snowmobile accidents in Minnesota and y represents the total number of snowmobile accidents in Minnesota. In what years is $x + 350 > y$? $\quad$ 1987–88, 1988–89, 1989–90

b. Graph $x + 350 > y$. See margin.

c. The snowmobiling season in Minnesota goes from November to March. The lack of snow during the 1994–1995 season caused many to resort to riding their snowmobiles on icy lakes instead of on land. As of December 1, there had been 7 fatal accidents, 1 water-related fatal accident, and 81 nonfatal accidents. Do you think that the 1994–1995 season's accident count was greater or less than the 1993–1994 season? Explain your answer. $\quad$ See students' work.

Mixed Review

49. See margin.

49. Statistics Make a box-and-whisker plot of the total number of snowmobiling accidents shown in the chart for Exercise 48. (Lesson 7–7)

50. Solve $5 - |2x - 7| > 2$. (Lesson 7–6) $\quad \{x \mid 2 < x < 5\}$

51. Write an equation in slope-intercept of the line that is parallel to the graph of $8x - 2y = 7$ and whose y-intercept is the same as the line whose equation is $2x - 9y = 18$. (Lesson 6–4) $\quad y = 4x - 2$

$52. \ -\dfrac{12}{5}$

52. Determine the value of r so that the line passing through $(r, 4)$ and $(-4, r)$ has a slope of 4. (Lesson 6–1)

53. Graph $-y + \dfrac{2}{7}x = 1$. (Lesson 5–4) $\quad$ See margin.

54. State the inverse of the relation $\{(4, -1), (3, 2), (-4, 0), (17, 9)\}$. (Lesson 5–2) $\quad \{(-1, 4), (2, 3), (0, -4), (9, 17)\}$

55. What is 98.5% of $140.32? (Lesson 4–4) $\quad \$138.22$

56. Find three consecutive integers whose sum is 87. (Lesson 3–3) $\quad$ 28, 29, 30

Lesson 7–8 Graphing Inequalities in Two Variables **441**

Extension

Connections Find the area of the triangular region determined as the intersection of the closed half-planes $x \le 4$, $y \ge -\dfrac{1}{2}x + 3$, and $y \le \dfrac{1}{2}x + 5$.

18 square units

Additional Answer

53.

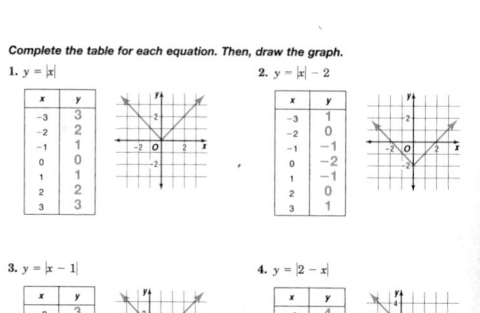

$-y + \dfrac{2}{7}x = 1$

Closing the Investigation

This activity provides students an opportunity to bring their work on the Investigation to a close. For each Investigation, students should present their findings to the class. Here are some ways students can display their work.

- Conduct and report on an interview or survey.
- Write a letter, proposal, or report.
- Write an article for the school or local paper.
- Make a display, including graphs and/or charts.
- Plan an activity.

Assessment

To assess students' understanding of the concepts and topics explored in the Investigation and its follow-up activities, you may wish to examine students' Investigation Folders.

The scoring guide provided in the *Investigations and Projects Masters,* p. 11, provides a means for you to score students' work on the Investigation.

Investigations and Projects Masters, p. 11

Scoring Guide
Chapters 6 and 7
Investigation

Level	Specific Criteria
3 Superior	• Shows thorough understanding of the concepts of *data collection and analysis, using formulas to calculate radius and volume of a sphere, making graphs, and writing equations and inequalities.* • Uses appropriate strategies to solve problems. • Computations are correct. • Written explanations are exemplary. • Charts, graphs, and report are appropriate and sensible. • Goes beyond the requirements of some or all problems.
2 Satisfactory, with Minor Flaws	• Shows understanding of the concepts of *data collection and analysis, using formulas to calculate radius and volume of a sphere, making graphs, and writing equations and inequalities.* • Uses appropriate strategies to solve problems. • Computations are mostly correct. • Written explanations are effective. • Charts, graphs, and report are appropriate and sensible. • Satisfies the requirements of problems.
1 Nearly Satisfactory, with Obvious Flaws	• Shows understanding of most of the concepts of *data collection and analysis, using formulas to calculate radius and volume of a sphere, making graphs, and writing equations and inequalities.* • May not use appropriate strategies to solve problems. • Computations are mostly correct. • Written explanations are satisfactory. • Charts, graphs, and report are appropriate and sensible. • Satisfies the requirements of problems.
0 Unsatisfactory	• Shows little or no understanding of the concepts of *data collection and analysis, using formulas to calculate radius and volume of a sphere, making graphs, and writing equations and inequalities.* • Does not use appropriate strategies to solve problems. • Computations are incorrect. • Written explanations are not satisfactory. • Charts, graphs, and report are not appropriate or sensible. • Does not satisfy the requirements of problems.

Smoke Gets In Your Eyes

Refer to the Investigation on pages 320–321. Add the results of your work below to your Investigation Folder.

The EPA's most recent long-term study on the effects of secondhand smoke shows that non-smokers married to smokers have a 19% increased risk of having lung cancer. Lung cancer is not the only danger of secondhand smoke. Twelve studies show that heart disease is another danger. Nonsmokers who are exposed to their spouses' smoke have a 30% increased chance of death from heart disease than do other nonsmokers. After reviewing a number of studies, the EPA's risk analysis has also concluded that secondhand smoke causes an extra 150,000 to 300,000 respiratory infections a year among the nations 5.5 million children under the age of 18 months.

Katharine Hammond, an environmental-health expert at the University of California, Berkeley, has also conducted a study on the *carcinogenic* components of secondhand smoke. The carcinogenic components are the parts of smoke that are known to cause cancer in humans. She found that "in the same room, at the same time, the nonsmoker is getting as much benzene (a chemical that is known to cause cancer in humans) as a smoker gets smoking six cigarettes."

James Repace and Alfred Lowery, two statistical researchers who study the effects of secondhand smoke, have concluded that a lifetime increase in lung-cancer risk of 1 in 1000 could be caused by long-term exposure to air containing more than 6.8 micrograms of nicotine per cubic meter of air.

442 Chapter 7 Investigation: Smoke Gets In Your Eyes

Analyze

You have conducted experiments and organized your data in various ways. It is now time to analyze your findings and state your conclusions.

PORTFOLIO ASSESSMENT

You may want to keep your work on this Investigation in your portfolio.

1. True secondhand smoke consists mostly of sidestream smoke. This is the smoke that comes from the smoldering cigarette. This smoke is much more toxic than inhaled smoke. How does this information affect your conclusions about the amount of smoke inhaled by nonsmokers in a room?

2. If you are a nonsmoker and live with a smoker, what are the cost factors involved? Explain your calculations.

3. Describe your personal experience with secondhand smoke.

Write

You want to inform people of the effects of secondhand smoke. You decide to write a letter to the editor of a local paper describing your investigation on the effects of secondhand smoke.

4. Use the information above and the results of your experiments and explorations to write a paper regarding the health risks of secondhand smoke.

5. You may want to do further research. The American Cancer Society and other agencies have information regarding the health risks of smoking and secondhand smoke.

6. Use data, charts, and graphs to justify your position. Use mathematics to help convince your readers of the conclusions you drew from this Investigation.

VOCABULARY

After completing this chapter, you should be able to define each term, property, or phrase and give an example or two of each.

Algebra

addition property for inequality (p. 385)
boundary (p. 437)
compound inequality (p. 405)
division property for inequality (p. 393)
half-plane (p. 437)
intersection (p. 406)
multiplication property for inequality (p. 393)
set-builder notation (p. 385)
subtraction property for inequality (p. 385)
union (p. 408)

Statistics

box-and-whisker plot (p. 427)
extreme values (p. 427)
whiskers (p. 428)

Probability

compound event (p. 414)
outcomes (p. 413)
simple events (p. 414)
tree diagram (p. 413)

Problem Solving

draw a diagram (p. 406)

UNDERSTANDING AND USING THE VOCABULARY

Choose the letter of the term that best matches each statement, algebraic expression, or algebraic sentence.

1. If $\frac{1}{2}x \le -5$, then $x \le -10$. h

2. If $8 > 4$, then $8 + 5 > 4 + 5$. b

3. $\{h \mid h > 43\}$ i

4. $x \ge -3$ or $x < -10$ c, k

5. $x \ge -4$ and $x < 2$ c, f

6. If $4x - 1 < 7$, then $4x - 4 < 4$. j

7. If $-3x < 9$, then $x > -3$. d

8. $>$ e

9. $<$ g

10. $7 > x > 1$ c, f

11. $|x + 6| > 12$ means $x + 6 > 12$ or $-(x + 6) > 12$. a

a. absolute value inequality
b. addition property for inequality
c. compound inequality
d. division property for inequality
e. greater than
f. intersection
g. less than
h. multiplication property for inequality
i. set builder notation
j. subtraction property for inequality
k. union

Chapter 7   Highlights   **443**

Instructional Resources

Three multiple-choice tests and three free-response tests are provided in the *Assessment and Evaluation Masters*. Forms 1A and 2A are for honors pacing, and Forms 1B, 1C, 2B, and 2C are for average pacing. Chapter 7 Test, Form 1B is shown at the right. Chapter 7 Test, Form 2B is shown on the next page.

Using the CHAPTER HIGHLIGHTS

The Chapter Highlights begins with a listing of the new terms, properties, and phrases that were introduced in this chapter. Have students define each term and provide an example or two of it, if appropriate.

Assessment and Evaluation Masters, pp. 171–172

---

7 NAME_____ DATE_____

Chapter 7 Test, Form 1B

Write the letter for the correct answer in the blank at the right of each problem.

1. Solve $2x - 7 \ge 3x$. **1. B**
   A. $\left\{x \mid x \le \frac{5}{7}\right\}$ B. $\{x \mid x \le -7\}$ C. $\{x \mid x \ge 7\}$ D. $\{x \mid x \ge -7\}$

2. Solve $-13 > w - (-12)$. **2. A**
   A. $\{w \mid w < -25\}$ B. $\{w \mid w > -25\}$ C. $\{w \mid w > -1\}$ D. $\{w \mid w < -1\}$

3. Solve $\frac{m}{5} < -3$. **3. B**
   A. $\{m \mid m > -15\}$ B. $\{m \mid m < -15\}$ C. $\{m \mid m < 15\}$ D. $\{m \mid m > 15\}$

4. Solve $-\frac{2}{3}s > 6$. **4. D**
   A. $\{s \mid s > -9\}$ B. $\{s \mid s > 9\}$ C. $\{s \mid s < 9\}$ D. $\{s \mid s < -9\}$

5. Solve $-1.1t \le 4.62$. **5. D**
   A. $\{t \mid t \le -5.06\}$ B. $\{t \mid t \ge -5.06\}$ C. $\{t \mid t \le -4.2\}$ D. $\{t \mid t \ge -4.2\}$

6. Solve $6d + 10 < 46$. **6. A**
   A. $\{d \mid d < 6\}$ B. $\{d \mid d > 6\}$ C. $\{d \mid d < -6\}$ D. $\{d \mid d > -6\}$

7. Solve $5z - 4 > 2z + 8$. **7. A**
   A. $\{z \mid z > 4\}$ B. $\{z \mid z < 1\}$ C. $\{z \mid z < 4\}$ D. $\{z \mid z > 1\}$

8. Solve $5w - (w - 8) > 9 + 3(2w - 3)$. **8. D**
   A. $\left\{w \mid w < \frac{11}{5}\right\}$ B. $\left\{w \mid w < -\frac{11}{5}\right\}$ C. $\{w \mid w < -4\}$ D. $\{w \mid w < 4\}$

9. Solve $(0.5)(r + 1) \le (0.6)(r - 2)$. **9. B**
   A. $\{r \mid r \ge 30\}$ B. $\{r \mid r \ge 17\}$ C. $\{r \mid r \ge 30\}$ D. $\{r \mid r \le 17\}$

10. The sum of two consecutive positive integers is at most 3. What is the greater integer? **10. D**
    A. 5 B. 1 C. 3 D. 2

11. If one coin and one die are tossed, what is P(one head and one 5)? **11. C**
    A. $\frac{1}{24}$ B. $\frac{1}{2}$ C. $\frac{1}{12}$ D. $\frac{1}{6}$

12. Renata's World Wide Web home page has 9 possible background colors and 3 possible text colors. How many combinations of background color and text color does the page have? **12. A**
    A. 27 B. 12 C. 6 D. 30

13. If 2 six-sided dice are tossed, what is the probability of getting two different numbers? **13. D**
    A. $\frac{1}{12}$ B. $\frac{1}{6}$ C. $\frac{1}{2}$ D. $\frac{5}{6}$

---

7 NAME_____ DATE_____

Chapter 7 Test, Form 1B (continued)

14. What percent of the data shown in the box-and-whisker plot at right is located between 30 and 70? **14. D**
    A. 25 B. 40 C. 50 D. 75

15. What compound sentence is graphed below? **15. D**
    A. $-2 < y < 3$ B. $-2 < y \le 3$ C. $y \ge -2$ or $y < 3$ D. $-2 \le y < 3$

16. Which of the following is a graph of the solution set of the compound sentence $x > 0$ or $x < -4$? **16. B**

17. Which of the following is the graph of the solution set of $-3 < 2x + 7 \le 13$? **17. A**

18. Which of the following is the solution set of $|x - 3| = 6$? **18. C**
    A. $\{9\}$ B. $\{-3\}$ C. $\{-3, 9\}$ D. $\{-9, 3\}$

19. Which inequality is graphed at the right? **19. A**
    A. $2x - 4y \le -6$
    B. $2x - 4y < -6$
    C. $2x - 4y \ge -6$
    D. $2x - 4y > -6$

20. Which of the following is a graph of the solution set of $|4x - 8| \le 16$? **20. C**

Bonus Graph $y = \begin{cases} 3 \text{ if } x \ge 1 \\ 2 \text{ if } x < 1 \end{cases}$. Use the coordinate plane provided.

Bonus

Using the STUDY GUIDE AND ASSESSMENT

Skills and Concepts Encourage students to refer to the objectives and examples on the left as they complete the review exercises on the right.

Assessment and Evaluation Masters, pp. 177–178

NAME _____ DATE _____

Chapter 7 Test, Form 2B

Solve each inequality.

1. $-14 \leq n + 5$ 1. $\{n \mid n \geq -19\}$

2. $3a < 6 + 4a$ 2. $\{a \mid a > -6\}$

3. $\frac{3y}{8} > -\frac{2}{5}$ 3. $\{y \mid y > \frac{16}{15}\}$

4. $-\frac{t}{6} \geq 14$ 4. $\{t \mid t \leq -84\}$

5. $-19.8 \geq 3.6y$ 5. $\{y \mid y \leq -5.5\}$

6. $4x - 5 < 2x + 11$ 6. $\{x \mid x < 8\}$

7. $1.3(c - 4) \leq 2.6 + 0.7c$ 7. $\{c \mid c \leq 13\}$

8. Write an inequality and solve: Forty less three times a number is no less than the number increased by 15.
    8. $40 - 3n \geq n + 15;\ n \leq 6\frac{1}{4}$

Define a variable, write an inequality, and solve.

9. Ray had scores of 75, 82, 94, and 77 on his first four science tests. What must he score on the next test to have an average of at least 85?   9. s = score needed; $\frac{75 + 82 + 94 + 77 + s}{5} \geq 85;$ **97 or greater**

10. Felicita's bank charges $2.25 a month plus $0.10 per check. How many checks does she write if her bank charges are always between $3.50 and $5.00?   10. c = no. of checks; $3.50 < 0.10c + 2.25 < 5.00;$ **between 13 and 27**

Solve each open sentence and graph the solution set on the number lines provided.

11. $3w < 6$ and $-5 < w$ 11. $-5 < w < 2$

12. $-4 \leq n$ or $3n + 1 < -2$ 12. {all numbers}

13. $-2 \leq x + 1 < 4$ 13. $-3 \leq x < 3$

14. $|1 - y| = 2$ 14. $\{-1, 3\}$

15. $|3 - 2x| \geq 1$ 15. $x \leq 1$ or $x \geq 2$

16. $|3w + 1| > -8$ 16. {all numbers}

NAME _____ DATE _____

Chapter 7 Test, Form 2B (continued)

17. Graph $2y - 4x < 8$. Use the coordinate plane provided.   17.

18. What percent of the data shown in the box-and-whisker plot below are between 20 and 70?   18. **75%**

(box-and-whisker plot: 10 20 30 40 50 60 70 80)

Bob, Ted, and Al each have to make a presentation in Mrs. Small's speech class. Mrs. Small chooses students for the three presentations at random. Each student speaks exactly once.

19. Draw a tree diagram to show the possible outcomes.   19.

20. What is the probability that Al speaks third?   20. $\frac{1}{3}$

21. One spinner is divided into 4 sections labeled 0–3. A second spinner is divided into 3 sections labeled 0, 1, and 5. Customers at a department store spin both spinners to determine the amount they will save on large purchases; the first spinner gives the tens digit and the second spinner gives the ones digit of the discount. What is the probability that a customer will save at least $10?   21. $\frac{3}{4}$

Use the following math quiz scores for exercises 22 and 23.

3, 3, 4, 5, 5, 6, 6, 6, 6, 7, 7, 7, 8, 8, 9, 10

22. Find the median.   22. **6**

23. Find the upper quartile.   23. **7.5**

Which ordered pairs from the given set are part of the solution set for each inequality?

24. $y > \frac{x}{2}$ {(−4, −2), (−6, −2), (0, 2), (0, 0), (6, 4), (6, 3)}   24. {(−6, −2), (0, 2), (6, 4)}

25. $2y - x > 5$ {(−1, 2), (−2, 4), (0, 2), (0, 3), (1, 3), (1, 4)}   25. {(−2, 4), (0, 3), (1, 4)}

Bonus Graph the solution set of the compound inequality $\frac{n+4}{5} < 3$ and $\frac{n}{5} + 4 > 3$.   **Bonus** $\frac{n}{-5} < n < 11$

OBJECTIVES AND EXAMPLES

Upon completing this chapter, you should be able to:

• solve inequalities by using addition and subtraction (Lesson 7–1)

$$56 > m + 16$$
$$56 - 16 > m + 16 - 16$$
$$40 > m$$
$$\{m \mid m < 40\}$$

12. $\{r \mid r > -12\}$ 13. $\{n \mid n > -35\}$
14. $\{t \mid t \leq 6\}$ 15. $\{p \mid p \leq -18\}$
16. $n - 3 \geq 2, \{n \mid n \geq 5\}$
17. $3n > 4n - 8, \{n \mid n < 8\}$

• solve inequalities by using multiplication and division (Lesson 7–2)

$$\frac{-5}{6}m > 25$$
$$\frac{-6}{5}\left(\frac{-5}{6}m\right) > \frac{-6}{5}(25)$$
$$m < -30$$
$$\{m \mid m < -30\}$$

20. $\left\{k \mid k \geq \frac{1}{5}\right\}$ 21. $\{x \mid x > 32\}$
22. $6n \leq 32.4, \{n \mid n \leq 5.4\}$
23. $-\frac{3}{4}n \leq 30, \{n \mid n \geq -40\}$

• solve linear inequalities involving more than one operation (Lesson 7–3)

$$15b - 12 > 7b + 60$$
$$15b - 7b - 12 > 7b - 7b + 60$$
$$8b - 12 > 60$$
$$8b - 12 + 12 > 60 + 12$$
$$8b > 72$$
$$\frac{8b}{8} > \frac{72}{8}$$
$$b > 9$$
$$\{b \mid b > 9\}$$

26. $\{r \mid r > 1.8\}$ 27. $\left\{y \mid y \leq -\frac{9}{2}\right\}$

REVIEW EXERCISES

Use these exercises to review and prepare for the chapter test.

Solve each inequality. Then check your solution.

12. $r + 7 > -5$ 13. $-35 + 6n < 7n$
14. $2t - 0.3 \leq 5.7 + t$ 15. $-14 + p \geq 4 - (-2p)$

Define a variable, write an inequality, and solve each problem. Then check your solution.

16. The difference of a number and 3 is at least 2.

17. Three times a number is greater than four times the number less eight.

Solve each inequality. Then check your solution. 18. $\{x \mid x \geq -8\}$ 19. $\{w \mid w \leq -15\}$

18. $7x \geq -56$ 19. $90 \leq -6w$
20. $\frac{2}{3}k \geq \frac{2}{15}$ 21. $9.6 < 0.3x$

Define a variable, write an inequality, and solve each problem. Then check your solution.

22. Six times a number is at most 32.4.

23. Negative three fourths of a number is no more than 30.

Find the solution set of each inequality if the replacement set for each variable is $\{-5, -4, -3, \ldots 3, 4, 5\}$.

24. $\frac{x - 5}{3} > -3$ $\{-3, -2, \ldots 4, 5\}$
25. $3 \leq -4x + 7$ $\{-5, -4, \ldots 0, 1\}$

Solve each inequality. Then check your solution.

26. $2r - 3.1 > 0.5$ 27. $4y - 11 \geq 8y + 7$
28. $-3(m - 2) > 12$ $\{m \mid m < -2\}$
29. $-5x + 3 < 3x + 23$ $\left\{x \mid x > -\frac{5}{2}\right\}$
30. $4(n - 1) < 7n + 8$ $\{n \mid n > -4\}$
31. $0.3(z - 4) \leq 0.8(0.2z + 2)$ $\{z \mid z \leq 20\}$

GLENCOE Technology

Test and Review Software

You may use this software, a combination of an item generator and item bank, to create your own tests or worksheets. Types of items include free response, multiple choice, short answer, and open ended.

For IBM & Macintosh

Additional Answers

33.
-4-3-2-1 0 1 2 3 4

OBJECTIVES AND EXAMPLES

- solve compound inequalities and graph their solution sets (Lesson 7-4)

$$2a > a - 3 \quad \text{and} \quad 3a < a + 6$$
$$2a - a > a - a - 3 \qquad 3a - a < a - a + 6$$
$$a > -3 \qquad\qquad 2a < 6$$
$$\qquad\qquad \frac{2a}{2} < \frac{6}{2}$$
$$\qquad\qquad a < 3$$

$$\{a \mid -3 < a < 3\}$$

-4 -3 -2 -1 0 1 2 3 4

- find the probability of a compound event (Lesson 7-5)

Draw a tree diagram to show the possibilities for boys and girls in a family of 3 children. Assume that the probabilities for girls and boys being born are the same.

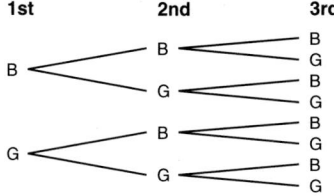

1st	2nd	3rd

The probability that the family has exactly 3 girls is $\frac{1}{8}$ or 0.125, because there is 1 way out of 8 for this to happen. The probability that the family has exactly 2 boys and 1 girl is $\frac{3}{8}$ or 0.375, because there are 3 ways out of 8 for this to happen.

- solve open sentences involving absolute value and graph the solutions (Lesson 7-6)

$$|2x + 1| > 1$$
$$2x + 1 > 1 \qquad \text{or} \qquad 2x + 1 < -1$$
$$2x + 1 - 1 > 1 - 1 \qquad 2x + 1 - 1 < -1 - 1$$
$$\frac{2x}{2} > \frac{0}{2} \qquad\qquad \frac{2x}{2} < \frac{-2}{2}$$
$$x > 0 \qquad\qquad x < -1$$

$$\{x \mid x > 0 \text{ or } x < -1\}$$

-4 -3 -2 -1 0 1 2 3

REVIEW EXERCISES

Solve each compound inequality. Then graph the solution set. 33–36. See margin for graphs.

32. $x - 5 < -2$ and $x - 5 > 2$ ∅

33. $2a + 5 \le 7$ or $2a \ge a - 3$ $\{a \mid a \text{ is a real number}\}$

34. $4r \ge 3r + 7$ and $3r + 7 < r + 29$ $\{r \mid 7 \le r < 11\}$

35. $-2b - 4 \ge 7$ or $-5 + 3b \le 10$ $\{b \mid b \le 5\}$

36. $a \ne 6$ and $3a + 1 > 10$ $\{a \mid a > 3 \text{ and } a \ne 6\}$

37. With each shrimp, salmon, or crab dinner at the Seafood Palace, you may have soup or salad. With shrimp, you may have broccoli or a baked potato. With salmon, you may have rice or broccoli. With crab, you may have rice, broccoli, or a potato. If all combinations are equally likely, find the probability of an order containing each item.

   a. salmon   **b.** soup

   c. rice   **d.** shrimp and rice

   e. salad and broccoli

   f. crab, soup, and rice

38. Matthew has 2 brown and 4 black socks in his dresser. While dressing one morning, he pulled out 2 socks without looking. What is the probability that he chose a matching pair?
$$\frac{7}{15} \approx 0.4\overline{6}$$

37a. $\frac{2}{7}$ **b.** $\frac{1}{2}$ **c.** $\frac{2}{7}$ **d.** 0   **e.** $\frac{3}{14}$ **f.** $\frac{1}{14}$

Solve each open sentence. Then graph the solution set. 39–44. See margin for graphs.

39. $|y + 5| > 0$ $\{y \mid y > -5 \text{ or } y < -5\}$

40. $|1 - n| \le 5$ $\{n \mid 6 \ge n \ge -4\}$

41. $|4k + 2| \le 14$ $\{k \mid 3 \ge k \ge -4\}$

42. $|3x - 12| < 12$ $\{x \mid 0 < x < 8\}$

43. $|13 - 5y| \ge 8$ $\left\{y \mid y \ge \frac{21}{5} \text{ or } y \le 1\right\}$

44. $\left|2p - \frac{1}{2}\right| > \frac{9}{2}$ $\left\{p \mid p > \frac{5}{2} \text{ or } p < -2\right\}$

34.
5 6 7 8 9 10 11 12 13

35.
-1 0 1 2 3 4 5 6 7

36.
1 2 3 4 5 6 7 8 9

39.
-9-8-7-6-5-4-3-2-1

40.
-8-6-4-2 0 2 4 6 8 10

41.
-5-4-3-2-1 0 1 2 3 4 5

42.
-4-2 0 2 4 6 8 10 12

43.
-2-1 0 1 2 3 4 5 6

44.
-4-3-2-1 0 1 2 3 4

Additional Answers

45.

50.

51.

52.

53.

CHAPTER 7   STUDY GUIDE AND ASSESSMENT

OBJECTIVES AND EXAMPLES

• display and interpret data on box-and-whisker plots (Lesson 7–7)

The following high temperatures were recorded during a two-week cold spell in St. Louis. Make a box-and-whisker plot of the temperatures.

20° 2° 12° 5° 4° 16° 17°

7° 6° 16° 5° 0° 5° 30°

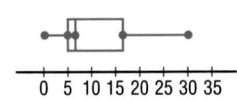

0 5 10 15 20 25 30 35

• graph inequalities in the coordinate plane (Lesson 7–8)

Graph $2x + 3y < 9$.

50–53. See margin for graphs.

REVIEW EXERCISES

The number of calories in a serving of french fries at 13 restaurants are 250, 240, 220, 348, 199, 200, 125, 230, 274, 239, 212, 240, and 327.   45. See margin.

45. Make a box-and-whisker plot of these data.

46. Are there any outliers? If so, name them.
   yes, 348

Find which ordered pairs from the given set are part of the solution set for each inequality.

47. $3x + 4y < 7$, $\{(1, 1), (2, -1), (-1, 1), (-2, 4)\}$

48. $4y - 8 \geq 0$, $\{(5, -1), (0, 2), (2, 5), (-2, 0)\}$

49. $-2x < 8 - y$, $\{(5, 10), (3, 6), (-4, 0), (-3, 6)\}$

47. $\{(2, -1), (-1, 1)\}$ **48.** $\{(0, 2), (2, 5)\}$

49. $\{(5, 10), (3, 6)\}$

Graph each inequality.

50. $x + 2y > 5$ **51.** $4x - y \leq 8$

52. $\frac{1}{2}y \geq x + 4$ **53.** $3x - 2y < 6$

APPLICATIONS AND PROBLEM SOLVING

54. Number Theory   The sum of three consecutive integers is less than 100. Find the three integers with the greatest sum. (Lesson 7–3)   **32, 33, 34**

56a. $\frac{1}{2}$ or 0.5   56b. $\frac{1}{4}$ or 0.25   56c. $\frac{1}{8}$ or 0.125

55. Shipping   An empty book crate weighs 30 pounds. The weight of a book is 1.5 pounds. For shipping, the crate must weigh at least 55 pounds and no more than 60 pounds. What is the acceptable number of books that can be packed in the crate? (Lesson 7–4)
17 to 20 books

56. Automobiles   An automobile dealer has cars available painted red or blue, with 4-cylinder or 6-cylinder engines, and with manual or automatic transmissions.   (Lesson 7–5)

   a. What is the probability of selecting a car with manual transmission?

   b. What is the probability of selecting a car with a 4-cylinder engine and a manual transmission?

   c. What is the probability of selecting a blue car with a 6-cylinder engine and an automatic transmission?

ALTERNATIVE ASSESSMENT

COOPERATIVE LEARNING PROJECT

Statistics In this chapter, you learned how to make and interpret a box-and-whisker plot. Making a box-and-whisker plot and interpreting the data from a box-and-whisker plot are two different skills, however. One can go through the routine of drawing the plot but not be able to use the data portrayed by the plot to answer appropriate questions.

For this project, suppose there are two companies that manufacture window glass. They have each submitted a bid to a contractor who is building a library. Since glass that varies in thickness can cause distortions, the contractor has decided to measure the thickness of panes of glass from each factory, at several locations on each pane. The table below shows measurements for the two panes, one from each manufacturer.

Glass Thickness (mm)	
Company A	**Company B**
10.2	9.4
12.0	13.0
11.6	8.2
10.1	14.9
11.2	12.6
9.7	7.7
10.7	13.2
11.6	12.2
10.4	10.2
9.8	9.5
10.6	9.9
10.3	9.7
8.5	11.5
10.2	11.5
9.7	10.5
9.2	10.6
8.6	6.4
11.3	13.5

Prepare a stem-and-leaf plot and a box-and-whisker plot to organize and compare the two sets of data.

Follow these steps to organize your data.

- Determine how to set up the stem-and-leaf plot using decimals.

- Determine if using a common stem would be useful in comparing the two sets of data.

- Compare the ranges of the two sets of data.

- For which company is there "bunching" or "spreading out evenly" of the data?

- Compare the stem-and-leaf plot shape with the box-and-whisker plot shape.

- Compare the middle half of the data for each company.

- Write a comparative description of the sets of data and determine, with support, which company's glass should have less distortion.

THINKING CRITICALLY

- Why are multiplication and division the only two out of the four operations for which it is necessary to distinguish between positive and negative numbers when solving linear inequalities?

- Under what conditions will the compound sentence $x < a$ and $-a < x$ have no solutions?

PORTFOLIO

Select one of the assignments from this chapter for which you felt organization of the problem and reevaluation of the answer were important in order to get an accurate answer. Revise your work as necessary and place it in your portfolio. Explain why organization and reevaluation were important.

SELF EVALUATION

Do you look beyond the obvious in your math answers? Many times math students will work through a math problem rather routinely and not evaluate or check their answer. An answer must make sense and be accurate.

Assess yourself. Do you take the obvious solution as the whole answer or do you evaluate your answers for accuracy and rationalness? List two problems in mathematics and/or your daily life whereby the obvious answer was incorrect, so you needed to evaluate your solution for accuracy.

Assessment and Evaluation Masters, pp. 182, 193

7 NAME_____ DATE _____

Chapter 7 Performance Assessment

Instructions: *Demonstrate your knowledge by giving a clear, concise solution to each problem. Be sure to include all relevant drawings and justify your answers. You may show your solution in more than one way or investigate beyond the requirements of the problem.*

1. An architect is designing a house for the Frazier family. In the design he must consider the desires of the family and the local building codes. The rectangular lot on which the house will be built has 91 feet of frontage on a lake and is 158 feet deep.

   a. The building codes state that one can build no closer than 10 feet to the lot line. Write an inequality and solve to see how long the front of the house facing the lake may be.

   b. The Fraziers requested that the house contain no less than 2800 ft² and no more than 3200 ft² of floor space. Write an inequality to represent the range of permissible widths for the house.

   c. The Fraziers have asked that the cost of the house be about $175,000 and are willing to deviate from this price no more than $20,000. Write an open sentence involving an absolute value and solve. Give the meaning of the answer.

2. a. Write a word problem involving an inequality with more than one operation.

   b. Solve and give the meaning of the answer.

3. Students on a band trip are given a soft drink and a sandwich chosen at random for lunch. They may then trade with other band members if they wish to do so. Students have an equal chance of getting root beer or orange drink and peanut butter and jelly, bologna, or ham sandwich. Use a tree diagram to find the probability of getting an orange drink and meat sandwich. Give your reasoning.

4. Describe the data set used to make this box-and-whisker plot. What can't you tell about the data by looking at the box-and-whisker plot?

Scoring Guide
Chapter 7
Performance Assessment

Level	Specific Criteria
3 Superior	· Shows thorough understanding of the concepts of *inequality, compound inequality, absolute value, solving inequalities, compound events,* and *box-and-whisker plot.* · Uses appropriate strategies to solve problems. · Computations are correct. · Written explanations are exemplary. · Word problem concerning inequality with more than one operation is appropriate and makes sense. · Goes beyond requirements of some or all problems.
2 Satisfactory, with Minor Flaws	· Shows understanding of the concepts of *inequality, compound inequality, absolute value, solving inequalities, compound events,* and *box-and-whisker plot.* · Uses appropriate strategies to solve problems. · Computations are mostly correct. · Written explanations are effective. · Word problem concerning inequality with more than one operation is appropriate and makes sense. · Satisfies all requirements of problems.
1 Nearly Satisfactory, with Serious Flaws	· Shows understanding of most of the concepts of *inequality, compound inequality, absolute value, solving inequalities, compound events,* and *box-and-whisker plot.* · May not use appropriate strategies to solve problems. · Computations are mostly correct. · Written explanations are satisfactory. · Word problem concerning inequality with more than one operation is mostly appropriate and sensible. · Satisfies most requirements of problems.
0 Unsatisfactory	· Shows little or no understanding of the concepts of *inequality, compound inequality, absolute value, solving inequalities, compound events,* and *box-and-whisker plot.* · May not use appropriate strategies to solve problems. · Computations are incorrect. · Written explanations are not satisfactory. · Word problem concerning inequality with more than one operation is not appropriate or sensible. · Does not satisfy requirements of problems.

Alternative Assessment

The Alternative Assessment section provides students with the opportunity to assess their own work by thinking critically, working with others, keeping a portfolio, and honestly evaluating their own progress. For more information on alternative forms of assessment, see *Alternative Assessment in the Mathematics Classroom*, one of the titles in the Glencoe Mathematics Professional Series.

Performance Assessment

Performance Assessment tasks for this chapter are included in the *Assessment and Evaluation Masters*. A scoring guide is also provided.

Ready, Set, Drop!

NCTM Standards: 1–6, 9

This Investigation is designed to be completed over several days or weeks. It may be considered optional. You may want to assign the Investigation and the follow-up activities to be completed at the same time.

Objective
Use mathematics to examine the relationship between the speed of descent and the size of a hang glider.

Mathematical Overview
This Investigation will use the following mathematical skills and concepts from Chapters 8 and 9.
- making charts and graphs
- solving systems of equations and inequalities
- adding and subtracting polynomials
- using the looking for a pattern strategy
- checking for hidden assumptions

Recommended Time

Part	Pages	Time
Investigation	448–449	1 class period
Working on the Investigation	461, 519, 541	20 minutes each
Closing the Investigation	548	1 class period

Instructional Resources

Investigations and Projects Masters, pp. 13–16

A recording sheet, teacher notes, and scoring guide are provided for each investigation in the *Investigations and Projects Masters.*

1 MOTIVATION

This Investigation uses simple items to investigate hang gliders. Ask students if any of them have ever seen a hang glider on TV or in person. Discuss the importance of math in the construction of a safe hang glider.

MATERIALS NEEDED

construction paper

metric ruler

paper clips

scissors

stopwatch

tape

tissue paper

washers

wire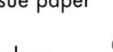

Hang gliding became popular in the United States in the early 1970s. In most states, a hang gliding certification is required before you are allowed to participate in the sport. The U.S. Hang Gliding Association is located in Los Angeles and certifies instructors and safety officers to train would-be hang gliders.

A hang glider looks like a manned kite. It consists of a triangular sail of synthetic fabric attached to an aluminum frame. The pilot hangs from a harness and steers the glider with a control bar that adjusts as the pilot shifts his or her body weight.

Hang gliders can be launched in several ways. The pilot can hold the glider and run down a hill until the glider is airborne. In areas with high cliffs, the pilot can run and jump from the cliff's edge, using the air currents below to fly. In flatter landscapes, the glider is often launched by towing it with a rope from a truck or boat and releasing it at an altitude of 400–500 feet.

Imagine that you are an engineer for an aeronautical engineering firm. A group of people who are interested in hang gliding have asked your firm to design a hang glider that can be used for recreational purposes. Your task is to design a hang glider that is as compact as possible, yet is safe for flight and landings. You have no previous experience designing hang gliders. You don't know what size hang glider is needed or whether or not the size of a hang glider depends on the size of its load. (The people range in size.)

With so many unknowns, you decide to conduct some tests to understand the principles involved. In this Investigation, you will use mathematics to examine the relationship between the speed of descent and the size of a hang glider. As part of a three-member research team, you will use tissue-paper triangles to study hang gliders.

Make an Investigation Folder in which you can store all of your work on this Investigation for future use.

448 *Investigation: Ready, Set, Drop!*

 ## Cooperative Learning

This Investigation offers an excellent opportunity for using cooperative learning groups. For more information on cooperative learning strategies and group management, see *Cooperative Learning in the Mathematics Classroom,* one of the titles in the Glencoe Mathematics Professional Series.

TRIANGLE TEST

Test	5 cm	10 cm	20 cm	35 cm
perimeter				
surface area				
1				
2				
3				
4				
5				

····· **THE EXPERIMENT**

1 Begin by copying the chart above.

2 Cut out four equilateral triangles from tissue paper. The sides of the triangles should be 5, 10, 20, and 35 centimeters long, respectively. These triangles will serve as models of hang gliders. Find the perimeter and surface area of each of these triangles and record them in your chart.

3 Measure a height of five feet on a wall. Mark this height with a piece of masking tape.

4 Hold the smallest triangle parallel to the ground at a height of 5 feet. Have a second person ready to use a stopwatch to time how long it takes the triangle to reach the floor. A third person should give the verbal command, "Ready, set, drop." At the drop command, the person holding the tissue paper glider should let go of the paper. The timer starts the stopwatch at the verbal command and stops it when the glider hits the ground. Repeat this process until five drops have been made, recording your data after each drop.

5 Repeat Step 4 for the other three gliders, recording the data for each drop.

6 Review the data that you collected. What observations can you make? Are there any relationships that you can see from the data?

Do the perimeter and surface area have a relationship with the time of the drop? Explain.

You will continue working on this Investigation throughout Chapters 8 and 9.

Be sure to keep your triangle models, charts, and other materials in your Investigation Folder.

Ready, Set, Drop! Investigation

Working on the Investigation
Lesson 8–1, p. 461
··················
Working on the Investigation
Lesson 9–4, p. 519
··················
Working on the Investigation
Lesson 9–7, p. 541
··················
Closing the Investigation
End of Chapter 9, p. 548
··················

Investigation: Ready, Set, Drop! **449**

You may wish to have a student read the first three paragraphs of the Investigation to provide information about hang gliders. Then read the next two paragraphs to introduce the activity. Discuss the activity with your students. Then separate the class into groups of four.

3 MANAGEMENT

Each group member should be responsible for a specific task.

Recorder Collects data.
Measurer Measures sides of triangles and height on the wall.
Triangles Builder Makes triangles and drops them from indicated height.
Timer Times how long it takes each triangle to reach the floor.

At the end of the activity, each member should turn in his or her respective equipment.

Sample Answers

Answers will vary as they are based on the size of the hang glider.

Investigations and Projects Masters, p. 16

4, 5   NAME_____ DATE _____

Investigation, Chapters 8 and 9   Student Edition Pages 448–449. 461, 519, 541, 548

Ready, Set, Drop!

Use this chart to record your data from Working on the Investigation, page 461.

	Launch Method	
Launch Trial	Launch Height	Horizontal Distance
1		
2		
3		
4		
5		
6		
7		
8		
9		
10		
Linear Equation:		
Slope:		

Use this chart to record your data from Working on the Investigation, page 541.

Triangle	Average glide time
1	
2	
3	
4	

Work with your group to answer the following questions and add others to be considered.

· You have investigated the effect of forward motion on the hang gliders. Using that information, what recommendation would you make to a hang glider in regards to their starting motions?

· What predictions about glide speed would you make for a hang glider that is larger than 32 feet wide?

Solving Systems of Linear Equations and Inequalities

PREVIEWING THE CHAPTER

This chapter introduces students to systems of linear equations by having them solve systems by graphing and classify systems as consistent or inconsistent and independent or dependent. Students learn to use algebraic methods, including the substitution method and the elimination method with addition or subtraction and with multiplication and addition. They also learn to assess systems of equations to decide which solution method to employ. Graphing calculator technology is integrated throughout to help students graph systems of equations. The chapter concludes with students solving systems of inequalities by graphing.

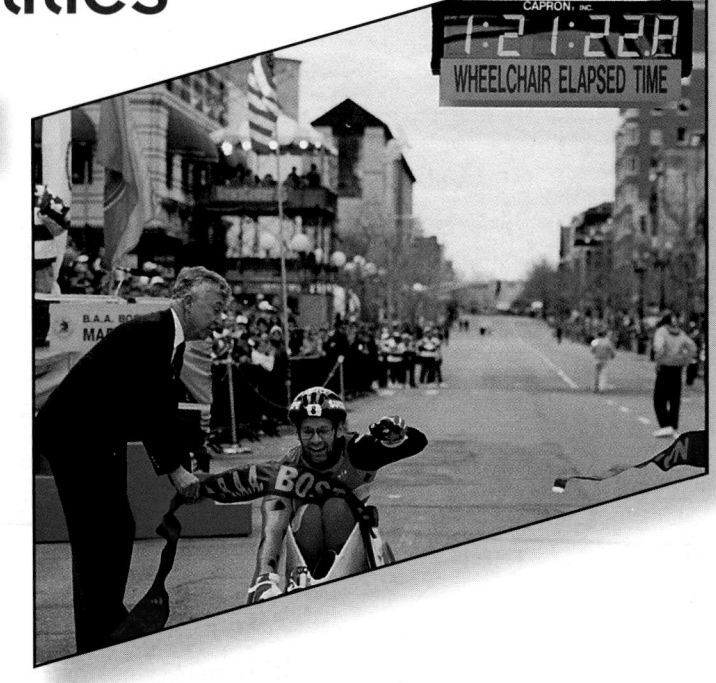

Lesson (Pages)	Lesson Objectives	NCTM Standards	State/Local Objectives
8-1A (452–453)	Use a graphing calculator to solve systems of equations.	1–5, 8	
8-1 (454–461)	Solve systems of equations by graphing. Determine by graphing whether a system of equations has one solution, no solutions, or infinitely many solutions.	1–5, 8	
8-2 (462–468)	Solve systems of equations by the substitution method. Organize data to solve problems.	1–5	
8-3 (469–474)	Solve systems of equations by the elimination method using addition or subtraction.	1–5	
8-4 (475–481)	Solve systems of equations by the elimination method using multiplication and addition.	1–5	
8-5 (482–486)	Solve systems of inequalities by graphing.	1–5, 8	

A complete, 1-page lesson plan is provided for each lesson in the *Lesson Planning Guide*. Answer keys for each lesson are available in the *Answer Key Masters*.

ORGANIZING THE CHAPTER

You may want to refer to the **Course Planning Calendar** on page T12 for detailed information on pacing.
PACING: Standard—9 days; **Honors**—8 days; **Block**—5 days; **Two Years**—14 days

LESSON PLANNING CHART

| Lesson (Pages) | Materials/ Manipulatives | Extra Practice (Student Edition) | BLACKLINE MASTERS | | | | | | | | | | Real-World Applications | Interactive Mathematics Tools Software | Teaching Transparencies |
			Study Guide	Practice	Enrichment	Assessment and Evaluation	Modeling Mathematics	Multicultural Activity	Tech Prep Applications	Graphing Calculator	Science and Math Lab Manual			
8-1A (452–453)	graphing calculator									pp. 28, 29				
8-1 (454–461)	geoboard rubber bands graphing calculator	p. 774	p. 56	p. 56	p. 56		p. 79		p. 15	p. 8		21	8-1	8-1A 8-1B
8-2 (462–468)	equation mat* cups and counters*	p. 774	p. 57	p. 57	p. 57	p. 212					pp. 35–38			8-2A 8-2B
8-3 (469–474)	graphing calculator	p. 775	p. 58	p. 58	p. 58	pp. 211, 212								8-3A 8-3B
8-4 (475–481)	graphing calculator	p. 775	p. 59	p. 59	p. 59	p. 213		p. 15	p. 16			22		8-4A 8-4B
8-5 (482–486)		p. 775	p. 60	p. 60	p. 60	p. 213	pp. 57–59		p. 16					8-5A 8-5B
Study Guide/ Assessment (487–491)						pp. 197– 210, 214– 216								

*Included in Glencoe's Student Manipulative Kit and Overhead Manipulative Resources.

ORGANIZING THE CHAPTER

OTHER CHAPTER RESOURCES

Student Edition
Investigation, pp. 448–449
Mathematics and Society, p. 481
Working on the Investigation,
   p. 461

Teacher's Classroom Resources
Investigations and Projects Masters,
   pp. 53–56
Algebra and Geometry Overhead
   Manipulative Resources,
   pp. 25–26

Technology
Test and Review Software (IBM
   and Macintosh)
CD-ROM Interactions (Windows
   and Macintosh)

Professional Publications
Block Scheduling Booklet
Glencoe Mathematics Professional
   Series

OUTSIDE RESOURCES

Books/Periodicals
Developing Skills in Algebra One, Book C, Dale
   Seymour Publications
Problem-Solving Experiences in Algebra,
   Addison-Wesley

Software
Alge-Blaster Plus, Davidson
Tools of Mathematics: Algebra, William K. Bradford
   Publishing Company

Videos/CD-ROMs
Solving Equations 2, The Wisconsin Foundation for
   Vocational, Technical, and Adult Education
*Solving Simultaneous Equations and Inequalities
   Algebraically and Geometrically,* Nasco

ASSESSMENT RESOURCES

Student Edition
Math Journal, pp. 478, 484
Mixed Review, pp. 461, 468,
   474, 481, 486
Self Test, p. 474
Chapter Highlights, p. 487
Chapter Study Guide and
   Assessment, pp. 488–490
Alternative Assessment, p. 491
   Portfolio, p.491

Cumulative Review, pp. 492–493

Teacher's Wraparound Edition
5-Minute Check, pp. 454, 462,
   469, 475, 482
Check for Understanding, pp. 458,
   466, 472, 478, 484
Closing Activity, pp. 461, 468,
   473, 481, 486
Cooperative Learning, pp. 471,
   477

Assessment and Evaluation Masters
Multiple-Choice Tests, Forms 1A
   (Honors), 1B (Average), 1C
   (Basic), pp. 197–202
Free-Response Tests, Forms 2A
   (Honors), 2B (Average), 2C
   (Basic), pp. 203–208
Calculator-Based Test, p. 209
Performance Assessment, p. 210
Mid-Chapter Test, p. 211
Quizzes A–D, pp. 212–213
Standardized Test Practice, p. 214
Cumulative Review, pp. 215–216

ENHANCING THE CHAPTER

Examples of some of the materials for enhancing Chapter 8 are shown below.

DIVERSITY
Multicultural Activity Masters, pp. 15, 16

APPLICATIONS
Real-World Applications, 21, 22

TECHNOLOGY
Graphing Calculator Masters, p. 8

TECH PREP
Tech Prep Applications Masters, pp. 15, 16

CONNECTIONS
Science and Math Lab Manual, pp. 35–38

PROBLEM SOLVING
Problem of the Week Cards, 22, 23, 24

CHAPTER

8

Solving Systems of Linear Equations and Inequalities

Objectives

In this chapter, you will:

- graph systems of equations,
- solve systems of equations using various methods,
- organize data to solve problems, and
- solve systems of inequalities by graphing.

Do you dream of starring in the NBA or playing for the New York Yankees? Do you devote all your leisure time to mainly one sport? Maybe you should try something new in the world of sports. Lacrosse, in-line skating, or judo, anyone?

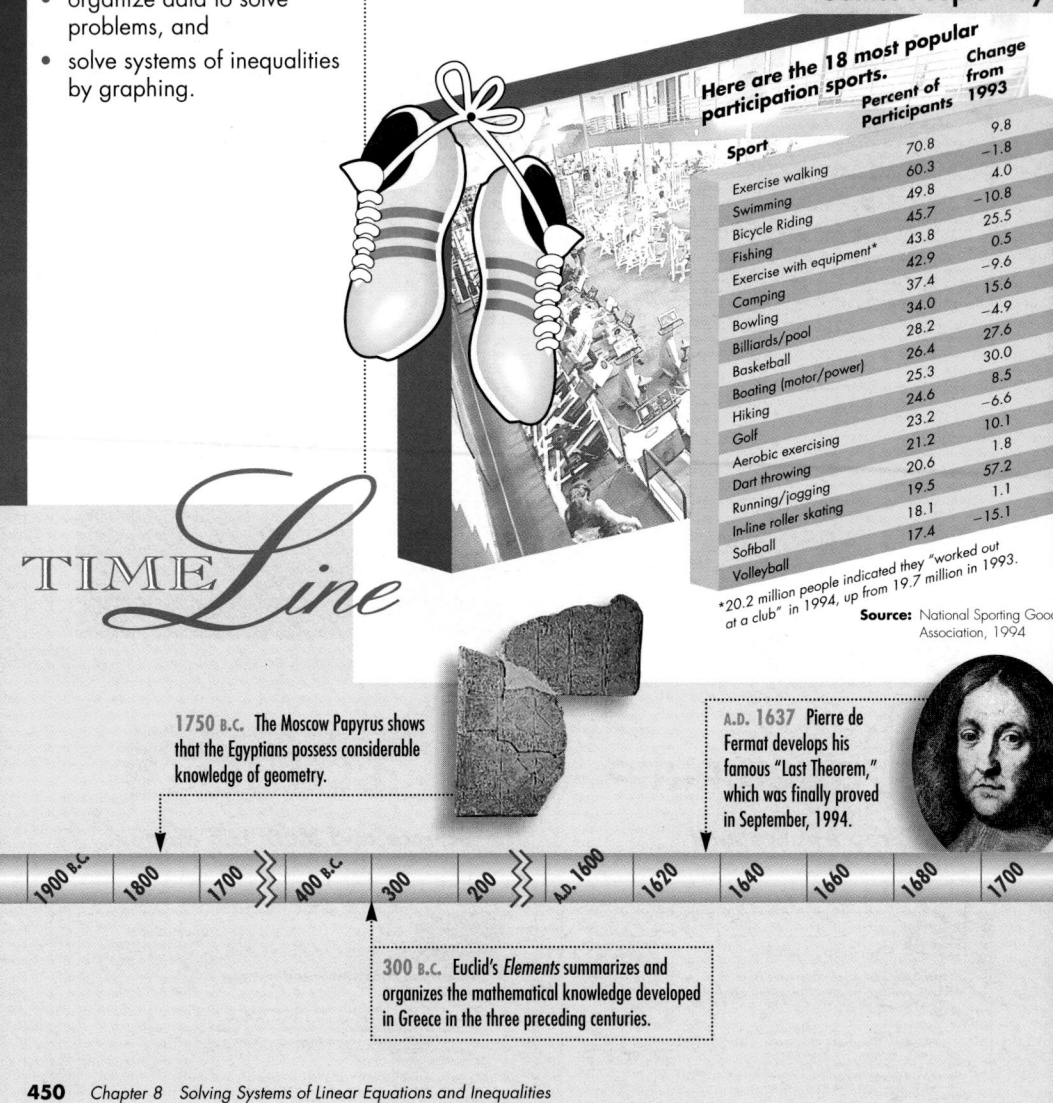

Games People Play

Here are the 18 most popular participation sports.

Sport	Percent of Participants	Change from 1993
Exercise walking	70.8	9.8
Swimming	60.3	–1.8
Bicycle Riding	49.8	4.0
Fishing	45.7	–10.8
Exercise with equipment*	43.8	25.5
Camping	42.9	0.5
Bowling	37.4	–9.6
Billiards/pool	34.0	15.6
Basketball	28.2	–4.9
Boating (motor/power)	26.4	27.6
Hiking	25.3	30.0
Golf	24.6	8.5
Aerobic exercising	23.2	–6.6
Dart throwing	21.2	10.1
Running/jogging	20.6	1.8
In-line roller skating	19.5	57.2
Softball	18.1	1.1
Volleyball	17.4	–15.1

*20.2 million people indicated they "worked out at a club" in 1994, up from 19.7 million in 1993.

Source: National Sporting Good Association, 1994

TIME *Line*

1750 B.C. The Moscow Papyrus shows that the Egyptians possess considerable knowledge of geometry.

A.D. 1637 Pierre de Fermat develops his famous "Last Theorem," which was finally proved in September, 1994.

| 1900 B.C. | 1800 | 1700 | 400 B.C. | 300 | 200 | A.D. 1600 | 1620 | 1640 | 1660 | 1680 | 1700 |

300 B.C. Euclid's *Elements* summarizes and organizes the mathematical knowledge developed in Greece in the three preceding centuries.

TIME *Line*

The Egyptians practiced an applied approach to mathematics. The Greeks employed what is known as pure mathematics. Have students research each approach and report on the influence of Egyptian and Greek mathematicians.

*inter*NET
CONNECTION

Use the Affirmative Fitness page on nutritional services, personal training, and other fitness links to learn more about physical fitness.

World Wide Web
http://student-www.uchicago.edu/users/tkhafen/affirmative.html

ChapterProject

Short-track speed skaters **Julie Goskowicz**, 15, and **Tony Goskowicz**, 18, are a brother-and-sister team aiming for the 1998 Olympics. They started skating eight years ago in their hometown of New Berlin, Wisconsin, when their father gave them each a pair of skates. Although both finished last in their first race, they enjoyed the sport and continued to train. Hard work has earned them a place at the U.S. Olympic Education Center in Marquette, Michigan, where they study and train while participating in the racing circuit.

In 1994, five-time Boston Marathon winner Jim Knaub tested his wheelchair's aerodynamics in the same wind tunnel that the Chrysler Corporation uses to test its car and truck designs. Knaub gained invaluable information concerning racing posture as well as helmet, wheel, and seat design. Earlier in the year, Knaub's Boston-Marathon-winning streak ended when he had to pull over twice during the race to make repairs. The winner was Heinz Frei of Switzerland.

- Suppose during a race, Frei's speed is 45 mph and Knaub is 264 feet ahead of him, racing at 36 mph.

- Write a system of equations to represent this situation. (*Hint:* Convert units from miles per hour to feet per second.)
  $y = 66x, y = 264 + 52.8x$

- If their speeds remained constant, when would Frei catch up with Knaub? Explain how you know using graphing. **After 20 seconds; the graphs intersect at (20, 1320); that is, after 20 seconds, Frei catches Knaub.**

Like most people, the Goskowiczes did not begin a new activity with instant success. In their first race, both Tony and Julie fell and finished last. "But we thought it was fun," Julie says, so they persevered and eventually mastered the sport.

ChapterProject

Cooperative Learning You may choose to have students pair off or arrange them in cooperative groups to discuss possible solutions for this activity. This project challenges students to examine the concepts of rate, time, and distance as they apply to a race situation and to represent that situation as a system of equations.

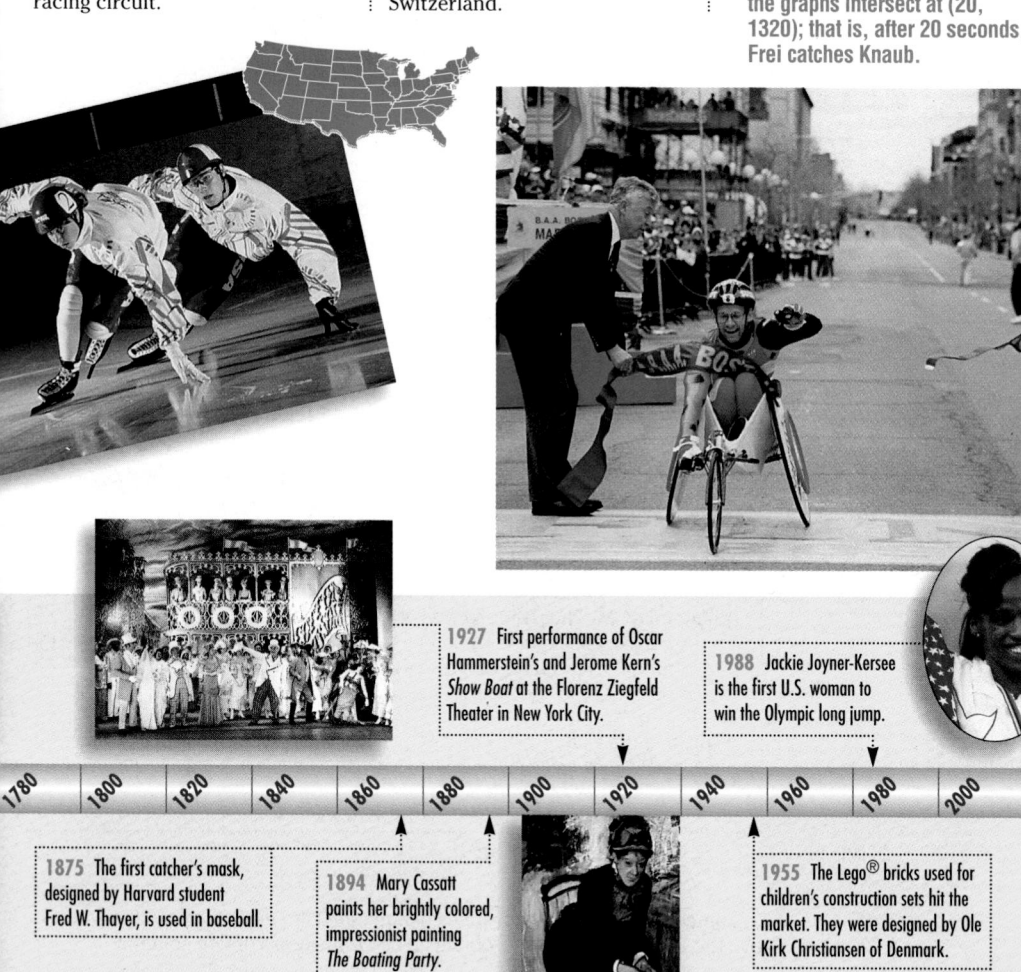

1927 First performance of Oscar Hammerstein's and Jerome Kern's *Show Boat* at the Florenz Ziegfeld Theater in New York City.

1988 Jackie Joyner-Kersee is the first U.S. woman to win the Olympic long jump.

1780 1800 1820 1840 1860 1880 1900 1920 1940 1960 1980 2000

1875 The first catcher's mask, designed by Harvard student Fred W. Thayer, is used in baseball.

1894 Mary Cassatt paints her brightly colored, impressionist painting *The Boating Party*.

1955 The Lego® bricks used for children's construction sets hit the market. They were designed by Ole Kirk Christiansen of Denmark.

Chapter 8 **451**

Alternative Chapter Projects

Two other chapter projects are included in the *Investigations and Projects Masters*. In Chapter 8 Project A, pp. 53–54, students extend the topic in the chapter opener. In Chapter 8 Project B, pp. 55–56, students explore and research linear programming.

Investigations and Projects Masters, p. 53

---

8 NAME_____ DATE _____
Chapter 8 Project A Student Edition Pages 450–48

Games People Play

1. Work with a partner. Select 50–100 people over the age of 7 in your neighborhood or family for a survey.

2. Ask the people you have selected to name any sports they have participated in at least once during the last year. (People who name swimming, exercise walking, exercising with equipment, running/jogging, or aerobics can be counted only if they have done one of these particular activities at least 6 times during the year.)

3. Determine the number of people who participated in each sport. What percent of the people you surveyed participated in each sport?

4. List the sports in order from greatest to least participation.

5. Compare your results with the table in the chapter opener. Are your results similar? If not, how do the results differ? Are you surprised at the results? Explain.

6. Complete one of the following activities based on your data:

   • Put your data together with that of another group or two. Find the percent of participation for each sport and make a new list of sports, ranked from greatest to least participation. Share your findings with the class by making a poster or giving an oral report.

   • Write an article for the school newspaper.

   • Make a pictograph for the school newspaper.

   • Do some library research and discover the participation in sports of past generations. What sporting pastimes were the most popular in 1850? 1900? 1950? What do you think will be the most popular sports in 2000? Write a report or make an oral presentation of your findings.

8-1A Graphing Technology
Systems of Equations

A Preview of Lesson 8–1

NCTM Standards: 1–5, 8

Objective

Use a graphing calculator to solve systems of equations.

Recommended Time

25 minutes

Instructional Resources

Graphing Calculator Masters, pp. 28 and 29

These masters provide keystroking instruction for this lesson for the TI-81 and Casio graphing calculators.

1 FOCUS

Motivating the Lesson

Ask students how many points of intersection are possible when two straight lines are drawn in the same plane. **0, 1, or infinitely many** Remind students that such pairs of lines are called *parallel*, *intersecting*, and *coincident*, respectively.

When solving systems of linear equations graphically, each equation is graphed on the same coordinate plane. The coordinates of the point at which the graphs intersect is the solution of the system. The graphing calculator permits us to graph several equations on the same coordinate plane and approximate the coordinates of the intersection point.

Example ① **Use a graphing calculator to solve the system of equations.**

$x + y = 9$
$2x - y = 15$

Begin by rewriting each equation in an equivalent form by solving for y.

$x + y = 9$
$\quad y = -x + 9$
$2x - y = 15$
$2x - 15 = y$

Graph each equation in the integer window $[-47, 47]$ by $[-31, 31]$. Recall that the integer window can be obtained by entering $\boxed{\text{ZOOM}}$ 6 $\boxed{\text{ZOOM}}$ 8 $\boxed{\text{ENTER}}$.

Enter: $\boxed{\text{Y=}}$ $\boxed{(-)}$ $\boxed{\text{X,T,}\theta}$ $\boxed{+}$

$\qquad$ 9 $\boxed{\text{ENTER}}$ 2 $\boxed{\text{X,T,}\theta}$

$\qquad$ $\boxed{-}$ 15 $\boxed{\text{ZOOM}}$ 6 $\boxed{\text{ZOOM}}$ 8 $\boxed{\text{ENTER}}$

The graphs intersect in one point. The coordinates of this point are the solution to the system of equations. Press the $\boxed{\text{TRACE}}$ key and use the arrow keys to move the cursor to the point of intersection. The coordinates of the point are $(8, 1)$. Thus, the solution is $(8, 1)$. We can check this solution by substituting these values into the two equations.

Check: $x + y = 9$ $\qquad\qquad$ $2x - y = 15$

$\qquad\quad 1 + 8 \overset{?}{=} 9$ $\qquad\qquad 2(8) - 1 \overset{?}{=} 15$

$\qquad\qquad\quad 9 = 9$ ✔ $\qquad\qquad\qquad 15 = 15$ ✔

Sometimes solutions to systems of equations are not integers. Then you can use the ZOOM IN process to obtain an accurate approximate solution.

Example Use a graphing calculator to solve this system of equations to the nearest hundredth.

$$y = 0.35x - 1.12$$
$$y = -2.25x - 4.05$$

Begin by graphing the equations in the standard viewing window.

Enter: Y= .35 X,T,θ − 1.12

ENTER (−) 2.25 X,T,θ

− 4.05 ZOOM 6

The graphs intersect at a point in the third quadrant. Use the TRACE function and the arrow keys to determine an approximation for the coordinates of the point of intersection. The ZOOM IN feature of the calculator is very useful for determining the coordinates of the intersection point with greater accuracy. Begin by placing the cursor on the intersection point and observing the coordinates, then press ZOOM 2 ENTER. Repeat this process as many times as necessary to get a more accurate answer.

You may want to use the INTERSECT feature to find the coordinates of the point of intersection. Press 2nd CALC 5 ENTER ENTER ENTER.

The solution is $(-1.13, -1.51)$.

X = −1.126923 Y = −1.514423

2 TEACH

Teaching Tip In both examples, the initial viewing window was specified. Sometimes, however, a point of intersection may lie outside the window. In this case, students would have to increase the window to locate the intersection.

3 PRACTICE/APPLY

Assignment Guide
Core: 1–10 **Enriched:** 1–10

4 ASSESS

Observing students working with technology is an excellent method of assessment.

EXERCISES

Use a graphing calculator to solve each system of equations. State each decimal solution to the nearest hundredth.

1. $y = x + 7$ (1, 8)
   $y = -x + 9$

2. $x + y = 27$ (17, 10)
   $3x - y = 41$

3. $y = 3x - 4$ (2.86, 4.57)
   $y = -0.5x + 6$

4. $x - y = 6$ (15, 9)
   $y = 9$

5. $x + y = 5.35$ (2.28, 3.08)
   $3x - y = 3.75$

6. $5x - 4y = 26$ (10.2, 6.25)
   $4x + 2y = 53.3$

7. $2x + 3y = 11$ (−2.9, 5.6)
   $4x + y = -6$

8. $2.93x + y = 6.08$ (0.91, 3.43)
   $8.32x - y = 4.11$

9. $125x - 200y = 800$ (1.14, −3.29)
   $65x - 20y = 140$

10. $0.22x + 0.15y = 0.30$ (−2.35, 5.44)
    $-0.33x + y = 6.22$

Lesson 8-1A Graphing Technology: Systems of Equations **453**

Using Technology
This lesson offers an excellent opportunity for using technology in your algebra classroom. For more information on using technology, see *Graphing Calculators in the Mathematics Classroom*, one of the titles in the Glencoe Mathematics Professional Series.

NCTM Standards: 1–5, 8

Instructional Resources

- Study Guide Master 8-1
- Practice Master 8-1
- Enrichment Master 8-1
- Graphing Calculator Masters, p. 8
- Modeling Mathematics Masters, p. 79
- Real-World Applications, 21
- Tech Prep Applications Masters, p. 15

Transparency 8-1A contains the 5-Minute Check for this lesson; **Transparency 8-1B** contains a teaching aid for this lesson.

Recommended Pacing	
Standard Pacing	Days 2 & 3 of 9
Honors Pacing	Day 2 of 8
Block Scheduling*	Day 1 of 5
Alg. 1 in Two Years*	Days 2 & 3 of 14

*For more information on pacing and possible lesson plans, refer to the *Block Scheduling Booklet* and *Algebra 1 in Two Years.*

1 FOCUS

5-Minute Check
(over Chapter 7)

Solve.

1. $6 + n > 40$ $n > 34$
2. $14p < 84$ $p < 6$
3. $12 + 4x < 20$ $x < 2$
4. $13r - 11 > 7r + 37$ $r > 8$
5. $3 + x < -4$ or $3 + x > 4$
   $x < -7$ or $x > 1$

Motivating the Lesson

Hands-On Activity Take two rulers and move them around in relation to each other. It should quickly become clear that they can intersect in, at most, one point. Then take two pieces of string and move them in relation to each other. Students should see that the strings can intersect many times.

Objectives
After studying this lesson, you should be able to:
- solve systems of equations by graphing, and
- determine whether a system of equations has one solution, no solution, or infinitely many solutions by graphing.

Fastest swimmers to cross the English Channel (hr:min)
1. Penny Lee Dean, 7:40
2. Philip Rush, 7:55
3. Richard Davey, 8:05
4. Irene van der Laan, 8:06
5. Paul Asmuth, 8:12

8-1

Graphing Systems of Equations

APPLICATION
World Records

Cape Verde is a group of islands located off the westernmost point of Africa. In December, 1994, Frenchman Guy Delage set off from these islands for a 2400-mile swim across the Atlantic Ocean. He arrived at Barbados in the West Indies eight weeks later. Every day he would swim a while and then rest while floating with the current on a huge raft equipped with a fax machine, a computer, and a two-way radio.

People are already considering trying to break Guy's record, but before a challenger makes the attempt, he or she should know what is required. Guy traveled approximately 44 miles per day. A good swimmer like Guy can swim about 3 miles per hour for an extended period, and the Atlantic currents will float a raft about 1 mile per hour. To match Guy's record, how many hours per day would one have to swim? How many hours would one be able to spend floating on the raft?

To solve this problem, let s represent the number of hours Guy swam, and let f represent the number of hours he floated. Then $3s$ represents the number of miles he traveled while swimming and $1f$ represents the number of miles he traveled while floating. You can write two equations to represent this situation.

number of hours swimming	*plus*	*number of hours floating*	*is*	*total number of hours in a day*	*miles traveled while swimming*	*plus*	*miles traveled while floating*	*is*	*total miles traveled in a day*
s	$+$	f	$=$	24	$3s$	$+$	$1f$	$=$	44

While the fastest human swimmer, the world freestyle champion, has a top speed of 5.37 miles per hour (mph), many animals are much faster. The fastest fish, the sailfish, can swim at 68 mph. The fastest mammal, the killer whale, can reach a speed of 34.5 mph.

The equations $s + f = 24$ and $3s + f = 44$ together are called a **system of equations**. The solution to this problem is the ordered pair of numbers that satisfies both of these equations.

One method for solving a system of equations is to carefully graph the equations on the same coordinate plane. The coordinates of the point at which the graphs intersect is the solution of the system.

Guy Delage's raft

With most graphs of systems of equations, we can only estimate the solution. In this case, the graphs of $s + f = 24$ and $3s + f = 44$ appear to intersect at the point with coordinates (10, 14).

Check: In each equation, replace s with 10 and f with 14.

$$s + f = 24 \qquad\qquad 3s + f = 44$$
$$10 + 14 \stackrel{?}{=} 24 \qquad 3(10) + 14 \stackrel{?}{=} 44$$
$$24 = 24 \quad\checkmark \qquad\qquad 44 = 44 \quad\checkmark$$

The solution of the system of equations $s + f = 24$ and $3s + f = 44$ is (10, 14). The ordered pair (10, 14) means that a person trying to match Guy Delage's record would have to spend approximately 10 hours a day swimming and 14 hours floating.

Example ❶ **Graph the system of equations to find the solution.**

$$x + 2y = 1$$
$$2x + y = 5$$

The graphs appear to intersect at the point with coordinates (3, −1). Check this estimate by replacing x with 3 and y with −1 in each equation.

Check:
$$x + 2y = 1 \qquad\qquad 2x + y = 5$$
$$3 + 2(-1) \stackrel{?}{=} 1 \qquad 2(3) + (-1) \stackrel{?}{=} 5$$
$$1 = 1 \quad\checkmark \qquad\qquad 5 = 5 \quad\checkmark$$

The solution is (3, −1).

A system of two linear equations has exactly one ordered pair as its solution when the graphs of the equations intersect at exactly one point. If the graphs coincide, they are the same line and have infinitely many points in common. In either case, the system of equations is said to be **consistent**. That is, it has *at least* one ordered pair that satisfies both equations.

In-Class Example

For Example 1
Graph the system of equations to find the solution.
$$y = -3x + 1$$
$$x + \frac{1}{2}y = -\frac{1}{2}$$
(2, −5)

Teaching Tip If necessary, review graphing linear equations using x- and y-intercepts or slope-intercept form before beginning this lesson.

Teaching Tip Point out that since there are two variables in Example 1, the solution, if it exists, will be an ordered pair or ordered pairs.

CAREER CHOICES

In the science of ecology, predator-prey relationships are often studied. If there are too few predators, the population of prey will increase. If there are too many predators, the prey population can vanish. In general, nature attempts to maintain a balance in the populations of these different groups. Systems of equations can be used to model these various relationships.

456 Chapter 8

In-Class Example

For Example 2
Graph each pair of equations. Then determine the number of solutions for the system of equations.

a. $x - 2y = 4$
 $x - 2y = -2$

no solution

b. $3x - y = 2$
 $12x - 4y = 8$

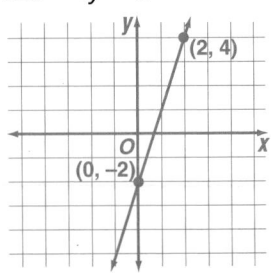

infinitely many solutions

Teaching Tip In Example 2b, explain that the graphs of these equations coincide.

It is also possible for the two graphs to be *parallel*. In this case, the system of equations is **inconsistent** because there is *no* ordered pair that satisfies both equations.

Another way to classify a system is by the number of solutions it has.

• If a system has exactly one solution, it is **independent**.

• If a system has an infinite number of solutions, it is **dependent**.

Thus, the system in Example 1 is said to be *consistent and independent*.

The chart below summarizes the possible solutions to systems of linear equations.

Graphs of Equations	Number of Solutions	Terminology
intersecting lines	exactly one	consistent and independent
same line	infinitely many	consistent and dependent
parallel lines	none	inconsistent

Example **Graph each system of equations to determine the number of solutions.**

a. $x + y = 4$
 $x + y = 1$

The graphs of the equations are parallel lines. Since they do not intersect, there is no solution to this system of equations. Notice that the two lines have the same slope but different y-intercepts.

Recall that a system of equations that has no solution is said to be <u>inconsistent</u>.

b. $x - y = 3$
 $2x - 2y = 6$

Each equation has the same graph. Any ordered pair on the graph will satisfy both equations. Therefore, there are infinitely many solutions of this system of equations. Notice that the graphs have the same slope and intercepts.

Recall that a system of equations that has infinitely many solutions is said to be <u>consistent and dependent</u>.

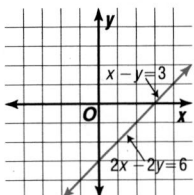

Check: Verify that the point at (4, 1) lies on both lines.

$$x - y = 3 \qquad\qquad 2x - 2y = 6$$
$$4 - 1 \overset{?}{=} 3 \qquad\qquad 2(4) - 2(1) \overset{?}{=} 6$$
$$3 = 3 \checkmark \qquad\qquad\quad 6 = 6 \checkmark$$

Alternative Learning Styles

Auditory Read to students the World Records application at the beginning of the lesson. At each stage, encourage students to think about how each part of the problem can be represented algebraically.

The methods you use to solve algebra problems are often useful in solving problems involving geometry.

Example **3**

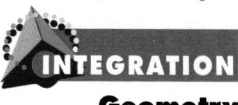

INTEGRATION

Geometry

The points $A(-1, 6)$, $B(4, 8)$, $C(8, 3)$ and $D(-2, -1)$ are vertices of a quadrilateral.

a. Use a graph to determine the point of intersection of the diagonals of quadrilateral $ABCD$.

b. Find the equations of the lines containing the diagonals to verify the solution.

a. Draw quadrilateral $ABCD$ with diagonals $\overline{AC}$ and $\overline{BD}$. The diagonals appear to intersect at the point $(2, 5)$.

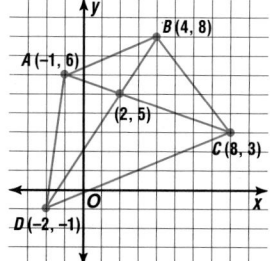

b. To check the solution, find the equations of lines AC and BD and then verify that $(2, 5)$ is a solution of both equations. First, find the slope of each line using

$$m = \frac{y_2 - y_1}{x_2 - x_1}.$$

Slope of $\overleftrightarrow{AC}$

$m = \dfrac{3 - 6}{8 - (-1)}$

$= \dfrac{-3}{9}$ or $-\dfrac{1}{3}$

Slope of $\overleftrightarrow{BD}$

$m = \dfrac{-1 - 8}{-2 - 4}$

$= \dfrac{-9}{-6}$ or $\dfrac{3}{2}$

Then use the slope-intercept form, $y = mx + b$, to determine the equations.

Equation for $\overleftrightarrow{AC}$

$y = mx + b$

$6 = -\dfrac{1}{3}(-1) + b$ *Replace m with $-\frac{1}{3}$ and (x, y) with (−1, 6).*

$\dfrac{17}{3} = b$ The equation for $\overleftrightarrow{AC}$ is $y = -\dfrac{1}{3}x + \dfrac{17}{3}$.

Equation for $\overleftrightarrow{BD}$

$y = mx + b$

$8 = \dfrac{3}{2}(4) + b$ *Replace m with $\frac{3}{2}$ and (x, y) with (4, 8).*

$2 = b$ The equation for $\overleftrightarrow{BD}$ is $y = \dfrac{3}{2}x + 2$.

Check that $(2, 5)$ is a solution to both equations.

$y = -\dfrac{1}{3}x + \dfrac{17}{3}$

$5 \stackrel{?}{=} -\dfrac{1}{3}(2) + \dfrac{17}{3}$ *(x, y) = (2, 5)*

$5 = 5$ ✔

$y = \dfrac{3}{2}x + 2$

$5 \stackrel{?}{=} \dfrac{3}{2}(2) + 2$ *(x, y) = (2, 5)*

$5 = 5$ ✔

The solution checks.

In-Class Example

For Example 3
The graphs of the equations $2x + y = 8$, $-x + y = 6$, and $7x + y = 3$ contain the sides of a triangle. Find the coordinates of the vertices of the triangle. $\left(\dfrac{2}{3}, 6\dfrac{2}{3}\right)$, $\left(-\dfrac{3}{8}, 5\dfrac{5}{8}\right)$, $(-1, 10)$

Lesson 8–1 Graphing Systems of Equations **457**

Chapter 8 **457**

Check for Understanding
Exercises 1–20 are designed to help you assess your students' understanding through reading, writing, speaking, and modeling. You should work through Exercises 1–7 with your students and then monitor their work on Exercises 8–20.

Error Analysis
Stress that students should check their answers in the original equations. Make sure it is in the original equation, too, because they could have made a mistake along the way.

Additional Answers

6.

14.

15.

16.

CHECK FOR UNDERSTANDING

Communicating Mathematics

1. (2, 4); The graphs intersect at the point (2, 4) and (2, 4) is a solution of both equations.

2. Find an ordered pair that satisifies both equations.

3. The graphs are the same line.

5. Sample answer:
$x + y = 2$
$x - y = -8$

MODELING MATHEMATICS

Study the lesson. Then complete the following.

1. **State** the solution of the system of equations shown in the graph at the right. Justify your answer.

2. **Explain** what it means to *solve* a system of linear equations.

3. **Describe** the graph of a linear system that has infinitely many solutions.

4. **Name** two of the solutions for the system of equations in Example 2b. Verify your answers algebraically. **Sample answer: (3, 0), (4, 1)**

5. **Write** a system of linear equations that has $(-3, 5)$ as its only solution.

6. **Sketch** the graph of a linear system that has *no* solution. **Graphs will vary, but lines must be parallel; see margin for sample graph.**

7. Use a geoboard and rubber bands to model a system of two equations that has the solution (3, 2). Let the lower left point on the geoboard represent the origin. **See students' work.**

Guided Practice

Use the graphs at the right to determine whether each system has *one* solution, *no* solution, or *infinitely many* solutions. If the system has one solution, name it.

9. no solution

11. one; $(-6, 2)$

8. $y = -x + 5$
$y = x - 3$ **one; (4, 1)**

9. $y = -x - 4$
$y = -x + 5$

10. $y = 2x + 14$
$y = -x + 5$ **one; $(-3, 8)$**

11. $y = -x - 4$
$y = 2x + 14$

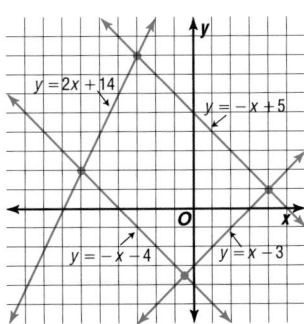

State whether the given ordered pair is a solution to each system. Write *yes* or *no*.

12. $x - y = 6$ **no**
$2x + y = 0$ $(-2, -4)$

13. $2x - y = 4$
$3x + y = 1$ $(1, -2)$ **yes**

Graph each system of equations. Then determine whether the system has *one* solution, *no* solution, or *infinitely many* solutions. If the system has one solution, name it. 14–19. See margin for graphs.

14. $y = 3x - 4$ $(0, -4)$
$y = -3x - 4$

15. $y = -x + 8$ $(3, 5)$
$y = 4x - 7$

16. $x + 2y = 5$ **no**
$2x + 4y = 2$ **solution**

17. $y = -6$ $(2, -6)$
$4x + y = 2$

18. $2x + 3y = 4$ **infinitely**
$-4x - 6y = -8$ **many**

19. $2x + y = -4$ $(-6, 8)$
$5x + 3y = -6$

20c. The graphs of the system are parallel lines. The system has no solution and is inconsistent.

20. **a.** Graph the line $y - x = 6$. a–b. See margin.
b. Slide the entire line four units to the right and down one unit. Draw the new line.
c. Describe this system of equations.

Reteaching

Using Alternative Methods Have half of the class determine the number of solutions for a system of equations by graphing. The other half can determine the number of solutions using the following method. They should write the equations in slope-intercept form. If the slopes and the y-intercepts are the same, the lines are identical. If the slopes are equal but the y-intercepts differ, the lines are parallel. If the equations have different slopes, the lines intersect in exactly one point. Have students discuss and compare the methods.

Practice

21. one, (3, −1)

23. no solution

25. one, (3, 3)
26. one, (9, 1)

Use the graphs below to determine whether each system has *one* solution, *no* solution, or *infinitely many* solutions. If the system has one solution, name it.

21. $y = x - 4$
$y = \frac{1}{3}x - 2$

22. $y = x - 4$ one, (6, 2)
$y = -\frac{1}{3}x + 4$

23. $y = \frac{1}{3}x + 2$
$y = \frac{1}{3}x - 2$

24. $y = x - 4$ one, (9, 5)
$y = \frac{1}{3}x + 2$

25. $y = -\frac{1}{3}x + 4$
$y = \frac{1}{3}x + 2$

26. $y = \frac{1}{3}x - 2$
$y = -\frac{1}{3}x + 4$

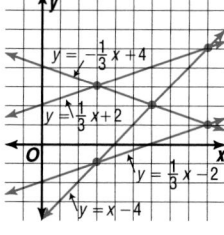

27–41. See Solutions Manual for graphs.

32. no solution

36. (−1, −5)

40. infinitely many

Graph each system of equations. Then determine whether the system has *one* solution, *no* solution, or *infinitely many* solutions. If the system has one solution, name it.

27. $y = -x$ (2, −2)
$y = 2x - 6$

28. $y = 2x + 6$ (−3, 0)
$y = -x - 3$

29. $x + y = 2$ (−1, 3)
$y = 4x + 7$

30. $2x + y = 10$ (4, 2)
$y = \frac{1}{2}x$

31. $x + y = 2$ (−2, 4)
$2y - x = 10$

32. $3x + 2y = 12$
$3x + 2y = 6$

33. $x - 2y = 2$ (2, 0)
$3x + y = 6$

34. $x - y = 2$ (3, 1)
$3y + 2x = 9$

35. $3x + y = 3$ infinitely
$2y = -6x + 6$ many

36. $2x + 3y = -17$
$y = x - 4$

37. $y = \frac{2}{3}x - 5$ no
$3y = 2x$ solution

38. $4x + 3y = 24$ (3, 4)
$5x - 8y = -17$

39. $\frac{1}{2}x + \frac{1}{3}y = 6$ (8, 6)
$y = \frac{1}{2}x + 2$

40. $6 - \frac{3}{8}y = x$
$\frac{2}{3}x + \frac{1}{4}y = 4$

41. $2x + 4y = 2$ infinitely
$3x + 6y = 3$ many

 INTEGRATION
Geometry

42. (2, 4), (0, 3), (−1, 10)

42. The graphs of the equations $-x + 2y = 6$, $7x + y = 3$, and $2x + y = 8$ contain the sides of a triangle. Find the coordinates of the vertices of the triangle.

43. Graph the system of equations below. Then find the area of the geometric figure.

$2x - 4 = 0$
$y = 8$
$x = 5$
$3y - 9 = 0$ See Solutions Manual for graph; 15 square units.

Graphing Calculator

Use a graphing calculator to solve each system of equations. Approximate the coordinates of the point of intersection to the nearest hundredth.

44. $y = x + 2$ (−1.50, 0.50)
$y = -x - 1$

45. $y = \frac{1}{4}x - 3$ (1.71, −2.57)
$y = -\frac{1}{3}x - 2$

46. $6x + y = 5$ (−0.44, 7.67)
$y = 9 + 3x$

47. $3 + y = x$ (−0.25, −3.25)
$2 + y = 5x$

Assignment Guide

Core: 21–47 odd, 48, 49, 51, 52–60
Enriched: 22–46 even, 48–60

For **Extra Practice,** see p. 774.

The red A, B, and C flags, printed only in the Teacher's Wraparound Edition, indicate the level of difficulty of the exercises.

Additional Answers

20a & 20b.

$y - x = 6$

Study Guide Masters, p. 56

8-1   NAME_____   DATE_____
Study Guide   Student Edition
Pages 454–461

Graphing Systems of Equations

Two or more linear equations involving the same variables form a **system of equations.** The solution set for the system is the set of ordered pairs that satisfy both equations. One method for solving a system of equations is to graph the equations on the same coordinate plane.

Examples: Solve each system of equations by graphing.

$x + y = 2$
$x - y = 4$ The point (3, −1) lies on both lines, thus (3, −1) is the solution set for the system of equations.

$3x + y = 2$
$3x + y = 4$
no solution

$y = 2x + 1$
$2y = 4x + 2$
infinitely many solutions

Use the graphs below to determine whether each system has one solution, no solution, or infinitely many solutions. If the system has one solution, name it.

1. $x - y = 1$
$x + y = -1$
one solution
(0, −1)

2. $x + 3y = 3$
$2x + y = 2$
one solution
$\left(\frac{3}{5}, \frac{4}{5}\right)$

3. $2x + 3y = 6$
$2x + 3y = -6$
no solution

4. $x - y = 2$
$2x - 2y = 4$
infinitely many solutions

Graph each system of equations on a separate sheet of paper. Then determine whether the system has one solution, no solution, or infinitely many solutions. If the system has one solution, name it.

5. $4x - 2y = 4$
$4x - 2y = 0$
no solution

6. $y = x + 2$
$y = 2x - 1$
one solution
(3, 5)

7. $x + y = 1$
$3x + 3y = 3$
infinitely many solutions

8. $2x - 2y = 2$
$y = x$
no solution

Additional Answers

17.

18.

19.

Critical Thinking

48. If (0, 0) and (2, 2) are known to be solutions of a system of two linear equations, does the system have any other solutions? Justify your answers. **Yes, the lines are coincident.**

49. The solution to the system of equations $Ax + y = 5$ and $Ax + By = 7$ is $(-1, 2)$. What are the values of A and B? **$A = -3$, $B = 2$**

Applications and Problem Solving

50a. (1000, 6000); See margin.

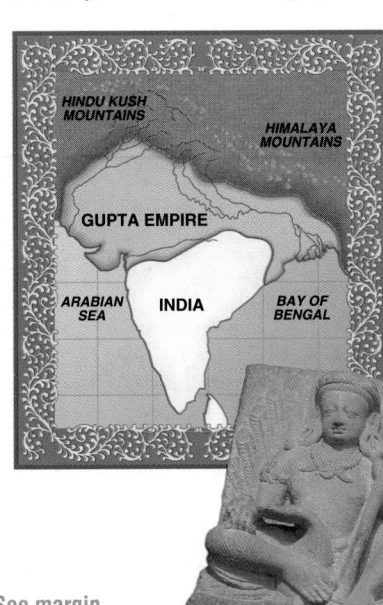

50. Business Mary Rodas is an 18-year-old toy specialist who tests and evaluates products at Catco, Inc., a company in New York City. She also helps design new toys, such as the Balzac Balloon Balls. Suppose the income from Balzac Balloon Balls is represented by the equation $d = 6b$ and the expenses are represented by the equation $d = b + 5000$. In both equations, b is the number of balls, and d is the number of dollars. Use the graph at the left to answer the following questions.

$d = b + 5000$

$d = 6b$

Dollars (thousands)

Number of Balls (thousands)

a. Find the solution to this system of equations. This solution is called the *break-even point.* What does this point represent? **a–c. See margin.**

b. A profit is made if income is greater than expenses. When is a profit made from the toys? How can you tell this from the graph?

c. Money is lost if expenses are greater than income. When is money lost from the Balzac Balloon Balls? How can you tell this from the graph?

Gupta doctors developed plastic surgery. Metallurgists made iron columns that are still free from rust after more than 1500 years. Gupta mathematicians developed a system of numbers that was later adopted by the Arabs.

51. World Cultures The Golden Age of India was during the expansion of the Gupta Empire, beginning in A.D. 320. India became a center of art, medicine, science, and mathematics. Suppose $P = \frac{1}{2}t + 22$ represents the percent of Indian people in the Gupta Empire, at time t.

Let $P = -\frac{1}{2}t + 78$ represent the percent of Indian people that were not Guptas. Graph the system of equations and estimate the year in which the percent of Guptas equaled the percent of Indians that were not Gupta. (*Hint:* Let $t = 0$ correspond to A.D. 320.)

A.D. 376; See margin for graph.

HINDU KUSH MOUNTAINS

HIMALAYA MOUNTAINS

GUPTA EMPIRE

ARABIAN SEA

INDIA

BAY OF BENGAL

Mixed Review

52. Graph $y - 7 > 3x$. (Lesson 7–8) **See margin.**

53. Solve $|2m + 15| = 12$. (Lesson 7–6) **$-1.5, -13.5$**

54. Solve $10p - 14 < 8p - 17$. (Lesson 7–3) $\left\{ p \mid p < -\frac{3}{2} \right\}$

55. Write an equation for the line that passes through the point at $(2, -2)$ and is parallel to $y = -2x + 21$. (Lesson 6–6) **$y = -2x + 2$**

Additional Answer

52.

$y - 7 = 3x$

56. Statistics Find the range, median, upper and lower quartiles, and interquartile range of the data in the stem-and-leaf plot at the right. (Lesson 5–7) **34; 446.5; 457; 439; 18**

Stem	Leaf
43	3 5 6 6 9
44	1 4 4 4 9 9
45	0 2 7 7 8
46	5 7 *44|9 = 449*

57. Finance Patricia invested $5000 for one year. Martin also invested $5000 for one year. Martin's account earned interest at a rate of 10% per year. At the end of the year, Martin's account had earned $125 more than Patricia's account. What was the annual interest rate on Patricia's account? (Lesson 4–4) $7\frac{1}{2}\%$

58. Solve $\frac{a-x}{-3} = \frac{-2}{b}$ for x. (Lesson 3–6) $-\frac{6}{b} + a$

59a. how many three- and four-bedroom homes will be built

59. Architecture Answer the related questions for the verbal problem below. A developer is designing a housing development. She proposes to have four times as many three-bedroom homes as four-bedroom homes. If the development is planned for 100 homes, how many three- and four-bedroom homes will be built? (Lesson 2–9)

  a. What does the problem ask?

  b. If h represents the number of four-bedroom homes that are planned, how many three-bedroom homes are planned? **$100 - h$ or $4h$**

  c. If 20 four-bedroom homes are planned, how many three-bedroom homes should be built? **80 homes**

60. Evaluate $\frac{6ab}{3x + 2y}$ if $a = 6$, $b = 4$, $x = 0.2$, and $y = 1.3$. (Lesson 1–3) **45**

WORKING ON THE

Refer to the Investigation on pages 448–449.

Ready, Set, Drop!

Your research team determines that hang gliders are not just dropped from a point as you did with your tissue paper triangles. They are always launched into forward motion before gliding. The team decides that scale models are needed in order to get a feel for the launching and landing aspects of a real hang glider.

1 Each team in your class will construct a hang glider model using tissue paper and wire. A table top will act as the top of the cliff from which the glider is to be launched.

2 Each team should discuss different types of methods for launching their hang glider models from the table top. They should present their ideas to the class, and the class should agree upon which method they prefer to use. Then each team tests their glider using the method that the class has chosen.

3 Launch the glider 10 times. For each trial, measure the horizontal distance (along the floor) from the table to the spot at which the glider lands. Record this measurement and the height of the launch site.

4 Use these data to write a linear equation that describes the path of your glider. What is the slope of the path for your glider?

5 Using the linear equations from each of the other teams' data, would your glider collide with any of the other teams' gliders if they were launched at the same time from cliffs that are opposite each other? Write a detailed report on your conclusions.

Add the results of your work to your Investigation Folder.

Lesson 8–1 Graphing Systems of Equations **461**

Extension

Connections The graphs of the equations $y = 2$, $3x + 2y = 1$, and $3x - 4y = -29$ contain the sides of a triangle. Find the measure of the area of this triangle. (*Hint:* Use the formula $A = \frac{1}{2}bh$.) **9**

In·ves·ti·ga·tion

Working on the Investigation

The Investigation on pages 448–449 is designed to be a long-term project that is completed over several days or weeks. Encourage students to keep their materials in their Investigation Folder as they work on the Investigation.

---

4 ASSESS

Closing Activity

Writing At the end of class, have students write down the key points to consider when solving a system of equations by graphing.

Enrichment Masters, p. 56

Chapter 8 **461**

NCTM Standards: 1–5

Instructional Resources

- Study Guide Master 8-2
- Practice Master 8-2
- Enrichment Master 8-2
- Assessment and Evaluation Masters, p. 212
- Science and Math Lab Manual, pp. 35–38

Transparency 8-2A contains the 5-Minute Check for this lesson; **Transparency 8-2B** contains a teaching aid for this lesson.

Recommended Pacing	
Standard Pacing	Day 4 of 9
Honors Pacing	Day 3 of 8
Block Scheduling*	Day 2 of 5 (along with Lesson 8-3)
Alg. 1 in Two Years*	Days 4 & 5 of 14

*For more information on pacing and possible lesson plans, refer to the *Block Scheduling Booklet* and *Algebra 1 in Two Years.*

1 FOCUS

5-Minute Check
(over Lesson 8-1)

Graph each system of equations and state its solution.

1. $x + y = 3$
   $y = x + 3$ **(0, 3)**
2. $y = x$
   $x + y = 4$ **(2, 2)**
3. $y = -x$
   $y = 2x$ **(0, 0)**
4. If two distinct lines intersect, how many solutions exist?
   one
5. If a system of two linear equations has no solution, what is true of their graphs?
   They are parallel lines.

8-2

Substitution

Objectives

After studying this lesson, you should be able to:
- solve systems of equations by using the substitution method, and
- organize data to solve problems.

FYI

California has 12% of the entire U.S. population.

Alaska, the largest state in the U.S., has the second smallest population.

CONNECTION
Geography

A recent article in *USA Today* reported that New York lost its position as the second most populous state when Texas slipped into the No. 2 spot at the end of 1994. Census Bureau projections show that New York will likely be pushed even further down the population ladder when Florida catches up early in the 21st century.

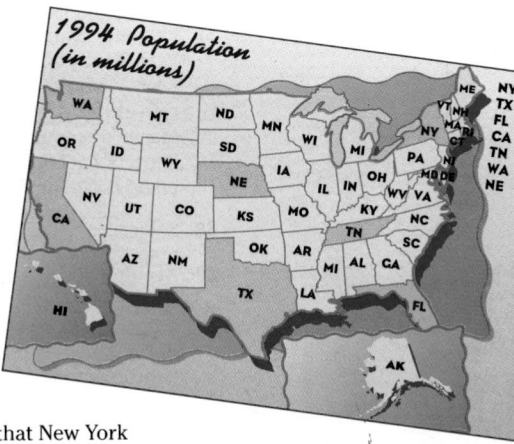

1994 Population (in millions)

If New York's population grows at a constant rate of 0.02 million people per year and Florida's population grows at a constant rate of 0.26 million people per year, when would Florida catch up to New York in population? What would their populations be then?

Let *P* represent the population in millions, and let *t* represent the amount of time in years. The information above can be described by the following system of equations.

$$P = 18.2 + 0.02t$$
$$P = 14 + 0.26t$$

You could try to solve this system of equations by graphing, as shown at the right. Notice that the *exact* coordinates of the point where the lines intersect cannot be easily determined from this graph. An estimate is (18, 18).

The exact solution of this system of equations can be found by using algebraic methods. One such method is called **substitution**.

From the first equation in the system, $P = 18.2 + 0.02t$, you know that *P* is equal to $18.2 + 0.02t$. Since *P* must have the same value in *both* equations, you can substitute $18.2 + 0.02t$ for *P* in the second equation $P = 14 + 0.26t$.

$$P = 14 + 0.26t$$
$$18.2 + 0.02t = 14 + 0.26t$$
$$0.02t = -4.2 + 0.26t$$
$$-0.24t = -4.2$$
$$t = 17.5$$

Substitute $18.2 + 0.02t$ for P so the equation will have only one variable.

462 Chapter 8 *Solving Systems of Linear Equations and Inequalities*

FYI

The population of the United States is not evenly distributed over its land area. The east and west coasts have the largest population centers. Current trends show the largest population growth occurring in the southern states, where the climate is warmer.

Now find the value of P by substituting 17.5 for t in either equation.

$P = 18.2 + 0.02t$
$ = 18.2 + 0.02(17.5)$ *You could also substitute 17.5 for t in P = 14 + 0.26t.*
$ = 18.55$

Check: In each equation, replace t with 17.5 and P with 18.55.

$$P = 18.2 + 0.02t \qquad\qquad P = 14 + 0.26t$$
$$18.55 \overset{?}{=} 18.2 + 0.02(17.5) \qquad 18.55 \overset{?}{=} 14 + 0.26(17.5)$$
$$18.55 = 18.55 \ \checkmark \qquad\qquad 18.55 = 18.55 \ \checkmark$$

The solution of the system of equations is (17.5, 18.55). Therefore after 17.5 years, or in 2011, the populations of New York and Florida would both be 18.55 million. *Compare this result to the estimate we obtained from the graph.*

You can use substitution to solve systems of equations even when the equations are more complex.

Example ❶ **Use substitution to solve each system of equations.**

a. $x + 4y = 1$
 $2x - 3y = -9$

Solve the first equation for x since the coefficient of x is 1.	Next, find the value of y by substituting $1 - 4y$ for x in the second equation.

$$x + 4y = 1$$
$$x = 1 - 4y$$

$$2x - 3y = -9$$
$$2(1 - 4y) - 3y = -9$$
$$2 - 8y - 3y = -9$$
$$-11y = -11$$
$$y = 1$$

Then substitute 1 for y in either of the original equations and find the value of x. *Choose the equation that is easier for you to solve.*

$$x + 4y = 1$$
$$x + 4(1) = 1$$
$$x + 4 = 1$$
$$x = -3$$

The solution of this system is $(-3, 1)$. *Use the graph at the left to verify this result.*

b. $\dfrac{5}{2}x + y = 4$

 $5x + 2y = 8$

Solve the first equation for y since the coefficient of y is 1.	Next, find the value of x by substituting $4 - \frac{5}{2}x$ for y in the second equation.

$$\frac{5}{2}x + y = 4$$
$$y = 4 - \frac{5}{2}x$$

$$5x + 2y = 8$$
$$5x + 2\left(4 - \frac{5}{2}x\right) = 8$$
$$5x + 8 - 5x = 8$$
$$8 = 8$$

The statement $8 = 8$ is true. This means that there are infinitely many solutions to the system of equations. This is true because the slope intercept form of both equations is $y = 4 - \frac{5}{2}x$. That is, the equations are equivalent, and both have the same graph.

Motivating the Lesson
Situational Problem Review solving equations with the following activity. Let $y = 3$. Find the value of x in each equation by substituting 3 for y and solving for x.

1. $y - x = 6$ **−3**
2. $4x - y = 9$ **3**
3. $2y - 3x = 6$ **0**

2 TEACH

In-Class Example

For Example 1
Use substitution to solve each system of equations.

a. $y = 2x$
 $2x + 5y = -12$ $(-1, -2)$
b. $2y = -3x$
 $4x + y = 5$ $(2, -3)$
c. $x + y = 6$
 $3x + y = 15$ $\left(\frac{9}{2}, \frac{3}{2}\right)$
d. $\dfrac{5}{2}x + 3y = 7$
 $5x + 6y = 9$ no solution
e. $3x + 4y = 7$
 $\dfrac{3}{2}x + 2y = 11$ no solution

Teaching Tip When an equation contains fractions, as in Example 1b, some students may find it easier to eliminate the fractions first. This can be done by multiplying each side of the equation by the LCM of the denominators.

MODELING MATHEMATICS The process students follow in this modeling activity mirrors the process of substitution. The arrangement of cups and counters on the mat represents the result of substitution: $4x + 3(x - 2) = 8$.

In-Class Example

For Example 2

A metal alloy is 25% copper. Another metal alloy is 50% copper. How much of each alloy should be used to make 1000 grams of a metal alloy that is 45% copper?

200 grams of the 25% copper alloy and 800 grams of the 50% copper alloy

---

In general, if you solve a system of linear equations and the result is a true statement (an identity such as $8 = 8$), the system has an infinite number of solutions; if the result is a false statement (for example, $8 = 12$), the system has no solution.

MODELING MATHEMATICS

Systems of Equations

Materials: cups and counters ▣ equation mat

Use a model to solve the system of equations.

$4x + 3y = 8$

$y = x - 2$

Your Turn   a. a cup and 2 negative counters

a. Let a cup represent the unknown value x. If $y = x - 2$, how can you represent y?

b. Represent $4x + 3y = 8$ on the equation mat. On one side of the mat, place four cups to represent $4x$ and three representations of y from step a. On the other side of the mat, place eight positive counters.

c. Use what you know about equation mats and zero pairs to solve the equation. What value of x is the solution of the system of equations?   2

d. Use the value of x from step c and the equation $y = x - 2$ to find the value of y.   0   e. (2, 0)

e. What is the solution of the system of equations?

Sometimes it is helpful to **organize data** before solving a problem. Some ways to organize data are to use tables, charts, different types of graphs, or diagrams.

Example

PROBLEM SOLVING

Organize Data

EJH Labs needs to make 1000 gallons of a 34% acid solution. The only solutions available are 25% acid and 50% acid. How many gallons of each solution should be mixed to make the 34% solution?

Explore   Let a represent the number of gallons of 25% acid.
Let b represent the number of gallons of 50% acid.

Make a table to organize the information in the problem.

	25% Acid	50% Acid	34% Acid
Total Gallons	a	b	1000
Gallons of Acid	$0.25a$	$0.50b$	$0.34(1000)$

Plan   The system of equations is $a + b = 1000$ and $0.25a + 0.50b = 0.34(1000)$. Use substitution to solve this system.

Solve   Since $a + b = 1000$, $a = 1000 - b$.

$0.25a + 0.50b = 0.34(1000)$

$0.25(1000 - b) + 0.50b = 340$ *Substitute $1000 - b$ for a.*

$250 - 0.25b + 0.50b = 340$ *Solve for b.*

$0.25b = 90$

$b = 360$

$a + b = 1000$

$a + 360 = 1000$ *Substitute 360 for b.*

$a = 640$ *Solve for a.*

Thus, 640 gallons of the 25% acid solution and 360 gallons of the 50% acid solution should be used.

Examine   The 34% acid solution contains $0.25(640) + 0.50(360) = 160 + 180$ or 340 gallons of acid. Since $0.34(1000) = 340$, the answer checks.

Systems of equations can be useful in representing real-life situations and solving real-life problems.

APPLICATION
Entertainment

The Williams family is going to the Johnstown Summer Carnival. They have two ticket options, as shown in the table below.

Ticket Option	Admission Price	Price Per Ride
A	$5	30¢
B	$3	80¢

a. Write an equation that represents the cost per person for each option.

b. Graph the equations and estimate a solution. Explain what the solution means.

c. Solve the system using substitution.

d. Write a short paragraph advising the Williams family which option to choose.

a. Let r represent the number of rides. The total cost C for each person will be the cost of admission plus the cost of the rides.

Option A: $C = 5 + 0.30r$ *The cost of the rides is the price per*

Option B: $C = 3 + 0.80r$ *ride × number of rides, r.*

b. We can estimate from the graph that the solution is about (4, 6). This means that when the number of rides equals 4, both ticket options cost about $6 per person.

c. Use substitution to solve this system.

$$C = 5 + 0.30r$$
$$3 + 0.80r = 5 + 0.30r \quad \textit{Replace C with 3 + 0.80r.}$$
$$0.50r = 2 \quad \textit{Solve for r.}$$
$$r = 4$$

$$C = 5 + 0.30r$$
$$= 5 + 0.30(4) \quad \textit{Replace r with 4.}$$
$$= 6.2 \quad \textit{Solve for C.}$$

The solution is (4, 6.2). This means that if a person rides 4 rides, both options cost the same, $6.20. From the graph, you can see that Option A tickets will cost less if a person rides more than 4 rides. Option B tickets will cost less if a person rides less than 4 rides.

d. You should advise the Williams family to purchase Option A tickets for those who plan to ride more than 4 rides and purchase Option B tickets for the rest of the family.

In-Class Example

For Example 3
Use the two ticket options below for Example 3 and advise when each option is appropriate.

A	$8	10
B	$1	1

For more than 8 rides, Option A is less expensive; for fewer than 8 rides, Option B is less expensive.

Classroom Vignette

"An alternative problem-solving strategy for part a of Example 3 is to have students observe a pattern like the one below and then determine the equation based on the pattern."

Cindy Boyd
Abilene High School
Abilene, Texas

Admission Price	Number of Rides @ 30¢	Cost
$5	1	5 + 0.30(1)
5	2	5 + 0.30(2)
•	3	5 + 0.30(3)
•	•	•
•	•	•
•	•	•
5	r	5 + 0.30(r)

Check for Understanding

Exercises 1–16 are designed to help you assess your students' understanding through reading, writing, speaking, and modeling. You should work through Exercises 1–6 with your students and then monitor their work on Exercises 7–16.

Additional Answers

1. From the first equation, y is equal to $2x - 4$, and y must have the same value in both equations.

5a. Yolanda is walking faster than Adele and catches up after about 5 seconds; 1 solution.

5b. Yolanda never catches up with Adele because they both are walking at the same rate, and Adele has a 30-ft headstart; no solution.

5c. Yolanda never catches up with Adele because Adele is walking at a faster rate and has a 30-ft headstart.

7. $x = 8 - 4y$; $y = 2 - \frac{1}{4}x$

8. $x = 4 + \frac{5}{3}y$; $y = -\frac{12}{5} + \frac{3}{5}x$

9. $x = -\frac{0.75}{0.8}y - 7.5$; $y = -\frac{0.8}{0.75}x - 8$

Study Guide Masters, p. 57

Communicating Mathematics

3. The graphs of both equations have the same slope, 9, but different y-intercepts, so the lines are parallel and the system has no solution.

Study the lesson. Then complete the following.

1. **Explain** why, when solving the system $y = 2x - 4$ and $4x - 2y = 0$, you can substitute $2x - 4$ for y in the second equation. **See margin.**

2. **State** what you would conclude if the solution to a system of linear equations yields the equation $8 = 0$. **There is no solution.**

3. **Describe** how you can tell just by looking at the equations $y = 9x + 2$ and $y = 9x - 5$ whether or not the system has a solution.

4. **Explain** why graphing a system of equations may not give you an exact solution. **Answers will vary.**

5. Yolanda is walking across campus when she sees Adele walking about 30 feet ahead of her. In each graph, t represents time in seconds and d represents distance in feet. Describe what happens in each case and how it relates to the solution. **See margin.**

a.

b.

c.

MODELING MATHEMATICS

6. Use cups and counters to model and solve the system of equations. **(3, 0)**

$$y = 2x - 6$$
$$3x + 2y = 9$$

Guided Practice

Solve each equation for x. Then, solve each equation for y.

7. $x + 4y = 8$ 8. $3x - 5y = 12$ 9. $0.8x + 6 = -0.75y$

7–9. See margin.

Use substitution to solve each system of equations. If the system does not have exactly one solution, state whether it has no solution or infinitely many solutions.

10. $y = 3x$ $(-3, -9)$
 $x + 2y = -21$

11. $x = 2y$
 $4x + 2y = 15$ $\left(3, \frac{3}{2}\right)$

12. $x + 5y = -3$ $(2, -1)$
 $3x - 2y = 8$

13. $8x + 2y = 13$ **no**
 $4x + y = 11$ **solution**

14. $2x - y = -4$ $(13, 30)$
 $-3x + y = -9$

15. $6x - 2y = -4$ infinitely
 $y = 3x + 2$ **many**

16. **Sales** Maria spent a long day working the cash register at Musicville during a sale on CDs. For this sale, all CDs in the store were marked either $12 or $10. Just when she thought she could go home, the store manager gave Maria the job of figuring out how many CDs they had sold at each price, so they could write the total in the store records. Maria doesn't want to sort through hundreds of sales slips, so she decided on an easier way. The counter at the exit of the store says that 500 people left with CDs (limit one per customer) during the sale, and the cash register contains $5750 from the day's sales. Maria wrote a system of equations for the number of $10 CDs and the number of $12 CDs.

a. What was the system of equations? $x + y = 500$, $12x + 10y = 5750$

b. How many CDs were sold at each price? $12: 375; $10: 125

Reteaching

Using Comparisons Give the class the following system of equations and have half the class use substitution by solving the first equation for y and the other half of the class use substitution by solving the second equation for x.

$$3x - 2y = 2$$
$$x + 4y = 10 \quad (2, 2)$$

Discuss whether it matters which equation is substituted into which equation and how to look for the "best" way.

For **Extra Practice**, see p. 774.

The red A, B, and C flags, printed only in the Teacher's Wraparound Edition, indicate the level of difficulty of the exercises.

Practice

Use substitution to solve each system of equations. If the system *does not* have exactly one solution, state whether it has *no solution* or *infinitely many* solutions.

17. $y = 3x - 8$ (3, 1)
$y = 4 - x$

18. $2x + 7y = 3$ (5, −1)
$x = 1 - 4y$

19. $x + y = 0$ (−4, 4)
$3x + y = -8$

20. $4c = 3d + 3$ (6, 7)
$c = d - 1$

21. $4x + 5y = 11$ (4, −1)
$y = 3x - 13$

22. $3x - 5y = 11$ (7, 2)
$x - 3y = 1$

24. $\left(\frac{9}{2}, \frac{3}{4}\right)$

23. $c - 5d = 2$ (2, 0)
$2c + d = 4$

24. $3x - 2y = 12$
$x + 2y = 6$

25. $x + 3y = 12$ (9, 1)
$x - y = 8$

26. $\left(\frac{21}{10}, \frac{7}{10}\right)$

26. $x - 3y = 0$
$3x + y = 7$

27. $5r - s = 5$ (2, 5)
$-4r + 5s = 17$

28. $2x + 3y = 1$ (−4, 3)
$-3x + y = 15$

31. (5, 2)

29. $8x + 6y = 44$ (4, 2)
$x - 8y = -12$

30. $0.5x - 2y = 17$ (50, 4)
$2x + y = 104$

31. $-0.3x + y = 0.5$
$0.5x - 0.3y = 1.9$

33. $\left(\frac{8}{3}, \frac{13}{3}\right)$

32. $x = \frac{1}{2}y + 3$ infinitely
$2x - y = 6$ many

33. $y = \frac{1}{2}x + 3$
$y = 2x - 1$

34. $y = \frac{3}{5}x$ no
$3x - 5y = 15$ solution

Use substitution to solve each system of equations. Write each solution as an ordered triple of the form (*x, y, z*).

35. $x + y + z = -54$
$x = -6y$
$z = 14y$ (36, −6, −84)

36. $2x + 3y - z = 17$
$y = -3z - 7$
$2x = z + 2$ (−1, 5, −4)

37. $12x - y + 7z = 99$
$x + 2z = 2$
$y + 3z = 9$

37. (14, 27, −6)

Critical Thinking

38. Number Theory If 36 is subtracted from certain two-digit positive integers, their digits are reversed. Find all integers for which this is true.
40, 51, 62, 73, 84, 95

Applications and Problem Solving

39. Entertainment American songwriter Cole Porter completed his first professional score in 1916 at age 23. At Harding High, this year's spring musical is *Anything Goes*, which Porter completed in 1934. The production is going to be part of a dinner theater; each ticket includes dinner and the show. The total cost of producing the show (stage, costumes, and so on) is $1000, and each dinner costs $5 to prepare. The drama club is going to sell tickets for $13 each.

a. Write a system of equations to represent the cost of and the income from the production. $y = 1000 + 5x$, $y = 13x$

b. How many tickets do they need to sell to break even? **125 tickets**

40. Humor Refer to the cartoon below. Solve the problem that is sending Peppermint Patty into a frenzy. Find how much cream and milk must be mixed together to obtain 50 gallons of cream containing $12\frac{1}{2}\%$ butterfat.
20.93 gal of cream, 29.07 gal of milk

Peanuts®

PEANUTS reprinted by permission of United Feature Syndicate, Inc.

Practice Masters, p. 57

Closing Activity

Modeling Have students create a flowchart diagram to model the steps used in solving a system of equations using the substitution method.

Chapter 8, Quiz A (Lessons 8-1 and 8-2), is available in the *Assessment and Evaluation Masters,* p. 212.

Additional Answers

43a.

	75% Gold (18-carat)	50% Gold (12-carat)	58% Gold (14-carat)
Total Grams	x	y	300
Grams of Pure Gold	$0.75x$	$0.50y$	$0.58(300)$

43b. $x + y = 300$
$0.75x + 0.50y = 0.58(300)$

44.

$y = 2x + 1$
$7y = 14x + 7$

Enrichment Masters, p. 57

---

41. **Athletes** According to *Health* magazine, top women athletes are narrowing the gap between their performances and those of their male counterparts. Speed skater Bonnie Blair's fastest time in the 500-meter would have won an Olympic gold medal in every men's 500-meter competition through 1976. The women's record time for the 500-meter in speed skating is 39.1 seconds, and the men's is 36.45 seconds. Suppose the women's record time decreases at an average rate of 0.20 second per year and the men's record time decreases at an average rate of 0.10 second per year.   **c. See students' work.**
    a. When would the women's record time equal the men's?   **26.5 years**
    b. What would the time be?   **33.8 seconds**
    c. Do you think this could actually happen? Why or why not?

42. **Accounting** Sometimes accountants must figure out how many stock shares to transfer from one person to another to reach a certain proportion of ownership. Suppose Rebeca Avila owns $3000 worth of stock in a new company that has no other stockholders. For tax purposes, the company is going to issue new stock to Muriel Eppick so that Ms. Avila owns 80%, rather than 100% of the total stock. Let S represent the new total value of company stock and let x represent the value of stock that Ms. Eppick is to receive. Use the equations below to find the value of stock to be issued to Ms. Eppick.   **$750**
    $S = 3000 + x$ *New total stock = Ms. Avila's share + Ms. Eppick's share.*
    $3000 = 0.80S$ *Ms. Avila's share is 80% of new total stock.*
    $x = 0.20S$ *Ms. Eppick's share is 20% of new total stock.*

43. **Organize Data** For thousands of years, gold has been considered one of Earth's most precious metals. When archaeologist Howard Carter discovered King Tutankhamun's tomb in 1922, he exclaimed that the tomb was filled with "strange animals, statues, and gold—everywhere the glint of gold." One hundred percent pure gold is 24-carat gold. If 18-carat gold is 75% gold and 12-carat gold is 50% gold, how much of each would be used to make a 14-carat gold bracelet weighing 300 grams? (*Hint:* 14-carat gold is about 58% gold.)
    a. Make a table to organize the data.   **See margin.**
    b. Write a system of equations that represents this problem.   **See margin.**
    c. How much 18-carat gold and 12-carat gold would it take to make a 14-carat gold bracelet weighing 300 grams?   **96 grams of 18-carat gold, 204 grams of 12-carat gold**

Mixed Review

44. Graph the system of equations below. Determine whether the system has *one* solution, *no* solutions, or *infinitely many* solutions. If the system has one solution, name it.   (Lesson 8–1)
    $y = 2x + 1$
    $7y = 14x + 7$ **Infinitely many; see margin for graph.**

45. **Finance** Michael uses at most 60% of his annual FlynnCo stock dividend to purchase more shares of FlynnCo stock. If his dividend last year was $885 and FlynnCo stock is selling for $14 per share, what is the greatest number of shares that he can purchase?   (Lesson 7–2)   **37 shares**

46. Graph $y = \frac{1}{5}x − 3$ using the slope and y-intercept.   (Lesson 6–5)   **See margin.**

47. $\{(−1, −7), (4, 8), (7, 17), (13, 35)\}$

47. Solve $3a − 4 = b$ if the domain is $\{−1, 4, 7, 13\}$.   (Lesson 5–3)

48. What is 25% less than 94?   (Lesson 4–5)   **70.5**

49. Solve $−8 − 12x = 28$.   (Lesson 3–3)   **−3**

50. See margin.

50. Graph the solution set of $n \le −2$ on a number line.   (Lesson 2–8)

51. Write an algebraic expression for *twelve less than m.*   (Lesson 1–1)   **m − 12**

Extension

Problem Solving The sum of the digits of a three-digit number is 15. The tens digit exceeds the units digit by the same amount that the hundreds digit exceeds the tens digit. If the digits are reversed, the new number is 76 times the original hundreds digit. Find the original number. **654**

Additional Answers

46.

$y = \frac{1}{5}x − 3$

50.
```
  +--+--+--⊖--+--+--+--+
 -5 -4 -3 -2 -1  0  1
```

Elimination Using Addition and Subtraction

Objective

After studying this lesson, you should be able to:

- solve systems of equations by using the elimination method with addition or subtraction.

APPLICATION

Entertainment

Disney cartoons are animated using an expensive computer process that makes the action flow smoothly and seem lifelike. In 1994, Disney's animated feature *The Lion King* was the top-grossing film of the year, making an estimated $300.4 million at the box office.

On a Saturday afternoon, the Johnson and Olivera families decided to go see *The Lion King* together. The Johnson family, two adults and four children, can afford to spend $30 from their entertainment budget this weekend for the movie tickets, while the Olivera family, two adults and two children, can afford to spend $21.50. Different theaters around town charge different amounts for adult and child tickets. What price can the Johnsons and Oliveras afford to pay for each adult and each child?

Let a represent the ticket price for one adult, and let c represent the ticket price for one child. Then the information in this problem can be represented by the following system of equations.

$$2a + 4c = 30$$
$$2a + 2c = 21.5$$

From the graph at the right, an estimated solution is ($7, $4). To get an exact solution, solve algebraically. You could solve this system by first solving either of the equations for a or c and then using substitution.

Recall that subtraction is the same as adding the opposite.

However, a simpler method of solution is to subtract one equation from the other since the coefficients of the variable a are the same. This method is called **elimination** because the subtraction eliminates one of the variables. First, write the equations in column form and subtract.

$$\begin{array}{r} 2a + 4c = 30 \\ (-)\,2a + 2c = 21.5 \end{array}$$

Multiply by −1.

$$\begin{array}{r} 2a + 4c = 30 \\ (+)\,-2a - 2c = -21.5 \\ \hline 2c = 8.5 \\ c = 4.25 \end{array}$$

Then, substitute 4.25 for c in either equation and find the value of a.

$$2a + 2c = 21.5$$
$$2a + 2(4.25) = 21.5 \quad \text{\textit{Substitute 4.25 for c.}}$$
$$2a + 8.5 = 21.5$$
$$2a = 13$$
$$a = 6.5 \quad \text{Is (6.5, 4.25) a solution of the system?}$$

NCTM Standards: 1–5

Instructional Resources

- Study Guide Master 8-3
- Practice Master 8-3
- Enrichment Master 8-3
- Assessment and Evaluation Masters, pp. 211–212

Transparency 8-3A contains the 5-Minute Check for this lesson; **Transparency 8-3B** contains a teaching aid for this lesson.

Recommended Pacing	
Standard Pacing	Day 5 of 9
Honors Pacing	Day 4 of 8
Block Scheduling*	Day 2 of 5 (along with Lesson 8-2)
Alg. 1 in Two Years*	Days 6 & 7 of 14

*For more information on pacing and possible lesson plans, refer to the *Block Scheduling Booklet* and *Algebra 1 in Two Years.*

1 FOCUS

5-Minute Check
(over Lesson 8-2)

Solve each equation for x. Then solve each equation for y.

1. $y + 1 = x$ $x = y + 1;$
   $y = x - 1$

2. $2x + 3y = 6$ $x = \frac{1}{2}(6 - 3y);$
   $y = \frac{1}{3}(6 - 2x)$

3. $y + 2x = 3$ $x = \frac{1}{2}(3 - y);$
   $y = 3 - 2x$

Motivating the Lesson

Situational Problem Tell students that you are thinking of two numbers whose sum is 30 and whose difference is 10. Ask them to identify the numbers. **20, 10** Have student volunteers make up additional problems like this one for the class to solve.

Teaching Tip To convince students that elimination is the easier way to solve the system of equations in Example 1, first have them solve the problem using substitution.

In-Class Examples

For Example 1
Use elimination to solve each system of equations.

a. $4x - 3y = 11$
$2x + 3y = 13$ $\left(4, \dfrac{5}{3}\right)$

b. $-3x + 4y = 12$
$3x - 6y = 18$ $(-24, -15)$

For Example 2
Use a system of equations to solve each problem.

a. The sum of two numbers is 27. Their difference is 5. Find the numbers. **11, 16**

b. The sum of the two numbers is 128. Their difference is 114. Find the numbers. **7, 121**

Check:

$2a + 4c = 30$	$2a + 2c = 21.5$
$2(6.5) + 4(4.25) \overset{?}{=} 30$	$2(6.5) + 2(4.25) \overset{?}{=} 21.5$
$30 = 30$ ✔	$21.5 = 21.5$ ✔

The solution of this system of equations is (6.5, 4.25). Thus, the Johnsons and Oliveras should look for a theater that charges $6.50 for each adult and $4.25 for each child.

In some systems of equations, the coefficients of terms containing the same variable are additive inverses. For these systems, the elimination method can be applied by adding the equations.

Example 1 **Use elimination to solve the system of equations.**

$3x - 2y = 4$
$4x + 2y = 10$

Since the coefficients of the y-terms, -2 and 2, are additive inverses, you can solve the system by adding the equations.

$$\begin{array}{rl} 3x - 2y = 4 & \textit{Write the equations in column form and add.} \\ (+)\ 4x + 2y = 10 & \textit{Notice that the variable y is eliminated.} \\ \hline 7x \qquad = 14 & \\ x = 2 & \end{array}$$

Now substitute 2 for x in either equation to find the value of y.

$$3x - 2y = 4$$
$$3(2) - 2y = 4$$
$$6 - 2y = 4$$
$$-2y = -2$$
$$y = 1$$

The solution of this system is (2, 1). *Check this result.*

Use subtraction to solve a system of two linear equations whenever one of the variables has the same coefficient in both equations.

Example 2 **The sum of two numbers is 18. The sum of the greater number and twice the smaller number is 25. Find the numbers.**

Let x = the greater number and let y = the lesser number. Since the sum of the numbers is 18, one equation is $x + y = 18$. Since the sum of the greater number and twice the smaller number is 25, the other equation is $x + 2y = 25$. Use elimination to solve this system.

$$\begin{array}{rl} x +\ y = 18 & \\ (-)\ x + 2y = 25 & \textit{Since the coefficients of the x terms are} \\ \hline -y = -7 & \textit{the same, use elimination by subtraction.} \\ y = 7 & \end{array}$$

Find x by substituting 7 for y in one of the equations.

$$x + y = 18$$
$$x + 7 = 18 \quad \textit{Substitute 7 for y.}$$
$$x = 11 \quad \textit{Solve for x.}$$

The solution is (11, 7), which means that the numbers are 7 and 11.
Check this result..

Alternative Learning Styles

Visual Any drawing can be approximated by a series of short straight lines. Students can draw, for example, a cat on a graph and write equations that describe each line segment.

Several software packages can be used to help you solve systems of equations.

EXPLORATION

EXPLORATION   GRAPHING SOFTWARE

The *Mathematics Exploration Toolkit (MET)* can be used to graph and solve systems of equations. Use the following CALC commands.

CLEAR F (clr f)   Removes previous graphs from the graphing window.

GRAPH (gra)   Graphs the most recent equation in the expression window.

SCALE (sca)   Sets limits on the *x*- and *y*-axes.

To set up the graphing window, enter clr f. Then enter sca 10. This sets limits on the axes at −10 to 10 for *x* and *y*. Since only two points are needed to graph a line, use the command gra 2.

Your Turn

a. Check the solutions of the examples using *MET*.   See students' work.

b. Graph the system $3x + 2y = 7$ and $5x + 2y = 17$ and find the solution using *MET*. $(5, -4)$

c. Describe the difference between solving a system of equations using graphing software and using a graphing calculator. Which do you prefer, and why?   See students' work.

EXPLORATION
This software offers an alternative method of graphing linear equations. It graphs a linear equation by specifying two points on the line. See the *Teacher's Guide for Software Resources* for information on how to obtain *MET*.

In-Class Example

For Example 3
Neal scored 1150 on his SAT. His math score was 250 points greater than his verbal score. What was his math score?   **700**

Example ❸

APPLICATION
Testing

Lina is preparing to take the Scholastic Assessment Test (SAT). She has been taking practice tests for a year, and her scores are steadily improving. She always scores about 150 points higher on the math test than on the verbal test. She needs a combined score of 1270 to get into the college she has chosen. If she assumes she will still have that 150-point difference between the two tests, how high does she need to score on each part?

Let *m* represent Lina's math score. Let *v* represent Lina's verbal score. Since the sum of her scores is 1270, one equation is $m + v = 1270$. Since the difference of her scores is 150, another equation is $m - v = 150$. Use elimination to solve this system.

$$\begin{array}{r} m + v = 1270 \\ (+)\ \underline{m - v = 150} \\ 2m\quad = 1420 \\ m = 710 \end{array}$$

Since the coefficients of the v term are additive inverses, use elimination by addition.

Find *v* by substituting 710 for *m* in one of the equations.

$$\begin{array}{r} m + v = 1270 \\ 710 + v = 1270 \\ v = 560 \end{array}$$

Substitute 710 for m.

The solution is (710, 560), which means that Lina must score 710 on the math portion and 560 on the verbal portion of the SAT.

Cooperative Learning

Think-Pair-Share Separate the class into pairs of students. Assign a system of equations and have one member of the pair solve the system using the elimination method while the other solves the system using a graphing calculator and the TRACE feature. Have the pairs check their answers together and then share them with the class. For more information on the think-pair-share strategy, see *Cooperative Learning in the Mathematics Classroom*, one of the titles in the Glencoe Mathematics Professional Series, pages 24–25.

Check for Understanding

Exercises 1–11 are designed to help you assess your students' understanding through reading, writing, speaking, and modeling. You should work through Exercises 1–3 with your students and then monitor their work on Exercises 4–11.

Error Analysis

When solving a system of equations by elimination using subtraction, have students change every sign of the equation they are subtracting in order to alleviate the problem of only subtracting the column of like variables and not the other columns too.

Additional Answers

1a. When the coefficients of one of the variables are the same.

1b. When the coefficients of one of the variables are additive inverses of each other.

Study Guide Masters, p. 58

Communicating Mathematics

Study the lesson. Then complete the following. 1–3. See margin.

1. **Explain** when it is easier to solve a system of equations in each way.
   a. by elimination using subtraction
   b. by elimination using addition

2. a. **State** the result when you add $3x - 8y = 29$ and $-3x + 8y = 16$. What does this result tell you about the system of equations?
   b. What does this result tell you about the graph of the system?

3. **You Decide** Maribela says that a system of equations has no solution if both variables are eliminated by addition or subtraction. Devin argues that there may be an infinite number of solutions. Who is correct? Explain your answer.

Guided Practice

4. Refer to the graph at the right.
   a. Estimate the solution of the system.
      Sample answer: $(-3, 4)$
   b. Use elimination to find the exact solution.
      $(-3.4, 4.2)$

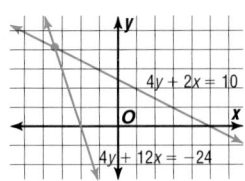

State whether addition, subtraction, or substitution would be most convenient to solve each system of equations. Then solve the system.

5. addition, $(1, 0)$
6. substitution, $(1, 2)$
7. subtraction, $\left(-\frac{5}{2}, -2\right)$
8. subtraction, $\left(\frac{3}{16}, -\frac{1}{2}\right)$

5. $3x - 5y = 3$
   $4x + 5y = 4$

6. $3x + 2y = 7$
   $y = 4x - 2$

7. $-4m + 2n = 6$
   $-4m + n = 8$

8. $8a + b = 1$
   $8a - 3b = 3$

9. $3x + y = 7$
   $2x + 5y = 22$
   substitution, $(1, 4)$

10. $2b + 4c = 8$
    $c - 2 = b$
    substitution, $(0, 2)$

11. **Statistics** The mean of two numbers is 28. Find the numbers if three times one of the numbers equals half the other number. 8, 48

Practice

For Exercises 12–14,
a. estimate the solution of each system of linear equations, and
b. use elimination to find the exact solution of each system.

A

12.

13.

14.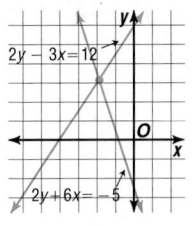

14. $(-2, 3)$; $\left(-\frac{17}{9}, \frac{19}{6}\right)$

$(-2, 1)$; $\left(-\frac{19}{9}, \frac{4}{3}\right)$

$(1, -4)$; $(1.29, -4.05)$

Reteaching

Using Models Let paper rectangles represent the coefficients of *x*, circles represent the coefficients of *y*, and squares represent constants. Use red for negatives and black for positives. Solve the system $x - 2y = 5$ and $2x + 2y = 7$. Since there are 3 rectangles after adding, separate the squares into groups of 3. Since there are 4 groups,

$x = 4$. Substitute 4 for *x* in either equation and solve for *y*. $\left(4, -\frac{1}{2}\right)$

State whether addition, subtraction, or substitution would be most convenient to solve each system of equations. Then solve the system.

18. –; (–5, 20)

19. –; (4, –7)

20. + or –; $\left(\frac{9}{2}, 2\right)$

22. –; (7, 4) **B**

23. sub; infinitely many

24. –; (–5, 4)

25. –; (–2, 3)

26. –; $\left(2, -\frac{3}{2}\right)$

27. –; $\left(\frac{1}{2}, 1\right)$

29. +; (10, –15)

 INTEGRATION
Number Theory

Critical Thinking

Applications and Problem Solving

41. Ling, 1.45 hours or 1 hour, 27 minutes; José, 1.15 hours, or 1 hour, 9 minutes

15. $x + y = 8$
    $x - y = 4$ +; (6, 2)

16. $2r + s = 5$
    $r - s = 1$ +; (2, 1)

17. $x - 3y = 7$
    $x + 2y = 2$ –; (4, –1)

18. $3x + y = 5$
    $2x + y = 10$

19. $5s + 2t = 6$
    $9s + 2t = 22$

20. $4x - 3y = 12$
    $4x + 3y = 24$

21. $2x + 3y = 13$
    $x - 3y = 2$ +; (5, 1)

22. $2m - 5n = -6$
    $2m - 7n = -14$

23. $x - 2y = 7$
    $-3x + 6y = -21$

24. $3r - 5s = -35$
    $2r - 5s = -30$

25. $13a + 5b = -11$
    $13a + 11b = 7$

26. $a - 2b - 5 = 0$
    $3a - 2b - 9 = 0$

27. $4x = 7 - 5y$
    $8x = 9 - 5y$

28. $\frac{2}{3}x + y = 7$
    $\frac{10}{3}x + 5y = 11$ sub; ∅

29. $\frac{3}{5}c - \frac{1}{5}d = 9$
    $\frac{7}{5}c + \frac{1}{5}d = 11$

30. $0.6m - 0.2n = 0.9$
    $0.3m = 0.45 - 0.1n$
    sub; (1.5, 0)

31. $1.44x - 3.24y = -5.58$
    $1.08x + 3.24y = 9.99$
    +; (1.75, 2.5)

32. $7.2m + 4.5n = 129.06$
    $7.2m + 6.7n = 136.54$
    –; (15.8, 3.4)

Use a system of equations and elimination to solve each problem.

33. Find two numbers whose sum is 64 and whose difference is 42.   11, 53

34. Find two numbers whose sum is 18 and whose difference is 22.   20, –2

35. Twice one number added to another number is 18. Four times the first number minus the other number is 12. Find the numbers.   5, 8

36. If $x + y = 11$ and $x - y = 5$, what does xy equal?   24

Use elimination twice to solve each system of equations. Write the solution as an ordered triple of the form (x, y, z).

C

37. $x + y = 5$
    $y + z = 10$
    $x + z = 9$ (2, 3, 7)

38. $2x + y + z = 13$
    $x - y + 2z = 8$
    $4x - 3z = 7$ (4, 2, 3)

39. $x + 2z = 2$
    $y + 3z = 9$
    $12x - y + 7z = 99$
    (14, 27, –6)

40. The graphs of $Ax + By = 7$ and $Ax - By = 9$ intersect at (4, –1). Find A and B.   (2, 1)

41. **On-Line Entertainment**   On June 27, 1994, Aerosmith became the first major rock band to release a song distributed exclusively in the U.S. through a computer on-line service. Users of the commercial service, CompuServe, were able to download the Aerosmith song *Head First* for free. However, it took a long time to download the song, which itself lasted only 3 minutes, 14 seconds, because of the high-fidelity sound. José and Ling share a personal computer, and one evening they each downloaded the song without realizing that the other had done it. Ling also wasted 18 minutes because he typed the wrong word and had to start over. At the end of the month, the bill from CompuServe said they had used a total of 2.6 hours of time that evening. How long did it take to download the song each time? (*Hint:* 18 minutes = 0.3 hour.)

Lesson 8–3 Elimination Using Addition and Subtraction **473**

Additional Answers

2a. The result is 0 = 45, which is false. Thus, the system has no solution.

2b. The graph is two parallel lines.

3. Both are correct. If the resulting statement is false, there is no solution. If the resulting statement is true, there is an infinite number of solutions.

Assignment Guide

Core: 13–39 odd, 40, 41, 43–49
Enriched: 12–38 even, 40–49
All: Self Test, 1–10

For **Extra Practice,** see p. 775.

The red A, B, and C flags, printed only in the Teacher's Wraparound Edition, indicate the level of difficulty of the exercises.

4 ASSESS

Closing Activity

Writing Have students write in outline form the steps to follow when solving a system of equations by the elimination method.

Chapter 8, Quiz B (Lesson 8-3), is available in the *Assessment and Evaluation Masters*, p. 212.

Mid-Chapter Test (Lessons 8-1 through 8-3) is available in the *Assessment and Evaluation Masters*, p. 211.

Practice Masters, p. 58

Additional Answer

46.

$6x - \frac{1}{2}y = -10$

42. World Cultures The ancient Egyptians believed that the pharaohs lived forever after death in their houses of eternity, the pyramids. Suppose the side of the pyramid containing the entrance is represented by the line $13x + 10y = 9600$, the opposite side of the pyramid, by the line $13x - 10y = 0$, and the descending corridor leading to the entrance, by the line $3x - 10y = 1500$, where x is the distance in feet and y is the height in feet.

    a. Find the coordinates of the entrance. **(693.75, 58.125)**

    b. Find the height of the pyramid. **480 ft**

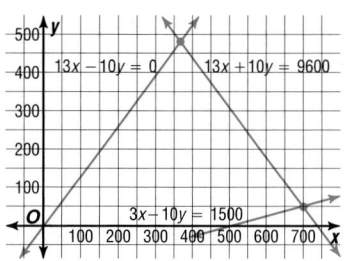

Mixed Review

43. Chemistry MX Labs needs to make 500 gallons of a 34% acid solution. The only solutions available are 25% acid and 50% acid. How many gallons of each solution should be mixed to make the 34% solution? Write and solve a system of equations by using substitution.
(Lesson 8–2) **320 gal of 25% and 180 gal of 50%**

44. Solve $5 - 8h \le 9$. (Lesson 7–1)

44. $\left\{ h \mid h \ge -\frac{1}{2} \right\}$

45. Determine the slope of the line that passes through the points at $(2, -9)$ and $(-1, 0)$. (Lesson 6–1) **−3**

46. Graph $6x - \frac{1}{2}y = -10$. (Lesson 5–4) **See margin.**

47. $-\frac{7}{6}$

47. Solve $-2(3t + 1) = 5$. (Lesson 3–3)

48. 3.87

48. Find an approximation to the nearest hundredth for $\sqrt{15}$. (Lesson 2–8)

49. Name the property illustrated by the following statement. (Lesson 1–6)
    If $6 = 2a$ and $a = 3$, then $6 = 2 \cdot 3$. **substitution (=)**

SELF TEST

Graph each system of equations. Then determine if the system has *one* solution, *no* solution, or *infinitely many* solutions. If the system has one solution, name it. (Lesson 8–1)

1. $x - y = 3$
    $3x + y = 1$ **(1, −2)**

2. $2x - 3y = 7$
    $3y = 7 + 2x$ **no solution**

3. $4x + y = 12$
    $x = 3 - \frac{1}{4}y$ **infinitely many**

1–3. See Solutions Manual for graphs.

Use substitution to solve each system of equations. (Lesson 8–2)

4. $y = 5x$
    $x + 2y = 22$ **(2, 10)**

5. $2y - x = -5$
    $y - 3x = 20$ **(−9, −7)**

6. $3x + 2y = 18$
    $x + \frac{8}{3}y = 12$ **(4, 3)**

Use elimination to solve each system of equations. (Lesson 8–3)

7. $x - y = -5$
    $x + y = 25$ **(10, 15)**

8. $3x + 5y = 14$
    $2x - 5y = 1$ **(3, 1)**

9. $5x + 4y = 12$
    $3x + 4y = 4$ **(4, −2)**

10. **Recreation** At a recreation and sports facility, 3 members and 3 nonmembers pay a total of $180 to take an aerobics class. A group of 5 members and 3 nonmembers pay $210 to take the same class. How much does it cost members and nonmembers to take an aerobics class? (Lesson 8–3)
$15 for members, $45 for nonmembers

474   Chapter 8   Solving Systems of Linear Equations and Inequalities

Extension

Connections The graphs of the equations $x - y = 1$, $x + y = 3$, and $-x + 5y = 19$ contain the sides of a triangle. Write the equation of the line that passes through the midpoint of the shortest side of the triangle and the vertex opposite the shortest side.
$x - 2y = -4$

SELF TEST

The Self Test provides students with a brief review of the concepts and skills in Lessons 8-1 through 8-3. Lesson numbers are given to the right of exercises or instruction lines so students can review concepts not yet mastered.

8-4 Elimination Using Multiplication

Objectives

After studying this lesson, you should be able to:

- solve systems of equations by using the elimination method with multiplication and addition, and
- determine the best method for solving systems of equations.

APPLICATION
Telecommunications

GBT Mobilnet provides monthly plans for cellular phone customers. Carla Ramos and Robert Johnson both selected Plan B for which monthly charges are based on per-minute rates of calls during peak and nonpeak hours. In one month, Carla made 75 minutes of peak calls and 30 minutes of nonpeak calls. Her bill was $40.05. During the same period, Robert made 50 minutes of peak calls and 60 minutes of nonpeak calls. His bill was $35.10. What is GBT Mobilnet's charge per minute for peak and nonpeak calls on Plan B?

Let p represent the rate per minute for peak calls, and let n represent the rate per minute for nonpeak calls. Then the information in this problem can be represented by the following system of equations.

$$75p + 30n = 40.05$$
$$50p + 60n = 35.10$$

So far, you have learned four methods for solving a system of two linear equations.

Method	The Best Time to Use
Graphing	if you want to estimate the solution, since graphing usually does not give an exact solution
Substitution	if one of the variables in either equation has a coefficient of 1 or −1
Addition	if one of the variables has opposite coefficients in the two equations
Subtraction	if one of the variables has the same coefficient in the two equations

The system above is not easily solved using any of these methods. However, there is an extension of the elimination method that can be used. Multiply one of the equations by some number so that adding or subtracting eliminates one of the variables.

For this system, multiply the first equation by -2 and add. Then the coefficient of n in both equations will be 60 or -60.

$$75p + 30n = 40.05 \quad \boxed{\text{Multiply by }-2.} \quad -150p - 60n = -80.10$$
$$50p + 60n = 35.10 \qquad\qquad\qquad \underline{(+)\ 50p + 60n = 35.10}$$
$$\qquad\qquad\qquad\qquad\qquad\qquad\qquad -100p \quad\quad = -45$$
$$\qquad\qquad\qquad\qquad\qquad\qquad\qquad\qquad p = 0.45$$

Now, solve for n by replacing p with 0.45.

$$75p + 30n = 40.05$$
$$75(0.45) + 30n = 40.05 \quad \textit{Substitute 0.45 for p.}$$
$$33.75 + 30n = 40.05 \quad \textit{Solve for n.}$$
$$30n = 6.3$$
$$n = 0.21 \quad \text{Is (0.45, 0.21) a solution?}$$

8-4 LESSON NOTES

NCTM Standards: 1–5

Instructional Resources

- Study Guide Master 8-4
- Practice Master 8-4
- Enrichment Master 8-4
- Assessment and Evaluation Masters, p. 213
- Multicultural Activity Masters, p. 15
- Real-World Applications, 22
- Tech Prep Applications Masters, p. 16

 Transparency 8-4A contains the 5-Minute Check for this lesson; **Transparency 8-4B** contains a teaching aid for this lesson.

Recommended Pacing

Standard Pacing	Day 6 of 9
Honors Pacing	Day 5 of 8
Block Scheduling*	Day 3 of 5 (along with Lesson 8-3)
Alg. 1 in Two Years*	Days 8 & 9 of 14

 *For more information on pacing and possible lesson plans, refer to the *Block Scheduling Booklet* and *Algebra 1 in Two Years*.

1 FOCUS

 5-Minute Check
(over Lesson 8-3)

Use elimination to solve each system of equations.

1. $2x + y = 8$
   $x - y = 3 \quad \left(\frac{11}{3}, \frac{2}{3}\right)$

2. $y + 5x = 9$
   $y - 5x = 7 \quad \left(\frac{1}{5}, 8\right)$

3. $y + 3x = 12$
   $3y - 3x = 6 \quad \left(\frac{5}{2}, \frac{9}{2}\right)$

4. The sum of two numbers is 20. Their difference is 4. Find the numbers. **12, 8**

5. Half the perimeter of a rectangle is 56 feet. The width of the rectangle is 8 feet less than the length. Find the dimensions of the rectangle. **24 ft by 32 ft**

Motivating the Lesson

Questioning Ask students if multiplying each side of an equation, such as $2x = 4$, by the same number will affect the value of the variable. For example, if each side of $2x = 4$ were multiplied by 2, the equation would read $4x = 8$. Students should recognize that the value of the variable is unaffected. Ask students if the same would hold true for an equation such as $2x + 3y = 8$. Explain that when simple addition or subtraction cannot be used to eliminate a variable in a system of equations, multiplication may be used in the elimination process.

2 TEACH

In-Class Example

For Example 1
Use elimination to solve each system of equations.

a. $3x + 5y = 11$
$2x + 3y = 7$ (2, 1)
b. $4x + 3y = 8$
$3x - 5y = -23$ (−1, 4)
c. $2x - 3y = 8$
$-5x + 2y = 13$ (−5, −6)

Teaching Tip In Example 1, emphasize that there are many other combinations of multipliers that can be used. You may want to have students suggest other multipliers.

Check:

$$75p + 30n = 40.05 \qquad\qquad 50p + 60n = 35.10$$
$$75(0.45) + 30(0.21) \overset{?}{=} 40.05 \qquad 50(0.45) + 60(0.21) \overset{?}{=} 35.10$$
$$40.05 = 40.05 \checkmark \qquad\qquad 35.10 = 35.10 \checkmark$$

The solution of this system is (0.45, 0.21). Thus, the per-minute rate for peak-hour calls is 45¢, and the per-minute rate for nonpeak calls is 21¢ on this plan.

For some systems of equations, it is necessary to multiply *each* equation by a different number in order to solve the system by elimination. You can choose to eliminate either variable.

Example **Use elimination to solve the system of equations in two different ways.**

$$2x + 3y = 5$$
$$5x + 4y = 16$$

Method 1
You can eliminate the variable x by multiplying the first equation by 5 and the second equation by -2 and then adding the resulting equations.

$$2x + 3y = 5 \qquad \text{Multiply by 5.} \qquad 10x + 15y = 25$$
$$5x + 4y = 16 \qquad \text{Multiply by } -2. \qquad \underline{(+) \ -10x - 8y = -32}$$
$$7y = -7$$
$$y = -1$$

Now find x using one of the original equations.

$$2x + 3y = 5$$
$$2x + 3(-1) = 5 \qquad \textit{Substitute } -1 \textit{ for y.}$$
$$2x - 3 = 5 \qquad \textit{Solve for x.}$$
$$2x = 8$$
$$x = 4$$

The solution of the system is (4, −1).

Method 2
You can also solve this system by eliminating the variable y. Multiply the first equation by -4 and the second equation by 3. Then add.

$$2x + 3y = 5 \qquad \text{Multiply by } -4. \qquad -8x - 12y = -20$$
$$5x + 4y = 16 \qquad \text{Multiply by 3.} \qquad \underline{(+) \ 15x + 12y = 48}$$
$$7x = 28$$
$$x = 4$$

Now find y.

$$2x + 3y = 5$$
$$2(4) + 3y = 5 \qquad \textit{Substitute 4 for x.}$$
$$8 + 3y = 5 \qquad \textit{Solve for y.}$$
$$3y = -3$$
$$y = -1$$

The solution is (4, −1), which matches the result obtained with Method 1.

Alternative Teaching Strategies

Reading Algebra The word *system* has many meanings. Have students research these meanings and determine which applies in the phrase "system of linear equations."

Example **2**

APPLICATION

Testing

Luis Diaz discovered while entering test scores into his computer that he had accidentally reversed the digits of a test and shorted a student 36 points. Mr. Diaz told the student that the sum of the digits was 14 and agreed to give the student his correct score plus extra credit if he could determine his actual score without looking at his test. What was his actual score on the test?

Explore Let t represent the tens digit of the score.
Let u represent the units digit.

The actual score on the test can be represented by $10t + u$. The amount entered in the computer can be represented by $10u + t$. *Why?*

Plan Since the sum of the digits is 14, one equation is $t + u = 14$. Since the teacher accidentally shorted the student by 36 points, another equation is $(10t + u) - (10u + t) = 36$ or $9t - 9u = 36$.

Solve

$t + u = 14$ Multiply by 9. ➤ $9t + 9u = 126$

$9t - 9u = 36$ $(+)\ 9t - 9u = 36$
                                          $\overline{ 18t = 162}$
                                                    $t = 9$

Now find u using one of the original equations.

$t + u = 14$

$9 + u = 14$ *Substitute 9 for t.*

$u = 5$ *Solve for u.*

The solution is $(9, 5)$, which means that the student's actual test score was $10(9) + 5$, or 95 points.

Examine The sum of the digits, $9 + 5$, is 14 and $95 - 59$ is 36.

You can use systems of equations to solve problems involving the distance formula, $rt = d$.

Example **3**

APPLICATION

Uniform Motion

F Y I

The world's largest riverboat is the 418-ft *American Queen*. Passengers paid up tp $9400 to take its first cruise in June, 1995.

A riverboat on the Mississippi River travels 48 miles upstream in 4 hours. The return trip takes the riverboat only 3 hours. Find the rate of the current.

Explore Let r represent the rate of the riverboat in still water. Let c represent the rate of the current.

Then $r + c$ represents the rate of the riverboat traveling downstream *with* the current and $r - c$ represents the rate of the riverboat traveling upstream *against* the current.

(continued on the next page)

Lesson 8–4 Elimination Using Multiplication **477**

In-Class Examples

For Example 2

A bank teller reversed the digits in the amount of a check and overpaid a customer by $9. The sum of the digits in the two-digit amount was 9. Find the amount of the check. **$45**

For Example 3

A boat is rowed 10 miles downstream in 2 hours, and then rowed the same distance upstream in $3\frac{1}{3}$ hours. Find the rate of the boat in still water and the rate of the current.

4 mph, 1 mph

Teaching Tip Discuss the chart in Example 3. What is the rate of the boat when it travels upstream? Why? **$r - c$; Because of the current, the boat travels upstream slower than in still water.** What is the rate of the boat when it travels downstream? Why? **$r + c$; Because of the current, the boat travels downstream faster than in still water.**

F Y I

The first commercially successful steamship, *The Clermont,* was created by Robert Fulton in 1807.

Bottom cooperative learning section

Cooperative Learning

Trade-A-Problem Separate the class into groups of four. One student selects a pair of equations for the other group members to solve. One group member should attempt to solve the equations by using graphing, another by substitution, another by elimination using addition and subtraction, and the last group member by elimination using multiplication. Group members should then discuss which methods worked the best. Have group members switch roles. For more information on the trade-a-problem strategy, see *Cooperative Learning in the Mathematics Classroom,* one of the titles in the Glencoe Mathematics Professional Series, pages 25–26.

Check for Understanding
Exercises 1–14 are designed to help you assess your students' understanding through reading, writing, speaking, and modeling. You should work through Exercises 1–4 with your students and then monitor their work on Exercises 5–14.

Error Analysis
When working a "digit problem," carefully explain the difference between the terms *digit* and *number*. Have students read the problem word for word and translate into mathematics. Example: tens digit = T
units digit = U
the number = $10T + U$
A digit has no value, but a number has a value.

Additional Answers

5. Multiply the first equation by −3, then add.
6. Multiply the first equation by 3, multiply the second equation by −2, then add.
7. Multiply the second equation by 5, then add.
8. Multiply the first equation by 5, multiply the second equation by −7, then add.
9. Multiply the first equation by 5, multiply the second equation by −8, then add.
10. Multiply the first equation by 4, multiply the second equation by 3, then add.

Plan Use the formula rate × time = distance, or $rt = d$, to write a system of equations. Then solve the system to find the value of c.

	r	t	d	$rt = d$
Downstream	$r + c$	3	48	$3r + 3c = 48$
Upstream	$r - c$	4	48	$4r - 4c = 48$

Solve

$3r + 3c = 48$ Multiply by 4. $12r + 12c = 192$

$4r - 4c = 48$ Multiply by −3. $(+) -12r + 12c = -144$

$$24c = 48$$
$$c = 2$$

The rate of the current is 2 miles per hour.

Examine Find the value of r for this system and then check the solution.

CHECK FOR UNDERSTANDING

Communicating Mathematics

Study the lesson. Then complete the following.

1. **Write** a problem about a real-life situation in which only an estimate of the solution is needed rather than the exact solution. The problem should involve a system of equations. **See students' work.**

2. To make either the *x*-term or *y*-term coefficients additive inverses.

2. **Explain** why you might need to multiply each equation by a different number when using elimination to solve a system of equations.

3. **Write** a system of equations that could best be solved by using multiplication and then elimination using addition or subtraction.
Sample answer: $3x + 5y = 7, 4x - 10y = 1$

 MATH JOURNAL

4. **Assess Yourself** Describe the method you like to use best when solving systems of linear equations. Explain your reasons. **See students' work.**

Guided Practice

Explain the steps you would follow to eliminate the variable x in each system of equations. Then solve the system. 5–7. See margin for steps.

5. $x + 5y = 4$
$3x - 7y = -10$ $(-1, 1)$

6. $2x - y = 6$
$3x + 4y = -2$ $(2, -2)$

7. $-5x + 3y = 6$
$x - y = 4$ $(-9, -13)$

Explain the steps you would follow to eliminate the variable y in each system of equations. Then solve the system. 8–10. See margin for steps.

8. $4x + 7y = 6$
$6x + 5y = 20$ $(5, -2)$

9. $3x - 8y = 13$
$4x - 5y = 6$ $(-1, -2)$

10. $2x - 3y = 2$
$5x + 4y = 28$ $(4, 2)$

Match each system of equations with the method that could be most efficiently used to solve it. Then solve the system.

11. $3x - 7y = 6$
$2x + 7y = 4$ b; (2, 0)

12. $y = 4x + 11$
$3x - 2y = -7$ a; (−3, −1)

13. $4x + 3y = 19$
$3x - 4y = 8$ c; (4, 1)

a. substitution
b. elimination using addition or subtraction
c. elimination using multiplication

Reteaching

Using Decision Making Have students go through a list of systems of equations and write which process would be best to use in order to solve (substitution, elimination using addition or subtraction, or elimination using multiplication). If elimination using multiplication is the most efficient process, have students use it to solve the problem.

14. Uniform Motion A riverboat travels 36 miles downstream in 2 hours. The return trip takes 3 hours.
a. Find the rate of the riverboat in still water. **15 mph**
b. Find the rate of the current. **3 mph**

EXERCISES

Practice

Use elimination to solve each system of equations. 23. $(-1, -2)$

15. $2x + y = 5$
$3x - 2y = 4$ **(2, 1)**

16. $4x - 3y = 12$
$x + 2y = 14$ **(6, 4)**

17. $3x - 2y = 19$
$5x + 4y = 17$ **(5, -2)**

18. $9x = 5y - 2$
$3x = 2y - 2$ **(2, 4)**

19. $7x + 3y = -1$
$4x + y = 3$ **(2, -5)**

20. $6x - 5y = 27$ **(2, -3)**
$3x + 10y = -24$

21. $8x - 3y = -11$
$2x - 5y = 27$ **(-4, -7)**

22. $11x - 5y = 80$
$9x - 15y = 120$ **(5, -5)**

23. $4x - 7y = 10$
$3x + 2y = -7$

24. $3x - \frac{1}{2}y = 10$
$5x + \frac{1}{4}y = 8$ **(2, -8)**

25. $2x + \frac{2}{3}y = 4$
$x - \frac{1}{2}y = 7$ **(4, -6)**

26. $\frac{2x + y}{3} = 15$
$\frac{3x - y}{5} = 1$ **(10, 25)**

27. $7x + 2y = 3(x + 16)$
$x + 16 = 5y + 3x$
(13, -2)

28. $0.4x + 0.5y = 2.5$
$1.2x - 3.5y = 2.5$
(5, 1)

29. $1.8x - 0.3y = 14.4$
$x - 0.6y = 2.8$
(10, 12)

INTEGRATION
Number Theory

Use a system of equations and elimination to solve each problem.

30. The sum of the digits of a two-digit number is 14. If the digits are reversed, the new number is 18 less than the original number. Find the original number. **86**

31. Three times one number equals twice a second number. Twice the first number is 3 more than the second number. Find the numbers. **6, 9**

32. The ratio of the tens digit to the units digit of a two-digit number is 1:4. If the digits are reversed, the sum of the new number and the original number is 110. Find the original number. **28**

Determine the best method to solve each system of equations. Then solve the system.

33. elim, addn; $\left(2, \frac{1}{8}\right)$

35. substitution or elim, mult.; infinitely many

39. (11, 12)

33. $9x - 8y = 17$
$4x + 8y = 9$

34. $3x - 4y = -10$ **elim, mult**
$5x + 8y = -2$ **(-2, 1)**

35. $x + 2y = -1$
$2x + 4y = -2$

36. $5x + 3y = 12$ **elim, mult**
$4x - 5y = 17$ **(3, -1)**

37. $\frac{2}{3}x - \frac{1}{2}y = 14$ **elim, subtr**
$\frac{5}{6}x - \frac{1}{2}y = 18$ **(24, 4)**

38. $\frac{1}{2}x - \frac{2}{3}y = \frac{7}{3}$ **elim, mult**
$\frac{3}{2}x + 2y = -25$ **(-6, -8)**

Use elimination to solve each system of equations.

39. $\frac{1}{x - 5} - \frac{3}{y + 6} = 0$
$\frac{2}{x + 7} - \frac{1}{y - 3} = 0$

40. $\frac{2}{x} + \frac{3}{y} = 16$
$\frac{1}{x} + \frac{1}{y} = 7$ $\left(\frac{1}{5}, \frac{1}{2}\right)$

41. $\frac{1}{x - y} = \frac{1}{y}$
$\frac{1}{x + y} = 2$ $\left(\frac{1}{3}, \frac{1}{6}\right)$

Lesson 8–4 Elimination Using Multiplication **479**

Assignment Guide

Core: 15–45 odd, 47–54
Enriched: 16–42 even, 43–54

For **Extra Practice,** see p. 775.

The red A, B, and C flags, printed only in the Teacher's Wraparound Edition, indicate the level of difficulty of the exercises.

Study Guide Masters, p. 59

NAME_____ DATE_____
8-4
Study Guide
Student Edition
Pages 475–481

Elimination Using Multiplication

Some systems of equations cannot be solved simply by adding or subtracting the equations. One or both equations must first be multiplied by a number before the system can be solved by elimination. Consider the following example.

Example: Use elimination to solve the system of equations
$x + 10y = 3$ and $4x + 5y = 5$.

$x + 10y = 3$ Multiply $x + 10y = 3$ $-4x - 40y = -12$
$4x + 5y = 5$ by -4. $4x + 5y = 5$
           Then add the $-35y = -7$
           equations. $y = \frac{1}{5}$

Substitute $\frac{1}{5}$ for y into either original equation and solve for x.

$x + 10\left(\frac{1}{5}\right) = 3$
$x + 2 = 3$
$x = 1$

The solution of the system is $\left(1, \frac{1}{5}\right)$.

Use elimination to solve each system of equations.

1. $3x + 2y = 0$
$x - 5y = 17$
(2, -3)

2. $2x + 3y = 6$
$x + 2y = 5$
(-3, 4)

3. $3x - y = 2$
$x + 2y = 3$
(1, 1)

4. $4x + 5y = 6$
$6x - 7y = -20$
(-1, 2)

Use a system of equations and elimination to solve each problem.

5. The length of Sally's garden is 4 meters greater than 3 times the width. The perimeter of her garden is 72 meters. What are the dimensions of Sally's garden?
28 meters by 8 meters

6. Anita is $4\frac{1}{2}$ years older than Basilio. Three times Anita's age added to six times Basilio's age is 36. How old are Anita and Basilio?
Basilio is $2\frac{1}{2}$ yr and Anita is 7 yr.

Programming

42. The graphing calculator program at the right finds the solution of two linear equations written in standard form.

$$ax + by = c$$
$$dx + ey = f$$

The formulas for the solution of this system are as follows.

$$x = \frac{ce - bf}{ae - bd}, \quad y = \frac{af - cd}{ae - bd}$$

Use the program to solve each system.

a. $8x + 2y = 0$ infinitely
   $12x + 3y = 0$ many

b. $x - 2y = 5$
   $3x - 5y = 8$ $(-9, -7)$

c. $5x + 5y = 16$ no
   $2x + 2y = 5$ solution

```
PROGRAM:SOLVE
: Disp "ENTER COEFFICIENTS"
: Prompt A, B, C, D, E, F
: If AE−BD = 0
: Then
: Goto 1
: End
: (CE−BF)/(AE−BD) → X
: (AF−CD)/(AE−BD) → Y
: Disp "THE SOLUTION IS"
: Disp "X= ", X
: Disp "Y= ", Y
: Stop
: Lbl 1
: If CE−BF=0 or AF−CD=0
: Then
: Disp "INFINITELY", "MANY"
: Else
: Disp "NO SOLUTION"
```

d. $7x - 3y = 5$ infinitely
   $14x - 6y = 10$ many

Critical Thinking

43. The graphs of the equations $5x + 4y = 18$, $2x + 9y = 59$, and $3x - 5y = -4$ contain the sides of a triangle. Determine the coordinates of the vertices of the triangle. $(-2, 7), (2, 2), (7, 5)$

Applications and Problem Solving

44. **Geography** Benjamin Banneker, a self-taught mathematician and astronomer, was the first African-American to publish an almanac. He is most noted for being the assistant surveyor on the team that designed the ten-mile square of Washington, D.C. The White House is located in the center of the square, at the intersection of Pennsylvania Avenue and New York Avenue. Let $-5x + 7y = 0$ represent New York Avenue and let $3x + 8y = 305$ represent Pennsylvania Avenue. Find the coordinates for the White House. $(35, 25)$

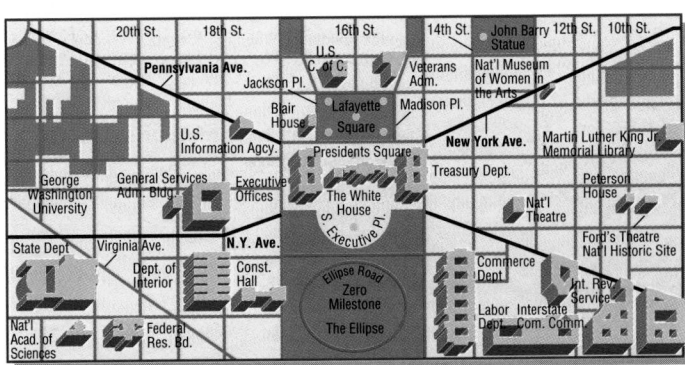

45. **Organize Data** At the new Cozy Inn Restaurant, which is still under construction, the owners have hired enough waiters and waitresses to handle 17 tables of customers. The fire marshall has looked at the plans for the restaurant and says he will approve it for a limit of 56 customers. The restaurant owners are now deciding how many two-seat tables and how many four-seat tables to buy for the restaurant. How many of each kind should they buy? 6 2-seat tables, 11 4-seat tables

480 Chapter 8 Solving Systems of Linear Equations and Inequalities

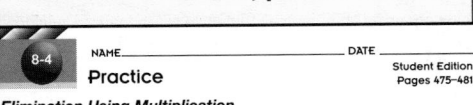
Extension ▬▬▬

Connections The graphs of the equations $5x + 4y = 18$, $2x + 9y = 59$, and $3x - 5y = -4$ contain the sides of a triangle. Determine the vertices of the triangle. $(-2, 7), (2, 2), (7, 5)$

Tech Prep

Surveyor Students who are interested in surveying may wish to do further research on the role of Benjamin Banneker in the building of Washington, D.C., as mentioned in Exercise 44, and on the role of surveyors in the construction of planned communities today. For more information on tech prep, see the *Teacher's Handbook*.

46. Information Highway The Mercury Center provides a reference and research service for on-line computer users based on peak and nonpeak usage. Miriam and Lesharo are both subscribers. The chart below displays the number of peak and nonpeak minutes each of them spent on-line in one month and how much it cost. Use the information to find the Mercury Center's rate per minute for its peak and nonpeak on-line research service. **$0.45 and $0.15**

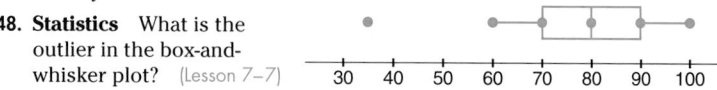

User	Number of Peak Minutes	Number of Nonpeak Minutes	Cost
Lesharo	45	50	$27.75
Miriam	70	30	$36

Mixed Review

47. Use elimination to solve the system of equations. (Lesson 8–3) **(3, −4)**
$$2x - y = 10$$
$$5x + 3y = 3$$

48. Statistics What is the outlier in the box-and-whisker plot? (Lesson 7–7)

48. 35
49. $\frac{1}{3}$

49. Probability If Bill, Raul, and Kenyatta each have an equal chance of winning a bicycle race, find the probability that Raul finishes last. (Lesson 7–5)

50. Find the coordinates of the midpoint of the line segment whose endpoints are at (1, 6) and (−3, 4). (Lesson 6–7) **(−1, 5)**

51. Track Alfonso runs a 440-yard race in 55 seconds, and Marcus runs it in 88 seconds. To have Alfonso and Marcus finish at the same time, how much of a head start should Alfonso give Marcus? (Lesson 4–7) **165 yd**

52. If $12m = 4$, then $3m =$ __?__ . (Lesson 3–2) **1**

53. 2 cups

53. Cooking If there are four sticks in a pound of butter and each stick is $\frac{1}{2}$ cup, how many cups of butter are in a pound of butter? (Lesson 2–6)

54. Evaluate $288 \div [3(9 + 3)]$. (Lesson 1–3) **8**

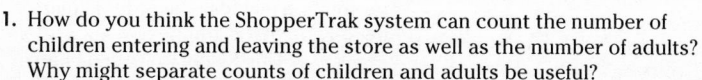

High-Tech Checkout Lanes   1–2. See margin.

The article below appeared in *Progressive Grocer* in February, 1994.

KMART...HAS INSTALLED A NEW technology in 48 stores that helps ensure that enough checklanes are open to serve customers in a store. The system counts the number of adults and children who are in a store at any given moment. The system, called ShopperTrak...uses infrared technology on door-mounted units to give a continuous count of shoppers entering and exiting a store...The data is channeled to software in a PC called FastLane, which uses it to calculate how many checklanes should be open during the next 20 minutes so that no more than two to three people are in line at each lane. Managers read the data at monitors stationed at the checkout area. ∎

1. How do you think the ShopperTrak system can count the number of children entering and leaving the store as well as the number of adults? Why might separate counts of children and adults be useful?

2. Do you think the average shopping times would differ between men and women, boys and girls, or senior citizens and young people? Why do you think the ShopperTrak system doesn't consider these factors?

Lesson 8–4 Elimination Using Multiplication **481**

Answers for Mathematics and Society

1. The infrared detection system can measure the heights of persons entering and leaving the store, and uses the height measurements to attempt to differentiate between adults and children. The higher the ratio of adults to children, the more checkout clerks would be needed.

2. Sample answer: The age and gender factors may not, by themselves, yield any useful data. Even if there were measurable differences, it may be too complicated and costly to include them in a system. For example, the device can count people in the ShopperTrak system, but it can't detect age or gender.

APPLICATION
Employment

Standards: 1–5, 8

Instructional Resources

- Study Guide Master 8-5
- Practice Master 8-5
- Enrichment Master 8-5
- Assessment and Evaluation Masters, p. 213
- Modeling Mathematics Masters, pp. 57–59
- Multicultural Activity Masters, p. 16

Transparency 8-5A contains the 5-Minute Check for this lesson; **Transparency 8-5B** contains a teaching aid for this lesson.

Recommended Pacing

Standard Pacing	Day 7 of 9
Honors Pacing	Day 6 of 8
Block Scheduling*	Day 4 of 5
Alg. 1 in Two Years*	Days 10 & 11 of 14

*For more information on pacing and possible lesson plans, refer to the *Block Scheduling Booklet* and *Algebra 1 in Two Years*.

1 FOCUS

5-Minute Check
(over Lesson 8-4)

Use elimination to solve each system of equations.

1. $\frac{1}{3}x - \frac{5}{6}y = -6$
   $4x + 7y = 30$ (−3, 6)
2. $2a + b = 19$
   $3a - 2b = -3$ (5, 9)
3. $x - y = 6$
   $x - 3y = 7$ $\left(\frac{11}{2}, -\frac{1}{2}\right)$
4. $x + y = 8$
   $2x - y = -6$ $\left(\frac{2}{3}, \frac{22}{3}\right)$
5. $y = 2x$
   $2x + y = 10$ $\left(\frac{5}{2}, 5\right)$

Objective

After studying this lesson, you should be able to:

- solve systems of inequalities by graphing.

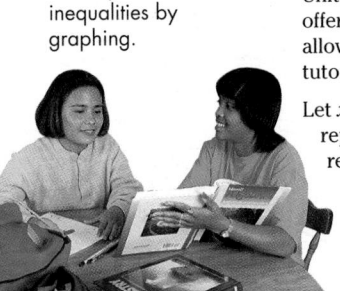

LOOK BACK

You can refer to Lesson 7-8 for information on graphing inequalities in two variables.

Unita likes her job as a baby-sitter, but it pays only $3 per hour. She has been offered a job as a tutor that pays $6 per hour. Because of school, her parents only allow her to work a maximum of 15 hours per week. How many hours can Unita tutor *and* baby-sit and still make at least $65 per week?

Let x represent the number of hours Unita can baby-sit each week. Let y represent the number of hours she can tutor each week. Since both x and y represent a number of hours, neither can be a negative number. Thus, $x \geq 0$ and $y \geq 0$. Then the following **system of inequalities** can be used to represent the conditions of this problem.

$$x \geq 0$$
$$y \geq 0$$
$$3x + 6y \geq 65 \quad \text{\textit{She wants to earn at least \$65.}}$$
$$x + y \leq 15 \quad \text{\textit{She can work up to 15 hours.}}$$

The solution of this system is the set of all ordered pairs that satisfies both inequalities and lies in the first quadrant. The solution can be determined by graphing each inequality on the same coordinate plane.

Recall that the graph of each inequality is called a *half-plane*. The intersection of the two half-planes represents the solution to the system of inequalities. This solution is a region that contains the graphs of an infinite number of ordered pairs. The boundary line of the half-plane is solid and is included in the graph if the inequality is ≤ or ≥. The boundary line of the half-plane is dashed and is not included in the graph if the inequality is < or >.

The graphs of $3x + 6y = 65$ and $x + y = 15$ are the boundaries of the region and are included in the graph of this system. This region is shown in green above. Only the portion in the first quadrant is shaded since $x \geq 0$ and $y \geq 0$. Every point in this region is a possible solution to the system. For example, since the graph of (5, 9) is a point in the region, Unita could baby-sit for 5 hours and tutor for 9 hours. In this case, she would make $3(5) + $6(9) or $69. *Does this meet her requirements of time and earnings?*

Example **Solve each system of inequalities by graphing.**

a. $y < 2x + 1$
   $y \geq -x + 3$

The solution includes the ordered pairs in the intersection of the graphs of $y < 2x + 1$ and $y \geq -x + 3$. This region is shaded in green at the right. The graphs of $y = 2x + 1$ and $y = -x + 3$ are the boundaries of this region. The graph of $y = 2x + 1$ is dashed and is *not* included in the graph of $y < 2x + 1$. The graph of $y = -x + 3$ is included in the graph of $y \geq -x + 3$.

GLOBAL CONNECTIONS

Europeans first learned of the Hawaiian Islands in 1778 when Captain James Cook travelled to the islands and named them the Sandwich Islands.

GLENCOE Technology

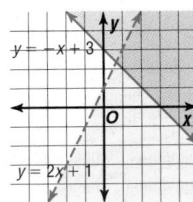
CD-ROM Interaction

A multimedia simulation allows students to use graphing to determine which cellular phone service is the most cost effective. A blackline master activity with teacher's notes provides a follow-up to the CD-ROM simulation.

For Windows & Macintosh

b. $2x + y \geq 4$
$y \leq -2x - 1$

The graphs of $2x + y = 4$ and $y = -2x - 1$ are parallel lines. Because the two regions have no points in common, the system of inequalities has no solution.

Sometimes in real-life problems involving systems, only whole-number solutions make sense.

Example 2

APPLICATION
Vacations

Elena Ayala wants to spend no more than $700 for hotels while vacationing in Hawaii. She wants to stay at the Hyatt Resort at least one night and at the Coral Reef Hotel for the remainder of her stay. The Hyatt Resort costs $130 per night, and the Coral Reef Hotel costs $85 per night.

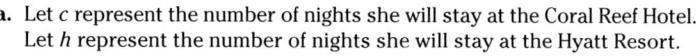

a. If she wants to stay in Hawaii at least 6 nights, how many nights could she spend at each hotel and still stay within her budget?

b. What advice might you give Elena concerning her options?

a. Let c represent the number of nights she will stay at the Coral Reef Hotel. Let h represent the number of nights she will stay at the Hyatt Resort.

Then the following system of inequalities can be used to represent the conditions of this problem.

$h + c \geq 6$ *Elena wants to stay at least 6 nights.*
$h \geq 1$ *She wants to stay at least 1 night at the Hyatt Resort.*
$130h + 85c \leq 700$ *She wants to spend no more than $700.*

The solution is the set of all ordered pairs whose graphs are in the intersection of the graphs of these inequalities. This region is shown in brown at the right.

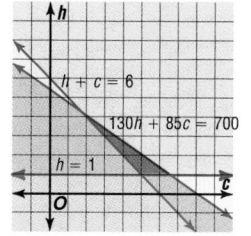

Any point in this region is a possible solution; however, only whole-number solutions make sense in this problem. *Why?* For example, since $(3, 3)$ is a point in the region, Elena could stay 3 nights at each hotel. In this case, she would spend 3($130) or $390 at the Hyatt Resort and 3($85) or $255 at the Coral Reef Hotel for a total of $645. The other solutions are $(5, 1)$, $(6, 1)$, $(4, 2)$, $(5, 2)$ and $(2, 4)$. *Check this result.*

b. You could advise Elena that she could stay in Hawaii a maximum of 7 nights if she stayed at the Hyatt Resort only 1 or 2 nights and stayed at the Coral Reef Hotel for the remainder of her vacation.

Lesson 8–5   Graphing Systems of Inequalities **483**

3 PRACTICE/APPLY

A graphing calculator is a useful tool for graphing systems of inequalities. It is important to enter the functions in the correct order, since this determines the shading.

EXPLORATION
GRAPHING CALCULATORS

You can use a graphing calculator to solve systems of inequalities. The TI-82 graphs functions and shades above the first function entered and below the second function entered. Select 7 on the DRAW menu to choose the SHADE feature. First, enter the function that is the lower boundary of the region to be shaded. (Note that inequalities that have > or $\ge$ are lower boundaries and inequalities that have < or $\le$ are upper boundaries.) Press [,]. Then enter the function that is the upper boundary of the region. Press [ )] [ENTER].

Your Turn   a. See margin.   b. See students' work.

a. Use a graphing calculator to graph the system of inequalities.
   $y \ge 4x - 3$
   $y \le -2x + 9$

b. Use a graphing calculator to work through the examples in this lesson. List and explain any disadvantages that you discovered when using the graphing calculator to graph systems of inequalities.

c. Describe the process of using a graphing calculator to solve systems of linear inequalities in your own words.   See students' work.

CHECK FOR UNDERSTANDING

Communicating Mathematics

1. A boundary line is included in the graph if the inequality is $\le$ or $\ge$ and not included if the inequality is < or >.

3. Sample answer: $y < x + 1$ and $y > x + 3$. There is no intersection.

MATH JOURNAL

Study the lesson. Then complete the following.

1. **Explain** how to determine whether boundary lines should be included in the graph of a system of inequalities.

2. **You Decide**   Joshua says that the intersection point of the boundary lines is always a solution of a system of inequalities. Rolanda says the point of intersection may not be part of the solution set. Explain who is correct and give an example to support your answer.   See margin.

3. **Write** a system of inequalities that has no solutions. Describe the graph of your system.

4. **State** which points are solutions to the system of inequalities graphed at the right. Explain how you know.
   a. $(0, 0)$ yes   b. $(-1, 4)$ yes
   c. $(2, 5)$ no   d. $(0.5, -1.7)$ no

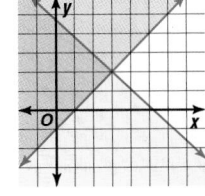

5. Describe a real-life situation that you can model using a system of linear inequalities.
   Answers will vary.

Reteaching

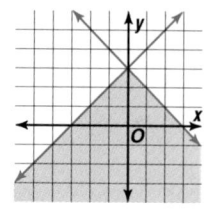

Solve each system of inequalities by graphing.

6. $x < 1$
   $x > -4$

7. $y \geq -2$
   $y - x < 1$

8. $y \geq 2x + 1$
   $y \leq -x + 1$

9. $y \geq 3x$
   $3y < 5x$

10. $y - x < 1$
    $y - x > 3$

11. $2x + y \leq 4$
    $3x - y \geq 6$

Write a system of inequalities for each graph.

12.

13. $y \leq -x + 3,$
    $y \leq x + 3$

    $y > x,$
    $y \leq x + 4$

14. **Sales** Ms. Johnson's homeroom class can order up to $90 of free pizzas from Angelino's Pizza as a reward for selling the most magazines during the magazine drive. They need to order at least 6 large pizzas in order to serve the entire class. If a pepperoni pizza costs $9.95 and a supreme pizza costs $12.95, how many of each type can they order? List three possible solutions. See margin.

EXERCISES

Solve each system of inequalities by graphing.

15. $x > 5$
    $y \leq 4$

16. $y < 0$
    $x \geq 0$

17. $y > 3$
    $y > -x + 4$

18. $x \leq 2$
    $y - 4 \geq 5$

19. $x \geq 2$
    $y + x \leq 5$

20. $y < -3$
    $x - y > 1$

21. $y \leq 2x + 3$
    $y < -x + 1$

22. $y - x < 3$
    $y - x \geq 2$

23. $y \geq 3x$
    $7y < 2x$

24. $x - y < -1$
    $x - y > 3$

25. $2y + x < 6$
    $3x - y > 4$

26. $3x - 4y < 1$
    $x + 2y \leq 7$

27. $y - 4 > x$
    $y + x < 4$

28. $5y \geq 3x + 10$
    $2y \leq 4x - 10$

29. $y + 2 \leq x$
    $2y - 3 > 2x$

30. $2x + y \geq -4$
    $-5x + 2y < 1$

31. $x + y > 4$
    $-2x + 3y < -12$

32. $-4x + 5y \leq 41$
    $x + y > -1$

Write a system of inequalities for each graph.

33. $y > -1, x \geq -2$
34. $3x - 5y \geq -25,$
    $y \geq 0, x \leq 0$
35. $y \leq x, y > x - 3$

33.

34.

35.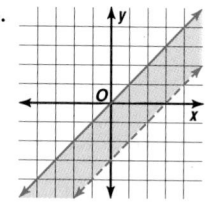

Alternative Teaching Strategies

Student Diversity When students are deciding which area in a graph to shade, have them think it through logically. Emphasize using a solution that makes sense rather than guessing.

Assignment Guide

Core: 15–47 odd, 48–56
Enriched: 16–44 even, 45–56

For **Extra Practice,** see p. 775.

The red A, B, and C flags, printed only in the Teacher's Wraparound Edition, indicate the level of difficulty of the exercises.

Additional Answers

2. Rolanda is correct; Sample answer: $y < 2x + 4$ and $y > x - 1$; the solution of $y = 2x + 4$ and $y = x - 1$, $(-5, -6)$, is not a solution to the system of inequalities.

14.

Sample answer:
7 pepperoni, 1 supreme;
5 pepperoni, 3 supreme;
2 pepperoni, 5 supreme

Practice Masters, p. 60

4 ASSESS

Closing Activity

Speaking Ask students to explain how to solve a system of inequalities by graphing.

Chapter 8, Quiz D (Lesson 8-5), is available in the *Assessment and Evaluation Masters*, p. 213.

Enrichment Masters, p. 60

Enrichment

NAME _____ DATE _____

Student Edition Pages 482–486

Describing Regions

The shaded region inside the triangle can be described with a system of three inequalities.

$y < -x + 1$
$y > \frac{1}{3}x - 3$
$y > -9x - 31$

Write systems of inequalities to describe each region. You may first need to divide a region into triangles or quadrilaterals.

1.
$y < \frac{5}{2}x + 12$
$y < -\frac{5}{2}x + 12$
$y < 2 \quad y > -3$

2.
$y > -x - 2 \qquad y < x + 2$
$y > x - 2 \qquad y > -x + 2$
$y < 5 \qquad y > -5$

3.
top: $y < \frac{3}{2}x + 7$, $y < -\frac{3}{2}x + 7$, $y > 4$
middle: $y < 4$, $y > 0$, $y > -\frac{4}{3}x - 4$, $y > \frac{4}{3}x - 4$
bottom left: $y < 4x + 12$, $y > \frac{1}{2}x - 2$, $y < 0$, $x < 0$
bottom right: $y < -4x + 12$, $y > -\frac{1}{2}x - 2$, $y < 0$, $x < 0$

486 Chapter 8

Write a system of inequalities for each graph.

C▷

36.
37.
38.

36. $2x - 3y \geq -6$, $x + y > -3$

37. $x \geq 0$, $y \geq 0$, $x + 2y \leq 6$

38. $4x - 3y \geq -12$, $2x + y \leq 4$, $x + 2y \geq 2$

Solve each system of inequalities by graphing.

39–41. See Solutions Manual.

39. $x - 2y \leq 2$
$3x + 4y \leq 12$
$x \geq 0$

40. $x - y \leq 5$
$5x + 3y \geq -6$
$y \leq 3$

41. $x < 2$
$4y > x$
$2x - y < -9$
$x + 3y < 9$

Graphing Calculator

Use a graphing calculator to solve each system of inequalities.

42. $y \geq 3x - 6$
$y \leq x + 1$

43. $y \leq x + 9$
$y > -x - 4$

44. $y < 2x + 10$
$y \geq 7x + 15$

42–44. See Solutions Manual.

Critical Thinking

45. Solve the inequality $|y| \leq 3$ by graphing. (*Hint:* Graph as a system of inequalities.) **See Solutions Manual.**

Applications and Problem Solving

Graph a system of inequalities to solve each problem.

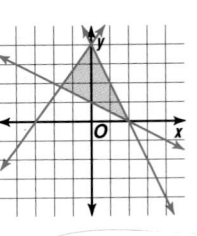

46. **Nutrition** Young people between the ages of 11 and 18 should get at least 1200 milligrams of calcium each day. One ounce of mozzarella cheese has 147 milligrams of calcium, and one ounce of Swiss cheese has 219 milligrams. If you wanted to eat no more than 8 ounces of cheese, how much of each type could you eat and still get your daily requirement of calcium? List three possible solutions. **See margin.**

47. See margin.

47. **Organize Data** Kenny Choung likes to exercise every day by walking and jogging at least 3 miles. Kenny walks at a rate of 4 mph and jogs at a rate of 8 mph. If he has only a half hour to exercise, how much time can he spend walking and jogging and cover at least 3 miles? List 3 possible solutions.

Mixed Review

48. **Number Theory** If the digits of a two-digit positive integer are reversed, the result is 6 less than twice the original number. Find all such integers for which this is true. (Lesson 8–4) **24**

49. **Organize Data** When Roberta cashed her check for $180, the bank teller gave her 12 bills, each one worth either $5 or $20. How many of each bill did she receive? (Lesson 8–2) **4 $5 bills, 8 $20 bills**

50. $\{a \mid -2 > a > 3\}$; See margin for graph.

50. Solve $4 > 4a + 12 > 24$ and graph the solution set. (Lesson 7–4)

51. $y = -\frac{1}{2}x + \frac{9}{2}$

51. Write an equation in slope-intercept form of a line that passes through the points at $(3, 3)$ and $(-1, 5)$. (Lesson 6–2)

52. Solve $y = -\frac{1}{2}x + 3$ if the domain is $\{2, 4, 6\}$. (Lesson 5–3) **{2, 1, 0}**

53. What number increased by 40% equals 14? (Lesson 4–5) **10**

54. **Travel** Paloma Rey drove to work on Wednesday at 40 miles per hour and arrived one minute late. She left home at the same time on Thursday, drove 45 miles per hour, and arrived one minute early. How far does Ms. Rey drive to work? (*Hint:* Convert hours to minutes.) (Lesson 3–5) **12 miles**

55. Let $y =$ the number of yards gained in both games; $y = 134 + (134 - 17)$

55. Define a variable, then write an equation for the following problem. Diego gained 134 yards running. This was 17 yards more than in the previous game. How many yards did he gain in both games? (Lesson 2–9)

56. assoc. (×)

56. Name the property illustrated by $(3 \cdot x) \cdot y = 3 \cdot (x \cdot y)$. (Lesson 1–8)

486 Chapter 8 *Solving Systems of Linear Equations and Inequalities*

Extension

Reasoning Solve the system of inequalities: $x + 4y \leq 13$, $x - 5y \leq 16$, $x + y \leq 4$, $x + 2y \geq -5$, and $2x - y \geq -10$.

Tech Prep

Nutritionist Nutritionists often plan meals to provide certain vitamins and minerals, such as the calcium mentioned in Exercise 46. Students who are interested in nutrition may wish to further explore the role of mathematics in this career. For more information on tech prep, see the *Teacher's Handbook*.

VOCABULARY

After completing this chapter, you should be able to define each term, property, or phrase and give an example or two of each.

Algebra

consistent (p. 455)
dependent (p. 456)
elimination (p. 469)
independent (p. 456)
inconsistent (p. 456)

substitution (p. 462)
system of equations (p. 455)
system of inequalities (p. 482)

Problem Solving

organize data (p. 464)

UNDERSTANDING AND USING THE VOCABULARY

Choose the correct term to complete each statement.

1. The method used in solving the following system of equations is (*elimination, substitution*).

$$x = 4y + 1 \}$$
$$x + y = 6 \}$$ $\rightarrow$

$(4y + 1) + y = 6$
$5y + 1 = 6$
$5y = 5$
$y = 1$

$x = 4(1) + 1$
$x = 4 + 1$
$x = 5$
solution: $(5, 1)$

substitution

2. If a system of equations has exactly one solution, it is (*dependent, independent*). **independent**

3. If the graph of a system of equations is parallel lines, the system of equations is said to be (*consistent, inconsistent*). **inconsistent**

4. A system of equations that has infinitely many solutions is (*dependent, independent*). **dependent**

5. The method used in solving the following system of equations is (*elimination, substitution*). **elimination**

$$-2c + b = 3 \}$$
$$-b - c = -6 \}$$ $\rightarrow$

$b - 2c = 3$
$(+) -b - c = -6$
$\overline{-3c = -3}$
$c = 1$

$-b - (1) = -6$
$-b - 1 = -6$
$-b = -5$
$b = 5$ solution: $(5, 1)$

6. If a system of equations has the same slope and different intercepts, the graph of the system is (*intersecting lines, parallel lines*). **parallel lines**

7. If a system of equations has the same slope and intercepts, the system has (*exactly one, infinitely many*) solutions. **infinitely many**

8. The solution to a system of equations is $(3, -5)$; therefore, this system is (*consistent, inconsistent*). **consistent**

9. The graph of a system of equations is shown at the right. This system has (*infinitely many, no*) solution. **no**

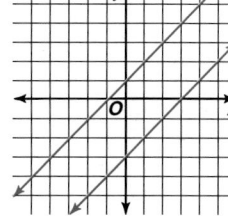

10. The solution to a system of inequalities is the (*intersection, union*) of two half-planes. **intersection**

11. A system of inequalities that includes $x < 0$ and $y > 0$ is in the (*second, fourth*) quadrant. **second**

Instructional Resources

Three multiple-choice tests and three free-response tests are provided in the *Assessment and Evaluation Masters*. Forms 1A and 2A are for honors pacing, and Forms 1B, 1C, 2B, and 2C are for average pacing. Chapter 8 Test, Form 1B is shown at the right. Chapter 8 Test, Form 2B is shown on the next page.

Using the CHAPTER HIGHLIGHTS

The Chapter Highlights begins with a listing of the new terms, properties, and phrases that were introduced in this chapter. Have students define each term and provide an example or two of it, if appropriate.

Assessment and Evaluation Masters, pp. 199–200

Using the STUDY GUIDE AND ASSESSMENT

Skills and Concepts Encourage students to refer to the objectives and examples on the left as they complete the review exercises on the right.

Assessment and Evaluation Masters, pp. 205–206

OBJECTIVES AND EXAMPLES

Upon completing this chapter, you should be able to:

• solve systems of equations by graphing (Lesson 8–1)

Graph $x + y = 6$ and $x - y = 2$. Then find the solution.

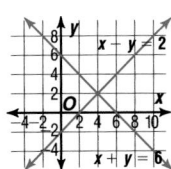

The solution is (4, 2).

• determine whether a system of equations has one solution, no solution, or infinitely many solutions by graphing (Lesson 8–1)

Graph $3x + y = -4$ and $6x + 2y = -8$. Then determine the number of solutions.

There are infinitely many solutions.

• solve systems of equations by using the substitution method (Lesson 8–2)

Use substitution to solve the system of equations.

$y = x - 1$
$4x - y = 19$

$$4x - y = 19 \qquad\qquad y = x - 1$$
$$4x - (x - 1) = 19 \qquad y = 6 - 1$$
$$4x - x + 1 = 19 \qquad\ y = 5$$
$$3x + 1 = 19$$
$$3x = 18$$
$$x = 6$$

The solution is (6, 5).

REVIEW EXERCISES

Use these exercises to review and prepare for the chapter test. **12–15. See margin for graphs.**

Graph each system of equations to find the solution.

12. $y = 2x - 7$ **(6, 5)**   13. $x + 2y = 6$
    $x + y = 11$ $2y - 8 = -x$

14. $3x + y = -8$ **(−3, 1)** 15. $5x - 3y = 11$
    $x + 6y = 3$ $2x + 3y = -25$

13. no solution   **(−2, −7)**

Graph each system of equations. Then determine whether the system of equations has *one* solution, *no* solution, or *infinitely many* solutions. If the system has one solution, name it.

16. $x - y = 9$ 17. $9x + 2 = 3y$
    $x + y = 11$ $y - 3x = 8$

18. $2x - 3y = 4$ 19. $3x - y = 8$
    $6y = 4x - 8$ $3x = 4 - y$

16. one, (10, 1)   17. no solution
18. infinitely many   19. one, (2, −2)
16–19. See margin for graphs.

Use substitution to solve each system of equations. If the system *does not* have exactly one solution, state whether it has *no* solution or *infinitely many* solutions.

20. $2m + n = 1$ **(3, −5)**   21. $3a - 2b = -4$ **(0, 2)**
    $m - n = 8$ $3a + b = 2$

22. $x = 3 - 2y$ 23. $3x - y = 1$
    $2x + 4y = 6$ $2x + 4y = 3$

    infinitely many $\left(\frac{1}{2}, \frac{1}{2}\right)$

GLENCOE Technology

Test and Review Software

You may use this software, a combination of an item generator and item bank, to create your own tests or worksheets. Types of items include free response, multiple choice, short answer, and open ended.

For IBM & Macintosh

Additional Answers

12.

13.
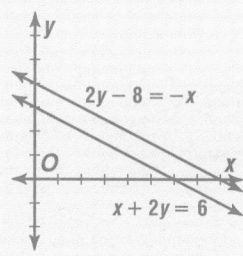

OBJECTIVES AND EXAMPLES

• solve systems of equations by using the elimination method with addition or subtraction (Lesson 8–3)

Use elimination to solve the system of equations.

$2m - n = 4$
$m + n = 2$

$$\begin{array}{c|c} 2m - n = 4 & m + n = 2 \\ (+)\ m + n = 2 & 2 + n = 2 \\ \hline 3m = 6 & n = 0 \\ m = 2 & \end{array}$$

The solution is $(2, 0)$.

REVIEW EXERCISES

Use elimination to solve each system of equations.

24. $x + 2y = 6$ $(2, 2)$
  $x - 3y = -4$

25. $2m - n = 5$ $(2, -1)$
  $2m + n = 3$

26. $3x - y = 11$ $(4, 1)$
  $x + y = 5$

27. $3s + 6r = 33$ $(5, 1)$
  $6r - 9s = 21$

28. $3x + 1 = -7y$
  $6x + 7y = 0$
  $\left(\frac{1}{3}, -\frac{2}{7}\right)$

29. $12x - 9y = 114$
  $7y + 12x = 82$
  $(8, -2)$

14.

• solve systems of equations by using the elimination method with multiplication and addition (Lesson 8–4)

Use elimination to solve the system of equations.

$3x - 4y = 7$
$2x + y = 1$

$$\begin{array}{cc} 3x - 4y = 7 & 3x - 4y = 7 \\ 2x + y = 1 \xrightarrow{\text{Multiply by 4.}} & (+)\ 8x + 4y = 4 \\ \hline & 11x\ \ \ \ = 11 \\ & x = 1 \end{array}$$

$2x + y = 1$
$2(1) + y = 1$
$y = -1$

The solution is $(1, -1)$.

Use elimination to solve each system of equations.

30. $x - 5y = 0$ $(5, 1)$
  $2x - 3y = 7$

31. $x - 2y = 5$
  $3x - 5y = 8$

32. $2x + 3y = 8$
  $x - y = 2$

33. $-5x + 8y = 21$
  $10x + 3y = 15$

34. $5m + 2n = -8$
  $4m + 3n = 2$

35. $6x + 7y = 5$ $(2, -1)$
  $2x - 3y = 7$

31. $(-9, -7)$

32. $\left(\frac{14}{5}, \frac{4}{5}\right)$

33. $\left(\frac{3}{5}, 3\right)$

34. $(-4, 6)$

15.

• determine the best method for solving systems of equations (Lesson 8–4)

Use the best method to solve the system of equations.

$x + 2y = 8$
$3x + 2y = 6$

$$\begin{array}{c|c} 3x + 2y = 6 & x + 2y = 8 \\ (-)\ x + 2y = 8 & -1 + 2y = 8 \\ \hline 2x = -2 & 2y = 9 \\ x = -1 & y = \frac{9}{2} \end{array}$$

The solution is $\left(-1, \frac{9}{2}\right)$.

Determine the best method to solve each system of equations. Then solve the system.

36. $y = 2x$
  $x + 2y = 8$

37. $9x + 8y = 7$
  $18x - 15y = 14$

38. $2x - y = 36$
  $3x - 0.5y = 26$

39. $3x + 5y = 2x$ $(0, 0)$
  $x + 3y = y$

40. $5x - 2y = 23$
  $5x + 2y = 17$

41. $2x + y = 3x - 15$
  $x + 5 = 4y + 2x$

36. $\left(\frac{8}{5}, \frac{16}{5}\right)$

37. $\left(\frac{7}{9}, 0\right)$

38. $(4, -28)$

40. $\left(4, -\frac{3}{2}\right)$

41. $(13, -2)$

16.

17.

Additional Answers

18.

19.

Applications and Problem Solving Encourage students to work through the exercises in the Applications and Problem Solving section to strengthen their problem-solving skills.

Additional Answers

42.

43.

44.

45.

46.

47.

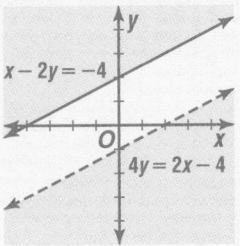

OBJECTIVES AND EXAMPLES

• solve systems of inequalities by graphing (Lesson 8–5)

Solve the system of inequalities.

$x \geq -3$
$y \leq x + 2$

REVIEW EXERCISES

Solve each system of inequalities by graphing.

42. $y < 3x$
$x + 2y \geq -21$

43. $y > -x - 1$
$y \leq 2x + 1$

44. $2x + y < 9$
$x + 11y < -6$

45. $x \geq 1$
$y + x \leq 3$

46. $y \geq x - 3$
$y \geq -x - 1$

47. $x - 2y \leq -4$
$4y < 2x - 4$

42–47. See margin for graphs.

APPLICATIONS AND PROBLEM SOLVING

48. **Ballooning** A hot-air balloon is 10 meters above the ground rising at a rate of 15 meters per minute. Another balloon is 150 meters above the ground descending at a rate of 20 meters per minute. (Lesson 8–1)
   a. After how long will the balloons be at the same height? **4 min**
   b. What is that height? **70 m**

49. **Number Theory** A two-digit number is 7 times its units digit. If 18 is added to the number, its digits are reversed. Find the original number. (Lesson 8–2) **35**

50. **Travel** While driving to Fullerton, Mrs. Sumner travels at an average speed of 40 mph. On the return trip, she travels at an average speed of 56 mph and saves two hours of travel time. How far does Mrs. Sumner live from Fullerton? (Lesson 8–4) **280 miles**

51. **Sales** The Beach Resort is offering two weekend specials. One includes a 2-night stay with 3 meals and costs $195. The other includes a 3-night stay with 5 meals and costs $300. (Lesson 8–4)
   a. What is the cost of a 1-night stay? **$75**
   b. What is the cost per meal? **$15**

52. **Organize Data** Abby plans to spend at most $24 to buy cashews and peanuts for her Fourth of July party. The Nut Shoppe sells peanuts for $3 a pound and cashews for $5 a pound. If Abby needs to have at least 5 pounds of nuts for the party, how many of each type can she buy? List three possible solutions. (Lesson 8–5)
   Answers will vary. Possible solutions: 1 lb of cashews, 6 lb of peanuts; 3 lb of cashews, 3 lb of peanuts; 4 lb of cashews, 1 lb of peanuts

A practice test for Chapter 8 is provided on page 794.

ALTERNATIVE ASSESSMENT

COOPERATIVE LEARNING PROJECT

Landscaping   Gavin Royse needs a layout of his yard. He is doing some landscaping and needs to have a grid of his yard for the landscape workers to use when putting in his fence and the two new trees that he bought. His house sits on an angle on the lot, which measures 80 feet by 80 feet. The boundaries of his house are described by the four equations below.

$3x - 4y = 55$
$3x + 2y = 85$
$3x - 4y = -125$
$3x + 2y = 175$

The house faces southwest.

The fence will run from the farthest north corner of the house to (5, 80) and then run along the north lot line to the corner of the lot (80, 80). It will then head south on the lot line to (80, 30). From there it will angle toward the house to (55, 5) and then head to the farthest east corner of the house.

A shade tree will be placed at the intersection of $3x - 4y = 55$ and $x + 3y = 170$, while an ornamental tree will be placed 15 feet south of the west corner of the house.

Prepare a graph for the landscape workers to use as a guide. Follow these steps to organize your data.

- Determine how to set up the graph and its scale.
- Construct a colorful and detailed graph that Gavin could give to the landscape workers.
- Create the system of inequalities that describes Gavin's fenced area.
- Determine the vertices of the house.
- Determine the vertices of the two trees.

Write a detailed description for the workers of the work that needs to be done and the area where the work is to be done.

THINKING CRITICALLY

- How does the elimination or substitution method show that a system of equations is inconsistent or that a system of equations is consistent and dependent?
- Create a system of equations for which there are no solutions. What were your criteria for creating this system?

PORTFOLIO

Select one of the systems of equations from this chapter that could be solved by various methods. Use this system of equations and solve it using each of the methods introduced in this chapter: graphing, substitution, elimination using addition and subtraction, and elimination using multiplication and addition. Write an explanation involving the pros and cons of using each of these methods for this problem.

SELF EVALUATION

Are you a team player? Do you pull your end of the load? Do you pull too much of the load? Since you will often need to work in cooperative groups, you need to be responsible for your actions and understand how they affect the group. A person who is too dominant can stifle the learning process of other students. On the other end of the spectrum, a person who is too passive can get left out and not experience the whole learning process.

Assess yourself. What kind of a group worker are you? Do you take the initiative to enhance the whole group or do you look out only for your own learning? Think about a group that you were recently a part of, either in mathematics or your daily life. List two positive actions that you observed in the group that helped the group. Then list two negative actions that you observed in the group that hindered the group.

Assessment and Evaluation Masters, pp. 210, 221

8   NAME_____ DATE _____

Chapter 8 Performance Assessment

Instructions: *Demonstrate your knowledge by giving a clear, concise solution to each problem. Be sure to include all relevant drawings and justify your answers. You may show your solution in more than one way or investigate beyond the requirements of the problem.*

1. **a.** Tell how you know that a system of two linear equations has no solutions.
   b. Write a system of two linear equations that has no solution.
   c. Tell how you know that a system of two linear equations has more than one solution. How many solutions will such a system have? Justify your answer.
   d. Write a system of two linear equations that has more than one solution.
   e. Write a system of two linear equations that has only one solution.
   f. Solve the system of equations in part e in at least two ways.
   g. Write a word problem that can be solved using a system of two linear equations. Solve the problem and give the meaning of the answer.

2. **a.** Write a system of inequalities whose solution is represented by the shaded region shown on the graph at the right.
   b. Name a point that is a solution of the system of inequalities.
   c. Name a point that is not a solution of the system of inequalities.

Alternative Assessment

The Alternative Assessment section provides students with the opportunity to assess their own work by thinking critically, working with others, keeping a portfolio, and honestly evaluating their own progress. For more information on alternative forms of assessment, see *Alternative Assessment in the Mathematics Classroom,* one of the titles in the Glencoe Mathematics Professional Series.

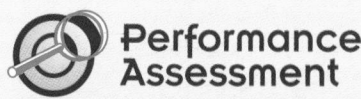

Performance Assessment

Performance Assessment tasks for this chapter are included in the *Assessment and Evaluation Masters.* A scoring guide is also provided.

Scoring Guide
Chapter 8
Performance Assessment

Level	Specific Criteria
3 Superior	• Shows thorough understanding of the concepts of *systems of linear equations and inequalities* and *solving such systems graphically and algebraically.* • Uses appropriate strategies to solve problems. • Computations are correct. • Written explanations are exemplary. • Word problem concerning a system of two linear equations is appropriate and makes sense. • Graphs are accurate and appropriate. • Goes beyond requirements of some or all problems.
2 Satisfactory, with Minor Flaws	• Shows understanding of the concepts of *systems of linear equations and inequalities* and *solving such systems graphically and algebraically.* • Uses appropriate strategies to solve problems. • Computations are mostly correct. • Written explanations are effective. • Word problem concerning a system of two linear equations is appropriate and makes sense. • Graphs are mostly accurate and appropriate. • Satisfies all requirements of problems.
1 Nearly Satisfactory, with Serious Flaws	• Shows understanding of most of the concepts of *systems of linear equations and inequalities* and *solving such systems graphically and algebraically.* • May not use appropriate strategies to solve problems. • Computations are mostly correct. • Written explanations are satisfactory. • Word problem concerning a system of two linear equations is mostly appropriate and sensible. • Graphs are mostly accurate and appropriate. • Satisfies most requirements of problems.
0 Unsatisfactory	• Shows little or no understanding of the concepts of *systems of linear equations and inequalities* and *solving such systems graphically and algebraically.* • May not use appropriate strategies to solve problems. • Computations are incorrect. • Written explanations are not satisfactory. • Word problem concerning a system of two linear equations is not appropriate or sensible. • Graphs are not accurate or appropriate. • Does not satisfy requirements of problems.

These two pages review the skills and concepts presented in Chapters 1–8. This review is formatted to reflect new trends in standardized testing.

A more traditional cumulative review is provided in the *Assessment and Evaluation Masters*, pp. 215–216.

Assessment and Evaluation Masters, pp. 215–216

8

NAME_____ DATE_____

Chapter 8 Cumulative Review

1. State the property shown in $3a + 7c = 7c + 3a$. (Lesson 1-6)
   1. ___commutative of addition___

2. Simplify $6(-4) + (-12)(-3)$. (Lesson 2-6)
   2. ___12___

Solve each equation. (Lessons 3-2, 3-3, and 3-5)
3. $-\frac{3}{4}x = 30$
   3. ___-40___
4. $2p - 3(p + 2) = 5(2p + 1)$
   4. ___-1___
5. $\frac{7}{10} = \frac{3}{x+1}$
   5. ___$3\frac{2}{7}$___

Solve each inequality. (Lessons 7-1 and 7-3)
6. $-6 + d > -14$
   6. ___$d > -8$___
7. $10y - 3(y + 4) \le 0$
   7. ___$y \le \frac{12}{7}$___
8. Solve the compound inequality $2 > -4$ and $2t - 4 \le 6$. Then graph the solution set. (Lesson 7-4)
   8. ___$\{t \mid -3 < t \le 5\}$___
   -5 -4 -3 -2 -1 0 1 2 3 4 5

9. A clothing store makes a profit of $5.50 on each tie sold. How many ties must the store sell to make a profit of at least $352.00? (Lesson 7-2)
   9. ___at least 64___

10. A price decreased from $85 to $72.25. Find the percent of decrease. (Lesson 4-5)
    10. ___15%___

11. Sara leaves home at 7 A.M. traveling at a rate of 45 mi/h. Her son discovers that she has forgotten her briefcase and starts out to catch up with her. Her son leaves at 7:30 A.M. traveling at a rate of 55 mi/h. At what time will he overtake his mother? (Lesson 4-7)
    11. ___9:45 A.M.___

12. Tracey Tierney invested $6,000 for one year, part at 10% annual interest and the balance at 13% annual interest. Her total interest for the year was $712.50. How much money did she invest at each rate? (Lesson 4-7)
    12. ___$2250 at 10%; $3750 at 13%___

Tell in which quadrant or on which axis the given point lies. (Lesson 5-1)
13. $(-2, 0)$
    13. ___x-axis___
14. $(5, -3)$
    14. ___Quadrant IV___

8

NAME_____ DATE_____

Chapter 8 Cumulative Review (continued)

For exercises 15 and 16, refer to the triangle shown below. (Lesson 4-3)
15. Find b to the nearest tenth.
    15. ___13.5___
16. Find c to the nearest tenth.
    16. ___3.7___

17. Determine the slope of the line passing through $(-2, 1)$ and $(-6, 5)$. (Lesson 6-1)
    17. ___-1___

18. Find the x- and y-intercepts for the graph of $x - 4y = 4$. (Lesson 6-4)
    18. ___4; -1___

19. Write an equation in slope-intercept form for the line that is perpendicular to the graph of $y - 3x = -4$ and passes through $(1, 2)$. (Lesson 6-6)
    19. ___$y = -\frac{1}{3}x + \frac{7}{3}$___

20. Find the coordinates of the midpoint of the line segment whose endpoints have the coordinates $(-1, 4)$ and $(2, 7)$. (Lesson 6-7)
    20. ___$\left(\frac{1}{2}, 5\frac{1}{2}\right)$___

21. Use the coordinate plane provided to graph the equations $x + y = 4$ and $y = x$. Then state the solution set. (Lesson 8-1)
    21. ___$\{(2, 2)\}$___

22. Use substitution to solve the system of equations $-x - 5y = 7$ and $x + y = 1$. (Lesson 8-2)
    22. ___$(3, -2)$___

Use elimination to solve each system of equations. (Lessons 8-3 and 8-4)
23. $2x + 4y = 1$
    $x - 4y = 5$
    23. ___$\left(2, -\frac{3}{4}\right)$___

24. $2x - 2y = 6$
    $x + y = 3$
    24. ___$(3, 0)$___

25. With the wind, an airplane travels 2200 miles in 4 hours. Against the wind, it takes 5 hours to travel 2350 miles. Find the rate of the wind and the rate of the plane in still air. (Lesson 8-4)
    25. ___40 mi/h; 510 mi/h___

CUMULATIVE REVIEW

CHAPTERS 1–8

SECTION ONE: MULTIPLE CHOICE

There are eight multiple-choice questions in this section. After working each problem, write the letter of the correct answer on your paper.

1. Choose the statement that is true for a system of two linear equations. **D**
   A. There are no solutions when the graphs of the equations are perpendicular lines.
   B. There is exactly one solution when the graphs of the equations are one line.
   C. A system can only be solved by graphing the equations.
   D. There are infinitely many solutions when the graphs of the equations have the same slope and intercepts.

2. The scale on a map is 2 centimeters to 5 kilometers. Doe Creek and Kent are 15.75 kilometers apart. How far apart are they on the map? **C**
   A. 7.88 cm
   B. 39.38 cm
   C. 6.3 cm
   D. 31.5 cm

3. Choose the graph that represents $2x - y < 6$. **C**

   A.
   B.
   C.
   D.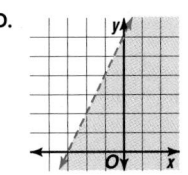

4. The units digit of a two-digit number exceeds twice the tens digit by 1. Find the number if the sum of its digits is 7. **A**
   A. 25   B. 16
   C. 34   D. 61

5. State which region in the graph shown below is the solution of the system. **B**
   $y \ge 2x + 2$
   $y \le -x - 1$

   A. Region A
   B. Region B
   C. Region C
   D. Region D

6. Abeytu's scores on the first four of five 100-point tests were 85, 89, 90, and 81. What score must she receive on the fifth test to have an average of at least 87 points for all of the tests? **B**
   A. at least 86 points
   B. at least 90 points
   C. at least 69 points
   D. at least 87 points

7. The frequency of a vibrating violin string is inversely proportional to its length. If a 10-inch violin string vibrates at a frequency of 512 cycles per second, find the frequency of an 8-inch violin string. **D**
   A. 284.4 cycles per second
   B. 409.6 cycles per second
   C. 514 cycles per second
   D. 640 cycles per second

8. Find the mean of $3\frac{1}{2}$, 5, $4\frac{1}{8}$, $7\frac{3}{4}$, 4, and $6\frac{5}{8}$. **A**
   A. $5\frac{1}{6}$
   B. 6
   C. $6\frac{1}{5}$
   D. 31

Standardized Test Practice Questions are also provided in the *Assessment and Evaluation Masters*, p. 214.

SECTION TWO: SHORT ANSWER

This section contains seven questions for which you will provide short answers. Write your answer on your paper.

9. Eric is preparing to run in the Bay Marathon. One day, he ran and walked a total of 16 miles. He ran the first mile, and after that walked one mile for every two miles he ran. How many miles did he run, and how many did he walk?

   11 miles, 5 miles

10. The number of Calories in a serving of French fries at 13 restaurants are 250, 240, 220, 348, 199, 200, 125, 230, 274, 239, 212, 240, and 327. Make a box-and-whisker plot of these data.

    See Solutions Manual.

11. Find three consecutive odd integers whose sum is 81.

    25, 27, 29

12. The concession stand sells hot dogs and soda during Beck High School football games. John bought 6 hot dogs and 4 sodas and paid $6.70. Jessica bought 4 hot dogs and 3 sodas and paid $4.65. At what prices are the hot dogs and sodas sold?

    hot dogs: $0.75, soda: $0.55

13. Draw a tree diagram to show the possible meals that could be created from the following choices.

    Meat: chicken, steak

    Vegetable: broccoli, baked potato, tossed salad, carrots

    Drink: milk, cola, juice **See Solutions Manual.**

14. Patricia is going to purchase graduation gifts for her friends, Sarah and Isabel. She wants to spend at least $5 more on Isabel's gift than on Sarah's. She can afford to spend at most a total of $56. Draw a graph showing the possible amounts she can spend on each gift.

    See Solutions manual.

15. The cost of a one-day car rental from Rossi Rentals is given by the formula $C(m) = 31 + 0.13m$, where m is the number of miles that the car is driven, $0.13 is the cost per mile driven, and $C(m)$ is the total cost. If Sheila drove a distance of 110 miles and back in one day, what is the cost of the car rental? **$45.30**

SECTION THREE: OPEN-ENDED

This section contains two open-ended problems. Demonstrate your knowledge by giving a clear, concise solution to each problem. Your score on these problems will depend on how well you do the following.

- Explain your reasoning.

- Show your understanding of the mathematics in an organized manner.

- Use charts, graphs, and diagrams in your explanation.

- Show the solution in more than one way or relate it to other situations.

- Investigate beyond the requirements of the problem.

16. Curtis has a coupon for 33% off any purchase of $50 or more at First Place Sports Shop. His grandmother has given him $75 for his birthday. Write a problem about how Curtis spends his money if he wants to buy several items, including a baseball hat for $18.75, a pair of cleats for $53.95, and a $26.95 sweatshirt discounted 15%. Then solve the problem.

    See students' work.

17. The graphs of the equations $y = x + 3$, $2x - 7y = 4$, and $2y + 3x = 6$ contain the sides of a triangle. Write a problem about these graphs that requires using systems of equations to solve. Then solve the problem.

    See students' work.

Cumulative Review Chapters 1–8 **493**

Exploring Polynomials

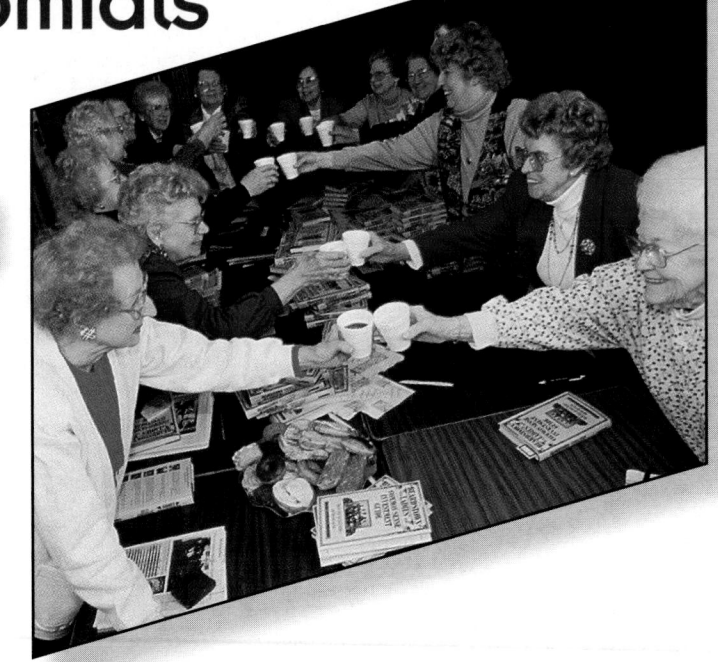

PREVIEWING THE CHAPTER

This chapter helps students master the skills necessary to perform operations with monomials and polynomials. Multiplication and division of monomials are developed first as students connect their knowledge of the powers of whole numbers with the concept of monomials containing variables and exponents. Students also learn to look for a pattern, an important problem-solving strategy. Polynomials are then defined, and students learn to add, subtract, and multiply polynomials, modeling the various operations with algebra tiles. The chapter concludes with a look at some special polynomial products.

Lesson (Pages)	Lesson Objectives	NCTM Standards	State/Local Objectives
9-1 (496–500)	Multiply monomials. Simplify expressions involving powers of monomials. Solve problems by looking for a pattern.	1–5	
9-2 (501–505)	Simplify expressions involving quotients of monomials. Simplify expressions containing negative exponents.	1–5	
9-3 (506–512)	Express numbers in scientific and standard notation. Find products and quotients of numbers expressed in scientific notation.	1–5	
9-4A (513)	Use algebra tiles to model polynomials.	1–5	
9-4 (514–519)	Find the degree of a polynomial. Arrange the terms of a polynomial so that the powers of a variable are in ascending or descending order.	1–5	
9-5A (520–521)	Use algebra tiles to add and subtract polynomials.	1–5	
9-5 (522–527)	Add and subtract polynomials.	1–5	
9-6A (528)	Use algebra tiles to model the product of simple polynomials.	1–5	
9-6 (529–533)	Multiply a polynomial by a monomial. Simplify expressions involving polynomials.	1–5	
9-7A (534–535)	Use algebra tiles to find the product of binomials.	1–5	
9-7 (536–541)	Use the FOIL method to multiply two binomials. Multiply any two polynomials by using the distributive property.	1–5	
9-8 (542–547)	Use patterns to find $(a + b)^2$, $(a - b)^2$, and $(a + b)(a - b)$.	1–5	

A complete, 1-page lesson plan is provided for each lesson in the *Lesson Planning Guide*. Answer keys for each lesson are available in the *Answer Key Masters*.

ORGANIZING THE CHAPTER

You may want to refer to the **Course Planning Calendar** on page T12 for detailed information on pacing.
PACING: **Standard**—15 days; **Honors**—14 days; **Block**—8 days; **Two Years**—26 days

LESSON PLANNING CHART

| Lesson (Pages) | Materials/ Manipulatives | Extra Practice (Student Edition) | BLACKLINE MASTERS | | | | | | | | | Real-World Applications | Interactive Mathematics Tools Software | Teaching Transparencies |
			Study Guide	Practice	Enrichment	Assessment and Evaluation	Modeling Mathematics	Multicultural Activity	Tech Prep Applications	Graphing Calculator	Science and Math Lab Manual			
9-1 (496–500)	calculator	p. 776	p. 61	p. 61	p. 61					p. 9		23	9-1	9-1A 9-1B
9-2 (501–505)	scientific calculator	p. 776	p. 62	p. 62	p. 62	p. 240		p. 17						9-2A 9-2B
9-3 (506–512)	graphing calculator	p. 776	p. 63	p. 63	p. 63				p. 17		pp. 39–42	24		9-3A 9-3B
9-4A (513)	algebra tiles*						p. 28						9-4A	
9-4 (514–519)		p. 777	p. 64	p. 64	p. 64	pp. 239, 240		p. 18						9-4A 9-4B
9-5A (520–521)	algebra tiles*						p. 29						9-5A.1 9-5A.2	
9-5 (522–527)		p. 777	p. 65	p. 65	p. 65									9-5A 9-5B
9-6A (528)	algebra tiles* product mat*						p. 30							
9-6 (529–533)		p. 777	p. 66	p. 66	p. 66	p. 241	p. 80		p. 18					9-6A 9-6B
9-7A (534–535)	algebra tiles* product mat*						p. 31						9-7A	
9-7 (536–541)		p. 778	p. 67	p. 67	p. 67		pp. 60–62							9-7A 9-7B
9-8 (542–547)	algebra tiles* product mat*	p. 778	p. 68	p. 68	p. 68	p. 241								9-8A 9-8B
Study Guide/ Assessment (549–553)						pp. 225– 238, 242– 244								

*Included in Glencoe's Student Manipulative Kit and Overhead Manipulative Resources.

ORGANIZING THE CHAPTER

OTHER CHAPTER RESOURCES

Student Edition
Chapter Opener, pp. 494–495
Mathematics and Society, p. 512
Working on the Investigation,
  pp. 519, 541
Closing the Investigation,
  p. 548

Teacher's Classroom Resources
Investigations and Projects Masters,
  pp. 57–60
Algebra and Geometry Overhead
  Manipulative Resources,
  pp. 26–31

Technology
Test and Review Software (IBM
  and Macintosh)
CD-ROM Interactions (Windows
  and Macintosh)

Professional Publications
Block Scheduling Booklet
Glencoe Mathematics Professional
  Series

OUTSIDE RESOURCES

Books/Periodicals
Developing Skills in Algebra One: Book B, Dale
  Seymour Publications

Software
Tools of Mathematics: Algebra, William K. Bradford
Visualizing Algebra: The Function Analyzer,
  Sunburst

Videos/CD-ROMs
*Adding and Subtracting Algebraic
  Expressions/Multiplying Polynomials,*
  Coronet/MTI
Polynomials, Dale Seymour Publications

ASSESSMENT RESOURCES

Student Edition
Math Journal, pp. 499, 504
Mixed Review, pp. 500, 505,
  511, 518, 526, 533, 540, 547
Self Test, p. 527
Chapter Highlights, p. 549
Chapter Study Guide and
  Assessment, pp. 550–552
Alternative Assessment, p. 553
  Portfolio, p. 553

Teacher's Wraparound Edition
5-Minute Check, pp. 496, 501,
  506, 514, 522, 529, 536, 542
Check for Understanding, pp. 499,
  503, 509, 516, 524, 531, 538,
  546
Closing Activity, pp. 500, 505,
  512, 519, 527, 533, 541, 547
Cooperative Learning, pp. 498,
  523

Assessment and Evaluation Masters
Multiple-Choice Tests, Forms 1A
  (Honors), 1B (Average), 1C
  (Basic), pp. 225–230
Free-Response Tests, Forms 2A
  (Honors), 2B (Average), 2C
  (Basic), pp. 231–236
Calculator-Based Test, p. 237
Performance Assessment, p. 238
Mid-Chapter Test, p. 239
Quizzes A–D, pp. 240–241
Standardized Test Practice, p. 242
Cumulative Review, pp. 243–244

ENHANCING THE CHAPTER

Examples of some of the materials for enhancing Chapter 9 are shown below.

DIVERSITY

Multicultural Activity Masters, pp. 17, 18

APPLICATIONS

Real-World Applications, 23, 24

TECHNOLOGY

Graphing Calculator Masters, p. 9

TECH PREP

Tech Prep Applications Masters, pp. 17, 18

CONNECTIONS

Science and Math Lab Manual, pp. 39–42

PROBLEM SOLVING

Problem of the Week Cards, 25, 26, 27

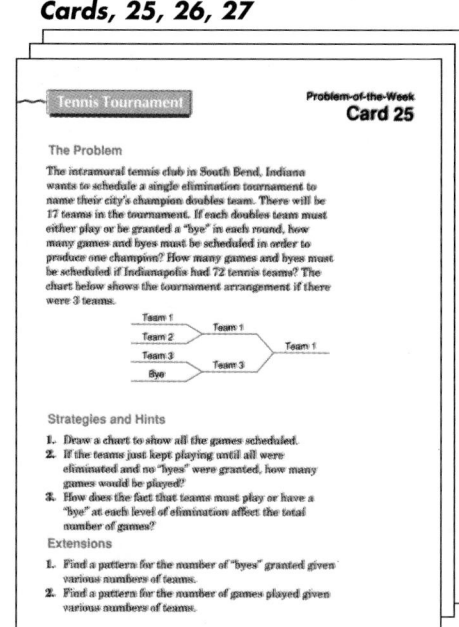

Exploring Polynomials

Objectives

In this chapter, you will:
- solve problems by looking for a pattern,
- multiply and divide monomials,
- express numbers in scientific notation, and
- add, subtract, and multiply polynomials.

Keeping Up With Technology

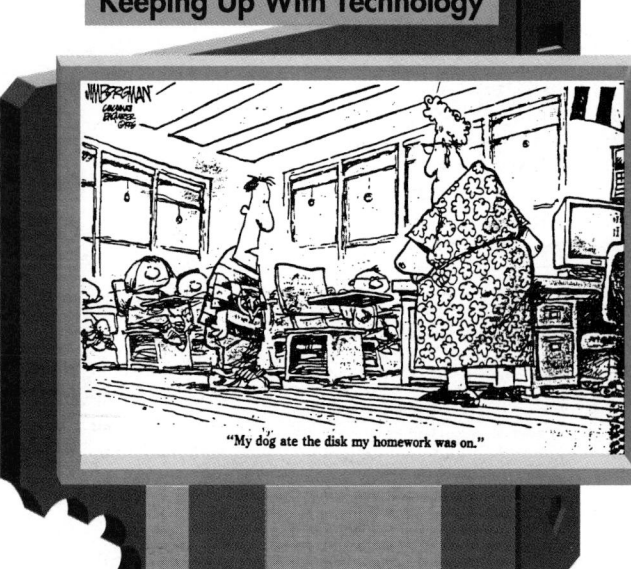

"My dog ate the disk my homework was on."

You've probably heard lots of excuses for not having homework done, but this one is probably a new one. Do you use a computer to do your homework? What are the advantages and disadvantages of doing your homework this way?

TIME *Line*

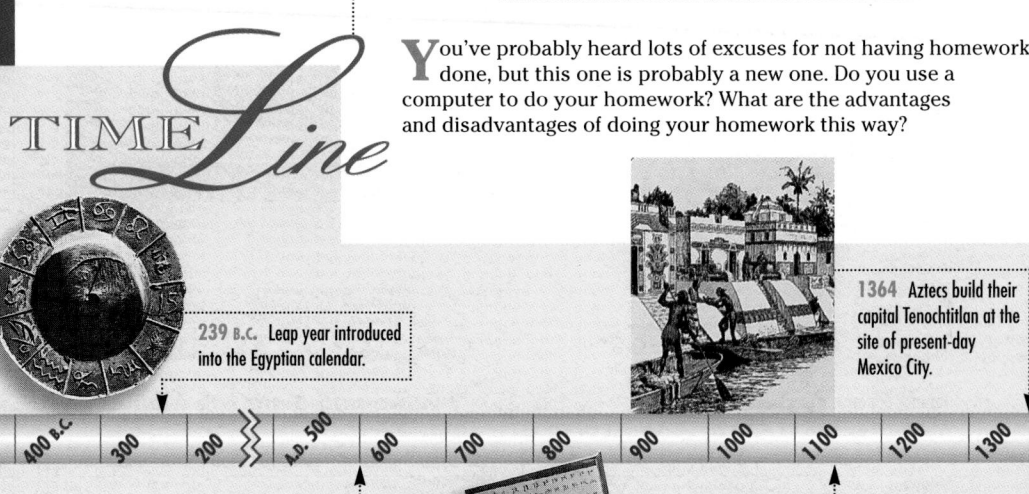

239 B.C. Leap year introduced into the Egyptian calendar.

A.D. 600 Zu Chong-zhi and his son Zu Geng-shi of China calculate π to be between 3.1415926 and 3.1415927.

1149 Hildegard von Bingen, a female medical practitioner, is named a supervisor at St. Ruperts on the Rhine.

1364 Aztecs build their capital Tenochtitlan at the site of present-day Mexico City.

400 B.C. | 300 | 200 | A.D. 500 | 600 | 700 | 800 | 900 | 1000 | 1100 | 1200 | 1300

494   Chapter 9   Exploring Polynomials

TIME *Line*

Earth revolves around the sun once every 365.25 days. Because a year must have a whole number of days, the accepted solution is to have three years of 365 days followed by a year with 366 days—a leap year.

inter**NET** CONNECTION

Explore technology and school reform, link to PBS's home page, and learn about child safety on-line.

World Wide Web
http://edweb.cnidr.org:90/
resource.cntnts.html

Jenny Slabaugh and Amy Gusfa help produce a weekly video show created by students at Dearborn High School in Michigan. When they were only 15 years old, their expertise in electronics and video won each a $1000 scholarship to the Sony Institute of Technology in Hollywood. They were the first females to receive a scholarship from the Sony Institute, and they spent a week there studying theoretical and applied electronics.

Their interest in video production came from Dearborn High's nationally honored video program. They are combining their interests in science, math, electronics, music, and art to create video programs. They both think that more girls should investigate video technology, and they see themselves contributing to this field in the future.

Chapter Project

A byte is a single unit of information such as a number or a letter processed by a computer. A megabyte is 1.048576×10^6 bytes, which is just over 1 million bytes.

* Research three personal computers and three laptop computers. Find out the amount of RAM and hard drive memory in megabytes for each computer. Write each number in standard notation and in scientific notation.

* Use scientific notation to write the amount of memory in bytes for each computer.

* Find the mean of the memory in megabytes for the three personal computers. Find the mean of the memory in megabytes for the three laptop computers. Write the answers in scientific notation.

* Write the ratio of the mean memory for the personal computers to the mean memory for the laptop computers. Write the ratio as a decimal. Discuss the significance of this ratio.

* Investigate the size of the memory of computers that are being developed for the future. How many bytes are in a gigabyte? How many times larger is a gigabyte than a megabyte?

Both Jenny and Amy feel that they were the first girls to receive the Sony Institute scholarship because many girls are intimidated by technology. Encourage a class discussion on this issue.

Chapter Project

Cooperative Learning You may choose to have students work in pairs or in cooperative groups to work on this project. One student, or several group members, can research personal computers. The other student or students can research laptop computers.

Investigations and Projects Masters, p. 57

1738 First cuckoo clocks are produced in the Black Forest district.

1881 Booker T. Washington establishes Tuskegee Institute.

1450 1500 1550 1600 1650 1700 1750 1800 1850 1900 1950 2000

1607 Italian composer Claudio Monteverdi's first opera, *L'Orfeo*, is performed.

1995 Microsoft spends billions of dollars to launch Windows '95.

Alternative Chapter Projects ▬

Two other chapter projects are included in the *Investigations and Projects Masters.* In Chapter 9 Project A, pp. 57–58, students extend the topic in the chapter opener. In Chapter 9 Project B, pp. 59–60, students investigate animal populations and endangered species.

Make History

1. Working in a small group, interview several adults to find out how their lives have been affected by the development of faster and more powerful computers over the years. If possible, make an audiotape or videotape recording of each interview.

2. Research important events in the history of computers—for example, the development of the semiconductor and the Internet. Be sure to find out how scientists and other researchers have achieved dramatic increases in the speed of computers (as measured by the number of calculations that can be performed in a certain amount of time) and in the amount of information that computers can store.

3. Decide how the group will present the information group members have gathered about the history of computers. You might consider doing one of the following.
   • Make an illustrated timeline.
   • Create an interactive museum exhibit.
   • Conduct additional interviews. Then write a newspaper article or put together a radio or TV documentary about people's attitudes toward computers.

4. Make your presentation to the rest of the class at a time designated by the teacher.

NCTM Standards: 1–5

Instructional Resources

- Study Guide Master 9-1
- Practice Master 9-1
- Enrichment Master 9-1
- Graphing Calculator Masters, p. 9
- Real-World Applications, 23

Transparency 9-1A contains the 5-Minute Check for this lesson; **Transparency 9-1B** contains a teaching aid for this lesson.

Recommended Pacing	
Standard Pacing	Days 1 & 2 of 15
Honors Pacing	Day 1 of 14
Block Scheduling*	Day 1 of 8 (along with Lesson 9-2)
Alg. 1 in Two Years*	Days 1, 2, & 3 of 26

*For more information on pacing and possible lesson plans, refer to the *Block Scheduling Booklet* and *Algebra 1 in Two Years.*

1 FOCUS

5-Minute Check
(over Chapter 8)

Solve each system of equations.

1. $3x + 5y = 11$
   $2x + 3y = 7$ **(2, 1)**
2. $x + y = 8$
   $2x - 3y = -9$ **(3, 5)**
3. $3x + 3y = 6$
   $2x - y = 1$ **(1, 1)**
4. $r - s = -5$
   $r + s = 25$ **(10, 15)**
5. $-3x + 2y = 16$
   $x - 2y = -12$ **(-2, 5)**

Motivating the Lesson

Questioning A certain number is multiplied by itself four times. The sum of the digits in the product is equal to the number. What is the number? **7**

Teaching Tip Point out that *monomial* is pronounced mah-NOH-mee-ul.

9–1 Multiplying Monomials

Objectives

After studying this lesson, you should be able to:
- multiply monomials,
- simplify expressions involving powers of monomials, and
- solve problems by looking for a pattern.

fabulous FIRSTS

Muriel Siebert (1932–)

The first woman to hold a seat on the New York Stock Exchange was Muriel ("Mickey") Siebert. She paid $445,000 and was admitted as a full member on December 28, 1967.

APPLICATION
Finance

Since 1983, the Beardstown Business and Professional Women's Investment Club ("Beardstown Ladies") has been able to earn enough of a profit in the stock market to make any market expert envious. The club was started when each of 16 women from a small town in Illinois contributed $100 to start an investment fund. Dividends and monthly dues of $25 are also invested. The Beardstown Ladies have earned an average annual profit of 23% on their investments. How has each woman's initial investment of $100 increased over the years?

At the end of the first year (1984), each initial investment would be worth $100(1 + 0.23)$, or $123. By the end of the second year, the initial investment would have grown to $100(1 + 0.23)(1 + 0.23)$, which is the same as $100(1 + 0.23)^2$ or $151.29. The table below shows the value of an initial investment of $100 for each of the first 13 years.

Year	Yearly Calculation	Value
1984	$100(1 + 0.23)$	$123.00
1985	$100(1 + 0.23)^2$	$151.29
1986	$100(1 + 0.23)^3$	$186.09
1987	$100(1 + 0.23)^4$	$228.89
1988	$100(1 + 0.23)^5$	$281.53
1989	$100(1 + 0.23)^6$	$346.28
1990	$100(1 + 0.23)^7$	$425.93
1991	$100(1 + 0.23)^8$	$523.89
1992	$100(1 + 0.23)^9$	$644.39
1993	$100(1 + 0.23)^{10}$	$792.59
1994	$100(1 + 0.23)^{11}$	$974.89
1995	$100(1 + 0.23)^{12}$	$1199.12
1996	$100(1 + 0.23)^{13}$	$1474.91

After 13 years, the initial $100 investment was worth $1474.91! If we let x equal the factor $(1 + 0.23)$, then the value of the initial investment after 13 years can be represented by $100x^{13}$.

An expression like $100x^{13}$ is called a **monomial**. A monomial is a number, a variable, or a product of a number and one or more variables. Monomials that are real numbers are called **constants**.

Monomials	Not Monomials
12	$a + b$
q	$\frac{a}{b}$
$4x^3$	$5 - 7d$
$11ab$	$\frac{5}{a^2}$
$\frac{1}{3}xyz^{12}$	$\frac{5a}{7b}$

fabulous FIRSTS

Full membership for Muriel Siebert brought with it the privilege of handling customers' orders on the floor.

Recall that an expression of the form x^n is a *power*. The base is x, and the exponent is n. A table of powers of 2 is shown below.

2^1	2^2	2^3	2^4	2^5	2^6	2^7	2^8	2^9	2^{10}
2	4	8	16	32	64	128	256	512	1024

In the following products, each number can be expressed as a power of 2. Study the pattern of the exponents.

Number	$8(32) = 256$	$8(64) = 512$	$4(16) = 64$	$16(32) = 512$
Power	$2^3(2^5) = 2^8$	$2^3(2^6) = 2^9$	$2^2(2^4) = 2^6$	$2^4(2^5) = 2^9$
Pattern of Exponents	$3 + 5 = 8$	$3 + 6 = 9$	$2 + 4 = 6$	$4 + 5 = 9$

These examples suggest that you can multiply powers with the same base by adding exponents.

Product of Powers	For any number a, and all integers m and n, $a^m \cdot a^n = a^{m+n}$.

Example **Simplify each expression.**

a. $(3a^6)(a^8)$

$(3a^6)(a^8) = 3a^{6+8}$
$\qquad = 3a^{14}$

b. $(8y^3)(-3x^2y^2)\left(\frac{3}{8}xy^4\right)$

$(8y^3)(-3x^2y^2)\left(\frac{3}{8}xy^4\right)$
$= \left(8 \cdot (-3) \cdot \frac{3}{8}\right)(x^2 \cdot x)(y^3 \cdot y^2 \cdot y^4)$
$= -9x^{2+1}y^{3+2+4}$
$= -9x^3y^9$

We looked for a pattern to discover the product of powers property. **Look for a pattern** is an important strategy in problem solving.

Example **Solve by extending the pattern.**

$4 \times 6 = 24$
$14 \times 16 = 224$
$24 \times 26 = 624$
$34 \times 36 = 1224$
$124 \times 126 = ?$

PROBLEM SOLVING
Look for a Pattern

Explore Look at the problem. You need to find a pattern to determine the product of 124 and 126.

Plan The last two digits of the product are always 24. To find the first digit(s) of the product, look at the tens place of each pair of factors. Notice that $0 \times 1 = 0$, $1 \times 2 = 2$, $2 \times 3 = 6$, and $3 \times 4 = 12$. Extend this pattern to find the product.

Solve $12 \times 13 = 156$

Therefore, $124 \times 126 = 15,624$

Examine Use a calculator to verify that the product is 15,624. The pattern remains true and the product is correct.

GLENCOE *Technology*

Interactive Mathematics Tools Software

This multimedia software provides an interactive lesson that uses real-life problems and patterns to use and find the Fibonacci sequence. A **Computer Journal** gives students an opportunity to write about what they have learned.

For Windows & Macintosh

Alternative Teaching Strategies

Student Diversity Review the definitions and properties discussed in the lesson. As you read a definition or property aloud, have students provide a number of examples using real numbers and variables, and write the examples on the board.

2 TEACH

Teaching Tip Emphasize the difference between a term and a monomial. A term may represent division of variables, but a monomial may not. The expressions in the text are not monomials for these reasons. $a + b$ shows addition, not multiplication. $\frac{a}{b}$, $\frac{5}{a^2}$, and $\frac{5a}{7b}$ show division, not multiplication. $5 - 7d$ shows subtraction, not multiplication.

Teaching Tip Point out that the product of powers rule applies to all integers m and n, but only positive integers will be treated in this lesson.

In-Class Examples

For Example 1
Simplify each expression.

a. $(21c^6)(c^7)$ $21c^{13}$
b. $(8x^4)(3x)$ $24x^5$
c. $(2a^4)(2a^3 b^2)(-3ab^3)$
   $-12a^8b^5$

For Example 2
Solve by extending the pattern.

$2 \times 3 = 6$
$12 \times 13 = 156$
$22 \times 23 = 506$
$32 \times 33 = 1056$
$122 \times 123 = ?$ $15,006$

Study the examples below.

$$(8^3)^5 = (8^3)(8^3)(8^3)(8^3)(8^3)$$
$$= 8^{3+3+3+3+3} \longleftarrow \text{Product of powers} \longrightarrow = y^{7+7+7}$$
$$= 8^{15}$$

$$(y^7)^3 = (y^7)(y^7)(y^7)$$
$$= y^{21}$$

Therefore, $(8^3)^5 = 8^{15}$ and $(y^7)^3 = y^{21}$. These examples suggest that you can find the power of a power by multiplying the exponents.

Power of a Power	For any number a, and all integers m and n, $(a^m)^n = a^{mn}$.

Look for a pattern in the examples below.

$$(ab)^4 = (ab)(ab)(ab)(ab)$$
$$= (a \cdot a \cdot a \cdot a)(b \cdot b \cdot b \cdot b)$$
$$= a^4b^4$$
$$(5pq)^5 = (5pq)(5pq)(5pq)(5pq)(5pq)$$
$$= (5 \cdot 5 \cdot 5 \cdot 5 \cdot 5)(p \cdot p \cdot p \cdot p \cdot p)(q \cdot q \cdot q \cdot q \cdot q)$$
$$= 5^5p^5q^5 \text{ or } 3125p^5q^5$$

LOOK BACK

You can refer to Lesson 1-1 for information on using a calculator to find a power of a number.

These examples suggest that the power of a product is the product of the powers.

Power of a Product	For all numbers a and b, and any integer m, $(ab)^m = a^mb^m$.

The power of a power property and the power of a product property can be combined into the following property.

Power of a Monomial	For all numbers a and b, and all integers m, n, and p, $(a^mb^n)^p = a^{mp}b^{np}$.

Example ③ Simplify $(2a^4b)^3[(-2b)^3]^2$.

$$(2a^4b)^3[(-2b)^3]^2 = (2a^4b)^3(-2b)^6 \quad \textit{Power of a power property}$$
$$= 2^3(a^4)^3b^3(-2)^6b^6 \quad \textit{Power of a product property}$$
$$= 8a^{12}b^3(64)b^6 \quad \textit{Power of a power property}$$
$$= 512a^{12}b^9 \quad \textit{Product of powers property}$$

To simplify an expression involving monomials, write an equivalent expression in which:

- there are no powers of powers,
- each base appears exactly once, and
- all fractions are in simplest form.

498 Chapter 9 Exploring Polynomials

 Cooperative Learning

Trade-A-Problem Separate the class into small groups. Have each group draw a square and a rectangular solid on a piece of grid paper and label the measure of one side of the square and three edges of the rectangular solid using monomials. Have groups switch papers and find the area of each square and volume of each rectangular solid.

Group members should all agree on their answers before checking them with other groups. For more information on the trade-a-problem strategy, see *Cooperative Learning in the Mathematics Classroom,* one of the titles in the Glencoe Mathematics Professional Series, pages 25–26.

CHECK FOR UNDERSTANDING

Communicating Mathematics

Study the lesson. Then complete the following.

1. **Write** in your own words. **See margin.**
   a. the product of powers property
   b. the power of a power property
   c. the power of product property

2. **Explain** why the product of powers property does not apply when the bases are different. **See Solutions Manual.**

3. **You Decide** Luisa says $10^4 \times 10^5 = 100^9$, but Taryn says that $10^4 \times 10^5 = 10^9$. Who is correct? Explain your answer. **See Solutions Manual.**

 MATH JOURNAL

4. **Write** 64 in six different ways using exponents; for example, $64 = (2^2)^3$. Answers will vary. Sample answers: 2^6, $(2^3)^2$, $2^2 \times 2^4$, 2×2^5, $2^3 \times 2^3$, $(2^2)^2 \times 2^2$

Guided Practice

Determine whether each pair of monomials is equivalent. Write *yes* or *no.*

5. $2d^3$ and $(2d)^3$ no

6. $(xy)^2$ and x^2y^2 yes

7. $-x^2$ and $(-x)^2$ no

8. $5(y^2)^2$ and $25y^4$ no

Simplify.

9. $a^4(a^7)(a)$ a^{12}

10. $(xy^4)(x^2y^3)$ x^3y^7

11. $[(3^2)^4]^2$ 3^{16} or 43,046,721

12. $(2a^2b)^2$ $4a^4b^2$

13. $(-27ay^3)\left(-\frac{1}{3}ay^3\right)$ $9a^2y^6$

14. $(2x^2)^2\left(\frac{1}{2}y^2\right)^2$ x^4y^4

15. **Geometry** Find the measure of the area of the rectangle at the right. $15a^4b^3$

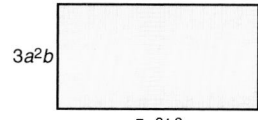

$3a^2b$

$5a^2b^2$

EXERCISES

Practice

A Simplify.

16. $b^3(b)(b^5)$ b^9

17. $(m^3n)(mn^2)$ m^4n^3

18. $(a^2b)(a^5b^4)$ a^7b^5

19. $[(2^3)^2]^2$ 2^{12} or 4096

20. $(3x^4y^3)(4x^4y)$ $12x^8y^4$

21. $(a^3x^2)^4$ $a^{12}x^8$

22. $m^7(m^3b^2)$ $m^{10}b^2$

23. $(3x^2y^2z)(2x^2y^2z^3)$ $6x^4y^4z^4$

B 24. $(0.6d)^3$ $0.216d^3$

25. $(ab)(ac)(bc)$ $a^2b^2c^2$

26. $-\frac{5}{6}c(12a^3)$ $-10a^3c$

27. $\left(\frac{2}{5}d\right)^2$ $\frac{4}{25}d^2$

28. $-3(ax^3y)^2$ $-3a^2x^6y^2$

29. $(0.3x^3y^2)^2$ $0.09x^6y^4$

30. $(-3ab)^3(2b^3)$ $-54a^3b^6$

31. $\left(\frac{3}{10}y^2\right)^2(10y^2)^3$ $90y^{10}$

32. $(3x^2)^2\left(\frac{1}{3}y^2\right)^2$ x^4y^4

33. $\left(\frac{2}{5}a\right)^2(25a)(13b)\left(\frac{1}{13}b^4\right)$ $4a^3b^5$

C 34. $(3a^2)^3 + 2(a^3)^2$ $29a^6$

35. $(-2x^3)^3 - (2x)^9$ $-520x^9$

Lesson 9-1 Multiplying Monomials **499**

Reteaching

Using Models Use circular markers or chips to illustrate the difference between the product of powers and a power of a power. For example, consider $2^3 \times 2^2$ and $(2^3)^2$. Let each marker represent the base 2. For the first example, lay out a group of 3 markers and a group of 2 markers. This is a total of 5 markers. Thus, $2^3 \times 2^2 = 2^5$. For the second example, lay out 2 groups of 3 markers. This is a total of 6. Thus, $(2^3)^2 = 2^6$. This model can be used with any examples in which the bases are the same.

Check for Understanding
Exercises 1–15 are designed to help you assess your students' understanding through reading, writing, speaking, and modeling. You should work through Exercises 1–4 with your students and then monitor their work on Exercises 5–15.

Assignment Guide

Core: 17–39 odd, 40–47
Enriched: 16–34 even, 36–47

For **Extra Practice,** see p. 776.

The red A, B, and C flags, printed only in the Teacher's Wraparound Edition, indicate the level of difficulty of the exercises.

Additional Answers

1a. When numbers with the same base are multiplied, the exponents are added.

1b. When a number is raised to a power and then raised to another power, the two exponents are multiplied.

1c. When two numbers are multiplied and the product is raised to a power, each number can be raised to the power before multiplying.

Practice Masters, p. 61

Closing Activity
Speaking Discuss each property of monomials described in this lesson. Have students provide examples to illustrate each property.

Additional Answers
36. For example, $(x + y)^2 = (x + y)(x + y)$ and not $x^2 + y^2$.

37. -2^4 equals $-(2)(2)(2)(2)$ or -16 and $(-2)^4$ equals $(-2)(-2)(-2)(-2)$ or 16.

41.

Enrichment Masters, p. 61

NAME _____ DATE _____
9-1
Enrichment   Student Edition
                                  Pages 496–500

Circular Areas and Volumes

Area of Circle	Volume of Cylinder	Volume of Cone
$A = \pi r^2$	$V = \pi r^2 h$	$V = \frac{1}{3}\pi r^2 h$

Write an algebraic expression for each shaded area. (Recall that the diameter of a circle is twice its radius.)

$\pi x^2 - \pi\left(\frac{x}{2}\right)^2 = \frac{3}{4}\pi x^2$ $\frac{\pi}{2}(y^2 + 2xy)$ $\frac{23\pi}{4}x^2$

Write an algebraic expression of the total volume of each figure.

4.   5.

$5\frac{2}{3}\pi x^3$ $\frac{\pi}{12}[13x^3 + (4a + 9b)x^2]$

Each figure has a cylindrical hole with a radius of 2 inches and a height of 5 inches. Find each volume.

6.   7.

$\frac{175\pi}{4}x^2 - 20\pi$ in³ $3\pi x^3 - 20\pi$ in³

Critical Thinking
36. Explain why $(x + y)^z$ does not equal $x^z + y^z$. **See margin.**
37. Explain why -2^4 does not equal $(-2)^4$. **See margin.**

Applications and Problem Solving
38. **Investments** Refer to the application at the beginning of the lesson. Each of the Beardstown Ladies added $25 to the investments each month. This amounts to $300 a year. Assume that each member invested the $300 at the beginning of each year starting in 1984. You can use the formula $T = p\left[\dfrac{(1 + r)^t - 1}{r}\right]$ to determine how each member's money grew. T represents the total amount, p represents the regular payment, r represents the annual interest rate, and t represents the time in years.
   a. How much money did each member make from their additional investments from 1984 to 1996? **$14,336.30**
   b. What was the total value of each member's investment in 1996? **$15,811.21**

39. **Look for a Pattern** The symbol used for a U.S. dollar is a capital S with a vertical line through it. The line separates the S into 4 parts, as shown at the right. How many parts would there be if the S had 100 vertical lines through it?
   301 parts

Mixed Review
40. Write a system of inequalities for the graph at the right. (Lesson 8–5)
   $x \leq 2, y \geq 2$

41. Graph the system of equations. Determine whether the system has *one* solution, *no* solution, or *infinitely many* solutions. If the system has one solution, name it. (Lesson 8–1) **See margin for graph; one; (−6, 3).**
   $x + 2y = 0$
   $y + 3 = -x$

42. **Business** Jorge Martinez has budgeted $150 to have business cards printed. A card printer charges $11 to set up each job and an additional $6 per box of 100 cards printed. What is the greatest number of cards Mr. Martinez can have printed? (Lesson 7–3) **2300 cards**

43. Write an equation from the relation shown in the chart below. Then copy and complete the chart. (Lesson 5–6)
   $n = 2m + 1$

m	−3	−2	−1	0	1
n	−5	−3	−1	1	3

44. **Travel** Tiffany wants to reach Dallas at 10 A.M. If she drives at 36 miles per hour, she would reach Dallas at 11 A.M. But if she drives 54 miles per hour, she would arrive at 9 A.M. At what average speed should she drive to reach Dallas exactly at 10 A.M.? (Lesson 4–8) **43.2 mph**

45. **Geometry** Find the supplement of 44°. (Lesson 3–4) **136°**

46. Simplify $-16 \div 8$. (Lesson 2–7) **−2**

47. Simplify $0.3(0.2 + 3y) + 0.21y$. (Lesson 1–8) **1.11y + 0.06**

500 Chapter 9 Exploring Polynomials

Extension
Reasoning Find a if $(3^{a+2})^2 \times 3^{3a-10} = 81$. (*Hint:* Write 81 as a power of 3.) $a = 2$

9-2

Dividing by Monomials

Objectives

After studying this lesson, you should be able to:

- simplify expressions involving quotients of monomials, and
- simplify expressions containing negative exponents.

Geometry

The volume of a cube with each side s units long is s^3 cubic units. So, the ratio of the measure of the volume of a cube to the measure of the length of each side is $\frac{s^3}{s}$. How can you express this ratio in simplest form?

Just as we used a pattern to discover the product of powers property, we can use a pattern to discover a property for a quotient of powers such as $\frac{s^3}{s}$. In the following quotients, each number can be expressed as a power of 2. Study the pattern of exponents.

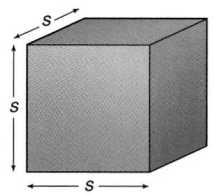

$V = s \cdot s \cdot s$ or s^3

Number	$\frac{64}{32} = 2$	$\frac{32}{8} = 4$	$\frac{64}{8} = 8$	$\frac{32}{2} = 16$
Power	$\frac{2^6}{2^5} = 2^1$	$\frac{2^5}{2^3} = 2^2$	$\frac{2^6}{2^3} = 2^3$	$\frac{2^5}{2^1} = 2^4$
Pattern of Exponents	$6 - 5 = 1$	$5 - 3 = 2$	$6 - 3 = 3$	$5 - 1 = 4$

These examples suggest that you can divide powers with the same base by subtracting exponents.

Quotient of Powers	For all integers m and n, and any nonzero number a, $\frac{a^m}{a^n} = a^{m-n}$.

To write $\frac{s^3}{s}$ in simplest form, subtract the exponents.

$\frac{s^3}{s} = s^{3-1}$ *Recall that $s = s^1$, and apply the quotient of powers property.*

$= s^2$

In simplest form, $\frac{s^3}{s} = s^2$.

Example ① Simplify $\frac{y^4 z^3}{y^2 z^2}$.

$\frac{y^4 z^3}{y^2 z^2} = \left(\frac{y^4}{y^2}\right)\left(\frac{z^3}{z^2}\right)$ *Group the powers with the same base, y^4 with y^2 and z^3 with z^2.*

$= (y^{4-2})(z^{3-2})$ *Quotient of powers property.*

$= y^2 z$

Lesson 9-2   Dividing by Monomials   **501**

9-2 LESSON NOTES

NCTM Standards: 1–5

Instructional Resources

- Study Guide Master 9-2
- Practice Master 9-2
- Enrichment Master 9-2
- Assessment and Evaluation Masters, p. 240
- Multicultural Activity Masters, p. 17

Transparency 9-2A contains the 5-Minute Check for this lesson; **Transparency 9-2B** contains a teaching aid for this lesson.

Recommended Pacing	
Standard Pacing	Day 3 of 15
Honors Pacing	Day 2 of 14
Block Scheduling*	Day 1 of 8 (along with Lesson 9-1)
Alg. 1 in Two Years*	Days 4, 5, & 6 of 26

*For more information on pacing and possible lesson plans, refer to the *Block Scheduling Booklet* and *Algebra 1 in Two Years*.

1 FOCUS

5-Minute Check
(over Lesson 9-1)

Simplify.

1. $(a^4)(a^7)$ a^{11}
2. $(2p^3)(5p)$ $10p^4$
3. $(x^4 y^5)^2$ $x^8 y^{10}$
4. $(2x^3 y^4)^2$ $4x^6 y^8$
5. $(-5mn)^2[(2mn^4)^2]^2$ $400m^6 n^{18}$

Motivating the Lesson

Questioning A number multiplied by itself five times divided by the number multiplied by itself twice is the number multiplied by itself three times. What is the number? **any positive whole integer**

In-Class Example

For Example 1
Simplify.

a. $\dfrac{m^7n^5}{mn^2}$ m^6n^3

b. $\dfrac{x^3y^7}{x^2y^6}$ xy

c. $\dfrac{3^4a^8b^3}{3^2a^5b}$ $9a^3b^2$

EXPLORATION

This Exploration provides students with a quick overview of the meaning of positive and negative exponents. Point out that a negative exponent does not yield a negative value; it yields a fraction.

Teaching Tip Reinforce the negative exponent concept with numerical examples like the one below.

$\dfrac{5^2}{5^4} = 5^{-2}$ and $\dfrac{5^2}{5^4} = \dfrac{1}{5^2}$,

so $5^{-2} = \dfrac{1}{5^2}$

Teaching Tip You may wish to point out that $\dfrac{r^3}{r^{-7}} = r^{3-(-7)}$ or r^{10}.

Also, point out that $\dfrac{1}{t^{-2}} = \dfrac{1}{\frac{1}{t^2}} = 1 \times t^2 = t^2$.

A calculator can be used to find the value of expressions with 0 as an exponent as well as expressions with negative exponents.

You can use the $\boxed{y^x}$ key to find the value of expressions with exponents.

Your Turn

a. Copy the following table. Use a scientific calculator to complete the table.

Exponential Expression	2^4	2^3	2^2	2^1	2^0	2^{-1}	2^{-2}	2^{-3}	2^{-4}
Decimal Form	16	8	4	2	1	0.5	0.25	0.125	0.0625
Fraction Form	$\frac{16}{1}$	$\frac{8}{1}$	$\frac{4}{1}$	$\frac{2}{1}$	$\frac{1}{1}$	$\frac{1}{2}$	$\frac{1}{4}$	$\frac{1}{8}$	$\frac{1}{16}$

b. According to your calculator display, what is the value of 2^0?   1
c. Compare the values of 2^2 and 2^{-2}.   They are reciprocals.
d. Compare the values of 2^4 and 2^{-4}.   They are reciprocals.
e. What happens when you evaluate 0^0?   An error message appears.

Study the following methods used to simplify $\dfrac{b^4}{b^4}$ where $b \neq 0$.

Method 1 **Definition of Powers or Expanded Form**	**Method 2** **Quotient of Powers**
$\dfrac{b^4}{b^4} = \dfrac{\overset{1}{\cancel{(b)}}\,\overset{1}{\cancel{(b)}}\,\overset{1}{\cancel{(b)}}\,\overset{1}{\cancel{(b)}}}{\underset{1}{\cancel{(b)}}\,\underset{1}{\cancel{(b)}}\,\underset{1}{\cancel{(b)}}\,\underset{1}{\cancel{(b)}}}$	$\dfrac{b^4}{b^4} = b^{4-4}$
$= 1$	$= b^0$

How does this value for b^0 compare with the value displayed on your calculator for 2^0?

Since $\dfrac{b^4}{b^4}$ cannot have two different values, we can conclude that $b^0 = 1$. This example and the Exploration suggest the following definition.

Zero Exponent	For any nonzero number a, $a^0 = 1$.

We can also simplify $\dfrac{r^3}{r^7}$ in two ways.

Did the display on your calculator for 2^{-4} equal $\dfrac{1}{2^4}$?

Method 1 **Definition of Powers or Expanded Form**	**Method 2** **Quotient of Powers**
$\dfrac{r^3}{r^7} = \dfrac{\overset{1}{\cancel{(r)}}\,\overset{1}{\cancel{(r)}}\,\overset{1}{\cancel{(r)}}}{\underset{1}{\cancel{(r)}}\,\underset{1}{\cancel{(r)}}\,\underset{1}{\cancel{(r)}}\,(r)\,(r)\,(r)\,(r)}$	$\dfrac{r^4}{r^7} = r^{3-7}$ *Quotient of powers*
$= \dfrac{1}{r^4}$	$= r^{-4}$

Alternative Learning Styles

Auditory Ask students to explain why 0^0 is undefined. 0^0 is derived from an expression such as $0^{x-x} = \dfrac{0^x}{0^x}$ which is not defined. Ask students to recall why division by zero is undefined.

Since $\frac{r^3}{r^7}$ cannot have two values, we conclude that $\frac{1}{r^4} = r^{-4}$. This example and the Exploration suggest the following definition.

Negative Exponents	For any nonzero number a and any integer n, $a^{-n} = \frac{1}{a^n}$.

Example **2** Simplify each expression.

a. $\dfrac{-9m^3n^5}{27m^{-2}n^5y^{-4}}$

$\dfrac{-9m^3n^5}{27m^{-2}n^5y^{-4}} = \left(\dfrac{-9}{27}\right)\left(\dfrac{m^3}{m^{-2}}\right)\left(\dfrac{n^5}{n^5}\right)\left(\dfrac{1}{y^{-4}}\right)$

$= \dfrac{-1}{3}m^{3-(-2)}n^{5-5}y^4 \qquad \frac{1}{y^{-4}} = y^4$

$= -\dfrac{1}{3}m^5n^0y^4 \qquad\qquad$ *Subtract the exponents.*

$= -\dfrac{m^5y^4}{3} \qquad\qquad\quad n^0 = 1$

b. $\dfrac{(5p^{-2})^{-2}}{(2p^3)^2}$

$\dfrac{(5p^{-2})^{-2}}{(2p^3)^2} = \dfrac{5^{-2}p^4}{2^2p^6} \qquad$ *Power of a monomial property*

$= \left(\dfrac{1}{2^2}\right)\left(\dfrac{1}{5^2}\right)\left(\dfrac{p^4}{p^6}\right)$

$= \left(\dfrac{1}{4}\right)\left(\dfrac{1}{25}\right)p^{4-6} \quad$ *Quotient of powers property*

$= \dfrac{1}{100}p^{-2}$

$= \dfrac{1}{100p^2} \qquad\qquad$ *Definition of negative exponents*

CHECK FOR UNDERSTANDING

Communicating Mathematics

2. Division by 0 is not defined.

3. Since each number is obtained by dividing the previous number by 3, $3^1 = 3$ and $3^0 = 1$.

Study the lesson. Then complete the following. 1. See margin.

1. **Explain** why 0^0 is not defined. (*Hint:* Think of computing $\frac{0^m}{0^m}$.)

2. **Explain** why a cannot equal zero in the negative exponents property.

3. **Write** a convincing argument to show that $3^0 = 1$ using the following pattern.
$$3^5 = 243,\ 3^4 = 81,\ 3^3 = 27,\ 3^2 = 9,\ \ldots$$

4. **Study** the pattern below.

5^5	5^4	5^3	5^2	5^1	5^0	5^{-1}	5^{-2}	5^{-3}	5^{-4}
3125	625	125	25	5	1	$\frac{1}{5}$	$\frac{1}{25}$	$\frac{1}{125}$	$\frac{1}{625}$

Copy the table and complete the pattern. Write an explanation of the pattern you used. **Divide the previous number by 5.**

Lesson 9-2 Dividing by Monomials **503**

Reteaching

Using Alternative Methods Describe the following simplified procedure for dividing a monomial by a monomial. Write the factors in the denominator in the same order in which they appear in the numerator. Think of the quotient as a product of simpler fractions in which the base in each fraction is the same in both the numerator and denominator.

Reduce each fraction. Have students solve several problems in this way. For example:

$\dfrac{4a^2b^3}{16ab^9} = \dfrac{2^2a^2b^3}{2^4ab^9}$

$= \dfrac{1}{2^2}\cdot a \cdot \dfrac{1}{b^6}$

$= \dfrac{a}{4b^6}$

In-Class Example

For Example 2
Simplify each expression.

a. $\dfrac{4a^8b^3c^2}{-28a^5b^{-6}c^2} \qquad -\dfrac{a^3b^9}{7}$

b. $\dfrac{(3c^{-2}d)^{-3}}{4c^3d^4} \qquad \dfrac{c^3}{108d^7}$

3 PRACTICE/APPLY

Check for Understanding

Exercises 1–13 are designed to help you assess your students' understanding through reading, writing, speaking, and modeling. You should work through Exercises 1–6 with your students and then monitor their work on Exercises 7–13.

Error Analysis

Students often evaluate -2^4 as $(-2)(-2)(-2)(-2)$ or 16. Does the exponent 4 refer to a base of 2 or (-2)? Compare -2^4 and $(-2)^4$. -2^4 means $-(2 \times 2 \times 2 \times 2)$ or -16. $(-2)^4$ means $(-2)(-2)(-2)(-2)$ or 16.

Additional Answer

1. $\dfrac{0^m}{0^m} \neq 0^{m-m}$, because $0^m = 0$ and division by 0 is not defined.

Study Guide Masters, p. 62

9-2 NAME_____ DATE_____

Study Guide

Student Edition
Pages 501–505

Dividing by Monomials

Study the following exponent rules.

	Rule	Example
Quotient of Powers	For all integers m and n, and any nonzero number a, $\frac{a^m}{a^n} = a^{m-n}$.	$q^a \div q^b = q^{a-b}$ $= q^5$
Zero Exponent	For any nonzero number a, $a^0 = 1$.	$4^0 = 1$ $6^0 = 1$
Negative Exponents	For any nonzero number a and any integer n, $a^{-n} = \frac{1}{a^n}$.	$\frac{r^3}{r^7} = \frac{r\cdot r\cdot r}{r\cdot r\cdot r\cdot r\cdot r\cdot r\cdot r}$ $= \frac{1}{r^4}$ or $\frac{r^3}{r^7} = r^{3-7} = r^{-4}$

Simplify. Assume no denominator is equal to zero.

1. $\frac{a^2}{a}$ a
2. $\frac{x^3y^2}{x^2y^2}$ y
3. $\frac{15a^2}{45a^2}$ $\frac{a}{3}$
4. $\frac{s^{-3}t^{-3}}{(s^2t^2)^{-1}}$ $\frac{1}{st^3}$

5. $\frac{a^4b^2}{a^2b^3}$ a^3b
6. $\frac{k^3}{k^7}$ $\frac{1}{k^4}$
7. $\frac{(6a^{-1}b)^2}{(b^2)^2}$ $\frac{36}{a^3b^6}$
8. $\frac{66w^3x^4y^6}{-22wxy^4}$ $-3w^2x^3y^2$

9. $\left(\frac{4m^2n^3}{8m^{-1}l}\right)^0$ 1
10. $\frac{15x^2}{5x^4}$ $3x^3$
11. $\frac{x^4}{x^3}$ x
12. $\frac{(3st^2)^2u^{-4}}{s^{-1}t^2u^2}$ $\frac{9s^3}{u^3}$

13. $\frac{b^6}{b^5}$ b
14. $\frac{x^9}{x^2}$ x^7
15. $\frac{24w^3t^4}{6w^3t^2}$ $4w^4t^2$
16. $\frac{9x^4z^5}{-3xz^3}$ $-3xz^2$

17. $\frac{(-x^{-3}y)^2}{4w^3y^3}$ $\frac{w}{4y^2}$
18. $\frac{w^3x}{wx}$ t^3
19. $\frac{w^3}{w}$ w
20. $\frac{(a^2b^3)^2}{(ab)^{-3}}$ a^6b^5

Chapter 9 **503**

Assignment Guide

Core: 15–41 odd, 43–50
Enriched: 14–34 even, 35–50

For **Extra Practice**, see p. 776.

The red A, B, and C flags, printed only in the Teacher's Wraparound Edition, indicate the level of difficulty of the exercises.

---

5. You Decide   Taigi and Isabel each simplified $\left(\dfrac{x^{-2}y^3}{x}\right)^{-2}$ correctly as shown below.

Taigi

$$\left(\frac{x^{-2}y^3}{x}\right)^{-2} = \frac{x^4y^{-6}}{x^{-2}}$$

$$= x^{4-(-2)}y^{-6}$$

$$= x^6y^{-6}$$

$$= \frac{x^6}{y^6}$$

Isabel

$$\left(\frac{x^{-2}y^3}{x}\right)^{-2} = (x^{-2-1}y^3)^{-2}$$

$$= (x^{-3}y^3)^{-2}$$

$$= x^6y^{-6}$$

$$= \frac{x^6}{y^6}$$

Whose method do you prefer? Why?   **See students' work.**

6. Assess Yourself   Which properties of monomials do you find easy to understand? Which properties do you need to study more?
See students' work.

Guided Practice   **Simplify. Assume that no denominator is equal to zero.**

7. 11^{-2} $\dfrac{1}{121}$

8. $(6^{-2})^2$ $\dfrac{1}{1296}$

9. $\left(\dfrac{1}{4} \cdot \dfrac{2}{3}\right)^{-2}$ 36

10. $a^4(a^{-7})(a^0)$ $\dfrac{1}{a^3}$

11. $\dfrac{6r^3}{r^7}$ $\dfrac{6}{r^4}$

12. $\dfrac{(a^7b^2)^2}{(a^{-2}b)^{-2}}$ $a^{10}b^6$

13. Geometry   Write the ratio of the area of the circle to the area of the square in simplest form.

$\dfrac{\pi}{4}$

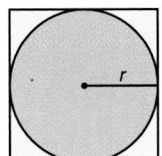

EXERCISES

Practice   **Simplify. Assume that no denominator is equal to zero.**

A

14. $a^0b^{-2}c^{-1}$ $\dfrac{1}{b^2c}$

15. $\dfrac{a^0}{a^{-2}}$ a^2

16. $\dfrac{5n^5}{n^8}$ $\dfrac{5}{n^3}$

17. $\dfrac{m^2}{m^{-4}}$ m^6

18. $\dfrac{b^5d^2}{b^3d^8}$ $\dfrac{b^2}{d^6}$

19. $\dfrac{10m^4}{30m}$ $\dfrac{m^3}{3}$

B

20. $\dfrac{(-y)^5m^8}{y^3m^{-7}}$ $-y^2m^{15}$

21. $\dfrac{b^6c^5}{b^{14}c^2}$ $\dfrac{c^3}{b^8}$

22. $\dfrac{22a^2b^5c^7}{-11abc^2}$ $-2ab^4c^5$

23. $\dfrac{(a^{-2}b^3)^2}{(a^2b)^{-2}}$ b^8

24. $\dfrac{7x^3z^5}{4z^{15}}$ $\dfrac{7x^3}{4z^{10}}$

25. $\dfrac{(-r)^5s^8}{r^5s^2}$ $-s^6$

26. $\dfrac{(r^{-4}k^2)^2}{(5k^2)^2}$ $\dfrac{1}{25r^8}$

27. $\dfrac{16b^4}{-4bc^3}$ $-\dfrac{4b^3}{c^3}$

28. $\dfrac{27a^4b^6c^9}{15a^3c^{15}}$ $\dfrac{9ab^6}{5c^6}$

29. $\dfrac{(4a^{-1})^{-2}}{(2a^4)^2}$ $\dfrac{1}{64a^6}$

30. $\left(\dfrac{3m^2n^2}{6m^{-1}k}\right)^0$ 1

31. $\dfrac{r^{-5}s^{-2}}{(r^2s^5)^{-1}}$ $\dfrac{s^3}{r^3}$

C

32. $\left(\dfrac{7m^{-1}n^3}{n^2r^{-1}}\right)^{-1}$ $\dfrac{m}{7rn}$

33. $\dfrac{(-b^{-1}c)^0}{4a^{-1}c^2}$ $\dfrac{a}{4c^2}$

34. $\left(\dfrac{3xy^{-2}z}{4x^{-2}y}\right)^{-2}$ $\dfrac{16y^6}{9x^6z^2}$

504   *Chapter 9   Exploring Polynomials*

Extension

Problem Solving   Simplify

$$\frac{a^0(a)(a^2)^2(a^3)^3}{(a^{-1})^{-1}(a^{-2})^{-2}(a^{-3})^{-3}} \cdot \quad 1$$

Simplify. Assume that no denominator is equal to zero.

35. $m^3(m^n)$ m^{3+n}

36. $y^{2c}(y^{5c})$ y^{7c}

37. $(3^{2x+1})(3^{2x-7})$ 3^{4x-6}

38. $\frac{r^{y-2}}{r^{y+3}}$ $\frac{1}{r^5}$

39. $\frac{(q^{y-7})^2}{(q^{y+2})^2}$ $\frac{1}{q^{18}}$

40. $\frac{y^x}{y^{a-x}}$ y^{2x-a}

Applications and Problem Solving

41. Finance You can use the formula

$$P = A\left[\frac{i}{1-(1+i)^{-n}}\right]$$ to determine the

monthly payment on a home. P represents the monthly payment, A represents the price of the home less the down payment, i represents the *monthly* interest rate (annual rate ÷ 12), and n is the total number of monthly payments. Find the monthly payment on a $180,000 home with 10% down and an *annual* interest rate of 8.6% over 30 years. $1257.14

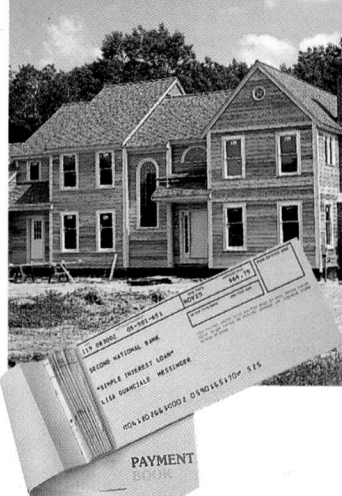

42. Finance You can use the formula

$$B = P\left[\frac{1-(1+i)^{k-n}}{i}\right]$$ to calculate

the balance due on a car loan after a certain number of payments have been made. B represents the balance due (or payoff), P represents the current monthly payment, i represents the *monthly* interest rate (annual rate ÷ 12), k represents the total number of monthly payments already made, and n is the total number of monthly payments. Find the balance due on a 48-month loan for $10,562 after 20 monthly payments of $265.86 have been made at an *annual* interest rate of 9.6%. $6645.57

Mixed Review

43. Simplify $(2a^3)(7ab^2)^2$.   (Lesson 9–1) $98a^5b^4$

44. Aviation Flying with the wind, a plane travels 300 miles in 40 minutes. Flying against the wind, it travels 300 miles in 45 minutes. Find the air speed of the plane.   (Lesson 8–4)   425 mph

45. Write an open sentence involving the absolute value for the graph at the right.   (Lesson 7–6) $|x+1| < 3$

$$\xleftarrow{\quad} \underset{-5\;-4\;-3\;-2\;-1\;\;0\;\;1\;\;2\;\;3}{\bullet\!-\!\!\!-\!\!\!-\!\!\!-\!\!\!-\!\!\!\bullet} \xrightarrow{\quad}$$

46. Solve $-\frac{2}{5} > \frac{4z}{7}$.   (Lesson 7–2) $-\frac{7}{10} > z$

47. Geometry Find the coordinates of the midpoint of the line segment whose endpoints are $(5, -3)$ and $(1, -7)$.   (Lesson 6–7) $(3, -5)$

48. Sales Latoya bought a new dress for $32.86. This included 6% sales tax. What was the cost of the dress before tax?   (Lesson 4–5) $31.00

49. Solve $\frac{4-x}{3+x} = \frac{16}{25}$.   (Lesson 4–1) $\frac{52}{41}$

50. Simplify $41y - (-41y)$.   (Lesson 2–3) $82y$

Lesson 9–2   Dividing by Monomials   **505**

Classroom Vignette

"To extend Exercise 42, I have students select a car they would like to buy and use the graphing calculator to determine the interest they would pay, the total cost of the loan, and the monthly payments. They also work with the remaining balance formula to discover how different numbers of payments affect the interest paid. This is a real eye-opener for students."

Harvey A. Johnecheck

Harvey A. Johnecheck
Gaylord High School
Gaylord, Michigan

---

4 ASSESS

Closing Activity

Writing Have students write down the key points to consider when dividing monomials using negative and zero exponents.

Chapter 9, Quiz A (Lessons 9-1 and 9-2), is available in the *Assessment and Evaluation Masters,* p. 240.

Enrichment Masters, p. 62

9-2	NAME	DATE	Student Edition
	Enrichment		Pages 501–505

Patterns with Powers

Use your calculator, if necessary, to complete each pattern.

	a.		b.		c.	
	$2^{10} =$	1024	$5^{10} =$	9,765,625	$4^{10} =$	1,048,576
	$2^9 =$	512	$5^9 =$	1,953,125	$4^9 =$	262,144
	$2^8 =$	256	$5^8 =$	390,625	$4^8 =$	65,536
	$2^7 =$	128	$5^7 =$	78,125	$4^7 =$	16,384
	$2^6 =$	64	$5^6 =$	15,625	$4^6 =$	4096
	$2^5 =$	32	$5^5 =$	3125	$4^5 =$	1024
	$2^4 =$	16	$5^4 =$	625	$4^4 =$	256
	$2^3 =$	8	$5^3 =$	125	$4^3 =$	64
	$2^2 =$	4	$5^2 =$	25	$4^2 =$	16
	$2^1 =$	2	$5^1 =$	5	$4^1 =$	4

Study the patterns for a, b, and c above. Then answer the questions.

1. Describe the pattern of the exponents from the top of each column to the bottom.   The exponents decrease by one from each row to the one below.

2. Describe the pattern of the powers from the top of the column to the bottom.   To get each power, divide the power on the row above by the base (2, 5, or 4).

3. What would you expect the following powers to be?
   2^0 1 5^0 1 4^0 1

4. Write a rule. Test it on patterns that you obtain using -2, -5, and -4 as bases.   Any nonzero number to the zero power equals one.

Study the pattern below. Then answer the questions.

$0^3 = 0$ $0^2 = 0$ $0^1 = 0$ $0^0 = \underline{\;?\;}$ 0^{-1} does not exist.
0^{-2} does not exist. 0^{-3} does not exist.

5. Why do 0^{-1}, 0^{-2}, and 0^{-3} not exist?   Negative exponents are not defined unless the base is nonzero.

6. Based upon the pattern, can you determine whether 0^0 exists?   No, since the pattern $0^n = 0$ breaks down for $n < 1$.

7. The symbol 0^0 is called an **indeterminate**, which means that it has no unique value. Thus it does not exist as a unique real number. Why do you think that 0^0 cannot equal 1?   Answers will vary. One answer is that if $0^0 = 1$, then $1 = \frac{1}{1} = \frac{1^0}{0^0} = \left(\frac{1}{0}\right)^0$, which is a false result, since division by zero is not allowed. Thus, 0^0 cannot equal 1.

NCTM Standards: 1–5

Instructional Resources

- Study Guide Master 9-3
- Practice Master 9-3
- Enrichment Master 9-3
- Real-World Applications, 24
- Science and Math Lab Manual, pp. 39–42
- Tech Prep Applications Masters, p. 17

 Transparency 9-3A contains the 5-Minute Check for this lesson; **Transparency 9-3B** contains a teaching aid for this lesson.

Recommended Pacing

Standard Pacing	Day 4 of 15
Honors Pacing	Day 3 of 14
Block Scheduling*	Day 2 of 8
Alg. 1 in Two Years*	Days 7 & 8 of 26

 *For more information on pacing and possible lesson plans, refer to the *Block Scheduling Booklet* and *Algebra 1 in Two Years*.

1 FOCUS

 ## 5-Minute Check
(over Lesson 9-2)

Simplify.

1. $\dfrac{b^5}{b^7}$ b^{-2}

2. $\dfrac{18m^8n^3}{36m^8n}$ $\dfrac{n^2}{2}$

3. $\dfrac{x^{-5}y^{-6}}{(9x^2y^6)^{-2}}$ $\dfrac{81y^6}{x}$

4. $\dfrac{36x^4y^2}{12x^6y}$ $\dfrac{3y}{x^2}$

5. $\dfrac{10t^6r}{40t^3r^5}$ $\dfrac{t^3}{4r^4}$

Motivating the Lesson

Situational Problem Begin the lesson by having students examine large numbers, such as the distance between Earth and the sun or the population of China. Ask students to describe problems that could arise when working with such large numbers.

 9-3

Scientific Notation

Objectives

After studying this lesson, you should be able to:
- express numbers in scientific and standard notation, and
- find products and quotients of numbers expressed in scientific notation.

APPLICATION
Transportation

The numbers of passengers arriving at and departing from some major U. S. airports for 1993 are listed in the table below.

Airport	City	Number of Passengers Arriving and Departing (nearest million)
O'Hare International	Chicago	65,000,000
Dallas/Ft. Worth International	Dallas/Ft. Worth	50,000,000
Los Angeles International	Los Angeles	48,000,000
Hartsfield Atlanta International	Atlanta	48,000,000
San Francisco International	San Francisco	32,000,000
Miami International	Miami	29,000,000
J. F. Kennedy International	New York	27,000,000
Newark International	Newark	26,000,000
Detroit Metropolitan Wayne County	Detroit	24,000,000
Logan International	Boston	24,000,000

Source: Air Transport Association of America

When dealing with very large numbers, keeping track of place value can be difficult. For this reason, it is not always desirable to express numbers in standard notation as shown in the chart. Large numbers such as these may be expressed in **scientific notation.**

Definition of Scientific Notation	A number is expressed in scientific notation when it is in the form $a \times 10^n$, where $1 \le a < 10$ and n is an integer.

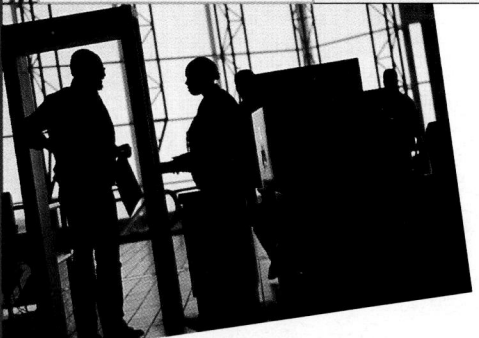

For example, the number of passengers arriving at and departing from Miami International was about 29,000,000. To write this number in scientific notation, express it as a product of a number greater than or equal to 1, but less than 10, and a power of 10.

$$29,000,000 = 2.9 \times 10,000,000$$
$$= 2.9 \times 10^7$$

Scientific notation is also used to express very small numbers. When numbers between zero and one are written in scientific notation, the exponent of 10 is negative.

Example **Express each number in scientific notation.**

a. **98,700,000,000**

b. **0.0000056**

$98,700,000,000$

$= 9.87 \times 10,000,000,000$

$= 9.87 \times 10^{10}$

$0.0000056 = 5.6 \times 0.000001$

$= 5.6 \times \frac{1}{1,000,000}$

$= 5.6 \times \frac{1}{10^6}$

$= 5.6 \times 10^{-6}$

Example ② **Express each number in standard notation.**

a. **3.45×10^5**

b. **9.72×10^{-4}**

$3.45 \times 10^5 = 3.45 \times 100,000$

$= 345,000$

$9.72 \times 10^{-4} = 9.72 \times \frac{1}{10^4}$

$= 9.72 \times \frac{1}{10,000}$

$= 9.72 \times 0.0001$

$= 0.000972$

You can use scientific notation to simplify computation with very large numbers and/or very small numbers.

EXPLORATION

GRAPHING CALCULATORS

You can use a graphing calculator to solve problems using scientific notation. First, put your calculator in scientific mode. To enter 3.5×10^9, enter 3.5 [X] 10 [∧] 9.

Your Turn

a. Use your calculator to find $(3.5 \times 10^9)(2.36 \times 10^{-3})$. **$8.26 \times 10^6$**

b. Explain how the calculator calculated the product in part a. **See margin.**

c. Write the product for part a in standard notation. **8,260,000**

d. Use your calculator to find $(5.544 \times 10^3) \div (1.54 \times 10^7)$. **$3.6 \times 10^{-4}$**

e. Explain how the calculator calculated the quotient in part d. **See margin.**

f. Write the quotient for part d in standard notation. **0.00036**

Example ③ **Use scientific notation to evaluate each expression.**

a. **(610)(2,500,000,000)**   *Estimate: 2.5 billion × 600 = 1.5 trillion*

$(610)(2,500,000,000) = (6.1 \times 10^2)(2.5 \times 10^9)$

$\qquad = (6.1 \times 2.5)(10^2 \times 10^9)$ *Associative property*

$\qquad = 15.25 \times 10^{11}$

$\qquad = 1.525 \times 10^{12}$ or $1,525,000,000,000$

b. **(0.000009)(3700)**   *Estimate: 0.00001 × 3700 = 0.037*

$(0.000009)(3700) = (9 \times 10^{-6})(3.7 \times 10^3)$

$\qquad = (9 \times 3.7)(10^{-6} \times 10^3)$ *Associative property*

$\qquad = 33.3 \times 10^{-3}$

$\qquad = 3.33 \times 10^{-2}$ or 0.0333

2 TEACH

Teaching Tip Emphasize the difference between scientific notation and standard notation. Students should feel comfortable with the terminology.

Teaching Tip When expressing numbers in scientific notation, such as those in Example 1, help students derive a method to recognize when base 10 will have a negative or a positive exponent.

In-Class Examples

For Example 1
Express each number in scientific notation.

a. 635,000,000 6.35×10^8
b. 4,000,000,000,000 4×10^{12}
c. 0.00234 2.34×10^{-3}
d. 0.000000028 2.8×10^{-8}

For Example 2
Express each number in standard notation.

a. 2.36×10^2 236
b. 8.04×10^{-7} 0.000000804

For Example 3
Use scientific notation to evaluate each expression.

a. (170) (475,000,000)
  8.075×10^{10}
b. (3,600,000,000) (0.0023)
  8.28×10^6
c. $\dfrac{4.473 \times 10^{12}}{4.26 \times 10^5}$ 1.05×10^7

EXPLORATION

It might be helpful to remind students that since multiplication is commutative, the expression in part a can be written as follows.
$(3.5 \times 2.36) \times (10^9 \times 10^{-3})$

Answers for the Exploration

b. Multiply 3.5 and 3.36 and then 10^9 and 10^{-3}.
e. Divide 5.544 by 1.54 and then 10^3 by 10^7.

In-Class Example

For Example 4
What would be the Schwarzschild radius of a star with a mass of 6.1×10^{31} kilograms?
9.0822×10^4 meters

Scientists believe that when a large star runs out of nuclear fuel, it begins to break down under the force of its own gravity. The gravity eventually becomes so intense that even light cannot escape, making the star look black.

Stephen W. Hawking is a theoretical physicist from Great Britain. Dr. Hawking has made many contributions to the field of science including extensive work dealing with black holes.

c. $\dfrac{2.0286 \times 10^8}{3.15 \times 10^3}$ Estimate: $\dfrac{210,000,000}{3000} = 70,000$

$$\dfrac{2.0286 \times 10^8}{3.15 \times 10^3} = \left(\dfrac{2.0286}{3.15}\right)\left(\dfrac{10^8}{10^3}\right)$$

$$= 0.644 \times 10^5$$

$$= 6.44 \times 10^4 \text{ or } 64{,}400$$

Scientific notation is extensively used by scientists in fields such as physics and astronomy.

Example 4

APPLICATION
Astronomy

A black hole is a region in space where matter seems to disappear. A star becomes a black hole when the radius of the star reaches a certain critical value called the *Schwarzschild radius*. The value is given by the equation $R_s = \dfrac{2GM}{c^2}$, where R_s is the Schwarzschild radius in meters, G is the gravitational constant (6.7×10^{-11}), M is the mass in kilograms, and c is the speed of light $(3 \times 10^8$ meters per second$)$.

a. The mass of the sun is 2×10^{30} kilograms. Find the Schwarzschild radius of the sun.

b. The actual radius of the sun is 700,000 kilometers. Is it in danger of becoming a black hole in the near future?

a. Use your knowledge of exponents and scientific notation to evaluate the expression for the sun's Schwarzschild radius.

$$R_s = \dfrac{2GM}{c^2}$$

$$= \dfrac{2(6.7 \times 10^{-11})(2 \times 10^{30})}{(3 \times 10^8)^2}$$

$$= \dfrac{2(6.7 \times 10^{-11})(2 \times 10^{30})}{3^2 \times 10^{16}} \qquad \textit{Power of a monomial property}$$

$$= \left(\dfrac{2(6.7)(2)}{3^2}\right)\left(\dfrac{10^{-11}(10^{30})}{10^{16}}\right)$$

$$= \left(\dfrac{26.8}{9}\right)10^{-11+30-16} \qquad \textit{Product and quotient of powers}$$

$$\approx 2.98 \times 10^3 \text{ or } 2980$$

The Schwarzschild radius for the sun is about 2980 meters.

b. Since the actual radius of the sun is 700,000 kilometers, it does not seem to be in danger of becoming a black hole in the near future.

508   *Chapter 9   Exploring Polynomials*

Classroom Vignette

"My students use the TI-82 graphing calculator throughout the year. I have the students write their earth science problems in standard scientific notation and then in TI-82 notation and compare the answers."

Frank Arcuri
Croton Harmon High School
Croton-on-Hudson, New York

Frank Arcuri

Communicating Mathematics

Study the lesson. Then complete the following. 1–4. See margin.

1. When do you use positive exponents in scientific notation?
2. When do you use negative exponents in scientific notation?
3. **Explain** how you can find the product of (1.2×10^5) and (4×10^8) without using pencil and paper or a calculator.
4. **Explain** how you can find the quotient of (4.4×10^4) and (4×10^7) without using pencil and paper or a calculator.

Guided Practice

Express each number in the second column in standard notation. Express each number in the third column in scientific notation.

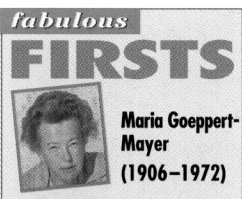

fabulous FIRSTS

Maria Goeppert-Mayer (1906–1972)

Maria Goeppert-Mayer was the first American woman to win the Nobel Prize for Physics. She won in 1963 for her work on the theory about the stability of atomic nuclei.

	Planet	Maximum Distance from Sun (miles)	Radius (miles)
5.	Mercury	4.34×10^7	1515
6.	Earth	9.46×10^7	3963
7.	Jupiter	5.07×10^8	44,419
8.	Uranus	1.8597×10^9	15,881
9.	Pluto	4.5514×10^9	714

5–9. See margin.

Express each number in scientific notation.

10. **Chemistry** The wavelength of cadmium's green line is 0.0000509 centimeters.
11. **Physics** The mass of a proton is 0.000000000000000000001672 milligrams.
12. **Health** The length of the AIDS virus is 0.00011 millimeters. 1.1×10^{-4} mm
13. **Biology** The diameter of an organism called the *Mycoplasma laidlawii* is 0.000004 inch. 4×10^{-6} in.

AIDS virus

Evaluate. Express each result in scientific and standard notation.

14. $(3.24 \times 10^3)(6.7 \times 10^4)$
15. $(0.2 \times 10^{-3})(31 \times 10^{-4})$
16. $\frac{8.1 \times 10^2}{2.7 \times 10^{-3}}$ $3 \times 10^5; 300,000$
17. $\frac{52,440,000,000}{(2.3 \times 10^6)(38 \times 10^{-5})}$ $6 \times 10^7; 60,000,000$

10. 5.09×10^{-5} cm
11. 1.672×10^{-21} mg
14. $2.1708 \times 10^8; 217,080,000$
15. $6.2 \times 10^{-7}; 0.00000062$

Practice

 A

Express each number in scientific notation.

18. 9500 9.5×10^3
19. 0.0095 9.5×10^{-3}
20. 56.9 5.69×10
21. 87,600,000,000 8.76×10^{10}
22. 0.000000000761 7.61×10^{-10}
23. 312,720,000 3.1272×10^8
24. 0.00000008 8×10^{-8}
25. 0.090909 9.0909×10^{-2}

 B

26. 355×10^7 3.55×10^9
27. 78.6×10^3 7.86×10^4
28. 112×10^{-8} 1.12×10^{-6}
29. 0.007×10^{-7} 7×10^{-10}
30. 7830×10^{-2} 7.83×10
31. 0.99×10^{-5} 9.9×10^{-6}

Lesson 9-3 Scientific Notation **509**

Check for Understanding

Exercises 1–17 are designed to help you assess your students' understanding through reading, writing, speaking, and modeling. You should work through Exercises 1–4 with your students and then monitor their work on Exercises 5–17.

fabulous FIRSTS

This theory, called the "shell model" of the atom, is still accepted today.

Assignment Guide

Core: 19–55 odd, 56–67
Enriched: 18–48 even, 49–67

For **Extra Practice,** see p. 776.

The red A, B, and C flags, printed only in the Teacher's Wraparound Edition, indicate the level of difficulty of the exercises.

Additional Answers

1. when the number is greater than or equal to 10
2. when the number is less than 1
3. In your head, first multiply 1.2 and 4 and then multiply 10^5 and 10^8. The answer is 4.8×10^{13}.
4. In your head, first divide 4.4 by 4 and then divide 10^4 by 10^7. The answer is 1.1×10^{-3}.
5. 43,400,000; 1.515×10^3
6. 94,600,000; 3.963×10^3
7. 507,000,000; 4.4419×10^4
8. 1,859,700,000; 1.5881×10^4
9. 4,551,400,000; 7.14×10^2

Reteaching

Using Cooperative Learning Have students work in groups of three. Student A will write a number in standard notation and give it to Student B. Student B will convert it to scientific notation and give it to Student C. Student C will convert it to standard notation and give it back to Student A, who will verify that it is the original number that he or she wrote.

32. 4.48×10^6; 4,480,000
33. 6×10^{-3}; 0.006
34. 2.088×10^{10}; 20,880,000,000
35. 8.992×10^{-7}; 0.0000008992
36. 4×10^7; 40,000,000
37. 4×10^{-2}; 0.04
38. 2.3×10^{-6}; 0.0000023
39. 6.5×10^{-6}; 0.0000065

Evaluate. Express each result in scientific and standard notation.

32. $(6.4 \times 10^3)(7 \times 10^2)$

33. $(4 \times 10^2)(15 \times 10^{-6})$

34. $360(5.8 \times 10^7)$

35. $(5.62 \times 10^{-3})(16 \times 10^{-5})$

36. $\dfrac{6.4 \times 10^9}{1.6 \times 10^2}$

37. $\dfrac{9.2 \times 10^3}{2.3 \times 10^5}$

38. $\dfrac{1.035 \times 10^{-3}}{4.5 \times 10^2}$

39. $\dfrac{2.795 \times 10^{-7}}{4.3 \times 10^{-2}}$

40. $\dfrac{3.6 \times 10^2}{1.2 \times 10^7}$

41. $\dfrac{5.412 \times 10^{-2}}{8.2 \times 10^3}$

42. $\dfrac{(35{,}921{,}000)(62 \times 10^3)}{3.1 \times 10^5}$

43. $\dfrac{1.6464 \times 10^5}{(98{,}000)(14 \times 10^3)}$

40. 3×10^{-5}; 0.00003
42. 7.1842×10^6; 7,184,200
41. 6.6×10^{-6}; 0.0000066
43. 1.2×10^{-4}; 0.00012

Graphing Calculator

Use a graphing calculator to evaluate each expression. Express each answer in scientific notation. 44. 2.7504×10^9 45. 2.4336×10^{-1}

44. $(4.8 \times 10^6)(5.73 \times 10^2)$

45. $(5.07 \times 10^{-4})(4.8 \times 10^2)$

46. $(9.1 \times 10^6) \div (2.6 \times 10^{10})$
3.5×10^{-4}

47. $(9.66 \times 10^3) \div (3.45 \times 10^{-2})$
2.8×10^5

Programming

48. The graphing calculator program at the right evaluates the expression $(2ab^2)^3$ for the values you input. It expresses the result in scientific notation. If $a = 9$ and $b = 10$, the result is 5.832E9.

```
PROGRAM:SCINOT
: Sci
: Prompt A, B, C
: Disp "(2AB^2)^3 =",
  (2AB^2)^3
```

Edit the program to evaluate each expression. Then evaluate for $a = 4$, $b = 6$, and $c = 8$.

a. $a^2b^3c^4$
1.416E7

b. $(-2a)^2(4b)^3$
8.84736E5

c. $(4a^2b^4)^3$
5.706E14

d. $(ac)^3 + (3b)^2$
3.3092E4

Critical Thinking

49a. Sample answer: overflow
49b. See margin.

49. Use a calculator to multiply 3.7×10^{112} and 5.6×10^{10}.
a. Describe what happens when you multiply these values.
b. Describe how you could find the product.
c. Write the product in scientific notation. 2.072×10^{123}

Applications and Problem Solving

50. about 6.57×10^6 times greater

50. **Biology** Seeds come in all sizes. The largest seed is the seed from the double coconut tree, which can have a mass as great as 23 kilograms. In contrast, the mass of the seed of an orchid is about 3.5×10^{-6} grams. Use a calculator to find how many times greater the mass of the seed of the double coconut tree is than the mass of the seed of the orchid. Express your answer in scientific notation.

Germination of wheat seedling

51. **Movies** In the movie *I.Q.*, Albert Einstein, played by Walter Matthau, tries to start a romance between his niece Catherine Boyd, played by Meg Ryan, and an auto mechanic named Ed Walters, played by Tim Robbins. When Ed asks Catherine to estimate the number of stars in the sky, she answers "$10^{12} + 1$." Write this number in standard notation. 1,000,000,000,001

510 Chapter 9 Exploring Polynomials

Study Guide Masters, p. 63

9-3

NAME_____ DATE _____

Study Guide

Student Edition
Pages 506–512

Scientific Notation

Scientists use scientific notation to express very large or very small numbers. A number is expressed in scientific notation when it is expressed as the product of a number between one and ten, and a power of ten.

Example 1: $37,000 = 3.7 \times 10^4$ *Example 2:* $0.00391 = 3.91 \times 10^{-3}$

You can find products or quotients of numbers that are expressed in scientific notation by following the rules for multiplying and dividing monomials and powers.

Example 3: $\dfrac{8.4 \times 10^4}{4.2 \times 10^5} = \dfrac{8.4}{4.2} \times \dfrac{10^4}{10^5}$ *Example 4:* $(6700)(0.00002)$
$= 2 \times 10^{-1}$, or 2000
$= (6.7 \times 10^3)(2 \times 10^{-5})$
$= (6.7 \times 2)(10^3 \times 10^{-5})$
$= 13.4 \times 10^{-2}$
$= 1.34 \times 10^{-1}$, or 0.134

Express each number in scientific notation.

1. 0.0000456
4.56×10^{-4}

2. 0.01
10^{-2}

3. 590,000,000
5.9×10^8

4. 640×10^5
6.4×10^7

Evaluate. Express each result in scientific and standard notation.

5. $\dfrac{1.4 \times 10^4}{0.2 \times 10^2}$
7×10^2; 700

6. $\dfrac{3.3 \times 10^{-12}}{1.1 \times 10^{-14}}$
3×10^2; 300

7. $\dfrac{0.000042}{600}$
7×10^{-8};
0.00000007

8. $(2 \times 10^4)(4 \times 10^{-5})$
8×10^{-1}; 0.8

9. $(77 \times 10^4)(0.02 \times 10^5)$
1.54×10^7;
15,400,000

10. $(3.2 \times 10^{-5})(2.0 \times 10^5)$
6.4×10^0; 6.4

11. $250,000 \div 0.0005$
5×10^8;
500,000,000

12. $(0.00004)(20,000)$
8×10^{-1}; 0.8

13. $(1.5 \times 10^5)(3.3 \times 10^{-4})$
4.95×10^1; 49.5

510 Chapter 9

52. **National Parks** Each year, millions of people visit our national parks and recreational areas. The five most popular locations are listed at the right. Write the number of visitors to each location in scientific notation. **See margin.**

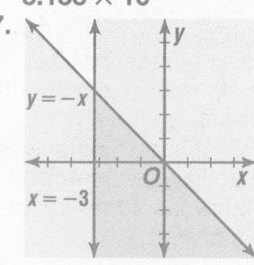

Location	State	Visitors in 1994 (nearest thousand)
Golden Gate National Recreation Area	California	15,309,000
Great Smokey Mountains National Park	Tennessee	9,227,000
Lake Mead National Recreational Area	Nevada	9,022,000
Gulf Islands National Seashore	Florida	5,460,000
Cape Cod National Seashore	Massachusetts	5,153,000

Source: *Good Housekeeping*, Sept., 1994

53. **Biology** There are an average of 25 billion red blood cells in the human body and about 270 million hemoglobin molecules in each red blood cell. Use a calculator to find the average number of hemoglobin molecules in the human body. Write your answer in scientific notation. 6.75×10^{18} molecules

54. **Health** Laboratory technicians look at bacteria through microscopes. A microscope set on 1000× makes an organism appear to be 1000 times larger than its actual size. Most bacteria are between 3×10^{-4} and 2×10^{-3} millimeters in diameter.

   a. How large would bacteria appear under a microscope set on 1000×? **0.3 mm to 2 mm**

   b. Do you think a microscope set on 1000× would allow the technician to see all the bacteria? Explain your answer.

55. **Economics** Suppose you try to feed all of the people on Earth using the 4.325×10^{11} kilograms of food produced each year in the United States and Canada. You need to divide this food among the population of the world so that each person receives the same amount of food each day. The population of Earth is about 4.8×10^9. **55a. about 0.25 kg**

   a. How much food will each person have each day?

   b. Do you think the amount in part a is enough to live on? Justify your answer. **Answers may vary.**

Mixed Review

56. Simplify $\dfrac{24x^2y^7z^3}{-6x^2y^3z}$. (Lesson 9–2) $-4y^4z^2$

57. Graph the system of inequalities. (Lesson 8–5) **See margin.**

   $y \leq -x$
   $x \geq -3$

58. Use substitution to solve the system of equations. (Lesson 8–2) **(-52, -20)**

   $x = 2y - 12$
   $x - 3y = 8$

59. **Statistics** What percent of the data represented at the right is between 15 and 25? (Lesson 7–7) **25%**

5 10 15 20 25 30 35 40 45 50 55

Lesson 9–3 Scientific Notation **511**

Extension

Reasoning Evaluate. Express the result in scientific notation.

$$\dfrac{(3.266 \times 10^3)^{-2}(0.815 \times 10^{-4})^{-2}}{(1.96 \times 10^{-5})^{-2}}$$

5.422×10^{-9}

Additional Answers

52. 1.5309×10^7;
   9.227×10^6;
   9.022×10^6;
   5.460×10^6;
   5.153×10^6

57.

Practice Masters, p. 63

Closing Activity

Modeling Have students identify three very large or very small numbers in the real world, such as the population of India or the mass of an element. Tell students to express each number in scientific and standard notation. Extend the activity by compiling a class list that ranks numbers from greatest to smallest.

Additional Answer

60.
-4 -3 -2 -1 0 1 2 3 4

Enrichment Masters, p. 63

NAME_____ DATE _____
Student Edition
Pages 506–512

Enrichment

Converting Metric Units

Scientific notation is convenient to use for unit conversions in the metric system.

Example 1: How many kilometers are there in 4,300,000 meters?

Divide the measure by the number of meters (1000) in one kilometer. Express both numbers in scientific notation.

$\frac{4.3 \times 10^6}{1 \times 10^3} = 4.3 \times 10^3$ The answer is 4.3×10^3 km.

Example 2: Convert 3700 grams into milligrams.

Multiply by the number of milligrams (1000) in 1 gram.

$(3.7 \times 10^3)(1 \times 10^3) = 3.7 \times 10^6$ There are 3.7×10^6 mg in 3700 g.

Complete the following. Express each answer in scientific notation.

1. 250,000 m = 2.5×10^2 km
2. 375 km = 3.75×10^5 m
3. 247 m = 2.47×10^4 cm
4. 5000 m = 5.0×10^6 mm
5. 0.0004 km = 4.0×10^{-1} m
6. 0.01 mm = 1.0×10^{-5} m
7. 6000 m = 6×10^{-6} mm
8. 340 cm = 3.4×10^{-3} km
9. 52,000 mg = 5.2×10^1 g
10. 420 kL = 4.2×10^5 L

Solve.

11. The planet Mars has a diameter of 6.76×10^3 km. What is the diameter of Mars in meters? Express the answer in both scientific and decimal notation.
6,760,000 m; 6.76×10^6 m

12. The distance of the earth from the sun is 149,590,000 km. Light travels 3.0×10^8 meters per second. How long does it take light from the sun to reach the earth in minutes?
8.31 min

13. A light-year is the distance that light travels in one year. (See Exercise 12.) How far is a light year in kilometers? Express your answer in scientific notation.
9.46×10^{12} km

---

60. Graph the solution set of $x < -3$ or $x \geq 1$. (Lesson 7–4) **See margin.**

61. **Geometry** Write an equation of the line that is perpendicular to the line $5x + 5y = 35$ and passes through the point at $(-3, 2)$. (Lesson 6–6) $y = x + 5$

62. **Aviation** An airplane passing over Sacramento at an elevation of 37,000 feet begins its descent to land at Reno, 140 miles away. If the elevation of Reno is 4500 feet, what should be the approximate slope of descent? (Lesson 6–1) **−232 ft/mi or about −0.044**

63. Is {(3, 4), (5, 4), (7, 5), (9, 5)} a function? (Lesson 5–5) **yes**

64. State the domain of {(4, 4), (0, 1), (21, 5), (13, 0), (3, 9)}. (Lesson 5–2) **{4, 0, 21, 13, 3}**

65. **Geometry** Suppose $\sin K = 0.4563$. Find the measure of $\angle K$ to the nearest degree. (Lesson 4–3) **27°**

66. **Swimming** Rosalinda swims the 50-yard freestyle for the Wachung High School swim team. Her times in the last six meets were 26.89 seconds, 26.27 seconds, 25.18 seconds, 25.63 seconds, 27.16 seconds, and 27.18 seconds. Find the mean and median of her times. (Lesson 3–7) **26.385, 26.58**

67. Solve $x - 44 = -207$. (Lesson 3–1) **−163**

Mathematics and SOCIETY

Measurements Great and Small

The excerpt below appeared in an article in the *New York Times* on September 12, 1993.

ONCE UPON A TIME, WHEN THE WORLD was simpler, a foot was really as long as someone's foot, and a cubit the distance from someone's elbow to the end of the middle finger. Now we measure things that are much smaller than shoes and larger than arms, but there is still a desire to put them into a human context. Take the micron, for example, a unit so small—one millionth of a meter—that it seems hard to cast in a human context. But that doesn't stop newspapers from trying; they inevitably hitch it to something else: a human hair. Earlier this year, describing the one-micron width of a computer circuit, one newspaper article said a hair, by comparison, is 100 microns across. But another article, on cancer-causing soot particles smaller than 10 microns, said a human hair was 75 microns in diameter. Last year, in an article on fiber-optic beams, a human hair was 70 microns; in 1982, in one on coatings for cutting tools, the hair was down to 25 microns. . . . It can be hard to grope with the big, too. *Strategically Speaking*, a marketing newsletter, said recently that 1.8 billion slices of frozen pizza are sold each year—enough to cover 511,366 square miles. How big is that? The newsletter had the sense not to give the answer in microns: enough to cover New York, California, Texas, Maine, Delaware, and Rhode Island combined. ■

1–2. See margin.

1. Based on the data above, what is the average size of one slice of frozen pizza? (State the answer in square feet.) Does this size seem reasonable to you? Explain.

2. Did you have any difficulty performing the calculations for Exercise 1? How is scientific notation useful in calculations with very large numbers?

3. See students' work.
3. When you can, do you check the numbers you see presented in newspapers or magazines to see if they are reasonable? Why or why not?

Mathematics and SOCIETY

Provide students with the area of your state in square miles. Have them determine how many of a certain item of their choice, for example, algebra textbooks or pizza boxes, are necessary to cover the state. Have students present their findings in scientific notation.

Answers for Mathematics and Society

1. 7920 ft² per slice; this is clearly not a reasonable answer. One or more of the figures stated in the article must be incorrect.

2. See students' work; the problem can be made more manageable by using scientific notation to arrive at the solution.

MODELING MATHEMATICS

A Preview of Lesson 9–4

9-4A Polynomials

Materials: algebra tiles

Algebra tiles can be used as a model for polynomials. A **polynomial** is a monomial or the sum of monomials. The diagram below shows the models.

Polynomial Models			
Polynomials are modeled using three types of tiles.	1	x	x^2
Each tile has an opposite.	-1	$-x$	$-x^2$

Activity   Use algebra tiles to model each polynomial.

a. $2x^2$

To model this monomial, you will need 2 blue x^2-tiles.

b. $x^2 - 3x$

To model this polynomial, you will need 1 blue x^2-tile and 3 red x-tiles.

c. $x^2 + 2x - 3$

To model this polynomial, you will need 1 blue x^2-tile, 2 green x-tiles, and 3 red 1-tiles.

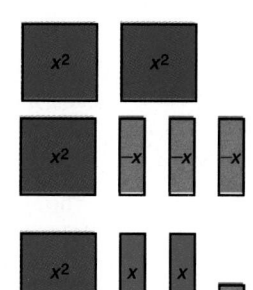

Model   Use algebra tiles to model each monomial or polynomial. Then draw a diagram of your model.   1–4. See Solutions Manual.

1. $-3x^2$ **2.** $2x^2 - 3x + 5$ **3.** $2x^2 - 7$ **4.** $6x - 4$

Write   Write each model as an algebraic expression.

5.

$x^2 - 3x + 2$

6.

$-2x^2 + 4x$

7.

$2x^2 - x - 5$

8.

$-x^2 + 3x - 1$

9. Write a few sentences giving reasons why algebra tiles are sometimes called *area tiles*.
x^2, x, and 1 represent the areas of the tiles.

---

4 ASSESS

Observing students working in cooperative groups is an excellent method of assessment.

NCTM Standards: 1–5

Objective
Use algebra tiles to model polynomials.

Recommended Time
Demonstration and discussion: 15 minutes; Exercises: 30 minutes

Instructional Resources
For each student or group of students
Student Manipulative Kit
• algebra tiles
Modeling Mathematics Masters
• p. 1 (algebra tiles)
• p. 28 (worksheet)
For teacher demonstration
Algebra and Geometry Overhead Manipulative Resources

1 FOCUS

Motivating the Lesson
Ask students the following questions.

• Does an x-tile need to be any particular length? Does its length matter? No; because it represents a variable, its length does not matter.

• Do the sides of an x^2-tile need to be a particular length? No

2 TEACH

Teaching Tip Be sure that students lay down all the x^2-tiles first, then the x-tiles, and finally the constant tiles.

3 PRACTICE/APPLY

Assignment Guide
Core: 1–9
Enriched: 1–9

NCTM Standards: 1–5

Instructional Resources

- Study Guide Master 9-4
- Practice Master 9-4
- Enrichment Master 9-4
- Assessment and Evaluation Masters, pp. 239–240
- Multicultural Activity Masters, p. 18

Transparency 9-4A contains the 5-Minute Check for this lesson; **Transparency 9-4B** contains a teaching aid for this lesson.

Recommended Pacing	
Standard Pacing	Day 6 of 15
Honors Pacing	Day 5 of 14
Block Scheduling*	Day 3 of 8
Alg. 1 in Two Years*	Days 10 & 11 of 26

*For more information on pacing and possible lesson plans, refer to the *Block Scheduling Booklet* and *Algebra 1 in Two Years.*

1 FOCUS

5-Minute Check
(over Lesson 9-3)

Express in scientific notation.

1. 42,345 4.2345×10^4
2. 0.000567 5.67×10^{-4}

Evaluate. Express each result in scientific and standard notation.

3. $(1.3 \times 10^5)(3 \times 10^4)$
   3.9×10^9; 3,900,000,000
4. $(8 \times 10^6)(2.6 \times 10^{-5})$
   2.08×10^2; 208
5. $\dfrac{2.416 \times 10^{-4}}{3.02 \times 10^5}$
   8×10^{-10}; 0.0000000008

Polynomials

9-4

Objectives

After studying this lesson, you should be able to:
- find the degree of a polynomial, and
- arrange the terms of a polynomial so that the powers of a variable are in ascending or descending order.

CONNECTION

Art

The picture at the right shows the painting *Composition with Red, Yellow, and Blue* by the Dutch painter Piet Mondrian. Mondrian preferred abstraction and simplification in his art. He liked to limit his palette to the primary colors and to use straight lines and right angles. The style of painting that Mondrian developed is called *neoplasticism,* and he is considered one of the most influential painters of the 20th century.

Selling Prices of Paintings

1. *Portrait du Dr. Gachet* by van Gogh, $75,000,000
2. *Au Moulin de la Galette* by Renoir, $71,000,000
3. *Les Noces de Pierrette* by Picasso, $51,700,000
4. *Irises* by van Gogh, $49,000,000
5. *Yo Picasso* by Picasso, $43,500,000

Consider a portion of the painting *Composition with Red, Yellow, and Blue.* The length of the sides and the area of each section are given.

We can find the area of this portion of the painting by adding the areas of each section.

$$rx + ry + rz + sx + sy + sz + tx + ty + tz$$

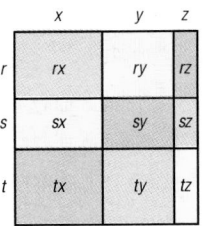

	x	y	z
r	rx	ry	rz
s	sx	sy	sz
t	tx	ty	tz

The expression representing the area of this portion of the painting is called a **polynomial**. A polynomial is a monomial or a sum of monomials. Recall that a monomial is a number, a variable, or a product of numbers and variables. The exponents of the variables of a monomial must be positive. A **binomial** is the sum of two monomials. A **trinomial** is the sum of three monomials. Here are some examples of each. *Polynomials with more than three terms have no special names.*

Monomial	Binomial	Trinomial
$3y^2$	$4x - 7$	$a + 2b + 4c$
$2abc^2$	$2x + 9y$	$x^2 + 8x + 9$
-9	$3x^2 - 11xy$	$x^2 + 2xy + y^2$
$14m$	$2 + 13x$	$3a - 7b^2 - 4c$

Portrait du Dr. Gachet by van Gogh

Irises by van Gogh

Au Moulin de la Galette by Renoir

During his lifetime (1853–1890), Vincent van Gogh was always desperately poor. Few exhibitions of his work were mounted while he was alive. Van Gogh's fame grew in the twentieth century.

Alternative Teaching Strategies

Reading Algebra Students often confuse the degree of a polynomial with the number of terms in the expression. Read the following polynomials to them and have them identify the degree and the number of terms.

$3x^3 + 2x^2 + 4$ 3, 3
$2x^3 + 5x^2 - x + 6$ 3, 4
$x^3 - 4x$ 3, 2
$3x^2 + x - 2$ 2, 3

Example State whether each expression is a polynomial. If it is a polynomial, identify it as either a *monomial, binomial,* or *trinomial.*

 a. $3a - 7bc$

  The expression $3a - 7bc$ can be written as $3a + (-7bc)$. Therefore, it is a polynomial. Since $3a - 7bc$ can be written as a sum of two monomials, $3a$ and $-7bc$, it is a binomial.

 b. $3x^2 + 7a - 2 + a$

  The expression $3x^2 + 7a - 2 + a$ can be written as $3x^2 + 8a + (-2)$. Therefore, it is a polynomial. Since it can be written as the sum of three monomials, $3x^2$, $8a$, and -2, it is a trinomial.

 c. $\frac{7}{2r^2} + 6$

  The expression $\frac{7}{2r^2} + 6$ is not a polynomial because $\frac{7}{2r^2}$ is not a monomial.

> **LOOK BACK**
>
> You can refer to Lesson 1-7 for information on simplifying expressions.

Polynomials can be used to represent savings accumulated over several years.

Example ② Each of the three summers before Li Chiang attends college, she plans to work as a lifeguard and save $2000 towards her college expenses. She plans to invest the money in a savings account at her local bank. The value of each year's savings can be found by the expression px^t where p represents the amount invested, x represents the sum of 1 and the annual interest rate, and t represents the time in years.

APPLICATION
Finance

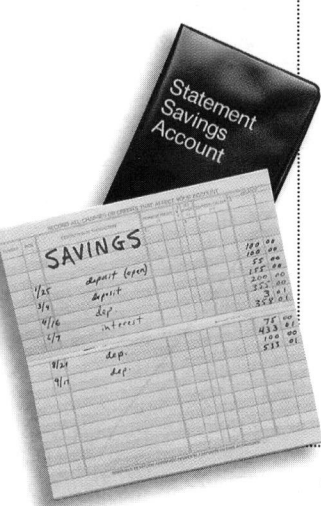

 a. Write a polynomial that represents the total amount of money Li will have when she starts college.

 b. Find the amount of money Li will have if the bank is paying 6% annually.

 a. By the time Li goes to college, the money she makes the last summer will be worth $2000. The money she makes the second summer will be worth $2000x$, and money she makes the first summer will be worth $2000x^2$. Altogether, she will have $2000 + 2000x + 2000x^2$.

 b. Replace x with 1.06 and calculate the total amount of money Li will have.

  $2000 + 2000(1.06) + 2000(1.06)^2 = 6367.20$ *$x = 1 + 6\%$ or 1.06*

  Li will have $6367.20 for college.

The **degree** of a monomial is the sum of the exponents of its variables.

Monomial	Degree
$8y^3$	3
$4y^2ab$	$2 + 1 + 1 = 4$
-14	0
$42abc$	$1 + 1 + 1 = 3$

Remember that $a = a^1$ and $b = b^1$.
Remember that $x^0 = 1$ and $-14 = -14x^0$.

Questioning Introduce the lesson with a discussion of the prefixes *mono-, bi-, tri-,* and *poly-.* Ask students to give examples of common terms that contain these prefixes. Develop the idea that *mono-* means "one," *bi-* means "two," *tri-* means "three," and *poly-* means "more than one."

2 TEACH

In-Class Examples

For Example 1
State whether each expression is a polynomial. If it is a polynomial, identify it as either a monomial, binomial, or trinomial.

a. $7y^3 - 4y^2 + 2$ trinomial
b. $10x^3y^2z$ monomial
c. $\frac{8}{r^6} + 11r$ not a polynomial

For Example 2
If Li's savings account pays 12% annually, how much will Li save for her first year of college? $6748.80

Teaching Tip Explain to students that the degree of a constant, such as -9, is 0 since -9 can be expressed as $-9x^0$. Emphasize that only a nonzero constant has a degree of 0. The number 0 and expressions such as $0x^2$ and $0a^2b^2$, which represent 0, have no degree.

In-Class Example

For Example 3
Find the degree of each polynomial.

a. $3x^2 + 5$ **2**
b. $y^7 + y^6 + 3x^4m^4$ **8**
c. $p^5 + p^3m^3 + 4m$ **6**
d. $3a^2 + 2ab + 4b^2$ **2**

Teaching Tip Explain to students that when placing terms in ascending or descending order (in x), the degree of y has no bearing on the placement of terms unless the degree of x is the same in more than one term. For $x^5y^3 + x^5y^2$, the terms are in descending powers of x. Even though the degree of x is the same in both terms, the first term is of degree 8, and the second term is of degree 7.

3 PRACTICE/APPLY

Check for Understanding

Exercises 1–13 are designed to help you assess your students' understanding through reading, writing, speaking, and modeling. You should work through Exercises 1–4 with your students and then monitor their work on Exercises 5–13.

Additional Answer

3. Sample answers:

   poly- more than one
   polygon, a plane closed figure formed by several line segments; *polygraph,* an instrument used to measure several different pulsations at one time

   mono- one
   monologue, a conversation by one person; *monotone,* a series of sounds in the same key or pitch

   bi- two
   bicycle, a vehicle with two wheels; *bifocal,* eyeglasses with two focal lengths

   tri- three
   triangle, a polygon with three sides; *tricycle,* vehicle with three wheels

---

To find the degree of a polynomial, you must find the degree of each term. The greatest degree of any term is the degree of the polynomial.

Polynomial	Terms	Degree of the Terms	Degree of the Polynomial
$3x^2 + 8a^2b - 4$	$3x^2, 8a^2b, -4$	2, 3, 0	3
$7x^4 - 9x^2y^7 + 4x$	$7x^4, -9x^2y^7, 4x$	4, 9, 1	9

Example ③ **Find the degree of each polynomial.**

a. $9xy + 2$

The degree of $9xy$ is 2.
The degree of 2 is 0.
Thus, the degree of $9xy + 2$ is 2.

b. $18x^2 + 21xy^2 + 13x - 2abc$

The degree of $18x^2$ is 2.
The degree of $21xy^2$ is 3.
The degree of $13x$ is 1.
The degree of $2abc$ is 3.
Thus, the degree of $18x^2 + 21xy^2 + 13x - 2abc$ is 3.

The terms of a polynomial are usually arranged so that the powers of one variable are in ascending or descending order. Later in this chapter, you will learn to add, subtract, and multiply polynomials. These operations are easier to perform if the polynomials are arranged in one of these orders.

Ascending Order	Descending Order
$4 + 5a - 6a^2 + 2a^3$	$2a^3 - 6a^2 + 5a + 4$
$-5 - 2x + 4x^5$	$4x^5 - 2x - 5$
(in x) $8xy - 3x^2y + x^5 - 2x^7y$	(in x) $-2x^7y + x^5 - 3x^2y + 8xy$
(in y) $2x^4 - 3x^3y + 2x^2y^2 - y^{12}$	(in y) $-y^{12} + 2x^2y^2 - 3x^3y + 2x^4$

CHECK FOR UNDERSTANDING

Communicating Mathematics

Study the lesson. Then complete the following.

1. **Explain** why the degree of an integer like -27 is 0. $-27 = -27x^0$

2. **Explain** why $m + \dfrac{34}{n}$ is not a binomial. $\dfrac{34}{n}$ is not a monomial.

3. In this lesson, you were introduced to the words *polynomial, monomial, binomial,* and *trinomial.* These words begin with the prefixes poly-, mono-, bi-, and tri- respectively. Find the meaning of each prefix. List two other words that begin with each prefix, and define each of these words. See margin.

4. The model below represents a polynomial.

Write this polynomial in simplest form. $2x^2 - 2x - 3$

Reteaching

Using Questioning If the expression is a polynomial, give the degree. If not, tell why.

1. $-3x + 7$ yes, 1
2. $3x^{-2} + 4x + 1$ no, not the sum of monomials
3. $4x^2y - 3xy^2$ yes, 3
4. $6x + 9 - 5x^2$ yes, 2

State whether each expression is a polynomial. If the expression is a polynomial, identify it as a *monomial*, a *binomial*, or a *trinomial*.

5. yes; trinomial

7. yes; binomial

5. $4x^3 - 11ab + 6$ 6. $x^3 - \frac{7}{4}x + \frac{y}{x^2}$ no   7. $4c + ab - c$

Find the degree of each polynomial.

8. $11d$ 1   9. 10 0   10. $42x^{12}y^3 - 23x^8y^6$ 15

11. Arrange the terms of $-11x + 5x^3 - 12x^6 + x^8$ so that the powers of x are in descending order. $x^8 - 12x^6 + 5x^3 - 11x$

12. Arrange the terms of $y^4x + y^5x^3 - x^2 + yx^5$ so that the powers of x are in ascending order. $y^4x - x^2 + y^5x^3 + yx^5$

13. **Geometry** The area of a rectangle equals the length times the width. The area of a square equals the square of the length of a side. The area of a circle equals the number pi (π) times the square of the radius. The figure at the right consists of a rectangle, a circle, and a square.

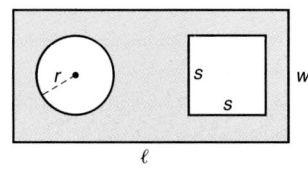

a. Write a polynomial expression that represents the area of the shaded region at the right. $\ell w - \pi r^2 - s^2$

b. Find the area of the shaded region if $w = 8$, $\ell = 15$, $r = 2$, and $s = 4$. about 91.43 square units

EXERCISES

State whether each expression is a polynomial. If the expression is a polynomial, identify it as a *monomial*, a *binomial*, or a *trinomial*.

15. yes; trinomial

17. yes; binomial

19. yes; trinomial

 14. $\frac{r^5}{26}$ yes; monomial 15. $x^2 - \frac{1}{3}x + \frac{y}{234}$ 16. $\frac{y^3}{12x}$ no

17. $5a - 6b - 3a$ 18. $\frac{7}{t} + t^2$ no   19. $9ag^2 + 1.5g^2 - 0.7ag$

Find the degree of each polynomial.

 20. $6a^2$ 2   21. $15t^3y^2$ 5   22. 24 0

23. $m^2 + n^3$ 3   24. $x^2y^3z - 4x^3z$ 6   25. $3x^2y^3z^4 - 18a^5f^3$ 9

26. $8r - 7y + 5d - 6h$ 1 27. $9 + t^2 - s^2t^2 + rs^2t$ 4 28. $-4yzw^4 + 10x^4z^2w$ 7

Arrange the terms of each polynomial so that the powers of *x* are in descending order.

30. $-2x^5 - 9x^2y + 8x + 5$

29. $5 + x^5 + 3x^3$ $x^5 + 3x^3 + 5$ 30. $8x - 9x^2y + 5 - 2x^5$

31. $abx^2 - bcx + 34 - x^7$ 32. $7a^3x + 9ax^2 - 14x^7 + \frac{12}{19}x^{12}$
    $-x^7 + abx^2 - bcx + 34$ $\frac{12}{19}x^{12} - 14x^7 + 9ax^2 + 7a^3x$

Arrange the terms of each polynomial so that the powers of *x* are in ascending order. 33. $1 + x^2 + x^3 + x^5$ 34. $y^4 + 3xy^4 - x^2y^3 + 4x^3y$

35. $7a^3x + \frac{2}{3}x^2 - 8a^3x^3 + \frac{1}{5}x^5$

33. $1 + x^3 + x^5 + x^2$ 34. $4x^3y + 3xy^4 - x^2y^3 + y^4$

36. $4 + \frac{2}{3}x - x^2 + \frac{3}{4}x^3y$ 35. $7a^3x - 8a^3x^3 + \frac{1}{5}x^5 + \frac{2}{3}x^2$ 36. $\frac{3}{4}x^3y - x^2 + 4 + \frac{2}{3}x$

Lesson 9-4 Polynomials **517**

Assignment Guide

Core: 15–43 odd, 45–52
Enriched: 14–40 even, 41–52

For **Extra Practice,** see p. 777.

For **Extra Practice,** see p. 777.

The red A, B, and C flags, printed only in the Teacher's Wraparound Edition, indicate the level of difficulty of the exercises.

Study Guide Masters, p. 64

Geometry

Write a polynomial to represent the area of each shaded region. Then find the area of each region if $a = 20$, $b = 6$, $c = 2$, $r = 5$, and $x = 1$.

C

37.
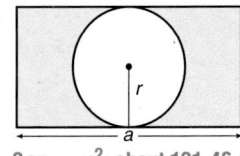

$2ab + \pi b^2$; about 353.10 square units

38.

$2ar - \pi r^2$; about 121.46 square units

39.

$ab - 4x^2$; 116 square units

40.

$2bc - \frac{1}{2}\pi c^2$; about 17.72 square units

Critical Thinking

41. You can write a numeral in base 10 in polynomial form. For example, $3892 = 3(10)^3 + 8(10)^2 + 9(10)^1 + 2(10)^0$.

  a. Write the year of your birth in polynomial form.   See students' work.

  b. Suppose 89435 is a numeral in base a. Write 89435 in polynomial form. $8a^4 + 9a^3 + 4a^2 + 3a^1 + 5a^0$

Applications and Problem Solving

42. Yes; after 6 years she has $53,070.58.

43. about 153 eggs

44a. $\pi r^2 h + \frac{2}{3}\pi r^3$

42. Banking Tawana Hodges inherited $25,000. She invested the money at an annual interest rate of 7.5%. Each year she added $2000 of her own money. Will her original investment of $25,000 double in 7 years? If not, how many years will it take to double the $25,000?

43. Biology The width of the abdomen of a certain type of female moth is useful in estimating the number of eggs that she can carry. The average number of eggs can be estimated by $14x^3 - 17x^2 - 16x + 34$, where x represents the width of the abdomen in millimeters. About how many eggs would you expect this type of moth to produce if her abdomen measures 2.75 millimeters?

44. Agriculture A diagram of a silo is shown at the left. The volume of a cylinder is the product of π, the square of the radius, and the height. The volume of a sphere is the product of $\frac{4}{3}$, π, and the cube of the radius.

  a. Write a polynomial that represents the volume of the silo.

  b. If the height is 40 feet and the radius is 8 feet, find the volume of the silo.   about 9114.81 ft^3

Mixed Review

45. Express 42,350 in scientific notation. (Lesson 9–3) 4.235×10^4

46. Use elimination to solve the system of equations. (Lesson 8–3) **(0, 25)**

$$\frac{3}{2}x + \frac{1}{5}y = 5$$

$$\frac{3}{4}x - \frac{1}{5}y = -5$$

47. Number Theory If 6.5 times an integer is increased by 11, the result is between 55 and 75. What is the integer? List all possible answers. (Lesson 7–4) **7, 8, 9**

Practice Masters, p. 64

48. **Manufacturing** At the end of January, Tanisha's Sporting Equipment had manufactured a total of 3250 tennis racket covers. Assuming that the planned production of tennis racket covers can be represented by a straight line, determine how many covers will be manufactured by the end of the year. (Lesson 6–4) **39,000 covers**

49. Solve $4x + 3y = 12$ if the domain is $\{-3, 1, 3, 9\}$. (Lesson 5–3) **See margin.**

50. $\frac{5}{31}$

50. **Probability** Suppose you meet someone whose birthday is in March. What is the probability that the day of the person's birthday has a 3 in the numeral? (Lesson 4–6)

51. $38.75

51. **Insurance** Insurance claims are usually paid based on the depreciated value of the item in question. An item stolen or destroyed with half its useful life remaining would bring a payment equal to half of its original cost. If a battery was stolen when it was 5 months old, find the depreciated value if the battery cost $45 new and was guaranteed for 36 months. (Lesson 4–1)

52. Solve $ax - by = 2cz$ for y. (Lesson 3–6) $\dfrac{ax - 2cz}{b}$

WORKING ON THE
In·ves·ti·ga·tion

Refer to the Investigation on pages 448–449.

Ready, Set, Drop!

You can determine an average drop time for each triangle and then use these data to predict glide times for triangles of different sizes.

1. Using either the mean, median, or mode, find an average glide time from the data for each of the four triangles. For each average, explain which measure you used and why you selected that measure to give you the best average or most typical time for the experiment. Are the average glide times best represented by scientific notation? Explain.

2. Describe a hypothetical case for which the mode would be the most appropriate average time, one for which the median would be the most appropriate average time, and one for which the mean would be most appropriate.

3. At this point, you have four average times—one for the typical glide time of each of the four triangles. Draw a scatter plot of the data.

Let the horizontal axis represent the perimeter, and let the vertical axis represent the average glide time.

4. Describe the graph. Describe any mathematical relationships that you see. What is the relationship of the perimeter of a triangular piece of tissue paper to the speed of the glide?

5. Draw a best-fit line for your data. Write a linear equation to model this best fit line.

6. Draw a second scatter plot in which the horizontal axis represents the area of the triangles and the vertical axis represents the average glide time.

7. Use your scatter plots to predict the glide time for a triangle with a perimeter of 36 centimeters. Use that glide time and the second scatter plot to predict the surface area of that triangle. How does your prediction compare with the actual area of an equilateral triangle whose perimeter is 36 centimeters? Explain your results.

Add the results of your work to your Investigation Folder.

Lesson 9-4 Polynomials **519**

Extension

Problem Solving Arrange the terms of the following polynomial so that the powers of y are in descending order. Then find the degree of the polynomial.

$16xy^4 - 8y^7z^5 - 2.8xyz + 14x^2y^3 + 3.9y^9z - 10$

$3.9y^9z - 8y^7z^5 + 16xy^4 + 14x^2y^3 - 2.8xyz - 10; \ 12$

In·ves·ti·ga·tion

Working on the Investigation

The Investigation on pages 448–449 is designed to be a long-term project that is completed over several days or weeks. Encourage students to keep their materials in their Investigation Folder as they work on the Investigation.

4 ASSESS

Closing Activity

Speaking Ask students to explain the relationship between a monomial and a polynomial. Then have them give examples of a monomial, a binomial, and a trinomial.

Chapter 9, Quiz B (Lessons 9-3 and 9-4), is available in the *Assessment and Evaluation Masters,* p. 240.

Mid-Chapter Test (Lessons 9-1 through 9-4) is available in the *Assessment and Evaluation Masters,* p. 239.

Additional Answer

49. $\left\{(-3,8), \left(1, \frac{8}{3}\right), (3,0), (9,-8)\right\}$

Enrichment Masters, p. 64

NCTM Standards: 1–5

Objective
Use algebra tiles to add and subtract polynomials.

Recommended Time
Demonstration and discussion: 30 minutes; Exercises: 30 minutes

Instructional Resources
For each student or group of students
Student Manipulative Kit
• algebra tiles
Modeling Mathematics Masters
• p. 1 (algebra tiles)
• p. 29 (worksheet)
For teacher demonstration
Algebra and Geometry Overhead Manipulative Resources

1 FOCUS

Motivating the Lesson
Review the concept of additive inverses with students. Present several numbers and have students name the inverses. Explain that students will use opposite, or inverse, tiles in some of the activities.

2 TEACH

Teaching Tip In Activity 1, stress that adding polynomials is like adding integers.

Teaching Tip Watch for students who have difficulty modeling the opposite of a polynomial. Point out that if they combine the original model with its opposite, they get all zero pairs.

Teaching Tip In Activity 3, another form of subtraction of polynomials is presented. This form is closer to the actual algebraic manipulation used when simplifying a difference.

MODELING MATHEMATICS

A Preview of Lesson 9–5

9–5A Adding and Subtracting Polynomials

Materials: algebra tiles

Monomials such as $4x^2$ and $-7x^2$ are called *like terms* because they have the same variable to the same power. When you use algebra tiles, you can recognize like terms because the individual tiles have the same size and shape.

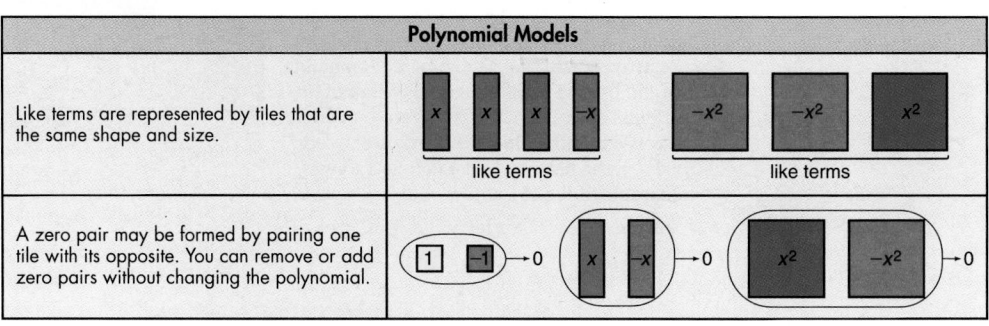

Polynomial Models	
Like terms are represented by tiles that are the same shape and size.	
A zero pair may be formed by pairing one tile with its opposite. You can remove or add zero pairs without changing the polynomial.	

Activity 1   Use algebra tiles to find $(2x^2 + 3x + 2) + (x^2 - 5x - 5)$.

Step 1   Model each polynomial. You may want to arrange like terms in columns for convenience.

Step 2   Combine like terms and remove all zero pairs.

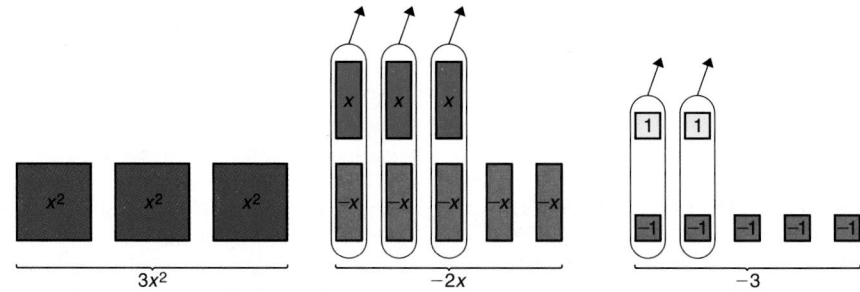

Step 3   Write the polynomial for the tiles that remain.
$$(2x^2 + 3x + 2) + (x^2 - 5x - 5) = 3x^2 - 2x - 3$$

Activity 2 Use algebra tiles to find (2x + 5) − (−3x + 2).

Step 1 Model the polynomial $2x + 5$.

 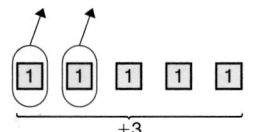

Step 2 To subtract $-3x + 2$, you must remove 3 red x-tiles and 2 yellow 1-tiles. You can remove the yellow 1-tiles, but there are no red x-tiles. Add 3 zero pairs of x-tiles. Then remove the 3 red x-tiles.

Remember that the value of a zero pair is 0.

Step 3 Write the polynomial for the tiles that remain.
$(2x + 5) − (−3x + 2) = 5x + 3$

Recall that you can subtract a number by adding its additive inverse or opposite. Similarly, you can subtract a polynomial by adding its opposite.

Activity 3 Use algebra tiles and the additive inverse, or opposite, to find (2x + 5) − (−3x + 2).

Step 1 To find the difference of $2x + 5$ and $-3x + 2$, add $2x + 5$ and the opposite of $-3x + 2$.

$2x + 5 \rightarrow$

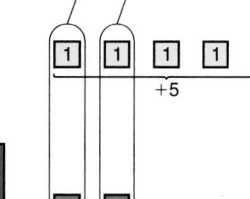

The opposite of $-3x + 2$ is $3x − 2$. $\rightarrow$

Step 2 Write the polynomial for the tiles that remain.
$(2x + 5) − (−3x + 2) = 5x + 3$ *This is the same answer as in Activity 2.*

Model Use algebra tiles to find each sum or difference. 1. $-x^2 + 6$ 2. $-2x^2 - 4x - 2$

1. $(2x^2 − 7x + 6) + (−3x^2 + 7x)$
2. $(−2x^2 + 3x) + (−7x − 2)$
3. $(x^2 − 4x) − (3x^2 + 2x)$
4. $(3x^2 − 5x − 2) − (x^2 − x + 1)$
5. $(x^2 + 2x) + (2x^2 − 3x + 4)$
6. $(2x^2 + 3x − 4) − (3x^2 − 4x + 1)$
   3. $-2x^2 − 6x$ 4. $2x^2 − 4x − 3$
   5. $3x^2 − x + 4$ 6. $-x^2 + 7x − 5$

Draw Is each statement *true* or *false*? Justify your answer with a drawing.

7–8. See margin for drawings.

7. $(3x^2 + 2x − 4) + (−x^2 + 2x − 3) = 2x^2 + 4x − 7$ **true**
8. $(x^2 − 2x) − (−3x^2 + 4x − 3) = -2x^2 − 6x − 3$ **false**

Write 9. Find $(x^2 − 2x + 4) − (4x + 3)$ using each method from Activity 2 and Activity 3. Illustrate with drawings and explain in writing how zero pairs are used in each case. **See margin.**

GLENCOE Technology

Interactive Mathematics Tools Software

This multimedia software provides an interactive lesson that uses algebra tiles to add and subtract like terms in polynomials. A **Computer Journal** gives students an opportunity to write about what they have learned.

For Windows & Macintosh

---

3 PRACTICE/APPLY

Assignment Guide

Core: 1–9
Enriched: 1–9

Additional Answers

7.

$3x^2 + 2x − 4$

$-x^2 + 2x − 3$

$2x^2 + 4x − 7$

8.
$x^2 − 2x$

opposite of $-3x^2 + 4x − 3$

$4x^2 − 6x + 3$

9. **Method from Activity 2:**

$x^2 − 6x + 1$

You need to add zero pairs so that you can remove 4 green x-tiles.

Method from Activity 3:
$x^2 − 2x + 4$

opposite of $4x + 3$

You remove all zero pairs to find the difference in simplest form.

4 ASSESS

Observing students working in cooperative groups is an excellent method of assessment.

Instructional Resources

- Study Guide Master 9-5
- Practice Master 9-5
- Enrichment Master 9-5

 Transparency 9-5A contains the 5-Minute Check for this lesson; **Transparency 9-5B** contains a teaching aid for this lesson.

Recommended Pacing

Standard Pacing	Day 8 of 15
Honors Pacing	Day 7 of 14
Block Scheduling*	Day 4 of 8
Alg. 1 in Two Years*	Days 13 & 14 of 26

 *For more information on pacing and possible lesson plans, refer to the *Block Scheduling Booklet* and *Algebra 1 in Two Years.*

1 FOCUS

 ## 5-Minute Check
(over Lesson 9-4)

State whether each expression is a polynomial. If the expression is a polynomial, identify it as either a monomial, binomial, or trinomial, and give its degree.

1. $8a^2 + 5ab$ binomial, 2

2. $\dfrac{3x^2 + 4x - 7}{x}$ not a polynomial

3. $6a^2b^2 + 7ab^5 - 6b^3$ trinomial, 6

4. $w^2x - \dfrac{7}{3}w + 6x$ trinomial, 3

Arrange the terms in the following polynomial so that the powers of *x* are in descending order.

5. $10x^2y + 6x^{10} + 5 + 5xy$
$6x^{10} + 10x^2y + 5xy + 5$

Adding and Subtracting Polynomials

Objective

After studying this lesson, you should be able to:

- add and subtract polynomials.

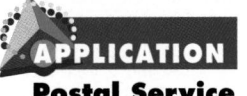 **APPLICATION**
Postal Service

The U.S. Postal Service has restrictions on the sizes of boxes that may be shipped by parcel post. The length plus the girth of the box must not exceed 108 inches. Girth is the shortest distance around the package. It is defined as follows.

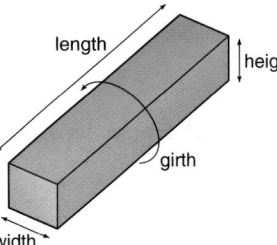

girth = twice the width + twice the height
or $w + w + h + h$

Mrs. Diaz wants to send a package to her daughter, who is a student at Auburn University. She can't find any cardboard boxes at home; all she has is a 60-by-40 inch rectangle of cardboard. She decides that she can make a box out of it by cutting squares out of each corner and folding up the flaps. Then she will use another rectangle of cardboard for the top of the box. She doesn't know how big the squares should be, though. For now, she is calling the side of each square *x*.

$$\text{Girth} = w + w + h + h$$
$$= (40 - 2x) + (40 - 2x) + x + x$$

In order to mail this box, the length plus the girth must be at most 108 inches. That is, $(60 - 2x) + (40 - 2x) + (40 - 2x) + x + x \le 108$. *You will solve this problem in Exercise 42.*

To add polynomials, you can group like terms and then find the sum, or you can write them in column form and then add.

Example Find $(4a^2 + 7a - 12) + (-9a^2 - 6 + 2a)$.

Method 1
Group the like terms together.

$(4a^2 + 7a - 12) + (-9a^2 - 6 + 2a)$

$= [4a^2 + (-9a^2)] + (7a + 2a) + [-12 + (-6)]$

$= [4 + (-9)]a^2 + (7 + 2)a + (-18)$ *Distributive property*

$= -5a^2 + 9a - 18$

Method 2
Arrange the like terms in column form and add.

$$\begin{array}{r} 4a^2 + 7a - 12 \\ (+)\ -9a^2 + 2a - 6 \\ \hline -5a^2 + 9a - 18 \end{array}$$ *Notice that terms are in descending order and like terms are aligned.*

Recall that you can subtract a rational number by adding its opposite or additive inverse. Similarly, you can subtract a polynomial by adding its additive inverse. To find the additive inverse of a polynomial, replace each term with its additive inverse or opposite.

Polynomial	Additive Inverse
$2a - 3b$	$-2a + 3b$
$4x^2 + 7x - 18$	$-4x^2 - 7x + 18$
$-9y + 4x - 2z$	$9y - 4x + 2z$
$7x^3 + 12x^2 + 21$	$-7x^3 - 12x^2 - 21$

Example Find $(6a^2 - 8a + 12b^3) - (-11a^2 + 6b^3)$.

Method 1
Find the additive inverse of $-11a^2 + 6b^3$. Then group the like terms and add.

The additive inverse of $-11a^2 + 6b^3$ is $11a^2 - 6b^3$.

$(6a^2 - 8a + 12b^3) - (-11a^2 + 6b^3)$

$= (6a^2 - 8a + 12b^3) + (11a^2 - 6b^3)$

$= (6a^2 + 11a^2) + (-8a) + [12b^3 + (-6b^3)]$

$= (6 + 11)a^2 - 8a + [12 + (-6)]b^3$

$= 17a^2 - 8a + 6b^3$

Method 2
Arrange like terms in column form and then subtract by adding the additive inverse.

$$\begin{array}{r} 6a^2 - 8a + 12b^3 \\ (-)\ -11a^2\quad\ +\ 6b^3 \\ \hline \end{array} \rightarrow \begin{array}{r} 6a^2 - 8a + 12b^3 \\ (+)\ 11a^2\quad\ -\ 6b^3 \\ \hline 17a^2 - 8a + 6b^3 \end{array}$$

Polynomials can be used to represent measures of geometric figures.

 Cooperative Learning

Trade-A-Problem Have students work in groups of four. Give students a piece of paper containing a list of six terms. Ask each student in the group to arrange the terms using parentheses so that they form the sum of two or more polynomials such that each has a different problem. Have them find the sum and compare their answers. For more information on the trade-a-problem strategy, see *Cooperative Learning in the Mathematics Classroom,* one of the titles in the Glencoe Mathematics Professional Series, pages 25–26.

Motivating the Lesson
Situational Problem Develop student understanding of polynomials through the following activity. The first student in a row identifies a monomial with a degree greater than 2. The next student adds a term to the monomial to create a binomial. The next student adds another term to create a trinomial. A fourth student then identifies the degree of the trinomial. Repeat the procedure until all class members have had a turn.

2 TEACH

Teaching Tip Stress to students that, as in Example 1, sometimes they will have to rearrange terms of a polynomial in order to use the column method.

In-Class Examples

For Example 1
Find each sum.

a. $(9y - 7x + 15a) + (-3y + 8x - 8a)$
$6y + x + 7a$

b. $(3a^2 + 3ab - b^2) + (4ab + 6b^2)$
$3a^2 + 7ab + 5b^2$

For Example 2
Find each difference.

a. $(7a - 10b) - (3a + 4b)$
$4a - 14b$

b. $(3y^2 + 7y + 8) - (2y^2 - 4y + 3)$
$y^2 + 11y + 5$

Teaching Tip For Example 2, point out that there is less chance for error if students subtract by adding the additive inverse.

Teaching Tip For Example 2, emphasize that the additive inverse of every term must be found.

3 PRACTICE/APPLY

Example 3

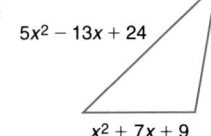

INTEGRATION
Geometry

The measure of the perimeter of the triangle at the right is represented by $11x^2 - 29x + 10$. Find the polynomial that represents the measure of the third side of the triangle.

Explore   Look at the problem. You know the perimeter of the triangle and the measures of two sides. You need to find the measure of the third side.

Plan   The perimeter of a triangle is the sum of the measures of the three sides. To find the measure of the missing side, subtract the two given measures from the perimeter.

Solve $(11x^2 - 29x + 10) - [(5x^2 - 13x + 24) + (x^2 + 7x + 9)]$

$= (11x^2 - 29x + 10) - (5x^2 - 13x + 24) - (x^2 + 7x + 9)$

$= (11x^2 - 29x + 10) + (-5x^2 + 13x - 24) + (-x^2 - 7x - 9)$

$= [11x^2 + (-5x^2) + (-x^2)] + [-29x + 13x + (-7x)] + [10 + (-24) + (-9)]$

$= 5x^2 - 23x - 23$

The measure of the third side of the triangle is $5x^2 - 23x - 23$.

Examine   The sum of the measures of the three sides should equal the perimeter.

$$\begin{array}{r} 5x^2 - 13x + 24 \\ x^2 + 7x + 9 \\ (+)\, 5x^2 - 23x - 23 \\ \hline 11x^2 - 29x + 10 \end{array}$$

The sum of the measures of the three sides of the triangle equals the measure of the perimeter. The answer is correct.

CHECK FOR UNDERSTANDING

Communicating Mathematics

Study the lesson. Then complete the following.

1. **Describe** the first step you take when you add or subtract polynomials in column form.   Align like terms.

2. **Write** three like terms containing powers of a and b. Find the sum of your three terms.   Sample answer: $2a^2b^3, -4a^2b^3, a^2b^3; -a^2b^3$

3. See margin.

3. **Explain** how to check your answer when you subtract two polynomials.

4. **Write** a paragraph explaining how to subtract a polynomial.   Find the additive inverse of the polynomial being subtracted and then add.

MODELING MATHEMATICS

5. Use algebra tiles to find each sum or difference.
   a. $(3x^2 + 2x - 7) + (-2x^2 + 15)$ **b.** $(4x + 1) - (x^2 - 2x + 3)$
     $x^2 + 2x + 8$ $-x^2 + 6x - 2$

Guided Practice

Find the additive inverse of each polynomial.
   6. $5y - 7z$ $-5y + 7z$ **7.** $-6a^2 + 3$ $6a^2 - 3$

9. $4x^2 + 3y^2 - 8y - 7x$ **8.** $7y^2 - 3x^2 + 2$ $-7y^2 + 3x^2 - 2$ **9.** $-4x^2 - 3y^2 + 8y + 7x$

Name the like terms in each group.

10. 3m and 6m; 8n and 5n

10. $3m, 8n, 4mn, 5n, 6m$

11. $-8y^2$ and $3y^2$; $2x$ and $4x$

11. $-8y^2, 2x, 3y^2, 4x, 2z$

12. $2x^3, 5xy, -x^2y, 14xy, 12xy$
    $5xy, 14xy,$ and $12xy$

13. $3p^3q, -2p, 10p^3q, 15pq, -p$
    $3p^3q$ and $10p^3q$; $-2p$ and $-p$

Find each sum or difference.

14.
$$5ax^2 + 3a^2x \quad - 5x$$
$$\underline{(+)\ 2ax^2 \quad\quad - 5ax + 7x}$$
$$7ax^2 + 3a^2x - 5ax + 2x$$

15.
$$11m^2n^2 + 2mn - 11$$
$$\underline{(-)\ 5m^2n^2 - 6mn + 17}$$
$$6m^2n^2 + 8mn - 28$$

16. $-3x^2 + 6x$

16. $(4x^2 + 5x) + (-7x^2 + x)$

17. $-4y^2 + 5y + 3$

17. $(3y^2 + 5y - 6) - (7y^2 - 9)$

18. $5b - 12ab + 12a$

18. $(5b - 7ab + 8a) - (5ab - 4a)$

19. $7p^3 - 3p^2 - 2p - 7$

19. $(6p^3 + 3p^2 - 7) + (p^3 - 6p^2 - 2p)$

20. **Geometry** The sum of the degree measures of the angles of a triangle is 180.

   a. Write a polynomial to represent the measure of the third angle of the triangle at the right. $(181 - 7x)°$

   b. If $x = 15$, find the measures of the three angles of the triangle. $31°, 73°, 76°$

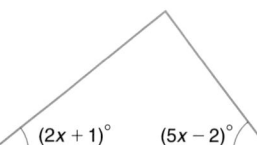
$(2x + 1)°$ $(5x - 2)°$

EXERCISES

Assignment Guide

Core: 21–43 odd, 45–52
Enriched: 22–40 even, 41–52
All: Self Test, 1–10

For **Extra Practice,** see p. 777.

The red A, B, and C flags, printed only in the Teacher's Wraparound Edition, indicate the level of difficulty of the exercises.

Practice

Find each sum or difference.

21.
$$4x^2 + 5xy - 3y^2$$
$$\underline{(+)\ 6x^2 + 8xy + 3y^2}$$
$$10x^2 + 13xy$$

22.
$$6x^2y^2 - 3xy - 7$$
$$\underline{(-)\ 5x^2y^2 + 2xy + 3}$$
$$x^2y^2 - 5xy - 10$$

23.
$$a^3 \quad\quad - b^3$$
$$\underline{(+)\ 3a^3 + 2a^2b - b^2 + 2b^3}$$
$$4a^3 + 2a^2b - b^2 + b^3$$

24.
$$3a^2 \quad - 8$$
$$\underline{(-)\ 5a^2 + 2a + 7}$$
$$-2a^2 - 2a - 15$$

25.
$$3a + 2b - 7c$$
$$-4a + 6b + 9c$$
$$\underline{(+)\ -3a - 2b - 7c}$$
$$-4a + 6b - 5c$$

26.
$$2x^2 - 5x + 7$$
$$5x^2 \quad - 3$$
$$\underline{(+)\ x^2 - x + 11}$$
$$8x^2 - 6x + 15$$

27. $3a - 11m$
28. $8 - 6a + 2a^2$
29. $-2n^2 + 7n + 5$
30. $2x^2 - 8x - 8$
32. $3ax^2 + 11ax - 4$
33. $-y^3 + 3y + 3$
34. $6p^2 - p - 18$

B

27. $(5a - 6m) - (2a + 5m)$

28. $(3 + 2a + a^2) + (5 - 8a + a^2)$

29. $(n^2 + 5n + 13) + (-3n^2 + 2n - 8)$

30. $(5x^2 - 4) - (3x^2 + 8x + 4)$

31. $(13x + 9y) - 11y$ $13x - 2y$

32. $(5ax^2 + 3ax) - (2ax^2 - 8ax + 4)$

33. $(3y^3 + 4y - 7) + (-4y^3 - y + 10)$

34. $(7p^2 - p - 7) - (p^2 + 11)$

35. $(4z^3 + 5z) + (-2z^2 - 4z)$
    $4z^3 - 2z^2 + z$

36. $(x^3 - 7x + 4x^2 - 2) - (2x^2 - 9x + 4)$
    $x^3 + 2x^2 + 2x - 6$

 INTEGRATION

Geometry

C

38. $10x^2 - 31x + 14$

The measures of two sides of a triangle are given. P represents the measure of the perimeter. Find the measure of the third side.

37. $P = 5x + 2y$ $2x + 3y$

38. $P = 13x^2 - 14x + 12$

$x + 2y$
$2x - 3y$

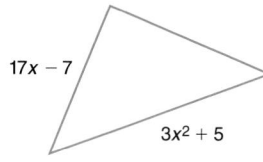
$17x - 7$
$3x^2 + 5$

Study Guide Masters, p. 65

9-5 NAME_____ DATE_____
Study Guide Student Edition Pages 522–527

Adding and Subtracting Polynomials

You can use the following methods when you add polynomials.

Example 1: Find $(4x + 6y) + (3x + 9y)$.
$(4x + 6y) + (3x + 9y) = (4x + 3x) + (6y + 9y)$
$= (4 + 3)x + (6 + 9)y$
$= 7x + 15y$

Example 2: Find $(3x^2 - 5xy + 8y^2) + (2x^2 + xy - 6y^2)$.
$3x^2 - 5xy + 8y^2$
$\underline{2x^2 + xy - 6y^2}$
$5x^2 - 4xy + 2y^2$

You can subtract a polynomial by adding its additive inverse.

Subtraction → additive inverses ← Addition
$-1 - (-4) = 3$ $-1 + 4 = 3$

Example 3: Find $(7x^2 - 8) - (-3x^2 + 1)$.

Method 1
$(7x^2 - 8) - (-3x^2 + 1)$
$= 7x^2 - 8 + 3x^2 - 1$
$= (7x^2 + 3x^2) + (-8 - 1)$
$= 10x^2 - 9$

Method 2
$7x^2 - 8$ $7x^2 - 8$
$\underline{(-)\ -3x^2 + 1}$ $\underline{(+)\ 3x^2 - 1}$
                    $10x^2 - 9$

Find each sum or difference.

1. $(4a - 5) + (3a + 6)$ $7a + 1$

2. $(3x^2 - 5xy^2 + y^3) + (-3x^2 + 5xy^2 - y^3)$ 0

3. $(3p^2 - 2p + 3) + (p^2 - 7p + 7)$
   $4p^2 - 9p + 10$

4. $(x^3 + y^3) - (-x^2 + y^3)$
   $2x^2$

5. $(2x^2 + 5xy + 4y^2) - (2x^2 + 5xy + 4y^2)$ 0

6. $(2a^2 - ab + b^2) + (3a^2 + 5ab - 7ab^2)$
   $5a^2 + 4ab - 7ab^2 + b^2$

7.
$$6x^2 + 3x$$
$$\underline{(+)\ x^2 - 4x - 3}$$
$$7x^2 - x - 3$$

8.
$$5x + 1\frac{1}{2}y + \frac{1}{4}$$
$$\underline{(+)\ 2x \quad - 7}$$
$$7x + 1\frac{1}{2}y - 6\frac{3}{4}$$

9.
$$x^2 + 2xy + y^2$$
$$x^2 - xy - y^2$$
$$\underline{(+)\ 3x^2 - 2xy - y^2}$$
$$5x^2 - xy - y^2$$

10.
$$10x^2 + 5x - 6$$
$$\underline{(-)\ 8x^2 - 2x + 7}$$
$$2x^2 + 7x - 13$$

11.
$$4.2x^2 \quad - 3.2$$
$$\underline{(-)\ 5.9x^2 + 2.6x - 1.9}$$
$$-1.7x^2 - 2.6x - 1.3$$

12.
$$7p^3q^2 - 8pq + 9$$
$$\underline{(-)\ p^3q^2 - 9pq - 10}$$
$$6p^3q^2 + pq + 19$$

The sum of the degree measures of the angles of a quadrilateral is 360. Find the measure of the fourth angle of each quadrilateral given the degree measures of the other three angles.

39. $4x + 12, 8x - 10, 6x + 5$
    $353 - 18x$

40. $x^2 - 2, x^2 + 5x - 9, -8x - 42$
    $413 - 2x^2 + 3x$

Critical Thinking

41. Your teacher gives you two polynomials. When the first polynomial is subtracted from the second polynomial, the difference is $2n^2 + n - 4$. What is the difference when the second polynomial is subtracted from the first? $-2n^2 - n + 4$

Applications and Problem Solving

42. **Postal Service**   Refer to the application at the beginning of the lesson.
    a. Solve the inequality to find the possible values of x Mrs. Diaz could use in designing her package. $x \geq 8$
    b. For reasons other than Postal Service regulation, what is the greatest value x could have?   20; The rectangle is only 40 inches wide, so 20 inches is the largest square that can be cut from each corner.

43. **Architecture**   The Sears Tower in Chicago is one of the tallest structures in the world. It is actually a building of varying heights as shown in the photo at the left. The diagram below indicates the height of each section in stories.

x	50 stories	89 stories	66 stories
x	110 stories	110 stories	89 stories
x	66 stories	89 stories	50 stories
	x	x	x

Use stories as a unit of measure. Assume that each section is x stories long and x stories wide. Write an expression for the volume of the Sears Tower.
$719x^2$ cubic stories

44. **Look for a Pattern**   At City Center Mall, there are 25 lockers numbered 1 through 25. Suppose a shopper opens every locker. Then a second shopper closes every second locker. Next a third shopper changes the state of every third locker. (If it's open, the shopper closes it. If it's closed, the shopper opens it.) Suppose this process continues until the 25th shopper changes the state of the 25th locker.
    a. Which lockers will still be open?   1, 4, 9, 16, 25
    b. Describe the numbers in your answer for part a.   square numbers

44c. 36, 49, 64
44d. square numbers less than or equal to n

    c. Give the next three numbers for the pattern you found in part a.
    d. If there were n lockers in this pattern, which lockers will be open?

Mixed Review

45. Arrange the terms of $-3x + 4x^5 - 2x^3$ so that the powers of x are in ascending order.   (Lesson 9-4) $-3x - 2x^3 + 4x^5$

46. **Finance**   The current balance on a car loan can be found by evaluating the expression $P\left[\dfrac{1-(1+r)^{k-n}}{r}\right]$, where P is the monthly payment, r is the monthly interest rate, k is the number of payments already made, and n is the total number of monthly payments. Find the current balance if $P = \$256, r = 0.01, k = 20$, and $n = 60$.   (Lesson 9-2) $8405.68

 Tech Prep

Postal Worker Students who are interested in a career in the postal service may wish to do further research on the use of mathematics within that occupation, as mentioned in Exercise 42. For more information on tech prep, see the *Teacher's Handbook*.

Practice Masters, p. 65

47. Solve the system of equations by graphing. Then state the solution of the system of equations. (Lesson 8–1)

$5x - 3y = 12$
$2x - 5y = 1$ **See margin for graph; (3, 1).**

48. Statistics What is the upper quartile of the data represented by the box-and-whisker plot at the right? (Lesson 7–7) **135**

49. Business Janet's Garage charges $83 for a two-hour repair job and $185 for a five-hour repair job. Define the variables and write a linear equation that Janet can use to bill customers for repair jobs of any length of time. (Lesson 6–4) **See margin.**

50. Finance The selling price of $145,000 for a home included a 6.5% commission for a real estate agent. How much money did the owners receive from the sale? (Lesson 4–5) **$136,150**

51. Solve $\frac{-3n - (-4)}{-6} = -9$. (Lesson 3–3) $-\frac{50}{3}$

52. Tennis The diameter of a circle is the distance across the circle. If the diameter of a tennis ball is $2\frac{1}{2}$ inches, how many tennis balls will fit in a can 12 inches high? (Lesson 2–7) **4 tennis balls**

SELF TEST

Simplify. Assume that no denominator is equal to zero. (Lessons 9–1 and 9–2)

1. $(-2n^4 y^3)(3ny^4)$
$-6n^5 y^7$

2. $(-3a^2 b^5)^2$
$9a^4 b^{10}$

3. $\frac{24a^3 b^6}{-2a^2 b^2}$ $-12ab^4$

4. $\frac{(5r^{-1} s)^3}{(s^2)^3}$ $\frac{125}{r^3 s^3}$

Express each number in scientific notation. (Lesson 9–3)

5. 5,670,000 5.67×10^6

6. 0.86×10^{-4} 8.6×10^{-5}

7. **Space Exploration** A space probe that is 2.85×10^9 miles away from Earth sends radio signals back to NASA. If the radio signals travel at the speed of light (186,000 miles per second), how long will it take the signals to reach NASA? (Lesson 9–3) **about 1.53×10^4 seconds or 4.25 hours**

8. Find the degree of the polynomial $11x^2 + 7ax^3 - 3x + 2a$. Then write the polynomial so that the powers of x are in ascending order. (Lesson 9–4) **4; $2a - 3x + 11x^2 + 7ax^3$**

Find each sum or difference. (Lesson 9–5)

9. $(x^2 + 3x - 5) + (4x^2 - 7x - 9)$
$5x^2 - 4x - 14$

10. $(2a - 7) - (2a^2 + 8a - 11)$
$-2a^2 - 6a + 4$

Lesson 9–5 Adding and Subtracting Polynomials **527**

Extension

Connections A pentagon is a geometric figure with five sides. Find its perimeter if the lengths of its sides are $(x^2 + 2x)$ ft, $(2x^2 - 5x)$ ft, $(4x + 3x^2)$ ft, $(6x + 2)$ ft, and $(7x^2 + 3x + 4)$ ft.
$(13x^2 + 10x + 6)$ ft

SELF TEST

The Self Test provides students with a brief review of the concepts and skills in Lessons 9-1 through 9-5. Lesson numbers are given to the right of exercises or instruction lines so students can review concepts not yet mastered.

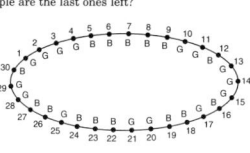
Chapter 9 **527**

NCTM Standards: 1–5

Objective
Use algebra tiles to model the product of simple polynomials.

Recommended Time
Demonstration and discussion: 15 minutes; Exercises: 30 minutes

Instructional Resources
For each student or group of students
Student Manipulative Kit
• algebra tiles
Modeling Mathematics Masters
• p. 1 (algebra tiles)
• p. 15 (product mat)
• p. 30 (worksheet)
For teacher demonstration
Algebra and Geometry Overhead Manipulative Resources

1 FOCUS

Motivating the Lesson
Encourage students to develop a method for using algebra tiles to model the multiplication of a polynomial by a monomial.

2 TEACH

Teaching Tip Students should recognize that this is an application of the distributive property. Emphasize that the top and side of the rectangle define the algebra tile to be used. That is, x by x requires an $x \times x$ or x^2-tile.

Teaching Tip Students may apply their integer knowledge to determine the appropriate tile to use. In Activity 1, $(-1)(1x) = -x$, which indicates that a red x-tile should be used.

3 PRACTICE/APPLY

Assignment Guide
Core: 1–9
Enriched: 1–9

MODELING MATHEMATICS

9-6A Multiplying a Polynomial by a Monomial

A Preview of Lesson 9–6

Materials: algebra tiles   product mat

You have used rectangles to model multiplication. In this activity, you will use algebra tiles to model the product of simple polynomials. The width and length of a rectangle will represent a monomial and a polynomial, respectively. The area of the rectangle will represent the product of the monomial and the polynomial.

Activity 1   Use algebra tiles to find $x(x - 4)$.

The rectangle will have a width of x units and a length of $(x - 4)$ units. Use your algebra tiles to mark off the dimensions on a product mat. Then make the rectangle with algebra tiles.

The rectangle consists of 1 blue x^2-tile and 4 red x-tiles. The area of the rectangle is $x^2 - 4x$. Therefore, $x(x - 4) = x^2 - 4x$.

Activity 2   Use algebra tiles to find $2x(x + 2)$.

The rectangle will have a width of $2x$ units and a length of $(x + 2)$ units. Make the rectangle with algebra tiles.

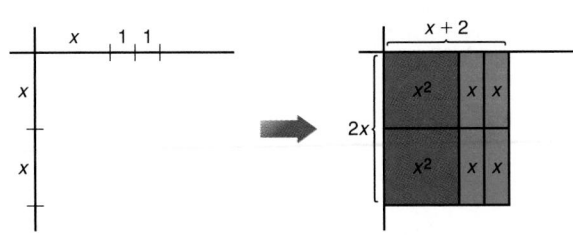

The rectangle consists of 2 blue x^2-tiles and 4 green x-tiles. The area of the rectangle is $2x^2 + 4x$. Therefore, $2x(x + 2) = 2x^2 + 4x$.

Model   **Use algebra tiles to find each product.**

1. $x(x + 2)$ $x^2 + 2x$
2. $x(x - 3)$ $x^2 - 3x$
3. $2x(x + 1)$ $2x^2 + 2x$
4. $2x(x - 3)$ $2x^2 - 6x$
5. $x(2x + 1)$ $2x^2 + x$
6. $3x(2x - 1)$ $6x^2 - 3x$

7–8. See margin for drawings.

Draw   **Is each statement *true* or *false*? Justify your answer with a drawing.**

7. $x(2x + 4) = 2x^2 + 4x$ true
8. $2x(3x - 4) = 6x^2 - 8$ false

Write   9. Suppose you have a square storage building that measures x feet on a side. You triple the length of the building and increase the width by 15 feet.

   a. What will be the dimensions of the new building? $3x$ and $x + 15$

   b. What is the area of the new building? Write your solution in paragraph form, complete with drawings. $(3x^2 + 45x)$ square feet; See Solutions Manual for drawing.

---

4 ASSESS

Observing students working in cooperative groups is an excellent method of assessment. You may wish to ask a student from each group to explain how to model and solve a problem. Also watch for and acknowledge students who are helping others understand the concept being taught.

Additional Answers

9-6

Multiplying a Polynomial by a Monomial

Objectives

After studying this lesson, you should be able to:

- multiply a polynomial by a monomial, and
- simplify expressions involving polynomials.

APPLICATION

Recreation

Have you ever played the game of hopscotch? Children from all over the world like to play some form of this game. The following diagrams show versions from different countries.

F Y I

Tourists can still see the hopscotch pattern carved into the floor of the ancient Forum in Rome.

Escargot (France)

Ta Galagala (Nigeria)

Gat Fei Gei (China)

Jumby (Trinidad)

On the Caribbean island of Trinidad, the children play Jumby. The pattern for this game is shown at the right. Suppose the dimensions of each rectangle are x and $x + 14$. To find the area of each rectangle, you must multiply its length by its width.

LOOK BACK

You can refer to Lesson 1-7 for information on the distributive property.

This diagram of one of the rectangles shows that the area is $x(x + 14)$.

This diagram of the same rectangle shows that the area is $x^2 + 14x$.

Since the areas are equal, $x(x + 14) = x^2 + 14x$.

The application above shows how the distributive property can be used to multiply a polynomial by a monomial.

Lesson 9-6 Multiplying a Polynomial by a Monomial **529**

F Y I

The "scotch" refers to the line drawn on the ground. The name therefore refers to the fact that the player must "hop" over the "scotch."

1 FOCUS

5-Minute Check
(over Lesson 9-5)

Find each sum or difference.

1. $(6a + 7b) + (11a + 4b)$
   $17a + 11b$
2. $(5x^2 - 7x + 3) + (16x + 4x^2 - 9)$
   $9x^2 + 9x - 6$
3. $(3a^2 + 7ab - 9b^2) + (4a^2 + ab - 11b^2)$
   $7a^2 + 8ab - 20b^2$
4. $(8d^2 - 3d - 7) - (d^2 - 13)$
   $7d^2 - 3d + 6$
5. $(6x^2 + 2x - 9) - (3x^2 - 8x + 11)$
   $3x^2 + 10x - 20$

Motivating the Lesson

Hands-On Activity Have groups of students draw hopscotch patterns on the pavement outside or on large sheets of paper. Have the groups measure all the dimensions and compute the total area for their hopscotch patterns.

2 TEACH

In-Class Examples

For Example 1
Find each product.

a. $5n(7n - 2)$ $35n^2 - 10n$

b. $\frac{3}{4}a(8a^2 + 12a + 24)$
$6a^3 + 9a^2 + 18a$

For Example 2
Find each product.

a. $6rs(r^2s - 3r + 2r^2s)$
$18r^3s^2 - 18r^2s$

b. $-4s^2t^2(5s^2t + 3s + 4t^2 - 6s^2t - st^2)$
$4s^4t^3 + 4s^3t^4 - 12s^3t^2 - 16s^2t^4$

For Example 3
If each lane is about 4 feet wide, how far apart should officials start the runners in the two inside lanes?
about 12.6 feet apart

Teaching Tip In Example 2, make sure that students are careful when distributing the negative term.

Example ① Find each product.

a. $7b(4b^2 - 18)$

You can multiply horizontally or vertically.

Method 1: Horizontal
$$7b(4b^2 - 18) = 7b(4b^2) - 7b(18)$$
$$= 28b^3 - 126b$$

Method 2: Vertical
$$4b^2 - 18$$
$$\underline{(\times) \quad\quad 7b}$$
$$28b^3 - 126b$$

b. $-3y^2(6y^2 - 8y + 12)$
$$-3y^2(6y^2 - 8y + 12) = -3y^2(6y^2) - (-3y^2)(8y) + (-3y^2)(12)$$
$$= -18y^4 + 24y^3 - 36y^2$$

Some expressions may contain like terms. In these cases, you will need to simplify by combining like terms.

Example ② Find $-3pq(p^2q + 2p - 3p^2q)$.

Method 1
Multiply first and then simplify by combining like terms.

$$-3pq(p^2q + 2p - 3p^2q) = -3pq(p^2q) + (-3pq)(2p) - (-3pq)(3p^2q)$$
$$= -3p^3q^2 - 6p^2q + 9p^3q^2$$
$$= 6p^3q^2 - 6p^2q$$

Method 2
Simplify by combining like terms and then multiply.

$$-3pq(p^2q + 2p - 3p^2q) = -3pq(-2p^2q + 2p)$$ *p^2q and $-3p^2q$*
$$= -3pq(-2p^2q) + (-3pq)(2p)$$ *are like terms.*
$$= 6p^3q^2 - 6p^2q$$

Example ③

Track

The runners in a 200-meter dash race around the curved part of a track. If the runners start and finish at the same line, the runner on the outside lane would run farther than the other runners. To compensate for this situation, the starting points of the runners are staggered. If the radius of the inside lane is x and each lane is 2.5 feet wide, how far apart should the officials start the runners in the two inside lanes?

The formula for the circumference C of a circle is $C = 2\pi r$, where r is the radius of the circle. The distance around half of a circle is πr. Use this information to find the distance around the curve for the two inside lanes and subtract the quantities to find the stagger distance.

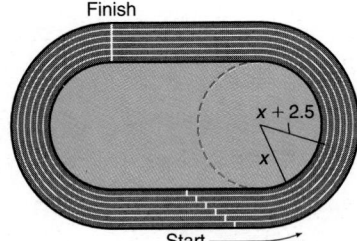

530 *Chapter 9 Exploring Polynomials*

Alternative Learning Styles

Kinesthetic Have students design a hopscotch pattern with a total area of 100 square feet. The pattern may contain only squares and rectangles. Once the pattern has been designed, students may draw it on the ground and measure it to verify the total area.

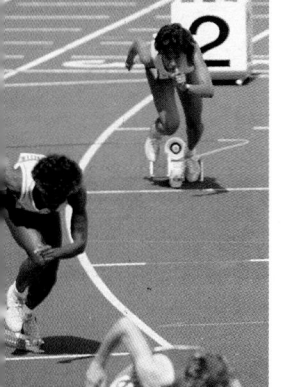

$$\underset{\substack{\text{outside} \\ \text{semicircle}}}{\pi(x + 2.5)} - \underset{\substack{\text{inside} \\ \text{semicircle}}}{\pi x} = \pi x + 2.5\pi - \pi x \qquad \textit{Distributive property.}$$
$$= 2.5\pi \qquad\qquad\qquad\quad \textit{Combine like terms.}$$

The two runners should start 2.5π, or about 7.9, feet apart.

Many equations contain polynomials that must be added, subtracted, or multiplied before the equation can be solved. The distributive property is frequently used as at least one of the steps in solving equations.

Example ④ **Solve** $x(x + 3) + 7x - 5 = x(8 + x) - 9x + 14$.

$$x(x + 3) + 7x - 5 = x(8 + x) - 9x + 14$$
$$x^2 + 3x + 7x - 5 = 8x + x^2 - 9x + 14 \qquad \textit{Distributive property.}$$
$$x^2 + 10x - 5 = x^2 - x + 14 \qquad\qquad \textit{Combine like terms.}$$
$$10x - 5 = -x + 14 \qquad\qquad\quad \textit{Subtract } x^2 \textit{ from each side.}$$
$$11x - 5 = 14 \qquad\qquad\qquad \textit{Add x to each side.}$$
$$11x = 19 \qquad\qquad\qquad\quad \textit{Add 5 to each side.}$$
$$x = \frac{19}{11} \qquad\qquad\qquad\quad \textit{Divide each side by 11.}$$

The solution is $\frac{19}{11}$.

CHECK FOR UNDERSTANDING

Communicating Mathematics

Study the lesson. Then complete the following. 1. distributive property

1. **Name** the property used to simplify $3a(5a^2 + 2b - 3c^2)$.

2. Refer to the application at the beginning of the lesson.
   a. **Describe** how you could find the area of the entire pattern used to play Jumby. **Multiply 7 times $x^2 + 14x$.** 2b. $7x^2 + 98x$
   b. **Write** an expression in simplest form for the area of this pattern.
   c. If x represents 8 inches, find the area of the pattern. **1232 in²**

3. Refer to Example 4. **3a. See margin.**
   a. **Explain** how you could check the solution to the equation.
   b. **Check** the solution. **See Solutions Manual.**

 MATH JOURNAL

4a. See margin.

4. A rectangular garden is $2x + 3$ units long and $3x$ units wide.
   a. Draw a model of the garden.
   b. Find the area of the garden. $6x^2 + 9x$

Guided Practice

Find each product. 6. $-32a^5c + 4a^2c - 44a^2$

5. $-7b(9b^3c + 1)$ $-63b^4c - 7b$

6. $4a^2(-8a^3c + c - 11)$

7. $5y - 13$
   (×) $2y$ $10y^2 - 26y$

8. $2ab - 5a$
   (×) $11ab$ $22a^2b^2 - 55a^2b$

Lesson 9–6 Multiplying a Polynomial by a Monomial **531**

Reteaching

Using Diagrams Have students examine the diagram below and answer the questions that follow.

a	a^2	a^2	a^2	ab	$3a$
a	a^2	a^2	a^2	ab	$3a$
	a	a	a	b	3

1. What is the width of the large rectangle? **2a**
2. What is the length? **3a + b + 3**
3. Find the area.
   $2a(3a + b + 3) = 6a^2 + 2ab + 6a$
4. Inside each square, write its area. **See diagram.**
5. Add the areas in Exercise 4. Compare with your answer in Exercise 3. **same**

In-Class Example

For Example 4
Solve $t(t - 5) + 2t - 1 = 7t + 3 + t(8 + t)$. $-\frac{2}{9}$

3 PRACTICE/APPLY

Check for Understanding
Exercises 1–13 are designed to help you assess your students' understanding through reading, writing, speaking, and modeling. You should work through Exercises 1–4 with your students and then monitor their work on Exercises 5–13.

Additional Answers

3a. Replace x with $\frac{19}{11}$ and check to see if both sides of the equation represent the same number.

4a.

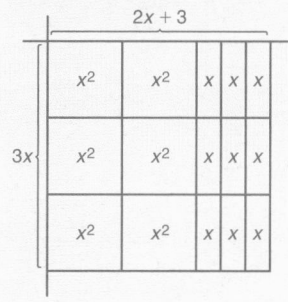

Study Guide Masters, p. 66

9-6 NAME_____ DATE_____
Study Guide Student Edition Pages 529–533

Multiplying a Polynomial by a Monomial

The example below shows how the distributive property can be used to multiply a polynomial by a monomial.

Example 1: $6a(a^2 + 5) = 6a(2a^2) + 6a(5)$
$= 12a^3 + 30a$

Many equations contain polynomials that must be added, subtracted, or multiplied before the equation can be solved.

Example 2:
Solve $-5 - (10 - 3p) = 12$.
$-5 - (10 - 3p) = 12$
$-5 - 10 + 3p = 12$
$-15 + 3p = 12$
$-15 + 15 + 3p = 12 + 15$
$3p = 27$
$\frac{3p}{3} = \frac{27}{3}$
$p = 9$

Example 3:
Solve $4(n - 2) + 5n = 6(3 - n) + 18$.
$4(n - 2) + 5n = 6(3 - n) + 18$
$4n - 8 + 5n = 18 - 6n + 18$
$9n - 8 = 36 - 6n$
$9n + 6n - 8 + 8 = 36 + 8 - 6n + 6n$
$15n = 44$
$\frac{15n}{15} = \frac{44}{15}$
$n = 2\frac{14}{15}$

Find each product.

1. $5(2a + 3b)$
   $10a + 15b$
2. $3x(4x - 2y)$
   $12x^2 - 6xy$
3. $-4(a^2 - b^2)$
   $-4a^2 + 4b^2$
4. $8b\left(b^3 - \frac{1}{2}b\right)$
   $8b^4 - 4b^2$
5. $-5t^5\left(\frac{1}{25}t^4 - \frac{1}{25}t + \frac{1}{5}\right)$
   $-\frac{1}{5}t^4 + \frac{1}{5}t^3 - t^2$
6. $\left(\frac{1}{2}x + y\right)\left(\frac{1}{2}y\right)$
   $\frac{1}{4}xy + \frac{1}{2}y^2$

Simplify.

7. $4r(2r^2 - 3r + 5) + 6r(4r^2 + 2r + 8)$
   $32r^3 + 68r$
8. $2b(b^2 + 4b + 8) - 3b(3b^2 + 9b - 18)$
   $-7b^3 - 19b^2 + 70b$

Solve each equation.

9. $3(x + 5) - 6 = 18$
   3
10. $3x(x - 5) - 3x^2 = -30$
    2
11. $(4y - 3) - 8y + 6 = 19$
    -4

12. Ria's corn field has a perimeter of 18,000 meters. The length of one side is 3 meters less than twice the width. What are the dimensions of the field?
    5999 m × 3001 m

13. Three sides of a triangle have measures that are consecutive even integers. What are the lengths of the sides if the perimeter is 114 meters?
    36 m, 38 m, 40 m

For **Extra Practice**, see p. 777.

The red A, B, and C flags, printed only in the Teacher's Wraparound Edition, indicate the level of difficulty of the exercises.

Additional Answers

43. Sample answer:
$1(8a^2b + 18ab)$
$a(8ab + 18b)$
$b(8a^2 + 18a)$
$2(4a^2b + 9ab)$
$(2a)(4ab + 9b)$
$(2b)(4a^2 + 9a)$
$(ab)(8a + 18)$
$(2ab)(4a + 9)$

47a. See page 533.

52.

$2x - y = 8$

Practice Masters, p. 66

Simplify.

9. $w(3w - 5) + 3w$
   $3w^2 - 2w$
10. $4y(2y^3 - 8y^2 + 2y + 9) - 3(y^2 + 8y)$
    $8y^4 - 32y^3 + 5y^2 + 12y$

Solve each equation.

11. $12(b + 14) - 20b = 11b + 65$ $\frac{103}{19}$
12. $x(x - 4) + 2x = x(x + 12) - 7$ $\frac{1}{2}$

13. **Number Theory** Suppose a is an even integer.   13a. $a^2 + a$
    a. Write the product, in simplest form, of a and the next integer after it.
    b. Write the product, in simplest form, of a and the next even integer after it. $a^2 + 2a$

EXERCISES

Practice A

Find each product.

14. $-7(2x + 9)$ $-14x - 63$
15. $\frac{1}{3}x(x - 27)$ $\frac{1}{3}x^2 - 9x$
16. $3st(5s^2 + 2st)$ $15s^3t + 6s^2t^2$
17. $-4m^3(5m^2 + 2m)$ $-20m^5 - 8m^4$
18. $3d(4d^2 - 8d - 15)$
19. $5m^3(6m^2 - 8mn + 12n^3)$
20. $7x^2y(5x^2 - 3xy + y)$
21. $-4d(7d^2 - 4d + 3)$
22. $2m^2(5m^2 - 7m + 8)$
23. $-8rs(4rs + 7r - 14s^2)$
24. $-\frac{3}{4}ab^2\left(\frac{1}{3}abc + \frac{4}{9}a - 6\right)$
25. $\frac{4}{5}x^2(9xy + \frac{5}{4}x - 30y)$

18. $12d^3 - 24d^2 - 45d$
19. $30m^5 - 40m^4n + 60m^3n^3$
20. $35x^4y - 21x^3y^2 + 7x^2y^2$
21. $-28d^3 + 16d^2 - 12d$
22. $10m^4 - 14m^3 + 16m^2$
23. $-32r^2s^2 - 56r^2s + 112rs^3$
24. $-\frac{1}{4}a^2b^3c - \frac{1}{3}a^2b^2 + \frac{9}{2}ab^2$
25. $\frac{36}{5}x^3y + x^3 - 24x^2y$

Simplify.

B

26. $b(4b - 1) + 10b$ $4b^2 + 9b$
27. $3t(2t - 4) + 6(5t^2 + 2t - 7)$ $36t^2 - 42$
28. $8m(-9m^2 + 2m - 6) + 11(2m^3 - 4m + 12)$ $-50m^3 + 16m^2 - 92m + 132$
29. $8y(11y^2 - 2y + 13) - 9(3y^3 - 7y + 2)$ $61y^3 - 16y^2 + 167y - 18$
30. $\frac{3}{4}t(8t^3 + 12t - 4) + \frac{3}{2}(8t^2 - 9t)$ $6t^4 + 21t^2 - \frac{33}{2}t$
31. $6a^2(3a - 4) + 5a(7a^2 - 6a + 5) - 3(a^2 + 6a)$ $53a^3 - 57a^2 + 7a$

Solve each equation.

32. $2(5w - 12) = 6(-2w + 3) + 2$ 2
33. $7(x - 12) = 13 + 5(3x - 4)$ $-\frac{77}{8}$
34. $\frac{1}{2}(2d - 34) = \frac{2}{3}(6d - 27)$ $\frac{1}{3}$
35. $p(p + 2) + 3p = p(p - 3)$ 0
36. $y(y + 12) - 8y = 14 + y(y - 4)$
37. $x(x - 3) - x(x + 4) = 17x - 23$
38. $a(a + 8) - a(a + 3) - 23 = 3a + 11$ 17
39. $t(t - 12) + t(t + 2) + 25 = 2t(t + 5) - 15$ 2

36. $\frac{7}{4}$

37. $\frac{23}{24}$

INTEGRATION
Geometry

C

Find the measure of the area of each shaded region in simplest terms.   40. $x^2 + 6x$ 41. $15p^2 + 32p$ 42. $12a^3 - 17a^2b + 15a^2 + 10b + 10$

40.

	3x	
	2x	
x + 2	x	

41.

4p + 8

4p   p, p

42.

4a + b + 5

3a² , 2a²b − b − 1

10

Critical Thinking

43. Write eight multiplication problems whose product is $8a^2b + 18ab$.
    See margin.

Applications and Problem Solving

44. Recreation In Honduras, children play a form of hopscotch called La Rayuela. The pattern for this game is shown at the right. Suppose that each rectangle is $2y + 1$ units long and y units wide.

La Rayuela
(Honduras)
cabeza
brazo casa brazo
tercera
segunda
primera

    a. Write an expression in simplest form for the area of the pattern. $14y^2 + 7y$

    b. If y represents 9 inches, find the area of the pattern. **1197 in²**

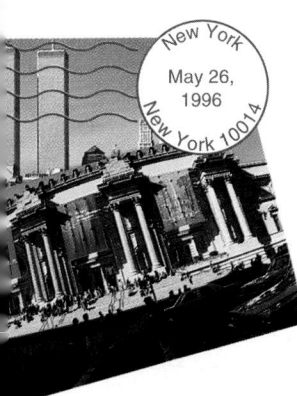
New York
May 26, 1996
New York 10014

45. Travel The Drama Club of Lincoln High School is visiting New York City. They plan to take taxis from the World Trade Center to the Metropolitan Museum of Art. The fare for a taxi is $2.75 for the first mile and $1.25 for each additional mile. Suppose the distance between the two locations is m miles and t taxis are needed to transport the entire group. Write an expression in simplest form for the cost to transport the group to the Metropolitan Museum of Art excluding the tip. $1.50t + 1.25mt$

46. Construction A landscaper is designing a rectangular garden for an office complex. There will be a concrete walkway on three sides of the garden, as shown at the right. The width of the garden will be 24 feet, and the length will be 42 feet. The width of the longer portion of the walkway will be 3 feet. The concrete will cost $20 per square yard, and the builders have told the landscaper that she can spend $820 on the concrete. How wide should the two remaining sides be? **4.5 ft**

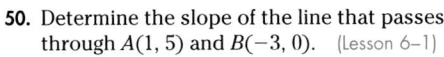
3 feet
42 feet
x 24 feet x

47. Geometry The number of diagonals that can be drawn for a polygon with n sides is represented by the expression $\frac{1}{2}n(n - 3)$.

    a. Draw polygons with 3, 4, 5, and 6 sides. Show that this expression is true for these polygons. **See margin.**

47b. $\frac{1}{2}n^2 - \frac{3}{2}n$

    b. Find the product of this expression.

    c. How many diagonals can be drawn for a polygon with 15 sides? **90 diagonals**

Mixed Review

49. 75 gal of 50%, 25 gal of 30%

50. $\frac{5}{4}$

51. $a^2 - 1$

52. See margin.

48. Find $(3a - 4ab + 7b) - (7a - 3b)$. (Lesson 9–5) $-4a - 4ab + 10b$

49. Chemistry One solution is 50% glycol, and another is 30% glycol. How much of each solution should be mixed to make a 100-gallon solution that is 45% glycol? (Lesson 8–2)

50. Determine the slope of the line that passes through $A(1, 5)$ and $B(-3, 0)$. (Lesson 6–1)

51. If $g(x) = x^2 + 2x$, find $g(a - 1)$. (Lesson 5–5)

52. Draw the graph of $2x - y = 8$. (Lesson 5–4)

53. Finance Antonio earned $340 in 4 days by mowing lawns and doing yard work. At this rate, how long will it take him to earn $935? (Lesson 4–8) **11 days**

54. Geometry Find the measure of an angle that is 44° less than its complement. (Lesson 3–4) **23°**

55. Evaluate $(15x)^3 - y$ if $x = 0.2$ and $y = 1.3$. (Lesson 1–3) **25.7**

Extension

Connections The perimeter of a square can be found by using the formula $P = 4s$, where s represents the length of a side. Find the perimeter of a square if its sides are each $(3x^2 + 2)$ cm long. $(12x^2 + 8)$ cm

---

Additional Answer (continued)

9 diagonals
$\frac{1}{2}(6)(6 - 3) = \frac{1}{2}(6)(3)$
$= 9$

NCTM Standards: 1–5

Objective
Use algebra tiles to find the product of binomials.

Recommended Time
Demonstration and discussion: 30 minutes; Exercises: 30 minutes

Instructional Resources
For each student or group of students
Student Manipulative Kit
• algebra tiles
Modeling Mathematics Masters
• p. 1 (algebra tiles)
• p. 15 (product mat)
• p. 31 (worksheet)
For teacher demonstration
Algebra and Geometry Overhead Manipulative Resources

1 FOCUS

Motivating the Lesson
Now that students have seen how the multiplication of a polynomial by a monomial is modeled by algebra tiles, encourage students to develop a method for using tiles to model the multiplication of a polynomial by another polynomial.

2 TEACH

Teaching Tip Stress that each multiplication is an extension of the distributive property.

Teaching Tip Some students may not require rearranging the tiles to determine the simplest form of the polynomial. You may want to encourage them to do so as a check of their mental math.

Teaching Tip After they have completed the activities, ask students to look for visual patterns in the models. When is a checkerboard pattern visible? When are exactly two different color rectangles visible?

MODELING MATHEMATICS

9–7A Multiplying Polynomials

Materials: ⬚ algebra tiles   ⬚ product mat

You can find the product of binomials by using algebra tiles.

A Preview of Lesson 9–7

Activity 1   Use algebra tiles to find $(x + 2)(x + 3)$.

The rectangle will have a width of $x + 2$ and a length of $x + 3$. Use your algebra tiles to mark off the dimensions on a product mat. Then make the rectangle with algebra tiles.

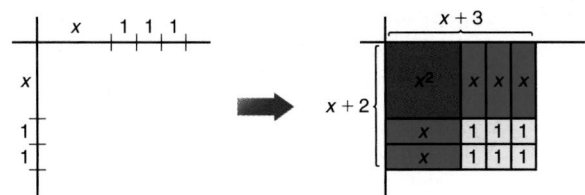

The rectangle consists of 1 blue x^2-tile, 5 green x-tiles, and 6 yellow 1-tiles.
The area of the rectangle is $x^2 + 5x + 6$. Therefore, $(x + 2)(x + 3) = x^2 + 5x + 6$.

Activity 2   Use algebra tiles to find $(x - 1)(x - 3)$.

Step 1 The rectangle will have a width of $(x - 1)$ units and a length of $(x - 3)$ units. Use your algebra tiles to mark off the dimensions on a product mat. Then begin to make the rectangle with algebra tiles.

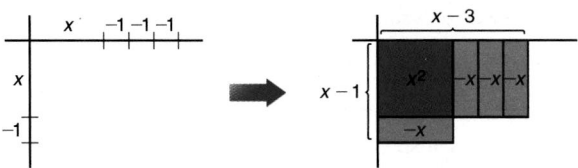

Step 2 Determine whether to use 3 yellow 1-tiles or 3 red 1-tiles to complete the rectangle. Remember that the numbers at the top and side give the dimensions of the tile needed. The area of each tile is the product of -1 and -1. This is represented by a yellow 1-tile. Fill in the space with 3 yellow 1-tiles to complete the rectangle.

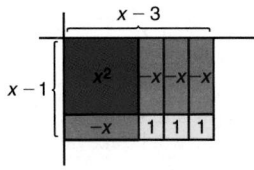

The rectangle consists of 1 blue x^2-tile, 4 red x-tiles, and 3 yellow 1-tiles.
The area of the rectangle is $x^2 - 4x + 3$. Therefore, $(x - 1)(x - 3) = x^2 - 4x + 3$.

534 *Chapter 9 Exploring Polynomials*

Activity 3 Use algebra tiles to find $(x + 1)(2x - 1)$.

Step 1 The rectangle will have a width of $(x + 1)$ units and a length of $(2x - 1)$. Use your algebra tiles to mark off the dimensions on a product mat. Then begin to make the rectangle with algebra tiles.

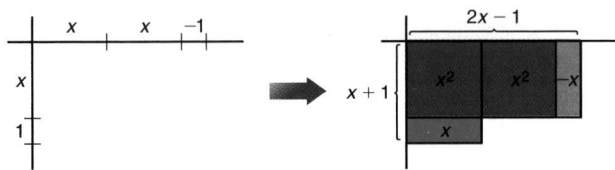

Step 2 Determine what color x-tile and what color 1-tile to use to complete the rectangle. The area of the x-tile is the product of x and 1. This is represented by a green x-tile. The area of the 1-tile is represented by the product of -1 and 1. This is represented by a red 1-tile.

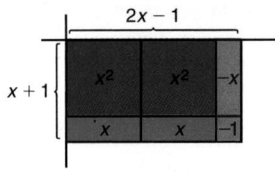

Step 3 Rearrange the tiles to simplify the polynomial you have formed. Notice that a zero pair is formed by the x-tiles.

There are 2 blue x^2-tiles, 1 green x-tile, and 1 red 1-tile left. In simplest form, $(x + 1)(2x - 1) = 2x^2 + x - 1$.

Model Use algebra tiles to find each product.

1. $x^2 + 3x + 2$ **1.** $(x + 1)(x + 2)$ **2.** $(x + 1)(x - 3)$ **3.** $(x - 2)(x - 4)$
2. $x^2 - 2x - 3$ **4.** $(x + 1)(2x + 2)$ **5.** $(x - 1)(2x + 2)$ **6.** $(x - 3)(2x - 1)$
3. $x^2 - 6x + 8$ $2x^2 + 4x + 2$ $2x^2 - 2$ $2x^2 - 7x + 3$

Draw Is each statement *true* or *false*? Justify your answer with a drawing.

     7. $(x + 4)(x + 6) = x^2 + 24$ false   **8.** $(x + 3)(x - 2) = x^2 + x - 6$ true

     9. $(x - 1)(x + 5) = x^2 - 4x - 5$ false   **10.** $(x - 2)(x - 3) = x^2 - 5x + 6$ true
     7–10. See margin for drawings.

Write 11. You can also use the distributive property to find the product of two binomials. The figure at the right shows the model for $(x + 3)(x + 2)$ separated into four parts. Write a paragraph explaining how this model shows the use of the distributive property.   By the distributive property, $(x + 3)(x + 2) = x(x + 2) + 3(x + 2)$. The top row represents $x(x + 2)$ or $x^2 + 2x$. The bottom row represents $3(x + 2)$ or $3x + 6$.

Assignment Guide

Core: 1–11
Enriched: 1–11

Additional Answers

7.

8.

9.

10.

4 ASSESS

Observing students working in cooperative groups is an excellent method of assessment.

Instructional Resources

- Study Guide Master 9-7
- Practice Master 9-7
- Enrichment Master 9-7
- Modeling Mathematics Masters, pp. 60–62

Transparency 9-7A contains the 5-Minute Check for this lesson; **Transparency 9-7B** contains a teaching aid for this lesson.

Recommended Pacing	
Standard Pacing	Day 12 of 15
Honors Pacing	Day 11 of 14
Block Scheduling*	Day 6 of 8
Alg. 1 in Two Years*	Days 19 & 20 of 26

*For more information on pacing and possible lesson plans, refer to the *Block Scheduling Booklet* and *Algebra 1 in Two Years.*

1 FOCUS

5-Minute Check
(over Lesson 9-6)

Simplify.

1. $3x(2x^2 - 7)$ $6x^3 - 21x$

2. $3c^2(4c^2 + 3c - 9)$
   $12c^4 + 9c^3 - 27c^2$

3. $5ab(4a^2b - 7ab^2 - 6b^3)$
   $20a^3b^2 - 35a^2b^3 - 30ab^4$

4. $2x(3x^2 - 4x + 7) -$
   $5(4x^2 + 3x - 9)$
   $6x^3 - 28x^2 - x + 45$

Solve.

5. $3a(a + 2b) + 7b(a + 2b)$
   $3a^2 + 13ab + 14b^2$

Motivating the Lesson

Questioning Introduce the lesson by having students describe possible ways of finding the product of $(2a + y)$ and $(2a + 5t - 6y + 5r)$.

Multiplying Polynomials

Objectives

After studying this lesson, you should be able to:

- use the FOIL method to multiply two binomials, and
- multiply any two polynomials by using the distributive property.

CONNECTION
Art

Have you ever flown over farm land and looked down? You probably saw various fields that gave the appearance of a patchwork quilt. However, if you had flown over a field designed by Stan Herd, you may have seen some sun flowers in a vase or a picture of Will Rogers. Since 1981, Stan Herd has been combining his interests in art and agriculture to form crop art. Most of Herd's work is harvested, and therefore is only visible for a short time.

In 1991, however, Herd created the picture above using native perennials. This picture is called *Little Girl in the Wind*. It depicts a Kickapoo Indian girl by the name of Carole Cadue. If you fly near Salina, Kansas, you may see this work of art.

GLOBAL CONNECTIONS

There is an historic work of art in Ohio that is best experienced from a bird's eye view. It is a mound built by the Adena Indians perhaps 30 centuries ago. The mound is in the shape of a snake and is about one-quarter mile long.

Suppose the measure of the length of the field used for *Little Girl in the Wind* can be represented by the polynomial $7x + 2$ units and the width can be represented by $5x + 1$. You know that the area of a rectangle is the product of its length and width. You can multiply $7x + 2$ and $5x + 1$ to find the area of the rectangle.

$$
\begin{aligned}
(7x + 2)(5x + 1) &= 7x(5x + 1) + 2(5x + 1) && \textit{Distributive property} \\
&= 7x(5x) + 7x(1) + 2(5x) + 2(1) && \textit{Distributive property} \\
&= 35x^2 + 7x + 10x + 2 && \textit{Substitution property} \\
&= 35x^2 + 17x + 2 && \textit{Combine like terms.}
\end{aligned}
$$

The area can also be determined by finding the sum of the areas of four smaller rectangles.

	$5x$	1
$7x$	$7x \cdot 5x$	$7x \cdot 1$
2	$2 \cdot 5x$	$2 \cdot 1$

$$
\begin{aligned}
(7x + 2)(5x + 1) &= 7x \cdot 5x + 7x \cdot 1 + 2 \cdot 5x + 2 \cdot 1 && \textit{Find the sum of the four areas.} \\
&= 35x^2 + 7x + 10x + 2 && \textit{Substitution property} \\
&= 35x^2 + 17x + 2 && \textit{Combine like terms.}
\end{aligned}
$$

GLOBAL CONNECTIONS

The serpent mound is twenty feet wide and averages five feet in height.

GLENCOE *Technology*

CD-ROM Interaction

A multimedia simulation connects polynomial functions with the construction of a swimming pool. A blackline master activity with teacher's notes provides a follow-up to the CD-ROM simulation.

For Windows & Macintosh

This example illustrates a shortcut of the distributive property called the **FOIL method**. You can use the FOIL method to multiply two binomials.

$$(7x + 2)(5x + 1) = (7x)(5x) + (7x)(1) + (2)(5x) + (2)(1)$$
$$\qquad\qquad\qquad\quad \uparrow F \qquad\quad \uparrow O \qquad\quad \uparrow I \qquad\quad \uparrow L$$
$$= 35x^2 + 7x + 10x + 2$$
$$= 35x^2 + 17x + 2$$

FOIL Method for Multiplying Two Binomials	**To multiply two binomials, find the sum of the products of**
	F the *First* terms,
	O the *Outer* terms,
	I the *Inner* terms, and
	L the *Last* terms.

Example **Find each product.**

a. $(x - 4)(x + 9)$

$$(x - 4)(x + 9) = (x)(x) + (x)(9) + (-4)(x) + (-4)(9)$$
$$\qquad\qquad\qquad F \qquad\quad O \qquad\quad\; I \qquad\quad\; L$$
$$= x^2 + 9x - 4x - 36$$
$$= x^2 + 5x - 36 \qquad \textit{Combine like terms.}$$

b. $(4x + 7)(3x - 8)$

$$(4x + 7)(3x - 8) = (4x)(3x) + (4x)(-8) + (7)(3x) + (7)(-8)$$
$$\qquad\qquad\qquad\qquad F \qquad\qquad O \qquad\qquad I \qquad\quad\; L$$
$$= 12x^2 - 32x + 21x - 56$$
$$= 12x^2 - 11x - 56 \qquad \textit{Combine like terms.}$$

The distributive property can be used to multiply any two polynomials.

Example **Find each product.**

a. $(2y + 5)(3y^2 - 8y + 7)$

$$(2y + 5)(3y^2 - 8y + 7)$$
$$= 2y(3y^2 - 8y + 7) + 5(3y^2 - 8y + 7) \qquad \textit{Distributive property}$$
$$= (6y^3 - 16y^2 + 14y) + (15y^2 - 40y + 35) \qquad \textit{Distributive property}$$
$$= 6y^3 - 16y^2 + 14y + 15y^2 - 40y + 35$$
$$= 6y^3 - y^2 - 26y + 35 \qquad \textit{Combine like terms.}$$

b. $(x^2 + 4x - 5)(3x^2 - 7x + 2)$

$$(x^2 + 4x - 5)(3x^2 - 7x + 2)$$
$$= x^2(3x^2 - 7x + 2) + 4x(3x^2 - 7x + 2) - 5(3x^2 - 7x + 2)$$
$$= (3x^4 - 7x^3 + 2x^2) + (12x^3 - 28x^2 + 8x) - (15x^2 - 35x + 10)$$
$$= 3x^4 - 7x^3 + 2x^2 + 12x^3 - 28x^2 + 8x - 15x^2 + 35x - 10$$
$$= 3x^4 + 5x^3 - 41x^2 + 43x - 10 \qquad \textit{Combine like terms.}$$

Teaching Tip Remind students that FOIL is a memory device, not a mathematical theorem or rule. Stress that this works only with two binomials.

In-Class Examples

For Example 1
Find each product.

a. $(m - 3)(m + 4)$
$m^2 + m - 12$

b. $(3a + 11)(5a - 2)$
$15a^2 + 49a - 22$

For Example 2
Find each product.

a. $(2x + 3)(x^2 + 3x + 8)$
$2x^3 + 9x^2 + 25x + 24$

b. $(4x^2 - 10x - 3)(2x^2 + 6x - 9)$
$8x^4 + 4x^3 - 102x^2 + 72x + 27$

Teaching Tip In Example 2, emphasize to students that they must multiply each term in the first polynomial by each term in the second polynomial.

Classroom Vignette

"I cut out colored transparencies of the rectangles in the diagram on page 536 and put them on the overhead. I ask the students to describe two different ways to find the total area. By doing this activity, they discover the 'FOIL' method on their own. Then I explain the two different methods to the entire class."

Linda Glover
Conway Junior High School
Conway, Arkansas

For Example 3
Find $(4x^3 - 2x^2 + 5)(5x - 6)$ using column form. $20x^4 - 34x^3 + 12x^2 + 25x - 30$

For Example 4
Write a polynomial expression that represents the volume of the prism shown here. Then find the volume if $y = 3$.

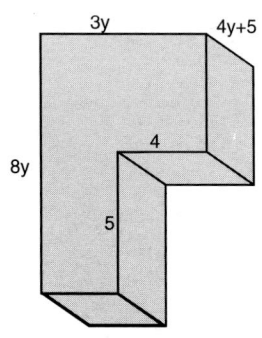

$(96y^3 + 120y^2 - 80y - 100)$ cubic units; 3332 cubic units

3 PRACTICE/APPLY

Check for Understanding
Exercises 1–13 are designed to help you assess your students' understanding through reading, writing, speaking, and modeling. You should work through Exercises 1–4 with your students and then monitor their work on Exercises 5–13.

Additional Answers

1a. Sample answer:
(42)(27)
$= (40 + 2)(20 + 7)$
$\quad$ F $\quad$ O $\quad$ I $\quad$ L
$= 40(20) + 40(7) + 2(20) + 2(7)$
$= 800 + 280 + 40 + 14$
$= 1134$

1b. Sample answer:
$4\frac{1}{2} \cdot 6\frac{3}{4}$
$= \left(4 + \frac{1}{2}\right)\left(6 + \frac{3}{4}\right)$
$\quad$ F $\quad$ O $\quad$ I $\quad$ L
$= 4(6) + 4\left(\frac{3}{4}\right) + \frac{1}{2}(6) + \frac{1}{2}\left(\frac{3}{4}\right)$
$= 24 + 3 + 3 + \frac{3}{8}$
$= 30\frac{3}{8}$

Polynomials can also be multiplied in column form. Be careful to align the like terms.

Example **3** Find $(x^3 - 8x^2 + 9)(3x + 4)$ using column form.

Since there is no x term in $x^3 - 8x^2 + 9$, $0x$ is used as a placeholder.

$$
\begin{array}{r}
x^3 - 8x^2 + 0x + 9 \\
(\times) \qquad\qquad 3x + 4 \\
\hline
4x^3 - 32x^2 + 0x + 36 \\
3x^4 - 24x^3 + 0x^2 + 27x \\
\hline
3x^4 - 20x^3 - 32x^2 + 27x + 36
\end{array}
$$

$\leftarrow$ *product of $x^3 - 8x^2 + 0x + 9$ and 4*
$\leftarrow$ *product of $x^3 - 8x^2 + 0x + 9$ and $3x$*
$\leftarrow$ *sum of the partial products*

Example **4** The volume V of a prism equals the area of the base B times the height h.

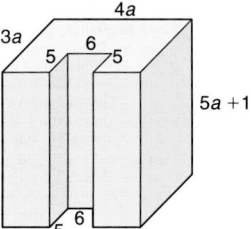

INTEGRATION
Geometry

a. **Write a polynomial expression that represents the volume of the prism shown at the right.**

b. **Find the volume if $a = 5$.**

a. A diagram of the base is shown below.

To find the area of the base, first find the area of a rectangle that is $3a$ by $4a$. Then subtract the area of a rectangle that is 6 by 5.

$B = 3a(4a) - 6(5)$
$\quad = 12a^2 - 30$

The volume of this prism equals the product of the base and the height, $12a^2 - 30$ and $5a + 1$, respectively. Use FOIL to find the product.

$$
\begin{array}{cccc}
& F & O & \qquad I \qquad L \\
(12a^2 - 30)(5a + 1) = & (12a^2)(5a) + (12a^2)(1) + (-30)(5a) + (-30)(1) \\
= & 60a^3 + 12a^2 - 150a - 30
\end{array}
$$

The volume of the prism is $(60a^3 + 12a^2 - 150a - 30)$ cubic units.

b. Substitute 5 for a and evaluate the expression.

$60(5^3) + 12(5^2) - 150(5) - 30 = 7500 + 300 - 750 - 30$
$\qquad\qquad\qquad\qquad\qquad\qquad = 7020$

If $a = 5$, the volume of the prism is 7020 cubic units.

CHECK FOR UNDERSTANDING

Communicating Mathematics

Study the lesson. Then complete the following.

1. Use the FOIL method to evaluate each product. **a–b. See margin.**
   a. 42(27) (*Hint:* Rewrite as (40 + 2) (20 + 7) or (40 + 2)(30 − 3).)
   b. $4\frac{1}{2} \cdot 6\frac{3}{4}$

 Alternative Learning Styles

Visual Have students draw a diagram of the garden in the following problem to illustrate its dimensions.
Pete has a rectangular garden 15 feet longer than it is wide. A cement divider that is 4 feet wide surrounds the garden. The total area of the divider is 632 square feet. What are the dimensions of the garden?
28 feet by 43 feet

2. You Decide Adita and Delbert used the following methods to find the product of $(t^3 - t^2 + 5t)$ and $(6t^2 + 8t - 7)$.

Adita:

$(t^3 - t^2 + 5t)(6t^2 + 8t - 7)$

$= t^3(6t^2 + 8t - 7) - t^2(6t^2 + 8t - 7) + 5t(6t^2 + 8t - 7)$

$= 6t^5 + 8t^4 - 7t^3 - 6t^4 - 8t^3 + 7t^2 + 30t^3 + 40t^2 - 35t$

$= 6t^5 + 2t^4 + 15t^3 + 47t^2 - 35t$

Delbert:

$$t^3 - t^2 + 5t$$
$$\underline{(\times)\ 6t^2 + 8t - 7}$$
$$-7t^3 + 7t^2 - 35t$$
$$8t^4 - 8t^3 + 40t^2$$
$$\underline{6t^5 - 6t^4 + 30t^3}$$
$$6t^5 + 2t^4 + 15t^3 + 47t^2 - 35t$$

Which method do you prefer? Why? **See students' work.**

3. Draw a diagram to show how you would use algebra tiles to find the product of $(2x - 3)$ and $(x + 2)$. **See margin.**

4. Write two binomials whose product is represented at the right.
Sample answer: $(a + x)(2x + 3)$

$2ax$	$3a$
$2x^2$	$3x$

MODELING MATHEMATICS

Guided Practice

5. $d^2 + 10d + 16$
6. $r^2 - 16r + 55$
7. $y^2 - 4y - 21$
8. $15p^2 - 19p - 10$
9. $2x^2 + 9x - 5$
10. $6m^2 - m - 40$
11. $10a^2 + 11ab - 6b^2$
12. $6x^3 - 25x^2 + 33x - 20$

Find each product.

5. $(d + 2)(d + 8)$ **6.** $(r - 5)(r - 11)$ **7.** $(y + 3)(y - 7)$
8. $(3p - 5)(5p + 2)$ **9.** $(2x - 1)(x + 5)$ **10.** $(2m + 5)(3m - 8)$
11. $(2a + 3b)(5a - 2b)$ **12.** $(2x - 5)(3x^2 - 5x + 4)$

13. a. Number Theory Find the product of three consecutive integers if the least integer is a. $a^3 + 3a^2 + 2a$

b. Choose an integer as the first of three consecutive integers. Find their product.   **See students' work.**

c. Evaluate the polynomial in part a for these integers. Describe the result.
See students' work. The result is the same as the product in part b.

EXERCISES

Practice

14. $y^2 + 12y + 35$
15. $c^2 - 10c + 21$
16. $x^2 - 4x - 32$
17. $w^2 - 6w - 27$
18. $2a^2 + 15a - 8$
19. $10b^2 - b - 3$
20. $132y^2 + 174y + 54$
21. $169x^2 - 9$
22. $24x^2 + 83xy + 63y^2$

Find each product. 23–33. See margin.

14. $(y + 5)(y + 7)$ **15.** $(c - 3)(c - 7)$ **16.** $(x + 4)(x - 8)$
17. $(w + 3)(w - 9)$ **18.** $(2a - 1)(a + 8)$ **19.** $(5b - 3)(2b + 1)$
20. $(11y + 9)(12y + 6)$ **21.** $(13x - 3)(13x + 3)$ **22.** $(8x + 9y)(3x + 7y)$
23. $(0.3v - 7)(0.5v + 2)$ **24.** $\left(3x + \frac{1}{3}\right)\left(2x - \frac{1}{9}\right)$ **25.** $\left(a - \frac{2}{3}b\right)\left(\frac{2}{3}a + \frac{1}{2}b\right)$
26. $(2r + 0.1)(5r - 0.3)$ **27.** $(0.7p + 2q)(0.9p + 3q)$
28. $(x + 7)(x^2 + 5x - 9)$ **29.** $(3x - 5)(2x^2 + 7x - 11)$

30. $\begin{array}{r} a^2 - 3a + 11 \\ \underline{(\times)\ 5a + 2} \end{array}$ **31.** $\begin{array}{r} 3x^2 - 7x + 2 \\ \underline{(\times)\ 3x - 8} \end{array}$

32. $\begin{array}{r} 5x^2 + 8x - 11 \\ \underline{(\times)\ x^2 - 2x - 1} \end{array}$ **33.** $\begin{array}{r} 5d^2 - 6d + 9 \\ \underline{(\times)\ 4d^2 + 3d + 11} \end{array}$

Lesson 9–7 Multiplying Polynomials **539**

Reteaching

Using Diagrams Have students use the diagram below to answer the following questions.

	a	a	b	b	b
a	a^2	a^2	ab	ab	ab
b	ab	ab	b^2	b^2	b^2
b	ab	ab	b^2	b^2	b^2

1. What is the width of the large rectangle? $a + 2b$
2. What is the length? $2a + 3b$
3. What is the area? $(a + 2b)(2a + 3b)$
   or $2a^2 + 7ab + 6b^2$
4. Inside each small square or rectangle, write its area.
   See diagram.
5. Add the areas in Exercise 4. Compare with your answer in Exercise 3. **same**

Assignment Guide

Core: 15–47 odd, 48–56
Enriched: 14–40 even, 42–56

For **Extra Practice**, see p. 778.

The red A, B, and C flags, printed only in the Teacher's Wraparound Edition, indicate the level of difficulty of the exercises.

Additional Answers

3.

23. $0.15v^2 - 2.9v - 14$
24. $6x^2 + \frac{1}{3}x - \frac{1}{27}$
25. $\frac{2}{3}a^2 + \frac{1}{18}ab - \frac{1}{3}b^2$
26. $10r^2 - 0.1r - 0.03$
27. $0.63p^2 + 3.9pq + 6q^2$
28. $x^3 + 12x^2 + 26x - 63$
29. $6x^3 + 11x^2 - 68x + 55$
30. $5a^3 - 13a^2 + 49a + 22$

(continued on p. 540)

Study Guide Masters, p. 67

9-7   NAME_____   DATE_____

Study Guide   Student Edition Pages 536–541

Multiplying Polynomials

The following example shows how the distributive property can be used to multiply any two polynomials.

Example 1: Find $(2x - 6)(3x + 1)$.

$$\begin{array}{cccc} & F & O & I & L \end{array}$$
$(2x - 6)(3x + 1) = 2x \cdot 3x + 2x \cdot 1 + (-6) \cdot 3x + (-6) \cdot 1$
$= 6x^2 + 2x - 18x - 6$
$= 6x^2 - 16x - 6$

You can also multiply polynomials vertically.

Example 2: Find $(3x^2 - x + 1)(5x + 2)$.

$$\begin{array}{rl} 3x^2 - x + 1 & \\ \underline{\times\quad\quad 5x + 2} & \\ 6x^2 - 2x + 2 & \text{Multiply } 3x^2 - x + 1 \text{ by 2.} \\ \underline{15x^3 - 5x^2 + 5x\quad} & \text{Multiply } 3x^2 - x + 1 \text{ by 5x.} \\ 15x^3 + x^2 + 3x + 2 & \text{Combine like terms.} \end{array}$$

Find each product.

1. $(5t + 4)(2t - 6)$ 2. $(5m - 3n)(4m - 2n)$ 3. $(a - 3b)(2a - 5b)$
   $10t^2 - 22t - 24$ $20m^2 - 22mn + 6n^2$ $2a^2 - 11ab + 15b^2$

4. $(3x - 0.1)(x + 0.1)$ 5. $(8x + 5)(8x - 5)$ 6. $(x + 5)(x + 2)$
   $3x^2 + 0.2x - 0.01$ $64x^2 - 25$ $x^2 + 7x + 10$

7. $(2x - 4)(2x + 5)$ 8. $\begin{array}{r} y^2 - 5y + 3 \\ \times\ 2y^2 + 7y - 4 \end{array}$ 9. $\begin{array}{r} 3b^3 - 2b^2 + b \\ \times\quad\ 2b - 3 \end{array}$
   $4x^2 + 2x - 20$ $2y^4 - 3y^3 - 33y^2 + 41y - 12$ $6b^4 - 13b^3 + 8b^2 - 3b$

Chapter 9 **539**

34. $2x^4 - 20x^3 + 39x^2 - 68x - 9$

Geometry
C

35. $10x^4 + 3x^3 + 51x^2 - 16x - 48$

36. $-35b^5 + 14b^4 - 18b^3 - 19b^2 + 14b - 12$

Critical Thinking

43. $-8x^4 - 6x^3 + 24x^2 - 12x + 80$

Applications and Problem Solving

Find each product.

34. $(x^2 - 8x - 1)(2x^2 - 4x + 9)$ **35.** $(5x^2 - x - 4)(2x^2 + x + 12)$

36. $(-7b^3 + 2b - 3)(5b^2 - 2b + 4)$ **37.** $(a^2 + 2a + 5)(a^2 - 3a - 7)$
$a^4 - a^3 - 8a^2 - 29a - 35$

Find the measure of the volume of each prism.

38.   **39.**   **40.**

$2a^3 + 10a^2 - 2a - 10$ $63y^3 - 57y^2 - 36y$ $15x^3 + 12x^2$

41. Geometry Refer to the prism in Exercise 38. Suppose a represents 15 centimeters.

   a. Find the length, width, and height of the prism.   **28 cm, 20 cm, 16 cm**

   b. Use the values in part a to find the volume of the prism.   **8960 cm³**

   c. Evaluate your answer for Exercise 38 if $a = 15$.   **8960**

   d. How do your answers for parts b and c compare?   **They are the same measure.**

If **A = 3x + 4, B = x² + 2, and C = x² + 3x − 2**, find each of the following.   **44.** $3x^5 + 13x^4 + 12x^3 + 18x^2 + 12x - 16$

42. $AC + B$ **43.** $2B(3A - 4C)$ **44.** ABC **45.** $(A + B)(B - C)$
$3x^3 + 14x^2 + 6x - 6$ $-3x^3 - 5x^2 - 6x + 24$

46. Construction A homeowner is considering installing a swimming pool in his backyard. He wants its length to be 5 yards longer than its width, to make room for a diving area at one end. Then he wants to surround it with a concrete walkway 4 yards wide. After finding out the price of concrete, he decides that he can afford 424 square yards of it for the walkway. What should the dimensions of the pool be?   **20 yd by 25 yd**

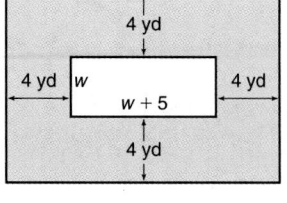

47a. Sample answer: $x - 2, x + 3$

47. Business Raul Agosto works for a company that has modular offices. His office space is presently a square. A new floor plan calls for his office to become 2 feet shorter in one direction and 3 feet longer in the other.   **47b. Sample answer:** $x^2 + x - 6$

   a. Write expressions that represent the new dimensions of Mr. Agosto's office.

   b. Find the area of his new office.

   c. Suppose his office is presently 8 feet by 8 feet. Will his new office be bigger or smaller than this office? by how much?   **larger, 2 sq ft**

Mixed Review

48. Find $\frac{3}{4}a(6a + 12)$.   (Lesson 9-6) $\frac{9}{2}a^2 + 9a$

49. Solve $6 - 9y < -10y$.   (Lesson 7-3) $\{y \mid y < -6\}$

Practice Masters, p. 67

9-7
NAME_____ DATE_____

Practice

Student Edition
Pages 536-541

Multiplying Polynomials

Find each product.

1. $(x + 2)(x + 7)$
$x^2 + 9x + 14$

2. $(y + 6)(y + 3)$
$y^2 + 9y + 18$

3. $(a - 3)(a - 8)$
$a^2 - 11a + 24$

4. $(r - 5)(r - 1)$
$r^2 - 6r + 5$

5. $(x - 10)(x + 2)$
$x^2 - 8x - 20$

6. $(3t + 2q)(7t - 9q)$
$21t^2 - 13qt - 18q^2$

7. $(7y - 4t)(2y + 5t)$
$14y^2 + 27ty - 20t^2$

8. $(5w - 3y)(w - 2y)$
$5w^2 - 13yw + 6y^2$

9. $(a + b)(2a - 3b)$
$2a^2 - ab - 3b^2$

10. $(0.3n + 5)(0.4n - 11)$
$0.12n^2 - 1.3n - 55$

11. $(1.3g + 3)(0.4g + 5)$
$0.52g^2 + 7.7g + 15$

12. $(x + y)(x + y)$
$x^2 + 2xy + y^2$

13. $(z - 2x)(z - 2x)$
$z^2 - 4xz + 4x^2$

14. $(3x + 2y)(2y + 3x)$
$4y^2 + 12xy + 9x^2$

15. $\begin{array}{r} y^2 + 3y - 6 \\ \times \quad 2y + 3 \\ \hline 2y^3 + 9y^2 - 3y - 18 \end{array}$

16. $\begin{array}{r} a^2 + 2a - 9 \\ \times \quad a + 3 \\ \hline a^3 + 5a^2 - 3a - 27 \end{array}$

17. $\begin{array}{r} e^2 - 2ef + f^2 \\ \times \quad e - f \\ \hline e^3 - 3e^2f + 3ef^2 - f^3 \end{array}$

18. $\begin{array}{r} 9m^2 - 12m + 4 \\ \times \quad 3m + 2 \\ \hline 27m^3 - 18m^2 - 12m + 8 \end{array}$

19. $(2a - 1)(a^2 + 3a + 5)$
$2a^3 + 5a^2 + 7a - 5$

20. $(2 + 3c)(4 + 6c + 9c^2)$
$8 + 24c + 36c^2 + 27c^3$

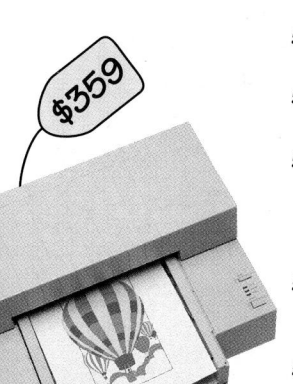

50. **National Landmarks** At the Royal Gorge in Colorado, an inclined railway takes visitors down to the Arkansas River. Suppose the slope is 50% and the vertical drop is 1015 feet. What is the horizontal change of the railway? (Lesson 6–1) **2030 ft**

51. Write an equation to represent the relation. (Lesson 5–6) $y = 2x + 1$
    $\{(-1, -1), (0, 1), (1, 3), (2, 5), (3, 7)\}$

52. Determine the domain, range, and inverse of the relation. (Lesson 5–2)
    $\{(8, 1), (4, 2), (6, -4), (5, -3), (6, 0)\}$ **See margin.**

53. **Geometry** $\triangle ABC$ and $\triangle XYZ$ are similar. Find the values of a and y. (Lesson 4–2) $a = 4, y = 9$

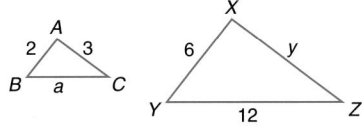

54. **Consumerism** The prices of six different models of printers in a computer store are $299, $369, $359, $228, $525, and $398. Find the mean and median prices for the printers. (Lesson 3–7) **$363; $364**

55. **Temperature** The formula for finding the Celsius temperature C when you know the Fahrenheit temperature F is $C = \frac{5}{9}(F - 32)$. Find the Celsius temperature when the Fahrenheit temperature is 59°. (Lesson 2–9) **15°C**

56. Replace the variable to make the sentence $\frac{3}{4}s = 6$ true. (Lesson 1–5) **8**

WORKING ON THE

Refer to the Investigation on pages 448–449.

Ready, Set, Drop!

You have experimented with various sizes of gliders to explore their flying abilities. You now need to investigate how the size of the glider and the weight of the load are related.

1 Cut out two equilateral triangles from construction paper so that the side of one is 6 centimeters long, and the side of the other is 12 centimeters long. What is the surface area and perimeter of each triangle?

2 Straighten two paper clips and bend them into the shape shown at the right. Punch the bent end of the paper clip through the center of the triangle, and tape it into place so that the hook-end hangs down under the other side of the triangle.

3 One at a time, drop each hang glider from the top of the bleachers or out of a second-floor window. (In order to get more accurate data, a greater height is needed than those used for previous experiments.) Record the glide time of each hang glider.

4 Add one washer onto the hook of each glider. Repeat the dropping procedure and record the times. How do those times compare with the first drop? Continue to add washers, one at a time, to each glider and record the glide times in a chart that compares the glide times with the number of washers carried by each glider.

5 Graph the relationship between the weight and the glide times. Let the horizontal axis represent the weight, and let the vertical axis represent the glide time. Analyze your findings.

6 Look at the weight, glide time, surface area, and perimeter in your data. Write a polynomial expression to relate some, if not all, of these measures.

Add the results of your work to your Investigation Folder.

Extension

Communication Have students find the product of $(4m^2 - m + 8)$ and $(m^3 + 2m^2 + 3m + 4)$ and then describe the process they used to solve the problem. $4m^5 + 7m^4 + 18m^3 + 29m^2 + 20m + 32$

Working on the Investigation
The Investigation on pages 448–449 is designed to be a long-term project that is completed over several days or weeks. Encourage students to keep their materials in their Investigation Folder as they work on the Investigation.

---

4 ASSESS

Closing Activity

Modeling The following diagram illustrates all the multiplications that must be performed when multiplying two polynomials.

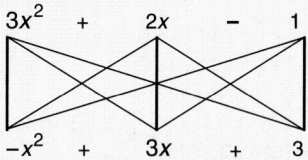

Have students draw such diagrams for several problems.

Additional Answer

52. $\{8, 4, 6, 5\}$; $\{1, 2, -4, -3, 0\}$; $\{(1, 8), (2, 4), (-4, 6), (-3, 5), (0, 6)\}$

Enrichment Masters, p. 67

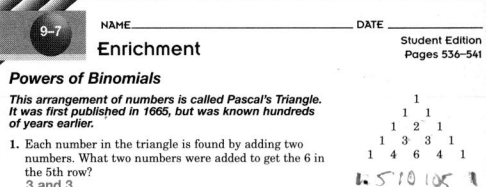

9-7 NAME _____ DATE _____
Enrichment Student Edition Pages 536–541

Powers of Binomials

This arrangement of numbers is called Pascal's Triangle. It was first published in 1665, but was known hundreds of years earlier.

1. Each number in the triangle is found by adding two numbers. What two numbers were added to get the 6 in the 5th row? **3 and 3**

2. Describe how to create the 6th row of Pascal's Triangle. The first and last numbers are 1. Evaluate $1 + 4$, $4 + 6$, $6 + 4$, and $4 + 1$ to find the other numbers.

3. Write the numbers for rows 6 through 10 of the triangle.
   Row 6: 1 5 10 10 5 1
   Row 7: 1 6 15 20 15 6 1
   Row 8: 1 7 21 35 35 21 7 1
   Row 9: 1 8 28 56 70 56 28 8 1
   Row 10: 1 9 37 84 126 126 84 37 9 1

Multiply to find the expanded form of each product.

4. $(a + b)^2$ $a^2 + 2ab + b^2$

5. $(a + b)^3$ $a^3 + 3a^2b + 3ab^2 + b^3$

6. $(a + b)^4$ $a^4 + 4a^3b + 6a^2b^2 + 4ab^3 + b^4$

Now compare the coefficients of the three products in Exercises 4–6 with Pascal's Triangle.

7. Describe the relationship between the expanded form of $(a + b)^n$ and Pascal's Triangle. The coefficients of the expanded form are found in row $n + 1$ of Pascal's Triangle.

8. Use Pascal's Triangle to write the expanded form of $(a + b)^6$. $a^6 + 6a^5b + 15a^4b^2 + 20a^3b^3 + 15a^2b^4 + 6ab^5 + b^6$

Instructional Resources

- Study Guide Master 9-8
- Practice Master 9-8
- Enrichment Master 9-8
- Assessment and Evaluation Masters, p. 241

 Transparency 9-8A contains the 5-Minute Check for this lesson; **Transparency 9-8B** contains a teaching aid for this lesson.

Recommended Pacing	
Standard Pacing	Day 13 of 15
Honors Pacing	Day 12 of 14
Block Scheduling*	Day 7 of 8
Alg. 1 in Two Years*	Days 21, 22, & 23 of 26

 *For more information on pacing and possible lesson plans, refer to the *Block Scheduling Booklet* and *Algebra 1 in Two Years.*

1 FOCUS

 ### 5-Minute Check
(over Lesson 9-7)

Find each product.

1. $(3w + 7)(2w + 5)$
   $6w^2 + 29w + 35$
2. $(a + 6)(a - 3)$
   $a^2 + 3a - 18$
3. $(3m - 4)(5m + 6)$
   $15m^2 - 2m - 24$
4. $(2b^2 - 5b - 3)(5b^2 + 3b - 2)$
   $10b^4 - 19b^3 - 34b^2 + b + 6$
5. $(8j^2 - 2j + 1)(2j^2 + 8j + 1)$
   $16j^4 + 60j^3 - 6j^2 + 6j + 1$

Motivating the Lesson

Hands-On Activity Have students use algebra tiles to represent the following perfect squares: 1, 4, 9, 16, 25, 36, 49, 64, and 81. Using the tile representations, students should determine the square root of each number.

1, 2, 3, 4, 5, 6, 7, 8, 9

9-8 Special Products

Objective
After studying this lesson, you should be able to:
- use patterns to find $(a + b)^2$, $(a - b)^2$, and $(a + b)(a - b)$.

Punnett squares are named after the English zoologist and geneticist Reginald Crundall Punnett (1875–1967). He held the first chair in genetics at Cambridge University and helped build the science of genetics in the 20th century.

CONNECTION
Biology

Punnett squares are diagrams that are used to show the possible ways that genes can combine at fertilization. In a Punnett square, *dominant* genes are shown with capital letters. Recessive genes are shown with lowercase letters. Letters representing the parents' genes are placed on two of the outer sides of the Punnett square. Letters inside the boxes of the square show the possible gene combinations of their offspring.

The Punnett square below represents a cross between tall pea plants and short pea plants. Let T represent the dominant gene for tallness. Let t represent the recessive gene for shortness. The parents are called *hybrids,* since they have one of each kind of gene.

Hybrid tall × Hybrid tall

Tall = T

Short = t

Offspring

$\frac{1}{4}$ or 25% pure tall (TT)

$\frac{2}{4}$ or 50% hybrid tall (Tt)

$\frac{1}{4}$ or 25% pure short (tt)

	T hybrid tall	**t**
T hybrid tall	**TT**	**Tt**
t	**Tt**	**tt**

Because the parent plants have both a dominant tall gene and a recessive short gene, biologists know that their offspring can be predicted by squaring the binomial $(0.5T + 0.5t)^2$. Therefore, the following must be true.

$(0.5T + 0.5t)^2 = (0.5T + 0.5t)(0.5T + 0.5t)$

$= 0.5T(0.5T) + 0.5T(0.5t) + 0.5t(0.5T) + 0.5t(0.5t)$

$= 0.25T^2 + 0.25Tt + 0.25Tt + 0.25t^2$

$= 0.25T^2 + 0.50Tt + 0.25t^2$ T^2 and t^2 represent TT and tt, respectively.

Squares very similar to Punnett squares are also used in logic, probability, and the design of digital computer circuits.

You can use the diagram below to derive a general form for the expression $(a + b)^2$.

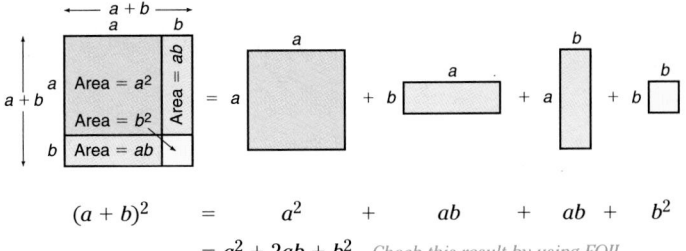

$$(a + b)^2 \quad = \quad a^2 \quad + \quad ab \quad + \quad ab \quad + \quad b^2$$
$$= a^2 + 2ab + b^2 \quad \textit{Check this result by using FOIL.}$$

In general, the square of a binomial that is a sum can be found by using the following rule.

Square of a Sum	$(a + b)^2 = (a + b)(a + b)$ $= a^2 + 2ab + b^2$

Example **1** Find each product.

a. $(y + 7)^2$

Method 1	Method 2
Use the square of a sum rule.	Use FOIL.
$(a + b)^2 = a^2 + 2ab + b^2$	$(y + 7)^2 = (y + 7)(y + 7)$
$(y + 7)^2 = y^2 + 2(y)(7) + 7^2$	$= y^2 + 7y + 7y + 49$
$= y^2 + 14y + 49$	$= y^2 + 14y + 49$

b. $(6p + 11q)^2$

$$(a + b)^2 = a^2 + 2ab + b^2$$
$$(6p + 11q)^2 = (6p)^2 + 2(6p)(11q) + (11q)^2 \quad a = 6p \text{ and } b = 11q$$
$$= 36p^2 + 132pq + 121q^2$$

The square of a sum rule can be used with other rules to simplify products of polynomials.

Example **2** Tourists to the southern part of England can visit the historic Gwennap Pit. In the 16th century, the pit of a tin mine was converted into an amphitheater. During the 18th century, John Wesley spoke to overflow crowds in this amphitheater. Gwennap Pit consists of a circular stage surrounded by circular levels used for seating. Each seating level is 1 meter wide. Suppose the radius of the stage is s meters. Find the area of the third seating level.

(continued on the next page)

Lesson 9–8   Special Products   **543**

In-Class Examples

For Example 1
Find each product.

a. $(3a + 2)^2$ $9a^2 + 12a + 4$
b. $(5b + 7c)^2$
    $25b^2 + 70bc + 49c^2$

For Example 2
What would be the area of the third seating level if each seating level were 2 meters wide?
$12.6s + 62.8$ square meters

Teaching Tip Emphasize to students that the square of a difference is another case of multiplying binomials by FOIL.

$(a - b)^2 = (a - b)(a - b)$

$\qquad\quad$ F $\quad$ O $\quad$ I $\quad$ L

$\qquad = a^2 - ab - ba + b^2$

$\qquad = a^2 - 2ab + b^2$

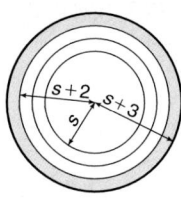

The area of a circle equals πr^2. The radius of the second seating level is $s + 2$ meters, and the radius of the third seating level is $s + 3$. The area of the third seating level can be found by subtracting the areas of two circles.

$$
\begin{aligned}
A &= \underbrace{\pi(s + 3)^2}_{\substack{\text{area of}\\\text{third level}}} - \underbrace{\pi(s + 2)^2}_{\substack{\text{area of}\\\text{second level}}} \\
&= \pi(s^2 + 6s + 9) - \pi(s^2 + 4s + 4) & \textit{Square of a sum rule} \\
&= (\pi s^2 + 6\pi s + 9\pi) - (\pi s^2 + 4\pi s + 4\pi) & \textit{Distributive property} \\
&= \pi s^2 + 6\pi s + 9\pi - \pi s^2 - 4\pi s - 4\pi \\
&= 2\pi s + 5\pi & \textit{Combine like terms.}
\end{aligned}
$$

The area of the third seating level is $2\pi s + 5\pi$, or about $6.3s + 15.7$ square meters.

To find $(a - b)^2$, write $(a - b)$ as $[a + (-b)]$ and square it.

$$
\begin{aligned}
(a - b)^2 &= [a + (-b)]^2 \\
&= a^2 + 2(a)(-b) + (-b)^2 \\
&= a^2 - 2ab + b^2
\end{aligned}
$$

$$
[a + (-b)][a + (-b)]
$$

In general, the square of a binomial that is a difference can be found by using the following rule.

Square of a Difference	$(a - b)^2 = (a - b)(a - b)$ $= a^2 - 2ab + b^2$

Example ❸ **Find each product.**

a. $(r - 6)^2$

Method 1

Use the square of a difference rule.

$$
\begin{aligned}
(a - b)^2 &= a^2 - 2ab + b^2 \\
(r - 6)^2 &= r^2 - 2(r)(6) + 6^2 \\
&= r^2 - 12r + 36
\end{aligned}
$$

Method 2

Use FOIL.

$$
\begin{aligned}
(r - 6)^2 &= (r - 6)(r - 6) \\
&= r^2 - 6r - 6r + 36 \\
&= r^2 - 12r + 36
\end{aligned}
$$

b. $(4x^2 - 7t)^2$

$$
\begin{aligned}
(a - b)^2 &= a^2 - 2ab + b^2 \\
(4x^2 - 7t)^2 &= (4x^2)^2 - 2(4x^2)(7t) + (7t)^2 & a = 4x^2 \text{ and } b = 7t \\
&= 16x^4 - 56x^2t + 49t^2
\end{aligned}
$$

Product of a Sum and a Difference

Materials: algebra tiles ☐ product mat

You have learned how to use algebra tiles to find the product of two binomials. In this activity, you will use algebra tiles to study a special situation.

Your Turn

a. Use algebra tiles to find each product.

$(x + 3)(x - 3)$ $x^2 - 9$ $(x + 5)(x - 5)$ $x^2 - 25$

b. They are the sum and difference of the same values.

$(x + 1)(x - 1)$ $x^2 - 1$ $(x + 2)(x - 2)$ $x^2 - 4$

$(x + 6)(x - 6)$ $x^2 - 36$ $(x + 4)(x - 4)$ $x^2 - 16$

b. What do you notice about the binomials used as factors in part a?

c. They are the differences of the square of the first number and the square of the second number.

c. What pattern do the products in part a have?

You can use the FOIL method to find the product of a sum and a difference of the same two numbers.

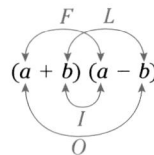

$$(a + b)(a - b) = a(a) + a(-b) + b(a) + b(-b)$$
$$= a^2 - ab + ab - b^2$$
$$= a^2 - b^2$$

The resulting product, $a^2 - b^2$, has a special name. It is called a **difference of squares.**

Difference of Squares	$(a + b)(a - b) = (a - b)(a + b)$ $= a^2 - b^2$

Example 4 Find each product.

a. $(m - 2n)(m + 2n)$

$$(a - b)(a + b) = a^2 - b^2$$
$$(m - 2n)(m + 2n) = m^2 - (2n)^2 \quad a = m \text{ and } b = 2n$$
$$= m^2 - 4n^2$$

b. $(0.3t + 0.25w^2)(0.3t - 0.25w^2)$

$$(a + b)(a - b) = a^2 - b^2$$
$$(0.3t + 0.25w^2)(0.3t - 0.25w^2) = (0.3t)^2 - (0.25w^2)^2 \quad a = 0.3t \text{ and } b = 0.25w^2$$
$$= 0.09t^2 - 0.0625w^4$$

Reteaching

Using Reasoning Use the pattern of the square of a sum, $(a + b)^2 = a^2 + 2ab + b^2$, to square numbers. For example, $(27)^2 = (20 + 7)^2 = 20^2 + 2 \cdot 140 + 49$ or 729.

Students must be careful not to confuse algebraic products with perfect squares in number theory:
$9 = 3 \times 3$, not $9 = 3 \times (-3)$;
$x^2 - 9 = (x + 3)(x - 3)$,
not $x^2 - 9 = (x + 3)(x + 3)$.

Teaching Tip Have students compare these expressions.
square of a difference
$(a - b)^2$
and
difference of two squares
$a^2 - b^2$
Point out that the difference of two squares is the only time the product of two binomials is a binomial.

In-Class Example

For Example 4
Find each product.

a. $(6x - 20y)(6x + 20y)$
$36x^2 - 400y^2$

b. $(14k + 9t^2)(14k - 9t^2)$
$196k^2 - 81t^4$

Study Guide Masters, p. 68

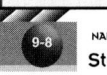

9-8

NAME _____ DATE _____

Student Edition
Pages 542–547

Study Guide

Special Products

You can use the FOIL method to find some special products.

Square of a Sum	$(a + b)^2 = (a + b)(a + b) = a^2 + 2ab + b^2$
Square of a Difference	$(a - b)^2 = (a - b)(a - b) = a^2 - 2ab + b^2$
Difference of Squares	$(a + b)(a - b) = (a - b)(a + b) = a^2 - b^2$

Study the examples below.

Binomials	Product
$(3n + 4)^2$	$9n^2 + 24n + 16$
$(2z - 9)^2$	$4z^2 - 36z + 81$
$(5x - 3y)(5x + 3y)$	$25x^2 - 9y^2$

Find each product.

1. $(x - 6)^2$
   $x^2 - 12x + 36$

2. $(3p + 4)^2$
   $9p^2 + 24p + 16$

3. $(x + 11)(x - 11)$
   $x^2 - 121$

4. $(2x + 3)(2x - 3)$
   $4x^2 - 9$

5. $(4x - 5)^2$
   $16x^2 - 40x + 25$

6. $(9x - y)(9x + y)$
   $81x^2 - y^2$

7. $(m + 5)^2$
   $m^2 + 10m + 25$

8. $(8a - 7b)(8a + 7b)$
   $64a^2 - 49b^2$

9. $(4a - 3b)^2$
   $16a^2 - 24ab + 9b^2$

10. $(3 - 5q)(3 + 5q)$
    $9 - 25q^2$

11. $(x^2 - 2)^2$
    $x^4 - 4x^2 + 4$

12. $(2.5 + q)^2$
    $6.25 + 5q + q^2$

13. $\left(\frac{3}{4}x + 1\right)\left(\frac{3}{4}x - 1\right)$
    $\frac{9}{16}x^2 - 1$

14. $(0.3p - 2q)^2$
    $0.09p^2 - 1.2pq + 4q^2$

15. $\left(\frac{1}{2}y + z\right)^2$
    $\frac{1}{4}y^2 + yz + z^2$

16. $(8 + x)^2$
    $64 + 16x + x^2$

17. $(6c - 10)(6c + 10)$
    $36c^2 - 100$

18. $(x^3 - 1)^2$
    $x^6 - 2x^3 + 1$

Check for Understanding

Exercises 1–12 are designed to help you assess your students' understanding through reading, writing, speaking, and modeling. You should work through Exercises 1–5 with your students and then monitor their work on Exercises 6–12.

Error Analysis

It is very common for students to assume that $(a + b)^2 = a^2 + b^2$. They reason that to square a quantity, it is acceptable to square its parts. Remind students to write out the multiplication. $(a + b)^2$ means $(a + b)(a + b)$ or $a^2 + 2ab + b^2$.

Assignment Guide
Core: 13–35 odd, 37–43
Enriched: 14–32 even, 33–43

For **Extra Practice,** see p. 778.

The red A, B, and C flags, printed only in the Teacher's Wraparound Edition, indicate the level of difficulty of the exercises.

Practice Masters, p. 68

546 Chapter 9

Communicating Mathematics

Study the lesson. Then complete the following.

1. **Explain** how the square of a difference and the square of a sum are different.   The middle terms have different signs.

2. **Compare and contrast** the square of a difference and the difference of two squares.   See margin.

3. **Explain** how you could mentally multiply 29×31. (*Hint:* $29 = 30 - 1$ and $31 = 30 + 1$) $(30 - 1)(30 + 1) = 900 - 1$ or 899

MODELING MATHEMATICS

4. Draw a diagram to represent each of the following.   a–b. See margin.
   a. $(x + y)^2$
   b. $(x - y)^2$

5. What does the diagram at the right represent if the shading represents regions to be removed or subtracted?
   $(a - b)^2 = a^2 - 2ab + b^2$

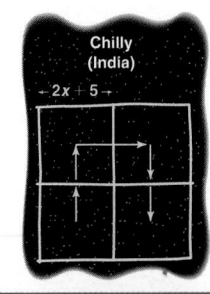

Guided Practice

Find each product.

6. $(2x + 3y)^2$ $4x^2 + 12xy + 9y^2$
7. $(m - 3n)^2$ $m^2 - 6mn + 9n^2$
8. $(2a + 3)(2a - 3)$ $4a^2 - 9$
9. $(m^2 + 4n)^2$ $m^4 + 8m^2n + 16n^2$
10. $(4y + 2z)(4y - 2z)$ $16y^2 - 4z^2$
11. $(5 - x)^2$ $25 - 10x + x^2$

12. **Recreation**   In India, children play a form of hopscotch called Chilly. One of the three possible patterns for this game is shown at the right. Suppose each side of the small squares is $2x + 5$ units long. Find the area of this Chilly pattern.
    $16x^2 + 80x + 100$ square units

Practice

Find each product.

A
13. $(x + 4y)^2$ $x^2 + 8xy + 16y^2$
14. $(m - 2n)^2$ $m^2 - 4mn + 4n^2$
15. $(3b - a)^2$ $9b^2 - 6ab + a^2$
16. $(3x + 5)(3x - 5)$ $9x^2 - 25$
17. $(9p - 2q)(9p + 2q)$ $81p^2 - 4q^2$
18. $(5s + 6t)^2$ $25s^2 + 60st + 36t^2$

B
19. $(5b - 12a)^2$ $25b^2 - 120ab + 144a^2$
20. $(2a + 0.5y)^2$ $4a^2 + 2ay + 0.25y^2$
21. $(x^3 + a^2)^2$ $x^6 + 2x^3a^2 + a^4$
22. $\left(\frac{1}{2}b^2 - a^2\right)^2$ $\frac{1}{4}b^4 - a^2b^2 + a^4$
23. $(8x^2 - 3y)(8x^2 + 3y)$ $64x^4 - 9y^2$
24. $(7c^2 + d^3)(7c^2 - d^3)$ $49c^4 - d^6$
25. $(1.1g + h^5)^2$ $1.21g^2 + 2.2gh^5 + h^{10}$
26. $(9 - z^9)(9 + z^9)$ $81 - z^{18}$
27. $\left(\frac{4}{3}x^2 - y\right)\left(\frac{4}{3}x^2 + y\right)$ $\frac{16}{9}x^4 - y^2$
28. $\left(\frac{1}{3}v^2 - \frac{1}{2}w^3\right)^2$

28. $\frac{1}{9}v^4 - \frac{1}{3}v^2w^3 + \frac{1}{4}w^6$

29. $9x^3 - 45x^2 - x + 5$

30. $x^4 - 29x^2 + 100$

33. $x^2 + y^2 + z^2 + 2xy + 2yz + 2xz$;
See margin for diagram.

C
29. $(3x + 1)(3x - 1)(x - 5)$
30. $(x - 2)(x + 5)(x + 2)(x - 5)$
31. $(a + 3b)^3$
    $a^3 + 9a^2b + 27ab^2 + 27b^3$
32. $(2m - n)^4$
    $16m^4 - 32m^3n + 24m^2n^2 - 8mn^3 + n^4$

Critical Thinking

33. Find $(x + y + z)^2$. Draw a diagram to show each term of the polynomial.

546   Chapter 9   Exploring Polynomials

Additional Answers

2. The square of a difference is $(a - b)^2$, which equals $a^2 - 2ab + b^2$.
   The difference of two squares is the product $(a + b)(a - b)$ or $a^2 - b^2$.

4a.

	x	y
x	x^2	xy
y	xy	y^2

4b.

	x	-y
x	x^2	$-xy$
-y	$-xy$	y^2

33.

	x	y	z
x	x^2	xy	xz
y	xy	y^2	yz
z	xz	yz	z^2

Applications and Problem Solving

34. Biology Refer to the application at the beginning of the lesson.
  a. Make a Punnett square for pea plants if one parent is pure short (*tt*) and the other parent is hybrid tall (*Tt*). **See margin.**
  b. What percent of the offspring will be pure short? **50%**
  c. What percent of the offspring will be hybrid tall? **50%**
  d. What percent of the offspring will be pure tall? **0%**

35. History Refer to Example 2.
  a. Write an expression for the area of the fourth seating level in the Gwennap Pit. $2\pi s + 7\pi$ **square meters**
  b. The radius of the stage level of the Gwennap Pit is 3 meters. Find the area of the stage. **about 28.27 square meters**
  c. Find the area of the fourth seating level in the Gwennap Pit. **about 40.84 square meters**

36. Photography Lenora cut off a 0.75-inch strip all around a square photograph so it would fit in an envelope she was mailing to her aunt. She decided to have the photo lab make a copy of the photo from the negative, but she forgot to measure how large the original photo was. All she had was the strips she cut off, whose area was 33.75 square inches. What were the original dimensions of the photograph? **12 in. by 12 in.**

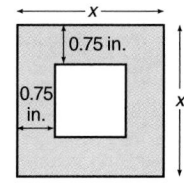

Mixed Review

37. Find $(3t - 3)(2t + 1)$. (Lesson 9–7) $6t^2 - 3t - 3$

38. Solve $-13z > -1.04$. (Lesson 7–2) $\{z \mid z < 0.08\}$

39. Statistics The table below shows the heights and weights of each of 12 players on a pro basketball team. (Lesson 6–3)

Height (in.)	75	82	75	74	80	80	75	79	80	78	76	81
Weight (lb)	180	235	184	185	230	205	185	230	221	195	205	215

39b. the taller the player, the greater the weight

  a. Make a scatter plot of these data. **See margin.**
  b. Describe the correlation between height and weight.

40. Write the standard form of the line that passes through (3, 1) and has a slope of $\frac{2}{7}$. (Lesson 6–2)
$2x - 7y = -1$

41. Graph the points $A(4, 2)$, $B(-3, 1)$, and $C(-2, -3)$. (Lesson 5–1) **See margin.**

42. about 18.26 amperes

42. Electricity The resistance R of a power circuit is 4.5 ohms. How much current I, in amperes, can the circuit generate if it can produce at most 1500 watts of power? Use $I^2R = P$. (Lesson 2–8)

43. Evaluate $5(9 \div 3^2)$. (Lesson 1–6) **5**

Lesson 9-8 Special Products **547**

Extension

Connections The formula for the volume of a cube is $V = e^3$, where e represents the length of an edge. Find the volume of a cube if the length of its edge is $a + b$.
$a^3 + 3a^2b + 3ab^2 + b^3$

Closing the Investigation

This activity provides students an opportunity to bring their work on the Investigation to a close. For each Investigation, students should present their findings to the class. Here are some ways students can display their work.

- Conduct and report on an interview or survey.
- Write a letter, proposal, or report.
- Write an article for the school or local paper.
- Make a display, including graphs and/or charts.
- Plan an activity.

Assessment

To assess students' understanding of the concepts and topics explored in this Investigation and its follow-up activities, you may wish to examine students' Investigation Folders.

The scoring guide provided in the *Investigations and Projects Masters,* p. 15, provides a means for you to score students' work on this Investigation.

Investigations and Projects Masters, p. 15

Scoring Guide
Chapters 8 and 9
Investigation

Level	Specific Criteria
3 Superior	· Shows thorough understanding of the concepts of *surface area, perimeter, linear equation, polynomial expression, scatter plot, slope, best-fit line,* and *using mean, median, or mode to determine an average.* · Uses appropriate strategies to solve problems. · Computations are correct. · Written explanations are exemplary. · Charts, graphs, and report are appropriate and sensible. · Goes beyond the requirements of some or all problems.
2 Satisfactory, with Minor Flaws	· Shows understanding of the concepts of *surface area, perimeter, linear equation, polynomial expression, scatter plot, slope, best-fit line,* and *using mean, median, or mode to determine an average.* · Uses appropriate strategies to solve problems. · Computations are mostly correct. · Written explanations are effective. · Charts, graphs, and report are appropriate and sensible. · Satisfies the requirements of problems.
1 Nearly Satisfactory, with Obvious Flaws	· Shows understanding of most of the concepts of *surface area, perimeter, linear equation, polynomial expression, scatter plot, slope, best-fit line,* and *using mean, median, or mode to determine an average.* · May not use appropriate strategies to solve problems. · Computations are mostly correct. · Written explanations are satisfactory. · Charts, graphs, and report are appropriate and sensible. · Satisfies the requirements of problems.
0 Unsatisfactory	· Shows little or no understanding of the concepts of *surface area, perimeter, linear equation, polynomial expression, scatter plot, slope, best-fit line,* and *using mean, median, or mode to determine an average.* · Does not use appropriate strategies to solve problems. · Computations are incorrect. · Written explanations are not satisfactory. · Charts, graphs, and report are not appropriate or sensible. · Does not satisfy the requirements of problems.

Ready, Set, Drop!

Refer to the Investigation on pages 448–449.

Analyze

You have conducted several experiments and organized your data in various ways. It is now time to analyze your findings and state your conclusions.

PORTFOLIO ASSESSMENT

You may want to keep your work on this Investigation in your portfolio.

1 Look over your data and organize it in such a way that the various relationships are obvious.

2 Describe the relationships in the data. What does weight have to do with glide time? Does perimeter or surface area have an effect on glide time? What other factors need to be considered?

Write

The report to the people interested in hang gliding should explain your process for investigating these hang glider models and what you found from your investigations.

3 Begin the report by stating the process you used to investigate the matter. Explain all the experiments conducted. State the purpose and findings for each.

4 Show the data you collected in tables, charts, and graphs. Explain your analysis of the data and conclusions you found.

5 Make a recommendation to the group about the size of hang glider(s) that is most suited for them. Include the weight of the object that the hang glider(s) should carry to be most efficient.

6 While you were conducting experiments, one of your team members found that the frames for most hang gliders are 32 feet wide. Explain how this information affects your generalizations regarding the weight and size of hang gliders.

7 Summarize your findings in a concluding statement to the group.

VOCABULARY

After completing this chapter, you should be able to define each term, property, or phrase and give an example or two of each.

Algebra

binomial (p. 514)

constants (p. 496)

degree of monomial (p. 515)

degree of polynomial (p. 516)

difference of squares (p. 545)

FOIL method (p. 537)

monomial (p. 496)

negative exponent (p. 503)

polynomial (pp. 513, 514)

power of a monomial (p. 498)

power of a power (p. 498)

power of a product (p. 498)

product of powers (p. 497)

quotient of powers (p. 501)

scientific notation (p. 506)

square of a difference (p. 544)

square of a sum (p. 543)

trinomial (p. 514)

zero exponent (p. 502)

Problem Solving

look for a pattern (p. 497)

UNDERSTANDING AND USING THE VOCABULARY

Choose the letter of the term that best matches each example.

1. $4^{-3} = \frac{1}{4^3}$ or $\frac{1}{64}$ e

2. $(x + 2y)(x - 2y) = x^2 - 4y^2$ b

3. $\frac{4x^2y}{8xy^3} = \frac{x}{2y^2}$ h

4. $4x^2$ d

5. $x^2 - 3x + 1$ i

6. $2^0 = 1$ j

7. $x^4 - 3x^3 + 2x^2 - 1$ f

8. $(x + 3)(x - 4) = x^2 - 4x + 3x - 12$ c

9. $x^2 + 2$ a

10. $(a^3b)(2ab^2) = 2a^4b^3$ g

a. binomial
b. difference of squares
c. FOIL method
d. monomial
e. negative exponent
f. polynomial
g. product of powers
h. quotient of powers
i. trinomial
j. zero exponent

Chapter 9 Highlights **549**

Instructional Resources

Three multiple-choice tests and three free-response tests are provided in the *Assessment and Evaluation Masters.* Forms 1A and 2A are for honors pacing, and Forms 1B, 1C, 2B, and 2C are for average pacing. Chapter 9 Test, Form 1B is shown at the right. Chapter 9 Test, Form 2B is shown on the next page.

Using the CHAPTER HIGHLIGHTS

The Chapter Highlights begins with a listing of the new terms, properties, and phrases that were introduced in this chapter. Have students define each term and provide an example or two of it, if appropriate.

Assessment and Evaluation Masters, pp. 227–228

---

9 NAME_____ DATE_____

Chapter 9 Test, Form 1B

Write the letter for the correct answer in the blank at the right of each problem.

1. What is the simplest form of $m^4 \cdot m^2$?
   A. m^2 B. m^8 C. m^6 D. $2m^8$ 1. ___C___

2. What is the simplest form of $(3w^2v)(-2w^5v^2)(4w^6v^5)$?
   A. $-24w^{13}v^8$ B. $5w^{60}v^{10}$ C. $-24w^{60}v^{10}$ D. $5w^{13}v^8$ 2. ___A___

3. What is the simplest form of $(y^3n^3)^4 + (3y^4n^6)^2$?
   A. $10y^8n^{12}$ B. $9y^{16}n^{27}$ C. $10y^8 + 2n^{12}$ D. $y^6n^7 + 9y^6n^8$ 3. ___A___

4. What is the simplest form of $\frac{m^6n^3}{m^2n^6}$? Assume the denominator is not equal to zero.
   A. $\frac{m^4}{n^3}$ B. $-\frac{m^4}{n^3}$ C. $-\frac{m^8}{n^3}$ D. $\frac{m^8}{n^3}$ 4. ___A___

5. What is the simplest form of $\frac{6n^{-3}yw^{-5}}{2n^{-1}y^{-3}w^2}$? Assume the denominator is not equal to zero.
   A. $\frac{4y^3}{n^2w^3}$ B. $\frac{3y^4}{n^2w^7}$ C. $\frac{3}{n^4w^3y^2}$ D. $\frac{3n^2}{y^4w^7}$ 5. ___B___

6. What is the simplest form of $\frac{(a^{-2}b^4)^{-6}}{(a^4b^{-8})^3}$? Assume the denominator is not equal to zero.
   A. ab^3 B. 1 C. $\frac{a^{24}}{b^{48}}$ D. $\frac{b^{48}}{a^{24}}$ 6. ___B___

7. What is 0.000543 expressed in scientific notation?
   A. 5.43×10^4 B. 5.43×10^{-3} C. 543×10^6 D. 5.43×10^{-4} 7. ___D___

8. What is the value of $\frac{2.88 \times 10^3}{2.4 \times 10^{-7}}$?
   A. 0.48×10^{-4} B. 1.2×10^{-4} C. 1.2×10^{10} D. 0.48×10^{10} 8. ___C___

9. What is the degree of $2x^3y - 4xy^2 + 9x^3y^2$?
   A. 4 B. 3 C. 12 D. 5   9. ___D___

10. Which of the following shows the terms of $x^2y^3 + 4xy^2 - 3x^3y + 6$ arranged so that the powers of x are in ascending order?
    A. $6 + 4xy^2 + x^2y^3 - 3x^3y$ B. $x^2y^3 - 3x^3y + 4xy^2 + 6$
    C. $6 + 4xy^2 - 3x^3y + x^2y^3$ D. $6 - 3x^3y + 4xy^2 + x^2y^3$ 10. ___A___

11. What is the simplest form of $(3c^2 - 8c + 5) + (c^2 - 8c - 6)$?
    A. $3c^2 - 1$ B. $4c^2 + 11$
    C. $4c^2 - 16c - 1$ D. $2c^2 - 16c - 1$ 11. ___C___

---

9 NAME_____ DATE_____

Chapter 9 Test, Form 1B (continued)

12. What is the simplest form of $(6w^2 - 5wz - 3z^2) - (4w^2 + 6wz - 8z^2)$?
    A. $10w^2 + wz - 11z^2$ B. $2w^2 - 11wz + 5z^2$
    C. $2w^2 + wz - 11z^2$ D. $2w^2 - 11wz - 5z^2$ 12. ___B___

13. What is the simplest form of $8a^2b(3a + 2ab^3)$?
    A. $24a^2b + 16a^2b^3$ B. $24a^3b + 2ab^3$
    C. $24a^3b + 16a^3b^4$ D. $11a^2b + 10a^2b^3$ 13. ___C___

14. What is the simplest form of $(x + 2)(x + 4)$?
    A. $x^2 + 8$ B. $2x + 6$ C. $2x + 8$ D. $x^2 + 6x + 8$ 14. ___D___

15. What is the simplest form of $(2n - 6)(4n + 2)$?
    A. $6n^2 + 28n - 12$ B. $8n^2 - 12$
    C. $8n^2 - 20n - 12$ D. $8n^2 + 20n - 12$ 15. ___C___

16. What is the simplest form of $(x + 6)^2$?
    A. $x^2 + 36$ B. $x^2 + 12x + 36$
    C. $x^2 + 12$ D. $2x + 12$ 16. ___B___

17. What is the simplest form of $(3y + 4z)(3y - 4z)$?
    A. $9y^2 - 16z^2$ B. $9y^2 - 24yz - 16z^2$
    C. $9y^2 + 16z^2$ D. $9y^2 - 24yz + 16z^2$ 17. ___A___

18. Find the solution of $-4(5 - 2n) = 8(-6 - 5n)$.
    A. $-\frac{1}{9}$ B. $-\frac{28}{3}$ C. $-\frac{7}{8}$ D. $-\frac{7}{12}$ 18. ___D___

19. Find the solution of $x(x + 3) - 2 = 2 + x(x + 1)$.
    A. 2 B. -2 C. 1 D. 0   19. ___A___

20. A picture is 4 inches longer than it is wide. It is surrounded by a mat 2 inches wide. The total area of the mat is 112 square inches. If w is the width of the picture, which equation is true?
    A. $(w + 4)(w + 8) + w(w + 4) = 112$
    B. $(w + 2)(w + 6) - w(w + 4) = 112$
    C. $(w + 4)(w + 8) - w(w + 4) = 112$
    D. $(w + 2)(w + 6) + w(w + 4) = 112$ 20. ___C___

Bonus Solve $3^{2n-1} \cdot 3^{5n} = (3^4)^{n+2}$ for n.   Bonus ___3___

Skills and Concepts Encourage students to refer to the objectives and examples on the left as they complete the review exercises on the right.

Assessment and Evaluation Masters, pp. 233–234

SKILLS AND CONCEPTS

OBJECTIVES AND EXAMPLES

Upon completing this chapter, you should be able to:

- multiply monomials and simplify expressions involving powers of monomials (Lesson 9–1)

$$(2ab^2)(3a^2b^3) = (2 \cdot 3)(a \cdot a^2)(b^2 \cdot b^3)$$
$$= 6a^3b^5$$

$$(2x^2y^3)^3 = 2^3(x^2)^3(y^3)^3$$
$$= 8x^6y^9$$

- simplify expressions involving quotients of monomials and negative exponents (Lesson 9–2)

$$\frac{2x^6y}{8x^2y^2} = \frac{2}{8} \cdot \frac{x^6}{x^2} \cdot \frac{y}{y^2}$$
$$= \frac{x^4}{4y}$$

$$\frac{3a^{-2}}{4a^6} = \frac{3}{4}(a^{-2-6})$$
$$= \frac{3}{4}(a^{-8}) \text{ or } \frac{3}{4a^8}$$

- express numbers in scientific and decimal notation (Lesson 9–3)

$$3,600,000 = 3.6 \times 1,000,000$$
$$= 3.6 \times 10^6$$

$$0.0021 = 2.1 \times 0.001$$
$$= 2.1 \times 10^{-3}$$

- find products and quotients of numbers expressed in scientific notation (Lesson 9–3)

$$(2 \times 10^2)(5.2 \times 10^6) = (2 \times 5.2)(10^2 \times 10^6)$$
$$= 10.4 \times 10^8$$
$$= 1.04 \times 10^9$$

$$\frac{1.2 \times 10^{-2}}{0.6 \times 10^3} = \frac{1.2}{0.6} \times \frac{10^{-2}}{10^3}$$
$$= 2 \times 10^{-5}$$

REVIEW EXERCISES

Use these exercises to review and prepare for the chapter test.   12. $-12a^3b^4$ 13. $20a^5x^5$

Simplify.   15. $576x^5y^2$ 16. $-432c^{14}d^4$

11. $y^3 \cdot y^3 \cdot y$ y^7 12. $(3ab)(-4a^2b^3)$

13. $(-4a^2x)(-5a^3x^4)$ 14. $(4a^2b)^3$ $64a^6b^3$

15. $(-3xy)^2(4x)^3$ 16. $(-2c^2d)^4(-3c^2)^3$

17. $-\frac{1}{2}(m^2n^4)^2$ 18. $(5a^2)^3 + 7(a^6)$

Simplify. Assume that no denominator is equal to zero.

19. $\frac{y^{10}}{y^6}$ y^4 20. $\frac{(3y)^0}{6a}$ $\frac{1}{6a}$

21. $\frac{42b^7}{14b^4}$ $3b^3$ 22. $\frac{27b^{-2}}{14b^{-3}}$ $\frac{27b}{14}$

23. $\frac{(3a^3bc^2)^2}{18a^2b^3c^4}$ $\frac{a^4}{2b}$ 24. $\frac{-16a^3b^2x^4y}{-48a^4bxy^3}$ $\frac{bx^3}{3ay^2}$

Express each number in scientific notation.

25. 240,000 2.4×10^5

26. 0.000314 3.14×10^{-4}

27. 4,880,000,000 4.88×10^9

28. 0.00000187 1.87×10^{-6}

29. 796×10^3 7.96×10^5

30. 0.03434×10^{-2} 3.434×10^{-4}

Evaluate. Express each result in scientific notation.

31. $(2 \times 10^5)(3 \times 10^6)$ 6×10^{11}

32. $(3 \times 10^3)(1.5 \times 10^6)$ 4.5×10^9

33. $\frac{5.4 \times 10^3}{0.9 \times 10^4}$ 6×10^{-1}

34. $\frac{8.4 \times 10^{-6}}{1.4 \times 10^{-9}}$ 6×10^3

35. $(3 \times 10^2)(5.6 \times 10^{-4})$ 1.68×10^{-1}

36. $34(4.7 \times 10^5)$ 1.598×10^7

9 NAME_____ DATE _____

Chapter 9 Test, Form 2B

1. Simplify $y^6 \cdot y^3$.   1. _____ y^9

2. Simplify $(9m^3n^5)(-2mn^2)(6m^2n^8)$.   2. _____ $-108m^6n^{15}$

3. Simplify $(w^5y^4)^3$.   3. _____ $w^{15}y^{12}$

4. Simplify $4a^3n^6 + 4(a^3n)^6 + 4(an^2)^3$.   4. _____ $8a^9n^6 + 4a^{18}n^6$

5. Simplify $\frac{p^6q^2}{p^5q}$. Assume the denominator is not equal to zero.   5. _____ p^9q

6. Simplify $\frac{16r^3s^{-5}}{4r^{-1}s^3}$. Assume the denominator is not equal to zero.   6. _____ $\frac{4r^4}{s^7}$

7. Simplify $\frac{(-8x^2y^2)^2}{(4x^2y^3)^3}$. Assume the denominator is not equal to zero.   7. _____ $\frac{y}{x^5}$

8. Express 0.000498 in scientific notation.   8. _____ 4.98×10^{-4}

9. Express 1.27×10^5 in decimal notation.   9. _____ 127,000

10. Evaluate $(2.5 \times 10^{-2})(4 \times 10^6)$. Express the result in scientific notation.   10. _____ 1×10^5

11. Evaluate $\frac{5.4 \times 10^{-8}}{1.5 \times 10^2}$. Express the result in scientific notation.   11. _____ 3.6×10^{-10}

12. Find the degree of $2x^3y^3 + 4xy - 10x^3y$.   12. _____ 6

13. Arrange the terms $4 + 3x^3y^3 - x^5y + xy$ so that the powers of x are in descending order.   13. _____ $-x^5y + 3x^3y^3 + xy + 4$

9 NAME_____ DATE _____

Chapter 9 Test, Form 2B (continued)

14. Find the difference. $\begin{aligned} 5n^2 - 2ny + 3y^2 \\ (-)\ 9n^2 - 8ny + 10y^2 \end{aligned}$ 14. _____ $-4n^2 + 6ny + 13y^2$

15. Simplify $(11m^2 - 2mn + 8n^2) + (8m^2 + 4mn - 2n^2)$.   15. _____ $19m^2 + 2mn + 6n^2$

16. Simplify $5hk^2(2h^3k - hk^3 + 4h^2k^2)$.   16. _____ $10h^4k^3 - 5h^2k^5 + 20h^4k^4$

17. Simplify $(x + 5) - (x + 6)$.   17. _____ -1

18. Simplify $(4x^2 + 2y^2)(2x^2 - y^2)$.   18. _____ $8x^4 - 2y^4$

19. Simplify $(3s + 5)(2s^2 - 8s + 6)$.   19. _____ $6s^3 - 14s^2 - 22s + 30$

20. Simplify $(5c - 4)^2$.   20. _____ $25c^2 - 40c + 16$

21. Simplify $(7a - 3b)(7a + 3b)$.   21. _____ $49a^2 - 9b^2$

22. Simplify $(4n + 1)^2$.   22. _____ $16n^2 + 8n + 1$

23. Solve $-6(3n - 2) = 4(-3 - 2n)$.   23. _____ 2.4

24. Solve $8n + 11 = 4 + 5(2n - 1)$.   24. _____ 6

25. The length of a rectangular garden is 8 feet longer than the width. The garden is surrounded by a 4-foot sidewalk. The sidewalk has an area of 320 square feet. Find the dimensions of the garden.   25. _____ 12 ft × 20 ft

Bonus If $(x + 1)^{20}$ is multiplied out, how many terms will there be? (*Hint*: Look at the smaller powers of $(x + 1)$. Find a pattern.)   **Bonus** _____ 21

GLENCOE Technology

Test and Review Software

You may use this software, a combination of an item generator and item bank, to create your own tests or worksheets. Types of items include free response, multiple choice, short answer, and open ended.

For IBM & Macintosh

OBJECTIVES AND EXAMPLES

find the degree of a polynomial   (Lesson 9–4)

Find the degree of $2xy^3 + x^2y$.

degree of $2xy^3$: $1 + 3$ or 4

degree of x^2y: $2 + 1$ or 3

degree of $2xy^3 + x^2y$: 4

REVIEW EXERCISES

Find the degree of each polynomial.

37. $n - 2p^2$ 2
38. $29n^2 + 17n^2t^2$ 4
39. $4xy + 9x^3z^2 + 17rs^3$ 5
40. $-6x^5y - 2y^4 + 4 - 8y^2$ 6
41. $3ab^3 - 5a^2b^2 + 4ab$ 4
42. $19m^3n^4 + 21m^5n^2$ 7

---

arrange the terms of a polynomial so that the powers of a variable are in ascending or descending order   (Lesson 9–4)

Arrange the terms of $4x^2 + 9x^3 - 2 - x$ in descending order.

$$9x^3 + 4x^2 - x - 2$$

Arrange the terms of each polynomial so that the powers of x are in descending order.

43. $3x^4 - x + x^2 - 5$ $3x^4 + x^2 - x - 5$
44. $-2x^2y^3 - 27 - 4x^4 + xy + 5x^3y^2$
    $-4x^4 + 5x^3y^2 - 2x^2y^3 + xy - 27$

---

add and subtract polynomials   (Lesson 9–5)

$$
\begin{array}{r}
4x^2 - 3x + 7 \\
(+)\ 2x^2 + 4x \quad\ \\
\hline
6x^2 +\ \ x + 7
\end{array}
$$

$(7r^2 + 9r) - (12r^2 - 4) = 7r^2 + 9r - 12r^2 + 4$
$\qquad\qquad\qquad = (7r^2 - 12r^2) + 9r + 4$
$\qquad\qquad\qquad = -5r^2 + 9r + 4$

45. $-3x^3 + x^2 - 5x + 5$
46. $4x^2 - 5xy + 6y^2$
49. $21m^4 - 10m - 1$

50. $-7p^2 - 2p + 25$

Find each sum or difference.

45. $(2x^2 - 5x + 7) - (3x^3 + x^2 + 2)$
46. $(x^2 - 6xy + 7y^2) + (3x^2 + xy - y^2)$

47.
$$
\begin{array}{r}
11m^2n^2 + 4mn - 6 \\
(+)\ 5m^2n^2 - 6mn + 17 \\
\hline
16m^2n^2 - 2mn + 11
\end{array}
$$

48.
$$
\begin{array}{r}
7z^2 \qquad\ \ + 4 \\
(-)\ 3z^2 + 2z - 6 \\
\hline
4z^2 - 2z + 10
\end{array}
$$

49.
$$
\begin{array}{r}
13m^4 - 7m - 10 \\
(+)\ 8m^4 - 3m + 9 \\
\hline
\end{array}
$$

50.
$$
\begin{array}{r}
-5p^2 + 3p + 49 \\
(-)\ 2p^2 + 5p + 24 \\
\hline
\end{array}
$$

---

multiply a polynomial by a monomial   (Lesson 9–6)

$ab(-3a^2 + 4ab - 7b^3) = -3a^3b + 4a^2b^2 - 7ab^4$

Find each product.

51. $4ab(3a^2 - 7b^2)$ $12a^3b - 28ab^3$
52. $7xy(x^2 + 4xy - 8y^2)$ $7x^3y + 28x^2y^2 - 56xy^3$
53. $4x^2y(2x^3 - 3x^2y^2 + y^4)$ $8x^5y - 12x^4y^3 + 4x^2y^5$
54. $5x^3(x^4 - 8x^2 + 16)$ $5x^7 - 40x^5 + 80x^3$

---

simplify expressions involving polynomials   (Lesson 9–6)

$x^2(x + 2) + 3(x^3 + 4x^2) = x^3 + 2x^2 + 3x^3 + 12x^2$
$\qquad\qquad\qquad\qquad\quad = 4x^3 + 14x^2$

Simplify.

55. $2x(x - y^2 + 5) - 5y^2(3x - 2)$
    $2x^2 - 17xy^2 + 10x + 10y^2$
56. $x(3x - 5) + 7(x^2 - 2x + 9)$ $10x^2 - 19x + 63$

Applications and Problem Solving Encourage students to work through the exercises in the Applications and Problem Solving section to strengthen their problem-solving skills.

OBJECTIVES AND EXAMPLES

• use the FOIL method to multiply two binomials and multiply any two polynomials by using the distributive property   (Lesson 9-7)

$$\begin{array}{cccc} F & O & I & L \end{array}$$
$$(3x + 2)(x - 2) = (3x)(x) + (3x)(-2) + (2)(x) + (2)(-2)$$
$$= 3x^2 - 6x + 2x - 4$$
$$= 3x^2 - 4x - 4$$

$$(4x - 3)(3x^2 - x + 2)$$
$$= 4x(3x^2 - x + 2) - 3(3x^2 - x + 2)$$
$$= (12x^3 - 4x^2 + 8x) - (9x^2 - 3x + 6)$$
$$= 12x^3 - 4x^2 + 8x - 9x^2 + 3x - 6$$
$$= 12x^3 - 13x^2 + 11x - 6$$

REVIEW EXERCISES

Find each product.

57. $(r - 3)(r + 7)$
$r^2 + 4r - 21$

58. $(x + 5)(3x - 2)$
$3x^2 + 13x - 10$

59. $(4x - 3)(x + 4)$
$4x^2 + 13x - 12$

60. $(2x + 5y)(3x - y)$
$6x^2 + 13xy - 5y^2$

61. $(3x + 0.25)(6x - 0.5)$
$18x^2 - 0.125$

62. $(5r - 7s)(4r + 3s)$
$20r^2 - 13rs - 21s^2$

63. $x^2 + 7x - 9$
$(\times)\ \ 2x + 1$

64. $a^2 - 17ab - 3b^2$
$(\times)\ \ \ \ \ 2a + b$

63. $2x^3 + 15x^2 - 11x - 9$

64. $2a^3 - 33a^2b - 23ab^2 - 3b^3$

• use patterns to find $(a + b)^2$, $(a - b)^2$, and $(a + b)(a - b)$ (Lesson 9-8)

$$(x + 4)^2 = x^2 + 2(4x) + 4^2$$
$$= x^2 + 8x + 16$$
$$(r - 5)^2 = r^2 - 2(5r) + 5^2$$
$$= r^2 - 10r + 25$$
$$(b + 9)(b - 9) = b^2 - 9^2$$
$$= b^2 - 81$$

Find each product.

65. $(x - 6)(x + 6)$
$x^2 - 36$

66. $(7 - 2x)(7 + 2x)$
$49 - 4x^2$

67. $(4x + 7)^2$
$16x^2 + 56x + 49$

68. $(8x - 5)^2$
$64x^2 - 80x + 25$

69. $(5x - 3y)(5x + 3y)$
$25x^2 - 9y^2$

70. $(a^2 + b)^2$
$a^4 + 2a^2b + b^2$

71. $(6a - 5b)^2$
$36a^2 - 60ab + 25b^2$

72. $(3m + 4n)^2$
$9m^2 + 24mn + 16n^2$

APPLICATIONS AND PROBLEM SOLVING

73. Finance   Find the current monthly payment on a 36-month car loan for $18,543. Twenty-five monthly payments have already been made at an annual interest rate of 8.7%. There is a balance due of $3216.27 at this time. Use the formula $B = P\left[\dfrac{1 - (1 + i)^{k-n}}{i}\right]$, where B represents the balance, P represents the current monthly payment, i represents the *monthly* interest rate (annual rate ÷ 12), k represents the total number of monthly payments already made, and n is the total number of monthly payments.   (Lesson 9-2)   **$305.26**

A practice test for Chapter 9 is provided on page 795.

74. Health   A radio station advertised the Columbus Marathon by saying that about 19,500,000 Calories would be burned in one day. If there were 6500 runners, about how many Calories did each runner burn? (Use scientific notation to solve.)   (Lesson 9-3)
about 3×10^3 Calories

75. Finance   Upon his graduation from college, Mark Price received $10,000 in a trust fund from his grandparents. If he invests this money in an account with an annual interest rate of 6% and adds $1000 of his own money to the account at the end of each year, will his money have doubled after 5 years? If not, when?
(Lesson 9-4)   **no; after 6 years**

ALTERNATIVE ASSESSMENT

COOPERATIVE LEARNING PROJECT

Saving for College In this chapter, you developed the concept of polynomials. You performed operations on polynomials, simplified polynomials, and solved polynomial equations. They were helpful in setting up a general formula to be used for inputting various data.

In this project, you will forecast a friend's finances. Jane has received $75 from her grandparents on every birthday since she was one year old. She has been saving the money in an account that pays 5% interest. She is saving her money to help pay for her college education, which she will start this fall after her 18th birthday. She also has been receiving birthday checks from her other relatives, but these didn't start until she was 12 years old. The amounts of these checks from her 12th birthday until her 18th birthday are $45, $45, $55, $50, $55, $60, and $65.

How much money will she have saved just from her birthdays by the time she starts college? Is this a reasonable amount to pay for a used car during her junior year in college? If she had invested her money in a different account that had earned 7% interest, how much more money would she have saved?

Follow these steps to accomplish your task.

- Construct a pattern for this situation.
- Develop a polynomial model to describe the amount of money she has each year.
- Determine the amount of money she received on birthdays 12 through 18.
- Determine what needs to be changed in your model when changing the interest rate.
- Write a paragraph describing the problem and your solution.

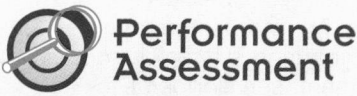

THINKING CRITICALLY

- Can $(-b)^2$ ever equal $-b^2$? Explain and give an example to support your answer.
- For all numbers a and b and any integer m, is $(a + b)^m = a^m + b^m$ a true sentence? Explain and give examples.

PORTFOLIO

Error analysis shows common mistakes that happen when performing an operation. Here is an example of an error when multiplying like bases.

$$4^3 \cdot 4^4 = 16^7$$

Actually, $4^3 \cdot 4^4 = 4^7$. The error of multiplying the bases while adding the exponents was incorrect. The base should stay the same while adding the exponent.

From the material in this chapter, find a problem that occurs often and write an error analysis for it. Describe the situation, give an example of the incorrect method, give the correct method for that example, and write a paragraph about it. Place this in your portfolio.

SELF EVALUATION

In this chapter, there are several words that have prefixes or suffixes that can be analyzed to determine what the word means. Do you break down words to find their meanings or do you just skip over those words and look for the meaning in the context of the sentence or paragraph? Maybe you go straight to the dictionary to get the meaning.

Assess yourself. How do you best learn new vocabulary words? After learning the meaning of the new word, do you then try to use that new word in your speaking and/or writing? Describe the plan that you use when learning a new word and how you accomplish it. Give an example of a new math-related word and a new word used in your daily life and explain how you found the meaning of each of these words.

Assessment and Evaluation Masters, pp. 238, 249

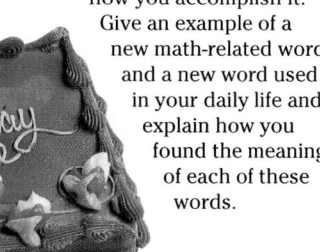

9   Chapter 9 Performance Assessment

Instructions: *Demonstrate your knowledge by giving a clear, concise solution to each problem. Be sure to include all relevant drawings and justify your answers. You may show your solution in more than one way or investigate beyond the requirements of the problem.*

1. Rectangular areas can be used to represent products of binomials.
   a. Find the product $(2x + 3)(x + 1)$. Tell how the product and the area at the right are related.
   b. Draw an area model demonstrating the product $(3x + 1)(x + 2)$. Find the product algebraically to verify your model.
   c. Write the two binomials whose product is demonstrated by the area model at the right. Find the product and show how it relates to the model.
   d. If the shaded regions represent a negative area and can be used to subtract or remove area from positive areas, tell how $x(2x - 1) = 2x^2 - x$ relates to the model.
   e. Draw an area model demonstrating the product $x(3x - 2)$. Find the product and tell how it relates to the model.
   f. The product of two binomials is $x^2 + 3x + 2$. Use the area models at the right to form a rectangle and find the two binomials. Check your answer.

2. Simplify $\frac{(-3a^3b^2)(4ab)}{2a^7b}$ in at least two ways. Explain each step.

3. Write a product of two numbers in scientific notation. Find the product and explain each step.

Scoring Guide
Chapter 9
Performance Assessment

Level	Specific Criteria
3 Superior	• Shows thorough understanding of the concepts of *multiplication of binomials, multiplication and division of monomials,* and *scientific notation.* • Computations are correct. • Written explanations are exemplary. • Diagrams are accurate and appropriate. • Goes beyond requirements of some or all problems.
2 Satisfactory, with Minor Flaws	• Shows understanding of the concepts of *multiplication of binomials, multiplication and division of monomials,* and *scientific notation.* • Computations are mostly correct. • Written explanations are effective. • Diagrams are mostly accurate and appropriate. • Satisfies all requirements of problems.
1 Nearly Satisfactory, with Serious Flaws	• Shows understanding of most of the concepts of *multiplication of binomials, multiplication and division of monomials,* and *scientific notation.* • Computations are mostly correct. • Written explanations are satisfactory. • Diagrams are mostly accurate and appropriate. • Satisfies most requirements of problems.
0 Unsatisfactory	• Shows little or no understanding of the concepts of *multiplication of binomials, multiplication and division of monomials,* and *scientific notation.* • Computations are incorrect. • Written explanations are not satisfactory. • Diagrams are not accurate or appropriate. • Does not satisfy requirements of problems.

⊘ Alternative Assessment

The Alternative Assessment section provides students with the opportunity to assess their own work by thinking critically, working with others, keeping a portfolio, and honestly evaluating their own progress. For more information on alternative forms of assessment, see *Alternative Assessment in the Mathematics Classroom,* one of the titles in the Glencoe Mathematics Professional Series.

⊘ Performance Assessment

Performance Assessment tasks for this chapter are included in the *Assessment and Evaluation Masters.* A scoring guide is also provided.

NCTM Standards: 1–5, 7, 9

This Investigation is designed to be completed over several days or weeks. It may be considered optional. You may want to assign the Investigation and the follow-up activities to be completed at the same time.

Objective
Design brick rectangular patios using small square bricks, large square bricks, and rectangular bricks.

Mathematical Overview
This Investigation will use the following mathematical skills and concepts from Chapters 10 and 11.

- making charts and tables
- identifying and factoring binomials that are the differences of squares
- using the zero product property to solve equations
- estimating roots of quadratic equations
- graphing quadratic functions

Recommended Time		
Part	**Pages**	**Time**
Investigation	554–555	1 class period
Working on the Investigation	586, 600, 617, 627	20 minutes each
Closing the Investigation	650	1 class period

Instructional Resources
Investigations and Projects Masters, pp. 17–20

A recording sheet, teacher notes, and scoring guide are provided for each investigation in the *Investigations and Projects Masters.*

1 MOTIVATION

This Investigation uses common materials to investigate the specifications given to design rectangular brick patios. Ask students if they have seen different designs of brick patios. Discuss the importance of having the correct measurements and the correct materials for each design.

the BRICKYARD

MATERIALS NEEDED

construction paper

scissors

ruler

You work for a construction company that specializes in brick patios. Your job is to create custom-designed patios. The company manufactures square and rectangular bricks. Recently, your manager sent your department the memo shown below.

In this Investigation, you must design brick patios that fit the specifications given in the memo. The design plans must be explicit and detailed, so that the construction crew can build them accordingly. Your design team consists of three people.

Make an Investigation Folder in which you can store all of your work on this Investigation for future use.

MEMO

To: Custom Design Department
From: Joanna Brown, Manager *JB*

We have a problem, and I need your help in solving it. We have an excess inventory of three types of bricks:
- small square bricks,
- large square bricks, and
- rectangular bricks that are as long as the large square brick and as wide as the small square brick.

We need to move this inventory, so I am asking you to investigate the possible patio design patterns using these three types of bricks.

I don't know if this is helpful, but the length of the larger brick is the same length as the diagonal of the smaller square brick.

Our other custom designs include using triangles, rectangles, and hexagons exclusively to create repeating patterns. However, for these surplus bricks, we need to concentrate on repeating rectangular patterns.

Please create several patio designs that will utilize these bricks. Submit at least three different plans, explaining the materials required for each patio. I am anxious to see the different ways in which these bricks can be arranged to form rectangular patios. Is there a general formula or pattern we can use to design these in the future? I look forward to your report on helping us solve our inventory problem.

 ## Cooperative Learning

This Investigation offers an excellent opportunity for using cooperative learning groups. For more information on cooperative learning strategies and group management, see *Cooperative Learning in the Mathematics Classroom,* one of the titles in the Glencoe Mathematics Professional Series.

PATIO SKETCH #____	DIMENSIONS	MATERIALS USED	number	total area
	small square: ____ × ____	bricks		
	large square: ____ × ____	sm. squares		
	rectangle: ____ × ____	lg. squares		
	design size: ____ × ____	rectangles		
		TOTAL		

CREATE MODELS

1 Copy the table above. You will use it, along with other tables, to record the data as you explore the designs that are possible.

2 Using the measurement requirements from Ms. Brown's memo, create a model of each of the three sizes of bricks.

3 After determining that your three models comply with the given specifications, use construction paper to make several copies of each model.

ANALYZE THE MODELS

4 Share the dimensions of your models with the other design teams in your class. Obviously the models of each design team will not be the same size, but each set of models must comply with the requirements Ms. Brown wrote in her memo.

5 Explain how the three sizes relate to each other. Make a chart listing the dimensions of each of the models from the other design teams. Do each set of dimensions relate in the same way that your dimensions do? Should they? Explain.

6 How do the lengths of the two square bricks relate? If you were to make one row of small squares and below it make one row of large squares, how many small squares would it take to match the exact length of the row of large squares? Explain your answer mathematically.

You will continue working on this Investigation throughout Chapters 10 and 11.

Be sure to keep your individual brick models, chart, and other materials in your Investigation Folder.

The Brickyard Investigation

Working on the Investigation
Lesson 10–4, p. 586

Working on the Investigation
Lesson 10–6, p. 600

Working on the Investigation
Lesson 11–1, p. 617

Working on the Investigation
Lesson 11–2, p. 627

Closing the Investigation
End of Chapter 11, p. 650

Investigation: The Brickyard **555**

2 SETUP

You may wish to have a student read the first two paragraphs of the Investigation to provide background information about the patio designs and specifications. You may then wish to read the memo to introduce the activity. Discuss the activity with your students. Then separate the class into groups of three.

3 MANAGEMENT

Each group member should be responsible for a specific task.

Recorder Collects data. Calculates dimensions.
Measurer Obtains measurement requirements from memo. Gathers materials.
Designer Creates models.

At the end of the activity, each member should turn in his or her respective equipment.

Sample Answers

Answers will vary as they are based on each design.

Investigations and Projects Masters, p. 20

10, 11 NAME_____ DATE_____

Investigation, Chapters 10 and 11 Student Edition Pages 554–555, 586, 600, 617, 627, 650

The Brickyard

Use this chart to record the data that you found from Working on the Investigation, page 627.

Situation 1

Dimensions			
Perimeter			
Area			

Situation 2

Dimensions			
Perimeter			
Area			

Situation 3

Dimensions			
Perimeter			
Area			

Work with your group to answer the following questions and add others to be considered.

- If you had two large square bricks, five small square bricks, and five rectangular bricks, what size rectangular patterns could you make?
- For each of the patterns that you designed, express the dimensions, perimeter, and area of each in terms of x and y.
- Which of the patterns would use up the inventory more quickly?

Using Factoring

PREVIEWING THE CHAPTER

This chapter covers factors and factoring. Students are introduced to factors and the greatest common factor. They then learn to factor using the distributive property. Students learn to factor trinomials and differences of squares. Finally, they learn to factor perfect square trinomials and apply all of their skills to solve quadratic equations.

Lesson (Pages)	Lesson Objectives	NCTM Standards	State/Local Objectives
10-1 (558–563)	Find prime factorizations of integers. Find greatest common factors (GCF) for sets of monomials.	1–5	
10-2A (564)	Use algebra tiles to factor binomials.	1–5	
10-2 (565–571)	Use the GCF and the distributive property to factor polynomials. Use grouping techniques to factor polynomials with four or more terms.	1–5	
10-3A (572–573)	Use algebra tiles to factor simple trinomials.	1–5	
10-3 (574–580)	Solve problems by using guess and check. Factor quadratic trinomials.	1–5	
10-4 (581–586)	Identify and factor binomials that are the differences of squares.	1–5	
10-5 (587–593)	Identify and factor perfect square trinomials.	1–5	
10-6 (594–600)	Use the zero product property to solve equations.	1–5	

ORGANIZING THE CHAPTER

You may want to refer to the **Course Planning Calendar** on page T12 for detailed information on pacing.
PACING: Standard—12 days; **Honors**—11 days; **Block**—6 days; **Two Years**—22 days

LESSON PLANNING CHART

| Lesson (Pages) | Materials/ Manipulatives | Extra Practice (Student Edition) | BLACKLINE MASTERS | | | | | | | | | | Real-World Applications | Interactive Mathematics Tools Software | Teaching Transparencies |
			Study Guide	Practice	Enrichment	Assessment and Evaluation	Modeling Mathematics	Multicultural Activity	Tech Prep Applications	Graphing Calculator	Science and Math Lab Manual			
10-1 (558–563)		p. 778	p. 69	p. 69	p. 69			p. 19					10-1	10-1A 10-1B
10-2A (564)	algebra tiles* product mat*						p. 32							
10-2 (565–571)		p. 779	p. 70	p. 70	p. 70	p. 268			p. 19					10-2A 10-2B
10-3A (572–573)	algebra tiles* product mat*						p. 33						10-3A	
10-3 (574–580)	graphing calculator	p. 779	p. 71	p. 71	p. 71	pp. 267, 268	pp. 63–65							10-3A 10-3B
10-4 (581–586)	algebra tiles* product mat*	p. 779	p. 72	p. 72	p. 72							25		10-4A 10-4B
10-5 (587–593)		p. 780	p. 73	p. 73	p. 73	p. 269	pp. 66–68, 81					26		10-5A 10-5B
10-6 (594–600)		p. 780	p. 74	p. 74	p. 74	p. 269		p. 20	p. 20	p. 10	pp. 43–48		10-6	10-6A 10-6B
Study Guide/ Assessment (601–605)						pp. 253–266, 270–272								

*Included in Glencoe's Student Manipulative Kit and Overhead Manipulative Resources.

ORGANIZING THE CHAPTER

OTHER CHAPTER RESOURCES

Student Edition
Investigation, pp. 554–555
Chapter Opener, pp. 556–557
Mathematics and Society,
    p. 563
Working on the Investigation,
    pp. 586, 600

Teacher's Classroom Resources
Investigations and Projects Masters,
    pp. 61–64
Algebra and Geometry Overhead
    Manipulative Resources,
    pp. 31–36

Technology
Test and Review Software (IBM
    and Macintosh)
CD-ROM Interactions (Windows
    and Macintosh)

Professional Publications
Block Scheduling Booklet
Glencoe Mathematics Professional
    Series

OUTSIDE RESOURCES

Books/Periodicals
The Ideas of Algebra, K–12, NCTM
Olson, Melfried, "A Geometric Look at Greatest
    Common Divisor," *Mathematics Teacher*, March
    1991

Software
Mathematics Curriculum and Teaching Program,
    NCTM

Videos/CD-ROMs
Mathematics Assessment: Alternative Approaches,
    NCTM

ASSESSMENT RESOURCES

Student Edition
Math Journal, pp. 561, 591
Mixed Review, pp. 563, 571,
    580, 585, 593, 600
Self Test, p. 580
Chapter Highlights, p. 601
Chapter Study Guide and
    Assessment, pp. 602–604
Alternative Assessment, p. 605
    Portfolio, p. 605

Cumulative Review, pp. 606–607

Teacher's Wraparound Edition
5-Minute Check, pp. 558, 565,
    574, 581, 587, 594
Check for Understanding, pp. 561,
    568, 578, 584, 591, 598
Closing Activity, pp. 562, 571,
    580, 586, 593, 600
Cooperative Learning, pp. 560,
    566

Assessment and Evaluation Masters
Multiple-Choice Tests, Forms 1A
    (Honors), 1B (Average), 1C
    (Basic), pp. 253–258
Free-Response Tests, Forms 2A
    (Honors), 2B (Average), 2C
    (Basic), pp. 259–264
Calculator-Based Test, p. 265
Performance Assessment, p. 266
Mid-Chapter Test, p. 267
Quizzes A–D, pp. 268–269
Standardized Test Practice, p. 270
Cumulative Review, pp. 271–272

Examples of some of the materials for enhancing Chapter 10 are shown below.

DIVERSITY

Multicultural Activity Masters, pp. 19, 20

Real-World Applications, 25, 26

TECHNOLOGY

Graphing Calculator Masters, p. 10

TECH PREP

Tech Prep Applications Masters, pp. 19, 20

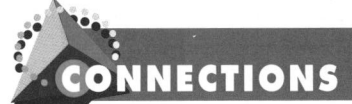

Science and Math Lab Manual, pp. 43–48

PROBLEM SOLVING

Problem of the Week Cards, 28, 29

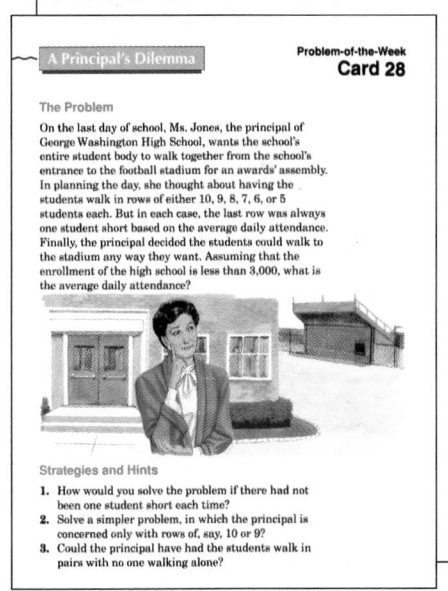

Using Factoring

The Impact of the Media

Objectives

In this chapter, you will:

- find the prime factorization of integers,
- find the greatest common factors (GCF) for sets of monomials,
- factor polynomials,
- solve problems by using guess and check, and
- use the zero product property to solve equations.

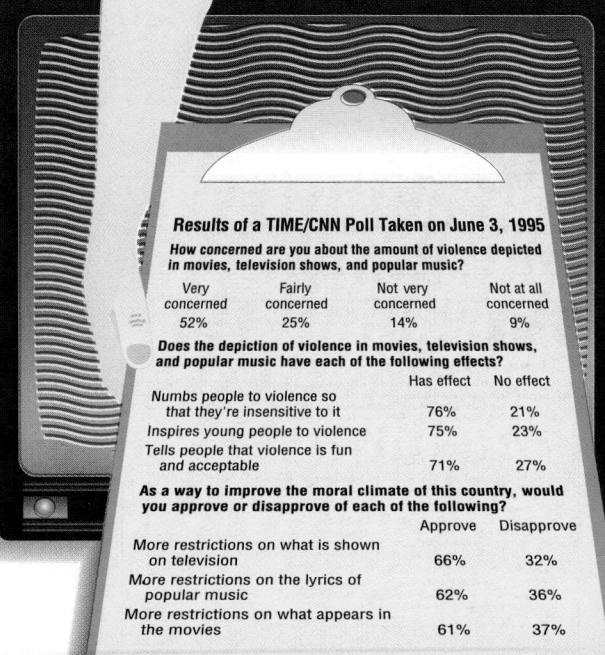

Results of a TIME/CNN Poll Taken on June 3, 1995

How concerned are you about the amount of violence depicted in movies, television shows, and popular music?

Very concerned	Fairly concerned	Not very concerned	Not at all concerned
52%	25%	14%	9%

Does the depiction of violence in movies, television shows, and popular music have each of the following effects?

	Has effect	No effect
Numbs people to violence so that they're insensitive to it	76%	21%
Inspires young people to violence	75%	23%
Tells people that violence is fun and acceptable	71%	27%

As a way to improve the moral climate of this country, would you approve or disapprove of each of the following?

	Approve	Disapprove
More restrictions on what is shown on television	66%	32%
More restrictions on the lyrics of popular music	62%	36%
More restrictions on what appears in the movies	61%	37%

Source: *Time Magazine,* June 12, 1995

Is American culture too violent? Do movies, television, magazines, and music reflect a true picture of America or do they contribute to the violence in our culture? What impact does the media have on American youth?

TIME*Line*

575 Ishtar Gate in Babylon is built.

1519 Ferdinand Magellan leaves Europe to sail around the world.

| 9000 B.C. | 8500 | 8000 | 7500 | 1000 | 500 | 0 | A.D.500 | 1000 | 1500 | 1600 | 1650 |

8000 B.C. People in Mesopotamia use clay tokens to record numbers of animals and amounts of grain.

A.D. 705 Ch'ang-an Pagoda in China is completed.

1663 Isaac Newton discovers the binomial theorem.

TIME*Line*

Students might want to study the 1663 discovery of the binomial theorem by John Newton. Like many mathematical discoveries, it was invented for just one purpose, but later many more applications were found.

interNET
CONNECTION

Learn about such topics as "Caught in the Crossfire" and "Healthy Children, Healthy Learning" on PBS's quarterly Merrow Report.

World Wide Web
http://www.merrow.oa.net:8003/

The title of **Robert Rodriquez's** new book is *Rebel Without a Crew: Or How a 23-Year-Old Filmmaker with $7,000 Became a Hollywood Player*. It is the story of how the University of Texas film student from Austin, Texas, made a feature film on a very small budget. He used friends as actors, wrote the script, directed the 14-day shoot, and handled the camera work as a one-man crew. The film, *El Mariachi*, went on to win the Audience Award at the Sundance Film Festival, was released by Columbia, and is now on video.

Robert's second film, a full budget production, was *Desperado*, released in 1995. He told Columbia that he would sign a contract if he could stay in Texas, near his family and his inspiration. His advice to future film makers, "Grab your camera and just do it."

Chapter Project

- Pick five of your favorite movies. List five hints for each movie that will help your classmates to guess the names of the movies.

- Exchange your list of hints with a classmate. Try to guess your classmate's favorite movies. Explain how each hint helped you to eliminate some movies and to concentrate on others. Did you guess your classmate's favorite movies correctly?

- Explain how guessing can help when factoring polynomials.

- List five polynomials for one of your classmates to factor. Make sure that one of your polynomials cannot be factored.

- Exchange your polynomials with a classmate and factor the polynomials on the list you receive.

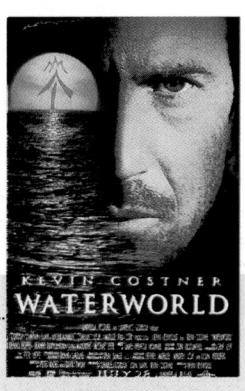

1995 *Waterworld*, which cost a record $175,000,000 to produce, is released.

1939 Marian Anderson gives a concert for 75,000 at the Lincoln Memorial.

1975 The VHS format (Video Home System) is launched by the Japanese company JVC.

Chapter 10 **557**

Alternative Chapter Projects

Two other chapter projects are included in the *Investigations and Projects Masters*. In Chapter 10 Project A, pp. 61–62, students extend the topic in the chapter opener. In Chapter 10 Project B, pp. 63–64, students research formulas and their use.

Rodriguez lists Alfred Hitchcock, Sergio Leone, and Sam Raimi as his mentors. The plot of *El Mariachi* is that of a typical action film, but the camera work (Rodriguez is the cameraman) is innovative. Because of the success of his films, Hollywood is interested in him. But, as is often the case, studio executives wanted to change his movie ideas to resemble other Hollywood blockbuster movies. Rodriguez plans to remain true to his own cinematic vision.

Chapter Project

There is a difference between guessing and intelligent guessing. Intelligent guessing requires knowing in advance what counts as a good guess and a bad guess. Factoring polynomials involves intelligent guessing.

Investigations and Projects Masters, p. 61

---

10   NAME_____ DATE_____

Chapter 10 Project A   Student Edition Pages 558–600

Movie Numbers

1. Numbers often appear in the titles of movies. Examples include *101 Dalmatians, Apollo 13,* and *Twelve Monkeys*. For this project, you will work in a small group to make a list of at least 20 such movies. Use newspaper movie listings for current films; visit a video store or consult a video catalogue for older titles.

2. Determine which of the numbers in your titles are prime and which are composite.

3. Find the prime factorization for each composite number in your list.

4. Create a board game in which players advance their game pieces by guessing movie titles with numbers in them. Write clues for each movie title you listed in exercise 1. For example, the clues for *Apollo 13* might be "Greek god of the sun" and "1 · 13," and those for *Twelve Monkeys* might be "2² · 3" and "small, lively primates that live together in social groups."

5. Exchange games with other groups and play them.

NCTM Standards: 1–5

Instructional Resources

- Study Guide Master 10-1
- Practice Master 10-1
- Enrichment Master 10-1
- Multicultural Activity Masters, p. 19

Transparency 10-1A contains the 5-Minute Check for this lesson; **Transparency 10-1B** contains a teaching aid for this lesson.

Recommended Pacing

Standard Pacing	Day 1 of 12
Honors Pacing	Day 1 of 11
Block Scheduling*	Day 1 of 6
Alg. 1 in Two Years*	Days 1, 2, & 3 of 22

*For more information on pacing and possible lesson plans, refer to the *Block Scheduling Booklet* and *Algebra 1 in Two Years*.

1 FOCUS

5-Minute Check
(over Chapter 9)

Find each product.

1. $(a + 2b)^2$ $a^2 + 4ab + 4b^2$
2. $(x - 3r)^2$ $x^2 - 6rx + 9r^2$
3. Simplify $\frac{63x^5}{9x^2}$. $7x^3$
4. Express $(2 \times 10^4)(1.5 \times 10^5)$ in scientific notation.
   3×10^9
5. Simplify $(8x^3 - 2x^2 + 7) - (3x^3 + 5x - 2)$.
   $5x^3 - 2x^2 - 5x + 9$

Motivating the Lesson

Questioning Introduce the lesson by having students identify factors of numbers such as 12, 36, 64, and 100.
$12 = 2 \times 2 \times 3$
$36 = 2 \times 2 \times 3 \times 3$
$64 = 2 \times 2 \times 2 \times 2 \times 2 \times 2$
$100 = 2 \times 2 \times 5 \times 5$

10-1

Factors and Greatest Common Factors

Objectives

After studying this lesson, you should be able to:
- find prime factorizations of integers, and
- find greatest common factors (GCF) for sets of monomials.

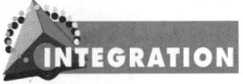
INTEGRATION
Geometry

Suppose you were asked to use grid paper to draw all of the possible rectangles with whole number dimensions that have areas of 12 square units each. The figure at the right shows one possible drawing.

Rectangles *A* and *B* are both 3 by 4 and, therefore, can be considered the same. Likewise, Rectangles *C* and *D* and Rectangles *E* and *F* are considered the same.

Recall that when two or more numbers are multiplied to form a product, each number is a *factor* of the product. In the example above, 12 is expressed as the product of different pairs of whole numbers.

$12 = 3 \times 4$	$12 = 2 \times 6$	$12 = 12 \times 1$
$12 = 4 \times 3$	$12 = 6 \times 2$	$12 = 1 \times 12$

The whole numbers 1, 2, 3, 4, 6, and 12 are factors of 12.

Example **Find the factors of 72.**

To find the factors of 72, list all the pairs of numbers whose product is 72.

1×72	2×36	3×24	4×18	6×12	8×9

Therefore, the factors of 72, in increasing order, are 1, 2, 3, 4, 6, 8, 9, 12, 18, 24, 36, and 72.

Some whole numbers have exactly two factors, the number itself and 1. These numbers are called **prime numbers**. Whole numbers that have more than two factors are called **composite numbers**.

Definitions of Prime and Composite Numbers	**A prime number is a whole number, greater than 1, whose only factors are 1 and itself. A composite number is a whole number, greater than 1, that is not prime.**

0 and 1 are neither prime nor composite.

F Y I

The greatest known prime number was discovered by computer scientists Paul Gage and David Slowinski in 1994, using a Cray supercomputer. The number has 258,716 digits and is expressed as $2^{859,433} - 1$.

The number 6 is a factor of 12, but not a *prime factor* of 12, since 6 is not a prime number. When a whole number is expressed as a product of factors that are all prime numbers, the expression is called the **prime factorization** of the number. Thus, the prime factorization of 12 is $2 \cdot 2 \cdot 3$ or $2^2 \cdot 3$.

The prime factorization of every number is unique except for the order in which the factors are written. For example, $2 \cdot 3 \cdot 2$ is also a prime factorization of 12, but it is the same as $2 \cdot 2 \cdot 3$. This property of numbers is called the **unique factorization theorem**.

558   *Chapter 10   Using Factoring*

F Y I

Investigating prime numbers belongs to a branch of mathematics called number theory. This field of research began around 600 B.C. At that time it was associated with numerology—a field that investigated the mystical properties of numbers.

GLENCOE *Technology*

Interactive Mathematics Tools Software

This multimedia software provides an interactive lesson that uses the arrangement of cells to find common factors and greatest common factors of two numbers. A **Computer Journal** gives students an opportunity to write about what they have learned.

For Windows & Macintosh

Example 2 Find the prime factorization of 140.

Method 1

$140 = 2 \cdot 70$ *The least prime factor of 140 is 2.*

$= 2 \cdot 2 \cdot 35$ *The least prime factor of 70 is 2.*

$= 2 \cdot 2 \cdot 5 \cdot 7$ *The least prime factor of 35 is 5.*

All the factors in the last row are prime. Thus, the prime factorization of 140 is $2 \cdot 2 \cdot 5 \cdot 7$ or $2^2 \cdot 5 \cdot 7$.

Method 2

Use a factor tree.

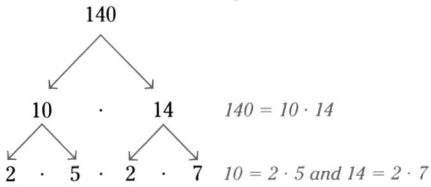

140 = 10 · 14

10 = 2 · 5 and 14 = 2 · 7

All of the factors in the last branch of the factor tree are prime. Thus, the prime factorization of 140 is $2 \cdot 2 \cdot 5 \cdot 7$ or $2^2 \cdot 5 \cdot 7$.

A negative integer is factored completely when it is expressed as the product of -1 and prime numbers.

Example 3 Factor -150 completely.

$-150 = -1 \cdot 150$ *Express -150 as -1 times 150.*

$= -1 \cdot 2 \cdot 75$ *Find the prime factors of 150.*

$= -1 \cdot 2 \cdot 3 \cdot 25$

$= -1 \cdot 2 \cdot 3 \cdot 5 \cdot 5$ or $-1 \cdot 2 \cdot 3 \cdot 5^2$

A monomial is in **factored form** when it is expressed as the product of prime numbers and variables and no variable has an exponent greater than 1.

Example 4 Factor $45x^3y^2$.

$45x^3y^2 = 3 \cdot 15 \cdot x \cdot x \cdot x \cdot y \cdot y$

$= 3 \cdot 3 \cdot 5 \cdot x \cdot x \cdot x \cdot y \cdot y$

Two or more numbers may have some common factors. Consider the numbers 84 and 70, for example.

Factors of 84: 1, 2, 3, 4, 6, 7, 12, 14, 21, 28, 42, 84

Factors of 70: 1, 2, 5, 7, 10, 14, 35, 70

There are some factors that appear on both lists. The greatest of these numbers is 14, which is called the **greatest common factor (GCF)** of 84 and 70.

Definition of Greatest Common Factor	The greatest common factor of two or more integers is the greatest number that is a factor of all of the integers.

Lesson 10–1 Factors and Greatest Common Factors **559**

2 TEACH

In-Class Examples

For Example 1
Find the factors of 84.
1, 2, 3, 4, 6, 7, 12, 14, 21, 42, 84

For Example 2
Find the prime factorization of each number.

a. 200 $2 \times 2 \times 2 \times 5 \times 5$
b. 650 $2 \times 5 \times 5 \times 13$
c. 420 $2 \times 2 \times 3 \times 5 \times 7$
d. 168 $2 \times 2 \times 2 \times 3 \times 7$

For Example 3
Factor each number completely.

a. -210 $-1 \times 2 \times 3 \times 5 \times 7$
b. -448 $-1 \times 2 \times 2 \times 2 \times 2 \times 2 \times 2 \times 7$

For Example 4
Factor each monomial.

a. $45a^2b^2$
   $3 \times 3 \times 5 \times a \times a \times b \times b$
b. $77ab^2$ $7 \times 11 \times a \times b \times b$

Teaching Tip The order of prime factors in a prime factorization is not critical. Usually the factors are ordered from the least prime factor to the greatest.

 Alternative Learning Styles

Kinesthetic Give students 72 marbles or counters. Their task is to group the marbles or counters into smaller groupings, such as 8 groups of 9 marbles or counters. Point out that 8 groups of 9 marbles represents 8×9. Therefore, 72 can be factored into 8×9. But 72 is not uniquely factorable. Have them explore other groupings.

CAREER CHOICES

In 1992 there were 3.1 million people working as chefs and cooks.

Short-order & fast-food cooks	714,000
Restaurant cooks	602,000
Institutional cooks	406,000
Bread and pastry bakers	146,000
Other kitchen workers	1,233,000

There is an easier way to find the GCF of numbers without having to find all of their factors. Look at the prime factorizations of the numbers, and multiply all the prime factors they have in common. *If there are no common prime factors, the GCF is 1.*

$$84 = 2 \cdot 2 \cdot 3 \cdot 7 \qquad 70 = 2 \cdot 5 \cdot 7$$

The integers 84 and 70 have 2 and 7 as common prime factors. The product of these common prime factors is 14, the GCF of 84 and 70.

Example 5 **Find the GCF of 54, 63, and 180.**

$54 = 2 \cdot 3 \cdot \textcircled{3} \cdot \textcircled{3}$ *Factor each number.*

$63 = \textcircled{3} \cdot \textcircled{3} \cdot 7$ *Circle the common factors.*

$180 = 2 \cdot 2 \cdot \textcircled{3} \cdot \textcircled{3} \cdot 5$

The GCF of 54, 63, and 180 is $3 \cdot 3$ or 9.

Example 6

APPLICATION
Packaging

A bakery packages its fat-free cookies in two sizes of boxes. One box contains 18 cookies, and the other contains 24 cookies. In order to keep the cookies fresh, the bakery plans to wrap a smaller number of cookies in cellophane before they are placed in the boxes. To save money, the bakery wants to use the same size cellophane packages for each box and to place the greatest possible number of cookies in each cellophane package.

a. How many cookies should the bakery place in each cellophane package?

b. How many cellophane packages will go in each size of box?

a. Find the GCF of 18 and 24.

$18 = \textcircled{2} \cdot 3 \cdot \textcircled{3}$

$24 = \textcircled{2} \cdot 2 \cdot 2 \cdot \textcircled{3}$

The bakery should put $2 \cdot 3$ or 6 cookies in each inner cellophane package.

b. The box of 18 cookies will contain $18 \div 6$ or 3 cellophane packages. The box of 24 cookies will contain $24 \div 6$ or 4 cellophane packages.

The GCF of two or more monomials is the product of their common factors, when each monomial is expressed in factored form.

Example 7 **Find the GCF of $12a^2b$ and $90a^2b^2c$.**

$12a^2b = \textcircled{2} \cdot 2 \cdot \textcircled{3} \cdot \textcircled{a} \cdot \textcircled{a} \cdot \textcircled{b}$ *Factor each monomial.*

$90a^2b^2c = \textcircled{2} \cdot 3 \cdot \textcircled{3} \cdot 5 \cdot \textcircled{a} \cdot \textcircled{a} \cdot \textcircled{b} \cdot b \cdot c$ *Circle the common factors.*

The GCF of $12a^2b$ and $90a^2b^2c$ is $2 \cdot 3 \cdot a \cdot a \cdot b$ or $6a^2b$.

560 *Chapter 10 Using Factoring*

Cooperative Learning

Pairs Check Separate the class into pairs. Give each pair 20 index cards. Students should write two numbers on each of 10 cards and the GCF of the numbers on separate index cards. Then ask them to exchange cards. Tell students to shuffle the cards and place them face down on a flat surface. Have students take turns flipping over two cards to match numbers and their GCF.

If a match is made, the student keeps the cards. The student holding the greatest number of cards once all the cards have been picked up is the winner. For more information on the pairs check strategy, see *Cooperative Learning in the Mathematics Classroom,* one of the titles in the Glencoe Mathematics Professional Series, pages 12–13.

Communicating Mathematics

Study this lesson. Then complete the following.

1. **Draw** and label as many rectangles as possible with whole number dimensions that have an area of 48 square inches.   See margin.

2. Is $2 \cdot 3^2 \cdot 4$ the prime factorization of 72? Why or why not?   No; 4 is not prime.

3. If the GCF of two numbers is 1, must the numbers be prime? Explain.

 MATH JOURNAL

4. How many prime numbers do you believe there are? Write a statement to support your opinion.   See students' work.

3. No; the GCF of 4 and 9 is 1, but neither number is prime.

Guided Practice

Find the factors of each number.

5. 4   1, 2, 4

6. 56   1, 2, 4, 7, 8, 14, 28, 56

State whether each number is *prime* or *composite*. If the number is composite, find its prime factorization.

7. 89   prime

8. 39   composite; $3 \cdot 13$

Factor each expression completely. Do not use exponents.

9. -30 $-1 \cdot 2 \cdot 3 \cdot 5$

10. $22m^2n$ $2 \cdot 11 \cdot m \cdot m \cdot n$

Find the GCF of the given monomials.   16. $6a^2$

11. 4, 12   4

12. 10, 15   5

13. $24d^2, 30c^2d$ $6d$

14. 18, 35   1

15. $-20gh, 36g^2h^2$ $4gh$

16. $30a^2, 42a^3, 54a^3b$

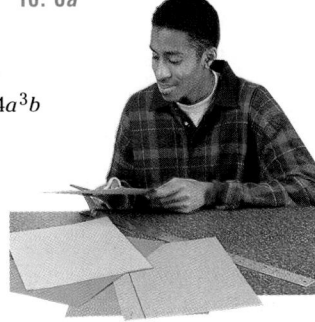

17. **Geometry** Suppose Terrell cuts out a rectangle that has an area of 96 square inches. If the length and width are both whole numbers, what is the minimum perimeter of the rectangle? Explain how you found the answer.   40 in.

20. 1, 2, 3, 4, 6, 9, 12, 18, 36

21. 1, 2, 4, 5, 8, 10, 16, 20, 40, 80

22. 1, 2, 4, 5, 8, 10, 16, 20, 25, 40, 50, 80, 100, 200, 400

Practice

A

Find the factors of each number.

18. 25   1, 5, 25

19. 67   1, 67

20. 36

21. 80

22. 400

23. 950

23. 1, 2, 5, 10, 19, 25, 38, 50, 95, 190, 475, 950

28. composite; $2^4 \cdot 19$

29. composite; $2^2 \cdot 5 \cdot 7 \cdot 11$

33. $2 \cdot 2 \cdot b \cdot b \cdot b \cdot d \cdot d$

34. $-1 \cdot 2 \cdot 3 \cdot 17 \cdot x \cdot x \cdot x \cdot y$

35. $-1 \cdot 2 \cdot 7 \cdot 7 \cdot a \cdot a \cdot b$

State whether each number is *prime* or *composite*. If the number is composite, find its prime factorization.

24. 17   prime

25. 63   composite; $3^2 \cdot 7$

26. 91   composite; $7 \cdot 13$

27. 97   prime

28. 304

29. 1540

Factor each expression completely. Do not use exponents.

30. -70 $-1 \cdot 2 \cdot 5 \cdot 7$

31. -117 $-1 \cdot 3 \cdot 3 \cdot 13$

32. $66z^2$ $2 \cdot 3 \cdot 11 \cdot z \cdot z$

33. $4b^3d^2$

34. $-102x^3y$

35. $-98a^2b$

B

3 PRACTICE/APPLY

Check for Understanding
Exercises 1–17 are designed to help you assess your students' understanding through reading, writing, speaking, and modeling. You should work through Exercises 1–4 with your students and then monitor their work on Exercises 5–17.

Assignment Guide
Core: 19–63 odd, 64–74 **Enriched:** 18–60 even, 61–74

For **Extra Practice,** see p. 778.

The red A, B, and C flags, printed only in the Teacher's Wraparound Edition, indicate the level of difficulty of the exercises.

Additional Answer

1.

Study Guide Masters, p. 69

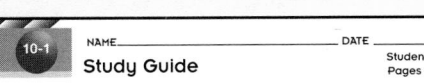

Reteaching

Using Alternative Methods To find the prime factorization of an integer, use "factor stairs." Divide the integer by its least prime factor. Repeat for the quotient. When you reach 1, the factorization is complete. The answer is on the steps.
$252 = 2 \cdot 2 \cdot 3 \cdot 3 \cdot 7$

Alternative Teaching Strategies

Student Diversity Many students confuse the GCF with the LCM. It may help them if they remember that any factor of *A* is *less* than *A*, even the *greatest* common factor. Any multiple of *A* is *greater* than *A*, even the *least* common multiple.

Closing Activity

Speaking Have students explain how they would determine whether 1237 is a prime number.

Using the Programming Exercises The program given in Exercise 60 is for use with a TI-82 graphing calculator. For other programmable calculators, have students consult their owner's manual for commands similar to those presented here.

Additional Answer

61.

1 cm, $8b^2$ cm; b cm, $8b$ cm; b^2 cm, 8 cm; $4b^2$ cm, 2 cm; $4b$ cm, $2b$ cm; $2b^2$ cm, 4 cm

---

Practice Masters, p. 69

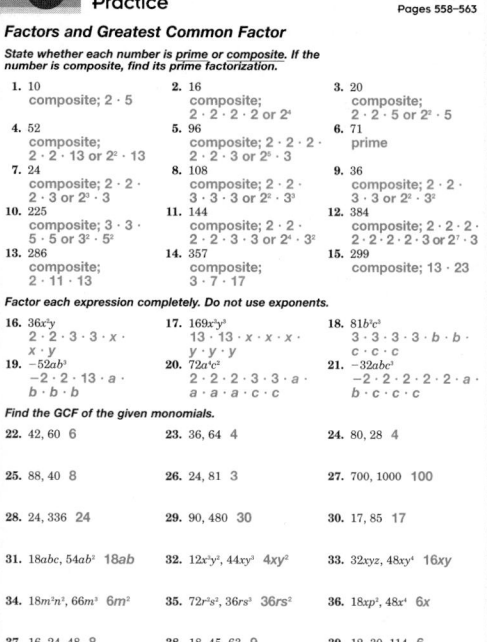

Find the GCF of the given monomials.

36. 18, 36   **18**
37. 18, 45   **9**
38. 84, 96   **12**
39. 28, 75   **1**
40. −34, 51   **17**
41. 95, −304   **19**
42. $17a, 34a^2$ **$17a$**
43. $21p^2q, 35pq^2$ **$7pq$**
44. $12an^2, 40a^4$ **$4a$**
45. $-60r^2s^2t^2, 45r^3t^3$ **$15r^2t^2$**
46. 18, 30, 54   **6**
47. 24, 84, 168   **12**
48. $14a^2b^3, 20a^3b^2c, 35ab^3c^2$ **ab^2**
49. $18x^2, 30x^3y^2, 54y^3$ **6**
50. $14a^2b^2, 18ab, 2a^3b^3$ **$2ab$**
51. $32m^2n^3, 8m^2n, 56m^3n^2$ **$8m^2n$**

Find each missing factor.

52. $42a^2b^5c = 7a^2b^3(\underline{\ ?\ })$ **$6b^2c$**
53. $-48x^4y^2z^3 = 4xyz(\underline{\ ?\ })$ **$-12x^3yz^2$**
54. $48a^5b^5 = 2ab^2(4ab)(\underline{\ ?\ })$ **$6a^3b^2$**
55. $36m^5n^7 = 2m^3n(6n^5)(\underline{\ ?\ })$ **$3m^2n$**

56. 1 in. by 116 in., 2 in. by 58 in., 4 in. by 29 in.

56. Geometry The area of a rectangle is 116 square inches. What are its possible whole number dimensions?

57. Geometry The area of a rectangle is 1363 square centimeters. If the measures of the length and width are both prime numbers, what are the dimensions of the rectangle?   **29 cm by 47 cm**

58. See students' work.

58. Number Theory Check to see if your house number is a prime number and if the last four digits in your telephone number form a prime number. Explain how you decided.

59. Number Theory *Twin primes* are two consecutive odd numbers that are prime, such as 11 and 13. List the twin primes where both primes are less than 100.   **3, 5; 5, 7; 11, 13; 17, 19; 29, 31; 41, 43; 59, 61; 71, 73**

Programming

60. The graphing calculator program below can be used to find the GCF of two numbers.

```
PROGRAM:GCF
: Input "INTEGER",A     : Goto 4
: Input "INTEGER",B     : A-B→A
: A→E                   : Goto R
: B→F                   : Lbl 4
: Lbl R                 : B-A→B
: If A=B                : Goto R
: Goto 5                : Lbl 5
: If A<B                : Disp "GCF IS", A
```

Use the program to find the GCF of each pair of numbers.

a. 896, 700   **28**
b. 1015, 3132   **29**
c. 567, 416   **1**
d. 486, 432   **54**
e. 891, 1701   **81**
f. 1105, 1445   **85**

Critical Thinking

61. Geometry Draw and label all the rectangles that satisfy the following conditions. The area of a rectangle is $8b^2$ square centimeters and the measure of each of its sides can be expressed as a monomial with integral coefficients. (*Hint:* There are 6.)   **See margin.**

562   Chapter 10   Using Factoring

Extension

Problem Solving Find the GCF of the given monomials so the second factor has integral coefficients.

$$\left(\frac{3}{8}\right)a^3b^3c^7, \left(\frac{7}{20}\right)a^2b^2c^6, \left(\frac{1}{12}\right)ab^6c^4$$

$$\frac{1}{120}ab^2c^4$$

Applications and Problem Solving

62. Gardening Marisela is planning to have 100 tomato plants in her garden. In what ways can she arrange them so that she has the same number of plants in each row, at least 5 rows of plants, and at least 5 plants in each row? **5 rows of 20 plants, 10 rows of 10 plants, 20 rows of 5 plants**

63. Sports A new athletic field is being sodded at Beck High School using 2-yard-by-2-yard squares of sod. If the length of the field is 70 yards longer than the width and its area is 6000 square yards, how many squares of sod will be needed? **1500 squares of sod**

Mixed Review

64. Find $(1.1x + y)^2$. (Lesson 9–8) **$1.21x^2 + 2.2xy + y^2$**

65. Simplify $\frac{12b^5}{4b^4}$. (Lesson 9–2) **$3b$**

66. Solve the system of equations by graphing. (Lesson 8–1) **(0, 0)**
$y = -x$
$y = 2x$

67. Graph the compound inequality $y > 2$ or $y < 1$. (Lesson 7–4) **See margin.**

68. Solve $16x < 96$. Check your solution. (Lesson 7–2) **$x < 6$**

69. Write the standard form of an equation of the line that passes through $(4, -2)$ and $(4, 8)$. (Lesson 6–2) **$x = 4$**

70. Graph $8x - y = 16$. (Lesson 5–4) **See margin.**

71. $9\frac{3}{5}$ feet

71. Physics Weights of 50 pounds and 75 pounds are placed on a lever. The two weights are 16 feet apart, and the lever is balanced. How far from the fulcrum is the 50-pound weight? (Lesson 4–8)

72. 37.8 feet

72. Waves The highest wave ever sighted and recorded was 112 feet high. This wave was brought on by a wind of 74 mph. Using ratios, determine how high a wave brought on by a 25-mph wind could reach. (Lesson 4–1)

74. the number z to the seventh power added to 2

73. Solve $9 = x + 13$. (Lesson 3–1) **-4**

74. Write a verbal expression for $z^7 + 2$. (Lesson 1–1)

Number Sieve

The article below appeared in *Science News* on October 1, 1994.

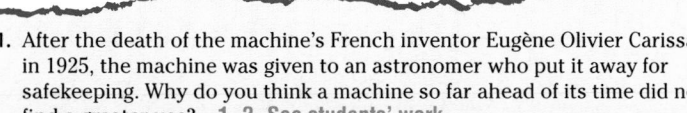

IT LOOKS LIKE A CROSS BETWEEN AN antique music box and an old-fashioned, hand-cranked phonograph. But no music emanates from the contraption. Instead, this ingenious mechanical device operates as a number sieve. It automatically sifts through arrays of numbers to identify certain patterns. From these data, mathematicians can determine whether a given number is a prime or the product of two or more primes multiplied together. Constructed 75 years ago, it also represents the first known, successful attempt to automate the factoring of whole numbers. Until three researchers tracked down the machine recently, few people knew of its existence. Now, this unique device can take its proper place in the history of computational number theory. ■

1. After the death of the machine's French inventor Eugène Olivier Carissan in 1925, the machine was given to an astronomer who put it away for safekeeping. Why do you think a machine so far ahead of its time did not find a greater use? **1–2. See students' work.**

2. One of the main uses of prime numbers today is in cryptography, the coding and decoding of data and messages. Why are more sophisticated codes needed today than they were in the 1920s?

Mathematics and SOCIETY

This machine, invented and built by Eugene Olivier Carissan, is about the size of a laptop computer. It consists of 14 brass gears of differing sizes. The machine works on a trial-and-error method, called the sieve method. It tries to divide the number by a smaller number to see whether it is or is not a factor.

Objective
Use algebra tiles to factor binomials.

Recommended Time
Demonstration and discussion: 15 minutes; Exercises: 30 minutes

Instructional Resources
For each student or group of students
Student Manipulative Kit
• algebra tiles
• product mat
Modeling Mathematics Masters
• p. 1 (algebra tiles)
• p. 15 (product mat)
• p. 32 (worksheet)
For teacher demonstration
Algebra and Geometry Overhead Manipulative Resources

1 FOCUS

Motivating the Lesson
The number 63 can be modeled by a rectangle of dimension 7 units by 9 units. Any product can be modeled by the area of a rectangle.

2 TEACH

Teaching Tip Students must form a rectangle with the given tiles. If a rectangle cannot be formed, the binomial is not factorable.

Teaching Tip Some students may want to add zero pairs to make the rectangle. Encourage this type of creative thinking.

3 PRACTICE/APPLY

Assignment Guide

Core: 1–9
Enriched: 1–9

MODELING MATHEMATICS

A Preview of Lesson 10–2

10–2A Factoring Using the Distributive Property

Materials: algebra tiles ☐ product mat

When two or more numbers are multiplied, these numbers are factors of the product. Sometimes you know the product of binomials and are asked to find the factors. This is called **factoring.** You can use algebra tiles to factor binomials.

Activity 1   Use algebra tiles to factor $2x + 8$.

Step 1 Model the polynomial $2x + 8$.

Step 2 Arrange the tiles into a rectangle. The total area of the tiles represents the product and its length and width represent the factors.

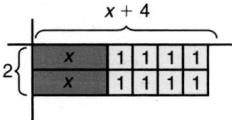

The rectangle has a width of 2 and a length of $x + 4$. Therefore, $2x + 8 = 2(x + 4)$.

Activity 2   Use algebra tiles to factor $x^2 - 3x$.

Step 1 Model the polynomial $x^2 - 3x$.

Step 2 Arrange the tiles into a rectangle.

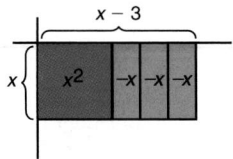

The rectangle has a width of x and a length of $x - 3$. Therefore, $x^2 - 3x = x(x - 3)$.

9. Binomials can be factored if they can be represented by a rectangle. Examples: $4x + 4$ can be factored and $4x + 3$ cannot be factored.

Model   Use algebra tiles to factor each binomial.

1. $3x + 9$
   $3(x + 3)$
2. $4x - 10$
   $2(2x - 5)$
3. $3x^2 + 4x$
   $x(3x + 4)$
4. $10 - 5x$
   $5(2 - x)$

Draw   Tell whether each binomial can be factored. Justify your answer with a drawing.

5. $2x + 3$ no
6. $3 - 9x$ yes
7. $x^2 - 5x$ yes
8. $3x^2 + 5$ no

5–8. See Solutions Manual for drawings.

Write   9. Write a paragraph that explains how you can determine whether a binomial can be factored. Include an example of one binomial that can be factored and one that cannot.

4 ASSESS

Observing students working in cooperative groups is an excellent method of assessment.

10-2

Factoring Using the Distributive Property

Objectives

After studying this lesson, you should be able to:

- use the greatest common factor (GCF) and the distributive property to factor polynomials, and
- use grouping techniques to factor polynomials with four or more terms.

APPLICATION
Sports

Rugby is a contact sport in which each of two teams tries to get an oval ball behind its opponent's goal line or kick it over its opponent's goal. It is similar to American football. However, the action in the game is almost nonstop, and the players wear little protective gear.

There are two versions of rugby—Rugby Union and Rugby League. Rugby Union is popular in Australia, Canada, England, France, Ireland, Japan, New Zealand, Scotland, South Africa, and Wales. It is played on a rectangular field.

LOOK BACK

You can refer to Lesson 1-7 for information on the distributive property.

If the width of this field is represented by x, its length can be represented by $x + 75$ and the area of the field by $x(x + 75)$, or $x^2 + 75x$. If $x(x + 75) = x^2 + 75x$, then $x^2 + 75x = x(x + 75)$.   *Why?*

The expression $x(x + 75)$ is called the *factored form* of $x^2 + 75x$. A polynomial is in factored form, or **factored,** when it is expressed as the product of monomials and polynomials.

In Chapter 9, you multiplied a polynomial by a monomial using the distributive property. You can also reverse this process and express a polynomial in factored form by using the distributive property.

GLOBAL CONNECTIONS

Rugby is named for Rugby School in England where the game was first played in 1823. The game has spread throughout the British Empire and into Asia. The Rugby World Cup, the World Championship played in 1995, was won by the South African team.

Model	Multiplying Polynomials	Factoring Polynomials
3 · [2a / 6a] [b / 3b]	$3(2a + b) = 6a + 3b$	$6a + 3b = 3(a + 2b)$
5x · [3x / 15x²] [−4y / −20xy]	$5x(3x - 4y) = 15x^2 - 20xy$	$15x^2 - 20xy = 5x(3x - 4y)$
3 · [x² / 3x²] [5x / 15x]	$3(x^2 + 5x) = 3x^2 + 15x$	$3x^2 + 15x = 3(x^2 + 5x)$

Lesson 10-2 Factoring Using the Distributive Property **565**

GLOBAL CONNECTIONS

Rugby was excluded from the Football Association, established in 1863. In 1871, the Rugby Football Union was formed to standardize the rules and regulate the game.

10-2 LESSON NOTES

NCTM Standards: 1–5

Instructional Resources

- Study Guide Master 10-2
- Practice Master 10-2
- Enrichment Master 10-2
- Assessment and Evaluation Masters, p. 268
- Tech Prep Applications Masters, p. 19

 Transparency 10-2A contains the 5-Minute Check for this lesson; **Transparency 10-2B** contains a teaching aid for this lesson.

Recommended Pacing	
Standard Pacing	Day 3 of 12
Honors Pacing	Day 3 of 11
Block Scheduling*	Day 2 of 6
Alg. 1 in Two Years*	Days 5 & 6 of 22

 *For more information on pacing and possible lesson plans, refer to the *Block Scheduling Booklet* and *Algebra 1 in Two Years*.

1 FOCUS

 5-Minute Check
(over Lesson 10-1)

Find the prime factorization of each number.

1. 86 2×43
2. 168 $2 \times 2 \times 2 \times 3 \times 7$

Find the GCF.

3. 49 and 77 7
4. 24 and 104 8
5. $40xy^2$ and $64yz^2$ $8y$

Motivating the Lesson

Situational Problem Show that 2(3) can be found by using the distributive property and only 1s. For example:
$$2(3) = (1 + 1)(1 + 1 + 1)$$
$$= (1 + 1)(1) + (1 + 1)(1) + (1 + 1)(1)$$
$$= 1 + 1 + 1 + 1 + 1 + 1$$
$$= 6$$
Have students find 3(5) using only 1s and 2s.

Factoring a polynomial or finding the factored form of a polynomial means to find its *completely* factored form. The expression $3(x^2 + 5x)$ above is not considered completely factored since the polynomial $x^2 + 5x$ can be factored as $x(x + 5)$. The completely factored form of $3x^2 + 15x$ is $3x(x + 5)$.

Example **Use the distributive property to factor each polynomial.**

a. $12mn^2 - 18m^2n^2$

First find the GCF for $12mn^2$ and $18m^2n^2$.

$12mn^2 = 2 \cdot 2 \cdot 3 \cdot m \cdot n \cdot n$
$18m^2n^2 = 2 \cdot 3 \cdot 3 \cdot m \cdot m \cdot n \cdot n$ *The GCF is $2 \cdot 3 \cdot m \cdot n \cdot n$ or $6mn^2$.*

Notice that $12mn^2 = 6mn^2(2)$ and $18m^2n^2 = 6mn^2(3m)$. Then use the distributive property to express the polynomial as the product of the GCF and the remaining factor of each term.

$12mn^2 - 18m^2n^2 = 6mn^2(2) - 6mn^2(3m)$
$\quad\quad\quad\quad\quad\quad = 6mn^2(2 - 3m)$ *Distributive property*

b. $20abc + 15a^2c - 5ac$

$20abc = 2 \cdot 2 \cdot 5 \cdot a \cdot b \cdot c$
$15a^2c = 3 \cdot 5 \cdot a \cdot a \cdot c$
$5ac = 5 \cdot a \cdot c$ *The GCF is $5ac$.*
$20abc + 15a^2c - 5ac = 5ac(4b) + 5ac(3a) - 5ac(1)$
$\quad\quad\quad\quad\quad\quad\quad\quad = 5ac(4b + 3a - 1)$

Factoring a polynomial can simplify computations.

Example **The Parkson family wants to build a swimming pool in the shape of the figure below. Although the family has not yet decided on the actual dimensions of the pool, they do know that they want to build a deck that is 4 feet wide around the pool.**

a. Write an equation for the area of the deck.

b. If they decide to let a be 24 feet long, b be 6 feet long, and c be 10 feet, find the area of the deck.

Cooperative Learning

Trade-A-Problem Factoring problems can be created by starting with the solution. For example, the solution $2x(a^2 + 4x)$ leads to the problem $2xa^2 + 8x^2$. Have students work in pairs. One student should create a factoring problem and give it to the other to solve. Then they should switch roles. For more information on the trade-a-problem strategy, see *Cooperative Learning in the Mathematics Classroom,* one of the titles in the Glencoe Mathematics Professional Series, pages 25–26.

a. You can find the area of the deck by finding the sum of the areas of the 6 rectangular sections shown in the figure. The resulting expression can be simplified by first using the distributive property and then factoring.

Section 1   Section 2   Section 3   Section 4   Section 5   Section 6

$A = 4(a + 4 + 4) + 4(b + 4) + 4c + 4(b + 4) + 4(a + 4 + 4) + 4(c + 4 + 4)$

$= 4a + 16 + 16 + 4b + 16 + 4c + 4b + 16 + 4a + 16 + 16 + 4c + 16 + 16$

$= 8a + 8b + 8c + 128$ *Combine like terms.*

$= 8(a + b + c + 16)$ *The GCF is 8.*

The area of the deck is $8(a + b + c + 16)$ square feet.
Would dividing the deck into different sections result in a different answer?

b. Replace a with 24, b with 6, and c with 10.

$A = 8(24 + 6 + 10 + 16)$

$= 8(56)$ or 448

The area of the deck would be 448 square feet.

Just as it is possible to use the distributive property to factor a polynomial into monomial and polynomial factors, it is possible to factor some polynomials containing four or more terms into the product of two polynomials. Consider $(3a + 2b)(4c + 7d) = 12ac + 21ad + 8bc + 14bd$. Here, the product of two binomials results in a polynomial with four terms. How can the process be reversed to factor the four-term polynomial into its two binomial factors?

Example ❸ **Factor $12ac + 21ad + 8bc + 14bd$.**

$12ac + 21ad + 8bc + 14bd$

$= (12ac + 21ad) + (8bc + 14bd)$ *Apply the associative property, since a common factor of 3a appears in the first two terms and a common factor of 2b appears in the last two terms.*

$= 3a(4c + 7d) + 2b(4c + 7d)$ *Factor the first two terms and the last two terms.*

$= (3a + 2b)(4c + 7d)$ *$4c + 7d$ is a common factor. Use the distributive property.*

Check by using FOIL.

$$\begin{array}{cccc} F & O & I & L \end{array}$$

$(3a + 2b)(4c + 7d) = (3a)(4c) + (3a)(7d) + (2b)(4c) + (2b)(7d)$

$= \quad 12ac \quad + \quad 21ad \quad + \quad 8bc \quad + \quad 14bd \checkmark$

> **LOOK BACK**
>
> You can refer to Lesson 9-6 for information on FOIL.

This method is called **factoring by grouping**. It is necessary to group the terms and factor each group separately so that the remaining polynomial factors of each group are the same. This allows the distributive property to be applied a second time with a polynomial as the common factor.

Lesson 10–2   Factoring Using the Distributive Property   **567**

In-Class Example

For Example 3
Factor.

a. $6x^2y + 9xyz + 4xy^2 + 6y^2z$
   $(3xy + 2y^2)(2x + 3z)$

b. $49m + 21n^2 + 35m^3n + 15m^2n^3$
   $(7m + 3n^2)(5m^2n + 7)$

3 PRACTICE/APPLY

Check for Understanding
Exercises 1–25 are designed to help you assess your students' understanding through reading, writing, speaking, and modeling. You should work through Exercises 1–6 with your students and then monitor their work on Exercises 7–25.

Error Analysis
Students may think they have factored by grouping when they factor problems like $3x + 3y + 4w + 4z$ as $3(x + y) + 4(x + z)$. Although some factoring was done in removing GCF from two binomials, the final form is a sum and not a product. For factoring by grouping to work, binomial factors must match.

Sometimes you can group the terms in more than one way when factoring a polynomial. For example, the polynomial in Example 3 could have been factored in the following way.

$$12ac + 21ad + 8bc + 14bd = (12ac + 8bc) + (21ad + 14bd)$$
$$= 4c(3a + 2b) + 7d(3a + 2b)$$
$$= (4c + 7d)(3a + 2b) \quad \textit{The result is the same as in Example 3.}$$

Recognizing binomials that are additive inverses is often helpful in factoring. For example, the binomials $3 - a$ and $a - 3$ are additive inverses since the sum of $3 - a$ and $a - 3$ is 0. Thus, $3 - a$ and $-a + 3$ are equivalent. What is the additive inverse of $5 - y$?

$-1(a - 3) = -a + 3$
$\qquad\qquad = 3 - a$

Example ④ **Factor $15x - 3xy + 4y - 20$.**

$$15x - 3xy + 4y - 20 = (15x - 3xy) + (4y - 20)$$
$$= 3x(5 - y) + 4(y - 5) \quad \textit{(5 - y) and (y - 5) are additive inverses.}$$
$$= 3x(-1)(y - 5) + 4(y - 5) \quad \textit{(5 - y) = (-1)(-5 + y) or (-1)(y - 5)}$$
$$= -3x(y - 5) + 4(y - 5)$$
$$= (-3x + 4)(y - 5)$$

Check: $(-3x + 4)(y - 5) = (-3x)(y) + (-3x)(-5) + 4(y) + 4(-5)$
$$= -3xy + 15x + 4y - 20$$
$$= 15x - 3xy + 4y - 20 \quad ✔$$

In summary, a polynomial can be factored by grouping if all of the following situations exist.

- There are four or more terms.
- Terms with common factors can be grouped together.
- The two common factors are identical or differ by a factor of -1.

CHECK FOR UNDERSTANDING

Communicating Mathematics

1a. $2(4d^2 - 7d)$,
$d(8d - 14)$, $2d(4d - 7)$

2a. $ac + ad + bc + bd$

2b. $(a + b)(c + d)$

2c. They are equal.

Study the lesson. Then complete the following.

1. **a.** Express $8d^2 - 14d$ as a product of factors in three different ways.
   b. Which of the three answers in part a is the completely factored form of $8d^2 - 14d$? Explain. $2d(4d - 7)$; $2d$ is the GCF of $8d^2$ and $14d$.

2. **a.** Express the area of the rectangle at the right by adding the areas of the smaller rectangles.
   b. Express the area as the product of the length and the width.
   c. What is the relationship between the expressions in parts a and b?

	c	d
a	ac	ad
b	bc	bd

Reteaching

Using Substeps The following format may be useful in helping students factor sums and differences of polynomials.
Factor $54x^2 + 72x$.

Terms	Prime Factors	GCF
$54x^2$	$2 \cdot 3 \cdot 3 \cdot 3 \cdot x \cdot x$	$2 \cdot 3 \cdot 3 \cdot x$ or
$72x$	$2 \cdot 2 \cdot 2 \cdot 3 \cdot 3 \cdot x$	$18x$

So, $54x^2 + 72x = 18x(3x + 4)$.

3. distributive, associative, commutative

4. $(4gh + 8h) + (3g + 6)$, $(4gh + 3g) + (8h + 6)$

Guided Practice

3. List the properties used to factor $4gh + 8h + 3g + 6$ by grouping.

4. Group the terms of $4gh + 8h + 3g + 6$ in pairs in two different ways so that the pairs of terms have a common monomial factor.

5. Write the additive inverse of $7p^2 - q$. $q - 7p^2$

6. Use algebra tiles to factor $2x^2 - x$. $x(2x - 1)$

Find the GCF of the terms in each expression.

7. $3y^2 + 12$ **3**   **8.** $5n - n^2$ **n**   **9.** $5a + 3b$ **1**

10. $6mn + 15m^2$ **3m**   **11.** $12x^2y^2 - 8xy^2$ **4xy²**   **12.** $4x^2y - 6xy^2$ **2xy**

Express each polynomial in factored form.   **13.** $(a + b)(x + y)$

14. $(3m + 5n)(a - 2b)$

13. $a(x + y) + b(x + y)$

14. $3m(a - 2b) + 5n(a - 2b)$

15. $x(3a + 4b) - y(3a + 4b)$ $(x - y)(3a + 4b)$

16. $x^2(a^2 + b^2) + (a^2 + b^2)$ $(x^2 + 1)(a^2 + b^2)$

Complete. In Exercise 18, both blanks represent the same expression.

17. $20s + 12t = 4(5s + \underline{\ ?\ })$ **3t**

18. $(6x^2 - 10xy) + (9x - 15y) = 2x(\underline{\ ?\ }) + 3(\underline{\ ?\ })$ **3x − 5y**

Factor each polynomial.

19. $29xy - 3x$ $x(29y - 3)$ **20.** $x^5y - x$ $x(x^4y - 1)$

21. $3c^2d - 6c^2d^2$ $3c^2d(1 - 2d)$ **22.** $ay - ab + cb - cy$ $(a - c)(y - b)$

23. $rx + 2ry + kx + 2ky$ **24.** $5a - 10a^2 + 2b - 4ab$

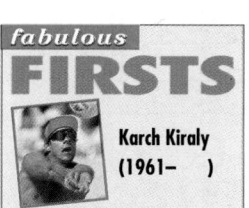
23. $(r + k)(x + 2y)$

24. $(5a + 2b)(1 - 2a)$

25a. $g = \frac{1}{2}n(n - 1)$

25. Volleyball Peta is scheduling the games for a volleyball league. To find the number of games she needs to schedule, she can use the equation $g = \frac{1}{2}n^2 - \frac{1}{2}n$, where g represents the number of games needed for each team to play each other team exactly once and n represents the number of teams.

  a. Write this equation in factored form.

  b. How many games are needed for 14 teams to play each other exactly once? **91 games**

  c. How many games are needed for 7 teams to play each other exactly 3 times? **63 games**

EXERCISES

Practice

Complete. In exercises with two blanks, both blanks represent the same expression.

 A

26. $10g - 15h = 5(\underline{\ ?\ } - 3h)$ **2g**

27. $8rst + 8rs^2 = \underline{\ ?\ }(t + s)$ **8rs**

28. $11p - 55p^2q = \underline{\ ?\ }(1 - 5pq)$ **11p**

29. $(6xy - 15x) + (-8y + 20) = 3x(\underline{\ ?\ }) - 4(\underline{\ ?\ })$ **2y − 5**

30. $(a^2 + 3ab) + (2ac + 6bc) = a(\underline{\ ?\ }) + 2c(\underline{\ ?\ })$ **a + 3b**

31. $(20k^2 - 28kp) + (7p^2 - 5kp) = 4k(\underline{\ ?\ }) - p(\underline{\ ?\ })$ **5k − 7p**

Lesson 10-2 Factoring Using the Distributive Property **569**

Assignment Guide

Core: 27–59 odd, 61–71
Enriched: 26–56 even, 57–71

For **Extra Practice,** see p. 779.

The red A, B, and C flags, printed only in the Teacher's Wraparound Edition, indicate the level of difficulty of the exercises.

Study Guide Masters, p. 70

Factor each polynomial.

32. $9t^2 + 36t$ $9t(t + 4)$

33. $14xz - 18xz^2$ $2xz(7 - 9z)$

34. $15xy^3 + y^4$ $y^3(15x + y)$

35. $17a - 41a^2b$ $a(17 - 41ab)$

36. $2ax + 6xc + ba + 3bc$

37. $2my + 7x + 7m + 2xy$

36. $(2x + b)(a + 3c)$

37. $(m + x)(2y + 7)$

38. $(m^2 + p^2)(3 - 5p)$

38. $3m^2 - 5m^2p + 3p^2 - 5p^3$

39. $3x^3y - 9xy^2 + 36xy$

40. $5a^2 - 4ab + 12b^3 - 15ab^2$

41. $2x^3 - 5xy^2 - 2x^2y + 5y^3$

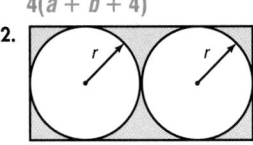

39. $3xy(x^2 - 3y + 12)$

42. $12ax + 20bx + 32cx$

43. $4ax - 14bx + 35by - 10ay$

44. $3my - ab + am - 3by$

45. $28a^2b^2c^2 + 21a^2bc^2 - 14abc$

46. $6a^2 - 6ab + 3bc - 3ca$

47. $12mx - 8m + 6rx - 4r$

48. $2ax + bx - 6ay - 3by - bz - 2az$

49. $7ax + 7bx + 3at + 3bt - 4a - 4b$

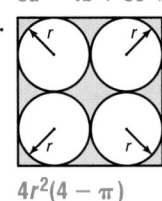
Geometry

Write an expression in factored form for the area of each shaded region.

40. $(a - 3b^2)(5a - 4b)$

41. $(2x^2 - 5y^2)(x - y)$

42. $4x(3a + 5b + 8c)$

43. $(2x - 5y)(2a - 7b)$

44. $(m - b)(3y + a)$

45. $7abc(4abc + 3ac - 2)$

46. $3(2a - c)(a - b)$

47. $2(2m + r)(3x - 2)$

48. $(x - 3y - z)(2a + b)$

49. $(7x + 3t - 4)(a + b)$

50.

$4(a + b + 4)$

51.

$8a - 4b + 8c + 16d + ab + 64$

52.

$2r^2(4 - \pi)$

53.

$4r^2(4 - \pi)$

Find the dimensions of a rectangle having the given area if its dimensions can be represented by binomials with integral coefficients.

54. $(5xy + 15x - 6y - 18)$ cm^2
$(5x - 6)$ cm by $(y + 3)$ cm

55. $(4z^2 - 24z - 18m + 3mz)$ cm^2
$(4z + 3m)$ cm by $(z - 6)$ cm

56. Geometry The perimeter of a square is $(12x + 20y)$ inches. Find the area of the square. $(9x^2 + 30xy + 25y^2)$ in^2

Critical Thinking

57. Geometry The perimeter of a rectangle is $(6a + 4b + 2ab + 12)$ centimeters. Find three possible expressions, in factored form, for the measure of its area.
Sample answer: $(3a + 2b)(ab + 6)$, $(3a + ab)(2b + 6)$, $(2b + ab)(3a + 6)$

Applications and Problem Solving

58. $w(2w + 13)$ ft^2

58. Gardening The length of Seth's garden is 5 feet more than twice its width w. This year, Seth decided to make the garden 4 feet longer and double its width. How much additional area did Seth add to his garden?

Tech Prep

Landscape Designer Students interested in gardening should look into landscape design as in Exercise 58, or horticulture and the actual growing of plants. For more information on tech prep, see the *Teacher's Handbook*.

Practice Masters, p. 70

NAME_____ DATE_____

10-2
Enrichment

Student Edition
Pages 565–571

Factoring Trinomials of Fourth Degree

Some trinomials of the form $a^4 + a^2b^2 + b^4$ can be written as the difference of two squares and then factored.

Example: Factor $4x^4 - 37x^2y^2 + 9y^4$.

Step 1: Find the square roots of the first and last terms.
$\sqrt{4x^4} = 2x^2$ $\sqrt{9y^4} = 3y^2$

Step 2: Find twice the product of the square roots.
$2(2x^2)(3y^2) = 12x^2y^2$

Step 3: Separate the middle term into two parts. One part is either your answer to Step 2 or its opposite. The other part should be the opposite of a perfect square.
$-37x^2y^2 = -12x^2y^2 - 25x^2y^2$

Step 4: Rewrite the trinomial as the difference of two squares and then factor.

$4x^4 - 37x^2y^2 + 9y^4 = (4x^4 - 12x^2y^2 + 9y^4) - 25x^2y^2$
$= (2x^2 - 3y^2)^2 - 25x^2y^2$
$= [(2x^2 - 3y^2) + 5xy][(2x^2 - 3y^2) - 5xy]$
$= (2x^2 + 5xy - 3y^2)(2x^2 - 5xy - 3y^2)$

Factor each trinomial.

1. $x^4 + x^2y^2 + y^4$
$(x^2 + xy + y^2)(x^2 - xy + y^2)$

2. $x^4 + x^2 + 1$
$(x^2 + x + 1)(x^2 - x + 1)$

3. $9a^4 - 15a^2 + 1$
$(3a^2 + 3a - 1)(3a^2 - 3a - 1)$

4. $16a^4 - 17a^2 + 1$
$(4a - 1)(a + 1)(4a + 1)(a - 1)$

5. $4a^4 - 13a^2 + 1$
$(2a^2 + 3a - 1)(2a^2 - 3a - 1)$

6. $9a^4 + 26a^2b^2 + 25b^4$
$(3a^2 + 2ab + 5b^2)(3a^2 - 2ab + 5b^2)$

7. $4x^4 - 21x^2y^2 + 9y^4$
$(2x^2 + 3xy - 3y^2)(2x^2 - 3xy - 3y^2)$

8. $4a^4 - 29a^2c^2 + 25c^4$
$(2a + 5c)(a - c)(2a - 5c)(a + c)$

59. Construction A 4-foot wide stone path is to be built along each of the longer sides of a rectangular flower garden. The length of the longer side of the garden is 3 feet less than twice the length of the shorter side s. Write an expression, in factored form, to represent the measure of the total area of the garden and path. $(2s - 3)(s + 8)$

60. Rugby Refer to the application at the beginning of the lesson.

 a. The width of a Rugby Union field is 69 meters. What is the length of the field? **144 m**

 b. What is the area of a Rugby Union field? **9936 m²**

 c. The length of a Rugby League field is 52 meters longer than its width. Write an expression for the area of the field. **$[w(w + 52)]$ m²**

 d. The width of a Rugby League field is 68 meters. Find the area of the field. **8160 m²**

 e. Which type of rugby field has a greater area? **Rugby Union**

Mixed Review

61. A and C sharp

61. Music Two musical notes played at the same time produce harmony. The closest harmony is produced by frequencies with the greatest GCF. A, C, and C sharp have frequencies of 220, 264, and 275, respectively. Which pair of these notes produces the closest harmony? (Lesson 10–1)

62. Government In 1990, the population of the United States was 248,200,000. The area of the United States is 3,540,000 square miles. (Lesson 9–3)

 a. If the population were equally spaced over the land, how many people would there be for each square mile? **about 70 people**

62b. $886.38

 b. In 1990, the federal budget deficit was $220,000,000,000. How much would each American have had to pay in 1990 to erase the deficit?

63. Geometry The graphs of $3x + 2y = 1$, $y = 2$, and $3x - 4y = -29$ contain the sides of a triangle. Find the measure of the area of the triangle. (Hint: Use the formula $A = \frac{1}{2}bh$.) (Lesson 8–2) **9**

64. Statistics The ten highest-paying occupations in America are shown in the table at the right. (Lesson 7–7)

 a. Make a box-and-whisker plot of the data. **See margin.**

 b. Name any outliers. **$148,000**

Occupation	Median Salary
Physician	$148,000
Dentist	93,000
Lobbyist	91,300
Management Consultant	61,900
Lawyer	60,500
Electrical Engineer	59,100
School Principal	57,300
Aeronautical Engineer	56,700
Airline Pilot	56,500
Civil Engineer	55,800

Source: Bureau of Labor Statistics

65. Solve $17.42 - 7.029z \geq 15.766 - 8.029z$. (Lesson 7–1) $\{z \mid z \geq -1.654\}$

66. $\left(6, \frac{1}{2}\right)$

66. Geometry Find the coordinates of the midpoint of the line segment whose endpoints are $A(5, -2)$ and $B(7, 3)$. (Lesson 6–7)

67. $\{(-3, 10),$ $\left(0, \frac{11}{2}\right), (1, 4),$ $\left(2, \frac{5}{2}\right), (5, -2)\}$

67. Solve $3a + 2b = 11$ if the domain is $\{-3, 0, 1, 2, 5\}$. (Lesson 5–3)

68. Fourteen is 50% less than what number? (Lesson 4–5) **28**

69. Solve $4x + 3y = 7$ for y. (Lesson 3–6)

69. $y = -\frac{4}{3}x + \frac{7}{3}$

70. Solve $\frac{5}{2}x = -25$. (Lesson 3–2) **−10**

71. Find the value of $(2^5 - 5^2) + (4^2 - 2^4)$. (Lesson 1–6) **7**

Lesson 10–2 Factoring Using the Distributive Property **571**

Extension

Connections The area of a triangle is represented by $7a^3 - 28a^2b$. What could be the measures of its base and altitude?
$14a^2$, $a - 4b$

Chapter 10 **571**

NCTM Standards: 1–5

Objective

Use algebra tiles to factor simple trinomials.

Recommended Time

Demonstration and discussion: 30 minutes; Exercises: 30 minutes

Instructional Resources

For each student or group of students
Student Manipulative Kit
• algebra tiles
• product mat
Modeling Mathematics Masters
• p. 1 (algebra tiles)
• p. 15 (product mat)
• p. 33 (worksheet)
For teacher demonstration
Algebra and Geometry Overhead Manipulative Resources

1 FOCUS

Motivating the Lesson

This modeling lesson extends the ideas of Lesson 10-2A to the case of trinomials. The strategy is to arrange the tiles in a rectangular shape.

2 TEACH

Teaching Tip In Activity 1, ask students to compare their models. Some may have factored $x^2 + 3x + 2$ as $(x + 2)(x + 1)$ instead of $(x + 1)(x + 2)$. Point out that these are both correct factorizations of the trinomial.

Teaching Tip Encourage students to experiment with different rectangles before they declare that a trinomial is not factorable. This is shown in Activity 2.

10-3A Factoring Trinomials

Materials: 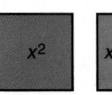 algebra tiles   [ ] product mat

You can use algebra tiles to factor trinomials. If a rectangle cannot be formed to represent the trinomial, then the trinomial is not factorable.

A Preview of Lesson 10–3

Activity 1   Use algebra tiles to factor $x^2 + 4x + 3$.

Step 1   Model the polynomial $x^2 + 4x + 3$.

Step 2   Place the x^2-tile at the corner of the product mat. Arrange the 1-tiles into a 1-by-3 rectangular array as shown.

Step 3   Complete the rectangle with the x-tiles.

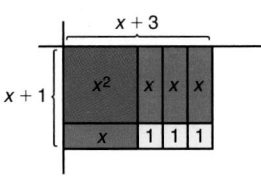

The rectangle has a width of $x + 1$ and a length of $x + 3$. Therefore, $x^2 + 4x + 3 = (x + 1)(x + 3)$.

You will need to use the guess-and-check strategy with many trinomials.

Activity 2   Use algebra tiles to factor $x^2 + 5x + 4$.

Step 1   Model the polynomial $x^2 + 5x + 4$.

Step 2   Place the x^2 tile at the corner of the product mat. Arrange the 1-tiles into a 2-by-2 rectangular array as shown. Try to complete the rectangle. Notice that there is an extra x-tile.

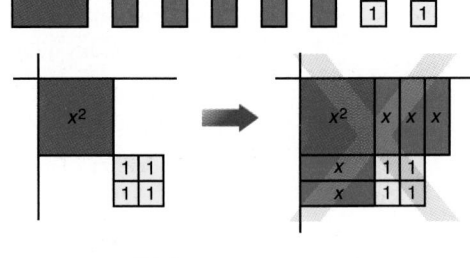

Step 3   Arrange the 1-tiles into a 1-by-4 rectangular array. This time you can complete the rectangle with the x-tiles.

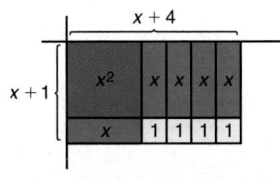

The rectangle has a width of $x + 1$ and a length of $x + 4$. Therefore, $x^2 + 5x + 4 = (x + 1)(x + 4)$.

Activity 3 Use algebra tiles to factor $x^2 - 4x + 4$.

Step 1 Model the polynomial $x^2 - 4x + 4$.

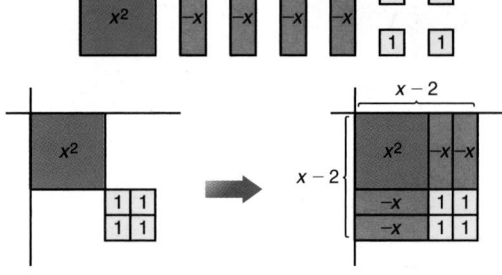

Step 2 Place the x^2-tile at the corner of the product mat. Arrange the 1-tiles into a 2-by-2 rectangular array as shown.

Step 3 Complete the rectangle with the x-tiles.

The rectangle has a width of $x - 2$ and a length of $x - 2$. Therefore, $x^2 - 4x + 4 = (x - 2)(x - 2)$.

Activity 4 Use algebra tiles to factor $x^2 - x - 2$.

Step 1 Model the polynomial $x^2 - x - 2$.

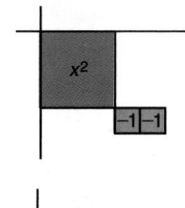

Step 2 Place the x^2-tile at the corner of the product mat. Arrange the 1-tiles into a 1-by-2 rectangular array as shown.

Step 3 Place the x-tile as shown. Recall that you can add zero-pairs without changing the value of the polynomial. In this case, add a zero pair of x-tiles.

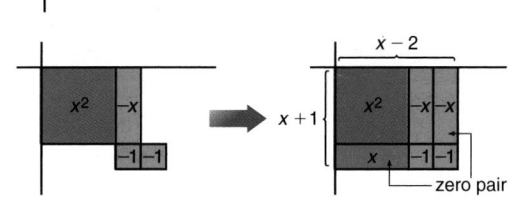

The rectangle has a width of $x + 1$ and a length of $x - 2$. Therefore, $x^2 - x - 2 = (x + 1)(x - 2)$.

1. $(x + 1)(x + 5)$ 2. $(x + 2)(x + 3)$ 3. $(x + 3)(x + 4)$ 4. $(x - 3)(x - 3)$
13. Trinomials can be factored if they can be represented by a rectangle. Sample answers: $x^2 + 4x + 4$ can be factored and $x^2 + 6x + 4$ cannot be factored.

Model **Use algebra tiles to factor each trinomial.**

1. $x^2 + 6x + 5$ 2. $x^2 + 5x + 6$ 3. $x^2 + 7x + 12$ 4. $x^2 - 6x + 9$
5. $x^2 - 3x + 2$ 6. $x^2 - 6x + 8$ 7. $x^2 + 4x - 5$ 8. $x^2 - x - 6$
   $(x - 1)(x - 2)$ $(x - 2)(x - 4)$ $(x - 1)(x + 5)$ $(x + 2)(x - 3)$

Draw **Tell whether each trinomial can be factored. Justify your answer with a drawing.** 9–12. See margin for drawings.

9. $x^2 + 7x + 10$ 10. $x^2 - 4x + 5$ 11. $x^2 + 5x - 4$ 12. $x^2 + 2x + 6$
   yes   no   no   no

Write 13. Write a paragraph that explains how you can determine whether a trinomial can be factored. Include an example of one trinomial that can be factored and one that cannot.

Teaching Tip Before proceeding with Activities 3 and 4, you may want students to review the visual patterns formed when multiplying binomials involving negatives.

Teaching Tip Before using Activity 4, remind students that after they try every arrangement they can think of, it may be necessary to add zero pairs to complete the rectangle. Remind them that adding zero pairs to any groups of tiles does not change the value of the groups.

3 PRACTICE/APPLY

Assignment Guide

Core: 1–13
Enriched: 1–13

4 ASSESS

Observing students working in cooperative groups is an excellent method of assessment.

Additional Answers
9.

10.

11.

12.

NCTM Standards: 1–5

Instructional Resources

- Study Guide Master 10-3
- Practice Master 10-3
- Enrichment Master 10-3
- Assessment and Evaluation Masters, pp. 267–268
- Modeling Mathematics Masters, pp. 63–65

Transparency 10-3A contains the 5-Minute Check for this lesson; **Transparency 10-3B** contains a teaching aid for this lesson.

Recommended Pacing	
Standard Pacing	Day 5 of 12
Honors Pacing	Day 5 of 11
Block Scheduling*	Day 3 of 6
Alg. 1 in Two Years*	Days 8, 9, & 10 of 22

*For more information on pacing and possible lesson plans, refer to the *Block Scheduling Booklet* and *Algebra 1 in Two Years.*

1 FOCUS

5-Minute Check
(over Lesson 10-2)

Factor by grouping.

1. $10xy + 25x - 14y - 35$
   $(5x - 7)(2y + 5)$

2. $18x^2y^2 + 21x^3y - 30y - 35x$
   $(3x^2y - 5)(6y + 7x)$

3. $49c - 7cd - 21 + 3d$
   $(7c - 3)(7 - d)$

Motivating the Lesson

Questioning Have students use FOIL and write out all four terms for the following products.
$(3x + 2)(5x + 3)$
$(t + 7)(3t - 5)$
$(2a - 5)(3a - 4)$
In groups, ask students to list all the relationships they notice relative to the coefficients. This can be used as a springboard to the sum-product method of factoring the general trinomial.

Factoring Trinomials

10-3

Objectives
After studying this lesson, you should be able to:
- solve problems by using guess and check, and
- factor quadratic trinomials.

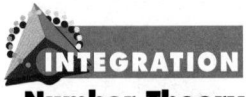
INTEGRATION
Number Theory

The product of two consecutive odd integers is 3363. What are the numbers?

One way to find the two numbers is to use a problem-solving strategy called **guess and check**. To use this strategy, guess the answer to the problem, and then check whether the guess is correct. If the first guess is incorrect, guess and check again until you find the correct answer. Often, the results of one guess can help you make a better guess. Always keep an organized record of your guesses so you don't make the same guess twice.

Guess: Try 41 and 43.

Check: $41 \times 43 = 1763$
This product is considerably less than 3363.

Guess: Try two numbers greater than 41 and 43, such as 61 and 63.

Check: $61 \times 63 = 3843$
This product is greater than 3363.

Guess: Try 51 and 53.

Check: $51 \times 53 = 2703$
This product is less than 3363. Since the ones digit is not 0 or 5, do not use 55 as one of the numbers. *Why?*

Guess: Try 57 and 59.

Check: $57 \times 59 = 3363$

The two numbers are 57 and 59.

In Lesson 10–1, you learned that when two numbers are multiplied, each number is a factor of the product. Similarly, when two binomials are multiplied, each binomial is a factor of the product. You can use the guess-and-check strategy to find the factors of a trinomial. Consider the binomials $5x + 3$ and $2x + 7$. You can use the FOIL method to find their product.

$$
\begin{aligned}
(5x + 3)(2x + 7) &= \overset{F}{(5x)(2x)} + \overset{O}{(5x)(7)} + \overset{I}{(3)(2x)} + \overset{L}{(3)(7)} \\
&= 10x^2 + 35x + 6x + 21 \\
&= 10x^2 + (35 + 6)x + 21 \quad \textit{Notice that } 10 \cdot 21 = 210 \\
&= 10x^2 + 41x + 21 \quad \textit{and } 35 \cdot 6 = 210.
\end{aligned}
$$

The binomials $5x + 3$ and $2x + 7$ are factors of $10x^2 + 41x + 21$.

When using the FOIL method above, look at the product of the coefficients of the first and last terms, 10 and 21. Notice that this product, 210, is the same as the product of the two middle terms, 35 and 6, whose sum is the coefficient of the middle term.

You can use this pattern to factor quadratic trinomials, such as $3y^2 + 10y + 8$.

$3y^2 + 10y + 8$ The product of 3 and 8 is 24.

$3y^2 + (\underline{\ ?\ } + \underline{\ ?\ })y + 8$

You need to find two integers whose *product is 24* and whose *sum is 10*.

Use the guess-and-check strategy to find these numbers.

Factors of 24	Sum of Factors	
1, 24	$1 + 24 = 25$	*no*
2, 12	$2 + 12 = 14$	*no*
3, 8	$3 + 8 = 11$	*no*
4, 6	$4 + 6 = 10$	*yes*

$3y^2 + 10y + 8$

$\quad = 3y^2 + (4 + 6)y + 8$ *Select the factors 4 and 6.*

$\quad = 3y^2 + 4y + 6y + 8$

$\quad = (3y^2 + 4y) + (6y + 8)$ *Group terms that have a common monomial factor.*

$\quad = y(3y + 4) + 2(3y + 4)$ *Factor.*

$\quad = (y + 2)(3y + 4)$ *Use the distributive property.*

Therefore, $3y^2 + 10y + 8 = (y + 2)(3y + 4)$.   *Check by using FOIL.*

Example **Factor $10x^2 - 27x + 18$.**

$10x^2 - 27x + 18$ The product of 10 and 18 is 180.

$10x^2 + (\underline{\ ?\ } + \underline{\ ?\ })x + 18$ Since the product is positive and the sum is negative, both factors of 180 must be negative. *Why?*

Factors of 180	Sum of Factors	
$-180, -1$	$-180 + (-1) = -181$	*no*
$-90, -2$	$-90 + (-2) = -92$	*no*
$-45, -4$	$-45 + (-4) = -49$	*no*
$-15, -12$	$-15 + (-12) = -27$	*yes* *You can stop listing factors when you find a pair that works.*

$10x^2 - 27x + 18$

$\quad = 10x^2 + [-15 + (-12)]x + 18$

$\quad = 10x^2 - 15x - 12x + 18$

$\quad = (10x^2 - 15x) + (-12x + 18)$

$\quad = 5x(2x - 3) + (-6)(2x - 3)$ *Factor the GCF from each group.*

$\quad = (5x - 6)(2x - 3)$ *Use the distributive property.*

Therefore, $10x^2 - 27x + 18 = (5x - 6)(2x - 3)$.   *Check by using FOIL.*

In-Class Example

For Example 1
Factor.

a. $2x^2 + 9x + 10$
    $(x + 2)(2x + 5)$

b. $3a^2 + 13a + 4$
    $(3a + 1)(a + 4)$

c. $6t^2 + 25t + 14$
    $(2t + 7)(3t + 2)$

d. $15y^2 + 34y + 15$
    $(3y + 5)(5y + 3)$

Teaching Tip Point out that when the sum of factors is positive, only positive factors need to be considered.

Teaching Tip Remind students to check for a GCF first.

Example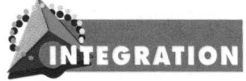

Geometry

The area of a rectangle is $(a^2 - 3a - 18)$ square inches. This area is increased by adding 5 inches to both the length and the width. If the dimensions of the original rectangle are represented by binomials with integral coefficients, find the area of the new rectangle.

To determine the area of the new rectangle, you must first find the dimensions of the original rectangle by factoring $a^2 - 3a - 18$. The coefficient of a^2 is 1. Thus, you must find two numbers whose product is $1 \cdot (-18)$ or -18 and whose sum is -3.

| Original Rectangle |
| Area = $(a^2 - 3a - 18)$ in² |

? in.

? in.

Factors of -18	Sum of Factors	
$-18, 1$	$-18 + 1 = -17$	no
$-9, 2$	$-9 + 2 = -7$	no
$-6, 3$	$-6 + 3 = -3$	yes

The factors of -18 should be chosen so that exactly one factor in each pair is negative and that factor has the greater absolute value. Why?

$$a^2 - 3a - 18 = a^2 + [(-6) + 3]a - 18$$
$$= a^2 - 6a + 3a - 18$$
$$= (a^2 - 6a) + (3a - 18)$$
$$= a(a - 6) + 3(a - 6)$$
$$= (a + 3)(a - 6) \quad \text{Check by using FOIL.}$$

The dimensions of the original rectangle are $(a + 3)$ inches and $(a - 6)$ inches. Therefore, the dimensions of the new rectangle are $(a + 3) + 5$ or $(a + 8)$ inches and $(a - 6) + 5$ or $(a - 1)$ inches. Now, find an expression for the area of the new rectangle.

$$(a + 8)(a - 1) = a^2 - a + 8a - 8$$
$$= a^2 + 7a - 8$$

The area of the new rectangle is $(a^2 + 7a - 8)$ square inches.

New Rectangle	$(a - 6) + 5$
Area = ? in²	or
	$(a - 1)$ in.

$(a + 3) + 5$
or
$(a + 8)$ in.

Let's study the factorization of $a^2 - 3a - 18$ from Example 2 more closely.

$$a^2 - 3a - 18 = (a + 3)(a - 6)$$

Notice that the sum of 3 and -6 is equal to -3, the coefficient of a in the trinomial. Also, the product of 3 and -6 is equal to -18, the constant term of the trinomial. This pattern holds for all trinomials whose quadratic term has a coefficient of 1.

Occasionally the terms of a trinomial will contain a common factor. In these cases, first use the distributive property to factor out the common factor, and then factor the trinomial.

Example **3** Factor $14t - 36 + 2t^2$.

First rewrite the trinomial so the terms are in descending order.

$$14t - 36 + 2t^2 = 2t^2 + 14t - 36$$
$$= 2(t^2 + 7t - 18) \quad \textit{The GCF of the terms is 2.}$$
$$\textit{Use the distributive property.}$$

Now factor $t^2 + 7t - 18$. Since the coefficient of t^2 is 1, we need to find two factors of -18 whose sum is 7.

Factors of -18	Sum of Factors	
18, -1	$18 + (-1) = 17$	*no*
9, -2	$9 + (-2) = 7$	*yes*

The desired factors are 9 and -2 and $t^2 + 7t - 18 = (t + 9)(t - 2)$.
Therefore, $2t^2 + 14t - 36 = 2(t + 9)(t - 2)$.

A polynomial that cannot be written as a product of two polynomials with integral coefficients is called a **prime polynomial**.

Example **4** Factor $2a^2 - 11a + 7$.

You must find two numbers whose product is $2 \cdot 7$ or 14 and whose sum is -11. Since the sum has to be negative, both factors of 14 have to be negative.

Factors of 14	Sum of Factors	
$-1, -14$	$-1 + (-14) = -15$	*no*
$-2, -7$	$-2 + (-7) = -9$	*no*

There are no factors of 14 whose sum is -11.

Therefore, $2a^2 - 11a + 7$ cannot be factored using integers.
Thus, $2a^2 - 11a + 7$ is a prime polynomial.

You can use your knowledge of factoring to write polynomials that can be factored using integers.

Example **5** Find all values of k so the trinomial $3x^2 + kx - 4$ can be factored using integers.

For $3x^2 + kx - 4$ to be factorable, k must equal the sum of the factors of $3(-4)$ or -12.

Factors of -12	Sum of Factors (k)
$-12, 1$	$-12 + 1 = -11$
$12, -1$	$12 + (-1) = 11$
$-6, 2$	$-6 + 2 = -4$
$6, -2$	$6 + (-2) = 4$
$-4, 3$	$-4 + 3 = -1$
$4, -3$	$4 + (-3) = 1$

Therefore, the values of k are $-11, 11, -4, 4, -1$, and 1.

3 PRACTICE/APPLY

Check for Understanding

Exercises 1–21 are designed to help you assess your students' understanding through reading, writing, speaking, and modeling. You should work through Exercises 1–4 with your students and then monitor their work on Exercises 5–21.

Additional Answers

1. So you do not waste time trying the same guess twice. It also helps you to make better guesses.

3a. It shows that the missing areas must have a sum of $8x$.

4.

Study Guide Masters, p. 71

NAME_____ DATE _____

Student Edition
Pages 574–580

10-3
Study Guide

Factoring Trinomials

To factor a trinomial of the form $ax^2 + bx + c$, follow Example 1 below.

Example 1: Factor $2d^2 + 15d + 18$.

The product of 2 and 18 is 36. You need to find two integers whose *product* is 36 and whose *sum* is 15.

Factors of 36	Sum of Factors
1, 36	1 + 36 = 37
2, 18	2 + 18 = 20
3, 12	3 + 12 = 15

$2d^2 + 15d + 18 = 2d^2 + (12 + 3)d + 18$
$= 2d^2 + 12d + 3d + 18$
$= (2d^2 + 12d) + (3d + 18)$
$= 2d(d + 6) + 3(d + 6)$ Factor the GCF from each group.
$= (2d + 3)(d + 6)$ Use the distributive property.

To factor a trinomial of the form given above when $a = 1$, you need to find only the factors of c whose sum is b.

Example 2: Factor $x^2 + 7x + 10$.

Since 2 and 5 are factors of 10 whose sum is 7,
$x^2 + 7x + 10 = (x + 2)(x + 5)$.

The same pattern can be used to factor a trinomial $ax^2 + bx + c$ when $a = 1$ and c is negative. When this occurs, the factors of the trinomial are a *difference* and a *sum*.

Complete.

1. $x^2 - 5x - 14 = (x + \underline{2})(x - 7)$
2. $a^2 + 13a + 36 = (a + 9)(a \underline{+} 4)$
3. $p^2 - 25 = (p + 5)(p \underline{- 5})$
4. $x^2 - 6xy - 16y^2 = (x \underline{- 8y})(x + 2y)$
5. $49 - n^2 = (7 \underline{+ n})(7 \underline{- n})$
6. $a^4 + 3xa^2 - 10x^2 = (a^2 \underline{- 2x})(a^2 \underline{+ 5x})$

Factor each trinomial, if possible. If the trinomial cannot be factored using integers, write prime.

7. $x^2 + 12y + 32$
$(y + 4)(y + 8)$
8. $x^2 - x - 6$
$(x - 3)(x + 2)$
9. $x^2 - 4x - 21$
$(x - 7)(x + 3)$
10. $y^2 + 22y + 121$
$(y + 11)(y + 11)$
11. $9 - 7n + n^2$
prime
12. $a^2 - 16a + 64$
$(a - 8)(a - 8)$
13. $3x^2 + 2x - 8$
$(3x - 4)(x + 2)$
14. $18h^2 - 27h - 5$
$(3h - 5)(6h + 1)$
15. $28x^2 + 60x - 25$
$(2x + 5)(14x - 5)$
16. $48x^2 + 22x - 15$
$(6x + 5)(8x - 3)$
17. $-4y^2 + 19y - 21$
$(4y - 7)(3 - y)$
18. $6a^2 - 7a + 18$
prime

578 Chapter 10

You can graph a polynomial and its factored form on the same axes to see if you have factored correctly. If the two graphs coincide, the factored form is probably correct.

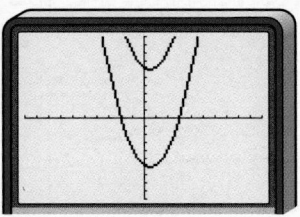
CHECK FOR UNDERSTANDING

Communicating Mathematics

Study this lesson. Then complete the following.   2. See students' work.

1. **Explain** why you should keep a record of your guesses when you are using the guess-and-check strategy.   See margin.

2. **Write** an example of a prime polynomial.

MODELING MATHEMATICS

3. **Study** the model at the right.
   a. **Explain** how this model could help to factor $x^2 + 8x + 12$.   See margin.
   b. **Factor** $x^2 + 8x + 12$. $(x + 6)(x + 2)$

4. Use algebra tiles to factor $x^2 + 3x - 4$. Make a drawing of the algebra tiles. $(x + 4)(x - 1)$; See margin for drawing.

Guided Practice

For each trinomial of the form $ax^2 + bx + c$, find two integers whose product is equal to ac and whose sum is equal to b.

5. $x^2 + 11x + 24$ 3, 8
6. $x^2 + 4x - 45$ $-5, 9$
7. $2x^2 + 13x + 20$ 8, 5
8. $3x^2 - 19x + 6$ $-1, -18$
9. $4x^2 - 8x + 3$ $-2, -6$
10. $5x^2 - 13x - 6$ $-15, 2$

Complete.

11. $r^2 - 5r - 14 = (r + 2)(r \underline{?} 7)$ $-$
12. $2g^2 + 5g - 12 = (2g - 3)(g + \underline{?})$ 4

Factor each trinomial, if possible. If the trinomial cannot be factored using integers, write *prime*.

13. $(t + 3)(t + 4)$
14. $(c - 4)(c - 9)$
15. $2(y + 2)(y - 3)$

13. $t^2 + 7t + 12$
14. $c^2 - 13c + 36$
15. $2y^2 - 2y - 12$
16. $3d^2 - 12d + 9$ $3(d - 3)(d - 1)$
17. $2x^2 + 5x - 2$ prime
18. $6p^2 + 15p - 9$ $3(2p - 1)(p + 3)$

Find all values of k so each trinomial can be factored using integers.

19. $x^2 + kx + 14$ $9, -9, 15, -15$
20. $2b^2 + kb - 3$ $1, -1, 5, -5$

Reteaching

Working Backward What values of b will make these trinomials factorable by unFOILing?

1. $x^2 + bx + 12$ $\{\pm 7, \pm 8, \pm 13\}$
2. $x^2 + bx - 20$ $\{\pm 19, \pm 8, \pm 1\}$
3. $x^2 + bx - 8$ $\{\pm 7, \pm 2\}$
4. $2x^2 + bx + 15$ $\{\pm 11, \pm 13, \pm 17, \pm 31\}$
5. $3x^2 + bx - 8$ $\{\pm 2, \pm 5, \pm 10, \pm 23\}$
6. $6x^2 + bx + 7$ $\{\pm 13, \pm 17, \pm 23, \pm 43\}$

21. Geometry The area of a rectangle is $(3x^2 + 14x + 15)$ square meters. This area is reduced by decreasing both the length and width by 3 meters. If the dimensions of the original rectangle are represented by binomials with integral coefficients, find the area of the new rectangle. $(3x^2 + 2x)$ m^2

 EXERCISES

Practice

A

28. $(b + 3)(b + 4)$
29. $(m - 4)(m - 10)$
30. $(z - 8)(z + 3)$
32. $(s - 12)(s + 15)$
33. $(2x + 7)(x - 3)$

B

34. $(7a + 1)(a + 3)$
35. $(2x + 3)(x - 4)$
37. $(2n - 7)(2n + 5)$
38. $2(9 - y)(4 - y)$
39. $(2 + 3m)(5 + 2m)$

C

42. $(5x - y)(3x - 2y)$
43. $2x(3x + 8)(2x - 5)$
47. $7, -7, 11, -11$
48. $10, -10, 11, -11,$
$14, -14, 25, -25$
49. $1, -1, 11, -11, 19,$
$-19, 41, -41$
50. $7, 12, 15, 16$
51. $6, 4$
53. r cm, $(5r + 6)$ cm,
$(3r - 7)$ cm
57. no; $(x - 3)(x - 3)$

Complete.

22. $a^2 + a - 30 = (a - 5)(a \underline{\,?\,} 6)$ +
 24. $4y^2 - y - 3 = (\underline{\,?\,} + 3)(y - 1)$ 4y
26. $4x^2 + 4x - 3 = (2x - 1)(\underline{\,?\,} + 3)$
 2x

23. $g^2 - 8g + 16 = (g - 4)(g \underline{\,?\,} 4)$ —
25. $6t^2 - 23t + 20 = (3t - 4)(2t - \underline{\,?\,})$ 5
27. $15g^2 + 34g + 15 = (5g + 3)(3g + \underline{\,?\,})$
5

Factor each trinomial, if possible. If the trinomial cannot be factored using integers, write *prime.*

28. $b^2 + 7b + 12$
29. $m^2 - 14m + 40$
30. $z^2 - 5z - 24$
31. $t^2 - 2t + 35$ prime
32. $s^2 + 3s - 180$
33. $2x^2 + x - 21$
34. $7a^2 + 22a + 3$
35. $2x^2 - 5x - 12$
36. $3c^2 - 3c - 5$ prime
37. $4n^2 - 4n - 35$
38. $72 - 26y + 2y^2$
39. $10 + 19m + 6m^2$
40. $a^2 + 2ab - 3b^2$ $(a + 3b)(a - b)$
41. $12r^2 - 11r + 3$ prime
42. $15x^2 - 13xy + 2y^2$
43. $12x^3 + 2x^2 - 80x$
44. $5a^3b^2 + 11a^2b^2 - 36ab^2$
$ab^2(5a - 9)(a + 4)$
45. $20a^4b - 58a^3b^2 + 42a^2b^3$
$2a^2b(5a - 7b)(2a - 3b)$

Find all values of k so each trinomial can be factored using integers.

46. $r^2 + kr - 13$ 12, −12
47. $x^2 + kx + 10$
48. $2c^2 + kc + 12$
49. $3s^2 + ks - 14$
50. $x^2 + 8x + k, k > 0$
51. $n^2 - 5n + k, k > 0$

52. **Geometry** The area of a rectangle is $(6x^2 - 31x + 35)$ square inches. If the dimensions of the rectangle are all whole numbers, what is the minimum possible area of the rectangle? 7 in^2

53. **Geometry** The volume of a rectangular prism is $(15r^3 - 17r^2 - 42r)$ cubic centimeters. If the dimensions of the prism are represented by polynomials with integral coefficients, find the dimensions of the prism.

Graphing Calculator

Use a graphing calculator to determine whether each equation is a true statement. If it is not correct, state the correct factorization of the trinomial. 54. yes 55. no; $(2x + 3)(x - 1)$ 56. no; $(3x + 2)(x - 2)$

54. $x^2 - 2x - 15 = (x - 5)(x + 3)$
55. $2x^2 + x - 3 = (2x - 1)(x + 3)$
56. $3x^2 - 4x - 4 = (3x - 2)(x + 2)$
57. $x^2 - 6x + 9 = (x + 3)(x - 3)$

Critical Thinking

58. Complete each polynomial in three different ways so that the resulting polynomial can be factored. Then factor each polynomial.
   a. $x^2 + 8x + \underline{\,?\,}$
   b. $x^2 + \underline{\,?\,} x - 10$ See margin.

Applications and Problem Solving

59. 27 ft^3

59. **Shipping** A shipping crate is to be built in the shape of a rectangular solid. The volume of the crate is $(45x^2 - 174x + 144)$ cubic feet where x is a positive integer. If the height of the crate is 3 feet, what is the minimum volume possible for this crate?

Lesson 10-3 Factoring Trinomials **579**

Additional Answers

58a. Sample answers:
$16, x^2 + 8x + 16 = (x + 4)(x + 4);$
$12, x^2 + 8x + 12 = (x + 6)(x + 2);$
$7, x^2 + 8x + 7 = (x + 7)(x + 1)$

58b. Sample answers:
$9, x^2 + 9x - 10 = (x + 10)(x - 1);$
$3, x^2 + 3x - 10 = (x + 5)(x - 2);$
$-3, x^2 + (-3)x - 10 = (x - 5)(x + 2)$

Assignment Guide

Core: 23–57 odd, 58, 59, 61–71
Enriched: 22–56 even, 58–71
All: Self Test, 1–10

For **Extra Practice,** see p. 779.

The red A, B, and C flags, printed only in the Teacher's Wraparound Edition, indicate the level of difficulty of the exercises.

Practice Masters, p. 71

10-3 NAME _____ DATE _____
Student Edition
Practice Pages 574–580

Factoring Trinomials
Factor each trinomial, if possible. If the trinomial cannot be factored using integers, write *prime.*

1. $t^2 + 8t + 12$
   $(t + 2)(t + 6)$
2. $w^2 + 24w + 144$
   $(w + 12)(w + 12)$
3. $m^2 - 7m + 12$
   $(m - 4)(m - 3)$
4. $n^2 + 3n - 18$
   $(n + 6)(n - 3)$
5. $v^2 - 18v + 80$
   $(v - 8)(v - 10)$
6. $p^2 - p - 56$
   $(p - 8)(p + 7)$
7. $b^2 + 8b - 22$
   prime
8. $x^2 + 7x - 44$
   $(x + 11)(x - 4)$
9. $y^2 - 5y - 84$
   $(y - 12)(y + 7)$
10. $32 + 18r + r^2$
    $(16 + r)(2 + r)$
11. $48 - 16e + e^2$
    $(12 - e)(4 - e)$
12. $s^2 + 17s + 52$
    $(s + 4)(s + 13)$
13. $102 - 23t + t^2$
    $(6 - t)(17 - t)$
14. $u^2 - 16u - 36$
    $(u - 18)(u + 2)$
15. $a^2b^2 + ab - 6$
    $(ab + 3)(ab - 2)$
16. $a^2b^2 + 5ab + 6$
    $(ab + 3)(ab + 2)$
17. $m^2 - mv - 56v^2$
    $(m - 8v)(m + 7v)$
18. $j^2 - 9jk - 10k^2$
    $(j - 10k)(j + k)$
19. $3h^2 + 2h - 16$
    $(3h + 8)(h - 2)$
20. $6c^2 + 7c + 2$
    $(3c + 2)(2c + 1)$
21. $5p^2 - 22p + 8$
    $(5p - 2)(p - 4)$
22. $8m^2 - 10m + 3$
    $(2m - 1)(4m - 3)$
23. $6z^2 - 5z - 4$
    $(3z - 4)(2z + 1)$
24. $15y^2 - y - 2$
    $(5y - 2)(3y + 1)$
25. $18x^2 + 9xz + z^2$
    $(6x + z)(3x + z)$
26. $20m^2 + 13mn + 2n^2$
    $(4m + n)(5m + 2n)$
27. $5\ell^2 - 26\ell x + 5x^2$
    $(5\ell - x)(\ell - 5x)$
28. $15s^2 - 16st + 4t^2$
    $(5s - 2t)(3s - 2t)$

Closing Activity

Speaking Have students explain the relationship between the sign of the last term in a quadratic trinomial and the signs of its binomial factors.

Chapter 10, Quiz B (Lesson 10-3), is available in the *Assessment and Evaluation Masters,* p. 268.

Mid-Chapter Test (Lessons 10-1 through 10-3) is available in the *Assessment and Evaluation Masters,* p. 267.

65. $y = -\frac{2}{3}x + \frac{14}{3}$

66. $-\frac{3}{2}$

67. D = {0, 1, 2};
R = {2, -2, 4}

Enrichment Masters, p. 71

60. **Guess and Check**   Place the digits 1, 2, 3, 4, 5, 6, 8, 9, 10, 12 on the dots at the right so that the sum of the integers on any line equals the sum on any other line.

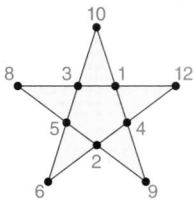

Mixed Review

61. **Finance**   During the first hour of trading, John Sugarman sold x shares of stock that cost $4 per share. During the next hour, he sold stock that cost $8 per share. He sold 5 more shares during the first hour than the second hour. If he had sold only the stock that cost $4 per share during the two hours, how many shares would he have needed to sell to have the same amount of total sales?   (Lesson 10-2)   **$(3x - 10)$ shares**

62. Find the degree of $7x^3 + 4xy + 3xz^3$.   (Lesson 9-4)   **4**

63. Use elimination to solve the system of equations.   (Lesson 8-3)   **(5, -2)**
$2x = 4 - 3y$
$3y - x = -11$

64. Solve $|2y - 7| \geq -6$.   (Lesson 7-6)   **{all numbers}**

65. Write an equation of the line that is parallel to the graph of $2x + 3y = 1$ and passes through (4, 2).   (Lesson 6-6)

66. Determine the slope of the line that passes through the points at $(-3, 6)$ and $(-5, 9)$.   (Lesson 6-1)

67. State the domain and range of {(0, 2), (1, -2), (2, 4)}.   (Lesson 5-2)

68. **Geography**   There is three times as much water as land on Earth's surface. What percent of Earth is covered by water?   (Lesson 4-4)   **75%**

69. Find the supplement of 90°.   (Lesson 3-4)   **90°**

70. **Time**   When it is noon in Richmond, Virginia, it is 2:00 A.M. the following morning in Kanagawa, Japan. Joel, who teaches English in Kanagawa, would like to call his mother in Richmond at 7:30 on the morning of her birthday, October 26. On what day and at what time would Joel have to call her?   (Lesson 2-3)   **Oct. 26 at 9:30 P.M.**

71. Simplify $3(x + 2y) - 2y$.   (Lesson 1-7)   **$3x + 4y$**

SELF TEST

Find the GCF of the given monomials.   (Lesson 10-1)

1. $50n^4, 40n^2p^2$ **$10n^2$**
2. $15abc, 35a^2c, 105a$ **$5a$**

Factor each polynomial, if possible. If the polynomial cannot be factored using integers, write *prime*.   (Lessons 10-2 and 10-3)   **4. $(2a - 1)(b + m)$ 8. $3mn(9mn - 25)$**

3. $18xy^2 - 24x^2y$ **$6xy(3y - 4x)$**
4. $2ab + 2am - b - m$
5. $2q^2 - 9q - 18$ **$(2q + 3)(q - 6)$**
6. $t^2 + 5t - 20$ **prime**
7. $3y^2 - 8y + 5$ **$(3y - 5)(y - 1)$**
8. $27m^2n^2 - 75mn$

9. **Guess and Check**   Write an eight-digit number using the digits 1, 2, 3, and 4 each twice so that the 1s are separated by 1 digit, the 2s are separated by 2 digits, the 3s are separated by 3 digits, and the 4s are separated by 4 digits.   (Lesson 10-3)   **41,312,432 or 23,421,314**

10. **Geometry**   The area of a rectangle is $(x^2 - x - 6)$ square meters. The length and width are each increased by 9 meters. If the dimensions of the original rectangle are binomials with integral coefficients, find the area of the new rectangle.   (Lesson 10-3)   **$(x^2 + 17x + 66)$ m^2**

Extension

Problem Solving   Factor
$24x^6 - 114x^4 - 225x^2$.
$3x^2(2x - 5)(2x + 5)(2x^2 + 3)$

SELF TEST

The Self Test provides students with a brief review of the concepts and skills in Lessons 10-1 through 10-3. Lesson numbers are given to the right of exercises or instruction lines so students can review concepts not yet mastered.

10-4 Factoring Differences of Squares

Objective

After studying this lesson, you should be able to:

- identify and factor binomials that are the differences of squares.

Modeling

You have used algebra tiles to factor trinomials. You can also use algebra tiles to factor some binomials.

MODELING MATHEMATICS

Difference of Squares

Materials: ⬛ algebra tiles ⬜ product mat

Factor $x^2 - 9$.

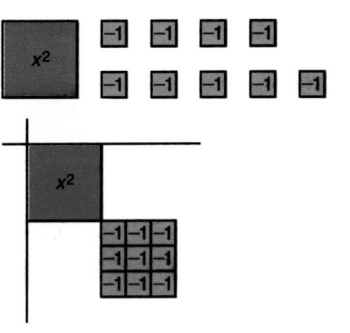

Step 1

Model the polynomial $x^2 - 9$.

Step 2

Place the x^2-tile at the corner of the product mat. Arrange the 1-tiles into a 3-by-3 square.

Step 3

Complete the rectangle using 3 zero pairs as shown.

The rectangle has a width of $x - 3$ and a length of $x + 3$.
Therefore, $x^2 - 9 = (x - 3)(x + 3)$.

Your Turn b, c, e. See margin.

a. Use algebra tiles to factor each binomial.

  $x^2 - 16$ $(x - 4)(x + 4)$ $x^2 - 4$ $(x - 2)(x + 2)$ $x^2 - 1$ $(x - 1)(x + 1)$
  $4x^2 - 9$ $(2x - 3)(2x + 3)$ $9x^2 - 4$ $(3x - 2)(3x + 2)$ $4x^2 - 1$ $(2x - 1)(2x + 1)$

b. Binomials such as those in part a are called the *difference of squares*. Explain why you think this term applies to these binomials.

c. Study the factors of the binomials in part a. What do you notice about the signs of the factors? about the terms of the factors?

d. Use the pattern that you observe to factor $x^2 - 100$. $(x - 10)(x + 10)$

e. Use FOIL to check your answer in part d. Was your answer correct?

Recall that the product of the sum and difference of two binomials such as $n + 8$ and $n - 8$ is called the *difference of squares*.

$$(n + 8)(n - 8) = n^2 - 8n + 8n - 64 \quad \textit{Use FOIL.}$$
$$= n^2 - 64$$

Note that this is the difference of two squares, n^2 and 64.

Lesson 10-4 Factoring Differences of Squares **581**

MODELING MATHEMATICS In this activity, students use factoring models developed earlier to study the problem of factoring a difference of squares. Emphasize to students that while in general the solution to a trinomial must be rectangular, the solution to a difference of squares must be square.

Answers for Modeling Mathematics

b. Each term is a square and the sign between them is always a subtraction sign.

c. One sign is minus and the other is plus. The terms are the same.

e. $(x - 10)(x + 10)$
  $= (x)(x) + (-10)(x) + (10)(x) + (-10)(10)$
  $= x^2 - 10x + 10x - 100$
  $= x^2 - 100$

10-4 LESSON NOTES

NCTM Standards: 1–5

Instructional Resources

- Study Guide Master 10-4
- Practice Master 10-4
- Enrichment Master 10-4
- Real-World Applications, 25

Transparency 10-4A contains the 5-Minute Check for this lesson; **Transparency 10-4B** contains a teaching aid for this lesson.

Recommended Pacing	
Standard Pacing	Days 6 & 7 of 12
Honors Pacing	Day 6 of 11
Block Scheduling*	Day 4 of 6 (along with Lesson 10-5)
Alg. 1 in Two Years*	Days 11, 12, & 13 of 22

*For more information on pacing and possible lesson plans, refer to the *Block Scheduling Booklet* and *Algebra 1 in Two Years*.

1 FOCUS

5-Minute Check
(over Lesson 10-3)

Factor each trinomial.

1. $10x^2 + 9x - 9$
   $(5x - 3)(2x + 3)$
2. $y^2 - 15y + 56$
   $(y - 7)(y - 8)$
3. $12t^2 + 35t + 8$
   $(4t + 1)(3t + 8)$
4. $p^2 + 6p - 27$
   $(p + 9)(p - 3)$
5. $6n^2 - 25n + 14$
   $(2n - 7)(3n - 2)$

Motivating the Lesson

Hands-On Activity Cut out pieces of paper as shown in the lesson to show how they fit to form a rectangle after a square has been removed from a larger square. This can be illustrated very well on an overhead projector. Students may want to cut out a square of any size and discover that the remaining "L" can always be cut to form a rectangle.

Chapter 10 **581**

In-Class Examples

For Example 1
Factor each binomial.

a. $x^2 - 25$ $(x - 5)(x + 5)$
b. $a^2 - 121$ $(a - 11)(a + 11)$
c. $16x^4 - z^4$
   $(4x^2 + z^2)(2x - z)(2x + z)$

For Example 2
Factor each polynomial.

a. $16x^2 - 25y^2$
   $(4x - 5y)(4x + 5y)$
b. $81a^2 - 16y^2$
   $(9a - 4y)(9a + 4y)$
c. $16x^2 - z^4$
   $(4x + z^2)(4x - z^2)$
d. $a^4 - 81b^8$
   $(a^2 + 9b^4)(a - 3b^2)(a + 3b^2)$

Teaching Tip Students should memorize the form of a difference of two squares.

Teaching Tip Review the FOIL method with students. For example, find $(x + 2)(x + 5)$.

$$\begin{array}{cccc} \text{F} & \text{O} & \text{I} & \text{L} \end{array}$$
$(x + 2)(x + 5) = x^2 + 5x + 2x + 10$
$\qquad\qquad\quad = x^2 + 7x + 10$

The Modeling Mathematics activity suggests the following rule for factoring the difference of squares.

Difference of Squares	$a^2 - b^2 = (a - b)(a + b) = (a + b)(a - b)$

You can use this rule to factor binomials that can be written in the form $a^2 - b^2$.

Example 1 Factor each binomial.

a. $m^2 - 81$

$$\begin{aligned} m^2 - 81 &= (m)^2 - (9)^2 && m \cdot m = m^2 \text{ and } 9 \cdot 9 = 81 \\ &= (m - 9)(m + 9) && \text{Use the difference of squares.} \end{aligned}$$

b. $100s^2 - 25t^2$

$$\begin{aligned} 100s^2 - 25t^2 &= 25(4s^2 - t^2) && \textit{25 is the GCF.} \\ &= 25[(2s)^2 - t^2] && 2s \cdot 2s = 4s^2 \text{ and } t \cdot t = t^2 \\ &= 25(2s - t)(2s + t) && \text{Use the difference of squares.} \end{aligned}$$

c. $\frac{1}{9}x^2 - \frac{4}{25}y^2$

$$\begin{aligned} \frac{1}{9}x^2 - \frac{4}{25}y^2 &= \left(\frac{1}{3}x\right)^2 - \left(\frac{2}{5}y\right)^2 && \text{Why?} \\ &= \left(\frac{1}{3}x - \frac{2}{5}y\right)\left(\frac{1}{3}x + \frac{2}{5}y\right) && \text{Check this result by using FOIL.} \end{aligned}$$

Sometimes the terms of a binomial have common factors. If so, the GCF should always be factored out first. Occasionally, the difference of squares needs to be applied more than once or along with grouping in order to completely factor a polynomial.

Example 2 Factor each polynomial.

a. $20cd^2 - 125c^5$

$$\begin{aligned} 20cd^2 - 125c^5 &= 5c(4d^2 - 25c^4) && \textit{The GCF of } 20cd^2 \text{ and } 125c^5 \text{ is } 5c. \\ &= 5c(2d - 5c^2)(2d + 5c^2) && 2d \cdot 2d = 4d^2 \text{ and } 5c^2 \cdot 5c^2 = 25c^4 \end{aligned}$$

b. $3k^4 - 48$

$$\begin{aligned} 3k^4 - 48 &= 3(k^4 - 16) && \text{Why?} \\ &= 3(k^2 - 4)(k^2 + 4) && k^2 \cdot k^2 = k^4 \\ &= 3(k - 2)(k + 2)(k^2 + 4) && k^2 + 4 \text{ cannot be factored. Why not?} \end{aligned}$$

c. $9x^5 + 11x^3y^2 - 100xy^4$

To factor the trinomial, we need two numbers whose product is -900 and whose sum is 11.

$$\begin{aligned} 9x^5 + 11x^3y^2 - 100xy^4 &= x(9x^4 + 11x^2y^2 - 100y^4) \\ &= x[(9x^4 - 25x^2y^2) + (36x^2y^2 - 100y^4)] \\ &= x[x^2(9x^2 - 25y^2) + 4y^2(9x^2 - 25y^2)] \\ &= x(x^2 + 4y^2)(9x^2 - 25y^2) \\ &= x(x^2 + 4y^2)(3x - 5y)(3x + 5y) \end{aligned}$$

Alternative Learning Styles

Visual The factorization $(a^2 - b^2) = (a - b)(a + b)$ can be modeled by drawing the rectangle for $(4^2 - 2^2) = (4 + 2)(4 - 2)$.

The difference of squares can be used to multiply numbers mentally.

Example **3** **Show a method for finding the product of 37 and 43 mentally.**

Since $37 = 40 - 3$ and $43 = 40 + 3$, the product of $(37)(43)$ can be expressed as $(40 - 3)(40 + 3)$.

$$(43)(37) = (40 + 3)(40 - 3)$$
$$= 40^2 - 3^2$$
$$= 1600 - 9 \text{ or } 1591$$

Pythagoras

According to the Pythagorean theorem, the sum of the squares of the measures of the legs of a right triangle equals the square of the measure of the hypotenuse.

$$a^2 + b^2 = c^2$$

A **Pythagorean triple** is a group of three whole numbers that satisfy the equation $a^2 + b^2 = c^2$. For example, the numbers 3, 4, and 5 form a Pythagorean triple.

$$3^2 + 4^2 \stackrel{?}{=} 5^2$$
$$9 + 16 \stackrel{?}{=} 25$$
$$25 = 25 \quad \checkmark$$

You can use the difference of squares to find Pythagorean triples.

Example **4** **Find a Pythagorean triple that includes 8 as one of its numbers.**

Number Theory

First find the square of 8. $8^2 = 64$

Factor 64 into two even factors or two odd factors. $64 = (2)(32)$

Find the mean of the two factors. $\frac{2 + 32}{2} = 17$

Complete the following statement.

Finding Pythagorean triples has interested mathematicians since ancient times. Some Pythagorean triples have been found on Babylonian cuneiform tablets.

$$(2)(32) = (17 - \underline{\ ?\ })(17 + \underline{\ ?\ })$$
$$= (17 - 15)(17 + 15)$$
$$= 17^2 - 15^2$$

Therefore, $8^2 = 17^2 - 15^2$ or $8^2 + 15^2 = 17^2$. The numbers 8, 15, and 17 form a Pythagorean triple.

Teaching Tip Point out that Example 3 shows a good method for doing multiplication without a calculator.

In-Class Examples

For Example 3
Find the product of 28 and 32 mentally.
$(28)(32) = (30 - 2)(30 + 2) = 30^2 - 2^2 = 900 - 4 = 896$

For Example 4
Find a Pythagorean triple that includes 6 as one of its numbers.
6, 8, 10

Teaching Tip A more in-depth discussion of the Pythagorean theorem is provided in Chapter 13.

F Y I

Pythagoras was born on the island of Samos in the Aegean Sea in about 580 B.C. In addition to being a brilliant mathematician, he founded a mystical cult based on relationships between numbers.

Check for Understanding

Exercises 1–21 are designed to help you assess your students' understanding through reading, writing, speaking, and modeling. You should work through Exercises 1–6 with your students and then monitor their work on Exercises 7–21.

Error Analysis

Here are two common errors in factoring.

1. $5a^2 - 20 = 5(a^2 - 4)$
   $= (a + 2)(a - 2)$
   The factor of 5 was omitted.

2. $16x^2 - 36 = (4x + 6)(4x - 6)$
   This was not factored completely. The GCF should be removed first.

Assignment Guide

Core: 23–53 odd, 54, 55, 57, 59–68
Enriched: 22–52 even, 54–68

For **Extra Practice,** see p. 779.

The red A, B, and C flags, printed only in the Teacher's Wraparound Edition, indicate the level of difficulty of the exercises.

Study Guide Masters, p. 72

Communicating Mathematics

5. $\dfrac{15}{16} \cdot \dfrac{17}{16} =$

$\dfrac{(16 - 1)(16 + 1)}{16^2} =$

$\dfrac{16^2 - 1^2}{16^2} = \dfrac{255}{256}$

Study this lesson. Then complete the following.

1. **Describe** a binomial that is the difference of two squares.   See margin.

2. **Write** a polynomial that is the difference of two squares. Factor your polynomial.   Sample answer: $a^2 - 25 = (a - 5)(a + 5)$

3. **Explain** how to factor a difference of squares by using the method for factoring trinomials presented in Lesson 10-3.   See margin.

4. **You Decide** Patsy says that $28f^2 - 7g^2$ can be factored using the difference of squares. Sally says it cannot. Who is correct? Explain.   See margin.

5. **Show** how to use the difference of squares to find $\frac{15}{16} \cdot \frac{17}{16}$.

MODELING MATHEMATICS

6. Use algebra tiles to factor $4 - x^2$. $(2 - x)(2 + x)$

Guided Practice

State whether each binomial can be factored as a difference of squares.

7. $p^2 - 49q^2$ yes
8. $25a^2 - 81b^4$ yes
9. $9x^2 + 16y^2$ no

Match each binomial with its factored form.

10. $4x^2 - 25$ c
11. $16x^2 - 4$ d
12. $25x^2 - 4$ b
13. $25x^2 - 25$ a

a. $25(x - 1)(x + 1)$
b. $(5x - 2)(5x + 2)$
c. $(2x - 5)(2x + 5)$
d. $4(2x - 1)(2x + 1)$

Factor each polynomial, if possible. If the polynomial cannot be factored, write prime.   14. $(t - 5)(t + 5)$ 15. $(1 - 4g)(1 + 4g)$

14. $t^2 - 25$
15. $1 - 16g^2$
16. $2a^2 - 25$ prime

17. $5(2m - 3n)(2m + 3n)$
17. $20m^2 - 45n^2$
18. $(a + b)^2 - c^2$
19. $x^4 - y^4$

18. $(a + b - c)(a + b + c)$

19. $(x - y)(x + y)(x^2 + y^2)$

20. Find the product of 17 and 23 mentally using difference of squares.   391

21. The difference of two numbers is 3. If the difference of their squares is 15, what is the sum of the numbers?   5

Practice

Factor each polynomial, if possible. If the polynomial cannot be factored, write prime.   22–45. See margin.

A
22. $w^2 - 81$
23. $4 - v^2$
24. $4q^2 - 9$
25. $100d^2 - 1$
26. $16a^2 - 25b^2$
27. $2z^2 - 98$

B
28. $9g^2 - 75$
29. $4t^2 - 27$
30. $8x^2 - 18$
31. $17 - 68k^2$
32. $25y^2 - 49z^4$
33. $36x^2 - 125y^2$
34. $-16 + 49h^2$
35. $16b^2c^4 + 25d^8$
36. $-9r^2 + 81$
37. $a^2x^2 - 0.64y^2$
38. $\frac{1}{16}x^2 - 25z^2$
39. $\frac{9}{2}a^2 - \frac{49}{2}b^2$

C
40. $(4p - 9q)^2 - 1$
41. $(a + b)^2 - (c + d)^2$
42. $25x^2 - (2y - 7z)^2$
43. $x^8 - 16y^4$
44. $a^6 - a^2b^4$
45. $a^4 + a^2b^2 - 20b^4$

Find each product mentally by using differences of squares.

46. 29×31 899
47. 24×26 624
48. 94×106 9964

Reteaching

Using Alternative Methods The difference of two squares can be used to find the product of two numbers that are the same difference from a multiple of ten. For example:
$(17)(23) = (20 - 3)(20 + 3)$
$= 20^2 - 3^2$
$= 400 - 9$
$= 391$

Additional Answers

1. Each term of the binomial is a perfect square, and the binomial can be written as a difference of terms.

3. Write the binomial as a trinomial where the coefficient of the middle term is 0, and then factor this trinomial.

4. Patsy; if 7 is factored from each term, the binomial factor is the difference of squares $4f^2 - g^2$.

Find the dimensions of a rectangle with the same area as the shaded region in each drawing. Assume that the dimensions of the rectangle must be represented by binomials with integral coefficients.

49. $(2a - b)$ in., $(2a + b)$ in.

50. $(\pi r - 5\pi)$ cm, $(r + 5)$ cm or $(r - 5)$ cm, $(\pi r + 5\pi)$ cm

51. $(x - 2)$ ft, $(x + 4)$ ft

52. $(p - r)$ cm, $(p + r)$ cm, $(7m + 2n)$ cm

53. $(a - 5b)$ in., $(a + 5b)$ in., $(5a + 3b)$ in.

Critical Thinking

Applications and Problem Solving

49.

50.
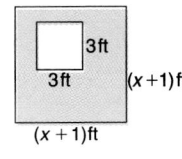

51.

Find the dimensions of a rectangular solid having the given volume if each dimension can be written as a binomial with integral coefficients.

52. $(7mp^2 + 2np^2 - 7mr^2 - 2nr^2)$ cubic centimeters

53. $(5a^3 - 125ab^2 - 75b^3 + 3a^2b)$ cubic inches

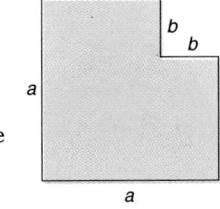

54. Show how to divide and rearrange the diagram at the right to show that $a^2 - b^2 = (a - b)(a + b)$. Make a diagram to show your reasoning.   See Solutions Manual.

55. Geometry   The side of a square is x centimeters long. The length of a rectangle is 5 centimeters longer than a side of the square, and the width of the rectangle is 5 centimeters shorter than the side of the square.
   a. Which has the greater area, the square or the rectangle?   square
   b. How much greater is that area?   25 cm^2

56. Number Theory   Find a Pythagorean triple that includes 7.   7, 24, 25

57. Number Theory   Find a Pythagorean triple that includes 9.   9, 12, 15

58. Geometry   Express the square of the length of the missing side of the triangle at the right as the product of two binomials. $(2a + 1)(4a + 3)$

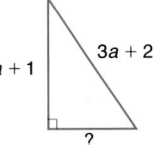

Mixed Review

59. Guess and Check   Julie went to the corner store to buy four items. The clerk at the store had to use a calculator to add the four prices, since her cash register was broken. When the clerk figured Julie's bill, she mistakenly hit the multiplication key each time, instead of the plus key. Julie had already mentally computed her sum, so, realizing that the total was actually correct, she paid the clerk the amount of $7.11. How much was each item that Julie bought?   (Lesson 10–3) $3.16, $1.50, $1.25, $1.20

60. Find $(n^2 + 5n + 3) + (2n^2 + 8n + 8)$.   (Lesson 9–5) $3n^2 + 13n + 11$

61. Use elimination to solve the system of equations.   (Lesson 8–4)   (4, 16)
   $x + y = 20$
   $0.4x + 0.15y = 4$

62. Probability   Use a tree diagram to find the probability of getting at least one tail when four fair coins are tossed.   (Lesson 7–5)

62. $\frac{15}{16}$

63. Graph $6x - 3y = 6$.   (Lesson 6–5)   **See margin.**

64. Write an equation for the relation given in the chart below.   (Lesson 5–6)

a	1	2	3	4	5	6
b	1	4	7	10	13	16

$b = 3a - 2$

Lesson 10–4   Factoring Differences of Squares   **585**

Closing Activity

Modeling Have students explore $a^3 - b^3$ by removing a cube from the corner of a larger cube. This can be done by building a cube from centimeter blocks and removing a cube from the corner. Students should write expressions for each of the remaining sections to illustrate the factoring of $a^3 - b^3$.

Additional Answer

65.

Enrichment Masters, p. 72

NAME_____ DATE _____

10-4
Enrichment

Student Edition
Pages 581–586

Writing Expressions of Area in Factored Form

Write an expression in factored form for the area A of the shaded region in each figure below.

1. $A = (x + y)(x - y)$
2. $A = 2r^2(4 - \pi)$
3. $A = \frac{\pi}{8}(a + b)(a - b)$
4. $A = r^2(8 + \pi)$
5. $A = \pi(R + 2r)(R - 2r)$
6. $A = a(3a + 2b)$
7. $A = x^2\left(8 - \frac{\pi}{2}\right)$
8. $A = r^2(\pi - 1)$

---

65. Graph $M(0, 3)$. (Lesson 5–1) **See margin.**

66. **Trigonometry** For the triangle at the right, find sin Y, cos Y, and tan Y to the nearest thousandth. (Lesson 4–3) **0.471; 0.882; 0.533**

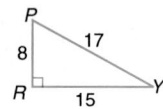

67. **Statistics** The populations in millions of the 50 states in 1993 are shown below. Find the mean, median, and mode population. (Lesson 3–7)

1.2	1.1	0.6	6.0	1.0	3.3	18.2	7.9	12.0
11.1	5.7	11.7	9.5	5.0	4.5	2.8	5.2	0.6
0.7	1.6	2.5	0.7	5.0	6.5	1.8	6.9	3.6
6.9	13.7	3.8	5.1	4.2	2.6	2.4	4.3	3.2
18.0	0.8	1.1	0.5	3.6	1.6	3.9	1.9	1.4
5.3	3.0	31.2	0.6	1.2	**5.14; 3.6; 0.6**			

68. In Mr. Tucker's algebra class, students can get extra points for finding the correct solution to the "Riddle of the Week." One week, Mr. Tucker posed this riddle. (Lesson 2–9)

Luis is 10 years older than his brother. Next year, he will be three times as old as his brother. How old is Luis now?

Josh's answer is: *Luis is 12 years old and his brother is 4.*

a. If the brother's age is represented by a, what is Luis' age? **$a + 10$**

b. What will be the brother's age next year? **$a + 1$**

c. Does Josh get the extra points for his answer? Why or why not? **no; $12 - 4 \neq 10$**

WORKING ON THE

In·ves·ti·ga·tion

Refer to the Investigation on pages 554–555.

the BRICKYARD

Suppose your manager, Ms. Brown, also gave you certain specifications as to which bricks to use and how many of each kind to use.

1. Using one of each type of brick, is there a rectangular pattern you can make with just three bricks? Justify your answer.

2. Select two of one type of brick and one each of the other two types. Is there a way to arrange these bricks into a rectangular pattern? If so, is there more than one pattern? Sketch a drawing of the rectangular pattern(s) you found and label the size of each brick.

3. If there is not a way to arrange the bricks into a rectangular pattern, explain why not. Is there more than one possible choice of four bricks that will make a rectangular pattern?

4. Now, using at least one brick of each type, find all the different patterns (if any) there are to arrange the bricks into a rectangular pattern of five bricks.

5. Now try designs using at least one of each type for a pattern of six bricks, seven bricks, eight bricks, nine bricks, and ten bricks.

6. Draw a diagram of all of the possible patterns. Label the size of each brick. Describe how you developed a process for finding the patterns. Are there generalizations you can make about the number of bricks and rectangular arrangements? Justify any generalizations. Explain how you know you have all of the possible patterns.

Add the results of your work to your Investigation Folder.

Extension

Problem Solving Factor $\frac{256}{6561}x^8 - 256$.

$\left(\frac{2}{3}x - 2\right)\left(\frac{2}{3}x + 2\right)\left(\frac{4}{9}x^2 + 4\right)\left(\frac{16}{81}x^4 + 16\right)$

Working on the Investigation

The Investigation on pages 554–555 is designed to be a long-term project that is completed over several days or weeks. Encourage students to keep their materials in their Investigation Folder as they work on the Investigation.

10-5

Perfect Squares and Factoring

Objective

After studying this lesson, you should be able to:

• identify and factor perfect square trinomials.

Number Theory

Recall that the numbers 1, 4, 9, 16, and 25 are called *perfect square* numbers, since they can each be expressed as the square of an integer.

The equation $5^2 = 25$ can be modeled as the area of a square having a side of length 5 as shown at the right.

Suppose $(3 + 2)$ is substituted for 5. Then $(3 + 2)^2$ can be modeled by using a 5-by-5 square and divided it into four regions as shown at the right. The sum of the areas of the four regions equals the area of the square.

$$(3 + 2)^2 = 3^2 + (2 \cdot 3) + (3 \cdot 2) + 2^2$$
$$= 3^2 + (3 \cdot 2) + (3 \cdot 2) + 2^2$$
$$= 3^2 + 2(3 \cdot 2) + 2^2$$

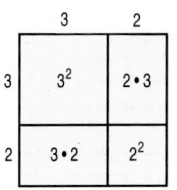

The last line shows a very interesting relationship.

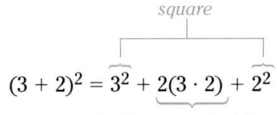

The square of this binomial is the sum of

• the square of the first term,
• twice the product of the first and second term, and
• the square of the second term.

This observation is generalized by the model at the right.

$$(a + b)^2 = a^2 + 2ab + b^2$$

To square $(a - b)$, write the binomial as $[a + (-b)]$ and then square this binomial.

$$(a - b)^2 = [a + (-b)]^2$$
$$= a^2 + 2a(-b) + (-b)^2$$
$$= a^2 - 2ab + b^2$$

Products of the form $(a + b)^2$ and $(a - b)^2$ are called perfect squares, and their expansions are **perfect square trinomials.**

Perfect Square Trinomials	$(a + b)^2 = a^2 + 2ab + b^2$ $(a - b)^2 = a^2 - 2ab + b^2$

These patterns can be used to factor trinomials.

LOOK BACK

You can refer to Lesson 9-8 for information on squares of sums and squares of differences.

Lesson 10–5 Perfect Squares and Factoring **587**

10-5 LESSON NOTES

NCTM Standards: 1–5

Instructional Resources

• Study Guide Master 10-5
• Practice Master 10-5
• Enrichment Master 10-5
• Assessment and Evaluation Masters, p. 269
• Modeling Mathematics Masters, pp. 66–68, 81
• Real-World Applications, 26

Transparency 10-5A contains the 5-Minute Check for this lesson; **Transparency 10-5B** contains a teaching aid for this lesson.

Recommended Pacing	
Standard Pacing	Day 8 of 12
Honors Pacing	Day 7 of 11
Block Scheduling*	Day 4 of 6 (along with Lesson 10-4)
Alg. 1 in Two Years*	Days 14, 15, & 16 of 22

*For more information on pacing and possible lesson plans, refer to the *Block Scheduling Booklet* and *Algebra 1 in Two Years.*

1 FOCUS

5-Minute Check
(over Lesson 10-4)

Factor each polynomial.

1. $n^2 - 81$ $(n - 9)(n + 9)$
2. $64p^2 - 25g^2$
   $(8p - 5g)(8p + 5g)$
3. $2a^5 - 162a$
   $2a(a^2 + 9)(a - 3)(a + 3)$
4. $75a^4 - 12a^2$
   $3a^2(5a - 2)(5a + 2)$

Motivating the Lesson

Situational Problem Winata said, "31^2 equals $30^2 + 1^2$." Do you agree or disagree? Emphasize the value of and need for the middle term when squaring a binomial.

Chapter 10 **587**

Model	Squaring a Binomial	Factoring a Perfect Square Trinomial
(model: v, 3 / v: v^2, $3v$ / 3: $3v$, 3^2)	$(v+3)^2 = v^2 + 2(v)(3) + 3^2$ $= v^2 + 6v + 9$	$v^2 + 6v + 9 = (v)^2 + 2(v)(3) + (3)^2$ $= (v+3)^2$
(model: $3p$, $-2q$ / $3p$: $(3p)^2$, $(-2q)(3p)$ / $-2q$: $(3p)(-2q)$, $(-2q)^2$)	$(3p-2q)^2 = (3p)^2 + 2(3p)(-2q) + (-2q)^2$ $= 9p^2 - 12pq + 4q^2$	$9p^2 - 12pq + 4q^2 = (3p)^2 - 2(3p)(2q) + (2q)^2$ $= (3p - 2q)^2$

Teaching Tip Caution students against confusing the difference of squares with the square of a difference.
$(a + b)(a - b) = a^2 - b^2$
$(a - b)^2 \neq (a^2 - b^2)$

Teaching Tip Point out that before factoring a trinomial, its terms should be arranged so that the powers of x are in descending or ascending order. For example, $8x + x^2 + 16$ should first be written as $x^2 + 8x + 16$ or $16 + 8x + x^2$.

Teaching Tip Answering the three questions gives students a system for identifying and factoring perfect square trinomials.

To determine whether a trinomial can be factored using these patterns, you must decide if it is a perfect square trinomial. In other words, you must determine if it can be written in the form $a^2 + 2ab + b^2$ or in the form $a^2 - 2ab + b^2$. For a trinomial to be in one of these forms, the following must be satisfied.

• The first term is a perfect square.
• The third term is a perfect square.
• The middle term is either 2 or -2 times the product of the square root of the first term and the square root of the last term.

Example Determine whether each trinomial is a perfect square trinomial. If so, factor it.

a. $4y^2 + 36yz + 81z^2$

To determine whether $4y^2 - 36yz + 81z^2$ is a perfect square trinomial, answer each question.

• Is the first term a perfect square? $4y^2 \stackrel{?}{=} (2y)^2$ *yes*

• Is the last term a perfect square? $81z^2 \stackrel{?}{=} (9z)^2$ *yes*

• Is the middle term twice the product of $2y$ and $9z$? $36yz \stackrel{?}{=} 2(2y)(9z)$ *yes*

$4y^2 + 36yz + 81z^2$ is a perfect square trinomial.
$4y^2 + 36yz + 81z^2 = (2y)^2 + 2(2y)(9z) + (9z)^2$
$= (2y + 9z)^2$

b. $9n^2 + 49 - 21n$

First arrange the terms of $9n^2 + 49 - 21n$ so that the powers of n are in descending order.

$9n^2 + 49 - 21n = 9n^2 - 21n + 49$

• Is the first term a perfect square? $9n^2 \stackrel{?}{=} (3n)^2$ *yes*

• Is the last term a perfect square? $49 \stackrel{?}{=} (7)^2$ *yes*

• Is the middle term the product of -2, $3n$, and 7? $-21n \stackrel{?}{=} -2(3n)(7)$ *no*

$9n^2 - 21n + 49$ is not a perfect square trinomial.

Example ❷

INTEGRATION
Geometry

Suppose the dimensions of a rectangle can be written as binomials with integral coefficients. Is the rectangle with the area of $(121x^2 - 198xy + 81y^2)$ square millimeters a square? If so, what is the measure of each side of the square?

Explore   You know that the dimensions of the rectangle can be written as binomials with integral coefficients. The problem gives the area of a rectangle and asks whether it is a square. If it is a square, you need to find the dimension of each side.

Plan   The rectangle is a square if $121x^2 - 198xy + 81y^2$ is a perfect square trinomial. You must answer three questions to determine if it is a perfect square trinomial. If it is a perfect square trinomial, you must factor it to find the measure of each side of the square.

Solve   • Is the first term a perfect square? $121x^2 \overset{?}{=} (11x)^2$ *yes*

    • Is the last term a perfect square? $81y^2 \overset{?}{=} (9y)^2$ *yes*

    • Is the middle term the product of -2, $11x$, and $9y$? $-198xy \overset{?}{=} -2(11x)(9y)$ *yes*

Since $121x^2 - 198xy + 81y^2$ is a perfect square trinomial, the rectangle is a square. To find the measure of each side, factor the trinomial.

$$121x^2 - 198xy + 81y^2 = (11x)^2 - 2(11x)(9y) + (9y)^2$$
$$= (11x - 9y)^2$$

The measure of each side is $(11x - 9y)$ millimeters.

Examine   If each side of the square is $(11x - 9y)$ millimeters, then the area of the square is $(11x - 9y)^2$ square millimeters. Use FOIL to see if $(11x + 9y)^2$ equals $(121x^2 - 198xy + 81y^2)$.

$$(11x - 9y)^2 = (11x - 9y)(11x - 9y)$$
$$= (11x)(11x) + (11x)(-9y) + (-9y)(11x) + (-9y)(-9y)$$
$$= 121x^2 - 99xy - 99xy + 81y^2$$
$$= 121x^2 - 198xy + 81y^2 \checkmark$$

As you continue your study of mathematics, you will find that forming a perfect square trinomial can sometimes be a useful tool for solving problems.

Example ❸

Determine all values of k that make $25x^2 + kx + 49$ a perfect square trinomial.

$$25x^2 + kx + 49 = (5x)^2 + kx + (7)^2$$

In order for this to be a perfect square trinomial, kx must equal either $2(5x)(7)$ or $-2(5x)(7)$. *Why?*

$kx = 2(5x)(7)$	$kx = -2(5x)(7)$
$kx = 70x$	$kx = -70x$
$k = 70$	$k = -70$

Check to see if $25x^2 + 70x + 49$ and $25x^2 - 70x + 49$ are perfect square trinomials.

In-Class Examples

For Example 2
Is it possible for $49a^2 - 42ab + 9b^2$ to be the measure of the area of a square? If so, what is the measure of each side of the square? $(7a - 3b)$

For Example 3
Determine all values of k that make each trinomial a perfect square trinomial.

a. $9x^2 + kx + 16$ ± 24

b. $49x^2 + kx + 9$ ± 42

Teaching Tip You may wish to suggest that students use FOIL to check their factorizations.

Teaching Tip In Example 4c, remind students that factoring completely often takes many steps.

In this chapter, you have learned various methods to factor different types of polynomials. The following chart summarizes these methods and can help you decide when to use a specific method.

Check for:	Number of Terms		
	Two	Three	Four or More
greatest common factor	✓	✓	✓
difference of squares	✓		
perfect square trinomials		✓	
trinomial that has two binomial factors		✓	
pairs of terms that have a common monomial factor			✓

Whenever there is a GCF other than 1, always factor it out first. Then, check the appropriate factoring methods in the order shown in the table. Use these methods to factor until all of the factors are prime.

Example **Factor each polynomial.**

a. $4k^2 - 100$

First check for a GCF. Then, since the polynomial has two terms, check for the difference of squares.

$$4k^2 - 100 = 4(k^2 - 25) \qquad \textit{The GCF is 4.}$$
$$= 4(k - 5)(k + 5) \qquad \textit{$k^2 - 25$ is the difference of squares}$$
$$\textit{since } k \cdot k = k^2 \textit{ and } 5 \cdot 5 = 25.$$

Therefore, $4k^2 - 25$ is completely factored as $4(k - 5)(k + 5)$.

b. $9x^2 - 3x - 20$

The polynomial has three terms. The GCF is 1. $9x^2 = (3x)^2$, but -20 is not a perfect square. The trinomial is not a perfect square trinomial.

Are there two numbers whose product is $9(-20)$ or -180 and whose sum is -3? Yes, the product of -15 and 12 is -180, and their sum is -3.

$$9x^2 - 3x - 20 = 9x^2 - 15x + 12x - 20$$
$$= (9x^2 - 15x) + (12x - 20)$$
$$= 3x(3x - 5) + 4(3x - 5)$$
$$= (3x + 4)(3x - 5)$$

Therefore, $9x^2 - 3x - 20$ is completely factored as $(3x + 4)(3x - 5)$.

c. $4m^4n + 6m^3n - 16m^2n^2 - 24mn^2$

Since the polynomial has four terms, first check for the GCF and then check for pairs of terms that have a common factor.

$$4m^4n + 6m^3n - 16m^2n^2 - 24mn^2 = 2mn(2m^3 + 3m^2 - 8mn - 12n)$$
$$= 2mn[(2m^3 + 3m^2) + (-8mn - 12n)]$$
$$= 2mn[m^2(2m + 3) + (-4n)(2m + 3)]$$
$$= 2mn(m^2 - 4n)(2m + 3)$$

Therefore, $4m^4n + 6m^3n - 16m^2n^2 - 24mn^2$ is completely factored as $2mn(m^2 - 4n)(2m + 3)$.

Communicating Mathematics

Study the lesson. Then complete the following.

1. a. **Draw** a rectangle to show how to factor $4x^2 + 12x + 9$. Label the dimensions and the area of the rectangle.   **See margin.**
   b. **Explain** why the name *perfect square trinomial* is appropriate for this trinomial.   **It is a trinomial that can be represented by a square.**

2a. See students' work.

2. a. **Write** a polynomial that is a perfect square trinomial.
   b. **Factor** your trinomial.   **Sample answer:** $x^2 + 2x + 1 = (x + 1)^2$

3. a. **Give** the first step in factoring any polynomial.   **See margin.**
   b. **Explain** why this step is important.   **See margin.**

4. **You Decide**   Robert says that $12a^4 - 8a^2 - 4$ is completely factored as $4(3a^2 + 1)(a^2 - 1)$. Samuel says that he can factor it further. Who is correct? Explain your answer.   **Samuel;** $(a^2 - 1)$ **can be factored as** $(a - 1)(a + 1)$.

 MATH JOURNAL

5. **Assess Yourself**   Describe the relationship between multiplying polynomials and factoring polynomials. Do you prefer to multiply polynomials or to factor polynomials? Explain.   **Multiplying and factoring polynomials are inverse operations; see students' work.**

Guided Practice

Complete.

6. $b^2 + 10b + 25 = (b + \underline{\ ?\ })^2$ **5**

7. $64a^2 - 16a + 1 = (\underline{\ ?\ } - 1)^2$ **8a**

8. $81n^2 + 36n + 4 = (\underline{\ ?\ } + 2)^2$ **9n**

9. $1 - 12c + 36c^2 = (1 - \underline{\ ?\ })^2$ **6c**

Determine whether each trinomial is a perfect square trinomial. If so, factor it.

10. $t^2 + 18t + 81$ **yes;** $(t + 9)^2$

11. $4n^2 - 28n + 49$ **yes;** $(2n - 7)^2$

12. $9y^2 + 30y - 25$ **no**

13. $16b^2 - 56bc + 49c^2$ **yes;** $(4b - 7c)^2$

Factor each polynomial, if possible. If the polynomial cannot be factored, write prime.

14. $15g^2 + 25$ **5**$(3g^2 + 5)$

15. $4a^2 - 36b^2$ **4**$(a - 3b)(a + 3b)$

16. $x^2 + 6x - 9$ **prime**

17. $50g^2 + 40g + 8$ **2**$(5g + 2)^2$

18. $9t^3 + 66t^2 - 48t$ **3**$t(3t - 2)(t + 8)$

19. (2a − 3b)(2a + 3b)
   (5x − y)

19. $20a^2x - 4a^2y - 45xb^2 + 9yb^2$

20. a. Find the missing value that makes the following a perfect square trinomial.

   $9x^2 + 24x + \underline{\ ?\ }$ **16**

   b. Copy and complete the model for this trinomial.

Practice

 A

Determine whether each trinomial is a perfect square trinomial. If so, factor it.

21. $r^2 - 8r + 16$ **yes;** $(r - 4)^2$

22. $d^2 + 50d + 225$ **no**

23. $49p^2 - 28p + 4$ **yes;** $(7p - 2)^2$

24. $4y^2 + 12yz + 9z^2$ **yes;** $(2y + 3z)^2$

25. $49s^2 - 42st + 36t^2$ **no**

26. $25y^2 + 20yz - 4z^2$ **no**

27. $4m^2 + 4mn + n^2$ **yes,** $(2m + n)^2$

28. $81t^2 - 180t + 100$ **yes;** $(9t - 10)^2$

Lesson 10–5   Perfect Squares and Factoring   **591**

Reteaching ▬▬▬▬

Using Manipulatives Have students use algebra tiles to model the trinomials in the exercises. Remind students that if the tiles form a square, then the trinomial is a perfect square. The length of each side represents each factor of the trinomial.

---

3 PRACTICE/APPLY

Check for Understanding

Exercises 1–20 are designed to help you assess your students' understanding through reading, writing, speaking, and modeling. You should work through Exercises 1–5 with your students and then monitor their work on Exercises 6–20.

Error Analysis

Students may overlook the sign of the constant term of a trinomial when factoring. For example, students might factor $x^2 - 12x - 36$ into $(x - 6)(x - 6)$. You may wish to review the difference between -6^2 and $(-6)^2$.
$-6^2 = -36$ and $(-6)^2 = 36$

Additional Answers

1.

3a. Factor out the GCF of the terms.

3b. The other factoring patterns will be more apparent after the GCF has been factored out.

Study Guide Masters, p. 73

The red A, B, and C flags, printed only in the Teacher's Wraparound Edition, indicate the level of difficulty of the exercises.

Determine whether each trinomial is a perfect square trinomial. If so, factor it.

29. $2g^2 - 10g + 25$ no

30. $1 + 100h^2 + 20h$ yes; $(10h + 1)^2$

31. $64b^2 - 72b + 81$ no

32. $9a^2 - 24a + 16$ yes; $(3a - 4)^2$

33. $\frac{1}{4}a^2 + 3a + 9$ yes; $\left(\frac{1}{2}a + 3\right)^2$

34. $\frac{4}{9}x^2 - \frac{16}{3}x + 16$ yes; $\left(\frac{2}{3}x - 4\right)^2$

Factor each polynomial, if possible. If the polynomial cannot be factored, write *prime*.

35. $45a^2 - 32ab$ $a(45a - 32b)$

36. $c^2 - 5c + 6$ $(c - 3)(c - 2)$

37. $v^2 - 30v + 225$ $(v - 15)^2$

38. $m^2 - p^4$ $(m - p^2)(m + p^2)$

40. $3ab(a + 2 + 3b)$

39. $9a^2 + 12a - 4$ prime

40. $3a^2b + 6ab + 9ab^2$

41. $3y^2 - 147$ $3(y - 7)(y + 7)$

42. $20n^2 + 34n + 6$ $2(5n + 1)(2n + 3)$

44. $3m(m + 8n)^2$

43. $18a^2 - 48a + 32$ $2(3a - 4)^2$

44. $3m^3 + 48m^2n + 192mn^2$

45. $(y^2 + z^2)(x - 1)(x + 1)$

45. $x^2y^2 - y^2 - z^2 + x^2z^2$

46. $5a^2 + 7a + 6b^2 - 4b$ prime

47. $4a^3 + 3a^2b^2 + 8a + 6b^2$

48. $(x + y)^2 - (w - z)^2$

47. $(a^2 + 2)(4a + 3b^2)$

49. $0.7p^2 - 3.5pq + 4.2q^2$

50. $(x + 2y)^2 - 3(x + 2y) + 2$

48. $(x + y - w + z)$
$(x + y + w - z)$

51. $g^4 + 6g^3 + 9g^2 - 3g^2h - 18gh - 27h$ $(g^2 - 3h)(g + 3)^2$

52. $12mp^2 - 15np^2 - 16m + 20np - 16mp + 20n$ $(4m - 5n)(3p + 2)(p - 2)$

49. $0.7(p - 3q)(p - 2q)$

Determine all values of k that make each of the following a perfect square trinomial.

50. $(x + 2y - 1)$
$(x + 2y - 2)$

53. $25t^2 - kt + 121$ $-110, 110$

54. $64x^2 - 16xy + k$ y^2

58. $(x - 3y)$ in.,
$(x + 3y)$ in., $(xy + 7)$ in.

55. $ka^2 - 72ab + 144b^2$ 9

56. $169n^2 + knp + 100p^2$ $-260, 260$

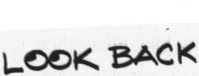
Geometry

57. The area of a circle is $(9y^2 + 78y + 169)\pi$ square centimeters. What is the diameter of the circle? $(6y + 26)$ cm

58. The volume of a rectangular prism is $(x^3y - 63y^2 + 7x^2 - 9xy^3)$ cubic inches. Find the dimensions of the prism, if its dimensions can be represented by binomials with integral coefficients.

LOOK BACK

You can refer to Lesson 2-8 to review square roots.

59. The length of a rectangle is 3 centimeters greater than the length of a side of a square. The width of the rectangle is one-half the length of the side of the square. If the area of the square is $(16x^2 - 56x + 49)$ square centimeters, what is the area of the rectangle? $(8x^2 - 22x + 14)$ cm^2

60. The area of a square is $(81 - 90x + 25x^2)$ square meters. If x is a positive integer, what is the least possible perimeter measure for the square?   4 m

Critical Thinking

61. Consider the value of $\sqrt{a^2 - 2ab + b^2}$.
   a. Under what circumstances does the value equal $a - b$? $a \geq b$
   b. Under what circumstances does the value equal $b - a$? $a \leq b$
   c. Under what circumstances does the value equal $a - b$ and $b - a$? $a = b$

Applications and Problem Solving

62. **Construction**   The builders of an office complex are looking for a square lot. They found a vacant lot that was long enough. However, its length was 60 yards more than its width w, so it was not a square. It also did not have enough area; they needed 900 additional square yards. So they are still looking for a lot. Write an expression for the length of a side of the square lot they should be looking for. $(w + 30)$ yd

---

Tech Prep

Construction Manager Students who are interested in construction management may wish to do further research on the data given in Exercise 62 and explore the potential growth of this career. For more information on tech prep, see the *Teacher's Handbook*.

Practice Masters, p. 73

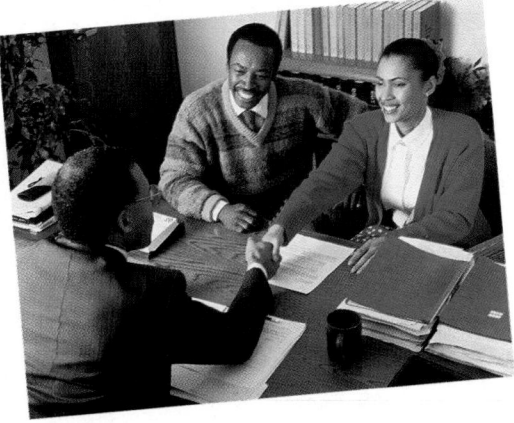

63. Investments Tamara plans to invest some money in a certificate of deposit. After 2 years, the value of the certificate will be $p + 2pr + pr^2$, where p represents the amount of money invested and r represents the annual interest rate.

   a. If Tamara invests $1000 at an annual interest rate of 8%, find the value of the certificate after 2 years. **$1166.40**

   b. Factor the expression that represents the value of the certificate after 2 years. $p(1 + r)^2$

   c. Suppose Tamara invests $1000 at 7%. Use your expression in part b to find the value of the certificate after 2 years. **$1144.90**

   d. Which form of the expression do you prefer to use to make your computations? Explain. **See students' work.**

Mixed Review

64. Factor $45x^2 - 20y^2z^2$. (Lesson 10–4) $5(3x - 2yz)(3x + 2yz)$

65. Simplify $2.5t(8t - 12) + 5.1(6t^2 + 10t - 20)$. (Lesson 9–6) $50.6t^2 + 21t - 102$

66. Simplify $(3a^2)(4a^3)$. (Lesson 9–1) $12a^5$

67. Employment Mike's parents allow him to work 30 hours a week. He would like to use this time to help out in his parents' hardware store, but it pays only $5 per hour. He could mow lawns for $7.50 per hour, but there is less than 20 hours of lawn work available. What is the maximum amount of time Mike can work in his parents' store and still make at least $175 per week? (Lesson 8–5) **20 hours**

68. $95° \le F \le 104°$

68. Physical Science A European-made hot tub is advertised to have a temperature of 35°C to 40°C, inclusive. What is the temperature range for the hot tub in degrees Fahrenheit? (*Hint:* Use $F = \frac{9}{5}C + 32$.) (Lesson 7–4)

Year	Median Income
1970	$ 8734
1975	11,800
1980	17,710
1981	19,074
1982	20,191
1983	21,018
1984	22,415
1985	23,618
1986	24,897
1987	26,061
1988	27,225
1989	28,905
1990	29,943
1991	30,126
1992	30,786

Source: U.S. Census Bureau

72. $b = \frac{70}{11}$, $c = \frac{30}{11}$

69. Statistics The median incomes of American families since 1970 are shown in the table at the left. (Lesson 6–3)

   a. Make a scatter plot of the data. **See margin.**

   b. Can the data be approximated by a straight line? If so, graph the line and write an equation of this line. **See margin.**

   c. Estimate the median family income for this year. **See students' work.**

70. Refer to Exercise 67 on page 586. Find the range and interquartile range of the state populations. (Lesson 5–7) **30.7, 4.6**

71. Probability If a card is selected at random from a deck of 52 cards, what are the odds of selecting a club? (Lesson 4–6) **1:3**

72. Geometry $\triangle ABC$ and $\triangle DEF$ are similar. If $a = 5$, $d = 11$, $f = 6$, and $e = 14$, find the missing measures. (Lesson 4–2)

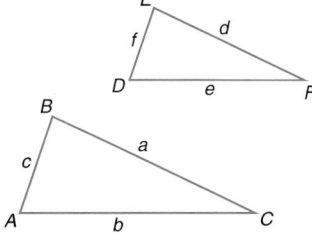

73. National Landmarks The Statue of Liberty and the pedestal on which it stands are 302 feet tall altogether. The pedestal is 2 feet shorter than the statue. How tall is the statue? (Lesson 3–3) **152 ft**

74. Basketball In 1962, Wilt Chamberlain set an NBA record by averaging 50.4 points per game for a few games. If he had been able to maintain this average over an 82-game season, how many total points would he have scored? Round to the nearest whole number. (Lesson 2–6) **4133 points**

Lesson 10–5 Perfect Squares and Factoring **593**

Extension

Reasoning If $x^n - y^m$ is the difference of two squares, what are some possible values of n and m?

any two positive, even integers

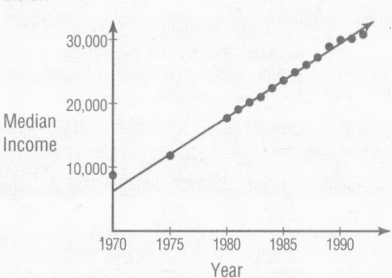

Instructional Resources

- Study Guide Master 10-6
- Practice Master 10-6
- Enrichment Master 10-6
- Assessment and Evaluation Masters, p. 269
- Graphing Calculator Masters, p. 10
- Multicultural Activity Masters, p. 20
- Science and Math Lab Manual, pp. 43–48
- Tech Prep Applications Masters, p. 20

 Transparency 10-6A contains the 5-Minute Check for this lesson; **Transparency 10-6B** contains a teaching aid for this lesson.

Recommended Pacing

Standard Pacing	Days 9 & 10 of 12
Honors Pacing	Days 8 & 9 of 11
Block Scheduling*	Day 5 of 6
Alg. 1 in Two Years*	Days 17, 18, & 19 of 22

 *For more information on pacing and possible lesson plans, refer to the *Block Scheduling Booklet* and *Algebra 1 in Two Years*.

1 FOCUS

 ### 5-Minute Check
(over Lesson 10-5)

Factor each polynomial.

1. $16x^4 - 36x^2y^2$
   $4x^2(2x - 3y)(2x + 3y)$

2. $20a^2 - 7a - 6$
   $(4a - 3)(5a + 2)$

3. $6t^3 - 21t^2 - 45t$
   $3t(t - 5)(2t + 3)$

4. $4n^3 - 20n^2 - 9n + 45$
   $(n - 5)(2n - 3)(2n + 3)$

5. The measure of the area of a rectangle is represented by $20a^2 - 13a - 15$. Find the measures of its dimensions.
   $5a + 3$ and $4a - 5$

Solving Equations by Factoring

Objective

After studying this lesson, you should be able to:

- use the zero product property to solve equations.

 ### Diving

In Acapulco, Mexico, divers leap from La Quebrada, the "Break in the Rocks," diving headfirst into the Pacific Ocean 87 feet below. Because the base rocks extend out 21 feet from the starting point, the divers must jump outward 27 feet. The equation $h = 87 + 8t - 16t^2$ describes the height h (in feet) of a diver t seconds after leaping.

To find the time at which the diver reaches the highest point of the dive, you can find the time at which he will be 87 feet above the ocean on his way down. The diver springs upward at the beginning of the dive and will be at the highest point halfway between this time and the start of the dive.

$$87 = 87 + 8t - 16t^2$$

$$0 = 8t - 16t^2 \qquad \textit{Subtract 87 from each side.}$$

$$0 = 8t(1 - 2t) \qquad \textit{Factor } 8t - 16t^2.$$

To solve this equation, you need to find the values of t that make the product $8t(1 - 2t)$ equal to 0. Consider the following products.

$$8(0) = 0 \qquad 0(-17)(0) = 0 \qquad (29 - 11)(0) = 0 \qquad 0(2a - 3) = 0$$

Notice that in each case, *at least one* of the factors is zero. These examples illustrate the **zero product property.**

Zero Product Property	**For all numbers a and b, if $ab = 0$, then $a = 0$, $b = 0$, or both a and b equal 0.**

Thus, if an equation can be written in the form $ab = 0$, then the zero product property can be applied to solve that equation.

Use this information to solve the equation about the diver. If $0 = 8t(1 - 2t)$, then either $8t = 0$ or $(1 - 2t) = 0$.

$$8t = 0 \qquad \text{or} \qquad 1 - 2t = 0$$

$$t = 0 \qquad\qquad\qquad 1 = 2t$$

$$\qquad\qquad\qquad\qquad \frac{1}{2} = t$$

The diver will be 87 feet above the ocean at the start of the dive ($t = 0$) and $\frac{1}{2}$ second later. He will be at the highest point of the dive $\frac{1}{4}$ second after starting the dive. *You will solve more problems about this dive in Example 5.*

594 *Chapter 10   Using Factoring*

 ### GLENCOE Technology

CD-ROM Interaction

A multimedia simulation allows students to use factoring and quadratic equations to determine the size of a farm. A blackline master activity with teacher's notes provides a follow-up to the CD-ROM simulation.

For Windows & Macintosh

Example **1** Solve each equation. Then check the solution.

a. $(p - 8)(2p + 7) = 0$

If $(p - 8)(2p + 7) = 0$, then $p - 8 = 0$ or $2p + 7 = 0$.   *Zero product property*

$$p - 8 = 0 \qquad \text{or} \qquad 2p + 7 = 0$$
$$p = 8 \qquad\qquad\qquad 2p = -7$$
$$\qquad\qquad\qquad\qquad p = -\frac{7}{2}$$

Check: $\qquad\qquad (p - 8)(2p + 7) = 0$

$$(8 - 8)[2(8) + 7] \overset{?}{=} 0 \qquad \text{or} \qquad \left(-\frac{7}{2} - 8\right)\left[2\left(-\frac{7}{2}\right) + 7\right] \overset{?}{=} 0$$
$$0(23) \overset{?}{=} 0 \qquad\qquad\qquad\qquad -\frac{23}{2}(0) \overset{?}{=} 0$$
$$0 = 0 \ \checkmark \qquad\qquad\qquad\qquad\qquad 0 = 0 \ \checkmark$$

The solution set is $\left\{8, -\frac{7}{2}\right\}$.

b. $t^2 = 9t$

Write the equation in the form $ab = 0$.

$$t^2 = 9t$$
$$t^2 - 9t = 0$$
$$t(t - 9) = 0 \quad \textit{Factor out the GCF, t.}$$
$$t = 0 \qquad \text{or} \qquad t - 9 = 0 \quad \textit{Zero product property}$$
$$\qquad\qquad\qquad\qquad t = 9$$

Check: $\qquad t^2 = 9t$

$$(0)^2 \overset{?}{=} 9(0) \qquad \text{or} \qquad (9)^2 \overset{?}{=} 9(9)$$
$$0 = 0 \ \checkmark \qquad\qquad\qquad 81 = 81 \ \checkmark$$

The solution set is $\{0, 9\}$.

To solve $t^2 = 9t$ in Example 1b, you may be tempted to divide each side of the equation by t. If you did, the solution would be 9. Since it is not possible for an equation to have two different solutions, which process is correct? Recall that division by 0 is undefined. But when you divide each side by t, t is unknown. Thus, you may actually be dividing by 0. In fact, 0 is one of the solutions of this equation. To avoid situations like this, keep in mind that you cannot divide each side of an equation by an expression containing a variable unless you know that the value of the expression is not 0.

Example **2** Solve $m^2 + 144 = 24m$. Then check the solution.

$$m^2 + 144 = 24m$$
$$m^2 - 24m + 144 = 0 \qquad \textit{Rewrite the equation.}$$
$$(m - 12)^2 = 0 \qquad \textit{Factor } m^2 - 24m + 144 \text{ as a perfect square trinomial.}$$
$$(m - 12)(m - 12) = 0$$
$$m - 12 = 0 \qquad \text{or} \qquad m - 12 = 0$$
$$m = 12 \qquad\qquad\qquad m = 12$$

(continued on the next page)

Lesson 10–6 Solving Equations by Factoring **595**

Hands-On Activity Use algebra tiles to solve this problem. Cosetta built a square pen in the corner of her square backyard. The side of her backyard is 24 feet longer than the side of the pen. Find the dimensions of the pen and Cosetta's backyard if the difference of their areas is 1152 square feet. **pen, 12 ft by 12 ft; backyard, 36 ft by 36 ft**

2 TEACH

In-Class Examples

For Example 1
Solve each equation.

a. $(x + 3)(x - 5) = 0$ $\{-3, 5\}$
b. $(2a + 4)(a + 7) = 0$ $\{-2, -7\}$
c. $x^2 - 36 = 5x$ $\{9, -4\}$

For Example 2
Solve each equation.

a. $a^2 - 24a = -144$ $\{12\}$
b. $4m^2 + 25 = 20m$ $\left\{\frac{5}{2}\right\}$

Teaching Tip Have students consider the equation $3(x + 7) = 0$. Since $3 \neq 0$, it is reasonable to set only the factor with the variable equal to zero. Therefore, $3(x + 7) = 0$ can be solved as follows.
$$3(x + 7) = 0$$
$$\frac{3(x + 7)}{3} = \frac{0}{3}$$
$$x + 7 = 0$$
$$x = -7$$

Check: $m^2 + 144 = 24m$

$$(12)^2 + 144 \stackrel{?}{=} 24(12)$$

$$144 + 144 \stackrel{?}{=} 288$$

$$288 = 288 \quad \checkmark$$

The solution set is $\{12\}$.

You can apply the zero product property to an equation that is written as the product of any number of factors equal to zero.

Example ③ Solve $5b^3 + 34b^2 = 7b$.

$$5b^3 + 34b^2 = 7b$$
$$5b^3 + 34b^2 - 7b = 0 \quad \textit{Arrange the terms so the powers of b are in descending order.}$$
$$b(5b^2 + 34b - 7) = 0 \quad \textit{Factor the GCF, b.}$$
$$b(5b - 1)(b + 7) = 0 \quad \textit{Factor } 5b^2 + 34b - 7.$$
$$b = 0 \qquad \text{or} \qquad 5b - 1 = 0 \qquad \text{or} \qquad b + 7 = 0$$
$$5b = 1 \qquad\qquad\qquad b = -7$$
$$b = \frac{1}{5}$$

The solution set is $\left\{0, \frac{1}{5}, -7\right\}$. *Check this result.*

If an object is launched from ground level, it reaches its maximum height in the air at the time halfway between the launch and impact times. Its height above the ground after t seconds is given by the formula $h = vt - 16t^2$. In this formula, h represents the height of the object in feet, and v represents the object's initial upward velocity in feet per second.

Example ④ **A flare is launched from a life raft with an initial upward velocity of 144 feet per second. How long will the flare stay aloft? What will be the maximum height attained by the flare?**

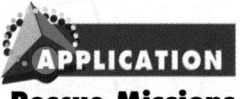
APPLICATION
Rescue Missions

Explore You know the initial upward velocity of the flare is 144 feet per second. You need to determine the length of time the flare will be in the air and how high the flare will go.

Plan The flare will be in the air until the height is 0. Use the general formula $h = vt - 16t^2$ to determine how long the flare will be in the air. The flare will reach its maximum height halfway between the launch and impact times. Use the formula again to determine the height at the middle of its flight.

Solve
$$h = vt - 16t^2$$
$$0 = 144t - 16t^2 \quad \textit{Replace v with 144.}$$
$$0 = 16t(9 - t)$$
$$16t = 0 \quad \text{or} \quad 9 - t = 0$$
$$t = 0 \qquad\qquad 9 = t$$

Since 0 seconds is the launch time, the landing time is 9 seconds. The flare will be aloft for 9 seconds. It will reach its maximum height halfway through its flight time at $\frac{1}{2}(9)$ or 4.5 seconds.

$h = 144t - 16t^2$

$\quad = 144(4.5) - 16(4.5)^2$ *Replace t with 4.5.*

$\quad = 648 - 324$

$\quad = 324$

The flare will reach its maximum height of 324 feet after 4.5 seconds.

Examine   Check to see if the flare will actually be at a height of 0 feet after 9 seconds.

$0 \overset{?}{=} 144(9) - 16(9)^2$

$0 \overset{?}{=} 1296 - 1296$

$0 = 0$ ✔

The flare will be aloft for 9 seconds and the maximum height will be reached after 4.5 seconds.

Example ⑤   **Refer to the application at the beginning of the lesson.**

APPLICATION

Diving

  a. What is the diver's maximum height?

  b. When will the diver enter the water?

  a. The diver will reach the maximum height after $\frac{1}{4}$ second.

$h = 87 + 8t - 16t^2$

$\quad = 87 + 8\left(\frac{1}{4}\right) - 16\left(\frac{1}{4}\right)^2$

$\quad = 87 + 2 - 1$

$\quad = 88$

The diver's maximum height is 88 feet. *Does this seem reasonable?*

  b. When the diver reaches 88 feet, his upward motion has stopped and the diver begins to fall to the sea below. His velocity at this time is 0. Therefore, the vt term in the general formula $h = vt - 16t^2$ is also 0.

$88 = 16t^2$ *Why does 88 equal $16t^2$ instead of $-16t^2$?*

$5.5 = t^2$

$2.35 \approx t$

It takes the diver $\frac{1}{4}$ or 0.25 second to reach the maximum height and about another 2.35 seconds to reach the water. The diver will reach the water about $0.25 + 2.35$ or 2.60 seconds after he starts his dive.

Teaching Tip Explain to students that velocity is positive when an object is going up and negative when it is going down. Velocity is zero when it changes direction from up to down.

GLENCOE *Technology*

 Interactive Mathematics Tools Software

This multimedia software provides an interactive lesson for solving an equation by factoring using the formula for the height of a falling object, $D = -16t^2 + vt + h$. A **Computer Journal** gives students an opportunity to write about what they have learned.

For Windows & Macintosh

Check for Understanding

Exercises 1–12 are designed to help you assess your students' understanding through reading, writing, speaking, and modeling. You should work through Exercises 1–4 with your students and then monitor their work on Exercises 5–12.

Error Analysis

Emphasize that an expression must be set equal to zero before the zero product property can be applied. Students may attempt to solve an equation such as $x(x - 4) = 3$ by saying that either $x = 3$ or $x - 4 = 3$. Neither 3 nor 7 is a correct solution.

Assignment Guide

Core: 13–41 odd, 42–48
Enriched: 14–32 even, 34–48

For **Extra Practice,** see p. 780.

The red A, B, and C flags, printed only in the Teacher's Wraparound Edition, indicate the level of difficulty of the exercises.

Study Guide Masters, p. 74

10-6 NAME_____ DATE _____
Study Guide Student Edition Pages 594–600

Solving Equations by Factoring

Factoring can be used to solve many kinds of problems.

Example: A rocket is fired with an initial velocity of 2288 feet per second. How many seconds will it take for the rocket to hit the ground?

Explore Use the formula $h = vt - 16t^2$.

Plan Substitute the appropriate values into the formula.

Solve $0 = 2288t - 16t^2$
$16t = 0$ or $143 - t = 0$ Zero product property
$t = 0$ $143 = t$

Examine An answer of 0 seconds is not a reasonable answer, so use only the value 143. The rocket returns to the ground in 143 seconds.

Solve each equation. Check your solutions.

1. $n^2 - 16 = 0$
{−4, 4}
2. $x^2 + 10x + 25 = 0$
{−5}
3. $9x^3 + x = 0$
$\left\{-\frac{1}{9}, 0\right\}$

4. $x^3 = 24 - 10x$
{−12, 2}
5. $x^3 - 18x = 7x^2$
{−2, 0, 9}
6. $x^3 - 36x = 9x^2$
{−3, 0, 12}

For each problem below, define a variable. Then use an equation to solve the problem. Disregard any unreasonable solutions.

7. The difference of the squares of two consecutive odd integers is 24. Find the integers. −5, −7
8. The length of a Charlotte, North Carolina, conservatory garden is 20 yards greater than its width. The area is 300 square yards. What are the dimensions?
30 yards by 10 yards

Use the formula $h = vt - 16t^2$ to solve each problem.

9. A punter can kick a football with an initial velocity of 48 feet per second. How many seconds will it take the ball to return to the ground?
3 seconds
10. If a rocket is launched at Houston, Texas, with an initial velocity of 1600 feet per second, when will the rocket be 14,400 feet high?
at 10 seconds and 90 seconds

Communicating Mathematics

Study the lesson. Then complete the following.

1. If the product of two or more factors is zero, what must be true of the factors? At least one of the factors equals 0.

2. **Describe** the type of equation that can be solved by using the zero-product property. equations that can be written as a product of factors that equal 0

3. Can $(x + 3)(x - 5) = 0$ be solved by dividing each side of the equation by $x + 3$? Explain. No; the division would eliminate −3 as a solution.

4. **You Decide** Diana says that if $(x + 2)(x - 3) = 8$, then $x + 2 = 8$ or $x - 3 = 8$. Caitlin disagrees. Who is correct? Caitlin; the zero product property only works for 0. If you multiply two numbers and get 8, that does not mean one of the numbers must have been 8.

Guided Practice

Solve each equation. Check your solutions.   6. {−2, 4}   7. $\left\{0, \frac{5}{3}\right\}$

5. $g(g + 5) = 0$ {0, −5}   6. $(n - 4)(n + 2) = 0$ 7. $5m = 3m^2$

8. $x^2 = 5x + 14$ {−2, 7}   9. $7r^2 = 70r - 175$ {5}   10. $a^3 - 29a^2 = -28a$
{0, 1, 28}

11. **Geometry** The dimensions of a rectangle are $(2x + 9)$ inches and $(2x - 1)$ inches. A square with side x inches is cut out of one of the corners. If the remaining area is 195 square inches, what is x? 6

(2x + 9) in.
(2x − 1) in.
x in.
x in.

12. **Baseball** Nolan Ryan, the greatest strike-out pitcher in the history of baseball, had a fastball clocked at 103 miles per hour, which is 151 feet per second. a. about 9.44 seconds

a. If he threw the ball directly upward with the same velocity, how many seconds would it take for the ball to return to his glove? (*Hint:* Use the general formula $h = vt - 16t^2$.)

b. How high above his glove would the ball travel? about 356 feet

Practice

15. $\left\{\frac{3}{2}, \frac{8}{3}\right\}$

16. $\left\{-\frac{5}{4}, \frac{7}{3}\right\}$

20. $\left\{0, \frac{5}{2}\right\}$

23. $\left\{-\frac{1}{3}, -\frac{5}{2}\right\}$

24. $\left\{-7, 0, \frac{1}{5}\right\}$

30. $\left\{-4, \frac{2}{3}\right\}$

A
B
C

Solve each equation. Check your solutions.

13. $x(x - 24) = 0$ {0, 24}
14. $(q + 4)(3q - 15) = 0$ {−4, 5}
15. $(2x - 3)(3x - 8) = 0$
16. $(4a + 5)(3a - 7) = 0$
17. $a^2 + 13a + 36 = 0$ {−9, −4}
18. $x^2 - x - 56 = 0$ {−7, 8}
19. $y^2 - 64 = 0$ {−8, 8}
20. $5s - 2s^2 = 0$
21. $3z^2 = 12z$ {0, 4}
22. $m^2 - 24m = -144$ {12}
23. $6q^2 + 5 = -17q$
24. $5b^3 + 34b^2 = 7b$
25. $\frac{x^2}{12} - \frac{2x}{3} - 4 = 0$ {12, −4}
26. $t^2 - \frac{t}{6} = \frac{35}{6}$ $\left\{-\frac{7}{3}, \frac{5}{2}\right\}$
27. $n^3 - 81n = 0$ {−9, 0, 9}
28. $(x + 8)(x + 1) = -12$ {−4, −5}
29. $(r - 1)(r - 1) = 36$ {−5, 7}
30. $(3y + 2)(y + 3) = y + 14$

Reteaching

Logical Thinking Abe adds up the number of runs that the Reds score in each game for the season. Bob multiplies together the number of runs that the Mets score in each game. Who will probably have the larger number at the end of the season? Why? Abe; think of the effects of a shutout.

First professional pitchers to throw a perfect game
1. Lee Richmond, June 12, 1880
2. Monte Ward, June 17, 1880
3. Cy Young, May 5, 1904
4. Adrian Joss, Oct. 2, 1908
5. Charlie Robertson, Apr. 30, 1922

There have been a total of 13 pitchers who have pitched perfect games. The most recent is Kenny Rogers on July 28, 1994.

31. Number Theory Find two consecutive even integers whose product is 168. **−14 and −12 or 12 and 14**

32. Number Theory Find two consecutive odd integers whose product is 1023. **−33 and −31 or 31 and 33**

33. Geometry The triangle at the right has an area of 40 square centimeters. Find the height h of the triangle. **5 cm**

h cm

(2*h* + 6) cm

Critical Thinking

34. Write an equation with integral coefficients that has $\{-3, 0, 7\}$ as its solution set. $x^3 - 4x^2 - 21x = 0$

35. Consider the equations $a^2 + 5a = 6$ and $|2x + 5| = 7$.
   a. Solve each equation. $\{-6, 1\}; \{-6, 1\}$
   b. What is the relationship between the two equations? What does that mean? **They are equivalent; they have the same solutions.**

Applications and Problem Solving

36. Gardening LaKeesha has enough bricks to make a 30-foot-long border around the rectangular vegetable garden she is planning. The booklet she got from the plant nursery when she bought the seeds says the plants will need space to grow, and it advises that the seeds should be planted in an area of 54 square feet. What should the dimensions of her garden be? **9 ft by 6 ft**

37. History During the late 16th century, Galileo was a professor at the University of Pisa, Italy. During this time, he demonstrated that objects of different weights fell at the same velocity by dropping two objects of different weights from the top of the Leaning Tower of Pisa.
   a. If he dropped the objects from a height of 180 feet, how long did it take them to hit the ground? **about 3.35 s**
   b. Research Galileo's life. Why was he criticized for experimenting with these weights? What other belief of his led to his trial by the Inquisition of 1633? **See margin.**

38. Fountains The tallest fountain in the world is at Fountain Hills, Arizona. When all three pumps are working, the nozzle speed of the water is 146.7 miles per hour (215.16 feet per second). It is claimed the water reaches a height of 625 feet. Do you agree or disagree with this claim? Explain. **See margin.**

39. Superheroes A meteorite is headed towards Metropolis when Superman intercepts it. He takes it to the top of the *Daily Planet* (180 feet high) and tosses the meteorite into space with an upward velocity of 2400 feet per second. What height will the meteorite attain before returning to Earth? **about 90,180 ft or 17 mi**

40. Bridges The chart at the right compares the highest bridge in the world at Royal Gorge, Colorado with the highest railroad bridge in the world at Kolasin, Yugoslavia. If Elva accidentally drops her keys from the Royal Gorge Bridge, will the keys hit the Arkansas River below within 8 seconds? **yes**

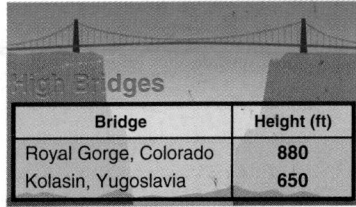

High Bridges

Bridge	Height (ft)
Royal Gorge, Colorado	880
Kolasin, Yugoslavia	650

Lesson 10–6 Solving Equations by Factoring **599**

Practice Masters, p. 74

10-6 Practice

NAME_____ DATE_____

Student Edition Pages 594–600

Solving Equations by Factoring
Solve each equation. Check your solutions.

1. $x(x - 7) = 0$ $\{0, 7\}$ 2. $4b(b + 4) = 0$ $\{0, -4\}$

3. $(y - 3)(y + 2) = 0$ $\{-2, 3\}$ 4. $(2y - 5)(y + 6) = 0$ $\{-6, \frac{5}{2}\}$

5. $(2z + 1)^2 = 0$ $\{-\frac{1}{2}\}$ 6. $(\frac{1}{2}t - 1)^2 = 0$ $\{2\}$

7. $3y^2 - 108y = 0$ $\{0, 36\}$ 8. $2v^2 + 20v = 0$ $\{0, -10\}$

9. $t^2 = 5t$ $\{0, 5\}$ 10. $\frac{1}{4}h^2 - \frac{3}{4}h = 0$ $\{0, 3\}$

11. $2u^2 - \frac{2}{3}u = 0$ $\{0, \frac{1}{3}\}$ 12. $5n^2 = 4n^2 - 4n$ $\{0, -4\}$

13. $3 - 13s = 3 - 8s + s^2$ $\{0, -5\}$ 14. $d^2 - 4d = 3d$ $\{0, 7\}$

Closing Activity

Writing Have students write about the advantage of factoring an equation in order to solve it.

Chapter 10, Quiz D (Lesson 10-6), is available in the *Assessment and Evaluation Masters,* p. 269.

Enrichment Masters, p. 74

10-6 NAME_____ DATE_____

Enrichment   Student Edition
                                        Pages 592–600

Using Factoring

The right side of each formula is a fraction. Factor the numerator and denominator of each fraction. Then reduce the fraction to simplest terms if possible. Use the result to answer the questions about each formula.

1. $t = \frac{\pi(d+r)^2w - \pi r^2w}{hws}$ where t = the approximate number of minutes of playing time remaining on a tape
   d = the depth of the tape in inches
   r = the radius of the core in inches
   w = the width of the tape in inches
   h = the thickness of the tape in inches
   and s = the speed at which the tape is played in inches per minute

Find the approximate number of minutes left on a tape where $d = \frac{5}{8}, r = \frac{7}{16}, w = \frac{3}{32}, h = 0.0015, s = 100,$ and $\pi = 3.14$.
19.6 minutes

2. $R = \frac{Mv^2 + 2eMv^2 + e^2Mv^2}{2gM^2 + 4gMm + 2gm^2}$ where m = the mass of a golf ball in grams
   M = the mass of a golf-club head in grams
   v = the velocity of the club head in meters per second
   e = the coefficient of restitution between the ball and the club head
   g = the acceleration due to gravity (9.8 meters per second per second)
   and R = the maximum range of the golf ball in meters

The value e, coefficient of restitution, is a measure of how elastic a golf ball is. As the temperature increases, the ball becomes more elastic. At 32°F, the value of e is about 0.64. At 80°F, the value of e is 0.75. The mass of a golf ball is about 50 grams. Assume the mass of the club head is 250 grams and the velocity of the club head is 45 meters/second.

a. Find the maximum range at 32°F in meters and convert it to yards (1 m = 1.093 yds). Round your answers to the nearest whole number. **211 yd**

b. Find the maximum range at 80°F in meters and yards to the nearest whole number. **219.72 m; 240 yd**

c. Why might a golfer want to buy a battery-heated glove to keep spare golf balls warm? **Increasing the temperature of the ball appears to increase the distance the ball can travel.**

---

41. **Oil Leases** A drilling company has the oil rights for a rectangular piece of land that is 6 kilometers by 5 kilometers. According to a lease agreement, the company can only dig their wells in the inner two-thirds of the land. A uniform strip around the edge of the land must remain untouched. What is the width of the strip of land that must remain untouched? **0.5 km**

Mixed Review

42. Factor $100x^2 + 20x + 1$. (Lesson 10–5) **$(10x + 1)^2$**

43. Find $(5q + 2r)(8q - 3r)$. (Lesson 9–7) **$40q^2 + rq - 6r^2$**

44. $\left\{x \mid x < \frac{3}{22}\right\}$

44. Solve $9x + 4 < 7 - 13x$. (Lesson 7–3)

45. 14; −4

45. Determine the x- and y-intercepts of the graph of $2x - 7y = 28$. (Lesson 6–4)

46. **Patterns** Copy and complete the table below. (Lesson 5–5)

s	4	2	0	−2	−4
r(s)	19	11	3	−5	−13

47. **World Records** The Huey P. Long Bridge in Metairie, Louisiana, is the longest railroad bridge in the world. If you were traveling on a train going 60 miles per hour across the 22,996-foot bridge, how long would it take you to cross it? (Lesson 4–7) **4.355 minutes or about 4 minutes 21 seconds**

48. **Consumerism** Julio is offered two payment plans when he buys a sofa. Under one plan, he pays $400 down and x dollars per month for 9 months. Under the other plan, he pays no money down and $x + 25$ dollars per month for 12 months. How much does the sofa cost? (Lesson 3–5) **$700**

WORKING ON THE

In·ves·ti·ga·tion

Refer to the Investigation on pages 554–555.

the BRICKYARD

The perimeter P of a square is given by the formula $P = 4s$, where s is the length of the side of the square. The formula for the area A of a square is $A = s^2$. The perimeter of a rectangle is found using $P = 2\ell + 2w$, where ℓ is the length of the rectangle and w is the width. The formula for the area of the rectangle is $A = \ell w$.

1 Suppose the large square brick has a side x units long and the small square brick has a side y units long. What are the dimensions of the rectangular brick? What is the area of each of the bricks? Draw a diagram to illustrate the dimensions of each brick.

2 Refer to the drawings of all the rectangular brick patterns you found using 4, 5, 6, 7, 8, 9, and 10 bricks. Calculate the perimeter and area in terms of x and y of all of the patterns you discovered.

3 Create a table to record the linear dimensions, the perimeter, and the area of the brick patterns you found. Use the table to record the dimensions, perimeter, and area in terms of x and y. The table should be set up like the one below.

Add the results of your work to your Investigation Folder.

# of bricks in pattern	length of pattern	width of pattern	perimeter of pattern	area of pattern
4				
5				
6				

Extension

Connections Two circles have radii that are consecutive numbers. Find their radii if their areas differ by 37π square feet. **18 and 19 feet**

In·ves·ti·ga·tion

Working on the Investigation

The Investigation on pages 554–555 is designed to be a long-term project that is completed over several days or weeks. Encourage students to keep their materials in their Investigation Folder as they work on the Investigation.

VOCABULARY

After completing this chapter, you should be able to define each term, property, or phrase and give an example or two of each.

Algebra

composite numbers   (p. 558)

difference of squares   (p. 581)

factored form   (p. 559)

factoring or factored   (pp. 564, 565)

factoring by grouping   (p. 567)

greatest common factor (GCF)   (p. 559)

perfect square trinomials   (p. 587)

prime factorization   (p. 558)

prime numbers   (p. 558)

prime polynomial   (p. 577)

Pythagorean triple   (p. 583)

unique factorization theorem   (p. 558)

zero product property   (p. 594)

Problem Solving

guess and check   (p. 574)

UNDERSTANDING AND USING THE VOCABULARY

State whether each sentence is *true* or *false*. If false, replace the underlined word or number to make a true sentence.

1. The number 27 is an example of a <u>prime</u> number.   false, composite

2. <u>$2x$</u> is the greatest common factor (GCF) of $12x^2$ and $14xy$.   true

3. <u>66</u> is an example of a perfect square.   false, sample answer: 64

4. 61 is a <u>factor</u> of 183.   true

5. The prime factorization for 48 is <u>$3 \cdot 4^2$</u>.   false, $2^4 \cdot 3$

6. $x^2 - 25$ is an example of a <u>perfect square trinomial</u>.   false, difference of squares

7. The number 35 is an example of a <u>composite</u> number.   true

8. <u>$x^2 - 3x - 70$</u> is an example of a prime polynomial.   false; sample answer: $x^2 + 2x + 2$

9. The <u>unique factorization theorem</u> allows you to solve equations.   false, zero product property

10. <u>$(b - 7)(b + 7)$</u> is the factorization of a difference of squares.   true

Chapter 10   Highlights   **601**

Using the CHAPTER HIGHLIGHTS

The Chapter Highlights begins with a listing of the new terms, properties, and phrases that were introduced in this chapter. Have students define each term and provide an example or two of it, if appropriate.

Assessment and Evaluation Masters, pp. 255–256

10   NAME_____ DATE_____
Chapter 10 Test, Form 1B

Write the letter for the correct answer in the blank at the right of each problem.

1. What is the prime factorization of 342?   1. __C__
   A. $2 \cdot 3 \cdot 57$ B. $1 \cdot 2 \cdot 3 \cdot 57$ C. $2 \cdot 3 \cdot 3 \cdot 19$ D. none of these

2. Find the GCF of $18abc^3$ and $54ab^2$.   2. __C__
   A. $18ab^2c^3$ B. $9ab$ C. $18ab$ D. $432a^2b^3c^3$

3. Find the GCF of $3ab$, $4bc$, and $5ac$.   3. __B__
   A. 0   B. 1   C. $2c$ D. $60a^2b^2c^2$

4. Factor $3x^2 + 75$.   4. __B__
   A. $3(x + 5)(x + 5)$ B. $3(x^2 + 25)$
   C. $3(x - 5)(x - 5)$ D. none of these

5. Factor $6y(y^2 - 6) + 5(6 - y^2)$.   5. __A__
   A. $(6y - 5)(y^2 - 6)$ B. $(6y + 5)(6 - y^2)$
   C. $(6y + 5)(y^2 - 6)$ D. none of these

6. If $x^3 - 7x^2 + 4x - 28$ is factored completely, one of the factors is   6. __D__
   A. $x + 2$.   B. $x + 7$.   C. $x^2 + 7$.   D. $x^2 + 4$.

7. Factor $5x^2 - 13x + 6$.   7. __D__
   A. $(x + 3)(5x - 2)$ B. $(x - 3)(5x + 2)$
   C. $(x + 2)(5x + 3)$ D. $(x - 2)(5x - 3)$

8. If $14a^2 - 15a + 4$ is factored completely, one of the factors is   8. __B__
   A. $7a + 2$.   B. $7a - 4$.   C. $7a - 1$.   D. $14a - 1$.

9. If $6x^2 + 11x - 2$ is factored completely, one of the factors is   9. __B__
   A. $3x - 1$.   B. $6x - 1$.   C. $2x + 1$.   D. $3x - 2$.

10. Which polynomial is prime?   10. __C__
    A. $x^2 + 13x - 114$ B. $n^2 - 2ny - 195y^2$
    C. $x^2 + 2x - 18$ D. More than one of these is prime.

11. Factor $100v^2 - 64y^2$.   11. __A__
    A. $(10v - 8y)(10v + 8y)$ B. $(10v - 8y)(10v - 8y)$
    C. $(10v + 8y)(10v + 8y)$ D. none of these

10   NAME_____ DATE_____
Chapter 10 Test, Form 1B (continued)

12. Factor $n^4 - 5n^2 - 36$.   12. __D__
    A. $(n + 2)(n - 2)(n + 3)(n - 3)$ B. $(n^2 + 9)(n + 2)(n - 2)$
    C. $(n^2 - 4)(n^2 + 9)$ D. none of these

13. How much less than 1600 is the product of 38 and 42?   13. __C__
    A. 1   B. 2   C. 4   D. 16

14. Each of the following is a perfect square except   14. __C__
    A. $4x^2 + 4x + 1$.   B. $x^2 - 6x + 9$.
    C. $x^2 + 10x - 25$.   D. All are perfect squares.

15. If $6x^2 + 48x + 96$ is factored completely, one of the factors is   15. __A__
    A. $x + 4$.   B. $3x + 8$.   C. $3x + 12$.   D. $6x + 16$.

16. Find the value of c that makes $25y^2 - 40y + c$ a perfect square trinomial.   16. __D__
    A. 64   B. 25   C. 20   D. 16

17. What is the solution set of the equation $(3w + 4)(2w - 7) = 0$?   17. __C__
    A. $\left\{-\frac{3}{4}, \frac{2}{7}\right\}$ B. $\left\{\frac{3}{4}, -\frac{2}{7}\right\}$ C. $\left\{-\frac{4}{3}, \frac{7}{2}\right\}$ D. $\left\{\frac{4}{3}, -\frac{7}{2}\right\}$

18. What is the solution set of the equation $4x^2 = 20x - 25$?   18. __D__
    A. $\left\{0, \frac{5}{4}\right\}$ B. $\left\{0, -\frac{5}{2}\right\}$ C. $\left\{-\frac{4}{5}\right\}$ D. $\left\{\frac{5}{2}\right\}$

19. What is the solution set of the equation $x^2 - 16x = 0$?   19. __C__
    A. $\{4, -4\}$ B. $\{0, 4, -4\}$ C. $\{0, 16\}$ D. $\{16\}$

20. The width of a rectangle is 6 centimeters less than the length. The area of the rectangle is 112 square centimeters. What is the length?   20. __D__
    A. 16 cm   B. 31 cm   C. 7 cm   D. 14 cm

Bonus   Insert parentheses so that a true equation results.   Bonus _____
$7 + 3 \cdot 5 - 6 - 2 \cdot 4 = 6$ $7 + 3 \cdot 5 - (6 - 2) \cdot 4 = 6$

Instructional Resources

Three multiple-choice tests and three free-response tests are provided in the *Assessment and Evaluation Masters*. Forms 1A and 2A are for honors pacing, and Forms 1B, 1C, 2B, and 2C are for average pacing. Chapter 10 Test, Form 1B is shown at the right. Chapter 10 Test, Form 2B is shown on the next page.

Skills and Concepts Encourage students to refer to the objectives and examples on the left as they complete the review exercises on the right.

Assessment and Evaluation Masters, pp. 261–262

CHAPTER **10** **STUDY GUIDE AND ASSESSMENT**

SKILLS AND CONCEPTS

OBJECTIVES AND EXAMPLES	REVIEW EXERCISES

Upon completing this chapter, you should be able to:

Use these exercises to review and prepare for the chapter test.

• find the prime factorization of integers (Lesson 10–1)

Find the prime factorization of 180.

$$180 = 2 \cdot 90$$
$$= 2 \cdot 2 \cdot 45$$
$$= 2 \cdot 2 \cdot 3 \cdot 15$$
$$= 2 \cdot 2 \cdot 3 \cdot 3 \cdot 5$$

The prime factorization of 180 is $2 \cdot 2 \cdot 3 \cdot 3 \cdot 5$ or $2^2 \cdot 3^2 \cdot 5$.

State whether each number is *prime* or *composite*. If the number is composite, find its prime factorization.

11. 28   composite, $2^2 \cdot 7$
12. 33   composite, $3 \cdot 11$
13. 150   composite, $2 \cdot 3 \cdot 5^2$
14. 301   prime
15. 83   prime
16. 378 $2 \cdot 3^3 \cdot 7$

• find the greatest common factors (GCF) for sets of monomials   (Lesson 10–1)

Find the GCF of $15x^2y$ and $45xy^2$.

$15x^2y = ③\cdot⑤\cdot⑥x\cdot x \cdot ⑥y$
$45xy^2 = ③\cdot 3 \cdot ⑤\cdot⑥x\cdot⑥y\cdot y$

The GCF is $3 \cdot 5 \cdot x \cdot y$ or $15xy$.

Find the GCF of the given monomials.

17. 35, 30   5
18. 12, 18, 40   2
19. $12ab, -4a^2b^2$ $4ab$
20. $16mrt, 30m^2r$ $2mr$
21. $20n^2, 25np^5$ $5n$
22. $60x^2y^2, 35xz^3$ $5x$
23. $56x^3y, 49ax^2$ $7x^2$
24. $6a^2, 18b^2, 9b^3$ 3

• use the greatest common factor (GCF) and the distributive property to factor polynomials (Lesson 10–2)

Factor $12a^2 - 8ab$.

$$12a^2 - 8ab = 4a(3a) - 4a(2b)$$
$$= 4a(3a - 2b)$$

Factor each polynomial.

25. $13x + 26y$ $13(x + 2y)$
26. $6x^2y + 12xy + 6$ $6(x^2y + 2xy + 1)$
27. $24a^2b^2 - 18ab$ $6ab(4ab - 3)$
28. $26ab + 18ac + 32a^2$ $2a(13b + 9c + 16a)$
29. $36p^2q^2 - 12pq$ $12pq(3pq - 1)$
30. $a + a^2b + a^3b^3$ $a(1 + ab + a^2b^3)$

• use grouping techniques to factor polynomials with four or more terms   (Lesson 10–2)

Factor $2x^2 - 3xz - 2xy + 3yz$.

$$2x^2 - 3xz - 2xy + 3yz = (2x^2 - 3xz) +$$
$$(- 2xy + 3yz)$$
$$= x(2x - 3z) - y(2x - 3z)$$
$$= (x - y)(2x - 3z)$$

Factor each polynomial.

31. $a^2 - 4ac + ab - 4bc$ $(a - 4c)(a + b)$
32. $4rs + 12ps + 2mr + 6mp$ $2(r + 3p)(2s + m)$
33. $16k^3 - 4k^2p^2 - 28kp + 7p^3$ $(4k - p^2)(4k^2 - 7p)$
34. $dm + mr + 7r + 7d$ $(d + r)(m + 7)$
35. $24am - 9an + 40bm - 15bn$
36. $a^3 - a^2b + ab^2 - b^3$ $(a - b)(a^2 + b^2)$

35. $(8m - 3n)(3a + 5b)$

602   *Chapter 10   Study Guide and Assessment*

---

10   NAME_____ DATE _____

Chapter 10 Test, Form 2B

State whether each number is prime or composite. If the number is composite, find its prime factorization.

1. 89

1. _____prime_____

2. 132

2. _____composite;_____
   $2 \cdot 2 \cdot 3 \cdot 11$

Find the GCF of the given monomials.

3. 305, 55

3. _____5_____

4. −80, 45

4. _____5_____

5. $12x^3y^2, 44xy^3$

5. _____$4xy^2$_____

6. $2a^2b^2, 5b^2c^2, 9a^2c^2$

6. _____1_____

Factor each polynomial.

7. $7x - 19xy^2$

7. _____$x(7 - 19y^2)$_____

8. $5s^3 + 45s$

8. _____$5s(s^2 + 9)$_____

9. $3x^2 + 21x + 27$

9. _____$3(x^2 + 7x + 9)$_____

10. $v^2n + 2v^2 - 4n - 8$

10. _____$(n + 2)(v - 2)(v + 2)$_____

Factor each polynomial, if possible. If the polynomial cannot be factored using integers, write prime.

11. $36m^2 - 64$

11. _____$4(3m + 4)(3m - 4)$_____

12. $u^4 - 15u^2 - 16$

12. _____$(u^2 + 1)(u + 4)(u - 4)$_____

13. $n^2 + nw - 42w^2$

13. _____$(n + 7w)(n - 6w)$_____

10   NAME_____ DATE _____

Chapter 10 Test, Form 2B (continued)

14. $54 + 3y - y^2$

14. _____$(9 - y)(6 + y)$_____

15. $b^2 - 6b + 10$

15. _____prime_____

16. $25w^2 - 60w + 36$

16. _____$(5w - 6)^2$_____

17. $8n^2y - 36ny + 40y$

17. _____$4y(2n - 5)(n - 2)$_____

18. $y^4 - v^4$

18. _____$(y^2 + v^2)(y - v)(y + v)$_____

19. $36x^3 - 84x^2 + 49x$

19. _____$x(6x - 7)^2$_____

20. The area of a square is $(49 - 56x + 16x^2)$ square feet. If x is a positive integer, what is the least possible perimeter measure for the square?

20. _____12 ft_____

21. Find the value of c that will make $9x^2 + 30x + c$ a perfect square trinomial.

21. _____25_____

Solve each equation.

22. $8y(2y - 1) = 0$

22. _____$0, \frac{1}{2}$_____

23. $9n^2 = 12n - 4$

23. _____$\frac{2}{3}$_____

For each problem, define a variable. Then use an equation to solve each problem.

24. The sum of two positive integers is 25. The difference of their squares is 275. Find the two integers.

24. _____7 and 18_____

25. The area of a rectangular room is 238 square feet. The width is 3 feet less than the length. What are the dimensions?

25. _____17 ft by 14 ft_____

Bonus Factor $v^2x^2 + 9n^2 - 9x^2 - v^2n^2$.

Bonus _____$(x + n)(x - n)(v + 3)(v - 3)$_____

GLENCOE Technology

Test and Review Software

You may use this software, a combination of an item generator and item bank, to create your own tests or worksheets. Types of items include free response, multiple choice, short answer, and open ended.

For IBM & Macintosh

OBJECTIVES AND EXAMPLES

• factor quadratic trinomials (Lesson 10–3)

$a^2 - 3a - 4 = (a + 1)(a - 4)$

$$4x^2 - 4xy - 15y^2 = 4x^2 + (-10 + 6)xy - 15y^2$$
$$= 4x^2 - 10xy + 6xy - 15y^2$$
$$= (4x^2 - 10xy) + (6xy - 15y^2)$$
$$= 2x(2x - 5y) + 3y(2x - 5y)$$
$$= (2x - 5y)(2x + 3y)$$

• identify and factor binomials that are the differences of squares (Lesson 10–4)

$a^2 - 9 = (a)^2 - (3)^2$
$$= (a - 3)(a + 3)$$

$3x^3 - 75x = 3x(x^2 - 25)$
$$= 3x(x - 5)(x + 5)$$

• identify and factor perfect square trinomials
(Lesson 10–5)

$16z^2 - 8z + 1$
$$= (4z)^2 - 2(4z)(1) + (1)^2$$
$$= (4z - 1)^2$$

$9x^2 + 24xy + 16y^2 = (3x)^2 - 2(3x)(4y) + (4y)^2$
$$= (3x + 4y)^2$$

REVIEW EXERCISES

Factor each trinomial, if possible. If the trinomial cannot be factored using integers, write *prime*.

37. $y^2 + 7y + 12$ $(y + 3)(y + 4)$

38. $x^2 - 9x - 36$ $(x - 12)(x + 3)$

39. $6z^2 + 7z + 3$ prime

40. $b^2 + 5b - 6$ $(b + 6)(b - 1)$

41. $2r^2 - 3r - 20$ $(r - 4)(2r + 5)$

42. $3a^2 - 13a + 14$ $(3a - 7)(a - 2)$

Factor each polynomial, if possible. If the polynomial cannot be factored using integers, write *prime*.

43. $b^2 - 16$ $(b - 4)(b + 4)$

44. $25 - 9y^2$ $(5 - 3y)(5 + 3y)$

45. $16a^2 - 81b^4$ $(4a - 9b^2)(4a + 9b^2)$

46. $2y^3 - 128y$ $2y(y - 8)(y + 8)$

47. $9b^2 - 20$ prime

48. $\frac{1}{4}n^2 - \frac{9}{16}r^2$ $\left(\frac{1}{2}n - \frac{3}{4}r\right)\left(\frac{1}{2}n + \frac{3}{4}r\right)$

Factor each polynomial, if possible. If the polynomial cannot be factored using integers, write *prime*.

49. $a^2 + 18a + 81$ $(a + 9)^2$

50. $9k^2 - 12k + 4$ $(3k - 2)^2$

51. $4 - 28r + 49r^2$ $(2 - 7r)^2$

52. $32n^2 - 80n + 50$ $2(4n - 5)^2$

53. $6b^3 - 24b^2g + 24bg^2$ $6b(b - 2g)^2$

54. $49m^2 - 126m + 81$ $(7m - 9)^2$

55. $25x^2 - 120x + 144$ $(5x - 12)^2$

CHAPTER 10   STUDY GUIDE AND ASSESSMENT

OBJECTIVES AND EXAMPLES	REVIEW EXERCISES

OBJECTIVES AND EXAMPLES

• use the zero product property to solve equations   (Lesson 10–6)

Solve $b^2 - b - 12 = 0$.

$$b^2 - b - 12 = 0$$

$$(b - 4)(b + 3) = 0$$

If $(b - 4)(b + 3) = 0$, then $(b - 4) = 0$ or $(b + 3) = 0$.

$$b - 4 = 0 \quad \text{or} \quad b + 3 = 0$$

$$b = 4 \qquad\qquad b = -3$$

The solution set is $\{4, -3\}$.

REVIEW EXERCISES

Solve each equation. Check your solution.

56. $y(y + 11) = 0$ $\{0, -11\}$

57. $(3x - 2)(4x + 7) = 0$ **57.** $\left\{\frac{2}{3}, -\frac{7}{4}\right\}$

58. $2a^2 - 9a = 0$ **58.** $\left\{0, \frac{9}{2}\right\}$

59. $n^2 = -17n$ $\{0, -17\}$

60. $\frac{3}{4}y = \frac{1}{2}y^2$ $\left\{0, \frac{3}{2}\right\}$

61. $y^2 + 13y + 40 = 0$ $\{-5, -8\}$

62. $2m^2 + 13m = 24$ **62.** $\left\{-8, \frac{3}{2}\right\}$

63. $25r^2 + 4 = -20r$ **63.** $\left\{-\frac{2}{5}\right\}$

APPLICATIONS AND PROBLEM SOLVING

64. Geometry   The measure of the area of a rectangle is $4m^2 - 3mp + 3p - 4m$. If the dimensions of the rectangle are represented by polynomials with integral coefficients, find the dimensions of the rectangle.   (Lesson 10–2) **$(4m - 3p)$ by $(m - 1)$**

65. Guess and Check   Numero Uno says, "I am thinking of a three-digit number. If you multiply the digits together and then multiply the result by 4, the answer is the number I'm thinking of. What is my number?"   (Lesson 10–3)   **384**

66. Photography   To get a square photograph to fit into a rectangular frame, Li-Chih had to trim a 1-inch strip from one pair of opposite sides of the photo and a 2-inch strip from the other two sides. In all, he trimmed off 64 square inches. What were the original dimensions of the photograph?   (Lesson 10–4)   **12 in. by 12 in.**

67. Guess and Check   Fill in each box below with a digit from 1 to 6 to make this multiplication work. Use each digit exactly once.   (Lesson 10–3)

$$\begin{array}{r} 3 \\ \times \; 5\;4 \\ \hline 1\;6\;2 \end{array}$$

68. Geometry   The measure of the area of a rectangle is $16x^2 - 9$. Find the measure of its perimeter.   (Lesson 10–4)   **16x**

69. Number Theory   The product of two consecutive odd integers is 99. Find the integers.   (Lesson 10–6)   **9, 11; −11, −9**

A practice test for Chapter 10 is provided on page 796.

ALTERNATIVE ASSESSMENT

COOPERATIVE LEARNING PROJECT

Art and Framing In this project, you will determine what size mat and frame will be the most visually stimulating for an abstract print that you just bought on your trip to The Chicago Institute of Art. The print is 9 inches by 12 inches and can be cropped an inch on either side and not spoil its effect.

You called a friend to get some advice on how to mat and frame your print. This friend majored in art and works in the art industry, but he also enjoys puzzling you. He gave you two problems and told you to solve them and decide which you think would be the most appropriate dimensions for your mat and frame. Here are the two problems.

- The inside rectangle has a width four inches less than its length. The length of the larger rectangle is twice its width. Find the dimensions of each rectangle if the matted area is twice as much as the inside rectangle area and the perimeter of the inside rectangle is 32 inches less than the perimeter of the larger rectangle.

- The inside rectangle has a width four inches less than its length. The length of the larger rectangle is twice its width. Find the dimensions of each rectangle if the matted area is seven times as much in square inches as the perimeter of the smaller rectangle is in inches, and the width of the larger rectangle is two inches more than the length of the smaller rectangle.

Which problem will give you the dimensions that will best suit your print? What are the dimensions? Will you have to crop your print? If so, how much?

Follow these steps to determine the appropriate dimensions for the mat and frame.

- Illustrate each of the problems.
- Label each of the drawings with as much information as possible.
- Develop an equation for each problem that describes the situation.
- Investigate your answers and determine the appropriate dimensions for the mat and frame.
- If possible, find a 9-inch by 12-inch photograph or advertisement from a magazine. Have two color photocopies made of it and frame each with a cardboard frame that meets the specifications.
- Write a paragraph describing the two problems and how you determined the solution.

THINKING CRITICALLY

- A *nasty* number is a positive integer with at least four different factors such that the difference between the numbers in one pair of factors equals the sum of the numbers in another pair. The first nasty number is 6 since $6 = 6 \cdot 1 = 2 \cdot 3$ and $6 - 1 = 2 + 3$. Find the next five nasty numbers. (*Hint*: They are all multiples of 6.)

- Write an equation with integral coefficients that has $\left\{\frac{2}{3}, -1\right\}$ as the solution.

PORTFOLIO

When using the guess-and-check strategy for solving a math problem, if a first guess does not work, you must make a second guess. How do you make a second guess when the first one is incorrect? In making this second guess, you must know how and why the first guess was incorrect. Find a problem from your work in this chapter in which you used the guess-and-check strategy. Write a step-by-step description of why you chose each successive guess after the previous guess for that problem. Place this in your portfolio.

SELF EVALUATION

Having a strategy before solving a problem is helpful. It helps you to focus on the problem and the solution and to be organized in the solution of your problem. Using strategies or check lists are fundamental skills used in critical thinking.

Assess yourself. Do you determine a strategy for solving a problem before you delve into it? Do you stay with the initial plan or do you reevaluate after a time period and try a new strategy? Do you keep a record of your strategy and attempts so that you can refer back to it to determine if you are still on track? Give an example of a math problem in which a strategy is beneficial in order to keep track of what has been attempted. Also, give an example of a daily life problem in which you used a strategy to help organize your plan for a solution.

Assessment and Evaluation Masters, pp. 266, 277

10   NAME_____ DATE _____

Chapter 10 Performance Assessment

Instructions: *Demonstrate your knowledge by giving a clear, concise solution to each problem. Be sure to include all relevant drawings and justify your answers. You may show your solution in more than one way or investigate beyond the requirements of the problem.*

1. One way to factor a trinomial such as $x^2 - 2x - 3$ is to assign a value such as 10 to x and evaluate the expression.

$$x^2 - 2x - 3 = 10^2 - 2(10) - 3$$
$$= 100 - 20 - 3$$
$$= 77$$

A factorization of 77 is 7×11. Since $x = 10$, $7 = (x - 3)$ and $11 = (x + 1)$. Multiply to see if $(x - 3)(x + 1) = x^2 - 2x - 3$.

   a. Try the above method to factor $x^2 - 8x + 15$. Show your work and explain each step. Did the method give you the correct factors?

   b. Try the above method to factor $2x^2 - 13x - 24$. *Hint: The factors will be of the form $(2x + a)(x + b)$.* Show your work and explain each step.

   c. Try the above method to factor $x^2 - 2x - 8$. Why is it difficult to find the correct factors for this trinomial using the above method?

   d. Evaluate the expression in part c for $x = 7$. Then use the above method to find the factors. Explain each step. Remember in finding your factors that $x = 7$.

2. a. Tell how to factor a polynomial with four terms by grouping.

   b. Use grouping to factor $10ax - 5ay + 2bx - by$. Explain each step.

   c. Factor $6x^3 + 3x^2y - 2xy - y^2$ in two different ways. Show your work.

Scoring Guide
Chapter 10
Performance Assessment

Level	Specific Criteria
3 Superior	• Shows thorough understanding of the concepts of *prime factorization of integers, factorization of polynomials, evaluating polynomials, and the guess-and-check strategy for problem solving*. • Computations are correct. • Written explanations are exemplary. • Goes beyond requirements of some or all problems.
2 Satisfactory, with Minor Flaws	• Shows understanding of the concepts of *prime factorization of integers, factorization of polynomials, evaluating polynomials, and the guess-and-check strategy for problem solving*. • Computations are mostly correct. • Written explanations are effective. • Satisfies all requirements of problems.
1 Nearly Satisfactory, with Serious Flaws	• Shows understanding of most of the concepts of *prime factorization of integers, factorization of polynomials, evaluating polynomials, and the guess-and-check strategy for problem solving*. • Computations are mostly correct. • Written explanations are satisfactory. • Satisfies most requirements of problems.
0 Unsatisfactory	• Shows little or no understanding of the concepts of *prime factorization of integers, factorization of polynomials, evaluating polynomials, and the guess-and-check strategy for problem solving*. • Computations are incorrect. • Written explanations are not satisfactory. • Does not satisfy requirements of problems.

Alternative Assessment

The Alternative Assessment section provides students with the opportunity to assess their own work by thinking critically, working with others, keeping a portfolio, and honestly evaluating their own progress. For more information on alternative forms of assessment, see *Alternative Assessment in the Mathematics Classroom,* one of the titles in the Glencoe Mathematics Professional Series.

Performance Assessment

Performance Assessment tasks for this chapter are included in the *Assessment and Evaluation Masters.* A scoring guide is also provided.

These two pages review the skills and concepts presented in Chapters 1–10. This review is formatted to reflect new trends in standardized testing.

A more traditional cumulative review, shown below, is provided in the *Assessment and Evaluation Masters*, pp. 271–272.

Assessment and Evaluation Masters, pp. 271–272

NAME _____ DATE _____

Chapter 10 Cumulative Review

1. Find $-\frac{3}{8} + \frac{5}{16}$. (Lesson 2-5)
1. $\frac{1}{16}$

2. Simplify $\frac{2}{3}(6a - 9b) - \frac{1}{2}(2a - 12b)$. (Lesson 2-6)
2. $3a$

3. Solve $\frac{3y-2}{5} = \frac{1}{10}y$. (Lesson 3-5)
3. $\frac{4}{5}$

4. Find two consecutive even integers whose sum is 74. (Lesson 3-3)
4. 36 and 38

5. Two airplanes leave Atlanta at the same time and fly in opposite directions. One plane travels 60 mi/h faster than the other. After $2\frac{1}{2}$ hours they are 1700 miles apart. What is the rate of each plane? (Lesson 4-7)
5. 310 mi/h; 370 mi/h

Find the probability of each outcome if two marbles are selected from a bag containing 3 red marbles, 4 blue marbles, and 2 green marbles. The first marble is not replaced before the second marble is selected. (Lesson 4-6)

6. selecting a blue marble first
6. $\frac{4}{9}$

7. selecting a yellow marble first
7. 0

8. selecting a red marble followed by a blue marble
8. $\frac{1}{6}$

9. selecting a green marble after selecting two marbles that are not green
9. $\frac{1}{6}$

10. Write an equation for the relationship between the variables in the chart at the right. (Lesson 5-6)

x	0	2	4	6
y	4	8	12	16

10. $y = 2x + 4$

11. If $f(x) = 2x^2 - 5x + 4$, find $f(-3)$. (Lesson 5-5)
11. 37

12. Graph $4x + 2y = 1$ using the slope and y-intercept. (Lesson 6-5)
12.

NAME _____ DATE _____

Chapter 10 Cumulative Review *(continued)*

13. Write an equation in slope-intercept form of the line that is parallel to the graph of $2y - 6x = 3$ and passes through $(-3, 2)$. (Lesson 6-6)
13. $y = 3x + 11$

14. Solve $\frac{2}{3}x > -10$. (Lesson 7-2)
14. $x > -15$

15. Graph the inequality $x + 2y < 4$. (Lesson 7-8)
15.

16. Solve the system of equations $3y - 2x = 4$ and $2y + 4x = 8$. (Lesson 8-4)
16. $(1, 2)$

17. Simplify $(-2x^2y^3)^4$. (Lesson 9-1)
17. $16x^8y^{12}$

18. Evaluate $\frac{3.6 \times 10^{-4}}{1.2 \times 10^{-8}}$. Express the result in scientific notation. (Lesson 9-3)
18. 3×10^4

19. Simplify $5(2y^2 + 3y - 2) - 8(3y^2 + 4y - 2)$. (Lesson 9-5)
19. $-14y^2 - 17y + 6$

20. Find $(a + 6)(a - 5)$. (Lesson 9-8)
20. $a^2 + a - 30$

Factor each polynomial, if possible. (Lessons 10-2 and 10-3)

21. $m^2 + 12m + 36$
21. $(m + 6)^2$

22. $2ar + 2rt - a - t$
22. $(2r - 1)(a + t)$

23. $36c^3 + 6c^2 - 6c$
23. $6c(3c - 1)(2c + 1)$

24. Solve $x(x + 12) = 0$. (Lesson 10-6)
24. $0, -12$

25. Find two consecutive odd integers whose product is 255. (Lesson 10-6)
25. 15 and 17 or -17 and -15

CUMULATIVE REVIEW

CHAPTERS 1–10

SECTION ONE: MULTIPLE CHOICE

There are nine multiple-choice questions in this section. After working each problem, write the letter of the correct answer on your paper.

1. The square of a number subtracted from 8 times the number is equal to twice the number. Find the number. **A**
   A. 0 or 6
   B. 3
   C. 2
   D. 0 or 10

2. Choose the open sentence that represents the range of acceptable diameters for a lawn mower bolt that will work properly only if its diameter differs from 2 cm by no more than 0.04 cm. **C**
   A. $|d - 0.04| \geq 2$
   B. $|d| < 1.96$
   C. $|d - 2| \leq 0.04$
   D. $|d| > 2.04$

3. Choose an expression for the area of the shaded region shown below. **D**

   A. $(4t)(t + 3)$
   B. $(4t - t)(t + 3) - (t + 1)$
   C. $(t + 1)(t + 3) - (4t)(t)$
   D. $(4t)(t + 3) - (t + 1)$

4. Choose the equivalent equation for $A = \frac{1}{2}h(a + b)$. **A**
   A. $h = \frac{2A}{a + b}$
   B. $h = 2A - (a + b)$
   C. $h = \frac{\frac{1}{2}(a + b)}{A}$
   D. $h = \frac{A - b}{2a}$

5. Bob and Vicki took a trip to Zuma Beach. On the way there, their average speed was 42 miles per hour. On the way home, their average speed was 56 miles per hour. If their total travel time was 7 hours, find the distance to the beach. **B**
   A. 126 miles
   B. 168 miles
   C. 98 miles
   D. 294 miles

6. Choose which statement is true. **B**
   A. The range is the difference between the greatest value and the lower quartile in a set of data.
   B. An outlier will only affect the mean of a set of data.
   C. Quartiles are values that divide a set of data into equal halves.
   D. The interquartile range is the sum of the upper and lower quartiles in a set of data.

7. Choose the prime polynomial. **C**
   A. $y^2 + 12y + 27$
   B. $6x^2 - 11x + 4$
   C. $h^2 + 5h - 8$
   D. $9k^2 + 30km + 25m^2$

8. Choose the equation of the line that passes through the points at $(9, 5)$ and $(-3, -4)$. **D**
   A. $y = -\frac{1}{12}x + \frac{23}{4}$
   B. $y = \frac{4}{3}x$
   C. $y = -\frac{7}{2}x - \frac{3}{4}$
   D. $y = \frac{3}{4}x - \frac{7}{4}$

9. Carrie's bowling scores for four games are $b + 2$, $b + 3$, $b - 2$, and $b - 1$. What must her score be on her fifth game to average $b + 2$? **A**
   A. $b + 8$
   B. b
   C. $b - 2$
   D. $b + 5$

Standardized Test Practice Questions are also provided in the *Assessment and Evaluation Masters*, p. 270.

SECTION TWO: SHORT ANSWER

This section contains ten questions for which you will provide short answers. Write your answer on your paper.

10. The measure of the perimeter of a square is $20m + 32p$. Find the measure of its area.
$$25m^2 + 80mp + 64p^2$$

11. At a bake sale, cakes cost twice as much as pies. Pies were $4 more than triple the price of cookies. Darin bought a cake, three pies, and four cookies for $24.75. What is the price of each item?
cake: $9.50, pie: $4.75, cookie: $0.25

12. Factor $3g^2 - 10gh - 8h^2$.
$$(3g + 2h)(g - 4h)$$

13. On the first day of school, 264 school note-books were sold. Some sold for 95¢ each, and the rest sold for $1.25 each. How many of each were sold if the total sales were $297?
110 at 95¢, 154 at $1.25

14. A rectangular photograph is 8 centimeters wide and 12 centimeters long. The photograph is enlarged by increasing the length and width by an equal amount. If the area of the new photograph is 69 square centimeters greater than the area of the original photograph, what are the dimensions of the new photograph?
11 cm by 15 cm

15. Linda plans to spend at most $50 on shorts and blouses. She bought 2 pairs of shorts for $14.20 each. How much can she spend on blouses?
$21.60 or less

16. The difference of two numbers is 3. If the difference of their squares is 15, what is the sum of the numbers?   5

17. Geometry The length of a rectangle is eight times its width. If the length was decreased by 10 meters and the width was decreased by 2 meters, the area would be decreased by 162 square meters. Find the original dimensions.   7 m by 56 m

18. Ben's car gets between 18 and 21 miles per gallon of gasoline. If his car's tank holds 15 gallons, what is the range of distance that Ben can drive his car on one tank of gasoline?
270 miles to 315 miles

19. Find the measure of the third side of a triangle if the perimeter of the triangle is $8x^2 + x + 15$ and the measures of two of its sides can be represented by the expressions $2x^2 - 5x + 7$ and $x^2 - x + 11$. $5x^2 + 7x - 3$

SECTION THREE: OPEN-ENDED

This section contains two open-ended problems. Demonstrate your knowledge by giving a clear, concise solution to each problem. Your score on these problems will depend on how well you do the following.

• Explain your reasoning.
• Show your understanding of the mathematics in an organized manner.
• Use charts, graphs, and diagrams in your explanation.
• Show the solution in more than one way or relate it to other situations.
• Investigate beyond the requirements of the problem.

20. A tinsmith is going to make a box out of tin by cutting a square from each corner and folding up the sides. The box needs to be 3 inches high (so he will be cutting out 3-inch squares), and it needs to be twice as long as it is wide, so that it can hold two smaller square cardboard boxes. Finally, its volume needs to be 1350 cubic inches. What should the dimensions be?
3 in. by 30 in. by 15 in.

21. The length of a rectangle is five times its width. If the length was increased by 7 meters and the width was decreased by 4 meters, the area would be decreased by 132 square meters. Find the original dimensions.   8 m by 40 m

Exploring Quadratic and Exponential Functions

PREVIEWING THE CHAPTER

This chapter introduces students to quadratic equations by having them find the equation of the axis of symmetry and the coordinates of the vertex. Students graph quadratic functions, compare different parabolas, and approximate roots of quadratic functions. Next, students solve quadratic equations using the quadratic formula, focus on the discriminant, and study how it can be used to determine the number of real roots of an equation. Students graph exponential functions on the graphing calculator and on paper. Finally, students analyze exponential functions and use problem-solving techniques to solve equations that have real exponents.

Lesson (Pages)	Lesson Objectives	NCTM Standards	State/Local Objectives
11-1A (610)	Use a graphing calculator to graph a quadratic function and find the coordinates of its vertex.	1–6	
11-1 (611–617)	Find the equation of the axis of symmetry and the coordinates of the vertex of a parabola. Graph quadratic functions.	1–6, 13	
11-1B (618–619)	Use a graphing calculator to study the characteristics of families of parabolas.	1–6	
11-2 (620–627)	Use estimation to find roots of quadratic equations. Find roots of quadratic equations by graphing.	1–6, 13	
11-3 (628–633)	Solve quadratic equations by using the quadratic formula.	1–6	
11-4A (634)	Use a graphing calculator to graph exponential functions.	1–6, 13	
11-4 (635–642)	Graph exponential functions. Determine if a set of data displays exponential behavior. Solve exponential equations.	1–6, 13	
11-5 (643–649)	Solve problems involving growth and decay.	1–6	

A complete, 1-page lesson plan is provided for each lesson in the *Lesson Planning Guide*. Answer keys for each lesson are available in the *Answer Key Masters*.

ORGANIZING THE CHAPTER

You may want to refer to the **Course Planning Calendar** on page T12 for detailed information on pacing.
PACING: Standard—11 days; **Honors**—10 days; **Block**—5 days; **Two Years**—19 days

LESSON PLANNING CHART

| Lesson (Pages) | Materials/ Manipulatives | Extra Practice (Student Edition) | BLACKLINE MASTERS | | | | | | | | | Real-World Applications | Interactive Mathematics Tools Software | Teaching Transparencies |
			Study Guide	Practice	Enrichment	Assessment and Evaluation	Modeling Mathematics	Multicultural Activity	Tech Prep Applications	Graphing Calculator	Science and Math Lab Manual			
11-1A (610)	graphing calculator									pp. 30, 31				
11-1 (611–617)	grid paper	p. 780	p. 75	p. 75	p. 75		p. 82			p. 11			11-1	11-1A 11-1B
11-1B (618–619)	graphing calculator									pp. 32, 33			11-1B	
11-2 (620–627)	graphing calculator	p. 781	p. 76	p. 76	p. 76	p. 296			p. 21					11-2A 11-2B
11-3 (628–633)	scientific calculator	p. 781	p. 77	p. 77	p. 77	pp. 295, 296		p. 21	p. 22			27	11-3.1 11-3.2	11-3A 11-3B
11-4A (634)	graphing calculator									pp. 34, 35				
11-4 (635–642)	large piece of paper graphing calculator	p. 781	p. 78	p. 78	p. 78	p. 297								11-4A 11-4B
11-5 (643–649)	spreadsheets	p. 782	p. 79	p. 79	p. 79	p. 297		p. 22			pp. 49–52	28		11-5A 11-5B
Study Guide/ Assessment (651–655)						pp. 281 –294, 298– 300								

ORGANIZING THE CHAPTER

OTHER CHAPTER RESOURCES

Student Edition
Chapter Opener, pp. 608–609
Mathematics and Society, p. 649
Working on the Investigation,
   pp. 617, 627
Closing the Investigation, p. 650

Teacher's Classroom Resources
Investigations and Projects Masters,
   pp. 65–68

Technology
Test and Review Software (IBM
   and Macintosh)
CD-ROM Interactions (Windows
   and Macintosh)

Professional Publications
Block Scheduling Booklet
Glencoe Mathematics Professional
   Series

OUTSIDE RESOURCES

Books/Periodicals
Isdell, Wendy, *A Gebra Named Al*, Free Spirit
   Publishing

Software
Algebra Xpresser, William K. Bradford

Videos/CD-ROMs
Algebra for Everyone, NCTM

ASSESSMENT RESOURCES

Student Edition
Math Journal, pp. 624, 631
Mixed Review, pp. 616, 627,
   633, 642, 649
Self Test, p. 633
Chapter Highlights, p. 651
Chapter Study Guide and
   Assessment, pp. 652–654
Alternative Assessment, p. 655
   Portfolio, p. 655

Teacher's Wraparound Edition
5-Minute Check, pp. 611, 620,
   628, 635, 643
Check for Understanding, pp. 614,
   624, 631, 640, 646
Closing Activity, pp. 617, 627,
   632, 641, 648
Cooperative Learning, pp. 616,
   630

Assessment and Evaluation Masters
Multiple-Choice Tests, Forms 1A
   (Honors), 1B (Average), 1C
   (Basic), pp. 281–286
Free-Response Tests, Forms 2A
   (Honors), 2B (Average), 2C
   (Basic), pp. 287–292
Calculator-Based Test, p. 293
Performance Assessment, p. 294
Mid-Chapter Test, p. 295
Quizzes A–D, pp. 296–297
Standardized Test Practice, p. 298
Cumulative Review, pp. 299–300

ENHANCING THE CHAPTER

Examples of some of the materials for enhancing Chapter 11 are shown below.

DIVERSITY

Multicultural Activity Masters, pp. 21, 22

11-3 NAME_____ DATE_____
Student Edition
Pages 628-632

Multicultural Activity

Emmy Noether

Emmy Noether (1882–1935) was known not only as a brilliant mathematician, but also as an exceptional teacher of mathematics. Born and raised in Germany, she taught at the University of Göttingen from 1916 until Hitler's rise to power, when she and several other of Germany's great scientists were forced to leave their country. She came to the United States in 1933, where she taught at Bryn Mawr College until her death in 1935.

Noether's greatest contributions to mathematics were in the field known as abstract algebra, in which problems are solved by a type of simplification; all extraneous information is eliminated, and only the underlying algebraic structure is examined. One of the most fundamental concepts of abstract algebra is **closure**.

A set of numbers is said to have **closure** under a given operation if, when you perform that operation on any numbers in the set, the result is also in the set. For instance, the set {−1, 0, 1} is closed under multiplication because, as you see at the right, any product of two numbers in the set is also in the set. However, this same set of numbers is *not closed* under addition. You need just one example to prove that this is true: the sum 1 + 1 is equal to 2, which is not in the given set.

{−1, 0, 1}
−1 × 0 = 0
−1 × 1 = −1
0 × 1 = 0
−1 × −1 = 1
0 × 0 = 0
1 × 1 = 1

Tell whether each set is closed or not closed under the given operation. If you think a set is not closed, give an example to support your answer.

1. {−1, 0, 1}; subtraction
   not closed; −1 − 1 = −2

2. {−1, 0, 1}; division
   not closed; $\frac{1}{0}$ is undefined

3. {−1, 1}; multiplication
   closed

4. {−1, 1}; division
   closed

5. {even integers}; addition
   closed

6. {odd integers}; addition
   not closed; 1 + 3 = 4

7. {positive even integers}; division
   not closed; $\frac{2}{4}$ = 0.5

8. {positive odd integers}; multiplication
   closed

9. {negative integers}; addition
   closed

10. {negative integers}; subtraction
    not closed; −1 − (−3) = 2

APPLICATIONS

Real-World Applications, 27, 28

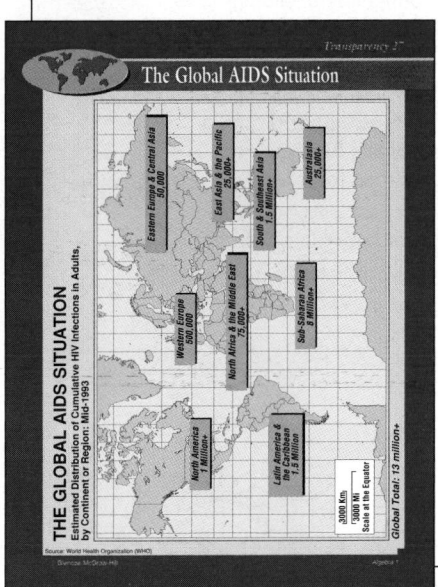

Transparency 27

The Global AIDS Situation

THE GLOBAL AIDS SITUATION
Estimated Distribution of Cumulative HIV Infections in Adults, by Continent or Region: Mid-1993

Eastern Europe & Central Asia 50,000
East Asia & the Pacific 25,000+
South & Southeast Asia 1.5 Million+
Australasia 25,000+
Western Europe 500,000
North Africa & the Middle East 75,000+
Sub-Saharan Africa 8 Million+
North America 1 Million+
Latin America & the Caribbean 1.5 Million

Global Total: 13 Million+

2000 Km
3000 Mi
Scale at the Equator

Source: World Health Organization (WHO)

TECHNOLOGY

Graphing Calculator Masters, p. 11

11-1 NAME_____ DATE_____
Student Edition
Pages 611-617

Graphing Calculator Activity

Maximum Area

Suppose Link Chainey wants to build a three-sided rectangular fence against the side of his barn. If x is the length of fence perpendicular to the barn, what is the maximum area of the rectangular enclosure he can build with 100 yards of fencing material? What are the dimensions of the enclosure?

The area y of this rectangle is given by the formula $y = x(100 − 2x)$. Use a graphing calculator to graph this function and then trace the coordinates of the highest point of the graph, the point where the area is a maximum.

a. **Select** convenient variables for the viewing window.

b. **Graph** the function $y = x(100 − 2x)$.

c. **Trace** the coordinates of the vertex of the parabola.

The coordinates of the vertex of the parabola are about (25, 1250). Therefore, the maximum area that 100 yards of fencing can enclose is 1250 square yards. The dimensions of the three-sided enclosure are 25 yd × 50 yd.

Use a graphing calculator to find the maximum area for each length of fence shown in the chart below. Give the dimensions of the rectangular enclosure.

	Length of Fence	Coordinates of Vertex	Maximum Area	Dimensions
1.	150 yd	(37.5, 2812.5)	2812.5 yd²	37.5 yd × 75 yd
2.	200 m	(50, 5000)	5000 m²	50 m × 100 m
3.	75 ft	(18.75, 703.125)	703.125 ft²	18.75 ft × 37.5 ft

4. Without graphing, predict the dimensions of the three-sided rectangular enclosure with the greatest area for a 50-meter length of fencing. 12.5 m × 25 m

TECH PREP

Tech Prep Applications Masters, pp. 21, 22

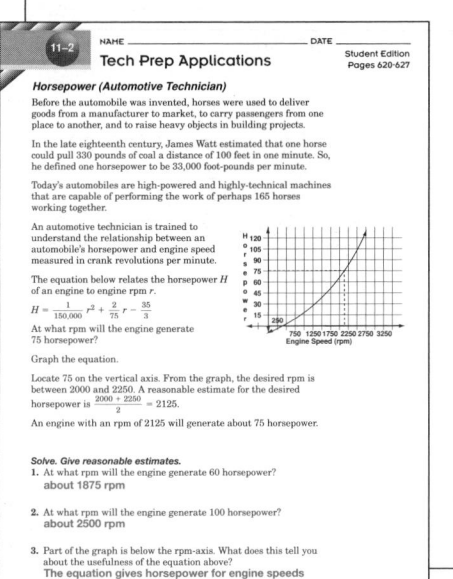

11-2 NAME_____ DATE_____
Student Edition
Pages 620-627

Tech Prep Applications

Horsepower (Automotive Technician)

Before the automobile was invented, horses were used to deliver goods from a manufacturer to market, to carry passengers from one place to another, and to raise heavy objects in building projects.

In the late eighteenth century, James Watt estimated that one horse could pull 330 pounds of coal a distance of 100 feet in one minute. So, he defined one horsepower to be 33,000 foot-pounds per minute.

Today's automobiles are high-powered and highly-technical machines that are capable of performing the work of perhaps 165 horses working together.

An automotive technician is trained to understand the relationship between an automobile's horsepower and engine speed measured in crank revolutions per minute.

The equation below relates the horsepower H of an engine to engine rpm r.

$H = \frac{1}{150,000} r^2 + \frac{2}{75} r − \frac{35}{3}$

At what rpm will the engine generate 75 horsepower?

Graph the equation.

Locate 75 on the vertical axis. From the graph, the desired rpm is between 2000 and 2250. A reasonable estimate for the desired horsepower is $\frac{2000 + 2250}{2}$ = 2125.

An engine with an rpm of 2125 will generate about 75 horsepower.

Solve. Give reasonable estimates.

1. At what rpm will the engine generate 60 horsepower?
   about 1875 rpm

2. At what rpm will the engine generate 100 horsepower?
   about 2500 rpm

3. Part of the graph is below the rpm-axis. What does this tell you about the usefulness of the equation above?
   The equation gives horsepower for engine speeds greater than about 375 rpm.

CONNECTIONS

Science and Math Lab Manual, pp. 49–52

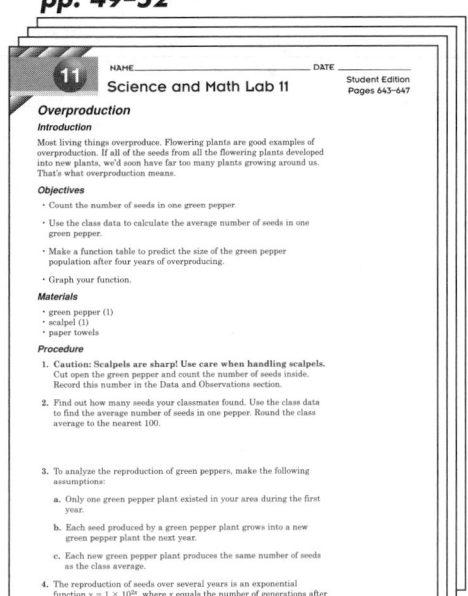

11 NAME_____ DATE_____
Student Edition
Pages 643-647

Science and Math Lab 11

Overproduction

Introduction

Most living things overproduce. Flowering plants are good examples of overproduction. If all of the seeds from all the flowering plants developed into new plants, we'd soon have far too many plants growing around us. That's what overproduction means.

Objectives

• Count the number of seeds in one green pepper.

• Use the class data to calculate the average number of seeds in one green pepper.

• Make a function table to predict the size of the green pepper population after four years of overproducing.

• Graph your function.

Materials

• green pepper (1)
• scalpel (1)
• paper towels

Procedure

1. **Caution: Scalpels are sharp! Use care when handling scalpels.** Cut open the green pepper and count the number of seeds inside. Record this number in the Data and Observations section.

2. Find out how many seeds your classmates found. Use the class data to find the average number of seeds in one pepper. Round the class average to the nearest 100.

3. To analyze the reproduction of green peppers, make the following assumptions:

   a. Only one green pepper plant existed in your area during the first year.

   b. Each seed produced by a green pepper plant grows into a new green pepper plant the next year.

   c. Each new green pepper plant produces the same number of seeds as the class average.

4. The reproduction of seeds over several years is an exponential function $y = 1 \times 10^{2x}$, where x equals the number of generations after the first plant. Use this formula to complete the Function Table.

PROBLEM SOLVING

Problem of the Week Cards, 30, 31

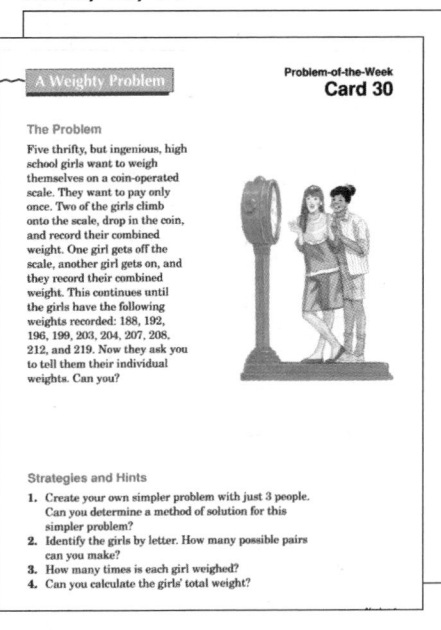

A Weighty Problem

Problem-of-the-Week
Card 30

The Problem

Five thrifty, but ingenious, high school girls want to weigh themselves on a coin-operated scale. They want to pay only once. Two of the girls climb onto the scale, drop in the coin, and record their combined weight. One girl gets off the scale, another girl gets on, and they record their combined weight. This continues until the girls have the following weights recorded: 188, 192, 196, 199, 203, 204, 207, 208, 212, and 219. Now they ask you to tell them their individual weights. Can you?

Strategies and Hints

1. Create your own simpler problem with just 3 people. Can you determine a method of solution for this simpler problem?

2. Identify the girls by letter. How many possible pairs can you make?

3. How many times is each girl weighed?

4. Can you calculate the girls' total weight?

Exploring Quadratic and Exponential Functions

Reaching Out to Help Others

Objectives

In this chapter, you will:

- find the equation of the axis of symmetry and the coordinates of the vertex of a parabola,
- graph quadratic and exponential functions,
- use estimation to find roots of quadratic equations by graphing,
- find roots of quadratic equations by using the quadratic formula,
- solve problems by looking for and using a pattern, and
- solve problems involving growth and decay.

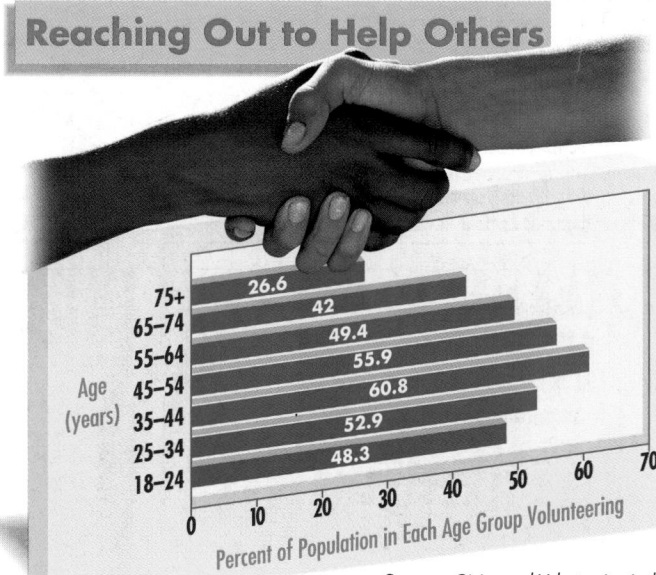

Age (years)	
75+	26.6
65–74	42
55–64	49.4
45–54	55.9
35–44	60.8
25–34	52.9
18–24	48.3

Percent of Population in Each Age Group Volunteering

Source: *Giving and Volunteering in the United States,* 1992 Edition

Dr. Robert Coles, a psychiatrist and Harvard professor, calls it "the call of service," men and women who volunteer their time to help others. Many teens would like to get involved in volunteer service, but may not know where to start. Some opportunities may be available in your own community—helping senior citizens, working as a Big Brother or Big Sister, or volunteering at a food pantry. Volunteering will help others and may help you feel great, too.

TIME *Line*

2100 B.C. Sumarian King Ur-Nammu builds the Ziggurat of Ur.

1482 Johannes Campanus's translation of Euclid's *Elements* is the first mathematical book of significance to be printed.

| 2300 BC | 2200 | 2100 | 2000 | 1900 | AD 1100 | 1200 | 1300 | 1400 | 1500 | 1600 | 1700 |

1245 Chin Chiu-Shao's *Mathematical treatise in nine sections* solves fourth-degree and higher-degree equations.

1654 Dutch inventor Christian Huygens develops the pendulum clock.

608 Chapter 11 Exploring Quadratic and Exponential Functions

TIME *Line*

interNET CONNECTION

Read the Internet NonProfit Center page for information and ideas about nonprofit organizations.

World Wide Web
http://www.human.com/inc/

ChapterProject

A perfect example of a teenage volunteer is **Liz Alvarez** of St. Petersburg, Florida. She has been a community volunteer at St. Anthony's Hospital, giving more than 600 hours of her time to help the staff. She answers phones in the pastoral care department, does filing as an administrative assistant, and carries messages for staff members. She helps patients by delivering items to their rooms, works in the gift shop, and assists other volunteers.

She enjoys her time spent volunteering, is happy to help others, and believes that the experience is very rewarding. She encourages other teens to investigate the opportunities in their communities and join a volunteer team.

O rganizations in every community study the characteristics of those who volunteer their time. In this manner, they know what population to target when seeking volunteers for a particular project or event. Survey companies, such as the Gallup Organization, Inc., often provide statistics for these organizations.

One of the characteristics the Gallup Organization studied was the income of volunteers. They found the following information on income brackets and the percent of the people in each bracket who volunteer.

Income ($)	Percent of Population Volunteering
Under 10,000	31.6
10,000–19,999	37.9
20,000–29,999	51.3
30,000–39,999	56.4
40,000–49,999	67.4
50,000–59,999	67.7
60,000–74,999	55.0
75,000–99,999	62.8
100,000 +	73.7

Analyze these findings.

• Make a graph of the data.

• What type of behavior do these data present?

• Why do you think this behavior exists? **Answers will vary.**

• Research in an almanac or statistical abstract to find the average income for the age groups listed in the graph on the previous page.

• Use your research, the graph on the previous page, and the table to make a conjecture about the characteristics of the average volunteer.

1940 U.S. photographer Helen Levitt chronicles life in the streets of New York City with her print *Children*.

1981 The Peace Corps becomes an independent agency, sending volunteers to improve living conditions in developing countries.

1945 | 1950 | 1955 | 1960 | 1965 | 1970 | 1975 | 1980 | 1985 | 1990 | 1995 | 2000

1961 American soprano Leontyne Price makes her Metropolitan Opera debut.

1991 Advances in music technology enable Natalie Cole to produce a duet album with her late father Nat "King" Cole.

Alternative Chapter Projects ▬

Two other chapter projects are included in the *Investigations and Projects Masters.* In Chapter 11 Project A, pp. 65–66, students extend the topic in the chapter opener. In Chapter 11 Project B, pp. 67–68, students investigate how engineers use parabolic curves in bridge designs.

Encourage students to talk about what volunteer work they have done. Encourage them to explain why they do it and what they get out of it. Would their work be as helpful if they were being paid? Would they feel differently?

ChapterProject

Cooperative Learning Students should work in groups of three for this project. One will be responsible for drawing the graphs. Another will be responsible for library research. The third member will present the results to the class. All should participate in a group discussion to answer the questions posed in this project.

Investigations and Projects Masters, p. 65

11

NAME_____ DATE_____

Chapter 11 Project A

Student Edition Pages 610–650

Volunteer Survey

1. Work with a partner to research and list organizations in your area that have volunteering opportunities. Possible places to check include health care organizations, schools, social service agencies, environmental programs, religious organizations, and museums.

2. Arrange the list you made in exercise 1 into different categories. Which categories have the most organizations listed? Which have the fewest?

3. Decide on a few organizations where you might be interested in volunteering. Interview a supervisor at one of these organizations to find out more about what the organization does. Find out how the organization uses its volunteers.

4. Find out how the organization uses the general equations for exponential growth and decay to make decisions about purchasing versus leasing, federal government funding, investments, population growth, and so on.

5. Write a newspaper article about volunteers and volunteering. At the end of your article, list different organizations that have volunteering opportunities and the phone numbers where people can get in touch with them.

1 FOCUS

NCTM Standards: 1–6

Objective
Use a graphing calculator to graph a quadratic function and find the coordinates of its vertex.

Recommended Time
15 minutes

Instructional Resources
Graphing Calculator Masters, pp. 30 and 31

These masters provide keystroking instruction for this lesson for the TI-81 and Casio graphing calculators.

Motivating the Lesson
Plot the points given by the coordinates in the table below.

x	y
0	0
1	2
2	4
3	6
4	8
5	10

Note that each value of *y* is a constant, increased over the preceding value of *y.* Ask students what the graph would look like if every increase in *y* value were greater than the preceding increase.

2 TEACH

Teaching Tip For some quadratic functions, the standard viewing window may need to be adjusted until the vertex is visible.

3 PRACTICE/APPLY

Assignment Guide
Core: 1–6
Enriched: 1–6

11–1A Graphing Technology
Quadratic Functions

A Preview of Lesson 11–1

Equations in the form $y = ax^2 + bx + c$ are called **quadratic functions,** and their graphs are called **parabolas.** A parabola is a U-shaped curve that can open upward or downward. The maximum or minimum point of a parabola is called its **vertex.** You can use a graphing calculator to graph a quadratic function and find the coordinates of its vertex.

Example

Graph $y = \frac{1}{4}x^2 - 4x - 2$ and locate its vertex. Use the integer window.

A quadratic function is entered into the Y= list in the same way that you entered linear functions.

Sometimes the coordinates of the vertex may not be integers. Use the ZOOM feature several times until you get coordinates that are fairly consistent (within three decimal places).

Enter: [Y=] .25 [X,T,θ] [x²] [−] 4 [X,T,θ] [−] 2 *Enters the function.*

[ZOOM] 6 [ZOOM] 8 [ENTER] *Selects the integer window.*

The parabola opens upward. The vertex is a minimum point.

Method 1: Use TRACE.

Press [TRACE] and use the left and right arrow keys to move the cursor to the vertex. Watch the coordinates at the bottom of the screen as you move the cursor. The vertex will be the point of the least *y* value. *Why?*

X = 8   Y = −18

Method 2: Use CALC.

Since the parabola opens upward, the vertex is a minimum point. Press [2nd] [CALC] 3.

• A "Lower Bound?" prompt appears. Move the cursor to a location left of where you think the vertex is and press [ENTER].

You can also press [ENTER] at the Guess? prompt without entering a guess.

• An "Upper Bound?" prompt appears. Move the cursor to a location right of where you think the vertex is and press [ENTER].

• When the "Guess?" prompt appears, move the cursor to your choice for the vertex and press [ENTER]. The screen will give you an approximation of the vertex coordinates.

The coordinates of the vertex are (8, −18).

Minimum
X = 8.0000006   Y = −18

1–6. See Solutions Manual for graphs.

EXERCISES

1. (−8, −5)
2. (−0.75, 3.25)
3. (5, 0) 4. (−2, 7)
5. (10, 14) 6. (−2, 5)

Graph each function. Make a sketch of the graph and note the ordered pair representing the vertex on the graph.

1. $y = x^2 + 16x + 59$ 2. $y = 12x^2 + 18x + 10$ 3. $y = x^2 - 10x + 25$

4. $y = -2x^2 - 8x - 1$ 5. $y = 2(x - 10)^2 + 14$ 6. $y = -0.5x^2 - 2x + 3$

4 ASSESS

Observing students working with technology is an excellent method of assessment.

Using Technology
This lesson offers an excellent opportunity for using technology in your algebra classroom. For more information on using technology, see *Graphing Calculators in the Mathematics Classroom,* one of the titles in the Glencoe Mathematics Professional Series.

Graphing Quadratic Functions

Objectives

After studying this lesson, you should be able to:

- find the equation of the axis of symmetry and the coordinates of the vertex of a parabola, and
- graph quadratic functions.

APPLICATION

Landmarks

The Gateway Arch of the Jefferson National Expansion Memorial in St. Louis, Missouri, is shaped like an upside-down U. This shape is actually a *catenary*, which resembles a geometric shape called a **parabola.** The shape of the arch can be approximated by the graph of the function $f(x) = -0.00635x^2 + 4.0005x - 0.07875$, where $f(x)$ is the height of the arch in feet and x is the horizontal distance from one base.

This type of function is an example of a **quadratic function.** A quadratic function can be written in the form $f(x) = ax^2 + bx + c$, where $a \neq 0$. Notice that this function has a degree of 2 and the exponents are positive.

Definition of a Quadratic Function	A quadratic function is a function that can be described by an equation of the form $y = ax^2 + bx + c$, where $a \neq 0$.

To graph a quadratic equation, you can use a table of values. The table at the right shows the distance (in 35-foot increments) from one base of the arch and the height of the arch at each increment (rounded to the nearest foot). Graph the ordered pairs and connect them with a smooth curve. The graph of $f(x) = -0.00635x^2 + 4.0005x - 0.07875$ is shown below.

x	f(x)
0	0
35	132
70	249
105	350
140	436
175	506
210	560
245	599
280	622
315	630
350	622
385	599
420	560
455	506
490	436
525	350
560	249
595	132
630	0

$f(x) = -0.00635x^2 + 4.0005x - 0.07875$

Height (feet)

Distance from Base (feet)

Notice that the value of a in this function is negative and the curve opens downward. The greatest value of $f(x)$ seems to be 630 feet, which occurs when the distance from the base is 315 feet. The point at (315, 630) would be the **vertex** of the parabola. For a parabola that opens downward, the vertex is a **maximum** point of the function. If the parabola opens upward, the vertex is a **minimum** point of the function.

Lesson 11-1   Graphing Quadratic Functions   **611**

11-1 LESSON NOTES

NCTM Standards: 1–6, 13

Instructional Resources

- Study Guide Master 11-1
- Practice Master 11-1
- Enrichment Master 11-1
- Graphing Calculator Masters, p. 11
- Modeling Mathematics Masters, p. 82

 Transparency 11-1A contains the 5-Minute Check for this lesson; **Transparency 11-1B** contains a teaching aid for this lesson.

Recommended Pacing	
Standard Pacing	Day 2 of 11
Honors Pacing	Day 2 of 10
Block Scheduling*	Day 1 of 5
Alg. 1 in Two Years*	Days 2 & 3 of 19

 *For more information on pacing and possible lesson plans, refer to the *Block Scheduling Booklet* and *Algebra 1 in Two Years.*

1 FOCUS

 5-Minute Check
(over Chapter 10)

Solve.

1. $x^3 + 2x^2 = 15x$ **0, −5, 3**
2. $a^3 - 13a^2 + 42a = 0$ **0, 6, 7**
3. $4r^3 - 9r = 0$ **0, $\frac{3}{2}$, $-\frac{3}{2}$**
4. The sum of the squares of two consecutive odd integers is 802. Find the integers.   **−21 and −19 or 19 and 21**
5. Find two consecutive odd integers such that the square of the greater decreased by the lesser is 274.   **15 and 17**

Motivating the Lesson

Situational Problem Ask students to describe a parabola. Have a volunteer draw a parabola on the chalkboard. Ask students to identify common objects that have this shape. Possible examples include a satellite dish and an automobile headlight.

This modeling activity illustrates that quadratic functions have a vertical axis of symmetry.

In-Class Example

For Example 1
Find the equation of the axis of symmetry and the coordinates of the vertex of the graph of each equation. Then graph each equation.

a. $y = x^2 - 4x - 5$
$x = 2$; $(2, -9)$

b. $y = -x^2 + 2x - 1$
$x = 1$; $(1, 0)$

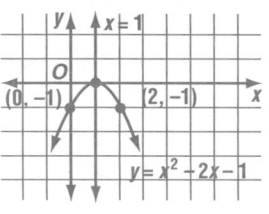

Teaching Tip The axis of symmetry is helpful in Example 1 for determining which elements of the domain to choose when graphing a function.

Parabolas possess a geometric property called **symmetry**. Symmetrical figures are those in which the figure can be folded and each half matches the other exactly. The following activity explores the symmetry of a parabola.

  Symmetry of Parabolas

Materials: grid paper

Your Turn

a. Graph $y = x^2 - 6x + 5$ on grid paper.

b. Hold your paper up to the light and fold the parabola in half so the two sides match exactly.

c. Unfold the paper. Which point on the parabola lies on the fold line? $(3, -4)$

d. Write an equation to describe the fold line $x = 3$

e. Write a few sentences to describe the symmetry of a parabola based on your findings in this activity. **See students' work.**

The fold line in the activity above is called the **axis of symmetry** for the parabola. Each point on the parabola that is on one side of the axis of symmetry has a corresponding point on the parabola on the other side of the axis. The vertex is the only point on the parabola that is on the axis of symmetry. The equation for the axis of symmetry can be determined from the equation of the parabola.

Equation of the Axis of Symmetry of a Parabola	The equation of the axis of symmetry for the graph of $y = ax^2 + bx + c$, where $a \neq 0$, is $x = -\dfrac{b}{2a}$.

You can determine a lot of information about a parabola from its equation.

Example Given the equation $y = x^2 - 4x + 5$,

  a. find the equation of the axis of symmetry,

  b. find the coordinates of the vertex of the parabola, and

  c. graph the equation.

  a. In the equation $y = x^2 - 4x + 5$, $a = 1$ and $b = -4$. Substitute these values into the equation of the axis of symmetry.

$$x = -\frac{b}{2a}$$

$$= -\frac{-4}{2(1)} \text{ or } 2$$

  b. Since the equation for the axis of symmetry is $x = 2$ and the vertex lies on the axis, the x-coordinate for the vertex is 2.

$$y = x^2 - 4x + 5$$

$$= 2^2 - 4(2) + 5 \quad \textit{Replace x with 2.}$$

$$= 4 - 8 + 5 \text{ or } 1$$

The coordinates of the vertex are $(2, 1)$.

 GLENCOE *Technology*

Interactive Mathematics Tools Software

This multimedia software helps students observe changes in graphs when the quadratic coefficient, linear coefficient, and constant term are manipulated. A **Computer Journal** gives students an opportunity to write about what they have learned.

For Windows & Macintosh

c. You can use the symmetry of the parabola to help you draw its graph. Draw a coordinate plane. Graph the vertex and axis of symmetry. Choose a value less than 2, say 0, and find the y-coordinate that satisfies the equation.

$$y = x^2 - 4x + 5$$
$$= 0^2 - 4(0) + 5 \text{ or } 5$$

Graph (0, 5). Since the graph is symmetrical, you can find another point on the other side of the axis of symmetry. The point at (0, 5) is 2 units left of the axis. Go 2 units right of the axis and plot a point at (4, 5). Repeat this for several other points. Then sketch the parabola.

Check: Does (4, 5) satisfy the equation?
$$y = x^2 - 4x + 5$$
$$5 \overset{?}{=} 4^2 - 4(4) + 5$$
$$5 \overset{?}{=} 16 - 16 + 5$$
$$5 = 5 \checkmark$$

The point (4, 5) satisfies the equation $y = x^2 - 4x + 5$ and is part of the graph.

The graph of an equation describing the height of an object propelled into the air can be an example of a quadratic function. Because gravity overtakes the initial force of the object, the object falls back to Earth.

Example 2

APPLICATION
Pyrotechnics

In Cincinnati, Ohio, radio station WEBN and Rozzi's Famous Fireworks, Inc. team together to develop a large fireworks display for Labor Day. The fireworks are coordinated with recorded music compiled by the radio station's DJs. Mr. Rozzi assumes that each rocket (firework) explodes at approximately its highest point after being propelled from a barge in the Ohio River. The formula that describes the height $H(t)$ of an object propelled into the air is $H(t) = v_0 t - \frac{1}{2}gt^2 + h_0$, where v_0 represents the initial velocity in m/s, t represents time in seconds, g represents the acceleration of gravity (about 9.8 m/s²), and h_0 is the initial height of the object at the time it is launched.

A certain rocket has an initial velocity of 39.2 m/s and is launched 1.6 meters above the surface of the water.

a. Graph the equation representing the height of the rocket.

b. What is the maximum height that the rocket achieves?

c. One of these rockets is scheduled to explode 2 minutes and 28 seconds into the program. When should the rocket be fired from the barge?

(continued on the next page)

Alternative Teaching Strategies

Reading Algebra The prefix *quad-* usually means "four." Warn students not to be misled. Quadratic functions are of degree 2 and have at most 2 roots.

Check for Understanding

Exercises 1–14 are designed to help you assess your students' understanding through reading, writing, speaking, and modeling. You should work through Exercises 1–6 with your students and then monitor their work on Exercises 7–14.

Additional Answers

2. Sample answer: Once you determine the axis of symmetry, there is a matching point on the other side of the axis for each point you find on one side. Suppose the axis is $x = 4$ and one point is $(-2, 3)$. -2 is 6 units left of the axis; another point is 6 units right of the axis with the same y-coordinate, that is, $(10, 3)$.

4. Angie is correct. The graph of $y = (-x)^2$ is the same graph as $y = x^2$, which opens upward from the origin. The graph of $y = -x^2$ opens downward from the origin.

6.

---

a. Use the given velocity and the force of gravity to determine the formula.

$$H(t) = v_0 t - \tfrac{1}{2} g t^2 + h_0$$
$$= 39.2t - \tfrac{1}{2}(9.8)t^2 + 1.6 \quad \textit{Replace } v_0, g, \textit{ and } h_0 \textit{ with the given values.}$$
$$= 39.2t - 4.9t^2 + 1.6 \quad \textit{Simplify.}$$

The values of a and b are -4.9 and 39.2, respectively.
$H(t)$ corresponds to $f(x)$ and t corresponds to x.

Find the equation of the axis of symmetry to help you locate the vertex.

$$t = -\frac{b}{2a}$$
$$= -\frac{39.2}{2(-4.9)} \text{ or 4 seconds}$$

Use a calculator to find $H(t)$ when $t = 4$.

$H(4) = 39.2(4) - 4.9(4)^2 + 1.6$ or 80 meters

The vertex of the graph is at $(4, 80)$.

We can use a table of values to find other points that satisfy the equation.

The graph shows the height of the rocket at any given time. It does not show the path that the rocket took.

t	$H(t)$
0	1.6
1	35.9
2	60.4
3	75.1
4	80.0
5	75.1
6	60.4
7	35.9
8	1.6
9	-42.5

Negative values are meaningless in this situation.

b. The maximum height is at the vertex. This height is 80 meters.

c. Since the maximum height is achieved 4 seconds after launch, the rocket must be fired 4 seconds before its scheduled explosion at 2 minutes 28 seconds into the program. The rocket should be fired at 2 minutes 24 seconds into the program.

2, 4. See margin.

3. If (p, q) is the vertex, $x = p$ is the equation of the axis of symmetry.

CHECK FOR UNDERSTANDING

Communicating Mathematics

5. If the coefficient of x^2 is positive, it is a minimum. If the coefficient is negative, it is a maximum.

6. It opens downward; see margin for graph; $x = 2$.

Study the lesson. Then complete the following.

1. **Determine** the distance between the bases of the arch in St. Louis.   630 ft

2. **Explain** how you can use symmetry to help you graph a parabola.

3. **Write** a sentence to explain how you can write the equation of the axis of symmetry if you know the coordinates of the vertex of a parabola.

4. **You Decide**   Lisa says that $y = -x^2$ is the same as $y = (-x)^2$. Angie says that they are different. Who is correct? Explain and include graphs.

5. How can you determine if the vertex is a maximum or minimum without graphing the equation first?

6. Graph $y = -x^2 + 4x - 4$. How does this graph differ from the one in the activity on page 612? Use paper folding to determine the axis of symmetry.

614   *Chapter 11   Exploring Quadratic and Exponential Functions*

Reteaching

Using Alternative Methods Have students graph the following equations on the same graph. Have them describe the changes that occur in the graph with each equation. Generalize the results and have students extend them to other equations.

1. $y = x^2$
2. $y = x^2 - 1$
3. $y = x^2 - 2x$
4. $y = x^2 - 2x - 1$

Guided Practice

Write the equation of the axis of symmetry and find the coordinates of the vertex of the graph of each equation. State if the vertex is a maximum or minimum. Then graph the equation.

7–12. See Solutions Manual for graphs.

9. $x = -2$, $(-2, -13)$, min.

10. $x = 7$, $(7, -36)$, min.

7. $y = x^2 + 2$ $x = 0$, $(0, 2)$, min.

8. $y = -2x^2$ $x = 0$, $(0, 0)$, max.

9. $y = x^2 + 4x - 9$

10. $y = x^2 - 14x + 13$

11. $y = -x^2 + 5x + 6$
$x = 2.5$, $(2.5, 12.25)$, max.

12. $y = -2x^2 + 4x + 6.5$
$x = 1$, $(1, 8.5)$, max.

13. Which equation describes the graph at the right?   c
   a. $f(x) = x^2 - 6x + 9$
   b. $f(x) = -x^2 + 6x + 9$
   c. $f(x) = x^2 + 6x + 9$

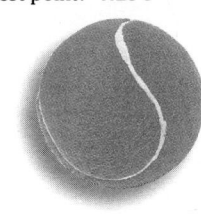

14. **Tennis** A tennis ball is propelled upward from the face of a racket at 40 ft/s. The racket face is 3 feet from the ground when it makes contact with the ball.
   a. If the force of gravity is 32 ft/s², at what time after hitting the ball is it at its highest point? 1.25 s
   b. How high does it go? 28 ft

15–29. See Solutions Manual for graphs.

EXERCISES

Practice

Write the equation of the axis of symmetry and find the coordinates of the vertex of the graph of each equation. State if the vertex is a maximum or minimum. Then graph the equation.

15. $x = 0$, $(0, 0)$, min.

16. $x = 2$, $(2, 3)$, max.   **A**

17. $x = -1$, $(-1, 17)$, min.

18. $x = \frac{3}{2}$,
$\left(\frac{3}{2}, -\frac{49}{4}\right)$, min.   **B**

19. $x = 0$, $(0, -5)$, min.

20. $x = 0$, $(0, 16)$, min.

21. $x = -3$, $(-3, -29)$, min.

22. $x = -4$, $(-4, 32)$, min.

23. $x = 0$, $(0, -25)$, min.

24. $x = -3$,
$(-3, 24)$, max.   **C**

25. $x = -1$, $(-1, 7)$, max.

26. $x = 4$, $(4, 37)$, max.

15. $y = 4x^2$

16. $y = -x^2 + 4x - 1$

17. $y = x^2 + 2x + 18$

18. $y = x^2 - 3x - 10$

19. $y = x^2 - 5$

20. $y = 4x^2 + 16$

21. $y = 2x^2 + 12x - 11$

22. $y = 3x^2 + 24x + 80$

23. $y = x^2 - 25$

24. $y = 15 - 6x - x^2$

25. $y = -3x^2 - 6x + 4$

26. $y = 5 + 16x - 2x^2$

27. $y = 3(x + 1)^2 - 20$
$x = -1$, $(-1, -20)$, min.

28. $y = -(x - 2)^2 + 1$
$x = 2$, $(2, 1)$, max.

29. $y = \frac{2}{3}(x + 1)^2 - 1$
$x = -1$, $(-1, -1)$, min.

Match each equation with its graph.

30. $f(x) = \frac{1}{2}x^2 + 1$ b

31. $f(x) = -\frac{1}{2}x^2 + 1$ c

32. $f(x) = \frac{1}{2}x^2 - 1$ a

      a.   b.   c.

Graph each equation.   33–36. See margin.

33. $y + 2 = x^2 - 10x + 25$

34. $y + 1 = 3x^2 + 12x + 12$

35. $y + 3 = -2(x - 4)^2$

36. $y - 5 = \frac{1}{3}(x + 2)^2$

37. What is the equation of the axis of symmetry of a parabola if its x-intercepts are -6 and 4? $x = -1$

Lesson 11-1   Graphing Quadratic Functions   **615**

Additional Answers

35.

$y + 3 = -2(x - 4)^2$

36.

$y - 5 = \frac{1}{3}(x + 2)^2$

Chapter 11   **615**

Assignment Guide

Core: 15–43 odd, 44, 45, 47–54
Enriched: 16–42 even, 44–54

For **Extra Practice,** see p. 780.

The red A, B, and C flags, printed only in the Teacher's Wraparound Edition, indicate the level of difficulty of the exercises.

Additional Answers

33.

$y + 2 = x^2 - 10x + 25$

34.

$y + 1 = 3x^2 + 12x + 12$

Study Guide Masters, p. 75

11-1   NAME_____   DATE_____

Study Guide   Student Edition
Pages 611–617

Graphing Quadratic Functions

An equation such as $y = x^2 - 4x + 1$ describes a type of function known as a **quadratic function.**

Definition of Quadratic Function
A quadratic function is a function that can be described by an equation of the form $y = ax^2 + bx + c$, where $a \neq 0$.

Graphs of quadratic functions have certain common characteristics. They have a general shape called a **parabola**. They have a minimum or maximum point. In general, a parabola will open upward and have a minimum point when the coefficient of y and x^2 have the same sign. It will open downward and have a maximum point when the coefficient of y and x^2 are opposite in sign. The vertical line containing the minimum or maximum point is the **axis of symmetry**. The minimum or maximum point is called the **vertex** of the parabola.

Equation of Axis of Symmetry
The equation of the axis of symmetry of $y = ax^2 + bx + c$, where $a \neq 0$ is $x = -\frac{b}{2a}$.

Write the equation of the axis of symmetry and find the coordinates of the vertex of the graph of each equation. Then graph the equation.

1. $y = x^2 + 3x$

$x = -\frac{3}{2}; \left(-\frac{3}{2}, -\frac{9}{4}\right)$

2. $y = 3x^2 + x + 1$

$x = -\frac{1}{6}; \left(-\frac{1}{6}, \frac{11}{12}\right)$

3. $y = -x^2 - 4$

$x = 0; (0, -4)$

4. $y = -\frac{1}{2}x^2 + x + \frac{5}{2}$

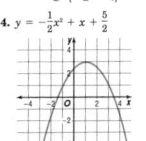

$x = 1; (1, 3)$

5. $y = -x^2 - 6x - 7$

$x = -3; (-3, 2)$

6. $y = 2x^2 + 12x + 9$

$x = -3; (-3, -9)$

Additional Answers

40.

41.

42.

43.

44.

Practice Masters, p. 75

 11-1
Practice

NAME_____ DATE _____

Student Edition
Pages 611–617

Graphing Quadratic Functions

Write the equation of the axis of symmetry and find the coordinates of the vertex of the graph of each quadratic function.

1. $y = x^2 - 2x - 8$
$x = 1; (1, -9)$

2. $y = -x^2 + x + 12$
$x = \frac{1}{2}; \left(\frac{1}{2}, \frac{49}{4}\right)$

3. $y = 8x^2 + 12x$
$x = -\frac{3}{4}; \left(-\frac{3}{4}, -\frac{9}{2}\right)$

4. $y = -2x^2 + x + 2$
$x = \frac{1}{4}; \left(\frac{1}{4}, \frac{17}{8}\right)$

Write the equation of the axis of symmetry and find the coordinates of the vertex of the graph of each quadratic function. State if the vertex is a maximum or minimum. Then graph the equation.

5. $y = x^2 - 3x + 2$

6. $y = -x^2 - 2x + 3$

7. $y = x^2 - 2$

8. $y = -\frac{1}{4}x^2$

---

38. The vertex of a parabola is at $(-4, -3)$. If one x-intercept is -11, what is the other x-intercept? **3**

39. Two points on a parabola are at $(-8, 7)$ and $(12, 7)$. What is the equation of the axis of symmetry? $x = 2$

Graphing Calculator

Sample answers:
40. $(-2.00, 12)$
41. $(-1.10, 125.8)$
42. $(-268.04, -1719.47)$
43. $(-0.15, 90.14)$
40–43. See margin for graphs.

You can draw the axis of symmetry on a graphed parabola by using the **VERTICAL** command on the **DRAW** menu. Press [2nd] [DRAW] 4 and use the arrow keys to move the line into place. To graph the axis of symmetry before the graph is drawn, enter the equation in the Y= list and press [2nd] [DRAW] 4 (the x value through which the line will pass) [ENTER]. Graph each equation and its axis of symmetry. Make a sketch of each graph on your paper, labeling the vertex of the graph.

40. $y = 8 - 4x - x^2$
41. $y = 20x^2 + 44x + 150$
42. $y = 0.023x^2 + 12.33x - 66.98$
43. $y = -78.23x^2 - 23.76x + 88.34$

Critical Thinking

44. Graph $y = x^2 + 2$ and $x + y = 8$ on the same coordinate plane. What are the coordinates of the points they have in common? Explain how you determined these points. **See margin for graph; $(-3, 11)$, $(2, 6)$; see students' work.**

Applications and Problem Solving

45c. See students' work.
46a. $H(t) = 80t - 16t^2 + 2$
46b. 66 ft, 98 ft, 98 ft

45. College The cost of a college education is increasing every year. Many parents are starting college funds for their children before they are even born. The average tuition and fees for public college during the years 1970–1993 can be estimated using the function $U(t) = 2.97t^2 - 6.78t + 329.96$, where $U(t)$ represents tuition and fees (in dollars) for one year and t represents the number of years after 1970.

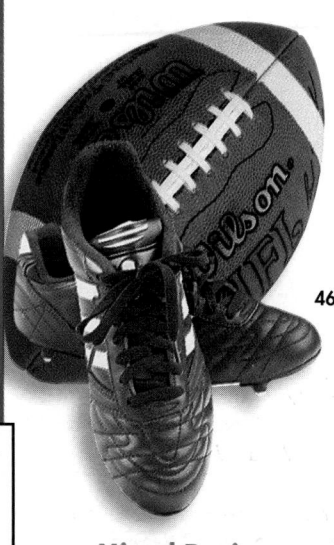

WHATSAMATA-U

Year	t	$U(t)$
1970	0	329.96
1975	5	370.31
1980	10	559.16
1985	15	896.51
1990	20	1382.36
1993	23	1745.15

a. Copy and complete the table at the right.
b. Graph this function. **See margin.**
c. Assume that this function is a model for all years after 1993. How much will the tuition and fees for your first year of college be if you attend a public college?

46. Football When a football player punts a football, he hopes for a long "hang time," the total amount of time the ball stays in the air. Any hang time of more than about 4.5 seconds is usually good. Manuel is the punter for his high school team. He can kick the ball with an upward velocity of 80 ft/s, and his foot meets the ball 2 feet off the ground.

a. Write a quadratic equation to describe the height of the football at any given time t. Use 32 ft/s² as the acceleration of gravity.
b. How high is the ball after 1 second? 2 seconds? 3 seconds?
c. What is Manuel's hang time? **about 5 seconds**

Mixed Review

47. Agriculture A field is 1.2 kilometers long and 0.9 kilometers wide. A farmer begins plowing the field by starting at the outer edge and going all the way around the field. When he stops for lunch, a strip of uniform width has been plowed on all sides of the field and half the field is plowed. What is the width of the strip? (Lesson 10-7) **0.15 km**

616 Chapter 11 Exploring Quadratic and Exponential Functions

 Cooperative Learning

Co-op Co-op You may want to separate students into groups of three. Have them graph $y = x^2$, $y = x^2 - 4$, and $y = (x + 2)^2 + 5$. They should identify the minimum points of each graph. **(0, 0), (0, -4), (-2, 5)** They should compare results and then predict the minimum point of $y = x^2 + 7$. **(0, 7)**

For more information on the co-op co-op strategy, see *Cooperative Learning in the Mathematics Classroom*, one of the titles in the Glencoe Mathematics Professional Series, page 31.

48. Find $(x - 4)(x - 8)$.   (Lesson 9-7) $x^2 - 12x + 32$

49. **Astronomy** Mars, located 227,920,000 kilometers from the sun, has a diameter of 6.79×10^3 kilometers.   (Lesson 9-3)   49b. 2.2792×10^8
   a. Write the diameter of Mars in decimal notation.   **6790**
   b. Write the distance Mars is from the sun in scientific notation.

50. Use elimination to solve the system of equations.   (Lesson 8-4)
   $3x + 4y = -25$
   $2x - 3y = 6$ $(-3, -4)$

51. **Meteorology** The table below lists the record 24-hour precipitation for each state as of 1990.   (Lesson 7-7)

State	Inches	State	Inches	State	Inches	State	Inches	State	Inches
AL	20.33	HI	38.00	MA	18.15	NM	11.28	SD	8.00
AK	15.20	ID	7.17	MI	9.78	NY	11.17	TN	11.00
AZ	11.40	IL	16.54	MN	10.84	NC	22.22	TX	43.00
AR	14.06	IN	10.50	MS	15.68	ND	8.10	UT	6.00
CA	26.12	IA	16.70	MO	18.18	OH	10.51	VT	8.77
CO	11.08	KS	12.59	MT	11.50	OK	15.50	VA	27.00
CT	12.77	KY	10.40	NE	13.15	OR	10.17	WA	12.00
DE	8.50	LA	22.00	NV	7.40	PA	34.50	WV	19.00
FL	38.70	ME	8.05	NH	10.38	RI	12.13	WI	11.72
GA	18.00	MD	14.75	NJ	14.81	SC	13.25	WY	6.06

Source: National Climatic Data Center

   a. Make a box-and-whisker plot of the data.   **See margin.**
   b. Are there any outliers? If so, list them.   **34.5, 38, 38.7, 43**

52. Graph $y = 3x + 4$.   (Lesson 5-5)   **See margin.**

53. Solve $3x = -15$.   (Lesson 3-2)   **−5**

54. **Patterns** Complete the pattern: 3, 6, 12, 24, _?_, _?_.   (Lesson 1-2)
   48, 96

WORKING ON THE
In·ves·ti·ga·tion

Refer to the Investigation on pages 554–555.

the BRICKYARD

Examine the table you created in Lesson 10-7 that included the length, width, perimeter, and area of the rectangular brick patterns. Measure each of your models in millimeters to determine the values of x and y and record the measures.

1 Graph the data, plotting the length of the pattern on the horizontal axis and the area of the pattern on the vertical axis. What kind of a graph is it? What relationship does it show?

2 Make another graph, plotting the perimeter of the pattern on the horizontal axis and the area of the pattern on the vertical axis. What kind of a graph is it? What kind of relationship does it show?

3 What is the relationship between the measures of the area, length, and width? Why are there some rectangular patterns that have the same area but different perimeters?

Add the results of your work to your Investigation Folder.

Lesson 11-1 Graphing Quadratic Functions   **617**

Extension

Problem Solving Graph the following pair of equations on the same coordinate plane. Then compare the graph of the second equation to that of the first equation.

$y = x^2$
$y = (x + 6)^2 - 8$

The second graph is 6 units to the left and 8 units below the first graph.

In·ves·ti·ga·tion

Working on the Investigation

The Investigation on pages 554–555 is designed to be a long-term project that is completed over several days or weeks. Encourage students to keep their materials in their Investigation Folder as they work on the Investigation.

---

4 ASSESS

Closing Activity
Speaking Have students compare the graph of a quadratic function whose coefficients of y^2 and y have the same sign and the graph of a quadratic function whose coefficients have opposite signs.

Additional Answers

45b.

$U(t) = 2.97t^2 - 6.78t + 329.96$

51a.

52.

$y = 3x + 4$

Enrichment Masters, p. 75

Chapter 11   **617**

11–1B Graphing Technology
Families of Parabolas

An Extension of Lesson 11–1

NCTM Standards: 1–6

Objective
Use a graphing calculator to study the characteristics of families of parabolas.

Recommended Time
25 minutes

Instructional Resources
Graphing Calculator Masters, pp. 32 and 33

These masters provide keystroking instruction for this lesson for the TI-81 and Casio graphing calculators.

1 FOCUS

Motivating the Lesson
In this lesson, students will explore the power of the graphing calculator to view a family of functions. They will study the effect of changing the coefficient of x^2 and the effect of changing the constant term.

2 TEACH

Teaching Tip Once students have graphed each equation, ask them if a better viewing window than the standard one would give them a better look at the parabolas being graphed.

A family of graphs is a group of graphs that have at least one characteristic in common. In Lesson 6–5A, you learned about families of linear graphs that shared the same slope or y-intercept. Families of parabolas often fall into two categories—those that have the same vertex and those that have the same shape. Graphing calculators make it easy to study the characteristics of families of parabolas.

Example Graph each group of equations on the same screen. Compare and contrast the graphs.

a. $y = x^2$, $y = 2x^2$, $y = 3x^2$

Each graph opens upward and has its vertex at the origin. The graphs of $y = 2x^2$ and $y = 3x^2$ are narrower than the graph of $y = x^2$.

b. $y = x^2$, $y = 0.5x^2$, $y = 0.3x^2$

Each graph opens upward and has its vertex at the origin. The graphs of $y = 0.5x^2$ and $y = 0.3x^2$ are wider than the graph of $y = x^2$.

How does the value of a in $y = ax^2$ affect the shape of the graph?

c. $y = x^2$, $y = x^2 + 2$,
$y = x^2 - 3$, $y = x^2 - 5$

Each graph opens upward and has the same shape as $y = x^2$. However, each parabola has a different vertex, located along the y-axis. *How does the value of the constant affect the placement of the graph?*

d. $y = x^2$, $y = (x - 2)^2$,
$y = (x + 3)^2$, $y = (x + 1)^2$

Each graph opens upward and has the same shape as $y = x^2$. However, each parabola has a different vertex, located along the x-axis. *How is the location of the vertex related to the equation of the graph?*

When analyzing or comparing the shapes of various graphs on different screens, it is important to compare the graphs using the same parameters. That is, the window used to compare the graphs should be the same, with the same scale factor. Suppose we graph the same equation using a different window for each. How will this affect the appearance of the graph?

Example **2** Graph $y = x^2 - 5$ in each viewing window. What conclusions can you draw about the appearance of a graph in the window used? *The scale is 1 unless otherwise noted.*

a. standard viewing window

b. $[-10, 10]$ by $[-100, 100]$ Yscl: 20

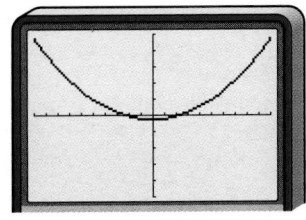

c. $[-50, 50]$ Xscl: 5 by $[-10, 10]$

d. $[-0.5, 0.5]$ Xscl: 0.1 by $[-10, 10]$

The window greatly affects the appearance of the parabola. Without knowing the window, graph b might be of the family $y = ax^2$, where $0 < a < 1$. Graph c makes the graph look like a member of $y = ax^2$, where $a > 1$. Graph d looks more like a line. However, all are graphs of the same equation.

EXERCISES

1–4. See margin.

Graph each group of equations on the same screen. Make a sketch of the screen on grid paper, and compare and contrast the graphs.

1. $y = -x^2$
   $y = -2x^2$
   $y = -5x^2$

2. $y = -x^2$
   $y = -0.3x^2$
   $y = -0.7x^2$

3. $y = -x^2$
   $y = -(x + 4)^2$
   $y = -(x - 8)^2$

4. $y = -x^2$
   $y = -x^2 + 6$
   $y = -x^2 - 4$

Use the families of graphs that have appeared in this lesson to predict the appearance of the graph of each equation. Then sketch the graph. 5–9. See Solutions Manual.

5. $y = 4x^2$
6. $y = x^2 - 6$
7. $y = -0.1x^2$
8. $y = (x + 1)^2$

9. Describe how each change in the equation of $y = x^2$ would affect the graph of $y = x^2$. Be sure to consider all values of a and b.
   a. $y = ax^2$
   b. $y = x^2 + a$
   c. $y = (x + a)^2$
   d. $y = (x + a)^2 + b$

Lesson 11–1B Graphing Technology: Families of Parabolas **619**

NCTM Standards: 1–6, 13

Instructional Resources

- Study Guide Master 11-2
- Practice Master 11-2
- Enrichment Master 11-2
- Assessment and Evaluation Masters, p. 296
- Tech Prep Applications Masters, p. 21

 Transparency 11-2A contains the 5-Minute Check for this lesson; **Transparency 11-2B** contains a teaching aid for this lesson.

Recommended Pacing	
Standard Pacing	Day 4 of 11
Honors Pacing	Day 4 of 10
Block Scheduling*	Day 2 of 5 (along with Lesson 11-3)
Alg. 1 in Two Years*	Days 5, 6, & 7 of 19

 *For more information on pacing and possible lesson plans, refer to the *Block Scheduling Booklet* and *Algebra 1 in Two Years.*

1 FOCUS

 ### 5-Minute Check
(over Lesson 11-1)

Write the equation of the axis of symmetry and find the coordinates of the vertex of the graph of each equation. State if the vertex is a maximum or a minimum. Then graph the equation.

1. $y = x^2 - 4x + 7$
minimum, (2, 3), $x = 2$

(continued at the right)

Objectives

After studying this lesson, you should be able to:
- use estimation to find roots of quadratic equations, and
- find roots of quadratic equations by graphing.

APPLICATION
Technology

In the United States, one of the fastest growing industries is the production of CD-ROMs for computers. CD-ROM stands for *Compact Disc-Read Only Memory.* CD-ROMs can hold text, music, photographic images, or combinations of any of these.

Any company that produces a product to sell finds that the profit they can make depends on the number of employees they have (among other things). The relationship between profit and the number of employees looks like the parabola drawn at right. The company does not make much of a profit if there are too few employees; as they hire more employees, the work can be done more efficiently, leading to higher profits. But if the company hires too many employees, it may not have room for them or enough work for them to do, yet they still have to be paid, causing lower profits.

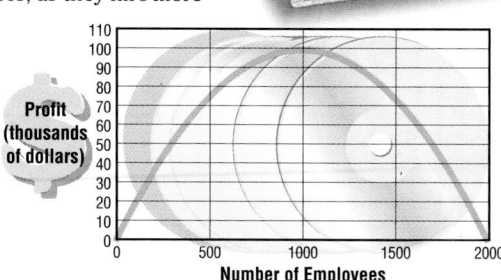

Suppose a company that produces CD-ROMs can express its profit as $P(x) = -0.1x^2 + 200x$, where x is the number of employees in the company. As the personnel manager of the company, it is your job to determine the least number of employees the company should have in order to reach its goal of having $75,000 in profit. *You will solve this problem in Exercise 4.*

A **quadratic equation** is an equation in which the value of the related quadratic function is 0. That is, for the quadratic equation $0 = x^2 + 6x - 7$, the related quadratic function is $f(x) = x^2 + 6x - 7$, where $f(x) = 0$. You have used factoring to solve equations like $x^2 + 6x - 7 = 0$. You can also use graphing to estimate the solutions to this equation.

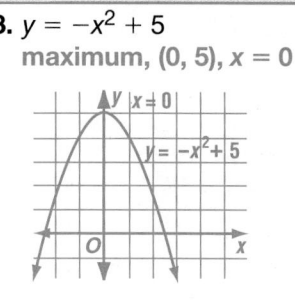

The solutions of a quadratic equation are called the **roots** of the equation. The roots of a quadratic equation can be found by finding the x-intercepts or **zeros** of the related quadratic function.

620 *Chapter 11 Exploring Quadratic and Exponential Functions*

---

2. $y = x^2 + 8x + 11$
minimum, (−4, −5), $x = −4$

3. $y = -x^2 + 5$
maximum, (0, 5), $x = 0$

Example ❶ Solve $x^2 - 2x - 3 = 0$ by graphing. Check by factoring.

Graph the related function $f(x) = x^2 - 2x - 3$. The equation of the axis of symmetry is $x = 1$, and the coordinates of the vertex are $(1, -4)$. Make a table of values to find other points to sketch the graph of $f(x) = x^2 - 2x - 3$.

x	f(x)
-1	0
0	-3
1	-4
2	-3
3	0

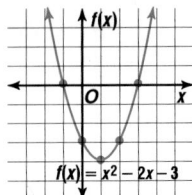

To solve the equation $x^2 - 2x - 3 = 0$, we need to know where the value of $f(x)$ is 0. On the graph, this occurs at the x-intercepts. The x-intercepts of the parabola appear to be -1 and 3.

LOOK BACK

You can refer to Lesson 10-6 for more information on the zero product property and solving equations by factoring.

Check: Solve by factoring.

$$x^2 - 2x - 3 = 0$$
$$(x - 3)(x + 1) = 0$$

Set each factor equal to 0.

$x - 3 = 0$	*Zero product property*
$x = 3$	*Add 3 to each side.*
$x + 1 = 0$	*Zero product property*
$x = -1$	*Add -1 to each side.*

The solutions of the equation are 3 and -1.

In Example 1, the zeros of the function were integers. Usually the zeros of a quadratic function are not integers. In these cases, use estimation to approximate the roots of the equation.

Example ❷ Solve $x^2 + 9x + 5 = 0$ by graphing. If integral roots cannot be found, estimate the roots by stating the consecutive integers between which the roots lie.

Use a table of values to graph the related function $f(x) = x^2 + 9x + 5$.

Notice in the table of values that the value of the function changes from negative to positive between the x values of -9 and -8 and -1 and 0.

x	f(x)
-9	5
-8	-3
-7	-9
-6	-13
-5	-15
-4	-15
-3	-13
-2	-9
-1	-3
0	5

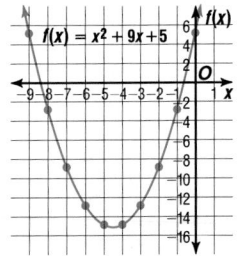

The x-intercepts of the graph are between -9 and -8 and between -1 and 0. So, one root of the equation is between -9 and -8, and the other root is between -1 and 0.

 Lesson 11–2 Solving Quadratic Equations by Graphing **621**

Motivating the Lesson

Hands-On Activity Toss a Ping-Pong™ ball and ask students to describe a curve that approximates the shape of its trajectory. Attach a Ping-Pong ball to each end of a piece of string and have the string hang in a parabolic shape. Remind students that the curve is not a parabola but a catenary, which is closely approximated by a parabola. Ask students to identify the axis of symmetry. Challenge them to determine if the axis of symmetry moves when the two Ping-Pong balls remain in the same place and a longer string is used. **Axis of symmetry does not move.**

2 TEACH

In-Class Examples

For Example 1
Solve $x^2 + 2x - 8 = 0$ by graphing. $x = -4, 2$

For Example 2
Solve $4x^2 - 4x - 3 = 0$ by graphing. If integral roots cannot be found, estimate the roots by stating the consecutive integers between which the roots lie. **One root is between 0 and -1. The other root is between 1 and 2.**

Teaching Tip Explain that the x-intercepts are the zeros of the function.

You can use a graphing calculator to find a more accurate estimate for the root of a quadratic equation than you can by using paper and pencil. One way to estimate the root is to ZOOM IN on the zero. Another method is to use the ROOT feature on the CALC menu.

The y-coordinate value may appear as a decimal in scientific notation, such as $-1E-12$ rather than 0.

Quadratic equations always have two roots. However, these roots may not be two distinct numbers.

Example   **Solve $x^2 - 12x + 36 = 0$ by graphing.**

Graph the related function $f(x) = x^2 - 12x + 36$.

x	f(x)
4	4
5	1
6	0
7	1
8	4

The equation of the axis of symmetry is $x = 6$. The vertex of the parabola is at $(6, 0)$.

$f(x) = x^2 - 12x + 36$

Notice that the vertex of the parabola is the *x*-intercept. Thus, one solution is 6. What is the other solution?

Try solving by factoring.

$x^2 - 12x + 36 = 0$

$(x - 6)(x - 6) = 0$

Set each factor equal to 0.

$x - 6 = 0$	$x - 6 = 0$	*Zero product property*
$x = 6$	$x = 6$	*Add 6 to each side.*

There are two identical roots to this equation. So, there is only one distinct root. The solution for $x^2 - 12x + 36 = 0$ is 6.

Thus far, we have seen that quadratic equations can have two distinct real roots or one distinct real root. Is it possible that there may be no real roots?

Example **4** **Solve $x^2 + 2x + 5 = 0$ by graphing.**

Graph the function $f(x) = x^2 + 2x + 5$.

x	f(x)
−3	8
−2	5
−1	4
0	5
1	8

The equation of the axis of symmetry is $x = -1$. The vertex of the parabola is at $(-1, 4)$.

$f(x) = x^2 + 2x + 5$

This graph has no *x*-intercept. Thus, there are no real number solutions for this equation.

The symbol ∅, indicating an empty set, is often used to represent no real solution.

Quadratic equations can be used to solve number problems.

Example **5** **Two numbers have a sum of 4. What are the numbers if their product is −12?**

Explore   Let *n* represent one of the numbers. Then the other number is $4 - n$.

Plan   A function that describes the product of these two numbers is $f(n) = n(4 - n)$ or $f(n) = -n^2 + 4n$. Find the value of *n* if $f(n)$ equals −12.

Solve   Solve $f(n) = -n^2 + 4n$ if $f(n) = -12$.

$f(n) = -n^2 + 4n$

$-12 = -n^2 + 4n$ $f(n) = -12$

$0 = -n^2 + 4n + 12$ *Rewrite the equation so one side is 0.*

Graph the related function $f(n) = -n^2 + 4n + 12$.

x	y
−3	−9
−2	0
−1	7
0	12
1	15
2	16
3	15
4	12
5	7
6	0
7	−9

The equation of the axis of symmetry is $n = 2$. The vertex of the parabola is at $(2, 16)$.

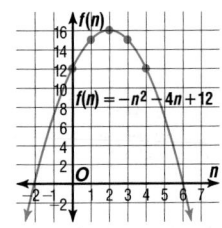

$f(n) = -n^2 + 4n + 12$

(continued on the next page)

Check for Understanding

Exercises 1–16 are designed to help you assess your students' understanding through reading, writing, speaking, and modeling. You should work through Exercises 1–5 with your students and then monitor their work on Exercises 6–16.

Additional Answers

10.

$f(x) = x^2 - 7x + 6$

11.

$f(c) = c^2 - 5c - 24$

12.

$f(n) = 5n^2 + 2n + 6$

13.

$f(w) = w^2 - 3w - 5$

14.

$f(b) = b^2 - b + 4$

The n-intercepts of the graph are -2 and 6. Use these values of n to find the value of the other number $4 - n$.

If $n = -2$, then $4 - n = 4 - (-2)$ or 6.

If $n = 6$, then $4 - n = 4 - 6$ or -2.

So the numbers are 6 and -2.

Examine   Test to see if the numbers satisfy the problem.

The sum of the numbers is 4.   The product of the numbers is -12.

$$-2 + 6 \stackrel{?}{=} 4 \qquad\qquad -2(6) \stackrel{?}{=} -12$$
$$4 = 4 \ \checkmark \qquad\qquad\qquad -12 = -12 \ \checkmark$$

The two numbers are -2 and 6.

CHECK FOR UNDERSTANDING

Communicating Mathematics

Study the lesson. Then complete the following.

1. **Explain** why the x-intercepts of a quadratic function can be used to solve a quadratic equation.   The x-intercept is where the function equals 0.

2. Hanna, because factoring only works when the polynomial making up the equation is factorable. Graphing will always give you an estimate of the solutions if they are real.

2. **You Decide**   Joshua says he likes to solve a quadratic equation by factoring rather than graphing. Hanna says that she likes graphing because she can always get an answer. Who is correct? Give examples to support your answer.

3. What is the related function you would use to solve $x^2 + 9x + 2 = 3x - 4$ by graphing? $f(x) = x^2 + 6x + 6$

4. Refer to the application at the beginning of the lesson. What is the least number of employees that would help the company make its goal of $75,000 profit?   **500 employees**

5. **Draw** an example of each type of situation that may occur when using graphing to solve a quadratic equation. Identify the number of real roots in each situation.   See students' work.

Guided Practice

Determine the number of real roots for each quadratic equation whose related function is graphed below.

6.

1 distinct root

7.

2 real roots

8.
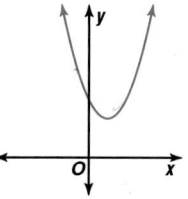
no real roots ($\varnothing$)

9. State the real roots of the quadratic equation whose related function is graphed at the right.
$-1, 2$

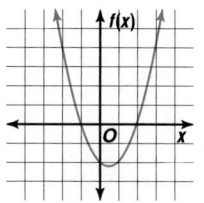

Reteaching

Decision Making   Give students a list of maximum and minimum points of a parabola and the value of a in $y = ax^2 + bx + c$. Then have them describe the number of roots of each equation and how they determined that number.

Example: Minimum point is $(-1, -1)$ and $a = 2$.   There are two roots because the minimum point is below the x-axis and the parabola will open up, therefore crossing the x-axis twice.

10–15. See margin for graphs.

10. 1, 6
11. −3, 8
12. ∅
13. −2 < w < −1, 4 < w < 5

Solve each equation by graphing. If integral roots cannot be found, state the consecutive integers between which the roots lie.

10. $x^2 - 7x + 6 = 0$ 11. $c^2 - 5c - 24 = 0$ 12. $5n^2 + 2n + 6 = 0$

13. $w^2 - 3w = 5$ 14. $b^2 - b + 4 = 0$ ∅   15. $a^2 - 10a = -25$ 5

16. **Number Theory** Use a quadratic equation to find two real numbers whose sum is 5 and whose product is −24. 8 and −3

EXERCISES

Practice

A

State the real roots of each quadratic equation whose related function is graphed below.

17.
18.
19.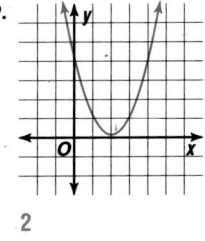

−2, −6   −2, 2   2

20–34. See Solutions Manual for graphs.

20. −3, −4
22. −5 < a < −4, −2 < a < −1
24. −6, 2
26. −8, −2
27. −4, 1
31. 8 < a < 9, −1 < a < 0
33. 3, −1 < n < 0
34. −4 < x < −3, 2 < x < 3

35–38. See Solutions Manual.

B

C

Solve each equation by graphing. If integral roots cannot be found, state the consecutive integers between which the roots lie.

20. $x^2 + 7x + 12 = 0$ 21. $x^2 - 16 = 0$ −4, 4   22. $a^2 + 6a + 7 = 0$

23. $x^2 + 6x + 9 = 0$ −3   24. $r^2 + 4r - 12 = 0$ 25. $c^2 + 3 = 0$ ∅

26. $2c^2 + 20c + 32 = 0$ 27. $3x^2 + 9x - 12 = 0$ 28. $2x^2 - 18 = 0$ −3, 3

29. $p^2 + 16 = 8p$ 4   30. $w^2 - 10w = -21$ 3, 7   31. $a^2 - 8a = 4$

32. $m^2 - 2m = -2$ ∅   33. $12n^2 - 26n = 30$ 34. $4x^2 - 35 = -4x$

The roots of a quadratic equation are given. Graph the related quadratic function if it has the indicated maximum or minimum point.

35. roots: 0, −6
   maximum point: (−3, 4)

36. roots: −2, −6
   minimum point : (−4, −2)

37. roots: no real roots
   minimum point: (2, 5)

38. roots: −4 < x < −3, 1 < x < 2
   maximum point: (−1, 6)

INTEGRATION
Number Theory

Use a quadratic equation to determine the two numbers that satisfy each situation.

39. Their difference is 4 and their product is 32. −4, −8 or 4, 8

40. Their sum is 9 and their product is 20. 4, 5

41. Their sum is 4 and their product is 5. no solution

42. They differ by 2. The sum of their squares is 130. 9 and 7 or −9 and −7

43–45. See margin for graphs.
43. −2, −6
44. 0 < y < 1, 3 < y < 4
45. no y-intercepts

Estimate the y-intercepts of each quadratic equation by graphing.

43. $x = -0.75y^2 - 6y - 9$ 44. $x = y^2 - 4y + 1$ 45. $x = 3y^2 + 2y + 4$

Lesson 11-2 Solving Quadratic Equations by Graphing **625**

Additional Answer

45.

$x = 3y^2 + 2y + 4$

Assignment Guide

Core: 17–53 odd, 54, 55, 57–66
Enriched: 18–52 even, 54–66

For **Extra Practice,** see p. 781.

The red A, B, and C flags, printed only in the Teacher's Wraparound Edition, indicate the level of difficulty of the exercises.

Additional Answers

15.

$f(a) = a^2 - 10a + 25$

43.

$x = -0.75y^2 - 6y - 9$

44.

$x = y^2 - 4y + 1$

Study Guide Masters, p. 76

NAME_____ DATE_____
11-2
Study Guide   Student Edition
Pages 620–627

Solving Quadratic Equations by Graphing

A **quadratic equation** is an equation in which the value of the related quadratic function is 0. The solutions of a quadratic equation are called the **roots** of the equation. The roots of a quadratic equation can be found by graphing the related quadratic function and finding the x-intercepts or **zeros** of the function.

Examples: Solve each equation by graphing. If exact roots cannot be found, estimate the roots by stating the consecutive integers between which the roots lie.

a. $x^2 + 7x + 12 = 0$ b. $x^2 + 6x + 6 = 0$ c. $x^2 - 4x + 5$

−3 and −4   between −5 and −4 and   no real roots
                between −2 and −1

Solve each equation by graphing. If exact roots cannot be found, state the consecutive integers between which the roots lie.

1. $x^2 - x - 12 = 0$ 4, −3   2. $x^2 + 4 = 0$ no real roots

3. $4x^2 - 12x + 3 = 0$ between 0 and 1 and 2 and 3   4. $4x^2 = 35 - 4x$ between −3 and −4 and 2 and 3

Practice Masters, p. 76

Graphing Calculator

Use a graphing calculator to solve each equation to the nearest hundredth. 46. −1.66, 1.66   47. no real roots   48. −11

46. $4x^2 - 11 = 0$ 47. $-2x^2 - x - 3 = 0$ 48. $x^2 + 22x + 121 = 0$
49. $6x^2 - 12x + 3 = 0$ 50. $5x^2 + 4x - 7 = 0$ 51. $-4x^2 + 7x + 8 = 0$
   0.29, 1.71   −1.6, 0.85   −0.79, 2.54

Use a graphing calculator to find values for k for each equation so that it will have (a) one distinct root, (b) two real roots, and (c) no real roots.

52. $x^2 + 3x + k = 0$ 53. $kx^2 + 4x - 2 = 0$
   2.25, $k < 2.25$, $k > 2.25$ −2, $k > -2$, $k < -2$

Critical Thinking

54. Suppose the value of a quadratic function is negative when $x = 10$ and positive when $x = 11$. Explain why it is reasonable to assume that the related equation has a root between 10 and 11. **See margin.**

Applications and Problem Solving

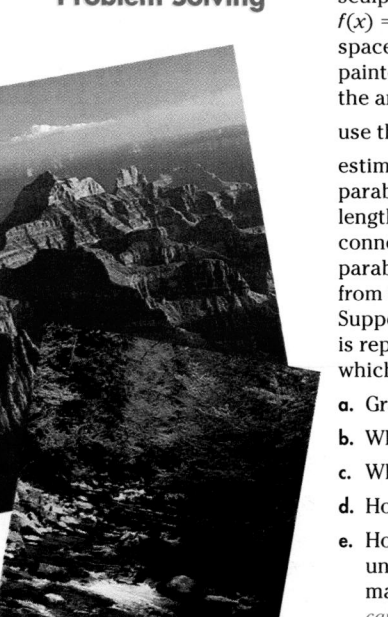

55. Architecture   A painter is hired to paint an art gallery whose walls are sculptured with arches that can be represented by the quadratic function $f(x) = -x^2 - 4x + 12$. The wall space under each arch is to be painted a different color from the arch itself. The painter can use the formula $A = \frac{2}{3}bh$ to estimate the area under a parabola, where b is the length of a horizontal segment connecting two points on the parabola and h is the height from that segment to the vertex. Suppose the horizontal segment is represented by the floor, which is the x-axis, and each unit represents 1 foot.

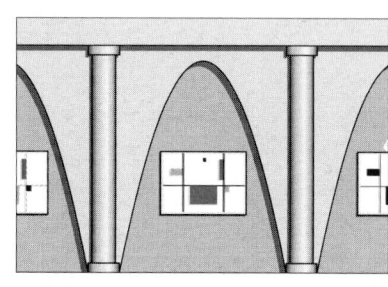

a. Graph the quadratic function and determine its x-intercepts.   **See margin.**
b. What is the length of the segment along the floor?   **8 feet**
c. What is the height of the arch?   **16 feet**
d. How much wall space is there under each arch? $85\frac{1}{3}$ ft²
e. How much would the paint cost to paint the walls under 12 arches if the paint is $27/gallon, she applies two coats, and the manufacturer states that each gallon will cover 200 ft²?   *Remember you cannot buy part of a gallon.*   **$297**

56. Make a Drawing   Banneker Park has set aside a section of the park as a nature preserve with an observation deck and telescopes so that people can observe the wildlife and plants up close. To accommodate the increased number of visitors they expect, the park commission has received funding to double the parking area. The current lot is 64 yards by 96 yards, and they are going to add strips of equal width to the end and side of the lot to create a larger rectangle that is twice the size of the original.

a. Make a drawing of the lot and the proposed additions.   **See margin.**
b. Write a quadratic equation to find x, the width of the strips, so that the area of the parking lot is doubled. $x^2 + 160x - 6144 = 0$
c. How wide are the strips to be added?   **32 yards**
d. What are the dimensions of the new parking lot?   **96 × 128 yards**

Most-Visited U.S. National Parks in 1993

Park	Visitors
1. Great Smoky Mountains	9,283,848
2. Grand Canyon	4,575,602
3. Yosemite	3,839,645
4. Yellowstone	2,912,193
5. Rocky Mountains	2,780,342

Listed below is the area in acres of each of the parks.

Great Smoky Mountains	520,004
Grand Canyon	1,218,375
Yosemite	748,542
Yellowstone	2,221,766
Rocky Mountains	265,668

. $x = 0$; (0, 4); See margin for graph.

. $\left\{0, -\frac{2}{9}\right\}$

. $(4x - 3)$ m by $x - 1)$ m

57. Find the equation of the axis of symmetry and the coordinates of the vertex of the graph of $y = -3x^2 + 4$. (Lesson 11–1)

58. Solve $81x^3 + 36x^2 = -4x$. (Lesson 10–6)

59. Geometry The area of a rectangle is $(8x^2 - 10x + 3)$ square meters. What are the dimensions of the rectangle? (Lesson 10–3)

60. Find $(x - 2y)^3$. (Lesson 9–8) $x^3 - 6x^2y + 12xy^2 - 8y^3$

61. Find the degree of $6x^2y + 5x^3y^2z - x + x^2y^2$. (Lesson 9–4)   **6**

62. Use graphing to solve the system of equations. (Lesson 8–1)
$x + y = 3$
$x + y = 4$ **no solution**

63. Solve $|3x + 4| < 8$. (Lesson 7–6) $-4 < x < \frac{4}{3}$

64. Graph $y = -x + 6$. (Lesson 6–5)   **See margin.**

65. Commerce Find the sale price of an item originally marked as $33 with 25% off. (Lesson 4–4)   **$24.75**

66. Find $-4 + 6 + (-10) + 8$. (Lesson 2–3)   **0**

WORKING ON THE
In·ves·ti·ga·tion

Refer to the Investigation on pages 554–555.

the BRICKYARD

order to best use the excess inventory of ricks, you are going to investigate the possible ombinations that will use up the inventory at a eady rate.

Suppose you have four large square bricks and three small square bricks. How many rectangular bricks would you need to create a rectangular pattern? Is there more than one pattern that can be made if you change the number of rectangular bricks?
• What are the dimensions, perimeters, and areas of the rectangular patterns formed?
• Use the variables x and y as the length of the large square and the length of the small square, respectively. List the dimensions, perimeter, and area of each pattern in terms of x and y.

• Explain your findings. How do the perimeters and areas of these patterns relate in this matter?

2 If you had three large square bricks, five rectangular bricks, and two small square bricks, what size rectangular patterns can you make? Express the dimensions, perimeter, and area of each pattern in terms of x and y.

3 If you had two large bricks, six rectangular bricks, and four small bricks, what size rectangular patterns can you make? Express the dimensions, perimeter, and area of each pattern in terms of x and y.

4 Explain the method you used to solve these problems and any generalizations you found.

Add the results of your work to your Investigation Folder.

Lesson 11–2 Solving Quadratic Equations by Graphing **627**

Extension

Connections You can show students that the solution to a system of equations involving one or more quadratics is the point or points at which the graphs intersect.

1. $y = x + 2$
$y = x^2$
(−1, 1), (2, 4)

2. $y = -3x - 8$
$y = x^2 - 4$
no real solution

3. $y = -x^2 + 2x$
$y = x^2 - 2x$
(0, 0), (2, 0)

In·ves·ti·ga·tion

Working on the Investigation
The Investigation on pages 554–555 is designed to be a long-term project that is completed over several days or weeks. Encourage students to keep their materials in their Investigation Folder as they work on the Investigation.

NCTM Standards: 1–6

Instructional Resources

- Study Guide Master 11-3
- Practice Master 11-3
- Enrichment Master 11-3
- Assessment and Evaluation Masters, pp. 295–296
- Multicultural Activity Masters, p. 21
- Real-World Applications, 27
- Tech Prep Applications Masters, p. 22

Transparency 11-3A contains the 5-Minute Check for this lesson; **Transparency 11-3B** contains a teaching aid for this lesson.

Recommended Pacing

Standard Pacing	Days 5 & 6 of 11
Honors Pacing	Day 5 of 10
Block Scheduling*	Day 2 of 5 (along with Lesson 11-2)
Alg. 1 in Two Years*	Days 8, 9, & 10 of 19

*For more information on pacing and possible lesson plans, refer to the *Block Scheduling Booklet* and *Algebra 1 in Two Years.*

1 FOCUS

5-Minute Check
(over Lesson 11-2)

Solve each equation by graphing.

1. $0 = x^2 - 8x + 12$
   2, 6
2. $x^2 + 6x + 9 = 0$
   −3
3. $x^2 - 3x + 7 = 0$
   no real roots
4. $x^2 - 4x + 5 = 0$
   no real roots
5. $x^2 - 2x + 1 = 0$
   1

Objective

After studying this lesson, you should be able to:

- solve quadratic equations by using the quadratic formula.

Civics

The number of citizens (in millions) voting in each presidential election since 1824 can be approximated by the quadratic function $V(t) = 0.0046t^2 - 0.185t + 3.30$, where t represents the number of years since 1824. For her history project, Marcela Ruiz needed to determine in what year the number of voters in a presidential election was approximately 55 million. She used the function above, replacing $V(t)$ with 55.

$$V(t) = 0.0046t^2 - 0.185t + 3.30$$
$$55 = 0.0046t^2 - 0.185t + 3.30$$
$$0 = 0.0046t^2 - 0.185t - 51.7$$

Marcela knew that she could use her graphing calculator to estimate the roots of this equation, but she wondered how she could get a good estimate of t if she didn't have a graphing calculator. *You will estimate these roots in Example 4.*

You can use the **quadratic formula** to solve any quadratic equation.

The Quadratic Formula	**The roots of a quadratic equation in the form $ax^2 + bx + c = 0$, where $a \neq 0$, are given by the formula** $$x = \frac{-b \pm \sqrt{b^2 - 4ac}}{2a}.$$

The quadratic formula can be used to solve any quadratic equation involving any variable.

Example **1** **Use the quadratic formula to solve each equation.**

a. $x^2 - 6x - 40 = 0$

In the equation, $a = 1$, $b = -6$, and $c = -40$. Substitute these values into the quadratic formula.

The symbol ± means to evaluate the expression first using + and then evaluate it again using −. This provides for the two roots of the equation.

$$x = \frac{-b \pm \sqrt{b^2 - 4ac}}{2a}$$
$$= \frac{-(-6) \pm \sqrt{(-6)^2 - 4(1)(-40)}}{2(1)} \qquad a = 1, b = -6, \text{ and } c = -40$$
$$= \frac{6 \pm \sqrt{36 + 160}}{2}$$
$$= \frac{6 \pm \sqrt{196}}{2}$$
$$= \frac{6 \pm 14}{2}$$

$$x = \frac{6 + 14}{2} \qquad \text{or} \qquad x = \frac{6 - 14}{2}$$

$$= \frac{20}{2} \text{ or } 10 \qquad\qquad = -\frac{8}{2} \text{ or } -4$$

You can also check the solution to any equation by substituting each value into the original equation.

Check: Solve by graphing the related function $f(x) = x^2 - 6x - 40$. The x-intercepts appear to be 10 and -4. This agrees with the algebraic solution.

The solutions are 10 and -4.

b. $y^2 - 6y + 9 = 0$

$$y = \frac{-b \pm \sqrt{b^2 - 4ac}}{2a}$$

$$= \frac{-(-6) \pm \sqrt{(-6)^2 - 4(1)(9)}}{2(1)} \qquad a = 1, b = -6, \text{ and } c = 9$$

$$= \frac{6 \pm \sqrt{36 - 36}}{2}$$

$$= \frac{6 \pm \sqrt{0}}{2}$$

$$y = \frac{6 + 0}{2} \quad \text{or} \quad y = \frac{6 - 0}{2}$$

$$= 3 \qquad\qquad = 3$$

Check: Solve by factoring.
$$y^2 - 6y + 9 = 0$$
$$(y - 3)(y - 3) = 0$$
$$y - 3 = 0 \qquad\qquad y - 3 = 0 \quad \textit{Zero product property}$$
$$y = 3 \qquad\qquad\quad y = 3 \quad \textit{Add 3 to each side.}$$

There is one distinct root, 3.

LOOK BACK

You can refer to Lesson 2-8 for more information on irrational numbers.

Sometimes when you use the quadratic formula, you find the roots are irrational numbers. It is helpful to use a calculator to estimate the values of the roots in this case.

Example **2** Use the quadratic formula to solve $2n^2 - 7n - 3 = 0$.

$$n = \frac{-b \pm \sqrt{b^2 - 4ac}}{2a}$$

$$= \frac{-(-7) \pm \sqrt{(-7)^2 - 4(2)(-3)}}{2(2)} \qquad a = 2, b = -7, \text{ and } c = -3$$

$$= \frac{7 \pm \sqrt{49 + 24}}{4}$$

$$= \frac{7 \pm \sqrt{73}}{4}$$

$\sqrt{73}$ is an irrational number. We can approximate the roots by using a calculator to find a decimal value for $\sqrt{73}$.

$$n = \frac{7 + \sqrt{73}}{4} \approx 3.886 \qquad\qquad n = \frac{7 - \sqrt{73}}{4} \approx -0.386$$

The two roots are approximately -0.386 and 3.886.

Lesson 11–3 Solving Quadratic Equations by Using the Quadratic Formula **629**

Alternative Teaching Strategies

Student Diversity Discuss the fact that the quadratic formula can be seen as a "cheat-sheet" to help students solve equations more easily. Any equation that can be rearranged and simplified to the form of $ax^2 + bx + c = 0$ can be solved using the quadratic formula.

Motivating the Lesson

Questioning Introduce the lesson with a review of the general form of a quadratic equation, $ax^2 + bx + c = 0$. Ask students to name a, b, and c for each quadratic equation.

1. $2x^2 - 4x + 3 = 0$ 2, -4, 3
2. $x^2 + x - 5 = 0$ 1, 1, -5
3. $7x^2 - 10x + 3 = 0$ 7, -10, 3
4. $-3x^2 + 4x - 12 = 0$ -3, 4, -12

2 TEACH

In-Class Examples

For Example 1
Use the quadratic formula to solve each equation.

a. $x^2 + 6x + 8 = 0$ $-2, -4$
b. $x^2 - 2x - 15 = 0$ $-3, 5$
c. $2x^2 + 7x - 15 = 0$ $-5, \frac{3}{2}$

For Example 2
Use the quadratic formula to solve $3n^2 - 8n - 2 = 0$. Use a calculator to approximate the roots.

$\frac{4 \pm \sqrt{22}}{3}$ or $-0.230, 2.897$

Teaching Tip For Examples 1 and 2, encourage students to memorize the quadratic formula. Emphasize that it can be used to solve any quadratic equation.

In-Class Examples

For Example 3
Use the quadratic formula to solve $3x^2 - x + 3 = 0$.
no real roots

For Example 4
Determine in what year the number of people voting in a presidential election was approximately 77 million. **1972**

Teaching Tip For Example 4, remind students that the square root of a negative number is not defined for the set of real numbers.

When we solved quadratic equations by graphing, we found that some quadratic equations have no real roots. How does the quadratic formula work in this situation?

Example ③ **Use the quadratic formula to solve $z^2 - 5z + 12 = 0$.**

$$z = \frac{-b \pm \sqrt{b^2 - 4ac}}{2a}$$

$$= \frac{-(-5) \pm \sqrt{(-5)^2 - 4(1)(12)}}{2(1)} \quad a = 1, b = -5, \text{ and } c = 12$$

$$= \frac{5 \pm \sqrt{25 - 48}}{2}$$

$$= \frac{5 \pm \sqrt{-23}}{2}$$

Since there is no real number that is the square root of a negative number, this equation has no real roots.

It is often helpful to use a calculator when using the quadratic formula to solve real-world problems. If there are no real roots for the equation, the calculator will give you an error message.

Example ④

Civics

Refer to the connection at the beginning of the lesson. Determine in what year the number of people voting in a presidential election was approximately 55 million.

Use a scientific calculator and the quadratic formula to find values for t. The values for a, b, and c are 0.0046, -0.185, and -51.7, respectively. Find the value of $\sqrt{b^2 - 4ac}$ and store it in the calculator's memory.

Enter: [( ] .185 [+/−] [x^2] [−] 4 [×] .0046 [×] 51.7 [+/−] [ )]

[√x] [STO] *0.992726044*

Now evaluate the quadratic formula.

Enter: [( ] [( ] .185 [+/−] [ )] [+/−] [+] [RCL] [ )] [÷] [( ] 2

[×] .0046 [ )] [=] *128.0137005*

Enter: [( ] [( ] .185 [+/−] [ )] [+/−] [−] [RCL] [ )] [÷] [( ] 2 [×]

.0046 [ )] [=] *−87.19630922*

The negative root has no meaning in this problem.

Since t represents the number of years since 1824, add 128 to 1824:
$128 + 1824 = 1952$.

So, 1952 was the year in which approximately 55 million people voted.

630 *Chapter 11 Exploring Quadratic and Exponential Functions*

Cooperative Learning

Pairs Check Separate students into pairs. One student will use the quadratic formula and a scientific calculator. The other student will use the graphing calculator to find the roots of each equation to the nearest hundredth.

a. $5.97x^2 - 17.1x + 2.54 = 0$
2.71, 0.16

b. $1.9y^2 - 11.5y - 6.2 = 0$
6.55, −0.50

c. $9.35t^2 + 1.97t - 7.42 = 0$
0.79, −1.00

For more information on the pairs check strategy, see *Cooperative Learning in the Mathematics Classroom*, one of the titles in the Glencoe Mathematics Professional Series, pages 12–13.

Communicating Mathematics

1. Evaluate the formula using + for one root; then reevaluate using − for the other.

MATH JOURNAL

Study the lesson. Then complete the following.

1. **Explain** how you get two roots when using the quadratic formula.

2. **Explain** what happens in the quadratic formula when there are no real roots of the equation.   **You get the square root of a negative number.**

3. Refer to the application at the beginning of the lesson.
   a. Estimate how many citizens voted in the 1992 election.   **102.05 million**
   b. Predict how many will vote in the year 2000.   **113.23 million**

4. **Assess Yourself** You have learned to use graphing, factoring, and the quadratic formula to solve quadratic equations. Which method do you prefer and why?   **See students' work.**

Guided Practice

6. −1, −5, −2; −4.56, −0.44

State the values of *a*, *b*, and *c* for each quadratic equation. Then solve the equation by using the quadratic formula. Approximate irrational roots to the nearest hundredth.

5. $x^2 + 3x - 18 = 0$ **1, 3, −18; 3, −6**   6. $14 = 12 - 5x - x^2$

7. $4x^2 - 2x + 15 = 0$
   4, −2, 15; no real roots

8. $x^2 = 25$ **1, 0, −25; ±5**

Solve each equation by using the quadratic formula. Approximate roots to the nearest hundredth if necessary.

9. $4x^2 + 2x - 17 = 0$ **1.83, −2.33**   10. $3b^2 + 5b + 11 = 0$ ∅

11. $x^2 + 7x + 6 = 0$ **−1, −6**   12. $z^2 - 13z = 32$ **−2.12, 15.12**

13. **Hydraulics**   Cox's formula for measuring the velocity of water escaping from a reservoir through a horizontal pipe is $4v^2 + 5v - 2 = \frac{1200HD}{L}$, where *v* represents the velocity of the water in feet per second, *H* the height of the reservoir in feet, *D* the diameter of the pipe in inches, and *L* the length of the pipe in feet. How fast is water flowing through a pipe 20 feet long with a diameter of 6 inches that is draining a swimming pool with a depth of 10 feet? Round your answer to the nearest tenth.   **about 29.4 ft/s**

EXERCISES

Practice

14. −4, 6   16. −10, −2

17. $-\frac{4}{5}$, 1   19. $-\frac{5}{3}$, 4

21. −2.91, 2.41

25. $-\frac{3}{2}$, $-\frac{5}{3}$

27. −6, 6

29. $-\frac{1}{3}$, 1

30. $-\frac{3}{4}$, $\frac{5}{6}$ 31. −3, $-\frac{1}{2}$

35. 0.5, 2

36. 0.6, 1.6

37. −0.2, 1.4

Solve each equation by using the quadratic formula. Approximate irrational roots to the nearest hundredth.   **15. −8.61, −1.39**

A

14. $x^2 - 2x - 24 = 0$ 15. $a^2 + 10a + 12 = 0$ 16. $c^2 + 12c + 20 = 0$

17. $5y^2 - y - 4 = 0$ 18. $r^2 + 25 = 0$ ∅   19. $3b^2 - 7b - 20 = 0$

20. $y^2 + 12y + 36 = 0$ **−6**   21. $2r^2 + r - 14 = 0$ 22. $2x^2 + 4x = 30$ **−5, 3**

23. $2x^2 - 28x + 98 = 0$ **7**   24. $24x^2 - 14x = 6$ **−0.29, 0.87**

25. $6x^2 + 15 = -19x$ 26. $12x^2 = 48$ **−2, 2**

27. $x^2 + 6x = 36 + 6x$ 28. $1.34a^2 - 1.1a = -1.02$ ∅

B

29. $3m^2 - 2m = 1$ 30. $24a^2 - 2a = 15$

31. $2w^2 = -(7w + 3)$ 32. $a^2 - \frac{3}{5}a + \frac{2}{25} = 0$ $\frac{2}{5}$, $\frac{1}{5}$

33. $-2x^2 + 0.7x = -0.3$ **0.60, −0.25**   34. $2y^2 - \frac{5}{4}y = \frac{1}{2}$ **−0.28, 0.90**

Without graphing, determine the *x*-intercepts of the graph of each function to the nearest tenth.

C

35. $f(x) = 2x^2 - 5x + 2$ 36. $f(x) = 4x^2 - 9x + 4$ 37. $f(x) = 13x^2 - 16x - 4$

Reteaching ▬▬▬

Using Discussion Have half the class solve a quadratic equation by using the quadratic formula and the other half of the class solve by graphing. Compare answers and discuss which method was better to use for that problem. Given a new problem, exchange tasks for each half of the class. Solve and discuss.

GLENCOE Technology

 Interactive Mathematics Tools Software

This multimedia software helps students observe the changes in maximum height of the graph of a projectile when the initial velocity and height are altered. A **Computer Journal** gives students an opportunity to write about what they have learned.

For Windows & Macintosh

---

3 PRACTICE/APPLY

Check for Understanding
Exercises 1–13 are designed to help you assess your students' understanding through reading, writing, speaking, and modeling. You should work through Exercises 1–4 with your students and then monitor their work on Exercises 5–13.

Error Analysis
Make sure students put quadratic equations in the form $ax^2 + bx + c = 0$. For Example, $2x^2 + 6x - 9 = (2)x^2 + (6)x + (-9)$ so $a = 2$, $b = 6$, and $c = -9$.

Assignment Guide

Core: 15–41 odd, 42, 43, 45–53
Enriched: 14–40 even, 41–53
All: Self Test, 1–10

For **Extra Practice,** see p. 781.

The red A, B, and C flags, printed only in the Teacher's Wraparound Edition, indicate the level of difficulty of the exercises.

Study Guide Masters, p. 77

The method of completing the square can be used to develop a general formula called the *quadratic formula* that can be used to solve any quadratic equation.

> **The Quadratic Formula**
> The roots of a quadratic equation of the form $ax^2 + bx + c = 0$, where $a \neq 0$, are given by the formula
> $x = \frac{-b \pm \sqrt{b^2 - 4ac}}{2a}$

In order to find a real value for $\sqrt{b^2 - 4ac}$, the value of $b^2 - 4ac$ must be nonnegative. If $b^2 - 4ac$ is negative, the equation has no real roots.

Example: Use the quadratic formula to solve $x^2 - 6x - 2 = 0$.

$$x = \frac{-b \pm \sqrt{b^2 - 4ac}}{2a}$$

$$= \frac{6 \pm \sqrt{(-6)^2 - 4(1)(-2)}}{2(1)}$$ $a = 1$, $b = -6$, and $c = -2$

$$= \frac{6 \pm \sqrt{44}}{2}$$

$x \approx 6.32$ or $x \approx 1.78$

The two roots are approximately 6.32 and 1.78.

Solve each equation by using the quadratic formula. Approximate irrational roots to the nearest hundredth.

1. $4t^2 = 144$ −6, 6
2. $2x^2 + 9x + 4 = 0$ -4, $-\frac{1}{2}$
3. $8x^2 + 17x + 2 = 0$ -2, $-\frac{1}{8}$

4. $3z^2 + 5z - 2 = 0$ -2, $\frac{1}{3}$
5. $-2m^2 + 8m + 4 = 0$ 4.45, −0.45
6. $x^2 + 3x - 2 = 0$ 0.56, −3.56

7. $2x^2 - 6x + 4 = 0$ 1, 2
8. $5x^2 + 10 = 15$ −1, 1
9. $2y^2 - 9y - 3 = 0$ 4.81, −0.31

38–40. Sample answers are given.

38. 1, 2, −2;
$x^2 + 2x - 2 = 0$

Programming

39. 4, 20, 23;
$4x^2 + 20x + 23 = 0$

40. 1, −4, $-\frac{13}{4}$;
$x^2 - 4x - \frac{13}{4} = 0$

Use the quadratic formula to determine values for *a*, *b*, and *c* if the given numbers are roots of a quadratic equation. Then write the equation.

38. $-1 \pm \sqrt{3}$

39. $\dfrac{-5 \pm \sqrt{2}}{2}$

40. $\dfrac{4 \pm \sqrt{29}}{2}$

41. The graphing calculator program at the right determines what type of roots a quadratic equation will have and then prints decimal approximations of the roots if they exist.

Use the program to find the roots of each equation.
a. $x^2 - 11x + 10 = 0$ 10, 1
b. $3x^2 - 2x + 1 = 0$ none
c. $4x^2 + 4x + 1 = 0$ −0.5
d. $7x^2 + 2x - 5 = 0$ −1, 0.7142857143

```
PROGRAM:ROOTS
: Prompt A, B, C
: B²−4AC→D
: If D < 0
: Then
: Disp "NO REAL ROOTS"
: Stop
: End
: If D = 0
: Then
: Disp "1 DISTINCT ROOT:"
   −B/2A
: Else
: Disp "2 REAL ROOTS",
   (−B+√D)/2A, (−B−√D)/2A
: End
```

Critical Thinking

42. The expression $b^2 - 4ac$ is called the **discriminant** of a quadratic equation. The discriminant can help you determine what type of roots to expect when you solve a quadratic equation. Copy and complete the table below.

Equation	$x^2 - 4x + 1 = 0$	$x^2 + 6x + 11 = 0$	$x^2 - 4x + 4 = 0$
Value of the Discriminant	12	−8	0
Graph of the Equation	See Solutions	Manual for	graphs.
Number of x-intercepts	2	none	1
Number of Real Roots	2	none	1 distinct root

43. Use the results of Examples 1–3 and the table above to describe the discriminant of a quadratic equation for each type of solution. See margin.
a. two irrational roots
b. two noninteger rational roots
c. two integral roots
d. 1 distinct integral root

Applications and Problem Solving

44. **Government** Between 1980 and 1993, the income (billions of dollars) received by the federal government can be modeled by the quadratic function $I(t) = 0.26t^2 + 49.94t + 511.4$, where *t* represents the number of years since 1980.
a. Determine the income in 1993. $1204.6 billion   b. $1614.2 billion
b. What is the projected federal income for the year 2000?
c. Determine in which year the federal income was $1000 billion or $1 trillion.   1989
d. In which year will the federal income be at least $2 trillion?   2007

Tech Prep

Accountant Students who are interested in accounting may wish to do further research on the information given in Exercise 44 and explore the potential growth of this career. For more information on tech prep, see the *Teacher's Handbook*.

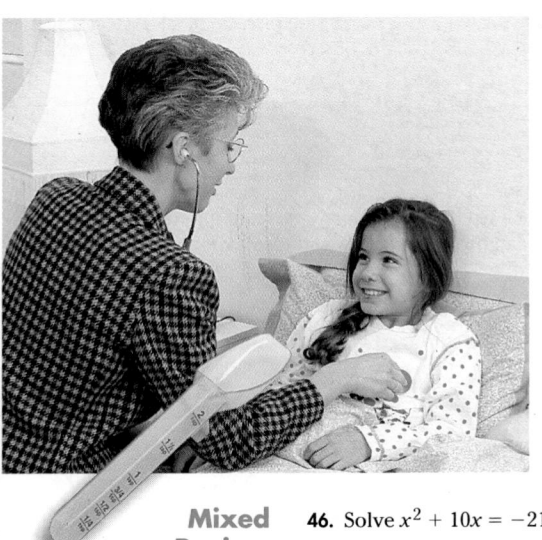

45. Medicine Two rules that govern the amount of medicine you should give a child if you know the adult dosage are Young's rule, $c = \frac{ad}{a + 12}$, and Cowling's rule, $c = \frac{(a + 1)d}{24}$. In both formulas, a represents the age of the child (in years), d represents the amount of the adult dosage, and c represents the amount of the child's dosage.

a. The adult dosage for a drug is 30 mg/day. Calculate the dosage for a 6-year-old using each rule. **Y, 10 mg; C, 8.75 mg**

b. Write a quadratic equation that represents the age(s) at which the two rules give the same dosage. Then solve the equation.
$0 = a^2 - 11a + 12$; 1.2 yr, 9.8 yr

Mixed Review

48. $21a^3 - 6a^2 - 46a + 28$

46. Solve $x^2 + 10x = -21$ by graphing. (Lesson 11–2) **−7, −3**

47. Solve $4s^2 = -36s$. (Lesson 10–6) **{0, −9}**

48. Find $3a^2(a - 4) + 6a(3a^2 + a - 7) - 4(a - 7)$. (Lesson 9–6)

49. Number Theory The sum of two numbers is 42. Their difference is 6. Find the numbers. (Lesson 8–3) **24, 18**

50. Graph the solution set for $b > 5$ or $b \le 0$. (Lesson 7–4) **See margin.**

51. Determine if the following relation is a function. (Lesson 5–5)
$\{(3, 2), (-3, -2), (-4, -2), (4, -2)\}$ **yes**

52. Solve $3x + 8 = 2x - 4$. (Lesson 3–5) **−12**

53. Weather The temperature at 8:00 A.M. was 36°F. A cold front went through that evening, and by 3:00 A.M. the next day, the temperature had fallen 40°. What was the temperature at 3:00 A.M.? (Lesson 2–1) **−4°F**

SELF TEST

Write the equation of the axis of symmetry and find the coordinates of the vertex of the graph of each equation. Then graph the equation. (Lesson 11–1) **1–3. See Solutions Manual for graphs.**

1. $y = x^2 - x - 6$
$x = \frac{1}{2}, \left(\frac{1}{2}, -\frac{25}{4}\right)$

2. $y = 2x^2 + 3$
$x = 0, (0, 3)$

3. $y = -x^2 + 7$
$x = 0, (0, 7)$

4. Physics The height h in feet of an experimental rocket t seconds after blast-off is given by the formula $h = -16t^2 + 2320t + 125$. (Lesson 11–1)

a. Approximately how long after blast-off does the rocket reach a height of 84,225 feet? **72.5 s**

b. How much longer from this height does it take the rocket to reach its maximum height?
0 seconds; this is the maximum height.

Solve each equation by graphing. If integral roots cannot be found, state the consecutive integers between which the roots lie. (Lesson 11–2) **5–7. See Solutions Manual for graphs.**

5. $x^2 = 81$ **−9, 9**

6. $4x^2 = 35 - 4x$
$2 < x < 3, -4 < x < -3$

7. $6x^2 + 36 = 0$ **∅**

Solve each equation by using the quadratic formula. Approximate irrational roots to the nearest hundredth. (Lesson 11–3) **10. −2.12, 15.12**

8. $a^2 + 7a = -6$ **−6, −1**

9. $y^2 + 6y + 10 = 0$ **∅**

10. $z^2 - 13z - 32 = 0$

Lesson 11–3 Solving Quadratic Equations by Using the Quadratic Formula **633**

Extension

Problem Solving Solve each equation. Express irrational roots in simplest radical form and in decimal form to the nearest hundredth.

1. $(2n - 3)^2 - 7 = 6n + 1$
$\frac{9 \pm \sqrt{77}}{4}$; 4.44, 0.06

2. $(3x + 2)^2 = (x - 1)^2 + 4$
$\frac{-7 \pm \sqrt{57}}{8}$; 0.07, −1.82

SELF TEST

The Self Test provides students with a brief review of the concepts and skills in Lessons 11-1 through 11-3. Lesson numbers are given to the right of exercises or instruction lines so students can review concepts not yet mastered.

4 ASSESS

Closing Activity

Speaking Ask students to explain how to use the quadratic formula to solve any quadratic equation.

Chapter 11, Quiz B (Lesson 3), is available in the *Assessment and Evaluation Masters*, p. 296.

Mid-Chapter Test (Lessons 11-1 through 11-3) is available in the *Assessment and Evaluation Masters*, p. 295.

Additional Answer

50.

Enrichment Masters, p. 77

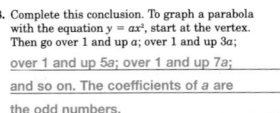

11-3 NAME _____ DATE _____
Enrichment
Student Edition Pages 628–633

Odd Numbers and Parabolas

The solid parabola and the dashed stair-step graph are related. The parabola intersects the stair steps at their inside corners.

Use the figure for Exercises 1–3.

1. What is the equation of the parabola?
$y = x^2$

2. Describe the horizontal sections of the stair-step graph.
Each is 1 unit wide.

3. Describe the vertical sections of the stair-step graph.
They form the sequence 1, 3, 5, 7.

Use the second figure for Exercises 4–6.

4. What is the equation of the parabola?
$y = \frac{1}{2}x^2$

5. Describe the horizontal sections of the stair steps.
Each is 1 unit wide.

6. Describe the vertical sections.
They form the sequence $\frac{1}{2}, \frac{3}{2}, \frac{5}{2}, \frac{7}{2}, \frac{9}{2}$.

7. How does the graph of $y = \frac{1}{2}x^2$ relate to the sequence of numbers $\frac{1}{2}, \frac{3}{2}, \frac{5}{2}, \frac{7}{2}, \ldots$?
If the x-values increase by 1, the y-values increase by the numbers in the sequence.

8. Complete this conclusion. To graph a parabola with the equation $y = ax^2$, start at the vertex. Then go over 1 and up a; over 1 and up $3a$;
over 1 and up 5a; over 1 and up 7a;
and so on. The coefficients of a are
the odd numbers.

Chapter 11 **633**

11–4A Graphing Technology
Exponential Functions

A Preview of Lesson 11–4

NCTM Standards: 1–6, 13

Objective
Use a graphing calculator to graph exponential functions of the form $y = a^x$, where $a > 0$ and $a \neq 1$.

Recommended Time
15 minutes

Instructional Resources
Graphing Calculator Masters, pp. 34 and 35

These masters provide keystroking instruction for this lesson for the TI-81 and Casio graphing calculators.

1 FOCUS

Motivating the Lesson
This activity introduces students to the graphs of simple exponential functions. The symmetry of n^x and $\left(\frac{1}{n}\right)^x$ are illustrated.

2 TEACH

Teaching Tip Set the *y*-range at [0, 1] to see how the graph approaches but never touches the *x*-axis.

3 PRACTICE/APPLY

Assignment Guide

Core: 1–10
Enriched: 1–10

4 ASSESS

Observing students working with technology is an excellent method of assessment.

Graphing calculators can be used to graph many types of functions easily so patterns in the functions can be studied. This includes **exponential functions** of the form $y = a^x$, where $a > 0$ and $a \neq 1$.

Example Graph each equation in the standard viewing window. Describe the graph.

a. $y = 3^x$

Enter the equation in the Y= list.

Enter: [Y=] 3 [∧] [X,T,θ] [ZOOM] 6

Notice that the graph increases rapidly as *x* becomes greater. The graph passes through the point at (0, 1). The domain of the function is all real numbers, and the range is all positive real numbers.

b. $y = \left(\frac{1}{3}\right)^x$

Enter: [Y=] [(] 1 [÷] 3 [)] [∧]
[X,T,θ] [GRAPH]

The graph decreases as *x* increases. The graph passes through the point at (0, 1). The domain is all real numbers, and the range is all positive real numbers.

1–9. See Solutions Manual.

EXERCISES

10. Sample answers:
(1) Graph $y = 1.2^x$ and $y = 10$ and find their intersection, (2) Graph $y = 1.2^x - 10$ and find its zero, (3) Graph $y = 1.2^x$ and use trace to find where $y = 10$.

Use a graphing calculator to graph each exponential equation. Sketch the graphs on a separate piece of paper.

1. $y = 2^x$
2. $y = 5^x$
3. $y = 0.1^x$
4. $y = \left(\frac{2}{3}\right)^x$
5. $y = 0.25^x$
6. $y = 1.6^x$
7. $y = 0.2^x$
8. $y = 0.5^{-x}$
9. $y = 10^x$

10. Solve $1.2^x = 10$ graphically. Explain how you solved the equation and write the solution accurately to the nearest hundredth. **12.63**

Using Technology
This lesson offers an excellent opportunity for using technology in your algebra classroom. For more information on using technology, see *Graphing Calculators in the Mathematics Classroom,* one of the titles in the Glencoe Mathematics Professional Series.

Exponential Functions

11-4

Objectives

After studying this lesson, you should be able to:

- graph exponential functions,
- determine if a set of data displays exponential behavior, and
- solve exponential equations.

APPLICATION
Folklore

A wise man asked his ruler to provide rice for feeding his people. Rather than receiving a constant daily supply of rice, the wise man asked the ruler to give him 2 grains of rice for the first square on a chessboard, 4 grains of rice for the second, 8 grains of rice for the third, 16 grains of rice for the fourth, and so on, doubling the amount of rice with each square of the board. How many grains of rice will he receive for the last (64th) square on the chessboard?

You could make a table and look for a pattern to determine how many grains of rice he received for each square of the chessboard.

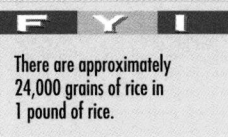

F Y I

There are approximately 24,000 grains of rice in 1 pound of rice.

Square	Grains	Pattern	Square	Grains	Pattern
1	2	2^1	19	524,288	2^{19}
2	4	2^2	20	1,048,576	2^{20}
3	8	2^3	21	2,097,152	2^{21}
4	16	2^4	22	4,194,304	2^{22}
5	32	2^5	23	8,388,608	2^{23}
6	64	2^6	24	16,777,216	2^{24}
7	128	2^7	25	33,554,432	2^{25}
8	256	2^8	26	67,108,864	2^{26}
9	512	2^9	27	134,217,728	2^{27}
10	1024	2^{10}	28	268,435,456	2^{28}
11	2048	2^{11}	29	536,870,912	2^{29}
12	4096	2^{12}	30	1,073,741,824	2^{30}
13	8192	2^{13}	31	2,147,483,648	2^{31}
14	16,384	2^{14}	32	4,294,967,296	2^{32}
15	32,768	2^{15}	33	8,589,934,592	2^{33}
16	65,536	2^{16}	34	17,179,869,184	2^{34}
17	131,072	2^{17}	35	34,359,738,368	2^{35}
18	262,144	2^{18}	36	68,719,476,736	2^{36}

Notice that when only 36 of the 64 squares are calculated, there are over 68 billion grains of rice. How many grains would there be for square 64? *You will answer this question in Exercise 1.*

Study the pattern column. Notice that the exponent number matches the number of the square on the chessboard. So we can write an equation to describe y, the number of grains of rice for any given square x as $y = 2^x$. This type of function, in which the variable is the exponent, is called an **exponential function**.

Definition of Exponential Function	An exponential function is a function that can be described by an equation of the form $y = a^x$, where $a > 0$ and $a \neq 1$.

F Y I

A simple computation on a scientific calculator reveals that the average grain of rice weighs 4.167×10^{-5} pounds.

NCTM Standards: 1–6, 13

Instructional Resources

- Study Guide Master 11-4
- Practice Master 11-4
- Enrichment Master 11-4
- Assessment and Evaluation Masters, p. 297

 Transparency 11-4A contains the 5-Minute Check for this lesson; **Transparency 11-4B** contains a teaching aid for this lesson.

Recommended Pacing	
Standard Pacing	Day 8 of 11
Honors Pacing	Day 7 of 10
Block Scheduling*	Day 3 of 5
Alg. 1 in Two Years*	Days 12 & 13 of 19

 *For more information on pacing and possible lesson plans, refer to the *Block Scheduling Booklet* and *Algebra 1 in Two Years*.

1 FOCUS

 5-Minute Check
(over Lesson 11-3)

Solve each equation by using the quadratic formula.

1. $y^2 + 8y + 15 = 0$ $\ -3, -5$
2. $2p^2 = 98$ $\ \pm 7$
3. $-2x^2 + 8x = 3$ $\ 0.42, 3.58$
4. $x^2 - 2x = 3$ $\ 3, -1$
5. $-6x^2 + 15x = 6$ $\ \frac{1}{2}, 2$

Motivating the Lesson

Situational Problem Have students look up the U.S. population during this century. Plot the data on a graph and point out that it is not a linear graph.

This hands-on activity illustrates how to generate the exponential function $y = 2^x$. Encourage students to figure out how to modify the activity to generate $y = 3^x$.

In-Class Example

For Example 1
Graph each function. State the y-intercept of each graph.

a. $y = 4^x$ y-intercept $= 1$

b. $y = \left(\frac{1}{4}\right)^x$ y-intercept $= 1$

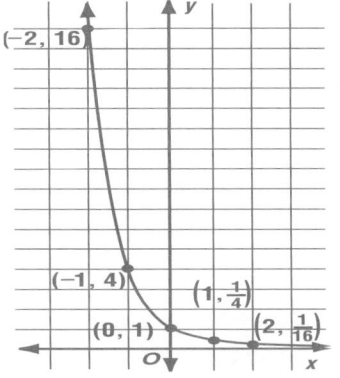

c. $y = 4^x - 8$ y-intercept $= -7$

You can use paper-folding to illustrate an exponential function.

MODELING MATHEMATICS

Modeling an Exponential Function

Materials: ☐ large piece of paper

Your Turn

a. Fold a large rectangular piece of paper in half. Unfold it and record how many sections are formed by the creases. Refold the paper.

b. Fold the paper in half again. Record how many sections are formed by the creases. Refold the paper.

c. Continue folding in half and recording the number of sections until you can no longer fold the paper.

d. How many folds could you make?   about 6

e. How many sections were formed?   64

f. What exponential function is modeled by the folds and sections created? $y = 2^x$

As with other functions, you can use ordered pairs to graph an exponential function. Use a table of values and a calculator to find ordered pairs that satisfy $y = 2^x$. While the negative values of x have no meaning in the rice problem, they should be included in the graph of the function. Connect the points to form a smooth curve.

x	y
−5	0.03125
−4	0.0625
−3	0.125
−2	0.25
−1	0.5
0	1
1	2
2	4
3	8
4	16

Notice that the graph has a y-intercept of 1. Does it have an x-intercept? *You will answer this question in Exercise 2.*

The graph shown above represents all real values of x and their corresponding values of y for $y = 2^x$.

You can use a scientific calculator to help you find ordered pairs to graph other exponential functions. For example, suppose $y = 3^x$ and $x = -2$.

Enter: 3 $\boxed{y^x}$ 2 $\boxed{+/-}$ $\boxed{=}$ *0.1111111111*

Example ❶ **Graph each function. State the y-intercept of each graph.**

a. $y = 3^x$

x	y
−3	0.037
−2	0.111
−1	0.333
0	1
1	3
2	9
3	27

The y-intercept is 1.

b. $y = \left(\frac{1}{3}\right)^x$

x	y
-3	27
-2	9
-1	3
0	1
1	0.333
2	0.111
3	0.037
4	0.012

The y-intercept is 1.

Notice that the graph of $y = \left(\frac{1}{a}\right)^x$ decreases rapidly as x increases.

c. $y = 3^x - 7$

x	y
-3	-6.96
-2	-6.89
-1	-6.67
0	-6
1	-4
2	2
3	20
4	74

The y-intercept is -6.

LOOK BACK
You can refer to Lesson 10-1 for more information on factors.

How do you know if a set of data is exponential? One method is to observe the shape of the graph. But the graph of an exponential function may resemble part of the graph of a quadratic function. Another way is to **look for a pattern** in the data.

Example 2 Determine whether each set of data displays exponential behavior.

a.

x	0	5	10	15	20	25
y	800	400	200	100	50	25

PROBLEM SOLVING
Look for a Pattern

TECHNOLOGY Tip
You could also use a graphing calculator to make a scatter plot of the data to observe the patterns.

Method 1: Look for a Pattern
The domain values are at regular intervals of 5. Let's see if there is a common factor among the range values.

800 400 200 100 50 25
   $\times\frac{1}{2}$ $\times\frac{1}{2}$ $\times\frac{1}{2}$ $\times\frac{1}{2}$ $\times\frac{1}{2}$

Since the domain values are at regular intervals and the range values have a common factor, the data are probably exponential. The equation for the data probably involves $\left(\frac{1}{2}\right)^x$.

Method 2: Graph the data.

The graph shows a rapidly decreasing value of y as x increases. This is a characteristic of exponential behavior.

In-Class Example

For Example 2
Determine whether each set of data displays exponential behavior.

a.

x	0	1	2	3	4	5
y	3	9	27	81	243	729

yes

b.

x	0	1	2	3	4
y	4	8	12	16	20

no

b.

x	0	5	10	15	20	25
y	3	6	9	12	15	18

Method 1: Look for a Pattern

The domain values are at regular intervals of 5. The range values have a common difference of 3.

3   6   9   12   15   18
  +3 +3 +3   +3   +3

These data do not display exponential behavior, but rather linear behavior.

Method 2: Graph the data.

This is the graph of a line, not an exponential function.

F Y I

Radiocarbon, or carbon-14, decays into nitrogen. It was first used in the late 1940s by Williard F. Libby to determine the age of an artifact. It has been used to date objects as old as 50,000 years.

Exponential functions are often used to describe real-life situations. In the late 1890s, Henri Becquerel discovered that fossils contain naturally-occurring radioactive atoms of carbon-14. When an organism is alive, it absorbs carbon-14 from the sun. When the organism dies, no more carbon-14 is absorbed, and the initial amount of carbon-14 begins to decay gradually into other elements.

The **half-life** of a radioactive element is defined as the time that it takes for one-half a quantity of the element to decay. Radioactive carbon-14 has a half-life of 5730 years. This means that after 5730 years, half the original amount of carbon-14 has decayed. In another 5730 years, half of the remaining half will decay, and so on. This pattern of decay can be described by $a = 0.5^t$, where a is the decay factor and t is the number of half-lives.

Study the chart to see how 64 grams of carbon-14 decays over several half-lives. Notice its relationship to the exponential function $a = 0.5^t$.

Half-life	Grams of Carbon-14 left	
0	64	$= 64 \times 0.5^0$
1	32	$= 64 \times 0.5^1$
2	16	$= 64 \times 0.5^2$
3	8	$= 64 \times 0.5^3$
4	4	$= 64 \times 0.5^4$
5	2	$= 64 \times 0.5^5$
6	1	$= 64 \times 0.5^6$
7	0.5	$= 64 \times 0.5^7$
8	0.25	$= 64 \times 0.5^8$
9	0.125	$= 64 \times 0.5^9$

Example 3

APPLICATION
Archaeology

LOOK BACK

You can refer to Lesson 9-3 for more information on numbers written in scientific notation.

If the original concentration of carbon-14 in a living organism was 256 grams, determine the concentration of carbon-14 remaining in a fossil for each situation.

a. The organism lived 1000 years ago.

b. The organism lived 10,000 years ago.

a. First determine how many carbon-14 half-lives there are in 1 thousand years.
$$\frac{1000}{5730} \approx 0.1745 \text{ half-lives}$$

Then determine the value of the disintegration factor when $t = 0.1745$.

$a = 0.5^t$

$= 0.5^{0.1745}$

≈ 0.886 *Use a calculator.*

 Alternative Learning Styles

Visual Using a graphing calculator, graph the functions $y = x$, $y = 2x^2$, and $y = 2^x$ in the same viewing window. Students will see that eventually 2^x surpasses x^2, which quickly surpasses x.

Multiply the original amount of carbon-14 by this factor.
256 grams $\times$ 0.886 = 226.8 grams

b. Find how many half-lives there are in 10,000 years.

$$\frac{10,000}{5730} \approx 1.75 \text{ half-lives}$$

Determine the value of the disintegration factor when $t = 1.75$.

$$a = 0.5^{1.75}$$

$$\approx 0.297$$

Multiply the original amount of carbon-14 by this factor.
256 grams $\times$ 0.297 $\approx$ 76.0 grams

You can use algebra to solve equations involving exponential expressions by using the following rule.

Property of Equality for Exponential Functions	**Suppose a is a positive number other than 1. Then $a^{x_1} = a^{x_2}$ if and only if $x_1 = x_2$.**

The skills you learned when solving quadratic equations can be helpful when solving some exponential equations.

Example Solve $64^3 = 4^{x^2}$.

Explore The two quantities do not have the same base, or value for a. However, 64 is a power of 4.

Plan In order to use the property of equality, we must rewrite the terms so that they have the same base. Then we can use the property of equality for exponential functions.

Solve

$$64^3 = 4^{x^2}$$

$$(4^3)^3 = 4^{x^2} \quad 64 = 4 \cdot 4 \cdot 4 \text{ or } 4^3$$

$$4^9 = 4^{x^2} \quad \textit{Product of powers}$$

$$9 = x^2 \quad \textit{Property of equality for exponential functions}$$

$$x^2 - 9 = 0 \quad \textit{Rewrite the equation in standard form.}$$

$$(x + 3)(x - 3) = 0 \quad \textit{Factor.}$$

$$x + 3 = 0 \qquad x - 3 = 0 \quad \textit{Zero product property}$$

$$x = -3 \qquad x = 3$$

Examine Check each solution by substituting it into the original equation.

$$64^3 = 4^{x^2} \qquad\qquad 64^3 = 4^{x^2}$$

$$64^3 \overset{?}{=} 4^{3^2} \quad x = 3 \qquad 64^3 \overset{?}{=} 4^{(-3)^2} \quad x = -3$$

$$262{,}144 \overset{?}{=} 4^9 \qquad\qquad 262{,}144 \overset{?}{=} 4^9$$

$$262{,}144 = 262{,}144 \quad \checkmark \qquad 262{,}144 = 262{,}144 \quad \checkmark$$

The solutions are 3 and -3.

LOOK BACK

You can refer to Lesson 9-1 for more information about properties of exponents.

Lesson 11–4 Exponential Functions **639**

F Y I

Willard F. Libby was an American chemist who received the Nobel Prize for Chemistry in 1960. In the 1940s he worked on the Manhattan Project, which developed the atomic bomb.

In-Class Example

For Example 4

Solve $9^2 = 3^{x^2}$.

2, -2

Check for Understanding
Exercises 1–16 are designed to help you assess your students' understanding through reading, writing, speaking, and modeling. You should work through Exercises 1–6 with your students and then monitor their work on Exercises 7–16.

Error Analysis
When looking at the graph of an exponential function, students may assume that the function is equal to zero for all x values below a certain point, while in fact it only approaches zero.

Assignment Guide
Core: 17–47 odd, 48, 49, 51–59 **Enriched:** 18–46 even, 48–59

For **Extra Practice,** see p. 781.

The red A, B, and C flags, printed only in the Teacher's Wraparound Edition, indicate the level of difficulty of the exercises.

Study Guide Masters, p. 78

CHECK FOR UNDERSTANDING

Communicating Mathematics

Study the lesson. Then complete the following.

1. **Refer** to the application at the beginning of the lesson.
   a. Use a calculator to determine how many grains there would be for the 64th square of the chessboard. $1.844674407 \times 10^{19}$
   b. How many tons of rice is this? (*Hint:* Recall that 1 T = 2000 lb.)

1b. over 3.8×10^{11} tons
2a. No, it does not.
2b. Sample answer: Use the graph of $y = 2^x$ and the TRACE feature on a graphing calculator to examine the y values.

2. a. **Determine** whether the graph of $y = 2^x$ has an x-intercept.
   b. **Describe** your method.
   c. Is this true of all exponential functions? no; only those of form $y = a^x$

3. a. **Determine** whether the graph of $y = 2^x$ has a vertex. No, it does not.
   b. **Describe** your method for answering part a. See students' work.
   c. Is this true for all exponential functions? Yes.

4. **Explain** why $a \neq 1$ in the definitions and properties involving exponential functions. If $a = 1$, $a^x = 1$ for any value of x.

5. **Write** a paragraph explaining why you think a graphing calculator might be a good tool to have when studying exponential functions. See students' work.

MODELING MATHEMATICS

6. Refer to the Modeling Mathematics activity on page 636.
   a. The area of the large rectangle is 1. Find the area of each section after each set of folds. Record your findings in a table. See margin.
   b. Compare the number of folds to the area of each section. What pattern do you see? See margin.
   c. Write an exponential function relating the number of folds to the area of each section. $y = 0.5^x$

Guided Practice

Use a calculator to determine the approximate value of each expression to the nearest hundredth.

7. $3^{1.5}$ 5.20
8. $3^{-0.9}$ 0.37
9. $3^{2.3}$ 12.51

Graph each function. State the y-intercept.

10–11. See Solutions Manual for graphs.

10. $y = 0.5^x$ 1
11. $y = 2^x + 6$ 7

12. Determine if the data in the table below displays exponential behavior. Describe the behavior.

x	0	1	2	3	4	5
y	1	6	36	216	1296	7776

Yes, increases by a factor of 6.

Solve each equation.

13. $5^{3y+4} = 5^y$ −2
14. $2^5 = 2^{2x-1}$ 3
15. $3^x = 9^{x+1}$ −2

16. **Biology** Suppose $B = 100 \cdot 2^t$ represents the number of bacteria B in a petri dish after t hours if you began with 100 bacteria. How long would it take to obtain 1000 bacteria? 3.32 hours

EXERCISES

Practice

A

Use a calculator to determine the approximate value of each expression to the nearest hundredth.

17. $4^{1.7}$ 10.56
18. $10^{-0.5}$ 0.32
19. $\left(\frac{2}{3}\right)^{-1.2}$ 1.63
20. $\left(\frac{1}{3}\right)^{4.1}$ 0.01

21. 25.86
24. 844.49

21. $50(3^{-0.6})$
22. $10(3^{-1.8})$ 1.38
23. $0.4(3^{0.7})$ 0.86
24. $20(0.25^{-2.7})$

Reteaching

Using Graphics Using a graphing calculator, graph a family of exponential functions $y = 2^x$, $y = 3^x$, $y = 5^x$, and $y = 11^x$. Note the effect of increasing the base.

Additional Answers

6a.
Fold	1	2	3	4	5	6
Area	$\frac{1}{2}$	$\frac{1}{4}$	$\frac{1}{8}$	$\frac{1}{16}$	$\frac{1}{32}$	$\frac{1}{64}$

6b. Each fold makes a rectangle whose area is $\frac{1}{2^x}$, where x is the number of folds.

Graph each function. State the y-intercept.

B

25. $y = 2^x + 4$ 5

26. $y = 2^{x+4}$ 16

27. $y = 3\left(\frac{1}{3}\right)^x$ 3

25–30. See Solutions Manual for graphs.

28. $y = 2 \cdot 3^x$ 2

29. $y = 4^x$ 1

30. $y = \left(\frac{1}{4}\right)^x$ 1

Determine if the data in each table display exponential behavior. Explain why or why not.

31.

x	y
−2	−5
−1	−2
0	1
1	4

no, linear

32.

x	y
0	1
1	0.5
2	0.25
3	0.125

yes, constant factor

33.

x	y
−1	−0.5
0	1.0
1	−2.0
2	4.0

no, no pattern

Solve each equation.

C

34. $5^{3x} = 5^{-3}$ −1

35. $2^{x+3} = 2^{-4}$ −7

36. $5^x = 5^{3x+1}$ −0.5

37. $10^x = 0.001$ −3

38. $2^{2x} = \frac{1}{8}$ $-\frac{3}{2}$

39. $\left(\frac{1}{6}\right)^q = 6^{q-6}$ 3

40. $16^{x-1} = 64^x$ −2

41. $81^x = 9^{x^2-3}$ 3, −1

42. $4^{x^2-2x} = 8^{x^2+1}$
−1, −3

Graphing Calculator

As with linear graphs and quadratic graphs, exponential graphs can form families of graphs. Graph each set of equations on the same screen. Sketch the graphs and discuss any similarities or differences.   43–46. See margin.

43. $y = 3^x$
$y = 3^{x+4}$
$y = 3^{x-2}$

44. $y = \left(\frac{1}{3}\right)^x$
$y = \left(\frac{1}{3}\right)^x + 5$
$y = \left(\frac{1}{3}\right)^x - 3$

45. $y = 2^x$
$y = 2^{x-7}$
$y = 2^{x-2}$

46. $y = 6^x$
$y = 6^{3x}$
$y = 6^{8x}$

47. **a.** Use a graphing calculator to solve $2.5^x = 10$. Write the solution to the nearest hundredth.   2.51

b. Explain why you cannot solve this equation algebraically like you did the equation in Exercise 41.   You cannot write 10 as a power of 2.5.

Critical Thinking

48. Refer to the equations in Example 1. Use a calculator to find additional values to complete the following.

a. For $y = 3^x$, as x decreases, the value of y approaches what number?   0

b. For $y = \left(\frac{1}{3}\right)^x$, as x increases, the value of y approaches what number?   0

48c. −7

c. For $y = 3^x - 7$, as x decreases, the value of y approaches what number?

d. For any equation $y = a^x + c$, where $a > 1$, what value does y approach as x decreases?   c

e. For any equation $y = a^x + c$, where $0 < a < 1$, what value does y approach as x increases?   c

Applications and Problem Solving

49. **Biology**   Mitosis is a process of cell reproduction in which one cell divides into two identical cells. *E. coli* is a fast-growing bacteria that is often responsible for food poisoning in uncooked meat. It can reproduce itself in 15 minutes. If you begin with 100 *e. coli* bacteria, how many bacteria will there be in 1 hour?   1600 bacteria

Lesson 11–4   Exponential Functions   **641**

Tech Prep

Medical Technician Students who are interested in medical laboratory assisting may wish to further investigate the data provided in Exercise 49 and explore the potential growth of this career. For more information on tech prep, see the *Teacher's Handbook.*

Additional Answers

43.

All have the same shape but different y-intercepts.

44.

All have the same shape but different y-intercepts.

45.

All have the same shape but are positioned at different places along the x-axis.

Practice Masters, p. 78

11-4

NAME_____   DATE_____

Practice

Student Edition
Pages 635–642

Exponential Functions

Use a calculator to determine the approximate value of each expression to the nearest hundredth.

1. $2^{0.7}$
1.62

2. $3^{-0.3}$
0.72

3. $4^{1.1}$
4.59

4. $5^{-4.1}$
0.00

Graph each function. State the y-intercept.

5. $y = 3^x - 2$

6. $y = 2^{x+1}$

7. $y = 3 \cdot 2^x$

8. $y = \left(\frac{1}{2}\right)^x$

Solve each equation.

9. $3^x = 3^{9x+1}$ $-\frac{1}{8}$

10. $3^x = 9^{x+1}$ −2

11. $8^{x-1} = 16^b$ $-\frac{1}{3}$

12. $9^{9b} = 27^{x+2}$ 2

13. $2^{2b-1} = 8^{x+1}$ −22

14. $4^{x-1} = 8^x$ −2

Additional Answer

46.

All have the same y-intercept, but each rises more steeply than the previous one.

Closing Activity

Modeling Begin with a jar full of pennies. Remove one penny, put it on the desk, remove 2 more, then 4 more, and so on, removing twice the number each time. Note how fast the jar is emptied.

Chapter 11, Quiz C (Lesson 11-4), is available in the *Assessment and Evaluation Masters,* p. 297.

fabulous
FIRSTS

Radioisotopes are used as tracers in biological research, in medicine to treat cancer, by geologists to date rocks, and for many uses in industry, including finding weak points in oil lines.

Additional Answer

52b. (4p − 5r) km
(4p − 5r) km

Enrichment Masters, p. 78

50a. 10, 40, 70, 90

51. −3, −$\frac{1}{2}$

55. 184

57a. See Solutions Manual.
57b. Sample answer:
y = 0.07x − 12

fabulous
FIRSTS

Clair C. Patterson (1922–1995)

Geochemist Clair Patterson, a professor at the California Institute of Technology, was the first scientist to establish Earth's age based on analysis of radioactive isotopes and their daughter products. In 1953, he established Earth's age to be 4.6 billion years.

50. **Currency** In the United States between 1910 and 1994, the amount of currency in circulation $M(t)$ (billions of dollars) can be approximated by the function $M(t) = 2.08(1.06)^t$, where t represents the number of years since 1910.
    a. Determine $M(t)$ for the years 1920, 1950, 1980, and 2000.
    b. Determine the amount of currency in circulation in the years 1920, 1950, 1980, and 2000. **$3.72 billion, $21.39 billion, $122.88 billion, $394.09 billion**

51. Solve $2x^2 + 3 = -7x$ by using the quadratic formula. Check your solution by factoring. (Lesson 11–3)

52. **Geometry** A rectangle has an area of $(16p^2 - 40pr + 25r^2)$ square kilometers. (Lesson 10–5)
    a. Find the dimensions of the rectangle. **(4p − 5r) km by (4p − 5r) km**
    b. Sketch the rectangle, labeling its dimensions. **See margin.**

53. Factor $\frac{4}{5}a^2b - \frac{3}{5}ab^2 - \frac{1}{5}ab$. (Lesson 10–2) $\frac{1}{5}ab(4a - 3b - 1)$

54. **City Planning** A section of Lithopolis is shaped like a trapezoid with an area of 81 square miles. The distance between Union Street and Lee Street is 9 miles. The length of Union Street is 14 miles less than 3 times the length of Lee Street. Find the length of Lee Street. Use $A = \frac{h(a + b)}{2}$. (Lesson 9–8) **8 miles**

55. **Basketball** On December 13, 1983, the Denver Nuggets and the Detroit Pistons broke the record for the highest score in a professional basketball game. The two teams scored a total of 370 points. If the Nuggets scored 2 points less than the Pistons, what was the Nuggets' final score? (Lesson 9–5)

56. Solve the system of inequalities by graphing. (Lesson 8–5)
    $y \leq 2x + 2$
    $y \geq -x - 1$ **See margin.**

57. **Astronomy** The table at the right shows the relationship between the distance from the sun, measured in millions of miles, and the time to complete an orbit, measured in Earth years. (Lesson 6–3)
    a. Make a scatter plot of these data.
    b. Draw a best-fit line and write an equation for the line.
    c. Suppose a tenth planet was discovered at a distance of 4.1 billion miles from the sun. Use the scatter plot to estimate how long it would take it to orbit the sun. **275 years**

Planet	Distance from Sun	Years per Orbit
Mercury	36.0	0.241
Venus	67.0	0.615
Earth	93.0	1.000
Mars	141.5	1.880
Jupiter	483.0	11.900
Saturn	886.0	29.500
Uranus	1782.0	84.000
Neptune	2793.0	165.000
Pluto	3670.0	248.000

58. **Geometry** Triangle ABC is similar to triangle ADE in the figure at the right. Find the value of s. (Lesson 4–2) **75 m**

59. Find $(-2)(3)(-10)$. (Lesson 2–6) **60**

Extension

Reasoning Do all functions of the form $y = (b \pm c)^x \pm d$ intersect the point (0, 1)? If $d = 0$, the graph goes through (0, 1). If $d \neq 0$, the graph does not.

Additional Answer

56.

Growth and Decay

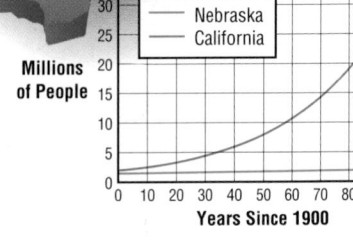

Objective

After studying this lesson, you should be able to:

- solve problems involving growth and decay.

APPLICATION
Demographics

How quickly has the population of your state grown in the 20th century? Both California and Nebraska have grown in population during this century by a constant percent. The graph at the right shows the population of each state where t represents the number of years since 1900. Since 1900, the population in Nebraska has grown at a rate of 0.4% per year, while the population of California has grown at a rate of 3% per year.

The exponential functions that model the growth of each state are given below.

California: $y = 1.77(1.03)^t$ *The population in 1900 was 1.77 million.*

Nebraska: $y = 1.14(1.004)^t$ *The population in 1900 was 1.14 million.*

Which state exhibits the more rapid growth? What will the population of each state be in the year 2000?   *You will answer these questions in Exercise 1.*

The equations for the two states' populations are variations of the equation $y = C(1 + r)^t$. This is the **general equation for exponential growth** in which the initial amount C increases by the same percent r over a given period of time t. This equation can be applied to many kinds of growth applications.

One of these applications is monetary growth. When solving problems involving compound interest, the growth equation becomes $A = P\left(1 + \frac{r}{n}\right)^{nt}$, where A is the amount of the investment over a period of time, P is the principal (initial amount of the investment), r is the annual rate of interest expressed as a decimal, n is the number of times that the interest is compounded each year, and t is the number of years that the money is invested.

CAREER CHOICES

Urban, or city, planners use statistics on demographics, traffic patterns, and economics to help officials make decisions on social, economic, and environmental issues in their communities. A bachelors degree in urban or regional planning is required, and a masters degree in civil engineering or planning is helpful for advancement.

For further information, contact:

American Planning Association
1776 Massachusetts Ave., NW
Washington, DC 20036

Example ①

APPLICATION
Finance

In the spring of 1994, Mr. and Mrs. Mitzu had $10,000 they wished to place in a bank certificate of deposit toward their retirement in the year 2004. The interest rate at that time was 2.5% compounded monthly. However, there were seven increases in the prime rate in a year so that in the spring of 1995, the interest rate had risen to 5.5%.

a. Determine the amount of their investment after 10 years if they invested the principal and let it remain at the 2.5% rate.

b. Determine the amount of the investment after 9 years if they waited and invested the principal at the 5.5% rate.

c. What are the best options for their investment?

(continued on the next page)

CAREER CHOICES

Urban planners develop a master plan of a city to show how land should be used in the future and how public facilities should be developed. Urban planners consult with many professionals with expertise in fields that include architecture, engineering, finance, air and water quality, law, economics, statistics, and sociology.

NCTM Standards: 1–6

Instructional Resources

- Study Guide Master 11-5
- Practice Master 11-5
- Enrichment Master 11-5
- Assessment and Evaluation Masters, p. 297
- Multicultural Activity Masters, p. 22
- Real-World Applications, 28
- Science and Math Lab Manual, pp. 49–52

Transparency 11-5A contains the 5-Minute Check for this lesson; **Transparency 11-5B** contains a teaching aid for this lesson.

Recommended Pacing	
Standard Pacing	Day 9 of 11
Honors Pacing	Day 8 of 10
Block Scheduling*	Day 4 of 5
Alg. 1 in Two Years*	Days 14, 15, & 16 of 19

*For more information on pacing and possible lesson plans, refer to the *Block Scheduling Booklet* and *Algebra 1 in Two Years*.

1 FOCUS

5-Minute Check
(over Lesson 11-4)

Use a calculator to determine the approximate value of each expression to the nearest hundredth.

1. $3^{2.6}$ **17.40**
2. $8^{-0.5}$ **0.35**
3. Graph and state the y-intercept for $y = 6^x$.
   y-intercept $= 1$

Solve each equation.

4. $2^{3x} = 2^{-3}$ -1
5. $7^{3x} = 7^{x-2}$ -1

Situational Problem In a small town, population 100, every married couple has two children. Everyone is married in this town. If this trend continues, what will the population be after 100 years?
100

2 TEACH

In-Class Examples

For Example 1

a. $5000 is invested at a rate of 4%. Determine the amount after five years. **$6104.98**

b. $5000 is invested at a rate of 8%. Determine the amount after five years. **$7449.23**

For Example 2
If a city with a population of 50,000 is decreasing at a rate of 1.16%, what will its population be after 10 years? 50 years? 100 years? **44,493; 27,900; 15,568**

Teaching Tip Use numbers for your town or a local city to make exercises more realistic for students.

---

a. The interest rate r as a decimal is 0.025 and $n = 12$. *Why?*

$$A = P\left(1 + \frac{r}{n}\right)^{nt}$$

$$= 10{,}000\left(1 + \frac{0.025}{12}\right)^{12 \cdot 10} \quad P = 10{,}000,\ r = 0.025,\ n = 12,\ and\ t = 10$$

Use a calculator.

Enter: 10000 ⊠ ⦅ 1 ⊞ ⦅ .025 ÷ 12 ⦆ ⦆
 120 ⊜ *12836.91542*

The amount of the account after 10 years at 2.5% is about $12,836.92.

b. The interest rate as a decimal is 0.055, and t is 9 years.

$$A = P\left(1 + \frac{r}{n}\right)^{nt}$$

$$= 10{,}000\left(1 + \frac{0.055}{12}\right)^{12 \cdot 9} \quad P = 10{,}000,\ r = 0.055,\ n = 12,\ and\ t = 9$$

$$\approx 16{,}386.44$$

c. If they waited until the rate went up, they would have had more money than leaving the money at the initial rate. However, if it is possible to reinvest the money each year, they could have invested the money for one year at 2.5% and then reinvested it the next year at 5.5%. That would yield even more money for their retirement in 2004.

A variation of the growth equation can be used as the **general equation for exponential decay.** In the formula $A = C(1 - r)^t$, A represents the final amount, C is the initial amount, r is the rate of decay, and t denotes time.

Example 2

APPLICATION
Demographics

The cities listed at the right have experienced declining populations since 1970.

City	1970 Population (thousands)	Annual Rate of Decrease
Baltimore, Maryland	894	1.029%
Cleveland, Ohio	733	1.055%

a. If t represents the number of years since 1970 and C represents the 1970 population, write an exponential decay equation for each city.

b. Assume that each city maintains the same decrease rate into the next century. Calculate the population of each city in the year 2070.

c. How do the projected populations of the cities in 2070 compare to the actual populations in 1970?

a. Baltimore

$C = 894$, $r = 0.01029$

$A = C(1 - r)^t$

$A = 894(1 - 0.01029)^t$

$A = 894(0.98971)^t$

Cleveland

$C = 733$, $r = 0.01055$

$A = C(1 - r)^t$

$A = 733(1 - 0.01055)^t$

$A = 733(0.98945)^t$

Alternative Learning Styles

Auditory Initiate a class discussion about the increasing world population. What are the consequences of an increasing population and decreasing nonrenewable natural resources?

b. The year 2070 is 100 years later. Evaluate each decay equation for $t = 100$.

Baltimore	Cleveland
$A = 894(0.98971)^t$	$A = 733(0.98945)^t$
$A = 894(0.98971)^{100}$	$A = 733(0.98945)^{100}$
$A \approx 317.78$	$A \approx 253.80$

c. In 1970, the difference in the populations of Baltimore and Cleveland was 161,000. In 2070, the difference is only about 64,000. If the populations would continue to decrease at the same rates, they would grow even closer together.

In-Class Example

For Example 3
If the car depreciates at 15% per year, how will the depreciated price compare with the buyout price of the lease? $12,205.19; $1436.81 less than the buyout price

Sometimes items decrease in value. For example, as equipment gets older, it *depreciates*. You can use the decay formula to determine the value of an item at a given time.

Example ③

APPLICATION
Consumerism

Hogan Blackburn is considering the purchase of a new car. He is faced with the decision to lease or buy. If he leases the car, he pays $369 a month for 2 years and then has the option to buy the car for $13,642. The price of the car now is $16,893.

a. If the car depreciates at 17% per year, how will the depreciated price compare with the buyout price of the lease?

b. At the end of the lease, the dealer offers a loan for 3 years at a rate of $435.79 per month. If he buys the car now, he will pay $464.56 monthly for 4 years. Which is the best way to go if he plans to keep the car for at least 5 years?

a. Use the decay formula.

$$A = C(1 - r)^t$$
$$= 16{,}893(1 - 0.17)^2 \quad \text{$r = 0.17$, $C = 16{,}893$, and $t = 2$}$$
$$= 11{,}637.59$$

The depreciated value is $2000 less than the buyout price.

b. Calculate the cost of each possibility over a 5-year period.

Lease and Buy: $369(24 months) + $435.79(36 months) = $24,544.44

Buy: $464.56(48 months) = $22,298.88

The best way to go may depend on Hogan's financial status.

- Overall, the buy option costs less and lasts only 4 years while the lease and buy option costs more and lasts for 5 years.

- However, if Hogan really wants a new car and cannot afford the larger monthly payment now, the lease option may be best. However, he will eventually have to make the larger payment if he keeps the car.

You can use a spreadsheet to quickly evaluate different values for any exponential growth or decay formula.

3 PRACTICE/APPLY

Recall that in a spreadsheet you can refer to each cell by its name (A1 means column A, row 1). Suppose we set up a spreadsheet to evaluate the formula for monetary exponential growth.

- You can enter labels into the first row to identify the data in each column. Let column A contain the values of P, column B the values of r, column C the values of n, column D the value of t, and column E the values of A.

> **LOOK BACK**
>
> Refer to the Exploration in Lesson 6-1 for more information about spreadsheets.

- You can enter a formula into cell E2 to evaluate the formula given the values entered into cells A2 through D2. The formula is A2 *(1 + (B2/C2))^(C2*D2).

- Copy this formula into the other cells in column E.

Your Turn

a. Use the spreadsheet to evaluate the data in Example 1.

b. How would you change the spreadsheet to evaluate the general formula for decay?   See margin.

CHECK FOR UNDERSTANDING

Communicating Mathematics

1b. CA: 34.02 million; NE: 1.70 million.

Study the lesson. Then complete the following.

1. Refer to the application at the beginning of the lesson.
   a. Which state exhibits more rapid growth?   California
   b. What will the population of each state be in the year 2000?

2. **Explain** how you could tell the difference between an exponential graph that shows 0.7% growth versus one that shows 7% growth.   The 7% graph would rise more steeply than the 0.7% graph when using the same scale.

Guided Practice

3. **State** the value of n in the monetary growth formula for each period of compounded interest.   3b. 2
   a. annually 1   b. semi-annually   c. quarterly 4   d. daily 365

Determine whether each exponential equation represents growth or decay.

4. $y = 10(1.03)^x$ growth   5. $y = 10(0.50)^x$ decay   6. $y = 10(0.75)^x$ decay

7. **History**   In 1626, Peter Minuit, governor of the colony of New Netherland, bought the island of Manhattan from the Indians for beads, cloth, and trinkets worth 60 Dutch guilders ($24). If that $24 had been invested at 6% per year compounded annually, how much money would there be in the year 2000?   about $70,000,000,000

8. **Farming** Many self-employed individuals such as farmers can depreciate the value of the machinery they buy as a part of their income tax returns. Suppose a tractor valued at $50,000 depreciates 10% per year. Make a table to determine in how many years the value of the tractor would be less than $25,000.   **about 7 years**

Assignment Guide

Core: 9–21 odd, 23–27
Enriched: 10–20 even, 21–27

For **Extra Practice,** see p. 782.

The red A, B, and C flags, printed only in the Teacher's Wraparound Edition, indicate the level of difficulty of the exercises.

Additional Answer

13c.

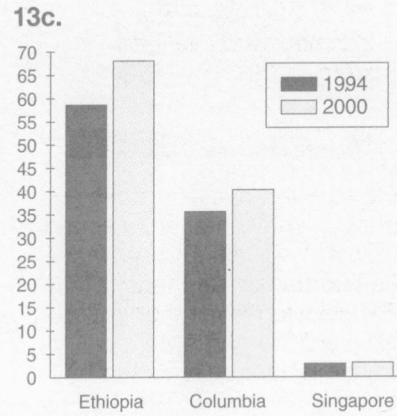

EXERCISES

Applications and Problem Solving

Finance Determine the final amount from the investment for each situation.

	Initial Amount	Annual Interest Rate	Time	Type of Compounding	
9.	$400	7.25%	7 years	quarterly	$661.44
10.	$500	5.75%	25 years	monthly	$2097.86
11.	$10,000	6.125%	18 months	daily	$10,962.19
12.	$250	10.3%	40 years	monthly	$15,121.61

13. **Demographics** The population in 1994 and the growth rate for four countries is given below.

Country	Continent	Growth Rate	1994 Population
Ethiopia	Africa	2.5%	58.7 million
India	Asia	1.9%	919.9 million
Colombia	S. America	2.1%	35.6 million
Singapore	Asia	1.3%	2.9 million

b. 68.1 million
1029.9 million
40.3 million
3.1 million

13a. Each equation represents growth and t is the number of years since 1994.

$y = 58.7(1.025)^t$,

$y = 919.9(1.019)^t$,

$y = 35.6(1.021)^t$,

$y = 2.9(1.013)^t$

a. Write an exponential equation for each country's growth.

b. Compute the estimated population for each country in the year 2000.

c. Exclude the data for India. Make a double-bar graph to show how the population in the other three countries in 1994 compares with the estimate for the year 2000.   See margin.

14. **Energy** The Environmental Protection Agency (EPA) has called for businesses to find cleaner sources of energy as we approach the 21st century. Coal is not considered to be a clean source of energy. In 1950, the use of coal by residential and commercial users was 114.6 million tons. Since then, the use of coal has decreased by 6.6% per year.

14a. $y = 114.6(0.934)^t$, t = years since 1950, decay

14b. It never reaches 0, but comes close in about 100 years.

a. Write an equation to represent the use of coal since 1950.

b. Suppose the use of coal continues to decrease at the same rate. Use a calculator to estimate in what year the use of coal will end.

Study Guide Masters, p. 79

Closing Activity

Writing Have students write a short essay explaining the difference between exponential growth and decay. Students need to make reference to the base of each function.

Chapter 11, Quiz D (Lesson 11-5), is available in the *Assessment and Evaluation Masters,* p. 297.

FM signals have less static and a truer representation of sound than AM signals. But AM waves can be transmitted farther than FM waves. Waves that travel skyward from an AM transmission are reflected by the ionosphere and can be received far from the antenna. Sky waves from the FM transmissions are not reflected by the atmosphere, so FM transmissions are limited to groundwaves, which cannot travel beyond the curve of Earth.

Practice Masters, p. 79

11-5

NAME_____ DATE _____

Practice

Student Edition
Pages 643–649

Growth and Decay

Solve.

1. Ten years ago, Mr. and Mrs. Boyce bought a house for $96,000. Their home is now worth $125,000. Assuming a steady rate of growth, what was the annual rate of appreciation? **about 3%**

2. The Greens bought a condominium for $83,000. Assuming that its value will appreciate 6% per year, how much will the condo be worth in five years when the Greens are ready to move? **$111,072.72**

3. Ten years ago, Tiffany's mother bought a new car for $9000. Tiffany is now going to buy the car for $900. Assuming a steady rate of depreciation, what was the approximate annual rate of depreciation? **21%**

4. A piece of machinery valued at $2500 depreciates at a steady rate of 10% yearly. The owner of the business plans to replace the equipment when its value has depreciated to $500. In how many years will the equipment be replaced? **$15\frac{1}{2}$ years**

5. Kyle has saved $500 of the money he earned working at Carousel Music. If he spends 10% of the money each week, after how many weeks will he have less than $1? **60 weeks**

6. Zeller Industries bought a piece of weaving equipment for $50,000. It is expected to depreciate at a steady rate of 10% per year. When will its value have depreciated to $25,000? **about 6.5 years**

7. The Fresh and Green Company has a savings plan for its employees. If the employee makes an initial contribution of $1000, the company pays 8% interest compounded quarterly.
   a. If an employee participating in the plan withdraws the balance of the account after five years, how much will the company have paid into the account? **$485.95**
   b. If an employee participating in the plan withdraws the balance of the account after thirty-five years, how much will the company have paid into the account? **$14,996.47**

16a. $y = 1.2(0.932)^t$, t = number of years since 1980, decay

16c. preservation efforts, conservation of land, domestic reproduction in zoos, poaching

FM stations began surpassing AM stations in popularity during the 1970s. This may have been due to the FM waves producing a clearer signal that could be heard over a longer distance.

---

15. **Insurance** The total amount of life insurance sold in 1950 was $231.5 billion. Since then, the annual increase has been estimated at 9.63%.
   a. Write an equation to represent the amount of insurance sold annually since 1950. $y = 231.5(1.0963)^t$, t = years since 1950, growth
   b. Find the estimated amount of life insurance that will be sold in the year 2010. **$57.6 trillion**

16. **Wildlife** In 1980, there were 1.2 million elephants living in Africa. Because the natural grazing lands for the elephant are disappearing due to increased population and cultivation of the land, the number of elephants in Africa has decreased by about 6.8% per year.
   a. Write an equation to represent the population of elephants in Africa.
   b. In what year did the population of African elephants drop to less than half of the number in 1980? **1990**
   c. What factors might affect the rate of decline in elephant population?

17. **Savings** Sheena is investing her $5000 inheritance in a saving certificate that matures in 4 years. The interest rate is 8.25% compounded quarterly.
   a. Determine the balance in the account after 4 years. **$6931.53**
   b. Her friend, LaDonna, invests the same amount of money at the same interest rate but her bank compounds daily. Determine how much she will have after 4 years. **$6954.58**
   c. What is the difference in the amount Sheena and LaDonna have after 4 years? **$23.05**
   d. Which type of compounding appears to be more profitable? **daily**

18. **Radio** FM broadcast frequencies range from 88 to 108 MHz in tenths. Before digital displays existed, you turned a knob and a bar slide from across the display or *dial* to find the station you wanted to hear. The equation that relates the MHz reading to position of the slide is $f(d) = 88(1.0137)^d$, where d is the distance from the left of the dial. Suppose someone didn't know the number of their favorite station, but knew it was about halfway across the dial. If the dial is 15 cm long, what station is their favorite? **97.5 MHz**

19. **Population** Research to find the population in your community for the past 50 years. **See students' work.**
   a. Make a graph to show the population change.
   b. Does the population of your community show growth or decay?
   c. Estimate what the population of your community will be when you are 50 years old.

Graphing Calculator

20. **Radioactivity** A formula for examining the decay of radioactive materials is $y = Ne^{kt}$, where N is the beginning amount in grams, $e \approx 2.72$, k is a negative constant for the substance, and t is the number of years. Use a graphing calculator to estimate each of the following.
   a. How long will it take 250 grams of a radioactive substance to reduce to 50 grams if $k = -0.08042$? **20 years**
   b. In 10 years, 200 grams of a radioactive substance is reduced to 100 grams. Find an estimate for the constant k for this substance. **−0.0693**

648 Chapter 11 *Exploring Quadratic and Exponential Functions*

Extension

Reasoning If interest were compounded daily instead of monthly, what would the formula for *A* look like?

$$A = P\left(1 + \frac{r}{365.25}\right)^{(365.25)t}$$

Critical Thinking

21. The general exponential formula for growth or decay is $y = Ca^x$. Determine what values of a describe growth or decay. What does x usually represent?

21. $a > 1$, growth; $0 < a < 1$, decay; x represents time

22. Consider the equation $y = C(1 + r)^x$. How do you determine the y-intercept of the graph of this equation without graphing or evaluating the function for values of x? The y-intercept is C.

Mixed Review

23. Solve $3^y = 3^{3y+1}$. (Lesson 11–4) $-\dfrac{1}{2}$

24. Find $(n^2 + 5n + 3) - (2n^2 + 8n + 8)$. (Lesson 9–5) $-n^2 - 3n - 5$

25. Simplify $\dfrac{-6r^3s^5}{18r^{-7}s^5t^{-2}}$. (Lesson 9–2) $-\dfrac{r^{10}t^2}{3}$

26. **Geometry** Find the volume of the cube shown at the right. (Lesson 9–1) y^3z^{12}

yz^4
yz^4
yz^4

27. **Geometry** Write the equation of a line that passes through $(-2, 7)$ and is perpendicular to the line whose equation is $2x - 5y = 3$. Use slope-intercept form. (Lesson 6–6) $y = -2.5x + 2$

Minimizing Computers

The following excerpt appeared in an article in *The New York Times* on November 22, 1994.

IN A BOLD EXPERIMENT THAT IS PROVOKING investigators to reconsider what a computer is and what it means to compute, a researcher has used DNA, the genetic material, as a sort of personal computer. Exploiting the extraordinary efficiency and speed of biological reactions, he translated a difficult mathematical problem into the language of molecular biology and solved it by carrying out a reaction in one-fiftieth of a teaspoon of solution in a test tube....Molecular computers can perform more than a trillion operations per second, which makes them a thousand times as fast as the fastest supercomputer. They are a billion times as energy efficient as conventional computers. And they can store information in a trillionth of the space required by ordinary computers. ∎

3. Sample answer: It could help to find the defective gene and perhaps eliminate the disease completely.

1. Does the idea of a molecular computer in a test tube surprise you? Why or why not? See students' work. 2. See margin for sample answer.

2. If biological systems can have computational abilities, what effects might this have on computer scientists, programmers, and mathematicians?

3. One class of problems that molecular computers could help solve involves finding one desired solution or path out of a huge number of possibilities. How could this be used if the problem was a defective gene causing the rate of a deadly disease to grow exponentially?

Lesson 11–5 Growth and Decay **649**

Mathematics and **SOCIETY**

The actual circuitry of today's computers is referred to as *hardware*. The program run on the hardware is referred to as *software*. The new biocomputers have been referred to as *wetware*.

Answer for Mathematics and Society

2. Sample answer: The methods used for rapid calculation in biological computers might be applied to similar problems in other real-world applications.

Closing the Investigation

This activity provides students an opportunity to bring their work on the Investigation to a close. For each Investigation, students should present their findings to the class. Here are some ways students can display their work.

- Conduct and report on an interview or survey.
- Write a letter, proposal, or report.
- Write an article for the school or local paper.
- Make a display, including graphs and/or charts.
- Plan an activity.

Assessment

To assess students' understanding of the concepts and topics explored in the Investigation and its follow-up activities, you may wish to examine students' Investigation Folders.

The scoring guide in the *Investigations and Projects Masters,* p. 19, provides a means for you to score students' work on the Investigation.

Investigations and Projects Masters, p. 19

Scoring Guide
Chapters 10 and 11
Investigation

Level	Specific Criteria
3 Superior	· Shows thorough understanding of the concepts of *creating different brick pattern designs; data collection and analysis; determining the dimensions, perimeter, and area of rectangles and squares in terms of x and y;* and *making graphs.* · Uses appropriate strategies to solve problems. · Computations are correct. · Written explanations are exemplary. · Charts, graphs, and report are appropriate and sensible. · Goes beyond the requirements of some or all problems.
2 Satisfactory, with Minor Flaws	· Shows understanding of the concepts of *creating different brick pattern designs; data collection and analysis; determining the dimensions, perimeter, and area of rectangles and squares in terms of x and y;* and *making graphs.* · Uses appropriate strategies to solve problems. · Computations are mostly correct. · Written explanations are effective. · Charts, graphs, and report are appropriate and sensible. · Satisfies the requirements of problems.
1 Nearly Satisfactory, with Obvious Flaws	· Shows understanding of most of the concepts of *creating different brick pattern designs; data collection and analysis; determining the dimensions, perimeter, and area of rectangles and squares in terms of x and y;* and *making graphs.* · May not use appropriate strategies to solve problems. · Computations are mostly correct. · Written explanations are satisfactory. · Charts, graphs, and report are appropriate and sensible. · Satisfies the requirements of problems.
0 Unsatisfactory	· Shows little or no understanding of the concepts of *creating different brick pattern designs; data collection and analysis; determining the dimensions, perimeter, and area of rectangles and squares in terms of x and y;* and *making graphs.* · Does not use appropriate strategies to solve problems. · Computations are incorrect. · Written explanations are not satisfactory. · Charts, graphs, and report are not appropriate or sensible. · Does not satisfy the requirements of problems.

---

the BRICKYARD

Refer to the Investigation on pages 554–555.

Review the knowledge you have gained from your experiments working with the bricks. Review the instructions given to you by your manager as you begin to close this Investigation.

> Please create several patio designs that will utilize these bricks. Submit at least three different plans, explaining the materials required for each patio. I am anxious to see the different ways in which these bricks can be arranged to form rectangular patios. Is there a general formula or pattern we can use to design these in the future? I look forward to your report on helping us solve our inventory problems.

Analyze

You have conducted experiments and organized your data in various ways. It is now time to analyze your findings and state your conclusions.

1 Look over your data and complete a chart like the one on page 555 for each design. Use actual measurements.

PORTFOLIO ASSESSMENT

You may want to keep your work on this Investigation in your portfolio.

2 What information does this chart reflect? Does it give information that can be used to generalize a method for forming future brick patio patterns? Explain.

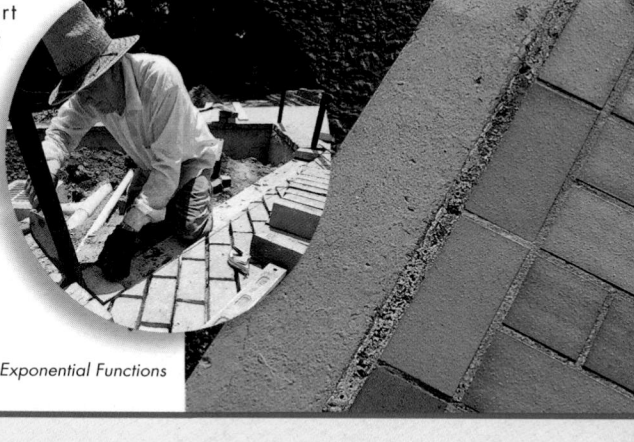

Write

The report to your manager should explain your process for investigating these rectangular brick patterns and what you found from your investigations.

3 What size rectangular brick patterns are possible? Draw sketches of the possible patterns. Describe the numbers of bricks used and include the dimensions, perimeter, and area of each pattern in terms of x and y.

4 Write procedures or generalizations that may be followed to find rectangular patterns in the following situations.
- You have a certain number of large squares and small squares. How many rectangular tiles are necessary to create a rectangular pattern?
- How can you find the dimensions of a pattern given the type and number of bricks available?
- How can you find the different possible patterns for any set number of bricks?
- How do you know that no possible rectangular pattern can be made given a set of the three types of bricks?

5 Summarize your findings and give recommendations for possible brick patterns. Explain methods for exploring more patterns in the future.

VOCABULARY

After completing this chapter, you should be able to define each term, property, or phrase and give an example or two of each.

Algebra

axis of symmetry   (p. 612)

discriminant   (p. 632)

exponential function   (pp. 634, 635)

general equation for exponential decay   (p. 644)

general equation for exponential growth   (p. 643)

half-life   (p. 638)

maximum   (p. 611)

minimum   (p. 611)

parabola   (pp. 610, 611)

quadratic equation   (p. 620)

quadratic formula   (p. 628)

quadratic function   (pp. 610, 611)

roots   (p. 620)

symmetry   (p. 612)

vertex   (pp. 610, 611)

zeros   (p. 620)

Problem Solving

look for a pattern   (p. 637)

UNDERSTANDING AND USING THE VOCABULARY

Choose the letter of the term that best matches each equation or phrase.

1. $y = C(1 + r)^t$ **d**

2. $f(x) = ax^2 + bx + c$ **g**

3. a geometric property of parabolas   **i**

4. $x = \dfrac{-b}{2a}$ **a**

5. $y = a^x$ **c**

6. maximum or minimum point of a parabola   **j**

7. $A = C(1 - r)^t$ **b**

8. solutions of a quadratic equation   **h**

9. $x = \dfrac{-b \pm \sqrt{b^2 - 4ac}}{2a}$ **f**

10. the graph of a quadratic function   **e**

a. equation of axis of symmetry

b. exponential decay formula

c. exponential function

d. exponential growth formula

e. parabola

f. quadratic formula

g. quadratic function

h. roots

i. symmetry

j. vertex

Chapter 11   Highlights   **651**

Instructional Resources

Three multiple-choice tests and three free-response tests are provided in the *Assessment and Evaluation Masters*. Forms 1A and 2A are for honors pacing, and Forms 1B, 1C, 2B, and 2C are for average pacing. Chapter 11 Test, Form 1B is shown at the right. Chapter 11 Test, Form 2B is shown on the next page.

11   NAME_____ DATE _____

Chapter 11 Test, Form 1B

Write the letter for the correct answer in the blank at the right of each problem.

1. What are the equation of the axis of symmetry and the coordinates of the vertex of the graph of $y = -x^2 - 10x + 17$?
A. $x = -5; (-5, -8)$ B. $x = -5; (-5, 42)$
C. $x = 5; (5, 92)$ D. $x = 5; (5, 32)$ ... 1. __B__

2. Find the equation of the axis of symmetry for the graph of $y = -2x^2 + x + 17$, and state whether the axis of symmetry contains the minimum or maximum point of the graph.
A. $x = \frac{1}{4}$; maximum   B. $x = -\frac{1}{4}$; maximum
C. $x = \frac{1}{4}$; minimum   D. $x = -\frac{1}{4}$; minimum   ... 2. __A__

3. The equation of the axis of symmetry of a parabola is $x = -1$. If the point $(0, 1)$ is on the parabola, which of the following points is also on the parabola?
A. $(-1, 0)$ B. $(0, -2)$ C. $(-2, 1)$ D. $(-1, -2)$ ... 3. __C__

4. Which of the following equations describes the graph at the right?
A. $y = x^2 - 1$ B. $y = x^2 + 1$
C. $y = -x^2 - 1$ D. $y = x^2$ ... 4. __A__

5. If a quadratic equation has exactly no real roots, then its graph intersects the x-axis at how many points?
A. two   B. one   C. none   D. It varies.   ... 5. __C__

6. If the roots of $x^2 - 3x - 3 = 0$ are located by graphing the related function, between which pair of integers does a root of the equation lie?
A. -1 and 0 B. 1 and 2   C. 2 and 3   D. -2 and -1 ... 6. __A__

7. What must be the value of c in the function $y = x^2 + 3x + c$ if its graph has a minimum point at $\left(-\frac{3}{2}, 0\right)$?
A. $-\frac{3}{2}$ B. $-\frac{9}{4}$ C. $\frac{9}{4}$ D. $\frac{3}{2}$ ... 7. __C__

8. How many real roots does $x^2 - 2x - 3 = 0$ have?
A. 0   B. 1   C. 2   D. cannot be determined   ... 8. __C__

9. Solve $9x^2 + 16 = 24x$.
A. 24   B. 0   C. $\frac{4}{3}$ D. $\frac{-8 \pm 2\sqrt{70}}{9}$ ... 9. __C__

10. Solve $3x^2 + 5x - 16 = 0$.
A. $\frac{-3 \pm 6\sqrt{3}}{5}$ B. $\frac{-5 \pm 6\sqrt{3}}{5}$ C. $\frac{-5 \pm \sqrt{167}}{6}$ D. $\frac{-5 \pm \sqrt{217}}{6}$ ... 10. __D__

11. Solve $d^2 - 6d + 4 = 0$.
A. $3 \pm \sqrt{5}$ B. $6 \pm 2\sqrt{5}$ C. $\frac{-6 \pm \sqrt{5}}{2}$ D. $\frac{3 \pm 2\sqrt{5}}{2}$ ... 11. __A__

11   NAME_____ DATE _____

Chapter 11 Test, Form 1B (continued)

12. If $3x^2 - \frac{2}{3}x = \frac{1}{2}$ is to be solved using the quadratic formula, which of these would be best to do first?
A. Add $\frac{1}{9}$ to each side.   B. Divide each side by 3.
C. Multiply each side by 3.   D. Multiply each side by 6.   ... 12. __D__

13. Use a calculator to determine the value of $0.5(2.5^{-2.5})$ to the nearest hundredth.
A. 0.10   B. -3.13 C. 0.05   D. 0.57   ... 13. __C__

14. Which of the following is the graph of $y = 3^x + 1$?
A.   B.   C.   D. none of them   ... 14. __C__

15. Which table of values displays exponential behavior?

A.
x	0	1	2	3
y	0	1	32	243

B.
x	0	1	2	3
y	1	5	25	125

C.
x	0	1	2	3
y	0	5	10	15

D. none of them   ... 15. __B__

16. Solve $4^x = 2^{3x-6}$.
A. $1\frac{1}{5}$ B. -6 C. 3   D. 6   ... 16. __D__

17. If $1000 is invested at an annual interest rate of 6% compounded monthly, which equation can you solve to find the value of the investment at the end of 2 years?
A. $A = 1000(1.005)^{24}$ B. $A = 1000(1.06)^2$
C. $A = 1000(1.005)^2$ D. $A = 1000(1.06)^{12}$ ... 17. __A__

18. Suppose a printing press valued at $100,000 depreciates 10% per year. After how many years will the value of the printing press be less than $50,000?
A. 5   B. 2   C. 7   D. 20   ... 18. __C__

19. In 1995, the annual rate of inflation for consumer goods was about 2.5%. Suppose the inflation rate remains constant until the year 2000. What would a bag of groceries that cost $25 in 1995 cost in 2000?
A. $28.29   B. $37.50   C. $76.29   D. $67.50   ... 19. __A__

20. In 1980, the population of Honolulu, Hawaii, was about 763,000. Since then, the population has been increasing at an average rate of 1.5% per year. What will be the population of Honolulu in 2030 if it continues to increase at the same rate?
A. about 1,335,000   B. about 826,831,000   C. about 358,000   D. about 1,606,000   ... 20. __D__

Bonus Solve $x + 7 + \frac{3}{x} = 0$.   **Bonus** $\dfrac{-7 \pm \sqrt{37}}{2}$

Skills and Concepts Encourage students to refer to the objectives and examples on the left as they complete the review exercises on the right.

Assessment and Evaluation Masters, pp. 289–290

---

11 NAME_____ DATE _____

Chapter 11 Test, Form 2B

1. Find the equation of the axis of symmetry and the coordinates of the vertex of the graph of $y = -2x^2 + 4x - 5$.
   1. ___$x = 1; (1, -3)$___

2. Find the equation of the axis of symmetry for the graph of $y = x^2 - 7x + 12$, and state whether the axis of symmetry contains a maximum point or a minimum point of the graph.
   2. ___$x = \frac{7}{2}$; minimum___

3. Find the coordinates of the vertex of the graph of $y = 6x^2 - 14x$.
   3. ___$\left(\frac{7}{6}, -\frac{49}{6}\right)$___

4. The equation of the axis of symmetry of a parabola is $x = 3$. The point $(10, 0)$ is on the parabola. Name another point that is also on the parabola.
   4. ___$(-4, 0)$___

5. For the graph shown, is the value of the coefficient of x^2 in the related quadratic equation positive or negative? How many real roots does the equation have?
   5. ___negative; none___

Solve each equation by graphing. If exact roots cannot be found, state the consecutive integers between which the roots lie.

6. $x^2 - 6x + 4 = 0$ 7. $x^2 - 4x + 5 = 0$
   6. ___$0 < x < 1, 5 < x < 6$___

   7. ___no real roots___

8. $x^2 - 6x + 9 = 0$
   8. ___3___

9. Suppose that the roots of a quadratic equation are given by $x = \frac{-5 \pm \sqrt{105}}{4}$. Approximate the roots of the equation to the nearest hundredth.
   9. ___$-3.81, 1.31$___

---

11 NAME_____ DATE _____

Chapter 11 Test, Form 2B (continued)

Solve using the quadratic formula.

10. $y^2 - 9y + 8 = 0$
    10. ___1, 8___

11. $3b^2 + 2b - 3 = 0$
    11. ___$\frac{-1 \pm \sqrt{10}}{3}$___

12. $3x^2 = 10x$
    12. ___$0, \frac{10}{3}$___

13. Use a calculator to determine the value of $10(0.4^{-1.5})$ to the nearest hundredth.
    13. ___39.53___

14. What is the y-intercept of the graph of $y = 2.5 \cdot 10^x$?
    14. ___2.5___

15. Graph $y = 0.4^x - 1$.
    15. [graph]

16. Solve $3^{2x+5} = 9^{2x-5}$.
    16. ___7.5___

17. Suppose $500 is invested at an annual interest rate of 8% compounded semiannually. Write an equation for finding the value of the investment at the end of 3 years.
    17. ___$A = 500(1.04)^6$___

18. A new car costing $16,500 depreciates 12% per year. What will be the value of the car after 10 years?
    18. ___$4595.27___

19. An oil painting that originally cost $2500 increases in value at an annual rate of 2%. What is the value of the painting after 10 years?
    19. ___$3047.49___

20. In 1980, the population of Rochester, New York, was about 1,030,000. Since then, the population has been increasing at an average rate of 0.25% per year. At this rate of increase, what would be the approximate population of Rochester in 2030?
    20. ___about 1,167,000___

Bonus Graph the solution set of $y \geq x^2 + 6x + 9$.
    Bonus [graph]

SKILLS AND CONCEPTS

OBJECTIVES AND EXAMPLES	REVIEW EXERCISES

Upon completing this chapter, you should be able to:

Use these exercises to review and prepare for the chapter test.

• find the equation of the axis of symmetry and the coordinates of the vertex of a parabola *(Lesson 11–1)*

In the equation $y = x^2 - 8x + 12$, $a = 1$ and $b = -8$.

The equation of the axis of symmetry is
$$x = -\frac{b}{2a} = -\frac{(-8)}{2(1)} \text{ or } 4.$$

Use the value $x = 4$ to find the coordinates of the vertex.

$$y = x^2 - 8x + 12$$
$$= (4)^2 - 8(4) + 12$$
$$= 16 - 32 + 12 \text{ or } -4$$

The coordinates of the vertex are $(4, -4)$.

Write the equation of the axis of symmetry and find the coordinates of the vertex of the graph of each equation.

11. $y = -3x^2 + 4$ $x = 0; (0, 4)$

12. $y = x^2 - 3x - 4$ $x = \frac{3}{2}; \left(\frac{3}{2}, -\frac{25}{4}\right)$

13. $y = 3x^2 + 6x - 17$ $x = -1; (-1, -20)$

14. $y = 3(x + 1)^2 - 20$ $x = -1; (-1, -20)$

15. $y = x^2 + 2x$ $x = -1; (-1, -1)$

• graph quadratic functions *(Lesson 11–1)*

Graph $y = x^2 - 8x + 12$. Use the information above.

Using the results from Exercises 11–15, graph each equation. 16–20. See margin.

16. $y = -3x^2 + 4$

17. $y = x^2 - 3x - 4$

18. $y = 3x^2 + 6x - 17$

19. $y = 3(x + 1)^2 - 20$

20. $y = x^2 + 2x$

• find roots of quadratic equations by graphing *(Lesson 11–2)*

Based on the graph of $y = x^2 - 8x + 12$ shown above, the roots of the equation $x^2 - 8x + 12 = 0$ are 2 and 6. This is because $y = 0$ at the x-intercepts, which appear to be at 2 and 6.

Substitute these values into the original equation.

$x^2 - 8x + 12 = 0$	$x^2 - 8x + 12 = 0$
$(2)^2 - 8(2) + 12 = 0$	$(6)^2 - 8(6) + 12 = 0$
$4 - 16 + 12 = 0$	$36 - 48 + 12 = 0$
$0 = 0 ✓$	$0 = 0 ✓$

The solutions of the equation are 2 and 6.

Solve each equation by graphing. If integral roots cannot be found, state the consecutive integers between which the roots lie.

21. $x^2 - x - 12 = 0$ $-3, 4$

22. $x^2 + 6x + 9 = 0$ -3

23. $x^2 + 4x - 3 = 0$ $-5 < x < -4, 0 < x < 1$

24. $2x^2 - 5x + 4 = 0$ $\varnothing$

25. $x^2 - 10x = -21$ $3, 7$

26. $6x^2 - 13x = 15$ $-1 < x < 0, 3$

21–26. See Solutions Manual for graphs.

GLENCOE Technology

Test and Review Software

You may use this software, a combination of an item generator and item bank, to create your own tests or worksheets. Types of items include free response, multiple choice, short answer, and open ended.

For IBM & Macintosh

16.

OBJECTIVES AND EXAMPLES

• solve quadratic equations by using the quadratic formula (Lesson 11–3)

Solve $2x^2 + 7x - 15 = 0$.

In the equation, $a = 2$, $b = 7$, and $c = -15$. Substitute these values into the quadratic formula.

$$x = \frac{-(7) \pm \sqrt{(7)^2 - 4(2)(-15)}}{2(2)}$$

$$= \frac{-7 \pm \sqrt{169}}{4}$$

$$x = \frac{-7 + 13}{4} \quad \text{or} \quad x = \frac{-7 - 13}{4}$$

$$= \frac{3}{2} \qquad\qquad = -5$$

REVIEW EXERCISES

Solve each equation by using the quadratic formula. Approximate irrational roots to the nearest hundredth.

27. $x^2 - 8x = 20$ $10, -2$

28. $r^2 + 10r + 9 = 0$ $-1, -9$

29. $4p^2 + 4p = 15$ $\frac{3}{2}, -\frac{5}{2}$

30. $2y^2 + 3 = -8y$ $-0.42, -3.58$

31. $9k^2 - 13k + 4 = 0$ $1, \frac{4}{9}$

32. $9a^2 + 25 = 30a$ $\frac{5}{3}$

33. $-a^2 + 5a - 6 = 0$ $2, 3$

34. $-2d^2 + 8d + 3 = 3$ $0, 4$

35. $21a^2 + 5a - 7 = 0$ $0.47, -0.71$

36. $2m^2 = \frac{17}{6}m - 1$ $\frac{3}{4}, \frac{2}{3}$

17.

18.

• graph exponential functions (Lesson 11–4)

Graph $y = 2^x - 3$.

x	y
-3	-2.875
-2	-2.75
-1	-2.5
0	-2
1	-1
2	1
3	5

The y-intercept is -2.

Graph each function. State the y-intercept.

37. $y = 3^x + 6$ 7

38. $y = 3^{x+2}$ 9

39. $y = 2^x$ 1

40. $y = 2\left(\frac{1}{2}\right)^x$ 2

37–40. See margin for graphs.

19.

20.

• solve exponential equations (Lesson 11–4)

Solve $25^{b+4} = \left(\frac{1}{5}\right)^{2b}$.

$$25^{b+4} = \left(\frac{1}{5}\right)^{2b}$$

$$(5^2)^{b+4} = (5^{-1})^{2b}$$

$$5^{2b+8} = 5^{-2b}$$

$$2b + 8 = -2b$$

$$8 = -4b$$

$$-2 = b$$

The solution is -2.

Solve each equation.

41. $3^{4x} = 3^{-12}$ -3

42. $7^x = 7^{4x+9}$ -3

43. $\left(\frac{1}{3}\right)^t = 27^{t+8}$ -6

44. $0.01 = \left(\frac{1}{10}\right)^{4r}$ $\frac{1}{2}$

45. $64^{y-3} = \left(\frac{1}{16}\right)^{y^2}$ $-3, \frac{3}{2}$

37.

Additional Answers

39.

40.

38.

CHAPTER 11   STUDY GUIDE AND ASSESSMENT

OBJECTIVES AND EXAMPLES

• solve problems involving growth and decay
(Lesson 11–5)

Find the final amount from an investment of $1500 invested at an interest rate of 7.5% compounded quarterly for 10 years.

$$A = P\left(1 + \frac{r}{n}\right)^{nt}$$

$$= 1500\left(1 + \frac{0.075}{4}\right)^{4 \cdot 10}$$

$$= 3153.523916$$

The amount of the account is about $3153.52.

REVIEW EXERCISES

Determine the final amount from the investment for each situation.

46. $3769.08
47. $12,067.68
48. $97,243.21
49. $24,688.37

	Initial Amount	Annual Interest Rate	Time	Type of Compounding
46.	$2000	8%	8 years	quarterly
47.	$5500	5.25%	15 years	monthly
48.	$15,000	7.5%	25 years	monthly
49.	$500	9.75%	40 years	daily

APPLICATIONS AND PROBLEM SOLVING

50. **Archery**   The height h, in feet, that a certain arrow will reach t seconds after being shot directly upward is given by the formula $h = 112t - 16t^2$. What is the maximum height for this arrow?   (Lesson 11–1)   **196 ft**

51. **Physics**   A projectile is shot vertically up in the air. Its distance s, in feet, after t seconds is given by the equation $s = 96t - 16t^2$. Find the values of t when s is 96 feet.   (Lesson 11–3)
1.3 second and 4.7 seconds

52. **Finance**   Kevin deposited $1400 for 8 years at $6\frac{1}{2}$% interest compounded quarterly. How much will he have at the end of 8 years?
(Lesson 11–5)   **$2345**

53. **Diving**   Wyatt is diving from a 10-meter platform. His height h in meters above the water when he is x meters away from the platform is given by the formula $h = -x^2 + 2x + 10$. Approximately how far away from the platform is he when he enters the water?

(Lesson 11–2)   **between 4 meters and 5 meters**

54. **Number Theory**   Find a number whose square is 168 greater than 2 times the number.
(Lesson 11–3)   **−12 or 14**

55. **Decision Making**   Juanita wants to buy a new computer but she only has $500. She decides to wait a year and invest her money. Should she put it in a 1-year CD with a rate of 8% compounded monthly or in a savings account with a rate of 6% compounded daily? Explain your answer.   (Lesson 11–5)   **CD, which yields $541.50 vs. savings at $530.92**

A practice test for Chapter 11 is provided on page 797.

654   *Chapter 11   Study Guide and Assessment*

ALTERNATIVE ASSESSMENT

COOPERATIVE LEARNING PROJECT

Sightseeing Tours   In this project, you will model a business's profit. The Wash Student Tour Company offers one week tours of Washington, D.C. in small groups. While some of Wash's cost per person go down as the number of people on the tour increases, other costs go up because they must reserve rooms in another motel and rent extra vans. Wash has a function that enables them to predict their profit per student. If x is the number of students on the tour, and $f(x)$ is the profit (in dollars) per student, then $f(x) = -0.6x^2 + 18x - 45$.

Write a summary, using this model that describes in detail the profit structure for the Tour Company. Find the number of students that will give Wash the largest profit per student. What is the maximum profit? What does this represent? The company will offer tours as long as they do not lose money. What is the least or greatest number of students they should accept?

Follow these steps to accomplish your task.

- Substitute various numbers of students into the function to determine the profit for each.
- Substitute various amounts of profits into the function to determine the number of students that must be on the tour.
- Graph the function.
- Using the model, discuss terms such as profit, loss, break even, and maximum or minimum.
- Write a summary.

THINKING CRITICALLY

- If the value of a quadratic function is negative when $x = 1$ and positive when $x = 2$, explain what this means in terms of the root and why.
- In the quadratic equation $ax^2 - bx + c = 0$, if $ac < 0$, what must be true about the nature of the roots of the equation?

PORTFOLIO

Use the quadratic formula and factoring to solve several quadratic equations. As you solve each quadratic equation in both ways, compare the similarities and the differences, if there are any. Write a description of how the quadratic formula can be used to determine whether or not a quadratic polynomial is factorable. Place this in your portfolio.

SELF EVALUATION

While the solution to a problem may be unforeseen, a good problem solver can use the details of the problem to plan a method of solution. Those details can often be part of the solution or preliminary steps needed to arrive at the solution.

Assess yourself. Do you pay attention to detail? Do you organize and evaluate as you go through a problem? Once you have a solution, do you analyze the solution by looking at all options or do you only look for the most obvious? Give an example of a math-related problem and a daily life problem in which detail was essential and explain how it helped.

Assessment and Evaluation Masters, pp. 294, 305

11   NAME_____ DATE _____

Chapter 11 Performance Assessment

Instructions: *Demonstrate your knowledge by giving a clear, concise solution to each problem. Be sure to include all relevant drawings and justify your answers. You may show your solution in more than one way or investigate beyond the requirements of the problem.*

1. a. Write the equation of a quadratic function.

   b. Determine what the axis of symmetry and the maximum or minimum point of the graph of the equation would be without graphing the equation.

   c. Graph the function.

   d. Tell how the graph indicates the nature of the function's roots.

2. a. Write a word problem involving finding two real numbers that could be solved with the equation $x(9 - x) = 14$.

   b. Show how to solve the equation in part a by graphing.

3. Describe two real-life situations that could be represented by the graph shown below.

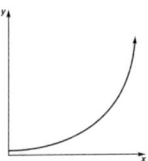

Scoring Guide
Chapter 11
Performance Assessment

Level	Specific Criteria
3 Superior	• Shows thorough understanding of the concepts of *quadratic functions, roots of quadratic equations,* and *exponential functions.* • Uses appropriate strategies to solve problems. • Computations are correct. • Written explanations are exemplary. • Word problem is appropriate and makes sense. • Graphs are accurate and appropriate. • Goes beyond requirements of some or all problems.
2 Satisfactory, with Minor Flaws	• Shows understanding of the concepts of *quadratic functions, roots of quadratic equations,* and *exponential functions.* • Uses appropriate strategies to solve problems. • Computations are mostly correct. • Written explanations are effective. • Word problem is appropriate and makes sense. • Graphs are mostly accurate and appropriate. • Satisfies all requirements of problems.
1 Nearly Satisfactory, with Serious Flaws	• Shows understanding of most of the concepts of *quadratic functions, roots of quadratic equations,* and *exponential functions.* • May not use appropriate strategies to solve problems. • Computations are mostly correct. • Written explanations are satisfactory. • Word problem is mostly appropriate and sensible. • Graphs are mostly accurate and appropriate. • Satisfies most requirements of problems.
0 Unsatisfactory	• Shows little or no understanding of the concepts of *quadratic functions, roots of quadratic equations,* and *exponential functions.* • May not use appropriate strategies to solve problems. • Computations are incorrect. • Written explanations are not satisfactory. • Word problem is not appropriate or sensible. • Graphs are not accurate or appropriate. • Does not satisfy requirements of problems.

 Alternative Assessment

The Alternative Assessment section provides students with the opportunity to assess their own work by thinking critically, working with others, keeping a portfolio, and honestly evaluating their own progress. For more information on alternative forms of assessment, see *Alternative Assessment in the Mathematics Classroom,* one of the titles in the Glencoe Mathematics Professional Series.

Performance Assessment

Performance Assessment tasks for this chapter are included in the *Assessment and Evaluation Masters.* A scoring guide is also provided.

NCTM Standards: 1–5, 7–8

This Investigation is designed to be completed over several days or weeks. It may be considered optional. You may want to assign the Investigation and the follow-up activities to be completed at the same time.

Objective
Create a model of a landscaping design and a bid for the Sanchez family.

Mathematical Overview
This Investigation will use the following mathematical skills and concepts from Chapters 12 and 13.

- simplifying rational expressions
- simplifying mixed expressions and complex functions
- using the Pythagorean theorem
- identifying subgoals
- completing squares

Recommended Time		
Part	Pages	Time
Investigation	656–657	1 class period
Working on the Investigation	665, 695, 718, 741	20 minutes each
Closing the Investigation	748	1 class period

Instructional Resources
Investigations and Projects Masters, pp. 21–24

A recording sheet, teacher notes, and scoring guide are provided for each Investigation in the *Investigations and Projects Masters.*

1 MOTIVATION

This Investigation uses common materials to investigate the specifications, materials, and costs involved in creating a model and a bid for a landscaping design. Discuss the importance of requesting bids from different contractors before making the final decision.

A Growing Concern

MATERIALS NEEDED

- calculator
- cardboard
- construction paper
- flashlight
- glue
- markers
- modeling clay
- paint
- ruler
- scissors
- duct tape

You have a small landscape company that specializes in residential landscape design. One of your clients is the Sanchez family. Mr. and Dr. Sanchez have three children, ages 5, 11, and 15. They have just moved into a home on a lot that is one third of an acre (1 acre = 43,560 ft²). Their one-story home and garage occupy 3425 square feet.

Dr. Sanchez loves to garden, and Mr. Sanchez likes to swim laps to keep in shape. So, the Sanchez family is interested in a backyard pool, hot tub, deck and/or patio, and a fairly good-sized lawn. They also want to leave room to later construct a play area for their youngest child and a garden for Dr. Sanchez. The front yard was fully landscaped by the construction company that built the house.

The family prefers a low-maintenance landscape, which consists mainly of lawn care. Because of the dry climate, daily watering must be done with a sprinkler system. They are also interested in concrete walkways, to match the driveway and the path to the front door.

For this job, the Sanchez family will request bids from several suppliers. They have asked your company to construct a model of your design, along with a bid, for them to view.

The Sanchezes are interested in a reasonably low price, but will choose a higher bid if they prefer the design and features. In either case, they plan to spend no more than $65,000.

When calculating bids, you must consider the cost of supplies, materials, labor, and a profit margin. Use the company's labor and material tables to estimate costs. Labor costs are determined either by the entire job or by the hour. This cost is dependent on the individual job. The profit margin is 20% of the total cost of materials, supplies, and labor.

Your design team consists of three people. Make an Investigation Folder in which you can store all of your work on this Investigation for future use.

 ## Cooperative Learning

This Investigation offers an excellent opportunity for using cooperative learning groups. For more information on cooperative learning strategies and group management, see *Cooperative Learning in the Mathematics Classroom,* one of the titles in the Glencoe Mathematics Professional Series.

THE PROPERTY

Use the dimensions given in the diagram below and your knowledge of geometry to design and create a *rough draft* of a landscaping plan for the Sanchez family's backyard. Be sure to include a pool, hot tub, lawn area, flower beds, trees, deck and/or patio, and concrete walkways in your design. Think about the dimensions of all of the features in your design. Be sure to leave ample space for a play area and garden to be developed later.

LABOR AND MATERIALS

POOL
Labor
digging: $1/cu ft
pool construction: $10/ft^2 of surface area
plumbing, filter, heating insulation: 8 h @ $35/h
Materials
pool construction: $18/ft^2 of surface area
plumbing and filter: $1250
heater: $1600

HOT TUB
Labor
digging: $1/cu ft
hot tub construction: $10/ft^2 of tub surface area
plumbing, filter, heating insulation: 8 h @ $35/h
   (3 hours if installed with a pool)
Materials
tub construction: $18/ft^2 of tub surface area
plumbing and filter: $1250
plumbing without filter: $250
heater: $1600 (pool and hot tub can share
   heater and filter)

DECK
Labor
8 ft^2 of deck per hour @ $25/h
Materials
redwood deck materials (4 ft × 1 ft): $12

PLANTS and TREES
Labor
plants/trees: 4 per hour @ $20/h
grass seed: 125 ft^2 per hour @ $20/h
sprinkler installation: 20 ft per hour @ $20/h
Materials
1 plant: $6.25
1 tree: $22.50
soil preparation: $1.75 per ft^2 of seeded area
sprinkler: 14 ft of sprinkler for every 10 ft^2 of lawn
   @ $1.50/ft

PATIO and WALKWAYS
Labor
12 ft^2 of patio or walkway per hour @ $22/h
Materials
concrete: $16/10 ft^2

You will continue working on this Investigation throughout Chapters 12 and 13.

Be sure to keep your designs, models, charts, and other materials in your Investigation Folder.

A Growing Concern Investigation

Working on the Investigation
Lesson 12–1, p. 665
....
Working on the Investigation
Lesson 12–7, p. 695
....
Working on the Investigation
Lesson 13–1, p. 718
....
Working on the Investigation
Lesson 13–5, p. 741
....
Closing the Investigation
End of Chapter 13, p. 748

Investigation: A Growing Concern **657**

2 SETUP

You may wish to have a student read the first six paragraphs of the Investigation to provide background information about creating the landscaping model and the bid. You may wish to read the next two paragraphs, which introduce the activity. Discuss the activity with your students. Then separate the class into groups of three.

3 MANAGEMENT

Each group member should be responsible for a specific task.

Recorder Collects data.
Measurer Sets dimensions of features for the model.
Contractor Calculates costs for the job.

At the end of the activity, each member should turn in his or her respective equipment.

Sample Answers

Answers will vary as they are based on the design chosen by each group.

Investigations and Projects Masters, p. 24

	NAME		DATE

Investigation, Chapters 12 and 13 Student Editi...
Pages 656–6...
665, 695, 718, 741, ...

A Growing Concern
Use this chart to record your calculations for the cost of materials, labor, and profit margin for each item in your plan.

Item	Cost of Materials	Labor	Profit

Work with your group to consider the following questions.

· How did the movement of the sun affect your decisions about plant choices and placement?

· If the Sanchez family was able to finance another $10,000, thus making the total amount that they can spend $75,000, what changes or additions would you make in your plan?

· List the elements that your group proposed to include for an effective sales presentation.

12

Exploring Rational Expressions and Equations

PREVIEWING THE CHAPTER

In this chapter, students build on their understanding of fractions to learn how to simplify, add, subtract, multiply, and divide rational expressions. The integration of graphing calculators helps students check the simplification of rational expressions. Students make a list of possible solutions to problems and determine which of the possibilities can be solutions. Students use their knowledge of mixed numbers and improper and complex fractions to study mixed expressions, rational expressions, and complex fractions containing variables. The chapter concludes with students applying their new knowledge to solve rational equations.

Lesson (Pages)	Lesson Objectives	NCTM Standards	State/Local Objectives
12-1 (660–665)	Simplify rational expressions. Identify values excluded from the domain of a rational expression.	1–5	
12-1B (666)	Use a graphing calculator to check the simplification of rational expressions.	1–6	
12-2 (667–670)	Multiply rational expressions.	1–5	
12-3 (671–674)	Divide rational expressions.	1–5	
12-4 (675–680)	Divide polynomials by monomials. Divide polynomials by binomials.	1–5	
12-5 (681–684)	Add and subtract rational expressions with like denominators.	1–5	
12-6 (685–689)	Add and subtract rational expressions with unlike denominators. Make an organized list of possibilities to solve problems.	1–5	
12-7 (690–695)	Simplify mixed expressions and complex fractions.	1–5	
12-8 (696–702)	Solve rational equations.	1–5	

A complete, 1-page lesson plan is provided for each lesson in the *Lesson Planning Guide*. Answer keys for each lesson are available in the *Answer Key Masters.*

ORGANIZING THE CHAPTER

You may want to refer to the **Course Planning Calendar** on page T12 for detailed information on pacing.
PACING: Standard—14 days; **Honors**—13 days; **Block**—6 days; **Two Years**—27 days

LESSON PLANNING CHART

| Lesson (Pages) | Materials/ Manipulatives | Extra Practice (Student Edition) | BLACKLINE MASTERS | | | | | | | | | | Real-World Applications | Interactive Mathematics Tools Software | Teaching Transparencies |
			Study Guide	Practice	Enrichment	Assessment and Evaluation	Modeling Mathematics	Multicultural Activity	Tech Prep Applications	Graphing Calculator	Science and Math Lab Manual			
12-1 (660–665)	scientific calculator graphing calculator	p. 782	p. 80	p. 80	p. 80						pp. 53–56			12-1A 12-1B
12-1B (666)	graphing calculator									pp. 36, 37				
12-2 (667–670)		p. 782	p. 81	p. 81	p. 81	p. 324								12-2A 12-2B
12-3 (671–674)		p. 783	p. 82	p. 82	p. 82									12-3A 12-3B
12-4 (675–680)	algebra tiles* product mat* graphing calculator	p. 783	p. 83	p. 83	p. 83	pp. 323, 324								12-4A 12-4B
12-5 (681–684)		p. 783	p. 84	p. 84	p. 84									12-5A 12-5B
12-6 (685–689)		p. 784	p. 85	p. 85	p. 85	p. 325		p. 23	p. 23					12-6A 12-6B
12-7 (690–695)	graphing calculator	p. 784	p. 86	p. 86	p. 86			p. 24	p. 24			29		12-7A 12-7B
12-8 (696–702)		p. 784	p. 87	p. 87	p. 87	p. 325	p. 83			p. 12		30	12-8	12-8A 12-8B
Study Guide/ Assessment (703–707)						pp. 309 –322, 326– 328								

*Included in Glencoe's Student Manipulative Kit and Overhead Manipulative Resources.

ORGANIZING THE CHAPTER

OTHER CHAPTER RESOURCES

Student Edition
Investigation, pp. 656–657
Chapter Opener, pp. 658–659
Mathematics and Society, p. 689
Working on the Investigation,
   pp. 665, 695

Teacher's Classroom Resources
Investigations and Projects Masters,
   pp. 69–72

Technology
Test and Review Software (IBM
   and Macintosh)
CD-ROM Interactions (Windows
   and Macintosh)

Professional Publications
Block Scheduling Booklet
Glencoe Mathematics Professional
   Series

OUTSIDE RESOURCES

Books/Periodicals
Marcy, Steve, *Algebra with Pizzazz!*, Creative
   Publications
Pedersen, Katherine, *Trivia Math: Algebra*,
   Creative Publications

Software
Algebra Concepts, Vol. 1, Ventura Educational
   Systems

Videos/CD-ROMs
Algebra for Everyone, NCTM

ASSESSMENT RESOURCES

Student Edition
Math Journal, pp. 668, 672, 699
Mixed Review, pp. 665, 670,
   674, 680, 684, 689, 695, 702
Self Test, p. 680
Chapter Highlights, p. 703
Chapter Study Guide and
   Assessment, pp. 704–706
Alternative Assessment, p. 707
 Portfolio, p. 707

Cumulative Review, pp. 708–709

Teacher's Wraparound Edition
5-Minute Check, pp. 660, 667,
   671, 675, 681, 685, 690, 696
Check for Understanding, pp. 663,
   669, 673, 678, 683, 687, 693,
   699
Closing Activity, pp. 665, 670,
   674, 680, 684, 689, 695, 702
Cooperative Learning, pp. 673,
   697

Assessment and Evaluation Masters
Multiple-Choice Tests, Forms 1A
   (Honors), 1B (Average), 1C
   (Basic), pp. 309–314
Free-Response Tests, Forms 2A
   (Honors), 2B (Average), 2C
   (Basic), pp. 315–320
Calculator-Based Test, p. 321
Performance Assessment, p. 322
Mid-Chapter Test, p. 323
Quizzes A–D, pp. 324–325
Standardized Test Practice, p. 326
Cumulative Review, pp. 327–328

ENHANCING THE CHAPTER

Examples of some of the materials for enhancing Chapter 12 are shown below.

DIVERSITY

Multicultural Activity Masters, pp. 23, 24

APPLICATIONS

Real-World Applications, 29, 30

TECHNOLOGY

Graphing Calculator Masters, p. 12

TECH PREP

Tech Prep Applications Masters, pp. 23, 24

CONNECTIONS

Science and Math Lab Manual, pp. 53–56

PROBLEM SOLVING

Problem of the Week Cards, 32, 33, 34

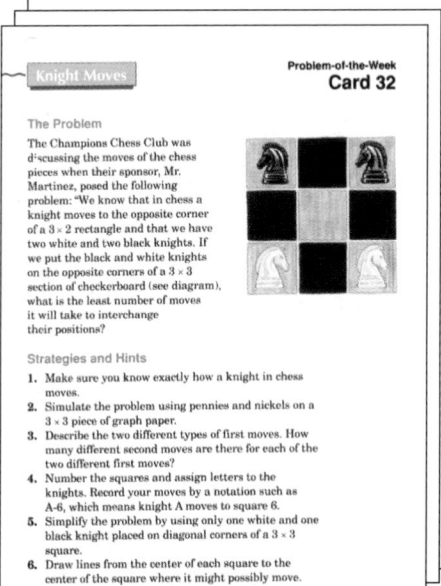

CHAPTER 12

Exploring Rational Expressions and Equations

Objectives

In this chapter, you will:

- simplify rational expressions,
- add, subtact, multiply, and divide rational expressions,
- divide polynomials, and
- make organized lists to solve problems.

National Endowment for the Arts

Funds Available ($ millions)

1985	171.7
1986	167.1
1987	170.9
1988	171.1
1989	166.7
1990	170.8
1991	166.5
1992	163
1993	159.7

Source: U.S. National Endowment for the Arts, *Annual Report*

The term "starving artist" is not without merit. Many artists often have other jobs to survive while trying to pursue their careers as artists. The National Endowment for the Arts is authorized to assist individuals and nonprofit organizations financially in a wide range of artistic endeavors. Some of the artistic forms that are funded are music, museums, theater, dance, media arts, and visual arts.

TIME Line

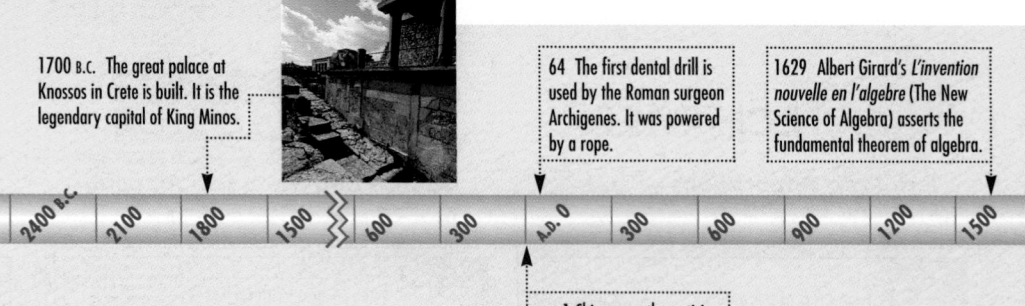

1700 B.C. The great palace at Knossos in Crete is built. It is the legendary capital of King Minos.

64 The first dental drill is used by the Roman surgeon Archigenes. It was powered by a rope.

1629 Albert Girard's *L'invention nouvelle en l'algebre* (The New Science of Algebra) asserts the fundamental theorem of algebra.

| 2400 B.C. | 2100 | 1800 | 1500 | 600 | 300 | A.D. 0 | 300 | 600 | 900 | 1200 | 1500 | |

A.D. 1 Chinese mathematician Liu Hsin is the first to use decimal fractions.

TIME Line

The fundamental theorem of algebra was not proven until 170 years after its discovery. Students might find it interesting to write a report on why mathematicians prove theorems that they often "know" to be true.

interNET CONNECTION

The Smithsonian Institution's National Museum of the American Indian covers languages, literature, history, and arts with links to Native American sites.

World Wide Web
http://www.si.edu/organiza/museums/amerind/nmai/

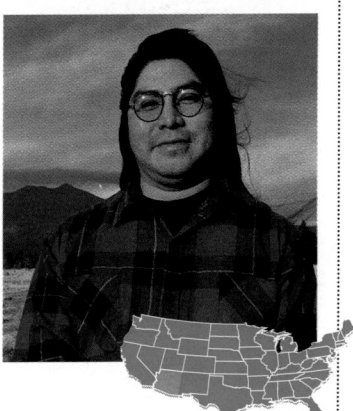

Chapter Project

Native American art also includes blanket weaving. Many of the designs include pictorial representations of everyday objects or characters from folklore about their ancestors.

- Design a 64″ × 80″ blanket. Choose either objects that represent you or some part of your personal history for the blanket's design.

- Make a drawing to represent your design. Include a scale expressed as a rational number to show the actual size of your blanket.

- Use yarn or construction paper to weave a portion of your design.

- Calculate how much yarn in each color you would need to create a real blanket.

- Estimate the cost of your blanket in materials and labor.

Suppose a blanket is priced at $375. How does this price compare with your estimate?

The expression of energy, vibrancy, and harmony are just some of the thoughts that flow through the creativity of **Joe Maktima** of Flagstaff, Arizona. The themes of his work in acrylic paints and mediums are deeply rooted in his pueblo culture. Joe began his artistic endeavors in high school as a hobby, but it was winning the Best of Show Award in a national competition of Native American high school art students that gave him the confidence to become a professional artist.

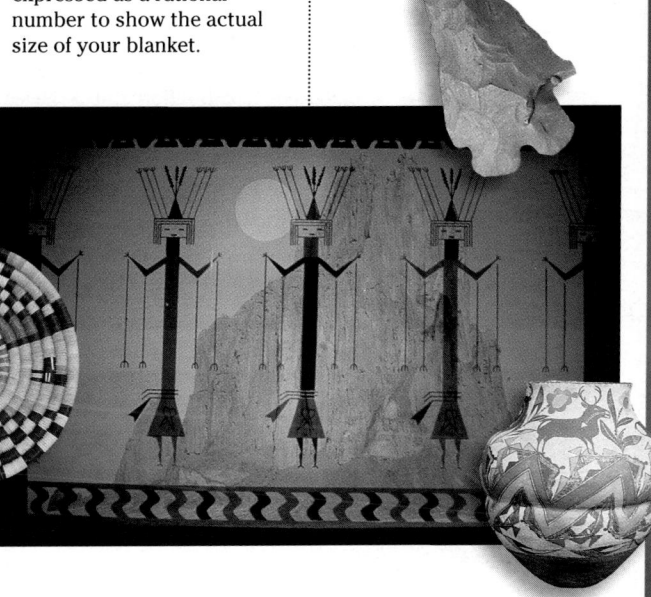

Joe Maktima has works on display in galleries in eight states and in Germany, Japan, and England. Mr. Maktima studied art at the Institute of American Indian Arts.

Chapter Project

Cooperative Learning Students can work on this project in groups of three. As a group, they can discuss and decide on the objects that will be represented in their design. One student will then draw the design, another student will estimate the cost, while the third student will produce a sample of the design in yarn.

1928 American portrait photographer Berenice Abbott takes her famous picture of writer James Joyce.

1983 A meter is officially redefined as the distance that light travels in $\frac{1}{299,792,458}$ of a second.

1920 1925 1930 1935 1965 1970 1975 1980 1985 1990 1995 2000

1971 Aretha Franklin releases her live album *Aretha Live at Fillmore West.*

1991 American artist Faith Ringgold writes and illustrates her children's book *Tar Beach.*

Chapter 12 **659**

Alternative Chapter Projects

Two other chapter projects are included in the *Investigations and Projects Masters.* In Chapter 12 Project A, pp. 69–70, students extend the topic in the chapter opener. In Chapter 12 Project B, pp. 71–72, students investigate focal length in relationship to a lens.

Investigations and Projects Masters, p. 69

12
NAME_____ DATE _____
Chapter 12 Project A
Student Edition
Pages 660–7

Mayan Time

1. The Maya were a Native American people whose complex civilization reached a peak during the years from about 250 to 900. For this project, you will work with a small group to research Mayan civilization. Investigate such topics as art and architecture, mathematics and astronomy, religion, and way of life.

2. Using careful astronomical observations, the Maya developed a solar calendar. They also developed a sacred calendar that ran simultaneously with it. Research these two calendars to find out the answers to the following questions.

   · How many days made up a year in the solar calendar?
   · How were the days of the solar calendar divided into months?
   · How many days made up a year in the sacred calendar?
   · How were the days of the sacred calendar named?

3. Use the information you found about the number of days in each calendar to determine how many days would pass before a given day of the sacred calendar coincided with the same day of the solar calendar again. How many years on each calendar would pass before the calendars reached the same point again? Explain how these problems relate to the concept of the least common multiple. Make up some other problems using your knowledge of the Mayan calendars and exchange them with other groups to solve.

4. With your group, design a museum exhibit about the Mayan civilization. Include a timeline that would show how the two calendars correspond. Share your timelines with other groups.

Simplifying Rational Expressions

NCTM Standards: 1–5

Instructional Resources

- Study Guide Master 12-1
- Practice Master 12-1
- Enrichment Master 12-1
- Science and Math Lab Manual, pp. 53–56

Transparency 12-1A contains the 5-Minute Check for this lesson; **Transparency 12-1B** contains a teaching aid for this lesson.

Recommended Pacing	
Standard Pacing	Day 1 of 14
Honors Pacing	Day 1 of 13
Block Scheduling*	Day 1 of 6
Alg. 1 in Two Years*	Days 1 & 2 of 27

*For more information on pacing and possible lesson plans, refer to the *Block Scheduling Booklet* and *Algebra 1 in Two Years*.

1 FOCUS

5-Minute Check
(over Chapter 11)

Solve each equation.

1. $n^2 + \frac{1}{6}n - \frac{1}{6} = 0$ $\left\{\frac{1}{3}, -\frac{1}{2}\right\}$

2. $t^2 - 14t + 49 = 0$ $\{7\}$

3. $y^2 + 5y = 0$ $\{-5, 0\}$

4. Find two numbers whose sum is 35 and whose product is 300. **15 and 20**

5. Find two numbers that differ by 7 and have a product of 144. **9 and 16, −9 and −16**

Motivating the Lesson

Hands-On Activity Set up a lever and fulcrum using a book and a ruler. One at a time, place objects of different weights and lift each by grasping the lever at different positions. Have students generalize their observations.

Teaching Tip In the examples at the bottom of the page, note that $x^2 - 5x + 6 = (x - 2)(x - 3)$. Use the zero product property to show that $x \neq 2$ and $x \neq 3$.

Objectives

After studying this lesson, you should be able to:
- simplify rational expressions, and
- identify values excluded from the domain of a rational expression.

LOOK BACK

Refer to Lesson 7-2 for more information on levers.

CONNECTION
Physical Science

Many bicyclists carry small tool kits in case they need to make adjustments to their bicycles while on the road. A wrench, for example, might be needed to tighten the bolts that keep the seat aligned. The force applied to one end of the wrench is multiplied so that the bolt at the other end of the wrench is made very secure.

A wrench is an example of a simple machine called a *lever*. In Lesson 7–2, you learned that to calculate the *mechanical advantage* (MA) of a lever, you find the ratio of the length of the effort arm to the length of the resistance arm. MA is usually expressed as a decimal value.

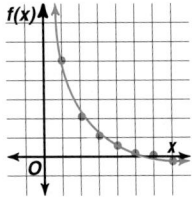

$$MA = \frac{\text{length of effort arm}}{\text{length of resistance arm}} = \frac{L_e}{L_r}$$

Suppose the total length of a lever is 6 feet. If the length of the resistance arm of the lever is x feet, then the length of the effort arm is $(6 - x)$ feet. The function $f(x) = \frac{6 - x}{x}$ represents the mechanical advantage of the lever. The table and graph at the right represent this function. They illustrate how varying the length of the resistance arm affects the mechanical advantage. Notice that, as x increases, the mechanical advantage $f(x)$ decreases. The expression that defines this function, $\frac{6-x}{x}$, is an example of a **rational expression**.

x	f(x)
0	undefined
1	5.0
2	2.0
3	1.0
4	0.5
5	0.2
6	0.0

Definition of a Rational Expression	**A rational expression is an algebraic fraction whose numerator and denominator are polynomials.**

$f(x) = \frac{6 - x}{x}$ is called a *rational function*.

Because a rational expression involves division, the denominator may not have a value of zero. Therefore, any values of a variable that result in a denominator of zero must be excluded from the domain of the variable. These are called **excluded values** of the rational expression.

For $\frac{6 - x}{x}$, exclude $x = 0$.

For $\frac{5m + 3}{m + 6}$, exclude $m = -6$, since $-6 + 6 = 0$.

For $\frac{x^2 - 5}{x^2 - 5x + 6}$, exclude $x = 2$ and $x = 3$. *Why?*

Alternative Learning Styles

Visual The rectangle at the right has length $4a^2 - 8a + 4$ and area $3a^2 - 4a + 1$. Express its width in simplest form. $w = \frac{3a - 1}{4a - 4}$

← $4a^2 - 8a + 4$ →

Area = $3a^2 - 4a + 1$

Example For each rational expression, state the values of the variable that must be excluded.

a. $\dfrac{7b}{b+5}$

Exclude the values for which $b + 5 = 0$.

$b + 5 = 0$

$b = -5$

Therefore, b cannot equal -5.

b. $\dfrac{r^2 + 32}{r^2 + 9r + 8}$

Exclude the values for which $r^2 + 9r + 8 = 0$.

$r^2 + 9r + 8 = 0$

$(r + 1)(r + 8) = 0$

$r = -1$ or $r = -8$ *Zero product property*

Therefore, r cannot equal -1 or -8.

To simplify a rational expression, you must eliminate any common factors of the numerator and denominator. To do this, use their greatest common factor (GCF). Remember that $\dfrac{ab}{ac} = \dfrac{a}{a} \cdot \dfrac{b}{c}$ and $\dfrac{a}{a} = 1$. So, $\dfrac{ab}{ac} = 1 \cdot \dfrac{b}{c}$ or $\dfrac{b}{c}$.

Example Simplify $\dfrac{9x^2yz}{24xyz^2}$. State the excluded values of x, y, and z.

$\dfrac{9x^2yz}{24xyz^2} = \dfrac{(3xyz)(3x)}{(3xyz)(8z)}$ *Factor; the GCF is 3xyz.*

$= \dfrac{(3xyz)(3x)}{(3xyz)(8z)}$ *Divide by the GCF.*

$= \dfrac{3x}{8z}$

Exclude the values for which $24xyz^2 = 0$: $x = 0$, $y = 0$, or $z = 0$. Therefore, neither x, y, nor z can equal 0.

You can use the same procedure to simplify a rational expression in which the numerator and denominator are polynomials.

Example **3** Simplify $\dfrac{a+3}{a^2 + 4a + 3}$. State the excluded values of a.

$\dfrac{a+3}{a^2 + 4a + 3} = \dfrac{a+3}{(a+1)(a+3)}$ *Factor the denominator.*

$= \dfrac{a+3}{(a+1)(a+3)}$ *The GCF is a + 3.*

$= \dfrac{1}{a+1}$

Exclude the values for which $a^2 + 4a + 3 = 0$.

$a^2 + 4a + 3 = 0$

$(a + 1)(a + 3) = 0$

$a = -1$ or $a = -3$

Therefore, a cannot equal -1 or -3.

Lesson 12–1 Simplifying Rational Expressions **661**

Alternative Teaching Strategies

Student Diversity Ask students to compare and contrast the expressions $(x - 2)$ and $\dfrac{x^2 - 4}{x + 2}$. Have them find the value for various values of x, including 2.

Teaching Tip For Example 1, it may be necessary to review factoring. You may want to draw a flowchart for factoring polynomials. Then have students test the flowchart with various examples.

In-Class Examples

For Example 1
For each rational expression, state the values of the variable that must be excluded.

a. $\dfrac{3r}{6-r}$ 6

b. $\dfrac{m+5}{m^2 - 4m + 3}$ 3, 1

c. $\dfrac{15k}{k^2 - 64}$ 8, −8

For Example 2
Simplify each rational expression. State the excluded values of the variables.

a. $\dfrac{3m^2 - 48}{3(m+4)}$

   $m - 4$, $m \neq -4$

b. $\dfrac{y^2 - 4}{y^2 + y - 2}$

   $\dfrac{y-2}{y-1}$, $y \neq -2, 1$

For Example 3
Simplify each rational expression. State the excluded values of the variables.

a. $\dfrac{b - 2m}{12m^2 - 3b^2}$

   $-\dfrac{1}{6m + 3b}$, $b \neq \pm 2m$

b. $\dfrac{2x^2 - 50}{5 - x}$

   $-2x + 10$, $x \neq 5$

Teaching Tip For Example 3, remind students that they must first factor the numerator and denominator in order to find the GCF. Stress that factors, not terms, can be canceled.

662 *Chapter 12*

This Exploration has two goals.

1. Students learn to use the calculator to compute values of rational functions.
2. Students learn what it means for a rational function to be undefined for certain values.

In-Class Example

For Example 4

A screwdriver that is 26.5 cm long is used to pry the lid off a paint can. It is placed so that 0.5 cm of its length extends inward from the rim of the can. The same 5 pounds of force is applied. What is the force applied on the lid? **260 pounds**

You can use a scientific or graphing calculator to evaluate rational expressions for given values of the variables. When you do this, you must be careful to use parentheses to group both the numerator and the denominator of the expression. For example, to evaluate $\frac{x^2 - 6x + 8}{x - 2}$ when $x = -3$, use a key sequence like this on a scientific calculator.

Enter: (3 +/- x^2 − 6 × 3 +/- + 8) ÷ (3 +/- − 2) =   −7

The key sequence for a graphing calculator is very similar.

Enter: (((−) 3) x^2 − 6 × (−) 3 + 8) ÷ ((−) 3 − 2) ENTER   −7

Your Turn

a. Copy and complete the table below.

x	−3	−2	−1	0	1	2	3
$\frac{x^2 - 6x + 8}{x - 2}$	−7	−6	−5	−4	−3	undef.	−1
$x - 4$	−7	−6	−5	−4	−3	−2	−1

b. For which value(s) of x is the value of $\frac{x^2 - 6x + 8}{x - 2}$ equal to the value of $x - 4$? Why does this result make sense? **all except $x = 2$**

c. For which value(s) of x is the value of $\frac{x^2 - 6x + 8}{x - 2}$ *not* equal to the value of $x - 4$? Why does this result make sense? **$x = 2$; It is undefined at $x = 2$.**

Example ④

To pry the lid off a paint can, a screwdriver that is 20.5 cm long is used as a lever. It is placed so that 0.5 cm of its length extends inward from the rim of the can. Then a force of 5 pounds is applied at the end of the screwdriver. What is the force placed on the lid?

Let s represent the total length of the screwdriver.

Let r represent the length that extends inward from the rim. This is the length of the resistance arm of the lever.

Then $s - r$ represents the length that extends outward from the rim. This is the length of the effort arm of the lever.

Use the formula given in the connection at the beginning of the lesson to write an expression for mechanical advantage.

$$MA = \frac{\text{length of effort arm}}{\text{length of resistance arm}} \text{ or } \frac{s - r}{r}$$

screwdriver
lid
rim of can (fulcrum of lever)
paint can

Classroom Vignette

"When factoring to simplify expressions, I teach my students that in $\frac{x + 3}{3}$, the $x + 3$ is like a baseball player trying out for the team. The coach will cut only a whole player, not part of a player. So the $x + 3$ can be cut only with another $x + 3$, not part of a player like an x or a 3."

Joyce Anne MacKenzie
Camdenton R-III High School
Camdenton, Missouri

Now evaluate the expression for the given values.

$$\frac{s - r}{r} = \frac{20.5 - 0.5}{0.5} \quad s = 20.5 \text{ and } r = 0.5$$

$$= \frac{20}{0.5} \text{ or } 40 \quad \textit{Simplify.}$$

The mechanical advantage is 40. The force placed on the lid is the product of this number and the force applied at the end of the screwdriver.

$40 \cdot 5$ pounds $= 200$ pounds

The force placed on the lid is 200 pounds.

CHECK FOR UNDERSTANDING

Communicating Mathematics

1. When $x = 2$, the denominator becomes zero. Division by zero is undefined.

4. Sample answer:
$$\frac{x - 2}{(x + 2)(x - 7)}$$

Guided Practice

10. $r - s$; $\frac{r + s}{r - s}$; $r - s \neq 0$

Study the lesson. Then complete the following.

1. **Explain** why $x = 2$ is excluded from the domain of $f(x) = \dfrac{x^2 + 7x + 12}{x - 2}$.

2. **Estimate** the mechanical advantage of the lever shown on page 660 using the graph if x is 1.5. about 3

3. **Explain** how you would determine the values to be excluded from the domain of $f(x) = \dfrac{x + 5}{x^2 + 6x + 5}$. See margin.

4. **Write** a rational expression involving one variable for which the excluded values are -2 and 7.

5. **Write** the meaning of the term *mechanical advantage* in your own words.
The multiplication effect of using a tool or lever to carry out a task.

For each expression, find the GCF of the numerator and the denominator. Then simplify. State the excluded values of the variables.

6. $\dfrac{13a}{14ay}$ a; $\dfrac{13}{14y}$; $a \neq 0$, $y \neq 0$

7. $\dfrac{-7a^2b^3}{21a^5b}$ $7a^2b$; $\dfrac{-b^2}{3a^3}$; $a \neq 0$, $b \neq 0$

8. $\dfrac{a(m + 3)}{a(m - 2)}$ a; $\dfrac{m + 3}{m - 2}$; $m \neq 2$, $a \neq 0$

9. $\dfrac{3b}{b(b + 5)}$ b; $\dfrac{3}{b + 5}$; $b \neq 0$, $b \neq -5$

10. $\dfrac{(r + s)(r - s)}{(r - s)(r - s)}$

11. $\dfrac{m - 3}{m^2 - 9}$ $m - 3$; $\dfrac{1}{m + 3}$; $m \neq \pm 3$

12. Evaluate $\dfrac{x^2 + 7x + 12}{x + 3}$ if $x = 2$. 6

13. **Landscaping** To clear land for a garden, Chang needs to move some large rocks. He plans to use a 6-foot-long pinch bar as a lever. He positions it next to each rock as shown at the right.
   a. Calculate the mechanical advantage. 5
   b. If Chang can apply a force of 150 pounds to the effort arm, what is the greatest weight he can lift? 750 lb

pinch bar
rock
5 feet
fulcrum

Lesson 12–1 Simplifying Rational Expressions **663**

Reteaching

Using Alternative Methods Write a rational expression with one variable and simplify it. Choose values for the variable and complete a table with headings *Variable, Value of Rational Expression,* and *Value of Simplified Form.* Write the table on the chalkboard. This may help students see how the rational expression is equivalent to its simplified form.

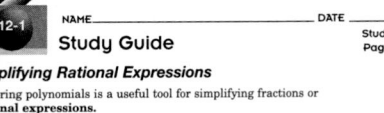
Chapter 12 **663**

Assignment Guide

Core: 15–35 odd, 36, 37, 39–46
Enriched: 14–34 even, 36–46

For **Extra Practice**, see p. 782.

The red A, B, and C flags, printed only in the Teacher's Wraparound Edition, indicate the level of difficulty of the exercises.

Additional Answers

14. $a \neq 0$
15. $y \neq 0, z \neq 0$
16. $a \neq 0, b \neq 0$
17. $x \neq 0, y \neq 0$
18. $a \neq -3$ or 0
19. $y \neq -\dfrac{1}{3}$
20. $x \neq -3$
21. $y \neq 7x$
22. $x \neq -5$ or 4
23. $a \neq 3$ or 4
24. $a \neq 2$ or 5
25. $x \neq -4$
26. $x \neq 4$ or -3
27. $a \neq -4$ or 2
28. $x \neq -6$ or 5
29. $x \neq 9$ or 4
30. $x \neq -1.5$ or -1
31. $x \neq -1$

Practice Masters, p. 80

EXERCISES

Practice

14–31. See margin for excluded values.

Simplify each rational expression. State the excluded values of the variables.

14. $\dfrac{15a}{39a^2}$ $\dfrac{5}{13a}$

15. $\dfrac{35y^2z}{14yz^2}$ $\dfrac{5y}{2z}$

16. $\dfrac{28a^2}{49ab}$ $\dfrac{4a}{7b}$

17. $\dfrac{56x^2y}{70x^3y}$ $\dfrac{4}{5x}$

18. $\dfrac{4a}{3a+a^2}$ $\dfrac{4}{3+a}$

19. $\dfrac{y+3y^2}{3y+1}$ y

20. $\dfrac{x^2-9}{2x+6}$ $\dfrac{x-3}{2}$

21. $\dfrac{y^2-49x^2}{y-7x}$ $y + 7x$

22. $\dfrac{x+5}{x^2+x-20}$ $\dfrac{1}{x-4}$

23. $\dfrac{a-3}{a^2-7a+12}$ $\dfrac{1}{a-4}$

24. $\dfrac{3x-15}{x^2-7x+10}$ $\dfrac{3}{x-2}$

25. $\dfrac{x+4}{x^2+8x+16}$ $\dfrac{1}{x+4}$

26. $\dfrac{x^2-2x-15}{x^2-x-12}$ $\dfrac{x-5}{x-4}$

27. $\dfrac{a^2+4a-12}{a^2+2a-8}$ $\dfrac{a+6}{a+4}$

28. $\dfrac{x^2-36}{x^2+x-30}$ $\dfrac{x-6}{x-5}$

29. $\dfrac{b^2-3b-4}{b^2-13b+36}$ $\dfrac{b+1}{b-9}$

30. $\dfrac{14x^2+35x+21}{12x^2+30x+18}$ $\dfrac{7}{6}$

31. $\dfrac{4x^2+8x+4}{5x^2+10x+5}$ $\dfrac{4}{5}$

Calculator

35. not possible, $y \neq -2$

Use a calculator to evaluate each expression for the given values.

32. $\dfrac{x^2-x}{3x}$, $x = -3$ $-\dfrac{4}{3}$

33. $\dfrac{x^4-16}{x^4-8x^2+16}$, $x = -1$ $-\dfrac{5}{3}$

34. $\dfrac{x+y}{x^2+2xy+y^2}$, $x = 3, y = 2$ $\dfrac{1}{5}$

35. $\dfrac{x^3y^3+5x^3y^2+6x^3y}{xy^5+5xy^4+6xy^3}$, $x = 1, y = -2$

Critical Thinking

Applications and Problem Solving

36. Because $m \neq -4$ in the original expression.

36. Explain why $\dfrac{m^2-16}{m+4}$ is not the same as $m - 4$.

37. **Carpentry** A house mover can lift a house from its foundation using a *jackscrew* like the one shown at the right. As shown, the effort force is applied in a circular motion. The vertical distance that the jackscrew moves in one turn is called its *pitch*. You can use the following formula to calculate the mechanical advantage of the jackscrew.

$$MA = \dfrac{\text{circumference of the circle}}{\text{pitch of the screw}}$$

Effort distance 48 in.

Effort force

Resistance distance, $\frac{1}{4}$ in.

Pitch

a. Calculate the mechanical advantage for the jackscrew shown above.   **192**

b. Suppose the effort force is 20 pounds. What is the greatest weight the jackscrew can lift?   **3840 lb**

38. **Aeronautics** Aircraft engineers can use the following formula to calculate the atmospheric pressure P in pounds per square inch when an aircraft is flying at an altitude a in feet.

$$P = \dfrac{-9.05\left[\left(\dfrac{a}{1000}\right)^2 - \dfrac{65a}{1000}\right]}{\left(\dfrac{a}{1000}\right)^2 + 40\left(\dfrac{a}{1000}\right)}$$

a. Calculate the atmospheric pressure outside a plane flying at an altitude of 20,000 feet.   **6.79 lb/in²**

b. Calculate the atmospheric pressure outside a plane flying at an altitude of 40,000 feet.   **2.83 lb/in²**

c. Is your answer to part b twice that of part a? By how much do they differ?   **no, 3.96 lb/in²**

Extension

Connections The measure of the area of a rectangle is $6x^2 - 7x - 20$, and the measure of one side is $2x - 5$. Find the measure of the other side of the rectangle. $3x + 4$

$39b. \ -\dfrac{95}{10} = -\dfrac{19}{2}$

39. Physics At sea level, the boiling point of water is 212°F. At the top of Mt. Everest, the boiling point is 159.8°F. The top of Mt. Everest is 29,002 ft above sea level.

  a. How many degrees does the boiling point drop for every mile up from sea level? Write the solution as a ratio involving units. **9.5°/mile**

  b. Simplify the expression you wrote in part a.

  c. Draw a graph that represents the boiling point in degrees Fahrenheit as a function of altitude in miles. **See margin.**

Mixed Review

40. Finance In the compound interest formula $A = P\left(1 + \dfrac{r}{n}\right)^{nt}$, A represents the value of the investment in the future, P is the amount of the original investment, r is the annual interest rate, t is the number of years of the investment, and n is the number of times the interest is compounded each year. Find the total amount after \$2500 is invested for 18 years at a rate of 6%, compounded quarterly. (Lesson 11–5) **\$7302.90**

41. Number Theory The square of a number decreased by 121 is 0. Find the number. (Lesson 10–7) **±11**

42. Find $3x(-5x^2 - 2x + 7)$. (Lesson 9–6) **$-15x^3 - 6x^2 + 21x$**

43. Graph the system of inequalities. (Lesson 8–5)
$y \geq x - 2$
$y \leq 2x - 1$ **See margin.**

44. Solve $4 + |x| = 12$. (Lesson 7–6) **{8, −8}**

45. Graph $y = 3x + 2$. (Lesson 5–4) **See margin.**

46. Solve $\dfrac{x}{4} + 7 = 6$. (Lesson 3–3) **−4**

WORKING ON THE

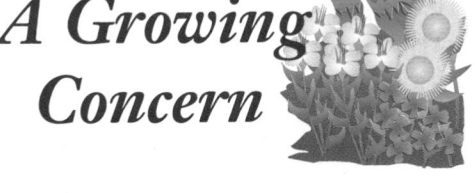

Refer to the Investigation on pages 656–657.

A Growing Concern

1 Use graph paper to make a scale drawing of the house, garage, driveway, boundary lines, and fences. This will serve as your template to which you will add the other details.

2 Develop a detailed plan for the pool and hot tub locations. Indicate the dimensions of both the pool and the hot tub. Explain why you

chose the size, location, and orientation for the pool.

3 Figure the costs for materials, labor, and profit margin for construction of the pool and hot tub. Justify your costs and make a case for building the pool and hot tub as you have designed them.

4 Add the pool and hot tub to the scale drawing, indicating their measurements.

Add the results of your work to your Investigation Folder.

Investigation

Working on the Investigation

The Investigation on pages 656–657 is designed to be a long-term project that is completed over several days or weeks. Encourage students to keep their materials in their Investigation Folder as they work on the Investigation.

---

4 ASSESS

Closing Activity

Writing At the end of class, have students write down the key points to consider in simplifying rational expressions.

Additional Answers

39c.

43.

45.

Enrichment Masters, p. 80

Enrichment

Student Edition
Pages 660–665

Continued Fractions

The following is an example of a continued fraction. By starting at the bottom you can simplify the expression to a rational number.

Example: Express $\dfrac{48}{19}$ as a continued fraction.

Write each continued fraction as a rational number.

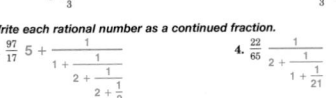

Write each rational number as a continued fraction.

NCTM Standards: 1–6

Objective
Use a graphing calculator to check the simplification of rational expressions.

Recommended Time
15 minutes

Instructional Resources
Graphing Calculator Masters, pp. 36 and 37

These masters provide keystroking instruction for this lesson for the TI-81 and Casio graphing calculators.

1 FOCUS

Motivating the Lesson
Have students complete the table below.

x	−1	0	1	2	3	4
$\dfrac{x^2 - 9}{x - 3}$	2	3	4	5	−	7
$x + 3$	2	3	4	5	6	7

2 TEACH

Teaching Tip Asymptotes are often difficult to detect on the graphing calculator. Reset the range of the viewing window and the x-scale to get a better view of a suspected asymptote.

3 PRACTICE/APPLY

Assignment Guide

Core: 1–9
Enriched: 1–9

4 ASSESS

Observing students working with technology is an excellent method of assessment.

12–1B Graphing Technology
Rational Expressions
An Extension of Lesson 12–1

When simplifying rational expressions, you can use a graphing calculator to support your answer. You can also use a calculator to find the excluded values.

Example ● **Simplify** $\dfrac{3x^2 - 8x + 5}{x^2 - 1}$.

$$\frac{3x^2 - 8x + 5}{x^2 - 1} = \frac{(3x - 5)(x - 1)}{(x + 1)(x - 1)} \qquad \textit{Factor the numerator and denominator.}$$

$$= \frac{3x - 5}{x + 1} \qquad \textit{Divide the common factor } (x - 1).$$

When $x = -1$ or $x = 1$, $x^2 - 1 = 0$. Therefore, x cannot equal -1 or 1.

Now use a graphing calculator. Graph $y = \dfrac{3x^2 - 8x + 5}{x^2 - 1}$ using the window $[-4.7, 4.7]$ by $[-10, 10]$ with scale factors of 1.

Enter: [Y=] [(] 3 [X,T,θ] [x^2] [−]

8 [X,T,θ] [+] 5 [)] [÷]

[(] [X,T,θ] [x^2] [−] 1 [)]

[GRAPH]

Use the TRACE key to observe the values of x and y. Notice that at $x = -1$ and $x = 1$, no value for y appears in the window. There are also "holes" in the graph at those points. These are the excluded values.

Now enter $y = \dfrac{3x - 5}{x + 1}$ as Y2 and observe the graphs.

Enter: [Y=] [(] 3 [X,T,θ] [−] 5 [)]

[÷] [(] [X,T,θ] [+] 1 [)]

[GRAPH]

Notice that the two graphs appear to be identical. This means that the simplified expression is probably correct.

Excluded values:
1. −5
2. −2, −5
3. 0, 4.5
4. 2, −3

EXERCISES

5. 8

6. 0, −16

7. 3

8. −1

9. −6, 5

Simplify each expression. Then verify your answer graphically. Name the excluded values.

1. $\dfrac{x^2 - 25}{x^2 + 10x + 25}$ $\dfrac{x - 5}{x + 5}$

2. $\dfrac{3x + 6}{x^2 + 7x + 10}$ $\dfrac{3}{x + 5}$

3. $\dfrac{2x - 9}{4x^2 - 18x}$ $\dfrac{1}{2x}$

4. $\dfrac{2x^2 - 4x}{x^2 + x - 6}$ $\dfrac{2x}{x + 3}$

5. $\dfrac{x^2 - 9x + 8}{x^2 - 16x + 64}$ $\dfrac{x - 1}{x - 8}$

6. $\dfrac{3x^2}{12x^2 + 192x}$ $\dfrac{x}{4(x + 16)}$

7. $\dfrac{-x^2 + 6x - 9}{x^2 - 6x + 9}$ -1

8. $\dfrac{5x^2 + 10x + 5}{3x^2 + 6x + 3}$ $\dfrac{5}{3}$

9. $\dfrac{25 - x^2}{x^2 + x - 30}$ $\dfrac{x + 5}{x + 6}$

Using Technology
This lesson offers an excellent opportunity for using technology in your algebra classroom. For more information on using technology, see *Graphing Calculators in the Mathematics Classroom*, one of the titles in the Glencoe Mathematics Professional Series.

Multiplying Rational Expressions

Objective

After studying this lesson, you should be able to:

- multiply rational expressions.

Traffic

On the Sunday after Thanksgiving in 1994, westbound traffic was backed up for 13 miles at one exit of the Massachusetts Turnpike. Assume that each vehicle occupied an average of 30 feet of space in a lane and that the turnpike has three lanes. You can use the expression below to estimate the number of vehicles involved in the backup.

$$3 \text{ lanes} \cdot \left(\frac{13 \text{ miles}}{\text{lane}} \right) \left(\frac{5280 \text{ feet}}{\text{mile}} \right) \left(\frac{1 \text{ vehicle}}{30 \text{ feet}} \right)$$

This expression will be simplified in Example 3.

LOOK BACK

You can refer to Lesson 2-6 for information about multiplying rational numbers.

This multiplication is similar to the multiplication of rational expressions. Recall that to multiply rational numbers that are expressed as fractions, you multiply numerators and multiply denominators. You can use this same method to multiply rational expressions. From this point on, you may assume that no denominator of a rational expression has a value of 0.

rational numbers

$$\frac{4}{5} \cdot \frac{3}{7} = \frac{4 \cdot 3}{5 \cdot 7}$$
$$= \frac{12}{35}$$

rational expressions

$$\frac{3}{m} \cdot \frac{k}{4} = \frac{3 \cdot k}{m \cdot 4}$$
$$= \frac{3k}{4m}$$

Example 1 Find $\dfrac{3x^2y}{2rs} \cdot \dfrac{24r^2s}{15xy^2}$.

Method 1: Divide by the common factors after multiplying.

$$\frac{3x^2y}{2rs} \cdot \frac{24r^2s}{15xy^2} = \frac{72r^2sx^2y}{30rsxy^2}$$

$$= \frac{\overset{1}{6rsxy}(12rx)}{\underset{1}{6rsxy}(5y)} \qquad \textit{The GCF is 6rsxy.}$$

$$= \frac{12rx}{5y}$$

Method 2: Divide by the common factors before multiplying.

$$\frac{3x^2y}{2rs} \cdot \frac{24r^2s}{15xy^2} = \frac{\overset{1\,x\,1}{3x^2y}}{\underset{1\,1\,1}{2rs}} \cdot \frac{\overset{12\ r\ 1}{24r^2s}}{\underset{5\ 1\,y}{15xy^2}}$$

$$= \frac{12rx}{5y}$$

Sometimes you must factor a quadratic expression before you can simplify a product of rational expressions.

12-2 LESSON NOTES

NCTM Standards: 1–5

Instructional Resources

- Study Guide Master 12-2
- Practice Master 12-2
- Enrichment Master 12-2
- Assessment and Evaluation Masters, p. 324

Transparency 12-2A contains the 5-Minute Check for this lesson; **Transparency 12-2B** contains a teaching aid for this lesson.

Recommended Pacing	
Standard Pacing	Day 3 of 14
Honors Pacing	Day 3 of 13
Block Scheduling*	Day 2 of 6 (along with Lesson 12-3)
Alg. 1 in Two Years*	Days 4, 5, & 6 of 27

*For more information on pacing and possible lesson plans, refer to the *Block Scheduling Booklet* and *Algebra 1 in Two Years*.

1 FOCUS

5-Minute Check
(over Lesson 12-1)

Simplify each rational expression. State the excluded values of the variable.

1. $\dfrac{(y+4)}{(y-4)(y+4)}$

   $\dfrac{1}{y-4}$; 4, −4

2. $\dfrac{4n^2-8}{4n-4}$

   $\dfrac{n^2-2}{n-1}$; 1

3. $\dfrac{x^2-49}{x^2-2x-35}$

   $\dfrac{x+7}{x+5}$; 7, −5

4. $\dfrac{m(m-1)}{m-1}$

   m; 1

5. $\dfrac{y-3}{y^2-9}$

   $\dfrac{1}{y+3}$; −3, 3

Motivating the Lesson

Questioning Review the multiplication of arithmetic fractions. Have students find the following products expressed in simplest form.

1. $\frac{1}{2} \times \frac{4}{5}$ $\frac{2}{5}$

2. $\frac{2}{7} \times \frac{49}{50}$ $\frac{7}{25}$

3. $\frac{3}{8} \times \frac{20}{12}$ $\frac{5}{8}$

4. $\frac{12}{15} \times \frac{25}{16}$ $\frac{5}{4}$

5. $\frac{10}{18} \times \frac{27}{30}$ $\frac{1}{2}$

2 TEACH

In-Class Examples

For Example 1

Find $\frac{6rst^2}{5s^2t} \cdot \frac{4r^2s}{3st^2} \cdot \frac{8r^3}{5st}$

For Example 2
Find each product.

a. $\frac{x + 6}{2x^2} \cdot \frac{8x^4}{x^2 - 36}$ $\frac{4x^2}{x - 6}$

b. $\frac{x^2 - 9}{3x - 9} \cdot \frac{6x - 12}{x + 3}$

$2x - 4$

For Example 3
A star is measured to be 24 kiloparsecs from Earth. A kiloparsec is 1000 parsecs. A parsec is 3.26 light years. How many light years away is this star? **78,240 light years**

Study Guide Masters, p. 81

Example 2 Find each product.

a. $\frac{m + 3}{5m} \cdot \frac{20m^2}{m^2 + 8m + 15}$

$$\frac{m + 3}{5m} \cdot \frac{20m^2}{m^2 + 8m + 15} = \frac{m + 3}{5m} \cdot \frac{20m^2}{(m + 3)(m + 5)} \quad \textit{Factor the denominator.}$$

$$= \frac{20m^2(m + 3)}{5m(m + 3)(m + 5)} \quad \textit{The GCF is } 5m(m + 3).$$

$$= \frac{4m}{m + 5}$$

b. $\frac{3(x + 5)}{4(2x^2 + 11x + 12)} \cdot (2x + 3)$

$$\frac{3(x + 5)}{4(2x^2 + 11x + 12)} \cdot (2x + 3) = \frac{3(x + 5)(2x + 3)}{4(2x^2 + 11x + 12)}$$

$$= \frac{3(x + 5)(2x + 3)}{4(2x + 3)(x + 4)} \quad \textit{Factor the denominator.}$$

$$= \frac{3(x + 5)(2x + 3)}{4(2x + 3)(x + 4)} \quad \textit{The GCF is } (2x + 3).$$

$$= \frac{3(x + 5)}{4(x + 4)}$$

$$= \frac{3x + 15}{4x + 16} \quad \begin{array}{l}\textit{Simplify the numerator}\\\textit{and the denominator.}\end{array}$$

When you multiply fractions that involve units of measure, you can divide by the units in the same way that you divide by variables. Recall that this process is called *dimensional analysis*.

Example 3 Refer to the application at the beginning of the lesson. About how many vehicles were involved in the backup?

APPLICATION
Traffic

$$3 \text{ lanes} \cdot \left(\frac{13 \text{ miles}}{\text{lane}}\right)\left(\frac{5280 \text{ feet}}{\text{mile}}\right)\left(\frac{1 \text{ vehicle}}{30 \text{ feet}}\right)$$

$$= \frac{3 \text{ lanes}}{1} \cdot \left(\frac{13 \text{ miles}}{\text{lane}}\right)\left(\frac{5280 \text{ feet}}{\text{mile}}\right)\left(\frac{1 \text{ vehicle}}{30 \text{ feet}}\right) \quad \textit{Divide by common units.}$$

$$= 6864 \text{ vehicles}$$

The backup involved about 6900 vehicles.

CHECK FOR UNDERSTANDING

Communicating Mathematics

Study the lesson. Then complete the following.

1. Write two rational expressions whose product is $\frac{xy}{x - 4} \cdot \frac{x}{x - 4}$, y

2. Angie; she divided out the GCF.

2. You Decide The work of two students who simplified $\frac{4x + 8}{4x - 8} \cdot \frac{x^2 + 2x - 8}{2x + 8}$ is shown below. Which one is correct, and why ?

Matt: $\frac{4x + 8}{4x - 8} \cdot \frac{x^2 + 2x - 8}{2x + 8} = x^2$

Angie: $\frac{4(x + 2)}{4(x - 2)} \cdot \frac{(x - 2)(x + 4)}{2(x + 4)} = \frac{x + 2}{2}$

MATH JOURNAL

3. Write a paragraph explaining which is better—to simplify an expression first and then multiply, or to multiply and then simplify. Include examples to support your answers. **See students' work.**

Teaching Tip Stress that each polynomial must be factored and then simplified.

Teaching Tip Caution students to write the negative sign before the fraction bar and not with the numerator. For example,

$\frac{-x + 4}{x + 2}$ is not the same as $-\frac{x + 4}{x + 2}$.

Guided Practice

Find each product. Assume that no denominator has a value of 0.

6. $\frac{r}{y+3}$

4. $\frac{12m^2y}{5r^2} \cdot \frac{r^2}{my} \cdot \frac{12m}{5}$

5. $\frac{16xy^2}{3m^2p} \cdot \frac{27m^3p}{32x^2y^3} \cdot \frac{9m}{2xy}$

6. $\frac{x-5}{r} \cdot \frac{r^2}{(x-5)(y+3)}$

7. $\frac{x+4}{4y} \cdot \frac{16y}{x^2+7x+12} \cdot \frac{4}{x+3}$

8. $\frac{a^2+7a+10}{a+1} \cdot \frac{3a+3}{a+2}$ $3(a+5)$

9. $\frac{4(x-7)}{3(x^2-10x+21)} \cdot (x-3)$ $\frac{4}{3}$

10. $3(x+6) \cdot \frac{x+3}{9(x^2+7x+6)}$ $\frac{x+3}{3(x+1)}$

11. **a.** Multiply $\frac{2.54 \text{ centimeters}}{1 \text{ inch}} \cdot \frac{12 \text{ inches}}{1 \text{ foot}} \cdot \frac{3 \text{ feet}}{1 \text{ yard}}$. Then simplify.   **91.44 cm/yd**

     b. What does the simplified expression represent?

12. **Bicycling** Marisa says that bicycling at the rate of 20 miles per hour is equivalent to a rate of 1760 feet per minute. Is she correct? Write a product involving units of measure to justify your answer.

EXERCISES

Practice

Find each product. Assume that no denominator has a value of 0.

Ⓐ

13. $\frac{7a^2}{5} \cdot \frac{15}{14a} \quad \frac{3a}{2}$

14. $\frac{3m^2}{2m} \cdot \frac{18m^2}{9m} \quad 3m^2$

15. $\frac{10r^3}{6x^3} \cdot \frac{42x^2}{35r^3} \quad \frac{2}{x}$

16. $\frac{7ab^3}{11r^2} \cdot \frac{44r^3}{21a^2b} \quad \frac{4b^2r}{3a}$

17. $\frac{64y^2}{5y} \cdot \frac{5y}{8y} \quad 8y$

18. $\frac{2a^2}{b} \cdot \frac{5bc}{6a} \quad \frac{5ac}{3}$

19. $\frac{m+4}{3m} \cdot \frac{4m^2}{m^2+9m+20} \quad \frac{4m}{3(m+5)}$

20. $\frac{m^2+8m+15}{a+b} \cdot \frac{7a+14b}{m+3}$

21. $\frac{5a+10}{10m^2} \cdot \frac{4m^3}{a^2+11a+18} \quad \frac{2m}{a+9}$

22. $\frac{6r+3}{r+6} \cdot \frac{r^2+9r+18}{2r+1} \quad 3(r+3)$

23. $2(x+1) \cdot \frac{x+4}{x^2+5x+4} \quad 2$

24. $4(a+7) \cdot \frac{12}{3(a^2+8a+7)} \quad \frac{16}{a+1}$

Ⓑ

25. $\frac{x^2-y^2}{12} \cdot \frac{36}{x+y} \quad 3(x-y)$

26. $\frac{3a+9}{a} \cdot \frac{a^2}{a^2-9} \quad \frac{3a}{a-3}$

27. $\frac{9}{3+2x} \cdot (12+8x) \quad 36$

28. $(3x+3) \cdot \frac{x+4}{x^2+5x+4} \quad 3$

29. $\frac{4x}{9x^2-25} \cdot (3x+5) \quad \frac{4x}{3x-5}$

30. $(b^2+12b+11) \cdot \frac{b+9}{b^2+20b+99}$

31. $\frac{4x+8}{x^2-25} \cdot \frac{x-5}{5x+10} \quad \frac{4}{5(x+5)}$

32. $\frac{a^2-a-6}{a^2-9} \cdot \frac{a^2+7a+12}{a^2+4a+4} \quad \frac{a+4}{a+2}$

Multiply. Explain what each expression represents.

Ⓒ

33. $\frac{32 \text{ feet}}{1 \text{ second}} \cdot \frac{60 \text{ seconds}}{1 \text{ minute}} \cdot \frac{60 \text{ minutes}}{1 \text{ hour}} \cdot \frac{1 \text{ mile}}{5280 \text{ feet}}$ **21.8 mph; converts ft/s to mph**

34. $10 \text{ feet} \cdot 18 \text{ feet} \cdot 3 \text{ feet} \cdot \frac{1 \text{ yard}^3}{27 \text{ feet}^3}$ **20 yd³; converts ft³ to yd³**

Simplify each expression.

35. **Heat** 20 grams at 540 Calories per gram   **10,800 Calories**

36. **Metal Refining** 4.025 grams per amp-hour at 2 amps for 5 hours

Critical Thinking

37. Find two different pairs of rational expressions whose product is $\frac{6x^2-6x-36}{x^2+3x-28}$. Sample answer: $\frac{3(x+2)}{x+7} \cdot \frac{2(x-3)}{x-4}$; $\frac{6}{x+7} \cdot \frac{x^2-x-1}{x-4}$

Lesson 12-2 Multiplying Rational Expressions   **669**

**Side notes (left margin):**

11b. changing centimeters to yards

12. yes; $\frac{20 \text{ mi}}{1 \text{ h}} \cdot \frac{5280 \text{ ft}}{1 \text{ mi}} \cdot \frac{1 \text{ h}}{60 \text{ min}}$

20. $\frac{7(a+2b)(m+5)}{a+b}$

30. $b+1$

36. 40.25 grams

Reteaching

Using Models Have students choose values for *a* and *b* and complete the table below. The ratio of opposites simplifies to −1.

a	*b*	*a* − *b*	*b* − *a*	$\frac{a-b}{b-a}$
1	4	−3	3	−1
3	−2	5	−5	−1

Pick *a* = *b* to see excluded values.

---

3 PRACTICE/APPLY

Check for Understanding

Exercises 1–12 are designed to help you assess your students' understanding through reading, writing, speaking, and modeling. You should work through Exercises 1–3 with your students and then monitor their work on Exercises 4–12.

Error Analysis

When all the factors in the numerator or denominator cancel, students may indicate the value of the expression as 0, rather than 1. For example,

$$\frac{4x(x-3)}{8x^2(x-3)} = \frac{0}{2x} \text{ rather than } \frac{1}{2x}.$$

Remind students to use small numbers when canceling to show that $\frac{a}{a} = 1$.

Assignment Guide

Core: 13–39 odd, 40–46
Enriched: 14–36 even, 37–46

For **Extra Practice**, see p. 782.

The red A, B, and C flags, printed only in the Teacher's Wraparound Edition, indicate the level of difficulty of the exercises.

Practice Masters, p. 81

12-2

NAME_____ DATE _____

Student Edition
Pages 667–670

Practice

Multiplying Rational Expressions

Find each product. Assume that no denominator has a value of 0.

1. $\frac{2a^3b}{b^3c} \cdot \frac{6}{a} \cdot \frac{2a}{c}$

2. $\frac{18x^5}{10y^2} \cdot \frac{15y^3}{24x} \quad \frac{9xy}{8}$

3. $\frac{24st^2}{8s^4t} \cdot \frac{12s^3t^4}{36s^2t} \quad \frac{1}{s^2}$

4. $\frac{12a^3b}{4} \cdot \frac{4a+8b}{20a^3b^2} \quad \frac{3a+6b}{5b^2}$

5. $\frac{12m-18}{18n} \cdot \frac{9n^2}{8m-12} \quad \frac{3n}{4}$

6. $\frac{y^2-9}{4} \cdot \frac{8}{y+3} \quad 2y-6$

7. $\frac{(x+2)^2}{8} \cdot \frac{72}{x^2-4} \quad \frac{9x+18}{x-2}$

8. $\frac{2a^2+a}{4a^2-1} \cdot \frac{6a-3}{4a} \quad \frac{3}{4}$

9. $\frac{5n+15}{8n+4} \cdot \frac{4n+2}{3n+9} \quad \frac{5}{6}$

10. $\frac{e^2-1}{2e-6} \cdot \frac{e^2-9}{3e-3} \quad \frac{e^2+4e+3}{6}$

11. $\frac{k^3-4}{8k^3+3k} \cdot \frac{16k+6}{k-2} \quad \frac{2k+4}{k}$

12. $\frac{4r+r^2}{8+2r} \cdot \frac{4}{2r^2} \quad \frac{1}{r^3}$

13. $\frac{25-c^2}{12} \cdot \frac{4}{5-c} \quad \frac{5+c}{3}$

14. $\frac{xy}{3x-3y} \cdot \frac{x^2-xy}{xy} \quad \frac{x}{3}$

15. $\frac{2c^3-5c-3}{c+d} \cdot \frac{c^2-d^2}{2c+1} \quad \frac{c^2-cd-3c+3d}{}$

16. $\frac{t^2+6t+9}{t^2-10t+25} \cdot \frac{t^2-t-20}{t^2+7t+12} \quad \frac{t+3}{t-5}$

Closing Activity

Modeling Use the formula $V = \ell wh$ to represent the volume of the rectangular solid shown. Write the expression in simplest form.

$$\frac{x^2 + 9x + 14}{2x^2 + x}$$

$$\frac{2x+1}{x}$$

$$\frac{x+7}{x-2}$$

$$\frac{x^2-4}{4x^2+4x+1}$$

Chapter 12, Quiz A (Lessons 12-1 and 12-2), is available in the *Assessment and Evaluation Masters*, p. 324.

Additional Answer

38a. $\dfrac{24 \text{ sec}}{1 \text{ car}} \cdot \dfrac{6864 \text{ cars}}{8 \text{ collectors}} \cdot \dfrac{1 \text{ min}}{60 \text{ sec}} \cdot$
$\dfrac{1h}{60 \text{ min}}$

Enrichment Masters, p. 81

NAME _____ **DATE** _____

Enrichment

Student Edition Pages 667–670

Surface Area of Solid Figures

Many solid objects are formed by rectangles and squares. A box is an example.

The dimensions of the box shown at the right are represented by letters. The length of the base is ℓ units, its width is w units, and the height of the box is h units.

Suppose the box is cut on the seams so that it can be spread out on a flattened surface as shown at the right. The area of this figure is the surface area of the box. Find a formula for the surface area of the box.

There are 6 rectangles in the figure. The surface area is the sum of the areas of the 6 rectangles.

$$S = hw + h\ell + \ell w + h\ell + hw + \ell w$$
$$S = 2\ell w + 2h\ell + 2hw$$

Find the surface area of a box with the given dimensions.

1. $\ell = 14$ cm, $w = 8$ cm, $h = 2$ cm
   312 cm²
2. $\ell = 40$ cm, $w = 30$ cm, $h = 25$ cm
   5900 cm²

3. $\ell = x$ cm, $w = (x - 3)$ cm,
   $h = (x + 3)$ cm 6($x^2 - 3$) cm²
4. $\ell = (s + 9)$ cm, $w = (s - 9)$ cm,
   $h = (s + 9)$ cm
   (6s^2 + 36s − 162) cm²

5. The surface area of a box is 142 cm². The length of the base is 2 cm longer than its width. The height of the box is 2 cm less than the width of the base. Find the dimensions of the box.
   $h = x = 3$ cm; $w = x + 2 = 5$ cm; $\ell = x + 4 = 7$ cm

6. Write an expression that represents the surface area of the figure shown at the right. Include the surface area of the base. 22x^2

Applications and Problem Solving

38. Traffic Refer to Example 3. Suppose there are eight toll collectors at the exit and it takes each an average of 24 seconds to collect the toll from one vehicle. **a. See margin.**
  a. Write a product involving units to estimate the time it would take in hours to collect tolls from all the vehicles in the backup.
  b. Simplify the product.

38b. 5.72 hours

39. Money The chart at the right shows a recent foreign exchange rate table. It indicates how much of another country's currency you would receive in exchange for one American dollar at the time these rates were in effect.
  a. Write a rational expression involving units of measure that indicates how many American dollars you would receive for one French franc. Write another expression that indicates how many French francs you would receive for one American dollar.

39a. $\dfrac{1 \text{ franc}}{0.1981 \text{ dollars}}$;
$\dfrac{1 \text{ dollar}}{5.05 \text{ francs}}$

39b. 12,500 pesos ·
$\dfrac{0.1597 \text{ dollars}}{1 \text{ peso}}$.
$\dfrac{1 \text{ franc}}{0.1981 \text{ dollars}}$

  b. Write a product involving units of measure that indicates how many French francs you would receive for 12,500 Mexican pesos. Then find the product. **about 10,077 francs**
  c. Explain how to determine an exchange rate between Canada's dollar and Hong Kong's dollar. **Convert to American dollars and then to Hong Kong's dollar.**

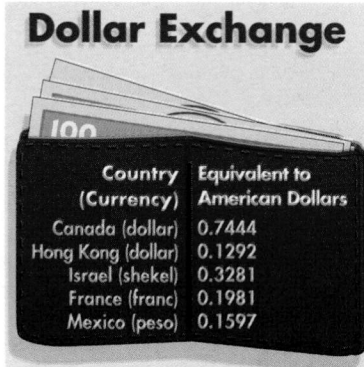

Dollar Exchange

Country (Currency)	Equivalent to American Dollars
Canada (dollar)	0.7444
Hong Kong (dollar)	0.1292
Israel (shekel)	0.3281
France (franc)	0.1981
Mexico (peso)	0.1597

Mixed Review

40. Cooking The formula $t = \dfrac{40(25 + 1.85a)}{50 - 1.85a}$ relates the time t in minutes that it takes to bake an average-size potato in an oven that is at an altitude of a thousands of feet. (Lesson 12–1)
  a. What is the value of a for an altitude of 4500 feet? **4.5**
  b. Calculate the time is takes to bake a potato at an altitude of 3500 feet.
  c. Calculate the time it takes to bake a potato at an altitude of 7000 feet.
  d. The altitude in part c is twice that of part b. How do your baking times compare for those two altitudes?

40b. about 29 minutes

40c. about 41 minutes

40d. The times are not doubled; the difference is 12 minutes.

41. Find the value of y for $y = 3^{2x}$ if $x = 3$. (Lesson 11–4) **729**

42. Solve $5x^2 + 30x + 45 = 0$ by factoring. (Lesson 10–6) **−3**

43. Find the degree of $6x^2yz + 5xyz - x^3$. (Lesson 9–4) **4**

44. Solve the system of equations. (Lesson 8–3)
$5 = 2x - 3y$
$-1 = -4x + 3y$ **(−2, −3)**

45. 0

45. Find the slope of the line that passes through $(-3, 5)$ and $(8, 5)$. (Lesson 5–1)

46. What is 45% of $1567 to the nearest dollar? (Lesson 4–4) **$705**

Tech Prep

Traffic Technician Students who are interested in traffic control may wish to do further research on the data given in Exercise 38 and explore the potential growth of this career. For more information on tech prep, see the *Teacher's Handbook*.

Extension

Connections Latest estimates suggest that the known universe is about 14 billion light years across. One light year equals 5.9×10^{12} miles. The top speed of our present-day spacecrafts is 25,000 mph. With this spacecraft, how long would it take to travel across the universe?
376,000,000,000,000 or 3.76×10^{14} or 376 trillion years

Dividing Rational Expressions

Objective

After studying this lesson, you should be able to:

• divide rational expressions.

APPLICATION
Parades

A crowd watching the Tournament of Roses Parade in Pasadena, California, fills the sidewalks along the route for about 5.5 miles on each side. Suppose these sidewalks are 10 feet wide and each person occupies an average of 4 square feet of space. To estimate the number of people along this part of the route, you can use dimensional analysis and the following expression.

$$\left(5.5 \text{ miles} \cdot \frac{5280 \text{ ft}}{1 \text{ mile}} \cdot 10 \text{ feet}\right) \div \frac{4 \text{ ft}^2}{1 \text{ person}}$$ *This expression will be simplified in Example 4.*

LOOK BACK

You can refer to Lesson 2-7 for information about dividing rational numbers.

This division is similar to the division of rational expressions. Recall that to divide rational numbers that are expressed as fractions you multiply by the reciprocal of the divisor. You can use this same method to divide algebraic rational expressions.

rational numbers

$$\frac{3}{2} \div \frac{2}{5} = \frac{3}{2} \cdot \frac{5}{2}$$ *The reciprocal of $\frac{2}{5}$ is $\frac{5}{2}$.*

$$= \frac{15}{4}$$

rational expressions

$$\frac{a}{b} \div \frac{c}{d} = \frac{a}{b} \cdot \frac{d}{c}$$ *The reciprocal of $\frac{c}{d}$ is $\frac{d}{c}$.*

$$= \frac{ad}{bc}$$

Example ① Find $\dfrac{2a}{a+3} \div \dfrac{a+7}{a+3}$.

$$\frac{2a}{a+3} \div \frac{a+7}{a+3} = \frac{2a}{a+3} \cdot \frac{a+3}{a+7}$$ *The reciprocal of $\frac{a+7}{a+3}$ is $\frac{a+3}{a+7}$.*

$$= \frac{2a}{a+3} \cdot \frac{\overset{1}{\cancel{a+3}}}{a+7}$$ *Divide by the common factor $a+3$.*

$$= \frac{2a}{a+7}$$

Sometimes a quotient of rational expressions involves a divisor that is a binomial.

Example ② Find $\dfrac{2m+6}{m+5} \div (m+3)$.

$$\frac{2m+6}{m+5} \div (m+3) = \frac{2m+6}{m+5} \cdot \frac{1}{m+3}$$ *The reciprocal of $m+3$ is $\frac{1}{m+3}$.*

$$= \frac{2(\cancel{m+3})^{1}}{(m+5)(\cancel{m+3})}$$ *Divide by the common factor $m+3$.*

$$= \frac{2}{m+5}$$

Sometimes you must factor before you can simplify a quotient of rational expressions.

NCTM Standards: 1–5

Instructional Resources

• Study Guide Master 12-3
• Practice Master 12-3
• Enrichment Master 12-3

Transparency 12-3A contains the 5-Minute Check for this lesson; **Transparency 12-3B** contains a teaching aid for this lesson.

Recommended Pacing	
Standard Pacing	Day 4 of 14
Honors Pacing	Day 4 of 13
Block Scheduling*	Day 2 of 6 (along with Lesson 12-2)
Alg. 1 in Two Years*	Days 7, 8, & 9 of 27

*For more information on pacing and possible lesson plans, refer to the *Block Scheduling Booklet* and *Algebra 1 in Two Years*.

1 FOCUS

5-Minute Check
(over Lesson 12-2)

Find each product. Assume that no denominator has a value of 0.

1. $\dfrac{3}{5} \cdot \dfrac{35}{99}$ $\dfrac{7}{33}$

2. $\dfrac{7a}{9a} \cdot \dfrac{3a^2}{6b^2}$ $\dfrac{7a^2}{18b^2}$

3. $\dfrac{a^2b}{b^3c} \cdot \dfrac{ac}{a^3}$ $\dfrac{1}{b^2}$

4. $\dfrac{5x-5}{4} \cdot \dfrac{12}{x-1}$ 15

5. $\dfrac{w-5}{w^2-7w+10} \cdot \dfrac{w-2}{12}$ $\dfrac{1}{12}$

Motivating the Lesson

Questioning Review dividing fractions. Ask students to find each quotient in simplest form.

1. $\dfrac{2}{3} \div \dfrac{8}{9}$ $\dfrac{3}{4}$ 2. $\dfrac{3}{4} \div \dfrac{5}{6}$ $\dfrac{9}{10}$

3. $\dfrac{10}{21} \div \dfrac{25}{42}$ $\dfrac{4}{5}$ 4. $\dfrac{5}{8} \div \dfrac{1}{60}$ $\dfrac{75}{2}$

In-Class Examples

For Example 1

Find $\dfrac{3x}{x+2} \div \dfrac{3x^2 - 3x}{x+2}$.

$\dfrac{1}{x-1}$

For Example 2

Find $\dfrac{3a}{a+2} \div (a-2)$.

$\dfrac{3a}{a^2 - 4}$

For Example 3

Find $\dfrac{y+5}{y^2 - 4y - 5} \div \dfrac{y^2 + 4y - 5}{y+1}$.

$\dfrac{1}{y^2 - 6y + 5}$

For Example 4

If the sidewalks are 18 feet wide and each person occupies an average of 3 square feet of space, estimate the number of people along this part of the parade route. **174,240 people**

Teaching Tip For Example 4, you may want to stress the idea of reciprocals and their importance in the division of rational expressions.

Study Guide Masters, p. 82

NAME_____ DATE _____

Study Guide

Student Edition
Pages 671–674

Dividing Rational Expressions

You should recall that two numbers whose product is 1 are called **multiplicative inverses** or **reciprocals**. To find the quotient of two rational expressions, you multiply the first rational expression by the reciprocal of the second rational expression.

> **Dividing Rational Expressions**
>
> For all rational numbers $\frac{a}{b}$ and $\frac{c}{d}$, where $b \neq 0$, $c \neq 0$ and $d \neq 0$,
> $\frac{a}{b} \div \frac{c}{d} = \frac{a}{b} \cdot \frac{d}{c}$.

Example 1: Find $\frac{2}{x} \div \frac{z}{y}$.

$\frac{2}{x} \div \frac{z}{y} = \frac{2}{x} \cdot \frac{y}{z}$

$= \frac{2y}{xz}$

Example 2: Find $\frac{x^2 + 6x - 27}{x^2 + 11x + 18} \div \frac{x-3}{x^2 + x - 2}$.

$\frac{x^2 + 6x - 27}{x^2 + 11x + 18} \div \frac{x-3}{x^2 + x - 2} = \frac{x^2 + 6x - 27}{x^2 + 11x + 18} \cdot \frac{x^2 + x - 2}{x-3}$

$= \frac{(x+9)(x-3)}{(x+9)(x+2)} \cdot \frac{(x+2)(x-1)}{x-3}$

$= x - 1$

Find each quotient. Assume that no denominator has a value of 0.

1. $\frac{5}{6} \div \frac{1}{6}$ 5
2. $1 \div \frac{1}{3}$ 3
3. $\frac{n}{4} \div \frac{m}{4}$ $\frac{n}{m}$
4. $\frac{6}{11} \div \frac{12}{7x}$ $\frac{7x}{22}$
5. $\frac{x}{y} \div \frac{2}{y}$ $\frac{x}{2}$
6. $\frac{a+b}{a-b} \div (a^2 + 2ab + b^2)$ $\frac{1}{a^2 - b^2}$
7. $\frac{3x^2y}{8} \div 6xy$ $\frac{x}{16}$
8. $\frac{y^2 - 36}{y^2 - 49} \div \frac{y+6}{y+7}$ $\frac{y-6}{y-7}$
9. $\frac{x^2 - 5x + 6}{5} \div \frac{x-3}{15}$ $3(x-2)$
10. $\frac{49 - m^2}{m} \div \frac{m^2 - 13m + 42}{3m^2}$ $\frac{-3m(m+7)}{m-6}$, or $\frac{-3m^2 - 21m}{m-6}$
11. $\frac{3a^2 - 27a}{2a^2 + 13a - 7} \div \frac{9a^3}{4a^2 - 1}$ $\frac{(a^2 - 9)(2a+1)}{3a(a+7)}$, or $\frac{2a^2 + a - 18a - 9}{3a^2 + 21a}$
12. $\frac{p^2 - 5p + 6}{p^2 + 3p} \div \frac{3-p}{4p+12}$ $\frac{4(2-p)}{p}$, or $\frac{8 - 4p}{p}$

Example ③ Find $\dfrac{x}{x+2} \div \dfrac{x^2}{x^2 + 5x + 6}$.

$\dfrac{x}{x+2} \div \dfrac{x^2}{x^2 + 5x + 6} = \dfrac{x}{x+2} \cdot \dfrac{x^2 + 5x + 6}{x^2}$

$= \dfrac{x}{x+2} \cdot \dfrac{(x+2)(x+3)}{x^2}$ *Divide by common factors.*

$= \dfrac{x+3}{x}$

Example ④

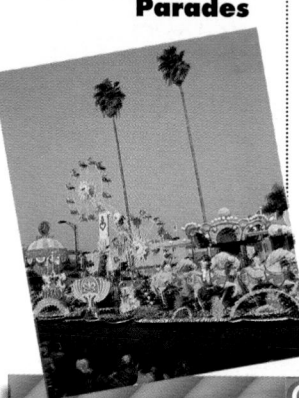

APPLICATION

Parades

Refer to the application at the beginning of the lesson. Estimate the number of people along the part of the parade route that is described.

$\left(5.5 \text{ miles} \cdot \dfrac{5280 \text{ ft}}{1 \text{ mile}} \cdot 10 \text{ feet} \right) \div \dfrac{4 \text{ ft}^2}{1 \text{ person}}$ *The reciprocal of*

$= \left(5.5 \text{ miles} \cdot \dfrac{5280 \text{ ft}}{1 \text{ mile}} \cdot 10 \text{ feet} \right) \cdot \dfrac{1 \text{ person}}{4 \text{ ft}^2}$ $\dfrac{4 \text{ ft}^2}{1 \text{ person}}$ *is* $\dfrac{1 \text{ person}}{4 \text{ ft}^2}$.

$= \dfrac{5.5 \text{ miles}}{1} \cdot \dfrac{\overset{1320}{5280 \text{ ft}}}{1 \text{ mile}} \cdot \dfrac{10 \text{ ft}}{1} \cdot \dfrac{1 \text{ person}}{4 \text{ ft}^2}$ *Divide by common units.*

$= 72{,}600 \text{ people}$

This means that there were about 72,600 people along *each side*. Since there were two sides, the total number of people along the route was about 2(72,600), or 145,200.

CHECK FOR UNDERSTANDING

Communicating Mathematics

Study the lesson. Then complete the following.

1. **Analyze** the solution shown below.

$\dfrac{a^2 - 9}{3a} \div \dfrac{a+3}{a-3} = \dfrac{3a}{a^2 - 9} \cdot \dfrac{a+3}{a-3}$

$= \dfrac{3a}{(a+3)(a-3)} \cdot \dfrac{(a+3)}{a-3}$

$= \dfrac{3a}{(a-3)^2}$

What error was made? The wrong reciprocal was used.

2. **Write** two rational expressions whose quotient is $\dfrac{10r}{d^2}$. $\dfrac{10}{d}, \dfrac{d}{r}$

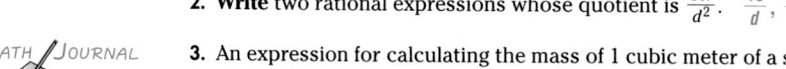

M_{ATH} J_{OURNAL}

3. An expression for calculating the mass of 1 cubic meter of a substance is given below. Write a procedure for simplifying the expression.

$\dfrac{5.96 \text{ grams}}{\text{centimeter}^3} \cdot \dfrac{1 \text{ kilogram}}{1000 \text{ grams}} \cdot \dfrac{100^3 \text{ centimeters}^3}{1 \text{ meter}^3} \cdot 1 \text{ meter}^3$ See students' work.

Guided Practice

Find the reciprocal of each expression.

4. $\dfrac{m^2}{3}$ $\dfrac{3}{m^2}$
5. $\dfrac{x}{5}$ $\dfrac{5}{x}$
6. $\dfrac{-9}{4y}$ $\dfrac{4y}{-9}$
7. $\dfrac{x^2 - 9}{y+3}$ $\dfrac{y+3}{x^2 - 9}$
8. $m - 3$ $\dfrac{1}{m-3}$
9. $x^2 + 2x + 5$ $\dfrac{1}{x^2 + 2x + 5}$

Reteaching

Logical Thinking Have each student create a division problem that has binomial factors that cancel. Take four arbitrary binomials (*A*, *B*, *C*, and *D*). The division problem is $\dfrac{E}{F} \div \dfrac{A}{C}$ where $AB = E$ and $CD = F$. The quotient is $\dfrac{B}{D}$. Have students exchange problems with a neighbor to check.

Find each quotient. Assume that no denominator has a value of 0.

10. $\frac{x}{x+7} \div \frac{x-5}{x+7}$ $\frac{x}{x-5}$

11. $\frac{m^2+3m+2}{4} \div \frac{m+1}{m+2}$ $\frac{(m+2)^2}{4}$

12. $\frac{5a+10}{a+5} \div (a+2)$ $\frac{5}{a+5}$

13. $\frac{x^2+7x+12}{x+6} \div (x+3)$ $\frac{x+4}{x+6}$

Simplify each dimensional expression. State what you think the expression represents.

14. $(8 \text{ feet} \cdot 3 \text{ feet} \cdot 12 \text{ feet}) \div \frac{27 \text{ feet}^3}{1 \text{ yard}^3}$ **10.7 cubic yards**

15. $(12 \text{ inches} \cdot 18 \text{ inches} \cdot 4 \text{ inches}) \div \frac{1728 \text{ inches}^3}{1 \text{ feet}^3}$ **0.5 cubic feet**

16. **Railroading** Jaheem is a railroad buff who likes to watch trains. At the railroad crossing near his house in Menominee Falls, Wisconsin, a Chicago & North Western freight train passes by at 40 miles per hour. Suppose each railroad car is 48 feet long.

16a. $\frac{40 \text{ mi}}{1 \text{ h}} \cdot \frac{5280 \text{ ft}}{1 \text{ mi}} \cdot$ $\frac{1 \text{ h}}{60 \text{ min}} \cdot \frac{1 \text{ car}}{48 \text{ ft}}$

a. Write an expression involving units that represents the number of cars that pass by him in one minute.

b. Find how many railroad cars would pass by him per minute. $73\frac{1}{3}$ **cars**

EXERCISES

Practice

Find each quotient. Assume that no denominator has a value of 0.

17. $\frac{a}{a+3} \div \frac{a+11}{a+3}$ $\frac{a}{a+11}$

18. $\frac{m+7}{m} \div \frac{m+7}{m+3}$ $\frac{m+3}{m}$

19. $\frac{a^2b^3c}{m^2y^2} \div \frac{a^2bc^3}{m^3y^2}$ $\frac{b^2m}{c^2}$

20. $\frac{5x^2}{7} \div \frac{10x^3}{21}$ $\frac{3}{2x}$

B

21. $\frac{3m+15}{m+4} \div \frac{3m}{m+4}$ $\frac{m+5}{m}$

22. $\frac{3x}{x+2} \div (x-1)$ $\frac{3x}{(x+2)(x-1)}$

23. $\frac{4z+8}{z+3} \div (z+2)$ $\frac{4}{z+3}$

24. $\frac{x+3}{x+1} \div (x^2+5x+6)$ $\frac{1}{(x+2)(x+1)}$

25. $\frac{2x+4}{x^2+11x+18} \div \frac{x+1}{x^2+14x+45}$

26. $\frac{k+3}{m^2+4m+4} \div \frac{2k+6}{m+2}$ $\frac{1}{2(m+2)}$

25. $\frac{2(x+5)}{x+1}$

27. What is the quotient when $\frac{2x+6}{x+5}$ is divided by $\frac{2}{x+5}$? $x+3$

28. Find the quotient when $\frac{m-8}{m+7}$ is divided by m^2-7m-8. $\frac{1}{(m+7)(m+1)}$

Find each quotient. Assume that no denominator has a value of 0.

29. $\frac{x^2+5x+6}{x^2-x-12} \div \frac{x+2}{x^2+x-20}$ $x+5$

30. $\frac{m^2+m-6}{m^2+8m+15} \div \frac{m^2-m-2}{m^2+9m+20}$ $\frac{m+4}{m+1}$

31. $\frac{2x^2+7x-15}{x+2} \div \frac{2x-3}{x^2+5x+6}$ $(x+5)(x+3)$

32. $\frac{t^2-2t-8}{w-3} \div \frac{t-4}{w^2-7w+12}$ $(t+2)(w-4)$

Simplify each expression. State what you think the expression represents.

33. $\left(\frac{60 \text{ miles}}{1 \text{ hour}} \cdot \frac{5280 \text{ feet}}{1 \text{ mile}} \div \frac{60 \text{ minutes}}{1 \text{ hour}} \right) \div \frac{60 \text{ seconds}}{1 \text{ minute}}$ **88 ft/s**

34. $\frac{23.75 \text{ inches}}{1 \text{ revolution}} \cdot \frac{33\frac{1}{3} \text{ revolutions}}{1 \text{ minute}} \cdot 16.5 \text{ minutes}$ **13,062.5 in.**

35. $(5 \text{ feet} \cdot 16.5 \text{ feet} \cdot 9 \text{ feet}) \div \frac{27 \text{ feet}^3}{1 \text{ yard}^3}$ **27.5 yd³**

36. $\left[\left(\frac{60 \text{ kilometers}}{1 \text{ hour}} \cdot \frac{1000 \text{ meters}}{1 \text{ kilometer}} \right) \div \frac{60 \text{ minutes}}{1 \text{ hour}} \right] \div \frac{60 \text{ seconds}}{1 \text{ minute}}$ **16.7 meters/s**

C

Sample answers:
33. changes 60 mph to ft/s
34. calculates how far a needle travels on a record for a song 16.5 minutes long
35. changes cubic feet to cubic yards
36. changes 60 km/h to m/s

Lesson 12–3 Dividing Rational Expressions **673**

---

3 PRACTICE/APPLY

Check for Understanding

Exercises 1–16 are designed to help you assess your students' understanding through reading, writing, speaking, and modeling. You should work through Exercises 1–3 with your students and then monitor their work on Exercises 4–16.

Assignment Guide
Core: 17–39 odd, 41–47
Enriched: 18–36 even, 37–47

For **Extra Practice**, see p. 783.

The red A, B, and C flags, printed only in the Teacher's Wraparound Edition, indicate the level of difficulty of the exercises.

Practice Masters, p. 82

Closing Activity

Writing Have students write key points to consider when dividing rational expressions.

Phonograph records were played at a speed of $33\frac{1}{3}$ RPMs, or revolutions per minute. This means that a typical 3-minute song required 100 revolutions.

Additional Answers

38a. $\dfrac{5 \text{ ft } (18 \text{ ft } + 15 \text{ ft})}{2} \cdot 9 \text{ ft} \div \dfrac{27 \text{ ft}^3}{1 \text{ yd}^3}$

40a. $\dfrac{2\pi\left(3\frac{3}{4}\text{ in.}\right)}{1 \text{ revolution}} \cdot \dfrac{33\frac{1}{3}\text{ revolutions}}{1 \text{ minute}} \cdot$

$\dfrac{16.5 \text{ minutes}}{1}$

Enrichment Masters, p. 82

Critical Thinking

Applications and Problem Solving

37. Geometry The area of a rectangle is $\dfrac{x^2 - y^2}{2}$, and its length is $2x + 2y$. Find the width. $\dfrac{x-y}{4}$

38. Construction A construction supervisor needs to determine how many truckloads of earth must be removed from a site before a foundation can be poured. The bed of the truck has the shape shown at the right.

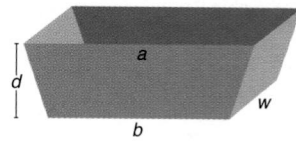

a. Use the formula $V = \dfrac{d(a+b)}{2} \cdot w$ to write an expression involving units that represents the volume of the truck bed in cubic yards if $a = 18$ ft, $b = 15$ ft, $w = 9$ ft, and $d = 5$ ft. **See margin. b. 727.27 truckloads**

b. There are 20,000 cubic yards of earth that must be removed from the excavation site. Write an expression involving units that represents the number of truckloads that will be required to remove all the earth.

Phonograph records are played on a turntable that rotates at a constant speed. Each side of the record has a single groove, or *track*. As the record turns, the stylus of a *tone arm* vibrates in response to variations in the track. The vibration is converted to an electrical signal, which in turn is converted into sound.

39. Railroads The table at the right shows data about railroads in the United States. **a. 5500 miles**

a. If one freight car is 48 feet long on average, how many miles long would a train without a locomotive be if all the freight cars for 1992 were connected?

b. How many trains whose length is the answer to part a would it take to occupy all of the track in 1992? **41.3 trains**

c. Repeat parts a and b for the year 1980. **10,618.2 miles; 27.3 trains**

Year	Miles of Track	Number of Freight Cars
1980	290,000	1,168,000
1985	257,000	867,000
1990	239,000	659,000
1992	227,000	605,000

40. Music Many old phonograph records turn at a rate of $33\frac{1}{3}$ revolutions per minute. Suppose a record of this type plays for 16.5 minutes and the average of the radii of the grooves on the record is $3\frac{3}{4}$ inches. **a. See margin.**

a. Write an expression involving units that represents how many inches the needle travels while playing the record.

b. Find the distance the needle travels. **about 1079.92 ft**

Mixed Review

41. Find $\dfrac{2m+3}{4} \cdot \dfrac{32}{(2m+3)(m-5)}$. (Lesson 12–2) $\dfrac{8}{m-5}$

42. Solve $x^2 + 6x + 8 = 0$ by graphing. (Lesson 11–2) $-4, -2$

43. Factor $16a^2 - 24ab^2 + 9b^4$. (Lesson 10–5) $(4a - 3b^2)^2$

44. Health If your heart beats once every second and you live to be 78 years old, your heart will have beat about 2,460,000,000 times. Write this number in scientific notation. (Lesson 9–3) 2.46×10^9

45. Solve $-3x + 6 > 12$. (Lesson 7–3) $x < -2$

46. Find $-3 + 4 - 10$. (Lesson 2–3) -9

47. Use the numbers 7 and 2 to write a mathematical sentence illustrating the commutative property of addition. (Lesson 1–8) $7 + 2 = 2 + 7$

674 Chapter 12 *Exploring Rational Expressions and Equations*

Extension

Problem Solving Perform the indicated operations and simplify

$$\dfrac{2y}{y^2 - 4} \div \dfrac{3}{y^2 - 4y + 4} \cdot \dfrac{y^2 + 4y + 4}{y^2 - y - 2}.$$

$$\dfrac{2y^2 + 4y}{3y + 3}$$

Dividing Polynomials

Objectives

After studying this lesson, you should be able to:

- divide polynomials by monomials, and
- divide polynomials by binomials.

APPLICATION
Interior Design

Tomi wants to put a decorative border waist-high around his dining room. The perimeter of the room is 52 feet, and the widths of the two windows and two doorways total $12\frac{3}{4}$ feet. To find the number of yards of border needed, he can use the expression $\frac{52 \text{ ft} - 12\frac{3}{4} \text{ ft}}{3 \text{ ft/yd}}$. The expression $\frac{52 \text{ ft}}{3 \text{ ft/yd}} - \frac{12\frac{3}{4} \text{ ft}}{3 \text{ ft/yd}}$ could also be used. In each case, each term of the numerator was divided by the denominator. *This expression will be simplified in Example 4.*

To divide a polynomial by a monomial, you divide each term of the polynomial by the monomial.

Example 1 Find each quotient.

a. $(3r^2 - 5) \div 12r$

$(3r^2 - 5) \div 12r = \frac{3r^2 - 5}{12r}$ *Write as a rational expression.*

$= \frac{3r^2}{12r} - \frac{5}{12r}$ *Divide each term by 12r.*

$= \frac{\overset{1}{\cancel{3r^2}}^{r}}{\underset{4}{\cancel{12r}}} - \frac{5}{12r}$ *Divide by the common factors.*

$= \frac{r}{4} - \frac{5}{12r}$

b. $(9n^2 - 15n + 24) \div 3n$

$(9n^2 - 15n + 24) \div 3n = \frac{9n^2 - 15n + 24}{3n}$

$= \frac{9n^2}{3n} - \frac{15n}{3n} + \frac{24}{3n}$

$= \frac{\overset{3}{\cancel{9n^2}}^{n}}{\underset{1}{\cancel{3n}}} - \frac{\overset{5}{\cancel{15n}}}{\underset{1}{\cancel{3n}}} + \frac{\overset{8}{\cancel{24}}}{\underset{1}{\cancel{3n}}}$ *Divide by the common factors.*

$= 3n - 5 + \frac{8}{n}$

Recall from Lesson 12–3 that, when you can factor, some divisions can be performed easily, as shown below.

$(a^2 + 7a + 12) \div (a + 3) = \frac{a^2 + 7a + 12}{(a + 3)}$

$= \frac{\overset{1}{\cancel{(a + 3)}}(a + 4)}{\underset{1}{\cancel{(a + 3)}}}$ *Factor the dividend.*

$= a + 4$

Lesson 12–4 Dividing Polynomials **675**

Instructional Resources

- Study Guide Master 12-4
- Practice Master 12-4
- Enrichment Master 12-4
- Assessment and Evaluation Masters, pp. 323–324

Transparency 12-4A contains the 5-Minute Check for this lesson; **Transparency 12-4B** contains a teaching aid for this lesson.

Recommended Pacing	
Standard Pacing	Days 5 & 6 of 14
Honors Pacing	Day 5 of 13
Block Scheduling*	Day 3 of 6
Alg. 1 in Two Years*	Days 10, 11, & 12 of 27

*For more information on pacing and possible lesson plans, refer to the *Block Scheduling Booklet* and *Algebra 1 in Two Years.*

1 FOCUS

5-Minute Check
(over Lesson 12-3)

Find each quotient. Assume that no denominator has a value of 0.

1. $\frac{5}{8} \div \frac{3}{4}$ $\frac{5}{6}$

2. $\frac{36}{64} \div \frac{27}{32}$ $\frac{2}{3}$

3. $\frac{5a}{6b} \div \frac{10d}{6c}$ $\frac{ac}{2bd}$

4. $\frac{a^2}{b} \div \frac{a^2}{b^2}$ b

5. $\frac{x^2 - 1}{x^2 + x - 2} \div \frac{x + 1}{x + 2}$ 1

Motivating the Lesson

Questioning Give each pair of students a worksheet containing several long-division problems to solve. Ask the first student to solve problem 1 and the other student to solve problem 2. Students should verify each other's answers and follow the same procedure to solve the remaining problems.

2 TEACH

In-Class Example

For Example 1
Find each quotient.

a. $(2x^2 - 8) \div 4x$

$\dfrac{x}{2} - \dfrac{2}{x}$

b. $(10x^2 - 8x + 20) \div 4x$

$\dfrac{5x}{2} - 2 + \dfrac{5}{x}$

 MODELING MATHEMATICS This activity applies what students know about algebra tiles to dividing polynomials. You may wish to compare it to the process of finding factors of polynomials.

 MODELING MATHEMATICS

Dividing Polynomials

Materials: algebra tiles ☐ product mat

Use algebra tiles to find $(x^2 + 2x - 8) \div (x + 4)$.

a. Model the polynomial $x^2 + 2x - 8$.

$$x^2 + 2x + (-8)$$

b. Place the x^2-tile at the corner of the mat. Arrange four of the 1-tiles as shown at the right, to make a length of $x + 4$.

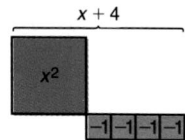

e. It doesn't work; there must be a remainder.

c. Use the remaining tiles to make a rectangular array. Recall that you can add zero-pairs without changing the value of the polynomial.

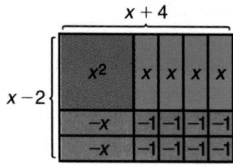

The width of the array, $x - 2$, is the quotient.

Your Turn
Use algebra tiles to find each quotient.
a. $(x^2 + 2x - 8) \div (x + 4)$ $x - 2$
b. $(x^2 + 3x - 10) \div (x + 5)$ $x - 2$
c. $(x^2 - 6x + 9) \div (x - 3)$ $x - 3$
d. $(x^2 - 9) \div (x + 3)$ $x - 3$
e. What happens when you try to model $(x^2 + 5x + 9) \div (x + 3)$? What do you think your result means?

When you cannot factor, you can use a long division process similar to the one you used in arithmetic. The division $(x^2 + 7x + 12) \div (x + 2)$ is shown below.

Step 1: To find the first term of the quotient, divide the first term of the dividend, x^2, by the first term of the divisor, x.

$$
\begin{array}{r}
x \\
x + 2 \overline{)x^2 + 7x + 12} \\
\underline{(-)\ x^2 + 2x} \\
5x
\end{array}
$$

$x^2 \div x = x$
Multiply x and x + 2.
Subtract.

Step 2: To find the next term of the quotient, divide the first term of the partial dividend, $5x$, by the first term of the divisor, x.

$$
\begin{array}{r}
x + 5 \\
x + 2 \overline{)x^2 + 7x + 12} \\
\underline{(-)\ x^2 + 2x} \\
5x + 12 \\
\underline{(-)\ 5x + 10} \\
2
\end{array}
$$

$5x \div x = 5.$
Bring down the 12.
Multiply 5 and x + 2.
Subtract.

Therefore, the quotient when $x^2 + 7x + 12$ is divided by $x + 2$ is $x + 5$ with a remainder of 2. Since there is a nonzero remainder, the divisor is not a factor of the dividend. The remainder can be expressed as shown below.

$$\underbrace{(x^2 + 7x + 12)}_{dividend} \div \underbrace{(x + 2)}_{divisor} = \underbrace{x + 5}_{quotient} + \frac{2}{x + 2}$$

Example Find $(3k^2 - 7k - 6) \div (3k + 2)$.

Method 1: Long Division

$$
\begin{array}{r}
k - 3 \\
3k + 2 \overline{)3k^2 - 7k - 6} \\
(-)\ \underline{3k^2 + 2k} \\
-9k - 6 \\
(-)\ \underline{-9k - 6} \\
0
\end{array}
$$

The quotient is $k - 3$ with a remainder of 0.

Method 2: Factoring

$$\frac{3k^2 - 7k - 6}{3k + 2} = \frac{\overset{1}{\cancel{(3k + 2)}}(k - 3)}{\cancel{3k + 2}}$$

$$= k - 3$$

When the dividend is an expression like $s^3 + 9$, there is no s^2 term or s term. In such situations, you must rename the dividend using 0 as the coefficient of the missing terms.

Example Find $(s^3 + 9) \div (s - 3)$.

$$
\begin{array}{r}
s^2 + 3s + 9 \\
s - 3 \overline{)s^3 + 0s^2 + 0s + 9} \quad s^3 + 9 = s^3 + 0s^2 + 0s + 9 \\
(-)\ \underline{s^3 - 3s^2} \\
3s^2 + 0s \\
(-)\ \underline{3s^2 - 9s} \\
9s + 9 \\
(-)\ \underline{9s - 27} \\
36
\end{array}
$$

Therefore, $(s^3 + 9) \div (s - 3) = s^2 + 3s + 9 + \dfrac{36}{s - 3}$.

Example **4**

APPLICATION
Interior Design

Refer to the application at the beginning of the lesson. If the border comes in 5-yard rolls, how many rolls of border should Tomi buy?

First find the total number of yards needed.

$$\frac{52\ \text{ft} - 12\frac{3}{4}\ \text{ft}}{3\ \text{ft/yd}} = \frac{52\ \text{ft}}{3\ \text{ft/yd}} - \frac{12\frac{3}{4}\ \text{ft}}{3\ \text{ft/yd}}$$

$$= \frac{52\ \text{ft}}{1} \cdot \frac{1\ \text{yd}}{3\ \text{ft}} - \frac{\overset{17}{\cancel{51}}}{4}\ \text{ft} \cdot \frac{1\ \text{yd}}{\underset{1}{\cancel{3}}\ \text{ft}}$$

$$= 17\frac{1}{3}\ \text{yd} - 4\frac{1}{4}\ \text{yd or } 13\frac{1}{12}\ \text{yd}$$

Two rolls are 10 yards, which is not enough. So, Tomi should buy 3 rolls of border.

Teaching Tip In Example 2, while presenting the technique of long division of polynomials, it may be helpful to present a parallel example of whole numbers to show the similarity of steps.

Teaching Tip In Example 2, you may want students to keep like terms in vertical columns as shown below.

$$
\begin{array}{r}
x + 3 \\
x + 5 \overline{)x^2 + 8x + 15} \\
\underline{x^2 + 5x} \\
3x + 15 \\
\underline{3x + 15} \\
0
\end{array}
$$

In-Class Examples

For Example 2
Find $(x^2 + 12x + 36) \div (x + 6)$.
$x + 6$

For Example 3
Find $(m^3 + 2m - 7) \div (m + 3)$.

$m^2 - 3m + 11 - \dfrac{40}{m + 3}$

For Example 4
A bulletin board is 3 yards long and 4 feet tall. One yard of fabric is 45 inches wide. How many yards of fabric are needed to cover the board?
3.2 yards

3 PRACTICE/APPLY

Check for Understanding

Exercises 1–11 are designed to help you assess your students' understanding through reading, writing, speaking, and modeling. You should work through Exercises 1–4 with your students and then monitor their work on Exercises 5–11.

Error Analysis

Students sometimes make errors when subtracting binomials by subtracting only the first term. Find the error below.

$$
\begin{array}{r}
4x \\
2x - 3\overline{)8x^2 + 14x - 39} \\
\underline{(-)8x^2 - 12x} \\
2x - 39
\end{array}
$$

The difference should be $26x - 39$, not $2x - 39$.

Study Guide Masters, p. 83

12-4	NAME_____ DATE_____

Study Guide

Student Edition
Pages 675–680

Dividing Polynomials

To divide a polynomial by a polynomial, you can use a long division process similar to that used for numbers.

Example: Find $(x^2 + 7x + 12) \div (x + 3)$.

Step 1 To find the first term of the quotient, divide the first term of the dividend (x^2) by the first term of the divisor (x).

$$
\begin{array}{r}
x \\
x + 3\overline{)x^2 + 7x + 12} \\
\underline{x^2 + 3x} \\
4x
\end{array}
$$

Step 2 To find the next term of the quotient, divide the first term of the partial dividend ($4x + 12$) by the first term of the divisor (x).

$$
\begin{array}{r}
x + 4 \\
x + 3\overline{)x^2 + 7x + 12} \\
\underline{x^2 + 3x} \\
4x + 12 \\
\underline{4x + 12} \\
0
\end{array}
$$

Therefore, $x^2 + 7x + 12$ divided by $x + 3$ is $x + 4$. This means that $(x + 3)(x + 4) = x^2 + 7x + 12$.

If the divisor is not a factor of the dividend, there will be a nonzero remainder. When there is a nonzero remainder, the quotient can be expressed as follows:

$$\text{quotient} = \text{partial quotient} + \frac{\text{remainder}}{\text{divisor}}$$

Find each quotient.

1. $(b^2 - 5b + 6) \div (b - 2)$
   $b - 3$
2. $(x^2 - x - 6) \div (x - 3)$
   $x + 2$
3. $(3x^3 + 15x^2 - 21x) \div 3x$
   $x^2 + 5x - 7$

4. $\frac{-6x - 9 + 8x^2}{2x - 3}$
   $4x + 3$
5. $\frac{x^3 + 1}{x - 2}$
   $x^2 + 2x + 4 + \frac{9}{x - 2}$
6. $\frac{6x^2 + 7x + 5}{2x + 5}$
   $3x - 4 + \frac{25}{2x + 5}$

7. $\frac{6x^3 + 11x^2 + 4x + 35}{2x + 5}$
   $3x^2 - 2x + 7$
8. $\frac{8x^3 + 27}{2x + 3}$
   $4x^2 - 6x + 9$
9. $\frac{15x^2 + 13x + 5}{5x + 1}$
   $3x + 2 + \frac{3}{5x + 1}$

You can use a graphing calculator to compare a rational expression with the quotient that results when you divide its numerator by its denominator.

Consider the following rational expression and its quotient.

rational expression   quotient
$$\frac{2x}{x + 5} = 2 - \frac{10}{x + 5}$$

Use a graphing calculator to graph
$y = \frac{2x}{x + 5}$ and $y = 2$ on the same screen.

Use the viewing window $[-15, 5]$ by $[-10, 10]$.

Enter: [Y=] 2 [X,T,θ] [÷] [(]
[X,T,θ] [+] 5 [)] [ENTER] 2 [GRAPH]

Your Turn b. As x increases, the value approaches 2.

a. What can you say about the two graphs from examining the calculator screen?   Sample answer: They approach each other but never intersect.

b. Change Xmin and Xmax to -47 and 47. What can you say about the value of $\frac{2x}{x + 5}$ and 2 as x gets larger and larger?

c. Use long division to find the quotient for $\frac{3x}{x - 5}$. $3 + \frac{15}{x - 5}$

d. What conclusion can you draw about the graph of $y = \frac{3x}{x - 5}$ and the graph of $y = 3$?   It approaches 3.

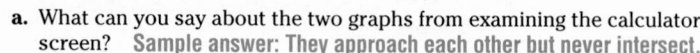

CHECK FOR UNDERSTANDING

Communicating Mathematics

Study the lesson. Then complete the following.

1. Copy the division below and label the dividend, divisor, quotient, and remainder.   See margin.

$$\frac{2x^2 - 11x - 20}{2x + 3} = x - 7 + \frac{1}{2x + 3}$$

2. It means the divisor goes evenly into the dividend or it is a factor of the dividend.

2. **Explain** the meaning of a remainder of 0 in a long division of a polynomial by a binomial.

3. Refer to the application at the beginning of the lesson. Suppose Tomi decided to place the border at the bottom of the room above the baseboard, but below the windows. How many rolls of border should he buy?   3 rolls

MODELING MATHEMATICS

4. a. Use algebra tiles to model $2x^2 - 9x + 9$. $(2x - 3)(x - 3)$
   b. Which of the following divisors of $2x^2 - 9x + 9$ results in a remainder of 0?
      $x + 3$ $x - 3$ $2x - 3$ $2x + 3$ $x - 3; 2x - 3$

Guided Practice

Find each quotient.

5. $(9b^2 - 15) \div 3$ $3b^2 - 5$
6. $(a^2 + 5a + 13) \div 5a$ $\frac{a}{5} + 1 + \frac{13}{5a}$
7. $(t^2 + 6t - 7) \div (t + 7)$ $t - 1$
8. $(s^2 + 11s + 18) \div (s + 2)$ $s + 9$
9. $\frac{2m^2 + 7m + 3}{m + 2}$ $2m + 3 + \frac{-3}{m + 2}$
10. $\frac{3r^2 + 11r + 7}{r + 5}$ $3r - 4 + \frac{27}{r + 5}$

Reteaching

Using Cooperative Learning Assign a division problem from the text to each student. After completing the division problem, have each student write the divisor and quotient with any remainder and exchange it with a neighbor. The neighbor computes the divisor times the quotient plus the remainder. Check this against the dividend.

Additional Answer

1. dividend, $2x^2 - 11x - 20$;
   divisor, $2x + 3$;
   quotient, $x - 7 + \frac{1}{2x + 3}$

11. **Geometry** Find the length of a rectangle if its area is $(10x^2 + 29x + 21)$ square meters and its width is $(5x + 7)$ meters. **$(2x + 3)$ meters**

EXERCISES

Assignment Guide

Core: 13–31 odd, 32, 35, 37, 39–45
Enriched: 12–30 even, 32–45
All: Self Test, 1–10

For **Extra Practice**, see p. 783.

The red A, B, and C flags, printed only in the Teacher's Wraparound Edition, indicate the level of difficulty of the exercises.

Using the Programming Exercises The program given in Exercise 32 is for use with a TI-82 graphing calculator. For other programmable calculators, have students consult their owner's manual for commands similar to those presented here.

Practice

14. $a + 2 + \frac{2}{3a}$

15. $\frac{m}{7} + 1 - \frac{4}{m}$

16. $3y - 5 + \frac{1}{xy}$

17. $a^2 + 2a + 12 + \frac{3}{a-2}$

18. $b + 2 - \frac{3}{2b-1}$

19. $m - 3 - \frac{2}{m+7}$

20. $x - 5 - \frac{1}{2x+3}$

21. $2x + 6 + \frac{1}{x-7}$

30. $x + 6 + \frac{-3}{x+3}$

A

Find each quotient.

12. $(x^3 + 2x^2 - 5) \div 2x$ $\quad \frac{x^2}{2} + x - \frac{5}{2x}$
13. $(b^2 + 9b - 7) \div 3b$ $\quad \frac{b}{3} + 3 - \frac{7}{3b}$

14. $(3a^2 + 6a + 2) \div 3a$

15. $(m^2 + 7m - 28) \div 7m$

16. $(9xy^2 - 15xy + 3) \div 3xy$

17. $(a^3 + 8a - 21) \div (a - 2)$

18. $(2b^2 + 3b - 5) \div (2b - 1)$

19. $(m^2 + 4m - 23) \div (m + 7)$

20. $(2x^2 - 7x - 16) \div (2x + 3)$

21. $(2x^2 - 8x - 41) \div (x - 7)$

B

22. $\frac{14a^2b^2 + 35ab^2 + 2a^2}{7a^2b^2}$ $\quad 2 + \frac{5}{a} + \frac{2}{7b^2}$
23. $\frac{12m^3k + 16mk^3 - 8mk}{4mk}$ $\quad 3m^2 + 4k^2 - 2$

24. $\frac{3r^2 + 20r + 11}{r + 6}$ $\quad 3r + 2 - \frac{1}{r+6}$
25. $\frac{a^2 + 10a + 20}{a + 3}$ $\quad a + 7 - \frac{1}{a+3}$

26. $\frac{4m^2 + 8m - 19}{2m + 7}$ $\quad 2m - 3 + \frac{2}{2m+7}$
27. $\frac{6x^2 + 5x + 15}{2x + 3}$ $\quad 3x - 2 + \frac{21}{2x+3}$

28. $\frac{y^2 - 19y + 9}{y - 4}$ $\quad y - 15 - \frac{51}{y-4}$
29. $\frac{4t^2 + 17t - 1}{4t + 1}$ $\quad t + 4 - \frac{5}{4t+1}$

C

30. Find the quotient when $x^2 + 9x + 15$ is divided by $x + 3$.

31. What is the quotient when $56x^3 + 32x^2 - 63x$ is divided by $7x$?
$\quad 8x^2 + \frac{32x}{7} - 9$

Programming

32a. $7x - 9 + \frac{15}{x+2}$

32b. $\frac{1}{3}x - \frac{46}{9} - \frac{4\frac{5}{9}}{3x+4}$

32. The graphing calculator program at the right will help you find the quotient and remainder when you divide a polynomial of the form $ax^2 + bx + c$ by a binomial of the form $x - r$. You enter the values of a, b, c, and r when prompted.

Run the program to find each quotient and remainder.

a. $(7x^2 + 5x - 3) \div (x + 2)$

b. $(x^2 - 14x - 25) \div (3x + 4)$

```
PROGRAM; POLYDIV
: Disp "ENTER A, B, C"
: Input A: Input B:
  Input C
: Disp "ENTER R"
: Input R
: Disp "COEFFICIENTS"
: Disp "OF QUOTIENT:"
: Disp A
: Disp B+A*R
: Disp "REMAINDER:"
: Disp C+B*R+A*R^2
```

33. Modify the graphing calculator program given above so that you can use it to find each quotient.

a. $(x^3 + 2x^2 - 4x - 8) \div (x - 2)$ $\quad x^2 + 4x + 4$

b. $(20t^3 - 27t^2 + t + 6) \div (4t - 3)$ $\quad 5t^2 - 3t - 2$

c. $(2a^3 + 9a^2 + 5a - 12) \div (a + 3)$ $\quad 2a^2 + 3a - 4$

Critical Thinking

34. Find the value of k if $x + 7$ is a factor of $x^2 - 2x - k$. **63**

35. Find the value of k if $2m - 3$ is a factor of $2m^2 + 7m + k$. **-15**

36. Find the value of k if the remainder is 15 when $x^3 - 7x^2 + 4x + k$ is divided by $x - 2$. **27**

Lesson 12–4 Dividing Polynomials **679**

Practice Masters, p. 83

Closing Activity

Speaking Have students identify similarities in the procedure of long division and that of dividing polynomials.

Chapter 12, Quiz B (Lessons 12-3 and 12-4), is in the *Assessment and Evaluation Masters*, p. 324.

Mid-Chapter Test (Lesson 12-1 through 12-4) is available in the *Assessment and Evaluation Masters*, p. 323.

Additional Answers

38a.

Material	Density
aluminum	2.7
gold	19.3
silver	10.5
steel	7.8
iron	7.8
copper	8.9
blood	1.1
lead	11.3
brass	8.6
concrete	2.0

38b.

Enrichment Masters, p. 83

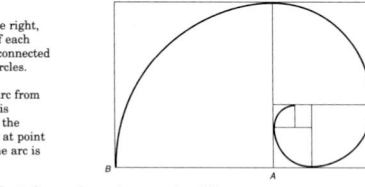

12-4

NAME_____ DATE_____
Student Edition
Pages 675–680

Enrichment

Golden Rectangles

A **golden rectangle** has the property that its sides satisfy the following proportion.

$$\frac{a+b}{a} = \frac{a}{b}$$

Two quadratic equations can be written from the proportion. These are sometimes called **golden quadratic** equations.

1. In the proportion, let $a = 1$. Use cross-multiplication to write a quadratic equation. $b^2 + b - 1 = 0$

2. Solve the equation in problem 1 for b. $b = \frac{-1 + \sqrt{5}}{2}$

3. In the proportion, let $b = 1$. Write a quadratic equation in a. $a^2 - a - 1 = 0$

4. Solve the equation in problem 3 for a. $a = \frac{1 + \sqrt{5}}{2}$

5. Explain why $\frac{1}{2}(\sqrt{5} + 1)$ and $\frac{1}{2}(\sqrt{5} - 1)$ are called golden ratios.

They are the ratios of the sides in a golden rectangle. The first is the ratio of the long side to the short side; the second is short side: long side.

Another property of golden rectangles is that a square drawn inside a golden rectangle creates another, smaller golden rectangle.

In the design at the right, opposite vertices of each square have been connected with quarters of circles.

For example, the arc from point B to point C is created by putting the point of a compass at point A. The radius of the arc is the length BA.

6. On a separate sheet of paper, draw a larger version of the design. Start with a golden rectangle with a long side of 10 inches.

The short side should be about $6\frac{3}{16}$ inches.

Applications and Problem Solving

silver

raw copper

37. Environment Due to concerns about air pollution, it is important to invest money on the reduction of pollutants. The equation $C = \frac{120,000p}{1 - p}$ models the expenditure C in dollars needed to reduce pollutants by p percent where p is the percent in decimal form. If a utility company wants to remove 80% of the pollutants their equipment emits, what expenditure must they make? **$480,000**

38. Science The *density* of a material is its mass per unit volume. A 2.48 g block of copper occupies 0.28 cm³. For example, the density of copper is found from the quotient

$$\frac{2.48\text{ g}}{0.28\text{ cm}^3} \approx 8.9 \text{ g/cm}^3.$$

Material	Mass(g)	Volume (cm³)
aluminum	4.15	1.54
gold	2.32	0.12
silver	6.30	0.60
steel	7.80	1.00
iron	15.20	1.95
copper	2.48	0.28
blood	4.35	4.10
lead	11.30	1.00
brass	17.90	2.08
concrete	40.00	20.00

a. Make a table of densities for the materials above.

b. Make a line plot of the densities computed in part a. Use densities rounded to the nearest whole number.

c. Interpret the line plot made in part b. The data's density is clustered around 9.

Mixed Review
38a–b. See margin.

39. Find $\frac{x^2 - 16}{16 - x^2} \div \frac{7}{x}$. (Lesson 12–3) $-\frac{x}{7}$

40. Graph $y = -x^2 + 2x + 3$. (Lesson 11–1) See Solutions Manual.

41. Factor $3x^2 - 6x - 105$. (Lesson 10–3) $3(x - 7)(x + 5)$ **gold**

42. Geometry Find the area of a rectangle if the length is $(2x + y)$ units and the width is $(x + y)$ units. (Lesson 9–7) $(2x^2 + 3xy + y^2)$ units²

43. Solve the system of equations by graphing. (Lesson 8–1)
$y = 2x + 1$
$y = -2x + 5$ **(1, 3)**

44. $y = 2x$

44. Write an equation for the relation $\{(2, 4), (3, 6), (-2, -4)\}$. (Lesson 5–6)

45. Graph $\{-2, -1, 4, 5\}$ on a number line. (Lesson 2–1) See Solutions Manual.

SELF TEST

Simplify each rational expression. (Lesson 12–1)

1. $\frac{25x^3y^4}{36x^2y^5}$ $\frac{25x}{36y}$

2. $\frac{4x^2 - 9}{2x^2 + 13x - 15}$ already simplified

Find each product or quotient. (Lessons 12–2 and 12–3)

3. $\frac{x^2 - 16}{x^2 + 5x + 6} \cdot \frac{4x^2 + 2x - 3}{x^2 - 5x + 4}$ $\frac{(x + 4)(4x^2 + 2x - 3)}{(x + 3)(x + 2)(x - 1)}$

4. $\frac{2x^2 - 5x + 2}{x^2 - 5x + 6} \div \frac{2x^2 + 9x - 5}{x^2 - 4x + 3}$ $\frac{x - 1}{x + 5}$

5. $(3x - 2) \cdot \frac{x - 5}{3x^2 + 10x - 8}$ $\frac{x - 5}{x + 4}$

6. $\frac{7x^2 + 36x + 5}{x - 5} \div (7x + 1)$ $\frac{x + 5}{x - 5}$

Find each quotient. (Lesson 12–4)

7. $(4x^2 - 18x + 20) \div (2x - 4)$ $2x - 5$

8. $\frac{3x^2 - 6x - 4}{x - 2}$ $3x + \frac{-4}{x - 2}$

9. Geometry A rectangular field has an area of $12x^2 + 20x - 8$ square units and a width of $x + 2$ units. Find its length in terms of x. (Lesson 12–4) $4(3x - 1)$ or $(12x - 4)$ units

10. Travel On the first day of a trip, Manuel drove 440 miles in 8 hours. The second day, he drove at the same speed but for only 6 hours. How far did he drive on the second day? (Lesson 12–2) **10. 330 miles**

680 Chapter 12 *Exploring Rational Expressions and Equations*

Extension

Problem Solving Find the quotient.

$$\frac{2x^4 + 3x^3 - 6x^2 + 7x - 6}{x^2 + 2x - 3}$$

$2x^2 - x + 2$

SELF TEST

The Self Test provides students with a brief review of the concepts and skills in Lessons 12-1 through 12-4. Lesson numbers are given to the right of exercises or instruction lines so students can review concepts not yet mastered.

Rational Expressions with Like Denominators

Objective

After studying this lesson, you should be able to:

- add and subtract rational expressions with like denominators.

CONNECTION
History

At 11:40 P.M. on the night of April 14, 1912, the lookout on the bridge of the *Titanic* spotted an iceberg directly ahead. Despite their heroic efforts, the crew could not steer clear of the iceberg, and a 300-foot gash was ripped in the side of the ship. The *Titanic* sank less than three hours later, and 1500 lives were lost.

The iceberg that the *Titanic* hit has been described as enormous. However, many people are surprised to learn that only about $\frac{1}{8}$ of any iceberg is visible above the surface of the water. This means the part of an iceberg that is submerged is about $1 - \frac{1}{8}$.

$$1 - \frac{1}{8} = \frac{8}{8} - \frac{1}{8} = \frac{8-1}{8} \text{ or } \frac{7}{8}$$

That is, about $\frac{7}{8}$, or 87.5% of the iceberg is submerged.

This example illustrates that, to add or subtract fractions with like denominators, you add or subtract the numerators and then write the sum or difference over the common denominator. You can use this same method to add or subtract rational expressions with like denominators.

rational numbers	*rational expressions*
$\frac{1}{9} + \frac{4}{9} = \frac{1+4}{9} = \frac{5}{9}$	$\frac{4}{y} + \frac{7}{y} = \frac{4+7}{y} = \frac{11}{y}$
$\frac{6}{7} - \frac{2}{7} = \frac{6-2}{7} = \frac{4}{7}$	$\frac{9}{5z} - \frac{6}{5z} = \frac{9-6}{5z} = \frac{3}{5z}$

Example Find $\frac{a}{15m} + \frac{2a}{15m}$.

$$\frac{a}{15m} + \frac{2a}{15m} = \frac{a+2a}{15m} \qquad \textit{The common denominator is 15m.}$$

$$= \frac{3a}{15m} \qquad \textit{Add numerators.}$$

$$= \frac{\overset{1}{\cancel{3a}}}{\underset{5}{\cancel{15m}}} \qquad \textit{Divide by the common factors.}$$

$$= \frac{a}{5m}$$

LOOK BACK

You can refer to Lesson 2-5 for information about adding and subtracting rational numbers.

Sometimes the denominators of rational expressions are binomials. As long as each rational expression in a sum or difference has exactly the same binomial as its denominator, the process of adding or subtracting is the same.

Example ❷ Find $\frac{4}{x+3} - \frac{1}{x+3}$.

$$\frac{4}{x+3} - \frac{1}{x+3} = \frac{4-1}{x+3} \qquad \textit{The common denominator is } x+3.$$

$$= \frac{3}{x+3} \qquad \textit{Subtract numerators.}$$

NCTM Standards: 1–5

Instructional Resources

- Study Guide Master 12-5
- Practice Master 12-5
- Enrichment Master 12-5

Transparency 12-5A contains the 5-Minute Check for this lesson; **Transparency 12-5B** contains a teaching aid for this lesson.

Recommended Pacing	
Standard Pacing	Day 7 of 14
Honors Pacing	Day 6 of 13
Block Scheduling*	Day 4 of 6 (along with Lesson 12-6)
Alg. 1 in Two Years*	Days 13, 14, & 15 of 27

*For more information on pacing and possible lesson plans, refer to the *Block Scheduling Booklet* and *Algebra 1 in Two Years.*

1 FOCUS

5-Minute Check
(over Lesson 12-4)

Find each quotient.

1. $\frac{d^2 + 3d + 2}{d + 1}$ $d + 2$

2. $\frac{s^2 - 5s + 4}{s - 1}$ $s - 4$

3. $\frac{x^2 + 12x + 11}{x + 11}$ $x + 1$

4. $\frac{x^2 + 9x + 20}{x + 5}$ $x + 4$

5. $\frac{a^2 - 2a - 35}{a - 7}$ $a + 5$

Motivating the Lesson

Questioning Find values for a, b, and c so that $\frac{a}{c} + \frac{b}{c} = 1$. What is the relationship among a, b, and c? Answers will vary. A sample answer is $\frac{2}{3} + \frac{1}{3} = \frac{3}{3}$ or 1. $a + b = c, c \neq 0$

Remember that, to subtract a polynomial, you add its additive inverse.

Example 3 Find $\frac{2m+3}{m-4} - \frac{m-2}{m-4}$.

$$\frac{2m+3}{m-4} - \frac{m-2}{m-4} = \frac{(2m+3)-(m-2)}{m-4}$$
$$= \frac{2m+3+[-(m-2)]}{m-4} \quad \textit{The additive inverse of } m-2 \textit{ is } -(m-2).$$
$$= \frac{2m+3-m+2}{m-4}$$
$$= \frac{m+5}{m-4}$$

Sometimes you must factor in order to simplify a sum or difference of rational expressions. Also a factor may need to be rewritten as its additive inverse to recognize the common denominator.

Example 4 Find $\frac{7k+2}{4k-3} + \frac{8-k}{3-4k}$.

The denominator $3 - 4k$ is the same as $-(-3 + 4k)$ or $-(4k - 3)$. Rewrite the second expression so that the denominator is the same as the first expression.

$$\frac{7k+2}{4k-3} + \frac{8-k}{3-4k} = \frac{7k+2}{4k-3} - \frac{8-k}{4k-3}$$
$$= \frac{7k+2-(8-k)}{4k-3}$$
$$= \frac{8k-6}{4k-3}$$
$$= \frac{2(4k-3)}{(4k-3)} \quad \textit{Factor the numerator.}$$
$$= \frac{\overset{1}{2(4k-3)}}{\underset{1}{(4k-3)}} \quad \textit{Divide by the common factors.}$$
$$= 2$$

Example 5 **Find an expression for the perimeter of rectangle ABCD.**

INTEGRATION
Geometry

$$P = 2\ell + 2w$$
$$= 2\left(\frac{9r}{2r+6s}\right) + 2\left(\frac{5r}{2r+6s}\right)$$
$$= \frac{2(9r) + 2(5r)}{2r+6s}$$
$$= \frac{18r + 10r}{2r+6s}$$
$$= \frac{28r}{2r+6s}$$
$$= \frac{28r}{2(r+3s)} \quad \textit{Factor the denominator.}$$
$$= \frac{\overset{14}{28r}}{\underset{1}{2(r+3s)}} \quad \textit{Divide by the common factors.}$$
$$= \frac{14r}{r+3s}$$

The perimeter of rectangle *ABCD* is $\left(\frac{14r}{r+3s}\right)$ cm.

Communicating Mathematics

Study the lesson. Then complete the following.

1. **Compare and contrast** two rational expressions whose sum is 0 with two rational expressions whose difference is 0.   See margin.

2. Combine the numerators, use their like denominator, and simplify.
3. She added the denominators.

2. **Summarize** the procedure for adding or subtracting two rational expressions whose denominators are the same.

3. **You Decide**   Abigail wrote $\frac{7x-3}{2x+11} + \frac{3x-9}{2x+11} = \frac{10x-12}{4x+22}$. What mistake did she make?

Guided Practice

Find each sum or difference. Express in simplest form.

4. $\frac{5x}{7} + \frac{2x}{7}$ x

5. $\frac{3}{x} + \frac{7}{x}$ $\frac{10}{x}$

6. $\frac{7}{3m} - \frac{4}{3m}$ $\frac{1}{m}$

7. $\frac{3}{a+2} + \frac{7}{a+2}$ $\frac{10}{a+2}$

8. $\frac{2m}{m+3} - \frac{-6}{m+3}$ 2

9. $\frac{3x}{x+4} - \frac{-12}{x+4}$ 3

10. 3x

10. If the sum of $\frac{2x-1}{3x+2}$ and another rational expression with denominator $3x+2$ is $\frac{5x-1}{3x+2}$, what is the numerator of the second rational expression?

Practice

A

Find each sum or difference. Express in simplest form.

11. $\frac{m}{3} + \frac{2m}{3}$ m

12. $\frac{3y}{11} - \frac{8y}{11}$ $-\frac{5y}{11}$

13. $\frac{5a}{12} - \frac{7a}{12}$ $-\frac{a}{6}$

14. $\frac{4}{3z} + \frac{-7}{3z}$ $-\frac{1}{z}$

15. $\frac{x}{2} - \frac{x-4}{2}$ 2

16. $\frac{a+3}{6} - \frac{a-3}{6}$ 1

17. $\frac{2}{x+7} + \frac{5}{x+7}$ $\frac{7}{x+7}$

18. $\frac{2x}{x+1} + \frac{2}{x+1}$ 2

19. $\frac{y}{y-2} + \frac{2}{2-y}$ 1

20. $\frac{4m+5}{2m+3}$

B

20. $\frac{4m}{2m+3} + \frac{5}{2m+3}$

21. $\frac{-5}{3x-5} + \frac{3x}{3x-5}$ 1

22. $\frac{3r}{r+5} + \frac{15}{r+5}$ 3

26. $\frac{3m-5}{m-2}$

23. $\frac{2x}{x+2} + \frac{2x}{x+2}$ $\frac{4x}{x+2}$

24. $\frac{2m}{m-9} + \frac{18}{9-m}$ 2

25. $\frac{2y}{y+3} + \frac{-6}{y+3}$ $\frac{2y-6}{y+3}$

28. $\frac{6t+2}{1-4t}$

26. $\frac{3m}{m-2} - \frac{5}{m-2}$

27. $\frac{4x}{2x+3} - \frac{-6}{2x+3}$ 2

28. $\frac{4t-1}{1-4t} + \frac{2t+3}{1-4t}$

29. $\frac{25x-93}{2x+5}$

C

29. If $\frac{3x-100}{2x+5}$ is added to the sum of $\frac{11x-5}{2x+5}$ and $\frac{11x+12}{2x+5}$, what is the result?

30. If $\frac{-3b+4}{2b+12}$ is subtracted from the sum of $\frac{b-15}{2b+12}$ and $\frac{-3b+12}{2b+12}$, what is the result? $\frac{b-7}{2b+12}$

 INTEGRATION

Geometry

Find an expression for the perimeter of each rectangle.

31.

$\frac{12x}{7x-2y}$

$\frac{24x+26y}{7x-2y}$

$\frac{13y}{7x-2y}$

32.

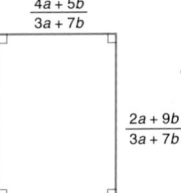

$\frac{4a+5b}{3a+7b}$

4

$\frac{2a+9b}{3a+7b}$

Lesson 12-5   Rational Expressions with Like Denominators   **683**

Reteaching ▬▬▬

Using Substeps   When working with rational expressions, answers must be given in simplest form. List the first two steps you would take to solve this rational expression.

$$\frac{7a}{a^2-b^2} - \frac{5a+4b}{a^2-b^2} + \frac{2b}{a^2-b^2}$$

Write common denominator; add/subtract like terms in numerator.

Additional Answer

1. The two expressions whose sum is zero have opposite signs but equal absolute values, and the two expressions whose difference is zero are equal and have the same sign.

3 PRACTICE/APPLY

Check for Understanding

Exercises 1–10 are designed to help you assess your students' understanding through reading, writing, speaking, and modeling. You should work through Exercises 1–3 with your students and then monitor their work on Exercises 4–10.

Assignment Guide

Core: 11–35 odd, 36–41
Enriched: 12–32 even, 33–41

For **Extra Practice**, see p. 783.

The red A, B, and C flags, printed only in the Teacher's Wraparound Edition, indicate the level of difficulty of the exercises.

Practice Masters, p. 84

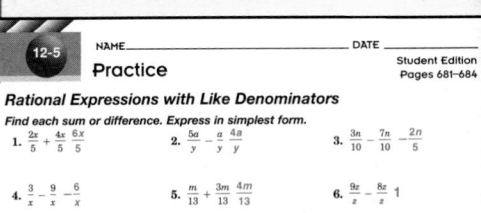

Closing Activity

Speaking Have students identify ways in which the addition and subtraction of fractions with like denominators is similar to the addition and subtraction of rational expressions with like denominators.

Additional Answer

40.

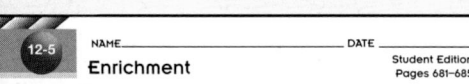

---

33. Which of the following rational numbers is not equivalent to the others? b
   a. $\frac{-3}{x - 2}$ b. $-\frac{3}{2 - x}$ c. $-\frac{3}{x - 2}$ d. $\frac{3}{2 - x}$

Applications and Problem Solving

34. **Icebergs** Water weighs 62.4 pounds per cubic foot. One cubic foot of water contains 7.48 gallons. Each cubic foot of ice yields 0.89 times as many gallons of water as a cubic foot of water.   a. 6.657 gallons
   a. How many gallons of water does one cubic foot of ice contain?
   b. Some people have suggested moving icebergs from the North Atlantic to parts of the world that have little fresh water. If an iceberg is 1 mile wide, 2 miles long, and 800 feet thick, about how much fresh water would such an iceberg yield?   296,938,414,100 gallons

35. **Utilities** A person in the United States on the average uses 168 gallons of water each day. Suppose that 25% of the iceberg in Exercise 34 is lost as it is being moved to another location. How many days would the iceberg sustain the Atlanta, Georgia, metropolitan area, which has a population of about 3 million people?   442 days

Mixed Review

36. 1680.7

36. **Engineering** The Canadian Pacific Railroad constructed a track in the form of a circle so the train will spiral over itself as it passes through a mountain tunnel. This reduces the degree of incline for the train to climb. If we assume that the spiral is a circle one mile long, what would be the diameter of the circle in feet? (Lesson 12–4)

37. $\pm\frac{6}{5}$

37. Solve $25x^2 = 36$ by factoring. (Lesson 10–6)

38. **Geometry** Find the perimeter of quadrilateral *MATH* shown at the right. (Lesson 9–5) $2x^2 + 8y + 4x + 4$

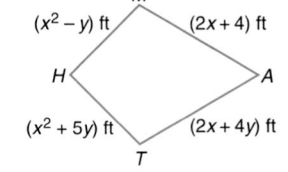

39. Use substitution to solve the system of equations. (Lesson 8–2)
   $x + 2y = 5$
   $x - 2y = -11$ $(-3, 4)$

40. **Astronomy** A *main sequence star* is a star that is in the neighborhood of the sun. The table below gives the names of the seven main sequence stars and the sun, their surface temperatures in thousands of °C, and their radii in multiples of the sun's radius. Draw a scatter plot with the temperatures on the horizontal axis and the radii on the vertical axis. (Lesson 6–3) **See margin.**

Stars	MU-1 Scorpii	Sirius A	Altair	Polycon A	Sun	61 Cygni A	Kueger 60	Barnard's Star
Surface Temp.	20	10.2	7.3	6.8	5.9	2.6	2.8	2.7
Radius	5.2	1.9	1.6	2.6	1.0	0.7	0.35	0.15

41. **Probability** A student rolls a die three times. What is the probability that each roll is a 1? (Lesson 4–6) $\frac{1}{216}$

Extension

Connections Find the perimeter of a rectangle with a width of $\frac{y}{2}$ and a length of $\frac{y - 6}{2}$. $2(y - 3)$

Rational Expressions with Unlike Denominators

Objectives

After studying this lesson, you should be able to:

- add and subtract rational expressions with unlike denominators, and
- make an organized list of possibilities to solve problems.

APPLICATION
Astronomy

Mars, Jupiter, and Saturn revolve around the sun approximately every 2 years, 12 years, and 30 years, respectively. The planets are said to be in *conjunction*, or alignment, when they appear close to one another in Earth's night sky. The last time this happened was in 1982. When will it happen again?

Mars will go through six revolutions for each one of Jupiter's. Mars will go through fifteen revolutions for each one of Saturn's. The times the planets align is related to the common multiples of 2, 12, and 30. In fact, the least number of years that will pass until the next alignment is the **least common multiple (LCM)** of these numbers. The least common multiple is the least number that is a common multiple of two or more numbers.

There are two methods you can use to find a least common multiple.

Method 1: Make an organized list.
You can often solve problems by listing the possibilities in an organized way. Use a systematic approach so you don't omit any important items.

Multiples of 2	Multiples of 12	Multiples of 30
$2 \cdot 0 = 0$	$12 \cdot 0 = 0$	$30 \cdot 0 = 0$
$2 \cdot 1 = 2$	$12 \cdot 1 = 12$	$30 \cdot 1 = 30$
$2 \cdot 2 = 4$	$12 \cdot 2 = 24$	$30 \cdot 2 = 60$
⋮	⋮	
$2 \cdot 30 = 60$	$12 \cdot 5 = 60$	

Compare the multiples in the table. Other than 0, the least number that is common to all three lists of multiples is 60. So, the LCM of 2, 12, and 30 is 60.

Method 2: Use prime factorization.
Find the prime factorization of each number.

$$2 = 2 \qquad 12 = 2 \cdot 2 \cdot 3 \qquad 30 = 2 \cdot 3 \cdot 5$$

Use each prime factor the greatest number of times it appears in any of the factorizations.

2 appears twice as a factor of 12. All other factors appear once.
Thus, the LCM of 2, 12, and 30 is $2 \cdot 2 \cdot 3 \cdot 5$ or 60.

Using either method, the LCM is 60. This means that the planets will be in alignment approximately 60 years after 1982, or about the year 2042.

You can use the same methods to find the LCM of two or more polynomials.

Example **Find the LCM of $15a^2b^2$ and $24a^2b$.**

List the multiples of each coefficient and each variable expression.

15: 15, 30, 45, 60, 75, 90, 105, 120, 135
24: 24, 48, 72, 96, 120, 144
a^2: a^2
b^2: b^2, b^4, b^6
b: b, b^2, b^3
LCM $= 120a^2b^2$

Alternative Learning Styles

Kinesthetic Place four common classroom objects in a bag. Ask students to predict in how many different orders the objects could be removed from the bag. Then have students remove the objects a number of times, noting the order, to check the accuracy of their predictions.

CAREER CHOICES

The Babylonians were the first astronomers, with the Greeks expanding the studies. Until the sixteenth century, it was thought that Earth was the center of the universe.

12-6 LESSON NOTES

NCTM Standards: 1–5

Instructional Resources

- Study Guide Master 12-6
- Practice Master 12-6
- Enrichment Master 12-6
- Assessment and Evaluation Masters, p. 325
- Multicultural Activity Masters, p. 23
- Tech Prep Applications Masters, p. 23

Transparency 12-6A contains the 5-Minute Check for this lesson; **Transparency 12-6B** contains a teaching aid for this lesson.

Recommended Pacing	
Standard Pacing	Days 8 & 9 of 14
Honors Pacing	Days 7 & 8 of 13
Block Scheduling*	Day 4 of 6 (along with Lesson 12-5)
Alg. 1 in Two Years*	Days 16, 17, & 18 of 27

*For more information on pacing and possible lesson plans, refer to the *Block Scheduling Booklet* and *Algebra 1 in Two Years*.

1 FOCUS

5-Minute Check
(over Lesson 12-5)

Find each sum or difference. Express in simplest form.

1. $\dfrac{3a}{16} - \dfrac{2a}{16} \quad \dfrac{a}{16}$

2. $\dfrac{8}{x} + \dfrac{7}{x} \quad \dfrac{15}{x}$

3. $\dfrac{y}{x+4} - \dfrac{3y}{x+4} \quad \dfrac{-2y}{x+4}$

4. $\dfrac{11}{r} - \dfrac{5}{r} \quad \dfrac{6}{r}$

5. $\dfrac{x^2}{x-y} - \dfrac{y^2}{x-y} \quad x+y$

Hands-On Activity Turn a bicycle upside-down. Have one student spin the back tire at a rate of 1 revolution every 2 seconds. Have another student spin the front tire at a rate of 1 revolution every 3 seconds. Before spinning, put a chalk mark on the top of each tire. Record how long it takes for the two tires to return to this same arrangement. **6 seconds**

2 TEACH

In-Class Examples

For Example 1
Find the LCM for each group of rational expressions.

a. $6x^2y$, $9xy^2$ 18^2y^2
b. $3a^2b$, $16ab^2$, $21ab$ $336a^2b^2$

For Example 2
Find the LCM for each pair of rational expressions.

a. $x^2 - 5x + 4$, $x^2 - 3x - 4$
$(x - 4)(x - 1)(x + 1)$
b. $x^2 - 9$, $x^2 + 6x + 9$
$(x - 3)(x + 3)^2$

For Example 3
Find each sum or difference.

a. $\dfrac{2}{3y^2} - \dfrac{1}{12y^3}$ $\dfrac{8y - 1}{12y^3}$
b. $\dfrac{4a}{mn} + \dfrac{2a}{np}$ $\dfrac{4ap + 2am}{mnp}$

For Example 4
Find each sum or difference.

a. $\dfrac{r}{r^2 - 16} - \dfrac{5}{r + 4}$
$\dfrac{-4r + 20}{r^2 - 16}$
b. $\dfrac{3}{y - 3} + \dfrac{2y}{y^2 - y - 6}$
$\dfrac{5y + 6}{y^2 - y - 6}$
c. $\dfrac{y}{(1 - y)(y + 1)} + \dfrac{y + 2}{(y + 1)^2}$
$\dfrac{-2}{(y - 1)(y + 1)^2}$
d. $\dfrac{a - 1}{(a + 3)^2} - \dfrac{a + 1}{(3 - a)(a + 3)}$
$\dfrac{2a^2 + 6}{(a - 3)(a + 3)^2}$

Teaching Tip Point out that the LCD of two rational expressions can be equal to the denominator of one of them if one denominator is a factor of the other.

Example ② Find the LCM of $x^2 - x - 6$ and $x^2 + 2x - 15$.

$x^2 - x - 6 = (x - 3)(x + 2)$ *Factor each expression.*
$x^2 + 2x - 15 = (x + 5)(x - 3)$
$\text{LCM} = (x - 3)(x + 2)(x + 5)$

To add or subtract fractions with unlike denominators, first rename the fractions so the denominators are alike. Any common denominator could be used. However, the computation usually is easier if you use the **least common denominator (LCD)**. Recall that the least common denominator is the LCM of the denominators.

You usually will use the following steps to add or subtract rational expressions with unlike denominators.

1. Find the LCD.
2. Change each rational expression into an equivalent expression with the LCD as the denominator.
3. Add or subtract as with rational expressions with like denominators.
4. Simplify if necessary.

Example ③ Find $\dfrac{7}{3m} + \dfrac{5}{6m^2}$.

$3m = 3 \cdot m$ *Factor each denominator.*
$6m^2 = 2 \cdot 3 \cdot m \cdot m$
$\text{LCM} = 2 \cdot 3 \cdot m \cdot m$, or $6m^2$

Since the denominator of $\dfrac{5}{6m^2}$ is already $6m^2$, only $\dfrac{7}{3m}$ needs to be renamed.

$\dfrac{7}{3m} + \dfrac{5}{6m^2} = \dfrac{7(2m)}{3m(2m)} + \dfrac{5}{6m^2}$ *Multiply $\dfrac{7}{3m}$ by $\dfrac{2m}{2m}$. Why?*

$= \dfrac{14m}{6m^2} + \dfrac{5}{6m^2}$

$= \dfrac{14m + 5}{6m^2}$ *Add the numerators.*

To combine rational expressions whose denominators are polynomials, follow the same procedure you use to combine expressions whose denominators are monomials.

Example ④ Find each sum or difference.

a. $\dfrac{s}{s + 3} + \dfrac{3}{s - 4}$

Since the denominators are $(s + 3)$ and $(s - 4)$, there are no common factors. The LCD is $(s + 3)(s - 4)$.

$\dfrac{s}{s + 3} + \dfrac{3}{s - 4} = \dfrac{s}{s + 3} \cdot \dfrac{s - 4}{s - 4} + \dfrac{3}{s - 4} \cdot \dfrac{s + 3}{s + 3}$

$= \dfrac{s(s - 4)}{(s + 3)(s - 4)} + \dfrac{3(s + 3)}{(s + 3)(s - 4)}$

$= \dfrac{s^2 - 4s}{(s + 3)(s - 4)} + \dfrac{3s + 9}{(s + 3)(s - 4)}$

$= \dfrac{s^2 - 4s + 3s + 9}{(s + 3)(s - 4)}$

$= \dfrac{s^2 - s + 9}{(s + 3)(s - 4)}$

686 *Chapter 12 Exploring Rational Expressions and Equations*

Teaching Tip Students should be aware of how the LCD is determined. Remind them that each factor is used the greatest number of times it appears in either factorization.

b. $\dfrac{n-4}{(2-n)^2} - \dfrac{n-5}{n^2+n-6}$

$\dfrac{n-4}{(2-n)^2} - \dfrac{n-5}{n^2+n-6}$

$= \dfrac{n-4}{(2-n)^2} - \dfrac{n-5}{(n-2)(n+3)}$ *Factor the denominators.*

$= \dfrac{n-4}{(n-2)(n-2)} - \dfrac{n-5}{(n-2)(n+3)}$ $(2-n)^2 = (n-2)^2$

$= \dfrac{n-4}{(n-2)(n-2)} \cdot \dfrac{(n+3)}{(n+3)} - \dfrac{n-5}{(n-2)(n+3)} \cdot \dfrac{(n-2)}{(n-2)}$ *Rename each fraction using the LCD.*

$= \dfrac{\left(n^2+3n-4n-12\right) - \left(n^2-2n-5n+10\right)}{(n-2)(n-2)(n+3)}$ *Subtract numerators*

$= \dfrac{n^2-n-12 - n^2+7n-10}{(n-2)^2(n+3)}$

$= \dfrac{6n-22}{(n-2)^2(n+3)}$

CHECK FOR UNDERSTANDING

Communicating Mathematics

1. Yes; no; no common factors were divided.

6. $(a+6)(a+7)$
7. $(x-3)(x+3)$
8. $2(x-4)$

Study the lesson. Then complete the following.

1. **You Decide**   Kaylee simplified $\dfrac{16}{64}$ as $\dfrac{1\!\!\!\!\diagup 6}{\diagup\!\!\!\!64} = \dfrac{1}{4}$. Is her answer correct? Was her method of simplifying correct? Explain.

2. **Describe** a situation in which the LCD of two or more rational expressions is equal to the denominator of one of the rational expressions. $\dfrac{3x}{4} + \dfrac{7}{8x}$; $8x$ is the LCD.

Guided Practice

Find the LCD for each pair of rational expressions.

3. $\dfrac{3}{x^2}, \dfrac{5}{x}$ x^2

4. $\dfrac{4}{a^2b}, \dfrac{3}{ab^2}$ a^2b^2

5. $\dfrac{4}{15m^2}, \dfrac{7}{18mb^2}$ $90m^2b^2$

6. $\dfrac{6}{a+6}, \dfrac{7}{a+7}$

7. $\dfrac{5}{x-3}, \dfrac{4}{x+3}$

8. $\dfrac{9}{2x-8}, \dfrac{10}{x-4}$

Find each sum or difference.   9–14. See margin.

9. $\dfrac{7}{15m^2} + \dfrac{3}{5m}$

10. $\dfrac{2}{x+3} + \dfrac{3}{x-2}$

11. $3x + 6 - \dfrac{9}{x+2}$

12. $\dfrac{m+2}{m^2+4m+3} - \dfrac{6}{m+3}$

13. $\dfrac{11}{3y^2} - \dfrac{7}{6y}$

14. $\dfrac{4}{2g-7} + \dfrac{5}{3g+1}$

15. **Parades**   At the Veteran's Day parade, the local members of the Veterans of Foreign Wars (VFW) found that they could arrange themselves in rows of 6, 7, or 8, with no one left over. What was the least number of VFW members in the parade?   **168 members**

EXERCISES

Practice
A

Find each sum or difference.

16. $\dfrac{m}{4} + \dfrac{3m}{5}$ $\dfrac{17m}{20}$

17. $\dfrac{x}{7} - \dfrac{2x}{9}$ $\dfrac{-5x}{63}$

18. $\dfrac{m+1}{m} + \dfrac{m-3}{3m}$ $\dfrac{4}{3}$

19. $\dfrac{7}{x} + \dfrac{3}{xyz}$ $\dfrac{7yz+3}{xyz}$

20. $\dfrac{7}{6a^2} - \dfrac{5}{3a}$ $\dfrac{7-10a}{6a^2}$

21. $\dfrac{2}{st^2} - \dfrac{3}{s^2t}$ $\dfrac{2s-3t}{s^2t^2}$

Lesson 12–6   Rational Expressions with Unlike Denominators   **687**

Reteaching

Using Problem Solving When adding rational expressions, a common denominator is needed. The LCD provides the shortest solution but any common denominator will do. Work the problem below twice, once using a common denominator of $(2x+6)(x+3)$ and once using the LCD.

$\dfrac{3x}{2x+6} - \dfrac{x-1}{x+3}$ $\dfrac{x+2}{2x+6}$

3 PRACTICE/APPLY

Check for Understanding
Exercises 1–15 are designed to help you assess your students' understanding through reading, writing, speaking, and modeling. You should work through Exercises 1–2 with your students and then monitor their work on Exercises 3–15.

Assignment Guide
Core: 17–43 odd, 44–50
Enriched: 16–38 even, 39–50

For **Extra Practice**, see p. 784.

The red A, B, and C flags, printed only in the Teacher's Wraparound Edition, indicate the level of difficulty of the exercises.

Additional Answers

9. $\dfrac{9m+7}{15m^2}$

10. $\dfrac{5x+5}{(x+3)(x-2)}$

11. $\dfrac{3(x^2+4x+1)}{x+2}$

12. $\dfrac{-5m-4}{(m+3)(m+1)}$

13. $\dfrac{22-7y}{6y^2}$

14. $\dfrac{22g-31}{(2g-7)(3g+1)}$

Study Guide Masters, p. 85

Find each sum or difference.

22. $\dfrac{3}{7m} + \dfrac{4}{5m^2}$ $\dfrac{15m + 28}{35m^2}$ **23.** $\dfrac{3}{z + 5} + \dfrac{4}{z - 4}$ **24.** $\dfrac{d}{d + 4} + \dfrac{3}{d + 3}$

25. $\dfrac{k}{k + 5} - \dfrac{2}{k + 3}$ **26.** $\dfrac{3}{y - 3} - \dfrac{y}{y + 4}$ **27.** $\dfrac{10}{3r - 2} - \dfrac{9}{r - 5}$

B

28. $\dfrac{4}{3a - 6} + \dfrac{a}{2 + a}$ **29.** $\dfrac{5}{2m - 3} - \dfrac{m}{6 - 4m}$ **30.** $\dfrac{b}{3b + 2} + \dfrac{2}{9b + 6}$ $\dfrac{1}{3}$

31. $\dfrac{w}{5w + 2} - \dfrac{4}{15w + 6}$ $\dfrac{3w - 4}{3(5w + 2)}$ **32.** $\dfrac{h - 2}{h^2 + 4h + 4} + \dfrac{h - 2}{h + 2}$ $\dfrac{h^2 + h - 6}{(h + 2)^2}$

33. $\dfrac{n + 2}{n^2 + 4n + 3} - \dfrac{6}{n + 3}$ $\dfrac{-5n - 4}{(n + 3)(n + 1)}$ **34.** $\dfrac{a}{5 - a} - \dfrac{3}{a^2 - 25}$ $\dfrac{-a^2 - 5a - 3}{(a + 5)(a - 5)}$

35. Find the difference when $\dfrac{2}{t + 3}$ is subtracted from $\dfrac{3}{10t - 9}$. $\dfrac{-17t + 27}{(t + 3)(10t - 9)}$

36. Find the sum of $\dfrac{2y}{y^2 + 7y + 12}$ and $\dfrac{y + 2}{y + 4}$. $\dfrac{y^2 + 7y + 6}{(y + 4)(y + 3)}$

C

37. Find the sum of $\dfrac{2}{v + 4}$, $\dfrac{v}{v - 1}$, and $\dfrac{5v}{v^2 + 3v - 4}$. $\dfrac{v^2 + 11v - 2}{(v + 4)(v - 1)}$

38. Find the difference when $\dfrac{2}{a + 1}$ is subtracted from $\dfrac{6}{a - 2}$. Then find the difference when $\dfrac{6}{a - 2}$ is subtracted from $\dfrac{2}{a + 1}$. How are the differences related?   **See margin.**

Critical Thinking

39. Copy and complete.
- **a.** $15 \cdot 24 = $ ___?___   **360**
  GCF of 15 and 24 = ___?___   **3**
  LCM of 15 and 24 = ___?___   **120**
  GCF · LCM = ___?___   **360**
- **b.** $18 \cdot 30 = $ ___?___   **540**
  GCF of 18 and 30 = ___?___   **6**
  LCM of 18 and 30 = ___?___   **90**
  GCF · LCM = ___?___   **540**
- **c.** Write a rule that describes the relationship between the GCF, the LCM, and the product of the two numbers.
- **d.** Use your rule to describe how to find the LCM of two numbers if you know their GCF.   **Divide the GCF into the product of the two numbers to find the LCM.**

Applications and Problem Solving

40. List the Possibilities Doug Paulsen, the choreographer of a Broadway musical, has asked the producer of the show to hire enough dancers so they can be arranged in groups of exactly three, six, or seven, with no dancer left out. What is the least number of dancers required?   **42**

41. Automobiles Car owners need to follow a regular maintenance schedule to keep their cars running smoothly. The table below shows several of the checkups that should be performed on a regular basis, according to the Spring, 1995, issue of *Know-How* magazine.

If all these inspections and services are performed on April 20, 1996, and the owner follows the recommendations shown in the table, what is the next date on which they all should be performed again?   **30 months later or Oct. 20, 1998**

Inspection or Service	Frequency
engine oil and oil filter change	every 3000 miles (about 3 months)
transmission fluid level check	every oil change
brake system inspection	every oil change
chassis lubrication	every other oil change
power steering pump fluid level check	twice a year
tire and wheel rotation and inspection	every 15,000 miles

42. Travel Jaheed Toliver spent, in this order, a third of his life in the United States, a sixth of his life in Kenya, twelve years in Saudi Arabia, half the remainder in Australia, and as long in Canada as he spent in Hong Kong. How many years did Jaheed live if he spent his 45th birthday in Saudi Arabia and he spent a whole number of years in each country? **66 years**

Mixed Review

43. Find $\frac{8z + 3}{3z + 4} - \frac{2z - 5}{3z + 4}$. (Lesson 12–5) **2**   44. $\frac{x + 6}{3x - 2}$; $-1, \frac{2}{3}$

44. Simplify $\frac{x^2 + 7x + 6}{3x^2 + x - 2}$. State the excluded values of x. (Lesson 12–1)

45. Solve $2x^2 - 3x - 4 = 0$ by using the quadratic formula. (Lesson 11–3)

45. $\frac{3 \pm \sqrt{41}}{4}$ or about 2.35 and -0.85

46. Factor $3x^2y + 6xy + 9y^2$. (Lesson 10–1) **$3y(x^2 + 2x + 3y)$**

47. Find $(3a^2b)(-5a^4b^2)$. (Lesson 9–1) **$-15a^6b^3$**

48. Graph the solution set of $3x > -15$ and $2x \le 6$. (Lesson 7–4) **See margin.**

49. Solve $3n - 12 = 5n - 20$. (Lesson 3–5) **4**

50. **Demographics** The populations of the capitals of some southern states are listed in the table below. Make a stem-and-leaf plot of the populations. (Lesson 1–4) **See margin.**

Capital	Population (thousands)	Capital	Population (thousands)
Atlanta, GA	394	Montgomery, AL	188
Austin, TX	466	Nashville, TN	488
Baton Rouge, LA	220	Oklahoma City, OK	445
Columbia, SC	98	Raleigh, NC	208
Frankfort, KY	26	Richmond, VA	203
Jackson, MS	197	Tallahassee, FL	125
Little Rock, AR	176		

Reading the Labels

The excerpt below appeared in an article in *Aging Magazine*, issue number 366, 1994.

BY THIS SUMMER, ALMOST ALL PACKAGED foods found in your local supermarket included new, improved labels that make it easier to understand the nutritional information contained on them and to compare the nutritional value of different products. This revolution in food labeling is the result of years of work by consumer advocates who pressed for more complete, accurate, and pertinent information on food labels....The percent of daily fat consumption should be no more than 30 percent of calories....The labels also give percentages of daily allowances for sodium, sugar, fiber, and protein. ∎

1. The food labels on items A, B, C, and D show fat contents of 25, 17, 4, and 33 grams per serving, respectively. If you are allowed a maximum daily allowance of 65 grams, which three-item combinations of these foods can you put together without exceeding the maximum allowance? **25, 17 and 4; 17, 4 and 33**

2. Numbers on the food labels are listed as "per serving" rather than per package, and the label also defines serving size. Do you think this is important? Explain your response. **Yes, because food intake is daily and per serving information better suits daily calculations.**

3. When you shop for food, do you use the nutritional data on the labels in deciding what to buy? Why or why not? **See students' work.**

Extension

Reasoning Find
$$\frac{2a^2 + 5a + 3}{2a^2 - a - 6} - \frac{a^2 - 4}{a^2 - a - 2}$$
and simplify.
$$\frac{2a + 5}{a^2 - a - 2}$$

Mathematics and SOCIETY

Even lower fat consumption has been shown to *reverse* heart disease. In 1990 Dr. Dean Ornish announced results of a study that showed that when fat consumption was kept to 10% of calories, plaque build-up around the heart had actually been reversed.

Instructional Resources

- Study Guide Master 12-7
- Practice Master 12-7
- Enrichment Master 12-7
- Multicultural Activity Masters, p. 24
- Real-World Applications, 29
- Tech Prep Applications Masters, p. 24

Transparency 12-7A contains the 5-Minute Check for this lesson; **Transparency 12-7B** contains a teaching aid for this lesson.

Recommended Pacing	
Standard Pacing	Day 10 of 14
Honors Pacing	Day 9 of 13
Block Scheduling*	Day 5 of 6 (along with Lesson 12-8)
Alg. 1 in Two Years*	Days 19, 20, & 21 of 27

*For more information on pacing and possible lesson plans, refer to the *Block Scheduling Booklet* and *Algebra 1 in Two Years.*

1 FOCUS

5-Minute Check
(over Lesson 12-6)

Find the LCD for each pair of rational expressions.

1. $\frac{7}{m^2n^3}$, $\frac{9}{mn^2}$ m^2n^3

2. $\frac{2x}{x+1}$, $\frac{5x}{x-1}$ $(x+1)(x-1)$

Find each sum or difference.

3. $\frac{5}{a} - \frac{3}{b}$ $\frac{5b-3a}{ab}$

4. $\frac{7}{5} + \frac{a}{5b}$ $\frac{7b+a}{5b}$

5. $\frac{6}{3x-9} + \frac{3}{x-3}$ $\frac{5}{x-3}$

Solve the problem by listing the possibilities.

6. There are 3 routes between Kelsey's house and the school. How many ways can Kelsey go from her house to the school and back home again? **9**

Mixed Expressions and Complex Fractions

12-7

Objective
After studying this lesson, you should be able to:
- simplify mixed expressions and complex fractions.

APPLICATION
Table Tennis

A Ping-Pong™ ball weighs about $\frac{1}{10}$ of an ounce. How many Ping-Pong balls weigh $1\frac{1}{2}$ pounds altogether? *This problem will be solved in Example 2.*

A number like $1\frac{1}{2}$ is a mixed number. Expressions like $a + \frac{b}{c}$ and $4 + \frac{x+y}{x-5}$ are **mixed expressions.** Changing mixed expressions to rational expressions is similar to changing mixed numbers to simple fractions (improper fractions).

mixed number to improper fraction

$$5\frac{4}{7} = 5 + \frac{4}{7}$$
$$= \frac{5(7) + 4}{7}$$
$$= \frac{35 + 4}{7}$$
$$= \frac{39}{7}$$

mixed expression to rational expression

$$4 + \frac{x+y}{x-5} = \frac{4(x-5)}{x-5} + \frac{x+y}{x-5}$$
$$= \frac{4(x-5) + (x+y)}{(x-5)}$$
$$= \frac{4x - 20 + x + y}{x-5}$$
$$= \frac{5x + y - 20}{x-5}$$

Example ① **Simplify $7 + \frac{y-3}{y+4}$.**

$$7 + \frac{y-3}{y+4} = \frac{7(y+4)}{y+4} + \frac{y-3}{y+4} \quad \text{The LCD is } y + 4.$$
$$= \frac{7(y+4) + y - 3}{y+4} \quad \text{Add the numerators.}$$
$$= \frac{7y + 28 + y - 3}{y+4}$$
$$= \frac{8y + 25}{y+4} \quad \text{Simplify.}$$

Now let's solve the Ping-Pong ball problem.

Example ② **Refer to the applications at the beginning of the lesson. How many Ping-Pong balls would weigh $1\frac{1}{2}$ pounds altogether?**

APPLICATION
Table Tennis

$$\frac{1\frac{1}{2} \text{ pounds}}{\frac{1}{10} \text{ ounce}} = \frac{\frac{3}{2} \text{ pounds}}{\frac{1}{10} \text{ ounce}}$$

$$= \frac{\frac{3}{2} \text{ pounds}}{\frac{1}{10} \text{ ounce}} \cdot \frac{16 \text{ ounces}}{1 \text{ pound}} \quad \begin{array}{l}\text{Convert pounds to ounces.}\\ \text{Divide by common units.}\end{array}$$

$$= \frac{24}{\frac{1}{10}} \text{ or } 240$$

It would take 240 Ping-Pong balls to weigh $1\frac{1}{2}$ pounds.

Alternative Learning Styles

Auditory Have the first student in a row name a mixed number such as $3\frac{1}{2}$. The next student in the row must name an equivalent improper fraction. A third student then names another mixed number. Continue the activity until all class members have had a turn speaking aloud.

Recall that if a fraction has one or more fractions in the numerator or denominator, it is called a *complex fraction*. Some complex fractions are shown below.

$$\frac{5\frac{1}{2}}{3\frac{3}{4}} \qquad \frac{9}{\frac{x}{y}} \qquad \frac{\frac{x+y}{y}}{\frac{x-y}{x}} \qquad \frac{\frac{1}{a}-\frac{1}{b}}{\frac{1}{a}+\frac{1}{b}}$$

You simplify an algebraic complex fraction in the same way you simplify a numerical complex fraction.

numerical

$$\frac{\frac{11}{2}}{\frac{15}{4}} = \frac{11}{2} \div \frac{15}{4}$$

$$= \frac{11}{2} \cdot \frac{4}{15} \quad \text{\textit{The reciprocal of}}$$
$$\qquad\qquad \textit{$\frac{15}{4}$ is $\frac{4}{15}$.}$$
$$= \frac{22}{15}$$

algebraic

$$\frac{\frac{a}{b}}{\frac{c}{d}} = \frac{a}{b} \div \frac{c}{d}$$

$$= \frac{a}{b} \cdot \frac{d}{c} \quad \text{\textit{The reciprocal of}}$$
$$\qquad\qquad \textit{$\frac{c}{d}$ is $\frac{d}{c}$.}$$
$$= \frac{ad}{bc}$$

Simplifying a Complex Fraction	Any complex fraction $\dfrac{\frac{a}{b}}{\frac{c}{d}}$, where $b \neq 0$, $c \neq 0$, and $d \neq 0$, can be expressed as $\dfrac{ad}{bc}$.

Example **Simplify each rational expression.**

a. $\dfrac{1 + \frac{4}{a}}{\frac{a}{6} + \frac{2}{3}}$

Simplify the numerator and denominator separately. Then divide.

$$\frac{1 + \frac{4}{a}}{\frac{a}{6} + \frac{2}{3}} = \frac{\frac{1}{1} \cdot \frac{a}{a} + \frac{4}{a}}{\frac{a}{6} + \frac{2}{3} \cdot \frac{2}{2}} \qquad \text{\textit{The LCD of the numerator is a.}}$$
$$\qquad\qquad\qquad\qquad\qquad\quad \text{\textit{The LCD of the denominator is 6.}}$$

$$= \frac{\frac{a+4}{a}}{\frac{a+4}{6}} \qquad \text{\textit{Add to simplify both numerator and denominator.}}$$

$$= \frac{a+4}{a} \div \frac{a+4}{6} \qquad \text{\textit{Rewrite as a division sentence.}}$$

$$= \frac{\overset{1}{\cancel{a+4}}}{a} \cdot \frac{6}{\underset{1}{\cancel{a+4}}} \qquad \text{\textit{The reciprocal of $\frac{a+4}{6}$ is $\frac{6}{a+4}$.}}$$

$$= \frac{6}{a} \qquad \text{\textit{Divide by common factors.}}$$

b. $\dfrac{m - \frac{m+5}{m-3}}{m+1}$

$$\frac{m - \frac{m+5}{m-3}}{m+1} = \frac{\frac{m(m-3)}{(m-3)} - \frac{m+5}{m-3}}{m+1} \qquad \text{\textit{The LCD of the numerator is $m-3$.}}$$
$$\qquad\qquad\qquad\qquad\qquad\qquad\quad \text{\textit{The LCD of the denominator is $m+1$.}}$$

$$= \frac{\frac{m^2 - 3m - m - 5}{m-3}}{m+1} \qquad \text{\textit{Subtract to simplify the numerator.}}$$

(continued on the next page)

Motivating the Lesson
Hands-On Activity Have students cut scrap paper to represent a mixed number. For example, cut three larger pieces and one half size to make $3\frac{1}{2}$. Have them cut the whole pieces to represent fractions such as $1 = \frac{2}{2}$. Then the students can name the improper fraction by counting the pieces.

2 TEACH

In-Class Examples

For Example 1
Simplify.

a. $4 + \dfrac{x-3}{x+7} \qquad \dfrac{5x+25}{x+7}$

b. $3 + \dfrac{r-4}{r^2-4} \qquad \dfrac{3r^2+r-16}{r^2-4}$

For Example 2
How many Ping-Pong balls will weigh $1\frac{7}{8}$ pounds altogether?
300

For Example 3
Simplify each rational expression.

a. $\dfrac{\frac{1}{a} - \frac{1}{b}}{\frac{1}{a} + \frac{1}{b}} \qquad \dfrac{b-a}{b+a}$

b. $\dfrac{r - \frac{r+5}{r-3}}{r+1} \qquad \dfrac{r-5}{r-3}$

Teaching Tip Help students to see that they are changing numbers with unlike denominators to numbers with like or common denominators. For example, consider $4 + \dfrac{x+y}{x-5}$ as $\dfrac{4}{1} + \dfrac{x+y}{x-5}$. Changing the mixed expression to an algebraic fraction requires that $\dfrac{4}{1}$ be renamed with the LCD $x - 5$ as its denominator.

$$= \dfrac{\dfrac{m^2 - 4m - 5}{m - 3}}{m + 1} \qquad \textit{Simplify.}$$

$$= \dfrac{\dfrac{(m + 1)(m - 5)}{m - 3}}{m + 1} \qquad \textit{Factor the numerator.}$$

$$= \dfrac{(m + 1)(m - 5)}{m - 3} \div (m + 1) \qquad \textit{Rewrite as a division sentence.}$$

$$= \dfrac{(m + 1)(m - 5)}{m - 3} \cdot \dfrac{1}{(m + 1)} \qquad \textit{The reciprocal of } m + 1 \textit{ is } \tfrac{1}{m + 1}.$$

$$= \dfrac{\overset{1}{\cancel{(m + 1)}}(m - 5)}{m - 3} \cdot \dfrac{1}{\underset{1}{\cancel{(m + 1)}}} \qquad \textit{Divide by the common factors.}$$

$$= \dfrac{m - 5}{m - 3}$$

You can use a graphing calculator to check the work you did in Example 3.

EXPLORATION

GRAPHING CALCULATORS

When you graph a function that involves rational expressions, remember to enclose every numerator and denominator in parentheses.

Your Turn   b. $x \neq 3, x \neq -1$ a, c. See margin.

a. Graph $y = \dfrac{x - \dfrac{x + 5}{x - 3}}{x + 1}$. Use the window $[-1.4, 8]$ by $[-5, 5]$.

b. What are the excluded values of x in the expression in part a? $3, -1$

c. Graph $y = \dfrac{x - 5}{x - 3}$ on the same display used in part a.

d. What are the excluded values of x in the expression in part c? $x \neq 3$

e. Except for excluded values, do the graphs appear the same?   yes

Example ④ **Simplify** $\dfrac{a - 2 + \dfrac{3}{a + 2}}{a + 1 - \dfrac{10}{a + 4}}$.

$$\dfrac{a - 2 + \dfrac{3}{a + 2}}{a + 1 - \dfrac{10}{a + 4}} = \dfrac{\dfrac{(a - 2)(a + 2)}{a + 2} + \dfrac{3}{a + 2}}{\dfrac{(a + 1)(a + 4)}{(a + 4)} - \dfrac{10}{a + 4}}$$

The LCD of the numerator is $a + 2$.

The LCD of the denominator is $a + 4$.

$$= \dfrac{\dfrac{a^2 - 4 + 3}{a + 2}}{\dfrac{a^2 + 5a + 4 - 10}{(a + 4)}}$$

Add to simplify the numerator.

Subtract to simplify the denominator.

$$= \dfrac{\dfrac{a^2 - 1}{(a + 2)}}{\dfrac{a^2 + 5a - 6}{(a + 4)}}$$

Simplify.

$$= \dfrac{\dfrac{(a + 1)(a - 1)}{(a + 2)}}{\dfrac{(a + 6)(a - 1)}{(a + 4)}}$$

Factor to simplify the numerator and denominator.

$$= \dfrac{(a + 1)\overset{1}{\cancel{(a - 1)}}}{(a + 2)} \cdot \dfrac{(a + 4)}{(a + 6)\underset{1}{\cancel{(a - 1)}}}$$

Multiply by the reciprocal.

$$= \dfrac{a^2 + 5a + 4}{a^2 + 8a + 12}$$

Multiply.

Communicating Mathematics

Study the lesson. Then complete the following.

1. **Determine** the simplified form of $\dfrac{\frac{3}{(x+1)(x+2)(x+3)}}{\frac{4}{(x+3)(x+2)(x+1)}}$ mentally. $\dfrac{3}{4}$

2. **a.** What is the LCD for the expression $3 + \dfrac{x}{2} + \dfrac{4}{x} + \dfrac{x^2 + 3x + 15}{x-2}$? $2x(x-2)$

   b. Write the expression in simplified form. $\dfrac{3x^3 + 10x^2 + 26x - 16}{2x(x-2)}$

Guided Practice

Write each mixed expression as a rational expression.

3. $8 + \dfrac{3}{x}$ $\dfrac{8x+3}{x}$

4. $5 + \dfrac{8}{3m}$ $\dfrac{15m+8}{3m}$

5. $3m + \dfrac{m+1}{2m}$ $\dfrac{6m^2 + m + 1}{2m}$

Simplify.

6. $\dfrac{4\frac{2}{3}}{5\frac{4}{7}}$ $\dfrac{7}{9}$

7. $\dfrac{6\frac{2}{5}}{3\frac{5}{9}}$ $\dfrac{9}{5}$

8. $\dfrac{\frac{3}{x}}{\frac{x}{3}}$ $\dfrac{9}{x^2}$

9. $\dfrac{\frac{5}{y}}{\frac{10}{y^2}}$ $\dfrac{y}{2}$

10. $\dfrac{\frac{x+4}{x-2}}{\frac{x+5}{x-2}}$ $\dfrac{x+4}{x+5}$

11. $\dfrac{\frac{a-b}{a+b}}{\frac{3}{a+b}}$ $\dfrac{a-b}{3}$

12. Nakita performed an operation on $\dfrac{x+2}{3x-1}$ and $\dfrac{2x^2-8}{3x-1}$ and got $\dfrac{1}{2(x-2)}$. What operation was it? Explain. **Division; see margin for explanation.**

Practice
A

Write each mixed expression as a rational expression.

13. $3 + \dfrac{6}{x+3}$ $\dfrac{3x+15}{x+3}$

14. $11 + \dfrac{a-b}{a+b}$ $\dfrac{12a+10b}{a+b}$

15. $3 - \dfrac{4}{2x+1}$ $\dfrac{6x-1}{2x+1}$

16. $3 + \dfrac{x-4}{x+y}$ $\dfrac{4x+3y-4}{x+y}$

17. $5 + \dfrac{r-3}{r^2-9}$ $\dfrac{5r^2 + r - 48}{r^2-9}$

18. $3 + \dfrac{x^2+y^2}{x^2-y^2}$ $\dfrac{4x^2 - 2y^2}{x^2-y^2}$

Simplify.

19. $\dfrac{7\frac{2}{3}}{5\frac{3}{4}}$ $\dfrac{4}{3}$

20. $\dfrac{6\frac{1}{7}}{8\frac{3}{5}}$ $\dfrac{5}{7}$

21. $\dfrac{\frac{a^3}{b}}{\frac{a^2}{b^2}}$ ab

22. $\dfrac{\frac{x^2y^2}{a}}{\frac{x^2y}{a^3}}$ a^2y

23. $\dfrac{2 + \frac{5}{x}}{\frac{x}{3} + \frac{5}{6}}$ $\dfrac{6}{x}$

24. $\dfrac{4 + \frac{3}{y}}{\frac{3}{8} + \frac{y}{2}}$ $\dfrac{8}{y}$

25. $\dfrac{a - \frac{15}{a-2}}{a+3}$ $\dfrac{a-5}{a-2}$

26. $\dfrac{x + \frac{35}{x+12}}{x+7}$ $\dfrac{x+5}{x+12}$

27. $\dfrac{\frac{x^2-4}{x^2+5x+6}}{x-2}$ $\dfrac{1}{x+3}$

28. $\dfrac{m + \frac{3m+7}{m+5}}{m+1}$ $\dfrac{m+7}{m+5}$ 2

30. $\dfrac{m + 5 + \frac{2}{m+2}}{m + 1 + \frac{6}{m+6}}$ $\dfrac{m+6}{m+2}$

30. $\dfrac{a + 1 + \frac{3}{a+5}}{a + 1 + \frac{3}{a-1}}$

31. $\dfrac{y + 6 + \frac{3}{y+2}}{y + 11 + \frac{48}{y-3}}$ $\dfrac{y-3}{y+2}$ 3

32. $\dfrac{(x+2)(x-1)(x+3)}{(x-2)(x^2+2x-5)}$

33. $\dfrac{x + 3 + \frac{4}{x-2}}{x - 1 - \frac{2}{x+3}}$

33. $\dfrac{t + 1 + \frac{1}{t+1}}{1 - t - \frac{1}{t+1}}$

33. $\dfrac{t^2 + 2t + 2}{-t^2}$

30. $\dfrac{(a+4)(a-1)(a+2)}{(a+5)(a^2+2)}$

Lesson 12-7 Mixed Expressions and Complex Fractions **693**

---

3 PRACTICE/APPLY

Check for Understanding
Exercises 1–12 are designed to help you assess your students' understanding through reading, writing, speaking, and modeling. You should work through Exercises 1–2 with your students and then monitor their work on Exercises 3–12.

Error Analysis
Since division is not commutative, identifying the major division bar in a complex fraction is crucial.

$$\dfrac{\frac{a}{b}}{c} = (a \div b) \div c \text{ or } a \div (b \div c)?$$

The major division bar should be indicated by some emphasis (bolder, longer, or wider).

Assignment Guide

Core: 13–37 odd, 38, 39, 41–48
Enriched: 14–36 even, 37–48

For **Extra Practice**, see p. 784.

The red A, B, and C flags, printed only in the Teacher's Wraparound Edition, indicate the level of difficulty of the exercises.

Study Guide Masters, p. 86

12-7 NAME_____ DATE_____
Study Guide Student Edition Pages 690–695

Mixed Expressions and Complex Fractions

Algebraic expressions such as $a + \frac{b}{c}$ and $5 + \frac{x+y}{x+3}$ are called **mixed expressions.** Changing mixed expressions to improper fractions is similar to changing mixed numbers to improper fractions.

If a fraction has one or more fractions in the numerator or denominator, it is called a **complex fraction.**

Example: Simplify $\dfrac{\frac{2+\frac{4}{a}}{a+2}}{3}$.

$\dfrac{\frac{2+\frac{4}{a}}{a+2}}{3} = \dfrac{\frac{2a+4}{a+2}}{\frac{a}{3}}$

$= \dfrac{\frac{2a+4}{a}}{3}$

$= \dfrac{2a+4}{a} \cdot \dfrac{3}{a+2}$

$= \dfrac{2(a+2)}{a} \cdot \dfrac{3}{a+2}$

$= \dfrac{6}{a}$

Simplifying Complex Fractions

Any complex fraction $\dfrac{\frac{a}{b}}{\frac{c}{d}}$, where $b \ne 0$, $c \ne 0$, and $d \ne 0$, may be expressed as $\dfrac{a}{b} \cdot \dfrac{d}{c}$ or $\dfrac{ad}{bc}$.

Simplify.

1. $10 + \dfrac{60}{x+5}$ $\dfrac{10x+110}{x+5}$

2. $12 + \dfrac{x-y}{x+y}$ $\dfrac{13x+11y}{x+y}$

3. $4 - \dfrac{4}{2x+1}$ $\dfrac{8x}{2x+1}$

4. $\dfrac{2\frac{2}{5}}{3\frac{3}{4}}$ $\dfrac{16}{25}$

5. $\dfrac{\frac{3}{x}}{\frac{4}{y}}$ $\dfrac{3y}{4x}$

6. $\dfrac{\frac{3}{y+2} - \frac{2}{y-2}}{\frac{1}{y+2} - \frac{2}{y-2}}$ $\dfrac{y-10}{-y-6}$

7. $\dfrac{1 - \frac{1}{x}}{1 - \frac{1}{x^2}}$ $\dfrac{x}{x+1}$

8. $\dfrac{\frac{1}{x-3}}{\frac{x+3}{x^2-9}}$ $\dfrac{x+3}{2}$

9. $\dfrac{\frac{x^2-25}{y}}{\frac{x+5}{x^2-5x^2}}$ $\dfrac{x+5}{x^3y}$

Reteaching

Using Models Have students complete the following table.

Single Fraction	Mixed Expression
$\dfrac{6x^2 + 4x + 1}{3}$	$2x^2 + \dfrac{4x+1}{3}$
$\dfrac{2x}{x+y}$	$1 + \dfrac{x-y}{x+y}$
$\dfrac{5x^2 + 5x + 6}{x+1}$	$5x + \dfrac{6}{x+1}$
$\dfrac{a + 3b}{a+b}$	$2 - \dfrac{a-b}{a+b}$

Additional Answer

12. $\dfrac{x+2}{3x-1} \div \dfrac{2x^2-8}{3x-1} =$

$\dfrac{x+2}{3x-1} \cdot \dfrac{3x-1}{2(x-2)(x+2)} = \dfrac{1}{2(x-2)}$

34. What is the quotient when $b + \frac{1}{b}$ is divided by $a + \frac{1}{a}$? $\frac{a(b^2+1)}{b(a^2+1)}$

35. $\frac{32b^2}{35}$

35. What is the product when $\frac{2b^2}{5c}$ is multiplied by the quotient of $\frac{4b^3}{2c}$ and $\frac{7b^3}{8c^2}$?

36. Write $1 + \cfrac{1}{1 + \cfrac{1}{1 + \cfrac{1}{1 + \frac{1}{x}}}}$ in simplest form. $\frac{5x+3}{3x+2}$

Graphing Calculator

37. a. Graph $y = \cfrac{\frac{x+3}{2x}}{\frac{3x+9}{4}}$. What are its excluded values? $x \neq 0, x \neq -3$

b. Graph $\frac{x+3}{2x} \cdot \frac{4}{3x+9}$ on the same display used in part a. What are its excluded values? $x \neq 0, x \neq -3$

c. Except for the excluded values, are the graphs the same? **yes**

Critical Thinking

38. Simplify $\cfrac{3}{1 - \frac{3}{3+y}} - \cfrac{3}{\frac{3}{3-y} - 1}$. **6**

Applications and Problem Solving

39. Acoustics If a train is moving toward you at v miles per hour and blowing its whistle at a frequency of f, then you hear it as though it were blowing its whistle with a frequency h, which is defined as $h = \cfrac{f}{1 - \frac{v}{s}}$, where s is the speed of sound.

a. Simplify the right side of this formula. $\frac{fs}{s-v}$

b. Suppose a train whistle blows at 370 cycles per second (at the same frequency as the first F sharp (F#) above middle C on the piano). The train is moving toward you at a speed of 80 miles per hour. The speed of sound is 760 miles per hour. Find the frequency of the sound as you hear it. **413.5**

c. The note F# and the notes to its right on the piano are listed with their frequencies in the table.

Note	F#	G	G#	A	A#	B	C	C#
Frequency	370.0	392.0	415.3	440.0	466.1	493.8	523.2	554.3

By approximately how many notes did the sound rise in part b? **2**

d. Find the frequency of the same whistle as you would hear it from an approaching TGV, the French train that is the fastest in the world ($v = 236$ miles per hour). By approximately how many notes would it rise? **6**

F Y I

Have you noticed that the pitch (frequency) of a train whistle or a horn from a speeding car gets higher as it approaches you and lower as it moves away from you? This effect is known as the *Doppler effect,* named for the Austrian scientist Christian Doppler, who discovered it in 1842.

Practice Masters, p. 86

12-7

NAME _____ DATE _____

Student Edition
Pages 690–695

Practice

Mixed Expressions and Complex Fractions

Simplify.

1. $\cfrac{2\frac{1}{2}}{4\frac{3}{4}}$ $\frac{10}{19}$

2. $\cfrac{\frac{a^2}{b}}{\frac{2a}{b^2}}$ $\frac{a^2b}{2}$

3. $\cfrac{\frac{m^2}{6n}}{\frac{3m}{n^2}}$ $\frac{nm}{18}$

4. $\cfrac{\frac{b-2}{b+3}}{\frac{b-2}{3}}$ $\frac{3}{b+3}$

5. $\cfrac{\frac{x^2-y^2}{x^2}}{\frac{x+y}{3x}}$ $\frac{3(x-y)}{x}$

6. $\cfrac{\frac{w+4}{w}}{\frac{w^2-16}{w}}$ $\frac{1}{w-4}$

7. $\cfrac{\frac{x^2-1}{x}}{\frac{x-1}{x^3}}$ $x(x+1)$

8. $\cfrac{\frac{a}{a-b}}{\frac{a^2}{a^2-b^2}}$ $\frac{a+b}{a}$

9. $\cfrac{\frac{r}{s}-\frac{s}{r}}{\frac{1}{s}+\frac{1}{r}}$ $r-s$

10. $\cfrac{\frac{4}{c^2-d^2}}{\frac{2}{c+d}}$ $\frac{2}{c-d}$

11. $\cfrac{\frac{m}{n}+1}{\frac{m}{n}-1}$ $\frac{m+n}{m-n}$

12. $\cfrac{\frac{3}{a}+\frac{4}{b}}{\frac{4}{a}-\frac{3}{b}}$ $\frac{3b+4a}{4b-3a}$

13. $\cfrac{\frac{x^2-y^2}{xy}}{\frac{x+y}{y}}$ $\frac{x-y}{x}$

14. $\cfrac{\frac{a^2+a-12}{a^2+3a-4}}{\frac{a-3}{a^2-a}}$ a

15. $\cfrac{x-\frac{6}{x+2}}{x+\frac{8}{x+5}}$ $\frac{x^3+7x^2+4x-30}{x^3+7x^2+18x+16}$

Extension

Reasoning

Simplify $\cfrac{\frac{a^2-a-1}{a-1}}{a - \frac{1}{a-1}}$. **1**

F Y I

The Doppler effect is used in astronomical measurements and in radar.

40. Statistics In 1993, New Jersey was the most densely populated state, and Alaska was the least densely populated. The population of New Jersey was 7,879,000, and the population of Alaska was 599,000. The land area of New Jersey is about 7419 square miles, and the land area of Alaska is about 570,374 square miles. How many more people were there per square mile in New Jersey than in Alaska in 1993? **approximately 1061**

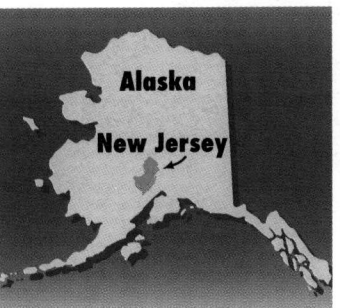

Mixed Review

41. Find $\frac{4x}{2x+6} + \frac{3}{x+3}$. (Lesson 12–6) $\frac{2x+3}{x+3}$

42. Find $\frac{b-5}{b^2-7b+10} \cdot \frac{b-2}{3}$. (Lesson 12–2) $\frac{1}{3}$

43. Factor $4x^2 - 1$. (Lesson 10–4) $(2x+1)(2x-1)$

44. Find $x^4y^5 \div xy^3$. (Lesson 9–2) x^3y^2

45. Solve the system of equations. (Lesson 8–4)
$3a + 4b = -25$
$2a - 3b = 6$ $(-3, -4)$

46. Geometry A quadrilaterial has vertices $A(-1, -4)$, $B(2, -1)$, $C(5, -4)$, and $D(2, -7)$. (Lesson 6–6)
a. Graph the quadrilateral and determine the relationship, if any, among its sides.
b. What type of quadrilaterial is $ABCD$? **square**

46a. See margin for graph; opposite sides are parallel, adjacent sides are perpendicular, perpendicular diagonals.

47. Solve $P = 2\ell + 2w$ for w. (Lesson 3–6)

48. Find $\sqrt{225}$. (Lesson 2–8) **15**

47. $w = \frac{P - 2\ell}{2}$

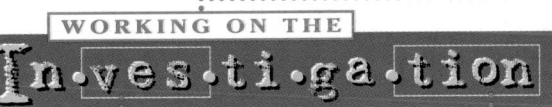

WORKING ON THE
In·ves·ti·ga·tion

Refer to the Investigation on pages 656–657.

A Growing Concern

1 Determine where the future play area and future garden should be placed. Mark off this space on your scale drawing. Justify your reasons for placing them where you did.

2 Develop a detailed plan for the deck and/or patio and walkways for the Sanchez family. Indicate the dimensions. Explain why you chose the placement and the type of materials for each place.

3 Figure the cost of materials, labor, and profit margin of the construction of the deck and/or patio. Justify your costs and make a case for building a deck and/or patio and the walkways as you designed them.

4 Be sure to add the deck and/or patio and the walkways to your drawing.

Add the results of your work to your Investigation Folder.

Lesson 12–7 Mixed Expressions and Complex Fractions **695**

Tech Prep

Statistician Students who are interested in statistics may wish to do further research on the data provided in Exercise 40 and explore the potential growth of this career. For more information on tech prep, see the *Teacher's Handbook*.

In·ves·ti·ga·tion

Working on the Investigation

The Investigation on pages 656–657 is designed to be a long-term project that is completed over several days or weeks. Encourage students to keep their materials in their Investigation Folder as they work on the Investigation.

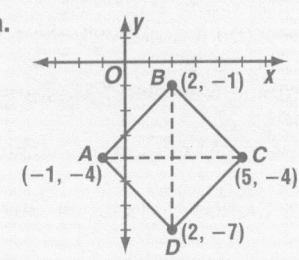

Solving Rational Equations

Instructional Resources

- Study Guide Master 12-8
- Practice Master 12-8
- Enrichment Master 12-8
- Assessment and Evaluation Masters, p. 325
- Graphing Calculator Masters, p. 12
- Modeling Mathematics Masters, p. 83
- Real-World Applications, 30

Transparency 12-8A contains the 5-Minute Check for this lesson; **Transparency 12-8B** contains a teaching aid for this lesson.

Recommended Pacing

Standard Pacing	Days 11 & 12 of 14
Honors Pacing	Days 10 & 11 of 13
Block Scheduling*	Day 5 of 6 (along with Lesson 12-7)
Alg. 1 in Two Years*	Days 22, 23, & 24 of 27

*For more information on pacing and possible lesson plans, refer to the *Block Scheduling Booklet* and *Algebra 1 in Two Years*.

1 FOCUS

5-Minute Check
(over Lesson 12-7)

Simplify.

1. $8 + \frac{5}{3y}$ $\frac{24y + 5}{3y}$

2. $3r^2 + \frac{4}{2r + 1}$ $\frac{6r^3 + 3r^2 + 4}{2r + 1}$

3. $\frac{\frac{b^2}{t}}{\frac{t}{b^3}}$ $\frac{b^5}{t^2}$

4. $\frac{\frac{j^3}{t^2}}{\frac{j+t}{j-t}}$ $\frac{j^3(j - t)}{t^2(j + t)}$

Motivating the Lesson

Situational Problem If Gene addresses 200 envelopes in 8 hours and Marty addresses 200 envelopes in 8 hours, how long will it take them to address 200 envelopes working together?

4 hours

Objective

After studying this lesson, you should be able to:
- solve rational equations.

LOOK BACK

You can refer to Lesson 3-3 for information on solving equations that involve multiple steps.

Work

Tiko and Julio have decided to start a lawn care service in their neighborhood. In order to schedule their clients in an organized manner, they compared notes on how long it took each of them to mow the same yard. Tiko could mow and trim Mrs. Harris's lawn in 3 hours, while it took Julio 2 hours. How long would it take if they both worked together?

You can answer this question by solving a **rational equation**. A rational equation is an equation that contains rational expressions.

Explore Since it takes Tiko 3 hours to do the yard, he can finish $\frac{1}{3}$ of the yard in 1 hour. At his rate, Julio can finish $\frac{1}{2}$ of the yard in an hour. Use the following formula.

$$\underbrace{rate\ of\ work}_{r} \cdot \underbrace{time}_{t} = \underbrace{work\ done}_{w}$$

Plan In t hours, Tiko can do $t \cdot \frac{1}{3}$ or $\frac{t}{3}$ of the job and Julio can do $t \cdot \frac{1}{2}$ or $\frac{t}{2}$ of the job. Thus, $\frac{t}{3} + \frac{t}{2} = 1$, where 1 represents the finished job.

Solve
$$\frac{t}{3} + \frac{t}{2} = 1$$
$$6\left(\frac{t}{3} + \frac{t}{2}\right) = 6(1) \qquad \textit{Multiply each side by the LCD, 3(2) or 6.}$$
$$2t + 3t = 6 \qquad \textit{Use the distributive property.}$$
$$5t = 6$$
$$t = \frac{6}{5} \text{ or } 1\frac{1}{5} \qquad \textit{Check this solution.}$$

Examine If Tiko and Julio work for $1\frac{1}{5}$ hours, will the whole yard get done?

$$\text{Tiko's } rt + \text{Julio's } rt \overset{?}{=} 1 \rightarrow \frac{1}{3} \cdot \frac{6}{5} + \frac{1}{2} \cdot \frac{6}{5} \overset{?}{=} 1$$
$$\frac{2}{5} + \frac{3}{5} \overset{?}{=} 1$$
$$1 = 1$$

Working together, Tiko and Julio can finish the yard in $1\frac{1}{5}$ hours.

Example ① Solve $\frac{10}{3y} - \frac{5}{2y} = \frac{1}{4}$.

$$\frac{10}{3y} - \frac{5}{2y} = \frac{1}{4}$$
$$12y\left(\frac{10}{3y} - \frac{5}{2y}\right) = 12y\left(\frac{1}{4}\right) \qquad \textit{Multiply each side by the LCD, 12y.}$$
$$12y \cdot \frac{10}{3y} - 12y \cdot \frac{5}{2y} = 12y \cdot \frac{1}{4} \qquad \textit{Use the distributive property.}$$
$$40 - 30 = 3y$$
$$10 = 3y$$
$$y = \frac{10}{3} \text{ or } 3\frac{1}{3}$$

GLENCOE Technology

CD-ROM Interaction

A multimedia simulation connects rational equations with underwater sonar devices. A blackline master activity with teacher's notes provides a follow-up to the CD-ROM simulation.

For Windows & Macintosh

Check: $\dfrac{10}{3\left(3\frac{1}{3}\right)} - \dfrac{5}{2\left(3\frac{1}{3}\right)} = \dfrac{1}{4}$

$$\dfrac{10}{10} - \dfrac{15}{20} \stackrel{?}{=} \dfrac{1}{4}$$

$$\dfrac{20}{20} - \dfrac{15}{20} \stackrel{?}{=} \dfrac{1}{4}$$

$$\dfrac{1}{4} = \dfrac{1}{4} \checkmark$$

The solution is $3\frac{1}{3}$.

Multiplying each side of an equation by the LCD of two rational expressions can yield results that are not solutions to the original equation. Such solutions are called **extraneous solutions** or "false" solutions.

Example ❷ Solve $\dfrac{x}{x-1} + \dfrac{2x-3}{x-1} = 2$.

$$\dfrac{x}{x-1} + \dfrac{2x-3}{x-1} = 2$$

$$(x-1)\left(\dfrac{x}{x-1} + \dfrac{2x-3}{x-1}\right) = (x-1)2 \quad \textit{Multiply each side by the LCD, } x-1.$$

$$(x-1)\left(\dfrac{x}{x-1}\right) + (x-1)\left(\dfrac{2x-3}{x-1}\right) = (x-1)2$$

$$x + 2x - 3 = 2x - 2$$

$$3x - 3 = 2x - 2$$

$$x = 1$$

The number 1 is not a solution, since 1 is an excluded value for x. Thus, the equation has no solution.

GL❂BAL CONNECTIONS

Algonquin Indians named the Mississippi River the "Father of Waters." Their phrase *Misi Sipi* meant big water. The Mississippi River drains one eighth of the North American continent, discharging 350 trillion gallons of water a day.

You can use the distance formula to solve real-world problems.

Example ❸

APPLICATION

Commerce

A grain barge operates between Minneapolis, Minnesota, and New Orleans, Louisiana, along the Mississippi River. The maximum speed of the barge in still water is 8 miles per hour. At this rate, a 30-mile trip downstream (with the current) takes as much time as an 18-mile trip upstream (against the current). What is the speed of the current?

Explore   Let c = the speed of the current. The speed of the barge traveling downstream is 8 miles per hour plus the speed of the current, that is, $(8 + c)$ miles per hour. The speed of the barge traveling upstream is 8 miles per hour minus the speed of the current, that is, $(8 - c)$ miles per hour.

Plan   To represent time t, solve $d = rt$ for t. Thus, $t = \dfrac{d}{r}$.

	d	r	$t = \dfrac{d}{r}$
downstream	30	$8 + c$	$\dfrac{30}{8+c}$
upstream	18	$8 - c$	$\dfrac{18}{8-c}$

(continued on the next page)

Longest Rivers in the World

River	Miles
1. Nile	4132
2. Amazon	4000
3. Yangtze	3915
4. Yenisey	3442
5. Huang Ho (Yellow)	3395

Longest Rivers in North America

River	Miles
1. Mackenzie-Peace	2635
2. Missouri-Red Rock	2540
3. Mississippi	2348
4. Missouri	2315
5. Yukon	1979

Solve

$$\frac{30}{8+c} = \frac{18}{8-c}$$

$$30(8-c) = 18(8+c) \quad \text{\textit{Find the cross products.}}$$

$$240 - 30c = 144 + 18c$$

$$96 = 48c$$

$$c = 2$$

Examine   Check the value to see if it makes sense.
The barge goes downstream at $(8 + 2)$ or 10 mph.
A 30-mile trip would take $30 \div 10$ or 3 hours.
The barge goes upstream at $(8 - 2)$ or 6 mph.
An 18-mile trip takes $18 \div 6$ or 3 hours.
Both trips take the same amount of time, so the speed of the current must be 2 miles per hour.

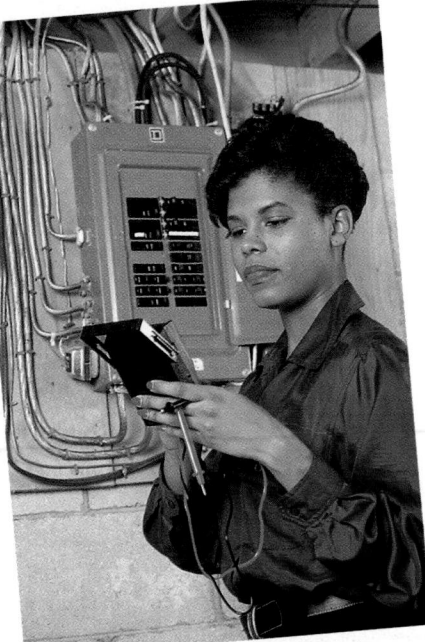

Electricity can be described as the flow of electrons through a conductor, such as a copper wire. Electricity flows more freely through some conductors than others. The force opposing the flow is called *resistance*. The unit of resistance commonly used is the *ohm*.

Resistances can occur one after another, that is, *in series*. Resistances can also occur in branches with the conductor going in the same direction, or *in parallel*.

Example **4**

APPLICATION
Electronics

Assume that $R_1 = 4$ ohms and $R_2 = 3$ ohms. Compute the total resistance of the conductor when the resistances are in series and in parallel.

series

$$R_T = R_1 + R_2$$

$$= 4 + 3 \text{ or } 7$$

parallel

$$\frac{1}{R_T} = \frac{1}{R_1} + \frac{1}{R_2}$$

$$\frac{1}{R_T} = \frac{1}{4} + \frac{1}{3}$$

$$\frac{1}{R_T} = \frac{7}{12}$$

$$12 = 7R_T \quad \text{\textit{Cross products}}$$

$$R_T = \frac{12}{7} \text{ or } 1\frac{5}{7}$$

698   *Chapter 12   Exploring Rational Expressions and Equations*

A circuit, or path for the flow of electrons, often has some resistances connected in series and others in parallel.

Example **5**

APPLICATION
Electronics

A parallel circuit has one branch in series as shown at the right. Given that the total resistance is 2.25 ohms, $R_1 = 3$ ohms, and $R_2 = 4$ ohms, find R_3.

$$\frac{1}{R_T} = \frac{1}{R_1} + \frac{1}{R_2 + R_3}$$ *The total resistance of the branch in series is $R_2 + R_3$.*

$$\frac{1}{2.25} = \frac{1}{3} + \frac{1}{4 + R_3}$$ *$R_T = 2.25$ and $R_1 = 3$*

$$\frac{1}{2.25} - \frac{1}{3} = \frac{1}{4 + R_3}$$

$$\frac{4}{9} - \frac{3}{9} = \frac{1}{4 + R_3}$$

$$\frac{1}{9} = \frac{1}{4 + R_3}$$

$$4 + R_3 = 9$$ *Find the cross products.*

$$R_3 = 5$$

Thus, R_3 is 5 ohms.

R_2 R_3 flow

R_1

In-Class Example

For Example 5
Solve each example for the circuit shown with Example 5.

a. Find R_3 if $R_T = \frac{14}{3}$ ohms, $R_1 = 6$ ohms, and $R_2 = 9$ ohms. **12 ohms**
b. Find R_1 if $R_T = 10$ ohms, $R_2 = 12$ ohms, and $R_3 = 8$ ohms. **20 ohms**

Teaching Tip For Example 5, remind students that when more than two resistances are connected in series, the total resistance is the sum of the individual resistances.
$R_T = R_1 + R_2 + R_3 + ...$
When more than two resistors are connected in parallel, the following formula is used for total resistance.
$$\frac{1}{R_T} = \frac{1}{R_1} + \frac{1}{R_2} + \frac{1}{R_3} + ...$$

CHECK FOR UNDERSTANDING

Communicating Mathematics

Study the lesson. Then complete the following.

1. Refer to the application at the beginning of the lesson. How would the solution be different if Julio takes 6 hours to complete the lawn? **2 hours**

2. **You Decide** Antoinette solved $\frac{2m}{1-m} + \frac{m+3}{m^2-1} = 1$ and claimed that 1 and $-\frac{4}{3}$ were the solutions. Joel says that $-\frac{4}{3}$ is the only solution. Who is correct, and why? **2–3. See margin.**

3. **Define** a rational equation and distinguish it from a linear equation.

MATH JOURNAL

4. **Assess Yourself** **a–b. See students' work.**
   a. Describe an activity that you and a friend do together that each can complete separately. Estimate the time it would take for each of you to complete it working alone.
   b. Use the math you have learned to find out how long it would take to complete the activity if the two of you worked together.

Guided Practice

Solve each equation.

5. $\frac{1}{4} + \frac{4}{x} = \frac{1}{x}$ **-12**

6. $\frac{1}{5} + \frac{3}{2y} = \frac{3}{3y}$ **$-\frac{5}{2}$**

7. $\frac{4}{x+5} = \frac{4}{3(x+2)}$ **$-\frac{1}{2}$**

8. $\frac{x}{2} = \frac{3}{x+1}$ **$-3, 2$**

9. $\frac{a-1}{a+1} - \frac{2a}{a-1} = -1$ **0**

10. $\frac{w-2}{w} - \frac{w-3}{w-6} = \frac{1}{w}$ **3**

11. **Work** Olivia can wash and wax her car, vacuum the interior, and wash the insides of the windows in 5 hours. What part of the job can she do in
    a. 1 hour? $\frac{1}{5}$
    b. 3 hours? $\frac{3}{5}$
    c. x hours? $\frac{x}{5}$

Lesson 12–8 Solving Rational Equations **699**

3 PRACTICE/APPLY

Check for Understanding
Exercises 1–15 are designed to help you assess your students' understanding through reading, writing, speaking, and modeling. You should work through Exercises 1–4 with your students and then monitor their work on Exercises 5–15.

Additional Answers
2. She correctly solved the equation, but $m = 1$ makes a denominator zero and it must therefore be discarded as a solution.
3. A rational equation contains rational expressions whereas a linear equation has integer coefficients.

Reteaching
Using Guess and Check Give a set of work and uniform motion problems with a proposed answer to each student. Students are to determine if the answers are correct without solving the problems. Students will have to identify and examine the conditions in each problem.

Assignment Guide

Core: 17–49 odd, 50–58
Enriched: 16–40 even, 41–58

For **Extra Practice**, see p. 784.

The red A, B, and C flags, printed only in the Teacher's Wraparound Edition, indicate the level of difficulty of the exercises.

---

12. **Recreation** Sally and her brother rented a boat to go fishing in Jones Creek. The maximum speed of the boat in still water was 3 miles per hour. At this rate, a 9-mile trip downstream (with the current) took the same amount of time as a 3-mile trip upstream. Let c = the speed of the current. Copy and complete the table below.

$9, 3 + c, \dfrac{9}{3 + c}$

$3, 3 - c, \dfrac{3}{3 - c}$

12a. $\dfrac{9}{3 + c} = \dfrac{3}{3 - c}$

a. Write an equation that represents the conditions in the problem.
b. Find the speed of the current. **1.5 mph**

Electronics: Exercises 13–15 refer to the diagram below.

13. Find the total resistance, R_T, given that $R_1 = 8$ ohms and $R_2 = 6$ ohms. **3.429 ohms**

14. **4 ohms**

14. Find R_1, given that R_T is $2.\overline{2}$ ohms and $R_2 = 5$ ohms.

15. Find R_1 and R_2, given that the total resistance is $2.\overline{6}$ ohms and R_1 is twice as great as R_2. **8 ohms, 4 ohms**

EXERCISES

Practice

A

Solve each equation.

16. $\dfrac{1}{4} + \dfrac{3}{x} = \dfrac{1}{x}$ -8

17. $\dfrac{1}{5} - \dfrac{4}{3m} = \dfrac{2}{m}$ $\dfrac{50}{3}$

18. $x + 3 = -\dfrac{2}{x}$ -1 or -2

19. $\dfrac{m + 1}{m} + \dfrac{m + 4}{m} = 6$ $\dfrac{5}{4}$

20. $\dfrac{x}{x + 1} + \dfrac{5}{x - 1} = 1$ $\dfrac{-3}{2}$

21. $\dfrac{m - 1}{m + 1} - \dfrac{2m}{m - 1} = -1$ 0

22. $\dfrac{-4}{a + 1} + \dfrac{3}{a} = 1$ 1 or -3

23. $\dfrac{3x}{10} - \dfrac{1}{5x} = \dfrac{1}{2}$ 2 or $-\dfrac{1}{3}$

24. $\dfrac{b}{4} + \dfrac{1}{b} = \dfrac{-5}{3}$ -6 or $-\dfrac{2}{3}$

25. $\dfrac{-4}{n} = 11 - 3n$ 4 or $-\dfrac{1}{3}$

26. $\dfrac{x - 3}{x} = \dfrac{x - 3}{x - 6}$ 3

27. $\dfrac{7}{a - 1} = \dfrac{5}{a + 3}$ -13

28. $\dfrac{3}{r + 4} - \dfrac{1}{r} = \dfrac{1}{r}$ 8

29. $\dfrac{3}{x} + \dfrac{4x}{x - 3} = 4$ $\dfrac{3}{5}$

30. $\dfrac{1}{4m} + \dfrac{2m}{m - 3} = 2$ $\dfrac{3}{25}$

31. $\dfrac{a - 2}{a} - \dfrac{a - 3}{a - 6} = \dfrac{1}{a}$ 3

B

32. $\dfrac{x + 3}{x + 5} + \dfrac{2}{x - 9} = \dfrac{5}{2x + 10}$ 3 or 1

33. $\dfrac{-1}{w + 2} = \dfrac{w^2 - 7w - 8}{3w^2 + 2w - 8}$ 6

Electronics: Refer to the diagram at the right.

34. Find R_T, given that $R_1 = 5$ ohms, $R_2 = 4$ ohms, and $R_3 = 3$ ohms. **2.92 ohms**

35. Find R_1, given that $R_T = 2\dfrac{10}{13}$ ohms, $R_2 = 3$ ohms, and $R_3 = 6$ ohms. **4 ohms**

36. Find R_2, given that $R_T = 3.5$ ohms, $R_1 = 5$ ohms, and $R_3 = 4$ ohms. **7.$\overline{6}$ ohms**

Electronics: Solve each formula for the variable indicated.

37. $R_1 = \dfrac{R_2 R_T}{R_2 - R_T}$

37. $\dfrac{1}{R_T} = \dfrac{1}{R_1} + \dfrac{1}{R_2}$, for R_1

38. $I = \dfrac{E}{r + R}$, for R $\quad R = \dfrac{E - Ir}{I}$

39. $I = \dfrac{nE}{nr + R}$, for n $\quad n = \dfrac{IR}{E - Ir}$

40. $I = \dfrac{E}{\frac{r}{n} + R}$, for r $\quad r = \dfrac{En - IRn}{I}$

Critical Thinking

41. What number would you add to both the numerator and denominator of $\dfrac{4}{11}$ to make a fraction equivalent to $\dfrac{2}{3}$? **10**

42a. $R_T = R_1 + \dfrac{R_2 R_3}{R_2 + R_3}$

42. Refer to the diagram at the right.
  a. Write an equation for the total resistance for the diagram.
  b. Find the total resistance, given that $R_1 = 5$ ohms, $R_2 = 4$ ohms, and $R_3 = 6$ ohms. **7.4 ohms**

Applications and Problem Solving

44. 12.63 ohms

43. **Electricity** Eight lights on a decorated tree are connected in series. Each has a resistance of 12 ohms. What is the total resistance? **96 ohms**

44. **Electricity** Three appliances are connected in parallel: a lamp with a resistance of 60 ohms, an iron with a resistance of 20 ohms, and a heating coil with a resistance of 80 ohms. Find the total resistance.

45. **Air Travel** The flying distance from Honolulu, Hawaii, to San Francisco, California, is approximately 2400 miles. The air speed of a 747-100 airplane is 520 miles per hour. A strong tailwind blows at 120 miles per hour. At 1000 miles into the trip (1400 miles from San Francisco), the aircraft loses power in one engine. As a navigator for the aircraft, you must answer the following questions.
  a. How long would it take to return to Honolulu? **2.5 hours**
  b. How long would it take to continue on to San Francisco? **2.19 hours**
  c. What is the point-of-no-return? That is, at what point into the flight (in miles) would it be quicker to continue the flight to San Francisco than return to Honolulu? **after 923 miles**

46a. $\dfrac{2}{3}$ h or 40 min, $10\dfrac{2}{3}$ miles from one end or $13\dfrac{1}{3}$ miles from the other

46b. 2 mph

46. **Bicycling** The Lake Pontchartrain Causeway in Louisiana is 24 miles long.
  a. Suppose Todd starts cycling at one end at 20 miles per hour and Kristie starts at the other end at 16 miles per hour. How long would it take them to meet? At what distance from either end would they meet?
  b. Todd's rate with no wind is 20 miles per hour, while Kristie's is 16 miles per hour. If Todd cycles against the wind, while Kristie cycles with the wind, and they meet at the midpoint of the bridge, what is the wind speed?
  c. The causeway consists of two parallel bridges. Assume Todd starts to bicycle from one end in one direction at 20 miles per hour. Kristie starts from the other end on the parallel bridge, bicycling at 16 miles per hour in the opposite direction. They pedal in a continuous loop on the two parallel causeways. How long does it take Todd to catch up with Kristie? (*Note:* She has a 24-mile head start.) **6 hours**

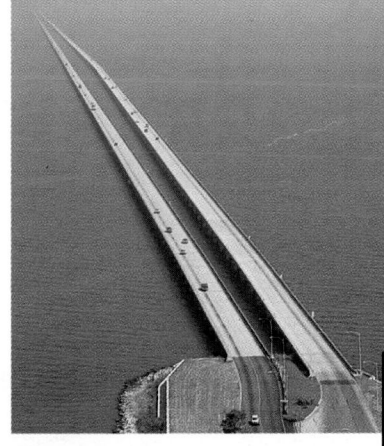

Lesson 12–8 Solving Rational Equations **701**

Closing Activity

Speaking Ask students to explain why 1 is not a solution to Example 4. **It makes the denominator 0.**

Chapter 12, Quiz D (Lessons 12-7 and 12-8), is available in the *Assessment and Evaluation Masters*, p. 325.

Additional Answer

53.

```
0  12 24 36 48 60 72 84 96 108 120
```

Enrichment Masters, p. 87

---

47. Anthropology The formula $c = \frac{100w}{\ell}$ provides a measure called the cephalic index c. Anthropologists use this index to identify skulls by their ethnic characteristics. You calculate the cephalic index by using the width w of a person's head, ear to ear, and the length ℓ of the head from face to back.

  a. Solve the formula for w in terms of c and ℓ.

  b. Solve the formula for ℓ in terms of c and w.

47a. $w = \frac{c\ell}{100}$ b. $\ell = \frac{100w}{c}$

48. Psychology The formula $i = \frac{100m}{c}$ provides a measure i of intelligence, called the intelligence quotient, or I.Q. In the formula, m represents the person's mental age, and c represents the person's chronological age.

  a. Solve the formula for m in terms of i and c.

  b. Solve the formula for c in terms of i and m.   48a. $m = \frac{ic}{100}$ b. $c = \frac{100m}{i}$

49. Baseball Chang has 32 hits in 128 times at bat. His current batting average is $\frac{32}{128} = 0.250$. How many consecutive hits must he get in his next x times at bat in order to get his average up to 0.300? **10**

Mixed Review

50. Simplify $\dfrac{\frac{x^2 - 5x}{x^2 + x - 30}}{\frac{x^2 + 2x}{x^2 + 9x + 18}}$. (Lesson 12–7) $\frac{x+3}{x+2}$

51. Simplify $\frac{a+2}{b^2 + 4b + 4} \div \frac{4a+8}{b+4}$. (Lesson 12–3) $\frac{b+4}{4(b+2)^2}$

52. Find $(0.5a + 0.25b)^2$. (Lesson 9–8) $0.25a^2 + 0.25ab + 0.0625b^2$

53.

53. Sales The amounts in the picture at the right are the cash register totals of 20 customers on the Wednesday before Thanksgiving. Make a box-and-whisker plot of these data. (Lesson 7–7) **See margin.**

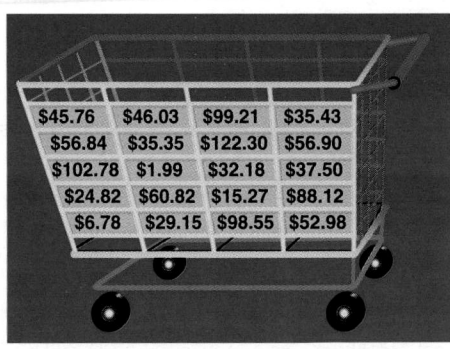

$45.76	$46.03	$99.21	$35.43
$56.84	$35.35	$122.30	$56.90
$102.78	$1.99	$32.18	$37.50
$24.82	$60.82	$15.27	$88.12
$6.78	$29.15	$98.55	$52.98

54. Statistics Refer to the data in Exercise 53. Find the range, median, upper quartile, lower quartile, and interquartile range of the data. Identify any outliers. (Lesson 5–7) **120.31, 45.90, 74.47, 30.67, 43.80; no outliers**

55. $\frac{48}{25}$

55. Write 12 pounds to 100 ounces as a fraction in simplest form. (Lesson 4–1)

56. Solve $3x = -15$. (Lesson 3–2) **−5**

57. Find $(-2)(3)(-3)$. (Lesson 2–6) **18**

58. Complete: $3(2 + x) = 6 + \underline{\ ?\ }$. (Lesson 1–7) **3x**

Extension

Connections A jet aircraft is flying from Honolulu to Los Angeles, a distance of 2570 miles. In still air, the plane flies at 600 mph. There is a 50 mph tailwind pushing the plane. At what point in the trip would it be quicker to go on to Los Angeles than to return to Honolulu? Give your answer in terms of distance from Honolulu. (*Hint:* Write expressions for equal times to continue or to return.) **1177.9 miles**

VOCABULARY

After completing this chapter, you should be able to define each term, property, or phrase and give an example or two of each.

Algebra

excluded values (p. 660)

extraneous solutions (p. 697)

least common denominator (LCD) (p. 686)

least common multiple (LCM) (p. 685)

mixed expressions (p. 690)

rational equation (p. 696)

rational expression (p. 660)

Problem Solving

make an organized list (p. 685)

UNDERSTANDING AND USING THE VOCABULARY

State whether each sentence is *true* or *false*. If false, replace the underlined word or number to make a true sentence.

1. A <u>mixed</u> expression is an algebraic fraction whose numerator and denominator are polynomials.
   false, rational

2. The complex fraction $\dfrac{\frac{4}{5}}{\frac{2}{3}}$ can be simplfied as $\dfrac{6}{5}$. **true**

3. The equation $\dfrac{x}{x-1} + \dfrac{2x-3}{x-1} = 2$ has an extraneous solution of <u>1</u>. **true**

4. The mixed expression $6 - \dfrac{a-2}{a+3}$ can be rewritten as the rational expression $\dfrac{5a+16}{a+3}$. **false,** $\dfrac{5a+20}{a+3}$

5. The least common multiple for $(x^2 - 144)$ and $(x + 12)$ is <u>$x + 12$</u>. **false, $x^2 - 144$**

6. The excluded values for $\dfrac{4x}{x^2 - x - 12}$ are <u>-3 and 4</u>. **true**

7. The least common denominator is the <u>greatest common factor</u> of the denominators. **false, least common multiple**

Instructional Resources

Three multiple-choice tests and three free-response tests are provided in the *Assessment and Evaluation Masters*. Forms 1A and 2A are for honors pacing, and Forms 1B, 1C, 2B, and 2C are for average pacing. Chapter 12 Test, Form 1B is shown at the right. Chapter 12 Test, Form 2B is shown on the next page.

Assessment and Evaluation Masters, pp. 311–312

12 NAME _____ DATE _____

Chapter 12 Test, Form 1B

Write the letter for the correct answer in the blank at the right of each problem.

1. Simplify $\dfrac{4x-4}{4x+4}$.
   A. $x-4$ B. $x+4$ C. $\dfrac{x-1}{x+1}$ D. -1
   1. __C__

2. Simplify $\dfrac{k^2-8k+16}{k^2-16}$.
   A. $k-4$ B. -1 C. $\dfrac{k-4}{k+4}$ D. $\dfrac{k+4}{k-4}$
   2. __C__

3. What are the excluded values of x in $\dfrac{x-3}{x^2-9}$?
   A. $x \neq 3$ B. $x \neq -3$ C. $x \neq 9$ D. $x \neq \pm 3$
   3. __D__

4. What are the excluded values of n in $\dfrac{n^2-3n-28}{n^2+3n-4}$?
   A. $n \neq -1, 4$ B. $n \neq 1$ C. $n \neq -1$ D. $n \neq 1, -4$
   4. __D__

5. Find $\dfrac{3a^2b}{b^2c^3} \cdot \dfrac{2bc^2}{15a}$.
   A. $\dfrac{ab}{3c}$ B. $\dfrac{2a^2b}{5c}$ C. $\dfrac{2a}{5c}$ D. $\dfrac{a^2}{3c^2}$
   5. __C__

6. Find $\dfrac{(y+2)^2}{8} \cdot \dfrac{72}{y^2-4}$.
   A. $\dfrac{9(y+2)}{y-2}$ B. $\dfrac{-9}{y+2}$ C. $\dfrac{9}{y-2}$ D. $\dfrac{-9}{8(y-2)}$
   6. __A__

7. Find $\dfrac{3}{x+2} + \dfrac{2x}{x+2}$.
   A. $\dfrac{5x}{x+2}$ B. $\dfrac{2x+3}{x+2}$ C. $\dfrac{2x+3}{2x+4}$ D. $\dfrac{5x}{2x+4}$
   7. __B__

8. Find $\dfrac{u+t}{w-6} + \dfrac{u+t}{6-w}$.
   A. 0 B. $\dfrac{2t}{6-w}$ C. $\dfrac{u+t}{w-6}$ D. $\dfrac{2u+2t}{w-6}$
   8. __A__

9. Find $\dfrac{16a^2-9w^2}{4a^2-3aw} \cdot \dfrac{a^2}{a+w}$.
   A. $\dfrac{4a-3}{a+w}$ B. $\dfrac{a(4a+3w)}{a+w}$ C. $\dfrac{a^2(4a-3w)}{(a+w)(2a-3w)(2a+w)}$ D. a
   9. __B__

10. Find $\dfrac{n^2+3n-10}{n^2+6n+8} \div \dfrac{n-2}{n^2+2n}$.
    A. $\dfrac{n(n+5)}{n+4}$ B. $\dfrac{(n+5)(n-2)^2}{(n+3)^3}$ C. $\dfrac{n+5}{n+2}$ D. $\dfrac{n(n-5)}{(n-2)^2}$
    10. __A__

11. Use long division to find the quotient $x + 3 \overline{)5x^2 - 9x + 20}$.
    A. $5x + 6 + \dfrac{38}{x+3}$ B. $5x + 6 + \dfrac{2}{x+3}$ C. $5x - 24 + \dfrac{92}{x+3}$ D. $5x - 24 + \dfrac{-52}{x+3}$
    11. __C__

12 NAME _____ DATE _____

Chapter 12 Test, Form 1B (continued)

12. What is the simplest form of $\dfrac{x - 11 + \frac{42}{x+2}}{x-5}$?
    A. $\dfrac{2x+31}{(x-5)(x+2)}$ B. $\dfrac{x^2+20}{(x-5)(x+2)}$ C. $\dfrac{x^2-9x+20}{x-5}$ D. $\dfrac{x-4}{x+2}$
    12. __D__

13. Solve $\dfrac{6}{y+2} + \dfrac{1}{y} = 3$.
    A. $-\dfrac{2}{3}$ and 1 B. $\dfrac{2}{3}$ C. 1 D. $\dfrac{1}{7}$
    13. __A__

14. Solve by making a list. In how many ways can the letters "C-A-T" be arranged to form a word? (A word is any arrangement of the letters.)
    A. 3 B. 4 C. 6 D. 1
    14. __C__

15. Find $\dfrac{2x+3}{x-4} + \dfrac{8}{x+1}$.
    A. $\dfrac{2x+11}{x^2-3x-4}$ B. $\dfrac{2x^2+5x+12}{x^2-3x-4}$ C. $\dfrac{10x-9}{x+1}$ D. $\dfrac{2x^2+13x-29}{x^2-3x-4}$
    15. __D__

16. Find $\dfrac{10}{a-b} - \dfrac{6b}{a^2-b^2}$.
    A. $\dfrac{4}{a-b}$ B. $\dfrac{10a-6b}{a^2-b^2}$ C. $\dfrac{4b}{a-b}$ D. $\dfrac{10a+4b}{a^3-b^2}$
    16. __D__

17. Solve $\dfrac{5x}{3x+1} - \dfrac{1}{9x+3} = \dfrac{7}{6}$.
    A. 1 B. -1 C. $-\dfrac{9}{11}$ D. $\dfrac{9}{11}$
    17. __A__

18. Solve $\dfrac{dn-n}{e} = f$ for n.
    A. $\dfrac{f-e}{d}$ B. $\dfrac{fe}{d-1}$ C. $\dfrac{fe}{d}$ D. $\dfrac{f-e}{d-1}$
    18. __B__

19. Simplify $\dfrac{1}{1 + \frac{1}{1 - \frac{1}{2}}}$.
    A. $\dfrac{1}{3}$ B. $\dfrac{3}{5}$ C. $\dfrac{2}{3}$ D. $\dfrac{1}{2}$
    19. __A__

20. The formula $\dfrac{1}{R_T} = \dfrac{1}{R_1} + \dfrac{1}{R_2} + \dfrac{1}{R_3}$ represents the total resistance of a circuit with 3 resistances connected in parallel. What is the resistance R_1 if $R_T = 2.5$ ohms, $R_2 = 10$ ohms, and $R_3 = 5$ ohms?
    A. 17.5 B. 2.5 C. 10 D. 7
    20. __C__

Bonus Find the value of k so that $x + 1$ is a factor of $3x^3 - 2x^2 + x + k$.
Bonus __6__

Using the STUDY GUIDE AND ASSESSMENT

Skills and Concepts Encourage students to refer to the objectives and examples on the left as they complete the review exercises on the right.

Assessment and Evaluation Masters, pp. 317–318

SKILLS AND CONCEPTS

OBJECTIVES AND EXAMPLES

Upon completing this chapter, you should be able to:

• simplify rational expressions (Lesson 12–1)

Simplify $\dfrac{x+y}{x^2+3xy+2y^2}$.

$$\dfrac{x+y}{x^2+3xy+2y^2} = \dfrac{\overset{1}{\cancel{x+y}}}{(x+y)(x+2y)}$$

$$= \dfrac{1}{x+2y}$$

• multiply rational expressions (Lesson 12–2)

$$\dfrac{1}{x^2+x-12}\cdot\dfrac{x-3}{x+5} = \dfrac{1}{(x+4)\cancel{(x-3)}}\cdot\dfrac{\overset{1}{\cancel{x-3}}}{x+5}$$

$$= \dfrac{1}{(x+4)(x+5)}$$

$$= \dfrac{1}{x^2+9x+20}$$

• divide rational expressions (Lesson 12–3)

$$\dfrac{y^2-16}{y^2-64}\div\dfrac{y+4}{y-8} = \dfrac{y^2-16}{y^2-64}\cdot\dfrac{y-8}{y+4}$$

$$= \dfrac{(y-4)\cancel{(y+4)}}{\cancel{(y-8)}(y+8)}\cdot\dfrac{\cancel{y-8}}{\cancel{y+4}}$$

$$= \dfrac{y-4}{y+8}$$

REVIEW EXERCISES

Use these exercises to review and prepare for the chapter test.

Simplify each rational expression. State the excluded values of the variables.

8. $\dfrac{3x^2y}{12xy^3z}$ $\dfrac{x}{4y^2z}$, $x, y, z \neq 0$

9. $\dfrac{z^2-3z}{z-3}$ $z, z \neq 3$

10. $\dfrac{a^2-25}{a^2+3a-10}$ $\dfrac{a-5}{a-2}$, $a \neq -5, 2$

11. $\dfrac{3a^3}{3a^3+6a^2}$ $\dfrac{a}{a+2}$, $a \neq 0, -2$

12. $\dfrac{x^2+10x+21}{x^3+x^2-42x}$ $\dfrac{x+3}{x(x-6)}$, $x \neq 0, -7, 6$

13. $\dfrac{b^2-5b+6}{b^4-13b^2+36}$ $\dfrac{1}{(b+3)(b+2)}$, $b \neq \pm 2, \pm 3$

Find each product. Assume that no denominator has a value of 0.

14. $\dfrac{7b^2}{9}\cdot\dfrac{6a^2}{b}$ $\dfrac{14a^2b}{3}$

15. $\dfrac{5x^2y}{8ab}\cdot\dfrac{12a^2b}{25x}$ $\dfrac{3axy}{10}$

16. $(3x+30)\cdot\dfrac{10}{x^2-100}$ $\dfrac{30}{x-10}$

17. $\dfrac{3a-6}{a^2-9}\cdot\dfrac{a+3}{a^2-2a}$ $\dfrac{3}{a^2-3a}$

18. $\dfrac{x^2+x-12}{x+2}\cdot\dfrac{x+4}{x^2-x-6}$ $\dfrac{(x+4)^2}{(x+2)^2}$

19. $\dfrac{b^2+19b+84}{b-3}\cdot\dfrac{b^2-9}{b^2+15b+36}$ $b+7$

Find each quotient. Assume that no denominator has a value of 0.

20. $\dfrac{p^3}{2q}\div\dfrac{p^2}{4q}$ $2p$

21. $\dfrac{y^2}{y+4}\div\dfrac{3y}{y^2-16}$ $\dfrac{y^2-4y}{3}$

22. $\dfrac{3y-12}{y+4}\div(y^2-6y+8)$ $\dfrac{3}{y^2+2y-8}$

23. $\dfrac{2m^2+7m-15}{m+5}\div\dfrac{9m^2-4}{3m+2}$ $\dfrac{2m-3}{3m-2}$

(Teacher edition reproduced test pages, left side:)

12 NAME_____ DATE_____

Chapter 12 Test, Form 2B

Perform the indicated operations.

1. $\dfrac{a^2-b^2}{a+b}$

2. $\dfrac{a^2-3a-28}{a^2+3a-4}$

3. $\dfrac{y^2-9}{4}\cdot\dfrac{8}{y+3}$

4. $\dfrac{s^2+2s-8}{s^2+3s-10}\cdot\dfrac{s+5}{s^2-16}$

5. $\dfrac{10-5x}{6+3x}\div\dfrac{5}{12+6x}$

6. $\dfrac{5m}{m+1}\div\dfrac{25m^2}{m^2+2m+1}$

7. $\dfrac{r}{r-5}-\dfrac{5}{r-5}$

8. $\dfrac{r^2}{r+s}-\dfrac{s^2}{r+s}$

9. $\dfrac{4n^2}{2n-3}+\dfrac{9}{3-2n}$

10. $\dfrac{7}{6x^3r}+\dfrac{10}{3xr^2}$

11. $\dfrac{2a+6}{a^2+6a+9}+\dfrac{1}{a+3}$

12. $\dfrac{8}{2a-2b}-\dfrac{3b}{a^2-b^2}$

13. $\dfrac{\frac{r^2+r-6}{r^2}}{\frac{r^2+3r}{r^2+2r}}$

14. $\dfrac{\frac{x-6}{x+2}}{\frac{8}{x+5}}$

15. What are the excluded values of y in the expression $\dfrac{y^2+y-2}{y^2-2y-15}\cdot\dfrac{y^2-5y+5?}{y^2+5y+6}$

16. Use long division to find the quotient $(t^2+4t-6)\div(t+5)$.

Answers column:
1. ___ $a-b$
2. ___ $\dfrac{a-7}{a-1}$
3. ___ $2y-6$
4. ___ $\dfrac{1}{s-4}$
5. ___ $4-2x$
6. ___ $\dfrac{m+1}{5m}$
7. ___ 1
8. ___ $r-s$
9. ___ $2n+3$
10. ___ $\dfrac{7r+20x}{6x^2r^2}$
11. ___ $\dfrac{3}{a+3}$
12. ___ $\dfrac{4a+b}{a^2-b^2}$
13. ___ $\dfrac{r^2-4}{r^2}$
14. ___ $\dfrac{(x+5)(x^2+2x-6)}{(x+2)(x^2+5x+8)}$
15. ___ $y \neq 5, -3, -2$
16. ___ $t-1-\dfrac{1}{t+5}$

12 NAME_____ DATE_____

Chapter 12 Test, Form 2B (continued)

17. Solve by making a tree diagram. For dinner, you can choose spaghetti, tacos, or egg foo yung. For a side dish, you can choose between a salad or soup. Cake, pie, and ice cream are offered for dessert. How many possible meals can you arrange?

17. ___ 18 meals

Solve each equation.

18. $\dfrac{a-3}{10}+\dfrac{a-5}{5}=\dfrac{1}{2}$

18. ___ 6

19. $\dfrac{2}{x+2}=\dfrac{9}{8}-\dfrac{5x}{4x+8}$

19. ___ 2

20. $\dfrac{2}{n-3}+\dfrac{2}{n+5}=\dfrac{5n-7}{n^2+2n-15}$

20. ___ 11

21. Solve $\dfrac{y}{c}+\dfrac{n}{d}=5$ for n.

21. ___ $n=\dfrac{5cd-dy}{c}$

22. Simplify $\left[\left(\dfrac{45\text{ miles}}{1\text{ hour}}\cdot\dfrac{5280\text{ feet}}{1\text{ mile}}\right)\div\dfrac{60\text{ minutes}}{1\text{ hour}}\right]\div\dfrac{60\text{ seconds}}{1\text{ minute}}$.

22. ___ 66 ft/sec

23. The volume of a box is $2x^3-11x^2+17x+6$. One dimension of the box is $x+2$. What are the other two dimensions if they are polynomials in x with integer coefficients?

23. ___ $x+3, 2x+1$

24. An airplane can fly at a rate of 525 mi/h in still air. It can fly 2800 miles with the wind in the same time that it can fly 2450 miles against the wind. Find the rate of the wind.

24. ___ 35 mi/h

25. Given the formula $\dfrac{1}{R_T}=\dfrac{1}{R_1}+\dfrac{1}{R_2}+\dfrac{1}{R_3}$ for the resistance of a circuit, determine the resistance R_2 if $R_T=2$ ohms, $R_1=3$ ohms, and $R_3=2\frac{3}{5}$ ohms.

25. ___ $3\frac{2}{5}$ ohms

Bonus Cardiss can do a job in 4 hours. Rachel can do the same job in 6 hours. Suppose that Rachel works on the job for 2 hours and then is joined by Cardiss. Find the time it will take both working together to finish the job.

Bonus ___ $1\frac{3}{5}$ h

GLENCOE Technology

Test and Review Software

You may use this software, a combination of an item generator and item bank, to create your own tests or worksheets. Types of items include free response, multiple choice, short answer, and open ended.

For IBM & Macintosh

OBJECTIVES AND EXAMPLES

- divide a polynomial by a binomial. (Lesson 12-4)

$$
\begin{array}{r}
x^2 + x - 19 \\
x - 3 \overline{)x^3 - 2x^2 - 22x + 21} \\
\underline{x^3 - 3x^2} \\
x^2 - 22x \\
\underline{x^2 - 3x} \\
- 19x + 21 \\
\underline{- 19x + 57} \\
- 36
\end{array}
$$

The quotient is $x^2 + x - 19 - \dfrac{36}{x-3}$.

REVIEW EXERCISES

Find each quotient.   24. $2ac^2 - 4a^2c + \dfrac{3c^2}{b}$

24. $(4a^2b^2c^2 - 8a^3b^2c + 6abc^2) \div (2ab^2)$

25. $(x^3 + 7x^2 + 10x - 6) \div (x + 3)$ $x^2 + 4x - 2$

26. $(x^3 - 7x + 6) \div (x - 2)$ $x^2 + 2x - 3$

27. $(x^4 + 3x^3 + 2x^2 - x + 6) \div (x - 2)$

28. $(48b^2 + 8b + 7) \div (12b - 1)$

27. $x^3 + 5x^2 + 12x + 23 + \dfrac{52}{x-2}$

28. $4b + 1 + \dfrac{8}{12b - 1}$

- add and subtract rational expressions with like denominators. (Lesson 12-5)

$$
\begin{aligned}
\frac{m^2}{m+4} - \frac{16}{m+4} &= \frac{m^2 - 16}{m+4} \\
&= \frac{(m-4)\cancel{(m+4)}^{\,1}}{\cancel{m+4}^{\,1}} \\
&= m - 4
\end{aligned}
$$

Find each sum or difference. Express in simplest form.

29. $\dfrac{7a}{m^2} - \dfrac{5a}{m^2}$ $\dfrac{2a}{m^2}$

30. $\dfrac{2x}{x-3} - \dfrac{6}{x-3}$ 2

31. $\dfrac{m+4}{5} + \dfrac{m-1}{5}$ $\dfrac{2m+3}{5}$

32. $\dfrac{-5}{2n-5} + \dfrac{2n}{2n-5}$ 1

33. $\dfrac{a^2}{a-b} + \dfrac{-b^2}{a-b}$ $a + b$

34. $\dfrac{m^2}{m-n} - \dfrac{2mn - n^2}{m-n}$ $m - n$

- add and subtract rational expressions with unlike denominators (Lesson 12-6)

$$
\begin{aligned}
\frac{x}{x+3} - \frac{5}{x-2} &= \frac{x}{x+3} \cdot \frac{x-2}{x-2} - \frac{5}{x-2} \cdot \frac{x+3}{x+3} \\
&= \frac{x(x-2)}{(x+3)(x-2)} - \frac{5(x+3)}{(x+3)(x-2)} \\
&= \frac{x^2 - 2x}{(x+3)(x-2)} - \frac{5x + 15}{(x+3)(x-2)} \\
&= \frac{x^2 - 2x - 5x - 15}{(x+3)(x-2)} \\
&= \frac{x^2 - 7x - 15}{x^2 + x - 6}
\end{aligned}
$$

Find each sum or difference.

35. $\dfrac{7n}{3} - \dfrac{9n}{7}$ $\dfrac{22n}{21}$

36. $\dfrac{7}{3a} - \dfrac{3}{6a^2}$ $\dfrac{14a - 3}{6a^2}$

37. $\dfrac{2c}{3d^2} + \dfrac{3}{2cd}$ $\dfrac{4c^2 + 9d}{6cd^2}$

38. $\dfrac{2a}{2a+8} - \dfrac{4}{5a+20}$ $\dfrac{5a - 4}{5a + 20}$

39. $\dfrac{r^2 + 21r}{r^2 - 9} + \dfrac{3r}{r+3}$ $\dfrac{4r}{r - 3}$

40. $\dfrac{3a}{a-2} + \dfrac{5a}{a+1}$ $\dfrac{8a^2 - 7a}{(a-2)(a+1)}$

Applications and Problem Solving Encourage students to work through the exercises in the Applications and Problem Solving section to strengthen their problem-solving skills.

OBJECTIVES AND EXAMPLES

- simplify mixed expressions and complex fractions. (Lesson 12–7)

$$\frac{y - \frac{40}{y-3}}{y+5} = \frac{\frac{y(y-3)}{(y-3)} - \frac{40}{y-3}}{y+5}$$

$$= \frac{\frac{y^2 - 3y - 40}{y-3}}{y+5}$$

$$= \frac{\frac{(y-8)(y+5)}{y-3}}{y+5}$$

$$= \frac{(y-8)(y+5)}{y-3} \div (y+5)$$

$$= \frac{(y-8)(\cancel{y+5})}{y-3} \cdot \frac{1}{\cancel{y+5}}$$

$$= \frac{y-8}{y-3}$$

REVIEW EXERCISES

Write each mixed expression as a rational expression.

41. $4 + \frac{m}{m-2}$ $\quad \frac{5m-8}{m-2}$ 42. $2 - \frac{x+2}{x^2-4}$ $\quad \frac{2x-5}{x-2}$

Simplify.

43. $\dfrac{\frac{x^2}{y^3}}{\frac{3x}{9y^2}}$ $\quad \frac{3x}{y}$ 44. $\dfrac{5 + \frac{4}{a}}{\frac{a}{2} - \frac{3}{4}}$ $\quad \frac{20a+16}{2a^2-3a}$

45. $\dfrac{x - \frac{35}{x+2}}{x + \frac{42}{x+13}}$ $\quad \frac{x^2+8x-65}{x^2+8x+12}$ 46. $\dfrac{y + 9 - \frac{6}{y+4}}{y + 4 + \frac{2}{y+1}}$ $\quad \frac{y^2+11y+10}{y^2+6y+8}$

- solve rational equations (Lesson 12–8)

$$\frac{3}{x} + \frac{1}{x-5} = \frac{1}{2x}$$

$$2x(x-5)\left(\frac{3}{x} + \frac{1}{x-5}\right) = \left(\frac{1}{2x}\right)2x(x-5)$$

$$6(x-5) + 2x = x - 5$$

$$6x - 30 + 2x = x - 5$$

$$7x = 25$$

$$x = \frac{25}{7}$$

Solve each equation. 47. −5

47. $\frac{4x}{3} + \frac{7}{2} = \frac{7x}{12} - \frac{1}{4}$ 48. $\frac{11}{2x} - \frac{2}{3x} = \frac{1}{6}$ $\quad$ 29

49. $\frac{2}{3r} - \frac{3r}{r-2} = -3$ $\quad -\frac{1}{4}$ 50. $\frac{x-2}{x} - \frac{x-3}{x-6} = \frac{1}{x}$ $\quad$ 3

51. $-\frac{5}{m} = 19 - 4m$ 52. $\frac{1}{h+1} + 2 = \frac{2h+3}{h-1}$

$\quad 5, -\frac{1}{4}$ $\qquad\qquad -\frac{3}{2}$

53. 6 bicycles, 24 tricycles, and 24 wagons

APPLICATIONS AND PROBLEM SOLVING

53. **List Possibilities** Wacky Wheels carries bicycles, tricycles, and wagons. They have an equal number of tricycles and wagons in stock. If there are 60 pedals and 180 wheels, how many bicycles, tricycles, and wagons are there in stock? (Lesson 12–6)

DENNIS THE MENACE

"MR. WILSON GAVE THEM TO ME! AND HE'S GOT LOTS MORE!"

54. **Electronics** Assume that $R_1 = 4$ ohms and $R_2 = 6$ ohms. What is the total resistance of the conductor if R_1 and R_2 are: (Lesson 12–8)

a. connected in series? **10 ohms**

b. connected in parallel? **2.4 ohms**

55. **Finance** Barrington High School is raising money to build a house for Habitat for Humanity by doing lawn work for friends and neighbors. Scott can rake a lawn and bag the leaves in 5 hours, while Kalyn can do it in 3 hours. If Scott and Kalyn work together, how long will it take them to rake a lawn and bag the leaves? (Lesson 12–8) $\quad 1\frac{7}{8}$ **hours**

A practice test for Chapter 12 is provided on page 798.

ALTERNATIVE ASSESSMENT

COOPERATIVE LEARNING PROJECT

Fun Puzzle   In this project, you will determine the mathematical equation that is used for a puzzle. Mara developed a puzzle in which the number that you start with ends up being the number that you end with. Edwin was intrigued by this and wanted to determine why it works and if it will work all the time.

Use the steps below to develop the algebraic equation for a puzzle. Then prove that it works by simplifying it. Will it work if the number chosen is zero? Will it work if the number chosen is a fraction? If not, find a counterexample. Will it always work?

1. Choose any whole number.
2. Multiply by three.
3. Add fifteen.
4. Multiply by four.
5. Subtract twice the chosen number.
6. Multiply by five.
7. Add ten times the chosen number.
8. Divide by six.
9. Subtract three times the chosen number.
10. Add thirteen.
11. Divide by seven.
12. Subtract nine.

Consider these ideas while accomplishing your task.

- Read the sequence and determine your variable.
- Develop the algebraic equation expressed by the sequence.
- Determine a plan for simplifying the algebraic equation.
- Simplify the equation.
- Determine when it works and when it doesn't work.

Write a report showing why it works and explaining if it works all the time.

THINKING CRITICALLY

- Write three different rational equations whose solution is the set of all real numbers except a.
- For all real numbers a, b, and x, tell whether each statement is *always true*, *sometimes true*, or *never true*. Give a reason.

$\dfrac{x}{x} = 1$ **Sometimes true, because if $x = 0$, it is not true.**

$\dfrac{ab^2}{b^2} = ab^3$ **Sometimes true, only when $b = 1$.**

$\dfrac{x^2 + 6x - 5}{2x + 2} = \dfrac{x + 5}{2}$ **never true**

PORTFOLIO

In the last chapter, you learned how to solve quadratic equations. Think about that process. Now choose a quadratic equation from your work in that lesson and solve it. Choose a rational equation from your work from this lesson and solve it right next to the quadratic equation that you solved. Refer to these two processes and compare and contrast solving quadratic equations and solving rational equations. Place this in your portfolio.

SELF EVALUATION

Applying a previous skill to help in developing a new skill is often a useful procedure. This can be refered to as building on a firm foundation. If a new idea is difficult to comprehend, stopping and relating it to a skill that has already been mastered can be helpful. The new skill is then "not so new" anymore and becomes more familiar.

Assess yourself. Do you build your skills? When developing a new idea, do you go back to something familiar and then build, or do you start from scratch? Give an example of when you could easily apply this method in learning a new skill in mathematics and in your daily life.

Assessment and Evaluation Masters, pp. 322, 333

12   NAME_____   DATE_____

Chapter 12 Performance Assessment

Instructions: *Demonstrate your knowledge by giving a clear, concise solution to each problem. Be sure to include all relevant drawings and justify your answers. You may show your solution in more than one way or investigate beyond the requirements of the problem.*

1. The mechanical advantage (*M.A.*) of a machine is the ratio of the force (R) put out of the machine to the force (r) put in. A machine with a mechanical advantage of $\frac{4}{3}$ can lift a 4-pound weight with a force of only 3 pounds. The mechanical advantage of the pulley system at right is $M.A. = \frac{R}{r}$.

   a. If $R = \frac{3}{4}$ foot and $r = \frac{1}{2}$ foot, tell how you find the mechanical advantage. Explain each step.

   b. If the weight (W) that can be lifted by a force (F) is given by $W = \frac{R}{r} \cdot F$, what is the weight of an object that can be lifted with a force of 5.5 pounds? Show your work.

   c. If $R = 3x^2y$, $r = 2xy^2$, and $F = 6x^3$, find the weight that can be lifted by the pulley system. Show your work.

   d. Study the pulley system shown at the right. If the mechanical advantage of the complete system is equal to the product of the mechanical advantages of the two smaller pulley systems, A and B, what is the mechanical advantage of the complete system?

   e. Draw your own pulley system and determine the mechanical advantage.

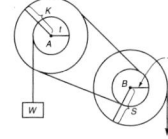

2. The lens maker's formula is $\frac{1}{f} = (\mu - 1)\left(\frac{1}{r_1} - \frac{1}{r_2}\right)$, where f is the focal length, μ is the refractive index, and r_1 and r_2 are the radii as light encounters the curvature of the lens surface.

   a. Solve the equation for f. Show your work.

   b. Find the focal length of a lens if the refractive index is 1.5, $r_1 = 0.8$ inches, and $r_2 = -0.6$ inches.

Scoring Guide
Chapter 12
Performance Assessment

Level	Specific Criteria
3 Superior	• Shows thorough understanding of the concepts of *simplifying rational expressions, multiplying rational expressions,* and *solving rational equations.* • Uses appropriate strategies to solve problems. • Computations are correct. • Written explanations are exemplary. • Diagrams are accurate and appropriate. • Goes beyond requirements of some or all problems.
2 Satisfactory, with Minor Flaws	• Shows understanding of the concepts of *simplifying rational expressions, multiplying rational expressions,* and *solving rational equations.* • Uses appropriate strategies to solve problems. • Computations are mostly correct. • Written explanations are effective. • Diagrams are mostly accurate and appropriate. • Satisfies all requirements of problems.
1 Nearly Satisfactory, with Serious Flaws	• Shows understanding of most of the concepts of *simplifying rational expressions, multiplying rational expressions,* and *solving rational equations.* • May not use appropriate strategies to solve problems. • Computations are mostly correct. • Written explanations are satisfactory. • Diagrams are mostly accurate and appropriate. • Satisfies most requirements of problems.
0 Unsatisfactory	• Shows little or no understanding of the concepts of *simplifying rational expressions, multiplying rational expressions,* and *solving rational equations.* • May not use appropriate strategies to solve problems. • Computations are incorrect. • Written explanations are not satisfactory. • Diagrams are not accurate or appropriate. • Does not satisfy requirements of problems.

Alternative Assessment

The Alternative Assessment section provides students with the opportunity to assess their own work by thinking critically, working with others, keeping a portfolio, and honestly evaluating their own progress. For more information on alternative forms of assessment, see *Alternative Assessment in the Mathematics Classroom,* one of the titles in the Glencoe Mathematics Professional Series.

Performance Assessment

Performance Assessment tasks for this chapter are included in the *Assessment and Evaluation Masters.* A scoring guide is also provided.

---

12   NAME_____ DATE _____

Chapter 12 Cumulative Review

1. State the property shown in $3 + 2 = 3 + 2$. (Lesson 1-6)   1. __reflexive__

2. Simplify $6x^2 + 9x^2 + 4x + 6x$. (Lesson 1-1)   2. __$15x^2 + 10x$__

3. Find $-2.349 + 1.456$. (Lesson 2-5)   3. __-0.893__

4. Use substitution to solve the system of equations $x + y = 2$ and $2x - y = 1$. (Lesson 8-2)   4. __$(1, 1)$__

5. Use elimination to solve the system of equations $2x + 2y = -4$ and $7x + 4 = 1$. (Lesson 8-4)   5. __$(3, -5)$__

Use the data 50, 46, 68, 62, 59, 74, 60, 70, and 64 for exercises 6–10. (Lessons 3-7, 5-7, and 7-7)

6. Find the mean to the nearest tenth.   6. __61.4__

7. Find the median.   7. __62__

8. Find the lower quartile.   8. __54.5__

9. Find the upper quartile.   9. __69__

10. Make a box-and-whisker plot for the data.   10. [box-and-whisker plot]

Simplify. (Lessons 9-1 and 9-2)

11. $(-4x^5y^3)(2xy^2)$ 11. __$-8x^6y^5$__

12. $\left(\frac{2}{3}c^2d^3\right)^4$ 12. __$\frac{16}{81}c^8d^{12}$__

13. $\frac{21hk^3j}{14h^5kj^8}$ 13. __$\frac{3k^2}{2h^4j^7}$__

14. Find $(2x^2y - 3xy + 4y^3) - (4xy - 3y^3 + x^2y)$. (Lesson 9-5)   14. __$x^2y - 7xy + 7y^3$__

15. Find $(3m - 2n)(3m + 2n)$. (Lesson 9-8)   15. __$9m^2 - 4n^2$__

Factor each polynomial, if possible. (Lessons 10-2 and 10-3)

16. $12c - 8d - 15rc + 10rd$ 16. __$(4 - 5r)(3c - 2d)$__

17. $3a^2 - 13a + 12$ 17. __$(3a - 4)(a - 3)$__

---

12   NAME_____ DATE _____

Chapter 12 Cumulative Review (continued)

Solve each equation. (Lessons 10-6 and 11-4)

18. $(x + 3)(x - 4) = 0$ 18. __$-3, 4$__

19. $4r^2 - 7r + 3 = 0$ 19. __$\frac{3}{4}, 1$__

20. $3^{5y + 4} = 3^y$ 20. __-1__

21. $4^{x-1} = 2^x$ 21. __2__

22. Simplify $\frac{2x^2 + 3x + 1}{4x^2 + 4x + 1}$. (Lesson 12-1)   22. __$\frac{x + 1}{2x + 1}$__

23. Simplify $\frac{4x^2 - 8x - 5}{2x^2 + x - 15}$. (Lesson 12-1)   23. __$\frac{2x + 1}{x + 3}$__

24. Find $\frac{a^3}{a - b} \cdot \frac{a^2 - b^2}{a}$. (Lesson 12-2)   24. __$a^2(a + b)$__

25. Find $(2x^2 + 3x - 5) \div (x - 3)$. (Lesson 12-4)   25. __$2x + 9 + \frac{22}{x - 3}$__

26. Find $\frac{2}{3k - 1} + \frac{k}{k + 3}$. (Lesson 12-6)   26. __$\frac{3k^2 + k + 6}{(3k - 1)(k + 3)}$__

27. Find $\frac{c^2}{c^2 - 4} - \frac{5c + 14}{c^2 - 4}$. (Lesson 12-5)   27. __$\frac{c - 7}{c - 2}$__

28. Simplify $\frac{z - \frac{3}{z - 2}}{z - \frac{6}{z - 1}}$. (Lesson 12-7)   28. __$\frac{(z + 1)(z - 1)}{(z + 2)(z - 2)}$__

29. Solve $wx - w = \frac{x}{y}$ for w. (Lesson 12-8)   29. __$w = \frac{x}{y(x - 1)}$__

30. Find three consecutive even integers such that the difference of the squares of the least and the greatest is 96. (Lesson 10-4)   30. __10, 12, and 14__

31. The length of a rectangular lot is 10 feet less than 3 times its width. The perimeter of the lot is 620 feet. Find the dimensions of the lot. (Lesson 3-3)   31. __80 ft by 230 ft__

32. Carlos paid $37.98 for some compact discs. This included $5\frac{1}{2}\%$ sales tax. What did the CDs cost without the tax? (Lesson 4-5)   32. __$36__

33. Find the dimensions of the rectangle whose width is 8 cm less than its length and whose area is 20 cm². (Lesson 11-3)   33. __2 cm and 10 cm__

CUMULATIVE REVIEW

CHAPTERS 1–12

SECTION ONE: MULTIPLE CHOICE

There are eight multiple-choice questions in this section. After working each problem, write the letter of the correct answer on your paper.

1. **Geometry** The measure of the area of a square is 129 square inches. What is the perimeter of this square, rounded to the nearest hundredth?   C
   - **A.** 11.36 in.
   - **B.** 32.25 in.
   - **C.** 45.43 in.
   - **D.** 1040.06 in.

2. Express the relation shown as a set of ordered pairs.   D

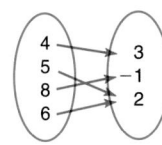

   - **A.** $\{(3, 4), (-1, 8), (2, 5), (2, -6)\}$
   - **B.** $\{(4, 3), (5, 2), (-1, 8), (-6, 2)\}$
   - **C.** $\{4, 5, 8, -6, 3, -1, 2\}$
   - **D.** $\{(4, 3), (5, 2), (8, -1), (-6, 2)\}$

3. **Geometry** Find the measure of the area of the rectangle shown below in simplest form.   B

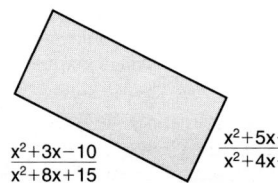

$\frac{x^2+3x-10}{x^2+8x+15}$ $\frac{x^2+5x+6}{x^2+4x+4}$

   - **A.** $\frac{15}{32}x^2 - 1$
   - **B.** $\frac{x - 2}{x + 2}$
   - **C.** $\frac{15x^2 - 32x - 60}{32x^2 + 92x + 60}$
   - **D.** $\frac{9x^2 - 42x + 49}{16x^2}$

4. **Probability** If a card is selected at random from a deck of 52 cards, what is the probability that it is not a face card?   D
   - **A.** $\frac{3}{13}$
   - **B.** $3:10$
   - **C.** $\frac{13}{25}$
   - **D.** $\frac{10}{13}$

5. **Physics** Rafael tossed a rock off the edge of a 10-meter-high cliff with an initial velocity of 15 meters per second. To the nearest tenth of a second, determine when the rock will hit the ground by using the formula $H = -4.9t^2 + vt + h$.   A
   - **A.** 3.6 s
   - **B.** 3.1 s
   - **C.** 4 s
   - **D.** 3.9 s

6. Simplify $\dfrac{\frac{x^2 + 8x + 15}{x^2 + x - 6}}{\frac{x^2 + 2x - 15}{x^2 - 2x - 3}}$.   D
   - **A.** $\frac{23}{30}$
   - **B.** $\frac{x^2 + 4x + 3}{x^2 + x - 6}$
   - **C.** $\frac{x^2 + 10x + 25}{x^2 - x - 2}$
   - **D.** $\frac{x + 1}{x - 2}$

7. **Geometry** What is the value of x if the perimeter of a square is 60 cm and its area is $4x^2 - 28x + 49$ cm²?   B
   - **A.** 4 only
   - **B.** 11 only
   - **C.** 4 and 11
   - **D.** -4 and 11

8. **Physics** The distance a force can move an object is $\frac{2a}{6a^2 - 17a - 3}$ yards. The distance a second force can move the same object is $\frac{a + 2}{a^2 - 9}$ yards. How much farther did the object move when the second force was applied than when the first force was applied?   A
   - **A.** $\frac{4a^2 + 7a + 2}{(a - 3)(a + 3)(6a + 1)}$
   - **B.** $\frac{a + 2}{5a^2 + 17a - 6}$
   - **C.** $\frac{4a^2 + 17a - 2}{(a - 3)(a + 3)(6a - 1)}$
   - **D.** $\frac{10a^2 + 11a + 1}{(3a - 2)(2a + 1)}$

SECTION TWO: SHORT ANSWER

This section contains nine questions for which you will provide short answers. Write your answer on your paper.

9. Evaluate
$42 \div 7 - 1 - 5 + 8 \cdot 2 + 14 \div 2 - 8.$ **15**

10. The rectangular penguin pond at the Bay Park Zoo is 12 meters long by 8 meters wide. The zoo wants to double the area of the pond by increasing the length and width by the same amount. By how much should the length and width be increased?   **4 meters**

11. Determine the slope, y-intercept, and x-intercept of the graph of $2x - 3y = 13.$ $\frac{2}{3}; -\frac{13}{3}; \frac{13}{2}$

12. Find the value of k if the remainder is 15 when $x^3 - 7x^2 + 4x + k$ is divided by $(x - 2)$.   **27**

13. The tens digit of a two-digit number exceeds twice its units digit by 1. If the digits are reversed, the number is 4 more than 3 times the sum of the digits. Find the number.   **52**

14. A long-distance cyclist pedaling at a steady rate travels 30 miles with the wind. He can travel only 18 miles against the wind in the same amount of time. If the rate of the wind is 3 miles per hour, what is the cyclist's rate without the wind?   **12 mph**

15. Solve $-2 \le 2x + 4 < 6$ and graph the solution set. $\{x \mid -3 \le x < 1\}$

16. **Biology** The 2-inch long hummingbird flaps its wings about forty to fifty times each second. At this rate, how many times does it flap its wings in half of an hour? Express your answer in scientific notation. 7.2×10^4 to 9×10^4

17. **Construction** Muturi has 120 meters of fence to make a rectangular pen for his rabbits. If a shed is used as one side of the pen, what would be the maximum area of the pen?   **1800 m²**

SECTION THREE: OPEN-ENDED

This section contains three open-ended problems. Demonstrate your knowledge by giving a clear, concise solution to each problem. Your score on these problems will depend on how well you do the following.

- Explain your reasoning.

- Show your understanding of the mathematics in an organized manner.

- Use charts, graphs, and diagrams in your explanation.

- Show the solution in more than one way or relate it to other situations.

- Investigate beyond the requirements of the problem.

18. Mrs. Bloom bought some impatiens and petunias for her landscaping business for $111.25. How many flats of each did she buy if each flat of impatiens was $10.00, each flat of petunias was $8.75, and if two fewer flats of impatiens than petunias were bought?
    5 flats of impatiens; 7 flats of petunias

19. Graph the quadratic function $y = -x^2 + 6x + 16.$ Include the equation of the axis of symmetry, the coordinates of the vertex, and the roots of the related quadratic equation.   **See margin.**

20. Solve the system of equations below by graphing. Explain how you determined the solution, and name at least three ordered pairs that satisfy the system.

$y \le x + 3$

$2x - 2y < 8$

$2y + 3x > 4$ **See margin.**

13

Exploring Radical Expressions and Equations

PREVIEWING THE CHAPTER

In this chapter, students apply the Pythagorean theorem to solve problems involving geometry. They simplify radical expressions with and without graphing technology. Students add and subtract radical expressions and apply this skill to solving radical equations. The distance formula is introduced as an integration with geometry. Students model completing the square and then apply this skill to solving equations.

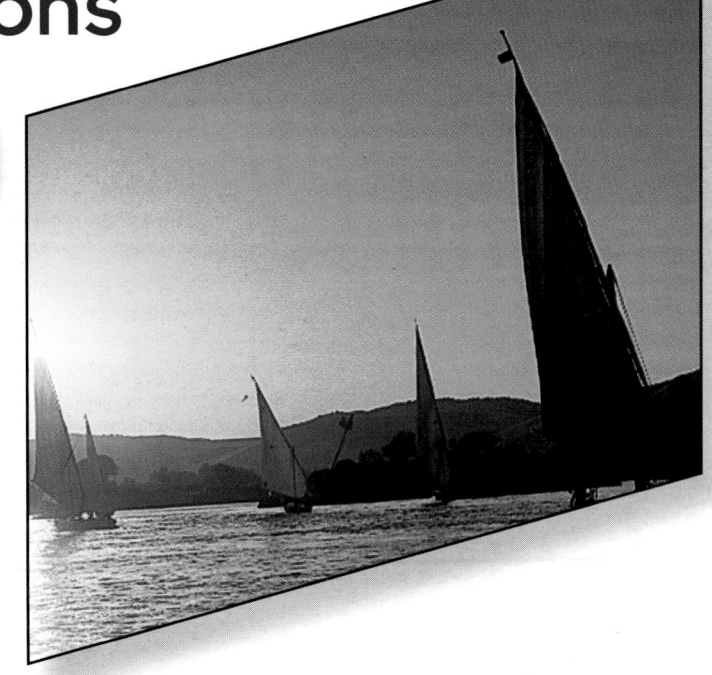

Lesson (Pages)	Lesson Objectives	NCTM Standards	State/Local Objectives
13-1A (712)	Use a geoboard to find areas by using the Pythagorean theorem.	1–5, 7	
13-1 (713–718)	Use the Pythagorean theorem to solve problems.	1–5, 7	
13-2 (719–725)	Simplify square roots. Simplify radical expressions.	1–5	
13-2B (726)	Use a graphing calculator to simplify and approximate values of expressions containing radicals.	1–5	
13-3 (727–731)	Simplify radical expressions involving addition, subtraction, and multiplication.	1–5	
13-4 (732–736)	Solve radical equations.	1–5	
13-5 (737–741)	Find the distance between two points in the coordinate plane.	1–5, 8	
13-6A (742)	Use algebra tiles as a model for completing the square.	1–5	
13-6 (743–747)	Solve problems by identifying subgoals. Solve quadratic equations by completing the square.	1–5	

A complete, 1-page lesson plan is provided for each lesson in the *Lesson Planning Guide*. Answer keys for each lesson are available in the *Answer Key Masters*.

ORGANIZING THE CHAPTER

You may want to refer to the **Course Planning Calendar** on page T12 for detailed information on pacing.
PACING: Honors—11 days; **Block**—7 days; **Two Years**—22 days

LESSON PLANNING CHART

Lesson (Pages)	Materials/ Manipulatives	Extra Practice (Student Edition)	Study Guide	Practice	Enrichment	Assessment and Evaluation	Modeling Mathematics	Multicultural Activity	Tech Prep Applications	Graphing Calculator	Science and Math Lab Manual	Real-World Applications	Interactive Mathematics Tools Software	Teaching Transparencies
							BLACKLINE MASTERS							
13-1A (712)	geoboard* dot paper						p. 34						13-1A	
13-1 (713–718)		p. 785	p. 88	p. 88	p. 88		pp. 69–71	p. 25	p. 25			31		13-1A 13-1B
13-2 (719–725)	graphing calculator	p. 785	p. 89	p. 89	p. 89	p. 352			p. 26				13-2	13-2A 13-2B
13-2B (726)	graphing calculator									pp. 38, 39				
13-3 (727–731)	calculator	p. 785	p. 90	p. 90	p. 90	pp. 351, 352	p. 84							13-3A 13-3B
13-4 (732–736)	graphing software	p. 786	p. 91	p. 91	p. 91									13-4A 13-4B
13-5 (737–741)		p. 786	p. 92	p. 92	p. 92	p. 353		p. 26		p. 13	pp. 57–62	13-5.1 13-5.2		13-5A 13-5B
13-6A (742)	algebra tiles* equation mat*						p. 35							
13-6 (743–747)	algebra tiles*	p. 786	p. 93	p. 93	p. 93	p. 353						32		13-6A 13-6B
Study Guide/ Assessment (749–753)						pp. 337 –350, 354 –356								

*Included in Glencoe's Student Manipulative Kit and Overhead Manipulative Resources.

ORGANIZING THE CHAPTER

OTHER CHAPTER RESOURCES

Student Edition
Chapter Opener, pp. 710–711
Mathematics and Society, p. 736
Working on the Investigation,
   pp. 718, 741
Closing the Investigation, p. 748

Teacher's Classroom Resources
Investigations and Projects Masters,
   pp. 73–76
Algebra and Geometry Overhead
Manipulative Resources,
   pp. 36–38

Technology
Test and Review Software (IBM
   and Macintosh)
CD-ROM Interactions (Windows
   and Macintosh)

Professional Publications
Block Scheduling Booklet
Glencoe Mathematics Professional
Series

OUTSIDE RESOURCES

Books/Periodicals
Downing, Douglas, *Algebra the Easy Way*, Dale
   Seymour Publications

Software
Connecting to Algebra, Prometheus Software,
   Macintosh

Videos/CD-ROMs
The Theorem of Pythagorus, California Institute
   of Technology

ASSESSMENT RESOURCES

Student Edition
Math Journal, pp. 723, 729, 734
Mixed Review, pp. 718, 725,
   731, 736, 741, 747
Self Test, p. 731
Chapter Highlights, p. 749
Chapter Study Guide and
   Assessment, pp. 750–752
Alternative Assessment, p. 753
    Portfolio, p. 753

Teacher's Wraparound Edition
5-Minute Check, pp. 713, 719,
   727, 732, 737, 743
Check for Understanding, pp. 715,
   723, 729, 734, 738, 745
Closing Activity, pp. 718, 725,
   731, 736, 741, 747
Cooperative Learning, pp. 733,
   744

Assessment and Evaluation Masters
Multiple-Choice Tests, Forms 1A
   (Honors), 1B (Average), 1C
   (Basic), pp. 337–342
Free-Response Tests, Forms 2A
   (Honors), 2B (Average), 2C
   (Basic), pp. 343–348
Calculator-Based Test, p. 349
Performance Assessment, p. 350
Mid-Chapter Test, p. 351
Quizzes A–D, pp. 352–353
Standardized Test Practice, p. 354
Cumulative Review, pp. 355–356

ENHANCING THE CHAPTER

Examples of some of the materials for enhancing Chapter 13 are shown below.

DIVERSITY

Multicultural Activity Masters, pp. 25, 26

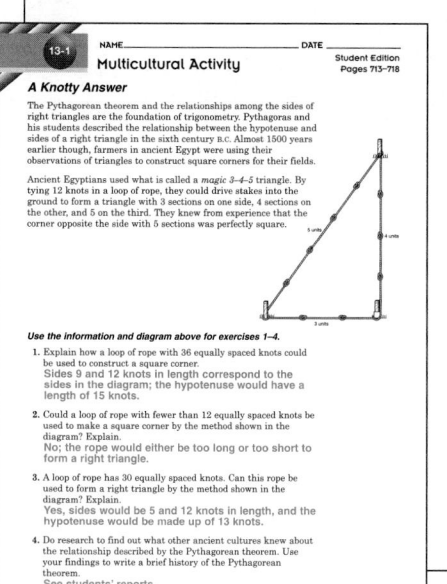

13-1   NAME_____   DATE_____

Multicultural Activity   Student Edition Pages 713–718

A Knotty Answer

The Pythagorean theorem and the relationships among the sides of right triangles are the foundation of trigonometry. Pythagoras and his students described the relationship between the hypotenuse and sides of a right triangle in the sixth century B.C. Almost 1500 years earlier though, farmers in ancient Egypt were using their observations of triangles to construct square corners for their fields.

Ancient Egyptians used what is called a *magic 3-4-5 triangle*. By tying 12 knots in a loop of rope, they could drive stakes into the ground to form a triangle with 3 sections on one side, 4 sections on the other, and 5 on the third. They knew from experience that the corner opposite the side with 5 sections was perfectly square.

Use the information and diagram above for exercises 1–4.

1. Explain how a loop of rope with 36 equally spaced knots could be used to construct a square corner.
Sides 9 and 12 knots in length correspond to the sides in the diagram; the hypotenuse would have a length of 15 knots.

2. Could a loop of rope with fewer than 12 equally spaced knots be used to make a square corner by the method shown in the diagram? Explain.
No; the rope would either be too long or too short to form a right triangle.

3. A loop of rope has 30 equally spaced knots. Can this rope be used to form a right triangle by the method shown in the diagram? Explain.
Yes, sides would be 5 and 12 knots in length, and the hypotenuse would be made up of 13 knots.

4. Do research to find out what other ancient cultures knew about the relationship described by the Pythagorean theorem. Use your findings to write a brief history of the Pythagorean theorem.
See students' reports.

APPLICATIONS

Real-World Applications, 31, 32

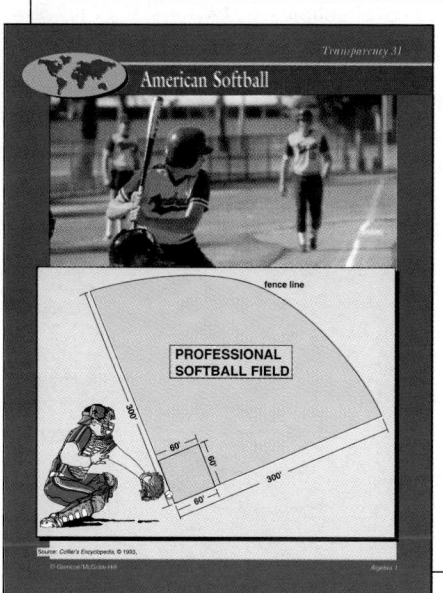

Transparency 31

American Softball

PROFESSIONAL SOFTBALL FIELD

fence line

TECHNOLOGY

Graphing Calculator Masters, p. 13

13-5   NAME_____   DATE_____

Graphing Calculator Activity   Student Edition Pages 737–741

Finding the Distance Between a Point and a Line

The following example uses a graphing calculator to find the distance between a point and a line.

Example:

Find the distance between the point at $(11, -1)$ and the line $y = 2x - 3$.

a. **Select the viewing window parameters.** Choose any values of x and y that are convenient.

b. **Graph** the point and the line.

c. Write the equation of the line through the given point that is perpendicular to the given line. The slope of the line $y = 2x - 3$ is 2. The slope of the line perpendicular to it must be $-\frac{1}{2}$ (so that the product of the two slopes is -1). Use the point-slope form of a linear equation.

Then an equation of the line through $(11, -1)$ that is perpendicular to the given line is $y + 1 = -\frac{1}{2}(x - 11)$, or $y = -\frac{1}{2}x + 4\frac{1}{2}$.

d. **Graph** $y = -0.5x + 4.5$.

e. **Trace** the coordinates of the intersection of the two lines (round to the nearest whole number). The intersection point is $(3, 3)$.

f. Use the distance formula to find the distance between the intersection point and the given point (round to the nearest tenth). This is the distance between the given point and the given line.

distance = $\sqrt{(11 - 3)^2 + (-1 - 3)^2}$
= $\sqrt{64 + 16}$
= $\sqrt{80} \approx 8.9$

Use a graphing calculator and follow the procedure above to find the distance between the given point and the line. Make a sketch of the calculator display. Give the equations of the two lines that you use. Students' graphs will vary.

1. point: $(7, 2)$
line: $y = x + 5$
distance ≈ 7
equation:
$y = -x + 9$

2. point: $(-8, -6)$
line: $y = \frac{2}{5}x + 3$
distance ≈ 5.4
equation:
$y = -\frac{5}{2}x - 26$

TECH PREP

Tech Prep Applications Masters, pp. 25, 26

13-1   NAME_____   DATE_____

Tech Prep Applications   Student Edition Pages 713–718

Roadways (Concrete Contractor)

Homeowners, business owners, school administrators, and city managers want sidewalks and roadways made of different kinds of material. Concrete is one of the most popular and serviceable. A concrete contractor is the one who mixes, delivers, and pours the concrete. First, however, the contractor must get some information about the geometry of the land.

Routes 82 and 51 intersect in the center of Pine City. Route 82 runs east and west and Route 51 runs north and south. To reduce the traffic in the center of town, government officials plan to build a road connecting a point 16 miles south of the center of town on Route 51 to a point 12 miles west of the center of town on Route 82. Before a contractor can give an estimate for the cost of the job, she must determine the length of the new road. Since the roads form a right triangle, the Pythagorean Theorem can be used to find the length of the new road.

$c^2 = a^2 + b^2$
$c^2 = 16^2 + 12^2$
$c^2 = 256 + 144$
$c^2 = 400$
$c = 20$

The new road will be 20 miles long.

Solve.

1. Suppose a person wishes to get from point S to point W, how many less miles will the person travel by taking the new road instead of Routes 51 and 82? 8 miles

2. If the new road is built from a point 14 miles south of the center of town to a point 10 miles west of the center of town, what is the length of the new road? about 17.2 miles

3. How much shorter would the road in Exercise 2 be than the road in the example? about 2.8 miles

CONNECTIONS

Science and Math Lab Manual, pp. 57–62

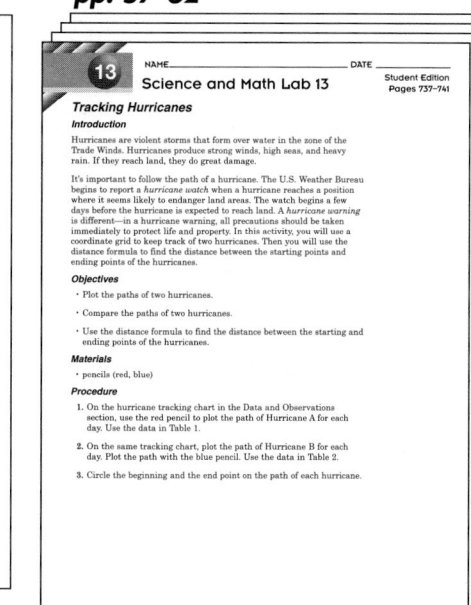

13   NAME_____   DATE_____

Science and Math Lab 13   Student Edition Pages 737–741

Tracking Hurricanes

Introduction

Hurricanes are violent storms that form over water in the zone of the Trade Winds. Hurricanes produce strong winds, high seas, and heavy rain. If they reach land, they do great damage.

It's important to follow the path of a hurricane. The U.S. Weather Bureau begins to report a *hurricane watch* when a hurricane reaches a position where it seems likely to endanger land areas. The watch begins a few days before the hurricane is expected to reach land. A *hurricane warning* is different—in a hurricane warning, all precautions should be taken immediately to protect life and property. In this activity, you will use a coordinate grid to keep track of two hurricanes. Then you will use the distance formula to find the distance between the starting points and ending points of the hurricanes.

Objectives

· Plot the paths of two hurricanes.

· Compare the paths of two hurricanes.

· Use the distance formula to find the distance between the starting and ending points of the hurricanes.

Materials

· pencils (red, blue)

Procedure

1. On the hurricane tracking chart in the Data and Observations section, use the red pencil to plot the path of Hurricane A for each day. Use the data in Table 1.

2. On the same tracking chart, plot the path of Hurricane B for each day. Plot the path with the blue pencil. Use the data in Table 2.

3. Circle the beginning and the end point on the path of each hurricane.

PROBLEM SOLVING

Problem of the Week Cards, 35, 36

No Time to Waste   Problem-of-the-Week Card 35

The Problem

Mr. Hansen wants to make 3 slices of French bread into Texas toast. Only 2 slices of bread will fit into his pan at a time. It takes him 5 seconds to put a slice of bread into the pan or take it out of the pan, and it takes 1 minute to fry each slice of bread. What is the shortest length of time Mr. Hansen needs to fry 3 slices of bread?

Strategies and Hints

1. Do you use the same units of measure throughout the problem?

2. On which action (toasting, turning, moving) should you concentrate?

3. All slices of bread can be toasted.

4. Flipping the toast is an important part of the problem.

Exploring Radical Expressions and Equations

Objectives

In this chapter, you will:

- use the Pythagorean theorem to solve problems,
- simplify radical expressions,
- solve problems involving radical equations,
- solve quadratic equations by completing the square, and
- solve problems by identifying subgoals.

Teens Talk to Parents

Teens rate the following things as important for parents to do for children who are 12 years old or younger: ○
For people their own age: ■

	Very important	Somewhat important	Not too important		Very important	Somewhat important	Not too important
Help them with their homework	○ 74%	22%	4%	Talk with their teachers regularly	○ 43%	44%	13%
	■ 42%	38%	20%		■ 32%	42%	25%
Give them a hug at least once a week	○ 70%	34%	8%	Be active in school activities like the PTA	○ 39%	43%	17%
	■ 46%	34%	19%		■ 24%	42%	34%

Source: *Oakland Press, 1995*

Almost everyone agrees that people who can communicate, who really listen to one another, have the healthiest relationships. This is certainly true when talking about teenagers and their parents. Kids can learn much from the experiences of their parents, and adults can benefit from the fresh ideas they share with their teens. What is the communication like in your home? What topics do you think are important to discuss with your parents?

TIME *Line*

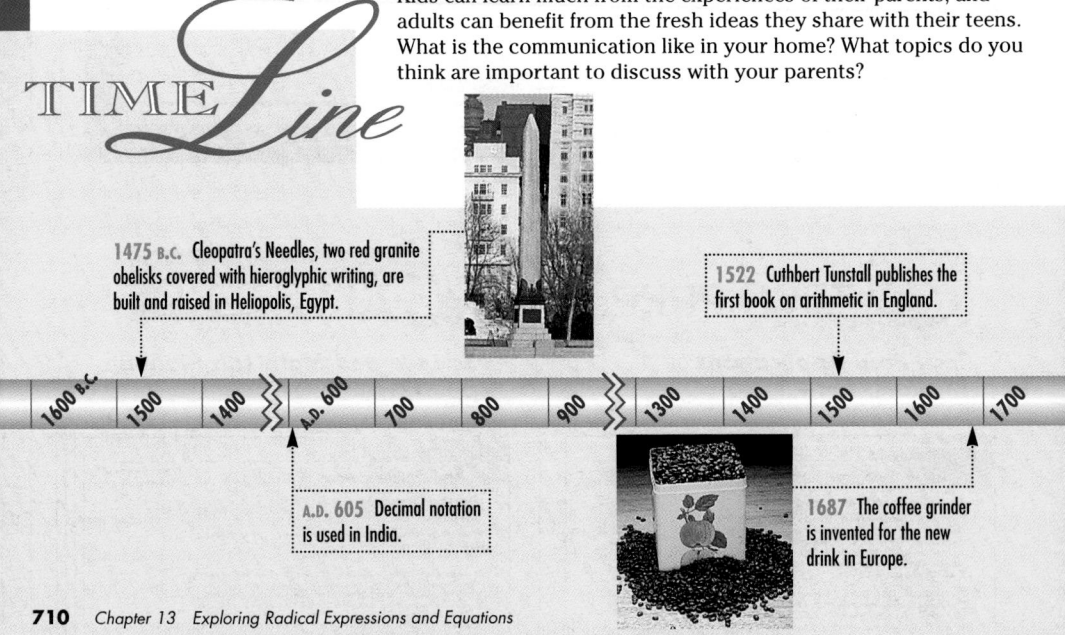

1475 B.C. Cleopatra's Needles, two red granite obelisks covered with hieroglyphic writing, are built and raised in Heliopolis, Egypt.

1522 Cuthbert Tunstall publishes the first book on arithmetic in England.

A.D. 605 Decimal notation is used in India.

1687 The coffee grinder is invented for the new drink in Europe.

Form teams and conduct your own survey of what teens want from their parents or guardians. You can use the questions from the chart on page 710 or add your own. Keep track of the age, gender, and number of people questioned and their responses. See if you can make conclusions about the attitude of teens toward their parents' communication with them. Create charts or graphs using your data and report to the class on your findings.

At age 12, **Ryan Holladay** of Arlington, Virginia, is already a published author. His book *What Preteens Want Their Parents to Know* contains about 180 tips for parents on subjects from discipline and homework to love and respect. "It's tough growing up from being a little kid to a teenager," says Holladay. "I noticed lots of advice books written by adults, but none by kids. So I began writing one myself." Holladay surveyed children aged 9 to 12 for more ideas for his book, which took two years to complete. Since his book was published, Holladay has been on TV and radio shows to talk about it. His best advice to parents and children: "Be flexible."

1937 French artist Marie Laurenein creates her ethereal painting *The Dancers*.

1987 The African-American experience is portrayed in the photograph *Basket of Millet* by artist Elisabeth Sunday.

1890 1900 1910 1920 1930 1940 1950 1960 1970 1980 1990 2000

1963 Charles Moore takes his famous photograph for *Life* magazine, *Birmingham Riots*, during the civil rights struggle in Alabama.

1994 Kim Campbell becomes Canada's first female prime minister.

Chapter 13 **711**

Alternative Chapter Projects

Two other chapter projects are included in the *Investigations and Projects Masters*. In Chapter 13 Project A, pp. 73–74, students extend the topic in the chapter opener. In Chapter 13 Project B, pp. 75–76, students construct spirals.

Ryan's book is available in many libraries. It was published in 1994 by McCracken Press, New York, NY.

Chapter Project

Cooperative Learning Groups should discuss what type of chart or graph would best represent the data collected. They might consider bar graphs, line graphs, or circle graphs. They should debate the pros and cons of each. They should conclude that bar graphs are the most descriptive.

Investigations and Projects Masters, p. 73

13 NAME_____ DATE_____
Chapter 13 Project A   Student Edition Pages 712–748

Radical Survey

1. Although you may not have thought about it, your parents and other adults you know probably spent time as teenagers studying math just as you are doing now. For this project, you will work in a small group to survey adults about their high school years and their experiences with math during and after high school. Begin by brainstorming a list of questions. Some possibilities are shown below.

   · What was high school like in your day?

   · What issues concerned you in high school?

   · What is your view of today's high schools?

   · Were your experiences with math in high school positive or negative? Explain your answer.

   · How do you or someone you work with use math on the job?

2. Design a questionnaire containing the five best questions from exercise 1. Distribute copies of your questionnaire to at least 20 adults. Be sure to explain the procedure for returning the completed questionnaires.

3. Make charts and graphs to organize and explain the data you and your group collected in exercise 2.

4. Use the data your group collected to prepare a newspaper article or a radio or TV report about adults and their experiences in high school.

NCTM Standards: 1–5, 7

Objective
Use a geoboard to find areas by using the Pythagorean theorem.

Recommended Time
Demonstration and discussion: 15 minutes; Exercises: 30 minutes

Instructional Resources
For each student or group of students
Student Manipulative Kit
• geoboard
Modeling Mathematics Masters
• p. 6 (geoboard pattern)
• p. 34 (worksheet)
For teacher demonstration
Algebra and Geometry Overhead Manipulative Resources

1 FOCUS

Motivating the Lesson
Discuss how triangles and squares relate to each other. Remind students of the meaning of *hypotenuse* and other terms related to triangles.

2 TEACH

Teaching Tip Students may be more willing to believe that the area of the square on the hypotenuse is 2 square units if they separate the square into right triangles as shown below.

3 PRACTICE/APPLY

Assignment Guide

Core: 1–10
Enriched: 1–10

MODELING MATHEMATICS

13-1A The Pythagorean Theorem

Materials: geoboard dot paper

In this activity, you will use the Pythagorean theorem to build squares on a geoboard or dot paper.

A Preview of Lesson 13–1

Activity Make a square with an area of 2 square units.

Step 1 Start with a right triangle like the one shown below.

Step 2 Build squares on the two legs. Each square has an area of 1 square unit.

Step 3 Now build a square on the hypotenuse.

You can find the area of the square on the hypotenuse by using the Pythagorean theorem. Let c represent the measure of the hypotenuse and a and b represent the measures of the legs.

$$c^2 = a^2 + b^2$$
$$= 1^2 + 1^2 \qquad \textit{Replace a with 1 and b with 1.}$$
$$= 1 + 1 \text{ or } 2$$

The area of the square on the hypotenuse is 2 square units.

10. The total area of the 4 triangles equals the area of the large square minus the area of the triangle in each corner minus the area of the small square, $49 - 4(6) - 16$ or 9 square units.

Model Build squares on each side of the triangles shown below using a geoboard or dot paper. Record the area of each square.

1.
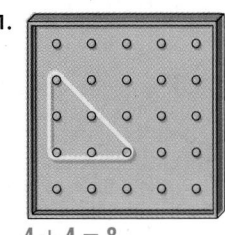
$4 + 4 = 8$

2.
$9 + 1 = 10$

3.
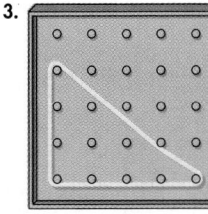
$16 + 9 = 25$

Draw Draw a square on dot paper having each area.

4–9. See Solutions Manual.

4. 4 square units
5. 9 square units
6. 8 square units
7. 13 square units
8. 17 square units
9. 32 square units

Write

10. Write a paragraph explaining how to find the total area of the shaded triangles in the drawing at the right.

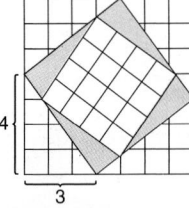

4 ASSESS

Observing students working in cooperative groups is an excellent method of assessment.

GLENCOE *Technology*

Interactive Mathematics Tools Software

This multimedia software provides an interactive lesson that has students manipulate a triangle and chart the observations of the relationship of the sides of the right angle. A **Computer Journal** gives students an opportunity to write about what they have learned.

For Windows & Macintosh

Integration: Geometry
The Pythagorean Theorem

Objective

After studying this lesson, you should be able to:

- use the Pythagorean theorem to solve problems.

APPLICATION
Sailing

One of the most prestigious sailboat races in the world is the America's Cup. In 1995, for the first time in the 144-year history of the race, one of the sailboats, *America*[3], had an all-woman crew. Theirs was one of three U.S. boats vying for the most prized trophy in sailing.

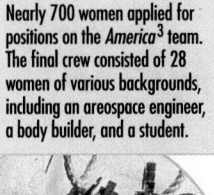

F Y I

Nearly 700 women applied for positions on the *America*[3] team. The final crew consisted of 28 women of various backgrounds, including an areospace engineer, a body builder, and a student.

A sailboat's *mast* and *boom* form a right angle. The sail itself, called a *mainsail*, is in the shape of a right triangle.

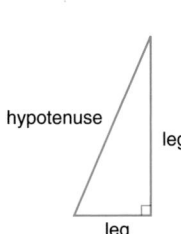

Recall that the side opposite the right angle in a right triangle is called the *hypotenuse*. This side is always the longest side of a right triangle. The other two sides are called the *legs* of the triangle.

To find the length of any side of a right triangle when the lengths of the other two are known, you can use a formula developed by the Greek mathematician Pythagoras.

The Pythagorean Theorem	If a and b are the measures of the legs of a right triangle and c is the measure of the hypotenuse, then $c^2 = a^2 + b^2$.

You can use the Pythagorean theorem to find the length of the hypotenuse of a right triangle when the lengths of the legs are known.

 Example **1** Find the length of the hypotenuse of a right triangle if $a = 12$ and $b = 5$.

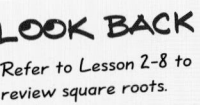
LOOK BACK

Refer to Lesson 2-8 to review square roots.

$$c^2 = a^2 + b^2 \quad \textit{Pythagorean theorem}$$
$$c^2 = 12^2 + 5^2 \quad \textit{a = 12 and b = 5}$$
$$c^2 = 144 + 25$$
$$c^2 = 169$$
$$c = \pm\sqrt{169}$$
$$c = \pm 13 \quad \textit{Disregard −13. Why?}$$

The length of the hypotenuse is 13 units.

F Y I

The America's Cup was first run in 1851 in Great Britain as the Hundred Guinea Cup. It was won by a New York schooner, the *America*. In 1857 the New York Yacht Club began running the competition. In 1987 the San Diego Yacht Club took over.

Motivating the Lesson

Hands-On Activity Open this lesson by measuring the length and width of the classroom. Then have students measure the length of a diagonal that cuts across the room. Write the three measurements on the board. Challenge students to determine a relationship among the numbers.

NCTM Standards: 1–5, 7

Instructional Resources

- Study Guide Master 13-1
- Practice Master 13-1
- Enrichment Master 13-1
- Modeling Mathematics Masters, pp. 69–71
- Multicultural Activity Masters, p. 25
- Real-World Applications, 31
- Tech Prep Applications Masters, p. 25

Transparency 13-1A contains the 5-Minute Check for this lesson; **Transparency 13-1B** contains a teaching aid for this lesson.

Recommended Pacing	
Honors Pacing	Day 2 of 11
Block Scheduling*	Day 1 of 7
Alg. 1 in Two Years*	Days 2 & 3 of 22

*For more information on pacing and possible lesson plans, refer to the *Block Scheduling Booklet* and *Algebra 1 in Two Years*.

1 FOCUS

5-Minute Check
(over Chapter 12)

1. Simplify and state the excluded values for

$$\frac{x^2 - 16}{x^2 + 6x + 8} \cdot \frac{x - 4}{x + 2}; \quad -4, -2$$

Find each quotient.

2. $\dfrac{2a + 4}{a + 1} \div \dfrac{3a^2 + 8a + 4}{2a^2 + 5a + 3}$

$\dfrac{4a + 12}{3a + 2}$

3. $\dfrac{3b^2 + 5b - 2}{b - 1}$ $3b + 8 + \dfrac{6}{b - 1}$

Find each difference.

4. $\dfrac{5x}{x - 2} - \dfrac{2x}{2 - x}$ $\dfrac{7x}{x - 2}$

5. $\dfrac{3}{a + 1} - \dfrac{2a}{2a + 3}$

$\dfrac{-2a^2 + 4a + 9}{(a + 1)(2a + 3)}$

In-Class Examples

For Example 1
Find the length of the hypotenuse of each triangle if the measures of the legs are given below.

a. $a = 12$, $b = 16$ **20**
b. $a = 11$, $b = \sqrt{23}$ **12**
c. $a = 5$, $b = 7$ $\sqrt{74}$ **or 8.60**

For Example 2
If c is the measure of the hypotenuse of a right triangle, find each missing measure.

a. $a = 4$, $c = 10$, $b = ?$
$\sqrt{84}$ or 9.17
b. $a = 5$, $c = 6$, $b = ?$
$\sqrt{11}$ or 3.32
c. $c = 10$, $b = 7$, $a = ?$
$\sqrt{51}$ or 7.14

For Example 3
Determine whether the following side measures would form right triangles.

a. $a = 3$, $b = 8$, $c = \sqrt{73}$ **yes**
b. $a = 8$, $b = 9$, $c = 11$ **no**
c. $a = 13$, $b = 84$, $c = 85$ **yes**

For Example 4
Find the length of the diagonal of a rectangle whose length is 8 meters and whose width is 5 meters.
$\sqrt{89}$ or 9.43 meters

Teaching Tip Point out that if c is the measure of the longest side of a triangle and $c^2 \neq a^2 + b^2$, then the triangle is not a right triangle. This is called the contrapositive of the Pythagorean theorem.

GLOBAL CONNECTIONS

Sails were originally made of animal hide. Then woven reeds were used. Sailcloth can be made of flax fiber, cotton, or other fabrics.

Example **Find the length of side a if $b = 9$ and $c = 21$. Round to the nearest hundredth.**

$$c^2 = a^2 + b^2 \quad \text{Pythagorean theorem}$$
$$21^2 = a^2 + 9^2 \quad b = 9 \text{ and } c = 21$$
$$441 = a^2 + 81$$
$$360 = a^2$$
$$\pm\sqrt{360} = a \quad \begin{array}{l}\text{Use a calculator to approximate}\\ \sqrt{360} \text{ to the nearest hundredth.}\end{array}$$
$$18.97 \approx a \quad \begin{array}{l}\text{Only the positive value of } a \text{ has}\\ \text{meaning in this situation.}\end{array}$$

The length of the leg, to the nearest hundredth, is 18.97 units.

The following corollary, based on the Pythagorean theorem, can be used to determine whether a triangle is a right triangle.

Corollary to the Pythagorean Theorem	If c is the measure of the longest side of a triangle and $c^2 \neq a^2 + b^2$, then the triangle is not a right triangle.

Example **Determine whether the following side measures would form right triangles.**

a. 6, 8, 10

Since the measure of the longest side is 10, let $c = 10$, $a = 6$, and $b = 8$. Then determine whether $c^2 = a^2 + b^2$.

$$10^2 \overset{?}{=} 6^2 + 8^2$$
$$100 \overset{?}{=} 36 + 64$$
$$100 = 100 \quad ✓$$

Since $c^2 = a^2 + b^2$, the triangle is a right triangle.

b. 7, 9, 12

Since the measure of the longest side is 12, let $c = 12$, $a = 7$, and $b = 9$. Then determine whether $c^2 = a^2 + b^2$.

$$12^2 \overset{?}{=} 7^2 + 9^2$$
$$144 \overset{?}{=} 49 + 81$$
$$144 \neq 130$$

Since $c^2 \neq a^2 + b^2$, the triangle is not a right triangle.

GLOBAL CONNECTIONS

One of the earliest inventions was the sail, and it was the most important means of powering boats for most of maritime history. Boats with sails appear in Egyptian rock carvings near the Nile, dating from 3300 B.C.

Example **Agriculture was very important in ancient Aztec culture. Aztec farmers kept records of their farms including calculations of the dimensions and area. Because the terrain was so rough, very few of the farms were rectangular. Yet, by using measuring ropes that measured length with a unit called a *quahuitl* (about 2.5 meters), they were able to make accurate calculations. In the farm shown at the right, the farmer measured three sides of his farm. He had trouble measuring the fourth side because it was located in a dense forest. Find the measure of the fourth side.**

APPLICATION
World Cultures

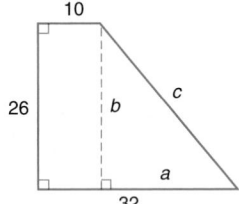

Alternative Learning Styles

Visual Challenge students to illustrate the Pythagorean theorem geometrically using other figures such as semicircles. Since $A = \pi r^2$ is the formula for the area of a circle, $A = \frac{1}{2}(\pi r^2)$ or $A = \frac{1}{2}\pi r^2$ is the formula for the area of a semicircle. Use this formula to illustrate the Pythagorean theorem using semicircles. The sides of the right triangle have lengths 3, 4, and 5 units.

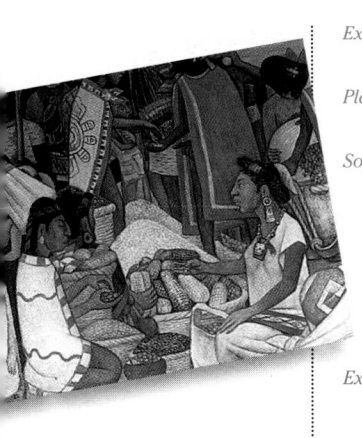

Explore   Let c represent the length of the fourth side of the farm. Note that it is the hypotenuse of a right triangle.

Plan   Use the Pythagorean theorem to find c. Let $a = 32 - 10$ or 22 and $b = 26$. Then solve the resulting equation.

Solve $c^2 = (22)^2 + (26)^2$

$c^2 = 484 + 676$

$c^2 = 1160$

$c \approx 34.06$

The length of the forest side of the farm is approximately 34.06 quahuitls.

Examine   Check the solution by substituting 34.06 for c in the Pythagorean theorem.

$$c^2 = a^2 + b^2$$

$$(34.06)^2 \stackrel{?}{=} (22)^2 + (26)^2$$

$$1160 = 1160 \checkmark$$

CHECK FOR UNDERSTANDING

Communicating Mathematics

Study the lesson. Then complete the following.   1–2. See margin.

1. **Draw** a right triangle and label each side with a letter.

2. **Explain** how you can determine whether a triangle is a right triangle if you know the lengths of the three sides.

3. In 1955, Greece issued the stamp shown at the left to honor the 2500th anniversary of the Pythagorean School. Notice that there is a triangle bordered on each side by a checkerboard pattern.
   a. Count the number of squares along each side of the triangle.   3, 4, 5
   b. Use the Pythagorean theorem to show that it is a right triangle.

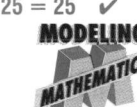
3b. $3^2 + 4^2 \stackrel{?}{=} 5^2$

$9 + 16 \stackrel{?}{=} 25$

$25 = 25$ ✔

MODELING MATHEMATICS

4. When taking the square root of a number, you can get a positive and a negative number. Why then, when using the Pythagorean theorem, is only the positive value used? Explain.   See margin.

5. Use a geoboard or dot paper to build squares on each side of the triangle at the right. Record the areas of the squares. $16 + 4 = 20$

Guided Practice

Solve each equation. Assume each variable represents a positive number.

6. $5^2 + 12^2 = c^2$ 13

7. $a^2 + 24^2 = 25^2$ 7

8. $16^2 + b^2 = 20^2$ 12

Find the length of each missing side. Round to the nearest hundredth.

9.

11.40

10.

14.14

Lesson 13–1 *Geometry*   The Pythagorean Theorem   **715**

3 PRACTICE/APPLY

Check for Understanding

Exercises 1–17 are designed to help you assess your students' understanding through reading, writing, speaking, and modeling. You should work through Exercises 1–5 with your students and then monitor their work on Exercises 6–17.

Additional Answers

1.

2. By squaring the two shorter sides and adding them; then, comparing their sum to the square of the third or longer side. If they are equal, it is a right triangle.

4. Since the Pythagorean theorem finds the length of a side, it has to be positive. Distance is not negative.

If *c* is the measure of the hypotenuse of a right triangle, find each missing measure. Round answers to the nearest hundredth, if necessary.

11. $a = 9, b = 12, c = ?$ **15**
12. $a = \sqrt{11}, c = 6, b = ?$ **5**
13. $b = \sqrt{30}, c = \sqrt{34}, a = ?$ **2**
14. $a = 7, b = 4, c = ?$ $\sqrt{65} \approx 8.06$

Determine whether the following side measures would form right triangles. Explain why or why not.

15. 12, 16, 20 **yes**
16. 2, 8, 8 **no; $2^2 + 8^2 \neq 8^2$**

17. **Baseball** A baseball scout uses many different tests to determine whether or not to draft a particular player. One test for catchers is to see how quickly they can throw a ball from home plate to second base. On a baseball diamond, the distance from one base to the next is 90 feet. What is the distance from home plate to second base? **127.28 ft**

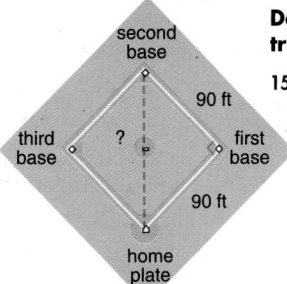

second base

third base

? P

first base

90 ft

90 ft

home plate

EXERCISES

Practice

Find the length of each missing side. Round to the nearest hundredth.

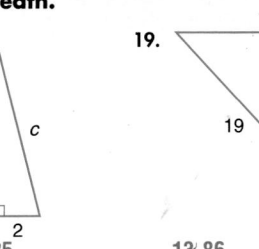

A

18.

8   c

2

8.25

19.

13

19   b

13.86

20.

16

a

5

15.20

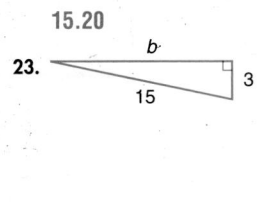

21.

5

14   a

13.08

22.

6

11   c

12.53

23.

b

15

3

14.70

If *c* is the measure of the hypotenuse of a right triangle, find each missing measure. Round answers to the nearest hundredth.

24. $a = 16, b = 30, c = ?$ **34**
25. $a = 11, c = 61, b = ?$ **60**
26. $b = 13, c = \sqrt{233}, a = ?$ **8**
27. $a = \sqrt{7}, b = \sqrt{9}, c = ?$ **4**

B

28. $a = 6, b = 3, c = ?$ $\sqrt{45} \approx 6.71$
29. $b = \sqrt{77}, c = 12, a = ?$

29. $\sqrt{67} \approx 8.19$

30. $b = 10, c = 11, a = ?$ $\sqrt{21} \approx 4.58$
31. $a = 4, b = \sqrt{11}, c = ?$

31. $\sqrt{27} \approx 5.20$

32. $a = 15, b = \sqrt{28}, c = ?$
33. $a = 12, c = 17, b = ?$

32. $\sqrt{253} \approx 15.91$

33. $\sqrt{145} \approx 12.04$

34. no; $6^2 + 9^2 \neq 12^2$

37. no; $11^2 + 12^2 \neq 15^2$

38. no; $16^2 + \left(\sqrt{32}\right)^2 \neq 20^2$

Determine whether the following side measures would form right triangles. Explain why or why not.

34. 6, 9, 12
35. 45, 60, 75 **yes**
36. 30, 40, 50 **yes**
37. 11, 12, 15
38. 16, $\sqrt{32}$, 20
39. 15, $\sqrt{31}$, 16 **yes**

716 Chapter 13 Exploring Radical Expressions and Equations

Alternative Teaching Strategies

Student Diversity The following is a proof of the Pythagorean theorem discovered by James Garfield in 1876.

H   B   D

C

A

In the figure, $BD = AH$, and $DC \parallel AH$ and equal to BH. Area of trapezoid $CDHA$ = area of ACB + 2 area of ABH.

$\frac{1}{2}(AH + CD) \times HD = \frac{1}{2} AB^2 + 2 \times \frac{1}{2}AH \times HB$ or $(AH + HB)^2 = AB^2 + 2AH \times HB$, so $AB^2 = BH^2 + AH^2$.

Therefore, $h^2 = a^2 + b^2$.

Use an equation to solve each problem. Round answers to the nearest hundredth.

41. $\sqrt{75}$ in. or about 8.66 in.

40. Find the length of the diagonal of a square if its area is 128 cm². **16 cm**

41. Find the length of the diagonal of a cube if each side of the cube is 5 inches long.

42a. $x^2 + (x+6)^2 = (30)^2$
42b. 18 cm, 24 cm

42. A right triangle has one leg that is 6 centimeters longer than the other. The hypotenuse is 30 centimeters long.

    a. Write an equation to find the length of the legs.
    b. Find the length of each leg of the triangle.

43. Look at the trapezoid at the right.
    a. Find the perimeter. (*Hint:* Drawing a second height will be helpful.) **44.49 m**
    b. Find the area. **78 m²**

Programming

44. The graphing calculator program at the right calculates the hypotenuse of a right triangle and displays the right triangle being measured.

    Find the hypotenuse *c* of each right triangle below, given the measures of sides *a* and *b*. Round to the nearest hundredth.

    a. $a = 2, b = 7$ **7.28**
    b. $a = 9, b = 12$ **15**
    c. $a = 13, b = 15$ **19.85**
    d. $a = 6, b = 9$ **10.82**
    e. $a = 12, b = 16$ **20**

```
PROGRAM:PYTH
:FnOff
:AxesOff
:-1→Xmin
:-1→Ymin
:2→Xscl
:2→Yscl
:ClrDraw
:ClrHome
:Split
:Input "SIDE A", A
:Line (0, A, 0, 0)
:Input "SIDE B", B
:Line (B, 0, 0, 0)
:√(A²+B²)→C
:Line (B, 0, 0, A)
:Text (1, 1,
 "HYPOTENUSE IS", C)
:Shade (0,
 (-A/B)X+A, 1, 0, B)
```

Critical Thinking

45. The window in Julia's attic was a square with an area of 1 square foot. After she remodeled her home, the width and height of the new attic window were the same as the original, but its area was half that of the original window. If the new window is also a square, explain how this is possible. Include a drawing. **See margin.**

Applications and Problem Solving

46. **Sailing** Refer to the application at the beginning of the lesson. If the edge of the mainsail that is attached to the mast is 100 feet long and the edge of the mainsail that is attached to the boom is 60 feet long, what is the length of the longest edge of the mainsail? **about 116.62 ft**

47. about 2.66 m

47. **Construction** The walls of the Downtown Recreation Center are being covered with paneling. The doorway into one room is 0.9 meters wide and 2.5 meters high. What is the length of the longest rectangular panel that can be taken through this doorway diagonally?

Using the Programming Exercises The program given in Exercise 44 is for use with a TI-82 graphing calculator. For other programmable calculators, have students consult their owner's manual for commands similar to those presented here.

Additional Answer

45.

1 ft
1 ft
Area = 1 ft²
or 144 in²

6 in.
6 in.
$(6\sqrt{2})$ in.
Area = $(6\sqrt{2})^2$
or 72 in²

Practice Masters, p. 88

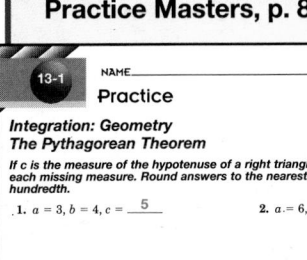

13-1
NAME _____ DATE _____
Practice
Student Edition
Pages 713–718

Integration: Geometry
The Pythagorean Theorem

If c is the measure of the hypotenuse of a right triangle, find each missing measure. Round answers to the nearest hundredth.

1. $a = 3, b = 4, c = $ __5__
2. $a = 6, c = 10, b = $ __8__
3. $b = 12, c = 13, a = $ __5__
4. $a = 6, c = 12, b = \sqrt{108} \approx 10.39$
5. $a = 8, b = 6, c = $ __10__
6. $a = 5, c = 13, b = $ __12__
7. $b = 0.8, c = 1.0, a = $ __0.6__
8. $a = 11, b = 4, c = \sqrt{137} \approx 11.70$
9. $a = \sqrt{12}, b = 6, c = \sqrt{48} \approx 6.93$
10. $b = 11, c = \sqrt{289}, a = \sqrt{168} \approx 12.96$
11. $a = 19, b = \sqrt{39}, c = $ __20__
12. $a = \sqrt{6}, b = \sqrt{19}, c = $ __5__

Determine whether the following side measures would form right triangles. Explain why or why not.

13. 20, 21, 29 yes
14. 15, 30, 34 no
15. 9, $\sqrt{40}$, 11 yes
16. 21, 72, 75 yes

Closing Activity

Writing At the end of class, have students write down the key points to consider when using the Pythagorean theorem.

Additional Answer

50.

$y = x^2 - 6x + 8$

TO ALL PARTS OF THE WORLD

48. Travel The cruise ship M.S. Starward has a right triangle-shaped dance floor on the cabaret deck. The lengths of the two shortest sides of the dance floor are equal, and the longest side next to the stage is 36 feet long. How long is one side of the dance floor? **about 25.5 ft**

Mixed Review

50. $x = 3$; $(3, -1)$; See margin for graph.

55. 41

49. Solve $\frac{a+2}{a} + \frac{a+5}{a} = 1$. (Lesson 12–8) **−7**

50. Find an equation of the axis of symmetry and the coordinates of the vertex of the graph of $y = x^2 - 6x + 8$. Then draw the graph. (Lesson 11–1)

51. Simplify $(4r^5)(3r^2)$. (Lesson 9–1) **$12r^7$**

52. Number Theory The sum of two numbers is 38. Their difference is 6. Find the numbers. (Lesson 8–3) **16, 22**

53. Solve $5c - 2 \geq c$. (Lesson 7–2) $\left\{ c \mid c \geq \frac{1}{2} \right\}$

54. State the slope and y-intercept of the line graphed at the right. Then write an equation of the line in slope-intercept form. (Lesson 6–4) **$-2, 3, y = -2x + 3$**

55. Given $f(x) = 2x^2 - 5x + 8$, find $f(-3)$. (Lesson 5–5)

56. Solve $-\frac{5}{6}y = 15$. (Lesson 3–2) **−18**

WORKING ON THE

In·ves·ti·ga·tion

Refer to the Investigation on pages 656–657.

A Growing Concern

1 Develop a detailed plan for planting the lawn, plants, and trees for the Sanchez family's yard. Indicate the dimensions of the lawn area as well as the location of the plants and trees you selected. Mark the length of each property line on the drawing.

2 Research the plants and trees that you feel would be good specimens for the areas in which you indicated that they would be planted. Explain why you chose those plants and

trees. Think about all four seasons and which plants best suit each season. Also consider the amount of sunlight each area receives during the day.

3 Calculate the cost of materials, labor, and profit margin of planting these items and laying the sprinkler system. Justify your costs and make a case for the quantity and location of the plants, trees, and lawn area in your design.

4 Be sure to add the plants, trees, and sprinkler system to your overall design.

Add the results of your work to your Investigation Folder.

Extension

Connections The length of a rectangle is 2 feet less than twice its width. Each diagonal is 34 feet. What are the dimensions of the rectangle? **16 ft, 30 ft**

In·ves·ti·ga·tion

Working on the Investigation

The Investigation on pages 656–657 is designed to be a long-term project that is completed over several days or weeks. Encourage students to keep their materials in their Investigation Folder as they work on the Investigation.

Enrichment Masters, p. 88

13-1

NAME_____ DATE_____

Enrichment

Student Edition
Pages 713–718

A Space-Saving Method

Two arrangements for cookies on a 32 cm by 40 cm cookie sheet are shown at the right. The cookies have 8-cm diameters after they are baked. The centers of the cookies are on the vertices of squares in the top arrangement. In the other, the centers are on the vertices of equilateral triangles. Which arrangement is more economical? *The triangle arrangement is more economical, because it contains one more cookie.*

8 cm

20 cookies

In the square arrangement, rows are placed every 8 cm. At what intervals are rows placed in the triangle arrangement?

Look at the right triangle labeled a, b, and c. A leg a of the triangle is the radius of a cookie, or 4 cm. The hypotenuse c is the sum of two radii, or 8 cm. Use the Pythagorean theorem to find b, the interval of the rows.

$b = ?$

20 cookies

$$c^2 = a^2 + b^2$$
$$8^2 = 4^2 + b^2$$
$$64 - 16 = b^2$$
$$\sqrt{48} = b$$
$$4\sqrt{3} = b$$
$$b = 4\sqrt{3} \approx 6.93$$

The rows are placed approximately every 6.93 cm.

Solve each problem.

1. Suppose cookies with 10-cm diameters are arranged in the triangular pattern shown above. What is the interval b of the rows? **8.66 cm**

2. Find the diameter of a cookie if the rows are placed in the triangular pattern every $3\sqrt{3}$ cm. **6 cm**

3. Describe other practical applications in which this kind of triangular pattern can be used to economize on space. **Answers will vary.**

Simplifying Radical Expressions

Objectives

After studying this lesson, you should be able to:
- simplify square roots, and
- simplify radical expressions.

Physics

The period of a pendulum is the time in seconds that it takes the pendulum to make one complete swing back and forth. The formula for the period P of a pendulum is $P = 2\pi\sqrt{\frac{\ell}{32}}$, where ℓ is the length of the pendulum in feet. Suppose a clock makes one "tick" after each complete swing back and forth of a 2-foot-long pendulum. How many ticks would the clock make in one minute? *This problem will be solved in Example 4.*

Can $2\pi\sqrt{\frac{\ell}{32}}$ be simplified? One rule used for simplifying radical expressions is that a radical expression is in *simplest form* if the **radicand,** the expression under the radical sign, contains no perfect square factors other than one. The following property can be used to simplify square roots.

Product Property of Square Roots	**For any numbers a and b, where $a \geq 0$ and $b \geq 0$,** $\sqrt{ab} = \sqrt{a} \cdot \sqrt{b}.$

The product property of square roots and prime factorization can be used to simplify radical expressions in which the radicand is not a perfect square.

Example **Simplify.**

a. $\sqrt{18}$

$\sqrt{18} = \sqrt{3 \cdot 3 \cdot 2}$ *Prime factorization of 18*

$= \sqrt{3^2} \cdot \sqrt{2}$ *Product property of square roots*

$= 3\sqrt{2}$

b. $\sqrt{140}$

$140 = \sqrt{2 \cdot 2 \cdot 5 \cdot 7}$ *Prime factorization of 140*

$= \sqrt{2^2} \cdot \sqrt{5 \cdot 7}$ *Product property of square roots*

$= 2\sqrt{35}$

When finding the principal square root of an expression containing variables, be sure that the result is not negative. Consider the expression $\sqrt{x^2}$. Its simplest form is not x since, for example, $\sqrt{(-4)^2} \neq -4$. For radical expressions like $\sqrt{x^2}$, use absolute value to ensure nonnegative results.

$$\sqrt{x^2} = |x| \qquad \sqrt{x^3} = x\sqrt{x} \qquad \sqrt{x^4} = x^2 \qquad \sqrt{x^5} = x^2\sqrt{x} \qquad \sqrt{x^6} = |x^3|$$

Lesson 13-2 Simplifying Radical Expressions **719**

NCTM Standards: 1–5

Instructional Resources

- Study Guide Master 13-2
- Practice Master 13-2
- Enrichment Master 13-2
- Assessment and Evaluation Masters, p. 352
- Tech Prep Applications Masters, p. 26

 Transparency 13-2A contains the 5-Minute Check for this lesson; **Transparency 13-2B** contains a teaching aid for this lesson.

Recommended Pacing	
Honors Pacing	Day 3 of 11
Block Scheduling*	Day 2 of 7
Alg. 1 in Two Years*	Days 4, 5, & 6 of 22

 *For more information on pacing and possible lesson plans, refer to the *Block Scheduling Booklet* and *Algebra 1 in Two Years*.

1 FOCUS

 ### 5-Minute Check
(over Lesson 13-1)

1. Find the length of the hypotenuse of a right triangle if the legs have measures of 7 and 11. $\sqrt{170}$
2. Find the measure of the other leg of a right triangle if one leg is 24 and the hypotenuse is 25. 7

The measures of three sides of a triangle are given. Determine whether each triangle is a right triangle.

3. 12, 13, 15 no
4. 6, 8, 10 yes
5. 9, 16, 20 no

Motivating the Lesson

Questioning You may want to open this lesson with a review of prime numbers and what it means to write the prime factorization of a number. You may want to use the Sieve of Eratosthenes to develop primes less than 100. Absolute value should also be reviewed.

2 TEACH

In-Class Examples

For Example 1
Simplify.

a. $\sqrt{147}$ $7\sqrt{3}$
b. $\sqrt{605}$ $11\sqrt{5}$

For Example 2
Simplify. Use absolute value symbols when necessary.

a. $\sqrt{x^6}$ $|x^3|$
b. $\sqrt{49x^2}$ $7|x|$

For Example 3
Simplify.

a. $\sqrt{6} \cdot \sqrt{10}$ $2\sqrt{15}$
b. $2\sqrt{14} \cdot \sqrt{21}$ $14\sqrt{6}$

Teaching Tip Point out that to simplify a square root with a variable, absolute value symbols are necessary when the variable has an even exponent and the exponent of the square root is odd. For example, in $\sqrt{x^4} = x^2$, since x is squared in the answer, it will automatically be positive. In $\sqrt{x^6} = x^3$, in order to guarantee that x^3 is positive, $|x^3|$ is necessary instead of x^3.

EXPLORATION

This activity gives students experience with real-world applications using radicals.

For $\sqrt{x^3}$, absolute value is not necessary. If x were negative, then x^3 would be negative, and $\sqrt{x^3}$ would not be defined as a real number. *Why is absolute value not necessary for $\sqrt{x^4}$?*

Example **Simplify $\sqrt{72x^3y^4z^5}$.**

$$\sqrt{72x^3y^4z^5} = \sqrt{2^3 \cdot 3^2 \cdot x^3 \cdot y^4 \cdot z^5} \quad \textit{Prime factorization}$$
$$= \sqrt{2^2} \cdot \sqrt{2} \cdot \sqrt{3^2} \cdot \sqrt{x^2} \cdot \sqrt{x} \cdot \sqrt{y^4} \cdot \sqrt{z^4} \cdot \sqrt{z} \quad \textit{Product property}$$
$$= 2 \cdot 3 \cdot |x| \cdot y^2 \cdot z^2 \cdot \sqrt{2xz} \quad \textit{Simplify.}$$
$$= 6|x|y^2z^2\sqrt{2xz} \quad \textit{The absolute value of x ensures a nonegative result.}$$

The product property can also be used to multiply square roots.

Example **Simplify $\sqrt{5} \cdot \sqrt{35}$.**

$$\sqrt{5} \cdot \sqrt{35} = \sqrt{5} \cdot \sqrt{5} \cdot \sqrt{7} \quad \textit{Product property of square roots}$$
$$= \sqrt{5^2 \cdot 7}$$
$$= 5\sqrt{7}$$

You can use a graphing calculator to explore and analyze radical expressions.

EXPLORATION
GRAPHING CALCULATORS

The formula $Y = 91.4 - (91.4 - T)[0.478 + 0.301(\sqrt{x} - 0.02x)]$ can be used to calculate the windchill factor. In this formula, Y represents the windchill, T represents the outside Fahrenheit temperature, and x represents the wind speed in miles per hour. When a meteorologist says that the temperature is 12 degrees, but it feels like 18 degrees below zero because of the wind, you can use the formula to find how fast the wind is blowing. Set the range at $[0, 40]$ by 2 and $[-50, 40]$ by 5.

Enter: `Y=` 91.4 `−` `(` 91.4 `−` `ALPHA` `T` `)` `(` .478
`+` .301 `(` `2nd` `√` `X,T,θ` `−` .02 `X,T,θ`
`)` `)` `ENTER`

Store 12 into T by pressing 12 `STO▸` `ALPHA` `T` `ENTER`. Graph the function and trace along the graph until Y = −18. At Y = −18, X = 10.2. So the wind is blowing at approximately 10 miles per hour.

Your Turn a. about 7.2 mph

a. Use the graph to find the wind speed if the temperature feels like −7°.
b. Enter the formula into Y1, Y2, and Y3 using different variables for T. Store three different temperatures for these variables and graph them simultaneously. **b–c. See students' work.**
c. Analyze and compare the graphs.

Alternative Learning Styles

Auditory Write $\sqrt{x^a} = |x^b|$ on the board. Have students come to the board one at a time and guess values for a and b that might work. When a student is wrong, the next student must explain why. When a student is right, the next student must try to guess the pattern and predict another correct answer. $a = 2, b = 1; a = 6, b = 3; a = 10, b = 5;$ and so on

CAREER CHOICES

Meteorologists study variations in temperature, humidity, air pressure, wind, cloud cover, and precipitation. Dynamic meteorology deals with motions of the atmosphere and airflow. Synoptic meteorology deals with the weather.

Example 4

Refer to the connection at the beginning of the lesson. How many ticks would the clock make in one minute? Use 3.14 for π and round to the nearest whole number.

Explore   The clock makes one "tick" after each complete swing back and forth of its 2-foot-long pendulum. The formula for the period P of a pendulum is $P = 2\pi\sqrt{\frac{\ell}{32}}$.

Plan   First find P, the number of seconds it takes for the pendulum to go back and forth one time. Then find $\frac{60}{P}$, the number of times the clock's pendulum swings back and forth in one minute.

Solve $P = 2\pi\sqrt{\frac{\ell}{32}}$

$\approx 2(3.14)\sqrt{\frac{2}{32}}$ $\pi \approx 3.14$

$\approx 6.28 \cdot \sqrt{\frac{1}{16}}$

$\approx 6.28 \cdot \frac{1}{4}$ or 1.57

So it takes about 1.57 seconds for the pendulum to go back and forth once.

$\frac{60}{1.57} \approx 38.22$

Thus, the clock makes about 38 ticks per minute.

Examine   Since 38.22 × 1.57 is about 40 × 1.5 or 60, the answer seems reasonable.

You can divide square roots and simplify radical expressions that involve division by using the quotient property of square roots.

Quotient Property of Square Roots	For any numbers a and b, where $a \geq 0$ and $b > 0$, $$\sqrt{\frac{a}{b}} = \frac{\sqrt{a}}{\sqrt{b}}.$$

A fraction containing radicals is in simplest form if no radicals are left in the denominator.

Example 5

a. Simplify $\frac{\sqrt{56}}{\sqrt{7}}$ and $\sqrt{\frac{34}{25}}$.

b. Compare the expressions using <, >, or =.

a. $\frac{\sqrt{56}}{\sqrt{7}} = \sqrt{\frac{56}{7}}$ *Quotient property of square roots* $\sqrt{\frac{34}{25}} = \frac{\sqrt{34}}{\sqrt{25}}$

$= \sqrt{8}$ $= \frac{\sqrt{34}}{5}$

$= \sqrt{4} \cdot \sqrt{2}$

$= 2\sqrt{2}$

Lesson 13–2 Simplifying Radical Expressions   **721**

GLENCOE *Technology*

Interactive Mathematics Tools Software

This multimedia software provides an interactive lesson that explores traveling upstream and downstream in a boat by manipulating the current and distance traveled. A **Computer Journal** gives students an opportunity to write about what they have learned.

For Windows & Macintosh

F Y I

Galileo determined that the period of a pendulum is constant by timing a swinging cathedral lamp using his pulse. Changing the length of the pendulum changes the period; changing the mass does not.

b. You can compare these expressions by estimating their values and then using a scientific calculator to find approximations for each simplified expression.

Estimate: $2\sqrt{2} \rightarrow$ *Since 2 is a little more than 1, $2\sqrt{2}$ will be a little more than 2.*

$\dfrac{\sqrt{34}}{5} \rightarrow \dfrac{\sqrt{36}}{5} = \dfrac{6}{5}$ This will be a little more than 1. So, $2\sqrt{2} > \dfrac{\sqrt{34}}{5}$.

Verify by using a calculator.

Enter: 2 $\boxed{\times}$ 2 $\boxed{\sqrt{x}}$ $\boxed{=}$ *2.828427125*

Enter: 34 $\boxed{\sqrt{x}}$ $\boxed{\div}$ 5 $\boxed{=}$ *1.166190379*

Since $2.8 > 1.2$, then $\dfrac{\sqrt{56}}{\sqrt{7}} > \sqrt{\dfrac{34}{25}}$.

PEANUTS®

PEANUTS reprinted by permission of United Feature Syndicate, Inc.

In the cartoon above, Woodstock simplified the radical expression by **rationalizing the denominator.** This method may be used to remove or eliminate radicals from the denominator of a fraction.

Example **6** **Simplify.**

a. $\dfrac{\sqrt{5}}{\sqrt{3}}$

$\dfrac{\sqrt{5}}{\sqrt{3}} = \dfrac{\sqrt{5}}{\sqrt{3}} \cdot \dfrac{\sqrt{3}}{\sqrt{3}}$ *Note that $\frac{\sqrt{3}}{\sqrt{3}} = 1$.*

$= \dfrac{\sqrt{15}}{3}$

b. $\dfrac{\sqrt{7}}{\sqrt{12}}$

$\dfrac{\sqrt{7}}{\sqrt{12}} = \dfrac{\sqrt{7}}{\sqrt{2 \cdot 2 \cdot 3}}$

$= \dfrac{\sqrt{7}}{\sqrt{2 \cdot 2 \cdot 3}} \cdot \dfrac{\sqrt{3}}{\sqrt{3}}$

$= \dfrac{\sqrt{7}\sqrt{3}}{2 \cdot 3}$

$= \dfrac{\sqrt{21}}{6}$

Binomials of the form $a\sqrt{b} + c\sqrt{d}$ and $a\sqrt{b} - c\sqrt{d}$ are called **conjugates** of each other. For example, $6 + \sqrt{2}$ and $6 - \sqrt{2}$ are conjugates. Conjugates are useful when simplifying radical expressions because their product is always a rational number with no radicals.

$(6 + \sqrt{2})(6 - \sqrt{2}) = 6^2 - (\sqrt{2})^2$ *Use the pattern $(a - b)$*
$= 36 - 2$ $\quad (a + b) = a^2 - b^2$ *to*
$= 34$ $\quad$ *simplify the product.*

This is true because of the following.

$(\sqrt{2})^2 = \sqrt{2} \cdot \sqrt{2}$
$= \sqrt{2 \cdot 2}$
$= \sqrt{2^2}$ or 2

Classroom Vignette

"In expressions such as the one in Example 7, I remind my students that $4 - \sqrt{3}$ is a binomial and they can't simplify it as follows.

$$\dfrac{\cancel{4}}{\cancel{4} - \sqrt{3}} = \dfrac{1}{-\sqrt{3}}$$ "

Barbara Owens
Davidson Middle School
Davidson, North Carolina

Conjugates are often used to rationalize the denominators of fractions containing square roots.

Example **Simplify** $\dfrac{4}{4-\sqrt{3}}$.

To rationalize the denominator, multiply both the numerator and denominator by $4+\sqrt{3}$, which is the conjugate of $4-\sqrt{3}$.

$$\frac{4}{4-\sqrt{3}} = \frac{4}{4-\sqrt{3}} \cdot \frac{4+\sqrt{3}}{4+\sqrt{3}} \quad \textit{Notice that } \frac{4+\sqrt{3}}{4+\sqrt{3}} = 1.$$

$$= \frac{4(4)+4\sqrt{3}}{4^2-\left(\sqrt{3}\right)^2} \quad \begin{array}{l}\textit{Use the distributive property to multiply numerators.}\\ \textit{Use the pattern } (a-b)(a+b) = a^2 - b^2 \textit{ to multiply}\\ \textit{denominators.}\end{array}$$

$$= \frac{16+4\sqrt{3}}{16-3}$$

$$= \frac{16+4\sqrt{3}}{13}$$

When simplifying radical expressions, check the following conditions to determine if the expression is in simplest form.

Simplest Radical Form

> **A radical expression is in simplest form when the following three conditions have been met.**
> **1. No radicands have perfect square factors other than 1.**
> **2. No radicands contain fractions.**
> **3. No radicals appear in the denominator of a fraction.**

CHECK FOR UNDERSTANDING

Communicating Mathematics

MATH JOURNAL

Study the lesson. Then complete the following. 2–4. See margin.

1. **Explain** why absolute values are sometimes needed when simplifying radical expressions containing variables. **to ensure nonnegative results**

2. **Describe** the steps you take to rationalize a denominator.

3. **You Decide** Niara showed the following equations to her friend Melanie and said, "I know that 6 can't equal 10, but all these steps make sense!" Melanie said, "One of the steps must be wrong." Who is correct? Can you explain the mistake?

$-60 = -60$	*Reflexive property of equality*
$36 - 96 = 100 - 160$	*Rewrite -60 as $36 - 96$ and $100 - 160$.*
$36 - 96 + 64 = 100 - 160 + 64$	*Add 64 to each side.*
$(6-8)^2 = (10-8)^2$	*Factor.*
$6 - 8 = 10 - 8$	*Take the square root of each side.*
$6 - 8 + 8 = 10 - 8 + 8$	*Add 8 to each side.*
$6 = 10$	*Simplify.*

4. Refer to the cartoon on page 722. Obviously, Woodstock realized that rationalizing the denominator is important. What are other steps that you may have to use to simplify a radical expression?

Lesson 13–2 Simplifying Radical Expressions **723**

Reteaching

Using Problem Solving To convince students that rationalizing the denominator produces equivalent results, have them use a calculator to find approximate values of the following expressions.

$$\frac{2}{\sqrt{3}} \qquad \frac{2 \cdot \sqrt{3}}{\sqrt{3} \cdot \sqrt{3}} \qquad \frac{2\sqrt{3}}{3}$$

The values all should be about 1.1547.

Alternative Teaching Strategies

Student Diversity Discuss the difference between $2\sqrt{2} \cdot 5$, $2\sqrt{2} \cdot \sqrt{5}$, and $2\sqrt{2} \cdot 5\sqrt{5}$. Have students write steps for each operation, trade with another student, and check the answers.

Chapter 13 **723**

Error Analysis

Students may make errors when squaring the product of a rational and an irrational number. For example, they may incorrectly evaluate $(3\sqrt{2})(3\sqrt{2})$ as $9 \cdot 4$ or $3 \cdot 2$. Point out that the rational numbers and the irrational numbers should be squared and their product should be found.

Assignment Guide

Core: 21–53 odd, 54, 55, 57–66
Enriched: 20–52 even, 54–66

For **Extra Practice,** see p. 785.

The red A, B, and C flags, printed only in the Teacher's Wraparound Edition, indicate the level of difficulty of the exercises.

Practice Masters, p. 89

State the conjugate of each expression. Then multiply the expression by its conjugate.

5. $5 + \sqrt{2}$ $5 - \sqrt{2}$; 23

6. $\sqrt{3} - \sqrt{7}$ $\sqrt{3} + \sqrt{7}$; -4

State the fraction by which each expression should be multiplied to rationalize the denominator.

7. $\dfrac{4}{\sqrt{7}}$ $\dfrac{\sqrt{7}}{\sqrt{7}}$

8. $\dfrac{2\sqrt{5}}{4 - \sqrt{3}}$ $\dfrac{4 + \sqrt{3}}{4 + \sqrt{3}}$

Simplify. Leave in radical form and use absolute value symbols when necessary.

9. $\sqrt{18}$ $3\sqrt{2}$

10. $\dfrac{\sqrt{20}}{\sqrt{5}}$ 2

11. $\sqrt{\dfrac{3}{7}}$ $\dfrac{\sqrt{21}}{7}$

12. $\sqrt{\dfrac{2}{3}} \cdot \sqrt{\dfrac{5}{2}}$ $\dfrac{\sqrt{15}}{3}$

13. $(\sqrt{2} + 4)(\sqrt{2} + 6)$ $10\sqrt{2} + 26$

14. $(y - \sqrt{5})(y + \sqrt{5})$ $y^2 - 5$

15. $\dfrac{6}{3 - \sqrt{2}}$ $\dfrac{18 + 6\sqrt{2}}{7}$

16. $\sqrt{80a^2b^3}$ $4|a|b\sqrt{5b}$

Compare each pair of expressions using <, >, or =.

17. $4\sqrt{3} \cdot \sqrt{3}$, $\sqrt{48} + \sqrt{8}$ >

18. $\sqrt{\dfrac{12}{7}}$, $\dfrac{\sqrt{18} \cdot \sqrt{2}}{\sqrt{7} \cdot \sqrt{3}}$ =

19. **Water Supply**   There is a relationship between a city's capacity to supply water to its citizens and the city's size. Suppose a city has a population P (in thousands). Then the number of gallons per minute that are required to assure water adequacy is given by the expression $1020\sqrt{P}(1 - 0.01\sqrt{P})$. If a city has a population of 55,000 people, how many gallons per minute must the city's pumping stations be able to supply?   **about 7003.5 gal/min**

EXERCISES

Practice

28. 150
29. $84\sqrt{5}$
30. $\dfrac{\sqrt{35}}{5}$
32. $2b^2\sqrt{10}$
33. $3|ab|\sqrt{6}$
34. $2|m|y^2\sqrt{15}$
35. $7x^2y^3\sqrt{3xy}$
38. $\dfrac{n^2\sqrt{5mn}}{2|m^3|}$
39. $\dfrac{x\sqrt{3xy}}{2y^3}$
40–47. See margin.
48. $2x - 6$
49. $x^2 - 5x\sqrt{3} + 12$

Simplify. Leave in radical form and use absolute value symbols when necessary.

A

20. $\sqrt{75}$ $5\sqrt{3}$

21. $\sqrt{80}$ $4\sqrt{5}$

22. $\sqrt{280}$ $2\sqrt{70}$

23. $\sqrt{500}$ $10\sqrt{5}$

24. $\dfrac{\sqrt{7}}{\sqrt{3}}$ $\dfrac{\sqrt{21}}{3}$

25. $\dfrac{\sqrt{5}}{\sqrt{10}}$ $\dfrac{\sqrt{2}}{2}$

26. $\sqrt{\dfrac{2}{7}}$ $\dfrac{\sqrt{14}}{7}$

27. $\sqrt{\dfrac{11}{32}}$ $\dfrac{\sqrt{22}}{8}$

28. $5\sqrt{10} \cdot 3\sqrt{10}$

29. $7\sqrt{30} \cdot 2\sqrt{6}$

30. $\sqrt{\dfrac{3}{5}} \cdot \sqrt{\dfrac{7}{3}}$

31. $\sqrt{\dfrac{1}{6}} \cdot \sqrt{\dfrac{6}{11}}$ $\dfrac{\sqrt{11}}{11}$

32. $\sqrt{40b^4}$

33. $\sqrt{54a^2b^2}$

34. $\sqrt{60m^2y^4}$

35. $\sqrt{147x^5y^7}$

B

36. $\sqrt{\dfrac{t}{8}}$ $\dfrac{\sqrt{2t}}{4}$

37. $\sqrt{\dfrac{27}{p^2}}$ $\dfrac{3\sqrt{3}}{|p|}$

38. $\sqrt{\dfrac{5n^5}{4m^5}}$

39. $\dfrac{\sqrt{9x^5y}}{\sqrt{12x^2y^6}}$

40. $(1 + 2\sqrt{5})^2$

41. $(y - \sqrt{7})^2$

42. $(\sqrt{m} + \sqrt{20})^2$

43. $\dfrac{14}{\sqrt{8} - \sqrt{5}}$

44. $\dfrac{9a}{6 + \sqrt{a}}$

45. $\dfrac{2\sqrt{5}}{-4 + \sqrt{8}}$

46. $\dfrac{3\sqrt{7}}{5\sqrt{3} + 3\sqrt{5}}$

47. $\dfrac{\sqrt{c} - \sqrt{d}}{\sqrt{c} + \sqrt{d}}$

48. $(\sqrt{2x} - \sqrt{6})(\sqrt{2x} + \sqrt{6})$

49. $(x - 4\sqrt{3})(x - \sqrt{3})$

40. $21 + 4\sqrt{5}$

41. $y^2 - 2y\sqrt{7} + 7$

42. $m + 4\sqrt{5m} + 20$

43. $\dfrac{28\sqrt{2} + 14\sqrt{5}}{3}$

44. $\dfrac{54a - 9a\sqrt{a}}{36 - a}$

45. $\dfrac{-2\sqrt{5} - \sqrt{10}}{2}$

46. $\dfrac{5\sqrt{21} - 3\sqrt{35}}{10}$

47. $\dfrac{c - 2\sqrt{cd} + d}{c - d}$

Compare each pair of expressions using <, >, or =.

50. $\sqrt{\dfrac{8}{9}} \cdot \dfrac{2}{\sqrt{8}}, \dfrac{2}{\sqrt{51}} \cdot \sqrt{\dfrac{17}{3}}$ =

51. $\sqrt{10} \cdot \sqrt{30}, \dfrac{10}{\sqrt{5}+9}$ >

52. $\dfrac{2}{\sqrt{6}-\sqrt{5}}, \dfrac{20}{6+\sqrt{3}}$ >

53. $\dfrac{3\sqrt{2}-\sqrt{7}}{2\sqrt{3}-5\sqrt{2}}, \dfrac{4\sqrt{5}-3\sqrt{7}}{\sqrt{6}}$ <

Critical Thinking

54. Determine whether $\sqrt{a \cdot b} = \sqrt{a} \cdot \sqrt{b}$ is true for negative real numbers. Give examples to support your answer.   **No, because square roots of negative numbers are not defined in the set of real numbers;**
$$\sqrt{(-2) \cdot (-3)} \neq \sqrt{-2} \cdot \sqrt{-3}$$

Applications and Problem Solving

55. **Racing**   In yacht racing from 1958 to 1987, 12-meter boats were not really 12 meters long, but the formula that governed their design contained numbers that equaled 12. In the expression $\dfrac{\sqrt{S}+L-F}{2.37}$, S is the area of the sails, L is the waterline length, and F is the distance from the deck of the boat to the waterline. The result must be less than or equal to 12 for a boat to be classified as a 12-meter boat. Determine if a boat for which $S = 158$ m², $L = 17.5$ m, and $F = 2$ m could be classified as a 12-meter boat.

56. **Electricity**   The voltage V required for a circuit is given by $V = \sqrt{PR}$ where P is the power in watts and R is the resistance in ohms. Find the volts needed to light a 75-watt bulb with a resistance of 110 ohms.   **about 90.83 volts**

Mixed Review

55. Yes, the result is about 11.84.

57. 18.44 cm

58. $\dfrac{2a-1}{a-8}$; $a \neq 8, -6$

59. $1, \dfrac{2}{3}$

57. **Geometry**   Find the length of the missing side of the triangle shown at the right. Round to the nearest hundredth.   (Lesson 13–1)

12 cm

14 cm

x

58. Simplify $\dfrac{2a^2 + 11a - 6}{a^2 - 2a - 48}$. State the excluded values of a.   (Lesson 12–1)

59. Solve $3x^2 - 5x + 2 = 0$ by using the quadratic formula.   (Lesson 11–3)

60. Factor $12a^2b^3 - 28ab^2c^2$.   (Lesson 10–2) $4ab^2(3ab - 7c^2)$

61. 2.7 × 10⁹ acres

61. **Forests**   The largest forested areas in the world are located in northern Russia. They cover 2,700,000,000 acres, and they make up 25% of the world's forests. Express the number of acres in scientific notation.   (Lesson 9–3)

62. Solve $4 - 2.3t < 17.8$.   (Lesson 7–3) $\{t \mid t > -6\}$

63. $y = \dfrac{1}{5}x + 6$

63. **Geometry**   Write an equation of the line that is perpendicular to the graph of $y = -5x + 2$ and passes through the point at $(0, 6)$.   (Lesson 6–6)

64. {(−1, 3), (−1, 4), (1, 4), (1, −3), (3, 5)}; {−1, 1, 3}; {3, 4, −3, 5}

64. Express the relation shown in the graph at the right as a set of ordered pairs. Then state the domain and range of the relation.   (Lesson 5–2)

65. Solve $6(x + 3) = 3x$.   (Lesson 3–5)   **−6**

66. Write an algebraic expression for the verbal expression *one-fourth the square of a number*.   (Lesson 1–1) $\dfrac{1}{4}x^2$

Extension

Problem Solving Simplify

$$\dfrac{2}{\sqrt{3} + \sqrt{5} + \sqrt{6}}.$$

$$\dfrac{4\sqrt{3} + 2\sqrt{5} + \sqrt{6} - 3\sqrt{10}}{14}$$

---

4 ASSESS

Closing Activity

Writing Have students summarize this section by writing a list of the key points that they learned.

Chapter 13, Quiz A (Lessons 13-1 and 13-2), is available in the *Assessment and Evaluation Masters,* p. 352.

Enrichment Masters, p. 89

13-2   NAME_____   DATE_____

Enrichment   Student Edition Pages 719–725

Rational Exponents

You have developed the following properties of powers when a is a positive real number and m and n are integers.

$a^m \cdot a^n = a^{m+n}$ $(ab)^n = a^n b^n$ $a^0 = 1$

$(a^m)^n = a^{mn}$ $\dfrac{a^m}{a^n} = a^{m-n}$ $a^{-n} = \dfrac{1}{a^n}$

Exponents need not be restricted to integers. We can define rational exponents so that operations involving them will be governed by the properties for integer exponents.

$\left(a^{\frac{1}{2}}\right)^2 = a^{\frac{1}{2} \cdot 2} = a$ $\left(a^{\frac{1}{3}}\right)^3 = a^{\frac{1}{3} \cdot 3} = a$ $\left(a^{\frac{1}{n}}\right)^n = a^{\frac{1}{n} \cdot n} = a$

$a^{\frac{1}{2}}$ squared is a. $a^{\frac{1}{3}}$ cubed is a. $a^{\frac{1}{n}}$ to the n power is a.

$a^{\frac{1}{2}}$ is a square root of a. $a^{\frac{1}{3}}$ is a cube root of a. $a^{\frac{1}{n}}$ is an nth root of a.

$a^{\frac{1}{2}} = \sqrt{a}$ $a^{\frac{1}{3}} = \sqrt[3]{a}$ $a^{\frac{1}{n}} = \sqrt[n]{a}$

Now let us investigate the meaning of $a^{\frac{m}{n}}$.

$a^{\frac{m}{n}} = a^{m \cdot \frac{1}{n}} = \left(a^m\right)^{\frac{1}{n}} = \sqrt[n]{a^m}$ $a^{\frac{m}{n}} = a^{\frac{1}{n} \cdot m} = \left(a^{\frac{1}{n}}\right)^m = (\sqrt[n]{a})^m$

Therefore, $a^{\frac{m}{n}} = \sqrt[n]{a^m}$ or $(\sqrt[n]{a})^m$.

Example 1: Write $\sqrt[4]{a^3}$ in exponential form.   **Example 2:** Write $a^{\frac{2}{3}}$ in radical form.
$\sqrt[4]{a^3} = a^{\frac{3}{4}}$ $a^{\frac{2}{3}} = \sqrt[3]{a^2}$

Example 3: Find $\dfrac{a^{\frac{2}{3}}}{a^{\frac{1}{3}}}$.

$\dfrac{a^{\frac{2}{3}}}{a^{\frac{1}{3}}} = a^{\frac{2}{3} - \frac{1}{3}} = a^{\frac{1}{3}}$ or $\sqrt[3]{a}$

Write each expression in radical form.

1. $b^{\frac{2}{3}}$ $\sqrt[3]{b^2}$ 2. $3c^{\frac{1}{3}}$ $3\sqrt[3]{c}$ 3. $(3c)^{\frac{1}{2}}$ $\sqrt{3c}$

Write each expression in exponential form.

4. $\sqrt[5]{b^2}$ $b^{\frac{2}{5}}$ 5. $\sqrt[3]{4a^3}$ $(4a^3)^{\frac{1}{3}} = 2a^{\frac{1}{3}}$ 6. $2 \cdot \sqrt[3]{b^2}$ $2b^{\frac{2}{3}}$

Perform the operation indicated. Answers should show positive exponents only.

7. $\left(a^3b^{\frac{1}{2}}\right)^{\frac{1}{3}}$ $a^9b^{\frac{1}{6}}$ 8. $\dfrac{-8a^{\frac{1}{2}}}{2a^{\frac{1}{4}}}$ $-4a^{\frac{1}{4}}$ 9. $\left(\dfrac{b^{\frac{1}{3}}}{b^{-\frac{2}{3}}}\right)^3$ b^3

10. $\sqrt[3]{a^3} \cdot \sqrt{a}$ a^2 11. $\left(a^3b^{-\frac{1}{3}}\right)^{-\frac{1}{3}}$ $\dfrac{b^{\frac{1}{9}}}{a}$ 12. $-2a^{\frac{1}{2}}b\left(5a^{\frac{1}{2}}b^{-\frac{3}{2}}\right)$ $\dfrac{-10a^2}{b^{\frac{1}{2}}}$

NCTM Standards: 1–5

Objective
Use the calculator to simplify and approximate values of expressions containing radicals.

Recommended Time
15 minutes

Instructional Resources
Graphing Calculator Masters, pp. 38 and 39

These masters provide keystroking instruction for this lesson for the TI-81 and Casio graphing calculators.

1 FOCUS

Motivating the Lesson

Have students estimate $\sqrt{\frac{2}{3}}$ without a calculator. They will have to guess and check to approximate $\sqrt{2}$ and $\sqrt{3}$. Then have students check their work with the graphing calculator.

2 TEACH

Teaching Tip Remind students that in most problems, a calculator's built-in square root function gives only an approximation.

3 PRACTICE/APPLY

Assignment Guide
Core: 1–11
Enriched: 1–11

4 ASSESS

Observing students working with technology is an excellent method of assessment.

The built-in square root feature of a graphing calculator allows us to simplify and approximate values of expressions containing radicals. In addition to obtaining approximate values for expressions, this feature is also useful for checking algebraic computations.

Example ● **Simplify each expression algebraically. Then check with a graphing calculator.**

a. $\sqrt{\frac{3}{5}}$

$$\sqrt{\frac{3}{5}} = \sqrt{\frac{3}{5}} \cdot \frac{\sqrt{5}}{\sqrt{5}}$$

$$= \frac{\sqrt{3 \cdot 5}}{5} \text{ or } \frac{\sqrt{15}}{5}$$

Verify with the calculator.

Enter: `2nd` `√` `(` `3` `÷` `5` `)`
`ENTER` .7745966692
`2nd` `√` `15` `÷` `5`
`ENTER` .7745966692

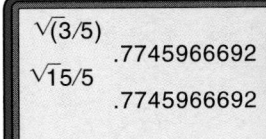

```
√(3/5)
        .7745966692
√15/5
        .7745966692
```

b. $\frac{1}{\sqrt{2} - 3}$

$$\frac{1}{\sqrt{2} - 3} = \frac{1}{\sqrt{2} - 3} \cdot \frac{\sqrt{2} + 3}{\sqrt{2} + 3}$$

$$= \frac{\sqrt{2} + 3}{(\sqrt{2})^2 - 3^2} \text{ or } \frac{\sqrt{2} + 3}{-7}$$

Verify with the calculator.

Enter: `1` `÷` `(` `2nd` `√` `2` `-`
`3` `)` `ENTER` -.6306019375
`(` `2nd` `√` `2` `+` `3` `)`
`÷` `(-)` `7` `ENTER`
-.6306019375

```
1/(√2 -3)
        -.6306019375
(√2+3)/-7
        -.6306019375
■
```

EXERCISES

Simplify each expression. Then check with a graphing calculator. Round answers to the nearest hundredth.

1. $\sqrt{1372}$ **2.** $\sqrt{32} \cdot \sqrt{12}$ **3.** $\sqrt{2}(\sqrt{6} + 3)$ **4.** $\sqrt{\frac{5}{6}}$

5. $\frac{4}{\sqrt{7}}$ **6.** $\frac{2}{\sqrt{11} + 8}$ **7.** $\sqrt{12} + \sqrt{3}$ **8.** $\frac{4}{5}\sqrt{2} + \frac{3}{5}\sqrt{2}$

9. $\frac{\sqrt{3}}{2} - \frac{\sqrt{5}}{3} + \sqrt{18}$ **10.** $\sqrt{18} + \sqrt{108} + \sqrt{50}$ **11.** $\sqrt{\frac{2}{3}} + \frac{\sqrt{6}}{3} - 6\sqrt{6}$

1. $14\sqrt{7}$ or 37.04
2. $8\sqrt{6}$ or 19.60
3. $2\sqrt{3} + 3\sqrt{2}$ or 7.71
4. $\frac{\sqrt{30}}{6}$ or 0.91
5. $\frac{4\sqrt{7}}{7}$ or 1.51
6. $-\frac{2\sqrt{11} - 16}{53}$ or 0.18
7. $3\sqrt{3}$ or 5.20
8. $\frac{7}{5}\sqrt{2}$ or 1.98
9. $\frac{\sqrt{3}}{2} - \frac{\sqrt{5}}{3} + 3\sqrt{2}$ or 4.36
10. $8\sqrt{2} + 6\sqrt{3}$ or 21.71
11. $-\frac{16\sqrt{6}}{3}$ or -13.06

Using Technology
This lesson offers an excellent opportunity for using technology in your algebra classroom. For more information on using technology, see *Graphing Calculators in the Mathematics Classroom*, one of the titles in the Glencoe Mathematics Professional Series.

Operations with Radical Expressions

Objective

After studying this lesson, you should be able to:

- simplify radical expressions involving addition, subtraction, and multiplication.

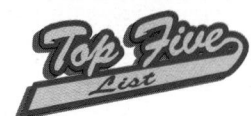

Largest Passenger Ships in the World (gross tonnage)

1. *Norway*, 76,049
2. *Majesty of the Seas*, 73,937
2. *Monarch of the Seas*, 73,937
4. *Sovereign of the Seas*, 73,192
5. *Sensation*, 70,367
5. *Ecstasy*, 70,367
5. *Fantasy*, 70,367

APPLICATION
Travel

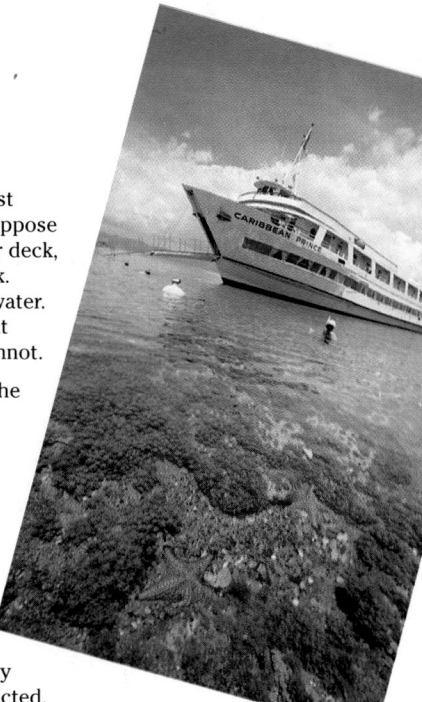

The *Norway* cruise ship is the largest passenger ship in the Caribbean. Suppose the captain of the ship is on the star deck, which is 48 feet above the pool deck. The pool deck is 72 feet above the water. The captain sees the next island, but the passengers on the pool deck cannot.

The equation $d = \sqrt{\dfrac{3h}{2}}$ represents the distance d in miles a person h feet high can see. So, $\sqrt{\dfrac{3(120)}{2}} - \sqrt{\dfrac{3(72)}{2}}$ describes how much farther the captain can see than the passengers. *You will find the value of this expression in Example 3.*

Radical expressions in which the radicands are alike can be added or subtracted in the same way that monomials are added or subtracted.

Monomials	Radical Expressions
$4x + 5x = (4 + 5)x$	$4\sqrt{5} + 5\sqrt{5} = (4 + 5)\sqrt{5}$
$\qquad = 9x$	$\qquad = 9\sqrt{5}$
$18y - 7y = (18 - 7)y$	$18\sqrt{2} - 7\sqrt{2} = (18 - 7)\sqrt{2}$
$\qquad = 11y$	$\qquad = 11\sqrt{2}$

Notice that the distributive property was used to simplify each radical expression.

Example ① Simplify each expression.

a. $6\sqrt{7} + 5\sqrt{7} - 3\sqrt{7}$

$$6\sqrt{7} + 5\sqrt{7} - 3\sqrt{7} = (6 + 5 - 3)\sqrt{7}$$
$$= 8\sqrt{7}$$

b. $5\sqrt{6} + 3\sqrt{7} + 4\sqrt{7} - 2\sqrt{6}$

$$5\sqrt{6} + 3\sqrt{7} + 4\sqrt{7} - 2\sqrt{6} = 5\sqrt{6} - 2\sqrt{6} + 3\sqrt{7} + 4\sqrt{7}$$
$$= (5 - 2)\sqrt{6} + (3 + 4)\sqrt{7}$$
$$= 3\sqrt{6} + 7\sqrt{7}$$

Lesson 13-3 Operations with Radical Expressions **727**

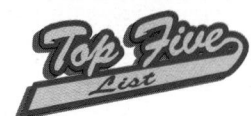

The first ocean liners were operated in the North Atlantic in 1840.

13-3 LESSON NOTES

NCTM Standards: 1–5

Instructional Resources

- Study Guide Master 13-3
- Practice Master 13-3
- Enrichment Master 13-3
- Assessment and Evaluation Masters, pp. 351–352
- Modeling Mathematics Masters, p. 84

Transparency 13-3A contains the 5-Minute Check for this lesson; **Transparency 13-3B** contains a teaching aid for this lesson.

Recommended Pacing	
Honors Pacing	Day 5 of 11
Block Scheduling*	Day 3 of 7
Alg. 1 in Two Years*	Days 8, 9, & 10 of 22

*For more information on pacing and possible lesson plans, refer to the *Block Scheduling Booklet* and *Algebra 1 in Two Years.*

1 FOCUS

5-Minute Check
(over Lesson 13-2)

Simplify.

1. $\sqrt{75}$ $5\sqrt{3}$
2. $\sqrt{48m^4}$ $4m^2\sqrt{3}$
3. $\sqrt{3} \cdot \sqrt{33}$ $3\sqrt{11}$
4. $\dfrac{\sqrt{7}}{\sqrt{5}}$ $\dfrac{\sqrt{35}}{5}$
5. $\dfrac{6}{3 - \sqrt{2}}$ $\dfrac{18 + 6\sqrt{2}}{7}$

Motivating the Lesson

Questioning Begin the lesson with a review of addition and subtraction of monomials. Ask students to simplify $4b + 7b$. **11b** Then have them simplify $4b^2 + 7b + 2b$. **$4b^2 + 9b$** Remind students that only like terms can be combined.

Teaching Tip If students can use the distributive property with ease, they should have no trouble adding and subtracting expressions with like radicands.

In-Class Examples

For Example 1
Simplify $6\sqrt{7} - 2\sqrt{7} - \sqrt{3} + 4\sqrt{3}$. $4\sqrt{7} + 3\sqrt{3}$

For Example 2
Simplify $3\sqrt{45} + 4\sqrt{80} - 2\sqrt{125}$. $15\sqrt{5}$

For Example 3
Find the exact measure of the perimeter of a rectangle whose width is $4\sqrt{6} + \sqrt{3}$ units and whose length is $2\sqrt{3} - 4$ units. $8\sqrt{6} + 6\sqrt{3} - 8$ sq units

Teaching Tip In Example 3, remind students that only when the radicands are alike can terms be added or subtracted.

In Example 1b, the expression $3\sqrt{6} + 7\sqrt{7}$ cannot be simplified further because the radicands are different. There are no common factors, and each radicand is in simplest form.

If the radicals in a radical expression are not in simplest form, simplify them first. Then use the distributive property wherever possible to further simplify the expression.

Example **Simplify $4\sqrt{27} + 5\sqrt{12} + 8\sqrt{75}$. Then use a scientific calculator to verify your answer.**

$$
\begin{aligned}
4\sqrt{27} + 5\sqrt{12} + 8\sqrt{75} &= 4\sqrt{3^2 \cdot 3} + 5\sqrt{2^2 \cdot 3} + 8\sqrt{5^2 \cdot 3} \\
&= 4(\sqrt{3^2} \cdot \sqrt{3}) + 5(\sqrt{2^2} \cdot \sqrt{3}) + 8(\sqrt{5^2} \cdot \sqrt{3}) \\
&= 4(3\sqrt{3}) + 5(2\sqrt{3}) + 8(5\sqrt{3}) \\
&= 12\sqrt{3} + 10\sqrt{3} + 40\sqrt{3} \\
&= 62\sqrt{3}
\end{aligned}
$$

The exact answer is $62\sqrt{3}$. Now, use a calculator to verify.

First, find a decimal approximation for the original expression.

Enter: 4 $\boxed{\times}$ 27 $\boxed{\sqrt{x}}$ $\boxed{+}$ 5 $\boxed{\times}$ 12 $\boxed{\sqrt{x}}$ $\boxed{+}$ 8 $\boxed{\times}$ 75 $\boxed{\sqrt{x}}$ $\boxed{=}$ *107.3871501*

Next, find a decimal approximation for the simplified expression.

Enter: 62 $\boxed{\times}$ 3 $\boxed{\sqrt{x}}$ $\boxed{=}$ *107.3871501*

Since the approximations are equal, the results have been verified.

Example **Refer to the application at the beginning of the lesson. How much farther is the captain able to see than the passengers on the pool deck?**

$$
\begin{aligned}
d &= \sqrt{\frac{3(120)}{2}} - \sqrt{\frac{3(72)}{2}} \\
&= \sqrt{\frac{360}{2}} - \sqrt{\frac{216}{2}} \\
&= \sqrt{180} - \sqrt{108} \\
&= \sqrt{6^2 \cdot 5} - \sqrt{6^2 \cdot 3} \\
&= 6\sqrt{5} - 6\sqrt{3} \\
&\approx 3.02
\end{aligned}
$$

The captain can see about 3 miles farther.

In the last lesson, you multiplied conjugates and expressions with like radicands. Multiplying two radical expressions with different radicands is similar to multiplying two binomials.

Example **4** Simplify $(2\sqrt{3} - \sqrt{5})(\sqrt{10} + 4\sqrt{6})$.

$(2\sqrt{3} - \sqrt{5})(\sqrt{10} + 4\sqrt{6})$

LOOK BACK

Refer to Lesson 9-7 to review multiplying polynomials.

First terms	Outer terms	Inner terms	Last terms

$= (2\sqrt{3})(\sqrt{10}) + (2\sqrt{3})(4\sqrt{6}) + (-\sqrt{5})(\sqrt{10}) + (-\sqrt{5})(4\sqrt{6})$

$= 2\sqrt{30} + 8\sqrt{18} - \sqrt{50} - 4\sqrt{30}$ *Multiply.*

$= 2\sqrt{30} + 24\sqrt{2} - 5\sqrt{2} - 4\sqrt{30}$ *Simplify each term.*

$= -2\sqrt{30} + 19\sqrt{2}$ *Combine like terms.*

CHECK FOR UNDERSTANDING

Communicating Mathematics

1. Sample answer: $4\sqrt{3}$, $2\sqrt{3}$, $6\sqrt{3}$

MATH JOURNAL

Study the lesson. Then complete the following.

1. **Write** three radical expressions that have the same radicand.

2. **Explain** why you should simplify each radical in a radical expression before adding or subtracting. to determine if there are any like radicands

3. **Explain** why $\sqrt{x} + \sqrt{y} \neq \sqrt{x + y}$. Give an example using numbers. See margin.

4. Explain how you use the distributive property to simplify like radicals that are added or subtracted. See margin.

Guided Practice

Name the expressions in each group that will have the same radicand after each expression is written in simplest form.

5. $3\sqrt{5}, 5\sqrt{6}, 3\sqrt{20}$

6. $-5\sqrt{7}, 2\sqrt{28}, 6\sqrt{14}$

7. $\sqrt{24}, \sqrt{12}, \sqrt{18}, \sqrt{28}$ none

8. $9\sqrt{32}, 2\sqrt{50}, \sqrt{48}, 3\sqrt{200}$

Simplify. 9. $13\sqrt{6}$ 10. in simplest form 11. $12\sqrt{7x}$

9. $3\sqrt{6} + 10\sqrt{6}$

10. $2\sqrt{5} - 5\sqrt{2}$

11. $8\sqrt{7x} + 4\sqrt{7x}$

Simplify. Then use a calculator to verify your answer.

13. $13\sqrt{3} + \sqrt{2}$; 23.93

12. $8\sqrt{5} + 3\sqrt{5}$ $11\sqrt{5}$; 24.60

13. $8\sqrt{3} - 2\sqrt{2} + 3\sqrt{2} + 5\sqrt{3}$

14. $2\sqrt{3} + \sqrt{12}$ $4\sqrt{3}$; 6.93

15. $\sqrt{7} + \sqrt{\frac{1}{7}}$ $\frac{8}{7}\sqrt{7}$; 3.02

Simplify.

16. $\sqrt{2}(\sqrt{18} + 4\sqrt{3})$ $6 + 4\sqrt{6}$

17. $(4 + \sqrt{5})(4 - \sqrt{5})$ 11

18. **Geometry** Find the exact measures of the perimeter and area in simplest form for the rectangle at the right.

$8\sqrt{7} - 6\sqrt{3}$; $4\sqrt{21} - 12$

 $4\sqrt{7} - 2\sqrt{12}$

$\sqrt{3}$

Reteaching

Logical Thinking Have students verify that the simplest form of a radical expression is equivalent to the radical expression for several problems. Use a calculator and round to 4 places. For example:

$$\sqrt{18} + \sqrt{50} = 8\sqrt{2}$$

$$4.243 + 7.071 = 8(1.414)$$

$$11.314 = 11.314$$

---

3 PRACTICE/APPLY

Check for Understanding

Exercises 1–18 are designed to help you assess your students' understanding through reading, writing, speaking, and modeling. You should work through Exercises 1–4 with your students and then monitor their work on Exercises 5–18.

Additional Answers

3. Sample answer:

$\sqrt{x} + \sqrt{y} \stackrel{?}{=} \sqrt{x + y}$

$\sqrt{3} + \sqrt{6} \stackrel{?}{=} \sqrt{9}$

4.1815 ≠ 3

4. The distributive property allows you to add like terms. Radicals with like radicands can be added or subtracted.

Study Guide Masters, p. 90

13-3 NAME_____ DATE_____
Study Guide Student Edition Pages 727–731

Operations with Radical Expressions

When adding or subtracting radical expressions, use the distributive and commutative properties to simplify the expressions. If radical expressions are not in simplest form, first simplify.

Example 1: Simplify $10\sqrt{6} - 5\sqrt{3} + 6\sqrt{3} - 4\sqrt{6}$.

$10\sqrt{6} - 5\sqrt{3} + 6\sqrt{3} - 4\sqrt{6} = 10\sqrt{6} - 4\sqrt{6} - 5\sqrt{3} + 6\sqrt{3}$
$= (10 - 4)\sqrt{6} + (-5 + 6)\sqrt{3}$
$= 6\sqrt{6} + \sqrt{3}$

Example 2: Simplify $3\sqrt{12} + 5\sqrt{75}$.

$3\sqrt{12} + 5\sqrt{75} = 3\sqrt{2^2 \cdot 3} + 5\sqrt{5^2 \cdot 3}$
$= 3 \cdot 2\sqrt{3} + 5 \cdot 5\sqrt{3}$
$= 6\sqrt{3} + 25\sqrt{3}$
$= (6 + 25)\sqrt{3}$
$= 31\sqrt{3}$

Simplify. Then use a calculator to verify your answer.

1. $8 + 3\sqrt{2}$
$8 + 3\sqrt{2}$;
$8 + 4.24 = 12.24$

2. $\sqrt{12} - \sqrt{27}$
$-\sqrt{3}$; 3.46 −
$5.20 \approx -1.73$

3. $\sqrt{27} - 2\sqrt{3}$
$\sqrt{3}$; 5.20 −
$3.46 \approx 1.73$

4. $\sqrt{20} + 2\sqrt{5} - 3\sqrt{5}$
$\sqrt{5}$; 4.47 + 4.47 −
$6.71 \approx 2.24$

5. $-5\sqrt{6} + 8\sqrt{6}$
$3\sqrt{6}$; −12.25 +
$19.60 = 7.35$

6. $\sqrt{200} - 3\sqrt{2}$
$7\sqrt{2}$; 14.14 −
$4.24 = 9.90$

7. $\sqrt{54} + \sqrt{24}$
$5\sqrt{6}$; 7.35 + 4.90 =
12.25

8. $\sqrt{18} - 3\sqrt{8} + \sqrt{50}$
$2\sqrt{2}$; 4.24 − 8.49 +
$7.07 \approx 2.83$

9. $\sqrt{80} - \sqrt{20} + \sqrt{180}$
$8\sqrt{5}$; 8.94 − 4.47 +
$13.42 = 17.89$

10. $2\sqrt{28} + 3\sqrt{63} - \sqrt{\frac{54}{3}}$
$13\sqrt{7} - 3\sqrt{2}$;
10.58 + 23.81 −
4.24 = 30.15 =
34.39 − 4.24

11. $\sqrt{12} + \sqrt{\frac{1}{3}}$
$\frac{7\sqrt{3}}{3}$; 3.46 +
$0.58 = 4.04$

12. $\sqrt{54} - \sqrt{\frac{1}{6}} + \sqrt{24}$
$\frac{29}{6}\sqrt{6}$; 7.35 −
$0.41 + 4.90 = 11.84$

Assignment Guide

Core: 19–45 odd, 46–53
Enriched: 18–42 even, 43–53
All: Self Test, 1–10

For **Extra Practice,** see p. 785.

The red A, B, and C flags, printed only in the Teacher's Wraparound Edition, indicate the level of difficulty of the exercises.

Practice Masters, p. 90

EXERCISES

Practice

A

Simplify.

19. $25\sqrt{13} + \sqrt{13}$ $26\sqrt{13}$

20. $7\sqrt{2} - 15\sqrt{2} + 8\sqrt{2}$ 0

21. $2\sqrt{6} - 8\sqrt{3}$ in simplest form

22. $2\sqrt{11} - 6\sqrt{11} - 3\sqrt{11}$ $-7\sqrt{11}$

23. $18\sqrt{2x} + 3\sqrt{2x}$ $21\sqrt{2x}$

24. $3\sqrt{5m} - 5\sqrt{5m}$ $-2\sqrt{5m}$

Simplify. Then use a calculator to verify your answer.

26. $-10\sqrt{5}$; -22.36

27. $\sqrt{6} + 2\sqrt{2} + \sqrt{10}$; 8.44

28. $4\sqrt{6} - 6\sqrt{2} + 5\sqrt{7}$; 14.54

32. $-2\sqrt{5} - 6\sqrt{6}$; -19.17

33. $4\sqrt{5} + 15\sqrt{2}$; 30.16

35. $4\sqrt{3} - 3\sqrt{5}$; 0.22

36. $\frac{53\sqrt{7}}{7}$; 20.03

39. $19\sqrt{5}$

41. $10\sqrt{3} + 16$

42. $15\sqrt{2} + 11\sqrt{5}$

B

25. $4\sqrt{3} + 7\sqrt{3} - 2\sqrt{3}$ $9\sqrt{3}$; 15.59

26. $5\sqrt{5} + 3\sqrt{5} - 18\sqrt{5}$

27. $\sqrt{6} + 2\sqrt{2} + \sqrt{10}$

28. $4\sqrt{6} + \sqrt{7} - 6\sqrt{2} + 4\sqrt{7}$

29. $3\sqrt{7} - 2\sqrt{28}$ $-\sqrt{7}$; -2.65

30. $2\sqrt{50} - 3\sqrt{32}$ $-2\sqrt{2}$; -2.83

31. $3\sqrt{27} + 5\sqrt{48}$ $29\sqrt{3}$; 50.23

32. $2\sqrt{20} - 3\sqrt{24} - \sqrt{180}$

33. $\sqrt{80} + \sqrt{98} + \sqrt{128}$

34. $\sqrt{10} - \sqrt{\frac{2}{5}}$ $\frac{4}{5}\sqrt{10}$; 2.53

35. $3\sqrt{3} - \sqrt{45} + 3\sqrt{\frac{1}{3}}$

36. $6\sqrt{\frac{7}{4}} + 3\sqrt{28} - 10\sqrt{\frac{1}{7}}$

Simplify.   37. $10\sqrt{2} + 3\sqrt{10}$

C

37. $\sqrt{5}(2\sqrt{10} + 3\sqrt{2})$

38. $\sqrt{6}(\sqrt{3} + 5\sqrt{2})$ $3\sqrt{2} + 10\sqrt{3}$

39. $(2\sqrt{10} + 3\sqrt{15})(3\sqrt{3} - 2\sqrt{2})$

40. $(\sqrt{5} - \sqrt{2})(\sqrt{14} + \sqrt{35})$ $3\sqrt{7}$

41. $(\sqrt{6} + \sqrt{8})(\sqrt{24} + \sqrt{2})$

42. $(5\sqrt{2} + 3\sqrt{5})(2\sqrt{10} - 3)$

Critical Thinking

43. Explain why the simplified form of $\sqrt{(x-5)^2}$ must have an absolute value sign, but $\sqrt{(x-5)^4}$ does not need one.   Both $(x-5)^2$ and $(x-5)^4$ must be nonnegative, but $x - 5$ may be negative.

Applications and Problem Solving

44. **Construction** *Slip forming* is the fastest method of erecting tall concrete buildings. With this method, the 1815-foot CN Tower in Toronto, Canada, was built at an average speed of 20 feet per day. At the beginning of the week, construction workers were 530 feet above the ground. After one week of construction, they were 670 feet above the ground. How many more miles could they see from the top of the building at the end of the week than at the beginning? Write your answer in exact form and as an approximation to the nearest hundredth. (*Hint:* Use the formula from the application at the beginning of the lesson.)
$\sqrt{1005} - \sqrt{795}$; 3.51 mi

45. **Construction** A wire is stretched from the top of a 12-foot pole to a stake in the ground and then to the base of the pole. If a total of 20 feet of wire is needed, how far is the stake from the pole? (*Hint:* In the figure, $a + b = 20$.) $6\frac{2}{5}$ ft

730   Chapter 13   *Exploring Radical Expressions and Equations*

Alternative Teaching Strategies

Reading Algebra Have a student read the expressions $2(\sqrt{6} + \sqrt{7})$ and $2\sqrt{6} + \sqrt{7}$ to the class. The student should read the first one as "2 times the quantity radical 6 plus radical 7." The student should read the second one as "2 times radical 6 plus radical 7."

Mixed Review

46. Simplify $\dfrac{\sqrt{3}}{\sqrt{6}}$. (Lesson 13-2) $\dfrac{\sqrt{2}}{2}$

47. Find $\dfrac{x^2 - y^2}{3} \cdot \dfrac{9}{x+y}$. (Lesson 12-2) $3x - 3y$

48. Factor $3a^2 + 19a - 14$. (Lesson 10-3) $(3a - 2)(a + 7)$

49. Find the degree of $16s^3t^2 + 3s^2t + 7s^6t$. (Lesson 9-4) 7

50. Statistics   Tim's scores on the first four of five 50-point quizzes were 47, 45, 48, and 45. What score must he receive on the fifth quiz to have an average of at least 46 points for all the quizzes? (Lesson 7-3)   **45**

51. Carpentry   When building a stairway, a carpenter considers the ratio of riser to tread. Write a ratio to describe the steepness of the stairs. (Lesson 6-1) $\dfrac{2}{3}$

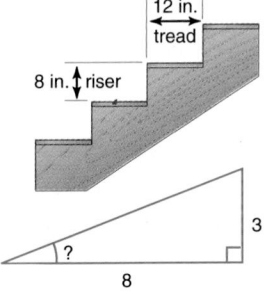

52. Geometry   Find the measure of the marked acute angle to the nearest degree. (Lesson 4-3)   **21°**

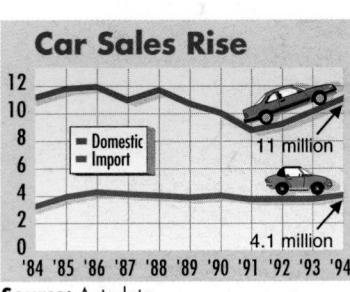

Car Sales Rise

11 million

4.1 million

Source: Autodata

53. Interpret Graphs   The graph at the left compares domestic and import brand auto sales from 1984 to 1994. (Lesson 1-9)

   a. During the ten-year period, when were domestic car sales the lowest? How many domestic cars were sold during that year?   **1991; about 9 million**

   b. In 1990, how many more domestic cars were sold than imports?   **6 million**

SELF TEST

If c is the measure of the hypotenuse of a right triangle, find each missing measure. Round answers to the nearest hundredth. (Lesson 13-1)

1. $a = 21, b = 28, c = ?$ **35**   **2.** $a = \sqrt{41}, c = 8, b = ?$ **4.80**   **3.** $b = 28, c = 54, a = ?$ **46.17**

Simplify. Leave in radical form and use absolute value symbols when necessary. (Lesson 13-2)

4. $\sqrt{20}$ $2\sqrt{5}$ **5.** $2\sqrt{5} \cdot \sqrt{5}$ **10**   **6.** $\dfrac{\sqrt{42x^2}}{\sqrt{6y^3}}$ $\dfrac{|x|\sqrt{7y}}{y^2}$

Simplify. (Lesson 13-3)

7. $8\sqrt{6} + 3\sqrt{6}$ $11\sqrt{6}$ **8.** $10\sqrt{17} + 9\sqrt{7} - 8\sqrt{17} + 6\sqrt{7}$ **9.** $(6 + \sqrt{3})(2\sqrt{5} - \sqrt{3})$
                                             $2\sqrt{17} + 15\sqrt{7}$ $12\sqrt{5} - 6\sqrt{3} + 2\sqrt{15} - 3$

10. Geometry   Find the perimeter and area of the figure at the right. Round answers to the nearest hundredth. (Lesson 13-1)
53.37 cm, 138 cm²

Lesson 13-3 Operations with Radical Expressions   **731**

Extension

Connections Simplify $\dfrac{\sqrt{a}}{\sqrt{b}} + \dfrac{\sqrt{b}}{\sqrt{a}}$.

$(a + b)\dfrac{\sqrt{ab}}{ab}$

---

4 ASSESS

Closing Activity

Speaking Have students explain how addition and subtraction of monomials is similar to that of radical expressions.

Chapter 13, Quiz B (Lesson 13-3), is available in the *Assessment and Evaluation Masters*, p. 352.

Mid-Chapter Test (Lessons 13-1 through 13-3) is available in the *Assessment and Evaluation Masters*, p. 351.

Enrichment Masters, p. 90

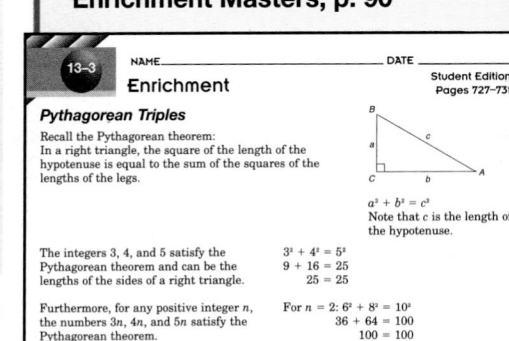

SELF TEST

The Self Test provides students with a brief review of the concepts and skills in Lessons 13-1 through 13-3. Lesson numbers are given to the right of exercises or instruction lines so students can review concepts not yet mastered.

Radical Equations

NCTM Standards: 1–5

Instructional Resources

- Study Guide Master 13-4
- Practice Master 13-4
- Enrichment Master 13-4

Transparency 13-4A contains the 5-Minute Check for this lesson; **Transparency 13-4B** contains a teaching aid for this lesson.

Recommended Pacing

Honors Pacing	Day 6 of 11
Block Scheduling*	Day 4 of 7
Alg. 1 in Two Years*	Days 11, 12, & 13 of 22

*For more information on pacing and possible lesson plans, refer to the *Block Scheduling Booklet* and *Algebra 1 in Two Years.*

1 FOCUS

5-Minute Check
(over Lesson 13-3)

Simplify. Use absolute value symbols when necessary.

1. $4\sqrt{2} + 5\sqrt{3} - \sqrt{2} + 3\sqrt{3}$
   $3\sqrt{2} + 8\sqrt{3}$

2. $8\sqrt{11} + 5\sqrt{11} - 16\sqrt{11}$
   $-3\sqrt{11}$

3. $3\sqrt{5} + 2\sqrt{20}$ $7\sqrt{5}$

4. $\left(3\sqrt{2} - 4\sqrt{5}\right)\left(5\sqrt{2} + 2\sqrt{5}\right)$
   $-10 - 14\sqrt{10}$

5. $\sqrt{3}\left(\sqrt{6} + 2\sqrt{3}\right)$ $3\sqrt{2} + 6$

Objective

After studying this lesson, you should be able to:
- solve radical equations.

F Y I

A tsunami may begin as a 2-foot high wave. After traveling hundreds of miles across the ocean at speeds of 450 to 500 miles per hour, it could approach shallow coastal waters as a towering 50-foot wall of water, capable of destroying anything in its path.

APPLICATION
Oceanography

Topographical map of the ocean surface

The Tonga Trench in the Pacific Ocean is a potential source for a *tsunami* (su-nom'-ee), a large ocean wave generated by an undersea earthquake. The formula for a tsunami's speed s in meters per second is $s = 3.1\sqrt{d}$, where d is the depth of the ocean in meters.

Equations like $s = 3.1\sqrt{d}$ that contain radicals with variables in the radicand are called **radical equations**. To solve these equations, first isolate the radical on one side of the equation. Then square each side of the equation to eliminate the radical.

Find the depth of the Tonga Trench if a tsunami's speed is 322 meters per second.

$322 = 3.1\sqrt{d}$ *Replace s with 322.*

$\dfrac{322}{3.1} = \dfrac{3.1\sqrt{d}}{3.1}$ *Divide each side by 3.1.*

$\left(\dfrac{322}{3.1}\right)^2 = \left(\sqrt{d}\right)^2$ *Square each side of the equation.*

$\left(\dfrac{322}{3.1}\right)^2 = d$ Use a scientific calculator to simplify $\left(\dfrac{322}{3.1}\right)^2$.

Enter: (322 ÷ 3.1) x^2 *10789.17194*

The depth of the Tonga Trench is approximately 10,789 meters.   *Check this result by substituting 10,789 for d into the original formula.*

Example **1** **Solve each equation.**

a. $\sqrt{x} + 4 = 7$

$\sqrt{x} + 4 = 7$

$\sqrt{x} + 4 - 4 = 7 - 4$ *Subtract 4 from each side.*

$\sqrt{x} = 3$ *Simplify.*

$\left(\sqrt{x}\right)^2 = 3^2$ *Square each side.*

$x = 9$ The solution is 9.

Check:

$\sqrt{x} + 4 = 7$

$\sqrt{9} + 4 \overset{?}{=} 7$ *x = 9*

$3 + 4 = 7$ ✔

b. $\sqrt{x + 3} + 5 = 9$

$\sqrt{x + 3} + 5 = 9$

$\sqrt{x + 3} + 5 - 5 = 9 - 5$ *Subtract 5 from each side.*

$\sqrt{x + 3} = 4$ *Simplify.*

$\left(\sqrt{x + 3}\right)^2 = 4^2$ *Square each side.*

$x + 3 = 16$

$x + 3 - 3 = 16 - 3$ *Subtract 3 from each side.*

$x = 13$ The solution is 13.   *Check this result.*

Squaring each side of an equation does not necessarily produce results that satisfy the original equation. Therefore, you must check all solutions when you solve radical equations.

F Y I

Tsunamis are usually caused by earthquakes or volcanic eruptions. When the ocean floor is quickly raised or lowered, it generates a seismic wave.

GLENCOE Technology

 CD-ROM Interaction

A multimedia simulation links radical expressions and the Pythagorean theorem with the installation of an electricity-generating windmill. A blackline master activity with teacher's notes provides a follow-up to the CD-ROM simulation.

For Windows & Macintosh

Example **2** Solve $\sqrt{3x-5}=x-5$.

$$\sqrt{3x-5}=x-5$$
$$(\sqrt{3x-5})^2=(x-5)^2 \qquad \textit{Square each side.}$$
$$3x-5=x^2-10x+25 \qquad \textit{Simplify.}$$
$$3x-3x-5+5=x^2-10x+25-3x+5 \qquad \textit{Add } -3x \textit{ and 5 to each side.}$$
$$0=x^2-13x+30 \qquad \textit{Simplify.}$$
$$0=(x-10)(x-3) \qquad \textit{Factor.}$$
$$x-10=0 \quad \text{or} \quad x-3=0 \qquad \textit{Use the zero product property.}$$
$$x=10 \qquad\qquad x=3$$

Check:

$$\sqrt{3x-5}=x-5 \qquad\qquad\qquad \sqrt{3x-5}=x-5$$
$$\sqrt{3(10)-5}\overset{?}{=}10-5 \qquad\qquad \sqrt{3(3)-5}\overset{?}{=}3-5$$
$$\sqrt{30-5}\overset{?}{=}5 \qquad\qquad\qquad \sqrt{9-5}\overset{?}{=}-2$$
$$\sqrt{25}\overset{?}{=}5 \qquad\qquad\qquad\qquad \sqrt{4}\overset{?}{=}-2$$
$$5=5 \checkmark \qquad\qquad\qquad\qquad 2\neq-2$$

Since 3 does not satisfy the original equation, 10 is the only solution.

You can use the *Mathematics Exploration Toolkit* (*MET*) to solve equations involving square roots.

EXPLORATION

GRAPHING SOFTWARE

The CALC commands below will be used.

ADD (add)	SUBTRACT (sub)	MULTIPLY (mult)
DIVIDE (div)	FACTOR (fac)	RAISETO (rai)
SIMPLIFY (simp)	STORE (sto)	SUBSTITUTE (subs)

To enter the square root symbol, type &.

Solve $\sqrt{x-2}=x-4$.

Enter: &(x − 2) = x − 4   **Result:** $\sqrt{x-2}=x-4$
     sto a   Saves the equation as *a*.
     rai 2 $(\sqrt{x-2})^2=(x-4)^2$
     simp $x-2=x^2-8x+16$
     sub x − 2 $x-2-(x-2)=x^2-8x+16-(x-2)$
     simp $0=x^2-9x+18$
     fac $0=(x-6)(x-3)$

By inspection, the solutions are $x=6$ or $x=3$. However, 3 does not satisfy the original equation. Therefore, 6 is the only solution.

Your Turn
Use CALC to solve each equation.

a. $3+\sqrt{2x}=7$ 8
b. $\sqrt{x+1}=x-1$ 3
c. $\sqrt{x}+6=1$ no solution
d. $\sqrt{3x-8}=5$ 11
e. $x+\sqrt{6-x}=4$ 2
f. $\sqrt{3x-9}=2x+6$ 15

Motivating the Lesson
Questioning Open the lesson by asking students to square each side of the following equations.

1. $\sqrt{x}=12$

2. $\sqrt{n}=8$

3. $\sqrt{2y}=-11$

Ask students to describe the results. Develop the idea that squaring each side of the equation eliminates the radical.

2 TEACH

Teaching Tip Emphasize the importance of isolating the radical by showing an example of what happens if it is not isolated.

$$\sqrt{y}+3=0$$
$$(\sqrt{y}+3)^2=0$$
$$y+6\sqrt{y}+9=0$$

The radical still remains.

Teaching Tip Stress that squaring each side of an equation often produces results that do not check. Thus, it is important to check all possible solutions. A result that does not check is called an *extraneous value*.

In-Class Examples

For Example 1
Solve each equation. Check your solution.

a. $3+\sqrt{m}=13$ $m=100$
b. $\sqrt{(4m+1)}-5=0$ $m=6$
c. $\sqrt{(6m-3)}-3=0$ $m=2$

For Example 2
Solve each equation. Check your solution.

a. $\sqrt{(y+2)}=y-4$ 7
b. $\sqrt{(3x-5)}=x-5$ 10

EXPLORATION

This activity makes use of the *MET* software. It may seem to relieve students of thinking. But, in fact, it makes them think about the role of each step in the program. Encourage students to explain in their own words what each built-in function does.

Check for Understanding

Exercises 1–15 are designed to help you assess your students' understanding through reading, writing, speaking, and modeling. You should work through Exercises 1–5 with your students and then monitor their work on Exercises 6–15.

Additional Answer

5a. Sample answer: Isolate the radical, square both sides of the equation, and simplify.

Example ③ **INTEGRATION**
Number Theory

The geometric mean of a and b is x if $\dfrac{a}{x}=\dfrac{x}{b}$. Find two numbers that have a geometric mean of 8 given that one number is 12 more than the other.

Explore   Let n represent the lesser number.
Then $n+12$ represents the greater number.

Plan   Use the equation $\dfrac{a}{x}=\dfrac{x}{b}$. Replace each variable with the appropriate value.

$$\frac{n}{8}=\frac{8}{n+12}\qquad\text{The geometric mean of the numbers is 8.}$$

Solve
$$n^2+12n=64$$
$$n^2+12n-64=0$$
$$(n+16)(n-4)=0\qquad\text{Factor.}$$
$$n+16=0\quad\text{or}\quad n-4=0\qquad\text{Zero product property}$$
$$n=-16\qquad\qquad n=4$$

If $n=-16$, then $n+12=-4$. If $n=4$, then $n+12=16$.

Thus, the numbers are -16 and -4, or 4 and 16.

Examine $\dfrac{-16}{8}\overset{?}{=}\dfrac{8}{-4}$ or $\dfrac{4}{8}\overset{?}{=}\dfrac{8}{16}$

$64=64$ ✓ $64=64$ ✓

CHECK FOR UNDERSTANDING

Communicating Mathematics

1. Isolate the radical.

3a. $d=\left(\dfrac{s}{3.1}\right)^2$
3b. about 16,649 m

Study the lesson. Then complete the following.

1. **Explain** the first step you should do when solving a radical equation.
2. **Write** an expression for the geometric mean of 7 and y. $\pm\sqrt{7y}$
3. Refer to the application at the beginning of the lesson.
   a. Solve $s=3.1\sqrt{d}$ for d.
   b. Use the equation you found in part a to find the depth of the ocean in meters if the speed of the tsunami is 400 meters per second.
4. **You Decide**   Alberto says that if you have an equation that contains a radical, you can always get a real solution by squaring each side of the equation. Ellen disagrees. Who is correct? Explain.   **Ellen is correct; for $\sqrt{x}=-4$, there is no real solution.**
5. a. **Assess Yourself**   Explain in your own words the process or steps needed to solve a radical equation.   **See margin.**
   b. Explain why it is important to check your answers when solving equations containing radicals.   **The solution may not satisfy the original equation.**

Math Journal

Guided Practice

Square each side of the following equations.

6. $\sqrt{x}=6$ $x=36$
7. $\sqrt{a+3}=2$ $a+3=4$
8. $13=\sqrt{2y-5}$ $169=2y-5$

Solve each equation. Check your solution. 10. no real solution

9. $\sqrt{m}=4$ **16**
10. $\sqrt{b}=-3$
11. $-\sqrt{x}=-6$ **36**
12. $\sqrt{7x}=7$ **7**
13. $\sqrt{-3a}=6$ **−12**
14. $\sqrt{y-2}=8$ **66**

Reteaching

Using Problem Solving Have students solve the following equation as an illustration of how combining like terms and factoring can be used together.

$$x+2=x\sqrt{3}$$
$$x-x\sqrt{3}=-2$$
$$x(1-\sqrt{3})=-2$$
$$x=\frac{-2}{1-\sqrt{3}}$$
$$x=1+\sqrt{3}$$

15. Engineering It is possible to measure the speed of water using an L-shaped tube. You can find the speed V of the water in miles per hour by measuring the height h of the column of water above the surface in inches and by using the formula $V = \sqrt{2.5h}$. If you take the tube into a river and the height of the column is 6 inches, what is the speed of the water to the nearest tenth of a mile per hour? **3.9 mph**

Assignment Guide

Core: 17–47 odd, 48–56
Enriched: 16–42 even, 43–56

For **Extra Practice,** see p. 786.

The red A, B, and C flags, printed only in the Teacher's Wraparound Edition, indicate the level of difficulty of the exercises.

EXERCISES

Practice

19. no real solution

21. $\frac{81}{25}$

24. no real solution

29. $\pm\frac{4}{3}\sqrt{3}$

36. $-\frac{16}{3}$, -27 or 9, 16

40. (9, 0)

41. $\left(\frac{1}{4}, \frac{25}{36}\right)$

Critical Thinking

Applications and Problem Solving

45. $54\frac{2}{3}$ mi

46a. 43.84°

Solve each equation. Check your solution.

16. $\sqrt{a} = 5\sqrt{2}$ **50**   17. $3\sqrt{7} = \sqrt{-x}$ **−63**   18. $\sqrt{m} - 4 = 0$ **16**

19. $\sqrt{2d} + 1 = 0$ 20. $10 - \sqrt{3y} = 1$ **27**   21. $3 + 5\sqrt{n} = 12$

22. $\sqrt{8s + 1} = 5$ **3**   23. $\sqrt{4b + 1} - 3 = 0$ **2**   24. $\sqrt{3r - 5} + 7 = 3$

25. $\sqrt{\frac{w}{6}} = 2$ **24**   26. $\sqrt{\frac{4x}{5}} - 9 = 3$ **180** 27. $5\sqrt{\frac{4t}{3}} - 2 = 0$ $\frac{3}{25}$

28. $\sqrt{2x^2 - 121} = x$ **11**   29. $7\sqrt{3z^2 - 15} = 7$ 30. $\sqrt{x + 2} = x - 4$ **7**

31. $\sqrt{5x^2 - 7} = 2x$ $\sqrt{7}$ 32. $\sqrt{1 - 2m} = 1 + m$ **0** 33. $4 + \sqrt{b - 2} = b$ **6**

INTEGRATION
Number Theory

34. The geometric mean of a certain number and 6 is 24. Find the number. **96**

35. Find two numbers with a geometric mean of $\sqrt{30}$ given that one number is 7 more than the other. **−3, −10 or 3, 10**

36. Find two numbers with a geometric mean of 12 given that one number is 11 less than three times the other.

Solve each equation. Check your solution.

37. $\sqrt{x - 12} = 6 - \sqrt{x}$ **16**   38. $\sqrt{x + 4} = \sqrt{x + 16}$ **0** 39. $\sqrt{x + 7} = 7 + \sqrt{x}$
                                                                                       no real solution

Solve each system of equations.

40. $2\sqrt{a} + 5\sqrt{b} = 6$ 41. $-3\sqrt{x} + 3\sqrt{y} = 1$ 42. $s = 4t$ **(16, 4)**
    $3\sqrt{a} - 5\sqrt{b} = 9$ $-4\sqrt{x} + 6\sqrt{y} = 3$ $\sqrt{s} - 5\sqrt{t} = -6$

43. Solve for x if $x + 2 = x\sqrt{3}$. $\sqrt{3} + 1$

44. Find two numbers such that the square root of their sum is 5 and the square root of their product is 12. **16 and 9**

45. Recreation The rangers at an aid station received a distress call from a group camping 60 miles east and 10 miles south of the station. A jeep sent to the campsite travels directly east for some number of miles and then turns and heads directly to the campsite. If the jeep traveled a total of 66 miles to get to the campsite, for how many miles did it travel due east?

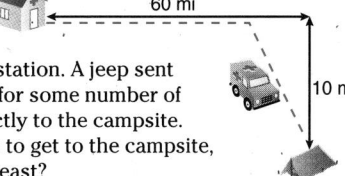

60 mi

10 mi

46. Sound The speed of sound near Earth's surface can be found with the equation $V = 20\sqrt{t + 273}$, where t is the surface temperature in degrees Celsius.

  a. Find the temperature if the speed of sound V is 356 meters per second.

  b. The speed of sound at Earth's surface is often given as 340 meters per second, but that's really only true at a certain temperature. On what temperature is the 340 m/s figure based? **16° C**

Extension

Problem Solving Five times the square root of a number is 6 less than the number. Find the number. **36**

Tech Prep

Park Ranger Students who are interested in recreational parks may wish to do further research on the use of mathematics in that occupation, as mentioned in Exercise 45. For more information on tech prep, see the *Teacher's Handbook*.

Practice Masters, p. 91

Closing Activity

Writing Have students develop new equations to solve. Collect the papers and then distribute randomly. Have students solve the new equations and return to the maker to be graded.

Additional Answer

50.

$y = x^2 - 2x - 8$

Mathematics and SOCIETY

Discuss the fact that in some problems, students were given data and then used graphing calculators to find a curve of best-fit. The r value was almost never 1. Did they settle for a poor model that they understood well? Why do models rarely fit the data perfectly? **Some of the factors include overly simple models, inaccuracy of measurements of data, other variables not occurring in the equation, and so on.**

Enrichment Masters, p. 91

---

47. **Travel** The speed s that a car is traveling in miles per hour, and the distance d in feet that it will skid when the brakes are applied, are related by the formula $s = \sqrt{30fd}$. In this formula, f is the coefficient of friction, which depends on the type and condition of the road. Sylvia Kwan told police she was traveling at about 30 miles per hour when she applied the brakes and skidded on a wet concrete road. The length of her skid marks was measured at 110 feet.

 a. If $f = 0.4$ for a wet concrete road, should Ms. Kwan's car have skidded that far when she applied the brakes?

 b. How fast was she traveling? **about 36 miles per hour**
 47a. No, at 30 mph, her car should have skidded 75 feet after the brakes were applied and not 110 feet.

Mixed Review

48. Simplify $5\sqrt{6} - 11\sqrt{3} - 8\sqrt{6} + \sqrt{27}$. (Lesson 13–3) $-3\sqrt{6} - 8\sqrt{3}$

49. Find $(6b^2 + 4b + 20) \div (b + 5)$. (Lesson 12–4)
49. $6b - 26 + \frac{150}{b + 5}$

50. Find the roots of $x^2 - 2x - 8 = 0$ by graphing its related function. (Lesson 11–2) $-2, 4$; See margin for graph.

51. Find the GCF of $12x^2y^3$ and $42xy^4$. (Lesson 10–1) $6xy^3$

52. Find $(6a - 2m) - (4a + 7m)$. (Lesson 9–5) $2a - 9m$

53. Use substitution to solve the system of equations. (Lesson 8–2) $(-4, 5)$
$x + 4y = 16$
$3x + 6y = 18$

54. Write an equation in slope-intercept form of the line that passes through the points at $(6, -1)$ and $(3, 2)$. (Lesson 6–4) $y = -x + 5$

55. **Probability** If the odds that an event will occur are 8:5, what is the probability that the event will occur? (Lesson 4–6)
55. $\frac{8}{13}$

56. Find $\frac{3}{7} + \left(-\frac{4}{9}\right)$. (Lesson 2–5) $-\frac{1}{63}$

Nonlinear Math

The excerpt below appeared in an article in *Business Week* on September 5, 1994.

ABOVE ALL ELSE, ENGINEERS ARE practical. If developing the perfect camera or oil refinery takes too long, they settle for a design that's "good enough." Increasingly, though, this no longer suffices. In companies driven by competition... engineers are being forced to dip into a new mathematics toolbox.... The new tools are called nonlinear equations, and the name says it all. These equations are used for precisely describing the behavior of things with an unpredictable facet. That's nearly everything—from the workings of car engines to the actions of DNA molecules. Even baking a cake is nonlinear: turning up the oven's temperature twice as high won't bake the cake twice as fast. And with some industrial recipes, such as those for making drugs and plastics, a tiny change in ingredients or processing conditions can mean a huge difference in the finished product. Nonlinear math can help explain such lopsided effects. ■

1. In nonlinear processes, you can't be sure how changes in input will affect the result. Therefore, what can you conclude about the number of variations you will need to enter into the equations you are solving?

2. Because of the huge number of variables that can be involved, solving nonlinear equations can require many millions of calculations. Why do you think the use of these equations has only recently begun to expand into many industries? 1–2. See margin.

Answers for Mathematics and Society

1. Sample answer: You need to enter all the possible variations. This could involve entering huge numbers of variables.

2. Sample answer: Advances in computing power and the widespread use of computers have enabled more organizations to solve nonlinear equations.

Integration: Geometry
The Distance Formula

Objective

After studying this lesson, you should be able to:
- find the distance between two points in the coordinate plane.

INTEGRATION
Geometry

In a coordinate plane, consider the points $A(-2, 6)$ and $B(5, 3)$. These two points do not lie on the same horizontal or vertical line. Therefore, you cannot find the distance between them by simply subtracting the x- or y-coordinates. A different method must be used.

Notice that a right triangle can be formed by drawing lines parallel to the axes through points at $(-2, 6)$ and $(5, 3)$. These lines intersect at $C(-2, 3)$. The measure of side b is the difference of the y-coordinates of the endpoints, $6 - 3$ or 3. The measure of side a is the difference of the x-coordinates of the endpoints, $5 - (-2)$ or 7.

Now the Pythagorean theorem can be used to find c, the distance between $A(-2, 6)$ and $B(5, 3)$.

$c^2 = a^2 + b^2$ *Pythagorean theorem*

$c^2 = 7^2 + 3^2$ *Replace a with 7 and b with 3.*

$c^2 = 49 + 9$

$c^2 = 58$

$c = \sqrt{58}$

$c \approx 7.62$

The distance between points A and B is approximately 7.62 units.

The method used for finding the distance between $A(-2, 6)$ and $B(5, 3)$ can also be used to find the distance between any two points in the coordinate plane. This method can be described by the following formula.

The Distance Formula	The distance d between any two points with coordinates (x_1, y_1) and (x_2, y_2) is given by the following formula. $$d = \sqrt{(x_2 - x_1)^2 + (y_2 - y_1)^2}$$

Example **1**   **Find the distance between the points with coordinates $(3, 5)$ and $(6, 4)$.**

$$d = \sqrt{(x_2 - x_1)^2 + (y_2 - y_1)^2}$$
$$= \sqrt{(6 - 3)^2 + (4 - 5)^2} \quad (x_1, y_1) = (3, 5) \text{ and } (x_2, y_2) = (6, 4)$$
$$= \sqrt{3^2 + (-1)^2}$$
$$= \sqrt{9 + 1}$$
$$= \sqrt{10} \text{ or about 3.16 units}$$

NCTM Standards: 1–5, 8

Instructional Resources

- Study Guide Master 13-5
- Practice Master 13-5
- Enrichment Master 13-5
- Assessment and Evaluation Masters, p. 353
- Graphing Calculator Masters, p. 13
- Multicultural Activity Masters, p. 26
- Science and Math Lab Manual, pp. 57–62

Transparency 13-5A contains the 5-Minute Check for this lesson; **Transparency 13-5B** contains a teaching aid for this lesson.

Recommended Pacing	
Honors Pacing	Day 7 of 11
Block Scheduling*	Day 5 of 7
Alg. 1 in Two Years*	Days 14, 15, & 16 of 22

*For more information on pacing and possible lesson plans, refer to the *Block Scheduling Booklet* and *Algebra 1 in Two Years.*

1 FOCUS

5-Minute Check
(over Lesson 13-4)

Simplify.

1. $\sqrt{x} = 4\sqrt{3}$ **48**
2. $5 - \sqrt{4a} = 2$ $\frac{9}{4}$
3. $\frac{\sqrt{3m}}{\sqrt{2}} - 7 = 1$ $\frac{128}{3}$
4. $\sqrt{4x^2} - 5 = 3x$ **no real solution**
5. Find two numbers that have a geometric mean of $10\sqrt{2}$ given that one number is 30 more than the other. **5, 35**

2 TEACH

In-Class Examples

For Example 1
Find the distance between each pair of points whose coordinates are given.

a. (2, 0), (3, 5) $\sqrt{26}$ or 5.099
b. (1, 1), (4, 4) $3\sqrt{2}$ or 4.243
c. (−1, −2), (−3, 0) $2\sqrt{2}$
  or 2.828

For Example 2
Determine if a triangle, *ABC*, with vertices *A*(2, 1), *B*(3, −2), and *C*(−3, −1) is an isosceles triangle. **no**

For Example 3
Find the value of *a* if the distance between the points at (*a*, 5) and (5, 1) is 5 units. **2**

Teaching Tip Encourage students to check the results in Example 3. Point out that the negative result does not need to be omitted. It represents the *y*-coordinate of a point in the plane, not a distance.

Additional Answer

1. The values that are subtracted are squared before being added, the square of a negative is always positive, and distances are never negative numbers.

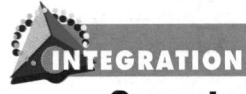

Example 2 Determine if triangle *ABC* with vertices *A*(−3, 4), *B*(5, 2), and *C*(−1, −5) is an isosceles triangle.

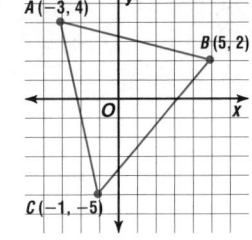

Geometry

A triangle is isosceles if at least two sides are congruent. Find AB, BC, and AC.

$$AB = \sqrt{[5 - (-3)]^2 + (2 - 4)^2}$$
$$= \sqrt{8^2 + (-2)^2} \text{ or } \sqrt{68}$$

$$BC = \sqrt{(-1 - 5)^2 + (-5 - 2)^2}$$
$$= \sqrt{(-6)^2 + (-7)^2} \text{ or } \sqrt{85}$$

$$AC = \sqrt{[-1 - (-3)]^2 + (-5 - 4)^2}$$
$$= \sqrt{2^2 + (-9)^2} \text{ or } \sqrt{85}$$

Since $\overline{BC}$ and $\overline{AC}$ have the same length, $\sqrt{85}$, they are congruent. So, triangle *ABC* is an isosceles triangle.

Suppose you know the coordinates of a point, one coordinate of another point, and the distance between the two points. You can use the distance formula to find the missing coordinate.

Example 3 Find the value of *a* if the distance between the points with coordinates (−3, −2) and (*a*, −5) is 5 units.

$$d = \sqrt{(x_2 - x_1)^2 + (y_2 - y_1)^2}$$
$$5 = \sqrt{[a - (-3)]^2 + [-5 - (-2)]^2} \qquad \text{Let } x_2 = a, x_1 = -3, y_2 = -5, y_1 = -2,$$
$$5 = \sqrt{(a + 3)^2 + (-3)^2} \qquad \text{and } d = 5.$$
$$5 = \sqrt{a^2 + 6a + 9 + 9}$$
$$5 = \sqrt{a^2 + 6a + 18}$$
$$(5)^2 = \left(\sqrt{a^2 + 6a + 18}\right)^2 \qquad \text{Square each side.}$$
$$25 = a^2 + 6a + 18$$
$$0 = a^2 + 6a - 7$$
$$0 = (a + 7)(a - 1) \qquad \text{Factor.}$$
$$a + 7 = 0 \quad \text{or} \quad a - 1 = 0 \qquad \text{Zero product property}$$
$$a = -7 \qquad\qquad a = 1$$

The value of *a* is −7 or 1.

CHECK FOR UNDERSTANDING

Communicating Mathematics

Study the lesson. Then complete the following.

1. **Explain** why the value calculated under the radical sign in the distance formula will never be negative. **See margin.**

2. **a. Write** two ordered pairs and label them $A(x_1, y_1)$ and $B(x_2, y_2)$. Does it matter which ordered pair is first when using the distance formula? Explain. **a–b. See students' work.**

   b. Find the distance between *A* and *B*.

738 Chapter 13 *Exploring Radical Expressions and Equations*

Reteaching

Using Alternative Methods

1. Show that a triangle with vertices *A*(−3, 4), *B*(5, 2), and *C*(−1, −5) is isosceles but not equilateral.
2. Show that a quadrilateral with vertices *A*(1, 1), *B*(4, −2), *C*(1, −3), and *D*(−2, 0) is a parallelogram.
3. Show that (2, 3) is the center of a circle through (5, 7), (−1, −1), and (−2, 0).

3. **a. Explain** how you can find the distance between $X(12, 4)$ and $Y(3, 4)$ without using the distance formula.   **a–b. See margin.**

   b. Explain how you can find the distance between $S(-2, 7)$ and $T(-2, -5)$ without using the distance formula.

4. Refer to Example 3. Check your answer by using the distance formula.
   See students' work.

Guided Practice

Find the distance between each pair of points whose coordinates are given. Express answers in simplest radical form and as decimal approximations rounded to the nearest hundredth if necessary.

5. $(6, 8), (3, 4)$ **5**   6. $(3, 7), (-2, -5)$ **13**

7. $(2, 2), (5, -1)$ $3\sqrt{2}$ **or 4.24**   8. $(2, 7), (10, -4)$ $\sqrt{185}$ **or 13.60**

Find the value of a if the points with the given coordinates are the indicated distance apart.

9. $(4, 7), (a, 3); d = 5$ **7 or 1**   10. $(5, a), (6, 1); d = \sqrt{10}$ **-2 or 4**

11. **Communication** Alpha Corporation is having a fiber optic cable system installed between two new offices. Alphatower I is 4 miles east and 5 miles north of Alpha Central. Alphatower II is 5 miles west and 2 miles north of Alpha Central. How many miles of cable will be needed to connect the new offices? (*Hint:* Alpha Central is located at $(0, 0)$.)
about 9.49 mi

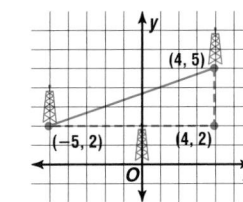

EXERCISES

Practice

Find the distance between each pair of points whose coordinates are given. Express answers in simplest radical form and as decimal approximations rounded to the nearest hundredth if necessary.

 12. $(5, -1), (11, 7)$ **10**   13. $(-4, 2), (4, 17)$ **17**

14. $(-3, 8), (5, 4)$ $4\sqrt{5}$ **or 8.94**   15. $(-8, -4), (-3, -8)$ $\sqrt{41}$ **or 6.40**

 16. $(9, -2), (3, -6)$ $2\sqrt{13}$ **or 7.21**   17. $(4, 2), \left(6, -\frac{2}{3}\right)$ $\frac{10}{3}$ **or 3.33**

18. $\left(3, \frac{3}{7}\right), \left(4, -\frac{2}{7}\right)$ $\frac{\sqrt{74}}{7}$ **or 1.23**   19. $\left(\frac{4}{5}, -1\right), \left(2, -\frac{1}{2}\right)$ $\frac{13}{10}$ **or 1.30**

20. $(4\sqrt{5}, 7), (6\sqrt{5}, 1)$ $2\sqrt{14}$ **or 7.48** 21. $(5\sqrt{2}, 8), (7\sqrt{2}, 10)$ $2\sqrt{3}$ **or 3.46**

Find the value of a if the points with the given coordinates are the indicated distance apart.

24. -2 or -12   22. $(3, -1), (a, 7); d = 10$ **9 or -3**   23. $(-4, a), (4, 2); d = 17$ **17 or -13**

25. -10 or 4   24. $(a, 5), (-7, 3); d = \sqrt{29}$ 25. $(6, -3), (-3, a); d = \sqrt{130}$

27. 8 or 32   26. $(10, a), (1, -6); d = \sqrt{145}$ **2 or -14** 27. $(20, -5), (a, 9); d = \sqrt{340}$

INTEGRATION
Geometry

Determine if the triangles with the following vertices are isosceles triangles.

28. $L(7, -4), M(-1, 2), N(5, -6)$ **yes** 29. $T(1, -8), U(3, 5), V(-1, 7)$ **no**

GLENCOE *Technology*

Interactive Mathematics Tools Software

This multimedia software provides an interactive lesson that uses the Pythagorean theorem to find the distance of a plane, if given the horizontal and vertical distances from the tracking station. A **Computer Journal** gives students an opportunity to write about what they have learned.

For Windows & Macintosh

Additional Answers

3a. The distance between them is the absolute value of the difference of their x-coordinates, $|12 - 3|$ or 9 units.

3b. The distance between them is the absolute value of the difference of their y-coordinates, $|7 - (-5)|$ or 12 units.

3 PRACTICE/APPLY

Check for Understanding
Exercises 1–11 are designed to help you assess your students' understanding through reading, writing, speaking, and modeling. You should work through Exercises 1–4 with your students and then monitor their work on Exercises 5–11.

Error Analysis
Make sure that students subtract the coordinates in corresponding order when using the distance formula. For example, some students may substitute as follows.

$d = \sqrt{(x_2 - x_1)^2 + (y_1 - y_2)^2}$, or

$d = \sqrt{(x_1 - x_2)^2 + (y_2 - y_1)^2}$

Assignment Guide

Core: 13–35 odd, 36–41
Enriched: 12–32 even, 33–41

For **Extra Practice,** see p. 786.

The red A, B, and C flags, printed only in the Teacher's Wraparound Edition, indicate the level of difficulty of the exercises.

Study Guide Masters, p. 92

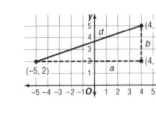

13-5 NAME_____ DATE_____
Study Guide   Student Edition
                          Pages 737–741

Integration: Geometry
The Distance Formula

The Pythagorean theorem can be used to derive the distance formula shown at the right. The distance formula can then be used to find the distance between any two points in the coordinate plane.

The Distance Formula
The distance between any two points with coordinates (x_1, y_1) and (x_2, y_2) is given by the following formula. $d = \sqrt{(x_2 - x_1)^2 + (y_2 - y_1)^2}$

Example: Find the distance between $(-5, 2)$ and $(4, 5)$.

In the diagram at the right below, the distance to be found is d. Use the distance formula.

$d = \sqrt{(x_2 - x_1)^2 + (y_2 - y_1)^2}$
$= \sqrt{[4 - (-5)]^2 + (5 - 2)^2}$
$= \sqrt{9^2 + 3^2}$
$= \sqrt{81 + 9}$
$= \sqrt{90} = 3\sqrt{10}$

The distance is $3\sqrt{10}$ units.

Find the distance between each pair of points whose coordinates are given. Express answers in simplest radical form and as decimal approximations rounded to the nearest hundredth.

1. $(1, 5), (3, 1)$ 2. $(0, 0), (6, 8)$ 3. $(-2, -8), (7, -3)$
   $2\sqrt{5}$; 4.47   10 $\sqrt{106}$; 10.30

4. $(6, -7), (-2, 8)$ 5. $(1, 5), (-8, 4)$ 6. $(3, -4), (-4, -4)$
   17                       $\sqrt{82}$; 9.06   7

Find the value of a if the points with the given coordinates are the indicated distance apart.

7. $(-3, -2), (a, -5); d = 5$ 8. $(2, -1), (a, 3); d = 5$ 9. $(a, 3), (2, -4); d = \sqrt{74}$
   -7 or 1   -1 or 5   -3 or 7

30. $12\sqrt{10}$ or 37.9 units

31. $\sqrt{157} \neq \sqrt{101}$; Trapezoid is not isosceles.

30. Find the perimeter of square *QRST* if two of the vertices are *Q*(6, 7) and *R*(−3, 4).

31. If the diagonals of a trapezoid have the same length, then the trapezoid is isosceles. Find the lengths of the diagonals of the trapezoid with vertices *A*(−2, 2), *B*(10, 6), *C*(9, 8), and *D*(0, 5) to determine if it is isosceles.

32. The program at the right calculates the distance between a pair of points whose coordinates are given.

Find the distance between each pair of points.

a. *A*(6, −3), *B*(12, 5) **10**
b. *M*(−3, 5), *N*(12, −2) ≈**16.55**
c. *S*(6.8, 9.9), *T*(−5.9, 4.3) ≈**13.88**

```
PROGRAM: DISTANCE
:ClrDraw
:Input "X1=", Q
:Input "Y1=", R
:Input "X2=", S
:Input "Y2=", T
:Line (Q, R, S, T)
:√((Q−S)² + (R−T)²)→D
:Text (5, 50, "DIST=", D)
```

Critical Thinking

33. Use the distance formula to show that the triangle with vertices at (3, −2), (−3, 7), and (−9, 3) is a right triangle. See margin.

Applications and Problem Solving

34. Art Egyptian artists about 5000 years ago decorated tombs of pharaohs by painting their pictures on the walls. The artists used small sketches on grids as a reference. In Egypt, the main standard of length was the cubit. It was the length of a man's forearm from the elbow to the tip of the outstretched fingers. Use the grid at the right to find the length of a cubit to the nearest inch if each unit represents 3.3 inches. **18 in.**

35. 109 miles

35. Telecommunications In order to set long distance rates, phone companies first superimpose an imaginary coordinate grid over the United States. Then the location of each exchange is represented by an ordered pair on the grid. The units of this grid are approximately equal to 0.316 miles. So, a distance of 3 units on the grid equals an actual distance of about 3(0.316) or 0.948 miles. Suppose the exchanges in two cities are at (132, 428) and (254, 105). Find the actual distance between these cities to the nearest mile.

740 Chapter 13 *Exploring Radical Expressions and Equations*

Extension

Connections The endpoints of $\overline{AB}$ are *A*(−2, 6) and *B*(4, 2). The endpoints of $\overline{CD}$ are *C*(3, 7) and *D*(1, 4). Show that $\overline{CD}$ is a perpendicular bisector of $\overline{AB}$.

a. Find the midpoint of *AB*. **(1, 4);** Since the midpoint of $\overline{AB}$ is point *D*, $\overline{CD}$ is a bisector of $\overline{AB}$.

b. Find the slopes of $\overline{AB}$ and $\overline{CD}$. slope of $\overline{AB} = -\frac{2}{3}$; slope of $\overline{CD} = \frac{3}{2}$; Since $\left(-\frac{2}{3}\right)\left(\frac{3}{2}\right) = -1$, $\overline{AB}$ and $\overline{CD}$ are perpendicular. Therefore, $\overline{CD}$ is a perpendicular bisector of $\overline{AB}$.

36. **Physics** The time t, in seconds, it takes an object to drop d feet is given by the formula $4t = \sqrt{d}$. Jessica and Lu-Chan each dropped a stone at the same time, but Jessica dropped hers from a spot higher than Lu-Chan's. Lu-Chan's stone hit the ground 1 second before Jessica's. If Jessica's stone dropped 112 feet farther than Lu-Chan's, how long did it take her stone to hit the ground? (Lesson 13–4) **4 seconds**

37. Simplify $\sqrt{\dfrac{8}{9}}$. (Lesson 13–2) $\dfrac{2\sqrt{2}}{3}$

38. Find $\dfrac{4p^3}{p-1} \div \dfrac{p^2}{p-1}$. (Lesson 12–3) **4p**

39. No solution; see margin for graph.

39. Graph the system of equations. Then determine whether the system has *one* solution, *no* solution, or *infinitely many* solutions. (Lesson 8–1)
$4x - y = 2$
$y - 4x = 4$

40. **Consumerism** Jackie wanted to buy a new coat that cost $145. If she waited until the coat went on sale for 30% off the original price, how much money did Jackie save? (Lesson 4–5) **$43.50**

41. **Air Conditioning** The formula for determining the BTU (British Thermal Units) rating of the air conditioner necessary to cool a room is BTU = Area (sq ft) × Exposure Factor × Climate Factor. Use this formula to determine the BTU necessary to cool each of the rooms described in the following chart. (Lesson 2–6)

41a. 7392 BTU
41b. 3705 BTU
41c. 6247.5 BTU
41d. 14,040 BTU
41e. 6831.45 BTU

	Room Dimensions (feet)	Exposure Factor	Climate Factor
a.	22 by 16	North: 20	Buffalo: 1.05
b.	13 by 12	West: 25	Portland: 0.95
c.	17 by 14	East: 25	Topeka: 1.05
d.	26 by 18	South: 30	San Diego: 1.00
e.	23.5 by 15.3	North: 20	Tacoma: 0.95

WORKING ON THE

In·ves·ti·ga·tion

Refer to the Investigation on pages 656–657.

A Growing Concern

1 Review your scale drawing. Make sure you have included everything that you think the Sanchez family wanted or will want in the future. Make any changes that you feel need to be made now that the plan is complete.

2 The Sanchez family had asked for a 3-dimensional model of the design. On a piece of cardboard, draw the boundary lines of the Sanchez property. Use modeling clay, construction paper, paint, markers, and whatever else you need to create a 3-dimensional model of your design.

3 Use a flashlight to model the sun's movement during the day. Note the patterns and the length of time certain areas are shaded.

Add the results of your work to your Investigation Folder.

Lesson 13–5 *INTEGRATION Geometry* The Distance Formula **741**

In·ves·ti·ga·tion

Working on the Investigation

The Investigation on pages 656–657 is designed to be a long-term project that is completed over several days or weeks. Encourage students to keep their materials in their Investigation Folder as they work on the Investigation.

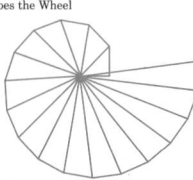

Objective
Use algebra tiles as a model for completing the square.

Recommended Time
Demonstration and discussion: 15 minutes; Exercises: 30 minutes

Instructional Resources
For each student or group of students
Student Manipulative Kit
• algebra tiles
• equation mat
Modeling Mathematics Masters
• p. 1 (algebra tiles)
• p. 14 (equation mat)
• p. 35 (worksheet)
For teacher demonstration
Algebra and Geometry Overhead Manipulative Resources

1 FOCUS

Motivating the Lesson
This modeling activity makes it clear that the phrase "completing the square" is literally true. The result of each modeling activity is a geometric square.

2 TEACH

Teaching Tip Point out that when completing the square, you can only add 1-tiles to each side of the mat.

3 PRACTICE/APPLY

Assignment Guide
Core: 1–8
Enriched: 1–8

4 ASSESS

Observing students working in cooperative groups is an excellent method of assessment.

MODELING MATHEMATICS

A Preview of Lesson 13–6

13–6A Completing the Square

Materials: algebra tiles   equation mat

One way to solve a quadratic equation is by **completing the square.** To use this method, the quadratic expression on one side of the equation must be a perfect square. You can use algebra tiles as a model for completing the square.

Activity Use algebra tiles to complete the square for the equation $x^2 + 4x + 1 = 0$.

Step 1 Subtract 1 from each side of the equation.

$$x^2 + 4x + 1 - 1 = 0 - 1$$
$$x^2 + 4x = -1$$

Then model the equation $x^2 + 4x = -1$.

Step 2 Begin to arrange the x^2-tile and the x-tiles into a square.

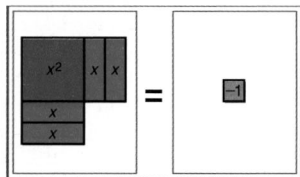

Step 3 In order to complete the square, you need to add 4 1-tiles to the left side of the mat. Since you are modeling an equation, add 4 1-tiles to the right side of the mat.

$$x^2 + 4x + 4 = -1 + 4$$

Step 4 Remove the zero pair on the right side of the mat. You have completed the square, and the equation is $x^2 + 4x + 4 = 3$ or $(x + 2)^2 = 3$.

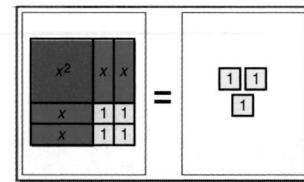

1. $(x + 2)^2 = 1$ 2. $(x - 3)^2 = 4$ 3. $(x + 2)^2 = 5$

Model Use algebra tiles to complete the square for each equation.

1. $x^2 + 4x + 3 = 0$
2. $x^2 - 6x + 5 = 0$
3. $x^2 + 4x - 1 = 0$
4. $x^2 - 2x + 5 = 3$
   $(x - 1)^2 = -1$
5. $x^2 - 4x + 7 = 8$
   $(x - 2)^2 = 5$
6. $0 = x^2 + 8x - 3$
   $(x + 4)^2 = 19$

Draw 7. In the equations shown above, the coefficient of x was always an even number. Sometimes you have an equation like $x^2 + 3x - 1 = 0$ in which the coefficient of x is an odd number. Complete the square by making a drawing. **See margin.**

Write 8. Write a paragraph explaining how you could complete the square with models without first rewriting the equation. Include a drawing. **See margin.**

Additional Answers

7. $(x + 1.5)^2 = 3.25$

8. Sample answer: You could model the expression on one side of the mat. Then add the appropriate 1-tiles needed to complete the square to each side of the mat.

Solving Quadratic Equations by Completing the Square

Objectives

After studying this lesson, you should be able to:

- solve problems by identifying subgoals, and
- solve quadratic equations by completing the square.

CONNECTION
Geography

The greatest flow of any river in the world is that of the Amazon, which discharges an average of 4.2 million cubic feet per second into the Atlantic Ocean. The rate at which water flows in a river varies depending on the distance from the shore.

Suppose the Castillos have a home on the bank of a river that is 40 yards wide. The rate of the river is given by the equation $y = -0.01x^2 + 0.4x$. Mr. and Mrs. Castillo do not want their children to wade in the water if the current is greater than 3 miles per hour. Find how many yards from shore the water flows at 3 miles per hour. *This problem will be solved in Example 4.*

Amazon River

You can solve some quadratic equations by taking the square root of each side.

Example **1** Solve $x^2 - 6x + 9 = 7$.

$$x^2 - 6x + 9 = 7$$
$$(x - 3)^2 = 7 \qquad x^2 - 6x + 9 \text{ is a perfect square trinomial.}$$
$$\sqrt{(x - 3)^2} = \sqrt{7} \qquad \text{Take the square root of each side.}$$
$$|x - 3| = \sqrt{7}$$
$$x - 3 = \pm\sqrt{7} \qquad \text{Why is this the case?}$$
$$x = 3 \pm\sqrt{7} \qquad \text{Add 3 to each side.}$$

The solution set is $\{3 + \sqrt{7}, 3 - \sqrt{7}\}$.

To use the method shown in Example 1, the quadratic expression on one side of the equation must be a perfect square. However, few quadratic expressions are perfect squares. To make any quadratic expression a perfect square, a method called **completing the square** may be used.

Consider the pattern for squaring a binomial such as $x + 5$.

$$(x + 5)^2 = x^2 + 2(5)x + 5^2$$
$$= x^2 + 10x + 25$$
$$\left(\frac{10}{2}\right)^2 \rightarrow 5^2 \qquad \text{Notice that one half of 10 is 5 and } 5^2 \text{ is 25.}$$

Lesson 13–6 Solving Quadratic Equations by Completing the Square **743**

Alternative Learning Styles

Kinesthetic Ask students to list important ideas, terminology, and problems that have been covered in the chapter. Have them write each item on the front of an index card. On the back, have them explain or define the term in their own words using complete sentences. They can use these cards to study alone or in pairs.

13-6 LESSON NOTES

NCTM Standards: 1–5

Instructional Resources

- Study Guide Master 13-6
- Practice Master 13-6
- Enrichment Master 13-6
- Assessment and Evaluation Masters, p. 353
- Real-World Applications, 32

 Transparency 13-6A contains the 5-Minute Check for this lesson; **Transparency 13-6B** contains a teaching aid for this lesson.

Recommended Pacing	
Standard Pacing	Day 9 of 11
Honors Pacing	Day 6 of 7
Block Scheduling*	Days 18 & 19 of 22

 *For more information on pacing and possible lesson plans, refer to the *Block Scheduling Booklet* and *Algebra 1 in Two Years.*

1 FOCUS

 5-Minute Check
(over Lesson 13-5)

Find the distance between each pair of points whose coordinates are given.

1. $(6, 0)(12, 8)$ **10**
2. $(-6, -4)(2, 15)$ $\sqrt{425}$
3. $(0, -11)(-6, -15)$ $2\sqrt{13}$

Determine if the triangles with the following vertices are isosceles triangles.

4. $A(0, 7)$, $B(8, 5)$, $C(2, -2)$ **yes**
5. $X(1, -3)$, $Y(3, 10)$, $Z(-1, 12)$ **no**

Motivating the Lesson

Questioning Introduce the lesson with a review of factoring. Ask students to factor each of the following.

1. $x^2 + 4x + 4$ $(x + 2)^2$
2. $x^2 - 6x + 9$ $(x - 3^2)$
3. $n^2 + 10n + 25$ $(n + 5)^2$
4. $y^2 - 2y + 1$ $(y - 1)^2$

Teaching Tip While students learned other ways of solving quadratic equations in Chapter 11, they needed to learn to work with radicals before being introduced to completing the square.

In-Class Examples

For Example 1
Solve $x^2 + 6x + 9 = 5$.
$\{-3 + \sqrt{5}, -3 - \sqrt{5}\}$

For Example 2
Find the value of c that makes each trinomial a perfect square.
a. $x^2 - 20x + c$ 100
b. $y^2 + y + c$ $\frac{1}{4}$
c. $m^2 - 3m + c$ $\frac{9}{4}$

For Example 3
Solve each equation by completing the square.
a. $x^2 - 6x + 8 = 0$ {2, 4}
b. $x^2 - 6x + 5 = 0$ {1, 5}
c. $m^2 + 4m - 9 = 0$
$\left\{-2 + \sqrt{13}, -2 - \sqrt{13}\right\}$
d. $x^2 - 6x + 7 = 0$
$\left\{3 + \sqrt{2}, 3 - \sqrt{2}\right\}$

For Example 4
How far from the shore is the river current 3.5 miles per hour?
20 ± 7.07

Teaching Tip A calculator may be used to aid in checking the roots of an equation.

To complete the square for a quadratic expression of the form $x^2 + bx$, you can follow the steps below.

Step 1 Find $\frac{1}{2}$ of b, the coefficient of x.

Step 2 Square the result of Step 1.

Step 3 Add the result of Step 2 to $x^2 + bx$, the original expression.

Example ② **Find the value of c that makes each trinomial a perfect square.**

a. $x^2 + 20x + c$

Step 1 Find $\frac{1}{2}$ of 20.	$\frac{20}{2} = 10$
Step 2 Square the result of Step 1.	$10^2 = 100$
Step 3 Add the result of Step 2 to $x^2 + 20x$.	$x^2 + 20x + 100$

Thus, $c = 100$. Notice that $x^2 + 20x + 100 = (x + 10)^2$.

b. $x^2 - 15x + c$

Step 1 Find $\frac{1}{2}$ of -15.	$\frac{-15}{2} = -7.5$
Step 2 Square the result of Step 1.	$(-7.5)^2 = 56.25$
Step 3 Add the result of Step 2 to $x^2 - 15x$.	$x^2 - 15x + 56.25$

Thus, $c = 56.25$. Notice that $x^2 - 15x + 56.25 = (x - 7.5)^2$.

Example ③ **Solve $x^2 + 8x - 18 = 0$ by completing the square.**

$$x^2 + 8x - 18 = 0 \quad \textit{Notice that } x^2 + 8x - 18 \textit{ is not a perfect square.}$$
$$x^2 + 8x = 18 \quad \textit{Add 18 to each side. Then complete the square.}$$
$$x^2 + 8x + 16 = 18 + 16 \quad \textit{Since } \left(\frac{8}{2}\right)^2 = 16, \textit{ add 16 to each side.}$$
$$(x + 4)^2 = 34 \quad \textit{Factor } x^2 + 8x + 16.$$
$$x + 4 = \pm\sqrt{34} \quad \textit{Take the square root of each side.}$$
$$x = -4 \pm \sqrt{34} \quad \textit{Subtract 4 from each side.}$$
$$x = -4 + \sqrt{34} \text{ or } x = -4 - \sqrt{34}$$

The solution set is $\{-4 + \sqrt{34}, -4 - \sqrt{34}\}$.   *Check this result.*

The method for solving quadratic equations cannot be used unless the coefficient of the first term is 1. To solve a quadratic equation in which the leading coefficient is not 1, divide each term by the coefficient.

Example ④ **Refer to the application at the beginning of the lesson. How far from shore will the rate of the current be 3 miles per hour?**

CONNECTION
Geography

$$y = -0.01x^2 + 0.4x$$
$$3 = -0.01x^2 + 0.4x \quad \textit{Replace y with 3 since the rate is 3 mph.}$$
$$-300 = x^2 - 40x \quad \textit{Divide each side by } -0.01.$$
$$-300 + 400 = x^2 - 40x + 400 \quad \textit{Complete the square. } \left(\frac{-40}{2}\right)^2 = 400$$
$$100 = (x - 20)^2 \quad \textit{Factor } x^2 - 40x + 400.$$
$$\pm 10 = x - 20 \quad \textit{Take the square root of each side.}$$
$$20 \pm 10 = x \quad \textit{Add 20 to each side.}$$

744   *Chapter 13 Exploring Radical Expressions and Equations*

Cooperative Learning

Pairs Check Have students work in pairs. One student will work an example of completing the square and then give it to the other student to check. The student doing the checking must either say what is wrong or, if correct, explain each step. Then switch roles. For more information on the pairs check strategy, see *Cooperative Learning in the Mathematics Classroom*, one of the titles in the Glencoe Mathematics Professional Series, pages 12–13.

The solutions are 20 + 10 or 30 and 20 − 10 or 10. Thus, the children should not be allowed to wade more than 10 yards from the shore; between 10 and 30 yards from the shore the water is flowing too fast.

Check this result with a graphing calculator by graphing the equation $y = -0.01x^2 + 0.4x$, and then estimating.

Sometimes finding the solution to a problem requires several steps. An important strategy for solving such problems is to **identify subgoals**. This strategy involves taking steps that will either produce part of the solution or make the problem easier to solve.

Example 5

PROBLEM SOLVING
Identify Subgoals

A square is extended in one direction by 14 centimeters. The resulting rectangle has an area of 51 square centimeters. What is the length of each side of the original square?

x cm   x cm   14 cm

Finding an equation to represent this problem will be easier if you develop the equation in steps rather than trying to write one directly from the given information.

Step 1   First, let x be the length of each side of the original square. Then the area of the square is x^2 cm.

Step 2   The extension is 14 cm long and x cm wide, so its area is $14x$ cm^2.

Step 3   Add the measures of the areas and set them equal to 51.
$$x^2 + 14x = 51$$

Step 4   Complete the square to find the value of x.

$$x^2 + 14x = 51$$
$$x^2 + 14x + 49 = 51 + 49 \quad \textit{Since } \left(\tfrac{14}{2}\right)^2 = 49, \textit{ add 49 to each side.}$$
$$(x + 7)^2 = 100 \quad \textit{Factor } x^2 + 14x + 49.$$
$$x + 7 = \pm 10 \quad \textit{Find the square root of each side.}$$
$$x = -7 \pm 10$$

$$x = -7 + 10 \qquad\qquad x = -7 - 10$$
$$= 3 \qquad\qquad\qquad = -17$$

Since lengths cannot be negative, the length of each side of the original square is 3 centimeters. *Check this result.*

CHECK FOR UNDERSTANDING

Communicating Mathematics

2–3. See margin.

Study the lesson. Then complete the following.

1. **Explain** which method for solving a quadratic equation always produces an exact solution, graphing or completing the square.   completing the square

2. **Explain** the three steps used to complete the square for the expression $x^2 + bx$.

3. **Write** a quadratic equation that has no real roots. After completing the square, explain how you could tell it had no real roots.

Lesson 13–6   Solving Quadratic Equations by Completing the Square   **745**

Reteaching

Using Discussion Discuss with students how to take half of a fractional or decimal coefficient. Does it change the method of completing the square? How do you square a fraction or a decimal?

Additional Answers

2. Step 1: Find one-half of b.
   Step 2: Square the result of Step 1.
   Step 3: Add the result of Step 2 to $x^2 + bx$.

3. Sample answer: $x^2 + 4x + 12 = 0$, $(x + 2)^2 = -8$; Since the number on the right side is negative, there are no real roots.

---

3 PRACTICE/APPLY

Check for Understanding
Exercises 1–13 are designed to help you assess your students' understanding through reading, writing, speaking, and modeling. You should work through Exercises 1–4 with your students and then monitor their work on Exercises 5–13.

Error Analysis
Students need to first look at the coefficient of the first term to be sure it is 1. Make sure they divide *all* terms by that coefficient if it is not 1.
Example: $3x^2 - 7x = 3$
$$x^2 - \frac{7x}{3} = 3 \qquad\qquad x^2 - \frac{7x}{3} = 1$$
no   yes

Study Guide Masters, p. 93

MODELING MATHEMATICS

4. Use algebra tiles to complete the square for the equation $x^2 + 6x + 2 = 0$. $(x+3)^2 = 7$

Guided Practice

Find the value of c that makes each trinomial a perfect square.

5. $x^2 + 16x + c$ 64

6. $a^2 - 7a + c$ $\frac{49}{4}$

Solve each equation by completing the square. Leave irrational roots in simplest radical form.

7. −1, −3
10. $\frac{5}{2}$ 12. $\frac{5}{2}, -4$
16. $\frac{25}{4}$ 17. $\frac{121}{4}$

7. $x^2 + 4x + 3 = 0$
8. $d^2 - 8d + 7 = 0$ 1, 7
9. $a^2 - 4a = 21$ 7, −3
10. $4x^2 - 20x + 25 = 0$
11. $r^2 - 4r = 2$ $2 \pm \sqrt{6}$
12. $2t^2 + 3t - 20 = 0$

13. **Sports** The dimensions of a regulation high school basketball court are 50 feet by 84 feet. The builders of an indoor sports arena can afford to construct an arena of 5600 square feet. They want it to have a regulation basketball court and walkways the same width around the court. Find the dimensions of the walkway.   4.9 ft

EXERCISES

Practice

23. $\frac{7}{3}$
24. $12 \pm 3\sqrt{15}$
26. $4 \pm 2\sqrt{5}$
27. $5 \pm 4\sqrt{3}$
28. $3, \frac{1}{2}$
29. $\frac{-5 \pm 2\sqrt{15}}{5}$
30. $-\frac{3}{2}, 4$ 31. $\frac{2}{3}, -1$
32. $-0.125 \pm \sqrt{0.515625}$
33. $\frac{7 \pm \sqrt{85}}{6}$
34. $\frac{5 \pm \sqrt{17}}{4}$
39. $2 \pm \sqrt{4 - c}$
40. $\frac{-b \pm \sqrt{b^2 - 4c}}{2}$
41. $b(-2 \pm \sqrt{3})$

A

Find the value of c that makes each trinomial a perfect square.

14. $x^2 - 6x + c$ 9
15. $b^2 + 8b + c$ 16
16. $m^2 - 5m + c$
17. $a^2 + 11a + c$
18. $9t^2 - 18t + c$ 9
19. $\frac{1}{2}x^2 - 4x + c$ 8

Solve each equation by completing the square. Leave irrational roots in simplest radical form.   20. −3, −4   21. 4, 1   22. 1, −15

20. $x^2 + 7x + 10 = -2$
21. $a^2 - 5a + 2 = -2$
22. $r^2 + 14r - 9 = 6$

B

23. $9b^2 - 42b + 49 = 0$
24. $x^2 - 24x + 9 = 0$
25. $t^2 + 4 = 6t$ $3 \pm \sqrt{5}$
26. $m^2 - 8m = 4$
27. $p^2 - 10p = 23$
28. $x^2 - \frac{7}{2}x + \frac{3}{2} = 0$
29. $5x^2 + 10x - 7 = 0$
30. $\frac{1}{2}d^2 - \frac{5}{4}d - 3 = 0$
31. $0.3t^2 + 0.1t = 0.2$
32. $b^2 + 0.25b = 0.5$
33. $3p^2 - 7p - 3 = 0$
34. $2r^2 - 5r + 8 = 7$

Find the value of c that makes each trinomial a perfect square.

35. $x^2 + cx + 81$ 18, −18
36. $4x^2 + cx + 225$ 60, −60
37. $cx^2 + 30x + 75$ 3
38. $cx^2 - 18x + 36$ $\frac{9}{4}$

C

Solve each equation by completing the square. Leave irrational roots in simplest radical form.

39. $x^2 - 4x + c = 0$
40. $x^2 + bx + c = 0$
41. $x^2 + 4bx + b^2 = 0$

Critical Thinking

42. **Geometry** Consider the quadratic function $y = x^2 - 8x + 15$.
   a. Write the function in the form $y = (x - h)^2 + k$. $y = (x-4)^2 - 1$
   b. Graph the function.   **See margin**
   c. What is the relationship of the point (h, k) to the graph?   It is the vertex of the graph.

746   Chapter 13   Exploring Radical Expressions and Equations

Tech Prep

Sportscaster Students who are interested in sportscasting may wish to do further research on the use of mathematics in that occupation, as mentioned in Exercise 13. For more information in tech prep, see the *Teacher's Handbook.*

Applications and Problem Solving

43. Identify Subgoals Two trains left the same station at the same time. One was traveling due north at a speed that was 10 mph faster than the other train, which was traveling due east. After one hour, the trains were 71 miles apart. How fast, to the nearest mile per hour, was each train traveling? (*Hint:* Use the Pythagorean theorem.) **45 mph, 55 mph**

44. Construction Arlando's Restaurant wants to add an outdoor café on the side of the restaurant. They are having a special water fountain shipped in that is 10 by 15 feet, and Arlando can afford to buy 1800 square feet of space next to his restaurant. He wishes to have a dining area around the fountain, of equal width all around.

a. Write an equation for *x*, the width of the dining area around the fountain and solve it. **(15 + 2x)(10 + 2x) = 1800; 15**

b. What should be the length and width of the piece of land Arlando buys for the café? **45 ft by 40 ft**

Mixed Review

45. Geometry Find the distance between points *A* and *B* graphed at the right. Round your answer to the nearest hundredth. (Lesson 13–5) **8.06**

46. Geometry The measures of the sides of a triangle are 5, 7, and 9. Determine whether this triangle is a right triangle. (Lesson 13–1) **no**

47. Astronomy Earth, Jupiter, and Saturn revolve around the Sun about once every 1, 12, and 30 years, respectively. The last time Jupiter and Saturn appeared close to each other in Earth's night sky was in 1982. When will this happen again? (Lesson 12–6) **2042**

48. Find $(5y - 3)(y + 2)$. (Lesson 9–7) **$5y^2 + 7y - 6$**

49. Use elimination to solve the system of equations. (Lesson 8–4) **(2, 2)**

$5y - 4x = 2$

$2y + x = 6$

50. If the graph of $P(x, y)$ satisfies the given conditions, name the quadrant in which point *P* is located. (Lesson 5–1)

a. $x > 0, y < 0$ **IV**   **b.** $x < 0, y = 3$ **II**   **c.** $x = -1, y < 0$ **III**

51. Food The chart at the right compares the Calorie content of regular tacos and the light tacos at Taco Bell®. (Lesson 3–7)

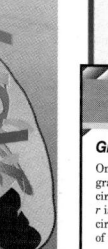

Item	Calories in Regular Tacos	Calories in Light Tacos
Taco	180	140
Soft Taco	220	180
Taco Supreme™	230	160
Soft Taco Supreme®	270	200
Chicken Soft Taco	223	180

a. Find the mean number of Calories for the regular tacos. **224.6**

b. Find the mean number of Calories for the light tacos. **172**

52. Evaluate $|a| + |4b|$ if $a = -3$ and $b = -6$. (Lesson 2–3) **27**

Lesson 13–6 Solving Quadratic Equations by Completing the Square **747**

Extension

Problem Solving Solve the equation for *x* in terms of *a* by completing the square. State the restrictions on *a* so that the equation will have real-number roots.

$ax^2 - 3x + 4 = 0$

$x = \dfrac{3 \pm \sqrt{9 - 16a}}{2a}; a \leq \dfrac{9}{16}, a \neq 0$

4 ASSESS

Closing Activity

Speaking Have students explain how to use the completing the square method to solve $x^2 - 36 = 0$.

Chapter 13, Quiz D (Lessons 13-5 and 13-6), is available in the *Assessment and Evaluation Masters,* p. 353.

Enrichment Masters, p. 93

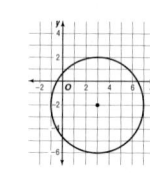

13-6   NAME_____   DATE_____
Enrichment   Student Edition
                               Pages 743–748

Graphing Circles by Completing Squares

One use for completing the square is to graph circles. The general equation for a circle with center at the origin and radius *r* is $x^2 + y^2 = r^2$. An equation represents a circle if it can be transformed into the sum of two squares.

$x^2 - 6x + y^2 + 4y - 3 = 0$
$(x^2 - 6x + \ ) + (y^2 + 4y + \ ) = 0$
$(x^2 - 6x + 9) + (y^2 + 4y + 4) = 3 + 9 + 4$
$(x - 3)^2 + (y + 2)^2 = 4^2$

Notice that the center of the circle is at the point $(3, -2)$.

Transform each equation into the sum of two squares. Then graph the circle represented by the equation. Use the coordinate plane provided at the bottom of the page.

1. $x^2 - 14x + y^2 + 6y + 49 = 0$
   $(x - 7)^2 + (y + 3)^2 = 3^2$

2. $x^2 + y^2 - 8y - 9 = 0$
   $x^2 + (y - 4)^2 = 5^2$

3. $x^2 + 10x + y^2 + 21 = 0$
   $(x + 5)^2 + y^2 = 2^2$

4. $x^2 + y^2 + 10y + 16 = 0$
   $x^2 + (y + 5)^2 = 3^2$

5. $x^2 - 30x + y^2 + 209 = 0$
   $(x - 15)^2 + y^2 = 4^2$

6. $x^2 - 18x + y^2 - 12y + 116 = 0$
   $(x - 9)^2 + (y - 6)^2 = 1^2$

7. $x^2 + 30x + y^2 - 4y + 193 = 0$
   $(x + 15)^2 + (y - 2)^2 = 6^2$

8. $x^2 + 38x + y^2 - 12y + 393 = 0$
   $(x + 19)^2 + (y - 6)^2 = 2^2$

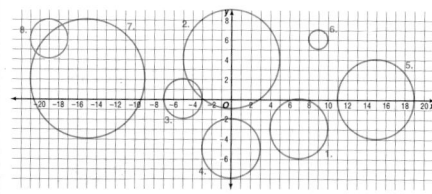

Chapter 13 **747**

In·ves·ti·ga·tion

TEACHER NOTES

Closing the Investigation

This activity provides students an opportunity to bring their work on the Investigation to a close. For each Investigation, students should present their findings to the class. Here are some ways students can display their work.

- Conduct and report on an interview or survey.
- Write a letter, proposal, or report.
- Write an article for the school or local paper.
- Make a display, including graphs and/or charts.
- Plan an activity.

Assessment

To assess students' understanding of the concepts and topics explored in the Investigation and its follow-up activities, you may wish to examine students' Investigation Folders.

The scoring guide provided in the *Investigations and Projects Masters*, p. 23, provides a means for you to score students' work on the Investigation.

Investigations and Projects Masters, p. 23

Scoring Guide
Chapters 12 and 13
Investigation

Level	Specific Criteria
3 Superior	• Shows thorough understanding of the concepts of *landscaping plans, profit margins, percents, scale drawings, models, calculating dimensions, and using a price chart to calculate costs.* • Uses appropriate strategies to solve problems. • Computations are correct. • Written explanations are exemplary. • Drawings, model, report, and presentation are appropriate and sensible. • Goes beyond the requirements of the investigation.
2 Satisfactory, with Minor Flaws	• Shows understanding of the concepts of *landscaping plans, profit margins, percents, scale drawings, models, calculating dimensions, and using a price chart to calculate costs.* • Uses appropriate strategies to solve problems. • Computations are mostly correct. • Written explanations are effective. • Drawings, model, report, and presentation are appropriate and sensible. • Satisfies the requirements of the investigation.
1 Nearly Satisfactory, with Obvious Flaws	• Shows understanding of most of the concepts of *landscaping plans, profit margins, percents, scale drawings, models, calculating dimensions, and using a price chart to calculate costs.* • May not use appropriate strategies to solve problems. • Computations are mostly correct. • Written explanations are satisfactory. • Drawings, model, report, and presentation are appropriate and sensible. • Satisfies the requirements of the investigation.
0 Unsatisfactory	• Shows little or no understanding of the concepts of *landscaping plans, profit margins, percents, scale drawings, models, calculating dimensions, and using a price chart to calculate costs.* • Does not use appropriate strategies to solve problems. • Computations are incorrect. • Written explanations are not satisfactory. • Drawings, model, report, and presentation are not appropriate or sensible. • Does not satisfy the requirements of the investigation.

A Growing Concern

Refer to the Investigation on pages 656–657.

When landscape businesses prepare a proposal for a client, they prepare all the specifications for the bid and include a sketch of the proposal and a photo display of other work they have done for individuals or companies. They also include a list of references that they give to prospective clients. These clients can then call these people and see other work the company has done in order to verify their credentials. What else might be good to include in a sales presentation to a prospective client?

Analyze

You have made a scale drawing of your design and organized your computations in various ways. It is now time to analyze your design and verify your conclusions.

PORTFOLIO ASSESSMENT

You may want to keep your work on this Investigation in your portfolio.

1 Look over your scale drawing of the landscape design and verify the dimensions and placement of each item.

2 Refer to your data on the shade patterns in the yard. Review your knowledge of the plants and trees that you suggested and verify the amount of sun and/or shade that they will get or that they need. Make any necessary changes.

3 Organize a detailed financial bid for the Sanchez family's backyard design. The bid must include the cost of supplies, materials, labor costs, and a profit margin for each phase of the project.

Present

Select members of your class to represent the Sanchez family. Present your proposal, scale drawing, model, and written report as you would for the real Sanchez family.

4 Begin the presentation by stating the requirements that the Sanchez family had specified that they wanted or needed for their backyard design.

5 Explain the process you used to develop your plan. Justify your design and fully explain your bid costs.

6 Prepare a detailed plan for the Sanchez family. Submit in this plan one scale drawing, the model, a written description of the backyard design, and a detailed financial bid.

7 Make sure that each aspect of the design is justified in writing. Include any options that may negate the terms of the proposal, such as a change in the types of plants used.

8 Summarize your plan with a sales statement that details why your plan is superior.

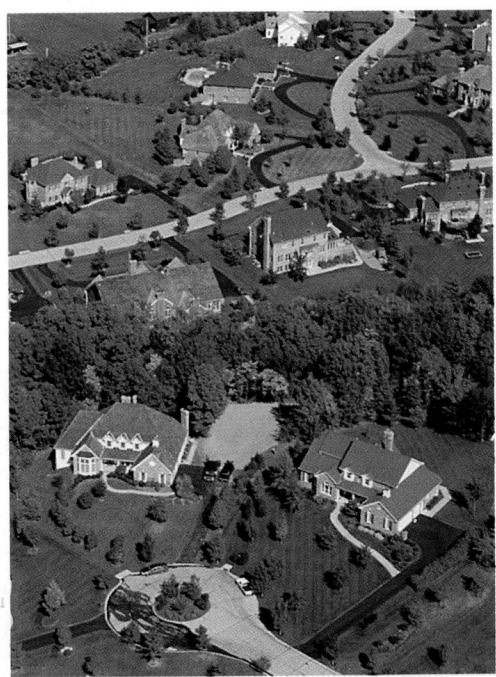

CHAPTER 13 HIGHLIGHTS

VOCABULARY

After completing this chapter, you should be able to define each term, property, or phrase and give an example or two of each.

Algebra
completing the square (pp. 742, 743)
conjugate (p. 722)
distance formula (p. 737)
product property of square roots (p. 719)
quotient property of square roots (p. 721)
radical equations (p. 732)
radicand (p. 719)

rationalizing the denominator (p. 722)
simplest radical form (p. 723)

Geometry
Pythagorean theorem (p. 713)

Problem Solving
identify subgoals (p. 745)

UNDERSTANDING AND USING THE VOCABULARY

State whether each sentence is *true* or *false*. If false, replace the underlined word or number to make a true sentence.

1. The binomials $-3 + \sqrt{7}$ and $\underline{3 - \sqrt{7}}$ are conjugates. false, $-3 - \sqrt{7}$

2. In the expression $-4\sqrt{5}$, the radicand is $\underline{5}$. true

3. The rational expression $\dfrac{1 + \sqrt{3}}{2 - \sqrt{5}}$ becomes $\underline{2 + 2\sqrt{3} + \sqrt{5} + \sqrt{15}}$ when the denominator is rationalized.
false, $-2 - 2\sqrt{3} - \sqrt{5} - \sqrt{15}$

4. The value of c that makes the trinomial $t^2 - 3t + c$ a perfect square is $\underline{9}$. false, $\dfrac{9}{4}$

5. The $\underline{\text{longest}}$ side of a right triangle is the hypotenuse. true

6. The distance formula can be expressed using the equation $\underline{d^2 = (x_2 - x_1)^2 + (y_2 - y_1)^2}$. true

7. After the first step in solving the rational equation $\sqrt{3x + 19} = x + 3$, you would have the equation $\underline{3x + 19 = x^2 + 9}$. false, $3x + 19 = x^2 + 6x + 9$

8. The two sides that form the right angle in a right triangle are called the $\underline{\text{legs}}$ of the triangle. true

9. The expression $\dfrac{2x\sqrt{3x}}{\sqrt{6y}}$ is in simplest radical form. false, $\dfrac{x\sqrt{2xy}}{y}$

10. To verify whether a triangle with sides having lengths of 25, 20, and 15 is a right triangle, the Pythagorean theorem would be used: $\underline{15^2 = 25^2 + 20^2}$. false, $25^2 = 15^2 + 20^2$

Chapter 13 Highlights **749**

Instructional Resources

Three multiple-choice tests and three free-response tests are provided in the *Assessment and Evaluation Masters*. Forms 1A and 2A are for honors pacing, and Forms 1B, 1C, 2B, and 2C are for average pacing. Chapter 13 Test, Form 1B is shown at the right. Chapter 13 Test, Form 2B is shown on the next page.

Using the CHAPTER HIGHLIGHTS

The Chapter Highlights begins with an alphabetical listing of the new terms, properties, and phrases that were introduced in this chapter. Have students define each term and provide an example or two of it, if appropriate.

Assessment and Evaluation Masters, pp. 339–340

13 NAME_____ DATE_____
Chapter 13 Test, Form 1B

Write the letter for the correct answer in the blank at the right of each problem.

1. What is the measure of the hypotenuse of a right triangle with side $a = \sqrt{6}$ and side $b = 8$?
A. $\sqrt{58}$ B. $\sqrt{70}$ C. $\sqrt{14}$ D. 2 — 1. __B__

2. Which of the following are the measures of three sides of a right triangle?
A. 6, 8, 10 B. 5, 9, 11 C. 11, 13, 16 D. 3, 8, 12 — 2. __A__

3. What is the length of the diagonal of a rectangle with a length of 12 m and width of 5 m?
A. 169 m B. 13 m C. $\sqrt{17}$ m D. $\sqrt{119}$ m — 3. __B__

4. Simplify $\sqrt{96}$.
A. $6\sqrt{4}$ B. $6\sqrt{16}$ C. $16\sqrt{6}$ D. $4\sqrt{6}$ — 4. __D__

5. Simplify $\sqrt{28x^2y^3}$.
A. $7x|y|\sqrt{2y}$ B. $2x|y|\sqrt{7y}$ C. $2|x|y\sqrt{7y}$ D. $7|x|y\sqrt{2y}$ — 5. __C__

6. Simplify $\sqrt{\dfrac{18}{40}}$.
A. $\dfrac{\sqrt{3}}{2}$ B. $\dfrac{3\sqrt{20}}{20}$ C. $\dfrac{3\sqrt{5}}{10}$ D. $\dfrac{\sqrt{45}}{10}$ — 6. __C__

7. Simplify $\dfrac{5}{4 + \sqrt{6}}$.
A. $\dfrac{4 - \sqrt{6}}{2}$ B. $\dfrac{20 - 5\sqrt{6}}{22}$ C. $\dfrac{20 + 5\sqrt{6}}{22}$ D. $\dfrac{-\sqrt{6}}{2}$ — 7. __A__

8. Simplify $5\sqrt{7} - \sqrt{7}$.
A. $4\sqrt{7}$ B. -35 C. 5 D. $-5\sqrt{7}$ — 8. __A__

9. Which expression cannot be simplified?
A. $5\sqrt{8} + 6\sqrt{2}$ B. $5\sqrt{10} + 3\sqrt{5}$
C. $7\sqrt{48} + \sqrt{108}$ D. $\sqrt{52} + \sqrt{13}$ — 9. __B__

10. Simplify $\sqrt{18} - \sqrt{54} + 2\sqrt{50}$.
A. $13\sqrt{2} - 3\sqrt{6}$ B. $-4\sqrt{3} + 4\sqrt{5}$
C. $-4\sqrt{3} - 4\sqrt{5}$ D. $8\sqrt{2} - 3\sqrt{6}$ — 10. __A__

13 NAME_____ DATE_____
Chapter 13 Test, Form 1B (continued)

11. Simplify $\sqrt{15} + \sqrt{\dfrac{3}{5}}$.
A. $\dfrac{2\sqrt{15}}{5}$ B. $\dfrac{6\sqrt{15}}{5}$ C. $2\sqrt{3}$ D. $\dfrac{\sqrt{15}}{5}$ — 11. __B__

12. Solve $\sqrt{3x - 2} = 4$.
A. {12} B. {6} C. $\left\{\dfrac{2}{3}\right\}$ D. $\left\{\dfrac{3}{2}\right\}$ — 12. __B__

13. Solve $\sqrt{3a + 28} = a$.
A. {−4} B. {7} C. {7, −4} D. no solution — 13. __B__

14. Solve $\sqrt{5n - 1} - n = 1$.
A. {1, 2} B. {−1, −2} C. $\left\{\dfrac{1}{4}\right\}$ D. {1} — 14. __C__

15. What is the distance between $(-3, 4)$ and $(2, 7)$?
A. $\sqrt{34}$ B. $\sqrt{74}$ C. $2\sqrt{30}$ D. $\sqrt{10}$ — 15. __A__

16. The distance between $(n, 4)$ and $(-2, 7)$ is $\sqrt{73}$. What is one possible value for n?
A. −6 B. 10 C. 8 D. −10 — 16. __D__

17. What is the distance between the origin and $(-6, 8)$?
A. 10 B. 2 C. 14 D. $\sqrt{28}$ — 17. __A__

18. What value of c makes $y^2 - y + c$ a perfect square?
A. $\dfrac{1}{2}$ B. $-\dfrac{1}{2}$ C. $\dfrac{1}{4}$ D. $-\dfrac{1}{4}$ — 18. __C__

19. If $b^2 + 6b - 10 = 0$ is to be solved by completing the square, what would be the best thing to do first?
A. Divide each side by 8. B. Add 10 to each side.
C. Subtract 8 from each side. D. Add 2 to each side. — 19. __B__

20. Solve $d^2 - 6d + 4 = 0$.
A. $3 \pm \sqrt{5}$ B. $6 \pm 2\sqrt{5}$ C. $\dfrac{-6 \pm \sqrt{5}}{2}$ D. $\dfrac{3 \pm 2\sqrt{5}}{2}$ — 20. __A__

Bonus Simplify $\sqrt{4x^2 + 12x + 9}$. **Bonus** $|2x + 3|$

Using the STUDY GUIDE AND ASSESSMENT

Skills and Concepts Encourage students to refer to the objectives and examples on the left as they complete the review exercises on the right.

Assessment and Evaluation Masters, pp. 345–346

---

13 NAME_____ DATE_____

Chapter 13 Test, Form 2B

1. Use the Pythagorean theorem to find the hypotenuse of a right triangle if side $a = 6$ and side $b = 10$.
   1. $2\sqrt{34}$, or 11.66

2. Find the width of a rectangle with a diagonal of 25 cm and a length of 24 cm.
   2. 7 cm

3. A rope from the top of a mast on a sailboat is attached to a point on the deck 5 feet from the base of the mast. If the rope is 13 feet long, how high is the mast?
   3. 12 ft

Simplify. Leave in radical form and use absolute value symbols when necessary.

4. $\sqrt{24}$
   4. $2\sqrt{6}$

5. $\sqrt{75y^4w^3}$
   5. $5y^2w\sqrt{3w}$

6. $\frac{\sqrt{14}}{\sqrt{45}}$
   6. $\frac{\sqrt{70}}{15}$

7. $\frac{6}{6-\sqrt{11}}$
   7. $\frac{36+6\sqrt{11}}{25}$

8. $3\sqrt{11}+2\sqrt{11}$
   8. $5\sqrt{11}$

9. $\sqrt{20}+2\sqrt{45}$
   9. $8\sqrt{5}$

10. $\sqrt{14}+\sqrt{\frac{2}{7}}$
    10. $\frac{8}{7}\sqrt{14}$

11. $\sqrt{6}(\sqrt{30}+4\sqrt{10})$
    11. $6\sqrt{5}+8\sqrt{15}$

---

13 NAME_____ DATE_____

Chapter 13 Test, Form 2B (continued)

Solve each equation. Check the solutions.

12. $\sqrt{x-8}=x-10$
    12. 12

13. $\sqrt{\frac{9a}{3}}-4=0$
    13. $\frac{16}{3}$

14. $\sqrt{m}=2\sqrt{3}$
    14. 12

15. Find the distance between $(0,-4)$ and $(5,2)$.
    15. $\sqrt{61}$, or 7.81

16. The distance between $(3,5)$ and $(7,a)$ is 5. Find the two possible values for a.
    16. 2, 8

17. Find the distance from the origin to $(5,-4)$.
    17. $\sqrt{41}$

Solve by completing the square.

18. $r^2+18r+9=0$
    18. $-9\pm6\sqrt{2}$

19. $b^2+8b+10=0$
    19. $-4\pm\sqrt{6}$

20. Find the value for c that makes x^2+3x+c a perfect square.
    20. $\frac{9}{4}$

Bonus For what values of x is $\sqrt{7x-4}$ a real number?
    Bonus $x\geq\frac{4}{7}$

SKILLS AND CONCEPTS

OBJECTIVES AND EXAMPLES

Upon completing this chapter, you should be able to:

- use the Pythagorean theorem to solve problems (Lesson 13–1)

Find the length of the missing side.

$$c^2 = a^2 + b^2$$
$$25^2 = 15^2 + b^2$$
$$625 = 225 + b^2$$
$$400 = b^2$$
$$20 = b$$

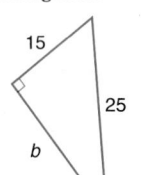

- simplify radical expressions (Lesson 13–2)

$$\sqrt{343x^2y^3} = \sqrt{7\cdot7^2\cdot x^2\cdot y\cdot y^2}$$
$$= \sqrt{7}\cdot\sqrt{7^2}\cdot\sqrt{x^2}\cdot\sqrt{y}\cdot\sqrt{y^2}$$
$$= 7|x|y\sqrt{7y}$$

$$\frac{3}{5-\sqrt{2}} = \frac{3}{5-\sqrt{2}}\cdot\frac{5+\sqrt{2}}{5+\sqrt{2}}$$
$$= \frac{3(5)+3\sqrt{2}}{5^2-(\sqrt{2})^2}$$
$$= \frac{15+3\sqrt{2}}{25-2}$$
$$= \frac{15+3\sqrt{2}}{23}$$

REVIEW EXERCISES

Use these exercises to review and prepare for the chapter test.

If c is the measure of the hypotenuse of a right triangle, find each missing measure. Round answers to the nearest hundredth.

11. $a = 30, b = 16, c = ?$ **34**
12. $a = 6, b = 10, c = ?$ **$2\sqrt{34}\approx11.66$**
13. $a = 10, c = 15, b = ?$ **$5\sqrt{5}\approx11.18$**
14. $b = 4, c = 56, a = ?$ **$4\sqrt{195}\approx55.86$**
15. $a = 18, c = 30, b = ?$ **24**
16. $a = 1.2, b = 1.6, c = ?$ **2**

Determine whether the following side measures would form right triangles. Explain why or why not.

17. $9, 16, 20$ **no; $9^2 + 16^2 \neq 20^2$**
18. $20, 21, 29$ **yes**
19. $9, 40, 41$ **yes**
20. $18, \sqrt{24}, 30$ **no; $18^2 + (\sqrt{24})^2 \neq 30^2$**

Simplify. Leave in radical form and use absolute value symbols when necessary.

21. $\sqrt{480}$ **$4\sqrt{30}$**
22. $\sqrt{\frac{60}{y^2}}$ **$\frac{2\sqrt{15}}{|y|}$**
23. $\sqrt{44a^2b^5}$ **$2|a|b^2\sqrt{11b}$**
24. $\sqrt{96x^4}$ **$4x^2\sqrt{6}$**
25. $(3-2\sqrt{12})^2$ **$57-24\sqrt{3}$**
26. $\frac{9}{3+\sqrt{2}}$ **$\frac{27-9\sqrt{2}}{7}$**
27. $\frac{2\sqrt{7}}{3\sqrt{5}+5\sqrt{3}}$ **$\frac{3\sqrt{35}-5\sqrt{21}}{-15}$**
28. $\frac{\sqrt{3a^3b^4}}{\sqrt{8ab^{10}}}$ **$\frac{|a|\sqrt{6}}{4|b^3|}$**

GLENCOE Technology

Test and Review Software

You may use this software, a combination of an item generator and item bank, to create your own tests or worksheets. Types of items include free response, multiple choice, short answer, and open ended.

For IBM & Macintosh

OBJECTIVES AND EXAMPLES

● simplify radical expressions involving addition, subtraction, and multiplication   (Lesson 13–3)

$\sqrt{6} - \sqrt{54} + 3\sqrt{12} + 5\sqrt{3}$

$= \sqrt{6} - \sqrt{3^2 \cdot 6} + 3\sqrt{2^2 \cdot 3} + 5\sqrt{3}$

$= \sqrt{6} - (\sqrt{3^2} \cdot \sqrt{6}) + 3(\sqrt{2^2} \cdot \sqrt{3}) + 5\sqrt{3}$

$= \sqrt{6} - 3\sqrt{6} + 3(2\sqrt{3}) + 5\sqrt{3}$

$= \sqrt{6} - 3\sqrt{6} + 6\sqrt{3} + 5\sqrt{3}$

$= -2\sqrt{6} + 11\sqrt{3}$

30. $5\sqrt{13} + 5\sqrt{15}$

REVIEW EXERCISES

Simplify. Then use a calculator to verify your answer.

29. $2\sqrt{6} - \sqrt{48}$ $2\sqrt{6} - 4\sqrt{3}$

30. $2\sqrt{13} + 8\sqrt{15} - 3\sqrt{15} + 3\sqrt{13}$

31. $4\sqrt{27} + 6\sqrt{48}$ $36\sqrt{3}$

32. $5\sqrt{18} - 3\sqrt{112} - 3\sqrt{98}$ $-6\sqrt{2} - 12\sqrt{7}$

33. $\sqrt{8} + \sqrt{\frac{1}{8}}$ $\frac{9\sqrt{2}}{4}$

34. $4\sqrt{7k} - 7\sqrt{7k} + 2\sqrt{7k}$ $-\sqrt{7k}$

● solve radical equations   (Lesson 13–4)

Solve $\sqrt{5 - 4x} - 6 = 7$.

$\sqrt{5 - 4x} - 6 = 7$

$\sqrt{5 - 4x} = 13$

$(\sqrt{5 - 4x})^2 = (13)^2$

$5 - 4x = 169$

$-4x = 164$

$x = -41$

Solve each equation. Check your solution.

35. $\sqrt{3x} = 6$ 12

36. $\sqrt{t} = 2\sqrt{6}$ 24

37. $\sqrt{7x - 1} = 5$ $\frac{26}{7}$

38. $\sqrt{x + 4} = x - 8$ 12

39. $\sqrt{r} = 3\sqrt{5}$ 45

40. $\sqrt{3x - 14} + x = 6$ 5

41. $\sqrt{\frac{4a}{3}} - 2 = 0$ 3

42. $9 = \sqrt{\frac{5n}{4}} - 1$ 80

43. $10 + 2\sqrt{b} = 0$ no solution

44. $\sqrt{a + 4} = 6$ 32

● find the distance between two points in the coordinate plane   (Lesson 13–5)

Find the distance between the pair of points with coordinates $(-5, 1)$ and $(1, 5)$.

$d = \sqrt{(x_2 - x_1)^2 + (y_2 - y_1)^2}$

$= \sqrt{(1 - (-5))^2 + (5 - 1)^2}$

$= \sqrt{6^2 + 4^2}$

$= \sqrt{36 + 16}$ or $\sqrt{52} \approx 7.21$

Find the distance between each pair of points whose coordinates are given.

45. $(9, -2), (1, 13)$ 17

46. $(4, 2), (7, -9)$ $\sqrt{130} \approx 11.40$

47. $(4, -6), (-2, 7)$ $\sqrt{205} \approx 14.32$

48. $(2\sqrt{5}, 9), (4\sqrt{5}, 3)$ $2\sqrt{14} \approx 7.48$

Find the value of a if the points with the given coordinates are the indicated distance apart.

49. $(-3, 2), (1, a); d = 5$ 5 or −1

50. $(5, -2), (a, -3); d = \sqrt{170}$ 18 or −8

51. $(1, 1), (4, a); d = 5$ 5 or −3

Applications and Problem Solving Encourage students to work through the exercises in the Applications and Problem Solving section to strengthen their problem-solving skills.

OBJECTIVES AND EXAMPLES

- solve quadratic equations by completing the square (Lesson 13–6)

Solve $y^2 + 6y + 2 = 0$ by completing the square.

$$y^2 + 6y + 2 = 0$$
$$y^2 + 6y = -2$$
$$y^2 + 6y + 9 = -2 + 9 \quad \textit{Complete the square.}$$
$$(y + 3)^2 = 7 \quad \textit{Since } \left(\frac{6}{2}\right)^2 = 9, \textit{ add 9 to}$$
$$y + 3 = \pm\sqrt{7} \quad \textit{each side}$$
$$y = -3 \pm \sqrt{7}$$

The solution set is $-3 + \sqrt{7}$ and $-3 - \sqrt{7}$.

53. $\frac{49}{4}$ 55. $\frac{1}{9}$ 57. $\frac{7 \pm \sqrt{69}}{2}$ 58. $-\frac{3}{2}, -\frac{5}{2}$

REVIEW EXERCISES

Find the value of c that makes each trinomial a perfect square.

52. $y^2 - 12y + c$ **36**   53. $m^2 + 7m + c$

54. $b^2 + 18b + c$ **81**   55. $p^2 - \frac{2}{3}p + c$

Solve each equation by completing the square. Leave irrational roots in simplest radical form.

56. $x^2 - 16x + 32 = 0$ $8 \pm 4\sqrt{2}$

57. $m^2 - 7m = 5$

58. $4a^2 + 16a + 15 = 0$

59. $\frac{1}{2}y^2 + 2y - 1 = 0$ $-2 \pm \sqrt{6}$

60. $n^2 - 3n + \frac{5}{4} = 0$ $\frac{1}{2}, \frac{5}{2}$

APPLICATIONS AND PROBLEM SOLVING

61. **Geometry** The sides of a triangle measure $4\sqrt{24}$ cm, $5\sqrt{6}$ cm, and $3\sqrt{54}$ cm. What is the perimeter of the triangle? (Lesson 13–3)
    $22\sqrt{6} \approx 53.9$ cm

62. **Sight Distance** In the movie *Angels in the Outfield*, Roger decides to climb a tree in order to be able to see the Angels game better. The formula $V = 3.5\sqrt{h}$ relates height and distance, where h is your height in meters above the ground and V is the distance in kilometers that you can see. (Lesson 13–4)   **62b. 256 m**

    a. How far could Roger see if he had climbed 9 meters to the top of a tree?   **10.5 km**

    b. How high would someone have to be if she wanted to be able to see 56 kilometers?

63. **Geometry** What kind of triangle has vertices at $(4, 2)$, $(-3, 1)$, and $(5, -4)$? (Lesson 13–5)
    scalene triangle

64. **Number Theory** When 6 is subtracted from 10 times a number, the result is equal to 4 times the square of the number. Find the numbers. (Lesson 13–6)   **1, 1.5**

65. **Nature** An 18-foot tall tree is broken by the wind. The top of the tree falls and touches the ground 12 feet from its base. How many feet from the base of the tree did the break occur? (Lesson 13–1)   **5 ft**

66. **Physics** The time T (in seconds) required for a pendulum of length L (in feet) to make one complete swing back and forth is given by the formula $T = 2\pi\sqrt{\frac{L}{32}}$. How long does it take a pendulum 4 feet long to make one complete swing? (Lesson 13–2) $\frac{\sqrt{2}}{2}\pi$ **or about 2.22 s**

67. **Law Enforcement** Lina told the police officer that she was traveling at 55 mph when she applied the brakes and skidded. The skid marks at the scene were 240 feet long. Should Lina's car have skidded that far if it was traveling at 55 mph? Use the formula $s = \sqrt{15d}$. (Lesson 13–4)   **No, it should skid about 201.7 ft.**

A practice test for Chapter 13 is provided on page 799.

ALTERNATIVE ASSESSMENT

COOPERATIVE LEARNING PROJECT

Battleship In this project, you will determine the placement of ships for a game. A game that Ryan and Nicholas enjoy playing is a form of the game Battleship®. They determine the lengths of five ships that they will draw on their grids. Each time a ship intersects a grid coordinate, that is one of the points where the ship can be attacked and hit. A ship can be placed on the grid horizontally, vertically, or diagonally. When all of the grid coordinates for a ship are hit, the ship is sunk. The goal is to sink all of the other person's ships first.

For their first game, Ryan and Nicholas decided on $2\sqrt{5}$ units, 3 units, $\sqrt{13}$ units, $\sqrt{2}$ units, and 2 units for their ship lengths. Place these five ships on the grid below making sure that they have the above stated lengths. How many hits would another player have to get in order to sink all of the ships on your grid? $3 + 4 + 2 + 2 + 3 = 14$

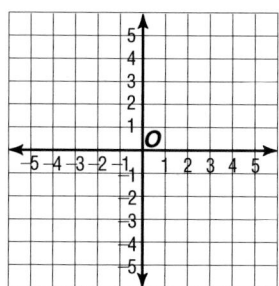

Follow these guidelines for your Battleship game.

- Determine whether each of the ships with the specified lengths have to be drawn horizontally, vertically, or diagonally.
- Devise a chart that will be helpful in organizing the data you will need; for example, length of ship, length of legs of triangle formed by diagonal ships, and number of grid coordinates for each ship.
- Create a grid that will incorporate the above specifications for the ships.
- Write a summary of the above game.
- Play the game with a fellow classmate.

THINKING CRITICALLY

- Is the sum of two irrational numbers always an irrational number? Explain.
- For any real number n, $\sqrt{n^2} = |n|$. What rule would be true for $\sqrt{n^t}$ if t were odd?

$$\sqrt{n^t} = n^{\frac{t-1}{2}}\sqrt{n}$$

PORTFOLIO

Visualizing a concept or skill is a good way to help understand a concept. Using colored pencils is one way to visualize steps that need to be remembered or aid in "seeing" items that are easy to forget. When simplifying a rational expression, write the perfect square number or variables using colored pencils. This will serve to remind you what needs to be written outside the radical sign and what needs to stay inside the radical sign. Use one of the exercises in this chapter and use colored pencils to simplify it. Write how you chose what to put in color and how it aided you in the problem-solving process. Place this in your portfolio.

SELF EVALUATION

When solving a problem, one may look for alternative methods to use. The same method may not work in every situation. Being open-minded about a different process is a positive value. Comparing methods can also be beneficial.

Assess yourself. Are you a person that sticks to the same method for a solution or do you also look for other means? Do you compare methods or not? Think of two problems that you have had, one from your daily life and one from your mathematical experiences, where you used two different methods to solve each of them.

Chapter 13   Study Guide and Assessment   **753**

Assessment and Evaluation Masters, pp. 350, 361

13 NAME_____ DATE_____

Chapter 13 Performance Assessment

Instructions: *Demonstrate your knowledge by giving a clear, concise solution to each problem. Be sure to include all relevant drawings and justify your answers. You may show your solution in more than one way or investigate beyond the requirements of the problem.*

1. Mrs. Raines is having a sprinkler system installed in her flower bed. The diagram below shows the location of sprinklers A, B, C, and D. The water source S is in the center of the bed.

   a. How much pipe would be required to hook each sprinkler directly to the water source? Explain your answer.
   b. Do you think she could save pipe by running pipe from S to A, from A to B, from B to C, and from C to D? Explain your answer.
   c. Would running pipe from A to B and from C to D before connecting the midpoint of each line to the water source S save pipe? Justify your answer.
   d. If A is connected to D and B is connected to C and the midpoints of these two lines connected to the water source, would Mrs. Raines save pipe over the arrangement in part c? Why?
   e. Would the arrangement in part d save pipe over the arrangement in part a? Justify your answer.
   f. What other considerations might Mrs. Raines consider in choosing the hook-up arrangement?
   g. Find the pipe arrangement that would use the least pipe for a square of side a feet. Justify your answer.

2. Solve $8\sqrt{4z^2 - 43} = 40$. Justify each step in your solution.

Scoring Guide
Chapter 13
Performance Assessment

Level	Specific Criteria
3 Superior	• Shows thorough understanding of the concepts of *finding square roots, using the Pythagorean theorem, simplifying radical expressions,* and *solving radical equations.* • Uses appropriate strategies to solve problems. • Computations are correct. • Written explanations are exemplary. • Diagrams are accurate and appropriate. • Goes beyond requirements of some or all problems.
2 Satisfactory, with Minor Flaws	• Shows understanding of the concepts of *finding square roots, using the Pythagorean theorem, simplifying radical expressions,* and *solving radical equations.* • Uses appropriate strategies to solve problems. • Computations are mostly correct. • Written explanations are effective. • Diagrams are mostly accurate and appropriate. • Satisfies all requirements of problems.
1 Nearly Satisfactory, with Serious Flaws	• Shows understanding of most of the concepts of *finding square roots, using the Pythagorean theorem, simplifying radical expressions,* and *solving radical equations.* • May not use appropriate strategies to solve problems. • Computations are mostly correct. • Written explanations are satisfactory. • Diagrams are mostly accurate and appropriate. • Satisfies most requirements of problems.
0 Unsatisfactory	• Shows little or no understanding of the concepts of *finding square roots, using the Pythagorean theorem, simplifying radical expressions,* and *solving radical equations.* • May not use appropriate strategies to solve problems. • Computations are incorrect. • Written explanations are not satisfactory. • Diagrams are not accurate or appropriate. • Does not satisfy requirements of problems.

 Alternative Assessment

The Alternative Assessment section provides students with the opportunity to assess their own work by thinking critically, working with others, keeping a portfolio, and honestly evaluating their own progress. For more information on alternative forms of assessment, see *Alternative Assessment in the Mathematics*

Classroom, one of the titles in the Glencoe Mathematics Professional Series.

Performance Assessment

Performance Assessment tasks for this chapter are included in the *Assessment and Evaluation Masters.* A scoring guide is also provided.

STUDENT HANDBOOK

For Additional Assessment

For Reference

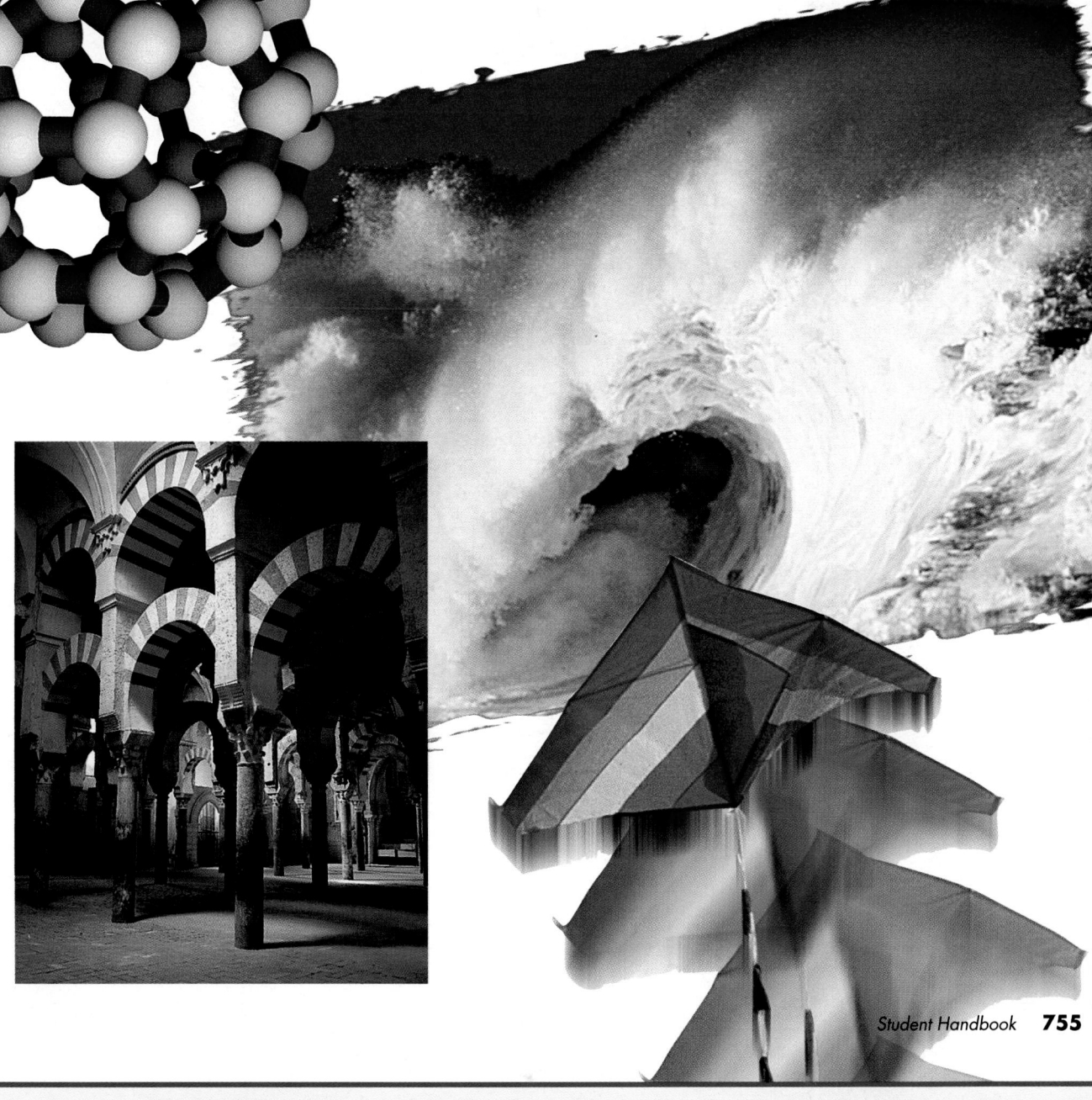

EXTRA PRACTICE

Lesson 1-1 Write an algebraic expression for each verbal expression.

1. the product of x and 7 $7x$

2. the quotient of r and s $\frac{r}{s}$

3. the sum of b and 21 $b + 21$

4. a number t decreased by 6
$t - 6$

5. a number a to the third power
a^3

6. sixteen squared 16^2

Write a verbal expression for each algebraic expression.

7. $n - 7$ n minus 7

8. xy the product of x and y

9. m^5 m to the fifth power

10. 8^4 8 to the fourth power

11. $6r^2$ 6 times r squared

12. $z^7 + 2$ z to the seventh power plus 2

Write each expression as an expression with exponents.

13. $5 \cdot 5 \cdot 5$ 5^3

14. $7 \cdot a \cdot a \cdot a \cdot a$ $7a^4$

15. $2(m)(m)(m)$ $2m^3$

16. $5 \cdot 5 \cdot 5 \cdot x \cdot x \cdot y$ 5^3x^2y

17. $p \cdot p \cdot p \cdot p \cdot p \cdot p$ p^6

18. $4 \cdot 4 \cdot 4 \cdot t \cdot t$ 4^3t^2

Evaluate each expression.

19. 2^4 16

20. 8^2 64

21. 7^3 343

22. 10^4 10,000

23. 3^6 729

24. 4^5 1024

Lesson 1-2 Give the next two items for each pattern.

1.

2.
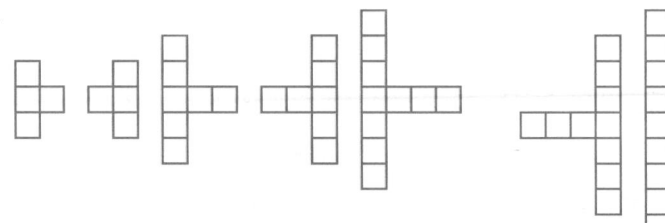

3. 12, 23, 34, 45, ... 56, 67

4. 39, 33, 27, 21, ... 15, 9

5. 6, 7.2, 8.4, 9.6, ... 10.8, 12

6. 86, 81.5, 77, 72.5, ... 68, 63.5

7. 4, 8, 16, 32, ... 64, 128

8. 3125, 625, 125, 25, ... 5, 1

9. 15, 16, 18, 21, 25, 30, ... 36, 43

10. $w - 2, w - 4, w - 6, w - 8, ...$
$w - 10, w - 12$

11. 13, 10, 11, 8, 9, 6, ... 7, 4

Lesson 1-3 Evaluate each expression.

1. $3 + 8 \div 2 - 5$ 2

2. $4 + 7 \cdot 2 + 8$ 26

3. $5(9 + 3) - 3 \cdot 4$ 48

4. $4(11 + 7) - 9 \cdot 8$ 0

5. $5^3 + 6^3 - 5^2$ 316

6. $16 \div 2 \cdot 5 \cdot 3 \div 6$ 20

7. $7(5^3 + 3^2)$ 938

8. $\frac{9 \cdot 4 + 2 \cdot 6}{7 \cdot 7}$ $\frac{48}{49}$

9. $25 - \frac{1}{3}(18 + 9)$ 16

Evaluate each expression when $a = 2$, $b = 5$, $x = 4$, and $n = 10$.

10. $8a + b$ 21

11. $12x + ab$ 58

12. $a(6 - 3n)$ -48

13. $bx + an$ 40

14. $x^2 - 4n$ -24

15. $3b + 16a - 9n$ -43

16. $n^2 + 3(a + 4)$ 118

17. $(2x)^2 + an - 5b$ 59

18. $[a + 8(b - 2)]^2 \div 4$ 169

Lesson 1-4 Suppose the number 16,782 is rounded to 16,800 and plotted using stem 16 and leaf 8. Write the stem and leaf for each number below if the numbers are part of the same set of data.

1. 24,640 **stem 24, leaf 6**
2. 35,788 **stem 35, leaf 8**
3. 4239 **stem 4, leaf 2**
4. 5865 **stem 5, leaf 9**
5. 611 **stem 0, leaf 6**
6. 17,903 **stem 17, leaf 9**

7. The stem-and-leaf plot at the right gives the average weekly earnings for various occupations in 1993. Use the plot to answer each question.

Average Weekly Earnings	
Stem	Leaf
2	6 7
3	1 2 8 9
4	1 3 6 8 8 8 9
5	0 1 6
6	
7	1 2 4

$3 \mid 8 = \$380-\$389$

  a. What were the highest weekly earnings? **$740–$749**
  b. What were the lowest weekly earnings? **$260–$269**
  c. How many occupations have weekly earnings of at least $500? **6**
  d. What does $5 \mid 1$ represent? **weekly earnings of $510–$519**

Lesson 1-5 State whether each equation is *true* or *false* for the value of the variable given.

1. $b + \frac{2}{3} = \frac{3}{4} + \frac{1}{3}, b = \frac{1}{2}$ **false**
2. $\frac{2 + 13}{y} = \frac{3}{5}y, y = 5$ **true**
3. $x^8 = 9^4, x = 3$ **true**
4. $4t^2 - 5(3) = 9, t = 7$ **false**
5. $\frac{3^2 - 5x}{3^2 - 1} \leq 2, x = 4$ **true**
6. $a^6 \div 4 \div a^3 \div a < 3, a = 2$ **true**

Find the solution set for each inequality if the replacement set for x is {4, 5, 6, 7, 8} and for y is {10, 12, 14, 16}.

7. $x + 2 > 7$ **{6, 7, 8}**
8. $x - 1 > 3$ **{5, 6, 7, 8}**
9. $2y - 15 \leq 17$ **{10, 12, 14, 16}**
10. $y + 12 < 25$ **{10, 12}**
11. $\frac{y + 12}{7} = 4$ **{16}**
12. $\frac{2(x - 2)}{3} = \frac{4}{7 - 5}$ **{4}**
13. $x - 4 > \frac{x + 2}{3}$ **{8}**
14. $y^2 - 100 \geq 4y$ **{14, 16}**
15. $9x - 20 = x^2$ **{4, 5}**
16. $0.3(x + 4) \leq 0.4(2x + 3)$ **{4, 5, 6, 7, 8}**
17. $1.3x - 12 < 0.9x + 4$ **{4, 5, 6, 7, 8}**
18. $1.2y - 8 \leq 0.7y - 3$ **{10}**

Solve each equation.

19. $x = \frac{17 + 9}{2}$ **13**
20. $3(8) + 4 = b$ **28**
21. $\frac{18 - 7}{13 - 2} = y$ **1**
22. $28 - (-14) = z$ **42**
23. $20.4 - 5.67 = t$ **14.73**
24. $t = 91.8 \div 27$ **3.4**
25. $-\frac{5}{8}\left(-\frac{4}{5}\right) = c$ **$\frac{1}{2}$**
26. $8\frac{1}{12} - 5\frac{5}{12} = e$ **$2\frac{2}{3}$**
27. $\frac{3}{4} - \frac{9}{16} = s$ **$\frac{3}{16}$**
28. $\frac{5}{8} + \frac{1}{4} = y$ **$\frac{7}{8}$**
29. $n = \frac{84 \div 7}{18 \div 9}$ **6**
30. $d = 3\frac{1}{2} \div 2$ **$\frac{7}{4}$**

Lesson 1-6 Name the property or properties illustrated by each statement.

1. If $8 \cdot 3 = 24$, then $24 = 8 \cdot 3$. **symmetric (=)**
2. $6 + (4 + 1) = 6 + 5$ **substitution (=)**
3. $(12 - 3)(7) = 9(7)$ **substitution (=)**
4. $qrs = 1qrs$ **multiplicative identity**
5. $\left(\frac{8}{9}\right)\left(\frac{9}{8}\right) = 1$ **multiplicative inverse**
6. $4\left(6^2 \cdot \frac{1}{36}\right) = 4$ **multiplicative inverse, multiplicative identity**
7. $0 + 45 = 45$ **additive identity**
8. If $5 = 9 - 4$, then $9 - 4 = 5$. **symmetric (=)**
9. $2(0) = 0$ **multiplicative property of 0**
10. $16 + 37 = 16 + 37$ **reflexive (=)**
11. $1(57) = 57$ **multiplicative identity**
12. $0 + h = h$ **additive identity**
13. If $9 + 1 = 10$ and $10 = 5(2)$, then $9 + 1 = 5(2)$. **transitive (=)**

Extra Practice **757**

1. multiplicative identity
2. commutative property (+)
3. commutative property (×)
4. additive identity
5. distributive property
6. commutative property (+)
7. associative property (×)
8. multiplicative property of zero
9. substitution (=)
10. associative property (+)
11. associative property (×)
12. associative property (+)

EXTRA PRACTICE

Lesson 1-7 Use the distributive property to rewrite each expression without parentheses.

1. $3(5 + w)$ $15 + 3w$
2. $(h - 8)7$ $7h - 56$
3. $6(y + 4)$ $6y + 24$
4. $9(3n + 5)$ $27n + 45$
5. $32\left(x - \frac{1}{8}\right)$ $32x - 4$
6. $c(7 - d)$ $7c - cd$

Use the distributive property to find each product.

7. $6 \cdot 55$ 330
8. $\left(4\frac{1}{18}\right) \times 18$ 73
9. $15(108)$ 1620
10. $14(3.7)$ 51.8
11. $689 \cdot 5$ 3445
12. 7×314 2198

Simplify each expression, if possible. If not possible, write *in simplest form*.

13. $13a + 5a$ $18a$
14. $21x - 10x$ $11x$
15. $8(3x + 7)$ $24x + 56$
16. $4m - 4n$ in simplest form
17. $3(5am - 4)$ $15am - 12$
18. $15x^2 + 7x^2$ $22x^2$
19. $9y^2 + 13y^2 + 3$ $22y^2 + 3$
20. $11a^2 - 11a^2 + 12a^2$ $12a^2$
21. $6a + 7a + 12b + 8b$
    $13a + 20b$

Lesson 1-8 Name the property illustrated by each statement.   1–12. See margin.

1. $1 \cdot a^2 = a^2$
2. $x^2 + (y + z) = x^2 + (z + y)$
3. $ax + 2b = xa + 2b$
4. $29 + 0 = 29$
5. $5(a + 3b) = 5a + 15b$
6. $5a + 3b = 3b + 5a$
7. $(4 \cdot c) \cdot d = 4 \cdot (c \cdot d)$
8. $(6x^3) \cdot 0 = 0$
9. $(4 + 1)x + 2 = 5x + 2$
10. $(a + b) + 3 = a + (b + 3)$
11. $5(ab) = (5a)b$
12. $5a + \left(\frac{1}{2}b + c\right) = \left(5a + \frac{1}{2}b\right) + c$

Simplify.   15. $3a + 13b + 2c$ 17. $-4p - 2q$ 22. $-11 + 3uv + u$ 23. $15x + 10y$ 24. $11.8a + 8.8b$

13. $5a + 6b + 7a$ $12a + 6b$
14. $8x + 4y + 9x$ $17x + 4y$
15. $3a + 5b + 2c + 8b$
16. $\frac{2}{3}x^2 + 5x + x^2$ $\frac{5}{3}x^2 + 5x$
17. $(4p - 7q) + (5q - 8p)$
18. $8q + 5r - 7q - 6r$ $q - r$
19. $4(2x + y) + 5x$ $13x + 4y$
20. $9r^5 + 2r^2 + r^5$ $10r^5 + 2r^2$
21. $12b^3 + 12 + 12b^3$ $24b^3 + 12$
22. $7 + 3(uv - 6) + u$
23. $3(x + 2y) + 4(3x + y)$
24. $6.2(a + b) + 2.6(a + b) + 3a$
25. $3 + 8(st + 3w) + 3st$
    $3 + 11st + 24w$
26. $5.4(s - 3t) + 3.6(s - 4)$
    $9s - 16.2t - 14.4$
27. $3[4 + 5(2x + 3y)]$
    $12 + 30x + 45y$

Lesson 1-9 Match each description with the most appropriate graph.

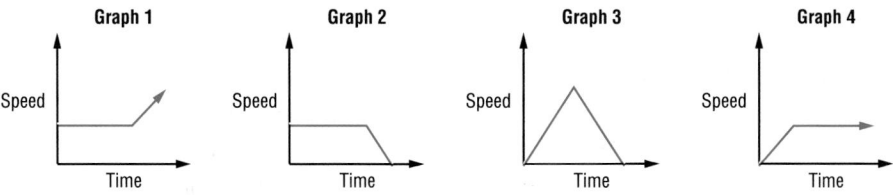

1. Jeremy picks up his speed while jogging down a hill. Then he slows down until he comes to a stop.   **Graph 3**
2. Ilene jogs up a hill at a steady speed. Then she runs down the hill and picks up her speed.   **Graph 1**
3. Casey runs down a hill and picks up his speed. Then he continues to jog at a steady pace.   **Graph 4**
4. Luisa jogs along a road at a steady speed. Then she slows down until she comes to a stop.   **Graph 2**

Lesson 2-1   Name the set of numbers graphed.   1–6. See margin.

1.
   -6 -5 -4 -3 -2 -1 0 1 2 3 4 5 6

2. ‹+++++++++++++›
   -4 -3 -2 -1 0 1 2 3 4 5 6 7 8

3. ‹+++++++++++++›
   -3 -2 -1 0 1 2 3 4 5 6 7 8 9

4. ‹+++++++++++++›
   4 5 6 7 8 9 10 11 12 13 14 15 16

5. ‹+++++++++++++›
   -10 -9 -8 -7 -6 -5 -4 -3 -2 -1 0 1 2

6. ‹+++++++++++++›
   -5 -4 -3 -2 -1 0 1 2 3 4 5 6 7

Graph each set of numbers on a number line.   7–12. See margin.

7. $\{-2, -4, -6\}$

8. $\{\ldots, -3, -2, -1, 0\}$

9. {integers greater than -1}

10. {integers less than -5 and greater than -10}

11. {integers less than or equal to 3}

12. {integers less than 0 and greater than or equal to -6}

Lesson 2-2   Use the line plot below to answer each question.

1. What was the highest score on the test?   50

2. What was the lowest score on the test?   22

3. How many students took the test?   31

4. How many students scored in the 40s?   8

5. What score was received by the most students?   26

```
                          x
            x      x      x                   x
            x  x   x  x             x                  x
x   xxx  x  xxxxxxx  x      xxx  xx  x   x
+--+--+--+--+--+--+--+--+--+--+--+--+--+--+--+--+--+--+
20 22 24 26 28 30 32 34 36 38 40 42 44 46 48 50 52 54
```

Make a line plot for each set of data.   6–9. See margin.

6. 134, 167, 137, 138, 120, 134, 145, 155, 152, 159, 164, 135, 144, 156

7. 19, 12, 11, 11, 7, 7, 8, 13, 12, 12, 9, 9, 8, 15, 11, 4, 12, 7, 7, 6

8. 66, 74, 72, 78, 68, 75, 80, 69, 62, 65, 63, 78, 81, 78, 76, 87, 80, 69, 81, 76, 79, 70, 62, 73, 85, 87, 70

9. 152, 156, 133, 154, 129, 146, 174, 138, 185, 141, 169, 176, 179, 168, 185, 154, 199, 200

Lesson 2-3   Find each sum or difference.

1. $-3 + 16$ 13

2. $27 - 19$ 8

3. $8 - 13$ -5

4. $14 + (-9)$ 5

5. $-18 + (-11)$ -29

6. $-25 + 47$ 22

7. $19m - 12m$ $7m$

8. $8h - 23h$ $-15h$

9. $24b - (-9b)$ $33b$

10. $97 + (-79)$ 18

11. $4 + (-12) + (-18)$ -26

12. $7 + (-11) + 32$ 28

13. $|-28 + (-67)|$ 95

14. $|-89 + 46|$ 43

15. $|-285 + (-641)|$ 926

16. $-35 - (-12)$ -23

17. $24 + (-15)$ 9

18. $-15 + (-13)$ -28

19. $-7 + (-21)$ -28

20. $8 - 17 + (-3)$ -12

21. $27 - 14 - (-19)$ 32

22. $|-9 + 15|$ 6

23. $\begin{bmatrix} 5 & -4 \\ 0 & 3 \end{bmatrix} + \begin{bmatrix} -3 & 4 \\ -2 & -4 \end{bmatrix}$ $\begin{bmatrix} 2 & 0 \\ -2 & -1 \end{bmatrix}$

24. $\begin{bmatrix} 1 & -4 \\ 5 & -6 \end{bmatrix} - \begin{bmatrix} 4 & -3 \\ 7 & -1 \end{bmatrix}$ $\begin{bmatrix} -3 & -1 \\ -2 & -5 \end{bmatrix}$

Extra Practice   **759**

Additional Answers for Lesson 2-1

1. $\{-3, -2, -1, 0, 1, 2, 3, 4\}$
2. $\{-2, 0, 2, 3, 6\}$
3. $\{2, 3, 4\}$
4. $\{7, 8, 9, 10, 11, 12, 13\}$
5. $\{-8, -6, -4, -2, 0\}$
6. $\{-2, -1, 0, 1, 2, 3, 4\}$

7. ‹++++++++++›
   -8 -7 -6 -5 -4 -3 -2 -1 0

8. ‹+++++++++›
   -4 -3 -2 -1 0 1 2 3 4

9. ‹+++++++++›
   -4 -3 -2 -1 0 1 2 3 4

10. ‹+++++++++›
    -12 -11 -10 -9 -8 -7 -6 -5 -4

11. ‹+++++++++›
    -1 0 1 2 3 4 5 6 7

12. ‹+++++++++›
    -7 -6 -5 -4 -3 -2 -1 0 1

Additional Answers for Lesson 2-2

6.
```
     x         xxxx   xx    x xx x     xx
+--+--+--+--+--+--+--+--+--+--+
118 124 130 136 142 148 154 160 166
```

7.
```
         x           x
         x         x x
         xxx       xx
x  xxxx  xxx  x        x
+--+--+--+--+
5       10      15      20
```

8.
```
                   x
   x        xx        x x xx        x
   xx xx  xxx  xxxxx  xxxx    x x
+--+--+--+--+--+--+
60    65    70    75    80    85
```

9.
```
              x              x
xx xx x  xxx       xx  xxx  x          xx
+--+--+--+--+--+--+--+--+--+--+--+--+
132 138 144 150 156 162 168 174 180 186 192 198
```

759

Lesson 2-4 Replace each __?__ with <, >, or = to make each sentence true.

1. 6 __?__ -4 $>$

2. 12 __?__ -21 $>$

3. -4 __?__ -10 $>$

4. 4 __?__ 14 $<$

5. -13 __?__ -8 $<$

6. $-5 + 2$ __?__ -3 $=$

7. 7 __?__ $13 - (-6)$ $<$

8. $3.4 - 5.7$ __?__ -2 $<$

9. $\frac{18}{-6}$ __?__ -3 $=$

10. $\frac{8}{13}$ __?__ $\frac{9}{14}$ $<$

11. $\frac{25}{-8}$ __?__ $\frac{-28}{7}$ $>$

12. $6\left(\frac{5}{3}\right)$ __?__ $\left(\frac{3}{2}\right)6$ $>$

13. $\frac{0.6}{7}$ __?__ $\frac{1.8}{12}$ $<$

14. 24.6 __?__ $13.8 - (-12.8)$ $<$

15. $-54 + 26.5$ __?__ 27.5 $<$

16. $\frac{5.4}{18}$ __?__ $-4 + 1$ $>$

17. $(4.1)(0.2)$ __?__ 8.4 $<$

18. $-\frac{12}{17}$ __?__ $-\frac{9}{14}$ $<$

Lesson 2-5 Find each sum or difference.

1. $-\frac{11}{9} + \left(-\frac{7}{9}\right)$ -2

2. $\frac{5}{11} - \frac{6}{11}$ $-\frac{1}{11}$

3. $\frac{2}{7} - \frac{3}{14}$ $\frac{1}{14}$

4. $-4.8 + 3.2$ -1.6

5. $-1.7 - 3.9$ -5.6

6. $-72.5 - 81.3$ -153.8

7. $-\frac{3}{5} + \frac{5}{6}$ $\frac{7}{30}$

8. $\frac{3}{8} + \left(-\frac{7}{12}\right)$ $-\frac{5}{24}$

9. $-\frac{7}{15} + \left(-\frac{5}{12}\right)$ $-\frac{53}{60}$

10. $-4.5 - 8.6$ -13.1

11. $89.3 - (-14.2)$ 103.5

12. $-0.007 + 0.06$ 0.053

13. $-\frac{2}{7} + \frac{3}{14} + \frac{3}{7}$ $\frac{5}{14}$

14. $-\frac{3}{5} + \frac{6}{7} + \left(-\frac{2}{35}\right)$ $\frac{1}{5}$

15. $\frac{7}{3} + \left(-\frac{5}{6}\right) + \left(-\frac{2}{3}\right)$ $\frac{5}{6}$

16. $-4.13 + (-5.18) + 9.63$ 0.32

17. $6.7 + (-8.1) + (-7.3)$ -8.7

18. $\frac{3}{4} + \left(-\frac{5}{8}\right) + \frac{3}{32}$ $\frac{7}{32}$

19. $1.9 - (-7)$ 8.9

20. $-1.8 - 3.7$ -5.5

21. $-18 - (-1.3)$ -16.7

Lesson 2-6 Find each product.

1. $5(12)$ 60

2. $(-6)(11)$ -66

3. $(-7)(-5)$ 35

4. $\left(-\frac{7}{8}\right)\left(-\frac{1}{3}\right)$ $\frac{7}{24}$

5. $(-5)\left(-\frac{2}{5}\right)$ 2

6. $(-6)(4)(-3)$ 72

7. $(4)(-2)(-1)(-3)$ -24

8. $(-6.8)(-5.415)(3.1)$ 114.1482

9. $(-5.34)(3.2)$ -17.088

10. $\left(\frac{3}{5}\right)\left(-\frac{5}{7}\right)$ $-\frac{3}{7}$

11. $-\frac{7}{15}\left(\frac{9}{14}\right)$ $-\frac{3}{10}$

12. $(4.2)(-5.1)(3.6)$ -77.112

13. $-6\left(\frac{5}{3}\right)\left(\frac{9}{10}\right)$ -9

14. $(3)(-6)(0)(-1)$ 0

15. $(-21)(-2)(-1)$ -42

Lesson 2-7 Simplify.

1. $\frac{-48}{8}$ -6

2. $-49 \div (-7)$ 7

3. $-64 \div 8$ -8

4. $-\frac{3}{4} \div 9$ $-\frac{1}{12}$

5. $-9 \div \left(-\frac{10}{17}\right)$ $\frac{153}{10} = 15\frac{3}{10}$

6. $\frac{-450n}{10}$ $-45n$

7. $\frac{-36a}{-6}$ $6a$

8. $\frac{63a}{-9}$ $-7a$

9. $8 \div \left(-\frac{5}{4}\right)$ $-\frac{32}{5} = -6\frac{2}{5}$

10. $\frac{\frac{7}{8}}{-10}$ $-\frac{7}{80}$

11. $\frac{12}{\frac{-8}{5}}$ $-\frac{15}{2} = -7\frac{1}{2}$

12. $\frac{6a + 24}{6}$ $a + 4$

13. $\frac{20a + 30b}{-2}$ $-10a - 15b$

14. $\frac{\frac{11}{5}}{-6}$ $-\frac{11}{30}$

15. $\frac{70a - 42b}{-14}$ $-5a + 3b$

16. $\frac{-32x + 12y}{-4}$ $8x - 3y$

17. $-\frac{7}{12} \div \frac{1}{18}$ $-\frac{21}{2} = -10\frac{1}{2}$

18. $\frac{\frac{5}{15}}{-7}$ $-\frac{7}{3} = -2\frac{1}{3}$

Lesson 2-8 Find each square root. Use a calculator if necessary. Round to the nearest hundredth if the result is not a whole number.

1. $-\sqrt{81}$ -9

2. $\sqrt{0.0016}$ 0.04

3. $\pm\sqrt{206}$ ± 14.35

4. $\pm\sqrt{\frac{81}{64}}$ $\pm\frac{9}{8}$

5. $\sqrt{85}$ 9.22

6. $-\sqrt{\frac{36}{196}}$ $-\frac{6}{14}$ or $-\frac{3}{7}$

7. $-\sqrt{149}$ -12.21

8. $\pm\sqrt{961}$ ± 31

9. $\sqrt{10.24}$ 3.2

Evaluate each expression. Use a calculator if necessary. Round to the nearest hundredth if the result is not a whole number.

10. $\sqrt{m}$, if $m = 529$ 23

11. $-\sqrt{c - d}$, if $c = 1.097$ and $d = 1.0171$ -0.28

12. $-\sqrt{ab}$, if $a = 1.2$ and $b = 2.7$ -1.8

13. $\pm\sqrt{\frac{x}{y}}$, if $x = 144$ and $y = 1521$ ± 0.31

Lesson 2-9 Translate each sentence into an equation, inequality, or formula.

1. The square of a decreased by the cube of b is equal to c. $a^2 - b^3 = c$

2. Twenty-nine decreased by the product of x and y is less than z. $29 - xy < z$

3. The perimeter P of a parallelogram is twice the sum of the lengths of two adjacent sides a and b. $P = 2(a + b)$

4. Four-fifths of the product of m, n, and the square of p is greater than 26. $\frac{4}{5}mnp^2 > 26$

5. Thirty increased by the quotient of s and t is equal to v. $30 + \frac{s}{t} = v$

6. The area A of a trapezoid is half the product of the height h and the sum of the two parallel bases a and b. $A = \frac{1}{2}h(a + b)$

Extra Practice   **761**

Lesson 3-1 Solve each equation. Then check your solution.

1. $-2 + g = 7$ 9
2. $9 + s = -5$ -14
3. $-4 + y = -9$ -5

4. $m + 6 = 2$ -4
5. $t + (-4) = 10$ 14
6. $v - 7 = -4$ 3

7. $a - (-6) = -5$ -11
8. $-2 - x = -8$ 6
9. $d + (-44) = -61$ -17

10. $e - (-26) = 41$ 15
11. $p - 47 = 22$ 69
12. $-63 - f = -82$ 19

13. $c + 5.4 = -11.33$ -16.73
14. $-6.11 + b = 14.321$ 20.431
15. $-5 = y - 22.7$ 17.7

16. $-5 - q = 1.19$ -6.19
17. $n + (-4.361) = 59.78$ 64.141
18. $t - (-46.1) = -3.673$ -49.77

19. $\frac{7}{10} - a = \frac{1}{2}$ $\frac{1}{5}$
20. $f - \left(-\frac{1}{8}\right) = \frac{3}{10}$ $\frac{7}{40}$
21. $-4\frac{5}{12} = t - \left(-10\frac{1}{36}\right)$ $-14\frac{4}{9}$

22. $x + \frac{3}{8} = \frac{1}{4}$ $-\frac{1}{8}$
23. $1\frac{7}{16} + s = \frac{9}{8}$ $-\frac{5}{16}$
24. $17\frac{8}{9} = d + \left(-2\frac{5}{6}\right)$ $20\frac{13}{18}$

Lesson 3-2 Solve each equation. Then check your solution.

1. $-5p = 35$ -7
2. $-3x = -24$ 8
3. $62y = -2356$ -38

4. $\frac{a}{-6} = -2$ 12
5. $\frac{c}{-59} = -7$ 413
6. $\frac{f}{14} = -63$ -882

7. $84 = \frac{x}{97}$ 8148
8. $\frac{w}{5} = 3$ 15
9. $\frac{q}{9} = -3$ -27

10. $\frac{2}{5}x = \frac{4}{7}$ $\frac{10}{7}$ or $1\frac{3}{7}$
11. $\frac{z}{6} = -\frac{5}{12}$ $-\frac{5}{2}$
12. $-\frac{5}{9}r = 7\frac{1}{2}$ $-13\frac{1}{2}$

13. $2\frac{1}{6}j = 5\frac{1}{5}$ $2\frac{2}{5}$
14. $3 = 1\frac{7}{11}q$ $1\frac{5}{6}$
15. $-1\frac{3}{4}p = -\frac{5}{8}$ $\frac{5}{14}$

16. $57k = 0.1824$ 0.0032
17. $0.0022b = 0.1958$ 89
18. $5j = -32.15$ -6.43

19. $\frac{w}{-2} = -2.48$ 4.96
20. $\frac{z}{2.8} = -6.2$ -17.36
21. $\frac{x}{-0.063} = 0.015$ -0.000945

22. $15\frac{3}{8} = -5.125p$ -3
23. $-7.25 = -3\frac{5}{8}g$ 2
24. $-18\frac{1}{4} = 2.50x$ -7.3

Lesson 3-3 Solve each equation. Then check your solution.

1. $2x - 5 = 3$ 4
2. $4t + 5 = 37$ 8
3. $7a + 6 = -36$ -6

4. $47 = -8g + 7$ -5
5. $-3c - 9 = -24$ 5
6. $5k - 7 = -52$ -9

7. $5s + 4s = -72$ -8
8. $3x - 7 = 2$ 3
9. $8 + 3x = 5$ -1

10. $-3y + 7.569 = 24.069$ -5.5
11. $7 - 9.1f = 137.585$ -14.35
12. $6.5 = 2.4m - 4.9$ 4.75

13. $\frac{e}{5} + 6 = -2$ -40
14. $\frac{d}{4} - 8 = -5$ 12
15. $-\frac{4}{13}y - 7 = 6$ $-42\frac{1}{4}$

16. $\frac{p + 10}{3} = 4$ 2
17. $\frac{h - 7}{6} = 1$ 13
18. $\frac{5f + 1}{8} = -3$ -5

19. $\frac{4n - 8}{-2} = 12$ -4
20. $\frac{2a}{7} + 9 = 3$ -21
21. $\frac{-3t - 4}{2} = 8$ $-6\frac{2}{3}$

Lesson 3-4 Find the complement of each angle measure.

1. $15°$ $75°$ 　　　　2. $79°$ $11°$ 　　　　3. $88°$ $2°$

4. $a°$ $(90 - a)°$ 　　5. $(3c)°$ $(90 - 3c)°$ 　　6. $(b - 15)°$ $(105 - b)°$

Find the supplement of each angle measure.

7. $156°$ $24°$ 　　　　8. $94°$ $86°$ 　　　　9. $21°$ $159°$

10. $a°$ $(180 - a)°$ 　11. $(3c)°$ $(180 - 3c)°$ 　12. $(b - 15)°$ $(195 - b)°$

Find the measure of the third angle of each triangle in which the measures of two angles of the triangle are given.

13. $90°, 2°$ $88°$ 　　　14. $34°, 132°$ $14°$ 　　15. $111°, 28°$ $41°$

16. $a°, b°$ $(180 - a - b)°$ 　17. $a°, (a - 15)°$ $(195 - 2a)°$ 　18. $b°, (3b - 2)°$ $(182 - 4b)°$

Lesson 3-5 Solve each equation. Then check your solution.

1. $6(y - 5) = 18$ 8 　　　　2. $-21 = 7(p - 10)$ 7 　　　3. $3(h + 2) = 12$ 2

4. $-3(x + 2) = -18$ 4 　　　5. $11.2n + 6 = 5.2n$ -1 　　6. $2m + 5 - 6m = 25$ -5

7. $3z - 1 = 23 - 3z$ 4 　　　8. $5a - 5 = 7a - 19$ 7 　　　9. $5b + 12 = 3b - 6$ -9

10. $3x - 5 = 7x + 7$ -3 　　11. $1.9s + 6 = 3.1 - s$ -1 　　12. $2.85y - 7 = 12.85y - 2$ $-\frac{1}{2}$

13. $2.9m + 1.7 = 3.5 + 2.3m$ 3 　14. $3(x + 1) - 5 = 3x - 2$ identity 　15. $4(2y - 1) = -10(y - 5)$ 3

16. $\frac{6v - 9}{3} = v$ 3 　　　　17. $\frac{3t + 1}{4} = \frac{3}{4}t - 5$ no solution 　18. $\frac{2}{5}y + \frac{y}{2} = 9$ 10

19. $3y - \frac{4}{5} = \frac{1}{3}y$ $\frac{3}{10}$ 　　　20. $\frac{3}{4}x - 4 = 7 + \frac{1}{2}x$ 44 　　21. $\frac{x}{2} - \frac{1}{3} = \frac{x}{3} - \frac{1}{2}$ -1

Lesson 3-6 Solve each equation for x.

1. $x + r = q$ $q - r$ 　　　2. $ax + 4 = 7$ $\frac{3}{a}$ 　　　3. $2bx - b = -5$ $\frac{-5 + b}{2b}$

4. $\frac{x - c}{c + a} = a$ $a^2 + ca + c$ 　5. $\frac{x + y}{c} = d$ $cd - y$ 　　6. $\frac{ax + 1}{2} = b$ $\frac{2b - 1}{a}$

7. $\frac{x + t}{4} = d$ $4d - t$ 　　8. $6x - 7 = -r$ $\frac{7 - r}{6}$ 　　9. $kx + 4y = 5z$ $\frac{5z - 4y}{k}$

10. $ax - 6 = t$ $\frac{t + 6}{a}$ 　　11. $\frac{2}{3}x + a = b$ $\frac{3}{2}(b - a)$ 　12. $q(x + 1) = 5$ $\frac{5}{q} - 1$

13. $\frac{x - y}{z} = 8$ $8z + y$ 　　14. $\frac{7a + b}{x} = 1$ $7a + b$ 　　15. $\frac{4cx + t}{7} = 2$ $\frac{14 - t}{4c}$

16. $\frac{9x - 4c}{z} = z$ $\frac{z^2 + 4c}{9}$ 　17. $cx + a = bx$ $\frac{-a}{c - b}$ 　　18. $\frac{12q - x}{5} = t$ $12q - 5t$

Lesson 3-7

1. Geography The areas in square miles of the 20 largest natural U.S. lakes are given below. Find the mean, median, and mode of the areas. **5166.6; 541.5; none**

31,700	1697	242	700	22,300	451	374	432	23,000	207
1000	1361	315	625	9910	215	458	360	7550	435

2. Work Each number below represents the number of days that each employee of the Cole Corporation was absent during 1996. Find the mean, median, and mode of the number of days absent. **4.45; 4; 3, 5**

0	10	8	5	8	9	3	3	2	9	7	0	4	2	4	6	2
9	13	3	1	5	5	7	2	6	5	3	4	7	1	1	5	3
3	1	4	5	1	2											

3. Football Tailback Michael Anderson, of the West High Bears, averaged 137.6 yards rushing per game for the first five games of the season. He rushes for 155 yards in the sixth game. What is his new rushing average? **140.5**

4. Olympic Games One of the events in the Winter Olympics is the men's 500-meter speed skating. The winning times for this event are shown at the right. Find the mean, median, and mode of the times. **40.164; 40.15; 40.2, 43.4**

Year	Time(s)	Year	Time(s)
1932	43.4	1968	40.3
1936	43.4	1972	39.4
1948	43.1	1976	39.2
1952	43.2	1980	38.0
1956	40.2	1984	38.2
1960	40.2	1988	36.5
1964	40.1	1992	37.1

Lesson 4-1 Solve each proportion.

1. $\frac{4}{5} = \frac{x}{20}$ **16**

2. $\frac{b}{63} = \frac{3}{7}$ **27**

3. $\frac{y}{5} = \frac{3}{4}$ $\frac{15}{4} = 3.75$

4. $\frac{7}{4} = \frac{3}{a}$ $\frac{12}{7} \approx 1.71$

5. $\frac{t-5}{4} = \frac{3}{2}$ **11**

6. $\frac{x}{9} = \frac{0.24}{3}$ **0.72**

7. $\frac{n}{3} = \frac{n+4}{7}$ **3**

8. $\frac{12q}{-7} = \frac{30}{14}$ $-\frac{5}{4} = -1.25$

9. $\frac{1}{y-3} = \frac{3}{y-5}$ **2**

10. $\frac{r-1}{r+1} = \frac{3}{5}$ **4**

11. $\frac{a-3}{8} = \frac{3}{4}$ **9**

12. $\frac{6p-2}{7} = \frac{5p+7}{8}$ **5**

13. $\frac{2}{9} = \frac{k+3}{2}$ $-\frac{23}{9} \approx -2.56$

14. $\frac{5m-3}{4} = \frac{5m+3}{6}$ **3**

15. $\frac{w-5}{4} = \frac{w+3}{3}$ **−27**

16. $\frac{96.8}{t} = \frac{12.1}{7}$ **56**

17. $\frac{x}{6.03} = \frac{4}{17.42}$ **1.385**

18. $\frac{4n+5}{5} = \frac{2n+7}{7}$ **0**

Lesson 4-2 Determine whether each pair of triangles is similar. Justify your answer.

1. 88° 60° 60° **no**

2. 50° 52° 78° 78° **yes**

3. 40° 40° 78° 65° **no**

4. 86° 46° 46° **no**

5. 60° 60° **yes**

6. 52° 55° 52° 52° **no**

Lesson 4-3 For each triangle, find sin N, cos N, and tan N to the nearest thousandth.   1–3. See margin.

1.

2.

3.

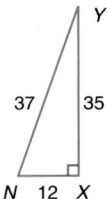

Use a calculator to find the value of each trigonometric ratio to the nearest ten thousandth.

4. cos 25° 0.9063   5. tan 31° 0.6009   6. sin 71° 0.9455

7. cos 64° 0.4384   8. tan 9° 0.1584   9. sin 2° 0.0349

Use a calculator to find the measure of each angle to the nearest degree.

10. tan B = 0.5427 28°   11. cos A = 0.8480 32°   12. sin J = 0.9654 75°

13. cos Q = 0.3645 69°   14. sin R = 0.2104 12°   15. tan V = 0.956 44°

Lesson 4-4 Write each ratio as a percent and then as a decimal.

1. $\frac{1}{4}$ 25%, 0.25   2. $\frac{34}{100}$ 34%, 0.34   3. $\frac{4}{25}$ 16%, 0.16

4. $\frac{3}{20}$ 15%, 0.15   5. $\frac{7}{8}$ $87\frac{1}{2}$%, 0.875   6. $\frac{9}{10}$ 90%, 0.9

7. $\frac{24}{40}$ 60%, 0.6   8. $\frac{4}{50}$ 8%, 0.08   9. $\frac{7}{15}$ $46\frac{2}{3}$%, $0.4\overline{6}$

10. $\frac{4}{9}$ $44\frac{4}{9}$%, $0.\overline{4}$ 11. $\frac{36}{15}$ 240%, 2.4   12. $\frac{18}{4}$ 450%, 4.5

Use a proportion to answer each question.

13. Twenty-four is what percent of 48? 50%   14. What percent of 70 is 14? 20%

15. Nine is what percent of 72? $12\frac{1}{2}$%; 12.5%   16. Fourteen is 17.5% of what number? 80

17. What percent of 16 is 5.12? 32%   18. What number is 25% of 64? 16

19. What percent of 80 is 2? 2.5%   20. Forty-five is what percent of 112.5? 40%

Lesson 4-5 State whether each percent of change is a percent of increase or a percent of decrease. Then find the percent of increase or decrease. Round to the nearest whole percent.

1. original: $100
   new: $67 D; 33%
2. original: 62 acres
   new: 98 acres I, 58%
3. original: 322 people
   new: 289 people D, 10%

4. original: 78 pennies
   new: 36 pennies D, 54%
5. original: $212
   new: $230 I, 8%
6. original: 35 mph
   new: 65 mph I, 86%

Find the final price of each item. When there is a discount and sales tax, first compute the discount price and then compute the sales tax and final price.

7. television: $299
   sales tax: 4.5%
   $312.46
8. boots: $49.99
   discount: 15%
   sales tax: 3.5% $43.98
9. backpack: $28.95
   discount: 10%
   sales tax: 5% $27.35

10. software: $36.99
    sales tax: 6.25%
    $39.30
11. jacket: $65
    discount: 30%
    sales tax: 4% $47.32
12. book: $15.95
    sales tax: 7%
    $17.07

Additional Answers for Lesson 4-3

1. sin N = 0.923, cos N = 0.385, tan N = 2.400
2. sin N = 0.280, cos N = 0.960, tan N = 0.292
3. sin N = 0.946, cos N = 0.324, tan N = 2.917

Lesson 4-6 Determine the probability of each event. 2. 1

1. a coin will land tails up $\frac{1}{2}$
2. There is a December 1 this year.
3. A baby will be a girl. $\frac{1}{2}$
4. Next year will have 400 days. 0
5. This is an algebra book. 1
6. Today is Wednesday. $\frac{1}{7}$

Find the probability of each outcome if a computer randomly chooses a letter in the word "success."

7. the letter e $\frac{1}{7}$
8. P(not c) $\frac{5}{7}$
9. the letter s $\frac{3}{7}$
10. the letter b 0
11. P(vowel) $\frac{2}{7}$
12. the letters u or c $\frac{3}{7}$

Find the odds of each outcome if a die is rolled.

13. a 4 1:5
14. a number greater than 3 1:1
15. a multiple of 3 1:2
16. a number less than 5 2:1
17. an odd number 1:1
18. not a 6 5:1

Lesson 4-7

1. **Advertising** An advertisement for an orange drink claims that the drink contains 10% orange juice. How much pure orange juice would have to be added to 5 quarts of the drink to obtain a mixture containing 40% orange juice? 2.5 quarts

2. **Finance** Jane Pham is investing $6000 in two accounts, part at 4.5% and the remainder at 6%. If the total annual interest earned from the two accounts is $279, how much did Jane deposit at each rate? $5400 at 4.5%, $600 at 6%

3. **Entertainment** At the Golden Oldies Theater, tickets for adults cost $5.50 and tickets for children cost $3.50. How many of each kind of ticket was purchased if 21 tickets were bought for $83.50? 5 adults, 16 children

4. **Automotives** A car radiator has a capacity of 14 quarts and is filled with a 20% antifreeze solution. How much must be drained off and replaced with pure antifreeze to obtain a 40% antifreeze solution? 3.5 quarts

5. **Nutrition** A liter of cream has 9.2% butterfat. How much skim milk containing 2% butterfat should be added to the cream to obtain a mixture with 6.4% butterfat? $0.6\overline{3}$ L

Lesson 4-8 Determine which equations represent inverse variations and which represent direct variations. Then find the constant of variation.

1. $ab = 6$ I, 6
2. $\frac{50}{y} = x$ I, 50
3. $\frac{1}{5}a = d$ D, $\frac{1}{5}$
4. $s = 3t$ D, 3
5. $14 = cd$ I, 14
6. $2x = y$ D, 2

Solve. Assume that y varies directly as x. 13. -12 16. -6

7. If $y = 45$ when $x = 9$, find y when $x = 7$. 35
8. If $y = 18$ when $x = 27$, find x when $y = 8$. 12
9. If $y = 450$ when $x = 6$, find y when $x = 10$. 750
10. If $y = 6$ when $x = 48$, find y when $x = 20$. 2.5
11. If $y = 25$ when $x = 20$, find x when $y = 35$. 28
12. If $y = 100$ when $x = 40$, find y when $x = 16$. 40
13. If $y = -7$ when $x = -1$, find x when $y = -84$.
14. If $y = 5$ when $x = -10$, find y when $x = 50$. -25
15. If $y = 24$ when $x = 6$, find y when $x = 14$. 56
16. If $y = -10$ when $x = -4$, find x when $y = -15$.

Solve. Assume that y varies inversely as x. 25. 11.7 26. 2.5

17. If $y = 54$ when $x = 4$, find x when $y = 27$. 8
18. If $y = 18$ when $x = 6$, find x when $y = 12$. 9
19. If $y = 2$ when $x = 26$, find x when $y = 4$. 13
20. If $y = 3$ when $x = 8$, find x when $y = 4$. 6
21. If $y = 12$ when $x = 24$, find x when $y = 9$. 32
22. If $y = 8$ when $x = -8$, find y when $x = -16$. 4
23. If $y = 3$ when $x = -8$, find y when $x = 4$. -6
24. If $y = 27$ when $x = \frac{1}{3}$, find y when $x = \frac{3}{4}$. 12
25. If $y = 19.5$ when $x = 6.3$, find x when $y = 10.5$.
26. If $y = 4.8$ when $x = 10$, find y when $x = 19.2$.

Lesson 5-1
Refer to the coordinate plane below. Write the ordered pair for each point. Name the quadrant in which the point is located.

1. B (1, 2), I
2. T (−5, 0), none
3. P (6, −2), IV
4. Q (0, 6), none
5. A (−2, −2), III
6. K (4, 5), I
7. J (2, −5), IV
8. L (4, 0), none
9. S (−3, 5), II
10. D (7, 3), I
11. M (−7, −5), III
12. N (−6, 4), II

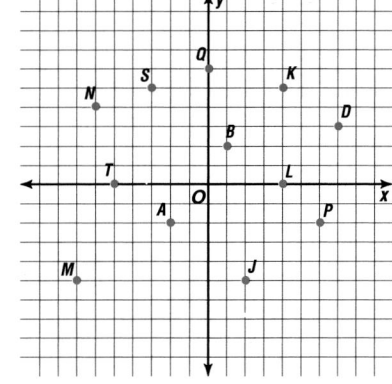

Graph each point.   13–20. See margin.

13. $A(2, -4)$
14. $B(3, 5)$
15. $C(-4, 0)$
16. $D(-4, 3)$
17. $E(-5, -5)$
18. $F(-1, 1)$
19. $G(0, -3)$
20. $H(2, 3)$

Lesson 5-2
State the domain and range of each relation.

1. {(5, 2), (0, 0), (−9, −1)}
   {5, 0, −9}; {2, 0, −1}
2. {(−4, 2), (−2, 0), (0, 2), (2, 4)}
   {−4, −2, 0, 2}; {2, 0, 4}
3. {(7, 5), (−2, −3), (4, 0), (5, −7), (−9, 2)}
   {7, −2, 4, 5, −9}; {5, −3, 0, −7, 2}
4. {(3.1, −1), (−4.7, 3.9), (2.4, −3.6), (−9, 12.12)}
   {3.1, −4.7, 2.4, −9}; {−1, 3.9, −3.6, 12.12}

Express the relation shown in each table, mapping, or graph as a set of ordered pairs. Then state the domain, range, and inverse of the relation.   5–10. See margin.

5.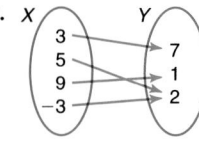

x	y
1	3
2	4
3	5
4	6
5	7

6.
x	y
−4	1
−2	3
0	1
2	3
4	1

7.

8.

9.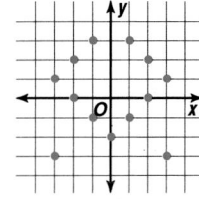

10.

Lesson 5-3
Which ordered pairs are solutions of each equation?

1. $3r = 8s - 4$ a, b, c
   a. $\left(\frac{2}{3}, \frac{3}{4}\right)$ b. $\left(0, \frac{1}{2}\right)$ c. (4, 2)   d. (2, 4)
2. $3y = x + 7$ c, d
   a. (2, 4)   b. (2, −1)   c. (2, 3)   d. (−1, 2)
3. $4x = 8 - 2y$ a
   a. (2, 0)   b. (0, 2)   c. (0.5, −3)   d. (1, −2)
4. $3n = 10 - 4m$ b, c
   a. (0, 3)   b. (−2, 6)   c. (1, 2)   d. (2, 1)

Solve each equation if the range is {−3, −1, 0, 2, 3}.   5–10. See margin.

5. $y = 2x$
6. $y = 5x + 1$
7. $2a + b = 4$
8. $4r + 3s = 13$
9. $5b = 8 - 4a$
10. $6m - n = -3$

767

Additional Answers for Lesson 5-1

13–20.

Additional Answers for Lesson 5-2

5. {(1, 3), (2, 4), (3, 5), (4, 6), (5, 7)}; {1, 2, 3, 4, 5}; {3, 4, 5, 6, 7}; {(3, 1), (4, 2), (5, 3), (6, 4), (7, 5)}
6. {(−4, 1), (−2, 3), (0, 1), (2, 3), (4, 1)}; {−4, −2, 0, 2, 4}; {1, 3}; {(1, −4), (3, −2), (1, 0), (3, 2), (1, 4)}
7. {(−1, 5), (−2, 5), (−2, 4), (−2, 1), (−6, 1)}; {−1, −2, −6}; {5, 4, 1}; {(5, −1), (5, −2), (4, −2), (1, −2), (1, −6)}
8. {(3, 7), (5, 2), (9, 1), (−3, 2)}; {3, 5, 9, −3}; {7, 1, 2}; {(7, 3), (2, 5), (1, 9), (2, −3)}
9. {(0, 0), (1, 1), (2, 3), (2, −1), (−2, 2), (−2, −1), (−4, −3)}; {0, 1, 2, −2, −4}; {0, 1, 3, −1, 2, −3}; {(0, 0), (1, 1), (3, 2), (−1, 2), (2, −2), (−1, −2), (−3, −4)}
10. {(−3, 1), (−3, −3), (−2, 0), (−2, 2), (−1, 3), (−1, −1), (0, −2), (1, 3), (1, −1), (2, 0), (2, 2), (3, 1), (3, −3)}; {−3, −2, −1, 0, 1, 2, 3}; {1, −3, 0, 2, 3, −1, −2}; {(1, −3), (−3, −3), (0, −2), (2, −2), (3, −1), (−1, −1), (−2, 0), (3, 1), (−1, 1), (0, 2), (2, 2), (1, 3), (−3, 3)}

Additional Answers for Lesson 5-3

5. $\left\{\left(-\frac{3}{2}, -3\right), \left(-\frac{1}{2}, -1\right), (0, 0), (1, 2), \left(\frac{3}{2}, 3\right)\right\}$
6. $\left\{\left(-\frac{4}{5}, -3\right), \left(-\frac{2}{5}, -1\right), \left(-\frac{1}{5}, 0\right), \left(\frac{1}{5}, 2\right), \left(\frac{2}{5}, 3\right)\right\}$
7. $\left\{\left(\frac{7}{2}, -3\right), \left(\frac{5}{2}, -1\right), (2, 0), (1, 2), \left(\frac{1}{2}, 3\right)\right\}$
8. $\left\{\left(\frac{11}{2}, -3\right), (4, -1), \left(\frac{13}{4}, 0\right), \left(\frac{7}{4}, 2\right), (1, 3)\right\}$
9. $\left\{\left(\frac{23}{4}, -3\right), \left(\frac{13}{4}, -1\right), (2, 0), \left(-\frac{1}{2}, 2\right), \left(-\frac{7}{4}, 3\right)\right\}$
10. $\left\{(-1, -3), \left(-\frac{2}{3}, -1\right), \left(-\frac{1}{2}, 0\right), \left(-\frac{1}{6}, 2\right), (0, 3)\right\}$

Additional Answers for Lesson 5-4

7.

$3x + y = 4$

8.

$y = 3x + 1$

9.

$3x - 2y = 12$

10.

$2x - y = 6$

11.

$3x - 2y = 8$

12.

$y = \frac{3}{4}$

13.

$y = 5x - 7$

768

EXTRA PRACTICE

Lesson 5-4 Determine whether each equation is a linear equation. If an equation is linear, rewrite it in the form $Ax + By = C$. 4. yes; $3x - 7y = -7$

1. $3x = 2y$ yes; $3x - 2y = 0$ **2.** $2x - 3 = y^2$ no   **3.** $3x - 2y = 8$ yes

4. $5x - 7y = 2x - 7$ **5.** $2x + 5x = 7y$ yes; $7x - 7y = 0$ **6.** $\frac{1}{x} + \frac{5}{y} = -4$ no

Graph each equation.   7–18. See margin.

7. $3x + y = 4$ **8.** $y = 3x + 1$ **9.** $3x - 2y = 12$

10. $2x - y = 6$ **11.** $3x - 2y = 8$ **12.** $y = \frac{3}{4}$

13. $y = 5x - 7$ **14.** $x + \frac{1}{3}y = 6$ **15.** $x = -\frac{5}{2}$

16. $5x - 2y = 8$ **17.** $4x + 2y = 9$ **18.** $4x + 3y = 12$

Lesson 5-5 Determine whether each relation is a function.

1.

x	y
1	3
2	5
1	-7
2	9
3	3

no

2.   no

3.   yes

4.   no

5.

a	b
1	-2
3	-4
5	-6
9	-4
10	1

yes

6. 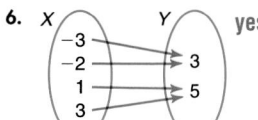 yes

7. $\{(-2, 4), (1, 3), (5, 2), (1, 4)\}$ no

8. $\{(5, 4), (-6, 5), (4, 5), (0, 4)\}$ yes

9. $\{(3, 1), (5, 1), (7, 1)\}$ yes

10. $\{(3, -2), (4, 7), (-2, 7), (4, 5)\}$ no

11. $y = 2$ yes

12. $x^2 + y = 11$ yes

Lesson 5-6 Write an equation for each relation.   9. $y = \frac{2}{3}x - 4$ 10. $y = -2x + 3$

1. $f(x) = \frac{1}{2}x - 5$

x	2	4	6	8	10
f(x)	-4	-3	-2	-1	0

2. $f(x) = x^2 - 4$

x	-2	-1	0	1	2
f(x)	0	-3	-4	-3	0

3. $f(x) = 4x + 3$

x	1	2	3	4	5
f(x)	7	11	15	19	23

4. $y = \frac{1}{3}x + 2$

5. $y = -\frac{3}{2}x - 2$

6. $y = 2x + 6$

7. $\{(3, 12), (4, 14), (5, 16), (6, 18), (7, 20)\}$ $y = 2x + 6$ **8.** $\{(1, 4), (2, 8), (3, 12), (-2, -8), (-3, -12)\}$ $y = 4x$

9. $\{(-6, -8), (-3, -6), (0, -4), (3, -2), (6, 0)\}$ **10.** $\{(-3, 9), (-2, 7), (1, 1), (4, -5), (6, -9)\}$

768 *Extra Practice*

14. $x + \frac{1}{3}y = 6$

15. $y = -\frac{5}{2}$

16. $5x - 2y = 8$

Lesson 5-7 Find the range, median, upper quartile, lower quartile, and interquartile range for each set of data. **4.** 8; 4; 7.1; 2.4; 4.7

1. 56, 45, 37, 43, 10, 34 46; 40; 45; 34; 11

2. 77, 78, 68, 96, 99, 84, 65 34; 78; 96; 68; 28

3. 30, 90, 40, 70, 50, 100, 80, 60 70; 65; 85; 45; 40

4. 4, 5.2, 1, 3, 2.4, 6, 3.7, 8, 1.3, 7.1, 9

5. 25°, 56°, 13°, 44°, 0°, 31°, 73°, 66°, 4°, 29°, 37°
73°; 31°; 56°; 13°; 43°

6. 234, 648, 369, 112, 527, 775, 406, 268, 400
663; 400; 587.5; 251; 336.5

7.

Stem	Leaf
0	0 2 3
1	1 7 9
2	2 3 5 6
3	3 4 4 5 9
4	0 7 8 8

$2|2 = 22$
48; 26; 39; 17; 22

8.

Stem	Leaf
7	3 4 7 8
8	0 0 3 5 7
9	4 6 8
10	0 1 8
11	1 9

$9|4 = 9.4$
4.6; 8.7; 10.05; 7.9; 2.15

9.

Stem	Leaf
25	0 3 7 9
26	1 3 4 5 5 6
27	1 5 6 6 9
28	1 2 3 5 8
29	2 5 6 9

$27|5 = 2750$
490; 2755; 2840; 2635; 205

Lesson 6-1 Determine the slope of each line.

1. a 3

2. b -1

3. c $\frac{3}{2}$

4. d 0

5. e undefined

6. f $-\frac{1}{5}$

7. g $\frac{1}{3}$

8. h -2

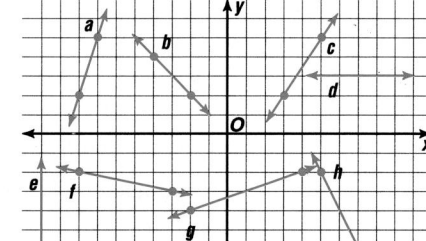

Determine the slope of the line that passes through each pair of points.

9. $(-2, 2), (3, -3)$ -1

10. $(-2, -8), (1, 4)$ 4

11. $(3, 4), (4, 6)$ 2

12. $(-5, 4), (-1, 11)$ $\frac{7}{4}$

13. $(18, -4), (6, -10)$ $\frac{1}{2}$

14. $(-4, -6), (-4, -8)$ undefined

Determine the value of r so the line that passes through each pair of points has the given slope.

15. $(-1, r), (1, -4), m = -5$ 6

16. $(-2, 1), (r, 4), m = \frac{3}{5}$ 3

17. $(-1, 3), (-3, r), m = -3$ 9

18. $(3, r), (7, -2), m = \frac{1}{2}$ -4

19. $(r, -2), (-7, -1), m = -\frac{1}{4}$ -3

20. $(-3, 2), (7, r), m = \frac{2}{3}$ $\frac{26}{3}$

Lesson 6-2 Write the point-slope form of an equation of the line that passes through the given point and has the given slope. **2.** $y - 4 = -5(x - 5)$ **3.** $y + 4 = \frac{3}{4}(x + 2)$

1. $(5, -2), m = 3$ $y + 2 = 3(x - 5)$

2. $(5, 4), m = -5$

3. $(-2, -4), m = \frac{3}{4}$

4. $(-3, 1), m = 0$ $y - 1 = 0$

5. $(-1, 0), m = \frac{2}{3}$ $y = \frac{2}{3}(x + 1)$

6. $(0, 6), m = -2$ $y - 6 = -2x$

Write the standard form of an equation of the line that passes through the given point and has the given slope. **7.** $x + 2y = -12$

7. $(-6, -3), m = -\frac{1}{2}$

8. $(4, -3), m = 2$ $2x - y = 11$

9. $(5, 4), m = -\frac{2}{3}$ $2x + 3y = 22$

10. $(1, 3), m = $ undefined $x = 1$

11. $(-2, 6), m = 0$ $y = 6$

12. $(6, -2), m = \frac{4}{3}$ $4x - 3y = 30$

Additional Answers for Lesson 5-4

17.

$4x + 2y = 9$

18.

$4x + 3y = 12$

Additional Answers for Lesson 6-4

7. $y = -\frac{2}{5}x + 2$; $2x + 5y = 10$

8. $y = 5x - 15$; $5x - y = 15$

9. $y = -\frac{7}{4}x + 2$; $7x + 4y = 8$

10. $y = -\frac{4}{3}x + \frac{5}{3}$; $4x + 3y = 5$

11. $y = -6x + 15$; $6x + y = 15$

12. $y = 12x - 24$;
$\quad 12x - y = 24$

Additional Answers for Lesson 6-5

1.
$4x + y = 8$

2.
$2x - y = 8$

3.
$3x - 2y = 6$

4.
$6x - 3y = 6$

5.
$x + \frac{1}{2}y = 4$

Lesson 6-3 Explain whether a scatter plot for each pair of variables would probably show a *positive*, *negative*, or *no* correlation between the variables.

1. playing time of a basketball game and points scored   positive

2. age of a car and its value   negative

3. heights of mothers and sons   positive

4. temperature of the water in an outdoor swimming pool and the temperature outside   positive

5. the weight of a person's car and the amount they pay to use a turnpike   no

6. the number of guests at a birthday party and the amount of food remaining after the party   negative

7. The scatter plot at the right compares the number of hours per week people watched television to the number of hours per week they spent doing some physical activity.

   a. As people watched more television, what happened to the number of hours they spend doing some physical activity?   It decreases.

   b. Is there a correlation between the variables? Is it positive or negative?   yes, negative

Hours Watching Television

Lesson 6-4 Find the *x*- and *y*-intercepts of the graph of each equation.

1. $3x + 2y = 6$ 2, 3

2. $5x + y = 10$ 2, 10

3. $2x + 5y = -11$ $-\frac{11}{2}, -\frac{11}{5}$

4. $3y = 12$ none, 4

5. $y - 6x = 5$ $-\frac{5}{6}, 5$

6. $x = -2$ −2, none

Write an equation in slope-intercept form of a line with the given slope and *y*-intercept. Then write the equation in standard form.   7–12. See margin.

7. $m = -\frac{2}{5}, b = 2$

8. $m = 5, b = -15$

9. $m = -\frac{7}{4}, b = 2$

10. $m = -\frac{4}{3}, b = \frac{5}{3}$

11. $m = -6, b = 15$

12. $m = 12, b = -24$

Find the slope and *y*-intercept of the graph of each equation.

13. $y - \frac{3}{5}x = -\frac{1}{4}$ $\frac{3}{5}, -\frac{1}{4}$

14. $y = 3x - 7$ 3, −7

15. $\frac{2}{3}x + \frac{1}{6}y = 2$ −4, 12

16. $2x + 3y = 5$ $-\frac{2}{3}, \frac{5}{3}$

17. $3y = 8x + 2$ $\frac{8}{3}, \frac{2}{3}$

18. $5y = -8x - 2$ $-\frac{8}{5}, -\frac{2}{5}$

Write an equation in standard form for a line that passes through each pair of points.

19. $(-1, 7), (8, -2)$ $x + y = 6$

20. $(6, 0), (0, 4)$ $2x + 3y = 12$

21. $(8, -1), (7, -1)$ $y = -1$

22. $(1, 0), (0, 1)$ $x + y = 1$

23. $(5, 7), (-1, 6)$ $x - 6y = -37$

24. $(-3, -5), (3, -15)$
$\quad 5x + 3y = -30$

Lesson 6-5 Graph each equation.   1–18. See Solutions Manual.

1. $4x + y = 8$

2. $2x - y = 8$

3. $3x - 2y = 6$

4. $6x - 3y = 6$

5. $x + \frac{1}{2}y = 4$

6. $4x + 5y = 20$

7. $y + 3 = -2(x + 4)$

8. $y - 1 = 3(x - 5)$

9. $y + 6 = -\frac{2}{3}(x + 1)$

10. $y - 5 = 4(x + 6)$

11. $y - 2 = (x + 7)$

12. $3(x - 1) = y + \frac{4}{5}$

13. $y = \frac{3}{4}x + 4$

14. $y = 4x - 1$

15. $-4x + y = 6$

16. $-2x + y = 3$

17. $5y - 6 = 3x$

18. $y = \frac{3}{2}x - 5$

6.
$4x + 5y = 20$

7.
$y + 3 = -2(x + 4)$

8.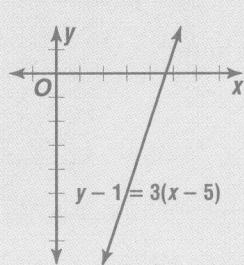
$y - 1 = 3(x - 5)$

Lesson 6-6 Determine whether the graphs of each pair of equations are *parallel, perpendicular,* or *neither.*

1. $2x + 3y = -12$
   $2x + 3y = 6$ parallel

2. $-4x + 3y = 12$
   $x + 3y = 12$ neither

3. $y = -3x + 9$
   $y + 3x = 14$ parallel

4. $y = 0.5x + 8$
   $2y = -8x - 3$ neither

5. $y = 7x + 2$
   $y = 2x + 7$ neither

6. $y + 5 = -9$
   $y + x = y - 6$ perpendicular

Write an equation in slope-intercept form of the line that passes through the given point and is parallel to the graph of each equation. 9. $y = \frac{2}{3}x + 2$ 11. $y = -\frac{3}{8}x + 4$

7. $(1, 6), y = 4x - 2$ $y = 4x + 2$

8. $(4, 6), y = 2x - 7$ $y = 2x - 2$

9. $(-3, 0), y = \frac{2}{3}x + 1$

10. $(2, 3), x - 5y = 7$ $y = \frac{1}{5}x + \frac{13}{5}$

11. $(0, 4), 3x + 8y = 4$

12. $(5, -2), y = -3x - 7$
    $y = -3x + 13$

Write an equation in slope-intercept form of the line that passes through the given point and is perpendicular to the graph of each equation. 13. $y = \frac{5}{3}x - 1$ 14. $y = \frac{1}{6}x + \frac{10}{3}$ 15. $y = -\frac{4}{3}x$

13. $(0, -1), y = -\frac{3}{5}x + 4$

14. $(-2, 3), 6x + y = 4$

15. $(0, 0), y = \frac{3}{4}x - 1$

16. $(4, 0), 4x - 3y = 2$
    $y = -\frac{3}{4}x + 3$

17. $(6, 7), 3x - 5y = 1$
    $y = -\frac{5}{3}x + 17$

18. $(5, -1), 8x + 4y = 15$
    $y = \frac{1}{2}x - \frac{7}{2}$

Lesson 6-7 Find the coordinates of the midpoint of a segment with each pair of endpoints.

1. $L(12, 2), M(8, 4)$ $(10, 3)$

2. $S(9, 5), T(17, 3)$ $(13, 4)$

3. $D(17, 9), E(11, -3)$ $(14, 3)$

4. $F(4, 2), G(8, -6)$ $(6, -2)$

5. $M(19, -3), N(11, 5)$ $(15, 1)$

6. $B(-6, 5), C(8, -11)$ $(1, -3)$

7. $T(-11, 6), U(13, 4)$ $(1, 5)$

8. $A(-6, 1), B(8, 9)$ $(1, 5)$

9. $J(6.4, -3), K(1.8, -3)$

10. $R(19, 5), S(7, 4)$ $\left(13, \frac{9}{2}\right)$

11. $G(8, 10), H(16, -6)$ $(12, 2)$

12. $C(7.6, 8.3), D(-5, 6.1)$
    $(1.3, 7.2)$

9. $(4.1, -3)$

Find the coordinates of the other endpoint of a segment given one endpoint and the midpoint M.

13. $P(9, 3), M(1, 2)$ $(-7, 1)$

14. $C(3, 5), M(5, -7)$ $(7, -19)$

15. $G(5, -9), M\left(8, -\frac{15}{2}\right)$ $(11, -6)$

16. $J(4, -7), M(-2, -3)$ $(-8, 1)$

17. $A(-3, 8), M(3, -5)$ $(9, -18)$

18. $F(5, 7), M(5, 6)$ $(5, 5)$

19. $T(-6, 12), M(4, 1)$ $(14, -10)$

20. $D(-8, 5), M\left(-\frac{1}{2}, 2\right)$ $(7, -1)$

21. $S(16, -9), M\left(\frac{3}{2}, -\frac{13}{2}\right)$

22. $U(-9, 14), M(0, 3)$ $(9, -8)$

23. $F(21, 18), M(19, 11)$ $(17, 4)$

24. $X\left(\frac{1}{4}, \frac{1}{3}\right), M\left(\frac{3}{16}, \frac{1}{3}\right)$
    $\left(\frac{1}{8}, \frac{1}{3}\right)$

21. $(-13, -4)$

Lesson 7-1 Solve each inequality. Then check your solution. 11. $\{b \mid b \geq -1\}$ 12. $\{s \mid s \geq -4.9\}$

1. $c + 9 \leq 3$ $\{c \mid c \leq -6\}$

2. $d - (-3) < 13$ $\{d \mid d < 10\}$

3. $z - 4 > 20$ $\{z \mid z > 24\}$

4. $h - (-7) > -2$ $\{h \mid h > -9\}$

5. $-11 > d - 4$ $\{d \mid d < -7\}$

6. $2x > x - 3$ $\{x \mid x > -3\}$

7. $2x - 3 \geq x$ $\{x \mid x \geq 3\}$

8. $16 + w < -20$ $\{w \mid w < -36\}$

9. $14p > 5 + 13p$ $\{p \mid p > 5\}$

10. $-7 < 16 - z$ $\{z \mid z < 23\}$

11. $-5 + 14b \leq -4 + 15b$

12. $2s - 6.5 \geq -11.4 + s$

13. $1.1v - 1 > 2.1v - 3$ $\{v \mid v < 2\}$

14. $\frac{1}{2}t + \frac{1}{4} \geq \frac{3}{2}t - \frac{2}{3}$ $\left\{t \mid t \leq \frac{11}{12}\right\}$

15. $9x < 8x - 2$ $\{x \mid x < -2\}$

16. $-2 + 9n \leq 10n$ $\{n \mid n \geq -2\}$

17. $a - 2.3 \geq -7.8$ $\{a \mid a \geq -5.5\}$

18. $5z - 6 > 4z$ $\{z \mid z > 6\}$

16.

$-2x + y = 3$

17.

$5y - 6 = 3x$

18.

$y = \frac{3}{2}x - 5$

Additional Answers for Lesson 6-5

9.

$y + 6 = -\frac{2}{3}(x + 1)$

10.

$y - 5 = 4(x + 6)$

11.

$y - 2 = (x + 7)$

12.

$3(x - 1) = y + \frac{4}{5}$

13.

$y = \frac{3}{4}x + 4$

14.

$y = 4x - 1$

15.

$-4x + y = 6$

771

1.
-8 -7 -6 -5 -4 -3 -2 -1 0 1 2 3 4

2.
-4 -3 -2 -1 0 1 2 3 4

3.
-3 -2 -1 0 1 2 3 4 5

4.
-3 -2 -1 0 1 2 3 4 5

5.
-5 -4 -3 -2 -1 0 1 2 3 4 5

6.
-9 -8 -7 -6 -5 -4 -3 -2 -1 0 1 2

7.
-5 -4 -3 -2 -1 0 1 2 3 4

8.
-4 -3 -2 -1 0 1 2 3 4

9.
-3 -2 -1 0 1 2 3 4 5

10.
2 1 0 1 2

13.
-4 -3 -2 -1 0 1 2 3 4

EXTRA PRACTICE

Lesson 7-2 Solve each inequality. Then check your solution.

1. $7b \geq -49$ $\{b \mid b \geq -7\}$

2. $-5j < -60$ $\{j \mid j > 12\}$

3. $\frac{w}{3} > -12$ $\{w \mid w > -36\}$

4. $\frac{p}{5} < 8$ $\{p \mid p < 40\}$

5. $-8f < 48$ $\{f \mid f > -6\}$

6. $\frac{t}{-4} \geq -10$ $\{t \mid t \leq 40\}$

7. $\frac{128}{-g} < 4$ $\{g \mid g > -32\}$

8. $-4.3x < -2.58$ $\{x \mid x > 0.6\}$

9. $4c \geq -6$ $\left\{c \mid c \geq -\frac{3}{2}\right\}$

10. $6 \leq 0.8n$ $\{n \mid n \geq 7.5\}$

11. $\frac{2}{3}m \geq -22$ $\{m \mid m \geq -33\}$

12. $-25 > \frac{a}{-6}$ $\{a \mid a > 150\}$

13. $-15a < -28$ $\left\{a \mid a > \frac{28}{15}\right\}$

14. $-\frac{7}{9}x < 42$ $\{x \mid x > -54\}$

15. $\frac{3y}{8} \leq 32$ $\left\{y \mid y \leq \frac{256}{3}\right\}$

16. $-7y \geq 91$ $\{y \mid y \leq -13\}$

17. $0.8t > 0.96$ $\{t \mid t > 1.2\}$

18. $\frac{4}{7}z \leq -\frac{2}{5}$ $\left\{z \mid z \leq -\frac{7}{10}\right\}$

Lesson 7-3 Solve each inequality. Then check your solution.   **18.** $\left\{y \mid y > \frac{11}{4}\right\}$

1. $3y - 4 > -37$ $\{y \mid y > -11\}$

2. $7s - 12 < 13$ $\left\{s \mid s < \frac{25}{7}\right\}$

3. $-5e + 9 > 24$ $\{e \mid e < -3\}$

4. $-6v - 3 \geq -33$ $\{v \mid v \leq 5\}$

5. $-2k + 12 < 30$ $\{k \mid k > -9\}$

6. $-2x + 1 < 16 - x$ $\{x \mid x > -15\}$

7. $15t - 4 > 11t - 16$ $\{t \mid t > -3\}$

8. $13 - y \leq 29 + 2y$ $\left\{y \mid y \geq -\frac{16}{3}\right\}$

9. $5q + 7 \leq 3(q + 1)$ $\{q \mid q \leq -2\}$

10. $2(w + 4) \geq 7(w - 1)$ $\{w \mid w \leq 3\}$

11. $-4t - 5 > 2t + 13$ $\{t \mid t < -3\}$

12. $\frac{2t + 5}{3} < -9$ $\{t \mid t < -16\}$

13. $\frac{z}{4} + 7 \geq -5$ $\{z \mid z \geq -48\}$

14. $13r - 11 > 7r + 37$ $\{r \mid r > 8\}$

15. $8c - (c - 5) > c + 17$ $\{c \mid c > 2\}$

16. $-5(k + 4) \geq 3(k - 4)$ $\{k \mid k \leq -1\}$

17. $9m + 7 < 2(4m - 1)$ $\{m \mid m < -9\}$

18. $3(3y + 1) < 13y - 8$

19. $5x \leq 10(3x + 4)$ $\left\{x \mid x \geq -\frac{8}{5}\right\}$

20. $3\left(a + \frac{2}{3}\right) \geq a - 1$ $\left\{a \mid a \geq -\frac{3}{2}\right\}$

21. $0.7(n - 3) \leq n - 0.6(n + 5)$ $\{n \mid n \leq -3\}$

1–14. See margin for graphs.

Lesson 7-4 Solve each compound inequality. Then graph the solution set.

1. $2 + x < -5$ or $2 + x > 5$ $\{x \mid x < -7 \text{ or } x > 3\}$

2. $-4 + t > -5$ or $-4 + t < 7$ {all numbers}

3. $3 \leq 2g + 7$ and $2g + 7 \leq 15$ $\{g \mid -2 \leq g \leq 4\}$

4. $2v - 2 \leq 3v$ and $4v - 1 \geq 3v$ $\{v \mid v \geq 1\}$

5. $3b - 4 \leq 7b + 12$ and $8b - 7 \leq 25$ $\{b \mid -4 \leq b \leq 4\}$

6. $-9 < 2z + 7 < 10$ $\left\{z \mid -8 < z < \frac{3}{2}\right\}$

7. $5m - 8 \geq 10 - m$ or $5m + 11 < -9$ $\{m \mid m \geq 3 \text{ or } m < -4\}$

8. $12c - 4 \leq 5c + 10$ or $-4c - 1 \leq c + 24$ {all numbers}

9. $2h - 2 \leq 3h \leq 4h - 1$ $\{h \mid h \geq 1\}$

10. $3p + 6 < 8 - p$ and $5p + 8 \geq p + 6$ $\left\{p \mid -\frac{1}{2} \leq p < \frac{1}{2}\right\}$

11. $2r + 8 > 16 - 2r$ and $7r + 21 < r - 9$ $\varnothing$

12. $4j + 3 < j + 22$ and $j - 3 < 2j - 15$ $\varnothing$

13. $2(q - 4) \leq 3(q + 2)$ or $q - 8 \leq 4 - q$ {all numbers}

14. $\frac{1}{2}w + 5 \geq w + 2 \geq \frac{1}{2}w + 9$ $\varnothing$

Lesson 7-5

1. **Food** For breakfast at Paul's Place, you can select one item from each of the following categories for $1.99.

Meat	Potato	Bread	Beverage
ham sausage bacon	hash browns country potatoes	toast muffin bagel biscuit	juice coffee

   a. What is the probability that a customer will have ham for breakfast? $\frac{1}{3}$

   b. What is the probability of selecting a biscuit and hash browns? $\frac{1}{8}$

   c. What is the probability of having a bagel, country potatoes, and coffee? $\frac{1}{16}$

2. **Music** In order to raise money for a trip to the opera, the music club has set up a lottery using two-digit numbers. The first digit will be a numeral from 1 to 4. The second digit will be a numeral from 3 to 8. The first digit in Trudy's lottery number is 2, but she can't remember the second digit. If only one two digit lottery number is drawn, and that number has a 2 as the first digit, what is the probability that Trudy will win? 2. $\frac{1}{6} = 0.1\overline{6}$

3. **Law** A three-judge panel is being used to settle a dispute. Both sides in the dispute have decided that a majority decision will be upheld. If each judge will render a favorable decision based on the evidence presented two-thirds of the time, what is the probability that the correct side will win the dispute? 3. $\frac{20}{27} \approx 0.741$

Lesson 7-6 Solve each open sentence. Then graph the solution set. 2, 5–15. See margin.

1. $|a - 5| = -3$ ∅
2. $|g + 6| > 8$
3. $|t - 5| \le 3$ $\{t \mid 2 \le t \le 8\}$
4. $|a + 5| \ge 0$ {all numbers}
5. $|14 - 2z| = 16$
6. $|y - 9| < 19$
7. $|2m - 5| > 13$
8. $|14 - w| \ge 20$
9. $|13 - 5y| = 8$
10. $|3p + 5| \le 23$
11. $|6b - 12| \le 36$
12. $|25 - 3x| < 5$
13. $|7 + 8x| > 39$
14. $|4c + 5| \ge 25$
15. $|4 - 5s| > 46$

Lesson 7-7

1. **Travel** Speeds of the fastest train runs in the U.S. and Canada are given below in miles per hour. Make a box-and-whisker plot of the data. See margin.

   93.5 82.5 89.3 83.8 81.8 86.8
   90.8 84.9 95.0 83.1 83.2 88.2

2. **Basketball** The numbers below represent the 20 highest points-scored-per-game averages for a season in the NBA from 1947 to 1990. Make a box-and-whisker plot of this data. See margin.

   35.0 33.5 32.5 37.9 31.2 38.4 34.5 34.7 32.9 44.8
   31.7 37.1 36.5 34.0 50.4 32.3 33.6 34.8 33.1 35.6

3. **Baseball** The stem-and-leaf plot at the right shows the number of home runs hit by the home run leaders in the National League in 1990.

   a. Find the median, upper quartile, lower quartile, and interquartile range. **26.5, 32.5, 24, 8.5**

   b. Are there any outliers? If so, name them. no

   c. Draw a box-and-whisker plot of the data. See margin.

Stem	Leaf
2	2 3 3 4 4 4 4 5 5 6 7 7 8
3	2 2 3 3 5 7
4	0

$3 \mid 3 = 33$

Additional Answers for Lesson 7-6

2. $\{g \mid g < -14 \text{ or } g > 2\}$

   -22 -18 -14 -10 -6 -2 2 6 10

3. $\{t \mid 2 \le t \le 8\}$

   1 2 3 4 5 6 7 8 9

4. {all numbers}

   -4 -3 -2 -1 0 1 2 3 4

5. $\{z \mid z = -1 \text{ or } z = 15\}$

   -2 0 2 4 6 8 10 12 14

6. $\{y \mid -10 < y < 28\}$

   -10   0   10   20   30

7. $\{m \mid m < -4 \text{ or } m > 9\}$

   -6   -2   2   6   10

8. $\{w \mid w \le -6 \text{ or } w \ge 34\}$

   -12 -6 0 6 12 18 24 30 36

9. $\left\{y \mid y = 1 \text{ or } y = \frac{21}{5}\right\}$

   -2 -1 0 1 2 3 4 5 6

10. $\left\{p \mid -\frac{28}{3} \le p \le 6\right\}$

    -10 -8 -6 -4 -2 0 2 4 6 8

11. $\{b \mid -4 \le b \le 8\}$

    -6 -4 -2 0 2 4 6 8 10

12. $\left\{x \mid \frac{20}{3} < x < 10\right\}$

    4 5 6 7 8 9 10 11 12

13. $\left\{x \mid x < -\frac{23}{4} \text{ or } x > 4\right\}$

    -10 -8 -6 -4 -2 0 2 4 6

14. $\left\{c \mid c \le -\frac{30}{4} \text{ or } c \ge 5\right\}$

    -8 -6 -4 -2 0 2 4 6 8

15. $\left\{s \mid s < -\frac{42}{5} \text{ or } s > 10\right\}$

    -14 -10 -6 -2 2 6 10 14 18

Additional Answers for Lesson 7-7

1.

2.

3c.

Lesson 7-8   Graph each inequality.   1–15. See Solutions Manual.

1. $y \leq -2$

2. $x < 4$

3. $x + y < -2$

4. $x + y > -4$

5. $y > 4x - 1$

6. $3x + y > 1$

7. $3y - 2x \leq 2$

8. $x < y$

9. $3x + y > 4$

10. $5x - y < 5$

11. $-4x + 3y \geq 12$

12. $-x + 3y \leq 9$

13. $y > -3x + 7$

14. $3x + 8y \leq 4$

15. $5x - 2y \geq 6$

1–15. See Solutions Manual for graphs.

Lesson 8-1   Graph each system of equations. Then determine whether the system has *one* solution, *no* solution, or *infinitely many* solutions. If the system has one solution, name it.

1. $y = 3x$
   $4x + 2y = 30$ (3, 9)

2. $x = -2y$
   $3x + 5y = 21$ (42, −21)

3. $y = x + 4$
   $3x + 2y = 18$ (2, 6)

4. $x + y = 6$
   $x - y = 2$ (4, 2)

5. $x + y = 6$
   $3x + 3y = 3$ no solution

6. $y = -3x$
   $4x + y = 2$ (2, −6)

7. $x + y = 8$
   $x - y = 2$ (5, 3)

8. $\frac{1}{5}x - y = \frac{12}{5}$
   $3x - 5y = 6$ (−3, −3)

9. $x + 2y = 0$
   $y + 3 = -x$ (−6, 3)

10. $x + 2y = -9$
   $x - y = 6$ (1, −5)

11. $x + \frac{1}{2}y = 3$
   $y = 3x - 4$ (2, 2)

12. $\frac{2}{3}x + \frac{1}{2}y = 2$
   $4x + 3y = 12$ infinitely many

13. $y = x - 4$
   $x + \frac{1}{2}y = \frac{5}{2}$ (3, −1)

14. $2x + y = 3$
   $4x + 2y = 6$ infinitely many

15. $12x - y = -21$
   $\frac{1}{2}x + \frac{2}{3}y = -3$ (−2, −3)

Lesson 8-2   Use substitution to solve each system of equations. If the system *does not* have exactly one solution, state whether it has *no* solution or *infinitely many* solutions.

1. $y = x$
   $5x = 12y$ (0, 0)

2. $y = 7 - x$
   $2x - y = 8$ (5, 2)

3. $x = 5 - y$
   $3y = 3x + 1$ $\left(\frac{7}{3}, \frac{8}{3}\right)$

4. $3x + y = 6$
   $y + 2 = x$ (2, 0)

5. $x - 3y = 3$
   $2x + 9y = 11$ $\left(4, \frac{1}{3}\right)$

6. $3x = -18 + 2y$
   $x + 3y = 4$ $\left(-\frac{46}{11}, \frac{30}{11}\right)$

7. $x + 2y = 10$
   $-x + y = 2$ (2, 4)

8. $2x = 3 - y$
   $2y = 12 - x$ (−2, 7)

9. $6y - x = -36$
   $y = -3x$ $\left(\frac{36}{19}, \frac{-108}{19}\right)$

10. $\frac{3}{4}x + \frac{1}{3}y = 1$
   $x - y = 10$ (4, −6)

11. $x + 6y = 1$
   $3x - 10y = 31$ (7, −1)

12. $3x - 2y = 12$
   $\frac{3}{2}x - y = 3$ no solution

13. $2x + 3y = 5$
   $4x - 9y = 9$ $\left(\frac{12}{5}, \frac{1}{15}\right)$

14. $x = 4 - 8y$
   $3x + 24y = 12$ infinitely many

15. $3x - 2y = -3$
   $25x + 10y = 215$ (5, 9)

Lesson 8-3 State whether addition, subtraction, or substitution would be most convenient to solve each system of equations. Then solve the system.

1. $x + y = 7$
$x - y = 9$ +; $(8, -1)$

2. $2x - y = 32$
$2x + y = 60$ +; $(23, 14)$

3. $-y + x = 6$
$y + x = 5$ +; $\left(\frac{11}{2}, -\frac{1}{2}\right)$

4. $s + 2t = 6$
$3s - 2t = 2$ +; $(2, 2)$

5. $x = y - 7$
$2x - 5y = -2$ sub; $(-11, -4)$

6. $3x + 5y = -16$
$3x - 2y = -2$ −; $(-2, -2)$

7. $x - y = 3$
$x + y = 3$ +; $(3, 0)$

8. $x + y = 8$
$2x - y = 6$ +; $\left(\frac{14}{3}, \frac{10}{3}\right)$

9. $2s - 3t = -4$
$s = 7 - 3t$ sub; $(1, 2)$

10. $-6x + 16y = -8$
$6x - 42 = 16y$ +; $\varnothing$

11. $3x + 0.2y = 7$
$3x = 0.4y + 4$ −; $(2, 5)$

12. $9x + 2y = 26$
$1.5x - 2y = 13$

13. $\frac{2}{3}x - \frac{1}{2}y = 14$
$\frac{5}{6}x - \frac{1}{2}y = 18$ −; $(24, 4)$

14. $4x - \frac{1}{3}y = 8$
$5x + \frac{1}{3}y = 6$ +; $\left(\frac{14}{9}, -\frac{16}{3}\right)$

15. $2x - y = 3$
$\frac{2}{3}x - y = -1$ −; $(3, 3)$

12. +; $\left(\frac{26}{7}, -\frac{26}{7}\right)$

Lesson 8-4 Use elimination to solve each system of equations.

1. $x + 8y = 3$
$4x - 2y = 7$ $\left(\frac{31}{17}, \frac{5}{34}\right)$

2. $4x - y = 4$
$x + 2y = 3$ $\left(\frac{11}{9}, \frac{8}{9}\right)$

3. $3y - 8x = 9$
$y - x = 2$ $\left(-\frac{3}{5}, \frac{7}{5}\right)$

4. $x + 4y = 30$
$2x - y = -6$ $\left(\frac{2}{3}, \frac{22}{3}\right)$

5. $3x - 2y = 0$
$4x + 4y = 5$ $\left(\frac{1}{2}, \frac{3}{4}\right)$

6. $9x - 3y = 5$
$x + y = 1$ $\left(\frac{2}{3}, \frac{1}{3}\right)$

7. $-3x + 2y = 10$
$-2x - y = -5$ $(0, 5)$

8. $2x + 5y = 13$
$4x - 3y = -13$ $(-1, 3)$

9. $5x + 3y = 4$
$-4x + 5y = -18$ $(2, -2)$

10. $2x - 7y = 9$
$-3x + 4y = 6$ $(-6, -3)$

11. $2x - 6y = -16$
$5x + 7y = -18$ $(-5, 1)$

12. $6x - 3y = -9$
$-8x + 2y = 4$ $\left(\frac{1}{2}, 4\right)$

13. $\frac{1}{3}x - y = -1$
$\frac{1}{5}x - \frac{2}{5}y = -1$ $(-9, -2)$

14. $3x - 5y = 8$
$4x - 7y = 10$ $(6, 2)$

15. $x - 0.5y = 1$
$0.4x + y = -2$ $(0, -2)$

Lesson 8-5 Solve each system of inequalities by graphing.   1–12. See Solutions Manual.

1. $x > 3$
$y < 6$

2. $y > 2$
$y > -x + 2$

3. $x \le 2$
$y - 3 \ge 5$

4. $x + y \le -1$
$2x + y \le 2$

5. $y \ge 2x + 2$
$y \ge -x - 1$

6. $y \le x + 3$
$y \ge x + 2$

7. $x + 3y \ge 4$
$2x - y < 5$

8. $y - x > 1$
$y + 2x \le 10$

9. $5x - 2y > 15$
$2x - 3y < 6$

10. $4x + 3y > 4$
$2x - y < 0$

11. $4x + 5y \ge 20$
$y \ge x + 1$

12. $-4x + 10y \le 5$
$-2x + 5y < -1$

Lesson 9-1 Simplify.

1. $a^5(a)(a^7)$ a^{13}
2. $(r^3t^4)(r^4t^4)$ r^7t^8
3. $(x^3y^4)(xy^3)$ x^4y^7

4. $(bc^3)(b^4c^3)$ b^5c^6
5. $(-3mn^2)(5m^3n^2)$ $-15m^4n^4$
6. $[(3^3)^2]^2$ 531,441

7. $(3s^3t^2)(-4s^3t^2)$ $-12s^6t^4$
8. $x^3(x^4y^3)$ x^7y^3
9. $(1.1g^2h^4)^3$ $1.331g^6h^{12}$

10. $-\frac{3}{4}a(a^2b^3c^4)$ $-\frac{3}{4}a^3b^3c^4$
11. $\left(\frac{1}{2}w^3\right)^2(w^4)^2$ $\frac{1}{4}w^{14}$
12. $\left(\frac{2}{3}y^3\right)(3y^2)^3$ $18y^9$

13. $[(-2^3)^3]^2$ 262,144
14. $(10s^3t)(-2s^2t^2)^3$ $-80s^9t^7$
15. $(-0.2u^3w^4)^3$ $-0.008u^9w^{12}$

Lesson 9-2 Simplify. Assume no denominator is equal to zero.

1. $\frac{b^6c^5}{b^3c^2}$ b^3c^3
2. $\frac{(-a)^4b^8}{a^4b^7}$ b
3. $\frac{(-x)^3y^3}{x^3y^6}$ $-\frac{1}{y^3}$

4. $\frac{12ab^5}{4a^4b^3}$ $\frac{3b^2}{a^3}$
5. $\frac{24x^5}{-8x^2}$ $-3x^3$
6. $\frac{-9h^2k^4}{18h^5j^3k^4}$ $\frac{-1}{2h^3j^3}$

7. $\frac{a^0}{2a^{-3}}$ $\frac{a^3}{2}$
8. $\frac{9a^2b^7c^3}{2a^5b^4c}$ $\frac{9b^3c^2}{2a^3}$
9. $\frac{-15xy^5z^7}{-10x^4y^6z^4}$ $\frac{3z^3}{2x^3y}$

10. $\frac{(u^{-3}v^3)^2}{(u^3v)^{-3}}$ u^3v^9
11. $\frac{(-r)s^5}{r^{-3}s^{-4}}$ $-r^4s^9$
12. $\frac{28a^{-4}b^0}{14a^3b^{-1}}$ $\frac{2b}{a^7}$

13. $\frac{(j^2k^3l)^4}{(jk^4)^{-1}}$ $j^9k^{16}l^4$
14. $\left(\frac{-2x^4y}{4y^2}\right)^0$ 1
15. $\frac{3m^7n^2p^4}{9m^2np^3}$ $\frac{m^5np}{3}$

Lesson 9-3 Express each number in scientific notation.

1. 6500 6.5×10^3
2. 953.56 9.5356×10^2
3. 0.697 6.97×10^{-1}

4. 843.5 8.435×10^2
5. 568,000 5.68×10^5
6. 0.0000269 2.69×10^{-5}

7. 0.121212 1.21212×10^{-1}
8. 543×10^4 5.43×10^6
9. 739.9×10^{-5} 7.399×10^{-3}

10. 6480×10^{-2} 6.48×10
11. 0.366×10^{-7} 3.66×10^{-8}
12. 167×10^3 1.67×10^5

19. 1.302×10^2; 130.2   20. 7.8×10^8; 780,000,000

Evaluate. Express each result in scientific and standard notation.

13. $(2 \times 10^5)(3 \times 10^{-8})$
    6×10^{-3}; 0.006
14. $\frac{4.8 \times 10^3}{1.6 \times 10^1}$ 3×10^2; 300
15. $(4 \times 10^2)(1.5 \times 10^6)$
    6×10^8; 600,000,000

16. $\frac{8.1 \times 10^2}{2.7 \times 10^{-3}}$ 3×10^5; 300,000
17. $\frac{7.8 \times 10^{-5}}{1.3 \times 10^{-7}}$ 6×10^2; 600
18. $(2.2 \times 10^{-2})(3.2 \times 10^5)$
    7.04×10^3; 7040

    6×10^{-3}; 0.006
19. $(3.1 \times 10^4)(4.2 \times 10^{-3})$
20. $(78 \times 10^6)(0.01 \times 10^3)$
21. $\frac{2.31 \times 10^{-2}}{3.3 \times 10^{-3}}$ 7×10^0; 7

Lesson 9-4 State whether each expression is a polynomial. If the expression is a polynomial, identify it as a *monomial,* a *binomial,* or a *trinomial* and find the degree of the polynomial.

1. $5x^2y + 3xy + 7$ yes, trinomial, 3
2. 0 yes, monomial, 0
3. $\frac{5}{k} - k^2y$ no

4. $3a^2x - 5a$ yes, binomial, 3
5. $a + \frac{5}{c}$ no
6. $14abcd - 6d^3$ yes, binomial, 4

7. $\frac{a^3}{3}$ yes, monomial, 3
8. $-4h^3$ yes, monomial, 3
9. $x^2 - \frac{x}{2} + \frac{1}{3}$ yes, trinomial, 2

Arrange the terms of each polynomial so that the powers of x are in descending order.   10–15. See margin.

10. $5x^2 - 3x^3 + 7 + 2x$
11. $-6x + x^5 + 4x^3 - 20$
12. $5b + b^3x^2 + \frac{2}{3}bx$

13. $21p^2x + 3px^3 + p^4$
14. $3ax^2 - 6a^2x^3 + 7a^3 - 8x$
15. $\frac{1}{3}s^2x^3 + 4x^4 - \frac{2}{5}s^4x^2 + \frac{1}{4}x$

Lesson 9-5   Find each sum or difference.

1.
$$\begin{array}{l} -7t^2 + 4ts - 6s^2 \\ \underline{(+) -5t^2 - 12ts + 3s^2} \end{array} \quad -12t^2 - 8ts - 3s^2$$

2.
$$\begin{array}{l} 6a^2 - 7ab - 4b^2 \\ \underline{(-) 2a^2 + 5ab + 6b^2} \end{array} \quad 4a^2 - 12ab - 10b^2$$

3.
$$\begin{array}{l} 4a^2 - 10b^2 + 7c^2 \\ -5a^2 \qquad + 2c^2 \qquad + 2b \\ \underline{(+) \qquad 7b^2 - 7c^2 + 7a} \\ -a^2 - 3b^2 + 2c^2 + 7a + 2b \end{array}$$

4.
$$\begin{array}{l} z^2 + 6z - 8 \\ \underline{(-) 4z^2 - 7z - 5} \quad -3z^2 + 13z - 3 \end{array}$$

5. $(4d + 3e - 8f) - (-3d + 10e - 5f + 6)$
$7d - 7e - 3f - 6$

6. $(7g + 8h - 9) + (-g - 3h - 6k)$
$6g + 5h - 6k - 9$

7. $(9x^2 - 11xy - 3y^2) - (x^2 - 16xy + 12y^2)$
$8x^2 + 5xy - 15y^2$

8. $(-3m + 9mn - 5n) + (14m - 5mn - 2n)$
$11m + 4mn - 7n$

9. $(4x^2 - 8y^2 - 3z^2) - (7x^2 - 14z^2 - 12)$
$-3x^2 - 8y^2 + 11z^2 + 12$

10. $(17z^4 - 5z^2 + 3z) - (4z^4 + 2z^3 + 3z)$
$13z^4 - 2z^3 - 5z^2$

11. $(6 - 7y + 3y^2) + (3 - 5y - 2y^2) + (-12 - 8y + y^2)$ $2y^2 - 20y - 3$
12. $(-3x^2 + 2x - 5) + (2x - 6) + (5x^2 + 3) + (-9x^2 - 7x + 4)$ $-7x^2 - 3x - 4$

8. $36m^4n + 4m^3n - 20m^2n^2$ 9. $-16s^3t^5 + 28s^6t^2 - 12s^2t^5$

Lesson 9-6   Find each product.

1. $-3(8x + 5)$ $-24x - 15$
2. $3b(5b + 8)$ $15b^2 + 24b$

3. $1.1a(2a + 7)$ $2.2a^2 + 7.7a$
4. $\frac{1}{2}x(8x - 6)$ $4x^2 - 3x$

5. $7xy(5x^2 - y^2)$ $35x^3y - 7xy^3$
6. $5y(y^2 - 3y + 6)$ $5y^3 - 15y^2 + 30y$

7. $-ab(3b^2 + 4ab - 6a^2)$ $-3ab^3 - 4a^2b^2 + 6a^3b$
8. $4m^2(9m^2n + mn - 5n^2)$

9. $4st^2(-4s^2t^3 + 7s^5 - 3st^3)$
10. $-\frac{1}{3}x(9x^2 + x - 5)$ $-3x^3 - \frac{1}{3}x^2 + \frac{5}{3}x$

11. $-2mn(8m^2 - 3mn + n^2)$
$-16m^3n + 6m^2n^2 - 2mn^3$

12. $-\frac{3}{4}ab^2\left(\frac{1}{3}b^2 - \frac{4}{9}b + 1\right)$
$-\frac{1}{4}ab^4 + \frac{1}{3}ab^3 - \frac{3}{4}ab^2$

Solve.

13. $-3(2a - 12) + 48 = 3a - 3$ $\frac{29}{3}$
14. $-6(12 - 2w) = 7(-2 - 3w)$ $\frac{58}{33}$

15. $a(a - 6) + 2a = 3 + a(a - 2)$ $-\frac{3}{2}$
16. $11(a - 3) + 5 = 2a + 44$ 8

17. $q(2q + 3) + 20 = 2q(q - 3)$ $-\frac{20}{9}$
18. $w(w + 12) = w(w + 14) + 12$ -6

19. $x(x + 8) - x(x + 3) - 23 = 3x + 11$ 17
20. $y(y - 12) + y(y + 2) + 25 = 2y(y + 5) - 15$ 2

21. $x(x - 3) + 4x - 3 = 8x + 4 + x(3 + x)$ $-\frac{7}{10}$
22. $c(c - 3) + 4(c - 2) = 12 - 2(4 + c) - c(1 - c)$ 3

Additional Answers for Lesson 9-4

10. $-3x^3 + 5x^2 + 2x + 7$
11. $x^5 + 4x^3 - 6x - 20$
12. $b^3x^2 + \frac{2}{3}bx + 5b$
13. $3px^3 + 21p^2x + p^4$
14. $-6a^2x^3 + 3ax^2 - 8x + 7a^3$
15. $4x^4 + \frac{1}{3}s^2x^3 - \frac{2}{5}s^4x^2 + \frac{1}{4}x$

Lesson 9–7   Find each product.

1. $(d + 2)(d + 3)$ $d^2 + 5d + 6$
2. $(z + 7)(z - 4)$ $z^2 + 3z - 28$
3. $(m - 8)(m - 5)$
4. $(2x - 5)(x + 6)$ $2x^2 + 7x - 30$
5. $(7a - 4)(2a - 5)$
6. $(4x + y)(2x - 3y)$
7. $(7v + 3)(v + 4)$ $7v^2 + 31v + 12$
8. $(7s - 8)(3s - 2)$
9. $(4g + 3h)(2g - 5h)$
10. $(4a + 3)(2a - 1)$ $8a^2 + 2a - 3$
11. $(7y - 1)(2y - 3)$ $14y^2 - 23y + 3$
12. $(2x + 3y)(5x + 2y)$ $10x^2 + 19xy + 6y^2$
13. $(12r - 4s)(5r + 8s)$ $60r^2 + 76rs - 32s^2$
14. $(x - 2)(x^2 + 2x + 4)$ $x^3 - 8$
15. $(3x + 5)(2x^2 - 5x + 11)$ $6x^3 - 5x^2 + 8x + 55$
16. $(4s + 5)(3s^2 + 8s - 9)$ $12s^3 + 47s^2 + 4s - 45$
17. $(3a + 5)(-8a^2 + 2a + 3)$
18. $(5x - 2)(-5x^2 + 2x + 7)$
19. $(x^2 - 7x + 4)(2x^2 - 3x - 6)$
20. $(a^2 + 2a + 5)(a^2 - 3a - 7)$
21. $(5x^4 - 2x^2 + 1)(x^2 - 5x + 3)$

3. $m^2 - 13m + 40$ 5. $14a^2 - 43a + 20$ 6. $8x^2 - 10xy - 3y^2$ 8. $21s^2 - 38s + 16$ 9. $8g^2 - 14gh - 15h^2$
17. $-24a^3 - 34a^2 + 19a + 15$ 18. $-25x^3 + 20x^2 + 31x - 14$ 19. $2x^4 - 17x^3 + 23x^2 + 30x - 24$
20. $a^4 - a^3 - 8a^2 - 29a - 35$ 21. $5x^6 - 25x^5 + 13x^4 + 10x^3 - 5x^2 - 5x + 3$

Lesson 9–8   Find each product.

1. $(t + 7)^2$ $t^2 + 14t + 49$
2. $(w - 12)(w + 12)$ $w^2 - 144$
3. $(q - 4h)^2$ $q^2 - 8qh + 16h^2$
4. $(10x + 11y)(10x - 11y)$
5. $(4e + 3)^2$ $16e^2 + 24e + 9$
6. $(2b - 4d)(2b + 4d)$
7. $(a + 2b)^2$ $a^2 + 4ab + 4b^2$
8. $(4x + y)^2$ $16x^2 + 8xy + y^2$
9. $(6m + 2n)^2$
10. $(5c - 2d)^2$ $25c^2 - 20cd + 4d^2$
11. $(5b - 6)(5b + 6)$ $25b^2 - 36$
12. $(1 + x)^2$ $1 + 2x + x^2$
13. $(4x - 9y)^2$ $16x^2 - 72xy + 81y^2$
14. $(8a - 2b)(8a + 2b)$ $64a^2 - 4b^2$
15. $\left(\frac{1}{2}a + b\right)^2$ $\frac{1}{4}a^2 + ab + b^2$
16. $(5a - 12b)^2$
17. $(a - 3b)^2$ $a^2 - 6ab + 9b^2$
18. $(7a^2 + b)(7a^2 - b)$ $49a^4 - b^2$
19. $(x + 2)(x - 2)(2x + 5)$
    $2x^3 + 5x^2 - 8x - 20$
20. $(4x - 1)(4x + 1)(x - 4)$
    $16x^3 - 64x^2 - x + 4$
21. $(x - 3)(x + 3)(x - 4)(x + 4)$
    $x^4 - 25x^2 + 144$

4. $100x^2 - 121y^2$ 6. $4b^2 - 16d^2$ 9. $36m^2 + 24mn + 4n^2$ 16. $25a^2 - 120ab + 144b^2$

Lesson 10–1   Find the factors of each number.

1. 17 1, 17
2. 21 1, 3, 7, 21
3. 81 1, 3, 9, 27, 81
4. 24 1, 2, 3, 4, 6, 8, 12, 24
5. 18 1, 2, 3, 6, 9, 18
6. 22 1, 2, 11, 22

State whether each number is *prime* or *composite*. If the number is composite, find its prime factorization.

7. 39 composite; $3 \cdot 13$
8. 89 prime
9. 72 composite; $2^3 \cdot 3^2$
10. 41 prime
11. 57 composite; $3 \cdot 19$
12. 60 composite; $2^2 \cdot 3 \cdot 5$

Factor each expression completely. Do not use exponents.

13. -64 $-1 \cdot 2 \cdot 2 \cdot 2 \cdot 2 \cdot 2 \cdot 2$
14. -26 $-1 \cdot 2 \cdot 13$
15. -240 $-1 \cdot 2 \cdot 2 \cdot 2 \cdot 2 \cdot 3 \cdot 5$
16. -231 $-1 \cdot 3 \cdot 7 \cdot 11$
17. $44rs^2t^3$
    $2 \cdot 2 \cdot 11 \cdot r \cdot s \cdot s \cdot t \cdot t \cdot t$
18. $756(mn)^2$
    $2 \cdot 2 \cdot 3 \cdot 3 \cdot 3 \cdot 7 \cdot m \cdot m \cdot n \cdot n$

Find the GCF of the given monomials.

19. 16, 60 4
20. 15, 50 5
21. $-80, 45$ 5
22. $29, -58$ 29
23. 305, 55 5
24. 252, 126 126
25. 128, 245 1
26. $7y^2, 14y^2$ $7y^2$
27. $4xy, -6x$ $2x$
28. $35t^2, 7t$ $7t$
29. $16pq^2, 12p^2q$ $4pq$
30. 5, 15, 10 5
31. $12mn, 10mn, 15mn$ mn
32. 14, 12, 20 2
33. $26jk^4, 16jk^3, 8j^2$ $2j$

Lesson 10-2 Complete. In exercises with two blanks, both blanks represent the same expression.

1. $6x + 3y = 3(\underline{\;?\;} + y)$ $2x$
2. $8x^2 - 4x = 4x(2x - \underline{\;?\;})$ 1
3. $12a^2b + 6a = 6a(\underline{\;?\;} + 1)$ $2ab$
4. $14r^2t - 42t = 14t(\underline{\;?\;} - 3)$ r^2
5. $24x^2 + 12y^2 = 12(\underline{\;?\;} + y^2)$ $2x^2$
6. $12xy + 12x^2 = \underline{\;?\;}(y + x)$ $12x$
7. $(bx + by) + (3ax + 3ay) = b(\underline{\;?\;}) + 3a(\underline{\;?\;})$ $x + y$
8. $(10x^2 - 6xy) + (15x - 9y) = 2x(\underline{\;?\;}) + 3(\underline{\;?\;})$ $5x - 3$
9. $(6x^3 + 6x) + (7x^2y + 7y) = 6x(\underline{\;?\;}) + 7y(\underline{\;?\;})$ $x^2 + 1$

Factor each polynomial.

10. $10a^2 + 40a$ $10a(a + 4)$
11. $15wx - 35wx^2$ $5wx(3 - 7x)$
12. $27a^2b + 9b^3$ $9b(3a^2 + b^2)$
13. $11x + 44x^2y$ $11x(1 + 4xy)$
14. $16y^2 + 8y$ $8y(2y + 1)$
15. $14mn^2 + 2mn$ $2mn(7n + 1)$
16. $25a^2b^2 + 30ab^3$ $5ab^2(5a + 6b)$
17. $2m^3n^2 - 16m^2n^3 + 8mn$ $2mn(m^2n - 8mn^2 + 4)$
18. $2ax + 6xc + ba + 3bc$ $(2x + b)(a + 3c)$
19. $6mx - 4m + 3rx - 2r$ $(2m + r)(3x - 2)$
20. $3ax - 6bx + 8b - 4a$ $(a - 2b)(3x - 4)$
21. $a^2 - 2ab + a - 2b$ $(a + 1)(a - 2b)$
22. $8ac - 2ad + 4bc - bd$ $(2a + b)(4c - d)$
23. $2e^2g + 2fg + 4e^2h + 4fh$ $2(e^2 + f)(g + 2h)$

Lesson 10-3 Complete.

1. $p^2 + 9p - 10 = (p + \underline{\;?\;})(p - 1)$ 10
2. $y^2 - 2y - 35 = (y + 5)(y - \underline{\;?\;})$ 7
3. $4a^2 + 4a - 63 = (2a - 7)(2a \underline{\;?\;} 9)$ $+$
4. $4r^2 - 25r + 6 = (r - 6)(\underline{\;?\;} - 1)$ $4r$
5. $b^2 + 12b + 35 = (b + 5)(b + \underline{\;?\;})$ 7
6. $3x^2 - 7x - 6 = (3x + 2)(x \underline{\;?\;} 7)$ $-$
7. $3a^2 - 2a - 21 = (a \underline{\;?\;} 3)(3a + 7)$ $-$
8. $4y^2 + 11y + 6 = (\underline{\;?\;} + 3)(y + 2)$ $4y$
9. $2z^2 - 11z + 15 = (\underline{\;?\;} - 5)(z - 3)$ $2z$
10. $6n^2 + 7n - 3 = (2n + \underline{\;?\;})(3n - 1)$ 3

Factor each trinomial, if possible. If the trinomial cannot be factored using integers, write *prime*.

11. $5x^2 - 17x + 14$ $(5x - 7)(x - 2)$
12. $a^2 - 9a - 36$ $(a - 12)(a + 3)$
13. $x^2 + 2x - 15$ $(x + 5)(x - 3)$
14. $n^2 - 8n + 15$ $(n - 3)(n - 5)$
15. $b^2 + 22b + 21$ $(b + 21)(b + 1)$
16. $c^2 + 2c - 3$ $(c + 3)(c - 1)$
17. $x^2 - 5x - 24$ $(x - 8)(x + 3)$
18. $2n^2 - 11n + 7$ prime
19. $8m^2 - 10m + 3$
20. $z^2 + 15z + 36$ $(z + 3)(z + 12)$
21. $s^2 - 13st - 30t^2$
22. $6y^2 + 2y - 2$ $2(3y^2 + y - 1)$
23. $2r^2 + 3r - 14$ $(2r + 7)(r - 2)$
24. $5x - 6 + x^2$ $(x - 1)(x + 6)$
25. $x^2 - 4xy - 5y^2$ $(x - 5y)(x + y)$
26. $5r^2 - 3r + 15$ prime
27. $18v^2 + 42v + 12$
28. $4k^2 + 2k - 12$ $2(2k - 3)(k + 2)$

19. $(4m - 3)(2m - 1)$ 21. $(s + 2t)(s - 15t)$ 27. $6(3v + 1)(v + 2)$

Lesson 10-4 Factor each polynomial, if possible. If the polynomial cannot be factored, write *prime*.
24, 26–30. See margin.   14. $(3x - 10y)(3x + 10y)$ 19. $-5(3m + 1)(3m - 1)$

1. $x^2 - 9$ $(x - 3)(x + 3)$
2. $a^2 - 64$ $(a - 8)(a + 8)$
3. $t^2 - 49$ $(t - 7)(t + 7)$
4. $4x^2 - 9y^2$ $(2x - 3y)(2x + 3y)$
5. $1 - 9z^2$ $(1 - 3z)(1 + 3z)$
6. $16a^2 - 9b^2$ $(4a - 3b)(4a + 3b)$
7. $8x^2 - 12y^2$ $4(2x^2 - 3y^2)$
8. $a^2 - 4b^2$ $(a - 2b)(a + 2b)$
9. $x^2 - y^2$ $(x - y)(x + y)$
10. $75r^2 - 48$ $3(5r + 4)(5r - 4)$
11. $x^2 - 36y^2$ $(x - 6y)(x + 6y)$
12. $3a^2 - 16$ prime
13. $12t^2 - 75$ $3(2t - 5)(2t + 5)$
14. $9x^2 - 100y^2$
15. $49 - a^2b^2$ $(7 - ab)(7 + ab)$
16. $12a^2 - 48$ $12(a - 2)(a + 2)$
17. $169 - 16t^2$ $(13 - 4t)(13 + 4t)$
18. $8r^2 - 4$ $4(2r^2 - 1)$
19. $-45m^2 + 5$
20. $9x^4 - 16y^2$ $(3x^2 - 4y)(3x^2 + 4y)$
21. $36b^2 - 64$ $4(3b - 4)(3b + 4)$
22. $5g^2 - 20h^2$ $5(g - 2h)(g + 2h)$
23. $\frac{1}{4}n^2 - 16$ $\left(\frac{1}{2}n - 4\right)\left(\frac{1}{2}n + 4\right)$
24. $\frac{1}{4}t^2 - \frac{4}{9}p^2$
25. $(r - t)^2 + t^2$ prime
26. $12x^3 - 27xy^2$
27. $0.01n^2 - 1.69r^2$
28. $0.04m^2 - 0.09n^2$
29. $(x - y)^2 - y^2$
30. $162m^4 - 32n^8$

Additional Answers for Lesson 10-4

24. $\left(\frac{1}{2}t - \frac{2}{3}p\right)\left(\frac{1}{2}t + \frac{2}{3}p\right)$
26. $3x(2x - 3y)(2x + 3y)$
27. $(0.1n - 1.3r)(0.1n + 1.3r)$
28. $(0.2m - 0.3n)(0.2m + 0.3n)$
29. $x(x - 2y)$
30. $2(3m - 2n^2)(3m + 2n^2)$
    $(9m^2 + 4n^4)$

EXTRA PRACTICE

Lesson 10-5 Determine whether each trinomial is a perfect square trinomial. If so, factor it.

1. $x^2 + 12x + 36$ yes; $(x + 6)^2$ **2.** $n^2 - 13n + 36$ no **3.** $a^2 + 4a + 4$ yes; $(a + 2)^2$

4. $b^2 - 14b + 49$ yes; $(b - 7)^2$ **5.** $x^2 + 20x - 100$ no **6.** $y^2 - 10y + 100$ no

7. $9b^2 - 6b + 1$ yes; $(3b - 1)^2$ **8.** $4x^2 + 4x + 1$ yes; $(2x + 1)^2$ **9.** $2n^2 + 17n + 21$ no

10. $9x^2 - 10x + 4$ no **11.** $9y^2 + 8y - 16$ no **12.** $4a^2 - 20a + 25$
yes; $(2a - 5)^2$

Factor each polynomial, if possible. If the polynomial cannot be factored, write *prime*.

13. $n^2 - 8n + 16$ $(n - 4)^2$ **14.** $4k^2 - 4k + 1$ $(2k - 1)^2$ **15.** $x^2 + 16x + 64$ $(x + 8)^2$

16. $t^2 - 4t + 1$ prime **17.** $x^2 + 22x + 121$ $(x + 11)^2$ **18.** $s^2 + 30s + 225$ $(s + 15)^2$

19. $1 - 10z + 25z^2$ $(1 - 5z)^2$ **20.** $9p^2 - 56p + 49$ prime **21.** $9n^2 - 36nm + 36m^2$

22. $16a^2 + 81 - 72a$ $(4a - 9)^2$ **23.** $9x^2 + 12xy + 4y^2$ $(3x + 2y)^2$ **24.** $m^2 + 16mn + 64n^2$ $(m + 8n)^2$

25. $8t^4 + 56t^3 + 98t^2$ $2t^2(2t + 7)^2$ **26.** $4p^2 + 12pr + 9r^2$ $(2p + 3r)^2$ **27.** $16m^4 - 72m^2n^2 + 81n^4$
21. $9(n - 2m)^2$ $(2m - 3n)^2(2m + 3n)^2$

Lesson 10-6 Solve each equation. Check your solutions.   12. $\{0, -2\}$

1. $y(y - 12) = 0$ $\{0, 12\}$ **2.** $2x(5x - 10) = 0$ $\{0, 2\}$ **3.** $7a(a + 6) = 0$ $\{0, -6\}$

4. $(b - 3)(b - 5) = 0$ $\{3, 5\}$ **5.** $(p - 5)(p + 5) = 0$ $\{5, -5\}$ **6.** $(4t + 4)(2t + 6) = 0$ $\{-1, -3\}$

7. $(3x - 5)^2 = 0$ $\left\{\dfrac{5}{3}\right\}$ **8.** $x^2 - 6x = 0$ $\{0, 6\}$ **9.** $n^2 + 36n = 0$ $\{0, -36\}$

10. $2x^2 + 4x = 0$ $\{0, -2\}$ **11.** $2x^2 = x^2 - 8x$ $\{0, -8\}$ **12.** $7y - 1 = -3y^2 + y - 1$

13. $\dfrac{1}{2}y^2 - \dfrac{1}{4}y = 0$ $\left\{0, \dfrac{1}{2}\right\}$ **14.** $\dfrac{5}{6}x^2 - \dfrac{1}{3}x = \dfrac{1}{3}x$ $\left\{0, \dfrac{4}{5}\right\}$ **15.** $\dfrac{2}{3}x = \dfrac{1}{3}x^2$ $\{0, 2\}$

16. $\dfrac{3}{4}a^2 + \dfrac{7}{8}a = a$ $\left\{0, \dfrac{1}{6}\right\}$ **17.** $n^2 - 3n = 0$ $\{0, 3\}$ **18.** $3x^2 - \dfrac{3}{4}x = 0$ $\left\{0, \dfrac{1}{4}\right\}$

19. $8a^2 = -4a$ $\left\{-\dfrac{1}{2}, 0\right\}$ **20.** $(2y + 8)(3y + 24) = 0$ $\{-4, -8\}$ **21.** $(4x - 7)(3x + 5) = 0$ $\left\{\dfrac{7}{4}, -\dfrac{5}{3}\right\}$

Lesson 11-1 Write the equation of the axis of symmetry and find the coordinates of the vertex of the graph of each equation. State if the vertex is a maximum or minimum. Then graph the equation.   1–24. See Solutions Manual.

1. $y = x^2 + 6x + 8$ **2.** $y = -x^2 + 3x$ **3.** $y = -x^2 + 7$

4. $y = x^2 + x + 3$ **5.** $y = -x^2 + 4x + 5$ **6.** $y = 3x^2 + 6x + 16$

7. $y = -x^2 + 2x - 3$ **8.** $y = 3x^2 + 24x + 80$ **9.** $y = x^2 - 4x - 4$

10. $y = 5x^2 - 20x + 37$ **11.** $y = 3x^2 + 6x + 3$ **12.** $y = 2x^2 + 12x$

13. $y = x^2 - 6x + 5$ **14.** $y = \dfrac{1}{2}x^2 + 3x + \dfrac{9}{2}$ **15.** $y = \dfrac{1}{4}x^2 - 4x + \dfrac{15}{4}$

16. $y = 4x^2 - 1$ **17.** $y = -2x^2 - 2x + 4$ **18.** $y = 6x^2 - 12x - 4$

19. $y = x^2 - 1$ **20.** $y = -x^2 + x + 1$ **21.** $y = -5x^2 - 3x + 2$

22. $y = x^2 - x - 6$ **23.** $y = 2x^2 + 5x - 2$ **24.** $y = -3x^2 - 18x - 15$

780 *Extra Practice*

780

Lesson 11-2 State the real roots of each quadratic equation whose related function is graphed below.

1.

$-1, 1$

2.

$0, 2$

3.

$\varnothing$

4.

2

Solve each equation by graphing. If exact roots cannot be found, state the consecutive integers between which the roots lie. 5–19. See Solutions Manual for graphs. 8. $0 < r < 1, 3 < r < 4$

5. $x^2 + 2x - 3 = 0$ $-3, 1$

6. $-x^2 + 6x - 5 = 0$ $1, 5$

7. $-a^2 - 2a + 3 = 0$ $-3, 1$

8. $2r^2 - 8r + 5 = 0$

9. $-3x^2 + 6x - 9 = 0$ $\varnothing$

10. $c^2 + c = 0$ $-1, 0$

11. $3t^2 + 2 = 0$ $\varnothing$

12. $-b^2 + 5b + 2 = 0$

13. $3x^2 + 7x = 1$

14. $x^2 + 5x - 24 = 0$ $-8, 3$

15. $8 - k^2 = 0$

16. $x^2 - 7x = 18$ $-2, 9$

17. $a^2 + 12a + 36 = 0$ -6

18. $64 - x^2 = 0$ $-8, 8$

19. $-4x^2 + 2x = -1$

12. $-1 < b < 0, 5 < b < 6$ 13. $-3 < x < -2, 0 < x < 1$ 15. $-3 < k < -2, 2 < k < 3$ 19. $-1 < x < 0, 0 < x < 1$

Lesson 11-3 Solve each equation by using the quadratic formula. Approximate irrational roots to the nearest hundredth.

1. $x^2 - 8x - 4 = 0$ $8.47, -0.47$

2. $x^2 + 7x + 6 = 0$ $-6, -1$

3. $x^2 + 5x - 6 = 0$ $-6, 1$

4. $y^2 - 7y - 8 = 0$ $-1, 8$

5. $m^2 - 2m = 35$ $-5, 7$

6. $4n^2 - 20n = 0$ $0, 5$

7. $m^2 + 4m + 2 = 0$ $-0.59, -3.41$

8. $2t^2 - t - 15 = 0$ $-2.5, 3$

9. $5t^2 = 125$ $-5, 5$

10. $t^2 + 16 = 0$ $\varnothing$

11. $-4x^2 + 8x = -3$ $2.32, -0.32$

12. $3k^2 + 2 = -8k$ $-2.39, -0.28$

13. $8t^2 + 10t + 3 = 0$ $-\dfrac{1}{2}, -\dfrac{3}{4}$

14. $3x^2 - \dfrac{5}{4}x - \dfrac{1}{2} = 0$ $\dfrac{2}{3}, -\dfrac{1}{4}$

15. $-5b^2 + 3b - 1 = 0$ $\varnothing$

16. $s^2 + 8s + 7 = 0$ $-7, -1$

17. $d^2 - 14d + 24 = 0$ $2, 12$

18. $3k^2 + 11k = 4$ $-4, \dfrac{1}{3}$

19. $n^2 - 3n + 1 = 0$ $2.62, 0.38$

20. $2z^2 + 5z - 1 = 0$ $0.19, -2.69$

21. $3h^2 = 27$ $3, -3$

22. $3f^2 + 2f = 6$ $1.12, -1.79$

23. $2x^2 = 0.7x + 0.3$ $0.6, -0.25$

24. $3w^2 - 8w + 2 = 0$ $2.39, 0.28$

25. $2r^2 - r - 3 = 0$ $-1, \dfrac{3}{2}$

26. $x^2 - 9x = 5$ $9.52, -0.52$

27. $6t^2 - 4t - 9 = 0$ $1.60, -0.94$

Lesson 11-4 Use a calculator to determine the approximate value of each expression to the nearest hundredth.

1. $3^{1.6}$ 5.80

2. $10^{-0.2}$ 0.63

3. $\left(\dfrac{1}{3}\right)^{-1.4}$ 4.66

4. $\left(\dfrac{2}{3}\right)^{5.1}$ 0.13

5. $40(2^{-0.5})$ 28.28

6. $10(2^{-1.6})$ 3.30

7. $0.3(4^{0.8})$ 0.91

8. $30(0.75^{-3.6})$ 84.51

9. $5^{1.75}$ 16.72

Graph each function. State the y-intercept. 10–21. See Solutions Manual for graphs.

10. $y = 3^x + 1$ 2

11. $y = 2^x - 5$ -4

12. $y = 2^{x+3}$ 8

13. $y = 3^{x+1}$ 3

14. $y = \left(\dfrac{1}{4}\right)^x$ 1

15. $y = 5\left(\dfrac{2}{5}\right)^x$ 5

16. $y = 3 \cdot 2^x$ 3

17. $y = 4 \cdot 5^x$ 4

18. $y = 6^x$ 1

19. $y = 3^x$ 1

20. $y = \left(\dfrac{1}{8}\right)^x$ 1

21. $y = \left(\dfrac{3}{4}\right)^x$ 1

Solve each equation.

22. $6^{3x-4} = 6^x$ 2

23. $3^4 = 3^{2x+2}$ 1

24. $4^x = 4^{5x+8}$ -2

25. $2^x = 4^{x+1}$ -2

26. $5^{4x} = 5^{-4}$ -1

27. $2^{x+3} = 2^{-5}$ -8

Lesson 11-5 Determine whether each exponential equation represents growth or decay.

1. $y = 3.89(1.05)^x$ **growth**   2. $y = 476(0.35)^x$ **decay**   3. $y = 19,520(0.98)^x$ **decay**

4. $y = 16(1.0432)^x$ **growth**   5. $y = 1.01(1.099)^x$ **growth**   6. $y = 84(0.03)^x$ **decay**

7. **Education** Marco withdrew all of the $2500 in his savings account to pay the tuition for his first semester at college. The account had earned 12% interest compounded monthly, and no withdrawals or additional deposits were made.

   a. If Marco's original deposit was $1250, how long ago did he open the account? **about 5 years, 10 months ago**

   b. If Marco's original deposit was $1500, how long ago did he open the account? **about 4 years, 3 months ago**

8. **Finance** Erin saved $500 of the money she earned working at the Dairy Dream last summer. She deposited the money in a certificate of deposit that earns 8.75% interest compounded monthly. If she rolls over the CD at the same rate each year, when will Erin's CD have a balance of $800? **5 years, 5 months**

9. **Demographics** In 1994, the metropolitan area of Pensacola, Florida, had a population of 371,000. The growth rate from 1990 to 1994 was 7.7%.

   a. Write an exponential equation for the area's growth. $y = 371,000(1.077)^t$

   b. Compute the estimated population for Pensacola in the year 2000. **578,986**

Lesson 12-1 Simplify each rational expression. State the excluded values of the variables.

1. $\dfrac{13a}{39a^2}$ $\dfrac{1}{3a}$; $a \neq 0$

2. $\dfrac{38x^2}{42xy}$ $\dfrac{19x}{21y}$; $x, y \neq 0$

3. $\dfrac{14y^2z}{49yz^3}$ $\dfrac{2y}{7z^2}$; $y, z \neq 0$

4. $\dfrac{p+5}{2(p+5)}$ $\dfrac{1}{2}$; $p \neq -5$

5. $\dfrac{79a^2b}{158a^3bc}$ $\dfrac{1}{2ac}$; $a, b, c \neq 0$

6. $\dfrac{a+b}{a^2-b^2}$ $\dfrac{1}{a-b}$; $a \neq \pm b$

7. $\dfrac{y+4}{(y-4)(y+4)}$ $\dfrac{1}{y-4}$; $y \neq 4, -4$

8. $\dfrac{c^2-4}{(c+2)^2}$ $\dfrac{c-2}{c+2}$; $c \neq -2$

9. $\dfrac{a^2-a}{a-1}$ a; $a \neq 1$

10. $\dfrac{(w-4)(w+4)}{(w-2)(w-4)}$ $\dfrac{w+4}{w-2}$; $w \neq 2, 4$

11. $\dfrac{m^2-2m}{m-2}$ m; $m \neq 2$

12. $\dfrac{x^2+4}{x^4-16}$ $\dfrac{1}{x^2-4}$; $x \neq \pm 2$

13. $\dfrac{r^3-r^2}{r-1}$ r^2; $r \neq 1$

14. $\dfrac{3m^3}{6m^2-3m}$ $\dfrac{m^2}{2m-1}$; $m \neq 0, \dfrac{1}{2}$

15. $\dfrac{4t^2-8}{4t-4}$ $\dfrac{t^2-2}{t-1}$; $t \neq 1$

16. $\dfrac{6y^3-12y^2}{12y^2-18}$ $\dfrac{y^2(y-2)}{2y^2-3}$; $y \neq \pm\dfrac{\sqrt{6}}{2}$

17. $\dfrac{x-3}{x^2+x-12}$ $\dfrac{1}{x+4}$; $x \neq -4, 3$

18. $\dfrac{5x^2+10x+5}{3x^2+6x+3}$ $\dfrac{5}{3}$; $x \neq -1$

Lesson 12-2 Find each product. Assume that no denominator has a value of 0.

1. $\dfrac{a^2b}{b^2c} \cdot \dfrac{c}{d}$ $\dfrac{a^2}{bd}$

2. $\dfrac{6a^2n}{8n^2} \cdot \dfrac{12n}{9a}$ a

3. $\dfrac{2a^2d}{3bc} \cdot \dfrac{9b^2c}{16ad^2}$ $\dfrac{3ab}{8d}$

4. $\dfrac{10n^3}{6x^3} \cdot \dfrac{12n^2x^4}{25n^2x^2}$ $\dfrac{4n^3}{5x}$

5. $\left(\dfrac{2a}{b}\right)^2 \cdot \dfrac{5c}{6a}$ $\dfrac{10ac}{3b^2}$

6. $\dfrac{6m^3n}{10a^2} \cdot \dfrac{4a^2m}{9n^3}$ $\dfrac{4m^4}{15n^2}$

7. $\dfrac{5n-5}{3} \cdot \dfrac{9}{n-1}$ 15

8. $\dfrac{a^2}{a-b} \cdot \dfrac{3a-3b}{a}$ $3a$

9. $\dfrac{2a+4b}{5} \cdot \dfrac{25}{6a+8b}$ $\dfrac{5a+10b}{3a+4b}$

10. $\dfrac{4t}{4t+40} \cdot \dfrac{3t+30}{2t}$ $\dfrac{3}{2}$

11. $\dfrac{3k+9}{k} \cdot \dfrac{k^2}{k^2-9}$ $\dfrac{3k}{k-3}$

12. $\dfrac{7xy^3}{11z^2} \cdot \dfrac{44z^3}{21x^2y}$ $\dfrac{4y^2z}{3x}$

13. $\dfrac{3}{x-y} \cdot \dfrac{(x-y)^2}{6}$ $\dfrac{x-y}{2}$

14. $\dfrac{x+5}{3x} \cdot \dfrac{12x^2}{x^2+7x+10}$ $\dfrac{4x}{x+2}$

15. $\dfrac{a^2-b^2}{4} \cdot \dfrac{16}{a+b}$ $4a-4b$

16. $\dfrac{4a+8}{a^2-25} \cdot \dfrac{a-5}{5a+10}$ $\dfrac{4}{5a+25}$

17. $\dfrac{r^2}{r-s} \cdot \dfrac{r^2-s^2}{s^2}$ $\dfrac{r^3+r^2s}{s^2}$

18. $\dfrac{a^2-b^2}{a-b} \cdot \dfrac{7}{a+b}$ 7

Lesson 12-3 Find each quotient. Assume that no denominator has a value of 0.

1. $\dfrac{5m^2n}{12a^2} \div \dfrac{30m^4}{18an}$ $\dfrac{n^2}{4am^2}$

2. $\dfrac{25g^7h}{28t^3} \div \dfrac{5g^5h^2}{42s^2t^3}$ $\dfrac{15g^2s^2}{2h}$

3. $\dfrac{6a + 3b}{36} \div \dfrac{3a + 2b}{45}$ $\dfrac{5}{2}$

4. $\dfrac{x^2y}{18z} \div \dfrac{2yz}{3x^2}$ $\dfrac{x^4}{12z^2}$

5. $\dfrac{p^2}{14qr^3} \div \dfrac{2r^2p}{7q}$ $\dfrac{p}{4r^5}$

6. $\dfrac{5e - f}{5e + f} \div (25e^2 - f^2)$ $\dfrac{1}{(5e + f)^2}$

7. $\dfrac{t^2 - 2t - 15}{t - 5} \div \dfrac{t + 3}{t + 5}$ $t + 5$

8. $\dfrac{5x + 10}{x + 2} \div (x + 2)$ $\dfrac{5}{x + 2}$

9. $\dfrac{3d}{2d^2 - 3d} \div \dfrac{9}{2d - 3}$ $\dfrac{1}{3}$

10. $\dfrac{3v^2 - 27}{15v} \div \dfrac{v + 3}{v^2}$ $\dfrac{v(v - 3)}{5}$

11. $\dfrac{3g^2 + 15g}{4} \div \dfrac{g + 5}{g^2}$ $\dfrac{3g^3}{4}$

12. $\dfrac{b^2 - 9}{4b} \div (b - 3)$ $\dfrac{b + 3}{4b}$

13. $\dfrac{p^2}{y^2 - 4} \div \dfrac{p}{2 - y}$ $-\dfrac{p}{y + 2}$

14. $\dfrac{k^2 - 81}{k^2 - 36} \div \dfrac{k - 9}{k + 6}$ $\dfrac{k + 9}{k - 6}$

15. $\dfrac{2a^3}{a + 1} \div \dfrac{a^2}{a + 1}$ $2a$

16. $\dfrac{x^2 - 16}{16 - x^2} \div \dfrac{7}{x}$ $-\dfrac{x}{7}$

17. $\dfrac{y}{5} \div \dfrac{y^2 - 25}{5 - y}$ $\dfrac{-y}{5y + 25}$

18. $\dfrac{3m}{m + 1} \div (m - 2)$ $\dfrac{3m}{m^2 - m - 2}$

Lesson 12-4 Find each quotient.

1. $(2x^2 - 11x - 20) \div (2x + 3)$ $x - 7 + \dfrac{1}{2x + 3}$

2. $(a^2 + 7a + 12) \div (a + 3)$ $a + 4$

3. $(m^2 + 9m + 20) \div (m + 5)$ $m + 4$

4. $(x^2 - 2x - 35) \div (x - 7)$ $x + 5$

5. $(c^2 + 12c + 36) \div (c + 9)$ $c + 3 + \dfrac{9}{c + 9}$

6. $(y^2 - 2y - 30) \div (y + 7)$ $y - 9 + \dfrac{33}{y + 7}$

7. $(3t^2 - 14t - 24) \div (3t + 4)$ $t - 6$

8. $(2r^2 - 3r - 35) \div (2r + 7)$ $r - 5$

9. $\dfrac{12n^2 + 36n + 15}{6n + 3}$ $2n + 5$

10. $\dfrac{10x^2 + 29x + 21}{5x + 7}$ $2x + 3$

11. $\dfrac{4t^3 + 17t^2 - 1}{4t + 1}$ $t^2 + 4t - 1$

12. $\dfrac{2a^3 + 9a^2 + 5a - 12}{a + 3}$ $2a^2 + 3a - 4$

13. $\dfrac{4m^3 + 5m - 21}{2m - 3}$ $2m^2 + 3m + 7$

14. $\dfrac{6t^3 + 5t^2 + 12}{2t + 3}$ $3t^2 - 2t + 3 + \dfrac{3}{2t + 3}$

15. $\dfrac{27c^2 - 24c + 8}{9c - 2}$ $3c - 2 + \dfrac{4}{9c - 2}$

16. $\dfrac{3b^3 + 8b^2 + b - 7}{b + 2}$ $3b^2 + 2b - 3 - \dfrac{1}{b + 2}$

17. $\dfrac{t^3 - 19t + 9}{t - 4}$ $t^2 + 4t - 3 - \dfrac{3}{t - 4}$

18. $\dfrac{9d^3 + 5d - 8}{3d - 2}$ $3d^2 + 2d + 3 - \dfrac{2}{3d - 2}$

Lesson 12-5 Find each sum or difference. Express in simplest form.

1. $\dfrac{4}{z} + \dfrac{3}{z}$ $\dfrac{7}{z}$

2. $\dfrac{a}{12} + \dfrac{2a}{12}$ $\dfrac{a}{4}$

3. $\dfrac{5}{2t} + \dfrac{-7}{2t}$ $-\dfrac{1}{t}$

4. $\dfrac{y}{2} + \dfrac{y}{2}$ y

5. $\dfrac{b}{x} + \dfrac{2}{x}$ $\dfrac{b + 2}{x}$

6. $\dfrac{5x}{24} - \dfrac{3x}{24}$ $\dfrac{x}{12}$

7. $\dfrac{7p}{p} - \dfrac{8p}{p}$ -1

8. $\dfrac{8k}{5m} - \dfrac{3k}{5m}$ $\dfrac{k}{m}$

9. $\dfrac{y}{2} + \dfrac{y - 6}{2}$ $y - 3$

10. $\dfrac{a + 2}{6} - \dfrac{a + 3}{6}$ $-\dfrac{1}{6}$

11. $\dfrac{8}{m - 2} - \dfrac{6}{m - 2}$ $\dfrac{2}{m - 2}$

12. $\dfrac{x}{x + 1} + \dfrac{1}{x + 1}$ 1

13. $\dfrac{2n}{2n - 5} + \dfrac{5}{5 - 2n}$ 1

14. $\dfrac{y}{b + 6} - \dfrac{2y}{b + 6}$ $\dfrac{-y}{b + 6}$

15. $\dfrac{x - y}{2 - y} + \dfrac{x + y}{y - 2}$ $\dfrac{2y}{y - 2}$

16. $\dfrac{r^2}{r - s} + \dfrac{s^2}{r - s}$ $\dfrac{r^2 + s^2}{r - s}$

17. $\dfrac{12n}{3n + 2} + \dfrac{8}{3n + 2}$ 4

18. $\dfrac{6x}{x + y} + \dfrac{6y}{x + y}$ 6

Lesson 12-6 Find each sum or difference.

1. $\dfrac{s}{3} + \dfrac{2s}{7}$ $\dfrac{13s}{21}$

2. $\dfrac{5}{2a} + \dfrac{-3}{6a}$ $\dfrac{2}{a}$

3. $\dfrac{2n}{5} - \dfrac{3m}{4}$ $\dfrac{8n - 15m}{20}$

4. $\dfrac{6}{5x} + \dfrac{7}{10x^2}$ $\dfrac{12x + 7}{10x^2}$

5. $\dfrac{3z}{7w^2} - \dfrac{2z}{w}$ $\dfrac{3z - 14wz}{7w^2}$

6. $\dfrac{s}{t^2} - \dfrac{r}{3t}$ $\dfrac{3s - rt}{3t^2}$

7. $\dfrac{5}{xy} + \dfrac{6}{yz}$ $\dfrac{5z + 6x}{xyz}$

8. $\dfrac{2}{t} + \dfrac{t + 3}{s}$ $\dfrac{2s + t^2 + 3t}{st}$

9. $\dfrac{a}{a - b} + \dfrac{b}{2b + 3a}$ $\dfrac{3a^2 + 3ab - b^2}{3a^2 - ab - 2b^2}$

10. $\dfrac{a}{a^2 - 4} - \dfrac{4}{a + 2}$ $\dfrac{-3a + 8}{a^2 - 4}$

11. $\dfrac{4a}{2a + 6} + \dfrac{3}{a + 3}$ $\dfrac{2a + 3}{a + 3}$

12. $\dfrac{m}{1(m - n)} - \dfrac{5}{m}$ $\dfrac{m^2 - 5m + 5n}{m(m - n)}$

13. $\dfrac{-3}{a - 5} + \dfrac{-6}{a^2 - 5a}$ $\dfrac{-3a - 6}{a^2 - 5a}$

14. $\dfrac{3t + 2}{3t - 6} - \dfrac{t + 2}{t^2 - 4}$ $\dfrac{3t - 1}{3t - 6}$

15. $\dfrac{y + 5}{y - 5} + \dfrac{2y}{y^2 - 25}$ $\dfrac{y^2 + 12y + 25}{y^2 - 25}$

16. $\dfrac{-18}{y^2 - 9} + \dfrac{7}{3 - y}$ $\dfrac{-7y - 39}{y^2 - 9}$

17. $\dfrac{c}{c^2 - 4c} - \dfrac{5c}{c - 4}$ $\dfrac{1 - 5c}{c - 4}$

18. $\dfrac{t + 10}{t^2 - 100} + \dfrac{1}{t - 10}$ $\dfrac{2}{t - 10}$

Lesson 12-7 Write each mixed expression as a rational expression.

1. $4 + \dfrac{2}{x}$ $\dfrac{4x + 2}{x}$

2. $8 + \dfrac{5}{3t}$ $\dfrac{24t + 5}{3t}$

3. $3b + \dfrac{b + 1}{2b}$ $\dfrac{6b^2 + b + 1}{2b}$

4. $2n + \dfrac{4 + n}{n}$ $\dfrac{2n^2 + n + 4}{n}$

5. $a^2 + \dfrac{2}{a - 2}$ $\dfrac{a^3 - 2a^2 + 2}{a - 2}$

6. $3r^2 + \dfrac{4}{2r + 1}$ $\dfrac{6r^3 + 3r^2 + 4}{2r + 1}$

Simplify.

7. $\dfrac{3\frac{1}{2}}{4\frac{3}{4}}$ $\dfrac{14}{19}$

8. $\dfrac{\frac{x^2}{y}}{\frac{y}{x^3}}$ $\dfrac{x^5}{y^2}$

9. $\dfrac{\frac{t^4}{u}}{\frac{t^3}{u^2}}$ tu

10. $\dfrac{\frac{x^3}{y^2}}{\frac{x + y}{x - y}}$ $\dfrac{x^3(x - y)}{y^2(x + y)}$

11. $\dfrac{\frac{y}{3} + \frac{5}{6}}{2 + \frac{5}{y}}$ $\dfrac{y}{6}$

12. $\dfrac{\frac{1}{x} + \frac{1}{y}}{\frac{1}{y} - \frac{1}{x}}$ $\dfrac{x + y}{x - y}$

13. $\dfrac{\frac{t - 2}{t^2 - 4}}{t^2 + 5t + 6}$ $t + 3$

14. $\dfrac{\frac{y^2 - 1}{y^2 + 3y - 4}}{y + 1}$ $\dfrac{1}{y + 4}$

Lesson 12-8 Solve each equation.

1. $\dfrac{k}{6} + \dfrac{2k}{3} = -\dfrac{5}{2}$ -3

2. $\dfrac{3x}{5} + \dfrac{3}{2} = \dfrac{7x}{10}$ 15

3. $\dfrac{18}{b} = \dfrac{3}{b} + 3$ 5

4. $\dfrac{3}{5x} + \dfrac{7}{2x} = 1$ $\dfrac{41}{10}$

5. $\dfrac{2a - 3}{6} = \dfrac{2a}{3} + \dfrac{1}{2}$ -3

6. $\dfrac{x + 1}{x} + \dfrac{x + 4}{x} = 6$ $\dfrac{5}{4}$

7. $\dfrac{2b - 3}{7} - \dfrac{b}{2} = \dfrac{b + 3}{14}$ $-\dfrac{9}{4}$

8. $\dfrac{2y}{y - 4} - \dfrac{3}{5} = 3$ 9

9. $\dfrac{2t}{t + 3} + \dfrac{3}{t} = 2$ 3

10. $\dfrac{5x}{x + 1} + \dfrac{1}{x} = 5$ $\dfrac{1}{4}$

11. $\dfrac{r - 1}{r + 1} - \dfrac{2r}{r - 1} = -1$ 0

12. $\dfrac{m}{m + 1} + \dfrac{5}{m - 1} = 1$ $-\dfrac{3}{2}$

13. $\dfrac{5}{5 - p} - \dfrac{p^2}{5 - p} = -2$ -5 or 3

14. $\dfrac{14}{b - 6} = \dfrac{1}{2} + \dfrac{6}{b - 8}$ 10 or 20

15. $\dfrac{r}{3r + 6} - \dfrac{r}{5r + 10} = \dfrac{2}{5}$ -3

16. $\dfrac{4x}{2x + 3} - \dfrac{2x}{2x - 3} = 1$ $\dfrac{1}{2}$

17. $\dfrac{2a - 3}{a - 3} - 2 = \dfrac{12}{a + 2}$ $\dfrac{14}{3}$

18. $\dfrac{z + 3}{z - 1} + \dfrac{z + 1}{z - 3} = 2$ 2

Lesson 13-1 If c is the measure of the hypotenuse of a right triangle, find each missing measure. Round answers to the nearest hundredth.

1. $b = 20, c = 29, a = ?$ 21
2. $a = 7, b = 24, c = ?$ 25
3. $a = 2, b = 6, c = ?$ $\sqrt{40} \approx 6.32$
4. $b = 10, c = \sqrt{200}, a = ?$ 10
5. $a = 3, c = 3\sqrt{2}, b = ?$ 3
6. $a = 6, c = 14, b = ?$ $\sqrt{160} \approx 12.65$
7. $a = \sqrt{11}, c = \sqrt{47}, b = ?$ 6
8. $a = \sqrt{13}, b = 6, c = ?$ 7
9. $a = \sqrt{6}, b = 3, c = ?$ $\sqrt{15} \approx 3.87$
10. $b = \sqrt{75}, c = 10, a = ?$ 5
11. $b = 9, c = \sqrt{130}, a = ?$ 7
12. $a = 9, c = 15, b = ?$ 12
13. $b = 5, c = 11, a = ?$ $\sqrt{96} \approx 9.80$
14. $a = \sqrt{33}, b = 4, c = ?$ 7

Determine whether the following side measures would form right triangles.

15. $14, 48, 50$ yes
16. $20, 30, 40$ no
17. $21, 72, 75$ yes
18. $5, 12, \sqrt{119}$ yes
19. $15, 39, 36$ yes
20. $\sqrt{5}, 12, 13$ no
21. $10, 12, \sqrt{22}$ no
22. $2, 3, 4$ no
23. $\sqrt{7}, 8, \sqrt{71}$ yes

Lesson 13-2 Simplify. Leave in radical form and use absolute value symbols when necessary.

1. $\sqrt{50}$ $5\sqrt{2}$
2. $\sqrt{20}$ $2\sqrt{5}$
3. $\sqrt{162}$ $9\sqrt{2}$
4. $\sqrt{700}$ $10\sqrt{7}$
5. $\dfrac{\sqrt{3}}{\sqrt{5}}$ $\dfrac{\sqrt{15}}{5}$
6. $\dfrac{\sqrt{72}}{\sqrt{6}}$ $2\sqrt{3}$
7. $\sqrt{\dfrac{8}{7}}$ $\dfrac{2\sqrt{14}}{7}$
8. $\sqrt{\dfrac{7}{32}}$ $\dfrac{\sqrt{14}}{8}$
9. $\sqrt{10} \cdot \sqrt{20}$ $10\sqrt{2}$
10. $\sqrt{7} \cdot \sqrt{3}$ $\sqrt{21}$
11. $6\sqrt{2} \cdot \sqrt{3}$ $6\sqrt{6}$
12. $5\sqrt{6} \cdot 2\sqrt{3}$ $30\sqrt{2}$
13. $\sqrt{4x^4y^3}$ $2x^2y\sqrt{y}$
14. $\sqrt{200m^2y^3}$
15. $\sqrt{12ts^3}$ $2s\sqrt{3ts}$
16. $\sqrt{175a^4b^6}$
17. $\sqrt{\dfrac{54}{g^2}}$ $\dfrac{3\sqrt{6}}{|g|}$
18. $\sqrt{99x^3y^7}$
19. $\sqrt{\dfrac{32c^5}{9d^2}}$ $\dfrac{4c^2\sqrt{2c}}{3|d|}$
20. $\sqrt{\dfrac{27p^4}{3p^2}}$ $3|p|$
21. $\dfrac{1}{3+\sqrt{5}}$ $\dfrac{3-\sqrt{5}}{4}$
22. $\dfrac{2}{\sqrt{3}-5}$ $\dfrac{\sqrt{3}+5}{-11}$
23. $\dfrac{\sqrt{3}}{\sqrt{3}-5}$ $\dfrac{3+5\sqrt{3}}{-22}$
24. $\dfrac{\sqrt{6}}{7-2\sqrt{3}}$ $\dfrac{7\sqrt{6}+6\sqrt{2}}{37}$
25. $(\sqrt{p}+\sqrt{10})^2$
26. $(2\sqrt{5}+\sqrt{7})(2\sqrt{5}-\sqrt{7})$ 13
27. $(t-2\sqrt{3})(t-\sqrt{3})$

14. $10|m|y\sqrt{2y}$ 16. $5a^2|b^3|\sqrt{7}$ 18. $3xy^3\sqrt{11xy}$ 25. $p+2\sqrt{10p}+10$ 27. $t^2-3t\sqrt{3}+6$

Lesson 13-3 Simplify. 4. $14\sqrt{7}-\sqrt{2}$ 10. $-6\sqrt{7}+24\sqrt{5}$ 15. $69\sqrt{2}-10\sqrt{3}$

1. $3\sqrt{11}+6\sqrt{11}-2\sqrt{11}$ $7\sqrt{11}$
2. $6\sqrt{13}+7\sqrt{13}$ $13\sqrt{13}$
3. $2\sqrt{12}+5\sqrt{3}$ $9\sqrt{3}$
4. $9\sqrt{7}-4\sqrt{2}+3\sqrt{2}+5\sqrt{7}$
5. $3\sqrt{5}-5\sqrt{3}$ in simplest form
6. $4\sqrt{8}-3\sqrt{5}$ $8\sqrt{2}-3\sqrt{5}$
7. $2\sqrt{27}-4\sqrt{12}$ $-2\sqrt{3}$
8. $8\sqrt{32}+4\sqrt{50}$ $52\sqrt{2}$
9. $\sqrt{45}+6\sqrt{20}$ $15\sqrt{5}$
10. $2\sqrt{63}-6\sqrt{28}+8\sqrt{45}$
11. $14\sqrt{3t}+8\sqrt{3t}$ $22\sqrt{3t}$
12. $7\sqrt{6x}-12\sqrt{6x}$ $-5\sqrt{6x}$
13. $5\sqrt{7}-3\sqrt{28}$ $-\sqrt{7}$
14. $7\sqrt{8}-\sqrt{18}$ $11\sqrt{2}$
15. $7\sqrt{98}+5\sqrt{32}-2\sqrt{75}$
16. $4\sqrt{6}+3\sqrt{2}-2\sqrt{5}$ in simplest form
17. $-3\sqrt{20}+2\sqrt{45}-\sqrt{7}$ $-\sqrt{7}$
18. $4\sqrt{75}+6\sqrt{27}$ $38\sqrt{3}$
19. $10\sqrt{\dfrac{1}{5}}-\sqrt{45}-12\sqrt{\dfrac{5}{9}}$ $-5\sqrt{5}$
20. $\sqrt{15}-\sqrt{\dfrac{3}{5}}$ $\dfrac{4\sqrt{15}}{5}$
21. $3\sqrt{\dfrac{1}{3}}-9\sqrt{\dfrac{1}{12}}+\sqrt{243}$ $\dfrac{17\sqrt{3}}{2}$

Lesson 13-4 Solve each equation. Check your solution.

1. $\sqrt{5x} = 5$ 5

2. $4\sqrt{7} = \sqrt{-m}$ -112

3. $\sqrt{t} - 5 = 0$ 25

4. $\sqrt{3b} + 2 = 0$ no real solution

5. $\sqrt{x-3} = 6$ 39

6. $5 - \sqrt{3x} = 1$ $\frac{16}{3}$

7. $2 + 3\sqrt{y} = 13$ $\frac{121}{9}$

8. $\sqrt{3g} = 6$ 12

9. $\sqrt{a} - 2 = 0$ 4

10. $\sqrt{2j} - 4 = 8$ 72

11. $5 + \sqrt{x} = 9$ 16

12. $\sqrt{5y+4} = 7$ 9

13. $7 + \sqrt{5c} = 9$ $\frac{4}{5}$

14. $2\sqrt{5t} = 10$ 5

15. $\sqrt{44} = 2\sqrt{p}$ 11

16. $4\sqrt{x-5} = 15$ $\frac{305}{16}$

17. $4 - \sqrt{x-3} = 9$ no real solution

18. $\sqrt{10x^2 - 5} = 3x$ $\sqrt{5}$

19. $\sqrt{2a^2 - 144} = a$ 12

20. $\sqrt{3y+1} = y - 3$ 8

21. $\sqrt{2x^2 - 12} = x$ $2\sqrt{3}$

22. $\sqrt{b^2 + 16} + 2b = 5b$ $\sqrt{2}$

23. $\sqrt{m+2} + m = 4$ 2

24. $\sqrt{3 - 2c} + 3 = 2c$ $\frac{3}{2}$

Lesson 13-5 Find the distance between each pair of points whose coordinates are given. Express answers in simplest radical form and as decimal approximations rounded to the nearest hundredth.

1. $(4, 2), (-2, 10)$ 10

2. $(-5, 1), (7, 6)$ 13

3. $(4, -2), (1, 2)$ 5

4. $(-2, 4), (4, -2)$ $6\sqrt{2}$ or 8.49

5. $(3, 1), (-2, -1)$ $\sqrt{29}$ or 5.39

6. $(-2, 4), (7, -8)$ 15

7. $(-5, 0), (-9, 6)$ $2\sqrt{13}$ or 7.21

8. $(5, -1), (5, 13)$ 14

9. $(2, -3), (10, 8)$ $\sqrt{185}$ or 13.60

10. $(-7, 5), (2, -7)$ 15

11. $(-6, -2), (-5, 4)$ $\sqrt{37}$ or 6.08

12. $(8, -10), (3, 2)$ 13

13. $(4, -3), (7, -9)$ $3\sqrt{5}$ or 6.71

14. $(6, 3), (9, 7)$ 5

15. $(10, 0), (9, 7)$ $5\sqrt{2}$ or 7.07

16. $(2, -1), (-3, 3)$ $\sqrt{41}$ or 6.40

17. $(-5, 4), (3, -2)$ 10

18. $(0, -9), (0, 7)$ 16

19. $(-1, 7), (8, 4)$ $3\sqrt{10}$ or 9.49

20. $(-9, 2), (3, -3)$ 13

21. $(3\sqrt{2}, 7), (5\sqrt{2}, 9)$

22. $(6, 3), (10, 0)$ 5

23. $(3, 6), (5, -5)$ $5\sqrt{5}$ or 11.18

24. $(-4, 2), (5, 4)$ $\sqrt{85}$ or 9.22

21. $2\sqrt{3}$ or 3.46

Lesson 13-6 Find the value of c that makes each trinomial a perfect square.

1. $a^2 + 6a + c$ 9

2. $x^2 + 10x + c$ 25

3. $t^2 + 12t + c$ 36

4. $y^2 - 9y + c$ $\frac{81}{4}$

5. $p^2 - 14p + c$ 49

6. $b^2 + 5b + c$ $\frac{25}{4}$

Solve each equation by completing the square. Leave irrational roots in simplest radical form.

7. $x^2 - 4x = 5$ $-1, 5$

8. $t^2 + 12t - 45 = 0$ $-15, 3$

9. $b^2 + 4b - 12 = 0$ $2, -6$

10. $a^2 - 8a - 84 = 0$ $14, -6$

11. $c^2 + 6 = -5c$ $-3, -2$

12. $t^2 - 7t = -10$ $2, 5$

13. $p^2 - 8p + 5 = 0$ $4 \pm \sqrt{11}$

14. $a^2 + 4a + 2 = 0$ $-2 \pm \sqrt{2}$

15. $2y^2 + 7y - 4 = 0$ $-4, \frac{1}{2}$

16. $t^2 + 3t = 40$ $5, -8$

17. $x^2 + 8x - 9 = 0$ $-9, 1$

18. $y^2 + 5y - 84 = 0$ $-12, 7$

19. $x^2 + 2x - 6 = 0$ $-1 \pm \sqrt{7}$

20. $t^2 + 12t + 32 = 0$ $-4, -8$

21. $2x - 3x^2 = -8$ $2, -\frac{4}{3}$

22. $2y^2 - y - 9 = 0$ $\frac{1 \pm \sqrt{73}}{4}$

23. $2z^2 - 5z - 4 = 0$ $\frac{5 \pm \sqrt{57}}{4}$

24. $4t^2 - 6t - \frac{1}{2} = 0$ $\frac{3 \pm \sqrt{11}}{4}$

Write an algebraic expression for each verbal expression.

1. the sum of a number x and 13 $x + 13$

2. the reciprocal of a number x squared $\frac{1}{x^2}$

3. the cube of a number x decreased by 7 $x^3 - 7$

4. the product of 5 and a number x squared $5x^2$

Find the next two items for each pattern.   5. See Solutions Manual.

5.

6. 4, 7, 10, 13, . . .   16, 19

7. 2, 5, 10, 17, . . .   26, 37

Evaluate each expression.

8. $5^2 - 12$ 13

9. $(0.5)^3 + 2 \cdot 7$ 14.125

10. $\frac{2}{5}(16 - 9)$ $2\frac{4}{5}$

Evaluate each expression when $a = 2$, $b = 0.5$, $c = 3$, and $d = \frac{4}{3}$.

11. $a^2b + c$ 5

12. $(cd)^3$ 64

13. $(a + d)c$ 10

Solve each equation.

14. $y = (4.5 + 0.8) - 3.2$ 2.1

15. $4^2 - 3(4 - 2) = y$ 10

16. $\frac{2^3 - 1^3}{2 + 1} = y$ $\frac{7}{3}$

Name the property illustrated by each statement.

17. $a = a + 0$ additive identity

18. $\frac{1}{a} \cdot a = 1$ multiplicative inverse

19. If $a = b$, and $b = c$, then $a = c$.   transitive

20. $a(bc) = (ab)c$ associative ($\times$)

21. $8(st) = 8(ts)$ commutative ($\times$)

22. $7y + 5x - 4y = 5x + 7y - 4y$ commutative ($+$)

Simplify each expression.

23. $2m + 3m$ $5m$

24. $4x + 2y - 2x + y$ $2x + 3y$

25. $3(2a + b) - 1.5a - 1.5b$
   $4.5a + 1.5b$

Use the stem-and-leaf plot below to complete Exercises 26–27.

Stem	Leaf
1	1 4 6 8 8
2	0 3 3 3 5 7 9
3	0 0 2 6 $2|4 = 24$

26. List the set of data represented by the plot on the left. 11, 14, 16, 18, 18, 20, 23, 23, 23, 25, 27, 29, 30, 30, 32, 36

27. Which number is used most frequently?   23

Sketch a reasonable graph for each situation.   28–30. See Solutions Manual.

28. A basketball is shot from the free throw line and falls through the net.

29. A slammer is dropped on a stack of pogs and bounces off.

30. An infant grows through adulthood.

Solve.   31. 250 miles

31. **Travel** If a car travels at an average speed of 50 miles per hour, how far can the car travel in 5 hours? Use the distance formula $d = r \cdot t$.

32. **Geometry** If the area of a circle is given by the equation $A = \pi r^2$, find the area if the radius is 4 in. (Use 3.14 for π.)   50.24 in^2

33. If a farmer wants to put a fence around his rectangular garden that measures 11 yards by 14 yards, how many yards of fencing will he need?   50 yd

CHAPTER 2 TEST

Find each sum or difference.

1. $12 - 19$ -7

2. $-21 + (-34)$ -55

3. $1.654 + (-2.367)$ -0.713

4. $-\frac{7}{16} - \frac{3}{8}$ $-\frac{13}{16}$

5. $18b + 13xy - 46b$ $-28b + 13xy$

6. $6.32 - (-7.41)$ 13.73

7. $\frac{5}{8} + \left(-\frac{3}{16}\right) + \left(-\frac{3}{4}\right)$ $-\frac{5}{16}$

8. $32y + (-73y)$ $-41y$

9. $|-28 + (-13)|$ 41

10. $\begin{bmatrix} -8 & 5 \\ 2 & -3 \end{bmatrix} + \begin{bmatrix} 3 & 6 \\ -4 & -7 \end{bmatrix}$ $\begin{bmatrix} -5 & 11 \\ -2 & -10 \end{bmatrix}$

11. $\begin{bmatrix} 1 & 0 \\ -9 & 3 \end{bmatrix} - \begin{bmatrix} 5 & 8 \\ -4 & 2 \end{bmatrix}$ $\begin{bmatrix} -4 & -8 \\ -5 & 1 \end{bmatrix}$

Evaluate each expression.

12. $-x - 38$, if $x = -2$ -36

13. $\left|-\frac{1}{2} + z\right|$, if $z = \frac{1}{4}$ $\frac{1}{4}$

14. $mp - k$, if $m = -12$, $p = 1.5$, and $k = -8$ -10

15. $w^2 - 15$, if $w = 5$ 10

Replace each __?__ with <, >, or = to make each sentence true.

16. $-14 \underline{\ ?\ } -15$ $>$

17. $\frac{9}{20} \underline{\ ?\ } \frac{7}{15}$ $<$

18. $-4.65 \underline{\ ?\ } -4.45$ $<$

Find a number between the given numbers. Answers will vary; sample answers given.

19. $-\frac{2}{3}$ and $-\frac{9}{14}$ $-\frac{55}{84}$

20. $\frac{4}{7}$ and $\frac{9}{4}$ $\frac{79}{56}$

21. $\frac{12}{7}$ and $\frac{15}{8}$ $\frac{99}{56}$

Simplify.

22. $\frac{8(-3)}{2}$ -12

23. $(-5)(-2)(-2) - (-6)(-3)$ -38

24. $\frac{2}{3}\left(\frac{1}{2}\right) - \left(-\frac{3}{2}\right)\left(-\frac{2}{3}\right)$ $-\frac{2}{3}$

25. $\frac{70x - 30y}{-5}$ $-14x + 6y$

26. $\frac{7}{\frac{-2}{5}}$ $-\frac{35}{2}$

27. $\frac{3}{4}(8x + 12y) - \frac{5}{7}(21x - 35y)$ $-9x + 34y$

Find each square root. Use a calculator if necessary. Round to the nearest hundredth if the result is not a whole number.

28. $\pm\sqrt{\frac{16}{81}}$ $\pm\frac{4}{9} \approx 0.44$

29. $\sqrt{40}$ 6.32

30. $\sqrt{2.89}$ 1.7

31. Statistics The height, in inches, of the students in a health class are 65, 63, 68, 66, 72, 61, 62, 63, 59, 58, 61, 74, 65, 63, 71, 60, 62, 63, 71, 70, 59, 66, 61, 62, 68, 69, 64, 63, 70, 61, 68, and 67.

a. Make a line plot of the data on the heights of the students in health class. See Solutions Manual.

b. What was the most common height for the students in this class? 63

32. Define a variable and write an equation for the following problem. Do *not* solve.
Each week for several weeks, Save-a-Buck stores reduced the price of a sofa by $18.25. The original price was $380.25. The final reduced price was $252.50. For how many weeks was the sofa on sale? Let w = number of weeks on sale; $380.25 - 18.25w = 252.50$.

33. Translate $r = (a - b)^3$ into a verbal sentence. The number r equals the cube of the difference of a and b.

Solve each equation. Then check your solution.

1. $-15 - k = 8$ $\quad -23$

2. $-1.2x = 7.2$ $\quad -6$

3. $\frac{3}{4}y = -27$ $\quad -36$

4. $\frac{t-7}{4} = 11$ $\quad 51$

5. $-12 = 7 - \frac{y}{3}$ $\quad 57$

6. $k - 16 = -21$ $\quad -5$

7. $t - (-3.4) = -5.3$ $\quad -8.7$

8. $-3(x+5) = 8x + 18$ $\quad -3$

9. $2 - \frac{1}{4}(b-12) = 9$ $\quad -16$

10. $\frac{r}{5} - 3 = \frac{2r}{5} + 16$ $\quad -95$

11. $25 - 7w = 46$ $\quad -3$

12. $-w + 11 = 4.6$ $\quad 6.4$

Find the mean, median, and mode for each set of data.

13. 67, 31, 15, 49, 31, 35, 42, 27

    37.125, 33, 31

14.

Stem	Leaf
18	0 5 8
19	3 3 4 4
20	8 8 9
21	4 5 5 5 9 $\quad 19|4 = 194$

    202, 208, 215

Define a variable, write an equation, and solve each problem. $\quad$ 16. $-\frac{2}{3}x = \frac{8}{5}$; $-\frac{12}{5}$

15. The sum of two integers is -23. One integer is -84. Find the other integer.
    $-84 + x = -23$; 61

16. Negative two thirds of a number is eight fifths. What is the number?

17. What number decreased by 37 is -65?
    $x - 37 = -65$; -28

18. The measures of two angles of a triangle are $23°$ and $121°$. Find the measure of the third angle. $\quad 180 = 23 + 121 + x$, $36°$

19. Find two consecutive odd integers whose sum is 172. $\quad x + (x+2) = 172$; 85, 87

Solve each equation or formula for the variable specified.

20. $h = at - 0.25vt^2$, for a $\quad a = \dfrac{h + 0.25vt^2}{t}$

21. $A = \frac{1}{2}(b + B)h$, for h $\quad h = \dfrac{2A}{b + B}$

Solve.

22. **Geometry** One of two supplementary angles measures $37°$ less than four times the other. Find the measure of each angle. $\quad 43.4°$, $136.6°$

23. **Grades** James had a cumulative score of 338 points after his first four tests, each worth 100 points. What is the least that he would need to score on his next chapter test in order to have a mean score of 82%? $\quad 72\%$

24. **Merchandise Stock** A store has 49 cartons of yogurt, some plain, some with blueberry flavoring. There are six times as many cartons of plain yogurt as flavored yogurt. How many cartons of plain yogurt are there?
    42 cartons

25. **Consumerism** Teri went to SuperValue to buy some groceries. Her total bill was $8.51. She bought 2 pounds of grapes at $0.99 per pound, a gallon of milk for $2.59, and three cans of orange juice. She used a coupon worth 50¢ for the three cans of orange juice. How much was each can of orange juice? $\quad \$1.48$

CHAPTER 4 TEST

Solve each proportion.

1. $\frac{2}{5} = \frac{x-3}{-2}$ $\frac{11}{5}$

2. $\frac{n}{4} = \frac{3.25}{52}$ $\frac{1}{4}$

3. $\frac{x-3}{x+5} = \frac{9}{11}$ 39

4. $\frac{x+1}{-3} = \frac{x-4}{5}$ $\frac{7}{8}$

Solve.

5. Find 6.5% of 80. **5.2**

6. 42 is what percent of 126? $33\frac{1}{3}\%$

7. 84 is 60% of what number? **140**

8. What number decreased by 20% is 16? **20**

9. 24 is what percent of 8? **300%**

10. 54 is 20% more than what number? **45**

11. A price decreased from $60 to $45. Find the percent of decrease. **25%**

12. The price in dollars p minus a 15% discount is $3.40. Find p. **$4**

△ABC and △JKH are similar. For each set of measures given, find the measures of the remaining sides.

13. $c = 20$, $h = 15$, $k = 16$, $j = 12$
    $a = 16$, $b = \frac{64}{3}$

14. $c = 12$, $b = 13$, $a = 6$, $h = 10$
    $j = 5$, $k = \frac{65}{6}$

15. $k = 5$, $c = 6.5$, $b = 7.5$, $a = 4.5$
    $h = 4.\overline{3}$, $j = 3$

16. $h = 1\frac{1}{2}$, $c = 4\frac{1}{2}$, $k = 2\frac{1}{4}$, $a = 3$
    $b = 6\frac{3}{4}$, $j = 1$

Solve each right triangle. State the side lengths to the nearest tenth and the angle measures to the nearest degree.

17.

18.
$m\angle A = 48°$
$a = 7.4$
$b = 6.7$

19. $m\angle A = 44°$
$m\angle B = 46°$
$a = 20$
$m\angle A = 36°$
$m\angle B = 54°$
$c = 25.5$

20. $a = 29.9$
$c = 30.9$
$m\angle A = 75°$

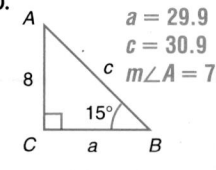

Solve.

21. **Music** During a 20-song sequence on a popular radio station, 8 soft-rock, 7 hard-rock, and 5 rap songs are played at random. You tune in to the station. **b. 1:3**

    a. What is the probability that a hard-rock song is playing? $\frac{7}{20}$

    b. What are the odds that a rap song is playing?

    c. What is the probability that a soft-rock song or a hard-rock song is playing? $\frac{3}{4}$

22. Two buildings are separated by an alley. Joe is looking out of a window 60 feet above the ground in one building. He observes the measurement of the angle of depression of the base of the second building to be 50° and the angle of elevation of the top to be 40°. How high is the second building? **about 102 feet**

23. **Finance** Gladys deposited an amount of money in the bank at 6.5% annual interest. After 6 months, she received $7.80 interest. How much money had Gladys deposited in the bank? **$240**

24. **Stereo** Larry is buying a stereo that costs $399. Since he is an employee of the store, he receives a 15% discount. He also has to pay 6% sales tax. If the discount is computed first, what is the total cost of Larry's stereo? **$359.50**

25. **Travel** At the same time Kris leaves Washington, D.C., for Detroit, Michigan, Amy leaves Detroit for Washington, D.C. The distance between the cities is 510 miles. Amy's average speed is 5 miles per hour faster than Kris's. How fast is Kris's average speed if they pass each other in 6 hours? **40 mph**

1. Graph $K(0, -5)$, $M(3, -5)$, and $N(-2, -3)$.   See margin.

2. Name the quadrant in which $P(-5, 1)$ is located.   II

Express the relations shown in each table, mapping, or graph as a set of ordered pairs. Then state the domain, range, and inverse of the relation.   3–5. See margin.

3.

x	f(x)
0	−1
2	4
4	5
6	10

4.

5.

Solve each equation if the domain is $\{-2, -1, 0, 2, 4\}$.

6. $y = -4x + 10$

7. $4 - 2x = 5y$

8. $-x + 3y = 1$

8. $\left\{\left(-2, \frac{-1}{3}\right), (-1, 0), \left(0, \frac{1}{3}\right), (2, 1), \left(4, \frac{5}{3}\right)\right\}$

6. $\{(-2, 18), (-1, 14), (0, 10), (2, 2), (4, -6)\}$

7. $\left\{\left(-2, \frac{8}{5}\right), \left(-1, \frac{6}{5}\right), \left(0, \frac{4}{5}\right), (2, 0), \left(4, \frac{-4}{5}\right)\right\}$.

Graph each equation.   9–11. See margin.

9. $x + 2y = -1$

10. $-3x = 5 - y$

11. $-4 = x - \frac{1}{2}y$

Determine whether each relation is a function.

12. $\{(2, 4), (3, 2), (4, 6), (5, 4)\}$ yes

13. $8y = 7 + 3x$ yes

If $f(x) = -2x + 5$ **and** $g(x) = x^2 - 4x + 1$, **find each value.**

14. $f\left(\frac{1}{2}\right)$ 4

15. $g(-2)$ 13

16. $-2g(3)$ 4

Write an equation for each relation.

17.

x	1	2	3	4	7
y	3	8	13	18	33

$y = 5x - 2$

18.

x	1	3	5	7	9	11	13
y	5	17	29	41	53	65	77

$y = 6x - 1$

Solve.   19c. Gina had more consistent scores since she had the smaller interquartile range.

19. School   Art and Gina's test scores in algebra are given below.
Art: 87, 54, 78, 97, 65, 82, 75, 68, 82, 73, 66, 75
Gina: 70, 80, 57, 100, 73, 74, 65, 77, 91, 69, 71, 76
  a. Find the range and interquartile range for each set of scores.
  b. Identify any outliers.   Gina's score of 100
  c. Which one had the more consistent scores?

20. Sales   When you use Jay's Taxi Service, a two-mile trip costs $6.30, a five-mile trip costs $11.25, and a ten-mile trip costs $19.50. Write an equation to describe this relationship and use it to find the cost of a one-mile trip. $y = 1.65x + 3$; $4.65
19a. Art: range, 43, interquartile range, 15;
Gina: range, 43, interquartile range, 9

CHAPTER 6 TEST

Determine the slope of the line that passes through each pair of points.

1. $(5, 8), (-3, 7)$ $\frac{1}{8}$

2. $(-2, 5), (2, 9)$ **1**

Find the slope and x- and y-intercepts of the graph of each equation.

3. $x - 8y = 3$ $\frac{1}{8}, 3, -\frac{3}{8}$

4. $3x - 2y = 9$ $\frac{3}{2}, 3, -\frac{9}{2}$

5. $y = 7$ **0, none, 7**

Graph each equation. **6–8. See Solutions Manual.**

6. $4x - 3y = 24$

7. $2x + 7y = 16$

8. $y = \frac{2}{3}x + 3$

Determine whether the graphs of each pair of equations are *parallel, perpendicular,* or *neither*.

9. $y = 4x - 11$ **parallel**
$2y + 1 = 8x$

10. $-7y = 4x + 14$ **perpendicular**
$7x - 4y = -12$

Write an equation in standard form of the line that satisfies the given conditions.

11. passes through $(2, 5)$ and $(8, -3)$
$4x + 3y = 23$

12. passes through $(-2, -1)$ and $(6, -4)$
$3x + 8y = -14$

13. has slope of 2 and y-intercept $= 3$
$2x - y = -3$

14. has y-intercept $= -4$ and passes through $(5, -3)$
$x - 5y = 20$

15. slope $= \frac{3}{4}$ and passes through $(6, -2)$
$3x - 4y = 26$

16. parallel to $6x - y = 7$ and passes through $(-2, 8)$
$6x - y = -20$

Write an equation in slope-intercept form of the line that satisfies the given conditions. **18.** $y = \frac{1}{5}x - \frac{23}{5}$

17. passes through $(4, -2)$ and the origin $y = -\frac{1}{2}x$

18. passes through $(-2, -5)$ and $(8, -3)$

19. passes through $(6, 4)$ with y-intercept $= -2$

20. slope $= -\frac{2}{3}$ and y-intercept $= 5$ $y = -\frac{2}{3}x + 5$

21. slope $= 6$ and passes through $(-3, -4)$
$y = 6x + 14$ **19.** $y = x - 2$

22. perpendicular to $5x - 3y = 9$ and passes through the origin $y = -\frac{3}{5}x$

23. parallel to $3x + 7y = 4$ and passes through $(5, -2)$
$y = -\frac{3}{7}x + \frac{1}{7}$

24. perpendicular to $x + 3y = 7$ and passes through $(5, 2)$ $y = 3x - 13$

25. Find the coordinates of the other endpoint of segment AB given $A(-2, -7)$ and midpoint $M(6, -5)$. $B(14, -3)$

26. The table below shows the number of students per computer in American classrooms since 1983.

School Year	Students per Computer	School Year	Students per Computer
1983–84	125	1989–90	22
1984–85	75	1990–91	20
1985–86	50	1991–92	18
1986–87	37	1992–93	16
1987–88	32	1993–94	14
1988–89	25	1994–95	12

a. Make a scatter plot of the data. **See Solutions Manual.**
b. Describe the correlation between the variables. **negative**
c. Are these data best modeled by a line? Explain your answer. **No, they are not linear.**

Solve each inequality. Then check your solution.

1. $-12 \le d + 7$ $\{d \mid d \ge -19\}$

2. $7x < 6x - 11$ $\{x \mid x < -11\}$

3. $z - 1 \ge 2z - 3$ $\{z \mid z \le 2\}$

4. $5 - 4b > -23$ $\{b \mid b < 7\}$

5. $-\frac{2}{3}r \le \frac{7}{12}$ $\left\{r \mid r \ge -\frac{7}{8}\right\}$

6. $8y + 3 < 13y - 9$ $\left\{y \mid y > \frac{12}{5}\right\}$

7. $8(1 - 2z) \le 25 + z$ $\{z \mid z \ge -1\}$

8. $0.3(m + 4) > 0.5(m - 4)$ $\{m \mid m < 16\}$

9. $\frac{2n - 3}{-7} \le 5$ $\{n \mid n \ge -16\}$

10. $y + \frac{5}{8} > \frac{11}{24}$ $\left\{y \mid y > -\frac{1}{6}\right\}$

Solve each compound inequality. Then graph the solution set. 11–16. See margin for graphs.

11. $x + 1 > -2$ and $3x < 6$ $\{x \mid -3 < x < 2\}$

12. $2n + 1 \ge 15$ or $2n + 1 \le -1$ $\{n \mid n \ge 7 \text{ or } n \le -1\}$

13. $8 + 3t > 2$ and $-12 > 11t - 1$ $\{t \mid -2 < t < -1\}$

14. $|2x - 1| < 5$ $\{x \mid -2 < x < 3\}$

15. $|5 - 3b| \ge 1$ $\left\{b \mid b \ge 2 \text{ or } b \le \frac{4}{3}\right\}$

16. $|3 - 5y| < 8$ $\left\{y \mid -1 < y < \frac{11}{5}\right\}$

Define a variable, write an inequality, and solve each problem. Then check your solution.

17. Twice a number subtracted from 12 is no less than the number increased by 27. $\{n \mid n \le -5\}$

18. Seven less than twice a number is between 71 and 83. $\{n \mid 39 < n < 45\}$

19. The product of two integers is no less than 30. One of the integers is 6. What is the other integer? **5 or more**

20. The average of four consecutive odd integers is less than 20. What are the greatest integers that satisfy this condition? **15, 17, 19, 21**

Graph each inequality. 21–23. See margin for graphs.

21. $y \ge 5x + 1$

22. $x - 2y > 8$

23. $3x - 2y < 6$

Solve.

24. **Business** Two men and three women are each waiting for a job interview. There is only enough time to interview two people before lunch. Two people are chosen at random.

    a. What is the probability that both people are women? $\frac{3}{10}$ or 0.3

    b. What is the probability that at least one person is a woman? $\frac{9}{10}$ or 0.9

    c. Which is more likely, one of the people is a woman and the other is a man, or both people are either men or women? **man and woman**

25. **Fire Safety** The city council of McBride is investigating the efficiency of the fire department. The time taken by the fire department to respond to a fire alarm was surveyed. It was found that the response times in minutes for 17 alarms were as follows.

    1, 3, 2, 2, 1, 9, 4, 6, 1, 10, 1, 4, 5, 10, 1, 3, 6

    a. Draw a box-and-whisker plot of the data. See margin.

    b. Between what two values of the data is the middle 50% of the data? **1 and 6**

CHAPTER 8 TEST

Graph each system of equations. Then determine whether the system has *one* solution, *no* solution, or *infinitely many* solutions. If the system has one solution, name it. 1–6. See Solutions Manual for graphs.

1. $y = x + 2$
   $y = 2x + 7$ one; $(-5, -3)$

2. $x + 2y = 11$
   $x = 14 - 2y$ no solution

3. $2x + 5y = 16$
   $5x - 2y = 11$ one; $(3, 2)$

4. $3x + y = 5$
   $2y - 10 = -6x$ infinitely many

5. $y + 2x = -1$
   $y - 4 = -2x$ no solution

6. $2x + y = -4$
   $5x + 3y = -6$ one; $(-6, 8)$

Use substitution or elimination to solve each system of equations.

7. $y = 7 - x$
   $x - y = -3$ $(2, 5)$

8. $x = 2y - 7$
   $y - 3x = -9$ $(5, 6)$

9. $x + y = 8$
   $x - y = 2$ $(5, 3)$

10. $3x - y = 11$
    $x + 2y = -36$ $(-2, -17)$

11. $3x + y = 10$
    $3x - 2y = 16$ $(4, -2)$

12. $5x - 3y = 12$
    $-2x + 3y = -3$ $(3, 1)$

13. $2x + 5y = 12$
    $x - 6y = -11$ $(1, 2)$

14. $x + y = 6$
    $3x - 3y = 13$ $\left(\frac{31}{6}, \frac{5}{6}\right)$

15. $3x + \frac{1}{3}y = 10$
    $2x - \frac{5}{3}y = 35$ $(5, -15)$

16. $8x - 6y = 14$
    $6x - 9y = 15$ $(1, -1)$

17. $5x - y = 1$
    $y = -3x + 1$ $\left(\frac{1}{4}, \frac{1}{4}\right)$

18. $7x + 3y = 13$
    $3x - 2y = -1$ $(1, 2)$

Solve each system of inequalities by graphing. 19–21. See Solutions Manual for graphs.

19. $y \le 3$
    $y > -x + 2$

20. $x \le 2y$
    $2x + 3y \le 7$

21. $x > y + 1$
    $2x + y \ge -4$

Solve.

22. **Number Theory** The units digit of a two-digit number exceeds twice the tens digit by 1. Find the number if the sum of its digits is 10. **37**

23. **Geometry** The difference between the length and width of a rectangle is 7 cm. Find the dimensions of the rectangle if its perimeter is 50 cm. **16 cm by 9 cm**

24. **Finance** Last year, Jodi invested $10,000, part at 6% annual interest and the rest at 8% annual interest. If she received $760 in interest at the end of the year, how much did she invest at each rate? **$2000 at 6%, $8000 at 8%**

25. **Organize Data** Joey sold 30 peaches from his fruit stand for a total of $7.50. He sold small ones for 20 cents each and large ones for 35 cents each. How many of each kind did he sell? **20 small and 10 large**

Simplify. Assume that no denominator is equal to zero.

1. $(a^2b^4)(a^3b^5)$ a^5b^9

2. $(-12abc)(4a^2b^4)$ $-48a^3b^5c$

3. $\left(\frac{3}{5}m\right)^2$ $\frac{9}{25}m^2$

4. $(-3a)^4(a^5b)^2$ $81a^{14}b^2$

5. $(-5a^2)(-6b^3)^2$ $-180a^2b^6$

6. $(5a)^2b + 7a^2b$ $32a^2b$

7. $\frac{y^{11}}{y^6}$ y^5

8. $\frac{mn^4}{m^3n^2}$ $\frac{n^2}{m^2}$

9. $\frac{9a^2bc^2}{63a^4bc}$ $\frac{c}{7a^2}$

10. $\frac{48a^2bc^5}{(3ab^3c^2)^2}$ $\frac{16c}{3b^5}$

11. $\frac{14ab^{-3}}{21a^2b^{-5}}$ $\frac{2b^2}{3a}$

12. $\frac{(10a^2bc^4)^{-2}}{(5^{-1}a^{-1}b^{-5})^2}$ $\frac{b^8}{4a^2c^8}$

Express each number in scientific notation.

13. $46,300$ 4.63×10^4

14. 0.003892 3.892×10^{-3}

15. 284×10^3 2.84×10^5

16. 0.0031×10^4 3.1×10

Evaluate. Express each result in scientific notation.

17. $(3 \times 10^3)(2 \times 10^4)$ 6×10^7

18. $\frac{2.5 \times 10^3}{5 \times 10^{-3}}$ 5×10^5

19. $\frac{14.72 \times 10^{-4}}{3.2 \times 10^{-3}}$ 4.6×10^{-1}

20. $(15 \times 10^{-7})(3.1 \times 10^4)$ 4.65×10^{-2}

21. Find the degree of $5ya^3 - 7 - y^2a^2 + 2y^3a$ and arrange the terms so that the powers of y are in descending order. $4; 2y^3a - y^2a^2 + 5ya^3 - 7$

Find each sum or difference.

22. $\begin{array}{r} 5ax^2 + 3a^2x - 7a^3 \\ (+)\ 2ax^2 - 8a^2x\ \ \ \ \ \ \ \ + 4 \\ \hline 7ax^2 - 5a^2x - 7a^3 + 4 \end{array}$

23. $\begin{array}{r} x^3 - 3x^2y + 4xy^2 + y^3 \\ (-)\ 7x^3 + x^2y - 9xy^2 + y^3 \\ \hline -6x^3 - 4x^2y + 13xy^2 \end{array}$

24. $(n^2 - 5n + 4) - (5n^2 + 3n - 1)$
$-4n^2 - 8n + 5$

25. $(ab^3 - 4a^2b^2 + ab - 7) + (-2ab^3 + 4ab^2 + 3ab + 2)$
$-ab^3 - 4a^2b^2 + 4ab^2 + 4ab - 5$

Simplify.   32. $3x^3 - 17x^2 - 9x$

26. $(h - 5)^2$ $h^2 - 10h + 25$

27. $(2x - 5)(7x + 3)$ $14x^2 - 29x - 15$

28. $(4x - y)(4x + y)$ $16x^2 - y^2$

29. $(2a^2b + b^2)^2$ $4a^4b^2 + 4a^2b^3 + b^4$

30. $3x^2y^3(2x - xy^2)$ $6x^3y^3 - 3x^3y^5$

31. $(4m + 3n)(2m - 5n)$ $8m^2 - 14mn - 15n^2$

32. $x^2(x - 8) - 3x(x^2 - 7x + 3) + 5(x^3 - 6x^2)$

33. $(x - 6)(x^2 - 4x + 5)$ $x^3 - 10x^2 + 29x - 30$

CHAPTER TEST

Find the GCF of the given monomials.

1. $48, 64$ 16

2. $18a^2b, 28a^3b^2$ $2a^2b$

3. $6x^2y^3, 12x^2y^2z, 15x^2y$ $3x^2y$

Factor each polynomial, if possible. If the polynomial cannot be factored using integers, write *prime*.

4. $25y^2 - 49w^2$ $(5y - 7w)(5y + 7w)$

5. $t^2 - 16t + 64$ $(t - 8)^2$

6. $x^2 + 14x + 24$ $(x + 12)(x + 2)$

7. $28m^2 + 18m$ $2m(14m + 9)$

8. $a^2 - 11ab + 18b^2$ $(a - 2b)(a - 9b)$

9. $12x^2 + 23x - 24$ $(3x + 8)(4x - 3)$

10. $2h^2 - 3h - 18$ prime

11. $6x^3 + 15x^2 - 9x$ $3x(x + 3)(2x - 1)$

12. $4my - 20m + 3py - 15p$ $(y - 5)(4m + 3p)$

13. $x^3 - 4x^2 - 9x + 36$ $(x - 4)(x - 3)(x + 3)$

14. $36a^2b^3 - 45ab^4$ $9ab^3(4a - 5b)$

15. $36m^2 + 60mn + 25n^2$ $(6m + 5n)^2$

16. $\frac{1}{4}a^2 - \frac{4}{9}$ $\left(\frac{1}{2}a - \frac{2}{3}\right)\left(\frac{1}{2}a + \frac{2}{3}\right)$

17. $64p^2 - 63p + 16$ prime

18. $15a^2b + 5a^2 - 10a$ $5a(3ab + a - 2)$

19. $6y^2 - 5y - 6$ $(2y - 3)(3y + 2)$

20. $4s^2 - 100t^2$ $4(s - 5t)(s + 5t)$

21. $2d^2 + d - 1$ $(2d - 1)(d + 1)$

22. $3g^2 + g + 1$ prime

23. $2xz + 2yz - x - y$ $(x + y)(2z - 1)$

Solve each equation. Check your solutions.

24. $(4x - 3)(3x + 2) = 0$ $\left\{\frac{3}{4}, -\frac{2}{3}\right\}$

25. $18s^2 + 72s = 0$ $\{0, -4\}$

26. $4x^2 = 36$ $\{-3, 3\}$

27. $t^2 + 25 = 10t$ $\{5\}$

28. $a^2 - 9a - 52 = 0$ $\{-4, 13\}$

29. $x^3 - 5x^2 - 66x = 0$ $\{-6, 0, 11\}$

30. $2x^2 = 9x + 5$ $\left\{-\frac{1}{2}, 5\right\}$

31. $3b^2 + 6 = 11b$ $\left\{\frac{2}{3}, 3\right\}$

Solve.

32. **Geometry**   A rectangle is 4 inches wide by 7 inches long. When the length and width are increased by the same amount, the area is increased by 26 square inches. What are the dimensions of the new rectangle?   6 in. by 9 in.

33. **Construction**   A rectangular lawn is 24 feet wide by 32 feet long. A sidewalk will be built along the inside edges of all four sides. The remaining lawn will have an area of 425 square feet. How wide will the walk be?   3.5 feet

CHAPTER 11 TEST

Write the equation of the axis of symmetry and find the coordinates of the vertex of the graph of each equation. State if the vertex is a maximum or minimum. Then graph the equation.

1. $y = x^2 - 4x + 13$ $x = 2; (2, 9)$; minimum

2. $y = -3x^2 - 6x + 4$ $x = -1; (-1, 7)$; maximum

3. $y = 2x^2 + 3$ $x = 0; (0, 3)$; minimum

4. $y = -1(x - 2)^2 + 1$ $x = 2; (2, 1)$; maximum

1–4. See Solutions Manual for graphs.

Solve each equation by graphing. If exact roots cannot be found, state the consecutive integers between which the roots lie.

5. $x^2 - 2x + 2 = 0$ $\varnothing$

6. $x^2 + 6x = -7$ $-5 < x < -4, -2 < x < -1$

7. $x^2 + 24x + 144 = 0$ -12

8. $2x^2 - 8x = 42$ $7, -3$

Solve each equation.

9. $x^2 + 7x + 6 = 0$ $-1, -6$

10. $2x^2 - 5x - 12 = 0$ $4, -\dfrac{3}{2}$

11. $6n^2 + 7n = 20$ $\dfrac{4}{3}, -\dfrac{5}{2}$

12. $3k^2 + 2k = 5$ $1, -\dfrac{5}{3}$

13. $y^2 - \dfrac{3y}{5} + \dfrac{2}{25} = 0$ $\dfrac{2}{5}, \dfrac{1}{5}$

14. $-3x^2 + 5 = 14x$ $\dfrac{1}{3}, -5$

15. $4^{x-2} = 16^{2x+5}$ -4

16. $1000^x = 10,000^{6x+4}$ $-\dfrac{16}{21}$

17. $5^{x^2} = 5^{15-2x}$ $-5, 3$

18. $\left(\dfrac{1}{2}\right)^{x-2} = 4^{5x}$ $\dfrac{2}{11}$

Graph each function. State the y-intercept.   19–21. See Solutions Manual for graphs.

19. $y = \left(\dfrac{1}{2}\right)^x$ 1

20. $y = 4 \cdot 2^x$ 4

21. $y = \left(\dfrac{1}{3}\right)^x - 3$ -2

Solve.

22. Automobile   Adina Ley needs to replace her car. If she leases a car, she will pay $410 a month for 2 years and then has the option to buy the car for $14, 458. The price of the car now is $17,369. If the car depreciates at 16% per year, how will the depreciated price compare with the buyout price of the lease?   **The depreciated value is $2202 less than the buyout price.**

23. Geometry   The area of a certain square is one-half the area of the rectangle formed if the length of one side of the square is increased by 2 cm and the length of an adjacent side is increased by 3 cm. What are the dimensions of the square?   **6 cm by 6 cm**

24. Number Theory   Find two integers whose sum is 21 and whose product is 90.   **6, 15**

25. Investment   After 6 years, a certain investment is worth $8479. If the money was invested at 9% interest compounded semiannually, find the original amount that was invested.   **$5000**

CHAPTER TEST

Simplify each rational expression. State the excluded values of the variables.

1. $\dfrac{5-2m}{6m-15}$ $-\dfrac{1}{3}, m \neq \dfrac{5}{2}$

2. $\dfrac{3+x}{2x^2+5x-3}$ $\dfrac{1}{2x-1}, x \neq \dfrac{1}{2}, -3$

3. $\dfrac{4c^2+12c+9}{2c^2-11c-21}$

   $\dfrac{2c+3}{c-7}, c \neq -\dfrac{3}{2}, 7$

Simplify each expression.

4. $\dfrac{1-\frac{9}{t}}{1-\frac{81}{t^2}}$ $\dfrac{t}{t+9}$

5. $\dfrac{\frac{5}{6}+\frac{u}{t}}{\frac{2u}{t}-3}$ $\dfrac{6u+5t}{12u-18t}$

6. $\dfrac{x+4+\frac{5}{x-2}}{x+6+\frac{15}{x-2}}$ $\dfrac{x-1}{x+1}$

Perform the indicated operations.

7. $\dfrac{2x}{x-7} - \dfrac{14}{x-7}$ 2

8. $\dfrac{n+3}{2n-8} \cdot \dfrac{6n-24}{2n+1}$ $\dfrac{3n+9}{2n+1}$

9. $(10m^2+9m-36) \div (2m-3)$ $5m+12$

10. $\dfrac{x^2+4x-32}{x+5} \cdot \dfrac{x-3}{x^2-7x+12}$ $\dfrac{x+8}{x+5}$

11. $\dfrac{z^2+2z-15}{z^2+9z+20} \div (z-3)$ $\dfrac{1}{z+4}$

12. $\dfrac{4x^2+11x+6}{x^2-x-6} \div \dfrac{x^2+8x+16}{x^2+x-12}$ $\dfrac{4x+3}{x+4}$

13. $(10z^4+5z^3-z^2) \div 5z^3$ $2z+1-\dfrac{1}{5z}$

14. $\dfrac{y}{7y+14} + \dfrac{6}{3y+6}$ $\dfrac{y+14}{7y+14}$

15. $\dfrac{x+5}{x+2} + 6$ $\dfrac{7x+17}{x+2}$

16. $\dfrac{x^2-1}{x+1} - \dfrac{x^2+1}{x-1}$ $\dfrac{-2x}{x-1}$

17. $\dfrac{-3}{a-5} + \dfrac{15}{a^2-5a}$ $\dfrac{-3}{a}$

18. $\dfrac{8}{m^2} \cdot \left(\dfrac{m^2}{2c}\right)^2$ $\dfrac{2m^2}{c^2}$

Solve each equation.

19. $\dfrac{2}{3t} + \dfrac{1}{2} = \dfrac{3}{4t}$ $\dfrac{1}{6}$

20. $\dfrac{2e}{e-4} - 2 = \dfrac{4}{e+5}$ -14

21. $\dfrac{4}{h-4} = \dfrac{3h}{h+3}$ $6, -\dfrac{2}{3}$

Solve each formula for the variable indicated.

22. $F = G\left(\dfrac{Mm}{d^2}\right)$, for G $G = \dfrac{Fd^2}{Mm}$

23. $\dfrac{1}{R_T} = \dfrac{1}{R_1} + \dfrac{1}{R_2}$, for R_2 $R_2 = \dfrac{R_T R_1}{R_1 - R_T}$

Solve. 25. $7\dfrac{1}{17}$ ohms

24. **Keyboarding** Willie can type a 200 word essay in 6 hours. Myra can type the same essay in $4\dfrac{1}{2}$ hours. If they work together, how long will it take them to type the essay? $2\dfrac{4}{7}$ **hours**

25. **Electronics** Three appliances are connected in parallel: a lamp of resistance 120 ohms, a toaster of resistance 20 ohms, and an iron of resistance 12 ohms. Find the total resistance.

Simplify. Leave in radical form and use absolute value symbols when necessary.

1. $\sqrt{480}$ $4\sqrt{30}$

2. $\sqrt{72} \cdot \sqrt{48}$ $24\sqrt{6}$

3. $\sqrt{54x^4y}$ $3x^2\sqrt{6y}$

4. $\sqrt{\frac{32}{25}}$ $\frac{4\sqrt{2}}{5}$

5. $\sqrt{\frac{3x^2}{4n^3}}$ $\frac{|x|\sqrt{3n}}{2n^2}$

6. $\sqrt{6} + \sqrt{\frac{2}{3}}$ $\frac{4\sqrt{6}}{3}$

7. $\left(x + \sqrt{3}\right)^2$ $x^2 + 2x\sqrt{3} + 3$

8. $\frac{7}{7 + \sqrt{5}}$ $\frac{49 - 7\sqrt{5}}{44}$

9. $3\sqrt{50} - 2\sqrt{8}$ $11\sqrt{2}$

10. $\sqrt{\frac{10}{3}} \cdot \sqrt{\frac{4}{30}}$ $\frac{2}{3}$

11. $2\sqrt{27} + \sqrt{63} - 4\sqrt{3}$ $2\sqrt{3} + 3\sqrt{7}$

12. $\left(1 - \sqrt{3}\right)\left(3 + \sqrt{2}\right)$ $3 + \sqrt{2} - 3\sqrt{3} - \sqrt{6}$

Find the distance between each pair of points whose coordinates are given. Express answers in simplest radical form.

13. $(4, 7), (4, -2)$ 9

14. $(-9, 2), \left(\frac{2}{3}, \frac{1}{2}\right)$ $\frac{\sqrt{3445}}{6}$

15. $(-1, 1), (1, -5)$ $2\sqrt{10}$

Find the length of each missing side. Round to the nearest hundredth.

16. $a = 8, b = 10, c = ?$ 12.81

17. $a = 12, c = 20, b = ?$ 16

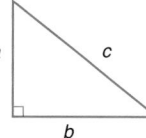

18. $a = 6\sqrt{2}, c = 12, b = ?$ 8.49

19. $b = 13, c = 17, a = ?$ 10.95

Solve each equation. Check your solution.

20. $\sqrt{4x + 1} = 5$ 6

21. $\sqrt{4x - 3} = 6 - x$ 3

22. $y^2 - 5 = -8y$ $-4 \pm \sqrt{21}$

23. $2x^2 - 10x - 3 = 0$ $\frac{5}{2} \pm \frac{\sqrt{31}}{2}$

Solve.

24. Geometry Find the measures of the perimeter and area, in simplest form, for the rectangle shown at the right. $16\sqrt{2} - 4\sqrt{6}, 16\sqrt{3} - 18$

25. Sports A hiker leaves her camp in the morning. How far is she from camp after walking 9 miles due west and then 12 miles due north? **15 miles**

GLOSSARY

A

absolute value (85)   The absolute value of a number is its distance from zero on a number line.

acute triangle (165)   In an acute triangle, all of the angles measure less than 90 degrees.

adding integers (86)

1. To add integers with the *same* sign, add their absolute values. Give the result the same sign as the integers.

2. To add integers with *different* signs, subtract the lesser absolute value from the greater absolute value. Give the result the same sign as the integer with the greater absolute value.

addition property for inequality (385)   For all numbers a, b, and c, the following are true:
1. if $a > b$, then $a + c > b + c$.

2. if $a < b$, then $a + c < b + c$.

addition property of equality (144)   For any numbers a, b, and c, if $a = b$, then $a + c = b + c$.

additive identity (37)   For any number a, $a + 0 = 0 + a = a$.

additive inverse property (87)   For any number a, $a + (-a) = 0$.

algebraic expression (6)   An expression consisting of one or more numbers and variables along with one or more arithmetic operations.

angle of depression (208)   An angle of depression is formed by a horizontal line and a line of sight below it.

angle of elevation (208)   An angle of elevation is formed by a horizontal line and a line of sight above it.

associative property (51)   For any numbers a, b, and c, $(a + b) + c = a + (b + c)$ and $(ab) c = a(bc)$.

axes (254)   Two perpendicular number lines that are used to locate points on a coordinate plane.

axis of symmetry (612)   The equation of the axis of symmetry for the graph of $y = ax^2 + bx + c$, where $a \neq 0$, is $x = -\frac{b}{2a}$.

B

back-to-back stem-and-leaf plot (27)   A back-to-back stem-and-leaf plot is used to compare two sets of data. The same stem is used for the leaves of both plots.

base   1. (7) In an expression of the form x^n, the base is x.   2. (215) The number that is divided into the percentage in the percent proportion.

best-fit line (341)   A line drawn on a scatter plot that passes close to most of the data points.

binomial (514)   The sum of two monomials.

boundary (437)   A boundary of an inequality is a line that separates the coordinate plane into half-planes.

box-and-whisker plot (427)   A type of diagram or graph that shows the quartiles and extreme values of data.

C

coefficient (47)   The numerical factor in a term.

commutative property (51)   For any numbers a and b, $a + b = b + a$ and $ab = ba$.

comparison property (94)   For any two numbers a and b, exactly one of the following sentences is true.
$$a < b \qquad a = b \qquad a > b$$

complementary angles (163)   Two angles are complementary if the sum of their measures is 90 degrees.

complete graph (278)   A complete graph shows the origin, the points at which the graph crosses the x- and y-axes, and other important characteristics of the graph.

completeness property (121)   Each real number corresponds to exactly one point on the number line. Each point on the number line corresponds to exactly one real number.

completeness property for points in the plane (256)

1. Exactly one point in the plane is named by a given ordered pair of numbers.

2. Exactly one ordered pair of numbers names a given point in the plane.

completing the square (742, 743)   To add a constant term to a binomial of the form $x^2 + bx$ so that the resulting trinomial is a perfect square.

complex fraction (114)   If a fraction has one or more fractions in the numerator or denominator, it is called a complex fraction.

composite numbers (558)   A whole number, greater than 1, that is not prime.

compound event (414)   A compound event consists of two or more simple events.

compound inequality (405)   Two inequalities connected by *and* or *or*.

congruent angles (164)   Angles that have the same measure.

conjugates (722)   Two binomials of the form $a\sqrt{b} + c\sqrt{d}$ and $a\sqrt{b} - c\sqrt{d}$.

consecutive integers (158)   Consecutive integers are integers in counting order.

consistent (455)   A system of equations is said to be consistent when it has at least one ordered pair that satisfies both equations.

constant of variation (239)   The number k in equations of the form $y = kx$ and $xy = k$.

constants (496)   Monomials that are real numbers.

coordinate (73)   The number that corresponds to a point on a number line.

coordinate plane (254)   The plane containing the x- and y-axes.

corresponding angles (201)   Matching angles in similar triangles, which have equal measures.

corresponding sides (201)   The sides opposite the corresponding angles in similar triangles.

cosine (206)   In a right triangle with acute angle A, the cosine of angle $A =$
$$\frac{\text{measure of leg adjacent to angle } A}{\text{measure of hypotenuse}}.$$

cross products (94)   When two fractions are compared, the cross products are the products of the terms on the diagonals.

D

---

data (25)   Numerical information.

defining the variable (127)   Choosing a variable to represent one of the unspecified numbers in a problem.

degree   **1.** (515) The degree of a monomial is the sum of the exponents of its variables. **2.** (516) The degree of a polynomial is the degree of the term of the greatest degree.

density property (96)   Between every pair of distinct rational numbers, there are infinitely many rational numbers.

dependent (456)   A system of equations that has an infinite number of solutions.

dependent variable (58)   The variable in a function whose value is determined by the independent variable.

difference of squares (545)   Two perfect squares separated by a subtraction sign, $a^2 - b^2 = (a + b)(a - b)$.

dimensional analysis (174)   The process of carrying units throughout a computation.

direct variation (239)   A direct variation is described by an equation of the form $y = kx$, where $k \neq 0$.

discrete mathematics (88)   A branch of mathematics that deals with finite or discontinuous quantities.

discriminant (632)   In the quadratic formula, the expression $b^2 - 4ac$.

distance formula (737)   The distance d between any two points with coordinates (x_1, y_1) and (x_2, y_2) is given by the following formula.
$$d = \sqrt{(x_2 - x_1)^2 + (y_2 - y_1)^2}$$

distributive property (46)   For any numbers a, b, and c:
1. $a(b + c) = ab + ac$ and $(b + c)a = ba + ca$.
2. $a(b - c) = ab - ac$ and $(b - c)a = ba - ca$.

dividing rational numbers (112)   The quotient of two rational numbers having the same sign is positive. The quotient of two rational numbers having different signs is negative.

division property for inequality (393)   For all numbers a, b, and c, the following are true:
1. If c is positive and $a < b$, then $\frac{a}{c} < \frac{b}{c}$, and if c is positive and $a > b$, then $\frac{a}{c} > \frac{b}{c}$.
2. If c is negative and $a < b$, then $\frac{a}{c} > \frac{b}{c}$, and if c is negative and $a > b$, then $\frac{a}{c} < \frac{b}{c}$.

division property of equality (151)   For any numbers a, b, and c, with $c \neq 0$, if $a = b$, then $\frac{a}{c} = \frac{b}{c}$.

domain (263)   The set of all first coordinates from the ordered pairs in a relation.

draw a diagram (406)   A problem-solving strategy that is often used as an organizational tool.

element (33)   A member of a set.

elimination (469)   The elimination method of solving a system of equations is a method that uses addition or subtraction to eliminate one of the variables to solve for the other variable.

equally likely (229)   Outcomes that have an equal chance of occurring.

equation (33)   A mathematical sentence that contains an equals sign, $=$.

equation in two variables (271)   An equation in two variables contains two unknown values.

equilateral triangle (164)   A triangle in which all the sides have the same length and all the angles have the same measure.

equivalent equation (144)   Equations that have the same solution.

equivalent expressions (47)   Expressions that denote the same number.

evaluate (8)   To find the value of an expression when the values of the variables are known.

excluded value (660)   A value is excluded from the domain of a variable because if that value were substituted for the variable, the result would have a denominator of zero.

exponent (7)   In an expression of the form x^n, the exponent is n.

exponential function (634, 635)   A function that can be described by an equation of the form $y = a^x$, where $a > 0$ and $a \neq 1$.

extraneous solutions (697)   Solutions derived from an equation that are not solutions of the original equation.

extreme values (427)   The least value and the greatest value in a set of data.

extremes (196)   *See* proportion.

factored form (559)   A monomial is written in factored form when it is expressed as the product of prime numbers and variables where no variable has an exponent greater than 1.

factoring (564)   To express a polynomial as the product of monomials and polynomials.

factoring by grouping (567)   A method of factoring polynomials with four or more terms.

factors (6)   In a multiplication expression, the quantities being multiplied are called factors.

family of graphs (354)   A family of graphs includes graphs and equations of graphs that have at least one characteristic in common.

FOIL method (537)   To multiply two binomials, find the sum of the products of

F   the first terms,
O   the outside terms,
I   the inside terms, and
L   the last terms.

formula (128)   An equation that states a rule for the relationship between certain quantities.

function   **1.** (56) A relationship between input and output in which the output depends on the input.   **2.** (287) A relation in which each element of the domain is paired with exactly one element of the range.

functional notation (289)   In functional notation, the equation $y = x + 5$ is written as $f(x) = x + 5$.

general equation for exponential decay (644)   The general equation for exponential decay is represented by the formula $A = C(1 - r)^t$.

general equation for exponential growth (643)   The general equation for exponential growth is represented by the formula $A = C(1 + r)^t$.

graph (73, 255)   To draw, or plot, the points named by certain numbers or ordered pairs on a number line or coordinate plane, respectively.

greatest common factor (GCF) (559)   The greatest common factor of two or more integers is the greatest number that is a factor of all the integers.

guess and check (574)   A problem-solving strategy in which several values or combinations of values are tried in order to find a solution to a problem.

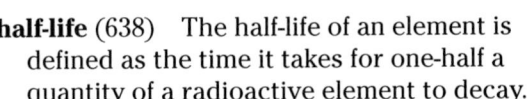

half-life (638)   The half-life of an element is defined as the time it takes for one-half a quantity of a radioactive element to decay.

half-plane (437)   The region of a graph on one side of a boundary is called a half-plane.

horizontal axis (56)   The horizontal line in a graph that represents the independent variable.

hypotenuse (206)   The side of a right triangle opposite the right angle.

identify subgoals (745)   A problem-solving strategy that uses a series of small steps, or subgoals.

identity (170)   An equation that is true for every value of the variable.

inconsistent (456)   A system of equations is said to be inconsistent when it has no ordered pair that satisfies both equations.

independent (456)   A system of equations is said to be independent if the system has exactly one solution.

independent variable (58)   The variable in a function whose value is subject to choice is the independent variable. The independent variable affects the value of the dependent variable.

inequality (33)   A mathematical sentence having the symbols $<$, $\leq$, $>$, or $\geq$.

integers (73)   The set of numbers represented as $\{..., -3, -2, -1, 0, 1, 2, 3, ...\}$.

interquartile range (304, 306)   The difference between the upper quartile and the lower quartile of a set of data. It represents the middle half, or 50%, of the data in the set.

intersection (406)   The intersection of two sets A and B is the set of elements common to both A and B.

inverse of a relation (264)   The inverse of any relation is obtained by switching the coordinates in each ordered pair.

inverse variation (241)   An inverse variation is described by an equation of the form $xy = k$, where $k \neq 0$.

irrational numbers (120)   A number that cannot be expressed in the form $\frac{a}{b}$, where a and b are integers and $b \neq 0$.

isosceles triangle (164)   In an isosceles triangle, at least two angles have the same measure and at least two sides have the same length.

least common denominator (LCD) (686)   The least common denominator is the least common multiple of the denominators of two or more fractions.

least common multiple (LCM) (685)   The least common multiple of two or more integers is the least positive integer that is divisible by each of the integers.

legs (206)   The sides of a right triangle that are not the hypotenuse.

like terms (47)   Terms that contain the same variables, with corresponding variables with the same power.

linear equation (280)   An equation whose graph is a line.

linear function (278)   An equation whose graph is a nonvertical line.

line plot (78)   Numerical data displayed on a number line.

look for a pattern (13, 497, 637)   A problem-solving strategy often involving the use of tables to organize information so that a pattern may be determined.

lower quartile (306)   The lower quartile divides the lower half of a set of data into two equal parts.

make an organized list (685)   A problem-solving strategy that uses an organized list to arrange and evaluate data in order to determine a solution.

mapping (263)   A mapping pairs one element in the domain with one element in the range.

matrix (88)   A matrix is a rectangular arrangement of elements in rows and columns.

maximum (611)   The highest point on the graph of a curve, such as a the vertex of parabola that opens downward.

mean (178)   The mean of a set of data is the sum of the numbers in the set divided by the number of numbers in the set.

Glossary   **803**

means (196)   *See* proportion.

measures of central tendency (178)   Numbers known as measures of central tendency are often used to describe sets of data because they represent a centralized, or middle, value.

measures of variation (306)   Measures of variation are used to describe the distribution of data.

median (178)   The median is the middle number of a set of data when the numbers are arranged in numerical order.

midpoint (369)   A point that is halfway between the endpoints of a segment.

minimum (611)   The lowest point on the graph of a curve, such as a the vertex of parabola that opens upward.

mixed expression (690)   An algebraic expression that contains a monomial and a rational expression.

mode (178)   The mode of a set of data is the number that occurs most often in the set.

monomial (496)   A monomial is a number, a variable, or a product of a number and one or more variables.

multi-step equations (157)   Multi-step equations are equations that need more than one operation to solve them.

multiplicative identity (38)   For any number a, $a \cdot 1 = 1 \cdot a = a$.

multiplicative inverse (38)   For every nonzero number $\frac{a}{b}$, where $a, b \neq 0$, there is exactly one number $\frac{b}{a}$ such that $a \cdot b = 1$.

multiplication property for inequality (393)   For all numbers a, b, and c, the following are true.

1. If c is positive and $a < b$, then $ac < bc$, $c \neq 0$, and if c is positive and $a > b$, then $ac > bc$, $c \neq 0$.

2. If c is negative and $a < b$, then $ac > bc$, $c \neq 0$, and if c is negative and $a > b$, then $ac < bc$, $c \neq 0$.

multiplicative property of −1 (107)   The product of any number and −1 is its additive inverse.

$$-1(a) = -a \text{ and } a(-1) = -a$$

multiplicative property of equality (150)   For any numbers a, b, and c, if $a = b$, then $a \cdot c = b \cdot c$.

multiplicative property of zero (38)   For any number a, $a \cdot 0 = 0 \cdot a = 0$.

804   *Glossary*

N

negative correlation (340)   There is a negative correlation between x and y if the values are related in opposite ways.

negative exponent (503)   For any nonzero number a and any integer n, $a^{-n} = \frac{1}{a^n}$.

negative number (73)   Any number that is less than zero.

number line (72)   A line with equal distances marked off to represent numbers.

number theory (158)   The study of numbers and the relationships between them.

O

obtuse triangle (165)   An obtuse triangle has one angle with measure greater than 90 degrees.

odds (229)   The odds of an event occurring is the ratio of the number of ways the event can occur (successes) to the number of ways the event cannot occur (failures).

open sentences (32)   Mathematical statements with one or more variables, or unknown numbers.

opposites (87)   The opposite of a number is its additive inverse.

order of operations (19)

1. Simplify the expressions inside grouping symbols, such as parentheses, brackets, and braces, and as indicated by fraction bars.
2. Evaluate all powers.
3. Do all multiplications and divisions from left to right.
4. Do all additions and subtractions from left to right.

ordered pair (57)   Pairs of numbers used to locate points in the coordinate plane.

organize data (464)   Organizing data is useful before solving a problem. Some ways to organize data are to use tables, charts, different types of graphs, or diagrams.

origin (254)   The point of intersection of the two axes in the coordinate plane.

outcomes (413)   Outcomes are all possible combinations of a counting problem.

outlier (307)   In a set of data, a value that is much greater or much less than the rest of the data can be called a outlier.

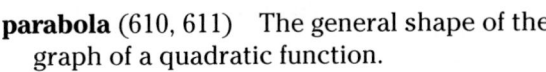

parabola (610, 611)   The general shape of the graph of a quadratic function.

parallel lines (362)   Lines in the plane that never intersect. Nonvertical parallel lines have the same slope.

parallelogram (362)   A quadrilateral in which opposite sides are parallel.

percent (215)   A percent is a ratio that compares a number to 100.

percentage (215)   The number that is divided by the base in a percent proportion.

percent of decrease (222)   The ratio of an amount of decrease to the previous amount, expressed as a percent.

percent of increase (222)   The ratio of an amount of increase to the previous amount, expressed as a percent.

percent proportion (215) $\frac{\text{Percentage}}{\text{Base}} = \frac{r}{100}$

perfect square (119)   A rational number whose square root is a rational number.

perfect square trinomial (587)   A trinomial which, when factored, has the form $(a + b)^2 = (a + b)(a + b)$ or $(a - b)^2 = (a - b)(a - b)$.

perpendicular lines (362)   Lines that meet to form right angles.

point-slope form (333)   For any point (x_1, y_1) on a nonvertical line having slope m, the point-slope form of a linear equation is as follows:
$$y - y_1 = m(x - x_1).$$

polynomial (513, 514)   A polynomial is a monomial or a sum of monomials.

positive correlation (340)   There is a positive correlation between x and y if the values are related in the same way.

power (7)   An expression of the form x^n is known as a power.

power of a monomial (498)   For any numbers a and b, and any integers m, n, and p,
$$(a^m b^n)^p = a^{mp} b^{np}.$$

power of a power (498)   For any number a, and all integers m and n, $(a^m)^n = a^{mn}$.

power of a product (498)   For all numbers a and b, and any integer m, $(ab)^m = a^m b^m$.

prime factorization (558)   A whole number expressed as a product of factors that are all prime numbers.

prime number (558)   A prime number is a whole number, greater than 1, whose only factors are 1 and itself.

prime polynomial (577)   A polynomial that cannot be written as a product of two polynomials with integral coefficients is called a prime polynomial.

principal square root (119)   The nonnegative square root of an expression.

probability (228)   The ratio that tells how likely it is that an event will take place.
$$P(\text{event}) = \frac{\text{number of favorable outcomes}}{\text{total number of possible outcomes}}$$

problem-solving plan (126)
1. Explore the problem.
2. Plan the solution.
3. Solve the problem.
4. Examine the solution.

product (6)   The result of multiplication.

product of powers (497)   For any number a, and all integers m and n, $a^m \cdot a^n = a^{m + n}$.

product property of square roots (719)   For any number a and b, where $a \geq 0$ and $b \geq 0$, $\sqrt{ab} = \sqrt{a} \cdot \sqrt{b}$.

proportion (195)   In a proportion, the product of the extremes is equal to the product of the means. If $\frac{a}{b} = \frac{c}{d}$, then $ad = bc$.

Pythagorean theorem (713)   If a and b are the measures of the legs of a right triangle and c is the measure of the hypotenuse, then $c^2 = a^2 + b^2$.

Pythagorean triple (583)   Three whole numbers a, b, and c such that $a^2 + b^2 = c^2$.

quadrant (254)   One of the four regions into which the x- and y-axes separate the coordinate plane.

quadratic equation (620)   A quadratic equation is one in which the value of the related quadratic function is 0.

quadratic formula (628)   The roots of a quadratic equation in the form $ax^2 + bx + c = 0$, where $a \neq 0$, are given by the formula $x = \frac{-b \pm \sqrt{b^2 - 4ac}}{2a}$.

quadratic function (610, 611)   A quadratic function is a function that can be described by an equation of the form $y = ax^2 + bx + c$, where $a \neq 0$.

quartiles (304, 306)   In a set of data, the quartiles are values that divide the data into four equal parts.

quotient of powers (501)   For all integers m and n and any nonzero number a, $\frac{a^m}{a^n} = a^{m-n}$.

quotient property of square roots (721)   For any numbers a and b, where $a > 0$ and $b > 0$, $\sqrt{\frac{a}{b}} = \frac{\sqrt{a}}{\sqrt{b}}$.

(R)

radical equations (732)   Equations that contain radicals with variables in the radicand.

radical sign (119)   The symbol $\sqrt{}$, indicating the principal or nonnegative root of an expression.

radicand (719)   The radicand is the expression under the radical sign.

random (229)   When an outcome is chosen without any preference, the outcome occurs at random.

range   **1.** (263) The set of all second coordinates from the ordered pairs in the relation. **2.** (306) The difference between the greatest and the least values of a set of data.

rate   **1.** (197) The ratio of two measurements having different units of measure.   **2.** (215) In the percent proportion, the rate is the fraction with a denominator of 100.

ratio (195)   A ratio is a comparison of two numbers by division.

rational equation (696)   A rational equation is an equation that contains rational expressions.

rational expression (660)   A rational expression is an algebraic fraction whose numerator and denominator are polynomials.

rational numbers (93)   A rational number is a number that can be expressed in the form $\frac{a}{b}$, where a and b are integers and $b \neq 0$.

rationalizing the denominator (722)   Rationalizing the denominator of a radical expression is a method used to remove or eliminate the radicals from the denominator of a fraction.

real numbers (121)   The set of rational numbers and the set of irrational numbers together form the set of real numbers.

reciprocal (38)   The multiplicative inverse of a number.

reflexive property of equality (39)   For any number a, $a = a$.

regression line (342)   The most accurate best-fit line for a set of data, and can be determined with a graphing calculator or computer.

relation (260, 263)   A relation is a set of ordered pairs.

replacement set (33)   A set of numbers from which replacements for a variable may be chosen.

right triangle (165)   A right triangle has one angle with a measure of 90 degrees.

rise (325)   The vertical change in a line.

roots (620)   The solutions of a quadratic equation.

run (325)   The horizontal change in a line.

(S)

scalar multiplication (108)   In scalar multiplication, each element of a matrix is multiplied by a constant.

scale (197)   A ratio called a scale is used when making a model to represent something that is too large or too small to be conveniently drawn at actual size.

scatter plot (339)   In a scatter plot, the two sets of data are plotted as ordered pairs in the coordinate plane.

scientific notation (506)   A number is expressed in scientific notation when it is in the form $a \times 10^n$, where $1 \leq a < 10$ and n is an integer.

set (33)   A collection of objects or numbers.

set-builder notation (385)   A notation used to describe the members of a set. For example, $\{y \mid y < 17\}$ represents the set of all numbers y such that y is less than 17.

sequence (13)   A set of numbers in a specific order.

similar triangles (201)   If two triangles are similar, the measures of their corresponding sides are proportional, and the measures of their corresponding angles are equal.

simple events (414)   A single event in a probability problem.

simple interest (217)   The amount paid or earned for the use of money. The formula $I = prt$ is used to solve simple interest problems.

simplest form (47)   An expression is in simplest form when it is replaced by an equivalent expression having no like terms and no parentheses.

simplest radical form (723)   A radical expression is in simplest radical form when the following three conditions have been met.

1. No radicands have perfect square factors other than one.
2. No radicands contain fractions.
3. No radicals appear in the denominator of a fraction.

sine (206)   In a right triangle with acute angle A, sine of angle

$$A = \frac{\text{measure of leg opposite angle } A}{\text{measure of the hypotenuse}}.$$

slope (324, 325)   The ratio of the rise to the run as you move from one point to another along a line.

slope-intercept form (347)   An equation of the form $y = mx + b$, where m is the slope and b is the y-intercept of a given line.

solution (32)   A replacement for the variable in an open sentence that results in a true sentence.

solution of an equation in two variables (271)   If a true statement results when the numbers in an ordered pair are substituted into an equation in two variables, then the ordered pair is a solution of the equation.

solution set (33)   The set of all replacements for the variable in an open sentence that result in a true sentence.

solve an equation (145)   To solve an equation means to isolate the variable having a coefficient of 1 on one side of the equation.

solving an open sentence (32)   Finding a replacement for the variable that results in a true sentence.

solving a triangle (208)   Finding the measures of all sides and angles of a right triangle.

square of a difference (544)   If a and b are any numbers, $(a - b)^2 = (a - b)(a - b) = a^2 - 2ab + b^2$.

square of a sum (543)   If a and b are any numbers, $(a + b)^2 = (a + b)(a + b) = a^2 + 2ab + b^2$.

square root (118, 119)   One of two identical factors of a number.

standard form (333)   The standard form of a linear equation is $Ax + By = C$, where A, B, and C are integers, $A \geq 0$, and A and B are not both zero.

statistics (25)   A branch of mathematics concerned with methods of collecting, organizing, and interpreting data.

stem-and-leaf plot (25)   In a stem-and-leaf plot, each piece of data is separated into two numbers that are used to form a stem and a leaf. The data are organized into two columns. The column on the left contains the stems and the column on the right contains the leaves.

substitution (462)   The substitution method of solving a system of equations is a method that uses substitution of one equation into the other equation to solve for the other variable.

substitution property of equality (39)   If $a = b$, then a may be replaced by b in any expression.

subtracting integers (87)   To subtract a number, add its additive inverse. For any numbers a and b, $a - b = a + (-b)$.

subtraction property for inequality (385)   For all numbers a, b, and c, the following are true:

1. if $a > b$, then $a - c > b - c$.
2. if $a < b$, then $a - c < b - c$.

subtraction property of equality (146)   For any numbers a, b, and c, if $a = b$, then $a - c = b - c$.

supplementary angles (162)   Two angles are supplementary if the sum of their measures is 180 degrees.

symmetric property of equality (39)   For any numbers a and b, if $a = b$, then $b = a$.

symmetry (612)   Symmetrical figures are those in which the figure can be folded and each half matches the other exactly.

system of equations (455)   A set of equations with the same variables.

system of inequalities (482)   A set of inequalities with the same variables.

tangent (206)   In a right triangle, the tangent of angle $A = \dfrac{\text{measure of leg opposite angle } A}{\text{measure of leg adjacent to angle } A}$.

term   **1.** (13) A number in a sequence.   **2.** (47) A number, a variable, or a product or quotient of numbers and variables.

transitive property of equality (39)   For any numbers a, b, and c, if $a = b$ and $b = c$, then $a = c$.

tree diagram (413)   A tree diagram is a diagram used to show the total number of possible outcomes.

triangle (163)   A triangle is a polygon with three sides and three angles.

trigonometric ratios (206)

$\sin A = \dfrac{a}{c}$ $\cos A = \dfrac{b}{c}$ $\tan A = \dfrac{a}{b}$

trinomials (514)   A trinomial is the sum of three monomials.

uniform motion (235)   When an object moves at a constant speed, or rate, it is said to be in uniform motion.

union (408)   The union of two sets A and B is the set of elements contained in both A or B.

unique factorization theorem (558)   The prime factorization of every number is unique except for the order in which the factors are written.

unit cost (95)   The cost of one unit of something.

upper quartile (306)   The upper quartile divides the upper half of a set of data into two equal parts.

use a model (339)   A problem-solving strategy that uses models, or simulations, of mathematical situations that are difficult to solve directly.

use a table (255)   A problem-solving strategy that uses tables to organize and solve problems.

variable (6)   Variables are symbols that are used to represent unspecified numbers.

Venn diagrams (73)   Venn diagrams are diagrams that use circles or ovals inside a rectangle to show relationships of sets.

vertex (610, 611)   The maximum or minimum point of a parabola.

vertical axis (56)   The vertical line in a graph that represents the dependent variable.

vertical line test (289)   If any vertical line passes through no more than one point of the graph of a relation, then the relation is a function.

weighted average (233)   The weighted average M of a set of data is the sum of the product of each number in the set and its weight divided by the sum of all the weights.

whiskers (428)   The whiskers of a box-and-whisker plot are the segments that are drawn from the lower quartile to the least value and from the upper quartile to the greatest value.

whole numbers (72)   The set of whole numbers is represented by {0, 1, 2, 3, ...}.

work backward (156)   A problem-solving strategy that uses inverse operations to determine an original value.

x-axis (254)   The horizontal number line.

x-coordinate (254)   The first number in an ordered pair.

x-intercept (346)   The coordinate at which a graph intersects the x-axis.

y-axis (254)   The vertical number line.

y-coordinate (254)   The second number in an ordered pair.

y-intercept (346)   The coordinate at which a graph intersects the y-axis.

Z

zero exponent (502)   For any nonzero number a, $a^0 = 1$.

zero product property (594)   For all numbers a and b, if $ab = 0$, then $a = 0$, $b = 0$, or both a and b equal 0.

zeros (620)   The zeros of a function are the roots, or x-intercepts, of the function.

absolute value/valor absoluto (85)   El valor absoluto de un número equivale al número de unidades que dicho número dista de cero en la recta numérica.

acute triangle/triángulo agudo (165)   En un triángulo agudo, todos los ángulos miden menos de 90°.

adding integers/suma de enteros (86)

1. Para sumar enteros del *mismo* signo, suma los valores absolutos de los números. Da al resultado el mismo signo de los números.

2. Para sumar enteros de *distinto* signo, resta el valor menor del valor mayor. Da al resultado el mismo signo que el número con el mayor valor absoluto.

addition property of equality/propiedad de adición de la igualdad (144)   Para cualquiera de los números a, b y c, si $a = b$, entonces $a + c = b + c$.

addition property of inequality/propiedad de adición de la desigualdad (385)   Para todos los números a, b y c:

1. si $a > b$, entonces $a + c > b + c$;

2. si $a < b$, entonces $a + c < b + c$.

additive identity/identidad aditiva (37)   Para cualquier número a, $a + 0 = 0 + a = a$.

additive inverse property/propiedad del inverso de la adición (87)   Para cualquier número a, $a + (-a) = 0$.

algebraic expression/expresión algebraica (6)   Una expresión que consiste de uno o más números y variables, además de una o más operaciones aritméticas.

angle of depression/ángulo de depresión (208)   El ángulo de depresión se forma por una línea horizontal y una línea visual por debajo de la misma.

angle of elevation/ángulo de elevación (208)   Un ángulo de elevación se forma por una línea horizontal y una línea visual por encima de la misma.

associative property/propiedad asociativa (51)   Para cualquiera de los números a, b, y c, $(a + b) + c = a + (b + c)$ y $(ab)c = a(bc)$.

axes/ejes (254)   Dos rectas numéricas perpendiculares que se usan para ubicar puntos en un plano de coordenadas.

axis of symmetry/eje de simetría (612)   La ecuación para el eje de simetría de la gráfica $y = ax^2 + bx + c$, en la cual $a \neq 0$ y $x = -\frac{b}{2a}$.

back-to-back stem-and-leaf plot/diagrama de tallo y hojas consecutivo (27)   Un diagrama de tallo y hojas consecutivo se usa para comparar dos conjuntos de datos. El mismo tallo se usa para las hojas de ambos diagramas.

base/base   1. (7) En una expresión de la forma x^n, la base es x.   **2.** (215) El número que se divide entre el porcentaje en una proporción de porcentaje.

best-fit line/línea de mejor encaje (341)   Es una línea que se dibuja en un diagrama de dispersión y que pasa cerca de la mayoría de los puntos de datos.

binomial/binomio (514)   La suma de dos monomios.

boundary/frontera (437)   La frontera de una desigualdad es una línea que separa el plano de coordenadas en dos mitades.

box-and-whisker plot/diagrama de caja y patillas (427)   Un tipo de diagrama o gráfica que muestra los valores cuartílicos y los extremos de los datos.

coefficient/coeficiente (47)   El factor numérico de un término.

commutative property/propiedad conmutativa (51)   Para cualquiera de los números a y b, $a + b = b + a$ y $ab = ba$.

comparison property/propiedad de comparación (94)   Para cualquier par de números a y b, exactamente una de las siguientes operaciones es válida.

$$a < b \qquad a = b \qquad a > b$$

complementary angles/ángulos complementarios (163)   Dos ángulos son complementarios si la suma de sus medidas es 90 grados.

complete graph/gráfica completa (278)   Una gráfica completa muestra el origen, los puntos en donde la gráfica cruza el eje de coordenadas x y el eje de coordenadas y, además de otras características importantes en la gráfica.

completeness property/propiedad del completo (121)   Cada número real corresponde exactamente a un punto en la recta numérica. Cada punto en la recta numérica corresponde exactamente a un número real.

completeness property for points in the plane/propiedad del completo de los puntos en el plano (256)

1. Exactamente un punto en el plano es nombrado por un par ordenado de números dado.

2. Exactamente un par ordenado de números nombra un punto dado en el plano.

completing the square/completar el cuadrado (742, 743)   Sumar un término constante a un binomio de la forma $x^2 = bx$, de modo que el trinomio resultante sea un cuadrado perfecto.

complex fraction/fracción compleja (114)   Si una fracción tiene uno o más fracciones en el numerador o denominador, entonces se llama una fracción compleja.

composite number/número compuesto (558)   Un número entero, mayor de 1, que no es un número primo.

compound event/evento compuesto (414)   Un evento compuesto consiste de dos o más eventos simples.

compound inequality/desigualdad compuesta (405)   Dos desigualdades conectadas por *y* u *o*.

congruent angles/ángulos congruentes (164)   Ángulos que tienen la misma medida.

conjugates/conjugados (722)   Dos binomios de la forma $a\sqrt{b} + c\sqrt{d}$ y $a\sqrt{b} - c\sqrt{d}$.

consecutive integers/números enteros consecutivos (158)   Son los números enteros en el orden de contar.

consistent/consistente (455)   Se dice que un sistema de ecuaciones es consistente cuando tiene por lo menos un par ordenado que satisface ambas ecuaciones.

constant of variation/constante de variación (239)   El número k en ecuaciones de la forma $y = kx$ y $xy = k$.

constants/constantes (469)   Los monomios que son números reales.

coordinate/coordenada (73)   El número que corresponde a un punto en una recta numérica.

coordinate plane/plano de coordenadas (254)   El plano que contiene el eje de coordenadas x y el eje de coordenadas y.

corresponding angles/ángulos correspondientes (201)   Ángulos que encajan en triángulos semejantes y que tienen medidas iguales.

corresponding sides/lados correspondientes (201)   Los lados opuestos a los ángulos correspondientes en triángulos semejantes.

cosine/coseno (206)   En un triángulo rectángulo con ángulo agudo A, el coseno del

ángulo $A = \dfrac{\text{medida del cateto adyacente al ángulo } A}{\text{medida de la hipotenusa}}$.

cross products/productos cruzados (94)   Cuando se comparan dos fracciones, los productos cruzados son los productos de los términos en diagonales.

data/datos (25)   La información numérica.

defining the variable/definir la variable (127)   El escoger una variable para representar uno de los números no especificados en un problema.

degree/grado   1. (515) El grado de un monomio es la suma de los exponentes de sus variables.   **2.** (516) El grado de un polinomio es el grado del término con el grado más alto.

density property/propiedad de la densidad (96)   Entre cada par de números racionales distintos, existe una infinidad de números racionales.

dependent/dependiente (456)   Un sistema de ecuaciones que tiene un número infinito de soluciones.

dependent variable/variable dependiente (58)   La variable en una función cuyo valor lo determina la variable independiente.

difference of squares/diferencia de cuadrados (545, 581)   Dos cuadrados perfectos separados por un signo de sustracción, $a^2 - b^2 = (a + b)(a - b)$.

dimensional analysis/análisis dimensional (174)   El proceso de llevar unidades a lo largo de un cómputo.

direct variation/variación directa (239)   Una función lineal descrita por una ecuación de la forma $y = kx$, en que $k \neq 0$.

discrete mathematics/matemáticas de números discretos (88)   Una rama de las matemáticas que estudia los conjuntos de números finitos o interrumpidos.

discriminant/discriminante (632)   En la fórmula cuadrática, la expresión $b^2 - 4ac$.

distance formula/fórmula de distancia (737)   La distancia d entre cualquier par de puntos con coordenadas (x_1, y_1) y (x_2, y_2) es dada por la siguiente fórmula.
$$d = \sqrt{(x_2 - x_1)^2 + (y_2 - y_1)^2}.$$

distributive property/propiedad distributiva (46)   Para cualquiera de los números a, b y c:

1. $a(b + c) = ab + ac$ y $(b + c)a = ba + ca$.

2. $a(b - c) = ab - ac$ y $(b - c)a = ba - ca$.

dividing rational numbers/división de números racionales (112)   El cociente de dos números racionales que tienen el mismo signo es positivo. El cociente de dos números racionales que tienen diferente signo es negativo.

division property of equality/propiedad de división de la igualdad (151)   Para cualquiera de los números a, b y c, en que $c \neq 0$, si $a = b$, entonces $\dfrac{a}{c} = \dfrac{b}{c}$.

division property of inequality/propiedad de división de la desigualdad (393)   Para cualquiera de los números reales a, b y c:

1. si c es positivo y $a < b$, entonces $\frac{a}{c} < \frac{b}{c}$ y

si c es positivo y $a > b$, entonces $\frac{a}{c} > \frac{b}{c}$.

2. si c es negativo y $a < b$, entonces $\frac{a}{c} > \frac{b}{c}$, y

si c es negativo y $a > b$, entonces $\frac{a}{c} < \frac{b}{c}$.

domain/dominio (263)   El conjunto de todas las primeras coordenadas de los pares ordenados de una relación.

draw a diagram/trazar un diagrama (406)   Una estrategia para resolver problemas que a menudo se usa como una herramienta organizadora.

element/elemento (33)   Un miembro de un conjunto.

elimination/eliminación (469)   El método de eliminación para resolver un sistema de ecuaciones es un método que usa adición o sustracción para eliminar una de las variables y así despejar la otra variable.

equally likely/igualmente verosímil (229)   Respuestas que tienen igual posibilidad de ocurrir.

equation/ecuación (33)   Un enunciado matemático que contiene un signo de igualdad, =.

equation in two variables/ecuación en dos variables (271)   Una ecuación en dos variables contiene dos valores desconocidos.

equilateral triangle/triángulo equilátero (164)   Un triángulo en el cual todos los lados y todos los ángulos tienen la misma medida.

equivalent equation/ecuación equivalente (144)   Ecuaciones que tienen la misma solución.

equivalente expressions/expresiones equivalentes (47)   Expresiones que denotan el mismo número.

evaluate/evaluar (8)   El método de hallar el valor de una expresión cuando se conocen los valores de las variables.

excluded value/valor excluido (660)   Se excluye un valor del dominio de una variable porque si se sustituyera ese valor por la variable, el resultado tendría un cero en el denominador.

exponent/exponente (7)   En una expresión de la forma x^n, el exponente es n.

exponential function/función exponencial (634, 635)   Una función que se puede describir por una ecuación de la forma $y = a^x$, en que $a > 0$ y $a \neq 1$.

extraneous solutions/soluciones extrañas (697)   Soluciones obtenidas de una ecuación y las cuales no son soluciones aceptables para la ecuación original.

extreme values/valores extremos (427)   Los valores extremos son el menor valor y el mayor valor en un conjunto de datos.

extremes/extremos (196)   *Ver* proporción.

factored form/forma factorial (559)   Un monomio está escrito en forma factorial cuando está expresado como el producto de números primos y variables y las variables no tienen un exponente mayor de 1.

factoring/factorizar (564)   Expresar un polinomio como el producto de monomios o polinomios.

factoring by grouping/factorizando por grupos (567)   Un método de factorización de polinomios de cuatro o mas términos.

factors/factores (6)   En una expresión de multiplicación, los factores son las cantidades que se multiplican.

family of graphs/familia de gráficas (354)   Una familia de gráficas incluye gráficas y ecuaciones de gráficas que tienen por lo menos una característica en común.

FOIL method/método FOIL (537)   Para multiplicar dos binomios, halla la suma de los productos de los primeros términos, los términos de afuera, los términos de adentro y los últimos términos.

formula/fórmula (128)   Una ecuación que enuncia una regla para la relación entre ciertas cantidades.

function/función   **1.** (56) Una relación entre los datos de entrada y de salida, en que los datos de salida dependen de los datos de entrada. **2.** (287) Una relación en que cada elemento del dominio se aparea exactamente con un elemento de la amplitud.

functional notation/notación funcional (289)   En notación funcional, la ecuación $y = x + 5$ se escribe $f(x) = x + 5$.

general equation for exponential decay/ ecuación general para la disminución exponencial (644)   La ecuación general para la disminución exponencial está representada por $A = C(1 - r)^t$.

general equation for exponential growth/ ecuación general para el crecimiento exponencial (643)   La ecuación general para el crecimiento exponencial está representada por $A = C(1 + r)^t$.

graph/gráfica (73, 255)   Consiste en dibujar, o trazar sobre una recta numérica o plano de coordenadas, los puntos nombrados por ciertos números o pares ordenados.

greatest common factor (GCF)/máximo común divisor (MCD) (559)   El máximo común divisor de dos o más números enteros es el número mayor que es un factor de todos los números.

guess and check/conjetura y cotejo (574)   Una estrategia para resolver problemas en la cual se prueban varios valores o combinaciones de valores para hallar una solución al problema.

half-life/media vida (638)   La media vida de un elemento se define como el tiempo que tarda en desintegrarse la mitad de un elemento radiactivo.

half-plane/medio plano (437)   La región de una gráfica en un lado de la frontera se llama el medio plano.

horizontal axis/eje horizontal (56)   La recta horizontal en una gráfica que representa la variable independiente.

hypotenuse/hipotenusa (206)   El lado de un triángulo rectángulo opuesto al ángulo recto.

identify subgoals/identificación de submetas (745)   Una estrategia para resolver problemas que utiliza una serie de pasos menores o submetas.

identity/identidad (170)   Una ecuación que es cierta para cualquier valor de la variable.

inconsistent/inconsistente (456)   Se dice que un sistema de ecuaciones es inconsistente cuando no tiene pares ordenados que satisfacen ambas ecuaciones.

independent/independiente (456)   Se dice que un sistema de ecuaciones es independiente si el sistema tiene exactamente una solución.

independent variable/variable independiente (58)   La variable en una función cuyo valor está sujeto a elección se dice que es una variable independiente. La variable independiente afecta el valor de la variable dependiente.

inequality/desigualdad (33)   Un enunciado matemático que contiene los símbolos $<$, $\leq$, $>$, o $\geq$.

integers/números enteros (73)   El conjunto de números enteros representados por $\{\ldots, -3, -2, -1, 0, 1, 2, 3, \ldots\}$.

interquartile range/amplitud intercuartílica (304, 306)   La diferencia entre el cuartillo superior y el inferior de un conjunto de datos. Representa la mitad inferior, ó 50%, de los datos en un conjunto.

intersection/intersección (406)   La intersección de dos conjuntos A y B es el conjunto de elementos comunes a ambos A y B.

inverse of a relation/inverso de una relación (264)   El inverso de cualquier relación se obtiene intercambiando las coordenadas en cada par ordenado.

inverse variation/variación inversa (241)   Una variación inversa se describe por una ecuación de la forma $xy = k$, en que $k \neq 0$.

irrational numbers/números irracionales (120)   Un número que no se puede expresar en la forma $\frac{a}{b}$, en que a y b son números enteros y $b \neq 0$.

isosceles triangle/triángulo isósceles (164)   En un triángulo isósceles, por lo menos dos ángulos tienen la misma medida y por lo menos dos lados tienen la misma longitud.

least common denominator (LCD)/mínimo común denominador (MCD) (686)   El mínimo común denominador es el mínimo común múltiplo de los denominadores de dos o más fracciones.

least common multiple (LCM)/mínimo común múltiplo (MCM) (685)   El mínimo común múltiplo de dos o más números enteros es el menor número entero positivo divisible entre cada uno de los números.

legs/catetos (206)   Los lados de un triángulo rectángulo que no son la hipotenusa.

like terms/términos semejantes (47)   Términos que contienen las mismas variables, con variables correspondientes con la misma potencia.

line plot/trazado gráfico lineal (78)   Datos numéricos desplegados sobre una recta numérica.

linear equation/ecuación lineal (280)   Una ecuación cuya gráfica es una recta.

linear function/función lineal (278)   Ecuación cuya gráfica es una recta no vertical.

look for a pattern/busca un patrón (13, 497, 637)   Una estrategia para resolver problemas que a menudo involucra el uso de tablas para organizar la información de manera que se pueda determinar un patrón.

lower quartile/cuartillo inferior (306)   El cuartillo inferior divide la mitad inferior de un conjunto de datos en dos partes iguales.

make an organized list/haz una lista organizada (685)   Una estrategia para resolver problemas que utiliza una lista organizada para arreglar y evaluar los datos y determinar una solución.

mapping/relación (263)   Una relación aparea un elemento en el dominio con un elemento en la amplitud.

matrix/matriz (88)   Una matriz es un arreglo rectangular de elementos en hileras y columnas.

maximum/máximo (611)   El punto más alto en la gráfica de una curva, tal como el vértice de la parábola que se abre hacia abajo.

mean/media (178)   La media de un conjunto de datos es la suma de los números en el conjunto dividida entre el número de números en el conjunto.

means/media proporcional (196)   *Ver* proporción.

measures of central tendency/medidas de tendencia central (178)   Los números conocidos como medidas de tendencia central se usan a menudo para describir conjuntos de datos porque estos representan un valor centralizado o en el medio.

measures of variation/medidas de variación (306) Las medidas de variación a menudo se usan para describir la distribución de los datos.

median/mediana (178) La mediana es el número en el centro de un conjunto de datos cuando los números se organizan en orden numérico.

midpoint/punto medio (369) El punto medio de un segmento es el punto equidistante entre los extremos del segmento.

minimum/mínimo (611) El punto más bajo en la gráfica de una curva, tal como el vértice de una parábola que se abre hacia arriba.

mixed expression/expresión mixta (690) Una expresión algebraica que contiene un monomio y una expresión racional.

mode/modal (178) La modal de un conjunto de datos es el número que ocurre con más frecuencia en un conjunto.

monomial/monomio (496) Un monomio es un número, una variable o el producto de un número y una o más variables.

multi-step equations/ecuaciones múltiples (157) Las ecuaciones múltiples son ecuaciones que requieren más de una operación para resolverlas.

multiplicative identity/identidad de multiplicación (38) Para cualquier número a, $a \cdot 1 = 1 \cdot a = a$.

multiplicative inverse/inverso multiplicativo (38) Para cada número no cero $\frac{a}{b}$, en que $a, b \neq 0$, hay exactamente un número $\frac{b}{a}$, tal que $a \cdot b = 1$.

multiplicative property of −1/propiedad multiplicativa de −1 (107) El producto de cualquier número y −1 es su inverso aditivo.

$$-1(a) = -a \text{ y } a(-1) = -a.$$

multiplicative property of equality/propiedad multiplicativa de la igualdad (150) Para cualquiera de los números a, b y c, si $a = b$, entonces $a \cdot c = b \cdot c$.

multiplication property of inequality/propiedad de multiplicación de la desigualdad (393) Para todos los números a, b y c, lo siguiente es cierto:

1. Si c es positivo y $a < b$, entonces $ac < bc$, $c \neq 0$, y si c es positivo y $a > b$, entonces $ac > bc$, $c \neq 0$.

2. Si c es negativo y $a < b$, entonces $ac > bc$, $c \neq 0$, y si c es negativo y $a > b$, entonces $ac < bc$, $c \neq 0$.

multiplicative property of zero/propiedad multiplicativa de cero (38) Para cualquier número a, $a \cdot 0 = 0 \cdot a = 0$.

negative correlation/correlación negativa (340) Existe una correlación negativa entre x y y si los valores están relacionados de maneras opuestas.

negative exponent/exponente negativo (503) Para cualquiera de los números no cero a y cualquier número entero n, $a^{-n} = \frac{1}{a^n}$

negative number/número negativo (73) Cualquier número que es menos de cero.

number line/recta numérica (72) Una recta con marcas equidistantes que se usa para representar números.

number theory/teoría de números (158) El estudio de los números y de las relaciones entre los mismos.

obtuse triangle/triángulo obtuso (165) Un triángulo obtuso tiene un ángulo cuya medida es mayor de 90 grados.

odds/posibilidades (229) Las posibilidades de que un evento ocurra son la proporción del número de formas en que el evento puede ocurrir (éxitos) comparada con el número de formas en que puede no ocurrir (fracasos).

open sentences/enunciados abiertos (32) Enunciado matemático que contiene una o más variables, o incógnitas.

opposites/opuestos (87) El opuesto de un número es su inverso aditivo.

order of operations/orden de operaciones (19)

1. Simplifica las expresiones dentro de los símbolos de agrupación, tales como paréntesis, corchetes y como lo indiquen las barras de fracción.

2. Evalúa todas las potencias.

3. Realiza todas las multiplicaciones y las divisiones de izquierda a derecha.

4. Realiza todas las sumas y las restas de izquierda a derecha.

ordered pair/par ordenado (57) Pares de números que se usan para ubicar puntos en el plano de coordenadas.

organize data/organizar datos (464) Una estrategia útil antes de resolver un problema. Algunas formas de organizar los datos son el uso de tablas, esquemas, diferentes tipos de gráficas o diagramas.

origin/origen (254) El punto de intersección de los dos ejes en el plano de coordenadas.

outcomes/resultados (413) Los resultados son todas las posibles combinaciones de un problema de cómputo.

outlier/valor atípico (307) En un conjunto de datos, un valor que es mucho mayor o mucho menor que el resto de los datos se puede llamar un valor atípico.

parabola/parábola (610, 611) Una parábola es la forma general de la gráfica de una función cuadrática.

parallel lines/rectas paralelas (362)   Rectas en el plano que nunca se intersecan. Las rectas paralelas no verticales tienen la misma pendiente.

parallelogram/paralelogramo (362)   Un cuadrilátero cuyos lados opuestos son paralelos.

percent/por ciento (215)   Un por ciento es una proporción que compara un número con 100.

percentage/porcentaje (215)   El número que se divide entre la base en un por ciento de proporción.

percent of decrease/porcentaje de disminución (222)   La proporción de una cantidad de disminución comparada con una cantidad previa, expresada en forma de por ciento.

percent of increase/porcentaje de aumento (222)   La proporción de una cantidad de aumento comparada con una cantidad previa, expresada en forma de por ciento.

percent proportion/proporción de porcentaje (215)
$$\frac{\text{Percentaje}}{\text{Base}} = \frac{r}{100}$$

perfect square/cuadrado perfecto (119)   Un número racional cuya raíz cuadrada es un número racional.

perfect square trinomial/cuadrado perfecto trinómico (587)   Un trinomio que al factorizarse tiene la forma $(a + b)^2 = (a + b)(a + b)$ o $(a - b)^2 = (a - b)(a - b)$.

perpendicular lines/rectas perpendiculares (362)   Rectas que se encuentran para formar ángulos rectos.

point-slope form/forma punto–pendiente (333)   Para cualquier punto (x_1, y_1) sobre una recta no vertical cuya pendiente es m, la forma punto–pendiente de una ecuación lineal es la siguiente:
$$y - y_1 = m(x - x_1).$$

polynomial/polinomio (513, 514)   Un polinomio es un monomio o la suma de monomios.

positive correlation/correlación positiva (340)   Existe una correlación positiva entre x y y si los valores están relacionados de la misma forma.

power/potencia (7)   Una expresión de la forma x^n se conoce como una potencia.

power of a monomial/potencia de un monomio (498)   Para cualquiera de los números a y b, y para todos los números enteros m, n y p, $(a^m b^n)^p = a^{mp} b^{np}$.

power of a power/potencia de una potencia (498)   Para cualquier número a y todos los números enteros m y n, $(a^m)^n = a^{mn}$.

power of a product/potencia de un producto (498)   Para todos los números a y b y cualquier número entero m, $(ab)^m = a^m b^m$.

prime factorization/factorización prima (558)   Un número entero expresado como un producto de factores que son todos números primos.

prime number/número primo (558)   Un número primo es un número entero mayor que 1 cuyos únicos factores son 1 y el número mismo.

prime polynomial/polinomio primo (577)   Un polinomio que no se puede escribir como el producto de dos polinomios con coeficientes integrales se llama un polinomio primo.

principal square root/raíz cuadrada principal (119)   La raíz cuadrada no negativa de una expresión.

probability/probabilidad (228)   Una razón que expresa la posibilidad de que algún evento suceda.
$$P(\text{evento}) = \frac{\text{número de resultados favorables}}{\text{número de resultados posibles}}$$

problem-solving plan/plan para solucionar problemas (126)

1. Explorar el problema.
2. Planificar la solución.
3. Resolver el problema.
4. Examinar la solución.

product/producto (6)   El resultado de la multiplicación.

product of powers/producto de potencias (497)   Para cualquiera de los números a y todos los números enteros m y n, $a^m \cdot a^n = a^{m + n}$.

product property of square roots/propiedad del producto de raíces cuadradas (719)   Para cualquiera de los números a y b, en que $a \geq 0$ y $b \geq 0$, $\sqrt{ab} = \sqrt{a} \cdot \sqrt{b}$.

proportion/proporción (195)   En una proporción, el producto de los extremos es igual al producto de las medias. Si $\frac{a}{b} = \frac{c}{d}$, entonces $ad = bc$.

Pythagorean theorem/teorema de Pitágoras (713)   Si a y b son las medidas de los catetos de un triángulo rectángulo y c es la medida de la hipotenusa, entonces, $c^2 = a^2 + b^2$.

Pythagorean triple/triplete de Pitágoras (583)   Tres números enteros a, b, y c, tales que $a^2 + b^2 = c^2$.

quadrant/cuadrante (254)   Una de las cuatro regiones en que el eje x y el eje y separan el plano de coordenadas.

quadratic equation/ecuación cuadrática (620)   Una ecuación cuadrática es una en que el valor de la función cuadrática relacionada es 0.

quadratic formula/fórmula cuadrática (628)   Las raíces de una ecuación cuadrática en la forma $ax^2 + bx + c = 0$, en la cual $a \neq 0$, son dadas por la fórmula
$$x = \frac{-b \pm \sqrt{b^2 - 4ac}}{2a}.$$

quadratic function/función cuadrática (610, 611)   Una función cuadrática es una función que se puede describir con una ecuación de la forma $ax^2 + bx + c = 0$, en la cual $a \neq 0$.

quartiles/cuartillos (304, 306)   En un conjunto de datos, los cuartillos son valores que dividen los datos en cuatro partes iguales.

quotient of powers/cociente de potencias (501)   Para todos los números enteros m y n y todo número no cero a, $\frac{a^m}{a^n} = a^{m-n}$.

quotient property of square roots/propiedad del cociente de raíces cuadradas (721)   Para cualquiera de los números a y b, en que $a > 0$ y $b > 0$, $\sqrt{\frac{a}{b}} = \frac{\sqrt{a}}{\sqrt{b}}$.

radical equations/ecuaciones radicales (732)   Ecuaciones que contienen radicales con variables en el radicando.

radical sign/signo radical (119)   El símbolo $\sqrt{\ }$ que se usa para indicar la raíz cuadrada principal no negativa de una expresión.

radicand/radicando (719)   El radicando es la expresión debajo del signo radical.

random/al azar (229)   Cuando se escoge un resultado sin ninguna preferencia, el resultado ocurre al azar.

range/amplitud   **1.** (263) El conjunto de todas las segundas coordenadas de los pares ordenados de una relación.   **2.** (306) La diferencia entre los valores mayor y menor en un conjunto de datos.

rate/razón (197)   La proporción de dos medidas que tienen diferentes unidades de medida.

rate/tasa (215)   En una proporción de por ciento, la tasa es la fracción con 100 como denominador.

ratio/razón (195)   Una razón es una comparación de dos números mediante división.

rational equation/ecuación racional (696)   Una ecuación racional es una ecuación que contiene expresiones racionales.

rational expression/expresión racional (660)   Una expresión racional es una fracción algebraica cuyo numerador y cuyo denominador son polinomios.

rational numbers/números racionales (93)   Un número racional es un número que se puede expresar en la forma $\frac{a}{b}$, en que a y b son números enteros y $b \neq 0$.

rationalizing the denominator/racionalizando el denominador (722)   Un proceso que se usa para quitar o eliminar radicales del denominador de una fracción, en una expresión radical.

real numbers/números reales (121)   El conjunto de números irracionales junto con los números racionales forman el conjunto de números reales.

reciprocal/recíproco (38)   El inverso multiplicativo de un número.

reflexive property of equality/propiedad reflexiva de la igualdad (39)   Para cualquier número a, $a = a$.

regression line/línea de regresión (342)   La línea más exacta de mejor ajuste para un conjunto de datos y la cual se puede determinar con una calculadora de graficar o una computadora.

relation/relación (260, 263)   Una relación es un conjunto de pares ordenados.

replacement set/conjunto de substitución (33)   Un conjunto de números de los cuales se pueden escoger números para reemplazar una variable.

right triangle/triángulo rectángulo (165)   Un triángulo rectángulo tiene un ángulo con una medida de 90 grados.

rise/altura (325)   El cambio vertical en una recta.

roots/raíces (620)   Las soluciones para una ecuación cuadrática.

run/carrera (325)   El cambio horizontal en una recta.

scalar multiplication/multiplicación escalar (108)   En multiplicación escalar, cada elemento de una matriz se multiplica por una constante.

scale/escala (197)   Una razón llamada una escala se usa en la construcción de un modelo para representar algo que es muy grande o muy pequeño para ser dibujado a su tamaño real.

scatter plot/diagrama de dispersión (339)   En un diagrama de dispersión, dos conjuntos de datos se trazan como pares ordenados en el plano de coordenadas.

scientific notation/notación científica (506)   Un número está expresado en notación científica cuando está en la forma de $a \times 10^n$, en que $1 \leq a < 10$ y n es un número entero.

set/conjunto (33)   Una colección de objetos o números.

set-builder notation/notación de construcción de conjuntos (385)   Una notación que se usa para describir los miembros de un conjunto. Por ejemplo, $\{y \mid y < 17\}$ representa el conjunto de todos los números y de modo que y es menos que 17.

sequence/secuencia (13)   Un conjunto de números en un orden específico.

similar triangles/triángulos similares (201)   Si dos triángulos son similares, las medidas de sus lados correspondientes son proporcionales y las medidas de sus ángulos correspondientes son iguales.

simple events/eventos simples (414)   Un evento simple es un evento sencillo en un problema de probabilidad.

simple interest/interés simple (217)   La cantidad pagada o ganada por el uso de una cantidad de dinero. La fórmula $I = prt$ se usa para resolver problemas de interés simple.

simplest form/forma reducida (47)   Una expresión está en su forma reducida cuando ha sido reemplazada por una expresión similar que no tiene términos semejantes ni paréntesis.

simplest radical form/forma radical reducida (723)   Una expresión radical se encuentra en forma reducida cuando se satisfacen las siguientes condiciones:

1. Ningún radicando tiene factores cuadrados perfectos además de uno.

2. Ningún radicando contiene fracciones.

3. No aparece ningún radical en el denominador de una fracción.

sine/seno (206)   En un triángulo rectángulo con ángulo agudo A, el seno del ángulo

$A = \dfrac{\text{medida del cateto opuesto al ángulo } A}{\text{medida de la hipotenusa}}$.

slope/pendiente (324, 325)   El cambio vertical (altura) al cambio horizontal (carrera) a medida que te mueves de un punto a otro a lo largo de la recta.

slope-intercept form/forma pendiente intersección (347)   Una ecuación de la forma $y = mx + b$, en que m es la pendiente y b es la intersección en y de una recta dada.

solution/solución (32)   Una sustitución por una variable en una ecuación que resulta en una ecuación válida.

solution of an equation in two variables/ solución de una ecuación de dos variables (271)   Si al sustituir los números en un par ordenado se satisface una ecuación de dos variables, entonces el par ordenado es una solución de la ecuación.

solution set/conjunto de solución (33)   El conjunto de todos los sustitutos para una variable en un enunciado abierto que satisfacen el enunciado.

solve an equation/resuelve una ecuación (145)   Resolver una ecuación quiere decir aislar la variable cuyo coeficiente es 1, en un lado de la ecuación.

solving an open sentence/resolviendo un enunciado abierto (32)   Hallar un sustituto que satisface la variable.

solving a triangle/resolviendo un triángulo (208)   El proceso de hallar las medidas de todos los lados y ángulos de un triángulo rectángulo.

square of a difference/cuadrado de una diferencia (544)   Si a y b son cualquier par de números, $(a - b)^2 = (a - b)(a - b) = a^2 - 2ab + b^2$.

square of a sum/cuadrado de una suma (543)   Si a y b son cualquier par de números, $(a + b)^2 = (a + b)(a + b) = a^2 + 2ab + b^2$.

square root/raíz cuadrada (118, 119)   Uno de los dos factores idénticos de un número.

standard form/forma estándar (333)   La forma estándar de una ecuación lineal es $Ax + By = C$, en la cual A, B y C son números enteros, $A \geq 0$ y A y B no son ceros ambos.

statistics/estadística (25)   Una rama de las matemáticas que tiene que ver con los métodos de recolección, organización e interpretación de datos.

stem-and-leaf plot/gráfica de tallo y hojas (25)   En una gráfica de tallo y hojas, cada dato se separa en dos números que se usan para formar un tallo y las hojas. Los datos se organizan en dos columnas. La columna de la izquierda contiene el tallo y la columna de la derecha las hojas.

substitution/sustitución (462)   El método de sustitución para resolver un sistema de ecuaciones es un método que sustituye una ecuación en la otra ecuación para despejar la otra variable.

substitution property of equality/propiedad de sustitución de la igualdad (39)   Si $a = b$, entonces, a se puede reemplazar por b en cualquier expresión.

subtracting integers/sustracción de enteros (87)   Para restar un entero, suma su inverso aditivo. Para cualquiera de los entras a y b, $a - b = a + (-b)$.

subtraction property of equality/propiedad de sustracción de la igualdad (146)   Para cualquiera de los números a, b y c, si $a = b$, entonces $a - c = b - c$.

subtraction property of inequality/propiedad de sustracción de la desigualdad (385)   Para todos los números a, b y c los siguientes son ciertos:

1. si $a > b$, entonces $a - c > b - c$;

2. si $a < b$, entonces $a - c < b - c$.

supplementary angles/ángulos suplementarios (162)   Dos ángulos son suplementarios si la suma de sus medidas es 180 grados.

symmetric property of equality/propiedad simétrica de la igualdad (39)   Para cualquiera de los números a y b, si $a = b$, entonces $b = a$.

symmetry/simetría (612)   Las figuras simétricas son aquellas en que la figura se puede doblar y cada mitad es exactamente igual a la otra.

system of equations/sistema de ecuaciones (455)   Un conjunto de ecuaciones con las mismas variables.

system of inequalities/sistema de desigualdades (482)   Un conjunto de desigualdades con las mismas variables.

tangent/tangente (206)   En un triángulo rectángulo con ángulo agudo A, la tangente del ángulo $A = \dfrac{\text{medida del cateto opuesto al ángulo } A}{\text{medida del cateto adyacente al ángulo } A}$.

term/término   1. (13) Un número en una secuencia. **2.** (47) Un número, una variable o un producto o cociente de números y variables.

transitive property of equality/propiedad transitiva de la igualdad (39)   Para cualquiera de los números a, b y c, si $a = b$ y $b = c$, entonces $a = c$.

tree diagram/diagrama de árbol (413)   Un diagrama de árbol es un diagrama que se usa para mostrar el número total de posibles resultados.

triangle/triángulo (163)   Un triángulo es un polígono con tres lados y tres ángulos.

trigonometric ratios/razones trigonométricas (206)   seno $A = \frac{a}{c}$ coseno $A = \frac{b}{c}$ tangente $A = \frac{a}{b}$

trinomials/trinomios (514)   Un trinomio es la suma de tres monomios.

uniform motion/movimiento uniforme (235)   Cuando un objeto se mueve a una velocidad, o a un ritmo constante, se dice que se mueve en movimiento uniforme.

union/unión (408)   La unión de dos conjuntos A y B es el conjunto de los elementos contenidos en ambos A o B.

unique factorization theorem/teorema de la factorización única (558)   La factorización prima de cada número es única excepto por el orden en que se escriben los factores.

unit cost/costo unitario (95)   El costo de una unidad de algo.

upper quartile/cuartillo superior (306)   El cuartillo superior divide la parte superior de un conjunto de datos en dos partes iguales.

use a model/usa un modelo (339)   Una estrategia para resolver problemas que usa modelos, o simulaciones, de situaciones matemáticas difíciles de resolver directamente.

use a table/usa una tabla (255)   Una estrategia para resolver problemas que usa tablas para organizar y resolver problemas.

variable/variable (6)   Las variables son símbolos que se usan para representar números desconocidos.

Venn diagrams/diagramas de Venn (73)   Los diagramas de Venn son diagramas que usan círculos u óvalos dentro de un rectángulo para mostrar relaciones de conjuntos.

vertex/vértice (610, 611)   El punto máximo o mínimo de una parábola.

vertical axis/eje vertical (56)   La recta vertical en una gráfica que representa la variable dependiente.

vertical line test/prueba de recta vertical (289)   Si cualquier recta vertical pasa por un solo punto de la gráfica de una relación, entonces la relación es una función.

weighted average/promedio ponderado (233)   El promedio ponderado M de un conjunto de datos es la suma del producto de cada número en el conjunto y su peso divididos entre la suma de todos los pesos.

whiskers/patillas (428)   Las patillas de un diagrama de caja y patillas son los segmentos que se dibujan desde el cuartillo inferior hasta el mínimo valor y desde el cuartillo superior hasta el máximo valor.

whole numbers/números enteros (72)   El conjunto de números enteros se representa por {0,1,2,3,...}.

work backward/trabaja al revés (156)   Una estrategia para resolver problemas que usa operaciones inversas para determinar un valor inicial.

x-axis/eje x (254)   La recta numérica horizontal.

x-coordinate/coordenada x (254)   El primer número en un par ordenado.

x-intercept/intersección con el eje x (346)   La coordenada en donde una gráfica interseca el eje x.

y-axis/eje y (254)   La recta numérica vertical.

y-coordinate/coordenada y (254)   El segundo número en un par ordenado.

y-intercept/intersección con el eje y (346)   La coordenada en donde una gráfica interseca el eje y.

zero exponent/exponente cero (502)   Para cualquier número no cero a, $a^0 = 1$.

zero product property/propiedad del producto cero (594)   Para todos los números a y b, si $ab = 0$, entonces $a = 0$, $b = 0$, o ambos a y b son iguales a cero.

zeros/ceros (620)   Los ceros de una función son las raíces, o intersecciones con el eje x de la función.

SELECTED ANSWERS

CHAPTER 1   EXPLORING EXPRESSIONS, EQUATIONS, AND FUNCTIONS

Pages 9–11   Lesson 1–1
7. $3y^2 - 6$ **9.** 3 times x squared increased by 4   **11.** a^7
13. 32   **15.** $k + 20$ **17.** a^7 **19.** $\frac{2x^2}{3}$ or $\frac{2}{3}x^2$ **21.** $b + 8$
23. 4 times m to the fifth power   **25.** c squared plus 23
27. 2 times 4 times 5 squared   **29.** 8^2 **31.** 4^7 **33.** z^5
35. 49   **37.** 64   **39.** 16   **41.** $3(55 - w^3)$ **43.** $a + b + \frac{a}{b}$
45. $y + 10x$ **47.** 16; 16   **47a.** They are equal; no; no.
47b. no; for example, $2^3 \neq 3^2$ **49a.** $3.5x$ **49b.** $3.5y$
49c. $3.5x + 3.5y$

Pages 15–18   Lesson 1–2
5.
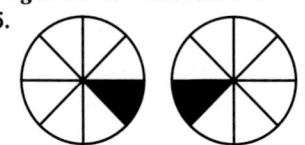
7. $5x + 1, 6x + 1$

9a.

4^1	4^2	4^3	4^4	4^5
4	16	64	256	1024

9b.   6; $4^6 = 4096$ **9c.** 4; When the exponent is odd, 4 is in the ones place.   **11.**
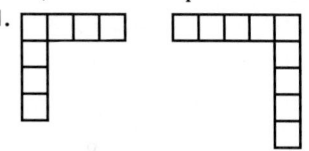
13. 48, 96   **15.** 25, 36   **17.** $a + 7, a + 9$
19a.

19b. White; even-numbered figures are white.
19c. 12 sides; The shapes come in pairs. Since $19 \div 2 = 9.5$, the 19th figure will be part of the 10th pair. The 10th pair will have $10 + 2$, or 12 sides.   **21a.** 4; 9; 16; 25
21b. The sums are perfect square numbers; that is 1^2, $2^2, 3^2, 4^2, 5^2, \ldots$ **21c.** 10,000   **21d.** x^2 **23a.** 1,999,998; 2,999,997; 3,999,996; 4,999,995   **23b.** 8,999,991
25. 4, 7, 10, 13, 16   **27.** 1:13 P.M.   **29a.** 100 cards
31. x cubed divided by 9   **33.** 4^3; 64 cubes
35. $x + \frac{1}{11}x$

Pages 22–24   Lesson 1–3
5. 173   **7.** 4   **9.** 25   **11.** $a^2 - a$ **13.** 14   **15.** 14
17. 9   **19.** 60   **21.** $\frac{11}{18}$ **23.** 6   **25.** 126   **27.** 147
29. 126   **31.** $r^2 + 3s$; 19   **33.** $(r + s)t^2$; $\frac{7}{4}$ **35.** 19.5 mm
37. 22 in.   **39.** 16.62   **41.** 3.77   **43a.** Sample answer:
$(4 - 2)5 \div (3 + 2)$ **43b.** $4 \times 2 \times 5 \times 3 \times 2$ **45a.** $\frac{1}{3}Bh$
45b. 198,450 m^3 **47.** 16, 19.5   **49.** June 18   **51.** $t + 3$
53. 9 more than two times y

Pages 28–31   Lesson 1–4
5. stem 12, leaf 2   **7.** stem 126, leaf 9   **9a.** 35 tickets sold one day   **9b.** 31 tickets   **9c.** 75 tickets   **9d.** 1043 tickets
11. stem 13, leaf 3   **13.** stem 44, leaf 3   **15.** stem 111, leaf 3   **17.** stem 14, leaf 3   **19.** stem 111, leaf 4
21. rounded to the nearest hundred: 9, 12, 24, 27, 38, 39, 40   **23a.** Possible answer: Both teens and young adults had similar distribution of responses.   **23b.** Sample answer: The market research shows that the new game is equally appealing to teens and young adults. Therefore, we should concentrate our marketing efforts towards both teens and young adults.

25a.

1980	Stem	1993
4	1	3
6	2	4
8 5 1	3	3 7 7
	4	6
1	5	

$3\,|\,7 = 3700$

25b. 3000 to 3900; 3000 to 3900   **25c.** Sample answer: Farm sizes seem to be shrinking.   **27.** 15   **29a.** $\frac{s}{5}$
29b. 2 miles; yes
31.

Pages 34–36   Lesson 1–5
7. false   **9.** {1, 3, 5}   **11.** 75   **13.** false   **15.** true   **17.** true
19. {5}   **21.** $\left\{\frac{1}{2}, \frac{3}{4}\right\}$ **23.** {10, 15, 20}   **25.** 3   **27.** 2
29. $4\frac{5}{6}$ **31.** Sample answer: $p = 1$ and $q = 2$, $p = 2$ and $q = 10$, $p = 3$ and $q = 8$, $p = 4$ and $q = 20$, and $p = 5$ and $q = 15$ **33a.** $C = \frac{3500 \cdot 4}{14}$ **33b.** 1000 Calories
35. 2, 5, 6, 7, 9   **37.** 8   **39.** 5 less than x to the fifth power

Page 36   Self Test
1. $3a + b^2$ **3.** 11:04, 11:08, 11:51, 11:55   **5.** 408

7.

Stem	Leaf
4	8
5	4
6	7
7	7
8	5 9

9. {6, 7, 8}

$6\,|\,7 = 67$

Pages 41–43   Lesson 1–6
7. $\frac{2}{9}$ **9.** c   **11.** e   **13.** d   **15.** f

17. $6(12 - 48 \div 4) + 9 \cdot 1$
$= 6(12 - 12) + 9 \cdot 1$ *Substitution* $(=)$
$= 6(0) + 9 \cdot 1$ *Substitution* $(=)$
$= 0 + 9 \cdot 1$ *Multiplicative property of 0*
$= 0 + 9$ *Multiplicative identity*
$= 9$ *Additive identity*
19a. $4(20) + 7$
19b. $4(20) + 7$
$= 80 + 7$ *Substitution* $(=)$
$= 87$ *Substitution* $(=)$

19c. 87 years **21.** 9 **23.** $\frac{1}{p}$ **25.** $\frac{2}{3}$ **27.** substitution ($=$)
29. multiplicative identity **31.** multiplicative inverse, multiplicative identity **33.** symmetric ($=$) **35.** reflexive ($=$) **37.** substitution ($=$); substitution ($=$); multiplicative identity; multiplicative inverse; substitution ($=$) **39.** substitution ($=$); substitution ($=$); substitution ($=$); multiplicative identity
41. $(15 - 8) \div 7 \cdot 25$

$$
\begin{aligned}
&= 7 \div 7 \cdot 25 \quad && \textit{Substitution}(=) \\
&= 1 \cdot 25 \quad && \textit{Substitution }(=) \\
&= 25 \quad && \textit{Multiplicative identity}
\end{aligned}
$$

43. $(2^5 - 5^2) + (4^2 - 2^4)$

$$
\begin{aligned}
&= (32 - 25) + (16 - 16) \quad && \textit{Substitution }(=) \\
&= 7 + 0 \quad && \textit{Substitution }(=) \\
&= 7 \quad && \textit{Additive identity}
\end{aligned}
$$

45. $5^3 + 9\left(\frac{1}{3}\right)^2$

$$
\begin{aligned}
&= 125 + \tfrac{1}{9}9\left(\ \right) \quad && \textit{Substitution }(=) \\
&= 125 + 1 \quad && \textit{Multiplicative inverse} \\
&= 126 \quad && \textit{Substitution }(=)
\end{aligned}
$$

47a. $[21(12 \cdot 2)] + [23(15 \cdot 2)] + [67(10 \cdot 2)]$
47b. $[21(12 \cdot 2)] + [23(15 \cdot 2)] + [67(10 \cdot 2)]$

$$
\begin{aligned}
&= [21(24)] + [23(30)] + 67(20)] \quad && \textit{Substitution }(=) \\
&= 504 + 690 + 1340 \quad && \textit{Substitution }(=) \\
&= 2534 \quad && \textit{Substitution }(=)
\end{aligned}
$$

47c. \$25.34 **49.** true **51.** false **55.** 36 **57.** $12y$

Page 44 Lesson 1–7A
1. $2x + 2$ **3.** $4x + 2$ **5.** false;

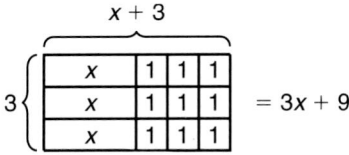

7a. Adita **7b.** Answers will vary. Answers should include the concept that multiplication by 3 is distributed over both terms in parentheses.

Pages 48–50 Lesson 1–7
5. b **7.** a **9.** c **11.** $2a - 2b$ **13.** 60 **15.** 7 **17.** $4y^4, y^4$
19. $3t^2 + 4t$ **21.** $23a^2b + 3ab^2$ **23.** $5.35(24) + 5.35(32)$; $5.35(24 + 32)$ **25.** $5g - 45$ **27.** $24m + 48$ **29.** $5a - ab$
31. 52 **33.** 60 **35.** 645 **37.** in simplest form **39.** $12a + 15b$
41. in simplest form **43.** $3x + 4y$ **45.** $1\frac{3}{5}a$ **47.** $9x + 5y$
49. no; sample counterexample: $2 + (4 \cdot 5) \neq (2 + 4)(2 + 5)$
51a. $2[x + (x + 14)] = 4x + 28$ **51b.** 96 **51c.** 527 ft^2
53. multiplicative property of 0 **55a.** $d = (1129)(2)$
55b. 2258 ft **57.** 2 **59.** 8 years

Pages 53–55 Lesson 1–8
7. commutative ($+$) **9.** associative ($\times$) **11.** $5a + 2b$
13. $14x + 3y$
15. $6z^2 + (7 + z^2 + 6)$

$$
\begin{aligned}
&= 6z^2 + (z^2 + 7 + 6) \quad && \textit{Commutative }(+) \\
&= (6z^2 + z^2) + (7 + 6) \quad && \textit{Associative }(+) \\
&= (6 + 1)z^2 + (7 + 6) \quad && \textit{Distributive property} \\
&= 7z^2 + 13 \quad && \textit{Substitution }(=)
\end{aligned}
$$

17. multiplicative identity **19.** associative ($\times$)
21. distributive property **23.** commutative ($\times$)
25. associative ($+$) **27.** $10x + 5y$ **29.** $7x + 10y$
31. $10x + 2y$ **33.** $32a^2 + 16$ **35.** $5x$ **37.** $\frac{3}{4} + \frac{5}{3}m + \frac{4}{3}n$

39. $2(s + t) - s$

$$
\begin{aligned}
&= 2s + 2t - s \quad && \textit{Distributive property} \\
&= 2t + 2s - s \quad && \textit{Commutative }(+) \\
&= 2t + (2s - s) \quad && \textit{Associative }(+) \\
&= 2t + (2s - 1s) \quad && \textit{Multiplicative identity} \\
&= 2t + (2 - 1)s \quad && \textit{Distributive property} \\
&= 2t + 1s \quad && \textit{Substitution }(=) \\
&= 2t + s \quad && \textit{Multiplicative identity}
\end{aligned}
$$

41. $5xy + 3xy$

$$
\begin{aligned}
&= (5 + 3)xy \quad && \textit{Distributive property} \\
&= 8xy \quad && \textit{Substitution }(=)
\end{aligned}
$$

43. $\frac{1}{100}$; Each denominator and the following numerator represent the number 1. The resulting expression is 1 in the numerator and 100 in the denominator multiplied by numerous 1s. Since 1 is the multiplicative identity, the product is $\frac{1}{100}$. **45.** no; Sample example: Let $a = 1$ and b $= 2$, then $1 * 2 = 1 + 2(2)$ or 5 and $2 * 1 = 2 + 2(1)$ or 4.
47a. $G = 3.73$ **49.** $100d + 80d + 8d$, $188d$ **51.** $\frac{3}{2}$

53.

Stem	Leaf
10	0 0 0 1 1 1 1 3
9	5 7 7 9 9 9

$10 | 3 = 103$

55.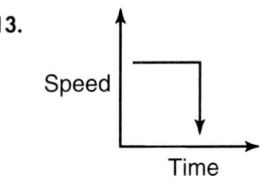

Pages 59–62 Lesson 1–9
5a. Graph 3 **5b.** Graph 4 **5c.** Graph 2 **5d.** Graph 1
7a. False, A is the younger player, but B runs the mile in less time. **7b.** False, A made more 3-point shots, but B made more 2-point shots. **7c.** True, B is the older player and B made more 2-point shots. **7d.** True, A is the younger player and A made more free-throw shots.
9. Graph a; An average person makes no money as a child, then his or her income rises for several years, and finally levels off. **11a.** Graph 5 **11b.** Graph 3 **11c.** Graph 1 **11d.** Graph 4 **11e.** Graph 6 **11f.** Graph 2

13.

15.

17a. horizontal axis: time in years; vertical axis: millions of dollars **17b.** In 1991 the telethon raised 45 million dollars. **17c.** Money raised in the telethon decreased, then increased steadily. **17d.** The money raised increased. **19.** $7p + 11q + 9$ **21.** 14 **23.** $\frac{1}{3}Bh$

Page 63 Chapter 1 Highlights
1. a **3.** e **5.** f **7.** g **9.** b

Pages 64–66 Chapter 1 Study Guide and Assessment
11. x^5 **13.** $5x^2$ **15.** three times a number m to the fifth power **17.** the difference of four times m squared and twice m **19.** 32, 64 **21.** $4x + y$, $5x + y$ **23.** $10^1, 10^2, 10^3, 10^4$ **25.** 11 **27.** 9 **29.** 26 **31.** 2.4 **33.** 9.2 **35.** 90

37. No; some former presidents are still living. **39.** false
41. true **43.** $13\frac{1}{2}$ **45.** $\frac{1}{3}$ **47.** 5 **49.** additive identity

51. $2 \times 4 + 2 \times 7$ **53.** $1 - 3p$ **55.** 294 **57.** $8m + 8n$
59. commutative (+) **61.** commutative ($\times$) **63.** $5x + 5y$
65. $3pq$ **67.** Graph c **69a.** 80s **69b.** 320 **69c.** 640
71a. $3 + 4 > 5, 4 + 5 > 3, 3 + 5 > 4$ **71b.** $x < 9$ ft and $x > 3$ ft

CHAPTER 2 EXPLORING RATIONAL NUMBERS

Pages 75–77 Lesson 2–1
7. $\{-1, 0, 1, 2, ...\}$
9.
$-4 \; -3 \; -2 \; -1 \; 0 \; 1 \; 2 \; 3 \; 4$

11. $-4 + (-3) = -7$ **13.** -5 **15.** 9-yard gain
17. $\{-4, -3, -2, -1\}$ **19.** $\{-7, -3\}$
21. $\{..., -5, -4, -3, -2, -1, 0\}$
23.
$-3 \; -2 \; -1 \; 0 \; 1 \; 2 \; 3 \; 4 \; 5$

25.
$-4 \; -3 \; -2 \; -1 \; 0 \; 1 \; 2 \; 3 \; 4$

27.
$-7 \; -6 \; -5 \; -4 \; -3 \; -2 \; -1 \; 0 \; 1$

29.
$-7 \; -6 \; -5 \; -4 \; -3 \; -2 \; -1 \; 0 \; 1$

31. 13 **33.** -5 **35.** -12 **37.** 6 **39.** -23
41. $-17 + 82 = 65; 65°F$ **43a.** 4°F
43b. $-31°F$ **43c.** $-15°F$ **43d.** 9°F

45.

Volume (vertical axis), *Time* (horizontal axis)

47. commutative ($\times$) **49.** substitution (=)
51. 13.9

Pages 80–83 Lesson 2–2
5. from 35 to 80;

x x   x   x x   x
35 40 45 50 55 60 65 70 75 80

7a.
x
x   x   x
x x x   x   x
x x x   x x x x   x
x x x   x xx x   x   x   x
30 35 40 45 50 55 60 65 70

7b. 70 mph; cheetah **7c.** 30 mph **7d.** 30 mph **7e.** 12 **7f.** 4
9a. from 20 to 75 by 5;

xxx x x   x   x   x   x
20 25 30 35 40 45 50 55 60 65 70 75

9b. yes; 25 and 42 **11a.** Ms. Martinez's **11b.** No; in Ms. Martinez's class, the hours that the students talked on the phone were close to each other (4 ± 2 hours). In Mr. Thomas' class, the students had a wider range of hours spent talking on the phone. The range was from 0 to 8 hours, with no hours clustered around a certain number.

13. Mr. Thomas', 3.6 hours; Ms. Martinez's, 3.7 hours; yes **15.** associative ($\times$) **17.** 2

Page 84 Lesson 2–3A
1. 6 **3.** -2 **5.** 2 **7.** 6 **9.** false **11.** true **13.** Answers should include using the number line.

Pages 90–92 Lesson 2–3
7. $-7, 7$ **9.** $0, 0$ **11.** -4 **13.** -11 **15.** 0 **17.** $40c$
19. -4 **21.** 3 **23.** $\begin{bmatrix} -1 & 4 \\ 6 & -6 \end{bmatrix}$ **25.** $-12, 12$ **27.** $302, 302$
29. 0 **31.** -32 **33.** -16 **35.** -70 **37.** -15 **39.** -5
41. $40b$ **43.** $-29p$ **45.** $26d$ **47.** -4 **49.** -15 **51.** 5
53. 8 **55.** 99 **57.** -59 **59.** $\begin{bmatrix} 0 & 3 \\ 2 & 2 \end{bmatrix}$ **61.** $\begin{bmatrix} -4 & -5 \\ -4 & -5 \\ 10 & -3 \end{bmatrix}$
63. 24th floor
65a.

Monday

	Sesame	Poppy	Blue	Plain
East Store	120	80	64	75
West Store	65	105	77	53

Tuesday

	Sesame	Poppy	Blue	Plain
East Store	112	76	56	74
West Store	69	95	82	50

65b. Monday + Tuesday

	Sesame	Poppy	Blue	Plain
East Store	232	156	120	149
West Store	134	200	159	103

65c. Monday − Tuesday

	Sesame	Poppy	Blue	Plain
East Store	8	4	8	1
West Store	−4	10	−5	3

This matrix represents the difference between Monday's and Tuesday's sales in each category.
67a. from 100 to 146
67b.
            x
            x
  x   xx   xxx xx x x   xx   x   x x   x x
100 105 110 115 120 125 130 135 140 145 150

67c. yes; 114 **67d.** Siberian Elm; American Elm
69.

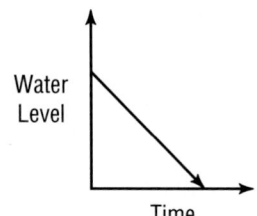

Water Level (vertical axis), *Time* (horizontal axis)

71. $55y^2$ **73.** symmetric (=) **75.** 14, 20, 29, 34, 37, 38, 43, 59, 64, 74, 84 **77.** 24

Pages 97–99 Lesson 2–4
7. $<$ **9.** $=$ **11.** $-0.5, \frac{3}{4}, \frac{7}{8}, 2.5$ **15.** $>$ **17.** $<$ **19.** $<$
21. $>$ **23.** $\frac{3}{8}, \frac{2}{3}, \frac{6}{7}$ **25.** $\frac{3}{23}, \frac{8}{42}, \frac{4}{14}$ **27.** $-\frac{2}{5}, -0.2, 0.2$
29. a 16-ounce drink for \$0.59 **31.** a package of 75 paper plates for \$3.29 **33.** Sample answer: $\frac{2}{3}$ **35.** Sample answer: $\frac{7}{20}$ **37.** Sample answer: $\frac{1}{6}$ **39.** $E = \frac{4}{14}$ or $\frac{2}{7}$; $G = \frac{10}{14}$ or $\frac{5}{7}$; $H = \frac{13}{14}$ **41.** 0.375 inch **43.** $-9, 9$

45.

47. $m + 2n + \frac{3}{2}$ 49. two times x squared plus six

Page 99   Self Test

1.

3. -17 5. 55

7a.

7b. yes; $270   9. $<$

Pages 102–104   Lesson 2–5

5. $-\frac{1}{9}$ 7. $-2\frac{5}{8}$ 9. 0.88   11. 5.75   13a. $+\frac{1}{2}$

13b.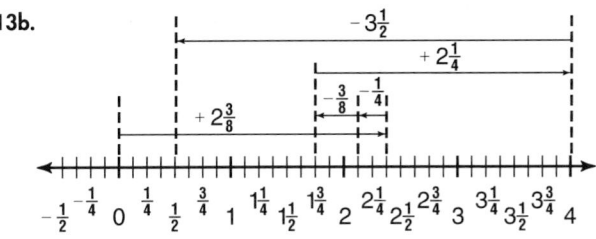

15. $-\frac{11}{16}$ 17. -1.3 19. $\frac{4}{9}$ 21. -0.2007 23. $-8\frac{5}{8}$

25. 0.0485   27. -22.94 29. -2.17 31. 1   33. -3.5

35. -16.7 37. $-\frac{52}{21}$ 39. $\begin{bmatrix} -3.8 & -0.1 \\ 1.7 & 2.9 \end{bmatrix}$ 41. $\begin{bmatrix} \frac{1}{4} & 11 \\ -5 & 8 \\ \frac{1}{2} & -12 \end{bmatrix}$

43. Answers will vary. Sample answer: $\begin{bmatrix} 5 & 3 \\ 1 & 7 \end{bmatrix}, \begin{bmatrix} 4 & -3 \\ 1 & 2 \end{bmatrix},$ $\begin{bmatrix} 9 & 0 \\ 2 & 9 \end{bmatrix}, \begin{bmatrix} 18 & 0 \\ 4 & 18 \end{bmatrix}; \begin{bmatrix} 5 & 3 \\ 1 & 7 \end{bmatrix} + \begin{bmatrix} 4 & -3 \\ 1 & 2 \end{bmatrix} = \begin{bmatrix} 9 & 0 \\ 2 & 9 \end{bmatrix} = \begin{bmatrix} 4 & -3 \\ 1 & 2 \end{bmatrix} +$ $\begin{bmatrix} 5 & 3 \\ 1 & 7 \end{bmatrix}; \left(\begin{bmatrix} 5 & 3 \\ 1 & 7 \end{bmatrix} + \begin{bmatrix} 4 & -3 \\ 1 & 2 \end{bmatrix} \right) + \begin{bmatrix} 9 & 0 \\ 2 & 9 \end{bmatrix} = \begin{bmatrix} 18 & 0 \\ 4 & 18 \end{bmatrix} =$ $\begin{bmatrix} 5 & 3 \\ 1 & 7 \end{bmatrix} + \left(\begin{bmatrix} 4 & -3 \\ 1 & 2 \end{bmatrix} + \begin{bmatrix} 9 & 0 \\ 2 & 9 \end{bmatrix} \right)$ 45a. 9.01, 10.13, 11.25

45b. 1.12   45c. $-2, -\frac{5}{4}, -\frac{1}{2}, \frac{1}{4}, 1; -2\frac{1}{2}$ 47. $<$

49a.

49b. no   51. multiplicative property of 0   53. $\frac{7}{24}$

Page 105   Lesson 2–6A

3. -10 5. -10 7. 10

Pages 109–111   Lesson 2–6

5. -18 7. 24   9. $-\frac{4}{5}$ 11. -4 13. $-46st$ 15. $\begin{bmatrix} -6 & 12 \\ -3 & 15 \end{bmatrix}$

17. $28.65   19. -60 21. -1 23. 2   25. 0.00879   27. $-\frac{6}{5}$

29. $-\frac{9}{17}$ 31. 85.7095   33. 3   35. -6 37. $\frac{13}{12}$ 39. $-\frac{179}{24}$

41. $\frac{25}{24}$ 43. $-30rt + 4s$ 45. $21x$ 47. $16.48x - 5.3y$

49. $\begin{bmatrix} 2 & 6 & 3 \\ \frac{5}{2} & 5 & 1 \end{bmatrix}$ 51. $\begin{bmatrix} -9 & 22.68 \\ -22.4 & -10 \\ 28.8 & 11.12 \end{bmatrix}$ 53. $\begin{bmatrix} 6 & 18 & 4 \\ 0 & 2 & \frac{8}{3} \end{bmatrix}$

55. It is negative.   57a. No; it meets the requirement that no dimensions can be less than 20 feet, but the minimum of 1250 square feet is not met because this yard would have 1216 square feet.   57b. No; it meets the requirement that yards must have a minimum of 1250 square feet because it would have 1330 square feet, but the requirement that no dimension can be less than 20 feet is not met because it has a side of length 19 feet.

57c. Answers will vary. Sample answer: 30 feet.

59. -2.2 61. -20 63. -7 65. $\{6, 7\}$ 67. 408

Pages 115–117   Lesson 2–7

5. -4 7. $-\frac{3}{32}$ 9. $-\frac{5}{48}$ 11. $-6x$ 13. about 2 minutes

15. 6   17. $-\frac{1}{18}$ 19. $-\frac{1}{16}$ 21. -6 23. $-\frac{1}{12}$ 25. $\frac{243}{10}$

27. $-\frac{35}{3}$ 29. $-65m$ 31. $r + 3$ 33. $-20a - 25b$

35. $-14c + 6d$ 37. $-a - 4b$ 39. -1.2 41. $-0.8\overline{3}$ 43. 4

45a.

	$1\frac{1}{2}$"
$\circ$	
	$\frac{3}{4}$"
$\circ$	
	$\frac{3}{4}$"
$\circ$	
	$\frac{3}{4}$"
$\circ$	
	4"

30"

45b. $5\frac{9}{16}$ inches   47. $4\frac{1}{20}$ inches   49. $\frac{13}{12}$ 51a. $\begin{bmatrix} -2 & 4 \\ 3 & 12 \end{bmatrix}$

51b. $\begin{bmatrix} 4 & 4 \\ 7 & 2 \end{bmatrix}$ 53. $20b + 24$

55a.

Stem	Leaf
13	0
15	4
17	2 3
19	4
20	0
22	0 2
24	5
25	1 4
27	9
29	0
30	4 7
33	0
34	6
52	5

$13 \mid 0 = 13,000$

55b. $13,000   55c. $52,500   55d. 9

Page 118   Lesson 2–8A

1. 4–5   3. 13–14   5. 1–2

Pages 123–125   Lesson 2–8

7. 8   9. 11.05   11. 16   13. Q   15. Q

17.

number line: 2 3 4 5 6 7 8 9 10

19.

number line: −2 −1 0 1 2 3 4 5 6

21. 13   **23.** $\frac{2}{3}$ **25.** 20.49   **27.** 15   **29.** −25   **31.** $\frac{3}{5}$ **33.** 9.33

35. −47   **37.** −20   **39.** Z, Q   **41.** Q   **43.** Q   **45.** Q   **47.** I

49. Q   **51.**

number line: −6 −5 −4 −3 −2 −1 0 1 2

53.

number line: −7 −6 −5 −4 −3 −2 −1 0 1

55.

number line: −15 −14 −13 −12 −11 −10 −9 −8 −7

57.

number line: −5.6 −5.5 −5.4 −5.3 −5.2 −5.1 −5.0 −4.9 −4.8

59.

number line: $4\frac{1}{2}$ $4\frac{3}{4}$ 5 $5\frac{1}{4}$ $5\frac{1}{2}$ $5\frac{3}{4}$ 6 $6\frac{1}{4}$ $6\frac{1}{2}$

61. Yes; it lies between 27 and 28.   **63.** about 50 miles
65. Fibonacci   **67.** −12   **69.** a 1-pound package of lunch
meat for $1.95   **71.** from 500 to 520 by 5s;

line plot: 500 505 510 515 520

73. b   **75a.** to the nearest ten thousand   **75b.** 10
75c. $690,000   **77.** 26, 32

Pages 130–132   Lesson 2–9
5a. 4 points   **5b.** 15 questions   **5c.** 86   **5d.** 6 points
5e. $15 - n$ **5f.** $4n$ **7.** $n + 5 \geq 48$ **9.** Let s = average
speed; $s = \frac{189}{3} = 63$.   **11.** The product of a times the sum
of y and 1 is b.   **13a.** 640 feet   **13b.** $y - 80$ **13c.** $\ell + \ell + w + w$ or $2\ell + 2w$ **13d.** 200 ft   **13e.** 24,000 ft^2 **13f.** Is
there enough space to build the playground as designed?
15. $a^2 + b^3 = 25$ **17.** $x < (a - 4)^2$ **19.** $\frac{7}{8}(a + b + c^2) = 48$

21. Sample answer: x = first number
$\qquad\qquad\qquad x + 25$ = second number
$\qquad\qquad\qquad 25 + 2x = 106$
23. Sample answer:
$\qquad y$ = how much money Hector saves
$\qquad y = 2(7)(50)$
25. V is equal to the quotient of a times h and three.
27. Sample answer: Three times the number of miles
home is from school is 36. How many miles away from
home is the school?   **29.** Sample answer: The number of
minutes spent waiting for a doctor in the waiting room is
5 minutes more than 6 times the number of appointments
she has. How long would you have to wait for a doctor
with d appointments?
31. $ab + cd + 2d - 8$

33. Sample answer: $4.5 + x + x = 24 + x$
$\qquad\qquad\qquad\qquad\qquad x = 19.5$

35a. $A = \frac{1}{2}h(a + b)$ **35b.** $A = 108\frac{1}{2}$ ft^2 **37.** −8

39. $\begin{bmatrix} 8 & -1.6 \\ 12 & -20 \end{bmatrix}$ **41.** −25, 25   **43.** {2, 3, 4, 5}

Page 133   Chapter 2 Highlights
1. true   **3.** true   **5.** false, sample answer: $\frac{1}{2}$ **7.** true
9. true

Pages 134–136   Chapter 2 Study Guide and Assessment
11.

number line: −4 −3 −2 −1 0 1 2 3 4 5 6

13.

number line: −3 −2 −1 0 1 2 3 4 5

15. −5   **17.** −4   **19.** −5   **21.** 56%   **23.** 8   **25.** −2
27. 4   **29.** −5   **31.** >   **33.** <   **35.** Sample answer: 0
37. −1   **39.** 12.37   **41.** −13.26   **43.** −99   **45.** $-\frac{3}{7}$
47. −90   **49.** −9   **51.** $\frac{-4}{35}$ **53.** −109   **55.** 14   **57.** −30

59a.

line plot: 5 10 15 20 25 30

59b. 14 detergents   **61.** 1.25 liters of soda for $1.31
63a. They both saw a gain of $\frac{1}{4}$ for the week.   **63b.** CompNet
63c. It changed $-\frac{3}{4}$ from Wednesday to Thursday.
65. Let n = the number; $3n - 21 = 57$.

CHAPTER 3   SOLVING LINEAR EQUATIONS

Page 143   Lesson 3–1A
1. 1   **3.** −11   **5.** 8   **7.** 4   **9.** no   **11.** yes

Pages 148–149   Lesson 3–1
7. −17   **9.** −12   **11.** −14   **13.** $n + (-56) = -82$; −26
15. −9   **17.** −1.3   **19.** 1.4   **21.** −3   **23.** −6   **25.** −7   **27.** 24
29. $1\frac{1}{4}$ **31.** $-\frac{1}{10}$ **33.** $n + 5 = 34$; 29   **35.** $n - (-23) = 35$;
12   **37.** $n + (-35) = 98$; 133   **39a.** $203,600 + s = 19,300,000$;
19,096,400 subscribers   **39b.** Sample answer: 25 million
41a. 24 years old   **41b.** 24 years   **41c.** 78 years   **41d.** 34
years old   **41e.** 24 years   **43.** −5   **45.** −0.73x **47.** 48

Pages 152–154   Lesson 3–2
7. 17   **9.** 48   **11.** $\frac{11}{15}$ **13.** $-7n = 1.477$; −0.211   **15.** −9
17. −17.33 or $-17\frac{1}{3}$ **19.** −14   **21.** 12.81   **23.** 600
25. −275   **27.** −25   **29.** $8\frac{6}{13}$ **31.** −8   **33.** $-12n = -156$; 13
35. $\frac{4}{3}n = 4.82$; 3.615   **37.** 45   **39.** 30   **41.** 12   **43a.** $7\ell = 350$, 50 people; $7\ell = 583$, ≈ 83 people   **43b.** $\frac{p}{7} = 65$; 455
people   **45a.** $0.10m = 2.30$; 23 minutes   **45b.** $0.10(18) = c$;
$1.80   **47.** $17 + 1\frac{1}{2}y = 33\frac{1}{2}$; 11 years   **49.** $\frac{19}{28}$ **51a.** The
number of bags increased during the day.   **51b.** The
machine was refilled.   **53.** 31

Page 155   Lesson 3–3A
1. 5   **3.** −2   **5.** 2   **7.** 4   **9.** −2

Pages 159–161   Lesson 3–3
9. −6   **11.** 35   **13.** −17   **15.** $n + (n + 2) + (n + 4) = 21$;

5, 7, 9   **17.** -1 **19.** -2 **21.** -42.72 **23.** $12\frac{2}{3}$ **25.** -6
27. -48 **29.** 30   **31.** 57   **33.** $4\frac{3}{7}$ **35.** $12 - 2n = -7; 9\frac{1}{2}$
37. $n + (n + 1) + (n + 2) + (n + 3) = 86$; 20, 21, 22, 23
39. $n + (n + 2) + (n + 4) = 39$; 11 m, 13 m, 15 m   **41.** 28
43. -38 **45.** $(3n - 1) + (3n + 1) + (3n + 3)$ **47.** $2a +$
$2 = 26$; 12 letters   **51.** 1   **53.** -5 **55.** 5

Page 161   Self Test
1. 44   **3.** 6   **5.** $-8\frac{1}{3}$ **7.** $23 - n = 42; -19$ **9.** $n + (n + 2) =$
126; 62, 64

Pages 165–167   Lesson 3–4
7. $79°, 169°$ **9.** $(90 - 3x)°, (180 - 3x)°$ **11.** $(110 - x)°,$
$(200 - x)°$ **13.** $85°$ **15.** $(c - 38) + c = 90; 26°, 64°$
17. $48°, 138°$ **19.** none, $55°$ **21.** $69°, 159°$ **23.** none, $81°$
25. $(90 - 3a)°, (180 - 3a)°$ **27.** $(128 - b)°, (218 - b)°$
29. $70°$ **31.** $105°$ **33.** $138°$ **35.** $(190 - 2p)°$ **37.** $45°$
39. $x + (x - 30) = 180; 75°$ **41.** $x + (30 + 3x) = 90; 15°,$
$75°$ **43.** $x + 3x + 4x = 180; 22.5°, 67.5°, 90°$ **45.** $60°$
47. $40°$ **49a.** $\$80$ **49b.** $\$74.75$ **51.** $7t$

53a.

53b. yes; 51, 54, 55, and 57

55.

Stem	Leaf
4	2 3 6 6 7 8 9 9
5	0 0 1 1 1 1 2 4
	4 4 4 5 5 5 5 6
	6 6 7 7 7 7 8
6	0 1 1 1 2 2 4 4
	5 8 9

$5 \,|\, 2 = 52$

Pages 170–172   Lesson 3–5
5. $-\frac{1}{2}$; Add $4x$ to each side, subtract 10 from each side,
then divide each side by 14.   **7.** -25; Subtract $\frac{1}{5}$ from
each side, subtract 3 from each side, then multiply by $\frac{5}{2}$.
9. all numbers   **11.** $2(n + 2) = 3n - 13$; 17, 19   **13.** $\frac{1}{2}n +$
$16 = \frac{2}{3}n - 4$; 120   **15.** no solution   **17.** -2 **19.** 5.6
21. -16 **23.** no solution   **25.** 4   **27.** 2.6 or $2\frac{3}{5}$ **29.** 8
31. 2   **33.** all numbers   **35.** $\frac{1}{5}n + 5n = 7n - 18$; 10
37. $n + (n + 2) + (n + 4) = 180; 58°, 60°, 62°$ **39a.** $4x +$
$6 = 4x + 6$; identity   **39b.** $5x - 7 = x + 3; 2.5$ **39c.** $-3x +$
$6 = 3x - 6; 2$ **39d.** $5.4x + 6.8 = 4.6x + 2.8; -5$ **39e.** $2x -$
$8 = 2x - 6$; no solution   **41a.** 6.827 years   **41b.** Air
conditioner sales are decreasing while fan sales are
increasing.   **43.** $148°$ **45.** -2.7 **47.** $\frac{11}{9}$ **49.** 11.05

Pages 175–177   Lesson 3–6
5. $\frac{3x - 7}{4}$ **7.** $\frac{c - b}{2}$ **9.** $\frac{2S - nt}{n}$ **11.** $\frac{b}{9}$ **13.** $3c - a$
15. $\frac{v - r}{t}$ **17.** $\frac{I}{pt}$ **19.** $\frac{H}{0.24I^2t}$ **21.** $\frac{-t + 5}{4}$ **23.** $\frac{4}{3}(c - b)$

25. $\frac{5x + y}{2}$ **27.** $2x + 12 = 3y - 31; \frac{3y - 43}{2}$ **29.** $\frac{2}{3}x + 5 =$
$\frac{1}{2}y - 3; \frac{3}{4}y - 12$ **31.** $\$900$ **33a.** 36 words per minute
33b. Clarence, with 74 words per minute   **35.** $5°$
37. $x \neq 2$ **39.** $\{8\}$

Pages 181–183   Lesson 3–7
11. 8.2; 8; 8   **13.** Sample answer: 20, 20, 30, 50, 90, 90
15a. 3.857; 3; 3   **15b.** mean   **17.** 96.8; 50; none   **19.** 9; 9;
none   **21.** 5.69; 4.56; none   **23.** 212.94; 218; 219   **25.** 14
27a. mean   **27b.** Yes, because the article indicates that the
median value is in the middle of the data set.   **27c.** No, he
should have used the term *mean* since the mean can be
affected by extremely high values, causing the majority to
be below the mean value.   **29.** $1660\frac{3}{5}; 1076\frac{1}{2}$; none
31. 37.5 mph   **33.** 16   **35.** -20 **37.** $5 + 9ac + 14b$

Page 185   Chapter 3 Highlights
1. b   **3.** i   **5.** d   **7.** k   **9.** a   **11.** g

Pages 186–188   Chapter 3 Study Guide and Assessment
13. -16 **15.** 21   **17.** -10 **19.** -32 **21.** -116 **23.** 7
25. $83\frac{1}{3}$ **27.** 10   **29.** -16 **31.** 3   **33.** -153 **35.** 11
37. 50   **39.** 16, 17   **41.** $21°; 111°$ **43.** $(70 - y)°; (160 - y)°$
45. $30°$ **47.** $15°$ **49.** $180° - (z + z - 30°)$ or $(210° - 2z)$
51. -30 **53.** 2   **55.** $\frac{y}{5}$ **57.** $\frac{a}{y - c}$ **59.** 21; 21; 21
61a. 25.5 years   **63.** $13\frac{3}{4}$ years

CHAPTER 4   USING PROPORTIONAL REASONING

Pages 198–200   Lesson 4–1
5. $=$ **7.** 12   **9.** 5   **11.** 4.62   **13.** 9.5 gallons   **15.** $=$ **17.** $\neq$
19. $=$ **21.** 9   **23.** $\frac{8}{5}$; 1.6   **25.** 0.84   **27.** $-\frac{149}{6}; -24.8$
29. 11   **31.** 2.28   **33.** 1.251

35a.

Louis' age	1	2	3	6	10	20	30
Mariah's age	9	10	11	14	18	28	38

35b. $9, 5, 3.\overline{6}, 2.\overline{3}, 1.8, 1.4, 1.2\overline{6}$ **35c.** $r = \frac{y + 8}{y}$ **35d.** The
ratio gets smaller.   **35e.** No; if the ratio equaled 1, Mariah
and Louis would be the same age.   **37.** 85 movies
39. 23; 19; 18   **41.** $4x - 2x = 100$; 50   **43a.** 93   **43b.** $\$9695$
43c. more than 40   **43d.** $93 - p$ **43e.** yes   **45.** 10

Pages 203–205   Lesson 4–2
5. $\triangle DEF$ **7.** yes   **9.** $\ell = 12, m = 6$ **11.** 27 feet high
13. $\triangle DFE$ **15.** no   **17.** yes   **19.** no   **21.** $a = \frac{55}{6}, b = \frac{22}{3}$
23. $d = \frac{51}{5}, c = 9$ **25.** $a = 2.78, c = 4.24$ **27.** $b = 16.2,$
$d = 6.3$ **29.** $c = \frac{7}{2}, d = \frac{17}{8}$
31. Sample answer:

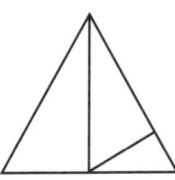

33a. $\frac{3}{9} = \frac{d}{d + 39}; d = 19.5$ ft; yes   **33b.** When the ball is served from 8 feet, $d = 23.4$ ft, and the serve is a fault. When the ball is served from 10 feet, $d = 16.7$ ft. Therefore, tall players have an easier time serving. **35.** $16\frac{1}{2}$ feet by 21 feet **37.** 160 **39.** 60 million; 420 million **41.** 23;

$$(19 - 12) \div 7 \cdot 23$$
$$= 7 \div 7 \cdot 23 \qquad \textit{Substitution } (=)$$
$$= 1 \cdot 23 \qquad \textit{Substitution } (=)$$
$$= 23 \qquad \textit{Multiplicative identity}$$

Pages 211–214   Lesson 4–3
7. $\sin Y = 0.600$, $\cos Y = 0.800$, $\tan Y = 0.750$ **9.** 0.8192 **11.** 0.1228 **13.** 46° **15.** 36° **17.** $m\angle B = 50°$, $AC = 12.3$ m, $BC = 10.3$ m **19.** $m\angle A = 30°$, $AC = 13.9$ m, $BC = 8$ m **21.** $\sin G = 0.6$, $\cos G = 0.8$, $\tan G = 0.75$ **23.** $\sin G = 0.471$, $\cos G = 0.882$, $\tan G = 0.533$ **25.** $\sin G = 0.923$, $\cos G = 0.385$, $\tan G = 2.4$ **27.** 0.3584 **29.** 0.9703 **31.** 0.3746 **33.** 33° **35.** 22° **37.** 62° **39.** 77° **41.** 58° **43.** 18° **45.** $m\angle B = 60°$, $AC = 12.1$ m, $BC = 7$ m **47.** $m\angle A = 45°$, $AC = 6$ ft, $AB = 8.5$ ft **49.** $m\angle A = 63°$, $AC = 9.1$ in., $BC = 17.8$ in. **51.** $m\angle A = 23°$, $m\angle B = 67°$, $AB \approx 12.8$ ft **53.** $m\angle A = 30°$, $m\angle B = 60°$, $AC = 5.2$ cm **55a.** true; $\sin C = \frac{c}{a}$, $\cos B = \frac{c}{a}$ **55b.** $\cos B = \frac{c}{a}$, $\frac{1}{\sin B} = \frac{a}{b}$ **55c.** $\tan C = \frac{c}{b}$, $\frac{\cos C}{\sin C} = \frac{b}{a} \div \frac{c}{a} = \frac{b}{c}$ **55d.** true; $\tan C = \frac{c}{b}$, $\frac{\sin C}{\cos C} = \frac{c}{a} \div \frac{b}{a} = \frac{c}{b}$ **55e.** true; $\sin B = \frac{b}{a}$, $(\tan B)(\cos B) = \frac{b}{c} \cdot \frac{c}{a} = \frac{b}{a}$ **57.** 2229 ft **59.** 3 **61.** 10 **63.** N, W, Z, Q **65.** $\frac{2}{15}$ **67.** $9x + 2y$ **69.** 125

Pages 218–221   Lesson 4–4
7. 43%, 0.43 **9.** 55% **11.** 15 **13.** $52.50 **15.** $1200 at 10%, $6000 at 14% **17.** 30%, 0.30 **19.** 70%, 0.70 **21.** 180%, 1.8 **23.** $62\frac{1}{2}$%; 0.625 **25.** 50% **27.** $12\frac{1}{2}$%, 12.5% **29.** 400% **31.** 2.5% **33.** 72.3 **35.** 702.4 **37.** 28% **39.** 25.92 **41.** 12% **43.** $3125 **45.** $2400 **47.** 39%, 34%, 15%, 12% **49.** 11.5% **51.** $5.45 **53a.** about 54 people; about 23 people **53b.** Respondents could choose more than one entree. **55.** about 277 feet **57.** $\frac{77}{4}$; 19.25 **59.** −4.7 **61.** −28y **63.** 2

Page 221   Self Test
1. 5 **3.** 5 **5.** $m\angle B = 34°$, $b \approx 11.5$, $c \approx 20.5$ **7.** no **9.** $62\frac{1}{2}$%, 62.5%

Pages 225–227   Lesson 4–5
5. I; 40% **7.** D; 50% **9.** $85.85 **11.** $196 **13.** I; 69% **15.** D; 55% **17.** D; 27% **19.** I; 2% **21.** $233.24 **23.** $19.90 **25.** $85.39 **27.** $16.59 **29.** $30.09 **31.** no **33.** 133% **35.** 8.6% **37.** an increase of more than 22% **39.** 11% **41.** 65 ft **43.** 11 **45.** heavy, 571; light, 286

Pages 230–232   Lesson 4–6
7. $\frac{1}{2}$ **9.** $\frac{2}{3}$ **11.** 5:1 **13.** $\frac{1}{2}$ **15.** $\frac{1}{2}$ **17.** $\frac{1}{2}$ **19.** $\frac{2}{11}$ **21.** 0 **23.** $\frac{8}{11}$ **25.** 1:5 **27.** 4:11 **29.** 23:7 **31.** $\frac{1}{2}$ **33.** 1:3 **35.** $\frac{8}{13}$ **37.** $\frac{1}{7}$ **39a.** 1:14 **39b.** $\frac{1}{27}$ **41a.** $\frac{3}{26}$ **41b.** $\frac{4}{13}$ **41c.** 3:23 **43.** 9° **45.** $50 **47.** $\frac{4}{21}$ **49.** 75

Pages 236–238   Lesson 4–7
5. 226 dozen chocolate chip; 311 dozen peanut butter **7.** 11:30 A.M. **9.** 5 quarters **11.** 3 hours **13.** 2.5 hours **15.** 46 mph **17.** 67 mph **19a.** 7 **19b.** 16 **21.** Sample answer: How much pure antifreeze must be mixed with a 20% solution to produce 40 quarts of a 28% solution? **23.** 0 **25.** $11,000 **27.** 49 **29.** associative (+)

Pages 243–244   Lesson 4–8
5. I, 5 **7.** 10 **9.** 99 **11.** 20 inches from the 8-ounce weight, 16 inches from the 10-ounce weight **13.** I, 15 **15.** I, 9 **17.** D, 4 **19.** $26\frac{1}{4}$ **21.** −8 **23.** 12 **25.** 6.075 **27.** 8.3875 **29.** $\frac{1}{4}$ **31.** 18 pounds **33.** 1.6 feet **35.** 2:1 **37.** $4300 at 5%, $7400 at 7% **39.** −12

Page 245   Chapter 4 Highlights
1. angle of elevation **3.** equal, proportional **5.** legs **7.** odds **9.** inverse variation

Pages 246–248   Chapter 4 Study Guide and Assessment
11. = **13.** ≠ **15.** 18 **17.** 16 **19.** $d = \frac{45}{8}$, $e = \frac{27}{4}$ **21.** $b = \frac{44}{3}$, $d = 6$ **23.** 0.528 **25.** 0.849 **27.** 1.607 **29.** 39° **31.** 80° **33.** $m\angle B = 45°$, $AB = 8.5$, $BC = 6$ **35.** $m\angle A \approx 66°$, $AB \approx 9.8$, $m\angle B \approx 24°$ **37.** 48 **39.** 87.5% **41.** 0.1881 **43.** decrease, 13% **45.** increase, 6% **47.** decrease, 10% **49.** $85.66 **51.** $9272.23 **53.** $\frac{1}{12}$ **55.** $\frac{5}{6}$ **57.** 10:39 **59.** 34:15 **61.** 21 **63.** 6 **65.** 21 **67.** 48 **69.** 450 mph and 530 mph

CHAPTER 5   GRAPHING RELATIONS AND FUNCTIONS

Pages 257–259   Lesson 5–1
5. $(-3, -1)$; III **7.** $(0, -2)$; none

8–11.

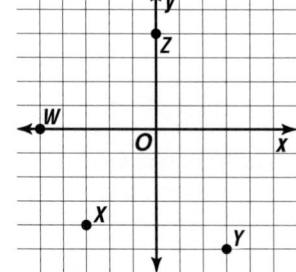

13. $(-1, -3)$; III **15.** $(0, 3)$; none **17.** $(0, 0)$; none **19.** $(3, -2)$; IV **21.** $(2, 2)$; I **23.** $(-5, 4)$; II

25–36.

37.

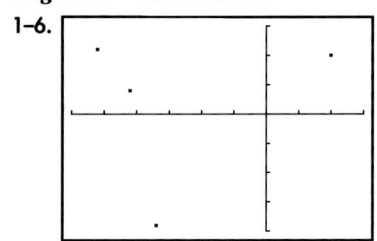

41a. New Orleans   **41b.** Oregon   **41c.** Sample answer: $(75°, 40°)$ **41d.** Honolulu, Hawaii   **41f.** Sample answer: The longitude lines are not the same distance apart; they meet at the poles.   **45.** $3\frac{1}{2}$ hours   **47.** 4; 3; none
49. $-a - 5$ **51.** $31a + 21b$

Page 261   Lesson 5–2A

1–6.

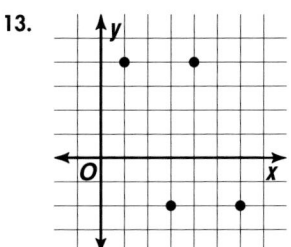

7. They are points on the axes.

Pages 266–269   Lesson 5–2

5. $D = \{0, 1, 2\}$; $R = \{2, -2, 4\}$ **7.** $\{(1, 3), (2, 4), (3, 5), (5, 7)\}$; $D = \{1, 2, 3, 5\}$; $R = \{3, 4, 5, 7\}$; Inv $= \{(3, 1), (4, 2), (5, 3), (7, 5)\}$ **9.** $\{(1, 3), (2, 2), (4, 9), (6, 5)\}$; $D = \{1, 2, 4, 6\}$; $R = \{2, 3, 5, 9\}$; Inv $= \{(3, 1), (2, 2), (9, 4), (5, 6)\}$
11. $\{(-2, 2), (-1, 1), (0, 1), (1, 1), (1, -1), (2, -1), (3, 1)\}$; $D = \{-2, -1, 0, 1, 2, 3\}$; $R = \{-1, 1, 2\}$; Inv $= \{(2, -2), (1, -1), (1, 0), (1, 1), (-1, 1), (-1, 2), (1, 3)\}$

13.

15b. 5.6%   **15c.** Except for the first 6 months of 1992, the unemployment rate seems to be decreasing.
17. $D = \{-5, -2, 1, 3\}$; $R = \{7\}$ **19.** $D = \left\{-5\frac{1}{4}, -3, \frac{1}{2},\right.$

$(1, -2)\}$; $D = \{1, 3, 4, 6\}$; $R = \{-2, 4\}$; Inv $= \{(4, 6), (-2, 4), (4, 3), (-2, 1)\}$ **23.** $\{(6, 0), (-3, 5), (2, -2), (-3, 3)\}$; $D = \{-3, 2, 6\}$, $R = \{-2, 0, 3, 5\}$; Inv $= \{(0, 6), (5, -3), (-2, 2), (3, -3)\}$ **25.** $(3, 4), (3, 2), (2, 9), (5, 4), (5, 8), (-7, 2)\}$; $D = \{-7, 2, 3, 5\}$; $R = \{2, 4, 8, 9\}$; Inv $= \{(4, 3), (2, 3), (9, 2), (4, 5), (8, 5), (2, -7)\}$ **27.** $\{(0, 25), (1, 50), (2, 75), (3, 100)\}$; $D = \{0, 1, 2, 3\}$; $R = \{25, 50, 75, 100\}$; Inv $= \{(25, 0), (50, 1), (75, 2), (100, 3)\}$ **29.** $\{(-3, 4), (-2, 2), (-1, -2), (2, 2)\}$; $D = \{-3, -2, -1, 2\}$; $R = \{-2, 2, 4\}$; Inv $= \{(4, -3), (2, -2), (-2, -1), (2, 2)\}$ **31.** $\{(-3, 3), (-3, -3), (3, 3), (3, -3), (0, 0)\}$; $D = \{-3, 0, 3\}$; $R = \{-3, 0, 3\}$; Inv $= \{(3, -3), (-3, -3), (3, 3), (-3, 3), (0, 0)\}$ **33.** $\{(-3, 1), (-1, 1), (2, 1), (3, 1), (4, 1)\}$; $D = \{-3, -1, 2, 3, 4\}$; $R = \{1\}$; Inv $= \{(1, -3), (1, -1), (1, 2), (1, 3), (1, 4)\}$

35.

37.

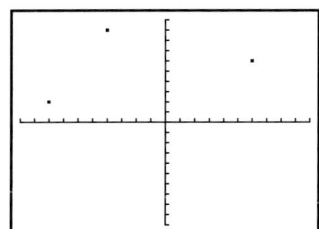

39a. $[-10, 10]$ by $[-10, 10]$

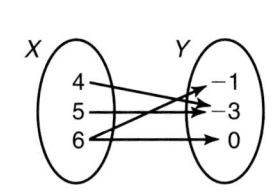

39b. $\{(10, 0), (-8, 2), (6, 6), (-4, 9)\}$

39c.

(x, y)	Quadrant	Inverse's Quadrant
(0, 10)	I	I
(2, -8)	IV	II
(6, 6)	I	I
(9, -4)	IV	II

41. If a point lies in Quadrants I or III, its inverse will lie in the same quadrant as the point. If a point lies in Quadrant II, its inverse lies in Quadrant IV, and vice versa. If a point lies on the x-axis, its inverse lies on the y-axis and vice versa.   **43.** Sample answers: **a.** 157 billion, 191 billion

b. Retail sales have increased from 1992 to 1994.
c. As unemployment decreases, retail sales increase, because people have more money to spend.

45a.

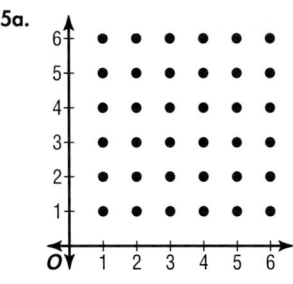

D = R = {1, 2, 3, 4, 5, 6}   **45b.** D = R = {1, 2, 3, 4, 5, 6}; relation = inverse

45c. 11 possible sums   **45d.**

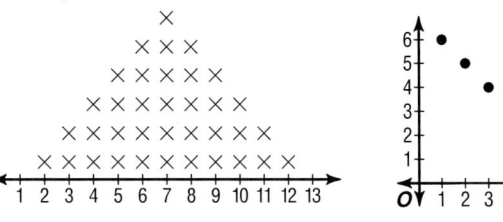

45e. $\frac{6}{36}$ or $\frac{1}{6}$; There are 6 out of 36 ways to roll a sum of 7.   **47.** 32   **49.** $480   **51.** 2214, 2290   **53.** $4 + 80x + 32y$

Page 270   Lesson 5–3A
1. {(−3, −19), (−2, −15), (−1, −11), (0, −7), (1, −3), (2, −1), (3, 5)}   **3.** {(−3, −10.4), (−2, −9.2), (−1, −8), (0, −6.8), (1, −5.6), (2, −4.4), (3, −3.2)}

Pages 274–277   Lesson 5–3
7. {(−2, −1), (−1, 1), (0, 3), (1, 5), (2, 7), (3, 9)}   **9.** a, d
11. {(−2, 8), (−1, 7.5), (0, 7), (1, 6.5), (2, 6)}

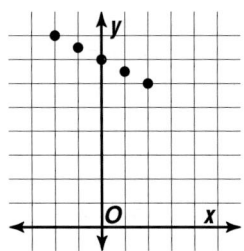

13. a, c   **15.** a, b   **17.** c   **19.** {(−3, −12), (−2, −8), (0, 0), (3, 12), (6, 24)}   **21.** {(−3, 10), (−2, 9), (0, 7), (3, 4), (6, 1)}   **23.** {(−3, 11), (−2, 9.5), (0, 6.5), (3, 2), (6, −2.5)}   **25.** {(−3, −12), (−2, −7), (0, 3), (3, 18), (6, 33)}   **27.** {(−3, 1.8), (−2, 1.4), (0, 0.6), (3, −0.6), (6, −1.8)}

29.

x	y
−5	−9
−3	−5
0	1
1	3
3	7
6	13

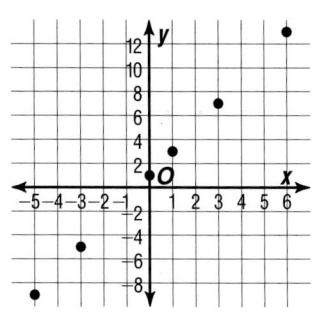

31.

a	b
−2	4.5
−1	3.25
0	2.00
1	0.75
3	−1.75
4	−3.00
5	−4.25

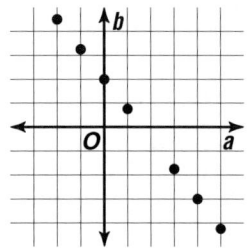

33a. $3x + 4y = 180$ **33b.** $y = 45 - \frac{3}{4}x$ **33c.** Sample answer:
(1, 44.25), (2, 43.5), (3, 42.75), (4, 42), (5, 41.25)
35. $\left\{-\frac{2}{3}, -\frac{1}{3}, 0, \frac{2}{3}, 1\right\}$ **37.** $\left\{-\frac{7}{4}, -\frac{1}{2}, 2, \frac{13}{4}, \frac{9}{2}\right\}$

39.

x	y
−2	−8
−1	−1
0	0
1	1
2	8

41. {(−2.5, −4.26), (−1.75, −3.21), (0, −0.76), (1.25, 0.99), (3.333, 3.902)}   **43.** [(−100, 350), (−30, 116.6̄), (0, 16.6̄), (120, −383.3̄), (360, −1183.3̄), (720, −2383.3̄)]
45a. {−6, −4, 0, 4, 6}   **45b.** {−13, −8, −4, 4, 8, 13}
45c. {−5, 0, 4, 8, 13}   **47a.** $D = \frac{m}{V}$ **47b.** silver and gasoline

49.   Women   Men

L	S	(L, S)	L	S	(L, S)
$9\frac{1}{3}$	6	$\left(9\frac{1}{3}, 6\right)$	$11\frac{1}{3}$	8	$\left(11\frac{1}{3}, 8\right)$
$9\frac{5}{6}$	$7\frac{1}{2}$	$\left(9\frac{5}{6}, 7\frac{1}{2}\right)$	$11\frac{5}{6}$	$9\frac{1}{2}$	$\left(11\frac{5}{6}, 9\frac{1}{2}\right)$
$10\frac{1}{6}$	$8\frac{1}{2}$	$\left(10\frac{1}{6}, 8\frac{1}{2}\right)$	$12\frac{1}{3}$	11	$\left(12\frac{1}{3}, 11\right)$
$10\frac{2}{3}$	10	$\left(10\frac{2}{3}, 10\right)$	$12\frac{5}{6}$	$12\frac{1}{2}$	$\left(12\frac{5}{6}, 12\frac{1}{2}\right)$

51a.

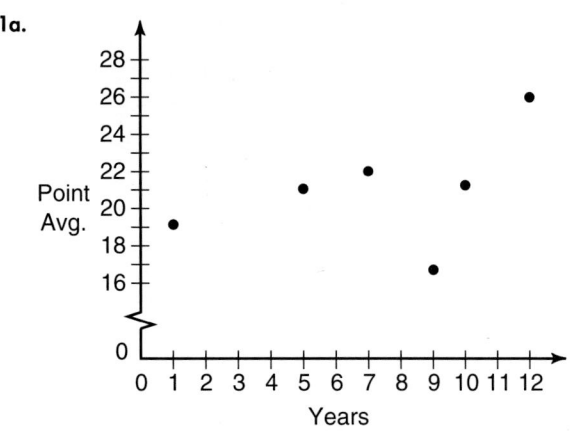

51b. The more years played, the higher the point per game average of the player.   **53.** $467.50   **55.** $18,000
57a. $2w + 2\ell = 148$ **57b.** w, 14.25 in.; ℓ, 59.75 in.
59. −7.976

Page 280   Lesson 5–4A

1.

3.

5.

7a. Sample answer:
$[-10, 110]$ by
$[-5, 15]$

7b.

7c. Sample answer:
$(0, 10)$, $(100, 0)$,
$(10, 9)$

9a. Sample answer:
$[-2.5, 0.5]$ by
$[-0.05, 0.05]$

9b.

9c. Sample answer:
$(-2, 0)$, $(0, 0.02)$,
$(1, 0.03)$

Pages 283–286   Lesson 5–4

7. yes; $2x + y = 6$ **9.** yes; $3x + 2y = 7$

11.

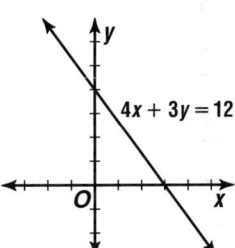

$4x + 3y = 12$

13.

$x = 6$

15.

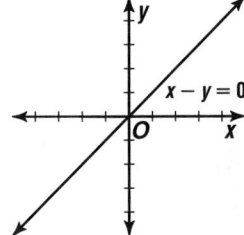

$x - y = 0$

17. yes; $\frac{3}{5}x - \frac{2}{3}y = 5$ **19.** no   **21.** yes; $3x - 2y = 8$ **23.** yes;
$7x - 7y = 0$ **25.** yes; $3m - 2n = 0$ **27.** yes; $6a - 7b = -5$

29.

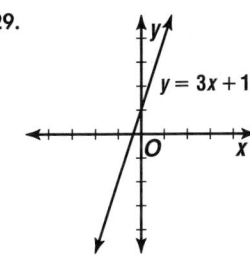

$y = 3x + 1$

33.

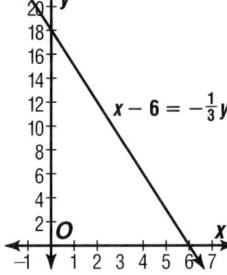

$x - 6 = -\frac{1}{3}y$

37.

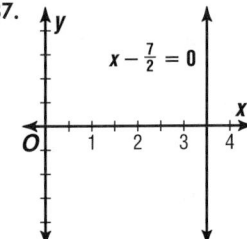

$x - \frac{7}{2} = 0$

41.

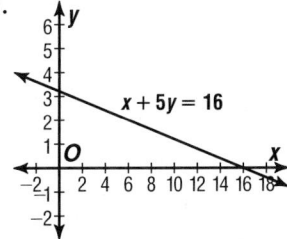

$x + 5y = 16$

45.

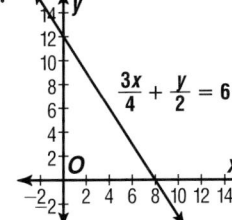

$\frac{3x}{4} + \frac{y}{2} = 6$

47. $-5, 7.5$ **49.** $6, 9, 10$

51. $[-5, 15]$ by $[-10, 10]$, Xscl: 1, Yscl: 1

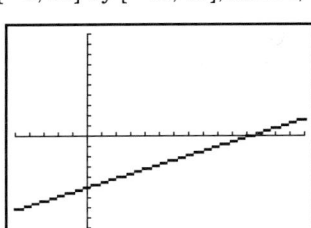

53. $[-10, 10]$ by $[-2, 2]$, Xscl: 1, Yscl: 0.25

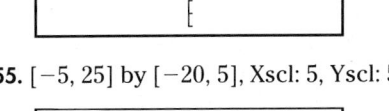

55. $[-5, 25]$ by $[-20, 5]$, Xscl: 5, Yscl: 5

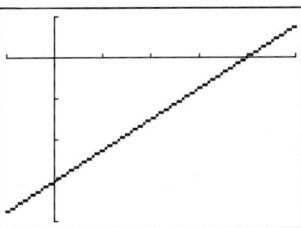

57a. Sample answer: Parallel lines that slant upward and intersect the x-axis at −7, −2.5, 0, and 4.5.

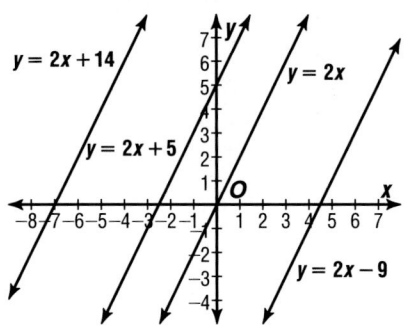

57b. Sample answer: Parallel lines that slant downward and intersect the x-axis at 0, $1\frac{1}{3}$, $2\frac{2}{3}$, and $-3\frac{1}{3}$.

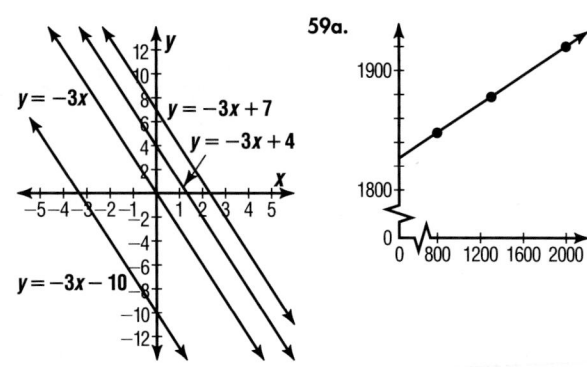

59b. yes, but only if her sales are $1300 or $2000 over target   **61.** $y = 3 - 4x$

63a.

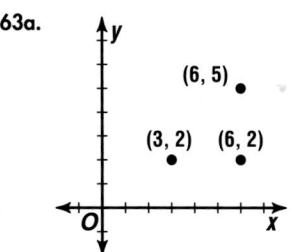

63b. (3, 5)   **63c.** 12 inches   **65.** 15%   **67.** 120°   **69.** $\frac{1}{48}$

Page 286   Self Test

1.

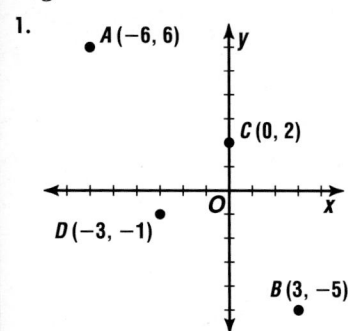

3. {(−5, −3), (−1, 4), (4, 4), (4, −3)}, D = {−5, −1, 4}, R = {−3, 4}, Inv = {(−3, −5), (4, −1), (4, 4), (−3, 4)}

5. $b = 3 - \frac{2}{3}a$: $\left\{\left(-2, 4\frac{1}{3}\right), \left(-1, 3\frac{2}{3}\right), (0, 3), \left(1, 2\frac{1}{3}\right), (3, 1)\right\}$

7.

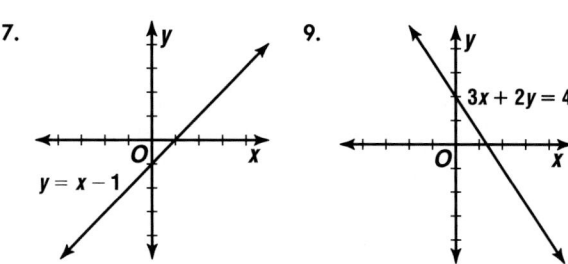

9.

Pages 291–294   Lesson 5–5

7. no   **9.** yes   **11.** yes   **13.** no   **15.** −10   **17.** $3w + 2$
19. yes   **21.** no   **23.** yes   **25.** yes   **27.** yes   **29.** yes
31. yes   **33.** no   **35.** −14   **37.** $-\frac{9}{25}$ **39.** 5.25   **41.** −18
43. $9b^2 - 6b$ **45.** 3   **47.** $5a^4 - 10a^2$ **49.** $24p - 36$
51a. Sample answer: $f(x) = x$.   **51b.** Sample answer: $f(x) = x^2$ **53.** a; sample answer: because they have a single price per shirt, and since you cannot have a fraction of a shirt, you must use points instead of lines. It is a function.

55a.

k	g
0	0
6	25
12	50
18	75
24	100

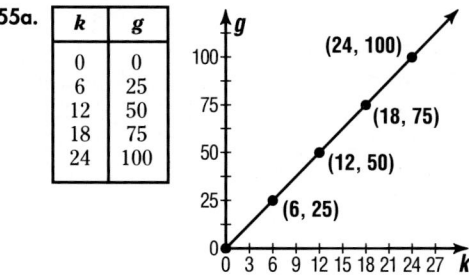

55b. It is a line.   **55c.** 24 karats   **57a.** $260.87
57b. Sample equation: $B = \frac{3}{23}P$ **59.** {−16, −13, 5, 11}
61. $\frac{3}{13}$ **63.** 45°   **65.** −48

67a.

67b. 149   **67c.** 91   **67d.** 121 and 125   **67e.** 13 players

Pages 299–302   Lesson 5–6

5. $f(x) = 2x + 6$ **7.** $y = \frac{1}{2}x - \frac{3}{2}$ **9.** −6   **11.** 48, 60
13. 8, −2   **15.** $f(x) = 5x$ **17.** $g(x) = 11 - x$ **19.** $h(x) = \frac{1}{3}x - 2$ **21.** $y = -3x$ **23.** $y = \frac{1}{2}x$ **25.** $y = 2x - 10$
27. $xy = -24$ **29.** $y = x^3$ **31.** y-intercept: $f(0)$, x-intercept: $f(x) = 0$ **33a.** $f(x) = 34x - 34$ **33b.** You must go deeper in fresh water to get the same pressure as in ocean water.
35a. 1514 C   **35b.** 11.04 C/min

35c.

Minutes	Calories
1	11.04
2	22.08
3	33.12
4	44.16
5	55.2
6	66.24
7	77.28
8	88.32
9	99.36
10	110.4

35d. $C(t) = 11.04t$; yes   **35e.** 1324.8 C burned; 189.2 C remaining   **37.** D = {1, 3, 5}; R = {2, 4, 5}   **39.** 5

41.

Page 304   Lesson 5–7A
1. Q1, 14; Med, 17; Q3, 20.5; R, 11; IQR, 6.5
3. Q1, 68; Med, 78; Q3, 96; R, 34; IQR, 28
5. Q1, 3.4; Med, 5.3, Q3, 21; R, 77; IQR, 17.6
7. At least 50% of the data is clustered around the median.

Pages 309–313   Lesson 5–7
7. 45, 40, 45, 34, 11; 11   **9.** 48, 26, 39, 17, 22
11. 34, 78, 96, 68, 28   **13.** 10, 5, 8, 2, 6   **15.** 1.1, 30.6, 30.9, 30.05, 0.85   **17.** 340, 1075, 1125, 1025, 100   **19.** 39, 218, 221, 202, 19   **21a.** 9,198,630; 11,750,000; 5,700,000; 24,000,000; 6,050,000   **21b.** No; the libraries would have accumulated books throughout the years.   **23.** 3760, 3224, 4201.5, 2708.5, 1421   **25.** 21,674; 9790; 12,194; 5475; 6719   **27a.** males: 36, 29, 37.5, 44, 15; females: 23, 29, 33, 39, 10   **27b.** There are no outliers.   **27c.** The ages of the top female golfers are less varied than those of the top male golfers.   **31.** c, d   **33a.** $6.2 + p = 9.4$; about 3.2 million people   **33b.** $6.0 + p = 6.9$; about 0.9 million people

Page 315   Chapter 5 Highlights
1. e   **3.** d   **5.** j   **7.** c   **9.** b

Pages 316–318   Chapter 5 Study Guide and Assessment
10–13.

15. (2, −1); IV   **17.** (1, 1); I   **19.** D = {−3, 4}, R = {5, 6}
21. D = {−3, −2, −1, 0}, R = {0, 1, 2}   **23.** {(2, 0), (−1, 3), (2, 2), (−1, −2)}   **25.** {(−4, −13), (−2, −11), (0, −9), (2, −7), (4, −5)}   **27.** $\left\{\left(-4, -5\frac{1}{3}\right), \left(-2, -2\frac{2}{3}\right), (0, 0), \left(2, 2\frac{2}{3}\right), \left(4, 5\frac{1}{3}\right)\right\}$

29.

x	y
−2	−7
0	3
2	13
4	23
6	33

31.   **33.**

35.

37. yes   **39.** no   **41.** yes   **43.** 3   **45.** $a^2 + a + 1$
47. $2a^2 - 14a + 26$ **49.** $y = x - 4$ **51.** 70, 65, 85, 45, 40
53. 37, 73, 77, 62, 15

55a.

55b. $56   **57a.** 9.6, 6, 7.05, 8.3, 2.3   **57b.** 12.6, 14.7

CHAPTER 6   ANALYZING LINEAR EQUATIONS

Page 324   Lesson 6–1A
1. $-\frac{1}{2}$ **3.** Sample answer: (3, 3)   **5.** Yes; sample answer: suppose the endpoints are $C(-4, -1)$ and $D(-2, -2)$. Let the upper right peg represent (0, 0). To go from C to D, the y value decreases by 1 and the x value increased by 2. The ratio is $\frac{-1}{2}$. If you use the rule from Exercise 4, $\frac{-2 - (-1)}{-2 - (-4)} = \frac{-1}{2}$. The results are the same.

Pages 329–331   Lesson 6–1
7. $\frac{5}{3}$ **9.** $\frac{3}{2}$ **11.** undefined   **13.** 2   **15.** $-\frac{1}{5}$ **17.** 0   **19.** 1

21. $\frac{4}{7}$ **23.** $-\frac{2}{3}$ **25.** undefined   **27.** undefined   **29.** $\frac{9}{5}$

31. -5 **33.** -1 **35.** 7

37.

(−2, −3)

39. Sample answer: (7, 6). The slope of the line containing A and B is -2. Use the slope to go 2 units down and 1 unit to the right of either point.
41. about 1478 feet
43. 11,160 feet   **45.** 77; 5; 8; 3.2; 4.8   **47.** 12%   **49.** -6

51a. 12 animals
51b.

```
      x   x
 x  x x x x x     x
x x x x x x xx x  x       x              x
+--+--+--+--+--+--+--+--+--+--+--+--+--+--+--+--+--+--+--+--+
0  2  4  6  8 10 12 14 16 18 20 22 24 26 28 30 32 34 36 38 40
```

51c. 15 years   **51d.** 4 animals

Pages 336–338   Lesson 6–2
5. 4; (2, −3)   **7.** 3; (−7, 1)   **9.** $y - 4 = -3(x + 2)$
11. $3x + 4y = -9$ **13.** $2x - y = -6$ **15.** $y + 2 = -\frac{4}{7}(x + 1)$
or $y - 2 = -\frac{4}{7}(x + 8)$; $4x + 7y = -18$ **17.** $3x - 5y = -2$
19. $y - 5 = 3(x - 4)$ **21.** $y - 1 = -4(x + 6)$ **23.** $y - 3 = -2(x - 1)$ **25.** $y + 3 = \frac{3}{4}(x - 8)$ **27.** $4x - y = -5$
29. $3x - 2y = -24$ **31.** $2x + 5y = 26$ **33.** $y - 2 = -\frac{1}{3}(x + 5)$ or $y + 1 = -\frac{1}{3}(x - 4)$ **35.** $y + 1 = \frac{3}{7}(x + 8)$ or $y - 5 = \frac{3}{7}(x - 6)$ **37.** $y + 2 = 0(x - 4)$ or $y + 2 = 0(x - 8)$
39. $4x + 3y = 39$ **41.** $11x + 8y = 17$ **43.** $36x - 102y = 61$
45. $\frac{5 - 1}{5 - 9} = \frac{4}{-4}$ or -1; An equation of the line is $(y - 1) = -1(x - 9)$. Let $y = 0$ in the equation and see if $x = 10$.
$$0 - 1 = -x + 9$$
$$-10 = -x$$
$$10 = x$$
(10, 0) lies on the line. Since (10, 0) is a point on the x-axis, the line intersects the x-axis at (10, 0).
47a. No; for a rise of 30 inches the ramp must be 30 feet long, but there only 18 feet available.
47b.

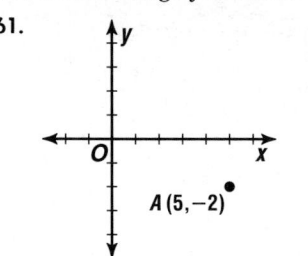

49a. $0.26, $0.09   **49b.** $0.32   **51.** 28   **53.** No; because there could be 2 pink and 1 white or 1 pink and 2 white.
55. associative property of addition

Pages 343–345   Lesson 6–3
7. positive   **9a.** It gets better.   **9b.** yes; negative
9c. (7, 18) and (8, 16)   **11.** positive   **13.** no   **15.** no

830   *Selected Answers*

17. positive   **19.** negative   **21.** Yes; sample reason: Dots are grouped in an upward diagonal pattern.   **23.** c; If 1 is correct then 19 are wrong, if 2 are correct, then 18 are wrong, and so on. Graph c shows these pairs of numbers.   **25a.** Sample answer: The more taxation increases, the more in debt the government becomes.
25b. Sample answer: You work harder and your grades go up.   **25c.** Sample answer: As more money is spent on research, fewer people die of cancer.   **25d.** Sample answer: Comparing the number of professional golfers with the number of holes-in-one.   **27a.** The correlation shown by the graph shows a slightly positive correlation between SAT scores and graduation rate.   **27b.** Sample equation: $y = 0.89x + 1142.17$

29a., c.

29b. positive   **29c.** Sample answer: $15x - 13y = 129$
31a. $7500x - y = 120,000$ **31b.** 15,000 feet   **31c.** No; it only describes the plane's path in that part of the flight.
33. $75°$ **35.** 1.45

Pages 350–353   Lesson 6–4
7a. 2   **7b.** 2, −4   **7c.** $y = 2x - 4$ **9.** $-\frac{28}{3}, \frac{7}{2}$
11. $y = \frac{2}{3}x - 10, 2x - 3y = 30$ **13.** −2, −4   **15.** $x + 5y = 13$
17. $y = \frac{11}{3}x$; 44   **19.** $-\frac{3}{2}$; 0, 0; $y = -\frac{3}{2}x$ **21.** $\frac{2}{5}$, 5, −2, $y = \frac{2}{5}x - 2$ **23.** 2, $\frac{8}{7}$ **25.** $-\frac{5}{6}$, 5   **27.** none, 6
29. $y = 7x - 2, 7x - y = 2$ **31.** $y = -1.5x + 3.75, 6x + 4y = 15$ **33.** $y = -7, y = -7$ **35.** $-\frac{5}{4}, \frac{5}{2}$ **37.** −4, 12
39. $\frac{4}{5}, \frac{4}{15}$ **41.** $y = -2$ **43.** $x = 3$ **45.** $y = 9$
47. $y = \frac{11}{24}x, \frac{33}{2}$ **49.** $y = \frac{2}{3}x - \frac{8}{3}$ **51.** (−3, −1)
53a. $y = 2.04x - 21.32$ **55a.** 42.24 ft^3 **55b.** 270 K
57a. $y = 0.1x + 3$ **57b.** Go to the other bank, since this one would charge you $5.50.   **59.** 0
61.

A (5, −2)

63. 13 ft 11 in.
65. −9
67. $14x + 14$

Page 353   Self Test
1. −1   **3.** undefined   **5.** $y - 4 = \frac{1}{2}(x + 6)$ **7.** $x + 5y = 17$
9. −6, −14

Pages 355   Lesson 6–5A

1. All graphs are of the family $y = -ax + 0$, where a represents different negative slopes.

3. All graphs have the same slope, -1, but have different y-intercepts.

5.

7.

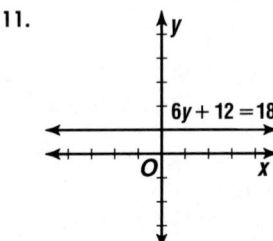

9. Sample answer: $y = -3x + 2$ **11.** Sample answer: The value of m tells how the line will slant, upward or downward, and how steep. The value of b tells where the line will intersect the y-axis.

Pages 359–360   Lesson 6–5

7.

$y + 5 = -2(x + 1)$

9.

$y = 2x - 3$

11.

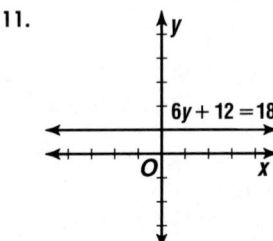

$6y + 12 = 18$

13a.

fingernail

toenail

in.

months

13b. $1\frac{3}{4}$ inches   **13c.** $\frac{1}{16}$ inch   **13d.** See 13a.

13e. the rate of growth per month

15.

$6x - y = 9$

19.

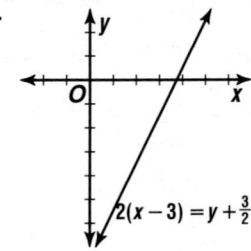

$2(x - 3) = y + \frac{3}{2}$

23.

$y = 3x - 5$

27.

$5x + 2y = 20$

31.

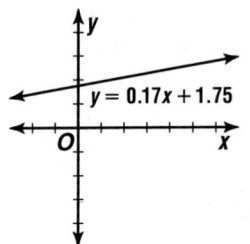

$y = 0.17x + 1.75$

35.

All lines have the same slope 4.

37a.

B

Breaths

Time (minutes)

37b. 122 breaths
39. $y = \frac{2}{5}x + 12$
41. 4:30 P.M.   **43.** 1.8
45. a number m minus 1

Pages 366–368   Lesson 6-6

7. $\frac{2}{3}, -\frac{3}{2}$ **9.** perpendicular   **11.** $y = \frac{5}{6}x - \frac{21}{2}$

13. $y = x - 9$ **15.** $y = \frac{9}{5}x$ **17.** perpendicular

19. parallel   **21.** perpendicular   **23.** perpendicular

25. $y = x + 1$ **27.** $y = \frac{8}{7}x + \frac{2}{7}$ **29.** $y = 2.5x - 5$

31. $y = \frac{2}{3}x + 4$ **33.** $y = -\frac{9}{2}x + 14$ **35.** $y = 3x - 19$

37. $y = -\frac{1}{5}x - 1$ **39.** $y = -\frac{7}{2}x + 11$ **41.** $y = -3$

43. $y = \frac{5}{4}x$ **45.** $y = \frac{1}{3}x - 6$ **47.** No, because the slope

of $\overline{AC}$ is $\frac{6}{7}$ and the slope of $\overline{BC}$ is $-\frac{2}{3}$. These slopes are
not negative reciprocals of each other, so the lines are
not perpendicular and the figure is not a rhombus.
49a. $y = \frac{1}{2}x + \frac{7}{2}, y = -2x + 11$ **49b.** right or 90° angle

51.

Miles Driven

53. $242.80   **55a.** $\frac{5}{12}; \frac{7}{12}$ **55b.** 16-karat gold
57. multiplicative identity

Pages 372–374   Lesson 6-7
5. $(1.5, 6)$ **7.** $(-6, 1)$ **9.** $(-11, 7)$ **11.** $(8, 9.8)$

13. $(12.5, 6)$ **15.** $(1, 5)$ **17.** $(3, 2)$ **19.** $\left(\frac{1}{2}, 1\right)$

21. $(0.8, 2.7)$ **23.** $(4x, 9y)$ **25.** $(-7, 0)$ **27.** $(21, -6)$
29. $(9, 10)$ **31.** $\left(\frac{5}{6}, \frac{1}{3}\right)$ **33.** $B(2.3, 6.8)$ **35.** $P(6.65, -1.85)$

37. $(-1, 5)$ **39.** $\left(1, \frac{5}{2}\right)$ **41a.** $N(6, 3), M(10, 3)$

41b. parallel, $MN = \frac{1}{2}AB$ **43a.** $P(-4, 1), Q(10, -1),$
$R(2, 9)$ **43b.** 62 square units; Sample answer: The area of
the smaller triangle is $\frac{1}{2}bh$. Since the base of the larger
triangle is twice that of the smaller one and the height is
also twice the length of the small one, the area of the
larger is $\frac{1}{2}(2b)(2h)$, or $2bh$. This is 4 times the area of the
small one.   **45.** $y = -\frac{7}{9}x - \frac{8}{3}$ **47.** yes; $9x - 6y = 7$

49. -20 **51.** $3.1x + 1.54$

Page 375   Chapter 6 Highlights
1. parallel   **3.** midpoint   **5.** perpendicular
7. slope-intercept   **9.** slope

Pages 376–378   Chapter 6 Study Guide and Assessment
11. $-\frac{1}{3}$ **13.** $\frac{2}{5}$ **15.** $\frac{25}{3}$ **17.** $y + 3 = -2(x - 4)$

19. $y - 7 = 0$ **21.** $y - 3 = \frac{3}{5}x$ **23.** $y - 1 = -\frac{6}{7}(x - 4)$

25. $3x - y = 18$ **27.** $3x - 4y = 22$ **29.** $y = 5$ **31.** $x = -2$
33a. Yes; it is positive   **33b.** Sample answer: 35 stories
33c. $y = 0.03x + 21$

35. $y = -3x$
37. $y = 5.5$
39. $-8, 4$
41. $-2, 12$

43.

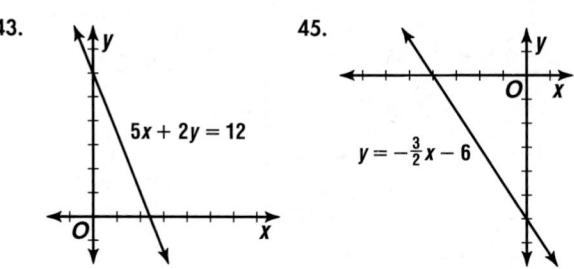

$5x + 2y = 12$

45.

$y = -\frac{3}{2}x - 6$

47.

$5x - \frac{1}{2}y = 2$

49.

$y + 5 = -\frac{3}{4}(x - 6)$

51. $y = -\frac{7}{2}x - 14$ **53.** $y = -\frac{3}{8}x + \frac{13}{2}$ **55.** $y = 5x - 15$

57. $\left(1, -\frac{5}{2}\right)$ **59.** $\left(5, \frac{11}{2}\right)$ **61.** $\left(\frac{7}{2}, -\frac{3}{2}\right)$ **63.** $(13, 11)$

65. $(-5, 20)$ **67.** $d = 45t - 10$

CHAPTER 7   SOLVING LINEAR EQUATIONS

Pages 388–390   Lesson 7-1
7. c   **9.** d   **11.** $\{x \mid x > -5\}$ **13.** $\{y \mid y < -5\}$
15. $x - 17 < -13, \{x \mid x < 4\}$
17. $\{a \mid a < 18\}$

19. $\{x \mid x \leq 1\}$

21. $\left\{x \mid x > \frac{11}{3}\right\}$

23. $\{x \mid x > 2\}$

25. $\left\{x \mid x < \frac{3}{8}\right\}$ **27.** $\{x \mid x \leq 15\}$ **29.** $\{x \mid x < 0.98\}$

31. $\{r \mid r < 10\}$ **33.** $x - (-4) \geq 9, \{x \mid x \geq 5\}$ **35.** $3x < 2x +$
$8, \{x \mid x < 8\}$ **37.** $20 + x < 53, \{x \mid x < 33\}$ **39.** $2x > x -$
$6, \{x \mid x > -6\}$ **41.** 12   **43.** -2 **45a.** no   **45b.** yes
45c. yes   **45d.** yes   **47.** The value of x falls between -2.4
and 3.6.   **49a.** $x \leq \$12.88$ **49b.** Sample answer: There
may be sales tax on his purchases.   **51.** $y = -3x + 3$
53. 42, 131, 145, 159, 28   **55.** 12   **57.** $<$

Page 391 Lesson 7–2A

1. Sample answer: The variable remains on the left, but the inequality symbol is reversed. **3.** When the coefficient of x is positive, you can solve the inequality like an equation and retain the same inequality symbol. If the coefficient of x is negative, you can solve like an equation, but the symbol must be reversed.

Pages 396–398 Lesson 7–2

9. multiply by $-\frac{1}{6}$ or divide by -6; yes; $\{y\,|\,y \le 4\}$

11. multiply by 4; no; $\{x\,|\,x < -20\}$ **13.** $\{x\,|\,x < 30\}$

15. $\{t\,|\,t \le -30\}$ **17.** $\frac{1}{5}x \le 4.025$; $\{x\,|\,x \le 20.125\}$ **19.** $s \ge 12$

21. $\{b\,|\,b > -12\}$ **23.** $\{x\,|\,x \ge -44\}$ **25.** $\{r\,|\,r < -6\}$

27. $\{t\,|\,t < 169\}$ **29.** $\{g\,|\,g \ge 7.5\}$ **31.** $\{x\,|\,x \ge -0.7\}$

33. $\left\{r\,\middle|\,r < -\frac{1}{20}\right\}$ **35.** $\{x\,|\,x < -27\}$ **37.** $\{m\,|\,m \ge -24\}$

39. $36 \ge \frac{1}{2}x$; $\{x\,|\,x \le 72\}$ **41.** $\frac{3}{4}x \le -24$; $\{x\,|\,x \le -32\}$

43. $-8x \le 144$; -18 or greater **45.** $y < 7.14$ meters **47.** $\ge$

49. $<$ **51.** up to 416 miles **53.** at least 5883 signatures

55. $(-1, 1)$ **57.** -2 **59.** \$155.64 **61.** 65 yd by 120 yd

Pages 402–404 Lesson 7–3

7. c **9.** $\{x\,|\,x > 2\}$ **11.** $\{d\,|\,d > -125\}$ **13.** $\{2, 3\}$

15a. $x + (x + 2) > 75$ **15b.** $x > 36.5$ **15c.** Sample answer: 38 and 40. **17.** $\{-10, -9, \ldots, 2, 3\}$ **19.** $\{-10, -9, \ldots, -5, -4\}$ **21.** $\{t\,|\,t > 3\}$ **23.** $\{w\,|\,w \le 15\}$ **25.** $\{n\,|\,n > -9\}$ **27.** $\{m\,|\,m < 15\}$ **29.** $\{x\,|\,x < -15\}$ **31.** $\left\{p\,\middle|\,p \le \frac{14}{3}\right\}$ **33.** $\{x\,|\,x > -10\}$ **35.** $\{k\,|\,k \le -1\}$ **37.** $\{y\,|\,y < -1\}$ **39.** $3(x + 7) > 5x - 13$; $\{x\,|\,x < 17\}$ **41.** $2x + 2 \le 18$ for $x > 0$; 7 and 9; 5 and 7; 3 and 5; 1 and 3 **43.** no solution $\{\varnothing\}$ **45a.** $x \le -8$ **45b.** $x > 8$ **45c.** $x > 2$ **45d.** $x \le -1$ **47.** $x + 0.04x + 0.15(x + 0.04x) \le \$50$, $x \le \$41.80$ **49.** at least \$571,428.57 **51a.** at most 2.9 weeks **51b.** no change **51c.** at most 4.1 weeks **53.** $\{y\,|\,y > 10\}$ **55.** $3x + 2y = 14$ **57.** $\{-5, -3, -2, 4, 16\}$ **59.** 25.1; 23.5; no mode

Pages 409–412 Lesson 7–4

7. $0 \le x \le 9$

9. $-3 < x \le 1$ **11.** The solution is the empty set. There are no numbers greater than 5 but less than -3.

13. $\{h\,|\,h \le -7 \text{ or } h \ge 1\}$

15. $\{w\,|\,1 > w \ge -5\}$

17. Drawings will vary; 16 pieces. **19.** $\varnothing$

21. **23.**

25. $-4 \le x \le 5$ **27.** $x \le -2$ or $x > 1$

29. $\{x\,|\,-1 < x < 5\}$

31. $\{x\,|\,x < -2 \text{ or } x > 3\}$

33. $\{c\,|\,c < 7\}$

35. $\varnothing$

37. $\{x\,|\,x \text{ is a real number.}\}$

39. $\{y\,|\,y > 3 \text{ and } y \ne 6\}$

41. $\{x\,|\,x \text{ is a real number.}\}$

43. $\{w\,|\,w < 4\}$

45. Sample answer: $x > 5$ and $x < -4$ **47.** $n + 2 \le 6$ or $n + 2 \ge 10$; $\{n\,|\,n \le 4 \text{ or } n \ge 8\}$ **49.** $31 \le 6n - 5 \le 37$; $\{n\,|\,6 \le n \le 7\}$

51. $\{m\,|\,-4 < m < 1\}$

53a. $\{x\,|\,x < -7 \text{ or } x > 1\}$ **53b.** $\{x\,|\,-5 \le x < 1\}$ **55.** $-4 \le x \le -1.5$ or $x \ge 2$ **57.** $4.4 < x < 6.7$ **59.** $\left\{m\,\middle|\,m \ge \frac{44}{3}\right\}$ **61.** -5

63.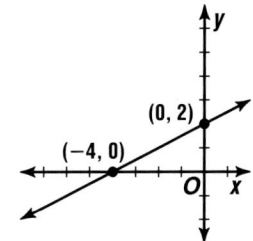

65. a little more than half a mile **67.** $18px - 15bg$

Page 412 Self Test

1. $\{y\,|\,y \ge -17\}$ **3.** $\{n\,|\,n < 4\}$ **5.** $\{g\,|\,g < -5\}$ **7.** c **9.** more than 17 points

Pages 415–419 Lesson 7–5

7a. outcomes from tree diagram: burger, soup, lemonade; burger, soup, soft drink; burger, salad, lemonade; burger, salad, soft drink; burger, french fries, lemonade; burger, french fries, soft drink; sandwich, soup, lemonade; sandwich, soup, soft drink; sandwich, salad, lemonade; sandwich, salad, soft drink; sandwich, french fries, lemonade; sandwich, french fries, soft drink; taco, soup, lemonade; taco, soup, soft drink; taco, salad, lemonade; taco, salad, soft drink; taco, french fries, lemonade; taco, french fries, soft drink; pizza, soup, lemonade; pizza, soup, soft drink; pizza, salad, lemonade; pizza, salad, soft drink; pizza, french fries, lemonade; pizza, french fries, soft drink **7b.** $\frac{1}{3}$ or $0.\overline{3}$ **7c.** $\frac{1}{12}$ or $0.08\overline{3}$ **7d.** $\frac{1}{24}$ or $0.041\overline{6}$ **9a.** 15 **9b.** $\frac{1}{5}$ or 0.2 **11.** $\frac{1}{3}$ or $0.\overline{3}$

13a. R3-G5, R3-R10, R3-B10, R3-G1, R3-Y14, B3-G5, B3-R10, B3-B10, B3-G1, B3-Y14, R5-G5, R5-R10, R5-B10, R5-G1, R5-Y14, R14-G5, R14-R10, R14-B10, R14-G1, R14-Y14, Y10-G5, Y10-R10, Y10-B10, Y10-G1, Y10-Y14 **13b.** $\frac{3}{25}$ or 0.12

13c. $\frac{2}{25}$ or 0.08 **13d.** 0 **13e.** $\frac{14}{25}$ or 0.56 **15a.** about 5.6% **15b.** about 26.3% **17.** 32% **19.** between 83 and 99, inclusive **21.** 3; -9 **23a.** $50°$ **23b.** $130°$ **23c.** yes

Pages 423–426 Lesson 7–6

5. c **7.** c **9.** d

11. $\{m\,|\,m \le -5 \text{ or } m \ge 5\}$

13. $\{r\,|\,-9 < r < 3\}$

15. $|x| = 2$

17. $\{-2, 6\}$

19. $\varnothing$

21. $\{y \mid 1 \le y \le 3\}$

23. $\varnothing$

25. $\left\{e \mid \frac{5}{3} < e < 3\right\}$

27. $\{y \mid y \text{ is a real number.}\}$

29. $\{w \mid 0 \le w \le 18\}$

31. $\{-2, 3\}$

33. $\left\{x \mid x \le -\frac{8}{3} \text{ or } x \ge 4\right\}$

35. $|p - 1| \le 0.01$ **37.** $|t - 50| > 50$
39. $|x + 1| = 3$ **41.** $|x - 1| \le 1$ **43.** $|x - 8| \ge 3$
45. $\{-2, -1, 0, 1, 2\}$ **47.** $2a + 1$ **49.** $a \ne 0$; never
51. $\frac{8}{13}$ or 0.61 **53.** no; $52 \le s \le 66$ **55.** $\$16{,}500 \le p \le$

$\$18{,}000$ **57a.** Outcomes from tree diagram: BBBB, BBBG, BBGB, BBGG, BGBB, BGBG, BGGB, BGGG, GBBB, GBBG, GBGB, GBGG, GGBB, GGBG, GGGB, GGGG
57b. $\frac{1}{16}$ or 0.0625, regardless of gender **57c.** $\frac{3}{8}$ or 0.375

59. $\{x \mid x \ge -1\}$ **61.** $\{k \mid k \ge -15\}$

63.   **65.** $m = -6 - \frac{n}{2}$

$2x - 9 = 2y$

Pages 430–432 Lesson 7-7
7a. A; 25, 65, 30, 60, 40; B; 20, 70, 40, 60, 45 **7b.** B **7c.** A
7d. B **9a.** Q2 = 6.5, Q3 = 16, Q1 = 5, IQR = 11 **9b.** no

9c.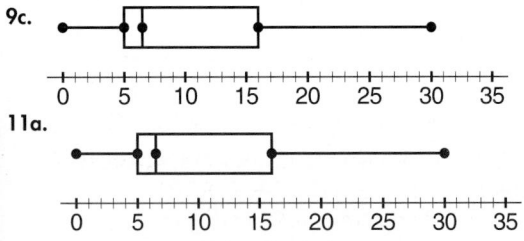

11a.

11b. clustered with lots of outliers **11c.** There are four western states that have more American Indian people than other states. **11d.** It is greater than the median.
13a.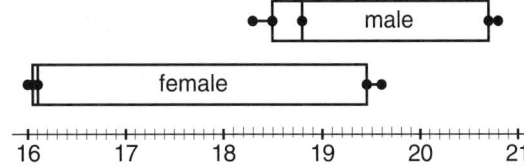

male

female

13b. 1990 **15.** Sample answer: Class A appears to be a more difficult class than B because the students don't do as well. **17.** $\{m \mid m > 1\}$

19.

(0, 6)

(−3, 2)

(8, −1)

Page 434 Lesson 7-7B
1. women identifying objects with their left hands
3. The left hand data are more clustered. **5.** males

Page 435 Lesson 7-8A
1.

3.

5.

7.

9.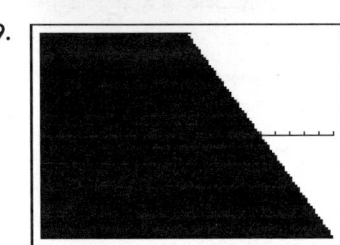

Page 439–441 Lesson 7-8
5. a **7.** b **9.** a, c; no

11.
$y = 3$

13.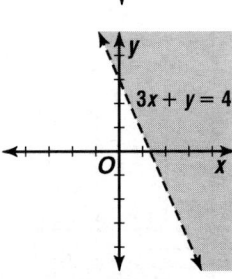
$2x + 3y = -2$

15.
$x = 4$

17.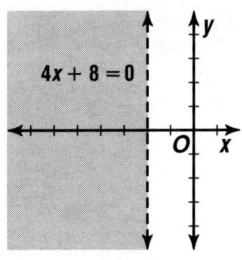
$3x + y = 4$

19. $\{(1, 1), (1, 2)\}$ **21.** $\varnothing$

23.
$y = -3$

25.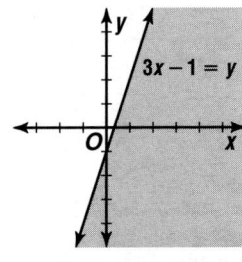
$4x + 8 = 0$

27.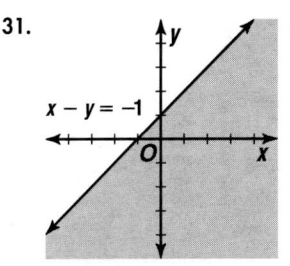
$x + y = 2$

29.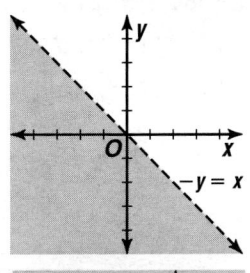
$3x - 1 = y$

31.
$x - y = -1$

33.
$-y = x$

35.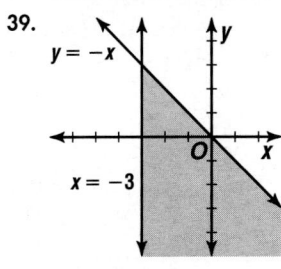
$8y + 3x = 16$

37.
$y = |x + 2|$

39.
$y = -x$
$x = -3$

41.

43. c **45.** a, b, d
47a. $0.7(220 - a) \leq z \leq 0.8(220 - a)$
47b. $32 \leq z \leq 37$
47c. improve cardiovascular conditioning

49.
250 300 350 400 450 500 550 600

51. $y = 4x - 2$

53.
$-y + \frac{2}{7}x = 1$

55. \$138.22

Page 443   Chapter 7 Highlights
1. h **3.** i **5.** c, f **7.** d **9.** g **11.** a

Pages 444–446   Chapter 7 Study Guide and Assessment
13. $\{n \mid n > -35\}$ **15.** $\{p \mid p \leq -18\}$ **17.** $3n > 4n - 8$, $\{n \mid n < 8\}$ **19.** $\{w \mid w \leq -15\}$ **21.** $\{x \mid x > 32\}$
23. $-\frac{3}{4}n \leq 30$, $\{n \mid n \geq -40\}$ **25.** $\{-5, -4, \ldots 0, 1\}$

27. $\left\{y \mid y \leq -\frac{9}{2}\right\}$ **29.** $\left\{x \mid x > -\frac{5}{2}\right\}$ **31.** $\{z \mid z \leq 20\}$

33. $\{a \mid a \text{ is a real number}\}$
-4 -3 -2 -1 0 1 2 3 4

35. $\{b \mid b \leq 5\}$
-1 0 1 2 3 4 5 6 7

37a. $\frac{2}{7}$ **37b.** $\frac{1}{2}$ **37c.** $\frac{2}{7}$ **37d.** 0 **37e.** $\frac{3}{14}$ **37f.** $\frac{1}{14}$
39. $\{y \mid y > -5 \text{ or } y < -5\}$
-9 -8 -7 -6 -5 -4 -3 -2 -1

41. $\{k \mid 3 \geq k \geq -4\}$
-5 -4 -3 -2 -1 0 1 2 3 4 5

43. $\left\{y \mid y \geq \frac{21}{5} \text{ or } y \leq 1\right\}$
-2 -1 0 1 2 3 4 5 6

45.
100 150 200 250 300 350

47. $\{(2, -1), (-1, 1)\}$ **49.** $\{(5, 10), (3, 6)\}$

51.
$4x - y = 8$

53.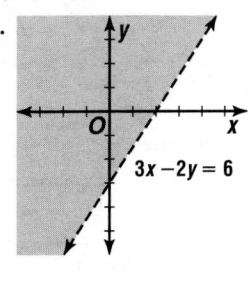
$3x - 2y = 6$

55. 17 to 20 books

Page 453   Lesson 8–1A
1. (1, 8)   **3.** (2.86, 4.57)   **5.** (2.28, 3.08)   **7.** (−2.9, 5.6)
9. (1.14, −3.29)

Pages 458–461   Lesson 8-1
9. no solution   **11.** one; (−6, 2)   **13.** yes
15. (3, 5)   **17.** (2, −6)

19. (−6, 8)

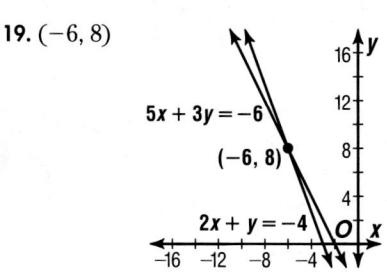

21. one, (3, −1)   **23.** no solution   **25.** one, (3, 3)

27. (2, −2)   **29.** (−1, 3)

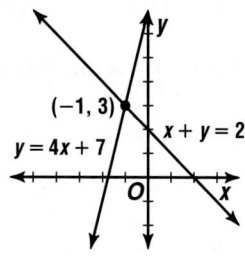

31. (−2, 4)   **33.** (2, 0)

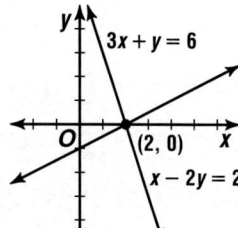

35. infinitely many   **37.** no solution

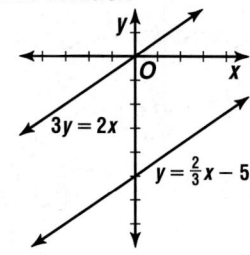

39. (8, 6)   **41.** infinitely many

43. 15 square units   **45.** (1.71, −2.57)
                                 47. (−0.25, −3.25)
                                 49. $A = −3, B = 2$

51. A.D. 376

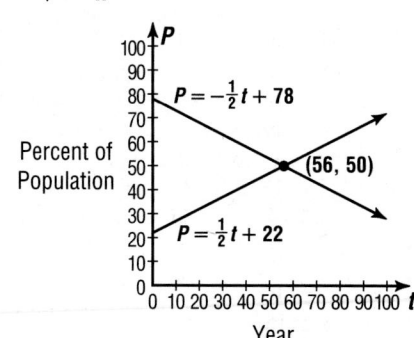

Year
($t = 0$ corresponds to A.D. 320)

53. −1.5, −13.5   **55.** $y = −2x + 2$ **57.** $7\frac{1}{2}$%   **59a.** how
many three- and four-bedroom homes will be built
59b. $100 − h$ **59c.** 80 homes

Pages 466–468   Lesson 8–2
7. $x = 8 − 4y, y = 2 − \frac{1}{4}x$ **9.** $x = −\frac{0.75}{0.8}y − 7.5; y = −\frac{0.8}{0.75}x − 8$
11. $\left(3, \frac{3}{2}\right)$ **13.** no solution   **15.** infinitely many   **17.** (3, 1)
19. (−4, 4)   **21.** (4, −1)   **23.** (2, 0)   **25.** (9, 1)   **27.** (2, 5)
29. (4, 2)   **31.** (5, 2)   **33.** $\left(\frac{8}{3}, \frac{13}{3}\right)$ **35.** (36, −6, −84)
37. (14, 27, −6)   **39a.** $y = 1000 + 5x, y = 13x$ **39b.** 125
tickets   **41a.** 26.5 years   **41b.** 33.8 seconds
43a.

	75% Gold (18-carat)	50% Gold (12-carat)	58% Gold (14-carat)
Total Grams	x	y	300
Grams of Pure Gold	0.75x	0.50y	0.58(300)

43b. $x + y = 300$; $0.75x + 0.50y = 0.58(300)$ **43c.** 96 grams
of 18-carat gold, 204 grams of 12-carat gold   **45.** 37 shares
47. {(−1, −7), (4, 8), (7, 17), (13, 35)}   **49.** −3   **51.** $m − 12$

Pages 472–474   Lesson 8–3
5. addition, (1, 0)   **7.** subtraction, $\left(−\frac{5}{2}, −2\right)$
9. substitution, (1, 4)   **11.** 8, 48   **13.** (1, −4), (1.29, −4.05)
15. +; (6, 2)   **17.** −; (4, −1)   **19.** −; (4, −7)   **21.** +; (5, 1)
23. sub; infinitely many   **25.** −; (−2, 3)   **27.** −; $\left(\frac{1}{2}, 1\right)$

29. +; $(10, -15)$ **31.** +; $(1.75, 2.5)$ **33.** 11, 53
35. 5, 8 **37.** $(2, 3, 7)$ **39.** $(14, 27, -6)$ **41.** Ling, 1.45 hours or 1 hour, 27 minutes; José, 1.15 hours, or 1 hour, 9 minutes **43.** 320 gal of 25% and 180 gal of 50% **45.** -3
47. $-\frac{7}{6}$ **49.** substitution (=)

Page 474 Self Test
1. $(1, -2)$ **3.** infinitely many

 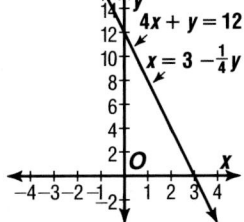

5. $(-9, -7)$ **7.** $(10, 15)$ **9.** $(4, -2)$

Pages 478–481 Lesson 8–4
5. $(-1, 1)$; Multiply the first equation by -3, then add.
7. $(-9, -13)$; Multiply the second equation by 5, then add.
9. $(-1, -2)$; Multiply the first equation by 5, multiply the second equation by -8, then add. **11.** b; $(2, 0)$ **13.** c; $(4, 1)$ **15.** $(2, 1)$ **17.** $(5, -2)$ **19.** $(2, -5)$ **21.** $(-4, -7)$
23. $(-1, -2)$ **25.** $(4, -6)$ **27.** $(13, -2)$ **29.** $(10, 12)$
31. 6, 9 **33.** elimination, addition; $\left(2, \frac{1}{8}\right)$ **35.** substitution or elimination, multiplication; infinitely many
37. elimination, subtraction; $(24, 4)$ **39.** $(11, 12)$
41. $\left(\frac{1}{3}, \frac{1}{6}\right)$ **43.** $(-2, 7), (2, 2), (7, 5)$ **45.** 6 2-seat tables, 11 4-seat tables **47.** $(3, -4)$ **49.** $\frac{1}{3}$ **51.** 165 yd **53.** 2 cups

Pages 485–486 Lesson 8–5
7. **9.**

11. 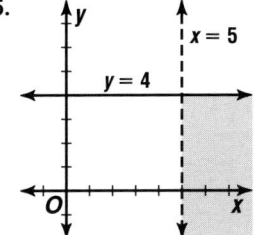 **13.** $y > x, y \le x + 4$

15.

17.

19. **21.**

23. **25.**

27. **29.**

31. 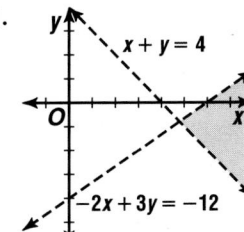   **33.** $y > -1, x \ge -2$
35. $y \le x, y > x - 3$
37. $x \ge 0, y \ge 0, x + 2y \le 6$

39.

41.

43.

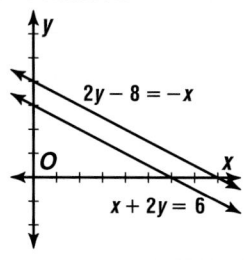

47. Sample answer: walk, 15 min, jog, 15 min; walk, 10 min, jog, 20 min; walk, 5 min, jog, 25 min.   49. 4 $5 bills, 8 $20 bills   51. $y = -\frac{1}{2}x + \frac{9}{2}$ 53. 10   55. Let $y =$ the number of yards gained in both games; $y = 134 + (134 - 17)$

Page 487   Chapter 8 Highlights
1. substitution   3. inconsistent   5. elimination
7. infinitely many   9. no   11. second

Pages 488–490   Chapter 8 Study Guide and Assessment
13. no solution   15. $(-2, -7)$

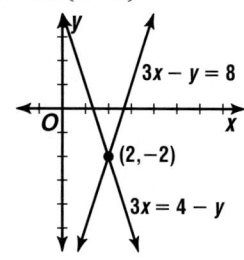

17. no solution   19. one, $(2, -2)$

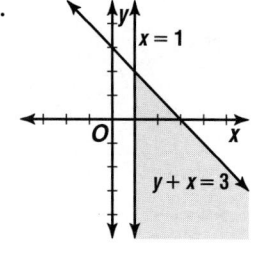

21. $(0, 2)$ 23. $\left(\frac{1}{2}, \frac{1}{2}\right)$ 25. $(2, -1)$ 27. $(5, 1)$ 29. $(8, -2)$
31. $(-9, -7)$ 33. $\left(\frac{3}{5}, 3\right)$ 35. $(2, -1)$ 37. $\left(\frac{7}{9}, 0\right)$ 39. $(0, 0)$
41. $(13, -2)$

43.

45.

47.

49. 35   51a. $75   51b. $15

Pages 499–500   Lesson 9–1
5. no   7. no   9. a^{12} 11. 3^{16} or 43,046,721   13. $9a^2y^6$
15. $15a^4b^3$ 17. m^4n^3 19. 2^{12} or 4096   21. $a^{12}x^8$
23. $6x^4y^4z^4$ 25. $a^2b^2c^2$ 27. $\frac{4}{25}d^2$ 29. $0.09x^6y^4$ 31. $90y^{10}$
33. $4a^3b^5$ 35. $-520x^9$ 37. -2^4 equals $-(2)(2)(2)(2)$ or -16 and $(-2)^4$ equals $(-2)(-2)(-2)(-2)$ or 16.
39. 301 parts
41. one; $(-6, 3)$

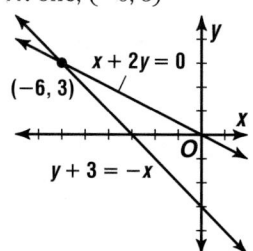

43. $n = 2m + 1$

m	−3	−2	−1	0	1
n	−5	−3	−1	1	3

45. $136°$ 47. $1.11y + 0.06$

Pages 504–505   Lesson 9–2
7. $\frac{1}{121}$ 9. 36   11. $\frac{6}{r^4}$ 13. $\frac{\pi}{4}$ 15. a^2 17. m^6 19. $\frac{m^3}{3}$
21. $\frac{c^3}{b^8}$ 23. b^8 25. $-s^6$ 27. $-\frac{4b^3}{c^3}$ 29. $\frac{1}{64a^6}$ 31. $\frac{s^3}{r^3}$
33. $\frac{a}{4c^2}$ 35. m^{3+n} 37. 3^{4x-6} 39. $\frac{1}{q^{18}}$ 41. $1257.14
43. $98a^5b^4$ 45. $|x + 1| < 3$ 47. $(3, -5)$ 49. $\frac{52}{41}$

Pages 509–512   Lesson 9–3
5. $43,400,000; 1.515 \times 10^3$ 7. $507,000,000; 4.4419 \times 10^4$
9. $4,551,400,000; 7.14 \times 10^2$ 11. 1.672×10^{-21} mg
13. 4×10^{-6} in.   15. $6.2 \times 10^{-7}; 0.00000062$ 17. $6 \times 10^7;$
$60,000,000$ 19. 9.5×10^{-3} 21. 8.76×10^{10} 23. $3.1272 \times$
10^8 25. 9.0909×10^{-2} 27. 7.86×10^4 29. 7×10^{-10}
31. 9.9×10^{-6} 33. $6 \times 10^{-3}; 0.006$ 35. $8.992 \times 10^{-7};$
0.0000008992 37. $4 \times 10^{-2}; 0.04$ 39. $6.5 \times 10^{-6};$
0.0000065 41. $6.6 \times 10^{-6}; 0.0000066$ 43. $1.2 \times 10^{-4};$
0.00012 45. 2.4336×10^{-1} 47. 2.8×10^5 49a. Sample
answer: overflow   49b. Multiply 3.7 and 5.6 and multiply
10^{112} and 10^{10}. Then write the product in scientific
notation.   49c. 2.072×10^{123} 51. $1,000,000,000,001$
53. 6.75×10^{18} molecules   55a. about 90.1 kg
57.

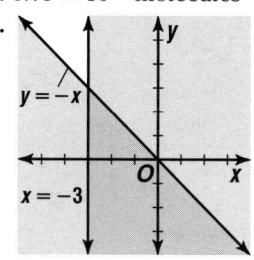

59. 25%   61. $y = x + 5$ 63. yes   65. $27°$ 67. -163

Page 513   Lesson 9–4A
1.

3.

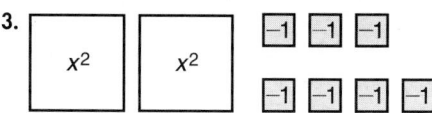

5. $x^2 - 3x + 2$ **7.** $2x^2 - x - 5$ **9.** x^2, x, and 1 represent the areas of the tiles.

Pages 517–519 Lesson 9–4
5. yes; trinomial **7.** yes; binomial **9.** 0 **11.** $x^8 - 12x^6 + 5x^3 - 11x$ **13a.** $\ell w - \pi r^2 - s^2$ **13b.** about 91.43 square units **15.** yes; trinomial **17.** yes; binomial **19.** yes; trinomial **21.** 5 **23.** 3 **25.** 9 **27.** 4 **29.** $x^5 + 3x^3 + 5$ **31.** $-x^7 + abx^2 - bcx + 34$ **33.** $1 + x^2 + x^3 + x^5$ **35.** $7a^3x + \frac{2}{3}x^2 - 8a^3x^3 + \frac{1}{5}x^5$ **37.** $2ab + \pi b^2$; about 353.10 square units **39.** $ab - 4x^2$; 116 square units **41b.** $8a^4 + 9a^3 + 4a^2 + 3a^1 + 5a^0$ **43.** about 153 eggs **45.** 4.235×10^4 **47.** 7, 8, 9 **49.** $\left\{(-3, 8), \left(1, \frac{8}{3}\right), (3, 0), (9, -8)\right\}$ **51.** \$38.75

Page 521 Lesson 9–5A
1. $-x^2 + 6$ **3.** $-2x^2 - 6x$ **5.** $3x^2 - x + 4$
7. true

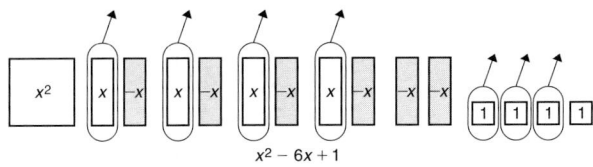

$2x^2 + 4x - 7$

9. Method from Activity 2:

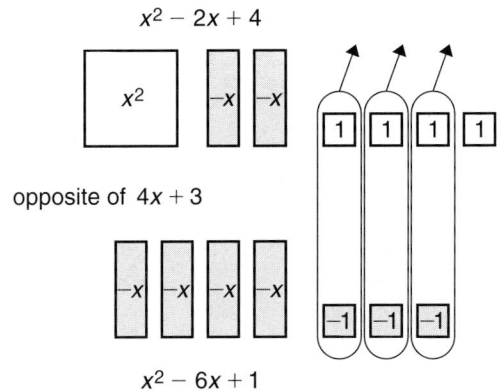

$x^2 - 6x + 1$

You need to add zero-pairs so that you can remove 4 green x-tiles.
  Method from Activity 4:

$$x^2 - 2x + 4$$

opposite of $4x + 3$

$x^2 - 6x + 1$

You remove all zero-pairs to find the difference in simplest form.

Pages 524–527 Lesson 9–5
7. $6a^2 - 3$ **9.** $4x^2 + 3y^2 - 8y - 7x$ **11.** $-8y^2$ and $3y^2$; $2x$ and $4x$ **13.** $3p^3q$ and $10p^3q$; $-2p$ and $-p$ **15.** $6m^2n^2 + 8mn - 28$ **17.** $-4y^2 + 5y + 3$ **19.** $7p^3 - 3p^2 - 2p - 7$ **21.** $10x^2 + 13xy$ **23.** $4a^3 + 2a^2b - b^2 + b^3$ **25.** $-4a + 6b - 5c$ **27.** $3a - 11m$ **29.** $-2n^2 + 7n + 5$ **31.** $13x - 2y$ **33.** $-y^3 + 3y + 3$ **35.** $4z^3 - 2z^2 + z$ **37.** $2x + 3y$

39. $353 - 18x$ **41.** $-2n^2 - n + 4$ **43.** $719x^2$ cubic stories **45.** $-3x - 2x^3 + 4x^5$
47. $(3, 1)$

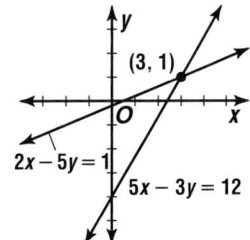

49. Sample answer: Let $h = $ the number of hours for the repair and let $c = $ the total charge; $c = 34h + 15$.

51. $-\frac{50}{3}$

Page 527 Self Test
1. $-6n^5y^7$ **3.** $-12ab^4$ **5.** 5.67×10^6 **7.** about 1.53×10^4 seconds or 4.25 hours **9.** $5x^2 - 4x - 14$

Page 528 Lesson 9–6A
1. $x^2 + 2x$ **3.** $2x^2 + 2x$ **5.** $2x^2 + x$
7. true

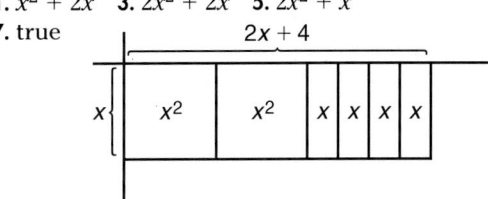

9a. $3x$ and $x + 15$
9b. $(3x^2 + 45x)$ square feet

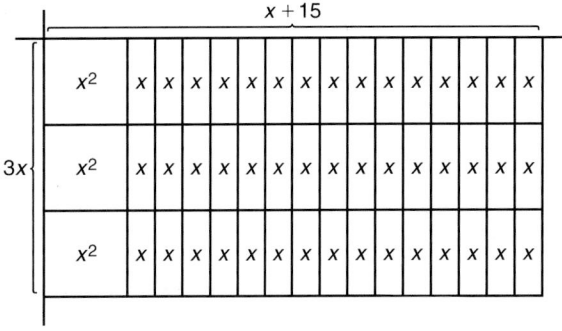

Pages 531–533 Lesson 9–6
5. $-63b^4c - 7b$ **7.** $10y^2 - 26y$ **9.** $3w^2 - 2w$ **11.** $\frac{103}{19}$

13a. $a^2 + a$ **13b.** $a^2 + 2a$ **15.** $\frac{1}{3}x^2 - 9x$ **17.** $-20m^5 - 8m^4$ **19.** $30m^5 - 40m^4n + 60m^3n^3$ **21.** $-28d^3 + 16d^2 - 12d$ **23.** $-32r^2s^2 - 56r^2s + 112rs^3$ **25.** $\frac{36}{5}x^3y + x^3 - 24x^2y$ **27.** $36t^2 - 42$ **29.** $61y^3 - 16y^2 + 167y - 18$ **31.** $53a^3 - 57a^2 + 7a$ **33.** $-\frac{77}{8}$ **35.** 0 **37.** $\frac{23}{24}$ **39.** 2

41. $15p^2 + 32p$ **43.** Sample answer: $1(8a^2b + 18ab)$, $a(8ab + 18b)$, $b(8a^2 + 18a)$, $2(4a^2b + 9ab)$, $(2a)(4ab + 9b)$, $(2b)(4a^2 + 9a)$, $(ab)(8a + 18)$, $(2ab)(4a + 9)$
45. $1.50t + 1.25mt$
47a.

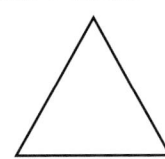

0 diagonals
$\frac{1}{2}(3)(3 - 3) = \frac{1}{2}(3)(0)$
$= 0$

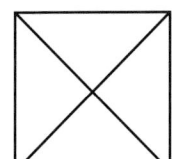

2 diagonals
$\frac{1}{2}(4)(4 - 3) = \frac{1}{2}(4)(1)$
$= 2$

5 diagonals $\qquad$ 9 diagonals

$\frac{1}{2}(5)(5-3) = \frac{1}{2}(5)(2)$ $\qquad$ $\frac{1}{2}(6)(6-3) = \frac{1}{2}(6)(3)$
$\phantom{\frac{1}{2}(5)(5-3)} = 5$ $\qquad$ $\phantom{\frac{1}{2}(6)(6-3)} = 9$

47b. $\frac{1}{2}n^2 - \frac{3}{2}n$ **47c.** 90 diagonals **49.** 75 gal of 50%,
25 gal of 30% **51.** $a^2 - 1$ **53.** 11 days **55.** 25.7

Page 535 Lesson 9–7A
1. $x^2 + 3x + 2$ **3.** $x^2 - 6x + 8$ **5.** $2x^2 - 2$
7. false

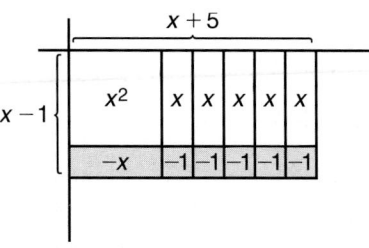

9. false

11. By the distributive property, $(x + 3)(x + 2) = x(x + 2)$ $+ 3(x + 2)$. The top row represents $x(x + 2)$ or $x^2 + 2x$. The bottom row represents $3(x + 2)$ or $3x + 6$.

Pages 539–541 Lesson 9–7
5. $d^2 + 10d + 16$ **7.** $y^2 - 4y - 21$ **9.** $2x^2 + 9x - 5$
11. $10a^2 + 11ab - 6b^2$ **13a.** $a^3 + 3a^2 + 2a$ **13c.** The result
is the same as the product in part b. **15.** $c^2 - 10c + 21$
17. $w^2 - 6w - 27$ **19.** $10b^2 - b - 3$ **21.** $169x^2 - 9$
23. $0.15v^2 - 2.9v - 14$ **25.** $\frac{2}{3}a^2 + \frac{1}{18}ab - \frac{1}{3}b^2$
27. $0.63p^2 + 3.9pq + 6q^2$ **29.** $6x^3 + 11x^2 - 68x + 55$
31. $9x^3 - 45x^2 + 62x - 16$ **33.** $20d^4 - 9d^3 + 73d^2 -$
$39d + 99$ **35.** $10x^4 + 3x^3 + 51x^2 - 16x - 48$ **37.** $a^4 -$
$a^3 - 8a^2 - 29a - 35$ **39.** $63y^3 - 57y^2 - 36y$ **41a.** 28 cm,
20 cm, 16 cm **41b.** 8960 cm³ **41c.** 8960 **41d.** They are
the same measure. **43.** $-8x^4 - 6x^3 + 24x^2 - 12x + 80$
45. $-3x^3 - 5x^2 - 6x + 24$ **47a.** Sample answer: $x - 2$,
$x + 3$ **47b.** Sample answer: $x^2 + x - 6$ **47c.** larger, 2 sq ft
49. $\{y\,|\,y < -6\}$ **51.** $y = 2x + 1$ **53.** $a = 4, y = 9$ **55.** 15°C

Pages 546–547 Lesson 9–8
7. $m^2 - 6mn + 9n^2$ **9.** $m^4 + 8m^2n + 16n^2$ **11.** $25 -$
$10x + x^2$ **13.** $x^2 + 8xy + 16y^2$ **15.** $9b^2 - 6ab + a^2$
17. $81p^2 - 4q^2$ **19.** $25b^2 - 120ab + 144a^2$ **21.** $x^6 +$
$2x^3a^2 + a^4$ **23.** $64x^4 - 9y^2$ **25.** $1.21g^2 + 2.2gh^5 + h^{10}$

840 *Selected Answers*

27. $\frac{16}{9}x^4 - y^2$ **29.** $9x^3 - 45x^2 - x + 5$ **31.** $a^3 + 9a^2b +$
$27ab^2 + 27b^3$ **33.** $x^2 + y^2 + z^2 + 2xy + 2yz + 2xz$

	x	y	z
x	x^2	xy	xz
y	xy	y^2	yz
z	xz	yz	z^2

35a. $2\pi s + 7\pi$ square meters **35b.** about 28.27 square
meters **35c.** about 40.84 square meters **37.** $6t^2 - 3t - 3$
39a.

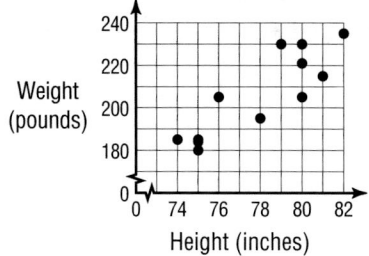

Weight (pounds) vs Height (inches)

39b. the taller the player, the greater the weight
41. $\qquad\qquad$ **43.** 5

$B(-3, 1)$ $A(4, 2)$
O
$C(-2, -3)$

Page 549 Chapter 9 Highlights
1. e **3.** h **5.** i **7.** f **9.** a

Pages 550–552 Chapter 9 Study Guide and Assessment
11. y^7 **13.** $20a^5x^5$ **15.** $576x^5y^2$ **17.** $-\frac{1}{2}m^4n^8$ **19.** y^4
21. $3b^3$ **23.** $\frac{a^4}{2b}$ **25.** 2.4×10^5 **27.** 4.88×10^9 **29.** 7.96
$\times 10^5$ **31.** 6×10^{11} **33.** 6×10^{-1} **35.** 1.68×10^{-1}
37. 2 **39.** 5 **41.** 4 **43.** $3x^4 + x^2 - x - 5$ **45.** $-3x^3 + x^2 -$
$5x + 5$ **47.** $16m^2n^2 - 2mn + 11$ **49.** $21m^4 - 10m - 1$
51. $12a^3b - 28ab^3$ **53.** $8x^5y - 12x^4y^3 + 4x^2y^5$ **55.** $2x^2 -$
$17xy^2 + 10x + 10y^2$ **57.** $r^2 + 4r - 21$ **59.** $4x^2 + 13x - 12$
61. $18x^2 - 0.125$ **63.** $2x^3 + 15x^2 - 11x - 9$ **65.** $x^2 - 36$
67. $16x^2 + 56x + 49$ **69.** $25x^2 - 9y^2$ **71.** $36a^2 - 60ab +$
$25b^2$ **73.** $\$305.26$ **75.** no; after 6 years

CHAPTER 10 USING FACTORING

Pages 561–563 Lesson 10–1
5. 1, 2, 4 **7.** prime **9.** $-1 \cdot 2 \cdot 3 \cdot 5$ **11.** 4 **13.** $6d$ **15.** $4gh$
17. 40 in. **19.** 1, 67 **21.** 1, 2, 4, 5, 8, 10, 16, 20, 40, 80
23. 1, 2, 5, 10, 19, 25, 38, 50, 95, 190, 475, 950 **25.** composite;
$3^2 \cdot 7$ **27.** prime **29.** composite; $2^2 \cdot 5 \cdot 7 \cdot 11$ **31.** $-1 \cdot 3$
$\cdot 3 \cdot 13$ **33.** $2 \cdot 2 \cdot b \cdot b \cdot b \cdot d \cdot d$ **35.** $-1 \cdot 2 \cdot 7 \cdot 7 \cdot a \cdot a \cdot b$
37. 9 **39.** 1 **41.** 19 **43.** $7pq$ **45.** $15r^2t^2$ **47.** 12 **49.** 6
51. $8m^2n$ **53.** $-12x^3yz^2$ **55.** $3m^2n$ **57.** 29 cm by 47 cm
59. 3, 5; 5, 7; 11, 13; 17, 19; 29, 31; 41, 43; 59, 61; 71, 73

61.

63. 1500 squares of sod **65.** $3b$
67. [number line from -1 to 5 with points marked]   **69.** $x = 4$ **71.** $9\frac{3}{5}$ ft
73. -4

Page 564   Lesson 10–2A
1. $3(x + 3)$ **3.** $x(3x + 4)$
5. no   **7.** yes
[algebra tile diagrams]

9. Binomials can be factored if they can be represented by a rectangle. Examples: $4x + 4$ can be factored and $4x + 3$ cannot be factored.

Pages 569–571   Lesson 10–2
7. 3   **9.** 1   **11.** $4xy^2$ **13.** $(a + b)(x + y)$ **15.** $(x - y)(3a + 4b)$ **17.** $3t$ **19.** $x(29y - 3)$ **21.** $3c^2d(1 - 2d)$
23. $(r + k)(x + 2y)$ **25a.** $g = \frac{1}{2}n(n - 1)$ **25b.** 91 games

25c. 63 games   **27.** $8rs$ **29.** $2y - 5$ **31.** $5k - 7p$
33. $2xz(7 - 9z)$ **35.** $a(17 - 41ab)$ **37.** $(m + x)(2y + 7)$
39. $3xy(x^2 - 3y + 12)$ **41.** $(2x^2 - 5y^2)(x - y)$ **43.** $(2x - 5y)(2a - 7b)$ **45.** $7abc(4abc + 3ac - 2)$ **47.** $2(2m + r)(3x - 2)$ **49.** $(7x + 3t - 4)(a + b)$ **51.** $8a - 4b + 8c + 16d + ab + 64$ **53.** $4r^2(4 - \pi)$ **55.** $(4z + 3m)$ cm by $(z - 6)$ cm   **57.** Sample answer: $(3a + 2b)(ab + 6)$, $(3a + ab)(2b + 6)$, $(2b + ab)(3a + 6)$ **59.** $(2s - 3)(s + 8)$ **61.** A and C sharp   **63.** 9   **65.** $\{z \mid z \geq -1.654\}$ **67.** $\{(-3, 10),$ $\left(0, \frac{11}{2}\right), (1, 4), \left(2, \frac{5}{2}\right), (5, -2)\}$ **69.** $y = -\frac{4}{3}x + \frac{7}{3}$ **71.** 7

Page 573   Lesson 10–3A
1. $(x + 1)(x + 5)$ **3.** $(x + 3)(x + 4)$ **5.** $(x - 1)(x - 2)$
7. $(x - 1)(x + 5)$
9. yes   **11.** no
[algebra tile diagrams]

13. Trinomials can be factored if they can be represented by a rectangle. Sample answers: $x^2 + 4x + 4$ can be factored and $x^2 + 6x + 4$ cannot be factored.

Pages 578–580   Lesson 10–3
5. 3, 8   **7.** 8, 5   **9.** $-2, -6$ **11.** $-$ **13.** $(t + 3)(t + 4)$
15. $2(y + 2)(y - 3)$ **17.** prime   **19.** 9, -9, 15, -15
21. $(3x^2 + 2x)$ m^2 **23.** $-$ **25.** 5   **27.** 5   **29.** $(m - 4)(m - 10)$

31. prime   **33.** $(2x + 7)(x - 3)$ **35.** $(2x + 3)(x - 4)$
37. $(2n - 7)(2n + 5)$ **39.** $(2 + 3m)(5 + 2m)$ **41.** prime
43. $2x(3x + 8)(2x - 5)$ **45.** $2a^2b(5a - 7b)(2a - 3b)$
47. 7, -7, 11, -11 **49.** 1, -1, 11, -11, 19, -19, 41, -41
51. 6, 4   **53.** r cm, $(5r + 6)$ cm, $(3r - 7)$ cm   **55.** no; $(2x + 3)(x - 1)$ **57.** no; $(x - 3)(x - 3)$ **59.** 27 ft^3
61. $(3x - 10)$ shares   **63.** $(5, -2)$ **65.** $y = -\frac{2}{3}x + \frac{14}{3}$
67. D = $\{0, 1, 2\}$; R = $\{2, -2, 4\}$ **69.** $90°$ **71.** $3x + 4y$

Page 580   Self Test
1. $10n^2$ **3.** $6xy(3y - 4x)$ **5.** $(2q + 3)(q - 6)$ **7.** $(3y - 5)(y - 1)$ **9.** 41,312,432 or 23,421,314

Pages 584–586   Lesson 10–4
7. yes   **9.** no   **11.** d   **13.** a   **15.** $(1 - 4g)(1 + 4g)$
17. $5(2m - 3n)(2m + 3n)$ **19.** $(x - y)(x + y)(x^2 + y^2)$
21. 5   **23.** $(2 - v)(2 + v)$ **25.** $(10d - 1)(10d + 1)$
27. $2(z - 7)(z + 7)$ **29.** prime   **31.** $17(1 - 2k)(1 + 2k)$
33. prime   **35.** prime   **37.** $(ax - 0.8y)(ax + 0.8y)$
39. $\frac{1}{2}(3a - 7b)(3a + 7b)$ **41.** $(a + b - c - d)(a + b + c + d)$ **43.** $(x^2 - 2y)(x^2 + 2y)(x^4 + 4y^2)$ **45.** $(a^2 + 5b^2)$ $(a - 2b)(a + 2b)$ **47.** 624   **49.** $(2a - b)$ in., $(2a + b)$ in.
51. $(x - 2)$ ft, $(x + 4)$ ft   **53.** $(a - 5b)$ in., $(a + 5b)$ in., $(5a + 3b)$ in.   **55a.** square   **55b.** 25 cm^2 **57.** 9, 12, 15
59. $3.16, $1.50, $1.25, $1.20   **61.** $(4, 16)$

63. 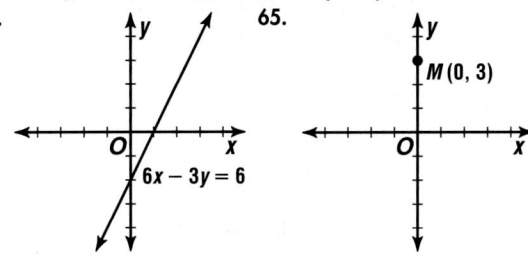   **65.**

67. 5.14, 3.6, 0.6

Pages 591–593   Lesson 10–5
7. 8a   **9.** 6c   **11.** yes; $(2n - 7)^2$ **13.** yes; $(4b - 7c)^2$
15. $4(a - 3b)(a + 3b)$ **17.** $2(5g + 2)^2$ **19.** $(2a - 3b)(2a + 3b)(5x - y)$ **21.** yes; $(r - 4)^2$ **23.** yes; $(7p - 2)^2$ **25.** no
27. yes; $(2m + n)^2$ **29.** no   **31.** no   **33.** yes; $\left(\frac{1}{2}a + 3\right)^2$

35. $a(45a - 32b)$ **37.** $(v - 15)^2$ **39.** prime   **41.** $3(y - 7)(y + 7)$ **43.** $2(3a - 4)^2$ **45.** $(y^2 + z^2)(x - 1)(x + 1)$
47. $(a^2 + 2)(4a + 3b^2)$ **49.** $0.7(p - 3q)(p - 2q)$
51. $(g^2 - 3h)(g + 3)^2$ **53.** $-110, 110$ **55.** 9   **57.** $(6y + 26)$ cm   **59.** $(8x^2 - 22x + 14)$ cm^2 **61a.** $a \geq b$ **61b.** $a \leq b$
61c. $a = b$ **63a.** $1166.40   **63b.** $p(1 + r)^2$ **63c.** $1144.90
65. $50.6t^2 + 21t - 102$ **67.** 20 hours
69a.

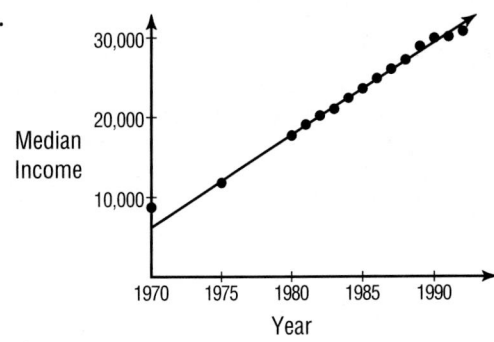

69b. Sample answer: Yes; $I = 1200y + 6000$, where I is the median income and y is the number of years since 1970.
71. 1:3   **73.** 152 ft

Pages 598–600   Lesson 10–6
5. $\{0, -5\}$ **7.** $\left\{0, \frac{5}{3}\right\}$ **9.** $\{5\}$ **11.** 6   **13.** $\{0, 24\}$ **15.** $\left\{\frac{3}{2},\right.$
$\left.\frac{8}{3}\right\}$ **17.** $\{-9, -4\}$ **19.** $\{-8, 8\}$ **21.** $\{0, 4\}$ **23.** $\left\{-\frac{1}{3}, -\frac{5}{2}\right\}$
25. $\{12, -4\}$ **27.** $\{-9, 0, 9\}$ **29.** $\{-5, 7\}$ **31.** -14 and
-12 or 12 and 14   **33.** 5 cm   **35a.** $\{-6, 1\}$; $\{-6, 1\}$
35b. They are equivalent; they have the same solution.
37a. about 3.35 s   **37b.** His ideas about falling objects
differed from what most people thought to be true. He
believed that Earth is a moving planet and that the sun
and planets do not revolve around Earth.   **39.** about
90,180 ft or 17 mi   **41.** 0.5 km   **43.** $40q^2 + rq - 6r^2$
45. $14; -4$ **47.** 4.355 minutes or about 4 minutes 21 seconds

Page 601   Chapter 10 Highlights
1. false; composite   **3.** false; sample answer: 64   **5.** false;
$2^4 \cdot 3$ **7.** true   **9.** false; zero product property

Pages 602–604   Chapter 10 Study Guide and Assessment
11. composite, $2^2 \cdot 7$ **13.** composite, $2 \cdot 3 \cdot 5^2$ **15.** prime
17. 5   **19.** $4ab$ **21.** $5n$ **23.** $7x^2$ **25.** $13(x + 2y)$
27. $6ab(4ab - 3)$ **29.** $12pq(3pq - 1)$ **31.** $(a - 4c)(a + b)$
33. $(4k - p^2)(4k^2 - 7p)$ **35.** $(8m - 3n)(3a + 5b)$
37. $(y + 3)(y + 4)$ **39.** prime   **41.** $(r - 4)(2r + 5)$
43. $(b - 4)(b + 4)$ **45.** $(4a - 9b^2)(4a + 9b^2)$ **47.** prime
49. $(a + 9)^2$ **51.** $(2 - 7r)^2$ **53.** $6b(b - 2g)^2$
55. $(5x - 12)^2$ **57.** $\left\{\frac{2}{3}, -\frac{7}{4}\right\}$ **59.** $\{0, -17\}$ **61.** $\{-5, -8\}$
63. $\left\{-\frac{2}{5}\right\}$ **65.** 384   **67.**

$$\begin{array}{ccc} & & \boxed{3} \\ \times & \boxed{5} & \boxed{4} \\ \hline \boxed{1} & \boxed{6} & \boxed{2} \end{array}$$

69. 9, 11; $-11, -9$

**CHAPTER 11   EXPLORING QUADRATIC AND
EXPONENTIAL FUNCTIONS**

Page 610   Lesson 11–1A
1. $(-8, -5)$

3. $(5, 0)$

5. $(10, 14)$
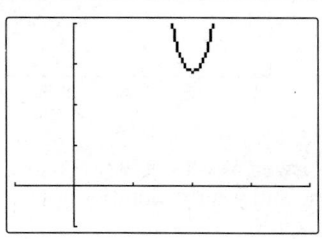

Pages 615–617   Lesson 11–1
7. $x = 0$, $(0, 2)$, min.
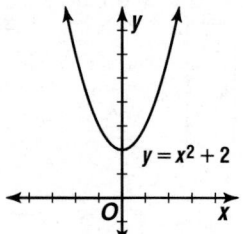

9. $x = -2$, $(-2, -13)$, min.
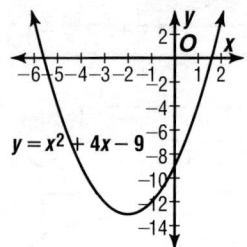

11. $x = 2.5$, $(2.5, 12.25)$, max.   **13.** c

15. $x = 0$, $(0, 0)$, min.
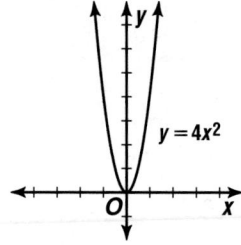

17. $x = -1$, $(-1, 17)$, min.

19. $x = 0$, $(0, -5)$, min.
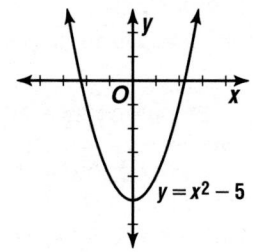

21. $x = -3$, $(-3, -29)$, min.

23. $x = 0$, $(0, -25)$, min.

25. $x = -1$, $(-1, 7)$, max.

27. $x = -1$, $(-1, -20)$, min.

29. $x = -1$, $(-1, -1)$, min.

31. c **33.**

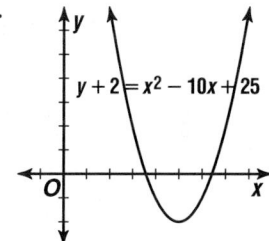

$y + 2 = x^2 - 10x + 25$

35.

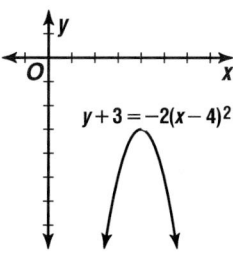

$y + 3 = -2(x - 4)^2$

37. $x = -1$ **39.** $x = 2$

41. $(-1.10, 125.8)$

43. $(-0.15, 90.14)$

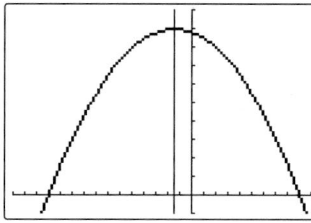

45a.

Year	t	$U(t)$
1970	0	329.96
1975	5	370.31
1980	10	559.16
1985	15	896.51
1990	20	1382.36
1993	23	1745.15

45b.

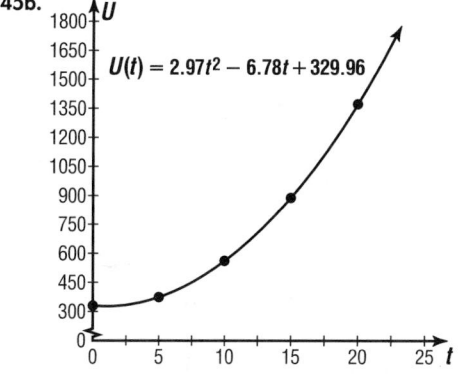

$U(t) = 2.97t^2 - 6.78t + 329.96$

47. 0.15 km **49a.** 6790 km **49b.** 2.2792×10^8 km

51a.

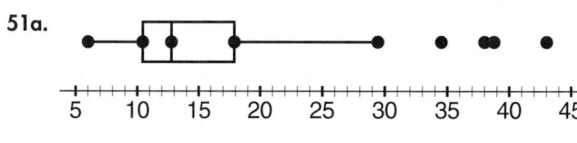

51b. 34.5, 38, 38.7, 43 **53.** -5

1. All the graphs open downward from the origin. $y = -2x^2$ is narrower than $y = -x^2$ and $y = -5x^2$ is the narrowest.

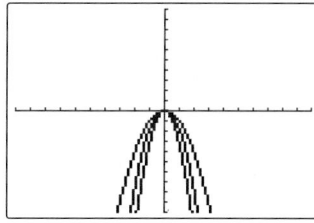

3. All open downward, have the same shape, and have vertices along the x-axis. However, each vertex is different.

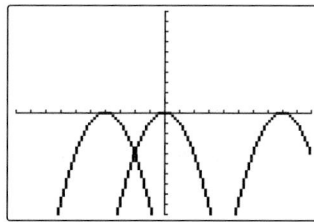

5. It will open upward, have a vertex at the origin, and be narrower than $y = x^2$.

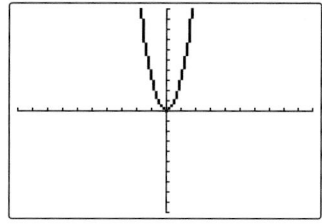

7. It will have vertex at the origin, open downward, and be wider than $y = x^2$.

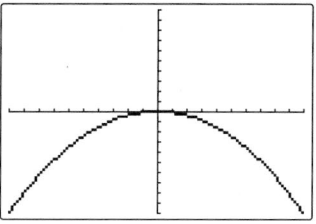

9a. If $|a| > 1$, the graph is narrower than the graph of $y = x^2$. If $0 < |a| < 1$, the graph is wider than the graph of $y = x^2$. If $a < 0$, it opens downward; if $a > 0$, it opens upward. **9b.** The graph has the same shape as $y = x^2$, but is shifted a units (up if $a > 0$, down if $a < 0$).
9c. The graph has the same shape as $y = x^2$, but is shifted a units (left if $a > 0$, right if $a < 0$). **9d.** The graph has the same shape as $y = x^2$ but is shifted a units left or right and b units up or down as prescribed in 9b and 9c.

Pages 624–627 Lesson 11–2
7. 2 real roots **9.** $-1, 2$
11. $-3, 8$ **13.** $-2 < w < -1, 4 < w < 5$

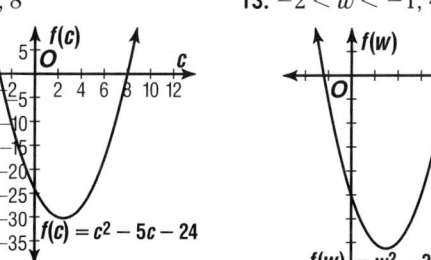

$f(c) = c^2 - 5c - 24$

$f(w) = w^2 - 3w - 5$

15. 5

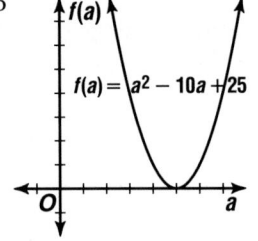

$f(a) = a^2 - 10a + 25$

17. $-2, -6$ **19.** 2   **21.** $-4, 4$ **23.** -3 **25.** $\varnothing$ **27.** $-4, 1$
29. 4   **31.** $8 < a < 9, -1 < a < 0$ **33.** $3, -1 < n < 0$

35.

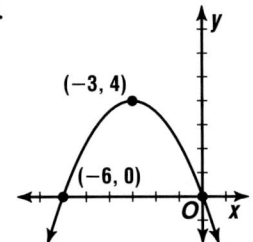

$(-3, 4)$
$(-6, 0)$

37.

$(2, 5)$

39. $-4, -8$ or $4, 8$ **41.** no solution
43. $-2, -6$ **45.** no y-intercepts   **47.** no real roots
49. $0.29, 1.71$ **51.** $-0.79, 2.54$ **53.** $-2, k > -2, k < -2$

55a.

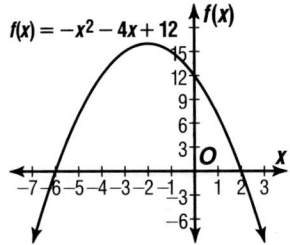

$f(x) = -x^2 - 4x + 12$

55b. 8 feet   **55c.** 16 feet   **55d.** $85\frac{1}{3}$ ft^2 **55e.** \$297

57. $x = 0; (0, 4)$

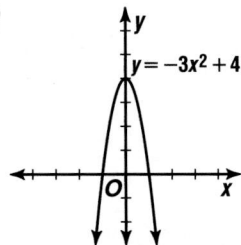

$y = -3x^2 + 4$

59. $(4x - 3)$ m by $(2x - 1)$ m   **61.** 6   **63.** $-4 < x < \frac{4}{3}$
65. \$24.75

Pages 631–633   Lesson 11–3
5. $1, 3, -18; 3, -6$ **7.** $4, -2, 15$; no real roots   **9.** $1.83,$
-2.33 **11.** $-1, -6$ **13.** about 29.4 ft/s   **15.** $-8.61, -1.39$
17. $-\frac{4}{5}, 1$ **19.** $-\frac{5}{3}, 4$ **21.** $-2.91, 2.41$ **23.** 7   **25.** $-\frac{3}{2}, -\frac{5}{3}$

27. $-6, 6$ **29.** $-\frac{1}{3}, 1$ **31.** $-3, -\frac{1}{2}$ **33.** $0.60, -0.25$

35. $0.5, 2$ **37.** $-0.2, 1.4$ **39.** Sample answer: 4, 20, 23;
$4x^2 + 20x + 23 = 0$ **41a.** 10, 1   **41b.** none   **41c.** -0.5
41d. $-1, 0.7142857143$ **43a.** Discriminant is not a perfect
square.   **43b.** Discriminant is a perfect square but the
expression is not an integer   **43c.** Discriminant is a perfect
square and the equation can be factored.
43d. Discriminant is 0 and the equation is a perfect square.
45a. Y, 10 mg; C, 8.75 mg   **45b.** $0 = a^2 - 11a + 12$; 1.2 yr,
9.8 yr   **47.** $\{0, -9\}$ **49.** 24, 18   **51.** yes   **53.** $-4°F$

Page 633   Self Test
1. $x = \frac{1}{2}, \left(\frac{1}{2}, -\frac{25}{4}\right)$

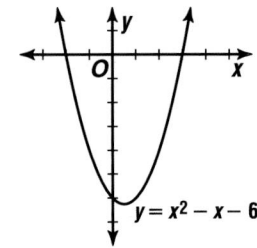

$y = x^2 - x - 6$

3. $x = 0, (0, 7)$

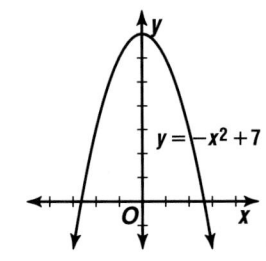

$y = -x^2 + 7$

5. $-9, 9$

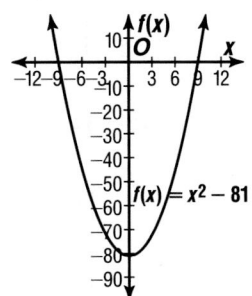

$f(x) = x^2 - 81$

7. $\varnothing$

$f(x) = 6x^2 + 36$

9. $\varnothing$

Page 634   Lesson 11–4A
1.

3.

5.

7.

9.

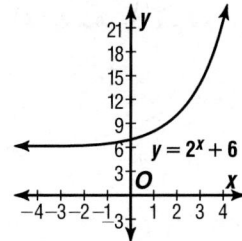

Pages 640–642   Lesson 11–4
7. 5.20   **9.** 12.51
11. 7

13. −2   **15.** −2   **17.** 10.56   **19.** 1.63   **21.** 25.86   **23.** 0.86
25. 5   **27.** 3

29. 1

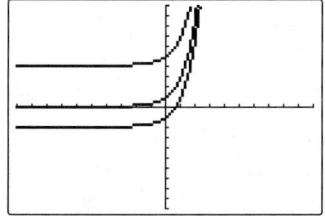

31. no, linear   **33.** no, no pattern   **35.** −7   **37.** −3
39. 3   **41.** 3, −1
43. All have the same shape but different y-intercepts.

45. All have the same shape but are positioned at different places along the x-axis.

47a. 2.51   **47b.** You cannot write 10 as a power or 2.5.
49. 1600 bacteria   **51.** −3, $-\frac{1}{2}$ **53.** $\frac{1}{5}ab(4a − 3b − 1)$
55. 184

57a–b.

Sample answer: $y = 0.07x − 12$ **57c.** 275 years   **59.** 60

Pages 646–649   Lesson 11–5
3a. 1   **3b.** 2   **3c.** 4   **3d.** 365   **5.** decay   **7.** about $70,000,000,000   **9.** $661.44   **11.** $10,962.19   **13a.** Each equation represents growth and t is the number of years since 1994. $y = 58.7(1.025)^t$, $y = 919.9(1.019)^t$, $y = 35.6(1.021)^t$, $y = 2.9(1.013)^t$ **13b.** 68.1 million, 1029.9 million, 40.3 million, 3.1 million

13c.

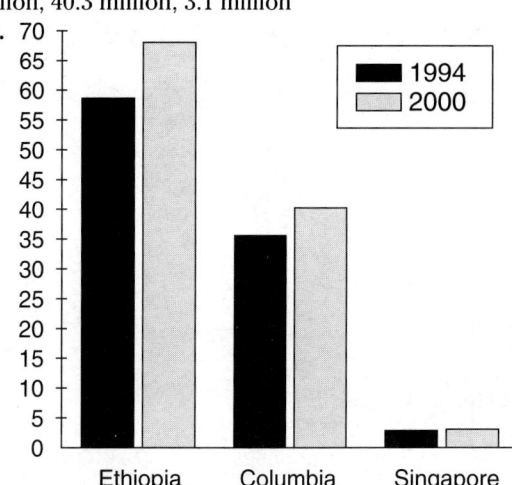

15a. $y = 231.5(1.0963)^t$, $t =$ years since 1950, growth
15b. 57.6 trillion   **17a.** $6931.53   **17b.** $6954.58
17c. $23.05   **17d.** daily   **21.** $a > 1$, growth; $0 < a < 1$, decay; x represents time   **23.** $-\frac{1}{2}$ **25.** $-\frac{r^{10}t^2}{3}$ **27.** $y = -2.5x + 2$

Page 651   Chapter 11 Highlights
1. d   **3.** i   **5.** c   **7.** b   **9.** f

Pages 652–654   Chapter 11 Study Guide and Assessment
11. $x = 0$; (0, 4)   **13.** $x = -1$; (−1, −20)   **15.** $x = -1$; (−1, −1)

17.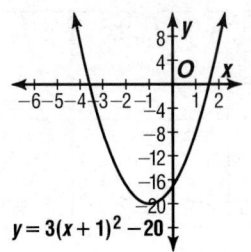

$y = x^2 - 3x - 4$

19.

$y = 3(x+1)^2 - 20$

21. $-3, 4$ **23.** $-5 < x < -4, 0 < x < 1$ **25.** $3, 7$
27. $10, -2$ **29.** $\frac{3}{2}, -\frac{5}{2}$ **31.** $1, \frac{4}{9}$ **33.** $2, 3$ **35.** $0.47, -0.71$

37. 7

$y = 3^x + 6$

39. 1

$y = 2^x$

41. -3 **43.** -6 **45.** $-3, \frac{3}{2}$ **47.** $\$12,067.68$ **49.** $\$24,688.37$

51. 1.3 seconds and 4.7 seconds   **53.** between 4 meters and 5 meters   **55.** CD, which yields $\$541.50$ vs. savings at $\$530.92$

CHAPTER 12   EXPLORING RATIONAL EXPRESSIONS AND EQUATIONS

Pages 663–665   Lesson 12–1
7. $7a^2b; \frac{-b^2}{3a^3}; a \neq 0, b \neq 0$ **9.** $b; \frac{3}{b+5}; b \neq 0,$
$b \neq -5$ **11.** $m - 3; \frac{1}{m+3}; m \neq \pm 3$ **13a.** 5 **13b.** 750 lb
15. $\frac{5y}{2z}; y \neq 0, z \neq 0$ **17.** $\frac{4}{5x}; x \neq 0, y \neq 0$ **19.** $y; y \neq -\frac{1}{3}$
21. $y + 7x; y \neq 7x$ **23.** $\frac{1}{a-4}; a \neq 4, -3$ **25.** $\frac{1}{x+4};$
$x \neq -4$ **27.** $\frac{a+6}{a+4}; a \neq -4, 2$ **29.** $\frac{b+1}{b-9}; b \neq 9, 4$
31. $\frac{4}{5}; x \neq -1$ **33.** $-\frac{5}{3}$ **35.** not possible, $y \neq -2$ **37a.** 192
37b. 3840 lb   **39a.** $9.5°$/mile   **39b.** $-\frac{95}{10} = -\frac{19}{2}$

39c.

Altitude (thousands of feet)

41. ± 11
43.

$y = 2x - 1$

$y = x - 2$

45.

$y = 3x + 2$

Page 666   Lesson 12–1B
1. $\frac{x-5}{x+5}; -5$ **3.** $\frac{1}{2x}; 0, 4.5$ **5.** $\frac{x-1}{x-8}; 8$ **7.** $-1; 3$
9. $-\frac{x+5}{x+6}; -6, 5$

Pages 669–670   Lesson 12–2
5. $\frac{9m}{2xy}$ **7.** $\frac{4}{x+3}$ **9.** $\frac{4}{3}$ **11a.** 91.44 cm/yd   **11b.** changing centimeters to yards   **13.** $\frac{3a}{2}$ **15.** $\frac{2}{x}$ **17.** $8y$
19. $\frac{4m}{3(m+5)}$ **21.** $\frac{2m}{a+9}$ **23.** 2 **25.** $3(x-y)$ **27.** 36
29. $\frac{4x}{3x-5}$ **31.** $\frac{4}{5(x+5)}$ **33.** 21.8 mph; converts ft/s to mph
35. 10,800 Calories   **37.** Sample answer: $\frac{3(x+2)}{x+7} \cdot \frac{2(x-3)}{x-4};$
$\frac{6}{x+7} \cdot \frac{x^2-x-1}{x-4}$ **39a.** $\frac{1 \text{ franc}}{0.1981 \text{ dollars}}; \frac{1 \text{ dollar}}{5.05 \text{ francs}}$
39b. 12,500 pesos $\cdot \frac{0.1597 \text{ dollars}}{1 \text{ peso}} \cdot \frac{1 \text{ franc}}{0.1981 \text{ dollars}};$
about 10,077 francs   **39c.** Convert to American dollars and then to Hong Kong's dollar.   **41.** 729   **43.** 4   **45.** 0

Pages 672–674   Lesson 12–3
5. $\frac{5}{x}$ **7.** $\frac{y+3}{x^2-9}$ **9.** $\frac{1}{x^2+2x+5}$ **11.** $\frac{(m+2)^2}{4}$ **13.** $\frac{x+4}{x+6}$
15. 0.5 ft^3 **17.** $\frac{a}{a+11}$ **19.** $\frac{b^2m}{c^2}$ **21.** $\frac{m}{m+5}$ **23.** $\frac{4}{z+3}$
25. $\frac{2(x+5)}{x+1}$ **27.** $x+3$ **29.** $x+5$ **31.** $(x+5)(x+3)$
33. 88 ft/s; sample answer: changes 60 mph to ft/s
35. 27.5 yd^3; changes ft^3 to yd^3 **37.** $\frac{x-y}{4}$ **39a.** 5500 miles   **39b.** 41.3 trains   **39c.** 10,618.2 miles; 27.3 trains
41. $\frac{8}{m-5}$ **43.** $(4a - 3b^2)^2$ **45.** $x < -2$ **47.** $7 + 2 = 2 + 7$

Pages 678–680   Lesson 12–4
5. $3b^2 - 5$ **7.** $t - 1$ **9.** $2m + 3 + \frac{-3}{m+2}$ **11.** $(2x+3)$ m
13. $\frac{b}{3} + 3 - \frac{7}{3b}$ **15.** $\frac{m}{7} + 1 - \frac{4}{m}$ **17.** $a^2 + 2a + 12 +$
$\frac{3}{a-2}$ **19.** $m - 3 - \frac{2}{m+7}$ **21.** $2x + 6 + \frac{1}{x-7}$
23. $3m^2 + 4k^2 - 2$ **25.** $a + 7 - \frac{1}{a+3}$ **27.** $3x - 2 + \frac{21}{2x+3}$
29. $t + 4 - \frac{5}{4t+1}$ **31.** $8x^2 + \frac{32x}{7} - 9$ **33a.** $x^2 + 4x + 4$
33b. $5t^2 - 3t - 2$ **33c.** $2a^2 + 3a - 4$ **35.** -15
37. $\$480,000$ **39.** $-\frac{x}{7}$ **41.** $3(x-7)(x+5)$ **43.** $(1, 3)$

45.

$-3 -2 -1 \ 0 \ 1 \ 2 \ 3 \ 4 \ 5 \ 6$

Page 680   Self Test
1. $\frac{25x}{36y}$ **3.** $\frac{(x+4)(4x^2+2x-3)}{(x+3)(x+2)(x-1)}$ **5.** $\frac{x-5}{x+4}$ **7.** $2x - 5$
9. $4(3x - 1)$ or $(12x - 4)$ units

Pages 683–684   Lesson 12–5
5. $\frac{10}{x}$ **7.** $\frac{10}{a+2}$ **9.** 3 **11.** m **13.** $-\frac{a}{6}$ **15.** 2 **17.** $\frac{7}{x+7}$
19. 1 **21.** 1 **23.** $\frac{4x}{x+2}$ **25.** $\frac{2y-6}{y+3}$ **27.** 2 **29.** $\frac{25x-93}{2x+5}$
31. $\frac{24x+26y}{7x-2y}$ **33.** b   **35.** 442 days   **37.** $\pm\frac{6}{5}$ **39.** $-3, 4$
41. $\frac{1}{216}$

Pages 687–689   Lesson 12–6

3. x^2 **5.** $90m^2b^2$ **7.** $(x-3)(x+3)$ **9.** $\dfrac{7+9m}{15m^2}$

11. $\dfrac{-20}{3(x+2)}$ **13.** $\dfrac{22-7y}{6y^2}$ **15.** 168 members   **17.** $\dfrac{-5x}{63}$

19. $\dfrac{7yz+3}{xyz}$ **21.** $\dfrac{2s-3t}{s^2t^2}$ **23.** $\dfrac{7z+8}{(z+5)(z-4)}$ **25.** $\dfrac{k^2+k-10}{(k+5)(k+3)}$

27. $\dfrac{-17r-32}{(3r-2)(r-5)}$ **29.** $\dfrac{10+m}{2(2m-3)}$ **31.** $\dfrac{3w-4}{3(5w+2)}$

33. $\dfrac{-5n-4}{(n+3)(n+1)}$ **35.** $\dfrac{-17t+27}{(t+3)(10t-9)}$ **37.** $\dfrac{v^2+11v-2}{(v+4)(v-1)}$

39a. 360, 3, 120, 360   **39b.** 540, 6, 90, 540   **39c.** The GCF times the LCM of two numbers is equal to the product of the two numbers.   **39d.** Divide the GCF into the product of the two numbers to find the LCM.   **41.** 30 months later or Oct. 20, 1998   **43.** 2   **45.** $\dfrac{3\pm\sqrt{41}}{4}\approx 2.35$ or -0.85 **47.** $-15a^6b^3$ **49.** 4

Pages 693–695   Lesson 12–7

3. $\dfrac{8x+3}{x}$ **5.** $\dfrac{6m^2+m+1}{2m}$ **7.** $\dfrac{9}{5}$ **9.** $\dfrac{y}{2}$ **11.** $\dfrac{a-b}{3}$

13. $\dfrac{3x+15}{x+3}$ **15.** $\dfrac{6x-1}{2x+1}$ **17.** $\dfrac{5r^2+r-48}{r^2-9}$ **19.** $\dfrac{4}{3}$ **21.** ab

23. $\dfrac{6}{x}$ **25.** $\dfrac{a-5}{a-2}$ **27.** $\dfrac{1}{x+3}$ **29.** $\dfrac{m+6}{m+2}$ **31.** $\dfrac{y-3}{y+2}$

33. $\dfrac{t^2+2t+2}{-t^2}$ **35.** $\dfrac{32b^2}{35}$ **37a.** $x\neq 0, x\neq -3$ **37b.** $x\neq 0,$ $x\neq -3$ **37c.** yes   **39a.** $\dfrac{fs}{s-v}$ **39b.** 413.5   **39c.** 2   **39d.** 6

41. $\dfrac{2x+3}{x+3}$ **43.** $(2x+1)(2x-1)$ **45.** $(-3, 4)$

47. $w=\dfrac{P-2\ell}{2}$

Pages 700–702   Lesson 12–8

5. -12 **7.** $-\dfrac{1}{2}$ **9.** 0   **11a.** $\dfrac{1}{5}$ **11b.** $\dfrac{3}{5}$ **11c.** $\dfrac{x}{5}$ **13.** 3.429 ohms   **15.** 8 ohms, 4 ohms   **17.** $\dfrac{50}{3}$ **19.** $\dfrac{5}{4}$ **21.** 0

23. 2 or $-\dfrac{1}{3}$ **25.** 4 or $-\dfrac{1}{3}$ **27.** -13 **29.** $\dfrac{3}{5}$ **31.** 3   **33.** 6

35. 4 ohms   **37.** $R_1=\dfrac{R_2R_T}{R_2-R_T}$ **39.** $n=\dfrac{IR}{E-Ir}$ **41.** 10

43. 96 ohms   **45a.** 2.5 hours   **45b.** 2.19 hours   **45c.** after 923 miles   **47a.** $w=\dfrac{c\ell}{100}$ **47b.** $\ell=\dfrac{100w}{c}$ **49.** 10

51. $\dfrac{b+4}{4(b+2)^2}$

53.

55. $\dfrac{48}{25}$ **57.** 36

Page 703   Chapter 12 Highlights

1. false, rational   **3.** true   **5.** false, x^2-144 **7.** false, least common multiple

Pages 704–706   Chapter 12 Study Guide and Assessment

9. $z, z\neq 3$ **11.** $\dfrac{a}{a+2}, a\neq 0, -2$ **13.** $\dfrac{1}{(b+3)(b+2)}$, $b\neq \pm 2, \pm 3$ **15.** $\dfrac{3axy}{10}$ **17.** $\dfrac{3}{a^2-3a}$ **19.** $b+7$

21. $\dfrac{y^2-4y}{3}$ **23.** $\dfrac{2m-3}{3m-2}$ **25.** x^2+4x-2 **27.** x^3+5x^2 $+12x+23+\dfrac{52}{x-2}$ **29.** $\dfrac{2a}{m^2}$ **31.** $\dfrac{2m+3}{5}$ **33.** $a+b$

35. $\dfrac{22n}{21}$ **37.** $\dfrac{4c^2+9d}{6cd^2}$ **39.** $\dfrac{4r}{r-3}$ **41.** $\dfrac{5m-8}{m-2}$ **43.** $\dfrac{3x}{y}$

45. $\dfrac{x^2+8x-65}{x^2+8x+12}$ **47.** -5 **49.** $-\dfrac{1}{4}$ **51.** 5, $-\dfrac{1}{4}$ **53.** 6 bicycles, 24 tricycles, and 24 wagons   **55.** $1\dfrac{7}{8}$ hours

CHAPTER 13   EXPLORING RADICAL EXPRESSIONS AND EQUATIONS

Page 712   Lesson 13–1A

1. $4+4=8$ **3.** $16+9=25$

5.

7.   **9.**

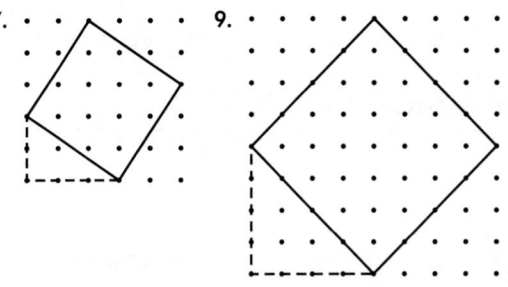

Pages 715–718   Lesson 13–1

7. 7   **9.** 11.40   **11.** 15   **13.** 2   **15.** yes   **17.** 127.28 ft **19.** 13.86   **21.** 13.08   **23.** 14.70   **25.** 60   **27.** 4 **29.** $\sqrt{67}\approx 8.19$ **31.** $\sqrt{27}\approx 5.20$ **33.** $\sqrt{145}\approx 12.04$ **35.** yes   **37.** no; $11^2+12^2\neq 15^2$ **39.** yes   **41.** $\sqrt{75}$ in. or about 8.66 in.   **43a.** 44.49 m   **43b.** 78 m²

45.

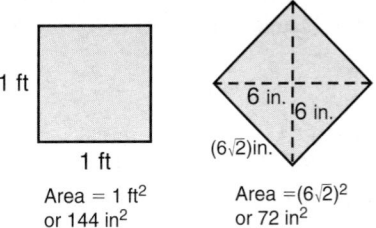

Area = 1 ft² or 144 in²   Area = $(6\sqrt{2})^2$ or 72 in²

47. about 2.66 m   **49.** -7 **51.** $12r^7$ **53.** $\left\{c\,\middle|\,c\geq\dfrac{1}{2}\right\}$ **55.** 41

Pages 724–725   Lesson 13–2

5. $5-\sqrt{2}; 23$ **7.** $\dfrac{\sqrt{7}}{\sqrt{7}}$ **9.** $3\sqrt{2}$ **11.** $\dfrac{\sqrt{21}}{7}$ **13.** $10\sqrt{2}+$ 26   **15.** $\dfrac{18+6\sqrt{2}}{7}$ **17.** $>$ **19.** about 7003.5 gal/min

21. $4\sqrt{5}$ **23.** $10\sqrt{5}$ **25.** $\dfrac{\sqrt{2}}{2}$ **27.** $\dfrac{\sqrt{22}}{8}$ **29.** $84\sqrt{5}$

31. $\dfrac{\sqrt{11}}{11}$ **33.** $3|ab|\sqrt{6}$ **35.** $7x^2y^3\sqrt{3xy}$ **37.** $\dfrac{3\sqrt{3}}{|p|}$

39. $\dfrac{x\sqrt{3xy}}{2y^3}$ **41.** $y^2-2y\sqrt{7}+7$ **43.** $\dfrac{28\sqrt{2}+14\sqrt{5}}{3}$

45. $\dfrac{-2\sqrt{5}-\sqrt{10}}{2}$ **47.** $\dfrac{c-2\sqrt{cd}+d}{c-d}$ **49.** $x^2-5x\sqrt{3}+12$

51. $>$ **53.** $<$ **55.** Yes; the result is about 11.84. **57.** 18.44 cm   **59.** 1, $\dfrac{2}{3}$ **61.** 2.7×10^9 acres

63. $y = \frac{1}{5}x + 6$ **65.** -6

Page 726   Lesson 13–2B
1. $14\sqrt{7}$ or 37.04   **3.** $2\sqrt{3} + 3\sqrt{2}$ or 7.71   **5.** $\frac{4\sqrt{7}}{7}$ or 1.51

7. $3\sqrt{3}$ or 5.20   **9.** $\frac{\sqrt{3}}{2} - \frac{\sqrt{5}}{3} + 3\sqrt{2}$ or 4.36

11. $\frac{-16\sqrt{6}}{3}$ or -13.06

Pages 729–731   Lesson 13–3
5. $3\sqrt{5}, 3\sqrt{20}$ **7.** none   **9.** $13\sqrt{6}$ **11.** $12\sqrt{7x}$
13. $13\sqrt{3} + \sqrt{2}$; 23.93   **15.** $\frac{8}{7}\sqrt{7}$; 3.02   **17.** 11
19. $26\sqrt{13}$ **21.** in simplest form   **23.** $21\sqrt{2x}$ **25.** $9\sqrt{3}$;
15.59   **27.** $\sqrt{6} + 2\sqrt{2} + \sqrt{10}$; 8.44   **29.** $-\sqrt{7}$; -2.65
31. $29\sqrt{3}$; 50.23   **33.** $4\sqrt{5} + 15\sqrt{2}$; 30.16   **35.** $4\sqrt{3} -$
$3\sqrt{5}$; 0.22   **37.** $10\sqrt{2} + 3\sqrt{10}$ **39.** $19\sqrt{5}$ **41.** $10\sqrt{3} +$
16   **43.** Both $(x - 5)^2$ and $(x - 5)^4$ must be nonnegative,
but $x - 5$ may be negative.   **45.** $6\frac{2}{5}$ ft   **47.** $3x - 3y$

49. 7   **51.** $\frac{2}{3}$ **53a.** 1991; about 9 million   **53b.** 6 million

Page 731   Self Test
1. 35   **3.** 46.17   **5.** 10   **7.** $11\sqrt{6}$ **9.** $12\sqrt{5} - 6\sqrt{3} +$
$2\sqrt{15} - 3$

Pages 734–736   Lesson 13–4
7. $a + 3 = 4$ **9.** 16   **11.** 36   **13.** -12 **15.** 3.9 mph
17. -63 **19.** no real solution   **21.** $\frac{81}{25}$ **23.** 2   **25.** 24

27. $\frac{3}{25}$ **29.** $\pm\frac{4}{3}\sqrt{3}$ **31.** $\sqrt{7}$ **33.** 6   **35.** $-3, -10$ or 3, 10

37. 16   **39.** no real solution   **41.** $\left(\frac{1}{4}, \frac{25}{36}\right)$ **43.** $\sqrt{3} + 1$

45. $54\frac{2}{3}$ mi   **47a.** No, at 30 mph, her car should have
skidded 75 feet after the breaks were applied and not
110 feet.   **47b.** about 36 miles per hour   **49.** $6b - 26 +$
$\frac{150}{b + 5}$ **51.** $6xy^3$ **53.** $(-4, 5)$ **55.** $\frac{8}{13}$

Pages 739–741   Lesson 13–5
5. 5   **7.** $3\sqrt{2}$ or 4.24   **9.** 7 or 1   **11.** about 9.49 mi   **13.** 17
15. $\sqrt{41}$ or 6.40   **17.** $\frac{10}{3}$ or 3.33   **19.** $\frac{13}{10}$ or 1.30

21. $2\sqrt{3}$ or 3.46   **23.** 17 or -13 **25.** -10 or 4   **27.** 8 or
32   **29.** no   **31.** $\sqrt{157} \neq \sqrt{101}$; Trapezoid is not
isosceles.   **33.** The distance between $(3, -2)$ and $(-3, 7)$
is $3\sqrt{13}$ units. The distance between $(-3, 7)$ and $(-9, 3)$
is $2\sqrt{13}$ units. The distance between $(3, -2)$ and $(-9, 3)$
is 13 units. Since $\left(3\sqrt{13}\right)^2 + \left(2\sqrt{13}\right)^2 = 13^2$, the triangle
is a right triangle.   **35.** 109 miles   **37.** $\frac{2\sqrt{2}}{3}$

39. no solution

41a. 7392 BTU   **41b.** 3705 BTU   **41c.** 6247.5 BTU
41d. 14,040 BTU   **41e.** 6831.45 BTU

Page 742   Lesson 13–6A
1. $(x + 2)^2 = 1$ **3.** $(x + 2)^2 = 5$ **5.** $(x - 2)^2 = 5$
7. $(x + 1.5)^2 = 3.25$

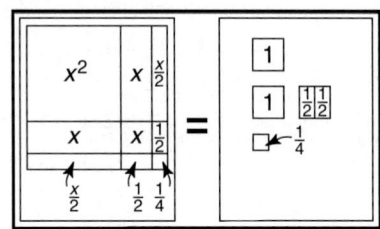

Pages 746–747   Lesson 13–6
5. 64   **7.** $-1, -3$ **9.** 7, -3 **11.** $2 \pm \sqrt{6}$ **13.** 4.9 ft
15. 16   **17.** $\frac{121}{4}$ **19.** 8   **21.** 4, 1   **23.** $\frac{7}{3}$ **25.** $3 \pm \sqrt{5}$

27. $5 \pm 4\sqrt{3}$ **29.** $\frac{-5 \pm 2\sqrt{15}}{5}$ **31.** $\frac{2}{3}, -1$ **33.** $\frac{7 \pm \sqrt{85}}{6}$

35. 18, -18 **37.** 3   **39.** $2 \pm \sqrt{4 - c}$ **41.** $b(-2 \pm \sqrt{3})$
43. 45 mph, 55 mph   **45.** 8.06   **47.** 2042   **49.** $(2, 2)$
51a. 224.6   **51b.** 172

Page 749   Chapter 13 Highlights
1. false, $-3 - \sqrt{7}$ **3.** false, $-2 - 2\sqrt{3} - \sqrt{5} - \sqrt{15}$
5. true   **7.** false, $3x + 19 = x^2 + 6x + 9$ **9.** false, $\frac{x\sqrt{2xy}}{y}$

Pages 750–752   Chapter 13 Study Guide and Assessment
11. 34   **13.** $5\sqrt{5} \approx 11.18$ **15.** 24   **17.** no; $9^2 + 16^2 \neq 20^2$
19. yes   **21.** $4\sqrt{30}$ **23.** $2|a|b^2\sqrt{11b}$ **25.** $57 - 24\sqrt{3}$
27. $\frac{3\sqrt{35} - 5\sqrt{21}}{-15}$ **29.** $2\sqrt{6} - 4\sqrt{3}$ **31.** $36\sqrt{3}$ **33.** $\frac{9\sqrt{2}}{4}$

35. 12   **37.** $\frac{26}{7}$ **39.** 45   **41.** 3   **43.** no solution   **45.** 17

47. $\sqrt{205} \approx 14.32$ **49.** 5 or -1 **51.** 5 or -3 **53.** $\frac{49}{4}$

55. $\frac{1}{9}$ **57.** $\frac{7 \pm \sqrt{69}}{2}$ **59.** $-2 \pm \sqrt{6}$ **61.** $22\sqrt{6} \approx 53.9$

63. scalene triangle   **65.** 5 ft   **67.** No, it should skid about
201.7 ft.

Cover (t) FPG International, (b) Nawrocki Stock Photo, (background) SuperStock; **iii** (t) courtesy Home of the Hamburger, Inc., (b) Aaron Haupt; **viii** (l) Terry Eiler, (r) Paul L. Ruben; **ix** (t) Geoff Butler, (b) Skip Comer; **x** (t) Stephen Webster, (bl) Aaron Haupt (br) Mark Peterson/Tony Stone Images; **xi** (t) Morton & White Photographic, (b) Tom Tietz; **xii** (t) Stan Fellerman, (b) Aaron Haupt; **xiii** (t) Tom McCarthy/The Stock Market, (c) Geoff Butler, (b) Charles H. Kuhn Collection, The Ohio State University Cartoon, Graphic & Photographic Arts Research Library/Reprinted with special permission of King Features Syndicate; **xiv** (t) Clark James Mishler/The Stock Market, (b) Julian Baum/Science Photo Library/Photo Researchers; **xv** (t) Jonathan Daniel/Allsport USA, (c) The Image Bank/Steve Niedorf, (b) The Image Bank/Jody Dole; **xvi** Tate Gallery, London/Art Resource, NY; **xvii** (t) Morton & White Photographic, (b) NCSA, University of IL/Science Photo Library/ Photo Researchers; **4** (r) David Ball/The Stock Market, (l) E. Hobson/ Ancient Art & Architecture Collection; **5** (tl) John Neubauer, (tr) Scott Cunningham, (bl) Ron Sheridan/Ancient Art & Architecture Collection, (br) Globe Photos; **6** Doug Martin; **10** (l) Aaron Haupt, (r) KS Studio; **11** (t) Ted Rice, (b) courtesy the Math Association Of America; **12** Terry Eiler; **18** Aaron Haupt; **19** Comnet/Westlight; **24** (t) Stephen Studd/Tony Stone Images, (b) Ken Reid/FPG International; **25** Aaron Haupt; **27** KS Studio; **30** (t) KS Studio, (b) Bruce Hands/Tony Stone Images; **31** Phil Degginger/Tony Stone Images; **32** courtesy Home of the Hamburger, Inc.; **36** Gail Shumwat/ FPG International; **37** (l) Duomo, (r) Duomo/Steven E. Sutton; **40** Doug Martin; **41** Mary Ann Evans; **43, 45** Aaron Haupt; **46** David Madison; **50** Eugene G. Schulz; **52** Kenji Kerins; **54** Doug Martin; **55** (l) KS Studio, (r) Paul L. Ruben; **57** The Image Bank/Ken Huang; **58** Aaron Haupt; **59** HERMAN © 1975 Jim Unger. Dist. by UNIVERSAL PRESS SYNDICATE. All rights reserved; **60** J. Parker/Westlight; **61** Aaron Haupt; **62** Sonderegger/Westlight; **67** (l) Aaron Haupt, (r) Jeff Bates **70** (t) Geoff Butler, (c) The Bettmann Archive, (b) Frank Lerner; **71** (tl) James Wasserman, (tr) Doug Martin, (bl) courtesy Nestle USA, (br) Archive Photos; **72–73** Geoff Butler; **75** KS Studio; **76** Geoff Butler; **77** Kenji Kerins; **78** James Avery/ Everett Collection; **79** Geoff Butler; **80** (l) R.B. Sanchez/The Stock Market, (r) Mark Petersen/Tony Stone Images; **82** Geoff Butler; **83** Focus on Sports; **85** AP/Space Shuttle/Wide World Photos; **88** Doug Martin; **89** Geoff Butler; **91** (t) American Photo Bank, (b) Geoff Butler; **92** William Clark/Tony Stone Images; **95** (t) Duomo/Paul Sutton, (b) Aaron Haupt; **97** Aaron Haupt; **98** (l) UPI/Bettmann, (r) Warren Morgan/Westlight; **99** Doug Martin; **100** (t) Westlight, (b) Stewart Cohen/Tony Stone Images; **102** Archive Photos; **103** James Marshall/The Stock Market, (c) Aaron Haupt, (c) Gary Mortimore/Allsport USA, (b) Geoff Butler; **106** (t) The Image Bank/Patti McConville, (b) R. Foulds/Washington Stock Photos; **108** Geoff Butler; **110** (b) Peter Saloutos/The Stock Market, (inset) Aaron Haupt; **112** (t) Holmes-Lebel/FPG International, (b) The Bettmann Archive; **113** (doctor scene) R.W. Jones/Westlight, (inset) Aaron Haupt; **116** Aaron Haupt; **117** (l) David Stoecklein/The Stock Market, (r) David Woods/The Stock Market; **119** Johnny Johnson/ Tony Stone Images; **122** CLOSE TO HOME © 1994 John McPherson/ Dist. by UNIVERSAL PRESS SYNDICATE. All rights reserved; **124** (t) Joe Towers/The Stock Market, (c) The Image Bank/Nick Nicholson, (b) Aaron Haupt; **125** (l) Aaron Haupt, (r) David Cannon/Allsport USA; **126** (t) Ian R. Howarth, (cl) Charles W. Melton, (cr) Roy Morsch/The Stock Market, (bl) Meryl Joseph/ The Stock Market, (br) William J. Weber, (background) Breck Kent; **127** (t) Elaine Shay, (b) John Neubauer; **129, 130** Aaron Haupt; **131** Bob Daemmrich; **132** Doug Martin; **136** Mark Burnett; **140** (t, chart) Mark Scott/FPG International, (b, chart) Frank Cezus, (c b) Ron Sheridan/Ancient Art & Architecture Collection; **141** (tl) courtesy Marianne Ragins, (tr) David Woods/The Stock Market, (cl br) The Bettmann Archive, (cr) UPI/Bettmann, (b) Ron Sheridan/ Ancient Art & Architecture Collection; **145** Michael Burr/NFLP; **149** Rick Weber; **150** AP/Wide World Photos; **153** Rick Weber; **154** Dave Lawrence/The Stock Market; **156** (t) Wide World Photos, (b) H.G. Ross/FPG International; **157** Archive Photos; **160** The Image Bank/ Benn Mitchell; **161** Skip Comer; **162** Ralph Cowan/Tony Stone Images; **163** Tim Davis/Photo Researchers; **167** (t) StudiOhio, (b)Rick Weber; **168** (l)UPI/Bettmann, (r) LPI/M. Yada/FPG International; **170** E. Alan McGee/FPG International; **172** Paul Barton/The Stock Market; **174** Rick Weber; **176** Aaron Haupt; **177** Bruce Ayres/Tony Stone Images; **178** David M. Grossman/ Photo Researchers; **180** Archive Photos/Martin Keene; **181** Focus On Sports; **183** Aaron Haupt; **184** (l) David M. Dennis, (r) NASA;

189 Rick Weber; **190** The Image Bank/James Meyer; **191** Morton & White Photographic; **192** (t) Tim Courlas, (bl) Archive Photos, (others) The Bettmann Archive; **193** (t) KS Studio, (c) John Karl Breun, (b) UPI/Bettmann; **195** Morton & White Photographic; **196** Larry Hamill; **197** (t) Phil Schofield/Tony Stone Images, (b) Universal Pictures/Shooting Star; **198** (l) David R. Frazier Photolibrary, (r) Glencoe photo; **199** (t) Biophoto Associates/ Photo Researchers, (b) Morton & White Photographic; **200** Kenji Kerins; **201** (t) Gus Chan, (c) TJ Collection/ Shooting Star, (b) AP/Wide World Photos; **202** Aaron Haupt; **203** James N. Westwater; **205** (t) Mike Yamashita/Westlight, (b) Scott Padgett/Tradd Street Stock; **206** (t) John Elk III, (b) The Image Bank/Jurgen Vogt; **209** Luc Novovitch/AP/Wide World Photos; **213** (t) The Image Bank/Ulf Wallin, (b) Doug Martin; **214** (t) Focus on Sports, (b) courtesy Ryan Morgan; **215** (t) Doug Martin, (b) Morton & White Photographic; **217** (t) Everett Collection, (bl br) Morton & White Photographic; **218** The Image Bank/Leo Nason; **219** KS Studio; **220** (t tc) Morton & White Photographic, (bc) Elaine Shay, (b) John Mead/Science Photo Library/Photo Researchers; **222** (t) Morton & White Photographic, (b) Charles Cangialosi/Westlight, (inset) Dr. E.R. Degginger/Color-Pic; **223–224** Morton & White Photographic; **226** (t) Morton & White Photographic, (b) Doug Martin; **227** Kenji Kerins; **228** (t) Roger Dons/People Weekly, (b) Morton & White Photographic; **229, 230, 231** Morton & White Photographic; **232** (t) Everett Collection, (c) The Image Bank/Chris Hackett; (b) Morton & White Photographic; **233** Everett Collection; **234** (t) Don Mason/ The Stock Market, (b) Morton & White Photographic; **235** Joe Bator/ The Stock Market; **236** Latent Image; **237** (t b) Morton & White Photographic, (c) Chuck Bankuti/ Shooting Star; **238** (t) Kenji Kerins, (b) Morton & White Photographic; **239** KS Studio; **240** Matt Meadows; **241** Morton & White Photographic; **242** Photo Researchers; **244** NASA; **248–249** KS Studio; **252** (t) Terry Vine/ Tony Stone Images, (c) The Bettmann Archive, (b) Ronald Sheridan/ Ancient Art & Architecture Collection; **253** (tl) Weldon McDougal, (tr) The Image Bank/David W. Hamilton, (bl) Jim Cornfield/ Westlight, (br) The Bettmann Archive; **254** (t) Mimi Ostendorf-Smith, (b) James Blank/The Stock Market; **262** (t) Clayton Fogle/Tony Stone Images, (cl) Daniel J. Cox/Tony Stone Images, (cr) Tom Tietz/Tony Stone Images, (b) Tim Davis/Tony Stone Images; **264** James P. Rowan/Tony Stone Images; **266** The Image Bank/Stephen Marks; **269** Stephen Webster; **271** (t) Doug Martin, (c) Brent Turner/BLT Productions,(b) Aaron Haupt; **273, 277** Aaron Haupt; **280** David Madison/Tony Stone Images; **281** (t) Doug Martin, (b) Tom McGuire; **282** KS Studio; **284** Ruth Dixon; **285** (t) Jim Gund/Sports Illustrated, (c) KS Studio, (b) Aaron Haupt; **286** Bob Daemmrich; **287** David O. Hill/Photo Researchers; **290** KS Studio; **293** The Image Bank/Stephen Marks; **294** (l) Duomo/Al Tielemans, (r) Duomo/Bryan Yablonsky; **295** Alese & Mort Pecher/ The Stock Market; **299** © 1995 Bill Watterson/Dist. by Universal Press Syndicate. All rights reserved; **301** (t) The Smithsonian Institution, (b) David Madison; **302** MAK-1; **304** KS Studio; **306** (t) Aaron Haupt, (b) Doug Martin; **308** David Barnes/The Stock Market; **309** KS Studio; **310** Aaron Haupt; **311** (t) Joan Marcus/Photofest (c bl) Photofest, (br) Aaron Haupt; **313** (t) Aaron Haupt, (b) The Image Bank/Harald Sund; **314** (t) Eric Lars Bakke/Allsport USA, (c b) In Fisherman Magazine; **318** Aaron Haupt; **319** Denis Valentine/The Stock Market; **321** (t) Matt Meadows, (b) Morton & White Photographic; **322** (t) The Bettmann Archive, (others) Culver Pictures; **323** (tl) courtesy Jorge Arturo Pineda Aguilar, (c) Archive Photos/Consolidated News, (tr) Morton & White Photographic, (b) People Weekly/Alan S. Weiner; **325** (t) David Ball/The Stock Market, (b) Morton & White Photographic: **328** Morton & White Photographic; **330** The Image Bank/J. Coolidge; **331** (tl) John M. Roberts/The Stock Market, (tc) The Image Bank/Andrea Pistolesi, (tr) Dallas & John Heaton/ Westlight, (cl) Gail Shumway/FPG International, (c) Tom Walker/ Tony Stone Images, (cr) Art Wolfe/Tony Stone Images, (b) UPI/ Bettmann; **332** (t) The Image Bank/Grant V. Faint, (cl) Van Bucher/ Photo Researchers, (cr) Scott Markewitz/FPG International, (b) Joyce Photographics/Photo Researchers; **335, 337** Doug Martin; **338, 339** Morton & White Photographic; **340** (t) Tom Stack/Tom Stack and Associates, (c) Bob Daemmrich, (b) Tracy I. Borland; **342** Stan Fellerman/The Stock Market; **343** (t) Doug Martin, (b) Morton & White Photographic; **344, 345** Morton & White Photographic; **349** The Bettmann Archive; **350** (t) John McDermott/ Tony Stone Images, (b) Morton & White Photographic; **353** Morton & White Photographic; **354** (t) Diane Padys/FPG International, (b) Dick Luria/FPG International; **356** Morton & White Photographic;

Academics. *See* Colleges; Education; School

Accounting, 23, 294, 468. *See also* Economics

Acoustics, 694

Advertising, 225, 404

Aeronautics, 664. *See also* Aviation; Space science

Agriculture, 31, 259, 269, 518, 616, 647, 714

Air conditioning, 172, 741

Airplanes
airline fares, 108, 287
atmospheric pressure on, 664–665
drilling holes in, 98
passenger statistics, 226, 506
pilots, 112, 507
routes, 416
slope of descent, 337, 512
speed of, 236, 248, 338, 505, 701
takeoff, 345

Anatomy, 276

Animals, 80, 172, 262, 264, 331, 341, 648

Anthropology, 702

Aquatics, 295, 300

Archaeology, 98, 213, 301, 413, 638–639

Architecture
art gallery, 626
housing development, 461
Leaning Tower of Pisa, 206, 208, 599
model home, 426
Sears Tower, 526
similar triangles, 201
slope of roof, 329, 331
tall buildings, 29, 183, 526

Art, 514, 536, 658, 659, 740

Astronomy, 508, 617, 642, 684, 685, 747. *See also* Space science

Automobiles, 330
antifreeze in, 237
braking distance for, 735
commuting in, 183, 486
depreciation of, 226
gas needed for, 198, 432
leasing vs. purchasing, 645
loans for, 505, 527, 552
maintenance schedule for, 688
MPG ratios for, 25, 26
options in purchasing, 446
popular, 311
rental of, 318, 397
repair charges for, 527
traffic and, 124, 667, 668, 670

Aviation, 98. *See also* Airplanes

Babies, 160, 418

Babysitting, 10

Balloons, 210, 490

Banking, 346, 352, 518. *See also* Finance

Birth rates, 429

Birth weight, 160

Book binding, 104

Book publishing, 4–5

Bridges, 205, 599

Budgeting, 294, 386–387, 398

Business and work
car repairs, 527
car wash, 404, 700
employment, 49, 100, 239–240, 266, 277, 285, 482, 593
keyboarding speeds, 176, 352
lawn care service, 696, 706
mileage reimbursement, 109
modular offices, 540
profit calculations, 40, 404
strikes, 404
toy evaluation, 460
T-shirt day, 416
See also Commerce; Sales

Carpentry, 116, 166, 330, 664, 731

Cartography, 256

Charities, 62, 178

City planning, 642

Civil engineering, 214

Climate. *See* Meteorology

Colleges, 238, 311, 382, 616
entrance exams for, 339, 345, 431, 471

Commerce, 308, 627, 697

Communication, 149, 153, 292, 475–476, 710, 739, 740

Community service, 608, 609

Commuting, 183, 486

Computers
CD-ROMs for, 620
chess and, 419
cost of, 353
graphing software for, 471, 733
keyboarding speed, 127
memory of, 495
molecular, 649
on-line services, 226
programming, 8, 49, 171
shipments for home use, 98
spreadsheets, 297, 327, 646
student use of, 418

Construction
building codes, 110
clearances, 325, 328
of deck, 566–567
of dining area with fountain, 747
excavation site, 674
of garden, 533
handrails for physically challenged persons, 422–423
ladders for, 227
lot size, 325, 328, 592
paneling, 717
ramps for people with disabilities, 150, 151
of roads, 330, 368
roof support, 353
slip forming, 730
slope of roof, 350
of swimming pool, 540, 566–567

Consumerism
cost comparisons, 95, 116, 136, 149
credit cards, 220, 302
discounting, 223, 224, 225, 226, 227
leasing vs. purchasing, 645
MPG ratios of automobiles, 25, 26
payment plans, 600
printer prices, 541
See also Sales, Economics

Contests, 228

Cooking, 481, 670

Cosmetics, 215

Cultures
Aztec agriculture, 714
Chinese calendar, 50
Chinese kite-making, 362
Egyptian pyramids, 24, 474
Gupta Empire (India), 460
Hawaiian alphabet, 160
Japanese royal wedding, 156

Currency. *See* Money

Decision making, 654

Demographics. *See* Population statistics

Diet, 70–71
for diabetic, 227
guidelines for, 384, 385
planning, 71
See also Food; Nutrition

Drafting, 116, 205

Driving. *See* Automobiles

Earthquakes, 30

Ecology, 205, 313. *See also* Environment

Economics
allowances, 50
cost comparisons, 95, 116, 136, 149
depreciation, 56–57, 226
farming, 31
feeding the world, 511
inflation, 161
money, 115, 220, 237, 641, 670
stock market, 19, 102, 103, 136, 468, 496, 500, 580
See also Finance; Consumerism

Education
ACT Program scores, 431
earnings potential and, 382
SAT scores, 339, 345, 471
See also Colleges; School

INDEX

Red type denotes items only in the Teacher's Wraparound Edition.

Index

Index

P